Challenges for Diadromous Fishes in a Dynamic Global Environment

Challenges for Diadromous Fishes in a Dynamic Global Environment

An International Symposium

Sponsored by

Northeastern Division, American Fisheries Society

Steering Committee

Alex J. Haro, Chair
Katherine L. Smith
Roger A. Rulifson
Paul R. Perra
Christine M. Moffitt
Ronald J. Klauda
Scott R. Decker
Michael J. Dadswell
Richard A. Cunjak
John E. Cooper
Kenneth L. Beal
Trevor S. Avery

Challenges for Diadromous Fishes in a Dynamic Global Environment

Edited by

Alex Haro, Chair

Katherine L. Smith

Roger A. Rulifson

Christine M. Moffitt

Ronald J. Klauda

Michael J. Dadswell

Richard A. Cunjak

John E. Cooper

Kenneth L. Beal

Trevor S. Avery

American Fisheries Society Symposium 69

Proceedings of the International Symposium
"Challenges for Diadromous Fishes in a Dynamic Global Environment"
Held in Halifax, Nova Scotia, Canada
June 18–21, 2007

American Fisheries Society
Bethesda, Maryland, USA
2009

The American Fisheries Society Symposium Series is a Registered Serial. Suggested citation formats are as follows.

Entire book

Haro, A. J., K. L. Smith, R. A. Rulifson, C. M. Moffitt, R. J. Klauda, M. J. Dadswell, R. A. Cunjak, J. E. Cooper, K. L. Beal, and T. S. Avery, editors. 2009. Challenges for diadromous fishes in a dynamic global environment. American Fisheries Society, Symposium 69, Bethesda, Maryland.

Chapter in book

Blaber, S. J. M. 2009. Socioeconomic and biopolitical linkages in the management of tropical shads. Pages 677–690 *in* A. J. Haro, K. L. Smith, R. A. Rulifson, C. M. Moffitt, R. J. Klauda, M. J. Dadswell, R. A. Cunjak, J. E. Cooper, K. L. Beal, and T. S. Avery, editors. Challenges for diadromous fishes in a dynamic global environment. American Fisheries Society, Symposium 69, Bethesda, Maryland.

Cover art and design by John E. Cooper. Front and bottom back cover fishes (from left): *Galaxias maculatus, Anguilla anguilla, Huso huso, Retropinna retropinna, Salangichthys microdon, Plecoglossus altivelus*. Back cover top: Logo from American Fisheries Society Symposium 1, *Common Strategies of Anadromous and Catadromous Fishes*.

Printed in the United States of America on acid-free paper.

Library of Congress Control Number 2009905527
ISBN 978-1-934874-08-0
ISSN 0892-2284

American Fisheries Society Web site: *www.fisheries.org*

American Fisheries Society
5410 Grosvenor Lane, Suite 110
Bethesda, Maryland 20814

Contents

Synopses

Poster Sessions

Poster Abstracts

Reviewer Acknowledgments

The editors thank the following additional referees for their contributions to the scientific quality of this book:

Paul Angermeier
Miran Aprahamian
Ken Ashley
Agnès Bardonnet
Kim Bell
Stephen Blaber
Sylvain Bonhommeau
Rod Bradford
Christopher Bunt
James Congleton
Charles Coutant
Willem Dekker
Brian Dempson
Walton Dickhoff
Ken Drinkwater
Panos Economidis
Mary Freeman
Irene Gregory-Eaves
Evelyn Habit
Dale Hanson
Melanie Harris
Gordon Hartman
Joseph Hightower
Morgan Hocking
Thomas Hoff
Arshad Hossain
Edward Houde
Midori Iida
James Irvine
Tim Jardine
Don Jellyman
Bror Jonsson
Brandon Kulik
Patrick Lambert
Benjamin Letcher
Karin Limburg
Martha Mather
David McGrath
John Michael
Brent Murry
Robert Naiman
Keith Nislow
John Olney
Tony Pitcher
Mark Plummer
Joe Rasmussen
David Reddin
James Reist
William Richkus
Stephen Rideout
Eric Rochard
David Rowe
Pekka Salmi
Daniel Schindler
Robin Schrock
David Secor
Mohammed Siddique
Michael Stokesbury
Donald Taphorn
Brad Taylor
James Terrell
Clive Trueman
John Waldman
Alan Whitfield
Fred Whoriskey

Cooperators

The Steering Committee expresses its sincere thanks to the following donors that contributed funds and in-kind services in support of the symposium:

Acadia University
AFS Marine Fisheries Section
AFS Atlantic International Chapter
AFS Bioengineering Section
AFS Canadian Aquatic Resources Section
AFS Estuaries Section
AFS Mid-Atlantic Chapter
AFS New York Chapter
AFS Pennsylvania Chapter
AFS Potomac Chapter
AFS Southern New England Chapter
AFS Tidewater Chapter
Alden Research Laboratory
AMIRIX Systems Inc., VEMCO Division
Atlantic Salmon Federation
Electric Power Research Institute
FishAmerica Foundation
Fisheries and Oceans Canada - Gulf Region
Floy Tag and Manufacturing, Inc.
HDR, Inc.
Halifax Marriott Harbourfront
Hudson River Foundation
Hydroacoustic Technology Inc.
Keith Campbell Foundation for the Environment
LOTEK Wireless Inc.
Maine Atlantic Salmon Commission
Maryland Department of Natural Resources
Massachusetts Division of Marine Fisheries
Miramichi Salmon Association
National Hydropower Association
Natural Sciences and Engineering Research Council-Atlantic/CRSNG-Atlantique
New York State Department of Environmental Conservation Bureau of Marine Resources and Hudson River Estuary Program
NOAA/NMFS Southeast Regional Office
NOAA/NMFS Office of Habitat Conservation
NOAA/NMFS Office of Habitat Conservation, Habitat Protection Division
NOAA/NMFS Office of International Affairs
NOAA/NMFS Office of Protected Resources, Endangered Species Division
Nova Scotia Department of Fisheries and Aquaculture
The Nature Conservancy
Trout Unlimited Canada
Trout Unlimited USA
U.S. Department of Agriculture
U.S. Environmental Protection Agency
U.S. Fish and Wildlife Service Headquarters
U.S. Geological Survey Biological Resources Discipline - Fisheries: Aquatic and Endangered Resources Program
U.S. Geological Survey Biological Resources Discipline - S.O. Conte Anadromous Fish Research Laboratory
Versar, Inc.

Symbols and Abbreviations

The following symbols and abbreviations may be found in this book without definition. Also undefined are standard mathematical and statistical symbols given in most dictionaries.

A	ampere
AC	alternating current
Bq	becquerel
C	coulomb
°C	degrees Celsius
cal	calorie
cd	candela
cm	centimeter
Co.	Company
Corp.	Corporation
cov	covariance
DC	direct current; District of Columbia
D	dextro (as a prefix)
d	day
d	dextrorotatory
df	degrees of freedom
dL	deciliter
E	east
E	expected value
e	base of natural logarithm (2.71828…)
e.g.	(exempli gratia) for example
eq	equivalent
et al.	(et alii) and others
etc.	et cetera
eV	electron volt
F	filial generation; Farad
°F	degrees Fahrenheit
fc	footcandle (0.0929 lx)
ft	foot (30.5 cm)
ft^3/s	cubic feet per second (0.0283 m^3/s)
g	gram
G	giga (10^9, as a prefix)
gal	gallon (3.79 L)
Gy	gray
h	hour
ha	hectare (2.47 acres)
hp	horsepower (746 W)
Hz	hertz
in	inch (2.54 cm)
Inc.	Incorporated
i.e.	(id est) that is
IU	international unit
J	joule
K	Kelvin (degrees above absolute zero)
k	kilo (10^3, as a prefix)
kg	kilogram
km	kilometer
l	levorotatory
L	levo (as a prefix)
L	liter (0.264 gal, 1.06 qt)
lb	pound (0.454 kg, 454g)
lm	lumen
log	logarithm
Ltd.	Limited
M	mega (10^6, as a prefix); molar (as a suffix or by itself)
m	meter (as a suffix or by itself); milli (10^{-3}, as a prefix)
mi	mile (1.61 km)
min	minute
mol	mole
N	normal (for chemistry); north (for geography); newton
N	sample size
NS	not significant
n	ploidy; nanno (10^{-9}, as a prefix)
o	ortho (as a chemical prefix)
oz	ounce (28.4 g)
P	probability
p	para (as a chemical prefix)
p	pico (10^{-12}, as a prefix)
Pa	pascal
pH	negative log of hydrogen ion activity
ppm	parts per million
qt	quart (0.946 L)
R	multiple correlation or regression coefficient
r	simple correlation or regression coefficient
rad	radian

S	siemens (for electrical conductance); south (for geography)
SD	standard deviation
SE	standard error
s	second
T	tesla
tris	tris(hydroxymethyl)-aminomethane (a buffer)
UK	United Kingdom
U.S.	United States (adjective)
USA	United States of America (noun)
V	volt
V, Var	variance (population)
var	variance (sample)
W	watt (for power); west (for geography)
Wb	weber
yd	yard (0.914 m, 91.4 cm)
α	probability of type I error (false rejection of null hypothesis)
β	probability of type II error (false acceptance of null hypothesis)
Ω	ohm
μ	micro (10^{-6}, as a prefix)
′	minute (angular)
″	second (angular)
°	degree (temperature as a prefix, angular as a suffix)
%	per cent (per hundred)
‰	per mille (per thousand)

Introduction

In 1983, members and associates of the American Fisheries Society (AFS) Northeastern Division began organizing an independent, international symposium on diadromous fishes. Their efforts resulted in the 1986 symposium *Common Strategies of Anadromous and Catadromous Fishes*, held in Boston, Massachusetts. The landmark symposium was one of the first scientific meetings to focus exclusively on diadromous fishes, uniquely from both a biological and management standpoint. The proceedings from that meeting comprised the highly successful American Fisheries Society's Symposium 1, a legacy publication that continues to be a highly desired source of information on diadromous fish biology and ecology. *Common Strategies* also brought together a diverse group of researchers, managers, and students from worldwide geographic locations. The attendees found commonality in their questions and problems, fostered new collaborations and information sharing, and (in a good way) conceived a couple of controversies. Many of those associations are still vibrant and productive today, and for many students (of whom I was then one), *Common Strategies* opened doors for investigation and career development at a global level that would not have been available otherwise.

During the intervening 23 years since the *Common Strategies* symposium, many of the participants of the original meeting discussed organizing a follow-up meeting, but it never materialized until 2005 when Steve Rideout, an organizer of the 1986 symposium, and I took the plunge and began to assemble a steering committee. Steve retired soon thereafter, and I was assigned the task of organizing a large-scale international meeting—a responsibility I had never tackled before! Fortunately, I had the assistance of a 10-member steering committee, six of whom had served in similar roles for the *Common Strategies* symposium almost a quarter century earlier, and they knew the process and potential pitfalls well. I am therefore eternally grateful to all the committee members, including the veterans (Roger Rulifson, Christine Moffitt, Ron Klauda, Mike Dadswell, John Cooper, and Ken Beal) and the new recruits (Rick Cunjak, Katherine Smith, and Trevor Avery), and symposium treasurers (Paul Perra and Scott Decker), for their advice and guidance. Very special thanks go to Ken Beal, who though recently retired, guided our symposium ship through some particularly rough fiscal seas.

It took some time for the committee to decide on a theme for this meeting; we wanted to bring the subject material into the 21st century and go beyond the original theme of life history and evolutionary strategies of diadromous fishes. We ultimately agreed that the main focus should be on a review of the scientific knowledge of the biology, ecology, human importance, management, and conservation of diadromous fishes, with the goal of providing a synergy to achieve innovative interpretations, increased scientific understanding, and opportunities for sustainable management. Because diadromous fishes use different environments and migration corridors to complete their life history in marine and freshwater environments, they are particularly vulnerable to direct and indirect consequences of human development and global climate change. This theme seemed to us particularly timely in that both development and climate change had moved to the forefront of concerns for both biodiversity and resource conservation. Therefore, it was agreed that the meeting would focus on the challenges that diadromous fishes face now and will likely face in the future, given the ever-changing impacts to their environments that are taking place at a global scale. The concept of another single stand-alone meeting was important to the committee and to the success of both events, to capitalize on developing questions and capture the entire group in informal and formal discussions and presentations.

*Challenges for Diadromous Fishes in a Dynamic Global Environmen*t brought an impressive response with more than 150 submissions for papers and posters from more than 15 countries, with representation of most groups of diadromous fishes. Selection of oral papers was carefully made to emphasize the diversity of fishes, geographies, and topics. The result was two plenary papers, 13 invited papers, 29 contributed papers, and 57 contributed posters within the symposium.

The success of *Challenges for Diadromous Fishes* built upon the sustained interest of many contributors, some of whom presented at the *Common Strategies* symposium and also presented papers at *Challenges for Diadromous Fishes* by revisiting their original topics or providing new perspectives on their interim research. We were especially fortunate to recruit Robert McDowall and Thomas Quinn, both presenters at *Common Strategies*, to serve as plenary speakers. Their reviews of the evolution, ecology, and life history of a broad spectrum of diadromous species and their evaluations of scientific and management progress in these areas over the past two decades were particularly insightful and provided a springboard for many presentations that followed.

We thank the AFS Northeastern Division membership for its major support of the meeting, including administration of the meeting, provision of publication finances that supported holding a membership vote to sponsor the meeting, signing for the venue, and covering a portion of the initial costs of the meeting. Northeastern Division (NED) Presidents C. Paola Ferreri, Margaret H. Murphy, Larry M. Miller, and Scott R. Decker championed the endeavor from its inception and throughout its 4-year development, while Secretary-Treasurers Paul Perra and Scott Decker handled the finances. We also appreciate the additional support in the form of funding and personal time from the NED chapters and their membership.

Finally, we are grateful to the many symposium donors (listed elsewhere in this proceedings) for their generous financial support, particularly the AFS Chapters and Sections, the U.S. National Oceanic and Atmospheric Administration Office of International Affairs, the U.S. National Marine Fisheries Service (Office of Habitat Conservation, Habitat Protection Division; Office of Protected Resources; and Southeast Regional Office), the U.S. Fish and Wildlife Service, the U.S. Geological Survey (Biological Resources Discipline, Fisheries: Aquatic and Endangered Resources Program), Acadia University, and Fisheries and Oceans Canada, who all made major and multiple donations to bring the meeting and publication into reality. Our thanks also go out to others who assisted in the endeavor at various levels, including Steve Rideout, Helen Beal, Ken Beal, Jr., Duane Currie, Tom Bigford, Melanie Harris, Kathy Pietryka, Abby Franklin, Aaron Spares, Eddie Halfyard, Philip Simpson, and Cindy Breau. We appreciate the effort of Aaron Lerner of AFS Publications for guiding us through the publication process. Our most sincere thanks go to Debby Lehman of AFS Publications for coordinating and implementing the extensive copy editing that was essential for the production of this book.

ALEX HARO
S.O. Conte Anadromous Fish Research Laboratory
U.S. Geological Survey, Biological Resources Discipline
Turners Falls, Massachusetts, USA

Major Symposium Contributors

The steering committee recognizes the following organizations for their major contributions to and support for the symposium:

Symposium Sponsor

National Oceanic and Atmospheric Administration – National Marine Fisheries Service

Session Sponsor

National Marine Fisheries Service – Office of Habitat Conservation, Habitat Protection Division

Benefactors

U.S. Geological Survey Biological Resources Discipline–Fisheries: Aquatic and Endangered Resources Program
U.S. Fish and Wildlife Service – Headquarters
Massachusetts Division of Marine Fisheries
American Fisheries Society International Fisheries Section, Carl R. Sullivan International Endowment Fund

Donors

National Marine Fisheries Service – Office of Protected Resources, Endangered Species Division
National Marine Fisheries Service – Southeast Regional Office
U.S. Department of Agriculture
Natural Sciences and Engineering Research Council – Atlantic/CRSNG-Atlantique
Fisheries and Oceans Canada/Pêches et Océans Canada, Gulf Region/Région du Golfe
Hudson River Foundation
New York State Department of Environmental Conservation Bureau of Marine Resources and Hudson River Estuary Program
Maine Atlantic Salmon Commission
Acadia University

Attendees of Challenges for *Diadromous Fishes in a Dynamic Global Environment* at the Halifax Mariott Harbourfront Hotel, June 2007. Photo credit: Michael J. Miller.

The steering committee and organizers of *Challenges for Diadromous Fishes in a Dynamic Global Environment* and the Northeastern Division of the American Fisheries Society wish to thank the Halifax Marriott Harbourfront for providing an excellent venue, accommodations, and services for our symposium.

Mariott Harbourfront Hotel
1919 Upper Water Street
Halifax, Nova Scotia B3J 3J5, Canada

American Fisheries Society Symposium 69:1–22, 2009

Making the Best of Two Worlds: Diadromy in the Evolution, Ecology, and Conservation of Aquatic Organisms

Robert M. McDowall*
National Institute of Water and Atmospheric Research
Post Office Box 8602, Christchurch, New Zealand

Abstract.—Diadromy is a life history attribute in a small number of fish species, and also some decapod crustaceans and gastropod mollusks, involving regular migrations at defined life history stages between freshwaters and the sea. Despite the relatively few species known to be diadromous, these are often of high fisheries importance owing in part to their migratory movements being spatially and temporally concentrated, making them vulnerable to exploitation. These migrations have important implications for a diverse array of aspects of biology: evolutionary, genetic, biogeographical, ecological, community, and conservation. There are also implications for the acclimatization of species into novel habitats, and deriving from these implications, diadromous fishes introduced into natural ecosystems may have seriously adverse impacts on the receiving ecosystems. As well, diadromy provides some species with the capacity to facilitate restoration of populations caused by adverse impacts on aquatic ecosystems across a broad range of spatial and temporal scales, from the effects of volcanism and glaciation to those of polluting discharges with anthropogenic sources. Allied with these questions is that relating to the importance of habitat connectivity and the vulnerability of diadromous species to events that interfere with up- and downstream migrations in fluvial ecosystems.

Introduction

Diadromy is a behavior that involves migrations between freshwater and marine biomes; because of differences in salinity between these biomes, diadromous migrations have osmoregulatory implications, so that it is substantially about animals being able to cope with the demanding physiological challenges involved in doing so. It is probably the ability to osmoregulate that limits the diversity of species that are diadromous, though many euryhaline species are not diadromous. Moreover, it is sometimes surprising where occasional diadromous species come from, in terms of phylogenetic relationships—diadromy is often unpredictably present. In this paper, I explore the significance of diadromy in the ecology and evolution of aquatic organisms. I focus to some extent on familiar groups like salmonids, and anguillid eels, but will also draw into the discussion some of the less widely known groups like the sicydiine gobies and the southern galaxiids, in part because they are just as interesting and in part because that is where I have done a lot of work. I refer in my title to aquatic organisms because, in addition to the well-known fishes, there are decapod crustaceans and gastropod mollusks that are just as diadromous as any fishes.

* Corresponding author: r.mcdowall@niwa.co.nz

Coping with the Physiological Challenges of Biome Shift

How various diadromous organisms deal with the physiological implications of diadromy varies across a broad continuum. At one extreme, some diadromous species are highly euryhaline and can make these interbiome movements at will and across a wide range of developmental stages. At the other extreme are amphihaline species, which can make the movements only at carefully mediated life stages when a temporally, sometimes quite transient, life stage-regular ability to osmoregulate develops (Fontaine 1975). I think that this is an important distinction that is too little recognized.

Thus, in terms of definitions, diadromous species are those that undertake seasonally and life stage-regular migrations between marine and freshwater biomes (Myers 1949; McDowall 1988; Metcalfe et al. 2002); I think that the presence of spatiotemporal and life stage predictability is an important part of the definition. Separation of diadromous species from what might be called "facultative wanderers" is far from explicit and often unclear. How far and how completely a species has to cross the marine/freshwater interface to be strictly diadromous is a matter for interpretation: does a species really, and always, have to move between fully marine waters and fully freshwaters to be regarded as diadromous? Whatever definitions we use, there are lots of species that are marginally diadromous, and there is another informal continuum.

The Occurrence of Diadromy across the Diversity of Fishes

Diadromy is best known in fishes, among which there are probably around 250 diadromous species (McDowall 1987, 1988), and they are thus a very small minority (<1%) of the c. 28,000 fish species known (Nelson 2006). It is not my intention to repeat, here, an account of their geographical and taxonomic diversity (McDowall 1987, 1988). In addition to these fishes, there is growing evidence that there are significant numbers of diadromous decapod crustaceans (families Atyidae and Palaemonidae: Fievet and Eppe 2002; March et al. 2002; Greathouse et al. 2005) and diadromous gastropod mollusks (family Neritidae: Ford and Kinzie 1982; Myers et al. 2000; Blanco and Scatena 2005) that make the same sorts of interbiome migrations as fishes do and, in every respect, fit easily into the diadromous category. I have little conceptual difficulty imagining small shrimps marching upstream from the sea into freshwaters, though imagining tiny, newly settled gastropods bravely doing so rather more stretches the mind—but they do it. At present, the numbers of crustaceans and mollusks involved in diadromous migrations have yet to be estimated, but there are perhaps 30–40 species. So, diadromy is a minority behavior, but not all that rare.

Knowledge of Diadromy

Diadromy involves some of the most valued fishes known. Diadromous fishes tend to assemble at high concentrations at predictable seasons of the year, and they typically move through habitat narrows (like river mouths), making them visually obvious and easily captured. Their great abundance, combined often with high palatability, generates high economic value, making them important for harvest. Their cultural popularity spans fishermen from artisanal/traditional to recreational and commercial, and they include such iconic human gastronomic delicacies as salmons, sturgeons, lampreys, eels, and shads. But it is not only these well-known and widely known species. Diadromy also implicates such highly valued fish as the ayu *Plecoglossus altivelis* of Japan; the tismiche (also known as lyre goby) *Evorthodus lyricus* of Mesoamerica and the Caribbean, the ipon *Sicyopterus lachrymosus* fishery of the Philippines, the bichique *Sicyopterus lagocephalus* of French tropical islands, all based on gobies; and the galaxiid whitebait of New Zealand and Chile (McDowall 1984). To place this in some perspective, you can at times expect to pay NZ$150/kg for whitebait in New Zealand, and it is so expensive that it is typically sold per 100 g. It is my understanding that the bichique is equally valued in French Polynesia and the Indian Ocean. So, these are really important fish, both culturally and economically, quite aside from their biodiversity interest.

Diadromy among a variety of these organisms gets very intensive attention by biologists, certainly out of all proportion to their number, and partly this is because of the high inherent interest in the phenomenon—migratory animals always attract a lot of interest, as mass migrations can be spectacular events. Dedication of this American Fisheries Society conference to just these groups illustrates something of that level of importance and interest. Equally, the International Council for the Exploration of the Sea has a specialist committee that addresses biological and fisheries aspects of diadromous fishes, and furthermore, there has been a series of conferences on the subject in recent years, including one in Bordeaux, France and another in Hilo, Hawai'i, both in 2005 (McDowall 2007b, 2008).

Also, the wide separation of disparate habitats that diadromous species occupy, to complete their life cycles, means that they need high habitat connectivity, an issue creating powerful conflicts between those, on the one hand, interested in exploitation, management, and conservation of these animals, and, on the other hand, those interested in alternative uses of river flows in which these animals live. These various features, fisheries importance, spatiotemporal concentration, and the need for connectivity, together make diadromous animals vulnerable to population decline, so that there are serious conservation concerns about many diadromous fishes (especially sturgeons [Birstein 1993], salmonids [Behnke 2002], and various other groups [McDowall 2006]), a rather higher proportion than conservation concern for fishes, generally.

Facultativeness in Diadromy

Biome shifts are mandatory for survival in many diadromous species, but in others, these shifts are variably facultative, and here is another continuum. This sort of variability can be seen among quite closely related taxa, as in the salmonid genus *Oncorhynchus*, in which, at one extreme, the chum salmon *O. keta* seems obligatorily diadromous, but at the other extreme, steelhead *O. mykiss* is greatly facultative (see Rounsefell 1958; Behnke 2002; Quinn and Myers 2004). There are similar examples among some of the southern families, as in galaxiids and retropinnids (McDowall 1988, 1990).

Although there is a surprising amount of facultativeness among fishes, concerning whether or not they migrate, in some taxa individual survival depends on each fish making the appropriate biome shift at the appropriate life stage/season (i.e., the biome shift is a physiological imperative: no migration, no survival). Thus,

- Diadromy is obligatory in some species, but there are often nondiadromous congenerics;
- Sometimes there are distinct diadromous and nondiadromous populations of a species;
- Or there are diadromous and nondiadromous individuals within populations of a species; and
- There are even sometimes diadromous and nondiadromous stages/phases within the lives of individual fish.

Facultativeness in diadromy has in some ways become most obvious in, and has had interesting implications for, the outcomes of projects involved with translocation of diadromous species. New Zealand has imported anadromous stocks of rainbow trout *O. mykiss*, Atlantic salmon *Salmo salar*, sockeye salmon *O. nerka*, and Chinook salmon *O. tshawytscha* (McDowall 1990, 1994a, 1994b). Despite the introduced stocks having a diadromous ancestry in all instances, only nondiadromous stocks have developed from the introductions of steelhead, Atlantic salmon, and sockeye salmon. However, with Chinook salmon, diadromous populations did become established, though there are also naturally landlocked populations despite this species not normally becoming landlocked in its natural range. With brown trout *S. trutta*, the opposite happened. Early imports were regarded as nonmigratory, but there were clearly some individuals with anadromous tendencies and diadromy developed rapidly, and the same is even clearer in the Falkland Islands (McDowall 1990, 1994a, 1994b; McDowall et al. 2001). Migratory behaviors may even vary ontogenetically in brown trout (i.e., individual fish may be diadromous in some years, going to sea to feed and mature, but in other years eschewing migration and remaining in freshwater throughout the year). Despite these difficulties, there is wide acceptance that the term diadromy is useful for focusing discussion on some types of migratory behaviors.

Different Sorts of Diadromy

Diadromy comes in three distinct types, as defined by Myers (1949): anadromy, catadromy, and amphidromy (Figure 1). I do not want to spend much time with the definitions of these terms, though some understanding is needed. Anadromy and amphidromy are often confused or combined, and yet, in many ways, amphidromy is more similar to catadromy than anadromy. Reproduction and most somatic growth take place in different biomes in both anadromy (freshwater and the sea, respectively) and catadromy (the sea and freshwater), and the return migration to the spawning biome, in both, is undertaken by mature, adult fish. They do little or no feeding and growing after initiating the return migration, which can therefore be described as reproductive. In contrast, most somatic growth and reproduction are in the same biome in amphidromous fishes (freshwater). The return migration to the spawning biome is undertaken by small juveniles, often only a month or two old. They return

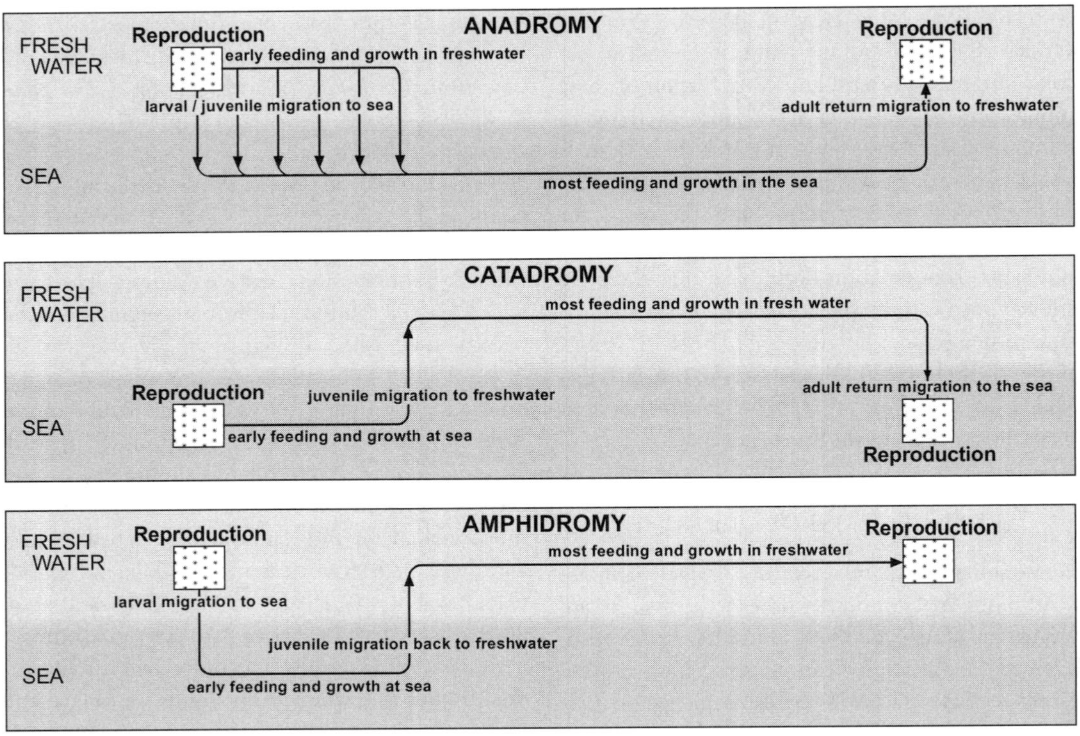

Figure 1.—The different types of diadromy.

to freshwater to feed and grow for many months, sometimes years, making their return migration a trophic one. They mature and spawn in freshwater (usually with no further migration). If you are having conceptual difficulty with this, contrast large, mature, adult anadromous fishes and very small, immature, amphidromous ones returning to freshwater from the sea. Strategically, they are very different life cycles.

So, in the end, definitions of these categories of diadromy depend substantially on where we place the emphasis: whether

- On where feeding and growth take place, or
- On where reproduction takes place, and
- Where in the life cycles of each species the interbiome migrations take place.

As with so many biological definitions, the separation of these phenomena is not always as explicit as their definitions might suggest (see McDowall 2007a). In some highly complex cases, there are instances where allocation to various of the categories differs between individuals of a population. Thus, there are populations of the southern smelt (also known as cumberfish) *Retropinna retropinna* that make an early return migration to freshwater as juveniles, making them amphidromous, but in other individuals, migration may be as mature adults, making them anadromous (McDowall 1990). But the several subcategories are useful and heuristic—they tell us something about the behavior of the various species described as diadromous, and they group species together that have analogous, sometimes homologous (commonly derived), behavioral attributes.

For those who insist that biological categories should be explicit, comprehensive, and exclusive, diadromy and its subcategories are likely to cause frustration. There are some conceptual and definitional difficulties in defining diadromy and the associated terms, just as is widely true in biology. But these difficulties are not insuperable: we cope with similar problems when we talk about animals that are carnivores, omnivores, and herbivores or are diurnal, nocturnal, and crepuscular or are benthic and pelagic. We need to ensure that these terms are our servants rather than our masters.

The Taxonomic Distribution of Diadromy

Diadromy occurs very widely across the diversity of fishes, in lots of disparate families, in occasional genera within these families, and sometimes in lots of species within genera. The numbers of species that are variously anadromous, catadromous, and amphidromous are not greatly different; all diadromous gastropods are amphidromous as are most diadromous decapod crustaceans. There is at least one catadromous decapod (the Asiatic mitten crab *Eriocheir japonica*: Kobayashi 1999) that may be found as much as 1,200 km upstream from the sea coast, but it seemingly returns to the sea for reproduction (and that is a lot of walking for a crab).

As far as I am aware, no new anadromous or catadromous species have been described in the past 20 years, though we have clarified the life history strategies of some long-recognized species. However, as Myers (1949) foresaw, the number of amphidromous species has risen substantially as a result of both the discovery of new species and the clarification of life histories of long-known species. In particular, the number of amphidromous sicydiine gobies has risen substantially (knowledge of these fishes is perhaps most accessible in a series of small books on the fish faunas of French-administered island territories across the tropics and subtropics of the Indo-West Pacific: Keith et al. 1999, 2002, 2006; Marquet et al. 2003).

Some Unexpected Diadromous Species

Sometimes regional freshwater fish faunas include occasional and often totally unpredictable species, typically belonging to primarily marine families that have become diadromous locally, like the Atlantic tomcod *Microgadus tomcod* (Scott and Crossman 1975), Tasmania's bovichtyid congolli *Pseudaphritis urvillii* (Andrews 1996), New Zealand's pinguipedid torrentfish *Cheimarrichthys fosteri* (McDowall 1990, 2000), and others (McDowall 1988).

The Implications of Diadromy for Evolution and Ecology

Diadromous migrations clearly have a whole series of implications for the evolution, biogeography, ecology, management, and conservation of these organisms, and these implications are evident at a very wide range of spatial and temporal scales.

A Role in Biogeographic Dispersal

At the broadest spatial scale, diadromy is believed to be closely associated with biogeographic dispersal. Some diadromous species have very broad distributions, especially when compared with other freshwater or terrestrial organisms. These broad distributions are clearly closely related to the existence of marine life stages that facilitate dispersal (McDowall 1988, 2002). Some diadromous species are among the most widespread of all aquatic organisms, such as the Indo-west Pacific amphidromous sicydiine goby bichique *Sicyopterus lagocephalus*, which is known across a range of c. 18,000 km, from the Marquesas Islands in eastern French Polynesia to the Seychelles and Comoros in the western Indian Ocean (Berrebi et al. 2005; Hoareau et al. 2007; note, however, that there is some ongoing and unresolved debate about the issue of whether one species is involved in populations across this vast range, or perhaps several—see Keith et al. 2005; Sparks and Smith 2006; Kottelat 2007; Smith and Sparks 2007).

Of similar magnitude are the ranges of two southern taxa, the anadromous southern pouched lamprey *Geotria australis* and the marginally catadromous inanga (also known as puye) *Galaxias maculatus* (McDowall 2002). Both are found in western and southeastern Australia, Tasmania, Lord Howe Island (inanga only), New Zealand, the Chatham Islands, Patagonian South America, and the Falkland Islands (though the Falklands record for *Geotria* is based on one observation only and is questionable—McDowall et al. 2005). Some of these huge distributions have prompted questions about whether all of the stocks do really represent just one, rather than several, species. Molecular studies are consistent with evidence from morphology that supports their species recognition across such wide ranges (McDowall 1990; Waters and Burridge 1999; Waters et al. 2000b; Gill et al. 2003; Berrebi et al. 2005; Keith et al. 2005). The presence of diadromy in these species is a key to their very wide ranges. These are not new ideas. I have been writing about them for more than 40 years (McDowall 1964, 2002), sometimes hotly criticized for doing so (Croizat et al. 1974; Rosen 1974; Craw 1979) but have been in good company—Darwin was writing about dispersal in inanga in 1872, but he knew nothing per se about diadromy.

Proving how these very broad geographical ranges developed is less easy and has to be inferential (but molecular evidence can be highly informative); nevertheless, it is parsimonious to think that these ranges are driven by the dispersal of long-lasting marine life stages. We should not be too seduced by these broad distributions, and recall that the nondiadromous northern pike *Esox lucius* is nearly pan-temperate in northern lands (Scott and Crossman 1975). So there are some distinctive, though not necessarily unique, broad-scale, distributions of nondiadromous species in the north.

When we look at southern lands, we see a similar pattern of very wide ranges in some diadromous taxa (McDowall 2002, 2006), but there are no comparably widespread nondiadromous southern taxa that could be seen as counterparts of the northern pike. This is probably because across the temperate/boreal northern hemisphere, there is now near-continuous land (and once it was continuous) so that the spread of species restricted to freshwaters has been relatively easy, including across what is now the Bering Strait. There is a deep contrast in southern latitudes, where there is nearly continuous sea, fragmented by quite large, but narrower, and very widely separated, lands that break the oceanic continuity—just the opposite of what is true of boreal lands, so we should expect northern and southern patterns to be different, and they are. These details aside, what is quite clear, is that diadromy facilitates the establishment of very wide geographical ranges of fish species in many parts of the globe, and I suggest that we are looking here at a causal relationship.

Regional Patterns of Occurrence

The presence of diadromous species at quite broad if more regional spatial/geographic scales is well exhibited in high latitudes, where some diadromous species are near circum-Arctic, such as Arctic lamprey *Lampetra japonica*; Arctic char *Salvelinus alpinus*, and rainbow smelt *Osmerus mordax*. There are also widespread northern diadromous taxa across the North Pacific, with most of the species of *Oncorhynchus* present in western catchments of North America and eastern catchments of northeastern Asia, as is also the osmerid pond smelt *Hypomesus olidus*. Similarly, the Atlantic salmon is widely present on both sides of the North Atlantic (Scott and Crossman 1975). Curiously, the equally diadromous brown trout is naturally absent from the western North Atlantic (though it was long ago introduced to catchments widely across North America).

Broad ranges in diadromous species are also seen in the tropics, as in the sicydiine Pacific river goby *Awaous guamensis*, which is widespread in the central/western tropical Pacific (Guam, Vanuatu, New Caledonia, Fiji, and Hawai'i: McDowall 2003), or at southern subtropical to cool-temperate latitudes, as in the southwestern Pacific shortfin eel *Anguilla australis* found in southeastern Australia, New Caledonia, Fiji, New Zealand, and perhaps other western Pacific islands or the koaro *Galaxias brevipinnis*, present in southeastern Australia, Tasmania, New Zealand, and the Chatham, Auckland, and Campbell Islands (McDowall 1990, 2002, 2004b).

Some Highly Localized Diadromous Species

A contrast with the very wide geographical ranges just discussed are species with much more local ranges, sometimes closely related to widespread diadromous congenerics, such as the western North Atlantic short-nosed sturgeon *Acipenser brevirostrum*, the Japanese masou salmon *O. masou*, the eastern North American brook char *Salvelinus fontinalis*, the eastern Pacific eulachon *Thaleichthys pacificus*, or the temperate East Asian ayu. There are distinct endemic species of amphidromous *Sicyopterus* in various individual islands across (within) the broad tropical Indo-Pacific range of bichique (Keith et al. 1999, 2002, 2006; Keith 2003; McDowall 2003), or, similarly, there are localized endemics in southern cool temperate latitudes, such as the Australian whitebait (also known as Tasmanian whitebait) *Lovettia sealii*, present in Tasmania, or New Zealand's Stokell's smelt *Stokellia anisodon*, which is found in only a few rivers along the east coast of New Zealand's South Island (McDowall 1990, 2006). All of these have relatively narrow ranges. Thus, whereas at times diadromy is an effective dispersal mechanism and enables species to become widespread, this is not true in all such species or in all places, and some species have much more local ranges. What causes these major differences in range is unknown, but it may relate to duration of the marine life stage or to local oceanographic processes and perhaps both.

Historical Dispersal Processes

Also, at a regional scale, there has been some interesting extension of the ranges of diadromous species in historic times, the best example being the arrival of the Australian/New Caledonian longfin eel (also known as speckled longfin eel) *Anguilla reinhardtii* in New Zealand, possibly for the first time in the 1970s (McDowall et al. 1998). It seemingly just began reaching New Zealand, perhaps associated with changes in oceanic current systems in the southwestern Pacific to the north of New Zealand. Note, however, that Phillipps (1925) did report this species from New Zealand in the 1920s. His identification is uncertain as there are no voucher specimens, but if Phillipps was right, perhaps we are looking at episodic arrival of this species in New Zealand. This idea is perhaps not quite as far-fetched as it might seem, as there are suggestions that the arrival of cohorts of other New Zealand anguillid species may also be episodic, though on a much smaller, decadal or narrower, time scale than would be true of Australian/New Caledonian longfin eel. The Tasmanian aplochitonine, Australian whitebait, has recently been recorded from the Australian mainland (T. Raadik, Department of Conservation and Heritage, Victoria, Australia, personal communication), but it is unclear whether this results from a recent dispersal event or is due just to more intensive study of the area's fish fauna.

The potential for dispersal of diadromous fishes is nicely exhibited following the successful translocation of American shad *Alosa sapidissima* to the Pacific coast of North America, after which it spread naturally to the western Pacific and now occurs along the coast of Siberia (Whitehead 1985). What this shad did is likely much the same as happened long ago in diadromous fishes now naturally present on either side of the boreal Pacific Ocean, spreading in one direction or the other, and it is not easy to determine in which direction these dispersals took place.

Diadromy and the Fish Faunas of Islands

Entire, or nearly entire, freshwater fish faunas are diadromous (McDowall 1988) in many islands of the globe, from Iceland and Greenland in the Arctic to Hawai'i, Guam, the Comoros, and Mauritius, across the tropics (McDowall 1988). The fish faunas of Arctic islands are largely salmonids. In the islands of the Indo-West Pacific and Caribbean tropics and subtropics, the fish faunas tend to be largely or entirely composed of diadromous species (McDowall 2003, 2004a, 2007a, 2007b), very often sicydiine gobies, often catadromous anguillid eels, sometimes catadromous gray (mugilid) mullets, and maybe kuhliids whose life histories are poorly known. The lack of anadromy on these islands is notable, and the presence of both amphidromy and catadromy probably reflects similarities in the way these fish use freshwater fluvial habitats for rearing and growth to maturity. And, as outlined above, often there are atyid or palaemonid shrimps and neritid gastropods as well—the faunas of remote island streams are often dominated by diadromous species.

Consistent with these island patterns, the fish faunas of the remote islands of the New Zealand region are also dominated by diadromous species. At the Chatham Islands, c. 800 km east of New Zealand, the freshwater fish fauna predominantly involves diadromous species present on mainland New Zealand from five families (McDowall 1990, 2002). Diadromy has clearly favored dispersal to these islands. Some geologists are suggesting that the Chathams disappeared completely beneath the sea during the mid-Cenozoic (Campbell 1998; Trewick et al. 2007), and if they are right, then everything that now lives there reached the islands by dispersal across the sea, including freshwater fishes. It is therefore no surprise that the Chatham Island's freshwater fish fauna is dominated by diadromous species. The only species present at the Auckland and Campbell Islands, far to the south of New Zealand, is the amphidromous koaro (McDowall 1970, 2004b), so the pattern there is the same. All of Hawai'i's five freshwater fish species are amphidromous gobioids (Devick et al. 1992; Chubb et al. 1998; McDowall 2003), and it is likely that these diadromous species have been able to reach remote Hawai'i's streams because they can disperse through the sea. Note that the five species belong to five different genera that are otherwise widespread on tropical and subtropical islands across the Indo-West Pacific, and so freshwater fishes in Hawai'i represent five distinct dispersal events to Hawai'i from elsewhere.

Patterns at the Local Scale

At the local scale, there is, within New Zealand at least, a deep dichotomy in geographical ranges between widespread, typically New Zealand-wide but often lower-elevation ranges in most of the diadromous species, whereas the ranges of nondiadromous species are narrower, are much more localized, and tend to be at higher elevations—there is a clear and simple dichotomy in the geographical distributions of diadromous and nondiadromous species. Similar patterns would probably be revealed in other lands, were there adequate data on distribution.

Diadromy Phylogenetic Relationships and Speciation

One significant evolutionary and ecological element that relates to facultativeness in diadromy is a related role in incipient speciation. As we look across the diversity of fishes there are many examples of nondiadromous (nonmigratory) stocks of otherwise diadromous species, and perhaps often deriving from this there are nondiadromous species that are regarded as being derived from a diadromous ancestry.

We often think of them as landlocked, though there are often no physical barriers preventing movements between freshwater and the sea.

The establishment of nondiadromous stocks derived from a diadromous parentage is common, for instance in anadromous sockeye salmon populations and numerous, independently derived stocks of kokanee (lacustrine sockeye salmon; Foote et al. 1994); the anadromous steelhead and nonanadromous populations of rainbow trout and redband trout in western North America (Behnke 2002); Scandinavian stocks of anadromous *Coregonus* species and landlocked derivatives (Svardson 1998); and lake populations derived from anadromous threespine sticklebacks *Gasterosteus aculeatus* (McPhail 1993). There are also similar problems in some more obscure taxa, such as a seeming flock of morphs/species in the Asian goby genus *Rhinogobius* (Iguchi and Mizuno 1990) and in Australian lineages of *Retropinna* (Hammer et al. 2007).

In various ways, these lineages are genetically distinct and they often have geographically discrete distributions, but their morphological diagnosis as species, and their identification either in the field or even in the laboratory is proving to be a serious headache, at the moment basically unresolvable. These complexities are an outcome of the capacity for diadromous species to spread widely through the sea and to then invade freshwater habitats and to independently establish localized, multimorph isolates in lakes and rivers, sometimes in sympatry. In the end, it seems to me more important that we study and understand the processes than that we are able to place the stocks in well-defined and clearly separated Linnaean species, though this makes it hard for identifications by field biologists and for all sorts of other issues: ecological, conservation, statutory, and management.

Herein lies a potential pitfall. There is a not inconsiderable number of examples in the literature where nondiadromous (landlocked) species have been described that appear to be derived from diadromous species; these so-called new species are based on real morphological differences. However, in some instances, these have proved to be no more than lacustrine stocks of otherwise diadromous species (McDowall 1970, 1972, 1979), and this probably has happened more often than we realize. The morphological divergence of derived, landlocked stocks of diadromous fishes is a serious trap for the unwary who can easily interpret this divergence as an indicator of speciation when it is merely ecophenotypic and may often happen in parallel and independently in separately derived landlocked populations. One such situation was exposed recently (Ilves and Taylor 2007) in the osmerid genus *Hypomesus*, where a lacustrine species had been described from several well-separated Sakhalin lakes (Saruwatari et al. 1997). The lake populations are independently derived, landlocked ecophenotypes of the anadromous Japanese smelt *H. nipponensis* in Sakhalin seas. And there are other, similar instances involving a range of diadromous species.

The Phylogenetic Significance of Loss of Diadromy

Some groups of diadromous species are so ecologically/behaviorally flexible that the loss of diadromy is commonplace, and this shows that such loss has little inherent value as an indicator of phylogenetic relationships. I could talk at length about the amphidromous koaro, in which there is a very large number of landlocked stocks in lakes widely across

New Zealand, mainland Australia, and Tasmania, all independently derived from diadromous populations (McDowall 1970, 1988, 2006). In these lake stocks, vertebral counts may have plummeted by as many as five to six vertebrae (McDowall 2004b), and there are other morphological differences. These, too, were once regarded as comprising a series of distinct lacustrine species (McDowall 1970), but it is clear that they are independently derived lacustrine stocks of a diadromous parental stock and the differences are merely ecophenotypic. This conclusion is supported by recent molecular studies (Waters and Wallis 2001). Similar circumstances apply in the southern smelt (genus *Retropinna*; McDowall 1979).

In addition, however, New Zealand has a flock of perhaps 10 nondiadromous, fluvial species derived from the same diadromous koaro (McDowall and Wallis 1996; Waters and Wallis 2001). At present, it is uncertain how many times these fluvial species have been derived from a diadromous ancestry. Clearly, much the same has happened in the osmerid smelts and salmonids (Svardson 1998; Behnke 2002).

Again, all of this leads to some really serious taxonomic difficulties, in which there are sometimes two or more sympatric, lacustrine or fluvial stocks of otherwise diadromous species and in which the complexity is really defying rational taxonomy. Much of the emerging clarity of these situations derives from the application of molecular methods for determining taxonomic identities and phylogenetic relationships among ostensibly freshwater fishes (McDowall and Wallis 1996; Waters and Wallis 2001). All this is providing new perspectives on the evolution and biogeography of diadromous species, particularly in determining patterns of speciation and phylogenetic relationships.

The frequency and apparent ease with which diadromy is lost casts serious doubt on its value for phylogenetic analysis. It cannot seriously be argued that loss of diadromy has equal probability to gain of diadromy in a phylogenetic analysis.

Diadromy and Genetic Structuring

Associated with the presence of diadromy is a quite frequent lack of genetic structuring, often across the broad ranges of diadromous species, which is a distinct contrast with the presence of substantial structuring in the often much narrower ranges of related nondiadromous derivative species (Allibone and Wallis 1993; Ward et al. 1994; Jerry and Baverstock 1998; DeWoody and Avise 2000). This is of particular interest in Hawai'i, where the same five species are present on all of the islands where any freshwater fish occur (Zink et al. 1996; Chubb et al. 1998). One of these five species is found widely across the western tropical Pacific, and we can imagine that it might be a relatively recent arrival in Hawai'i. Or it is possible that there is continuing dispersal of this species among Hawai'i and other islands. The other species that are local endemics have perhaps been there longer. Larval life in the sea is clearly promoting gene flow, so that there are no single-island endemics or even local genetic structuring among the freshwater fishes across these islands (Zink et al. 1996; Chubb et al. 1998), a distinct contrast to the common development of local island endemics in Hawai'i's terrestrial biota (Wagner and Funk 1995).

The French are exploring the distributions of some amphidromous sicydiine goby species, whose broad ranges are clearly facilitated by early larval life spent at sea. Molecular studies are showing that there is little or no divergence among populations of bichique across its huge range of 18,000 km, from French Polynesia in the east to the Comoro Islands in the western Indian Ocean (Berrebi et al. 2005). They are arguing for continuing, contemporary, gene flow and are even exploring the prospect that populations in the western Indian Ocean are being derived, year by year, at the cohort scale, from populations far to the east across the Indian Ocean, in Indonesia (Hoareau et al. 2007).

Viewed broadly across the diversity of diadromous species, there is what seems to be a paradoxical negative correlation between the extent of species' ranges and the amount of genetic divergence across those ranges—a positive correlation would seem more likely, with deviation related to how far apart the populations are. Were it not for our understanding of diadromy, this pattern would be counterintuitive. The reason, of course, is simply that the same dispersal through the sea that generates the broad ranges of these diadromous species also fosters continuing gene flow that hinders local genetic divergence and speciation. It is likely that these hugely broad patterns by so-called freshwater fishes would not be believed were it not for clarity brought by molecular studies—though some biogeographers remain unbelieving.

These processes have interesting interactions with the question of homing and the existence of separate stocks within diadromous species across their geographical ranges. The very intensively studied western North American species of *Oncorhynchus* are an interesting exception to the general rule that there tends to be little genetic structuring across the geographic ranges of diadromous species (and there may be others, as in alosid shads: Waters et al. 2000a). Salmonid biologists are arguing strongly for the formation of independent salmon stocks peculiar to individual river systems and even to individual subcatchments in such big rivers as the Fraser and Columbia. Here we are seeing the interesting interactions between widespread ranges driven by diadromy, but the development of local stocks all the time being protected by homing—an interesting ecological, behavioral, and evolutionary tension (McDowall 2001).

A Role for Diadromy in Fish Community Assembly

If we extend the processes discussed above, in which there is invasion of streams from the sea by diadromous species, we can discern a role for diadromy in community assembly, in fish faunas dominated by diadromous species, with a series of consequential implications in the structure of fluvial populations and communities. It is becoming increasingly clear that community assembly and the structuring of riverine fish faunas are in part driven by the invasion processes involved with diadromy. This has long been studied in some salmonids and alosid shads, perhaps to the point that the connection to diadromy is all but lost, but it is a key element in the community and population ecology of diadromous fishes all over the world. Invasion processes play a key role in determining species composition, and the age and size structure of fish populations in which diadromy is an important attribute (McDowall 1993, 1996a; Joy and Death 2004; Leathwick et al. 2005). There are largely untouched questions of how invading individuals of diadromous species establish themselves in existing fluvial stream communities, especially the small juveniles of amphidromous and catadromous species, whether already populated by other diadromous species by nondiadromous ones.

In fluvial populations where diadromous species play a prominent role, there may be a shift from a predominance of diadromous species at low elevations and close to the sea towards a higher proportion of nondiadromous species at higher elevations and farther upstream. Why this is so is unclear, but it could be due to a combination of several processes. Fairly obviously, one of these processes is upstream penetration and how it is influenced by several subprocesses working together:

- Just the swimming effort in getting upstream, and the time it takes;
- Changing channel gradients and parallel habitat change—substrate characteristics, flow velocities, water temperatures;
- Perhaps channel size;
- Possibly effects of riparian features like forest cover and composition;
- These may interact with species' properties like habitat preferences, and swimming instincts and abilities; as well as
- The role of the fish fauna already living in the streams, whether diadromous or not.

In addition, there may be influences from an upstream changing biota, perhaps changing competitive environments. Nondiadromous species may fail to hold their place in stream communities where diadromous species are invading. All of these influences are likely to be highly local, rather than regional, or general in character and need to be addressed at a hierarchy of microhabitat, habitat reach, and catchment-wide scales.

Deriving from these, there are likely to be altitudinal differences in the species composition, individual species' abundance, and age/size structure of the populations that relate to the invasion process. Thus, there may be a tendency for fish communities composed mainly of diadromous species to be more taxonomically diverse at lower elevations, with diversity declining with distance upstream and elevation. There will also often be downstream/upstream differences in age- and size-classes, with these varying according to whether each species is, on the one hand, anadromous or, on the other hand, catadromous or amphidromous, these falling together because in both, the invasion process involves emigration from the sea of small juveniles. Climbing ability may favor the latter, though they may be at a competitive advantage owing to their small size.

Moreover, perhaps their slow upstream penetration is adaptive as it gives them time to grow.

In populations of anadromous species, and particularly in the fish communities that are dominated by anadromous species, as is true of catchments of temperate to boreal North America and the northern Pacific coasts of eastern Asia, there are two distinct, seasonally variable gradients. One of these involves a downstream–upstream gradient in which invading adult anadromous fish are less sexually mature in downstream reaches, with increasing sexual maturity with distance upstream, the fully ripe adults occupying the uppermost reaches of the species ranges where they will spawn and often die. What follows for those postspawners that may not die in iteroparous species, is so variable as to be difficult to summarize. There may be postspawning adults anywhere from the upstream spawning sites down to the river mouth—or even at sea, since sometimes the spent adults return to sea to feed and regain condition before repeating spawning.

There is a second age/size gradient that involves new cohorts of anadromous populations, with the youngest, smallest individuals in the populations typically being at the uppermost limits of the species' range, close to the spawning habitats. A progressive downstream gradient may include some very recently hatched juveniles that emigrate rapidly towards the sea, as well as often some rather older/larger juveniles that may be feeding and growing as they move downstream either continuously or in pulsed movements. In some anadromous species, of course, the newly hatched individuals move straight to sea so that the upstream/downstream, age/size juvenile gradient will be generated only across a time scale of hours or days. And so, there is a complete continuum from newly hatched larvae, as in some populations of some Pacific salmons, osmerid smelts, or alosid shads, which migrate very rapidly and undertake no or very little feeding before migrating to sea, to Atlantic salmon in which the age structure of the catchments' population of juveniles, heading downstream on their way to sea, may include individuals up to several years old. So here again, we are dealing with broad variation in detail across a wide taxonomic variety of anadromous species.

When we examine similar issues in catadromous and amphidromous species, we encounter a very different pattern. In both groups, the individuals invading freshwater are small, varying in size from just 15–20 mm to larger ones that may be 60–70 mm long, but in all instances, they are juveniles. In these species, there tends to be a downstream–upstream gradient in both fish size and age. The typical pattern is for juveniles to migrate into freshwater, often in large shoals. They tend to settle in the lower reaches of rivers, and they may undertake a less explicitly coordinated upstream migration that takes weeks, months, or even years. So in the lower reaches, there may be a wide size range of individuals from very small juveniles to adults, and often there is an upstream gradient in which the age structure becomes narrower, with fewer younger, smaller fish present with increasing distance up stream. The slope of this gradient also varies, depending, in any river system, on the individualistic behavioral attributes of each species and the nature of the catchment. In some species with weak penetration, the entire size and age structure of the populations may be found in just the first few kilometers up from the tidal zone of the river. But here is another continuum across the diversity of catadromous and amphidromous species in the extent of their inland penetration, and this is paralleled by a change in the age and size structure of the populations. At least in the New Zealand context, the New Zealand longfin eel *Anguilla dieffenbachii* may penetrate as much as 350 km, and the more upstream populations will consist of mostly larger adults, and in this instance, there is, as well, a change in sex ratio, with the more upstream individuals being largely or even entirely females. A similar change in age and size structure applies to the amphidromous koaro, a smallish galaxiid (to 280 mm), but one that is a very powerful upstream mover, known to penetrate as much as 400 km upstream, and there again, the upstream populations will consist of larger, older, adult fish, but of both sexes.

Looking at riverine fish populations at the community scale, there are also likely to be gradients in species richness, with a tendency for richness to be greater at lower elevations. If we turn to the amphidromous pinguipedid torrentfish *Cheimarrichthys fosteri*, the same increasing age/size distribution will apply across a distance of more than 200 km, but for reasons that, at the moment, elude explanation, the females are largely upstream and the males downstream, with a broad zone of overlap in the midreaches (McDowall 1990). What movements happen for the sexes to get together, at spawning time, is undiscovered.

We need to avoid getting overwhelmed by the details and their explanations. What it is really important to recognize is that these species exhibit downstream/upstream gradients that involve a broad range of community/population metrics involving species composition, species richness, age and size structures, and sex ratios. Moreover, these gradients are likely to be highly idiosyncratic in each catchment, varying with species composition and the slope and distance inland/habitat characteristics of each river system. The processes generating these patterns are likely to be broadly similar across a wide geographical front, and the generalized patterns are likely to be similar, but the details will be highly local in character and probably need to be addressed on a catchment-by-catchment basis.

With these come some problems, especially relating to riverine connectivity and natural impediments to upstream migration such as torrents and falls. These have severely negative impacts on fish communities and populations. Species may be excluded from upstream reaches of river systems, despite proximate habitat conditions being suitable, because they cannot penetrate upstream, especially in catadromous and amphidromous species that invade rivers from the sea as small juveniles.

Climbing Ability in Diadromous Fishes

There is, at least in some diadromous taxa, a sometimes quite astonishing ability to penetrate upstream past barriers. These abilities are well recognized in the leaping ability of salmons at falls, and one can see photos or film footage of fish struggling upstream through torrential flows in a manner that defies the imagination. Many diadromous species, especially those that are catadromous or amphidromous, exhibit an equally astonishing ability to climb—as is well known for elvers of anguillid eels. The sicydiines, which are becoming known as "sucker gobies," have their pelvic fins fused to form a cup that the fish use to help them climb falls (Schoenfuss and Blob 2003)—and they reach astonishing heights. The Hawai'ian *Lentipes concolor* is clearly the champion: it has been found upstream of falls of a cumulative height of 600 m, including one vertical drop of 300 m (Englund and Filbert 1997). Some of these gobies actually live on the surfaces of torrents. Some galaxiids that are similarly impressive climbers have distinctive rugosities on the ventral surfaces of their ventral-facing pectoral and pelvic fins that may well function in a similar sort of way. Anguillid eels climb as juveniles, based simply on small size, anguilliform swimming movements, and surface tension on wetted surfaces.

These fish get out of the main flows of the falls, and climb up the spray-wetted margins. This climbing has substantial energetics implications. The sicydiine gobies are herbivores, and we can imagine them spending as long as they need on the faces of some of these falls, perhaps because they can feed on periphyton on the way up—and so herbivory may be a useful preadaptation for successful climbing. Juvenile galaxiids and anguillid eels are pure carnivores (McDowall 1990), so that feeding on the way up falls is probably less feasible and energetics issues become more serious. So, here we see some of the processes of diadromous fish invasion that are key elements of the distributions and community ecologies of stream ecosystems.

In a discussion on the ecology of anadromous species in the rivers of the Atlantic coast of North America, Saunders et al. (2006) have suggested that the various species there may have coevolved, that the cycles of abundance in some species may be dependent to some extent on the cycles in others. The idea is that Atlantic salmon populations may depend on the abundance of various shads *Alosa* spp. This is a novel and interesting idea that warrants further exploration. It might be very difficult to prove but is certainly worth thinking about and may have important implications for restoring salmon populations.

Diadromy, Habitat Perturbation, Restoration Ecology, and Conservation Issues

Quite closely associated with the issues discussed above, diadromy has a key and distinctive role in conservation, especially in restoration ecology and recovery from perturbation, which may occur at a very broad range of temporal scales that might be described as geological to historical to contemporary.

The broad-scale role may be seen in retreat of glacial ice sheets or massive regional historic volcanism that have caused broad-scale extinction. The role of diadromy has clearly been very important at a high temporal scale (thousands of years) for the reinvasion of freshwater habitats after the retreat of Pleistocene glacial ice sheets across North America and in the Baltic and Scandinavia (Legendre and Legendre 1984; Black et al. 1986; Hammar 1987; Bernatchez and Wilson 1998; McDowall 1999; Curry 2007). This historical response to the retreat of the ice is still happening for contemporary invasion of Alaskan streams by salmon species (Milner 1987; Milner and Bailey 1989), as habitats become available and congenial for fish. There are similar patterns in New Zealand where the only freshwater fish now present in streams that were completely ice-covered in the Pleistocene are diadromous (Main 1989; McDowall 1996b), and this is because nondiadromous species have difficulty reoccupying streams as they emerge from ice cover. That the first fish colonists are diadromous species is not surprising. With global warming, we may expect to see more of this as temperatures ameliorate in the Arctic. This is likely to be a general phenomenon, and were the details available, I suspect that we could show that diadromous species dominate in many places where there was major local/regional perturbation by glaciation or volcanism.

At a rather more localized spatiotemporal scale, diadromy clearly has had a role in restoration after major perturbations resulting from volcanism, as with the eruption of Mt. St. Helens in western North America in 1980 (Martin et al. 1986; Bisson et al. 1988); a similar pattern is evident in New Zealand following the largest volcanic eruption in human history in c. 186 AD, when clearly there was total defaunation of river ecosystems over a broad area of volcanic deposition (McDowall 1996b). These rivers were repopulated by immigration through coastal seas, and only diadromous fish are now found across the zones of ash deposition. Nondiadromous species are found in river systems around the fringes of the ash-covered zone but have not yet reinvaded the ash-fall zone. Returning to Hawai'i, active volcanism may be a mechanism for local, periodic defaunation of streams, and diadromy important for restoration.

It is, no doubt, the reason why all fishes in Iceland rivers are diadromous—there would have been extirpation of all life during cover by Pleistocene ice, and possibly, there has also been subsequent local extirpation from volcanism. Whichever were crucial, only diadromous species have been able to reinvade Icelandic streams, and they have clearly done so through the sea.

The process of local extirpation, and then restoration through diadromy, applies at the highly localized/recent spatiotemporal scale, as when fish kills defaunate an entire river. Recovery is most rapid by diadromous species. There are islands in the Caribbean in which the entire fish and crustacean fauna is amphidromous and where shrimps are harvested using chlorine bleach (Greathouse et al. 2005). This is a practice that has no long-term future, though diadromous shrimp populations (and also those of diadromous mullets and gobies that probably also suffer) can become reestablished in these streams by invasion by their marine life stages. This at least makes recovery in streams possible—as long as the entire region's streams are not treated in the same uncouth way. Thus, diadromy provides a mechanism for the restoration of the faunas by dispersal through coastal seas. Similar sorts of events are important, globally. All of us know of fish kills caused by petrochemical, or pulp and paper, industries, point-source pollution associated with cities, and so forth. There is a serious risk that resource managers can just shrug and say "They'll come back!" And in a way, diadromous species do have a remarkable ability to do so. Dealing with events like these is a continual nightmare for fisheries managers and conservation agencies.

Looking more generally, diadromy has a key role in restoration ecology and recovery from anthropogenic perturbation. The behavioral/migratory processes that make diadromy possible renders these fishes highly vulnerable to events that prevent migration, leading to loss of what has come to be called connectivity—causing local extirpation where access is impeded. There are two distinct, perhaps contrary aspects of this. The fundamental need for the sort of stream connectivity that makes it possible for diadromous species to populate stream systems by moving upstream from the sea fairly obviously also makes these species highly vulnerable to local extirpation where access is impeded. The dual roles of diadromy in making for vulnerability to perturbation and facilitating recovery from perturbation are different sides of the same coin.

Where impoundment interferes with connectivity, we naturally turn to construction of fish passes, and I want to make only one comment on this question and that is to plead for the distinctive needs of small amphidromous species to be recognized with appropriate fish passes being constructed. Small upstream-migrating climbing gobies and galaxiids need very different kinds of fish passes from large, powerful, jumping salmonids and shads.

Diadromy and the Application of Indices of Biotic Integrity

Let me explore just one more issue that is distinctive to the community/population ecology of diadromous fishes. One of the emerging ways of quantifying the conservation status of streams is through the use of indices of biological integrity (IBI) (Hughes and Oberdorff 1999). In these indices, the fish populations of an area of interest are compared with fish populations in known, pristine, ideal habitats of the same general type. Use of IBIs for diadromous fishes is likely to fail because there is a strong tendency for the relationship between habitat quality and fish populations to change with increasing distance upstream. This is simply because of the downstream/upstream gradients in various fish population metrics, discussed earlier—changes in fish populations that involve a downstream–upstream shift in (1) age structure with more juveniles in downstream reaches; (2) size of individual fish, driven simply by the changing age structure; (3) species diversity, as some fish penetrate upstream further than others; and (4) absolute abundance. It is a question of access, distance, stream gradients, time, and species' instinctive migratory drives and swimming/climbing abilities. Addressing this question in fine detail, for the moment, pairs of stream reaches that have exactly the same critical habitat qualities, as to substrate, depth, velocity, water quality/chemistry, food organisms, riparian characteristics, and so on, are likely to have very different fish populations if they are at very different distances upstream from the sea, for the simple reason that habitats farther downstream are much easier for diadromous fishes to reach. Moreover, if there is plenty of habitat space (low population densities), it is possible that, as they migrate upstream, individual fish may stop migrating as soon as they find plentiful, suitable habitat, and this may result in habitats farther upstream having few or none of the species, despite being entirely suitable for the species and being accessible to them. I have shown, elsewhere, that the New Zealand torrentfish may penetrate long distances inland in rivers systems that have gentle slopes, but they may reach higher elevations in very steep streams (McDowall 2000). This seems counterintuitive, but it is possibly because in low-gradient streams, there is less riffle habitat available per meter of elevation and so fish have to penetrate well upstream to find suitable habitat. Again, we need to avoid getting trapped by the detail but, rather, need to recognize the principle that the relationship between habitat quality and fish abundance may break down with increasing distance upstream (McDowall 1998).

There are interesting parallels in the patterns of distribution, abundance, species diversity, and other population metrics in diadromous fish communities in river systems to the river continuum concept (RCC; Vannote et al. 1980), which points to accumulating upstream–downstream vectors of change in river systems. These ostensibly relate to the changing nature of river systems and their biotas, as water flows downstream and waterways enlarge. There are parallel changes, driven by diadromy, that are downstream–upstream vectors that have nothing to do with the predictions of the RCC.

To the extent that indices of biological integrity break down, special approaches are needed to permit comparisons of habitat quality across a river catchment or across a region. Recent modeling studies that factor in the effect of distance inland and elevation on a species' frequency of occurrence are making progress towards solving this problem using neural networks (Joy and Death 2004), boosted regression trees (Leathwick et al. 2005), and geographic information systems technologies (Eikaas et al. 2005; Eikaas and McIntosh 2006). These studies allow comparisons of what is found at sites with what might be expected from the best sites, using modeling studies. These methods are perhaps not quite as good as IBIs, but there is little alternative when dealing with diadromous species. There is a need for a substantial database on the fauna, which we now have in New Zealand (Figure 2).

Maintenance of Fish Populations on Small Islands

The fish faunas of the smaller islands of the globe, especially in the tropics and subtropics, are strongly

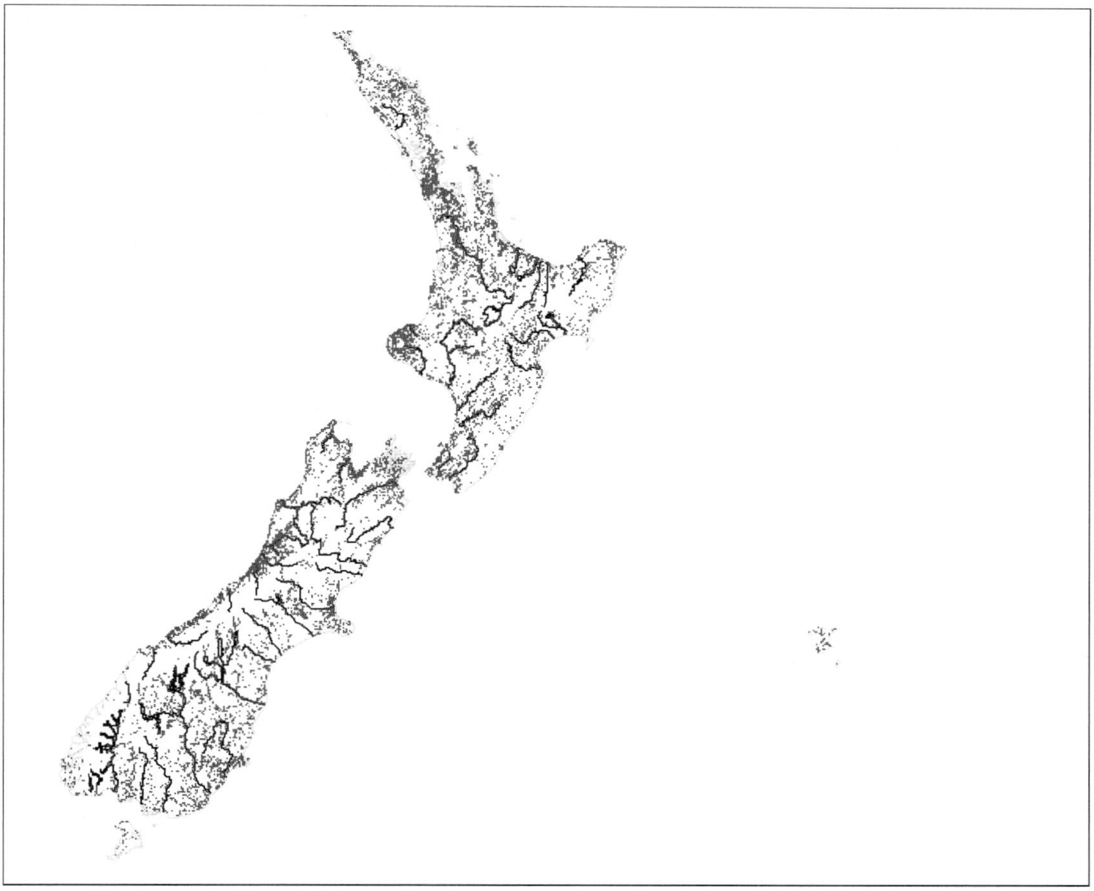

Figure 2.—The distribution of sites for which there are data on New Zealand freshwater fish communities, from the NIWA New Zealand Freshwater Fish Database (>26,000 sites).

dominated by amphidromous fishes—often gobies of diverse sorts. It seems to me likely that amphidromy is a highly adaptive life history strategy that promotes the invasion of these remote island streams in the first place. The same process facilitates repopulation of the small streams on islands prone to dewatering on account of the small size of their catchments and the risk that there is not enough rain to ensure that they flow continuously.

This may be of high significance to Hawai'i (McDowall 2003) and is especially true on the leeward sides of these often quite small and steep islands where rainfall is less in both volume and reliability. Stream fish populations may be extirpated repeatedly by loss of flows and then reestablished by spread through coastal seas from elsewhere among the islands. The same processes that are leading to lack of genetic divergence and structuring across the islands (discussed above) are probably also ensuring that streams continue to support fishes across the medium/short time scale.

Population Dispersion, Sources, and Sinks

I have been thinking about the sources/sinks mechanism (Pulliam 1988; Dunning et al. 1992) and wondering whether some geographical fringe populations of diadromous species may function as continual sinks. Is it not possible that the same spread of diadromous fishes through the sea, which provides the ability to provide backup for reinvasion of streams that have been defaunated, is also a strategy that continually enables species to be always probing out into the distant unknown, providing the capacity for newly available stream systems to obtain fish populations? Imagine Iceland, emerging from the Pleistocene and from permanent ice cover (as northern Alaska is doing today). Straying of salmonids, which might otherwise have been dispersed beyond the possibility of return to freshwater habitats to spawn (expatrial dispersal), might have been the elemental key to natural restocking Icelandic streams as they became available.

The French have been considering how bichique can populate the rivers of the Comoro Islands in the far western Indian Ocean. Given the patterns of oceanic currents there, they cannot see how the populations self-recruit. This species has a larval life of several months, and maybe the Comoro Islands are perpetual sinks for this goby, derived from farther east (Hoareau et al. 2007).

Is this a more general pattern? If diadromy has an important role in facilitating the spread of species into habitats where they are not present (for whatever reason), a corollary seems, to me, to be their likely expatrial dispersal into reproductively neutral or negative habitats. How was it possible for the Australian longfin eel (also known as the speckled longfin eel) *Anguilla reinhardtii* to begin arriving in New Zealand (McDowall et al. 1998)? I think we may need to envisage an almost constant process of oceanic straying/spread of propagules into areas where a species either does not, at present, exist (including long-distance waif dispersal to far-off places) or their routine spread into stream habitats where populations become established but cannot effectively contribute to the species' reproduction. If this is so, there is probably a continual process of source/sink population dynamics, with at least some populations across the ranges of diadromous species not contributing to the recruitment pool but being population sinks. This might well be a significant aspect of the population ecology and biogeography of diadromous species, especially on islands, that there are populations, perhaps mostly around the fringes of their geographical ranges (though not necessarily just there), that release larvae or juveniles into coastal seas in areas where the ocean currents carry them beyond the reach of life stages ready to return to freshwater. Are diadromous species the fisheries counterpart of Mayr and Diamond's (2001) ornithological "tramps"? Is this one of the population costs/benefits of being diadromous?

Energetics Risks from Diadromous Migrations

Japanese work is suggesting that there may be risks from diadromy that relate to possible nonrecoverable starvation of newly hatched larvae of amphidromous species during their downstream migrations to sea from upstream spawning sites, if it takes too long (Moriyama et al. 1998; Iguchi and Mizuno 1999).

If we have a look at the life histories of sicydiine gobies, which are very often amphidromous, we can see some potentially interesting aspects. Typically, it seems, the sicydiine gobies hatch as very tiny larvae (<2 mm) after less than two days' development. I suspect that this may be an adaptation to provide the tiny larvae enough time to migrate downstream in river systems from the sea and to reach the trophically rich marine plankton before they run out of the endogenous energy sources provided in the egg (McDowall 2009). Thus, they are undertaking part of their development while migrating downstream This may well be a strategy for facilitating long-distance downstream dispersal of the larvae (i.e., they may do much of their embryonic development as an already-hatched prolarva, growing "on the move" downstream, instead of within the egg prehatching). This is unlikely to be of much significance on the many small oceanic islands where there are lots of amphidromous sicydiines because their streams are usually short, steep, and fast-flowing, but it may be an issue for these species in the much longer river systems on larger islands or continents.

A more adventurous life history strategy might have been for them to evolve pelagic eggs, which can develop as they disperse downstream (as is true of the North American, anadromous striped bass *Morone saxatilis*—Rulifson 1982), but seemingly, gobies are strongly locked into a life history where there is reproductive pairing of the sexes and deposition of the eggs on a substrate boulder.

Also, in at least some groups, there is a difference in egg size between diadromous species and nondiadromous congenerics (McDowall 1990). Diadromous species may have smaller eggs, but more of them, producing larger numbers of smaller larvae, perhaps compensating for both the higher availability of small planktonic foods in the sea and the risk of loss by expatrial dispersal in ocean currents—in contrast with fewer larger larvae, having more difficulty finding small foods in fluvial environments but needing greater swimming powers to cope with flow velocities in streams. Thus, there is a series of additional trade-offs that relate also to the amount of yolk/energy in eggs, age at hatch, and survival during downstream transport.

In a group of diadromous galaxiids, spawning takes place during elevated streamflows. This means that when flows abate, the developing eggs end up on streambanks, not immersed in water, and develop in the humid atmosphere, scattered among leaf litter from riparian forest or among the coarse cobble substrates along stream margins. The eggs hatch when a later event of elevated flows reimmerses them in water and they are agitated by the flood flows. The need for agitation is probably important as it reduces the risk of rainfall stimulating hatch, leaving the larvae stranded among the substrate materials. Timing of hatch during elevated flows may have other adaptive advantages: (1) it might mean that the larvae are less vulnerable to predation, and (2) it might result in a quicker trip to sea caused by augmented flows (McDowall and Charteris 2006), thus minimizing the prospect of the irreversible starvation discussed by the Japanese. And as some of these species spawn several hundred kilometers upstream from the sea, the prospect of larval mortality from both predation and starvation is real.

The Question of Facultativeness and Conservation

We need to understand the role of facultativeness in diadromous behaviors across a wide range of metrics. Some diadromous species may be more facultative in their diadromy than we presently realize. In species that are facultatively diadromous, survival of stocks in freshwater lakes may provide limited protection of stocks that are threatened, but we need to be wary of using this as a surrogate for real conservation of the species' full behavioral and genetic diversity.

Diadromy and Invasive Species

There are some serious conservation issues with invasive diadromous species. In general, there has been quite poor success in human attempts to translocated diadromous species into new regions. Many anadromous salmonids, in particular, have been translocated all over the globe, and most often there is a failure for the translocated fishes to establish diadromous populations (McDowall 1988), but there are some important exceptions. In particular, brown trout have established anadromous populations in many areas. These trout have caused substantial conservation chaos in receiving ecosystems creating real concern (McDowall 2006). Where there are sea-migratory populations of introduced brown trout, it is almost impossible to limit the river systems where this species is found. The establishment of brown trout has been substantially driven by spread of anadromous fish through coastal seas in southern New Zealand, also at the Falklands and probably in Tasmania and southern Patagonian South America. Where this happens and where the native fish fauna is vulnerable to trout predation, there are serious conservation issues that are very difficult to manage. Indigenous fish faunas are safe from trout predation only above falls that prevent trout from penetrating upstream. And even there, the reckless behavior of some trout fisheries managers or trout anglers is leading to trout being released upstream of these falls, and so they continue their depredations on the indigenous faunas (Cadwallader 1996). At least in New Zealand, there is very strong evidence for major ecosystem changes that result from the establishment of alien brown trout (McDowall 2006).

Results of a survey of freshwater fishes across the Falkland Islands suggested that the zebra trout *Aplochiton zebra* has suffered severely from brown trout invasion. In the nearly 150 sites sampled over a month at the islands, it was found that brown trout had become very widespread and had clearly spread around the islands through the sea. Zebra trout had disappeared across a wide expanse of countryside and were generally present only at sites where brown trout were not found. Zebra trout and brown trout were found together at only two sites. The brown trout fishery at the Falklands is based largely on anadromous fish living in coastal seas and moving into river estuaries to spawn, and it is there that the anglers can catch big trout. Decline in zebra trout is clearly due to brown trout invasion to the extent that the future of the zebra trout is a matter of serious conservation concern (McDowall et al. 2001, 2005). This scenario is representative of what has probably happened widely across southern lands (McDowall 2006). What is happening in Tierra del

Fuego is virtually unknown, but we can guess! Perhaps the clarity of this situation in the Falklands is a warning to us about what might be happening more widely but is not so sharply etched in more speciose, complex, and lesser known freshwater ecosystems.

There is some evidence that additional anadromous salmonids are becoming established in Patagonia, with the prospect that steelhead/rainbow trout, Chinook salmon, and perhaps other species are establishing anadromous stocks in Patagonia, in addition to the brown trout already there. The establishment of these species may be driven by massive escapes of fish from sea-cage salmon farms—huge escapes of hundreds of thousands of fish are recorded (Soto et al. 2001). The implications for the indigenous fishes may be serious, though the extent to which the invasions of river systems across Patagonia are being driven by these farm escapes is presently unknown. Establishment of alien salmonids in southern lands is creating an unfortunate conflict of interests owing to the economic/tourism value of recreational angling fisheries to these countries (Cadwallader 1996; Pascual et al. 2001, 2002; McDowall 2006).

References

Allibone, R. M., and G. P. Wallis. 1993. Genetic variation and diadromy in some native New Zealand galaxiids (Teleostei: Galaxiidae). Biological Journal of the Linnean Society 50:19–-33.

Andrews, A. P. 1996. Family Bovicthidae. Pages 198–199 *in* R. M. McDowall, editor. Freshwater fishes of south-eastern Australia. Reed, Sydney, Australia.

Behnke, R. J. 2002. Trout and salmon of North America. Chanticleer Press, New York.

Bernatchez, L., and C. C. Wilson. 1998. Comparative phylogeography of Nearctic and Palearctic fishes. Molecular Ecology 7:431–452.

Berrebi, P., G. Cattaneo-Berrebi, P. Valade, J. F. Ricou, and T. Hoareau. 2005. Genetic homogeneity in eight freshwater populations of *Sicyopterus lagocephalus*, an amphidromous gobiid of La Reunion Island. Marine Biology 148:179–188.

Birstein, V. J. 1993. Sturgeons and paddlefishes: threatened fishes in need of conservation. Conservation Biology 7:773–-787.

Bisson, P. A., J. L. Neilsen, and J. W. Ward. 1988. Summer production of coho salmon stocked in Mount St. Helens streams 3–6 years after the 1980 eruption. Transactions of the American Fisheries Society 117:322–335.

Black, G. A., J. B. Dempson, and W. J. Bruce. 1986. Distribution and postglacial dispersal of freshwater fishes of Labrador. Canadian Journal of Zoology 64:21–31.

Blanco, J. F., and F. N. Scatena. 2005. Floods, habitat hydraulics and upstream migration of *Neritina virginea* (Gastropoda: Neritidae) in northeastern Puerto Rico. Caribbean Journal of Science 41:55–74.

Cadwallader, P. L. 1996. Overview of the impacts of introduced salmonids on Australian native fauna. Australian Nature Conservation Agency, Canberra.

Campbell, H. J. 1998. Fauna and flora of the Chatham Islands: less than 4 m.y. old? Geological Society of New Zealand Miscellaneous Publication 97:15–16.

Chubb, A. L., R. M. Zink, and J. M. Fitzsimons. 1998. Patterns of mtDNA variation in Hawaiian freshwater fishes: phylogeographic consequences of amphidromy. Journal of Heredity 89:8–16.

Craw, R. C. 1979. Generalized tracks and dispersal in biogeography: a response to R. M. McDowall. Systematic Zoology 28:99–107.

Croizat, L., G. J. Nelson, and D. E. Rosen. 1974. Centers of origin and related concepts. Systematic Zoology 23:265–287.

Curry, R. A. 2007. Late glacial impacts on dispersal and colonization of Atlantic Canada and Maine by freshwater fishes. Quaternary Research 67:225–233.

Darwin, C. 1872. On the origin of species by means of natural selection, or the preservation of favored races in the struggle for life. 6th edition. Murray, London.

Devick, W. S., J. M. Fitzsimons, and R. T. Nishimoto. 1992. Conservation of Hawaiian freshwater fishes. Department of Land and Natural Resources, Division of Aquatic Resources, Honolulu.

DeWoody, J. A., and J. C. Avise. 2000. Microsatellite variation in marine, freshwater and anadromous fishes compared with other animals. Journal of Fish Biology 56:461–473.

Dunning, J. B., B. J. Danielson, and H. R. Pulliam. 1992. Ecological processes that affect populations in complex landscapes. Oikos 65:169–175.

Eikaas, H., and A. R. McIntosh. 2006. Habitat loss through disruption of constrained dispersal networks. Ecological Applications 16:987–998.

Eikaas, H., A. Kliskey, and A. R. McIntosh. 2005. Spatial modelling and habitat quantification for two diadromous fish in New Zealand streams: a GIS-based approach with application for conservation management. Environmental Management 36:726–740.

Englund, R., and R. Filbert. 1997. Discovery of the native stream goby, *Lentipes concolor*, above Hawaii's highest waterfall, Hiilawe Falls. Bishop Museum Occasional Papers 49:62–64.

Fievet, E., and R. Eppe. 2002. Genetic differentiation among populations of an amphidromous shrimp

Atya innocuous (Herbst) and obstacles to their upstream migration. Archiv für Hydrobiologie 153:287–300.

Fontaine, M. 1975. Physiological mechanisms in the migrations of marine and amphihaline fish. Advances in Marine Biology 13:241–335.

Ford, J. I., and R. A. Kinzie. 1982. Life crawls upstream. Natural History (New York) 91:60–67.

Foote, C. J., I. Mayer, C. C. Wood, W. C. Clarke, and J. Blackburn. 1994. On the developmental pathway to nonanadromy in sockeye salmon, *Oncorhynchus nerka*. Canadian Journal of Zoology 72:397–405.

Gill, H. S., C. B. Renaud, F. Chapleau, R. L. Mayden, and I. C. Potter. 2003. Phylogeny of living parasitic lampreys (Petromyzontiformes) based on morphological data. Copeia 2003:687–703.

Greathouse, E. A., J. G. March, and C. M. Pringle. 2005. Recovery of a tropical stream after a harvest-related chlorine poisoning event. Freshwater Biology 50:603–615.

Hammar, J. 1987. Zoogeographical zonation of fish communities in insular Newfoundland: a preliminary attempt to use the Arctic char population ecology to describe early postglacial colonization interactions. International Society of Arctic Char Fanatics Information Series 4:31–38.

Hammer, M. P., M. Adams, P. J. Unmack, and K. F. Walker. 2007. A rethink on *Retropinna*: conservation implications of new taxa and significant genetic sub-structure in Australian smelts (Pisces: Retropinnidae). Marine and Freshwater Research 58:327–341.

Hoareau, T. B., R. Lecomte-Finiger, H. P. Grondin, C. Conand, and P. Berrebi. 2007. Oceanic larval life of La Reunion 'bichiques', amphidromous gobiid post-larvae. Marine Ecology Progress Series 333:303–308.

Hughes, R., and T. Oberdorff. 1999. Application of IBI concepts and metrics to waters outside of the United States. Assessing the sustainability and biological integrity of water resources using fish communities. Pages 79–93 *in* T. P. Simon, editor. Assessing the sustainability and biological integrity of water resources using fish communities. CRC Press, Boca Raton, Florida.

Iguchi, K., and N. Mizuno. 1990. Diel changes in larval drift among amphidromous gobies in Japan, especially *Rhinogobius brunneus*. Journal of Fish Biology 37:255–264.

Iguchi, K., and N. Mizuno. 1999. Early starvation limits survival in amphidromous fishes. Journal of Fish Biology 54:705–712.

Ilves, K. L., and E. B. Taylor. 2007. Are *Hypomesus chishimaensis* and *H. nipponensis* (Osmeridae) distinct species? A molecular assessment using comparative sequence data from five genes. Copeia 1997:180–185.

Jerry, D. R., and P. R. Baverstock. 1998. Consequences of a catadromous life-strategy for levels of mitochondrial DNA differentiation among populations of the Australian bass, *Macquaria novemaculeata*. Molecular Ecology 7:1003–1013.

Joy, M. K., and R. G. Death. 2004. Application of the index of biotic integrity methodology to New Zealand freshwater fish communities. Environmental Management 34:415–428.

Keith, P. 2003. Biology and ecology of amphidromous Gobiidae of the Indo-Pacific and Caribbean regions. Journal of Fish Biology 63:831–847.

Keith, P., E. Vigneux, and P. Bosc. 1999. Atlas des poissons et des crustaces d'eau douce de la Reunion. Patrimoines Naturels 39:1–136.

Keith, P., E. Vigneux, and G. Marquet. 2002. Atlas des poissons et des crustaces d'eau douce de Polynesie Francaise. Patrimoines Naturels 55:1–177.

Keith, P., T. Galewski, G. Cattaneo-Berrebi, T. Hoareau, and P. Berrebi. 2005. Ubiquity of *Sicyopterus lagocephalus* (Teleostei: Gobioidei) and phylogeography of the genus *Sicyopterus* in the Indo-Pacific area inferred from mitochondrial cyctochrome b gene. Molecular Phylogenetics and Evolution 37:721–732.

Keith, P., G. Marquet, P. Valade, P. Bosc, and E. Vigneux. 2006. Atlas des poisons et des crustaces d'eau douce des Comores, Mascareignes et Seychelles. Patrimoines Naturels 65:1–250.

Kobayashi, S. 1999. Reproductive ecology of the Japanese mitten crab *Eriocheir japonica* (de Haan): a review. Japanese Journal of Benthology 54:24–35.

Kottelat, M. 2007. Nomenclatural status and identity of *Gobius lagocephalus* (Teleostei: Gobiidae). Molecular Phylogenetics and Evolution 43:693–695.

Leathwick, J. R., D. Rowe, J. Richardson, J. Elith, and T. Hastie. 2005. Using multivariate adaptive regression spines to predict the distributions of New Zealand's freshwater diadromous fish. Freshwater Biology 50:2034–2052.

Legendre, P., and V. Legendre. 1984. Postglacial dispersal of freshwater fishes in the Quebec Peninsula. Canadian Journal of Fisheries and Aquatic Sciences 41:1781–1802.

Main, M. R. 1989. Distribution and post-glacial dispersal of freshwater fishes in South Westland, New Zealand. Journal of the Royal Society of New Zealand 19:161–169.

March, J. G., C. M. Pringle, M. J. Townsend, and A. I. Wilson. 2002. Effects of freshwater shrimp assemblages on benthic communities along an altitudinal gradient of a tropical island stream. Freshwater Biology 47:377–390.

Marquet, G., P. Keith, and E. Vigneux. 2003. Atlas des poissons et des crustaces d'eau douce de Nouvelle Caledonie. Patrimoines Naturels 58:1–284.

Martin, D. F., L. J. Wasserman, and V. H. Dale. 1986. Influence of riparian vegetation on posteruption survival of coho salmon fingerlings on the west-side streams of Mount St. Helens, Washington. North American Journal of Fisheries Management 6:1–8.

Mayr, E., and J. Diamond. 2001. The birds of northern Melanesia: speciation, ecology and biogeography. Oxford University Press, Oxford, UK.

McDowall, R. M. 1964. The affinities and derivation of the New Zealand fresh-water fish fauna. Tuatara 12:59–67.

McDowall, R. M. 1970. The galaxiid fishes of New Zealand. Bulletin of the Museum of Comparative Zoology at Harvard University 139:341–431.

McDowall, R. M. 1972. The species problem in freshwater fishes and the taxonomy of diadromous and lacustrine populations of *Galaxias maculatus* (Jenyns). Journal of the Royal Society of New Zealand 2:325–367.

McDowall, R. M. 1979. Fishes of the family Retropinnidae (Pisces: Salmoniformes)—a taxonomic revision and synopsis. Journal of the Royal Society of New Zealand 9:85–121.

McDowall, R. M. 1984. The New Zealand whitebait book. Reed, Wellington, New Zealand.

McDowall, R. M. 1987. The occurrence and distribution of diadromy among fishes. Pages 1–13 *in* M. J. Dadswell, R. J. Klauda, C. M. Moffitt, R. L. Saunders, R. A. Rulifson, and J. E. Cooper, editors. American Fisheries Society, Symposium 1, Bethesda, Maryland.

McDowall, R. M. 1988. Diadromy in fishes: migrations between freshwater and marine environments. Croom Helm, London.

McDowall, R. M. 1990. New Zealand freshwater fishes: a natural history and guide. Heinemann Reed, Auckland, New Zealand.

McDowall, R. M. 1993. Implications of diadromy for the structuring and modelling of riverine fish communities in New Zealand. New Zealand Journal of Marine and Freshwater Research 27:453–462.

McDowall, R. M. 1994a. The origins of New Zealand's Chinook salmon, *Oncorhynchus tshawytscha*. Marine Fisheries Review 56:1–7.

McDowall, R. M. 1994b. Gamekeepers for the nation: the story of New Zealand's acclimatisation societies, 1861–1990. Canterbury University, Christchurch, New Zealand.

McDowall, R. M. 1996a. Diadromy and the assembly and restoration of riverine fish communities: a downstream view. Canadian Journal of Fisheries and Aquatic Sciences 53(Supplement 1):219–236.

McDowall, R. M. 1996b. Volcanism and freshwater fish biogeography in the northeastern North Island of New Zealand. Journal of Biogeography 23:139–148.

McDowall, R. M. 1998. Fighting the flow: downstream–upstream linkages in the ecology of diadromous fish faunas in west coast New Zealand rivers. Freshwater Biology 40:111–122.

McDowall, R. M. 1999. Diadromy and genetic diversity in Nearctic and Palearctic fishes. Molecular Ecology 8:527–528.

McDowall, R. M. 2000. Biogeography of the New Zealand torrentfish, *Cheimarrichthys fosteri* (Teleostei: Pinguipedidae): a distribution driven mostly by ecology and behaviour. Environmental Biology of Fishes 58:119–131.

McDowall, R. M. 2001. Anadromy and homing: two life history traits with adaptive synergies in salmonid fishes. Fish and Fisheries 2:78–85.

McDowall, R. M. 2002. Accumulating evidence for a dispersal biogeography of southern cool temperate freshwater fishes. Journal of Biogeography 29:207–220.

McDowall, R. M. 2003. Hawaiian biogeography and the islands' freshwater fish fauna. Journal of Biogeography 30:703–710.

McDowall, R. M. 2004a. Ancestry and amphidromy in island freshwater fish faunas. Fish and Fisheries 5:75–85.

McDowall, R. M. 2004b. Variation in vertebral number in galaxiid fishes, how fish swim and a possible reason for pleomerism. Reviews in Fish Biology and Fisheries 13:246–263.

McDowall, R. M. 2006. Crying wolf, crying foul, or crying shame: alien salmonids and a biodiversity crisis in the southern cool-temperate galaxioid fishes. Reviews in Fish Biology and Fisheries 16:233–422.

McDowall, R. M. 2007a. On amphidromy, a distinct form of diadromy in aquatic organisms. Fish and Fisheries 8:1–13.

McDowall, R. M. 2007b. Hawaiian stream fishes: the role of amphidromy in history, ecology and conservation biology. Bishop Museum Bulletin in Cultural and Environmental Studies 3:3–9.

McDowall, R. M. 2008. Diadromy, history and ecology: a question of scale. Hydrobiologia 602:5–14

McDowall, R. M. 2009. Early hatch: a strategy for safe downstream transport in amphidromous gobies? Reviews in Fish Biology and Fisheries 19:1–8.

McDowall, R. M., and S. C. Charteris. 2006. The possible adaptive advantages of terrestrial egg deposition in some fluvial diadromous galaxiid fishes (Teleostei: Galaxiidae). Fish and Fisheries 7:153–164.

McDowall, R. M., and G. P. Wallis. 1996. Description and redescription of *Galaxias* species (Teleostei: Galaxiidae) from Otago and Southland. Journal of the Royal Society of New Zealand 26:401–427.

McDowall, R. M., D. J. Jellyman, and L. C. Dijkstra. 1998. Arrival of an Australian anguillid eel in New

Zealand: an example of transoceanic dispersal. Environmental Biology of Fishes 51:1–6.

McDowall, R. M., R. M. Allibone, and W. L. Chadderton. 2001. Issues for the conservation and management of Falkland Islands freshwater fishes. Aquatic Conservation Marine and Freshwater Ecosystems 11:473–486.

McDowall, R. M., R. M. Allibone, and W. L. Chadderton. 2005. Falkland Islands freshwater fishes: a natural history. Falklands Conservation, London.

McPhail, J. D. 1993. Ecology and evolution of sympatric sticklebacks (*Gasterosteus*): origin of the species pairs. Canadian Journal of Zoology 71:515–523.

Metcalfe, J., G. Arnold, and R. M. McDowall. 2002. Migration. Pages 175–199 *in* P. J. B. Reynolds and J. D. Reynolds, editors. Handbook of fish biology and fisheries, volume 1. Blackwell Scientific Publications, Oxford, UK.

Milner, A. M. 1987. Colonization and ecological development of new streams in Glacier Bay National Park, Alaska. Freshwater Biology 18:53–70.

Milner, A. M., and R. G. Bailey. 1989. Salmonid colonization of new streams in Glacier Bay National Park, Alaska. Aquaculture and Fisheries Management 20:179–192.

Moriyama, A., Y. Yanagisawa, N. Mizuno, and K. Omori. 1998. Starvation of drifting goby larvae due to retention of free embryos in upstream reaches. Environmental Biology of Fishes 52:321–329.

Myers, G. S. 1949. Usage of anadromous, catadromous and allied terms for migratory fishes. Copeia 1949:89–97.

Myers, M. J., C. P. Meyer, and V. H. Resh. 2000. Neritid and thiarid gastropods from French Polynesian streams: how reproduction (sexual, parthenogenetic) and dispersal (active/passive) affect population structure. Freshwater Biology 44:535–545.

Nelson, J. S. 2006. Fishes of the world. Wiley, Hoboken, New Jersey.

Pascual, M., P. Bentzen, S. R. Rossi, G. Mackey, M. T. Kinnison, and R. Walker. 2001. First documented case of anadromy in a population of introduced rainbow trout in Patagonia, Argentina. Transactions of the American Fisheries Society 130:53–67.

Pascual, M., P. Macchi, J. Urbanski, F. Marcos, C. Riva Rossi, M. Novara, and P. Dell'Arcipetre. 2002. Evaluating potential effects of exotic freshwater fish from incomplete species presence–absence data. Biological Invasions 4:101–113.

Phillipps, W. J. 1925. New Zealand eels. New Zealand Journal of Science and Technology 8:28–30.

Pulliam, H. R. 1988. Sources and sinks and population regulation. The American Naturalist 43:187–193.

Quinn, T. P., and K. W. Myers. 2004. Anadromy and the marine migrations of Pacific salmon and trout: Rounsefell revisited. Reviews in Fish Biology and Fisheries 14:421–442.

Rosen, D. E. 1974. Phylogeny and zoogeography of salmoniform fishes and relationships of *Lepidogalaxias salamandroides*. Bulletin of the American Museum of Natural History 153:265–326.

Rounsefell, G. A. 1958. Anadromy in North American Salmonidae. U.S. Fish and Wildlife Service Fishery Bulletin 62:171–185.

Rulifson, R. A. 1982. Anadromous fish in the southeastern United States and recommendations for a management plan. U.S. Fish and Wildlife Service, Fishery Resource Region 4, Atlanta, Georgia.

Saruwatari, T., J. A. Lopez, and T. W. Pietsch. 1997. A revision of the osmerid genus *Hypomesus* Gill (Teleostei: Salmoniformes) with the description of a new species from the southern Kuril Islands. Species Diversity 2:59–82.

Saunders, R., M. A. Hachey, and C. W. Fay. 2006. Maine's diadromous fish community: past, present and implications for Atlantic salmon recovery. Fisheries 31:537–547.

Schoenfuss, H. L., and R. W. Blob. 2003. Kinematics of waterfall climbing in Hawaiian freshwater fishes (Gobiidae); vertical propulsion at the aquatic-terrestrial interface. Journal of Zoology, London 261:191–206.

Scott, W. B., and E. J. Crossman 1975. Freshwater fishes of Canada. Fisheries Research Board of Canada Bulletin 184:1–996.

Smith, W. L. and J. S. Sparks. 2007. *Gobius lagocephalus*: the world's most widespread nomen dubium. Molecular Phylogenetics and Evolution 43:696–698.

Soto, D., F. Jara, and C. Moreno. 2001. Escaped salmon in inner seas, southern Chile: facing ecological and social conflicts. Ecological Applications 11:1750–1762.

Sparks, J. S. and W. L. Smith. 2006. *Sicyopterus lagocephalus*: widespread species, species complex, or neither? A critique on the use of molecular data for species identification. Molecular Phylogenetics and Evolution 40:900–902.

Svardson, G. 1998. Postglacial dispersal and reticulate evolution in Nordic coregonids. Nordic Journal of Freshwater Research 78:3–32.

Trewick, S. A., A. M. Paterson, and H. J. Campbell. 2007. Hello New Zealand. Journal of Biogeography 34:1–6.

Vannote, R. L., G. W. Minshall, C. W. Cummins, J. R. Seddell, and C. E. Cushing. 1980. The river continuum concept. Canadian Journal of Fisheries and Aquatic Sciences 37:130–137.

Wagner, W. L., and V. A. Funk. 1995. Hawaiian biogeography: evolution on a hot-spot archipelago. Smithsonian Institution Press, Washington, D.C.

Ward, R. D., M. Woodwark, and D. O. F. Skibinski. 1994. A comparison of diversity levels in marine, freshwater, and anadromous fishes. Journal of Fish Biology 44:213–232.

Waters, J. M., and C. P. Burridge. 1999. Extreme intraspecific mitochondrial DNA sequence divergence in *Galaxias maculatus* (Osteichthys: Galaxiidae), one of the world's most widespread freshwater fish. Molecular Phylogenetics and Evolution 11:1–12.

Waters, J. M., and G. P. Wallis. 2001. Cladogenesis and loss of the marine life-history phase in freshwater galaxiid fishes (Osmeriformes: Galaxiidae). Evolution 55:587–597.

Waters, J. M., J. M. Epifanio, T. Gunter, and B. L. Brown. 2000a. Homing behaviour facilitates subtle genetic differentiation among river populations of *Alosa sapidissima*: microsatellites and mtDNA. Journal of Fish Biology 56:622–636.

Waters, J. M., L. H. Dijkstra, and G. P. Wallis. 2000b. Biogeography of a southern hemisphere freshwater fish: how important is marine dispersal? Molecular Ecology 9:1815–1821.

Whitehead, P. J. P. 1985. Clupeoid fishes of the world (suborder Clupeiodei): an annotated and illustrated catalogue of the herrings, sardines, pilchards, sprats, shads, anchovies and wolf-herrings. Part 1. Chirocentridae, Clupeidae and Pristigasteridae. FAO Fisheries Synopsis 125(7):1–303.

Zink, R. M., J. M. Fitzsimons, D. L. Dittman, D. R. Reynolds, and R. T. Nishimoto. 1996. Evolutionary genetics of Hawaiian freshwater fish. Copeia 1996:330–335.

American Fisheries Society Symposium 69:23–42, 2009

Diadromy and the Life History of Sockeye Salmon: Nature, Nurture, and the Hand of Man

Thomas P. Quinn*, Katy Doctor, Neala Kendall, and Harry B. Rich, Jr.
School of Aquatic and Fishery Sciences
Box 355020, University of Washington, Seattle, Washington 98195, USA

Abstract.—All diadromous fishes share some common themes in their life histories that are linked to the two migratory transitions between freshwater and marine habitats that define this group. We explore these linkages using Pacific salmon and specifically the sockeye salmon *Oncorhynchus nerka* populations from Bristol Bay, Alaska as a model system to illustrate concepts that are broadly applicable to diadromous fishes. In general, large size is beneficial for survival at sea or reproduction, but because the seasonal migration window is often narrow, the main option in many cases is to go in 1 year or delay until the next. The age at which fish make this transition is related to growth, but it reflects population-specific norms of reaction rather than species-wide rules. Common growing conditions have different effects on discrete populations that are shaped by the trade-offs between the benefits of large size in the new environment and the risks of mortality in the present environment associated with another year of growth. Within-season variation in timing is also related to body size (larger fish generally migrating earlier), reflecting a balance between the need to grow as large as possible before making the transition and the need to migrate at the optimal date. These complex life history connections are affected by natural variation in growing conditions driven by biotic and abiotic processes. In addition to the natural evolutionary and ecological processes affecting life history transitions, humans can also have profound effects. Fishing affects the abundance of prespawning adults, hence the levels of intraspecific competition and growth of their juvenile offspring in freshwater prior to seaward migration. In addition, fishing is often selective with respect to the body size and timing of migration, and the linkage between these traits further complicates the effects of selection.

Introduction

The essential feature of migratory animals, regularly scheduled movement between habitats, compels them to decide when to go on a series of time scales (this year or next, this season or another, today or tomorrow, etc.). The migrations of diadromous fishes between freshwater and saltwater allow them to take full advantage of the differences in growing conditions and breeding habitats (Northcote 1978; Gross et al. 1988) in the two environments. Diadromy is comparatively rare among fishes, being found in fewer than 1% of the species (McDowall 1988), but it is widely distributed among fish taxa. All such fishes are confronted, at least twice in their lives, with the trade-offs of staying where they are or migrating to the other environment. However, for many species, the details of these trade-offs are poorly understood, as are their implications for conservation.

The purpose of the paper is not to review the scientific literature on diadromous fishes; the tremendous quantity and diversity of research, and the complex linkages between life history, migration and conservation, are too daunting. Rather, our purpose is to use as a case history a well-studied (though not necessarily typical) group of diadromous fishes to illustrate important themes in life history and conservation. Specifically, we review information on sockeye salmon *Oncorhynchus nerka* from Bristol Bay, Alaska, revealing (1) the genetic controls and phenotypic plasticity controlling the transitions to and from the ocean, (2) the scope for selection in

* Corresponding author: tquinn@u.washington.edu

fisheries with respect to body size and migration timing, and (3) the complex challenges for conservation that arise from mismatches between population structure of the fish and the spatial scales at which fishery management operates. We take it as given that the species and populations in question are diadromous; the fascinating and related subject of diadromous and nondiadromous populations within species is beyond our scope here. We include selected references to work on other diadromous fishes to illustrate the generality of our points, rather than as a comprehensive review, and make no effort to review the vast literature on these transitions in other salmonids (e.g., quantitative genetics, connections to aquaculture, etc.). We also note that the relationship between growth and life history transitions is a very general one (Day and Rowe 2002) but do not attempt to review the theoretical literature on the subject.

The Bristol Bay System

Bristol Bay has a number of large lakes in several major drainage systems (Figure 1), and these drainages are the fundamental units of salmon population structure, ecology, and management (Minard and Meacham 1987; Hilborn et al. 2003). Juvenile sockeye salmon typically (though not invariably) rear in lakes for their first 1 or 2 years of life (Burgner 1991), so this region has an especially high abundance and diversity of this species. Togiak Lake ($40\ km^2$), and its tributaries, is the western-most system, contributing the fewest returning salmon. Clockwise around the bay from the Togiak man-

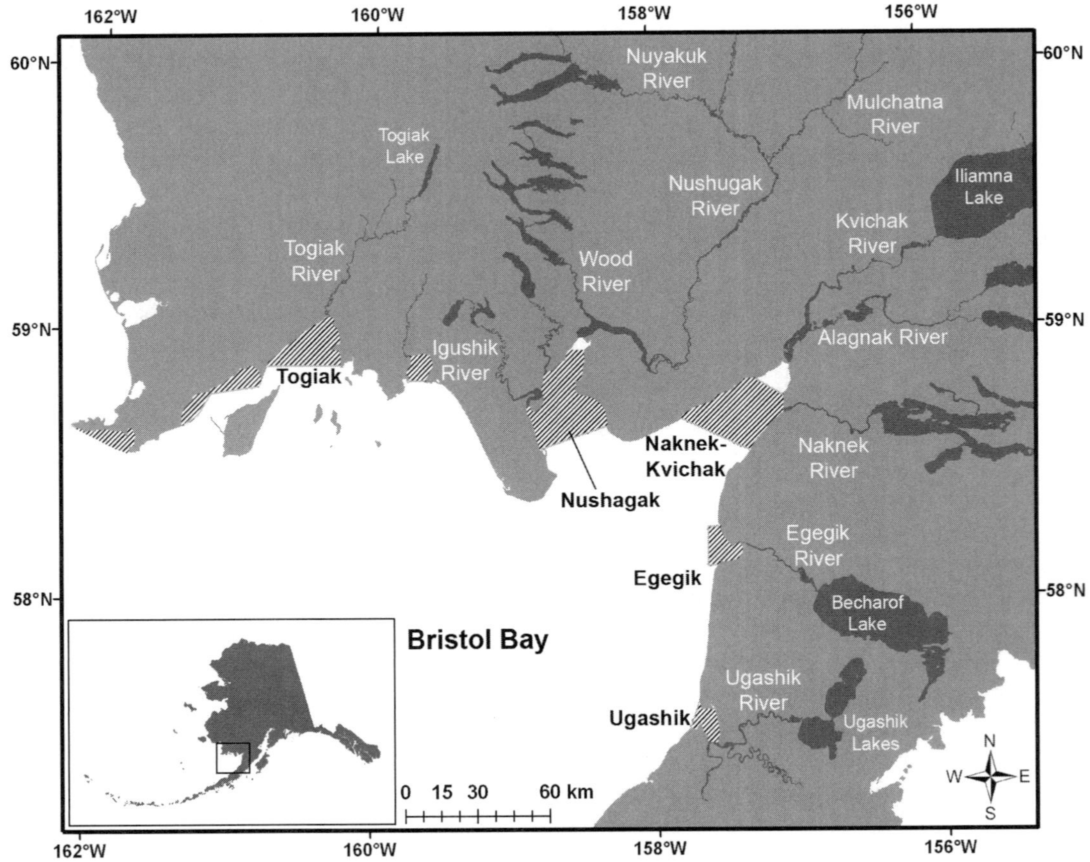

Figure 1.—Map of southwestern Alaska showing the locations of the five commercial fishing districts of Bristol Bay (indicated by diagonal lines) and the major rivers and lakes.

agement district, we find the Nushagak District, including the Igushik River system (two lakes with a combined area of 74 km^2), the Snake River system (one lake, 89 km^2), the Wood River system (four main lakes and several smaller ones, totaling 425 km^2), the large Nushagak River and its major tributaries, the Mulchatna River and the Nuyakuk (Tikchik) system (three lakes: 279 km^2). The Naknek–Kvichak District is the most complex, as it includes the Kvichak River system, draining Iliamna Lake (2,622 km^2), Lake Clark (267 km^2), and other smaller lakes and also two other drainages, the Alagnak and the Naknek. The Alagnak River flows into the Kvichak River just above Kvichak Bay, and it drains two large lakes, Nonvianuk (121 km^2) and Kukaklek (176 km^2), and several smaller ones. The Naknek River flows into Bristol Bay near the mouth of the Kvichak River, and it drains Naknek Lake (610 km^2) and three smaller lakes above it. Farther down the Alaska Peninsula, we find the Egegik District, composed of Lake Becharof (1,132 km^2) and the Egegik River, and finally the Ugashik District, composed of upper (208 km^2) and lower (177 km^2) Ugashik lakes and the Ugashik River.

All five species of Pacific salmon found in North America occur in Bristol Bay, but sockeye salmon constituted 95% of the catch from 1985 to 2005 by number (Alaska Department of Fish and Game [ADF&G] data). Sockeye are primarily caught in drift gill nets (85% of the catch), but fixed gill nets (set nets) take 15%, and a small fraction (<1%) are taken in subsistence, personal use, and recreational fisheries (1985–2005 averages). In addition to the semelparous salmon, the Bristol Bay region also has healthy populations of rainbow trout *O. mykiss*, Dolly Varden *Salvelinus malma*, Arctic char *S. alpinus*, and a variety of other fishes in a community that is entirely native. The habitat is in excellent condition; there is no commercial logging, agriculture or other land-use practices to degrade the streams, and there are no dams to block or impede fish passage. The University of Washington began a program of investigations into the ecology and management of sockeye salmon in this region in 1946, with emphasis on the Wood River system and, in the 1950s, the Kvichak River system. Many of these data sets have been maintained by the University of Washington or ADF&G, and they are the basis for many of the examples in this paper.

Genetic Controls and Phenotypic Plasticity: When to Go to Sea and Back Again

Seaward Migration

Age and size of migrants.—The defining trait of diadromous fishes, migration between marine and freshwater environments, presents them with a recurring dilemma: do I go this year or wait until next year?" In some species, the life history simplifies the decision process because there is only one option; all migrants are the same age (McDowall 1988). However, many species vary in the number of years spent prior to one or both transitions and also in the date within the season when migration occurs. The sockeye salmon in Bristol Bay exemplify these patterns and may shed light on some of the underlying processes.

Pacific salmon spawn in the fall and for Bristol Bay sockeye salmon the period is primarily from late July to early September. A protracted period of embryonic development follows, and fry emerge from the gravel in streams and lake beaches around the time when ice breaks up the following May and June (Abrey 2005). The most common life history pattern is for juveniles to rear in a lake for 1 or 2 years prior to seaward migration, but a number of populations over the range of the species migrate to sea in their first year of life or rear in a river (Burgner 1991; Wood 1995; Gustafson and Winans 1999). Within Bristol Bay, ocean-type sockeye (i.e., those migrating to sea in their first year of life) are uncommon and largely confined to the Nushagak District (Table 1). Within the Nushagak District, they are almost exclusively in the Nushagak River itself rather than the Wood, Igushik, or Nuyakuk River systems (Table 2). Presumably, this is related to the fact that the Nushagak River has suitable spawning areas for salmon but there is no lake for the fry to rear in.

Sockeye salmon are smaller than the other salmon species at emergence (Quinn 2005), and their life history (relatively slow growth for a year or two in a lake) reflects the trade-off between the size-selective mortality at sea (favoring a long period in freshwater) and the risk of mortality in freshwater (favoring a short period there). We built a simple model to illustrate this trade-off between the risk

Table 1.—Variation among sockeye salmon from different Bristol Bay districts in the number of years spent in freshwater prior to seaward migration (data on male salmon in samples of the escapements from 1979 through 2006).

District	% freshwater age 0	% freshwater age 1	% freshwater age 2
Nushagak	5.08	88.39	6.53
Togiak	1.81	86.89	11.30
Ugashik	0.51	51.15	48.34
Naknek/Kvichak	0.16	53.47	46.37
Egegik	0.08	27.40	72.53

of mortality, which accumulates with time spent in freshwater, and the benefits of size-biased survival once the fish go to sea (Figure 2). In this model, based on generalized data from Bristol Bay, age-1 smolts have the highest overall survival, followed by age-2 smolts, and age-0 smolts fare poorly.

Age at seaward migration varies among systems (Tables 1 and 2) and one might expect that this reflects differences in growth rates among lakes. This is true to some extent, but each population apparently has its own norm of reaction whereby growth rate or size at the end of the first summer's growing season determines the probability of seaward migration the following spring. In Iliamna Lake, large average size at the end of the summer is positively correlated with the proportion of age-1 smolts from that cohort, whereas in Lake Aleknagik, the great majority of smolts leave after 1 year regardless of mean size (Figure 3). Thus the lake-specific growth and mortality rates, combined with the relationship between size and survival at sea for each system, have apparently resulted in norms of reaction that are adaptations to the biotic and abiotic features of each system.

Variation in growth among years within a given lake is largely a function of temperature and food availability (Burgner 1987). Temperature regimes during the growing season in northern lakes are affected by the date when the ice leaves, air temperatures, and wind. In general, ice has been breaking up earlier in the season recently than in the past (e.g., Iliamna Lake; Figure 4), and summer air temperatures have been warmer. In addition to the direct effect of climate, there is an indirect effect of climate through density-dependent processes. Survival rates of sockeye salmon in Bristol Bay have generally (but not universally) improved during the recent warming period that began around 1977 (Hilborn et al. 2003), and the density of juvenile sockeye salmon tends to depress growth of both that cohort (Schindler et al. 2005) and the cohort to follow in cases where a large fraction of the fish spend 2 years in the lake (Rich et al. 2009). In Lake Aleknagik, density of sockeye salmon exerts a stronger influence on their growth than temperature (Schindler et al. 2005) but the opposite pattern occurs in Iliamna Lake (Rich et al. 2009). Overall, the size of fry in Iliamna Lake reflects the combined effects of temperature and density, and both factors have contributed to the reduction in age-2 smolts from an average of 52% from 1963 to 1992 to only 28% from 1993 to 2000 (Rich et al. 2009).

Table 2.—Variation among sockeye salmon populations within the Nushagak District in the number of years spent in freshwater prior to seaward migration (data from the escapement, 1979 through 2006).

	% age 0 freshwater		% age 1 freshwater		% age 2 freshwater	
Location	females	males	females	males	females	males
Nushagak River	22.0	24.2	75.1	73.6	2.9	2.3
Nuyakuk River	2.4	4.1	90.8	90.0	6.9	5.8
Wood River	0.3	0.6	93.6	93.7	6.2	5.7
Igushik River	0.2	0.2	92.7	93.6	7.1	6.1

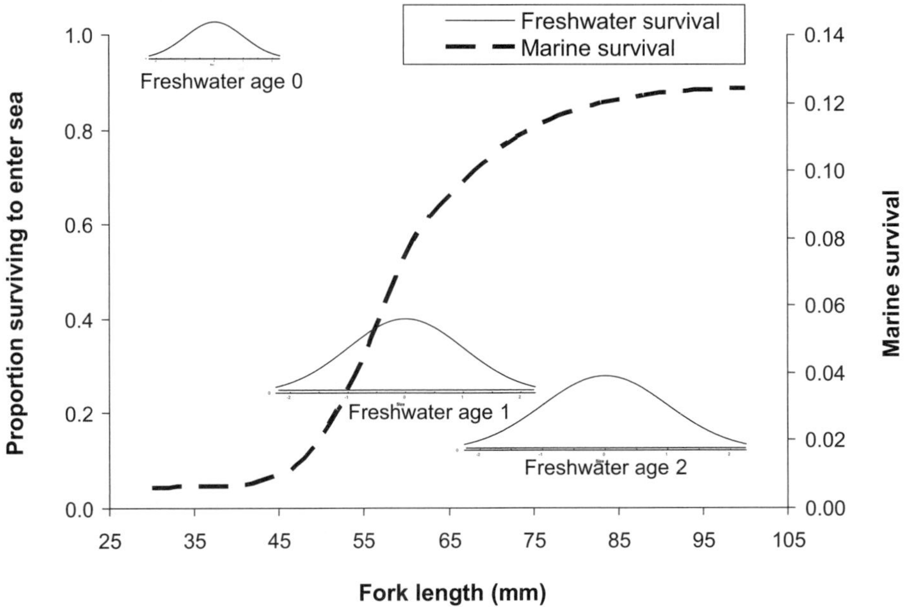

Figure 2.—Model illustrating the tradeoff between size-biased survival at sea (which favors large, old smolts) and the risks of mortality in freshwater needed to achieve such sizes. The values are hypothetical but are based on generalized data for Bristol Bay sockeye salmon. We assume high survival from emergence to seaward migration of age-0 migrants (0.95), 0.25 survival rate to age 1, and 0.5 survival rate from age 1 to age 2 in freshwater. With average growth rates in freshwater for each age-group, we assume survival rates at sea of 0.006 for age-0 smolts, 0.08 for age-1 smolts, and 0.12 for age-2 smolts.

Data collected from the Kvichak River system (1962–2000 by ADF&G) indicated that age-1 smolts were 18% shorter and 43% lighter than age-2 smolts and had lower survival rates at sea. Age-1 smolts were 51% (mean annual value from brood years 1955–1999) of the out-migrants, but only 42% of the returning adults had spent a single year in the lake. However, the age-1 smolts have avoided the mortality that might have taken place during the second year in the lake, so the net effects on productivity depend on the balance between mortality in the lake and the strength of size-selective mortality at sea (Figure 2).

The timing of migration.—In addition to the decision regarding how many years to spend in freshwater before migrating, the fish also need to decide when during the year to migrate. In various species of salmon, including sockeye, the early part of the smolt migration is dominated by larger or older fish, and those migrating later are smaller, even though they may have an extra month in freshwater to grow. Thus, the late migrants must have been quite a bit smaller than the early migrants at a common date at the beginning of the season (Figure 5). This phenomenon is not unique to sockeye salmon but is seen in other salmonids as well (Irvine and Ward 1989). The time of seawater transition is extremely important for the survival of salmon at sea, as indicated by studies in which smolt size and release date were systematically manipulated (Bilton et al. 1982) and also studies in which wild smolts were differentially marked by migration date (Lum 2003). It is not surprising that the distributions of prey, predators, and favorable physical conditions for survival vary seasonally, and so there may be some long-term average optimum date to enter the estuarine and marine waters. Interestingly, the same pattern of large migrants at the beginning of the run and smaller ones later is seen in larval *Galaxias* (McDowall et al. 1994) and *Anguilla* (Jellyman 1977) migrating from sea into freshwater in New Zealand.

Salmon have to initiate migration without any direct opportunity to sample the conditions and so

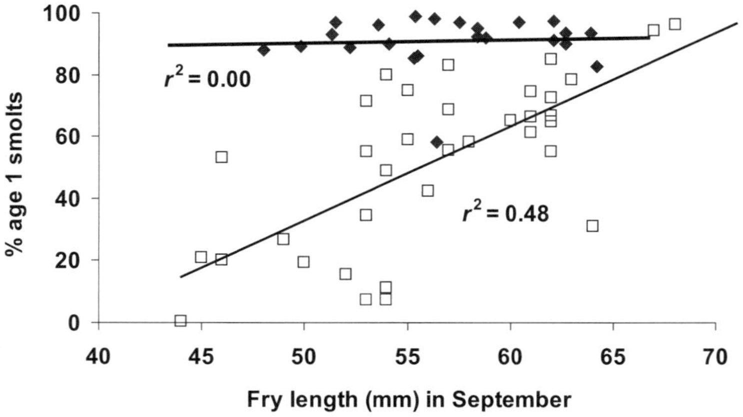

Figure 3.—Relationship between size of sockeye salmon fry at the end of the growing season (September 1) based on samples from the lake and the percentage of the cohort that migrated to sea the following spring, based on samples of smolts at the outlet, for lakes Aleknagik (solid diamonds) and Iliamna (open squares).

respond to changes in photoperiod in a population-specific manner (Stewart et al. 2006) to get them to the ocean at the right time of year. However, survival at sea is also positively related to smolt size within a cohort for many salmonid species (Reimers 1971; Healey 1982; Ward et al. 1989; Holtby et al. 1990; Henderson and Cass 1991; Yamamoto et al. 1999), so what is a small individual to do? The very smallest may postpone seaward migration for another year, but those of intermediate size delay migration until later in the season, presumably to grow somewhat larger and increase their probability of survival, despite the fact that they are giving up the opportunity to grow much more rapidly at sea than they could in freshwater. As indicated in the section below, smaller smolts often spend an extra year at sea, compared to larger ones, so this is also part of the complex trade-off that they experience.

Finally, as with everything else, there is environmental modulation of the date of seaward migration. In general, warm spring temperatures tend to advance the date of migration and in northern systems the date when ice leaves can be important as either an index of spring conditions or as a direct causal influence (Burgner 1962; Quinn 2005). Early spring warming is likely to affect not only the

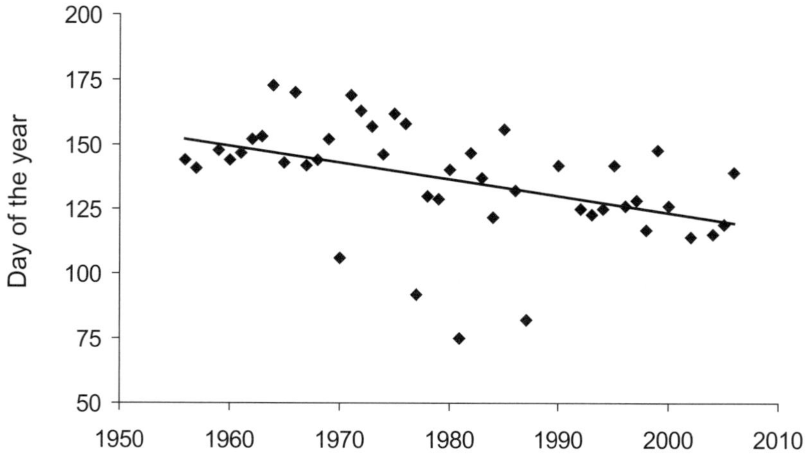

Figure 4.—Trend in the date of ice-off on Iliamna Lake, 1956–2006.

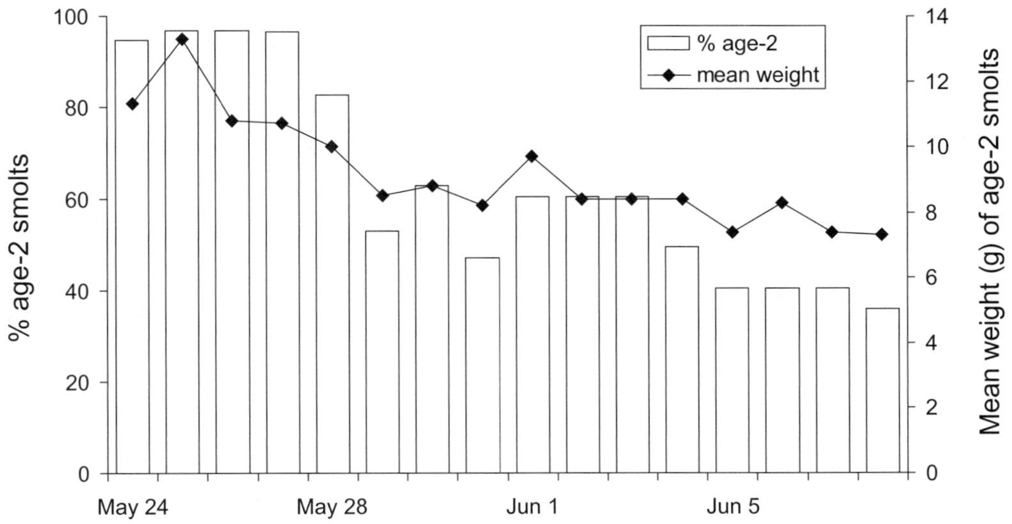

Figure 5.—Shift in size and age of sockeye salmon smolts migrating down the Kvichak River (data from 1992 as an example), showing the greater proportion of older (age 2 rather than age 1) smolts at the beginning of the season, and the tendency for larger age-2 smolts to migrate earlier than smaller ones of the same age (Crawford and Cross 1994).

freshwater environment where the fish are residing prior to seaward migration, but also the conditions along the migratory corridor and in the estuarine environment as well, so environmental modulation of migration timing would seem to have substantial adaptive value. However, when environmental conditions affecting the fish in freshwater are not linked to conditions at sea (e.g., increased temperatures as a result of logging operations) the resulting mismatch between migration and favorable ocean conditions can reduce survival (Holtby and Scrivener 1989).

Return Migration from the Sea

Age and size of migrants.—The processes by which juvenile salmon in freshwater make the commitment to migrate to sea in a given year or delay to the next year are mirrored by those made by immature salmon at sea—to mature and return this year or remain at sea. The benefits of staying at sea are only obtained after the fish leave the ocean to reproduce. Females benefit from large size by having larger and more numerous eggs (Beacham and Murray 1993; Quinn et al. 1995) and by having the ability to obtain nesting sites (Foote 1990) and to bury their eggs deeper, hence safer from disturbance (Steen and Quinn 1999). However, all of these relationships are highly variable, and early models indicating great benefits of large size (van den Berghe and Gross 1989) have not been supported by DNA parentage analysis (Dickerson et al. 2005; Seamons et al. 2007). Nevertheless, some balance between reproductive benefits and marine mortality (Healey 1987) would seem to be the appropriate way to consider age and size at maturity in female salmon. Likewise, large males tend to dominate smaller males in competition (Fleming and Gross 1994; Quinn and Foote 1994) and are favored as mates by females (Foote 1989; Berejikian et al. 2000). As with females, the realized benefits of large size in males are less than one might expect, but all other things being equal, larger males generally achieve greater reproductive success than smaller males (Seamons et al. 2007).

Male and female salmon differ in age at maturity in almost all species. Some anadromous males, known as jacks, mature a year younger than the youngest females of the population, thus reducing their risk of mortality at sea in exchange for a less competitive status (Gross 1985; Allen et al. 2007) and rejection by females (Foote 1989). The phenomenon of variation in age at maturity between sexes is common in fishes, including many diadromous species, and females are often older and larger than males: Atlantic sturgeon *Acipenser oxyrinchus* (Smith 1985), striped bass *Morone saxatilis* (Rulifson and Dadswell 1995), American shad *Alosa sapidissima* (Carscadden and Leggett 1975; Jessop 1975; Saunders 1984) and most notably in anguillid eels (Helfman et al. 1987; Jellyman 2003).

The growth of sockeye salmon at sea varies among years, reduced by density of conspecifics and enhanced by warmer temperatures (Rogers 1980; Rogers and Ruggerone 1993; Pyper and Peterman 1999). Correlations among populations are strongest within regions (e.g., Bristol Bay), but the broad areas over which growth and age at maturity are correlated indicate either extensive spatial overlap among populations or the influence of large-scale environmental processes (Pyper et al. 1999). In addition, there is evidence that competition for food with pink salmon *Oncorhynchus gorbuscha*, the most abundant species of salmon, also depresses the growth of Bristol Bay sockeye salmon (Ruggerone et al. 2003, 2005; Ruggerone and Nielsen 2004). So, broad-scale climate- and density-driven processes affect growth of sockeye salmon during their period of foraging at sea. However, these factors do little to explain the long-term variation in age at maturity among populations, nor the variation within a population in a given year.

In general, the sockeye salmon of a given age at sea that are maturing are larger than immature fish of the same age (French et al. 1976), indicating a link between size or growth and maturity. Among years, there is a negative correlation between length at age and age at maturity in sockeye salmon (Pyper et al. 1999), though the patterns are not very strong and the processes determining them seen to occur at different time periods.

> [I]nterannual variation in adult length of sockeye salmon appears to be largely determined by conditions during the final year at sea. In contrast, environmentally induced variation in age at maturity likely occurs within the first 2 years of marine life. (Pyper et al. 1999)

Variation among years in length at age was more closely correlated among populations than age at maturity (Pyper et al. 1999), suggesting stronger environmental (e.g., temperature and competition) controls over growth and stronger genetic controls over age at maturity. Examination of age composition records (e.g., Table 3) reveals that the number of years spent at sea is inversely related to the number of years spent in freshwater. That is, individuals that mature early (after 2 years and especially after 1 year at sea) are primarily those that went to sea after 2 years in freshwater rather than one. This is consistent with the finding that early maturity in male Chinook salmon *O. tshawytscha* and coho salmon *O. kisutch* was unrelated to growth opportunity at sea, but rather was a function of smolt size; cohorts of larger smolts produced a higher proportion of early maturing adults than smaller smolts (Vøllestad et al. 2004). However, a progressive shift towards older age at maturity has been noted for sockeye salmon from Bristol Bay and other regions, suggesting that broad-scale processes such as density-dependent growth at sea may be operating on age at maturity (Holt and Peterman 2004).

In addition to the roles of growth at sea (mediated by competition, temperature, and other environmental factors) and the size of the salmon when they left freshwater, the age at maturity is also strongly affected by genetics. This can be readily seen in wild populations that experience common environmental conditions. For example, Lake Aleknagik, in the Wood River system, is the shared rearing habitat for juvenile sockeye salmon spawned in various rivers and streams flowing into it. After a

Table 3.—Relationship between the number of years spent by sockeye salmon in freshwater and the number of years they spent at sea, expressed as a percent of the sample for each freshwater age-group (data from the Egegik District escapement samples of males from 1979 to 2005; Alaska Department of Fish and Game; data on the relatively scarce age-0 smolts were excluded from analysis).

	Years spent at sea			
Years in freshwater	1	2	3	sample
1	0.92	38.05	61.03	8,142
2	5.07	56.87	38.05	27,385

year in the lake, these fish all go to sea and return synchronously, affording them the same growing opportunities. Notwithstanding these common environments, there is a clear tendency for the populations spawning in the large rivers to be older at maturity than those spawning in smaller streams (Rogers 1987; Quinn et al. 2001; Table 4). Thus, the oldest (hence largest) fish tend to spawn in the deepest, fastest rivers. This may reflect the selective effects of bear predation against large fish in small creeks (Quinn and Buck 2001; Quinn et al. 2001) or the advantages that large fish may have in holding position and spawning in a large river with fast water and large gravel.

Based on the general inverse relationship between growth and year of transition between freshwater and marine environments, we would expect that salmon from populations characterized by delayed maturity would be smaller for their age than fish of the same age from populations that mature at a younger age. Paradoxically, the opposite is the case. In the populations where selection seems to favor small fish, the salmon are both young and small for their age. For example, Hansen Creek, a narrow and shallow stream, has 6.6 times more jacks (male sockeye salmon that spent only 1 year at sea) than nearby but larger Bear Creek (1.91 versus 0.29% of the males based on annual samples from 1956 to 2006). This higher proportion of jacks in Hansen Creek does not result from faster growth; Bear Creek fish are heavier than Hansen Creek fish at a given age at maturity (2.04 versus 1.62 kg after 2 years at sea and 3.19 versus 2.52 kg after 3 years). Thus, there are different norms of reaction controlling the relationship between growth and age at maturity among populations, though it is unclear how some fish grow so much faster than others, given the same opportunities.

Ultimately, the decision as to whether a salmon should stay at sea or begin the process of maturation and return to freshwater depends on the balance between the risk of mortality during the extra year at sea and the reproductive benefits associated with the increase in size that the year will bring. It is not possible to determine with certainty the rate of mortality experienced by salmon during specific periods at sea, but all the models (Mathews 1968; Ricker 1976) assume and data (Parker 1968; Pearcy 1992) indicate that mortality rate is a declining function of size and duration at sea. Nevertheless, the mortality does not go to zero, and even large salmon experience some risk at sea. We built a simple model (Figure 6) in parallel to that built for seaward migration (Figure 2), using length at age data from Wood River and Iliamna Lake sockeye salmon, length—fecundity data for sockeye salmon and kokanee (lacustrine sockeye salmon) (Quinn et al. 1998), an annual mortality rate of 0.5 from marine age 1–2, and 0.42 from ages 2–3 and 3–4 (Mathews 1968). By multiplying the probability of surviving to a given age times the reproductive output (simplistically measured as egg production) at the length the fish might achieve at that age, we can see that maturation by females after 1 year at sea is unsuccessful because they produce too few eggs for replacement, and spending 4 years at sea is too risky to compensate for the increase in fecundity. Marine ages 2 and 3 are optimal, with the balance being determined by rates of growth and mortality and the fecundity–length relationship. The similar process pertains, in reverse, to catadromous fish "deciding" when to leave freshwater, where they are presumably comparatively safe, and return to sea to breed (Helfman et al. 1987; Jellyman 2001).

The timing of migration.—The relatively young age at maturity of most salmon indicates that the combination of declining growth rate and accumu-

Table 4.—Age at maturity of sockeye salmon from a series of populations in tributaries of the Wood River system, Alaska, expressed as the percent of the adults that had spent 1, 2, or 3 years at sea. Samples of about 100 males and 100 females were collected annually from 1946 to 2006.

Spawning site	Mean width (m)	% 1 ocean	% 2 ocean	% 3 ocean
A & C creeks	1.8	0.3	80.7	19.0
Hansen Creek	3.8	0.8	70.5	28.7
Bear Creek	5.1	0.1	49.0	50.9
Ice Creek	15.7	0.1	35.7	64.2
Agulowok River	77.3	0.01	21.7	78.3

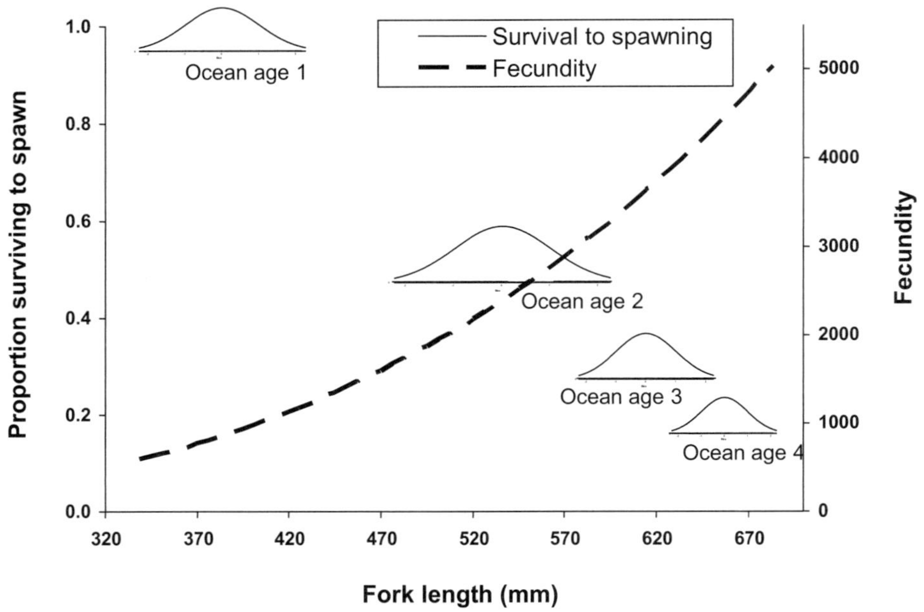

Figure 6.—Model illustrating the trade-off between survival at sea from year to year and the egg production associated with the average lengths achieved at those ages. The values are based on generalized data for Bristol Bay sockeye salmon. We assume that salmon have already survived 1 year at sea and can mature at marine ages 1, 2, 3, or 4. High survival rate (0.95) is assumed for fish maturing at marine age 1, with 0.5 survival to age 2 and 0.58 annual survival rate thereafter (Mathews 1968). Fecundity is a power function of length, based on data for sockeye salmon and kokanee (Quinn et al. 1998).

lating risk of mortality generally favor a short stay at sea, in terms of the number of years. However, the time within the year that the salmon leave the ocean holds fascinating mysteries. One might think that because they went to sea to grow, the salmon would stay at sea as long as possible and only return to freshwater with sufficient time to migrate to their spawning grounds, complete final maturation, spawn, and then die or return to sea. In some cases, this is the pattern, and many pink salmon and chum salmon *O. keta*, for example, are in an advanced state of maturity when they leave saltwater, and they spawn only a few days later (Heard 1991; Salo 1991). On the other hand, some Chinook salmon leave the ocean in spring and spawn in fall (Healey 1991), whereas others leave the ocean only about a month prior to spawning. Likewise, stream-maturing steelhead (anadromous rainbow trout) migrate in late summer and fall but do not spawn until the following spring, whereas ocean-maturing steelhead migrate in late winter and spawn in spring (Busby et al. 1996). The timing of return from the ocean is among the most important, defining traits of populations and tends to be quite consistent among years. In some cases, the early arrival of the population seems to be an adaptation to avoid stressful thermal conditions. So, for example, some sockeye salmon populations enter freshwater in early summer, hold below the thermocline in a lake, and ascend spawning streams in fall (Hodgson and Quinn 2002; Newell and Quinn 2005). If they tried to migrate up the outlet of the lake in late summer or early fall, the water might be too warm, and so they sacrifice the opportunity to grow at sea in order to reach favorable breeding grounds at the appropriate time to spawn. This pattern of premature migration is seen primarily at the southern end of the range of sockeye salmon.

The migration timing of Bristol Bay sockeye salmon is noteworthy (and somewhat atypical of the species and of salmon in general) because a large number of populations migrate very rapidly from the ocean through coastal and estuarine waters within a very narrow window in time.

> From 1956–1976, for example, 80% of the Bristol Bay run passed through the fishery during a 12.9 d period (SD 1.58 d) with a mean of 4 July … and standard deviation of only 2.92 d. (Burgner 1980)

This compression in migration timing is especially remarkable because the several hundred discrete populations (totaling tens of millions of adult salmon) do not spawn synchronously; each population has a specific timing that might be as early as late July or as late as September. Consistent timing of spawning by each population is characteristic of salmon and is generally correlated with the temperatures to be experienced by developing embryos, with earlier spawning in colder water (Brannon 1987; Webb and McLay 1996; Hodgson and Quinn 2002). What is unusual is the fact that the timing of migration does not closely correspond to the timing of spawning (Jensen and Mathisen 1987), unlike the orderly sequence that is seen in other systems such as the Fraser River (Killick 1955; Woodey 1987). Why do the populations that spawn in late August and early September not stay at sea for an extra month to grow larger, rather than migrating from sea at the same time as populations that spawn in July?

In some cases, there are obvious environmental factors that would seem to select for premature migration by the population, such as thermal barriers or difficulties in migrating past rapids at certain flow conditions. However, in other cases, it is simply not obvious why the salmon migrate into freshwater so long before they spawn. In some rivers in Scotland (Shearer 1990) and Ireland (Went 1964; Quinn et al. 2006), the salmon run at virtually all months of the year. Some of this variation may result from the amalgamation of data from many populations, each with discrete but different timing, but in other cases, a discrete population may indeed have a huge range of migration dates (Quinn and Myers 2004). Thus, there is variation among populations in both how early the salmon migrate into freshwater, relative to their spawning date (i.e., the length of the holding period), and also in the extent to which the migration and spawning periods are compressed or protracted (Mathisen 1994). There does not seem to have been a systematic study of these phenomena, but the patterns of variation presumably reflect the balance between the benefits of growth at sea and the difficulties in reaching the spawning grounds and surviving until the spawning season, and the extent to which conditions experienced by adults and progeny vary from year to year. Rivers with predictable regimes of temperature and flow during the migration and spawning periods, and predictable, narrow windows for optimizing juvenile emergence and growth the following spring, should select for salmon to migrate and spawn at the optimal dates. Variation in conditions experienced by migrants and spawners should disrupt the selection and result in protracted timing because in some years, the early fish are successful, but in other years, the later ones are.

Interestingly, within a population, the timing of return is linked to age, size, and sex of the mature fish. Just as the older and larger smolts tended to migrate before the younger and smaller ones, the early migrating salmon are often older and larger than the later ones. This pattern, well known and often very dramatic in Atlantic salmon *Salmo salar* (Shearer 1990; McKinnell et al. 1994; Stewart et al. 2002; Quinn et al. 2006), is common (but more subtle) in Pacific salmon as well, including those in Bristol Bay (Figure 7). The fascinating pattern of earlier arrival by larger salmon has not been adequately explained, and there are three classes of explanations: those related to migration performance itself, those related to processes operating at sea, and those related to processes operating on the spawning grounds. In the first instance, if large and small salmon departed from the same areas at sea at the same time, we would expect earlier arrival by larger fish because they swim faster. This might explain the comparatively subtle differences in timing seen in some species, but in others, the larger, older salmon arrive months prior to the arrival of smaller, younger salmon, so it is not plausible that migration speed alone is the explanation. Second, the benefits and risks of staying at sea to forage may depend on fish size. For example, a salmon that is already large may grow comparatively less than a small salmon if each stays an extra month at sea. To the extent that these gains in size translate into reproductive success, the smaller salmon gains more by staying at sea than the larger salmon. There is some risk of mortality during the period at sea, so the large salmon may strike a different balance between foraging benefits and mortality risks than small salmon. Finally, the patterns of migration timing and body size may reflect processes taking place on spawning grounds. Large salmon commonly breed before small salmon. In the case of females, this may be because

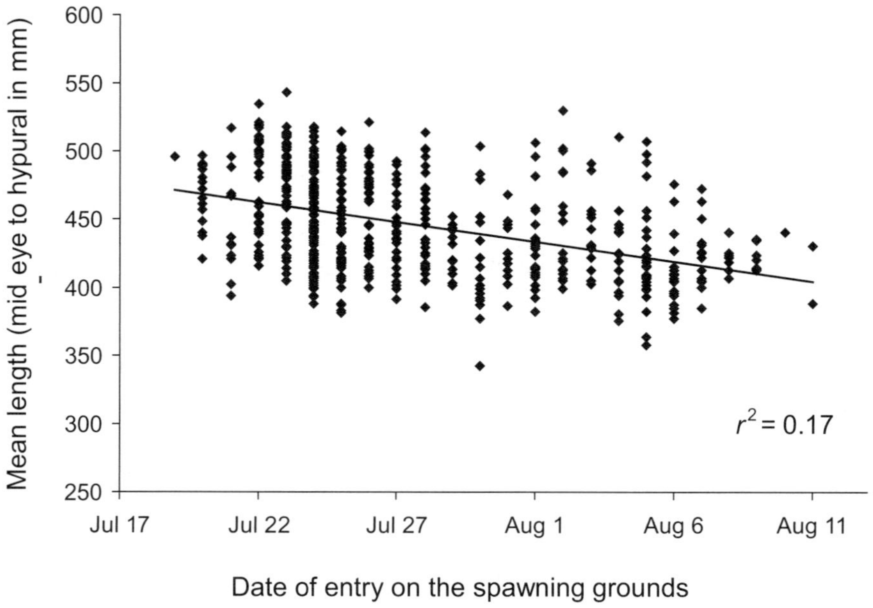

Figure 7.—Date of entry of female sockeye salmon on the spawning grounds at Hansen Creek as a function of body size. Data based on individuals measured and tagged in Lake Aleknagik and monitored on arrival into the stream, 1997–2006.

large females can breed at the optimal date but smaller females cannot risk nest disturbance by the larger females (van den Berghe and Gross 1984; Steen and Quinn 1999) and so are constrained to breed later. The decline in male size over the season might be linked to that in females. Each of these three types of explanations has merit, and they are not mutually exclusive. Comparisons to species without parental care such as striped bass (Grant 1974) and alewife *Alosa pseudoharengus* and blueback herring *A. aestivalis* (Loesch 1987) that also show this size-biased arrival pattern might help elucidate the reasons for it.

In addition for the tendency for older salmon to return earlier than younger salmon, males tend to migrate earlier than females through coastal waters (McKinstry 1993), up large rivers, and onto spawning grounds (Morbey 2000) (Figure 8). This seems puzzling at first because all nest site selection and preparation are done by females. However, the females complete spawning soon after they arrive, and the operational sex ratio soon becomes very male-biased because all males are courting the few remaining ripe females. Thus males lose more in reproductive opportunities by being a bit late rather than a bit early, and this presumably drives the pattern that is often seen (Quinn 2005). Similar shifts in sex ratio from males to females are seen in catadromous eels (Todd 1981) and also anadromous fishes such as river herring (Loesch 1987) and striped bass. Indeed, in the latter species, this pattern is used to regulate fisheries, reducing the catch of females (Kornegay and Thomas 2004).

The timing of migration and reproduction are under strong genetic control in salmon, as indicated by the very regular patterns seen among populations (Ricker 1972; Hodgson and Quinn 2002) and the results of controlled breeding studies (Hansen and Jonsson 1991; Smoker et al. 1998; Quinn et al. 2000; Sato et al. 2000). There is some interannual variation in timing that is correlated with environmental conditions in the ocean and the rivers (Hodgson et al. 2006), but in general, the maturation process in salmonids seems to be less flexible than that of other teleosts. The very large eggs may necessitate this from a physiological standpoint, and the long period of incubation also means that the conditions experienced by the migrating adults are only weakly correlated to those experienced by their offspring, so the best strategy is to spawn at the date determined by long-term average conditions rather than rely on

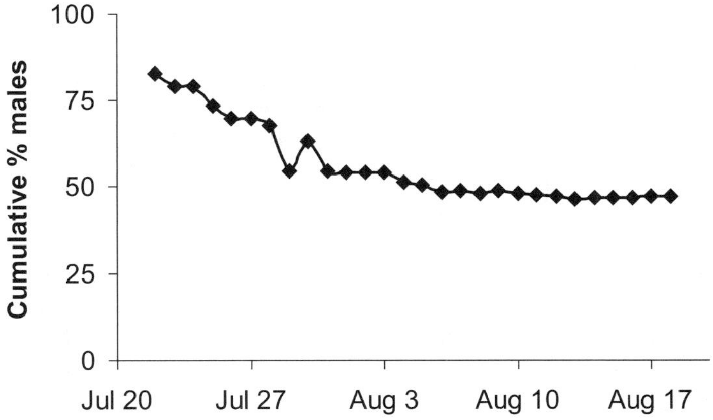

Figure 8.—Cumulative percent male sockeye salmon in Hansen Creek, Alaska in 1998, as an example of the general tendency for males to arrive on spawning grounds before females. The data are based on daily foot surveys of the entire creek on which live and dead salmon were counted and identified to sex.

proximate conditions to advance or delay migration and spawning (Quinn and Adams 1996). Such strong genetic control over migrating timing is not necessarily typical of anadromous fishes; American shad *Alosa sapidissima* migrations, for example, are strongly influenced by interannual variation in temperature (Leggett and Whitney 1972; Quinn and Adams 1996).

Scope for selection in fisheries.—The transitions to and from the ocean, common to all diadromous fishes, are sculpted by combinations of natural and sexual selection interacting with variable environmental conditions. As with many diadromous fishes, salmon are subjected to intensive fishing during their migration, and this constitutes an important source of mortality at what would otherwise be a comparatively safe period in their lives. The fishing mortality decreases the abundance of adults, and the general goal is to bring the spawning population down to the level at which sustainable yield is optimized by reducing density dependence (Ricker 1954; Hilborn et al. 1995). The density of juveniles affects growth, and growth affects smolt age, size, and timing, and these traits affect the duration of marine residence and survival as described above. Fishing can thus have profound phenotypic effects on the population, but it can also have important selective effects because all individuals are not equally likely to get caught.

Exploitation of immature fish increases the mortality rate and so decreases the proportion of fish living to older ages, reducing the age and size distribution of the population. Extensive exploitation of both maturing and immature salmon took place on the high seas during part of the 20th century (Harris 1987) and continues to occur in coastal waters. To the extent that the age at maturity is a heritable trait, we have been selecting against fish that were predisposed to mature at an old age, and this is one of the possible causes of the declines in size of salmon (Ricker 1981). In the case of Bristol Bay sockeye salmon, exploitation of immature fish that had spent two winters at sea (Harris 1987) would have selected against older fish. This presumably resulted in lower exploitation rates of populations that were genetically disposed to spend 2 years at sea and also favored younger fish within populations.

As salmon mature and migrate homeward, they are subjected to various forms of fishing gear, and Bristol Bay sockeye salmon are primarily taken in gill nets. These nets can be highly size-selective on salmon (Todd and Larkin 1971) and other migratory fishes (Mansueti 1961). Depending on the size of the mesh and the size distribution of the fish, they may be more likely to catch fish of large, small, or intermediate sizes. Detailed analysis of the salmon caught in commercial gill nets and those escaping to spawn over many decades in the Wood River system revealed a general tendency for the fishery to exploit large fish (Kendall 2007). However, the populations in this system vary in age distribution, size at age, and body depth (Quinn et al. 2001; Rogers 1987), and fish that are deep-bodied on the spawning grounds were also greater in girth when they passed through the fisheries (Hamon et al. 2000). There is considerable variation in exploitation rate among populations, as well as selection for fish with certain phenotypes within populations (Kendall and Quinn, in press). It would be tempting to test the hypothesis that the selection has changed the size distribution of the salmon run as a whole by simply looking at a time series of length and age data. However, the selective effects of the fishery would have to be disentangled from the effects of climate and density (McKinnell 1995; Cox and Hinch 1997; Pyper and Peterman 1999; Ruggerone et al. 2005), and the effects of high seas fishing (Harris 1987), greatly complicating this kind of analysis.

In addition to the fact that the Bristol Bay gillnet fisheries are size-selective, there is also evidence for selection with respect to the timing of migration. In this area, the fisheries are managed on the basis of escapement goals. That is, estimates are made of the number of adults needed to reach the carrying capacity of each system, and the manager opens and closes the fishery on a daily basis to ensure that this goal is met if at all possible, using both preseason forecasts, test fisheries, and in-season assessment of abundance (Minard and Meacham 1987). Given the uncertainties surrounding forecasts the variation in timing among years, cautious managers may restrict the fishing during the beginning of the run until it is clear that the overall escapement goal will be met and then will allow higher exploitation rates to permit fishermen to take the surplus. Records from the Egegik District illustrated this pattern clearly, and there is evidence that the overall timing is getting earlier, as would be predicted from this kind of selection (Quinn et al. 2007).

Therefore, the fisheries tend to be selective, taking larger than average fish and taking a higher proportion of the later-migrating fish. However, as was noted earlier, the late migrants tend to be smaller than the earlier ones, so there is a complex interaction between size, migration timing, and vulnerability to the fishery. If all the discrete breeding populations within a system (e.g., the fishery management district) migrated simultaneously, the size-selective fishery would tend to exploit the populations with large-bodied fish more than the populations with smaller fish (in average age, length at age, and girth). The temporal selection would tend to remove the later migrants from all populations if they migrated simultaneously, but if they migrated sequentially, then they would experience different exploitation rates. Tributary-specific migration timing is well known in other salmonids (Boatright et al. 2004; Killick 1955; Stewart et al. 2002), but initial research indicated that all breeding populations migrated at the same time, even though they spawned at different times in the Kvichak River (Jensen and Mathisen 1987; Smith 1964) and Wood River (Rowse 1985). However, a large-scale tagging study in 2006 indicated that salmon spawning in creeks tended to migrate earlier than those spawning in large rivers, consistent with the tendency for spawning to occur later in the rivers (Doctor 2008). Moreover, in one intensively monitored creek, the early migrants also tended to enter the spawning grounds earlier than later migrants, though sockeye salmon elsewhere did not show this pattern (Newell et al. 2007).

Conclusions

The sockeye salmon of Bristol Bay, whose biology and management have been briefly reviewed here, are not necessarily typical of Pacific salmon, much less all diadromous fishes. Nevertheless, many of their life history transitions and conservation challenges are shared by taxonomically distant fishes with different migration patterns, so they may serve as a model against which others can be contrasted. The option of growing in one environment, where spawning took place, prior to making the first transition to the other rearing environment is common to all diadromous fishes (McDowall 1988). In some species, the duration of residence prior to migration is fixed, but often there is some variation, and presumably the same ecological factors (growth rate and risk of mortality in the present environment and size-selective mortality in the other environment) determine how long the fish stay where they are before migrating. Likewise, immature adults must migrate in order to spawn, and this requires a parallel set of life history decisions, involving growth rate and mortality risk in the present environment and the benefits of size for migration performance and reproduction in the other environment. The question of how many years an eel should spend in a river prior to seaward migration, and why it differs between males and females (Helfman et al. 1987;

Jellyman 2003), is not fundamentally different from the question of how long a salmon should spend in the ocean, despite the obvious differences between salmon and eels.

The two life history transitions of diadromous fishes to and from the ocean are controlled by a combination of genetic and environmental processes. In some species, the nature of larval development, metamorphic changes, and maturation processes may dictate that the timing and size vary a great deal from year to year, whereas in other species, genetic controls may predominate. However, not only will there be both genetic and environmental influences, but we can expect that the norms of reaction will differ among populations or closely related species, such that common growth rates produce different effects. The details about growth rate, smolt size and transformation date, duration of marine residence, egg size and fecundity, timing of return migration and spawning, and other life history traits in salmon serve as examples of things that may also be true of other fishes or are different from other fishes in interesting ways.

Finally, like many migratory fishes, the sockeye salmon in Bristol Bay are subjected to fisheries that are not completely matched to the structure of the populations. The salmon migrate rapidly from vast areas of the open ocean, and hundreds of thousands or even millions of them stream through nearshore and estuarine waters in various districts each day at the height of the season. The fishery takes fish that are homing to different breeding sites (e.g., one creek or another) and different habitats (e.g., a beach, creek, or river) at different rates, depending on the interaction between their body size and migratory timing and the mesh size of the nets and the dates when fishing is open and closed. It is simply impractical for the fisheries to be managed on a creek-by-creek basis because there are so many different creeks at remote locations that the processing would be impossible and the fish are of low commercial value by the time they reach the spawning grounds. Therefore, the management must compromise between the fine-scale structure of the populations and the practicalities of monitoring huge runs while the salmon are still migrating and not yet on the specific spawning grounds and managing the fisheries in a manner that is conservative yet recognizes the financial implications of all decisions (Minard and Meacham 1987). The Alaska Department of Fish and Game biologists do remarkably well with this exceptionally challenging balance between the need to maintain sufficient number of spawning salmon and the expectations of the fishermen and the processors.

Mixed stock fisheries are a common problem for management, especially when the fish migrate across governmental boundaries in coastal areas or enter international waters. The very precise timing patterns that characterize many diadromous fishes only make them more vulnerable to fishing. The most obvious effect of fishing, the reduction in density, necessarily attracts the most attention, but there is increasing awareness of the possible evolutionary effects of size-selective fishing (Conover and Munch 2002; Conover et al. 2005). The long-term effects of size-selective fishing will depend on the balance between genetic and environmental effects on growth and age at maturation in the population and the intensity of selection (Hard 2004). Likewise, many fisheries are selective with respect to migratory timing because of exploitation patterns (Consuegra et al. 2005) and management regulations, though this has not attracted the same level of scientific attention as size-selection. Documenting such selection and then considering the possible long-term consequences will provide scientists working on such species with plenty of work in the future. All of these processes will be playing out in the 21st century as we also experience changing climate conditions, posing even greater challenges for the fishes and those working on them.

Acknowledgments

The program of research in Bristol Bay was initiated decades ago, and we gratefully acknowledge the pioneering scientists for their intellectual vision and hard work, especially Robert Burgner, the late Donald Rogers, the late Ole Mathisen, and the many staff and students working with the program over many decades. More recently, Ray Hilborn and Daniel Schindler have energized the program, and the results presented in this paper reflect many of their ideas and efforts. Financial support for the program has come from many sources, including the commercial salmon processing industry, the National Science Foundation, the Gordon and Betty Moore Foundation, and the University of Washington's School of Aquatic and Fishery Sciences. We also thank the Alaska Department of Fish and Game for access to data and for cooperation in many aspects of our program.

References

Abrey, C. A. 2005. Variation in the early life history of sockeye salmon (*Oncorhynchus nerka*): emergence timing, an ontogenetic shift, and population productivity. Doctoral dissertation. University of Washington, Seattle.

Allen, C. S., H. B. Rich, and T. P. Quinn. 2007. Condition-dependent reproductive tactics by large and small anadromous male sockeye salmon *Oncorhynchus nerka*. Journal of Fish Biology 70:1302–1307.

Beacham, T. D., and C. B. Murray. 1993. Fecundity and egg size variation in North American Pacific salmon (*Oncorhynchus*). Journal of Fish Biology 42:485–508.

Berejikian, B. A., E. P. Tezak, and A. L. LaRae. 2000. Female mate choice and spawning behaviour of Chinook salmon under experimental conditions. Journal of Fish Biology 57:647–661.

Bilton, H. T., D. F. Alderdice, and J. T. Schnute. 1982. Influence of time at release of juvenile coho salmon (*Oncorhynchus kisutch*) on returns at maturity. Canadian Journal of Fisheries and Aquatic Sciences 39:426–447.

Boatright, C., T. P. Quinn, and R. Hilborn. 2004. Timing of adult migration and stock structure of sockeye salmon in Bear Lake, Alaska. Transactions of the American Fisheries Society 133:911–921.

Brannon, E. L. 1987. Mechanisms stabilizing salmonid fry emergence timing. Canadian Special Publication of Fisheries and Aquatic Sciences 96:120–124.

Burgner, R. L. 1962. Studies of red salmon smolts from the Wood River Lakes, Alaska. Pages 247–314 *in* T. S. Y. Koo, editor. Studies of Alaska red salmon. University of Washington Press, Seattle.

Burgner, R. L. 1980. Some features of ocean migrations and timing of Pacific salmon. Pages 153–164 *in* W. J. McNeil, and D. C. Himsworth, editors. Salmonid ecosystems of the North Pacific. Oregon State University Press, Corvallis.

Burgner, R. L. 1987. Factors influencing age and growth of juvenile sockeye salmon (*Oncorhynchus nerka*) in lakes. Canadian Special Publication of Fisheries and Aquatic Sciences 96:129–142.

Burgner, R. L. 1991. Life history of sockeye salmon (*Oncorhynchus nerka*). Pages 1–117 *in* C. Groot, and L. Margolis, editors. Pacific salmon life histories. University of British Columbia Press, Vancouver.

Busby, P. J., T. C. Wainwright, G. J. Bryant, L. J. Lierheimer, R. S. Waples, F. W. Waknitz, and I. V. Lagomarsino. 1996. Status review of West Coast steelhead from Washington, Idaho, Oregon and California. National Marine Fisheries Service, Technical Memorandum NWFSC-27, Seattle.

Carscadden, J. E., and W. C. Leggett. 1975. Life history variations in populations of American shad, *Alosa sapidissima* (Wilson), spawning in tributaries of the St John River, New Brunswick. Journal of Fish Biology 7:595–609.

Conover, D. O., S. A. Arnott, M. R. Walsh, and S. B. Munch. 2005. Darwinian fishery science: lessons from the Atlantic silverside (*Menidia menidia*). Canadian Journal of Fisheries and Aquatic Sciences 62:730–737.

Conover, D. O., and S. B. Munch. 2002. Sustaining fisheries yields over evolutionary time scales. Science 297:94–96.

Consuegra, S., C. García de Leaniz, A. Serdio, and E. Verspoor. 2005. Selective exploitation of early running fish may induce genetic and phenotypic changes in Atlantic salmon. Journal of Fish Biology 67(Supplement A):129–145.

Cox, S. P., and S. G. Hinch. 1997. Changes in size at maturity of Fraser River sockeye salmon (*Oncorhynchus nerka*) (1952–1993) and associations with temperature. Canadian Journal of Fisheries and Aquatic Sciences 54:1159–1165.

Crawford, D. L., and B. A. Cross. 1994. Bristol Bay sockeye salmon smolt studies for 1992. Alaska Department of Fish and Game, Technical Fishery Report 94–19, Juneau.

Day, T., and L. Rowe. 2002. Developmental thresholds and the evolution of reaction norms for age and size at life-history transitions. American Naturalist 159:338–350.

Dickerson, B. R., K. W. Brinck, M. F. Willson, P. Bentzen, and T. P. Quinn. 2005. Relative importance of body size and timing of arrival to breeding grounds in reproductive success of pink salmon. Ecology 86:347–352.

Doctor, K. 2008. Fishing out evolution? Spatial and temporal patterns of migration in sockeye salmon. Master's thesis. University of Washington, Seattle.

Fleming, I. A., and M. R. Gross. 1994. Breeding competition in a Pacific salmon (coho: *Oncorhynchus kisutch*): measures of natural and sexual selection. Evolution 48:637–657.

Foote, C. J. 1989. Female mate preference in Pacific salmon. Animal Behaviour 38:721–722.

Foote, C. J. 1990. An experimental comparison of male and female spawning territoriality in a Pacific salmon. Behaviour 115(3–4):283–313.

French, R., H. Bilton, M. Osako, and A. Hartt. 1976. Distribution and origin of sockeye salmon (*Oncorhynchus nerka*) in offshore waters of the North Pacific Ocean. International North Pacific Fisheries Commission Bulletin 34:1–113.

Grant, G. C. 1974. The age composition of striped bass catches in Virginia rivers, 1967–1971, and a description of the fishery. Fishery Bulletin 72:193–199.

Gross, M. R. 1985. Disruptive selection for alternative life histories in salmon. Nature (London) 313:47–48.

Gross, M. R., R. M. Coleman, and R. M. McDowall. 1988. Aquatic productivity and the evolution of diadromous fish migration. Science 239:1291–1293.

Gustafson, R. G., and G. A. Winans. 1999. Distribution and population genetic structure of river- and sea-type sockeye salmon in western North America. Ecology of Freshwater Fish 8:181–193.

Hamon, T. R., C. J. Foote, R. Hilborn, and D. E. Rogers. 2000. Selection on morphology of spawning wild sockeye salmon by a gill-net fishery. Transactions of the American Fisheries Society 129:1300–1315.

Hansen, L. P., and B. Jonsson. 1991. Evidence of a genetic component in the seasonal return pattern of Atlantic salmon, *Salmo salar* L. Journal of Fish Biology 38:251–258.

Hard, J. J. 2004. Evolution of chinook salmon life history under size-selective harvest. Pages 315–337 *in* A. P. Hendry, and S. C. Stearns, editors. Evolution illuminated: salmon and their relatives. Oxford University Press, Oxford, UK.

Harris, C. K. 1987. Catches of North American sockeye salmon (*Oncorhynchus nerka*) by the Japanese high seas salmon fisheries, 1972–84. Canadian Special Publication of Fisheries and Aquatic Sciences 96:458–479.

Healey, M. C. 1982. Timing and relative intensity of size-selective mortality of juvenile chum salmon (*Oncorhynchus keta*) during early sea life. Canadian Journal of Fisheries and Aquatic Sciences 39:952–957.

Healey, M. C. 1987. The adaptive significance of age and size at maturity in female sockeye salmon (*Oncorhynchus nerka*). Canadian Special Publication of Fisheries and Aquatic Sciences 96:110–117.

Healey, M. C. 1991. Life history of chinook salmon (*Oncorhynchus tshawytscha*) Pages 311–393 *in* C. Groot, and L. Margolis, editors. Pacific salmon life histories. University of British Columbia Press, Vancouver.

Heard, W. R. 1991. Life history of pink salmon (*Oncorhynchus gorbuscha*). Pages 119–230 *in* C. Groot, and L. Margolis, editors. Pacific salmon life histories. University of British Columbia Press, Vancouver.

Helfman, G. S., D. E. Facey, L. S. Hales, and E. L. Bozeman. 1987. Reproductive ecology of the American eel. Pages 42–56 *in* M. J. Dadswell, R. J. Klauda, C. M. Moffitt, R. L. Saunders, R. A. Rulifson, and J. E. Cooper, editors. Common strategies of anadromous and catadromous fishes. American Fisheries Society, Symposium 1, Bethesda, Maryland.

Henderson, M. A., and A. J. Cass. 1991. Effect of smolt size on smolt-to-adult survival for Chilko Lake sockeye salmon (*Oncorhynchus nerka*). Canadian Journal of Fisheries and Aquatic Sciences 48:988–994.

Hilborn, R., T. P. Quinn, D. E. Schindler, and D. E. Rogers. 2003. Biocomplexity and fisheries sustainability. Proceedings of the National Academy of Science 100:6564–6568.

Hilborn, R., C. J. Walters, and D. Ludwig. 1995. Sustainable exploitation of renewable resources. Annual Review of Ecology Systems 26:45–67.

Hodgson, S., and T. P. Quinn. 2002. The timing of adult sockeye salmon migration into fresh water: adaptations by populations to prevailing thermal regimes. Canadian Journal of Zoology 80:542–555.

Hodgson, S., T. P. Quinn, R. Hilborn, R. C. Francis, and D. E. Rogers. 2006. Marine and freshwater climatic factors affecting interannual variation in the timing of return migration to fresh water of sockeye salmon (*Oncorhynchus nerka*). Fisheries Oceanography 15:1–24.

Holt, C. A., and R. M. Peterman. 2004. Long-term trends in age-specific recruitment of sockeye salmon (*Oncorhynchus nerka*) in a changing environment. Canadian Journal of Fisheries and Aquatic Sciences 61:2455–2470.

Holtby, L. B., B. C. Anderson, and R. K. Kadowaki. 1990. Importance of smolt size and early ocean growth to interannual variability in marine survival of coho salmon (*Oncorhynchus kisutch*). Canadian Journal of Fisheries and Aquatic Sciences 47:2181–2194.

Holtby, L. B., and J. C. Scrivener. 1989. Observed and simulated effects of climatic variability, clear-cut logging, and fishing on the numbers of chum salmon (*Oncorhynchus keta*) and coho salmon (*O. kisutch*) returning to Carnation Creek, British Columbia. Canadian Special Publication of Fisheries and Aquatic Sciences 105:62–81.

Irvine, J. R., and B. R. Ward. 1989. Patterns of timing and size of wild coho salmon (*Oncorhynchus kisutch*) smolts migrating from the Keogh River Watershed on northern Vancouver Island. Canadian Journal of Fisheries and Aquatic Sciences 46:1086–1094.

Jellyman, D. J. 1977. Invasion of a New Zealand freshwater stream by glass-eels of two *Anguilla* spp. New Zealand Journal of Marine and Freshwater Research 11:193–209.

Jellyman, D. J. 2001. The influence of growth rate on the size of migrating female eels in Lake Ellesmere, New Zealand. Journal of Fish Biology 58:725–736.

Jellyman, D. J. 2003. The distribution and biology of the South Pacific species of *Anguilla*. Pages 275–292 *in* K. Aida, K. Tsukamoto, and K. Yamauchi, editors. Eel biology. Springer-Verlag, Berlin.

Jensen, K. A., and O. A. Mathisen. 1987. Migratory

structure of the Kvichak River sockeye salmon (*Oncorhynchus nerka*) escapement, 1983. Canadian Special Publication of Fisheries and Aquatic Sciences 96:101–109.

Jessop, B. M. 1975. A review of the American shad (*Alosa sapidissima*) stocks of the Saint John River, New Brunswick, with particular reference to the adverse effects of hydroelectric developments. Environment Canada, Fisheries and Marine Service, Technical Report Series MAR/T-75–6, Halifax.

Kendall, N. 2007. Long term fishery selection on age and size at maturity of Bristol Bay, Alaska sockeye salmon. Master's thesis. University of Washington, Seattle.

Kendall, N. W., and T. P. Quinn. In press. Effects of population-specific variation in age and length on fishery selection and exploitation rates of sockeye salmon. Canadian Journal of Fisheries and Aquatic Sciences.

Killick, S. R. 1955. The chronological order of Fraser River sockeye salmon during migration, spawning and death. International Pacific Salmon Fisheries Commission Bulletin 12.

Kornegay, J. W., and C. D. Thomas. 2004. Review of striped bass fisheries and monitoring programs in North Carolina—2003. North Carolina Division of Marine Fisheries and North Carolina Wildlife Resources Commission, Raleigh.

Leggett, W. C., and R. R. Whitney. 1972. Water temperature and the migrations of American shad. Fishery Bulletin 70:659–670.

Loesch, J. G. 1987. Overview of life history aspects of anadromous alewife and blueback herring in freshwater habitats. Pages 89–103 *in* M. J. Dadswell, R. J. Klauda, C. M. Moffitt, R. L. Saunders, R. A. Rulifson, and J. E. Cooper, editors. Common strategies of anadromous and catadromous fishes. American Fisheries Society, Symposium 1, Bethesda, Maryland.

Lum, J. L. 2003. Effects of smolt length and emigration timing on marine survival and age at maturity of wild coho salmon (*Oncorhynchus kisutch*) at Auke Creek, Southeast Alaska. Master's thesis. University of Alaska, Juneau.

Mansueti, R. J. 1961. Age, growth, and movements of the striped bass, *Roccus saxatilis*, taken in size-selective fishing gear in Maryland. Chesapeake Science 2:9–36.

Mathews, S. B. 1968. An estimate of ocean mortality of Bristol Bay sockeye salmon three years at sea. Journal of the Fisheries Research Board of Canada 25:1219–1227.

Mathisen, O. A. 1994. Spawning characteristics of the pink salmon (*Oncorhynchus gorbuscha*) in the eastern North Pacific Ocean. Aquaculture and Fisheries Management 25(Supplement 2):147–156.

McDowall, R. M. 1988. Diadromy in fishes. Timber Press, Portland, Oregon.

McDowall, R. M., C. P. Mitchell, and E. B. Brothers. 1994. Age at migration from the sea of juvenile *Galaxias* in New Zealand (Pisces: Galaxiidae). Bulletin of Marine Science 54:385–402.

McKinnell, S. 1995. Age-specific effects of sockeye abundance on adult body size of selected British Columbia sockeye stocks. Canadian Journal of Fisheries and Aquatic Sciences 52:1050–1063.

McKinnell, S., H. Lundqvist, and H. Johansson. 1994. Biological characteristics of the upstream migration of naturally and hatchery-reared Baltic salmon, *Salmo salar* L. Aquaculture and Fisheries Management 25(Supplement 2):45–63.

McKinstry, C. A. 1993. Forecasting migratory timing and abundance of pink salmon (*Oncorhynchus gorbuscha*) runs using sex-ratio information. Master's thesis. University of Washington, Seattle.

Minard, R. E., and C. P. Meacham. 1987. Sockeye salmon (*Oncorhynchus nerka*) management in Bristol Bay, Alaska. Canadian Special Publication of Fisheries and Aquatic Sciences 96:336–342.

Morbey, Y. 2000. Protandry in Pacific salmon. Canadian Journal of Fisheries and Aquatic Sciences 57:1252–1257.

Newell, J., and T. P. Quinn. 2005. Behavioral thermoregulation by maturing adult sockeye salmon (*Oncorhynchus nerka*) in a stratified lake prior to spawning. Canadian Journal of Zoology 83:1232–1239.

Newell, J. C., K. L. Fresh, and T. P. Quinn. 2007. Arrival patterns and movements of adult sockeye salmon (*Oncorhynchus nerka*) in Lake Washington: implications for management of an urban fishery. North American Journal of Fisheries Management 27:908–917.

Northcote, T. G. 1978. Migratory strategies and production in freshwater fishes. Pages 326–359 *in* S. D. Gerking, editor. Ecology of freshwater fish production. Wiley, New York.

Parker, R. R. 1968. Marine mortality schedules of pink salmon of the Bella Coola River, central British Columbia. Journal of the Fisheries Research Board of Canada 25:757–794.

Pearcy, W. G. 1992. Ocean ecology of North Pacific salmonids. University of Washington, Washington Sea Grant Program, Seattle.

Pyper, B. J., and R. M. Peterman. 1999. Relationship among adult body length, abundance, and ocean temperature for British Columbia and Alaska sockeye salmon (*Oncorhynchus nerka*), 1967–1997. Canadian Journal of Fisheries and Aquatic Sciences 56:1716–1720.

Pyper, B. J., R. M. Peterman, M. F. Lapointe, and C. J. Walters. 1999. Patterns of covariation in length and age at maturity of British Columbia and Alaska sockeye salmon (*Oncorhynchus nerka*) stocks. Canadian Journal of Fisheries and Aquatic Sciences 56:1046–1057.

Quinn, T. P. 2005. The behavior and ecology of Pacific salmon and trout. University of Washington Press, Seattle.

Quinn, T. P., and D. J. Adams. 1996. Environmental changes affecting the migratory timing of American shad and sockeye salmon. Ecology 77:1151–1162.

Quinn, T. P., and G. B. Buck. 2001. Size and sex selective mortality on adult Pacific salmon: bears, gulls and fish out of water. Transactions of the American Fisheries Society 130:995–1005.

Quinn, T. P., and C. J. Foote. 1994. The effects of body size and sexual dimorphism on the reproductive behaviour of sockeye salmon (*Oncorhynchus nerka*). Animal Behaviour 48:751–761.

Quinn, T. P., E. Graynoth, C. C. Wood, and C. J. Foote. 1998. Genotypic and phenotypic divergence of sockeye salmon in New Zealand from their ancestral British Columbia populations. Transactions of the American Fisheries Society 127:517–534.

Quinn, T. P., A. P. Hendry, and L. A. Wetzel. 1995. The influence of life history trade-offs and the size of incubation gravels on egg size variation in sockeye salmon (*Oncorhynchus nerka*). Oikos 74:425–438.

Quinn, T. P., S. Hodgson, L. Flynn, R. Hilborn, and D. E. Rogers. 2007. Directional selection by commercial fisheries and the timing of sockeye salmon (*Oncorhynchus nerka*) migrations. Ecological Applications 17:731–739.

Quinn, T. P., P. McGinnity, and T. F. Cross. 2006. Long-term declines in body size and shifts in run timing of Atlantic salmon in Ireland. Journal of Fish Biology 68:1713–1730.

Quinn, T. P., and K. W. Myers. 2004. Anadromy and the marine migrations of Pacific salmon and trout: Rounsefell revisited. Reviews in Fish Biology and Fisheries 14:421–442.

Quinn, T. P., M. J. Unwin, and M. T. Kinnison. 2000. Evolution of temporal isolation in the wild: genetic divergence in timing of migration and breeding by introduced Chinook salmon populations. Evolution 54:1372–1385.

Quinn, T. P., L. Wetzel, S. Bishop, K. Overberg, and D. E. Rogers. 2001. Influence of breeding habitat on bear predation, and age at maturity and sexual dimorphism of sockeye salmon populations. Canadian Journal of Zoology 79:1782–1793.

Reimers, P. E. 1971. The length of residence of juvenile fall chinook salmon in Sixes River, Oregon. Doctoral dissertation. Oregon State University, Corvallis.

Rich, H. B. 2006. Effects of climate and density on the distribution, growth, and life history of juvenile sockeye salmon (*Oncorhynchus nerka*) in Iliamna Lake, Alaska. Master's thesis. University of Washington, Seattle.

Rich, H. B., Jr., T. P. Quinn, M. D. Scheuerell, and D. E. Schindler. 2009. Climate and intra-specific competition control the growth and life history of juvenile sockeye salmon (*Oncorhynchus nerka*) in Iliamna Lake, Alaska. Canadian Journal of Fisheries and Aquatic Sciences 66:238–246.

Ricker, W. E. 1954. Stock and recruitment. Journal of the Fisheries Research Board of Canada 11:559–623.

Ricker, W. E. 1972. Hereditary and environmental factors affecting certain salmonid populations. Pages 19–160 *in* R. C. Simon, and P. A. Larkin, editors. The stock concept in Pacific salmon. University of British Columbia Press, H. R. MacMillan Lectures in Fisheries, Vancouver.

Ricker, W. E. 1976. Review of the rate of growth and mortality of Pacific salmon in salt water, and noncatch mortality caused by fishing. Journal of the Fisheries Research Board of Canada 33:1483–1525.

Ricker, W. E. 1981. Changes in the average size and average age of Pacific salmon. Canadian Journal of Fisheries and Aquatic Sciences 38:1636–1656.

Rogers, D. E. 1980. Density-dependent growth of Bristol Bay sockeye salmon. Pages 267–283 *in* W. J. McNeil, and D. C. Himsworth, editors. Salmonid ecosystems of the North Pacific. Oregon State University Press, Corvallis.

Rogers, D. E. 1987. The regulation of age at maturity in Wood River sockeye salmon (*Oncorhynchus nerka*). Canadian Special Publication of Fisheries and Aquatic Sciences 96:78–89.

Rogers, D. E., and G. T. Ruggerone. 1993. Factors affecting marine growth of Bristol Bay sockeye salmon. Fisheries Research 18:89–103.

Rowse, M. L. 1985. Migration timing of sockeye salmon (*Oncorhynchus nerka*) in the Wood River lake system, Alaska. Master's thesis. University of Washington, Seattle.

Ruggerone, G. T., E. Farley, J. Nielsen, and P. Hagen. 2005. Seasonal marine growth of Bristol Bay sockeye salmon (*Oncorhynchus nerka*) in relation to competition with Asian pink salmon (*O. gorbuscha*) and the 1977 ocean regime shift. Fishery Bulletin 103:355–370.

Ruggerone, G. T., and J. L. Nielsen. 2004. Evidence for competitive dominance of pink salmon (*Oncorhynchus gorbuscha*) over other salmonids in the North Pacific Ocean. Reviews in Fish Biology and Fisheries 14:371–390.

Ruggerone, G. T., M. Zimmermann, K. W. Myers, J. L. Nielsen, and D. E. Rogers. 2003. Competition between Asian pink salmon (*Oncorhynchus gorbuscha*) and Alaskan sockeye salmon (*O. nerka*) in the North Pacific Ocean. Fisheries Oceanography 12:209–219.

Rulifson, R. A., and M. J. Dadswell. 1995. Life history and population characteristics of striped bass in At-

lantic Canada. Transactions of the American Fisheries Society 124:477–507.
Salo, E. O. 1991. Life history of chum salmon (*Oncorhynchus keta*). Pages 231–309 *in* C. Groot, and L. Margolis, editors. Pacific salmon life histories. University of British Columbia Press, Vancouver.
Sato, H., A. Amagaya, M. Ube, N. Ono, and H. Kudo. 2000. Manipulating the timing of a chum salmon (*Oncorhynchus keta*) run using preserved sperm. North Pacific Anadromous Fish Commission Bulletin 2:353–357.
Saunders, W. P., Jr. 1984. Effects of anadromous fish restoration programs on the population dynamics of American shad (*Alosa sapidissima*) in the Connecticut River. Master's thesis. University of Massachusetts, Amherst.
Schindler, D. E., D. E. Rogers, M. D. Scheuerell, and C. A. Abrey. 2005. Effects of changing climate on zooplankton and juvenile sockeye salmon growth in southwestern Alaska. Ecology 86:198–209.
Seamons, T. R., P. Bentzen, and T. P. Quinn. 2007. DNA parentage analysis reveals inter-annual variation in selection: results from 19 consecutive brood years in steelhead trout. Evolutionary Ecology Research 9:409–431.
Shearer, W. M. 1990. The Atlantic salmon (*Salmo salar* L.) of the North Esk with particular reference to the relationship between river and sea age and time of return to home waters. Fisheries Research 10:93–123.
Smith, H. D. 1964. The segregation of red salmon in the escapements to the Kvichak River system, Alaska. U.S. Fish and Wildlife Service Special Scientific Report Fisheries 470.
Smith, T. I. J. 1985. The fishery, biology, and management of Atlantic sturgeon, *Acipenser oxyrinchus*, in North America. Environmental Biology of Fishes 14:61–72.
Smoker, W. W., A. J. Gharrett, and M. S. Stekoll. 1998. Genetic variation of return date in a population of pink salmon: a consequence of fluctuating environment and dispersive selection? Alaska Fishery Research Bulletin 5:46–54.
Steen, R. P., and T. P. Quinn. 1999. Egg burial depth by sockeye salmon (*Oncorhynchus nerka*): implications for survival of embryos and natural selection on female body size. Canadian Journal of Zoology 77:836–841.
Stewart, D. C., S. J. Middlemas, and A. F. Youngson. 2006. Population structuring in Atlantic salmon (*Salmo salar*): evidence of genetic influence on the timing of smolt migration in sub-catchment stocks. Ecology of Freshwater Fish 15:552–558.
Stewart, D. C., G. W. Smith, and A. F. Youngson. 2002. Tributary-specific variation in timing of return of adult Atlantic salmon (*Salmo salar*) to fresh water has a genetic component. Canadian Journal of Fisheries and Aquatic Sciences 59:276–281.
Todd, I. S. P., and P. A. Larkin. 1971. Gillnet selectivity on sockeye (*Oncorhynchus nerka*) and pink salmon (*O. gorbuscha*) of the Skeena River system, British Columbia Journal of the Fisheries Research Board of Canada 28:821–842.
Todd, P. R. 1981. Timing and periodicity of migrating New Zealand freshwater eels (*Anguilla* spp.). New Zealand Journal of Marine and Freshwater Research 15:225–235.
van den Berghe, E. P., and M. R. Gross. 1984. Female size and nest depth in coho salmon (*Oncorhynchus kisutch*). Canadian Journal of Fisheries and Aquatic Sciences 41:204–206.
van den Berghe, E. P., and M. R. Gross. 1989. Natural selection resulting from female breeding competition in a Pacific salmon (coho: *Oncorhynchus kisutch*). Evolution 43:125–140.
Vøllestad, L. A., J. Peterson, and T. P. Quinn. 2004. Effects of fresh water and marine growth rates on early maturity in male coho and Chinook salmon. Transactions of the American Fisheries Society 133:495–503.
Ward, B. R., P. A. Slaney, A. R. Facchin, and R. W. Land. 1989. Size-based survival in steelhead trout (*Oncorhynchus mykiss*): back-calculated lengths from adults' scales compared to migrating smolts at the Keogh River, British Columbia. Canadian Journal of Fisheries and Aquatic Sciences 46:1853–1858.
Webb, J. H., and H. A. McLay. 1996. Variations in the time of spawning of Atlantic salmon (*Salmo salar*) and its relationship to temperature in the Aberdeenshire Dee, Scotland. Canadian Journal of Fisheries and Aquatic Sciences 53:2739–2744.
Went, A. E. J. 1964. Irish salmon—a review of investigations up to 1963. Scientific Proceedings of the Royal Dublin Society, Series A 1(15):365–412.
Wood, C. C. 1995. Life history variation and population structure in sockeye salmon. Pages 195–216 *in* J. L. Nielsen, editor. Evolution and the aquatic ecosystem: defining unique units in population conservation. American Fisheries Society, Symposium 17, Bethesda, Maryland.
Woodey, J. C. 1987. In-season management of Fraser River sockeye salmon (*Oncorhynchus nerka*): meeting multiple objectives. Canadian Special Publication of Fisheries and Aquatic Sciences 96:367–374.
Yamamoto, S., K. Morita, and A. Goto. 1999. Marine growth and survival of white-spotted charr, *Salvelinus leucomaenis*, in relation to smolt size. Ichthyological Research 46:85–92.

American Fisheries Society Symposium 69:43–44, 2009

The Dynamic Nature of Diadromy
Preamble

Roger A. Rulifson
Institute for Coastal Science and Policy, and Department of Biology
East Carolina University Greenville, North Carolina 27858, USA

Diadromy is a behavior that involves physiological changes required to accommodate the extremes in habitat shifts between freshwaters and seawaters at a number of different life cycle stages, depending on the species. The physiological responses to changing environmental conditions such as temperature and day length trigger or are otherwise modulated by important behavioral changes that influence migration, environmental preferences, and other factors.

What makes diadromous behavior so interesting and difficult to understand is the inconsistency of diadromy among related species and within one species. Closely related species within a family may or may not exhibit diadromy. For one species, there may be latitudinal differences in populations exhibiting diadromy rather than residency. For some species, the diadromy-residency life histories differ between watersheds, and for others, both residency and diadromy occur within the same watershed. There also can be great individual variability within a species as to when diadromy begins and also whether diadromy continues through the life of the fish or ceases temporarily or permanently at some point during the fish's life. Thus, there is great variability in the general concept of diadromy, and a diadromous life history itself can be extremely plastic or dynamic.

At the first diadromy symposium held in 1986 (Dadswell et al. 1987), the prevailing paradigm for diadromy was the concept of fish populations exhibiting a uniform behavior. Now, researchers are beginning to identify migratory contingents within a single watershed and even differences in migratory behaviors between individuals within a contingent. Individuals may become anadromous or resident or "change their minds" and their lifestyles later in life (e.g., Secor 1999; Gemperline et al. 2001; Paramore and Rulifson 2001).

This session addresses these inconsistencies to bring further understanding to the causative mechanisms involved in facilitating the life histories of diadromous fishes. At the first diadromy symposium, Gross (1987) proposed a paradigm for how anadromy and catadromy might have evolved. In the current symposium, this paradigm is challenged by Dodson et al. (2009, this volume) and by Tsukamoto et al. (2009, this volume). Dodson et al. (2009) use a comparison of families in the Salmoniformes and Osmeriformes to hypothesize the existence of contrasting evolutionary pathways that explain the current observed patterns of diadromy in these groups. Tsukamoto et al. (2009) offer an explanation of how upstream migration of fish might be initiated and how it prevails through generations, through positive responses of individual fish in schools to fish density, endocrinological condition, and hunger level.

Patterns of diadromy that do not hold across species within families or subpopulations within a single watershed continue to be discovered. Bradbury et al. (2009, this volume) examine how anadromous smelt populations in postglacial temperate latitudes may have poor gene flow, resulting in different migration patterns and morphological phenotypes. Howland et al. (2009, this volume) examine how one coregonid species in northern Canada, the inconnu *Stenodus leucichthys*, exhibits residency, anadromy, and potadromy (e.g., estuarine anadromy) within a single watershed. Finally, Grothues et al. (2009, this volume) review how nonnatal habitats are used by migrating species.

Understanding these differences is important for fisheries managers and greatly increases the complexity of establishing effective measures for fishery management plans. It has become apparent since Dadswell et al. (1987) that diadromy itself has evolutionarily significant plasticity, which is an important attribute for adapting populations to ever-changing environmental stressors over generations. This session contributes important information and challenges old concepts, but we must continue pushing the envelope of knowledge about the dynamic nature of diadromy in order to deal with management issues about anthropogenic changes and global warming.

References

Bradbury, I. R., M. W. Coulson, S. E. Campana, E. Baggs, and P. Bentzen. 2009. Postglacial recolonization and the loss of anadromy in rainbow smelt from coastal Newfoundland. Pages 79–96 *in* A. J. Haro, K. L. Smith, R. A. Rulifson, C. M. Moffitt, R. J. Klauda, M. J. Dadswell, R. A. Cunjak, J. E. Cooper, K. L. Beal, and T. S. Avery, editors. Challenges for diadromous fishes in a dynamic global environment. American Fisheries Society, Symposium 69, Bethesda, Maryland.

Dadswell, M. J., R. J. Klauda, C. M. Moffitt, R. L. Saunders, R. A. Rulifson, and J. E. Cooper, editors. 1987. Common strategies of anadromous and catadromous fishes. American Fisheries Society, Symposium 1, Bethesda, Maryland.

Dodson, J. J., J. Laroche, and F. Lécomte. 2009. Contrasting evolutionary pathways of anadromy in euteleostean fishes. Pages 63–77 *in* A. J. Haro, K. L. Smith, R. A. Rulifson, C. M. Moffitt, R. J. Klauda, M. J. Dadswell, R. A. Cunjak, J. E. Cooper, K. L. Beal, and T. S. Avery, editors. Challenges for diadromous fishes in a dynamic global environment. American Fisheries Society, Symposium 69, Bethesda, Maryland.

Gemperline, P. J., R. A. Rulifson, and L. M. Paramore. 2001. Multi-way analysis of trace elements in fish otoliths to track migratory patterns. Chemometrics and Intelligent Laboratory Systems 60(2002):135–146.

Gross, M. R. 1987. Evolution of diadromy in fishes. Pages 14–25 *in* M. J. Dadswell, R. J. Klauda, C. M. Moffitt, R. L. Saunders, R. A. Rulifson, and J. E. Cooper, editors. Common strategies of anadromous and catadromous fishes. American Fisheries Society, Symposium 1, Bethesda, Maryland.

Grothues, T. M., K. W. Able, J. Carter, and T. Arienti. 2009. Migration patterns of striped bass through nonnatal estuaries of the U.S. Atlantic coast. Pages 135–150 *in* A. J. Haro, K. L. Smith, R. A. Rulifson, C. M. Moffitt, R. J. Klauda, M. J. Dadswell, R. A. Cunjak, J. E. Cooper, K. L. Beal, and T. S. Avery, editors. Challenges for diadromous fishes in a dynamic global environment. American Fisheries Society, Symposium 69, Bethesda, Maryland.

Howland, K. L., J. A. Babaluk, D. Chiperzak, R. F. Tallman, and G. Low. 2009. Variability in diadromous behavior of a northern coregonid based on scanning proton microprobe analysis of otolith strontium. Pages 121–134 *in* A. J. Haro, K. L. Smith, R. A. Rulifson, C. M. Moffitt, R. J. Klauda, M. J. Dadswell, R. A. Cunjak, J. E. Cooper, K. L. Beal, and T. S. Avery, editors. Challenges for diadromous fishes in a dynamic global environment. American Fisheries Society, Symposium 69, Bethesda, Maryland.

Paramore, L. M., and R. A. Rulifson. 2001. Dorsal coloration as an indicator of different life history patterns for striped bass within a single watershed of Atlantic Canada. Transactions of the American Fisheries Society 130:663–674.

Secor, D. H. 1999. Specifying divergent migrations in the concept of stock: the contingent hypothesis. Fisheries Research 43:13–34.

Tsukamoto, K., M. J. Miller, A. Kotake, Jun Aoyama, and Kazuo Uchida. 2009. The origin of fish migration: the random escapement hypotheses. Pages 45–61 *in* A. J. Haro, K. L. Smith, R. A. Rulifson, C. M. Moffitt, R. J. Klauda, M. J. Dadswell, R. A. Cunjak, J. E. Cooper, K. L. Beal, and T. S. Avery, editors. Challenges for diadromous fishes in a dynamic global environment. American Fisheries Society, Symposium 69, Bethesda, Maryland.

American Fisheries Society Symposium 69:45–61, 2009

The Origin of Fish Migration: The Random Escapement Hypothesis

Katsumi Tsukamoto*, Michael J. Miller, Aya Kotake, Jun Aoyama, and Kazuo Uchida
Ocean Research Institute, The University of Tokyo
Minamidai, Nakano, Tokyo 164-8639, Japan

Abstract.—The remarkable migrations of some fishes are still the subject of active research, but there has been much less attention on the evolutionary or behavioral origins of diadromous migrations. Diadromy likely evolved as a result of the adaptive advantage of using higher productivity habitats for growth, but the reason why the first individuals left their habitat to enter a new one is unknown. The objective of this paper is to examine the possible evolutionary and behavioral origin of diadromous fish migration and to explore the possible factors that could have led to the development of migratory behavior, using a behavioral model for a triggering mechanism of fish migration. Eels likely evolved in the tropics from a marine ancestor, and salmon evolved in temperate regions from a freshwater ancestor. Evidence of the ancestral life history states of these fishes can be seen in geographic clines of occurrence of nonmigratory residents of the sympatric Japanese eel *Anguilla japonica*, a catadromous species, and masu salmon *Oncorhynchus masou*, an anadromous species, with a higher proportion of resident type "sea eels" at higher latitudes, whereas masu salmon have more river residents at lower latitudes. The amphidromous ayu *Plecoglossus altivelis* was used as a model species for exploring the behavioral origins of migration. The triggering mechanism of upstream migration of ayu could be explained by a three-step model of necessary conditions for starting migration that include age and body size, endocrinological condition, and psychological processes. A behavioral model of drive was proposed to explain the last step that is influenced by exogenous/endogenous factors such as water temperature, fish density, and hunger level. The drive of upstream migration behavior appears to be a psychological repulsion that occurs in stressed fish when the behavioral rule of optimum interindividual distance is broken, since fish with stronger repulsion and a larger interindividual distance showed more active jumping behavior and stronger upstream migration. A vacuum activity of jumping behavior of ayu contained in a limited space occurred in random directions without any behavioral stimuli when water temperature increased. This vacuum activity can be interpreted as an escapement behavior of fish that bolt out of unfavorable environments, and this escapement might be a behavioral origin of fish migration that may also help to understand the evolutionary origin of migration (random escapement hypothesis).

Introduction

Diadromous fishes such as eels and salmon are some of the most famous species of migrating animals. These types of migrations have always fascinated biologists because of the amazing ability of various species to move thousands of kilometers with surprising regularity and accuracy. How they are able to achieve these remarkable migrations, however (Bingman and Cheng 2005; Alerstam 2006; Papi 2006; Åkesson and Hedenström 2007), and why they evolved this ability (Gross 1987; Roff 1988; Salewski and Bruderer 2007) has remained mysterious.

Historically, migrations were studied using various tagging techniques in birds, fishes, sea turtles, and other animals. Recently, there has been use of

* Corresponding author: ktpc@ori.u-tokyo.ac.jp

tags that can communicate with satellites to obtain data on marine animals that make remarkably long migrations. For species that migrate within or above the ocean, these techniques have provided some of the first information on the routes of their migrations. These types of studies have been made on the spawning migrations of eels for which there is no information available about their particular spawning area (Jellyman and Tsukamoto 2002), and a range of valuable new insight has been gained about the movements of tuna (Block et al. 2005; Wilson et al. 2005; Teo et al. 2007), billfishes (Sippel et al. 2007), sharks (Eckert and Stewart 2001; Bonfil et al. 2005; Weng et al. 2005), sea turtles (Luschi et al. 2003, 2006; James et al. 2005; Hatase et al. 2006), whales (Zerbini et al. 2006) and seabirds (Croxall et al. 2005; Shaffer et al. 2006). In addition to these studies on the process of migration by tagging techniques, behavioral and evolutionary approaches are important for research on migration. Except for terrestrial birds though, there have been few studies on why animals evolved these migrations or how these types of movements began in ancient times.

When animals migrate from one habitat to another, three conditions are needed for their migration (Tsukamoto 1994): (1) locomotory ability and energy storage to move to the destination, (2) orientation ability to know the direction of the destination, and (3) a psychological factor, termed "drive" or "motivation" that urges an animal to behave in a certain way (Berridge 2004). These conditions are all indispensable for animals to complete their migration, and the lack of even one of the three may result in migration failure. Although there have been a substantial number of physiological and ecological studies on locomotion and orientation, psychological research is scarce. To understand the evolutionary origin of migration and the triggering mechanism of present-day migratory behavior, behavioral and evolutionary approaches are needed.

In the case of terrestrial birds, which mostly migrate between high-latitude breeding areas in summer and low-latitude feeding areas in winter, a variety of hypotheses have been proposed for why birds evolved this ability. The migratory behavior of birds may be related to ecological processes such as food limitation, exploitation of seasonal peaks in local food availability, direct climatic effects on physiological function, and risk of nest predation (Salewski and Bruderer 2007). Birds have been the subjects of a number of studies and theoretical examinations of the evolution of migration (e.g., Bell 2000; Zink 2002; Rappole et al. 2003; Alerstam et al. 2003; Boyle and Conway 2007), probably because they are easier to observe in nature and to maintain in the laboratory for experiments on migratory motivation than most migratory marine organisms in the ocean. Some insects also make long-distance migrations, but little is known about why their migrations evolved (Holland et al. 2006) or how they start migration.

In fishes, there are several types of migrations that are undertaken for similar reasons as in other types of animals. One type is potamodromy of freshwater fishes such as catfish, carp, and some sturgeon that migrate only within rivers or lakes (e.g., Bemis and Kynard 1997), and another type is oceanodromy of widely ranging marine fishes such as tunas, billfishes, and sharks that can move great distances in the ocean between spawning and feeding areas (Eckert and Stewart 2001; Block et al. 2005; Bonfil et al. 2005; Weng et al. 2005; Wilson et al. 2005; Sippel et al. 2007; Teo et al. 2007). The other types of migrations are those of diadromous fishes that move between freshwater and marine environments at regular periods in their life histories (McDowall 1987, 1997). These fishes include catadromous species such as eels, mugilids and sculpins, anadromous species such as lampreys, salmon, clupeids and sturgeon, and amphidromous fishes such as ayu *Plecoglossus altivelis*, gobies, and galaxiids (Dadswell et al. 1987; McDowall 1987; Bemis and Kynard 1997; Keith 2003; Quinn and Myers 2004; Bradbury et al. 2009 and Iida et al. 2009, both this volume). In each of these forms of diadromy, various species use either freshwater or marine environments for particular parts of their life histories, such as spawning, larval and juvenile growth and maturation.

Life history strategies that involve migration between different types of environments appear to have evolved to gain a greater fitness by using alternate habitats during each stage of development (Gross 1987; Gross et al. 1988). Although many species of salmon use freshwater during their early juvenile stages, in some species such as pink salmon *Oncorhynchus gorbuscha*, the freshwater feeding stage has almost been completely lost (Quinn and Myers 2004). In some amphidromous species such as gobies and ayu, spawning occurs in freshwater and the larvae drift directly into marine habitats where they grow larger before returning to freshwater for further growth and eventual maturation (Tsukamoto et al. 1987; Keith 2003; Iida et al. 2009).

But it is much less clear how the actual behavior of moving between freshwater and marine habitats first evolved in the wide range of species that developed diadromous life histories independently in different lineages of fishes. At some point, individuals had to make the transition from one habitat to the other and gain some adaptive advantage that gave those individuals greater fitness than those that remained within the single habitat. In this paper, we consider the definition of migration using the "migration loop" concept and discuss the evolutionary origin of diadromy by exploring the possible factors that could have led to the evolutionary development of migratory behavior in diadromous fishes. Furthermore, we provide a behavioral model of a triggering mechanism for migration based on experimental data from amphidromous ayu and suggest that a "random escapement" behavior could be a behavioral origin of diadromous fish migrations in their evolutionary history.

What Is Migration?

Definition

Animals show movements on a daily, seasonal, or life history basis when they are foraging, seeking shelter, or meeting reproductive mates. Among many types of movements, migrations are regular long-distance movements that are made at a certain time in the life history of a species, such as for reproduction or growth (Figure 1; Tsukamoto 1994; Tsukamoto et al. 2002). These migrations typically include a habitat transition like the seasonal migrations of birds or the movements between fresh and salt water by diadromous fishes. It is generally considered that simple movements of different individuals to various areas within their geographic range while searching for food or a suitable place to live are not migrations because it is not an established part of the life history of the species. The types of migrations in fishes have been examined previously (e.g., McKeown 1984; Rankin 1985), and a broad perspective on the definition of what is migration has been discussed recently by Dingle and Alistair (2007), who characterize migration as an adaptation to resources that fluctuate spatiotemporally either seasonally or less predictably. In this paper, however, we merely define migration as regular movement of an individual at a particular life history stage between two different habitats (Figure 1).

Salmon species show an accurate return migration to their natal river from their marine growth habitat, and anguillid eels return to the same spawning area where they were born. However, the amphidromous ayu *Plecoglossus altivelis* and many amphidromous gobies do not necessarily show homing migration to their natal rivers (Tsukamoto 1990). It seems that some migratory fish do not migrate back to the exact natal habitat where they were born, so salmon and eels with their very specific spawning areas are rather exceptional. Such accurate homing appears to be an evolutionarily obtained ability in individual fish that originally consisted of a basic movement between the seawater and freshwater habitats. Therefore, when considering the origins of fish migration, the habitats between which fishes migrate back and forth should be categorized not as specific locations like a natal river for individuals, but as a broader concept of ecological habitats for species with different environmental properties such as freshwater and saltwater, coastal areas and offshore, pelagic and demersal layers, or high and low latitudes.

Migration Loop

When individuals return to the habitat where they were before, the migrations involve a round trip or loop (Figure 1). The established migrations of species can be visualized in terms of "migration loops" that connect two different habitats typically involved in growth and reproduction (Tsukamoto

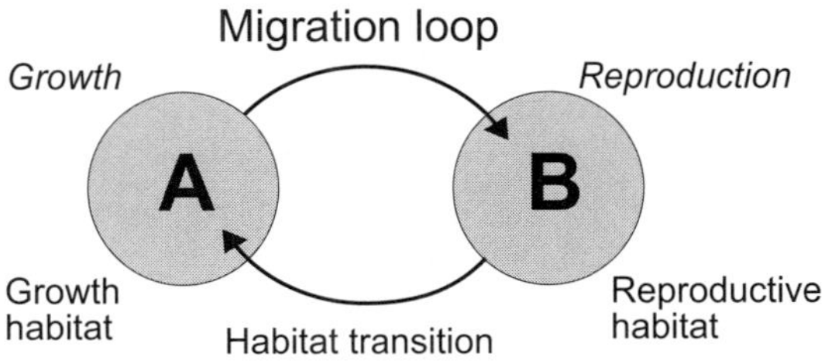

Figure 1.—Migration is defined as regular movement between different habitats, and this can be conceptualized as a migration loop of a species or population.

and Aoyama 1998). Accordingly, it can be said that a migration loop is a species specific or population specific condition that consists of "one species, one loop" or "one population, one loop."

Typically, the reproductive habitat is a much smaller, well-discriminated area compared with the larger region of the growth habitat that has less distinct margins (Tsukamoto et al. 2002). Salmon are famous for their ability to return to the small stream in which they were born after covering thousands of kilometers during their growth phase in the ocean (Dittman and Quinn 1996). Similarly, anguillid eels return to the same area of the open ocean in which they were born (Tsukamoto 1992). The Japanese eels *Anguilla japonica* return to a spawning area in the North Equatorial Current near three seamounts to the west of the Mariana Islands (Tsukamoto et al. 2003; Tsukamoto 2006). In the Atlantic, the European eel *A. anguilla* and American eel *A. rostrata* return to the southern Sargasso Sea for spawning (Schmidt 1922; McCleave et al. 1987) after being widely dispersed on both sides of the ocean. The Australasian shortfin eel *A. australis* appears to return to the South Equatorial Current region of the western South Pacific to spawn (see Miller et al. 2009, this volume). Thus, the individuals of both of these diadromous fishes all converge back to their natal habitats during the reproductive phase of their migration loops (Tsukamoto et al. 2002). Sea turtles and other types of migratory animals also return to the same place where they were born for reproduction (Bowen et al. 2005; Lee et al. 2007).

All of these species also show a period of individual dispersal during their oceanic larval development and a passive transport period in the case of eel leptocephali and the ocean growth phase of juvenile salmon and turtles, during which individuals can disperse over a wide area. Therefore, the migration loops of many species include a phase of "convergence" from a large growth habitat to a small reproductive habitat and "divergence" as they migrate or disperse from the reproductive habitat to their growth habitat.

Evolutionary Origin of Diadromy

Evolutionary Process of Migration

Most aquatic animals live only in either marine or freshwater habitats, so the original diadromous species of most taxa likely evolved from a single habitat species that expanded into a new type of habitat. Gross (1987) hypothesized that diadromy evolved from euryhaline "wandering" to amphidromy and then to anadromy or catadromy. These life history changes can be conceptualized as a gradual expansion of the migration loop of each species until eventually a completely new diadromous migration has been established (Figure 2).

In the case of anguillid eels and salmon, their different types of diadromous migrations appear to reflect their origins in different habitats. Except for anguillid eels and a few rare ophichthid or muraenid eels that may enter freshwater to some degree in the tropics, all other eels of the Anguilliformes are completely marine species. Because of this, and the phylogenetic position of anguillid eels within the Anguilliformes (Inoue et al. 2004), the freshwater eels of the genus *Anguilla* likely evolved from a marine ancestor (Tsukamoto et al. 2002). In contrast, since a recent molecular phylogenetic tree suggests a freshwater ancestry of the Salmoniformes (Ishiguro et al. 2003), salmon and trout may have originated in freshwater in temperate regions and then they gained the ability to migrate into the ocean (Gross et al. 1988; McDowall 2002; Dodson et al. 2009, this volume). Salmon would likely have originated as freshwater fishes with a potamodromous migration only within freshwater habitats. Their occasional visit to estuaries might

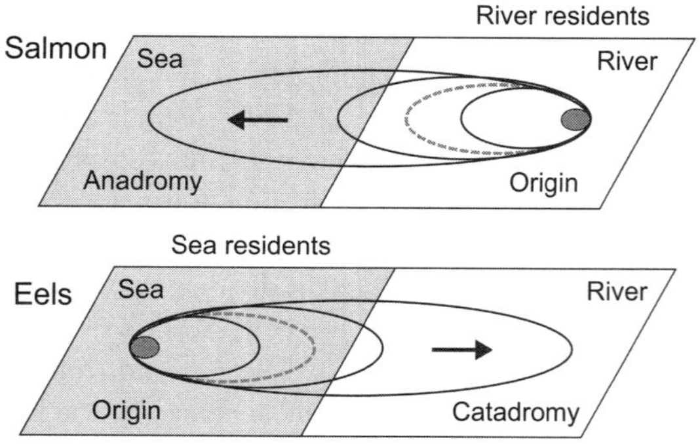

Figure 2.—The origin and evolution of diadromous migration in eels and salmon (Tsukamoto et al. 2002). Expansion of the migration loops of salmon and eels show the progression of the development of their diadromous migrations between marine and freshwater habitats.

have caused an amphidromous migration (Gross et al. 1988), and then, they finally developed the present anadromous migration as a result of the expansion of their migration loop from freshwater to the ocean (Figure 2).

To help evaluate this hypothesis, some evidence can be found in the phylogenetic relationships of the salmonid genera *Hucho*, *Salvelinus*, *Salmo*, and *Oncorhynchus* (Murata et al. 1993) and in their increasing dependencies on marine habitats during their juvenile growth phase. The most basal taxon was *Hucho*, which have a mainly freshwater life history with wandering into coastal waters during their growth phase in winter, and the more diverged species have the more regular migration into the ocean. In addition, there is also a cline within the genus *Oncorhynchus*, in the dependency of each species on marine growth habitat among the masu salmon *O. masou*, coho salmon *O. kisutch*, Chinook salmon *O. tschawytscha*, sockeye salmon *O. nerka*, chum salmon *O. keta*, and pink salmon *O. gorbuscha*, in order from low to high marine dependency, which also roughly reflects their phylogenetic relationships (Kaeriyama 1994). Because pink salmon can spawn in the estuary and their eggs develop even in brackish water, this species may eventually evolve into a marine species with a completely marine migration loop. However, diadromy is an adaptive behavior, and therefore, it has clearly evolved several times in fishes such as gobies and sculpins, and its occurrence is polyphyletic, with very widely separated taxa having similar migration patterns (McDowall 1987, 1997).

How a species establishes a new migration loop to move across the sea/freshwater boundary to establish a diadromous life history is likely related to an adaptive advantage gained from entering the new habitat. Gross (1987) proposed that diadromous migration occurs when the gain in fitness from using a second habitat, minus the migration costs, is greater than the fitness from staying in only one habitat. It was hypothesized that the difference in productivity between north and south and between sea and freshwater habitats may drive fishes to migrate to another habitat (Gross 1987; Gross et al. 1988). In the case of catadromous eels, the tropical freshwater habitats represented a highly productive environment with no other eel species to compete with, in comparison to the marine environment that likely had many other families of eels utilizing similar habitats. Therefore, the catadromous life history of anguillid eels may have evolved to take advantage of the greater fitness associated with the freshwater environments of tropical areas, while retaining their ancestral spawning areas in the ocean. The anadromous migrations of salmon probably originated in a freshwater species living at high latitudes where their fitness could be greatly enhanced by a migration to the ocean for feeding and growth, while retaining their spawning areas in freshwater habitats that provide relatively safe larval development and early growth areas for their offspring in comparison to the marine environments. These scenarios then depend on some sort of conditions being present where the first individuals or "pioneers" enter the new habitat and gain an adaptive advantage within the population. Eventually, as the genotypes of the pioneers spread in the population, more and more individuals enter the new habitat, until the behavioral trait is fixed in the population.

Speciation within existing diadromous species could also be caused by pioneers that have new migration patterns. After these individuals cause some aspect of their migration loop to change, such as a latitudinal or longitudinal expansion of species range (Tsukamoto et al. 2002), eventually a new migration loop may be created that could lead to speciation. If there is a shift in the spawning area or season to facilitate successful recruitment to the new area, this type of reproductive isolation could cause speciation. The Atlantic anguillid eels are a good example of this because as the Atlantic basin expanded longitudinally due to continental drift, eventually the ancestral Atlantic eel species had to diverge into two species that have different patterns of larval migrations and timing of metamorphosis that enabled them to utilize an overlapping spawning area, yet recruit to opposite sides of the North Atlantic after different distances of larval migration (Tsukamoto and Aoyama 1998; Tsukamoto et al. 2002; Kuroki et al. 2008). The Australasian shortfin eel in the western South Pacific may also be an example of an early stage of divergence, since a longer larval duration is needed to recruit to New Zealand compared to Australia, and genetic and morphological differences have been detected between the two subspecies (Watanabe et al. 2006; Shen and Tzeng 2007; Minegishi et al. 2009, this volume).

Latitudinal Cline of Migrants Versus Nonmigrants

Some evidence of the original life history traits of several diadromous species can be inferred from the variations in their migratory behaviors. Eels have been shown to have a wide range of habitat use patterns during their yellow eel growth phase, with some following the typical catadromous migration pattern of entering freshwater as glass eels and elvers (termed "river eel" in Tsukamoto et al. 1998) and others appearing to never enter freshwater and remain in estuarine or coastal waters until maturation ("estuarine eel" and "sea eel," respectively) (Tsukamoto et al. 1998; Tsukamoto and Arai 2001; Daverat et al. 2006). Some individuals appear to even move back and forth between freshwater and saltwater for feeding one or more times during their life (Tsukamoto and Arai 2001; Tzeng et al. 2002; Arai et al. 2004; Jessop et al. 2004; McCleave and Edeline 2009, this volume). These discoveries were made using analyses of the Sr/Ca ratios in the otoliths of eels, which provide an estimate of the salinity in the environments inhabited by each individual. Studies using these techniques have found that sea eels and estuarine eels of the Japanese eel primarily use estuarine habitats (Tsukamoto and Arai 2001; Tzeng et al. 2002; Arai et al. 2003; Kotake et al. 2003, 2005), and other temperate species such as the American eel (Jessop et al. 2002, 2004; Morrison et al. 2003) and the two temperate species in New Zealand (Arai et al. 2004) also show evidence of similar behaviors. However, there appears to be a latitudinal cline in the degree to which individuals of the Japanese eel enter freshwater (Figure 3), with a higher proportion of eels entering freshwater in the southern regions where productivity would be high compared to the freshwater habitats in the north (Kotake et al. 2003, 2005). In contrast, fewer eels appear to enter freshwater in the north where productivity is much higher in marine habitats (Arai et al. 2003). This type of cline may also be present in the other northern temperate anguillid species, the European and American eels (Daverat et al. 2006).

Similarly, some anadromous and amphidromous species have life history variants that do not make the typical diadromous migrations that are made by the rest of the individuals of the species. Landlocked and conditional river resident forms are well known in some anadromous species such as sockeye salmon, char, and trout that never migrate to the sea (Groot and Margolis 1991; Kaeriyama 1994; Quinn and Myers 2004; Chernoff and Curry 2007). Nonmigrant forms are also known for some amphidromous species such as ayu, sculpins, and gobies, which normally spawn in freshwater, but complete their larval development and early juvenile stage in coastal marine areas (Tsukamoto et al. 1987; Keith 2003; Yokoyama and Goto 2005; Yokoi and Hosoya 2006).

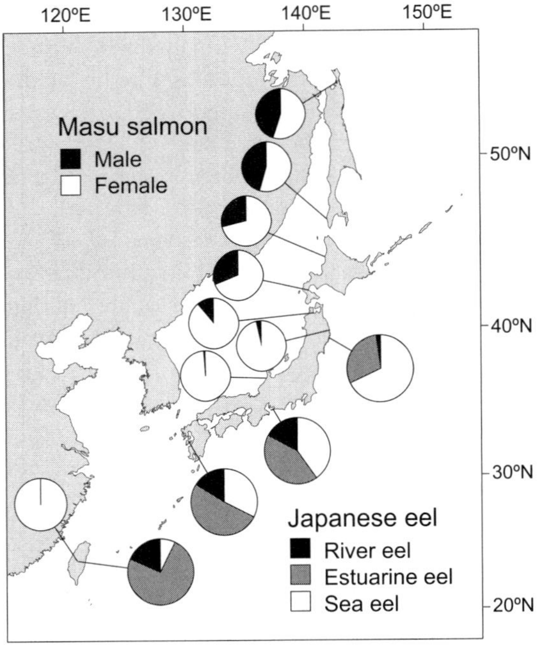

Figure 3.—Latitudinal cline of the proportion of nonmigrants (residents) in the Japanese eel based on Sr:Ca ratios (right side), and masu salmon based on sex ratios (left side). The proportion of nonmigrants in the masu salmon is shown by the sex ratio of migrants since it roughly reflects the ratio of migrants and nonmigrants in the population.

Anadromous masu salmon have a latitudinally wide species range that overlaps sympatrically with the Japanese eel, but in addition to having many of the typical salmonid life history traits, they also show regional variation in migratory behaviors (Kubo 1980; Kato 1991). The masu salmon are born in freshwater, but they show variation in their proportion of males in each area that perform an anadromous migration to the sea along with the females (Figure 3). This species has a complicated

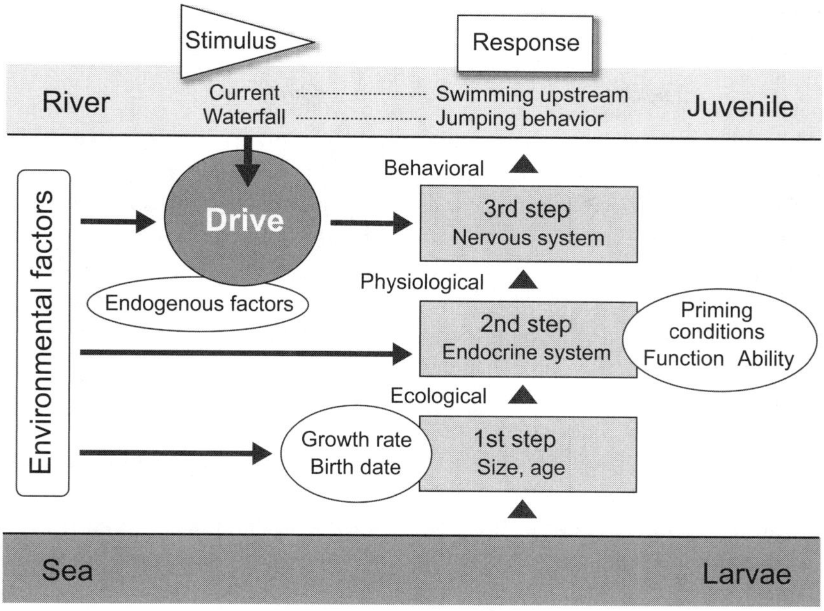

Figure 4.—Three-step model for the releasing mechanism of upstream migration from the sea to freshwater in the juvenile ayu. Environmental factors can influence all three steps, and drive can be influenced by endogenous and exogenous factors such as hunger, temperature, and fish density. The stimuli of current and a waterfall release the upstream migration as the third step through nervous system.

life history because the generally small males with a faster growth rate stay in rivers to become residents or nonmigrants and mature earlier (<2 years old) and mate with large females that migrated to the sea and grew to a large size before returning to freshwater for spawning (Kubo 1980; Tamate and Maekawa 2002; Yamamoto 2004). Other salmonid species such as Atlantic salmon *Salmo salar*, char, and trout have also similar conditional life history patterns with resident mature male parr (Fleming 1998; Quinn and Myers 2004). However, in the northern part of their distribution range, most males and females appear to be anadromous and migrate out to the sea, with the males returning at a larger size than the freshwater male form. But further to the south, the proportion of males that go to sea is much less, with a greater number of males staying in freshwater. This is illustrated by Figure 3 showing the higher proportion of males migrating back to freshwater in the northern regions after entering the sea along with the females. Although there is no direct data on the exact proportion of nonmigrants that stay in freshwater compared to migrants from the sea, the sex ratio of the returning salmon gives an indirect indication of the rate of nonmigrant occurrence in each area, as long as there is no difference in the overall sex ratio in each area and no mixing of different stocks.

The higher proportion of males returning from the sea along with the females in northern areas where the productivity in freshwater is much lower than in southern areas indicates that there is an apparent latitudinal cline in the proportion of males migrating to the sea instead of remaining in freshwater. At the northern limit of their range, approximately equal numbers of males and females migrate to the sea (Kato 1991), with almost no individuals remaining in freshwater. The proportion of males decreases in the more southern areas, suggesting that more males stay in freshwater along with some females. This indicates an increase in the percentage of nonmigrants in a population in an area because the overall sex ratio of the species does not vary greatly in each area. At the other extreme, in Taiwan, at the southern limit of the species range, both males and females stay in freshwater, with no individuals migrating into the sea. An increased river residency as nonmigrants in the southern areas appears to be caused by higher productivity in freshwater to allow for growth and reproduction.

The geographic variation in migratory behavior of the sympatric anadromous masu salmon is a mirror image of that exhibited by catadromous eels. Nonmigratory residents of both diadromous species reflect the respective origins of migration by reverting back their ancestral behavioral state. The variation in migratory pattern of both species appears to be related to the temperate freshwater origin of salmon and the tropical marine origin of eels, and the evolutionary direction of diadromous migration can be explained by the cline of food abundance between freshwater and marine habitat in different latitudes. In addition, this latitudinal cline may suggest the direction of expansion of their geographic range: salmon from the north to south, and eels from south to north. As is shown in Figure 3, the present proportion of nonmigrants representing the ancestral state in a population of a species likely reflects the evolution of either catadromy or anadromy. These types of factors can help explain why these diadromous migrations may have evolved from a fitness and reproductive success perspective, but to try to better understand how they first developed each migration, a behavioral approach is needed.

Behavioral Origin of Fish Migration

Model Species

Compared to eels and salmon, other types of shorter migration loops are seen in amphidromous fishes (McDowall 2007) and invertebrates, such as crustaceans (Kuribayashi et al. 2006) and mollusks (Blanco and Scatena 2007). These species reproduce in freshwater and have newly hatched larvae that drift downstream to marine habitats where they have a period of feeding and growth of usually at least a few months before they return to freshwater as juveniles (McDowall 2007). They then have a period of additional feeding and growth before they reproduce in freshwater. These migrations differ from anadromous and catadromous migrations because they include periods of feeding and growth after each migration, so that reproduction is not the immediate result of each migration as it is in the spawning migrations of eels and salmon. However, the migrations facilitate rapid growth, which leads to early and successful reproduction.

The amphidromous life history of the ayu that lives in the rivers, estuaries, and coastal regions of Japan and other parts of East Asia (Tsukamoto et al. 1987; Tsukamoto 1994) is a good model for studying migration because their annual life cycle and the size of their juveniles make them easy to handle in the laboratory. The ayu spawn in the lower reaches of rivers in autumn, and just after hatching, larvae drift downstream to spend several months in coastal waters during winter. In spring, the juveniles that are about 60 mm long (standard length) migrate from the sea and into the rivers, showing vigorous jumping and upstream swimming behavior that is followed by further growth in the middle reaches of rivers in summer where fish reach about 200 mm. When they start maturation in autumn, they migrate downstream to spawn in the lower reaches of a river and then die, so the life span of the species is just 1 year. Artificial production techniques have been well established for this species, which make it available for research. In addition, the upstream migration of ayu present clear and more reliable behavioral indices of migratory motivation, such as jumping behavior and swimming upstream, compared to the downstream migration of other salmonid juveniles that sometimes could not be separated between an active behavior of downstream migration and a passive downward drift with the current. Roff (1988) hypothesized that schooling behavior may be related to the evolution of migration in marine fishes, so the ayu, which displays schooling behavior in its larval and early juvenile stages (Tsukamoto et al. 1987), may be a good model species for examining the behavioral basis of migration.

Three-Step Model

How fish start migration can be conceptualized from a three-step model based on field observations and laboratory experiments on amphidromous ayu. The first step for starting their upstream migration is an age or body size threshold, which is reached depending on the hatching date and the growth rate of the individual (Figure 4). The second step is to reach a particular physiological condition in preparation for migration that includes smoltification, osmoregulation, and energy accumulation, which are regulated by the endocrine system. In the case of the upstream migration of ayu, it has been reported that thyroid hormone (T4) played an important role for the initiation of migration (Tsukamoto et al. 1988). In an experimental stream, fish with a high level of plasma T4 swam upstream (migrant), while fish with a low level of T4 swam downstream or stayed in the stream where they were released (nonmigrant). T4 is known to affect smolification and migration behavior in ayu and other salmonids (Tsukamoto et al. 1988; Iwata 1995; Dickhoff et al. 1997) and to affect the migratory behavior of anguillid glass eels (Edeline et al. 2004, 2005). However, migration does not necessarily occur in fish that attain the age and size and endocrinological conditions needed for migration without some additional trigger for migration. As the last step, they need to initiate a behavioral process to release the actual migratory behavior by receiving a trigger from particular endogenous or exogenous factors.

A laboratory experiment suggested that the triggering mechanisms may include many factors such as an increase in water temperature, fish density, or hunger, which raised the level of behavioral "drive" (Berridge 2004) and led fish to respond to the stimuli of water current or a waterfall to which fish had not responded before (described in detail below) (Tsukamoto et al. 1987; Tsukamoto 1988; Tsukamoto and Uchida 1992). The fish can start migration only when these three steps are cleared in order (Tsukamoto et al. 1987). These general steps in preparation for migration are likely similar in most diadromous fishes as they approach the times in their life history when they must initiate an active migration between different habitats.

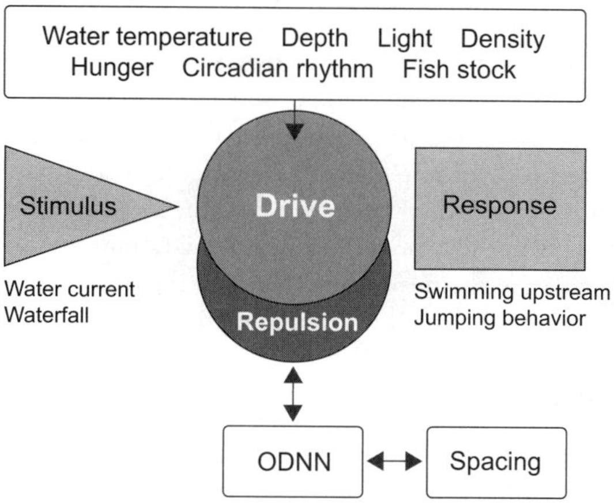

Figure 5.—Behavioral model for the upstream migration of the juvenile ayu. The psychological drive for behaviors relevant to upstream migration of the ayu may be a repulsion that occurs between individuals when their ODNN (optimum distance to the nearest neighbor) is broken in a limited space. In the presence of the appropriate stimuli of current or a waterfall and the elevated level of drive, the response of jumping behavior or upstream migration would be released.

Drive

Animals show behavioral responses to very specific types of stimuli (Tsukamoto 1988; Tsukamoto and Uchida 1992). A stimulus is an external factor that causes a specific behavior to occur in response to the stimulus given to the animal (Figure 5). The behavioral response is usually directional and occurs towards or away from the stimulus. It is well known that the response increases when stimulus intensity increases. However, even if a stimulus is constant, the response sometimes may vary greatly under certain conditions, so we have to assume drive or motivation as an intermediate function between stimulus and response (Figure 5). Drive has been conceptualized as an endogenous factor that urges an animal to perform a behavior and has the potential to amplify or reduce the intensity of the expression of the behavior (Berridge 2004).

In the case of upstream migration of amphidromous ayu, which includes "jumping behavior" and "swimming upstream" as behavioral elements (Tsukamoto 1988; Uchida et al. 1990), the stimuli for jumping are waterfalls and current in the river for swimming upstream (Figure 5). A waterfall is an overall stimulus for jumping behavior that involves three elements: sound waves of about 200–300 Hz, flickering light, and water current from upstream to downstream. All are effective in releasing jumping behavior towards each stimulus. However, experimental results on the migratory behaviors of ayu showed that even when the same stimuli associated with a waterfall and current speed were held constant, an increased jumping behavior or swimming upstream would occur when water temperature was raised, there was decreased water depth, there was increased fish density, there was hunger, or when light conditions were drastically changed (Tsukamoto 1988; Uchida et al. 1990; Tsukamoto and Uchida 1992). In addition, ayu never show upstream migration at night, even when the same level of stimuli (waterfall, currents) continuously exist in a river, which indicates they have a clear daily rhythm. These endogenous and exogenous factors likely affect the drive level of ayu and modify the magnitude of response at a fixed level of stimulus. This suggests that there must be some sort of a drive that is needed for the upstream migration of ayu, regulating the relationship between the stimuli and migratory behavior (Tsukamoto 1988).

Spacing Behavior

Spacing is another important behavior closely related to the upstream migration of ayu (Tsukamoto 1988; Tsukamoto and Uchida 1990). Animals tend to keep their interindividual distance constant under a fixed set of conditions, and this distance is defined as the "optimum distance to the nearest neighbor" (ODNN), which is an equilibrium point in their spacing behavior. Juvenile ayu of a particular stock also show a specific ODNN under certain environmental and physiological conditions (Figure 6A; Tsukamoto and Uchida 1990, 1992). When they spread out beyond their ODNN, they come back close to each other like an attraction, and contrarily, when they approach too close inside the ODNN, they swim away from each other by repulsion (Figure 6B). It was found that increased water temperatures and hunger levels expanded the ODNN of a school or aggregation and that decreased levels would decrease the ODNN (Figure 7; Tsukamoto and Uchida 1990, 1992). It was also confirmed that a group of ayu with a larger ODNN showed more active jumping behavior and swimming upstream. Furthermore, when they were released into a natural river, they showed a stronger tendency to migrate upstream (Tsukamoto et al. 1990; Tsukamoto and Uchida 1992). These observations showed that all these behaviors were closely linked to one another and appeared to characterize the upstream migration of the ayu. Thus, spacing behavior appears to share a common causal relationship with other behaviors related to upstream migration such as jumping behavior and swimming upstream.

Behavioral Model and Triggering Mechanism

Assuming that a drive for upstream migration of ayu is repulsion that occurs between individuals, all the experimental results of the migratory behaviors of the ayu can be comprehensively explained (Figure 5). When fish aggregate at high density below a waterfall, their ODNN is compressed and a repulsive force is generated. This condition causes an elevated drive level in the ayu, which in turn exhibits a stronger response even at a fixed stimulus level of a waterfall. Fish with a large ODNN, showing wider spacing, have a stronger

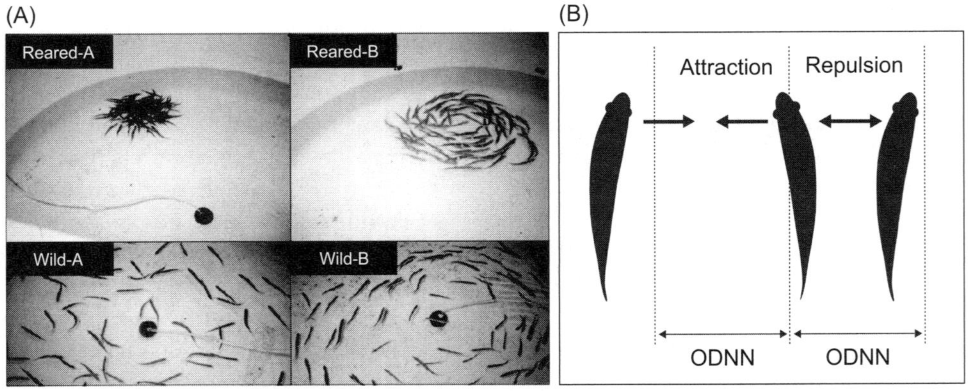

Figure 6.—Spacing behavior of the amphidromous ayu. (A) Spacing of four different stocks of ayu: reared fishes from the hatcheries A (Reared-A) and B (Reared-B), wild fishes from Lake Biwa (Wild-A, landlocked population) and the Yabe River (Wild-B, amphidromous population). (B) A representation of how juvenile ayu react with attraction when the interindividual distance exceeds their ODNN (optimum distance to the nearest neighbor) and with repulsion when the distance is less than the ODNN.

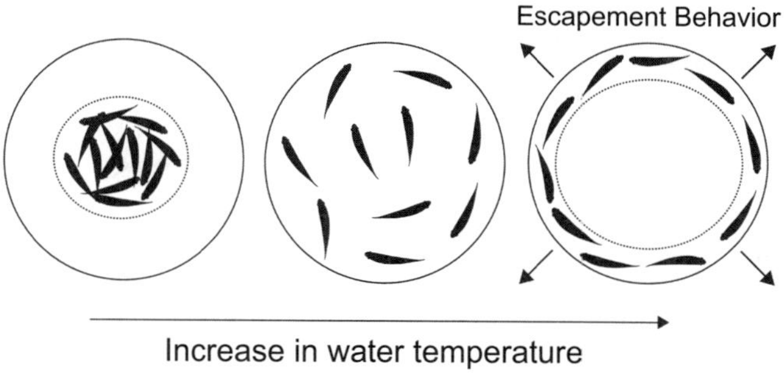

Figure 7.—Change in spacing behavior and jumping-out motion in random directions by juvenile ayu, along the edge of the tank (escapement behavior) when water temperature is increased.

repulsion in a limited space and show a stronger tendency for upstream migration. Increased water temperature causes an elevated level of repulsion, with an expanded spacing and results in a greater manifestation of jumping behavior and swimming upstream, even when exposed to a fixed stimulus. Hunger level, water depth, change in light intensity, and circadian rhythm also directly affect drive level and modify the response (Figure 5).

In the three-step model of the triggering mechanism of juvenile ayu (Figure 4), the final step is triggered by the behavioral process explained above (Figure 5). In the early morning at low water temperature, even fish with sufficient body size and age (first step) and suitable endocrinological condition (second step) cannot react to stimuli in a river such as a waterfall and or current. However, in the afternoon at an elevated water temperature, fish have their drive elevated and can respond to the same level of stimuli to finally start their upstream migration. Other factors affecting drive level could cause upstream migration in the same manner.

Random Escapement Hypothesis

Vacuum Activity of Ayu

One more important observation is that of a "jumping-out" motion observed in juvenile ayu that were contained in a small tank (e.g., 1 m diameter, 20 cm water depth) (Figure 7). They showed a small aggregation at low water temperature (e.g., 15°C) and had a larger spacing at higher temperature. When temperature further increased to about 25°C, at which temperature the ayu feel unfavorable near their upper limit, they showed a doughnut-shaped distribution in the tank and suddenly started strong jumping motions in various directions along the walls of the tank (Figure 7). The motion of this activity is essentially the same as that of the jumping behavior that is released towards the waterfall (stimulus), but is different in the point of direction of the motion: fish jump toward a stimulus in jumping behavior, while in the jumping-out motion that was observed, they jump in random directions because there is no specific stimulus. This jumping-out may be a "vacuum activity" with no appropriate stimulus (Berridge 2004), which was caused only by an extremely elevated drive resulting from an increased water temperature (Figure 7). Other factors such as an increase in fish density or hunger level can likely raise drive level and cause vacuum activity without a stimulus. This jumping-out can be interpreted as "random escapement from unfavorable environments" because too much of an increase in water temperature, fish density, or hunger level are all unfavorable conditions for the ayu.

Escapement Behavior

We propose that random escapement behavior may be the basic behavioral element underlying upstream migration of the ayu. If this is the case for the present-day migration of ayu, a pioneer migrant of the ancestral ayu might have presented this type of random escapement behavior in an estuary at the very beginning of the prototype migration in the past. Consequently, it is possible that the evolutionary process of migration of ayu started with this random escapement behavior at some point in the past and that the migrants moving upstream profited from better growth by feeding on algae on the rocks and thus gained an adaptive advantage that was consequently favored by natural selection. The first migrating ayu with the highest levels of drive, or the lowest threshold for the vacuum activity, may have shown random escapement behavior that resulted in upstream migration as the first step of a regular migration.

The random escapement hypothesis may also be applicable as the most primitive origin of migration for other types of migration in both anadromous and catadromous fishes (e.g., downstream migration by salmonid smolts or upstream migration of anguillid glass eels) and possibly for potamodromy and oceanodromy because the initial movements of any type of migration are essentially the same. First, they leave their habitat (Habitat A, Figure 8–1) by random escapement because they feel unfavorable there, and some of them successfully find a new habitat (Habitat B, Figure 8–2). After spending a period of time there, they feel that Habitat B is unfavorable and escape from the habitat (Figure 8–3). When some of them can successfully migrate back to Habitat A (Figure 8–4), a migration loop is established and eventually obtains stability and regularity of migration if it is advantageous (Figure 8–5).

Various factors could cause unfavorable conditions for fish within a habitat that could drive them away through escapement behavior, such as food availability, predators, fish density, water temperature, light condition, water depth, salinity, pH, or any other environmental stressors. In fact, as Liem (1987) showed experimentally and from field observations, that unfavorable interindividual competition for food in the aquatic habitat and the resulting hunger could cause an overland excursion of mudskippers (*Monopterus* and *Anabas* spp.). He further concluded that the evolutionary cause of invasion from water to land in teleosts would not be a shortage of water during the dry season, but was the severe competition among individuals under sufficiently wet conditions. This observation of mudskippers that escaped from unfavorable aquatic environment to terrestrial habitat is consistent with the random escapement hypothesis as a possible origin of migration.

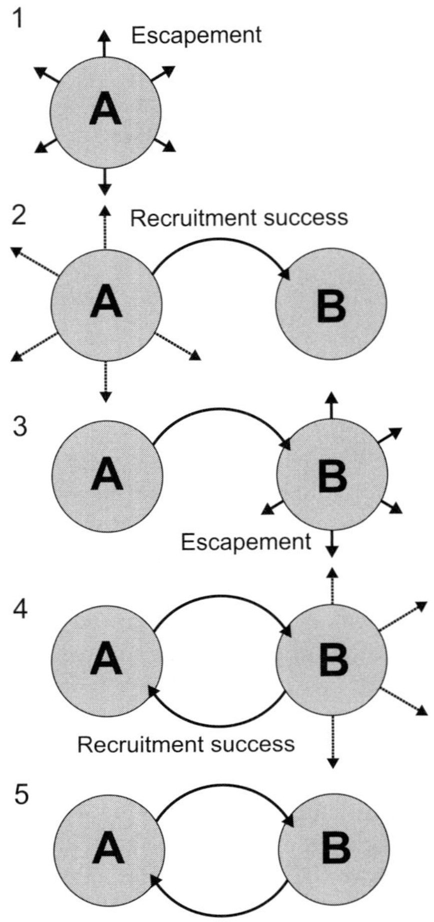

Figure 8.—A schematic representation of how progressive steps of random escapement from one habitat (1) and successful recruitment into another (2–4) could lead to the establishment of a new migration loop (5).

Summary and Conclusions

Migration can be defined as regularly timed habitat transitions that are part of the life history of a species, which are typically used for growth and reproduction. A migration loop model is useful to help conceptualize the process of speciation in diadromous fishes, with an expansion or shift of a migration loop potentially resulting in diadromy or speciation. Phylogenetic relationships suggest a freshwater origin of salmonids with an increasing sea dependency through anadromous migrations and a marine origin of anguillid eels, which appear to be facultatively catadromous in temperate regions, with some individuals not entering freshwater for growth. These patterns appear to have resulted from differences in productivity between the marine and freshwater habitats at higher and lower latitudes. Evidence of the ancestral life history states of these fishes can be seen in geographic clines in the occurrence of nonmigratory residents of the sympatric Japanese eel and the masu salmon.

A behavioral approach for understanding the triggering mechanism of fish migration using the upstream migration of amphidromous ayu as a behavioral model suggests that endogenous and exogenous factors, such as an increase in water temperature, fish density, or hunger level, can raise drive level and release behavior associated with upstream migration. Experimental observations of ayu suggest that random escapement behavior from unfavorable environments may result from an increased drive level. If an adaptive advantage was gained by entering a new habitat as a result of random escapement behavior, it could lead to the establishment of a new diadromous migration pattern. It is hypothesized that this type of escapement behavior might be an origin of fish migration, including diadromy between freshwater and marine habitats. This was likely later accompanied by the more sophisticated abilities such as locomotion, osmoregulation, and the other remarkable abilities of very precise homing that is seen in some diadromous fishes.

References

Åkesson, S., and A. Hedenström. 2007. How migrants get there: migratory performance and orientation. Bioscience 57:123–133.

Alerstam, T. 2006. Conflicting evidence about long-distance animal navigation. Science 313:791–793.

Alerstam, T., A. Hedenström, and S. Åkesson. 2003. Long-distance migration: evolution and determinants. Oikos 103:247–260.

Arai, T., A. Kotake, P. M. Lokman, M. J. Miller, and K. Tsukamoto. 2004. Evidence of different habitat use by New Zealand freshwater eels, *Anguilla australis* and *A. dieffenbachii*, as revealed by otolith microchemistry. Marine Ecology Progress Series 266:213–225.

Arai, T., A. Kotake, M. Ohji, N. Miyazaki, and K. Tsukamoto. 2003. Migratory history and habitat use of Japanese eel *Anguilla japonica* in the Sanriku coast of Japan. Fisheries Science 69:813–818.

Bell, C. P. 2000. Process in the evolution of bird migration and pattern in avian ecogeography. Journal of Avian Biology 31:258–265.

Bemis, W. E., and B. Kynard. 1997. Sturgeon rivers: an introduction to acipenseriform biogeography and life history. Environmental Biology of Fishes 48:167–183.

Berridge, K. C. 2004. Motivation concepts in behavioral neuroscience. Physiology and Behavior 81:179–209.

Bingman, V. P., and K. Cheng. 2005. Mechanisms of animal global navigation: comparative perspectives and enduring challenges. Ethology, Ecology and Evolution 17:295–318.

Blanco, J. F., and F. N. Scatena. 2007. Floods, habitat hydraulics and upstream migration of *Neritina virginea* (Gastropoda: Neritidae) in northeastern Puerto Rico. Caribbean Journal of Science 41:55–74.

Block, B. A., S. L. H. Teo, A. Walli, A. Boustany, M. J. W. Stokesbury, C. J. Farwell, K. C. Weng, H. Dewar, and T. D. Williams. 2005. Electronic tagging and population structure of Atlantic bluefin tuna. Nature (London) 434:1121–1127.

Bonfil, R., M. Meÿer, M. C. Scholl, R. Johnson, S. O'Brien, H. Oosthuizen, S. Swanson, D. Kotze, and M. Paterson. 2005. Transoceanic migration, spatial dynamics, and population linkages of white sharks. Science 310:100–103.

Bowen, B. W., A. L. Bass, L. Soares, and R. J. Toonen. 2005. Conservation implications of complex population structure: lessons from the loggerhead turtle (*Caretta caretta*). Molecular Ecology 14:2389–2402.

Boyle, W. A., and C. J. Conway. 2007. Why migrate? A test of the evolutionary precursor hypothesis. American Naturalist 169:344–359.

Bradbury, I. R., M. W. Coulson, S. E. Campana, E. Baggs, and P. Bentzen. 2009. Postglacial recolonization and the loss of anadromy in rainbow smelt from coastal Newfoundland. Pages 79–96 *in* A. J. Haro, K. L. Smith, R. A. Rulifson, C. M. Moffitt, R. J. Klauda, M. J. Dadswell, R. A. Cunjak, J. E. Cooper, K. L. Beal, and T. S. Avery, editors. Challenges for diadromous fishes in a dynamic global environment. American Fisheries Society, Symposium 69, Bethesda, Maryland.

Chernoff, E., and R. A. Curry. 2007. First summer growth predetermined in anadromous and resident brook charr. Journal of Fish Biology 70:334–346.

Croxall, J. P., J. R. D. Silk, R. A. Phillips, V. Afanasyev, and D. R. Briggs. 2005. Global circumnavigations: Tracking year-round ranges of nonbreeding albatrosses. Science 307:249–250.

Dadswell, M. J., R. J. Klauda, C. M. Moffitt, R. L. Saunders, R. A. Rulifson, and J. E. Cooper, editors. 1987. Common strategies of anadromous and catadromous fishes. American Fisheries Society, Symposium 1, Bethesda, Maryland.

Daverat, F., K. E. Limburg, I. Thibault, J. C. Shiao, J. J. Dodson, F. Caron, W. N. Tzeng, Y. Iizuka, and H. Wickström. 2006. Phenotypic plasticity of habitat use by three temperate eel species, *Anguilla anguilla*, *A. japonica* and *A. rostrata*. Marine Ecology Progress Series 308:231–241.

Dickhoff, W. W., B. R. Beckman, D. A. Larsen, C. Duan, and S. Moriyama. 1997. The role of growth in endocrine regulation of salmon smoltification. Fish Physiology and Biochemistry 17:231–236.

Dingle, H., and D. V. Alistair. 2007. What is migration? Bioscience 57:113–121.

Dittman, A. H., and T. P. Quinn. 1996. Homing in Pacific salmon: mechanisms and ecological basis. Journal of Experimental Biology 199:83–91.

Dodson, J. J., J. Laroche, and F. Lecomte. 2009. Contrasting evolutionary pathways of anadromy in euteleostean fishes. Pages 63–77 *in* A. J. Haro, K. L. Smith, R. A. Rulifson, C. M. Moffitt, R. J. Klauda, M. J. Dadswell, R. A. Cunjak, J. E. Cooper, K. L. Beal, and T. S. Avery, editors. Challenges for diadromous fishes in a dynamic global environment. American Fisheries Society, Symposium 69, Bethesda, Maryland.

Eckert, S., and B. Stewart. 2001. Telemetry and satellite tracking of whale sharks, *Rhincodon typus*, in the Sea of Cortez, Mexico, and North Pacific Ocean. Environmental Biology of Fishes 60:299–308.

Edeline, E., A. Bardonnet, V. Bolliet, S. Dufour, and P. Elie. 2005. Endocrine control of *Anguilla anguilla* glass eel dispersal: effect of thyroid hormones on locomotor activity and rheotactic behavior. Hormones and Behavior 48:53–63.

Edeline, E., S. Dufour, C. Briand, D. Fatin, and P. Elie. 2004. Thyroid status is related to migratory behavior in *Anguilla anguilla* glass eels. Marine Ecology Progress Series 282:261–270.

Fleming, I. A. 1998. Pattern and variability in the breeding system of Atlantic salmon (*Salmo salar*), with comparisons to other salmonids. Canadian Journal Fisheries and Aquatic Science 55:59–76.

Groot, C., and L. Margolis, editors. 1991. Pacific salmon life histories. University of British Columbia Press, Vancouver.

Gross, M. R. 1987. Evolution of diadromy in fishes. Pages 12–25 *in* M. J. Dadswell, R. J. Klauda, C. M. Moffitt, R. L. Saunders, R. A. Rulifson, and J. E. Cooper, editors. American Fisheries Society, Symposium 1, Bethesda, Maryland.

Gross, M. R., R. M. Coleman, R. M. McDowall. 1988. Aquatic productivity and the evolution of diadromous fish migration. Science 239:1291–1293.

Hatase, H., K. Sato, M. Yamaguchi, K. Takahashi, and K. Tsukamoto. 2006. Individual variation in feeding habitat use by adult female green sea turtles (*Chelonia mydas*): are they obligately neritic herbivores? Oecologia 149:52–64.

Holland, R. A., M. Wilkelski, and D. S. Wilcove. 2006. How and why do insects migrate? Science 313:794–796.

Iida, M., S. Watanabe, and K. Tsukamoto. 2009. Life history characteristics of a Sicydiinae goby in Japan, compared with its relatives and other amphidromous fishes. Pages 355–373 *in* A. J. Haro, K. L. Smith, R. A. Rulifson, C. M. Moffitt, R. J. Klauda, M. J. Dadswell, R. A. Cunjak, J. E. Cooper, K. L. Beal, and T. S. Avery, editors. Challenges for diadromous fishes in a dynamic global environment. American Fisheries Society, Symposium 69, Bethesda, Maryland.

Inoue, J. G., M. Miya, K. Tsukamoto, and M. Nishida. 2004. Mitogenomic evidence for the monophyly of Elopomorph fishes (Teleostei) and the evolutionary origin of the leptocephalus larva. Molecular Phylogenetics and Evolution 32:274–286.

Ishiguro, N. B., M. Miya, and M. Nishida. 2003. Basal euteleostean relationships: a mitogenomic perspective on the phylogenetic reality of the "Protacanthopterygii." Molecular Phylogenetics and Evolution 27:476–488.

Iwata, M. 1995. Downstream migratory behavior of salmonids and its relationship with cortisol and thyroid hormones: a review. Aquaculture 135:131–139.

James, M. C., R. A. Myers, and C. A. Ottensmeyer. 2005. Behaviour of leatherback sea turtles, *Dermochelys coriacea*, during the migratory cycle. Proceedings of the Royal Society of London Series B 272:1547–1555.

Jellyman, D., and K. Tsukamoto. 2002. First use of archival transmitters to track migrating freshwater eels *Anguilla dieffenbachii* at sea. Marine Ecology Progress Series 233:207–215.

Jessop, B. M., J. C. Shiao, Y. Iizuka, and W. N. Tzeng. 2002. Migratory behaviour and habitat use by American eels *Anguilla rostrata* as revealed by otolith microchemistry. Marine Ecology Progress Series 233:217–229.

Jessop, B .M., J. C. Shiao, Y. Iizuka, and W. N. Tzeng. 2004. Variation in the annual growth, by sex and migration history, of silver American eels *Anguilla rostrata*. Marine Ecology Progress Series 272:231–244.

Kaeriyama, M. 1994. Life history strategy of the sockeye salmon. Pages 111–113 *in* A. Goto, K. Tsukamoto, and K. Maekawa, editors. Diadromous fishes. Toukai University Press, Tokyo.

Kato, F. 1991. Life histories of masu and amago salmon (*Oncorhynchus masou* and *Oncorhynchus rhodurus*). Pages 447–520 *in* C. Groot and L. Margolis, editors. Pacific salmon life histories. University of British Columbia Press, Vancouver.

Keith, P. 2003. Biology and ecology of amphidromous Gobiidae of the Indo-Pacific and the Caribbean regions. Journal of Fish Biology 63:831–847.

Kotake, A., T. Arai, T. Ozawa, S. Nojima, M. J. Miller, and K. Tsukamoto. 2003. Variation in migratory history of Japanese eels, *Anguilla japonica*, collected in coastal waters of the Amakusa Islands, Japan, inferred from otolith Sr/Ca ratios. Marine Biology 142:849–854.

Kotake, A., A. Okamura, Y. Yamada, T. Uto, T. Arai, M. J. Miller, H. P. Oka, and K. Tsukamoto. 2005. Seasonal variation in migratory history of the Japanese eel, *Anguilla japonica*, in Mikawa Bay, Japan. Marine Ecology Progress Series 293:213–221.

Kubo, T. 1980. Studies on the life history of the masu salmon (*Oncorhynchus masou*) in Hokkaido. Scientific Reports of the Hokkaido Salmon Hatchery 34:1–95. (In Japanese with English abstract.)

Kuribayashi, K., H. Katakura, M. Kyono, M. H. Dick, and S. F. Mawatari. 2006. Round-trip catadromous migration in a Japanese amphipod, *Sternomoera rhyaca* (Gammaridea: Eusiridae). Zoological Science 23:763–774.

Kuroki, M., M. Kawai, B. Jónsson, J. Aoyama, M. J. Miller, D. L.G. Noakes, and K. Tsukamoto. 2008. Inshore migration and otolith microstructure/microchemistry of anguillid glass eels recruited to Iceland. Environmental Biology of Fishes 83:309–325.

Lee, P. L. M., P. Luschi, and G. C. Hays. 2007. Detecting female precise natal philopatry in green turtles using assignment methods. Molecular Ecology 16:61–74.

Liem, K. F. 1987. Functional design of the air ventilation apparatus and overland excursions by teleosts. Fieldiana Zoology 37:1–29.

Luschi, P., G. C. Hays, and F. Papi. 2003. A review of long distance movements by marine turtles, and the possible role of ocean currents. Oikos 103:293–302.

Luschi, P., J. R. E. Lutjeharms, P. Lambardi, R. Mencacci, G. R. Hughes, and G. C. Hays. 2006. A review of migratory behaviour of sea turtles off southeastern Africa. South African Journal of Science 102:51–58.

McCleave, J. D., and E. Edeline. 2009. Diadromy as a conditional strategy: patterns and drivers of eel movements in continental habitats. Pages 97–120 *in* A. J. Haro, K. L. Smith, R. A. Rulifson, C. M. Moffitt, R. J. Klauda, M. J. Dadswell, R. A. Cunjak, J. E. Cooper, K. L. Beal, and T. S. Avery, editors. Challenges for diadromous fishes in a dynamic global environment. American Fisheries Society, Symposium 69, Bethesda, Maryland.

McCleave, J. D., R. C. Kleckner, and M. Castonguay. 1987. Reproductive sympatry of American and European eels and implications for migration and taxonomy. Pages 286–297 *in* M. J. Dadswell, R. J. Klauda, C. M. Moffitt, R. L. Saunders, R. A. Rulifson, and J. E. Cooper, editors. American Fisheries Society, Symposium 1, Bethesda, Maryland.

McDowall, R. M. 1987. The occurrence and distribution of diadromy among fishes. Pages 1–13 *in* M. J. Dadswell, R. J. Klauda, C. M. Moffitt, R. L. Saunders, R. A. Rulifson, and J. E. Cooper, editors. American Fisheries Society, Symposium 1, Bethesda, Maryland.

McDowall, R. M. 1997. The evolution of diadromy in fishes (revisited) and its place in phylogenetic analysis. Reviews in Fish Biology and Fisheries 7:443–462.

McDowall, R. M. 2002. The origin of the salmonid fishes: marine, freshwater ... or neither? Reviews in Fish Biology and Fisheries 11:171–179.

McDowall, R. M. 2007. On amphidromy, a distinct form of diadromy in aquatic organisms. Fish and Fisheries 8:1–13.

McKeown B. A. 1984. Fish migration. Croom Helm, London.

Miller, M. J., S. Kimura, K. D. Friedland, B. Knights, H.-Y. Kim, D. J. Jellyman, and K. Tsukamoto. 2009. Review of ocean-atmospheric factors in the Atlantic and Pacific oceans influencing spawning and recruitment of anguillid eels. Pages 231–249 *in* A. J. Haro, K. L. Smith, R. A. Rulifson, C. M. Moffitt, R. J. Klauda, M. J. Dadswell, R. A. Cunjak, J. E. Cooper, K. L. Beal, and T. S. Avery, editors. Challenges for diadromous fishes in a dynamic global environment. American Fisheries Society, Symposium 69, Bethesda, Maryland.

Minegishi, Y., J. Aoyama, and K. Tsukamoto. 2009. Migratory characteristics as a key factor for determining the population structure of a diadromous species: eels tend to be panmictic while salmon easily divide populations. Pages 851–853 *in* A. J. Haro, K. L. Smith, R. A. Rulifson, C. M. Moffitt, R. J. Klauda, M. J. Dadswell, R. A. Cunjak, J. E. Cooper, K. L. Beal, and T. S. Avery, editors. Challenges for diadromous fishes in a dynamic global environment. American Fisheries Society, Symposium 69, Bethesda, Maryland.

Morrison, W. E., D. H. Secor, and P. M. Piccoli. 2003. Estuarine habitat use by Hudson River American eels as determined by otolith strontium:calcium ratios. Pages 87–99 *in* D. A. Dixon editor. Biology, management, and protection of catadromous eels. American Fisheries Society, Symposium 33, Bethesda, Maryland.

Murata, S., N. Takasaki, M. Saitoh, and N. Okada. 1993. Determination of the phylogenetic relationships among Pacific salmonids by using short interspersed elements (SINEs) as temporal landmarks of evolution. Proceedings of the National Academy of Sciences of the United States of America 90:6995–6999.

Papi, F. 2006. Navigation of marine, freshwater and coastal animals: concepts and current problems. Marine and Freshwater Behaviour and Physiology 39:3–12.

Quinn, T. P., and K. W. Myers. 2004. Anadromy and the marine migrations of Pacific salmon and trout: Rounsefell revisited. Reviews in Fish Biology and Fisheries 14:421–442.

Rankin, M. A. editor. 1985. Migrations: mechanisms and adaptive significance. Contributions in Marine Science 27(Supplement).

Rappole, J. H., B. Helm, and M. A. Ramos. 2003. An integrative framework for understanding the origin and evolution of avian migration. Journal of Avian Biology 34:124–128.

Roff, D. A. 1988. The evolution of migration and some life history parameters in marine fishes. Environmental Biology of Fishes 22:133–146.

Salewski, V., and B. Bruderer. 2007. The evolution of bird migration—a synthesis. Naturwissenschaften 94:268–279.

Schmidt, J. 1922. The breeding places of the eel. Philosophical Transactions of the Royal Society B 211:179–208.

Shaffer, S. A., Y. Tremblay, H. Weimerskirch, D. Scott, D. R. Thompson, P. M. Sagar, H. Moller, G. A. Taylor, D. G. Foley, B. A. Block, and D. P. Costa. 2006. Migratory shearwaters integrate oceanic resources across the Pacific Ocean in an endless summer. Proceedings of the National Academy of Sciences of the United States of America 103:12799–12802.

Shen, K. N., and W. N. Tzeng. 2007. Genetic differentiation among populations of the shortfinned eel *Anguilla australis* from East Australia and New Zealand. Journal of Fish Biology 70:177–190.

Sippel, T. J., P. S. Davie, J. C. Holdsworth, and B. A. Block. 2007. Striped marlin (*Tetrapturus audax*) movements and habitat utilization during a summer and autumn in the southwest Pacific Ocean. Fisheries Oceanography 16:459–472.

Tamate, T., and K. Maekawa. 2002. Individual growth and phase differentiation of lacustrine masu salmon, *Oncorhynchus masou*, under artificial rearing conditions. Ichthyological Research 49:397–400.

Teo, S. L. H., A. Boustany H. Dewar, M. J. W. Stokesbury, K. C. Weng, S. Beemer, A. C. Seitz, C. J. Farwell, E. D. Prince, and B. A. Block. 2007. Annual migrations, diving behavior, and thermal biology of Atlantic bluefin tuna, *Thunnus thynnus*, on their Gulf of Mexico breeding grounds. Marine Biology 151:1–18.

Tsukamoto, K. 1988. Migratory mechanisms and behavioral characateristics in ayu. Pages 100–133 *in* T. Ueno and M. Okiyama, editors. Ichthyology currents. Asakurashoten, Tokyo.

Tsukamoto, K. 1990. Ecological aspects and resources of ayu, in the marine habitat. Suisanzoshoku 38:208–209.

Tsukamoto, K. 1992. Discovery of the spawning area for the Japanese eel. Nature(London) 356:789–791.

Tsukamoto, K. 1994. Origin of diadromous fishes and mechanism of migration. Pages 2–17 *in* A. Goto, K. Tsukamoto, and K. Maekawa, editors. Freshwater fishes migrating between river and the sea. Tokai University Press, Tokyo.

Tsukamoto, K. 2006. Spawning of eels near a seamount. Nature(London) 439:929.

Tsukamoto, K., K. Aida, and T. Otake. 1988. Plasma thyroxine concentration and upstream migratory behavior of juvenile ayu. Nippon Suisan Gakkaishi 54:1687–1693.

Tsukamoto, K., and J. Aoyama. 1998. Evolution of freshwater eels of the genus *Anguilla*: a probable scenario. Environmental Biology of Fishes 52:139–148.

Tsukamoto, K., J. Aoyama, and M. J. Miller. 2002. Migration, speciation and the evolution of diadromy in anguillid eels. Canadian Journal of Fisheries and Aquatic Sciences 59:1989–1998.

Tsukamoto, K., and T. Arai. 2001. Facultative catadromy of the eel *Anguilla japonica* between freshwater and seawater habitats. Marine Ecology Progress Series 220:265–276.

Tsukamoto, K., R. Ishida, K. Naka, and T. Kajihara. 1987. Switching of size and migratory pattern in successive generations of landlocked ayu. Pages 492–506 *in* M. J. Dadswell, R. J. Klauda, C. M. Moffitt, R. L. Saunders, R. A. Rulifson, and J. E. Cooper, editiors. Common strategies of anadromous and catadromous fishes. American Fisheries Society, Symposium 1, Bethesda, Maryland.

Tsukamoto, K., S. Masuda, M. Endo, and T. Otake. 1990. Behavioural characteristics of the ayu, *Plecoglossus altivelis*, as predictive indices for stocking effectiveness in a river. Nippon Suisan Gakkaishi 56:1177–1186.

Tsukamoto, K., I. Nakai, and F. W. Tesch. 1998. Do all freshwater eels migrate? Nature (London) 396:635–636.

Tsukamoto, K., T. Otake, N. Mochioka, T. W. Lee, H. Fricke, T. Inagaki, J. Aoyama, S. Ishikawa, S. Kimura, M. J. Miller, H. Hasumoto, M. Oya, and Y. Suzuki. 2003. Seamounts, new moon and eel spawning: the search for the spawning site of the Japanese eel. Environmental Biology of Fishes 66:221–229.

Tsukamoto, K., and K. Uchida. 1990. Spacing and jumping behaviours of the ayu, *Plecoglossus altivelis*. Nippon Suisan Gakkaishi 56:1383–1392.

Tsukamoto, K., and K. Uchida. 1992. Migration mechanism of the ayu. Pages 145–172 *in* V. I. Ilyichev and V. V. Anikiev, editors. Oceanic and anthropogenic controls of life in the Pacific Ocean. Kluwer Academic Publishers, Amsterdam.

Tzeng, W. N., J. C. Shiao, and Y. Iizuka. 2002. Use of otolith Sr:Ca ratios to study the riverine migratory behaviors of Japanese eel *Anguilla japonica*. Marine Ecology Progress Series 245:213–221.

Uchida, K., K. Tsukamoto, and T. Kajihara. 1990. Effects of environmental factors on jumping behaviour of the juvenile ayu *Plecoglossus altivelis* with special reference to their upstream migration. Nippon Suisan Gakkaishi 56:1393–1399.

Watanabe, S., J. Aoyama, and K. Tsukamoto. 2006. Confirmation of morphological differences between *Anguilla australis australis* and *Anguilla australis schmidtii*. New Zealand Journal of Marine and Freshwater Research 40:325–331.

Weng, K. C., P. C. Castilho, J. M. Morrissette, A. M. Landeira-Fernandez, D. B. Holts, R. J. Schallert, K. J. Goldman, and B. A. Block. 2005. Satellite tagging and cardiac physiology reveal niche expansion in salmon sharks. Science 310:104–106.

Wilson, S. G., M. E. Lutcavage, R. W. Brill, M. P. Genovese, A. B. Cooper, and A. W. Everly. 2005. Movements of bluefin tuna (*Thunnus thynnus*) in the northwestern Atlantic Ocean recorded by pop-up satellite archival tags. Marine Biology 146:409–423.

Yamamoto, T. 2004. 2004. Sex-specific growth pattern during early life history in masu salmon, *Oncorhynchus masou*. Ecology of Freshwater Fish 13:203–207.

Yokoi, K. I., and K. Hosoya. 2006. Early development of the endangered freshwater goby, *Rhinogobius* sp BI (Gobiidae). Ichthyological Research 53:160–165.

Yokoyama, R., and A. Goto. 2005. Evolutionary history of freshwater sculpins, genus *Cottus* (Teleostei: Cottidae) and related taxa, as inferred from mitochondrial DNA phylogeny. Molecular Phylogenetics and Evolution 36:654–668.

Zerbini, A. N., A. Andriolo, M. P. Heide-Jørgensen, J. L. Pizzorno, Y. G. Maia, G. R. VanBlaricom, D. P. DeMaster, P. C. Simões-Lopes, S. Moreira, and C. Bethlem. 2006. Satellite-monitored movements of humpback whales *Megaptera novaeangliae* in the southwest Atlantic Ocean. Marine Ecology Progress Series 313:295–304.

Zink, R. M. 2002. Towards a framework for understanding the evolution of avian migration. Journal of Avian Biology 33:433–436.

American Fisheries Society Symposium 69:63–77, 2009

Contrasting Evolutionary Pathways of Anadromy in Euteleostean Fishes

Julian J. Dodson*
Département de biologie, Université Laval, Pavillon Vachon, Québec G1V 0A6, Canada

Jérôme Laroche
Centre de bioinformatique et de biologie computationnelle, Université Laval Pavillon Marchand, Québec G1V 0A6, Canada

Frédéric Lecomte[1]
Département de biologie, Université Laval, Pavillon Vachon, Québec G1V 0A6, Canada

Abstract.—We develop the view, based on life cycle differences and recently published sister group relationships, that the freshwater life cycle was the ancestral character state leading to anadromy among salmoniforms, whereas the marine life cycle was the ancestral character state leading to anadromy among osmeriforms. In contrast to most salmonid fishes, the reproductive migrations of smelts are generally characterized by brief excursions to spawn in freshwater, and larvae may spend no more than 24 h in freshwater before being transported to coastal marine or estuarine environments. We reconstructed the phylogeny of the suborder Osmeroidei to establish the phylogenetic relationships among anadromous, marine, and freshwater species of this taxon. We mapped these life cycles onto phylogenetic trees of osmeriforms and salmoniforms and applied character-reconstruction methodology based on simple parsimony and likelihood methodologies. A freshwater origin of salmonids was supported by our analyses, whereas either marine or anadromous life cycles characterized the evolution of osmeroids. The possibility that the evolution of anadromy in salmonids and osmeroids followed separate paths requires a reconsideration of some generalizations concerning anadromy. We hypothesize that anadromy in osmeroids may be first and foremost an adaptation to place embryos and the early larval stages in reproductive safe sites to maximize their survival. The evolution of exclusive freshwater species of osmeriforms has occurred via anadromy through the various processes associated with landlocking. Freshwater amphidromy in osmeroids is most likely a consequence of anadromy rather than a precursor and may be contingent upon the availability of food resources in freshwater. Finally, marine osmeroids have been derived from anadromous ancestors and are "safe-site" specialists, exploiting principally the upper intertidal zone for reproduction. We also suggest that such contrasting evolutionary pathways to anadromy may provide insight into the evolution of partial migration, observed uniquely in salmonids, and the nature and extent of population genetic structure found in the two groups of fishes.

Introduction

Fishes that migrate regularly between freshwater and saltwater for the purposes of reproduction and feeding (diadromy) exhibit three migratory patterns (Myers 1949). Anadromous fishes such as salmon move from the sea to breed in freshwater, whereas catadromous fishes such as eels move from freshwater to breed in the sea. The third category, amphidromy, is sometimes included within either anadromy or catadromy but has been considered

* Corresponding author: julian.dodson@bio.ulaval.ca
[1] Current address: INRA, UMR ECOBIOP, Quartier Ibarron, 64310 St Pée-sur-Nivelle, France

as a distinct strategy (McDowall 1988). McDowall (2004) restricted the definition of amphidromy to fishes that reproduce in freshwater and drift downstream following hatching to feed at sea for variable periods of time before returning to freshwater to feed and grow to maturity. We refer to this life cycle more specifically as freshwater amphidromy to distinguish it from those diadromous fishes that reproduce at sea, migrate to freshwater early in life to feed for variable periods of time, and return to sea to feed and grow to maturity (marine amphidromy, Gross 1987). Diadromy is shown by at least 227 fish species, of which 110 are anadromous (McDowall 1988; Nelson 1994). Anadromy appears to have evolved early among fishes, as 90% of the 110 species identified by McDowall (1988) as anadromous are among either the most primitive living fishes (e.g., lampreys, sturgeons; 20 species) or the basal clupeocephalan teleosts (79 species.). However, its presence among some neoteleosts (e.g., Gasterosteiformes, Gobioidei) suggests more than one independent origin of anadromy.

One of the most influential contributions to understanding the adaptive significance and evolution of diadromy was formulated by Gross (1987). To evolve, diadromy (or any other migratory pattern) must result in a level of lifetime reproductive success (reproductive success × survivorship) exceeding that of individuals who do not migrate. The most important biological variable in explaining the presence and direction of diadromy was identified as the relative availability of food in marine and freshwater habitats (Gross et al. 1988). Marine habitats are more productive at high latitudes and less so towards the tropics, whereas the inverse occurs in freshwater. As such, anadromy is favored at high latitudes, whereas catadromy is favored at low latitudes. Gross (1987) envisioned amphidromy as the ancestral state in the evolution of anadromy and catadromy. Finally, the evolutionary scenarios proposed by Gross (1987) were based on the assumption that the ancestral state of diadromous fishes involves the reproductive environment. As such, catadromous fishes are derived from marine species that continue to exploit the ancestral reproductive environment. The ultimate derived state involves a completely freshwater life cycle. On the other hand, anadromous fishes are derived from freshwater species that continue to exploit the ancestral reproductive environment and the ultimate derived state involves a completely marine life cycle.

Since they were first published, these comprehensive evolutionary scenarios have generated a great deal of discussion and ensuing confusion. McDowall (1997) objected to the contention that amphidromy is an intermediate stage in the evolution of anadromy on the basis that there are few documented cases of freshwater fishes becoming facultative marine wanderers, the initial stage in Gross' evolutionary pathway to anadromy. However, McDowall (1988) defined a category of freshwater fishes that occasionally move into marine waters (euryhaline wanderers) as being distinct from anadromous fishes. The well-documented ability of many freshwater species to exploit the oligohaline waters of estuaries (Moyle and Cech 2004) seems to fit the definition of euryhaline wanderers and represents a plausible evolutionary route to an anadromous life style. McDowall (1997) also stated that there is little empirical evidence that marine amphidromy is a precursor to the evolution of catadromy in formerly marine species as there are virtually no cases of amphidromous species that spawn at sea. On the contrary, recent observations reveal that marine amphidromy is more common than previously suggested, even among the catadromous anguillid eels (Daverat et al. 2006 and references therein). Despite numerous publications on the subject, we are no closer to formulating a comprehensive evolutionary framework to understand the development of diadromy beyond that provided by Gross (1987) 20 years ago. In the present paper, we aim to provide an alternative, yet complementary, model of the evolution of anadromy.

The starting point of this argument is to challenge the assumption that anadromy has been uniquely derived from freshwater origins. Several anadromous fishes such as the Atlantic tomcod *Microgadus tomcod* (Walbaum) and the clupeid shads (*Alosa* spp. and *Hilsa* spp.) appear to have been derived from marine families (McDowall 1993, 1997; Dodson 1997). Secor (2002) hypothesized that anadromy in striped bass *Morone saxatilis* (Walbaum) evolved from a marine-spawning ancestor. Gobies may represent another, more complex, example. The suborder Gobioidei is sister to the widespread, primarily marine family Apogonidae. The phylogenetic analysis of Thacker and Hardman (2005) supports the hypothesis that gobioids arose in freshwater, from a marine ancestor, then returned to marine habitats on several occasions. Of the 42 valid genera of basal gobioid families identified by Thacker and Hardman (2005), 17 are classified as having freshwater life cycles, 17 as salt-tolerant, occurring in freshwater and estuaries, and 8 as having marine life cycles. Exclusively freshwater ecology is optimized as the condition at the root of Gobioidei, and invasion of marine habitats is hypothesized to have occurred independently on several occasions. The major radiation of Gobioidei, the gobiine gobiids, is known primarily from marine habitats. The occurrence of the anadromous life history strategy (particularly freshwater amphidromy) within this taxon thus represents a derived character state within Gobioidae. These observations also illustrate that the transition among marine, freshwater, euryhaline, and diadromous life cycles across phylogenetic history appears to be unconstrained, at least within the Gobioidae.

A serious impediment to clarifying the origins of anadromy resides in the phylogenetic reconstructions of the basal teleost fishes (Ramsden et al. 2003). The Clupeocephalan teleosts include the Tselfatiiformes, the Otocephala (Clupeomorpha and Ostariophysi), and the Euteleostei (Taverne and Gayet 2005). Without a well-supported phylogeny of these taxa, it is virtually impossible to track the evolutionary origins of anadromy. McDowall's argument that salmonids probably evolved from a diadromous ancestor (McDowall 2001) was based on the assumption that the sister group of salmonids is the family Osmeridae. However, the study by Ishiguro et al. (2003), based on mitochondrial genome sequences, concluded a sister group relationship between salmoniforms and esociforms, a group that inhabits freshwater almost exclusively. In addition, the sister of Salmoninae (*Eosalmo driftwoodensis*) included in the analysis of Ramsden et al. (2003) is considered a freshwater taxon, as fossil evidence indicates that all life history stages occurred in freshwater. These observations appear to add weight to the argument that the ancestor of the salmoniforms was a freshwater species and that anadromy evolved from freshwater origins (Ramsden et al. 2003).

The phylogenetic reconstruction of Ishiguro et al. (2003) also sheds light on the evolution of anadromy in osmeriforms. The superfamily Argentinoidae (herring smelts, deep-sea smelts, and spookfishes), a group of strictly marine fishes (Nelson 1994), and the suborder Osmeroidei (the diadromous smelts, galaxiids, and icefishes) are the two major clades peripheral to the basal neoteleosts, all of which are marine species. The maximum parsimony analysis of Ishiguro et al. (2003) revealed that Argentoidae is the sister group to Osmeroidei, and together, they form the sister group of neoteleosts. It thus appears plausible that anadromy among osmeroids may have evolved from marine origins. In contrast to salmoniforms, reproduction in freshwater of anadromous osmeroids may be considered the derived state. The emergence of anadromy may thus have followed very different evolutionary pathways in salmoniforms and osmeriforms.

The hypothesis presented in this paper is developed in three stages. We first reconstructed the phylogenetic history of the suborder Osmeroidae to establish the phylogenetic relationships among anadromous, marine, and freshwater species of this important group of fishes. Unlike salmonids, the osmeroids exhibit the complete range of anadromous (including amphidromous), marine, and freshwater life cycles. We mapped life cycles onto our phylogenetic reconstruction of osmeroids and onto the previously published phylogenetic reconstructions of salmoniforms (see below). We then applied ancestral character-state reconstruction methodology (see below) to test alternative hypotheses concerning the evolution of anadromy in salmonids and osmeroids. Second, we formulate an alternative hypothesis concerning the evolution of anadromy, based on the possibility that anadromy may have evolved from marine origins as well as from freshwater origins. We refer to this alternative view as the safe-site hypothesis and contrast it with the model presented by Gross (1987). Finally, we consider some important biological differences between anadromous osmerids and salmonids and discuss how these differences may be related to the hypothesized origins of anadromy in salmoniforms and osmeriforms.

Methods

Phylogenetic Reconstruction

The overall classification of euteleosts is based on Nelson (1994), with exceptions as noted (Table 1). To assess the evolution of anadromy in osmeroids, we conducted a phylogenetic reconstruction of this group based on the mitochondrial cytochrome *b*

Table 1.—Classification and common names used in this study. The nomenclature follows Nelson (1994), Ishiguro et al. (2003), and Waters et al. (2002).

Classification	Common names used
Protacanthopterygii	
Order Esociformes	Esociforms: pikes, pickerels, mudminnows
Order Osmeriformes	Osmeriforms
Superfamily Argentinoidea	Argentinoids: herring smelts, deep-sea smelts, spookfishes
Suborder Osmeroidei	Osmeroids
Family Osmeridae	Osmerids: northern smelts
Family Galaxiidae	Galaxiids
Family Retropinnidae	Retropinnids: southern smelts
Family Salangidae	Salangids: icefishes
Order Salmoniformes	Salmoniforms
Family Salmonidae	Salmonids

gene. All sequences were published in GenBank (NCBI 2005), and accession numbers of the various sequences used are listed in Table 2. All known sequences available for the family Osmeridae were used. We did not include the numerous sequences available for the galaxiid fishes. Waters and Wallis (2001) have argued that the freshwater, nonmigratory lineages among galaxiids represent independent radiations from diadromous lineages with repeated loss of diadromy. Thus, galaxiid fishes were represented here only by the anadromous Tasmanian whitebait *Lovettia sealii* (Johnston). As reported by Nelson (1994), 13 species of Osmeridae are recognized. However, taking into account the recent literature (Nishida 1988; Wilson and Williams 1991; Taylor and Bentzen 1993; Taylor and Dodson 1994; Johnson and Patterson 1996; Saruwatari et al. 1997; Ishiguro et al. 2001; Waters et al. 2002; Ishiguro et al. 2003; Skurikhina et al. 2004; Ilves and Taylor 2007), we recognized 15 valid osmerid species (excluding subspecies). Within the Osmeridae, we obtained cyt *b* sequences for 11 species. We also obtained sequences for Salangidae (icefishes) and Retropinnidae (southern smelts) (Table 2). The trees were rooted with representative species of the superfamily Argentinoidae, the sister group of suborder Osmeroidei in the maximum parsimony analysis of Ishiguro et al. (2003). Sequences were aligned using Clustal W (Thompson et al. 1994), and multiple alignments were corrected by eye using Seaview (Galtier et al. 1996). Maximum likelihood trees were generated by dnaml (PHYLIP) (Felsenstein 2005), and the robustness of tree branches was assessed using 500 bootstrap replications. Only values above 50% were reported for basal and intermediate branches. We indicated whether the species were anadromous, marine, or exclusively freshwater based on published accounts of their ecology (see McAllister 1963; FishBase, www.fishbase.org, version 12/2004).

The phylogenetic tree of the salmonid fishes used was the one obtained by Crespi and Fulton (2004; Figure 2C) from the analysis of seven nuclear DNA genes (7,530 DNA positions), involving 21 species. We also included the phylogenetic reconstruction of the esociforms (10 species of pikes, pickerels, and mudminnows; Lopez et al. 2004), as this group was identified as the sister of the salmoniforms by these authors and by Ishiguro et al. (2003). The concatenation was done by hand to produce one phylogenetic reconstruction upon which we mapped freshwater and anadromous life cycle states (there are no salmoniforms with uniquely marine life cycles). This general topology excluded branch lengths, as different sequence data were available for the two groups. Huchen *Hucho hucho* (Linnaeus) was classified as a uniquely freshwater taxon (Holcik et al. 1988). The movements of European and Siberian Hucho are quite limited and they are found in the large rivers of piedmont and upland areas. East Asian Hucho are also found far inland (M. Kottelat, independent consultant, Cornol, Switzerland, personal communication).

Reconstructing Ancestral Character States

It is well known that anadromy encompasses a wide array of behaviors. Rounsefell (1958) and Quinn

Table 2.—Species and family names and GenBank accession numbers relative to the mitochondrial cytochrome *b* (cyt *b*) gene sequences of species used to reconstruct the phylogeny of osmeriforms (Figure 1). F = freshwater life cycle; M = marine life cycle; A = anadromy/amphidromy.

Family - species	cyt *b*	Life cycle
Salangidae		
Hemisalanx brachyrostralis [Fang]	AY279372	F
Leucosoma (=*Salanx*) *reevesii* [Gray]	AY279375	F
Neosalanx jordani [Wakiya and Takahashi]	AY279374	A
Protosalanx chinensis [Basilewsky]	AY279370	F
Japanese icefish *Salangichthys microdon* [Bleeker]	NC_004599	A
Salanx ariakensis [Kishinouye]	NC_006918	A
Noodlefish *S. cuvieri* [Valenciennes]	AB196913	A
Osmeridae		
Japanese smelt *Hypomesus japonicus* [Brevoort]	AB049019	A
Wakasagi *H. nipponensis* [McAllister][a]	DQ010190	A
Pond smelt *H. olidus* [Pallas]	DQ010196	A
Delta smelt *H. transpacificus* [McAllister]	AB089609	A
Capelin *Mallotus villosus* [Müller]	AB049022	M
Arctic rainbow smelt *Osmerus dentex* [Steindachner and Kner]	AB114911	A
European smelt *O. eperlanus* [Linnaeus]	U05667	A
Rainbow smelt *O. mordax* [Mitchill]	U05666	A
Ayu *Plecoglossus altivelis* [Temminck and Schlegel]	NC_002734	A
Spirinchus lanceolatus [Hikita]	AB094410	A
Eulachon *Thaleichthys pacificus* [Richardson]	AY279378	A
Retropinnidae		
Cucumberfish *Retropinna retropinna* [Richardson]	NC_004598	A
Tasmanian smelt *R. tasmanica* [McCulloch]	AF112321	A
Galaxiidae		
Tasmanian whitebait *Lovettia sealii* [Johnston]	AF112320	A
Outgroups		
Deepsea smelt *Glossanodon semifasciatus* [Kishinouye]	NC_004595	M
Eared blacksmelt *Lipolagus* (=*Bathylagus*) *ochotensis* [Schmidt]	NC_004591	M
Robust smallmouth *Nansenia ardesiaca* [Jordan and Thompson]	NC_004596	M
Barrel-eye *Opisthoproctus soleatus* [Vaillant]	NC_004600	M
Stout blacksmelt *Pseudobathylagus milleri* [Jordan and Gilbert]	AY973046	M

[a] Identical to *Hypomesus chishimaensis* [Saruwatari, López and Pietsch] (see Ilves and Taylor 2007).

and Myers (2004) have argued that anadromy in salmonids is not a single trait, but rather represents a suite of life history traits that are expressed along a continuum. The freshwater portion of the life cycle may be limited to spawning and only several days following hatching. Anadromous individuals may coexist with freshwater residents within some or all populations of some species. One salmonid genus, *Oncorhynchus* [Suckley], illustrates the entire range of such behaviors (Quinn and Myers 2004). Freshwater residency and feeding prior to spawning, typical of freshwater amphidromy, may vary from several years (e.g., *Galaxias fasciatus* [Gray], McDowall 1988) to several weeks or months in some populations of southern smelts (Eldon and Greager 1983). Despite this diversity of life cycle characteristics, we here consider anadromy and its variants together as one evolutionary condition, the physiological capacity to exploit freshwaters to reproduce yet tolerate saline waters for varying periods of time for feeding. We thus limited our reconstruction of ancestral character states to three general categories of life cycles: freshwater, marine, and anadromous. We broadly defined the latter group to include spe-

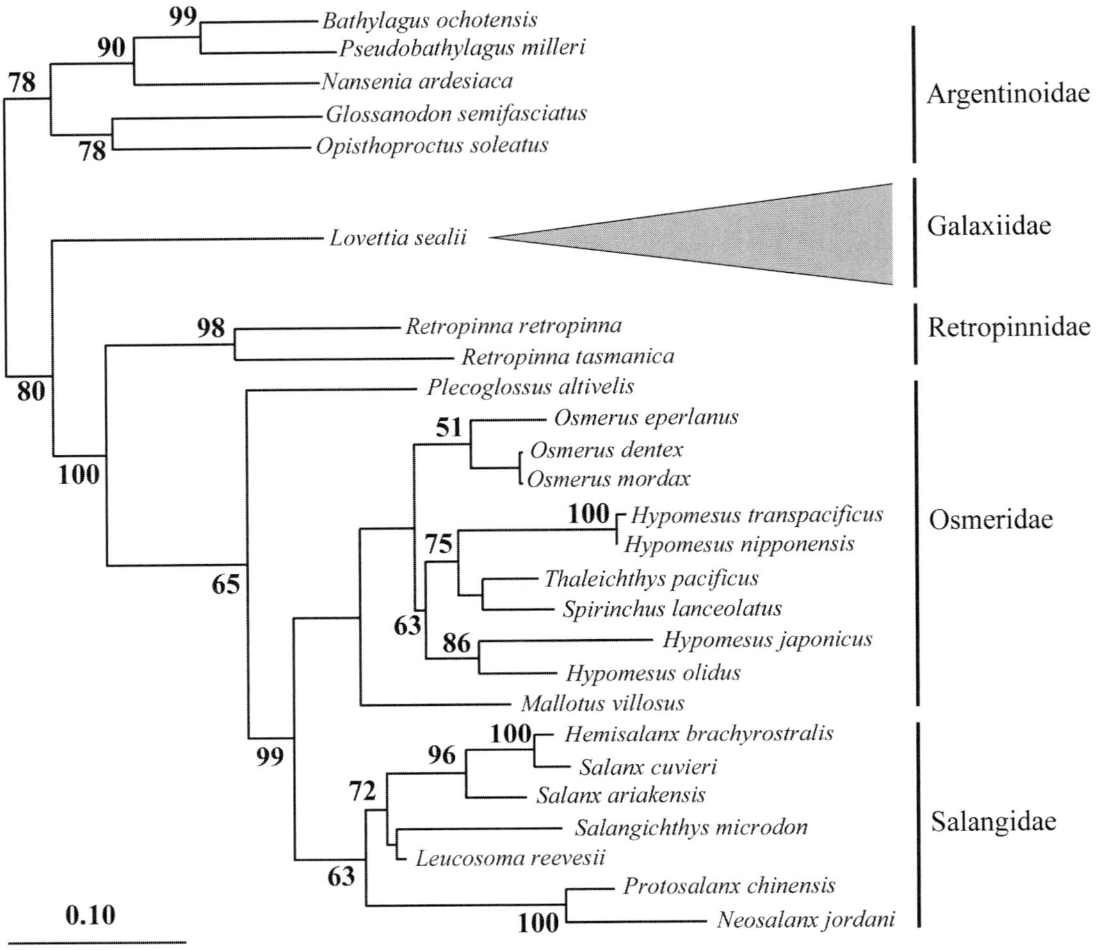

Figure 1.—Phylogenetic reconstruction of osmeriforms based on the cytochrome *b* mitochondrial gene. A total of 1,043 nucleotides were obtained from GenBank to construct a consensus tree based on 500 bootstraps. Numbers refer to the percent bootstrap support. Scale bar indicates number of changes per site.

cies with populations or population components that reproduce in freshwater and migrate to sea to feed during at least part of their life cycle.

We first reconstructed ancestral character states using maximum parsimony methods. Parsimony reconstruction finds the ancestral states that minimize the number of steps of character change given a phylogenetic tree and the observed character distribution across the terminal branches (Maddison et al. 1984). Parsimony analysis was based on simple parsimony: reversals and convergence were counted equally. Whereas parsimony implicitly assumes an equal probability of character change on all branches, Waters et al. (2002) have argued in favor of character weighting (unequal probabilities of character change) in cases involving reductive characters and when a priori evidence suggests that homoplasy is wide-spread (e.g., losses of anadromy in galaxiid fishes, which repeatedly lose their marine life history phase; McDowall 1988). We have avoided such character weighting as counseled by Kitching (1992), Swofford et al. (1996), and Grant and Kluge (2003). In the present analysis, character weighting would necessarily be a subjective exercise. As phylogenetic hypotheses based on subjective weighting schemes run the risk of becoming self-fulfilling prophesies, we adopt the view that equal weighting of all transformations in parsimony and

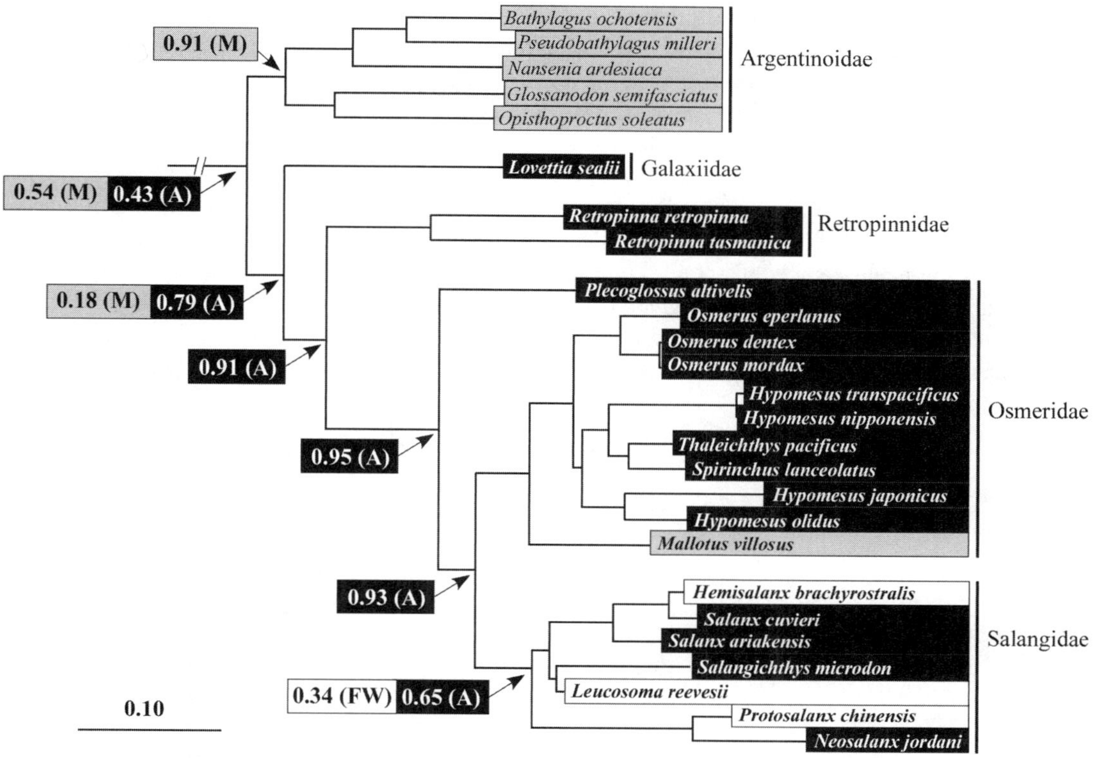

Figure 2.—Ancestral character state reconstruction of anadromous (A), marine (M), and freshwater (F) life cycles in osmeriforms based on the phylogenetic reconstruction depicted in Figure 1. Numbers at nodes represent significant liklihoods reported as proportional likelihoods. gray boxes—marine species, open boxes—freshwater species, black boxes—anadromous species.

likelihood analyses provides the least-biased test of competing hypotheses.

A fundamental problem with parsimony is that it does not account for the uncertainty in the process of character change (Huelsenbeck et al. 2003). There is always the chance that histories other than the most parsimonious one produced the observed distribution of character states. To deal with this problem, we used likelihood reconstruction methods to find ancestral states that maximize the probability that the observed states would evolve under a stochastic model of evolution (Schluter et al. 1997). For each node, the reconstruction finds the state assignment that maximizes the probability of arriving at the observed states in the terminal taxa and allowing the states at all other nodes to vary. We used the Markov k-state parameter model of evolution (Maddison and Maddison 2006). The single parameter is the rate of change and any particular change is equally probable. All analyses were conducted using the program Mesquite (Maddison and Maddison 2006).

Results

Phylogenetic Reconstruction and the Evolution of Anadromy in Osmeroids and Salmonids

The phylogenetic reconstruction of osmeroids, based on 1,043 nucleotides of the cyt *b* gene (Figure 1), strongly supported a sister group relationship between southern smelts (Retropinnidae) and northern smelts (Osmeridae) + icefishes (Salangidae), as initially shown by Waters et al. (2002) and López et al. (2004). Poor bootstrap support was obtained for the basal nodes of Osmeridae, largely because of uncertainty concerning the placement of capelin *Mallotus villosus* (Müller). Both northern and southern smelts are considered as anadromous. Salangidae (icefishes) are also anadromous, with some freshwater neotonic forms (Johnson and Patterson 1996; Froese and Pauly 2004). Within osmerids, four species are considered marine, but cyt *b* sequences were only available for capelin. Based on both parsimony and likelihood methodologies, anadromy is the likeliest ancestral state leading to the Osmeroidae, but there is a high degree of uncertainty in the reconstruction (likelihood ratio of 4.4:1, anadromous versus marine character state; Figure 2). The marine life cycle is the likeliest ancestral state leading to the Argentoidae (likelihood of 0.91 for a marine origin) (Figure 2). The life history cycle of the common ancestor of the two groups is ambiguous, with a likelihood ratio of nearly 1:1 for marine or diadromous origins.

Ancestral nodes in the salmoniform reconstruction exhibited exclusively freshwater life cycles based on both parsimony and likelihood procedures. The freshwater life cycle is supported at a likelihood level of 0.91 at the root of the tree (Figure 3). Moving into the tree, the freshwater life cycle is again

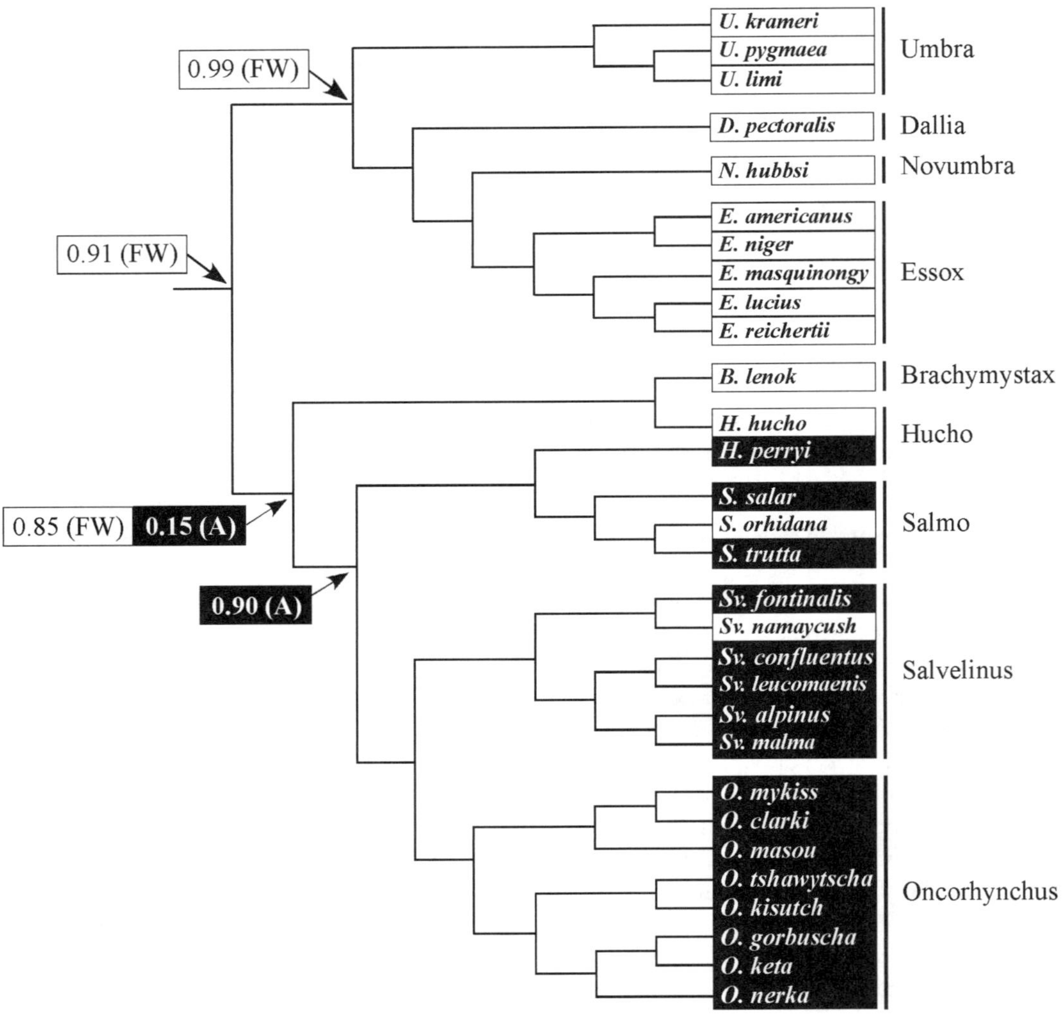

Figure 3.—Ancestral character state reconstruction of anadromous and freshwater life cycles in salmonids and esociformes. Numbers at nodes represent significant likelihoods reported as proportional likelihoods. Open boxes—freshwater species; black boxes—anadromous species.

strongly supported with a likelihood ratio of 5.6 to 1. Finally, the anadromous life cycle is strongly supported at the ancestral node of Japanese huchen *Hucho perryi*, *Salmo*, *Salvelinus*, and *Oncorhynchus*. Two species reverted independently to the freshwater life cycle.

Discussion

The Safe-Site Hypothesis

We initially proposed that the freshwater life cycle was the ancestral character state leading to salmoniform anadromy and that the marine life cycle was the ancestral character state leading to osmeriform anadromy. This prediction was partially supported by the character-state reconstruction reported here. The freshwater life cycle characterized the last common ancestor of salmoniforms and esociforms using both parsimony and likelihood reconstruction methods. In contrast, either anadromy or a marine life cycle was the ancestral state of the last common ancestor of osmeroids and argentines. Although we may neither accept nor reject a marine ancestry for osmeriform anadromy, it appears evident that anadromy among osmeriforms may have been derived without the precursor of a freshwater life cycle. The emergence of anadromy may thus have followed different evolutionary pathways in salmoniforms and osmeriforms.

The hypothesis that anadromy was derived from freshwater species to exploit the rich marine feeding grounds at high latitudes (Gross 1987; Gross et al. 1988) is consistent with the freshwater origins of salmoniforms as they appear to have experienced a strictly freshwater phase in their evolutionary history. If we assume that osmeriform anadromy was derived from a marine ancestry, we hypothesize that this evolutionary pathway represents a means of exploiting freshwater as a reproductive environment to protect the most vulnerable, early life history stages from marine predators (Figure 4). Many populations of plants and animals are controlled by heavy mortality at some life cycle stage and only individuals present in special "safe" sites may escape catastrophic mortality (Harper et al. 1961). For example, initiation of emergence and drift in beach-spawning capelin is associated with rapid wind induced water-mass exchange, which in turn results in the replacement of cold, high-saline, predator-laden waters with warmer, less-saline waters in which the abundance of predators was reduced 3–20-fold and zooplankton prey increased 2–3-fold (Frank and Leggett 1982). The abundance of eggs and larvae of 11 other marine fish species was found to be associated with the safe site water mass (Frank and Leggett 1982). The attraction to freshwater for spawning by a variety of marine fishes may be more widespread than generally believed. The groundwater seepage hypothesis (Harden Jones 1981) proposes that assembly areas, spawning areas, and spawning grounds of fishes with marine life cycles as diverse as plaice, cod, herring, pilchard, and mackerel could be identified by reference to groundwater seepage. Although Harden Jones (1981) postulated that the functional significance of groundwater seepage in spawning areas was associated with imprinting and homing to the chemicals associated with such seepages, we suggest that such seepage would also contribute to reducing the incidence of marine predators such as chaetognaths, jellyfish, and ctenophores, thus creating localized safe sites for embryos and early larvae. Given the massive mortality that occurs in marine fishes during the earliest life history stages, the transition to exploiting freshwater as a safe haven for embryos and the early larval stages would result in important fitness advantages by reducing mortality during the most vulnerable stages. Finally, the importance of estuaries to marine fish as refuges from predation has been well-established in the literature for many years (Moyle and Cech 2004).

We propose that the exploitation of low-salinity waters by some marine species led to freshwater reproduction and possibly prolonged feeding in freshwaters and the development of the amphidromous life cycle. It is no coincidence that amphidromy in osmeroids is dominated by southern hemisphere taxa. To the best of our knowledge, no salmonids have been unequivocally characterized as amphidromous. The Australian zoogeographic region has long been isolated from evolutionary events taking place elsewhere, and this is reflected in its depauperate primary freshwater fish fauna. The vast majority of this region's fish fauna is diadromous or derived from marine ancestors, reflecting the availability of freshwater niches for marine invaders. Similarly, in Patagonia, only three genera (*Aplochiton* [Jenyns], *Brachygalaxias* [Eigenmann], and *Galaxias* [Cuvier]) and eight freshwater species are present. The Andean lakes of Patagonia were formed by the last

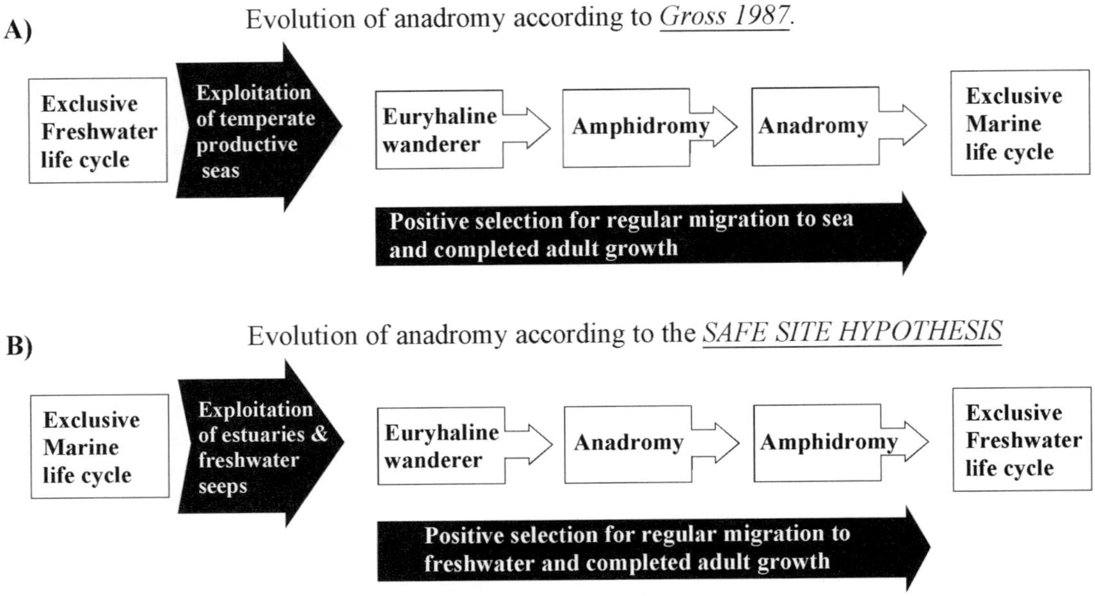

Figure 4.—(A) Hypothesized origins of anadromy as presented by Gross (1987) and applicable to the evolution of of salmoniform anadromy from freshwater origins. (B) The safe-site hypothesis as applied to the marine origins of osmeriform anadromy.

retreat of the Pleistocene ice sheets, some 10,000–20,000 years ago and were colonized by previously displaced gondwanic fishes (silurids and galaxiids), marine fishes (percichthiids and aterinopsiids), and marine dispersants (galaxiids) (Cussac et al. 2004). Thus, the presence of amphidromous and freshwater populations of galaxiids in Patagonia also appears to be related to a depauperate freshwater fish fauna and the availability of freshwater niches for a variety of anadromous species. This interpretation is consistent with the view of Waters and Wallis (2001) that cladogenesis of freshwater, nonmigratory, lineages among galaxiids represent independent radiations from migratory lineages with repeated loss of amphidromy.

Ecological Correlates of Contrasting Evolutionary Pathways to Anadromy

In contrast to most salmonids, the reproductive migrations of both northern and southern smelts are generally characterized by brief excursions to spawn in freshwater. Larval smelt of the subfamily Osmerinea (including *Osmerus*, *Allosmerus*, *Spirinchus*, and *Thaleichthys*) may spend no more than 24 h in freshwater before being transported to coastal marine or estuarine environments (Hart and McHugh 1944; Ouellet and Dodson 1985; Mclean et al. 1999). In both northern and southern smelts, the establishment of landlocked populations is widespread (McDowall, 1988; Nishida 1988; Nellbring 1989; Taylor and Bentzen 1993; Bernatchez 1997). Within osmerids, four species are considered marine, but most retain some connection to fresh or brackish water. The cyt *b* sequence was available for only one marine osmerid, the capelin, a coastal marine species that spawns on the sea bottom or on beaches. Spawning beaches in Newfoundland are located at the heads of embayments having small freshwater streams percolating through the sediments (Carscadden et al. 1988). Other marine Osmerinae include the surf smelt *Hypomesus pretiosus* (Girard), a coastal species spawning in the upper surf zone (Martin and Swiderski 2001); whitebait smelt *Allosmerus elongates*, a species that inhabits freshwater at times but appears to be neither anadromous nor beach spawning (Eschmeyer 1983); and night smelt *Spirinchus starksi*, also reported to spawn in the surf zone (McAllister, 1963). The almost complete abandonment of anadromy for a marine life cycle thus appears to have several evolutionary origins within osmerids. Most cases involve seeking safe reproductive havens in the upper intertidal reaches that may be characterized by freshwater percolation.

Another striking difference between osmeroid and salmoniform life histories concerns the phenomenon of partial migration. Salmonids are known for the occurrence in sympatry of two life history forms, one that undergoes migration to sea before returning to freshwater to reproduce (anadromous) and one that inhabits freshwater without a migration phase (resident). Moreover, experimental crosses and transplant studies have shown that juveniles from "pure" anadromous or resident crosses can either become one form or the other and that transplanted resident fish have given rise to anadromous stock or vice versa (e.g., Olsson and Greenberg 2004; Schreiber and Diefenbach 2005). By using highly polymorphic microsatellite loci and parentage analysis in a natural population of sympatric anadromous and resident brook char *Salvelinus fontinalis*, Thériault et al. (2007) found that extensive gene flow occurred between the two forms and was mediated by resident males mating with both resident and anadromous females. Furthermore, although no anadromous brook char occur further south than the Gulf of Maine, coastal resident brook char still persist (Castric and Bernatchez 2003). Assuming a reduction in the resource gradient from freshwater to the sea, the species at the southern limit of its distribution has adopted the freshwater life cycle, even though their fluvial habitats are connected to the sea. The conditional nature of anadromy in salmonids is most probably ancestral (Crespi and Teo 2002) and represents an alternative way to maximize fitness according to the metabolic status of an ancestral freshwater fish. Although osmeriforms readily establish freshwater populations through the process of landlocking, we know of no example within osmeriform fishes of the sympatric occurrence of diadromous and freshwater forms or the existence of freshwater forms that have abandoned anadromy other than by the process of landlocking. One possible exception to this is the recent demonstration that the lake-dwelling *Hypomesus chishimaensis* and and the anadromous wakasagi are not genetically distinct and do not merit separate species status (Ilves and Taylor 2007). However, similarity in mitochondrial and nuclear sequence data cannot be used to establish if life histories are under the control of a conditional strategy. Although *Hypomesus chishimaensis* may very well be an alternative ecotype of *H. nipponensis*, more work is required to demonstrate the sharing of a common gene pool in sympatry.

There also appear to be differences in the occurrence and the adaptive significance of homing to natal spawning sites between salmoniforms and osmeriforms. The cooccurrence of anadromy and homing is common in salmonid fishes and is generally believed to have evolved to facilitate adaptations to local, population-specific conditions (Leggett 1977). Recent molecular studies of population structure in salmonids and osmerids suggest that homing is not as well developed in osmerids. Although considerable population structure can be detected among populations of salmon and trout species (e.g., Fontaine et al. 1997; Shaklee et al. 1999; Castric et al. 2001), such structure is far less evident among anadromous osmerids. Studies of rainbow smelt *Osmerus mordax* (Lecomte F and J.J. Dodson unpublished data) and eulachon *Thaleichthys pacificus* (Mclean et al. 1999; Mclean and Taylor 2001; Beacham et al. 2005) reveal little genetic structuring among local populations. These local groups, or demes, may be considered to form metapopulations (Kritzer and Sale 2004) whereby extensive gene flow occurs among the demes forming the metapopulation. Anadromous salmonids bear the imprint of genetically structured freshwater species characterized by high levels of cladogenesis associated with homing to freshwater spawning sites, whereas anadromous osmerids bear the imprint of genetically less heterogeneous marine species with relatively low levels of cladogenesis associated with marine dispersal. If this assertion is correct, we would predict that allelic richness and the slope of isolation-by-distance relationships should be lower, and the extent of population differentiation (F_{st}) should be higher among anadromous salmonids relative to anadromous osmerids. However, it is difficult to find the data to test this idea. As gene flow is strongly regulated by geographic features (Bradbury et al. 2006), any comparisons between anadromous salmonids and osmerids must be carried out over the same geographical region and, ideally, in the same rivers. For example, the work of Beacham and his colleagues on the west coast of Canada has revealed that the genetic differentiation observed among steelhead *O. mykiss* populations was approximately

10 times greater than that observed among eulachon populations observed over the same coastline but in different rivers (Beacham et al. 2004; Beacham et al. 2005). Although far more research is required to evaluate the genetic structure of a variety of anadromous osmerids, we speculate that the lower levels of cladogenesis observed among anadromous osmerids studied to date is a manifestation of an ancestry that is more associated with marine than freshwater origins.

Concluding Remarks

This change in perspective regarding the evolution of anadromy identifies several promising research avenues. The impact of anadromous life cycle diversity on the comparative genetic population structure of salmonids and osmerids merits attention. Such studies may provide insights into the trade-off between homing and dispersal and its impact on the population structure and life history evolution of fishes. Far more basic biological information is required to better describe the life cycles of euryhaline species. Although there is much evidence showing the ease by which anadromous osmerids and salmonids may establish freshwater populations through the process of landlocking, little effort has been invested in exploring the ability of anadromous species to establish marine populations or the ability of marine and freshwater species to establish euryhaline populations, the most likely starting point in the evolution of anadromy. The analysis of strontium and calcium concentrations in the otoliths of some species is changing this situation and revealing brackish water life history cycles in what were previously presumed to be uniquely anadromous (ninespine stickleback *Pungitius pungitius* [Linnaeus]; Arai and Goto 2005) or even freshwater (northern pike *Esox lucius* [Linnaeus]; Westin and Limburg 2002) species. There is also evidence that estuarine environments impose distinct selective regimes, generating physiologically adapted populations divergent from their marine counterparts, and the potential for in situ speciation in complete or partial isolation (Bilton et al. 2002 and references therein). In a similar vein, the sympatric anadromous rainbow smelt populations of the St. Lawrence estuary exhibit different salinity tolerances, with the smelt population occupying the north shore of the estuary spending more time in freshwater and occupying lower salinities in the estuarine transition zone than the more marine south-shore population (Lecomte and Dodson 2005). The cryptic nature of many of these evolutionary partitions requires a much closer examination of anadromous and so-called "euryhaline" life cycles.

Acknowledgments

This work was supported by grants from NSERC (Discovery program) to J. J. Dodson and scholarships from the Natural Sciences and Engineering Research Council of Canada (NSERC) and Fonds pour la Formation des Chercheurs et l'Aide à la Recherche (FCAR, Québec) to F. Lecomte. We thank Bernard Crespi for kindly providing us the database needed to conduct the reconstruction of salmonid phylogeny presented here, Le Centre de bioinformatique et de biologie computationelle for the use of their supercomputer and software, and R. M. McDowall, Richard Hedger, and anonymous referees for reviewing earlier drafts of this paper.

References

Arai, T., and A. Goto. 2005. Flexible life history strategies of ninespine sticklebacks, genus *Pungitius*. Environmental Biology of Fishes 74:43–50.

Beacham, T. D., E. H. Douglas, and D. L. Khai. 2005. Population structure and stock identification of eulachon (*Thaleichthyes pacificus*), an anadromous smelt, in the Pacific Northwest. Marine Biotechnology 7:363–372.

Beacham, T. D., K. D. Le, and J. R. Candy. 2004. Population structure and stock identification of steelhead trout (*Oncorhynchus mykiss*) in British Columbia and the Columbia River based on microsatellite variation. Environmental Biology of Fishes 69:95–109.

Bernatchez, L. 1997. Mitochondrial DNA analysis confirms the existence of two glacial races of rainbow smelt (*Osmerus mordax*) and their reproductive isolation in the St-Lawrence R. estuary (Québec, Canada). Molecular Ecology 6:73–83.

Bilton, D. J., J. Paula, and J. D. D. Bishop. 2002. Dispersal, genetic differentiation and speciation in estuarine organisms. Estuarine, Coastal and Shelf Science 55:937–952.

Bradbury, I. R., M. W. Coulson, S. E. Campana, and P. Bentzen. 2006. Morphological and genetic differentiation in anadromous smelt *Osmerus mordax* (Mitchill): disentangling the effects of geography and morphology on gene flow. Journal of Fish Biology 69(Supplement C):95–114.

Carscadden, J. E., K. T. Frank, and D. S. Miller. 1988. Distribution of capelin (*Mallotus villosus*) in relation to physical features of the Southeast Shoal. North Atlantic Fisheries Organization, Scientific Council Research Document 88/90, Dartmouth, Nova Scotia.

Castric, V., and L. Bernatchez. 2003. The rise and fall of isolation by distance in the anadromous brook charr (*Salvelinis fontinalis* Mitchill). Genetics 163:983–996.

Castric, V., F. Bonney, and L. Bernatchez. 2001. Landscape structure and hierarchical genetic diversity in the brook charr, *Salvelinus fontinalis*. Evolution 55:1016–1028.

Crespi, B. J., and M. J. Fulton. 2004. Molecular systematics of Salmonidae: combined nuclear data yields a robust phylogeny. Molecular Phylogenetics and Evolution 31:658–679.

Crespi, B. J., and R. Teo. 2002. Comparative phylogenetic analysis of the evolution of semelparity and life-history in salmonid fishes. Evolution 56:1008–1020.

Cussac, V., S. Ortubay, G. Iglesias, D. Milano, M. E. Lattuca, J. P. Barriga, M. Battini, and M. Gross. 2004. The distribution of South American galaxiid fishes: the role of biological traits and post-glacial history. Journal of Biogeography 31:103–121.

Daverat, F., K. Limburg, I. Thibault, J-C. Shiao, J. J. Dodson, F. Caron, W-N. Tzeng, Y. Iizuka, and H. Wickstöm. 2006. Plasticity of habitat use in temperate eel species. Marine Ecology Progress Series 308:231–241.

Dodson, J. J. 1997. Fish migration: an evolutionary perspective. Pages 10–36 *in* J.-G. Godin, editor. Behavioural ecology of teleost fishes. Oxford University Press, Oxford, UK.

Eldon, G. A., and A. J. Greager. 1983. Fishes of the Rakaia Lagoon. New Zealand Ministry of Agriculture and Fisheries, Fisheries Environmental Report 30, Wellington.

Eschmeyer, W. N. 1983. A field guide to Pacific Coast fishes of North America. Houghton Mifflin, Boston.

Felsenstein, J. 2005. PHYLIP (Phylogeny Inference Package) version 3.6. University of Washington, Department of Genome Sciences, Seattle.

Fontaine, P.-M., J. J. Dodson, L. Bernatchez, and A. Slettan. 1997. A genetic test of metapopulation structure in Atlantic salmon (*Salmo salar*) using microsatellites. Canadian Journal of Fisheries and Aquatic Sciences 54:2434–2442.

Frank, K. T., and W. C. Leggett. 1982. Coastal water mass replacement: its effect on zooplankton dynamics and the predator prey complex associated with larval capelin (*Mallotus villosus*). Canadian Journal of Fisheries and Aquatic Sciences 39:991–1003.

Galtier, N., M. Gouy, and C. Gautier. 1996. SEAVIEW and PHYLO_WIN: two graphic tools for sequence alignment and molecular phylogeny. Computational Applications in the Biosciences 12:543–548.

Grant, T., and A. G. Kluge. 2003. Data exploration in phylogenetic inference: scientific, heuristic or neither. Cladistics 19:379–418.

Gross, M. R. 1987. Evolution of diadromy in fishes. Pages 14–24 *in* M. J. Dadswell, R. J. Klauda, C. M. Moffitt, and R. L. Saunders, editors. Common strategies of anadromous and catadromous fishes. American Fisheries Society, Symposium 1, Bethesda, Maryland.

Gross, M. R., R. M. Coleman, and R. M. McDowall. 1988. Aquatic productivity and the evolution of diadromous fish migration. Science 239:1291–1293.

Harden Jones, F. R. 1981. Fish migration: strategy and tactics. Cambridge University Press, New York.

Harper, J. L., J. N. Clatworthy, I. H. McNaughten, and G. R. Sagar. 1961. The evolution and ecology of closely related species living in the same area. Evolution 15:209–227.

Hart, J. L., and J. L. McHugh. 1944. The smelts (Osmeridae) of British Columbia. Research Board of Canada Bulletin 64:1–27.

Holcik, J., K. Hensel, J. Nieslanik, and L. Skacel. 1988. The Eurasian huche. Junk, Dordrecht, Netherlands.

Huelsenbeck, J. P., R. Nielson, and J. P. Bollback. 2003. Stochastic mapping of morphological characters. Systematic Biology 52:131–158.

Ilves, K. L., and E. B. Taylor. 2007. Are *Hypomesus chishimaensis* and *H. nipponensis* (Osmeridae) distinct species? A molecular assessment using comparative sequence data from five genes. Copeia 2007:180–185.

Ishiguro, N., M. Miya, and M. Nishida. 2001. Complete mitochondrial DNA sequence of ayu *Plecoglossus altivelis*. Fisheries Science 67:474–481.

Ishiguro, N. B., M. Miya, and M. Nishida. 2003. Basal euteleostean relationships: a mitogenomic perspective on the phylogenetic reality of the "Protacanthopterygii." Molecular Phylogenetics and Evolution 27:476–88.

Johnson, G. D., and C. Patterson. 1996. Relationships of lower euteleostean fishes. Pages 251–332 *in* M. L. J. Stiassny, L. R. Parenti, and G. D. Johnson, editors. Interrelationships of fishes. Academic Press, San Diego, California.

Kitching, I. J. 1992. Tree-building techniques. Pages 44–71 *in* P. L. Forey, C. J. Humphries, I. J. Kitching, R. W. Scotland, D. J. Siebert, and D. M. Williams, editors. Cladistics, a practical course in systematics. Clarendon Press, Oxford, UK.

Kritzer, J. P., and P. F. Sale. 2004. Metapopulation ecology in the sea: from Levin's model to marine ecology and fisheries science. Fish and Fisheries 5:131–140.

Lecomte, F., and J. J. Dodson. 2005. Distinguishing trophic and habitat partitioning among sympatric populations of an estuarine fish (*Osmerus mordax* Mitchill). Journal of Fish Biology 66:1601–1623.

Leggett, W. C. 1977. The ecology of fish migration. Annual Review of Ecology and Systematics 8:285–308.

López, J. A., W-J. Chen, and G. Orti. 2004. Esociform phylogeny. Copeia 2004:449–464.

Maddison, W. P., M. J. Donoghue, and D. R. Maddison. 1984. Outgroup analysis and parsimony. Systematic Zoology 33:83–104.

Maddison, W. P. and D. R. Maddison. 2006. Mesquite: a modular system for evolutionary analysis. Version 1.12. Available: http://mesquiteproject.org (January 2009).

Martin, K. L. M., and D. L. Swiderski. 2001. Beach spawning in fishes: phylogenetic tests of hypotheses. American Zoologist 41:526–537.

McAllister, D. E. 1963. A revision of the smelt family, Osmeridae. National Museum of Canada Bulletin 191:1–53.

McDowall, R. M. 1988. Diadromy in fishes. Timber Press, Portland, Oregon.

McDowall, R. M. 1993. A recent marine ancestry for diadromous fishes? Sometimes yes but mostly no. Environmental Biology of Fishes 37:329–335.

McDowall, R. M. 1997. The evolution of diadromy in fishes (revisited) and its place in phylogenetic analysis. Reviews in Fish Biology and Fisheries 7:443–462.

McDowall, R. M. 2001. Anadromy and homing: two life-history traits with adaptive synergies in salmonid fishes? Fish and Fisheries 2:78–85.

McDowall, R. M. 2004. Ancestry and amphidromy in island freshwater fish faunas. Fish and Fisheries 5:75–85.

Mclean, J. E., D. E. Hay, and E. B. Taylor. 1999. Marine population structure in an anadromous fish: life-history influences patterns of mitochondrial DNA variation in the eulachon, *Thaleichthys pacificus*. Molecular Ecology 8:S143–S158.

Mclean, J. E., and E. B. Taylor. 2001. Resolution of population structure in a species with high gene flow: microsatellite variation in the eulachon, (Osmeridae: *Thaleichthys pacificus*). Marine Biology 139:411–420.

Moyle, P. B., and J. J. Cech, Jr. 2004. Fishes: an introduction to ichthyology, 5th edition. Prentice-Hall, Upper Saddle River, New Jersey.

Myers, G. S. 1949. Usage of anadromous, catadromous and allied terms for migratory fishes. Copeia 1949:89–97.

NCBI (National Center for Biotechnology Information). 2005. Taxonomy browser. National Center for Biotechnology Information databases. Available: www.ncbi.nlm.nih.gov/Taxonomy/taxonomyhome.html (March 2006).

Nellbring, S. 1989. The ecology of smelts (genus *Osmerus*): a literature review. Nordic Journal of Freshwater Research 65:116–145.

Nelson, J. S. 1994. Fishes of the world, 3rd edition. Wiley, New York.

Nishida, M. 1988. A new subspecies of the ayu, *Plecoglossus altivelis*, (Plecoglossidae) from the Ryukyu Islands. Japanese Journal of Ichthyology 35:236–242.

Olsson, I. C., and L. A. Greenberg. 2004. Partial migration in a landlocked brown trout population. Journal of Fish Biology 65:106–121.

Ouellet, P., and J. J. Dodson. 1985. Tidal exchange of anadromous rainbow smelt (*Osmerus mordax*) larvae between a shallow spawning tributary and the St. Lawrence estuary. Canadian Journal of Fisheries and Aquatic Sciences 42:1352–1358.

Quinn, T. P., and K. W. Myers. 2004. Anadromy and the marine migrations of Pacific salmon and trout: Rounsefell revisited. Reviews in Fish Biology and Fisheries 14:421–442.

Ramsden, S. D., H. Brinkmann, C. W. Hawryshyn, and J. S. Taylor. 2003. Mitogenomics and the sister of Salmonidae. Trends in Ecology and Evolution 18:607–610.

Rounsefell, G. A. 1958. Anadromy in North American Salmonidae. Fishery Bulletin 131:171–185.

Saruwatari, T., J. A. López, and T. W. Pietsch. 1997. A revision of the osmerid genus *Hypomesus* Gill (Teleostei: Salmoniforms), with the description of a new species from the southern Kuril Islands. Species Diversity 2:59–82.

Schluter, D., A. Price, A.O. Mooers, and D. S. Ludwig. 1997. Likelihood of ancestor states in adaptive radiation. Evolution 51:1699–1711.

Secor, D. H. 2002. Estuarine dependency and life history evolution in temperate sea basses. Fisheries Science 68(Supplement 1):178–181

Schreiber, A., and G. Diefenbach. 2005. Population genetics of the European trout (*Salmo trutta* L.) migration system in the river Rhine: recolonisation by sea trout. Ecology of Freshwater Fishes 14:1–13.

Shaklee, J. B., T. D. Beacham, L. Seeb, and B. A. White. 1999. Managing fisheries using genetic data: case studies from four species of Pacific salmon. Fisheries Research 43:45–78.

Skurikhina, L. A., A. G. Oleinik, and M. V. Pan'kova. 2004. Comparative analysis of mitochondrial DNA diversity in smelts. Russian Journal of Marine Biology 30:259–265.

Swofford, D. L., G. J. Olsen, P. J. Wadell, and D. M. Hillis. 1996. Phylogenetic inference in molecular systematics. Pages 407–514 *in* D. M. Hillis, C. Moritz and B. K. Mable, editors. Molecular systematics, 2nd edition. Sinauer Associates, Sunderland, Massachusetts.

Taverne, L., and M. Gayet. 2005. Phylogenetical relationships and palaeozoogeography of the marine Cretaceous Tselfatiiformes (Teleostei, Clupeocephala). Cybium 29:65–87.

Taylor, E. B., and P. Bentzen. 1993. Evidence for multiple origins and sympatric divergence of trophic ecotypes of smelt (*Osmerus*) in northeastern North America. Evolution 47:813–832.

Taylor, E. B., and J. J. Dodson. 1994. A molecular analysis of relationships and biogeography within a species complex of Holarctic fish (genus *Osmerus*). Molecular Ecology 3:235–248.

Thacker, C. E., and M. A. Hardman. 2005. Molecular phylogeny of basal gobioid fishes: Rhyacichthyidae, Odontobutidae, Xenisthmidae, Eleotridae (Teleostei: Perciformes: Gobioidei). Molecular Phylogenetics and Evolution 37:858–871.

Thériault, V., L. Bernatchez, and J. J. Dodson. 2007. Mating system and individual reproductive success of sympatric anadromous and resident brook charr, *Salvelinus fontinalis*, under natural conditions. Behavioural Ecology and Sociobiology 62:51–65.

Thompson, J. D., D. G. Higgins, and T. J. Gibson. 1994. CLUSTAL W: improving the sensitivity of progressive multiple sequence alignment through sequence weighting, position-specific gap penalties and weight matrix choice. Nucleic Acids Research 22:4673–4680.

Waters, J. M., T. Saruwatari, T. Kobayashi, I. Oohara, R. M. McDowall, and G. P. Wallis. 2002. Phylogenetic placement of retropinnid fishes: data set incongruence can be reduced by using asymmetric character state transformation costs. Systematic Biology 51:432–449.

Waters, J. M., and G. P. Wallis. 2001. Cladogenesis and loss of the marine life history phase in freshwater galaxiid fishes (Osmeriformes: Galaxiidae). Evolution 55:587–597.

Westin, L., and K. E. Limburg. 2002. Newly discovered reproductive isolation reveals sympatric populations of *Esox lucius* in the Baltic. Journal of Fish Biology 61:1647–1652.

Wilson, M. V. H., and R. R. G. Williams. 1991. New Paleocene genus and species of smelt (Teleostei: Osmeridae) from freshwater deposits of the Paskapoo Formation, Alberta, Canada, and comments on osmerid phylogeny. Journal of Vertebrate Paleontology 11:434–451.

American Fisheries Society Symposium 69:79–96, 2009

Postglacial Recolonization and the Loss of Anadromy in Rainbow Smelt from Coastal Newfoundland

Ian R. Bradbury* and Mark W. Coulson
Marine Gene Probe Laboratory, Biology Department, Life Sciences Centre
Dalhousie University, Halifax, Nova Scotia B3H 4J1, Canada

Steven E. Campana
Population Ecology Division, Bedford Institute of Oceanography
Post Office Box 1006, Dartmouth, Nova Scotia B2Y 4A2, Canada

Eric Baggs
Biology Department, Memorial University of Newfoundland
St. John's, Newfoundland A1C 5S7, Canada

Paul Bentzen
Marine Gene Probe Laboratory, Biology Department, Life Sciences Centre
Dalhousie University, Halifax, Nova Scotia B3H 4J2, Canada

Abstract.—Postglacial colonization of temperate latitudes by anadromous fishes has frequently resulted in radiation into multiple migration and morphological phenotypes, among which evolutionary relationships are often poorly understood. Freshwater phenotypes may be derived from freshwater forms or independent postglacial colonization by a marine ancestor (i.e., raceme structure). We examined genetic and morphological variation associated with the transition from anadromy to freshwater residency in rainbow smelt *Osmerus mordax*. Colonization pathways were reconstructed using nine microsatellite loci and mtDNA sequence data, between three pairs of anadromous and adjacent landlocked populations in southeastern Newfoundland. Microsatellite data indicate strong associations between adjacent anadromous and freshwater populations in two of three pairs consistent with independent parallel colonization. In contrast mtDNA sequence analysis suggests little variation and colonization of most of the region by a single glacial race. To explore the independent parallel transition from anadromous to nonanadromous populations, we document a novel estuarine dwarf phenotype inhabiting a partially landlocked fjord (Holyrood Pond) and examine the hypothesis that this estuarine dwarf represents a transitional form between the commonly observed anadromous and freshwater morphs. Habitat reconstructions using LA-ICP-MS and otolith elemental composition from each of the anadromous and freshwater locations as well as Holyrood Pond indicate that Holyrood Pond smelt exist primarily in a stable estuarine environment. Likewise, microsatellite loci show similar levels of diversity and only moderate divergence between nearby anadromous locations and Holyrood Pond. In contrast, morphological and shape variation suggests some similarities between Holyrood Pond and freshwater dwarfs. Comparison of phenotypic and genetic differentiation (i.e., P_{ST} and F_{ST}) suggest phenotypic convergence among freshwater forms and a greater phenotypic similarity between freshwater and Holyrood Pond dwarfs than between Holyrood Pond and anadromous locations. Estimates of P_{ST} were twice those of F_{ST} between anadromous and Holyrood Pond smelt consistent with divergence with gene flow. We conclude that postglacial colonization in smelt primarily oc-

* Corresponding author: ibradbur@dal.ca

curs independently through parallel radiations, and may occur even in the presence of gene flow likely resulting in large variation in migration phenotype and life history.

Introduction

Adaptive radiation, the generation of diversity in a lineage resulting primarily from natural selection, is commonly associated with glacial isolation and subsequent colonization of vacant niches and habitats (e.g., Rundle et al. 2000; Carstens and Knowles 2006). Accordingly, recolonization of temperate freshwater habitats by anadromous fishes following the Pleistocene glaciation has been associated with rapid radiations of many species into multiple ecotypes (e.g., Claytor and Verspoor 1991; Taylor and Bentzen 1993; Bernatchez et al. 1999). In fact, postglacially derived populations of many northern fish species display phenotypic divergence equal in magnitude to that separating other species and families of fishes (e.g., Bell 1976; Lanteigne and McAllister 1983). This divergence is often repeated geographically, and the parallel evolution of similar ecotypes in anadromous/freshwater systems (e.g., Rundle et al. 2000; Colosimo et al. 2005) provides independent replication of these radiations. Nonetheless, the mechanisms, the rate of divergence, and role of selection driving these radiations often remain illusive.

An increasing number of studies support the role of Pleistocene glacial isolation and postglacial radiation as a major evolutionary force in anadromous and freshwater fish species located in the northern hemisphere (e.g., Durand et al. 1999; Mäkinen et al. 2006). The late Wisconsinian maximum, which occurred between 22,000 and 16,000 BP, saw much of eastern Canada ice covered (Catto 1998). The retreat of Laurentide ice, which began approximately 15,000–14,000 BP, likely caused the opening of coastal and inland habitat for colonization. In many temperate environments, the opening of vacant habitats permitted the rapid radiation of anadromous fish species into unexploited niches (Bernatchez and Wilson 1998), resulting in a loss or modification of diadromous migration behavior. In some species, this radiation has been hypothesized to follow a phylogenetic raceme model, resulting from independent colonizations and radiation from an ancestral anadromous source (Williams 1992; Bell and Foster 1994; McPhail 1994). Most notably, examinations of threespine stickleback *Gasterosteus aculeatus* postglacial divergence suggest parallel evolution of a low-plated freshwater morph following deglaciation and independent recolonization of drainages (e.g., Raeymaeker et al. 2005). Two general predictions should hold if a raceme model is supported: source populations should be closely related, though not always if gene flow is restricted, and derived freshwater forms should be independent.

The rainbow smelt *Osmerus mordax* (Mitchill) is a small pelagic fish found in coastal and freshwater systems throughout northeastern North America from New Jersey to Labrador (Nellbring 1989). Anadromous smelt spawn near the head of the tide in coastal rivers and streams and the larvae develop in downstream estuaries (e.g., Ouellet and Dodson 1985; Bradbury et al. 2004). Phylogeographic studies suggest postglacial recolonization of two major mtDNA clades present throughout Canada's east coast, forming secondary contact zones within the St. Lawrence estuary (Baby et al. 1991; Taylor and Bentzen 1993; Bernatchez 1997) and in eastern Newfoundland (M. W. Coulson, unpublished data). The two clades are hypothesized to be derived from different Pleistocene refugia, one along the Atlantic coastal plains and a second on the Grand Banks off Newfoundland (Bernatchez 1997). However recent examinations of microsatellite (Bradbury et al. 2008) and mtDNA (Coulson, unpublished data) data suggest that hypotheses of simple colonization from these two refugia require reexamination.

Smelt are characterized by high levels of phenotypic diversity, life history variation, and habitat types (Nellbring 1989). Freshwater "dwarf" forms are common throughout the range and are usually characterized by a smaller size (60–150 mm), relatively larger eyes, and a higher number of gill rakers compared to anadromous forms (Kendall 1927; Copeman 1977; Copeman and McAllister 1978). In fact, several dwarf occurrences have been historically accorded species status (pigmy smelt *Osmerus spectrum*, Cope 1870; Lanteigne and McAllister 1983), though this remains contentious. In contrast to the dwarf phenotype, anadromous and so-called "normal" freshwater forms typically mature at a larger size (150–300 mm). Both morphs may exist in sympatry in freshwater, and previous works suggest that freshwater ecotypes are polyphyletic, resulting from independent divergence from ancestral anadromous forms (Baby et al. 1991; Taylor and Bentzen 1993). Given the high divergence and recent colonization of freshwater habitats, smelt seem an appropriate model for which to examine the interaction between colonization pathways and postglacial radiation.

In the present study, we examine postglacial colonization and radiation of rainbow smelt in eastern Newfoundland using several approaches. First, we tested for parallel postglacial colonization of lacustrine habitats using microsatellite and mtDNA variation associated with three pairs of adjacent anadromous and freshwater landlocked populations. Second, we document a novel anadromous dwarf phenotype inhabiting a partially landlocked fjord and examine the hypothesis that this estuarine dwarf represents a transitional form between the commonly observed anadromous and dwarf freshwater morphs. This hypothesis is examined using a combination of molecular, otolith elemental, and morphological approaches.

Methods

Mature adult smelt were collected by dip and fyke netting during the spawning and postspawning period in 2003–2006. Pectoral or caudal fin clips were taken and immediately placed in 95% ethanol, and fish were immediately frozen for later otolith extraction and morphological examination. The sampling sites included three anadromous spawning locations, Holyrood Pond, as well as four freshwater locations upstream of these spawning locations all of which were encompassing St. Mary's and Placentia Bay (Figure 1). Due to difficulty sampling some of the freshwater locations, only otoliths and fin clips were sampled from Gull Pond, and only fin clips were available from Hawco's Pond.

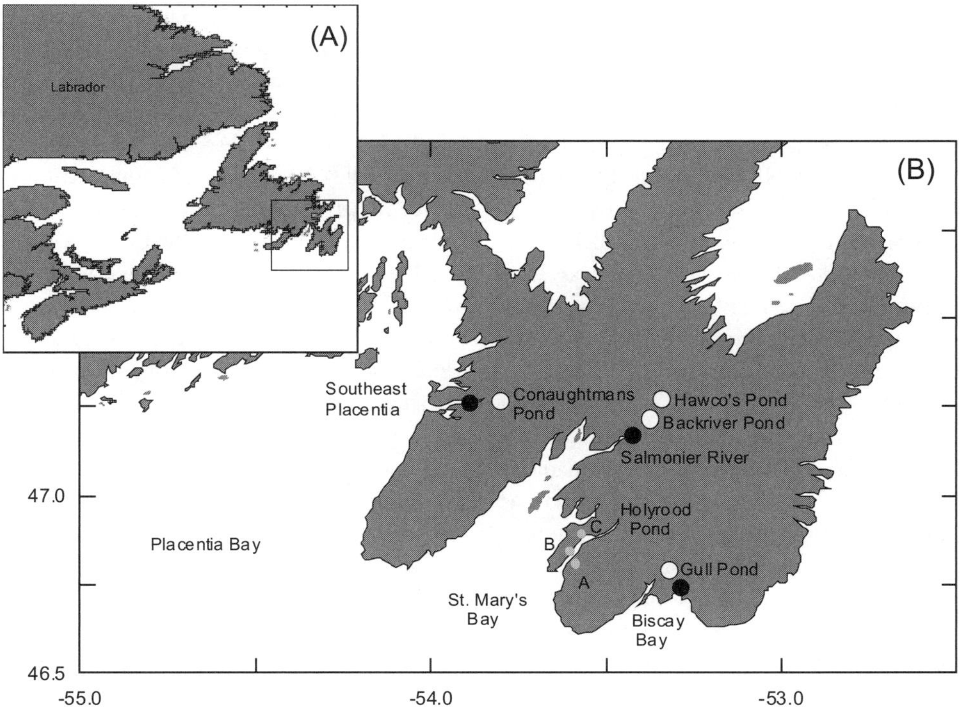

Figure 1.—Map of sample locations, inset (A) shows location of southeast coast of Newfoundland (B) with respect to eastern Canada. Open circles represent freshwater sample locations. Black circles represent sampled anadromous smelt spawning locations, and gray represent locations sampled within Holyrood Pond. A, B, and C refer to locations within Holyrood Pond: Deer Pond Brook, Holyrood Pond Park Brook, and Pathend Brook, respectively.

Genetic Analysis

DNA was extracted following the protocol of Elphinstone et al. (2003), modified to work with a 96-well filter plate, and automated on a robotic liquid handling system (Perkin Elmer). Nine tetranucelotide microsatellite loci were used following Coulson et al. (2006). Individuals were genotyped using polymerase chain reaction conditions of 5- or 10-μL volumes containing 20–100 ng DNA, 1.5 mM MgCl2, 80 μM each dNTP, 0.5 U Taq DNA polymerase (New England Biolabs), 0.3 μM each primer (forward primers were end-labeled with Hex, or Rox dye), and 10× PCR buffer (10 mM tris-HCl, pH 8.3; 50 mM KCl). Two temperature profiles were used for touchdown in order to allow for the possibility of multiplex PCRs. Touchdown PCR conditions were as follows: 94°C for 2 min, followed by 4–5 cycles of 94°C for 30 s, program specific touchdown annealing temperatures minus 1°C per cycle for 30 s, 72°C for 30 s, followed by 25–26 cycles where the T_a was held constant at 4°C below the starting temperature. A final extension was held at 72°C for 5 min. Reactions were run on Eppendorf thermocyclers and imaged on an FMBioII system (Hitachi Genetic Systems). See Coulson et al. (2006) for further details. Data were checked for the presence of null alleles and scoring errors using MICRO-CHECKER (van Oosterhout et al. 2004). Genetic polymorphism was quantified by examination of the number of alleles, and observed and expected heterozygosities using GENETIX (version 4.05.2, Belkhir et al. 2004). Tests for Hardy-Weinberg equilibrium (HWE) and linkage disequilibrium were done using FSTAT (version 2.9.3.3, Goudet 1995). F-statistics and significance were calculated using FSTAT and ARLEQUIN (Schneider et al. 2000). Contemporary estimates of gene flow were made using a Bayesian method as implemented in BAYESASS (Wilson and Rannala 2003). BAYESASS uses a Bayesian approach and Markov chain Monte Carlo (MCMC) methods to estimate whether an individual is an immigrant or of immigrant ancestry (one generation). The analysis was run with 10^7 MCMC iterations and a burn-in period of 10^6.

For mitochondrial DNA analysis, a portion of the ND5 gene was amplified in 25 μL PCRs as follows: 1 μL of template DNA, 2.5 μL of PCR buffer, 2.5 μL of 2 mM dNTPs, 0.75 μL of 10 μM forward and reverse primers (modified from Pigeon et al. 1998), and 1 U Taq DNA polymerase. Due to suspected primer binding mismatch and poor PCR amplification, internal primers were designed and utilized on the Back River Pond fish. PCR products were visualized on 0.8% agarose gels stained with Gel Green (Biotium, Inc.) and cleaned with the Exo-Sap protocol (USB Corporation) prior to sequencing. Sequences were run on an ABI3730XL (Macrogen, Inc.). An average of 12 individuals were sequenced per location. Phylogenetic relationships among individuals were constructed with an unrooted neighbor-joining tree on the unique haplotypes in PAUP* (Swofford 1998).

Otolith Elemental Composition

Sagittal otoliths were extracted from 10 randomly selected adults from each of the anadromous, freshwater, and Holyrood Pond locations (Figure 1). Otoliths were cleaned using Super Q water and stored in acid washed vials. Otoliths were embedded in epoxy, sectioned using an Isomet saw to approximately 400 μm, and sections were imaged and aged in years. Otolith elemental composition was measured using laser ablation inductively coupled plasma mass spectrometry (LA-ICP-MS). All assays were conducted at WRL laboratories at Arkansa State University. We used a CETAC LSX500 (260 nm) laser to ablate otolith material coupled to a Perkin-Elmer DRCII Dynamic Reaction Cell ICP-MS for detection and measurement. All assays were done with a 25 μm spot size and using the Digi-Laz software (Perkin-Elmer). Two assays were taken in triplicate in each winter and summer growth zone. A standard reference material, micro-analytical carbonate standard 1 (MACS-1, U.S. Geological Survey), was used to calibrate and control for drift every five to nine samples. Concentrations of each of 35 analytes were estimated from each otolith sample using the Geo-Pro software (Cetac Technologies). ^{42}Ca was used as an internal standard as Ca is 38% by weight in MACS-1 and otoliths and we measured ^{43}Ca to estimate total Ca. The limit of quantification was estimated as 3× the standard deviation for seven blank replicates for each of the 35 analytes. Analysis revealed multiple detectable elements, but Sr and Ba were chosen as they have been shown to be most related to salinity (e.g., Elsdon and Gillanders 2005). Limits of detection for each element were as follows: Sr—12 parts per billion (ppb); Ba—37 ppb; Ca—102 ppm. Data analysis followed Campana (2005). To control for the amount of ablated material, all concentrations were expressed in comparison to Ca in parts per thousand. Discriminant Function Analysis (DFA) assigned samples to either a freshwater, estuarine, or marine baseline. See Coghlan et al. (2007) for details regarding protocols or Bradbury et al. (2008) for discussion of baseline sample details and DFA analysis.

Morphological Measurements

Morphological and meristic measurements were made on ~50 individuals from each of the sampled locations. All sampled fish were frozen to facilitate rapid collection and preservation. Within six months of collection, 14 morphological and meristic traits were measured (see Figure 2), including fork length (tip of snout to caudal fork), eye diameter (diameter of orbit along body axis), upper jaw length (tip of jaw to end of maxilla), head depth (depth at opercular insert), head length (tip of head to posterior edge of operculum), predorsal length (tip of upper jaw to anterior insert of dorsal fin), prepelvic length (tip of upper jaw to anterior insert of pelvic fin) prepectoral length (tip of upper jaw to anterior insert of pectoral fin), pelvic-pectoral fin distance (distance from pelvic insert to pectoral fin insert), gill raker count (number of rakers on first arch), length of first gill arch, length of longest raker on long arm, and length of longest raker on small arm of first gill arch. All linear measurements were made with Vernier calipers. Covariation due to size was removed following Lecomte and Dodson (2005). Pairwise comparisons were made for size and gill raker counts among habitats using analysis of variance (ANOVA). A "size" variable was calculated as the first factor in a factorial principle component analysis of all traits. Individual observations were plotted against this size variable and residuals were then used in further analysis. DFA (Systat v11) was used to examine how well differing migration phenotypes/habitats self-assigned based on morphology.

Phenotypic variation was also explored using all linear and meristic traits and a pairwise measure of phenotypic divergence (i.e., Q_{ST}), which was used as a measure of phenotypic variation between populations (e.g., Saint-Laurent et al. 2003). Following Bradbury et al. (2006), this was defined here as P_{ST}, emphasizing that phenotypic measures were utilized and no estimates of genetic variance were made. The distinction made here between P_{ST} and Q_{ST} seems appropriate, as environmental influences on these traits are unknown. The phenotypic variance was equal to twice the variance of individuals within populations ($2\sigma^2_w$), and the phenotypic variance between populations was used as σ^2_B. P_{ST} was calculated as

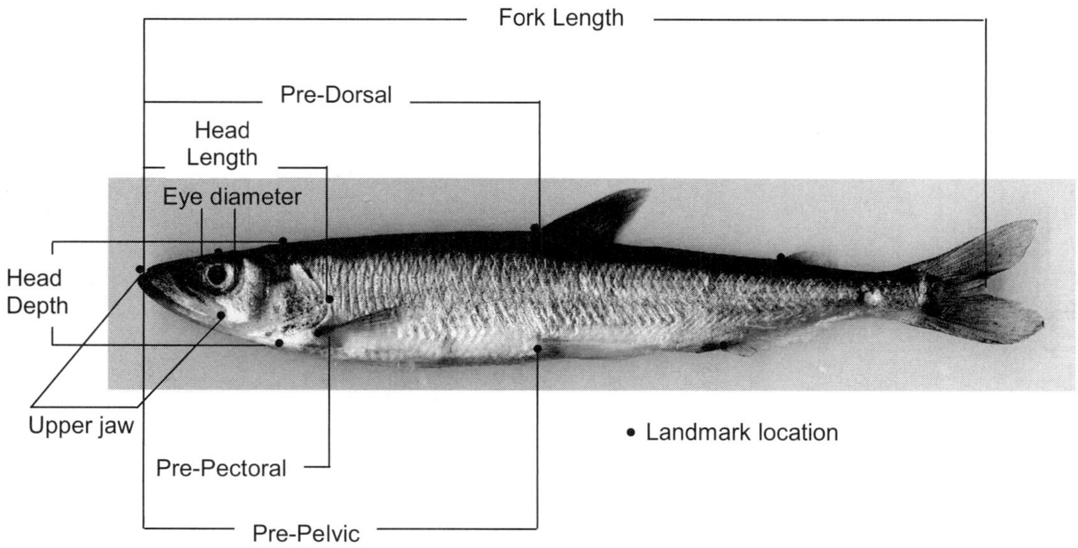

Figure 2.—Linear morphometric and landmark (closed circles) measures used in analysis. See Methods for details and explanation. Note: four gill raker characters not shown.

$$P_{ST} = \sigma^2_B/(\sigma^2_B + 2\sigma^2_w),$$

following Spitze (1993).

In addition to linear measurements, geometric morphometric analysis (e.g., Zelditch et al. 2004) and relative warp analysis (Rohlf and Marcus 1993) was used to examine variation in shape between each of the habitat types. All fish were digitally photographed using a digital camera placed overhead, and coordinates of 13 landmarks (Figure 2) were obtained using TPSDIG (Rohlf 2001a). Coordinates were imported into the program TPSRELW (Rohlf 2001b), which applies a Procrustes superimposition approach generating a consensus against which each individual is compared using a thin-plate-spline analysis. This method uses a principle component analysis of the "partial warps" to generate "relative warps," which explain the major trends in shape change of which the first two were used to document shape change between habitats.

Results

Genetic Variation

Microsatellites.—Average heterozygosities were lower in freshwater locations (0.49–0.59) than anadromous sites (0.66–0.74) and in Holyrood Pond (0.71–0.72; Table 1). Data were checked for the presence of null alleles and scoring errors using MICRO-CHECKER (van Oosterhout et al. 2004). Deviations from HWE were observed at two locations and primarily associated with *Omo*3 and *Omo*16. Instances of null alleles were rare (<5%) and primarily associated with *Omo*3 and *Omo*16, which also displayed evidence of linkage in a subset of populations. Yet, as removal of either locus had no significant effect on observed trends, they were included in further analyses.

Average F_{ST} varied from 0.38 among freshwater sites to 0.08 among estuarine sites and 0.002 among sites within Holyrood Pond. Differentiation between habitat types was lower between Holyrood Pond and anadromous populations (average F_{ST}= 0.06) than between anadromous and freshwater locations (average F_{ST} > 0.20) or Holyrood Pond and freshwater locations (average F_{ST} = 0.21). Multidimensional scaling of F_{ST} values revealed clear associations between adjacent freshwater and anadromous sites at both Biscay Bay and Southeast Placentia (Figure 3). Back River Pond and Hawco's Pond were distinct with the first axis separating these sites from all others (Figure 3). Bayesian estimates of contemporary gene flow suggested limited dispersal with migration rates of 0.001 or less between all sites with the exception of the nearby freshwater sites of Hawco's Pond and Back River Pond (Figure 3).

Table 1.—Locations, details, and summary microsatellite genetic statistics for rainbow smelt samples from southeastern Newfoundland 2003-2006. Samples encompassed estuarine (EST), freshwater (FW) and Holyrood Pond (HRP) locations. Obs hz = observed heterozygosity.

Site	*N*	# loci	Obs hz	Obs hz SD	No alleles	No alleles SD
Hawco's Pond (FW)	34.0	9	0.5046	0.0297	4.89	1.36
Gull Pond (FW)	37.0	9	0.5920	0.0293	9.67	5.15
Backriver Pond (FW)	80.0	9	0.5121	0.0196	6.56	2.83
Conaughtman's Pond (FW)	96.0	9	0.4924	0.0177	6.11	3.44
Average	61.8		0.5253	0.0240	6.80	3.19
Salmonier River (EST)	93.0	9	0.7398	0.0156	10.20	4.15
Biscay Bay River (EST)	93.0	9	0.6682	0.0171	9.00	3.94
Southeast Placentia (EST)	96.0	9	0.6636	0.0173	9.33	4.21
Average	94.0		0.6905	0.0167	9.52	4.10
Holyrood Pond Brook (HRP)	96.0	9	0.7143	0.0165	9.89	3.44
Deer Pond Brook (HRP)	41.0	9	0.7119	0.0240	8.44	3.78
Pathend Brook (HRP)	96.0	9	0.7211	0.0156	10.00	3.64
Average	77.7		0.7158	0.01869	9.44	3.62

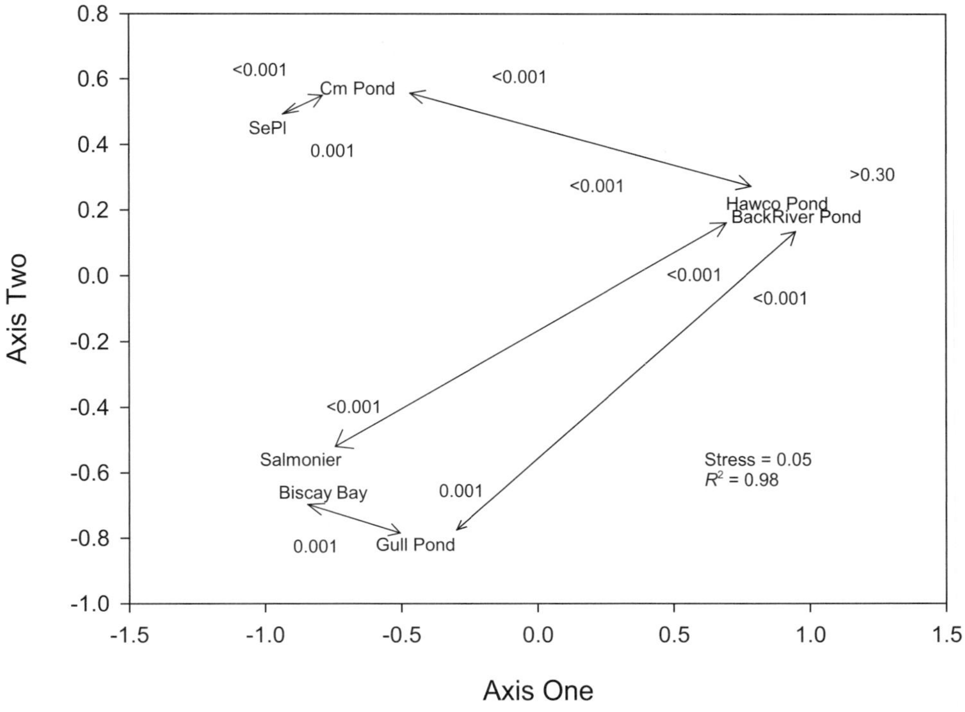

Figure 3.—Multidimensional scaling of genetic differentiation based on nine microsatellite loci for seven anadromous and freshwater populations of rainbow smelt from eastern Newfoundland. Sample and site details in Table 1 and Figure 1. Estimates of contemporary migration rate based on Bayesian estimates using BayesAss.

mtDNA.—This portion of the ND5 has previously been used in smelt to classify individuals into either of two glacial races (e.g., Bernatchez 1997). The complete alignment (730 base pairs) defined six unique haplotypes and included individuals of both races though one accounted for the majority of haplotypes. DNA from Hawco's Pond smelt did not successfully amplify and was excluded from this analysis. Haplotype 1 was the predominant haplotype found in 62 of 84 individuals (73.8%) (Figure 4). Three other haplotypes differed from haplotype 1 by a single mutational event and were each found in 1–3 individuals (Figure 4). The second most common haplotype was unique to all 14 Gull Pond smelt and differed from haplotype 1 at three sites. Finally, a single individual from Salmonier had a haplotype that differed from Gull Pond at a single base.

Otolith Elemental

Frequency distributions of assays taken across the otoliths of individuals from the various habitats suggest significant variation both among and within regions (Figure 5). Holyrood Pond displayed the least variation with the majority of Sr:Ca values between 0.005 and 0.0055. Freshwater locations displayed very limited overlap with other sites with the majority of Sr:Ca values below 0.005 or Ba:Ca values above 0.00001 (Figure 5). Discriminant function analysis (DFA) using marine, estuarine, and freshwater baselines correctly classified 97% of baseline samples supporting its utility in differentiation habitats (See Bradbury et al. 2008 for further details). Accordingly, the DFA classified more than 95% of Holyrood Pond samples as indicative of estuarine residency.

Morphological Variation

Large significant differences in length (ANOVA $P < 0.001$) were observed among samples from the three habitats (Figure 6), and all pairwise comparisons among the habitats were significant ($P < 0.001$). All anadromous fish were 140 mm in fork

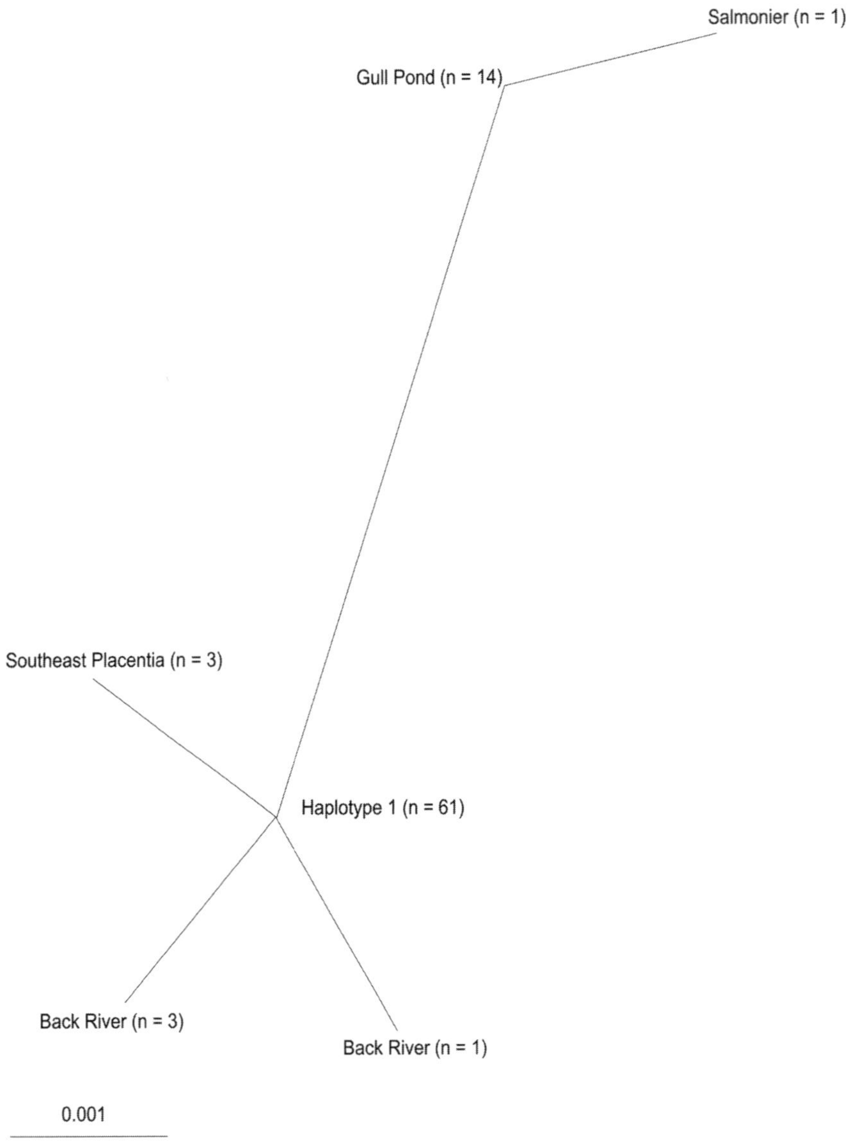

Figure 4.—Unrooted neighbor-joining tree on the unique haplotypes for 730 base pairs of the ND5 gene in rainbow smelt from southeastern Newfoundland carried out in PAUP* (Swofford 1998). See Methods for site details.

length or larger. In contrast, the freshwater and Holyrood Pond fish were between 80 and 120 mm in length (Figure 6B, 6C). Similarly, gill raker counts were generally lower in anadromous fish than in the freshwater and Holyrood Pond samples (Figure 7). Again, significant differences in gill raker count were observed overall (ANOVA $P < 0.001$) among samples from the three habitats (Figure 7), and all pairwise comparisons among the three habitats were significant ($P < 0.001$). Interestingly, Southeast Placentia displayed the lowest variation in gill raker counts and approached freshwater values in some cases.

Multivariate analysis of morphology used both DFA of linear measurements and relative warp analysis of shape variables (principle warps). Discriminant function analysis of linear measurements correctly assigned individuals to habitat with an average accuracy of 78% (Figure 8A). The first DFA

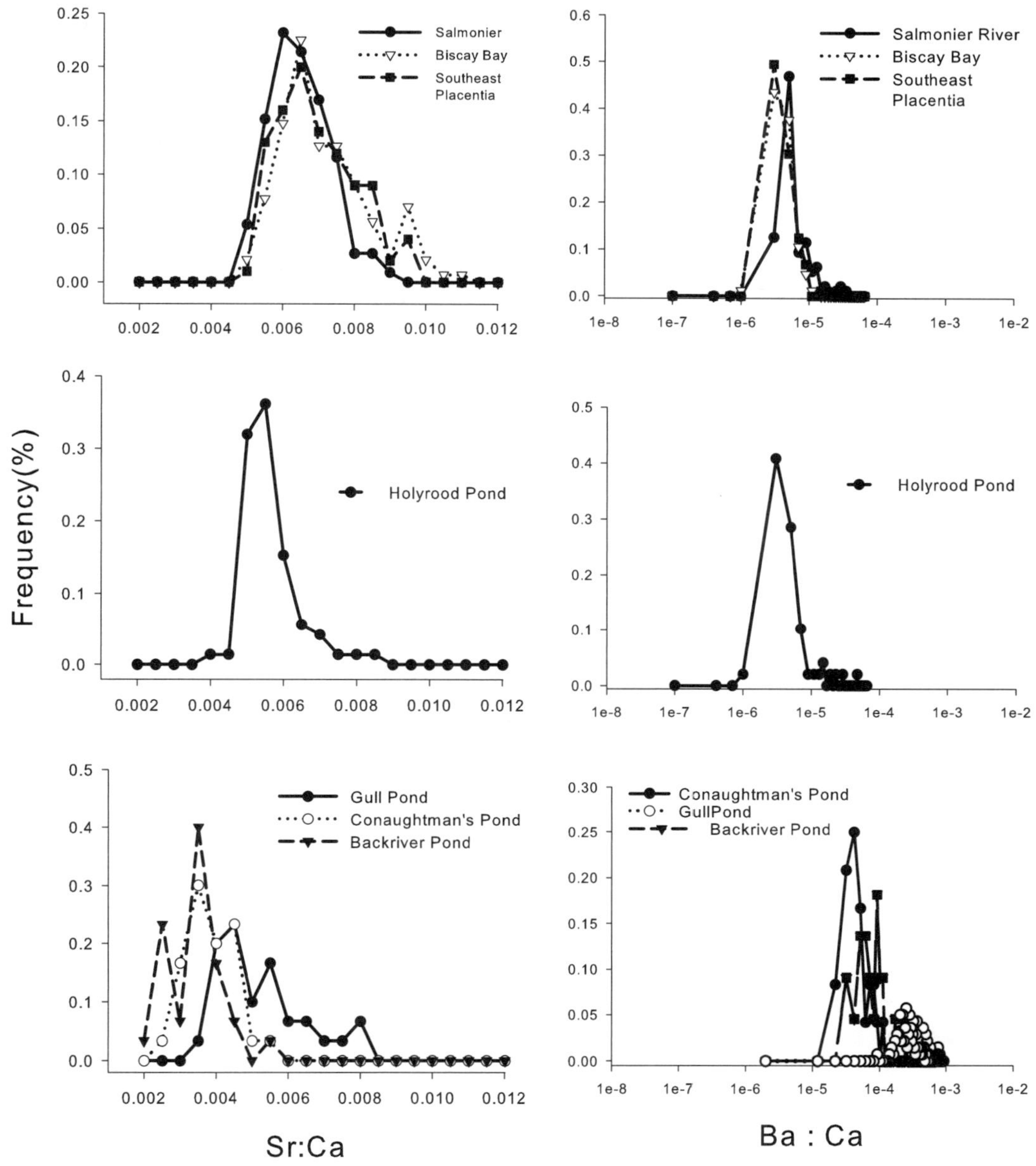

Figure 5.—Frequency distributions of sample specific values of otolith Sr:Ca (A) and Ba:Ca (B) measured using LA-ICP-MS for seven populations of smelt from anadromous (i), Holyrood Pond (ii), and freshwater locations. See Methods for details regarding assays and otolith handling procedures.

axis differentiated Holyrood Pond from remaining samples and the second DFA axis separated freshwater from anadromous habitats on which Holyrood Pond was intermediate (Figure 8A). DFA classification success was primarily driven by gill raker count, eye diameter, and upper jaw length. Geometric morphometric analysis produced a first axis, which was uninformative, but the second relative warp clearly differentiated all dwarf samples (i.e., Holyrood Pond and freshwater) from the large anadromous samples (Figure 8B).

P_{ST} analysis indicated significant morphological divergence between habitats (Figures 9 and 10). Multidimensional scaling of P_{ST} values grouped

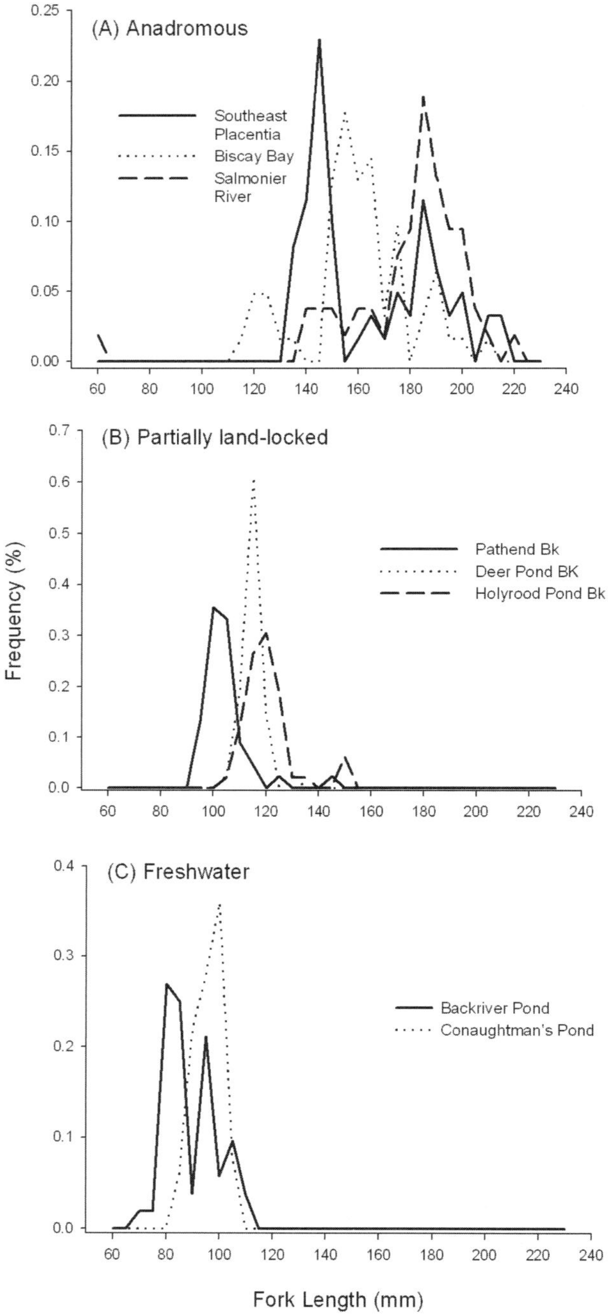

Figure 6.—Length-frequency distributions of smelt sampled in three habitat types: (A) Anadromous, (B) Holyrood Pond (partially landlocked), and (C) Freshwater.

freshwater and Holyrood Pond samples together along the first axis with the anadromous locations not grouping and at greater distances (Figure 9). Estimates of P_{ST} among sites within a habitat type generally displayed low differentiation (Figure 10A), although this was greatest among anadromous loca-

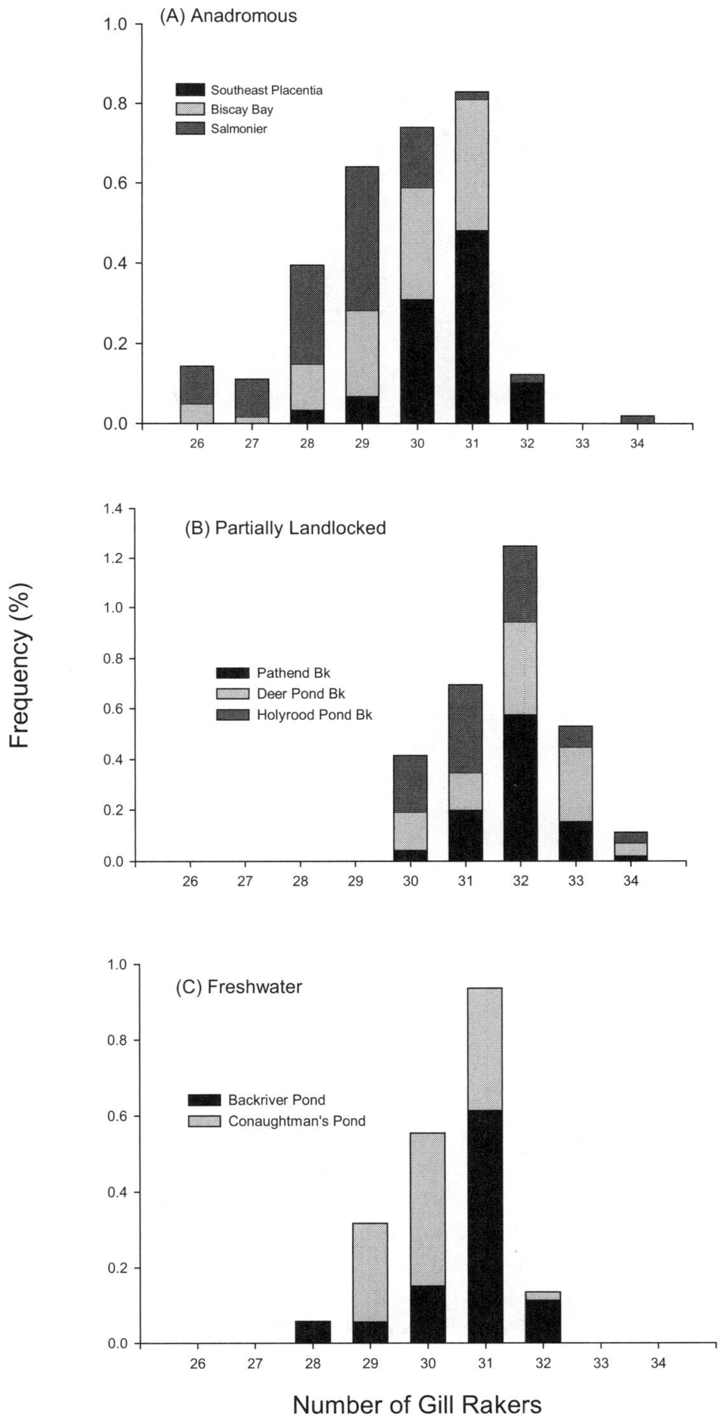

Figure 7.—Frequency distributions of gill raker counts of smelt sampled in three habitat types: (A) Anadromous, (B) Holyrood Pond (partially landlocked), and (C) Freshwater. See Methods for details.

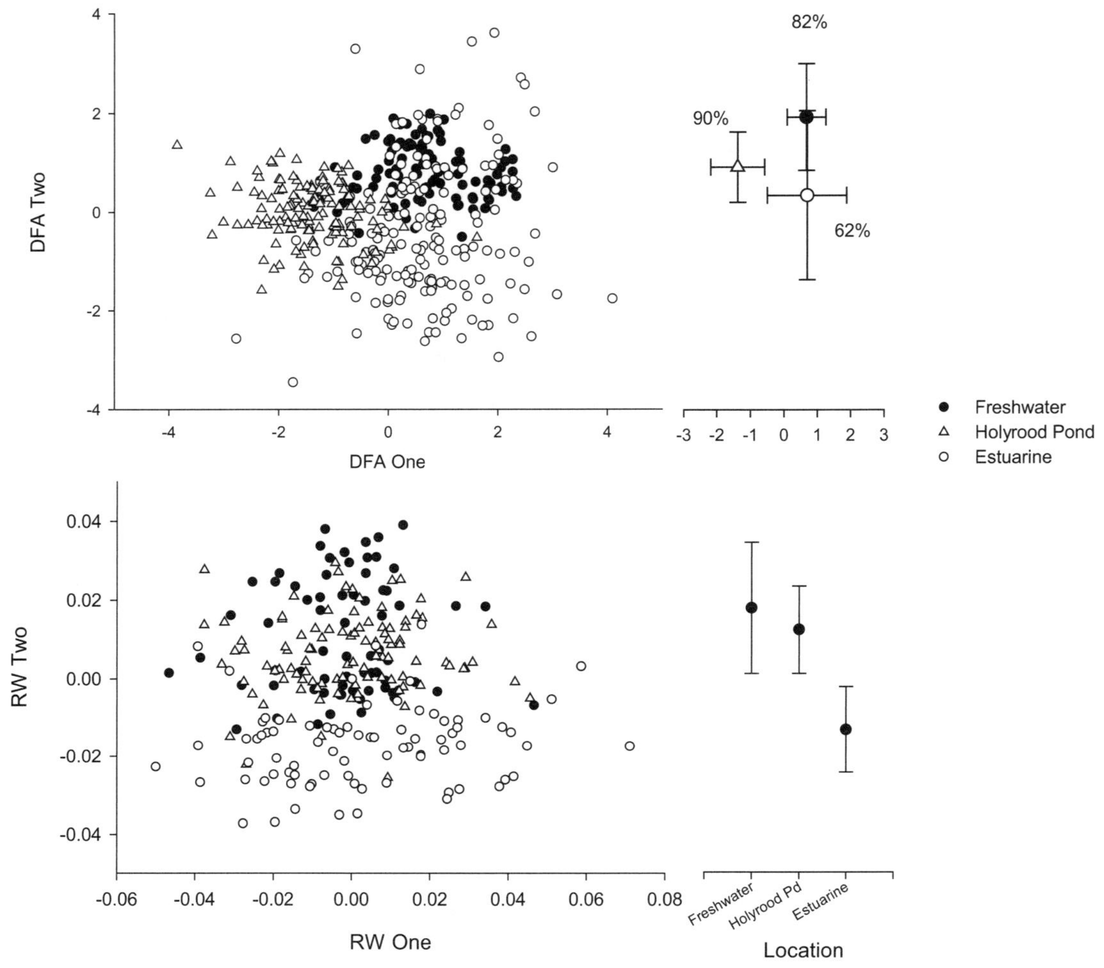

Figure 8.—Morphological divergence of habitat types inhabited by smelt in eastern Newfoundland examined using discriminant function analysis of linear measurements (A) and relative warp scores of shape analysis (B). See methods for details regarding each analysis.

tions. Between habitat comparisons indicated similar levels of divergence between the anadromous samples and all others (Holyrood Pond and freshwater) and lower differentiation between freshwater and Holyrood Pond samples (Figure 10B). Comparable estimates of F_{ST} between Holyrood Pond and anadromous locations were on average less than half that of P_{ST} values for the same comparisons (Figure 10B).

Discussion

Postglacial temperate fish populations illustrate the role of radiation associated with recolonization of vacant habitat following glacial recedence (Bernatchez and Wilson 1998). Accordingly, contemporary spatial and phenotypic diversity is the culmination of colonization history and local selective pressures, which are often difficult to disentangle (e.g., Mäkinen et al. 2006; Raeymaekers et al. 2007). Our results suggest that in rainbow smelt in Newfoundland, primary colonization of lacustrine habitats has occurred in parallel from adjacent independent estuarine sources. Consistent with this inference, morphological analysis supports the hypothesis of intermediate status of the estuarine-dwarf morph located in Holyrood Pond. Moreover, it seems that the transition into the freshwater ecotype may oc-

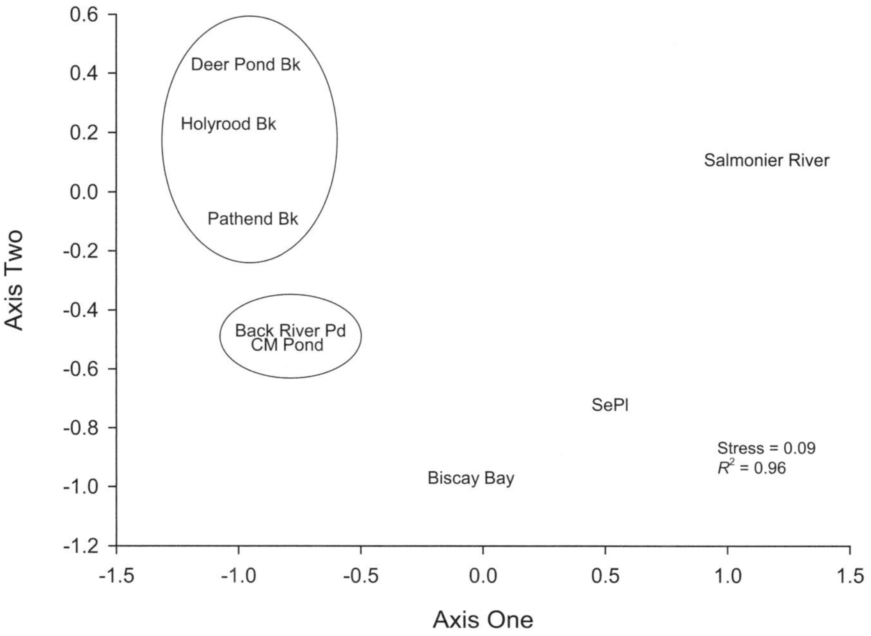

Figure 9.—Multidimensional scaling of morphological divergence (P_{ST}) for eight populations of rainbow smelt from eastern Newfoundland from contrasting habitats. Sample and site details in Table 1 and Figure 1.

cur before complete isolation is present, likely in the presence of moderate gene flow consistent with contemporary local selection. Our comparison of colonization history with a contemporary intermediate example (Holyrood Pond) provides a uniquely complete mechanistic view of how these radiations into freshwater may occur in rainbow smelt and perhaps other anadromous fishes.

The raceme model of divergence proposes independent colonization of freshwater habitats from a common ancestral marine or anadromous source (Bell and Foster 1994). Such a model predicts a loss of genetic diversity and that a random assortment of anadromous alleles will be present in freshwater habitats (Raeymaekers et al. 2005). As predicted, we observed a significant loss of genetic diversity in freshwater smelt populations relative to anadromous populations. However, unlike in other species, which are characterized by genetically homogeneous marine ancestral populations (e.g., McKinnon et al. 2004; Mäkinen et al. 2006; Rayemaekers et al. 2007), regional population structure within the anadromous morph seems reflected in derived freshwater populations such that two of three pairs of freshwater and anadromous source locations are closely related. This result is consistent with a hypothesis of regional diversification among anadromous populations prior to the colonization of freshwater habitats.

Comparison among molecular markers suggests primary postglacial colonization of this area by a single glacial race and supports a hypothesis of subsequent parallel colonization of freshwater habitats from a common ancestor and contemporary isolation. This is consistent with previous work examining only anadromous locations (Coulson, unpublished data), suggesting that the southern Avalon Peninsula is characterized primarily by a single glacial race with transition zones along the south and north coasts of the peninsula. Interestingly, several instances of clear disagreement between markers were observed. We observed a single location, Gull Pond, which was characterized as the second glacial race (i.e., Acadian race, c.f. Bernatchez 1997) and, yet, displayed only moderate contemporary isolation observed through the microsatellite data clustering with its adjacent anadromous location, which is indicative of some gene flow. Such a discrepancy between marker types suggests either a nonneutral or sex-biased dispersal explanation, although at this

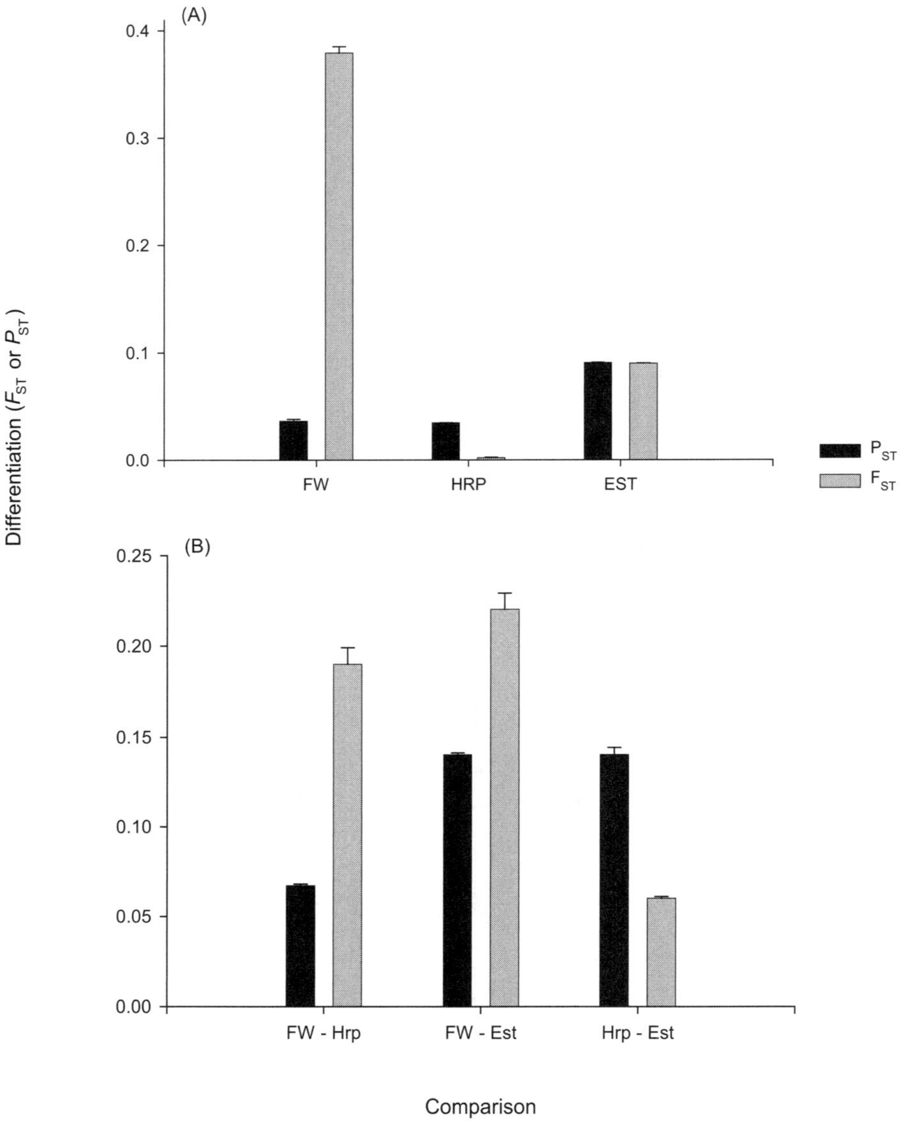

Figure 10.—Genetic divergence (F_{ST}) based on nine microsatellite loci and morphological divergence both within each habitat (A) and among habitat types (B) as measured using P_{ST} nine populations of rainbow smelt from Canada's east coast. See methods for details of genetic and morphological traits.

point, the exact mechanism is unclear. Additionally, in contrast to the mtDNA data, the microsatellite data suggest strong contemporary isolation of Back River Pond and Hawcos Pond from the adjacent anadromous location. Geographic isolation from the south coast may play a role, as Salmonier River is separated from these freshwater ponds by a series of small waterfalls, which likely form a barrier to upstream dispersal, a feature absent from other locations; however, this fails to explain why downstream gene flow is not possible over such short distances (<5 km) and perhaps suggests another isolating mechanism such as selection against downstream migrants.

Newfoundland was completely glaciated during the last glacial maximum (Catto 1998), and several possible mechanistic hypotheses exist for the colonization of freshwater habitats by anadromous fishes.

Long inland migrations could result in significant numbers of adults or offspring becoming either physically or behaviorally landlocked, or alternatively, lowering sea level may have isolated coastal estuaries, transforming them into coastal ponds. Given that smelt migrations are largely limited to lower reaches of most coastal watersheds rarely extending beyond the influence of tidal forcing (McKenzie 1964; Nellbring 1989), it seems likely that isolation due to lowering sea level was the primary source of freshwater populations with perhaps upstream secondary colonization within larger watersheds. Our observation of the estuarine-dwarf ecotype in Holyrood Pond supports this hypothesis, as Holyrood Pond is partially isolated and characterized by irregular periods of connection to surrounding coastal environment for periods of weeks to months each year. Although large-scale movements and gene flow of other marine species have been observed into Holyrood pond (O'Connell et al., Department of Fisheries and Oceans, unpublished data), recent work (Bradbury et al. 2007) observed significant genetic structure associated with the pond in several large marine fish species (e.g., Atlantic cod *Gadus morhua*, white hake *Urophycis tenuis*).

Our results suggest that the affinity of Holyrood Pond smelt to either the anadromous or freshwater group depends strongly on the particular attributes (i.e., morphological, genetic, or otolith composition) examined. Elemental reconstructions of dispersal history and habitat of anadromous smelt from coastal Newfoundland have been done elsewhere (see Bradbury et al. 2008) and suggest limited marine excursions in anadromous fish inhabiting coastal estuaries. Comparison of elemental composition of Holyrood Pond smelt otoliths to both nearby anadromous and freshwater locations clearly indicate estuarine residency. These trends are apparent in both single element concentrations (i.e., Sr:Ca and Ba:Ca), as well as in the multivariate DFA approach. Similarly, heterozygosities observed at the nine microsatellite loci examined suggest similarities between Holyrood Pond smelt and anadromous populations, with both significantly exceeding observed heterozygosities in freshwater populations. Moreover, estimates of genetic divergence between Holyrood Pond and the surrounding spawning locations are comparable to other coastal Newfoundland comparisons (Bradbury et al. 2006), indicating an absence of restricted gene flow associated with the isolation of Holyrood Pond. Accordingly, both otolith elemental and molecular genetic data consistently support a strong similarity between Holyrood Pond and anadromous populations.

Repeated parallel convergent evolution to a freshwater morph has also been documented in three-spine stickleback where freshwater habitats select for a small, low-plated morph (Bell and Foster 1994). Similarly, we observed a common dwarf phenotype in our freshwater samples and Holyrood Pond contrasting the similarities in genetic structure and otolith composition between Holyrood Pond and anadromous samples. Overlap between the freshwater and Holyrood Pond samples was observed in fork length, gill raker count, and relative warp scores. Moreover, estimates of morphological divergence (i.e., P_{ST}) among groups were lowest between freshwater and Holyrood Pond samples and relatively high between anadromous samples and all dwarfs.

The contrast between morphological observations and otolith and molecular data are consistent with an intermediate evolutionary and ecological status of Holyrood Pond smelt. Moreover, the large morphological divergence suggests the presence of strong selection, allowing for morphological differentiation, despite gene flow likely associated with trophic strategy. Previous work has both suggested evidence for selection against dispersal in these habitats and the morphological independence of anadromous populations (see Bradbury et al. 2006). This is further supported by the comparison of P_{ST} and F_{ST} estimates between Holyrood Pond and adjacent anadromous samples where phenotypic divergence significantly exceeded genetic divergence. Similar conclusions were reached by Saint-Laurent et al. (2003) in an examination of sympatric dwarf-normal freshwater smelt where large phenotypic differentiation was not accompanied by large genetic differentiation. Such conclusions are tentative in the absence of heritability information for the traits in question, although similar traits in sticklebacks display heritabilities of 0.30–0.50 consistent with a significant genetic component and support a hypothesis of selection acting on morphology in Holyrood Pond.

Interestingly, anadromous samples displayed larger phenotypic variation among spawning sites than either of the other groups. Morphological diversity among anadromous populations within coastal Newfoundland has been shown to be independent of gene flow (Bradbury et al. 2006, 2008). Extensive phenotypic variation was observed among the anadromous populations and was particularly evident in gill raker count, where Southeast Placentia (and to some degree Biscay Bay) displayed an unusually high count for an anadromous population. Such exceptions and independence among anadromous sites are likely linked to population-specific habitat differences. Southeast Placentia and Biscay Bay are both barred estuaries with permanent access to coastal habitats, and previous analysis of otolith composition has suggested movements between high- and low-salinity regimes in these habitats (Bradbury et al. 2008).

Rapid evolution of freshwater morphs has been noted elsewhere, though usually in the absence of gene flow. Kristjansson (2005) documented rapid adaptation to freshwater habitats in an introduced stickleback population and concluded that phenotypic plasticity and strong selection resulted in substantial phenotypic divergence in only a few generations. Indeed, many examples of rapid evolution of stickleback phenotype have been observed (Bell and Foster 1994). Similar rapid modifications of smelt phenotype have been noted following introductions (e.g., Rupp and Redmond 1966), but the equilibrium phenotype seems similar to the ancestral form consistent with a strong genetic component (Copeman and McAllister 1978). If, indeed, selection were acting on the Holyrood Pond population, the question remains as to the nature of the selective mechanism. Associations between trophic characters and Holyrood Pond suggest that trophic ecology may be more similar between the pond and freshwater locations. In addition, body size seems an obvious candidate for selection as it differed substantially between Holyrood Pond smelt and neighboring anadromous locations, and size has been implicated in other species undergoing anadromous–freshwater diversification. For example, McKinnon et al. (2004) suggest that divergence in stickleback populations may arise as a byproduct of ecological differences and selection on only a few phenotypic traits such as size. Nonetheless, despite clear trends in size and trophic characters, the exact mechanism of colonization and radiation remains unclear.

In conclusion, the radiation of anadromous fish into deglaciated freshwater habitats is a major source of diversification at temperate latitudes, yet the dynamics of this diversification are often poorly understood. This study suggests that postglacial colonization in smelt primarily occurs independently following a raceme model, and the radiation into freshwater may occur even in the presence of gene flow likely associated with strong selection.

Acknowledgments

We thank Ian Paterson, and all members of the Marine Gene Probe Laboratory. We are grateful to various summer students who assisted with sample collection and two anonymous reviewers for constructive comments on an earlier version of this manuscript. Funding for this project was provided by a National Sciences and Engineering Research Coucil Stategic Grant, The Department of Fisheries and Oceans, and The Killam Trusts.

References

Baby, M. C., L. Bernatchez, and J. J. Dodson. 1991. Genetic structure and relationships among anadromous and landlocked populations of rainbow smelt, *Osmerus mordax*, Mitchill, as revealed by mtDNA restriction analysis. Journal of Fish Biology 39:61–68.

Belkhir, K., P. Borsa, L. Chikhi, N. Raufaste, and F. Bonhomme. 2004. Genetix 4.02, logiciel sous Windows pour la genetique des populations. Laboratoire Genome, Populations, Interactions. CNRS UMR 5000. Universite de Montpellier II, Montpellier, France.

Bell, M. A. 1976. Reproductive character displacement in threespine sticklebacks. Evolution 30:847–848.

Bell, M. A., and S. A. Foster. 1994. The evolutionary biology of the threespine stickleback. Oxford University Press, Oxford, UK.

Bernatchez, L. 1997. Mitochondrial DNA analysis confirms the existence of two glacial races of rainbow smelt *Osmerus mordax* and their reproductive isolation in the St Lawrence River estuary (Quebec, Canada). Molecular Ecology 6(1):73–83.

Bernatchez, L., A. Chouinard, and G. Lu. 1999. Integrating molecular genetics and ecology in studies of adaptive radiation: whitefish, *Coregonus* sp., as a case study. Biological Journal of the Linnean Society 68(1):173–194.

Bernatchez, L., and C. C. Wilson. 1998. Comparative phylogeography of nearctic and palearctic fishes. Molecular Ecology 7:431–452.

Bradbury, I. R., S. E. Campana, and P. Bentzen. 2008. Integrating molecular and otolith elemental estimates of dispersal in an estuarine fish, rainbow smelt (*Osmerus mordax*) from coastal Newfoundland. Molecular Ecology 17:1438–1450.

Bradbury, I. R., S. E. Campana, P. Bentzen, and P. V. R. Snelgrove. 2004. Synchronized hatch and its ecological significance in rainbow smelt *Osmerus mordax* in St. Mary's Bay, Newfoundland. Limnology and Oceanography 49(6):2310–2315.

Bradbury, I. R., M. Coulson, S. E. Campana, and P. Bentzen. 2006. Morphological and genetic differentiation in anadromous rainbow smelt: disentangling the effects of geography and morphology on gene flow. Journal of Fish Biology 69:95–114.

Bradbury, I. R., P. V. R. Snelgrove, P. Bentzen, and B. DeYoung. 2007. Structural and functional connectivity of marine fish within a semi-enclosed fjord from coastal Newfoundland. Report of the Harris Center of Memorial University of Newfoundland, St. John's.

Carstens, B. C., and L. Knowles. 2006. Shifting distributions and speciation: species divergence during rapid climate change. Molecular Ecology 16 619–627.

Catto, N. R. 1998. The pattern of glaciation on the Avalon Peninsula of Newfoundland. Géographie physique et Quaternaire 52(1):1–24.

Claytor, R. R., and E. Verspoor. 1991. Discordant phenotypic variation in sympatric resident and anadromous Atlantic salmon (*Salmo salar*) populations. Canadian Journal of Zoology 69:2846–2852.

Coghlan, S. M., Jr., M. S. Lyerly, T. R. Bly, J. S. Williams, D. Bowmand, and R. Hannigan. 2007. Otolith chemistry discriminates among hatchery-reared and tributary spawned salmonines in a tailwater system. North American Journal of Fisheries Management 27(2):531–541.

Colosimo, P. F., K. E. Hosemann, S. Balabhadra, G. Villarreal, M. Dickson, J. Grimwood, J. Schmutz, R. M. Myers, D. Schluter, and D. M. Kingsley. 2005. Widespread parallel evolution in sticklebacks by repeated fixation of Ectodysplasin alleles. Science 307(5717):1928–1933.

Cope, E. D. 1870. A partial synopsis of the fishes of fresh waters of North Carolina. Proceedings of the American Philosophical Society 11:448–495.

Copeman, D. G. 1977. Population differences in rainbow smelt, *Osmerus mordax*: multivariate analysis of mensural and meristic data. Journal of the Fisheries Research Board of Canada 34(8):1220–1229.

Copeman, D. G., and D. E. McAllister. 1978. Analysis of the effect of transplantation on morphometric and meristic characters in lake populations of the rainbow smelt, *Osmerus mordax* (Mitchill). Environmental Biology of Fishes 3(3):253–259.

Coulson, M. W., I. G. Paterson, A. Green, R. Kepkay, and P. Bentzen. 2006. Characterization of di- and tetranucleotide microsatellite markers in rainbow smelt (*Osmerus mordax*). Molecular Ecology Notes 6:942–944.

Durand, J. D., H. Persat, and Y. Bouvet. 1999. Phylogeography and postglacial dispersion of the chub (*Leuciscus cephalus*) in Europe. Molecular Ecology 8(6):989–997.

Elsdon, T. S., and B. M. Gillanders. 2005. Alternative life-history patterns of estuarine fish: barium in otoliths elucidates freshwater residency. Canadian Journal of Fisheries and Aquatic Sciences 62:1143–1152.

Elphinstone, M. S., G. N. Hinten, M. J. Anderson, and C. J. Nock. 2003. An inexpensive and high-throughput procedure to extract and purify total genomic DNA for population studies. Molecular Ecology 3:317–320.

Goudet, J. 1995. Fstat version 1.2: a computer program to calculate F-statistics 86:485.

Kendall, W. C. 1927. The smelts. Fishery Bulletin of the Fish and Wildlife Service 42:217–325.

Kristjansson, B. K. 2005. Rapid morphological changes in threespine stickleback, *Gasterostues aculeatus*, in freshwater. Environmental Biology of Fishes 74:357–363.

Lanteigne, J., and D. E. McAllister. 1983. The pygmy smelt, *Osmerus spectrum* Cope, 1870, a forgotten sibling species of eastern North American fish. Syllogeus 45:1–32.

Lecomte, F., and J. J. Dodson. 2005. Distinguishing trophic and habitat partitioning among sympatric populations of the estuarine fish *Osmerus mordax* Mitchill. Journal of Fish Biology 66(6):1601–1623.

Mäkinen, H. S., J. M. Cano, and J. Merila. 2006. Genetic relationships among marine and freshwater populations of the european three-spined stickleback (*Gasterosteus aculeatus*) revealed by microsatellites. Molecular Ecology 15:1519–1534.

McKenzie, R. A. 1964. Smelt life history and fishery in the Miramichi, New Brunswick. Bulletin of the Fisheries Research Board of Canada 144.

McKinnon, J. S., S. Mori, B. Blackman, L. David, D. M. Kingsley, L. Jamieson, J. Chou, and D. Schluter. 2004. Evidence for ecology's role in speciation. Nature (London) 429:294–298.

McPhail, J. D. 1994. Speciation and the evolution of reproductive isolation in the sticklebacks (Gasterosteus) of southwestern British Columbia. Pages 399–437 *in* M. A. Bell, and S. A. Foster, editors. The evolutionary biology of the threespined stickleback. Oxford University Press, Oxford, UK.

Nellbring, S. 1989. The ecology of smelts (genus *Osmerus*): a literature review. Nordic Journal of Freshwater Research 65:116–145.

Ouellet, P., and J. J. Dodson. 1985. Tidal exchange of anadromous rainbow smelt (*Osmerus mordax*) larvae between a shallow spawning tributary and the St. Lawrence estuary. Canadian Journal of Fisheries and Aquatic Sciences 42(8):1352–1358.

Pigeon, D., J. J. Dodson, and L. Bernatchez. 1998. A mtDNA analysis of spatiotemporal distribution of two sympatric larval populations of rainbow smelt (*Osmerus mordax*) in the St. Lawrence River estuary, Quebec, Canada. Canadian Journal of Fisheries and Aquatic Sciences 55(7):1739.

Raeymaekers, J., J. Van Houdt, M. Larmuseau, S. Geldof, and F. Volckaert. 2007. Divergent selection as revealed by PST and QTL-based FST in three-spined stickleback (*Gasterosteus aculeatus*) populations along a coastal-inland gradient. Molecular Ecology 16:891–905.

Raeymaekers, J. A. M., G. E. Maes, E. Audenaert, and F. A. M. Volckaert. 2005. Detecting holocene divergence in the anadromous-freshwater three-spined stickleback (*Gasterosteus aculeatus*) system. Molecular Ecology 14:1001–1014.

Rohlf, F. J. 2001a. tpsDIG. State University of New York, Department of Ecology and Evolution, Stony Brook.

Rohlf, F. J. 2001b. tpsRelw. State University of New York, Department of Ecology and Evolution, Stony Brook.

Rohlf, F. J., and L. F. Marcus. 1993. A revolution in morphometrics. Trends in Ecology & Evolution 8:129–132.

Rundle, H. D., L. Nagel, J. W. Boughman, and D. Schluter. 2000. Natural selection and parallel speciation in sympatric speciation in sympatric sticklebacks. Science 287:306–308.

Rupp, R. S., and M. A. Redmond. 1966. Transfer studies of ecologic and genetic variation in American smelt. Ecology 47:253–259.

Saint-Laurent, R., M. Legault, and L. Bernatchez. 2003. Divergent selection maintains adaptive differentiation despite high gene flow between sympatric rainbow smelt ecotypes (*Osmerus mordax* Mitchill). Molecular Ecology 2:315–330.

Schneider, S., D. Roessli, and L. Excoffier. 2000. ARLEQUIN, version 2.0: a software for population genetic analysis. University of Geneva, Genetics and Biometry Laboratory, Geneva, Switzerland.

Spitze, K. 1993. Population structure in *Daphnia obtusa*: quantitative genetic and allozymic variation. Genetics 135:367–374.

Swofford, D. L. 1998. PAUP*: phylogenetic analysis using parsimony, version 4.0. Sinauer, Sunderland, Massachusetts..

Taylor, E. B., and P. Bentzen. 1993. Evidence for multiple origins and sympatric divergence of trophic ecotypes of smelt (*Osmerus*) in northeastern North America. Evolution 47(3):813–832.

van Oosterhout, C., W. F. Hutchison, and D. M. P. Wills. 2004. MICROCHECKER: software for identifying and correcting genotyping errors in microsatellite data. Molecular Ecology Notes 4:535.

Williams, G. C. 1992. Natural selection: domains, levels, and applications. Oxford University Press, Oxford, UK.

Wilson, G. A., and B. Rannala. 2003. Bayesian inference of recent migration rates using multilocus genotypes. Genetics 163:1177–1191.

Zelditch, M. L., D. L. Swiderski, H. D. Sheets, and W. L. Fink. 2004. Geometric morphometrics for biologists. Elsevier Academic Press, New York.

American Fisheries Society Symposium 69:97–119, 2009

Diadromy As a Conditional Strategy: Patterns and Drivers of Eel Movements in Continental Habitats

JAMES D. MCCLEAVE*
School of Marine Sciences, University of Maine, 5741 Libby Hall, Orono, Maine 04469, USA

ERIC EDELINE[1]
Centre for Ecological and Evolutionary Synthesis (CEES), Department of Biology, University of Oslo, Post Office Box 1066, Blindern, Oslo N-0316, Norway

Abstract.— Diadromy, the change of habitat salinity, is found in many fishes and is often facultative (i.e., not all individuals of a population are diadromous). In some *Anguilla* spp., Sr:Ca ratios along transects from core to periphery of otoliths have revealed that, within a river basin, some eels enter freshwaters as glass eels and remain there, some remain in coastal or estuarine waters, some move between the habitats after some years, and some move between habitats annually or irregularly. These various patterns of diadromy reflect two fundamentally different behaviors: migration of individuals arriving from the sea and settling at different levels of the watershed, and foraging that is initiated after settlement and may lead to habitat relocations. Migration is an internally initiated process while foraging is essentially environmentally initiated. We depict diadromy as a conditional strategy (i.e., a threshold reaction norm responding to internal cues during migration and to environmental cues during foraging). During the migratory phase, internal cues for being a migrant versus a settler are provided by individual energetic and ontogenetic status while, during foraging, environmental cues for dispersal are provided by agonistic interactions and resource availability. During both phases, the switchpoint for alternative movement decisions is underlain by significant genetic variation so that the conditional strategy may evolve to an evolutionary stable state. We provide a speculative endocrine mechanism for proximate mediation of alternative migratory tactics, which is based on an antagonism between thyroid hormones and a growth hormone-cortisol group. Changes in selective pressures during the past decades (decreased densities in saltwater habitats and increased selection against upstream migrants) may have displaced the stable state of the conditional strategy and thus induced an adaptive drop in the proportion of diadromous individuals among populations.

Introduction

The quintessential catadromous fishes, eels of the genus *Anguilla*, have lost some of their quintessence with the widespread findings that their catadromy is facultative rather than obligatory (e.g., Tsukamoto et al. 1998; Tsukamoto and Arai 2001; Morrison et al. 2003; Daverat et al. 2006; Thibault et al. 2007a, 2007b). Diadromy is a general term referring to physiologically mediated reciprocal migrations between the sea and freshwaters, occurring at certain life history stages (Myers 1949; McDowall 1997). Such migrations usually have been considered obligatory within a population of a species (McDowall 1997). For sake of generality, we will refer here to any cross of salinity boundaries as diadromy.

Catadromy is a life history characteristic in which reproduction occurs in the sea, where little somatic growth of larvae or juveniles occurs, followed by migration to the primary growth biome (freshwaters), followed by return to the reproductive biome (the sea) of maturing adults (McDowall 1997). Ana-

* Corresponding author: mccleave@maine.edu
[1] Current address: UMPC-Paris 6, UMR (CNRS) 7618, Laboratoire Biogéochimie et Ecologie des Milieux Continentaux, 46 rue d'Ulm, F-75230 Paris Cedex 05, France

dromy is the reciprocal life history characteristic. Amphidromy refers to migrations between freshwaters and the sea, occurring at regular times in the life history not associated with reproduction (Myers 1949; McDowall 1988). However, McDowall has come to restrict the term amphidromy to species that spawn in freshwater and have a larval or juvenile migration to sea for early feeding followed by a migration into freshwater for a period of extended feeding, growth, and then reproduction (McDowall 1997, 2004, 2007, 2009, this volume).

Facultative catadromy, or more generally facultative diadromy, as considered here, is defined as the condition in which various individuals within a population have different patterns of migration between the reproductive and growth biomes or may not migrate at all from the reproductive biome. By considering the variation in movement patterns within populations, one can consider controlling factors for migration or dispersal operating at the level of the individual.

This paper has two objectives. First, we review the literature describing patterns of migration, dispersal, and habitat use in continental waters in *Anguilla* spp. Second, we consider ultimate and proximate factors controlling the various patterns seen in individuals in the context of conditional evolutionarily stable strategies (ESS; Edeline 2005, 2007; Edeline et al. 2009).

The development of methods for measuring the concentrations of stable and radioactive isotopes in otoliths has led to insights about stock structure, age validation, and habitat use in fishes. Measurements of strontium and calcium (and other elements) concentrations along transects from core to periphery of otoliths has facilitated the determination of habitat use and the movement history of individual fish (Kalish 1990; Radtke et al. 1990; Limburg 1995; Secor et al. 1995; Tzeng 1996; Secor and Rooker 2000; Gemperline et al. 2002). Sr and Ca determinations usually are made with a wavelength dispersive electron microprobe or sometimes with a scanning proton microprobe, which can measure concentrations in spots of approximately 5 μm^2 or less on sectioned and polished otoliths. Because Sr is generally more abundant relative to Ca in the sea than in freshwater (Kraus and Secor 2004), the ratio of the two can be used to determine whether and when an individual fish has lived in the sea, in the estuary, and in freshwater (Figure 1). Environmental temperature, diet, and physiological processes also influence the Sr:Ca to some degree, so the actual salinity of residency cannot be determined unless calibrated for a population/species by rearing experiments (Secor et al. 1995; Secor and Piccoli 1996; Kraus and Secor 2003; Daverat et al. 2005; Zimmerman 2005). Recent evidence also suggests that the Sr:Ca ratio in eel otoliths is sensitive to the respective proportions of vaterite and aragonite, two different crystalline morphs of calcium carbonate (Tzeng et al. 2007). Finally, Fablet et al. (2007) cautioned that the temporal resolution of Sr:Ca ratio measurement in ottoliths decreases with eel age.

Use of inductively coupled plasma mass spectrometry to measure isotopic composition of Sr and, for example, Sr:Ba, Ba:Ca, or Cr:Ca can allow distinctions among rivers or estuaries of residence (McCulloch et al. 2005), as well as assisting in distinguishing marine and freshwater habitats (Arai and Hirata 2006). The freshwater to marine gradients of carbon and nitrogen stable isotopes ($\delta^{13}C$ and $\delta^{15}N$) have also been used in conjunction with Sr:Ca to discriminate fish of unknown origin (Limburg 1998; Harrod et al. 2005). Tag-and-recapture and telemetric methods (Deelder and Tesch 1970; Parker 1995; Thibault et al. in 2007a) are also useful, though labor intensive, in determining movement patterns of individual eels.

Patterns of Movement of *Anguilla*

Six or seven (or more) patterns of movement occur in some or all of the six species of *Anguilla* that have been examined (Table 1), the actual number of patterns depending on how fine a detail one puts on them (Daverat et al. 2006; Fablet et al. 2007). Most or all patterns may exist in the same catchment. (For additional detail, see Appendix 1, Table A.1.)

In the classic catadromous pattern, glass eels ascend into freshwater and remain there until approaching maturity, when the seaward migration of silver eels occurs. Microprobe transects across otoliths of eels having this life history pattern show a peak value of Sr:Ca at the time of metamorphosis from leptocephalus to glass eel, followed by a dramatic decline to levels associated with freshwater residence (Sr:Ca generally < 0.002) maintained until the time of capture (Figure 2, Type 2). Some individuals of all studied species show this pattern (Table 1).

In contrast, some individual eels never enter freshwater, remaining in coastal or estuarine waters

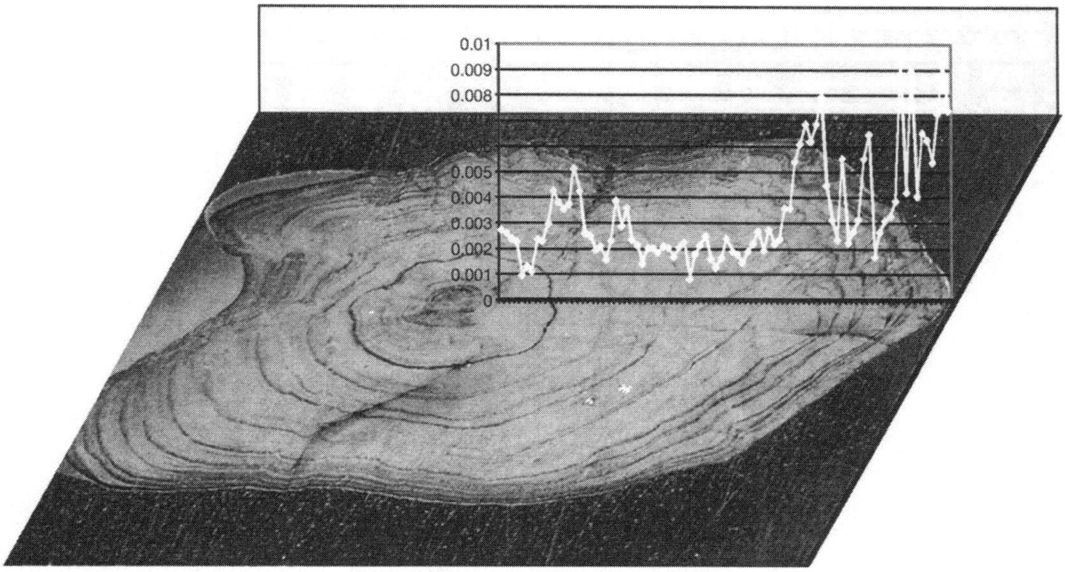

Figure 1.—Variation in Sr:Ca ratio along a radius of an otolith of a European eel. This eel entered freshwater probably as a glass eel (as indicated by low Sr:Ca), then moved back and forth twice between freshwater and the estuary, then resided in freshwater for a few years, and then moved back and forth several times between high salinity water (high Sr:Ca) and low-salinity water (low Sr:Ca). From Daverat and Tomás (2006), with permission.

during the entire growth phase. These have been termed sea eels and estuarine eels (Tsukamoto and Arai 2001; Kotake et al. 2003, 2005) or marine residents and estuarine residents (Daverat and Tomás 2006) (Figure 2, Types 1, 3). Estuarine type eels have been documented in all six studied species, but truly marine eel have only been documented for American eels *Anguilla rostrata*, European eels *A. anguilla*, and Japanese eels *A. japonica* (Table 1). Sr:Ca ratios in marine residents usually exceed 0.006, while those of estuarine residents vary considerably between 0.003 and 0.005 depending on salinity (Tsukamoto and Arai 2001; Kotake et al. 2003, 2005; Daverat et al. 2006). Perhaps these individuals should be termed oceanodromous (Metcalfe et al. 2002) in recognition of their extensive migrations between growth and reproductive areas of the sea. This life cycle pattern was called semicatadromous by Tzeng et al. (2000) and Daverat et al. (2005).

There are two principal variants on strict catadromy, sensu McDowall (1997). One involves individuals that remain in the sea or estuary for a period of years and then disperse into freshwater instead of migrating as glass eels (Figure 2, Type 5). This variant has been shown in five of the six studied species, all except the giant mottled eel *A. marmorata* (Table 1). The other variant involves individuals that enter freshwaters as glass eels and return to the estuary or coastal waters after some years (Figure 2, Type 4), as demonstrated for the Japanese eel, the American eel, and the European eel. These movement patterns have been termed nomadic (Daverat and Tomás 2006; Daverat et al. 2006). However, that term implies wandering with no fixed home range, which does not seem to be the case here. Rather, it seems as if these individuals have changed from a lengthy period of residence in one habitat for a lengthy period in a contrasting habitat. A pattern also exists in *Anguilla* where movements back and forth between fresh and estuarine waters occur either seasonally or irregularly (Figure 2, Type 6), a pattern demonstrated for all studied species except the giant mottled eel (Table 1). The back and forth movements, inferred from Sr:Ca ratios, need to be interpreted with caution, as Sr:Ca ratios at a given

Table 1.—Patterns of habitat use by six species of *Anguilla* as illustrated by Daverat et al. (2006). Marine and estuarine types of eels spend their entire continental lives in coastal or brackish waters. Catadromous types ascend into freshwater for their entire continental sojourn. Downstream and upstream types dwell in freshwater or salt/brackish water for one to several years and then switch habitats. Back-and-forth types switch habitats seasonally or occasionally. For additional detail, see Appendix 1, Table A.1.

Pattern type	Eel species	Example references	Pattern type	Eel species	Example references
1. Marine	American	Lamson et al. 2006	4. Downstream	American	Jessop et al. 2002
	European	Daverat and Tomás 2006		European	Daverat and Tomás 2006
	Japanese	Kotake et al. 2005		Japanese	Tzeng et al. 2003b
2. Catadromous	American	Lamson et al. 2006	5. Upstream	American	Thibault et al. 2007b
	Australian shortfin	Arai et al. 2004		Australian shortfin	Arai et al. 2004
	European	Arai et al. 2006		European	Daverat et al. 2006
	Giant mottled	Shiao et al. 2003		Japanese	Daverat et al. 2006
	Japanese	Tzeng et al. 2003a		New Zealand longfin	Arai et al. 2004
	New Zealand longfin	Arai et al. 2004	6. Back and forth	American	Jessop et al. 2002
3. Estuarine	American	Thibault et al. 2007b		Australian shortfin	Arai et al. 2004
	Australian shortfin	Arai et al. 2004		European	Daverat et al. 2006
	European	Daverat and Tomás 2006		Japanese	Arai et al. 2003a
	Giant mottled	Shiao et al. 2003		New Zealand longfin	Arai et al. 2003b
	Japanese	Tsukamoto and Arai 2001			
	New Zealand longfin	Arai et al. 2004			

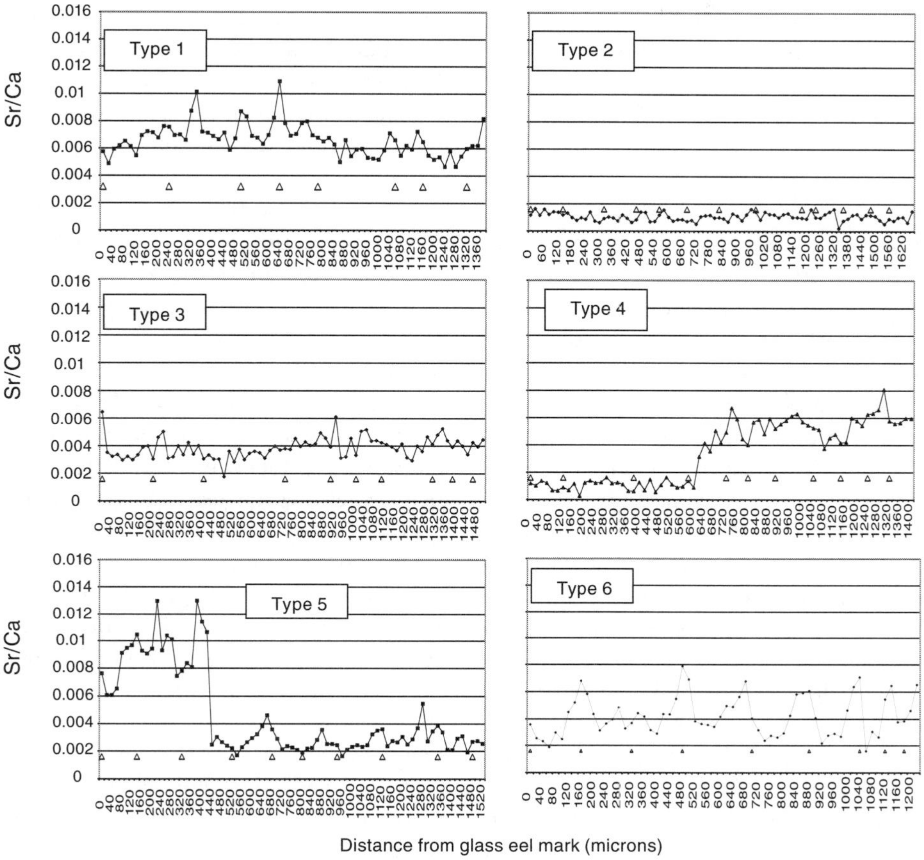

Type	Name	Interpretation
1	Marine resident	Residence in coastal compartment
2	Riverine resident	Residence in freshwater
3	Estuarine resident	Residence in the estuary
4	Downstream nomad	Residence in freshwater, movement to the estuary
5	Upstream nomad	Residence in marine compartment, movement to the estuary
6	Seasonal nomad	Seasonal movements between the lower and upper part of the estuary

Figure 2.—Life history variation revealed by Sr:Ca ratios along radii of otoliths of six European eels collected in the Gironde catchment, France and having 8–13 years of continental residence. Type 1: marine resident; Type 2: freshwater resident; Type 3: estuarine resident; Type 4: downstream mover (freshwater, then estuarine resident); Type 5: upstream mover (marine, then estuarine resident); Type 6: seasonal mover. Triangles indicate annuli. From Daverat and Tomás (2006), with permission.

location in an estuary can vary from very low to very high as salinity changes with seasonal freshets and flood or drought conditions. All these variations on strict catadromy have been termed amphidromy, with appropriate modifiers by Thibault et al. (2007b). Individuals such as these have been called facultative marine/freshwater wanderers (McDowall 2009).

Fablet et al. (2007), using eels from Daverat et al. (2005), conducted an unsupervised Bayesian reconstruction of individual European eel life histories. Patterns of habitat use were not determined a priori, as they are in most studies. The analysis of 192 eels resulted in 36 movement patterns, most of which reflected simply the number and directions of interhabitat movements. However, most eels had the simpler patterns described above.

Caution should be used in assigning eels to estuarine or marine residency, as the distinction based on Sr:Ca ratio may be less accurate than implied in many studies. Although the concentrations of Sr and Ca in the water are both linearly related to salinity, the Sr:Ca ratio in water is curvilinearly related to salinity, with most of the variation occurring at salinities below about 8‰ (Surge and Lohmann 2002; Kraus and Secor 2004). In rearing experiments with young Japanese and American eels, significant linear relations were found between salinity and the Sr:Ca ratio in otoliths (Tzeng 1996; Kraus and Secor 2003), but a given eel could be misclassified with respect to habitat of residency. Knowledge of the Sr and Ca concentrations in the water of estuaries being studied would assist in interpretation, as Kraus and Secor (2004) found in laboratory and field studies with young white perch *Morone americana*. The stable isotope $\delta^{13}C$ concentration in otoliths provided better classification of white perch to known salinity habitats than did Sr:Ca ratios (Kerr et al. 2007). Harrod et al. (2005) could correctly classify about 85% of European eels as marine, brackish, or freshwater residents in a catchment using discriminant analysis of $\delta^{13}C$, $\delta^{15}N$, and C:N in muscle tissue.

Tzeng et al. (2007) cautioned that the salinity history of eels might be misidentified because of the inclusion of areas of the vaterite form of $CaCO_3$ within the normal aragonite form in otoliths. They found vaterite inclusions in 48% of 108 European eels collected for the study of Shiao et al. (2006) in Lithuania on the Baltic Sea. Vaterite has a much lower Sr:Ca ratio than aragonite, and Tzeng et al. (2007) illustrated how one individual estuarine eel could be misclassified as a freshwater resident or an occasional visitor to freshwater. However, the life history patterns reported by Shiao et al. (2006) seem consistent with expectations. Many eels collected in freshwaters showed only low Sr:Ca beyond the elver stage, as expected given elver stocking programs in those waters. Others had patterns of higher Sr:Ca for some years before the ratio fell, consistent with natural recruitment through the brackish waters of the Baltic Sea. Shiao et al. (2006) were able to divide Sr:Ca patterns of eels into phases corresponding to migration through the North Sea and then through the Baltic Sea, quite as expected. Nevertheless, it is possible that some of the "noise" in Sr:Ca transects in this and other studies is due to vaterite inclusions rather than habitat shift.

Fortuitous experiments suggest misidentification due to vaterite inclusions has not been a serious problem. European eels stocked into three Swedish lakes had the expected life history patterns as interpreted from Sr:Ca transects at ages 8–18+ years (Tzeng et al. 1997). Eels from two lakes stocked with elvers showed patterns that would be interpreted as catadromy. Eels caught in brackish coastal waters and stocked as yellow eels had patterns that would be interpreted as upstream movers. Silver eels from a stocking of elvers in a fourth lake also showed the expected pattern (Tzeng et al. 2000).

The proportion of eels exhibiting various life history patterns undoubtedly varies widely among catchments, depending on a number of environmental factors. Factors might include comparative surface areas of coastal, estuarine, and freshwater habitat; comparative water temperatures and productivities; and comparative population densities of conspecifics and competitors (see section on evolutionary ecology of eel diadromy). The only way to estimate proportions of life history patterns for an entire catchment would be to sample silver eels at sea leaving coastal waters, a difficult or impossible task. Thus, caution in interpretation of proportions given in the literature is necessary.

The most complete data are for European eels sampled in the Gironde watershed in southwestern France (Daverat and Tomás 2006). Eels captured in coastal and estuarine waters had varied dispersal histories, with a large component of those captured in the estuary having had some period of freshwater residence (56%) (Table 2). Eels with an upstream-movement history (Figure 2, Type 5) that were captured subsequently in coastal and estuarine waters were rare (7% and 2%, respectively), which explains the low percentage of eels caught in freshwater with previous estuarine residency (3%). Those few were caught in freshwater that was still under tidal influence, while none was caught upstream of tidal influence. Daverat and Tomás (2006) suggested that the higher proportion of downstream-moving eels than upstream-moving eels caught later in the estuarine and coastal waters may reflect the need to reside in

Table 2.—Percentages of European eels collected in the Gironde watershed, France, having various migratory histories prior to capture (data from Daverat and Tomás 2006).

	Location of capture		
Migratory history	Coastal (n = 45)	Estuary (n = 162)	River (n = 63)
Marine resident	57	0	0
Estuarine resident	16	44	3
Upstream migrant	7	2	0
Downstream migrant	16	27	0
Seasonal migrant	4	12	0
Riverine resident	0	15	97

freshwater for a time (until about 30 cm long) to reduce competition with other glass eels and yellow eels.

The variation in proportions of eels exhibiting different patterns of life histories in different localities emphasizes the high variety of habitat use in the genus *Anguilla*. Surprisingly, high proportions (22–85% in some locations) of American eels in Nova Scotia and Prince Edward Island, Canada would be classified as habitat shifters (i.e., eels that made one or more shifts between freshwater and saltwater other than at the glass eel stage [Jessop et al. 2006; Lamson et al. 2006]). Although proportions of habitat shifters, per se, were not given, the presence of freshwater and estuarine classified Japanese eels in the sea, marine eels in estuaries, and estuarine eels in freshwater (Tsukamoto and Arai 2001) implies habitat shifting. Tsukamoto and Arai (2001) did report such eels.

Salinity History and Somatic Growth

One of the assumptions of conceptual arguments we present later in the paper is that primary productivity is higher overall in saltwater habitats than in freshwater habitats in temperate latitudes while the reverse is true in tropical latitudes. One way to test this assumption is to examine whether growth rates of temperate *Anguilla* spp. are higher in marine and brackish habitats than in freshwater habitats; unfortunately, no data are available for tropical eels. Many studies of eels in the wild, based either on length-at-age back-calculation from otoliths or on mark–recapture experiments, are equivocal but tend to support the assumption of higher food availability in saltwater habitats. In the European eel, Arai et al. (2006) reported greater growth rate among undifferentiated eels collected in an Irish bay (42 mm/year) than older eels collected from six rivers (19–41 mm/year). This result must be interpreted with caution as young eels generally have greater growth rates than older eels. Freshwater resident European eels in a French river basin had a significantly lower growth rate (53 mm/year) than those eels with histories, including some or all marine or estuarine residence (61 mm/year) (Daverat and Tomás 2006). Data from 17 catchments also suggest that eels grow faster in brackish than in freshwater (Fernandez-Delgado et al. 1989). Silver Japanese eels, collected in bays and freshwaters in Japan, were classified on the basis of Sr:Ca ratios in otoliths into freshwater, estuarine, and marine residents. The marine- and estuarine-resident females had significantly greater mean lengths at age than freshwater residents for the first 5 years, after which the gap narrowed (Tzeng et al. 2003a); males of the three types did not differ significantly in length at age. These age and sex effects on the salinity–growth interaction are most likely the result of a biased sampling. Indeed, because they grow faster, saltwater female eels likely silvered and left habitats at a younger age than freshwater female eels. Therefore, a sample of old female eels includes both fast- and slow-growing eels in freshwater but only slow-growing eels in saltwater. The effect is probably even stronger for males because they silver earlier than females; fast-growing saltwater males were thus probably underrepresented in the whole sample. Such sampling biases must be accounted for when comparing growth rates between different groups of eels. Growth rates of mostly silver-phase male and female eels in the same Japanese bay were significantly greater in marine and estuarine eels than freshwater eels, if several outliers were removed (Kotake et al. 2005). In contrast, lengths at age of similarly sized female Japanese eels collected in a Taiwanese estuary were not significantly different among the three life history patterns, but the sample size of marine and freshwater types was small (Tzeng et al. 2002).

American eels in a Prince Edward Island catchment with a marine- or brackish-water history had an otolith radius at age (proxy for growth rate) two times greater than freshwater residents after the first 4 years of continental life (Cairns et al. 2004). In a study in a river basin in Nova Scotia, Jessop et al. (2006) categorized yellow-phase eels after otolith Sr:Ca analysis into those where less than 50%, 50–75%, and more than 75% of growth had occurred in freshwater. The less than 50% freshwater (more marine) group had greater growth rates in years 3–5 than the other two groups, after which growth rates were similar for all groups (see comment above about biased sampling). This gave greater lengths at age for the more marine group that persisted to at least age 9. In the same location, Jessop et al. (2004) also found, for silver eels, that eels with more than 50% residence in marine or brackish waters had grown faster than those with less than 50% residence time there. In this case, the length-at-age differences persisted through most of years 2–16. Greater growth rate of saltwater eels than of freshwater American eels was also reported for the St. Jean River, Quebec (Thibault et al. 2007b). In the Hudson River, New York, growth rates of eels captured at two brackish water sites were two to three times greater than those of eels captured at four freshwater sites, as measured both by mark–recapture and length-at-age data back-calculated from otoliths (Morrison and Secor 2003). Annual growth rates of female yellow Australian longfin eels *Anguilla reinhardtii* were greater in brackish than in freshwater habitats in three rivers in southeastern Australia, but with large variation among individuals in both habitats (Walsh et al. 2006). In brackish habitats, females grew faster than males. Finally, Arai et al. (2004) also found higher growth rates in eels that experienced saline habitats in New Zealand shortfin eel *A. australis* and New Zealand longfin eel *A. dieffenbachii*.

The fact that temperate eel growth is often faster in saltwater than in freshwater supports the hypothesis that there is more food in saltwater than in freshwater habitats in temperate latitudes. However, the large variation in lengths at age of eels within a single location (e.g., Morrison and Secor 2003; Walsh et al. 2006), among nearby locations (e.g., Matthews et al. 2003), and among distant locations (e.g., Jellyman 1997), points out that somatic growth is a complex process that is influenced by a number of environmental parameters. Competitors and predators may have contrasted effects on growth. Higher competitor density may deplete food resources, while predators may hinder foraging, both effects resulting in slower growth rates. However, competition may be coupled to cannibalism, and both cannibals and predators often target small, slow-growing prey, which are easier to catch and ingest. Hence, cannibalism and predation may result in higher growth in the prey population by removing slow growers. Finally, temperature strongly affects fish growth (optimal is around 23°C in the eel), and in teleosts, salinity may directly influence somatic growth by altering the endocrine system (see Edeline et al. 2005b and references therein). Under experimentally controlled conditions, salinity stimulates feeding behavior and cannibalism in juvenile European eels (Edeline et al. 2005b; but see Tzeng 1996 for contrasting results). Therefore, further studies are needed to clarify the respective contributions of food availability, competition, cannibalism, predation, temperature, and salinity (nonexhaustive list) to the growth variations observed in eels inhabiting different habitats.

Evolutionary Ecology of Eel Diadromy

Two Types of Eel Movements in Continental Habitats

In order to interpret the patterns observed in otolith Sr:Ca studies correctly, it is important to realize that eel movements in continental habitats (and thus diadromy) result from two profoundly distinct behaviors: migration and foraging (Edeline et al. 2007, 2009). Precise distinctions between migration and foraging are provided by Dingle (2006). Migration, as used in our paper, is a directed, long-distance movement between the breeding and feeding areas. Migration should be seen as a syndrome initiated by a number of morphological, physiological, and behavioral components, and a migrant is thus not responsive to local variations in habitat resources (Dingle 2006). For instance, glass eels may migrate across coastal habitats and up-estuary, even if resources are abundant there. Hence, we consider that migration is an internally cued behavior, although the environment (salinity, temperature, etc.) may plastically modulate internal cues. Foraging movements occur after settlement (i.e., after the cessation of migration, which drove the individual onto the foraging grounds). In contrast to migration, foraging is not directed over long distances but is a habitat exploration driven by resource search and thus ceases when food or a new suitable habitat is found. Therefore, we consider that foraging is an environmentally cued behavior, although there may be internal variation for sensitivity thresholds to environmental cues. While migration is essentially a density-independent behavior (but evolution of migration is density-dependent; see next section), foraging behavior leads to density-dependent dispersal (Sutherland et al. 2002). Despite the fact that density-dependent dispersal has not been explicitly demonstrated in the eel, several studies strongly support the view that river colonization by yellow eels is equivalent to a process of random diffusion of particles from a point source, the point of the watershed where most of individuals settle (Smogor et al. 1995; Ibbotson et al. 2002). Based on these definitions of migration and foraging, we consider that migration can pilot a first cross of the saltwater–freshwater interface only, while foraging can pilot both a first cross and subsequent crosses of the interface as recorded in eel otoliths.

Recent experimental and field data suggest that in European eels, a majority of individuals settle (i.e., shift from migration to foraging) during the glass–yellow eel transition (stage 7 as defined by Elie et al. 1982), that is, with acquisition of fully developed feeding capacity. Edeline et al. (2007) have investigated the distribution of glass eels and elvers (partially pigmented eels) relative to the distribution of newly transformed yellow eels (fully pigmented and shorter than 15 cm) at the tidal limit of a large river. They found that glass eels and elvers accumulated at the break point of tidal streams, reflecting an upstream migration that was slowed by the loss of tidal streams available to help movement. In contrast, newly transformed yellow eels were homogeneously distributed around the point where they initially accumulated at the glass eel and elver stages, suggesting that transformation into a yellow eel induced settlement and shift to density-dependent dispersal. Accordingly, when kept in an aquarium, late metamorphic glass eels (i.e., elvers) shift from a pelagic to a benthic behavior as they complete their metamorphosis into a yellow eel (Jegstrup and Rosenkilde 2003). In a flume-tank experiment, transformation into a yellow eel induced a drop in upstream climbing activity (Imbert et al. 2008). Additionally, upstream climbers among elvers had higher body condition than inactive ones, but upstream climbers among yellow eels were smaller and physiologically more stressed than inactive ones, suggesting that the triggering factors for movements were profoundly different between elvers and yellow eels (Imbert et al. 2008). We interpret this result as a shift from internally initiated to environmentally initiated movements (i.e., from migration to foraging). Indeed, high energy load triggered climbing in elvers because they were migratory and migration requires much energy (see the following section), while small size triggered climbing in yellow eels to escape agonistic interaction (and resultant stress) from larger conspecifics. Yellow eels are indeed strongly agonistic and cannibalistic under artificial rearing conditions (Peters et al. 1980; Degani and Levanon 1983; Knights 1987). Field data from a variety of habitats show that movement probability in yellow eels is inversely proportional to their body size (Feunteun et al. 2003), further supporting the view that competitive and agonistic interactions are major drivers of movement in yellow eels.

Therefore, the typical ontogenetic sequence for the European eel is an upstream migration until transformation into a yellow eel, which triggers settlement and foraging. However, settlement is a complex trait embedding multiple physiological and behavioral changes, and complex traits generally show high variation in populations. Indeed, occurrence of strictly marine eels, as revealed by otolith Sr:Ca studies, suggests that settlement may occur at the glass eel and elver stages (for instance, as migrants run out of energy stores; see next section). Accordingly, some glass eels are strongly saltwater seeking, exhibit a low upstream swimming behavior, and seem to be programmed for saltwater residency. Such individual glass eels show low positive rheotaxis and consistently prefer saltwater to freshwater in experimental behavioral tanks. They subsequently grow faster under 2 months excess feeding conditions in saltwater (but not freshwater) compared to glass eels that initially showed weak or no preference for saltwater (Edeline et al. 2005b). In contrast to early settlers, some small yellow eels may maintain upstream migration during episodes often termed "runs" (Tesch 1977; Moriarty 1986). Feunteun et al. (2003) also suggest occurrence of "pioneer type" small yellow eels, which continue moving upstream even if habitat is suitable and densities are low. Postsettlement, seasonal, or nonperiodic changes in environmental conditions (food, space, temperature, oxygen) or predator or competitor numbers may necessitate habitat relocations following an ideal free distribution (i.e., fitness-maximizing movement; see Fretwell and Lucas (1970)). After Dingle (2006), we suggest that foraging movements occur at two scales: within (patch scale), and between home ranges (habitat scale), depending on habitat stability. For instance, during growth seasons, yellow eels seem to be restricted to small foraging patches, but winter conditions may drive them from their home range, inducing habitat relocations to warmer places (Tesch 1977). Some such movements may cross the freshwater–saltwater interface, resulting in some of the life history patterns documented above in otolith microchemical studies.

Evolutionary Constraints on Eel Movements

Bioenergetics.— Migration and dispersal are selected against due to their high energetic and mortality costs (Bernatchez and Dodson 1987; Gandon and Michalakis 2001; Alerstam et al. 2003). For instance, during long distance migrations, exhaustion may directly lead to death or to increased sensitivity to infections and predators. Additionally, movement increases encounter rate with agonistic conspecifics, predators, and decreases competitive ability during the colonization of a new habitat compared to conspecifics already holding the ground. Finally, locomotion directly conflicts with foraging and digestion for energy and oxygen allocation (Owen 2001), indicating that a migrant cannot grow optimally or cannot grow at all. The hypothesis of a physiological trade-off between migration and growth is supported by experimental data showing inverse correlation between glass eel migratory intensity (salinity preference and positive rheotaxis) and subsequent juvenile growth performance over a period of 2 months (Edeline et al. 2005b). These bioenergetic limitations are expected to induce adaptive behavioral responses in migrants. Accordingly, both experimental and field results suggest that migratory behavior of glass eels and elvers responds plastically to variations of individuals' energetic status. European glass eel and elver salinity preferences shift from freshwater to salt water as their body condition decreases (Edeline et al. 2006), and swimmers have a higher energetic status (as measured by body length, weight, and energy content) than settlers when tested in different types of large experimental flumes (Bardonnet et al. 2003; Bureau du Colombier et al. 2007; Imbert et al. 2008). Furthermore, Edeline et al. (2004) sampled different groups of glass eels in the Vilaine River estuary where a dam equipped with an eel ladder blocks the tides. They found that settlers, caught on the bottom of the estuary during flood tide, were lighter than upstream migrants caught climbing the eel ladder.

Trade-off between habitat productivity and competition and predator avoidance.—Migration and dispersal in general are selected for when the habitat is heterogeneous in space or time. Indeed, moving gives access to more resources or allows escape from unsuitable conditions, predators, and competitors (Gandon and Michalakis 2001; Weisser 2001; Alerstam et al. 2003). Presumably, diadromy has evolved from differences in primary productivity between saltwater and freshwater ecosystems (Gross et al. 1988). In tropical latitudes, marine species tend to evolve a growth phase in the more productive freshwater systems. Hence, catadromy is probably an ancestral trait in the genus *Anguilla* that originates from tropical marine waters (Aoyama et al. 2001). Shift of several *Anguilla* species to European, American, and Asian temperate latitudes has favored evolution of facultative diadromy, presumably because the saltwater:freshwater productivity ratio increases in temperate areas (Tsukamoto and Arai 2001). The view that eel species evolve migratory propensity in relation to the productivity pattern prevailing in their main geographical range is further supported by data from Taiwanese rivers where the tropical giant mottled eel and the temperate Japanese eel co-occur. The giant mottled eel colonizes upper river reaches, while the Japanese eel tends to stay in the estuary and the lower river reaches (Shiao et al. 2003).

As noted above, social interactions, competition, and predation are also strong drivers of the evolution of migration and dispersal. We suggest that social interactions are crucial drivers for evolution of migration and dispersal in eel species because maintenance of diadromy at a high energetic cost by temperate eels may only be explained by fitness gain from avoiding competition and predation in saltwater habitats. Indeed, the available data suggest that eel density decreases with increasing distance from the river mouth (Naismith and Knights 1993; Smogor et al. 1995; Ibbotson et al. 2002), and fisheries location and landings suggest that eel densities are higher in marine and brackish waters than in freshwater (Moriarty 2003). Also, predation is generally considered to be higher in marine than in freshwaters (Jonsson and Jonsson 1993). In particular, predation by conger eels *Conger* spp. (among other predators) is thought to favor diadromy in temperate eels (Moriarty 1987). Therefore, evolution of eel diadromy may result from a complex interaction of opposing selective forces. The benefits of colonizing productive habitats balance against avoidance of competition, which favors shift to less productive habitats. Bioenergetic constraints hinder movements in both cases. We suggest that anguillid eels have evolved a conditional strategy to cope with these selective pressures.

Conditional ESS model for eel facultative diadromy.—Conditional strategies provide a game-theoretic framework to explain the evolution of discrete, alternative phenotypes within populations (Hazel et al. 1990; Gross 1996; Roff 1996; Hazel et al. 2004). They define a special case of phenotypic plasticity in which the alternative phenotypes, termed the tactics of the strategy, involve a fitness-maximizing decision by the individual. The strategy results in a threshold reaction norm, which responds to reliable cues given by the environment or by status of the individual, which is composed of one or more phenotypic traits (Hazel et al. 1990; Schlichting and Pigliucci 1995; Gross 1996; Roff 1996; Hazel et al. 2004). A conditional strategy is presumably superior to a strict genetic polymorphism as a means of adapting to environmental heterogeneity and unpredictability. In accordance with this prediction, conditional strategies are widespread in nature (Hazel et al. 1990; Gross 1996; Roff 1996). Conditional strategies can evolve to a stable equilibrium (i.e., can be ESS) even when the fitnesses of the alternative tactics are unequal and when the strategy is heritable (Hazel and Smock 2000).

Based on the work of Gross (1996), Edeline (2005) developed the idea that eel diadromy is a conditional strategy internally cued by energetic status of an individual. Then, Edeline (2007) provided a more complete conditional ESS model for eel diadromy by combining the strategic model of Gross (1996) and the quantitative genetic model of Hazel et al. (2004). The model, presented in Figure 3, illustrates how an individual eel may adaptively choose between being a migrant or a settler, depending on whether its energetic status is above or below an internal switchpoint. Importantly, in this model, both the switchpoint between tactics and the cueing trait are polygenic traits. In other words, for an individual eel, the cueing trait and the switchpoint are determined by a large number of genes, whose effects may be modulated by environmental influences. For instance, a glass eel's energetic status (cueing trait) may be affected by the amount of food encountered during the larval stage, and the switchpoint for being a migrant or not is affected by temperature (Edeline et al. 2006). The combination of environmental and genetic variability for the cueing trait and the switchpoint results in normal distributions among the population (Figure 3), and evolution may act on the mean and the variance of these distributions. Following Hazel et al. (2004), the model allows a combination of conditional and unconditional individuals in the population if, at equilibrium, variance in the switchpoint distribution is higher than variance in the cueing trait distribution. Conditional individuals are those having a switchpoint lying in the range of possible values of the cueing trait, while unconditional individuals have a switchpoint lying outside the limits of the cueing trait and thus will be necessarily nondiadromous (extremely high switchpoint) or diadromous (extremely low switchpoint). It must be noted that even in the absence of, or large changes in, selective pressures, environmental variation in space and time affects the exact position of the fitness functions (e.g., diadromy does not return the same fitness in every watershed). Thus, at equilibrium, selection does not favor a single switchpoint in the population, adding to the variability in the distribution of switchpoints and cueing traits.

Reliable internal cues for migration should be good predictors of future fitness gain because, in contrast to a forager, a migrant does not explore the habitat and thus cannot directly evaluate its quality. The cue for settlement should thus be in-

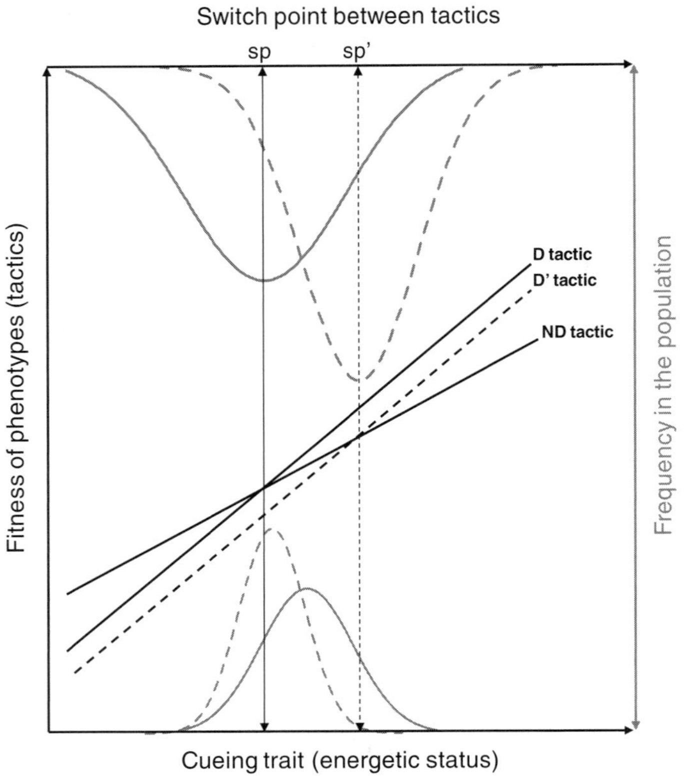

Figure 3.—Conditional ESS model for alternative migratory tactics in the eel (D: diadromy; ND: nondiadromy), modified from Edeline (2007). First, the conditional strategy at the individual level. Consider black solid lines referring to the fitness-maximizing decision in an individual (left-hand vertical axis). Fitness functions of the tactics D and ND are status-dependent (i.e., they depend on the value of the cueing trait—here energetic status). Migration is selected for (i.e., D higher than ND) in an eel with a high energetic status because competition and predation are lower in rivers than in marine and estuarine habitats. In contrast, migration is selected against (i.e., D lower than ND) in an eel with a low energetic status because further energy expense due to migration seriously impairs immediate survival and competitive ability during settlement. The fitness functions of the D and ND tactics intersect at the ESS switchpoint (sp). Both the cueing trait and the switchpoint are polygenic traits. The individual migrates if its energetic status lies above sp but settles if its energetic status decreases below sp (see also Figure 4). Second, the conditional strategy at the population level. Consider gray solid curves referring to trait distributions in the population at equilibrium (right-hand vertical axis). Environmental and genetic variation for the cueing trait and the switchpoint result in normal distributions of both traits in the population. The sp figured here is therefore a mean switchpoint, and the proportion of each phenotype (diadromous versus nondiadromous) in the population is determined by the distribution of energetic status (lower horizontal axis) relative to the distribution of switchpoints (upper horizontal axis). In this example, the mean switchpoint is lower than the mean energetic status, resulting in a majority of diadromous individuals in the population. Variation for the switchpoint is larger than variation for the energetic status in the population, leading to existence of unconditional migrants and unconditional sedentary individuals. Third, anthropogenic change in the equilibrium distributions. Consider dashed lines and curves simulating increased selection against diadromy (i.e., increased selection against low switchpoint and high energetic status). The fitness function of the D tactic is lowered to D', displacing sp to sp' and thus decreasing the proportion of eels migrating to freshwater. At the same time, the distribution of the cueing trait is shifted to the left, further decreasing the proportion of diadromous individuals in the population. Here, we assume that increased selection against migrants overall strengthens stabilizing selection and thus reduces variance in the distributions, but strict directional selection is also possible. Because the surface area below the curve is held constant, the peak value of the curve increases.

ternally programmed (ontogeny) and be responsive to energetic status in order to avoid exhaustion. The conditional ESS model may also apply to foraging movements, provided that the cue is not internal but environmental (Hazel et al. 1990, 2004). The environmental cue may be an abiotic factor. For instance, foragers may disperse to a warmer habitat when temperature drops in winter, and there could be interindividual variation for the sensitivity threshold to temperature drop. The environmental cue may also be a biotic factor, for instance, opponent's body size relative to the size of the focal individual. In such cases of agonistic interaction, one can hypothesize that variation in the switchpoint is related to variation in boldness (i.e., willingness to fight or avoid confrontation). Because foraging is environmentally cued, evolution of foraging behavior may proceed through changes in gene frequencies for the switchpoint only, while evolution of migration may proceed through changes in gene frequencies for both the switchpoint and the cueing trait (see section on changes in evolutionary equilibrium).

Endocrine Control of the Alternative Migratory Tactics

Hormone production is controlled by genes, integrates environmental influences, and regulates gene transcription through complex feedbacks. During diadromous migrations, hormones orchestrate the suite of morphological, physiological, and behavioral adaptations that are functionally connected to form a migratory syndrome. Hence, hormones may be seen as the key mediators of adaptive phenotypic plasticity during eel migration. So far, study of the endocrine control of eel movements has been done only relative to thyroid hormones, but we speculate that growth hormone and cortisol could also be involved in the mediation of the alternative migratory tactics.

Thyroid Hormones

Thyroid hormones (THs; thyroxine T4 and triiodothyronine T3) have a crucial role in controlling leptocephalus metamorphosis (Yamano et al. 1991; Ozaki et al. 2000), body pigmentation, and gut transformations (Vilter 1946; Jegstrup and Rosenkilde 2003). In glass eels and elvers, THs stimulate migratory behavior (Edeline et al. 2004, 2005a; Imbert et al. 2008). Thyroid gland activity decreases at the yellow stage (Callamand and Fontaine 1942), likely promoting settlement during transformation of elvers into small yellow eels. Individual yellow eels caught climbing waterfalls had higher plasma T4 levels and locomotor activity compared to sedentary individuals (Castonguay et al. 1990), suggesting that maintenance of high T4 levels at the yellow stage could delay settlement. Our knowledge of the mechanisms through which THs regulate eel migration remains fragmented. THs activate locomotor activity and could also affect rheotaxis (see Edeline et al. 2005a and references therein). However, it is likely that THs are also involved in the regulation of eel olfactory sensitivity, salinity preference, and muscle physiology, as suggested by results obtained in other teleosts.

Olfaction plays a central role in the orientation of salmon migrations, at least in coastal waters and freshwaters (Døving and Stabell 2002). In smolting coho salmon *Oncorhynchus kisutch*, THs induce cell proliferation in the olfactory epithelium and could thus be involved in the imprinting process (Plate et al. 2002; Lema and Nevitt 2004). Migrating glass eels and elvers orient upstream following increasing gradients of earthy odorants (Sorensen 1986; Tosi and Sola 1993; Sola 1995). High TH levels from metamorphosis to settlement could activate olfactory orientation in the eel. Salinity is also a major orienting signal for migratory fishes. In European glass eels, preference for freshwater is related to high locomotor activity, a behavioral syndrome presumably favoring migration to rivers. High TH levels, together with other hormones, could participate in mediating this link between salinity preference and locomotor activity. Indeed, THs stimulate locomotor activity in glass eels (Edeline et al. 2005a) and influence salinity preference during diadromous migration in Pacific salmon *Oncorhynchus* spp. and threespine sticklebacks *Gasterosteus aculeatus* (Baggerman 1960, 1962). In eels, THs could further reduce somatic growth because they inhibit secretion of growth hormone (GH) (Rousseau et al. 2002), a hormone activating appetite and growth in teleosts (Le Bail and Boeuf 1997; Dufour et al. 2001). Finally, THs have been shown to change muscle physiology towards increased anaerobic swimming capacity in juvenile coho salmon (Katzman and Cech 2001). One can hypothesize that, in the eel also, THs may increase burst swimming capacity and thus favor upstream colonization. Maintenance of a high thyroid gland activity (indicated by high T4 levels and low T3:T4 ratios, see Edeline et al. [2004]) at the yellow stage could favor colonization far upstream and mediate slow growth rate leading to female sex differentiation and to an old age at maturation.

Growth Hormone and Cortisol

The effects of GH and cortisol on eel behavior have not been studied in any detail, but some data suggest that these two hormones may play an important role in the control of eel migration. Both influence locomotor activity in salmonids (Øverli et al. 2002; Johansson et al. 2004) and act synergistically to promote adaptation to saltwater in teleosts (McCormick 2001). In glass eels and elvers, shift from diadromy to nondiadromy may be regulated by a shift from a "TH-dominated" to a "GH/cortisol-dominated" physiology, as controlled by energetic status in the conditional strategy (Figure 4). Indeed, in teleosts, a decreased caloric state lowers thyroid gland activity (Eales 1988). In the eel, fasting and physiological stress activate cortisol and GH secretion (Marchelidon et al. 1996), and cortisol inhibits production of THs (Redding et al. 1986). Finally, as

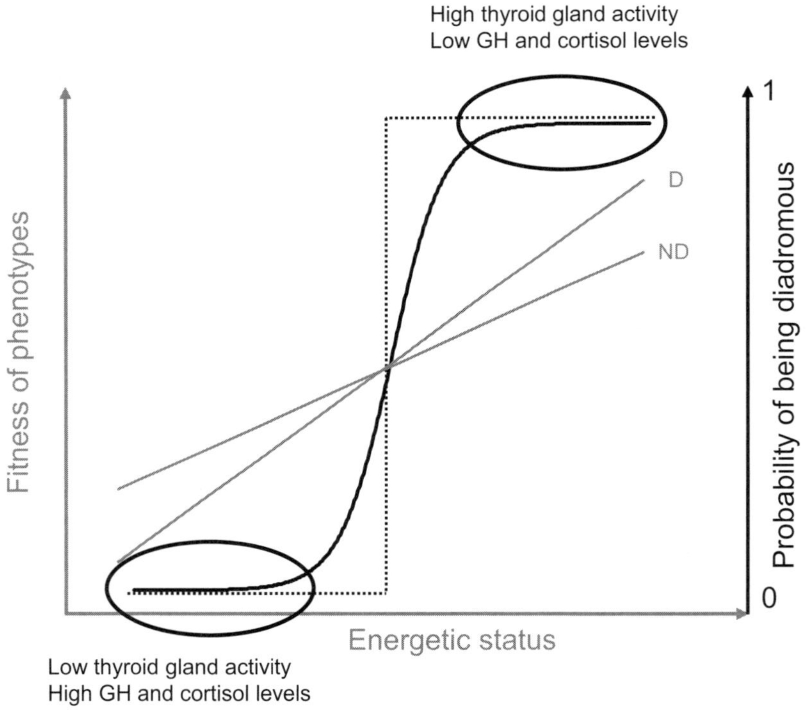

Figure 4.—Speculative endocrine mediation of the alternative migratory tactics in an individual eel. The threshold migratory reaction norm is here represented in two ways. A reaction norm (in dashed black) with abrupt behavioral change from migration to settlement depicts the genetically programmed response (i.e., response in the absence of environmental influence on the switchpoint). However, the reaction norm is more realistically depicted (in solid black) as a logistic function of the cueing trait with 50% probability of being diadromous located on the evolutionary switchpoint because of environmental influences modulating the effects of the genes encoding the switchpoint. Adaptive switch from diadromy to nondiadromy is suggested to be mediated by a switch from a "THs-dominated" to a "GH/cortisol-dominated" physiology. High thyroid gland activity promotes upstream migration and induces low GH levels through a negative feedback. Low THs levels, as induced by a decreased energetic status, favor settlement. At the same time, high GH and cortisol levels could induce preference for saltwater, enhance appetite, and induce aggressive behavior (see text for detailed explanation). THs: thyroid hormones; GH: growth hormone.

mentioned earlier, THs exert a negative feedback on GH production by the pituitary. Hence, shift from upstream migration to settlement in saltwater may be controlled by an endocrine cascade triggered by energetic stress followed by increased cortisol production, decreased thyroid status, and a rise in GH secretion. Field data support this hypothesis. In European glass eels and elvers, decreasing body condition is related to increasing GH secretion (Lambert et al. 2003), and settlement in saltwater is associated with higher T3:T4 ratios, suggesting high GH levels (Edeline et al. 2004) because GH activates the deiodination of T4 into T3 (De Luze and Leloup 1984). The same endocrine pattern indicative of physiological stress (high T3:T4 ratio) is found in dispersing yellow eels (Imbert et al. 2008), suggesting that physiological stress inhibits migration in glass eels and elvers but stimulates dispersal in yellow eels (probably in response to agonistic interactions). In other words, the effects of GH and cortisol on eel behavior could be stage-specific. Further studies are needed to test this hypothesis.

Changes in the Evolutionary Equilibrium

Alterations of selective pressures experienced by eels in continental habitats are expected to modify the evolutionary equilibrium in the conditional strategy (i.e., are expected to change the distributions of the cueing trait and the switchpoint in the population). For instance, anthropogenic pressures like overharvesting, habitat fragmentation and loss, or introduction of species are strong agents of selection that may induce rapid evolution (Stockwell et al. 2003). In the eel, anthropogenic pressures exert more strongly on freshwater habitats than on saltwater habitats and thus select against diadromy. For instance, fisheries for both glass eels and silver eels are located on rivers and estuaries only and thus selectively target the diadromous fraction of the population. Exploitation rates by fisheries may be severe as suggested by the results of Briand et al. (2003) in the Vilaine River estuary where the fishery may deplete up to 99% of the glass eel stock because a dam blocks migrants. Obstructions to migration further induce habitat fragmentation and loss, while silver eels can be killed in turbines during their downstream migration (McCleave 2001; Durif et al. 2003). Additionally, wetland reclamation and drying presumably reduced surface area of these high-value habitats by 50–90% during the past century in Europe (Feunteun 2002). Introduction of strong freshwater predators such as the Wels catfish *Silurus glanis* in western Europe could seriously impair the benefits from colonizing the remaining accessible habitats. Also, prevalence of parasites may be higher in fresh than in saltwater, as observed with the swim bladder nematode *Anguillicola crassus* (Sauvaget et al. 2003). Finally, pollution impacts fitness through decreased individual condition and genetic erosion (Maes et al. 2005), and it probably also disrupts complex behaviors through various endocrine and metabolic pathways (Robinet and Feunteun 2002; Scott and Sloman 2004). We suggest that this increase in selective pressures acting against migrants induced an evolutionary drop in the proportion of migrants through a mechanism presented in Figure 3.

Equilibrium displacement in the ESS results not only from selection against migrants but also from a density-dependent increase in fitness gain from being a saltwater eel. Indeed, global warming affects oceanic gyres, reduces nutrient availability, and slows oceanic migration and could thus increase larval mortality through enhanced starvation and predation and reduced transport (Knights 2003; Friedland et al. 2007). A reduction in the number of recruits reaching the coastline necessarily induces decreased densities in saltwater habitats and thus, by increasing fitness gain from the nondiadromous tactic, induces a further reduction in the proportion of diadromous individuals in the population (Edeline 2007). Anthropogenic changes in evolutionary equilibrium, such as described above, may explain the collapse of eel recruitment to freshwater systems monitored during the last decades in temperate zones of the northern hemisphere (Dekker et al. 2003). If this assumption is correct, then population decline has been less steep in saltwater than in freshwater. Evolutionary game theory predicts that adaptive loss of the diadromous tactic in the conditional strategy may be hindered by frequency-dependent selection (i.e., as the proportion of the diadromous phenotype decreases, its fitness increases). Frequency-dependent selection is likely to occur in the eel due to strong competitive interactions and could then be seen as the only selection pressure preventing the loss of diadromous eels.

General Discussion

With respect to migration and movement, conditional strategies are widespread. For instance, in populations of many bird species, some individuals migrate from the breeding areas while others do not—a conditional strategy (Pulido et al. 1996). In salmonids, the decisions for both maturation and migration are conditional strategies controlled by body size and the amount of surplus energy gained during a critical period (Hazel et al. 1990; Gross 1996; Thorpe et al. 1998; Hutchings 2002; Thériault and Dodson 2003). More generally, dispersal in a variety of invertebrate and vertebrate taxa is often condition-dependent (Ims and Hjermann 2001).

Therefore, we suggest that many of the alternative migratory patterns observed in fishes are conditional strategies. With respect to *Anguilla*, one might consider Gross et al.'s (1988) ideas about increased catadromy in the tropics, where freshwaters are more productive than marine waters, in light of the EES ideas developed in this paper. It would be instructive to compare the salinity and growth histories of tropical eels with temperate eels. Shiao et al. (2003) showed that the giant mottled eel mostly inhabited freshwater portions of rivers, while Japanese eels were spread among marine, estuarine, and freshwater habitats in Taiwan. However, it is unclear whether the contrast in distribution results from alternative migratory behaviors or from interspecific competition after settlement or from both. Behavioral experiments of migratory propensity in temperate versus tropical eels could provide a reliable test for Gross et al.'s (1988) habitat productivity hypothesis. Also, facultatively diadromous eel species should keep higher osmotic capacity at the yellow stage than more strictly diadromous species. Appropriate physiological experiments could investigate this question. In general, much could be learned from comparing life histories of tropical and temperate eel species.

Temperate eels are facing a crisis with strong decline of population numbers, together with a speculated drop in the proportion of the diadromous phenotype among populations. The consequences of such profound changes are not known. Maybe temperate eels have not yet reached evolutionary equilibrium and the proportion of upstream migrants will continue decreasing in the future. One might think that marine-resident eels are enough to prevent the extinctions of these species. Such speculation would be hazardous and risk prone. Further loss of freshwater eels may have serious demographic impacts because freshwater eels, by silvering at a larger size than sea eels, have higher fecundity. A recent genetic study suggests that effective population sizes in European and American eels are surprisingly low, 5,000–10,000 (Wirth and Bernatchez 2003). Smaller populations are generally considered to have reduced potential for adaptive evolution due to reduced genetic variability. Hence, small populations are more sensitive to demographic and environmental stochasticity than larger populations. Estimates of the effective population size necessary for long-term adaptive evolution vary from 500 to 5,000 (Stockwell et al. 2003 and references therein), and therefore, European and American eels are close to dangerous limits. Under strong selective pressure, such as climate change and human-induced selection against migrants, small populations may decline to extinction while larger populations would have adapted to the new prevailing conditions (Figure 3). In all evidence, a precautionary approach is necessary for the management of temperate eel populations. Tropical eels, by relying on freshwater habitats more strongly than do temperate eels, are probably even more fragile and thus need particular attention.

References

Alerstam, T., A. Hedenström, and S. Åkesson. 2003. Long-distance migration: evolution and determinants. Oikos 103:247–260.

Aoyama, J., M. Nishida, and K. Tsukamoto. 2001. Molecular phylogeny and evolution of the freshwater eel, genus *Anguilla*. Molecular Phylogenetics and Evolution 20:459–459.

Arai, T., and T. Hirata. 2006. Differences in the trace element deposition in otoliths between marine- and freshwater-resident Japanese eels, *Anguilla japonica*, as determined by laser ablation ICPMS. Environmental Biology of Fishes 75:173–182.

Arai, T., A. Kotake, P. M. Lokman, M. J. Miller, and K. Tsukamoto. 2004. Evidence of different habitat use by New Zealand freshwater eels *Anguilla australis* and *A. dieffenbachi*, as revealed by otolith microchemistry. Marine Ecology Progress Series 266:213–225.

Arai, T., A. Kotake, P. M. Lokman, and K. Tsukamoto. 2003b. Migratory history and habitat use in New Zealand freshwater eels *Anguilla dieffenbachii* and *A.*

australis, as revealed by otolith microchemistry. Ichthyological Research 50:190–194.

Arai, T., A. Kotake, and T. K. McCarthy. 2006. Habitat use by the European eel *Anguilla anguilla* in Irish waters. Estuarine, Coastal and Shelf Science 67:569–578.

Arai, T., A. Kotake, M. Ohji, M. J. Miller, K. Tsukamoto, and N. Miyazaki. 2003a. Occurrence of sea eels of *Anguilla japonica* along the Sanriku coast of Japan. Ichthyological Research 50:78–81.

Baggerman, B. 1960. Salinity preference, thyroid activity and the seaward migration of four species of Pacific salmon (*Oncorhynchus*). Journal of the Fisheries Research Board of Canada 17:295–322.

Baggerman, B. 1962. Some endocrine aspects of fish migration. General and Comparative Endocrinology Supplement 1:188–205.

Bardonnet, A., S. Dasse, M. Parade, and M. Heland. 2003. Study of glass-eels movements in a flume in relation to nycthemeral changes. Bulletin Français de la Pêche et de la Pisciculture 368:9–19.

Bernatchez, L., and J. J. Dodson. 1987. Relationship between bioenergetics and behavior in anadromous fish populations. Canadian Journal of Fisheries and Aquatic Sciences 44:399–407.

Briand, C., D. Fatin, G. Fontenelle, and E. Feunteun. 2003. Estuarine and fluvial recruitment of the European glass eel, *Anguilla anguilla*, in an exploited Atlantic estuary. Fisheries Management and Ecology 10:377–384.

Bureau du Colombier, S., V. Bolliet, P. Lambert, and A. Bardonnet. 2007. Energy and migratory behavior in glass eels (*Anguilla anguilla*). Physiology and Behavior 92:684–690.

Callamand, O., and M. Fontaine. 1942. L'activité thyroïdienne de l'anguille au cours de son développement. Archives de Zoologie Expérimentale et Générale 82:129–136.

Cairns, D. K., J. C. Shiao, Y. Iizuka, W.-N. Tzeng, and C. D. MacPherson. 2004. Movement patterns of American eels in an impounded watercourse, as indicated by otolith microchemistry. North American Journal of Fisheries Management 24:452–458.

Castonguay, M., J. D. Dutil, C. Audet, and R. Miller. 1990. Locomotor activity and concentration of thyroid hormones in migratory and sedentary juvenile American eels. Transactions of the American Fisheries Society 119:946–956.

Daverat, F., K. E. Limburg, I. Thibault, J.-C. Shiao, J. J. Dodson, F. Caron, W.-N. Tzeng, Y. Iizuka, and H. Wickström. 2006. Phenotypic plasticity of habitat use by three temperate eel species, *Anguilla anguilla*, *A. japonica* and *A. rostrata*. Marine Ecology Progress Series 308:231–241.

Daverat, F., and J. Tomás. 2006. Tactics and demographic attributes in the European eel *Anguilla anguilla* in the Gironde watershed, SW France. Marine Ecology Progress Series 307:247–257.

Daverat, F., J. Tomás, M. Lahaye, M. Palmer, and P. Elie. 2005. Tracking continental habitat shifts of eels using otolith Sr/Ca ratios: validation and application to the coastal, estuarine and riverine eels of the Gironde-Garonne-Dordogne watershed. Marine and Freshwater Research 56:619–627.

Deelder, C. L., and F. W. Tesch. 1970. Heimfindevermögen van Aalen (*Anguilla anguilla*), die über große Entfernungen verpflanzt worden waren. Marine Biology 6:81–92.

De Luze, A., and J. Leloup. 1984. Fish growth hormone enhances peripheral conversion of thyroxine to triiodothyronine in the eel (*Anguilla anguilla* L.). General and Comparative Endocrinology 56:308–312.

Degani, G., and D. Levanon. 1983. The influence of low density on food adaptation, cannibalism and growth of eels (*Anguilla anguilla* (L.)). Bamidgeh 35:53–60.

Dekker, W., J. M. Casselman, D. K. Cairns, K. Tsukamoto, D. Jellyman, and H. Lickers. 2003. Worldwide decline of eel resources necessitates immediate action. Fisheries 28:28–30.

Dingle, H. 2006. Animal migration: is there a common migratory syndrome? Journal of Ornithology 147:212–220.

Døving, K. B., and O. B. Stabell. 2002. Trails in open waters: sensory cues in salmon migration. Pages 39–52 *in* S. P. Collin, and N. J. Marshall, editors. Sensory processing in aquatic environments. Springer, New York.

Dufour, S., K. Rousseau, M. Sbaihi, N. Le Belle, B. Vidal, J. Marchelidon, and M. Schmitz. 2001. Multiple neuroendocrine regulation of growth hormone in the European eel *Anguilla anguilla*. Journal of Taiwan Fisheries Research 9:77–92.

Durif, C., P. Elie, C. Gosset, J. Rives, and F. Travade. 2003. Behavioral study of downstream migrating eels by radio-telemetry at a small hydroelectric power plant. Pages 343–356 *in* D. A. Dixon, editor. Biology, management, and protection of catadromous eels. American Fisheries Society, Symposium 33, Bethesda, Maryland.

Eales, J. G. 1988. The influence of nutritional state on thyroid function in various vertebrates. American Zoologist 28:351–362.

Edeline, E. 2005. Facteurs du contrôle de la dispersion continentale chez l'anguille. Doctoral dissertation. CEMAGREF/University of Toulouse III, Toulouse, France.

Edeline, E. 2007. Adaptive phenotypic plasticity of eel diadromy. Marine Ecology Progress Series 341:229–232.

Edeline, E., A. Bardonnet, V. Bolliet, S. Dufour, and P. Elie. 2005a. Endocrine control of *Anguilla anguilla* glass eel dispersal: effect of thyroid hormones on locomotor activity and rheotactic behavior. Hormones and Behavior 48:53–63.

Edeline, E., L. Beaulaton, R. Le Barh, and P. Elie. 2007. Dispersal in metamorphosing *Anguilla anguilla* juvenile eel. Marine Ecology Progress Series 344:213–218.

Edeline, E., S. Dufour, C. Briand, D. Fatin, and P. Elie. 2004. Thyroid status is related to migratory behavior in *Anguilla anguilla* glass eels. Marine Ecology Progress Series 282:261–270.

Edeline, E., S. Dufour, and P. Elie. 2005b. Role of glass eel salinity preference in the control of habitat selection and growth plasticity in *Anguilla anguilla*. Marine Ecology Progress Series 304:191–199.

Edeline, E., S. Dufour, and P. Elie. 2009. Proximate and ultimate control of eel continental dispersal. Pages 433–461 *in* G. van den Thillart, S. Dufour, and J. C. Rankin, editors. Spawning migration of the European eel. Springer Netherlands, Dordrecht.

Edeline, E., P. Lambert, C. Rigaud, and P. Elie. 2006. Effects of body condition and water temperature on *Anguilla anguilla* glass eels migratory behavior. Journal of Experimental Marine Biology and Ecology 331:217–225.

Elie, P., R. Lecomte-Finiger, I. Cantrelle, and N. Charlon. 1982. Définition des limites des différents stades pigmentaires durant la phase civelle d'*Anguilla anguilla* L. (poisson téléostéen anguilliforme). Vie et Milieu 32:149–157.

Fablet, R., F. Daverat, and H. de Pontual. 2007. Unsupervised Bayesian reconstruction of individual life histories from otolith signatures: case study of Sr:Ca transects of European eel (*Anguilla anguilla*) otoliths. Canadian Journal of Fisheries and Aquatic Sciences 64:125–165.

Fernandez-Delgado, C., J. A. Hernando, M. Herrera, and M. Bellido. 1989. Age and growth of yellow eels, *Anguilla anguilla*, in the estuary of the Guadalquivir River (south-west Spain). Journal of Fish Biology 34:561–570.

Feunteun, E. 2002. Management and restoration of European eel population (*Anguilla anguilla*): an impossible bargain. Ecological Engineering 18:575–591.

Feunteun, E., P. Laffaille, T. Robinet, C. Briand, A. Baisez, J. M. Olivier, and A. Acou. 2003. A review of upstream migration and movements in inland waters by anguillid eels. Towards a general theory. Pages 191–213 *in* K. Aida, K. Tsukamoto, and K. Yamauchi, editors. Eel biology. Springer, Tokyo.

Fretwell, S. D., and H. L. Lucas. 1970. On territorial behavior and other factors influencing habitat distribution in birds. Acta Biotheoretica 19:16–36.

Friedland, K. D., M. J. Miller, and B. Knights. 2007. Oceanic changes in the Sargasso Sea and declines in recruitment of the European eel. ICES Journal of Marine Science 64:519–530.

Gandon, S., and Y. Michalakis. 2001. Multiple causes of the evolution of dispersal. Pages 155–167 *in* J. Clobert, E. Danchin, A. A. Dhondt, and J. D. Nichols, editors. Dispersal. Oxford University Press, Oxford, UK.

Gemplerine, P. J., R. A. Rulifson, and L. Paramore. 2002. Multi-way analysis of trace elements in fish otoliths to rack migratory patterns. Chemometrics and Intelligent Laboratory Systems 60:135–146.

Gross, M. R. 1996. Alternative reproductive strategies and tactics: diversity within sexes. Trends in Ecology and Evolution 11:92–98.

Gross, M. R., R. M. Coleman, and R. M. McDowall. 1988. Aquatic productivity and the evolution of diadromous fish migration. Science 139:1291–1293.

Harrod, C., J. Grey, T. K. McCarthy, and M. Morrissey. 2005. Stable isotope analyses provide new insights into ecological plasticity in a mixohaline population of European eel. Oecologia 144:673–683.

Hazel, W., and R. Smock. 2000. Inheritance in the conditional strategy revisited. Journal of Theoretical Biology 204:307–309.

Hazel, W. N., R. Smock, and M. D. Johnson. 1990. A polygenic model for the evolution and maintenance of conditional strategies. Proceedings of the Royal Society B 242:181–187.

Hazel, W., R. Smock, and C. M. Lively. 2004. The ecological genetics of conditional strategies. American Naturalist 163:888–900.

Hutchings, J. A. 2002. Life histories of fish. Pages 149–174 *in* P. J. B. Hart and J. D. Reynolds, editors. Handbook of fish biology and fisheries, volume 1, fish biology. Blackwell Scientific Publications, Malden, Massachusetts.

Ibbotson, A., J. Smith, P. Scarlett, and M. W. Aprahamian. 2002. Colonisation of freshwater habitats by the European eel *Anguilla anguilla*. Freshwater Biology 47:1696–1706.

Imbert, H., R. Arrowsmith, S. Dufour, and P. Elie. 2008. Relationships between locomotor behavior, morphometric characters and thyroid hormone levels give evidence of stage-dependent mechanisms in European eel upstream migration. Hormones and Behavior 53:69–81.

Ims, R. A., and D. O. Hjermann. 2001. Condition-dependent dispersal. Pages 203–216 *in* J. Clobert, E. Danchin, A. A. Dhondt, and J. D. Nichols, editors. Dispersal. Oxford University Press, Oxford, UK.

Jegstrup, I. M., and P. Rosenkilde. 2003. Regulation of post-larval development in the European eel: thyroid hormone level, progress of pigmentation and changes in behaviour. Journal of Fish Biology 63:168–175.

Jellyman, D. J. 1997. Variability in growth rates of freshwater eels (*Anguilla* spp.) in New Zealand. Ecology of Freshwater Fish 6:108–115.

Jessop, B. M., J. C. Shiao, Y. Iizuka, and W. N. Tzeng. 2004. Variation in the annual growth rate, by sex and migration history, of silver American eels *Anguilla rostrata*. Marine Ecology Progress Series 272:231–244.

Jessop, B. M., J.-C. Shiao, Y. Iizuka, and W.-N. Tzeng. 2002. Migratory behaviour and habitat use by American eels *Anguilla rostrata* as revealed by otolith microchemistry. Marine Ecology Progress Series 233:217–229.

Jessop, B. M., J.-C. Shiao, Y. Iizuka, and W.-N. Tzeng. 2006. Migration of juvenile American eels *Anguilla rostrata* between freshwater and estuary, as revealed by otolith microchemistry. Marine Ecology Progress Series 310:219–233.

Johansson, V., S. Winberg, E. Jönsson, D. Hall, and B. T. Björnsson. 2004. Peripherally administered growth hormone increases brain dopaminergic activity and swimming in rainbow trout. Hormones and Behavior 46:436–443.

Jonsson, B., and N. Jonsson. 1993. Partial migration: niche shift versus sexual maturation in fishes. Reviews in Fish Biology and Fisheries 3:348–365.

Kalish, J. M. 1990. Use of otolith microchemistry to distinguish progeny of sympatric anadromous and non-anadromous salmonids. Fishery Bulletin 88:657–666.

Katzman, S., and J. J. Cech. 2001. Juvenile coho salmon locomotion and mosaic muscle are modified by 3',3',5'-tri-iodo-l-thyronine (T(3)). Journal of Experimental Biology 204:1711–1717.

Kerr, L. A., D. H. Secor, and R. T. Kraus. 2007. Stable isotope ($\delta^{13}C$ and $\delta^{18}O$) and Sr/Ca composition of otoliths as proxies for environmental salinity experienced by an estuarine fish. Marine Ecology Progress Series 349:245–253.

Knights, B. 1987. Agonistic behaviour and growth in the European eel, *Anguilla anguilla* L., in relation to warm-water aquaculture. Journal of Fish Biology 31:265–276.

Knights, B. 2003. A review of the possible impacts of long-term oceanic and climate changes and fishing mortality on recruitment of anguillid eels of the northern hemisphere. Science of the Total Environment 310:237–244.

Kotake, A., T. Arai, M. Ohji, S. Yamane, N. Miyazaki, and K. Tsukamoto. 2004. Application of otolith microchemistry to estimate the migratory history of Japanese eel *Anguilla japonica* on the Sanriku coast of Japan. Journal of Applied Ichthyology 20:150–153.

Kotake, A., T. Arai, T. Ozawa, S. Nojima, M. J. Miller, and K. Tsukamoto. 2003. Variation in migratory history of Japanese eels, *Anguilla japonica*, collected in coastal waters of the Amakusa Islands, Japan, inferred from otolith Sr/Ca ratios. Marine Biology 142:849–854.

Kotake, A., A. Okamura, Y. Yamada, T. Utoh, T. Arai, M. J. Miller, H. P. Oka, and K. Tsukamoto. 2005. Seasonal variation in the migratory history of the Japanese eel *Anguilla japonica* in Mikawa Bay, Japan. Marine Ecology Progress Series 293:213–221.

Kraus, R. T., and D. H. Secor. 2003. Response of otolith Sr:Ca to a manipulated environment in young American eels. Pages 79–85 *in* D. A. Dixon, editor. Biology, management, and protection of catadromous eels. American Fisheries Society, Symposium 33, Bethesda, Maryland.

Kraus, R. T., and D. H. Secor. 2004. Incorporation of strontium into otoliths of an estuarine fish. Journal of Experimental Marine Biology and Ecology 302:85–106.

Lambert, P., M. Sbaihi, E. Rochard, J. Marchelidon, S. Dufour, and P. Elie. 2003. Morphological and growth hormone variabilities of European glass-eel (*Anguilla anguilla*) in Gironde estuary during 1997–1998 season. Bulletin Français de la Pêche et de la Pisciculture 368:69–83.

Lamson, H. M., J.-C. Shiao, Y. Iizuka, W.-N. Tzeng, and D. K. Cairns. 2006. Movement patterns of American eels (*Anguilla rostrata*) between salt- and freshwater in a coastal watershed, based on otolith microchemistry. Marine Biology 149:1567–1576.

Le Bail, P. Y., and G. Boeuf. 1997. What hormones may regulate food intake in fish? Aquatic Living Resources 10:371–379.

Lema, S. C., and G. A. Nevitt. 2004. Evidence that thyroid hormone induces olfactory cellular proliferation in salmon during a sensitive period for imprinting. Journal of Experimental Biology 207:3317–3327.

Limburg, K. E. 1995. Otolith strontium traces environmental history of subyearling American shad *Alosa sapidissima*. Marine Ecology Progress Series 119:25–35.

Limburg, K. E. 1998. Anomalous migrations of anadromous herrings revealed with chemical tracers. Canadian Journal of Fisheries and Aquatic Sciences 55:431–437.

Limburg, K. E., H. Wickström, H. Svedäng, M. Elfman, and P. Kristiansson. 2003. Do stocked freshwater eels migrate? Evidence from the Baltic suggests "yes." Pages 275–284 *in* D. A. Dixon, editor. Biology, management, and protection of catadromous eels. American Fisheries Society, Symposium 33, Bethesda, Maryland.

Maes, G. E., J. A. M. Raeymaekers, C. Pampoulie, A. Seynaeve, G. Goemans, C. Belpaire, and F. A. Volckaert. 2005. The catadromous European eel *Anguilla anguilla* (L.) as a model for freshwater evolutionary ecotoxicology: relationship between heavy metal bioaccumulation, condition and genetic variability. Aquatic Toxicology 73:99–114.

Marchelidon, J., M. Schmitz, L. M. Houdebine, B. Vidal, N. Le Belle, and S. Dufour. 1996. Development of a radioimmunoassay for European eel growth hormone and application to the study of silvering and experimental fasting. General and Comparative Endocrinology 102:360–369.

Matthews, M. A., D. W. Evans, C. A. McClintock, and C. Moriarty. 2003. Age, growth, and catch-related data of yellow eel *Anguilla anguilla* (L.) from lakes of the Erne catchment, Ireland. Pages 207–215 in D. A. Dixon, editor. Biology, management, and protection of catadromous eels. American Fisheries Society, Symposium 33, Bethesda, Maryland.

McCleave, J. D. 2001. Simulation of the impact of dams and fishing weirs on reproductive potential of silver-phase American eels in the Kennebec River basin, Maine. North American Journal of Fisheries Management 21:592–605.

McCormick, S. D. 2001. Endocrine control of osmoregulation in teleost fish. American Zoologist 41:781–794.

McCulloch, M., M. Cappo, and M. Aumend, W. 2005. Tracing the life history of individual barramundi using laser ablation MC-ICP-MS Sr-isotopic and Sr/BA ratios in otoliths. Marine and Freshwarer Research 56:637–644.

McDowall, R. M. 1988. Diadromy in fishes: migrations between freshwater and marine environments. Croom-Helm, London.

McDowall, R. M. 1997. The evolution of diadromy in fishes (revisited) and its place in phylogenetic analysis. Reviews in Fish Biology and Fisheries 7:443–462.

McDowall, R. M. 2004. Ancestry and amphidromy in island freshwater fish faunas. Fish and Fisheries 5:75–85.

McDowall, R. M. 2007. On amphidromy, a distinct form of diadromy in aquatic organisms. Fish and Fisheries 8:1–13.

McDowall, R. M. 2009. Making the best of two worlds: diadromy in the evolution, ecology, and conservation of aquatic organisms. Pages 1–22 *in* A. J. Haro, K. L. Smith, R. A. Rulifson, C. M. Moffitt, R. J. Klauda, M. J. Dadswell, R. A. Cunjak, J. E. Cooper, K. L. Beal, and T. S. Avery, editors. Challenges for diadromous fishes in a dynamic global environment. American Fisheries Society, Symposium 69, Bethesda, Maryland.

Metcalfe, J., G. P. Arnold, and R. M. McDowall. 2002. Migration. Pages 175–199 *in* P. J. B. Hart and J. D. Reynolds, editors. Handbook of fish biology and fisheries, volume 1, fish biology. Blackwell Scientific Publications, Oxford, UK.

Moriarty, C. 1986. Riverine migrations of young eels *Anguilla anguilla* (L.). Fisheries Research 4:43–58.

Moriarty, C. 1987. Factors influencing recruitment of the Atlantic species of anguillid eels. Pages 483–491 *in* M. J. Dadswell, R. J. Klauda, C. M. Moffitt, R. L. Saunders, R. A. Rulifson, and J. E. Cooper, editors. Common strategies of anadromous and catadromous fishes. American Fisheries Society, Symposium 1, Bethesda, Maryland.

Moriarty, C. 2003. The yellow eel. Pages 89–105 *in* K. Aida, K. Tsukamoto, and K. Yamauchi, editors. Eel biology. Springer, Tokyo.

Morrison, W. E., and D. H. Secor. 2003. Demographic attributes of yellow-phase American eels (*Anguilla rostrata*) in the Hudson River estuary. Canadian Journal of Fisheries and Aquatic Sciences 60:1487–1501.

Morrison, W. E., D. H. Secor, and P. M. Piccoli. 2003. Estuarine habitat use by Hudson River American eels as determined by otolith strontium:calcium ratios. Pages 87–99 *in* D. A. Dixon, editor. Biology, management, and protection of catadromous eels. American Fisheries Society, Symposium 33, Bethesda, Maryland.

Myers, G. S. 1949. Usage of anadromous, catadromous and allied terms for migratory fishes. Copeia 1949:89–97.

Naismith, I. A., and B. Knights. 1993. The distribution, density and growth of the European eel, *Anguilla anguilla*, in the freshwater catchment of the River Thames. Journal of Fish Biology 42:217–226.

Øverli, Ø., S. Kotzian, and S. Winberg. 2002. Effects of cortisol on aggression and locomotor activity in rainbow trout. Hormones and Behavior 42:53–61.

Owen, S. F. 2001. Meeting energy budgets by modulation of behaviour and physiology in the eel (*Anguilla anguilla* L.). Comparative Biochemistry and Physiology Part A: Molecular & Integrative Physiology 128:629–642.

Ozaki, Y., H. Okamura, Y. Kazeto, T. Ikeuchi, S. Ijiri, M. Nagae, S. Adachi, and K. Yamauchi. 2000. Developmental changes in pituitary-thyroid axis, and formation of gonads in leptocephali and glass eels of *Anguilla* spp. Fisheries Science 66:1115–1122.

Parker, S. J. 1995. Homing ability and home range of yellow-phase American eels in a tidally dominated estuary. Journal of the Marine Biological Association of the UK 75:127–140.

Peters, G., H. Delventhal, and H. Klinger. 1980. Physiological and morphological effects of social stress in the eel, (*Anguilla anguilla* L.). Archiv für Fischereiwissenschaft 30:157–180.

Plate, E. M., B. A. Adams, W. T. Allison, G. Martens, C. W. Hawryshyn, and J. G. Eales. 2002. The effects of thyroxine or a GnRH analogue on thyroid hormone deiodination in the olfactory epithelium and retina of rainbow trout, *Oncorhynchus mykiss*, and sockeye salmon, *Oncorhynchus nerka*. General and Comparative Endocrinology 127:59–65.

Pulido, F., P. Berthold, and A. J. Van Noordwijk. 1996.

Frequency of migrants and migratory activity are genetically correlated in a bird population: evolutionary implications. Proceedings of the National Academy of Sciences of the United States of America 93:14642–14647.

Radtke, R. L., D. W. Townsend, S. D. Folsom, and M. A. Morrison. 1990. Strontium:calcium concentratiuon ratios in otoliths of herring larvae as indicators of environmental histories. Environmental Biology of Fishes 27:51–61.

Redding, J. M., A. Deluze, J. Leloup-Hâtey, and J. Leloup. 1986. Suppression of plasma thyroid hormone concentrations by cortisol in the European eel *Anguilla anguilla*. Comparative Biochemistry and Physiology Part A: Physiology 83:409–413.

Robinet, T., and E. Feunteun. 2002. Sublethal effects of exposure to chemical compounds: a cause for the decline in Atlantic eels? Ecotoxicology 11:265–277.

Roff, D. A. 1996. The evolution of threshold traits in animals. Quaterly Review of Biology 71:3–35.

Rousseau, K., N. Le Belle, M. Sbaihi, J. Marchelidon, M. Schmitz, and S. Dufour. 2002. Evidence for a negative feedback in the control of eel growth hormone by thyroid hormones. Journal of Endocrinology 175:605–613.

Sauvaget, B., D. Fatin, and C. Briand. 2003. Contamination by *Anguillicola crassus* of five coastal eel populations (*Anguilla anguilla*) in Brittany (France). Bulletin Français de la Pêche et de la Pisciculture 368:21–26.

Schlichting, C. D., and M. Pigliucci. 1995. Gene regulation, quantitative genetics and the evolution of reaction norms. Evolutionary Ecology 9:154–168.

Scott, G. R., and K. A. Sloman. 2004. The effects of environmental pollutants on complex fish behaviour: integrating behavioural and physiological indicators of toxicity. Aquatic Toxicology 68:369–392.

Secor, D. H., A. Henderson-Arzapalo, and P. M. Piccoli. 1995. Can otolith microchemistry chart patterns of migration and habitat utilization in anadromous fishes? Journal of Experimental Marine Biology and Ecology 192:15–23.

Secor, D. H., and P. M. Piccoli. 1996. Age- and sex-dependent migrations of striped bass in the Hudson River as determined by chemical microanalysis of otoliths. Estuaries 19:778–793.

Secor, D. H., and J. R. Rooker. 2000. Is otolith strontium a useful scalar of life cycles in estuarine fishes. Fisheries Research 46:359–371.

Shiao, J. C., Y. Iizuka, C. W. Chang, and W. N. Tzeng. 2003. Disparities in habitat use and migratory behavior between tropical eel *Anguilla marmorata* and temperate eel *A. japonica* in four Taiwanese rivers. Marine Ecology Progress Series 261:233–242.

Shiao, J. C., L. Ložys, Y. Iizuka, and W. N. Tzeng. 2006. Migratory patterns and contributions of stocking to the population of European eel in Lithuanian waters as indicated by otolith Sr:Ca ratios. Journal of Fish Biology 69:749–769.

Smogor, R. A., P. L. Angermeier, and C. K. Gaylord. 1995. Distribution and abundance of American eels in Virginia streams: test of null models across spatial scales. Transactions of the American Fisheries Society 124:789–803.

Sola, C. 1995. Chemoattraction of upstream migrating glass eels *Anguilla anguilla* to earthy and green odorants. Environmental Biology of Fishes 43:179–185.

Sorensen, P. W. 1986. Origins of the freshwater attractant(s) of migrating elvers of the American eel, *Anguilla rostrata*. Environmental Biology of Fishes 17:185–200.

Stockwell, C. A., A. P. Hendry, and M. T. Kinnison. 2003. Contemporary evolution meets conservation biology. Trends in Ecology and Evolution 18:94–101.

Surge, D. M., and K. C. Lohmann. 2002. Temporal and spatial differences in salinity and water chemistry in SW Florida estuaries: effects of human-impacted watersheds. Estuaries 25:393–408.

Sutherland, W. J., J. A. Gill, and K. Norris. 2002. Density-dependent dispersal in animals: concepts, evidence, mechanisms and consequences. Pages 134–151 *in* J. M. Bollock, R. E. Kenward, and R. S. Hails, editors. Dispersal ecology. Blackwell Scientific Publications, Oxford, UK.

Tesch, F. W. 1977. The eel. Chapman and Hall, London.

Thériault, V., and J. J. Dodson. 2003. Body size and the adoption of a migratory tactic in brook charr. Journal of Fish Biology 63:1144–1159.

Thibault, I., J. J. Dodson, and F. Caron. 2007a. Yellow stage American eel movements determined by microtagging and acoustic telemetry in the St. Jean River watershed, Gaspé (Quebec, Canada). Journal of Fish Biology 71:1095–1112.

Thibault, I., J. J. Dodson, F. Caron, W.-N. Tzeng, Y. Iizuka, and J.-C. Shiao. 2007b. Facultative catadromy in American eels: testing the conditional strategy hypothesis. Marine Ecology Progress Series 344:219–229.

Thorpe, J. E., M. Mangel, N. B. Metcalfe, and F. A. Huntingford. 1998. Modelling the proximate basis of salmonid life-history variation, with application to Atlantic salmon, *Salmo salar* L. Evolutionary Ecology 12:581–599.

Tosi, L., and C. Sola. 1993. Role of geosmin, a typical inland water odour, in guiding glass eel *Anguilla anguilla* (L.) migration. Ethology 95:177–195.

Tsukamoto, K., and T. Arai. 2001. Facultative catadromy of the eel *Anguilla japonica* between freshwater and seawater habitats. Marine Ecology Progress Series 220:265–276.

Tsukamoto, K., I. Nakai, and F.-W. Tesch. 1998. Do all freshwater eels migrate? Nature 396:635.

Tzeng, W.-N. 1996. Effects of salinity and ontogenetic movements on strontium:calcium ratios in the otoliths of the Japanese eel, *Anguilla japonica* Temminck and Schlegel. Journal of Experimental Marine Biology and Ecology 199:111–122.

Tzeng, W.-N., C.-W. Chang, C.-H. Wang, J.-C. Shiao, Y. Iizuka, Y.-J. Yang, C.-F. You, and L. Ložys. 2007. Misidentification of the migratory history of anguillid eels by Sr/Ca ratios of vaterite otoliths. Marine Ecology Progress Series 348:285–295.

Tzeng, W.-N., Y. Iizuka, J.-C. Shiao, Y. Yamada, and H. P. Oka. 2003a. Identification and growth rates comparison of divergent migratory contingents of Japanese eel (*Anguilla japonica*). Aquaculture 216:77–86.

Tzeng, W. N., K. P. Severin, and H. Wickström. 1997. Use of otolith microchemistry to investigate the environmental history of European eel *Anguilla anguilla*. Marine Ecology Progress Series 149:73–81.

Tzeng, W. N., J. C. Shiao, and Y. Iizuka. 2002. Use of otolith Sr:Ca ratios to study the riverine migratory behaviors of Japanese eel *Anguilla japonica*. Marine Ecology Progress Series 245:213–221.

Tzeng, W.-N., J.-C. Shiao, Y. Yamada, and H. P. Oka. 2003b. Life history patterns of Japanese eel *Anguilla japonica* in Mikawa Bay, Japan. Pages 285–293 *in* D. A. Dixon, editor. Biology, management, and protection of catadromous eels. American Fisheries Society, Symposium 33, Bethesda, Maryland.

Tzeng, W. N., C. H. Wang, H. Wickström, and M. Reizenstein. 2000. Occurrence of the semi-catadromous European eel *Anguilla anguilla* in the Baltic Sea. Marine Biology 137:93–98.

Vilter, V. 1946. Action de la thyroxine sur la métamorphose larvaire de l'anguille. Comptes Rendus des Séances de la Société de Biologie et de ses Filiales 140:783–785.

Walsh, C. T., B. C. Pease, S. D. Hoyle, and D. J. Booth. 2006. Variability in growth of longfinned eels among coastal catchments of south-eastern Australia. Journal of Fish Biology 68:1693–1706.

Weisser, W. W. 2001. The effects of predation on dispersal. Pages 180–188 *in* J. Clobert, E. Danchin, A. A. Dhondt, and J. D. Nichols, editors. Dispersal. Oxford University Press, Oxford, UK.

Wirth, T., and L. Bernatchez. 2003. Decline of North Atlantic eels: a fatal synergy? Proceedings of the Royal Society B 270:681–688.

Yamano, K., M. Tagawa, E. G. De Jesus, T. Hirano, S. Miwa, and Y. Inui. 1991. Changes in whole body concentrations of thyroid-hormones and cortisol in metamorphosing conger eel. Journal of Comparative Physiology, B: Biochemical, Systematic and Environmental Physiology 161:371–375.

Zimmerman, C. E. 2005. Relationship of otolith strontium-to-calcium ratios and salinity: experimental validation for juvenile salmonids. Canadian Journal of Fisheries and Aquatic Sciences 62:88–97.

Appendix 1: Literature on Habitat Patterns in *Anguilla*

Table A.1.—Patterns of habitat use by natural stocks of six species of *Anguilla* and literature documenting them. Habitat patterns and terminology are our interpretations and may differ from those of the authors. Cat = catadromous (freshwater residents); est = estuarine or brackish-water residents; mar = marine residents; down = downstream movers after freshwater or estuarine residence; up = upstream movers after marine or estuarine residence; b-f = multiple movements between habitats of differing salinity; seas = annual movements between habitats of differing salinity; mover = direction and frequency of habitat shift not determined. NA = not applicable. All studies were based on transects of Sr:Ca across otoliths, except three as footnoted.

Species	Habitat patterns	Location	References
European eel	est[a]	Sweden	Tzeng et al. (1997)
	est[a], up	Baltic Sea	Tzeng et al. (2000)
	cat, est[a], mar, up	Baltic Sea, North Sea	Limburg et al. (2003)
	cat, est, mar, down, up, seas	France	Daverat et al. (2005)
	cat, est, mar, mover[b]	Ireland	Harrod et al. (2005)
	cat, mar, b-f	Ireland	Arai et al. (2006)
	cat, est, mar, down, up, b-f	France	Daverat and Tomás (2006)

Table A.1.—Continued.

Species	Habitat patterns	Location	References
European eel	est, down, up, b-f, seas	Lithuania, Baltic Sea	Shiao et al. (2006)
	37 patterns	NA	Fablet et al. (2007)
Australian shortfin eel	est, up	New Zealand	Arai et al. (2003b)
	cat, est, up, b-f	New Zealand	Arai et al. (2004)
New Zealand longfin eel	cat, b-f	New Zealand	Arai et al. (2003b)
	cat, est, up, b-f	New Zealand	Arai et al. (2004)
Japanese eel	cat, mar	Japan, East China Sea	Tsukamoto et al. (1998)
	cat, est, mar, down, b-f	Japan	Tsukamoto and Arai (2001)
	cat, est, mar, down	Taiwan	Tzeng et al. (2002)
	mar, down, b-f	Japan	Arai et al. (2003a)
	cat, est, mar, down	Japan	Kotake et al. (2003)
	cat, mar, down	Taiwan	Shiao et al. (2003)
	cat, mar, down	China, Japan, Taiwan	Tzeng et al. (2003a)
	cat, mar, down, b-f	Japan	Tzeng et al. (2003b)
	mar, down	Japan	Kotake et al. (2004)
	cat, est, mar	Japan	Kotake et al. (2005)
	cat, mar[c]	Japan	Arai and Hirata (2006)
	cat, mar, down, up, b-f	China, Japan, Taiwan	Daverat et al. (2006)
Giant mottled eel	cat, est	Taiwan	Shiao et al (2003)
American eel	cat, est, b-f[d]	Canada	Jessop et al. (2002)
	cat, up, b-f[e]	USA	Jessop et al. (2002, 2004)
	cat, est, down	USA	Morrison et al. (2003)
	cat, est, down	Canada	Cairns et al. (2004)
	cat, b-f	Canada	Jessop et al. (2006)
	cat, est, mar, b-f	Canada	Lamson et al. (2006)
	cat, est, down, up[f]	Canada	Thibault et al. (2007a)
	cat, est, down, up, b-f	Canada	Thibault et al. (2007b)

[a] Brackish waters of the Baltic Sea.
[b] Used $\delta^{13}O$, δ^{15}, and C:N in muscle tissue.
[c] Used Ba:Ca, Cr:Ca, Mg:Ca, Mn:Ca, and Sr:Ca.
[d] All $\leq$ 4 years old.
[e] Silver eels $\geq$ 10 years old.
[f] Used a 3-year tagging and 1-year telemetry study.

American Fisheries Society Symposium 69:121–134, 2009

Variability in Diadromous Behavior of a Northern Coregonid Based on Scanning Proton Microprobe Analysis of Otolith Strontium

Kimberly L. Howland* and John A. Babaluk
Department of Fisheries and Oceans
501 University Crescent, Winnipeg, Manitoba R3T 2N6, Canada

Doug Chiperzak
Department of Fisheries and Oceans
200 Kent Street, Ottawa, Ontario K1A 0E6, Canada

Ross F. Tallman
Department of Fisheries and Oceans
501 University Crescent, Winnipeg, Manitoba R3T 2N6, Canada

George Low
Department of Fisheries and Oceans
42043 Mackenzie Highway, Hay River, Northwest Territories X0E 0R9, Canada

Abstract.—Several coregonid species in northern latitudes exhibit diadromous behavior and may share river/lake systems with conspecifics who are freshwater residents. Although both life history strategies have been reported in a variety of systems, little information on the extent and variability in diadromous behavior both within and among populations exists. We addressed this issue using inconnu *Stenodus leucichthys* from the Mackenzie River drainage (Northwest Territories, Canada) as a model. At the extremes of their distribution within this system, inconnu have been characterized by alternative life history strategies. Those in the north exhibit diadromous behavior, while those in the south demonstrate life history closure within the upper reaches of the system (i.e., freshwater residency). We examined the otolith strontium profiles of inconnu collected from 12 regions throughout the Mackenzie watershed to (1) determine the geographical range over which diadromy occurs in this system, (2) characterize the variability in diadromous behavior within and among populations, and (3) identify physical/environmental factors that may influence choice between a diadromous or freshwater life history. Our results demonstrate that the degree of diadromy within populations of inconnu tends to gradually decrease with distance from marine waters, rather than showing an abrupt transition from diadromy to freshwater residency. We conclude that distance from a marine food source plays an important role in determining whether fish remain in freshwater or migrate to the sea to feed. Within diadromous populations of inconnu, there is a substantial level of flexibility in the extent and frequency of marine migration by individuals.

Introduction

Diadromy is common for many northern coregonids. However, the degree to which species exhibit this behavior varies considerably. For example, in North America, the Arctic cisco *Coregonus autumnalis* and Bering cisco *C. laurettae* appear to be fully anadromous (Dillinger et al. 1992; Brown et al. 2007), while species such as inconnu *Stenodus leucichthys*, broad whitefish *C. nasus*, lake whitefish

* Corresponding author: kimberly.howland@dfo-mpo.gc.ca

C. clupeaformis, and least cisco *C. sardinella* vary in the extent to which they utilize saltwater habitat and typically have freshwater and anadromous populations (Reist and Bond 1988; Howland et al. 2001a, 2001b; Tallman et al. 2002). In contrast, round whitefish *Prosopium cylindraceum* are rarely reported from coastal collections and appear to occur mainly in freshwater (e.g., Craig 1984; Lawrence et al. 1984; Brown et al. 2007). The co-occurrence of both anadromous and freshwater populations within river systems has been documented for a number of these species in both North America and Eurasia (Berg 1948; Nikol'skii 1954; Behnke 1972). Although both life history strategies have been reported for northern coregonids in a variety of systems, there is little information on the extent and variability in diadromous behavior both within and among populations.

We addressed this question using inconnu in the Mackenzie River drainage. The inconnu is a highly migratory species and has both anadromous and freshwater populations throughout its range in North America and Siberia (Berg 1948; Alt 1977; Brown 2000; Howland et al. 2001b). The anadromous form typically spawns in the headwaters of large rivers, and young are thought to move downstream with the spring freshet (Berg 1948; Reist and Bond 1988). After spending the early part of their lives in freshwater, anadromous inconnu typically spend the summer months feeding in estuarine and nearshore coastal areas and regularly move back into freshwater to overwinter (most likely in deeper areas of rivers and deltas) (Berg 1948; Reist and Bond 1988; Howland et al. 2001b). Freshwater forms may also be migratory, moving shorter distances between rivers and lakes (adfluvial) or within rivers only (Howland et al. 2000). In the Canadian Arctic, they form an important component of both aboriginal subsistence and commercial fisheries. Within the Mackenzie River system, the distribution of inconnu spans several land claim areas. Given the highly migratory nature of this species, particularly the anadromous form, there are management concerns among resource users, with the potential for transboundary movements across different land claim areas necessitating cooperation by multiple stakeholders. Thus, there is a need to document the extent and nature of anadromous migrations by harvested inconnu populations.

Historical tag and recapture data from various studies conducted in the Mackenzie River system over the past 30 years suggest three possible types of inconnu in the Mackenzie River System: anadromous (lower Mackenzie River), freshwater adfluvial (Great Slave Lake area), and freshwater riverine (upper Mackenzie River) (Figure 1; Howland et al. 2000). Recent genetic work and radio telemetry provide support for a separation between inconnu in the Great Slave Lake area and lower Mackenzie (Howland et al. 2000; Howland 2005), and initial otolith microchemistry work has confirmed that inconnu at the extreme southern extent of the river system (Great Slave Lake area) are freshwater, whereas those at the northern end of the system (closer to the sea) are primarily anadromous (Howland et al. 2001b).

Although there are clear differences in the life histories of inconnu populations at either end of their distribution in the Mackenzie River system, there is some evidence to suggest that the degree of diadromy may vary within and among populations. For example, a high level of intrapopulation variability in migratory behavior was noted among anadromous individuals for the above mentioned otolith microchemistry study (Howland et al. 2001b). Similarly, a recent radio telemetry study on the Liard River (upper Mackenzie River system) found that a small proportion of individuals tagged in the upper Liard were later recaptured in the Mackenzie Delta (2 of 44 fish); these individuals were later shown to be anadromous (Stephenson et al. 2005), suggesting that the migratory behavior of upper river populations may be more complex than previously thought.

Analysis of trace elements in the otoliths of teleost fish has proven useful in reconstructing the environmental histories of fish (e.g., Secor 1999; Thresher 1999). The analysis of changes in otolith strontium (Sr) has been particularly effective in determining the migratory patterns of diadromous species (Babaluk et al. 1997; Howland et al. 2001b; Brown et al. 2007). The Sr concentration of water is directly proportional to salinity (Ingram and Sloan 1992); thus, Sr concentration within the otoliths is hypothesized to reflect the salinities encountered by a fish during its lifetime (Secor 1992; Babaluk et al. 1997). A variety of studies provide good support for this hypothesis (see Howland et al. 2001b for a review).

In this paper, we examined the otolith Sr profiles of inconnu collected from a variety of popu-

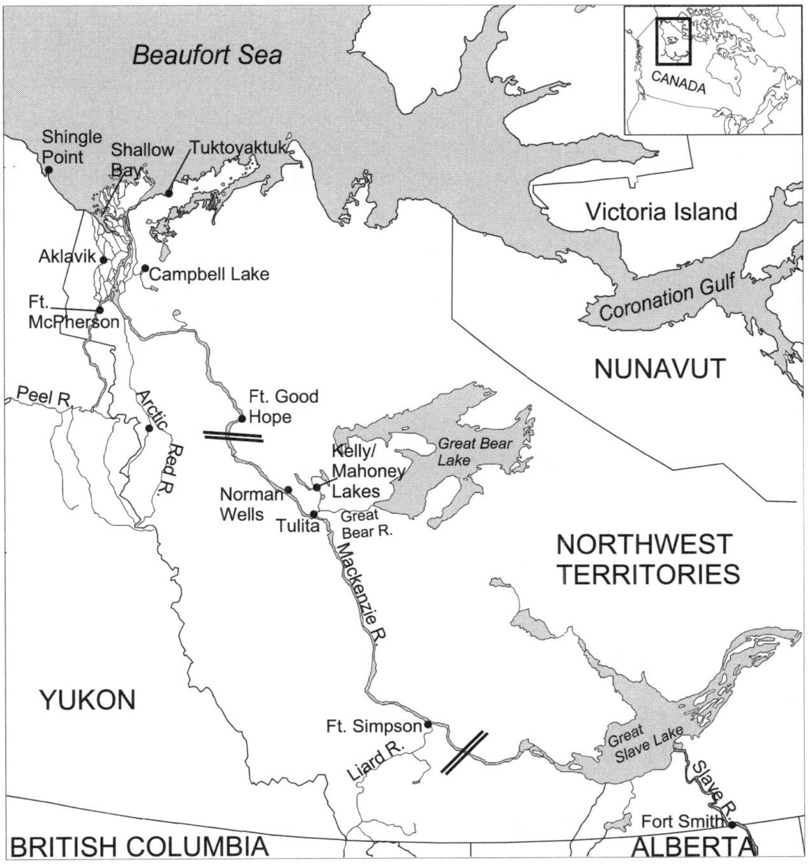

Figure 1.—The Mackenzie River system showing inconnu collection sites for analysis of otolith Sr and other locations mentioned in the text. Double bars delineate different regions of the Mackenzie River system (lower, upper and Great Slave Lake) referred to in the text.

lations throughout the Mackenzie watershed with the objectives of (1) determining the geographical range over which diadromy occurs in this system, (2) characterizing the variability in diadromous behavior within and among populations (with respect to prevalence of diadromy, age at first migration, and frequency of migration to saltwater habitats), and (3) identifying physical and environmental factors that could be involved in determining whether populations or individuals undertake a diadromous or freshwater life history.

Study Area

The Mackenzie River runs 1,650 km from the west end of Great Slave Lake to the Beaufort Sea coast, receiving outflow from numerous tributaries, the largest of which are the Arctic Red, Peel, and Liard rivers (Figure 1). Before entering the Beaufort Sea, the river forms the Mackenzie Delta, a 12,170 km^2 network of lakes and channels (Brunskill 1986). Once the Mackenzie River enters the Beaufort Sea, its freshwater plume extends northeast along the Tuktoyaktuk Peninsula, west to Shingle Point, and encircles the numerous islands in the outer delta, creating a narrow zone of freshened water (Reist and Bond 1988). Inconnu are distributed from the lower reaches of tributaries draining into Great Slave Lake, including the Slave River, north to the lower Mackenzie River and associated tributaries, the Mackenzie Delta, and the Beaufort Sea coast from Shingle Point to the Tuktoyaktuk Peninsula (Figure 1; Scott and Crossman 1973). We analyzed archived sagittal otoliths of inconnu collected from

12 locations throughout the Mackenzie River system, ranging from the Beaufort Sea coast to 2,212 km upstream (Table 1). Sample locations were from three regions of the lower Mackenzie (lower, upper, and Slave River–Great Slave Lake), delineated based on suggested distributions from earlier tagging and genetic studies and included various reaches of the Mackenzie River proper, large tributaries such as the Arctic Red, Peel, Liard, and Slave rivers, and two connected lakes (Campbell and Kelly lakes) (Table 1; Figure 1).

Methods

All otoliths were stored dry in coin envelopes prior to analysis. We tried to obtain a sample size of at least 10 individuals per location with approximately equal representation for both sexes. Most fish were sexually mature at the time of capture; however, for some locations where insufficient numbers of sexually mature fish were available, older juveniles (>450 mm fork length) were included in the sample. Ages of fish ranged from 5 to 31 years and sizes ranged from 459 to 1,122 mm (Table 1).

Otoliths were prepared for scanning proton microprobe (SPM) analysis by embedding them in epoxy resin and sectioning in a dorsoventral transverse plane through the core. The posterior portion of each otolith was then re-embedded in a standard 25.4-mm epoxy resin disk probe. The exposed surface of each otolith was ground (20 μm diamond grit), polished (0.5 μm followed by 0.2 μm diamond or aluminum paste), ultrasonically cleaned and coated with carbon (Howland et al. 2001b).

We analyzed otolith Sr using the SPM at the University of Guelph. To infer life history and migratory patterns of inconnu, we determined Sr levels at 4-μm intervals along the longitudinal growth axis, from the core to the otolith edge, using one-dimensional vertical line scans. Digital images of the otoliths, made prior to the microprobe analyses, were used as reference maps for the line scans. Strontium profiles were produced by rastering a 5 μm (high) × 12 μm (wide) proton beam (3 MeV, 8.5 nA) over each specimen 50 times (ca. 10–15 min/otolith). The proton beam entered each otolith at a 45° angle to the surface, and otoliths were oriented such that the growth zones were horizontal, while the longitudinal growth axis was vertical to the beam. The limit of detection was 1–2 mg/kg (Campbell et al. 1995), and the 90% depth yield (D_{90}) for Sr was ca. 30 μm. All X-ray spectra generated from line scans were analyzed using the GUPIX software package (Maxwell et al. 1989). Other details regarding operating procedures, spectrum fitting, and standardization for this SPM system are outlined in Czamanske et al. (1993) and Campbell et al. (1995).

Following microprobe analyses, otoliths were lightly polished to remove the carbon coating, stained with hematoxylin (with acetic acid), and rephotographed under reflected light. Ages were determined from digital images of stained otoliths, following Howland et al. (2001b). To relate Sr distribution patterns from microprobe analyses to specific periods within the lifetime of each fish, we overlayed Sr distribution profiles and otolith images, following Morris et al. (2005). The residual stain mark left from the proton beam path allowed for the accurate alignment of distribution profiles with images, thus permitting us to determine age at first migration and frequency of migration for inconnu with anadromous life histories.

We developed criteria for classifying life history patterns using a combination of methods from Howland et al. (2001b) and Brown et al. (2007). Howland et al. (2001b) determined the range and maxima within annual increments and classified the patterns of individual fish using a multivariate approach. Brown et al. (2007) simplified this method by examining the maximum and a measure of the coefficient of variation (a surrogate for range) over the entire Sr profile and classifying fish using threshold values determined from analyses of known anadromous and resident samples. They found distinct groupings by life history type, with a clear threshold separating known anadromous and freshwater forms. Due to the large number of samples that we were interested in classifying, we chose to adopt this simplified approach but used a combination of the maximum and range in Sr concentration over individual profiles to categorize life histories of inconnu. We plotted the Sr concentration range versus maxima for all samples (including samples of known life history—anadromous and freshwater) and looked for distinct groupings that were likely to represent individuals with different life histories. A Kruskall-Wallis analysis of variance (ANOVA) on ranks with post hoc Dunn's test was used to compare the mean range and maxima of suggested groupings, and in-

Table 1.—Summary of river reach and upstream distance along with the biological characteristics, prevalence of anadromy, mean age at first migration, and mean frequency of anadromous migration for inconnu by sample location.

Location	Region of Mackenzie system	Distance upstream (km)	Size range	Age range	Sample size	% anadromous	Age at first migration (mean ± SD)	Migration frequency (Mean ± SD)
1. Beaufort Sea Coast (Tuktoyaktuk)	Lower	0	488–923	8–21	10	100	1.9 ± 1.5	0.80 ± .25
2. Beaufort Sea Coast (Shingle Point)	Lower	0	616–940	13–22	10	90	2.5 ± 1.3	0.85 ± .16
3. Mackenzie Delta (Aklavik)	Lower	130	585–820	9–16	20	50	2.9 ± 1.0	0.64 ± .24
4. Peel River (Ft. McPherson)	Lower	331	710–927	10–18	10	100	3.0 ± 1.9	0.82 ± .22
5. Upper Arctic Red River	Lower	530	700–1,122	11–31	16	88	2.7 ± 1.8	0.87 ± .30
6. Mackenzie River (Fort Good Hope)	Lower	646	507–995	8–25	10	80	2.9 ± 1.1	0.72 ± .32
7. Mackenzie River (Norman Wells)	Upper	881	556–889	7–20	10	60	2.8 ± 1.2	0.69 ± .25
8. Mackenzie River (Tulita)	Upper	916	575–866	6–16	10	50	5.2 ± 2.0	0.37 ± .65
9. Liard River (Ft. Simpson)	Upper	1,411	589–735	5–12	8	13	2.0 (*n* = 1 anadromous)	0.50 (*n* = 1 anadromous)
10. Slave River (Ft. Smith)	Slave River–Great Slave Lake	2,212	677–1,100	8–25	11	0	–	–
11. Campbell Lake	Lake	235	459–768	7–15	10	0	–	–
12. Kelly/Mahoney lakes	Lake	1,006	Not available	10–28	7	0?	–	–

dividual fish were classified based upon their grouping. A threshold value for determining age at first migration and frequency of migration for individual fish was based on the groupings generated by this plot.

The incidence of anadromy was summarized by sampling location and region (lower Mackenzie, upper Mackenzie, Slave River–Great Slave Lake), and comparisons among regions were made using Chi-squared tests with Bonferroni correction. Mean age at first migration was determined by location and region for all inconnu classified as anadromous. An index of migration frequency was calculated for each fish as follows:

Index of migration frequency = number of migrations/(age – age at first migration + 1).

This index is a ratio of the of the number of migrations to the number of annuli from age at first migration until capture; a fish that migrated annually following age at first migration would have an index of 1 whereas a fish that migrated less frequently would have an index less than 1. A mean index of migration frequency was calculated by location and region. We compared mean age at first migration and index of migration frequency among locations using Kruskall-Wallis ANOVA on ranks with post hoc Dunn's tests and between regions (lower and upper Mackenzie) using a Mann-Whitney rank sum test. Means are reported with plus or minus one standard deviation. Differences were considered significant at a probability level of 0.05.

Results

A comparison of Sr range to Sr concentration revealed three distinct groups of inconnu within the Mackenzie River drainage (Figure 2). The mean range (group 1: 816 ± 147 mg/kg, group 2: 2,582 ± 598 mg/kg; group 3: 7,713 ± 1,356 mg/kg) and maxima (group 1: 1,373 ± 174 mg/kg, group 2: 3,216 ± 600 mg/kg; group 3: 10,146 ± 794 mg/kg) for these groups were significantly different (Kruskall Wallis ANOVA on ranks with post hoc Dunn's test; $P < 0.001$ for all pairwise comparisons).

Fish in group 1, which we interpret as having freshwater life histories, included individuals with Sr concentration ranges of less than 1,150 mg/kg and maxima of less than 1,700 mg/kg. All 11 individuals from the Slave River of confirmed freshwater origin (Howland et al. 2001b) fell within this group. Profiles in this group were characteristically

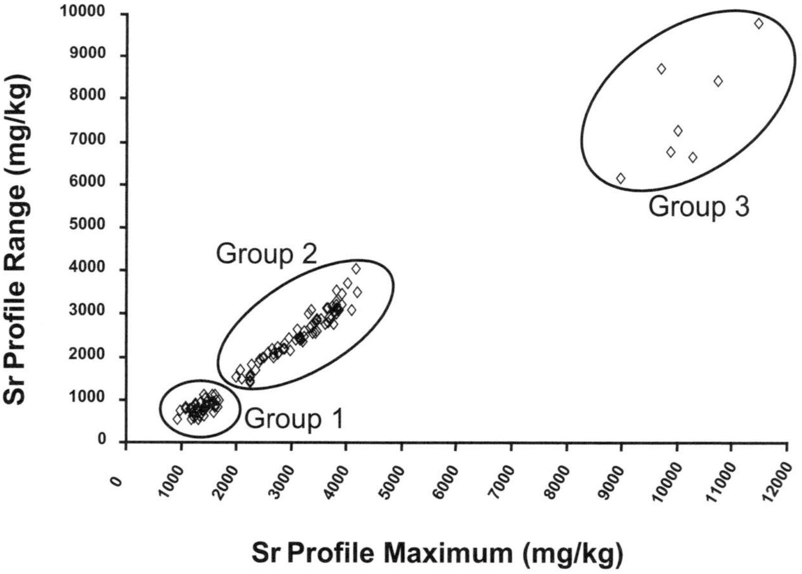

Figure 2.—Strontium profile range versus maximum Sr concentration for all sampled inconnu. Ovals denote distinct groupings referred to in the text: 1 = freshwater inconnu, 2 = anadromous inconnu, 3 = inconnu from Kelly Lake, a high Sr freshwater lake.

low and flat, although some showed a slight oscillatory pattern (Figure 3).

Fish in group 2, which we interpret as anadromous or representing individuals that made at least one migration during their lifetime, had otoliths with Sr concentration ranges of 1,545–4,030 mg/kg and maxima of 2,000–4,200 mg/kg (Figure 2). This group encompassed all individuals from the Arctic Red River previously classified as anadromous and estuarine anadromous using the earlier multivariate methods of Howland et al. (2001b). Most fish in this group had profiles showing the typical pattern of low Sr near the core of the otolith followed by annual oscillations in Sr concentration, indicating an early life history in freshwater followed by regular migrations into a higher Sr environment, most likely the Beaufort Sea (Howland et al. 2001b); however, some individuals also showed irregular or infrequent migrations (Figure 4).

The third group of fish identified in our analysis were all from a single location in the Mackenzie River system, the Kelly/Mahoney Lakes system and had profiles with unusually high Sr concentration ranges (6,170–9,950 mg/kg) and maxima (8,950–11,460 mg/kg) that far exceeded levels found in any of the other populations sampled in our analysis (Figure 2). These profiles were also atypical with respect to the baseline Sr levels at the core of the otolith, which ranged from ~3,000 to 6,000 mg/kg, three to four times higher than baseline levels seen for typical inconnu Sr profiles (less than 1,200 mg/kg) (Figure 5). Most Sr profiles for fish from this location showed some oscillation; however, levels rarely dropped below baseline and, in most cases, gradually increased over the lifetime of the fish. The high Sr concentrations in the otoliths of inconnu from Kelly/Mahoney Lakes system were consistent with the unusually high freshwater Sr levels at the outflow of these water bodies (Loche River, Table 2). Because fish from this location had anomalous profiles and are likely lake residents, we treated them as a distinct grouping and excluded them from subsequent analyses of migratory/life history patterns with the Mackenzie River system.

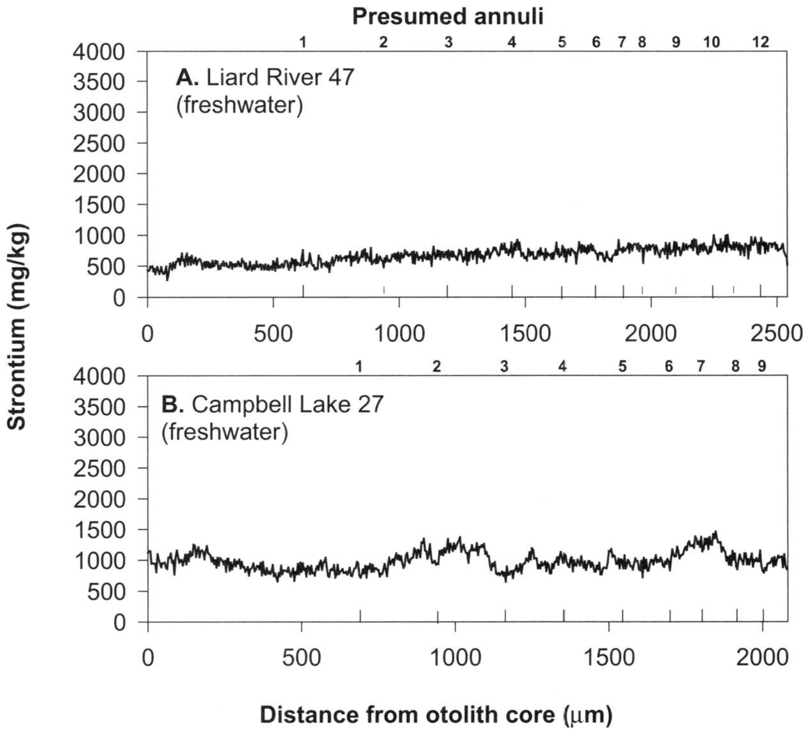

Figure 3.—Sr profiles from scanning proton microprobe line scans of freshwater inconnu from (A) the Liard River and (B) Campbell Lake.

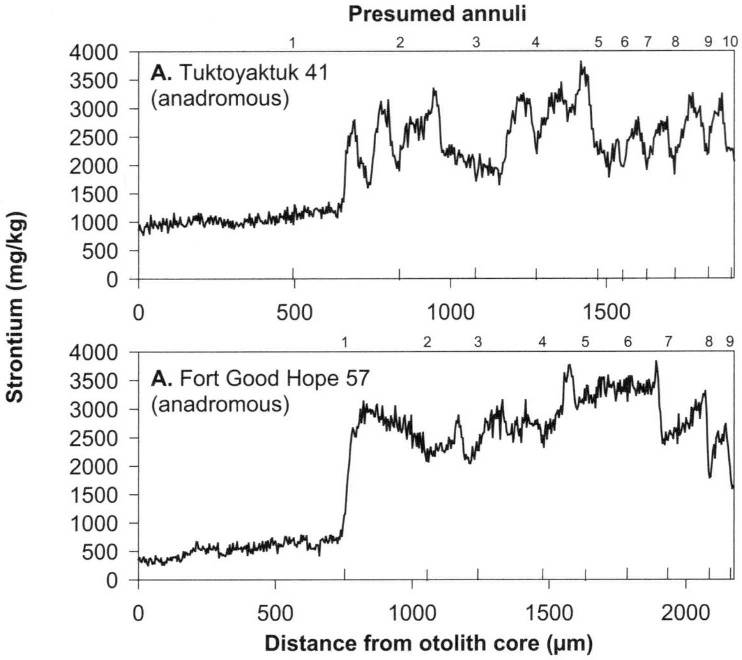

Figure 4.— Sr profiles from scanning proton microprobe line scans of anadromous inconnu from (A) Tuktoyaktuk and (B) Fort Good Hope.

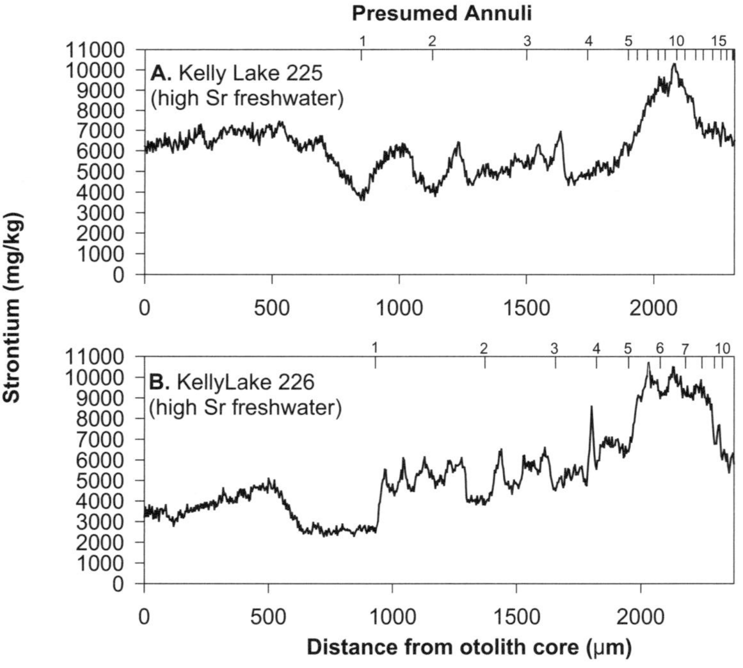

FIGURE 5.—Sr profiles from scanning proton microprobe line-scans of two inconnu from a high Sr freshwater system: (A) Kelly/Mahoney Lakes 225 and (B) Kelly/Mahoney Lakes 226.

TABLE 2.—Strontium levels of water from various locations in the Mackenzie River system and in coastal areas of the Northwest Territories.

Location	Salinity	Sr (mg/L)	Source
Coronation Gulf	Seawater	4.740	Babaluk, unpublished data
Coast of Victoria Island	Seawater	3.226	Babaluk, unpublished data
Peel River	Freshwater	0.13	Reeder et al. 1972
Arctic Red River	Freshwater	0.15	Reeder et al. 1972
Mackenzie River (various locations)	Freshwater	0.13–0.21	Reeder et al. 1972
Mackenzie River at Norman Wells	Freshwater	0.195	J. A. Babaluk and G. Low, unpublished data
Great Bear River	Freshwater	0.09	Reeder et al. 1972
Loche River (outlet of Kelly/Mahoney Lake)	Freshwater	1.46	Babaluk and Low, unpublished data
Liard R. (various locations)	Freshwater	0.10–0.13	Reeder et al. 1972
Great Slave Lake	Freshwater	0.13	Reeder et al. 1972
Slave River (various locations)	Freshwater	0.13	Reeder et al. 1972

The threshold separating groups 1 and 2 matched the levels established by Brown et al. (2007) using known freshwater and andromous fishes from a variety of species. The range in maximum otolith Sr concentrations for inconnu in the anadromous group was also within levels reported for the estuarine/marine phase of migration in other species analyzed by SPM (e.g., sockeye salmon *Oncorhynchus nerka* ~1,800–3,300 mg/kg, Babaluk et al. 2000; striped bass, *Morone saxatilis* ~4,000–8,000 mg/kg, Morris et al. 2005; Arctic char *Salvelinus alpinus* ~1,800–2,000 mg/kg, Babaluk et al. 1997). Based on the concurrence with other studies and the clarity of our results, we were confident that the above-described groupings accurately distinguished fish that had made at least one anadromous migration from those that remained in freshwater throughout their life history.

Using groups 1 and 2 from Figure 2, we determined the proportion of anadromous fish by sample location (Table 1). Anadromous inconnu occurred throughout the Mackenzie River system up to the Liard River, although only one of eight sampled individuals from this location was anadromous. We observed a gradual decrease in the incidence of anadromy as distance from the sea increased, with 100% of inconnu in the Slave River–Great Slave Lake area exhibiting a freshwater life history (Table 1). Although this general pattern was observed, two locations in the lower Mackenzie region had unusually high numbers of freshwater individuals relative to other locations in the same vicinity. One was a population in Campbell Lake, which has access to Mackenzie River but, based on our results, appear to be lake resident (this population was excluded from subsequent analyses since our focus was on understanding the extent of anadromy in riverine migratory populations). The second location was in the Aklavik area of the Mackenzie Delta. When we compared different reaches of the Mackenzie River system (excluding the Campbell and Kelly Lake samples), we found significant differences in the incidence of anadromy for fish in the lower (53/66 anadromous), upper (12/28 anadromous), and Slave River–Great Slave Lake (0/11 anadromous) reaches of the Mackenzie River system (Table 1; Chi-squared test $P < 0.001$).

Based on the distribution of Sr profile maxima for inconnu in groups 1 and 2 (Figure 2), we considered a migration into saline waters to have occurred if the maximum within a given growth increment exceeded 1,700 mg/kg. The average age at which inconnu first began to migrate to the sea was similar among locations, with most in the range of 1.9–3 years (Kruskall Wallis ANOVA on ranks with post hoc Dunn's test; $P > 0.05$ for all pairwise comparisons; Table 1); an exception was the sample from Tulita, which had an average age of 5.2 ± 2.0 years and was significantly different from Tuktoyaktuk, the location with the youngest average age at first migration (1.9 ± 1.5 years) (Kruskall Wallis ANOVA on ranks with post hoc Dunn's test; $P =$

0.019) (Table 1). Although average age at first migration tended to be similar across locations, when the data were pooled and compared between regions, the average age at first migration was greater for inconnu in the upper (3.6 ± 1.9 years) as compared to the lower (2.5 ± 1.5 years) reaches of the Mackenzie River system (Mann-Whitney rank sum test; $P = 0.034$).

The average index of migration frequency, a measure of the tendency for anadromous inconnu to make regular marine migrations, showed a decreasing trend with distance from sea, ranging from a high of 0.87 ± 0.30 in the lower Mackenzie region to a low of 0.37 ± 0.65 in the upper Mackenzie (Table 1); however, no significant differences were found in our comparisons among locations or between regions (Kruskall Wallis ANOVA on ranks with post hoc Dunn's test, $P > 0.05$ for all pairwise comparisons; Mann-Whitney rank sum test; $P = 0.07$). The Aklavik area was again unique; in addition to the above-mentioned observation of a high number of freshwater individuals, we also found a decreased tendency for anadromous individuals to make regular seaward migrations relative to other populations in the same vicinity (Table 1). When this location was excluded from our comparison between regions, we found there was a significant difference in the index of migration frequency between the lower (0.82 ± 0.26) and upper Mackenzie regions (0.66 ± 0.27) (Mann-Whitney rank sum test; $P = 0.032$).

Discussion

Anadromous inconnu occurred throughout the Mackenzie River upstream to the mouth of the Liard River at Fort Simpson. However, their prevalence decreased with distance from the sea, demonstrating a lower incidence in upper river locations (between Norman Wells and Fort Simpson) and a complete absence from the Slave River–Great Slave Lake area. This is consistent with earlier Floy tagging studies, which suggest a low degree of movement between lower (presumably anadromous) and upper river populations and no mixing or movement between Slave River–Great Slave Lake and either upper or lower Mackenzie River populations (summarized in Howland et al. 2000). These observations concur with the recent findings of Stephenson et al. (2005) who demonstrated that a only a small number of inconnu tagged in the upper Liard River were anadromous and exhibited movement into the lower Mackenzie Delta region.

Interpopulation Variability in Diadromy

The populations examined in our study showed a gradual decrease in the incidence of anadromy with increasing distance from the sea rather than an abrupt shift from one life history strategy to the other. This was accompanied by a trend of increased age at first migration and decrease in frequency of migration with distance from the sea for the anadromous component of populations. The combination of these patterns resulted in a high degree of both among and within population variability in diadromous behavior.

The Mackenzie River system flows in a south–north direction for more than 2,000 km, creating a gradient in both distance and latitude. Thus, a variety of factors may drive the system-wide patterns of variation in diadromous behavior we observed. Variability among populations may be the result of trade-offs between the costs of long distance migration and the benefits of feeding in more productive coastal areas following the classic hypothesis by Gross et al. (1988) that in cases where the nutritional benefits of feeding in a marine environment outweigh the costs of long distance migration, fish should opt for an anadromous life history. For inconnu populations in the Mackenzie River system, the gradual decline in anadromy with distance from the sea may be explained by the increased cost of migrating further upstream.

The decision of whether or not to migrate appears to be closely linked with growth and energetic requirements. However, the mechanisms leading to divergence between freshwater and diadromous life histories remains unclear. In some species, faster growing individuals, typically males, opt to remain in freshwater and become residents (e.g., coho salmon *Oncorhynchus kisutch* and Atlantic salmon *Salmo salar*, Gross 1991). However, in other species such as brown tout *S. trutta* (Arai et al. 2002) and Arctic char (Radtke et al. 1996), the faster growing individuals appear to adopt an anadromous lifestyle and the slower growing individuals become freshwater-resident. Given the length of the Mackenzie River system, we suggest that there may be energetic and/or physiological limitations to migration for inconnu populations in the upper Mackenzie. The inverse relationship between body size and migration cost is well demonstrated (e.g., Glebe and Leggett 1981; Roff 1991), and the upper reaches of northern rivers such as the Mackenzie likely provide limited growth opportunities for fishes due to the low productivity of these areas. Therefore, a higher proportion of individuals may have difficulty reaching an optimal size for migration. We hypothesize that for upper Mackenzie River inconnu populations, anadromous migrations may only occur in individuals that are faster growing and thus able to reach a "threshold size." In our study, a higher proportion of inconnu in upstream locations appeared to make occasional or infrequent anadromous migrations, and there was a trend towards a later age at first migration. These characteristics may be related to the increased time required to build up sufficient reserves prior to migrating. Alternatively, upper Mackenzie River populations may be composed of a mixture of riverine and anadromous stocks that overlap at certain times of the year. This scenario is less likely given that a high degree of gene flow and evidence of mating between sympatric freshwater and anadromous morphs is frequently observed for other salmonids (e.g., Arctic char, Jonsson and Jonsson 1993; Atlantic salmon, Hutchings and Myers 1985).

Diadromy has also been found to vary with latitude in a number of salmonids, showing a gradual decrease from north to south in the northern hemisphere with some species becoming strictly freshwater at southern latitudes (e.g., brown trout, McDowall 1997; white-spotted char *Salvelinus leucomaenis*, Arai et al. 2005). Most examples from the literature involve comparisons among populations that have similar access to saltwater, but at different latitudes. Our study is different in that it involves a single river system that flows in a south to north direction. Thus, it is difficult to tease apart the effects of latitude versus distance. As previously discussed in Howland (2005), the combination of increased distance from the sea (making migration more costly) and decreased latitude (conferring a growth advantage due to a longer growing season) may produce conditions that favor freshwater residency in upstream populations. There may be further selection for a freshwater lifestyle in the Slave River–Great Slave Lake area since the shallow west basin provides productive feeding habitat for inconnu populations that spawn in the various tributaries flowing into Great Slave Lake (Brunskill 1986). Increasing productivity in freshwater systems can lead to a reversion towards freshwater residency in anadromous individuals (Nordeng 1983), which may explain why inconnu are strictly freshwater resident in this region of the Mackenzie. In contrast, the upper reaches of the Mackenzie River are less productive and are likely to have a more restrictive carrying capacity, which may promote the movement of some individuals to more productive feeding habitats (Secor 1999), resulting in the small proportion of anadromous individuals observed in these populations.

Although there was a general reduction in the incidence of anadromy with distance from the sea, some populations deviated from this pattern. Inconnu from both connected lake populations in this study (Campbell and Kelly lakes) had Sr profiles, which suggested freshwater residency. Additionally, inconnu from the Aklavik area of the Mackenize delta (in close proximity to the sea) had an unusually high proportion of individuals with freshwater Sr profiles. In these locations, inconnu have access to large bodies of freshwater that likely provide a sufficient food base and access to suitable overwintering habitat, making freshwater residency a viable life history tactic. Campbell and Kelly lakes are both relatively large freshwater bodies with deep basins suitable for overwintering. Inconnu in the Aklavik area are on the east channel of the Mackenzie delta, in close proximity to Shallow Bay—a large coastal bay with strong freshwater influence from the Mackenzie River. Inconnu from the Aklavik area may feed in Shallow Bay and then utilize deeper areas of the Mackenzie delta for overwintering (Figure 1).

Inconnu from the Kelly/Mahoney lakes population had anomalously high levels of Sr in their otoliths despite being from a freshwater system. We attribute these elevated Sr levels to the naturally high Sr levels in these water bodies (which we were not aware of when these samples were included in the study). Most areas of the Mackenzie River system range from 0.13 to 0.2 mg/L Sr, whereas the water sample from the outlet of these lakes had an Sr concentration of 1.6 mg/L (Loche River, Table 2). The unusually high Sr levels in these waters likely stem from the surface geology of the area (large amounts of fossilized limestone and dolomite predominate in the region) (Reeder et al. 1972). Most Sr profiles for fish from this location showed some oscillations, suggesting movements between areas of varying Sr levels or perhaps seasonal variation in Sr incorporation. Strontium levels rarely dropped below baseline and, in most cases, gradually increased over the lifetime of the fish, suggesting that they remained in a high Sr environment throughout life rather than migrating to the lower Sr environment of the Mackenzie River. We also noted an increase in Sr levels with age for all individuals from this location. This observation raises some interesting thoughts regarding how Sr incorporation changes with age. Assuming these fish were permanent residents of the Kelly Lake system and assuming Sr levels within the system were relatively stable, these profiles suggest that Sr incorporation or apparent concentration in the otoliths may increase with age independent of Sr levels in the environment.

Intrapopulation Variability in Diadromy

Most anadromous inconnu showed the typical pattern of low Sr levels at the core, indicating that they spent their early life history in freshwater (average of 2–5 years), followed by regular oscillations (usually peaking in the summer growth phase), indicating annual migrations between freshwater and waters of higher salinity (estuarine or coastal). However, similar to previous otolith microchemistry studies that focused on inconnu in the Arctic Red River region of the lower Mackenzie system only (Howland et al. 2001b), we found a high degree of within population variability in the Sr profiles of anadromous inconnu throughout the Mackenzie. We observed variation in the magnitude, frequency, and onset of diadromy, suggesting substantial flexibility in migratory strategies. The unusually late onset and extreme irregularity of migration in some individuals (particularly in the upper Mackenzie populations) suggests that inconnu may have the option of switching between freshwater and anadromous life histories and may not necessarily be locked into a single strategy early in life. Similar patterns have been observed in other salmonids where individuals have been shown to become anadromous later in life (Nordeng 1983; Jonsson and Jonsson 1993) or, in some cases, to become resident for periods following migration (Naslund 1990). The variability in anadromous behavior in inconnu as well as other northern coregonids (Brown et al. 2007; J. A. Babaluk, unpublished data) may be facilitated by the wide range of habitat and environmental conditions available in the large river systems they inhabit.

Conclusions

Inconnu are a highly dynamic species with respect to diadromous behavior, a characteristic they appear to share with other coregonids. We observed a high degree of variability in the migratory strategies both within and among populations. Anadromous migration is likely facultative in coregonids and age at first migration, frequency of migration, and the magnitude of diadromous migration may be flexible at an individual level depending on a number of factors, including condition of fish, body size, sex, physiological state, and environmental conditions.

Acknowledgments

This study was cooperatively funded through the Inuvialuit (Fisheries Joint Management Committee), Gwich'in (Gwich'in Renewable Resources Board), and Sahtu (Sahtu Renewable Resources Board) aboriginal groups who live along this river system. We thank R. Wastle and S. H. Wiley for completing age reading, otolith preparation, and image overlays. Thanks are extended to DFO Inuvik, DFO Hay River, and RL&L for providing archived samples. For advice and use of equipment and otolith preparation, we thank W. J. Teesdale and J. L. Campbell, University of Guelph, and the Geology Departments at University of Alberta and University of Manitoba.

References

Alt, K. T. 1977. Inconnu, *Stenodus leucichthys*, migration studies in Alaska. Journal of the Fisheries Research Board of Canada 34:129–133.

Arai, T., A. Kotake, T. Aoyama, H. Hayano, and N. Miyazaki. 2002. Identifying sea run brown trout, *Salmo trutta*, using Sr:Ca ratios of otoliths. Ichthyological Research 49:380.

Arai, T., A. Kotake, and T. Kitamura. 2005. Migration of anadromous white-spotted charr *Salvelinus leucomaenis*, as determined by otolith strontium:calcium ratios. Fisheries Science 71:731–737.

Babaluk, J. A., N. M. Halden, J. D. Reist, A. H. Kristofferson, J. L Campbell, and W. J. Teesdale. 1997. Evidence for non-anadromous behaviour of Arcic charr (*Salvelinus alpinus*) from Lake Hazen, Ellesmere Island, Northwest Territories, Canada, based on scanning proton microprobe analysis of otolith strontium distribution. Arctic 50:224–233.

Babaluk, J. A., J. D. Reist, and G. Low. 2000. First record of Kokanee salmon, *Oncorhynchus nerka*, in Great Slave Lake, Northwest Territories. Canadian Field-Naturalist 114:680–684.

Behnke, R. J. 1972. The systematics of salmonid fishes of recently glaciated lakes. Journal of the Fisheries Research Board of Canada 29:639–671.

Berg, L. S. 1948. Freshwater fishes of the U.S.S.R. and adjacent countries. Volume 1. 4th edition. Academy of Science, Moscow. (Translated from Russian by Isreal Program for Scientific Translations, Jerusalem.)

Brown, R. J. 2000. Migratory patterns of Yukon River inconnu as determined with otolith microchemistry and radio telemetry. Master's thesis. University of Alaska, Fairbanks.

Brown, R. J., N. Bickford, and K. Severin. 2007. Otolith trace element chemistry as an idicator of anadromy in Yukon River drainage Coregonine fishes. Transactions of the American Fisheries Society 136:678–690.

Brunskill, G. J. 1986. Environmental features of the Mackenzie system. Pages 435–471 *in* B. R. Davies and K. F. Walker, editors. The ecology of river systems. Dr. W. Junk Publishers, Dordrecht, Netherlands.

Campbell, J. L., W. J. Teesdale, and N. M. Halden. 1995. Theory, practice and application of micro-PIXE analysis and element-distribution maps. The Canadian Mineralogist 33:279–292.

Craig, P. C. 1984. Fish use of coastal waters of the Alaskan Beaufort Sea: a review. Transactions of the American Fisheries Society 113:265–282.

Czamanske, G. K., T. W. Sisson, J. L. Campbell, and W. J. Teesdale. 1993. Micro-PIXE analysis of silicate reference standards. The American Mineralogist 78:893–903.

Dillinger, R. E., Jr., T.P. Birt, and J. M. Green. 1992. Arctic cisco, *Coregonus autumnalis*, distribution, migration and spawning in the Mackenzie River. Canadian Field-Naturalist 106:175–180.

Glebe, B. D., and W. C. Leggett. 1981. Latitudinal differences in energy allocation and use during the freshwater migrations of American shad (*Alosa sapidissima*) and their life history consequences. Canadian Journal of Fisheries and Aquatic Sciences 38:806–820.

Gross, M. R., R. M. Coleman, and R. M. McDowall. 1988. Aquatic productivity and the evolution of diadromous fish migration. Science 239:1291–1293.

Gross, M. R. 1991. Salmon breeding behaviour and life history evolution in changing environments. Ecology 72:1180–1186.

Howland, K. L., R. F. Tallman, and W. M. Tonn. 2000. Migration patterns of freshwater and anadromous inconnu, *Stenodus leucichthys*, in the Mackenzie River system. Transactions of the American Fisheries Society 129:41–59.

Howland, K. L., W. M. Tonn, and G. Goss. 2001a. Contrasts in the hypo-osmoregulatory abilities of a freshwater and an anadromous population of inconnu. Journal of Fish Biology 59:916–927.

Howland, K. L., W. M. Tonn, J. A. Babaluk and R. F. Tallman. 2001b. Identification of freshwater and anadromous inconnu in the Mackenzie River system by analysis of otolith strontium. Transactions of the American Fisheries Society 130:725–741.

Howland, K.L. 2005. Population differentiation of inconnu, *Stenodus leucichthys*, in the Mackenzie River system. Doctoral thesis. University of Alberta, Edmonton.

Hutchings, J. A., and R. A. Myers. 1985. Mating between anadromous and non-anadromous Atlantic salmon, *Salmo salar*. Canadian Journal of Zoology 63:2219–2221.

Ingram, B. L., and D. Sloan. 1992. Strontium isotopic composition of estuarine sediments as paleosalinity-paleoclimate indicators. Science 255:68–72.

Jonsson, B., and N. Jonsson. 1993. Partial migration: niche shift versus sexual maturation in fishes. Reviews in Fish Biology and Fisheries 3:348–365.

Lawrence, M. J., G. Lacho, and S. Davies. 1984. A survey of the coastal fishes of the southeastern Beaufort Sea. Canadian Technical Report of Fisheries and Aquatic Sciences 1220.

Maxwell, J. L., J. L. Campbell, and W. J. Teesdale. 1989. The Guelph PIXE software package. Nuclear Instruments and Methods in Physics Research B 43:218–230.

McDowall, R. M. 1997. The evolution of diadromy in fishes (revisited) and its place in phylogenetic analysis. Reviews in Fish Biology and Fisheries 7:443–462.

Morris, J. A. Jr., R. A. Rulifson, J. A. Babaluk, P. G. May, and J. L. Campbell. 2005. Use of micro-PIXE to investigate otolith Sr distributions of the anadromous striped bass, *Morone saxatilis*. X-Ray Spectrometry 34:301–305.

Naslund, I. 1990. The development of regular seasonal habitat shifts in landlocked Arctic charr, *Salvelinus alpinus* L. population. Journal of Fish Biology 36:401–414.

Nikol'skii, G. V. 1954. Special ichthyology (2nd edition revised). (Translated from Russian by Israel Program for Scientific Translations, Jerusalem.)

Nordeng, H. 1983. Solution to the "char problem" based on Arctic char (*Salvelinus alpinus*) in Norway. Canadian Journal of Fisheries and Aquatic Sciences 40:1372–1387.

Radtke, R., M. Svenning, D. Malone, A. Klementson,

J. Ruzicka, and D. Fey. 1996. Migrations in an extreme northern populations of Arctic charr *Salvelinus aplinus*: insights from otolith microchemistry. Marine Ecology Progress Series 136:13–23.

Reeder, S. W., B. Hitchon, and A. A. Levinson. 1972. Hydrogeochemistry of the surface waters of the Mackenzie River drainage basin, Canada—I. Factors controlling inorganic composition. Geochimica et Cosmochimica Acta 36:825–865.

Reist, J. D., and W. A. Bond. 1988. Life history characteristics of migratory coregonids of the lower Mackenzie River, Northwest Territories, Canada. Finnish Fisheries Research 9:133–144.

Roff, D. A. 1991. Life history consequences of bioenergetic and biomechanical constraints on migration. American Zoologist 31:205–215.

Scott, W. B., and E. J. Crossman. 1973. Freshwater fishes of Canada. Fisheries Research Board of Canada Bulletin 184.

Secor, D.H. 1992. Application of otolith microchemistry analysis to investigate anadromy in Chesapeake Bay striped bass, *Morone saxatilis*. Fishery Bulletin 90:798–806.

Secor, D. H. 1999. Specifying divergent migrations in the concept of stock: the contingent hypothesis. Fisheries Research 43:13–34.

Stephenson, S. A., J. A. Burrows, and J. A. Babaluk. 2005. Long-distance migrations by inconnu (*Stenodus leucichthys*) in the Mackenzie River system. Arctic 58:21–25.

Tallman, R. F., M. V. Abrahams, and D. H. Chudobiak. 2002. Migration and life history alternatives in a high latitude species, the broad whitefish, *Coregonus nasus* Pallas. Ecology of Freshwater Fish 11:101–111.

Thresher, R. E. 1999. Elemental composition of otoliths as a stock delineator in fishes. Fisheries Research 43:165–204.

American Fisheries Society Symposium 69:135–150, 2009

Migration Patterns of Striped Bass through Nonnatal Estuaries of the U.S. Atlantic Coast

THOMAS M. GROTHUES* AND KENNETH W. ABLE
Marine Field Station, Institute of Marine and Coastal Sciences, Rutgers University 800 c/o 132 Great Bay Boulevard, Tuckerton, New Jersey 08087, USA

JACQUE CARTER AND TIMOTHY W. ARIENTI
University of New England, University Campus 11 Hills Beach Road, Biddeford, Maine 04005, USA

Abstract.—Telemetered adult striped bass *Morone saxatilis* (n = 170) in two small nonnatal U.S. Atlantic coast estuaries, the Mullica River–Great Bay in New Jersey and the Saco River in Maine, displayed a variety of movements relative to migration and habitat use. Individual presence in both systems ranged from hours to many months from spring through fall but seldom during the winter. Some made upriver runs during the presumed spawning season. An absence of eggs, larvae, or juvenile stage striped bass or suitable spawning grounds suggests that fish utilizing both systems are members of migrant contingents originating elsewhere. In both systems, some seasonal residents occupied individual "home" ranges throughout the salinity gradient, but others were plastic in their behavior and utilized several sites or visited the estuary either briefly or for a whole season in different years. Movement of fish between study sites and recaptures away from them indicated wide dispersal during time away from the study estuaries. The above supports the idea that migratory and seasonal residence behaviors of migrant striped bass are not compulsive or predictive but reactive or learned. Despite long seasonal occurrence in these small estuaries by some individuals, none appeared to become full-time residents. Small estuaries may lack the year-round resources for spawning and feeding to enable this. However, the observed behaviors could promote recolonization of spawning stocks with residents in larger restored rivers and estuaries where they have been previously extirpated. Localized fishery depletion of small estuaries could occur quickly but would be temporary in the absence of stock wide depletion.

Introduction

Striped bass *Morone saxatilis* exploit marine, estuarine, riverine, and lacustrine habitats but must spawn in freshwater. They naturally occupy coastal marine environments as migrants along most of eastern North America from North Carolina into Canada (Collette and Klein-MacPhee 2002). Permanent residency of most of the population in open systems at the fringes of their range (along the Gulf Coast, in central Florida and North Carolina and perhaps Nova Scotia) may be related to recent climatic oceanic thermal barriers that inhibit coastal ranging (Raney 1957; Dudley et al. 1977; Haeseker et al. 1996, Bjorgo et al. 2000), but individual nonmigrants (residents) also occur among stocks in the central part of the range (Secor et al. 2001; Secor and Piccoli 2007).

Diversity in residency and migration is evident even at local scales. The presence of telemetered striped bass across the salinity gradient within one river/estuary is due to not just the residence of individuals in microhabitats along the gradient, but also the presence of some individuals that moved along the gradient (Able and Grothues 2007; Ng et al. 2007). Within-stock diversity in migratory behavior is captured in the concept of "contingents" (Clark 1968; Secor 1999; Secor et al. 2001). As first described by mark–recapture patterns (Clark 1968), and later on the basis of otolith

* Corresponding author: grothues@marine.rutgers.edu

microchemistry (Secor et al. 2001; Morris et al. 2003, Zlokovitz et al. 2003), adults sort into retentive (nonmigratory) freshwater residents, retentive estuarine residents (remaining in the estuary of a natal system), and ocean-going migrants. Segregation along habitat salinity gradients is also evidenced by color and fatty acid analysis (Paramore and Rulifson 2001). Movement into marine environments is facultative and hypothesized to arise from learned experience during ranging or deterioration in the value of the natal habitat (Gross 1987; Secor 1999; Paramore and Rulifson 2001). The ability and need for larger fish to exploit oceanic prey may explain a propensity for both an ontogenetic shift and female sex bias towards ocean migration (Secor and Piccoli 2007) and may be a key selective pressure to the evolution of diadromy (Gross 1987).

What is not yet well understood is the degree of behavioral plasticity related to the migrant individuals visiting nonnatal estuaries. Do they express retentive behavior in different environments (i.e., recruit as new residents)? These would necessarily be excluded from fish collected for otolith chemistry analysis on natal spawning grounds. Different from diversity (among individuals), behavioral plasticity is an individual's ability to choose or change behavior (as opposed to compulsion). Plasticity provides much faster population-wide response to environmental perturbation than diversity and can happen within a generation (Sutherland 1996). For example, a generational response in migrant behavior to a change in food resources has been demonstrated for anadromous fishes. Fertilization of natal system lakes to improve growth of juvenile sockeye salmon *Oncorhynchus nerka* for better sea survival led instead to the establishment of freshwater residents (Hyatt and Stockner 1985). Also, high feeding rates decreased anadromy in crossbreeding experiments with Arctic char *Salvelinus alpinus* (Nordeng 1983). The extent to which anadromy or other forms of specific habitat fidelity is plastic probably differs among taxa. The most direct genetic control of its expression appears to be in semelparous fish such as Pacific salmon *Onchorhynchus* spp., where it is compulsive and may be readily acted upon in natural selection (Carl and Healy 1984; Saxton et al. 1984; Gross 1987; Randall et al. 1987).

However, while striped bass share an anadromous life history with salmonines, they differ in being iteroparous and in having less arduous spawning runs unaccompanied by morphometric changes or feeding cessation; in fact, spawning runs coincide with those of favored anadromous prey, river herring *Alosa* spp. (Collette and Klein-MacPhee 2002). Therefore, striped bass face different selective pressure than semelparous salmon (Rago and Goodyear 1987). Annual reproductive payoff for striped bass is a gamble and usually dismal (Rutherford and Houde 1995; Houde 1996, Secor 2000). Environmental uncertainty on spawning grounds should select for more plasticity in the way of straying or skipped or repeated spawning (Rago and Goodyear 1987; Secor 2007 and 2009, this volume). This is important because unlike most marine fishes, anadromous fishes do not disperse to disparate habitats as larvae. The role of recruitment (addition to or colonization of new or nonnatal spawning populations) must be primarily fulfilled by migrating adults or subadults, although exceptions of age-0 striped bass dispersal along shallow beaches in the Gulf of Saint Lawrence has been documented between close rivers (Robinson et al. 2004).

Here, we investigate the plasticity of habitat use and migration of striped bass from two small rivers and their estuaries as observed with fixed and mobile acoustic telemetry. The study sites, Mullica River–Great Bay estuary in southern New Jersey and the Saco River in southern Maine, are particularly interesting because they are not recognized as functional spawning sites. This means that striped bass using these habitats migrated to them. Both of these river systems are much smaller and shallower than the major recognized spawning sites along the central east coast of the United States (the Chesapeake and its tributaries and the Delaware and Hudson rivers). By applying common data analysis procedures to the Maine and New Jersey data sets and synthesizing both fine-scale and estuarine-scale habitat use from earlier works with new findings, we compare and contrast results to begin addressing such questions as do individuals retain the same patterns of movement across years? Are behaviors plastic enough to allow "recruitment" of "new" residents into small nonnatal systems from a pool of migrants? In concert with other factors, the answers to these questions are useful to predicting the result of management decisions. For example, how might stock structure be affected by selective pressure (harvest or habitat alteration) on striped bass given the observed level of plasticity? The understanding of this specie's characteristic is also important to understanding the evolution of anadromy.

Materials and Methods

Study Areas

The Saco River is a small river in southern Maine, generally less than 0.5 km wide and stretching approximately 7 river kilometers from the mouth to a high head dam (Figure 1A). A fish lift is used to transport fish above the dam, but striped bass are targeted for removal from the lift to bolster the success of Atlantic salmon *Salmo salar* restoration. Salinity varies from freshwater to 5 practical salinity units (psu) at the Cataract Dam and 10 psu and 30 psu near the river mouth. Thus, it is dammed below suitable spawning grounds for striped bass (Davis 1966). The polyhaline river mouth is shallow with a rocky or sand bottom. A salt marsh border in low-

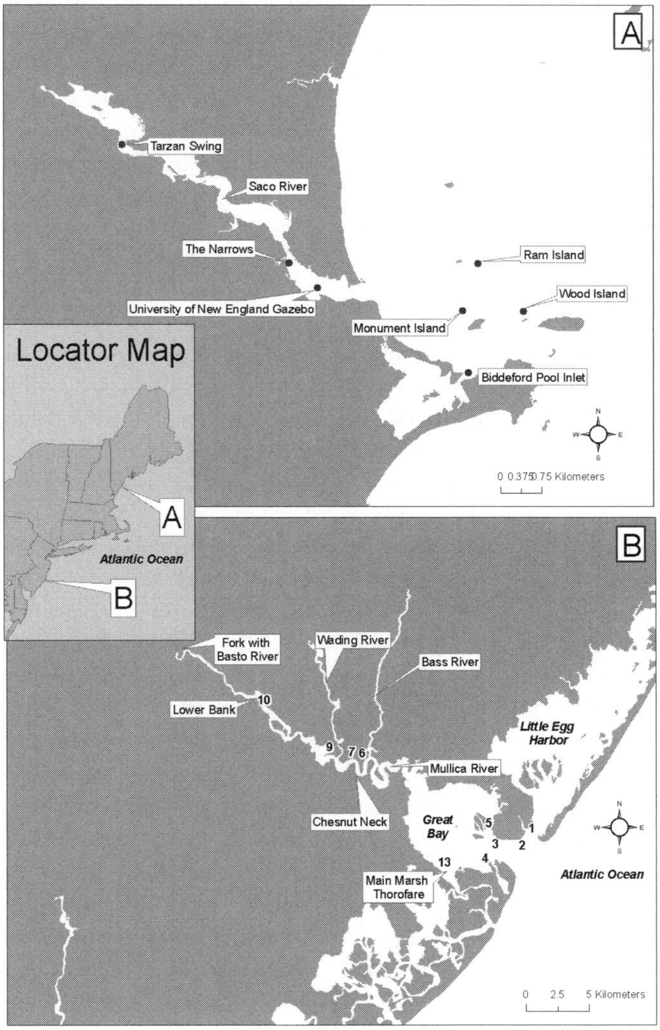

Figure 1.—Study sites in southern (A) Maine (Saco River) and (B) New Jersey (Mullica River–Great Bay). Compatible tags and hydrophones were used in both sites. Details of hydrophone function are given in Grothues et al. (2005). Study areas differ in morphology, but hydrophones near the mouth of the Saco River (Ram Island, Wood Island, Monument Island, and Biddeford Pool), Maine are within the same salinity regime as those in Little Egg Inlet and lower Great Bay, New Jersey. The farthest upriver hydrophone site in the Maine array (Tarzan Swing) is in freshwater just below Cataract Dam.

er reaches turns to cobble beach or bluffs upriver. The upper Saco River currently serves as spawning grounds for anadromous river herring and Atlantic salmon, which obtain access via fish passages at the first three dams. A fish lift at the lowest dam, Cataract Dam, moves only about 10 adult striped bass annually into the upper reaches and no young of year are seen to move downriver from there despite intense monitoring for river herring, Atlantic salmon, and brown trout *Salmo trutta* (Kelhir 2005; FLP Maine Hydro, unpublished data). The Saco River is 600 km north of the Hudson River (around Cape Cod), which is thought to be the major contributor to the stock of fish from there and northwards (Waldman et al. 1990). However, striped bass young of year are frequently collected by surveys in the Kennebec River, Maine (57 km to the northeast; State of Maine Marine Resources Department, unpublished data). The Maine and New Jersey study sites are separated by approximately 700 km, following the coast.

The Mullica River–Great Bay estuary is a major component of the Jacques Cousteau National Estuarine Research Reserve in southern New Jersey (Figure 1B). Euhaline and polyhaline portions are bordered by *Spartina*-dominated salt marsh. It is approximately 40 river kilometers from Little Egg Inlet to the Mullica River fork with the Batsto River. In the upper 10 km, the river is shallow with frequent gravel bars and bordered by cedar and pine forest that lower pH to as low as 5 (Kennish 2004). The habitat, its fauna, and seasonal gradients are relatively well known (Able and Fahay 1998; Martino and Able 2003). The Great Bay estuary is 90 km northeast of the Delaware Bay and 120 km southwest of the Hudson River estuary, both important striped bass spawning sites (Waldman et al. 1990). Our own surveys of the Mullica River system over the course of 18 years (Able and Fahay 1998; Martino and Able 2003; K. W. Able, unpublished data) as well as those of Stockton College (S. Evert, Richard Stockton College of New Jersey, personal communication) since the early 1970s have collected less than 20 striped bass within the young-of-year size range (<100 mm total length) in any region of the river. A study directed at describing striped bass reproduction in the river collected 142 eggs, 7 young of year, and a single yearling out of 524 specimens (Hoff 1976). Thus, it is likely that spawning occurs here but that it is rarely successful, possibly owing to lethally low ambient pH (Tatum et al. 1965; Mehrle et al. 1984, Martino and Able 2003; Kennish 2004). Thus, the Mullica River–Great Bay estuary is farther south than Saco River, warms earlier, fluctuates a meter less in tide level, and has lower pH.

Telemetry

Striped bass in both river systems were surgically implanted (tagged) with acoustic transmitters (76.8 kHz, CAFT-series, Lotek Wireless Inc., St. Johns, Canada) following Grothues et al. (2005) and Able and Grothues (2007). Most transmitters were of 2-year duration, but some fish received smaller transmitters of 1-year duration. Fish were secondarily marked with external tags to encourage release or notification. Implantation was conducted under the Rutgers University Animal Use Protocol # 04–020 and the University of New England Protocol #UNE08–2007.

Fish were detected when they came within range of wireless hydrophones (WHS_1100, Lotek Wireless, Inc.) moored in the estuary inlets and other constrained areas (Figure 1). Hydrophone detection areas divided the estuaries and rivers approximately at euhaline/polyhaline, polyhaline/oligohaline, or freshwater interfaces (Figure 1). See Grothues et al. (2005) for a complete description of the hydrophone array and its function in the Mullica River–Great Bay estuary. Mobile tracking was also undertaken from a boat throughout the Mullica River–Great Bay, covering 113 preset listening points with overlapping radii of detection (Ng et al. 2007). Mobile tracking was used to examine habitat use of striped bass down to the meter scale twice weekly or more often in spring, summer, and fall, but was only occasionally undertaken when the wireless hydrophone data indicated that all tagged fish had left the system in winter. Mobile telemetry was accompanied by physicochemical measures using a handheld CTD (conductivity, temperature, and depth). In the Saco River, four hydrophones monitored the polyhaline bay outside of the inlet and the polyhaline Biddeford Pool, two hydrophones gated passage to the lower oligohaline reach, and one monitored freshwater use near the Cataract Dam (this one in 2005 only). Mobile telemetry in the Saco River was less frequent, occurring primarily in 2006 to help monitor use of the upper reaches in the absence of a moored hydrophone there.

Data Analysis

We examined the movement of fish through and among the study sites on an individual basis to comparatively describe habitat use and migration. We compared the incidence of absence, revisitation, residence time, and recaptures inside and away from the study sites as indications of movement outside of the study sites. Further, we statistically described gradients in habitat use and visitation using ranked scores. We calculated a similarity matrix from a compiled score matrix of individual striped bass presence in various regions of each estuary on a biweekly scale. For the first and second half of each month, each fish was scored on the basis of its presence in the polyhaline (1), oligohaline (2) both (3), fresh (4), oligohaline and fresh (5), or all segments (6) of an estuary, separately for the New Jersey and Maine study areas. A score of 7 was given for a period in which a fish tagged in one estuary was detected in the other estuary and 8 if it entered the river in that estuary. Similarity was presented using cluster analysis, which represents ranked order of similarity between pair groups of individuals as a dendrogram (Legendre and Legendre 1998). It is therefore useful in delineating natural groups based on similar sets of behavior. Natural groups are expected when control of behavior owes to few or strong factors, such as tight coupling to genotype.

Cluster analysis required the standardization of both timing of tagging and "time at liberty." As used here, this indicates the time between when a tag was surgically implanted and the time during which the battery expired or the fish was harvested, whichever came first. To standardize, we classified the time periods simply as the first, second, or third repetition of the first or second half of a particular month at liberty without reference to year. Then, we truncated both the beginning and end periods of the matrix so that all fish would have had the same time at liberty. While this forced the deletion of much of the data on a given fish (although only for the purpose of clustering), it did allow the inclusion of some fish with shorter-lived tags and some that had been harvested by anglers before their tags would have expired. Fish that were not detected subsequent to their tagging but were not reported as harvested were still included. For the Mullica River–Great Bay study site, this yielded a matrix of 19 months and 68 fish. A 12-month period was used for the Saco River for 35 fish. Calculation of the distance matrix was based on the nonnormalized Euclidian metric. The unweighted pair-group arithmetic mean algorithm was used to construct cluster dendrograms (Primer E. Limited, Plymouth, UK).

Standardization of time at liberty had the side effect of removing year as a variable from calculation of the distance matrix. To address this, we analyzed the effect of year of capture as an independent variable using a one-way nonparametric analysis of similarity (global and pairwise analysis of similarities). Additionally, the effect of fish size at capture on Euclidean distance was explored with a Mantel family test using the Spearman's ranked correlation statistic (both also in Primer E).

Results

Characteristics of Sampled Fish

Striped bass were tagged in both the Mullica River–Great Bay, New Jersey (n = 115, 324–952 mm fork length [FL]) and the Saco River, Maine (n = 55, 406–889 mm FL). The timing of initial capture and tagging spanned both a wider annual and monthly range in New Jersey (November 2002 to June 2007) than in Maine (October 2003 to September 2006). Both larger and smaller fish were tagged in New Jersey, but the average size of fish (605 mm FL) sampled in Maine was significantly smaller than in New Jersey (683 mm FL) ($P < 0.0001$, Kolmogorov-Smirnov test, $\alpha = 0.05$; Figure 2). Fish in both estuaries were tagged throughout the river to ocean gradient, and in both estuaries, several fish were captured and released outside of the telemetered study area. In Maine, five fish were tagged in the Kennebec River; one subsequently entered the Saco River study area, although it stayed in the coastal ocean. In New Jersey, four fish were tagged in Little Egg Harbor (Figure 1), an estuary adjacent to and sharing one of two inlets with the study area; three of these eventually entered the New Jersey study area. Also, six fish were tagged in the ocean outside of the study area.

Migrations Between New Jersey and Maine and Elsewhere

Exchanges occurred between study sites primarily as fish tagged in Maine and subsequently detected in New Jersey. Of two exceptions, one was tagged in

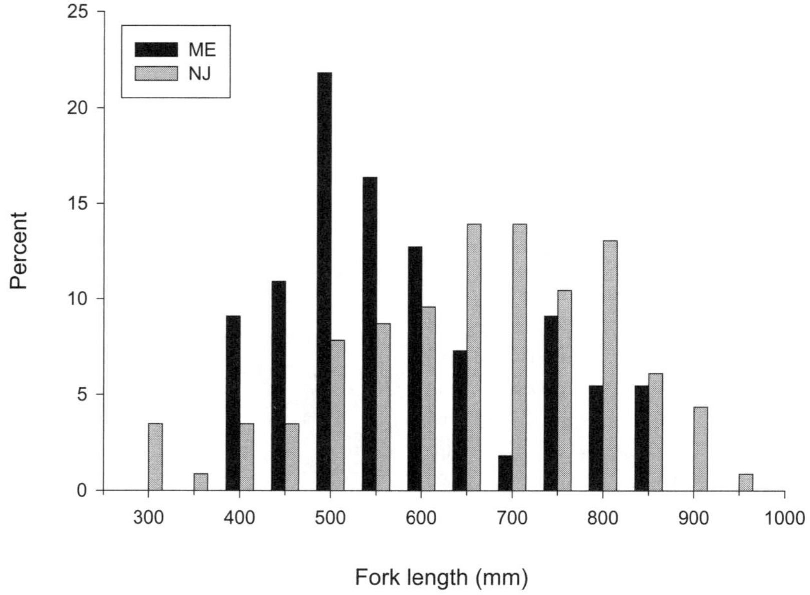

Figure 2.—Size distribution of striped bass tagged with acoustic transmitters in New Jersey and Maine between 2002 and 2007.

New Jersey in May 2003 when it ascended the Mullica River. It was detected briefly in Maine in August 2004 but did not enter the Saco River. It returned to the New Jersey study area in April 2005 but did not ascend the river, left, and returned to Maine where it was harvested by an angler in May 2005. Thus, it was not resident at either study site for more than a few days. Transit time between the second New Jersey visit and its harvest in Maine was 43 d. Another fish, tagged in November 2002, moved to the Saco River in June 2004 and thereafter was occasionally detected in the bay from August to early October 2004 before moving again to New Jersey in late November, taking 54 d for the return trip. Of fish that were tagged in Maine, four were seen in New Jersey in the same season of their initial departure from the Maine study site (southerly fall migration), taking 45, 52, 63, and 80 d in transit. Of these, one resided in the Saco River for more than 2 months in each year before and after migration. Tags on several of the other fish that migrated through New Jersey were of short duration and may not have been able to document return migration. One individual was tagged in October 2004 in Maine but was only detected in New Jersey in April 2005, and another tagged in Maine in August 2004 was detected in a Chesapeake Bay tributary (York River) hydrophone array (for American shad *Alosa sapidissima* study) for several days in late March 2005 (J. Olney and B.Watkins, Virginia Institute of Marine Science, personal communication); thus, they were not detected in the New Jersey study area during their southern migration.

Recapture location outside of the study areas by anglers was consistent with spring/summer movement northward and fall movement to the south (Table 1). Thirty-four (20%) tagged striped bass were captured again; 30 were from New Jersey samples (26% recapture rate) and 4 from the Maine samples (7% recapture rate) (Table 1). Two fish were recaptured twice. Eleven fish were recaptured while still inside the study area where tagged (one twice), only one of these in Maine. Fish that were recaptured away from the study areas included fish that spent as much as a month in the estuary of tagging (seasonal residents) and some that stayed only briefly. Fish that were recaptured within the array included primarily individuals that spent many consecutive months inside the study site as seasonal residents.

Interannual and Seasonal Differences in Estuarine Movement

When the duration and timing of tagging allowed, we determined interannual differences in

Table 1.—Recaptures of fish tagged with acoustic transmitters. Unless otherwise noted in the "Fish ID" column by "(ME)", fish were originally tagged in the New Jersey study site. Some tag identifications were reused after battery expiration and are identified with "(2)" following the identification number. All dates are MM/DD/YY.

Fish ID	Date tagged	Date recaptured	Last date in array	Days from array exit to recapture	Approximate distance (km)	Location of recapture
2	11/14/02	08/01/04	3/27/04	127	680	NH
165	11/25/02	12/06/03	11/25/03	11	–	Little Egg Inlet, Great Bay, NJ
151	03/25/03	04/05/05	04/26/03	711	–	Little Egg Inlet, Great Bay, NJ
153	03/27/03	12/03/03	11/06/03	27	–	Barnegat Bay, NJ
149	04/02/03	11/25/05	04/14/04	590	90	Delaware Bay
124	05/02/03	05/30/05	04/17/05	43	700	Two Lights, Cape Elizabeth State Park, ME
126	05/02/03	12/10/05	05/17/04	572	140	Raritan Bay, NJ
108	05/05/03	06/05/04	05/07/04	29	110	Sandy Hook, NJ
134	05/09/03	10/10/03	05/17/03	146	8	Absecon Inlet, NJ
136	05/09/03	08/05/03	05/20/03	77	240	Shinecock Inlet, Long Island, NY
163	05/22/03	06/22/03	05/22/03	31	8	Absecon Bay, NJ
127	06/10/03	05/09/04	–	–	–	Holgate tip, Great Bay, NJ
119	09/03/03	10/03/04	–	–	–	Mullica River, NJ
119	09/03/03	Fall 2005	–	–	–	Mullica River, NJ
141	09/26/03	09/18/05	05/19/04	487	130	Southern Long Island, NY
180	10/16/03	10/22/04	10/24/03	364	17	Brandt Beach , NJ
15	10/24/03	06/17/05	–	–	–	Great Bay, NJ
196	04/22/04	11/05/06	–	–	–	Great Bay, NJ
132	04/29/04	08/06/05	05/14/04	449	470	Monomoy Island, MA
90	05/14/04	10/30/04	–	–	–	Mouth of Mullica River, NJ
122	05/21/04	06/18/05	–	–	–	Great Bay, NJ
37	06/04/04	02/20/05	06/11/04	254	460	Croatan Sound, NC
186	06/04/04	06/25/06	–	–	–	Little Egg Harbor., NJ
205 (ME)	9/23/04	05/11/05	–	230	650	Barnegat Inlet, NJ
205 (ME)	9/23/04	11/12/06	–	550	540	Sandy Hook, NJ
2 (2)	04/07/05	05/10/07	10/15/05	527	310	MD
168	05/27/05	06/18/05	–	–	–	Great Bay, NJ
22 (ME)	06/13/05	11/30/05	07/16/05	137	810	Bowers Beach, DE
122 (2)	06/29/05	11/08/05	10/24/05	15	–	Brigantine, NJ
210 (ME)	07/21/05	08/23/06	07/29/05	390	85	Merrimack River, MA
130	07/26/05	11/05/05	08/01/05	96	460	Fire Island Inlet, Long Island, NY
59	10/24/05	05/28/07	–	–	–	Little Egg Inlet, Great Bay, NJ
106	05/02/06	06/06/06	05/30/06	7	17	Atlantic City, NJ
203	06/21/06	07/02/06	–	–	–	Little Egg Inlet. Great Bay, NJ
4 (ME)	07/27/06	08/18/06	–	–	–	Biddeford Pool, ME
204	11/01/06	04/28/07	–	178	330	Chesapeake Bay, ME

the proportion of fish moving upstream during the spring spawning period (Figure 3), as well as in the timing of estuarine entrance and egress. In New Jersey, in 2003, 82% of the fish tagged in the spring and previous fall proceeded upriver at least to the polyhaline/oligohaline interface at Chestnut Neck and 36% as far as the freshwater boundary at Lower Bank. These included indi-

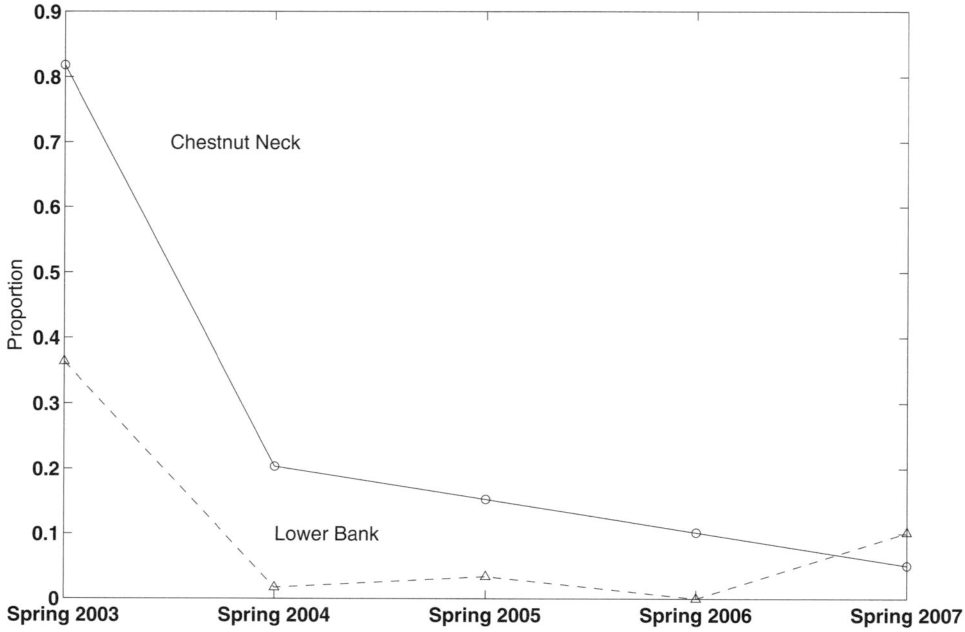

Figure 3.—Proportion of the total striped bass with active tags in each year (i.e., batteries not expired or fish not harvested before each spring) that utilized river stretches at Chestnut Neck or Lower Bank in New Jersey (see Figure 1B) between April 1 and June 30. The proportion of fish at Lower Bank is higher than downstream (Chestnut Neck) in 2007 because fish tagged between the sites in that year ran upstream and did not descend past Chestnut Neck until after June 30.

viduals that left the estuary within days of moving upriver as well as individuals that remained in the river and bay for much of the ensuing summer. Few of the fish tagged in early spring of 2004 ascended the river. Also, few of the fish returning to the estuary from previous tagging moved into the river. Therefore, it is unlikely that tagging efforts from 2004 simply missed this movement. This trend continued in 2005 and 2006. In 2007, newly tagged fish made an upstream movement in spring, but fewer of the total tagged fish from previous years returned to the estuary by the end of June and so the proportion moving upstream remained low.

Analysis for consistency in upstream movement into freshwater was based on 74 individuals, including 5 fish from Maine that had 2-year duration tags and that were not recaptured before their second spring. Of these, 25 (32.7%) were never detected again after their initial departure from the study area (this may include undocumented mortality). Of those subsequently detected, 19 fish (25.6% of the total) followed the same pattern in both years (either stayed in polyhaline water or ascended the river), 21 (28.4%) switched from ascending the river in the first year to staying in the polyhaline bay the second year, 3 (4%) switched from only polyhaline use to oligohaline or freshwater in the second year, and 6 (8.1%) utilized part of the study area in one spring period and were absent in the other but did enter the estuary in summer or fall of the second year and thus differ from the group with potential mortalities. One fish stayed in the Mullica River–Great Bay for a full year but left in the second spring and did not return. Two other fish overwintered after entering the estuary in late fall. Otherwise, striped bass used both study areas only seasonally. The season of use was longer in New Jersey, between April and December, than in Maine where it was generally between June and November. A number of individuals stayed for the duration of this period in both estuaries while others were present for only a few hours or days (Figure 4).

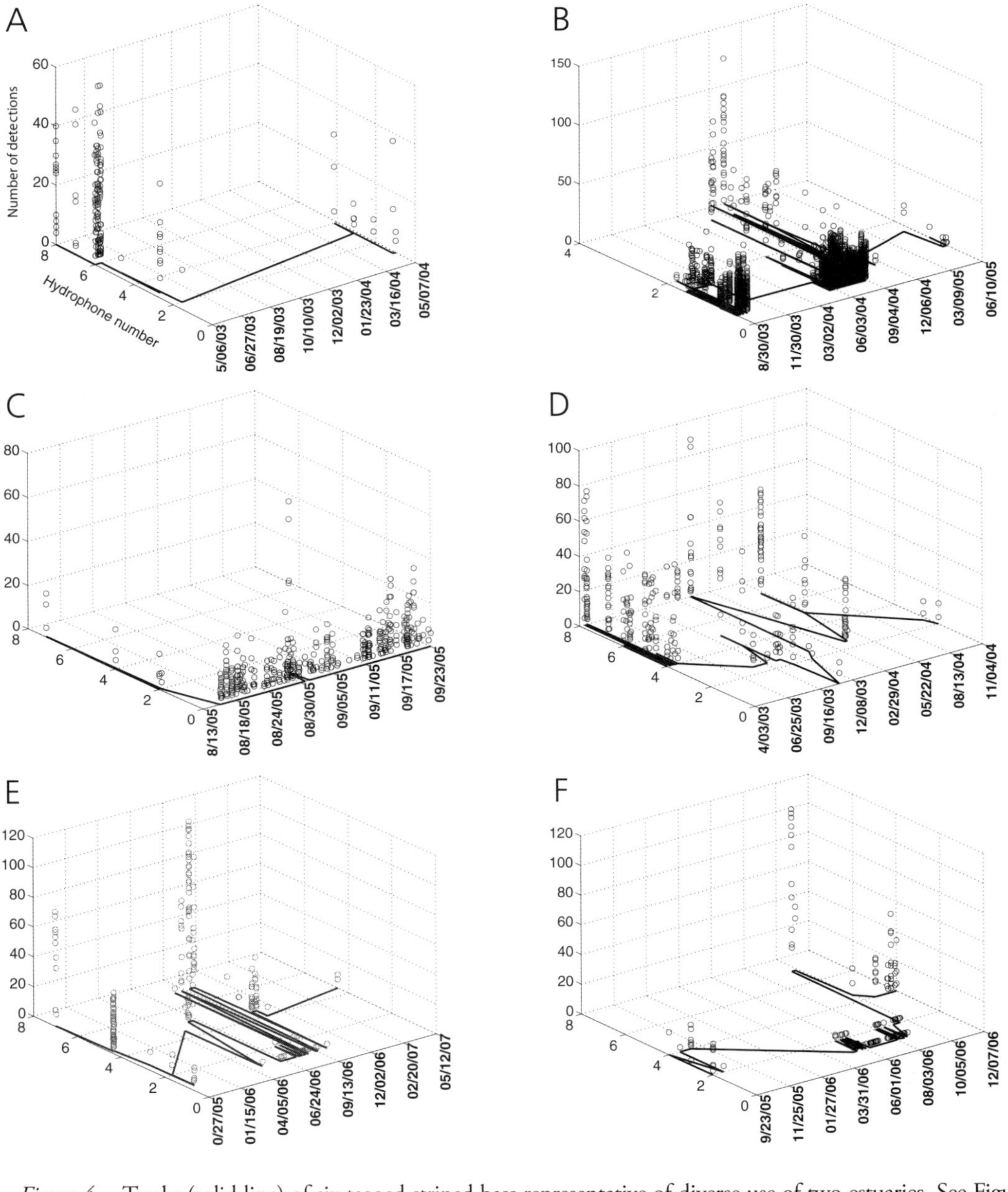

Figure 4.—Tracks (solid line) of six tagged striped bass representative of diverse use of two estuaries. See Figure 1 for locations of hydrophones. Height of bubbles along the *y*-axis indicates the number of tag detections in 15-min increments (time along *x*-axis) for a given hydrophone (*z*-axis). Examples are from (A) a fish that briefly ascended the Mullica River, New Jersey to freshwater in spring of 2003 returned briefly in 2004 but did not ascend the river; (B) a fish that resided in Little Egg Inlet, New Jersey, from capture in 2003 through late summer, returned to the site in late spring 2004 and made several short excursions upriver in summer and fall, left, and returned in summer of 2005 before tag expiration; (C) a fish that ascended the Saco River, Maine in mid-August and retreated to the bay where it resided near Biddeford Pool for the following month; (D) A fish that made three excursions between the inlet and brackish or fresh reaches of the Mullica River, New Jersey in early spring 2003, remained in the lower estuary in fall 2003, overwintered in fresh reaches, returned to the bay, and repeated the spring run upriver as well as summer and fall residency in 2004 prior to egress in November; (E) a fish that entered the oligohaline stretch of the Mullica River, New Jersey in fall 2005 and 2006 but only the euhaline bay in spring (2006 and 2007); and (F) a fish that utilized the oligohaline Mullica River, New Jersey in fall of 2005 and returned in 2006 and overwintered in polyhaline stretches.

Pattern of Estuarine and Habitat Use

In both estuaries, the dominant habitat use was of the euhaline portion of the estuary, although these estuaries differed in geomorphology. However, classification did not fall into distinct groups at high rank levels of similarity (Figure 5) because of diversity in the length of stay, rate of return, and use of the estuarine gradient throughout the years. At one end of the spectrum, a low proportion of tagged fish in both estuaries spent a majority of their time, relative to the length of the season, well into the river; others made repeated excursions between the inlet and freshwater regions on the order of days near both the beginning and end of the season but spent a majority of time in the inlet. Others used only the polyhaline regions just outside (Maine) or just inside the inlet (New Jersey).

A common behavioral trait among striped bass in New Jersey that stayed for a period of months, and for fishes with 2-year tags that returned to the estuaries in which they were tagged, was a high fidelity to "home" ranges. This was captured on the broad scale by the pattern of detection on fixed hydrophones and confirmed by mobile tracking to be oriented to prominent physical features such as deep cuts (Ng et al. 2007). In cases where fall tagging allowed observations of three spring returns, site fidelity was observed to extend over 3 years. Still others utilized two primary locations between which they switched sea-

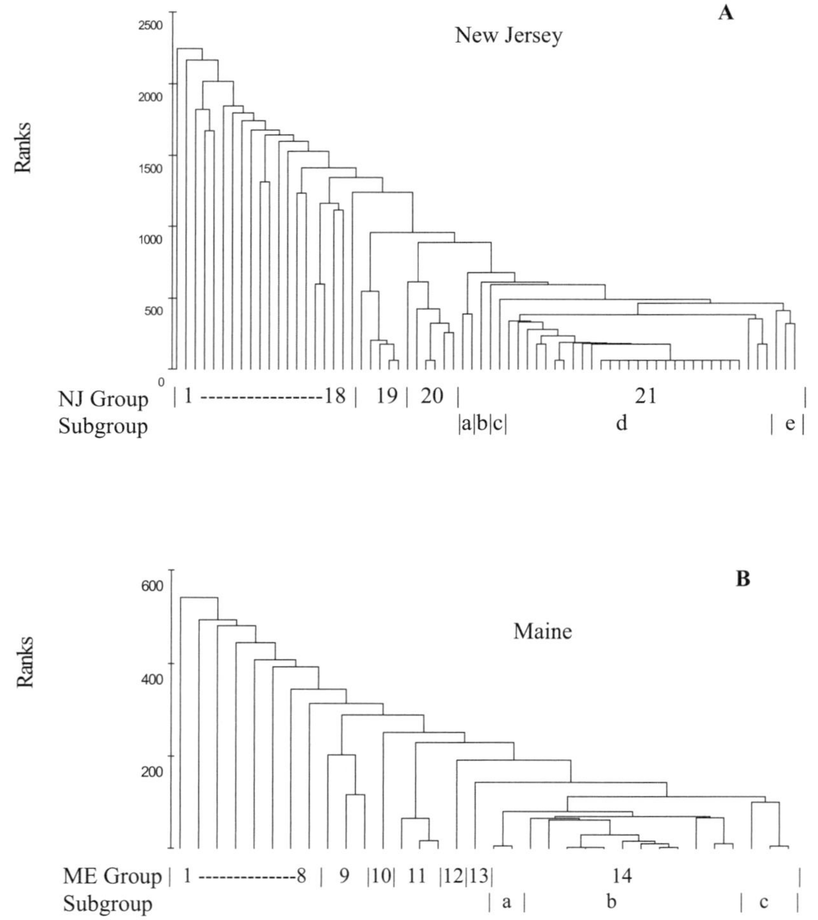

Figure 5.—Dendrograms of ranked Euclidean distance between individual striped bass based on their use of fresh, oligohaline, and euhaline stretches in the (A) Mullica River–Great Bay, New Jersey and (B) Saco River, Maine. *X*-axis labels identify cluster groups defined in the text.

sonally, even repeating this bimodal fidelity among years. This pattern was evident, though less so, in the Saco River, but was observed in the polyhaline bay (J. Carter and T. Arienti, unpublished data).

The diversity of habitat use patterns in both estuaries was captured in the cluster diagrams (Figure 5). It was possible to identify 18 singletons and pair groups (New Jersey group1–18) in a linkage cascade above the 800 rank level of similarity. These differed widely in habitat use, from frequent but short return visits to season-long residency in the river or bay and appearance in Maine. Four clusters grouped fish below this level based on repetition and timing of short spring visits. Individuals in New Jersey group 19 (n = 5) made spring runs upriver in the first spring after tagging but revisited the estuary in a second spring season without ascending the river. Individuals in New Jersey group 20, (n = 6) ascended the river briefly in their first spring but never reentered the study area. Individuals from New Jersey group 21 (n = 37) never ascended the river as far as Chestnut Neck, but differed in their use of the bay by timing, duration, and repetition. New Jersey group 21d included fish with brief or no visits in the inlet during the standardized time series. Fish from New Jersey group 21a and 21b spent much of their first full spring in the inlet or lower bay (beginning April or May) and remained there until October without ascending the river. Several of these also visited the estuary in the beginning of their second spring but did not ascend the river then either.

Some of the cluster assignments at the lower level, especially between New Jersey groups 19/20 and 21 appeared to be an effect of the difference in year of tagging (Table 2). A global test for difference, based on the year of their capture (and therefore the year in which they could have been documented to ascend the river), was significant (Spearmans Statistic R = 0.059, P = 0.044, 6,000 permutations). Pairwise tests proved significant for differences between fish tagged in 2002 (n = 3 fish) and those from each other year (Table 2). Use of the estuarine and river habitat as scored here did not correlate with fish size (ρ = –0.071, P = 0.896, 2000 permutations).

Clustering of fish tagged in Maine also did not resolve groups at a high level of similarity (Figure 5). A cascade of 10 singletons (Maine groups 1–8, 10, and 13) and two triplets Maine group 9 and 11) accounted for all structure above rank 180. Of the singletons, five were distinguished by their appearance in the New Jersey study site in late fall, winter, or early April. However, use of the Maine study site among all of these varied in length of residence and timing of repeated upriver excursions. Below this rank level, three groups were resolved (Maine 14a–c). One pair (Maine group 14a) briefly entered the oligohaline Saco River in their first summer only. A group of three (Maine group 14c) entered the oligohaline stretch in summer and retreated to the bay for a period of weeks to months but did not return after departing. The third subgroup (Maine 14b, n = 12) had short visits (2 weeks or less) or no appearances in the hydrophone array during this period and varied in the timing of their visit. Of those, two entered the oligohaline reach, but none entered freshwater. Fish size did not correlate with ranked similarity distance

Table 2.—Test for effect of tagging year on fish rank in similarity matrix. *P* values were obtained using a Monte Carlo random permutation (n = 4000 permutations). Significant values are in bold.

Global test
Sample statistic (Global *R*): 0.059
P = 0.044

Years	*R*	*P*	Possible permutations	Actual permutations	Number observed
2002, 2003	0.571	**0.017**	3,276	3,276	56
2002, 2004	0.693	**0.008**	1,771	1,771	15
2002, 2005	0.631	**0.012**	1,771	1,771	21
2003, 2004	–0.008	0.510	Too many	6,000	3,058
2003, 2005	–0.024	0.788	Too many	6,000	4,730
2004, 2005	–0.012	0.808	Too many	6,000	4,846

for Maine samples (Spearman Rank Correlation ρ = –0.163, P = 0.97.9, 2000 permutations).

Discussion

It is clear from this and previous work (Able and Grothues 2007; Ng et al. 2007) that individual striped bass are using nonnatal estuaries along the U.S. mid-Atlantic coast both as destinations (following Ramenofsky and Wingfield 2007) and as habitats that they only briefly enter during extended coastal movements. All of the fish that we observed were migrants and thus members of contingents that mingle along the coast (Boreman and Lewis 1987; Fabrizio 1987, Wirgin et al. 1997, Rulifson et al. 2008); thus, our observations did not include any true natal freshwater or natal estuarine residents. However, we have shown that some migratory fish also use primarily polyhaline or oligohaline waters and that some of the migratory fish are not continuous ocean water users. Observed behaviors included seasonal maintenance of small ranges and periodic weeks-long excursions indicative of exploration or exploitation of temporary resources. These also included movements into freshwater during the spawning season for periods that would exclude spawning elsewhere and thus may be failed reproductive attempts or skipped spawning opportunities, as documented in other species (Rideout et al. 2005; Jorgensen et al. 2006). These behaviors sometimes changed within individuals among years.

Several caveats must be addressed before discussing the implications of these results, especially as they relate to a perspective of life history strategies coming from earlier works (i.e., Secor et al. 2001; Morris et al. 2003). First, the spatial and temporal scale of our observations are finer than those on which a hypothesis of divergent life history is formulated (otolith microchemistry), but like those, they indicate whether and for how long individual fish were resident in or absent from the estuary. Classification on the basis of 2-week observation blocks may be fine enough to smear the boundaries of segregation into natural groups similar to contingents. However, these fine-scale observations show that individuals differentiate on habitat choice along the salinity gradient within a single estuary, as has been used to define contingents in juvenile white perch *Morone americana* (Kraus and Secor 2004). Since every fish participated in migration at some time after tagging, we can say that all fish remained migratory.

Second, there is a potential size bias in our understanding of whether striped bass may be recruited to residency. Unlike studies that are conducted in larger natal systems, we observed no large adult fish remaining in the study systems overwinter after summer/fall residency, although several smaller individuals did remain and others entered the Mullica River from the ocean to overwinter. Thus, we would predict that smaller striped bass that did make their way into the estuary would be more likely to stay the winter and following spring than larger fish. This was not born out in our analysis, but it is quite possible that the smaller fish we sampled were above a threshold size or age where this might be observed. Telemetry of even smaller (<300 mm FL) subadults, which do occur in these rivers and estuaries despite the virtual absence of young of year, remains an important direction for further work. The failure to detect true residents (as opposed to seasonal residents) also agrees with the idea that true residency is learned early in life (Secor 1999), and it may, therefore, not become readily established in a nonnatal estuary. However, the fidelity of some fish to particular sites, as well as the plasticity in the behavior of some other individuals, demonstrates that learning and habitat evaluation and selection continues for larger fish.

About half of all striped bass sampled in larger systems of the United States eventually join coastal migrants, but the timing of this varies among individuals and with sex (Secor and Piccoli 1996; Secor et al. 2001; Secor and Piccoli 2007). Striped bass in Atlantic Canada are also observed (through mark–recapture methods) after spawning to segregate into a resident group and a migrant group of larger fish (>390 mm at release) that mix with fish of U.S. origins (Rulifson and Dadswell 1995; Rulifson et al. 2008). The variability and plasticity that we observed in habitat choice agrees with Secor (1999) that migration is more reactive to local conditions or is learned (e.g., as for Atlantic cod *Gadus morhua*, Windle and Rose 2005) and less compulsive than in some salmoninae. This provides for the possibility that extirpated or newly restored areas (such as following dam removal) could quickly recruit adults or subadults to establish new spawning populations if that new habitat proved suitable over the whole year. Several lines of inquiry follow from this: Is it even possible for smaller river systems north of Cape Hatteras (i.e., other than the Chesapeake, Delaware, Hudson, and Kennebec rivers) to offer suitable habitat for adult fish year-round? Do smaller systems lack the thermohaline refugia of larger water bodies (e.g., Hurst and Conover 2002) in the current climate and have the scale for stable predator–prey dynamics (e.g., Hartman 2003; Uphoff 2003)? Establishment of new resident populations or recovery of extirpated populations such as the St. Lawrence River (Magnin and Beaulieu 1967; Bradford et al. 2001) might also require a critical number of individuals. A California coastal fishery for striped bass became self-sufficient and yielded annual harvests of 450,000 kg just 10 years after stocking of 332 juveniles into the San Francisco Bay (Stevens et al. 1987), but this is a large system and founders could have benefited by being imprinted at an early age. Numerous other introductions of striped bass in the United States and worldwide failed to establish self-sustaining populations (Westin and Rogers 1978). These questions are addressable through continued telemetry, particularly with compatible listening stations in natal and nonnatal river systems.

Our results also imply that intense fishing pressure in smaller systems can result in annual local depletion with little effect on the size or structure of a healthy parent stock. Seasonal feeding habitats vacated by capture of striped bass in fisheries or from natural mortality should remain unfilled until replenishment in the ensuing year from a migrant stock. Conversely, reserve designation of one or more of these small estuaries would offer only marginal stock protection. The potential occurrence of seasonal destination sites in the ocean needs to be examined and compared to seasonal residence habitat in estuaries to fully understand habitat use and migration in response to natural and anthropogenic perturbation in stock size, community change, and climate change.

Acknowledgments

We thank R. Nichols, J. Turnure, R. Vohden, M. Greany, M. Holon, C. Ng, and D. Sackett in New Jersey and N. Hamacher, J. Kyne, A. Deventitis, J. Carter, W. Gunn, and M. Meserve in Maine who helped with fixed and mobile telemetry. Many individuals supported this effort through the "adoption" of fish in donations for the transmitters, especially P. Gieseler and J. Wilson of the BassBarn.com cooperative effort and the students and faculty of the Eisenhower School District in New Jersey. The local knowledge and angling skills of D. Messerschmid, L. Webb in New Jersey and Rippin Lips Charters, Trina Lyn Charters, and Obsession Charters in Maine were crucial to the success of our tagging efforts. J. Oddi, J. and G. McDonald, and B. Flannery provided property access for equipment siting, and J. Olney and B. Watkins loaned additional equipment. The manuscript was greatly improved with the help of S. Haag and several reviewers. This project was supported by the Maine Department of Marine Resources, the Maine Technology Institute's Maine Marine Research Fund, the Maine Outdoor Heritage Fund, the Coastal Conservation Association of Maine, the Bluefish/Striped Bass Dynamics Research Program of Rutgers University and NOAA/National Marine Fisheries Service, and a generous gift from Jules Plangere, Jr. This is contribution no. 2008-13 of the Rutgers University Institute of Marine and Coastal Sciences and contribution no. 6 from the University of New England's Marine Science Center.

References

Able, K. W., and M. P. Fahay. 1998. The first year in the life of estuarine fishes in the Middle Atlantic Bight. Rutgers University Press, New Brunswick, New Jersey.

Able, K. W., and T. M. Grothues. 2007. Diversity of striped bass (*Morone saxatilis*) estuarine movements: synoptic examination in a passive, gated estuarine observatory. Fisheries Bulletin 105:426–435.

Bjorgo, K. A., J. J. Isely, C. S. Thomason. 2000. Seasonal movement and habitat use by striped bass in the Combahee River, South Carolina. Transactions of the American Fisheries Society 129(6):1281–1287.

Boreman, J., and R. R. Lewis. 1987. Atlantic coastal migration of striped bass. Pages 331–339 *in* M. J. Dadswell, R. J. Klauda, C. M. Moffitt, R. L. Saunders, R. A. Rulifson, and J. E. Cooper, editors. Common strategies of anadromous and catadromous fishes. American Fisheries Society, Symposium 1, Bethesda, Maryland.

Bradford, R. G., D. Cairns, and B. Jessop. 2001. Update on the status of striped bass (*Morone saxatilis*) in eastern Canada in 1998. Department of Fisheries and Oceans, Canadian Stock Assessment Secretariat Research Document 2001/007, Ottawa.

Carl, L. M., and M. C. Healy. 1984. Differences in enzyme frequency and body morphology among three juvenile life history traits of Chinook salmon (*Onchyrynchus tshawyscha*) in the Naniamo River, British Columbia. Canadian Journal of Fisheries and Aquatic Sciences 41:1979–1077.

Clark, J. 1968. Seasonal movements of striped bass contingents of Long Island Sound and the New York Bight. Transactions of the American Fisheries Society 97(4):320–343.

Collette, B. B., and G. Klein-MacPhee. 2002. Bigelow and Schroeder's Fishes of the Gulf of Maine. Smithsonian Institution Press, Washington, D.C.

Davis, R. H., Jr. 1966. Population studies of striped bass *Roccus saxatilis* (Walbaum), in Maine based on age distribution and growth rate. Master's thesis. University of Maine, Orono.

Dudley, R., A. Mullis, and J. Terrel. 1977. Movements of striped bass in the Savannah River, Georgia. Transactions of the American Fisheries Society 106:314–322.

Fabrizio, M. C. 1987. Contribution of Chesapeake Bay and Hudson River stocks of striped bass to Rhode Island coastal waters as estimated by isoelectric focusing of eye lens proteins. Transactions of the American Fisheries Society 116:588–593.

Gross, M. R. 1987. Evolution of diadromy in fishes. Pages 14–25 *in* M. J. Dadswell, R. J. Klauda, C. M. Moffitt, R. L. Saunders, R. A. Rulifson, and J. E. Cooper, editors. Common strategies of anadromous and catadromous fishes. American Fisheries Society, Symposium 1, Bethesda, Maryland.

Grothues, T. M., K. W. Able, J. McDonnell, and M. M. Sisak. 2005. An estuarine observatory for real-time telemetry of migrant macrofauna: design, performance, and constraints. Limnology and Oceanography Methods 3:275–289.

Haeseker, S. L., J. T. Carmichael, and J. E. Hightower. 1996. Summer distribution and condition of striped bass within Albemarle Sound, North Carolina. Transactions of the American Fisheries Society 125:690–704.

Hartman, J. K. 2003. Population-level consumption by Atlantic coastal striped bass and the influence of population recovery upon prey communities. Fisheries Management and Ecology 10(5):281–288.

Hoff, H. K. 1976. The life history of striped bass (*Morone saxatilis* Walbaum), in the Great Bay–Mullica River estuary and the vicinity of Little Egg Inlet. Pages 43–53 *in* C. B. Milstein and D. L. Thomas, editors. Ecological studies of the bays and other waterways near Little Egg Inlet and the ocean in the vicinity of the proposed site for the Atlantic generating station. New Jersey Progress Report Jan–Dec 1975. Ichthyological Associates, Inc., Lansing, New York.

Houde, E. D. 1996. Evaluating stage-specific survival during the early life of fishes. Pages 51–66 *in* Y. Watanabe, Y. Yamashita, and Y. Oozeki, editors. Survival strategies of early life stages of marine resources. A. A. Balkema, Rotterdam, Netherlands.

Hurst, T. P., and D. O. Conover. 2002. Effects of temperature and salinity on survival of young-of-the-year Hudson River striped bass (*Morone saxatilis*): implications for optimal overwintering habitats. Canadian Journal of Fisheries and Aquatic Sciences 59:787–795.

Hyatt, K. D., and J. D. Stockner. 1985. Response of sockeye salmon (*Oncorhynchus nerka*) to fertilization of British Columbia coastal lakes. Canadian Journal of Fisheries and Aquatic Sciences 42:320–331.

Jorgensen, C., B. Ernande, O. Fiksen, and U. Dieckmann. 2006. The logic of skipped spawning in fish. Canadian Journal of Fisheries and Aquatic Sciences 63:200–211.

Kelhir, P. 2005. 2004 Annual Report of the Maine Atlantic Salmon Commission. Maine Atlantic Salmon Commission, Augusta.

Kennish, M. J. 2004. Jacques Cousteau National Estuarine Research Reserve. Pages 59–115 *in* M. J. Kennish, editor. Estuarine research monitoring, and resource protection. CRC Press, Baton Raton, Florida.

Kraus, R. T., and D. H. Secor. 2004. The dynamics of white perch (*Morone americana* Gmelin) population contingents in the Patuxent River estuary, Maryland, USA. Marine Ecology Progress Series 279:247–259.

Legendre, P. and L. Legendre. 1998. Numerical ecology. Elsevier, Amsterdam.

Magnin, E., and G. Beaulieu. 1967. Striped bass of the St. Lawrence River. Naturaliste Canadien (Quebec) 94:539–555.

Martino, E., and K. W. Able. 2003. Fish assemblages across the marine to low salinity transition zone of a temperate estuary. Estuarine Coastal and Shelf Science 56(5–6):967–985.

Mehrle, P. M., D. Buckler, S. Finger, and L. Ludke. 1984. Impact of contaminants on striped bass. U.S. Fish and Wildlife Service, Columbia National Fisheries Research Library, Columbia, Missouri.

Morris, J. A., R. A. Rulifson, and L. H. Toburen. 2003. Life history strategies of striped bass, *Morone saxatilis*, populations inferred from otolith microchemistry. Fisheries Research 62:53–63.

Ng, C. L., K. W. Able, and T. M. Grothues. 2007. Habitat use, site fidelity, and movement of adult striped bass in a southern New Jersey estuary based on mobile acoustic telemetry. Transactions of the American Fisheries Society 136:1344–1355.

Nordeng, H. 1983. Solution to the "char" problem based on Arctic char (*Salvelinus alpinus*) in Norway. Canadian Journal of Fisheries and Aquatic Sciences 40:1372–1387.

Paramore, L. M., and R. A. Rulifson. 2001. Dorsal coloration as an indicator of different life history patterns for striped bass within a single watershed of Atlantic

Canada. Transactions of the American Fisheries Society 130:663–674.

Rago, P. J., and C. P. Goodyear. 1987. Recruitment mechanisms of striped bass and Atlantic salmon: comparative liabilities of alternate life histories. Pages 402–416 *in* M. J. Dadswell, R. J. Klauda, C. M. Moffitt, R. L. Saunders, R. A. Rulifson, and J. E. Cooper, editors. Common strategies of anadromous and catadromous fishes. American Fisheries Society, Symposium 1, Bethesda, Maryland.

Ramenofsky, M., and J. C. Wingfield. 2007. Regulation of migration. Bioscience 57(2):135–143.

Randall, R. G., M. C. Healey, J. B. Dempsey. 1987. Variability in length of freshwater residence of salmon, trout, and char. Pages 27–41 *in* M. J. Dadswell, R. J. Klauda, C. M. Moffitt, R. L. Saunders, R. A. Rulifson, and J. E. Cooper, editors. Common strategies of anadromous and catadromous fishes. American Fisheries Society, Symposium 1, Bethesda, Maryland.

Raney, E. C. 1957. Subpopulations of the striped bass *Roccus saxatilis* (Walbaum), in tributaries of Chesapeake Bay. Pages 85–107 *in* J. C. Marr, coordinator. Contributions to the study of subpopulations of fishes. U.S. Fish Wildlife Service Special Scientific Report Fisheries 208.

Rideout, R. M., G. A. Rose, and M. P. M. Burton. 2005. Skipped spawning in female iteroparous fishes. Fish and Fisheries 6:50–72.

Robinson, M., S. Courtnay, T. Benfey, L. Maceda, I. Wirgin. 2004. Origin and movements of young-of-the-year striped bass in the southern Gulf of St. Lawrence, New Brunswick. Transactions of the American Fisheries Society 133:412–426.

Rulifson, R. A., and M. J. Dadswell. 1995. Life history and population characteristics of striped bass in Atlantic Canada. Transactions of the American Fisheries Society 124:477–507.

Rulifson, R. A., S. A. McKenna, and M. J. Dadswell. 2008. Intertidal habitat use, population characteristics, movement, and exploitation of striped bass in the inner Bay of Fundy. Transactions of the American Fisheries Society 137:23–32.

Rutherford, E. S., and E. D. Houde. 1995. The influence of temperature on cohort-specific growth, survival and recruitment of striped bass *Morone saxatilis* larvae in Chesapeake Bay. Fisheries Bulletin 93:315–332.

Saxton, A. M. W. K. Hershberger, and R. N. Iwamoto. 1984. Smoltification in the net pen culture of coho salmon: quantitative genetic analysis. Transactions of the American Fisheries Society 113:339–347.

Secor, D. H. 1999. Specifying divergent migrations in the concept of stock: the contingent hypothesis. Fisheries Research 43:13–34.

Secor, D. H. 2000.Spawning in the nick of time? Effect of adult demographics on spawning behavior and recruitment of Chesapeake Bay striped bass. ICES Journal of Marine Science 57:403–411.

Secor, D. H. 2007. The year-class phenomenon and the storage effect in fishes. Journal of Sea Research 57(2–3):91–103.

Secor, D. H. 2009. Lexicon of life cycle diversity in diadromous and other fishes. Pages 537–556 *in* A. J. Haro, K. L. Smith, R. A. Rulifson, C. M. Moffitt, R. J. Klauda, M. J. Dadswell, R. A. Cunjak, J. E. Cooper, K. L. Beal, and T. S. Avery, editors. Challenges for diadromous fishes in a dynamic global environment. American Fisheries Society, Symposium 69, Bethesda, Maryland.

Secor, D. H., and P. M. Piccoli. 1996. Age- and sex-dependent migrations of the Hudson River striped bass population determined from otolith microanalysis. Estuaries 19:778–793.

Secor, D. H., and P. M. Piccoli. 2007. Oceanic migration rates of upper Chesapeake Bay striped bass (*Morone saxatilis*) determined by otolith microchemical analysis. Fisheries Bulletin 105:62–73.

Secor, D. H., J. R. Rooker, E. Zlokovitz, V. S. Zdanovwcz. 2001. Identification of riverine, estuarine and coastal contingents of Hudson River striped bass based on elemental fingerprints. Marine Ecology Progress Series 211:245–253.

Stevens, D. E., H. K. Chadwick, and R. E. Painter. 1987. American shad and striped bass in California's Sacramento-San Joaquin River System. Pages 66–78 *in* M. J. Dadswell, R. J. Klauda, C. M. Moffitt, R. L. Saunders, R. A. Rulifson, and J. E. Cooper, editors. Common strategies of anadromous and catadromous fishes. American Fisheries Society, Symposium 1, Bethesda, Maryland.

Sutherland, W. J. 1996. From individual behavior to population biology. Oxford University Press, New York.

Tatum, B. L., J. D. Bayless, E. G. McCoy, and W. B. Smith. 1965. Preliminary experiments in the artificial propagation of striped bass, *Roccus saxatilis*. North Carolina Wildlife Research Commission, Division of Inland Fisheries, Raleigh.

Uphoff, J. H., Jr. 2003. Predator–prey analysis of striped bass and Atlantic menhaden in upper Chesapeake Bay. Fisheries Management and Ecology 10(5):313–322.

Waldman, J. R., D. J. Dunning, Q. E. Ross, and M. T. Mattson. 1990. Range dynamics of Hudson River striped bass along the Atlantic coast. Transactions of the American Fisheries Society 119:910–919.

Westin D. T. and B. A. Rogers. 1978. Synopsis of the biological data on the striped bass *Morone saxatilis* (Walbaum 1792). University of Rhode Island, Marine Technical Report 67, Kingston.

Windle, M. J. S., and G. A. Rose. 2005. Migration route familiarity and homing of transplanted Atlantic cod (*Gadus morhua*). Fisheries Research 75(1):193–199.

Wirgin, I. I., J. R. Waldman, L. Maceda, J. Stabile, and V. J. Vecchio. 1997. Mixed-stock analysis of Atlantic coast striped bass (*Morone saxatilis*) using nuclear DNA and mitochondrial DNA markers. Canadian Journal of Fisheries and Aquatic Sciences 54:2814–2826.

Zlokovitz, E. R., D. H. Secor, and P. M. Piccoli. 2003. Patterns of migration in Hudson River striped bass as determined by otolith microchemistry. Fisheries Research 63:245–259.

American Fisheries Society Symposium 69:151–153, 2009

Climate Change and Anthropogenic Influences Preamble

CHRISTINE M. MOFFITT
Idaho Cooperative Fish and Wildlife Research Unit, U.S. Geological Survey
University of Idaho, Moscow, Idaho 83844, USA

Our understanding of the challenges for diadromous fishes has increased manyfold since our first symposium in 1986. In our first symposium, the discussion of anthropogenic influences mainly focused on factors controlling smolt physiology and differences between hatchery and wild fish stocks. Our concerns regarding human modifications of the environment were flow modification and migration barriers, although we discussed the risks of toxic compounds and changes in nutrient/trophic relationships in the freshwater or estuarine habitats. Stock harvest was considered as an appropriate concern within the marine environment. In 1986, we were absorbed in hypotheses of mechanistic relationships for single or limited factors. Many of us were young, and our science stayed in our biological comfort zone. We were optimistic that we could improve the survival and management of diadromous fishes by simply understanding the intrinsic life history components of these fishes.

During the 21 years between the two conferences, many factors contributed to a changed perspective. We now have many more sophisticated technical and molecular tools with which to interrogate and understand mechanistic genetic, physiological, and behavioral factors affecting fish stocks. Also revolutionary is the increased capability for us to model and approach synthetic multifactor aspects of stocks or environments through advanced statistical modeling and data analyses. Even scientists at remote sites have access to sophisticated computer-assisted tools. Finally, we have changed the pace and methods for communication to allow for incorporation of real time and historical information in computer databases for use in adaptive management and decision analysis tools.

Today, we recognize the many changes in the physical environment of the world and its biota. In 1986, slightly more than 4.8 billion people were using global resources; today, the Earth is supporting 6.7 billion people. In 1986, scientists warned of global climate changes and UNESCO had convened work groups to address global challenges, but few of us suspected the amplification and threshold effects we would see with these changes, as well as potential interaction with fishing effects on the oceans. In years since our first conference, the Food and Agriculture Organization of the United Nations began discussions of the effects of overfishing, and the United Nations increased efforts to regulate high seas fishing. The U.S. government, among others, promoted buy backs of fishing fleets in efforts to reduce the overcapitalization that occurred decades before. At the same time, the growth in economics and industrialization across the world exploded demands and production of goods and food products and increased global consumption of fossil fuels. These effects of global economic development are felt throughout the world, and these effects have widespread influence on diadromous fish stocks. The introductions of xenobiotic chemicals have accumulated in the food chain, and trophic alterations from fishing down the food web have affected marine stocks throughout the world. With this section, we seek to explore some of these interrelationships with papers from scientists from the United States, Canada, France, England, Korea, Japan, Australia, and New Zealand.

The opening paper by Beamish et al. (2009, this volume) challenges us to consider the surprises and adaptive changes in relative abundance of populations of Pacific salmon. These abundance shifts are associated with natural- and anthropogenic-caused climate shifts that affect ocean conditions. Beamish et al. (2009) discuss how large-scale atmospheric processes driving ocean conditions are related to astronomical processes and speculate on the likely challenges ahead.

Lassalle et al. (2009, this volume) provide a synthetic model approach using generalized additive models to explore the effects of climate scenarios on a suite of more than 20 diadromous species occupying river basins of Europe and North Africa. Their models of habitat association were based on historical distribution documented before the industrialized era, and the authors relate presence or absence of these species with landscape and environmental factors in these drainages. Using scenarios predicted for climate change, they explore the likely future presence for these fish species in their study area by the year 2100. Lassalle et al. (2009) predict northward expansion for some species with disappearance or decline of others in the south.

Model approaches are used by Monk and Curry (2009, this volume) who provide and test the validity of several models to predict likely future habitat conditions for freshwater phases of Atlantic salmon *Salmo salar* in 17 sites in northeastern North America. Their models were developed using historical data sets, present day information, and projected future conditions under two scenarios for climate change based on predicted air temperatures. Monk and Curry (2009) approached this analysis by classifying the freshwater habitat at these sites into one of three conditions based on maximum water temperatures: cool, intermediate, and warm. Their models predicted that the duration and the number of sites experiencing mean weekly maximum stream temperatures greater than the 20°C were projected to increase under both climate change scenarios.

McCormick et al. (2009, this volume) provide a review and synthesis of contaminants, including estrogenic compounds that affect developmental physiology or underlying endocrine control mechanisms and thus can affect marine survival. They synthesize considerable information on acid and aluminum exposure and seawater tolerance in salmon and provide compelling data documenting reduced survival and return rates of exposed fish. The authors discuss ways that elevated water temperatures and stresses during downstream migration can affect survival directly and indirectly through suppressed immune competency.

Scientists studying European eels *Anguilla anguilla*, American eels *A. rostrata*, Japanese temperate eels *A. japonica*, Australasian shortfin eels *A. australis*, and New Zealand longfin eels *A. dieffenbachii* collaborate in a thorough review of the factors affecting recruitment in marine environments. Miller et al. (2009, this volume) review and test hypotheses that major ocean-atmospheric associations in the western Pacific and North Atlantic oceans are factors key to understanding the recruitment of eel populations. The scientists clearly show that changes in ocean productivity must be considered in context with the cumulative effects of freshwater habitat loss and degradations for adequate management of these transnational stocks in the future.

In this section, the international scientists studying these diverse stocks illustrate that the challenges for diadromous fish are linked to global environmental factors. It is our collective hope that subsequent scientific discussions, analyses, and management entities use these global intellectual resources and diverse analytical tools to improve future fisheries management.

References

Beamish, R. J., R. M. Sweeting, and C. M. Neville. 2009. Planning the management of Pacific salmon in a changing climate. Pages 155–173 *in* A. J. Haro, K. L. Smith, R. A. Rulifson, C. M. Moffitt, R. J. Klauda, M. J. Dadswell, R. A. Cunjak, J. E. Cooper, K. L. Beal, and T. S. Avery, editors. Challenges for diadromous fishes in a dynamic global environment. American Fisheries Society, Symposium 69, Bethesda, Maryland.

Lassalle, G., M. Béguer, L. Beaulaton, and E. Rochard. 2009. Learning from the past to predict the future: responses of European diadromous fish to climate change. Pages 175–193 *in* A. J. Haro, K. L. Smith, R. A. Rulifson, C. M. Moffitt, R. J. Klauda, M. J. Dadswell, R. A. Cunjak, J. E. Cooper, K. L. Beal, and T. S. Avery, editors. Challenges for diadromous fishes in a dynamic global environment. American Fisheries Society, Symposium 69, Bethesda, Maryland.

McCormick, S. D., D. T. Lerner, M. Y. Monette, K. Nieves-Puigdoller, J. T. Kelly, and B. T. Bjönsson. 2009. Taking it with you when you go: how perturbations to the freshwater environment, including temperatures, dams, and contaminants, affect marine survival of salmon. Pages 195–214 *in* A. J. Haro, K. L. Smith, R. A. Rulifson, C. M. Moffitt, R. J. Klauda, M. J. Dadswell, R. A. Cunjak, J. E. Cooper, K. L. Beal, and T. S. Avery, editors. Challenges for diadromous fishes in a dynamic global environment. American Fisheries Society, Symposium 69, Bethesda, Maryland.

Miller, M. J., S. Kimura, K. D. Friedland, B. Knights, H. Kim, D. J. Jellyman, and K. Tsukamoto. 2009. Review of ocean-atmospheric factors in the Atlantic and Pacific oceans influencing spawning and recruitment of anguillid eels. Pages 231–249 *in* A. J. Haro, K. L. Smith, R. A. Rulifson, C. M. Moffitt, R. J. Klauda, M. J. Dadswell, R. A. Cunjak, J. E. Cooper, K. L. Beal, and T. S. Avery, editors. Challenges for diadromous fishes in a dynamic global environment. American Fisheries Society, Symposium 69, Bethesda, Maryland.

Monk, W. A., and R. A. Curry. 2009. Models of past, present, and future stream temperatures for selected Atlantic salmon rivers in northeastern North America. Pages 215–230 *in* A. J. Haro, K. L. Smith, R. A. Rulifson, C. M. Moffitt, R. J. Klauda, M. J. Dadswell, R. A. Cunjak, J. E. Cooper, K. L. Beal, and T. S. Avery, editors. Challenges for diadromous fishes in a dynamic global environment. American Fisheries Society, Symposium 69, Bethesda, Maryland.

American Fisheries Society Symposium 69:155–173, 2009

Planning the Management of Pacific Salmon in a Changing Climate

Richard J. Beamish*, Ruston M. Sweeting and Chrys M. Neville
Fisheries and Oceans Canada, Pacific Biological Station
3190 Hammond Bay Road, Nanaimo, British Columbia V9T 6N7, Canada

Abstract.—Pacific salmon *Oncorhynchus* spp. catches are at historic high levels. It is significant that one of the world's major fisheries for a group of species that dominates the surface waters of the subarctic Pacific is actually very healthy. Natural trends in climate are now recognized to cause large fluctuations in Pacific salmon production, as shown in historical records of catch and recent changes probably have been affected by greenhouse gas induced climate changes. Pink salmon *O. gorbuscha* and chum salmon *O. keta* production and catch has increased in the past 30 years and may continue in a similar trend for for the next few decades. Coho salmon *O. kisutch* and Chinook salmon *O. tshawytscha* catches have been declining for several decades, particularly at the southern end of their range, and they may continue to decline. In the 1970s, hatcheries were considered to be a method of adding to the wild production of coho and Chinook salmon because the ocean capacity to produce these species was assumed to be underutilized. Large-scale changes in Pacific salmon abundances are linked to changes in large-scale atmospheric processes. These large-scale atmospheric processes are also linked to planetary energy transfers, and there is a decadal scale pattern to these relationships. Pacific salmon production in general is higher in decades of intense Aleutian lows than in periods of weak Aleutian lows. Key to understanding the impact of climate change on Pacific salmon is understanding how the Aleutian low will change. Chinook and coho salmon are minor species in the total commercial catch, but important socially and economically in North America. A wise use of hatcheries may be needed to maintain abundances of these species in future decades.

Introduction

It is now widely recognized that climate has always profoundly altered the abundance of Pacific salmon *Oncorhynchus* spp. (Beamish and Bouillon 1993; Francis and Hare 1994; Mantua et al. 1997; Beamish et al. 2000; Finney et al. 2000, 2002; Ruggerone et al. 2003; Ruggerone and Goetz 2004; Briscoe et al. 2005). The new information is not that climate affects the production of Pacific salmon, but that the effects occur in trends. The assumption that climate effects did not occur in trends facilitated the initial development of models of stock and recruitment relationships. The world famous Ricker curve (Ricker 1954) was developed using an assumption that the effects of climate were random (Ricker 1958). This initial assumption was probably based on available climate data, which were sparse and did not appear to show trends. There now is clear evidence of trends in climate (Trenberth 1990; Trenberth and Hurrell 1994; Hurrell 1995; Polovina et al. 1995; Latif and Barnett 1996; Mantua et al. 1997; Thompson and Wallace 1998; Overland et al. 1999; Minobe 2000; Yasunaka and Hanawa 2002), and there is evidence that Ricker-type stock and recruitment relationships can improve when the analysis is restricted to the particular climate regime (Beamish et al. 2004a). The relationship between trends in climate and fish production has also been identified for other commercial species in the North Pacific and the North Atlantic (Kawasaki 1991; Beamish 1993; Bailey 2000; Hollowed et al. 2001; Rothschild 2007).

This recognition of the importance of climate and climate trends in the regulation of the production of Pacific salmon and the realization of the poor understanding of the mechanisms puts the manage-

* Corresponding author: richard.beamish@dfo-mpo.gc.ca

ment of Pacific salmon in a rather perilous position. There is a poor understanding of the linkages between climate and the dynamics of the life history strategies of the various species of Pacific salmon at a time when climate is being altered by accumulations of greenhouse gases in the atmosphere.

In this paper, we examine the current state of Pacific salmon and the recent pattern in the total commercial catch of all species to determine if the trends provide information about future effects of climate. Historical catches are an indicator of abundance for Pacific salmon because of the high exploitation rates commonly range between 60% and 80%, indicating that large abundances are captured in the fisheries (Beamish et al. 2004a) We look at the major indicators of climate on Pacific salmon to show that natural fluctuations in abundance should be expected to continue and that understanding future climate effects requires recognition of when natural shifts in climate occur so that natural changes can be distinguished from global warming-related changes. We show that the two popular North American species, coho salmon *O. kisutch* and Chinook salmon *O. tshawytscha*, are not doing as well as pink salmon *O. gorbuscha* and chum salmon *O. keta*, particularly in southern British Columbia. If there are social requirements to provide fishing opportunities for coho and Chinook salmon, then we suggest a wise use of hatcheries, which requires a management that monitors wild and hatchery salmon and manages them differently.

Trends in the Total Catch of Pacific Salmon

The total commercial catch of Pacific salmon by all countries (Canada, Japan, Korea, Russia, and the United States)(Figure 1, www.npafc.org) is at historic high levels, contrary to a popular belief. The total commercial catches for the five major species of Pacific salmon (pink salmon, chum salmon, sockeye salmon *O. nerka*, coho salmon, and Chinook salmon) represent more than 99% of the catch of all species of Pacific salmon. The highest catch of 985,000 metric tons (mt) was in 1995, the second highest catch was in 2005, and the third highest catch in 2003 (Figure 1). Preliminary estimates of the Russian catch in 2006 of 199,700 mt are the highest returns of pink salmon ever recorded; thus, the high catch levels appear to be continuing (Anonymous 2007).

Pink and chum salmon now dominate the total salmon catch (Figure 2). In 2003–2005, these two species made up an average 82% of the total catch (Figure 3A), an increase from an average of about 70% in the late 1970s. In contrast, however, since the late 1970s, the catches of sockeye salmon decreased (Figures 3B). At the same time, there has been a dramatic decline in the percentage of the catch

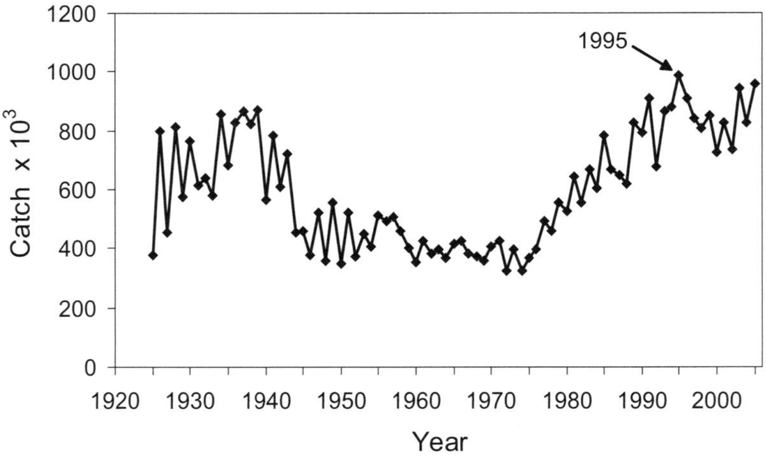

Figure 1.—Total catch of Pacific salmon from 1925 to 2005 (for 1925–1992 data, e-mail nevillec@pac.dfo-mpo.gc.ca; for 1993–2005 data, see www.npafc.org/new/pub_statistics.html). The largest catches occurred in 1995, 2003, and 2005 with 985,100, 942,400, and 959,000 metric tons, respectively.

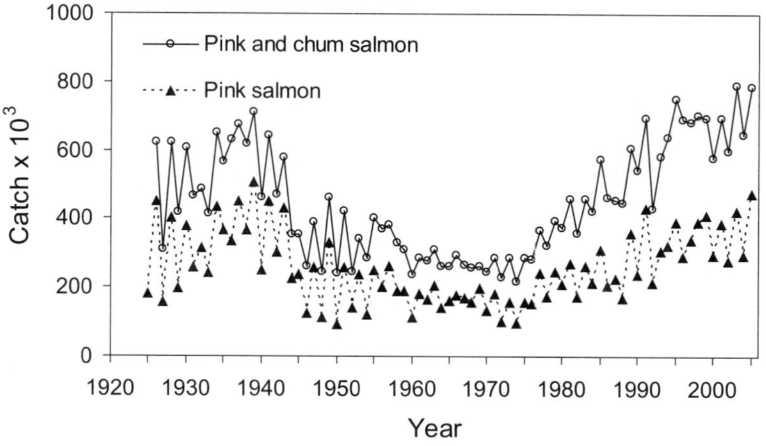

Figure 2.—Total catch of pink salmon and pink and chum salmon from 1925 to 2005. Data updated from North Pacific Anadromous Fish Commission Statistical Yearbook, 2002, and historical data, www.npafc.org.

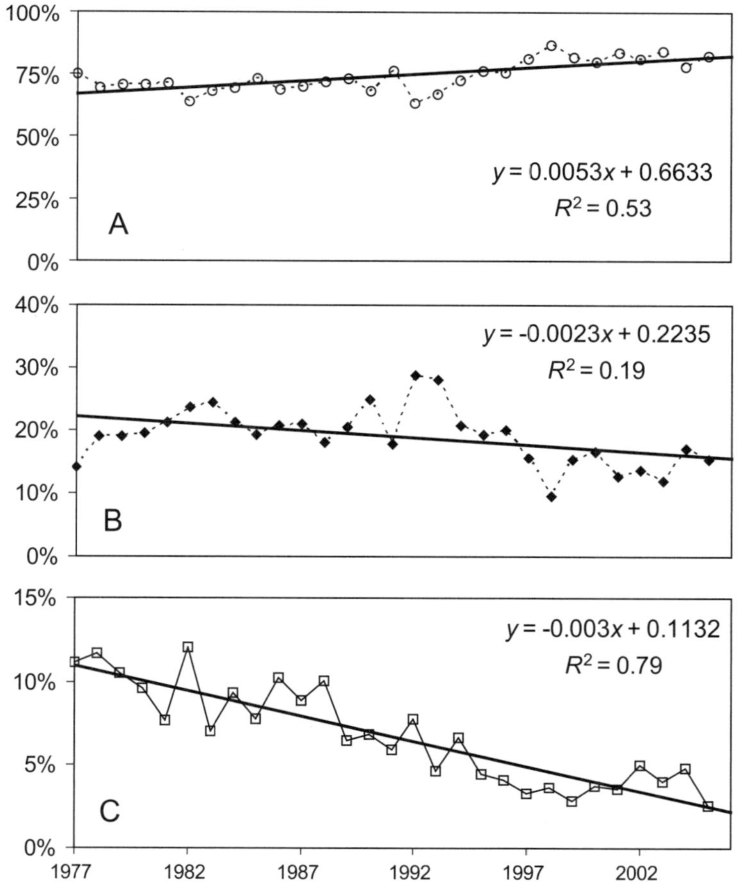

Figure 3.—Percentage of total catch from 1977 to 2005 of (A) pink and chum salmon, (B) sockeye salmon, and (C) coho and Chinook salmon.

of coho and Chinook salmon from about 11% in the late 1970s to approximately 2.5% at the present time (Figure 3C). Pink and chum salmon may continue to benefit from climate conditions and dynamics of the Aleutian low. Coho and Chinook salmon appear to be negatively impacted by the current climate, possibly because of their later entry times into the ocean.

The contribution to the total catch by Canada declined from levels of 19% in the late 1960s and early 1970s to about 3% in recent years (Figure 4). The reasons for the declines are a combination of management actions, poor markets, and reduced abundances of some stocks. It is noteworthy, however, that an assessment of the state of pink salmon concluded that in recent years, there has been an increase in escapement for most areas (Riddell and Beamish 2003). The largest pink salmon population in British Columbia is in the Fraser River (Figure 5), where recent production is at historic high levels and the recent number of spawning pink salmon (escapement) is about 16 times historic levels (Figure 6). Thus, the reduced total Canadian catch is not related to reduced abundances of pink salmon, which is currently the most abundant of Pacific salmon in British Columbia.

Climate Regimes and Production of Pacific Salmon

Beamish and Bouillon (1993) may have been the first to show a close relationship between climate and patterns in total Pacific salmon catch (Figure 7A). These decadal-scale trends are generally termed regimes, and the rapid shifts are regime shifts (see Benson and Trites 2002 for a review). We consider a regime shift to be a climate forced, abrupt, and persistent change in the productivity of an ecosystem. The validity of regimes and regime shifts in relation to the 1998 regime shift was reviewed by an international panel of experts (King 2005) who concluded that regimes were real. The generally accepted regime shift years are 1925, 1947, 1977, 1989, and 1998 (Mantua et al. 1997; Minobe 1997; Hare and Mantua 2000), and most mark changes in the Pacific salmon catch trends (Figure 7A). Perhaps the most significant regime shift in recent times occurred in 1977 (Overland et al. 1999). There are literally dozens of studies that documented a wide variety of changes that occurred at the time of the 1977 regime shift. It is intriguing that the mechanism that caused this dramatic shift in climate and species abundances remains to be discovered. The relationship between catch trends and regimes is not perfect, but it is close. Regimes and regime shifts reliably describe the production of sockeye salmon production from the Fraser River, one of the few data sets that has scientifically based estimates of total production (catch and escapement)(Beamish et al. 1997b, 2004c).

Pacific salmon abundance and the corresponding catches tended to be large during periods

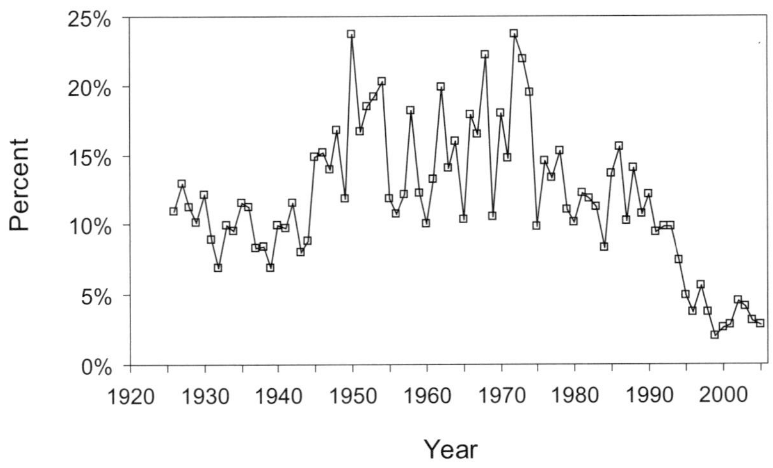

Figure 4.—Pacific salmon catch in Canada shown as a percentage of the total Pacific salmon catch from 1925 to 2005.

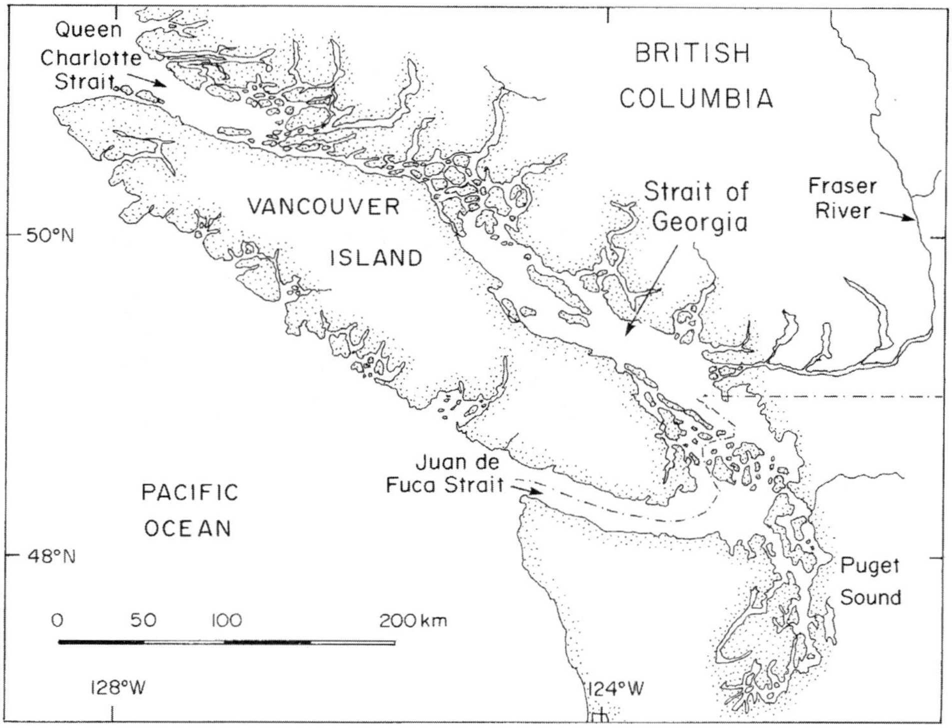

Figure 5.—Location of the Strait of Georgia and Fraser River in southern British Columbia, Canada.

of strong Aleutian lows in the regimes from 1925 to 1947, 1977 to 1989, and 1998 to the present (Figure 7B). The mechanism that links stormier winters caused by a strong Aleutian low in the North Pacific to improved Pacific salmon production appears to result from increased mid-ocean upwelling (Polovina et al. 1995; Deser et al. 1996, Gargett 1997; Bromirski et al. 2003; Goes et al. 2004; Miller et al. 2004). Thus, it is the dynamics of the Aleutian low that appears to be the key to understanding the decadal-scale natural variability of Pacific salmon production, and Rauthe and Pueth (2004) showed that it is the Aleutian low that dominates the climate of the North Pacific Ocean. Climate conditions in the subarctic Pacific have been commonly indexed by the Pacific Decadal Oscillation (Mantua et al. 1997), the Aleutian low pressure index (Beamish et al. 1997b), the Pacific-North America pattern, and other indices. It is the Pacific Decadal Oscillation that is the most commonly used index. The Pacific Decadal Oscillation is the first mode of decadal variability in the sea surface temperature of the northern North Pacific (Mantua et al. 1997). The Aleutian low pressure index is the anomaly of the winter mean area of the northern North Pacific that is covered by the Aleutian low that is less than 100.5 kPa (Beamish et al. 1997b). A positive Aleutian low pressure index results in cooler water being upwelled in the central North Pacific, contributing to a positive Pacific Decadal Oscillation. There is a close relationship between the Pacific Decadal Oscillation and the Aleutian low pressure index (Figure 8) up to about 1998 when Aleutian low pressure index shifted to a positive mode, but the Pacific Decadal Oscillation as originally estimated appears to follow an opposite trend.

A new index of winter atmospheric circulation over the North Pacific identifies the dominant atmospheric circulation processes (King et al. 1998, 2006; McFarlane et al. 2000). The index categorizes atmospheric processes from 1900 to 2004 into zonal or west–east flows of air masses and two meridional or wavy flow patterns. One meridional process results in a southwesterly transfer of air mass along the coast of the western North Pacific and a northwest-

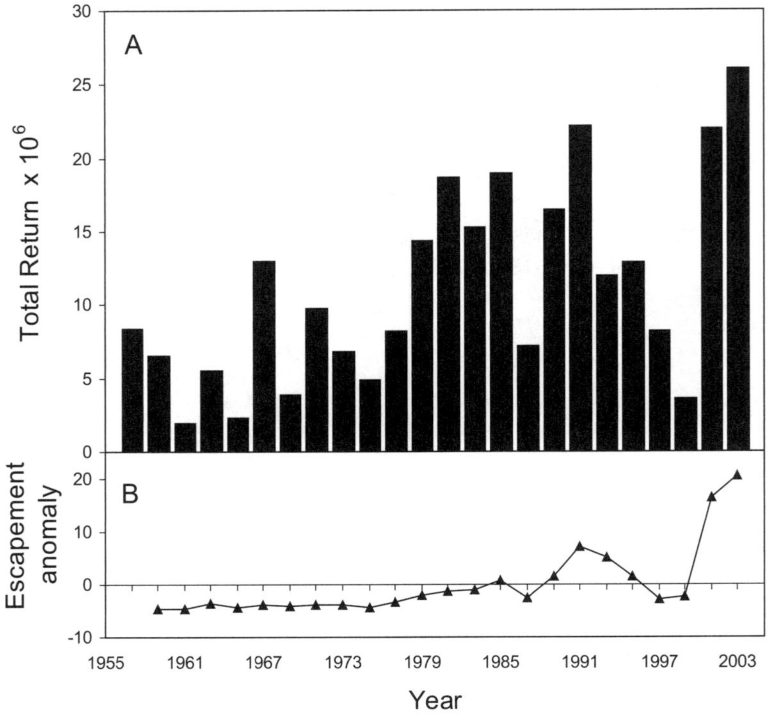

Figure 6.—(A) Total return of pink salmon to the Fraser River in Canada showing the total return (catch and escapement) since reliable data were available in the early 1950s, and (B) the total escapement of pink salmon 1960 to 2003 calculated as an anomaly from the 1960s to 2003 average escapement, showing the exceptional number of spawning fish in 2001 and 2003. Virtually all pink salmon spawn in the Fraser River in odd-numbered years.

erly transfer along the eastern North Pacific (Figure 9, from King et al. 2006).

The other meridional process is opposite with northwesterly flow circulation along the western North Pacific and southwesterly flow in the eastern North Pacific. The method used to develop the index was described by Girs (1971) and was used to produce a similar index for the Atlantic–Eurasia region, the atmospheric circulation index (see Beamish et al. 1999b for summary of atmospheric circulation index). The winter (December to March) frequencies for each process were expressed as anomalies from the long-term mean (1900–1966). The trend of these anomalies identified periods when one of the three processes was above average. The anomalies of each process will sum to zero, such that the negative of one meridional anomaly will equal the sum of the other two. The cumulative sum (CuSum; Murdoch 1979; Noakes and Campbell 1992) of this negative anomaly is used to identify trends and change points. A CuSum plot shows trends in data by including information from past data points and giving less weight to single points. A positive slope indicates an above average or increasing trend, a negative slope indicates a decreasing trend, and a horizontal trend identifies the long-term mean. The new Pacific circulation index is the winter (December to March) pattern. When the Pacific circulation index is compared to the atmospheric circulation index (Figure 10), the trends are similar, but the Pacific circulation index more closely captures the years when regimes shift. The Pacific circulation index also is very closely related to the Aleutian low pressure index when Aleutian low pressure index is expressed as a CuSum (Figure 11). It is important to note that Pacific circulation index is the winter atmospheric pattern over the subarctic Pacific and atmospheric circulation index is an annual pattern over an area that approximates the location of Europe. We include the atmospheric circulation index

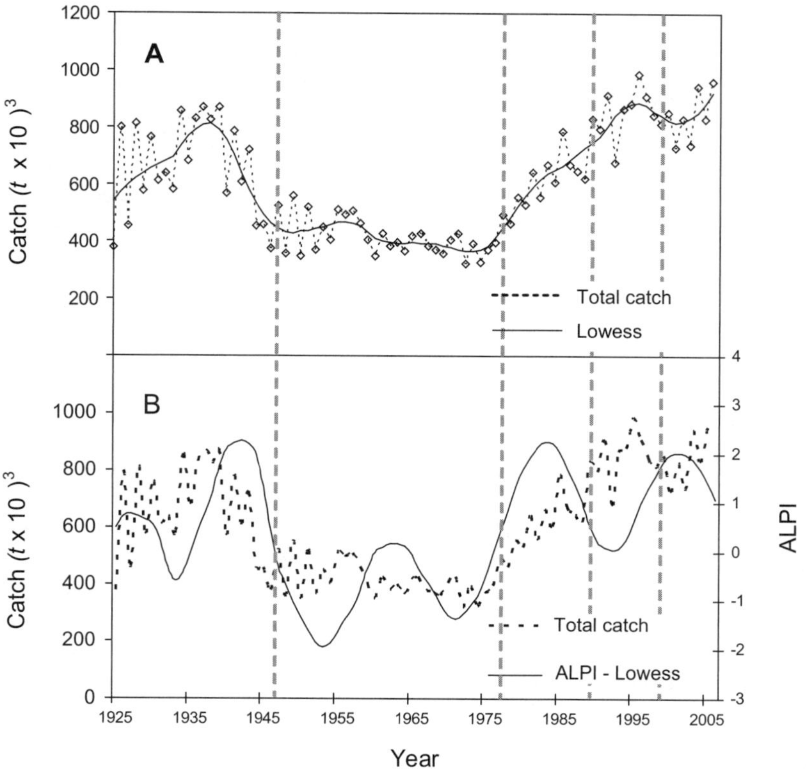

Figure 7.—(A) Total Pacific salmon catch smoothed using lowess (band width, f, 0.2, Cleveland 1985). Vertical dashed lines indicate regime shift years. (B) Total Pacific salmon catch and Aleutian low pressure index (ALPI) smoothed with lowess (0.2). Vertical dashed lines indicate regime shift years.

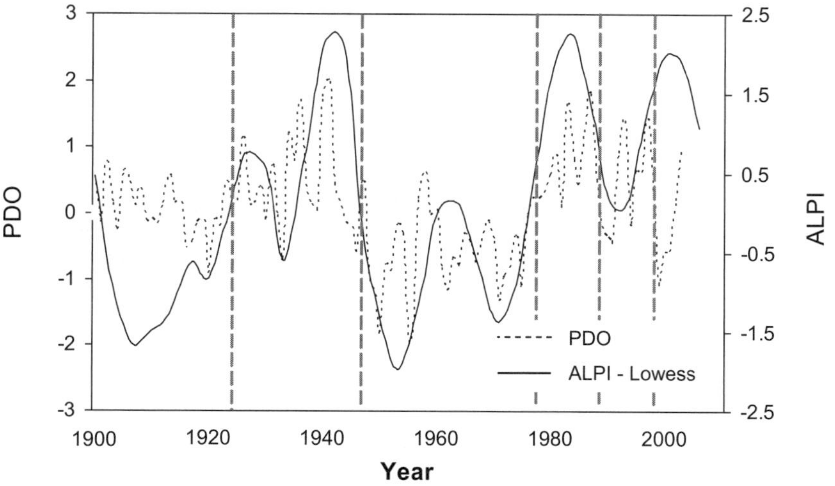

Figure 8.—The Pacific Decadal Oscillation (PDO) and Aleutian low pressure index (ALPI) smoothed with lowess (band width, f, 0.2). Vertical dashed lines indicate regime shift years.

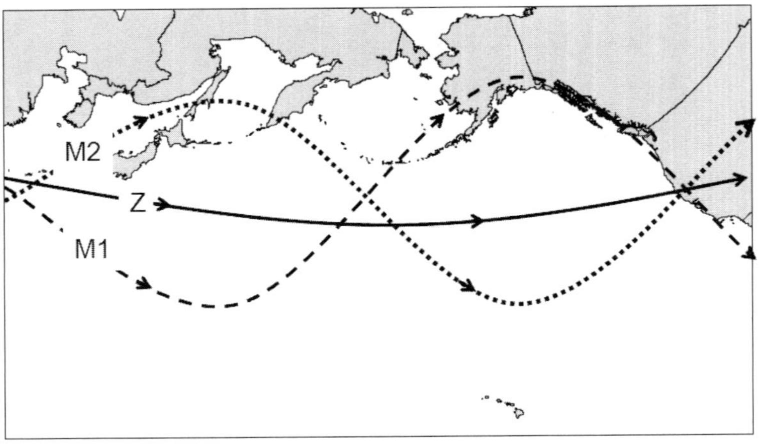

Figure 9.—Atmospheric processes over the North Pacific (from King et al. 2006). This is the classification scheme used to estimate the Pacific circulation index. Atmospheric circulation over the northern North Pacific is characterized by three processes: Z-type for westerly zonal flow, M2-type for southwesterly meridional flow, and M1-type for northwesterly meridional flow (Girs 1971). The atmospheric circulation for each day was categorized as only one of the three processes. The number of days for each type was then expressed as anomalies from a long-term mean. The anomalies were then used to identify periods when one of the three processes generally remained above average while the other two were generally below average.

in this paper because it is the index that has been used for North Pacific Ocean species (Klayshtorin 1998). Pacific Ocean species should be indexed by atmospheric processes over the North Pacific, and this index is now available.

The Pacific circulation index and the atmospheric circulation index are also related to decadal-scale changes in the index of the earth's rotational speed or length of day (LOD; Figure 12). The LOD is the difference between the astronomically determined duration of the day and the standard LOD which was established as exactly 86,400 s on January 1, 1958. Changes in the LOD are expressed as the difference between the measured LOD relative

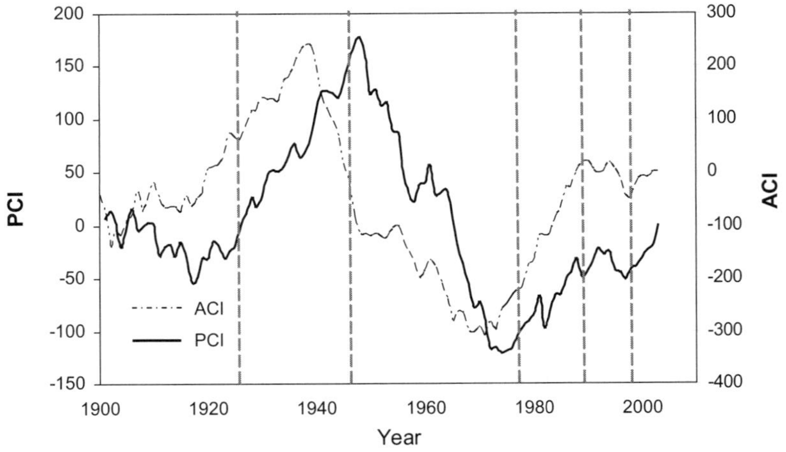

Figure 10.—The atmospheric circulation index (ACI) and Pacific circulation index (PCI) from 1900 to 2003. Vertical dashed lines indicate regime shift years.

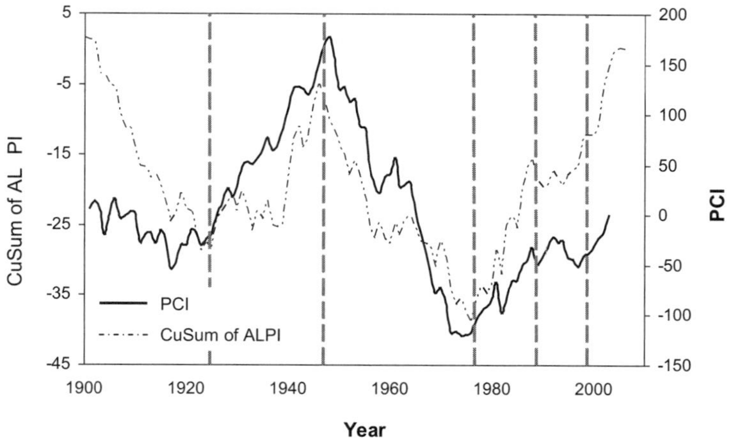

Figure 11.—Annual Pacific circulation index (PCI) and cumulative sum (CuSum) of the Aleutian low pressure index (ALPI) from 1900 to 2003. Vertical dashed lines indicate regime shift years.

to the standard LOD (Figure 13A). Seasonal changes in LOD (Figure 13B) are closely associated with changes in the atmosphere, but decadal changes in LOD (Figure 13A) are believed to be linked to energy transfers between the solid earth and the core. As energy in a body rotating in a frictionless environment is conserved and because the four shells of the planet (atmosphere, hydrosphere or oceans, solid earth, and core) rotate at different speeds, the energy lost from one shell must be transferred to one of the other three shells. For example, Eubanks (1993) noted that from about 1895 to 1905, LOD increased (a slowing of the spinning of the solid Earth) by 3–4 milliseconds (Figure 13). The energy lost from the solid Earth would be extremely large (equivalent to the output of 3–4 million power sta-

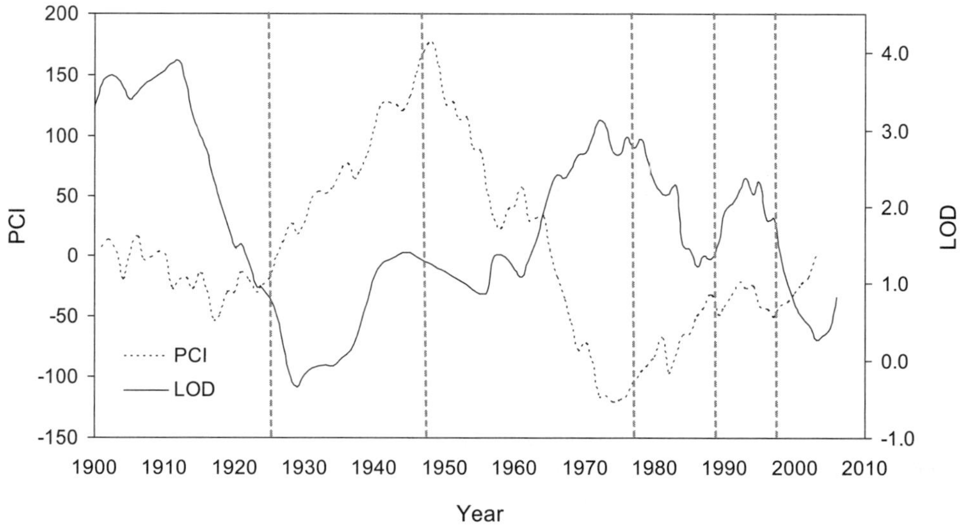

Figure 12.—A comparison of the trends in (A) the Pacific circulation index (PCI) and annual length of day (LOD) and (B) the atmospheric circulation index (ACI) and annual LOD from 1900 to 2005. Vertical dashed lines indicate regime shift years.

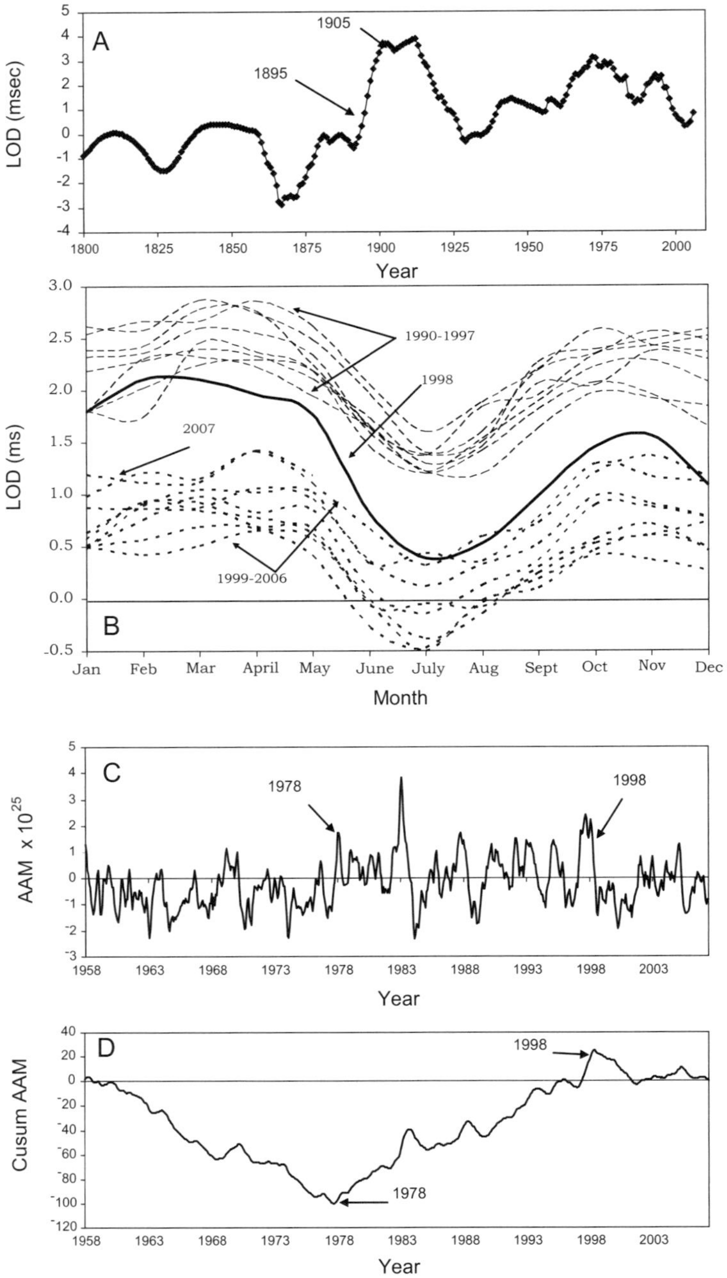

Figure 13.—(A) The average annual length of day (LOD) from 1800 to 2006 showing the slowing of the solid Earth from 1895 to 1905. (B) The average monthly LOD showing the trends in seasonal changes shifting in 1998. The dashed lines represent the years after the 1989 regime shift (1990–1997) and after the 1998 regime shift (1999–2005). The solid line shows the change in 1998. (C) Global atmospheric angular momentum (AAM) showing the increasing trend in intensity and variation in 1978 and a decreasing trend beginning in 1998. Monthly atmospheric angular momentum values (in kg m^2/s, scaled to 10^{25}) from 1958–2007, obtained from www.cdc.noaa.gov and derived from National Centers for Environmental Predictions and the National Center for Atmospheric Research reanalysis project (Kalnay et al. 1996). (D) cumulative sum (CuSum) of data in Figure 13C showing the years in which trends change.

tions of 100 MW each). Eubanks (1993) showed that if the energy lost from the solid Earth from 1895 to 1905 went to the atmosphere, there would have been a doubling of the mean zonal wind velocity, which did not happen. If the energy went to the oceans, the sea level would have increased by about 20 cm. As this did not occur, Eubanks (1993) points out that it is generally assumed that decadal fluctuations in LOD reflect core–mantle transfers. The almost inverse relationship between the Pacific circulation index and LOD (Figure 12) identified a relationship between LOD and decadal-scale patterns in the atmospheric circulation (Sidorendkov and Svirenko 1991). Beamish et al. (1999b) suggested that periods of decreasing LOD or a speeding up of the solid earth were associated with increased Pacific salmon production and a deeper Aleutian low. These conditions also match a positive Pacific Decadal Oscillation and an above average Pacific circulation index. An above average Pacific Circulation index identifies above average winter zonal and southwest winds. Although the mechanisms that cause these decadal transfers in energy are unknown, it is important to recognize that they are naturally occurring planetary events that link to atmospheric circulation, which is linked to Pacific salmon production. The large fluctuations in Figure 13A and the study of Beamish et al. (1999b, Figure 7) would suggest that in the past, large, natural fluctuations might have occurred in Pacific salmon abundance. Finney et al. (2002) showed that fluctuatiosn of this magnitude have occurred over the past 2,200 years for sockeye salmon returning to a lake in Alaska. Thus, it is probable that there have been large, natural fluctuations in Pacific salmon production and that these fluctuations will continue into the future.

The dynamic interaction between the solid Earth and the atmosphere is also reflected in the atmospheric angular momentum. A wide range of processes affect atmospheric angular momentum over a wide range of time scales (Barnes et al. 1983; Gross et al. 2005). There is a strong hemispheric component in the atmospheric angular momentum; hence, the zonal angular momentum is larger in the northern hemisphere compared to the southern hemisphere (Salstein 2003). This is reflective of the greater land mass area with high mountains in the northern hemisphere interacting with ocean and atmospheric circulation.

In 1991, the National Centers for Environmental Prediction and the National Center for Atmospheric Research began a reanalysis of 40 years (1957–1996) of global analyses of atmospheric fields (Rosen and Salstein 2000). This effort was undertaken to assimilate data from a number of wide ranging data sets into a single system to provide reliable atmospheric data (Kalnay et al. 1996). Currently, the Special Bureau for the Atmosphere of the International Earth Rotation Service collects data from several of the world's major weather centers (U.S. National Centers for Environmental Prediction, European Center for Medium-Range Weather Forecasts, the Japan Meteorology Agency, and the United Kingdom Meteorological Office) and provides this data at 6-h intervals. Langley et al. (1981) reported a close, inverse relationship in the total global atmospheric angular momentum and LOD. The data in Figure 13C show a change in the strength and variability of the pattern. A CuSum analysis shows that the change in trends occurred about in 1978 and 1998 (Figure 13D). It is noteworthy that a global measurement of wind intensity shifts at the time of the 1977 and 1998 regime shifts. We note, however, that the 1989 regime shift is not evident in the global atmospheric angular momentum.

Future Climate Changes with Addition of Greenhouse Gasses

Once it is determined that natural shifts in atmospheric circulation or climate are associated with astronomical processes profoundly affecting Pacific salmon abundance, it should be apparent that these changes would continue with or without the additions of greenhouse gases to the atmosphere. Thus, energy would be expected to be transferred into the atmosphere and out of the atmosphere, resulting in natural changes in the Aleutian low in response to the mechanism causing these shifts. Periods of strong Aleutian lows would be expected to increase the carrying capacity for the aggregate species of Pacific salmon, while periods of weak Aleutian lows would correspond to years of reduced production.

Thus, it will be important to be able to identify when a regime has changed or better to be able to forecast when change will occur. We suggest that natural large-scale climate shifts might be detected in the energy transfers identified by the LOD (Figure 13B) or in the measurements of atmospheric angular momentum (Figure 13C), as evidenced by the clear and abrupt change in the LOD early in 1998 (Figure 13B) that appears to be associated with the 1998 regime shift. We noted previously that there also were changes in the intensity and variability of the planet's atmospheric angular momentum at the time of the 1977 and 1998 regime shifts (Figure 13B). The associations between these energy transfers and the astronomical mechanisms that precipitated the shifts remain to be discovered. However, it appears to be a promising area of study.

It is the effect of increasing greenhouse gases on the intensity of the Aleutian low that probably will most affect on Pacific salmon production. Changes in the temperature of freshwater and ocean habitats will be important in some areas, but the stocks of the species of Pacific salmon occupy broad temperature ranges, and in aggregate, we suggest that it is the impact on the Aleutian low that in the near term will cause the largest total abundance changes. Studies by Mote et al. (1999) showed that the intensity of the Aleutian low would increase early in the 21st century using the Canadian Climate Centre model or later in the 21st century using the Hadley Centre model. Graham and Diaz (2001) provided evidence of an intensification of North Pacific winter cyclones since 1948, indicating that there has been a trend towards increasing intense Aleutian lows. Raible et al. (2005) also showed a trend to more intense Aleutian lows from 1951 to 2000. They speculated that this trend to stronger Aleutian lows may be influenced by external forcing such as global warming, volcanism, and solar activity change. A recent study by Salathé (2006) may confirm that global warming will cause the Aleutian low to intensify and move northward.

Global warming could also affect the factors that regulate the timing of the beginning of the plankton bloom in the spring. There is evidence of an earlier spring on land (Brown et al. 1999; Walther et al. 2002; Nemani et al. 2003; Parmesan and Yohe 2003; Parmesan 2006) and in the North Pacific Ocean (Stabeno and Overland 2001; Grebmeier et al. 2006) that appear to be linked to global warming impacts. An earlier bloom of production could increase the carrying capacity of marine ecosystems for pink and chum salmon species as they enter the ocean before the other species. As Arctic ice melts, Pacific salmon may move into the ice-free waters in the summer. Scientists from the National Marine Fisheries Service Ocean Carrying Capacity Program conducted a survey during the fall 2007 within the Chukchi Sea and eastern Bering Sea to provide ecological data on the pelagic ecosystem and salmon stocks during their juvenile and immature life history stage. The preliminary results of this cruise indicated a relatively large abundance of juvenile Pacific salmon (J. Moss, personal communication, Alaska Fisheries Science Center, Auke Bay Laboratory, Juneau). All five major species were present as far north as 70°N, but pink and chum salmon catches were the largest. This is the first indication of significant numbers of juvenile Pacific salmon north of the Arctic Circle.

The mechanism linking climate and Pacific salmon production may be in the first few months or even weeks after juvenile salmon enter the ocean. We speculate that juvenile Pacific salmon must grow quickly to a critical size by a critical period so that they can store lipids (Post and Parkinson 2001) to improve survival during the winter. There is evidence that this critical size–critical period hypothesis is valid (Beamish et al. 2004c; Moss et al. 2005; Morita et al. 2006; see Farley et al. 2007 for a review), but the idea remains to be tested. We also speculate that the critical period could be as early as the summer solstice when the day length begins to shorten. Again, this linkage with the summer solstice remains to be investigated.

An earlier spring plankton bloom would favor pink and chum salmon production, as juveniles of these species enter the ocean first (Groot and Margolis 1992). According to this interpretation, Chinook and coho salmon would progressively have reduced productivity as they would not grow to critical sizes or accumulate the necessary energy reserves by the critical period. Improved abundances of pink and chum salmon would also result in increased competition for Chinook and coho salmon in this early marine period (Ruggerone et al. 2003; Ruggerone and Goetz 2004; Beamish et al. 2008). Any additional competition from artificially reared pink and chum or even coho and Chinook salmon could add to a reduced rate of growth, affecting lipid accumulation and survival over the first marine winter (Beamish et al. 2008). Thus, according to this speculation, pink and chum production will continue to fluctuate, but the trend would be to continued increases in abundances into the near future. At the same time, the opposite may be true for coho and Chinook salmon, that is, on average, marine survivals of wild coho and Chinook salmon would continue to decline. Beamish and Noakes (2002), using the model output previously identified, suggested that total Pacific salmon catch could increase during the next 100 years. Radchenko et al. (2007) found a significant relationship ($r = 0.59$, $p < 0.001$) between pink salmon catches in Russia and the world ocean heat content for the top 700 m (Levitus et al. 2005). Relationships between ocean heat content and pink salmon catches off the North American coast were also significant. The authors viewed the increasing heat content as a noncyclic component of the climate impacts on pink salmon that was increasing carrying capacity for pink salmon, possibly by increased abundances of prey that also were a function of warmer temperatures (Huntley and Lopez 1992). Thus, if warmer oceans and stormier winters are a future possibility, then pink and chum salmon abundances could become even larger. However, we caution that all biological organisms have limits and when the limits are exceeded, Pacific salmon production would begin to decline.

Evidence from Studies of Coho Salmon

Our speculation relating to coho and Chinook salmon in this paper is based, in part, from evidence from our studies of coho salmon. We studied the effects of climate changes on the dynamics of coho salmon in the Strait of Georgia from 1997 until the present (Beamish et al. 2000, 2004b, 2008).

The commercial and recreational fisheries for coho salmon needed minimal attention until the early 1990s (DFO 1992; Beamish et al. 2000). The total catch of coho salmon started to decline in the late 1980s, despite almost constant hatchery production (Figure 14A). Declines in catch and the ultimate collapse of the commercial and recreational fisheries resulted from a precipitous decline in marine survival (Figure 14A; Beamish et al. 2000, 2008) and a change in behavior that resulted in an absence of ocean age 1+ coho salmon in the Strait of Georgia, except during their return to spawning areas (Beamish et al. 1999a). All of these changes were opposite to the expectations of the Salmon Enhancement Program (Fisheries and Environment Canada 1978), which proposed that the addition of hatchery coho salmon into the Strait of Georgia would about double the catches.

We determined the percentages of hatchery and wild juvenile coho salmon in a 10-year study in the Strait of Georgia (Beamish et al. 2008). We observed that there has been a steady decline in the percentage of juvenile hatchery coho salmon (Figure 14C). The reasons for the decline are not easy to study, but we determined that hatchery coho salmon have been released at a constant average time around about the third week of May (Figure 14D). We proposed that the gradual decline in hatchery percentages resulted from an earlier production of prey that better matched the earlier entry time of wild coho salmon. There is only one reliable, long-term data set of the timing of ocean entry of wild coho salmon, and it supports the interpretation that wild coho salmon are entering the ocean earlier (Beamish et al. 2008). Radchenko et al. (2007) also found that the stocks of juvenile pink salmon that entered the ocean earlier were increasing in abundance relative to stocks that entered the ocean later.

In managed fisheries, it may be the very early marine period in the life history of Pacific salmon that is the most immediate indicator of large changes in productivity. After more than 10 years of studying the early marine ecology of juvenile Pacific salmon, we know that our studies suffer from not sampling in May and June, not having identifications of the stocks of juveniles, not having lipid measurements, not having reliable escapement estimates, and not being able to determine the major causes of mortality. Acquiring the information is expensive and time consuming. It also requires a change in the priorities of agencies providing research support. We suggest that until more effort is directed at the early marine period of Pacific salmon, progress will be slow in understanding the impacts of global warming on the sustainability of fisheries and the stewardship of threatened stocks.

Management Needs and Use of Hatchery Fish

Virtually all chum salmon in Japan are produced in hatcheries, and the recent reported chum salmon

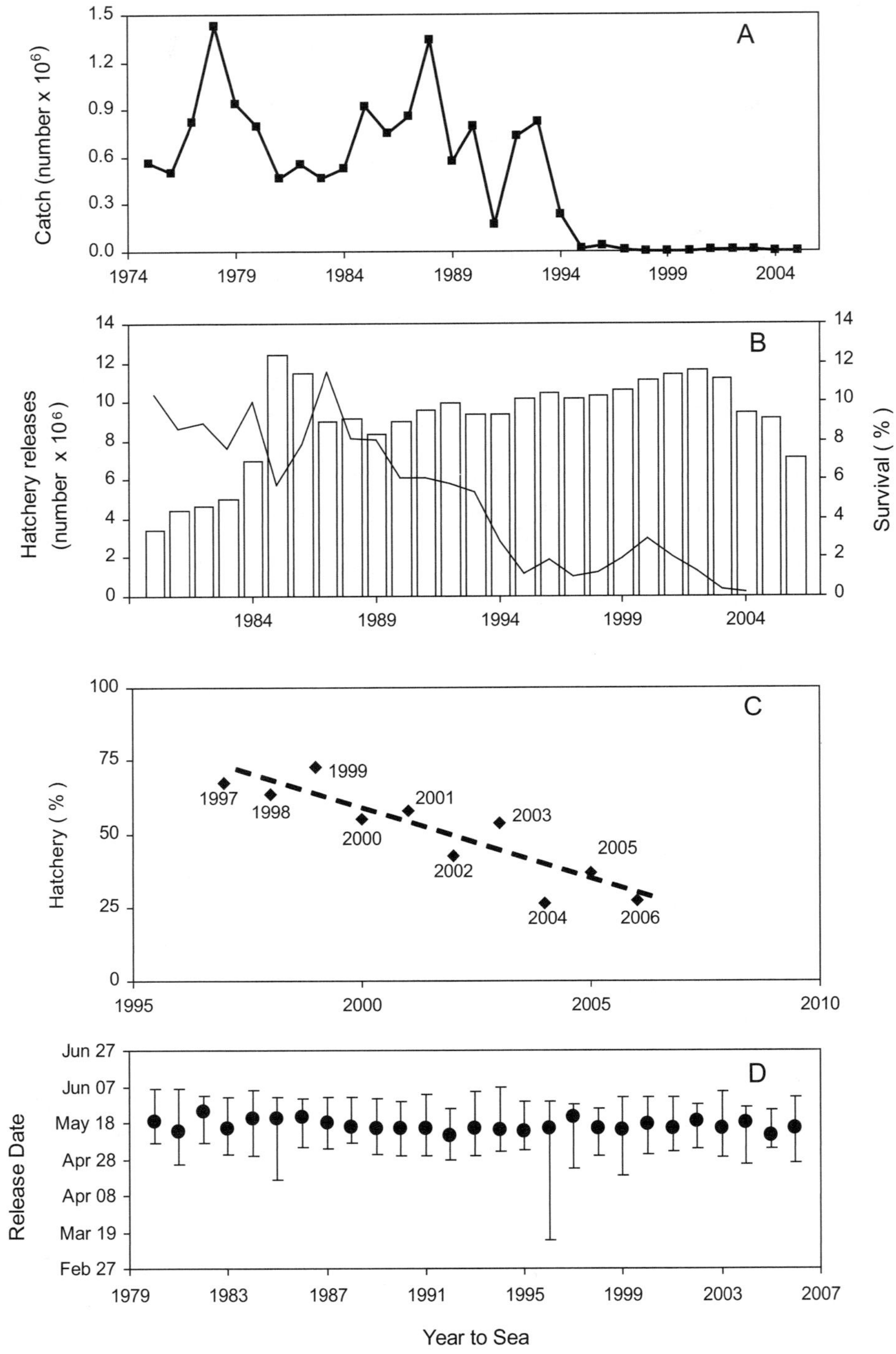

Figure 14.—(A) The commercial and recreational catch of coho salmon in the Strait of Georgia from 1975 to 2005. (B) The number of coho salmon released from hatcheries into the Strait of Georgia (bars) and corresponding marine survival lagged to year of ocean entry from 1980 to 2004. (C) The percentage of hatchery coho salmon in the Strait of Georgia in September, 1997–2006. (D) The average date of release of 50% of all hatchery coho salmon in the Strait of Georgia from 1980 to 2006.

production by Japan represents 69% of the total Pacific salmon catch and 26% of the total catch of all chum salmon. Beamish et al. (1997a) estimated that 84% of all chum salmon catches were hatchery fish. Thus, hatcheries are successful. Coho and Chinook salmon are more highly esteemed in some areas in North America and by some individuals than are pink and chum salmon. If our speculation that coho and Chinook salmon will be adversely affected by a global warming induced trend to an earlier spring in the ocean is correct, then the maintenance of fisheries for coho and Chinook salmon probably would require the continued use of hatcheries. However, the use of hatcheries remains controversial (Hilborn and Eggers 2000; Wertheimer et al. 2001) because of the unknown impact on wild stocks and the general desire to protect wild stocks (Fisheries and Oceans Canada 2005). The use of hatcheries to mitigate impacts of a warming climate, therefore, would have to be carefully managed. It may be appropriate to consider that hatchery fish of a particular species represent a distinct form of the species because it is not known if hatchery fish compete with wild fish in the ocean. Taxonomically, the forms would be identical, but management would assume that one form could compete with the other form until better information is available. Wise use of hatcheries would require recognition of the possible competition and a requirement to monitor the dynamics of both wild and hatchery forms.

Conclusions

Catches of Pacific salmon by all countries were at historic high levels in the past 15 years. Most of the catch consists of pink and chum salmon. These species enter the ocean earlier than other species of Pacific salmon, and we propose that it is this early entry and improved prey availability that is the reason for the very large catches. Large catches could result from very high exploitation rates, but the relatively short life span of pink salmon (2 years) and chum salmon (3–5 years) indicates that the catches have been sustainable (see Beamish and Noakes [2002] and Radchenko et al. [2007] for additional discussion on the use of catch data). We propose that climate and ocean conditions are responsible for the sustainability of these catches. The large numbers of juvenile pink and chum salmon that are produced in hatcheries also contribute to the catch, probably because the capacity to produce these fish in the ocean is expanding.

It is now recognized that natural changes in climate result in trends in Pacific salmon production and the decadal-scale changes cause large fluctuations in abundance and in catch. The natural changes in climate will continue as they are related to large-scale atmospheric processes, which are related to astronomical processes. In fact, it is the relationship between the strengthening and weakening of the Aleutian low and energy transfers among the rotating shells of the planet that may provide an accurate method of determining when regimes have shifted. It may even be possible to forecast when regimes will shift by using astronomical linkages. It appears that global warming will increase the intensity of the Aleutian low. Recent trends show an increase in the intensity, and this increase may continue. Future studies should consider how any resulting energy transfers will affect the planet's ecosystems as energy required to increase the Aleutian low must come from another of the shells of the planet. There is accumulating evidence of an earlier spring on land and perhaps an earlier beginning of plankton production in the ocean. If global warming effects were to result in an intensification of the Aleutian low and an earlier spring plankton bloom, then the trend of increased catches of pink and chum salmon should continue. At the same time, wild coho and Chinook production could continue to decline. If there are social reasons to maintain fisheries for these two species, then hatchery production may be needed, but it should be used wisely until more information is available relating to competition between the wild and hatchery fish. We suggest that managing Pacific salmon in the future will be more difficult until there is a better understanding of the processes affecting marine carrying capacity and unless individual efforts can be integrated among investigators and among countries.

Acknowledgments

The lead author thanks the organizers and sponsors (Northeastern Division of the American Fisheries Society) for support to attend the conference "Challenges for Diadromous Fishes in a Dynamic Global Environment," an international symposium held in Halifax, Nova Scotia on June 18–21, 2007. We thank David Stern for the calculation of the energy represented when there are millisecond changes in LOD. We also thank Lana Fitzpatrick for her assistance with the manuscript preparation.

References

Anonymous. 2007. Biostatistical information on salmon catches and escapement in Russia in 2006. North Pacific Anadromous Fish Commission Document 1063. Available: www.npafc.org (September 2007).

Bailey, K. M. 2000. Shifting control of recruitment of walleye pollock (*Theragra chalcogramma*) after a major climate and ecosystem change. Marine Ecological Progress Series 198:215–224.

Barnes, R., R. Hide, A. White, and C. R. Wilson. 1983. Atmospheric angular momentum functions, length-of-day changes and polar motion. Proceedings of the Royal Society of London 387:31–73.

Beamish, R. J. 1993. Climate and exceptional fish production off the west coast of North America. Canadian Journal of Fisheries and Aquatic Sciences 50:2270–2291.

Beamish, R. J., Benson, A. J., Sweeting, R. M., and Neville, C. M. 2004a. Regimes and the history of the major fisheries off Canada's west coast. Progress in Oceanography 60:355–385.

Beamish, R. J., and D. R. Bouillon. 1993. Pacific salmon production trends in relation to climate. Canadian Journal of Fisheries and Aquatic Sciences 50:1002–1016.

Beamish, R. J., C. Mahnken, and C. M. Neville. 2004c. Evidence that reduced early marine growth is associated with lower marine survival of coho salmon. Transactions of the American Fisheries Society 133:26–33.

Beamish, R. J., C. Mahnken, and C. M. Neville. 1997b. Hatchery and wild production of Pacific salmon in relation to large-scale, natural shifts in the productivity of the marine environment. ICES Journal of Marine Science 54:1200–1215.

Beamish, R. J., G. A. McFarlane, and R. E. Thomson. 1999b. Recent declines in the recreational catch of coho salmon (*Oncorhynchus kisutch*) in the Strait of Georgia are related to climate. Canadian Journal of Fisheries and Aquatic Sciences 56:506–515

Beamish, R. J., C.-E. M. Neville, and A. J. Cass. 1997a. Production of Fraser River sockeye salmon (*Oncorhynchus nerka*) in relation to decadal-scale changes in the climate and the ocean. Canadian Journal of Fisheries and Aquatic Sciences 54:543–554.

Beamish, R. J., and D. J. Noakes. 2002. The role of climate in the past, present and future of Pacific salmon fisheries off the west coast of Canada. Pages 231–244 *in* N. A. McGinn, editor. Fisheries in a changing climate. American Fisheries Society, Symposium 32, Bethesda, Maryland.

Beamish, R. J., D. J. Noakes, G. A. McFarlane, L. Klyashtorin, V. V. Ivanov, and V. Kurashov. 1999a. The regime concept and natural trends in the production of Pacific salmon. Canadian Journal of Fisheries and Aquatic Sciences 56:516–526.

Beamish, R. J., D. J. Noakes, G. A. McFarlane, W. Pinnix, R. Sweeting, and J. King. 2000. Trends in coho marine survival in relation to the regime concept. Fisheries Oceanography 9:114–119.

Beamish, R. J., J. T. Schnute, A. J. Cass, C. M. Neville, and R. M. Sweeting. 2004b. The influence of climate on the stock and recruitment of pink and sockeye salmon from the Fraser River, British Columbia, Canada. Transactions of the American Fisheries Society 133:1396–1412.

Beamish, R. J., R. M. Sweeting, K. L. Lange, and C. M. Neville. 2008. Changes in the population ecology of hatchery and wild coho salmon in the Strait of Georgia. Transactions of the American Fisheries Society 137(2):503–520.

Benson, A. J., and Trites, A. W. 2002. Ecological effects of regime shifts in the Bering Sea and eastern North Pacific. Fish and Fisheries 3:95–113.

Briscoe, R. J., M. D. Adkison, A. Wertheimer, and S. G. Taylor. 2005. Biophysical factors associated with the marine survival of Auke Creek, Alaska, coho salmon. Transactions of the American Fisheries Society 134:817–828.

Bromirski, P., R. Flick, and D. R. Cayan. 2003. Storminess variability along the California coast: 1858–2000. Journal of Climate 16:982–993.

Brown, J. L., S.-H. Li, and N. Bhagabati. 1999. Long-term trend toward earlier breeding in an American bird: a response to global warming? Proceedings of the National Academy of Sciences of the United States of America 96:5565–5569.

Cleveland, W.J. 1985. The elements of graphing data. Wadsworth Inc., Monterey, California.

Deser, C., M. A. Alexander, and M. S. Timlin. 1996. Upper-ocean thermal variations in the North Pacific during 1970–1991. Journal of Climate 9:1840–1855.

DFO (Department of Fisheries and Oceans Canada). 1992. Strait of Georgia coho planning process and recommendations: south coast coho initiative final report. Department of Fisheries and Oceans Canada, Coho Steering Committee Final Report, Vancouver.

Eubanks, T. M. 1993. Variations in the orientation of the Earth. Pages 1–54 *in* D. E. Smith and D. L. Turcotte, editors. Contributions of space geodesy to geodynamics: Earth dynamics. Geodynamic Series, volume 24. American Geophysical Union, Washington, D.C.

Farley, E. V., Jr., J. H. Moss, and R. J. Beamish. 2007. A review of the critical size, critical period hypothesis for juvenile Pacific salmon. North Pacific Anadromous Fish Commission Bulletin 4:311–317.

Finney, B. P., I. Gregory-Eaves, M. S. V. Douglas, and J. P. Smol. 2002. Fisheries productivity in the northeastern Pacific Ocean over the past 2000 years. Nature (London) 416:729–733.

Finney, B. P., I. Gregory-Eaves, J. Sweetman, M. S. V. Douglas, and J. P. Smol. 2000. Impacts of climate change and fishing on Pacific salmon abundance over the past 300 years. Science 290:795–799.

Fisheries and Environment Canada. 1978. The salmonid enhancement program. A public discussion paper. Information Branch, Fisheries and Marine Science, Vancouver.

Fisheries and Oceans Canada. 2005. Canada's policy for conservation of wild Pacific salmon. Fisheries and Oceans Canada, Vancouver.

Francis, R. C., and S. R. Hare. 1994. Decadal-scale regime shifts in the large marine ecosystem of the northeast Pacific: a case for historical science. Fisheries Oceanography 3:249–291.

Gargett, A. E. 1997. The optimal stability "window": a mechanism underlying decadal fluctuations in North Pacific salmon stocks? Fisheries Oceanography 6:109–117.

Girs, A.A. 1971. Long-term fluctuations of the atmospheric circulation and hydrometeorological forecasts. Hydrometeorological Monographs, St. Petersburg, Russia.

Goes, J. I., H. R. Gomes, A. Limsakul, and T. Saino. 2004. The influence of large-scale environmental changes on carbon export in the North Pacific Ocean using satellite and shipboard data. Deep Sea Research II 51:247–279.

Graham, N. E., and H. F. Diaz. 2001. Evidence for intensification of North Pacific winter cyclones since 1948. Bulletin of the American Meteorological Society 82:1869–1893.

Grebmeier, J. M., J. E. Overland, S. E. Moore, E. V. Farley, E. C. Carmack, L. W. Cooper, K. E. Frey, J. E. Helle, F. A. McLaughlin, and S. L. McNutt. 2006. A major ecosystem shift observed in the northern Bering Sea. Science 311:1461–1464.

Groot, C., and L. Margolis, editors. 1992. Pacific salmon life histories. University of British Columbia Press, Vancouver.

Gross, R. S., I. Fukumori, and D. Menemenlis. 2005. Atmospheric and oceanic excitation of decadal-scale Earth orientation variations. Journal of Geophysical Research, 110B, doi:10.10299/2004JB003565.

Hare, S. R., and N. J. Mantua. 2000. Empirical evidence for North Pacific regime shifts in 1977 and 1989. Progress in Oceanography 47:103–145.

Hilborn, R., and D. Eggers. 2000. A review of the hatchery programs for pink salmon in Prince William Sound and Kodiak Island, Alaska. Transactions of the American Fisheries Society 129:333–350.

Hollowed, A. B., S. R. Hare, and W. S. Wooster. 2001. Pacific basing climate variability and patterns of Northeast Pacific marine fish production. Progress in Oceanography 49:257–282.

Huntley, M. E., and M. D. G. Lopez. 1992. Temperature-dependent production of marine copepods: a global analysis. American Naturalist 140:201–242.

Hurrell, J. W. 1995. Decadal trends in the North Atlantic Oscillation: regional temperatures and precipitation. Science 269:676–679.

Kalnay, E., M. Kanamitsu, R. Kistler, W. Collins, D.Deaven, L. Gandin, M. Iredell, S. Saha, G. White, J. Woollen, Y. Zhu, M. Chelliah, W. Ebisuzaki, W. Higgins, J. Janowiak, K. C. Mo, C. Ropelewski, J. Wang, A. Leetmaa, R. Reynolds, R. Jenne, and D. Joseph. 1996. The National Centers for Environmental Predictions and the National Center for Atmospheric Research 40-Year Reanalysis Project. Bulletin of the American Meteorological Society 77(3):437–471.

Kawasaki, K. 1991. Long-term variability in the pelagic fish populations. Pages 47–60 *in* T. Kawasaka, S. Tanaka, Y. Toba, and A. Taniguch, editors. Proceedings of the International Symposium on Long-Term Variability of Pelagic Fish Populations and Their Environment. November 1989, Pergamon, Tokyo.

King, J. R., editor. 2005. Report of the study group on fisheries and ecosystem responses to recent regime shifts. North Pacific Marine Science Organization (PICES) Scientific Report No. 28, Sidney, British Columbia.

King, J. R., V. V. Ivanov, V. Kurashov, R.J. Beamish, and G. A. McFarlane. 2006. An index of general circulation of the atmosphere over the North Pacific. Izvestiya TINRO 145:220–227.

King, J. R., V. V. Ivanov, V. Kurashov, R. J. Beamish, and G. A. McFarlane. 1998. General circulation of the atmosphere over the North Pacific and its relationship to the Aleutian low. Department of Fisheries and Oceans, Sciences Branch–Pacific Region, Pacific Biological Station, North Pacific Anadromous Fish Commission Document 318, Nanaimo, British Columbia.

Klayshtorin, L. B. 1998. Long-term climate change and main commercial fish production in the Atlantic and Pacific. Fisheries Research 37:115–125.

Langley, R. B., R. W. King, I.I. Shapiro, R.D. Rosen, and D. A. Salstein. 1981. Atmospheric angular momentum and the length of day: a common fluctuation with a period near 50 days. Nature (London) 294:730–732.

Latif, M., and T. P. Barnett. 1996. Decadal climate variability over the North Pacific and North America: dynamics and predictability. Journal of Climate 9:2407–2423.

Levitus, S. J. Antonov, and T. Boyer. 2005. Warming of the world ocean, 1955–2003. Geophysical Research Letters 32:L02604, doi:10.1029/2004GL021592.

Mantua, N. J., S. R. Hare, Y. Zhang, J. M. Wallace, and R. C. Francis. 1997. A Pacific interdecadal climate oscillation with impacts on salmon production. Bulletin of the American Meteorological Society 78:1069–1079.

McFarlane, G. A., J. R. King, and R. J. Beamish. 2000. Have there been recent changes in climate? Ask the fish. Progress in Oceanography 47:147–169.

Miller, A. J., F. Chair, S. Chiba, J.R. Moisan, and D. J. Neilson. 2004. Decadal-scale climate and ecosystem interactions in the North Pacific Ocean. Journal of Oceanography 60:163–188.

Minobe, S. 1997. A 50–70 year climatic oscillation over the North Pacific and North America. Geophysical Research Letters 24:683–686.

Minobe, S. 2000. Spatio-temporal structure of the pentadecadal variability over the North Pacific. Progress in Oceanography 47:381–408.

Morita, K., S. H. Morita, and M. Fukuwaka. 2006. Population dynamics of Japanese pink salmon (*Oncorhynchus gorbuscha*): are recent increases explained by hatchery programs or climatic variations? Canadian Journal of Fisheries and Aquatic Sciences 63:55–62.

Moss, J. H., D. A. Beauchamp, A. D. Cross, K. W. Myers, E. V. Farley, J. M. Murphy, and J. H. Helle. 2005. Evidence for size-selective mortality after the first summer of ocean growth by pink salmon. Transactions of the American Fisheries Society 134:1313–1322.

Mote, P., D. Canning, D. Fluharty, R. Francis, J. Franklin, A. Hamlet, M. Hershman, M. Holmberg, K. Gray-Ideker, W.S. Keeton, D. Lettenmaier, R. Leung, N Mantua, E. Miles, B. Noble, H. Parandvash, D.W. Peterson, A. Snover, and S. Willard. 1999. Climate variability and change, Pacific Northwest. National Atmospheric and Oceanic Administration, Office of Global Programs, and JISAO/SMA Climate Impacts Group, Seattle.

Murdoch, J. 1979. Control charts. MacMillan Press, London.

Nemani, R. R., C. D. Keeling, H. Hashimoto, W. M. Jolly, S. C. Piper, C. J. Tucker, R. B. Myneni, and S. W. Running. 2003. Climate-driven increases in global terrestrial net primary production from 1982 to 1999. Science 300:1560–1563.

Noakes, D. J., and A. Campbell. 1992. Use of geoduck clams to indicate changes in the marine environment of Ladysmith Harbour, British Columbia. Environmetrics 3:81–97.

Overland, J. E., J. M. Adams, and N. A. Bond. 1999. Decadal variability of the Aleutian low and its relation to high-latitude circulation. Journal of Climate 12:1542–1548.

Parmesan, C. 2006. Ecological and evolutionary responses to recent climate change. Annual Review of Ecology, Evolution, and Systematics 37:637–669.

Parmesan, C., and G. Yohe. 2003. A globally coherent fingerprint of climate change impacts across natural systems. Nature (London) 421:37–42.

Polovina, J. J., G. T. Mitchum, and G. T. Evans. 1995. Decadal and basin-scale variation in mixed layer depth and the impact on biological production in the Central and North Pacific 1960–88. Deep Sea Research 42:1701–1716.

Post, J. R., and E. A. Parkinson. 2001. Energy allocation strategy in young fish: allometry and survival. Ecology 82:1040–1051.

Radchenko, V.I., O.S. Temnykh, and V.V. Lapko. 2007. Trends in abundance and biological characteristics of pink salmon (*Oncorhynchus gorbuscha*) in the North Pacific Ocean. North Pacific Anadromous Fish Commission Bulletin 4:7–21.

Raible, C. C., T. F. Stocker, M. Yoshimori, M. Renold, U. Beyerle, C. Casty, and J. Luterbacher. 2005. Northern hemispheric trends of pressure indices and atmospheric circulation patterns in observations, reconstructions, and coupled GCM simulations. Journal of Climate 18:3968–3982.

Rauthe, M., and H. Paeth. 2004. Relative importance of northern hemisphere circulation modes in predicting regional climate change. Journal of Climate 17:4180–4189.

Ricker, W. E. 1954. Stock and recruitment. Journal of the Fisheries Research Board of Canada 11:559–623.

Ricker, W. E. 1958. Maximum sustained yields from fluctuating environments and mixed stocks. Journal of the Fisheries Research Board of Canada 15:991–1006.

Riddell, B. E., and R. J. Beamish. 2003. Distribution and monitoring of pink salmon (*Oncorhynchus gorbuscha*) in British Columbia, Canada—a contribution to the development of a pink salmon watch. Department of Fisheries and Oceans, North Pacific Anadromous Fish Commission, Science Branch, Pacific Region, Document 707, Nanaimo, British Columbia.

Rosen, R. D., and D. A. Salstein. 2000. Multidecadal signals in the interannual variability of atmospheric angular momentum. Climate Dynamics 16:693–700.

Rothschild, B. J. 2007. Coherence of Atlantic cod stock dynamics in the northwest Atlantic Ocean. Transactions of the American Fisheries Society 136:858–874.

Ruggerone, G. T., and F. A. Goetz. 2004. Survival of Puget Sound chinook salmon (*Oncorhynchus tshawytscha*) in response to climate-induced competition with pink salmon (*Oncorhynchus gorbuscha*). Canadian Journal of Fisheries and Aquatic Sciences 61:1756–1770.

Ruggerone, G. T., M. Zimmermann, K. W. Myers, J. L. Nielsen, and D. E. Rogers. 2003. Competition between Asian pink salmon (*Oncorhynchus gorbuscha*) and Alaskan sockeye salmon (*O. nerka*) in the North Pacific Ocean. Fisheries Oceanography 12(3):209–219.

Salathé, E. P. 2006. Influences of a shift in North Pacific storm tracks on western North American precipitation under global warming. Geophysical Research Letters 33, L19820, doi: 10.1029/2006GL026882.

Salstein, D. A. 2003. Atmospheric mass and motion signals in the Earth's orientation and other properties. Extended abstract, 14th Symposium on Global Change and Climate Variations, Long Beach California, February 2003.

Sidorendkov, I. S., and P. I. Svirenko. 1991. Long-term changes in atmospheric circulation and fluctuations in the primary natural synoptic region. Pages 93–104 *in* Atmospheric processes of the Earth, volume 316. The Hydrometeorological Scientific Research Centre of Russia, Moscow.

Stabeno, P. J., and J. E. Overland. 2001. Bering Sea shifts towards an earlier spring transition. Eos, Transactions of the American Geophysical Union 82:317–321.

Thompson, D. W. J., and J. M. Wallace. 1998. The Arctic Oscillation signature in the winter time geopotential height and temperature fields. Geophysical Research Letters 25:1297–1300.

Trenberth, K. E. 1990. Recent observed interdecadal climate changes in the northern hemisphere. Bulletin of the American Meteorological Society 71:988–993.

Trenberth, K. E., and J. W. Hurrell. 1994. Decadal ocean-atmosphere variations in the Pacific. Climate Dynamics 9:303–319.

Walther, G. R., E. Post, P. Convey, A. Menzel, C. Parmesan, T. J. C. Beebee, J. M. Fromentin, O. Hoegh-Guldberg, and F. Bairlein. 2002. Ecological responses to recent climate change. Nature (London) 416:389–395.

Wertheimer, A. C., W. W. Smoker, T. L. Joyce, and W. R. Heard. 2001. Comment: a review of the hatchery programs for pink salmon in Prince William Sound and Kodiak Island, Alaska. Transactions of the American Fisheries Society 130:712–720.

Yasunaka, S., and K. Hanawa. 2002. Regime shifts found in the northern hemisphere SST field. Journal of the Meteorological Society of Japan 80:119–135.

American Fisheries Society Symposium 69:175–193, 2009

Learning from the Past to Predict the Future: Responses of European Diadromous Fish to Climate Change

Géraldine Lassalle* and Mélanie Béguer
Cemagref: Estuarine ecosystems and diadromous fish research unit
50 av. de Verdun F 33612 CESTAS Cedex, France

Laurent Beaulaton
ONEMA: Direction de l'action scientifique et technique
"Le Nadar" Hall C. 5 square Félix Nadar F 94300 Vincennes, France

Éric Rochard
Cemagref: Estuarine ecosystems and diadromous fish research unit
50 av. de Verdun F 33612 CESTAS Cedex, France

Abstract.—Climate change can have an effect on species distributions. The 1900 distribution and potential future distribution of diadromous fish in Europe, North Africa, and the Middle East were explored using generalized additive models (GAMs) and selected habitat characteristics of 196 basins. Robust presence–absence models were built for 20 of the 28 diadromous species in the study area using longitude, annual temperature, drainage surface area, annual precipitation, and source elevation as explanatory variables. Inspection of the relationship between each variable and species presence–absence revealed that the GAMs were generally interpretable and plausible. Given the predicted rise in annual temperature in climate models ranging between 1°C and 7°C by 2100, the fish species were classified according to those losing suitable basins, those gaining suitable basins, and those showing little or no change. It was found that the climate envelopes based on temperature and precipitation for diadromous species would, in general, be shifted farther northeastwards by 2100, and these shifting ranges were comparable with those assessed in other studies. The uncertain future of some species was highlighted, and it was concluded that conservation policy and management plans will need to be revised in the face of climate change.

Introduction

The past decade was the warmest of the last millennium, and temperatures will probably continue to rise in the future (IPCC 2007). The different scenarios outlined by the Intergovernmental Panel on Climate Change project an increase in global mean surface air temperature, relative to 1980–1999, of 1.1–6.4°C by 2100. Many studies covering a wide range of ecosystems have already attempted to understand and estimate the ecological consequences of climate change in various organisms. These studies have typically highlighted a change in distribution and community structure (Thomas and Lennon 1999; Walther et al. 2002; MacKenzie et al. 2007).

Anadromous and catadromous fish rely on freshwater, estuarine, and marine ecosystems to complete their life cycles (McDowall 2009; Quinn et al. 2009). This unique life history characteristic has enabled them to adjust their distribution to cope with environmental changes and could allow them to do the same in a dynamic changing environment. Moreover, diadromous species are economically important in many parts of the world. The native range of the European eel *Anguilla anguilla* covers all of Europe and North Africa, and this species is exploited in all the countries

* Corresponding author: geraldine.lassalle@cemagref.fr

in this area (Dekker 2003). Allis shad *Alosa alosa* are harvested by commercial fisheries throughout their range, extending historically from Morocco to Norway, in estuaries, in the middle sections of rivers, or in their resident form in lakes (Baglinière et al. 2003). Lampreys have long been considered a gastronomic delicacy in Europe, encouraging the development of commercial fisheries for these species (Kelly and King 2001). The river lamprey *Lampetra fluviatilis* is an important source of income for many fishermen in Sweden and Finland (Sjöberg 1980; Tuunainen et al. 1980), and the sea lamprey *Petromyzon marinus* is also a commercially important species in Spain and Portugal (Almeida et al. 2002).

Static modeling techniques can be used to assess the impact of global change on biodiversity distribution by predicting current species distributions and applying statistical models from current distribution to project future distributions under global change scenarios (i.e., Bakkenes et al. [2002] for plants, Hilbert et al. [2004] for birds, Chu et al. [2005] for fish, and Thuiller et al. [2006] for mammals). The model relates the current observed species distribution to a set of available environmental variables. Several modeling methods are available such as generalized linear models (GLMs), generalized additive models (GAMs), classification tree analysis, artificial neural network, or multiple adaptive regression splines. Some authors have compared the performance of these techniques in predicting species distribution (Olden and Jackson 2002; Thuiller 2003; Leathwick et al. 2006; Heikkinen et al. 2007). In most cases, GAMs provided a good stability and performance because of their flexible nonparametric or semiparametric framework to model the relationship between a response and one or more predictor variables (Hastie and Tibshirani 1990) and were easily interpretable.

The objectives of the work reported here were to develop a model that would relate the 1900 diadromous fish distribution to major climatic and physical factors using GAMs and then to project the distribution at the end of the 21st century under climate change. Our work focused on suitable climatic conditions without taking into account regional anthropic pressures such as the erection of dams in main stems (Gowans et al. 1999; Larinier 2001; Hill 2009), the deterioration in water quality in industrialized areas (Breine et al. 2007; Maes et al. 2007; Woodland et al. 2009), or unsustainable fisheries (Masters et al. 2006). We chose to consider the distribution of diadromous fish species at the beginning of the 20th century before the main biodiversity erosion. The first warming period occurred worldwide between 1910 and 1945, with a temperature increase of 0.15°C per decade (IPCC 2007). Moreover, in Europe, the large dams policy (above 15 m) was started in the 1950s (CIGB/ICOLD 2003). Nevertheless, Hoffmann (1996, 2005) demonstrated that during the Middle Ages, increasing human populations and demand had already produced great change in many European fisheries and altered the freshwater biodiversity.

Methods

Study Area

We selected 196 river basins in Europe, North Africa, and the Middle East in a two-steps process: (1) we selected all the large rivers in the area according to the list of "major European river catchments" (31 basins exceeding 50,000 km^2 and draining two-thirds of the continent) provided by the European Environment Agency (www.eea.europa.eu), and (2) we selected medium-sized rivers to obtain a balanced geographical coverage (Figure 1). Basins ranged from Greenland in the west (Kapisillit River: 51.76°W) to Iran in the east (Gorgan Rud River: 54.04°E). Basins from Morocco, Algeria, Tunisia, and Egypt, as well as Greenland and Iceland, were added to our study to fully cover the distribution area of the diadromous species considered in this work. No other diadromous species were encountered in these regions. In North Africa, the river network is very scattered, especially in the desert country, Libya. We selected all the major basins (seven in all) in this region.

Species Data

All over Europe, a pool of 28 diadromous fish species was encountered in the basins (Table 1; i.e., 25 anadromous species and 3 catadromous species). This list was based on the heuristic checklist of the freshwater fish of Europe (Kottelat 1997; Kottelat and Freyhof 2007) and the list of diadromous fish species established by McDowall (1988). Because of taxonomical uncertainties, all Baltic *Coregonus* species were pooled under the appellation *Corego-*

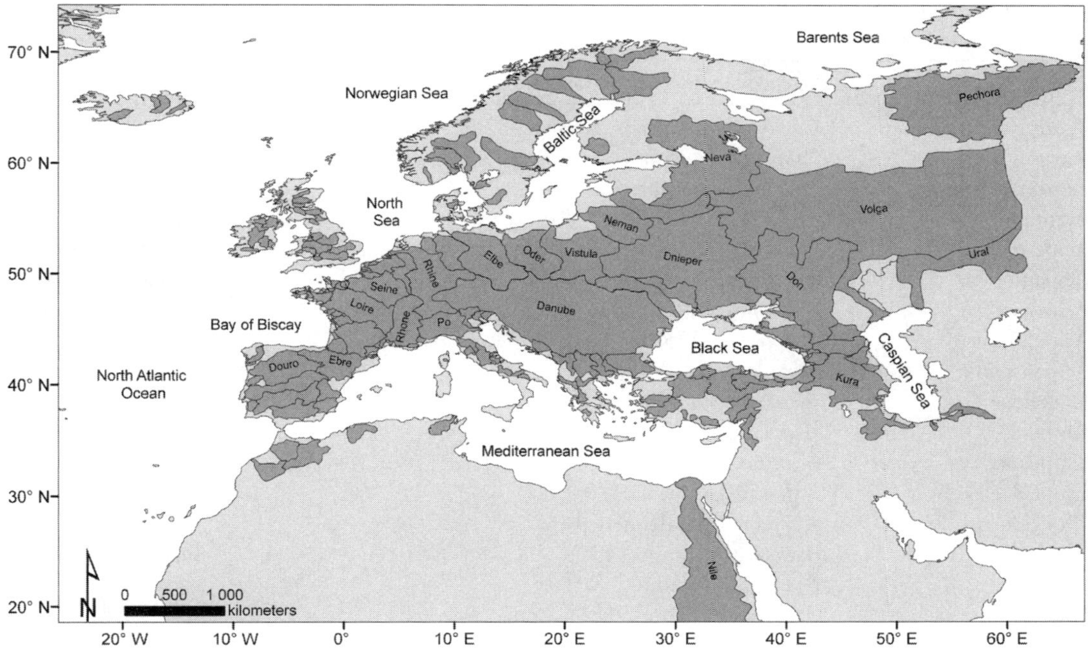

Figure 1.—The study area covering the whole of Europe, North Africa, and the Middle East. The names of the largest basins were provided. Nile (Egypt) and Kapisillit (Greenland) basins were truncated to offer a better visualization of the whole study area.

nus spp. Many species present high morphological variations. Thus, some taxonomists grouped similar species in groups or complexes while others recognized different species or subspecies for some local or regional variants. All the selected species demonstrated physiological and ecological capabilities to express diadromous tactics (McDowall 1988). In some cases, these tactics are not currently expressed (i.e., resident forms exist for some species,).

A presence–absence database for these species at the beginning of the 20th century in the selected basins (EuroDiad 2.0 database) was produced from a literature review and was checked by national experts from the Diadfish network (www.diadfish.org). Béguer et al. (2007) used the first version of EuroDiad centered on Western Europe and provided the related reference list in their paper. This list constitutes a major overview of the literature types (thesis, scientific papers or officials reports about diadromous species distribution) that we used in our study. Brotons et al. (2004) demonstrated that models based on presence–absence information were more relevant than presence-only modeling methods for mapping current species distributions under different conditions.

Explanatory Variables

Five types of variables (temperature, precipitation, longitude, basin surface area, and source elevation) were selected according to the amount of data available on a large scale and in accordance with theories regarding species response to environmental gradients and resources (Austin 2007). The description of the model variables and the references for where these data were located were provided in Table 2. Because of the amount of data, only the major references were cited. If data originated from Web sites, we searched for confirmation in peer review articles.

We selected three variables providing a physical description of the basins. Longitude at the mouth (Long) can be related to biogeographical history with two major events that profoundly affected fish extinction, dispersal, and speciation

Table 1.—Diadromous fish species studied. The list is taxonomically ordered according to Kottelat (1997) and Kottelat and Freyhof (2007). * represents the catadromous species compared to the anadromous species.

Family	Scientific name and authority	Common name
Petromyzontidae	*Caspiomyzon wagneri* (Kessler, 1870)	Caspian lamprey
Petromyzontidae	*Lampetra fluviatilis* (L., 1758)	River lamprey
Petromyzontidae	*Petromyzon marinus* L., 1758	Sea lamprey
Anguillidae*	*Anguilla anguilla* (L., 1758)	European eel
Acipenseridae	*Acipenser gueldenstaedtii* Brandt and Ratzeburg, 1833	Russian sturgeon
Acipenseridae	*A. naccarii* Bonaparte, 1836	Adriatic sturgeon
Acipenseridae	*A. stellatus* Pallas, 1871	Stellate sturgeon
Acipenseridae	*A. sturio* L., 1758	European sturgeon
Acipenseridae	*Huso huso* (L., 1758)	Beluga
Clupeidae	*Alosa algeriensis* Regan, 1916	
Clupeidae	*A. alosa* (L., 1758)	Allis shad
Clupeidae	*A. caspia* (Eichwald, 1838)	Caspian shad
Clupeidae	*A. fallax* (Lacépède, 1803)	Twaite shad
Clupeidae	*A. immaculata* Bennett, 1835	Pontic shad
Clupeidae	*A. kessleri* (Grimm, 1837)	Blackspined herring
Clupeidae	*A. tanaica* (Grimm, 1901)	
Clupeidae	*A. volgensis* (Berg, 1913)	Azon shad
Cyprinidae	*Vimba vimba* (L., 1758)	Vimba
Osmeridae	*Osmerus eperlanus* (L., 1758)	European smelt
Coregonidae	*Coregonus oxyrinchus* (L.,1758)	Houting
Coregonidae	*Coregonus* spp.	
Salmonidae	*Salmo caspius* Kessler, 1877	Caspian trout
Salmonidae	*S. labrax* Pallas, 1814	Black Sea salmon
Salmonidae	*S. salar* L., 1758	Atlantic salmon
Salmonidae	*S. trutta* L., 1758	Brown trout
Salmonidae	*Salvelinus alpinus* (L., 1758)	Arctic char
Mugilidae*	*Liza ramada* (Risso, 1826)	Thin-lipped mullet
Pleuronectidae*	*Platichthys flesus* (L., 1758)	European flounder

after the end of the Miocene: the Messinian salinity crisis (5 million years ago [Ma]) of the Mediterranean and the last glacial maximum (18,000 years ago) (Keith 1998; Reyjol et al. 2007). To test longitude as a predictor variable in our models, coordinates of the western basins were converted to negative values.

According to the species–area theory (MacArthur 1984; Angermeier and Schlosser 1989), the surface area of the drainage basin (Surf) should influence the number of species present in a basin. Increasing the sample area includes additional habitat types with groups of different species.

The source elevation of a basin (Elv) is an element of topography informing about the altitude range encountered in a basin (Leopold et al. 1995). Combined with precipitation, it represents the hydraulic potential of the river. This has consequences for habitat characteristics and how demanding are the swimming skills required.

We selected two factors to define the climate experienced in the basins: air temperature at the mouth and precipitation across the whole basin. Air temperature in a certain range is a good proxy for water temperature. A linear relationship was found to exist between air temperature above 0°C and corresponding water temperature (Preud'homme and Stefan 1993; Pilgrim et al. 1998; Erickson and Stefan 2000). However, at low air temperature, stream temperatures often reach 0°C as an asymptote, and at high air temperature, strong evaporative cooling causes a flatter stream temperature–air temperature relationship (Mohseni and Stefan 1999). Precipitation informs about the amount of water a basin receives (Leopold et al. 1995). Discharge data are

Table 2.—Predictor variables, unit, range, and source data used in generalized additive model training.

Variable name and description	Unit	Mean	Minimum	Maximum	References
Climatic variables averaged over 1901–1910					Mitchell and Jones 2005 (CRU TS 2.1 database)
At the outlet					
TempAnn—annual air temperature	°C	11	4.8	20.2	
TempSum—summer air temperature	°C	18.9	6.2	30.1	
TempWin—winter air temperature	°C	3.5	–18	14	
On the whole basin					
PrecAnn—annual precipitation	mm/month	63.3	8.2	156.1	
PrecSum—summer precipitation	mm/month	55.9	0	172.1	
PrecWin—winter precipitation	mm/month	65.9	19.2	148.4	
Physical variables					
Long—longitude at the outlet	° (decimals)	14.1	–51.8	54.0	http://earth.google.com
Surf—surface drainage area	km^2	57,600	52	3,350,000	www.rivernet.org
Elv—elevation at the source	m	1,181	3	4,014	The Time atlas of the world, comprehensive edition 1975

difficult to obtain on so large a scale and models enabling precipitation to be converted into discharge are finalized at the local scale only and cannot currently be applied to regions.

We compiled historic data from the CRU TS 2.1 (Mitchell and Jones 2005) database, which consists of 1,200 monthly grids of observed climate for the period 1901–2002, covering the global land surface at 0.5-degree resolution. Temperature and precipitation data used in our models were averaged over 10 years between 1901 and 1910 to smooth interannual variability. As stated in the introduction, this decade was before the first warming period registered during the 20th century (IPCC 2007). Climate data were not available on a European scale before 1900. Precipitation across the whole basin corresponded to the average for all the grids contained in the basin. This calculation was performed with ArcGIS 9.0 and the Zonal Statistics tool (ESRI 2004).

Annual air temperature/precipitation (TempAnn/PrecAnn) at the river mouth and across the whole basin, respectively; the mean over the three summer months (June, July, and August—TempSum/PrecSum) and the mean over the three winter months (December, January, and February—TempWin/PrecWin) were tested separately.

Distribution Models of Diadromous Fish

Model building.—From among the various methods for predicting and explaining the presence–absence of species in relation to environmental variables (Guisan and Zimmermann 2000), we used GAMs (Hastie and Tibshirani 1990), which are semiparametric extensions of GLMs. Generalized additive models are more flexible; they permit both linear and complex additive response shapes, as well as combinations of the two within the same model (Wood and Augustin 2002).

In order to evaluate the quality of predictions for each species, we divided the original database into two subsets: calibration and validation. The first, a random sample from 75% of the total database, was used to calibrate (train) the models, whereas the second, constituted with the remaining data (25%), was used to evaluate (test) model predictions.

We used R software (R foundation core team 2005), the gam package (Hastie 2005), and some custom functions to implement the GAMs.

For each species, we examined all possible combinations of the five environmental variables, without posing and testing only plausible approaches, and selected the best models using the Akaike information criterion (Burnham and Anderson 2002). Interactions between variables were not relevant and were therefore not included in the analysis. To circumvent predictor colinearity, we added the three temperature/precipitation expressions (annual, summer, and winter) to the model separately.

We applied a binomial distribution of error via a logistic link function. A cubic smoothing spline method was chosen to smooth the studied variables, using a maximum of two degrees of freedom to avoid overfitting.

To obtain the relative importance of the predictors (their rank) in our models, we assessed the residual deviance change after excluding each variable in turn from the final model (Virkkala et al. 2005).

Model evaluation.—To assess the overall classification success of the model, we first derived a confusion matrix to tabulate the observed and predicted presence–absence patterns in order to provide a summary of the number of correct and incorrect classifications from the model (Fielding and Bell 1997). Using this matrix, five prediction success metrics were produced: (1) the percentage of sites where the model correctly predicted the presence–absence of each species was assessed (% well classified), (2) the percentage of sites where presences were correctly predicted was defined as model sensibility, (3) the percentage of sites where absences were correctly predicted was defined as model specificity; (4) the kappa coefficient measures the correct classification rate (proportion of correctly classified presences and absences) after the probability of chance agreement has been removed (Cohen 1960). Landis et al. (1977) proposed a scale to describe degree of concordance: 0.81–1.00, almost perfect; 0.61–0.80, substantial; 0.41–0.60, moderate; 0.21–0.40, fair, and <0.20, fail; (5) area under the curve (AUC) relates true positive proportion (sensibility) and false positive proportion (1-specificity) over a wide and continuous range of threshold levels, which make it a threshold-independent measure (Fielding and Bell 1997). The AUC ranges from 0.5 for models with no discrimination ability to 1 for models with perfect discrimination. An appropriate guide for classifying the accuracy of AUC is Swets (1988): 0.90–1.00, excellent; 0.80–0.90, good; 0.70–0.80, fair; 0.60–0.70, poor; 0.50–0.60, fail. Last, we used the percentage of deviance, explained (DevExp) by our final models (Virkkala et al. 2005),

$$\text{DevExp} = \frac{\text{Null model defiance} - \text{final modual residual deviance}}{\text{Null model deviance}} \times 100$$

as the sixth evaluation method.

Model Predictions under Climate Change Scenario

To examine future species distributions, we used the TYN SC 2.03 data set (Mitchell et al. 2004), which comprises monthly grids of modeled climate for the period 2001–2100 and covers the global land surface at 0.5-degree resolution. The control scenario represented the evolution of surface climate over the 21st century under the assumption that the mean climate remained fixed at 1961–1990 levels. The 20 climate change scenarios were made up of all permutations of five global climate models (GCMs) with four emission scenarios (IPCC 2000) (A1FI, A2, B2, B1).

Potential future distributions were projected by changing climate as predicted by the HadCM3 GCM under one of the most pessimistic and frequently used scenarios (A2). Temperatures and precipitation were averaged over 30 years from 2070 to 2099 in order to smooth interannual variability. The climate model resulted in a rise in annual temperature in our study area of between 1°C and 7°C and a reduction in annual precipitation for the southern basins in particular of up to 1% and 47%. Annual precipitation, on the contrary, increased in the same range for the northern basins of our set.

To transform probability values for each model into presence–absence, we used the threshold T, maximizing kappa statistics (Fielding and Bell 1997).

Results

Successful Model Building and Evaluation

For 20 species, we built robust models, including climatic variables and allowing predictions into the future. During calibration, all the models presented high AUC values (0.86–1), indicating their good accuracy. In the same way, kappa values (0.62–1) indicated substantial to almost perfect model discrimination ability. On average, final models explained 72% of deviance (Table 3).

The validation of single species models received high AUC values (0.70–1). The AUC values did not differ greatly between the calibration and the evaluation data set in the different species. However, kappa values were always weaker during the validation phase. Models were almost perfect for three species (sea lamprey, Adriatic sturgeon, and Atlantic salmon), substantial for eight species, moderate for seven species, and fair for two species (Table 3). All the selected variables were highly significant ($P < 0.001$). Longitude was included in all the GAMs models built for the 20 species. However, other variables also emerged as important predictors of species distribution. Temperature was retained in 15 models and was in fact the most explanatory variable for six species (rank 1). Surface drainage area was selected in 12 models mainly in rank 2 (eight species); precipitation was included in seven models and source elevation in one model. Relationships between species presence–absence and each predictor variable varied from simple increasing or decreasing trends to more complex unimodal trends. Unimodal characterizes a symmetrical or nonsymmetrical distribution that has only one mode (Table 4).

Successful Model Predictions under Climate Change Scenario

Predictions for the 20 modeled species revealed responses that were species-dependent. Nevertheless, these species could be placed into three categories: those gaining suitable basins with an expansion of the 1900 presence area of between 10% to 30%, those losing suitable basins with a contraction of between 16% and 92%, and those showing little or no change (Figure 2).

The expected trend of northward expansion was apparent for twaite shad and thin-lipped mullet, with no decreasing trend in the southern part of their distribution area and a gain of suitable basins in the southern Baltic as well as in Ireland, Scotland, and Iceland (Figure 3a).

Caspian lamprey distribution decreased with the disappearance of this lamprey in the southern Caspian Sea (in Iran and Turkmenistan). Changes in the distribution of Russian sturgeon and stellate sturgeon tended to show the eradication of these species from small basins in the Black Sea and Caspian Sea. Potential suitable basins for Acipenseridae were recorded in Scandinavia, such as the Neva River near the Gulf of Finland or the Vistula River in Poland. However, these basins were a long way from the present distribution of these sturgeons. *Coregonus* spp. remained in the extreme northern part of its distribution area (Gulf of Bothnia and northern Norway). Future potential distribution for vimba *Vimba vimba* indicated a regression in small basins of the Black Sea and Caspian Sea and a northward movement in the Baltic. Arctic char *Salvelinus alpinus* remained present only in the four basins studied in Iceland and in the Altä River in northern Norway (Figure 3b). In the same way, the European smelt *Osmerus eperlanus* model tended towards a high rate of disappearance of this species across its entire distribution area with a potential new refuge in Iceland. The model for Allis shad indicated a major decrease in suitable conditions in Morocco, Portugal, and Spain but a northward expansion into Scotland and Iceland. For the European river lamprey distribution area, a decrease mainly in southern France around the Bay of Biscay was predicted. Next, the model projected a stable presence area, particularly in the British Isles. Model predictions for Atlantic salmon highlighted a disappearance from the Iberian Peninsula, the basins around the Bay of Biscay, and the small basins bordering the English Channel. In the same way, future predictions for sea lamprey highlighted a decrease in presence of this lamprey in Italian basins and in the Iberian Peninsula (Figure 3c). For brown trout *Salmo trutta*, the resident populations were predicted to disappear for the most part in all the countries bordering the Mediterranean up to the Charente River. The European flounder *Platichthys flesus* model tended towards a disappearance of this species in all the countries bordering the Black Sea and the Mediterranean (Figure 3d).

Adriatic sturgeon, Azov shad *Alosa tanaica*, Black Sea salmon *S. labrax*, Caspian trout *S. caspius*,

Table 3.—Summary of the final generalized additive models selected to explain the distribution of the diadromous fish. Calibration: training set (147 basins) considered to compute the model; validation: testing set (49 basins) considered to validate the model. *T* is threshold. Kappa values were calculated using the optimal threshold (*T*) that yielded maximum values for each of the models based on the model calibration data. The list is ordered taxonomically according to Kottelat (1997) and Kottelat and Freyhof (2007). AUC = area under the curve; DevExp = percent of deviance, explained.

		Calibration						Validation				
Species	*T*	% well classified	Sensibility	Specificity	Kappa index	AUC	DevExp (%)	% well classified	Sensibility	Specificity	Kappa index	AUC
Caspian lamprey	0.50	99	100	99	0.92	0.99	85	94	50	98	0.54	0.74
River lamprey	0.41	92	96	88	0.83	0.96	66	80	78	81	0.59	0.91
Sea lamprey	0.50	97	100	94	0.94	0.99	75	94	90	100	0.87	0.95
European eel	0.72	99	100	93	0.96	0.99	80	94	95	83	0.73	0.90
Russian sturgeon	0.40	98	90	100	0.94	0.99	82	90	50	93	0.39	0.72
Adriatic sturgeon	0.32	99	83	99	0.83	0.99	83	100	100	100	1	1
Stellate sturgeon	0.37	94	87	96	0.81	0.97	65	94	50	100	0.64	0.74
European sturgeon	0.37	77	80	77	0.54	0.82	25	77	53	91	0.47	0.72
Beluga	0.51	98	96	98	0.92	0.98	78	94	67	96	0.54	0.80
Allis shad	0.59	90	85	92	0.78	0.94	51	82	71	88	0.59	0.94
Twaite shad	0.40	81	99	62	0.62	0.86	37	75	78	72	0.50	0.81
Azov shad	0.16	95	100	95	0.67	0.98	64	90	33	98	0.40	0.73
Vimba	0.36	93	85	95	0.80	0.97	65	87	40	100	0.51	0.70
European smelt	0.61	96	97	99	0.90	0.99	77	86	67	93	0.63	0.90
Coregonus spp.	0.13	100	100	100	1	1	96	96	75	98	0.73	0.86
Caspian trout	0.10	100	100	100	1	1	99	96	100	96	0.73	0.98
Black Sea salmon	0.46	97	87	98	0.86	0.99	71	98	100	98	0.66	0.98
Atlantic salmon	0.20	99	99	96	0.94	0.99	88	94	100	87	0.88	0.98
Brown trout	0.69	86	83	93	0.69	0.92	43	79	83	67	0.48	0.86
Arctic char	0.25	99	100	99	0.97	0.99	91	94	100	93	0.79	0.96
Thin-lipped mullet	0.59	91	93	87	0.80	0.95	49	83	79	92	0.61	0.91
European founder	0.48	95	99	82	0.85	0.97	66	86	93	67	0.63	0.88

Table 4.—Variables selected in final models showing nature of the relationship between each continuous variable and diadromous fish presence–absence. For each variable, numbers indicate its relative contribution (Rank) to the model. (s) represents smooth term of generalized additive models. Effect represents with a letter (P, positive; N, negative; U, unimodal with an intermediate optimum) the direction of the effect. For climatic variables, Type informs on the expression (Annual, Summer, or Winter) retained in the final model. The list is taxonomically ordered according to Kottelat (1997) and Kottelat and Freyhof (2007).

	Long		TempAnn TempSum TempWin			Surf		PrecAnn PrecSum PrecWin			Elv	
Species	Rank	Effect	Rank	Type	Effect	Rank	Effect	Rank	Type	Effect	Rank	Effect
Caspian lamprey	1	P						2(s)	Annual	U		
European river lamprey	1(s)	U	2(s)	Annual	U						3	N
Sea lamprey	1(s)	U	3(s)	Summer	U			2(s)	Annual	U		
European eel	1(s)	U										
Russian sturgeon	1(s)	P	3(s)	Annual	U	2	P					
Adriatic sturgeon	1(s)	U						2(s)	Annual	P		
Stellate sturgeon	1	P	3(s)	Annual	U	2	P					
European sturgeon	2(s)	U				1(s)	U	3	Annual	P		
Beluga	1	P				2(s)	P	3	Annual	P		
Allis shad	1(s)	N	2(s)	Summer	U	3(s)	U					
Twaite shad	1(s)	N	3(s)	Annual	P	2(s)	U					
Azov shad	1(s)	P	3(s)	Annual	U	2(s)	P					
Vimba	1(s)	P	3(s)	Winter	U	2(s)	P					
European smelt	3(s)	U	1(s)	Summer	U	2(s)	U					
Coregonus spp.	2(s)	U	1(s)	Annual	N	3(s)	U					
Caspian trout	1(s)	P						2(s)	Annual	U		
Black Sea salmon	1(s)	P				2(s)	P	3(s)	Annual	U		
Atlantic salmon	2(s)	U	1	Annual	N	3(s)	U					
Brown trout	1(s)	U	2(s)	Winter	N	3	P					
Arctic char	2	N	1(s)	Winter	N							
Thin-lipped mullet	2(s)	U	1(s)	Annual	P							
European flounder	2(s)	U	1(s)	Summer		U		3(s)	Annual	U		

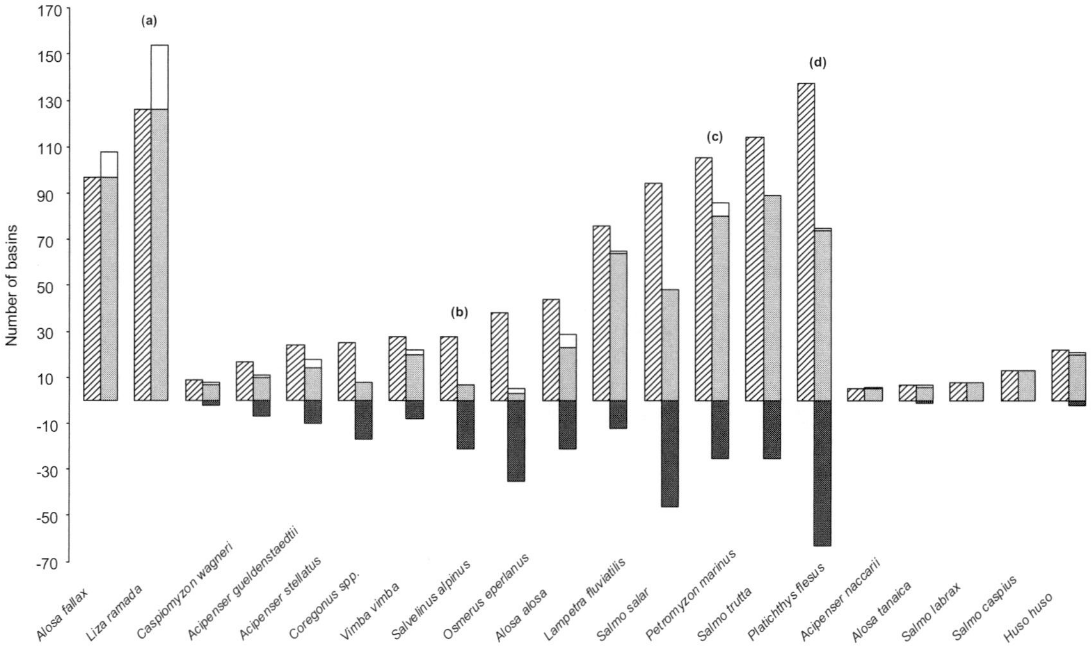

Figure 2.—Changes in distribution predicted for each species by generalized additive models. Hatched bars represent the number of basins where the species occurred in 1900. Light gray bars represent basins where the species remained in 2100 under climate change. White and dark gray bars represent respectively the number of basins gained and lost by a species in 2100 under climate change. Letters indicate the species for which a map was produced in Figure 3. The species were sorted to highlight groups according to response to climate change. Within a group, the species were sorted by increasing frequency of occurrence in the study area.

and beluga remained stable without clear expansion or contraction of their distribution area under global change.

Species Failing during the Modeling Process

First, no models were established for *Alosa algeriensis*, Caspian shad, blackspined shad, Volga herring, and houting because of their too narrow distribution range (presence in only two to four basins over the 196 investigated). The Pontic shad model was not mentioned because of poor results during the validation phase. The kappa value inferior to 0.20 indicated that the model failed especially in the prediction of presences.

Second, the model for European sturgeon *Acipenser sturio* was not sufficiently relevant to perform an accurate prediction, with only 25% of deviance explained by this model and the weakest kappa and AUC values during the calibration phase (Table 3). Finally, climatic variables were not significant in the European eel model; thus prediction under climate change scenario could not be performed (Table 4).

Discussion

The Model Approaches and the Use of GAMs

Presence–absence models performed well (i.e., discriminated successfully between presences and absences for 20 species out of 28). The level of discrimination in these models was good to excellent (AUC values 0.86–1) during the calibration phase. The high calibration kappa values confirmed the good discrimination ability of the models. Furthermore, predictions made for the independent validation data set successfully predicted occurrence for these species. Both validation AUC values and validation kappa values ranged from fair to excellent. Our overall con-

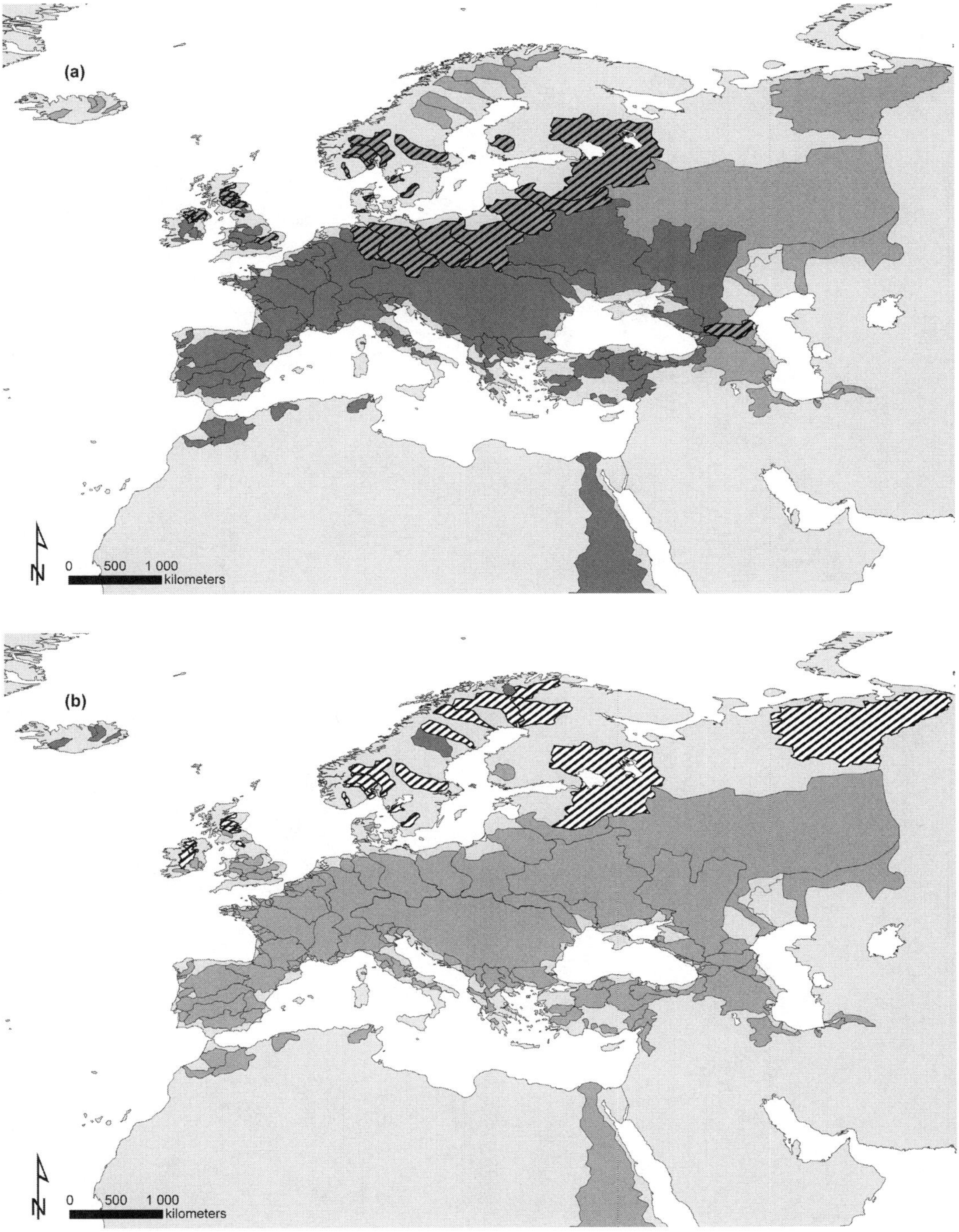

Figure 3.—Projected future potential distribution for (a) thin-lipped mullet, (b) Arctic char, (c) sea lamprey, and (d) European flounder. Dark gray color corresponds to stable suitable basins, clear gray to stable unsuitable basins, hatched with a white background to loss of suitable basins, and hatched with a gray background to gain of suitable basins. Nile (Egypt) and Kapisillit (Greenland) basins were truncated to offer a better visualization of the whole study area.

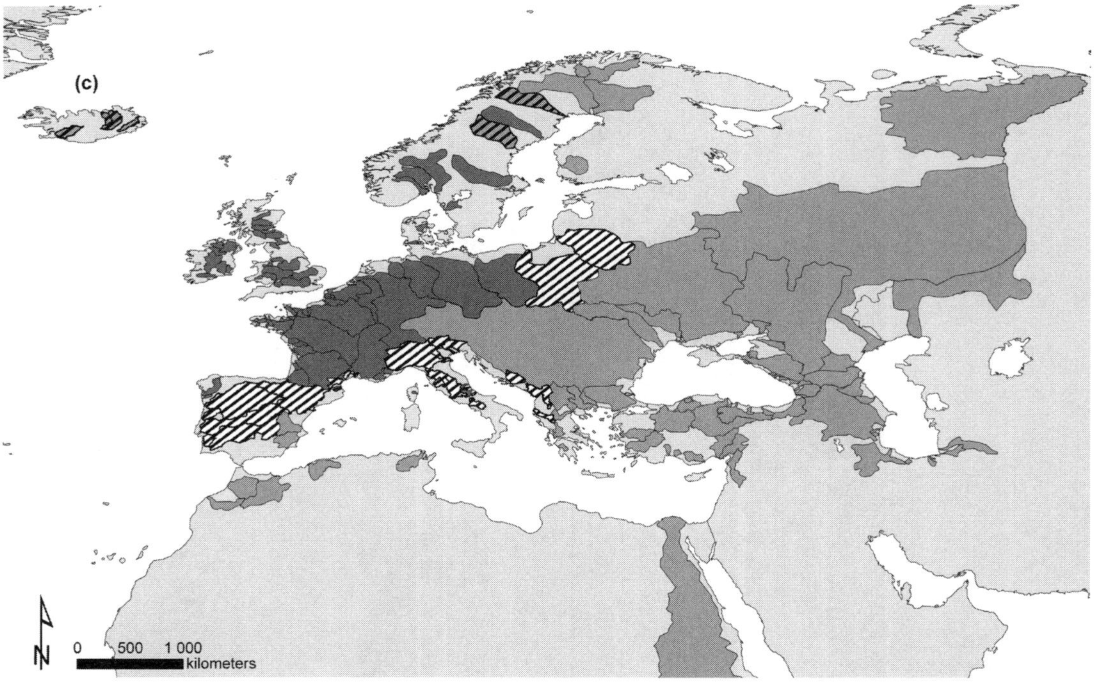

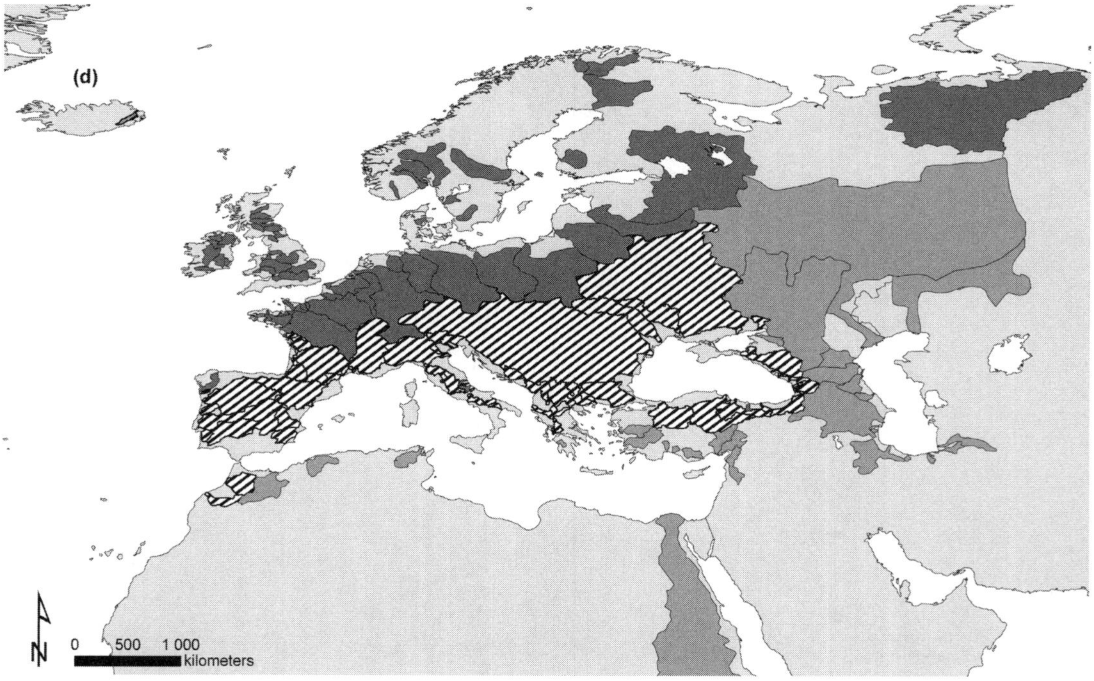

Figure 3.—Continued.

clusion is that GAMs were successful in describing and predicting the occurrence of most species.

No models could be established for *Alosa algeriensis*, Caspian shad, blackspined shad, Volga herring, and houting because of the rarity of these species in the study area (i.e., only one to four basins with these species in the study area). Indeed, in our database EuroDiad 2.0, *A. algeriensis* is only recorded in North Africa, Caspian shad is endemic to the southern Caspian Sea, blackspined shad and Volga herring spawn in the Caspian Sea, mainly in two northern rivers, and houting is recorded in a few basins in Germany and the Netherlands. To establish models with high discrimination abilities for these species, we would have to enlarge and detail our database in these specific regions.

The Pontic shad model failed during the validation phase with a null kappa value. Source elevation was retained as a predictor with a bimodal response curve (low and high optima) indicating a variable that was too distal with no direct physiological effect. A more ecologically relevant variable related to source elevation has to be found to build a distribution model for pontic shad. The European eel model included only longitude as an explanatory variable, mainly because the species was recorded as present in most of the basins. In further study, an abundance index should be used.

Environmental Variables and Their Relevance in the Modeling Process

The individual response functions for the continuous variables presented in Table 4 are u-shaped (bimodal) as well as bell-shaped (unimodal) and truncated versions of these functions. These results are in agreement with Austin's (2007) conclusions. Inspection of the relationship between each variable and species presence–absence revealed that the GAMs were generally interpretable and plausible.

Longitude was retained in each model, mainly in first rank. The eight species endemic to the Caspian Sea had a positive correlation between their probability of presence and longitude. This is in accordance with the biogeographical history and particularly with the formation of the present Ponto-Caspian region 5 Ma. It is now widely accepted that during the Messinian (7.1–5.3 Ma; Berggren et al. 1995), the Mediterranean (Tethys) was at least partially dried up (Hsü et al. 1973, 1977). Hsü (1972) has suggested that Paratethys, a northern arm of the Mediterranean extending from the Rhone basin to the Caucasus, dried up too. This phenomenon resulted during the Pliocene (5.3–1.8 Ma; Berggren et al. 1995) in the formation of the Black Sea and Azov Sea, the Balaton Lake in Hungary, the Caspian and the Aral seas, and the development of a highly endemic fauna in this area. However, under the climate change scenario used in our study, Baltic basins were predicted to become suitable for Beluga, Russian, and stellate sturgeons. These basins, which are in the same range of longitude as the Ponto-Caspian region, are not connected to this region. This result underlined the necessity to find a more relevant proxy for endemism and biogeographical history (Oberdorff et al. 1997, 1999).

Temperature was used in 15 models. For the species described as coldwater species in the literature, Atlantic salmon, brown trout, Arctic char (Klemetsen et al. 2003), and *Coregonus* spp. (Svärdson 1970), models were mainly established using the winter temperature as an explanatory variable with a decreasing trend for this variable. Two species retained the annual temperature with an increasing trend. Indeed, for twaite shad, temperature appears to play an important role in the onset of spawning, with a range of 18–22°C (Aprahamian et al. 2003). Thin-lipped mullet is recorded in all the temperate and tropical seas entering lagoons and estuaries and rivers between temperatures of 8–24°C (Keith and Allardi 2001). Temperature appeared to have an influence for nine other species with bell-shaped response curves. Both annual and summer temperatures were retained equally in these models. Optimal temperatures registered for each species were in accordance with ecological observations (European river lamprey and sea lamprey [Maitland 1980], Russian sturgeon and stellate sturgeon [Holcik 1989], Allis shad and Azov shad [Hoestlandt 1991], vimba [Zayanchkauskas 1970], European smelt [Nellbring 1989], and European flounder [Fonds et al. 1992]).

Surface of the drainage basin appeared to have an influence on continental diadromous fish distribution. This variable was selected with an increasing trend in 12 models out of 20. This could be related to the species–area theory (MacArthur 1984; Angermeier and Schlosser 1989) and more precisely to environmental heterogeneity. Increasing the sample area includes additional available habitat types

(Williams 1964). Nevertheless, probability of presence decreased for five species for very large basins more than 200,000 km^2. There may be a trade-off between habitat availability and stream hydraulicity. Moreover, species presenting this decrease were among the smallest ones, such as Allis shad, twaite shad, European smelt, and *Coregonus* spp.. For example, Allis shad cannot swim upstream if the speed of the current exceeds 2 m/s, and they cannot jump (Cassou-Leins 1981). The spawning sites of twaite shad are homogeneous, sandy, and muddy, indicating a low stream velocity (Hoestlandt 1991).

Precipitation was included in 7 models out of 20, and annual precipitation was the only selected expression. This variable is easily available on a large temporal and spatial scale and acts as surrogate for a more proximal variable such as depth or a linked factor like current speed or associated granulometry. Indeed, the two species that presented an increasing trend with this variable are sensitive to stream depth during spawning periods. During both the seaward and the spawning migration, the beluga normally travels in the deepest parts of the riverbed (Holcik 1989; Vecsei et al. 2002). In rivers, Adriatic sturgeon seeks out deep pools with high flow rates (Bronzi et al. 2005).

Elevation acted only for European river lamprey, which exhibited a decreasing trend with this variable. Ammocoetes live in rivers where they burrow in the silt bottom either in the main channel or along the bank. Moreover, in Scandinavian rivers, average flow rates over the burrows of *Lampetra* sp. have been recorded at 0.08–0.1 m/s (Hardisty 1986). These elements of their ecology suggest a preference for rivers with a low hydraulic potential.

Limitations of the Approach

Accurate absence data in particular are often difficult to obtain (Hirzel et al. 2002). A species may be undetected even though it was present, especially for inconspicuous species. Nevertheless, most diadromous fish species are not very small in size and all of them can be caught by fishermen; therefore, the existence of cryptic populations is very unlikely. A mobile species could be absent from the site only temporarily. However, for each basin in our data set, we collected several references to its fish fauna from different dates. In some cases, authors neglect rare species or species whose distribution is limited to the lower reaches of basins. Thus, in the case of contradictions between data sources of the same quality, data indicating presence were favored in our study. Furthermore, difficulties might arise when working with historical data. As historical records mostly originate from opportunistic sampling, they show sampling biases (e.g., certain localities have been investigated more frequently than others or no information is available on the exact locality of the survey) (Lütolf et al. 2006). Thus, we chose to investigate species presence–absence on a large scale (basin scale) to reduce the sources of error in our data. To go further in the predictions of diadromous fish distributions, an abundance index should be used to split the presences into several classes such as rare, moderate, or abundant levels. In this way, a population predicted to be still present but in strong decline could be detected, which was not possible with presence–absence data.

It is important to remember that the protocol developed in our study corresponds to projections of a pristine situation in a changing global environment, focusing on the climatic suitability of the habitat. Our models show habitat changes and infer changes in fish distributions. However, the potential success of these species depends on other crucial factors such as the distance to cover and their dispersion tactics (Solomon and Shugart 1993; Bakkenes et al. 2002). Further studies on variations in life history traits along a latitudinal gradient could address the issue more precisely and could provide information on the possibility of the predicted distribution being filled by the species according to the dynamic of its populations (Lassalle et al. 2008). Furthermore, this study does not include any estimates of the regional anthropic pressures that are placed on aquatic ecosystems, such as the erection of dams in main stems (Gowans et al. 1999; Larinier 2001; Hill 2009), the deterioration in water quality in industrialized areas (Breine et al. 2007; Maes et al. 2007; Woodland et al. 2009), or unsustainable fisheries (Masters et al. 2006). However, there are two main limitations to the inclusion of these variables into the modeling process: first, finding relevant proxies and then, developing evolution scenarios for these variables in the future.

There are at least three sources of uncertainty inherent in the modeling process presented in this paper and other published studies (Thuiller 2003): uncertainty based on climate change scenarios, uncertainty based on the modeling techniques, and uncertainty based on the selection of the threshold to transform probability values into presence–absence form. We tried to minimize the two sources of error that we could control. First, the modeling technique used in this work (GAMs) allowed us to fit realistic descriptions of relationships between species and environmental predictors with complex and nonlinear response curves (Hastie and Tibshirani 1990; Austin 2007). Moreover, rather than simply following the conventional decision threshold of 0.5 (Oberdorff et al. 2001), we examined thresholds with 0.01 accuracy for each of our models and selected the probability level that gave the highest Cohen's kappa values (Fielding and Bell 1997). To show the range of possibilities, it would be necessary to test all the emission scenarios from the most optimistic to the most pessimistic.

Implications for Management

Our results showed that basins that were suitable in 1900 could become unsuitable under climate change. There are species for which the existence of suitable space is predicted to be sufficiently reduced to cause serious concern and thus further support their conservation. This category would include Nordic species such as Arctic char and European smelt, which are currently classified as LR (low risk) and DD (data deficient), respectively, in the International Union for the Conservation of Nature (IUCN) World Red Data Book (IUCN 2006). These Nordic species are at the border with the Arctic region and thus do not have new territories to colonize if they were to face global warming (Figure 3b). More generally, 20 European diadromous fish species are listed under the 2006 IUCN red list and the most threatened are also protected by European and national laws. Consequently, restoration and management plans have been set up to reduce harmful impacts on species, to sustain populations by stocking, or to reintroduce an extinct species into a basin (Lyle and Maitland 1997; Prignon et al. 1999; Kirschbaum and Gessner 2000). Action plans aim to reverse the trend for the remaining populations, but also to restore the species to their historical range (i.e., 1900). Therefore, in the light of the distribution changes predicted by our models, we argue that global warming issues should be included in diadromous fish conservation or restoration plans (Rochard et al. 2009).

In conclusion, models were established for almost all diadromous fish species recorded in Europe, North Africa, and the Middle East. In accordance with ecological theories, five variables related to physical habitat and climate were retained as explanatory variables in GAMs. Predictions at the end of this century under the global warming assumption revealed a shift in distribution for many diadromous species. The modeled species could be placed mainly into three categories: those losing suitable basins, those gaining suitable basins, and those showing little or no change. This first assessment of future diadromous fish distributions provides useful guidelines for their conservation and management.

Acknowledgments

We would like to thank colleagues in the Diadfish network who checked the EuroDiad 2.0 database: Miran Aprahamian from the Environment Agency, United Kingdom; Pier Giorgio Bianco from The University of Napoli, Biology Department, Italy; Panos Stavros Economidis from the Aristotle University, Zoology Department, Greece; Mejdeddine Kraïem from the National Institute of Marine Sciences and Technologies, Tunisia; Ion Navodaru from the Danube Delta National Institute for Research and Development, Romania; and Panu Orell from the Finnish Game and Fisheries Research Institute, Finland.

This study was carried out with financial support from the Adour-Garonne Water Agency, "Programme Seine-Aval," the French Ministry of Ecology and Sustainable Development "Programme GICC, Gestion et Impacts du Changement Climatique," and the National Office of Water and the Aquatic Environments (ONEMA).

References

Almeida, P. R., B. R. Quintella, N. M. Dias, and N. Andrade. 2002. The anadromous sea lamprey in Portugal: biology and conservation perspectives. Pages 49–58 *in* M. Moser, J. Bayer, and D. MacKinlay, editors. International Congress on the Biology of Fish: the biology of lampreys. American Fisheries Society, Physiology Section, Bethesda, Maryland.

Angermeier, P., and I. J. Schlosser. 1989. Species–area relationships for stream fishes. Ecology 70:1450–1462.

Aprahamian, M. W., C. D. Aprahamian, J. L. Baglinière, R. Sabatié, and P. Alexandrino. 2003. *Alosa alosa* and

Alosa fallax spp. literature review and bibliography. Environment Agency, Warrington, UK.

Austin, M. 2007. Species distribution models and ecological theory: a critical assessment and some possible new approaches. Ecological Modelling 200:1–19.

Baglinière, J. L., M. R. Sabatié, E. Rochard, P. J. Alexandrino, and M. W. Aprahamian. 2003. The Allis shad (*Alosa alosa* Linnaeus, 1758): biology, ecology, range and status of populations. Pages 85–102 *in* K. E. Limburg, and J. R. Waldman, editors. Biodiversity, status, and conservation of the world's shads. American Fisheries Society, Bethesda, Maryland.

Bakkenes, M., J. R. M. Alkemade, F. Ihle, R. Leemans, and J. B. Latour. 2002. Assessing effects of forecasted climate change on the diversity and distribution of European higher plants for 2050. Global Change Biology 8(4):390–407.

Béguer, M., L. Beaulaton, and E. Rochard. 2007. Distribution and richness of diadromous fish assemblages in Western Europe: large scale explanatory factors. Ecology of Freshwater Fish 16:221–237.

Berggren, W. A., D. V. Kent, C. C. Swisher, and M. P. Aubry. 1995. A revised Cenozoic geochronology and chronostratigraphy. Pages 129–212 *in* W. A. Berggren, D. V. Kent, M. P. Aubry, and J. Hardenbol, editors. Geochronology, time scales and global stratigraphic correlation. Society of Economic Paleontologists and Mineralogists Special Publication 54.

Breine, J. J., J. Maes, P. Quataert, E. Van der Bergh, I. Simoens, G. Van Thuyne, and C. Belpaire. 2007. A fish-based assessment tool for the ecological quality of the brackish Schelde estuary in Flanders (Belgium). Hydrobiologia 575:141–159.

Bronzi, P., P. Vecsei, and G. Arlati. 2005. Threatened fishes of the world: *Acipenser naccarii* Bonaparte, 1836 (Acipenseridae). Environmental Biology of Fishes 72(1):66.

Brotons, L., W. Thuiller, M. B. Araujo, and A. H. Hirzel. 2004. Presence–absence versus presence-only modelling methods for predicting bird habitat suitability. Ecography 27:437–448.

Burnham, K. P., and D. R. Anderson. 2002. Model selection and multimodel inference: a practical information-theoretic approach, 2nd edition. Springer–Verlag, New-York.

Cassou-Leins, F. 1981. Recherches sur la biologie et l'halieutique des migrateurs de la Garonne et principalement de l'alose: *Alosa alosa* L. Thèse de doctorat. Institut National Polytechnique de Toulouse, Toulouse, France.

Chu, C., N. E. Mandrak, and C. K. Minns. 2005. Potential impacts of climate change on the distributions of several common and rare freshwater fishes in Canada. Diversity and Distributions 11(4):299–310.

Cohen, J. 1960. A coefficient of agreement for nominal scales. Educational and Psychological Measurement 20:27–46.

Dekker, W. 2003. On the distribution of the European eel (*Anguilla anguilla*) and its fisheries. Canadian Journal of Fisheries and Aquatic Sciences 60:787–799.

Dill, W. A. 1993. Inland fisheries of Europe. Food and Agriculture Organization of the United Nations, European Inland Fisheries Advisory Technical Paper 52, Rome.

ESRI (Environmental System Research Institute). 2004. ArcGIS, version 9.0. Environmental System Research Institute, Redlands, California.

Erickson, T. R., and H. G. Stefan. 2000. Linear air/water temperature correlations for streams during open water periods. Journal of Hydrologic Engineering 5(3):317–321.

Fielding, A. H., and J. F. Bell. 1997. A review of methods for the assessment of predictor error in conservation presence/absence models. Environmental Conservation 24:38–49.

Fonds, M., R. Cronie, A. D. Vethaak, and P. van der Puyl. 1992. Metabolism, food consumption and growth of plaice (*Pleuronectes platessa*) and flounder (*Platichthys flesus*) in relation to fish size and temperature. Netherlands Journal of Sea Research 29:127–143.

CIGB/ICOLD. 2003. World register of dams. Registre mondial des barrages, D. 8.3 100 EEA11784, Paris.

Gowans, A. R. D., J. D. Armstrong, and I. G. Priede. 1999. Movements of adult Atlantic salmon in relation to a hydroelectric dam and fish ladder. Journal of Fish Biology 54(4):713–726.

Guisan, A., and N. E. Zimmermann. 2000. Predictive habitat distribution models in ecology. Ecological Modelling 135(2–3):147–186.

Hardisty, M. W. 1986. *Lampetra fluviatilis* (Linnaeus, 1758). Pages 249–278 *in* J. Holcik, editor. The freshwater fishes of Europe, volume 1, part 1. Petromyzontiformes. Aula-Verlag, Wiesbaden, Germany.

Hastie, T. J. 2005. gam: generalized additive models. R package version 0.94.

Hastie, T. J., and R. J. Tibshirani. 1990. Generalized additive models. Chapman and Hall.

Heikkinen, R. K., M. Luoto, M. Kuussaari, and T. Toivonen. 2007. Modelling the spatial distribution of a threatened butterfly: impacts of scale and statistical technique. Landscape and Urban Planning 79(3–4):347–357.

Hilbert, D. W., M. Bradford, T. Parker, and D. A. Westcott. 2004. Golden bowerbird (*Prionodura newtonia*) habitat in past, present and future climates: predicted extinction of a vertebrate in tropical highlands due to global warming. Biological Conservation 116(3):367–377.

Hill, A. K. 2009. The Santee River basin accord: restoring diadromous fish through prioritization of subbasins. Pages 769–782 *in* A. J. Haro, K. L. Smith, R. A. Rulifson, C. M. Moffitt, R. J. Klauda, M. J. Dadswell, R. A. Cunjak, J. E. Cooper, K. L. Beal, and T. S. Avery, editors. Challenges for diadromous fishes in a dynamic global environment. American Fisheries Society, Symposium 69, Bethesda, Maryland.

Hirzel, A. H., J. Hausser, D. Chessel, and N. Perrin. 2002. Ecological-niche factor analysis: how to compute habitat-suitability maps without absence data? Ecology 83(7):2027–2036.

Hoestlandt, H., editor. 1991. Clupeidae - Anguillidae, The freshwater fishes of Europe, volume 2. Aula-Verlag, Wiesbaden.

Hoffmann, R. C. 1996. Economic development and aquatic ecosystems in medieval Europe. American Historical Review 101:631–669.

Hoffmann, R. C. 2005. A brief history of aquatic resource use in medieval Europe. Helgoland Marine Research 59(1):22–30.

Holcik, J., editor. 1989. General introduction to fishes Acipenseriformes, The freshwater fishes of Europe, volume 1/II. Aula-Verlag, Wiesbaden.

Hsü, K. 1972. When the Mediterranean dried up. Scientific American 227:26–36.

Hsü, K., L. Montadert, D. Bernouilli, M. B. Cita, A. Erickson, R. E. Garrison, R.B. Kidd, F. Melieres, and R. C. Müller. 1977. History of the Mediterranean salinity crisis. Nature (London) 267:399–403.

Hsü, K. J., W. B. F. Ryan, and M. B. Cita. 1973. Late Miocene desiccation of the Mediterranean Sea. Nature (London) 242:240–244.

IPCC (Intergovernmental Panel on Climate Change). 2000. Emissions scenarios, Special Reports of the IPCC. Cambridge University Press, Cambridge, UK.

IPCC (Intergovernmental Panel on Climate Change). 2007. Climate change 2007: the physical science basis. Contribution of working group I to the fourth assessment report of the Intergovernmental Panel on Climate Change. Cambridge University Press, Cambridge, UK.

IUCN (Intergovernmental Panel on Climate Change). 2006. 2006 IUCN red list of threatened species. Intergovernmental Panel on Climate Change, Gland, Switzerland.

Keith, P. 1998. Evolution des peuplements ichtyologiques de France et stratégies de conservation. Thèse de doctorat. Université de Rennes I, Rennes, France.

Keith, P., and J. Allardi. 2001. Atlas des poissons d'eau douce de France, patrimoines naturels, volume 47. SPN/IEGB/MNHN, Paris.

Kelly, F. L., and J. J. King. 2001. A review of the ecology and distribution of three lamprey species, *Lampetra fluviatilis* (L.), *Lampetra planeri* (Bloch) and *Petromyzon marinus* (L.): a context for conservation and biodiversity considerations in Ireland. Biology and Environment: Proceedings of the Royal Irish Academy 101B(3):165–185.

Kirschbaum, F., and J. Gessner. 2000. Re-establishment programme for *Acipenser sturio* L., 1758: the German approach. Boletin Instituto Espanol de Oceanografia 16(1–4):149–156.

Klemetsen, A., P. A. Amundsen, J. B. Dempson, B. Jonsson, N. Jonsson, M. F. O'Connell, and E. Mortensen. 2003. Atlantic salmon *Salmo salar* L., brown trout *Salmo trutta* L. and Arctic charr *Salvelinus alpinus* (L.): a review of aspects of their life histories. Ecology of Freshwater Fish 12:1–59.

Kottelat, M. 1997. European freshwater fishes: an heuristic checklist of the freshwater fishes of Europe (exclusive of former USSR), with an introduction for non-systematists and comments on nomenclature and conservation. Biologia (Bratislava) 52(Supplement 5):1–271.

Kottelat, M., and J. Freyhof. 2007. Handbook of European freshwater fishes. Kottelat, Cornol, Switzerland and Freyhof, Berlin, Germany.

Landis, J. R., G. G. Koch, and J. Lawton. 1977. The measurements of observer agreement for categorical data. Biometrics 33:159–174.

Larinier, M. 2001. Environmental issues, dams and fish migration. Pages 45–90 *in* G. Marmulla, editor. Dams, fish and fisheries: opportunities, challenges and conflict resolution. Food and Agriculture Organization of the United Nations, FAO Fisheries Technical Paper 419, Rome.

Lassalle, G., T. Trancart, P. Lambert, and E. Rochard. 2008. Latitudinal variations in age and size at maturity among allis shad populations. Journal of Fish Biology 73:1799–1809.

Leathwick, J. R., J. Elith, and T. Hastie. 2006. Comparative performance of generalized additive models and multivariate adaptive regression splines for statistical modelling of species distribution. Ecological Modelling 199:188–196.

Leopold, L. B., M. G. Wolman, and J. P. Miller. 1995. Fluvial processes in geomorphology. Courier Dover Publications, New York.

Lütolf, M., F. Kienast, and A. Guisan. 2006. The ghost of past species occurrence: improving species distribution models for presence-only data. Journal of Applied Ecology 43:802–815.

Lyle, A. A., and P. S. Maitland. 1997. The spawning migration and conservation of smelt *Osmerus eperlanus* in the River Cree, southwest Scotland. Biological Conservation 80(3):303–311.

MacArthur, R. H. 1984. Geographical ecology: patterns in the distribution of species. Princeton University Press, Princeton, New Jersey.

MacKenzie, B. R., H. Gislason, C. Möllmann, and F. W. Köster. 2007. Impact of 21st century climate change on the Baltic Sea fish community and fisheries. Global Change Biology 13:1348–1367.

Maes, J., M. Stevens, and J. Breine. 2007. Modelling the migration opportunities of diadromous fish species along a gradient of dissolved oxygen concentration in a European tidal watershed. Estuarine, Coastal and Shelf Science 75:151–162.

Maitland, P. S. 1980. Review of the ecology of lampreys in northern Europe. Canadian Journal of Fisheries and Aquatic Sciences 37:1944–1952.

Masters, J. E. G., M. H. Jang, K. Ha, P. D. Bird, P. A. Frear, and M. C. Lucas. 2006. The commercial exploitation of a protected anadromous species, the river lamprey (*Lampetra fluviatilis* (L.)), in the tidal River Ouse, north-east England. Aquatic Conservation-Marine and Freshwater Ecosystems 16(1):77–92.

McDowall, R. M. 1988. Diadromy in fishes: migrations between freshwater and marine environments. Croom Helm, London.

McDowall, R. M. 2009. Making the best of two worlds: diadromy in the evolution, ecology, and conservation of aquatic organisms. Pages 1–22 *in* A. J. Haro, K. L. Smith, R. A. Rulifson, C. M. Moffitt, R. J. Klauda, M. J. Dadswell, R. A. Cunjak, J. E. Cooper, K. L. Beal, and T. S. Avery, editors. Challenges for diadromous fishes in a dynamic global environment. American Fisheries Society, Symposium 69, Bethesda, Maryland.

Mitchell, T. D., T. R. Carter, P. D. Jones, M. Hulme, and M. G. New. 2004. A comprehensive set of high-resolution grids of monthly climate for Europe and the globe: the observed record (1901–2000) and 16 scenarios (2001–2100) Tyndall Centre, Norwich, UK.

Mitchell, T. D., and P. D. Jones. 2005. An improved method of constructing a database of monthly climate observations and associated high-resolution grids. International Journal of Climatology 25(6):693–712.

Mohseni, O., and H. G. Stefan. 1999. Stream temperature/air temperature relationship: a physical interpretation. Journal of Hydrology 218(3–4):128–141.

Nellbring, S. 1989. The ecology of smelts (Genus *Osmerus*): a literature review. Nordic Journal of Freshwater Research 65:116–145.

Oberdorff, T., B. Hugueny, and J. F. Guégan. 1997. Is there an influence of historical events on contemporary fish species richness in rivers? Comparisons between Western Europe and North America. Journal of Biogeography 24(4):461–467.

Oberdorff, T., S. Lek, and J. F. Guégan. 1999. Patterns of endemism in riverine fish of the northern hemisphere. Ecology Letters 2(2):75–81.

Oberdorff, T., D. Pont, B. Hugueny, and J. Porcher. 2001. A probabilistic model characterizing fish assemblages of French rivers: a framework for environmental assessment. Freshwater Biology 46:399–415.

Olden, J. D., and D. A. Jackson. 2002. A comparison of statistical approaches for modelling fish species distributions. Freshwater Biology 47:1976–1995.

Pilgrim, J. M., X. Fang, and H. G. Stefan. 1998. Stream temperature correlations with air temperatures in Minnesota: implications for climate warming. Journal of the American Water Resources Association 34(5):1109–1121.

Preud'homme, E. B., and H. G. Stefan. 1993. Stream temperature estimate from air temperature. Water Resource Bulletin 29(1):27–45.

Prignon, C., J. C. Micha, G. Rimbaud, and J. C. Philippart. 1999. Rehabilitation efforts for Atlantic salmon in the Meuse basin area: synthesis 1983–1998. Hydrobiologia 410:69–77.

Quinn, T. P., K. Doctor, N. Kendall, and H. B. Rich. 2009. Diadromy and the life history of sockeye salmon: nature, nurture, and the hand of man. Pages 23–42 *in* A. J. Haro, K. L. Smith, R. A. Rulifson, C. M. Moffitt, R. J. Klauda, M. J. Dadswell, R. A. Cunjak, J. E. Cooper, K. L. Beal, and T. S. Avery, editors. Challenges for diadromous fishes in a dynamic global environment. American Fisheries Society, Symposium 69, Bethesda, Maryland.

R foundation core team. 2005. R: a language and environment for statistical computing. R Foundation for Statistical Computing, Vienna, Austria.

Reyjol, Y., B. Hugueny, D. Pont, P. G. Bianco, U. Beier, N. Ciola, F. Casals, I. Cowx, A. Economou, T. Ferreira, G. Haidvogl, R. Noble, A. Sostoa, T. Vigneron, and T. Virbickas. 2007. Patterns in species richness and endemism of European freshwater fish. Global Ecology and Biogeography 16:65–75.

Rochard, E., P. Pellegrini, J. Marchal, M. Béguer, D. Ombredane, G. Lassalle, E. Menvielle, and J.-L. Baglinière. 2009. Identification of diadromous fish species on which to focus river restoration: an example using an eco-anthropological approach (the Seine basin, France). Pages 691–711 *in* A. J. Haro, K. L. Smith, R. A. Rulifson, C. M. Moffitt, R. J. Klauda, M. J. Dadswell, R. A. Cunjak, J. E. Cooper, K. L. Beal, and T. S. Avery, editors. Challenges for diadromous fishes in a dynamic global environment. American Fisheries Society, Symposium 69, Bethesda, Maryland.

Sjöberg, K. 1980. Ecology of the European River lamprey (*Lampetra fluviatilis*) in northern Sweden. Canadian Journal of Fisheries and Aquatic Sciences 37:1974–1980.

Solomon, A. M., and H. H. Shugart, editors. 1993. Vegetation dynamics and global change. Chapman and Hall, London.

Svärdson, G. 1970. Significance of introgression in coregonid evolution. Pages 33–59 *in* C. C. Lindsey, and C. S. Woods, editors. Biology of coregonid fishes. University of Manitoba Press, Winnipeg.

Swets, K. 1988. Measuring the accuracy of diagnostic systems. Science 240:1285–1293.

The Times atlas of the world, comprehensive edition. 1975. 5th edition. Times Books & John Bartholomew and Son, London.

Thomas, C. D., and J. J. Lennon. 1999. Birds extend their ranges northwards. Nature (London) 399(6733):213.

Thuiller, W. 2003. BIOMOD: optimizing predictions of species distributions and projecting potential future shifts under global change. Global Change Biology 9:1353–1362.

Thuiller, W., O. Broennimann, G. Hughes, J. R. M. Alkemade, G. F. Midgley, and F. Corsi. 2006. Vulnerability of African mammals to anthropogenic climate change under conservative land transformation assumptions. Global Change Biology 12(3):424–440.

Tuunainen, P., E. Ikonen, and H. Auvinen. 1980. Lampreys and lamprey fisheries in Finland. Canadian Journal of Fisheries and Aquatic Sciences 37:1953–1959.

Vecsei, P., R. Sucui, and D. Peterson. 2002. Threatened fishes of the world: *Huso huso* (Linnaeus, 1758) (Acipenseridae). Environmental Biology of Fishes 65:363–365.

Virkkala, R., M. Luoto, K. Heikkinen, and N. Leikola. 2005. Distribution patterns of boreal marshland birds: modelling the relationships to land cover and climate. Journal of Biogeography 32:1957–1970.

Walther, G. R., E. Post, P. Convey, A. Menzel, C. Parmesan, T. J. C. Beebee, J.-M. Fromentin, O. Hoegh-Guldberg, and F. Bairlein. 2002. Ecological responses to recent climate change. Nature (London) 416:389–395.

Williams, C. B. 1964. Patterns in the balance of nature. Academic Press, London.

Wood, S., and N. Augustin. 2002. GAMs with integrated model selection using penalized regression splines and applications to environmental modelling. Ecological Modelling 157:157–177.

Woodland, R. J., D. H. Secor, and E. J. Niklitschek. 2009. Past and future habitat suitability for the Hudson River population of shornose sturgeon: a bioenergetic approach to modeling habitat suitability for an endangered species. Pages 589–604 *in* A. J. Haro, K. L. Smith, R. A. Rulifson, C. M. Moffitt, R. J. Klauda, M. J. Dadswell, R. A. Cunjak, J. E. Cooper, K. L. Beal, and T. S. Avery, editors. Challenges for diadromous fishes in a dynamic global environment. American Fisheries Society, Symposium 69, Bethesda, Maryland.

Zayanchkauskas, P., editor. 1970. Biology and fisheries of vimba in Europe. Mintis Press, Vilnius, Lithuania.

American Fisheries Society Symposium 69:195–214, 2009

Taking It with You When You Go: How Perturbations to the Freshwater Environment, Including Temperature, Dams, and Contaminants, Affect Marine Survival of Salmon

STEPHEN D. MCCORMICK*
U.S. Geological Survey, Conte Anadromous Fish Research Center
Turners Falls, Massachusetts 01376, USA

Program in Organismic and Evolutionary Biology, University of Massachusetts
Amherst, Massachusetts 01003, USA

DARREN T. LERNER, MICHELLE Y. MONETTE, AND
KATHERINE NIEVES-PUIGDOLLER
U.S. Geological Survey, Conte Anadromous Fish Research Center
Turners Falls, Massachusetts 01376, USA

Program in Organismic and Evolutionary Biology, University of Massachusetts
Amherst, Massachusetts 01003, USA

JOHN T. KELLY
U.S. Geological Survey, Conte Anadromous Fish Research Center
Turners Falls, Massachusetts 01376, USA

Department of Natural Resource Conservation, University of Massachusetts
Amherst, Massachusetts 01003, USA

BJÖRN THRANDUR BJÖRNSSON
Department of Zoophysiology, University of Göteborg, Göteborg 405 30, Sweden

Abstract.—Most anadromous fish undergo physiological and behavioral changes that are preparatory and adaptive for seawater entry. In anadromous salmonids, these preparatory changes are collectively known as smolting. Smolt development is regulated by environmental factors such as photoperiod and temperature and mediated by the neuroendocrine system. In this paper, we review evidence that a variety of anthropogenic factors can influence smolt development and affect marine survival. Hatchery rearing can affect the size of smolts and the extent and timing of smolt development. Smolt development is reversible, and the period of peak physiological preparedness in salmon smolts is limited by time and temperature. By influencing temperature and the duration of the migratory period, climate change and dams will have negative effects on smolt survival beyond direct lethal impacts. Contaminants acting on developmental physiology or underlying endocrine control mechanisms can also reduce marine survival. Exposure to estrogenic compounds prior to or during smolt development can reduce seawater tolerance and preference. Acid and aluminum exposure can reverse the development of seawater tolerance and reduce adult return rates. We conclude that environmental conditions in freshwater can affect physiological development, estuarine and ocean behavior, early seawater survival, and long-term seawater growth and homing, thus having influences on adult returns and long-term population sustainability of anadromous fishes.

* Corresponding author: steve_mccormick@usgs.gov

Introduction

The freshwater and marine survival of anadromous fishes is critically intertwined, though our understanding of the relationship between the two is limited. Availability of favorable freshwater habitat can affect the number of juvenile anadromous fish that enter seawater from any given river. The number of emigrating juveniles will thus be determined by both the number of spawning adults and the ability of the freshwater habitat to produce and sustain the entire freshwater life cycle (Jonsson et al. 1998; Pess et al. 2002). In addition to the number of fish that leave the freshwater environment, the quality of these fish and their capacity to enter the marine environment will also be important. In this paper, we review the adaptive value of smolt development in salmon and its relation to marine survival. We also examine the environmental factors, both natural and anthropogenic, that influence smolt development and seawater performance. Although most of the information on preparatory physiological adaptations for seawater entry and residence comes from research on salmonids, we hypothesize that similar environmental influences will affect the seawater performance of other anadromous species, particularly those in which preparatory physiological adaptations also occur. The effect of environmental factors in freshwater on seawater performance may have far-reaching impacts on the long-term sustainability of anadromous fish populations.

Smolt Development and the Adaptive Value of Smolting

Smolts are the juvenile downstream-migratory stage of salmon. In addition to the behavioral changes that are required for downstream migration, the transition from parr to smolt is accompanied by physiological and morphological changes that are generally recognized to be adaptive for migration and seawater entry. Although there are some variations in smolt development among salmon, particularly with regard to the timing of the event, most species share many common features. Several excellent reviews have summarized the major features of smolt development (Hoar 1988; Boeuf 1993; McCormick et al. 1998), and we will limit this review to those that are likely to have the greatest influence on seawater performance.

There is substantial variation in the timing of downstream migration and seawater entry among anadromous salmonids (Rounsefell 1958; Hoar 1988; McCormick 1994). Pink salmon *Oncorhynchus gorbuscha* and chum salmon *O. keta* enter seawater soon after hatching, whereas Chinook salmon *O. tshawytscha*, coho salmon *O. kisutch*, steelhead *O. mykiss*, and Atlantic salmon *Salmo salar* spend more substantial periods in freshwater as parr. The earliest life history stages (eggs and yolk sac larvae in the case of pink and chum salmon; eggs, yolk-sac larvae, and parr in the others) have relatively poor salinity tolerance (Weisbart 1968). In contrast, smolts have the capacity to withstand a direct transfer to seawater (30–35 parts per thousand [ppt]), with little perturbation in plasma ions (Figure 1). This feat is accomplished through a preparatory reorganization of the osmoregulatory tissues: the gill, gut, kidney, and urinary bladder. The gill is the site of sodium and chloride uptake in freshwater and secretion in seawater (Evans et al. 2005), involving specialized cells known as chloride cells or mitochondrion-rich cells. Both the number and size of chloride cells increase during smolt development, and this is accompanied by an increase in the amount and activity of Na+,K+-ATPase (NKA; Figure 1). As the key enzyme for salt secretion by the gill, its branchial activity has been widely used as a marker for the acquisition of salinity tolerance and smolt development in salmon.

An inherent challenge facing smolts is that while the mechanisms for salt secretion develop and the hypoosmoregulatory capacity of the fish increases, the capacity to osmoregulate in freshwater can be compromised. This makes smolts vulnerable to external factors that may perturb ion regulation in freshwater. Although loss of plasma ions does not appear to be a necessary feature of smolt development or migration (McCormick and Saunders 1987), ion flux rates in freshwater are greater in Atlantic salmon smolts than in parr (Primmett et al. 1988), and decreased plasma osmolality has been observed during coho salmon smoltification (Björnsson et al. 1987). The respiratory function of the gill requires increased blood flow and decreased diffusion distance in the face of increased oxygen demand (for instance during prolonged swimming), which results in a net loss of plasma ions (Nilsson and Sundin 1998). Known as the "osmorespiratory compromise," this may be part of the reason that smolts in freshwater experience greater decreases in plasma ions in response to

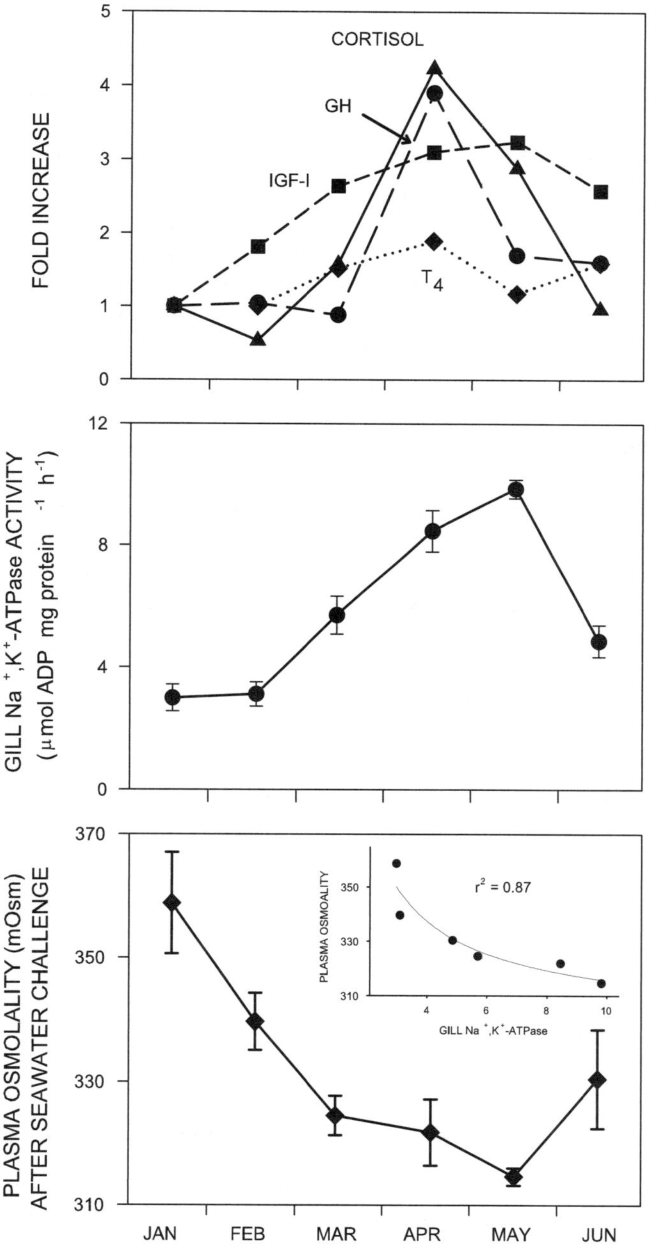

Figure 1.—Plasma hormone levels, gill Na^+,K^+-ATPase activity, and salinity tolerance of Atlantic salmon during the parr–smolt transformation. Abbreviations: GH = growth hormone, IGF-I = insulin-like growth factor I, T4 = thyroxine. Increase in hormone levels were expressed by setting the mean value of the January sample (February in the case of thyroxine) to 1 and expressing all subsequent changes relative to that value. Details of rearing regime and analytical methods used can be found in McCormick et al. (2000).

a handling stress than do parr (Carey and McCormick 1998).

Although there have been no direct comparisons of maximum growth rate or scope for growth in seawater between parr and smolt, Brett (1979) reviewed several different studies and species and concluded that smolts had a greater maximum growth capacity in seawater than parr. It should be noted that these conclusions are circumstantial and are confounded by the inherent size difference between parr and smolt. McCormick et al. (1987) found that continuous light (L24) applied at the time of first feeding in May resulted in good growth compared to fish on natural photoperiod, but prevented normal development of gill NKA activity and salinity tolerance the following spring. These L24 fish grew to the same size as smolts reared under normal photoperiod but, when transferred to sea cages of 28–30 ppt seawater for 3 months, had a 60% lower linear growth rate. Similarly, (Handeland and Stefansson 2001) found that juvenile Atlantic salmon maintained under continuous light had increased growth in freshwater, lower gill NKA activity and salinity tolerance, and reduced growth rates following transfer to seawater when compared to fish reared under normal photoperiod. Further, a number of studies indicate that the growth hormone system, involved both in promoting growth and hypoosmoregulatory capacity, is activated following exposure of salmonids to seawater (see Björnsson 1997). Thus, there is strong indirect evidence that smolt development is accompanied by an increase in the capacity for growth in seawater.

Other physiological and behavioral changes that occur during smolt development are also likely to be adaptive for marine conditions. Loss of parr marks, increased silvering, and darkened fin margins of smolts may be important to the development of schooling behavior (Johnston and Eales 1967), which itself may be adaptive for food acquisition and predator avoidance. Increased muscle phosphofructokinase (Leonard and McCormick 2001) and buffering capacity (Ogata et al. 1998) of smolt relative to parr may provide the former a greater capacity for repeated sprint performance necessary in an open ocean environment. Peake and McKinley (1998) have dispelled the notion that smolts have reduced swimming capacity relative to parr, indicating that initiation of downstream migration is unlikely to be a passive response to increased flow. Additionally, although some aspects of imprinting may occur throughout the juvenile phase, there is a critical period of imprinting during smolting that can override previous experience (Dittman and Quinn 1996). It should be noted, however, that relatively little is known about the behavior of smolts once they enter seawater.

It is likely that virtually all of the features of smolt development are under the control of the neuroendocrine system (see reviews by Hoar 1988; McCormick et al. 1998). Although hormonal control of salinity tolerance has received substantial attention, relatively little is known about the hormonal control of other physiological and behavioral changes during smolting. Changes in circulating levels (Figure 1) and exogenous treatment with hormones indicate that salinity tolerance is under the positive control of cortisol, growth hormone (GH), and insulin-like growth factor I (IGF-I) (McCormick 2001). Given its importance in maintaining ion uptake in freshwater fish (Pickford and Phillips 1959), prolactin is likely to have a negative affect on smolt development (Thorpe 1982; Young et al. 1989), though surprisingly little research has been done in this area. Thyroid hormones (thyroxine and triiodo-L-thyronine) may play only a supportive role in development of salinity tolerance, but are more directly involved in the morphological changes (such as silvering and retinal pigmentation), behavioral changes, and imprinting (Dickhoff and Sullivan 1987; Dittman and Quinn 1996). Given the known role of the GH-IGF-I system in regulating growth and salt secretion in fish, and their increase during smolting and seawater entry, it seems likely that this endocrine axis will play a key role in the increased capacity for growth in seawater, though to date, this has not been substantiated. It has been suggested that early growth in seawater is a critical factor in determining overall seawater survival (Friedland et al. 1996).

The Physiological Smolt Window

For all anadromous salmonids, the salinity tolerance and other aspects of development that occur during smolting (or as fry in the case of pink and chum salmon) are lost if fish remain in freshwater (Figure 1; Hoar 1988). Smolting is therefore a reversible process, and there is a limited period of time during which the fish are at peak preparedness for seawater entry, known as the “physiological smolt window.” The temporal limits of the physiological smolt window will be determined by the environmental factors that control both the stimulation and loss of smolt characters. By way of example, smolt development in Atlantic salmon occurs in the spring and is stimulated primarily by photoperiod and secondarily by temperature (Björnsson 1997). Both the timing of development and the factors that control it differ among salmonids, driven by the evolution of their life histories (McCormick 1994). Smolt development in coho salmon and steelhead are also driven by changes in increased day length, but photoperiod may be less important in pink and chum salmon that develop salinity tolerance soon after hatching.

In Atlantic and coho salmon and steelhead, it appears that the loss of salinity tolerance and other aspects of smolt development are driven primarily by temperature and time (Zaugg and McLain 1976; Duston et al. 1991). This relationship has been investigated in detail in Atlantic salmon, where there is a dome-shaped relationship between degree-days (summation of average daily temperatures experienced after the peak of smolt development) and the loss of salinity tolerance and/or gill NKA activity (Figure 2). After peak smolt development is reached, fish that experience 500 degree-days will have completely reverted to presmolt or parr levels (Figure 2). It also appears that the regulation of smolt behavior is similar to salinity tolerance in that higher temperatures lead to an earlier cessation of downstream migratory behavior (Wagner 1974; Zydlewski et al. 2005).

Many studies have shown that adult return rates—perhaps the most important index of seawater performance—of hatchery-reared salmon smolts are strongly dependent on the timing of release. Hatchery-reared Atlantic salmon released in late May during the normal migratory period of wild fish have higher adult return rates than fish released in April and June (Staurnes et al. 1993b; Lundqvist et al. 1994). Similarly, Bilton et al. (1982) found substantially higher return rates in hatchery-reared coho salmon released in April and May than in June and July. In the studies where it was measured, the peak of physiological smolt development corresponds with the timing of release that yielded the highest adult return rates (Virtanen et al. 1991). This correspondence is likely dependent on both the physiological smolt window and coincidence of release with optimum environmental conditions for smolt survival (ecological smolt window) (McCormick et al. 1998). It seems likely that natural selection will have shaped the physiological smolt window to coincide with the ecological smolt window. Stock-specific differences in the timing of downstream migration (Orciari and Leonard 1996) or increases in salinity tolerance (Handeland et al. 2004) under the same environmental cues are a likely example of selection acting on the timing of smolt development.

In addition to the timing of release, freshwater rearing conditions in hatcheries will affect the overall seawater performance of smolts. Beckman et al. (1999) found that the same strain of spring-run Chinook salmon reared at three different hatcheries had different smolt-to-adult return rates and that these were most strongly correlated with spring growth, plasma IGF-I levels, and gill NKA activity. Fish density and raceway flows experienced by juveniles in freshwater affect the adult return rates of spring-run Chinook salmon (Banks 1994). Several studies have found a positive correlation between size of released smolts and smolt survival and adult return rates (Miyakoshi et al. 2001; Connor 2004). The effect of size appears to be more pronounced in years with poor overall marine survival (Saloniemi et al. 2004). These effects are generally smaller than the effects of time of release and some other factors discussed, but nonetheless indicate that growth conditions in freshwater will influence marine survival. While these studies have focused on hatcheries, they underscore the principal that variations in environmental conditions in freshwater impact smolt development and seawater performance.

The mechanisms through which smolt development affects seawater survival and growth are likely to be more complicated than that of simple acute mortality, due to osmoregulatory failure following seawater entry. Minor compromises to physiological and/or behavioral preparations that are sublethal in the laboratory may have dire consequences in the wild. For example, Dieperink et al. (2001) reported daily predation rates as high as 34% percent in brown trout *S. trutta* smolts during the first 2 d of seawater entry. Atlantic salmon smolts directly transferred to seawater suffered greater experimental predation in seawater than smolts that had been acclimated to seawater for 30 d (Jarvi 1990). Similarly, Handeland et al. (1996) determined that the greatest predation rates on smolts occurred during the first days of exposure to seawater when osmotic perturbations were greatest. We have observed that smolts that perform poorly in seawater challenge tests often swim at the

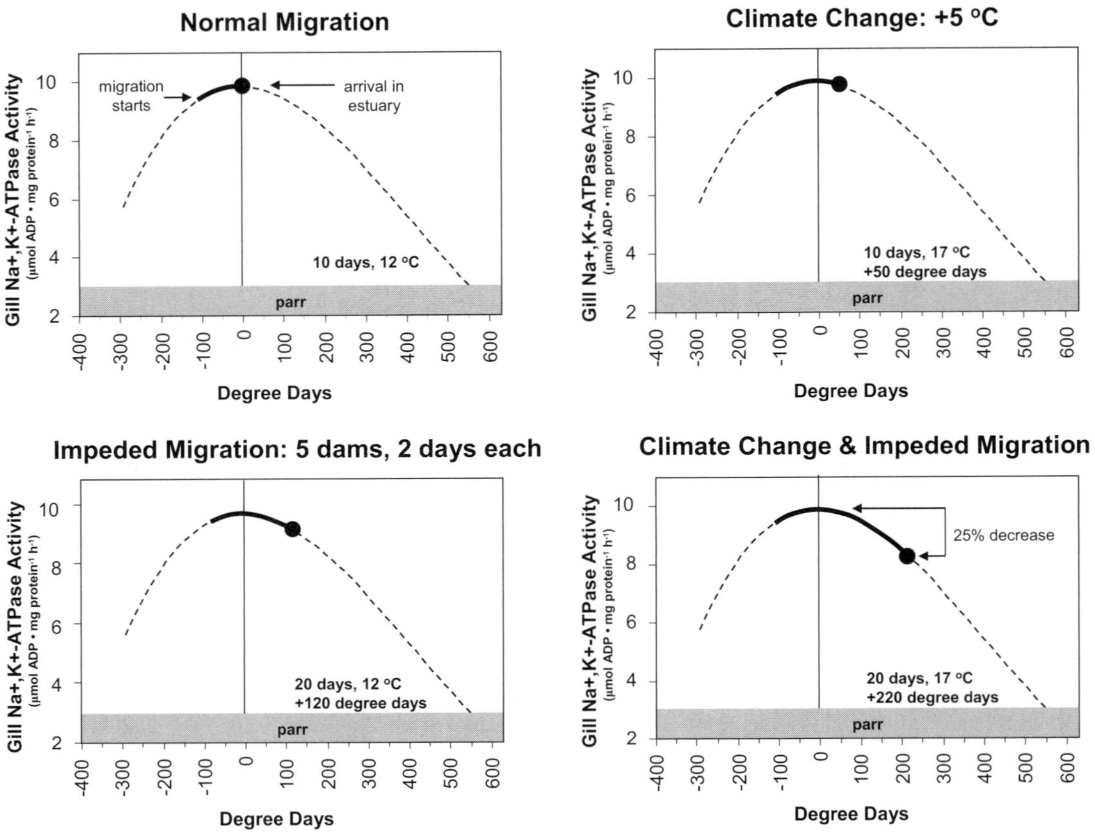

Figure 2.—Hypothetical impacts of increased spring temperature and dams on seawater preparedness of Atlantic salmon smolts. The dashed line in each panel is the response of gill Na^+,K^+-ATPase (NKA) activity (µmol ADP mg/protein/h) to degree-days (cumulative daily temperature) since March 30 (McCormick et al. 1997; Handeland et al. 2004). Gill NKA and salinity tolerance are tightly correlated (Figure 1). Degree-days initially increased gill NKA activity (and salinity tolerance) and then caused a decline. Vertical line represents the peak levels of gill NKA activity. In this model, the average fish (bold line, upper left graph) has a 10-d migratory period at 12°C and arrives at the estuary (dot symbol) with peak levels of gill NKA activity (i.e., highest seawater preparedness). An increase in average temperature of 5°C (upper right) may be due to annual differences in spring temperature, global climate change, or a combination and results in slight (1.4%) decreases in gill Na,K-ATPase (a 100% decrease would be reversion to parr levels of 2–3 µmol ADP mg/protein/h, shown in the gray area). Imposition of delays in migration of 10 d (e.g., five dams with 2 d of migratory delay each; lower left) also results in minor decreases (8.3%). Combined temperature increases (5°C) and delay in migration (10 d) result in synergistic decreases in seawater preparedness of 25% (lower right). It should be noted that this represents the average individual and that late migrants or smolts in river systems with warmer temperatures will likely experience even greater impacts (McCormick et al. 1999).

surface of a tank and are less responsive to external stimuli such as netting. In the wild, Chinook salmon smolts subjected to a handling stress spent more time in the upper, freshwater layers of a simulated estuary than did unstressed fish (Price and Schreck 2003), which the authors suggest would expose them to greater avian predation. Furthermore, lower levels of gill NKA activity are associated with greater levels of avian predation in estuaries during downstream migration of Chinook salmon (Schreck et al. 2006). Osmotic and other stressors may also result in increased susceptibility to diseases that reduce estuarine and marine survival (Loge et al. 2005). Adult return rates of coho salmon subjected to transportation stress just prior to release were reduced compared to fish released directly from the hatchery or those allowed 6 weeks to recover from transportation stress (Schreck et al. 1989). These studies indicate that factors negatively influencing the physiological and behavioral capacity of smolts to make the transition from freshwater to seawater are likely to result in an overall decrease in marine survival, even if final cause of mortality is predation or disease rather than osmoregulatory failure.

Climate Change, Dams, and the Smolt Window

As outlined above, laboratory and hatchery studies indicate that increased temperature and time in freshwater decrease the amount of time that smolts are at the highest levels of seawater preparedness. Global climate change scenarios suggest that an increase in temperature will occur in many temperate and arctic regions, the natural distribution zone of anadromous salmonids (Battin et al. 2007; Hayhoe et al. 2007). Such an increase in temperature is likely to decrease the width of the smolt window and thus have negative impacts on overall seawater preparedness of many smolt populations. Populations already close to the upper temperature tolerance limit should be especially vulnerable, as demonstrated by McCormick et al. (1999), who found that decreased gill NKA activity of smolts at the end of the migratory period was seen only in southern populations that experienced warm spring temperatures and not in northern populations that experienced cool spring temperatures. Furthermore, the proportion of the smolt population with reduced gill NKA activity was greater in years with warmer spring temperatures. Temperature increases associated with global climate change are likely to result in more years with high numbers of smolts that experience loss of seawater preparedness (Figure 2). Higher freshwater temperatures may also result in asynchrony between migration timing and ocean conditions, which could decrease marine survival (McCormick et al. 1998). Recent comprehensive assessment of the geographical distributions of all *Oncorhynchus* species (Augerot 2007) and Atlantic salmon (Parrish et al. 1998) shows that stock disappearances have largely taken place at the southern-most edge of their natural distribution. To what extent these are due to climate change or other anthropogenic disturbances changes, such as habitat loss, dams, and pollution, remains unclear.

Global climate change may have other effects beyond the impact of temperature. Both the amount of rainfall and its seasonal distribution are likely to change, though impacts will vary by region (Hayhoe et al. 2007). Areas with reduced snowmelt or spring rain will have lower water velocity and water levels in spring, which may have profound impacts on in-river smolt mortality. Given the variation that exists in the timing of migration especially among Pacific salmon, the importance of flow and seasonal hydrographs is likely to vary among species. Although a full review is outside the scope of this paper, there is evidence that flow has a substantial impact on smolt survival both in river and after fish reach the ocean. A relation between freshwater discharge and adult survival has been shown for hatchery-reared Atlantic salmon, with greatest survival occurring when smolts are released into high levels of freshwater discharge (Hvidsten and Hansen 1988). Flow has been shown to be a positive factor for in-river survival of steelhead and spring-migrating sockeye salmon, but less important for yearling and summer-migrating Chinook salmon in the mid-Columbia River basin (Giorgi et al. 1998). In the Snake River, the in-river survival of migrating juvenile fall Chinook salmon increases with increasing flow and decreasing temperature (Connor et al. 2003). In addition to providing correlative evidence of a link between flow during out-migration and marine survival of coho salmon, Lawson et al. (2004) demonstrate that annual variations in climate that improve freshwater survival also improve marine survival. In addition to the direct effects of reduced flow on in-river survival, by reducing the speed of migration, reduced water flow may affect smolt survival through loss of smolt development or altered timing of seawater entry (Budy et al. 2002). There are other climate change issues that may have profound effects on seawater entry and marine performance of anadromous fish, including increased salinity intrusion in estuaries due to higher sea levels, as well as a projected decrease of seawater pH (Orr et al. 2005).

Dams and associated hydropower facilities are known to have direct negative impacts on smolt survival. Turbines and spillways can be directly lethal to smolts, with mortality rates of 6–20% for passage through a single turbine or deep plunge spillway (Bickford and Skalski 2000). Dams may also have indirect effects by delaying migration and altering the migratory paths of smolts, often funneling them into narrow corridors with substantial numbers of predators (Venditti et al. 2000). As noted above, there is generally a positive relationship between flow and survival of salmon; dams that decrease or substantially alter flow are therefore likely to decrease smolt survival. The long reservoirs and/or highly altered hydrographs caused by dams can impose substantial delays in migration (Bentley and Raymond 1976) that will affect whether the physiological smolt window is still open by the time fish arrive at the estuary. There is surprisingly little information on the impact of physical stressors, such as those that might occur during passage through turbines, spillways, or over dams on salinity tolerance or other aspects of smolt performance. Experimental descaling, similar to that which occurs during passage through turbines, can decrease salinity tolerance of Atlantic salmon smolts (G. Zydlewski, University of Maine, Orono, personal communication). Pinpointing the specific impacts of dams on delayed mortality or overall seawater performance of smolts in a "natural" setting is difficult. Because the impact of stress and delays in migration are cumulative and often delayed (Barton et al. 1986), measurement of the impact of a single dam will underestimate the impacts of multiple dams. For instance, Budy et al. (2002) concluded that the accumulated stress and altered timing of moving through multiple dams on the Snake and Columbia rivers result in higher delayed (marine) mortality than would occur in this river system without dams. Changes in management of hydropower systems that reduced in-river mortality for Chinook salmon did not improve adult returns (Williams et al. 2001). Although this could be due to a variety of factors, including variation in ocean conditions, one possibility is that dams continued to have an impact on delayed or early marine mortality.

A hypothetical scenario of the impact of the interaction between increased temperature experienced by smolts under global climate change and delays in migration due to dams is shown in Figure 2. Temperature increases associated with global climate change are likely to result in more years with increasing numbers of smolts that have experienced loss of seawater preparedness. Delays in migration appear to have a greater impact than temperature increases, though it should be noted that the effect of temperature does not take into account those river systems in which temperature will rise above the levels that are lethal to smolts. Perhaps most importantly, there is likely to be a synergistic interaction with global climate change, such that the presence of dams and increased temperature will have more than an additive negative effect on seawater performance (Figure 2).

Acid Rain and Aluminum Impacts

Acid rain resulting from sulfur and nitrogen oxide emissions from coal-burning power plants is a known cause of the decline of Atlantic salmon in eastern Canada (Watt 1987; Lacroix 1989) and southern Norway (Hesthagen 1989; Hesthagen and Hansen 1991). As a result of their underlying geology, rivers and streams in these regions have poor buffering capacity and are thus vulnerable to chronic (year-round) acidification. As a consequence of acidification, aluminum (Al), one of the most abundant elements in the soil, is leached into the surrounding watershed (Driscoll 1984). In addition, the solubility of Al increases as a direct result of decreased pH leading to the increased presence of inorganic Al (Al_i), the form most toxic to fish (Gensemer and Playle 1999). Recently, episodic (days to weeks) acidification and its associated Al toxicity has been identified as a possible cause of salmon decline in the northeastern United States, including Maine, where many salmon rivers have been listed as federally endangered (National Academy of Science 2004). In this region, rivers and streams are not chronically acidified, but instead experience episodic decreases in pH during acidification events such as spring snowmelts and fall storms.

During exposure to acid and aluminum (acid/Al), Al accumulates both on the surface and within the gill (Youson and Neville 1987; Wilkinson and Campbell 1993; Lacroix et al. 1993; Teien et al. 2004) damaging the branchial epithelium and ultimately leading to ion regulatory disturbance. Loss of ion regulatory ability is likely due to both an increase in branchial permeability and an inhibition of active ion uptake (Booth et al. 1988; McDonald et al. 1991). Increased permeability may be caused by the displacement of Ca^{2+} ions from anionic gill binding sites by Al, resulting in weakening of intercellular tight junctions (Booth et al. 1988; Freda et al. 1991), whereas inhibition of ion uptake may result from damage to or alteration of gill chloride cells (Jagoe and Haines 1997), and decreased NKA activity (Staurnes et al. 1993a, 1996; Kroglund and Staurnes 1999; Magee et al. 2003).

Smolts appear to be the most sensitive of the salmon life stages to the effects of acid/Al exposure. In Atlantic salmon, smolts are more prone than parr to plasma ion losses and inhibition of gill NKA activity, and they suffer greater acute mortality during exposure to both short-term and chronic acid and aluminum (Rosseland and Skogheim 1984; Leivestad et al. 1987; Rosseland et al. 2001). Parr and smolts accumulate similar levels of aluminum in the gill following the same acid and aluminum exposure levels, indicating that increased aluminum alone is not the reason smolts are more sensitive than parr (Monette and McCormick 2008). The increased sensitivity of smolts may be due to the higher ion flux rates they experience in freshwater (Primmett et al. 1988), putting them at greater risk of any factor that affects ion uptake.

Previous research has established that long-term (weeks–months) exposure to acid/Al during the par–smolt transformation can compromise survival, growth, and the smolting process. Thus, Atlantic salmon reared under low pH (4.2–4.7) from February to June had impaired ion regulation in freshwater, reduced seawater tolerance and lower gill NKA activity levels than control fish reared at high pH (6.4–6.7) (Saunders et al. 1983). Smolting Atlantic salmon kept for more than 1 month in soft water of pH 5.0 and 50 μg/L Al had increased mortality, decreased condition factor, impaired ion regulation in both freshwater and seawater, and substantial declines in gill NKA and carbonic anhydrase activities (Staurnes et al. 1993a). Exposure of Atlantic salmon smolts to pH 5.5 and 158 μg/L Al_i for ≥16 d significantly altered gill morphology, as indicated by changes in the number, size, and shape of chloride cells (Jagoe and Haines 1997).

There is an increasing body of evidence that short-term or episodic (days–week) exposure to acid/Al can also impair smolt development. Atlantic salmon smolts exposed to several episodic pulses of acid/Al (pH ~5.2, 59 μg/L Al_i) over the course of 31 d and then exposed to seawater experienced reductions in hematocrit and increases in plasma ions and suffered higher mortalities than controls (Magee et al. 2003). Staurnes et al. (1996) found that smolts placed in floating cages in the river Lygna (pH 4.7–4.9, 30–120 Al_i) for only 12 h had completely lost seawater tolerance, as indicated by 100% mortality during a 24 h, 30 ppt seawater challenge. Interestingly, during short-term exposure to acid/Al, impaired seawater tolerance can be seen in the absence of any impact on ion regulation in freshwater. Monette (2007) determined that plasma Cl^- levels of smolts exposed to pH 5.4, 42 μg/L Al_i for as little as 2 d did not differ from controls in freshwater, but were significantly elevated relative to controls after 24 h in 35 ppt seawater, indicating loss of seawater tolerance in this group. Also in this study, smolts exposed to pH 5.3, 11 μg/L Al_i for 6 d exhibited impaired seawater tolerance, demonstrating that the hypoosmoregulatory system of smolts is highly sensitive to acid and very low levels of Al_i.

Several studies have begun to make the important link between the osmoregulatory status of smolts after acid/Al exposure and overall seawater performance (Staurnes et al. 1996; Magee et al. 2001; Kroglund and Finstad 2003). Staurnes et al. (1996) Carlin-tagged over 5,000 Atlantic salmon smolts, released them into acid (Lygna) and limed (Audna) rivers in southern Norway, and monitored adult return rates over 2 years. They found a significant correlation between adult recapture rates and the osmoregulatory status of smolts held in floating cages at the respective release sites. Kroglund and Finstad (2003) determined that exposure of smolts to very low levels of Al_i (6 μg/L) impaired freshwater growth and reduced marine survival by 20–30%. Following exposure of Atlantic salmon smolts to sublethal acid/Al conditions, Kroglund et al. (2007) found a strong correlation of both gill NKA activity and salinity tolerance (plasma chloride after 24 h seawater [SW] challenge) and adult return rates. In their study, exposure of smolts to sublethal acid/Al reduced smolt-to-adult survival by 50% relative to controls. Further evidence for the impact of acid and aluminum on seawater performance is provided by the response of Atlantic salmon populations in eastern Norway to long-term liming. These programs target a pH of 6.5 in order to protect smolt development and have resulted in recoveries of Atlantic salmon populations that had been extirpated due to the effects of acid rain (Hesthagen and Larsen 2003).

To date, most research on acid impacts on salmon smolts has been done with Atlantic salmon, due to the regional nature of acidification impacts. Given the rise of industrialization in China coupled with global weather patterns, future impacts on Pacific salmon are possible, and it seems likely that Pacific salmon smolts (and other life stages) will have sensitivity similar to Atlantic salmon. In addition, heavy metals have a similar mode of action and impact on the osmoregulatory physiology of fish as aluminum, though heavy metal impacts are not necessarily dependent on low pH. Coho salmon have a lower lethal threshold (i.e., were more sensitive) to copper and zinc in November as parr than in May as smolts (Lorz and McPherson 1976), and although zinc did not affect seawater tolerance, copper at one-tenth the lethal level was sufficient to reduce gill NKA activity and seawater tolerance.

Environmental Estrogens

Exposure to environmental estrogens can adversely affect reproductive development and overall population stability (Kidd et al. 2007). From the mid-1970s to the mid-1980s, the pesticide Matacil 1.8D, which contained the xenoestrogen nonylphenol (NP) as a surfactant, was applied to forests in the Atlantic coast watersheds of Canada to control damage from spruce budworm *Choristoneura fumiferana*. Application of this insecticide coincided with spring smolting and downstream migration of Atlantic salmon. In a retrospective study, Fairchild et al. (1999) calculated that the resulting concentrations of NP after pesticide application were within ranges found in municipal sewage outfalls and industrial effluents. Importantly, they demonstrated that for 16 rivers exposed to spraying between 1973 and 1990, the lowest adult salmon returns occurred in rivers with the largest exposure to Maticil 1.8D (Fairchild et al. 1999). These authors conclude that NP entered the aqueous environment at levels sufficient to have estrogenic actions, specifically impacting the capacity of smolts to enter seawater.

Previous research has established that administration of both androgens and estrogens are inhibitory to the development of salinity tolerance in salmonids (Ikuta et al. 1987; Schmitz and Mayer 1993). Madsen and Korsgaard (1989) found that estradiol (E_2) inhibited silvering and reduced gill NKA activity in Atlantic salmon smolts, while inducing vitellogenesis, indicated by increased total plasma calcium, hepatosomatic index and, total liver RNA. Further work by this laboratory confirmed parallels in the physiological impacts of NP and E2, with both compounds inhibiting smolting by reducing gill NKA activity and gill chloride cell density, as well as reducing hypoosmoregulatory ability of smolts (Madsen et al. 1997). Further, E_2 and NP exposure just prior to smolting delayed smolt development and downstream migration (Madsen et al. (2004). In an effort to define the endocrine pathways involved in the physiological impairment of SW tolerance, McCormick et al. (2005) demonstrated that E_2 and NP suppressed plasma concentrations of IGF-I and thyroid hormones. Aqueous exposure of Atlantic salmon smolts to E_2 and NP also reduced gill NKA activity and plasma concentrations of IGF-I, in addition to decreasing growth and increasing stress sensitivity (Arsenault et al. 2004; Lerner et al. 2007b).

While providing useful insight into the potential impacts of estrogenic compounds on smolt development, many of the studies mentioned above have applied NP levels that were higher than those likely to be present in most salmon rivers (Kolpin et al. 2002), including those that experienced Matacil spraying in eastern Canada. It is possible that exposure to environmentally relevant concentrations at an earlier developmental stage has long term impacts on smolt development. To address this, Lerner et al. (2007c) exposed Atlantic salmon yolk sac larvae to two environmentally relevant aqueous levels of NP or E2 for 21 d. Nonylphenol at 80 μg/L resulted in 50% mortality during the 21-d exposure period, reaching 100% 120 d after treatment was terminated. Yolk sac larvae treated with the lower dose of NP (7 μg/L) or E2 (2 μg/L) were not affected during the exposure period but exhibited delayed mortality (four-fold greater than controls) 31–60 d after treatment was terminated. A full year after exposure, at the peak of smolting, yolk sac larvae that had been exposed to NP or E_2 had reduced plasma IGF-I levels, gill NKA activity, and seawater tolerance. Behavioral impacts were also detected, with NP and E_2 substantially reducing behavioral preference for seawater. These studies establish that early exposure to contaminants, well before smolt development, can have long-term impacts on seawater preparedness. They also provide a possible explanation for reduced marine survival of fish exposed to nonylphenol in nature (Fairchild et al. 1999).

Environmental disruption of SW performance during smolting has been linked to negative effects of estrogenic compounds on the GH-IGF-I system. The impacts of E_2 on circulating levels of GH in fishes appear to vary among species, developmental stage, and environmental salinity (Melamed et al.

1998; McCormick et al. 2005; Lerner et al. 2007b, 2007c). On the other hand, a reduction in plasma levels of IGF-I is consistent among fish treated with estrogenic compounds (Arsenault et al. 2004; McCormick et al. 2005; Lerner et al. 2007b, 2007c). Insulin-like growth factor-I is a mediator of many of the effects of GH on osmoregulatory physiology and probably includes both endocrine action (from GH binding to its hepatic receptor and release of IGF-I into circulation) and paracrine/autocrine action (GH induction of IGF-I release at the target tissue) (Duan 1998). In mammals, E_2 reduces GH receptor (GHR) density in the liver (Domene et al. 1994), and there is evidence that E_2 and some environmental estrogens reduce hepatic and branchial GHR in Atlantic salmon (Lerner 2007). Taken together, these findings suggest that modification of GH and/or IGF-I levels and their receptors may provide a likely mechanistic route for mediating the negative impacts of estrogenic compounds on the maintenance of osmoregulatory homeostasis. Exposure to estrogenic compounds also decreases the levels of thyroid hormones during smolt development (McCormick et al. 2005). Although thyroid hormones appear to have an indirect role in controlling osmoregulation in salmonids, they have a more direct role in other aspects of smolt development such as migratory behavior and imprinting (Hoar 1988). In addition to affecting seawater tolerance, estrogenic compounds may therefore inhibit other aspects of smolt development through their impact on thyroid hormones. The inhibitory effect of environmental estrogens on seawater performance may be related to the anadromous life history of salmon; endogenous sex steroids are low during early development (Sower et al. 1984), and when they increase during reproductive maturation, they may act as a signal for both migration into freshwater and the accompanying loss of hyperosmoregulatory ability (Uchida et al. 1997; Shrimpton et al. 2005; Makino et al. 2007).

Other Contaminants and Interactions

In agricultural and managed forestry areas, river contamination through rain runoff can occur due to the high application rates and water solubility of many pesticides. A number of pesticides are often present in salmon habitats, and spring pesticides application is especially problematic because the timing of spring application of pesticides coincides with smolt development. Widely used herbicides such as such hexazinone (HEX) and atrazine (ATZ), nonselective herbicides in the triazine family, are employed to control undesirable plants in crops such as corn, alfalfa, sugar cane, pineapples, and blueberries. ATZ can cause damage to gill epithelium and kidneys, increasing the renal excretion of ions and proteins of rainbow trout (nonanadromous *O. mykiss*) (Neskovic et al. 1993). Exposure of Atlantic salmon smolts to concentrations up to 100 μg/L ATZ compromised osmoregulatory ability in freshwater (FW) and after 24 h SW challenge (Waring and Moore 2004; Nieves-Puigdoller et al. 2007). Both studies found a slight decrease in survival and growth and elevated plasma cortisol during the exposure period. Furthermore, Nieves-Puigdoller (2007) found that prior exposure to ATZ in FW resulted in reduced growth of Atlantic salmon smolts reared in clean seawater (free of any contaminants) for 1 month. Given the evidence that early growth in seawater is critical to overall marine survival (Beamish et al. 2004), this impact of contaminant exposure in freshwater on seawater growth indicates a detrimental effect on marine survival.

In salmon, the yolk sac larval stage coincides with pesticide application in the spring. Exposure of yolk sac larvae to 100 μg/L ATZ for 21 d resulted in significantly smaller smolts 1 year after exposure (Nieves-Puigdoller 2007). This is somewhat surprising given the capacity of fish to exhibit compensatory growth and suggests that some permanent physiological effect may have occurred. Plasma cortisol levels were higher in ATZ-exposed fish a year after exposure, which suggests a permanent alteration in the hypothalamic-pituitary-interrenal axis (Nieves-Puigdoller 2007). Increased cortisol may be valuable in mobilizing energy for responding to and repairing damage caused by contaminants, but it can have negative consequences for growth and disease resistance (Wendelaar Bonga 1997).

In addition to herbicides, fungicides are widely used in numerous applications. The use of didecyldimethylammonium chloride (DDAC) in wood preservation may be a problem for anadromous fish, as most lumber storage facilities are near large rivers. Exposure of coho salmon smolts to 0.75 mg/L DDAC for 24 h followed by 24 h SW challenge resulted in increased plasma cortisol, reduced gill

NKA activity, and lower salinity tolerance (Johnston et al. 1998). Chlorothalonil (CTL) is the most commonly used fungicide in the United States, but to date, there are no studies that have examined the impact of this compound on smolt development. However, Nieves-Puigdoller (2007) found that exposure of yolk sac larvae to 10 μg/L CTL for 12 d resulted in increased whole-larvae NKA activity and Na+ ion, which may be a compensatory response to moderate damage of ion regulatory tissues.

Organophosphate insecticides act through inhibition of acetylcholinesterase (AChE) activity, affecting normal cholinergic nerve transmission. The inhibition of AChE causes accumulation of acetylcholine in synapses and neuromuscular junctions, resulting in overstimulation of cholinergic receptors (Colombo et al. 2005). Exposure of juvenile cutthroat trout *O. clarkii clarkii* to environmentally relevant concentrations (750 μg/L) of the insecticide carbaryl for 6 h decreased brain and muscle AChE activity (Labenia et al. 2007). In trout species such as cutthroat (Labenia et al. 2007) and rainbow (Little et al. 1990), decreased AChE activity resulted in reduced swimming performance and increased experimental predation. Furthermore, decreases in AChE activity in juvenile coho salmon following exposure to another common insecticide, chlorpyrifos, were associated with reduced swimming and feeding activities (Sandahl et al. 2005). The specific impact of AChE inhibitors on seawater performance of anadromous fish has yet to be examined.

Polychlorinated biphenyls (PCBs) are highly stable anthropogenic chemicals classified as "persistent organic pollutants" (POPs). Polychlorinated biphenyls are ubiquitous in the aquatic environment and are present in significant amounts in downstream migrating smolts (Johnson et al. 2007) and adult farmed and wild salmon (Hites et al. 2004). Polychlorinated biphenyl mixtures and their parent congeners are known to reduce plasma thyroid hormone levels in mammals, which may have significant impact on brain development (Zoeller 2005). Thyroid hormones are thought to play a role in controlling downstream migratory behavior and SW preference of anadromous salmonids (Iwata et al. 1990; Iwata 1995). There is evidence that exposure to the PCB mixture Aroclor 1254 can negatively affect mechanisms involved in smolt development and migration of Arctic char *Salvelinus alpinus* (Jorgensen et al. 2002, 2004). Exposure of Atlantic salmon smolts to 10 μg/L aqueous PCB resulted in loss of gill NKA activity and reductions of plasma ions in freshwater (Lerner et al. 2007a). Behavioral preference for seawater was also decreased, but salinity tolerance was not detectably impacted. There is also evidence that exposure can reduce time spent in SW in striped bass *Morone saxatilis* (Zlokovitz and Secor 1999). These data indicate that exposure to sublethal, environmentally relevant concentrations of A1254 can affect critical aspects of behavior and physiology, which in turn could affect marine survival and population sustainability.

Oil spills present another frequent source of aquatic pollution. Exposure of pink salmon eggs to doses of polynuclear aromatic hydrocarbon (PAHs, 1–45 μg/L) resulted in a dose-dependent decrease of growth rate in sea cages 10 months after the exposure period (Heintz et al. 2000). Furthermore, PAH-exposure of embryos that were subsequently released as downstream migrating fry decreased marine survival from 1.3% in controls to 0.8% in fish exposed to 19 μg/L PAH. Interestingly, these authors found that the impact of PAH was greater (1.3% to 0.8%) in a year with poor marine survival, compared to a year when marine survival was high (2.0% reduced to 1.7% by PAH). The study indicates that freshwater exposure to a contaminant can have substantial impact on marine survival and suggests that the level of this impact will be greater when marine conditions are unfavorable.

Most aquatic toxicology studies have focused on single compounds. However, aquatic organisms are usually exposed to mixtures of contaminants that vary with land use. A mixture of pesticides may increase the toxicity to aquatic fauna, and only recently have efforts been made to understand the effects and mechanisms of pesticide interactions (Anderson and Lydy 2002). For example, exposure of Atlantic salmon smolts to a mixture of 4-nonylphenol and ATZ at environmentally relevant levels causes increased mortality and compromises osmoregulatory ability in SW (Moore et al. 2003); however, neither of the compounds alone affected osmoregulation in SW. In addition, Nieves-Puigdoller (2007) found that body size was smaller in yolk sac larvae exposed to a combination of pesticides than when fish were only exposed to a single compound. Moreover, exposure to contaminants and other stressors may interact with diseases to increase mortality (Loge et al. 2005).

Most studies to date have examined only a limited number of features of the preparatory adaptations for seawater entry in anadromous fish, and many have focused on the development of salinity tolerance. While this is obviously a critical factor in marine survival, other features of preparatory development may also be impacted by exposure to contaminants. Given the demonstrated sensitivity of the olfactory system to a variety of anthropogenic factors, including low pH (Leduc et al. 2007), carbofuran (Waring and Moore 1997), cypermethrine (Moore and Waring 2001), and atrazine (Moore and Waring 1998), it would not be surprising to discover that the process of imprinting is sensitive to perturbation by contaminants.

Perspectives

Developmental increases in salinity tolerance probably occur in all anadromous species; however, how this is integrated into the overall osmoregulatory capacity of the animal during its life history varies with species. In some species, such as the striped bass, the developmental acquisition of hypoosmoregulatory ability increases the osmoregulatory range of the species, leaving it highly euryhaline through most of its life cycle (Hurst and Conover 2002). In other species, such as American shad *Alosa sapidissima*, salinity tolerance is acquired and maintained just prior to and during the downstream migratory phase, and freshwater tolerance is lost at the end of the migratory period (Zydlewski and McCormick 1997a, 1997b). While the developmental processes and strategies that govern the transition from a freshwater to marine environment varies among species, some basic aspects, such as the development of salinity tolerance (and its underlying physiology), behavioral changes during downstream migration, and the underlying neuroendocrine regulation of these changes, are likely to be shared to some degree. Thus, the environmental factors and contaminants that affect smolt performance may also affect the seawater performance of other anadromous species.

Although we are only at the beginning of focused research in this area, it appears likely that even subtle, sublethal effects during key developmental stages may have profound effects during subsequent life stages. In this review, we have presented evidence that exposure to poor environmental conditions in freshwater such as high temperature, physical obstacles (dams), and contaminants such as acid rain disrupt smolt development and salinity tolerance (Figure 3). In the wild, these impacts may compromise growth, competitive ability, migration, disease

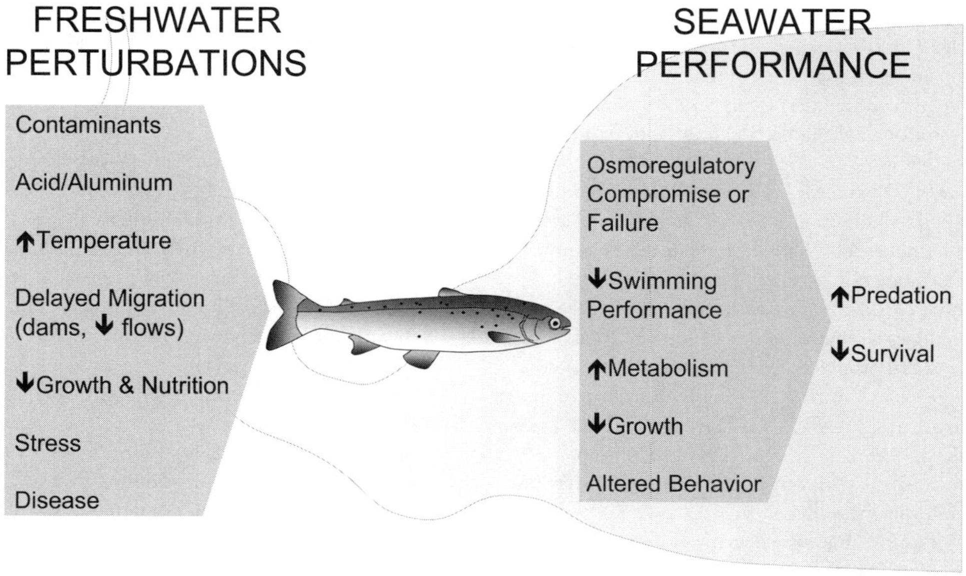

Figure 3.—Environmental impacts on salmon in freshwater and the organismal outcomes that affect survival and growth in the marine environment.

resistance, and predator avoidance of smolts, affecting their survival and population sustainability. There may also be important interactions between performance capacities affected in freshwater and their expression under various marine conditions: poor performance may be masked when ocean conditions are excellent but accentuated when marine conditions are poor. Poor conditions for salmon survival and growth in the marine environment may become more common with global warming (Friedland et al. 2003; Lawson et al. 2004), thus increasing the impacts of environmental perturbations in freshwater. The accumulated evidence summarized here indicates that decreased marine returns are due not only to changes in oceanic conditions, but are also adversely affected by anthropogenic influences during their freshwater phase. Anadromous salmonids take it with them when they go, having lasting impacts on marine survival and population sustainability.

Acknowledgments

We thank Theodore Castro-Santos and Gayle Zydlewski for their suggestions for improving the manuscript. This paper is dedicated to Harold Raymond "Hap" McCormick (1956–2008): fisherman, conservationist, brother.

References

Alexander, G., R. Sweeting, and B. Mckeown. 1994. The shift in visual pigment dominance in the retinae of juvenile coho salmon (*Oncorhynchus kisutch*): an indicator of smolt status. The Journal of Experimental Biology 195:185–197.

Anderson, T. D., and M. J. Lydy. 2002. Increased toxicity to invertebrates associated with a mixture of atrazine and organophosphate insecticides. Environmental Toxicology & Chemistry 21:1507–1514.

Arsenault, J. T. M., W. L. Fairchild, D. L. MacLatchy, L. Burtidge, K. Haya, and S. B. Brown. 2004. Effects of water-borne 4-nonylphenol and 17 beta-estradiol exposures during par–smolt transformation on growth and plasma IGF-I of Atlantic salmon (*Salmo salar* L.). Aquatic Toxicology 66(3):255–265.

Augerot, X. 2007. Atlas of Pacific salmon. University of California Press, Berkeley.

Banks, J. L. 1994. Raceway density and water flow as factors affecting spring chinook salmon (*Oncorhynchus tshawytscha*) during rearing and after release. Aquaculture 119:201–217.

Barton, B. A., C. B. Schreck, and L. A. Sigismondi. 1986. Multiple acute disturbances evoke cumulative physiological stress responses in juvenile Chinook salmon. Transactions of the American Fisheries Society 115:245–251.

Battin, J., M. W. Wiley, M. H. Ruckelshaus, R. N. Palmer, E. Korb, K. K. Bartz, and H. Imaki. 2007. Projected impacts of climate change on salmon habitat restoration. Proceedings of the National Academy of Sciences of the United States of America 104:6720–6725.

Beamish, R. J., C. Mahnken, and C. M. Neville. 2004. Evidence that reduced early marine growth is associated with lower marine survival of coho salmon. Transactions of the American Fisheries Society 133(1):26–33.

Beckman, B. R., W. W. Dickhoff, W. S. Zaugg, C. Sharpe, S. Hirtzel, R. Schrock, D. A. Larsen, R. D. Ewing, A. Palmisano, C. B. Schreck, and C. W. Mahnken. 1999. Growth, smoltification, and smolt-to-adult return of spring Chinook salmon from hatcheries on the Deschutes River, Oregon [Review]. Transactions of the American Fisheries Society 128(6):1125–1150.

Bentley, W. W., and H. L. Raymond. 1976. Delayed migrations of yearling Chinook salmon since the completion of Lower Monumental and Little Goose dams on the Snake River. Transactions of the American Fisheries Society 105:422–424.

Bickford, S. A., and J. R. Skalski. 2000. Reanalysis and interpretation of 25 years of Snake–Columbia River juvenile salmonid survival studies. North American Journal of Fisheries Management 20:53–68.

Bilton, H. T., D. F. Alderdice, and J. T. Schnute. 1982. Influence of time and size at release of juvenile coho salmon (*Oncorhynchus kisutch*) on returns at maturity. Canadian Journal of Fisheries and Aquatic Sciences 39:426–447.

Björnsson, B. Th. 1997. The biology of salmon growth hormone: from daylight to dominance. Fish Physiology & Biochemistry 17(1–6):9–24.

Björnsson, B. Th., K. Yamauchi, R. S. Nishioka, L. J. Deftos, and H. A. Bern. 1987. Effects of hypophysectomy and subsequent hormonal replacement therapy on hormonal and osmoregulatory status of coho salmon, *Oncorhynchus kisutch*. General and Comparative Endocrinology 68:421–430.

Boeuf, G. 1993. Salmonid smolting: a preadaptation to the oceanic environment. Pages 105–135 *in* J. C. Rankin and F. B. Jensen, editors. Fish ecophysiology. Chapman and Hall, London.

Booth, C. E., M. D. McDonald, B. P. Simons, and C. M. Wood. 1988. Effects of aluminum and low pH on net ion fluxes and ion balance in the brook trout (*Salvelinus fontinalis*). Canadian Journal of Fisheries and Aquatic Sciences 45(9):1563–1574.

Brett, J. R. 1979. Environmental factors and growth. Pages 599–675 *in* W. S. Hoar, D. J. Randall, and J. R. Brett, editors. Fish physiology, volume VIII, bioenergetics and growth. Academic Press, New York.

Budy, P., G. P. Thiede, N. Bouwes, C. E. Petrosky, and H. Schaller. 2002. Evidence linking delayed mortality of snake river salmon to their earlier hydrosystem experience [Review]. North American Journal of Fisheries Management 22(1):35–51.

Carey, J. B., and S. D. McCormick. 1998. Atlantic salmon smolts are more responsive to an acute handling and confinement stress than parr. Aquaculture 168(1–4):237–253.

Colombo, A., F. Orsi, and P. Bonfanti. 2005. Exposure to the organophosphate pesticide chlorpyrifos inhibits acetylcholinesterase activity and affects muscular integrity in *Xenopus laevis* larvae. Chemosphere 61:1665–1667.

Connor, W. P., H. L. Burge, J. R. Yearsley, and T. C. Bjornn. 2003. Influence of flow and temperature on survival of wild subyearling fall Chinook salmon in the Snake River. North American Journal of Fisheries Management 23(2):362–375.

Connor, W. P., S. G. Smith, T. Andersen, S. M. Bradbury, D. C. Burum, E. E. Hockersmith, M. L. Schuck, G. W. Mendel, and R. M. Bugert. 2004. Postrelease performance of hatchery yearling and subyearling fall Chinook salmon released into the Snake River. North American Journal of Fisheries Management 24:545–560.

Dickhoff, W. W., and C. V. Sullivan. 1987. Involvement of the thyroid gland in smoltification, with special reference to metabolic and developmental processes. Pages 197–210 *in* M. J. Dadswell, R. J. Klauda, C. M. Moffitt, R. L. Saunders, R. A. Rulifson, and J. E. Cooper, editors. American Fisheries Society, Symposium 1, Bethesda, Maryland.

Dieperink, C., S. Pedersen, and M. I. Pedersen. 2001. Estuarine predation on radiotagged wild and domesticated sea trout (*Salmo trutta* L.) smolts. Ecology of Freshwater Fish 10(3):177–183.

Dittman, A. H., and T. P. Quinn. 1996. Homing in Pacific salmon: mechanisms and ecological basis. The Journal of Experimental Biology 199:83–91.

Domene, H. M., G. Marin, J. Sztein, Y. M. Yu, J. Baron, and F. G. Cassorla. 1994. Estradiol inhibits growth hormone receptor gene expression in rabbit liver. Molecular and Cellular Endocrinology 103:81–87.

Driscoll, C. T. 1984. A procedure for the fractionation of aqueous aluminum in dilute acidic waters. International Journal of Environmental Analytical Chemistry 16(4):267–283.

Duan, C. M. 1998. Nutritional and developmental regulation of insulin-like growth factors in fish. Journal of Nutrition 128(2):306–314.

Duston, J., R. L. Saunders, and D. E. Knox. 1991. Effects of increases in freshwater temperature on loss of smolt characteristics in Atlantic salmon (*Salmo salar*). Canadian Journal of Fisheries and Aquatic Sciences 48:164–169.

Evans, D. H., P. M. Piermarini, and K. P. Choe. 2005. The multifunctional fish gill: dominant site of gas exchange, osmoregulation, acid–base regulation, and excretion of nitrogenous waste. Physiological Reviews 85(1):97–177.

Fairchild, W. L., E. O. Swansburg, J. T. Arsenault, and S. B. Brown. 1999. Does an association between pesticide use and subsequent declines in catch of Atlantic salmon (*Salmo salar*) represent a case of endocrine disruption? Environmental Health Perspectives 107(5):349–357.

Freda, J., D. A. Sanchez, and H. L. Bergman. 1991. Shortening of branchial tight junctions in acid exposed rainbow trout (*Oncorhynchus mykiss*). Canadian Journal of Fisheries and Aquatic Sciences 48(10):2028–2033.

Friedland, K. D., R. E. Haas, and T. F. Sheehan. 1996. Postsmolt growth, maturation, and survival of two stocks of Atlantic salmon. Fishery Bulletin 94:654–663.

Friedland, K. D., D. G. Reddin, J. R. McMenemy, and K. F. 2003. Drinkwater. Multidecadal trends in North American Atlantic salmon (*Salmo salar*) stocks and climate trends relevant to juvenile survival. Canadian Journal of Fisheries & Aquatic Sciences 60(5):563–583.

Gensemer, R. W. and R. C. Playle. 1999. The bioavailability and toxicity of aluminum in aquatic environments. Critical Reviews in Environmental Science and Technology 29(4):315–450.

Giorgi, A. E., T. W. Hillman, J. R. Stevenson, S. G. Hays, and C. M. Peven. 1998. Factors that influence the downstream migration rates of juvenile salmon and steelhead through the hydroelectric system in the mid-Columbia River basin. North American Journal of Fisheries Management 17:268–282.

Handeland, S. O., T. Jarvi, A. Ferno, and S. O. Stefansson. 1996. Osmotic stress, antipredator behaviour, and mortality of Atlantic salmon (*Salmo salar*) smolts. Canadian Journal of Fisheries & Aquatic Sciences 53:2673–2680.

Handeland, S. O., and S. O. Stefansson. 2001. Photoperiod control and influence of body size on off-season par–smolt transformation and postsmolt growth. Aquaculture 192(2–4):291–307.

Handeland, S. O., E. Wilkinson, B. Sveinsbo, SD. McCormick, and S. O. Stefansson. 2004. Temperature influence on the development and loss of seawater tolerance in two fast-growing strains of Atlantic salmon. Aquaculture 233(1–4):513–529.

Hayhoe, K., C. P. Wake, T. G. Huntington, L. F. Luo, M. D. Schwartz, J. Sheffield, E. Wood, B. Anderson, J. Bradbury, A. DeGaetano, T. J. Troy, and D. Wolfe. 2007. Past and future changes in climate and hydrological indicators in the U.S. Northeast. Climate Dynamics 28(4):381–407.

Heintz, R. A., SD. Rice, A. C. Wertheimer, R. F. Bradshaw, F. P. Thrower, J. E. Joyce, and J. W. Short. 2000. Delayed effects on growth and marine survival of pink salmon *Oncorhynchus gorbuscha* after exposure to crude oil during embryonic development. Marine Ecology Progress Series 208:205–216.

Hesthagen, T. 1989. Episodic fish kills in an acidified salmon river in southwestern Norway. Fisheries 14(3):10–17.

Hesthagen, T., and L. P. Hansen. 1991. Estimates of the annual loss of Atlantic salmon (*Salmo salar* L.) in Norway due to acidification. Aquaculture and Fisheries Management 22:85–91.

Hesthagen, T., and B. M. Larsen. 2003. Recovery and re-establishment of Atlantic salmon, *Salmo salar*, in limed Norwegian rivers. Fisheries Management & Ecology 10(2):87–95.

Hites, R. A., J. A. Foran, D. O. Carpenter, M. C. Hamilton, B. A. Knuth, and S. J. Schwager. 2004. Global assessment of organic contaminants in farmed salmon. Science 303(5655):226–229.

Hoar, W. S. 1988. The physiology of smolting salmonids. Pages 275–343 *in* W. S. Hoar and D. Randall, editors. Fish physiology, volume XIB. Academic Press, New York.

Hurst, T. P. and D. O. Conover. 2002. Effects of temperature and salinity on survival of young-of-the-year Hudson River striped bass (*Morone saxatilis*): implications for optimal overwintering habitats. Canadian Journal of Fisheries & Aquatic Sciences 59(5):787–795.

Hvidsten, N. A., and L. P. Hansen. 1988. Increased recapture rate of adult Atlantic salmon, *Salmo salar* L., stocked as smolts at high water discharge. Journal of Fish Biology 32:153–154.

Ikuta, K., K. Aida, N. Okumoto, and I. Hanyu. 1987. Effects of sex steroids on the smoltification of masu salmon, *Oncorhynchus masou*. General and Comparative Endocrinology 65:99–110.

Iwata, M. 1995. Downstream migratory behavior of salmonids and its relationship with cortisol and thyroid hormones: a review. Aquaculture 135(1–3):131–139.

Iwata, M., K. Yamauchi, R. S. Nishioka, R. Lin, and H. A. Bern. 1990. Effects of thyroxine, growth hormone and cortisol on salinity preference of juvenile coho salmon (*Oncorhynchus kisutch*). Marine Behavior and Physiology 17:191–201.

Jagoe, C. H. and T. A. Haines. 1997. Changes in gill morphology of Atlantic salmon (*Salmo salar*) smolts due to addition of acid and aluminum to stream water. Environmental Pollution 97(1–2):137–146.

Jarvi, T. 1990. Cumulative acute physiological stress in Atlantic salmon smolts: the effect of osmotic imbalance and the presence of predators. Aquaculture 89:337–350.

Johnson, L. L., G. M. Ylitalo, M. R. Arkoosh, A. N. Kagley, C. Stafford, J. L. Bolton, J. Buzitis, B. F. Anulacion, and T. K. Collier. 2007. Contaminant exposure in out-migrant juvenile salmon from Pacific Northwest estuaries of the United States. Environmental Monitoring and Assessment 124(1–3):167–194.

Johnston, B. D., J. M. Seubert, and C. J. Kennedy. 1998. Biochemical effects of didecyldimethylammonium chloride (DDAC) exposure and osmoregulatory stress on juvenile coho salmon, *Oncorhynchus kisutch*. Archives of Environmental Contamination & Toxicology 34(3):275–279.

Johnston, C. E. and J. G. Eales. 1967. Purines in the integument of the Atlantic salmon (*Salmo salar*) during par–smolt transformation. Journal of the Fisheries Research Board of Canada 24:955–964.

Jonsson, N., B. Jonsson, and L. P. Hansen. 1998. The relative role of density-dependent and density-independent survival in the life cycle of Atlantic salmon *Salmo salar*. Journal of Animal Ecology 67(5):751–762.

Jorgensen, E. H., O. Aas-Hansen, A. G. Maule, J. E. T. Strand, and M. M. Vijayan. 2004. PCB impairs smoltification and seawater performance in anadromous Arctic char (*Salvelinus alpinus*). Comparative Biochemistry and Physiology C-Toxicology & Pharmacology 138(2):203–212.

Jorgensen, E. H., M. M. Vijayan, N. Aluru, and A. G. Maule. 2002. Fasting modifies Aroclor 1254 impact on plasma cortisol, glucose and lactate responses to a handling disturbance in Arctic char. Comparative Biochemistry and Physiology C 132(2):235–245.

Kidd, K. A., P. J. Blanchfield, K. H. Mills, V. P. Palace, R. E. Evans, J. M. Lazorchak, and R. W. Flick. 2007. Collapse of a fish population after exposure to a synthetic estrogen. Proceedings of the National Academy of Sciences of the United States of America 104:8897–8901.

Koch, H. J. A. 1982. Hemoglobin changes with size in the Atlantic salmon (*Salmo salar* L.). Aquaculture 28:231–240.

Kolpin, D. W., E. T. Furlong, M. T. Meyer, E. M. Thurman, S. D. Zaugg, L. B. Barber, and H. T. Buxton. 2002. Pharmaceuticals, hormones, and other organic wastewater contaminants in U.S. streams, 1999–2000: a national reconnaissance. Environmental Science & Technology 36(6):1202–1211.

Kroglund, F., and B. Finstad. 2003. Low concentrations of inorganic monomeric aluminum impair physiological status and marine survival of Atlantic salmon. Aquaculture 222(1–4):119–133.

Kroglund, F., B. Finstad, S. O. Stefansson, T. O. Nilsen, T. Kristensen, B. O. Rosseland, H. C. Teien, and B. Salbu. 2007. Exposure to moderate acid water and aluminum reduces Atlantic salmon postsmolt survival. Aquaculture 273:360–373.

Kroglund, F. and M. Staurnes. 1999. Water quality requirements of smolting Atlantic salmon (*Salmo salar*) in limed acid rivers. Canadian Journal of Fisheries and Aquatic Sciences 56(11):2078–2086.

Labenia, J. S., D. H. Baldwin, B. L. French, J. W. Davis, and N. L. Scholz. 2007. Behavioral impairment and increased predation mortality in cutthroat trout exposed to carbaryl. Marine Ecology Progress Series 329:1–11.

Lacroix, G. L. 1989. Ecological and physiological responses of Atlantic salmon in acidic organic rivers of Nova Scotia, Canada. Water Air and Soil Pollution 46(1–4):375–386.

Lacroix, G. L., R. H. Peterson, C. S. Belfry, and D. J. Martinrobichaud. 1993. Aluminum dynamics on gills of Atlantic salmon fry in the presence of citrate and effects on integrity of gill structures. Aquatic Toxicology 27(3–4):373–401.

Lawson, P. W., E. A. Logerwell, N. J. Mantua, R. C. Francis, and V. N. Agostini. 2004. Environmental factors influencing freshwater survival and smolt production in Pacific Northwest coho salmon (*Oncorhynchus kisutch*). Canadian Journal of Fisheries and Aquatic Sciences 61(3):360–373.

Leduc, A. O. H. C., E. Roh, M. C. Harvey, and G. E. Brown. 2007. Impaired detection of chemical alarm cues by juvenile wild Atlantic salmon (*Salmo salar*) in a weakly acidic environment. Canadian Journal of Fisheries and Aquatic Sciences 63:2356–2363.

Leivestad, H., E. Jensen, H. Kjartansson, and L. Xingfu. 1987. Aqueous speciation of aluminum and toxic effects on Atlantic salmon. Annales de la Societe Royale Zoologique de Belgique 117:387–398.

Leonard, J. B. K., and S. D. McCormick. 2001. Metabolic enzyme activity during smolting in stream- and hatchery-related Atlantic salmon (*Salmo salar*). Canadian Journal of Fisheries and Aquatic Sciences 58:1585–1593.

Lerner, D. T. 2007. Endocrine control and disruption of physiological and behavioral smolt development of Atlantic salmon *Salmo salar*. Doctoral dissertation. University of Massachusetts, Amherst.

Lerner, D. T., B. T. Bjornsson, and S. D. McCormick. 2007a. Effects of aqueous exposure to polychlorinated biphenyls (Aroclor 1254) on physiology and behavior of smolt development of Atlantic salmon. Aquatic Toxicology 81(3):329–336.

Lerner, D. T., B. T. Björnsson, and S. D. McCormick. 2007b. Aqueous exposure to 4-nonylphenol and 17β-estradiol increases stress sensitivity and disrupts ion regulatory ability of juvenile Atlantic salmon. Environmental Toxicology and Chemistry 26(7):1433–1440.

Lerner, D. T., B. T. Björnsson, and S. D. McCormick. 2007c. Larval exposure to 4-nonylphenol and 17β-estradiol affects physiological and behavioral development of seawater adaptation in Atlantic salmon smolts. Environmental Science & Technology 41(12):4479–4485.

Little, E. E., R. D. Archeski, B. A. Flerov, and V. I. Kozlovskaya. 1990. Behavioral indicators of sublethal toxicity in rainbow trout. Archives of Environmental Contamination & Toxicology 19:380–385.

Loge, F. J., M. R. Arkoosh, T. R. Ginn, L. L. Johnson, and T. K. Collier. 2005. Impact of environmental stressors on the dynamics of disease transmission. Environmental Science & Technology 39:7329–7336.

Lorz, H. W., and B. P. McPherson. 1976. Effects of copper or zinc in freshwater on the adaptation to seawater and ATPase activity, and the effects of copper on migratory disposition of coho salmon (*Oncorhynchus kisutch*). Journal of the Fisheries Research Board of Canada 33:2023–2030.

Lundqvist, H., S. Mckinnell, H. Fangstam, and I. Berglund. 1994. The effect of time, size and sex on recapture rates and yield after river releases of *Salmo salar* smolts. Aquaculture 121:245–257.

Madsen, S. S., and B. Korsgaard. 1989. Time-course effects of repetitive estradiol-17β and thyroxine injections on the natural spring smolting of Atlantic salmon, *Salmo salar* L. Journal of Fish Biology 35:119–128.

Madsen, S. S., A. B. Mathiesen, and B. Korsgaard. 1997. Effects of 17 beta-estradiol and 4-nonylphenol on smoltification and vitellogenesis in Atlantic salmon (*Salmo salar*). Fish Physiology and Biochemistry 17(1–6):303–312.

Madsen, S. S., S. Skovbolling, C. Nielsen, and B. Korsgaard. 2004. 17-beta estradiol and 4-nonylphenol delay smolt development and downstream migration in Atlantic salmon, *Salmo salar*. Aquatic Toxicology 68(2):109–120.

Magee, J. A., T. A. Haines, J. F. Kocik, K. F. Beland, and S. D. McCormick. 2001. Effects of acidity and aluminum on the physiology and migratory behavior of Atlantic salmon smolts in Maine, USA. Water Air and Soil Pollution 130(1–4):881–886.

Magee, J. A., M. Obedzinski, S. D. McCormick, and J. F. Kocik. 2003. Effects of episodic acidification on Atlantic salmon (*Salmo salar*) smolts. Canadian Journal of Fisheries and Aquatic Sciences 60(2):214–221.

Makino, K., T. A. Onuma, T. Kitahashi, H. Ando, M. Ban, A. Urano, and A. Urano. 2007. Expression

of hormone genes and osmoregulation in homing chum salmon: a minireview. General and Comparative Endocrinology 152(2–3):304–309.
McCormick, S. D. 1994. Ontogeny and evolution of salinity tolerance in anadromous salmonids: hormones and heterochrony. Estuaries 17:26–33.
McCormick, S. D. 2001. Endocrine control of osmoregulation in teleost fish. American Zoologist 41(4):781–794.
McCormick, S. D, R. A. Cunjak, B. Dempson, M. F. O'Dea, and J. Carey. 1999. Temperature-related loss of smolt characteristics in Atlantic salmon in the wild. Canadian Journal of Fisheries and Aquatic Sciences 56:1649–1658.
McCormick, S. D., L. P. Hansen, T. P. Quinn, and R. L. Saunders. 1998. Movement, migration, and smolting of Atlantic salmon (*Salmo salar*). Canadian Journal of Fisheries and Aquatic Sciences 55(Supplement 1):77–92.
McCormick, S. D., S. Moriyama, and B. T. Bjornsson. 2000. Low temperature limits photoperiod control of smolting in Atlantic salmon through endocrine mechanisms. American Journal of Physiology 278(5):R1352–R1361.
McCormick, S. D., M. F. O'Dea, A. M. Moeckel, D. T. Lerner, and B. T. Bjornsson. 2005. Endocrine disruption of par–smolt transformation and seawater tolerance of Atlantic salmon by 4-nonylphenol and 17 beta-estradiol. General and Comparative Endocrinology 142(3):280–288.
McCormick, S. D., and R. L. Saunders. 1987. Preparatory physiological adaptations for marine life in salmonids: osmoregulation, growth and metabolism. Pages 211–229 *in* M. J. Dadswell, R. J. Klauda, C. M. Moffitt, R. L. Saunders, R. A. Rulifson, and J. E. Cooper, editors. American Fisheries Society, Symposium 1, Bethesda, Maryland.
McCormick, S. D., R. L. Saunders, E. B. Henderson, and P. R. Harmon. 1987. Photoperiod control of par–smolt transformation in Atlantic salmon (Salmo salar): changes in salinity tolerance, gill Na+,K+-ATPase activity and plasma thryoid hormones. Canadian Journal of Fisheries and Aquatic Sciences 45:1462–1468.
McCormick, S. D., J. M. Shrimpton, and J. D. Zydlewski. 1997. Temperature effects on osmoregulatory physiology of juvenile anadromous fish. In C. M. Wood and D. G. Mcdonald, editors. Global warming: implications for freshwater and marine fish. Cambridge University Press, Cambridge, UK.
McDonald, D. G., C. M. Wood, and R. G. Rhem. 1991. Nature and time course of acclimation to aluminum in juvenile brook trout (*Salvelinus fontinalis*). I. Physiology. Canadian Journal of Fisheries and Aquatic Sciences 48:2006–2015.
Melamed, P., H. Rosenfeld, A. Elizur, and Z. Yaron. 1998. Endocrine regulation of gonadotropin and growth hormone gene transcription in fish. Comparative Biochemistry and Physiology C 119(3):325–338.
Miyakoshi, Y., M. Nagata, and S. Kitada. 2001. Effect of smolt size on postrelease survival of hatchery-reared masu salmon *Oncorhynchus masou*. Fisheries Science 67(1):134–137.
Monette, M. 2007. Impacts of episodic acid and aluminum exposure on the physiology of Atlantic salmon, *Salmo salar*, smolt development. Doctoral dissertation. University of Massachusetts, Amherst.
Monette, M. Y., and S. D. McCormick. 2008. Impacts of short-term acid and aluminum exposure on Atlantic salmon (*Salmo salar*) physiology: a direct comparison of parr and smolts. Aquatic Toxicology 86(2):216–226.
Moore, A., A. P. Scott, N. Lower, I. Katsiadaki, and L. Greenwood. 2003. The effects of 4-nonylphenol and atrazine on Atlantic salmon (*Salmo salar* L) smolts. Aquaculture 222(1–4):253–263.
Moore, A., and C. P. Waring. 1998. Mechanistic effects of a triazine pesticide on reproductive endocrine function in mature male Atlantic salmon (*Salmo salar* L.) parr. Pesticide Biochemistry & Physiology 62(1):41–50.
Moore, A., and C. P. Waring. 2001. The effects of a synthetic pyrethroid pesticide on some aspects of reproduction in Atlantic salmon (*Salmo salar* L.). Aquatic Toxicology 52(1):1–12.
National Academy of Science. 2004. Atlantic salmon in Maine. Report of the National Research Council of the National Academies. The National Academies Press, Washington, D.C.
Neskovic, N. K., I. Elezovic, V. Karan, V. Poleksic, and M. Budimir. 1993. Acute and sub-acute toxicity of atrazine to carp (*Cyprinus carpio* L.). Ecotoxicology & Environmental Safety 25:173–182.
Nieves-Puigdoller, K. 2007. Effect of contaminants on sensitive life history stages of Atlantic salmon.
Nieves-Puigdoller, K., B. T. Bjornsson, and S. D. McCormick. 2007. Effects of hexazinone and atrazine on the physiology and endocrinology of smolt development in Atlantic salmon. Aquatic Toxicology 84(1):27–37.
Nilsson, S. and L. Sundin. 1998. Gill blood flow control. Comparative Biochemistry and Physiology A 119(1):137–147.
Ogata, H. Y., S. Konnno, and J. T. Silverstein. 1998. Muscular buffering capacity of the parr and smolts in *Oncorhynchus masou*. Aquaculture 168:303–310.
Orciari, R. D., and G. H. Leonard. 1996. Length characteristics of smolts and timing of downstream migration among three strains of Atlantic salmon in

a southern New England stream. North American Journal of Fisheries Management 16:851–860.

Orr, J. C., V. J. Fabry, O. Aumont, L. Bopp, S. C. Doney, and R. A. Feely. 2005. Anthropogenic ocean acidification over the twenty-first century and its impact on calcifying organisms. Nature 437:681–686.

Parrish, D. L., R. J. Behnke, S. R. Gephard, S. D. McCormick, and G. H. Reeves. 1998. Why aren't there more Atlantic salmon (*Salmo salar*)? Canadian Journal of Fisheries and Aquatic Sciences 55(Supplement 1):281–287.

Peake, S., and R. S. McKinley. 1998. A re-evaluation of swimming performance in juvenile salmonids relative to downstream migration. Canadian Journal of Fisheries and Aquatic Sciences 55(3):682–687.

Pess, G. R., D. R. Montgomery, E. A. Steel, R. E. Bilby, B. E. Feist, and H. M. Greenberg. 2002. Landscape characteristics, land use, and coho salmon abundance, Snohomish River, Wash., USA. Canadian Journal of Fisheries and Aquatic Sciences 59:613–623.

Pickford, G. E., and J. G. Phillips. 1959. Prolactin, a factor promoting survival of hypophysectomized killifish in freshwater. Science 130:454–455.

Price, C. S., and C. B. Schreck. 2003. Stress and saltwater-entry behavior of juvenile chinook salmon (*Oncorhynchus tshawytscha*): conflicts in physiological motivation. Canadian Journal of Fisheries and Aquatic Sciences 60(8):910–918.

Primmett, D. R. N., F. B. Eddy, M. S. Miles, C. Talbot, and J. E. Thorpe. 1988. Transepithelial ion exhange in smolting Atlantic salmon (*Salmo salar* L.). Fish Physiology and Biochemistry 5:181–186.

Rosseland, B. O., F. Kroglund, M. Staurnes, K. Hindar, and A. Kvellestad. 2001. Tolerance to acid water among strains and life stages of Atlantic salmon (*Salmo salar* L.). Water Air and Soil Pollution 130(1–4):899–904.

Rosseland, B. O., and O. K. Skogheim. 1984. A comparative study on salmonid fish species in acid aluminum-rich water II. Physiological stress and mortality of one- and two-year-old fish. Institute of Freshwater Research Drottningholm Report 61:186–194.

Rounsefell, G. A. 1958. Anadromy in North American salmonidae. Fishery Bulletin 58:171–185.

Saloniemi, I., E. Jokikokko, I. Kallio-Nyberg, E. Jutila, and P. Pasanen. 2004. Survival of reared and wild Atlantic salmon smolts: size matters more in bad years. ICES Journal of Marine Science 61(5):782–787.

Sandahl, J. F., D. H. Baldwin, J. J. Jenkins, and N. L. Scholz. 2005. Comparative thresholds for acetylcholinesterase inhibition and behavioral impairment in coho salmon exposed to chlorpyrifos. Environmental Toxicology & Chemistry 24:136–145.

Saunders, R. L. 1965. Adjustment of buoyancy in young Atlantic salmon and brook trout by changes in swim bladder volume. Journal of the Fisheries Research Board of Canada 22:335–352.

Saunders, R. L., E. B. Henderson, P. R. Harmon, C. E. Johnston, and J. G. Eales. 1983. Effects of low environmental pH on smolting of Atlantic salmon (*Salmo salar*). Canadian Journal of Fisheries and Aquatic Sciences 40(8):1203–1211.

Schmitz, M. and I. Mayer. 1993. Effect of androgens on seawater adaptation in Arctic char, *Salvelinus alpinus*. Fish Physiology and Biochemistry 12:11–20.

Schreck, C. B., M. F. Solazzi, S. L. Johnson, and T. E. Nickelson. 1989. Transportation stress affects performance of coho salmon, *Oncorhynchus kisutch*. Aquaculture 82:15–20.

Schreck, C. B., T. P. Stahl, L. E. Davis, D. D. Roby, and B. J. Clemens. 2006. Mortality estimates of juvenile spring-summer Chinook salmon in the lower Columbia River and estuary, 1992–1998: evidence for delayed mortality? Transactions of the American Fisheries Society 135:457–475.

Shrimpton, J. M., D. A. Patterson, J. G. Richards, S. J. Cooke, P. M. Schulte, S. G. Hinch, and A. P. Farrell. 2005. Ionoregulatory changes in different populations of maturing sockeye salmon *Oncorhynchus nerka* during ocean and river migration. Journal of Experimental Biology 208(21):4069–4078.

Sower, S. A., C. V. Sullivan, and A. Gorbman. 1984. Changes in plasma estradiol and effects of triiodothyronine on plasma estradiol during smoltification of coho salmon (*Oncorhynchus kisutch*). General and Comparative Endocrinology 54:486–492.

Staurnes, M., P. Blix, and O. B. Reite. 1993a. Effects of acid water and aluminum on par–smolt transformation and seawater tolerance in Atlantic Salmon, *Salmo salar*. Canadian Journal of Fisheries and Aquatic Sciences 50(9):1816–1827.

Staurnes, M., L. P. Hansen, K. Fugelli, and O. Haraldstad. 1996. Short-term exposure to acid water impairs osmoregulation, seawater tolerance, and subsequent marine survival of smelts of Atlantic salmon (*Salmo salar* L). Canadian Journal of Fisheries and Aquatic Sciences 53(8):1695–1704.

Staurnes, M., G. Lysfjord, L. P. Hansen, and T. G. Heggberget. 1993b. Recapture rates of hatchery-reared Atlantic salmon (*Salmo salar*) related to smolt development and time of release. Aquaculture 118:327–337.

Teien, H. C., B. Salbu, F. Kroglund, and B. O. Rosseland. 2004. Transformation of positively charged aluminum-species in unstable mixing zones following liming. Science of the Total Environment 330(1–3):217–232.

Thorpe, J. E. 1982. Migration in salmonids, with special

reference to juvenile movements in freshwater. Pages 86–97 *in* E. L. Brannon and E. O. Salo, editors. Proceedings of the Salmon and Trout Migratory Behavior Symposium. University of Washington, Seattle.

Uchida, K., T. Kaneko, A. Yamaguchi, T. Ogasawara, and T. Hirano. 1997. Reduced hypoosmoregulatory ability and alteration in gill chloride cell distribution in mature chum salmon (*Oncorhynchus keta*) migrating upstream for spawning. Marine Biology 129(2):247–253.

Venditti, D. A., D. W. Rondorf, and J. M. Kraut. 2000. Migratory behavior and forebay delay of radio-tagged juvenile fall Chinook salmon in a lower Snake River impoundment. North American Journal of Fisheries Management 20:41–52.

Virtanen, E., L. Soderholmtana, A. Soivio, L. Forsman, and M. Muona. 1991. Effect of physiological condition and smoltification status at smolt release on subsequent catches of adult salmon. Aquaculture 97:231–257.

Wagner, H. H. 1974. Photoperiod and temperature regulation of smolting in steelhead trout. Canadian Journal of Zoology 52:219–234.

Waring, C. P., and A. Moore. 1997. Sublethal effects of a carbamate pesticide on pheromonal mediated endocrine function in mature male Atlantic salmon (*Salmo salar* L.) parr. Fish Physiology & Biochemistry 17(1–6):203–211.

Waring, C. P., and A. Moore. 2004. The effect of atrazine on Atlantic salmon (*Salmo salar*) smolts in freshwater and after seawater transfer. Aquatic Toxicology 66(1):93–104.

Watt, W. D. 1987. A summary of the impact of acid rain on Atlantic salmon (*Salmo salar*) in Canada. Water Air and Soil Pollution 35:27–35.

Weisbart, M. 1968. Osmotic and ionic regulation in embryos, alevins, and fry of the five species of Pacific salmon. Canadian Journal of Zoology 46:385–397.

Wendelaar Bonga, S. E. 1997. The stress response in fish. Physiological Reviews 77:591–625.

Wilkinson, K. J., and P. G. C. Campbell. 1993. Aluminum bioconcentration at the gill surface of juvenile Atlantic salmon in acidic media. Environmental Toxicology and Chemistry 12(11):2083–2095.

Williams, J. G., S. G. Smith, and W. D. Muir. 2001. Survival estimates for downstream migrant yearling juvenile salmonids through the Snake and Columbia rivers hydropower system, 1966–1980 and 1993–1999. North American Journal of Fisheries Management 21:310–317.

Young, G., B. Th. Björnsson, P. Prunet, R. J. Lin, and H. A. Bern. 1989. Smoltification and seawater adaptation in coho salmon (*Oncorhynchus kisutch*): plasma prolactin, growth hormone, thyroid hormones, and cortisol. General and Comparative Endocrinology 74:335–345.

Youson, J. H. and C. M. Neville. 1987. Deposition of aluminum in the gill epithelium of rainbow trout (*Salmo gairdneri* Richardson) subjected to sublethal concentrations of the metal. Canadian Journal of Zoology 65(3):647–656.

Zaugg, W. S., and L. R. McLain. 1976. Influence of water temperature on gill sodium, potassium-stimulated ATPase activity in juvenile coho salmon (*Oncorhynchus kisutch*). Comparative Biochemistry and Physiology A 54:419–421.

Zlokovitz, E. R., and D. H. Secor. 1999. Effect of habitat use on PCB body burden in Hudson River striped bass (*Morone saxatilis*). Canadian Journal of Fisheries and Aquatic Sciences 56:86–93.

Zoeller, R. T. 2005. Environmental chemicals as thyroid hormone analogues: new studies indicate that thyroid hormone receptors are targets of industrial chemicals? Molecular and Cellular Endocrinology 242(1–2):10–15.

Zydlewski, G. B., A. Haro, and S. D. McCormick. 2005. Evidence for cumulative temperature as an initiating and terminating factor in downstream migratory behavior of Atlantic salmon (*Salmo salar*) smolts. Canadian Journal of Fisheries and Aquatic Sciences 62(1):68–78.

Zydlewski, J. D., and S. D. McCormick. 1997a. The loss of hyperosmoregulatory ability in migrating juvenile American shad, *Alosa sapidissima*. Canadian Journal of Fisheries and Aquatic Sciences 54:2377–2387.

Zydlewski, J. D., and S. D. McCormick. 1997b. The ontogeny of salinity tolerance in American shad, *Alosa sappidissima*. Canadian Journal of Fisheries and Aquatic Sciences 54:182–189.

American Fisheries Society Symposium 69:215–230, 2009

Models of Past, Present, and Future Stream Temperatures for Selected Atlantic Salmon Rivers in Northeastern North America

Wendy A. Monk* and R. Allen Curry
Post Office Box 4400, University of New Brunswick
Fredericton, New Brunswick, E3B 5A3, Canada

Abstract.—In this paper, we develop logistic stream temperature models for 17 selected sites in northeastern North America and evaluate the potential changes from warming climate under two scenarios (low and medium-high emissions). Classification of the magnitude of the (1) long-term (1980–2002) and (2) annual thermal regimes allowed examination of the relative spatial and temporal patterns of instream thermal variability across the 17 sites. At the regional scale, the classification identified three broad groups of rivers (cool, intermediate, and warm) reflecting geographical location and moderated by site-specific factors. The interannual classification identified four thermal year types reflecting increasing magnitude and variability in the annual thermal regime. The dominance of thermal year types and the frequency of occurrence indicated significant variability between years for all sites and within thermal regions. Under the two climate change scenarios, stream temperatures in the 17 sites are projected to increase by 2050. However, there are regional differences with intermediate and warm region rivers projected to be more affected, particularly under the medium-high emissions scenario. More significantly, the duration of weeks when temperatures exceed 20°C (taken as a threshold of thermal stress for Atlantic salmon *Salmo salar*) is projected to increase with variability in response between river groups. We comment on the ecological significance of these potential future increases in stream temperature and duration for Atlantic salmon in the region.

Introduction

Stream temperature is a key component in determining the health and suitability of instream habitats for fishes (i.e., temperature regulates biochemistry and physiological processes of animals; Fry 1971). Controlled by multiple factors, instream thermal regimes demonstrate naturally high levels of spatial and temporal variability across different scales (Torgersen et al. 1999; Poole and Berman 2001). However, fishes have evolved adaptations for surviving in present-day thermal habitats and therefore, alterations of these thermal regimes within river systems can impact species survival, invasions rates of new species, and the overall diversity of biological communities (Jensen 2003).

* Corresponding author: wmonk@unb.ca

In temperate freshwaters, fishes are generally classified as coldwater, coolwater, and warmwater species, which reflects species-specific adaptations. Cool- and coldwater fishes are predicted to be most negatively impacted by the continuing increase in global temperatures due to reductions in the availability of suitable thermal habitats (e.g., Chatters et al. 1991; Eaton and Scheller 1996; Magnuson et al. 1997; Mohseni et al. 2003). The Atlantic salmon *Salmo salar* of the north Atlantic region are a coldwater species that are likely susceptible to factors related to increasing global temperatures (e.g., Friedland 1998; Dempson et al. 2001). Changes in both freshwater and ocean thermal regimes will affect life history stages in different ways. For example, mature Atlantic salmon returning from the sea to freshwater rivers for reproduction can spend months, including the warmest period of the summer, in rivers. Their embryos incubate in coarse gravel from late fall to spring when they emerge, grow for 1–3 years, and then smoltify before migrating to the sea in spring. Each of these life history events is strongly regulated by temperature (e.g., Fry 1971; Fleming and Jensen 2002; Zydlewski et al. 2005), including direct impacts from changing global temperatures that have been predicted for growth (Swansburg et al. 2002), timing of smolt emigration (McCormick et al. 1999), and timing of adult runs (e.g., Juanes et al. 2004; Swansburg et al. 2004).

Climate change scenarios predict an increase in atmospheric carbon dioxide, suggesting that there is a high probability of warming air temperatures and therefore instream temperatures. Air temperature is well simulated by general circulation models and offers an ability to assess the potential effects of future climate change on river ecosystems. A number of studies have attempted to explore the potential effects of a changing climate on river systems and their components (e.g., Magnuson et al. 1997). For example, predictions of stream temperatures using a modified logistic function for sites across the continental United States suggested that maximum river temperatures associated with a 2 × CO_2 condition are unlikely to exceed the thermal tolerances of warmwater fishes, while there will be moderate decreases in the available thermal habitat for cold- and coolwater fishes (Mohseni et al. 2003). Effects are predicted to be less direct for coolwater and coldwater fishes (Ficke et al. 2007) (e.g., altered behavior or habitats related to warmer temperatures and reduced water levels; Chatters et al. 1991) or the cumulative effects of sublethal stress in warmer environments (e.g., Zydlewski et al. 2005).

While air temperatures are predicted to increase, very little is known about the concurrent changes anticipated for thermal regimes of rivers within the northeastern region of North America, as discussed by Caissie (2006). Under future climate scenarios, published research suggests that this region will have decreased snow cover amount and duration with possible large reductions in stream discharge leading to changes in magnitude, timing of ice freezing/break up with consequences on spring flooding, which may cause shifts in fish species distributions and migration patterns (Moore et al. 1997; Hodgkins et al. 2005; Hayhoe et al. 2007). Therefore, we would predict higher maximum stream temperatures and longer durations of these higher instream temperatures and thus direct impacts on thermal habitats for Atlantic salmon. In this study, we examine 17 rivers with historical and current Atlantic salmon populations across northeastern North America to (1) develop and apply logistic stream temperature models for predicting current weekly instream thermal maxima, (2) quantify spatial (regional) and temporal (interannual) patterns in the thermal regimes of these rivers (1980–2002), and (3) examine the potential variability in future instream thermal regimes under B1 and A1B climate change scenarios (2046–2055) using Canadian Climate Centre for modeling and analysis (CCCma) models and then comment on the implications for Atlantic salmon rivers in the region.

Methods

Stream and Air Temperature Data

The historical range of Atlantic salmon in northeastern North America falls between the Connecticut River, Connecticut in the south through to populations in Labrador and Northern Quebec in Canada (MacCrimmon and Gots 1979). The geographical distribution in this region reflects the optimal temperature ranges for normal feeding and growth (Jensen et al. 1989). Seventeen rivers with current Atlantic salmon populations have been selected from Maine, New Brunswick, and Newfoundland (Figure 1).

Daily maximum stream temperature records at monitoring stations for the 17 rivers were extracted from the U.S. Geological Survey (USGS 2007), Water Resources Management Division of the Government of Newfoundland and Labrador, and the New Brunswick Aquatic Data Warehouse (New Brunswick Aquatic Data Warehouse 2007; Table 1). The spatial distribution of the analysis sites was limited by the availability of accurate and consistent water temperature records (e.g., recorded using comparable data loggers and equipment; Figure 1). Sites had a minimum of 3 years of stream temperature data allowing for models to be developed and tested on an independent set of data. The majority of temperature records were available for years within the period 1999–2003. Each stream temperature gauging site was paired with a nearby weather station (between 2.7 and 67.2 km from the stream monitoring site) with long-term high qual-

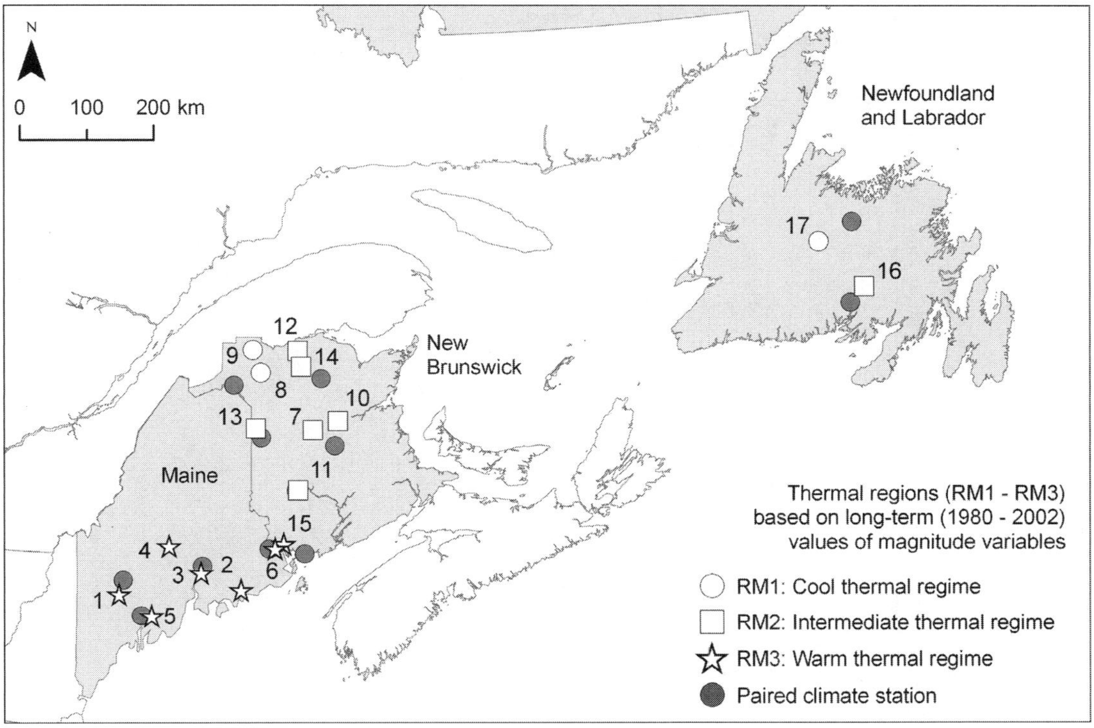

Figure 1.—Location map of northeastern North America showing the 17 stream temperature analysis sites with their paired weather stations and the distribution of the identified thermal regions (RM1–RM3). Site codes: 1 = Androscoggin, 2 = Narraguagus, 3 = Penobscot, 4 = Piscataquis, 5 = Sheepscot, 6 = St. Croix, 7 = Dungarvon, 8 = Kedgwick, 9 = Little Main Restigouche, 10 = Little Southwest Miramichi, 11 = Saint John River at Mactaquac, 12 = Northwest Upsalquitch, 13 = Tobique, 14 = Upsalquitch, 15 = Waweig, 16 = Conne, and 17 = Exploits.

ity records. Long-term daily mean air temperature data were extracted from the U.S. Historical Climatological Network (Williams et al. 2006) and Environment Canada historical data portal (Environment Canada 2008). Sites required a minimum of 23 years continuous air temperature data between 1980 and 2002 in addition to records overlapping with the stream temperature records. Sites with less than 10% missing data were interpolated using a standard linear interpolation method.

Analyses were based on standard weeks with week 1 reflecting the mean of 7 d maximum stream temperature from January 1 of that calendar year. Additional days (e.g., week 53) were combined with the preceding week to allow for standardized methods and leap years were included in the standard week at the end of February. Daily stream and air temperatures were extracted from standard week 17 through week 45. Ideally, complete annual stream temperature records would be explored; however, because of ice presence and the potential complications during the annual thaw, the majority of the temperature gauges are removed from the rivers between late autumn and early spring. The period between standard weeks 17 and 45 represents a common monitoring period for all of the selected temperature gauging stations. Weekly stream and air temperatures were calculated by taking the arithmetic mean of daily maximum stream and daily mean air temperatures, respectively. The week-long averaging period masks some of the extreme, daily temperature events but absorbs much of the expected lag between the air and water temperatures (Stefan and Preud'homme 1993). The optimal temperatures for Atlantic salmon are between 13°C and 17°C, with an upper incipient lethal limit of 25°C (Wallace 1993). Therefore, a threshold

Table 1.—Summary information for each of the seventeen stream gauging stations, their paired climate stations and associated catchment information.

Map reference	Site code	Stream temperature station	Province/ state	Latitude	Longitude	Catchment area (km²)	Paired weather station	Distance between weather and stream temperature stations (km)
1	AND	Androscoggin	ME	44.502464	–70.224375	2,488	Farmington	21.5
2	NAR	Narraguagus	ME	44.608056	–67.935278	227	Orono	67.2
3	PEN	Penobscot at Orono	ME	44.826667	–68.696667	7,764	Orono	9.5
4	PIS	Piscataquis	ME	45.175000	–69.314722	298	Orono	58.7
5	SHE	Sheepscot	ME	44.222778	–69.593889	145	Gardiner	15.7
6	STC	St. Croix	ME	45.169906	–67.296775	1,455	Woodland	9.0
7	DUN	Dungarvon	NB	46.762992	–66.618903	583.3	Doaktown	40.5
8	KED	Kedgwick	NB	47.843839	–67.799381	799.7	Edmundston	60.3
9	LMR	Little Main Restigouche	NB	47.520118	–67.641327	914.5	Edmundston	44.2
10	LSW	Little Southwest Miramichi	NB	46.874280	–66.114913	1,230.9	Doaktown	37.2
11	MAC	Saint John River at Mactaquac	NB	45.969083	–66.879755	40,527.3	Mactaquac	2.8
12	NWU	Northwest Upsalquitch	NB	47.586285	–66.831154	756.8	Upsalquitch Lake	34.7
13	TOB	Tobique	NB	46.807471	–67.715123	4,184.2	Bon Accord	16.9
14	UPS	Upsalquitch	NB	47.809728	–66.908024	2,249.6	Upsalquitch Lake	54.4
15	WAW	Waweig	NB	45.235407	–67.162733	140.4	Pennfield	23.1
16	CON	Conne	NL	48.165417	–55.483333	602	Bay d'Espoir	29.1
17	EXP	Exploits	NL	48.838589	–56.276121	1,100	Wooddale	56.2

of 20°C was chosen as this temperature represents a stressful temperature for free-swimming life stages of Atlantic salmon in rivers (e.g., Alabaster 1990; Breau et al. 2007).

Modeling Stream Temperatures

Stream temperatures increase linearly at moderate air temperatures (i.e., between approximately 5°C and 25°C; Figure 2). As air temperatures rise above 25°C, stream temperatures stabilize because of extensive evaporative cooling and back radiation, while below 5°C, stream temperatures are constrained by the lower bound of around 0°C (Mohseni and Stefan 1999). This relationship can be best represented by an S-shaped function as a linear extrapolation of the stream temperature/air temperature relationship beyond 25°C can overestimate stream temperatures (equation 1; Mohseni et al. 1998).

$$T_{sim} = \mu + \frac{\alpha - \mu}{1 + e^{\gamma(\beta - T_a)}} \quad (1)$$

T_{sim} = estimated weekly stream temperature
T_a = weekly air temperature measured at or near the stream gauging site
α = upper bound stream temperature
μ = minimum estimated stream temperature
β = air temperature at the inflection point
γ = measure of the steepest slope of the function.

The component α is an estimate of the probable maximum stream temperature at a gauging station approximated using a universal standard deviate method proposed by Mohseni et al. (2002). The values of β, γ, and μ were determined by a least squares regression method. A separate logistic model was developed for each site to allow site-specific variations to be incorporated into the model.

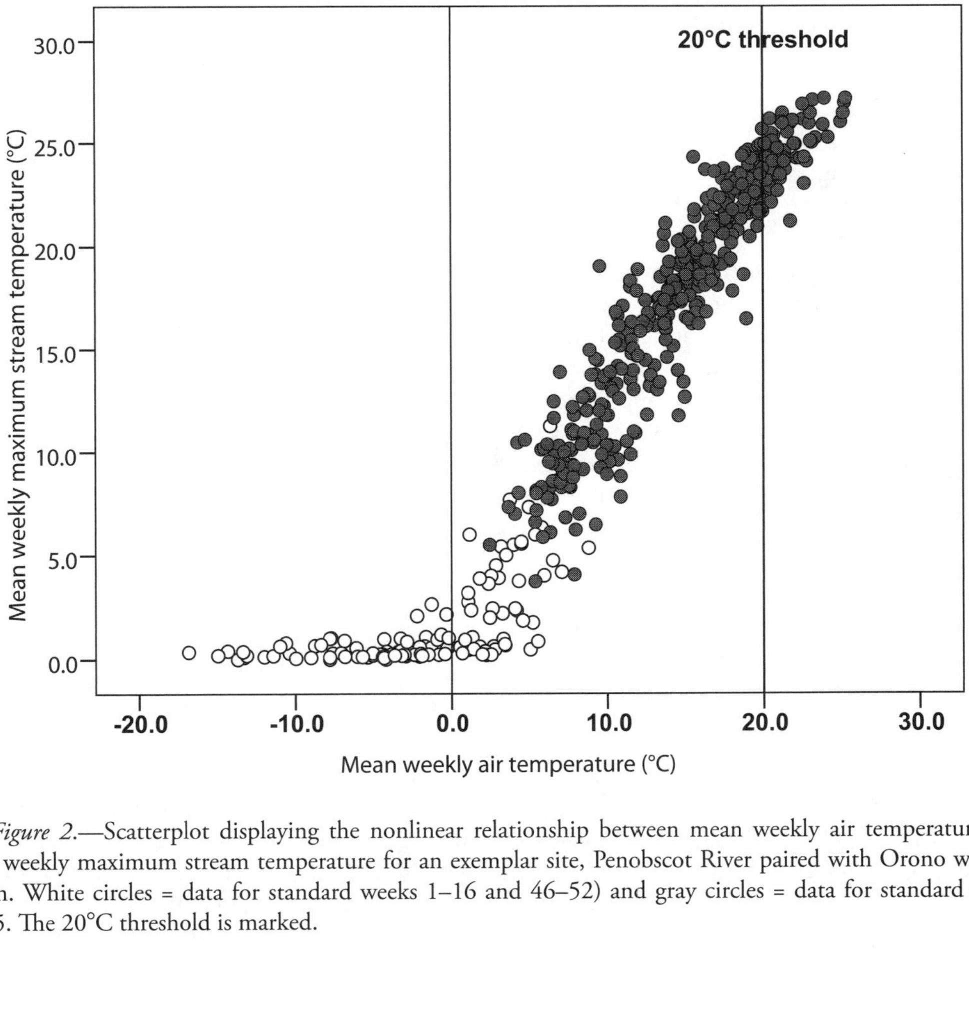

Figure 2.—Scatterplot displaying the nonlinear relationship between mean weekly air temperature and mean weekly maximum stream temperature for an exemplar site, Penobscot River paired with Orono weather station. White circles = data for standard weeks 1–16 and 46–52) and gray circles = data for standard weeks 17–45. The 20°C threshold is marked.

The fit between the model and the available stream temperature data for individual streams was measured by the Nash-Sutcliffe coefficient (NSC; Nash and Sutcliffe 1970) (equation 2) and the root mean square error (RMSE) method. The NSC varies between 0 and 1, with higher values indicating greater concordance between the observed and simulated values. Only sites with high NSC and low RMSE values were retained as these stations were the most responsive to air temperatures. A threshold NSC value of 0.85 was originally proposed by Mohseni et al. (2003), as this was suggested to be high enough to exclude all stream gauging stations adversely affected by anthropogenic effects.

$$NSC = 1 - \frac{\sum_{i=1}^{n} (T_{sim_i} - T_{obs_i})^2}{\sum_{i=1}^{n} (\overline{T}_{obs} - T_{obs_i})^2} \quad (2)$$

T_{sim} = estimated weekly stream temperature
T_{obs} = observed weekly stream temperature
$\overline{T}_{obs}$ = mean weekly observed stream temperatures

A seasonal effect (hysteresis) (Webb and Nobilis 1994; Mohseni et al. 1998; Webb et al. 2003) in the stream and air temperature relationship was identified. Therefore, two models were developed for each site, one using the complete data series and the second using two separate functions fitted to the data representing the rising limb and falling limb of the stream temperature/air temperature relationship. The limbs were distinguished using associated week numbers, with the first week of analysis being standard week 17. The minimum of the mean weekly air temperatures was used to set the starting week of the rising limb (i.e., ending week of the falling limb) and the maximum of mean weekly air temperatures was used to set the ending week of the rising limb (i.e., the starting week of the falling limb). The two separate temperature models (standard and two-functions) were developed and compared for each of the 17 sites, with the most statistically robust in terms of error retained for the analysis.

Cluster Analysis

A hierarchical agglomerative clustering analytical procedure (Ward's method) was utilized to examine the spatial and temporal variability of the stream temperature regimes of the sites using the Statistical Package for the Social Sciences (SPSS; SPSS Inc. 2006). The number of clusters was selected by examining the agglomeration schedule plot in addition to assessing the type of representation needed from the data. The approach was originally developed by Hannah et al. (2000) and modified by Harris et al. (2000) for examining and quantifying relative variation in hydrological regimes. The method utilized herein has been adapted to use standard weekly data involving a classification procedure for the 17 sites to separately identify thermally homogenous: (1) regions (spatially) consisting of sites with similar long-term (1980–2002) values in the magnitude of the thermal regime magnitude, and (2) interannual (temporal) grouping years with a similar average annual magnitude of the thermal regime.

The magnitude (size) of the thermal regime was quantified by four indices: mean, maximum, minimum, and standard deviation of the long-term (1980–2002 mean; thermal regions) or annual values (single year; thermal years) of the stream temperature series, regardless of timing of annual peaks. The temperature time-series for each site were divided into thermal years commencing in standard week 17 through to standard week 45. Prior to analysis, stream temperatures were standardized using *z*-scores (mean = 0, standard deviation = 1) to allow unbiased comparison of geographically different stations. The identified thermal regions will be used to structure further analyses and examination of the response of these rivers to potential future temperature scenarios.

Temperature Change Scenarios

Used in the International Panel on Climate Change Fourth Assessment Report (Meehl et al. 2007), the third generation coupled global climate model (CGCM3.1/T63) developed by the Canadian Centre of Climate Modeling and Analysis (CCCma) was used in the analysis as it is representative of other climate models developed through other climate research centers. The CGCM3.1 model, having a spatial resolution of 2.81 degrees latitude/longitude, is based on the earlier CGCM2 model (Flato and Boer 2001), with an updated atmospheric component as described in Scinocca and McFarlane (2004). Within this model, two climate change scenarios reflecting both medium-high (A1B, 720 ppm CO_2 stabilization) and low (B1, 550 ppm CO_2 stabilization) emissions scenarios were explored. Although all equally likely, each climate change scenario reflects potential economic, technological, social, and environmental changes (Nakićenović and Swart 2000). The A1 scenario describes a future world of very rapid economic growth, low population growth, and the rapid introduction of new and more efficient technologies. The A1B scenario represents a subfamily of the A1 scenario and is based on a balanced reliance on fossil intensive and nonfossil energy sources (Nakićenović and Swart 2000). The B1 scenario describes a convergent world with the same low population growth as in the A1 scenario but with rapid changes in economic structures toward a service and information economy, with reductions in material intensity, and the introduction of clean and resource-efficient technologies (Nakićenović and Swart 2000). Mohseni et al. (2003) suggests that air temperatures do not vary significantly within a grid cell on a weekly time series, and therefore, no downscaling method was employed for air temperature projections. Therefore, air temperatures simulated by the CCCma model were taken from the grid point nearest to the weather station using the freeware program, Panoply (Schmunk 2006).

Baseline stream temperature models were developed by applying the site-specific logistic model to a long-term mean series of weekly mean air temperature values averaged over the 23-year period (1980–2002). Future stream temperature regimes were predicted by applying the water–air temperature relationship to the baseline air temperature data at each station incremented by increases in air temperatures projected by CCCma for 2050 (average 2046–2055) for both a low and medium-high emissions scenario as defined by the B1 and A1B, respectively. Errors associated with the projection of stream temperatures were assessed using a simple two-tailed paired *t*-test with a 10% significance level between the mean of the errors, assuming that errors associated with the stream temperature model simulations would not change under new climate condition (Mohseni et al. 2003).

Results

Stream Temperature Modeling

The 17 rivers were paired with the nearest weather station with suitable long-term data, with minimal missing data meaning some rivers were paired with the same weather station. Logistic functions between observed weekly air and weekly maxima stream temperature data with minimal error and high concordance between observed and simulated data on independent data sets were developed for 17 rivers. Five of the 17 sites displayed significant hysteresis in their thermal regimes, and therefore, two-function models (functions developed separately for the rising and falling limbs) were developed for these rivers. The NSC values ranged between 0.85 and 1.0. The distance between the stream temperature monitoring station and the corresponding weather station was from 2.7 to 67.2 km. These distances did not have a significant effect ($p > 0.05$) on the goodness-of-fit of the models as quantified by the NSC and RMSE.

Using standard weekly mean air temperature data for standard weeks 17–45 for 1980–2002, baseline stream temperature models were developed for each site and provide the basis for further modeling. Current baseline weekly maxima stream temperatures range between 1.5°C and 28.7°C with a mean of 15.3°C, with sites at the southern geographical extent already experiencing thermal conditions known to be stressful to coldwater fishes.

Identification of Thermal Regions

Results of the regional thermal regime classification based on the four magnitude variables derived from baseline data (mean of 1980–2002 standard weekly data) suggest that three significantly different thermal regions exist within the analysis sites (Figure 1). The thermal regions (RM1–RM3) can be arranged in order of increasing magnitude of the values of the four thermal regime variables (mean, standard deviation, minimum, and maximum) (Table 2 and Figure 1):

- Region 1 (RM1)—Cool, with the lowest values for the mean, maximum, minimum, and stan dard deviation indices (three sites);
- Region 2 (RM2)—Intermediate, with the in termediate values for each of the variables (sev en sites);
- Region 3 (RM3)—Warm, with the highest val ues for the mean, maximum, minimum, and standard deviation variables (seven sites).

Table 2.—Mean values (variability within class as quantified by SD) of weekly maximum stream temperature (°C) for the four cluster variables used in the analysis to discriminate among each thermal region (17 sites; 1980–2002)

Stream temperature indices	Class average (°C) 1. Cool	2. Intermediate	3. Warm	Average (all stations)
Number of stations	3	7	7	17
Mean	11.30 (0.348)	14.45 (1.185)	17.86 (0.557)	15.29 (2.615)
Standard deviation	4.04 (0.025)	4.96 (0.562)	5.72 (0.520)	5.11 (0.779)
Minimum	4.28 (0.588)	4.94 (1.633)	6.92 (1.238)	5.60 (1.708)
Maximum	16.75 (0.539)	20.47 (0.936)	24.65 (0.664)	21.53 (3.090)

The spatial distribution of the clusters reflects geography and the natural site-specific factors of these sites. The coolest temperatures (Region 1) occur in two sites northeastern New Brunswick (within the Restigouche watershed) and a single site in northern Newfoundland. Region 2 sites are located across New Brunswick and one site in southern Newfoundland. Warm long-term thermal regimes (Region 3) were confined to rivers within Maine with the exception of the Waweig River, located near the southern New Brunswick border with Maine (Figure 1). The identified thermal regions are used to structure further analysis and examine whether rivers with different thermal regimes could respond differently to potential projected future changes in air temperatures.

Identification of Thermal Year Types

Thermal year "types" were identified by quantifying the natural variability in the annual thermal regimes (1980–2002) of the 17 sites. Examination of the agglomeration schedule plot and significance analysis suggested that four annual thermal magnitude (size) clusters were present relative to the input data. These reflect increasing values and variability in the magnitude of the mean weekly stream temperatures, as represented by the four magnitude variables (IA1: cool; IA2: intermediate with low variability; IA3: intermediate with high variability; and IA4: warm) (three example sites for each of the thermal regions (RM1—RM3) presented in Table 3). Examining the frequency of the thermal year types over the baseline period (1980–2002) identifies years with a dominant thermal year type (Table 4). Eight of the 23 years exhibited dominant thermal year types across all sites. In addition, there appears to be natural variability and switching between year types with the exception of 1981–1984, which was mostly dominated by a stable IA2 regime (Table 4).

Patterns were clearer when regional groups were considered, although all three regions (RM1—RM3) were dominated by both of the intermediate annual thermal regimes with the cool and warm annual thermal regimes occurring less regularly (Table 4). Sites within the warm RM3 region appeared to display greater similarity in thermal year types as opposed to sites within the intermediate RM2 region, which were more variable. The long-term (1980–2002 average) regime magnitude regionalization highlighted that distinct regime types (regions) can be identified across the study area, and perhaps more significantly, the interannual thermal regime variability was relatively similar among regions (Table 4; Figure 2).

Temperature Change Scenarios

Application of the two climate change scenarios for the year 2050 (represented by the 2046–2055 average) relative to the baseline mean weekly maximum thermal regimes for the analysis period suggest an average projected increase in mean weekly maximum stream temperatures of 0.56°C and 1.44°C for the low and medium-high emissions scenarios, respectively, for the 17 analysis sites. Results of one-way analysis of variance indicate that the projected increases in modeled mean weekly maximum stream temperatures under the two scenarios were not significantly different ($p > 0.05$) from the baseline conditions (1980–2002 average) when all sites were considered. However, there are regional variations with sites within the cool RM1 region ($p < 0.05$) and warm RM3 region ($p < 0.001$) project-

Table 3.—Average values for indices within each class for exemplar stations selected from long-term thermal magnitude regions.

Thermal region	Exemplar station	Thermal indices	Class average (°C)			
			IA1: cool	IA2: Intermediate, low variability	IA3: Intermediate, high variability	IA4: warm
RM1: Cool	Kedgwick, NB	Mean	10.383	11.201	11.267	10.954
		Maximum	15.994	16.945	17.835	18.213
		Minimum	2.508	3.992	2.578	3.165
		SD	4.064	4.065	4.373	4.606
		Number of station years	3	9	6	5
RM2: Intermediate	Dungarvon, NB	Mean	13.308	14.054	14.274	13.599
		Maximum	19.057	20.350	20.876	20.922
		Minimum	3.771	5.821	4.269	4.398
		SD	4.599	4.474	4.696	4.970
		Number of station years	3	10	6	4
RM3: Warm	Penobscot, ME	Mean	16.278	18.333	18.263	17.290
		Maximum	24.865	26.617	26.910	27.121
		Minimum	4.893	7.474	5.861	6.069
		SD	5.986	5.697	6.082	6.580
		Number of station years	6	6	6	5

ed to experience significantly higher mean weekly maximum stream temperatures for the year 2050 under the medium-high emissions scenario relative to the baseline 1980–2002 average (Figure 1; Table 5). In addition, all of the thermal regions are projected to experience significantly (RM1 $p < 0.05$; RM2 and RM3 $p < 0.001$) higher annual weekly maximum stream temperatures for the year 2050 under the medium-high emissions scenario relative to the baseline 1980–2002 average. In particular, sites within the warm RM3 region are projected to have annual maximum weekly stream temperatures warmer than the upper thermal threshold of 25°C suggested by Wallace (1993).

The consecutive numbers of weeks with mean weekly maximum temperatures exceeding 20°C are projected to increase under the projected future climate scenarios from an average baseline duration of 7.41 weeks rising to an average of 8.65 weeks under a low emissions scenario and 10.00 weeks under a medium-high emissions scenario (Table 5). However, spatial variations within thermal regions are apparent with statistically significant ($p < 0.01$) projected increases under the medium-high emissions scenario for sites within the intermediate RM2 region (Table 5). Sites within the cool RM1 region demonstrate mean weekly maximum stream temperatures throughout the analysis period (weeks 17–45) well below the 20°C thermal threshold under both temperature change scenarios. Sites grouped within intermediate RM2 region currently experience annual weekly maximum stream temperatures greater than the 20°C threshold, and the duration of these are projected to increase by an average of 1.83 weeks and 4.17 weeks under the B1 and A1B scenarios, respectively, relative to the

Table 4.—Interannual thermal regime classes [● = IA1 (cool thermal year); ■ = IA2 (intermediate thermal year—low variability); ✚ = IA3 (intermediate thermal year—high variability); and ▲ = IA4 (warm thermal year)]: matrix of spatial and temporal variability. For site codes, refer to Table 1.

Thermal region	Site	Year																						
		1980	1981	1982	1983	1984	1985	1986	1987	1988	1989	1990	1991	1992	1993	1994	1995	1996	1997	1998	1999	2000	2001	2002
RM1: Cool	EXP	■	✚	●	■	▲	▲	●	✚	■	■	▲	●	✚	✚	✚	■	■	●	■	✚	■	■	▲
	KED	▲	▲	■	■	■	■	●	✚	▲	■	✚	✚	●	▲	■	✚	■	●	✚	✚	■	■	▲
	LMR	▲	▲	■	■	■	■	●	✚	▲	■	✚	✚	●	▲	■	✚	■	●	✚	✚	■	■	▲
RM2: Intermediate	CON	●	✚	●	■	▲	▲	●	✚	✚	■	▲	●	●	▲	✚	■	■	■	✚	✚	■	■	▲
	DUN	▲	■	■	■	■	■	●	✚	▲	■	✚	✚	●	▲	■	✚	■	●	✚	✚	■	■	▲
	LSW	▲	■	■	■	■	■	●	✚	▲	■	✚	✚	●	▲	■	✚	■	●	✚	✚	■	■	▲
	MAC	▲	✚	■	■	■	■	■	✚	✚	✚	■	■	●	▲	✚	✚	■	●	✚	■	●	■	▲
	NWU	■	■	■	■	■	✚	●	✚	▲	●	▲	■	●	▲	■	✚	✚	●	✚	✚	■	■	▲
	TOB	▲	✚	■	■	■	■	■	✚	✚	✚	▲	■	●	▲	✚	✚	■	▲	✚	✚	●	■	▲
	UPS	■	■	■	■	■	✚	●	✚	▲	●	▲	■	●	▲	■	✚	✚	●	✚	✚	■	■	▲
RM3: Warm	AND	▲	▲	■	✚	✚	✚	●	▲	▲	■	■	✚	●	▲	■	✚	■	●	✚	✚	■	■	▲
	NAR	▲	■	■	✚	▲	●	●	✚	▲	●	●	✚	●	▲	■	✚	✚	▲	■	✚	■	■	▲
	PEN	▲	■	■	✚	▲	●	●	✚	▲	●	●	✚	●	▲	■	✚	✚	●	■	✚	■	■	▲
	PIS	▲	■	■	✚	▲	●	●	✚	▲	●	●	✚	●	▲	■	✚	✚	●	■	✚	■	■	▲
	SHE	▲	✚	■	✚	■	■	●	▲	▲	■	■	✚	●	▲	■	✚	■	●	✚	✚	■	■	▲
	STC	■	■	■	✚	▲	✚	●	✚	▲	■	■	✚	●	▲	▲	✚	●	▲	✚	✚	■	■	▲

Table 5.—Mean and range of mean weekly maximum stream temperatures (°C) and mean duration of weeks greater than 20°C for all sites and for sites within regional magnitude cluster (RM1–RM3) for (1) baseline conditions (1980–2002), (2) under low emissions B1 scenario (2046–2055), and (3) under medium-high emissions A1B scenario (2046–2055).

		Baseline (1980–2002)	B1 (2046–2050)	A1B (2046–2050)
All sites	Mean	15.4	15.96	16.84
(n = 17 sites)	Range	22.28	22.33	23.56
	Mean duration of weeks > 20°C	7.41	8.65	10
RM1: Cool				
(n = 3 sites)	Mean	11.3	11.88	12.51
	Range	13.47	13.96	15.64
	Mean duration of weeks > 20°C	0	0	0
RM2: Intermediate				
(n = 7 sites)	Mean	14.71	15.25	16.04
	Range	18.3	18.42	19.86
	Mean duration of weeks > 20°C	4.71	7	9.29
RM3: Warm				
(n = 7 sites)	Mean	17.86	18.42	19.49
	Range	19.78	19.93	20.59
	Mean duration of weeks > 20°C	13.29	14.00	15.00

baseline data. These sites are projected to experience temperatures greater than this threshold an average of 1.5 weeks (B1) and 2.3 weeks (A1B) earlier relative to the baseline period. Although the average weekly maximum stream temperatures for the analysis period (weeks 17–45) for sites within the warm RM3 region do not exceed 20°C, the duration of weekly maximum stream temperatures are projected to increase by 0.71 and 1.71 weeks for the two scenarios relative to the baseline data. These increases in duration are projected to occur an average of 0.14 weeks (B1) and 0.71 weeks (A1B) earlier relative to the baseline period. For sites that currently experience mean weekly maximum stream temperatures greater than 20°C (all of the warm RM3 region sites and six of the seven intermediate RM2 sites), they are projected to experience these temperatures earlier under the climate change scenarios with an average 0.8 weeks earlier under the B1 scenario and an average of 1.5 weeks earlier under the A1B scenario. This pattern breaks down by region with intermediate sites projected to experience more significant earlier timing of an average of 1.5 weeks earlier under the B1 and 2.33 weeks under the A1B scenario. Warm region sites are projected to 0.14 weeks under the B1 and 0.71 under the A1B scenario, possibly reflecting the current susceptibility of these sites to warmer temperatures.

Discussion

We modeled stream temperatures in 17 Atlantic salmon rivers of northeastern North America, and using existing global climate models for a low (B1) and medium-high (A1B) scenario, demonstrated that warming summer habitats are predicted for these rivers, although the magnitude of change varies between systems. Three thermal regions of cool (RM1), intermediate (RM2), and warm (RM3) were identified from the clustering of four variables quantifying the magnitude of the long-term (1980–2002) thermal regime for each river site. Examination of the between-year variability in the annual thermal regimes suggests that four significantly different thermal year types over the 1980–2002 period could be identified. These thermal years reflect an increasing magnitude and variability in the regime and can be characterized as cool (IA1), intermediate but with low variability (IA2), intermediate but with high variability (IA3), and warm (IA4). The identification of these different thermal year types suggests that these systems do not have stable thermal regimes and demonstrate natural variability between years. The natural variability between the regions may reflect their geographic location; for example, the spatial range in the intermediate RM2 sites could account for the greater level of variability between sites for an individual year. However, these site-specific factors were obviously overridden in years with highly dominant air temperature events. For example, 16 of the 17 sites exhibited an IA1 (cool) year type in 1992, which was one of the coolest years on record with a –0.9°C departure from the regional 1948–2007 mean for the Atlantic region (Meteorological Service of Canada 2007). The dominance of intermediate and warm thermal year types between 1998 and 2002 reflects warmer than average air temperature years in the Atlantic region relative to the 1948–2007 region (Meteorological Service of Canada 2007).

Our models of stream temperature based on air temperature were robust because we applied site-specific logistic models to incorporate local variations in solar insolation, riparian vegetation, groundwater inflow, and so forth. (Mohseni and Stefan 1999; Caissie 2006). The classification of sites demonstrated spatial and temporal patterns at regional (among-site) and local (within-site) scales. However, there are factors that can affect local thermal regimes that we cannot yet model (e.g., the hydrogeological setting of a river). Groundwater discharging to rivers can be localized but are also significant for survival in many populations of coldwater fishes (Curry et al. 1995; Power et al. 1999). Salmonids can significantly depend on groundwater discharge for cool water refugia during high temperature events (Breau et al. 2007). However, the availability and effectiveness of cold groundwater refugia is predicted to decrease as temperatures increase concurrently with mean global temperatures (Schindler 1997).

Temporal and spatial variability in river flow is similarly recognized to fundamentally control instream habitat and availability and ultimately riverine ecosystem biodiversity and integrity (e.g., Beitinger et al. 2000; Harris et al. 2000). The ecological significance of flow regime variability in structuring instream macroinvertebrate communities has been demonstrated at both a regional (Monk et al. 2006) and interannual (Monk et al. 2008) scale. Steel and Lange (2007) identified the significance of reservoirs and dams that significantly mute short temporal scale variations in stream temperatures. Many of the southern and warm RM3 region Atlantic salmon rivers have such impoundments (e.g., the Penobscot and Saint John rivers). Temperatures associated with impoundment releases will probably increase under a warmer environment (Sinokrot et al. 1995), and thus our predictions may be conservative in these rivers. All of these factors affecting the instream thermal habitats for salmon have yet to be explored and incorporated into our models of future changes under climate change scenarios. We need more analyses of each to move from predicting changes in the physical environment to predicting changes at the population scale where the real impacts will occur.

The duration and number of sites experiencing mean weekly maximum stream temperatures greater than the suggested 20°C threshold is projected to increase under the two climate change scenarios with regional differences in response between the identified thermal regions. Mean weekly maximum stream temperatures for sites within the cool (RM1) region are projected to increase but will not exceed this thermal threshold. However, the number of sites in the intermediate (RM2) and warm (RM3) experiencing warmer temperatures is projected to increase. Perhaps more significantly for the Atlantic salmon populations in these systems, the duration of the temperatures greater than the thermal threshold are projected to be extended and occur earlier in the season, particularly for intermediate (RM2) rivers. These predicted changes in summer thermal regimes would be hypothesized to impact the habitat availability, biology, and therefore survival of the salmon in these rivers.

Freshwater fishes are constrained by the range of temperatures available in the environment (e.g., increased stream temperatures will alter physiological functions affecting acclimatization, growth, and reproductive success; Fry 1971). With the projected increases in mean weekly maximum stream temperature, Atlantic salmon populations in this region may increase or decrease in abundance, experience range expansions or contractions, or face extinction. Projected increases in stream temperatures could result in a decrease of nearly 50% in the suitable cold- and coolwater fish habitat in North American streams (as defined by exceeding the maximum thermal tolerances of these thermal guilds) (Eaton and Scheller 1996). Salmonids are likely to experience range shifts under even conservative increases in stream temperatures because of the loss of suitable thermal habitats (Webb and Walsh 2004; Pederson and Sand-Jensen 2007). These species exist within narrow coldwater ranges and have limited abilities to adjust to changing biotic pressures (Van Der Kraak and Pankhurst 1997) or abiotic conditions (e.g., changes in stream temperatures and river flow). We would predict response for all life history stages (e.g., cumulative exposure to stream temperatures appears to control the timing of salmon smolt out-migration; Antonsson and Gurjonsson 2002; Zydlewski et al. 2005), as suggested in the River Imsa, Norway where smolt descent started earlier in years when the stream temperature was high or temperature increased rapidly (Jonsson and Ruud-Hansen 1985). Species may experience range contraction or extension to shift northwards to cooler thermal habitats. However, this assumes that species are able to adapt successfully and are not impeded by a boundary that would impede migration, such as a lack of viable migration routes or warmer waters that can isolate cool- and coldwater fish in headwater areas (Vander Zanden et al. 1999). Additional stress could occur with the invasion of nonnative species, such as the migration of southern warmwater species to regions currently occupied by cool- and coldwater fishes.

The uncertainties within temperature predictions are strongly dependent on the reliability of the global climate scenario predictions. While the development of air/stream temperature models for stream types are statistically reliable, predictions of future stream temperatures should always be viewed with caution. In examining the potential impacts of a changing climate on river systems and the potential implications for freshwater fishes, we need to separate the complicating effects of anthropogenic activities, such as land use changes, which can sometimes mask the true source of change in these systems (e.g., the reservoir effect). Future studies should focus on providing analyses of additional environmental and anthropogenic factors to better predict changes in the local river's physical environment in order to allow the prediction of changes at the population scale where the real impacts on Atlantic salmon and other fishes will occur.

Acknowledgments

We would like to thank C. Moffitt and an anonymous reviewer for their comments and suggestions that have improved the clarity and organization of this manuscript. The authors would like to acknowledge the opportunity to present this research at the ANACAT conference in Haifax, Nova Scotia in June 2007. We would like to thank C. Breau and G. Chaput for their comments on this research project. This research is data intensive and therefore we would like to acknowledge the National Oceanic and Atmospheric Administration (via the U.S. Historical Climatology Network) and Environment Canada for long-term air temperature records; U.S. Geological Survey, Newfoundland and Labrador Department of Environment, and sources (e.g., Department of Fisheries and Oceans Canada (DFO), Department of Natural Resources, and New Brunswick Department of Environment) within the New Brunswick Aquatic Data Warehouse (NBADW) for stream temperature records; NBADW (and sources within); and DFO Newfoundland and Labrador for adult salmon and smolt count/run data. Additional geographic information system layers were provided through the NBADW repository.

References

Alabaster, J. S. 1990. The temperature requirements of adult Atlantic salmon, *Salmo salar* L., during their upstream migration in the River Dee. Journal of Fish Biology 37:659–661.

Antonsson, T., and S. Gurjonsson. 2002. Variability in timing and characteristics of Atlantic salmon smolt in Icelandic rivers. Transactions of the American Fisheries Society 131:643–655.

Beitinger, T. L., W. A. Bennett, and R. W. McCauley. 2000. Temperature tolerances of North American freshwater fishes exposed to dynamic changes in temperature. Environmental Biology of Fishes 58:237–275.

Breau, C., R. A. Cunjak, and G. Bremset. 2007. Age-specific aggregation of wild juvenile Atlantic salmon *Salmo salar* at cool water sources during high temperature events. Journal of Fish Biology 71:1179–1191.

Caissie, D. 2006. The thermal regime of rivers: a review. Freshwater Biology 51:1389–1406.

Chatters, J. C., D. A. Neitzel, M. J. Scott, and S. A. Shankle. 1991. Potential impacts of global climate change on Pacific Northwest spring Chinook salmon (*Oncorhynchus tshawytscha*): an exploratory case study. Northwest Environmental Journal 7:71–92.

Curry, R. A., D. L. G. Noakes, and G. E. Morgan. 1995. Groundwater and incubation habitats, embryo survival, and alevin emergence of brook charr (*Salvelinus fontinalis*). Canadian Journal of Fisheries and Aquatic Sciences 52:1741–1749.

Dempson, J. B., M. F. O'Connell, and N. M. Cochrane. 2001. Potential impact of climate warming on recreational fishing opportunities for Atlantic salmon (*Salmo salar* L.) in Newfoundland, Canada. Fisheries Management and Ecology 8:69–82.

Eaton, J. G., and R. M. Scheller. 1996. Effects of climate warming on fish thermal habitat in streams of the United States. Limnology and Oceanography 41:1109–1115.

Environment Canada. 2008. Daily climate data. Available: www.climate.weatheroffice.ec.gc.ca/climateData/dailydata_e.html (January 2008).

Ficke, A. D., C. A. Myrick, and L. J. Hansen. 2007. Potential impacts of global climate change on freshwater fisheries. Reviews in Fish Biology and Fisheries 17:581–613.

Flato, G. M., and G. J. Boer. 2001. Warming asymmetry in climate change simulations. Geophysical Research Letters 28:195–198.

Fleming, I. A., and A. Jensen. 2002. Fisheries: effects of climate change on the life cycles of salmon. Pages 309–312 *in* I. Douglas, editor. Encyclopedia of global environmental change. Volume 3, causes and consequences of global environmental change. Wiley, Chichester, UK.

Friedland, K. 1998. Ocean climate influences on critical Atlantic salmon (*Salmo salar*) life history events. Canadian Journal of Fisheries and Aquatic Sciences 55(Supplement 1):119–130.

Fry, F. E. J. 1971. The effect of environmental factors on the physiology of fish. Pages 1–98 *in* W. S. Hoar and D. J. Randall, editors. Fish physiology. Academic Press, New York.

Hannah, D. M., B. P. G. Smith, A. M. Gurnell, and G. R. McGregor. 2000. An approach to hydrograph classification. Hydrological Processes 14:317–338.

Harris, N. M., A. M. Gurnell, D. M. Hannah, and G. E. Petts. 2000. Classification of river regimes: a context for hydroecology. Hydrological Processes 14:2831–2848.

Hayhoe, K., C. P. Wake, T. G. Huntington, L. Luo, M. D. Schwartz, J. Sheffield, E. Wood, B. Anderson, J. Bradbury, A. DeGaetano, T. J. Troy, and D. Wolfe. 2007. Past and future changes in climate and hydrological indicators in the U.S. Northeast. Climate Dyamics 28:381–407.

Hodgkins, G. A., R. W. Dudley, and T. G. Huntington. 2005. Changes in the number and timing of days of ice-affected flow on northern New England rivers, 1930–2000. Climate Change 71:319–340.

Jensen, A. J. 2003. Atlantic salmon (*Salmo salar*) in the regulated River Alta: effects of altered water temperature on parr growth. River Research and Applications 19:733–747.

Jensen, A. J., B. O. Johsen, and L. Saksgard. 1989. Temperature requirements in Atlantic salmon (*Salmo salar*), brown trout (*Salmo trutta*), and Arctic char (*Salvelinus alpinus*) from hatching to initial feeding compared with geographic distribution. Canadian Journal of Fisheries and Aquatic Sciences 46:786–789.

Jonsson, B., and J. Ruud-Hansen. 1985. Water temperature as the primary influence on timing of seaward migrations of Atlantic salmon (*Salmo salar*) smolts. Canadian Journal of Fisheries and Aquatic Sciences 42:593–595.

Juanes, F., S. Gephard., and K. F. Beland. 2004. Long-term changes in migration timing of adult Atlantic salmon (*Salmo salar*) at the southern edge of the species distribution. Canadian Journal of Fisheries and Aquatic Sciences 61:2392–2400.

MacCrimmon, H. R., and B. L. Gots. 1979. World distribution of Atlantic salmon, *Salmo salar*. Journal of the Fisheries Research Board of Canada 36:422–457.

Magnuson, J. J., K. E. Webster, R. A. Assel, C. J. Bowser, P. J. Dillon, J. G. Eaton, H. E . Evans, E. J. Fee, R. I. Hall, L. R. Mortsch, D. W. Schindler, and F. H. Quinn. 1997. Potential effects of climate changes on aquatic systems: Laurentian Great Lakes and Precambrian shield region. Hydrological Processes 11:825–871.

McCormick, S. D., R. A. Cunjak, B. Dempson, M. F. O'Dea, and J. B. Carey. 1999. Temperature-related loss of smolt characteristics in Atlantic salmon (*Salmo salar*) in the wild. Canadian Journal of Fisheries and Aquatic Sciences 56:1649–1658.

Meehl, G. A., T. F. Stocker, W. D. Collins, P. Friedlingstein, A. T. Gaye, J. M. Gregory, A. Kitoh, R. Knutti, J. M. Murphy, A. Noda, S. C. B. Raper, I. G. Watterson, A. J. Weaver, and Z.-C Zhao. 2007. Global climate projections. Pages 748–845 *in* S. Soloman, D. Qin, M. Manning, Z. Chen, M. Marquis, K. B. Averyt, M. Tignor, and H. L. Miller, editors. Climate change 2007: the physical science basis. Contribution of Working Group I to the Fourth

Assessment. Report of the Intergovernmental Panel on Climate Change. Cambridge University Press, Cambridge, UK.

Meteorological Service of Canada (2007). Climate trends and variations bulletin: regional temperature. Available: www.msc-smc.ec.gc.ca/ccrm/bulletin/regional_e.cfm (January 2008).

Mohseni, O., and H. G. Stefan. 1999. Stream temperature/air temperature relationship: a physical interaction. Journal of Hydrology 218:128–141.

Mohseni, O., T. R. Erickson, and H. G. Stefan. 2002. Upper bounds for stream temperatures in the contiguous United States. Journal of Environmental Engineering January:4–11.

Mohseni, O., H. G. Stefan, and J. G. Eaton. 2003. Global warming and potential changes in fish habitat in U.S. streams. Climatic Change 59:389–409.

Mohseni, O., H. G. Stefan, and D. L. Erickson. 1998. A non-linear regression model for weekly stream temperatures. Water Resources Research 34:2685–2692.

Monk, W. A., P. J., Wood, D. M. Hannah, and D. A. Wilson. 2008. Macroinvertebrate community response to inter-annual and regional river flow regime dynamics. Rivers Research and Application 24(7):988–1001.

Monk, W. A., P. J. Wood, D. M. Hannah, D. A. Wilson, C. A. Extence, and R. P. Chadd. 2006. Flow variability and macroinvertebrate community response within riverine systems in England and Wales. Rivers Research and Applications 22:595–615.

Moore, M. V., M. L. Pace, J. R. Mather, P. S. Murdoch, R. W. Howarth, C. L. Folt, C. Y. Chen, H. F. Hemond, P. A. Flebbe, and C. T. Driscoll. 1997. Potential effects of climate change on freshwater ecosystems of the New England/Mid-Atlantic region. Hydrological Processes 11:925–947.

Nakićenović, N., and R. Swart, editors. 2000: special report on emissions scenarios. A Special Report of Working Group III of the Intergovernmental Panel on Climate Change. Cambridge University Press, Cambridge, UK.

Nash, J. E., and J. V. Sutcliffe. 1970. River flow forecasting through conceptual models. Part I-A. Discussion of principles Journal of Hydrology 10:282–290.

New Brunswick Aquatic Data Warehouse. 2007. Daily stream temperature and fish count records. Available: http://nbwaters.unb.ca (June 2007).

Pederson, N. L., and K. Sand-Jensen. 2007. Temperature in lowland Danish streams: contemporary patterns, empirical models and future scenarios. Hydrological Processes 21:348–358.

Poole, G. C., and C. H. Berman. 2001. An ecological perspective on instream temperature: natural heat dynamics and mechanisms of human-caused thermal degradation. Environmental Management 27:787–802.

Power, G., R. S. Brown, and J. G. Imhof. 1999. Groundwater and fish: insights from northern North America. Hydrological Processes 13:401–422.

Schindler, D. W. 1997. Widespread effects of climatic warming on freshwater ecosystems in North America. Hydrological Processes 11:1043–1067.

Schmunk, R. B. 2006. Panoply. Version 2.1.4. NASA Goddard Institute for Space Studies, New York.

Scinocca, J. F., and N. A. McFarlane. 2004. The variability of modelled tropical precipitation. Journal of Atmospheric Sciences 61:1993–2015.

Sinokrot, B. A., H. G. Stefan, J. H. McCormick, and J. G. Eaton. 1995. Modeling of climate change effects on stream temperatures and fish habitats below dams and near groundwater inputs. Climatic Change 30:181–200.

SPSS Inc. 2006. SPSS for Windows. Version 15.0.0. SPSS Inc., Chicago.

Steel, E. A., and I. A. Lange. 2007. Using wavelet analysis to detect changes in water temperature regimes at multiple scales: effects of multi-purpose dams in the Willamette River basin. River Research and Applications 23:351–359.

Stefan, H. G., and E. B. Preud'homme. 1993. Stream temperature estimation from air temperature. Water Resources Bulletin 29:27–45.

Swansburg, E., G. Chaput, D. Moore, D. Caissie, and N. El-Jabi. 2002. Size variability of juvenile Atlantic salmon: links to environmental conditions. Journal of Fish Biology 61:661–683.

Swansburg, E., N. El-Jabi, D. Caissie, and G. Chaput. 2004. Hydrometeorological trends in the Miramichi river, Canada: implications for Atlantic salmon growth. North American Journal of Fisheries Management 24:561–576.

Torgersen, C. E., D. M. Price, H. W. Li, and B. A. McIntosh. 1999. Multiscale thermal refugia and stream habitat associations of Chinook salmon in northeastern Oregon. Ecological Applications 9:301–319.

USGS (U.S. Geological Survey). 2007. Water-quality for the nation. Available: http://waterdata.usgs.gov/nwis/qw (June 2007).

Van Der Kraak G., and N. W. Pankhurst. 1997. Temperature effects on the reproductive performance of fish. Pages 159–176 *in* C. M. Wood and D. G. McDonald, editors. Global warming: implications for freshwater and marine fish. University Press, Cambridge, UK.

Vander Zanden, M. J., J. M. Cassleman, and J. B. Rasmussen. 1999. Stable isotope evidence for the food web consequences of species invasions in lakes. Nature (London) 401:464–467.

Wallace J. 1993. Environmental considerations. Pages

127–143 *in* K. Heen, R. Monahan, and F. Utter, editors. Salmon aquaculture. Fishing News Books, Oxford, UK.

Webb, B. W., and F. Nobilis. 1994. Water temperature behaviour in the River Danube during the twentieth century. Hydrobiologia 291:105–113.

Webb, B. W., and A. J. Walsh. 2004. Changing UK river temperatures and their impact on fish populations. Pages 177–191 *in* B. Webb, M. Acreman, C. Maksimovic, H. Smithers, and C. Kirby, editors. Hydrology: science and practice for the 21st century, volume II (Proceedings of the British Hydrological Society International Conference, Imperial College, London, July 2004). British Hydrological Society, London.

Webb, B. W., P. D. Clack, and D. E. Walling. 2003. Water-air temperature relationships in a Devon river system and the role of flow. Hydrological Processes 17:3069–3084.

Williams, C. N., Jr., M. J. Menne, R. S. Vose, and D. R. Easterling. 2006. United States Historical Climatology Network daily temperature, precipitation, and snow data ORNL/CDIAC-118, NDP-070. U.S. Department of Energy, Oak Ridge National Laboratory, Carbon Dioxide Information Analysis Center, Oak Ridge, Tennessee. Available: http://cdiac.ornl.gov/epubs/ndp/ushcn/usa.html (January 2008).

Zydlewski, G. B., A. Haro, and S. D. McCormick. 2005. Evidence for cumulative temperature as an initiating and terminating factor in downstream migratory behavior of Atlantic salmon (*Salmo salar*) smolts. Canadian Journal of Fisheries and Aquatic Sciences 62:68–78.

American Fisheries Society Symposium 69:231–249, 2009

Review of Ocean-Atmospheric Factors in the Atlantic and Pacific Oceans Influencing Spawning and Recruitment of Anguillid Eels

MICHAEL J. MILLER* AND SHINGO KIMURA
Ocean Research Institute, The University of Tokyo,, Minamidai, Nakano, Tokyo 164-8639, Japan

KEVIN D. FRIEDLAND
National Marine Fisheries Service, 28 Tarzwell Drive, Narragansett, Rhode Island 02882, USA

BRIAN KNIGHTS[1]
University of Westminster, 115 New Cavendish Street, London W1W 6UW, UK

HEEYONG KIM[2]
Ocean Research Institute, The University of Tokyo, Minamidai, Nakano, Tokyo 164-8639, Japan

DONALD J. JELLYMAN
National Institute of Water and Atmosphere, Post Office Box 8602, Christchurch, New Zealand

KATSUMI TSUKAMOTO
Ocean Research Institute, The University of Tokyo, Minamidai, Nakano, Tokyo 164-8639, Japan

Abstract.—Declines in recruitment of temperate anguillid eels have occurred in the past 30 years in many areas of their species ranges. The cumulative effects of anthropogenic changes to their freshwater growth habitats are likely contributors to reductions in population sizes, but changes in ocean-atmospheric conditions in the ocean also appear to be contributing to the declines. This paper reviews how changes in the ocean may contribute to recruitment declines by affecting the spawning location of silver eels, larval feeding success, or the transport of their leptocephalus larvae by ocean currents. Recruitment of European eels *Anguilla anguillla* has shown correlations with the North Atlantic Oscillation and specific changes in ocean conditions in the Sargasso Sea where spawning and development occurs. The American eel *A. rostrata* spawns in an area that overlaps with the European eel and so could also be affected by these types of changes. Recruitment of Japanese eels *A. japonica* appears to be correlated to the El Niño Southern Oscillation index and latitudinal changes in salinity fronts in the western North Pacific. The general spawning and recruitment patterns of the temperate Australasian shortfin eels *A. australis* and New Zealand longfin eels *A. dieffenbachii* in the western South Pacific are similar to those of the northern temperate anguillids and also may be affected by El Niño-related factors. The changes in ocean conditions associated with atmospheric forcing or a warming of the ocean could alter the biological characteristics of the surface layer where leptocephali feed, due to changes in productivity or community structure, in addition to having possible effects on larval

* Corresponding author: miller@ori.u-tokyo.ac.jp
[1] Current address: King's College London, Norfolk Building, Surrey Street, London WC2R 2LS, UK
[2] Current address: National Fisheries Research and Development Institute, 408-1 Sirang-ri, Gijang-eup, Gjiang-kunm Busan, 619-705, Korea

transport and location of the spawning areas by silver eels. Changes in ocean-atmospheric conditions could result in lower feeding success and survival of leptocephali, or increased retention in offshore areas due to changes in the location of spawning areas, resulting in reductions in recruitment.

Introduction

The most important fisheries species of anguillid eels are those found at the higher latitudes of the North Atlantic and western Pacific, and major declines in recruitment particularly at the margins of the species ranges have raised serious concerns among scientists (Dekker et al. 2003). These temperate species include the European eel *Anguilla anguilla* and American eel *A. rostrata* in the North Atlantic, the Japanese eel *A. japonica* in the western North Pacific, and the Australasian shortfin eel *A. australis* and the New Zealand longfin eel *A. dieffenbachii* in the western South Pacific. Concern about these species has increased recently because recruitment of the northern temperate species has declined markedly from the most recent peak levels in the late 1970s to early 1980s (Figure 1; Dekker 1998; Haro et al. 2000; McCleave 2001; Casselman 2003; Tatsukawa 2003). There also are some indications that the two southern temperate anguillid species that recruit to New Zealand may be showing signs of declines (Jellyman 2007, 2009; Jellyman et al. 2000). However, because the catadromous life histories of anguillid eels are so complex and difficult to monitor during their oceanic stages, the specific causes of these declines have been difficult to determine.

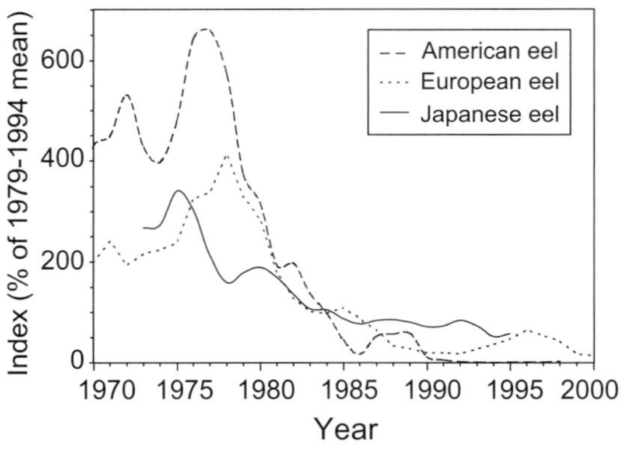

Figure 1.—Trends in recruitment indices of northern temperate anguillid species (% of 1979–1994 means, 3-year running averages) for the American eel in the St. Laurence River in Canada (Moses-Saunders Dam trap, assuming age of 6+ of yellow eels at time of trapping in the 1970–1980s), the European eel in the Netherlands (Den Oever glass eel fishery-independent surveys), and the Japanese eel in Japan (total commercial catches of glass eels). Data are from Casselman (2003), ICES (2006, 2007), and Kimura (2003).

The amount of freshwater and estuarine growth habitats available for the production of silver eels has been reduced on decadal scales by anthropogenic factors such as land reclamation, construction of dams, and pollution throughout much of their ranges (Haro et al. 2000; Feunteun 2002; Knights 2003; Dekker 2004). Other anthropogenic factors in addition to fishing mortality that have been proposed as influencing recruitment include parasite introductions and toxins in the body of silver eels that could cause physiological disruptions during their migration or spawning (e.g., Robinet and Feunteun 2002; Kirk 2003; Palstra et al. 2007). Recent research also has suggested that virus infections may reduce the ability of silver eels to reach their spawning areas (van Ginneken et al. 2005). The effects of such impacts on entire populations have not been determined, but scientists and management agencies have expressed concerns (e.g., ICES 2001, 2006, 2007; Dekker et al. 2003).

In addition to the effects of anthropogenic factors, it has been hypothesized that changes in ocean-climate conditions could have had important impacts on the northern temperate species (Castonguay et al. 1994; Kimura et al. 2001; Knights 2003; Friedland et al. 2007; Bonhommeau et al. 2008a, 2008b; Kettle et al. 2008). Even for the European eels though, only a few studies have clarified exactly how oceanic factors might affect larval migrations or their feeding success (Friedland et al. 2007; Bonhommeau et al. 2008a, 2008b). Because silver eels have never been seen in their spawning areas, it is not known if the factors leading to reduced recruitment are due to fewer silver eels migrating to spawn each year, a reduced survival of these migrants, changes in oceanic conditions affecting spawning success, or poor larval feeding, transport, or survival.

In this paper, we review the possible ways that changes in ocean-atmospheric conditions could affect the recruitment of temperate anguillid eels. To help evaluate these potential factors, the literature on the spawning areas and larval distributions of temperate anguillid eels and their relationship to patterns of ocean-atmospheric conditions in the western Pacific and North Atlantic oceans are overviewed. There is increasing understanding of how climatic changes and ocean regime shifts are affecting physical and biological conditions in regions of the world's oceans (Weijerman et al. 2005; Drinkwater 2006; Lehodey et al. 2006; Beamish et al. 2009, this volume), so the objective of this review is to examine the possible relationships between specific life history events of anguillid eels and changes in ocean conditions.

In the North Atlantic, changes in ocean-atmospheric factors have been found to occur in association with the North Atlantic Oscillation (NAO; Hurrell 1995; Hurrell and Dickson 2004) that have been linked to changes in physical (Curry and McCartney 2001) and biological (Hays et al. 2005; Lehodey et al. 2006; Stige et al. 2006) conditions. In the Pacific, the El Niño Southern Oscillation (ENSO) is known to affect sea surface temperatures and ocean current patterns in the eastern Pacific and for having major influences on climatic conditions worldwide (McPhaden 1999). These effects of ENSO have major effects on the equatorial Pacific and the biological characteristics of the eastern Pacific (Sugimoto et al. 2001; Lehodey et al. 2006). Another longer-term regime shift in ocean-atmospheric factors in the Pacific Ocean has been termed the Pacific Decadal Oscillation (PDO), which has major affects on the ecosystem of the northeastern Pacific (Hare and Mantua 2000; Lehodey et al. 2006; Beamish et al. 2009). In addition to these types of regime shifts, there is also evidence that there has been a gradual warming of the world's oceans (Levitus et al. 2000) that may be affecting productivity in the surface layer (Behrenfeld et al. 2006) where leptocephali feed.

One obvious way that various types of changes in oceanic conditions could be affecting anguillid eels is by reducing their feeding success during their long larval migrations from their offshore spawning areas to their freshwater and estuarine growth habitats where they recruit. However, evaluating the possible effects of oceanic factors on the feeding of leptocephali is particularly complicated because their morphology and physiology are very different from other fish larvae. Leptocephali appear to feed on marine snow or other particulate organic matter such as discarded larvacean houses or even dissolved organic matter (Otake et al. 1993; Mochioka and Iwamizu 1996), which are not easily measured. The availability of marine snow and other types of particulate matter in the surface layer where leptocephali live is thought to depend on primary and secondary production and is coupled to phytoplankton blooms (Turner 2002). In the open ocean, primary production is controlled by, among other factors, nutrient availability and surface mixing dynamics (Palter et al. 2005), thus providing possible linkages between ocean-atmospheric changes and the feeding success of leptocephali.

In the North Atlantic, the possibility that ocean-atmospheric changes could affect anguillid eels led to hypotheses that changes associated with the NAO may be affecting the recruitment of the European eel and possibly the American eel (Castonguay et al. 1994; ICES 2001; Knights 2003; Friedland et al. 2007; Kettle et al. 2008). Both species spawn at the southern edge of the Gulf Stream recirculation gyre in the Sargasso Sea (Figure 2), but the American eel only needs to move west and across the Gulf Stream, whereas the European eel must move eastward all the way across the North Atlantic basin. The possibility that the long migration of the European eel could be influenced by ocean-atmospheric changes has been supported by correlations between glass eel recruitment data and the NAO index and other ocean-atmospheric parameters in the Sargasso Sea (Knights 2003; Friedland et al. 2007; Bonhommeau et al. 2008b).

The Japanese eel spawns in a similar westward flowing current in the western North Pacific (Tsukamoto 1992, 2006), and ocean-atmospheric changes also have been hypothesized to affect its recruitment (Kimura et al. 2001). This species is distributed in East Asia from Taiwan to Japan and spawning occurs in the North Equatorial Current, which bifurcates into both a northward flow that transports larvae to their recruitment areas, and a southward flow to areas outside of its range (Kimura et al. 1994). This transport situation is also complicated by north–south movements of the salinity

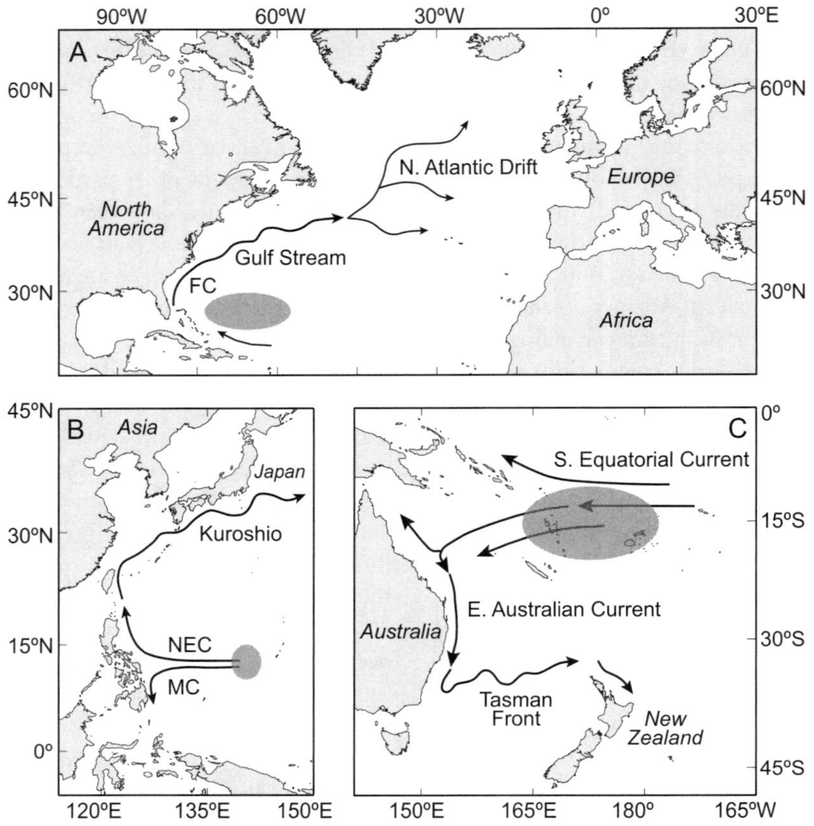

Figure 2.—The spawning areas of the northern temperate species of anguillid eels in the subtropical gyres of the western North Atlantic showing the Florida Current (FC), Gulf Stream, and North Atlantic Drift in panel (A) and in the western North Pacific showing the North Equatorial Current (NEC), Mindanao Current (MC), and Kuroshio Current in panel (B). The region where spawning likely occurs by the southern temperate anguillid species in the western South Pacific is also shown along with the South Equatorial Current and the East Australian Current in panel (C).

front typically present at the northern margin of the spawning area. A southward movement of the front during El Niño could contribute to larvae entering the southward flow (Kimura and Tsukamoto 2006), and correlations between the El Niño Southern Oscillation index (SOI) and recruitment to southern Japan have been found (Kimura et al. 2001).

Changes in the equatorial Pacific from ENSO also could have an effect on the anguillid spawning and larval development areas in the western South Pacific where there are two species of temperate anguillid eels that recruit to eastern Australia and throughout New Zealand (Jellyman 2003). Based on the presence of their leptocephali in the region (Jellyman 1987, 2003; Aoyama et al. 1999), it is probable that Australasian shortfin eels spawn somewhere in the westward flowing South Equatorial Current. Their leptocephali presumably move westward and then southward in the East Australian Current where some move westward and recruit to eastern Australia. Other larvae recruit to New Zealand, although the actual migration route has not been determined. Another species, the New Zealand longfin eel, also recruits to New Zealand, but its leptocephali have never been distinguished so its spawning area remains a mystery (Jellyman and Tsukamoto 2002). Teleconnections of ENSO reach into the western South Pacific (Gouriou and Delcroix 2002), so the spawning area and larval transport of these South Pacific temperate eels could be affected by ocean-atmospheric factors.

In the following sections, each species of temperate anguillid eel and their relationship to patterns of ocean-atmospheric conditions in the three subtropical gyres of the western Pacific and North Atlantic oceans are examined. The objective of this review is to provide a general perspective on how temperate eel populations may be responding to ocean-atmospheric changes that can help to facilitate future research and efforts to manage and conserve these important catadromous fishes.

Atlantic Eels

Spawning Area and Oceanic Conditions

The spawning area of the anguillid eels in the Sargasso Sea was discovered by the Danish fisheries scientist Johannes Schmidt in the early part of the 20th century after collections of leptocephali were made in many regions of the Atlantic and Mediterranean Sea (Schmidt 1922). The smallest leptocephali of both species were found in the southern Sargasso Sea, and larvae smaller than 10 mm were later shown to be widely distributed in overlapping areas south of about 30°N (Figure 2), with the European eel appearing to spawn slightly to the east of the American eel (McCleave et al. 1987). The Sargasso Sea is the western part of the North Atlantic subtropical gyre that has the northward flowing Florida Current to the west and its continuation as the Gulf Stream to the north (Schmitz and McCartney 1993). The anguillid eels spawn in the southern Sargasso Sea, where there is westward current flow that transports their leptocephali towards the Florida Current.

The general location of the main spawning area of the Atlantic eels has been determined from small recently hatched leptocephali 4–7 mm total length (TL) that were collected in several surveys over a wide region (Schoth and Tesch 1982; Boëtius and Harding 1985; McCleave et al. 1987; Kleckner and McCleave 1988). It has been shown that spawning was occurring south of distinct temperature fronts that are consistently present during the spawning season in late winter and early spring (Kleckner and McCleave 1988). These fronts were also found to cause discontinuities in the assemblages of leptocephali of various families, and the associated frontal jets likely transport some leptocephali eastward (Miller and McCleave 1994).

After the spawning season, the leptocephali of both anguillid species appear to become widely distributed as they are transported by the currents and eddies in the region (Boëtius and Harding 1985; Kleckner and McCleave 1985; McCleave and Kleckner 1987; Kettle and Haines 2006). Various sizes were present in the summer and early fall across the southern Sargasso, and leptocephali of both species have been documented in the Florida Current (Kleckner and McCleave 1982; McCleave and Kleckner 1987). One of the more direct transport pathways out of the southern Sargasso Sea may be the flow of the Antilles Current off the northeastern edge of the northern Bahamas. American eel leptocephali must cross the Florida Current and Gulf Stream to recruit to the east coast of North America, but those of the European eel must continue being transported to the east by the Gulf Stream and North Atlantic Drift. The actual pathway and transport time of European eel leptocephali remains poorly understood, however, due to the limited spatial and temporal coverage of sampling surveys across the eastern part of the basin (McCleave et al. 1998).

Studies of leptocephalus distribution and modeling of drift patterns suggest that most European eel leptocephali probably use the Gulf Stream–North Atlantic Drift system and, possibly, the Azores Current after moving west through the southern Sargasso (McCleave and Kleckner 1987; Kettle and Haines 2006). However, direct eastward or northeastward movement of some leptocephali appears to be a possible second route due to flows associated with the frontal jets that form in the subtropical convergence zone each year (Miller and McCleave 1994; Ullman et al. 2007). Regardless of how they achieve it or how long it takes, the migration of the European eel is probably the longest of any anguillid species, making its recruitment especially vulnerable to oceanic changes.

Possible Oceanic Factors Affecting Recruitment

Oceanic changes associated with shifts in the NAO in the North Atlantic have been hypothesized to contribute to changes in Atlantic eel recruitment due the variety of physical and biological factors affected by this atmospheric phenomenon. The NAO winter index (Hurrell 1995) is defined as the difference in sea level pressure between Iceland and Portugal in the North Atlantic. Correlations between the NAO and the index of catches of glass eels at Den Oever in the Netherlands (DOI), which is based on standardized fishery-independent sampling (Dekker 1998, 2002), led to the hypothesis that changes in the ocean related to the NAO may be influencing recruitment fluctuations (ICES 2001; Knights 2003). Westerberg (ICES 2001), Knights (2003), and Friedland et al. (2007) have demonstrated significant inverse correlations between the NAO and DOI. However, the strength of these correlations has been questioned by Dekker (2004) because trends since the 1990s in the NAO towards zero have not resulted in clear increases in the DOI. It appears that the effects of global warming that have occurred since the 1980s, reflected in a steep and continual increase in the northern hemisphere temperature and Sargasso Sea surface temperature anomalies, have overridden the effects of the NAO (B. Knights and S. Bonhommeau, Agrocampus Rennes, Rennes, France, unpublished analysis). However, another more recent study has also found correlations between the NAO and catch data of both glass eels and adults of the European eel from many different parts of its range (see Kettle et al. 2008).

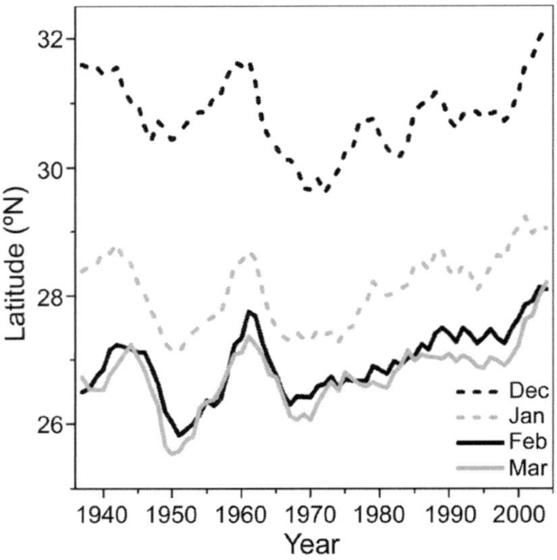

Figure 3.—Changes in the latitude of the 22.5°C isotherm in the Sargasso Sea in the 4 months just before and during the early spawning season of the Atlantic eels (adapted from Friedland et al. 2007). This isotherm is associated with the temperature front that forms the northern boundary of the spawning area of both the American and European eels.

To clarify possible cause–effect relationships between oceanic changes and European eel recruitment, Friedland et al. (2007) examined the correlations between several specific ocean-climate parameters in the Sargasso Sea with the DOI. Because temperature changes have occurred in the surface layer of the North Atlantic and southern Sargasso Sea spawning area (Grey et al. 2000; Polyakov et al. 2005), an isotherm-latitude analysis was performed. This showed that the latitude of the 22.5°C isotherm in January and February has moved from as far south as 26–27°N in the 1950s to as far north as 28–29°N in recent years (Figure 3). Peak negative correlations with the DOI were found in January and February in an area centered at 60°W, which is the longitude of the eastern region of the European eel spawning area (Friedland et al. 2007). The 22.5°C isotherm was used because it is linked to the northernmost temperature front that appears to form at the northern boundary of anguillid spawning in the Sargasso Sea (Kleckner and McCleave 1988). This front has been observed to form at the same water density in all previous studies in the region prior to 1994 (Kleckner and McCleave 1988; Miller and McCleave 1994), and migrating silver eels may use this front to identify the spawning area. Changes in the latitude of these fronts could affect both the spawning location and early larval transport towards the Florida Current because currents in this region are likely complex (Reverdin et al. 2003). Length–frequency data for American eel leptocephali in the summer and fall suggest that some leptocephali are retained offshore to the east of the Florida Current (McCleave and Kleckner 1987). Therefore, a northward movement of the spawning area could affect the larval transport out of the Sargasso Sea if the strongest westward flow is in the southern areas, such as off the northeast edge of the northern Bahamas where leptocephali of both species were abundant (McCleave and Kleckner 1987).

Friedland et al. (2007) also identified potentially important changes in wind-driven factors and mixed layer depth (MLD) in the Sargasso Sea that could affect larval transport or feeding success. Spring wind speeds and southward Ekman transports between 1949 and 2003 were found to have decreased in the northern Sargasso Sea region, possibly contributing to increased larval retention in the Sargasso gyre.

Ekman depths were about 30–50 m during the first three decades of the time series but have shallowed to 10–20 m in recent years. Spring MLD in the northern Sargasso Sea has also shallowed during February to May since 1985 and for the spring months of March, April, and May during the peak spawning season. This decrease in MLD was observed to have been about 50–100 m, using two different methods of calculation (Figure 4). Because wind stress, eddies, and MLD can affect nutrient levels and primary productivity in the surface layer (Palter et al. 2005), the changes observed could affect the feeding success of both species of Atlantic anguillid leptocephali, as hypothesized by Knights (2003).

The hypothesis that recruitment fluctuations of the European eel may be related to feeding success was supported recently by negative correlations between ocean surface layer temperature (0–100 m) in the northern Sargasso Sea and recruitment to Europe (Bonhommeau et al. 2008b). The ocean temperature in the northern Sargasso Sea has increased since about 1970, and an apparent regime shift in sea surface temperature corresponded to sharp decreases in recruitment at seven different sites in Europe from Spain to Germany (Bonhommeau et al. 2008b). Increases in temperature cause lower productivity (Behrenfeld et al. 2006), so correlations between recruitment and ocean temperature may be reflecting reduced feeding success of leptocephali. Direct measurements of primary productivity in the northern Sargasso Sea were also found to be correlated with a 3-year lag to the Loire River recruitment time series in France (Figure 5), but not those at the other locations (Bonhommeau et al. 2008b). A similar, but more recent study has also found correlations between primary productivity in the northern Sargasso Sea and recruitment of both European and American eels over short periods of 11 or 13 years, at time lags of 2.5 and 1.5 years, for the European and American eels, respectively (see Bonhommeau et al. 2008a). Changes in ocean productivity may also be associated with changes in the length and condition of glass eels recruiting to Europe (Désaunay and Guérault 1997; Dekker 1998, 2004).

Changes in the North Atlantic gyres could also affect recruitment success. For example, fluctuations in the strength or location of the Gulf Stream have been proposed as possible factors affecting larval transport (Castonguay et al. 1994). Curry and McCartney (2001) showed that Gulf Stream transport increased during the positive phase of the NAO in recent decades compared to the negative phase that ended in the early 1970s. This resulted in a strength-

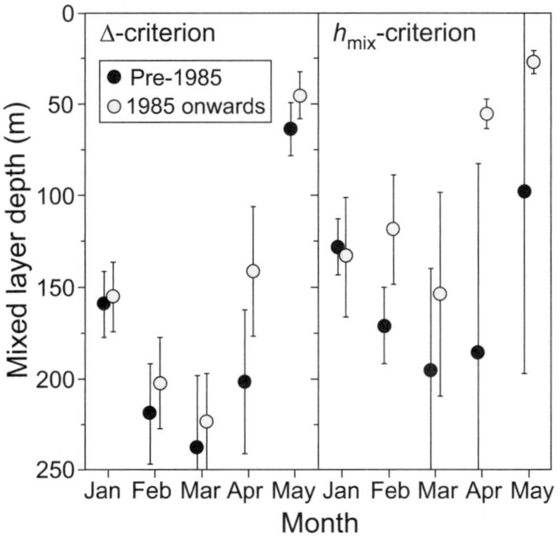

Figure 4.—Changes in the mixed layer depth (MLD) in the northern Sargasso Sea near Bermuda in the 4 months during the spawning season measured using two calculation techniques (adapted from Friedland et al. 2007).

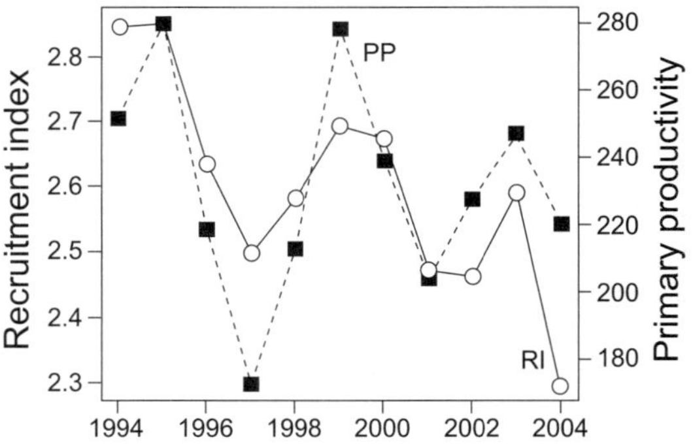

Figure 5.—Time-series of a log-transformed recruitment index (RI) of glass eels of the European eel in the Loire River, France, from the 1994 to 2004 recruitment seasons (solid line with circles), and 3-years lagged integrated primary production (PP) in the upper 140 m (mg C/m^2; dashed line with squares) in the northern Sargasso Sea near Bermuda from 1991 to 2001 (adapted from Bonhommeau et al. 2008b).

ening of the subtropical gyre with increased transport, a weakening of the subpolar gyre, and a slight northward shift in the location of the Gulf Stream. Although no correlation was detected between the transport index of Curry and McCartney (2001) and recruitment of glass eels to Europe (Bonhommeau et al. 2008b), this type of change could affect the proportional transport of leptocephali into the northern and southern branches of the Gulf Stream flow or the rates of transport towards Europe.

The NAO also appears to be associated with changes in patterns of the northern branch of the Gulf Stream that enters the subpolar gyre, the North Atlantic Drift, as it reaches the northeast Atlantic (Bersch 2002). This could affect leptocephalus transport to the northern part of the range of the European eel. Alternatively, the southern branch flows into the Azores Current, which also has shown apparent changes related to the NAO, such as a slight southward shift and apparent increased recirculation south or west back into the Gulf Stream (Reverdin et al. 2003; Siedler et al. 2005). Therefore, although strengthening of the Gulf Stream might be expected to facilitate successful recruitment to Europe, this increased flow may have caused a strengthening and expansion of Gulf Stream recirculation in recent decades (Curry and McCartney 2001). This could have resulted in increased retention of leptocephali in the western half of the gyre.

The various trends mentioned above may reflect changes acting not only in the Sargasso Sea, but also later in the larval migration phase. Significant regime shifts in the physical and biological structure of the northeast Atlantic and North Sea have been observed, affecting a wide range of marine organisms from zooplankton to fishes (Richardson and Schoeman 2004; Hays et al. 2005; Weijerman et al. 2005; Lehodey et al. 2006; Stige et al. 2006). More local factors could be acting on nonfeeding eels as they cross the continental shelf after metamorphosis from the leptocephalus stage. For example, although Atlantic inflow to the North Sea may have increased in association with changes in the NAO, wind-driven currents in the North Sea and inflows of seawater through the Skagerrak-Kattegat straits could have been less favorable for transporting glass and yellow eel into the Baltic Sea (Westerberg 1998; Knights 2003 and unpublished results).

The Japanese Eel

Spawning Area and Oceanic Conditions

The Japanese eel spawns in the westward flowing North Equatorial Current region of the western North Pacific (Tsukamoto 1992), so this species has a general spawning and larval recruitment pattern very similar to that of the American eel in the Atlantic. The North Pacific is also a large subtropical gyre, with the powerful Kuroshio Current being the western boundary current and recirculation occurring in the western part of the gyre. It differs in that the southern border of the Philippine Sea is much further south of the spawning area than it is in the Sargasso Sea, which has the Bahamas and the Greater Antilles along its southern margin (Figure 2). This is important because although the westward flow of the North Equatorial Current is much stronger (Reverdin et al. 1994) than the flow in the southern Sargasso Sea, the North Equatorial Current bifurcates into northward and southward flows at a latitude that can change both seasonally and annually (Kim et al. 2004).

The spawning area in the North Equatorial Current may typically be located from about 13–15°N to the west of seamounts of the West Mariana Ridge at about 143°E because no leptocephali have been collected further east (Tsukamoto et al. 2003; Tsukamoto 2006). Small 4–5-mm larvae have been collected recently in an area just west of one of these seamounts. All larval records in the past 15 years indicate a limited spawning area (Tsukamoto 2006) just south of a salinity front in the region (Tsukamoto 1992; Kimura and Tsukamoto 2006). To the north of this front, there are many eddies and alternating east and west flows with several countercurrents apparently present (Kaneko et al. 1998; Kobashi and Kawamura 2001), so that region is not good for the transport of leptocephali.

Because of the poor transport conditions to the north and the bifurcation of the North Equatorial Current at the western margin of the North Pacific, it appears that the Japanese eel spawning area has ended up being located in a much more precise area than the spawning area of the Atlantic eels (Tsukamoto 2006). Japanese eel leptocephali must avoid becoming trapped in eddies and avoid being entrained from the North Equatorial Current bifurcation zone into the southward transport of the Mindanao Current (Figure 2). This would carry their leptocephali into the Celebes Sea or into the North Equatorial Countercurrent that flows back to the east (Kimura et al. 1994, 2001). The Japanese eel is not distributed in northern Indonesia or the southern Philippines, although there are several other tropical anguillid species living there.

Behavioral mechanisms used by leptocephali, such as diel vertical migration into the Ekman layer, have been proposed as possibly facilitating entrainment into the northward flowing branch of the North Equatorial Current that enters the Kuroshio (Kimura et al. 1994). Once in the northward flow, some larvae may be transported into the South China Sea to the south of Taiwan to recruit to the southern parts of their species range, and others are transported north into the Kuroshio in the East China Sea for eventual recruitment to Korea and Japan at the northern extent of their range (Kimura et al. 1999).

Possible Oceanic Factors Affecting Recruitment

Although the basic geography of spawning and recruitment of the Japanese eel is similar to that of the American eel in the western North Atlantic, there are several different factors that also appear to make the Japanese eel vulnerable to ocean-atmospheric changes. The Japanese eel has a much narrower pathway leading to successful recruitment because entrainment into the southward flow of the Mindanao Current or into a large persistent eddy east of Taiwan near the beginning of the Kuroshio are two possible factors that could reduce recruitment (Figure 2). Kimura et al. (2001) presented evidence that changes in the location of the salinity front typically found at the northern margin of the spawning area could influence the effect of these factors. One possibility is that a steady northward movement of this front in recent decades could have caused more leptocephali to be retained offshore in the large eddy east of Taiwan (Kimura et al. 2001), and consequently contributed to a decline in recruitment since the 1970s.

Interannual recruitment fluctuations, however, may be more related to the southward movement of the salinity front in some years, which appeared to correlate to fluctuations of glass eel catches in southern Japan (Kimura et al. 2001). The movement of the front occurs primarily in association with El Niño events (Kimura et al. 2001; Figure 6), which affects the amount of rainfall in the region and therefore the location of the salinity front. The southward movement of the salinity front and presumably its associated eastward countercurrent could contribute to lower recruitment success in several ways. If salinity is an important cue used by silver eels to locate the spawning area, this could result in spawning further south than normal (Kimura et al. 2001; Kimura and Tsukamoto 2006). This would likely result in more leptocephali drifting southward in the Mindanao Current. Alternatively, it could disrupt the westward flow at the appropriate latitude for spawning and reduce the success of entrainment into the Kuroshio due to increased entrainment into eddies or countercurrents. Recent modeling research indicates that a southward displacement of the spawning area from 15°N (Figure 7, top panels) during El Niño events to 13.8°N and 12.4°N (Figure 7, bottom panels) may cause a much higher percentage of leptocephali to enter the Mindanao Current (see Kim et al. 2007).

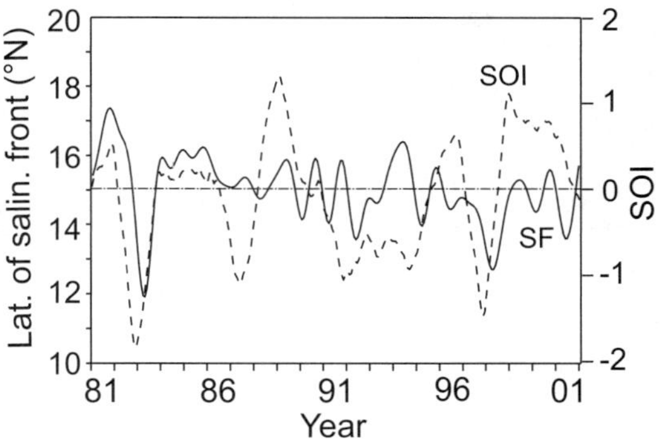

Figure 6.—Plots of the latitude of the salinity front (SF) along 137°E in the western North Pacific and of the Southern Oscillation index (SOI) from 1981 to 2001.

Support for the hypothesis that a southward movement of the salinity front could affect the location of the Japanese eel spawning area was seen in 2002. The salinity front was further south than usual at about 12°N in 2002 (Kimura and Tsukamoto 2006), which was a mild El Niño year (Lagerloef et al. 2003) compared to the 1997–1998 El Niño (McPhaden 1999). The 1997–1998 El Niño appeared to cause disruptions in the westward flow in the spawn-

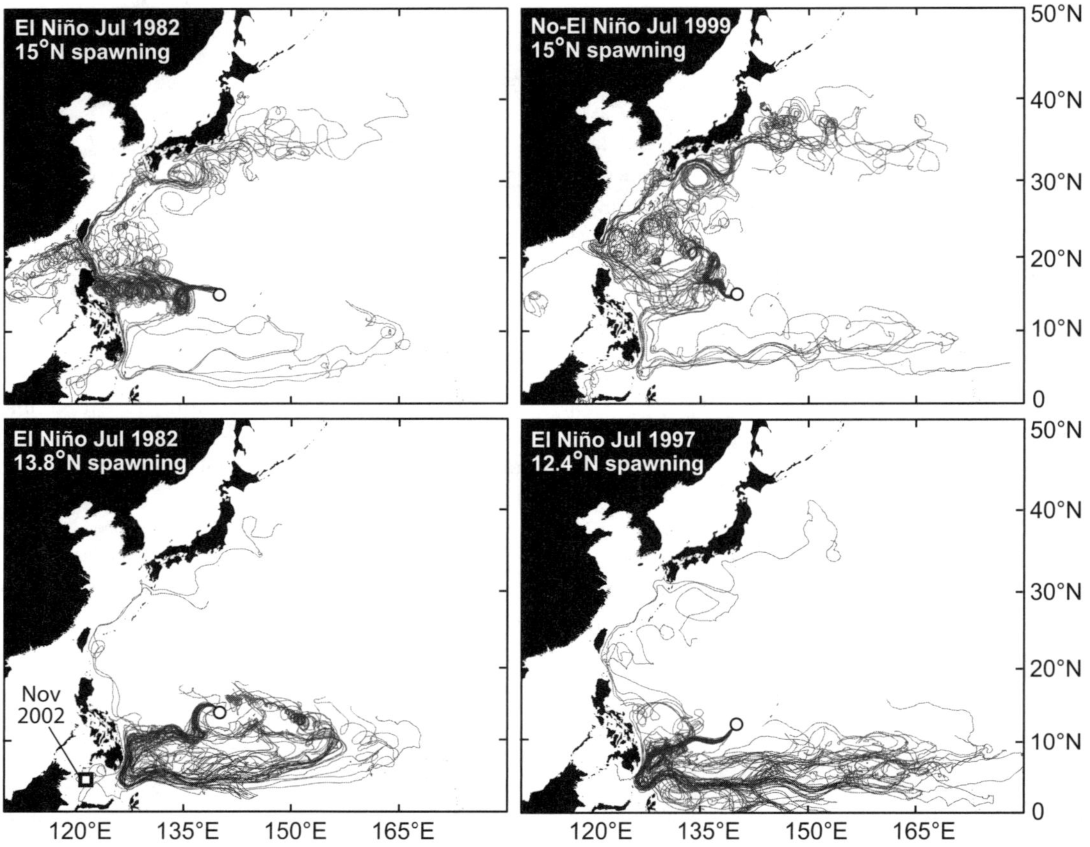

Figure 7.—Plots of modeled trajectories of particles/leptocephali of the Japanese eel in the western North Pacific after being released/spawned at different spawning locations (white circles) in July of various years and latitudes, which show abnormal larval transport resulting from spawning at the more southern locations. Top panels show trajectories when spawning occurs near the northern margin of the spawning area (i.e., salinity front at normal location) during El Niño and normal years. Bottom panels show trajectories when the spawning area was shifted to the south (i.e., salinity front further south than usual) during El Niño conditions (see Kim et al. 2007 for details about modeling procedures). The location where a 42-mm Japanese eel leptocephalus was collected in November in the Celebes Sea outside of the normal species range during the mild El Niño year of 2002 is shown with the white square in the bottom left.

ing area in 1998 (Tsukamoto et al. 2003), but the southward movement of the salinity front in 2002 may have caused an increase in entrainment of leptocephali into the Mindanao Current. Direct evidence of entrainment into the southward flow of the Mindanao Current came later in 2002, when a 42.8-mm Japanese eel leptocephalus was collected at 4.9°N, 120.9°E in the Celebes Sea on November 19 (Figure 7). The specimen was identified genetically, and its otolith microstructure indicated it had likely hatched on about July 22, 2002 (A. Shinoda, Ocean Research Institute, University of Tokyo, personal communication). The collection of 8–11-mm leptocephali just south of the front in 2002 at a lower latitude than in non-El Niño years (Kimura and Tsukamoto 2006), and the observation that drifters released at that time at 12°N and 10°N moved south into the Mindanao Current, suggested that the leptocephali located south of the front had a high probability of being entrained into the Mindanao Current (Kimura and Tsukamoto 2006). At least part of the flow of this current typically enters the Celebes Sea (Lukas et al. 1991), so leptocephali were likely transported from the southern region of the North Equatorial Current (NEC) into the Mindanao Current and then into the Celebes Sea. The collection of a Japanese eel leptocephalus at this location was the first confirmation of the transport of a larva outside of the range of this species and represents clear evidence of the potential for the southward movement of the salinity front to disrupt larval transport of this species.

In addition, it is also possible that changes in the productivity or community structure of the surface layer of the western North Pacific have occurred that affect the survival or growth of Japanese eel leptocephali, as has been hypothesized for the Atlantic eels. Major physical and biological changes have occurred in the northwestern and northeastern Pacific after the PDO shifted state in 1977 (Hare and Mantua 2000; Sugimoto et al. 2001; Lehodey et al. 2006). However, fluctuating but decreasing trends in MLD, nutrients, chlorophyll *a*, and net community production concentrations also have been detected in the region just to the west of the spawning area along 137°E between 1971 and 2000 (Sugimoto and Tadokoro 1998; Watanabe et al. 2005). Bonhommeau et al. (2008a) also showed evidence of increasing sea surface temperatures in this region, and found a correlation with temperature fluctuations and glass eel recruitment to Japan (Tatsukawa 2003) using a 0.5-year tag. These changes occurred when the Japanese eel showed decreasing recruitment and at latitudes where leptocephali are located, so their early larval migration feeding success may have been affected. In comparison to the changes in the more northern areas of the North Pacific, the impact of the PDO on the conditions in the NEC spawning and larval development areas of the Japanese eel appear to be much less. However, the frequency of El Niño events appear to have increased since the PDO regime shift in the mid-1970s (Kimura and Tsukamoto 2006), which may have had a substantial influence on interannual recruitment fluctuation in the Japanese eel in the past three decades. The effects of regime shifts can interact differently with diadromous species with contrasting life histories, however, because catches of pink salmon *Oncorhynchus gorbuscha* and chum salmon *O. keta* in the North Pacific have increased since the late 1970s (Beamish et al. 2009).

South Pacific Eels

Spawning Area and Oceanic Conditions

The spawning areas of the two species of temperate anguillid eels in the western South Pacific have not been determined as clearly as the northern hemisphere species, but they have been presumed to be in the westward-flowing South Equatorial Current (Figures 2 and 8) (Jellyman 1987, 2003). The genetic identification of Australasian shortfin eel leptocephali (22–32 mm TL) in the South Equatorial Current region (Figure 8) to the south of the Solomon Islands supports this hypothesis (Aoyama et al. 1999), as do other recent catches of anguillid leptocephali in the region (Kuroki et al. 2008). Other smaller anguillids were collected just to the northwest of Fiji in the Equatorial Pacific "Fresh Pool," a distinct lens of low salinity water in the upper 100 m (Delcroix and Picaut 1998; Hénin et al. 1998) that was present at about 10–17°S in August of 1995 (Miller et al. 2006). This hydrographic feature may provide salinity fronts to act as landmarks of the spawning area, as fronts seem to in the northern hemisphere anguillid spawning areas.

Spawning by the Australasian shortfin eel in this region would represent a typical location compared to the northern temperate anguillid species because the South Equatorial Current would transport leptocephali westward, and the western boundary current of the western South Pacific subtropical gyre, the East Australian Current, would transport them southward (Jellyman 1987, 2003; Arai et al. 1999; Shiao et al. 2001). The spawning area is likely in the southern part of the South Equatorial Current, which usually has a two-core structure with westward flow in the upper 300–400 m from the equator to about 18°S (Delcroix et al. 1987; Kessler and Taft 1987; Roden 1998). Between the two branches of westward flow, there is often a narrow band of eastward flow of the South Equatorial Countercurrent, which is typically between 5°S and 10°S (Delcroix et al. 1987; Kessler and Taft 1987; Qiu and Chen 2004). The strength of the South Equatorial Countercurrent appears to be greatest in March and weakest in August (Chen and Qiu 2004; Qiu and Chen 2004). At the western margin of the western South Pacific, the South Equatorial Current bifurcates at about

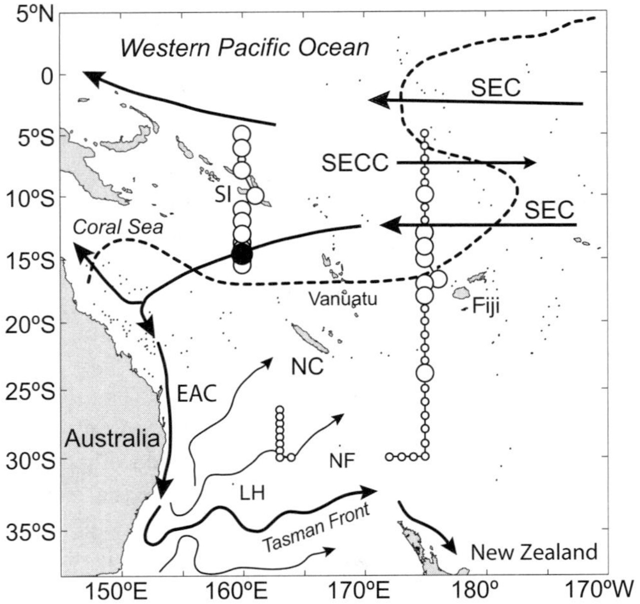

Figure 8.—Map of the western South Pacific showing the location where genetically identified Australasian shortfin eel leptocephali (large black circle) were collected on September 4, 1995 (N = 7; 22–32 mm) and where other anguillid leptocephali (large white circles) were collected in August and September 1995 during the KH-95-2 survey for leptocephali based on Aoyama et al. (1999) and Miller et al. (2006). The small white circles show the locations of other stations during the survey where no anguillid leptocephali were collected. The average location of the low salinity fresh pool (salinity < 35.0) is on the northwest side of the dotted line based on Delcroix and Picaut (1998). The major branches of eastward flow originating from the East Australian Current (EAC) are shown based on Ridgway and Dunn (2003). The South Equatorial Current (SEC), South Equatorial Countercurrent (SECC), and the Solomon Islands (SI), are shown. New Caledonia (NC), Norfolk Island (NF), and Lord Howe Island (LH) where Australasian shortfin eels also recruit, in addition to Australia and New Zealand, are labeled.

18°S, with the southward branch becoming the East Australian Current (Qu and Lindstrom 2002; Ridgway and Dunn 2003). Such oceanographic features would suggest that that the spawning area of Australasian shortfin eels is located to the south of the typical latitude of the South Equatorial Countercurrent and north of the southern edge of the South Equatorial Current. Spawning in this location would maximize the chance of entering the East Australian Current, which would transport leptocephali southward to where they recruit. However, recent larval dispersal modeling indicates that likely spawning areas of Australasian shortfin eels and New Zealand longfin eels may extend further south than the southern boundary of the South Equatorial Current (Jellyman and Bowen 2009, this volume).

Once in the East Australian Current, those leptocephali that recruit to Australia must detrain from the current and move westward. It has previously been suggested that larvae that will recruit to New Zealand must stay in the flow of the East Australian Current that turns eastward and moves across the Tasman Sea (Aoyama et al. 1999). The recent modeling effort by Jellyman and Bowen (2009) indicates that this route is questionable, as the ages at arrival in New Zealand may be too young to accommodate a trans-Tasman migration. However, because the geostrophic flow to the north of New Zealand and up to 15°S near Fiji is typically to the east (Morris et al. 1996; Qiu and Chen 2004), the leptocephali recruiting to New Zealand may have to move along the flow of the Tasman Front (Tilburg et al. 2001) after being transported southward from the South Equatorial Current by the East Australian Current. The flow of the East Australian Current has been recently shown to have several branches of eastward flow

(Ridgway and Dunn 2003), which would explain how this species can recruit to regions ranging from New Caledonia to New Zealand (Figure 8). These possibilities and the results of the recent modeling should be considered in future surveys to find the spawning areas of these eels.

The spawning area of the other New Zealand species, the New Zealand longfin eel, is unknown because no leptocephali have ever been identified. Larval duration estimated from otolithometry in glass eels showed that New Zealand longfin eels have a longer leptocephalus stage than Australasian shortfin eels and most other anguillids (Marui et al. 2001; Jellyman 2003). However, females of this species appear to leave their freshwater growth habitats at a more advanced state of reproductive maturation, suggesting that the spawning migration is shorter (Jellyman 1987, 2003). The recent use of satellite pop-up tagging techniques with migrating silver eels has shown potential to assist in understanding their migrations and identifying potential spawning sites (Jellyman and Tsukamoto 2002).

Possible Oceanic Factors Affecting Recruitment

Because the precise location of the spawning area of Australasian shortfin eels in the South Equatorial Current region is unknown, it is difficult to determine how spawning behavior or early larval development may be affected by changes in ocean-atmospheric conditions in the western South Pacific. However, the collection of small anguillid leptocephali in a narrow band of latitude just to the northwest of Fiji and to the south of the Solomon Islands suggests that alterations to the physical environment of the region of the South Equatorial Current on either side of Fiji could affect the spawning area and early larval development or transport of Australasian shortfin eels. Analyses of the effects of ENSO forcing on the upper ocean conditions in the western Pacific have focused primarily on the equatorial region (i.e., Johnson and McPhaden 2000), but the likely spawning area in the southern branch of the South Equatorial Current also appears to be affected by these events.

The western South Equatorial Current region is likely the part of the western South Pacific most affected by ENSO events, based on the changes documented in association with the 1997–1998 El Niño. The 1997 ENSO had major effects on the salinity, temperature and current structure of the equatorial Pacific (Delcroix and Picaut 1998; Hénin et al. 1998; Johnson and McPhaden 2000). The surface salinity of this region increased drastically during the 1997–1998 El Niño and the longitudinal fluctuations in the extent of the low salinity fresh pool appear to correspond to changes in the SOI (Gouriou and Delcroix 2002). These effects extended into the northwestern South Pacific, primarily as changes to the fresh pool region to the north and northeast of Fiji. The changes in 1997–1998 were also observed in measures of sea surface height variations in this area, with a lower sea surface height being present as a result of an increase in salinity (Bowen et al. 2006; Qiu and Chen 2006). A smaller reduction of sea surface height also appears to have occurred in association with the 2002 El Niño event (Lagerloef et al. 2003). Variations in the salinity and temperature structure between 10°S and 20°S also have been seen in earlier multiyear data sets, along with apparent variations in the strength or latitudinal location of the South Equatorial Current (Kessler and Taft 1987; Delcroix and Picaut 1998; Hénin et al. 1998).

A change in the surface salinity in this area could be important if the salinity front that likely forms along the edges of the fresh pool plays a role in defining the spawning area in the western South Pacific, as fronts appear to do in the western North Pacific and in the Sargasso Sea (Kleckner and McCleave 1988; Kimura and Tsukamoto 2006). Apparent salinity fronts were present at the southern edge of the fresh pool north or northwest of Fiji in August of 1993 (Roden 1998) and in August 1995 when anguillid leptocephali were collected in the low-salinity water (Figure 8; Miller et al. 2006). In contrast to the Sargasso Sea where primarily temperature varies across the fronts and the western North South Pacific where salinity varies, in the western South Pacific, both temperature and salinity vary at this latitude and a distinct density front was observed north of Fiji in August by Roden (1998). A westward or northwestward shift in the location of this front in association with ENSO events (Gouriou and Delcroix 2002) could result in a corresponding shift in the location of the spawning area or disruptions in locating the spawning area. A shift in the location of spawning could then affect the transport pattern of the leptocephali. Chisnall et al. (2002) found variations in recruitment patterns and sizes of glass eels of both species that appeared to be related to El Niño in 1997 and 1998, so it is possible that one or more aspects of the early life history or larval transport were affected by ENSO related changes in the region.

Some evidence that ENSO changes in ocean conditions may influence recruitment in the western South Pacific has also been found in a preliminary analysis of glass eel catch data in New Zealand in relation to the SOI (Jellyman and Bowen 2009). Since 1995, the annual recruitment of glass eels to New Zealand has been monitored in six small streams from four geographic regions (Jellyman et al. 1999, 2002). From these data, the mean density of newly arrived glass eels was investigated for any relationship with the mean SOI for a 3-month period from November to January (representing the possible spawning or early larval transport period 8–10 months before recruitment). Recruitment for both species in all four regions was positively correlated with the mean SOI during this period, with correlation coefficients ranging from 0.06 to 0.66. However, confidence intervals for all but one species × region combination were not significant, so these results should be interpreted as indicative rather than predictive. Because of a suggested relationship between the SOI and glass eel arrival patterns to New Zealand (Chisnall et al. 2002), a second time period representing the time immediately preceding arrival in New Zealand waters (July–September) was investigated, but there was no evidence of any consistent relationship between recruitment and the prevailing SOI at the time of arrival.

Fluctuations in wind patterns in the region could affect the flow of the South Equatorial Current in the region just north of Fiji, which in turn could affect the transport of recently spawned leptocephali. The South Equatorial Current does not appear to be as consistently strong as the North Equatorial Current north of the equator in direct temporal comparisons (Kessler and Taft 1987; Reverdin et al. 1994), and the South Equatorial Current appears to have major fluctuations in its strength and latitudinal locations (Delcroix et al. 1987; Morris et al. 1996; Hénin et al. 1998). In addition, there appears to be a seasonal cycle in the presence and strength of the South Equatorial Countercurrent and eddy variability in the region just south of the South Equatorial Current (Chen and Qiu 2004; Qiu and Chen 2004) that may affect westward transport in the region. Future studies using longer term data sets such as those briefly mentioned here, and modeling research (Jellyman and Bowman 2009) used in combination with life history studies of glass eels recruiting to Australia and New Zealand (Arai et al. 1999; Marui et al. 2001; Shiao et al. 2001; Chisnall et al. 2002; Jellyman et al. 2003) may be useful to determine how ocean-atmosphere changes may affect the western South Pacific anguillid species.

Summary and Conclusions

This examination of the potential effects of ocean-atmospheric changes on temperate anguillid eel recruitment suggests that there are at least three ways that these species could be affected: (1) changes in oceanic frontal features could shift the locations of their spawning areas resulting in reduced transport to recruitment areas, (2) changes in current flow patterns along the larval migration routes could be altered and reduce successful transport and recruitment, and (3) changes in productivity along the larval migration routes could reduce survival due to reduced feeding success. This review has shown that spawning and larval migration patterns of each species appear similar in most respects, and thus these factors may act upon all of them. However, there are distinct differences in oceanographic features in the three ocean basins that may respond somewhat differently to ocean-atmospheric changes. The general lack of information about many important aspects of the biology of each species makes it difficult though to establish clear links to changes in oceanic conditions that may affect their eventual recruitment.

In the North Atlantic, the NAO has been associated with changes in the physical and biological environment of the northeast Atlantic and the Sargasso Sea, and ocean warming may have reduced productivity. These changes appear to have the potential to both modify the transport patterns of Atlantic eel leptocephali or affect their feeding success, and similar processes may be involved in the western Pacific regions. In the western North Pacific, teleconnections with the PDO may have had an effect on the productivity in the surface layer and have caused an increase in the number or severity of El Niño events that appear to have a negative affect on the transport pattern of Japanese eel leptocephali. Links between ocean-atmospheric events and recruitment fluctuations in temperate anguillid eels may also be present in the western South Pacific where ENSO events appear to have an effect on the oceanographic conditions of the spawning area that could potentially affect recruitment.

Decadal-scale cycles in the ocean-climate of the North Atlantic and other areas likely have occurred in the past though (Hare and Mantua 2000; Hurrell and Dickson 2004; Drinkwater 2006), so it is possible that recruitment and stock levels have stabilized in more recent years, although at lower levels than in the 1970s–1980s. Anguillid eel populations can likely survive wide-ranging changes in oceanic and continental climates, considering that the Atlantic eel species have survived extreme changes such as ice ages since their evolution millions of years ago. However, especially in the case of the European eel whose silver eel migrations are so long, and whose recruitment remains dangerously low in some areas, other factors such as fishing pressure, accumulation of pollutants, and parasite or virus infections that may disrupt the ability of spawners to reach the Sargasso Sea must also be considered as possible contributors to the recruitment declines. The cumulative effects of habitat loss or degradation due to changes in agriculture and industrialization also must have reduced the effective population sizes of each species during the decades leading up to the declines, which may reduce the ability of populations to recover quickly. If increases in temperature reduce productivity enough to affect the feeding success of leptocephali, then a continued global warming trend is an additional concern. This mosaic of potential factors, including anthropogenic and oceanic changes, necessitate further research and action to prevent further declines in these mysterious catadromous fishes that live in both the oceans and in freshwater.

References

Aoyama, J., N. Mochioka, T. Otake, S. Ishikawa, Y. Kawakami, P. H. J. Castle, M. Nishida, and K. Tsukamoto. 1999. Distribution and dispersal of anguillid leptocephali in the western Pacific Ocean revealed by molecular analysis. Marine Ecology Progress Series 188:193–200.

Arai, T., T. Otake, D. J. Jellyman, and K. Tsukamoto. 1999. Differences in the early life history of the Australasian shortfinned eel, *Anguilla australis* from Australia and New Zealand, as revealed by otolith microstructure and microchemistry. Marine Biology 135:381–389.

Beamish R. J., R. M. Sweeting and C. M. Neville. 2009. Planning the management of Pacific salmon in a changing climate. Pages 155–173 *in* A. J. Haro, K. L. Smith, R. A. Rulifson, C. M. Moffitt, R. J. Klauda, M. J. Dadswell, R. A. Cunjak, J. E. Cooper, K. L. Beal, and T. S. Avery, editors. Challenges for diadromous fishes in a dynamic global environment. American Fisheries Society, Symposium 69, Bethesda, Maryland.

Behrenfeld, M. J., R. T. O'Malley, D. A. Siegel, C. R. McClain, J. L. Sarmiento, G. C. Feldman, A. J. Milligan, P. G. Falkowski, R. M. Letelier, and E. S. Boss. 2006. Climate-driven trends in contemporary ocean productivity. Nature (London) 444:752–755.

Bersch, M. 2002. North Atlantic Oscillation-induced changes of the upper layer circulation in the northern North Atlantic Ocean. Journal of Geophysical Research 107:3156, doi:10.1029/2001JC000901.

Boëtius, J., and E. F. Harding. 1985. A re-examination of Johannes Schmidt's Atlantic eel investigations. Dana 4:129–162.

Bonnhommeau, S., E. Chassot, B. Planque, E. Rivot, A. H. Knap, and O. Le Pape. 2008a. Impact of climate on eel populations of the Northern Hemisphere. Marine Ecology Progress Series 373:71–80.

Bonhommeau, S., E. Chassot, and E. Rivot. 2008b. Fluctuations in European eel (*Anguilla anguilla*) recruitment resulting from environmental changes in the Sargasso Sea. Fisheries Oceanography 17:32–44.

Bowen, M. M., P. J. H. Sutton, and D. Roemmich. 2006. Wind-driven and steric fluctuations of sea surface height in the southwest Pacific. Geophysical Research Letters 33:L14617, doi:10.1029/2006GL026160.

Casselman, J. M. 2003. Dynamics of resources of the American eel, *Anguilla rostrata*: declining abundance in the 1990s. Pages 255–274 *in* K. Aida, K. Tsukamoto, and K. Yamauchi, editors. Eel biology, Springer Verlag, Tokyo.

Castonguay, M., P. V. Hodson, C. Moriarty, K. F. Drinkwater, and B. M. Jessop. 1994. Is there a role of ocean environment in American and European eel decline? Fisheries Oceanography 3:197–203.

Chen, S., and B. Qiu. 2004. Seasonal variability of the South

Equatorial Countercurrent. Journal of Geophysical Research 109:C08003, doi:10.1029/2003JC002243.

Chisnall, B. L., D. J. Jellyman, M. L. Bonnett, and J. R. Sykes. 2002. Spatial and temporal variability in length of glass eels (*Anguilla* spp.) in New Zealand. Zealand Journal of Marine and Freshwater Research 36:89–104.

Curry, R., and M. McCartney. 2001. Ocean gyre circulation changes associated with the North Atlantic Oscillation. Journal of Physical Oceanography 31:3374–3400.

Dekker, W. 1998. Long-term trends in the glass eels immigrating at Den Oever, the Netherlands. Bulletin Francais de la Pêche et de la Pisciculture 349:199–214.

Dekker, W. 2002. Monitoring of glass eel recruitment. Netherlands Institute of Fisheries Research, Report C007/02-WD, IJmuiden.

Dekker, W. 2004. Synthesis and analyses. Pages 127–145 *in* Slipping through our hands: population dynamics of the European eel. Doctoral dissertation. Faculteit der Natutuurweten Schappen, Wiskunde an Informatica, The Netherlands.

Dekker, W., J. M. Casselman, D. K. Cairns, K. Tsukamoto, D. Jellyman, and H. Lickers. 2003. Worldwide decline of eel resources necessitates immediate action: Quebec declaration of concern. Fisheries 28:2830.

Delcroix, T. G., G. Eldin, and C. Hénin. 1987. Upper ocean water masses and transports in the western tropical Pacific (165°E). Journal Physical Oceanography 17:2248–2262.

Delcroix, T., and J. Picaut. 1998. Zonal displacement of western equatorial Pacific "fresh pool." Journal of Geophysical Research 103:1087–1098.

Désaunay, Y., and D. Guérault. 1997. Seasonal and long-term changes in biometrics of eel larvae: a possible relationship between recruitment variation and North Atlantic ecosystem productivity. Journal of Fish Biology 51:317–339.

Drinkwater, K. F. 2006. The regime shift of the 1920s and 1930s in the North Atlantic. Progress in Oceanography 68:134–151.

Feunteun, E. 2002. Management and restoration of European eel population (*Anguilla anguilla*): an impossible bargain. Ecological Engineering 18:575–591.

Friedland, K. D., M. J. Miller, and B. Knights. 2007. Oceanic changes in the Sargasso Sea and declines in recruitment of the European eel. ICES Journal of Marine Science 64:519–530.

Gouriou, Y., and T. Delcroix. 2002. Seasonal and ENSO variations of sea surface salinity and temperature in the South Pacific Convergence Zone during 1976–2000. Journal of Geophysical Research 107:8011, doi:10.1029/2001JC000830.

Grey, M., H. Haines, and A.Troccoli. 2000. A study of temperature changes in the upper North Atlantic: 1950–94. Journal of Climate Change 13:2697–2711.

Hare, S. R., and N. J. Mantua. 2000. Empirical evidence for North Pacific regime shifts in 1977 and 1989. Progress in Oceanography 47:381–408.

Haro A., W. Richkus, K. Whalen, A. Hoar, W-D. Busch, S. Lary, T. Brush, and D. Dixon. 2000. Population decline of the American eel: implications for research and management. Fisheries 25:7–16.

Hays, G. C., A. J. Richardson, and C. Robinson. 2005. Climate change and marine plankton. Trends in Ecology and Evolution 20:337–344.

Hénin, C., Y. du Penhoat, and M. Ioualalen. 1998. Observations of sea surface salinity in the western Pacific fresh pool: large-scale changes in 1992–1995. Journal of Geophysical Research 103:7523–7536.

Hurrell, J. W. 1995. Decadal trends in the North Atlantic Oscillation: regional temperatures and precipitations. Science 269:676–679.

Hurrell, J. W., and R. R. Dickson. 2004. Climate variability over the North Atlantic. Pages 15–31 *in* N. C., Stenseth, G. Ottersen, J. W. Hurrell, and A. Belgrano, editors. Marine ecosystems and climate variation: the North Atlantic. Oxford University Press, New York.

ICES (International Council for the Exploration of the Sea). 2001. International Council for the Exploration of the Sea. Report of the ICES/EIFAC Working Group on Eels. International Council for the Exploration of the Sea, Advisory Committee on Fisheries Management, ICES CM 2001/ACFM:03, Copenhagen, Denmark.

ICES (International Council for the Exploration of the Sea). 2006. Report of the 2006 Session of the Joint EIFAC/ICES Working Group on Eels. International Council for the Exploration of the Sea, Advisory Committee on Fisheries Management, ICES CM. 2006/ACFM:16, Copenhagen, Denmark.

ICES (International Council for the Exploration of the Sea). 2007. Report of the 2007 session of the Joint EIFAC/ICES Working Group on Eels. International Council for the Exploration of the Sea, Advisory Committee on Fisheries Management, ICES CM 2007/ACFM:23, Copenhagen, Denmark.

Jellyman, D. J. 1987. Review of the marine life history of Australian temperate species of *Anguilla*. Pages 276–285 *in* M. J. Dadswell, R. J. Klauda, C. M. Moffitt, R. L. Saunders, R. A. Rulifson, and J. E. Cooper. American Fisheries Society, Symposium 1, Bethesda, Maryland.

Jellyman, D. J. 2003. The distribution and biology of the South Pacific species of *Anguilla*. Pages 275–292 *in* K. Aida, K. Tsukamoto, and K. Yamauchi, editors. Eel biology. Springer Verlag, Tokyo.

Jellyman, D. J. 2007. Status of New Zealand freshwater eel stocks and management initiatives. ICES Journal of Marine Science 64:1379–1386.

Jellyman, D. J. 2009. Forty years on—the impact of commercial fishing on stocks of New Zealand freshwater eels. Pages 36–56 *in* J. Casselman and D. K. Cairns, editors. Eels at the edge: science, status, and conservation concerns. American Fisheries Society, Symposium 58, Bethesda, Maryland.

Jellyman, D. J., and M. Bowen. 2009. Modelling larval migration routes and spawning areas of anguillid eels of New Zealand and Australia. Pages 255–274 *in* A. J. Haro, K. L. Smith, R. A. Rulifson, C. M. Moffitt, R. J. Klauda, M. J. Dadswell, R. A. Cunjak, J. E. Cooper, K. L. Beal, and T. S. Avery, editors. Challenges for diadromous fishes in a dynamic global environment. American Fisheries Society, Symposium 69, Bethesda, Maryland.

Jellyman, D. J., B. L. Chisnall, M. L. Bonnett, and J. R. E. Sykes. 1999. Seasonal arrival patterns of juvenile freshwater eels (*Anguilla* spp.) in New Zealand. New Zealand Journal of Marine and Freshwater Research 33:249–261.

Jellyman, D. J., B. L. Chisnall, J. R. Sykes, and M. L. Bonnett. 2002. Variability in spatial and temporal abundance of glass eels (*Anguilla* spp.) in New Zealand waterways. New Zealand Journal of Marine and Freshwater Research 36:511–517.

Jellyman, D. J., E. Graynoth, R. I. C. C. Francis, B. L. Chisnall, and M. P. Beentjes. 2000. A review of the evidence for a decline in the abundance of longfinned eels (*Anguilla dieffenbachii*) in New Zealand. Final Research Report for Ministry of Fisheries Research, Project EEL9802, Wellington, New Zealand.

Jellyman, D. J., and K. Tsukamoto. 2002. First use of archival transmitters to track migrating freshwater eels *Anguilla dieffenbachii* at sea. Marine Ecology Progress Series 233:207–215.

Johnson, G. C., and M. J. McPhaden. 2000. Upper equatorial Pacific Ocean current and salinity variability during the 1996–1998 El Niño–La Niña cycle. Journal of Geophysical Research 105:1037–1053.

Kaneko, I., Y. Takatsuki, H. Kamiya, and S. Kawae. 1998. Water property and current distributions along the WHP-P9 section (137°–142°E) in the western North Pacific. Journal of Geophysical Research 103:12,959– 12:984.

Kessler, W. S., and B. A. Taft. 1987. Dynamic heights and zonal geostrophic transports in the central tropical Pacific during 1979–1984. Journal of Physical Oceanography 17:97–122.

Kettle, A. J., D. C. E. Bakker, and K. Haines. 2008. Impact of the North Atlantic Oscillation on the trans-Atlantic migrations of the European eel (*Anguilla anguilla*). Journal of Geophysical Research 113:G03004, doi:10.1029/2007JG000589.

Kettle, A. J., and K. Haines. 2006. How does the European eel (*Anguilla anguilla*) retain its population structure during its larval migration across the North Atlantic Ocean? Canadian Journal of Fisheries and Aquatic Sciences 63:90–106.

Kim, H., S. Kimura, A. Shinoda, T. Kitagawa, Y. Sasai and H. Sasaki. 2007. Effect of El Niño on migration and larval transport of the Japanese eel, *Anguilla japonica*. ICES Journal of Marine Science 64:1387–1395.

Kim, Y. Y., T. Qu,. T. Jensen, T. Miyama, H. Mitsudera, H. W. Kang, and A. Ishida. 2004. Seasonal and interannual variations of the North Equatorial Current bifurcation in a high-resolution OGCM. Journal of Geophysical Research 109:C03040.

Kimura, S. 2003. Larval transport of the Japanese eel. Pages 169–179 *in* K. Aida, K. Tsukamoto, and K. Yamauchi, editors. Eel biology. Springer Verlag, Tokyo.

Kimura, S., K. Döös, and A. C. Coward. 1999. Numerical simulation to resolve the downstream migration of the Japanese eel. Marine Ecology Progress Series 186:303–306.

Kimura, S., T. Inoue, and T. Sugimoto. 2001. Fluctuation in the distribution of low-salinity water in the North Equatorial Current and its effect on the larval transport of the Japanese eel. Fisheries Oceanography 10:51–60.

Kimura, S., and K. Tsukamoto. 2006. The salinity front in the North Equatorial Current: a landmark for the spawning migration of the Japanese eel (*Anguilla japonica*) related to the stock recruitment. Deep-Sea Research II 53:315–325.

Kimura, S., K. Tsukamoto, and T. Sugimoto. 1994. A model for the larval migration of the Japanese eel: roles of the trade winds and salinity front. Marine Biology 119:185–190.

Kirk, R. S. 2003. The impact of *Anguillicola crassus* on European eels. Fisheries Management and Ecology 10:385–394.

Kleckner, R. C., and J. D. McCleave. 1982. Entry of migrating American eel leptocephali into the Gulf Stream system. Helgoländer wiss Meeresunters 35:329–339.

Kleckner, R. C., and J. D. McCleave. 1985. Spatial and temporal distribution of American eel larvae in relation to North Atlantic Ocean current systems. Dana 4:67–92.

Kleckner, R. C., and J. D. McCleave. 1988. The northern limit of spawning by Atlantic eels (*Anguilla* spp.) in the Sargasso Sea in relation to thermal fronts and surface water masses. Journal of Marine Research 46:647–667.

Knights, B. 2003. A review of the possible impacts of

long-term oceanic and climate changes and fishing mortality on recruitment of anguillid eels of the northern hemisphere. Science of the Total Environment 310:237–244.

Kobashi, F., and H. Kawamura. 2001. Variation of sea surface height at periods of 65–220 days in the subtropical gyre of the North Pacific. Journal of Geophysical Research 106:26,817– 26,813.

Kuroki, M., J. Aoyama, M. J. Miller, S. Watanabe, A. Shinoda, D. J. Jellyman, E. Feunteun, and K. Tsukamoto. 2008. Distribution and early life history characteristics of anguillid leptocephali in the western South Pacific. Marine and Freshwater Research 59:1035–1047.

Lagerloef, G. S. E., R. Lukas, F. Bonjean, J. T. Gunn, G. T. Mitchum, M. Bourassa, and A. J. Busalacchi. 2003. El Niño Tropical Pacific Ocean surface current and temperature evolution in 2002 and outlook for early 2003. Geophysical Research Letters 30:1514, doi:10.1029/2003GL017096.

Lehodey, P., J. Alheit, M. Barange, T. Baumgartner, G. Heaugrand, K. Drinkwater, J.-M. Fromentin, S. R. Hare, G. Ottersen, R. I. Perry, C. Roy, C. D. van der Lingen, and F. Werner. 2006. Climate variability, fish, and fisheries. Journal of Climate 19:5009–5030.

Levitus, S., J. I. Antonov, and T. P. Boyer. 2000. Warming of the world ocean. Science 287:2225–2229.

Lukas, R., E. R. Firing, P. Hacker, P. L. Richardson, C. A. Collins, R. Fine, and R. Gammon. 1991. Observations of the Mindanao Current during the western Equatorial Pacific Ocean Circulation Study. Journal of Geophysical Research 96:7089–7104.

Marui, M., T. Arai, M. J. Miller, D. J. Jellyman, and K. Tsukamoto. 2001. Comparison of early life history between New Zealand temperate eels and Pacific tropical eels revealed by otolith microstructure and microchemistry. Marine Ecology Progress Series 213:273–284.

McCleave, J. D. 2001. Simulation of the impact of dams and fishing weirs on reproductive potential of silver-phase American eels in the Kennebec River basin, Maine. North American Journal of Fisheries Management 21:592–605.

McCleave, J. D., P. J. Brickley, K. M. O'Brien, D. A. Kistner-Morris, M. W. Wong, M. Gallagher, and S. M. Watson. 1998. Do leptocephali of the European eel swim to reach continental waters? Status of the question. Journal of the Marine Biological Association of the UK 78:285–306.

McCleave, J. D., and R. C. Kleckner. 1987. Distribution of leptocephali of the catadromous *Anguilla* species in the western Sargasso Sea in relation to water circulation and migration. Bulletin of Marine Science 41:789–806.

McCleave, J. D., R. C. Kleckner, and M. Castonguay. 1987. Reproductive sympatry of American and European eels and implications for migration and taxonomy. Pages 286–297 *in* M. J. Dadswell, R. J. Klauda, C. M. Moffitt, R. L. Saunders, R. A. Rulifson, and J. E. Cooper. American Fisheries Society, Symposium 1, Bethesda, Maryland.

McPhaden, M. J. 1999. Genesis and evolution of the 1997–98 El Niño. Science 283:950–954.

Miller, M. J., J. Aoyama, N. Mochioka, T. Otake, P. H. J. Castle, G. Minagawa, T. Inagaki, and K. Tsukamoto. 2006. Geographic variation in the assemblages of leptocephali in the western South Pacific. Deep-Sea Research I 53:776–794.

Miller, M. J., and J. D. McCleave. 1994. Species assemblages of leptocephali in the subtropical convergence zone of the Sargasso Sea. Journal of Marine Research 52:743–772.

Mochioka, N., and M. Iwamizu. 1996. Diet of anguillid larvae: leptocephali feed selectively on larvacean houses and fecal pellets. Marine Biology 125:447–452.

Morris, M., D. Roemmich, and B. Cornuelle. 1996. Observations of variability in the South Pacific subtropical gyre. Journal Physical Oceanography 26:2359–2360.

Otake, T., K. Nogami, and K. Maruyama. 1993. Dissolved and particulate organic matter as possible food sources for eel leptocephali. Marine Ecology Progress Series 92:27–34.

Palstra, A. P., D. F. M. Heppener, V. J. T. van Ginneken, C. Székely, and G. E. E. J. M. van den Thillart. 2007. Swimming performance of silver eels is severely impaired by the swim-bladder parasite *Anguillicola crassus*. Journal of Experimental Marine Biology and Ecology 352:244–256.

Palter, J. B., Lozier, M. S. and Barber, R. T. 2005. The effect of advection on the nutrient reservoir in the North Atlantic subtropical gyre. Nature (London) 437:687–692.

Polyakov, I. V., U. S. Bhatt, H. L. Simmons, D. Walsh, J. E. Walsh, and X. Zhang. 2005. Multidecadal variability of North Atlantic temperature and salinity during the twentieth century. Journal of Climate 18:4562–4581.

Qiu, B., and S. Chen. 2004. Seasonal modulations in the eddy field of the South Pacific Ocean. Journal of Physical Oceanography 34:1515–1527.

Qiu B., and S. M. Chen. 2006. Decadal variability in the large-scale sea surface height field of the South Pacific Ocean: observations and causes. Journal Physical Oceanography 36:1751–1762.

Qu, T., and E. J. Lindstrom. 2002. A climatological interpretation of the circulation in the western South Pacific. Journal Physical Oceanography 32:2492–2508.

Reverdin, G., C. Frankignoul, E. Kestenare, and M .J. McPhaden. 1994. Seasonal variability in the surface currents of the equatorial Pacific. Journal of Geophysical Research 99:20323–20344.

Reverdin, G., P. P. Niiler, and H. Valdimarsson. 2003. North Atlantic Ocean surface currents. Journal of Geophysical Research 108:3002, doi:10.1029/2001JC001020.

Richardson A. J., and D. S. Schoeman. 2004. Climate impact on plankton ecosystems in the Northeast Atlantic. Science 305:1609–1612.

Ridgway, K. R., and J. R. Dunn. 2003. Mesoscale structure of the mean East Australian Current System and its relationship with topography. Progress in Oceanography 56:189–222.

Robinet, T., and E. Feunteun. 2002. Sublethal effects of exposure to chemical compounds: a cause for the decline in Atlantic eels? Ecotoxicology 11:265–277.

Roden, G. 1998. Upper ocean thermohaline, oxygen, nutrient, and flow structure near the date line in the summer of 1993. Journal of Geophysical Research 103:12,919– 12:939.

Schmidt, J. 1922. The breeding places of the eel. Philosophical Transactions of the Royal Society B 211:179–208.

Schmitz, W. J., and M. S. McCartney. 1993. On the North Atlantic circulation. Reviews of Geophysics 31:29–49.

Schoth, M., and Tesch, F.-W. 1982. Spatial distribution of 0-group eel larvae (*Anguilla* sp.) in the Sargasso Sea. Helgoländer wiss Meeresunters 35:309–320.

Shiao, J. C., W. N. Tzeng, A. Collins, and D. J. Jellyman. 2001. Dispersal pattern of glass eel stage of *Anguilla australis* revealed by otolith growth increments. Marine Ecology Progress Series 219:241–250.

Siedler, G. S., L. Armi, and T. J. Müller. 2005. Meddies and decadal changes at the Azores Front from 1980 to 2000. Deep-Sea Research II 52:583–604.

Stige, L. C., G. Ottersen, K. Brander, K.-S. Chan, and N. C. Stenseth. 2006. Cod and climate: effect of the North Atlantic Oscillation on recruitment in the North Atlantic. Marine Ecology Progress Series 325:227–241.

Sugimoto, T., S. Kimura, and K. Tadokoro. 2001. Impact of El Niño events and climate regime shift on living resources in the western North Pacific. Progress in Oceanography 49:113–127.

Sugimoto, T., and K. Tadokoro. 1998. Interdecadal variations of plankton biomass and physical environment in the North Pacific. Fisheries Oceanography 7:289–299.

Tatsukawa, T. 2003. Eel resources in East Asia. Pages 293–298 *in* K. Aida, K. Tsukamoto, and K. Yamauchi, editors. Eel biology. Springer Verlag, Tokyo.

Tilburg, C. E., H. E. Hurlburt, J. J. O'Brien, J. F. Shriver. 2001. The dynamics of the East Australian Current System: the Tasman front, the East Auckland Current, and the East Cape Current. Journal Physical Oceanography 31:2917–2943.

Tsukamoto, K. 1992. Discovery of the spawning area for Japanese eel. Nature (London) 356:789–791.

Tsukamoto, K. 2006. Spawning of eels near a seamount. Nature (London) 439:929.

Tsukamoto, K., T. Otake, N. Mochioka, T. W. Lee, H. Fricke, T. Inagaki, J. Aoyama, S. Ishikawa, S. Kimura, M. J. Miller, H. Hasumoto, M. Oya, and Y. Suzuki. 2003. Seamounts, new moon and eel spawning: the search for the spawning site of the Japanese eel. Environmental Biology of Fishes 66:221–229.

Turner, J. T. 2002. Zooplankton fecal pellets, marine snow and sinking phytoplankton blooms. Aquatic Microbial Ecology 27:57–102.

Ullman D. S., P. C. Cornillon, and Z. Shan. 2007. On the characteristics of subtropical fronts in the North Atlantic. Journal of Geophysical Research 112:C01010, doi:10.1029/2006JC003601.

van Ginneken, V., T, B. Ballieux, R. Willemze, K. Coldenhoff , E. Lentjes, E. Antonissen, O. Haenen, and G. van den Thillart. 2005. Hematology patterns of migrating European eels and the role of EVEX virus. Comparative Biochemistry and Physiology Part C 140:97–102.

Watanabe, Y. W., H. Ishida, T. Nakano, and N. Naga. 2005. Spatiotemporal decreases of nutrients and chlorophyll-a in the surface mixed layer of the Western North Pacific from 1971 to 2000. Journal of Oceanography 61:1011–1016.

Weijerman, M., H. Lindeboom, and A. F. Zuur. 2005. Regime shifts in marine ecosystems of the North Sea and Wadden Sea. Marine Ecology Progress Series 298:21–39.

Westerberg, H. 1998. Oceanographic aspects of the recruitment of eels to the Baltic Sea. Bulletin Francais de la Pêche et de la Pisciculture 349:177–185.

American Fisheries Society Symposium 69:251–253, 2009

Ocean Migration of Diadromous Fishes in a Changing Global Environment

Preamble

Michael J. Dadswell
Department of Biology, Acadia University
Wolfville, Nova Scotia B4P 2R6, Canada

The ocean migratory stage of many diadromous fishes remains one of the most poorly known portions of their life cycles. For all the work done on North American stocks of Atlantic salmon over the past 60 years, we still only have sketchy details of their ocean migration. Atlantic salmon *Salmo salar* migrate to sea and virtually vanish into the North Atlantic for one or more years, except for popping up off west Greenland for six or more weeks in late summer, if they stay at sea for a second or more summers. As you examine the papers in the following section, this aspect of the biology of diadromous fishes will present itself numerous times. In my opinion, future work must focus on describing these marine migrations. Our development of knowledge concerning the marine phase of life cycles is moving too slowly in the face of possible global climate change. Advances have been made since our first international symposium in Boston in 1986, but we must do better.

Anguillid eels are an excellent example. Of the 15 described species, the ocean spawning grounds are known for only four (Jellyman and Bowen 2009, this volume). Many eel populations, particularly European and American, are declining rapidly for unknown reasons (MacGregor et al. 2009, this volume), and we may have little time to discover the problem. In this session, Jellyman and Bowen have combined modeling and tracking ocean migrant adults with pop-up tags in an attempt to suggest oceanic regions north of New Zealand and Australia with the greatest potential for spawning sites of the three eel species they studied, but much work needs to be done to find the actual sites. Perhaps now with their work as a guide, researchers will be successful in the search.

During our talks in Halifax, we discovered that virtually all species of amphidromous fishes that go to sea as small larvae (whitebait, tritri, etc) have ocean nursery grounds that are unknown to science (McDowell 2009, this volume). Papers presented on Sicydiine gobies from the Caribbean and Japan clearly expressed this lack of knowledge. These studies examined more or less mutually exclusive, alternate hypotheses that suggest that larval gobies either (1) remain close to their natal streams in nearshore, mesohaline environments (Bell 2009, this volume), or (2) because their genetic structure indicates poor stock definition, larvae mix while moving long distances in ocean currents (Iida et al. 2009, this volume). Rowe and Kelley (2009, this volume) examined the ocean larval duration of galaxiids from New Zealand and obtained much new information on growth rates at sea, which they have used to develop some interesting hypotheses regarding the ocean growth phase. However, they were unable to resolve differences in growth between the two studied populations because ocean nursery sites were unknown. Since whitebait is an important commercial fishery in New Zealand and many populations have declined (Rowe and Kelley 2009), perhaps studies will be funded in the future that will complete the knowledge of galaxiid life cycles by discovering the ocean nursery sites.

On the other hand, ocean tracking of diadromous fishes has reached a new level of sophistication that we heard nothing about during the 1986 symposium, the advent of acoustic tagging. Acoustic systems and their uses have grown dramatically since the 1980s as the technology has steadily improved. Besides the use of pop-up tags to study the ocean migration of adult New Zealand longfin silver eels *Anguilla dieffenbachi*, we heard about acoustic studies on the early phase of ocean migration of American silver eels *A. rostrata* (Bradford et al. 2009, this volume), the first few weeks of Atlantic salmon smolt migration in the marine environment (Kocik et al. 2009, this volume), and the development of extensive ocean shelf acoustic listening chains (Stokesbury et al. 2009, this volume) that will expand our knowledge many fold concerning the marine migrations of numerous diadromous species. Acoustic studies have for the first time opened a window into almost the minute to minute movements of fishes in the marine environment, rather than just the beginning and endpoint knowledge obtained from most conventional tag studies. The ocean listening chains, when fully developed, will extend our knowledge of diadromous fish migrations over the shelf regions of the worlds continents, and anticipated technology advances should finally open up many of the secrets of the open ocean. With the development of integrated acoustic and archival tags as part of a calling card approach and combined with large, living, mobile receiver platforms (sturgeon, sharks, seals; Stokesbury et al. 2009), we will finally have real time ecological information, interactions with other members of the community, and movements of multiple targets, including the platform.

Of course, since they are so small, it will probably be a long time (I think) before acoustic technology will develop to allow tracking of larval amphidromous species to their nursery grounds, which means there remains the need for more conventional studies at sea. But finally, we may be able to learn where those elusive Atlantic salmon are in the North Atlantic during the winter. Hopefully, at our next international symposium, some of these questions will have been resolved.

References

Bell, K. N. I. 2009. What comes down must go up: the migration cycle of juvenile-return anadromous taxa. Pages 321–341 *in* A. J. Haro, K. L. Smith, R. A. Rulifson, C. M. Moffitt, R. J. Klauda, M. J. Dadswell, R. A. Cunjak, J. E. Cooper, K. L. Beal, and T. S. Avery, editors. Challenges for diadromous fishes in a dynamic global environment. American Fisheries Society, Symposium 69, Bethesda, Maryland.

Bradford, R. G., J. W. Carr, F. H. Page, and Fred Whoriskey. 2009. Migration of silver American eels through a macrotidal estuary and bay. Pages 275–292 *in* A. J. Haro, K. L. Smith, R. A. Rulifson, C. M. Moffitt, R. J. Klauda, M. J. Dadswell, R. A. Cunjak, J. E. Cooper, K. L. Beal, and T. S. Avery, editors. Challenges for diadromous fishes in a dynamic global environment. American Fisheries Society, Symposium 69, Bethesda, Maryland.

Iida, M., S. Watanabe, and K. Tsukamoto. 2009. Life history characteristics of a Sicydiinae goby in Japan, compared with its relatives and other amphidromous fishes. Pages 355–373 *in* A. J. Haro, K. L. Smith, R. A. Rulifson, C. M. Moffitt, R. J. Klauda, M. J. Dadswell, R. A. Cunjak, J. E. Cooper, K. L. Beal, and T. S. Avery, editors. Challenges for diadromous fishes in a dynamic global environment. American Fisheries Society, Symposium 69, Bethesda, Maryland.

Jellyman, D. J., and M. M. Bowen. 2009. Modeling larval migration routes and spawning areas of anguillid eels of New Zealand and Australia. Pages 255–274 *in* A. J. Haro, K. L. Smith, R. A. Rulifson, C. M. Moffitt, R. J. Klauda, M. J. Dadswell, R. A. Cunjak, J. E. Cooper, K. L. Beal, and T. S. Avery, editors. Challenges for diadromous fishes in a dynamic global environment. American Fisheries Society, Symposium 69, Bethesda, Maryland.

Kocik, J. F., J. P. Hawkes, T. F. Sheehan, P. A. Music, and K. F. Beland. 2009. Assessing estuarine and coastal migration and survival of wild Atlantic salmon smolts from the Narraguagus River, Maine using ultrasonic telemetry. Pages 293–310 *in* A. J. Haro, K. L. Smith, R. A. Rulifson, C. M. Moffitt, R. J. Klauda, M. J. Dadswell, R. A. Cunjak, J. E. Cooper, K. L. Beal, and T. S. Avery, editors. Challenges for diadromous fishes in a dynamic global environment. American Fisheries Society, Symposium 69, Bethesda, Maryland.

MacGregor, R., J. M. Casselman, W. A. Allen, T. Haxton, J. M. Dettmers, A. Mathers, S. LaPan, T. C. Pratt, P. Thompson, M. Stanfield, L. Marcogliese, and J-D. Dutil. 2009. Natural heritage, anthropogenic impacts, and biopolitical issues related to the status and sustainable management of American eel: a retrospective analysis and management perspective at the population level. Pages 713–740 *in* A. J. Haro, K. L. Smith, R. A. Rulifson, C. M. Moffitt, R. J. Klauda, M. J. Dadswell, R. A. Cunjak, J. E. Cooper, K. L. Beal, and T. S. Avery, editors. Challenges for diadromous fishes in a dynamic global environment. American Fisheries Society, Symposium 69, Bethesda, Maryland.

Rowe, D. K., and G. Kelly. 2009. Duration of the oceanic phase for inanga whitebait (Galaxiidae) is inversely related to growth rate at sea. Pages 343–354 *in* A. J. Haro, K. L. Smith, R. A. Rulifson, C. M. Moffitt, R. J. Klauda, M. J. Dadswell, R. A. Cunjak, J. E. Cooper, K. L. Beal, and T. S. Avery, editors. Challenges for diadromous fishes in a dynamic global environment. American Fisheries Society, Symposium 69, Bethesda, Maryland.

Stokesbury, M. J. W., M. J. Dadswell, K. N. Holland, G. D. Jackson, W. D. Bowen, and R. K. O'Dor. 2009. Tracking diadromous fishes at sea: the electronic future using hybrid acoustic and archival tags. Pages 311–320 *in* A. J. Haro, K. L. Smith, R. A. Rulifson, C. M. Moffitt, R. J. Klauda, M. J. Dadswell, R. A. Cunjak, J. E. Cooper, K. L. Beal, and T. S. Avery, editors. Challenges for diadromous fishes in a dynamic global environment. American Fisheries Society, Symposium 69, Bethesda, Maryland.

American Fisheries Society Symposium 69:255–274, 2009

Modelling Larval Migration Routes and Spawning Areas of Anguillid Eels of New Zealand and Australia

Donald J. Jellyman*
National Institute of Water and Atmospheric Research
Post Office Box 8602, Riccarton, Christchurch 8440, New Zealand

Melissa M. Bowen
National Institute of Water and Atmospheric Research,
Post Office Box 14–091, Wellington 6241, New Zealand

Abstract.—The location of the spawning grounds of the three species of *Anguilla* that occur in New Zealand and Australia, shortfin eel *A. australis*, Australian longfin eel *A. dieffenbachii*, and Australian longfin eel (also known as speckled longfin eel) *A. reinhardtii*, are unknown. No larvae of New Zealand longfin eels have been collected, and too few shortfin eel and speckled longfin eel larvae have been collected to use conventional back-tracking of progressively smaller larvae to determine likely spawning areas. The limited larval material together with results from satellite tracking pop-up tags from New Zealand longfin eels indicate that spawning of all three species will be in the tropics, and possible areas were further demarcated by developing a Lagrangian trajectory model based on surface currents derived from hydrography, satellite altimetry, and wind stress. The initial model assumed passive drift of larvae, a third of the total time spent in near-surface layers, and arrival within the larval lifetimes indicated by ages of metamorphosing glass eels. The proportion of successful trajectories enabling arrival offshore of New Zealand or Australia was substantially improved by addition to the model of directed swimming of the larvae towards a destination. The model indicated that possible spawning areas for all three species would be in the northeast of New Caledonia, perhaps within the North Fiji basin between Vanuatu and Fiji. Spawning within this region is consistent with the locations of known larvae, probable migration routes, and the distribution of adult eels in both countries.

Introduction

Of the 15 species of freshwater eel (Tsukamoto and Aoyama 1998; Watanabe et al. 2005) the oceanic spawning grounds of only four are known with any certainty: European eel *Anguilla anguilla* and American eel *A. rostrata* (Schmidt 1922), Japanese eel *A. japonica* (Tsukamoto 1992), and Celebes longfin eel *A. celebesensis* (Aoyama et al. 2003). While it is likely that spawning areas of tropical eels are not far distant from their freshwater habitats (Aoyama et al. 2003; Kuroki et al. 2006), the temperate species are considered to undertake extensive migrations using gyre current systems. However, as evidenced by the known spawning grounds of European eel and Japanese eel, together with tracking information of tagged New Zealand longfin eel *A. dieffenbachii* Gray 1842 (Jellyman and Tsukamoto 2005), spawning areas of temperate species are also likely to be in the tropics (Lecomte-Finiger 2003). This conclusion is consistent with results from molecular phylogeny that propose ancestral eels originated in the tropics near present-day Indonesia (Aoyama and Tsukamoto 1997; Aoyama et al. 2001).

New Zealand has two main species of *Anguilla*—the shortfin eel *Anguilla australis* Richardson, 1848 and the New Zealand longfin eel. The shortfin is found throughout New Zealand and offshore islands and also on the east coast of Australia, Tasmania, Lord Howe Island, New Caledonia, and possibly Fiji.

* Corresponding author: d.jellyman@niwa.co.nz

The New Zealand longfin is endemic to New Zealand and is found throughout the country and offshore islands. The Australian longfin eel (also known as speckled longfin eel) *Anguilla reinhardtii* Steindachner, 1867 is a tropical species that was first recorded from New Zealand waterways in 1996 (Jellyman et al. 1996), but now seems to be a regular arrival to the upper part of the North Island. In Australia, this species occurs along the entire eastern coast, including Tasmania, but is particularly common in the tropical rivers of Queensland (Pusey et al. 2004); it is also found on Lord Howe Island, New Caledonia, and New Guinea.

The classical method of determining spawning areas of freshwater eels has been through the collection of progressively smaller leptocephali (larvae) at sea (Schmidt 1922). Unfortunately, collections of leptocephali of Australasian eels are few and consist of a total of 13 larvae of shortfin eel (Jespersen 1942; Castle 1963; Aoyama et al. 1999) and 14 larvae of Australian longfin eel (Castle 1963; Aoyama et al. 1999). New Zealand longfin eel remains the only species of *Anguilla* for which no larvae have been collected (Jellyman 2003). The recent advent of pop-up tags provides another method of determining likely spawning grounds by tracking seaward migrating silver eels, (Jellyman and Tsukamoto 2002, 2005).

The ability to determine surface current directions and velocities by remote sensing has made it possible to simulate larval trajectories based either on measurements (Polovina et al. 1998; Chiswell et al. 2003) or with output from numerical models tuned or driven by observations (Kimura et al. 1999; Kettle and Haines 2006). Such models have proved useful in verifying likely larval transport vectors of northern hemisphere species of *Anguilla* (Power and McCleave 1983; Kimura et al. 1999; Kettle and Haines 2006). The objectives of this study were to determine likely migration routes of larvae and possible spawning areas for the three Australasian species—shortfin eel, New Zealand longfin eel, and Australian longfin eel, including whether larvae arriving in New Zealand needed to traverse the east coast of Australia or arrive via a more direct route.

Methods

Model Assumptions and Constraints

Model outcomes needed to be capable of providing pathways that resulted in a high probability of shortfin eel arriving in both Australia and New Zealand (Jellyman 1987; Shiao et al. 2001), New Zealand longfin eel arriving only in New Zealand and Australian longfin eel arriving mainly in Australia but occasionally in New Zealand. Further, likely spawning areas and subsequent larval dispersal should accommodate the limited data on larvae of shortfin eel and Australian longfin eel and the ascent locations of satellite tracking tags from migrating New Zealand longfin eel females (Jellyman and Tsukamoto 2005).

A further requirement of the model was that transport to within 150 km offshore (an average width of the continental shelf) of mainland Australia or northern New Zealand must occur within the estimated duration of larval life (as estimated from daily growth ring studies). Given that transport on the East Australian Current (EAC; Figure 1) has often been assumed as the primary vector to both Australia and New Zealand (Castle 1963; Aoyama et al. 1999), we investigated its likelihood also.

The initial model tested passive transport only (i.e., larvae did not actively swim), although it incorporated diel vertical movements, meaning that larvae would spend a proportion of their time within the shallow, directly wind-driven depths. The model required values for several parameters, including likely diel swimming depths (and hence time larvae spent in the wind-driven depths), the duration of larval life, and the probable age of larvae. We adopted a depth of 50 m for the wind-driven layer and assumed that larvae would spend up to 8 h/d within this zone. For the duration of larval life, we used age at metamorphosis rather than age at recruitment, as age at metamorphosis represents arrival off the continental shelf where metamorphosis is presumed to take place. We used average larval lifetimes of 140, 185, and 250 d and maximum ages of 175, 270, and 300 d for Australian longfin eels, shortfin eels, and New Zealand longfin eels, respectively (Table 1).

The smallest larvae of shortfin eels (Table 2) range from 20.5 to 32.2 mm; the size range for Australian longfin eels is very similar (20.0–34.0 mm). To estimate the ages of these larvae, we used growth rates estimated for shortfin eel larvae in Australia of 0.25 mm/d (Shiao et al. 2002). These rates were similar to those estimated for American eel (Kleckner and McCleave 1985). Locations of captured larvae, together with ascent locations of pop-up tags released from silver eels, corroborated the approxi-

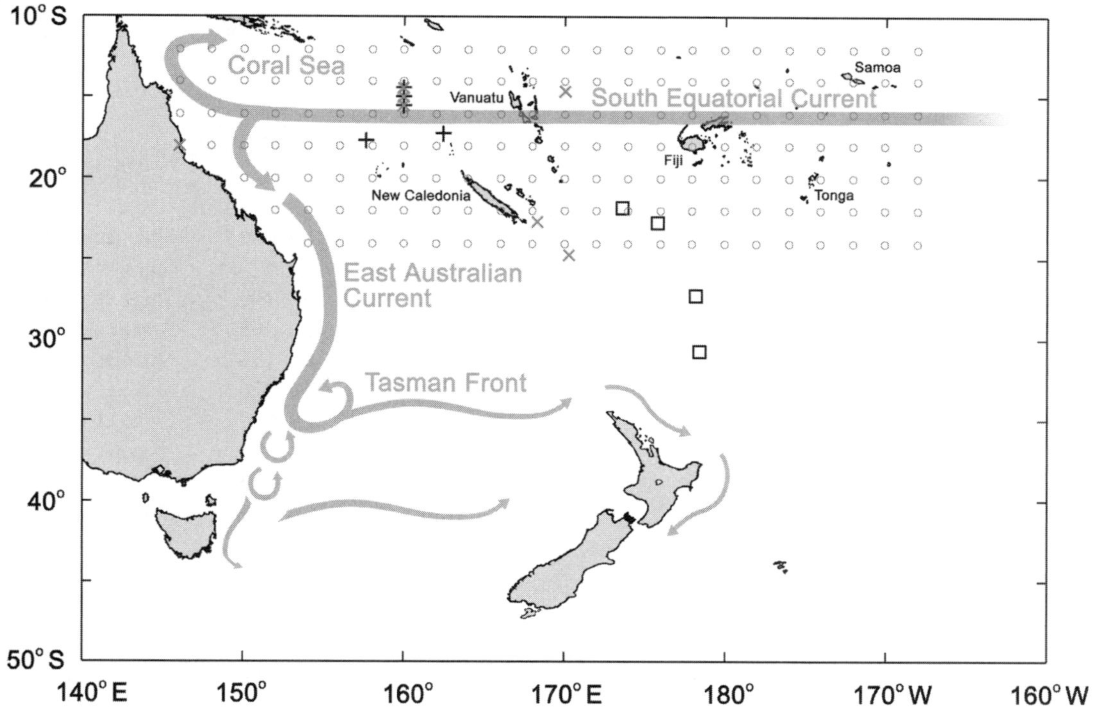

Figure 1.—The main surface currents of the South Pacific circulation. Indicated are the starting locations of the model trajectories (small circles), distribution of collections of larvae of shortfin eels (crosses) and Australilan longfin eel (pluses), and the ascent locations of pop-up tags from adult New Zealand longfin eel (squares).

mate start point of larval return trajectories (Figure 1; Jellyman and Tsukamoto 2005; National Institute of Water and Atmospheric Research [NIWA], unpublished data)

Two modifications were subsequently made to the model—first, an allowance was made for the likelihood that larvae could actively swim horizontally, and hence dispersal was not entirely dependent upon passive drift in surface currents; second, that larvae had an innate sense of direction and the ability to navigate.

For most simulations, swimming speed (u_s) was assumed it be positively related to length and increased logarithmically with time (t) from 0 cm/s at hatching to a maximum speed at the average larval lifetime (T).

$$u_s = 10\,(1 - e^{-}\frac{t}{0.2T})\ \text{cm/s} \qquad (1)$$

In recognition that swimming of newly hatched larvae would be, at best, weak and possibly nonexistent, we also tested a hyperbolic tangent swimming curve that assumed larvae did not commence swimming until they reached an age equivalent to more than 20% of their total larval lifetime and achieved 50% of their maximum swimming ability by half their larval lifetime.

For most simulations, we assumed a maximum swimming speed of 10 cm/s, although for sensitivity testing of the model, we also examined the affect of both 5 and 15 cm/s. The direction of swimming was updated throughout the simulation. For simulations with swimming towards New Zealand, swimming was aimed towards the center of the North Island of New Zealand. For simulations swimming towards Australia, swimming of each simulated larva was aimed at all times towards the coastline closest to that larva.

Trajectory Simulations

Larval trajectories were simulated using data on ocean currents and estimated swimming speeds. Position along each trajectory (x, y) was incremented

Table 1.—Age in days at metamorphosis of Australian and New Zealand glass eels. WC = West Coast; SI = South Island; EC = East Coast ; NI = North Island.

Country	Location	No.	Age at metamorphosis (d) Mean	Range	Reference
Shortfin eels					
Australia	Fitzroy R (~23°S)	29	160 ± 14	130–180	Shiao et al. 2002
	Kolan R (~25°S)	33	160 ± 12		Shiao et al. 2001
	Albert R (~27°S)	27	161 ± 13	144–190	Shiao et al. 2002
	Albert R (~27°S)	10	164 ± 19	138–198	Arai et al. 1999
	Port Hacking (~34°S)	26	168 ± 15	130–196	Shiao et al. 2002
	Brodribb R (~37°S)	31	184 ± 22	150–226	Shiao et al. 2002
	Tarwin R (~38°S)	37	189 ± 17	160–245	Shiao et al. 2002
New Zealand	Arahura (WC, SI)	22	188 ± 12		Shiao et al. 2001
	Arahura (WC, SI)	10	185 ± 17	161–208	Arai et al. 1999
	Purau St (EC, SI)	19	180 ± 9		Shiao et al. 2001
	Pigeon Bay St (EC, SI)	12	187 ± 19	161–211	Arai et al. 1999
	Makara St (WC, NI)	10	204 ± 39	151–265	Marui et al. 2001
New Zealand longfin eels					
	Arahura (WC, SI)	10	248 ± 31	189–282	Marui et al. 2001
	Pigeon Bay St (EC, SI)	10	251 ± 32	204–294	Marui et al. 2001
Australian longfin eels					
	Barron R (~16°S)	30	143 ± 11	130–171	Shiao et al. 2002
	Fitzroy R (~23°S)	31	148 ± 11	126–177	Shiao et al. 2002
	Albert R (~27°S)	90	136 ± 14	96–175	Shiao et al. 2002
	Port Hacking (~34°S)	29	143 ± 8	130–157	Shiao et al. 2002

Table 2.—Distributions of known larvae of shortfin eels and Australian longfin eels. * = based on growth rates for shortfin eel larvae reaching Australia of 0.25 mm/d (Shiao et al. 2002).

	Year	Latitude	Longitude	No.	Length (mm)	Age (d)*	Reference
Shortfin eels							
	1928	24°46′	170°18′	1	43.0	172	Jespersen 1942
	1956	14°37′	170°03′	1	24.6	98	Castle 1963
	1960	22°38’	168°20′	1	47.6	190	Castle 1963
	?	17–19°	~146°W	1	75.7		Castle 1963
	1995	14°30′ to 15°30′	160°00′ to 1603′	9	20.5–32.2	82–129	Aoyama et al. 1999
Australian longfin eels							
	1960	17°13′	162°30′	1	27.3	109	Castle 1963
	1960	17°40′	157°40′	1	27.6	110	Castle 1963
	1995	14°30′ to 15°30′	160°00′ to 160°03′	12	20.0–34.0	80–136	Aoyama et al.1999

at each time step (Δt) using the velocity (U) at that location and time.

$$x_{n+1} = x_n + U_x \Delta t; y_{n+1} = y_n + U_y \Delta t \quad (2)$$

The velocities were the sum of the mean currents (U_m), the time-dependent geostrophic currents (U_e), the directly wind-driven currents (U_w), and a swimming speed (U_s). The mean flow (U_m) was estimated by calculating the dynamic height relative to 2,000 m from temperature and salinity climatologies in CARS (the CSIRO Atlas of Regional Seas; Ridgway and Dunn 2003). Time-varying geostrophic velocities (U_e) were estimated using weekly mapped sea surface height anomalies derived from satellite altimeter observations distributed by AVISO (Ducet et al. 2000). Geostrophic velocities were derived from sea surface height gradients, assuming a geostrophic balance. Altimeter data used were from the start of the time series October 15, 1992 until June 30, 2006.

The directly wind-driven flow was computed from the Ekman transport (Tomczak and Godfrey 1994) using the NCEP Reanalysis surface ocean wind stress (τ) divided by an Ekman depth (δ) of 50 m with the coriolis parameter (f) and density (ρ). Assuming that the larvae are in the wind-driven layer a third of the time, the magnitude of the directly wind-driven flow is given by the expression below and is 90 degrees to the left of the wind.

$$u_e = \frac{1}{3} \left| \frac{\tau}{\rho f \delta} \right| \quad (3)$$

Starting locations for the simulations (Figure 1) were chosen to cover the range of spawning locations previously suggested in the literature. Trajectories were started every two degrees of latitude and longitude between 12°S and 24°S and 146°E and 192°E, providing 155 locations. Simulations started 10 d apart were regarded as being independent; more than 470 trajectories were carried out for species at each location and each of the three swimming speeds. Trajectories were determined by integrating velocities (all interpolated to the AVISO time and space grid) with a fourth-order Runge-Kutta method (Press et al. 1989), using a time step of 3 h over the region between latitude 10–50°S (Figure 1).

A trajectory that came within 150 km of the coast (i.e., the edge of the continental shelf) within the defined larval lifetime for that species was defined as "successful"; the total number of successful trajectories for any given start point was then expressed as the overall likelihood of success of that point. Start points that resulted in trajectories that reached the coast in less than or greater than the larval lifetime (Table 1, Australian longfin eels 96–177 d, shortfin eels (New Zealand) 151–265 d, shortfin eels (Australia) 130–245 d, New Zealand longfin eel 189–294 d) were eliminated as a potential spawning location. No attempt was made to model subsequent dispersal of metamorphosed glass eels (the stage of recruitment into freshwater) around coastlines, as currents for this would be coastal rather than oceanic.

Sensitivity tests were performed to assess the dependence of the trajectories on the mean flow, the velocities derived from the altimeter data, and the fraction of time spent in the wind-driven currents. An independent measure of the distances particles can travel with surface currents was obtained from the tracks from ocean drifters released in the South Pacific between 1979 and 2005 (www.meds-sdmm.dfo-mpo.gc.ca/meds/Home_e.htm). The drifting buoys ride higher in the water column (15 m) than the eel larvae and hence their tracks are indicative of the fastest velocities and more directly wind-driven flow.

In addition to the above simulations, we investigated the likelihood of "accidental" arrivals of shortfin eel and Australian longfin eel in both Australia and New Zealand. For this, the likelihood of either species bound for one country but accidentally arriving in the other was determined to see whether this provided a plausible mechanism for recruitment of relatively small numbers of larvae to the accidental country.

Near-Surface Oceanography of the South Pacific

Flow is westward in the South Equatorial Current (SEC) along the northern edge of the South Pacific subtropical gyre (Figure 1), with maximum flows near 17°S (Ridgway and Dunn 2003). The SEC bifurcates near the Australian coast at about 18°S, with some flow turning north and the southward flow feeding the EAC. The EAC is the fastest current in the region with sustained speeds of 60 cm/s (Mata et al. 2000), with instances of currents reaching nearly 2 m/s (Cresswell and Legeckis 1986).

Near 33°S, the EAC separates from the coast and turns eastward to form the Tasman Front (TF), a feature which extends from the Australian coast to the northern tip of New Zealand (Tomczak and Godfrey 1994). A portion of the flow in the EAC continues southward along the Australian coast.

To investigate the likelihood of larvae of all three species using the EAC to reach New Zealand, the expression for swimming speeds (equation 1) and the mean currents quoted above were used to back-calculate where the larvae of each species could originate if they travel along the EAC pathway. Assuming the larvae travel a distance D_x from the spawning grounds to the entrance of the EAC and then swim with the fastest mean currents to northern New Zealand, the distances can be written as an integral of the velocities along the path:

$$\int_0^{t_1} u_{swim}\,dt = D_x; \int_{t_1}^{t_2} (U_{eac} + U_{swim})\,dt = D_{eac}; \int_{t_2}^{t_3} (U_{tf} + U_{swim})\,dt = D_{tf} \quad (4)$$

where U_{swim} is given by equation (4), U_{eac} = 0.4 m/s, U_{tf} = 0.1 m/s, D_{eac} = 1,000 km, D_{tf} = 1,800 km, t_1 = the time between spawning and reaching the entry of the EAC, t_2 = the time at which the larvae reach the end of the EAC/beginning of the tf, and t_3 = the time at which larvae reach New Zealand. The distance between the entrance of the East Australian Current and the hatching location (D_x) could then be estimated for the range of larval lifetimes.

Results

Passive Drift

Initially, simulations were run using only the ocean currents to advect particles. Very few of the simulations produced trajectories that arrived in Australia or New Zealand within the measured larval lifetimes (Figure 2). For all species, the number of trajectories at any successful location was small (less than 3%).

For shortfin eels and New Zealand longfin eels, only a few of trajectories made it to New Zealand from locations near the entrance of the EAC. Shortfin eels were able to reach Australia if they started at locations between New Caledonia and Vanuatu and northeast of Vanuatu. Simulations did not result in shortfin eels reaching both Australia and New Zealand from a common spawning region. Australian longfin eels could reach Australia from locations west of New Caledonia and Vanuatu and north of Vanuatu, but passive drift did not allow them to reach New Zealand within the duration of their larval life.

Results from ocean drifters were similar to the trajectory simulations. None of the trajectories from

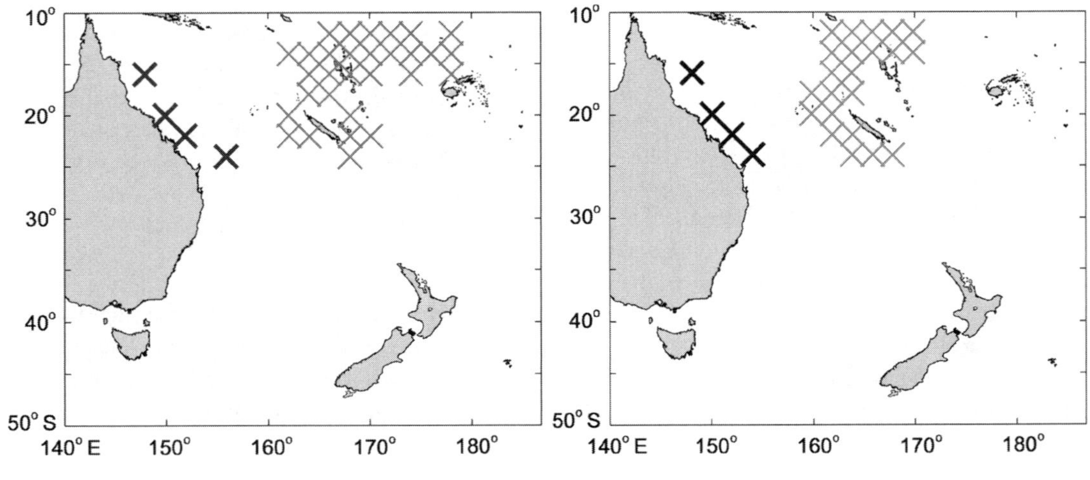

Figure 2.—Passive drift simulations. Left: starting locations of trajectories that arrive in New Zealand (bold crosses) and Australia (crosses) within larval lifetimes of shortfin eels. Right: starting locations of trajectories that arrive in New Zealand within New Zealand longfin eel larval lifetimes (bold crosses) and trajectories that arrive in Australian waters within Australian longfin eel larval lifetimes (crosses).

the ocean drifters that pass near the New Zealand coast were in the tropics 300 d beforehand. Drifters arrive near the Australian coast from tropical locations as far east as the date line (in 250 d), between New Caledonia and Fiji (within 185 d) and all regions of the Coral Sea within 140 d.

Addition of Larval Swimming

Addition of larval swimming at higher velocities to the model significantly improved the incidence of successful trajectories. Assuming swimming ability increased logarithmically with size; if the larvae could swim in a calm ocean, they would reach Australia and New Zealand from locations within calculated distances of 640–2,540 km, depending on species (Table 3). Shortfin eels would be able to reach both Australia and New Zealand from a region around New Caledonia. New Zealand longfin eels could swim to New Zealand from a semicircular region west of New Caledonia, northward to Vanuatu, and east to southern Fiji and Tonga. Australian longfin eels could swim across the Coral Sea to reach Australia; however, to reach New Zealand, a common spawning ground would need to be near New Caledonia.

The mean and variable ocean currents can assist the larvae in reaching coastal waters from some starting locations and reduce the likelihood of arrival from other locations. With a maximum swimming speed of 5 cm/s, shortfin eels have less than a 10% chance of arrival in either Australia or New Zealand from any of the potential spawning locations (Figure 3). Sites east of New Caledonia allowed some larvae to reach both countries. With a swimming speed of 10 cm/s, larvae could reach both countries starting from locations between 166°E and 172°E, with more reaching New Zealand from the southern region and more reaching Australia from the northern region. A swimming speed of 15 cm/s extended the possible spawning region further east and north. Chances of reaching both countries were greatest if the spawning areas lie between Vanuatu and Fiji and north of Fiji. It was only possible for larvae to accidentally arrive in New Zealand from starting locations near 24°S (not shown). Most accidental arrivals in Australia required starting locations near Australia.

With a swimming speed of 5 cm/s, less than 10% of the New Zealand longfin eel larvae starting at any location could reach New Zealand within the larval lifetime (Figure 4). A swimming speed of 10 cm/s enabled greater percentages of larvae to reach New Zealand from locations originating south of Tonga, between Fiji and New Caledonia, and around Vanuatu and in the Coral Sea. Swimming at 15 cm/s, larvae needed to start well north of New Caledonia and Tonga. Locations with the highest probability are north of Vanuatu and the central Coral Sea, north of Fiji, and between Tonga, Samoa, and Niue. However, the location of the pop-up tags from New Zealand longfin eel adults (Figure 1) suggests that spawning sites between New Caledonia and Fiji are more likely than locations further east or west.

The simulations were also examined for New Zealand longfin eels accidentally arriving in Australia while swimming towards New Zealand in order to see whether these simulations would help restrict possible spawning areas for this species. Results (Figure 4, gray symbols) indicated that if the spawning ground was east of 160°E, there was some likelihood that New Zealand longfin eel larvae would arrive in Australian waters, something that has not been recorded.

Few trajectories of Australian longfin eel larvae were successful with a swimming speed of 5 cm/s (Figure 5). At 10 cm/s, larvae could reach Australia from just east of Vanuatu. Many of these locations, especially closer to New Caledonia, also allowed a

Table 3.—The total distance (km) that each eel species could swim during their average larval lifetime, at three continuous maximum swimming speeds.

Maximum swimming speed	Shortfin eels	New Zealand longfin eels	Australian longfin eels
5 cm/s	640	860	480
10 cm/s	1,280	1,730	970
15 cm/s	1,920	2,590	1,450

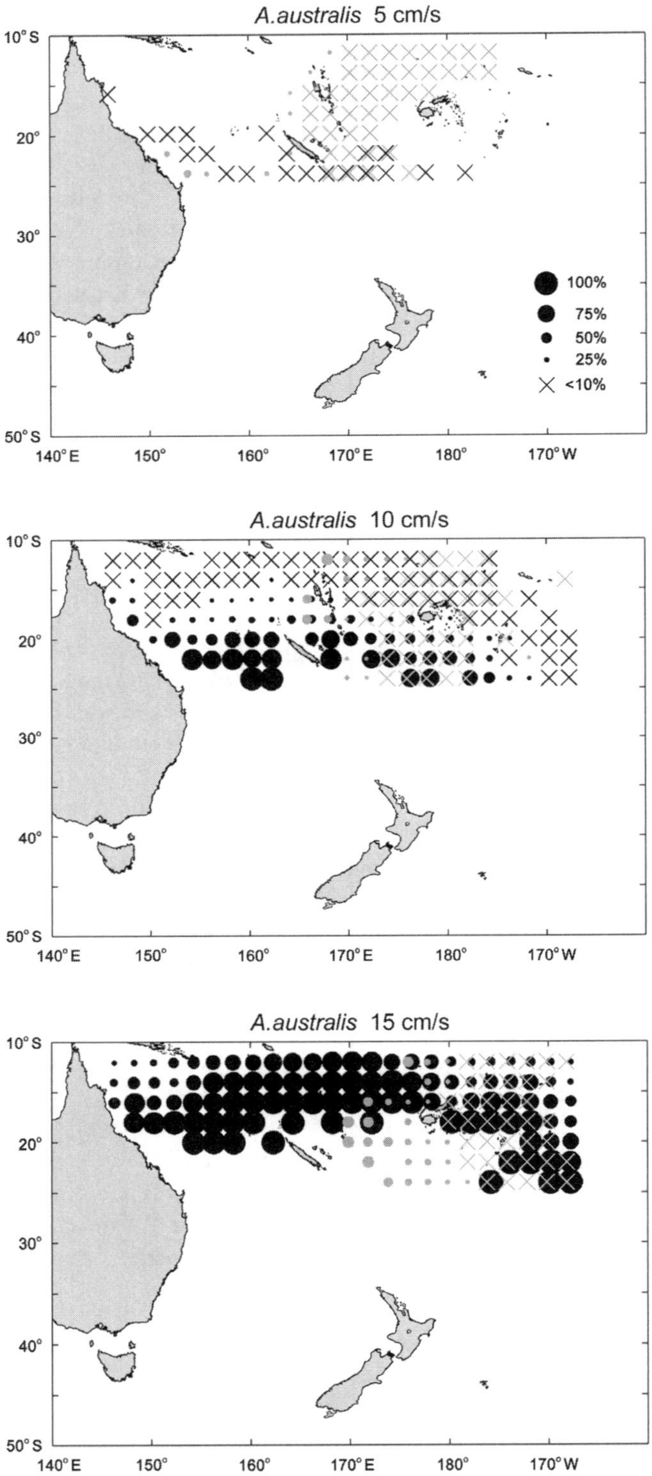

Figure 3.—Percentage of successful trajectories for starting locations of shortfin eels swimming towards Australia (gray) and New Zealand (black) for three maximum swimming velocities.

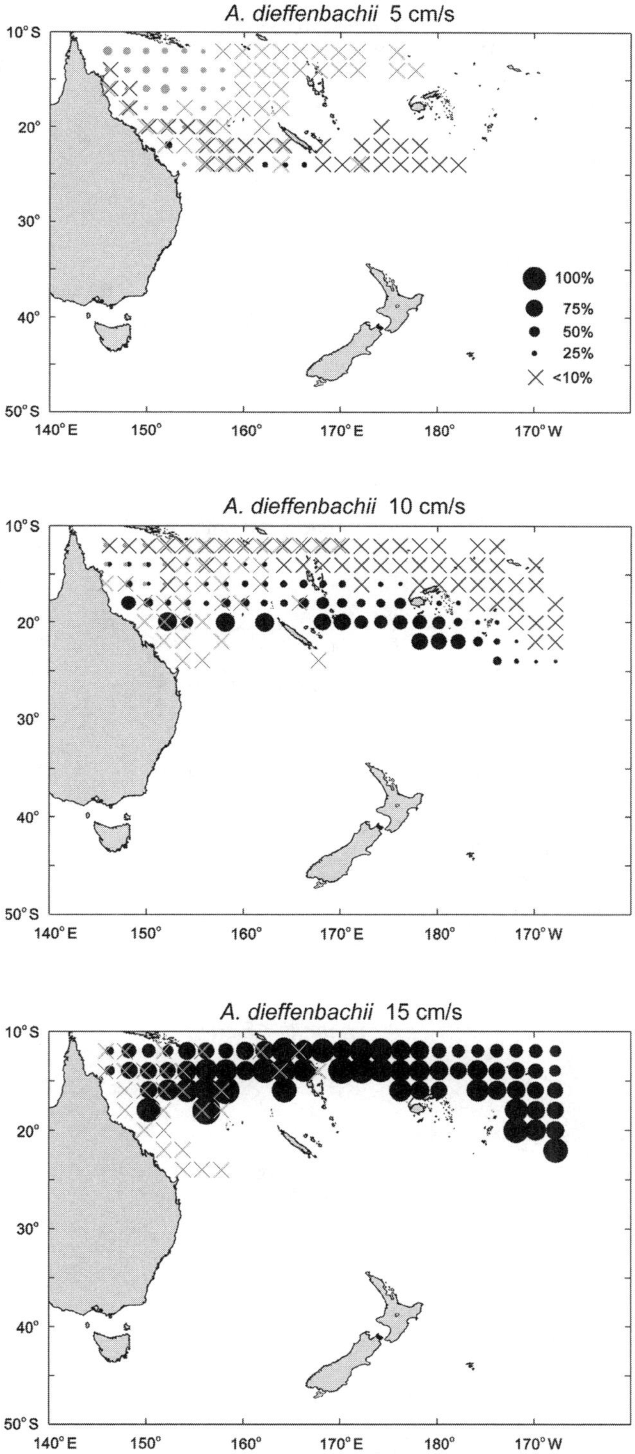

Figure 4.—The percentage of successful trajectories found for starting locations of New Zealand longfin eels swimming towards New Zealand (black) for three maximum swimming velocities. Indicated also are the locations (gray) that would allow arrival in Australia while swimming towards New Zealand.

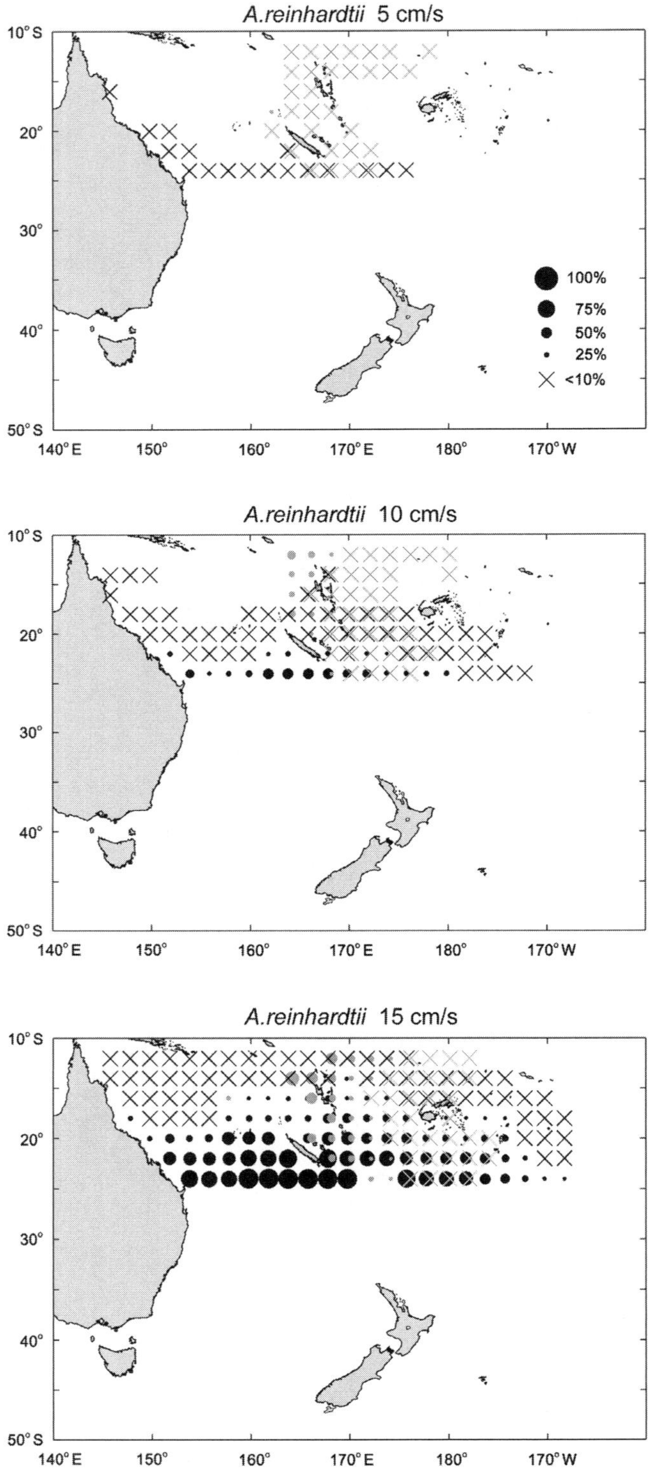

Figure 5.—Percentage of successful trajectories found for starting locations of Australian longfin eels swimming towards Australia (gray) and New Zealand (black) for three maximum swimming velocities.

small percentage of arrivals in New Zealand. A few accidental arrivals in New Zealand were only possible from a site far south of Fiji. This suggests that Australian longfin eels are unlikely to accidentally arrive in New Zealand while swimming towards Australia.

Results for 15 cm/s for Australian longfin eels were generally similar to those for 10 cm/s, with the largest number of successful trajectories arriving in Australia from spawning sites in northern Vanuatu. These sites also allowed a small percentage of larvae to reach New Zealand, New Guinea, and Lord Howe Island.

Overall, the start locations where eels can most successfully reach Australia and New Zealand are not locations where they can easily accidentally arrive in the other country. The start locations of accidental arrivals were different from those of the successful trajectories, which suggest that swimming direction is a deliberate choice.

Sensitivity Tests

There was little change in the number or distribution of successful trajectories using mean flows derived from the World Ocean Atlas climatology, doubling the velocities derived from the altimeter, or assuming that larvae remain either below or within the wind-driven flow at all times. Whether a trajectory reached land was far more dependent on the swimming ability than variations in the ocean currents. Preliminary assessments of the effect of interannual variability indicated that total recruitment could vary by a factor of five over the 14 years of oceanographic record available. The impact of such variability should be explored. The use of the hyperbolic tangent swimming speed curve gave possible spawning locations (not shown) that were not as distant as those generated by the equivalent logarithmic curve. For example, a hyperbolic swimming curve with a maximum swimming speed of 10 cm/s gave possible spawning areas that were approximately midway between those produced from the logarithmic swimming speed curve for swimming speeds of 5 cm/s and 10 cm/s.

The East Australian Current Pathway

At the mean current speed of the EAC (40 cm/s), it would take passively drifting larvae about 240 d and swimming larvae about 130 d to traverse the distance from the inflow to the EAC (at about 25°S on the Australian coast) to the northern tip of New Zealand. Results from the simulation of swimming larvae using the EAC to arrive at New Zealand showed that for all three species, the spawning locations would need to be confined to the Coral Sea (i.e., Australian longfin eels would need to originate very near the point of entry of the EAC while shortfin eels and New Zealand longfin eels could have originated within a radius of 225 km and 650 km, respectively.

Consistency of Proposed Spawning Areas with Larval Distributions

Starting locations for shortfin eels and Australian longfin eels, where trajectories passed less than 150 km from each larval collection site within ±10 d of the larval age at that site, were assessed with the model (Figure 6). Locations for Australian longfin eels and shortfin eels swimming at 15 cm/s (results not shown) are consistent with the areas suggested by the directed swimming model as possible spawning areas for these species. However, the consistent locations for shortfin eels at 10 cm/s were further north and east than the trajectory results indicated (Figure 3).

Possible Spawning Areas

Overall conclusions were that the three eel species could spawn in the same general area, north or east of Vanuatu and between Vanuatu and Fiji (Figure 7). For shortfin eels, the area suggested by the 15 cm/s maximum swimming speed may be more likely than the alternative area (10 cm/s) further west, as it better accommodates the distribution of the larval samples. As spawning west of Vanuatu would mean some likelihood of New Zealand longfin eels being present in Australia north of 25°S, it seems more likely that this species would spawn east of Vanuatu, an area partially confirmed by the ascent locations of the pop-up tags (Figure 1).

Discussion

Advantages of Modelling

Although oceanic expeditions are still used to investigate likely spawning grounds of anguillid eels (Aoyama et al. 1999), the high cost of mounting

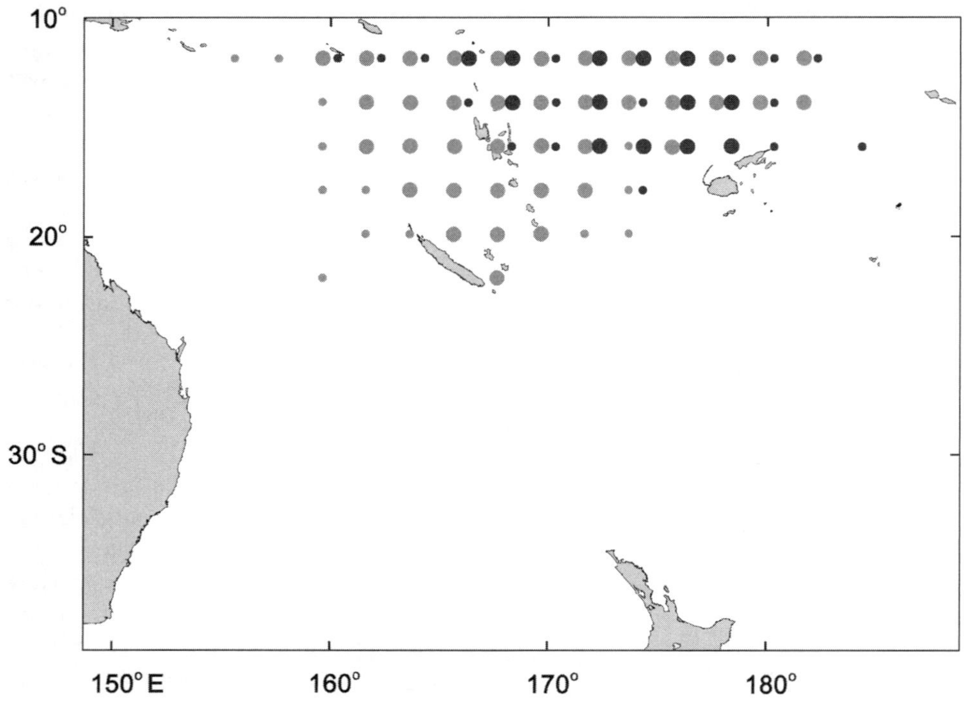

Figure 6.—The percentage of successful trajectories for starting locations of shortfin eels (black) and Australian longfin eels (gray) that were consistent with the distributions of larval collections of each species. Large circles indicate 50% success of trajectories, smaller circles 10% success. Simulations are for maximum swimming speed of 10 cm/s.

such expeditions and their often lack of success mean that alternative methods should be investigated. More than a decade of high-quality ocean observations enables simulations of near-surface drift trajectories over wide areas and under varying conditions. Previous models of anguillid larvae dispersal were based on advection-diffusion equations (Power and McCleave 1983) or particle drift models (Kimura et al. 1999; Kimura 2003). A recent Lagrangian model (Kettle and Haines 2006) seeded virtual drifters in a numerical model over 4 years of ocean circulation data and assumed passive drift and no vertical movement of larvae. Our model differed by carrying out simulations over 14 years of ocean currents and incorporating both vertical and directed horizontal swimming.

Model Assumptions

The directly wind-driven layer is relatively shallow, and other researchers have assumed depths of 50 m (Chiswell et al. 2003; Griffin et al. 2001) and 70 m (Kimura et al. 1994). The proportion of time that larvae spend within this region depends on their diel vertical migrations. Knowledge of vertical migrations is limited (Tsukamoto et al. 2003a), but swimming depths of larvae more than 5 mm are thought to be 80–250 m during the day, and 30–100 m during the night (Castonguay and McCleave 1987; Otake et al. 1998). For the present model, we used a 50 m depth and assumed that larvae would be present within this for 8 h/d.

For *Anguilla* spp., there has been uncertainty whether the putative daily growth rings are actually deposited daily (McCleave et al. 1998). Also, whether deposition continues during metamorphosis (Antunes and Tesch 1997) or there is some resorption during this phase, is a question (Cieri and McCleave 2000). The formation of daily growth rings has been validated for glass eels of both temperate (American eel, Martin 1995; Cieri and McCleave 2001), and tropical *Anguilla* spp.

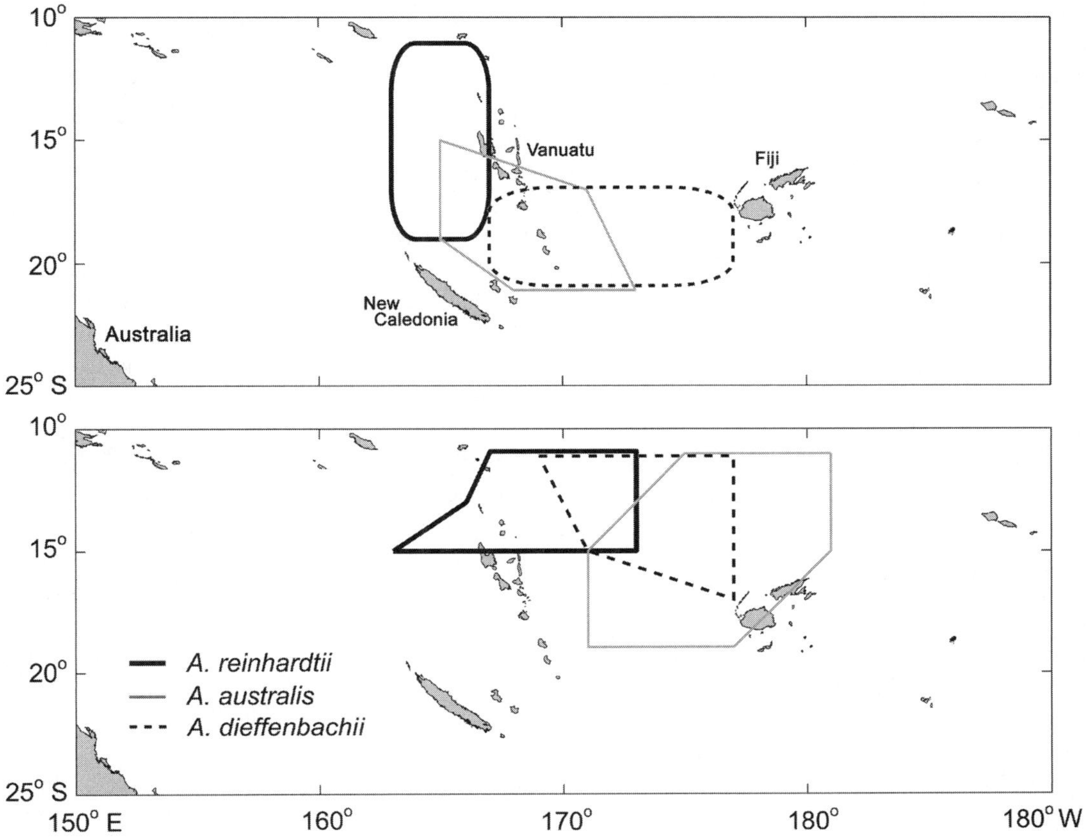

Figure 7.—Possible spawning locations for Australian longfin eels, shortfin eels, and New Zealand longfin eels suggested by the logarithmic swimming speed model for maximum swimming speed.

(marbled eel *A. marmorata*, Sugeha et al. 2001; Celebes longfin eel). Thus, we felt warranted in assuming that the sum of the (putative) daily rings was a reasonable indicator of duration of larval life, and the use of age at metamorphosis avoided any uncertainty about ring deposition during this phase.

Early accounts of larval transport assumed passive drift within near-surface ocean currents (Schmidt 1922). Subsequent information has shown that larvae make significant diel vertical movements (Castonguay and McCleave 1987; Otake et al. 1998), and American (and European) eel larvae must actively swim to detrain from the Gulf Stream. While some researchers suggest that active swimming is required to transport larvae across the Atlantic Ocean at faster rates than accounted for by passive drift (e.g., Lecomte-Finiger 1992; Arai et al. 2000b), other authors remain skeptical whether active swimming plays a significant role in larval transport (McCleave et al. 1998).

To enable a significant proportion of trajectories to achieve landfall, our model indicated that active swimming was necessary. However, there is considerable uncertainty about possible speed and duration of larval swimming. Tesch et al. (1986) assumed that large larvae could swim 20 cm/s and smaller larvae at two body lengths/s (approximately 7 cm/s for half-grown larvae). Based on the small quantity of red muscle that leptocephali have, McCleave et al. (1998) considered these speeds too high. Glass eels are active swimmers, with Australasian glass eels having sustained swimming speeds of more than 50 cm/s (Langdon and Collins 2000). The only measurements of Anguilliform larval swimming speeds are those for late-stage leptocephali of a marine conger eel *Conger oceanicus*, which have sustained swimming speeds of approximately 9 cm/s, equivalent to approximately one body length/s (M. Wuenschel, Rutgers University, personal communication). In the absence of verified data, we assumed a length-related swimming speed, with speeds ranging from 0 km/s at hatching to maximums of 5, 10, or 15 cm/s for full grown larvae.

The model requirement of an innate sense of swimming direction is the first time such a mechanism has been proposed for migrating eel larvae. However, it is known that both glass eels and adult eels are magnetosensitive (Tesch et al. 1992; Nishi and Kawamura 2005; van Ginneken et al. 2005), and it has been suggested (Nishi and Kawamura 2005) that migration of silver eels may follow a similar path to that which they followed as larvae by reversing the goemagnetic cues "memorized" during the larval stage. Irrespective of this, the fact that glass eels and presumably leptocephali are sensitive to magnetic fields could provide a means of navigating, although the choice of direction must be innate.

Such a mechanism could explain why larval American eels detrain from the Gulf Stream to invade North America, while coexisting larvae of European eels continue towards Europe. Previous authors have postulated that vertical migration differences might enable differential dispersal of larvae of both species and subsequent detrainment of American eels (Castonguay and McCleave 1987; Kettle and Haines 2006), although no such vertical differences have been observed. Another suggestion has been different development rates of both species (Tesch 2003), but given the spatial and temporal overlaps in spawning (McCleave and Kleckner 1987; McCleave 2003), this explanation is also problematic. While unproven, our explanation of directed migration is known from studies of demersal fish larvae (Leis 2006) and from migratory birds where naïve fledglings use magnetic cues to migrate extensive distances unaccompanied by adults (Muheim et al. 2006; Shaffer et al. 2006). Unlike other *Anguilla* species where "missing" of the target landmass could still result in larvae making landfall in nearby countries, eels destined for New Zealand must show a high degree of navigational accuracy or they would miss New Zealand and presumably perish in the open ocean of the South Pacific. The ability to navigate to a predetermined destination would greatly enhance recruitment success.

Possible Spawning Areas

Results from the present study confirm that all three species (shortfin eel, New Zealand longfin eel, and Australian longfin eel) spawn in the tropics, a feature assumed to be common to all *Anguilla* species (Lecomte-Finiger 2003). Further, all three species are likely to spawn between 10°S and 25°S and between 165°W and 180°W, possibly in geographically overlapping areas. Overlaps in spawning areas occur for European eels and American eels (Schmidt 1922; McCleave 2003), and although it seems likely that there are some temporal differences in peak spawning times, the spawning of both species is at least partially sympatric in space and time (McCleave and Kleckner 1987).

For all three species, the use of the faster maximum swimming speed of 15 cm/s has the effect of shifting possible spawning areas slightly north and east of those suggested by using 10 cm/s. For convenience, the suggested spawning areas for 10 and 15 cm/s are drawn with distinct boundaries (Figure 7). In reality, the areas are unlikely to be geographically fixed and actual spawning areas may lie on a continuum between the suggested areas.

A number of authors have suggested possible locations of spawning areas of Australasian eels. Schmidt (1928) assumed that the New Caledonia submarine ridge formed a physical barrier between the spawning areas of Australian and New Zealand shortfins. Other researchers have suggested locations further east, with Castle (1963) and Arai et al. (1999) suggesting an area between Fiji and Tahiti, although Jellyman (1987) considered that the area might be further north, in the vicinity of 150–170°W and 5–15°S. Since New Zealand longfin eel is confined to New Zealand, and compared with the coexisting species, shortfin eel, it is larger, more sexually advanced at migration, and migrates later, Jellyman (1987) considered a spawning area might lie closer to New Zealand to the east of Tonga. Jespersen (1942) considered that the spawning ground for Australian longfin eel would be within the Coral Sea, with the EAC providing larval transport to the eastern coast of Australia. The overlapping distributions of small numbers of larvae of this species and shortfin eels caught northwest of New Caledonia led Aoyama et al. (1999) to conclude that the same dispersal routes were utilized by both species. More recently, Shiao et al. (2002) concluded that the spawning site for Australian longfin eels would be further west than that of shortfin eels, possibly near or west of Fiji, and Miller et al. (2006) recorded greatest catches of anguillid larvae at the latitudes of the so-called fresh pool within the SEC. Our results indicate that possible spawning areas are considerably further west than those proposed by Castle (1963), Jellyman (1987), and Arai et al. (1999), further east than areas suggested by Schmidt (1928) for shortfin eel and by Jespersen (1942) for Australian longfin eel, but in the vicinity of the fresh pool suggested by Miller et al. (2009, this volume).

Similar to the spawning areas of European eels and American eels, which are not geographically fixed but bounded by thermal fronts (Kleckner et al. 1983; Kleckner and McCleave 1985, 1988; Tesch and Wegner 1990), it is likely that spawning areas of Australian and New Zealand eels will also be associated with hydrological rather than physical features. The spawning area of the Japanese eel may also be bounded to the north by a marked salinity front at the northern margin of the North Equatorial Current (Tsukamoto 1992; Tsukamoto et al. 2003a), although the presence of seamounts in this vicinity has led to suggestions that spawning might occur at these mounts (Fricke and Tsukamoto 1998; Tsukamoto et al. 2003b). Given the ecological plasticity that typifies the genus *Anguilla* (Helfman et al. 1987), it would seem unlikely that one species would adopt an inflexible spawning strategy of spawning at a geographically fixed location that could make spawning success vulnerable to shifts in oceanic conditions.

There is still some uncertainty about the subspecies status of shortfin eels. From examination of mtDNA or allozymes, several researchers concluded that the species is genetically homogeneous (Dijkstra and Jellyman 1999; Aoyama et al. 2000; Smith et al. 2001), inferring a single spawning area (panmixia) or, at least, incomplete spawning segregation of eastern (New Zealand) and western stocks (Australian). In contrast, Watanabe et al. (2005) and Shen and Tzeng (2007) recorded considerable genetic variability for a different mtDNA sequence and microsatellites, respectively. Shen and Tzeng (2007) concluded that the two populations are reproductively isolated, although it is possible that some genetic exchange occurs between them.

Irrespective of these divergent findings, there are small but consistent differences in morphological features between Australian and New Zealand populations of shortfin eels (Schmidt 1928; Ege 1939; Jellyman 1987; Watanabe et al. 2006). Jellyman (1987) suggested that such differences could be explained by small temperature differences across the spawning ground(s), as there is some lability in vertebral number during larval development (Blaxter 1969), with temperature being influential (Tåning 1952; Pavolv and Shadrin 1998). For the present paper, we have adopted the suggestion of Jellyman (1987) of a single spawning area, with partial segregation into eastern and western stocks. While the possibility of separate spawning sites cannot be ruled out, only the most widely distributed of the 15 Anguillid species, marbled eel, is known to have multiple (five) spawning areas (Ishikawa et al. 2004). Accordingly, there would seem little likelihood of shortfin eels having multiple spawning areas.

There seems little doubt that the SEC is the primary means of transport of shortfin eel larvae to Australia, but there has been some conjecture whether shortfin eels arriving in New Zealand have come via the EAC and TF (Castle 1963; Aoyama et al. 1999) or via a separate and more direct route from north of New Zealand (Jellyman 1987; Shiao et al. 2001). In support of the latter, Shiao et al. (2001) noted that the differences in average ages of glass eels at metamorphosis between Australia and New Zealand was only 8 d, too short for trans-Tasman transport. Results from the present study support the hypothesis that recruitment of shortfin eel in New Zealand is not via the EAC for two reasons: first, using the mean swimming speed of 10 cm/s, it would take about 100 d for larvae to traverse the Tasman Sea, which is outside the range of ages for metamorphosing glass eels from both countries. Second, back-calculations of successful trajectories of larvae transported via the EAC show that the most distant spawning grounds for shortfin eels would be in the Coral Sea, a location at variance with the more substantive modelling results of the present study and opinions of other authors (Castle 1963; Jellyman 1987; Aoyama et al. 1999; Shiao et al. 2001).

Conclusions and Future Directions

Our model was successful in identifying possible spawning areas that are compatible with larval distribution and provide differential migratory routes to Australia and New Zealand consistent with adult distributions. It also indicated the likelihood of adjacent or overlapping spawning areas for the three species investigated. The areas demarcated as possible spawning locations are broadly of similar size to those given by previous researchers for spawning grounds of European eels and American eels (McCleave et al. 1987). Given the assumptions and uncertainties in the model, "forcing" it to define smaller areas is unwarranted and may also be biologically unrealistic as actual locations of the spawning areas may change with oceanic variability (Friedland et al. 2007).

Recruitment of northern hemisphere *Anguilla* species is characterized by both high interannual variability (Moriarty and Dekker 1997), but also an alarming decline in recent decades (Dekker 2003). While a variety of anthropogenic impacts have been suggested as influencing recruitment (Castonguay et al. 1994; Feunteun 2002; Dekker 2003), the role of climate change and oceanographic variability has also been implicated (Knights 2003; Friedland et al. 2007; Bonhommeau et al. 2008). Similar variability in the South Pacific would be expected to influence recruitment success to Australia and New Zealand. For instance, the El Niño Southern Oscillation may influence both the direction and magnitude of recruitment of glass eels to New Zealand, as there is some evidence that strongest recruitment periods are associated with the positive phase La Niña, which is characterized by more persistent northeasterly winds that might enhance larval transport to New Zealand (Chisnall et al. 2002; Miller et al. 2009). There are concerns about the reductions in recruitment of New Zealand species (Jellyman 2007), and being able to distinguish between environmental and anthropogenic causes would be of considerable interest.

At present, the interannual variability in larval arrival from our model is unlikely to explain the interannual variability in recruitment, which can be 10-fold in adjacent years (Jellyman 1979). This suggests that variability in ocean currents alone does not account for variability in recruitment, and factors like sea temperatures and food availability should be included in any subsequent investigations.

Capture of further larvae of all species would provide additional inputs to our model(s), while the temporal variability in ages of shortfin eel glass eels arriving in the north of their range in Australia and of Australian longfin eels arriving in northern New Zealand would help elucidate actual pathways of these species. Given the influence that swimming speed has on determining likely spawning areas, observations on the actual speed and duration of larval swimming should be a priority for research. Future advances in the miniaturization of pop-up tags should enable the tagging of emigrating silver eels of smaller species like shortfin eel. Convergence of pop-up tags from adults of this species tagged in New Zealand and Australia would be a strong indicator of likely spawning areas and could resolve any differences in the spatial and temporal distribution of spawning.

Acknowledgments

We thank Phil Sutton and Steve Chiswell (NIWA, Wellington) for assistance with development of the model. We also thank Mike Miller (The University of Tokyo) for helpful discussions about swimming speeds of leptocephali, and Doug Booker, NIWA, provided many helpful comments on an earlier version of the manuscript. Comments from Brian Knights and an anonymous referee further improved the text.

References

Antunes, C., and F.-W. Tesch. 1997. A critical consideration of the metamorphosis zones when identifying daily rings in otoliths of European eel, *Anguilla anguilla*. Ecology of Freshwater Fish 6:102–107.

Aoyama, J., and K. Tsukamoto. 1997. Evolution of the freshwater eels. Naturwissenschaften 84:17–21.

Aoyama, J., N. Mochioka, T. Otake, S. Ishikawa, Y. Kawakami, P. Castle, M. Nishida, and K. Tsukamoto. 1999. Distribution and dispersal of anguillid leptocephali in the western Pacific Ocean revealed by molecular analysis. Marine Ecology Progress Series 188:193–200.

Aoyama, J., S. Watanabe, M. Nishida, and K. Tsukamoto. 2000. Discrimination of catadromous eels of the genus *Anguilla* using polymerase chain reaction—restriction fragment length polymorphism analysis of the mitochnondrial 16S ribosomal RNA domain. Transactions of the American Fisheries Society 129:873–878.

Aoyama, J., M. Nishida, and K. Tsukamoto. 2001. Molecular phylogeny and evolution of the freshwater eel, genus *Anguilla*. Molecular Phylogenetics and Evolution 20:450–459.

Aoyama, J., S. Wouthuyzen, M. J. Miller, T. Inagaki, and K. Tsukamoto. 2003. Short-distance spawning migration of tropical freshwater eels. Biological Bulletin 204:104–108.

Arai, T., T. Otake, D. J. Jellyman, and K. Tsukamoto. 1999. Differences in the early life history of the Australasian shortfinned eel *Anguilla australis* from Australia and New Zealand, as revealed by otolith microstructure and microchemistry. Marine Biology 135:381–389.

Arai, T., D. Limbong, and K. Tsukamoto. 2000a. Validation of otolith daily increments in the tropical eel *Anguilla celebesensis*. Canadian Journal of Zoology 78:1078–1084.

Arai, T., T. Otake, and K. Tsukamoto. 2000b. Timing of metamorphosis and larval segregation of the Atlantic eels *Anguilla rostrata* and *A. anguilla*, as revealed by otolith microstructure and microchemistry. Marine Biology 137:39–45.

Blaxter, J. H. S. 1969. Development: eggs and larvae. Pages 152–177 *in* W. S. Hoar, and D. J. Randall, editors. Fish physiology 3. Academic, New York.

Bonhommeau, S., Chassot, E., and Rivot, E. 2008. Fluctuations in European eel (*Anguilla anguilla*) recruitment resulting from environmental changes in the Sargasso Sea. Fisheries Oceanography 17:32–44.

Castle, P. H. J. 1963. Anguillid leptocephali in the southwest Pacific. Zoology Publications from Victoria University of Wellington 33.

Castonguay, M., and J. D. McCleave. 1987. Vertical distributions, diel and ontogenetic vertical migrations and net avoidance of leptocephali of *Anguilla* and other common species in the Sargasso Sea. Journal of Plankton Research 9:195–214.

Castonguay, M., P. V. Hodson, C. M. Couillard, M. J. Eckersley, J-D. Dutil, and G. Verreault. 1994. Why is recruitment of the American eel, *Anguilla rostrata*, declining in the St Lawrence River and Gulf? Canadian Journal of Fisheries and Aquatic Sciences 51:479–488.

Chisnall, B. L., D. J. Jellyman, M. L. Bonnett, and J. R. Sykes. 2002. Spatial and temporal variability in length of glass eels (*Anguilla* spp.) in New Zealand. New Zealand Journal of Marine and Freshwater Research 36:89–104.

Chiswell, S. M., J. Wilkin, J. D. Booth, and B. Stanton. 2003. Trans-Tasman Sea larval transport: is Australia a source for New Zealand rock lobsters? Marine Ecology Progress Series 247:173–182.

Cieri, M. D., and J. D. McCleave. 2000. Discrepancies between otoliths of larvae and juveniles of the American eel: is something fishy happening at metamorphosis? Journal of Fish Biology 57:1189–1198.

Cieri, M. D., and J. D. McCleave. 2001. Validation of daily otolith increments in glass-phase American eels *Anguilla rostrata* (Lesueur) during estuarine residency. Journal of Experimental Marine Biology and Ecology 257:219–227.

Cresswell, G. R. and R, Legeckis. 1986. Eddies off southeastern Australia. Deep-Sea Research Part A 33:1527–1562.

Dekker, W. 2003. Status of the European eel stocks and fisheries. Pages 237–254 *in* K. Aida, K. Tsukamoto, and K. Yamauchi, editors. Eel biology. Springer-Verlag, Tokyo.

Dijkstra, L. H., and D. J. Jellyman. 1999. Is the subspecies classification of the freshwater eels *Anguilla australis australis* Richardson and *A. a. schmidtii* Phillipps still valid? Marine and Freshwater Research 50:261–263.

Ducet, N., P. Y. LeTraon and G. Reverdin. 2000. Global high-resolution mapping of ocean circulation from TOPEX/Poseidon and ERS-1/2. Journal of Geophysical Research 105:19477–19498.

Ege, V. 1939. A revision of the genus *Anguilla Shaw*: a systematic, phylogenetic and geographical study. Dana Report 3:1–256.

Feunteun, E. 2002. Management and restoration of European eel population (*Anguilla anguilla*): an impossible bargain. Ecological Engineering 18:575–591.

Fricke, H., and K. Tsukamoto. 1998. Seamounts and the mystery of eel spawning. Naturwissenschaften 85:290–291.

Friedland, K. D., M. J. Miller and B. Knights, B. 2007. Oceanic changes in the Sargasso Sea and declines in recruitment of the European eel. ICES Journal of Marine Science 64:519–530.

van Ginneken, V., B. Muusze, J. K. Breteler, D. Jansma, and G. van de Thillart. 2005. Microelectronic detection of activity level and magnetic orientation of yellow European eel, *Anguilla anguilla* L., in a pond. Environmental Biology of Fishes 72:313–320.

Griffin, D. A., J. L. Wilkin, C. F. Chubb, A. F. Pearce, and N. Caputi. 2001. Ocean currents and the larval phase of Australian western rock lobster, *Panulirus cygnus*. Marine and Freshwater Research 52:1187–1199.

Helfman, G. S., D. J. Facey, J. L. S. Hales, and J. E. L. Bozeman. 1987. Reproductive ecology of the American eel. Pages 42–56 *in* M. J. Dadswell, R. J. Klauda, C. M. Moffitt, R. L. Saunders, R. A. Rulifson, and J. E. Cooper. American Fisheries Society, Symposium 1, Bethesda, Maryland.

Ishikawa, S., K. Tsukamoto, and M. Nisida. 2004. Genetic evidence for multiple geographic populations of the giant mottled eel *Anguilla marmorata* in the Pacific and Indian oceans. Ichthyological Research 51:343–353.

Jellyman, D. J. 1979. Upstream migration of glass-eels (*Anguilla* spp.) in the Waikato River. New Zealand

Journal of Marine and Freshwater Research 13:13–22.

Jellyman, D. J. 1987. Review of the marine life history of Australasian temperate species of *Anguilla*. Pages 276–285 *in* M. J. Dadswell, R. J. Klauda, C. M. Moffitt, R. L. Saunders, R. A. Rulifson, and J. E. Cooper. American Fisheries Society, Symposium 1, Bethesda, Maryland.

Jellyman, D. J. 2003. The distribution and biology of the South Pacific species of Anguilla. Pages 275–292 *in* K. Aida, K. Tsukamoto, and Yamauchi, editors. Eel biology. Springer-Verlag, Tokyo.

Jellyman, D. J. 2007. Status of New Zealand fresh-water eel stocks and management initiatives. ICES Journal of Marine Science 64:1379–1386.

Jellyman, D. J. and K. Tsukamoto. 2002. First use of archival transmitters to track migrating freshwater eels *Anguilla dieffenbachii* at sea. Marine Ecology Progress Series 233:207. –215.

Jellyman, D. J., and K. Tsukamoto. 2005. Swimming depths of offshore migrating longfin eels *Anguilla dieffenbachii*. Marine Ecology Progress Series 286:261–267.

Jellyman, D. J., B. L. Chisnall, L. H. Dijkstra, and J. A. T. Boubée. 1996. First record of the Australian longfinned eel, *Anguilla reinhardtii*, in New Zealand. Marine and Freshwater Research 47:1037–1040.

Jespersen, P. 1942. Indo-Pacific leptocephalids of the genus *Anguilla*. Dana Report 22:1–128.

Kettle, A. J., and K. Haines. 2006. How does the European eel (*Anguilla anguilla*) retain its population structure during its larval migration across the North Atlantic Ocean? Canadian Journal of Fisheries and Aquatic Sciences 63:90–106.

Kimura, S. 2003. Larval transport of the Japanese eel. Pages 169–179 *in* K. Aida, K. Tsukamoto, and K. Yamauchi editors. Eel biology. Springer-Verlag, Tokyo.

Kimura, S., K. Tsukamoto, and T. Sugimoto. 1994. A model for the larval migration of the Japanese eel: roles of the trade winds and salinity fronts. Marine Biology 119:185–190.

Kimura, S., K. Doos, and A. C. Coward. 1999. Numerical simulation to resolve the issue of downstream migration of the Japanese eel. Marine Ecology Progress Series 186:303–306.

Kleckner, R. C., and J. D. McCleave. 1985. Spatial and temporal distribution of American eel larvae in relation to North Atlantic Ocean current systems. Dana 4:67–92.

Kleckner, R. C., and J. D. McCleave. 1988. The northern limit of spawning by Atlantic eels (*Anguilla* spp.) in the Sargasso Sea in relation to thermal fronts and surface water masses. Journal of Marine Research 46:647–667.

Kleckner, R. C., J. D. McCleave, and G. S. Wipplehauser. 1983. Spawning of American eel, *Anguilla rostrata*, relative to thermal fronts in the Sargasso Sea. Environmental Biology of Fishes 9:289–293.

Knights, B. 2003. A review of the possible impacts of long-term oceanic and climate changes and fishing mortality on recruitment of anguillid eels of the northern hemisphere. Science of the Total Environment 310:237–244.

Kuroki, M., J. Aoyama, M. J. Miller, S. Wouthuyzen, T, Arai and K. Tsukamoto. 2006. Contrasting patterns of growth and migration of tropical anguillid leptocephali in the western Pacific and Indonesian seas. Marine Ecology Progress Series 309:233–246.

Langdon, S. A., and A. L. Collins. 2000. Quantification of the maximal swimming performance of Australasian glass eels, *Anguilla australis* and *Anguilla reinhardtii*, using a hydraulic flume swimming chamber New Zealand. Journal of Marine and Freshwater Research 34:629–636.

Lecomte-Finiger, R. 1992. Growth history and age at recruitment of European glass eels (*Anguilla anguilla*) as revealed by otolith microstructure. Marine Biology 114:205–210.

Lecomte-Finiger, R. 2003. The genus *Anguilla* Schrank, 1798: current state of knowledge and questions. Reviews in Fish Biology and Fisheries 13:265–279.

Leis, J. M. 2006. Are larvae of demersal fishes plankton or nekton? Advances in Marine Biology 51:59–141.

Martin, M. H. 1995. Validation of daily growth increments in otoliths of *Anguilla rostrata* (Lesueur) elvers. Canadian Journal of Zoology 73:208–211.

Marui, M., T. Arai, M. J. Miller, D. J. Jellyman, and K. Tsukamoto. 2001. Comparison of early life history between New Zealand temperate eels and Pacific tropical eels revealed by otolith microstructure and microchemistry. Marine Ecology Progress Series 213:273–284.

Mata, M. M., M. Tomczhak, S. Wijffels, and J. A. Church. 2000. East Australian Current volume transports at 30S: estimates from the World Ocean Circulation Experiment hydrographic sections PR11/PR6 and the PCM3 current meter array. Journal of Geophysical Research 105:28509–28526.

McCleave, J. D. 2003. Spawning areas of Atlantic eels. Pages 141–155 *in* K. Aida, K. Tsukamoto, and K. Yamauchi editors. Eel biology. Springer-Verlag, Tokyo.

McCleave, J. D., and R. C. Kleckner. 1987. Reproductive sympatry of American and European eels and implications for migration and taxonomy. Pages 286–297 *in* M. J. Dadswell, R. J. Klauda, C. M. Moffitt, R. L. Saunders, R. A. Rulifson, and J. E. Cooper. American Fisheries Society, Symposium 1, Bethesda, Maryland.

McCleave, J. D., P. J. Brickley, K. M. O'Brien, M. W. Kistner, D.A. Wong. M. Gallagher, and S. M. Watson. 1998. Do leptocephali of the European eel swim to reach continental waters? Status of the question. Journal of the Marine Biological Association of the U.K 78:285–306.

Miller, M. J., J. Aoyama, N. Mochioka, T. Otake, P. H. J. Castle, G. Minagawa, T. Inagaki and K. Tsukamoto, K. 2006. Geographic variation in the assemblages of leptocephali in the western South Pacific. Deep-Sea Research 53: 776–794.

Miller, M., S., Kimura, K. D. Friedland, B. Knights, D.J. Jellyman and K. Tsukamoto. 2009. Review of ocean-atmospheric factors in the Atlantic and Pacific oceans influencing spawning and recruitment of anguillid eels. Pages 231–249 *in* A. J. Haro, K. L. Smith, R. A. Rulifson, C. M. Moffitt, R. J. Klauda, M. J. Dadswell, R. A. Cunjak, J. E. Cooper, K. L. Beal, and T. S. Avery, editors. Challenges for diadromous fishes in a dynamic global environment. American Fisheries Society, Symposium 69, Bethesda, Maryland.

Moriarty, C., and W. Dekker. 1997. Management of the European eel. Fisheries Bulletin (Dublin) 15.

Muheim, R., F. R. Moore, and J. B. Phillips. 2006. Calibration of magnetic and celestial compass cues in migratory birds: a review of cue-conflicting experiments. The Journal of Experimental Biology 209:2–17.

Nishi, T., and G. Kawamura. 2005. *Anguilla japonica* is already magnetosensitive at the glass eel phase. Journal of Fish Biology 67:1213–1224.

Otake, T., T. Inagaki, H. Hasumoto, N. Mochioka, and K. Tsukamoto. 1998. Diel vertical distribution of *Anguilla japonica* leptocephali. Ichthyological Research 45:208–211.

Pavolv, D. A., and A. M. Shadrin. 1998. Development of variation in the number of myomeres and vertebrae in the White Sea herring, *Clupea pallasi marisalbi*, under the influence of temperature. Journal of Ichthyology 39:251–261.

Polovina, J. J., P. Kleiber and D. R. Kobayashi. 1998. Application of TOPEX-POSEIDON satellite altimetry to simulate transport dynamics of larvae of spiny lobster, Panulirus marginatus, in the northwestern Hawaiian islands. Fishery Bulletin 97:132–143.

Power, J. H. and J. D. McCleave. 1983. Simulation of the North Atlantic Ocean drift of *Anguilla leptocephali*. Fishery Bulletin 81:483–500.

Press, W. H., B. P. Flannery, S. A. Teukolsky and W. T. Vetterling. 1989. Numerical recipes: the art of scientific computing. Cambridge University Press, Cambridge, UK.

Pusey, B., M. Kennard, and A. Arthington. 2004. Freshwater fishes of north-eastern Australia. CSIRO Publishing, Collingwood, Victoria, Australia.

Ridgway, K. R., and J. R. Dunn. 2003. Mesoscale structure of the mean East Australian Current System and its relationship with topography. Progress in Oceanography 56:189–222.

Schmidt, J. 1922. The breeding places of the eel. Philosophical Transactions of the Royal Society of London B 211:179–208.

Schmidt, J. 1928. The fresh-water eels of Australia with some remarks on the shortfinned species of *Anguilla*. Records of the Australian Museum 16:179–201.

Shaffer, S. A., Y. Tremblay, H. Weimerskirch, D. Scott, R. R. Thompson, P. M. Sagar, H. Moller, G. A. Taylor, D. G. Foley, B. A. Block, and D. P. Costa. 2006. Migratory shearwaters integrate oceanic resources across the Pacific Ocean in an endless summer. Proceedings of the National Academy of Sciences of the United States of America 103:12799–12802.

Shen, K. N., and W. N. Tzeng. 2007. Genetic differentiation among populations of the shortfinned eel *Anguilla australis* from East Australia and New Zealand. Journal of Fish Biology 70(Supplement B):177–190.

Shiao, J. C., W. N. Tzeng, A. Collins and D. J. Jellyman. 2001. Dispersal pattern of glass eel stage of *Anguilla australis* revealed by otolith growth increments. Marine Ecology Progress Series 219:241–250.

Shiao, J. C., W. N. Tzeng, A. Collins, and Y. Iizuka. 2002. Role of marine larval duration and growth rate of glass eels in determining the distribution of *Anguilla reinhardtii* and *A. australis* on Australian eastern coasts. Marine and Freshwater Research 53:687–695.

Smith, P. J., P. G. Benson, C. Stanger, B. L. Chisnall and D. J. Jellyman. 2001. Genetic structure of New Zealand eels *Anguilla dieffenbachii* and *A. australis* with allozyme markers. Ecology of Freshwater Fish 10:132—137.

Sugeha, H. Y., A. Shinoda, M. Marui, T. Arai, and K. Tsukamoto. 2001. Validation of otolith daily increments in the tropical eel *Anguilla marmorata*. Marine Ecology Progress Series 220:291–294.

Tåning, A. V. 1952. Experimental study of meristic characters in fishes. Biological Review 271:169–193.

Tesch, F-W. 2003. The eel, 3rd edition. Blackwell Scientific Publications, Oxford. UK.

Tesch, F-W., and G. Wegner. 1990. The distribution of small larvae of *Anguilla* sp. related to hydrographic conditions 1981 between Bermuda and Porto Rico. Internationale Revue der gesamten Hydrobiologie 75:845–858.

Tesch, F-W., U. Niermann, and A. Plaga. 1986. Differences in development stage and stock density of larval *Anguilla anguilla* off the west coast of Europe. Vie Milieu 36:255–260.

Tesch, F-W., T. Wendt, and L. Karlsson. 1992. Influence of geomagnetism on the activity and orientation of the eel, *Anguilla anguilla* (L.), as evident from laboratory experiments. Ecology of Freshwater Fish 1:52–60.

Tomczak, M. and J. S. Godfrey. 1994. Regional oceanography: an introduction. Pergamon, London.

Tsukamoto, K. 1992. Discovery of the spawning area for Japanese eel. Nature (London) 356:789–790.

Tsukamoto, K., and J. Aoyama. 1998. Evolution of freshwater eels of the genus *Anguilla*: a probable scenario. Environmental Biology of Fishes 52:139–148.

Tsukamoto, K., T-W. Lee and H. Fricke. 2003a. Spawning area of the Japanese eel. Pages 121–140 *in* K. Aida, K, Tsukamoto, and K. Yamauchi, editors. Eel biology. Springer-Verlag, Tokyo.

Tsukamoto, K., T. Otake, N. Mochioka, T. W. Lee, H. Fricke, T. Inagaki, J. Aoyama, S. Ishikawa, S. Kimura, M. J. Miller, H. Hasumoto, M. Oya, and Y. I. Suzuki. 2003b. Seamounts, new moon and eel spawning: the search for the spawning site of the Japanese eel. Environmental Biology of Fishes 66:221–229.

Watanabe, S., J. Aoyama and K. T. Mishida. 2005. A molecular genetic evaluation of the taxonomy of eels of the genus *Anguilla* (Pisces: Anguilliformes). Bulletin of Marine Science 76:675–690.

Watanabe, S., J. Aoyama, and K. Tsukamoto. 2006. Confirmation of morphological differences between *Anguilla australis australis* and *A. australis schmidtii.* New Zealand Journal of Marine and Freshwater Research 40:325–331.

American Fisheries Society Symposium 69:275–292, 2009

Migration of Silver American Eels through a Macrotidal Estuary and Bay

RODNEY G. BRADFORD*
Department of Fisheries and Oceans, Population Ecology Division, Bedford Institute of Oceanography Post Office Box 1006, Dartmouth, Nova Scotia B2Y 4A2, Canada

JONATHAN W. CARR
Atlantic Salmon Federation
Post Office Box 5200, St. Andrews, New Brunswick E5B 3S8, Canada

FRED H. PAGE
Department of Fisheries and Oceans, St. Andrews Biological Station
531 Brandy Cove Road, St. Andrews, New Brunswick E5B 2L9, Canada

FRED WHORISKEY
Atlantic Salmon Federation,
Post Office Box 5200, St. Andrews, New Brunswick E5B 3S8, Canada

Abstract.—The horizontal and vertical movements of 20 silver American eels *Anguilla rostrata* migrating through a macrotidal estuary and bay were tracked with hydroacoustic telemetry. Eels initiated their seaward migrations around or shortly after sunset, mostly migrated at night, exhibited no bias for migration on ebb versus flood tides, and swam both with and against tidal currents with little preference for depth. Profiles of eel vertical distributions during transit through the estuary, the bay, and the tidal passages that connected the bay to the open sea differed and differed from analogous profiles for silver European eels *A. anguilla* transiting the Baltic Sea where tidal forcing is negligible. The underlying mechanism(s) associated with their direction of net displacement appears to be innate, although local circulation features may influence transit speeds and exit routes to the open sea. Our results in combination with published information for both the silver American and the European eel suggest that the horizontal migration speed for both species is relatively invariant, around 0.5 body lengths per second. The rate of ascent/descent of eels exhibiting a change in swimming depth of greater than 5 m could not be estimated with precision from the available records of eel swimming depths. The observed average rate of 0.11 m/s was, however, within the relatively narrow range of the average vertical swimming speeds (0.11–0.18 m/s) reported in the literature for silver eels executing vertical dives.

Introduction

Decadal scale declines in abundance within portions of the Canadian range of the American eel *Anguilla rostrata* Lesueur, 1817 have elevated the concern for the current status of the species (Dekker et al. 2003; COSEWIC 2006). In response, a management framework is under development to guide the eel conservation activities under the jurisdictions of the Canadian federal and provincial governments and to link these activities to the jurisdictions within the United States (MacGregor et al. 2009, this volume). Measures to reduce the effects of both existing and proposed human activities on the escapement of sexually maturing adult silver eels are a core element of many of the recommended activities (MacGregor 2009).

Evaluations of conservation measures to in-

* Corresponding author: bradfordr@mar.dfo-mpo.gc.ca

crease spawner escapement should extend beyond considerations of direct mortality (e.g., from fishing, hydroelectric turbine mortality, complete barriers to downstream migration). These sublethal effects may include physiological injuries incurred either during passage through turbines (Montén 1985), falling over dams, bioaccumulation of nonnatural chemicals (Castonguay et al. 1994), or infection by parasites such as the nonnative (exotic) swim bladder nematode *Anguillicola crassus* (Fries et al. 1996; Moser et al. 2001). Emerging technologies such as those proposed to harness tidal energy on a large scale (Whitford 2008) will also require consideration if facilities are constructed within the migratory corridors for adult eels or within the rearing habitat of eels that reside for a portion of their subadult lives in coastal habitats (Jessop et al. 2002). The potential for increased spawning escapement via the translocation of elvers over large distance from areas of secure status into vacant habitat is also under discussion (MacGregor et al. 2009). In all of the above cases, the choice, design, and implementation of assessment and mitigation protocols will benefit from a better understanding of the biological and environmental conditions influencing the migratory behavior of adult American eels.

The study of eel migration patterns at sea is often limited by logistic and financial restrictions to the coastal environment, which represents a very small proportion of the total spawning migration. Published studies of direct, telemetry-based, research on the horizontal and vertical movements of silver American eel migration through coastal areas are limited to those of Stasko and Rommel (1974) in Passamaquoddy Bay, Canada and Parker and McCleave (1997) in Penobscot Bay, USA. Both studies, conducted in areas of pronounced tidal activity, were based on small numbers of eels (five and six, respectively) tracked over relatively brief periods (4–40 h and 55–93 h, respectively). Parker and McCleave (1997) showed that most migration activity occurs during darkness, and both studies demonstrated that silver eels execute vertical dives of varying depths relative to water depth.

The underlying bases for movements documented by Stasko and Rommel (1974) remains to be determined. The posited mechanistic basis, selective tidal stream transport (STST) of Parker and McCleave (1997) has not been retested for the American eel, either with a larger sample size or within a different study area. Studies conducted on silver European eels *A. anguilla* (Linneaus, 1758) migration offer only partial (McCleave and Arnold 1999) or no support (Tesch 1992) for STST. Whether possible additional mechanisms or cues for the migration of silver European eels might apply to the spawning migration of the silver American eel is not clear. The potential cues include orientation and directed swimming through geomagnetic reference (Tesch 1978, 1989; Hanson et al. 1984; Tesch et al. 1992; van Ginneken et al. 2005b; Westerberg et al. 2007), celestial navigation (Westerberg et al. 2007), imprinting of the migratory route during recruitment to continental waters (Westin 1998), directional information from currents (Westerberg 1979), and possibly olfactory cues as has been shown for yellow phase American eels (Barbin 1998).

The purpose of our study was to identify the behavioral and external, physical factors associated with silver American eel migration through the coastal zone. The physical character of the study area, a macrotidal embayment and estuary in the Bay of Fundy, allowed for retesting of hypotheses in a previously studied migratory corridor for adult American eels (Stasko and Rommel 1974). The possibility that the observed timing of the horizontal and vertical movements of silver American eels have underlying obligate, facultative, or innate origins is discussed.

Materials and Methods

Study Area

Passamaquoddy Bay and Passages.—Passamaquoddy Bay is a relatively large (264 km^2 in area) macrotidal embayment located in the southwestern portion of the province of New Brunswick, Canada (Figure 1). The tidal range, which averages 6 m and can be as great as 8.3 m on spring tides (Gregory et al. 1993), is the dominant feature of the oceanography of the bay. Temporal variation in tidal height is dominated by the lunar, semidiurnal tide (M_2), which generates a tidal cycle every 12.4 h or approximately two tidal cycles per day. The difference in range of these two cycles is small (Trites and Garrett 1983). Tidally generated water currents are typically greater than 1 m/s through the four passages (Letete Passage, Little Letete Passage, Pendelton Passage,

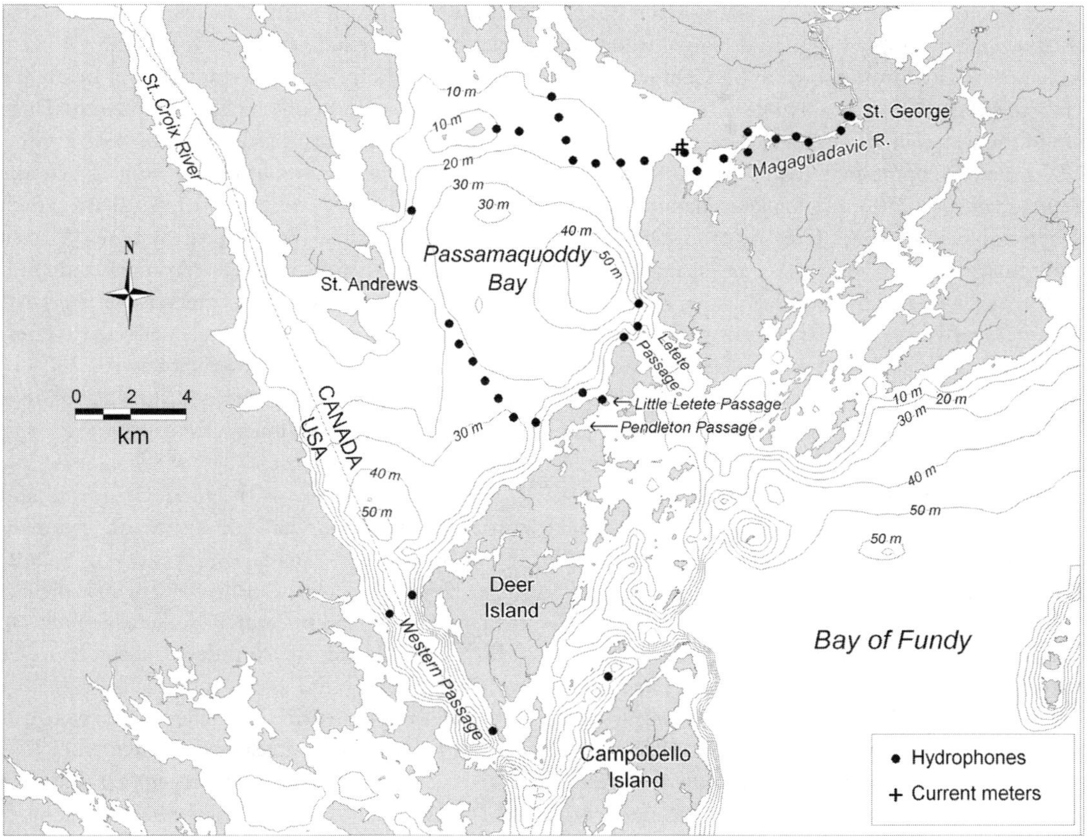

Figure 1.—Map of the Magaguadavic River estuary, Passamaquoddy Bay, and associated passages showing the locations of fixed acoustic receivers and current meter deployments. Depth contours are in meters.

and Western Passage; Figure 1) that connect Passamaquoddy Bay to the Bay of Fundy. Tidal current speeds are much less within Passamaquoddy Bay, with values less than 0.1 m/s in some locations (Trites and Garrett 1983; Greenberg et al. 2005; Chang et al. 2006).

The total flow of water through the passages connecting Passamaquoddy Bay to the Bay of Fundy during each half-tidal cycle (6.2 h) varies from ~1 × 10^9 m^3 during neap tides to ~2 × 10^9 m^3 during spring tides (Trites and MacGregor 1959). The average tidal volume represents ~34% of the total volume of the bay (~3.2 × 10^9 m^3, Gregory et al. 1993). The majority of the flow (95%) occurs through the Western (61%) and Letete passages (34%) (Trites and MacGregor 1959). The large exchange through these two passages at the western and eastern entrances, respectively, to Passmaquoddy Bay produces short (15 min or less) phase shifts in the timing of both high tide and low tide at the extreme locations of the study area. The freshwater component of this flow, while seasonally variable, is small, the ratio of tidal volume to freshwater volume is on the order of 330 (Trites and MacGregor 1959; Gregory et al. 1993).

Numerous release and recovery drift bottle experiments (Chevrier and Trites 1960) suggest a counterclockwise surface circulation within Passamaquoddy Bay, particularly during the summer when winds are relatively weak. During the autumn months when adult eels migrate through the bay, the prevailing winds are northerly and stronger and this circulation pattern is not apparent. Instead, the surface circulation appears to be through Western Passage toward the Bay of Fundy (Chevrier and Trites 1960). Unfortunately, the release and recover drift bottle experiments did not provide reliable information on the rate of transit of drift bottles across the bay.

Magaguadavic River and Estuary.—The Magaguadavic River estuary begins at a natural water fall and gorge located in the town of St. George (Figure 1). The river, which has a hydroelectric dam at the top of the waterfall, drains an area of approximately 1,860 km^2 or approximately 27% of the watershed drainage area (6,950 km^2) for Passamaquoddy Bay (Trites and Garrett 1983; Gregory et al. 1993). River discharge (annual mean 33.1 m^3/s; years 1917–2006) varies seasonally (April, 91.4 m^3/s; August, 12.1 m^3/s). Average discharge during October and November when silver eels migrate seaward are 23.4 and 37.5 m^3/s, respectively.

The estuary consists of a narrow (less than 1 km in width), relatively short (less than 10 km in length) rocky shore channel that deepens gradually from less than 3 m (relative to mean low water) at the head of tide to ~20 m where it joins Passamaquoddy Bay. Vertical profiles of salinity acquired on several occasions during autumn 2006 (F. H. Page, unpublished data) indicated that during the time of the eel tracking study, surface salinities within the 1 km section below the head of tide typically varied between 7 and 25 parts per thousand (ppt) depending upon river discharge. Surface salinities at the mouth of the estuary varied from 30 to 32 ppt. Bottom salinities generally exceeded 25 ppt throughout the estuary, being lowest within the inner portions. Bottom salinities at locations more than 2 km downstream of the head of tide were always greater than 30 ppt.

Water flushing times within both the estuary and bay vary seasonally in response to changes in river discharge and both the strength and direction of winds. The flushing time of the Magaguadavic River estuary has been estimated as 2 d during June (Forgeron 1959), a month of relatively moderate to low river discharge. The flushing time for Passamaquoddy Bay during August has been estimated as 15 d by Ketchum and Keen (1953) and as 17 and 8 d by Forgeron (1959) for the summer and autumn periods, respectively.

Environmental Monitoring of the Estuary

Current meters.—An acoustic Doppler current profiler (ADCP: 300 kHz RDI Workhorse Sentinel) and an Anadaarra RCM9 current meter were moored ~8 km below the head of tide and within the mouth of the estuary (Figure 1) to record currents every 15 min. Water velocity data collected from the ADCP moorings was extracted as a time series of velocity data at constant depths of 4.22 m above the bottom and a constant 2.75 m below the surface. The latter approach accommodates change in tidal height at the mooring location with time.

Surface currents.—Clusters of four or five surface (upper 1 m of water column) drift barrels each containing a Global Positioning System (Garmin model GPSmap 60CS) set to record latitude and longitude positions at 1-min intervals were released on ebbing tides within the Magaguadavic River estuary on five occasions from October 10, 2006 to November 1, 2006. The barrels drifted for periods of 0.6–3.9 h. Shorter times were usually the result of stranding on the shore.

Sunrise and sunset.—The times of sunrise and sunset (Atlantic standard time) for St. Andrews (New Brunswick, Canada; longitude = 67° 49.2′ West, latitude = 45° 9′ North) were acquired from the National Research Council of Canada's Web site (NRC 2007b) and are considered accurate to be within ±2 min (NRC 2007a).

Tide height predictions.—Tide heights were calculated for St. Andrews using the Foreman (1977) tidal prediction software with recent estimates of tidal constituent amplitudes and phases (Chang et al. 2006).

Silver eel collection, transmitter attachment, release, tracking.—Twenty large (87.7 ± 4.8 cm total length; 1.34 ± 0.16 kg total weight, Table 1) silver eels exhibiting no external injuries were selected over a 3-d period beginning October 8, 2006 from the collection chamber of a downstream fish bypass facility installed in the hydroelectric dam located just above the head of tide (Figure 1). Coded depth-recording (maximum 100 m, resolution ~0.3 m) hydroacoustic transmitters (VEMCO V9P-6 L-R64K) 38 mm in length and 9 mm in diameter and weighing 2.2 g in water were surgically implanted into the visceral cavities of the eels, as recommended by Cottrill et al. (2006), to increase the likelihood of transmitter retention without affecting swimming performance. All transmitters had an expected battery life of 25 d, operated at a frequency of 69 kHz, had 15–45 s off time, and had a power output of 137 Db.

Eels, anesthetized with clove oil (100–120 mg/L), were placed ventral side up in a v-shaped trough lined with soft foam to stabilize the eel's body during surgery. The transmitters were inserted into

Table 1.—Total weight (kg), total length (cm) and the time, hour, and tidal stage (L = low water, H = high water, E = ebb, F = flood) of release for 20 silver American eels surgically implanted with hydro acoustic transmitters.

Fish no.	Total weight (kg)	Total length (cm)	Day	Hour	Tide stage
1	1.18	83.3	Oct 12	0400	L
2	1.45	90.2	Oct 12	0400	L
3	1.45	89.1	Oct 12	0400	L
4	1.30	92.8	Oct 12	0400	L
5	1.60	94.5	Oct 11	2145	H
6	1.16	82.6	Oct 11	2145	H
7	1.40	89.3	Oct 11	2145	H
8	1.70	97.1	Oct 11	2145	H
9	1.20	79.5	Oct 12	0400	L
10	1.14	81.4	Oct 12	0400	L
11	1.30	82.2	Oct 12	0400	L
12	1.60	95.5	Oct 12	0400	L
13	1.26	88.0	Oct 12	0400	L
14	1.40	89.3	Oct 12	0400	L
15	1.14	87.0	Oct 11	2145	H
16	1.50	86.6	Oct 11	2145	H
17	1.35	87.0	Oct 11	2145	H
18	1.24	84.4	Oct 11	2145	H
19	1.22	89.3	Oct 11	2145	H
20	1.30	85.7	Oct 11	2145	H
Mean	1.34	87.7			
SD	0.16	4.8			

the visceral cavity through a 20-mm incision on the mid-ventral line 50 mm anterior to the anus. Usually only one, and no more than two, sutures (3–0 Ethilon black monofilament nylon) were required to close the incisions. Furacin was used to disinfect the ventral surface of the fish prior to the making of the incision and following suturing. All 20 eels were processed in less than 2 h; individual surgeries were completed in less than 2 min (1.25 min average/eel). Eels resumed a dorsal side up orientation in 10 min or less when placed in a holding pen receiving fresh, running river water, and all were liberated within 6–18 h following surgery.

Moonlight has been shown to influence silver European eel migratory activity (Tesch 1978, 1989; Tesch et al. 1992). To reduce the influence of moon phase on eel behavior and attempt to evaluate the relative influence of behavioral versus oceanographic factors on silver eel migration, releases were timed to allow for the first 2 weeks of the experiment to occur during a period of low or no moon light. Therefore, on October 11, 2 d before the lunar third quarter, 10 silver eels were liberated at night during low tide (2145 hours) approximately 500 m below the head of tide. The remaining 10 silver eels were released the same night during the following high tide (October 12 at 0400 hours) at the same location.

Detection.—The locations of the 20 liberated eels were monitored with anchored, submersible receivers (VEMCO VR2 and VR2W) possessing omnidirectional hydrophones and data logging components programmed to (1) decode and identify individual transmitters, and (2) log swimming depth. The receivers were organized spatially relative to the release site as follows: a linear series along the axis of the Magaguadavic River estuary, a linear series across the northern portion of Passamaquoddy Bay near the confluence of the estuary with the bay, a linear series that roughly bisected Passamaquoddy Bay, and two or more receivers across the entrances to the tidal passages (Figure 1). Other receivers were deployed around the northerly portions of the bay in order to detect if any eels moved north, away from the channels (Figure 1). Tracking studies of Atlantic cod *Gadus morhua* indicated that the detection radius of the receivers in the estuary, the bay, and the passages was somewhat greater than 0.5 km (Brooking et al. 2006). The spacing between adjacent receivers organized as linear series was therefore 1 km or less. All receivers were deployed prior to the release of the silver eels. The experiment was terminated on November 6.

Definition of Migration

European and American studies of silver eel migrations have reported among-animal variability in time to first movement relative to release time (Parker and McCleave 1997), migration speeds (Parker and McCleave 1997; Tesch 1978; McCleave and Arnold 1999; Westerberg et al. 2007), tidal cycle (Tesch 1978; Parker and McCleave 1997; McCleave and Arnold 1999), migration depth (Stasko and Rommel Jr. 1974; Parker and McCleave 1997), direction of movement (Parker and McCleave 1997; McCleave and Arnold 1999), and movements relative to the day/night cycle (Tesch 1978; Tesch 1989; Parker and McCleave 1997; McCleave and Arnold 1999; Westerberg et al. 2007). In many cases, interpretation of the results included the caveat that at least some of the observed among-animal variability could be the result of the act of handling the test animals, including, for example, stress from capture and holding, possible loss of continuous sensory input following exposure to an anesthetic, and recovery from the act of surgery. We therefore defined the onset of migration as the time at which an irreversible progression away from the head of tide has occurred, while allowing for interruption of net seaward displacement during periods of daylight.

Discrete Observations versus Continuous Observations

Studies concerning either silver American or European eel migrations are based upon continuous records of eel movements either through active tracking or the release–recapture of specimens carrying data storage tags. The present study was supported with data downloaded from stationary receivers that did not provide 100% coverage of the study area, particularly in Passamaquoddy Bay, and the data records were therefore of a more discrete nature.

In order to relate our results to those of previous studies, we have incorporated information acquired from a recent data-storage-tag based study of silver European eels exiting the Baltic Sea along the coast of Sweden (Westerberg et al. 2007). These data were generously provided to us by the investigators. Briefly, the Baltic study yielded high frequency (every minute) information on the depth of swimming by the eels between their time of release and recapture and relative to the day–night cycle (Westerberg et al. 2007). We resampled the data series at n = 100, the approximate minimum number of observations available to us at any of the study sites (i.e., estuary, bay, passages) in order to assess whether our discrete observations were consistent with continuous records of behavior.

Results

The time of onset of continuous migration varied from virtually the time of release to 10 d following release (Table 2; Figure 2). Four eels (Fish 4, 12, 13, 14) departed immediately following release, and two others (Fish 17, 19) departed the following day, possibly a consequence of the interruption of their migration by daylight. Residency until the onset of migration was limited to the inner (1.1 ± 1.1 km) portions of the estuary except for a single eel (Fish 10) that moved 4.3 km down the estuary over a period of about 6.5 d before committing to continuous migration (Table 2).

The onset of migration was initiated by all 20 eels, either at or shortly following sundown, with no overall preference exhibited for either flood or ebb tides (Table 2; Figure 2). Once continuous migration started, departure from the estuary tended to be rapid, with 15 of the 20 eels exiting to the bay in less than 0.25 d, the approximate time scale of one-half a tidal cycle. Five eels (Fish 1, 7, 12, 16, 19) negotiated the estuary to the bay within a single ebb tide; an additional five eels (Fish 5, 8, 10, 15, 20) transited the estuary on a single flood tide. Migration was interrupted by daylight for three eels (Fish 4, 13, 14), with the result that their departure from the estuary required 17–22 h (Table 2).

Table 2.—The day, hour, tidal stage (E = ebb, F = flood, H = high tide), and estuary location (km) at onset of continued migration. Tidal stage at time of departure from the estuary, the duration of the estuary migration and estuary migration speed are expressed in m/sec and body lengths (BL) per second. The duration (days) of each eels residence within Passamaquoddy Bay is shown along with the total number of days each eel remained within the study area, and their location, and swimming depth (m) at the time of last detection.

			Migration through											
			Estuary							Bay	Last detection			
Fish	Onset		Tide stage		Estuary kilometer		Duration			Duration	Days since		Tide	Depth
no.	Day	Hour	Begin	End	Begin	End	(days)	m/s	BL/s	(days)	release	Passage	stage	(m)
1	Oct 13	1955	E	E	2.7	9.6	0.11	0.73	0.61	1.98	3.75	Western	E	0.0
2	Oct 13	1956	E	F	2.7	9.6	0.25	0.32	0.29	1.78	3.68	Western	E	9.0
3	Oct 18	2003	F	E	0.3	8.3	2.25	0.04	0.04	27.83	37.02	Letete	H/E	4.9
4	Oct 12	0410	F	E	0.3	8.3	0.70	0.13	0.12	3.10	3.80	Letete	E	13.9
5	Oct 21	1947	F	F	2.7	9.6	0.14	0.56	0.53	1.21	11.27	Western	F/E	42.0
6	Oct 18	2116	F	E	0.3	8.3	0.19	0.48	0.40	8.36	15.53	Western	F/E	0.0
7	Oct 16	2053	E	E	0.5	6.9	0.13	0.57	0.50	3.00	8.10	Western	F/E	5.5
8	Oct 21	1913	F	F	0.5	9.6	0.22	0.48	0.46	1.11	11.23	Western	E	49.5
9	Oct 16	1942	F	E	0.5	6.9	0.12	0.60	0.47	2.09	6.86	Western	E	21.5
10	Oct 18	1857	F	F	4.3	8.3	0.09	0.53	0.43	0.14	6.85	Letete	E	59.0
11	Oct 18	2118	F	E	1.0	9.6	0.16	0.63	0.52	3.83	10.70	Western	F/E	55.0
12	Oct 12	0439	E	E	0.5	8.3	0.11	0.86	0.82	2.58	2.68	Western	E	0.0
13	Oct 12	0411	E	F	0.5	8.3	0.94	0.10	0.08	2.92	3.86	Western	E	67.5
14	Oct 12	0404	E	F	0.3	8.3	0.91	0.10	0.09	0.86	1.77	Letete	E	0.6
15	Oct 21	2024	F	F	0.5	8.3	0.18	0.51	0.44	8.77	18.89	Western	E	0.0
16	Oct 13	1907	E	E	0.5	9.6	0.17	0.63	0.55	0.90	2.96	Western	E	6.2
17	Oct 12	1752	E	E	0.5	9.6	8.86	0.01	0.01	8.65	18.09	Western	F/E	4.9
18	Oct 18	2134	F	E	1.4	6.9	0.12	0.53	0.45	1.15	8.26	Unknown	E	55.2
19	Oct 21	1929	E	E	0.5	8.3	0.13	0.68	0.60	1.88	2.92	Letete	E	17.0
20	Oct 21	1935	F	F	0.5	8.3	0.19	0.47	0.40	8.28	18.39	Western	E	48.5
Mean					1.1		0.80	0.45	0.39	4.52	9.83			23.0
SD					1.1		1.97	0.25	0.22	6.18	8.53			24.3

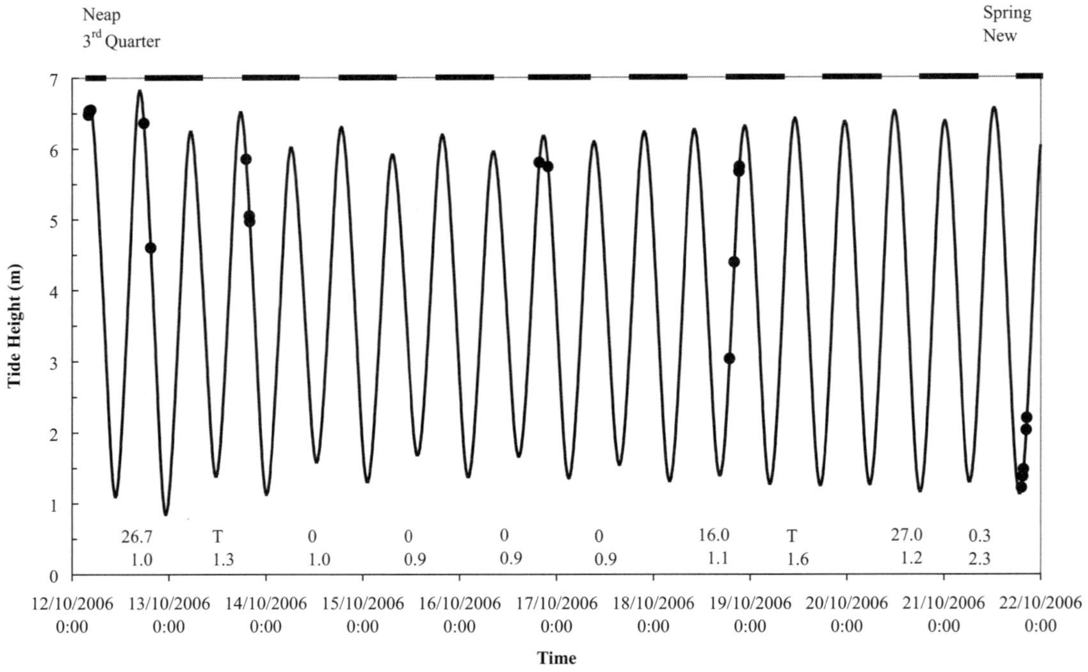

Figure 2.—Onset of continued migration from the Magaguadavic River estuary for each silver eel (n = 20) relative to tide height (m), neap to spring tide, lunar quarter, daily precipitation (mm, upper row), and relative river discharge (nondimensional, scaled to 1 for October 12, lower row). Light and dark bars along the top of the figure represent day and night, respectively.

Although silver eels migrated on both flood and ebb tides in approximately equal proportions, a temporal shift in the timing of migration relative to tidal phase was evident. All migrations initiated on ebb tides occurred before October 17, whereas all but one migration on a flood tide occurred after October 18 (Figure 2).

Fifteen of the 20 eels initiated their final down estuary movements during evenings with measurable precipitation (Figure 2). However, not every precipitation event resulted in down estuary movements. There was no apparent trend in number of migrants with amount of precipitation. For example, 6 of 20 (30%) eels migrated October 12 (26.7 mm precipitation), 5 (36%) of the remaining 14 eels migrated on October 19 (16.0 mm precipitation, none (0%) of the remaining nine eels migrated on October 20 (27 mm precipitation), and four (44%) of the remaining nine eels migrated on 21 October 21, (0.3 mm precipitation). Eighteen of the 20 eels initiated migrations on evenings with river discharge greater than the previous day (Figure 2; note: mean daily river discharge was 4.8 m^3/s on October 11), but only during 4 of the 7d of increased discharge and without a trend in number of migrants with proportional increase in discharge.

Average transit speed ($\pm$1 SD) through the estuary as calculated between the start and end distances and times for the 15 eels that exited to the bay within 0.25 d was 0.57 $\pm$ 0.12 m/s or 0.50 $\pm$0.12 body lengths (BL)/s (Table 2). Velocity for eels (n = 5) that completed the migration on a single ebb tide (0.69 $\pm$ 0.11 m/s, 0.62 $\pm$ 0.12 BL/s) were faster than those for eels (n = 5) that completed the migration on a single flood tide (0.50 $\pm$ 0.04 m/s, 0.45 $\pm$ 0.05 BL/s) (analysis of variances [ANOVAs]: df = 1,9, Fm/s = 12.52, p < 0.01; df = 1, 9, FBL = 7.88, p = 0.02).

Regression slopes for the relationship between transit distance (m) versus transit time (s) for eels detected passing through the estuarial array of receivers (Figure 3A) were significantly faster on ebb tides (0.61 m/s) than on flood tides (0.44 m/s) (df = 1, 250; T = 5.20, p < 0.001; SAS 1999; Proce-

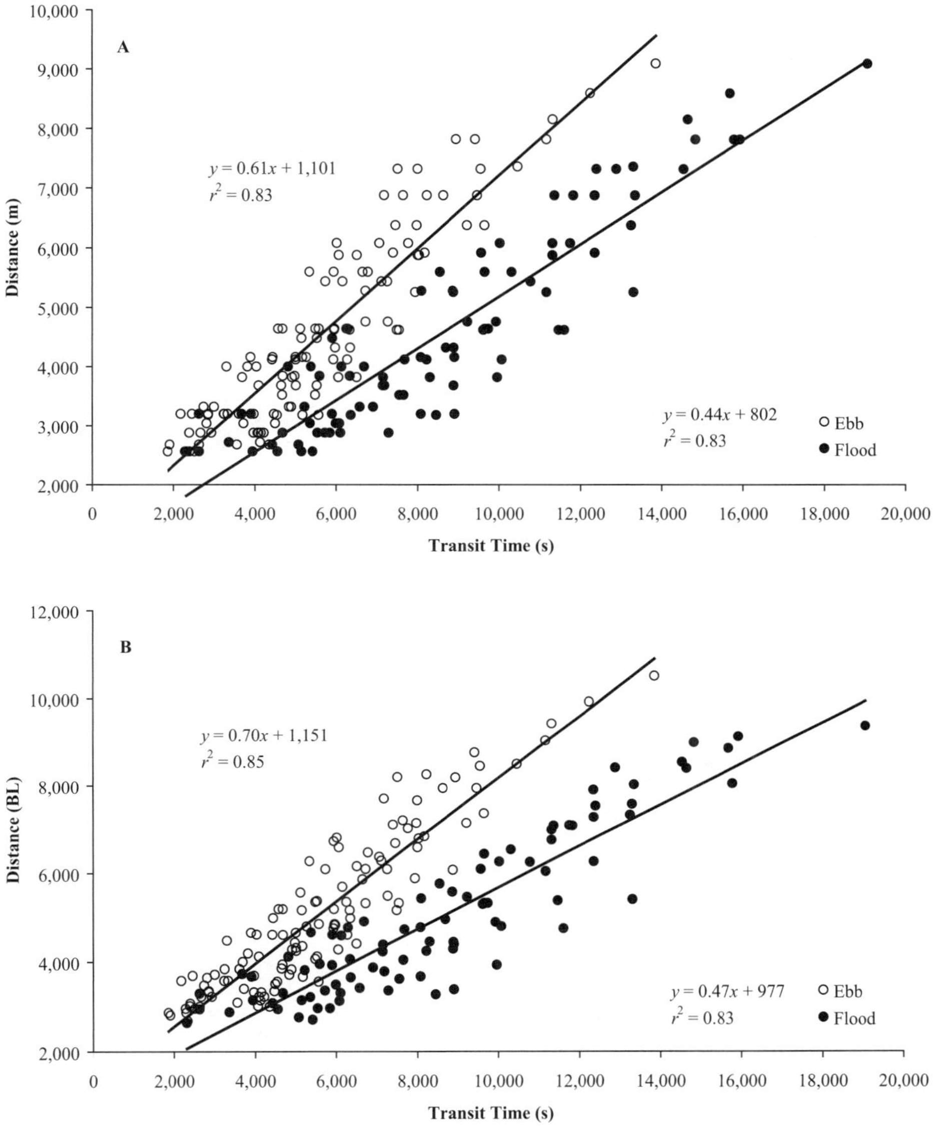

Figure 3.—Transit distance versus transit time (s) plots for the silver eels that exited the Magaguadavic River estuary on either a single ebb tide (open circles, n = 5) or a single flood tide (closed circles, n = 5). Panel A shows distance in meters; panel B shows distance in body lengths (BL).

dure REG). The intercepts (1,101, ebb; 802, flood) did not statistically differ (df = 2, 250; T = 1.11, p = 0.27). Similarly, regression slopes for the relationship between transit distances expressed as body lengths and transit times (Figure 3B) were faster on ebb tides (0.70 BL/s) than on flood tides (0.47 BL/s) (df = 2, 250, T = 6.44, p < 0.001). The intercepts (1,151, ebb; 977, flood) did not statistically differ (df = 2, 250, T = 0.52, p = 0.60). These estimates were based on transit times between receivers more than 2.5 km apart, such that the detection radius of 0.5 km was greater than 0.2 of the transit distance.

Surface (2.7 m below) and bottom (4.2 m above) currents, measured at the mouth of the estuary between September 28, 2006 and November 2, 2006, were generally bidirectional, either into or out of the estuary depending on tidal stage (Figure 4B, 4D). However, progressive displacement of the

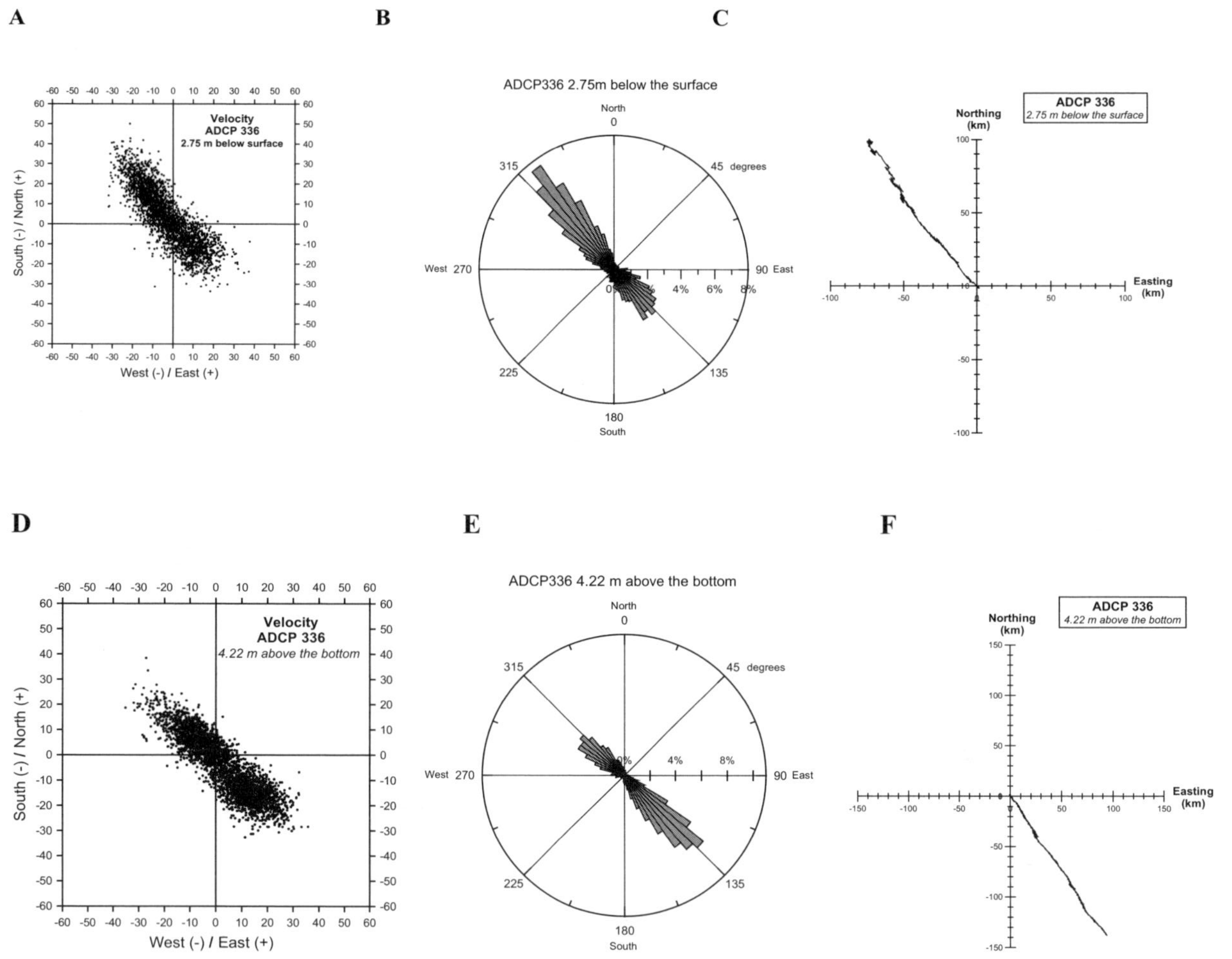

Figure 4.—Scatter plots of the east–west and north–south components of water current velocity (cm/s), radial frequency distributions of the water velocity (cm/s), and the direction of progressive displacement of water plot (km) as estimated 2.75 m below the surface (panels A, B, and C, respectively) and 4.22 m above the bottom (panels D, E, and F, respectively) at the mouth of the Magaguadavic River estuary.

flow (i.e., the residual flow) was seaward at the surface and landward near the bottom (Figure 4C, 4F). Eel migrations within 2.5 m of the surface on ebb tides would have occurred in the same direction as currents traveling an average of 0.20 m/s (maximum ~0.54 m/s during peak flow), whereas migrations on flood tides would have been against incoming currents averaging 0.16 m/s (maximum ~0.44 m/s; Figure 4A, 4D).

None of the silver eels exhibited a depth preference while migrating at night through the estuary, either during the flood or the ebb tides. The records of swimming depth versus time during the overnight from sunset to sunrise are therefore displayed as a composite (Figure 5A). Average swimming depth on ebb tides (n = 531, mean = 4.55 ± 4.23 m) did not statistically differ from average swimming depth during flood tides (n = 388, mean = 4.42 ± 4.54 m) (ANOVA df = 1,918, F = 0.216, p = 0.64).

Silver eels were more active during the night than during the day once within Passamaquoddy Bay. Total daylight detections through the receiver arrays was 708 (393 on ebb, 316 on flood) versus 2,038 (1,115 on ebb, 923 on flood) between sunset and sunrise. Only two eels (Fish 12, 17) exhibited changes in position during daylight; the remainder of the daylight detections represented eels that apparently were resting on the bottom within the detection radius of a receiver.

As was the case within the estuary, none of the silver eels exhibited a depth preference while migrating at night through the bay, either during the flood or the ebb tides (Figure 5B). Average swimming depth on ebb tides (n = 129; mean = 11.64 ± 14.38 m) did not statistically differ from average swimming depth during flood tides (n = 143, mean = 11.78 ± 12.39 m; ANOVA: df = 1,270, F = 0.008, p = 0.93).

Nineteen of the 20 eels were last detected either within Western Passage or Letete Passage (Table 2) and were considered to have exited the study area. The remaining eel (Fish 18), although last detected less than 1 km inside of Letete Passage, could not be definitively assessed as having departed the study area. Fourteen of the 19 eels departed in 4 d or less (Table 2); the effective transit times are likely less for the 17 eels that were inactive during daylight, perhaps by as much as 40%. No clear association between transit times through the estuary or the bay were evident, with the exception that the two eels with the greatest estuarial transit times (Fish 3, 17) also exhibited lengthy residencies within Passamaquoddy Bay (Table 2). The average transit time through the bay (4.5 ± 6.8 d) was less than the range of flushing times that could be anticipated during the late summer–autumn (8–17 d; Forgeron 1959).

Fourteen of the 19 eels last detected at the seaward extremes of the study area were detected in the Western Passage; the remaining five eels were last detected at Letete Passage (Table 2; Figure 1). These proportions are consistent with expectations based on proportional water exchange of 0.61 and 0.34 for Western and Letete passages, respectively, as shown relative to normalized probability (Poisson) distributions (Figure 6).

The last detections for all of the eels in the area of the passages occurred at night and either during ebbing tides or at the transition from flood to ebb (Table 2). Observed swimming depths within the passages although variable (Figure 5C) were shallower on ebb tides (n = 90; mean = 29.81 ± 22.69 m) than during flood tides (n = 90; mean 40.73 ± 25.35 m; ANOVA: df = 1, 179, F = 9.724, p < 0.01). Swimming depths at time of last detection varied between 0 and 67.5 m (mean 23.0 ± 24.3 m; Table 2).

The data set provided by Westerberg et al.(2007) for a single silver European eel migrating through the Baltic yielded 6,940 observations between sunset and sunrise over 12 nights. Cumulative frequency distributions (n = 12) derived from random resampling at n = 100 observations per sample reproduced the overall results of Westerberg et al. (2007) that the eels remained within the upper 1 m of the water column more than 90% of the time (Figure 7). In comparison, silver American eels in our study occurred within the upper 1 m less than 40% of the time while transiting both the estuary and bay (Figure 7) and less than 15% of the time while in the passages. The depths of 50% occurrence while the eel were transiting the estuary, bay, and channels were 3.5 m, 9.5 m, and greater than 30 m, respectively.

Discussion

Onset of Migration.

Silver American eels, apparently like silver European eels (Westerberg et al. 2007), initiated their seaward

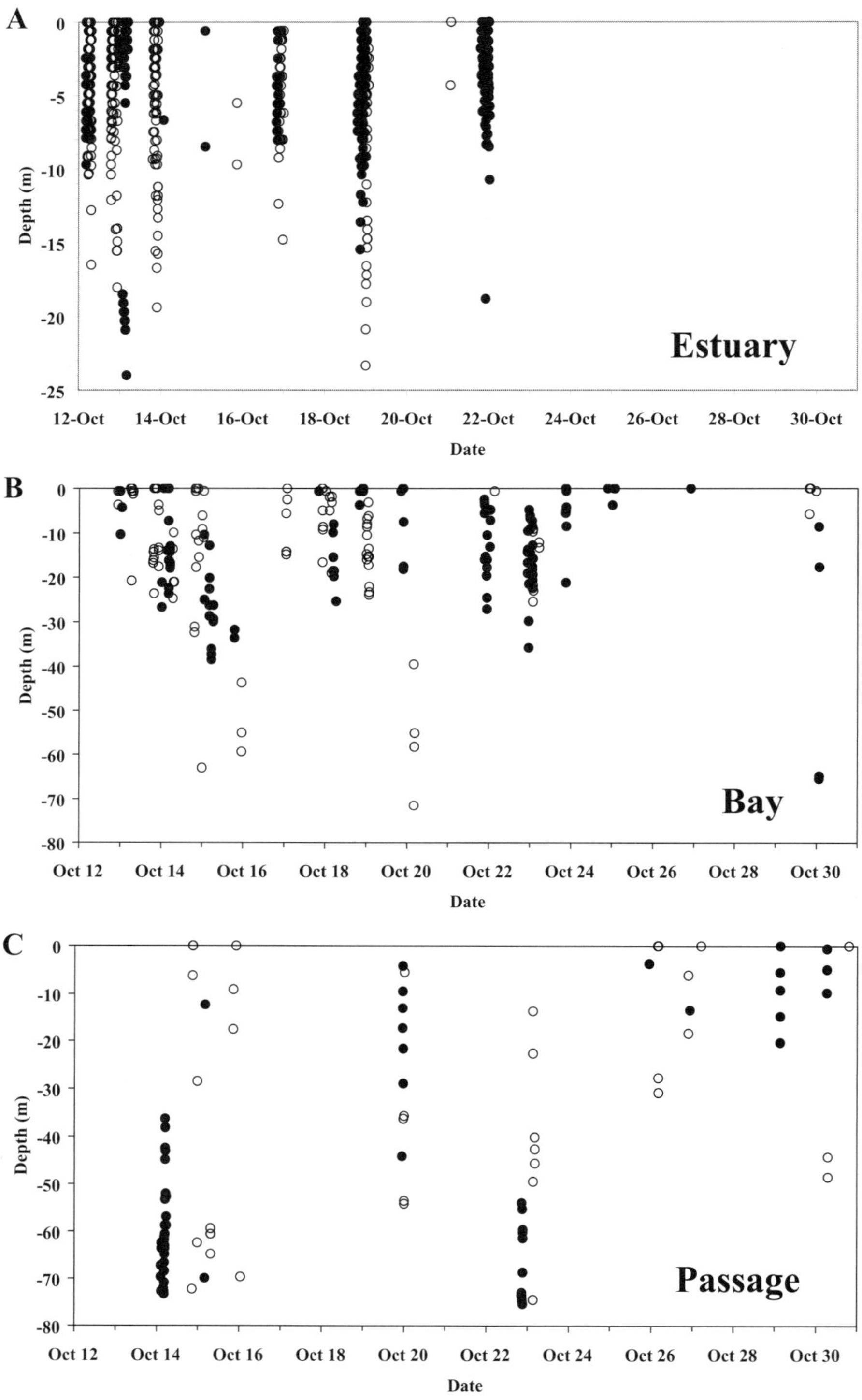

Figure 5.—Composite plots of swimming depth (m) of all silver eels detected during their transit of (A) the Magauadavic River estuary, (B) Passmaquoddy Bay, and (C) the passages on ebb tides (open circles) and flood tides (closed circles) between October 12 and 30, 2006.

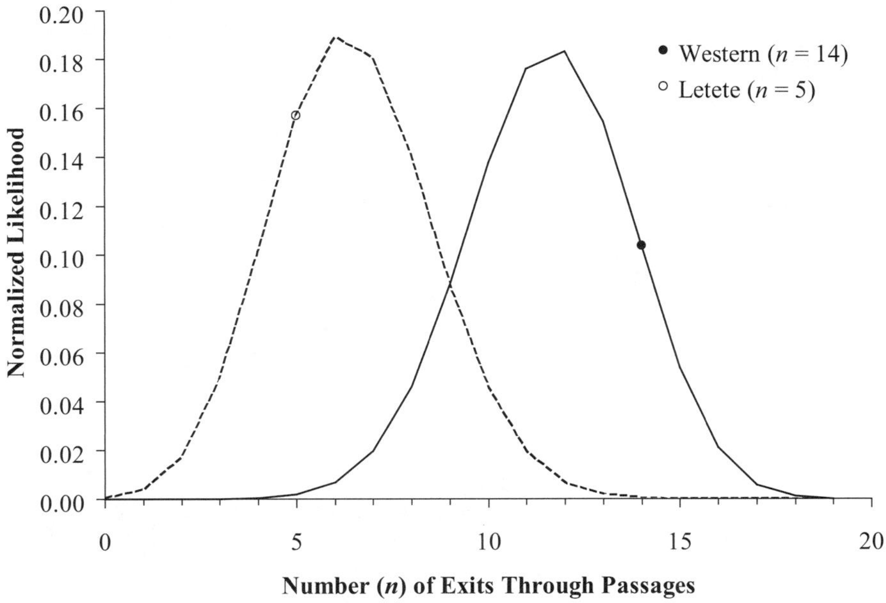

Figure 6.—Proportion of the 19 silver eels last detected in either the Western (closed circle, n = 14) or Letete (open circle, n = 5) Passage exits from Passamaquoddy Bay relative to the normalized probability (Poisson) distribution expected for n = 19 and proportional water volume exchanges of 0.61 and 0.34 for each passage, respectively.

migrations around or a few hours following sunset, and most migratory activity was at night. The migrations resulted in a net movement down the Magaguadavic River estuary on both flood and ebb tides. These results differed from those of Parker and McCleave (1997) that indicated that significant seaward progressions for silver American eels occurred on ebb tides through Penobscot Bay. The immediate departure of four eels to Passamaquoddy Bay (32 ppt) from the release site at the head of tide (prior to which they were held in freshwater), in combination with early departures of others with migrations interrupted by daylight, eliminates acclimation to seawater as a significant limiting factor to the onset of migration. Silver American eels, like immature yellow American eels, are evidently not impeded in their movements from freshwater to seawater by an osmotic barrier (Parker 1995).

No external physical cue was associated with the specific day of departure for each eel, with the possible exception of rainy weather. However, the potential role of weather is not clear for several reasons. First, total rainfall on days of migration did not result in appreciable increases in river discharge in all cases. Second, moonrise occurred well after sundown in the third quarter and darkness was continuous during the new moon. Therefore, cloud cover would not be expected to greatly influence light intensity. Third, it is not clear how eels, recently released into tidal waters, could discriminate changes in atmospheric pressure in response to changing weather when the background change in hydrostatic pressure associated with diurnal swings in tidal height of nearly 6 m would be much greater.

Migrations were initiated on both flood tides and ebb tides in approximately equal proportion, a result not previously reported for silver American eels. Initial migratory movements occurred during the early, mid, and late stages of both tides, except for the late ebb tide when no departures were recorded. However, the phasing of the tides relative to local sunset times would have resulted in a late stage of the ebb tide at sunset on only one occasion, October 20. The null observation may therefore be a chance event. None of the five eels remaining in the Magaguadavic estuary migrated that evening.

Variability among the eels in physiological readiness for marine migration at their time of capture might partially explain the observed differences in timing of the onset of migration. However, it

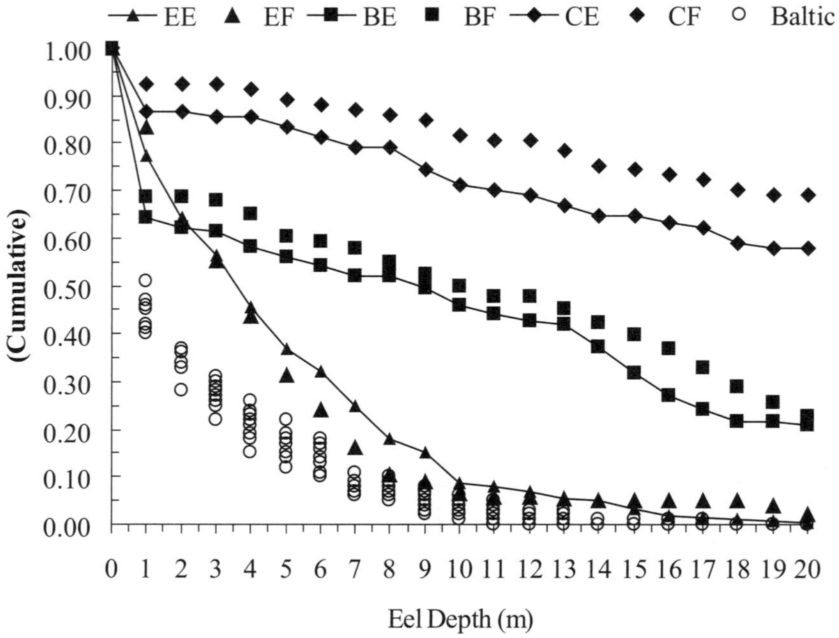

Figure 7.—Cumulative frequency distributions of silver eel swimming depths during transit through the estuary on ebb tides (EE, triangle and line) and flood tides (EF, triangle), Passamaquoddy Bay on ebb tides (BE, square and line) and flood tides (BF, square), and within the tidal passages on ebb tides (CE, diamond and line) and flood tides (CF, diamond). Relational data are shown for a silver European eel (open circles) transiting the Baltic Sea, an area of low tidal forcing.

does not offer a ready explanation for the nonrandom distribution of the departure dates, specifically the tendency for migrations to occur on ebb tides early in the experiment and on flood tides late in the experiment. Further study with releases distributed across the spring-neap tidal cycle may help explain these observations.

Higher silver eel activity levels during the new moon phase are well known from field-based data, including fisheries catches for both the American and European eel. Experimental evidence indicates that changes in activity levels are in response to the variability in the earth's geomagnetic field associated with the lunar cycle (Tesch et al. 1992). Both yellow- (van Ginneken et al. 2005b), and silver-phase eels (Tesch et al. 1992) can detect and orient to weak geomagnetic fields. The onset of migration by 9 of the 20 silver eels within ±3 d of the new moon, the expected period of highest activity (Tesch et al. 1992), may represent a response to a cue received from a change in the earth's geomagnetic field. It, of course, does not explain the departure of the other 11 eels; although Tesch et al. (1992) report that a secondary peak in activity can be expected in advance of the new moon, no time is given relative to the new moon at which this might occur.

Swimming Speed and Direction

The net horizontal speed of 0.45 BL/s for silver eels down estuary while swimming against the current and 0.7 BL/s while swimming with the current are consistent with the estimated velocities of migrating silver European eels over relatively short (0.45–0.75 BL/s; McCleave and Arnold 1999) and long (0.31–0.89 BL/s; Westerberg 1979; 0.36–0.54 BL/s; Westerberg et al. 2007) distances. Generally, the horizontal swimming speeds reported for silver eels when averaged over distances of at least several kilometers and once adjusted for the effects of local circulation appear to be relatively invariant around 0.5 BL/s, irrespective of species. Silver American eels have apparently evolved the same energetically efficient sustained swimming speed as a migration tactic as the European eel (van den Thillart et al. 2004; van Ginneken et al. 2005a). The demonstration by Gillis (1998) that silver American eels recruit only a small percentage of their swimming musculature, all located in the posterior region of the body, for distance swimming lends support to this interpretation.

The successful, usually rapid, negotiation of the estuary irrespective of the tidal cycle, in combination with relatively rapid silver eel departures from Passamaquoddy Bay, indicate that innate cues (e.g., geomagnetism) may be of relatively greater importance than external cues (e.g., current reversals, changing hydrostatic pressure, turbulence, sound patterns, astronomical cycles; Parker and McCleave 1997) early in the estuarine migration phase. Additional support is offered by the inferred movements of the eels through the tidal passages where depth averaged current velocities can exceed 0.5 m/s (Forrester 1959). Although all 19 of the eels determined to have departed the study area did so on ebb tides, 10 of the eels were also recorded as present in the passages on flood tides. The receivers recorded an average of six more hits for these eels on flood tides than on ebb tides. Average residency of each fish within the detection radius of the receivers was longer by 22 min on flood tides than ebb tides. The most plausible explanation is that the eels attempted to negotiate the passages without reference or cue to tidal stage but were more successful in doing so only on ebb tides.

Our results cannot address whether imprinting of the migratory route of the silver American eels during the recruitment stage might explain the observed migration behaviors (Westin 1998). Also, the physical boundary conditions of the site may be relevant. The east–west axis of the estuary is nearly perpendicular to the north–south axis of the exit route from Passamaquoddy Bay through Letete Passage and roughly 45 degrees to the south and west through Western Passage (Figure 1). The length scales of the Passamaquoddy Bay study are short (0.2 degrees of latitude, 0.3 degrees of longitude). If navigating the estuary only by compass and not by route, the silver eels might then exhibit a left-lateral bias (i.e., to the south) while progressing to the west. The relatively large 0.5-km detection radius of the receivers could not resolve lateral movements relative to net direction of travel within the estuary (<1 km in width). The number of receiver arrays, and the spatial density of the arrays deployed in Passamaquoddy Bay, was low relative to requirements to estimate direction with sufficient accuracy. It was thus not possible to evaluate whether a "map" might be used by seaward migrating silver eels.

Vertical Movements

Neither pronounced variability in swimming depth nor variability in swimming depths among the estuary, bay, and passages, respectively (Figure 8), were expected based on the observed regularity with which the silver European eels swam at the surface during their seaward migration through the Baltic Sea (Westerberg et al. 2007). Water depths were similar in both studies. In contrast, average swimming depths were similar for silver American eel migrations down the tidally dominated Penobscot River (~4 m, Parker and McCleave [1997]) and Magaguadavic River estuaries (~3.5 m, this study). The variability around the mean swimming depth within a given area, whether estuary or embayment, and so forth, and in mean swimming depths among areas differing in character, such as tidal energy, suggests that silver eels are responsive to their ambient environment during migration.

The reasons why swimming depth might vary with coastal circulation are not clear, although they may be linked to another migration behavior, the execution of vertical dives. These have been documented for both silver American (Stasko and Rommel Jr. 1974; Parker and McCleave 1997) and European eels (Tesch 1978, 1989; Westerberg 1979; McCleave and Arnold 1999; Westerberg et al. 2007). The reported speeds of ascent or descent are similar irrespective of species, water depths, or study area (0.11–0.18 m/s).

Our exploration of vertical dive behavior was limited by records of swimming depth only at the time of last detection of each eel at the individual receivers. It was therefore not possible to know if the direction of change in vertical position (either up or down) was constant between consecutive records. It can however be noted that the average vertical speed over changes in depth of ≥ 5 m was estimated as 0.11 ± 0.12 m/s (217 observations, data not shown). This estimate is consistent with the rates of ascent/descent reported elsewhere, albeit with considerable variability around the estimated average. Future investigations of silver eel migratory behavior based on deployed stationary receivers may benefit from incorporation of active tracking into the survey design.

The variability around the mean swimming depth within a given area, whether estuary, bay, or passage, and in mean swimming depths among areas differing in water circulation (Baltic Sea versus Passamaquoddy Bay; Figure 7), and indications that ascent/descent speeds may be relatively invariant around about 0.15 m/s suggest that the role of the vertical dives is facultative, to enable progress along a specific direction of travel, for example. An angled dive of constant speed while maintaining directed movement on the horizontal (e.g., a geomagnetic reference) might be sufficient to detect vertical water velocity shear at the length scale of the receptor, in the case of the entire body as the receptor, variations in water velocity occurring on length scales of 0.5 to 1 m. The fact that silver American eels detected within the passages where tidal current speeds are greatest occurred at greater depths (i.e., where the presence of a bottom boundary layer might be expected) on flood tides than on ebb tides lends some support to this interpretation.

Vertical dives as a means to enable progress along a specific direction of travel may also explain the equivocal support for STST as a mechanism for silver eel migrations, if STST is considered in the context of a learned rather than an innate mechanism of migration. For example, in areas of continental shelves where tidal flows are bidirectional and strong (on the order of the body length-specific sustained swimming speed or greater) and prevail over extensive sections of the migration corridor, silver eels might use vertical dives to resolve over time the bottom boundary layer flow. A greater prevalence of silver eels at the range of depths that offer expedient transport might be anticipated, as may have been the case for the silver European eels tracked within in the North Sea by McCleave and Arnold (1999).

Similarly, entrainment of resident eels to predictable bidirectional currents in estuaries could be anticipated, as was shown for yellow American eels in the Penobscot River estuary (Parker and McCleave 1997). As well, eels with a protracted or periodic residency in estuaries as yellow eels (Jessop et al. 2002; Cairns et al. 2004) may be more likely to exhibit STST following silvering than eels that may have resided in nontidal waters, if STST is a learned behavior. This may explain the prevalence of departure of silver American eels from the Penobscot River estuary on ebb tides (Parker and McCleave 1997), whereas silver eels departed the equally tidally dominated Magaguadavic River on both flood and ebb tides. Information concerning the level of facultative catadromy exhibited by these two river populations would be of interest.

Acknowledgments

We thank Ron Farquharson, Gerald Burke, and Tony Craig of JD Irving Ltd. for their logistic support during the course of the study and for the flexibility shown in accommodating this study within the ongoing operations of the hydroelectric facility. We thank Mike Best, Matthew Beck, James Connop, Paul Brooking, and Ellen Merrill of the Atlantic Salmon Federation for their efficient and essential field support and general management of the project. Field work and circulation modeling by Randy Losier, data analyses by Blythe Chang, and field work and data processing by Paul McCurdy, all with the St. Andrews Biological Station, contributed to the successful execution of the project. Håkan Westerberg, Ingvar Lagenfelt, and Henrik Svedäng with the Swedish Board of Fisheries provided important and greatly appreciated relational European eel migration data. Comments provided by three anonymous reviewers on a previous draft substantively improved the manuscript. The project was supported financially by Fisheries and Oceans Canada, Atlantic Elver Fishery, and JD Irving Ltd.

References

Barbin, G. P. 1998. The role of olfaction in homing and estuarine migratory behavior of yellow-phase American eels. Canadian Journal of Fisheries and Aquatic Sciences 55:564–575.

Brooking, P., G. Doucette, S. Tinker, and F. Whoriskey. 2006. Sonic tracking of wild cod, *Gadus morhua*, in an inshore region of the Bay of Fundy: a contribution to understanding the impact of cod farming for wild cod and endangered salmon populations. ICES Journal of Marine Science 63:1364–1371.

Cairns, D. K., J. C. Shiao, Y. Iizuka, W. N. Tzeng, and C. D. MacPherson. 2004. Movement patterns of American eels in an impounded watercourse, as indicated by otolith microchemistry. North American Journal of Fisheries Management 24:452–458.

Castonguay, M., P. V. Hodson, C. M. Couillard, M. J.

Eckersley, J. D. Dutil, and G. Verreault. 1994. Why is recruitment of the American eel, *Anguilla rostrata*, declining in the St Lawrence River and Gulf? Canadian Journal of Fisheries and Aquatic Sciences 51:479–488.

Chang, B.D., F. H. Page, R. J. Losier, D. A. Greenberg, J. D. Chaffey, and E. P. McCurdy. 2006. Water circulation and management of infectious salmon anemia in the salmon aquaculture industry in Passamaquoddy Bay, southwestern New Brunswick. Canadian Technical Report of Fisheries and Aquatic Sciences 2622.

Chevrier, J. R., and R. W. Trites. 1960. Drift-bottle experiments in the Quoddy Region, Bay of Fundy. Journal of the Fisheries Research Board of Canada 17:743–762.

COSEWIC (Committee on the Status of Endangered Wildlife in Canada). 2006. COSEWIC assessment and status report on the American eel *Anguilla rostrata* in Canada. Committee on the Status of Endangered Wildlife in Canada, Ottawa.

Cottrill, R. A., F. Økland, K. Aerestrup, N. Jepsen, A. Koed, K.J. Hunter, K. G. Butterworth, and R. S. McKinley. 2006. Evaluation of three transmitter attachment methods for female silver-phase American eel (*Anguilla rostrata* Lesueur). Journal of Great Lakes Research 32:502–511.

Dekker, W., J. M. Casselman, D. K. Cairns, K. Tsukamoto, D. Jellyman, and H. Lickers. 2003. Worldwide decline of eel resources necessitates immediate action. Quebec declaration of concern. Fisheries 28(12):28–30.

Foreman, M. G. G. 1977. Manual for tidal heights analysis and prediction. Institute of Ocean Sciences, Pacific Marine Science Report 77–10, Patricia Bay, Sidney, British Columbia.

Forgeron, F. D. 1959. Temperature and salinity in the Quoddy Region. Chapter 1 in The International Passamaquoddy Fisheries Board Report to the International Joint Commission, Washington, D.C.

Forrester, W. D. 1959. Current measurements in Passamaquoddy Bay and the Bay of Fundy 1957 and 1958. Chapter 3 in The International Passamaquoddy Fisheries Board Report to the International Joint Commission, Washington, D.C.

Fries, L. T., D. J. Williams, and S. K. Johnson. 1996. Occurrence of *Anguillicola crassus*, an exotic parasitic swim bladder nematode of eels, in the southeastern United States. Transactions of the American Fisheries Society 125:794–797.

Gillis, G. B. 1998. Neuromuscular control of anguilliform locomotion: patterns of red and white muscle activity during swimming in the American eel, *Anguilla rostrata*. Journal of Experimental Biology 201:3245–3256.

Greenberg, D. A., J. A. Shore, F. H. Page, and M. Dowd. 2005. Modelling embayments with drying intertidal areas for application to the Quoddy region of the Bay of Fundy. Ocean Modeling 10(2005):211–231.

Gregory, D., B. Petrie, F. Jordan, and P. Langille. 1993. Oceanographic, geographic and hydrological parameters of Scotia-Fundy and southern Gulf of St. Lawrence inlets. Canadian Technical Report of Hydrography and Ocean Sciences 143.

Hanson, L., L. Karlsson, and H. Westerberg. 1984. Magnetic material in European eel (*Anguilla anguilla* L.). Comparative Biochemistry and Physiology 77A:221–224.

Jessop, B. M., J. C. Shiao, Y. Iizuka, and W. N. Tzeng. 2002. Migratory behaviour and habitat use by American eels *Anguilla rostrata* as revealed by otolith microchemistry. Marine Ecology Progress Series 233:217–229.

Ketchum, B. H., and D. J. Keen. 1953. The changes of fresh and salt waters in the Bay of Fundy and in Passamaquoddy Bay. Journal of the Fisheries Research Board of Canada 10:97–124.

MacGregor, R., J. M. Casselman, W. A. Allen, T. Haxton, J. M. Dettmers, A. Mathers, T. C. Pratt, P. Thompson, M. Stanfield, and L. Marcogliese. 2009. Natural heritage, anthropogenic impacts, and biopolitical issues related to the status and sustainable management of American eel: a retrospective analysis and management perspective at the population level. Pages 713–740 *in* A. J. Haro, K. L. Smith, R. A. Rulifson, C. M. Moffitt, R. J. Klauda, M. J. Dadswell, R. A. Cunjak, J. E. Cooper, K. L. Beal, and T. S. Avery, editors. Challenges for diadromous fishes in a dynamic global environment. American Fisheries Society, Symposium 69, Bethesda, Maryland.

McCleave, J. D., and G. P. Arnold. 1999. Movements of yellow- and silver-phase European eels (*Anguilla anguilla* L.) tracked in the western North Sea. ICES Journal of Maine Science 56:510–536.

Montén, E. 1985. Fish and turbines: fish injuries during passage through power station turbines. Vattenfall, Stockholm, Sweden.

Moser, M. L., W. S. Patrick, and J. U. Grutchfield, Jr. 2001. Infection of American eels, *Anguilla rostrata*, by an introduced Nematode parasite, *Anguillicola crassus*, in North Carolina. Copeia 2001(3):848–853.

NRC (National Research Coucil). 2007a. Fee, accuracy and confirmation of sunrise/sunset data Web page of the National Research Council of Canada Herzberg Institute of Astrophysics. Available: http://hia-iha.nrc-cnrc.gc.ca/sunrise_e.html (April 2007).

NRC (National Research Council). 2007b. Standard times of solar rise/set for Saint Andrews NB 2006. Output generated from the National Research Council of Canada Herzberg Institute of Astrophysics Sunrise/Sunset Calculator. Available: http://hia-iha.nrc-cnrc.gc.ca/sunrise_e.html (April 2007).

Parker, S. J. 1995. Homing ability and home range of yellow-phase American eels in a tidally dominated estuary. Journal of the Marine Biological Association of the U.K 75:127–140.

Parker, S. J., and J. D. McCleave. 1997. Selective tidal stream transport by American eels during homing movements and estuarine migration. Journal of the Marine Biological Association of the U.K. 77:871–889.

SAS. 1999. SAS/STAT users guide, 4th edition. SAS Institute Inc., Cary, North Carolina.

Stasko, A. B., and S. A. Rommel, Jr. 1974. Swimming depth of adult American eels (*Anguilla rostrata*) in a saltwater bay as determined by ultrasonic tracking. Journal of the Fisheries Research Board of Canada 31:1148–1150.

Tesch, F.-W. 1978. Telemetric observations on the spawning migration of the eel (*Anguilla anguilla*) west of the European continental shelf. Environmental Biology of Fishes 3:203–209.

Tesch, F. -W. 1989. Changes in swimming depths and direction of silver eels (*Anguilla anguilla* L.) from the continental shelf of the deep sea. Aquatic Living Resources/Resources vivantes aquaticues. Nantes 2(1):9–20.

Tesch, F.-W. 1992. Insignificance of tidal currents for silver eel migration as studied by eel trackings and current measurements. Irish Fisheries Investigations A (Freshwater) 36:105–109.

Tesch, F.-W., T. Wendt, and L. Karlsson. 1992. Influence of geomagnetism on the activity and orientation of eel, *Anguilla anguilla*, as evident from laboratory experiments. Ecology of Freshwater Fish 1:52–60.

Trites, R. W., and C. J. R. Garrett. 1983. Physical oceanography of the Quoddy Region. Pages 9–34 *in* M. L. H. Thomas, editor. Marine and coastal systems of the Quoddy Region, New Brunswick. Canadian Special Publication of Fisheries and Aquatic Sciences 64.

Trites, R. W. and D. G. MacGregor. 1959. Flow of water in the passages of Passamaquoddy Bay measured by the electromagnetic method. Chapter 4 *in* The International Passamaquoddy Fisheries Board Report to the International Joint Commission, Washington, D.C.

van den Thillart, G., V. van Ginneken , F. Körner, R. Heijmans, R. van der Linden, and A. Gluvers. 2004. Endurance swimming of European eel. Journal of Fish Biology 65:1–7.

van Ginneken, V., E. Antoissen, U.K. Müller, R. Booms, E. Eding, J. Verreth, and G. van den Thillart. 2005a. Eel migration to the Sargasso: remarkably high swimming efficiency and low energy costs. Journal of Experimental Biology 208:1329–1335.

van Ginneken, V., B. Muusze, J.K. Breteler, D. Jansma, and G. van den Thillart. 2005b. Microelectronic detection of activity level and magnetic orientation of yellow European eel, *Anguilla anguilla* L., in a pond. Environmental Biology of Fishes 72:313–320.

van Ginneken, V., and G. van den Thillart. 2000. Eel fat stores are enough to reach the Sargasso. Nature (London) 403:156–157.

Westerberg, H. 1979. Counter-current orientation in the migration of the European eel. Rapport Proces Verbeaux Reunion Conseil International l'Exploration de la Mer 174:134–143.

Westerberg, H., I. Lagenfelt, and Svendang, H. 2007. Silver eel migration behavior in the Baltic. ICES Journal of Marine Science 64:1457–1462.

Westin, L. 1998. The spawning migration of European silver eel (*Anguilla anguilla* L.) with particular reference to stocked eel in the Baltic. Fisheries Research 38:257–270.

Whitford, J. 2008. Final Report: background report for the Fundy Tidal Energy Strategic Environmental Assessment. Offshore Energy Environmental Research Association, Project No. 1028476, Dartmouth, Nova Scotia.

American Fisheries Society Symposium 69:293–310, 2009

Assessing Estuarine and Coastal Migration and Survival of Wild Atlantic Salmon Smolts from the Narraguagus River, Maine Using Ultrasonic Telemetry

John F. Kocik* and James P. Hawkes
National Oceanic and Atmospheric Administration, National Marine Fisheries Service Northeast Fisheries Science Center, 17 Godfrey Drive, Suite 1, Orono, Maine 04473, USA

Timothy F. Sheehan
National Oceanic and Atmospheric Administration, National Marine Fisheries Service Northeast Fisheries Science Center, 166 Water Street, Woods Hole, Massachusetts 02543, USA

Paul A. Music
National Oceanic and Atmospheric Administration, National Marine Fisheries Service Northeast Fisheries Science Center, 17 Godfrey Drive, Suite 1, Orono, Maine 04473, USA

Kenneth F. Beland
Maine Atlantic Salmon Commission, 650 State Street, Bangor, Maine 04401, USA

Abstract.—Understanding estuarine and early marine ecology of Atlantic salmon *Salmo salar* smolts within the relatively large and spatially dynamic environments of eastern Maine is challenging. Using ultrasonic transmitters and a large network of fixed receivers, we monitored natural smolt migration in the Narraguagus River, Narraguagus Bay, and coastal environment of the western Gulf of Maine. Our 30-km-long study area began in the lower river and extended 8 km downstream to head-of-tide, 7 km through the estuary, then fanned out seaward 15 km into the western gulf along the interface with Maine Coastal Current. From 1997 to 2004, we increased sampling network density in the estuary and expanded marine arrays further into the Gulf of Maine. We designed receiver networks to monitor all smolt exit routes and were able to (1) estimate smolt survival to the Gulf of Maine, (2) map primary migration paths, and (3) document emigration timing. Survival ranged from 36% to 47% to our outer receiver array. Median migration rates were 0.7 km/h in the estuary to middle bay and 1.0 km/h in the outer bay. Smolts generally traveled with the tides and upon entering saltwater most commonly used the western 6 km of a 23-km-wide embayment. These are among the first quantitative data to estimate survival during early marine migration of wild Atlantic salmon smolts. A Cormack-Jolly-Seber model estimated site efficiency and smolt survival simultaneously, providing a useful methodology and information benchmarks for other studies to better understand emigration dynamics and to help identify mortality factors at sea.

Introduction

Atlantic salmon *Salmo salar* Linneaus, 1758 populations are declining throughout their range, particularly at the southern end of their distribution where many populations are extirpated or at historically low abundance (Parrish et al. 1998; National Research Council 2004). Historic factors attributed to industrialization (dams, pollution, and overharvest) impacted all U.S. populations and resulted in the extirpation of U.S. populations south of Maine by 1890 (National Research Council 2004; Fay et al. 2006). Since the early 1990s, remaining populations have undergone a notable decline hypothesized to be driven by a second factor—a marine re-

* Corresponding author: john.kocik@noaa.gov

gime shift—that broadly affected North American Atlantic salmon populations (Chaput et al. 2005). Several groups have investigated the reasons for Atlantic salmon declines and identified research needs (Mather et al. 1998; O'Neil et al. 2000; Cairns 2001; National Research Council 2004). They found that while freshwater production may limit some stocks, marine survival should be a top research priority. Marine survival is an important factor constraining U.S. Atlantic salmon recovery.

The last remaining U.S. Atlantic salmon populations with notable natural reproduction spawn in Maine rivers are identified collectively as the Gulf of Maine distinct population segment (GoM DPS; Colligan et al. 1999; Fay et al. 2006). Because of low freshwater production and high marine mortality, GoM DPS populations are low and have been listed under the Endangered Species Act (U.S. Department of Commerce 2000). Determining the population dynamics of these populations and understanding mortality is important to conservation management.

We studied the Atlantic salmon population in the Narraguagus River ecosystem in eastern Maine because of a long time series of baseline data (U.S. Atlantic Salmon Assessment Committee 2007). Narraguagus River smolt production averaged 2,300 fish (0.5 smolt per production habitat unit), and marine survival in this system averaged only 0.68% in the past decade (U.S. Atlantic Salmon Assessment Committee 2007). There has been no significant increase in spawning adults despite a decade of extensive fry stocking that averaged 340,000/year and was designed to supplement the limited natural reproduction. Understanding the general ecology and dynamics of smolt migration in lower river and coastal environments is important because it may identify coastal population bottlenecks.

Past information on U.S. salmon stocks at sea was derived from marking, tagging, and scale studies of hatchery fish released in large rivers (Meister 1984; Baum and Spencer 1990). This research has improved our understanding of the general movements and exploitation of these stocks at sea. Corresponding work has not been done in smaller rivers still producing wild salmon, nor have those studies spatially partitioned emigration mortality. Recent advances in fish telemetry make it an increasingly effective approach for determining migration success. Telemetry studies of Atlantic salmon focused on behavior of hatchery fish and several were done in large river systems (Fried et al. 1978; LaBar et al. 1978; Lacroix and McCurdy 1996; Lacroix and Knox 2005; Lacroix et al. 2005). Less common are studies of wild smolts that focus on survival during emigration (Riley et al. 2002; Økland et al. 2006). Because many coastal rivers that GoM DPS Atlantic salmon occupy are of moderate size (<8,000-ha watersheds) and smolts are primarily naturally reared (wild spawned or stocked as fry), additional research is needed. Our work will provide useful comparisons to studies of hatchery smolts and larger ecosystems.

Our study used small ultrasonic transmitters and automated receivers (Moore et al. 1990; Lacroix and McCurdy 1996; Voegeli et al. 1998) to establish a broad spatial network with continual monitoring capability to study naturally reared smolts. This approach offers the potential to assess other marine ecology factors such as predation and migratory corridors, generating information that is useful in isolating mortality spatially and temporally. Our goal was to obtain quantitative information on smolt emigration mortality, movements, and dispersal to allow a better assessment of early marine survival and management opportunities. Specific objectives were to (1) measure smolt emigration timing, (2) quantify migration success, (3) identify ecological zones where mortalities occur, and (4) describe general behavior patterns.

Methods

Overview

Methodology focused on (1) establishing a network of passive ultrasonic receivers to detect smolt emigration at ecological transition zones, (2) tagging smolts with ultrasonic transmitters throughout the entire emigration period, (3) using receiver data to determine location and potential disposition of smolts within the study area, and (4) quantitatively evaluating smolt migration ecology. Our study had two phases (Figure 1)—Phase I (1997–1999) evaluated smolts in the lower river, estuary, and inner to middle bay. Phase II (2002–2004) built upon this information to study smolt behavior more extensively in the estuary and expand the study area to the outer bay interface with the Gulf of Maine (GoM). During both phases, technological innovations to

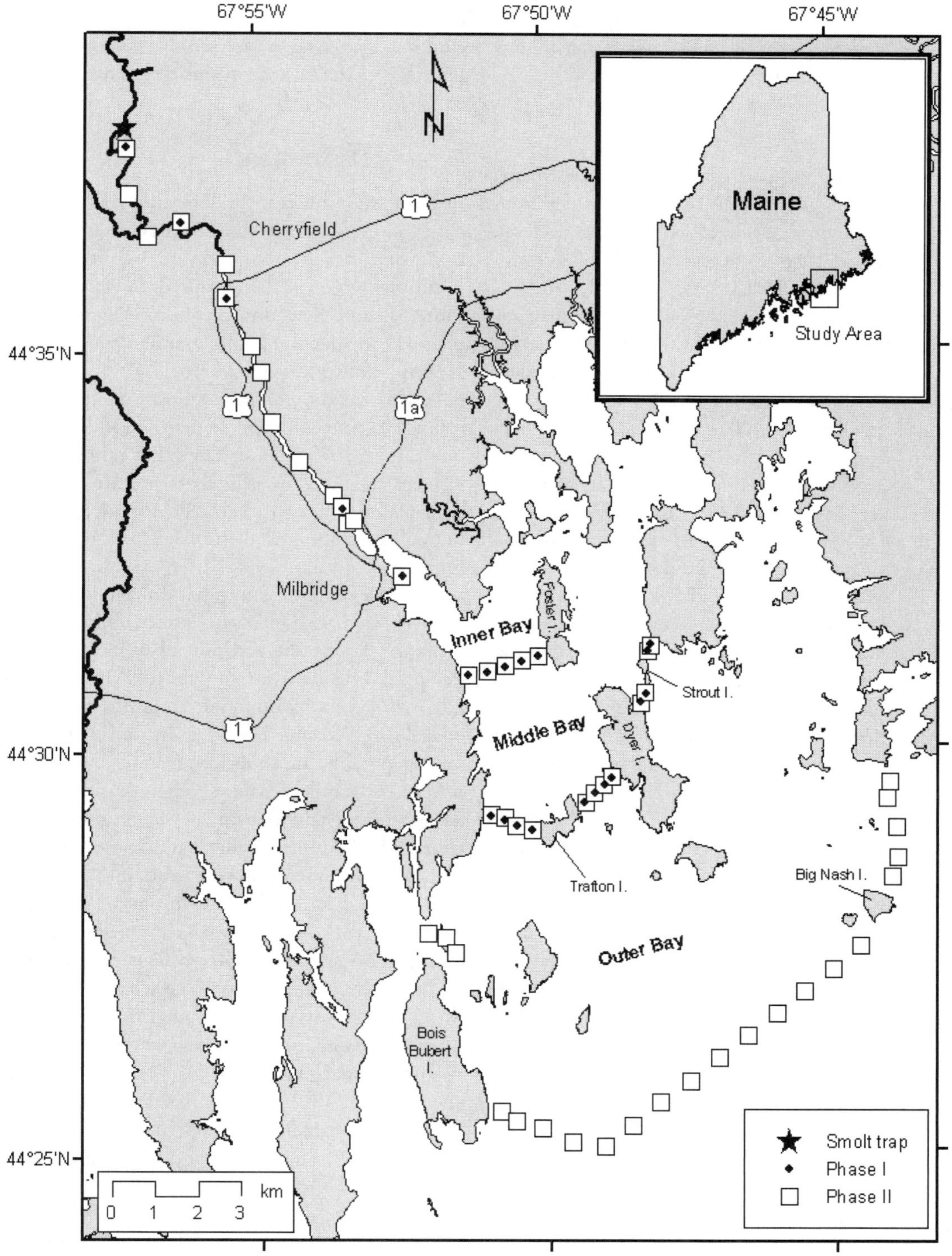

Figure 1.—Study area in Narraguagus River and Bay, Maine, illustrating location of rotary screw trapping and release site as well as locations of hydroacoustic receivers during phase I (VR-20 receivers) and phase II (VR-2 receivers). Inset indicates extent of study area and Narraguagus watershed.

gear occurred that made expansion of networks more cost-effective (lower transmitter and receiver prices) and less complicated (coded transmitters eliminated monitoring multiple frequencies).

Study Site

The Narraguagus River located in eastern Maine, USA (44.67°N, 68.00°W) flows southeasterly from Eagle Lake at an elevation of 140 m above sea level, a distance of 70 km to head of tide in Cherryfield, Maine (Baum and Jordan 1982; Figure 1). The watershed drains about 588 km^2 and has a median annual discharge of 14 m^3/s. Most salmon production habitat is in the upper two-thirds of the watershed and totals about 60 ha (Abbott 2006). The estuary begins approximately 400 m upstream of U.S. Route 1 in Cherryfield and flows southeasterly approximately 6 km where it empties into Narraguagus Bay. The estuary is only 50–120 m wide and depth is 3 m at mean low water with 3.4-m tidal fluctuation (Baum and Jordan 1982). The tidal freshwater zone is 260 ha and the estuary/mixing zone is 1,300 ha (Strategic Assessments Branch 1985). Narraguagus Bay has flora and fauna typical of eastern Maine and southern Atlantic Canada and encompasses 16,600 ha with tidal fluctuations of 4.8 m (Strategic Assessments Branch 1985; Jury et al. 1994). As smolts exit Narraguagus Bay, the first oceanographic feature they encounter in the GoM is the Eastern Maine Coastal Current (EMCC)—a cold coastal band that flows southwest from the Bay of Fundy to Penobscot Bay (Pettigrew et al. 2005).

We divided our study area into separate ecological zones to understand migration dynamics (Figure 1). The riverine zone of our study area began at the smolt trap site and extended downstream 8.6 km to head of tide in Cherryfield. The estuarine zone extended 6.3 km downstream from the Route 1 bridge in Cherryfield to just above the Route 1A bridges in Milbridge. The marine zone has three subzones—inner, middle, and outer bay. The estuary empties into the inner bay and, in 5 km, widens from 0.4 to 2 km at the Foster Island Receiver Array. The middle bay was delineated by the Trafton-Dyer-Strout Receiver Array that was sited among islands to both take advantage of and better understand usage of distinct corridors. The middle bay was 2.8 km seaward and spanned a breadth of almost 8 km. The Nash Island Receiver Array averaged 6.5 km seaward in a 23-km arc delineating the outer bay. At the apex, the Nash Island Receiver Array transected the western edge of the EMCC and passage of this array characterized entry into the GoM.

Receiver Deployment

In both study phases, the longitudinal migration path was monitored at discrete sites; each site consisted of either a node (one receiver) or an array (multiple receivers), as required to achieve full shore-to-shore detection capability (Figure 1). In phase I, we deployed and maintained a passive acoustic network with Vemco VR-20 receivers bolted to custom concrete moorings—freshwater (16 kg), estuary (23 kg), or marine (36 kg). Each mooring had a four-post frame that protected the receiver and acted as a purchase point to attach a 0.95-cm buoy line to a 5 × 28 cm spongex lobster buoy. Line scope varied from 1.5–3:1 to account for tides.

In phase II of the study, bottom-mounted VR2 receivers were used for river, estuary, and inner bay locations. In middle and outer bay locations, we deployed VR2 receivers suspended 10 m deep on a buoy line in a hydrophone up position (Voegeli et al. 1998). We attached receivers on (1) an active lobster pot line in cooperation with commercial fishermen (trap and ballast weights varied) or (2) a mushroom-chain mooring (11 kg mushroom anchor and 11 kg of anchor chain). Lines on both platforms were attached with weak links (breaking strain of 272 kg) and splices to ensure that gear was whale-safe. Depending on conditions, we used a single or double 20 × 38 cm float, with a 15 × 28 cm pick-up toggle float. The scope of the line was 2–3:1, and typically, we used 0.95 cm nylon or polypropylene sink line, but where there was a risk of line snagging, we used 0.87 cm polypropylene float line. VR2 receivers were attached to lines using plastic cable ties at 2–3 anchor points that were passed through the stranded line to prevent movement of the receiver up or down the buoy line (Voegeli et al. 1998; Moore et al. 1990).

For both phases, we tested receivers for water tightness (24-h soak test) and function and then initialized and synched them to an atomic clock. Receivers were deployed at predetermined locations where each receiver was checked with a test transmitter to confirm function and mark the start of the deployment in the receiver memory. Boats were used to deploy receivers. The vessels maneuvered to target coordinates using a Global Positioning System (GPS). We deployed receivers by hand or pot hauler.

Receiver Retrieval and Tends

During smolt migration, we tended receivers at 3–4-week intervals to (1) replace batteries (VR-20 only), (2) test receiver operation, and (3) collect data in-season (a data safeguard). Receivers were located using the coordinates taken during deployment, and if the receiver had moved due to current, storm, gear interactions, or other factors, the new and original coordinates were averaged to determine position. Receivers were checked with a test transmitter to mark the retrieval, data downloaded to receiver memory, and batteries replaced or receivers exchanged. Data were downloaded using a field laptop computer and Vemco interface cables and software. If the receiver was functioning and had sufficient memory, it was redeployed; otherwise, the receiver was replaced. Redeployment was at original target coordinates.

Active Searches

Active searches determined the fate and location of some deployed transmitters that did not pass successfully into the GoM. Vemco VR28, VR60, and VR100 receivers were used to locate transmitters that remained in the study area. We used a boat with hydrophones deployed at the bow to search using the center of the river and estuary or transects in the inner bay. We used the GPS to record the track of the boat along the search transect and the location of any transmitter detections. A directional hydrophone pinpointed the location of the transmitter and whether it was stationary or moving. These data were used to supplement fixed receiver network data.

Smolt Collection

We collected smolts using two 1.5-m-diameter rotary screw traps (E.G. Solutions, Inc., Corvallis, Oregon) at river km 7.6 (7.6 km from the river mouth). The first smolts handled daily that met the size threshold (see surgery section) were selected for surgical implantation of transmitters. There was a target distribution for days: 1 (one transmitter); 2 (two transmitters); 3 (five transmitters); 4–6 (six transmitters); and 7–10 (eight transmitters), and this was mirrored for the next 10 d to spread transmitters throughout the emigration period.

Surgical Transmitter Implantation

We surgically implanted acoustic transmitters, which varied in frequencies, weight, and size (Table 1). We typically selected smolts greater than 145 mm fork length to minimize transmitter-size affects. Smolts were sedated using buffered tricaine methanesulfonate (MS-222) or clove oil (1999 only) until the fish was immobilized. Before surgery, we sterilized transmitters and surgical apparatus with Benz-All followed with a distilled water rinse. We placed sedated smolts ventral side up in a

Table 1.—Number and type of transmitters and receivers deployed each year.

Year	Receivers deployed					Transmitters			
	Type	River	Estuary	Marine	Total	Tag type	Pingers	Frequency (kHz)	Weight (g)
1997	VR-20	5	1	11	17	V8SC	109	67.3, 69, 72, 75, 76.8, 78	4
1998	VR-20	4	1	18	23	V8SC	95	65.5, 67.3, 69, 72, 75, 76.8, 78	4
1999	VR-20	5	1	19	25	V8SC	102	65.5, 67.3, 69, 72, 75, 76.8	4
2002	VR-2	5	5	35	45	V8SC	100	69	3.1
2003	VR-2	6	8	40	54	V8SC	101	69	3.1
2004	VR-2	6	8	46	60	V8SC	74	69	3.1

parr couch (wet chamois cloth draped over a slotted sponge drenched with river water in a surgical tray) to stabilize the fish. Smolt gills were irrigated intermittently with anesthetic river water. We wiped the incision area with sterile gauze, then used a scalpel to make an incision (three-fourths of the transmitter length; ~15 mm) anterior to the pelvic fins at the midline. The transmitter was inserted and a 4/0 Ethilon nylon suture used to tie two or three simple, interrupted square knots equally spaced to close the incision. We dried the suture area with gauze and sealed it with a thin layer of BioGlue, then applied Taro Pharmaceuticals triple antibiotic ointment (Bacitracin Zinc, Gramicidin, and Polymyxin B Sulfate). We placed the smolt into a river-water recovery bucket until it regained equilibrium and was released. Typical handling durations were 2–3 min to anesthetize, 2–3 min for surgery, and 3–5 min for recovery.

Survival and Efficiency Analysis

We used StatTools (Palisade Corporation) for statistical analysis of biological traits and emigration timing data. We modeled survivorship (ϕ) and efficiency (ρ) probabilities between monitoring sites using the Cormack-Jolly-Seber (CJS) release–recapture model (Seber 1982). Parameters were estimated in Program MARK from complete capture histories at multiple sites with removals (White et al. 1982). We used the full CJS model, where both ϕ and ρ are site-dependent, for each year. Because outermost array efficiencies cannot be calculated in the CJS model due to their terminal location, we performed an efficiency test on the Nash Island Receiver Array in 2004 by adding receivers at the midpoint between the seven westernmost receivers to measure fish passing this section undetected by typical array spacing.

Results

Receiver Deployment, Retrieval, and Active Searches

During phase I, we deployed between 17 and 25 VR-20 receivers in the Narraguagus River system from mid-April until mid-June 1997–1999 (Figure 1). Moorings performed well and receivers were upright with exception of two receivers that became entangled in aquatic vegetation in the inner bay (1998); receivers remained functional, but efficiency was likely reduced (see below). During phase I, we released 306 naturally reared smolts with transmitter implants (Table 1). In phase II, we deployed between 45 and 60 VR-2 receivers in the expanded study area from April 20 until final retrieval in mid to late June (Figure 1; Table 1). Because of longer battery life and modified mooring design, receivers remained in place but were checked and data downloaded in mid- to late May and at retrieval. During phase II, 175 smolts were released from the trapping site. Active searches were conducted in the river and estuary study zones at least twice annually, one time during the course of migration and once within 10 d after the last release. Data from active searches were used to corroborate detection data as well as to document that some transmitters stranded between fixed receivers but still active.

Temporal and Biological Characterization of Study Population

We collected smolts from rotary screw traps as they migrated to sea in late April through early June (Figure 2). Total catch data centered on a median of standard day 131, with 90% of fish moving between day 117 and 145. Our sample population implanted with transmitters had only slight differences—centered on day 132 with 90% of study fish released between days 119 and 146. Thus, our sampling protocols successfully captured the range of natural emigration dates (Figure 2).

Smolts that we collected in rotary screw traps (N = 6,701) had an average fork length (FL) of 165 mm (N = 6,673, SD ± 15) and weight of 46 g (N = 2,832, SD ± 13) with a condition factor of 0.96 (N = 2,832, SD ± 0.11). Study fish (N = 581) were larger because we maintained a size minimum of 145–150 mm FL for surgeries—171 mm FL (N = 565, SD ± 13), 49 g (N = 535, SD ± 12.8), and a condition factor of 0.97 (N = 535, SD ± 0.10). Upon reaching the size threshold, the size distribution of fish implanted with transmitters followed the trend of the overall catch peaking at 165–175 mm FL and then decreasing to about 200 mm FL. As such, study fish size distributions tracked those of total catch and represented about 90% of the population.

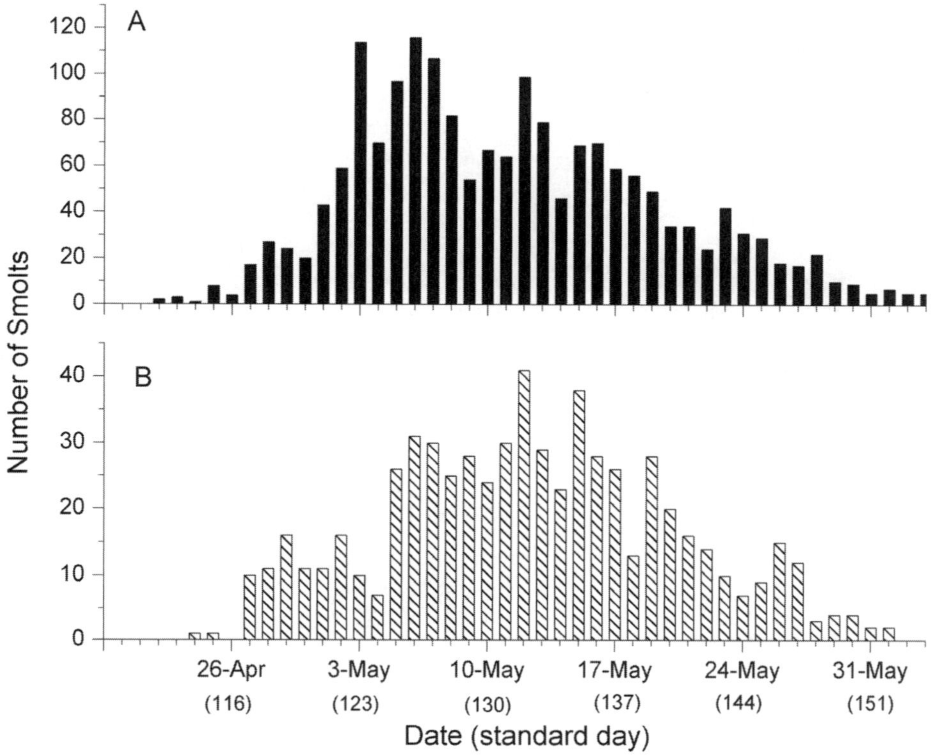

Figure 2.—Number of total smolts caught (panel A) and transmitters deployed daily (panel B) for entire study period.

Smolt Emigration Timing and Behavior

We calculated smolt emigration speeds arithmetically: Speed (km/h) = $(r\text{Km}_u - r\text{Km}_d)/(T_{\text{lu}} - T_{\text{fd}})$, where

$r\text{Km}_u$ = distance along centerline to nearest 0.1 km of upstream receiver

$r\text{Km}_d$ = distance along centerline to nearest 0.1 km of more seaward receiver

T_{lu} = time of last detection at upstream receiver

T_{fd} = time of first detection at seaward receiver

Emigration timing trends showed median speeds were similar across years with notable differences between individual fish illustrated by percentile ranges (Figure 3, panels A–D). Migration speed in the river was typically the lowest, averaging 0.38 km/h during the 6 years. Migration speed increased in the estuary zone and averaged 1.07 km/h, but slowed in the middle bay to 0.69 km/h. Speed once again increased to 1.05 km/h in the outer bay.

Total time from freshwater to exiting either the middle bay or outer bay varied but provided an understanding of the smolt migration window (Figure 3, panels E–F). Within 75 h, about 50% of migrants had exited the middle bay, and by 125 h, 75% exited. In fact, more than 95% of successful migrants exited the middle bay in less than 200 h (8.3 d). Similar trends were seen in outer bay exit times from phase II—within 75 h, 55% of smolts had entered the GoM, 95% of smolts had entered the GoM in less than 200 h, and all successful smolts entered the GoM in less than 14 d.

Smolt behavior in each ecological zone of the migration route affected overall migration time. This is most notable in the riverine and the estuary zones where unique behaviors affected overall transit time. In freshwater reaches, the preference for nighttime movement was greater than in estuaries and bays, as illustrated in enumeration of initial detections as fish actively enter detection fields at individual receivers (Figure 4). These data found that smolts were less likely to move during the day, particularly in freshwater. As such, total speeds and

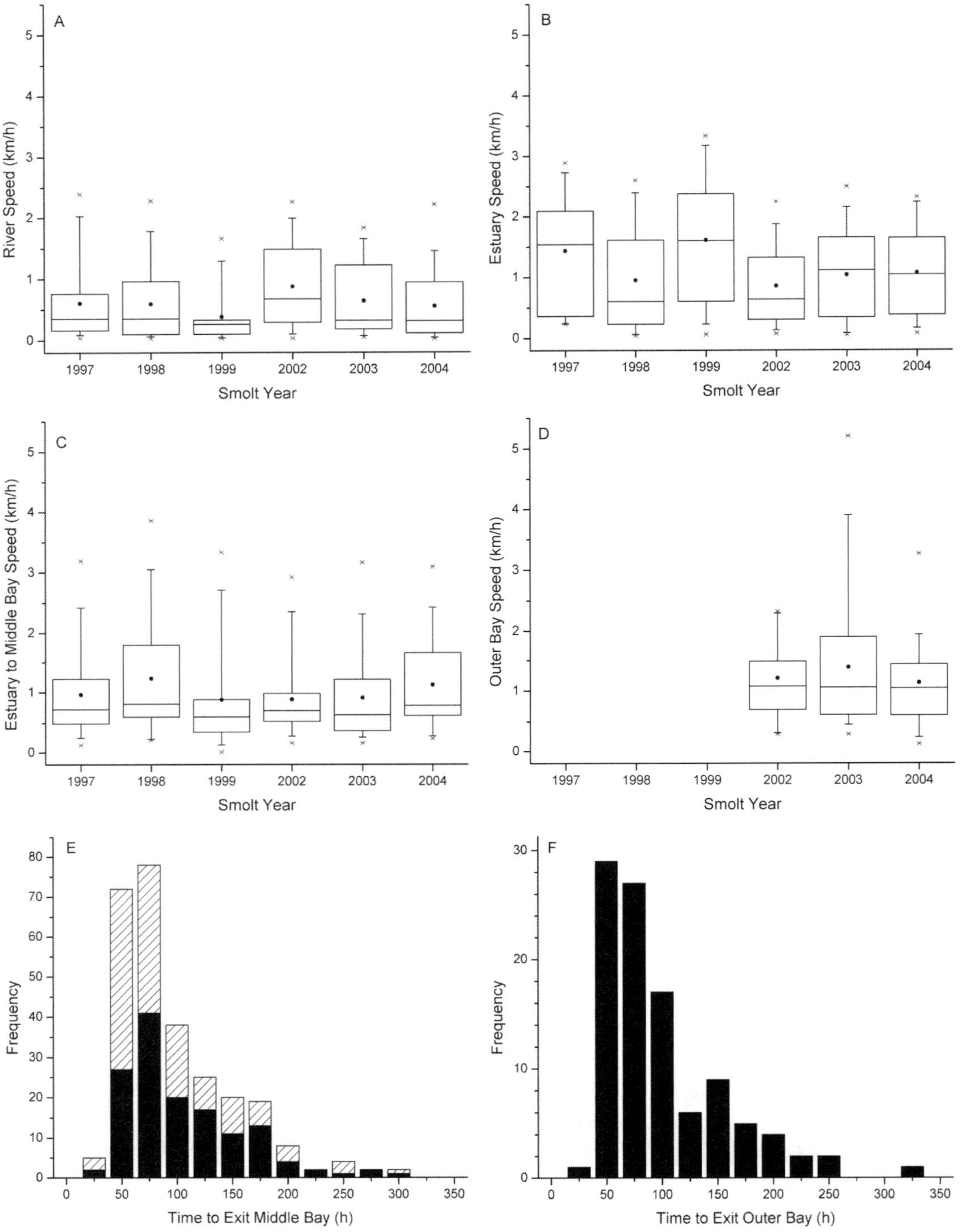

Figure 3.—Atlantic salmon smolt migration speed by ecological zone: river (panel A), estuary (panel B), middle bay (panel C), and outer bay (panel D). For each panel, the box denotes 25th and 75th percentiles, whiskers denote the 5th and 95th percentiles, X denotes the 1st and 99th percentiles, the solid circle is the mean, and the horizontal bar in the box is the median. Panels E and F represents frequency distributions of total migration time to middle (panel E) or outer (panel F) bay for phase I (striped bar) and phase 2 (solid bar).

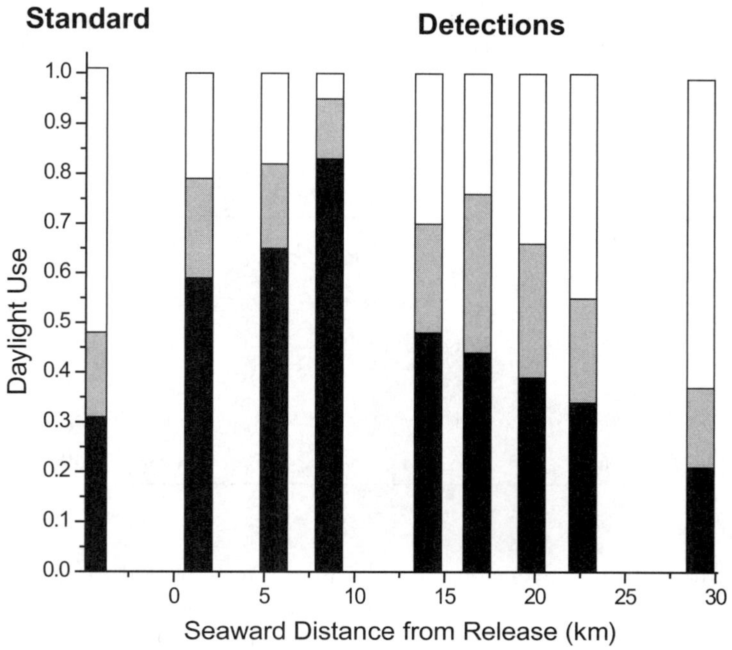

Figure 4.—Temporal distribution of initial detections indicating active movement of smolts at selected transects compared to the seasonal standard of available hours (standard) of daylight (white), twilight (gray), and night (black) at the midpoint of the smolt run. Distance from release site illustrates changes in behavior between freshwater, estuarine, and marine zones.

residence times include periods when fish may not be actively moving, but holding within the river. Similarly, smolt behavior in the estuary zone of the migration route accounted for additional variability, not only in speeds in that zone, but in the overall transit time from the release site to the GoM. In phase I, we noted estuary losses and relatively high variability in estuary speeds (Figure 3).

During phase II, in order to better understand estuary dynamics, we increased receiver coverage from one mid-estuary receiver to eight receivers in the zone. This level of resolution allowed us to analyze individual fish movements and identify reversals in migration (Table 2). We documented unidirectional outward migration behavior (no reversals) in 142 (56%) of all smolts tracked. Defining success as detection at the Nash Island receiver array, 98 of these fish were unsuccessful (65% of the unsuccessful group). Smolts that reversed directions at least once totaled 112–44% of all smolts tracked. The 59 successful smolts that reversed direction comprised 57% of the successful category. Examination of the success ratio within the two general behavior categories reveals success for 31% of the unidirectional behavior and 53% for the reversal behavior. Of fish that reversed both successful and nonsuccessful, we found that a single reversal was the most common and that frequency dropped for two and more reversals. However, success ratio increased with increased reversals (Table 2). As such, unidirectional outward migration was the most commonly documented behavior, but reversal behavior had a higher success rate. Additionally, while there is still high individual variability, these reversals help explain some of the variation in overall time to exit the river and enter the GoM when categories of behavior are taken into account.

Emigration Pathways

As smolts enter Narraguagus Bay, a wider selection of migratory options are available with wider channels and different passageways to exit to the GoM. Because inner, middle, and outer bay sites contained receiver arrays generally running east to west, trends in detections indicate preferred passageways.

Table 2.—Summary of the fate of emigrating Atlantic salmon smolts based on estuary behavior category—unidirectional outward (uni) migration or reversal migration (reverse), up-river movement(s). Reversal detail provides information for subgroups. Success ratio summarizes percentage of each behavior column category that reaches the Gulf of Maine.

	Behavior category		Reversal detail			
Fate	Uni	Reverse	1	2	>2	Total
Successful	44	59	31	19	9	103
	43%	57%	30%	18%	9%	
Unsuccessful	98	53	33	16	4	151
	65%	35%	22%	11%	3%	
Overall	142	112	64	35	13	254
	56%	44%	25%	14%	5%	
Success ratio	31%	53%	48%	54%	69%	

In general, smolts were detected most frequently on the western shore of all three marine subzones (Figure 5). In the inner bay, smolts were found most commonly passing the two western most receivers that were west of the dredged navigation channel, but fish were found across the array. The middle bay is divided into two main passages, as it is bisected by Dyer Island; the western passage is also bisected by Trafton Island and ledges and the eastern passage bisected by Strout Island. Smolts most commonly used the western passage and are primarily detected at the three receivers west and the two receivers east of Trafton Island (Figure 5). As smolts enter the outer bay, there is a small passage to the west through Pigeon Hill Bay, but the mouth opens to about a 100° scope of choice. Most smolts enter the GoM through the main portion of the outer bay to the east of Bois Bubert Island, with most detections occurring at the receivers within 6 km of the island. As such, despite the wide opening to the sea, most smolts use only 5% of the available waters, choosing a restricted pathway with limited dispersal to the east (Figure 5).

Emigration Success and Ecological Zone-Specific Losses

We used detection (1) and no detection (0) binary coding to classify individual migration at each site or array and summarized these data by ecological areas—river, estuary, inner bay, middle bay, and outer bay. Using the program MARK with a full CJS model, we estimated both ϕ and ρ for each site as well as associated variance (Figure 6; Table 3). Detection efficiencies wre generally greater than 0.80, with the exception of the Trafton-Dyer-Strout receiver array (middle bay) during 1998 and the Foster Island receiver array that was consistently the most variable and lowest efficiency area (Table 3). In fact, site efficiency was most often greater than 0.90. The modeling-based approach used in MARK provides estimates of site efficiency and allows calculation of SE for both survival and efficiency. One caveat is that at the last detection site, both ϕ and ρ are estimated at the same value since no additional data are available to assess efficiency and error. In an effort to understand the impact of this on our model, we supplemented these estimates with grid efficiency tests on our Nash Island receiver array in 2004. Our analysis indicated that typical ρ in this section was 0.773, lower than estimated by MARK (0.84–0.90). This suggests that because this site efficiency is overestimated, survival in the Outer Bay may be higher.

Our active searches provide ancillary but informative data because they were conducted in the freshwater and estuary only as time permitted. For smolts that did not exit the receiver network, we located an average of 12% (annual rates ranged from 3% to 22%) of them in river or estuarine searches of variable effort. These supplemental data suggest that transmitters not located either exited the water (through either avian or mammalian predators or transmitters became stranded on stream mar-

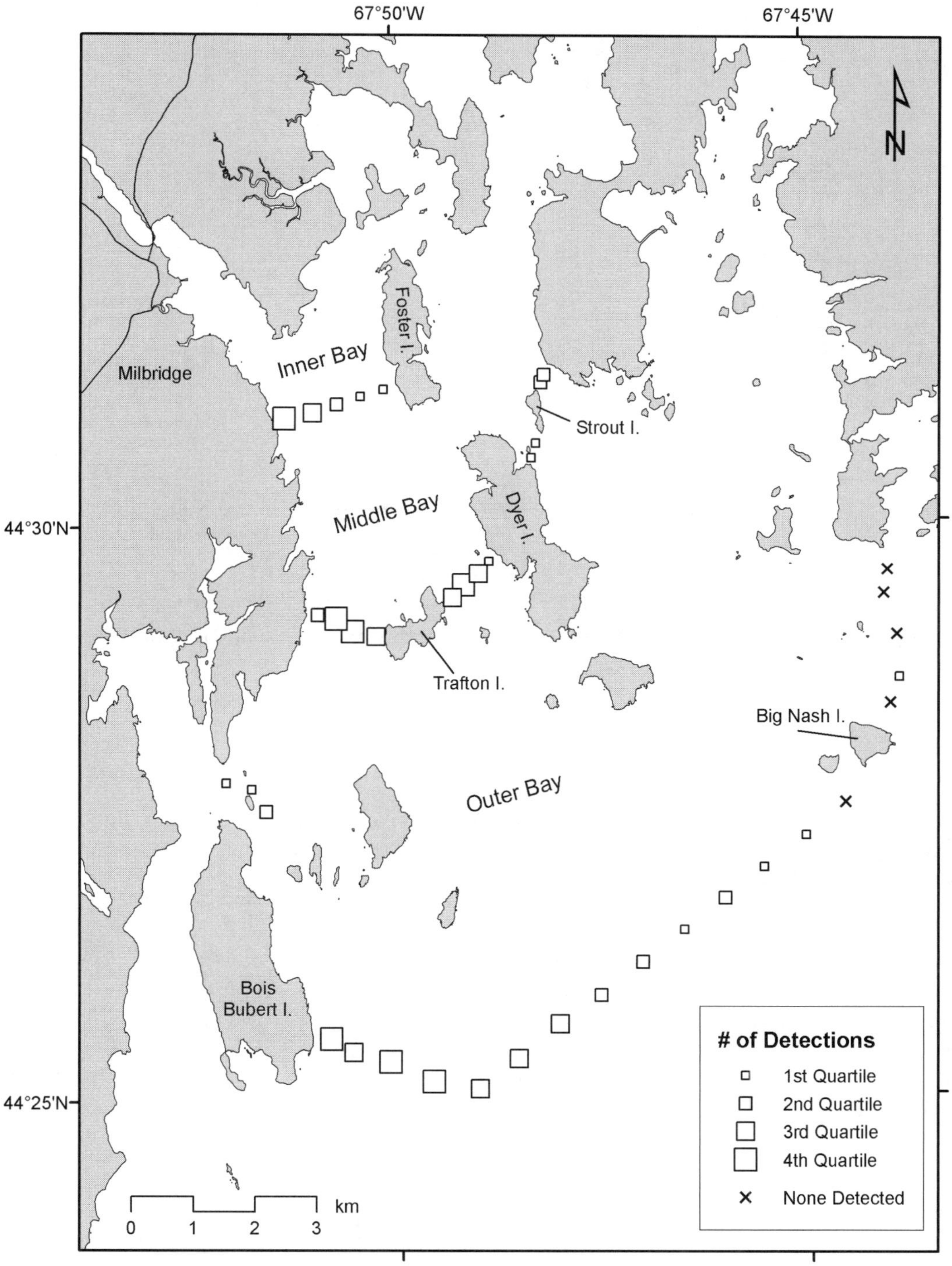

Figure 5.—Map of Narraguagus Bay marine study area with receiver locations identifying quartile ranking of detection numbers. Independent quartile breaks were made for each array and are represented by different sized squares: quartile 1 (<25%), quartile 2 (25% to <50%), quartile 3 (50% to <75%), and quartile 4 (>75%), and locations with no detections were represented with an X. This figure illustrates the most frequently used pathways that favored the western channels and exit from the study area.

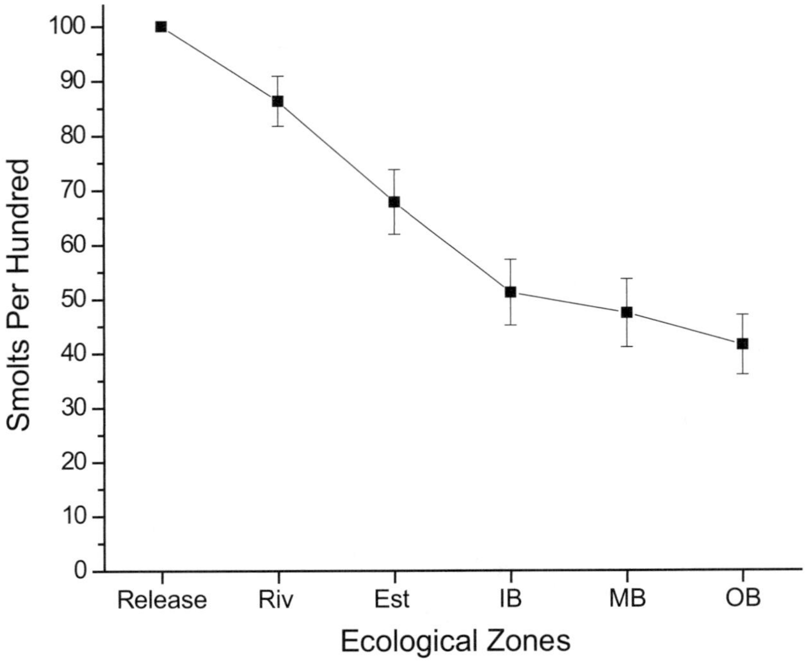

Figure 6.—Results of a simulation that sampled from each year and location independently to illustrate composite trends and smolt survival for all years combined.

Table 3.—Survivorship (ϕ) and receiver site efficiency (ρ) probabilities (SE) calculated using a full Cormack-Jolly-Seber release-recapture model (ϕ and ρ are site- and year-dependent). Note that terminal ϕ and ρ values and SE are identical due to calculation methods.

	Survival ϕ				
Year	River	Estuary	Inner Bay	Middle Bay	Outer Bay
1997	0.68 (0.054)	0.88 (0.047)	–	0.85 (0.000)	–
1998	0.79 (0.044)	0.90 (0.038)	1.00 (0.000)	0.39 (0.000)	–
1999	0.84 (0.038)	0.89 (0.037)	0.86 (0.044)	0.97 (0.000)	–
2002	0.91 (0.031)	0.81 (0.044)	0.74 (0.055)	0.94 (0.038)	0.84 (0.000)
2003	0.84 (0.043)	0.85 (0.045)	0.81 (0.052)	0.98 (0.028)	0.88 (0.000)
2004	0.91 (0.042)	0.75 (0.058)	0.77 (0.061)	1.00 (0.000)	0.90 (0.000)
	Site efficiency ρ				
Year	River	Estuary	Inner Bay	Middle Bay	Outer Bay
1997	0.91 (0.041)	1.00 (0.000)	–	0.85 (0.000)	–
1998	0.98 (0.016)	0.98 (0.020)	0.78 (0.053)	0.39 (0.000)	–
1999	0.99 (0.014)	0.98 (0.017)	0.82 (0.051)	0.97 (0.000)	–
2002	0.94 (0.027)	0.96 (0.026)	0.67 (0.066)	1.00 (0.000)	0.84 (0.000)
2003	0.88 (0.041)	1.00 (0.000)	0.74 (0.064)	0.97 (0.027)	0.88 (0.000)
2004	0.88 (0.048)	1.00 (0.000)	0.89 (0.051)	1.00 (0.000)	0.90 (0.000)

gins that dewatered as spring flows subsided) or stopped functioning. Active searches in this study did find two tags in locations that would have been dewatered under lower flow conditions. A transmitter from this study was also found in a cormorant nest.

MARK results indicated that zone-specific survival generally exceeded 80% and approached 100% in the middle bay (Table 3). Survival trajectories show increased losses in the estuary and inner bay zones and lower losses in the middle and outer bay zones (Figure 6). Our simulations of overall survival using zone specific data from all years indicated that for every 100 smolts exiting the river, 62–74 reached the inner bay, 41–54 reached the middle bay, and 36–47 reached the outer bay (Figure 6). While mortality decreases in the marine environment, our analysis indicates that less than half the smolts survive the approximately 10-d period of migration from the river to the GoM. The survival trajectory also illustrates the increased losses that occur transiting the estuary and inner bay zones (Figure 6). Measured survival increased in the outer bay zone; thus, the survival curve seems to flatten somewhat within this relatively short period of time.

Discussion

Tagging studies, scale imaging, telemetry, and postsmolt trawls are allowing fisheries scientists to better understand Atlantic salmon marine ecology. These tools allow a glimpse into the life of the GoM Atlantic salmon stock complex as they emigrate as smolts and transit various freshwater and marine zones. The results from our telemetry studies have added to this understanding and demonstrate that estuaries and nearshore waters are zones of high mortality and complex behavior during a relatively brief period of time. Our analysis shows that Narraguagus River smolts entered the GoM through relatively narrow migration corridor while catches of fish in surface trawl surveys have documented that within 45–60 d, Maine postsmolts have dispersed widely in the Gulf of Maine south of the Bay of Fundy (Lacroix and Knox 2005). Tagging and scale analyses indicate that later in the summer postsmolts reach the southern Labrador Sea and the Gulf of St. Lawrence (Friedland et al. 1999). These areas are marine habitat for Atlantic salmon postsmolts from across New England and Atlantic Canada. By their second summer at sea, GoM salmon are found off the coast of Greenland intermixed with stocks from North America and Europe (WGNAS 2005).

Despite a better understanding of ocean ecology, there are still many unknowns about Atlantic salmon at sea, and given high marine mortality, additional information is critical to conservation (O'Neil et al. 2000; Cairns 2001; National Research Council 2004). For GoM stocks, understanding migration routes and ocean distribution will enhance our understanding of their ecology—particularly how they use the Eastern Maine Coastal Current, the Labrador Current, and other features for coastal and offshore movements. Even abundant Atlantic salmon populations are a challenge to study because they would still be a relatively rare species in pelagic environments dominated by marine species, but new technologies such as ultrasonic tracking and pelagic trawling have allowed additional insights (Holm et al. 2000; Lacroix and Knox 2005). Integrated programs of ultrasonic tracking and postsmolt trawling in the GoM and further offshore would provide more information than either platform is capable of alone.

This study highlights the utility of telemetry work and how, in combination with mark–recapture models, quantitative population dynamics information can be combined with ecological observations to better understand the interplay between migration behavior and survival. In the lower Narraguagus River system, the smolt migration window to the GoM was short, taking only 3–10 d in most instances. Combined with an overall river emigration window of about 25 d for the entire population, this relatively short 28–35-d period is one of high losses with cohort mortality of 53–64%. Mortality was notably higher in the estuary and inner bay, suggesting that losses are somehow associated with affects of saltwater transition, the ecology of these zones, or a combination of the two (Tytler et al. 1978; Lacroix and McCurdy 1996; Stefansson et al. 2003). To gain insights into common strategies, we compare our observations with other Atlantic salmon populations and Pacific salmon *Oncorhynchus* spp. smolts that are ecologically similar.

River discharge and tidal flow both appear to influence migration speed, but active behaviors observed in our study show that migration to the ocean was not a totally passive event. Studies on hatchery smolts in another GoM river, the Penobscot River, found that movement patterns were regulated nearly exclusively by tidal forces (Fried et al. 1978; LaBar et al. 1978). Fish behavior was thought to play a more important role in other regions (Moser et al. 1991; Moore et al. 1995; Lacroix and McCurdy 1996) and for other salmonid species (Quinn and Brannon 1982; Wood et al. 1993). In some of these studies, smolts actively swam against tides and/or held position in flooding or ebbing tides to maintain downstream gains (Macdonald et al. 1987; Greenstreet 1992; Lacroix and McCurdy 1996). Differences in these conclusions are likely due to differences between estuary structure across geography and the adaptive traits developed across systems by species and even stocks, making generalizations difficult (Tytler et al. 1978; Greenstreet 1992; Lacroix and McCurdy 1996). Because smolts in this study and the Penobscot River studies (Fried et al. 1978; LaBar et al. 1978) are part of the same stock complex (Fay et al. 2006), explaining mechanisms that might determine passive versus active migration behavior could help with understanding regulating factors. Smolts from the earlier Penobscot River studies were all reared in hatcheries, and differences between those fish and Narraguagus River stocks may be due to hatchery influences on behavior. Likewise, differing water quality, estuary structure, and temperatures between systems could be an important regulator of seawater entry behavioral ecology (Staurnes et al. 1993; Magee et al. 2001; Zydlewski et al. 2005). Penobscot stock hatchery smolts released in the Narraguagus River as part of a pH exposure experiment appear to have behaved more similarly to the native stock (Magee et al. 2001). However, this study was conducted using the phase I network that lacked fine-scale detail of smolt movements in the estuary. A study in Norway found that survival was similar between hatchery and wild smolts and suggested that larger wild fish were more successful and that early migration speed (size dependent) may be an influence (Thorstad et al. 2007). More research is needed to compare and contrast the estuary ecology of GoM Atlantic salmon to determine the contributions of rearing (stream or hatchery), stock, and environment to smolt migration speed.

Our observations indicated that smolts used specific migration corridors. This information is important for managers since it indicates that not all coastal pathways are used equally. From a system-specific perspective, this is essential to help conserve and protect habitat and populations. To expand this information to other systems, understanding the factors that influence route choice is equally important. In conjunction with this study, a hydrodynamic model of Narraguagus Bay was developed that provides spatially and temporally explicit temperatures, salinities, and water direction/velocity (D. Williamson, Baird Associates, Ottawa, personal communication; Incze 2000). At a population level, our comparisons of smolt migration corridors with dynamic visualizations are suggestive that smolts use high tidal movement areas preferentially. Modeled salinities in the middle and outer bays generally range from 29.2 to 30.5 parts per thousand across migration corridors, and variation does not seem linked to spatial choice at a population scale. Salinity preferences have been suggested to be a cue for seaward migration in other studies (Holm et al. 2000; Lacroix and Knox 2005). Across our marine arrays, salinity was typically within the range Lacroix and Knox (2005) suggesting that salmon may be seeking a buffer as they transition to full strength seawater. In the outer bay, channels become less distinct and tidal influences more similar across broad geographic areas. Here, compass bearings of smolts suggest a southwesterly tendency as the migration scope of choice expands; this direction would be along the edge of the EMCC a cold coastal band that flows southwest towards Penobscot Bay (Pettigrew et al. 2005). Additionally, celestial and magnetic cues have been found important in sockeye salmon orientation during migration (Quinn and Brannon 1982) and olfaction is another potential tool for orientation since it is important in homing (Brannon et al. 1984; Johannesson 1987). The above aids to navigation in salmonids combined with both immediate and evolved responses to tides and currents could facilitate directed movements to the GoM and perhaps to specific oceanographic features. While this study was designed to investigate spatial ecology in estuary and coastal waters and not migration cues, finer scale analyses of smolt movements facilitated by telemetry could be a powerful tool for investigating active versus passive migration hypotheses and biological aids to navigation. Further, examining smolt migration at an individual level with associated environmental choices can provide insights into route selection of emigrating smolts. This might be particularly important in this population given differential behavior upon encountering saltwater in the estuary (reversing and nonreversing groups) and should be pursued.

Differences in data interpretation between phase I and phase II demonstrated the benefits of increasingly fine-scale continuous monitoring relative to smolt behavior. With the wider spacing of receivers in phase I, the migration times appeared quite variable in the estuary–inner bay transition zones. Fine-scale longitudinal data in phase II identified that this variation was due to differences in individual behavior, with distinct unidirectional or reversal patterns for each smolt. Other studies mention the role of tides in estuary transit (Fried et al. 1978; LaBar et al. 1978; Lacroix and McCurdy 1996; Lacroix et al. 2005), but to our knowledge, none categorized movements of successful and unsuccessful individuals. Our findings suggest that two very different behaviors can lead to successful smolt emigration to the GoM. Further studies are needed to more fully understand these behaviors and environmental conditions and triggers: how this affects cohort success and if patterns are part of a suite of naturally adaptive behaviors or symptomatic of dysfunction in altered ecosystems. Given that environmental stressors can impact both physiological readiness for smoltification and ability to sense chemical gradients such as salt wedges (Staurnes et al. 1993; Magee et al. 2001; McCormick et al. 2005; Zydlewski et al. 2005), understanding behavior should be considered essential to salmon conservation given historically low marine survival (Mather et al. 1998; O'Neil et al. 2000; Cairns 2001; National Research Council 2004; Chaput et al. 2005).

For some species of Pacific salmon, particularly species that emigrate without much freshwater rearing, the estuary is an important nursery area (Healey 1981; Kocik and Taylor 1987). Only one of the large-smolt species, coho salmon *O. kisutch*, uses estuaries for extended periods of time (McMahon and Holtby 1992). For Atlantic salmon, our results support observations that natal estuaries and coastal waters appear to be regions that larger smolt-types transit quickly en route to other environments (Fried et al. 1978; LaBar et al. 1978; Lacroix and McCurdy 1996; Lacroix and Knox 2005; Lacroix et al. 2005). The emigration window is relatively short temporally and spatially for GoM stocks that spend 2 years at sea and travel more than 1,200 km. However, losses during this time period are relatively heavy. Estimates of the monthly marine mortality for Narraguagus River population at sea range from 30% to 42% (WGNAS 2005). Our estimates of first month mortality during emigration to offshore waters range from 53% to 64%. Comparative data with quantitative estimates of initial smolt survival is limited. In inner Bay of Fundy stocks, initial marine mortality was lower, ranging from 0% to 10% (Lacroix and Stokesbury 2004; Lacroix et al. 2005). Recent work on northern New Brunswick Atlantic salmon stocks found that mortality there was comparable to our findings (F. Whoriskey, Atlantic Salmon Federation, St. Andrews, New Brunswick, personal communication). Again, this variability could be due to inherent differences in estuary environments, stocks, and environment–stock interactions.

One criticism of telemetry-based survival estimates that we addressed was lack of quantitative efficiency estimates and associated description of the variance of survival estimates. We believe variance estimates strengthen conclusions and indicate where improvements could be made for future work. This topic is currently a focus of survival estimate analysis for Pacific salmon, and such advances should increase our understanding of marine survival dynamics. Given the limited time series of other studies, the inherent variability in estuaries, and the relatively limited amount of data overall, it is difficult to determine if rates measured for the Narraguagus stock are within norms for the region or are particularly low. Additionally, it is impossible to put these data into a longer time context when overall marine survival was higher, since this technology was not available, so it is therefore important to repeat such studies during high marine survival. Given ecological theory that mortality is highest when fish are smaller, it is possible that these early marine survival patterns cascade through Atlantic salmon life in the ocean and influence overall marine survival.

Acknowledgments

We gratefully acknowledge the assistance of many individuals from the National Oceanic and Atmospheric Administration, the Atlantic Salmon Commission, and the U.S. Fish and Wildlife Service that assisted in field studies. Particular thanks are extended to Christopher Holbrook and Gregg Horton for their assistance in using MARK, Brian Cullum and Daniel Skall for their assistance in literature searches and documentation, and Chad Keith for assistance with maps. We also thank Mike Dadswell and anonymous reviewers for comments that enhanced this paper greatly.

References

Abbott, A. 2006. Maine Atlantic salmon habitat atlas. 3rd edition. Dennys, Ducktrap, E. Machias, Machias, Narraguagus, Pleasant, and Sheepscot rivers. U.S. Fish and Wildlife Service, Falmouth, Maine.

Baum, E. T., and R. M. Jordan. 1982. The Narraguagus River: an Atlantic salmon river management report. Pages 1–42 *in* A river management report by the Atlantic Sea Run Salmon Commission: Narraguagus and Pleasant. Atlantic Sea Run Salmon Commission, Bangor, Maine.

Baum, E. T., and R. C. Spencer. 1990. Homing of adult Atlantic salmon released as hatchery-reared smolts in Maine rivers. International Council for the Exploration of the Sea, Working Group on North Atlantic Salmon, Bangor, Maine.

Brannon, E. L., R. P. Whitman, and T. P. Quinn. 1984. Responses of returning adult coho salmon to home water and population-specific odors. Transactions of the American Fisheries Society 113:374–377.

Cairns, D. K. 2001. An evaluation of possible causes of the decline in pre-fishery abundance of North American Atlantic salmon. Canadian Technical Report of Fisheries and Aquatic Sciences 2358.

Chaput, G., C. M. Legault, D. G. Reddin, F. Caron, and P. G. Amiro. 2005. Provision of catch advice taking account of non-stationarity in productivity of Atlantic salmon (*Salmo salar* L.) in the northwest Atlantic. ICES Journal of Marine Science 62:131–143.

Colligan, M. A., J. F. Kocik, D. C. Kimball, G. Marancik, J. F. McKeon, and P. R. Nickerson. 1999. Status review for anadromous Atlantic salmon in the United States. National Marine Fisheries Service/U.S. Fish and Wildlife Service Joint Publication, Gloucester, Massachusetts.

U.S. Department of Commerce. 2000. Endangered and threatened species; final endangered status for a distinct population segment of anadromous Atlantic salmon (*Salmo salar*) in the Gulf of Maine. Federal Register Notice 63(223):69459–69483.

Fay, C., M. Bartron, S. Craig, A. Hecht, J. Pruden, R. Saunders, T. Sheehan, and J. Trial. 2006. Status review for anadromous Atlantic salmon (*Salmo salar*) in the United States. Report to the National Marine Fisheries Service and U.S. Fish and Wildlife Service, Washington, D.C.

Fried, S. M., J. D. McCleave, and G. W. LaBar. 1978. Seaward migration of hatchery-reared Atlantic salmon, *Salmo salar*, smolts in the Penobscot River estuary, Maine: riverine movements. Journal of Fisheries Research Board of Canada 35:76–87.

Friedland, K. D., J. Dutil, and T. Sadusky. 1999. Growth patterns in postsmolts and the nature of the marine juvenile nursery for Atlantic salmon, *Salmo salar*. Fisheries Bulletin 97: 472–481.

Greenstreet, S. P. R. 1992. Migration of hatchery reared juvenile Atlantic salmon, *Salmo salar* L., smolts down a release ladder. 1. Environmental effects on migratory activity. Journal of Fish Biology 40:655–666.

Healey, M. 1981. Juvenile Pacific salmon in estuaries: the life support system. Estuaries 4:285.

Holm, M., J. C. Holst, and L. P. Hansen. 2000. Spatial and temporal distribution of post-smolts of Atlantic salmon (*Salmo salar* L.) in the Norwegian Sea and adjacent areas. ICES Journal of Marine Science 57:955–964.

Incze, L. S. 2000. A preliminary survey of hydrography and plankton in the Narraguagus–Pleasant Bay system of eastern Maine. National Marine Fisheries Service, Final Report, Grant NA97FL0215, West Boothbay Harbor, Maine.

Johannesson, B. 1987. Observations related to the homing instinct of Atlantic salmon (*Salmo salar* L.). Aquaculture 64:339–341.

Jury, S. H., D. Field, S. L. Stone, D. M. Nelson, and M. E. Monaco. 1994. Distribution and abundance of fishes and invertebrates in North Atlantic estuaries. National Oceanic and Atmospheric Administration, National Ocean Service, Strategic Environmental Assessments Division, ELMR Report 13, Silver Spring, Maryland.

Kocik, J. F., and W. W. Taylor. 1987. Effect of fall and winter instream flow on year-class strength of Pacific salmon evolutionarily adapted to early fry outmigration: a Great Lakes perspective. Pages 430–440 *in* M. J. Dadswell, R. J. Klauda, C. M. Moffitt, R. L. Saunders, R. A. Rulifson, and J. E. Cooper, editors. Common strategies of anadromous and catadromous fishes. American Fisheries Society, Symposium 1, Bethesda, Maryland.

LaBar, G. W., J. D. McCleave, and S. M. Fried. 1978. Seaward migration of hatchery-reared Atlantic salmon (*Salmo salar*) smolts in the Penobscot River estuary, Maine: open-water movements. Journal du Conseil International Council for the Exploration of the Sea 38:257–269.

Lacroix, G. L., and P. McCurdy. 1996. Migratory behaviour of post-smolt Atlantic salmon during initial stages of seaward migration. Journal of Fish Biology 49:1086–1101.

Lacroix, G. L., and M. J. W. Stokesbury. 2004. Adult return of farmed Atlantic salmon escaped as juveniles into freshwater. Transactions of the American Fisheries Society 133:484–490.

Lacroix, G. L., and D. Knox. 2005. Distribution of Atlantic salmon (*Salmo salar*) postsmolts of different origins in the Bay of Fundy and Gulf of Maine and evaluation of factors affecting migration, growth, and survival. Canadian Journal of Fisheries and Aquatic Sciences 62:1363–1376.

Lacroix, G. L., D. Knox, and M. J. Stokesbury. 2005. Survival and behaviour of post-smolt Atlantic salmon in coastal habitat with extreme tides. Journal of Fish Biology 66:485–498.

Macdonald, J. S., I. K. Birtwell, and G. M. Kruzynski. 1987. Food and habitat utilization by juvenile salmonids in the Campbell River estuary. Canadian Journal of Fisheries and Aquatic Sciences 44:1233–1246.

Magee, J., T. A. Haines, J. F. Kocik, K. F. Beland, and S. D. McCormick. 2001. Effects of acidity and aluminum on the physiology and migratory behavior of Atlantic salmon smolts in Maine, USA. Water, Air and Soil Pollution 130:881–886.

Mather, M. E., D. L. Parrish, C. L. Folt, and R. M. DeGraaf. 1998. Integrating across scales: effectively applying science for the successful conservation of Atlantic salmon (*Salmo salar*). Canadian Journal of Fisheries and Aquatic Sciences 55:1–8.

McCormick, S. D., M. F. O'Dea, A. M. Moeckel, D. T. Lerner, and B. T. Bjornsson. 2005. Endocrine disruption of parr-smolt transformation and seawater tolerance of Atlantic salmon by 4-nonylphenol and 17 beta -estradiol. General and Comparative Endocrinology 142:280–288.

McMahon, T. E., and L. B. Holtby. 1992. Behavior, habitat use, and movements of coho salmon (*Oncorhynchus kisutch*) smolts during seaward migration. Canadian Journal of Fisheries and Aquatic Sciences 49:1478–1485.

Meister, A. L. 1984. The marine migrations of tagged Atlantic salmon (*Salmo salar* L.) of USA origin. Anadromous and Catadromous Fish Committee, ICES CM 1984:M27, Copenhagen, Denmark.

Moore, A., I. C. Russell, and E. C. E. Potter. 1990. Preliminary results from the use of a new technique for tracking the estuarine movements of Atlantic salmon, *Salmo salar* L. smolts. Aquaculture and Fisheries Management 21:369–371.

Moore, A., E.C.E. Potter, New Jersey Milner, and S. Bamber. 1995. The migratory behaviour of wild Atlantic salmon (*Salmo salar*) smolts in the estuary of the River Conwy, North Wales. Canadian Journal of Fisheries and Aquatic Sciences 52:1923–1935.

Moser, M. L., A. F. Olson, and T. P. Quinn. 1991. Riverine and estuarine migratory behavior of coho salmon (*Oncorhynchus kisutch*) smolts. Canadian Journal of Fisheries and Aquatic Sciences 48:1670–1678.

National Research Council. 2004. Atlantic salmon in Maine. National Academy Press, Washington, D.C.

O'Neil, S., J. Ritter, and K. Robichaud-LeBlanc, editors. 2000. Proceedings of a workshop on research strategies into the causes of declining Atlantic salmon returns to North American rivers. Department of Fisheries and Oceans Canada, Canadian Stock Assessment Secretariat Proceedings Series 2000/18, Ottawa.

Økland, F., E. B. Thorstad, B. Finstad, R. Sivertsgård, N. Plantalech, N. Jepsen, and R. S. McKinley. 2006. Swimming speeds and orientation of wild Atlantic salmon post-smolts during the first stage of the marine migration. Fisheries Management and Ecology 13:271–274.

Parrish, D.L., R. J. Behnke, S.R. Gephard, S. D. McCormick, and G.H. Reeves. 1998. Why aren't there more Atlantic salmon (*Salmo salar*)? Canadian Journal of Fisheries and Aquatic Sciences 55:281–287.

Pettigrew, N. R., J. H. Churchill, C. D. Janzen, L. J. Mangum, R. P. Signell, A. C. Thomas, D. W. Townsend, J. P. Wallinga, and H. Xue. 2005. The kinematic and hydrographic structure of the Gulf of Maine Coastal Current. Deep Sea Research II 52:2369–2391.

Quinn, T. P., and E. L. Brannon. 1982. The use of celestial and magnetic cues by orienting sockeye salmon smolts. Journal of Comparative Physiology A 147:547–552.

Riley, W. D., M. O. Eagle, and S. J. Ives. 2002. The onset of downstream movement of juvenile Atlantic salmon, *Salmo salar* L., in a chalk stream. Fisheries Management and Ecology 9:87–94.

Seber, G. A. F. 1982. The estimation of animal abundance and related parameters. 2nd edition. Macmillan Publishing, New York.

Staurnes, M., P. Blix, and O. B. Reite. 1993. Effects of acid water and aluminum on parr-smolt transformation and seawater tolerance in Atlantic salmon, *Salmo salar*. Canadian Journal of Fisheries and Aquatic Sciences 50:1816–1827.

Stefansson, S. O., P. McGinnity, B. T. Bjornsson, C. B. Schreck, and S. D. McCormick. 2003. The importance of smolt development to salmon conservation, culture, and management: perspectives from the 6th international workshop on salmonid smoltification. Aquaculture 222:1–14.

Strategic Assessments Branch. 1985. National estuarine inventory data atlas. National Oceanic and Atmospheric Administration, National Ocean Service, Ocean Assessments Division, Rockville, Maryland.

Thorstad, E. B., F. Økland, B. Finstad, R. Sivertsgård, N. Plantalech, P. A. Bjorn, and R. S. McKinley. 2007. Fjord migration and survival of wild and hatchery-reared Atlantic salmon and wild brown trout post-smolts. Hydrobiologica 582:99–107.

Tytler, P., J. E. Thorpe, and W. M. Shearer. 1978. Ultrasonic tracking of the movements of Atlantic salmon smolts (*Salmo salar* L.) in the estuaries of two Scottish rivers. Journal of Fish Biology 12:575–586.

U.S. Atlantic Salmon Assessment Committee. 2007. Annual Report of the U.S. Atlantic Salmon Assessment Committee, Report 19—2006 activities. U.S. Atlantic Salmon Assessment Committee, Gloucester, Massachusetts.

Voegeli, F. A., G. L. Lacroix, and J. M. Anderson. 1998. Development of miniature pingers for tracking Atlantic salmon smolts at sea. Hydrobiologia 371/372:35–46.

Working Group on North Atlantic Salmon. 2005. Report of the Working Group on North Atlantic Salmon. International Council for the Exploration of the Sea, Copenhagen, Denmark.

White, G. C., D. R. Anderson, K. P. Burnham, and D. L. Otis. 1982. Capture-recapture and removal methods for sampling closed populations. Los Alamos National Laboratory, Los Alamos, New Mexico.

Wood, C. C., N. B. Hargreaves, D. T. Rutherford, and B. T. Emmett. 1993. Downstream and early marine migratory behaviour of sockeye salmon (*Oncorhynchus nerka*) smolts entering Barkley Sound, Vancouver Island Canadian Journal of Fisheries and Aquatic Sciences 50:1329–1337.

Zydlewski, G. B., A. Haro, and S. D. McCormick. 2005. Evidence for cumulative temperature as an initiating and terminating factor in downstream migratory behavior of Atlantic salmon (*Salmo salar*) smolts. Canadian Journal of Fisheries and Aquatic Sciences 62:68–78.

American Fisheries Society Symposium 69:311–320, 2009

Tracking Diadromous Fishes at Sea: The Electronic Future Using Hybrid Acoustic and Archival Tags

Michael J. W. Stokesbury*
The Ocean Tracking Network, Faculty of Science, Dalhousie University
Halifax, Nova Scotia B3H 4J1, Canada

Michael J. Dadswell
Biology Department, Acadia University, Wolfville, Nova Scotia B4P 2R6, Canada

Kim N. Holland
University of Hawaii, HIMB, Kaneohe, Hawaii 96744, USA

George D. Jackson
Institute of Antarctic and Southern Ocean Studies, University of Tasmania
Private Bag 77, Hobart, Tasmania 7001, Australia

W. Don Bowen
Fisheries and Oceans Canada, Bedford Institute of Oceanography
Marine Fish Division Dartmouth, Nova Scotia B2Y 4A2, Canada

Ronald K. O'Dor
Consortium for Ocean Leadership
4th Floor, New York Avenue NW, Washington, D.C. 20005, USA

Abstract.—Tagging fish with electronic tags can provide information on movement, migration, behavior, and stock structure while diadromous species are at sea. The state of the art technology for tracking fishes in the marine environment includes two families of tags. Archival tags store data and either relay them to satellites or require recapture for interrogation. Low return rates for diadromous species make these tags very expensive to use. A second type, acoustic tags, sends signals to passive receivers. Information is collected from the fish only when it is within range of a receiver. Technology is now being developed to mesh these tags into a fully integrated tag that will permit archived data to be transmitted acoustically over multiple frequencies to receivers allowing data retrieval without recapturing the animal. The new technology includes a "business card" tag that is a miniaturized receiver coupled with a coded pulse transmitter. These tags will exchange and record individual-specific codes when two animals carrying them come within acoustic range of each other, which will allow data from many animals to be moved ashore through few animals. These devices would be ideal for quantifying the degree of school fidelity (or, conversely, mixing) or the degree of at sea interaction of fishes from different river systems and provide ecological information to enhance management in an ecosystem approach to fisheries.

* Corresponding author: mstokesb@dal.ca

Introduction

Recent advances in tagging of aquatic animals with electronic tags has led to critical information in the areas of animal movement and migration (Block et al. 2001, 2005; Boustany et al. 2002; Lacroix et al. 2005), behavior (Austin et al. 2006; James et al. 2006a; Teo et al. 2007), natural history (Stokesbury et al. 2005; Weng et al. 2005, James et al. 2006b), management (Comeau et al. 2002; Welch et al. 2004; Block et al. 2005), and conservation (Block et al. 2005; Heide-Jørgensen and Reeves 2006; James et al. 2006c). Electronic tag technology used to track animals in the marine environment includes two families of tags: archival tags and acoustic tags.

Archival tags store data collected by sensors in the tag. To access the stored information, the data can be relayed from the tag to a satellite system such as ARGOS or Iridium. Data are transmitted directly, if the animal is at the surface of the water for long enough (Block et al. 2005; Weng et al. 2005; James et al. 2006a), or by disengagement of the tag, which then ascends to the surface and downloads its information (Block et al. 1998; Lutcavage et al. 1999; Stokesbury et al. 2004). Data from archival tags can also be downloaded if the animal is captured and the tag removed and returned to the researcher (Block et al. 2005). A new type of archival sonic tag downloads data when cued by an acoustic receiver. This tag is called a communicating histogram archiving transmitter (CHAT), and the concept has been tested successfully with tiger sharks *Galeocerdo cuvier* (Holland et al. 2001). However, in that prototype, data download over the acoustic link required a long time and mobile animals could move out of the receiver range prematurely.

Other than CHAT tags, acoustic tags do not store data but continuously transmit acoustic signals that are received only when the animal is within range (approximately 500 m) of submerged hydrophones—either mounted on a tracking vessel or on a mooring. Consequently, data are collected only while the fish is within range of a receiver (Lacroix and McCurdy 1996; Lacroix et al. 2005, Stokesbury et al. 2005). This technology is more cost-effective in some designs and, due to the generally smaller size of the tags, is often more useful for smaller fishes than archival technology. It is used mostly for animals that are not at the surface long enough to contact satellites.

Large-Scale Acoustic Tracking on the East Coast of North America

Acoustic receivers can be deployed on a matrix of moorings such as fish aggregation buoys (Dagorn et al. 2007) or as "curtain arrays" (Lacroix and McCurdy 1996; Heupel et al. 2006). In this configuration, a line of receivers is deployed across an area in such a way that if an acoustically tagged fish passes through, it will be detected (Figure 1). This procedure was first developed and deployed in the Bay of Fundy (Lacroix and McCurdy 1996), and the array design has been used in many areas, including forming the basic array geometry of the Pacific Ocean Shelf Tracking system (POST; Welch et al. 2003) and the Ocean Tracking Network (OTN; O'Dor et al. 2007). The OTN will deploy acoustic curtains in 14 ocean areas between the years 2008 and 2012 (Figure 2). There will be several installations in the northwest Atlantic Ocean, including arrays that stretch from shore to the continental shelf slope off of New Jersey, connected to the Long-Term Ecosystem Observatory at 15 m (LEO 15), a cabled system operated by Rutgers University and off of Nova Scotia on the "Halifax line." The Halifax line has been traditionally sampled for oceanographic parameters by the Department of Fisheries and Oceans Canada. Also, curtain arrays will be deployed across the Cabot Strait, the Strait of Belle Isle, the Bay of Fundy, and individual receiver deployments on the Gulf of Maine Ocean Observing System (GoMOOS) buoys (Figure 3). As a result, an extensive system of curtain arrays will be in place to provide information on acoustically tagged fishes migrating on the continental shelf of eastern North America.

Argument for Using Electronic Tags to Recover Unbiased Fish Movement Data

In recent decades, the use of conventional external and internal tags has proven less effective. Fishers have become reluctant to return capture information because of perceived threats of fishery closures and the bycatch of protected species. A case in point is the West Greenland fishery for Atlantic salmon *Salmo salar,* which took ocean-feeding salmon from virtually every salmon-producing na-

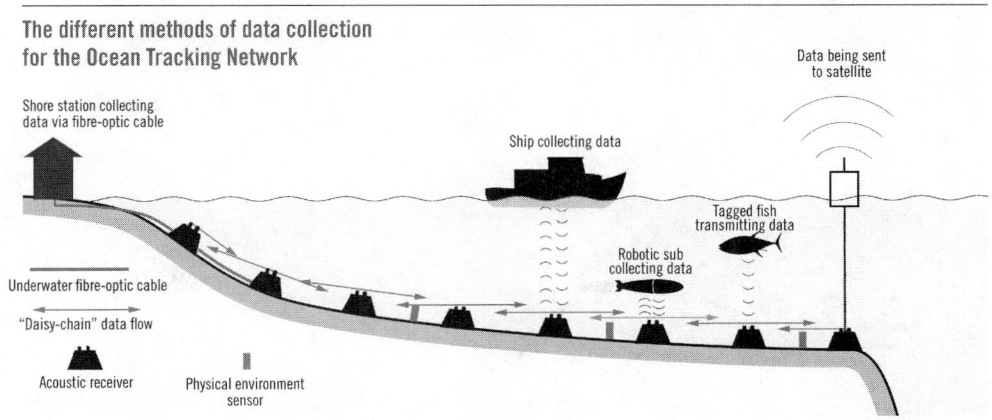

Figure 1.—Proposed OTN arrays will create an integrated system for collecting physical oceanographic and biological information from oceans throughout the world.

tion, data for which was provided by the return of external tags by the fishers (Jensen 1990). The fishery was initiated about 1960, and by 1970, landings had grown to more than 2,000 metric tons (mt)/year as it was exploited by both Greenland and international vessels (Shearer 1992). Because of the real threat that this mixed-stock fishery was collapsing home-returns of Atlantic salmon (Paloheimo and Elson 1974) international pressure was applied to first remove foreign vessels from the fleet (after 1975), impose quotas (1976), and, finally in 2002, close it completely (Atlantic Salmon Federation 2007).

The fishers responded to these threats by reducing the number of external tags they returned, and tag return data after 1980 was biased. Between 1966 and 1992, 1.5 million Atlantic salmon smolts were tagged in Maine using external Carlin tags, and 2,155 tags were returned from the West Greenland fishery (Baum 1997). Data from the return of these tags demonstrate that tags were being held back by fishers after management changes to the fishery (Table 1). Between 1969 and 1976 average landings were 2,077 mt/year or about an estimated 700,000 salmon/year, of which an average of 42% were of North America origin (Reddin and Friedland

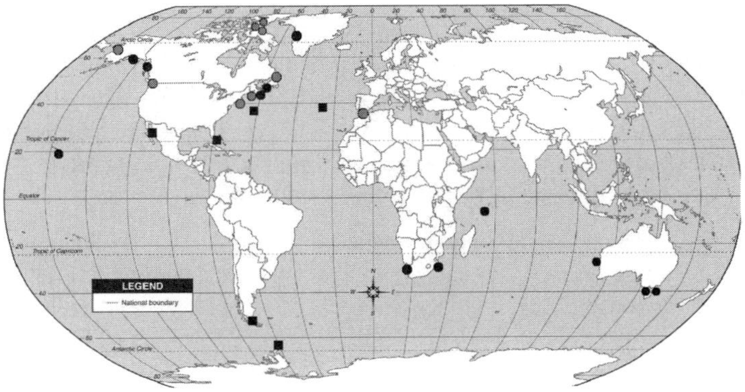

Figure 2.—The OTN's scheduled deployments in 14 ocean regions giving a global perspective on ocean and animal movements; arrays deployed in 2009 (•) 2010 (•), 2011–2012 (■).

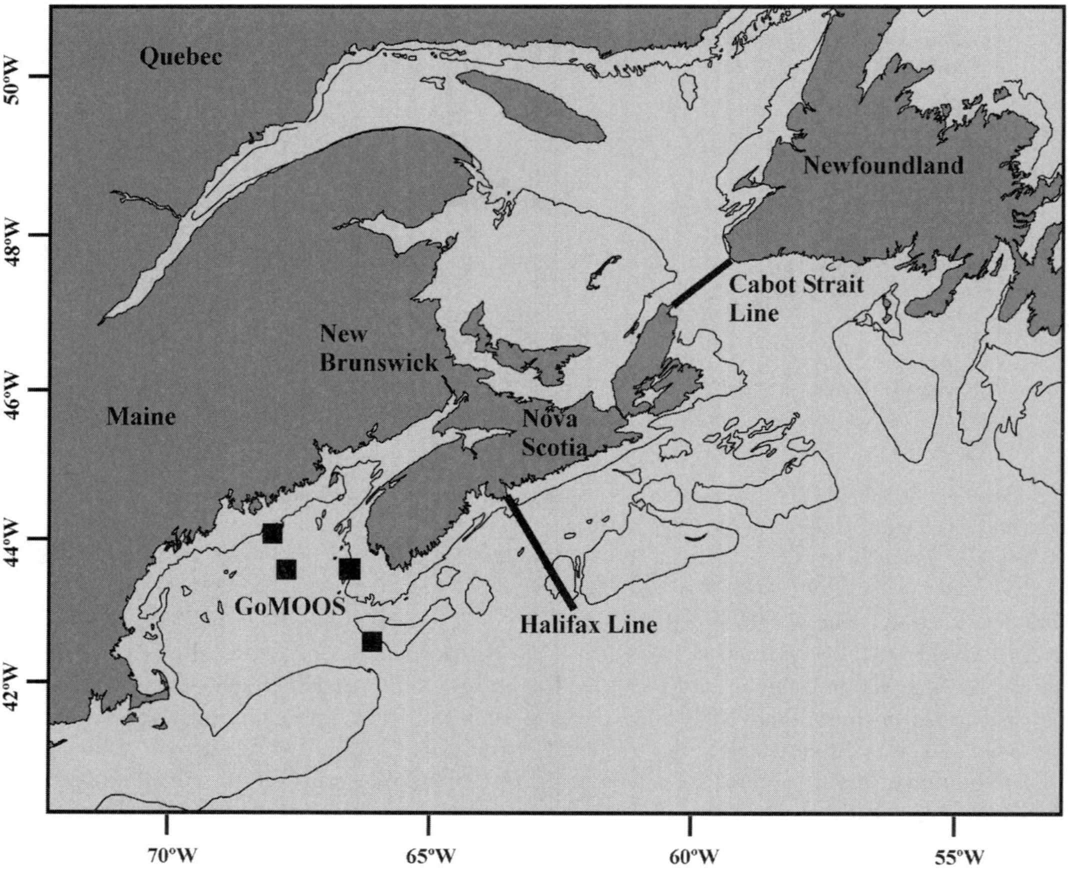

Figure 3.—Arrays of acoustic receivers to be deployed by the Ocean Tracking Network in the Gulf of Maine, the Gulf of St. Lawrence and on the Scotian Shelf between the years of 2008 and 20012.

Table l.—Return rates for external Carlin tagged Atlantic salmon smolts from Maine, USA from the commercial fishery off West Greenland during 1969–1976 and 1979–1986 (after Baum 1997). Between 1966 and 1992 a total of 1,585,335 smolts were released from Maine and 2,155 tags were returned from off West Greenland. Separation of data was decided on because no tags were applied in 1978 (Baum 1997), foreign vessels were restricted after 1975, and a quota was set in 1976 (Shearer 1992). mt = metric tons.

Period	Average landings/year (mt)	Continental origin (mean %) (NA)[b]	(E)[b]	Total tags applied	Total tags returned	Return rate %
1969–1976	2,077	42	58	340,895	1,255	0.37
1979–1986[a]	920	52	48	557,170	562	0.10

[a] Average quota/year 1979–1986 was 1,090 mt/year (Shearer 1992).

[b] NA = North American origin; E = European origin. Mean cross-validated, misclassified from known origin salmon is 15.9% ± 4.3% (Reddin and Friedland 1999).

1999). The average return rate of Maine tags during this period was 0.37% (Table 1). During the period 1979–1986, after fishery restrictions had been applied and the annual catch was reduced by half to an average of 920 mt/year, the tag return rate fell to 0.10%, almost a fourfold decrease. The change in return rate happened even though the number of tags released from Maine nearly doubled and the average proportion of North American salmon off West Greenland during the period increased to a mean of 52% a year (Table 1; Reddin and Friedland 1999). For example, of 48,210 tagged smolts released from Maine in 1970, a total of 403 tags were returned from West Greenland (0.83%), compared to 1980 when 49,760 were released but only 49 tags were returned (0.10%; Baum 1997).

Increased use of internal (coded wire) tags for diadromous fishes has occurred because of their homing behavior, but these methods are only useful if large-scale monitoring programs and the necessary detection systems are available (Russell et al. 1991). Also, internal tags are not often discovered by fishers or the public and valuable data on movements is lost. For these reasons, as well as the concurrent technical advances for miniaturizing and cost decrease of electronic tags, electronic methods are becoming widely accepted and used (Rassam 2008). The development of new electronic tags has the ability to provide unbiased information on fish migrations, enhance our ability to observe a wide range of fish behavior, determine critical habitat use, and discover sources of exploitation in real time.

Fish Tagging and Population Ecology

Electronic tags provide data only on an individual level. It is inherently difficult to combine data from individuals that were tagged at different times, often in different areas, and with mixed ages to test a population level hypothesis. Many animals need to be tagged so that statistically rigorous methods may be used to derive population level characteristics from tagging data (Sibert et al. 2006). This is a lengthy and expensive process.

A move toward an ecosystem approach to fisheries (Garcia et al. 2003) presents new challenges to the animal researcher (Cury 2004). In addition to the need for population level answers from data derived on the individual level, it is required that researchers link ecological processes to ecosystem level patterns (Cury 2004). To obtain this objective, a procedure is required that acquires data from individuals when they interact with members of their population. It also requires recording intraspecific interactions and relationships between animals and the physical properties of their habitat. For this, we need to combine some already existing technologies and to develop wholly new ways of acquiring data.

The Fully Integrated Tag

Several studies have combined both archival and acoustic tag technology (Jackson et al. 2005; Stokesbury et al. 2005). Building upon this idea, the OTN has proposed to combine acoustic and archival technologies with a business card (BC) tag to create a fully integrated tag (FIT). The BC tag includes a miniaturized acoustic receiver that records an acoustic code when a fish carrying it comes near another tagged fish. This tag also transmits its own individual code. These tags may help quantify the degree of school fidelity (or conversely, mixing) in schooling fish such as Atlantic salmon, American shad *Alosa sapidissima*, or striped bass *Morone saxatilis*. The tagged animal becomes a mobile receiver "bioprobe," moving and recording proximity (within 400 m) to others in the ecosystem. Basic research and development of BC tags is being developed in a joint venture between Amirix Systems Inc. and the Pelagic Fisheries Research Program at the University of Hawaii.

The FIT will permit retrieval of high resolution archival data on migrating animals in the ocean, including records of proximity with other tagged animals and information on geolocation that can be downloaded to hydrophones over a very fast spread spectrum acoustic link (Figure 1). Consequently, the tag does not have to be physically recovered and does not have to be downloaded over a slow and/or expensive satellite link. Development of the FIT tag is outlined in O'Dor et al. (2007).

Using FIT Tags to Study Diadromous Fish Interactions and Life Histories

As we move toward managing systems on the basis of large marine ecosystems (Sissenwine and Murawski 2004), we need a more complete vision of the ecological interactions between animals at the same and different trophic levels. This is important both to management and understanding system ecology in terms of food webs and energy flows. We envision two arenas where mobile and widely distributed animals move through coastal and oceanic waters and report movements, environmental preferences, and interactions: the continental shelf region of the northwest Atlantic and the inner Bay of Fundy.

Species that are natural fits for testing the FIT technology in these Arenas are gray seals *Halichoerus grypus* and Atlantic sturgeon *Acipenser oxyrinchus*. Gray seals return from marine migrations to Sable Island where they move to the land and are accessible to researchers (Bowen and Harrison 1994). These animals have been the focus of a large, electronic-tagging research program (n = 70 tagged individuals) conducted by researchers at the Bedford Institute of Oceanography and Dalhousie University (Austin et al. 2006). During the project, researchers deployed tags that link to satellites and archive data.

Gray seals disperse widely along shelf and oceanic water from the Gulf of Maine to southern Labrador, including the Gulf of St. Lawrence (Figure 4; Mansfield and Beck 1977). We can expect considerable overlap in range between gray seals and many of the diadromous species that will be electronically tagged along the Atlantic seaboard. Numerous wild Atlantic salmon stocks are endangered on the Atlantic coast of North America (Amiro 2003; Legault 2005). Salmon from North America

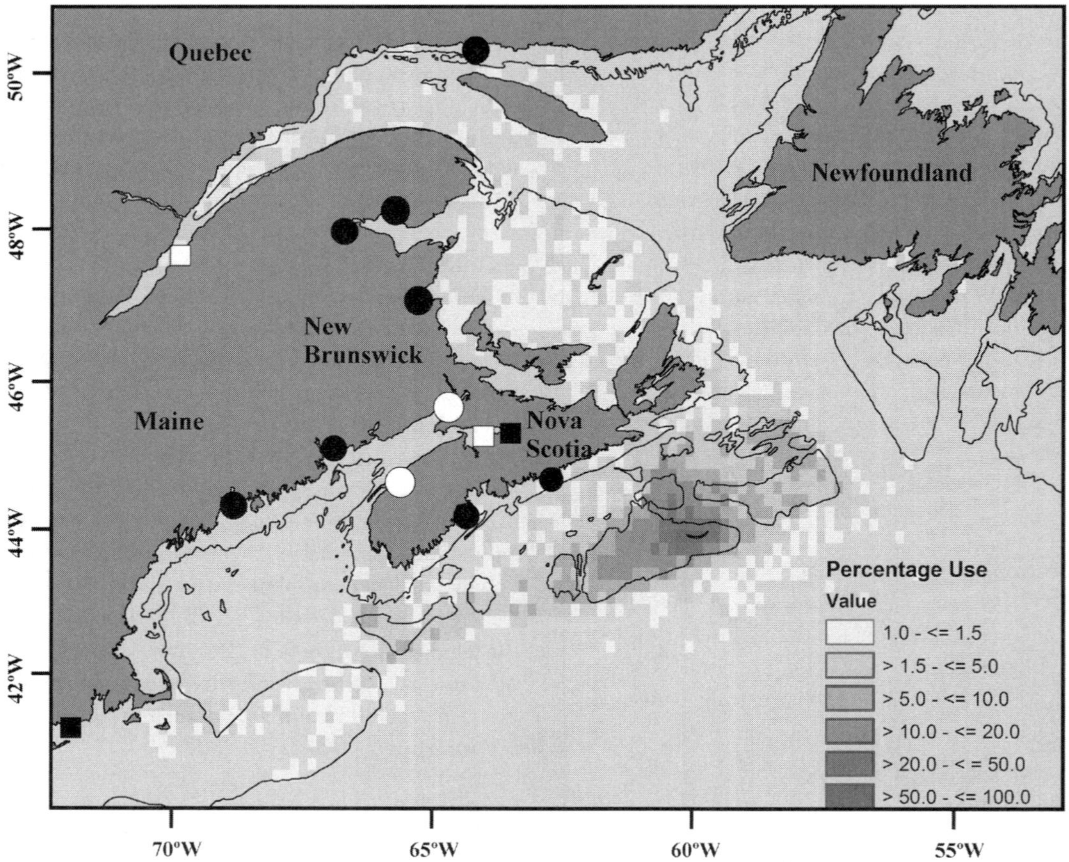

Figure 4.—Grey seal (n = 70) movement and percentage use of habitat along the Atlantic seaboard and electronic tagging sites of diadromous species in this region, including Atlantic salmon (●), American shad (○) Atlantic sturgeon (□), and striped bass (■).

migrate into the North Atlantic and most do not return (Anderson et al. 2000; Spares et al. 2007). Seals from rookeries in North America also swim into the North Atlantic searching for salmon and other prey (Austin et al. 2004). The Atlantic coast of North America has a very wide continental shelf, and in some places it is not practical to construct a POST-type continental-scale acoustic receiver array (Welch et al. 2003). However, the new BC and FIT technology can overcome this problem by using seals as mobile acoustic receivers to record where tagged salmon are in the North Atlantic. Besides Atlantic salmon we expect interactions between gray seals and other tagged diadromous species including American shad, alewife *A. pseudoharengus*, American eel *Anguilla rostrata*, and striped bass, many of which are eaten by this seal (Mansfield and Beck 1977). These tagged seals would provide time-stamped, georeferrenced data on interactions with fishes and measures of the physical environment in which these interactions occurred.

Gray seals, however, forage mainly along the eastern shelf region of Nova Scotia (Figure 4). In order to gain a more inclusive picture of ecological interactions in the Atlantic and the inner Bay of Fundy arenas, we need to complement gray seal tagging with the tagging of a species that is widely distributed along the North American Atlantic coast. Atlantic sturgeon is a species whose life history lends itself to tagging and also to providing information on other animals that occupy the same environments. Since Atlantic sturgeon are a large, long-lived species with adult weights commonly 200 kg, they represent an excellent chance to utilize a diadromous fish that is capable of performing as a mobile platform for multiple and/or FIT tagging.

Electronic tagging of sturgeon is well established, and the survivorship of sturgeons from the tagging process is high (Welch et al. 2006; Erickson and Hightower 2007). They are large enough to carry multiple tags with minimal effect on behavior. They migrate on a seasonal basis in a north–south pattern along the eastern seaboard from North Carolina to the Bay of Fundy and possibly Labrador, with many diadromous fishes such as other sturgeon, Atlantic salmon, American shad, alewife, American eel, and striped bass (Dadswell 2006; Laney et al. 2007). Migration depths are apparently restricted to ≤200 m, and they enter estuaries and rivers as well as ranging over the continental shelf (Secor et al. 2000; Stein et al. 2004). As a result, they are targeting a variety of environments that provide crucial habitat for many diadromous species. In some larger river systems, such as the St. Lawrence (Caron et al. 2002) and the Saint John (Dadswell et al. 1984), their spawning distribution is known. Recapturing Atlantic sturgeon in these rivers, especially if they are carrying acoustic tags that will reveal precisely where in the river system they are located, should not be difficult. Also, there are commercial fisheries for Atlantic sturgeon in the St. Lawrence and Saint John rivers (Dadswell 2006), and some tags may be returned through recapture in the fishery.

The Minas Basin and Chignecto Bay provide a unique marine habitat within the inner Bay of Fundy arena. Although these basins have a large tidal range (10–12 m) that is similar to other areas in the Bay of Fundy, they are summer warm enclaves due to shallow water and extensive tide flats (Bousfield and Leim 1958). Both contain seasonal fish faunal elements that are a mix of marine and diadromous species and which migrate long distances on the Atlantic coast, with their summer migration terminus in the inner Bay of Fundy (Dadswell et al. 1987; Rulifson and Dadswell 1995; Moore 1998; Dadswell 2006; Rulifson et al. 2008). Additionally, the fishes in this region are under continual threat of the development of tidal power, an occurrence that, by virtue of their extensive migrations, could harm diadromous fish stocks and protected species both from Canada and the United States over a wide area (Dadswell and Rulifson 1994; Cada et al. 2007).

Atlantic sturgeons are an important part of the inner Bay of Fundy ecosystem (Armitage and Gingras 2003), and they have been tagged with both conventional dart tags and acoustic tags in this region when captured as bycatch in local fisheries (Wehrell 2005: M. J. Dadswell, personal communication). For many diadromous species, details of their migration pathways, stock composition, seasonal distribution, and interactions with other fishes in the region are poorly known. Because of the constrained topography of the inner Bay of Fundy with its numerous narrow passageways (Minas Channel, Economy Point, etc.), access to the fish and potential for tracking with a limited number of acoustic receivers is feasible.

We could also test a number of hypotheses on the stock specific nature of the coastal migration of different Atlantic sturgeon populations by tagging these fish with FITs in two or three widely separated Canadian or U.S. rivers (Saint John, St. Lawrence, Hudson, etc.) and in the Bay of Fundy. FIT tagged sturgeons would interact with each other, other acoustically tagged fishes, and the various OTN receiver chains that are proposed for the Atlantic coast (Figure 4). Since the Atlantic sturgeon fishery in the United States is currently under a moratorium and sturgeon fishing in Canada is partially restricted (Dadswell 2006), tagged sturgeon will be allowed freedom of movement and protection for themselves and their tags.

References

Amiro, P. G. 2003. Population status of inner Bay Atlantic salmon (*Salmo salar*), to 1999. Canadian Technical Report of Fisheries and Aquatic Sciences 2488.

Anderson, J. M., F. G. Whoriskey, and A. Goode. 2000. Atlantic salmon on the brink. Endangered Species Update 17:15–21.

Armitage, I. A., and M. K. Gingras. 2003. Sedimentologic and environmental implications of Atlantic sturgeon (*Acipenser oxyrhynchus*) feeding traces, Bay of Fundy, New Brunswick, Canada. Geological Society of America Abstracts with Programs 35(6)578, Boulder, Colorado.

Atlantic Salmon Federation. 2007. ASF news: new Atlantic salmon conservation agreement. Available: www.asf.ca/issues.php?id=14 (January 2008).

Austin, D., W. D. Bowen, and J. I. McMillan. 2004. Intraspecific variation in movement patterns: modelling individual behaviour in a large marine predator. Oikos 105:1–15.

Austin, D., W. D. Bowen, J. I. McMillan, and S. J. Iverson. 2006. Linking movement, diving and habitat to foraging success in a large marine predator. Ecology 87:3095–3108.

Baum, E. 1997. Maine Atlantic salmon: a national treasure. Atlantic Salmon Unlimited, Hermon, Maine.

Block, B. A., H. Dewar, C. Farwell, and E. Prince. 1998. A new satellite technology for tracking the movements of Atlantic bluefin tuna. Proceedings of the National Academy of Sciences of the United States of America 95:9384–9389.

Block, B. A., H. Dewar, S. B. Blackwell, T. D. Williams, E. D. Prince, C. J. Farwell, A. Boustany, S. L. H. Teo, A. Seitz, A. Walli, and D. Fudge. 2001. Migratory movements, depth preferences, and thermal biology of Atlantic bluefin tuna. Science 293:1310–1314.

Block, B. A., S. L. H. Teo, A. Walli, A. Boustany, M. J. W. Stokesbury, C. J. Farwell, K. C. Weng, H. Dewar, and T. D. Williams. 2005. Electronic tagging and population structure of Atlantic bluefin tuna. Nature (London) 434:1121–1127.

Bousfield, E. L., and A. H. Leim. 1958. The fauna of Minas basin and Minas Channel. National Museum of Canada Bulletin 16.

Boustany, A. M., S. F. Davis, P. Pyle, South Dakota Anderson, B.J. Le Boeuf, and B. A. Block. 2002. Expanded niche for white sharks. Nature (London) 415:35–36.

Bowen, W. D., and G. D. Harrison. 1994. Offshore diet of grey seals, *Halichoerus grypus*, near Sable Island, Canada. Marine Ecology Progress Series 112:1–11.

Cada, G., J. Ahlgrimm, M. Bahleda, T. Bigford, South Dakota Stavrakas, D. Hall, R. Moursund, and M. Sale. 2007. Potential impacts of hydrokinetic and wave energy conversion technologies on aquatic environments. Fisheries 32:174–181.

Caron, F., D. Hatin, and R. Fortin. 2002. Biological characteristics of adult Atlantic sturgeon (*Acipenser oxyrinchus*) in the St. Lawrence estuary and the effectiveness of management rules. Journal of Applied Icthyology 18:580–585.

Comeau L. A., S. E. Campana, and M. Castonguay. 2002. Automated monitoring of a large-scale cod (*Gadus morhua*) migration in the open ocean. Canadian Journal of Fisheries and Aquatic Sciences 59:1845–1850.

Cury, P. M. 2004. Turning the ecoscope for the ecosystem approach to fisheries. Marine Ecology Progress Series 274:272–275.

Dadswell, M. J. 2006. A review of the status of Atlantic sturgeon in Canada, with comparisons to populations in the United States and Europe. Fisheries 31:218-229.

Dadswell, M.J. and R.A. Rulifson. 1994. Macrotidal estuaries: a region of collision between migratory marine animals and tidal power development. Biological Journal of the Linnean Society 51:93–113.

Dadswell, M. J., R. Bradford, A. H. Leim, G. D. Melvin, R. G. Appy, and D. J. Scarratt. 1984. A review of fish and fisheries research in the Bay of Fundy between 1976 and 1983. Canadian Technical Report of Fisheries and Aquatic Sciences 1256:163–294.

Dadswell, M. J., G. D. Melvin, P. J. Williams, and D. E. Themelis. 1987. Influences of origin, life history, and chance on the Atlantic coast migration of American shad. Pages 313–330 *in* M. J. Dadswell, R. J. Klauda, C. M. Moffitt, R. L. Saunders, R. A. Rulifson, and J. E. Cooper, editors. Common strategies of anadromous and catadromous fishes. American Fisheries Society, Symposium 1, Bethesda, Maryland.

Erickson, D. L., and J. E. Hightower. 2007. Oceanic distribution and behavior of green sturgeon. Pages

197–211 *in* J. Munro, D. Hatin, J. E. Hightower, K. McKown, K. J. Sulak, A. W. Kahnle, and F. Caron, editors. Anadromous sturgeons: habitats, threats, and management. American Fisheries Society, Symposium 56, Bethesda, Maryland.

Dagorn, L. C., K. N. Holland, and D. G. Itano. 2007. Behavior of yellowfin (*Thunnus albacares*) and bigeye (*Thunnus obesus*) tuna in a network of fish aggregating devices (FADs). Marine Biology 151:595–606.

Garcia, S. M., A. Zerbi, C. Aliaume, T. Do Chi, and G. Lasserre. 2003. The ecosystem approach to fisheries. Issues, terminology, principles, institutional foundations, implementation and outlook. FAO Fisheries Technical Paper 443.

Heide-Jørgenson, M-P., and R. R. Reeves. 2006. Evidence of a decline in beluga, *Delphinapterus leucas*, abundance off West Greenland. ICES Journal of Marine Sciences 53:61–72.

Heupel, M. R., J. M. Semmens, and A. J. Hobday. 2006. Automated acoustic tracking of aquatic animals: scales, design and deployment of listening station arrays. Marine and Freshwater Research 57:1–13.

Holland, K. N., A. Bush, S. M. Kajiura, C. G. Meyer, B. M. Wetherbee, and C. G. Lowe. 2001. Five tags applied to a single species in a single location: the tiger shark experience. Pages 237–247 *in* J. R. Sibert and J. Nielsen, editors. Electronic tagging and tracking in marine fisheries. Kluver Academic Publishers, Dordrecht, Netherlands.

Jackson, G. D., R. K. O'Dor, and Y. Andrade. 2005. First tests of hybrid acoustic/archival tags on squid and cuttlefish. Marine and Freshwater Research 56:425–430.

James, M. C., C. A. Ottensmeyer, S. A. Eckert, and R. A. Myers. 2006a. Changes in diel diving patterns accompany skirts between northern foraging and southward migration in leatherback turtles. Canadian Journal of Zoology 84:754–765.

James, M. C., S. A. Sherrill-Mix, K. Martin, and R. A. Myers. 2006b. Canadian waters provide critical foraging habitat for leatherback sea turtles. Biological Conservation 133:347–357.

James, M. C., J. Davenport, and G. C. Hays. 2006c. Expanded thermal niche for a diving vertebrate: a leatherback turtle diving into near-freezing water. Journal of Experimental Marine Biology and Ecology 335:221–226.

Jensen, J. M. 1990. Atlantic salmon at Greenland. Fisheries Research 10:29–52.

Lacroix, G. L., and P. McCurdy. 1996. Migratory behaviour of post-smolt Atlantic salmon during initial stages of seaward migration. Journal of Fish Biology 49:1086–1101.

Lacroix, G. L., D. Knox, and M. J. W. Stokesbury. 2005. Influence of tidal flow on Atlantic salmon post-smolt migration and behaviour in coastal habitat. Journal of Fish Biology 66:485–498.

Laney, R. W., J. E. Hightower, B. R. Versak, M. F. Mangold, W. W. Cole, Jr., and S. E. Winslow. 2007. Distribution, habitat use, and size of Atlantic sturgeon captured during cooperative winter tagging cruises, 1988–2006. Pages 167–182 *in* J. Munro, D. Hatin, J. E. Hightower, K. McKown, K. J. Sulak, A. W. Kahnle, and F. Caron, editors. Anadromous sturgeons: habitats, threats, and management. American Fisheries Society, Symposium 56, Bethesda, Maryland.

Legault, C. M. 2005. Population viability analysis of Atlantic salmon in Maine, USA. Transactions of the American Fisheries Society 134:549–562.

Lutcavage, M., R.Brill, G. Skomal, B. Chase, and P. Howey. 1999. Results of pop-up satellite tagging on spawning size class fish in the Gulf of Maine. Do North Atlantic bluefin tuna spawn in the mid-Atlantic? Canadian Journal of Fisheries and Aquatic Sciences 56:173–177.

Mansfield, A. W., and B. Beck. 1977. The grey seal in eastern Canada. Fisheries and Marine Services Technical Report 704.

Moore, T. M. 1998. Population characteristics of the spiny dogfish, *Squalus acanthias* Linneaus, 1758, from geographically distinct locations in Atlantic Canada during the summer and fall of 1996. Master's thesis. Acadia University, Wolfville, Nova Scotia.

O'Dor, R. K., M. J. W. Stokesbury, and G. D. Jackson. 2007. Tracking marine species—taking the next steps. Pages 6–12 *in* J. M. Lyle, D. M. Furlani, and C. D. Buxton, editors. Cutting-edge technologies in fish and fisheries science. Australian Society for Fish Biology Workshop Proceedings, Hobart, Tasmania.

Paloheimo, J. E., and P. F. Elson. 1974. Reduction of Atlantic salmon (*Salmo salar*) catches in Canada attributed to the Greenland fishery. Journal of the Fisheries Research Board of Canada 31:1467–1480.

Rassam, G. 2008. Fish tagging symposium—a successful international model for the future. Fisheries 33:129.

Reddin, D. G. and K.D. Friedland. 1999. A history of identification to continent of origin of Atlantic salmon (*Salmo salar* L.) at west Greenland, 1969–1997. Fisheries Research 43:221–235.

Rulifson, R. A., and M. J. Dadswell. 1995. Population characteristics of striped bass in Atlantic Canada. Transactions of the American Fisheries Society 124:477–507.

Rulifson, R. A., McKenna, S. A., and M. J. Dadswell. 2008. Intertidal habitat use, population characteristics, movement, and exploitation of striped bass in the Inner Bay of Fundy, Canada. Transactions of the American Fisheries Society 137:23–32.

Russell, I. C., E. C. E. Potter, D. G. Reddin, and K. D. Friedland. 1991. Recoveries of coded-wire microtags from salmon caught at west Greenland in 1990. ICES CM 1991/M:11, Copenhagen, Denmark.

Secor, D. H., E. J. Niklitschek, J. T. Stevenson, T. E. Gunderson, S. P. Minkkinen, B. Richardson, B. Florence, M. Mangold, J. Skjeveland, and A. Henderson-Arzapalo. 2000. Dispersal and growth of yearling Atlantic sturgeon, *Acipenser oxyrinchus*, released into Chesapeake Bay. Fisheries Bulletin 98: 800–810.

Shearer, W. M. 1992. The Atlantic salmon: natural history, exploitation and future management. Fishing News Books, Oxford, UK.

Sibert, J. R., M. E. Lutcavage, A. Nielson, R. Brill, and S. G. Wilson. 2006. Interannual variation in large-sclae movement of Atlantic bluefin tuna (*Thunnus thynnus*) determined from pop-up satellite archival tags. Canadian Journal of Fisheries and Aquatic Sciences 63:2154–2166.

Sissenwine, M. and S. Murawski. 2004. Moving beyond "intelligent tinkering": advancing an ecosystem approach to fisheries. Marine Ecology Progress Series 274: 291295.

Spares, A. D., J. M. Reader, M. J. W. Stokesbury, T. McDermott, L. Zikovsky, T. S. Avery, and M. J. Dadswell. 2007. Inferring marine distribution of Canadian and Irish Atlantic salmon (*Salmo salar*) in the North Atlantic from tissue concentrations of bioaccumulated 137Cs. ICES Journal of Marine Science 64:394–404.

Stein, A. B., K. D. Freidland, and M. Sutherland. 2004. Atlantic sturgeon marine distribution and habitat use along the northeastern coast of the United States. Transactions of the American Fisheries Society 133: 527–537.

Stokesbury, M. J. W., S. L. H. Teo, A. Seitz, R. K. O'Dor, and B. A. Block. 2004. Movement of Atlantic bluefin tuna (*Thunnus thynnus*) as determined by satellite tagging experiments initiated off New England. Canadian Journal of Fisheries and Aquatic Sciences 61:1976–1987.

Stokesbury, M. J. W., C. Harvey-Clark, J. Gallant, B. A. Block, and R. A. Myers. 2005. Movement and environmental preferences of Greenland sharks (*Somniosus microcephalus*) electronically tagged in the St. Lawrence estuary, Canada. Marine Biology 148:159–165.

Teo, S. L. H., A. Boustany, H. Dewar, M. J.W. Stokesbury, K. C. Weng, S. Beemer, A.C. Seitz, C. J. Farwell, E. D. Prince, and B. A. Block. 2007. Annual migrations, diving behaviour, and thermal biology of Atlantic bluefin tuna, *Thunnus thynnus*, on their Gulf of Mexico breeding grounds. Marine Biology 151:1–18.

Wehrell, S. 2005. A survey of the groundfish caught by the summer trawl fishery in Minas basin and Scots Bay. Master's thesis. Acadia University, Wolfville, Nova Scotia.

Welch, D. W., G. W. Boehlert, and B. R. Ward. 2003. POST—the Pacific Ocean salmon tracking project. Oceanologia Acta 25:243–253.

Welch, D. W., B. R. Ward, and S. D. Batten. 2004. Early ocean survival and marine movements of hatchery and wild steelhead trout (*Oncorhynchus mykiss*) determined by an acoustic array: Queen Charlotte Strait, British Columbia. Deep Sea Research Part II 51:897–909.

Welch, D. W., S. Turo, and S. D. Batten. 2006. Large-scale marine and freshwater movements of white sturgeon. Transactions of the American Fisheries Society 135:140–143.

Weng, K. C., P. C. Castilho, J. M. Morrissette, A. M. Landeira-Fernandez, D. B. Holts, R. J. Schallert, K. J. Goldman, and B. A. Block. 2005. Satellite tagging and cardiac physiology reveal niche expansion in salmon sharks. Science 310:103–106.

American Fisheries Society Symposium 69:321–341, 2009

What Comes Down Must Go Up: The Migration Cycle of Juvenile-Return Anadromous Taxa

Kim N. I. Bell*
Department of Ichthyology and Fisheries Science, Rhodes University
Grahamstown 6140, South Africa

Abstract.—The migration cycle of juvenile-return anadromous (a.k.a. amphidromous) taxa is overviewed using the goby *Sicydium punctatum* as representative. The first migration, a physiologically obligate drift to the sea, has a cost of extreme mortality, frequently exceeding 0.5/h, effectively a 1-h half-life. Stream drift may be the most hazardous known natural habitat. A nest sited 5 km upstream reduces larval drift survival to $0.5^5 = 0.0313$ or is 97% lower than a downstream nest. The conservation implication is strong: the lower reaches are disproportionately important for population maintenance. Mortality implies two new metrics for conservation and for understanding evolution of habitat choice: the mainstay habitat and point of expatriation. The second migration is oceanic, of which little is known beyond its duration until return to freshwater (recruitment), but salinity choice experiments demonstrated a requirement for intermediate salinities. Consistent with that and with others' stable isotope work, observations suggest that the combination of rain and onshore wind creates intermediate-salinity habitats on windward coasts of high islands. The third migration is the upstream migration of juveniles and adults. Drift (first migration) mortality, however, has strong implications for adult habitat choice: it challenges the notion that this goby can be successful by migrating far upstream because that entails a large reproductive penalty. The assumption that populations can be maintained through conservation of adults at high altitude or far from the sea is challenged.

Introduction

This paper will attempt to explore some new directions and place them in context of the juvenile-return anadromous migration cycle, using the goby *Sicydium punctatum* Perugia, 1896 (sensu Brockmann 1965) on the Caribbean island of Dominica as the main example. The migration cycle is the same as that for other coinhabitanting juvenile-return anadromous taxa: gobies, neritid snails, and decapod shrimps. *Sicydium punctatum* is the ideal exemplar because we have more key information about it than for other taxa. The issues here are broadly relevant to most juvenile-return anadromous fishes and form a background against which differences among them will be more informative. The relationships between parts of the migration cycle are useful in visualizing evolutionary and conservation issues. I will overview and integrate previously published work with some newer work to improve the understanding of the life cycle, extract implications for conservation and management, and raise hypotheses that may help account for the worldwide declines documented for Sicydiine gobies.

A note is necessary regarding anadromy versus amphidromy. Despite that *Sicydium* was Myers's (1949) type genus for his definition of amphidromy, the ambiguous and teleological term was not justifiable as a term on the same level as anadromous and catadromous. Instead, there is good reason to treat it as a form of anadromy, from which it is distinguished solely (McDowall 2009, this volume) by a difference in the stage that migrates back to freshwaters. Other differences have been conscripted to support distinctness of amphidromy, but they are unavoidable correlates following on stage at return and therefore not independent, thus incapable of supporting distinctness. Amphidromy is for this

* Corresponding author: kbell@mun.ca

and other reasons most logically and constructively considered a form of anadromy, here termed "juvenile-return anadromy." That makes anadromy the including term available to communicate with general audiences that may otherwise miss messages associated with the word amphidromous. The new term is synonymous with "amphidromy" but is unencumbered by its history.

Juvenile-return anadromous taxa are widespread: in the tropics gobies, eleotrids, gastropods, and decapod shrimps occur along virtually all volcanic coastlines and on islands. Many gobies support fisheries, which generally have declined (Bell 1999), such as the Philippines (Manacop 1953) where in the 1930s, the commercial fishery in northern Luzon alone approached 20,000 metric tons (mt)/year, or Reunion where landings of 65 mt were recorded in 1966 (Aboussouan 1969). In Dominica, the fishery at Layou, a principal fishing site, had, by 1990, declined to a few metric tons per year from yields credibly recalled to have been "two or three canoes [skiffs of about 5–7 m] full" and large enough to saturate the market, requiring it to be transported across the island to sell it all. By 1990, it was said that "tritri" (local name for postlarval gobies) used to be plentiful, but now to get any, "you have to know someone." Declines worldwide were not simultaneous, but plausibly followed land-use changes or development in each location.

The goby fry fishery in Dominica (15°30'N, 61°25'W) is dominated by *S. punctatum*, accounting for 95% or more of the catch (Bell 1994). There are two kinds of fishery for returning *Sicydium* postlarvae in Dominica: at a river mouth using a beach seine or in-river traps. The trap fishery is conducted by women and sometimes children, while the beach seine fishery at river mouths (only those with sandy substrates) is conducted by fishermen preferentially targeting the brief (1–2 d) but lucrative goby runs. The trap fishery begins a little later and diminishes over several days.

The beach seine is made of fine net material, commonly mosquito netting, about 8 × 3 m, with rope loops at the corners for holding with feet and hands. It is walked through the water and lifted to the surface, and the catch (which at Layou typically includes tritri, shrimps, and leaves) is coaxed into a mass at one end. The tritri are induced to self-sort by pulling the net in the direction away from the mass, stimulating a rheotactic response in the tritri, which rush to the pulled end of the net. The net is then gathered around them for transfer to buckets while the undesired leaves, flotsam, and shrimps are discarded.

The in-river fishery uses a "bag-and-stones" trap to catch postlarvae after they have become benthic. It is reminiscent of the fishing baskets briefly mentioned by Atwood (1791), which may have been not ordinary baskets but basketwork traps, as are well known from the Philippines, and the Carib baskets may well have been similarly refined for fishing. The present trap is now a fertilizer or woven bag, slit up one side, placed with the open end facing downstream, and stabilized with stones (not mentioned by Atwood). The benthic postlarvae advance between the stones until they encounter the fabric and they remain trapped by their own rheotaxis; the bag is later lifted and the stones removed, leaving the tritri.

Life Cycle

In Dominica, the juvenile-anadromous life cycle (Figure 1) is seen in sicydiine gobies, some Eleotridae, some decapod shrimps, and Neritid gastropods. The *S. punctatum* life cycle (Figure 1) is rapid with immediate flush of hatched larvae to the sea (Bell 1994; Bell et al. 1995). After 50–150 d at sea, they return as transparent, pelagic, schooling, terminally mouthed postlarvae ~20 mm standard length (SL). When they reach hard substrates in rivers, they settle (with loss of a little SL) and become cryptically pigmented, benthic, inferior-mouthed, and nonschooling juveniles. They can mature (developing adult sexually dimorphic coloration) at ~30 mm SL and are iteroparous and long-lived. Spawning is typically in caves excavated in gravels below rocks. Reproduction is all year (as for all of Dominica's juvenile-return anadromous taxa) at the coast, but seasonally interrupted at higher altitudes.

Laboratory studies revealed spawning by temporary pairs, with males tending eggs. Females ~45 mm SL had clutch sizes of 20,000–30,000 eggs. Females could repeat spawn at intervals of about 14 d. Adult growth appears slow: a female collected in 1991 (likely hatched in December 1990) was, in 2006 and at 16.5 years old, only 67 mm SL (author's personal observation). If growth in rivers is comparable, then biologists should use restraint when sampling adults because they may well be

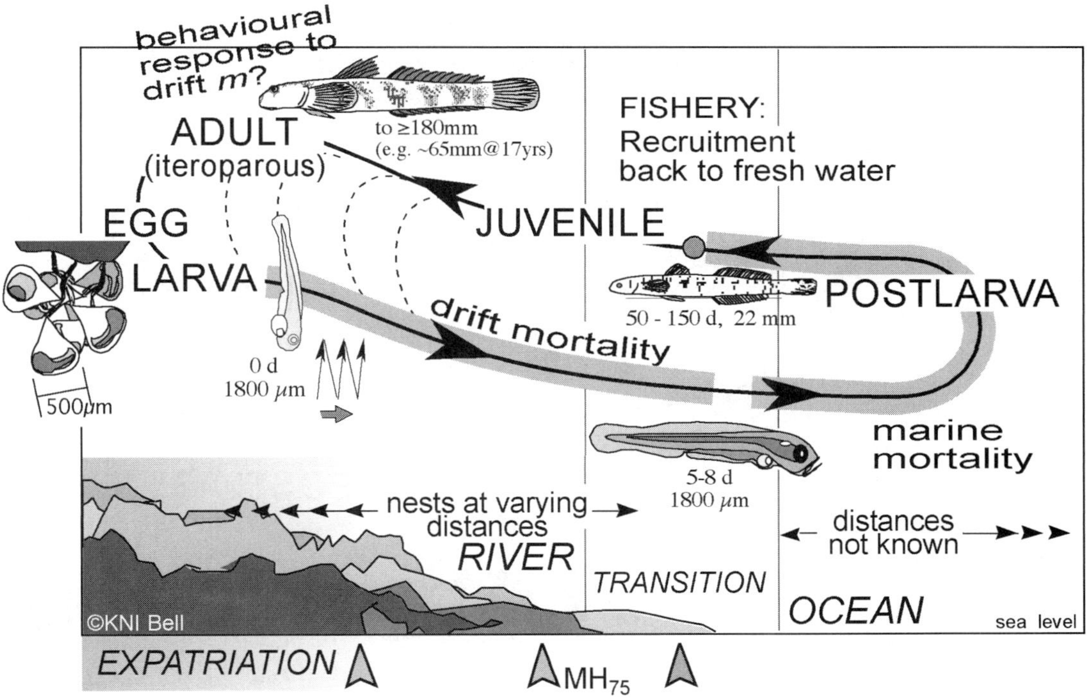

Figure 1.—Life cycle of *Sicydium punctatum* in Dominica, West Indies. MH_{75} = mainstay habitat below which (from point to river mouth) 75% of the river's drift survivors originate. PEX_x illustrates point of expatriation, above which drift survivorship falls below a critical level chosen (e.g., 5%); this metric is independent of spatial distribution of reproduction.

one to many decades old, and not quickly replaced. Laboratory studies also revealed one female using a "sneaky" strategy of intruding on a spawning in progress, displaying the female spawning motions without laying eggs, and resisting the male's effort to drive her away until the male in presumed confusion drove the actual spawning female away. The sneaker then ate the other female's eggs. Such behavior may permit the female sneaker to acquire extra nutrients to spawn more often or could result in better-provisioned larvae.

Disturbance (even by illumination) causes premature hatching of larvae of 1.3 mm total length (TL); even earlier hatching releases larvae that are for some time unable to uncurl or swim properly. In contrast, larvae of undisturbed nests hatch at 1.8 mm TL and larvae retain that length for several days (I suggest that the marker of proper hatching time is a stable TL, reported by many authors, in the first few days). Eggs hatch as quickly as a few hours to more than a day; however, the entire nest does not seem to hatch simultaneously but rather over a period of a day.

Drift is obligate since larvae cannot live long in freshwater (Bell and Brown 1995). Larvae begin drifting immediately following hatch, swimming vertically to keep suspended in the flow (Figure 2). Despite many experiments with larvae, no routine settling on the bottom was seen. The journey begins, however, with a shock: stream drift mortality is extreme and may be the most hazardous naturally occupied habitat we know of with survival only ~0.5/h or less. This has profound implications for conservation and habitat choice.

Methods

Otolith methods and protocols are given in previous papers (Bell 1994; Bell et al. 1995). The recruitment data available come from the Layou River fishery or from purpose-designed traps (Bell 1994; Figure 3) that face downstream and include a boul-

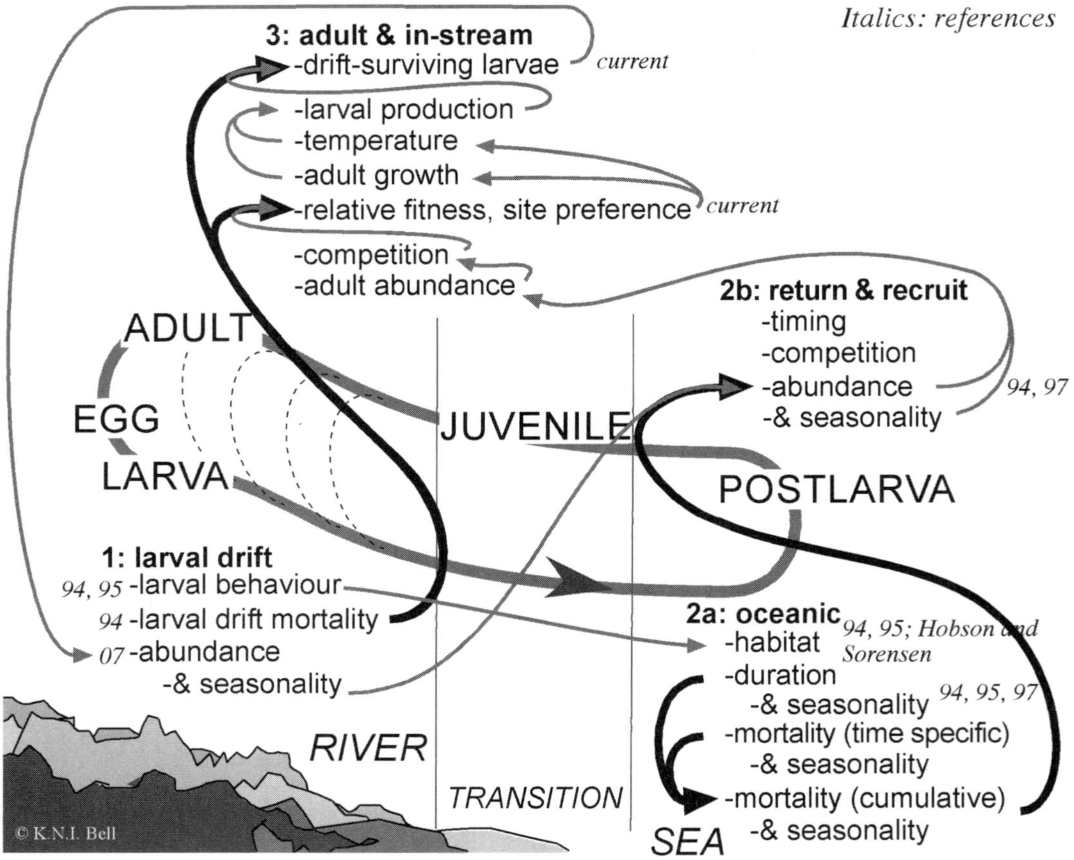

Figure 2.—Relationships among aspects associated with different of stages of the migration cycle. Heavier lines indicate key links. Italicized numbers (94, 95, 97, 07) refer to, respectively, Bell 1994, Bell and Brown 1995, Bell et al. 1995, Bell et al. 1997, and Bell 2007), and Sorensen and Hobson refer to Sorensen and Hobson (2005) and to Hobson et al. (2007). Larval drift mortality (first migration, downstream) interacts with upstream distance to directly determine larval survivorship and abundance. It has profound implications for adult habitat choice and, thence, adults' fitness through net production of drift survivors.

der-supported mesh (1 × 1 mm) fence that leads to a mesh cone, then to a removable accumulator. Methods for plankton sampling referred to here are given in Bell (1994) and Bell (2007).

Periodic regression (Bliss 1958, 1970; Batschelet 1981; Bell et al. 1995; Bell 2008) is a powerful and robust means of analyzing cycles. Significances are generally reported as exact P-values from parametric distributions (with exception of the novel D_{95}).

Remotely sensed oceanographic data maps were obtained from National Oceanic and Atmospheric Administration's CoastWatch Program and NASA's Goddard Space Flight Center, OceanColor Web (http://las.pfeg.noaa.gov/oceanWatch/oceanwatch_safari.php).

Comparing Cycles with Multiple Confidence Ellipses

Phasing can be compared by plotting as multiple confidence ellipses (MCEs), an extension of the single descriptive confidence ellipse (Bliss 1958). In this extension, if a straight line from the origin can intersect two ellipses, then they are not significantly different. If a centered circle can intersect two ellipses, their amplitudes (their effects on *Y*) are not significantly different.

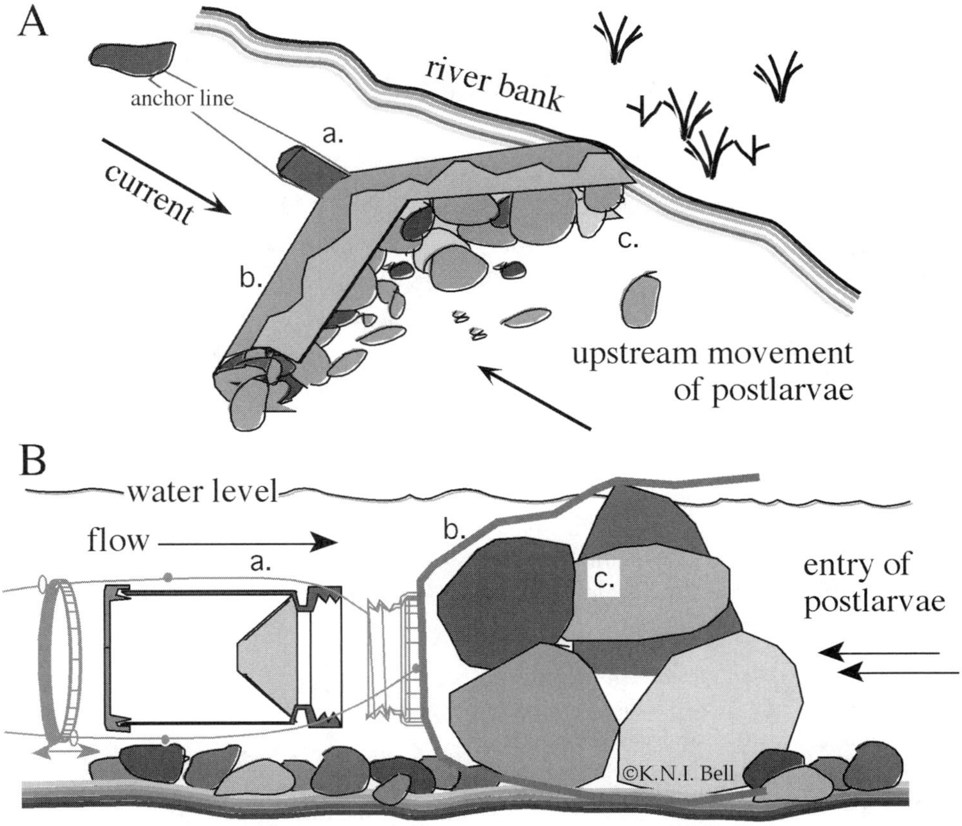

Figure 3.—Trap for upstream-migrating fauna, as installed (A) and in section (B). The trap includes a removable collector jar (a) having mesh windows to admit water flow, and a guide fence of ~1 mm mesh (b), which is supported by boulders (c).

Bliss (1970) used a superior method of calculating the ellipse; the approximation here is a computational economy more simply based on the 95% confidence bounds for a cycle's (sin, cos) coefficents in the periodic regression, which are plotted as rectangles in (sin,cos) space and the ellipse inscribed. I believe this computational economy results in fractionally larger ellipses than Bliss's method and they would therefore be conservative for purposes of declaring significance of cycles and significance of difference between them.

Drift Mortality: Estimation Method

Mortality is a key demographic parameter in the stream drift. Drift mortality estimation contains some traps and subtleties, and there is limited space to present it all here. The basic method (Bell 1994) uses paired samples (Figure 4) in the before/after sense, as has been applied to marine larvae. It had never been applied to stream drift, perhaps because the adjacency of adults and the drifting stages being sampled leads to ambiguities that can be quite daunting. Nevertheless, the stream situation virtually invites the challenge because, unlike the marine situation, the population is confined by stream banks.

Principles

Estimation can be complex and problematic, but for larvae that drift without settling and in a location where reproduction is rare, it becomes tractable. Survival (S) and mortality (M) sum to 1, and it is necessary to calculate S first. Survival is calculated from paired abundances (A, units/m^3) upstream

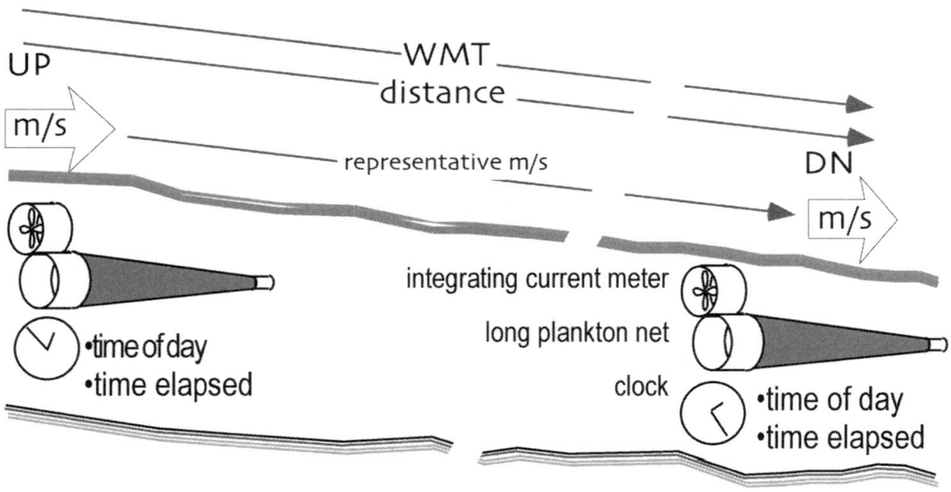

Figure 4.—Paired-net estimation of drift mortality. WMT = time for water mass to transit the sample reach, calculated from current speed and intersample distance.

(A_0) and downstream (A_t) and from exposure time t (here in hours, from water mass to transit (Figure 4) or distance (kilometers, which in calculation functions like t). S is expressed per hour or per kilometer, as S_h or S_{km}. S, A, and t are related as

$$A_t = A_0 S^t \text{ or } S = (A_t/A_0)^{1/t}, \tag{1}$$

which assumes that S dominates over settling and inputs of larvae (the no-reproduction assumption). For juvenile-return anadromous species like *Sicydium* that remain in the water column during drift, the key ambiguity of reproduction can be eliminated in reaches unsuitable for reproduction. Once eliminated, or tolerably reduced, in situ mortality can be estimated.

Generally, there is a nonsensicality to $S > 1.0$, so it is preferable to express a difference rather than a ratio like S because larvae do not reproduce (i.e., larvae cannot come from larvae, which are counted, but they can come from adults, which are not counted and are therefore an externality). The exception is where $S > 1.0$ is believed to result from sample error—instead of from hatching in the sample reach—and expressing the data point as S lets it be used with other estimates to generate fair means or medians. Objective disambiguation of error from spawning is possible using the criterion D_{95} below.

Drift survivorship, S_{drift}, is the drift survival probability or the proportion of larvae, from a given distance d upstream, that survives drift and reach the sea:

$$S_{drift} = S_{km}{}^{d}. \tag{2}$$

Sites and Sampling

In the Canefield and Layou rivers, Dominica, I sampled reaches within 1 km of the mouth. The upper and lower stations were separated by about 400 m in Layou River and 150 m at Canefield River. There were no tributaries to these reaches. The key to reducing ambiguities is to choose a reach with poor nesting conditions in order to reduce role of reproduction on the data and to assume that reproduction = 0.

The following are axiomatic but easily missed:

1. Where $S < 1.0$, S is a conservative estimate of mortality.
2. Where $S > 1.0$, $A_t - A_0$ is a conservative estimate of production.

Even in a reach generally unsuitable for reproduction, there may be some rock or object that provides a little grazing and nesting habitat, and one hatching of 30,000 larvae (within the capability of a 45-mm adult as frequently as every 10 d) can occasionally elevate local abundance and generate negative mortalities. Negative mortalities are a key issue in any mortality estimation and must be handled in a fair way. It is not justifiable to simply delete instances of $S \geq 1.0$ as nonsensical—even though in a limited sense they are—because they represent either sampling error (in which case they are a valid part of the survival data and must be included in aggregate statistics) or violation of the no-reproduction assumption (in which case they are not part of the mortality data). Only when mortality is dominated by reproduction are negative mortalities not part of the survival data. We need a way of deciding which.

To objectively disambiguate sample error from violation of the assumption that reproduction = 0, for each taxon, I obtained the D_{95} criterion, representing the 95th percentile of differences expected under the null hypothesis, using a resampling approach with a larger body of real plankton data from the same locations. Abundances A_i from the same locations were detrended for effects of time (seasonal, lunar, and daily cycles) by periodic regression (like equation 3, except that Y is rheoplanktonic larvae not recruits). Temporal detrending was required because mortality samples pairs are near-simultaneous, while the A_i used for D_{95} were not. The residuals were used to generate all possible differences (simulating $[A_t - A_0]$ in the null situation), comprising the null distribution from which D_{95} is identified for the taxon. D_{95} is applied to each case i of the downstream-minus-upstream difference x_i, and if $x_i = (A_t - A_0)_i > D_{95}$, then we say x_i is significantly different from the null and inconsistent with the no-reproduction assumption, and we treat it as production and not mortality. It is then possible to estimate S without contamination from reproduction.

Estimating Mainstay Habitat and Point of Expatriation

Mainstay habitat (MH_c; Figure 1), the most productive habitats accounting for a chosen c% of drift survivors, is calculated by integrating $S_{km}{}^x$ over a downriver range 0 (min) to x km (using the definite integral $[S_{km}\hat{}x]/\ln[S_{km}] - [S_{km}\hat{}\min]/\ln[S_{km}]$), and over an upriver range x km to 40 (max) km ($[S_{km}\hat{}\max]/\ln[S_{km}] - [S_{km}\hat{}x]/\ln[S_{km}]$), and then iteratively solving for the value of x giving downriver integral equal to c% of the total. The truncation of river length to 40 km is an economy justified by the fact that at survival values in the range of the present results, 40 km is well past the point where drift survivorship is nearly nil: about one drift survivor per thousand larvae at $S_{km} = 0.8$ and about one in a million at $S_{km} = 0.5$.

Point of expatriation (PEX; Figure 1) is the point above which the population cannot be self-sustaining. Conceptually simple, it requires either knowledge of all survivals through the life cycle to identify the point where one adult's entire reproduction results in one replacement adult or an assumption about the minimum drift survivorship required for that to pertain. As drift survivorship $S_{drift} = S_{km}{}^d$, where d is a distance in kilometers, an assumed minimal necessary S_{drift} allows d_{PEX} to be calculated as $\ln(S_{drift\_minimal})/\ln(S_{km})$.

Results and Discussion

In this overview, I will discuss the first migration (drift to the sea and mortality and implications), the time at sea, and the return to freshwaters (age at return, synchrony within and among taxa and its implications).

Drift Mortality

Mortality in the stream drift is important to trophic relationships, demographics, and dispersal; yet, despite many mentions in the literature (Elliott 1970; McLay 1970; Allan 1978; Allan and Russek 1985), the first direct estimate was by Bell (1994). There is not space here to treat this issue fully, so this brief overview uses data of Bell (1994), augmented in 1997 from Dominica.

Drift mortalities are extreme (Figure 5). While there are complexities in mortality analysis, including how to handle so-called negative mortalities, the data simply do not allow room to think that mortalities are trivial: stream drift mortalities far exceed any known mortalities for fish larvae and may reflect the most hazardous naturally occupied habitat yet found. Published data (Pepin et al. 1987), available only for marine fish larvae, show survivals generally higher than 0.97/d. In contrast, the present data show such extreme mortality that survival (S) has to be expressed on an hourly basis (it is near zero on a per-day basis). I use in discussion, conservatively and for simplicity, a value of S = 0.5/h, noting that the implications are more severe with higher mortalities.

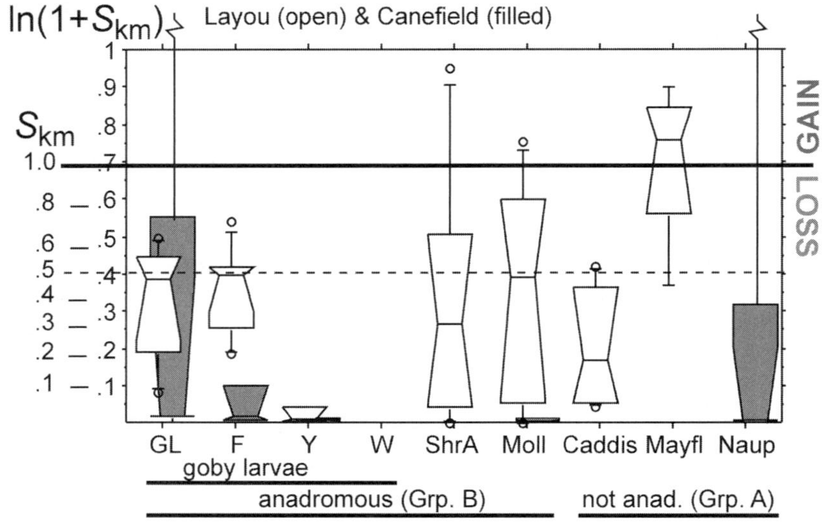

Figure 5.—Per-kilometer survival rates (S_{km}) in lower Layou (open symbols), lower Canefield (filled), filtered by D_{95} criterion, which identifies whether high outliers can be considered part of the mortality data population or not. Plots of survival for four juvenile-return anadromous taxa and three others for comparison: two insect (nonanadromous) taxa and nauplii (unknown life cycle, not expected to be anadromous). GL = all goby larvae; F = *Sicydium punctatum*; Y and W are unknown goby larvae (W data all were above the D_{95} critical level, indicating net production rather than sample error). ShrA is a complex of decapod shrimps; Moll is Neritid gastropods; Caddis are caddis-like larvae; Mayfl are mayfly-like larvae; Naup are nauplii. Box plots show median, 95% confidence interval for median (notches), the 25th and 75th percentile estimates (boxes), 10th and 90th percentiles (T-bars), and outer data (circles).

Mortality rates were far more severe in Canefield River than in Layou River. Although Canefield is visibly impacted by siltation from light industry (gravel works), the difference is not usefully amenable to statistical generalization (i.e., we could distinguish Canefield from Layou, but that is trite because we do not have the degrees of freedom for river condition to associate that difference with the impacts).

The high rate of mortality is supported by the stage distribution of larvae retrieved in drift plankton samples, which are predominantly at very early stages of development (eye at outline stage, no lens function, no eye pigmentation). Though we cannot identify sources of mortality; I suggest that predation by fish dominates. I have seen an adult *Sicydium* (SL ≈ 70 mm) in aquaria picking off larvae continuously, at the rate of one every few seconds for 8 h, despite that the mouth of an adult *Sicydium* is subterminal and adapted for grazing aufwuchs and despite that mouth–eye coordination was clearly a little clumsy for capturing swimming prey. Also likely as predators are mountain mullet *Agonostomus monticola*, abundant in the lower reaches of Dominican rivers.

If drift takes a day, then survival to the sea would be $S_{24\ h} = 0.5^{24} = 5.96E-08$, and at that level, a lifetime of spawning would likely yield zero drift survivors. That would certainly imply expatriation. That S_h^{24} is nearly zero does not mean an error in calculation or judgment; it simply means that average drift time cannot possibly be that long. Plausible drift time is a few hours at most. It is useful to consider the rates on a per-kilometer as well as a per-hour basis. Per-hour rates allow comparison with the literature; per-kilometer rates allow us to develop the implications. Rates/h approximates rates of per kilometer because typical river flow at my study sites approximated 1 km/h.

Conservation Implications

Conservation implications are profound. Distance-specific survival data let us, in principle, calculate the extent of (1) the most productive nesting habitats (those accounting for most of the larvae reaching the sea), and (2) the point above which reproduction is insufficient to sustain the local adult population. We can name these two useful concepts: (a) mainstay habitat (MH_x), and (b) PEX.

Drift mortality implies mainstay habitat.—Let MH_x denote the habitat accounting for *x*% of the larvae reaching the sea. MH_x integrates production and drift survivorship beginning at the lowest-altitude reproductive habitat, extending upstream to include the habitat accounting for *x*% of the river's drift survivors. MH can accommodate spatial variations in production, but we rarely have that information, so we assume uniform reproductive output (eggs/*t*/m, eggs per unit time per unit river length [and across all tributaries]) along a river. Reproduction, however, is likely greater downstream, so MH_{75} would be even smaller than estimated under the simplifying assumption. With reasonable assumptions or historic data on substrate, stream cover, and so forth, the approach can be used to estimate production already lost due to previous land-use changes. Higher *S* (lower mortality) extends MH_x over a greater area, while lower *S* reduces MH_x. We can choose *x*, and here $MH_x = MH_{75}$. $S \approx 0.5$/km implies that 75% of drift-surviving larvae originate in the first 2 km of river, so the 75% mainstay habitat or MH_{75} is 2 km (Figure 1).

Unavoidably, and consistently with the findings of Iguchi and Mizuno (1999), lower reaches are disproportionately responsible for population maintenance and therefore critical for conservation. We waste our time if we think a few adults 10 km upstream constitute conservation (yes, those habitats are important but mostly as water source). Thus, conservation of the coastal zone, the most contended for by development, is a critical conservation target.

Selection and behavior implications also generate interesting and testable hypotheses about adult choices and behaviors. The mathematically unavoidable decline in drift survivorship (the portion of larvae that survive to reach the sea) with upstream distance implies that adults should compete for downstream habitats and that the upstream adults are those that earlier in life were subordinate and excluded and that their upstream journey was not a desire for altitude as some (including I) have sometimes supposed, but instead a forced exile.

Drift mortality implies a point of expatriation.—Any mortality ($M_{km} = 1 - S_{km}$) implies a point (the PEX) upstream of which adults are ecologically nonreproductive (i.e., do not replace themselves). The very word "reproduction" implies success. In contrast, unsuccessful reproduction is typical of expatriate populations whose adults may be healthy and reproductive but whose larvae cannot access the nursery grounds from which they can recruit back to the parental population. Expatriate populations (Haedrich and Judkins 1979) are those sustained by exogenous reproduction but may not look any different. The challenge in the context of conservation of stream gobies is that there are often healthy looking populations far upstream; yet, even at 5 km, the drift survivorship (S_{drift}) is only 0.03%.

We cannot easily know life cycle's survival values well enough to precisely identify the point at which the population replacement rate falls below two recruits per parent lifetime, so we must assume a drift survivorship level, for instance $S_{drift} \leq 0.01$, below which we should assume expatriation. Then, at $S_{km} \approx 0.5$, $S_{drift} \leq 0.01$ occurs just 6.64 km upstream ($0.5^{[6.64]} = 0.01$).

Upstream adults may be expatriate or ecologically sterile because of low drift survivorship of their larvae. Upstream fish are often noted to be larger than the downstream adults, but the low drift survivorship of their larvae makes us wonder why they are there at all. The answer may be that they were out-competed for superior space downstream once those habitats became saturated. Do they have some remaining possibilities; for instance, do their larvae hatch during a flood so that higher current speeds will shorten drift time? We know nothing of drift mortality during those conditions however, so we have no basis to deny that floods leave drift survivorship unimproved. Reliance on floods also means relying on a state that is both unpredictable and brief, thereby reducing the effective spawning season.

Do upstream adults know where they are?—Adult Sicydiines have been reported to be faithful to home ranges or sites for long periods (Nishimoto and Fitzsimons 1986; Fitzsimons and Nishimoto 1990). This is appealing, logical, and plausible for a dominant individual in the best habitat it has access to, but what about nondominant individuals? We need to know more about competition for habitats, about which habitats are "choice'" from the goby point of view on spatial scales both small (within-pool) and large (along-stream) and about male–female differences in territoriality and habitat choice.

Questions are raised by the very fact of any nonnegligible drift mortality rate. Do upstream adults recognize that their larvae face a very low drift survivorship? Cues could be temperature (usually the upstream reaches are cooler) and the fish's own travel history. Are there circumstances in which drift survival increases (e.g., during floods?), and if so, do they take advantage of these?

Could large upstream adults drop down waterfalls to return to the more desirable lower reaches after they attain competitive size? The goboid sucker has a trailing form, suggesting it would be more difficult for the fish to establish a seal when moving backwards, and the sucker is situated well anterior of the center of mass and drag, so a headfirst controlled descent seems implausible because it would be opposed by forces that would tend to rotate the fish back to a head-upstream position. Those factors have tended, until recently, to make me regard upstream migration as a one-way trip. But if the lower reaches afford sufficiently greater larval drift survivorship, and if the larger sizes acquired by adults having grown in cooler and less desirable habitats would give them advantage in lower reaches, the question becomes not whether but when the risk of a relatively uncontrolled descent would be balanced by the rewards (Figure 1). A large adult may profitably accept a substantial risk of injury or death in descending downstream for a chance to be the bigger fish in a better pool and realize the advantage of higher drift survivorship.

The possibility for an adult downstream journey, even if risky, invites questions about the partitioning of energy to growth and reproduction and whether partitioning is contingent on location so as to support a strategy of exile to upper reaches followed by return at advantageous size. The choices faced by males and females may be qualitatively different because it seems that males are territorial and females are not (author's personal observation), so it is possible that the sex distribution of upper-reaches gobies would be biased toward males. Microchemical composition of otoliths might help show whether some downstream adults have an upstream history.

Marine Stage of Migration Cycle

We know very little about the marine journeys undertaken by sicydiine gobies or other juvenile-return anadromous taxa. We have some information on the time spent in the marine environment by juvenile-return anadromous taxa: some gobies (Iida et al. 2009, this volume) and Galaxiids (Rowe and Kelly 2009, this volume), but not for neritid gastropods (Hau 2007) or decapod shrimps (Hunte and Mahon 1983). We have some strong indications (for sicydiine gobies) of the required habitat for the first week of life and, beyond that, indications that the journeys might be small in extent (Sorensen and Hobson 2005) and some indications to the contrary (Maeda et al. 2007; Iida et al. 2009) but not necessarily for the same situations, so both may prevail.

The time spent at sea, from soon after hatch to recruitment, was first examined for diadromous gobies in Hawaii (Radtke et al. 1988). This period is referred to in the literature variously as PLD for postlarval duration or AAR for age at recruitment or OLD/MLD for oceanic/marine larval duration (Iida et al. 2009; Rowe and Kelley 2009). Because the freshwater period for juvenile-return-anadromous gobies before entry to the ocean is almost certainly mere hours, there is little scope for any of these terms to be much different.

Until the work in Dominica, variation in OLD had been ignored as error, but Bell et al. (1995) showed that this was incorrect: temporal structure was instead apparent as a highly significant seasonal cycle in OLD shown by periodic regression (Bliss 1958, 1970; Batschelet 1981; Bell et al. 1995; Bell 2008) for *S. punctatum* ($r^2 = 0.28, p \leq 0.0001$). The regression equation was

$$OLD = 83.835 + 10.1 \sin`DOY - 0.213\cos`DOY,$$

where the grave accent mark (`) indicates a proper transform of a variable (here, day of year [DOY]) to the correct units for taking sine and cosine. The intercept—or periodic mean, called a "mesor" in the periodic regression context to show its special utility as the expected mean value (correcting for or detrending with respect to) through the cycle(s)—is 83 d. The amplitude (*A*, contribution of the cycle to the value of OLD through the cycle), calculated by Pythagoras' theorem from the sine and cosine coefficients (10.1^2 + (–0.213)^2^0.5 = 102.055^0.5) = 10.102, meaning a seasonally associated fluctuation of ±10.1 d around the mesor; the regression indicates a seasonal acrophase (timing of peak or maximum contribution of the cycle to *Y*) near April 2 each year. The "error" contained temporal structure: a cycle that was informative for a new source of recruitment variation (Bell 1997), which was a plausible cause of the observed fishery seasonality. Regrettably, however, temporal structure remains largely ignored in otolith aging studies.

The cycle held a key to explaining seasonal structure in recruitment via two routes. One was as an exponent of time-specific survival. The second was a variation in the lag from hatch to recruitment, which forms the first application of the Doppler effect (Doppler 1842, cited by Hujer 1955) outside sound or electromagnetic radiation. Both routes were combined in Bell-Doppler theory (Bell 1997), which can be numerically simulated or explicitly estimated via an equation. It can mimic (potentially explain) a fairly complex recruitment pattern, so it is no longer justified to assume that a recruitment pattern is a lagged copy of the reproduction pattern; that is shown in the case of *S. punctatum* where the spawning pattern is not a tidy preview of the recruitment pattern, but recruitment well matched by a model using the observed OLD cycle and constant reproduction over the year.

Recruitment to Freshwater: The Journey Home—Synchrony within and among Taxa

Many juvenile-return anadromous taxa are reported to recruit at some stage of the lunar cycle (Bell 1999). In Dominica, recruitment (several taxa) is phased to the fourth day following the last quarter moon (Bell 1994).

Periodic regression (Bliss 1958; Batschelet 1981; Bell et al. 1995; Bell 2008) can, by definition, confirm (according to significance of a given cycle) synchrony within a (statistical) population. For example, for temporal recruitment of *S. punctatum*, the periodic regression equation found for the instream trap data (ln[CPUE]) was equation (3)

Y = ln(1 + CPUE) = mesor + f(day cycle, second harmonic of day cycle, seasonal cycle, second harmonic of seasonal cycle, lunar cycle, second harmonic of lunar cycle, linear trend)

$$Y = \ln(1 + \text{CPUE}) = 6.985 - 0.64{*}\sin`\text{DOY} + 0.97{*}\cos`\text{DOY} + 0.669{*}\sin 2\text{DOY} + 0.602{*}\cos`2\text{DOY} + 1.008{*}\sin`\text{L} - 0.858{*}\cos`\text{L} - 0.363{*}\sin`2\text{L} - 1.125{*}\cos`2\text{L} - 0.011{*}\text{d}, \quad (3)$$

where the grave symbol (`) functions as above. This regression was significant at $p \leq 0.0001$ (and highly significant in first and/or second harmonics of the day, lunar, and seasonal cycles), $R^2 = 0.625$, $n = 186$ (Bell et al. 1995).

Regressions for all other juvenile-return anadromous taxa in Dominica were also significant ($p \leq 0.0084$). All showed significance in either primary or secondary lunar cycles or both. None showed a significant tendency to depart from the same phasing.

Synchrony among Taxa

Lunar phasing does not significantly differ among juvenile-return anadromous taxa in Dominica, either with regard to the first harmonic (29-d cycle) or the second harmonic (≈14.5-d cycle). Phasing was compared by plotting as MCEs (Figures 6 and 7). Note that because the second harmonic completes its full cycle in one-half of the full cycle, it "begins twice" in the full cycle. There are peaks about 8 d after the first quarter and the last quarter. The second peak is within ~3 d of the main or first harmonic; the recruitment we see at that time is the sum of the first and second harmonic peaks, whereas the first peak of the second harmonic acts alone about two weeks later at ~8 d after the first (not last) lunar quarter.

Hypothesis: Passive Tidal Transport (Eliminated)

Tidal transport leaps to mind because of significance of response of recruitment to second lunar component. However, (1) tides in the region are small, partly due to the presence of an amphidromic point south of Puerto Rico, (2) tidal currents are diminished by the great water depth, and (3) use of reversing currents in open water requires vertical migration through most of the depth, so it is not advantageous in deep water. These reasons appear sufficient to eliminate passive tidal transport.

That suggests that the lunar response is merely a cue to enable synchrony. Consistent with that, lunar phasing varies worldwide (Figure 8): the data are incomplete but show clearly that there is a considerable variation in lunar phasing, which is not what we would expect if tidal transport was generally important. Predator swamping may be the sole

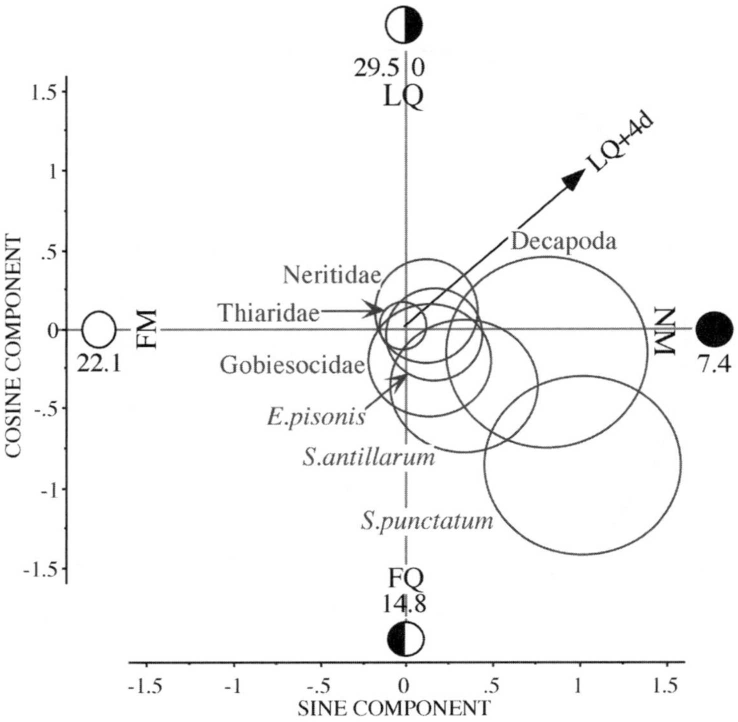

Figure 6.—Lunar recruitment and synchrony across taxa in Dominica: multiple confidence ellipses for the amplitude (distance from origin) and phase (angle from the origin) on the first harmonic cycle. The Layou beach seine fishery occurs on the fourth to fifth day after the last quarter, but the in-river traps catch slightly later and longer, hence the later in-river peak for *Sicydium punctatum* indicated at about 8–12 d.

plausible advantage to account for similar lunar phasing among taxa.

Where Are Early Growth Habitats?

The obvious constraint on early growth habitats is that larvae must be able to reach them and return. Recent work (Leis and Carson-Ewart 1997; Bellwood and Fisher 2001; Leis 2006) frees us from the constraint that larval fish are helpless against the currents. Larvae can sustain speeds of 10–50 body lengths (bl) per second, which means that larvae of 5–10 mm can generally counter the currents typical of the eastern Caribbean (rarely exceeding 0.07 m/s except in the channels between the Windward Islands). Leis and his colleagues (1997) and Leis (2006) have not denied that larvae can be plankton, nor claimed to be first to say they can be nekton, but what is of great value is that they have dissociated swimming incompetence with the word "larva." Next we can acknowledge—and this harkens back to the classic migration triangle—that whether a fish is plankton or nekton depends both on the fish (size or age) and the currents it is in (i.e. that it is conditional and that there is a transition of which the timing is also conditional). It remains reasonable for us to think of transport of newly hatched larvae as being mainly due to currents, but with the contribution or constraint of currents diminishing as the larva grows. That also would not be a knife-edged transition, but certainly if, for the later half of its pelagic larval period, a fish could sustain speeds of more than double the current speed, then wherever it drifted during the first-half of the pelagic period would be within range of return. Excess sustained swimming capacity could be allocated to foraging; indeed, what we see as sustained swimming ability probably has its origins in foraging and the need for it.

We can use swimming speeds and current speeds to make an order-of-magnitude estimate of the ability of a postlarva to move from an early

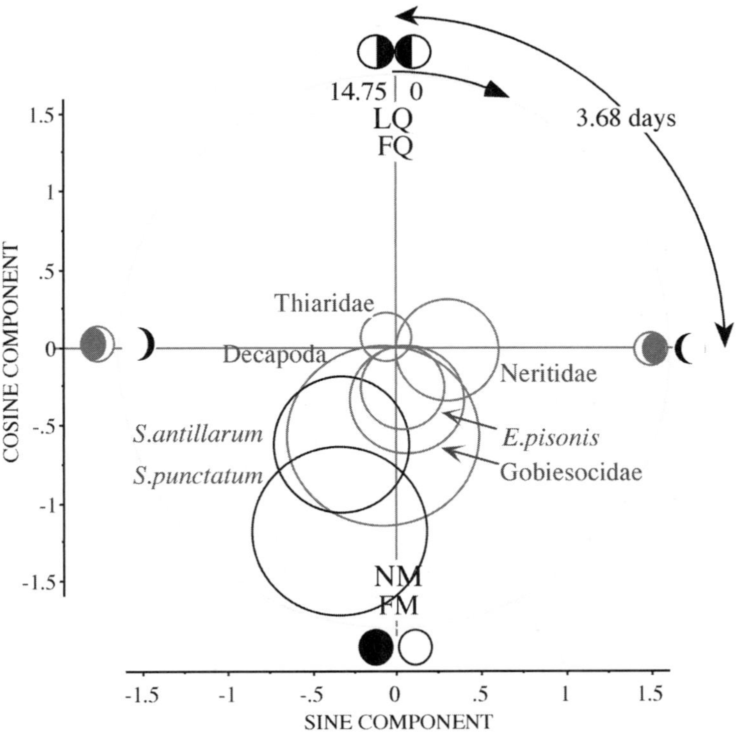

Figure 7.—Lunar recruitment and synchrony across taxa in Dominica: multiple confidence ellipses for amplitude (distance from origin) and phase (angle from the origin) on the second harmonic cycle. The second harmonic peak strikes twice in the first harmonic cycle, and one of those peaks occurs at 7–9 d following the last quarter. In other words, a portion of the recruitment seen 8 (traps) or 4 (seine fishery at Layou) d after the LQ is due to the second harmonic.

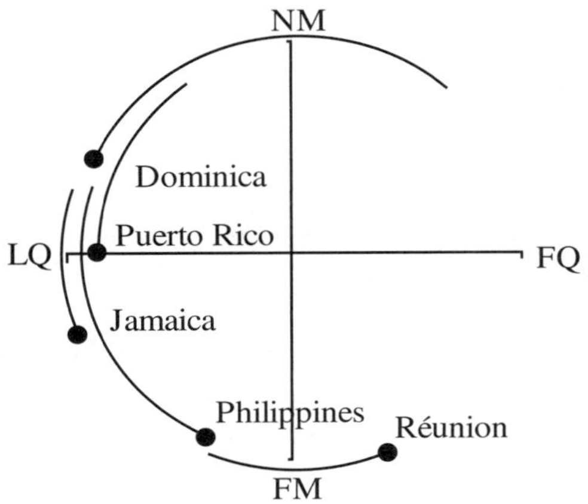

Figure 8.—Recruitment timing on the lunar scale, worldwide.

growth habitat back to the Windward Islands. This will give us some idea of the (maximum) distances it might travel during its marine stage. Consider an average of in situ estimates of sustained swimming speeds published for larvae of a range of fish taxa; sicydiine gobies are not included in the taxa reported, but let us suppose that as a fusiform larva, it would have a capability of at least average. Assume a larva of length 15 mm and a sustained swimming speed of 10 bl/s (Leis and Miller 1976) or 15 cm/s, less say 5 cm/s for opposing currents, for a net sustained progress of 10 cm/s, which equates to 8.64 km/d. Assume that the larva is three-fourths of the way through its marine period and has 20 d to reach its destination river or island: during that time, it can cover a distance (net, after allowing for opposing currents) of about 170 km. If, however, favorable currents exist and behaviors enable larvae to use them, then the potential travel distance could be twice as far (i.e., [15 + 5]/[15 − 5] = 2). Other considerations would include productivity, the ability of a larva to continue feeding and growing on its way home, the best path through the currents, the availability of suitable cues to prompt appropriate behavioral responses, and so forth. At the low precision of data we have, we should not take this as anything more than a reasonable indication that an early growth habitat for postlarvae could be, or could extend, hundreds of kilometers away from recruitment sites.

Population independence—often a key conservation question—is related to whether EGHs are one or many for a region or are isolated or contiguous. Lack of genetic/molecular difference between sites may imply a single population, but certainty with negatives is notoriously difficult and we rely on exhaustiveness for confirmation: differences may exist, but outside the molecular markers investigated. It is also possible for populations to be functionally independent (and subject to local extirpation), even without genetic distinctiveness.

Larval Active Salinity Choice Shows Early Growth Habitat Intermediate Salinity Requirement

Sicydium punctatum larvae actively chose intermediate salinities for the first 10 d of life (Bell and Brown 1995). The experiments involved setting up haloclines in experimental containers and placing larvae in them. In a halocline, larvae show an impressively uniform response, such that although one cannot see a halocline unless it is sharp enough to generate an optical discontinuity, the point where sinking larvae turn and swim upward is so uniform that they appear to be bouncing off a flat surface. Drift mortality rates do not permit larvae to spend more than a few hours in rivers, so those 10 d must reflect EGH, which therefore must have intermediate salinities.

Oceanography and Likely Intermediate-Salinity Regions

Visible photography from space shows the island wakes of the Windward Islands (Figure 9). The photo is useful for showing the extent of influence of the islands, but these visible components are atmospheric and may not correlate with oceanic wake components. The problem is that visible photography has a far higher resolution than present remotely sensed current or chlorophyll data, and such oceanic properties do not show well in visible photography. To visualize wakes, eddies, and fronts, a resolution of about 1 km is necessary, but only 27-km resolution data are available (Anonymous 2007) so inferences will have to be limited.

From first principles, a sheltered lee zone is not doubtable. Freshwater outflows would then be slower to mix on the leeward side because of reduced wind velocity, so suitable halocline conditions are plausible here. However, these low-salinity lenses might not be stable if subject to intermittent mixing and dissipation by anomalous winds or atmospheric systems passing through, and the lack of mixing in calm times may mean limited productivity. They could also be subject to runoff, and runoff regimes can be affected (made less moderated, more noisy) by deforestation and land-use changes (including area draining systems) that reduce soil water storage. If freshwater lenses in the island lee are important as EGHs, then such locally anthropogenic hydrographic change would affect population dynamics and recruitment.

Wake eddies are known in the literature (Coutis and Middleton 2002; Hasegawa et al. 2004). Eddies seem probable in the oceanic island wake of Dominica: the island's calculated Reynolds number Re ≈ 3E10 is vastly greater than turbulence-threshold Re of 500 (Chen and Jirka 1995) and therefore implies a turbulent oceanic wake. It is not clear to what extent eddies might successfully transport nutrients to the photic zone, but there are cases where enrichment is known or argued (Hasegawa et al. 2004). We do not know to what extent eddies are consistent with moderately stable intermediate salinity conditions and with production that would be needed to sustain larvae.

So at present, we still have questions about eddies. If there are eddies in a turbulent oceanic wake, they may exist perennially or seasonally. Von Karman vortex streets are a possibility and intriguing for transport because given appropriate behaviors by larvae, they could supply currents that could help transport larvae both down and up the wake.

As Leis (2006) said, fish larvae are "nekton, not plankton" because their sustained swimming speeds can be several tens of bl/s, while the oceanic current speeds in the Caribbean west of Dominica are generally of the order of 5 cm/s (Anonymous 2007). Island

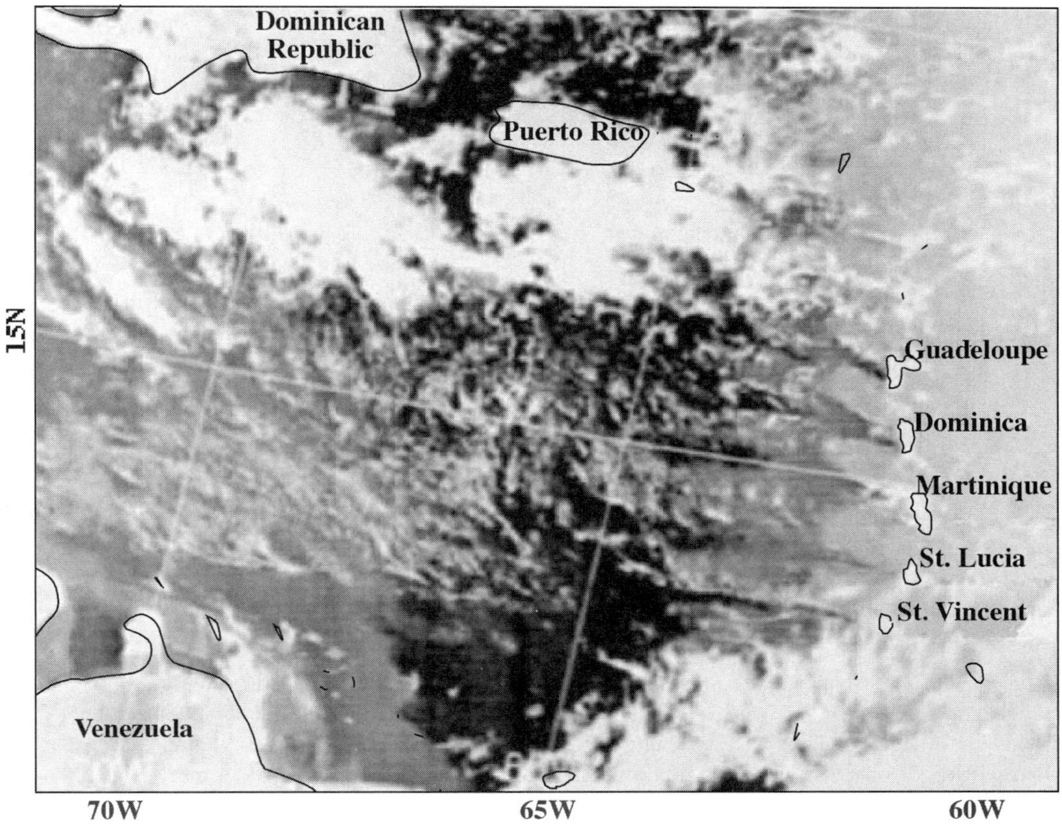

Figure 9.—Island wakes (atmospheric) of the Windward Islands, May 10 1979, photographed from space (Defense Meterological Satellite Program). Note that the low islands north of Guadeloupe have qualitatively different wakes.

wakes would be weaker currents than the surrounding waters. If larvae use wakes, they should not have trouble moving up the wakes to regain the islands. The strongest currents are those in the channels between the Windward Islands, reportedly averaging about 50 km/d or slightly more than 0.5 m/s. Those currents would likely challenge most larvae swimming against them, but the boundary and turbulent layers near the coasts might well allow fish to move from the lee to the windward coasts. The textbook image of the islands situated in a steady westward current is of course simpler than reality, but just how much so becomes apparent from detailed information (Figure 10). Available satellite data (Anonymous 2007) on currents is too coarse to detect oceanic wake eddies. Over time, however, there may be variation in the channel currents, just as there is temporal variation in the current field surrounding the islands.

Currents to the west of the Windward Islands show considerable temporal variations. Using the data (27-km resolution, 14-d averages) from January 2007 to June 2007 (all that is available from this source), the earlier months show currents tending to be more directed to the west or southwest, while by May, more meandering and features that could be interpreted as vortices appear. Unfortunately a full year cycle is not available from this source, but when it becomes available, it will be interesting to compare the current fields associated with low and high recruitment to the fishery.

Lee and Windward Systems of Island Arcs (e.g., Eastern Caribbean)

Chlorophyll-*a* distributions from imagery described in this paper are difficult to relate to currents in the

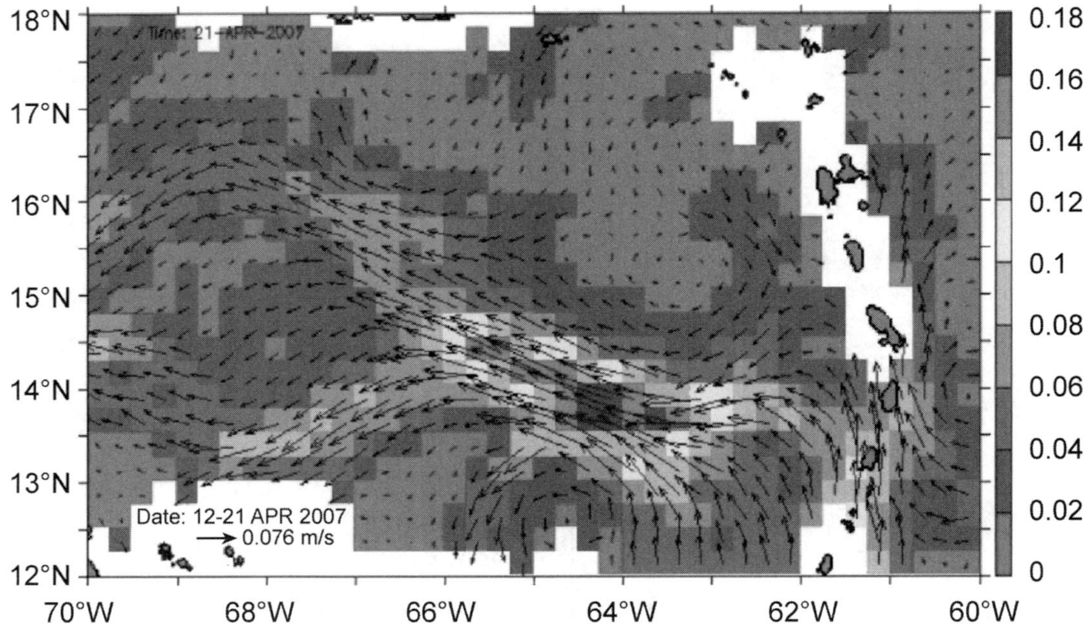

Figure 10.—Currents (geostrophic) on April 21, 2007, 27-km resolution, 14-d averages. The current structures are complex and are highly variable over time. Note that the current vectors are symbolic and not to map scale, and most of the currents are of the order of 5 cm/s or less, so that if sustained swimming speeds are only 10× body length, a larva as small as 5 mm could overcome most of the currents on this map.

remotely sensed data maps, except that the southeast Caribbean often shows higher productivity than surrounding areas and that seems to be related to flow from the southeast. I discussed above the ability of larvae to cover significant distances. Larvae may not, however, need to venture far to find high-productivity regions. Remotely sensed data (Anonymous 2007) suggests (persistently) higher chlorophyll *a* off windward coasts of the high islands (Figure 11). The high Hawaiian Islands also tend to show windward enhancement. There is little indication of high leeward productivity except in the low islands, (e.g., Antigua, St. Croix), suggesting a qualitatively different behavior in high and low islands. High islands tend to have higher rainfall (Dominica has up to 900 cm of rain per year) and a more sheltered (less well mixed, probably less upwelling) zone to the leeward. These could account for the difference by island type. Why should the productivity be highest off the windward coasts of high islands? Terrigenous nutrient outwash is likely highest on the windward coasts, but by no means absent from the leeward coast. In a well-mixed sea at low latitudes (where the Coriolis force is weak), an onshore wind will cause shoreward surface currents and downwelling at the shore—so far is not promising for productivity—but the high rainfall and freshwater input could add complexity to that circulation pattern.

Consistent with the observations of regionally elevated chlorophyll *a* on windward coasts of the high islands, first principles and simple shore-based observations suggest a windward circulation system that could provide the required intermediate salinities and the mixing that could support high productivity for growth. Bell and Brown (1995) reported low salinities all along the beach of Castle Bruce despite constant onshore wind and breaking waves. After heavy rain, I photographed a visible front on the order of 1 km out from Rosalie beach, with the inshore water made visible by the mud carried in a brief flash flood (Figure 12). In fair weather a characteristic lighter color of inshore water can also be seen, as photographed near Bibier on the windward coast (Figure 13). These indicate a circulation more complex than that expected in a well-mixed sea. The key is that mixing takes energy and time: there must

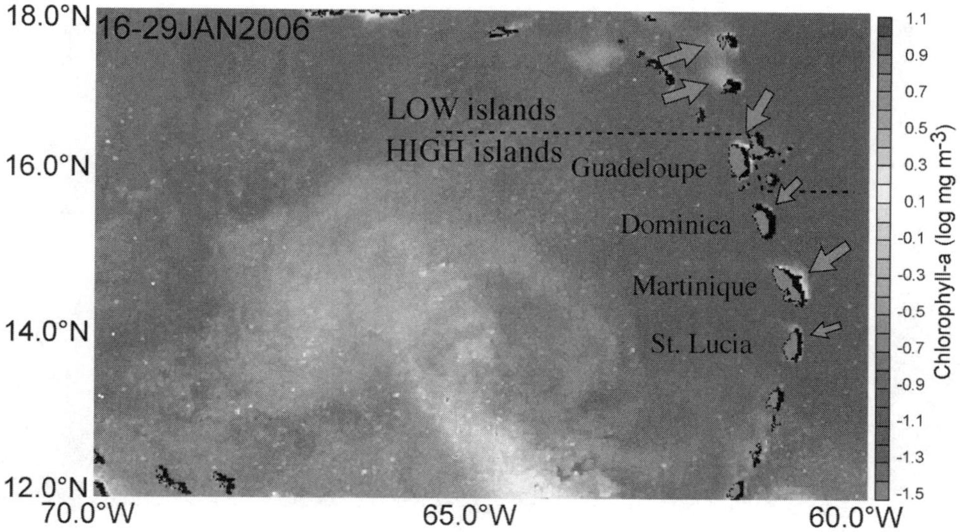

Figure 11.—Chlorophyll-*a* distribution on January 29, 2006 in the eastern Caribbean (NASA: MODIS on Aqua 2.5 km). Resolution is coarse, and high productivity (light gray) could be hidden by land-mask pixels (black). Arrows show high-productivity areas. High islands show enhanced productivity only on windward coasts. High islands generate orogenic rainfall, most heavily on windward coasts where enhanced productivity is apparent (clearest in Martinique), and have a sheltered lee with low productivity; the low islands generate little orogenic rainfall, lack a sheltered lee, and seem to show higher neritic productivity to the leeward.

Figure 12.—Rosalie beach, windward coast of Dominica, November 24, 1989. Turbid flash flood waters mixed through most of the bay in less than an hour. The demarcation at the outer edge (highlighted with four short lines) is interpreted as the outer edge of the circulation cell that comprises a nearshore system featuring intermediate-salinity water.

Figure 13.—Bibier (a few kilometers south of Rosalie), windward coast of Dominica, February 4, 1990. The different color of nearshore water (pointed to by two short lines) suggests an inshore system that is different from oceanic water.

always be, from fresh to salt, a gradient; its extent determined by freshwater input balanced against the energy available to mix it.

I propose that we can envision the windward circulation system of high-rainfall coasts being first a lower-salinity water mass (Figure 14), which develops its own circulation due to wind stress, with the upper layer moving landward, then downwelling along the bottom to some depth, then rising again at a front edge where oceanic water also moving landward under wind stress dips below it, creating a mixing zone with opposed upwelling (low salinity) and downwelling (oceanic salinity). The freshwater is typically several degrees cooler than seawater in Dominica, which adds an interesting feature. This circulation pattern explains the salinity and color observations and is consistent with the high chlorophyll *a* seen to the windward in the tall islands of the Lesser Antilles.

Such a circulation system must be dependent on a density difference sustained by high freshwater runoff. The system would be charged by runoff while being eroded by mixing at the frontal zone. Its extent could be seasonally variable (with rainfall and wind) in some locations or subject to breakdown if rainfall anomalously diminishes or if climate change reduces runoff or increases variance in it. Likewise, as commented in the context of leeward low-salinity lenses formed by runoff, development and land-use changes can make runoff noisy and river flows more variable (lower on average with excess water flushing in flash floods). If my hypothesized "windward system" circulation exists, and if it is important as an EGH for juvenile-return anadromous gobies, then disruptions could affect the population dynamics and recruitment dynamics of these species. It would also mean a longer exposure of larvae to terrigenous pollutants that persist in the windward system circulation than might be the case if larvae merely transited this system and went elsewhere.

The role of high freshwater runoff in early growth habitats is consistent with larval behavior

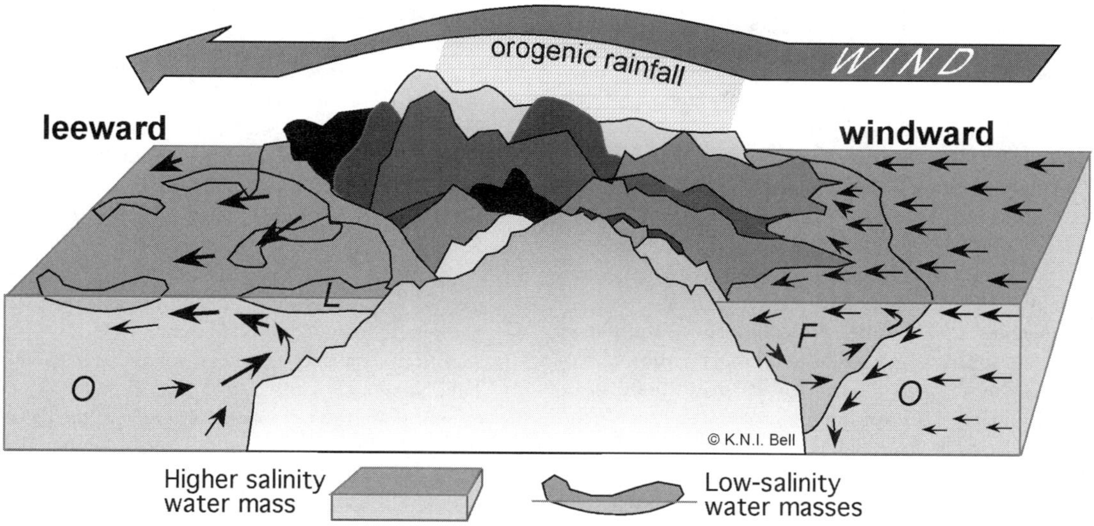

Figure 14.—An illustration of the proposed circulation model for the nearshore of an island like Dominica. Leeward freshwater lenses (L) sheltered from prevailing winds may persist and mix slowly. The model has a low-salinity water mass (F) kept to the shore by prevailing winds and has a shoreward current at the surface, a downwelling at the shore, and an upwelling at the oceanic front, where oceanic water (O) must downwell and mix.

(Bell and Brown 1995) and the implications of stable isotope analyses for Hawaiian gobies (Sorensen and Hobson 2005; Hobson et al. 2007), which help answer the question for these fishes of whether the life history period that we have generally been referring to as marine is not fully marine but considerably influenced by freshwater. The authors comment on interannual variability and on future extensions of the work (including different isotopes) that we all hope takes place, and I would suggest that to include samples randomized over seasons could also add even more interest to the results.

Conclusions

A core motivation of this work has been to supply information and ideas relevant to conservation of juvenile-return anadromous gobies and other juvenile-return anadromous fauna. There have been considerable declines in juvenile-return anadromous goby fisheries (Bell 1999). The declines did not all occur simultaneously. This paper provides at least two new routes by which anthropogenic changes could have played a role in the decline: (1) stream drift mortality implies a disproportionate reliance on downstream mainstay habitats, and degradation of mainstay habitats has certainly occurred due to deforestation, canalization, and coastal zone development; (2) reduced or more noisy runoff pattern, which could affect the stability and persistence of two candidate EGHs. Both of these could have occurred due to local effects in the regions where we have reports of fishery declines, and the hypothesis does not require them to have occurred at the same time worldwide.

Acknowledgments

I thank IDRC Young Canadian Researchers Award; Rhodes, Dalhousie, and Memorial Universities; Nigel Lawrence and the Dominica Fisheries Division; Clemson University and the Archbold Tropical Research Center; NOAA, NASA, CIT (imagery via www.las.pfeg.noaa.gov/oceanWatch); Peter Bell, Einar Wide, Don Erdman, Lieda and Jane Robin Bell, Pierre Pepin, Sammy Wyche, Bob and Edna Weisel, John D. Archbold, Moida Kelly, Jennifer D. Archbold, Anne and Cuthbert Jno. Baptiste of Papillote, Mona Dill, David Schneider, W.G. Warren, Katherine Bridger Ayranto, the Deorksen Institute, Horst Kaiser, Tony Booth, many others, and three anonymous reviewers. Heartfelt thanks to the ANACAT symposium organizers and attendees for a remarkable experience, to Mike Dadswell for his thoughtful editing, and to Debby Lehman for her patience.

References

Aboussouan, A. 1969. Note sur les «bichiques» de l'île de la réunion. Recordes de Travaille de Station Marine Endoume 1969(9):25–31.

Allan, J. D. 1978. Trout predation and the size composition of stream drift. Limnology and Oceanography 23:1231–1237.

Allan, J. D., and E. Russek. 1985. The quantification of stream drift. Canadian Journal of Fisheries and Aquatic Sciences 42:210–215.

Anonymous. 2007. National Oceanic and Atmospheric Administration's CoastWatch Program and NASA's Goddard Space Flight Center, OceanColor Web. Available: http://las.pfeg.noaa.gov/oceanWatch/oceanwatch_safari.php (December 2007).

Atwood, T. 1791. The history of the island of Dominica. (Orig. pub: J. Johnson, No. 72, St. Paul's Churchyard, London MDCCXCI.) Reprint facsimile edition 1971, Cass Library of West Indian Studies 27, Frank Cass and Company Ltd., 67 Great Russell Street, London.

Batschelet, E. 1981. Circular statistics in biology. Mathematics in biology. R. Sibson and J. Cohen, editors. Academic Press, London.

Bell, K. N. I. 1994. Life cycle, early life history, fisheries and recruitment dynamics of diadromous gobies of Dominica, W. I., emphasizing *Sicydium punctatum* Perugia. Doctoral dissertation. Memorial University of Newfoundland, St. John's.

Bell, K. N. I. 1997. Complex recruitment dynamics with Doppler-like effects caused by shifts and cycles in age-at-recruitment. Canadian Journal of Fisheries and Aquatic Sciences 54:1668–1681.

Bell, K. N. I. 1999. An overview of goby-fry fisheries. NAGA—the ICLARM quarterly 22:30–36.

Bell, K. N. I. 2007. Opportunities in stream drift: methods, goby larval types, temporal cycles, in situ mortality estimation, and conservation implications. Pages 35–61 *in* N. L. Evenhuis and J. M. Fitzsimons, editors. Bishop Museum Bulletin in Cultural and Environmental Studies 3:35–61.

Bell, K. N. I. 2008. Analyzing cycles in biology and medicine—a practical introduction to circular variables and periodic regression. Razorbill Press, St. John's, Newfoundland.

Bell, K. N. I., and J. A. Brown. 1995. Active salinity choice and enhanced swimming endurance in 0 to 8-d-old larvae of diadromous gobies, with emphasis on *Sicydium punctatum* (Pisces), in Dominica, West Indies. Marine Biology 121:409–417.

Bell, K. N. I., P. Pepin, and J. A. Brown. 1995. Seasonal, inverse cycling of length- and age-at-recruitment in the diadromous gobies *Sicydium punctatum* and *Sicydium antillarum* (Pisces) in Dominica, West Indies. Canadian Journal of Fisheries and Aquatic Sciences 52:1535–1545.

Bell, K. N. I., P. Pepin, and J. A. Brown. 1997. Variation in age-at-recruitment can drive recruitment dynamics: the example of *Sicydium* spp. (Pisces:Gobiidae) in Dominica, West Indies. Micronesica 30:25.

Bellwood, D. R., and R. Fisher. 2001. Relative swimming speeds in reef fish larvae. Marine Ecology Progress Series 211:299–303.

Bliss, C. I. 1958. Periodic regression in biology and climatology. Connecticut Agricultural Experimental Station, Bulletin 615, New Haven.

Bliss, C. I. 1970. Statistics in biology, volume 2. McGraw-Hill, New York.

Brockmann, F. W. 1965. A taxonomic study of the Panamanian freshwater gobies of the genera Awaous and Sicydium. Master's dissertation. Florida State University, Tallahasee, Florida.

Chen, D., and G. H. Jirka. 1995. Experimental study of plane turbulent wakes in a shallow water layer. Fluid Dynamics Research 16:11–41.

Coutis, P. F., and J. H. Middleton. 2002. The physical and biological impact of a small island wake in the deep ocean. Deep-Sea Research Part I Oceanographic Research Papers 49:1341–1361.

Doppler, J. C. 1842. Ueber das Farbige Licht der Doppelsterne und einiger anderer Gestirne des Himmels—Versuch einer das Bradley'sche Aberrations—Theorem als integrirenden Theil in sich schliessenden allgemeineren Theorie (On the coloured light of the binary stars and some other stars of the heavens—attempt at a general theory including Bradley's theorem as an integral part. Abhandlung der Konigl. Böhmischen Gesellschaft der Wissenschaften, 5th ser. II:465–482.

Elliott, J. M. 1970. Diel changes in invertebrate drift and the food of trout. Journal of Fish Biology 2:161–165.

Fitzsimons, J. M., and R. T. Nishimoto. 1990. Territories and site tenacity in males of the Hawaiian stream goby *Lentipes concolor* (Pisces: Gobiidae). Ichthyological Explorations in Freshwaters 1:185–189.

Haedrich, R. L. and D. C. Judkins. 1979. Macrozooplankton and its environment. Pages 4–28 *in* S. van der Spoel and A. C. Pierrot-Bults, editors. Zoogeography and diversity of plankton. Halsted Press, New York.

Hasegawa, D., H. Yamazaki, R. G. Lueck, and L. Seuront. 2004. How islands stir and fertilize the upper ocean. Geophysical Research Letters 31(16):L16303.

Hau, S. 2007. Hīhīwai (Neritina granosa Sowerby) recruitment in ʻĪao and Honomanū streams on the island of Maui, Hawaiʻi. N. L. Evenhuis and J. M. Fitzsimons, editors. Bishop Museum Bulletin in Cultural and Environmental Studies 3:171–181.

Hobson, K. A., R. J. F. Smith, and P. Sorensen. 2007. Applications of stable isotope analysis to tracing nutrient sources to Hawaiian gobioid fishes and other stream organisms. N. L. Evenhuis and J. M. Fitzsimons, editors. Bishop Museum Bulletin in Cultural and Environmental Studies 3:99–111.

Hujer, K. 1955. Sesquicentennial of Christian Doppler. American Journal of Physics 23:51–54.

Hunte, W., and R. Mahon. 1983. Life history and exploitation of Macrobrachium faustinum in a tropical high-gradient river. Fisheries Bulletin 81:654–660.

Iguchi, K., and N. Mizuno. 1999. Early starvation limiits survival in amphidromous fishes. Journal of Fish Biology 54:705–712.

Iida, M., S. Watanabe, and K. Tsukamoto. 2009. Life history characteristics of a Sicydiinae goby in Japan, compared with its relatives and other amphidromous fishes. Pages 355–373 *in* A. J. Haro, K. L. Smith, R. A. Rulifson, C. M. Moffitt, R. J. Klauda, M. J. Dadswell, R. A. Cunjak, J. E. Cooper, K. L. Beal, and T. S. Avery, editors. Challenges for diadromous fishes in a dynamic global environment. American Fisheries Society, Symposium 69, Bethesda, Maryland.

Leis, J. M. 2006. Are larvae of demersal fishes plankton or nekton? Advances in Marine Biology 51:57–141.

Leis, J. M., and J. M. Miller. 1976. Offshore distributional patterns of Hawaiian fish larvae. Mar. Biol 36:359–367.

Leis, J. M., and B. M. Carson-Ewart. 1997. In situ swimming speeds of the late pelagic larvae of some Indo-Pacific coral reef fishes. Marine Ecology Progress Series 159:165–174.

Maeda, K., N. Yamasaki, and K. Tachihara. 2007. Size and age at recruitment and spawning season of sleeper, genus *Eleotris* (Teleostei: Eleotridae) on Okinawa island, southern Japan. Raffles Bulletin of Zoology Supplement 14: 199–207.

Manacop, P. R. 1953. The life history and habits of the goby, *Sicyopterus extraneus* Herre (añga) Gobiidae with an account of the goby-fry fishery of Cagayan River, Oriental Misamis [Province, Mindanao, Philippines]. Philippine Journal of Fisheries 2:1–60.

McDowall, R. M. 2009. Making the best of two worlds: diadromy in the evolution, ecology, and conservation of aquatic organisms. Pages 1–22 *in* A. J. Haro, K. L. Smith, R. A. Rulifson, C. M. Moffitt, R. J. Klauda, M. J. Dadswell, R. A. Cunjak, J. E. Cooper, K. L. Beal, and T. S. Avery, editors. Challenges for diadromous fishes in a dynamic global environment. American Fisheries Society, Symposium 69, Bethesda, Maryland.

McLay, C. 1970. A theory concerning the distance travelled by animals entering the drift of a stream. Journal of the Fisheries Research Board of Canada 27:359–370.

Myers, G. S. 1949. Usage of anadromous, catadromous and allied terms for migratory fishes. Copeia 1949:89–97. Annual Report of the Smithsonian Institute for 1937:339–364.

Nishimoto, R. T. and J. M. Fitzsimons. 1986. Courtship, territoriality, and coloration in the endemic Hawaiian freshwater goby, *Lentipes concolor*. Pages 811–817 *in* T. Uyeno, R. Arai, T. Taniuchi, and K. Matsuura editors, Indo-Pacific fish biology: Proceedings of the second international conference on Indo-Pacific fishes. Ichthyological Society of Japan, Tokyo.

Pepin, P., S. J. Pearre, and J. A. Koslow. 1987. Predation on larval fish by Atlantic mackerel, *Scomber scombrus*, with a comparison of predation by zooplankton Canadian Journal of Fisheries and Aquatic Sciences 44:2012–2018.

Radtke, R. L., R. A. Kinzie III, and S. D. Folsom. 1988. Age at recruitment of Hawaiian freshwater gobies. Environmental Biology of Fishes 23:205–213.

Rowe, D. K., and G. Kelly. 2009. Duration of the oceanic phase for inanga whitebait (Galaxiidae) is inversely related to growth rate at sea. Pages 343–354 *in* A. J. Haro, K. L. Smith, R. A. Rulifson, C. M. Moffitt, R. J. Klauda, M. J. Dadswell, R. A. Cunjak, J. E. Cooper, K. L. Beal, and T. S. Avery, editors. Challenges for diadromous fishes in a dynamic global environment. American Fisheries Society, Symposium 69, Bethesda, Maryland.

Sorensen, P. W., and K. A. Hobson. 2005. Stable isotope analysis of amphidromous Hawaiian gobies suggests their larvae spend a substantial period of time in freshwater river plumes. Environmental Biology of Fishes 74:31–42.

American Fisheries Society Symposium 69:343–354, 2009

Duration of the Oceanic Phase for Inanga Whitebait (Galaxiidae) Is Inversely Related to Growth Rate at Sea

David K. Rowe*
National Institute for Water and Atmospheric Research Limited
Post Office Box 11–115, Hamilton, New Zealand 3251

Greg Kelly
National Institute of Water and Atmospheric Research Limited
Post Office Box 8602, Christchurch, New Zealand 8011

Abstract.— Juvenile inanga *Galaxias maculatus* are the main species in the New Zealand whitebait fishery. Whitebait were sampled throughout the spring fishing season (August–November 2002) as they entered the mouths of two New Zealand rivers, one in the North Island and the other more than 500 km south in the South Island. Fish length, date of arrival at the river mouth, and the number of daily growth rings in otoliths were used to determine age, hatch date, and mean growth rate at sea. Backdating indicated that whitebait hatched between mid-March and late July 2002, spent 104–168 d at sea, and experienced overall growth rates ranging from 0.24 to 0.41 mm/d. Mean length on return to the river mouth ranged from 45 to 55 mm total length and varied among months as well as between rivers. The duration of the oceanic rearing period (in days) was inversely related to growth rate, implying that the time of return to freshwater is endogenously determined. However, average size at river entry was larger for the South Island than for the North Island fish because, on average, the South Island fish remained at sea for a further 10 d. A longer journey from marine feeding grounds back to the coast rather than a delay at the river mouth before entry would account for this difference, but colder oceanic waters in the south may also play a role.

Introduction

Important fisheries for the juvenile (40–60 mm total length [TL]), migrant stage of diadromous galaxiids occur during spring months at the mouths of many rivers throughout New Zealand. At this time, the juvenile galaxiids migrate en masse into rivers on their way to adult habitats upstream (McDowall 1984). The catch can consist of up to five species of galaxiid, which are collectively known as whitebait, and these fish can fetch up to NZ$100/kg in city markets. The dominant species (>90% of the catch) in most rivers is now inanga *Galaxias maculatus* (Jenyns, 1842), but banded kokopu *G. fasciatus* Gray,1842 and koaro *G. brevipinnis* Günther,1866 can also comprise a significant component of the catch (up to 30% in some rivers), and they can dominate the catch in rivers where inanga habitat is minimal (Rowe et al. 1992).

The whitebait fishery has declined throughout New Zealand since the early 1900s (McDowall 1984), and some of this decline can be attributed to deforestation, which has reduced adult habitat for both koaro and banded kokopu (McDowall 1990; Rowe et al. 1999). For example, models of these species' adult habitats developed by Leathwick et al. (2005) were used to calculate the total length of stream habitat for each species in New Zealand both before the arrival of Europeans to New Zealand (i.e., before widespread deforestation) and today. The models indicated an overall, nationwide drop in stream habitat of 6.6% for koaro and 35.3% for banded kokopu, but a 28.8% increase in habitat for inanga. Land-use changes will therefore have reduced the proportions of koaro and especially banded kokopu in the catch while increasing the

* Corresponding author: d.rowe@niwa.co.nz

relative contribution of inanga. However, the increase in stream habitat for inanga caused by deforestation does not necessarily imply an increase in its overall abundance. This species spawns at the base of terrestrial vegetation on riverbanks in the tidally influenced regions of rivers when this vegetation is periodically inundated by spring high tides (Burnet 1965; Benzie 1968a). Loss of this spatially restricted spawning habitat is thought to be a major factor responsible for the decline of inanga and hence for the decline of whitebait fisheries throughout New Zealand.

Although management to ensure the sustainability of inanga stocks involves fishery regulation (to allow some escapement), fish passage provision (to allow access past culverts), and habitat restoration in streams (to improve adult habitat), considerable effort is also being applied to the problem of identifying and restoring spawning habitat (Richardson and Taylor 2002). A major limitation with the restoration of habitats is the difficulty in measuring its success because large variations in whitebait catches occur throughout both time and between different rivers, which means in some years and/or rivers, whitebait fishers experience very good catches, whereas in other years and/or rivers, catches are poor. Riverine factors such as overfishing, loss of habitat, high egg mortality, and changed river flows are all blamed when catches are poor (McDowall 1984) and are then raised as key issues for management to address. However, marine factors such as larval mortality rates, larval growth rates, and changes in the strength (and direction) of coastal current systems that can influence the time of return to freshwaters may also influence the level of recruitment into rivers. Such marine factors may overwhelm the effects of river-based restoration strategies and mask any effects of improvements in adult habitat or spawning grounds. It is clear that knowledge of the marine phase of inanga is needed to underpin better management of the whitebait fisheries and that a lack of information on its marine life history will hamper future management. We seek to address this gap by exploring some of the characteristics of the life history of inanga at sea.

McDowall et al. (1994) found that, as with many other fishes, the otoliths of juvenile inanga contained daily growth rings. Their study validated the daily periodicity of these rings and indicated the oceanic larval duration (OLD, sensu Iida et al. [2009, this volume] and akin to the pelagic larval duration used by Bell et al. [1995]), ranged from 103 to 202 d for inanga collected from 12 river mouths (10 fish per river) around the New Zealand coastline. McDowall et al. (1994) found that inanga entering rivers later in the season were smaller than those returning earlier in the season and that the former fish were younger rather than slower growing. However, large-scale geographical variations in size and age were not addressed and the sample size for each river was small and taken on a single occasion during the migration season. Large-scale geographic and temporal variations could well occur in the marine life history of inanga if marine conditions affecting growth and survival differ. For example, McDowall and Eldon (1980) found distinct differences in the size of inanga between South Island and North Island rivers, suggesting that some geographic differentiation in either fish stocks or marine rearing environments occurs. Because geographical and temporal variations in marine rearing environments could influence the size and number of inanga whitebait entering river mouths, our study focused on this aspect of marine life history. In addition, we examined whitebait length at river entry in relation to both the duration of the OLD and marine growth rates.

Methods

Samples of fresh-run whitebait containing 100–700 individual fish, depending on the size of the catch, were collected by experienced whitebait fishers from the mouths of the Mokau and Hokitika rivers (Figure 1). Samples were obtained at approximately weekly intervals from late August to mid-November 2002, encompassing the main whitebait migration and fishing seasons. Whitebait samples were preserved in 10% buffered formalin and all fish were later identified to species and their TL measured. Inanga were separated from other whitebait species (e.g., koaro and banded kokopu), using the methods developed by McDowall and Eldon (1980). Mean total lengths were plotted to determine variations in size between both months and rivers and significant differences determined using one-way analysis of variances (ANOVAs).

A subsample of inanga was taken from near the start (early September), middle (mid-October), and end (mid-November) of the peak migration season in each river, and the sagittal otoliths of 20–25 fish from each month and river were collected. Otoliths (approximately 0.5 mm in diameter) were cleaned

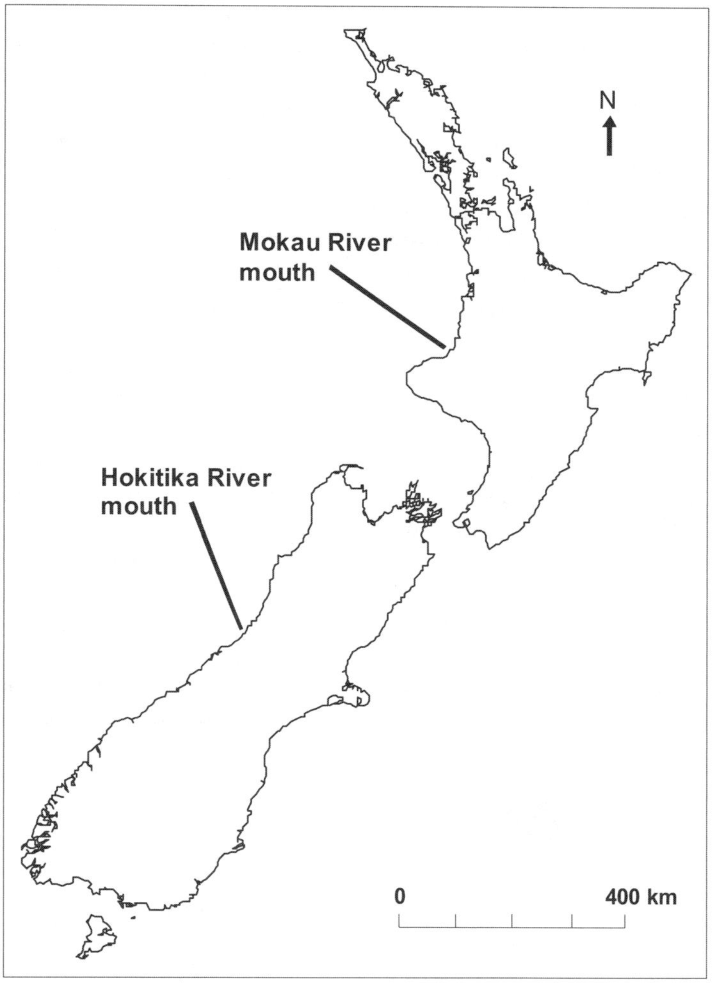

Figure 1.—Location of the Mokau and Hokitika River mouths on the west coast of New Zealand.

and mounted on a glass slide using Crystalbond. They were repeatedly ground on each side (using 1,000 and then 2,000 grit paper) and examined under a microscope until a thin cross section, including the nucleus, was obtained. The number of daily growth rings beyond the hatch zone (as defined by McDowall et al. 1994) was counted along the main axis using a graticule and 600× magnification without oil immersion. Counts of the number of daily growth rings were made by two independent readers, and these confirmed that there was no bias in the counts and that the error in ring counts was negligible.

The oceanic larval duration measured in days at sea, the estimated hatch date, and the overall mean growth rate at sea (assuming a standard size at hatch of 7 mm) was calculated for each fish. The standard hatch size of 7 mm TL was based on the range of 7–8 mm reported by McDowall (1990). Relationships between fish size at river entry, the duration of the marine growth phase, and the mean growth rate at sea were then examined and compared between both months and rivers using one-way ANOVA and regression analysis.

Differences in the marine rearing environments experienced by these fish off the coast of the Hokitika and Mokau River mouths were respectively determined using data collected by NOAA (U.S. National Oceanic and Atmospheric Administration) satellites and subsequently analyzed by the National Institute for Water and Atmospheric Research (NIWA; Uddstrom and Oien 1999). Maps portraying average monthly sea-surface temperature isotherms and chlorophyll-*a* concentration isoclines for each month in 2002 were downloaded from the NIWA Web site (www.niwa.co.nz) and mean temperatures and chlorophyll-*a* concentrations recorded for the areas within 50 km of the coast in the vicinity of both the Mokau and Hokitika River mouths for each month. Mean monthly sea temperatures were then plotted against chlorophyll-*a* concentrations for each region to identify differences in marine growing environments.

The maximum heights of spring tides near the mouths of the Hokitika and Mokau rivers were obtained for the months January to July 2002 from the New Zealand NIWA tide forecaster (www.niwa-science.co.nz/services/free/tides), which is based on the tide model developed by Goring (2001).

Results

Marine rearing conditions for fish larvae within an area approximately 50 km out from the coastline and within 100 km both north and south of each of the Mokau and Hokitika River mouths differed in terms of both mean monthly sea surface temperature and chlorophyll-*a* concentration (Figure 2). Water temperatures in both regions declined between February and July and increased between August and January. However, sea temperatures were generally warmer in the vicinity of the Mokau River than off the more southerly Hokitika River. Mean chlorophyll-*a* concentrations in the vicinity of the Mokau River increased as water temperature declined, but in the vicinity of the Hokitika River, they remained high throughout the year.

The mean length of inanga at river arrival was 46–54 mm TL and varied both among months and between rivers (Figure 3). In general, mean length increased at the start of the migration season and then declined towards the end of the season. Inanga whitebait entering the Hokitika River were generally larger than those entering the more northern Mokau River (Figure 3).

The duration of OLD for inanga ranged from 104 to 168 d (Figure 4), and the mean OLD var-

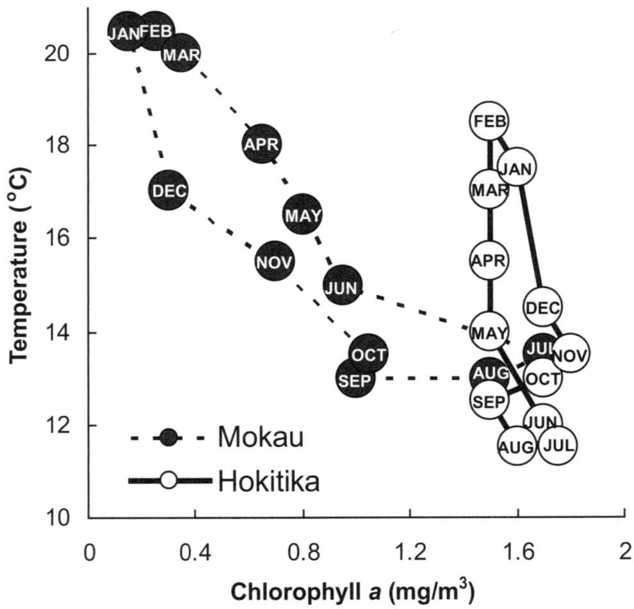

Figure 2.—Variations in mean monthly sea-surface primary productivity (indicated by chlorophyll *a*) and water temperature as determined from National Oceanic and Atmospheric Administration satellite imagery over the area within 50 km of the shoreline in the vicinity of the Hokitika and Mokau River mouths, respectively, from January to December 2002. Juvenile inanga develop at sea mainly during autumn and winter months (i.e., March to August) as marine water temperatures decline.

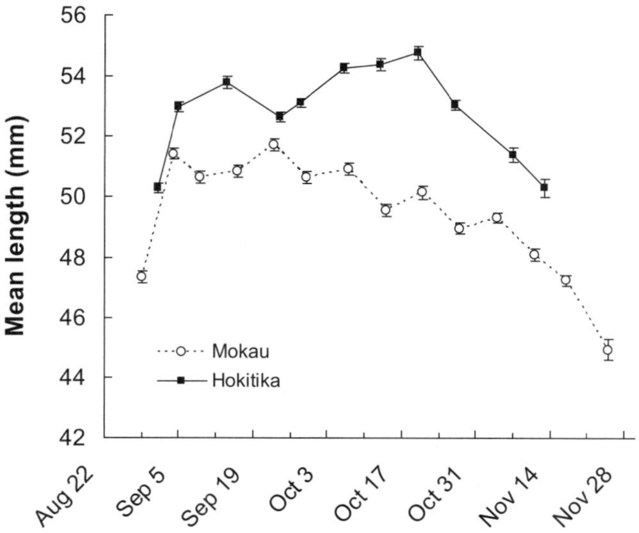

Figure 3.—Temporal variation in the mean size (±SE) of juvenile inanga as they entered the mouths of the Mokau and Hokitika rivers in 2002.

ied among months for both the Mokau (ANOVA, $F = 15.65$, $P < 0.001$, df = 74) and the Hokitika River ($F = 14.56$, $P < 0.001$, df = 59). Although the maximum OLD was very similar for both rivers (i.e., 167 d in the Hokitika River and 168 d in the Mokau River), the overall mean of 135 d for fish entering the Mokau River was significantly fewer than the 145 d for fish entering the Hokitika River ($F = 16.72$, $P < 0.001$, df = 134).

The back-calculated hatch dates for inanga ranged from March 13 to July 27, 2002 for fish from the Mokau River and from March 27 to July 14, 2002 for fish entering the Hokitika River (Figure 5). In general, inanga that entered the river mouth at the start of the migration season (i.e., early September) had earlier hatch dates than fish entering the river mouth at the end of the season and vice versa. There was very little difference in the earliest and latest hatch dates for inanga returning to either the Mokau or Hokitika rivers, indicating that spawning times are similar in both rivers.

Marine growth rates for individual fish ranged

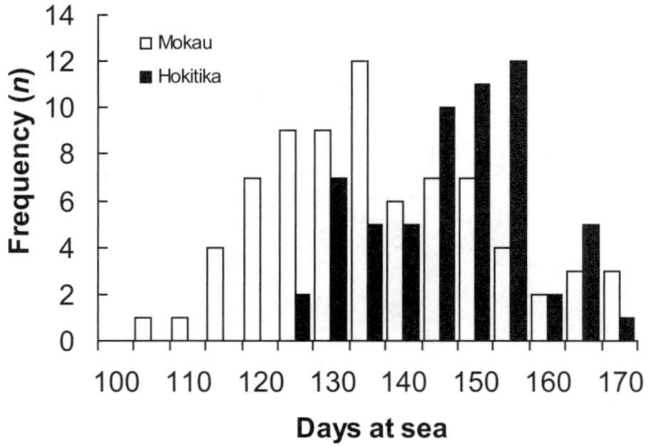

Figure 4.—Differences in the frequency distributions for time in days spent at sea before juvenile inanga entered the Mokau and Hokitika River mouths in 2002.

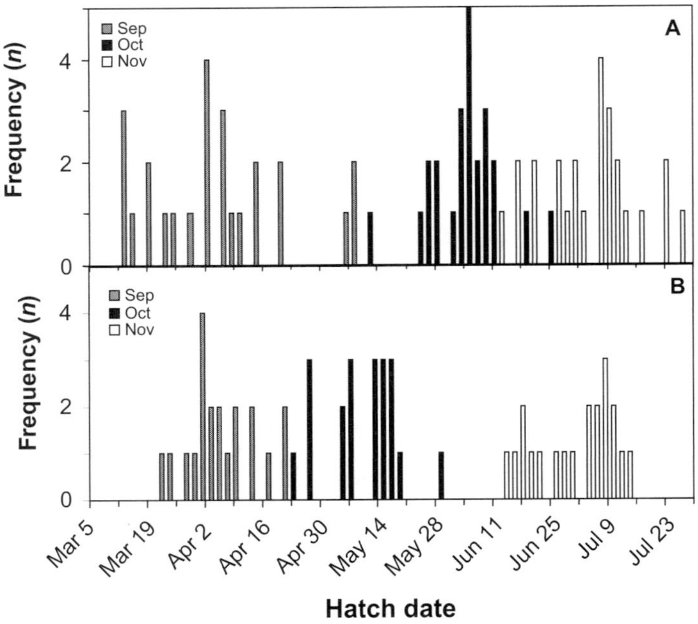

Figure 5.—Hatch dates calculated for juvenile inanga returning to (A) the Mokau River mouth, and (B) the Hokitika River mouth in September, October, and November 2002, respectively.

from 0.24 to 0.41 mm/d, and although mean growth rate varied significantly among months (Table 1) for both the Mokau (ANOVA, F = 29.2, P < 0.001, df = 74) and Hokitika rivers (F = 25.1, P < 0.001, df = 59), there was no difference in the overall mean of 0.31 mm/d between rivers (F = 0.08, P > 0.50, df = 134).

Fish length at river entry was not related to time at sea (i.e., OLD values) nor to marine growth rates (Table 2), but there was a strong inverse relationship between OLD and growth rate for all months in both rivers (Table 2; Figure 6). When the relationship between mean growth rate and the duration of OLD was compared between rivers, it was apparent that inanga from the Hokitika River spent about 10 more days at sea on average than fish entering the Mokau River, irrespective of the monthly differences in growth rate (Figure 7).

The range in hatch dates for fish entering the

Table 1.—The mean (±SE) duration of oceanic larval development (OLD), growth rate, and length at river entry for inanga entering the Mokau and Hokitika River mouths in September, October, and November 2002.

	Month of river entry		
	September	October	November
Mokau River			
OLD (days at sea)	145.0 (3.3)	124.8 (1.8)	136.0 (2.3)
Growth rate (mm/d)	0.28 (0.007)	0.35 (0.005)	0.30 (0.006)
Length (mm)	47.5 (0.25)	50.1 (0.36)	48.1 (0.32)
Hokitika River			
OLD (days at sea)	149.1 (1.9)	149.3 (2.1)	135.8 (2.1)
Growth rate (mm/d)	0.29 (0.004)	0.31 (0.004)	0.34 (0.005)
Length (mm)	50.2 (0.15)	53.7 (0.34)	52.4 (0.35)

Table 2.—Correlations between length at river entry, the duration of oceanic larval development (OLD) and growth rate for inanga entering the Mokau and Hokitika rivers in September, October, and November 2002. (Correlations with significance levels less than $P = 0.05$ are shown in bold).

	Mokau River			Hokitika River		
Variables	Sept	Oct	Nov	Sept	Oct	Nov
Length vs. OLD	0.34	0.22	–0.10	0.10	0.29	0.20
Length vs. growth rate	–0.09	0.36	**0.47**	0.18	0.26	0.29
Growth rate vs. OLD	**–0.96**	**–0.83**	**–0.92**	**0.96**	**–0.85**	**–0.88**

Mokau and Hokitika rivers (mid-March to late July 2002) coincided with a period of declining spring tide heights in both rivers, with the overall monthly maximum height (full moon maximum) dropping by more than 10 cm a month (Figure 8).

Discussion

The pattern of declining spring tide height during the whitebait spawning season is an annual one and is typical for the west coast of New Zealand.

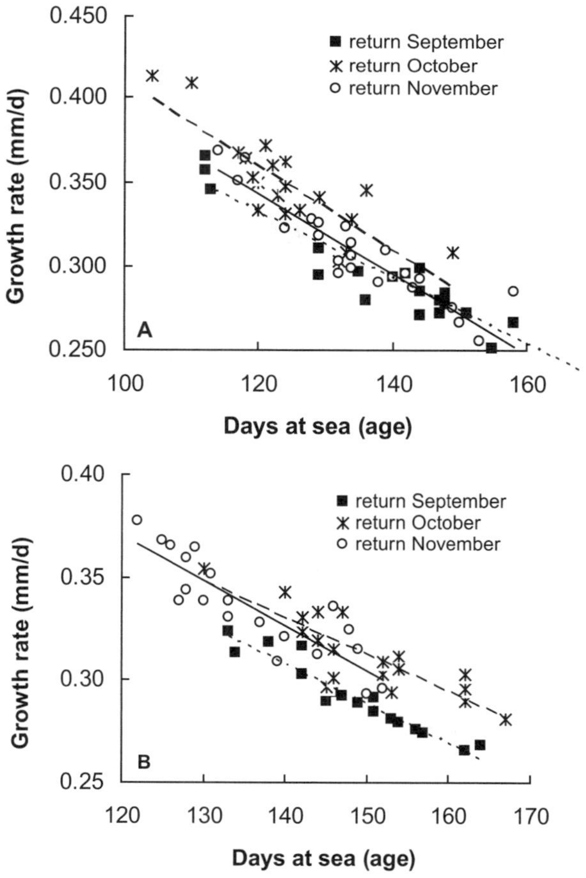

Figure 6.—Relationships between growth rate and the duration of the oceanic larval development phase (OLD) for individual inanga returning to (A) the Mokau River, and (B) the Hokitika River during different months in 2002. Lines are the least squares regressions for September (dotted line), October (dashed line), and November (solid line).

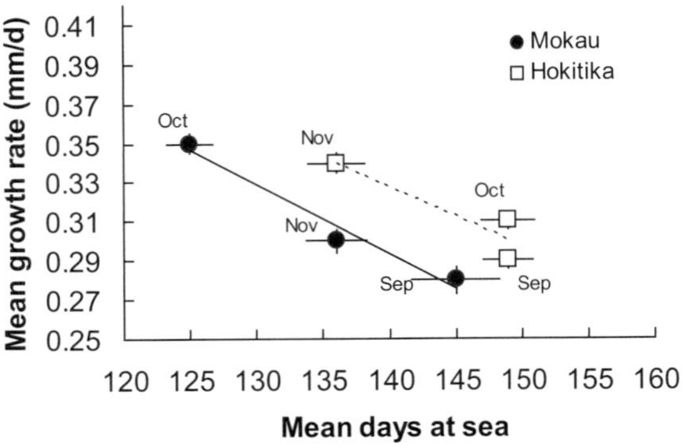

Figure 7.—Relationships between mean growth rate and the duration of the oceanic larval development phase (OLD) for inanga returning to the Mokau and Hokitika rivers at different times in 2002. Straight lines are the least squares regressions, horizontal, and vertical bars on mean values represent standard errors.

It was somewhat unexpected and counterintuitive as a declining spring high tide height would reduce the chances of inanga eggs being reinundated by a subsequent spring tide.

Three explanations are possible for this apparent contradiction and all require some rethinking of the current model of inanga spawning behavior. It should be noted that two spring high tides occur at each of these river mouths each month. The first and smaller high is associated with the new moon and the second larger high is associated with the full moon. Inanga could therefore spawn mainly on the lower spring high with eggs hatching approximately 14 d later on the higher one. Direct observations of inanga spawnings are too rare to permit any evaluation of this explanation at present. However, it is consistent with the minimum incubation time of 10 d for inanga eggs at 17°C (Benzie 1968a). Alternatively, inanga may spawn on the highest spring high tide (as described in McDowall 1990) but deposit eggs either 5 cm or more below the maximum water level so that eggs are inundated by the spring tide occurring a fortnight later. However, they would need to be deposited more than 10 cm below the maximum water level to hatch on the spring tide 28 d later. While this explanation is plausible, it

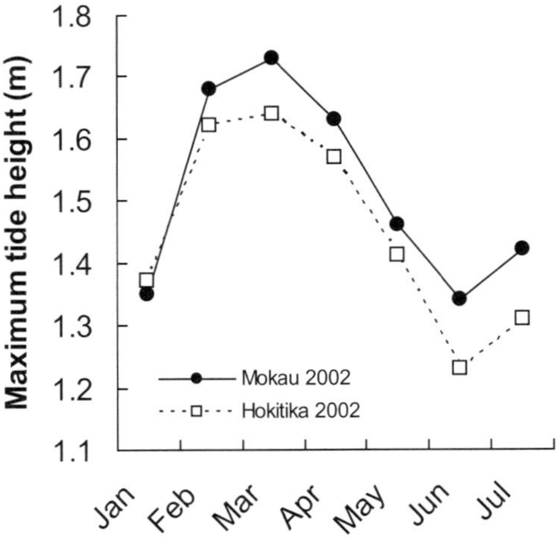

Figure 8.—Maximum tide heights (spring peak per month) for the Mokau and Hokitika River mouths over the inanga spawning season of March to July 2002.

conflicts with the few direct observations of inanga egg deposition available to date. These fish tend to lay their eggs at the very edge of the tidally flooded zone of riparian vegetation "almost out of the water itself" and "in the inch or two of water covering the ground" (McDowall 1990). Such observations are supported by direct monitoring of inanga spawning by one of us (D. Rowe), which indicated that eggs are deposited within 5 cm of the peak high tide level. Hence, they would not be reimmersed by future spring tides in the Hokitika or Mokau rivers over the remainder of the spawning season unless river water levels were increased by higher flow. The third explanation is that inanga do not spawn on the exact day of the spring high tide but a few days later, as indicated by McDowall and Eldon (1980). Even so, egg deposition over a period of declining spring tide heights implies that hatching may depend more on flood flows, large enough to raise river water levels above the previous spring tide height, than on subsequent high tides.

McDowall et al. (1994) found some evidence for a role of flood flows on the recruitment of inanga when they examined the temporal distribution of inanga hatch dates for fish returning to a small North Island stream. The timing of the main hatching peak in this stream coincided more with a flood event than with a spring tide. Such an egg hatching strategy is consistent with the behavior of other closely related galaxiids such as koaro, banded kokopu, and shortjaw kokopu *G. postvectis* (Mitchell and Penlington 1982; O'Connor and Koehn 1998; Charteris et al. 2003), which all spawn near adult habitats in the upper reaches of rivers and so rely solely on floods for both spawning and hatching.

There may be a strong selective advantage for inanga to adopt a recruitment strategy not cued precisely to spring high tides. Egg hatching at predictable times such as spring high tides can be expected to maximize predation of newly hatched larvae because water levels are high at such times and velocities are low or static. In contrast, egg hatching at irregular times related to flood events is more likely to maximize larval survival because floods create higher turbidity levels, providing better visual protection against predators (McDowall and Charteris 2006) and the increased flow to transport larvae out of rivers faster and to a greater distance offshore. However, dependence on flood flows also carries the risk of prolonged drought and failure to hatch. The long spawning period of 5 months no doubt provides inanga with some protection against this possibility. Should flood flows prove to be important for inanga hatching, damming and flow regulations that reduce flood peaks at river mouths may affect the hatching success of inanga.

McDowall et al. (1994) indicated that the small inanga found towards the end of the whitebait migration season were younger than their earlier returning and larger counterparts. But they were not slower growing, indicating that they experienced a shorter time at sea compared with the larger, earlier migrants. We also found that the smaller whitebait present towards the end of the migration season in 2002 were younger than those occurring at the start or in the middle of the season. But this only occurred in the Hokitika River and not in the Mokau River. Size at river entry for inanga is influenced by both growth rate and the OLD period, but the relative importance of these variables can clearly change with both time and location. For example, the generally smaller size of inanga entering the Mokau River in November was related to both the relatively short period of OLD and the low growth rate these fish experienced. However, in both rivers, inanga were also smaller in September at the start of the migration season, and for these fish, size at river entry was related to a low growth rate, not a shorter period of OLD.

Monthly and riverine variation aside, it was clear that OLD varied inversely and strongly with growth rate at sea. Such a relationship could occur if some whitebait congregated around river mouths and experienced a delay before river entry. However, this is unlikely since whitebait migrating from the sea into rivers do not feed (McDowall and Eldon 1980) and a delay in river entry would result in an apparent slow growth rate at sea (i.e., an increased age at river entry for little commensurate increase in length). It would also produce a zone of narrow growth rings on the outer edge of the otoliths of such fish. The last 20 rings of all but one of the slower-growing fish were more widely spaced than others, so a period of reduced growth just before river entry cannot account for their slower growth. Some shrinkage in body length (up to 3 mm) can occur on exposure to freshwater (McDowall and Eldon 1980), which could contribute to an inverse relationship between growth rate and the duration of OLD. However, shrinkage would only produce an inverse relationship between OLD and growth rate if some, and not all fish, were affected, which seems implausible. The relationship we found indicates that those fish that experienced a fast growth rate at sea returned to river mouths earlier than their slower-growing counterparts (and vice versa), irrespective of month and river location.

An inverse relationship between growth rate and OLD implies that fish return to rivers at a similar size and helps explain the small seasonal and geographic variation in size observed. Hypothetically, an optimal size at river entry can be expected to maximize key components of fitness such as survival and growth. For example, migration into rivers at too small a size may be a disadvantage because it would incur an increased mortality risk if body reserves were too low to support the energetic costs of migration into rivers and adaptation to freshwater. Furthermore, migrants entering rivers need to have the size and swimming power to move upstream against the river flow towards adult habitats. Any delay in upstream movement in rivers related to slower upstream migration could increase exposure to predation and increase mortality rates. Similarly, if inanga juveniles remained in marine waters until they were larger, they may experience a higher mortality rate from marine predators.

An optimal size for migration by inanga back to rivers is therefore likely to be selected for, but how do fish at sea know that they are at the appropriate size for return to freshwater? We may presume an endogenous, physiologically mediated test of whether they are at the right size for return to freshwater (sensu Benzie 1968b), similar to that used by salmonids. In the Salmonidae, the decision to migrate back to spawning grounds is associated with maturation, and maturation is endogenously triggered by changes in hormonal levels related to the accumulation of sufficient fat reserves by a certain time of year (Rowe et al. 1991; Silverstein et al. 1998; Shearer and Swanson 2000). The seasonal window for this decision seems to be determined primarily by a circadian clock mechanism set by changes in day length (Porter et al. 1999; Bromage et al. 2001). In inanga, maturation and spawning is not the immediate goal of migration, so the level of fat reserves may be less important than a measure of size as determined from physiological correlates of growth rate.

The observed inverse relationship between growth rate and OLD for inanga means that a faster growth rate at sea results in an earlier return to rivers and vice versa. This mechanism constrains whitebait size at river entry and implies that runs of inanga into rivers will occur earlier in the season when growing conditions at sea are optimal and later when growing conditions are poor. The rate of marine growth experienced by juvenile inanga at sea can therefore be expected to affect the timing of migration back to river mouths and so influence the temporal distribution of recruitment. Bell (1997) showed, through the modeling of recruitment as applied to *Sicydium punctatum*, that the overall level of recruitment can be related to both the temporal distribution of size at return and marine mortality rates. Our results help support this model.

Inanga entering the Hokitika River were larger than Mokau River fish, which was caused by a longer OLD (by 10 d), irrespective of growth rate and month. Either inanga entering the Hokitika need to be larger or, if not, they take a longer time to return from marine feeding areas, resulting in their larger size. The latter is considered more likely as there are no obvious biological or physical reasons why inanga should be larger on entry to the Hokitika River than the Mokau River. However, sea surface conditions off the coast of the Hokitika River are somewhat colder than off the coast of the Mokau River and this difference in the marine growth environment may influence size at river entry through effects of sea temperature on growth and migration. For example, colder water may reduce fish swimming speeds, resulting in delayed river entry, or it may reduce growth rate and hence delay the physiologically mediated decision to return to freshwater. Conversely, an increase in sea temperature may hasten migration back to freshwater. Clearly, a better understanding of the role of temperature in controlling the growth rate and migratory behavior of juvenile inanga in saltwater will help unravel the main factors influencing the marine life history phase of this species. This will lead to a better understanding of factors influencing temporal and geographic variations in recruitment.

More knowledge of the marine phase of inanga is clearly needed to underpin better management of the whitebait fisheries because a lack of information on factors affecting recruitment will hamper management initiatives to improve riverine habitat. Our study has provided new insights into inanga spawning in rivers and on the effects of marine conditions on the timing of migrations back to river mouths.

The results help identify the role of some key riverine and marine factors affecting inanga recruitment and so contribute to the goal of better fishery management.

Acknowledgments

This project depended on the careful and regular collection of whitebait samples by whitebaiters in the Mokau and Hokitika rivers and we thank Ray and Bev Christiansen as well as Faye Gordon for their help and persistence with this. J. Smith assisted with the counting of daily growth rings and the manuscript benefited from the careful scrutiny and comments of my colleague Dr. C. Baker and two referees, particularly Dr. K. N. I. Bell. The study was partially funded by the New Zealand Foundation for Research Science and Technology under contract CO1X020.

References

Bell, K.N.I. 1997. Complex recruitment dynamics with Doppler-like effects caused by shifts and cycles in age-at-recruitment. Canadian Journal of Fisheries and Aquatic Sciences 52:1668–1681.

Bell, K. N. I., P. Peppin, and J. A. Brown. 1995. Seasonal inverse cycling of length and age-at-recruitment in the diadromous species *Sicydium punctatum* and *Sicydium antillarum* in Dominica, West Indies. Canadian Journal of Fisheries and Aquatic Sciences 52:1535–1545.

Benzie, V. L. 1968a. Stages in the normal development of *G. maculatus attenuatus* (Jenyns). New Zealand Journal of Marine and Freshwater Research 2:606–627.

Benzie, V. L. 1968b. A consideration of the whitebait stage of *Galaxias maculatus attenuatus* (Jenyns). New Zealand Journal of Marine and Freshwater Research 2:559–573.

Bromage, N., M. Porter, and C. Randall. 2001. The environmental regulation of maturation in farmed finfish with special reference to the role of photoperiod and melatonin. Aquaculture 197:63–98.

Burnet, A. M. R. 1965. Observations on the spawning migrations of *G. attenuatus* (Jenyns). New Zealand Journal of Science 8:79–87.

Charteris, S. C., R. M. Allibone, and R. Death. 2003. Spawning site selection, egg development and larval drift of *Galaxias postvectis* and *G. fasciatus* in a New Zealand stream. New Zealand Journal of Marine and Freshwater Research 37:493–506.

Goring, D. 2001. Computer models define tide variability. The Industrial Physicist October/November 2001:14–17.

Iida, M., S. Watanabe and K. Tsukamoto. 2009. Life history traits of a temperate sicydiine goby in Japan, compared with its relatives and other amphidromous fishes. Pages 355–373 *in* A. J. Haro, K. L. Smith, R. A. Rulifson, C. M. Moffitt, R. J. Klauda, M. J. Dadswell, R. A. Cunjak, J. E. Cooper, K. L. Beal, and T. S. Avery, editors. Challenges for diadromous fishes in a dynamic global environment. American Fisheries Society, Symposium 69, Bethesda, Maryland.

Leathwick, J., D. K. Rowe, J. Richardson, J. Elith, and T. Hastie. 2005. Using multivariate regression splines to predict the distribution of New Zealand's freshwater diadromous fish. Freshwater Biology 50:2034–2052.

McDowall, R. M. 1984. The New Zealand whitebait book. A.H. & A.W. Reed, Wellington, New Zealand.

McDowall, R.M. 1990. New Zealand freshwater fishes: a natural history and guide. Heinemann Press, Auckland, New Zealand.

McDowall, R. M., and T. Eldon. 1980. The ecology of whitebait migrations (Galaxiidae: *Galaxias* spp.). New Zealand Ministry of Agriculture and Fisheries, Fisheries Research Bulletin 20, Wellington.

McDowall, R. M., C. P. Mitchell, and E. B. Brothers. 1994. Age at migration from the sea of juvenile galaxias in New Zealand (Pisces: Galaxiidae). Bulletin of Marine Science 54:385–402.

McDowall, R. M., and S. C. Charteris. 2006. The possible adaptive advantages of terrestrial egg deposition in some fluvial diadromous galaxiid fish (Teleostei: Galaxiidae). Fish and Fisheries 7:153–164.

Mitchell, C. P., and Penlington, P. B. 1982. Spawning of *Galaxias fasciatus* Gray (Salmoniformes: Galaxiidae). New Zealand Journal of Marine and Freshwater Research 16:131–133.

O'Connor, W. G., and J. D. Koehn. 1998. Spawning of the broad-finned galaxias *Galaxias brevipinnis* Günther (Pisces: Galaxiidae) in coastal streams of southeastern Australia. Ecology of Freshwater Fishes 7:95–100.

Porter, M. J. R., N. J. Duncan, D. Mitchell, and N. R Bromage. 1999. The use of cage lighting to reduce plasma melatonin in Atlantic salmon (*Salmo salar*) and its effect on the inhibition of grilsing. Aquaculture 176:237–244.

Richardson, J., and M. Taylor. 2002. A guide to restoring inanga habitat. National Institute of Water and Atmospheric Research Ltd., NIWA Science and Technology Series 40, Wellington.

Rowe, D. K., J. E. Thorpe, and A. M. Shanks. 1991. Role of fat stores in the maturation of male Atlantic salmon (*Salmo salar*) parr. Canadian Journal of Fisheries and Aquatic Sciences 48:405–413.

Rowe, D. K., B. A. Saxton, and A. G. Stancliff. 1992. Species composition of whitebait (Galaxiidae) fisheries in 12 Bay of Plenty rivers, New Zealand: evidence for river mouth selection by juvenile *Galaxias brevipinnis* (Günther). New Zealand Journal of Marine and Freshwater Research 26:219–228.

Rowe, D. K., B. L. Chisnall, T. Dean, and J. Richardson. 1999. Effects of land use on native fish communities in east coast streams of the North Island of New Zealand. New Zealand Journal of Marine and Freshwater Research 33:141–151.

Shearer, K. D., and P. Swanson. 2000. The effect of whole body lipid on early sexual maturation of 1+ age male Chinook salmon (*Oncorhynchus tshawytscha*) Aquaculture 190:343–367.

Silverstein, J. T., K. D. Shearer, W. W. Dickoff, and E. M. Plisetskaya. 1998. Effects of growth and fatness on sexual development of Chinook salmon (*Oncorhynchus tshawytscha*) parr. Canadian Journal of Fisheries and Aquatic Sciences 55:2376–2382.

Uddstrom, M. J., and N. A Oien. 1999. On the use of high resolution satellite data to describe the spatial and temporal variability of sea surface temperatures in the New Zealand region. Journal of Geophysical Research (Oceans) 104(C9):20729–20751.

American Fisheries Society Symposium 69:355–373, 2009

Life History Characteristics of a Sicydiinae Goby in Japan, Compared with Its Relatives and Other Amphidromous Fishes

Midori Iida*, Shun Watanabe, and Katsumi Tsukamoto
Ocean Research Institute, The University of Tokyo
Minamidai, Nakano, Tokyo 164–8639, Japan

Abstract—*Sicyopterus japonicus* is unique because it is the only temperate amphidromous goby of the subfamily Sicydiinae. Life history and migration characteristics of *S. japonicus*, including seasonal changes of condition factor, spawning season, hatching size, oceanic larval duration, recruitment season, and size at recruitment, were examined in the temperate region of western Japan and were compared with those of other Sicydiinae species, all of which inhabit the tropics and subtropics. The condition factor varied seasonally, with peaks in July and November. The spawning season of *S. japonicus* ranged from July to September, with a peak in July–August, and was shorter than that of other species that normally spawn from 7 to 12 months/year. The hatching size (1.5 mm total length) was similar to other Sicydiinae, but oceanic larval duration (range: 173–253 d) was longer than most other species. The long larval duration and single annual reproduction period suggest that it is adapted to long-distance oceanic dispersal. The recruitment season (about 4 months) is the shortest known among the Sicydiinae, but its size at river entry (mean: 26.3 mm standard length) was similar to other species. The shorter reproductive and recruitment seasons for *S. japonicus* probably indicate that its migration strategy is determined by seasonal changes in the temperate region. Sicydiinae species have larger clutch sizes and smaller hatching sizes compared to other temperate amphidromous families such as the Galaxiidae, Osmeridae, and Cottidae. Sicydiinae species have no nondiadromous forms, suggesting that they may have a migration strategy with a high oceanic dependency that facilitates oceanic dispersion.

Introduction

Amphidromous migrations from freshwater to the sea, or vice versa, are not for the purpose of breeding, but occur regularly at some definite stage in the life cycle (Myers 1949; McDowall 1988). This type of migration is a specialization known in several groups of freshwater biota but is most commonly seen in fishes, although it also occurs in some decapod crustaceans and gastropod mollusks (Lee and Fielder 1979; Starmuhlner 1979; McDowall 1988, 1997, 2004, 2007; Hamano and Hayashi 1992; Hodges and Allendorf 1998; Fièvet and Eppe 2002; March et al. 2002). In this paper, we limit our discussion about amphidromy in fishes to freshwater amphidromy, where spawning is in freshwater and the larvae/juveniles are temporarily marine before returning to freshwater to grow and mature.

Amphidromous fishes are distributed mainly in the high latitude temperate regions of the northern and southern hemispheres (McDowall 1988) and include species in families such as the Osmeridae in East Asia, the Galaxiidae in southern temperate regions, the Retropinnidae in Australia and New Zealand, the Cheimarrichthyidae in New Zealand, and the Cottidae in northern temperate regions (Figure 1). Amphidromous fishes in tropical regions were not known in detail until McDowall (1988) first showed that amphidromous gobies were present in both temperate and tropical regions. Since then, much effort has been made to study the biology and ecology of tropical amphidromous gobies, includ-

* Corresponding author: midori@ori.u-tokyo.ac.jp

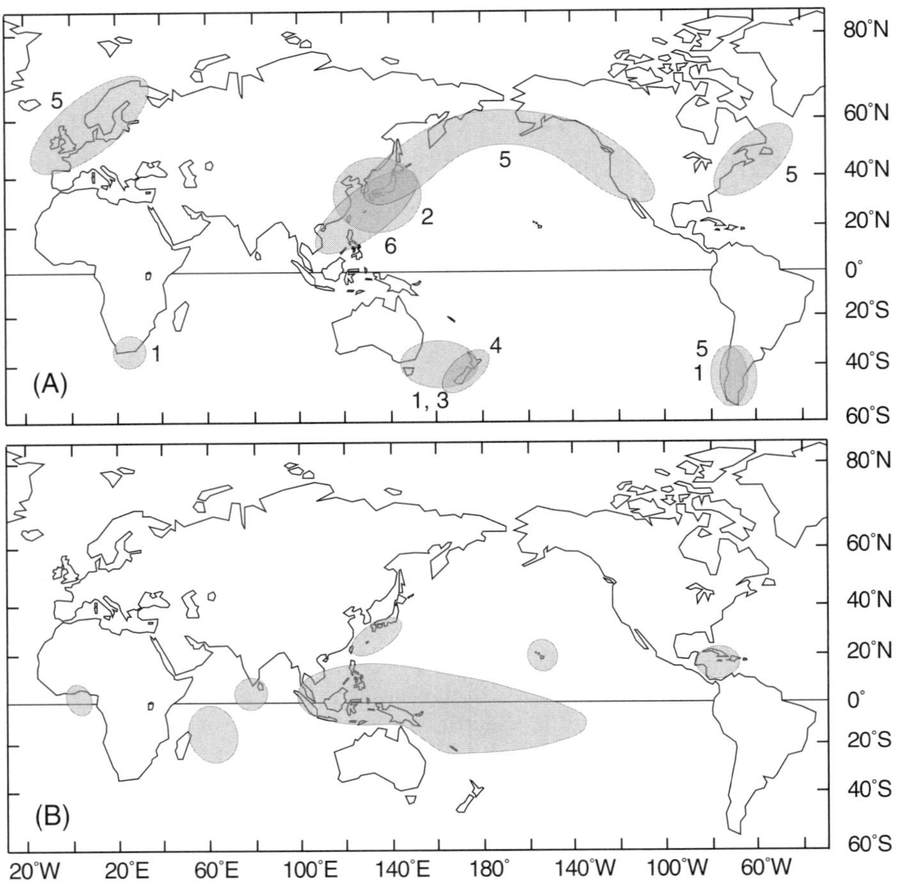

Figure 1.—Distribution of (A) the main temperate amphidromous families (following McDowall 1988) 1: Galaxiidae (spotted mountain trout *Galaxias truttaceus*, kaoro *G. brevipinnis*, Tasmanian mudfish *G. cleaveri*, giant kokopu *G. argenteus*, banded kokopu *G. fasciatus*, shortjaw kokopu *G. postvectis*, *Aplochiton taeniatus*), 2: Osmeridae (ayu *Plecoglossus altivelis*), 3: Retropinnidae (Australian grayling *Prototroctes maraena*, New Zealand grayling *P. oxyrhynchus*), 4: Cheimarrichthyidae (torrent fish *Cheimarrichthys fosteri*), 5: Cottidae (Pacific staghorn sculpin *Leptocottus armatus*, *Cottus hangiongensis*), 6: Gobiidae (*Rhinogobius* spp.); and (B) Sicydiinae species.

ing the Sicydiinae (Fitzsimons et al. 1993; Bell et al. 1995; Radtke et al. 2001; Keith 2003). The Sicydiinae includes about 100 species that all have amphidromous life histories (Froese and Pauly 2007). They are widely distributed mainly in tropical oceanic islands from the western Indian Ocean through Indonesia, the eastern Pacific Ocean, and the Caribbean Sea to western Africa (Harrison 1993; Bell 1999; Keith 2003). Myers (1949) used the genus *Sicydium* as an example of amphidromy when he first defined the term. McDowall (2003a, 2004) suggested that the oceanic life stages of the Sicydiinae and other amphidromous gobies, including the Rhyacichthyidae, Eleotoridae, the genus *Stenogobius* and the genus *Awaous*, could be regarded as the key element in explaining their distributions. However, no studies have investigated the relationships between the migration strategies of the Sicydiinae with prolonged oceanic life stages and their distributions.

Sicyopterus japonicus (Tanaka 1909) is the only known species of the Sicydiinae that lives in a temperate region, and we postulated that it possibly has different life history characteristics than the other species with more tropical distributions. The distribution of *S. japonicus* ranges along the southwestern margin of the North Pacific from Taiwan to Fukushima Prefecture in Japan (Figure

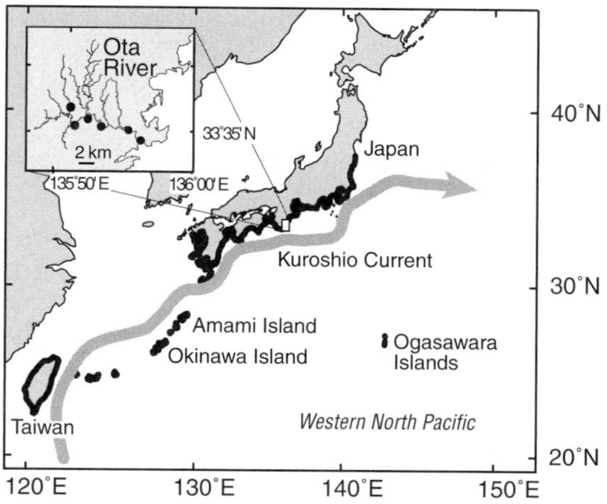

Figure 2.—Map of the East Asia region showing the geographic distribution of *Sicyopterus japonicus* (thick black lines) and the study area in the Ota River, Wakayama Prefecture, Japan. Solid circles in the inset map indicate the sampling stations.

2; Akihito et al. 2000). Water temperature regimes in this region vary seasonally compared with the almost constant conditions throughout the year in tropical regions. There are many reports of differences in life history parameters in association with latitude or with environmental temperature (Lowe-McConnell 1979; Huggett et al. 2003; Watanabe et al. 2003; Kokita 2004); consequently, the life history parameters of *S. japonicus* would be expected to differ from those of tropical sicydiine gobies.

The objectives of this study were to (1) describe the life history characteristics of *S. japonicus* using specimens collected from the Ota River, in western Japan, (2) characterize the migration strategy of *S. japonicus*, (3) compare migration strategies of Sicydiinae species in the tropics and subtropics with the strategy of *S. japonicus* in the temperate zone, and (4) evaluate the amphidromy of the Sicydiinae in comparison with those of other amphidromous families found in temperate regions, such as the Osmeridae, Galaxiidae, and Cottidae.

Methods

Sampling and Measuring

Sampling of *S. japonicus* was conducted in the Ota River, Wakayama, western Japan (34°N) at six sites from 0.9 to 20 km from the river mouth (Figure 2). Juveniles and adults were collected monthly from November 2003 to October 2004 (except June 2004) using an electric shocker and a hand net. At each sampling, water temperature was measured to the nearest 0.1°C using a mercury thermometer. Standard length (SL) and body weight (BW) were measured to the nearest 0.1 mm and 0.01 g, respectively, to calculate condition factor (CF). The CF of *S. japonicus* (n = 377) was calculated as CF = 1,000 $\times$ BW/SL^3. Gonads were removed from 71 subsampled females (3–10 specimens per month, 46.0–70.0 mm SL) and weighed to the nearest 0.01 g. The gonad somatic index (GSI) of females was calculated as: GSI = 100 $\times$ GW/BW, where GW is the gonad weight (g). An egg clutch was collected in the Ota River in August 2004 and hatching was observed in the laboratory.

Otolith Analysis

Newly recruited larvae of *S. japonicus* ($n \geq 200$) were collected by a square lift net near the river mouth of the Ota River on April 30, 2005. After collection, the specimens were frozen. The SL of 30 randomly selected specimens was measured to the nearest 0.1 mm, and the sagittal otoliths were extracted from each specimen for age determination. Otoliths were mounted on glass slides with euparal (Chroma-Gesellschaft) and examined and photographed at 50–500$\times$ under transmitted light. Otolith increments were counted on the photographs. The presence or absence of a metamorphosis check (MC) in the otoliths was confirmed in the specimens examined. The MC was previously reported to have been deposited in *S. japonicus* just after recruitment in association with a change in head morphology (Shen and Tzeng 2002). It was tentatively assumed that this species has daily periodicity in the deposition of otolith growth increments, as has been validated in *S. lagocephalus* at Réunion Island (Hoareau et al. 2007) and in *Stiphodon percnopterygionus* at Okinawa Island (Yamasaki et al. 2007).

Recruitment

The recruitment season of *S. japonicus* was determined by sampling upstream migrating larvae using a set net (7.8 m wide × 0.9 m high, with 1.5 mm mesh) in the estuary of the Ota River at a site 0.9 km upstream from the river mouth. The net was set near the riverbank with the mouth facing downstream so larvae migrating upstream could be caught. Because larvae were not observed in the river from fall to early spring (September–March) in an earlier study of the life history of this species (Iida 2005), larval sampling was from April to August. Sampling was conducted 24 h/d from April 6 to August 27, 2006 and the net was emptied of larvae every day, except on days of heavy rain when strong river flow made it impossible to leave the net in the river. Collected specimens were frozen, later blotted dry with a paper towel, and the SL of randomly selected subsamples of up to 30 specimens from each day were measured to the nearest 0.1 mm. Upstream migratory behavior of recruited larvae was observed by snorkeling in the estuary of the Ota River.

Results

Condition Factor and Reproduction

Reproductive characteristics of *S. japonicus* were examined during 2003 to 2004 when the mean water temperature at the six sites in the Ota River ranged from 9°C to 24°C through the four seasons (Figure 3). The CF of females (n = 192) and males (n = 185) ranged from 12.6 to 29.6 (mean: 18.7) and 11.8–26.0 (mean: 18.7), respectively. The CF of both female and male *S. japonicus* increased from May to a peak in July, decreased greatly in August, increased again in November, and then declined until April (Figure 3).

The GSI data from 2004 and other observations of this species in the Ota River (Iida 2005) indicated that *S. japonicus* has a very short spawning season in late summer (Table 1). The GSI was high (up to 20.5) from July to September and was low (0.0–1.2) in other months (Figure 3). Based on these data and the timing of collections of downstream migrating larvae (July–September; Iida 2005), the spawning season of *S. japonicus* in this river was from July to September, with a peak in July–August.

Eggs from the clutch that was taken to the laboratory were almost spherical with a long axis diameter of 0.5 mm. The mean hatchling size of the larvae from this clutch was 1.5 mm total length (TL).

Oceanic Larval Duration

The otoliths of the recruiting larvae of *S. japonicus* collected in the estuary of the Ota River had a wide range of ages. Their oceanic larval duration (OLD), here equivalent to pelagic larval duration or age at recruitment, ranged from 173 to 253 d (mean: 208) for larvae 24.5–29.5 mm SL. Among the sampled larvae, there was MC at the edge of the otoliths of 12 (40%), while 18 larvae (60%) had no MC.

Recruitment Season

The recruitment season began April 18 and lasted until August 26. The total number of recruiting larvae collected was 12,766 individuals. Large catches of 500 or more larvae/d were observed six times during the recruitment season, on April 22 and 23 and on May 16, 17, 21, and 22 (Figure 4). All larvae collected were judged as newly recruited based on their short SL (mean: 26.3 mm, range 23.5–30.0 mm). Underwater observations revealed that larvae displayed a head-down posture (a headstand) when they remained in one place in slow water.

Discussion

Reproduction of Sicyopterus japonicus

The spawning season of *S. japonicus* in the Ota River in Wakayama, Japan was short, and like many fishes, the reproductive season of this species appears to

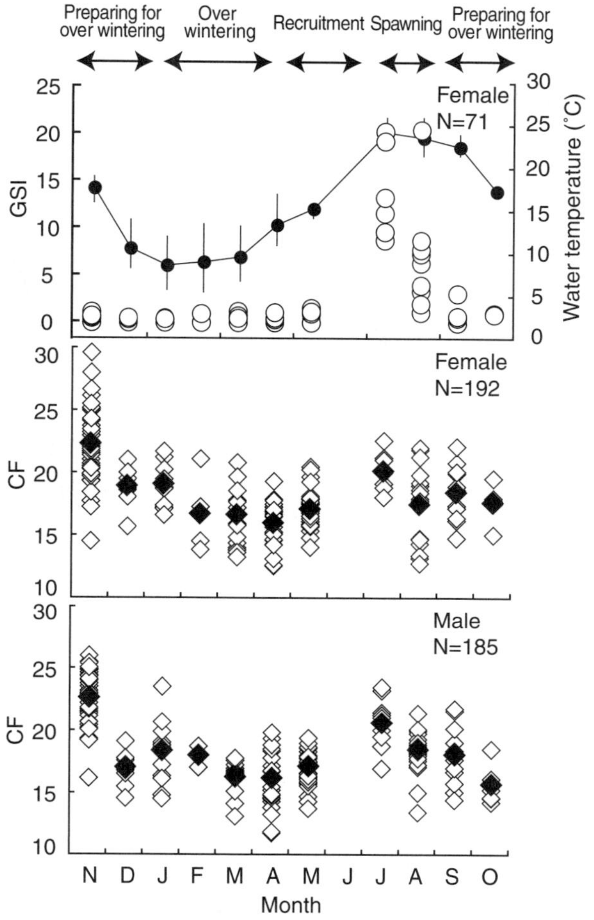

Figure 3.—The monthly changes of gonad somatic index (GSI) and condition factor (CF) of *Sicyopterus japonicus* collected in the Ota River in relation to water temperature. Open circles indicate individual GSI, and solid circles and line (bars show range) indicate water temperature in the Ota River each month (top panel). Open diamonds indicate individual CF and solid diamonds mean CF each month (bottom two panels). The life history periods (overwintering, recruitment season, spawning season, preparing for overwintering) of this species are shown at the top.

shorten with increasing latitude. The spawning season of *S. japonicus* ranges from July to September (3 months) in Wakayama (34°N), from May to August (4 months) in Okinawa (26°N; Yamasaki 2006) and from July to October (4 months) in Taiwan (23°N; Ju 2001), suggesting that the spawning season of this species is slightly longer at lower latitudes or at warmer average water temperatures. Since there are many examples to support the effect of photoperiod on fish reproduction in temperate regions (e.g., Kadmon et al. 1985; Tominaga et al. 2005), day length could also affect the spawning season of *S. japonicus*.

We made general observations on various aspects of the reproductive biology of *S. japonicus*. Adhesive eggs laid on the underside of stones were observed visually from July to September by diving. The males of this species were seen to lead the females to a nest following courtship display. Similar behavior has been observed in Stimpson's goby *Sicyopterus stimpsoni* (Fitzsimons et al. 1993). Males then cared for and guarded the eggs as reported for many other species of gobies (e.g., Gibson 1970; Marconato et al. 1989; Takegaki 2000). We found that each male would protect several clutches in his nest under a stone.

Table 1.—The life history characteristics of gobies of the subfamily Sicydiinae, other amphidromous fishes, and fluvial nonamphidromous species. The latitudes, spawning seasons, clutch sizes, hatching sizes, oceanic larval durations, recruitment seasons, recruitment sizes, and river residents are given for each locality that was studied. The sources of each study are shown by the numbers. See sources in footnotes. OLD = oceanic larval duration.

Species	Localities (latitude)	Spawning season (duration) <peak>	Clutch size	Hatching size (mm TL)	OLD (days) range (mean)	Recruitment season (duration) <peak>	Recruitment size (mm) range (mean)	River resident	Sources
Sicydiinae:									
Sicyopterus japonicus	Japan (34°N)	Jul–Sep (3) <Aug> (1)	–	1.5	173–253 (208)	Apr–Aug (5) <Apr–May>	23.5–30.0 (26.3) SL	–	1
	Japan (32°N)	Jul–Sep? (3?)	225,000	1.5	–	Mar–May? (3?)	(33) TL	–	2
S. lagocephalus	Philippines (8°N)	all year (12)	–	1.1	–	all year (12) <Jan–Mar>	21–33 (25.9) TL	–	3
	Réunion (19°S)	Nov–May (7)	50,000–70,000	–	133–266 (199)	–	(32.7) TL	–	4–6
Stimpson's goby *S. stimpsoni*	Hawai'i (18°N)	all year (12)	160,000	1.8	–	all year (12)	(23.6) SL	–	7–10
Stiphodon percnopterygionus	Okinawa (26°N)	May–Dec (8)	1,000–10,000	1.3	78–146 (99)	Jun–Oct, Apr (?)	12.2–13.6 SL	–	11,12
Lentipes concolor	Hawai'i (18°N)	Oct–Jun (9)	12,000–14,000	1.8	63–106 (86)	all year (12)	14.1–17.9 (16.0) TL	–	8, 10, 13
Sicydium punctatum	Dominica (15°N)	all year (12)	50,000–60,000	1.8	54–136 (83)	all year (12) <Aug–Nov?>	16–25 (19.5) SL	–	5, 14, 15
S. antillarum	Dominica (15°N)	–	–	–	63–139 (87)	all year (12) <Aug–Nov?>	18.8–25.5 (21.8) SL	–	14
Cotylopus acutipinnis	Réunion (19°S)	–	–	–	78–150 (101)	–	(22.5) TL	–	6
Other amphidromous fishes:									
Koaro *Galaxias brevipinnis*	N. Zealand (37°S)	Apr–May (2)	–	8.6	110–155	<Sep, Oct> (2)	36.5–47.5 SL	+	16–20
Banded kokopu *G. fasciatus*	N. Zealand (37°S)	May–Jun (2)	3,800–23,600	9.0	93–128	<Oct> (1)	30.5–37.4 SL	+	16–18, 21, 22

Table 1.—Continued.

Species	Localities (latitude)	Spawning season (duration) <peak>	Clutch size	Hatching size (mm TL)	OLD (days) range (mean)	Recruitment season (duration) <peak>	Recruitment size (mm) range (mean)	River resident	Sources
Giant kokopu *Galaxias argenteus*	N. Zealand (37°S)	Jun–Aug (3)	–	–	113–137	<Nov> (1)	36.0–39.6 (38.4) SL	+	16–18, 23
Shortjaw kokopu *G. postvectis*	N. Zealand (37°S)	May–Jun (2)	–	8.8	120–136	–	30.0–34.5 (33.5) SL	–	17, 18, 22
* Dwarf galaxias *G. divergens*	N. Zealand (37°S)	Sep–Dec (4)	145–252	7.8	/	/	/	/	24
Ayu *Plecoglossus altivelis*	Japan (34°N)	Sep–Nov (3)	5,000–15,000	6.0	93–164	Mar–May (3)	37–63 SL	+	25–28
Japanese fluvial sculpin *Cottus pollux* ME	Japan (42°N)	–	400–1,000	–	–	–	–	+	29, 30
C. hangiongensis	Japan (42°N)	Apr–Jun (3)	400–1,700	8.0	–	–	–	+	29, 31, 32
* *Cottus pollux* LE	Japan (35°N)	Feb–May (4)	150–300	8.0–9.5	/	/	/	/	29, 33
Rhinogobius sp. CB	Japan (34°N)	May–Jul (3)	475–4,817	3.8–4.3	–	–	15–20 TL	+	34–36
Rhinogobius sp. DA	Japan (33°N)	Apr–Jul (4)	490–4,420	3.8	–	–	15–20 TL	+	34, 37, 38
Rhinogobius sp. OR	Japan (35°N)	May–Jul (3)	–	–	–	Aug–?	15 TL	+	34, 39
Rhinogobius sp. BI	Ogasawara I. (27°N)	–	–	3.5	–	–	–	–	40
* *Rhinogobius flumineus*	Japan (35°N)	May–Aug (4)	–	–	/	/	/	/	34
Eleotris oxycephala	Japan (33°N)	–	15,000<	0.94–1.11	–	Jul–Aug? (2?)	–	?	41
Dusky sleeper *E. fusca*	Okinawa (26°N)	Mar,Aug,Nov, Dec (?)	–	1.0–1.4	84–111	May–Dec (8)	15.5–19.6 SL	–	42, 43

Table 1.—Continued.

Species	Localities (latitude)	Spawning season (duration) <peak>	Clutch size	Hatching size (mm TL)	OLD (days) range (mean)	Recruitment season (duration) <peak>	Recruitment size (mm) range (mean)	River resident	Sources
Broadhead sleeper *Eleotris melanosoma*	Okinawa (26°N)	Nov, Dec (2)	–	–	81–84	–	11.2–12.8 SL	–	42, 43
E. acanthopoma	Okinawa (26°N)	all year (12)	–	1.0–1.4	62–125	May–Dec (8)	9.7–13.4 SL	–	42, 43
	Taiwan (23°N)	–	–	–	113, 129	Jun–Aug (3)	–	+	44, 45
Sandwich Island sleeper *E. sandwicensis*	Hawai'i (18°N)	–	–	1.4	–	all year (12) <Oct–Mar?>	13.5 SL	–	9, 10, 46
Spinycheek sleeper *E. pisonis*	Dominica (15°N)	–	–	–	85?–170?	–	–	–	14
	Costa Rica (11°N)	all year? (12?)	–	–	–	–	–	–	47
Large-scaled spinycheek sleeper *E. amblyopsis*	Costa Rica (11°N)	all year? (12?)	–	–	–	–	–	–	47

* fluvial nonamphidromous species

Sources: (1) this study; (2) Dotu and Mito 1955; (3) Manacop 1953; (4) Bielsa et al. 2003; (5) Keith 2003; (6) Hoareau et al. 2007; (7) Fitzsimons et al. 1993; (8) Kinzie 1993; (9) Nishimoto and Kuamo'o 1997; (10) Lindstrom 1999; (11) Yamasaki and Tachihara 2006; (12) Yamasaki et al. 2007; (13) Radtke et al. 2001; (14) Bell et al. 1995; (15) Bell and Brown 1995; (16) McDowall 1978; (17) McDowall 1988; (18) McDowall et al. 1994; (19) O'Connor and Koehn 1998; (20) Allibone and Caskey 2000; (21) Hopkins 1979; (22) Charteris et al. 2003; (23) McDowall and Kelly 1999; (24) Hopkins 1971; (25) Matsuyama and Matsuura 1982; (26) Hasegawa et al. 1983; (27) Tsukamoto 1991; (28) Takahashi 2005; (29) Goto and Iguchi 2001; (30) Goto et al. 2002; (31) Goto 1987; (32) Goto 1993; (33) Natsumeda et al. 1997; (34) Mizuno et al. 1995; (35) Tamada and Iwata 2005; (36) Sakai and Yasuda 1978; (37) Moriyama et al. 1998; (38) Takahashi and Yanagisawa 1999; (39) Maruyama et al. 2001; (40) Yokoi and Hosoya 2006; (41) Dotu and Fujita 1959; (42) Maeda and Tachihara 2005; (43) Maeda et al. 2007; (44) Shen et al. 1998; (45) Wang et al. 2001; (46) Benbow et al. 2004; (47) Winemiller and Ponwith 1998.

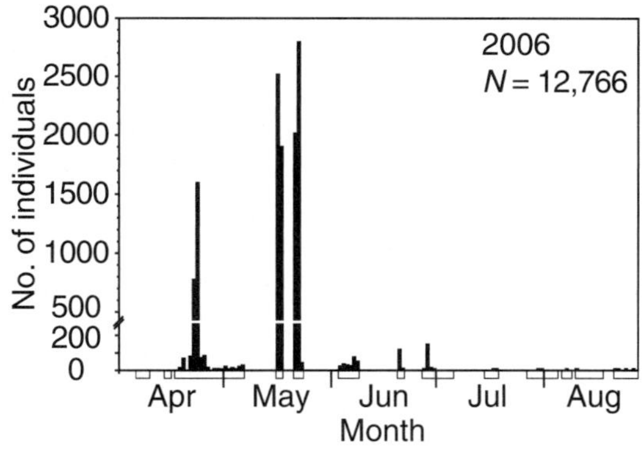

Figure 4.—Number of *Sicyopterus japonicus* larvae collected by set net in the estuary of the Ota River from April 6 to August 27, 2006 (adapted from Iida et al. 2008). Standard sized black bars above the *x*-axis indicate collections of less than 10 larvae on a particular date, and white boxes below the *x*-axis, the sampling periods.

Oceanic Larval Duration of Sicyopterus japonicus

Similar to latitudinal differences in spawning seasons, there also appear to be differences in the OLD for *S. japonicus* that successfully recruit in different parts of its range. The OLD of *S. japonicus* ranged from 173 to 253 d at Wakayama (34°N), from 182 to 215 d in Okinawa (26°N; Yamasaki 2006), and from 125 to 186 d in Taiwan (23°N; Shen et al. 1998), which suggests a latitudinal cline with the OLD increasing at higher latitudes. Alternatively, there may be differences in the oceanographic conditions affecting the recruitment patterns in each region.

The distribution range of *S. japonicus* is located along the path of the Kuroshio Current, and because of its long OLD, it is possible that this species represents a single population. Watanabe et al. (2006) showed that *S. japonicus* had no spatial heterogeneity in genetic diversity among 77 specimens from the four localities in Okinawa, Kochi, Wakayama, and Shizuoka in Japan (448 sites of the control region of the mitochondrial DNA). Furthermore, there was no difference in population structure between specimens from Amami Island in Japan and those from Taiwan using mtDNA D-loop and 16S rRNA (Ju 2001). These studies provided no evidence for different stocks of *S. japonicus*, also suggesting a single population for this species (approximate distribution range: 2,500 km).

Other species of gobies also demonstrate little population heterogeneity, but some species have definite stocks. There was no genetic structuring (mtDNA fragment and cyt *b* of the mtDNA) among discrete Hawaiian island populations of five species of amphidromous gobies (*Awaous guamensis*, *Stenogobius hawaiiensis*, *Lentipes concolor*, Stimpson's goby *Sicyopterus stimpsoni*, and Sandwich Island sleeper *Eleotris sandwicensis*, approximate species ranges: 1,500 km; Zink et al. 1996; Chubb et al. 1998). Comparative phylogeographic studies using the control region of mtDNA of the sympatric sister species, arrow goby *Clevelandia ios* (approximate species range: 3,200 km), and tidewater goby *Eucyclogobius newberryi*, along the coast of California, also indicated that there was no geographic, genetic differentiation in arrow goby; however, there was for tidewater goby (Dawson et al. 2001, 2002). Genetic studies using mitochondrial 12S and 16S rRNA sequences of *Pandaka* gobies and *Eutaeniichthys gilli*, distributed on both the main islands of Japan and the Ryukyu Islands to the south, found evidence of genetic differentiation between the main islands and the Ryukyu Islands (Mukai et al. 2003, 2004). However, there was no genetic differentiation between these two regions in *S. japonicus*.

A possible factor that would help explain these various patterns of population structure is the OLD of each particular species of goby. The OLD for *S. japonicus* ranges from 125 to 253 d (Shen et al. 1998; Yamasaki 2006; Iida et al. 2008). The OLD range for several Hawaiian gobies was 150–169 d for *A. guamensis*, 119–151 d for *S. hawaiiensis*, and 63–106 d for *L. concolor* (Radtke et al. 1988, 2001). It is possible that the OLD of these species is sufficiently long enough for them to disperse widely among the Japanese and Hawaiian Islands depending on current patterns.

Conceivably the OLD of gobies may be roughly correlated with an adaptation for long oceanic transport. Dawson et al. (2002) suggested that arrow goby has a greater dispersal ability along the coast of California than its sister taxon tidewater goby based on habitat structure and life history differences such as larval duration, which is 2–4 weeks in arrow goby and is possibly as little as a few days in tidewater goby. Species such as arrow goby, the Hawaiian amphidromous gobies, and *S. japonicus* that have a long OLD are likely to disperse over longer distances in the sea, which may lead to single population structures. Larvae of *S. japonicus* could disperse widely because their

OLD is relatively long and the powerful Kuroshio Current passes along most of its geographic range.

Recruitment of Sicyopterus japonicus

The recruitment season of *S. japonicus* in the Ota River occurred mainly during a 2–month period (April–May) when several catches of large, inward migrant schools were made. The SL at recruitment of *S. japonicus* is relatively constant, suggesting that size may be a more important factor than OLD for determining the ability of larvae to recruit. Similar observations have been made among diadromous galaxiids (Rowe and Kelley 2009, this volume).

Larvae were collected in the Ota River before metamorphosis, when they were transparent without external pigmentation. Minimal red pigmentation on parts of the dorsal fin and caudal fin base suggested that these larvae had just come into the river from the sea. The postlarvae of this species and other Sicydiinae undergo abrupt morphological changes in pigmentation, fin shape, mouth shape, jaw structure, and digestive system during upstream migration (Manacop 1953; Erdman 1986; Schoenfuss and Blanchard 1997).

Life in the River of Sicyopterus japonicus

Seasonal changes in CF for *S. japonicus* observed in the Ota River possibly reflected important annual life history events of the adults. Mean CF demonstrated two peaks, one in summer and one in fall. The increase during spring–summer appeared to correspond to preparation for spawning since females developed ripe gonads comprising 5–20% of their body weight during the spawning season (Iida 2005). The increase in CF of males from spring to summer may indicate weight gain in preparation for reduced foraging during egg guarding.

A rapid decrease of CF in summer probably resulted from female spawning and possibly reduced feeding associated with parental care by the males. The CF of male amphidromous gobies, *Rhinogobius* sp., in Japan were found to decrease during the spawning season probably because of a lack of feeding (Takahashi and Yanagisawa 1999). Somatic weight loss of males during the spawning season is widely known in fishes with male paternal care (Unger 1983; Okuda and Yanagisawa 1996). Therefore, it is implied that male *S. japonicus*, as well as other goby males, do not feed, or feed less, or have a higher energetic output not compensated for by higher feeding during the spawning season, resulting in low CF by fall.

Grazing behavior and activity was observed in spring, summer, and fall in both the rapids and the pools of the river, but *S. japonicus* disappeared in winter when water temperatures are low, and they are assumed to hide under large rocks in the slow pools (Iida 2005). It appeared that the activities of *S. japonicus* are suppressed by low water temperatures in winter, and this species shows overwintering behavior in the Ota River. A low CF value in winter probably resulted from reduction in feeding. The increase of CF in fall by active feeding was possibly energy sequestering for overwintering. The life history of *S. japonicus* in the Ota River appeared to be controlled by seasonal changes in water temperature.

Life History Characteristics of *Sicyopterus japonicus* among the Sicydiinae

Reproduction of the Sicydiinae

The spawning season of *S. japonicus* in the Ota River had one peak in July–August when water temperature was highest for the year (24°C). Water temperature probably limits the duration of the spawning season of *S. japonicus* in the temperate region. Water temperature is known affect the spawning activity of gobies in other localities (Gill et al. 1996; Mazzoldi and Rasotto 2001).

The spawning season of most Sicydiinae species in the tropics is all year (*S. lagocephalus* in the Philippines [Manacop 1953], Stimpson's goby in Hawai'i [Fitzsimons et al. 1993], and *Sicydium punctatum* in Dominica [Bell et al. 1995]). Other Sicydiinae in the tropics and subtropics apparently spawn over 7–9 months (*Sicyopterus lagocephalus* on Réunion Island [Bielsa et al. 2003], *Lentipes concolor* in Hawai'i [Kinzie 1993], and *Stiphodon percnopterygionus* in Okinawa [Yamasaki and Tachihara 2006]).

The spawning season of *Sicyopterus japonicus* is the shortest known among the Sicydiinae, and it possibly varies in duration with latitude (Table 1; Ju 2001; Yamasaki 2006). The duration of Sicydiinae spawning seasons are significantly correlated with latitude of the sampling localities (Figure 5; Spearman's rank correlation, $n = 7$, $p = 0.012$) and the spawning seasons of most gobies appear to follow this general trend of a shortened spawning season with increasing latitude (Lowe-McConnell 1979; Ebeling and Hixon 1991; Privitera 2002).

The ovarian clutch size of *S. japonicus* of about 225,000 eggs (Dotu and Mito 1955) is larger than that of other Sicydiinae, which range from 1,000 to 160,000 eggs (Table 1; Manacop 1953; Kinzie 1993; Keith 2003; Yamasaki and Tachihara 2006). Although clutch sizes among fishes are variable because of a general correlation with female body size, the larger clutch size of *S. japonicus* might be correlated with their concentrated spawning season in comparison with almost year-round reproduction of other Sicydiinae in the tropics.

The eggs of *S. japonicus* are very small (0.45–0.5 mm; this study; Dotu and Mito 1955), and other species of the Sicydiinae also have small eggs (0.34–0.58 mm; Manacop 1953; Fitzsimons et al. 1993; Kinzie 1993; Bell and Brown 1995; Yamasaki and Tachihara 2006). The hatching size of *S. japonicus* larvae is also small (1.5 mm TL) but similar to other Sicydiinae (1.1–1.8 mm TL; Table 1; Manacop 1953; Bell and Brown 1995; Lindstrom 1999; Yamasaki and Tachihara 2006).

Oceanic Larval Duration of the Sicydiinae

Oceanic larval duration of *S. japonicus* was longer than most other Sicydiinae except for *S. lagocephalus*. *Sicyopterus japonicus* had an OLD of 125–253 d, which was similar to *S. lagocephalus*, of Réunion Island, with an OLD of 133–266 d (Figure 6; Hoareau et al. 2007). Other Sicydiinae have an OLD that is about half as long (Table 1; Bell et al. 1995; Radtke et al. 2001; Hoareau et al. 2007; Yamasaki et al. 2007). The longer OLD for *S. japonicus* and *S. lagocephalus* may be correlated with their wider distribution and is possibly related to wider dispersal in ocean currents. The latter species has the widest geographic distribution of any Sicydiinae (Watson et al. 2000).

Recruitment of the Sicydiinae

The recruitment period of the Sicydiinae may be related to environmental conditions such as water temperature. The recruitment period for most

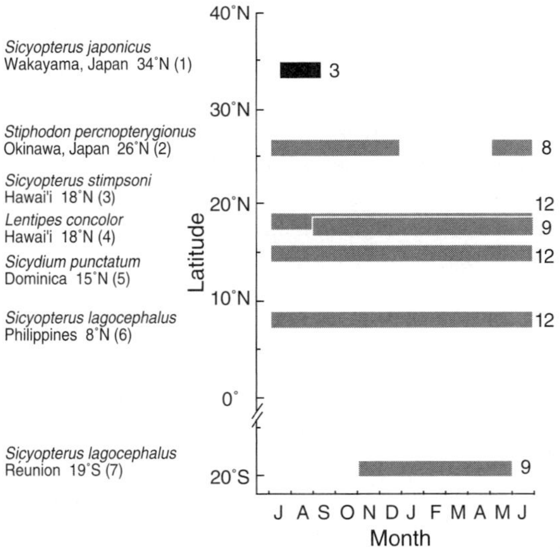

Figure 5.—Comparison of spawning season and latitude of Sicydiinae species among *Sicyopterus japonicus* in Wakayama Japan (1—this study), *Stiphodon percnopterygionus* in Okinawa Japan (2—Yamasaki and Tachihara 2006), Stimpson's goby *Sicyopterus stimpsoni* in Hawai'i (3—Fitzsimons et al. 1993), *Lentipes concolor* in Hawai'i (4—Kinzie 1993), *Sicydium punctatum* in Dominica (5—Bell et al. 1995), and *Sicyopterus lagocephalus* in the Philippines and Réunion Island (6—Manacop 1953; 7—Bielsa et al. 2003). Numbers to the right of the bars indicate the duration of the spawning season in months. Numbers in parentheses indicate the various sources of the data used.

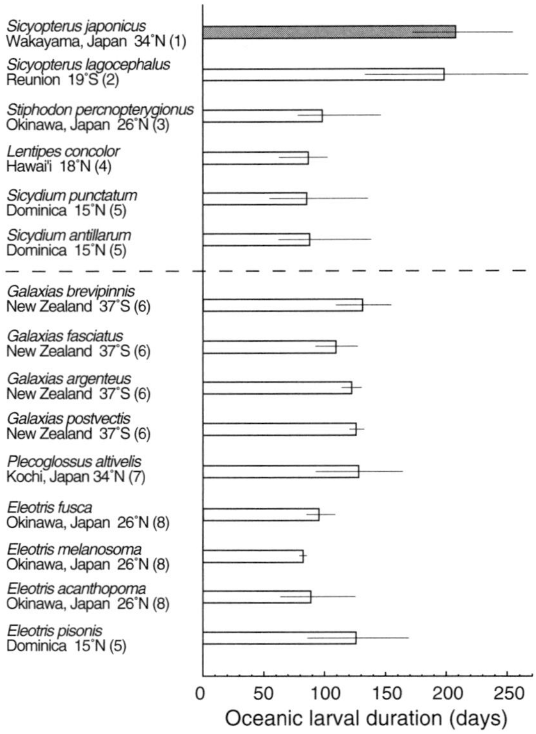

Figure 6.—Comparison of the oceanic larval durations (OLD) of Sicydiinae species, including *Sicyopterus japonicus* in Wakayama Japan (1—this study), *S. lagocephalus* on Réunion Island (2—Hoareau et al. 2007), *Stiphodon percnopterygionus* in Okinawa Japan (3—Yamasaki et al. 2007), *Lentipes concolor* in Hawai'i (4—Radtke et al. 2001) and *Sicydium punctatum* and *S. antillarum* in Dominica (5—Bell et al. 1995), and other amphidromous fishes, including kaoro *Galaxias brevippinis*, banded kokopu *G. fasciatus*, giant kokopu *G. argenteus*, and shortjaw kokopu *G. postvectis* in New Zealand (6—McDowall et al. 1994), ayu *Plecoglossus altivelis* in Kochi Japan (7—Takahashi 2005), dusky sleeper *Eleotris fusca*, broadhead sleeper *E. melanosoma*, *E. acanthopoma* in Okinawa Japan (8—Maeda et al. 2007), and spinycheek sleeper *E. pisonis* in Dominica (5—Bell et al. 1995). Horizontal bars indicate means of OLD when available and lines indicate the range. When no mean value was provided in the references, the middle values of the range are shown instead of means. Numbers in parentheses indicate the various sources of the data. The dashed line separates the Sicydiinae from other amphidromous fish species.

tropical and subtropical Sicydiinae appears to be all year for several species such as Stimpson's goby in Hawai'i (Nishimoto and Kuamo'o 1997), *S. lagocephalus* in the Philippines (Manacop 1953), *Lentipes concolor* in Hawai'i (Radtke et al. 2001), and *Sicydium punctatum* and *S. antillarum* in Dominica (Bell et al. 1995; Table 1). The 4-month recruitment period of *Sicyopterus japonicus* is the shortest known among the Sicydiinae. Since temperate regions have greater seasonal changes of water temperature, the window for recruitment is probably short. Seawater temperatures in Hawai'i ranged from 22°C to 27°C all year (NODC 2007), whereas water temperature in the Ota River ranged from 9°C to 24°C, and low water temperature in winter probably suppressed recruitment of *S. japonicus*. Similar restriction of recruitment to spring was reported for the amphidromous osmerid fish, the ayu *Plecoglossus altivelis*, in Japan (Tsukamoto 1988).

Body Length at Recruitment of the Sicydiinae

The body length at recruitment of *S. japonicus* (26.3 mm SL) was similar to *S. lagocephalus* but larger than other tropical Sicydiinae, most of which are

12–21 mm SL at recruitment (Table 1). The slightly larger body length at recruitment of *S. japonicus* and *S. lagocephalus* may be correlated with their longer OLD (Figure 6; Table 1) and further studies are needed to reveal any correlation between OLD and body length at recruitment.

Life History and Migration of Sicyopterus japonicus among the Sicydiinae

There are numerous differences in the life history characteristics between *S. japonicus* inhabiting the temperate zone and other Sicydiinae inhabiting tropical and subtropical regions. These differences are probably based on the differences in the environmental factors characterizing their habitats at high and low latitudes. Except for *S. japonicus*, the other Sicydiinae all inhabit tropical and subtropical regions where there are no major seasonal changes in environmental conditions. This would result in longer spawning and recruitment seasons among tropical species. The same phenomenon occurs in anguillid eels, where the tropical species apparently spawn and recruit throughout much of the year while temperate species spawn and recruit during a more limited period (Arai et al. 2001). Water temperature variation is well known to control life history characteristics such as spawning, recruitment, and activity, and every life history event in temperate regions must be performed within a narrower time window compared with species inhabiting tropical regions.

Amphidromy in the Sicydiinae and other Temperate Fishes

Clutch and Hatching Size

One obvious way that the Sicydiinae differ from other temperate amphidromous fishes is their clutch sizes. The clutch sizes of the Sicydiinae, which typically range from 1,000 (Yamasaki and Tachihara 2006) to 225,000 eggs (Dotu and Mito 1955), are larger than most temperate amphidromous fishes that usually range from 400 to 15,000 eggs (Table 1; Dotu and Fujita 1959; Matsuyama and Matsuura 1982; Goto and Iguchi 2001; Tamada and Iwata 2005).

Another characteristic of the Sicydiinae that appears to differ from other amphidromous fishes is hatchling size. Hatching larvae of the Sicydiinae range from 1.1 (Manacop 1953) to 1.8 mm TL (Bell and Brown 1995; Lindstrom 1999) and are smaller than most other temperate amphidromous fishes. Other gobies hatch at a body size of 3.5–3.8 mm TL (Moriyama et al. 1998; Yokoi and Hosoya 2006), galaxiids at 8.6–9.0 mm TL (O'Connor and Koehn 1998; Charteris et al. 2003), and osmerids and cottids at 6.0–8.0 mm TL (Tsukamoto 1991; Goto and Iguchi 2001). Conversely, hatchling sleepers of the genus *Eleotris* are similar in size to the Sicydiinae (0.9–1.4 mm TL; Dotu and Fujita 1959; Maeda et al. 2007).

These comparisons indicate that clutch sizes of the Sicydiinae are larger and hatching size smaller than most other amphidromous fishes. Since larvae hatched from large eggs are larger in general (Helfman et al. 1997), the small hatchling size of Sicydiinae larvae are possibly a direct result of large clutch sizes and small egg size.

Oceanic Larval Duration

Although there are some reports of collections of Sicydiinae larvae in the ocean (Dotu and Mito 1955), their oceanic ecology is unknown except for what has been learned from studies of the otolith microstructure of the recruited larvae. Although the OLD of *S. japonicus* and *S. lagocephalus* are nearly twice as long as other Sicydiinae, these latter species have an OLD very similar to other amphidromous fishes, which have oceanic larval durations of 62–155 d (Table 1; McDowall et al. 1994; Shen et al. 1998; Takahashi 2005; Maeda et al. 2007).

In addition to studies on OLD, some indirect information about the potential for oceanic dispersal can be gained from studies on the genetic population structure of amphidromous families. Although the species range of *S. japonicus* is wider than that of the ayu (Akihito et al. 2000), mtDNA studies indicate that the ayu also has little genetic structuring (Iwata et al. 2006). Evidence of marine dispersal of kaoro *Galaxias brevipinnis* between Tasmanian and New Zealand was reported using a mitochondrial control region (Waters and Wallis 2001), and Berrebi et al. (2005) using allozyme electrophoresis suggested that *S. lagocephalus* has a single population in the Indo-Pacific region, although there are other opinions about the population structure of that widespread species (Keith et al. 2005; Berrebi et al. 2006; Sparks and Smith 2006). Even a low level of population structure over such a large region might suggest that there is a tendency for wide dispersal of *S. lagocephalus* larvae. Hoareau et al. (2007) postulated that OLD is one of the crucial factors influencing the dispersal process and the distribution range of *S. lagocephalus*.

These data suggest that some Sicydiinae gobies such as *S. japonicus* and *S. lagocephalus* have large-scale dispersal compared to most other amphidromous fishes. The amphidromous sleepers of the genus *Eleotris* may have a similar tendency (Maeda et al. 2007), but more research is needed to evaluate this hypothesis. Although a long OLD might be one factor influencing species with wide geographic distribution ranges, it does not always determine the species range. For example, temperate amphidromous Galaxiidae with a relatively short OLD in comparison to *S. japonicus* and *S. lagocephalus* also have wide distributions (McDowall 1988). Currents systems and growth of larvae may also influence the duration of oceanic life (Rowe and Kelley 2009) and hence affect dispersal.

River Residents

There are no reports of nondiadromous forms among the Sicydiinae, although other temperate amphidromous families frequently have river resident forms (Table 1). Species with river resident forms include galaxiids in New Zealand (McDowall 1988), ayu in Japan (Hasegawa et al. 1983) and the Japanese fluvial sculpin in Japan (Goto et al. 2002). Strontium and calcium concentrations in otoliths indicate that some gallaxiids exhibit nondiadromous recruitment despite open access to an estuary and living only 3 km from the sea, (A. Hicks, University of Otago, personal communication). Genetic studies using mitochondrial DNA 12SrRNA and control region sequences of amphidromous-fluvial species pairs of the genus *Cottus* indicated that pairing occurred independently in different lineages (Yokoyama and Goto 2005), which suggests that fluvial forms evolved from amphidromous species with an oceanic larval phase, to fluvial species without an oceanic phase.

The family Galaxiidae is widely distributed in southern circum-temperate regions with about 45 species (McDowall 1969, 2003b), including 7 diadromous species and many nondiadromous (freshwater) species. Phylogenetic relationships are not described rigorously for the Galaxiidae as a whole, but there are hypotheses of relationships in some species groups based on cytotaxonomic and genetic studies. Diadromy is probably an ancient trait among galaxiid fishes (McDowall 1988, 1993) and speciation often takes place after formerly diadromous stocks become restricted to freshwater (Johnson and White 1981; Johnson et al. 1981, 1983; McDowall and Frankenberg 1981; Ovenden and White 1990; Ovenden et al. 1993; Allibone et al. 1996; Waters and White 1997; Waters and Burridge 1999; Waters and Wallis 2001; McDowall 2003b).

On the other hand, larvae of *Sicydium punctatum* clearly showed a physiological obligation for marine occupation in laboratory experiments (Bell and Brown 1995). Therefore, it appears that the Sicydiinae have a greater ocean dependency in comparison with other temperate amphidromous families such as Osmeridae, Galaxiidae, and Cottidae.

Migration Strategy of the Sicydiinae

We hypothesize that *Sicyopterus japonicus* has a migration strategy involving high oceanic dependency with a great dispersal tendency, which might be unique in comparison with other temperate amphidromous fishes. Although most Sicydiinae have a similar OLD in comparison to other amphidromous fishes, all Sicydiinae have larger clutch sizes and a smaller hatching size, and they do not have any river resident forms. Large clutch size could obviously compensate for large-scale dispersive loss, or failure to find recruitment habitats.

The Sicydiinae have high oceanic dependency and long OLD, which promotes a wide geographic distribution in oceanic islands where there are variable environments. Variable environments increase the chance of population extinction and would lead to a greater need for wide dispersal. We suggest that this type of life history might be defined as "dispersal amphidromy" in comparison with other temperate amphidromous fishes that could be defined as having "retention amphidromy." Research also suggests that the same characteristics are found in the genus *Eleotris* (Maeda et al.

2007), but further studies are needed to clarify this type of amphidromy. Further consideration of the evolutionary basis for different life history traits in amphidromous fishes using phylogenetic analysis is needed to better understand these interesting diadromous fishes.

Acknowledgments

We sincerely thank S. Fukui, M. Itou, and S. Takino for their kind help during sampling. We are grateful to S. Harada, T. Gosho, T. Yamada, and N. Horiki for special permission to sample in the river. We especially thank N. Yamasaki and K. Maeda for their useful suggestions. We are grateful to A. Hicks of University of Otago for permission to refer to his unpublished data. This paper benefited from suggestions by M. J. Miller, and we thank J. Aoyama and M. Oya for assistance with many aspects of the study. M. Iida was partly supported by the Sasakawa Scientific Research Grant from The Japan Science Society. This study was partially supported by a Research Fellowship of the Japan Society for the Promotion of Science for Young Scientists (No. 18780143) and Grants-in-Aid from the Ministry of Education, Culture, Sports, Science, and Technology, Japan (No. 19208019).

References

Akihito, K. Sakamoto, Y. Ikeda, and A. Iwata. 2000. Gobioidei. Pages 1139–1310 *in* T. Nakabo, editor. Fishes of Japan with pictorial keys to the species, 2nd edition. Tokai University Press, Tokyo.

Allibone, R. M., and D. Caskey. 2000. Timing and habitat of koaro (*Galaxias brevipinnis*) spawning in streams draining Mt Taranaki, New Zealand. New Zealand Journal of Marine and Freshwater Research 34:593–595.

Allibone, R. M., T. A. Crowl, J. M. Holmes, T. M. King, R. M. McDowall, C. R. Townsend, and G. P. Wallis. 1996. Isozyme analysis of *Galaxias* species (Teleostei: Galaxiidae) from the Taieri River, South Island, New Zealand: a species complex revealed. Biological Journal of the Linnaean Society 57:107–127.

Arai, T., D. Limbong, T. Otake, and K. Tsukamoto. 2001. Recruitment mechanisms of tropical eels *Anguilla* spp. and implications for the evolution of oceanic migration in the genus *Anguilla*. Marine Ecology Progress Series 216:253–264.

Bell, K. N. I. 1999. An overview of goby-fry fisheries. Naga 22:30–36.

Bell, K. N. I., and J. A. Brown. 1995. Active salinity choice and enhanced swimming endurance in 0-d-old to 8-d-old larvae of diadromous gobies, including *Sicydium punctatum* (Pisces), in Dominica, West Indies. Marine Biology 121:409–417.

Bell, K. N. I., P. Pepin, and J. A. Brown. 1995. Seasonal, inverse cycling of length and age-at-recruitment in the diadromous gobies *Sicydium punctatum* and *Sicydium antillarum* in Dominica, West Indies. Canadian Journal of Fisheries and Aquatic Sciences 52:1535–1545.

Benbow, M. E., A. J. Burky, and C. M. Way. 2004. Morphological characteristics and species separation of Hawaiian postlarval amphidromous fishes. Micronesica 37:129–144.

Berrebi, P., G. Cattaneo-Berrebi, P. Valade, J. F. Ricou, and T. Hoareau. 2005. Genetic homogeneity in eight freshwater populations of *Sicyopterus lagocephalus*, an amphidromous gobiid of La Réunion Island. Marine Biology 148:179–188.

Berrebi, P., T. Galewski, and P. Keith. 2006. *Sicyopterus* "*lagocephalus*", a unique widespread taxon confirmed by mtDNA sequences. Molecular Phylogenetics and Evolution 40:903–904.

Bielsa, S., P. Francisco, S. Mastrorillo, and J. P. Parent. 2003. Seasonal changes of periphytic nutritive quality for *Sicyopterus lagocephalus* (Pallas, 1770) (gobiidae) in three streams of Reunion Island. Annales De Limnologie-International Journal of Limnology 39:115–127.

Charteris, S. C., R. M. Allibone, and R. G. Death. 2003. Spawning site selection, egg development, and larval drift of *Galaxias postvectis* and *G. fasciatus* in a New Zealand stream. New Zealand Journal of Marine and Freshwater Research 37:493–505.

Chubb, A. L., R. M. Zink, and J. M. Fitzsimons. 1998. Patterns of mtDNA variation in Hawaiian freshwater fishes: the phylogeographic consequences of amphidromy. Journal of Heredity 89:8–16.

Dawson, M. N., J. L. Staton, and D. K. Jacobs. 2001. Phylogeography of the tidewater goby, *Eucyclogobius newberryi* (Teleostei, Gobiidae) in coastal California. Evolution 55:1167–1179.

Dawson, M. N., K. D. Louie, M. Barlow, D. K. Jacobs, and C. C. Swift. 2002. Comparative phylogeography of sympatric sister species, *Clevelandia ios* and *Eucyclogobius newberryi* (Teleostei, Gobiidae), across the California Transition Zone. Molecular Ecology 11:1065–1075.

Dotu, Y., and S. Mito. 1955. Life history of the gobioid fish, *Sicydium japonicum* Tanaka. Scientific Bulletin of the Faculty of Agriculture Kyushu University 10:120–126.

Dotu, Y., and S. Fujita. 1959. The bionomics and life history of the eleotrid fish, *Eleotris oxycephala* Temminck et Schlegel. Bulletin of the Faculty of Fisheries Nagasaki University 8:191–195.

Ebeling, A. W., and M. A. Hixon. 1991. Tropical and temperate reef fishes: comparison of community structures. Pages 509–563 *in* P. F. Sale, editor. The ecology of fishes on coral reefs. Academic Press, San Diego, California.

Erdman, D. S. 1986. The green stream goby, *Sicydium plumieri*, in Puerto Rico. Tropical Fish Hobbyist 2:70–74.

Fièvet, E., and R. Eppe. 2002. Genetic differentiation among populations of the amphidromous shrimp *Atya innocuous* (Hebst) and obstacles to their upstream migration. Archiv für Hydrobiologie 153:287–300.

Fitzsimons, J. M., R. T. Nishimoto, and A. R. Yuen. 1993. Courtship and territorial behavior in the native Hawaiian stream goby, *Sicyopterus stimpsoni*. Ichthyological Exploration of Freshwaters 4:1–10.

Froese, R., and D. Pauly. 2007. FishBase. World Wide Web electronic publication. Version (Apr 2007). Available: www.fishbase.org/search (Jun 2007).

Gibson, R. N. 1970. Observations on the biology of the giant goby *Gobius cobitis* Pallas. Journal of Fish Biology 2:281–288.

Gill, H. S., B. S. Wise, I. C. Potter, and J. A. Chaplin. 1996. Biannual spawning periods and resultant divergent patterns of growth in the estuarine goby *Pseudogobius olorum*: temperature-induced? Marine Biology 125:453–466.

Goto, A. 1987. Polygyny in the river sculpin, *Cottus hangiongensis* (Pisces, Cottidae), with special reference to male mating success. Copeia 1987:32–40.

Goto, A. 1993. Clinical life-history variation in the river-sculpin, *Cottus hangiongensis*: an example of phenotypic plasticity. Japanese Journal of Ichthyology 39:363–370.

Goto, A. and K. Iguchi. 2001. Evolutionary biology of egg size in aquatic animals. Kaiyusha, Tokyo.

Goto, A., R. Yokoyama, and M. Yamada. 2002. A fluvial population of *Cottus pollux* (middle-egg type) from the Honmyo River, Kyushu Island, Japan. Ichthyological Research 49:318–323.

Hamano, T., and K. I. Hayashi. 1992. Ecology of an atyid shrimp *Caridina japonica* (De Man, 1892) migrating to upstream habitats in the Shiwagi Rivulet, Tokushima Prefecture. Researches on Crustacea 21:1–12.

Harrison, I. J. 1993. The West African sicydiine fishes, with notes on the genus *Lentipes* (Teleostei: Gobiidae). Ichthyological Exploration of Freshwaters 4:201–232.

Hasegawa, S., T. Hirano, N. Kuniya, Y. Abe, and K. Suzuki. 1983. Seawater adaptability of anadromous and landlocked forms of the ayu *Plecoglossus altivelis*. Bulletin of the Japanese Society of Scientific Fisheries 49:569–575.

Helfman, G. S., B. B. Collette, and D. E. Facey. 1997. The diversity of fishes. Blackwell Scientific Publications, Malden, Massachusetts.

Hoareau, T. B., R. Lecomte-Finiger, H.-P. Grondin, C. Conand, and P. Berrebi. 2007. Oceanic larval life of La Réunion 'bichiques', amphidromous gobiid post-larvae. Marine Ecology Progress Series 333:303–308.

Hodges, M. H., and F. W. Allendorf. 1998. Population genetics and pattern of larval dispersal of the endemic Hawaiian freshwater amphidromous gastropod *Neritina granosa* (Prosobranchia: Neritidae). Pacific Science 52:237–249.

Hopkins, C. L. 1971. Life history of *Galaxias divergens* (Salmonoidea: Galaxiidae). New Zealand Journal of Marine and Freshwater Research 5:41–57.

Hopkins, C. L. 1979. Reproduction in *Galaxias fasciatus* Gray (Salmoniformes, Galaxiidae). New Zealand Journal of Marine and Freshwater Research 13:225–230.

Huggett, J., P. Freon, C. Mullon, and P. Penven. 2003. Modelling the transport success of anchovy *Engraulis encrasicolus* eggs and larvae in the southern Benguela: the effect of spatio-temporal spawning patterns. Marine Ecology Progress Series 250:247–262.

Iida, M. 2005. Life history of *Sicyopterus japonicus* in the Ota River, Wakayama, Japan. Master's thesis. The University of Tokyo, Tokyo.

Iida, M., S. Watanabe, A. Shinoda, and K. Tsukamoto. 2008. Recruitment of the amphidromous goby *Sicyopterus japonicus* to the estuary of the Ota River, Wakayama, Japan. Environmental Biology of Fishes 83:331–341

Iwata, H., K. Watanabe, H. Takeshima, K. Iguchi, and M. Nishida. 2006. Isolation and characterization of 49 polymorphic microsatellite loci in the ayu, *Plecoglossus altivelis*. Molecular Ecology Notes 6:1076–1079.

Johnson, C. R., and R. W. G. White. 1981. The variation in nonspecific esterases in the white muscles of 2 species of *Galaxias* in Tasmania. Biochemical Systematics and Ecology 9:221–223.

Johnson, C. R., R. W. G. White, and Y. A. E. Bick. 1981. Cyto-taxonomy of 7 species of *Glaxias* (Pisces, Galaxiidae) in Tasmania. Genetica 56:17–21.

Johnson, C. R., D. A. Ratkowsky, and R. W. G. White. 1983. Multivariate analysis of the phenotypic relationships of the species of *Paragalaxias* and *Galaxias* (Pisces, Galaxiidae) in Tasmania. Journal of Fish Biology 23:49–63.

Ju, Y. M. 2001. The studies of molecular evolution of genera *Sicyopterus* and reproduction biology of *Sicy-*

opterus japonicus in Taiwan. Master's thesis. National Sun Yat-Sen University, Kaohsiung, Taiwan.

Kadmon, G., Z. Yaron, and H. Gordin. 1985. Sequence of gonadal events and oestradiol levels in *Sparus aurata* (L.) under two photoperiod regimes. Journal of Fish Biology 26:609–620.

Keith, P. 2003. Biology and ecology of amphidromous Gobiidae of the Indo-Pacific and the Caribbean regions. Journal of Fish Biology 63:831–847.

Keith, P., T. Galewski, G. Cattaneo-Berrebi, T. Hoareau, and P. Berrebi. 2005. Ubiquity of *Sicyopterus lagocephalus* (Teleostei : Gobioidei) and phylogeography of the genus *Sicyopterus* in the Indo-Pacific area inferred from mitochondrial cytochrome *b* gene. Molecular Phylogenetics and Evolution 37:721–732.

Kinzie, R. A. 1993. Reproductive biology of an endemic, amphidromous goby *Lentipes concolor* in Hawaiian streams. Environmental Biology of Fishes 37:257–268.

Kokita, T. 2004. Latitudinal compensation in female reproductive rate of a geographically widespread reef fish. Environmental Biology of Fishes 71:213–224.

Lee, C. L., and D. R. Fielder. 1979. A mass migration of the freshwater prawn, *Macrobrachium australiense* Holthuis, 1950 (Decapoda: Palamonidae). Crustaceana 37:219–222.

Lindstrom, D. P. 1999. Molecular species identification of newly hatched Hawaiian amphidromous gobioid larvae. Marine Biotechnology 1:167–174.

Lowe-McConnell, R. H. 1979. Ecological aspects of seasonality in fishes of tropical waters. Symposia of the Zoological Society of London.

Maeda, K., and K. Tachihara. 2005. Recruitment of amphidromous sleepers *Eleotris melanosoma*, and *Eleotris fusca* into the Teima River, Okinawa Island. Ichthyological Research 52:325–335.

Maeda, K., N. Yamasaki, and K. Tachihara. 2007. Size and age at recruitment and spawning season of sleeper, genus *Eleotris* (Teleostei: Eleotriidae) on Okinawa Island, southern Japan. The Raffles Bulletin of Zoology Supplement 14:199–207.

Manacop, P. R. 1953. The life history and habits of the goby, *Sicyopterus extraneus* Herre (Anga) gobiidae with an account of the goby-fry fishery of Cagayan River, Oriental Misamis. The Philippine Journal of Fisheries 2:1–58.

March, J. G., C. M. Pringle, M. J. Townsend, and A. I. Wilson. 2002. Effects of freshwater shrimp assemblages on benthic communities along an altitudinal gradient of a tropical island stream. Freshwater Biology 47:377–390.

Marconato, A., A. Bisazza, and G. Marin. 1989. Correlates of male reproductive success in *Padogobius martensi* (Gobiidae). Journal of Fish Biology 34:889–899.

Maruyama, A., Y. Yamada, M. Yuma, and B. Rusuwa. 2001. Stable nitrogen and carbon isotope ratios as migration tracers of a landlocked goby, *Rhinogobius* sp. (the orange form), in the Lake Biwa water system. Ecological Research 16:697–703.

Matsuyama, M., and S. Matsuura. 1982. Ovarian maturation and ovulation of the amphidromous type ayu *Plecoglossus altivelis* in Chikugo River based on histological observations. Bulletin of the Japanese Society of Scientific Fisheries 48:1573–1580.

Mazzoldi, C., and M. B. Rasotto. 2001. Extended breeding season in the marbled goby, *Pomatoschistus marmoratus* (Teleostei: Gobiidae), in the Venetian Lagoon. Environmental Biology of Fishes 61:175–183.

McDowall, R. M. 1969. Relationships of galaxioid fishes with a further discussion of salmoniform classification. Copeia 1969:796–824.

McDowall, R. M. 1978. New Zealand freshwater fishes. Heinemann Educational Books, Auckland, New Zealand.

McDowall, R. M. 1988. Diadromy in fishes: migrations between freshwater and marine environments. Croom Helm, London.

McDowall, R. M. 1993. A recent marine ancestry for diadromous fishes: sometimes yes, but mostly no. Environmental Biology of Fishes 37:329–335.

McDowall, R. M. 1997. Is there such a thing as amphidromy? Micronesica 30:3–14.

McDowall, R. M. 2003a. Hawaiian biogeography and the islands' freshwater fish fauna. Journal of Biogeography 30:703–710.

McDowall, R. M. 2003b. Variation in vertebral number in galaxiid fishes (Teleostei: Galaxiidae): a legacy of life history, latitude and length. Environmental Biology of Fishes 66:361–381.

McDowall, R. M. 2004. Ancestry and amphidromy in island freshwater fish faunas. Fish and Fisheries 5:75–85.

McDowall, R. M. 2007. On amphidromy, a distinct form of diadromy in aquatic organisms. Fish and Fisheries 8:1–13.

McDowall, R. M., and R. S. Frankenberg. 1981. The galaxiid fishes of Australia. Records of the Australian Museum 33:443–605.

McDowall, R. M., and G. R. Kelly. 1999. Date and age at migration in juvenile giant kokopu, *Galaxias argenteus* (Gmelin) (Teleostei: Galaxiidae) and estimation of spawning season. New Zealand Journal of Marine and Freshwater Research 33:263–270.

McDowall, R. M., C. P. Mitchell, and E. B. Brothers. 1994. Age at migration from the sea of juvenile *Galaxias* in New Zealand (Pisces, Galaxiidae). Bulletin of Marine Science 54:385–402.

Mizuno, N., K. Tsuji, T. Suzuki, T. Koshikawa, and A. Iwata. 1995. Genus *Rhinogobius*. Pages 584–603 *in* H. Kawanabe and N. Mizuno, editors. Freshwater fishes of Japan, 2nd edition. Yamatokeikoku-sha, Tokyo.

Moriyama, A., Y. Yanagisawa, N. Mizuno, and K. Omori. 1998. Starvation of drifting goby larvae due to retention of free embryos in upstream reaches. Environmental Biology of Fishes 52:321–329.

Mukai, T., T. Suzuki, and M. Nishida. 2003. Genetic differentiation of the brackish water goby, *Eutaeniichthys gilli* (Perciformes, Gobiidae), between the Japanese and the Ryukyu archipelagos. Biogeography 5:49–53.

Mukai, T., T. Suzuki, and M. Nishida. 2004. Genetic and geographical differentiation of *Pandaka* gobies in Japan. Ichthyological Research 51:222–227.

Myers, G. S. 1949. Usage of anadromous, catadromous and allied terms for migratory fishes. Copeia 1949:89–97.

Natsumeda, T., S. Kimura, and Y. Nagata. 1997. Sexual size dimorphism, growth and maturity of the Japanese fluvial sculpin, *Cottus pollux* (large egg type), in the Inabe River, Mie Prefecture, central Japan. Ichthyological Research 44:43–50.

Nishimoto, R. T., and D. G. K. Kuamo'o. 1997. Recruitment of goby postlarvae into Hakalau Stream, Hawai'i Island. Micronesica 30:41–49.

NODC (The National Oceanographic Data Center). 2007. Coastal water temperature guide. U.S. Department of Commerce, National Oceanic and Atmospheric Administration. Available: www.nodc.noaa.gov. (June 2007).

O'Connor, W. G., and J. D. Koehn. 1998. Spawning of the broad-finned Galaxias, *Galaxias brevipinnis* Günther (Pisces: Galaxiidae) in coastal streams of southeastern Australia. Ecology of Freshwater Fish 7:95–100.

Okuda, N., and Y. Yanagisawa. 1996. Filial cannibalism in a paternal mouthbrooding fish in relation to mate availability. Animal Behaviour 52:307–314.

Ovenden, J. R., and R. W. G. White. 1990. Mitochondrial and allozyme genetics of incipient speciation in a landlocked population of *Galaxias truttaceus* (Pisces, Galaxiidae). Genetics 124:701–716.

Ovenden, J. R., R. W. G. White, and M. Adams. 1993. Mitochondrial and allozyme genetics of two Tasmanian galaxiids (*Galaxias auratus* and *G. tanycephalus*, Pisces, Galaxiidae) with restricted lacustrine distributions. Heredity 70:223–230.

Privitera, L. A. 2002. Reproductive biology of the coral-reef goby, *Asterropteryx semipunctata*, in Kaneohe Bay, Hawaii. Environmental Biology of Fishes 65:289–310.

Radtke, R. L., R. A. Kinzie, and S. D. Folsom. 1988. Age at recruitment of Hawaiian freshwater gobies. Environmental Biology of Fishes 23:205–213.

Radtke, R. L., R. A. Kinzie, and D. J. Shafer. 2001. Temporal and spatial variation in length of larval life and size at settlement of the Hawaiian amphidromous goby *Lentipes concolor*. Journal of Fish Biology 59:928–938.

Rowe, D. K., and G. Kelley. 2009. Duration of the oceanic phase for inanga whitebait (Galaxiidae) is inversely related to growth rate at sea. Pages 343–354 *in* A. J. Haro, K. L. Smith, R. A. Rulifson, C. M. Moffitt, R. J. Klauda, M. J. Dadswell, R. A. Cunjak, J. E. Cooper, K. L. Beal, and T. S. Avery, editors. Challenges for diadromous fishes in a dynamic global environment. American Fisheries Society, Symposium 69, Bethesda, Maryland.

Sakai, H., and F. Yasuda. 1978. Development of eggs and larvae of freshwater goby, *Rhinogobius brunneus*. Japanese Journal of Ichthyology 25:92–100.

Schoenfuss, H. L., and T. A. Blanchard. 1997. Metamorphosis in the cranium of postlarval *Sicyopterus stimpsoni*, an endemic Hawaiian stream goby. Micronesica 30:93–104.

Shen, K. N., and W. N. Tzeng. 2002. Formation of a metamorphosis check in otoliths of the amphidromous goby *Sicyopterus japonicus*. Marine Ecology Progress Series 228:205–211.

Shen, K. N., Y. C. Lee, and W. N. Tzeng. 1998. Use of otolith microchemistry to investigate the life history pattern of gobies in a Taiwanese stream. Zoological Studies 37:322–329.

Sparks, J. S. and W. L. Smith. 2006. *Sicyopterus lagocephalus*: widespread species, species complex, or neither? A critique on the use of molecular data for species identification. Molecular Phylogenetics and Evolution 40:900–902.

Starmuhlner, F. 1979. Distribution of freshwater molluscs in mountain streams of tropical Indo-Pacific islands. Malacologia 18:245–255.

Takahashi, D., and Y. Yanagisawa. 1999. Breeding ecology of an amphidromous goby of the genus *Rhinogobius*. Ichthyological Research 46:185–191.

Takahashi, I. 2005. Study on the early life history of ayu in the Shimanto estuary, Japan. Bulletin of Marine Science and Fisheries Kochi University 23:113–173.

Takegaki, T. 2000. Monogamous mating system and spawning cycle in the gobiid fish, *Amblygobius phalaena* (Gobiidae). Environmental Biology of Fishes 59:61–67.

Tamada, K., and K. Iwata. 2005. Intra-specific variations of egg size, clutch size and larval survival related to maternal size in amphidromous *Rhinogobius* goby. Environmental Biology of Fishes 73:379–389.

Tominaga, O., M. Inoue, M. Kamata, and T. Seikai. 2005. Reproductive cycle of yellow sea bream *Dentex tumifrons* in Wakasa Bay, the Sea of Japan off central Honshu. Fisheries Science 71:1069–1077.

Tsukamoto, K. 1988. Migratory mechanisms and behavioral characteristics in ayu. Pages 100–133 *in* T. Ueno and M. Okiyama, editors. Ichthyology currents. Asakura-Shobo, Tokyo.

Tsukamoto, K. 1991. Age and growth of ayu larvae *Plecoglossus altivelis* collected in the Nagara, Kiso and Tone River during the downstream migration. Nippon Suisan Gakkaishi 57:2013–2022.

Unger, L. M. 1983. Nest defense by deceit in the fathead minnow, *Pimephales promelas*. Behavioral Ecology and Sociobiology 13:125–130.

Wang, H. Y., C. F. Weng, M. C. Tu, and S. C. Lee. 2001. Synchronization of plasma sexual steroid concentrations and gonadal cycles in the sleeper, *Eleotris acanthopoma*. Zoological Studies 40:14–20.

Watanabe, S., M. Iida, Y. Kimura, E. Feunteun, and K. Tsukamoto. 2006. Genetic diversity of *Sicyopterus japonicus* as revealed by mitochondrial DNA sequencing. Coastal Marine Science 30:473–479.

Watanabe, Y., Y. Kurita, M. Noto, Y. Oozeki, and D. Kitagawa. 2003. Growth and survival of Pacific saury *Cololabis saira* in the Kuroshio-Oyashio transitional waters. Journal of Oceanography 59:403–414.

Waters, J. M., and R. W. G. White. 1997. Molecular phylogeny and biogeography of the Tasmanian and New Zealand mudfishes (Salmoniformes: Galaxiidae). Australian Journal of Zoology 45:39–48.

Waters, J. M., and C. P. Burridge. 1999. Extreme intraspecific mitochondrial DNA sequence divergence in *Galaxias maculatus* (Osteichthys : Galaxiidae). One of the world's most widespread freshwater fish. Molecular Phylogenetics and Evolution 11:1–12.

Waters, J. M., and G. P. Wallis. 2001. Cladogenesis and loss of the marine life-history phase in freshwater galaxiid fishes (Osmeriformes: Galaxiidae). Evolution 55:587–597.

Watson, R. E., G. Marquet, and C. Pollabauer. 2000. New Caledonia fish species of the genus *Sicyopterus* (Teleostei: Gobioidei: Sicydiinae). Aqua 4:5–34.

Winemiller, K. O., and B. J. Ponwith. 1998. Comparative ecology of eleotrid fishes in Central American coastal streams. Environmental Biology of Fishes 53:373–384.

Yamasaki, N. 2006. Life history of sicydiine gobies, *Stenogobius* sp., and *Awaous melanocephalus* on Okinawa Island. Master's thesis. University of the Ryukyus, Okinawa, Japan.

Yamasaki, N., and K. Tachihara. 2006. Reproductive biology and morphology of eggs and larvae of *Stiphodon percnopterygionus* (Gobiidae: Sicydiinae) collected from Okinawa Island. Ichthyological Research 53:13–18.

Yamasaki, N., K. Maeda, and K. Tachihara. 2007. Pelagic larval duration and morphology at recruitment of *Stiphodon percnopterygionus* (Gobiidae: Sicydiinae). The Raffles Bulletin of Zoology Supplement 14:209–214.

Yokoi, K. I., and K. Hosoya. 2006. Early development of the endangered freshwater goby, *Rhinogobius* sp. BI (Gobiidae). Ichthyological Research 53:160–165.

Yokoyama, R., and A. Goto. 2005. Evolutionary history of freshwater sculpins, genus *Cottus* (Teleostei: Cottidae) and related taxa, as inferred from mitochondrial DNA phylogeny. Molecular Phylogenetics and Evolution 36:654–668.

Zink, R. M., J. M. Fitzsimons, D. L. Dittmann, D. R. Reynods, and R. T. Nishimoto. 1996. Evolutionary genetics of Hawaiian freshwater fish. Copeia 1996:330–335.

American Fisheries Society Symposium 69:375–377, 2009

Linkages with Ecosystem Energetics
Preamble

RICHARD A. CUNJAK
Canadian Rivers Institute and the Department of Biology, University of New Brunswick
Fredericton, New Brunswick, E3B 6K9, Canada

The original intent of this session was to focus on ecosystem-level impacts of diadromous fishes, specifically the historic versus current extent of diadromy, linkages between diadromous fish and managed fisheries, the keystone role of some of these species, and the nutrient contributions of diadromous fish to freshwater and terrestrial ecosystems (Willson et al. 1998; Naiman and Latterell 2005). The six papers herein certainly demonstrate such breadth. Geographically, the work encompasses the rich foraging regions of the North Atlantic Ocean: freshwater systems of two continents (Europe and North America), rivers running through coastal rainforests, temperate and boreal zones, as well as northern (Arctic) ecosystems. Taxonomically, the papers focus on a variety of diadromous species that include salmonids, alosids, osmerids, and petromyzontids. In terms of temporal range, the research reflects trends from the past decade to millennial time scales.

Unlike the situation 20 years ago when the first diadromous symposium was held, scientists now commonly use novel technologies and approaches such as stable isotope analysis (SIA) to provide insight into the dynamics of marine-derived nutrients in freshwater systems (Wipfli et al. 2004; Hicks et al. 2005). This was obvious during our session where five of the six papers relied directly on stable isotope information to demonstrate the ecological processes central to their research objectives. For example, Sinnatamby et al. (2009, this volume) used SIA to determine if unsustainable exploitation by north Atlantic fisheries may have led to changes in marine food webs, forcing a trophic feeding shift in adult Atlantic salmon. Stable isotope signatures of the marine growth portion of scales showed no consistent patterns over time to support the hypothesis that the mean trophic level of adult Atlantic salmon *Salmo salar* has changed over the past several decades.

The specific mechanisms and pathways for ecosystem enrichment from marine-derived nutrients (MDNs) delivered by diadromous fishes remains a complex puzzle. The very thorough review by Naiman et al. (2009, this volume) clearly demonstrates the importance of MDNs to freshwater, estuarine, and riparian ecosystems in western North America. The influence of MDN on terrestrial plants and wildlife is especially fascinating. For example, the authors cite the work of Jauquet et al. (2003) who reported consumption of salmon by 10 bird species and four mammal species, including cougar *Felis concolor*, for a single stream in western Washington. For this reason, salmon have been described as keystone species significantly affecting coastal ecosystems (Willson et al. 1998).

The review paper by Gregory-Eaves et al (2009, this volume) is impressive in that diadromous fish (in this case, sockeye salmon *Oncorhynchus nerka*) population dynamics are reconstructed over millennial timescales using a multiproxy approach involving stable isotope analyses of lake sediments as well as paleolimnological indicators such as zooplankton and algal (mainly diatom) assemblages as proxies for sockeye salmon abundance in dated sediment cores. Such data have allowed paleolimnologists and fisheries scientists and resource managers to collaborate to test if climate-induced oceanographic changes may explain recent variation in salmon stock abundance by relating salmon historic salmon abundance to past climate regimes prior to anthropogenic impacts that would otherwise confound the interpretation.

By contrast with the Pacific coast, the importance of MDN subsidies to freshwater ecosystems of the Atlantic coast is poorly documented (but see Durbin et al. 1979). The paper by Jardine et al (2009, this volume) presents strong evidence, from several rivers in eastern Canada, for direct predation of osmerid and alosid eggs by predatory macroinvertebrates (e.g., perlid stoneflies), but the evidence of an indirect MDN uptake by epilithic flora and fauna (biofilm) is ambiguous at best. How important such marine nutrient subsidies are for maintaining freshwater productivity presently and historically (when diadromous species were much more abundant in Atlantic river systems) still needs to be determined.

Swanson and Kidd (2009, this volume) also studied how diadromous fishes might influence freshwater ecosystems, but from the perspective of an iteroparous species, the Arctic char *Salvelinus alpinus*, with a very complex life history. Unlike the situation for semelparous Pacific salmon (especially sockeye salmon), the import of marine-derived nutrients (N, C, and P) to lakes in the central Canadian Arctic was negligible and largely offset by the export of nutrients via the emigrant smolt stages of char. Nislow and Kynard (2009, this volume) also measured the transport of nutrients between freshwater and marine environments—this time for the semelparous sea lamprey *Petromyzon marinus* in rivers of eastern North America. For one New England river, they estimated that adult diadromous sea lamprey may import as much as of 20% of the total annual phosphorous loading to streams where spawning and juvenile rearing habitat are suitable. However, as was the case for Arctic char, the marine-derived nutrient loading for lamprey is probably largely offset by an extended juvenile stage (and adult survival for which there are no reliable data). Both papers highlight the importance (and difficulty) of understanding the complex life-histories of diadromous fishes in order to adequately quantify the importance of MDNs to freshwater ecosystems.

Collectively, the research in the papers from this session and the other sessions covered in this volume provide valuable information for scientists and resource managers to recognize the contributions of diadromous fishes to complex ecosystem function. The research also provides evidence that despite the evidence of resilience and the natural cycling of diadromous populations, recent stock collapses and large-scale food web shifts underscore the critical stage where we now find ourselves. We cannot wait another 20 years to address the environmental problems that plague aquatic ecosystems, for which diadromous fishes are simply one component. Surely, the time for determined, science-based action is now.

References

Durbin, A. G., S. W. Nixon, and C. A. Oviatt. 1979. Effects of the spawning migration of the alewife, *Alosa pseudoharengus*, on freshwater ecosystems. Ecology 60:8–17.

Gregory-Eaves, I., D. T. Selbie, J. N. Sweetman, B. P. Finney, and J. P. Smol. 2009. Tracking sockeye salmon population dynamics from lake sediment cores: a review and synthesis. Pages 379–393 *in* A. J. Haro, K. L. Smith, R. A. Rulifson, C. M. Moffitt, R. J. Klauda, M. J. Dadswell, R. A. Cunjak, J. E. Cooper, K. L. Beal, and T. S. Avery, editors. Challenges for diadromous fishes in a dynamic global environment. American Fisheries Society, Symposium 69, Bethesda, Maryland.

Hicks, B. J., M. S. Wipfli, D. W. Lang, and M. E. Lang. 2005. Marine-derived nitrogen and carbon in freshwater-riparian food webs of the Copper River delta, southcentral Alaska. Oecologia 144:558-569.

Jardine, T. D., J.-M. Roussel, S. Mitchell and R. A. Cunjak. 2009. Detecting marine nutrient and organic matter inputs into multiple trophic levels in streams of Atlantic Canada and France. Pages 427–445 *in* A. J. Haro, K. L. Smith, R. A. Rulifson, C. M. Moffitt, R. J. Klauda, M. J. Dadswell, R. A. Cunjak, J. E. Cooper, K. L. Beal, and T. S. Avery, editors. Challenges for diadromous fishes in a dynamic global environment. American Fisheries Society, Symposium 69, Bethesda, Maryland.

Jauquet, J., N. Pittman, J. A. Heinis, S. Thompson, N. Tatyama, and C. J. Cederholm. 2003. Observations of chum salmon consumption by wildlife and changes in water chemistry at Kennedy Creek during 1997–2000. Pages 71–88 *in* J. G. Stockner, editor. Nutrients in salmonid ecosystems: sustaining production and biodiversity. American Fisheries Society, Symposium 34, Bethesda, Maryland.

Naiman, R. J., and J. J. Latterell. 2005. Principles for linking fish habitat to fisheries management and conservation. Journal of Fish Biology 67:166–185.

Naiman, R. J., J. M. Helfield, K. K. Bartz, D. C. Drake, and J. M. Honea. 2009. Pacific salmon, marine-derived nutrients and the dynamics of aquatic and riparian ecosystems. Pages 395–425 *in* A. J. Haro, K. L. Smith, R. A. Rulifson, C. M. Moffitt, R. J.

Klauda, M. J. Dadswell, R. A. Cunjak, J. E. Cooper, K. L. Beal, and T. S. Avery, editors. Challenges for diadromous fishes in a dynamic global environment. American Fisheries Society, Symposium 69, Bethesda, Maryland.

Nislow, K. H., and B. E. Kynard. 2009. The role of anadromous sea lamprey (*Petromyzon marinus*) in nutrient and material transport between marine and freshwater environments. Pages 485–494 *in* A. J. Haro, K. L. Smith, R. A. Rulifson, C. M. Moffitt, R. J. Klauda, M. J. Dadswell, R. A. Cunjak, J. E. Cooper, K. L. Beal, and T. S. Avery, editors. Challenges for diadromous fishes in a dynamic global environment. American Fisheries Society, Symposium 69, Bethesda, Maryland.

Sinnatamby, R. N., J. B. Dempson, G. Chaput, F. Caron, E. Niemelä, J. Erkinaro, and M. Power. 2009. Spatial and temporal variability in the trophic ecology of Atlantic salmon in the North Atlantic inferred from analyses of stable isotope signatures. Pages 447–463 *in* A. J. Haro, K. L. Smith, R. A. Rulifson, C. M. Moffitt, R. J. Klauda, M. J. Dadswell, R. A. Cunjak, J. E. Cooper, K. L. Beal, and T. S. Avery, editors. Challenges for diadromous fishes in a dynamic global environment. American Fisheries Society, Symposium 69, Bethesda, Maryland.

Swanson, H. K., and K. A. Kidd. 2009. A preliminary investigation of the effects of anadromous Arctic char on food web structures and nutrient transport in coastal Arctic lakes. Pages 465–483 *in* A. J. Haro, K. L. Smith, R. A. Rulifson, C. M. Moffitt, R. J. Klauda, M. J. Dadswell, R. A. Cunjak, J. E. Cooper, K. L. Beal, and T. S. Avery, editors. Challenges for diadromous fishes in a dynamic global environment. American Fisheries Society, Symposium 69, Bethesda, Maryland.

Willson, M. F., Gende, S. M., and B. H. Marston. 1998. Fishes and the Forest. Bioscience 48:455–462.

Wipfli, M. S., J. P. Hudson, and J. P. Caouette. 2004. Restoring productivity of salmon-based food webs: contrasting effects of salmon carcass and salmon analog additions on stream-resident salmonids. Transactions of the American Fisheries Society 133:1440–1454.

American Fisheries Society Symposium 69:379–393, 2009

Tracking Sockeye Salmon Population Dynamics from Lake Sediment Cores: A Review and Synthesis

IRENE GREGORY-EAVES*
Department of Biology, McGill University
1205 Avenue Docteur Penfield, Montreal, Quebec H3A 1B1, Canada

DANIEL T. SELBIE[1]
Department of Biology, McGill University
1205 Avenue Docteur Penfield, Montreal, Quebec H3A 1B1, Canada

and

Paleoecological Environmental Assessment and Research Laboratory (PEARL)
Department of Biology, Queen's University, Kingston, Ontario K7L 3N6, Canada

JON N. SWEETMAN
Paleoecological Environmental Assessment and Research Laboratory (PEARL)
Department of Biology, Queen's University, Kingston, Ontario K7L 3N6, Canada

Western and Northern Service Centre, Parks Canada
145 McDermot Avenue, Winnipeg, Manitoba R3B 0R9, Canada

BRUCE P. FINNEY[2]
Institute of Marine Science, University of Alaska Fairbanks
Fairbanks, Alaska 99775, U.S.A

JOHN P. SMOL
Paleoecological Environmental Assessment and Research Laboratory (PEARL)
Department of Biology, Queen's University, Kingston, Ontario K7L 3N6, Canada

Abstract.—Pacific salmon *Oncorhynchus* spp. play a central role in coastal ecosystems that rim the North Pacific Ocean. Given the ecological, cultural, and economic importance of Pacific salmon, there is great interest in defining the magnitude and frequency of change in these fish stocks. Fisheries scientists, through analyzing harvest records, have demonstrated pronounced salmon production variability. The causes underlying such marked fluctuations are currently debated. Collating harvest records across a broad geographic range over the past ~80 years, fisheries scientists have advanced a plausible argument that climate-induced oceanographic changes explain a significant fraction of the variation in salmon catch records. However, without data that predate the introduction of large-scale human interventions (e.g., commercial harvesting, dams, hatchery releases), it is difficult to isolate the role of climate in shaping fish stock

* Corresponding author: irene.gregory-eaves@mcgill.ca
[1] Current address: Fisheries and Oceans Canada, Science Branch, Salmon and Freshwater Ecosystems Division, Cultus Lake Salmon Research Laboratory, Cultus Lake, British Columbia V2R 5B6, Canada
[2] Current address: Department of Biological Sciences, Idaho State University, Pocatello, Idaho 83209, USA

dynamics. Within the past decade, however, we have developed a paleolimnological approach for tracking past sockeye salmon *Oncorhynchus nerka* population abundances, and numerous papers have applied this approach to infer changes in these fish over the past hundreds to thousands of years. Here, we provide an overview of the approach and a synthesis of the work that has been conducted in this field to date. It is clear that numerous sockeye salmon populations have undergone pronounced changes, even prior to human interventions. Furthermore, tracking salmon populations over millennial timescales with paleolimnology has revealed modes of change that were previously never imagined possible. Such long-term perspectives indicate that sockeye salmon is a resilient fish species. We note, however, that when natural environmental changes are compounded by intense human impacts, populations have been particularly susceptible to extirpation.

Introduction

Pacific salmon *Oncorhynchus* spp. are an integral part of the ecology, economy, and culture of the North Pacific. Ecologically, Pacific salmon represent significant nutrient and food sources for biological communities. Despite the importance of this genus, we only have a limited understanding of the primary factors influencing their population dynamics. Analyses of harvest data have highlighted dramatic interannual and interdecadal variability in Pacific salmon abundances, and considerable controversy exists over the causes for such fluctuations (e.g., Mantua et al. 1997; Collie et al. 2000; Mann and Plummer 2000). This debate is heightened and perhaps more critical at the southern end of the range of Pacific salmon where numerous populations have been extirpated or are at risk (Nehlsen 1997).

A major stumbling block in determining the relative importance of the myriad of stressors acting on Pacific salmon populations, including commercial fishing, dam construction, habitat destruction, hatcheries, and climatic change, is that monitoring data do not predate the period of large-scale human interventions. Without knowledge of baseline conditions, it is difficult to assess the impact of individual stressors (Smol 2008). However, paleolimnological approaches have recently been developed to reconstruct past sockeye salmon *O. nerka* population dynamics using information preserved in nursery lake sediments (Finney et al. 2000). As sockeye salmon populations can markedly influence the limnological characteristics of their nursery lake systems (e.g., Naiman et al. 2009, this volume), changes in the numbers of sockeye salmon spawners are recorded as shifts in chemical and biological components of lake sediments. The objectives of this paper are to provide an overview of this paleolimnological approach, highlight some key findings from recent studies, and discuss the major challenges in the field.

We begin with a synopsis of the current state of knowledge regarding Pacific salmon population trends, based on historical monitoring records, as this forms the key impetus for this paleolimnological research. We then review the paleolimnological approach and highlight some of the key findings. Overall, we show that paleolimnological analyses of nursery lake sediments provide many important and otherwise missing data on the population dynamics of sockeye salmon, as well as crucial information on the changes in freshwater ecosystem where sockeye salmon rear as young.

Historical Trends in Pacific Salmon Abundances

Much of what we know about population trends for Pacific salmon comes from catch data. The commercial fishery, marked by the opening of the first cannery in California, began in 1864 (Lichatowich 1999), and not long after, commercial operations were opened up at more northern and remote locations (i.e., in 1878 in Alaska; Wertheimer 1997). Whereas one might think commercial catch records could be biased by socioeconomic factors (e.g., decreases in catches during times of war), fisheries scientists have argued that catch records since the 1920s are a reasonable proxy for total salmon production, as these fish have been targeted fairly consistently over this time (Beamish and Bouillon 1993; Hare et al. 1999).

At least on a localized scale, there is evidence that declines in numerous stocks of Pacific salmon have been linked to anthropogenic activities. One of the most striking examples of this is the dramatic loss of Fraser River (BC) pink salmon *O. gorbuscha* and sockeye salmon stocks originating from lakes and rivers upstream of Hell's Gate. A severe landslide in 1913, triggered by the construction of the Canadian National Railway, greatly restricted access to upstream spawning grounds. This event, combined with subsequent mismanagement of Pacific salmon fisheries until 1937, exterminated pink salmon stocks upriver of Hell's Gate (Northcote and Atagi 1997) and depressed sockeye salmon stocks to very low abundances for several decades (Mundy 1997).

Analyses of catch records on a broader geographical scale, however, demonstrate regionally synchronous patterns, providing evidence to suggest that Pacific salmon stocks are also influenced by changes in the basin-scale, coupled ocean-atmospheric system. Beamish and Bouillon (1993) were among the first fisheries scientists to analyze salmon catch records over the 20th century and demonstrate a broadly coherent, interregional pattern across Canada, the United States, and Russia.

By examining the salmon catch records across a north–south gradient, fisheries scientists have recognized that interdecadal regime changes have both temporal and spatial components (Figure 1). Southern and northern populations of salmon from the North Pacific have shown synchronous but 180 degrees out-of-phase changes in 1947, 1977, and 1989 (Mantua et al. 1997; Hare et al. 1999; Hare and Mantua 2000). For example, Alaskan salmon stocks increased in abundance after 1977, while southern stocks showed lower than average returns (Figure 1; Mantua et al. 1997). Dramatic changes have also been identified around 1976–1977 in physical parameters of the North Pacific, primary and secondary productivity measures (e.g., Ebbesmeyer et al. 1991; Brodeur and Ware 1992), and abundances of other strictly marine fish (e.g., walleye pollock *Theragra chalcogramma*, Pacific cod *Gadus macrocephalus*, and Pacific halibut *Hippoglossus stenolepis*; e.g. Beamish

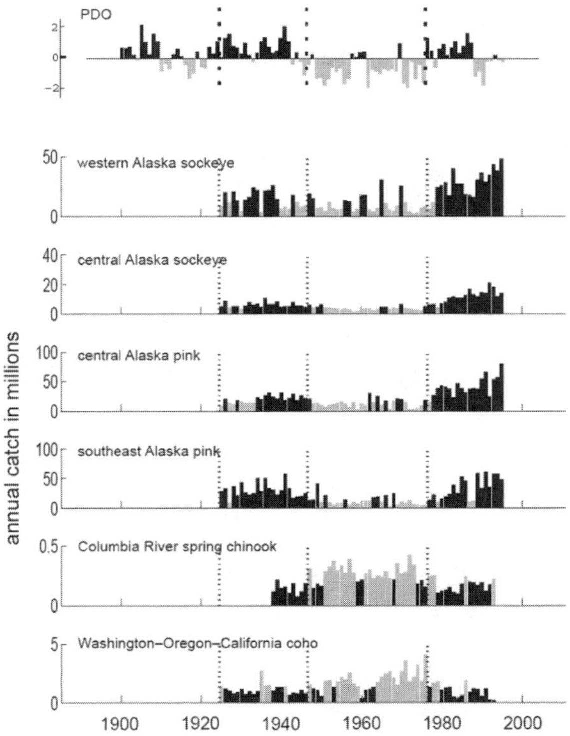

Figure 1.—Pacific salmon catch records from a north–south gradient along the coast of North America demonstrating changes that are synchronous with changes in the Pacific Decadal Oscillation (from Mantua et al. 1997).

1993; Clark and Hare 2002). These broad ecosystem-level changes have led numerous investigators to suggest that decadal-scale shifts in salmon abundances are likely driven by environmental changes in the North Pacific (e.g., Beamish and Bouillon 1993; Francis and Hare 1994; Mantua et al. 1997; Hare et al. 1999). The term "regime shift" was coined to describe the change in average characteristics of the North Pacific ecosystem that have persistent patterns at decadal scales, but then show abrupt transformations in states within a year or so (Hare and Mantua 2000). Further analyses of tree-ring data have revealed that the North Pacific climate has undergone repeated regime shifts with a periodicity of 50–70 years over the past few centuries (e.g., Minobe 1997; MacDonald and Case 2005).

The mechanism(s) through which physical changes in the North Pacific affect salmon (and other marine organisms) is still an area of active research, but several investigators have suggested that the mechanism(s) are not direct, as there are poor relationships between physical and biological variables on interannual time scales (Beamish and Bouillon 1993; Francis and Hare 1994). In an attempt to explain inverse salmon production regimes, Gargett (1997) proposed a comprehensive hypothesis that relates salmon abundances to environmental changes in the North Pacific. Her interpretation, which has since been widely cited by the scientific community, is centered on changes in the intensity of the Aleutian Low (a low pressure system that develops over the Aleutian Islands in winter). The Aleutian Low strongly couples the physical properties of the atmosphere with those of the North Pacific Ocean and is believed to be the key driving force in interdecadal patterns of variability in sea surface temperatures (i.e., the Pacific Decadal Oscillation; Mantua et al. 1997). In years when there is a strong Aleutian Low, temperatures along the coast of North America tend to be warmer and the coastal waters more strongly stratified. Coastal conditions are believed to be of prime importance to Pacific salmon because it is the early marine phase, when the smolts are in the coastal areas, that salmon are most sensitive to environmental changes (e.g., sea surface temperatures (SST); Pyper et al. 2005). Greater coastal stability has opposite effects on phytoplankton populations in the northern (Alaska) and southern (Pacific Northwest) regions of the northeast Pacific. These changes in phytoplankton production are significant to all marine organisms because they form the base of the food web (Miller et al. 2004). In the Gulf of Alaska, phytoplankton is believed to be light-limited, and thus greater water column stability allows for phytoplankton to remain in the photic zone longer and achieve higher production rates (Polovina et al. 1995). Meanwhile, further south, greater coastal stability is unfavorable for nutrient-limited phytoplankton because stronger water column stability deters the upwelling of deep, nutrient-rich waters with surface waters (McGowan et al. 2003). Gargett (1997) suggests that salmon abundances are largely affected by the availability of their forage, which in turn is driven by primary production, ultimately influenced by nutrient availability and water column stability.

While there is a reasonable case to be made that climatic shifts are driving marine production in the North Pacific, it is hard to isolate climate from other potentially important variables influencing Pacific salmon, given contemporary data sets. As stated above, numerous anthropogenic stressors have been introduced over the past century. In order to tease apart the impact of multiple stressors, it is necessary to extend records of Pacific salmon abundances back in time so that the predisturbance production variability (i.e., premid- to late 1800s) can be assessed. Hindcasts of past salmon abundances on these temporal scales will allow investigators to uncover natural salmon population dynamics and the role of climatic change, as well as constrain influences of anthropogenic impacts.

The Paleolimnological Approach for Reconstructing Sockeye Salmon Population Dynamics

Paleolimnology, the study of the biological, chemical, and physical indicators preserved in lake and river sediments, is now a well-established science (Smol 2008). Lake sediments represent natural environmental archives of local and regional conditions as autochthonous (i.e., derived from within the lake) and allochthonous (i.e., coming from outside the lake) materials are continuously deposited at the bottom of lakes. To date, a vast array of paleolimnological indicators have been studied to define changes in a wide range of environmental variables (e.g., nutrients, pH, temperature; Smol 2008). Unfortunately, reconstructing past fish abundances has always been a challenge for paleolimnologists because fish bones and scales are rarely preserved in sufficient numbers in deepwater lake sediments (Smol 2008). However, Pacific salmon, particularly sockeye salmon, have a somewhat unique life history that enables paleolimnologists to infer past population dynamics through changes in proxy indicators preserved in nursery lake sediments.

Pacific salmon have an anadromous life cycle (i.e., organisms that migrate from the ocean to freshwaters to spawn) and most are semelparous (spawn once and then die). Sockeye salmon differ from other Pacific salmon in that they make use of nursery lakes and their tributary streams to spawn (other species use streams and rivers exclusively). The annual return and subsequent death of thousands to millions of sockeye salmon spawners in nursery lake environments may represent a significant source of nutrient loading (Krohkin 1975; Koenings and Burkett 1987; Kline et al. 1997). As a result, changes in spawner abundances can greatly alter nursery lake nutrient conditions (Koenings and Burkett 1987; Gregory-Eaves et al. 2003). The technique described herein could be extended to other anadromous lake-rearing fish that have a similar influence on the water chemistry of their nursery environment (e.g., alewife *Alosa pseudoharengus*; Durbin et al. 1979). More often than not, however, other anadromous lake-rearing fishes contribute only a small fraction of the total nutrient pool (e.g., Swanson and Kidd 2009, this volume).

Finney (1998) demonstrated that alterations in salmon-derived nutrient loading (and thus sockeye salmon spawner abundances) could be inferred by analyzing the stable isotopic signature of lake sediments. Finney et al. (2000) were the first to use multiple paleolimnological indicators to track sockeye salmon population dynamics in Alaskan nursery lakes over the past ~300 years. Below, we summarize key aspects of the main indicators used in these reconstructions.

Stable Nitrogen Isotope Ratio $\delta^{15}N$

The stable isotope ratio $\delta^{15}N$ has repeatedly been shown to be a valuable tracer of anadromous salmon in both food web and paleolimnological research (e.g., Kline et al. 1993; Finney 1998). Salmon are enriched in ^{15}N (relative to the much more common isotope ^{14}N) in part because they feed relatively high on the food web and in part because they accumulate most of their biomass in the marine environment (i.e., >95%, Burgner 1991). Adult salmon feed on a diet of small pelagic fish, squid, and small crustaceans (Burgner 1991), which results in ^{15}N enrichment in sockeye salmon relative to lower trophic levels. As was first demonstrated by Minagawa and Wada (1984), $\delta^{15}N$ progressively increases in organisms of higher trophic levels because the lighter ^{14}N isotope is preferentially excreted. Feeding in the marine environment further contributes to the heavy $\delta^{15}N$ signature of adult salmon. The $\delta^{15}N$ signature of the marine environment (based on analyses of nitrate, particulate matter, and organisms) is relatively enriched because the primary source of nitrogen is upwelled nitrate, which has been enriched in ^{15}N via denitrification (a process that occurs in oxygen-depleted waters; reviewed in Michener and Schell 1994). Finney's (1998) paper demonstrated that the sedimentary $\delta^{15}N$ profile from Karluk Lake, Alaska, was significantly correlated to the historical time series of sockeye salmon spawners, which had been monitored since 1922. Finney et al. (2000) went on to demonstrate that sedimentary $\delta^{15}N$ was not only related to escapement through time, but that similar patterns were present from a wide spectrum of Alaskan lakes.

There are, however, numerous factors that can also influence $\delta^{15}N$ in lakes, including changes in organic matter source, nitrogen utilization, nitrogen fixation, denitrification, and ammonia volatilization (Finney et al. 2000). Careful consideration of these factors is necessary when interpreting $\delta^{15}N$ profiles. For example, a sudden input of terrestrial organic matter (e.g., as a result of a forest fire or earthquake) would dilute the salmon-derived nutrient signal and could lead an investigator to conclude that sockeye salmon populations decreased. Such an interpretation, however, can be constrained by plotting ratios of sedimentary carbon to nitrogen (C:N; e.g., Holtham et al. 2004), as terrestrial organic matter sources have higher C:N than those of aquatic organic matter due to higher cellulose concentrations in land-based plants (Meyers and Ishiwatari 1993). Sedimentary charcoal analysis could also be used to isolate forest fire events and evaluate whether deviations in $\delta^{15}N$ were influenced by fires (Whitlock and Larsen 2001). Depletion of a lake's nitrogen supply and the shift in the organic matter source of nitrogen to that fixed by cyanobacteria (i.e., atmospheric N with a $\delta^{15}N$ of 0 ‰) is less likely in most sockeye salmon nursery lakes, as these systems tend to be P-limited more than N-limited (Edmundson and Carlson 1998; Shortreed et al. 2001). However, data regarding the water chemistry (e.g., TN:TP, POC:PON), as well as analysis of algal pigments in the sediments, can help to confirm that changes in the N source in the lake are minimal. In contrast to the processes described above, which can depress the $\delta^{15}N$ signal, denitrification, and ammonia volatilization, are two pathways through which sedimentary $\delta^{15}N$ can be enriched without the influence of salmon-derived nutrients. These pathways are, however, not likely to be prevalent in sockeye salmon nursery lakes, as they only operate with any magnitude under rather severe environmental conditions (i.e., low oxygen for denitrification and high pH for ammonia volatilization). Along with evaluating the relative importance of such conflicting factors in a paleolimnological analysis, a robust approach includes the analysis of several, independent proxies. For example, biological microfossil indicators (as discussed below) are less susceptible to postdepositional alterations than geochemical tracers (e.g., $\delta^{15}N$) and have proven to be a valuable complement (Gregory-Eaves et al. 2003; Gregory-Eaves et al. 2004; Holtham et al. 2004).

Diatoms and Other Algal Indicators

A second proxy group demonstrated to indirectly track past sockeye salmon abundances through salmon-derived nutrient loading are algal indicators (e.g., Gregory-Eaves et al. 2003; Schindler et al. 2005). Numerous investigators have observed the presence of dense algal mats in areas near decomposing salmon carcasses (e.g., Mathisen 1972). Because algal growth is often limited by nutrients, salmon-derived nutrient loading can influence both algal production (e.g., Fisher-Wold and Hershey 1999) and species composition (e.g., Gregory-Eaves et al. 2003, 2004).

Diatoms (Bacillariophyceae) have been widely applied in paleolimnology because (1) they are a ubiquitous group, present in large abundances in many aquatic habitats; (2) they are responsive indicators with short generation times; (3) they are identifiable to the species level or lower; (4) they are a very diverse group, with different species having specific optima and tolerances to numerous environmental variables; and (5) their siliceous cell walls preserve well in most lake sediments (Smol 2008). A common application of diatoms has been to trace past lake trophic status, as nutrient concentrations have been shown to be a significant factor in influencing diatom species assemblages (reviewed by Hall and Smol 1999). Other variables, however, are also known to influence diatom community composition (e.g., pH and salinity), and therefore it is important to select lakes where salmon-nutrient loading is likely to be the strongest signal. In particular, this technique is only likely to be effective where salmon-derived nutrients (SDN) make up a significant fraction of the total nutrient budget (or at least did in the past). The diatom analysis should be done in conjunction with a marine tracer such as $\delta^{15}N$, as the investigator is unable to determine the nutrient source from the diatom community composition alone. Numerous lakes on the west coast of North America satisfy these conditions, as many salmon nursery systems are remote and thus are less influenced by large nutrient inputs from human activities or by other sources of anthropogenic impact. In such lakes, Finney et al. (2000) and Gregory-Eaves et al. (2003) have demonstrated that fossil diatom assemblages can be strongly correlated to historical numbers of returning adult spawners (i.e., escapement).

Sedimentary algal pigments have also been used for tracking past algal responses to salmon-derived nutrients (Schindler et al. 2005; Brock et al. 2006, 2007). The preservation of chlorophylls and carotenoids in the sediment record allows investigators to quantify the magnitude of change in algal production and to infer how different algal groups (e.g., siliceous algae, chlorophytes, and cyanobacteria) have varied over the course of the time (Leavitt and Hodgson 2001). Since the development of the use of algal pigments in paleolimnology, there have been some concerns about the preservation of pigments (e.g., poor preservation under alkaline conditions and changes in preservation with lake depth), but salmon lakes tend to be circumneutral in pH and deep, open drainage systems (and thus relatively insensitive to changes in atmospheric moisture balance; Shortreed et al. 2001).

Biogenic silica is another geochemical indicator of algal production and is easily measured from sediment samples. Biogenic silica is often used as a surrogate measure for diatom production as diatoms are commonly the most abundant siliceous algae (e.g., Conley and Schelske 2001). Total diatom production, in turn, has been shown to be a reasonable proxy for inferring changes in phosphorus loading in many lakes until silica becomes a limiting nutrient (Conley and Schelske 2001). Therefore, in lakes where the sockeye salmon returns are relatively high and silica is not a limiting nutrient, biogenic silica can track changes in SDN loading. This approach was used successfully in Packers Lake, Alaska, where Gregory-Eaves et al. (2004) demonstrated that biogenic silica was strongly correlated to $\delta^{15}N$ and the changes in diatom species assemblages.

Zooplankton

The study of zooplankton remains can provide insight into the extent to which salmon nutrients are incorporated into freshwater food webs as well as information on the predation pressure by juvenile sockeye salmon. Cladocerans, and to varying degrees copepods and aquatic insect larvae, are primary prey items of juvenile sockeye salmon, that will spend up to 3 years in the nursery lake (Burgner 1991). The chitinous exoskeletons of cladocerans, though disarticulated into several parts in the sediments, are well preserved and are used to reconstruct a key component of the zooplankton assemblage (Korhola and Rautio 2001). Both the species composition and the size structure of the cladocerans provide insightful metrics.

Sweetman and Finney (2003) demonstrated the influence of both nutrient-stimulated production and predation pressure on zooplankton populations from two lakes (Karluk and Frazer) in Alaska. The two study lakes were selected because they both had long-term monitoring records of escapement and were similar in morphometries and geographic location but differed substantially in their histories. Due to a steep waterfall at the outlet of Frazer Lake, this system had no resident sockeye salmon population until two fish passes were constructed in the 1960s and 1970s. In contrast, Karluk Lake is a naturally producing sockeye salmon lake.

Interestingly, the sediment record from both lakes showed a positive relationship between the flux of the dominant zooplankton species *Bosmina longirostris* and historical sockeye salmon escapement, highlighting the significance of changes in salmon-derived nutrients in the production of zooplankton over broad temporal scales. Morphometric data for *B. longirostris*, however, showed different responses in each of the lakes to variation in sockeye salmon escapement, which could be explained in terms of the importance of different predators. In Frazer Lake, the size of *B. longirostris* was inversely related to the abundance of sockeye salmon, whereby strong predation pressure by introduced juvenile sockeye elicited an adaptive response in the zooplankton to become smaller and more evasive. Karluk Lake provides a contrasting picture, however, as a positive relationship was documented between the size of *B. longirostris* and the abundance of sockeye salmon. Given that Karluk Lake is more productive (which is likely related to its long history of strong sockeye salmon returns) and has a much more abundant zooplankton community than Frazer Lake, it is perhaps not surprising that the juvenile sockeye salmon are a less effective selection pressure than the impact of enhanced nutrient loading from salmon-derived nutrients. The relatively strong presence of the zooplanktivorous zooplankton *Cyclops* sp. in Karluk Lake may also have played a role in size increases of *B. longirostris*, as *Cyclops* sp. are unable to consume the larger sized zooplankton. Ultimately, the Sweetman and Finney (2003) study was instrumental in demonstrating the influence of salmon-derived nutrients on the density and size of sockeye salmon forage.

What Have We Learned so Far?

The use of paleolimnological approaches to track past salmon abundances is a relatively new area of research. However, several important questions have already been answered using information preserved in nursery lake sediment profiles. For example, the ability of paleolimnology to extend these historical records beyond the period of pronounced anthropogenic impact has allowed us to reconstruct natural population dynamics and bring forth understanding on the effects of anthropogenic stressors. The major findings in the area of sockeye salmon reconstructions can be summarized into the following categories: (1) broad-scale synchrony in northern salmon production over the past ~300 years, (2) pronounced and sustained variation in some populations over the past ~2,200 years, (3) sustained low population sizes in other systems for the past ~1,500 years, and (4) the importance of site characteristics in selecting a suitable lake for salmon reconstruction.

In the first paper to use a multiproxy approach to reconstruct past sockeye salmon abundances, Finney et al. (2000) showed evidence of broad regional synchrony in $\delta^{15}N$, a proxy for changes in sockeye salmon populations, over the past ~300 years (Figure 2). By

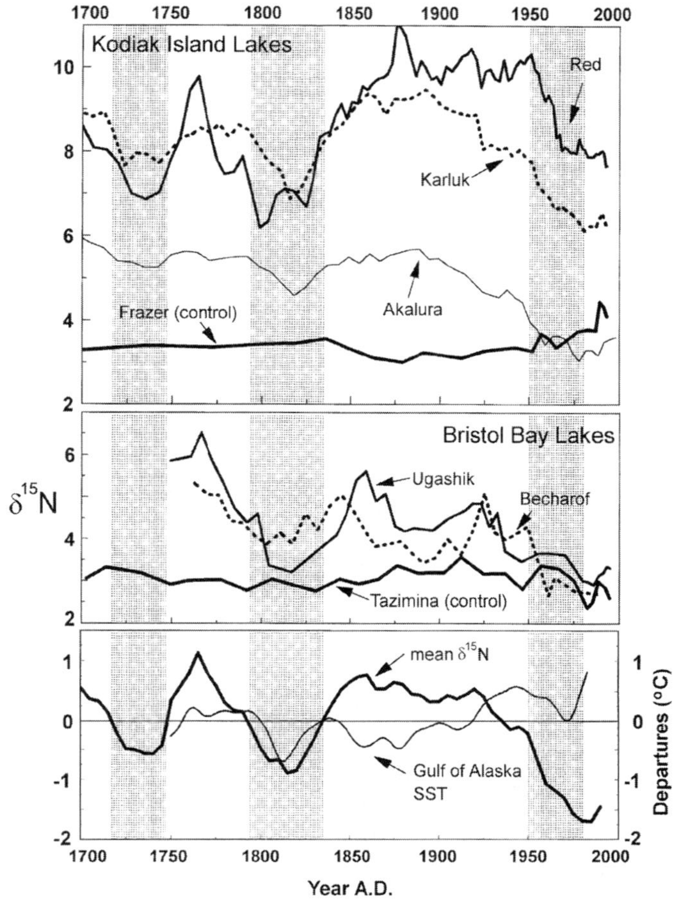

Figure 2.—Reconstructions of sockeye salmon population dynamics, as inferred from $\delta^{15}N$ profiles, from three nursery lakes on Kodiak Island and two nursery lakes in the Bristol Bay region (from Finney et al. 2000). Plotting the 250-years reconstruction of sea surface temperatures from the Gulf of Alaska (based on dendrochronological analysis; D'Arrigo et al. 1999) alongside our composite salmon index (bottom panel) illustrates that the early 1800 deviation in the salmon records corresponds to the coldest period on record.

correlating the $\delta^{15}N$ profiles from sediment cores from five nursery lakes, many of the series have strong relationships (i.e., $r > 0.50$; Table 1). In contrast, the relationship between the composite $\delta^{15}N$ index from the pre-fishing era (Figure 2), based on the five salmon lakes and the reconstruction of SST from the Gulf of Alaska (based on tree ring data; D'Arrigo et al. 1999), is weak and nonsignificant ($r = 0.23$, $p = 0.37$, $n = 19$). The coldest period on record, however, corresponds to a marked deviation in the mean $\delta^{15}N$. This finding is similar to analyses of historical time series where correlations between catch records and climate variables are weak at an interannual time scale. Climate regime changes in these historical records, however, have corresponded to marked deviations in catch (Francis and Hare 1994; Mantua et al. 1997). The direction of change, where colder conditions correspond to lower than average sockeye salmon abundances in northern populations, is also the same between the catch and paleolimnological time series. Thus, the paleolimnological time series confirmed that regime changes had occurred before the advent of intensive anthropogenic disturbances, supporting the notion that the 20th century regime changes were also likely climate-driven.

Table 1.—A spearkman rank correlation matrix of the $\delta^{15}N$ values from the pre-fishing era (pre-1880 AD) presented in Figure 2. The data were first interpolated such that they would be on the same time scale and resolution (n = 19) and then transformed into Z-scores. * indicates $p < 0.05$; ** indicates $p < 0.01$.

	Red	Karluk	Akalura	Becharof	Ugashik
Red	1.00				
Karluk	0.84**	1.00			
Akalura	0.72**	0.64**	1.00		
Becharof	0.05	-0.13	0.26	1.00	
Ugashik	0.52*	0.50*	0.72**	0.49*	1.00

Note: As there was significant first-order autocorrelation in all data series, the p-values are likely underestimated.

The Finney et al. (2000) study also presented striking declines in $\delta^{15}N$ (reflecting declines in the number of spawners) in all five sockeye salmon populations over the past century, which was inferred to be, at least in part, a product of the commercial fishery. The impact of the declines in the number of spawners, however, appears to have different consequences in different lakes (Finney et al. 2000). To address this issue more closely, Schindler et al. (2005) developed a mixing model formula to quantitatively infer past sockeye salmon escapement. When the historical catch record was added to the inferred escapement profile from Lake Nerka, Alaska, they showed that despite at least 50 years of commercial fishing and declines in the inferred escapement record, the total production of the system over the past few decades has been at a record high (Figure 3A). An ecological argument supporting this finding is that there was increased freshwater production of juvenile sockeye salmon as a result of less density dependence for spawning habitat and freshwater forage. In contrast, when we apply the same mixing model to Karluk Lake and add the historical catch record to the inferred sockeye salmon escapement profile, we can see that the current total production is at least three times lower than what is was at the turn of the 20th century (Figure 3B). This result from Karluk Lake, however, is not surprising, as increasing the returns of

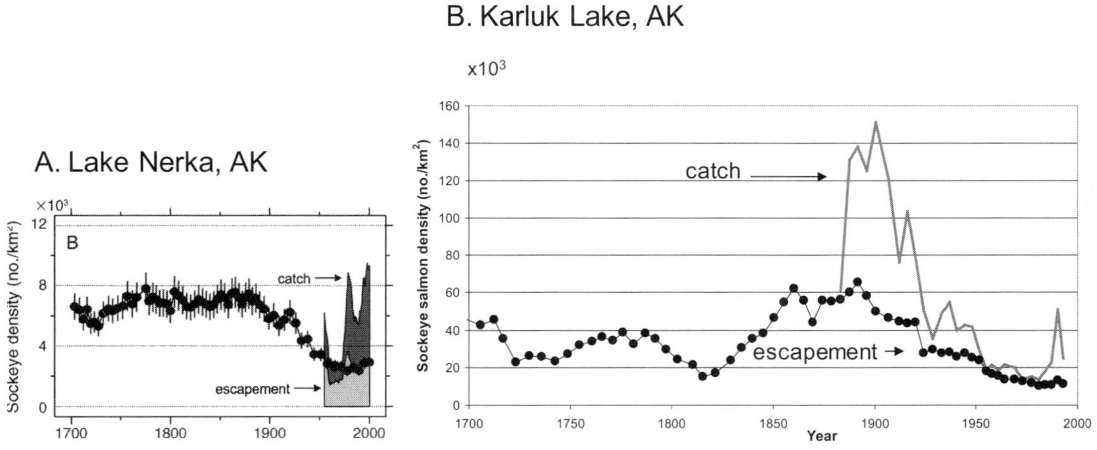

Figure 3.—The history of sockeye salmon total production (i.e., sum of escapement and catch) from two Alaskan systems that have responded very differently to the commercial harvest. (A) Lake Nerka has experienced the greatest total production in the past few decades (modified from Schindler et al. 2005). (B) The Karluk Lake time series shows that total production was greatest around the time that commercial harvesting began. Note: in both profiles, the catch series has been added to the escapement and thus reflects the best estimate of total production trends.

sockeye salmon spawners to the system have resulted in substantial increases in surface water total phosphorus concentrations (Koenings and Burkett 1987; Gregory-Eaves et al. 2003). Given that sockeye salmon nursery lakes represent a diversity of habitats (including glacial, meromictic, and clearwater systems; Edmundson and Carlson 1998), we reiterate that the importance of sockeye salmon in fertilizing these systems will vary greatly (e.g., Finney et al. 2000; Hyatt et al. 2004).

While the ~300 year records have demonstrated that decadal-scale variability existed in salmon populations long before the commercial harvest, Finney et al. (2002) extended these time series over the past ~2,200 years and showed much more prolonged and pronounced changes. The ~2,200-year $\delta^{15}N$ records illustrated that sockeye salmon production could be depressed for hundreds of years and then return to peak abundances without large-scale interventions by humans. We never suspected, nor did many colleagues in the fisheries community, that relatively unperturbed sockeye salmon populations would be this variable. By comparing the Karluk Lake $\delta^{15}N$ profile to reconstructions of other marine fish to the south (i.e., southern British Columbia and California), Finney et al. (2002) suggested that northern and southern fish populations within the North Pacific region are broadly synchronous, but that southern fish are 180 degrees out of phase with northern populations. As previously discussed, similar patterns in north–south sockeye salmon populations have been observed historically (e.g., Mantua et al. 1997; Hare et al. 1999) but have been complicated by the presence of multiple stressors. These results show that we are just beginning to unravel the true range of spatiotemporal variability in the magnitude and persistence of sockeye salmon population dynamics.

Although the Alaskan paleolimnological work has revealed that some populations have undergone large changes in size (i.e., changes on the order of greater than 10,000 spawners/km^2), there have been a few studies where the $\delta^{15}N$ profiles show subtle to no variations over the time frames under study (Holtham et al. 2004; Selbie et al. 2007; Hobbs and Wolfe 2007, 2008). There are two possible reasons for this: (1) the sockeye salmon population has been naturally low and invariable over the course of history, and/or (2) the conditions of the lake and watershed are such that other factors swamp any signal that might have been recorded by salmon in the nursery lake sediments. An example of the former is highlighted in a recent paper by Selbie et al. (2007), who developed paleolimnological records from current (Redfish Lake) and past spawning habitat (Pettit Lake) of the federally endangered Snake River sockeye salmon in the Sawtooth Valley of Idaho. The key questions addressed in this study were (1) How abundant and variable were sockeye salmon populations in these systems prehistorically? (2) How important were salmon-derived nutrients to the rearing system ecology? (3) When did the population declines occur and what historical influences were most likely responsible? (4) What are and have been the potential impediments to stock recovery efforts? The $\delta^{15}N$ profiles from the two sockeye salmon nursery systems showed that although they are higher than the control lake profile in the pre-1950 sediments, the values were not as high or variable as in Alaskan nursery systems (mostly less than 4 ‰). Using the predictive equation from Finney et al. (2000) that would allow the approximate conversion of $\delta^{15}N$ values into spawner density numbers, we estimate that sockeye salmon spawner densities in Redfish Lake were generally less than 6,500/km^2. A pattern that was coherent across the Idaho salmon systems was a unidirectional decline in $\delta^{15}N$ beginning ~1860, which corresponds to the onset of commercial fishing on the Columbia River. A further negative inflection in the $\delta^{15}N$ profile was evident and coincident with completion of the Sunbeam Dam, a small hydroelectric project that represented a significant barrier to the passage of sockeye salmon from 1910 to 1934 (Selbie et al. 2007). Thus, although the Idaho study lakes have experienced substantial declines coincident with fishing and dam construction, mitigation targets of escapement of more than 6,500/km^2 in Redfish Lake would likely be unrealistic because the system likely has not supported larger densities over the past ~1,400 years.

By adopting a multiproxy approach in the Idaho study, Selbie et al. (2007) also demonstrated that there was a pronounced reorganization of the nursery lake food web following the decline in sockeye salmon and subsequent fisheries management interventions. Historical fisheries "enhancement" practices, consisting of the introduction of hundreds of thousands to millions of nonnative sockeye salmon and kokanee (lacustrine sockeye salmon) juveniles and eggs to Sawtooth Valley rearing systems, potentially augmented by changes in nutrient loading (i.e., atmospheric nitrogen deposition and climatically altered biogeochemical cycling), shifted the zooplankton community to a smaller, less energetically beneficial forage (from *Daphnia* to *Bosmina*) for juvenile Snake River sockeye salmon. Selbie et al. (2007) identified this as a likely factor impeding the success of restoration efforts in rebuilding the sockeye salmon run.

Paleolimnological analyses of sockeye salmon nursery lake sediments have also shown that the salmon signal preserved in sediments may be harder to detect in systems with high flushing rates (i.e., low water residence times), with relatively high loads of terrestrial organic matter, and/or where the nutrient loading from sockeye salmon is low. Holtham et al. (2004) addressed these issues with their paleolimnological investigations of coastal nursery lakes. Although two of their study lakes currently have high densities of sockeye salmon spawners (i.e., greater than 10,000 spawers/km^2), the sedimentary $\delta^{15}N$ profiles from two of the lakes were relatively low and invariant. Through a close examination of the limnology of their study systems, Holtham et al. (2004) identified that both lakes had very low water residence times (i.e., <0.5 years). Given that such low water residence times result in the swift flow through of nutrients and plankton (Stockner et al. 2005), the limnic conditions prevent planktonic communities from fully exploiting salmon-derived nutrients. The fast movement of plankton through the system also prevents this material from being deposited at the lake bottom in the form of sediments. Terrestrial organic matter can also be more important in such systems as watersheds receiving relatively high amounts of precipitation also receive greater loads of nutrients from the watershed through precipitation-fed streams and surface runoff (Gross et al. 1998). When terrestrial organic matter loading is very high, the sediment may be dominated by cellulose-rich matter, which has a high C:N ratio (as discussed above). This was particularly evident in Hobiton Lake, BC, where a prominent peak in the C:N profile was detected around the time of a major earthquake in the region (Holtham et al. 2004). Even when the C:N peaks were excluded, the $\delta^{15}N$ profile was consistently low, suggesting that sockeye salmon loading to the system was never substantial. Given that Hobiton Lake's water residence time is not extremely low (i.e., ~1 year), the most parsimonious hypothesis is that this site did not experience high sockeye salmon spawner densities over the past ~200 years. Overall, the Holtham et al. (2004) study emphasized the importance of site selection in developing a reconstruction of past sockeye salmon dynamics from lake sediments. To summarize the key factors that must be considered in selecting a suitable nursery lake for paleolimnological analysis, we have developed a conceptual model (Figure 4). This flow diagram not only highlights the factors that influence the strength of the salmon signal, but also suggests how these factors may be evaluated a priori.

Emerging Areas of Study and Conclusions

After approximately a decade of research, the field has made significant advances—perhaps most notable has been identifying the widespread role of natural factors in regulating population dynamics. When embarking on this kind of research, however, it is crucial that scientists evaluate the potential of a site to produce a high-quality salmon reconstruction. Our work to date provides a number of guiding principles that should be addressed before investing large amounts of resources into developing paleolimnological analyses.

Promising topics for future applications of the approach described herein include tracking sockeye salmon population dynamics in (1) the Pacific Northwest and California, (2) Russia, and (3) over longer term time scales such as over the course of the entire Holocene and into the full glacial period. There have also been some interesting developments that have shown that Pacific salmon are important vectors of contaminants (e.g., Ewald et al. 1998; Krümmel et al. 2003; Gregory-Eaves et al. 2007).

Paleolimnological research has and will continue to play a central role in defining sockeye salmon population variability. Given that monitoring data are lacking in many areas and, where they exist, are often of insufficient length or quality to determine how large-scale human interventions have impacted Pacific salmon, paleolimnological research is clearly a critical tool in developing sustainable fisheries management plans. This

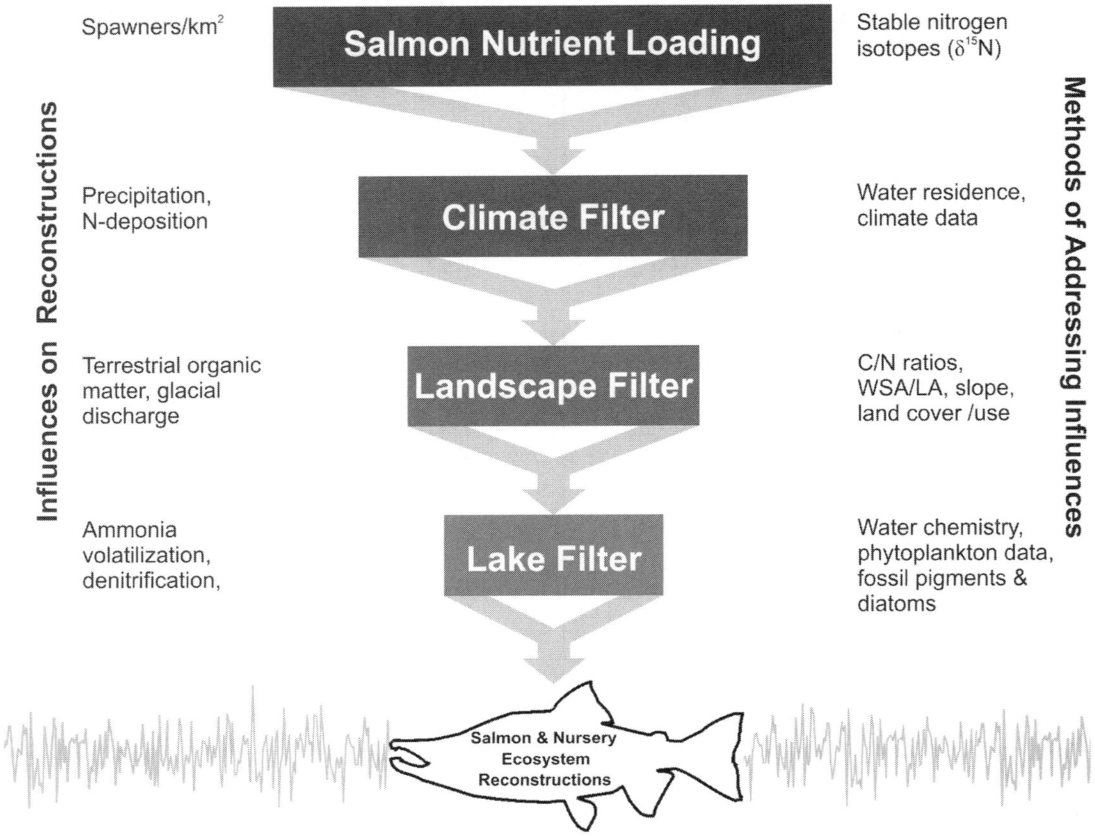

Figure 4.—A conceptual model that illustrates the hierarchy of decision making an investigator should consider when selecting a site for paleolimnological salmon reconstruction. Along the right of the diagram are methods that can be used to guide each level of decision making. (Modified from Selbie 2008)

need is further heightened by the importance of Pacific salmon in the ecology, economy, and culture of North Pacific communities and the threatened status of many stocks.

Acknowledgments

The Gordon and Betty Moore Foundation, the Natural Science and Engineering Research Council, the U.S. Global Ocean Ecosystem Dynamics program, Alaska Sea Grant program, and the NOAA Auke Bay Laboratory Ocean Carrying Capacity program provided funding for this research. We would also like to acknowledge collaborators (past and present) with whom we have had stimulating discussions: Jules Blais, Guangjie Chen, Brian Cumming, Marianne Douglas, Eva Krümmel, Peter Leavitt, Robie MacDonald, and Daniel Schindler.

References

Beamish, R. J. 1993. Climate and exceptional fish production off the west coast of North America. Canadian Journal of Fisheries and Aquatic Sciences 50:2270–2291.

Beamish, R. J., and D. R. Bouillon. 1993. Pacific salmon production trends in relation to climate. Canadian Journal of Fisheries and Aquatic Sciences 50:1002–1016.

Brock, C. S., P. R. Leavitt, D. E. Schindler, S. P. Johnson, and J. W. Moore. 2006. Spatial variability of stable isotopes and fossil pigments in surface sediments of Alaskan coastal lakes: constraints on quantitative estimates of past salmon abundance. Limnology and Oceanography 51:1637–1647.

Brock, C. S., P. R. Leavitt, D. E. Schindler, and P. D. Quay. 2007. Variable effects of marine-derived nutrients on algal production in salmon nursery lakes of Alaska during the past 300 years. Limnology and Oceanography 52:1588–1598.

Brodeur, R. D., and D. M. Ware. 1992. Long-term variability in zooplankton biomass in the subarctic Pacific Ocean. Fisheries Oceanography 1:32–38.

Burgner, R. L. 1991. Life history of sockeye salmon. Pages 1–117 *in* C. Groot and L. Margolis, editors. Pacific salmon life histories. UBC Press, Vancouver.

Clark, W. G., and S. R. Hare. 2002. Effects of climate and stock size on recruitment and growth of Pacific halibut. North American Journal of Fisheries Management 22:852–862.

Collie, J., S. Saila, C. Walters, and S. Carpenter. 2000. Of salmon and dams. Science 290:933–934.

Conley, D. J., and C. L. Schelske. 2001. Biogenic silica. Pages 1–13 *in* J. P. Smol, H. J. B. Birks, and W. M. Last, editors. Tracking environmental change using lake sediments, volume 3: terrestrial, algal and siliceous indicators. Kluwer Academic Publishers, Dordrecht, Netherlands.

D'Arrigo, R. D., G. Wiles, G. Jacoby, and R. Villalba. 1999. North Pacific sea surface temperatures: past variations inferred from tree rings. Geophysical Research Letters 26:2757–2760.

Durbin, A. G., S. W. Nixon, and C. A. Oviatt. 1979. Effects of the spawning migration of the alewife, *Alosa pseudoharengus*, on freshwater ecosystems. Ecology 60:8–17.

Ebbesmeyer, C. C., D. R. Cayan, D. R. McLain, F. H. Nichols, D. H. Peterson, and K. T. Redmond. 1991. 1976 step in the Pacific climate: forty environmental changes between 1968–1975 and 1977–1984. Pages 115–126 *in* J. L. Betancourt and V. L. Tharp, editors. Proceedings of the Seventh Annual Pacific Climate (PACLIM) Workshop. California Department of Water Resources, Interagency Ecological Studies Program Technical Report 26, Sacramento, California.

Edmundson, J. A., and S. R. Carlson. 1998. Lake typology influences on the phosphorus–chlorophyll relationship in subarctic, Alaskan lakes. Journal of Lake and Reservoir Management 14: 440–450.

Ewald, G., P. Larsson, H. Linge, L. Okla, and N. Szarzi. 1998. Biotransport of organic pollutants to an inland Alaska lake by migrating sockeye salmon (*Oncorhynchus nerka*). Arctic 51:40–47.

Finney, B. P. 1998. Long-term variability of Alaskan sockeye salmon abundance determined by analysis of sediment cores. North Pacific Anadromous Fisheries Council Bulletin 1:388–395.

Finney, B. P., I. Gregory-Eaves, M. S. V. Douglas, and J. P. Smol. 2002. Fisheries productivity in the northeast Pacific over the past 2,200 years. Nature (London) 416:729–733.

Finney, B. P., I. Gregory-Eaves, J. Sweetman, M. S. V. Douglas, and J. P. Smol. 2000. Impacts of climatic change and fishing on Pacific salmon abundance over the past three hundred years. Science 290:795–799.

Fisher-Wold, A. K., and A. E. Hershey. 1999. Effects of salmon carcass decomposition on biofilm growth and wood decomposition. Canadian Journal of Fisheries and Aquatic Sciences 56:767–773.

Francis, B., and S. R. Hare. 1994. Decadal-scale regime shifts in the large marine ecosystems of the northeast Pacific: a case for historical science. Fisheries Oceanography 3:279–291.

Gargett, A. E. 1997. The optimal stability 'window': a mechanism underlying decadal fluctuations in North Pacific salmon stocks? Fisheries Oceanography 6:109–117.

Gregory-Eaves, I., M. J. Demers, M. J., L. Kimpe, E. M. Krümmel, R. W. Macdonald, B. P. Finney, and J. M. Blais. 2007. Tracing salmon-derived nutrients and contaminants in freshwater food webs across a pronounced spawner density gradient. Environmental Toxicology and Chemistry 26:135–143.

Gregory-Eaves, I., B. P. Finney, M. S. V. Douglas, and J. P. Smol. 2004. Inferring sockeye salmon (*Oncorhynchus nerka*) population dynamics and water-quality changes in a stained nursery lake over the past ~500 years. Canadian Journal of Fisheries and Aquatic Sciences 61:1235–1246.

Gregory-Eaves, I., J. P. Smol, M. S. V. Douglas, and B. P. Finney. 2003. Diatoms and sockeye salmon (*Oncorhynchus nerka*) population dynamics: reconstructions of salmon-derived nutrients in two lakes from Kodiak Island, Alaska. Journal of Paleolimnology 30:35–53.

Gross, H. P., W. A. Wurtsbaugh, and C. Luecke. 1998. The role of anadromous sockeye salmon in the nutrient loading and productivity of Redfish Lake, Idaho. Transactions of the American Fisheries Society 127:1–18.

Hall, R. I., and J. P. Smol. 1999. Diatoms as indicators of lake eutrophication. Pages 128–168 *in* E. F. Stoermer and J. P. Smol, editors. The diatoms: application for the environmental and earth sciences. Cambridge University Press, Cambridge, UK.

Hare, S. R., and N. J. Mantua. 2000. Empirical evidence for North Pacific regime shifts in 1977 and 1989. Progress in Oceanography 47:103–145.

Hare, S. R., N. J. Mantua, and R. C. Francis. 1999. Inverse production regimes: Alaska and West Coast Pacific salmon. Fisheries 24:6–14.

Hobbs W. O., and A. P. Wolfe. 2007. Caveats on the use of paleolimnology to infer Pacific salmon returns. Limnology and Oceanography 52:2053–2061.

Hobbs W. O., and A. P. Wolfe. 2008. Recent paleolimnology of three lakes in the Fraser River basin (BC, Canada): no response to the collapse of sockeye salmon stocks following the Hells Gate landslides. Journal of Paleolimnology 40:295–308.

Holtham, A. J., I. Gregory-Eaves, M. Pellatt, D. T. Selbie, L. Stewart, B. P. Finney, and J. P. Smol. 2004. The influence of flushing rates, terrestrial input and low salmon escapement densities on paleolimnological reconstructions sockeye salmon (*Oncorhynchus nerka*) nutrient dynamics in Alaska and British Columbia. Journal of Paleolimnology 32:255–271.

Hyatt, K. D., D. J. McQueen, K. S. Shortreed, and D. P. Rankin. 2004. Sockeye salmon (*Oncorhynchus nerka*) nursery lake fertilization: review and summary of results. Environmental Reviews 12:133–162.

Kline, T. C., J. J. Goering, O. A. Mathiesen, P. H. Poe, P. L. Parker, and R. S. Scalan. 1993. Recycling of elements transported upstream by runs of Pacific salmon. II. $\delta^{15}N$ and $\delta^{13}C$ evidence in the Kvichak River watershed, Bristol Bay, Southwestern Alaska. Canadian Journal of Fisheries and Aquatic Sciences 50:2350–2365.

Kline, T. C., Jr., J. J. Goering, and R. J. Piorkowski. 1997. The effect of salmon carcasses on Alaskan freshwaters. Pages 205–227 *in* A. M. Milner and M. W. Oswood, editors. Freshwaters of Alaska, Ecological Studies 119. Springer, New York.

Koenings, J. P., and R. D. Burkett. 1987. An aquatic Rubic's cube: restoration of the Karluk Lake sockeye salmon (*Oncorhynchus nerka*). Pages 419–434 *in* H. D. Smith, L. Margolis, and C. C. Wood, editors. Sockeye salmon (*Oncorhyncus nerka*) population biology and future management. Canadian Special Publication of Fisheries and Aquatic Sciences 96.

Korhola, A., and M. Rautio. 2001. Cladocera and other branchiopod crustaceans. Pages 5–41 *in* J. P. Smol, H. J. B. Birks, and W. M. Last, editors. Tracking environmental change using lake sediments, volume 4: zoological indicators. Kluwer Academic Publishers, Dordrecht, Netherlands.

Krohkin, E. M. 1975. Transport of nutrients by salmon migrating from the sea into lakes. Pages 153–156 *in* A. D. Hasler, editor. Coupling of land and water systems. Springer-Verlag, New York

Krümmel, E., R. MacDonald, L. Kimpe, I. Gregory-Eaves, M. Demers, J. P. Smol, B. Finney, and J. M. Blais. 2003. Delivery of pollutants by spawning salmon. Nature 425:255–256.

Leavitt, P. R., and D. A. Hodgson. 2001. Sedimentary pigments. Pages 295–325 *in* J. P. Smol, H. J. B Birks, and W. M. Last, editors. Tracking environmental changes using lake sediments, volume 3: terrestrial, algal, and siliceous indicators. Kluwer Academic Publishers, Dordrecht, Netherlands.

Lichatowich, J. 1999. Salmon without rivers: a history of the Pacific salmon crisis. Island Press, Washington, D.C.

MacDonald, G. M., and R. A. Case. 2005. Variations in the Pacific Decadal Oscillation over the past millennium. Geophysical Research Letters 32:L08703.

Mann, C. C., and M. L. Plummer. 2000. Can science rescue salmon? Science 289:716–719.

Mantua, N. J., S. R. Hare, Y. Zhang, J. M. Wallace, and R. C. Francis. 1997. A Pacific interdecadal climate oscillation with impacts on salmon production. Bulletin of the American Meteorological Society 78(6):1069–1079.

Mathisen, O. A. 1972. Biogenic enrichment of sockeye salmon lakes and stock productivity. Verhandlungen der Internationalen Vereinigung Limnologen 18:1089–1095.

McGowan, J. A., S. J. Bograd, R.J. Lynn, and A. J. Miller. 2003. The biological response to the 1977 regime shift in the California Current. Deep Sea Research Part II 50:2567–2582.

Meyers, P. A., and R. Ishiwatari. 1993. Lacustrine organic geochemistry: an overview of indicators of organic matter sources and diagenesis in lake sediments. Organic Geochemistry 20:867–900.

Michener, R. H., and D. M. Schell. 1994. Stable isotope ratios as tracers in marine aquatic food webs. Pages 138–157 *in* K. Lajtha and R. H. Michener, editors. Stable isotopes in ecology and environmental science. Blackwell Scientific Publications, Oxford, UK.

Miller, A. J., F. Chai, S. Chiba, J. R. Moisan, and D. J. Neilson. 2004. Decadal-scale climate and ecosystem interactions in the North Pacific Ocean. Journal of Oceanography 60:163–188.

Minagawa, M., and E. Wada. 1984. Stepwise enrichment of ^{15}N along food chains: further evidence and the relation between $\delta^{15}N$ and animal age. Geochimica et Cosmochimica 48:1135–1140.

Minobe, S. 1997. A 50–70 year climatic oscillation over the North Pacific and North America. Geophysical Research Letters 24:683–686.

Mundy, P. R. 1997. The role of harvest management in the future of Pacific salmon populations: shaping human behavior to enable the persistence of salmon. Pages 315–329 *in* D. J. Stouder, P. A. Bisson, and R. J. Naiman, editors. Pacific salmon and their ecosystems: status and future options. Chapman and Hall, New York.

Naiman, R. J., J. M. Helfield, K. K. Bartz, D. C. Drake, and J. M. Honea. 2009. Pacific salmon, marine-derived nutrients and the dynamics of aquatic ecosystems. Pages 395–425 *in* A. J. Haro, K. L. Smith, R. A. Rulifson, C. M. Moffitt, R. J. Klauda, M. J. Dadswell, R. A. Cunjak, J. E. Cooper, K. L. Beal, and T. S. Avery, editors. Challenges for diadromous fishes in a dynamic global environment. American Fisheries Society, Symposium 69, Bethesda, Maryland.

Nehlsen, W. 1997. Pacific salmon status and trends: a

coastwide perspective. Pages 41–50 *in* D. J. Stouder, P. A. Bisson, and R. J. Naiman, editors. Pacific salmon and their ecosystems: status and future options. Chapman and Hall, New York.

Northcote, T. G., and D. Y. Atagi. 1997. Pacific salmon abundance trends in the Fraser River Watershed compared to other British Columbia systems. Pages 199–219 *in* D. J. Stouder, P. A. Bisson, and R. J. Naiman, editors. Pacific salmon and their ecosystems: status and future options. Chapman and Hall, New York.

Polovina, J. J., G. T. Mitchum, and G. T. Evans. 1995. Decadal and basin-scale variation in mixed layer depth and the impact on biological production in the Central and North Pacific, 1960–88. Deep Sea Research Part 1 42:1701–1716.

Pyper, B. J., F. J. Mueter, and R. M. Peterman. 2005. Across-species comparisons of spatial scales of environmental effects on survival rates of northeast Pacific salmon. Transactions of the American Fisheries Society 134:86–104.

Schindler, D. E., P. R. Leavitt, C. S. Brock, S. P. Johnson, and P. D. Quay. 2005. Marine-derived nutrients, commercial fisheries, and production of salmon and lake algae in Alaska. Ecology 86:3225–3231.

Selbie, D. T. 2008. Large-scale exogenous forcing of long-term Pacific salmon production and ecosystem interactions in western North America. Ph.D thesis. Queen's University, Kingston, Ontario.

Selbie, D. T., B. Lewis, J. P. Smol, and B. P. Finney. 2007. Long-term population dynamics of endangered Snake River sockeye salmon: evidence of past influences on stock decline and impediments to recovery. Transactions of the American Fisheries Society 136:800–821.

Shortreed, K. S., K. F. Morton, K. Malange, and J. M. B. Hume. 2001. Factors limiting juvenile sockeye production and enhancement potential for selected B.C. nursery lakes. Canadian Science Advisory Secretariat, Research Document 2001/098, Ottawa.

Smol, J. P. 2008. Pollution of lakes and rivers: a paleolimnological perspective, 2nd edition. Blackwell Scientific Publications Publishing, Oxford, UK.

Stockner J., A. Langston, D. Sebastian, and G. Wilson. 2005. The limnology of Williston Reservoir: British Columbia's largest lacustrine ecosystem. Water Quality Research Journal of Canada 40:28–50.

Swanson, H. K., and K. A. Kidd. 2009. A preliminary investigation of the effects of anadromous Arctic char on food web structure and nutrient transport in coastal Arctic lakes. Pages 465–483 *in* A. J. Haro, K. L. Smith, R. A. Rulifson, C. M. Moffitt, R. J. Klauda, M. J. Dadswell, R. A. Cunjak, J. E. Cooper, K. L. Beal, and T. S. Avery, editors. Challenges for diadromous fishes in a dynamic global environment. American Fisheries Society, Symposium 69, Bethesda, Maryland.

Sweetman, J. N., and B. P. Finney. 2003. Differential responses of zooplankton populations (*Bosmina longirostris*) to fish predation and nutrient-loading in an introduced and a natural sockeye salmon nursery lake on Kodiak Island, Alaska, USA. Journal of Paleolimnology 30:183–193.

Wertheimer, A. C. 1997. Status of Alaska salmon. Pages 179–197 *in* D. J. Stouder, P. A. Bisson, and R. J. Naiman, editors. Pacific salmon and their ecosystems: status and future options. Chapman and Hall, New York.

Whitlock, C., and C. P. S. Larsen. 2001. Charcoal as a fire proxy. Pages 75–98 *in* J. P. Smol, H. J. B. Birks, and W. M. Last, editors. Tracking environmental change using lake dediments, volume 3: terrestrial, slgal and diliceous indicators. Kluwer Academic Publishers, Dordrecht, Netherlands.

American Fisheries Society Symposium 69:395–425, 2009

Pacific Salmon, Marine-Derived Nutrients, and the Characteristics of Aquatic and Riparian Ecosystems

Robert J. Naiman*
School of Aquatic and Fishery Sciences, University of Washington Box 355020, Seattle, Washington 98195, USA

James M. Helfield
Department of Environmental Sciences, Huxley College of the Environment Western Washington University, Bellingham, Washington 98225, USA

Krista K. Bartz
National Marine Fisheries Service, Northwest Fisheries Science Center 2725 Montlake Boulevard East, Seattle, Washington 98112, USA

Deanne C. Drake[1]
The Ecosystem Center, Marine Biological Laboratory Woods Hole, Massachusetts 02543, USA

Jon M. Honea
National Marine Fisheries Service, Northwest Fisheries Science Center 2725 Montlake Boulevard East, Seattle, Washington 98112, USA

Abstract.—We review the current understanding of major pathways, mechanisms, and consequences of salmon-borne marine-derived nutrients (MDN) in estuarine, freshwater, and riparian ecosystems. Semelparous Pacific salmon *Oncorhynchus* spp. acquire most of their body mass while at sea before returning to spawn and die in natal streams. The annual spawning migrations transport substantial quantities of MDN from the fertile North Pacific Ocean to relatively nutrient-poor coastal freshwater and terrestrial ecosystems. People have been long aware of the importance of salmon-borne MDN for the productivity of freshwater ecosystems in western North America, and the rapidly increasing knowledge base supports this notion. Nevertheless, many details associated with nutrient pathways, cycling processes, and the ecosystem-scale consequences of MDN transfer remain to be elucidated. The collective data suggest that freshwater portions of the salmon production system, as well as the dynamics of local terrestrial plant and animal communities, are intimately linked to MDN in complex ways. At the same time, the ecological importance of MDN, relative to other major nutrient sources, is temporally and spatially dependent and influenced by the life histories and abundances of salmonid stocks. Although interactions among climate cycles, salmon, riparian vegetation, predators, and MDN flowpaths and feedbacks are complex, they also form a wonderfully integrated ecological system with a high degree of resilience and productivity. Understanding this complex system and its inherent temporal and spatial variability requires a holistic scientific perspective that values important interactions among the salmonid life cycle, the physical setting, and the numerous linkages to other ecosystem components.

* Corresponding author: naiman@u.washington.edu
[1] Current address: School of Animal, Plant and Environmental Sciences, University of the Witwatersrand, Private Bag 3, Johannesburg 2050, South Africa

Introduction

Anadromous Pacific salmon *Oncorhynchus* spp. spend more than half of their lives at sea before returning to spawn and die in natal streams. Consequently, annual spawning migrations provide a mechanism for transporting nutrients from the fertile northern Pacific Ocean to freshwater and terrestrial ecosystems. This phenomenon has been recognized for centuries. Aboriginal peoples marked and celebrated the initial return of adult salmon as tangible symbols of health, replenishment, and vigor. Returning salmon were widely regarded as vessels of nutrition, and it was accepted that nature would be fundamentally different, and poorer, without them (Roche and McHutchison 1998).

These basic insights have not changed. Salmon continue to play a central role in the cultures, economies, and ecosystems of the Pacific Coastal Ecoregion. Returning salmon replenish and help sustain aquatic ecosystems, although in recent decades, their numbers have diminished greatly in many areas (Stouder et al. 1997; Gresh et al. 2000). These declines have brought increased attention to the management and restoration of salmon stocks, fueling an intense debate about salmon–ecosystem interactions. Researchers are discovering an amazingly complex process that is surprisingly sensitive to climate, latitude, land use, species, and individual stock characteristics. This exciting story has captured the attention of modern naturalists in much the same way it captured the imagination of historic generations. However, with continued exploitation of Pacific salmon, and with projected changes in climate and land use, it is not yet clear how the story ultimately will unfold.

The accumulation of knowledge about salmon and ecosystem processes in recent years is impressive (e.g., Willson et al. 1998; Cederholm et al. 2000; Gende et al. 2002; Naiman et al. 2002; Schindler et al. 2003). However, the accumulation of recent insights suggests it is timely to provide an updated synthesis, especially concerning the salmon as vectors for nutrient movements and the fate and consequences of those nutrients in aquatic and riparian communities. Here, we review the major pathways, mechanisms, and consequences of salmon-borne marine-derived nutrients (MDN) in estuarine, freshwater, and riparian ecosystems. In so doing, we seek to identify gaps in understanding and elucidate the breadth of MDN influence on the long-term viability of salmon populations and on the structure and dynamics of coupled salmon-aquatic-riparian ecosystems. We start by examining the use of stable isotopes in tracking MDN. We then consider the major roles, processes and effects of MDN in aquatic and riparian ecosystems before identifying key emerging issues and concepts.

The Use of Stable Isotopes in Tracking Marine-Derived Nutrients

The fate of marine-derived carbon (C), nitrogen (N), and sulfur (S) may be quantified through analyses of naturally occurring stable isotopes. Over time, the biogeochemical cycling of elements has resulted in a predictable distribution of heavy and light isotopes in nature. For example, atmospheric C (CO_2) is enriched in the heavier isotope (a higher $^{13}C/^{12}C$ ratio relative to carbon in the oceans), but marine C is also enriched relative to terrestrial organic matter in plants and soils. Marine SO_4^{2-} is enriched in the heavier isotope (^{34}S), while freshwater SO_4^{2-} is highly variable, encompassing enriched marine values and much more depleted values. The heavier isotope of nitrogen (^{15}N) is present in higher abundance in the oceans and also accumulates as trophic level increases (Peterson and Fry 1987; Fry 2006). As a result of these natural distributions and trophic accumulation, C, N, and S in returning anadromous salmon tissues are isotopically distinct from most components of freshwater and riparian ecosystems (Mathisen et al. 1988; Kline et al. 1990, Scheuerell et al. 2005). N isotopes have been the most widely used natural tracer of fate of salmon tissues because N is often a limiting nutrient, so there is greater confidence that it is taken up when available. N is also more abundant in biological pools (salmon consumers, soils) than sulfur, facilitating mass spectroscopy analyses. Carbon isotopes are used to determine the importance of salmon in the diets of consumers (invertebrates, fish; Bilby et al. 1996) but are not applicable for the fate of MDN in primary producers, which derive C from the atmosphere. Sulfur isotopes are not used widely in salmon research, but a few studies have shown promise (e.g., Hesslein et al. 1991; MacAvoy et al. 1998).

Isotopic ratios, typically expressed as δ values, represent the parts per thousand deviation in a sample from an international isotope standard. For example, the international N standard is atmospheric N_2, which contains 0.3663% ^{15}N, or $\delta^{15}N$ = 0‰. N in salmon tissues contain slightly more of the heavier isotope (^{15}N), frequently in the range of 0.3707% ^{15}N or $\delta^{15}N$ = 12‰. A common approach is to compare $\delta^{13}C$, $\delta^{15}N$, or $\delta^{34}S$ values observed at sites with spawning salmon to those observed at comparable sites to which anadromous access is blocked (e.g., Helfield and Naiman 2001). δ values closer to those of salmon than reference site values are indicative of salmon nutrients. Observed δ values may be converted to proportions of marine-derived C, N, or S through use of linear two-source mixing models (Kline et al. 1990; Bilby et al. 1996), but two-source models are generally oversimplifications of the complex nutrient cycles in natural systems. Fortunately, there is a growing body of work on multisource models (Phillips and Gregg 2003; Phillips et al. 2005) yielding new ecological insights.

Use of stable isotopes to quantify MDN contributions to freshwater or riparian ecosystems is predicated on the assumption that MDN (C, N, and S) are isotopically distinct from other nutrient sources. All inferences are based on differences between the fractions being examined, and subtle environmental influences such as resorption of nutrients from senescing leaves or age of the plant can result in biases (Fry 2006). Isotopic fractionation associated with various biogeochemical processes alters isotopic signals and can confound inferences about MDN. In freshwater ecosystems, such confounding effects are generally considered insignificant. In terrestrial ecosystems, biogeochemical processes that have high fluxes generally have slight fractionations (N mineralization, nitrification, N_2-fixation), whereas those with potentially large fractionations (ammonia volatilization, denitrification) tend to have relatively small fluxes and are generally confined to specific areas within the landscape (Nadelhoffer and Fry 1994). Riparian zones are among those areas within the landscape where processes with high fractionations are most likely to occur. For example, studies of riparian soils adjacent to spawning streams suggest that under anaerobic conditions with sufficient influx of nitrate (NO_3^-), denitrification could result in ^{15}N enrichment (e.g., Pinay et al. 2003); however, there is no evidence to date to suggest that MDN enrichments are significantly process-based. Nevertheless, metabolic modifications by scavengers and decomposers in transferring MDN from salmon to riparian ecosystems may result in fractionations that, if not accounted for, can cause errors in mixing model estimates (Morris et al. 2005). Additionally, there are issues with propagation of error in mixing models (Fry 2006).

Recent technological advances have made stable isotope analyses more accessible, allowing for investigators to use a wide range of organic matter types. This has contributed greatly to understanding the fate of MDN, but studies are limited to elements for which naturally occurring stable isotopes exist (e.g., C, N). This is often to the exclusion of other potentially important elements for which no such tracers exist (e.g., P) or, if present, are expensive to analyze (e.g., S). As a result, comparatively few studies have examined the pathways and effects of marine-derived P, Mg, or Ca in stream and riparian ecosystems. Fortunately, isotopes of Si, Ca, and Fe are starting to be used (Clementz et al. 2003; Varele et al. 2004; Rouxel et al. 2005) and may eventually have applications in better understanding MDN cycling processes and rates.

Major Roles and Processes of Marine-Derived Nutrients in Aquatic and Riparian Ecosystems

Effects on Estuary Ecosystems

Relatively little is known about the potential impacts of salmon-borne MDN on rearing habitat, trophic interactions, or salmonid growth in estuarine ecosystems. The few studies that have included estuarine systems report mixed results regarding the importance of MDN. On one hand, this is not surprising considering the tidal influences and the variability in estuarine residence times among salmon species. On the other hand, one might expect a clear MDN impact, given that pink salmon *O. gorbuscha* and chum salmon *O. keta* salmon frequently spawn short distances upstream and that salmon-borne nutrients and postspawn carcasses are washed downstream (Brickell and Goering 1970; Goering and Brickell 1972). Furthermore, estuaries are used extensively by juvenile chum salmon and ocean-type Chinook salmon *O. tshawytscha*, and all anadromous salmon must spend some portion of their life cycle there (Groot and Margolis 1991).

There is some documentation that estuarine invertebrates utilize MDN, based on their isotopic signatures. Two species of clams located in the high intertidal zone near a river with abundant spawners, relative to the same species located in the mid-intertidal zone, exhibited elevated $\delta^{15}N$ values (Chow 2007). Sedimentary organic matter, an indirect measure of the clams' food source, was also enriched in the high intertidal zone. Further, in Seldovia Bay, Alaska harpacticoid copepods consume the macroalga *Ulva* sp. as a primary food source, and *Ulva* utilizes salmon-borne MDN when available (Fujiwara and Highsmith 1997). Since the stomach contents of juvenile chum salmon rearing in the estuary were dominated by harpacticoid copepods, an indirect link was proposed between adult and juvenile chum, whereby MDN inputs enhance algal and harpacticoid production, ultimately increasing food availability and smolt production. According to other studies, nutrient inputs appear to have little effect on growth of juvenile salmonids in estuaries. For example, decomposition of adult salmon carcasses resulted in increased concentrations of organic N in Iliuliuk Bay, Alaska, but no beneficial effects were observed for juvenile salmon (Brickell and Goering 1970). Similarly, the application of urea fertilizer to the Yakoun River estuary on the Queen Charlotte Islands, British Columbia resulted in no appreciable enhancement of salmonid growth or survival (Stockner and Levings 1982).

Several factors may account for the lack of influence of salmon-borne MDN on juvenile salmon in estuaries. In some systems, smolt production may be limited by factors unrelated to estuarine habitat (e.g., egg to fry survival in streams). Where estuarine production is important, estuaries might be less prone to nutrient limitation because they receive inputs from the entire catchment, as well as the ocean, and are typically among the most nutrient-rich environments within the catchment. Moreover, nutrient inputs might be sequestered by estuarine sediments before they can be incorporated into the food web. In the Brickell and Goering (1970) study, salmon carcass decomposition resulted in increased dissolved organic N concentrations but no change in NH_4^+, as estuarine sediments sequestered organic N before it could be mineralized. This process might be exacerbated by the fact that estuaries typically have fine sediments with high capacities for adsorption of organic compounds or by periodic drying and inundation, which results in alternating oxidizing and reducing conditions causing specific adsorption of dissolved elements onto hydrous oxides of iron (Fe) and aluminum (Al), thereby transforming dissolved nutrients into less soluble and less bioavailable forms (Faulkner and Richardson 1989). Similarly, aquatic macrophytes tend to be more abundant in estuaries than in riverine or lacustrine habitats, and macrophytic sequesteration may divert some nutrient inputs from the planktonic trophic web and render them inaccessible to juvenile salmon. However, there is some evidence to suggest that vascular macrophytes such as eelgrass *Zostera marina* do not use MDN inputs to the same extent as macroalgae (e.g., *Ulva*; Fujiwara and Highsmith 1997). In estuaries where MDN inputs enhance algal and harpacticoid production, beneficial effects on juvenile salmon might be offset by detrimental effects of adult salmon carcasses. For example, carcass decomposition in estuarine sediment may result in decreased dissolved oxygen, which may affect egg survival in spawning streams near the intertidal zone (Brickell and Goering 1970; Goering and Brickell 1972), as well as attract scavengers or predators, which might also prey on fry and smolts (Eastman 1996; Bilby et al. 1998).

Effects on Lake Ecosystems

Nutrient status and primary producers.—Nutrient inputs from returning salmon have been long known to affect the N and P status of oligotrophic lakes (e.g., Juday et al. 1932; Barnaby 1944; Donaldson 1967). Lake studies have focused primarily on sockeye salmon *O. nerka*, a species that typically spawns in dense aggregations in tributary streams or along lake beaches (Burgner 1991; Quinn 2005). Marine-derived nutrients are transferred to lakes via downstream movement from spawning tributaries and via in situ decomposition of beach spawners. Sockeye-producing lakes are typically oligotrophic, and primary production is often severely nutrient-limited (Goldman 1960; Hyatt and Stockner 1985; Wurtsbaugh et al. 1997). Consequently, MDN inputs, depending on the physical setting, have significant potential to stimulate overall lake primary productivity.

Since flushing rates tend to be low in lakes, MDN inputs may be comparatively high in relation to losses. For example, in Iliamna Lake, Alaska, annual salmon-borne inputs are equivalent to 50–100% of total annual dissolved N losses via flushing (Kline et al. 1993) and comprise more than 50% of total lake P (Donaldson 1967). Similarly, on the Kamchatka Peninsula of northeastern Russia, spawning sockeye contribute an estimated 30–40% of total annual P inputs (Krohkin 1975). In contrast, simulation models of sockeye-producing lakes in Idaho indicate that MDN historically has been of minimal importance, contributing only 3% of annual P budgets, much of which is quickly lost due to high flushing rates and P export by smolts (Gross et al. 1998). Regardless of flushing rates, the significance of MDN inputs may be reduced by dilution in large volume lakes (Krohkin 1967; Durbin et al. 1979). For example, Krohkin (1967) estimated a minimum of 500 salmon per 106 m^3 of water would be required for P inputs to be significant in lakes. However, effective concentrations may be greater in localized habitats such as near nesting colonies of avian scavengers (Payne and Moore 2006), the mouths of spawning tributaries, or in the vicinity of beach spawning. Beach spawning typically occurs in areas of groundwater or hyporheic upwelling (Burgner 1991; Quinn 2005), where increased water circulation may promote mixing. In general, MDN inputs are more likely to influence production in lakes where spawning densities and contributions of salmon derived nutrients are high relative to lake nutrient pools and where flushing rates are low.

Studies of Alaskan and Russian lakes have demonstrated a link between MDN and primary production. Temporal analyses of sockeye-producing lakes indicate increased phytoplankton biomass and primary production in years of high salmon escapement (Krohkin 1975; Schmidt et al. 1998; Schindler et al. 2005), while comparisons with lakes to which anadromous access is blocked indicate significantly decreased P concentrations and phytoplankton standing stocks (Kyle 1996). These latter results, however, are confounded as most non-salmon lakes sit high in the hydrologic landscape and salmon lakes are generally low in the landscape. Marine-derived nutrient inputs might also influence species composition as well as productivity and biomass in lake algal communities. Analyses of fossilized diatoms and $\delta^{15}N$ signals in sediment cores indicate strong historical correlations between diatom assemblages and salmon spawning densities in Alaskan lakes (Finney et al. 2000, 2002; Gregory-Eaves et al. 2003, 2008). Periods of high salmon abundance correlate with increased abundance of mesotrophic to eutrophic indicators such as *Stephanodiscus minutulus/parvus* and *Fragilaria crotonensis*, whereas periods of low salmon abundance are marked by more oligotrophic taxa such as *Cyclotella comensis* and *C. ocellata*. These responses are similar to those seen following lake fertilization in other coastal systems (Ennis et al. 1983). Comparisons with nonsockeye reference lakes suggest that these responses reflect the influence of MDN inputs on diatom competitive dynamics rather than a correlated response of both MDN and diatom assemblages to changes in climate.

Zooplankton.—Responses of zooplankton to MDN inputs are less consistent than those of phytoplankton. These inconsistencies are most likely due to bottom-up effects of MDN fertilization being confounded by top-down effects of foraging by juvenile sockeye, who feed primarily on zooplankton (Burgner 1991) and who also increase in abundance with increased spawner returns. Some of the increases in algal production seen in sockeye lakes may be due to reductions in grazing by zooplankton (cf. Carpenter et al. 1985). While some studies have reported increased zooplankton production associated with high escapements (Krohkin 1975; Sweetman and Finney 2003) and positive correlations between spawner density and zooplankton $\delta^{15}N$ (Finney et al. 2000), others have reported that foraging by juvenile sockeye minimizes the effects of increased algal production (Hardy et al. 1986) or results in decreased zooplankton biomass (Schmidt et al. 1998).

Resident lake fishes.—Responses of other components of lake food webs to MDN inputs are poorly understood. Marine-derived nutrients may be incorporated into lake food webs via autotrophic sequestration of dissolved nutrients and subsequent consumption by zooplankton and planktivorous fishes or by direct consumption of salmon eggs and carcass tissues by fishes and benthic invertebrates. Direct consumption is likely not as important in lakes due to their typically large benthic areas and the fact that a significant proportion of MDN inputs are delivered as dissolved nutrients from spawning tributaries. For example, direct consumption of eggs and fry does not represent a significant pathway for incorporation of MDN in Iliamna Lake fishes, suggesting that lake food webs are enriched primarily through mineralization of MDN by autotrophic production (Kline et al. 1993; but see Foote and Brown 1998). Isotopic analyses demonstrate that MDN are used by benthic invertebrates and resident fishes in systems with large spawning runs (Mathisen et al. 1988; Kline et al. 1993), but relatively little is known about consequences of MDN for the structure and dynamics of lake food webs. These will likely vary according to the competitive dynamics and number of trophic levels present (Lienesch et al. 2005).

Feedback effects on salmon production.—Since juvenile sockeye production is correlated with lake primary production and forage resources (Burgner 1987; Koenings and Burkett 1987; Hume et al. 1996), it has been suggested that MDN from returning spawners may provide an important feedback for future stocks (Mathisen 1972; Mathisen et al. 1988). Accordingly, N and P fertilization projects have been undertaken by various management agencies to enhance production of sockeye (LeBrasseur et al. 1979; Stockner 1981, 1987; Hyatt and Stockner 1985; Griswold et al. 2003) and other salmon species that may forage as juveniles on zooplankton in oligotrophic lakes (Kyle 1994). Nonetheless, it is difficult to establish an unequivocal link between MDN and salmon production in lakes. Studies of coastal lakes in Alaska and Russia have demonstrated positive correlations between spawning density and smolt $\delta^{15}N$ values (Finney et al. 2000), and between spawner abundance and total primary and secondary production, including smolt production (Krohkin 1975; Schmidt et al. 1998). Lake to lake comparisons suggest that juvenile salmon obtain MDN through the lake food web and that enhanced lake productivity is associated with increased smolt production, but it is not clear that increased recruitment is a result of MDN enrichment of nursery habitat or if it is a result of increased spawning. Further, within a single lake, the pattern is not clear either. Over a minimum threshold, all salmon show strong negative density dependence in growth and survival. As MDN increase, so does primary production, but it does not necessarily produce more fish in a single lake (Schindler et al. 2005). Additionally, even where MDN inputs enhance primary and secondary production in nursery lakes, these effects will not be necessarily translated to increased smolt production where other factors (e.g., spawning habitat availability) are limiting.

Effects on River Ecosystems

Nutrient concentrations and primary production.—Spawning and subsequent decomposition of salmon cause seasonal pulses of dissolved and particulate nutrients in streams. These pulses typically are measured as changes in concentrations of N (NH_4^+, NO_3^-, total N) or phosphorus (PO_4^{-2}, soluble reactive P, total P), relative to background levels. More studies report elevated NH_4^+ in response to salmon than other forms of N, such as NO_3^-, which may be driven by discharge rather than spawning (Table 1). NH_4^+ is the form of nitrogen excreted by salmon as metabolic waste and is also the primary nitrogenous product of microbial decay of salmon carcasses (Drake et al. 2005). Conversely, studies report elevated concentrations of various forms of P, with no single form dominating the measurements (Table 1). These nutrient pulses generally increase with run size or carcass abundance (e.g., Johnston et al. 2004; Mitchell and Lamberti 2005; Hood et al. 2007). Their duration varies; most studies report concentrations returning to background levels within 1–3 months of the pulse onset (Sugai and Burrell 1984; Johnston et al. 2004; Claeson et al. 2006).

Several mechanisms have been proposed to explain how salmon alter stream nutrient concentrations during and after spawning (summarized in Moore et al. 2007). First, the excretion of dissolved nutrients from dead and dying salmon could boost concentrations in the water column. Alternatively, the creation of redds could release antecedent dissolved nutrient pools trapped in the interstitial spaces below the sediment surface. The creation of redds also could resuspend nutrient-laden sediments and organic matter. Finally, redd creation could scour or bury periphyton that would otherwise sequester dissolved nutrients during primary production. It is likely that a combination of these mechanisms contributes to the seasonal pulse of nutrients (Moore et al. 2007), although observed increases in concentrations following the addition of carcasses to mesocosms and nonsalmon-bearing streams suggest that the first mechanism may be sufficient to cause significant changes in water chemistry on its own (Mitchell and Lamberti 2005; Claeson et al. 2006).

The response of freshwater primary producers to the annual influx of MDN is highly variable. It is often quantified as a change in chlorophyll *a* or ash-free dry mass per unit area, and it may be as

Table 1.—Summary of changes in water column nutrients and epilithic primary production during and after salmon spawning or carcass addition in various studies. Changes relative to reference sites are indicated as + (increases), – (decreases), 0 (no significant change), or blank (variable was not measured).

	Carbon	Nitrogen					Phosphorus				Primary production	
Source	DOC	DON	NH_4^+	NO_3^-	NO_2^{-2}	TN	PO_4^{-2}	SRP	TDP	TP	Chl *a*	AFDM
Brickell and Goering 1970			+			+						
Richey et al. 1975			+	+			+				+	
Sugai and Burrell 1984	0		+	0	0		+					
Schuldt and Hershey 1995			0	0	0	0		+		+	+	
Bilby et al. 1996			0	+					0			
Wipfli et al. 1998												+ , 0
Fisher Wold and Hershey 1999											+	+
Minakawa and Gara 1999			+						+		–	
Wipfli et al. 1999											+	+
Chaloner et al. 2004, 2007	0		+	0				+			+	+
Johnston et al. 2004			+	0		0		+	+	+	+	
Moore et al. 2004											–	
Sarica et al. 2004	+		+	+						+		
Mitchell and Lamberti 2005	0		+	0				+			+ , 0	0
Claeson et al. 2006	0	0	+	0	0			0			0	
Hood et al. 2007	+	+	+	0				+				
Moore et al. 2007						+				+		

Notes: DOC = dissolved organic carbon; DON = dissolved organic nitrogen; TN = total nitrogen; SRP = soluble reactive phosphorus; TDP = total dissolved phosphorus; TP = total phosphorus; Chl *a* = Chlorophyll *a*; AFDM = ash-free dry mass.

great as a 20-fold increase in chlorophyll a (Mitchell and Lamberti 2005) and a 15-fold increase in ash-free dry mass (Wipfli et al. 1998; Table 1). In other cases, redd creation decreases primary production (e.g., Moore et al. 2004) or no significant change in production parameters is observed (e.g., Claeson et al. 2006). This variability in response may be due, in part, to differences in study design, since artificial streams bypass any negative effects of redd excavation on periphyton abundance. The variability may also be due to other biotic and abiotic factors, including (1) run timing, magnitude, and distribution; (2) stream temperature, discharge, and shading; (3) background nutrient levels; (4) carcass retention capacity; (5) nutrient storage capacity; and (6) composition of the biological community (Wipfli et al. 1999; Ambrose et al. 2004). The influence of MDN on primary production likely will be greatest where spawning runs are dense, where uptake processes occur at rates greater than the rates at which nutrients are flushed from the system, and where nutrients are limiting production at the time of spawning.

Salmon also may be an important source of dissolved organic nutrients in streams (Table 1). For example, concentrations of dissolved organic C and N were significantly increased during spawning in a salmon-bearing humic-stained stream in southeast Alaska, relative to an upstream reference site (Hood et al. 2007). This study used fluorescence spectroscopy to trace salmon-derived amino acids (e.g., tyrosine), a technique which may be a valuable complement to stable isotope analyses for distinguishing MDN from other sources of organic nutrients. Other studies relying on different techniques typically reported no relationship between spawning salmon and dissolved organic nutrients (Table 1; but see Sarica et al. 2004).

Aquatic macroinvertebrate communities.—In addition to MDN contributions, salmon impact stream macroinvertebrate communities through disturbance of benthic habitats by upstream migration and by redd excavation. This includes altering sediment composition and removal of periphyton (Field-Dodgson 1987; Montgomery et al. 1996; Moore et al. 2004, 2007), as well as increasing rates of macroinvertebrate drift (Peterson and Foote 2000). Macroinvertebrate densities generally decrease (Field-Dodgson 1987; Minakawa and Gara 2003; Moore et al. 2004), although some taxa may increase in abundance (Moore et al. 2004), and disturbed sediments are recolonized by the end of the spawning season (Field-Dodgson 1987).

As spawners die, macroinvertebrates rapidly colonize the carcasses (Minakawa and Gara 1999; Chaloner et al. 2002b) and the redds (Claire and Phillips 1968; Nicola 1968; Ellis 1970; Elliot and Bartoo 1981), with densities of some taxa increasing relative to areas without MDN (Wipfli et al. 1998; Minakawa and Gara 2003; Walter et al. 2006). In their attraction to salmon, macroinvertebrates are selective, choosing higher quality carcasses where available. For example, caddisflies prefer bear-killed carcasses of prespawn salmon over senescent post-spawn carcasses, which are relatively depleted in energy-rich lipids (Winder et al. 2005).

Analyses of $\delta^{13}C$ and $\delta^{15}N$ confirm that macroinvertebrates consume and assimilate MDN, either directly from salmon tissue or indirectly via grazing of algae and fungi containing MDN or via predation on other MDN-enriched invertebrate taxa (Kline et al. 1990; Bilby et al. 1996; MacKenzie 2001; Chaloner et al. 2002a; Claeson et al. 2006). Evidence from natural streams and artificial microcosms indicates that macroinvertebrate density, standing stock, and individual growth rate are increased in the presence of salmon carcasses (Chaloner and Wipfli 2002; Minakawa et al. 2002). Responses vary by species, likely as a result of differences in feeding ecology. The most significant responses are seen in taxa belonging to the shredder and collector functional feeding groups, with collectors potentially playing an important role in transferring MDN to the benthic food web (Chaloner and Wipfli 2002). However, we note that some of these results are from mesocosm experiments in which the effects of redd excavation are excluded.

It is not clear to what extent observed increases in macroinvertebrate growth and density are indicative of MDN-enhancement. Increased secondary production due to MDN uptake may not make up for decreased standing stock due to benthic disturbances associated with spawning (Minakawa and Gara 2003). Furthermore, it may be that overall secondary production is unchanged as macroinvertebrates are drawn away from areas lacking MDN. For example, individuals of the caddisfly *Ecclisomyia conspersa* colonizing carcasses are older than those found elsewhere, suggesting that the larger caddisflies are drawn from adjacent areas without carcasses

(Walter et al. 2006). Further, there are reduced rates of alder *Alnus rubra* leaf decomposition as otherwise detritivorous caddisflies are drawn to more energy-rich salmon carcasses (Zhang et al. 2003). In this latter case, although effects on secondary production are unclear, the effects of MDN on macroinvertebrate distribution and organic matter cycling are substantial. A better understanding of the role of macroinvertebrates in transferring MDN to higher trophic levels will require research into secondary production with the use of stable isotope or other tracers to link production responses to differences in nutrient uptake.

Resident freshwater fishes.—Stable isotope analyses show that MDN are incorporated into resident freshwater fishes through several trophic pathways (Kline et al. 1990, 1993; Bilby et al. 1996). Marine-derived nutrient subsidies may affect the growth, condition, biomass, density, and lipid content of resident freshwater fishes (Bilby et al. 1998; Wipfli et al. 2003, 2004; Table 2). The duration of effects varies; some persist only as long as carcasses are available, while others last for months after decomposition. These analyses tend to have small sample sizes (n = 1 or 2 streams), necessitating the use of individual fish as pseudoreplicates (Wipfli et al. 2004). In one truly replicated study, carcass addition did not significantly affect the growth, biomass, or density of resident trout (Wilzbach et al. 2005), although a posthoc power analysis revealed little capacity to detect a response. During and after spawning, salmon eggs and flesh comprise a significant proportion of the diet of resident fishes, such as Arctic char *Salvelinus alpinus* (Eastman 1996), sculpin *Cottus* spp. (Foote and Brown 1998), and juvenile steelhead *O. mykiss* (Bilby et al. 1998). Before salmon eggs and carcasses are available for consumption, digging by female spawners disturbs benthic invertebrate communities, increasing drift densities and availability to Arctic char, rainbow trout *O. mykiss*, and Arctic grayling *Thymallus arcticus* (Peterson and Foote 2000).

Feedback effects on salmon populations via enhanced productivity within streams.—Each run of spawning salmon delivers a pulse of energy and nutrients and actively modifies the physical stream bed, all of which influence stream productivity. Local densities of juvenile salmonids increase during spawning (Giannico and Hinch 2007), and as juvenile and resident salmonids consume spawner tissue and eggs, their lipid content, condition factor, growth rate, and survivorship also increase (Bilby et al. 1998; Wipfli et al. 2003; Heintz et al. 2004). These feedbacks often occur in the winter, a critical period when other food sources are limited, and likely influence life histories since larger, more energy-dense fish have a greater propensity to out-migrate to marine systems (Jonsson and Jonsson 2005).

Spawning salmon affect their progeny and other salmonids indirectly as well. Redd excavation resets succession among epilithon communities at a time when spawner eggs, tissues, and remineralized nutrients become available to stimulate recolonization and sustain organisms throughout the trophic web (Hicks et al. 2005), including other salmonid species (Michael 1995; Wipfli et al. 2003). The link between stream productivity and salmonid production, as evidenced by the results of river fertilization experiments (e.g., Johnston et al. 1990; Deegan and Peterson 1992), suggests that depleted populations risk spiraling toward further depletion, exacerbating the challenge of recovery efforts (Larkin and Slaney 1997; Gresh et al. 2000; Scheuerell et al. 2005), and necessitating the development of nutrient restoration programs (e.g., Pearsons et al. 2007; Kohler et al. 2008). For this reason, the effects of MDN on stream productivity should be considered in setting fisheries escapement numbers, for example to a level above which an increase in MDN transfer is no longer observed (Bilby et al. 2001).

Potential adverse effects.—Salmon transport not only nutrients from marine to freshwater systems, but also persistent industrial pollutants. These pollutants include methylmercury (Zhang et al. 2001), polybrominated diphenyl ethers (Hites et al. 2004a), and various organochlorine contaminants such as polychlorinated biphenyls (PCBs; Krümmel et al. 2003; Hites et al. 2004b; Kelly et al. 2007) and polychlorinated dibenzodioxin and dibenzofuran (Kelly et al. 2007). Such compounds are lipophilic and biomagnified in relatively fatty fish that occupy higher trophic levels, such as salmon (Thomann et al. 1992). Because salmon species differ in marine distribution and diet, they also differ in their contaminant levels (O'Neill et al. 2006), with higher tissue concentrations typically found in species with coastal distributions (e.g., Chinook salmon) than in those with more oceanic distributions (e.g., pink salmon; Figure 1). Pollutants transported upstream

Table 2.—Responses of resident freshwater fish species to carcass or carcass analog additions, measured according to several parameters: + indicates that the parameter was higher in the carcass-enriched reach than in the control reach, – indicates that the parameter was lower in the carcass-enriched reach than in the control reach, and 0 indicates that the parameter did not differ significantly between the carcass-enriched and control reaches.

Species	Parameter	Units	Response	Duration	Location	Carcass load	Source
Cutthroat trout[a]	Biomass	% Δ in mg/m²/d	0	Aug–Sep	SE AK	0.54 kg/m²	Wipfli et al. 2004
Cutthroat trout	Biomass	% Δ in mg/m²/d	+	Aug–Sep	SE AK	[b]	Wipfli et al. 2004
Cutthroat trout	Condition	% Δ in mass/length³	+	Aug–Sep	SE AK	0.54 kg/m²	Wipfli et al. 2004
Cutthroat trout	Condition	% Δ in mass/length³	+	Aug–Sep	SE AK	[b]	Wipfli et al. 2004
Cutthroat trout	Growth	% Δ	+	Sep–May	SE AK	0.82 kg/m²	Wipfli et al. 2003
Cutthroat trout	Growth	% Δ	+	Sep–Oct	SE AK	0.82 kg/m²	Wipfli et al. 2003
Cutthroat trout	Growth	% Δ	–	Oct–May	SE AK	0.82 kg/m²	Wipfli et al. 2003
Cutthroat trout	Lipids	% total lipid	+	Aug–Sep	SE AK	0.54 kg/m²	Wipfli et al. 2004
Cutthroat trout	Lipids	% total lipid	+	Aug–Sep	SE AK	[b]	Wipfli et al. 2004
Trout species[c]	Biomass	Δ in g/m²	0	[d]	NW CA	0.94 kg/m²	Wilzbach et al. 2005
Trout species[c]	Density	Δ in fish/m²	0	[d]	NW CA	0.94 kg/m²	Wilzbach et al. 2005
Trout species[c]	Growth	%/d	0	[d]	NW CA	0.94 kg/m²	Wilzbach et al. 2005
Dolly Varden[e]	Growth	% Δ	+	Sep–Oct	SE AK	0.82 kg/m²	Wipfli et al. 2003
Steelhead (age 0+)	Condition	100/mass/length³	+	Jan	SW WA	0.56 kg/m²	Bilby et al. 1998
Steelhead (age 0+)	Condition	100/mass/length³	–	Feb	SW WA	0.56 kg/m²	Bilby et al. 1998
Steelhead (age 0+)	Density	Fish/m²	+	Oct–Mar	SW WA	0.71 kg/m²	Bilby et al. 1998
Steelhead (age 1+)	Density	Fish/m²	+ then 0	Oct–Mar	SW WA	0.71 kg/m2	Bilby et al. 1998

[a] *Oncorhynchus clarkii*
[b] 4 g phosphorus/m² from carcass analogs.
[c] Cutthroat trout and rainbow trout.
[d] All dates tested (Jun 2002, Oct 2002, Jun 2003, Oct 2003) relative to the pretreatment date (Oct 2001).
[e] *Salvelinus malma*

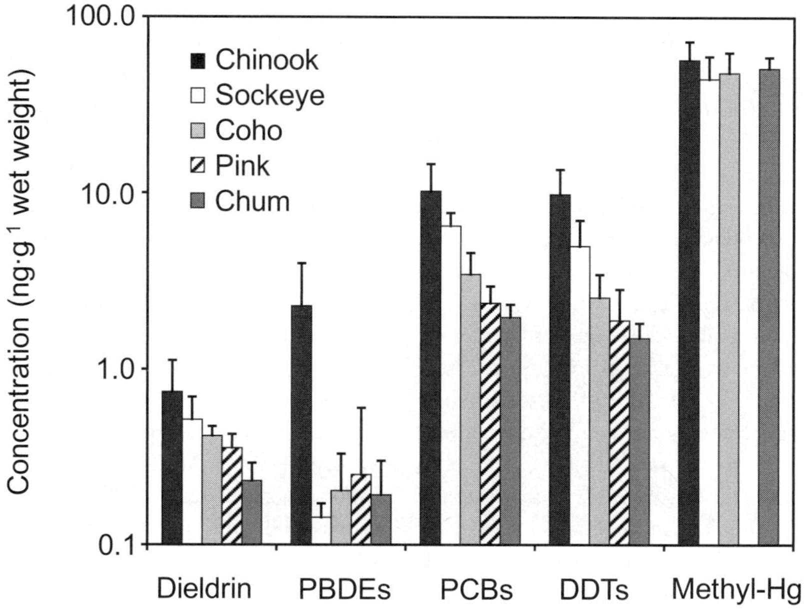

Figure 1.—Concentrations of pollutants measured in wild adult salmon tissues from various species, as reported by Zhang et al. (2001), Hites et al. (2004a, 2004b), and Kelly et al. (2007). Bars indicate average concentrations (+SD; n = 9 fish per bar), with two exceptions. First, the PCB concentration for sockeye is an average of two mean values (one from Hites et al. 2004a, the other from Kelly et al. 2007). Second, all methyl-Hg concentrations represent averages of 10–13 mean values. Note that the y-axis is depicted on a log scale.

by salmon can exceed background levels from atmospheric deposition (Zhang et al. 2001; Krümmel et al. 2003; Kelly et al. 2007). Even in heavily urbanized and industrial watersheds such as the Fraser River near Vancouver, British Columbia, the majority of contaminants may originate from marine sources (Kelly et al. 2007). Not surprisingly, pollutant concentrations in lake sediments near spawning grounds correlate with spawner densities (Krümmel et al. 2003).

Pollutants have also been measured in salmon eggs, carcass biofilm, and both vertebrate and invertebrate consumers of spawning salmon, eggs, and carcasses (Ewald et al. 1998; Zhang et al. 2001, Sarica et al. 2004; Christensen et al. 2005, Gregory-Eaves et al. 2007). The consequences of salmon-borne pollutants for stream ecosystems or subsequent salmon generations are unknown. Polychlorinated biphenyls and other contaminants have been detected at high levels in juvenile salmon (Johnson et al. 2007), where they may increase susceptibility to disease and, ultimately, cause mortality (Arkoosh et al. 1998). Thus, the positive feedback of nutrient cycling between spawning adults and their progeny may be counterbalanced by a negative feedback of pollution, a complication that is relatively recent in their evolutionary history.

Nutrient export.—Instream retention of MDN occurs via three potential pathways: biotic uptake by primary producers, abiotic uptake by streambed substrates, and direct consumption by stream consumers. However, not all MDN stays where it is deposited. What proportion of MDN imported by spawners is retained in streams and what proportion is exported? A nutrient budget model indicates that only a small percentage of the MDN imported by sockeye salmon (<11% of N, and <37% of P) is retained in a British Columbia spawning stream, while the majority is exported downstream or removed to terrestrial habitats (Johnston et al. 2004). Likewise, nutrient mass balance estimates for a stream in southeast Alaska indicate that 46% of the inorganic N and 60% of the inorganic P introduced by spawning salmon are exported downstream to the estuary (Mitchell and Lamberti 2005). In 10 southwest Alaskan streams where total N and P were measured before, during, and after spawning for 4 years, spawning sockeye exported 55% and 60% of the total amount of N and P imported in their bodies when all streams and years were averaged and export unrelated to salmon was accounted for (Moore et al. 2007). Furthermore, in some streams and years, total salmon-induced nutrient export exceeded salmon-borne MDN import.

Nutrient export via out-migrating juvenile salmon may also be significant. Moore and Schindler (2004) report that sockeye smolts from four major river systems in Alaska export, on average, 12% of the N and 16% of the P imported by their parents. These percentages are context dependent, varying across space and time, and range from less than 1–47% for N and 1–65% for P. In a threatened population of Chinook salmon, smolts export proportionally more P with decreasing spawner abundance (Scheuerell et al. 2005). In this case, smolts export, on average, 24% of the P imported by their parents, with smolt exports exceeding spawner imports in 12% of the years analyzed.

Effects on Terrestrial Wildlife

Returning salmon provide a seasonal high-quality food source used by numerous predators and scavengers. Coastal brown bear *Ursus arctos*, river otter *Lontra canadensis*, and mink *Mustela vison* feed regularly on salmon (Ben-David et al. 1997b; Blundell et al. 2002, Hilderbrand et al. 1999b, 2004; Reimchen 2000), while marten *Martes americana* and wolf *Canis lupus* feed on salmon in times of low abundance of principal prey species (Ben-David et al. 1997a; Szepanski et al. 1999; Darimont and Reimchen 2002; Darimont et al. 2003). Bald eagle *Haliaeetus leucephalus* and gulls *Larus* spp. scavenge and occasionally kill salmon in Alaskan streams (Quinn and Buck 2000; Christie and Reimchen 2005). Gulls then redistribute salmon nutrients to distant habitats, increasing biological productivity in areas otherwise isolated from MDN (Payne and Moore 2006). On the Olympic Peninsula of Washington, where brown bear are absent and black bear *U. americanus* are relatively scarce, there are 22 species of mammals and birds known to consume salmon carcasses (Cederholm et al. 1989). Over a 3-year period on a single stream in western Washington, Jauquet et al. (2003) reported consumption of live spawners, carcass flesh, and eggs by 10 bird species and four mammal species, including cougar *Felis concolor*. For this reason, salmon have been described as keystone species significantly affecting terrestrial and aquatic food webs in coastal ecosystems (Willson and Halupka 1995; Willson et al. 1998).

Through their feeding activities, bear are agents of natural selection within salmon populations. For example, bears have been found to be more likely to kill male salmon and larger-bodied salmon of both sexes, which may exert selective pressures affecting size distributions and body morphology, as well as operational sex ratios, paternity relationships, and genetic variance in salmon populations (Quinn and Kinnison 1999; Quinn and Buck 2000; Reimchen 2000; Ruggerone et al. 2000; Carlson et al. 2004). Conversely, salmon represent an important seasonal food source for bear (Hilderbrand et al. 1996, 1999b), and variations in salmon abundance influence patterns of aggregation, social dominance, infanticide, and reproductive success within bear populations (Klinka and Reimchen 2002; Ben-David et al. 2004; Gende and Quinn 2004).

Salmon carcasses also contribute significant amounts of nutrients to terrestrial invertebrates in riparian zones. Depending on taxa and site, up to ~70% of the total N in invertebrate tissues may be derived from salmon in coastal British Columbia (Hocking and Reimchen 2002). Some terrestrial invertebrates feed and breed directly on carcasses deposited by vertebrate predators or scavengers in the riparian zone (Meehan et al. 2005; Hocking et al. 2006). For example, fly larvae may number up to 50,000 individuals per carcass (Meehan et al. 2005), and a majority of the larvae of burying beetles *Nicrophorus investigator* in the riparian zone of a salmon-bearing watershed were observed to feed on salmon (Hocking et al. 2007). Other invertebrate taxa acquire MDN indirectly via litter and vegetation, as indicated by uptake of salmon-derived N but not C (Hocking and Reimchen 2002; Reimchen et al. 2003). As spawner numbers increase, so does the rate of nutrient transfer to invertebrates as large vertebrate predators and scavengers consume smaller portions of each fish (Gende et al. 2001; Meehan et al. 2005). Thus, salmon influence terrestrial invertebrates by subsidizing the productivity of specific components through various pathways; pathways that depend on run size, hydrology, and facilitation by vertebrates. These salmon-subsidized invertebrates contribute, in turn, to the diet of other riparian and stream organisms (Hocking and Reimchen 2002, 2006).

Marine-derived nutrients might also have indirect influences on terrestrial browsers. Plants with higher foliar N contents are generally more palatable and nutritious for mammalian browsers such as moose *Alces alces* and snowshoe hare *Lepus americanus* (Bryant 1987; Pastor et al. 1988). To the extent that riparian plants receiving MDN subsidies have higher foliar N contents (see below), MDN might therefore influence patterns of browsing, which in turn affects nutrient cycling and plant species composition (Kielland and Bryant 1998; Suominen et al. 1999). Ben-David et al. (1998) reported increased $\delta^{15}N$ but not $\delta^{13}C$ values in muscle tissues of small mammals (*Permyscus keeni, Microtus longicaudas, Sorex cinereus, Tamiasciurus hudsonicus*) near spawning streams in southeast Alaska, suggesting indirect assimilation of MDN through terrestrial vegetation rather than direct consumption of salmon carcasses. Since these animals can be insectivorous or omnivorous as well as granivorous, it is possible that MDN might be obtained not only via primary consumption of MDN-enriched vegetation, but also via secondary consumption of invertebrates that have been feeding on MDN-enriched vegetation. There is still much to learn about this linkage.

Effects on Riparian Ecosystems

Pathways for riparian MDN delivery.—Large animals play an important role in mediating MDN transfer from spawning streams to riparian ecosystems. Bear in particular can consume a large proportion of total spawner biomass (Quinn and Kinnison 1999; Quinn and Buck 2000; Reimchen 2000, Ruggerone et al. 2000; Gende et al. 2001, Quinn et al. 2001), with the number of salmon killed determined largely by spawner density (Quinn et al. 2003). Marine-derived nutrients are subsequently transferred to riparian soils via dissemination of partially eaten salmon carcasses and salmon-enriched wastes. On the Kenai Peninsula, Alaska, each adult female brown bear passes 0.046–0.051 kg/ha/year of salmon-derived N annually via urine and, to a lesser extent, feces and decomposition of body tissues after death (Hilderbrand et al. 1999a). This pathway accounts for 83–84% of salmon-borne N detected in white spruce *Picea glauca* foliage within 500 m of spawning streams. These findings are consistent with other studies demonstrating spatial correlations between bear activity and $\delta^{15}N$ signals in riparian foliage (Ben-David et al. 1998; Helfield and Naiman 2002; Wilkinson et al. 2005). Most salmon consumption occurs within the immediate vicinity of the stream channel, although bears may carry carcasses up to 200 m from the stream. This distance will depend on spawner density and social dominance interactions among bears (Gende and Quinn 2004). Helfield and Naiman (2006) used a mass balance model to estimate N flux in the riparian zone of a southwestern Alaskan stream, finding that total annual N inputs were significantly increased by the presence of salmon and bear together, but not by either species individually. Since nutrient inputs exert significant influences on the structure and dynamics of riparian ecosystems (see below), salmon and bear have been described as "mobile links" (Lundberg and Moberg 2003) and "keystone" interactors (Helfield and Naiman 2006) essential to the development and resilience of coastal ecosystems.

Although bear are the most visible consumers of salmon, MDN are also disseminated by other salmon-eating mammals and birds (e.g., Ben-David et al. 1998; Hilderbrand et al. 2004; Christie and Reimchen 2005; Payne and Moore 2006). Nutrients are also carried from streams to adjacent riparian areas via the metamorphosis of aquatic insects into winged adults and their subsequent death and consumption by terrestrial insectivores (Jackson and Fisher 1986), although studies in southwestern Alaska suggest that this represents a relatively minor vector for MDN transfer (Francis et al. 2006). Nonetheless, terrestrial flies and carrion beetles might play an important role in distributing and diffusing MDN via colonization of salmon carcasses and subsequent dispersal within the riparian zone (Meehan et al. 2005; Hocking and Reimchen 2006). The relative importance of different species as MDN vectors, as well as the spatial extent of their MDN distribution, varies according to abundance, mobility, and per capita rates of salmon consumption.

Marine-derived nutrients are also transferred to riparian habitats through abiotic processes. In larger rivers, flooding deposits salmon carcasses on streambanks (Cederholm et al. 1989; Ben-David et al. 1998). In streams with hydraulically conductive substrates, dissolved nutrients from decomposing carcasses may downwell into hyporheic flowpaths, where they may be accessible to the roots of floodplain plants. For example, increased concentrations of NH_4^+ and soluble reactive P have been observed in a southwestern Alaskan stream following entry of spawning sockeye salmon in peak run years, and this nutrient-enriched surface water subsequently downwells into the hyporheic zone beneath the riparian forest (O'Keefe and Edwards 2003). The relative importance of these pathways varies in time and space with salmon escapement and predator abundance, as well as climate, channel morphology, and vegetative characteristics (Helfield and Naiman 2006).

Soil nutrient processes.—Most pathways for the movement of salmon-derived nutrients in riparian areas are mediated through soils, and lead to one of three fates: short-term retention, long-term retention, or export (Table 3). Riparian soils are especially heterogeneous and their distribution is driven by frequent fluvial disturbance—soils along unconstrained alluvial rivers are often in early stages of development (Décamps 1996; Latterell et al. 2006; Bechtold 2007), with relatively thin organic soil layers overlying mineral alluvium and sediment. In most soils, decaying organic matter is the main source of nutrients to biota (Killham 1994), but soils receiving annual inputs of salmon-derived nutrients and energy present a special case.

Riparian soils adjacent to streams show significant ^{15}N enrichment when spawning densities are high (Reimchen et al. 2003; Bartz and Naiman 2005). Although ^{15}N enrichment can be potentially caused by biogeochemical processes unrelated to

Table 3.—Fates (export, retention) and major known pathways of salmon-derived nutrients (SDN) in riparian soils.

Mechanism	Chemical/biophysical process	Conditions when important
Export		
Leaching	Nutrient ions (especially anions) dissolved in water move into ground or surface waters	Transition from dry to wet seasons, flushed from soils by rains.
Biological export	Consumed by animals (e.g., insects, bears) that leave the area.	Direct and indirect (e.g., browsing of plants containing SDN) consumption
Volatilization	NH_4^+ oxidized to gaseous NH_3.	Warm weather and high pH soils
Denitrification	NO_3^- is converted into gaseous N (N_2, NO, NO_2) and released into the atmosphere.	Saturated, anoxic conditions (e.g. wetlands bordering rivers)
Disturbance	Fire, forestry, erosion during floods	
Retention (short-term)		Turnover time
Microbial pools		days–weeks
Nonmicrobial soil biota	Longer-lived organisms such as fungi, worms	days–months
Loose binding in soils	Especially cations such as Ca+, Mg+, NH_4^+	days +
Short-lived primary producers	Periphyton, annual plants	days–months
Animal processing	Nutrients not assimilated (i.e., excreted)	days
Retention (long-term)		
Long-lived primary producers	Trees and other perennial plants, some fungi	years +
Immobilization in soils		years +
Animal assimilation		months +

MDN (e.g., Pinay et al. 2003), the inference that MDN inputs contribute to soil nutrient pools is supported by the results of experimental additions of salmon carcasses to riparian soils. In the coastal rainforests of western Washington, NH_4^+ and NO_3^-, bound mainly in the soft tissues of salmon, enter extractable, bioavailable soil pools within weeks to months of death, whereas calcium (Ca^{+2}) and P, bound largely in slowly decaying bone, are released over months to years (Drake et al. 2005). During the course of soft tissue decay, NH_4^+ concentrations in soils within 20 cm of carcasses increased by a factor of greater than 200, while phosphate (PO_4^{-2}) and sulfate (SO_4^{-2}) concentrations increased by a factor of ~5. Lateral movement of NH_4^+ and NO_3^- was evident in soils further (~1 m) from carcasses. Maximum availability of NH_4^+ and NO_3^- in soils occurs at ~2 months and ~4.5 months, respectively, after carcass deposition. As expected, MDN increase soil nutrient availability, at least within the immediate vicinity of salmon carcasses, suggesting a localized mosaic pattern of MDN concentrations.

Microbial communities largely control rates of turnover and mineralization of soil nutrients (Killham 1994). The supply of soil carbon to microbial pools regulates the cycling of other nutrients with more than 90% of total soil C in refractory forms. It is generally accepted that carbon quality limits soil microorganisms (Fontaine et al. 2003). Salmon (as carcasses or excreta) contribute all elements important to biological processes (Table 4), including a labile carbon supply. The capacity of the riparian area for nutrient uptake may, in some cases, be temporarily exceeded by the large, patchy, short-term contributions of salmon nutrients.

How do salmon-borne MDN inputs compare with nutrients from riparian litterfall? Natural N loading rates from litterfall vary from 20 to 80 kg/ha/year in Pacific Northwest coniferous forests but can be as much as ~125 kg/ha/year in 50–100 year-old stands dominated by alder (O'Keefe and Naiman 2006). The lower part of this range (near 20 kg per year) probably results in N deficiency in trees (Cole 1995). Salmon-derived nutrients in the forms of carcasses and excreta from predators and scavengers add ~0.6–3.3 kg N/ha/year or more, depending on salmon and predator densities (Helfield and Naiman 2006). In contrast to litterfall, nutrient contributions from salmon carcasses are readily bioavailable (Drake et al. 2005) and therefore become an important contribution to soils and eventually to plants.

Plant species composition and diversity.—Foliar ^{15}N enrichment in riparian plants growing near spawning streams is well documented (Bilby et al.

Table 4.—Nutrient content of a 3.58-kg spawned-out whole salmon, and the skeleton alone, which weighed 44.16 g after ashing. The total and skeletal weights from a specific fish were used because its skeleton was the most complete at the end of the experiment. Table reproduced from Drake et al. (2005) with permission from Northwest Science.

Element	% by weight of whole salmon (spawned-out)	mg in 3.58 kg salmon	% by weight in bone	% (of whole fish) in bone
N	2.50[a]	89.50	6.25[d]	3.08
P	0.43[a]	15.39	17.30[e]	49.63
Ca	0.50[b]	17.90	35.60[e]	87.83
K	0.14[b]	5.01	0.36[e]	3.30
S	0.10[c]	3.58	0.10[c]	1.23
Mg	0.03[b]	1.07	0.61[e]	44.16
Fe	0.003[b]	0.11	0.007[e]	44.00
Mn	0.0002[b]	0.01	0.003[e]	43.17

[a] Value from Larkin and Slaney (1997)
[b] estimated from Shearer et al. (1994)
[c] estimated from published values of other vertebrate tissues
[d] values based on N:P ratios from Sterner and Elser (2002)
[e] values estimated from Hamada et al. (1994).

1996, 2003; Ben-David et al. 1998; Hilderbrand et al. 1999a; Helfield and Naiman 2001, 2002). Although patterns of foliar isotopic composition may be influenced by edaphic factors unrelated to MDN (Vitousek et al. 1989; Garten 1993; Schulze et al. 1994; Handley and Scrimgeour 1997), the weight of evidence suggests that most riparian plants assimilate MDN. Consistent ^{15}N enrichment has been quantified at sites with high spawning densities relative to comparable sites without salmon in coastal rainforests (Bilby et al. 1996, 2003; Ben-David et al. 1998; Helfield and Naiman 2001; Mathewson et al. 2003; Reimchen et al. 2003), in boreal forests (Hilderbrand et al. 1999a; Helfield and Naiman 2002), and in interior forests in North America (Koyama et al. 2005) as well as in Asia (Nagasaka et al. 2006). The relationship between MDN and foliar ^{15}N enrichment has been corroborated by experimental fertilization studies (e.g., Koyama et al. 2005) and by empirical observations above and below salmon-blocking waterfalls (Bartz and Naiman 2005). Additionally, an N isotope tracer study examining the fate of labeled NH_4^+ (the primary product of carcass decay) in riparian soils and trees on a salmon river showed that ~37% of labeled NH_4^+ was found in riparian trees within 6 months of addition (Drake et al. 2006).

Foliar nutrient concentrations are often higher in plants receiving MDN inputs (Helfield and Naiman 2001, 2002; Bilby et al. 2003; Mathewson et al. 2003; Wilkinson et al. 2005), although this is not always the case (Ben-David et al. 1998; Bartz and Naiman 2005; Drake et al. 2006). Marine-derived nutrient subsidies might not enhance foliar nutrient concentrations in cases where increased growth rates result in increased foliar biomass and diluted nutrient concentrations or where nutrients are translocated from foliage (Nambiar and Fife 1991; Drake et al. 2006). Alternatively, at sites where soil nutrients are abundant, MDN additions might have little effect on the nutrient status of plants. For example, salmon-derived N subsidies appear to be less important to riparian ecosystems where symbiotic N_2-fixation by alder *Alnus* spp. is prevalent (Helfield and Naiman 2002).

Marine-derived nutrient inputs may also influence the species composition and diversity of riparian plant communities (Table 5). MDN-influenced sites in coastal British Columbia feature increased bryophyte species richness (Wilkinson et al. 2005), and there are significantly increased overstory densities at MDN-influenced sites in southwestern Alaska, with correspondingly increased densities of understory plants at sites without MDN (Bartz and Naiman 2005). These differences in density were driven primarily by the abundance of willows *Salix* spp. In the same region, decreased alder abundances at sites adjacent to spawning streams have been observed and it is hypothesized that consistent annual MDN inputs might decrease the competitive advantage of N_2-fixing species (Helfield and Naiman 2002). Alternatively, alder and salmon abundance may be inversely correlated because alder are more likely to dominate sites subjected to frequent fluvial disturbance, which entails less preferable spawning habitat (Helfield and Naiman 2002).

Many plants can be classified as either N-poor or N-rich soil indicator species, according to their preference for soil nutrient conditions (Klinka et al. 1989). N-rich indicators tend to be more abundant than N-poor indicators adjacent to spawning streams, whereas the opposite is true at reference sites, where N-poor indicators are more abundant. These differences in riparian composition were observed at two streams in British Columbia (Mathewson et al. 2003; Wilkinson et al. 2005). Comparable differences did not occur consistently among 10 streams in southwest Alaska (Bartz 2002; Figure 2); however, average spawning densities at both British Columbia sites (1.6 and 1.9 salmon per square meter) were an order of magnitude greater than that at the densest Alaska site (0.15 salmon per square meter).

Forest growth rates.—Dendrochronological analyses indicate that MDN inputs may enhance the productivity of riparian trees. In coastal rainforests of southeast Alaska, enhanced growth of Sitka spruce *Picea sitchensis* has been reported in areas where MDN inputs are greatest (Helfield and Naiman 2001). Among trees within 25 m of the stream channel, basal area growth rates are more than three times greater at sites with spawning salmon relative to comparable sites without spawning salmon. As a result, there is a disproportionate occurrence of large productive forest in close proximity to spawning streams (Helfield and Naiman 2003; Kirchhoff et al. 2003). Similar patterns are observed for white spruce *P. glauca* in

Table 5.—Effects of MDN inputs on riparian plant community characteristics, as reported by various studies. Effects are characterized as positive (+), negative (–), or not significant (0), generally at the 0.05 level.

Plant type and layer	Characteristic	Units	Effect	Source[a]
Vascular				
Overstory	Stem density	Stems/ha	+ , – , 0	1, 2, 3
	Basal area	m^2/ha	0	1, 3
	Aboveground biomass	kg/ha	0	1
	Canopy cover	%	0	2, 3
	Alder cover	%	– , 0	2, 3
Understory	Stem density	Stems/ha	–	1
	Basal area	m^2/ha	0	1
	Aboveground biomass	kg/ha	0	1
	Cover	%	+ , 0	2
	Alder cover	%	0	3
	N-rich indicator species cover	%	+	4
	Diversity	Species/transect	+ , 0	2
Ground layer	Stem density	Plants/plot	+	2
	Cover	%	+ , 0	2
	Graminoid cover	%	0	1
	Dwarf shrub cover	%	–	1
	N-rich indicator species cover	%	+	4
	Diversity	Species/transect	+	2
	Forb diversity	Species/transect	0	1
Nonvascular				
Ground layer	Cover	%	–	1
	N-rich indicator species cover	%	+	5
	N-poor indicator species cover	%	–	5
	Diversity	Species/transect	+	5
	Shannon diversity index	H'	0	5

[a] 1 = Bartz and Naiman 2005; 2 = Bilby et al. 2003; 3 = Helfield and Naiman 2002; 4 = Mathewson et al. 2003 and M. Hocking, University of Victoria, Victoria, British Columbia, personal communication; 5 = Wilkinson et al. 2005.

boreal forests of southwestern Alaska, although, as with foliar $\delta^{15}N$, the effects of MDN on spruce growth appear to be confounded in the presence of alder (Helfield and Naiman 2002).

These studies are largely inferential (see Helfield and Naiman 2003; Kirchhoff et al. 2003), and although subsequent studies of long-term (50-year) riparian tree productivity are equivocal, short-term (annual) relationships between salmon escapement and tree-ring growth have been demonstrated for various salmon and tree species throughout the Pacific Northwest (Drake et al. 2002; Drake and Naiman 2007). Although trees receiving MDN might not always grow faster than their counterparts at reference sites, tree-ring chronologies suggest that trees grow faster when salmon are abundant than they would in the absence of salmon. Additionally, in one Columbia River tributary, riparian tree-ring growth rates decreased concomitantly with the construction of a dam and consequent extirpation of salmon (Table 6; Drake 2005). This trend was statistically significant at only one of tributaries examined, but the technique (i.e., intervention analysis) holds promise for assessing the importance of MDN to riparian growth and the system-scale consequences of declining salmon runs.

Feedback effects on salmon populations via enhanced riparian functioning.—The influence of riparian vegetation on the quality of salmonid habitat

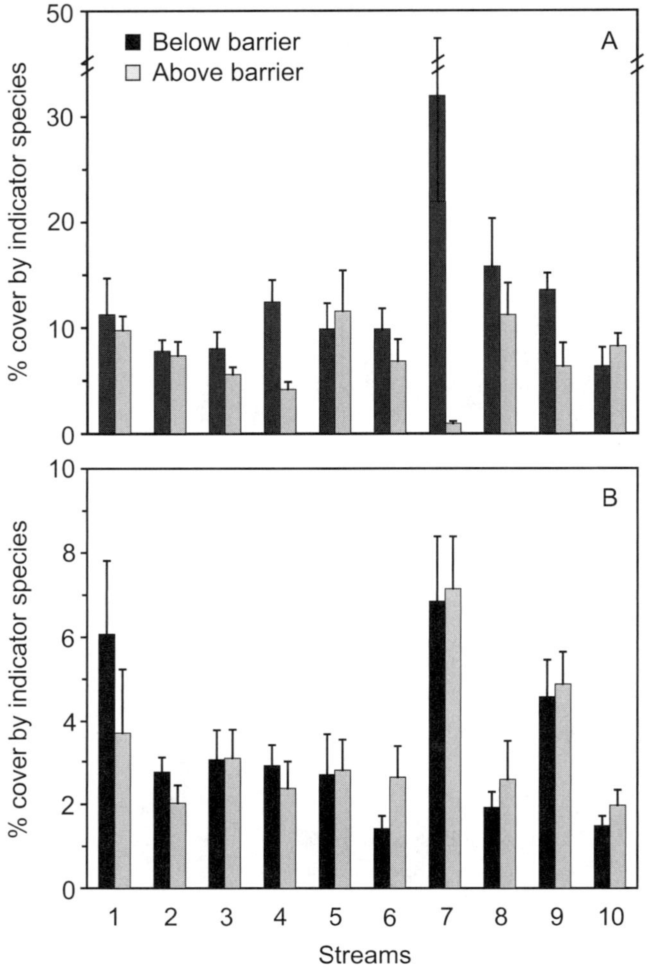

Figure 2.—Average covers of riparian plant species indicative of N-rich (A) and N-poor (B) soils near 10 streams in southwest Alaska. At each stream, percent cover was measured by species along 12 transects: 6 downstream and 6 upstream of a barrier to salmon migration. Averages were calculated first for all species within a given category (i.e., N-rich or N-poor) on a single transect and then among all transects within a given location (i.e., below or above the barrier). In this figure, streams are ordered according to their mean annual below-barrier spawning density, with densities increasing to the right from less than 0.01–0.15 sockeye per square meter. Error bars indicate standard errors. According to paired t-tests among streams, covers of indicator species did not differ significantly above and below barriers (p = 0.09 in A and 0.86 in B; n = 10 streams).

has been well documented. Riparian forests regulate instream primary production through shading, while providing allochthonous organic matter in the form of litter inputs (Naiman and Décamps 1997; Naiman et al. 2005). These processes affect the species composition and production of aquatic invertebrates, which provide an essential food source for juvenile salmon (Meehan et al. 1977). In some systems, riparian canopy effects may exert a greater influence on salmonid biomass, density, and growth than do direct inputs of nutrients and organic matter in the form of salmon carcasses (Wilzbach et al. 2005). Marine-derived nutrient enrichment of riparian invertebrates might also affect juvenile salmon production, as terrestrial invertebrates often support a substantial proportion of the diet of salmonids (Wipfli 1997; Baxter et al. 2005).

Table 6.—Intervention analysis of deviation from predicted tree-ring growth associated with salmon extirpation. Trees at the Metolius River, Oregon grew significantly slower than predicted after the local decline of salmon (1930 cutoff date). Trees at the Deschutes River reference site grew at about the expected rate after 1930. Trees at the Cowlitz River salmon site (salmon extirpated in 1968), and the Coal Creek reference site grew at about the predicted rate after 1968. Table reproduced from Drake (2005).

	Intervention sites		Reference sites
	Metolius River, OR Intervention 1930		Deschutes River, OR Reference
Average annual deviation from predicted growth (mm)	–0.879	$P < 0.005$	0.155
	Cowlitz River, WA Intervention 1967		Coal Creek, WA Reference
Average annual deviation from predicted growth (mm)	–0.002	NS	0.102

Riparian forests also affect salmon populations by enhancing physical habitat (Meehan et al. 1977; Naiman and Latterell 2005). Marine-derived nutrient subsidies can improve the growth of some riparian trees that eventually provide large wood (LW) to stream channels. Larger pieces typically persist longer in streams as they take longer to decompose and are more difficult to flush downstream (Murphy and Koski 1989; Latterell and Naiman 2007). The height of riparian trees also affects LW recruitment as the probability of a tree entering the stream increases with the height of the tree relative to its distance from the channel's edge (Van Sickle and Gregory 1990). Among other functions, LW traps sediment and increases the structural complexity of stream channels (Keller and Swanson 1979; Harmon et al. 1986), creating preferred habitat for spawning and rearing (Bjornn and Reiser 1991). Large wood also creates areas of low flow velocity and shear stress, providing shelter from winter high flows and bed scour (Murphy et al. 1985; McMahon and Hartman 1989), which are important causes of mortality in incubating embryos and overwintering fry (McNeil 1964). Overall, the presence of LW in spawning streams enhances production of salmonids (Murphy et al. 1985; Fausch and Northcote 1992; Crispin et al. 1993). By enhancing the growth of riparian trees and the production of LW, MDN inputs to the riparian zone might therefore serve as a positive feedback mechanism by which spawning salmon help to enhance the survivorship of subsequent generations.

Emerging Issues and Concepts

Reconstruction of Past Salmon Abundances

Analyses of $\delta^{15}N$ values and indicators of productivity in lake sediment cores have been used successfully to reconstruct historical abundances of sockeye salmon (Finney et al. 2000, 2002; Gregory-Eaves et al. 2003, 2004; Schindler et al. 2005; Selbie et al. 2007). In rivers, however, where use of sediment cores is not viable, an analogous approach has been to examine ^{15}N abundance in core samples from riparian trees. In theory, if salmon provide a large portion of the N taken up and used by riparian trees, $\delta^{15}N$ values should be higher in rings corresponding to years of high salmon abundance. Yet, despite widespread interest and investment in this approach, no clear quantitative relationship between salmon abundance and tree-ring $\delta^{15}N$ has been reported in the peer-review literature. There is one report of decreasing xylem $\delta^{15}N$ with distance from salmon-bearing rivers (Reimchen et al. 2003), but this study did not isolate $\delta^{15}N$ signals in individual growth rings or groups of rings so as to create a time series that could be used for retrospective analyses. The difficulties in detecting salmon-related $\delta^{15}N$ signals in tree rings are likely due to the mobility of N within trees. Where N limits growth, trees conserve N through tight internal cycling. For example, N may be withdrawn from the tissue of a given ring and redistributed to rings grown in subsequent growth years. Additionally, up to 80% of the N found in wood is contained in sap, which is highly mobile among rings (Sheppard and Topa 2002; Drake 2005). Therefore, sap must be removed to analyze N bound in the structure of wood, a labor intensive process that leaves the wood N-poor, thereby requiring large, carbon-rich samples to be analyzed via mass spectroscopy with a resulting decrease in analytical precision.

In contrast, the relationship between salmon escapement and riparian nonclimatic tree-ring growth has been used successfully in reconstructing past salmon abundances in rivers (Drake et al. 2002; Drake and Naiman 2007). The approach controls for large-scale climate patterns such as the Pacific Decadal Oscillation, which can influence patterns in tree growth and are also correlated with salmon abundance over large geographical areas (Mantua et al. 1997). But the approach also has stringent data and site requirements and therefore cannot be applied everywhere; nevertheless, the locations where it has been successfully applied show reconstructions capturing lower-frequency cycles of abundance and is most useful for determination and comparison of relative salmon abundance, cycles, and the effects of interventions (e.g., dam construction; see Table 6). The results provide the only river-specific, century-scale perspectives on salmon abundance and cycles and also account for climate variation that is independent of salmon variation.

Effects of Mass Spawners versus Other Species

Nearly all results summarized here are from Pacific salmon species spawning in large and dense aggregations (i.e., pink, chum, sockeye), and the effects of MDN are readily apparent at the ecosystem scale. Other species of Pacific salmon (i.e., coho, Chinook, and steelhead) often spawn in lesser numbers, even though they may be locally dense. The ecosystem-scale effects of MDN from these species still require additional quantification. Marine-derived nutrients from these species can be detected in riparian vegetation (Bilby et al. 1996), but it is not clear if the ecological responses to the added nutrients are locally significant or are spatially extensive.

This latter point emphasizes that the importance of MDN is context-dependent. It may not only depend on the life history strategy of an individual species, but also on temporal and spatial variations in climate, channel morphology, escapement, bear density, and other local biophysical characteristics. This is well illustrated by the mass balance model developed by Helfield and Naiman (2006), which shows how MDN varies in relative importance in to other nutrient sources as well as in time and space.

Marine-Derived Nutrient Effects on Smolt Production and Adult Returns

The overarching concept that MDN enhances smolt size and production appears to be well established. However, it is not clear how that translates to increased numbers of returning adults, especially given uncertainties regarding future climate change and other factors affecting salmon survivorship in the ocean. Year-to-year variation in marine survival is only weakly correlated with smolt size and often not at all (Quinn 2005). Indeed, marine survival may be better predicted by ocean conditions, such as the intensity of coastal upwelling (Scheuerell and Williams 2005).

Marine-Derived Nutrient Effects on Forest Conditions

Most studies relating MDN to riparian soils and plants rely on comparisons of salmon and non-salmon sites at a small number of regionally restricted locations. Consequently, a key question arises: how general are the findings regarding the effects of MDN on riparian forest conditions? The current range of Pacific salmon encompasses seven nations and 66 ecoregions containing numerous forest types (Augerot 2005). Thus, it is conceivable, or even likely, that conclusions drawn from a single region may not be extrapolated easily across the full range of salmon. One technique that might help broaden the spatial scale of future studies is the analysis of remotely sensed imagery via geographic information systems (GIS). Other studies have used GIS to estimate salmon abundance and redd density from geospatial habitat data, such as terrestrial land cover (e.g., Pess et al. 2002; Steel et al. 2004). It would be informative to try the reverse, to see if coarse scale patterns emerge in the terrestrial land cover data based on salmon abundance or spawning distribution.

Conclusions

Conceptually, there is general understanding of the importance and complexity of salmon-borne MDN in shaping ecosystem processes (Figure 3). Collectively, the data suggest that MDN provide significant links between marine, freshwater, and riparian ecosystems, so much so that the long-term viability of salmon populations, riparian plant communities, and terrestrial predators and scavengers may be mutually dependent. Such findings clearly illustrate the complexity of interactions in freshwater and riparian ecosystems and the importance of linkages across ecosystem boundaries (Naiman and Latterell 2005).

These results have significant management implications, especially in using MDN to set salmonid escapement goals or system-scale recovery goals. Should salmon-borne MDN or riparian tree growth (as a proxy of historic MDN availabilities) be used to help establish appropriate ecosystem-scale salmon escapements? Clearly, managing salmon populations

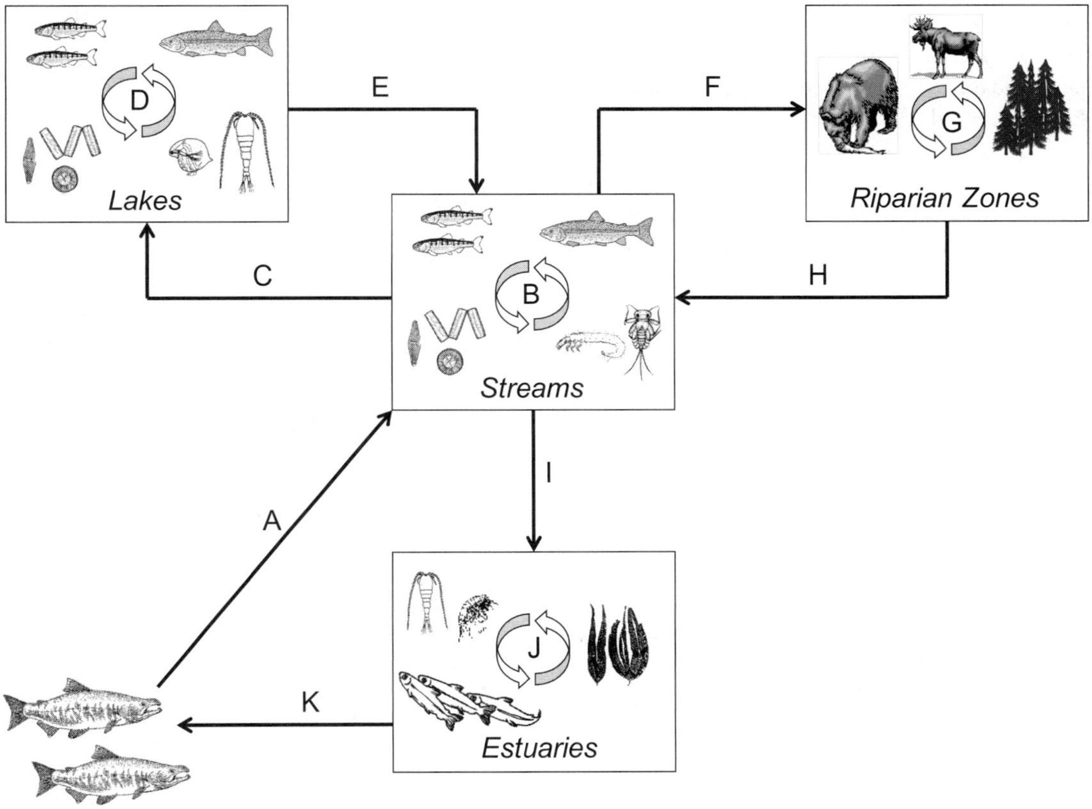

Figure 3.—Cycling of marine-derived nutrients (MDN) in coastal ecosystems: (A) spawning salmon transport MDN upstream; (B) MDN are incorporated into stream food webs (i.e., periphyton, aquatic macroinvertebrates, resident stream fishes, and salmon fry); (C) MDN are delivered to lakes via beach spawners and discharge from spawning tributaries; (D) MDN are incorporated into lake food webs (i.e., phytoplankton, zooplankton, resident lake fishes, and sockeye salmon fry); (E) MDN are washed from lakes into receiving streams; (F) MDN are delivered to riparian zones via salmon-eating fauna, flooding, or hyporheic exchange; (G) MDN is taken up by riparian vegetation and incorporated into riparian food webs; (H) Riparian trees provide shade, bank stabilization, allochthonous organic matter, and large woody debris, enhancing the quality of instream habitat for salmonid fishes; (I) MDN washes down from spawning streams to estuaries; (J) MDN is incorporated into estuarine food webs (i.e., macroalgae, zooplankton, and smolts; and (K) Enhanced smolt production potentially leads to increased returns of adult spawners.

according to traditional stock–recruitment models needs to be expanded to account for the diversity of MDN effects on ecosystems (e.g., food to bear and resident fish, nutrients for riparian communities; Bilby et al. 2003). Sustainable management of salmon fisheries requires a holistic approach that treats salmon population dynamics, ocean climatic cycles, riparian vegetation, predators, and scavengers as an integrated system. Failure to maintain the MDN requirements of the entire system will eventually result in subtle but persistent declines in system-scale integrity that, eventually, will lead to loss of the fundamental ecological characteristics that one is trying to conserve. We suggest that it would be prudent to explore these possibilities in geographically dispersed adaptive management scenarios, especially in light of the widespread loss of salmon under present management protocols.

Finally, one must consider that there are potentially many nutrient sources available for aquatic systems. The relative importance of MDN with respect to other nutrient sources remains to be quantified. In this review, we have focused on salmon-borne MDN—and it is certainly an important nutrient source, especially so because of its immediate bioavailability. Nonetheless, one should not ignore several other nutrient sources for freshwater and riparian systems that also underpin system productivity. These include riparian litterfall that may add many times more in total amounts of nutrients than MDN per unit area (O'Keefe and Naiman 2006), riparian epiphytes that fix N_2 and capture of marine micronutrients in aerosols and precipitation (R. Van Pelt and R. J. Naiman, University of Washington, unpublished data), long range atmospheric transport of nutrients (e.g., N), and translocation of nutrients from upslope regions of the catchment dissolved in subsurface water flows (e.g., N from N_2-fixation by upslope alder and nutrients associated with river-borne silts; Bechtold and Naiman 2006). Collectively, the importance of other nutrient sources relative to MDN also require further quantification in terms of their impacts on the overall productivity of freshwater and riparian habitats. Nevertheless, the marine environment is a highly important nutrient source, and with major alterations to salmon abundance throughout the North Pacific, the consequences for continental systems may be problematic for the continued long-term vitality of many oligotrophic freshwater systems.

Acknowledgments

Research support by the National Science Foundation, Pacific Northwest Research Station of the U.S. Forest Service, University of Washington Sea Grant Program, University of Washington School of Aquatic and Fishery Sciences, NOAA Fisheries, and the Pacific Seafood Processors Association is gratefully acknowledged. We thank Ronald A. Hites for his generous sharing of data for Figure 1, Thomas P. Quinn and Daniel E. Schindler for insightful comments on an early version of the manuscript, and I. Gregory-Eaves and M. D. Hocking whose thoughtful suggestions improved the final version.

References

Ambrose, H. E., M. A. Wilzbach, and K. W. Cummins. 2004. Periphyton response to increased light and salmon carcass introduction in northern California streams. Journal of the North American Benthological Society 23:701–712.

Arkoosh, M. R., E. Casillas, E. R. Clemons, A. N. Kagley, R. Olson, P. Reno, and J. E. Stein. 1998. Effect of pollution on fish diseases: potential impacts on salmonid populations. Journal of Aquatic Animal Health 10:182–190.

Augerot, X. 2005. Atlas of Pacific salmon. University of California Press, Berkeley.

Barnaby, J. T. 1944. Fluctuations in abundance of red salmon, *Oncorhynchus nerka* (Walbaum), of the Karluk River, Alaska. Fishery Bulletin 50:237–295.

Bartz, K. K. 2002. Impacts of salmon-borne nutrients on riparian soils and vegetation in southwest Alaska. Master's thesis. University of Washington, Seattle.

Bartz, K. K., and R. J. Naiman. 2005. Effects of salmon-borne nutrients on riparian soils and vegetation in southwest Alaska. Ecosystems 8:529–545.

Baxter, C. V., K. D. Fausch, and C. Saunders. 2005. Tangled webs: reciprocal flows of invertebrate prey link streams and riparian zones. Freshwater Biology 50:201–220.

Bechtold, J. S. 2007. Influence of fluvial sediments on floodplain soil biogeochemistry. Doctoral dissertation. University of Washington, Seattle.

Bechtold, J. S., and R. J. Naiman. 2006. Soil nitrogen mineralization potential across a riparian toposequence in a semi-arid savanna. Soil Biology and Biogeochemistry 38:1325–1333.

Ben-David, M., R. W. Flynn, and D. M. Schell. 1997a. Annual and seasonal changes in diets of martens: evidence from stable isotope analysis. Oecologia 111:280–291.

Ben-David, M., T. A. Hanley, D.R. Klein, and D. M. Schell. 1997b. Seasonal changes in diets of coastal and riverine mink: the role of spawning Pacific salmon. Canadian Journal of Zoology 75:803–811.

Ben-David, M., T. A. Hanley, and S. M. Schell. 1998. Fertilization of terrestrial vegetation by spawning Pacific salmon: the role of flooding and predator activity. Oikos 83:47–55.

Ben-David, M., K. Titus, and L. R. Beier. 2004. Consumption of salmon by Alaskan brown bears: a trade-off between nutritional requirements and the risk of infanticide? Oecologia 138:465–474.

Bilby, R. E., E. W. Beach, B. R. Fransen, J. K. Walter, and P. A. Bisson. 2003. Transfer of nutrients from spawning salmon to riparian vegetation in western Washington. Transactions of the American Fisheries Society 132:733–745.

Bilby, R. E., B. R. Fransen, and P. A. Bisson. 1996. Incorporation of nitrogen and carbon from spawning coho salmon into the trophic system of small streams: evidence from stable isotopes. Canadian Journal of Fisheries and Aquatic Sciences 53:164–173.

Bilby, R. E., B. R. Fransen, P. A. Bisson., and J. K. Walter. 1998. Response of juvenile coho salmon (*Oncorhynchus kisutch*) and steelhead (*O. mykiss*) to the addition of salmon carcasses to two streams in southwestern Washington, U.S.A. Canadian Journal of Fisheries and Aquatic Sciences 55:1909–1918.

Bilby, R. E., B. R. Fransen, J. K. Walter, C. J. Cederholm, and W. J. Scarlett. 2001. Preliminary evaluation of the use of nitrogen stable isotope ratios to establish escapement levels for Pacific salmon. Fisheries 26:6–14.

Bjornn, T. C., and D. W. Reiser. 1991. Habitat requirements of salmonids in streams. Pages 83–138 *in* W. R. Meehan, editor. Influences of forest and rangeland management on salmonid fishes and their habitats. American Fisheries Society, Special Publication 19, Bethesda, Maryland.

Blundell, G. M., M. Ben-David, and R. T. Bowyer. 2002. Sociality in river otters: cooperative foraging or reproductive strategies? Behavioural Ecology 13:134–141.

Brickell, D. C., and J. J. Goering. 1970. Chemical effects of salmon decomposition on aquatic ecosystems. Pages 125–138 *in* R. S. Murphy, editor. First international symposium on water pollution control in cold climates. U.S. Government Printing Office, Washington, D.C.

Bryant, J. P. 1987. Feltleaf willow-snowshoe hare interactions: plant carbon/nutrient balance and floodplain succession. Ecology 68:1319–1327.

Burgner, R.L. 1987. Factors influencing age and growth of juvenile sockeye salmon (*Oncorhynchus nerka*) in lakes. Pages 129–142 *in* H. D. Smith, L. Margolis, and C. C. Wood, editors. Sockeye salmon (*Oncorhynchus nerka*) population biology and future management. Government of Canada Department of Fisheries and Oceans, Ottawa.

Burgner, R. L. 1991. Life history of sockeye salmon (*Oncorhynchus nerka*). Pages 1–117 *in* C. Groot and L. Margolis, editors. Pacific salmon life histories. University of British Columbia Press/Government of Canada Department of Fisheries and Oceans, Vancouver.

Carlson, S. M., H. B. Rich, and T. P. Quinn. 2004. Reproductive life-span and sources of mortality for alternative male life-history strategies in sockeye salmon, *Oncorhynchus nerka*. Canadian Journal of Zoology 82:1878–1885.

Carpenter, S. R., J. F. Kitchell, and J. R. Hodgson. 1985. Cascading trophic interactions and lake productivity. BioScience 35:634–639.

Cederholm, C. J., D. B. Houston, D.L. Cole, and W. J. Scarlett. 1989. Fate of coho salmon (*Oncorhynchus kisutch*) carcasses in spawning streams. Canadian Journal of Fisheries and Aquatic Sciences 46:1347–1355.

Cederholm, C. J., D. H. Johnson, R. E. Bilby, L. G. Dominguez, A. M. Garrett, W. H. Graeber, E. L. Greda, M. D. Kunze, B. G. Marcot, J. F. Palmisano, R. W. Plotnikoff, W. G. Pearcy, C. A. Simenstad, and P. C. Trotter. 2000. Pacific salmon and wildlife: ecological contexts, relationships, and implications for management. Washington Department of Fish and Wildlife, Olympia.

Chaloner, D. T., G. A. Lamberti, A. D. Cak, N. L. Blair, and R. T. Edwards. 2007. Inter-annual variation in responses of water chemistry and epilithon to Pacific salmon spawners in an Alaskan stream. Freshwater Biology 52:478–490.

Chaloner, D. T., G. A. Lamberti, R. W. Merritt, N. L. Mitchell, P. H. Ostrum, and M. S. Wipfli. 2004. Variation in responses to spawning Pacific salmon among three southeastern Alaska streams. Freshwater Biology 49:587–599.

Chaloner, D. T., K. M. Martin, M. Wipfli, P. H. Ostrom, and G. A. Lamberti. 2002a. Marine carbon and nitrogen in southeastern Alaska stream food webs: evidence from artificial and natural streams. Canadian Journal of Fisheries and Aquatic Sciences 59:1257–1265.

Chaloner, D. T., and M. S. Wipfli. 2002. Influence of decomposing Pacific salmon carcasses on macroinvertebrate growth and standing stock in southeastern Alaska streams. Journal of the North American Benthological Society 21:430–442.

Chaloner, D. T., M. S. Wipfli, and J. P. Caouette. 2002b. Mass loss and macroinvertebrate colonisation of Pacific salmon carcasses in south-eastern Alaskan streams. Freshwater Biology 47:263–273.

Chow, J. K. 2007. Nutrient linkages between freshwater

and marine ecosystems: uptake of salmon-derived nutrients in estuaries. Master's thesis. University of Victoria, Victoria, British Columbia.

Christensen, J. R., M. MacDuffee, R.W. Macdonald, M. Whiticar, and P. S. Ross. 2005. Persistent organic pollutants in British Columbia grizzly bears: consequence of divergent diets. Environmental Science and Technology 39:6952–6960.

Christie, K. S., and T. E. Reimchen. 2005. Post-reproductive Pacific salmon, *Oncorhynchus* spp., as a major nutrient source for large aggregations of gulls, *Larus* spp. Canadian Field Naturalist 119:202–207.

Claeson, S. M., J. L. Li, J. E. Compton, and P. A. Bisson. 2006. Response of nutrients, biofilm, and benthic insects to salmon carcass addition. Canadian Journal of Fisheries and Aquatic Sciences 63:1230–1241.

Claire, E. W., and R. W. Phillips. 1968. The stonefly *Acroneuria pacifica* as a potential predator on salmonid embryos. Transactions of the American Fisheries Society 97:50–52.

Clementz, M. T., P. Holden, and P. L. Koch. 2003. Are calcium isotopes a reliable monitor of trophic level in marine settings? International Journal of Osteoarchaeology 13:29–36.

Cole, D. W. 1995. Soil nutrient supply in natural and managed forests. Plant and Soil. 168:43–53

Crispin, V., R. House, and D. Roberts. 1993. Change in instream habitat, large woody debris, and salmon habitat after the restructuring of a coastal Oregon stream. North American Journal of Fisheries Management 13:96–102.

Darimont, C. T., and T. E. Reimchen. 2002. Intra-hair stable isotope analysis implies seasonal shift to salmon in gray wolf diet. Canadian Journal of Zoology 80:1638–1642.

Darimont, C. T., T. E. Reimchen, and P. C. Paquet. 2003. Foraging behaviour by gray wolves on salmon streams in coastal British Columbia. Canadian Journal of Zoology 81:349–353.

Décamps, H. 1996. The renewal of floodplain forests along rivers: a landscape perspective. Verhandlungen der Internationalen Vereinigung für Theoretische und Angewandte Limnologie 26:35–59.

Deegan, L. A., and B. J. Peterson. 1992. Whole-river fertilization stimulates fish production in an Arctic tundra river. Canadian Journal of Fisheries and Aquatic Sciences 49:1890–1901.

Donaldson, J.R. 1967. The phosphorus budget of Iliamna Lake, Alaska, as related to the cyclic abundance of sockeye salmon. Doctoral dissertation. University of Washington, Seattle.

Drake, D. C. 2005. The fate and influence of salmon nutrients in riparian soils and trees: seasonal to multicentury patterns. Doctoral dissertation. University of Washington, Seattle.

Drake, D. C., and R. J. Naiman. 2007. Reconstruction of Pacific salmon abundance from riparian tree ring growth. Ecological Applications 17:1523–1542.

Drake, D. C., R. J. Naiman, and J. S. Bechtold. 2006. Fate of nitrogen in riparian forest soils and trees: an ^{15}N tracer study simulating salmon decay. Ecology 87:1256–1266.

Drake, D. C., R. J. Naiman, and J. M. Helfield. 2002. Reconstructing salmon abundance in rivers: an initial dendrochronological evaluation. Ecology 83:2971–2977.

Drake, D. C., J. V. Smith, and R. J. Naiman. 2005. Salmon decay and nutrient contributions to riparian forest soils. Northwest Science 79:60–70.

Durbin, A. G., S. W. Nixon, and C. A. Oviatt. 1979. Effects of the spawning migration of the alewife, *Alosa pseudoharengus*, on freshwater ecosystems. Ecology 60:8–17.

Eastman, D. 1996. Response of freshwater fish communities to spawning sockeye salmon. Master's thesis. University of Washington, Seattle.

Elliot, S. T., and R. Bartoo. 1981. Relation of larval *Polypedilum* (Diptera, Chironomidae) to pink salmon eggs and alevins in an Alaskan stream Progressive Fish-Culturist 43:220–221.

Ellis, R. J. 1970. Alloperla stonefly nymphs: predators or scavengers on salmon eggs and alevins. Transactions of the American Fisheries Society 4:677–682.

Ennis, G. L., T. G. Northcote, and J. G. Stockner. 1983. Recent trophic changes in Kootenay Lake, British Columbia, as recorded by fossil diatoms. Canadian Journal of Botany 61:1983–1992.

Ewald, G., P. Larsson, H. Linge, L. Okla, and N. Szarzi. 1998. Biotransport of organic pollutants to an inland Alaska lake by migrating sockeye salmon (*Oncorhynchus nerka*). Arctic 51:40–7.

Faulkner, S. P., and C. J. Richardson. 1989. Physical and chemical characteristics of freshwater wetland soils. Pages 41–72 *in* D. A. Hammer, editor. Constructed wetlands for wastewater treatment: municipal, industrial, and agricultural. Lewis Publishers, Chelsea, Michigan.

Fausch, K. D., and T. G. Northcote. 1992. Large woody debris and salmonid habitat in a small coastal British Columbia stream. Canadian Journal of Fisheries and Aquatic Sciences 49:682–693.

Field-Dodgson, M. S. 1987. The effect of salmon redd excavation on stream substrate and benthic community of two salmon spawing streams in Canterbury, New Zealand. Hydrobiologia 154:3–11.

Finney, B. P., I. Gregory-Eaves, M. S. V. Douglas, and J. P. Smol. 2002. Fisheries productivity in the northeastern Pacific Ocean over the past 2,200 years. Nature (London) 416:729–733.

Finney, B. P., I. Gregory-Eaves, J. Sweetman, M. S. V.

Douglas, and J. P. Smol. 2000. Impacts of climatic change and fishing on Pacific salmon abundance over the past 300 years. Science 290:795–799.

Fisher Wold, A. K., and A. E. Hershey. 1999. Effects of salmon carcass decomposition on biofilm growth and wood decomposition. Canadian Journal of Fisheries and Aquatic Sciences 56:767–773.

Fontaine, S., A. Mariotti, and L. Abbadie. 2003. The priming effect of organic matter: a question of microbial competition? Soil Biology and Biochemistry 35:837–843.

Foote, C. J., and G. S. Brown. 1998. Ecological relationship between freshwater sculpins (genus *Cottus*) and beach-spawning sockeye salmon (*Oncorhynchus nerka*) in Iliamna Lake, Alaska. Canadian Journal of Fisheries and Aquatic Sciences 55:1524–1533.

Francis, T. B., D. E. Schindler, and J. W. Moore. 2006. Aquatic insects play a minor role in dispersing salmon-derived nutrients into riparian forests in southwestern Alaska. Canadian Journal of Fisheries and Aquatic Sciences 63:2543–2552.

Fry, B. 2006. Mixing. Pages 119–182 *in* B. Fry, editor. Stable isotope ecology. Springer, New York.

Fujiwara, M., and R. C. Highsmith. 1997. Harpacticoid copepods: potential link between inbound adult salmon and outbound juvenile salmon. Marine Ecology Progress Series 158:205–16.

Garten, C.T. Jr. 1993. Variation in foliar ^{15}N abundance and the availability of soil nitrogen on Walker Branch watershed. Ecology 74:2098–113.

Gende, S. M., R. T. Edwards, M. F. Willson, and M. S. Wipfli. 2002. Pacific salmon in aquatic and terrestrial ecosystems. BioScience 52:917–928.

Gende, S. M., and T. P. Quinn. 2004. The relative importance of prey density and social dominance in determining energy intake by bears feeding on Pacific salmon. Canadian Journal of Zoology 82:75–85.

Gende, S. M., T. P. Quinn, and M. F. Willson. 2001. Consumption choice by bears feeding on salmon. Oecologia 127:372–382.

Giannico, G. R., and S. G. Hinch. 2007. Juvenile coho salmon (*Oncorhynchus kisutch*) responses to salmon carcasses and in-stream wood manipulations during winter and spring. Canadian Journal of Fisheries and Aquatic Sciences 64:324–335.

Goering, J. J., and D. C. Brickell. 1972. Distribution of organics from salmon decomposition. Alaska Institute of Water Resources, Publication No. IWR-34, Fairbanks.

Goldman, C. R. 1960. Primary productivity and limiting factors in three lakes of the Alaskan Peninsula. Ecological Monographs 30:207–270.

Gregory-Eaves, I., M. J. Demers, M. J., L. Kimpe, E. M. Krümmel, R. W. Macdonald, B. P. Finney, and J. M. Blais. 2007. Tracing salmon-derived nutrients and contaminants in freshwater food webs across a pronounced spawner density gradient. Environmental Toxicology and Chemistry 26:1100–1108.

Gregory-Eaves, I., B. P. Finney, M. S. V. Douglas, and J. P. Smol. 2004. Inferring sockeye salmon (*Oncorhynchus nerka*) population dynamics and water-quality changes in a stained nursery lake over the past ~500 years. Canadian Journal of Fisheries and Aquatic Sciences 61:1235–1246.

Gregory-Eaves,I., D. T. Selbie, J. N. Sweetman, B. P. Finney, and J. P. Smol. 2009. Tracking sockeye salmon population dynamics from lake sediment cores: a review and synthesis. Pages 379–393 *in* A. J. Haro, K. L. Smith, R. A. Rulifson, C. M. Moffitt, R. J. Klauda, M. J. Dadswell, R. A. Cunjak, J. E. Cooper, K. L. Beal, and T. S. Avery, editors. Challenges for diadromous fishes in a dynamic global environment. American Fisheries Society, Symposium 69, Bethesda, Maryland.

Gregory-Eaves, I., J. P. Smol, M. S. V. Douglas, and B. P. Finney. 2003. Diatoms and sockeye salmon (*Oncorhynchus nerka*) population dynamics: reconstructions of salmon-derived nutrients over the past 2,200 years in two lakes from Kodiak Island, Alaska. Journal of Paleolimnology 30:35–53.

Gresh, T., J. Lichatowich, and P. Schoonmaker. 2000. An estimation of historic and current levels of salmon production in the northeast Pacific ecosystem: evidence of a nutrient deficit in the freshwater systems of the Pacific Northwest. Fisheries 25:15–21.

Griswold, R. G., D. Taki, and J. G. Stockner. 2003. Redfish Lake sockeye salmon: nutrient supplementation as a means of restoration. Pages 197–211 *in* J. Stockner, editor. Nutrients in salmonid ecosystems: sustaining production and biodiversity. American Fisheries Society, Symposium 34, Bethesda, Maryland.

Groot, C., and L. Margolis, editors. 1991. Pacific salmon life histories. University of British Columbia Press, Vancouver.

Gross, H. P., W. A. Wurtsbaugh, and C. Luecke. 1998. The role of anadromous sockeye salmon (*Oncorhynchus nerka*) in the nutrient loading and productivity of Redfish Lake, Idaho. Transactions of the American Fisheries Society 127:1–18.

Handley, L. L., and C. M. Scrimgeour. 1997. Terrestrial plant ecology and ^{15}N natural abundance: the present limits to interpretation for uncultivated systems with data from a Scottish old field. Advances in Ecological Research 27:133–212.

Hardy, F. J., K. S. Shortread, and J. G. Stockner. 1986. Bacterioplankton, phytoplankton, and zooplankton communities in a British Columbia coastal lake before and after nutrient reduction. Canadian Journal of Fisheries and Aquatic Sciences 43:1504–1514.

Harmon, M. E., J. F. Franklin, F. J. Swanson, P. Sollins,

S.V. Gregory, J. D. Lattin, N. H. Anderson, S. P. Cline, N. G. Aumen, J. R. Sedell, G. W. Linkaemper, K. Cromack, Jr., and K. W. Cummins. 1986. Ecology of coarse woody debris in temperate ecosystems. Pages 133–302 *in* A. Macfadyen and E. D. Ford editors. Advances in ecological research. Volume 15. Academic Press, New York.

Heintz, R. A., B. D. Nelson, J. Hudson, M. Larsen, and L. Holland. 2004. Marine subsidies in freshwater: effects of salmon carcasses on lipid class and fatty acid composition of juvenile coho salmon. Transactions of the American Fisheries Society 133:559–567.

Helfield, J. M., and R. J. Naiman. 2001. Effects of salmon-derived nitrogen on riparian forest growth and implications for stream productivity. Ecology 82:2403–2409.

Helfield, J. H., and R. J. Naiman. 2002. Salmon and alder as nitrogen sources to riparian forests in a boreal Alaskan watershed. Oecologia 133:573–582.

Helfield, J. M., and R. J. Naiman. 2003. Effects of salmon-derived nitrogen on riparian forest growth and implications for stream productivity: response. Ecology 84:3399–3401.

Helfield, J. M., and R. J. Naiman. 2006. Keystone interactions: salmon and bear in riparian forests of Alaska. Ecosystems 9:167–180.

Hesslein, R. H., M. J. Capel, D. E. Fox, and K. A. Hallard. 1991. Stable isotopes of sulfur, carbon and nitrogen as indicators of trophic level and fish migration in the lower Mackenzie River basin, Canada. Canadian Journal of Fisheries and Aquatic Sciences 48:2258–2265.

Hicks, B. J., M. S. Wipfli, D. W. Lang, and M. E. Lang. 2005. Marine-derived nitrogen and carbon in freshwater-riparian food webs of the Copper River Delta, southcentral Alaska. Oecologia 144:558.

Hilderbrand, G. V., S. D. Farley, C. T. Robbins, T. A. Hanley, K. Titus, and C. Servheen. 1996. Use of stable isotopes to determine diets of living and extinct bears. Canadian Journal of Zoology 74:2080–2088.

Hilderbrand, G. V., S. D. Farley, C. C. Schwartz, and C. T. Robbins. 2004. Importance of salmon to wildlife: implications for integrated management. Ursus 15:1–9.

Hilderbrand, G. V., T. A. Hanley, C. T. Robbins, and C. C. Schwartz. 1999a. Role of brown bears (*Ursus arctos*) in the flow of marine nitrogen into a terrestrial ecosystem. Oecologia 121:546–550.

Hilderbrand, G. V., C. C. Schwartz, C. T. Robbins, M. E. Jacoby, T. A. Hanley, S. M. Arthur, and C. Servheen. 1999b. The importance of meat, particularly salmon, to body size, population productivity, and conservation of North American brown bears. Canadian Journal of Zoology 77:132–138.

Hites, R. A., J. A. Foran, D. O. Carpenter, M. C. Hamilton, B. A. Knuth, and S. J. Schwager. 2004a. Global assessment of organic contaminants in farmed salmon. Science 303:226–229.

Hites, R. A., J. A. Foran, S. J. Schwager, B. A. Knuth, M. C. Hamilton, and D. O. Carpenter. 2004b. Global assessment of polybrominated diphenyl ethers in farmed and wild salmon. Environmental Science and Technology 38:4945–4949.

Hocking, M. D., C. T. Darimont, K. S. Christie, and T. E. Reimchen. 2007. Niche variation in burying beetles (*Nicrophorus* spp.) associated with marine and terrestrial carrion. Canadian Journal of Zoology-Revue Canadienne de Zoologie 85:437–442.

Hocking, M. D., and T. E. Reimchen. 2002. Salmon-derived nitrogen in terrestrial invertebrates from coniferous forests of the Pacific Northwest. BMC Ecology 2:4.

Hocking, M. D., and T. E. Reimchen. 2006. Consumption and distribution of salmon (*Oncorhynchus* spp.) nutrients and energy by terrestrial flies. Canadian Journal of Fisheries and Aquatic Sciences 63:2076–2086.

Hocking, M. D., R. A. Ring, and T. E. Reimchen. 2006. Burying beetle *Nicrophorus investigator* reproduction on Pacific salmon carcasses. Ecological Entomology 31:5–12.

Hood, E., J. Fellman, and R. T. Edwards. 2007. Salmon influences on dissolved organic matter in a coastal temperate brown-water stream: an application of fluorescence spectroscopy. Limnology and Oceanography 52:1580–1587.

Hume, J. M. B., K. S. Shortreed, and K. F. Morton. 1996. Juvenile sockeye rearing capacity of three lakes in the Fraser River system. Canadian Journal of Fisheries and Aquatic Sciences 53:719–733.

Hyatt, K. D., and J. G. Stockner. 1985. Responses of sockeye salmon (*Oncorhynchus nerka*) to fertilization of British Columbia coastal lakes. Canadian Journal of Fisheries and Aquatic Sciences 42:320–331.

Jackson, J. K., and S. G. Fisher. 1986. Secondary production, emergence and export of aquatic insects of a Sonoran Desert stream. Ecology 67:629–638.

Jauquet, J., N. Pittman, J. A. Heinis, S. Thompson, N. Tatyama, and C. J. Cederholm. 2003. Observations of chum salmon consumption by wildlife and changes in water chemistry at Kennedy Creek during 1997–2000. Pages 71–88 *in* J. Stockner, editor. Nutrients in salmonid ecosystems: sustaining production and biodiversity. American Fisheries Society, Symposium 34, Bethesda, Maryland.

Johnson, L. L., G. M. Ylitalo, M. R. Arkoosh, A. N. Kagley, C. L. Stafford, J. L. Bolton, J. Buzitis, B. F. Anulacion, and T. K. Collier. 2007. Contaminant exposure in outmigrant juvenile salmon from Pacific

Northwest estuaries. Environmental Monitoring and Assessment 124:167–194.

Johnston, N. T., C. J. Perrin, P. A. Slaney, and B. R. Ward. 1990. Increased juvenile salmonid growth by whole-river fertilization. Canadian Journal of Fisheries and Aquatic Sciences 47:862–872.

Johnston, N. T., E. A. MacIsaac, P. J. Tschaplinski, and K. J. Hall. 2004. Effects of the abundance of spawning sockeye salmon (*Oncorhynchus nerka*) on nutrients and algal biomass in forested streams. Canadian Journal of Fisheries and Aquatic Sciences 61:384–403.

Jonsson, B., and N. Jonsson. 2005. Lipid energy reserves influence life-history decision of Atlantic salmon (*Salmo salar*) and brown trout (*S. trutta*) in fresh water. Ecology of Freshwater Fish 14:296–301.

Juday, C., W. H. Rich, G. I. Kemmerer, and A. Mann. 1932. Limnological studies of Karluk Lake, Alaska 1926–1930. Fishery Bulletin 47:407–436.

Keller, E. A., and F. J. Swanson. 1979. Effects of large organic material on channel form and fluvial processes. Earth Surface Processes 4:361–380.

Kelly, B. C., S. L. Gray, M. G. Ikonomou, J. S. Macdonald, S. M. Bandiera, and E. G. Hrycay. 2007. Lipid reserve dynamics and magnification of persistent organic pollutants in spawning sockeye salmon (*Oncorhynchus nerka*) from the Fraser River, British Columbia. Environmental Science and Technology 41:3083–3089.

Kielland K., and J. P. Bryant. 1998. Moose herbivory in taiga: effects on biogeochemistry and vegetation dynamics in primary succession. Oikos 82:377–383.

Killham, K. 1994. The soil biota. Pages 35–61 *in* K. Killham, editors. Soil ecology. Cambridge University Press, Cambridge, UK.

Kirchhoff, M. D. 2003. Effects of salmon-derived nitrogen on riparian forest growth and implications for stream productivity: comment. Ecology 84:3396–3399.

Kline, T. C., Jr., J. J. Goering, O. A. Mathisen, P. H. Poe, and P. P. Parker. 1990. Recycling of elements transported upstream by runs of Pacific salmon: I. $\delta^{15}N$ and $\delta^{13}C$ evidence in Sashin Creek, southeastern Alaska. Canadian Journal of Fisheries and Aquatic Sciences 47:136–144.

Kline, T. C., Jr., J. J. Goering, O. A. Mathisen, P. H. Poe, P. L. Parker, and R. S. Scanlan. 1993. Recycling of elements transported upstream by runs of Pacific salmon: II. $\delta^{15}N$ and $\delta^{13}C$ evidence in the Kvichak river watershed, Bristol Bay, southwestern Alaska. Canadian Journal of Fisheries and Aquatic Sciences 50:2350–2365.

Klinka, D. R., and T. E. Reimchen. 2002. Nocturnal and diurnal foraging behaviour of brown bears (*Ursus arctos*) on a salmon stream in coastal British Columbia. Canadian Journal of Zoology 80:1317–1322.

Klinka, K. M., V. J. Krajina, A. Ceska, and A. M. Scagel. 1989. Indicator plants of coastal British Columbia. University of British Columbia Press, Vancouver.

Koenings, J. P., and R. D. Burkett. 1987. An aquatic Rubic's cube: restoration of the Karluk lake sockeye salmon (*Oncorhynchus nerka*). Pages 419–434 *in* H. D. Smith, L. Margolis, and C. C. Wood, editors. Sockeye salmon (*Oncorhynchus nerka*) population biology and future management. Canadian Special Publication of Fisheries and Aquatic Sciences 96.

Kohler, A. E., A. Rugenski, and D. Taki. 2008. Stream food web response to a salmon carcass analogue addition in two central Idaho, U.S.A. streams. Freshwater Biology 53:446–460.

Koyama, A., K. Kavanagh, and A. Robinson. 2005. Marine nitrogen in central Idaho riparian forests: evidence from stable isotopes. Canadian Journal of Fisheries and Aquatic Sciences 62:518–526.

Krohkin, E. M. 1967. Influence of the intensity of passage of the sockeye salmon *Oncorhynchus nerka* (Wald.) on the phosphate content of spawning lakes. Izdanija "Nauka," Lenningrad. 15:26–31 (Translation from Russian by Fisheries Research Board of Canada Translation Series 1273, 1968).

Krohkin E. M. 1975. Transport of nutrients by salmon migrating from the sea into lakes. Pages 153–156 *in* A. D. Hasler, editor. Coupling of land and water systems. Springer-Verlag, New York.

Krümmel, E., R. W. Macdonald, L. E. Kimpe, I. Gregory-Eaves, M. Demers, J. P. Smol, B. Finney, and J. M. Blais. 2003. Delivery of pollutants by spawning salmon. Nature (London) 425:255–256.

Kyle, G. B. 1994. Assessment of trophic-level responses and coho salmon (*Oncorhynchus kisutch*) production following nutrient treatment (1981–1986) of Bear Lake, Alaska. Fisheries Research 20:243–261.

Kyle, G. B. 1996. Stocking sockeye salmon (*Oncorhynchus nerka*) in barren lakes of Alaska: effects on the macrozooplankton. Fisheries Research 28:29–44.

Larkin, G. A., and P. A. Slaney. 1997. Implications of trends in marine-derived nutrient influx to south coastal British Columbia salmonid production. Fisheries 22:16–24.

Latterell, J. J., J. S. Bechtold, T. C. O'Keefe, R. Van Pelt, and R. J. Naiman. 2006. Dynamic patch mosaics and channel movement in an unconfined river valley of the Olympic Mountains. Freshwater Biology 51:523–544.

Latterell, J. J., and R. J. Naiman. 2007. Sources and dynamics of large logs in a temperate floodplain river. Ecological Applications 17:1127–1141.

LeBrasseur, R. J., C. D. McAllister, and T. R. Parsons. 1979. Addition of nutrients to a lake leads to greatly increased catch of salmon. Environmental Conservation 6:187–190.

Lienesch, P. W., M. E. McDonald, A. E. Hershey, W. J, O'Brien, and N. D. Bettez. 2005. Effects of a whole-lake, experimental fertilization on lake trout in a small oligotrophic arctic lake. Hydrobiologia 548:51–66.

Lundberg, J., and F. Moberg. 2003. Mobile link organisms and ecosystem functioning: implications for ecosystem resilience and management. Ecosystems 6:87–98.

MacAvoy, S. E., S. A. Macko, and G. C. Garman. 1998. Tracing marine biomass into tidal freshwater ecosystems using stable sulfur isotopes Naturwissenschaften 85:544–546.

MacKenzie, G. 2001. Trophic relations between coho salmon carcasses, oomycetes, and select caddisfly larvae. Master's thesis. University of Washington, Seattle.

Mantua, N. J., S. R. Hare, Y. Zhang, J. M. Wallace, and R. C. Francis. 1997. A Pacific interdecadal climate oscillation with impacts on salmon production. Bulletin of the American Meterological Society 78:1969–1079.

Mathewson, D. D., M. D. Hocking, and T. E. Reimchen. 2003. Nitrogen uptake in riparian plant communities across a sharp ecological boundary of salmon density. BMC Ecology 3:4–15.

Mathisen, O. A. 1972. Biogenic enrichment of sockeye salmon lakes and stock productivity. Verhandlungen Internationale Vereinigung Limnologie 18:1089–1095.

Mathisen, O. A., P. L. Parker, J. J. Goering, T. C. Kline, P. H. Poe, and R. S. Scanlan. 1988. Recycling of marine elements transported into freshwater by anadromous salmon. Verhandlungen Internationale Vereinigung Limnologie 23:2249–2258.

McMahon, T. E., and G. F. Hartman. 1989. Influence of cover complexity and current velocity on winter habitat use by juvenile coho salmon (*Oncorhynchus kisutch*). Canadian Journal of Fisheries and Aquatic Sciences 46:1551–1557.

McNeil, W. J. 1964. Effect of the spawning bed environment on reproduction of pink and chum salmon. Fishery Bulletin 65:495–523.

Meehan, E. P., E. E. Seminet-Reneau, and T. P. Quinn. 2005. Bear predation on Pacific salmon facilitates colonization of carcasses by fly maggots. American Midland Naturalist 153:142–151.

Meehan, W. R., F. J. Swanson, and J. R. Sedell. 1977. Influences of riparian vegetation on aquatic ecosystems with particular reference to salmonid fishes and their food supply. Pages 137–145 *in* J. R. Roy and D. A. Jones, technical coordinators. Importance, preservation, and management of riparian habitat: a symposium. U.S. Forest Service, Pacific Northwest Forest and Range Experiment Station, Corvallis, Oregon.

Michael, J. H. 1995. Enhancement effects of spawning pink salmon on stream rearing juvenile coho salmon: managing one resource to benefit another. Northwest Science 69:228–233.

Minakawa, N., and R. I. Gara. 1999. Ecological effects of a chum salmon (*Oncorhynchus keta*) spawning run in a small stream of the Pacific Northwest. Journal of Freshwater Ecology 14:327–335.

Minakawa, N., and R. I. Gara. 2003. Effects of chum salmon redd excavation on benthic communities in a stream in the Pacific Northwest. Transactions of the American Fisheries Society 132:598–604.

Minakawa, N., R. I. Gara, and J. M. Honea. 2002. Increased individual growth rate and community biomass of stream insects associated with salmon carcasses. Journal of the North American Benthological Society 21:651–659.

Mitchell, N. L., and G. A. Lamberti. 2005. Responses in dissolved nutrients and epilithon abundance to spawning salmon in southeast Alaska streams. Limnology and Oceanography 50:217–227.

Montgomery, D. R., J. M. Buffington, N. P. Peterson, D. Schuett-Hames, and T. P. Quinn. 1996. Stream-bed scour, egg burial depths, and the influence of salmonid spawning on bed surface mobility and embryo survival. Canadian Journal of Fisheries and Aquatic Sciences 53:1061–1070.

Moore, J. W., and D. E. Schindler. 2004. Nutrient export from freshwater ecosystems by anadromous sockeye salmon (*Oncorhynchus nerka*). Canadian Journal of Fisheries and Aquatic Sciences 61:1582–1589.

Moore, J. W., D. E. Schindler, J. L. Carter, J. Fox, J. Griffiths, and G. W. Holtgrieve. 2007. Biotic control of stream fluxes: spawning salmon drive nutrient and matter export. Ecology 88:1278–1291.

Moore, J. W., D. E. Schindler, and M. D. Scheuerell. 2004. Disturbance of freshwater habitats by anadromous salmon in Alaska. Oecologia 139:298–308.

Morris, A. E. L., J. M. Stark, and B. K. Gilbert. 2005. Evaluation of isotopic fractionation error on calculations of marine-derived nitrogen in terrestrial ecosystems. Canadian Journal of Forest Research 35:1604–1616.

Murphy, M. L., and K. V. Koski. 1989. Input and depletion of woody debris in Alaska streams and implications for streamside management. North American Journal of Fisheries Management 9:427–436.

Murphy, M. L., K. V. Koski, J. Heifetz, S. W. Johnson, D. Kirchofer, and J. F. Thedinga. 1985. Role of large organic debris as winter habitat for juvenile salmonids in Alaska streams. Proceedings of the Western Association of Fish and Wildlife Agencies 64:251–262, Victoria, British Columbia.

Nadelhoffer, K. J., and B. Fry. 1994. Nitrogen isotope

studies in forest ecosystems. Pages 22–44 *in* K. Lajtha and R. H. Michener, editors. Stable isotopes in ecology and environmental science. Blackwell Scientific Publications, London.

Nagasaka, A., Y. Nagasaka, K. Ito, T. Mano, M. Yamanaka, A. Katayama, Y. Sato, A. L. Grankin, A. I. Zdorikov, and G. A. Boronov. 2006. Contributions of salmon-derived nitrogen to riparian vegetation in the northwest Pacific region. Journal of Forest Research 11:377–382.

Naiman, R. J., R. E. Bilby, D. E. Schindler, and J. M. Helfield. 2002. Pacific salmon, nutrients, and the dynamics of freshwater and riparian ecosystems. Ecosystems 5:399–417.

Naiman, R. J., and H. Décamps. 1997. The ecology of interfaces: riparian zones. Annual Review of Ecology and Systematics 28:621–58.

Naiman, R. J., H. Décamps, and M. E. McClain. 2005. Riparia: ecology, conservation and management of streamside communities. Elsevier/Academic Press, San Diego, California.

Naiman, R. J., and J. J. Latterell. 2005. Principles for linking fish habitat to fisheries management and conservation. Journal of Fish Biology 67:166–185.

Nambiar, E. K. S., and D. N. Fife. 1991. Nutrient retranslocation in temperate conifers. Tree Physiology 9:185–207.

Nicola, S. J. 1968. Scavenging by *Alloperla* (Plecoptera: Chloroperlidae) nymphs on dead pink (*Oncorhynchus gorbuscha*) and chum (*O. keta*) salmon embryos. Canadian Journal Zoology 46:787–796.

O'Keefe, T. C., and R. T. Edwards. 2003. Evidence for hyporheic transfer and storage of marine-derived nutrients in sockeye streams in southwest Alaska. Pages 99–107 *in* J. Stockner, editor. Nutrients in salmonid ecosystems: sustaining production and biodiversity. American Fisheries Society, Symposium 34, Bethesda, Maryland.

O'Keefe, T. C., and R. J. Naiman. 2006. The influence of forest structure on riparian litterfall in a Pacific coastal rainforest. Canadian Journal of Forest Research 36:2852–2863.

O'Neill, S. M., G. M. Ylitalo, J. E. West, J. Bolton, C. A. Sloan, and M. M. Krahn. 2006. Regional patterns of persistent organic pollutants in five Pacific salmon species (*Oncorhynchus* spp.) April 3–5, 2006. Extended abstract from the 2006 Southern Resident Killer Whale Symposium, April 3–5, 2006, Seattle, Washington.

Pastor J. P., R. J. Naiman, B. Dewey, and P. McInnes. 1988. Moose, microbes and the boreal forest. BioScience 38:770–777.

Payne, L. X., and J. W. Moore. 2006. Mobile scavengers create hotspots of freshwater productivity. Oikos 115:69–80.

Pearsons, T. N., D. D. Roley, and C. L. Johnson. 2007. Development of a carcass analog for nutrient restoration in streams. Fisheries 32:114–124.

Pess, G. R., D. R. Montgomery, E. A. Steel, R. E. Bilby, B. E. Feist, and H. M. Greenberg. 2002. Landscape characteristics, land use, and coho salmon (*Oncorhynchus kisutch*) abundance, Snohomish River, Wash., USA. Canadian Journal of Fisheries and Aquatic Sciences 59:613–623.

Peterson, B. J., and B. Fry. 1987. Stable isotopes in ecosystem studies. Annual Review of Ecology and Systematics 18:293–320.

Peterson, D. P., and C. J. Foote. 2000. Disturbance of small-stream habitat by spawning sockeye salmon in Alaska. Transactions of the American Fisheries Society 129:924–934.

Phillips, D. L., and J. W. Gregg. 2003. Source partitioning using stable isotopes: coping with too many sources. Oecologia 136:261–269.

Phillips, D. L., S. D. Newsome, and J. W. Gregg. 2005. Combining sources in stable isotope mixing models: alternative methods. Oecologia 144:520.

Pinay, G., T. C. O'Keefe, R. T. Edwards, and R. J. Naiman. 2003. Potential denitrification activity in the landscape of a western Alaska drainage basin. Ecosystems 6:336–343.

Quinn, T. P. 2005. The behavior and ecology of Pacific salmon and trout. American Fisheries Society, Bethesda, Maryland and University of Washington Press, Seattle.

Quinn, T. P., and G. B. Buck. 2000. Scavenging by brown bears, *Ursus arctos*, and glaucous-winged gulls, *Larus glaucescens*, on adult sockeye salmon, *Oncorhynchus nerka*. Canadian Field-Naturalist 114:217–223.

Quinn, T. P., S. M. Gende, and D. E. Rogers. 2003. Density-dependent predation by brown bear (*Ursus arctos*) on sockeye salmon (*Oncorhynchus nerka*). Canadian Journal of Fisheries and Aquatic Sciences 60:553–562.

Quinn, T. P., and M. T. Kinnison. 1999. Size-selective predation by brown bears on sockeye salmon. Oecologia 121:273–282.

Quinn, T. P., L. Wetzel, S. Bishop, K. Overberg, and D. E. Rogers. 2001. Influence of breeding habitat on bear predation and age at maturity and sexual dimorphism of sockeye salmon populations. Canadian Journal of Zoology 79:1782–1793.

Reimchen, T. E. 2000. Some ecological and evolutionary aspects of bear-salmon interactions in coastal British Columbia. Canadian Journal of Zoology 78:448–457.

Reimchen, T. E., D. Mathewson, M. D. Hocking, J. Moran, and D. Harris. 2003. Isotopic evidence for enrichment of salmon-derived nutrients in vegetation, soil, and insects in riparian zones in coastal British Columbia. Pages 59–69 *in* J. Stockner, editor. Nutrients in salmonid ecosystems: sustaining production and biodiversity. American Fisheries Society, Symposium 34, Bethesda, Maryland.

Richey, J. E., M. A. Perkins, and C. R. Goldman. 1975. Effects of kokanee salmon (*Oncorhynchus nerka*) decomposition on the ecology of a subalpine stream. Journal of the Fisheries Research Board of Canada 32:817–820.

Roche, J., and M. McHutchison. 1998. First fish/first people: salmon tales of the North Pacific Rim. University of Washington Press, Seattle.

Rouxel, O. J., A. Bekker, and K. J. Edwards. 2005. Iron isotope constraints Archaen and Paleoproterozoic ocean redox state. Science 307:1088–1091.

Ruggerone, G. T., R. Hanson, and D. E. Rogers. 2000. Selective predation by brown bears (*Ursus arctos*) foraging on spawning sockeye salmon (*Oncorhynchus nerka*). Canadian Journal of Zoology 78:974–981.

Sarica, J., M. Amyot, L. Hare, M.-R. Doyon, and L. W. Stanfield. 2004. Salmon-derived mercury and nutrients in a Lake Ontario spawning stream. Limnology and Oceanography 49:891–899.

Scheuerell, M. D., P. S. Levin, R. W. Zabel, J. G. Williams, and B. L. Sanderson. 2005. A new perspective on the importance of marine-derived nutrients to threatened stocks of Pacific salmon (*Oncorhynchus* spp.). Canadian Journal of Fisheries and Aquatic Sciences 62:961–964.

Scheuerell, M. D., and J. G. Williams. 2005. Forecasting climate-induced changes in the survival of Snake River spring/summer Chinook salmon. Fisheries Oceanography 14:448–457.

Schindler, D. E., P. R. Leavitt, C. S. Brock, S. P. Johnson, and P. D. Quay. 2005. Marine-derived nutrients, commercial fisheries, and marine production of salmon and lake algae in Alaska. Ecology 86:3225–3231.

Schindler, D. E., M. D. Scheuerell, J. W. Moore, S. M. Gende, T. B. Francis, and W. J. Palen. 2003. Pacific salmon and the ecology of coastal ecosystems. Frontiers in Ecology and the Environment 1:31–37.

Schmidt, D. C., S. R. Carlson, G. B. Kyle, and B. P. Finney. 1998. Influence of carcass-derived nutrients on sockeye salmon productivity of Karluk Lake, Alaska: importance in the assessment of an escapement goal. North American Journal of Fisheries Management 18:743–763.

Schuldt, J. A., and A. E. Hershey. 1995. Effect of salmon carcass decomposition on Lake Superior tributary streams. Journal of the North American Benthological Society 14:259–268.

Schulze, E. D., F. S. Chapin, III, and G. Gebauer. 1994. Nitrogen nutrition and isotope differences among life forms at the northern treeline of Alaska. Oecologia 100:406–12.

Selbie, D. T., B. Lewis, J. P. Smol, and B. P. Finney. 2007. Long-term population dynamics of endangered Snake River sockeye salmon: evidence of past influences on stock decline and impediments to recovery. Transactions of the American Fisheries Society 136:800–821.

Sheppard, P. R., and M. A. Topa. 2002. Physical-chemical pretreatment of wood for measuring tree-ring nitrogen. Pages 303–304 *in* Y. Bégin, editor. Dendrochronology, environmental change and human history. Sixth International Conference on Dendrochronology. Université Laval, Centre d'Études Nordiques, Québec.

Steel, E. A., B. E. Feist, D. Jenson, G. R. Pess, M. B. Sheer, J. Brauner, and R. E. Bilby. 2004. Landscape models to understand steelhead (*Oncorhynchus mykiss*) distribution and help prioritize barrier removals in the Willamette Basin, OR, U.S.A. Canadian Journal of Fisheries and Aquatic Sciences 61:999–1011.

Stockner, J. G. 1981. Whole-lake fertilization for the enhancement of sockeye salmon (*Oncorhynchus nerka*) in British Columbia, Canada. Verhandlungen Internationale Vereinigung Limnologie 21:293–299.

Stockner, J.G. 1987. Lake fertilization: the enrichment cycle and lake sockeye salmon (*Oncorhynchus nerka*) production. Pages 198–215 *in* H. D. Smith., L. Margolis, and C. C. Wood, editors. Sockeye salmon (*Oncorhynchus nerka*) population biology and future management. Canadian Special Publication of Fisheries and Aquatic Sciences 96.

Stockner, J. G. and C. D. Levings. 1982. Biological reconnaissance of the Yakoun River estuary, Queen Charlotte Islands, and results of a trial fertilization with urea. Canadian Technical Report of Fisheries and Aquatic Sciences 1132.

Stouder, D. J., P. A. Bisson, and R. J. Naiman editors. 1997. Pacific salmon and their ecosystems: status and future options. Chapman and Hall, New York.

Sugai, S. F., and D. C. Burrell. 1984. Transport of dissolved organic carbon, nutrients, and trace metals from the Wilson and Blossom rivers to Smeaton Bay, southeast Alaska. Canadian Journal of Fisheries and Aquatic Sciences 41:180–190.

Suominen, O., K. Danell, and J. P. Bryant. 1999. Indirect effects of mammalian browsers on vegetation and ground-dwelling insects in an Alaskan floodplain. Ecoscience 6:505–510.

Sweetman, J. N., and B. P. Finney. 2003. Differential responses of zooplankton populations (*Bosmina longirostris*) to fish predation and nutrient-loading in an introduced and a natural sockeye salmon nursery lake on Kodiak Island, Alaska, USA. Journal of Paleolimnology 30:183–193.

Szepanski, M. M., M. Ben-David, and V. Van Ballenberghe. 1999. Assessment of anadromous salmon resources in the diet of the Alexander Archipelago wolf using stable isotope analysis. Oecologia 120:327–335.

Thomann, R. V., J. P. Connolly, and T. F. Parkerton. 1992. An equilibrium model of organic chemical accumulation in aquatic food webs with sediment interaction. Environmental Toxicology and Chemistry 11:615–629.

Van Sickle, J., and S. V. Gregory. 1990. Modeling inputs of large woody debris to streams from falling trees. Canadian Journal of Forest Research 20:1593–1601.

Varele, D. E., C. J. Pride, and M. A. Brzezinski. 2004. Biological fractionation of silicon isotopes in Southern Ocean surface waters. Global Biogeochemical Cycles. 18:GB01047, doi:10.1029/2003GB002140.

Vitousek, P. M., G. Shearer, and D. H. Kohl. 1989. Foliar ^{15}N natural abundance in Hawaiian rainforest: patterns and possible mechanisms. Oecologia 78:383–388.

Walter, J. K., R. E. Bilby, and B. R. Fransen. 2006. Effects of Pacific salmon spawning and carcass availability on the caddisfly *Ecclisomyia conspersa* (Trichoptera: Limnephilidae). Freshwater Biology 51:1211–1218.

Willson, M. F., S. M. Gende, and B. H. Marston. 1998. Fishes and the forest: expanding perspectives on fish-wildlife interactions. BioScience 48:455–62.

Willson, M. F., and K. C. Halupka. 1995. Anadromous fish as keystone species in vertebrate communities. Conservation Biology 9:489–497.

Wilkinson, C. E., M. D. Hocking, and T. E. Reimchen. 2005. Uptake of salmon-derived nitrogen by mosses and liverworts in coastal British Columbia. Oikos 108:85–98.

Wilzbach, M. A., B. C. Harvey, J. L. White, and R. J. Nakamoto. 2005. Effects of riparian canopy opening and salmon carcass addition on the abundance and growth of resident salmonids. Canadian Journal of Fisheries and Aquatic Sciences 62:58–67.

Winder, M., D. E. Schindler, J. W. Moore, S. P. Johnson, and W. J. Palen. 2005. Do bears facilitate transfer of salmon resources to aquatic macroinvertebrates? Canadian Journal of Fisheries and Aquatic Sciences 62:2285–2293.

Wipfli, M. S. 1997. Terrestrial invertebrates as salmonid prey and nitrogen sources in streams: contrasting old-growth and young-growth riparian forests in southeastern Alaska, U.S.A. Canadian Journal of Fisheries and Aquatic Sciences 54:1259–1269.

Wipfli, M. S., J. Hudson, and J. Caouette. 1998. Influence of salmon carcasses on stream productivity: response of biofilm and benthic macroinvertebrates in southeastern Alaska, USA. Canadian Journal of Fisheries and Aquatic Sciences 55:1503–1511.

Wipfli, M. S., J. P. Hudson, J. P. Caouette, and D. T. Chaloner. 2003. Marine subsidies in freshwater ecosystems: salmon carcasses increase the growth rates of stream-resident salmonids. Transactions of the American Fisheries Society 132:371–381.

Wipfli, M. S., J. P. Hudson, and J. P. Caouette. 2004. Restoring productivity of salmon-based food webs: contrasting effects of salmon carcass and salmon carcass analog additions on stream-resident salmonids. Transactions of the American Fisheries Society 133:1440–1454.

Wipfli, M. S., J. P. Hudson, D. T. Chaloner, and J. P. Caouette. 1999. The influence of salmon spawner densities on stream productivity in southeast Alaska. Canadian Journal of Fisheries and Aquatic Sciences 56:1600–1611.

Wurtsbaugh, W. A., H. P. Gross, C. Luecke, and P. Budy. 1997. Nutrient limitation of oligotrophic sockeye salmon lakes of Idaho (USA). Verhandlungen Internationale Vereinigung für Theoretische und Angewandte Limnologie 26:413–419.

Zhang, X., A. S. Naidu, J. J. Kelley, S. C. Jewett, D. Dasher, and L. K. Duffy. 2001. Baseline concentrations of total mercury and methylmercury in salmon returning via the Bering Sea (1999–2000). Marine Pollution Bulletin 42:993–997.

Zhang, Y. X., J. N. Negishi, J. S. Richardson, and R. Kolodziejczyk. 2003. Impacts of marine-derived nutrients on stream ecosystem functioning. Proceedings of the Royal Society of London Series B 270:2117–2123.

American Fisheries Society Symposium 69:427–445, 2009

Detecting Marine Nutrient and Organic Matter Inputs into Multiple Trophic Levels in Streams of Atlantic Canada and France

Timothy D. Jardine*
Canadian Rivers Institute and Department of Biology, University of New Brunswick Fredericton, New Brunswick E3B 5A3, Canada

Jean-Marc Roussel
INRA, Agrocampus Rennes, UMR 985, Laboratoire d'Ecologie Aquatique Rennes, France

Sean C. Mitchell
Canadian Rivers Institute and Department of Biology, St. Francis Xavier University Antigonish, Nova Scotia, Canada

Richard A. Cunjak
Canadian Rivers Institute and Department of Biology, University of New Brunswick Fredericton, New Brunswick E3B 5A3, Canada

Abstract.—We used stable isotope analysis in an attempt to detect marine subsidies from anadromous fish to freshwater benthos in four river systems draining to the Atlantic Ocean. Benthic invertebrates in the West River, Nova Scotia, Canada, had elevated $\delta^{13}C$, $\delta^{15}N$, and $\delta^{34}S$ values in a downstream reach that suggested consumption of marine-derived organic matter from spawning blueback herring *Alosa aestivalis*. In Doctor's Brook, Nova Scotia, the arrival of rainbow smelt *Osmerus mordax* to spawn led to rapid increases in the $\delta^{13}C$ and $\delta^{15}N$ of a predatory stonefly (Perlidae), but lower trophic levels (mayflies and biofilm) showed inconsistent responses. Sculpin *Cottus* sp. showed no evidence of predation on Atlantic salmon *Salmo salar* eggs in Catamaran Brook, New Brunswick, Canada or the Scorff River, Brittany, France. These analyses suggest that marine organic matter subsidies, in the form of direct consumption of eggs and/or carcasses, are important in streams with concentrated spawning activity such as by alosid and osmerid species, whereas carbon and nitrogen contributions from more sparse spawning species such as by Atlantic salmon may be minimal.

Introduction

The long-term stability of food webs in systems with limited primary productivity may be dependent on inputs of nutrients and organic matter from adjacent or connected productive systems (Polis et al. 1997). One such pairing in temperate regions is oligotrophic streams and rivers with the ocean (Gross et al. 1988). Many rivers have a limited supply of phosphorus, nitrogen, or both (Dodds et al. 2002; Tank and Dodds 2003), and autotrophs and heterotrophs residing within these watercourses may therefore be reliant on seasonal inputs of nutrients and organic matter from beyond the stream boundaries. For example, on the west coast of North America, spawning Pacific salmon *Oncorhynchus* sp. annually deliver more than 10,000 metric tons (mt) of biomass, more than 40 mt of phosphorus, and 400 mt of nitrogen to oligotrophic streams, rivers, and lakes (Gresh et al. 2000). While these num-

* Corresponding author: tim.jardine@unb.ca

bers are small relative to historic levels (Gresh et al. 2000), they likely contribute to sustaining productivity levels higher than those in nearby systems lacking salmon (Wipfli et al. 1999; Chaloner et al. 2004; Naiman et al. 2009, this volume).

While perhaps not as dramatic as the effects observed in Pacific coast systems (Naiman et al. 2009), previous studies have suggested that the spawning run of anadromous fishes may contribute to the energy budget of rivers draining to the Atlantic Ocean (Durbin et al. 1979; Jonsson and Jonsson 2003; Nislow et al. 2004). Anadromous species such as Atlantic salmon *Salmo salar*, anadromous brown trout *S. trutta*, rainbow smelt *Osmerus mordax*, and several members of the family Alosidae, including American shad *Alosa sapidissima*, allis shad *A. alosa*, twaid shad *A. fallax*, blueback herring *A. aestivalis*, and alewife *A. pseudoharengus* occur in sufficient numbers to suggest that despite their generally iteroparous life histories, they provide considerable quantities of nutrients and organic matter to coastal rivers and streams through excretion and release of gametes and partial die-off.

Many studies have used stable isotope analysis (SIA) to estimate the relative importance of marine nutrient and organic matter subsidies carried by salmon and other anadromous fishes into freshwater (e.g., Bilby et al. 1996; MacAvoy et al. 2000). Due to predictable patterns in isotope fractionation by primary producers between freshwater and the sea (Fry and Sherr 1984; France 1995a), enriched ^{13}C, ^{15}N, and ^{34}S values are often deemed as evidence of marine subsidies to a variety of trophic levels (Kline et al. 1990; Harding et al. 2004). The recent explosion in popularity of SIA as a research tool (Kelly 2000) coupled with modeling efforts that suggest marine nutrients can be important in Atlantic coast streams (Durbin et al. 1979; Jonsson and Jonsson 2003; Nislow et al. 2004) together beg the question, why have we not seen stable isotope evidence for such subsidies in the benthos of Atlantic coast streams (Garman and Macko 1998)?

We initiated a program designed to determine if marine subsidies were present in four systems from both sides of the Atlantic Ocean, two studies related to Atlantic salmon, one related to blueback herring, and one related to rainbow smelt. We attempted to distinguish between nutrient subsidies and organic matter subsidies, discriminating the former as dissolved carbon and nitrogen that would be incorporated by growing algae and the latter as a direct food resource for consumers in the form of eggs and/or carcasses. Any evidence of marine isotope ratios in the benthos of these catchments would provide future studies with baseline data on the significance of such subsidies in Atlantic drainage streams and would allow for an assessment of the importance of anadromous fishes to their overall energy budgets.

In the Miramichi River drainage, New Brunswick, Canada, Atlantic salmon spawn in several tributaries, including Catamaran Brook (Cunjak et al. 1993). Similarly, in the Scorff River, Brittany, France, Atlantic salmon populations spawn in certain tributaries, including the Scave River and Kerlégan Brook (Baglinière and Champigneulle 1986). Also present in many salmon spawning tributaries are fishes of the family Cottidae (e.g., slimy sculpin *Cottus cognatus* and bullhead *C. gobio*) hereafter collectively referred to as sculpin) that have long been suspected of eating the eggs of salmonid fishes (Moyle 1977; Foote and Brown 1998). Sculpin are spring spawners and females invest large amounts of energy into gonad development during the winter and early spring (Petrosky and Waters 1975; Gray and Munkittrick 2005), with likely high rates of food consumption (Jardine et al. 2005). We therefore sampled both sexes but expected a greater isotopic response in females with the prediction that if salmon eggs were indeed being consumed by sculpin, marine isotope ratios would most likely be measured in muscle and gonad tissue.

In the West River, Nova Scotia, Canada, we sampled both long-lived (dragonflies and stoneflies) and short-lived (mayflies) aquatic macroinvertebrates to assess patterns in stable isotope ratios at a location where blueback herring spawn and at an upstream location above the limit of spawning. We predicted that scavenging of eggs and carcasses by invertebrates in the downstream reach would result in marine isotopic ratios (Chaloner et al. 2002) that were highest in summer following the spawning run and low or absent in the spring prior to the run. We expected that long-lived invertebrates would show limited seasonal variation in response to marine inputs because of their larger body size and lower turnover, while short-lived invertebrates would show rapid and dramatic responses.

In Doctor's Brook, Nova Scotia, Canada, rainbow smelt typically enter the stream in May and deposit eggs and sperm on the stream substrate. We sampled biofilm and invertebrate grazers and predators in the spawning reach and in an upstream reach not accessible to smelt. These sites were sampled several times over the course of one growing season to determine the timing and extent of marine inputs and to differentiate between direct consumption of marine organic matter (e.g., eggs) and the use of dissolved marine-derived nutrients by primary producers.

Methods

Study Sites and Sample Collection

West River, Nova Scotia, Canada.—The West River (Figure 1) drains an area of approximately 350 km^2, flowing directly into the Gulf of St. Lawrence (N 45°37′W 61°58′). The landscape of the drainage basin has been greatly modified from natural hardwood forest. Now, the lower West River, where this study was conducted, is largely under agricultural and rural development with the consequent loss of forest associated with these activities. However, riparian buffer strips in these areas remain intact. Bank-full river discharge is estimated at between 30 and 40 m^3/s (MacLaren Plansearch 1983). There are 13 lakes within the drainage, ranging in size between less than 5 ha and 65 ha. Gaspereau Lake, the largest, lies off of the main-stem West River and is connected to the main stem by a short (<2.0 km) tributary.

Freshwater fish species known to be present in the West River drainage include yellow perch *Perca fluviatilis*, blueback herring, alewife, white sucker *Catostomus commersonii*, northern redbelly dace *Phoxinus eos* (also known as *Chrosomus eos*), American eel *Anguilla rostrata*, creek chub *Semotilus atromaculatus*, and sticklebacks (threespine stickleback *Gasterosteus aculeatus* and ninespine stickleback *Pungitius pungitius*) (S. C. Mitchell, unpublished data; C. MacInnis, Fisheries and Oceans Canada, Antigonish, Nova Scotia, personal communication). The river also contains species with known or suspected sea-run life histories (Atlantic salmon, brown trout, and brook trout *Salvelinus fontinalis*), but these occur in low abundance and spawn in late fall; thus they are unlikely to affect isotope ratios of benthic invertebrates in our study. Jamieson et al. (1982) report the alewife run in the West River occurring generally between May 15 and June 15, with spawning primarily in lakes for this species and the blueback herring between June 15 and July 15. On June 30, 2003 the lower 5 km of the West River was surveyed during the time of the *Alosa* spawning run to determine distribution and sites of abundance for subsequent sampling. The first sign of *Alosa* (scales on the bank from predation) was found approximately 3 km upstream of the mouth, suggesting that most were remaining downstream near the mouth. An earlier study on osprey *Pandion haliaetus* (Jamieson et al. 1982) found that the birds were feeding heavily on alewife in Gaspereau Lake, indicating that this lake is a significant area for spawning. Therefore, we assume that the principal spawning areas are Gasperau Lake for alewife and the lower West River for blueback herring. The downstream sample site was selected as it was reasonably certain to include the spawning fish and the location was one of few riffles in the lower reach of the stream. An upstream site (N 45°34′W 62°06′) was selected to act as reference as we were reasonably certain that this site was not used for *Alosa* spawning. We base our decision of the lack of *Alosa* spawning at this site on the following: this tributary is (1) upstream of the confluence of Gaspereau Lake (presumably where much of the alewife spawning in the system occurs) and West River, (2) has no lakes and so would not be attractive to lake-spawning alewife, and (3) undergoes a significant increase in gradient within 1.5 km of its confluence, making it less attractive to these fish.

Invertebrates were collected using D-frame kick nets and identified to family (Merritt and Cummins 1996) at the two sites on several occasions (May 12, 2005, July 16–18 in 2003, 2004, 2005; September 3, 2003; September 22, 2004; November 13, 2004;). We also returned to the two sites in May and July of 2006 to collect water samples that were analyzed for nutrients (nitrate nitrogen, total nitrogen, and total phosphorus). Given that the first run of *Alosa* occurs between June 15 and July 15, our water and biotic samples collected in May (aka spring samples) represent samples that are not directly impacted by marine-derived nutrients. Whenever possible, invertebrates were analyzed individually, but in some instances (e.g., S analysis), pooled samples of 2–10 individuals were used. Additional C and N isotope data were obtained for muscle tissue of blueback herring from the southern Gulf of St. Lawrence (Pastershank 2001). *Alosa* that

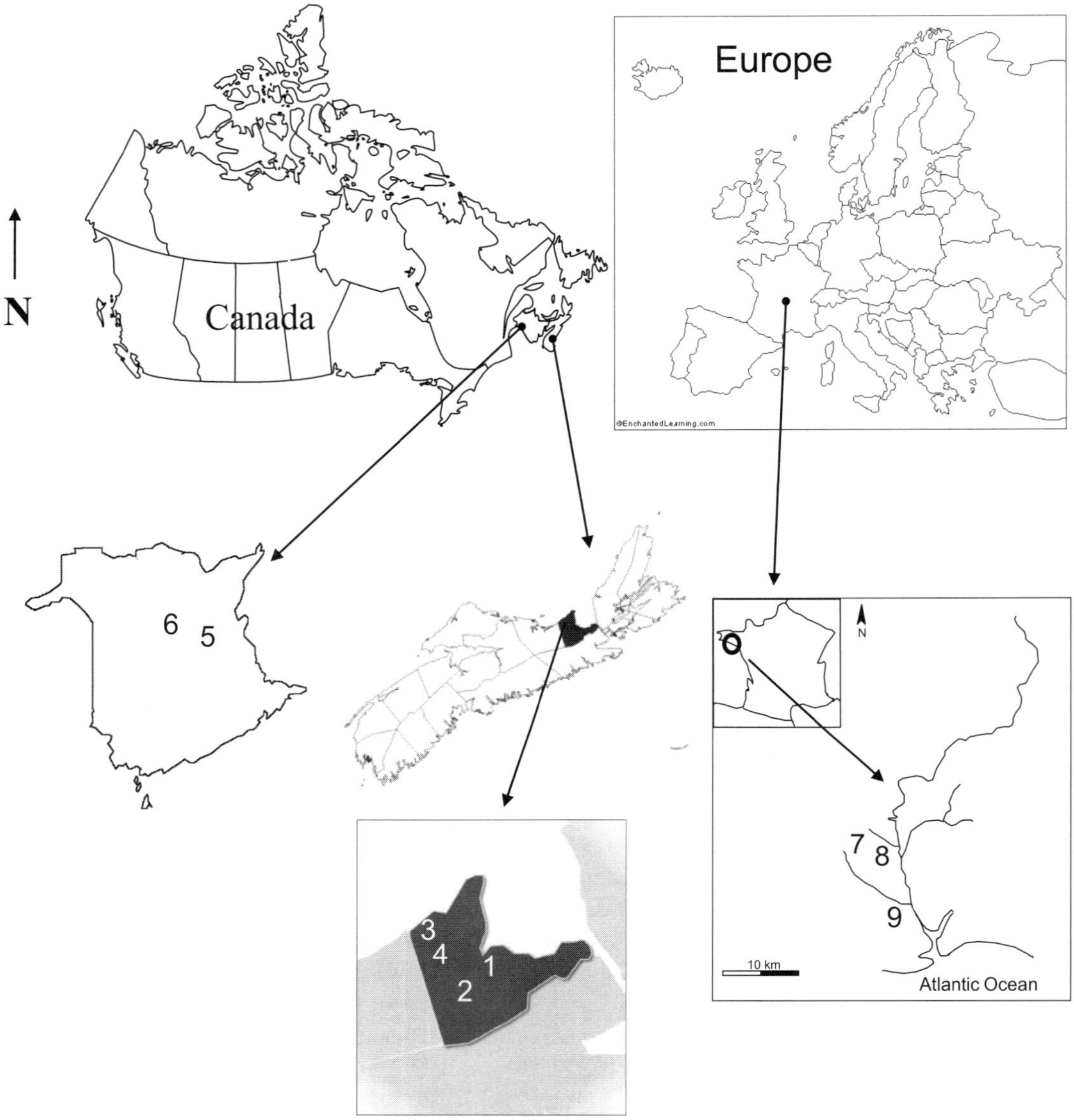

Figure 1.—Sampling locations in this study: West River (1—downstream and 2—upstream) and Doctor's Brook (3—downstream and 4—upstream) in Nova Scotia, Canada; Stewart Brook (5) and Catamaran Brook (6) in New Brunswick, Canada; and Kerlegan Brook (7—upstream and 8—downstream) and Scave Brook (9) in Brittany, France.

spawn in the West River would have similar marine isotope ratios as those fish and were considered as representative of the marine isotope signal for this system (Pastershank 2001).

Doctor's Brook, Nova Scotia, Canada.—Doctor's Brook (Figure 1) is a third-order coastal stream draining an area of 25.3 km^2 and flowing 10 km to meet the Northumberland Strait of the Gulf of St. Lawrence in northeastern Nova Scotia (N 45°47′W 62°07′). The lower 6 km is relatively low gradient (<0.05%) but, above this point, drains rugged, high-gradient (~2%) terrain. Forest cover of the catchment is largely intact with some roading, rural development, and small-scale forest harvesting. Riparian vegetation is a mix of softwoods and hardwoods. The hydrologic regime of this small brook is unknown. Fish species present include brook char, Atlantic salmon, and rainbow smelt (MacInnis, personal communication). Sea-run char and salmon are both rare in the system. Two sites were sampled in this brook, a downstream site approximately 600 m upstream of the brook mouth and an upstream (reference) site about 5 km upstream of the mouth and above known smelt spawning areas.

All samples in Doctor's Brook were collected in 2006. Our first sample occurred at the beginning of the spawning period for smelt (May 8), as evidenced by adults in the river (observed visually and captured in kick nets). Both the downstream reach with spawners and the upstream reach that is inaccessible to spawners were sampled that same day. We targeted three groups for our analysis: predatory stoneflies (Perlidae), herbivorous mayflies (Heptageniidae), and biofilm (rock scrapings). Invertebrates were sampled with D-frame kick nets, and biofilm was removed from the surfaces of stones with a toothbrush (n = 3 samples per site consisting of a minimum of three stones each). The downstream reach was sampled more frequently over the course of the growing season (May 8, 15, 22, 29, June 4, 18, July 7, August 7, September 10) than the upstream reach (May 8, 29, June 18, July 7, August 7, September 10). Smelt eggs were also picked from stones at the downstream reach on May 8 and 15 to obtain stable isotope ratios for this marine food source.

Catamaran Brook, New Brunswick, Canada.— Catamaran Brook (Figure 1) is the location of a long-term biological research program (Cunjak et al. 1993). It is a third-order, 20.5-km-long brook with a drainage area of 52 km^2 that flows to the Little Southwest Miramichi River in central New Brunswick (N 46°52′W 66°06′). Sampling was conducted in the middle reach (km 5–7), which has a low gradient and a mean annual discharge of 0.69 m^3/s (Cunjak et al. 1993). A fish trap located 500 m from the mouth of the brook has been operated since 1990 (Cunjak et al. 1993). Adult Atlantic salmon enter Catamaran Brook in late September; spawning occurs in late October and early November, with young-of-the-year emergence typically in April and May of the following year. The brook also supports several other coolwater fish species, the most common being brook trout, eastern blacknose dace *Rhinichthys atratulus*, lake chub *Couesius plumbeus*, and slimy sculpin (Mitchell et al. 2004). Some sea-run char enter the brook, but their abundance is low (R. A. Cunjak, unpublished data). Samples of slimy sculpin muscle tissue were collected by electrofishing in May 2004 within the same reach where approximately 20 salmon redds were identified the preceding autumn (L. Weir, Concordia University, Montreal, Canada, personal communication). The reference site for this study was a tributary of the Northwest Miramichi River, Stewart Brook (N 46°58′W 65°39′), which has an impassable barrier to fish migration. Sculpin samples were collected from Stewart Brook on the same day to compare isotope ratios with those from Catamaran Brook. We also used data from a previous study (Gray et al. 2004) as reference values for sculpin collected above an impassable waterfall in the St. John River, New Brunswick.

Scorff River, Brittany, France.—The Scorff River (Figure 1) is a 62-km-long stream located in Brittany, France (N 47°50′W 3°23′) that flows into the Atlantic Ocean. It rises at low altitude (spring source at 230 m) and drains an area of 480 km^2. Urban areas, woods, and farming (both arable lands and pasture) represent 10%, 31%, and 59% of the drainage area, respectively. The mean annual discharge is 5.01 m^3/s, and dissolved nitrogen levels are generally high (from 20 to 30 mg/L of nitrate, Giovanni 1996). A fish trap located 1 km upstream from the estuary has been operated since 1994 to monitor Atlantic salmon population abundance and movements in the Scorff River. Adult salmon spawn in December; young-of-the-year start to emerge from redds in March of the following year, and smolts (mostly age 1) migrate to sea in March and April of the year following emergence. The Scorff River and tributaries also support other freshwater fish species, including brown trout, European eel *Anguilla anguilla*, European brook lamprey *Lampetra planeri* and European river lamprey *L. fluviatilis*, sea lamprey *Petromyzon marinus*, stone loach *Barbatula barbatula*, European minnow *Phoxinus phoxinus*, and sculpin. Seatrout are rare in the system and sea lamprey spawn in the lower reaches far from our study sites.

The study was carried out in two streams of the lower third of the Scorff River drainage: Kerlégan Brook, a 6.7-km-long tributary and Scave Brook, a 13-km-long stream that flows into the Scorff River estuary. Kerlégan Brook is disrupted by an impassable dam, and only the lower kilometer is accessible to Atlantic salmon. Sculpin were electrofished in November 2003 and January 2004 (i.e., before and after salmon spawning, at three different sites in Scave Brook (about 1 km upstream from the estuary), Kerlégan Brook downstream of the dam, and Kerlégan Brook upstream of the dam (reference site). Benthic invertebrates representing various foraging strategies were sampled using kick nets at each site in February 2005, together with additional sculpin samples collected in Kerlégan Brook (downstream site). Atlantic salmon ova (~10–15) were taken by manual stripping of three mature females caught in the Scorff River to establish reference marine isotope ratios. In the laboratory, sculpin were killed in a solution of phenoxyethanol (0.1%) as per regulations for ethical treatment of wild animals. White muscle and gonadal tissues from fish, and pooled invertebrate samples (~2–10 individuals), were hermetically stored in vials and frozen until analysis. The abundance (or absence) of Atlantic salmon redds at each site was visually assessed in December 2003 and 2004 (N. Jeannot, INRA Rennes, personal communication) and confirmed by an electrofishing census of young-of-the-year salmon performed in 2004 and 2005.

Stable Isotope Analysis

All fish and invertebrate tissue samples were oven-dried (50–60°C for 24–48 h) and ground to a homogenous powder. For C & N isotope analysis at the Stable Isotopes in Nature Laboratory, Fredericton, New Brunswick, samples were weighed into tin cups and combusted in a Carlo Erba NC2500 elemental analyzer at high temperature. Resultant gases were delivered via continuous flow to a Finnigan Mat Delta XP isotope ratio mass spectrometer (Thermo Finnigan, Bremen, Germany). Isotope ratios are reported in delta notation relative to international standards Peedee Belemnite Carbonate (C) and atmospheric nitrogen (N). Data were corrected using working standards (bass muscle, bovine liver, nicotinamide) that were previously calibrated against International Atomic Energy Agency (IAEA) standards CH6, CH7, N1, and N2. A commercially available standard (acetanilide, Elemental Microanalysis, Ltd.) yielded mean $\delta^{13}C$ and $\delta^{15}N$ of $-33.60 \pm 0.13‰$ SD and $-3.13 \pm 0.19‰$ SD, respectively ($n = 62$). Standard deviations of samples analyzed in duplicate averaged 0.11‰ and 0.10‰ for carbon and nitrogen, respectively.

A subset of *Drunella* sp. samples from West River was analyzed for sulfur isotope ratios at the Colorado Plateau Analytical Laboratory, Flagstaff, Arizona using a Thermo Finnigan Delta Advantage Mass Spectrometer. Isotope ratios are reported in delta notation relative to the international standard Canyon Diablo Troilite. Data were corrected using IAEA standards S1, S2, and S3. A commercially available standard (bovine liver, National Institute of Standards) yielded $\delta^{34}S = 7.61 \pm 0.25‰$ SD ($n = 6$).

Statistical Analysis

For West River invertebrates, differences in stable isotope ratios between sites at different sampling times were tested for those taxa collected at both sites, using a single factor analysis of variance (ANOVA). These included dragonfly larvae (Gomphidae and Corduliidae), stonefly nymphs (Perlidae and Perlodidae), and mayflies *Drunella* sp. For Doctor's Brook, differences in stable isotope ratios over time were tested among groups (Perlidae, Heptageniidae, biofilm) separately for both isotopes and at the upstream and downstream sites, using ANOVA.

For all sculpin, we grouped isotope values for male and female muscle tissues together based on a high degree of similarity within sites that was consistent with previous work (Jardine et al. 2005). For sculpin samples in Catamaran Brook, we tested for differences in isotope ratios between our spawning site and the reference site. For sculpin in the Scorff River, we tested for differences in stable isotope ratios over time in relation to the availability of salmon eggs. For sculpin gonad samples, we compared differences between muscle and gonad $\delta^{13}C$ and $\delta^{15}N$ with previously measured values for sculpin in the absence of potential marine inputs (Jardine et al. 2005). Due to heteroscedasticity of variance, all statistical testing was done using a Kruskal-Wallis one-way ANOVA on ranks, with tests considered significant at $\alpha < 0.05$.

Given that we sampled sites repeatedly over time, we recognize the lack of independence of data within each system. However, we were primarily interested in patterns within systems that served as our level of inference (e.g., within the West River, do sites differ in isotope ratios and are differences dependent on season?). We viewed this as the first step in assessing the importance of marine nutrients and organic matter to Atlantic coast streams. Inferring general patterns across the landscape will be dependent on observing similar trends in several different systems.

Results

Benthic Invertebrates in the West River, Nova Scotia

Consistent with the delivery of marine derived organic matter, invertebrates sampled in the summer following the spawning of blueback herring at the downstream site in the West River were enriched in ^{13}C and ^{15}N relative to those from the upstream site at the same time (Figure 2). However, given that land-use can significantly affect baseline isotope ratios, to account for these differences between sites and to ensure that these observations of enriched ^{13}C and ^{15}N were not artiacts of other factors, we used two strategies: temporal sampling and stable sulfur isotope analysis.

For the mayflies *Drunella* sp., individuals at the downstream site were enriched in ^{15}N at all times (spring 2005 and summer 2004 and 2005, Figure 3A) and enriched in ^{13}C in both summer samples (2004 and 2005, $p < 0.01$) but not the spring ($p = 0.873$, Figure 3A). For dragonfly larvae (Odonata: Corduliidae) sampled from the two sites at different times, individuals from the downstream (spawning) site were enriched in ^{15}N at all times compared with the upstream site (spring, summer, and fall, $p < 0.01$, Figure 3B). For $\delta^{13}C$, individuals at the downstream site were enriched in ^{13}C compared to the upstream site in the summer and fall ($p < 0.01$) but not in the spring ($p = 0.109$, Figure 3B). Similarly, for stonefly larvae (Plecoptera), individuals at the downstream site were enriched in ^{15}N at all times (Figure 3C) and enriched in ^{13}C in the summer and fall ($p < 0.01$) but not the spring ($p = 0.071$, Figure 3C).

For the subset of *Drunella* sp. samples analyzed for $\delta^{34}S$ (Table 1), individuals from the upstream site in the spring had the lowest $\delta^{34}S$ while values at the upstream site in the summer were similar to those at the downstream site in the spring. Samples from the downstream site in consecutive summers had the highest $\delta^{34}S$ values.

In streamwater samples, nitrogen and phosphorus concentrations were slightly higher at the downstream site compared to the upstream site (upstream May: nitrate < 0.05 mg/L, total nitrogen (TN) < 0.3 mg/L, total phosphorus (TP) < 0.005 mg/L; July: nitrate = 0.07 mg/L, TN = 0.18 mg/L, TP = 0.007 mg/L; downstream May: nitrate <0.05 mg/L, TN < 0.3 mg/L, TP = 0.006 mg/L; July: nitrate = 0.11 mg/L, TN = 0.32 mg/L, TP = 0.012 mg/L). Both sites, however, had concen-

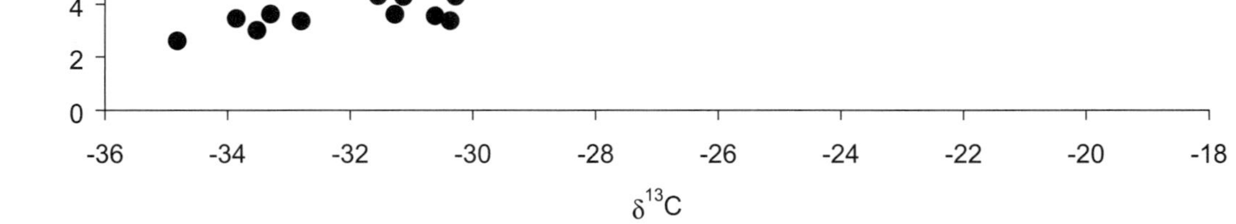

Figure 2.—Stable carbon ($\delta^{13}C$) and nitrogen ($\delta^{15}N$) isotope ratios of all invertebrates sampled during the summer in the West River, Nova Scotia, Canada. Samples from the upstream site (no blueback herring spawning) are indicated by solid symbols; from the downstream site (blueback herring spawning), by open symbols. Taxa include *Drunella* sp. (circles), Perlidae (triangles), Corduliidae (diamonds), Megaloptera (squares), Psephenidae (X's, only present at the downstream site), and Pteronarcyidae (+'s, only present at the downstream site). Isotope ratios for *Alosa* sp. in the upper right are shown for reference (data from Pastershank 2001).

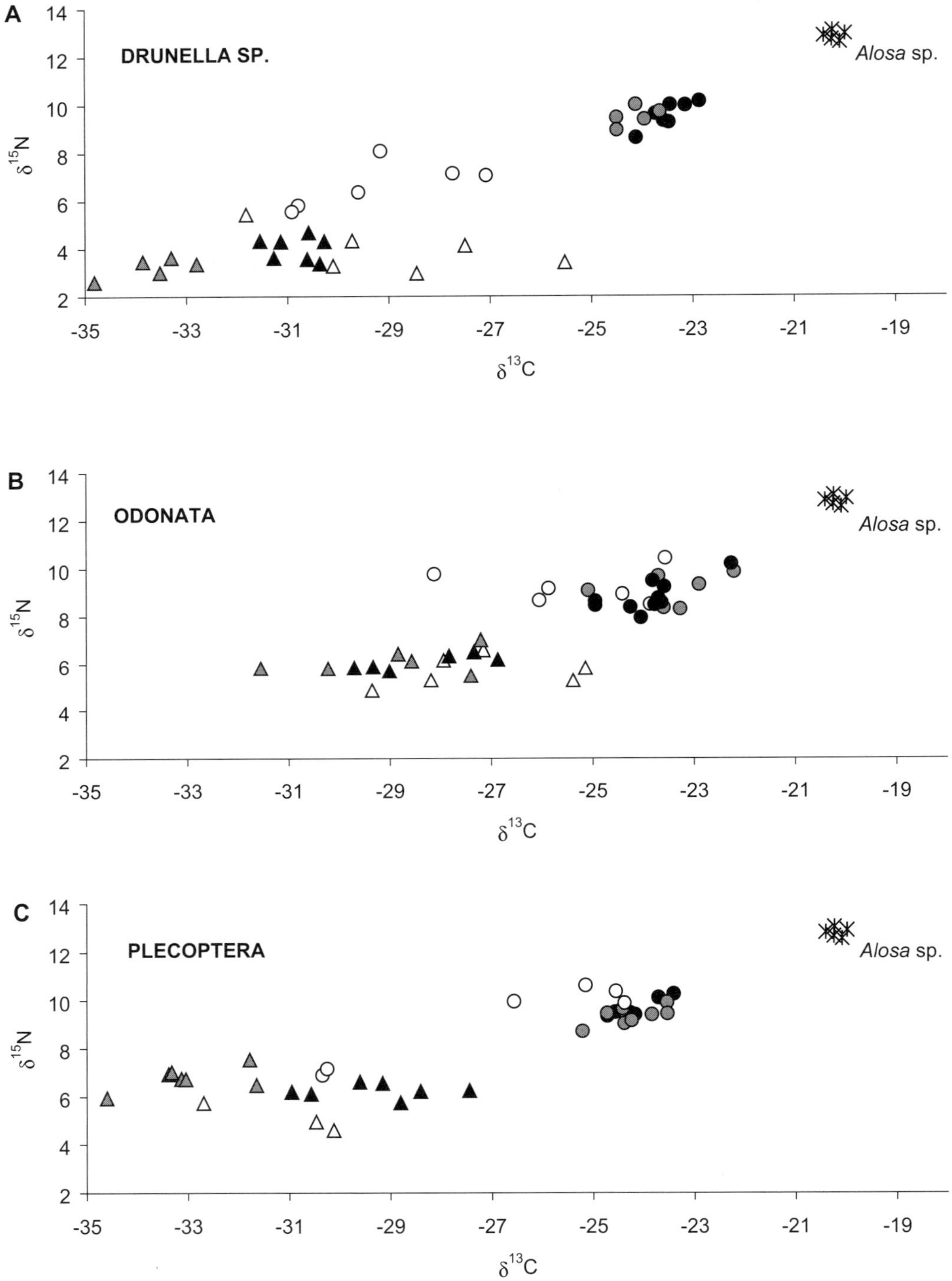

Figure 3.—Scatterplot of *Drunella* sp. (A), Odonata (B), and Plecoptera (C) stable carbon and nitrogen isotope ratios from the upstream (reference—triangles) and downstream (spawning—circles) sites in the West River, Nova Scotia, Canada in spring prior to blueback herring spawning (open symbols) and following blueback herring spawning (shaded and solid symbols). Isotope ratios for *Alosa* sp. in the upper right are shown for reference (data from Pastershank 2001).

Table 1.—Stable sulfur isotope ratios ($\delta^{34}S \pm SD$, n = 2–3 pooled samples per site and time) of *Drunella* sp. at the upstream and downstream sites in the West River, Nova Scotia, Canada prior to blueback herring spawning (spring) and for each of 2 years following spawning (summer). Note: summer 2005 data from the upstream site were unavailable because of limited remaining sample material.

Season (year)	Site	$\delta^{34}S$
Summer (2004)	Upstream	12.5 ± 0.1
	Downstream	14.1 ± 0.3
Spring (2005)	Upstream	9.8 ± 0.2
	Downstream	12.2 ± 0.1
Summer (2005)	Downstream	14.1 ± 0.7

trations below levels expected to be found at sites impacted by agriculture (e.g., nitrate > 0.15 mg/L, Harrington et al. 1998) and are in the range of values considered to have ideal or acceptable water quality (TN = 0.47 mg/L, TP = 0.06 mg/L, Dodds and Welch 2000).

Benthic Invertebrates in Doctor's Brook, Nova Scotia

Smelt eggs were enriched in ^{13}C and ^{15}N relative to available food sources for consumers in Doctor's Brook (Figure 4). Consumers at the downstream (spawning) site had elevated $\delta^{15}N$ but similar or lower $\delta^{13}C$ compared to the upstream (reference) site. Both time and group significantly affected both isotopes at both sites, with some interaction between terms ($p < 0.05$, Figures 5 and 6). At the downstream site, all three groups (Perlidae, Heptageniidae, biofilm) showed a distinct increase in $\delta^{13}C$ coincident with smelt spawning (Figure 5A). Perlidae showed the most pronounced increase, rising ~4‰ between the first sample and its peak on Julian day 142. Peaks in $\delta^{13}C$ of biofilm (Julian day 155) and Heptageniidae (Julian day 169) occurred later than those observed in Perlidae and were not accompanied by peaks in $\delta^{15}N$. The upstream site also had variable $\delta^{13}C$ in the three groups, but changes over time were not consistent in any direction (Figure 5B). Also, the range of mean $\delta^{13}C$ values observed at the downstream site was far greater compared with the upstream site for all three groups (Perlidae downstream range = 5.5‰, upstream = 2.3‰; biofilm downstream range = 3.1‰, upstream = 1.8‰; Heptageniidae downstream range = 5.3‰, upstream = 4.3‰). Perlidae at the downstream site had $\delta^{15}N$ values that changed over time suggestive of a contribution of marine organic matter to their diet from smelt (Figure 6A). The peak in Perlidae $\delta^{15}N$ oc-

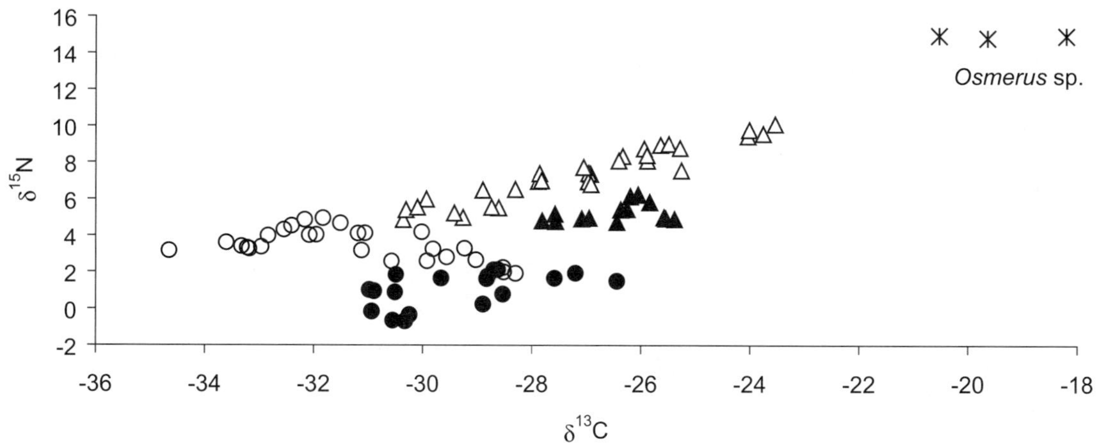

Figure 4.—Stable carbon ($\delta^{13}C$) and nitrogen ($\delta^{15}N$) isotope ratios of all invertebrates sampled in Doctor's Brook, Nova Scotia, Canada. Samples from the upstream site (no smelt spawning) are indicated by solid symbols for Perlidae (triangles) and Heptageniidae (circles). Samples from the downstream site (smelt spawning) are indicated by open symbols for Perlidae (triangles) and Heptageniidae (circles). Isotope ratios for smelt eggs in the upper right are shown for reference.

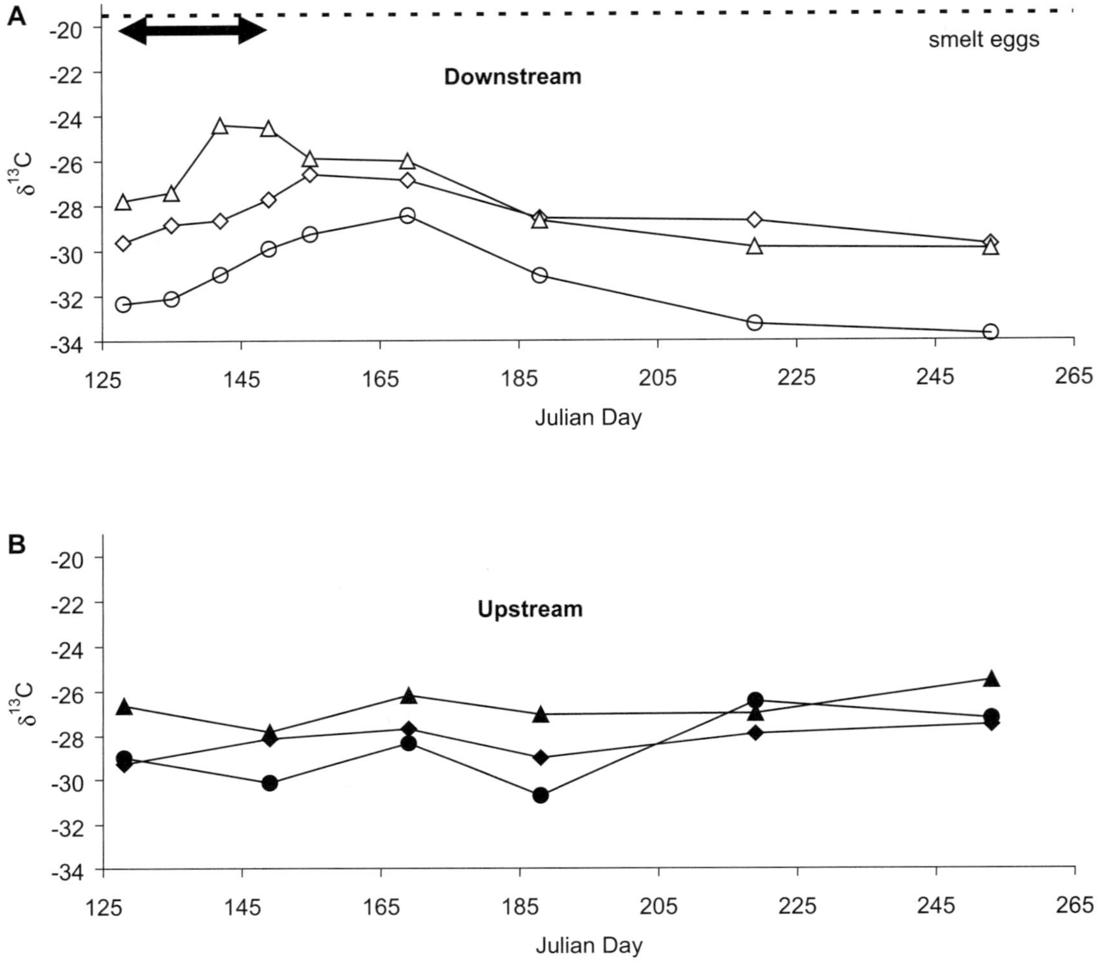

Figure 5.—Stable carbon isotope ratios in Doctor's Brook, Nova Scotia, Canada over time in a downstream reach with smelt spawning (A) and an upstream reach that is inaccessible to smelt (B). The hatched line indicates $\delta^{13}C$ of smelt eggs, and the arrow indicates the duration when eggs were deposited on the stream bottom and available as an organic matter and nutrient source to Perlidae (triangles), Heptageniidae (circles) and biofilm (diamonds).

curred on the same day (Julian Day = 142) as the peak in Perlidae $\delta^{13}C$, and overall Perlidae $\delta^{13}C$ was strongly correlated with Perlidae $\delta^{15}N$ at the downstream site (Figure 4). Biofilm and Heptageniidae $\delta^{15}N$ values did not show a consistent response to smelt spawning, but were enriched in ^{15}N relative to the upstream site (Figure 6B). The range of values in $\delta^{15}N$ was greater in Perlidae at the downstream site (4.2‰ versus 1.1‰), but ranges in biofilm (3.2‰, 3.9‰) and Heptageniidae (2.5‰, 1.9‰) $\delta^{15}N$ were similar between the two sites.

Sculpin in Catamaran Brook and the Scorff River

Salmon eggs were highly enriched in ^{13}C (–21.1 ± 0.5‰ SD) compared with values for invertebrates from all Scorff River sites measured in the current study (–29.5 ± 2.2‰ SD, Table 2) and those from Catamaran Brook's Middle Reach (reported by Doucett 1999; Table 2). For $\delta^{15}N$, salmon eggs (12.7 ± 0.3‰ SD) were indistinguishable from invertebrates from the Scave River and the upstream site in Kerlégan Brook (Table 2), enriched in ^{15}N

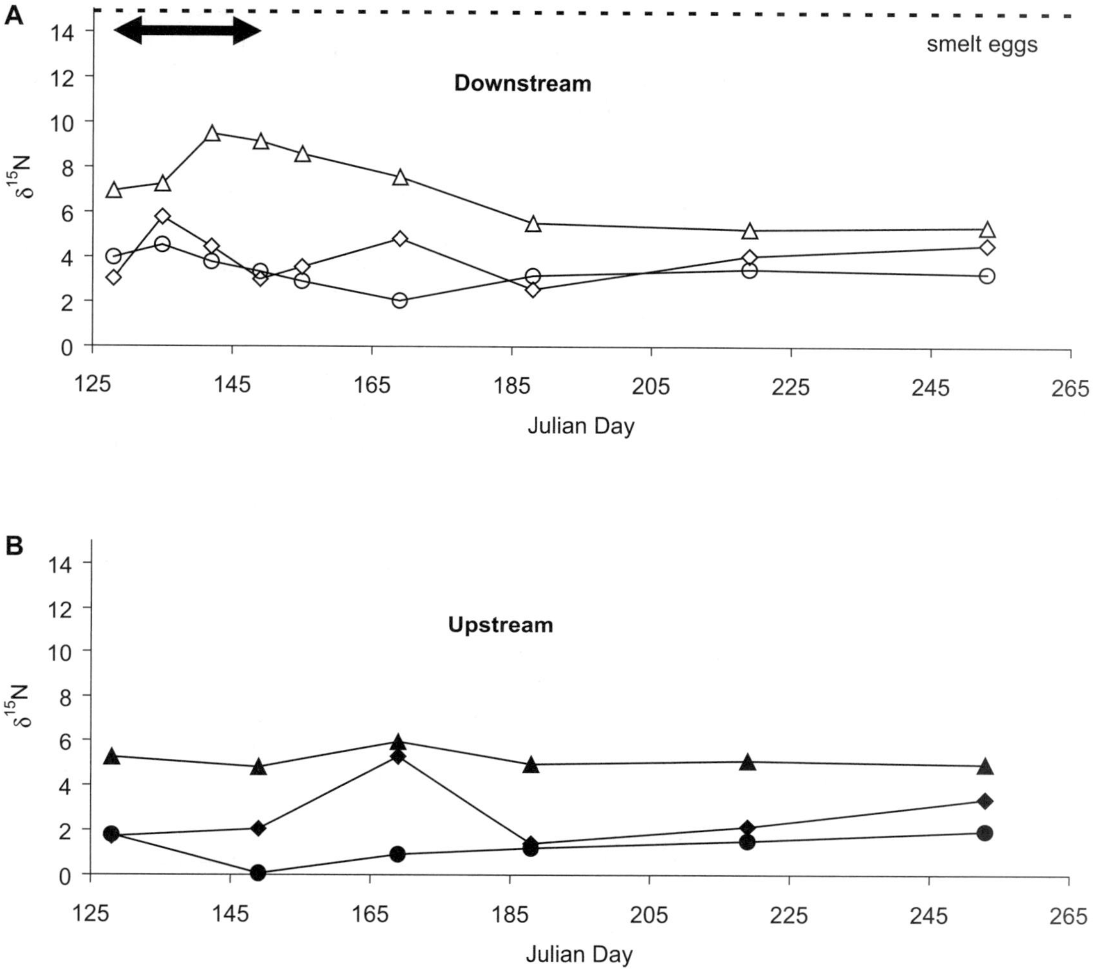

Figure 6.—Stable nitrogen isotope ratios in Doctor's Brook, Nova Scotia, Canada over time in a downstream reach with smelt spawning (A) and an upstream reach that is inaccessible to smelt (B). The hatched line indicates $\delta^{15}N$ of smelt eggs, and the arrow indicates the duration when eggs were deposited on the stream bottom and available as an organic matter and nutrient source to Perlidae (triangles), Heptageniidae (circles) and biofilm (diamonds).

relative to invertebrates from the downstream site in Kerlégan Brook (Table 2), and enriched in ^{15}N compared with invertebrates from Catamaran's Middle Reach (Doucett 1999; Table 2).

Contrary to our expectations, there was very little within-site variation in sculpin muscle $\delta^{13}C$ (Catamaran Brook SD = 0.6‰, Stewart Brook SD = 0.6‰, Kerlégan Brook downstream SD = 0.7‰, Kerlégan Brook upstream SD = 0.6‰, Scave River SD = 0.6‰, Table 2) and $\delta^{15}N$ (Catamaran Brook SD = 0.3‰, Stewart Brook SD = 0.4‰, Kerlégan Brook downstream SD = 0.4‰, Kerlégan Brook upstream SD = 0.5‰, Scave River SD = 0.8‰, Table 2) at all sites on both sides of the Atlantic Ocean. No individuals had isotope ratios suggestive of a marine contribution to the diet, and at most sites, sculpin were strongly linked to local aquatic invertebrates with diet-tissue differences consistent with previous studies (Catamaran sculpin $\delta^{13}C$ – invertebrate $\delta^{13}C$ = –0.1‰, sculpin $\delta^{15}N$ – invertebrate $\delta^{15}N$ = 5.9‰; Kerlégan downstream sculpin $\delta^{13}C$ – invertebrate $\delta^{13}C$ = 0.6‰, sculpin $\delta^{15}N$ – invertebrate $\delta^{15}N$ = 2.0‰; Scave sculpin $\delta^{13}C$ – invertebrate $\delta^{13}C$ = 1.5‰, sculpin

Table 2.—Stable carbon ($\delta^{13}C$ ± SD) and nitrogen ($\delta^{15}N$ ± S.D.) isotope ratios of invertebrates (inverts) and *Cottus* sp. muscle and gonad tissues relative to availability of Atlantic salmon eggs ($\delta^{13}C$ = –21.1 ± 0.5‰, $\delta^{15}N$ = 12.7 ± 0.3‰) in Canadian and French rivers. Note: invert data for Catamaran Brook are from Doucett (1999); no invert data were available for Stewart Brook.

Site	Egg availability	Date	Sex	Tissue (*n*)	$\delta^{13}C$	$\delta^{15}N$
Catamaran Brook	High	May 2004	M + F	muscle (11)	–28.2 ± 0.6	8.9 ± 0.3
		June 1998		inverts (14)	–28.1 ± 2.1	4.0 ± 1.4
Stewart Brook	Absent	May 2004	M + F	muscle (15)	–27.4 ± 0.6	11.1 ± 0.4
Kerlegan Brook (downstream)	Absent	November 2003	M + F	muscle (5)	–26.6 ± 0.6	10.8 ± 0.4
			F	gonad (4)	–27.7 ± 0.6	10.7 ± 0.5
	Low	January 2004	M + F	muscle (6)	–26.5 ± 0.4	10.5 ± 0.4
			F	gonad (3)	–27.9 ± 0.4	11.8 ± 0.6
	Low	May 2004	M + F	muscle (6)	–26.9 ± 1.1	10.5 ± 0.5
			F	gonad (2)	–29.6 ± 4.0	9.4 ± 0.1
	High	February 2005	M + F	muscle (14)	–27.0 ± 0.5	10.3 ± 0.2
			F	gonad (10)	–28.1 ± 0.4	11.3 ± 0.3
		February 2005		inverts (5)	–27.4 ± 1.0	8.5 ± 2.0
Kerlegan Brook (upstream)	Absent	November 2003	M + F	muscle (6)	–27.7 ± 0.6	10.9 ± 0.4
			F	gonad (4)	–28.1 ± 0.6	9.7 ± 0.5
	Absent	January 2004	M + F	muscle (6)	–27.4 ± 0.3	10.7 ± 0.3
			F	gonad (3)	–28.4 ± 0.3	10.9 ± 0.9
	Absent	May 2004	M + F	muscle (6)	–27.5 ± 0.8	10.3 ± 0.5
			F	gonad (5)	–27.7 ± 0.6	9.6 ± 0.4
		February 2005		inverts (4)	–31.4 ± 2.0	12.9 ± 2.7
Scave River	Absent	November 2003	M + F	muscle (7)	–28.5 ± 0.5	17.5 ± 0.6
			F	gonad (2)	–29.0 ± 0.3	18.8 ± 0.2
	High	January 2004	M + F	muscle (6)	–28.6 ± 0.5	17.4 ± 0.3
			F	gonad (3)	–28.5 ± 0.0	17.7 ± 0.2
	High	May 2004	M + F	muscle (6)	–29.2 ± 0.5	17.4 ± 1.3
			F	gonad (3)	–29.0 ± 0.6	17.1 ± 0.4
		February 2005		inverts (5)	–30.2 ± 1.2	11.8 ± 4.1

$\delta^{15}N$ – invertebrate $\delta^{15}N$ = 5.7‰). The exception was the upstream site in Kerlégan Brook where sculpin were depleted in ^{15}N (10.6 ± 0.4‰ SD) relative to both local invertebrates ($\delta^{15}N$ = 12.9 ± 2.7‰ SD) and salmon eggs, suggesting an alternative ^{15}N-depleted food source or that our $\delta^{15}N$ value for invertebrates was artificially high at the time of sampling due to unpredictable N inputs to the system. Over time, sculpin showed no evidence of an isotopic shift in relation to the availability of salmon eggs ($p > 0.05$, Table 2), and often sites that were inaccessible to salmon had sculpin with higher $\delta^{13}C$ and $\delta^{15}N$ than those from spawning sites. Female sculpin gonads similarly showed no indication of a shift towards a diet of salmon eggs at sites where salmon spawners were present (Table 2). Sculpin gonads were depleted in ^{13}C compared with muscle, similar to previous estimates for sculpin with no opportunity to forage on salmon eggs (Jardine et al. 2005).

Discussion

We found considerable evidence for the use of marine-derived organic matter by benthic invertebrates in the West River and Doctor's Brook but no evidence of consumption of Atlantic salmon eggs by sculpin in either Catamaran Brook or the Scorff River. These data indicate that the importance of marine carbon and nitrogen carried by anadromous fishes may be more apparent in streams with concentrated spawning activity, such as by *Alosa* sp. and *Osmerus* sp., and less obvious in areas with more scattered spawning activity by reduced populations of Atlantic salmon.

In the West River, we observed considerable differences in isotope ratios between the upstream and downstream sites and between pre- and post-spawning by blueback herring. While $\delta^{15}N$ values were higher in all taxa at the downstream site at all times, $\delta^{13}C$ values were similar between sites in the spring prior to the arrival of blueback herring. This may be related to nutrient limitation, in that benthic algae at the downstream site could be N-limited and hence incorporate marine nitrogen leading to high $\delta^{15}N$ throughout the growing season (Chaloner et al. 2002), while marine carbon is consumed directly by invertebrate scavenging of eggs and carcasses that are only available for a short duration. While we did not measure isotope ratios in benthic algae directly, known herbivores at the downstream site (water pennies Psephenidae, mayflies Heptageniidae) collected after the spawning run all had elevated $\delta^{15}N$. Herbivore isotope ratios have been shown to approximate those of their algal diet (Doucett 1999; Finlay 2001). On the west coast of North America, streams are often nitrogen limited, and hence, the $\delta^{15}N$ signal from spawning Pacific salmon can be more apparent in benthic food webs than $\delta^{13}C$ (Kline et al. 1990). Stream benthic communities are known to efficiently retain nitrogen, especially when it is in particulate form (Dodds et al. 2004), as would be expected under marine subsidy scenarios. Marine carbon, meanwhile, may appear briefly in a variety of taxa through direct egg/carcass consumption, but its relative importance diminishes over the growing season as that food source is consumed (Kline et al. 1990).

Similar to the West River, in Doctor's Brook, the downstream site had consistently higher $\delta^{15}N$ than the upstream site. Kline et al. (1993) also found that salmon-influenced sites had benthic algae with higher $\delta^{15}N$ than control sites regardless of the timing of sampling. In Doctor's Brook, all three of our study groups showed a $\delta^{13}C$ response coincident with smelt spawning, but only the predatory stoneflies (Perlidae) indicated a response in their $\delta^{15}N$. This suggests that direct consumption of eggs is likely the major pathway for marine inputs to enter this system, and the increases in $\delta^{13}C$ observed in biofilm and its direct consumer (Heptageniidae) may have simply reflected seasonal variation (Finlay 2004) in the absence of a marine influence or contamination of biofilm samples by broken and decomposing eggs. Alternatively, the system may be phosphorus-limited, and the addition of P by decomposing smelt eggs stimulated primary production in the downstream reach, leading to reduced dissolved CO_2 availability, less discrimination against ^{13}C by algae, and subsequent elevations in primary producer $\delta^{13}C$ that was tracked by grazing Heptageniidae (Finlay 2004). A third and less likely scenario is that benthic algae, fungi, and bacteria present in biofilm are carbon-limited and the increase in $\delta^{13}C$ in the downstream reach following smelt spawning was a result of the incorporation of marine C that is enriched in ^{13}C. The resolution of these questions will require detailed analyses of stream water nutrient concentrations over time. Temporal analyses of water nutrients (and their isotope ratios) would also aid in unraveling the relative influence of direct (organic matter) and indirect (dissolved nutrients) pathways of entry into stream food webs. Kline et al. (1990) proposed that the latter pathway would be made apparent by increased $\delta^{15}N$ in all trophic levels at spawning sites while the former would be indicated by increased $\delta^{13}C$ in consumers at those sites. Our results from the West River and Doctor's Brook suggest that both of these pathways may be important.

Slimy sculpin in Catamaran Brook where salmon spawners were present had $\delta^{13}C$ and $\delta^{15}N$ values that differed little from sculpin collected above a barrier impassable to anadromous fishes in both the current study and an earlier study (Gray et al. 2004). Similarly, in the Scorff River, sculpin showed no evidence of salmon egg consumption since stable isotope values did not differ between sites or before and after salmon spawning. Contrasting results were obtained on the Pacific coast by Kline et al. (1993) who found a significant contribution of sockeye salmon *Oncorhynchus nerka* eggs and emergent fry to the diet of the coastrange sculpin *Cottus aleuticus*. Other studies, however, have found egg consumption by sculpin to be extremely rare, with eggs occurring in only 45 of 7,785 stomachs (0.6%, summarized in Moyle 1977). Given the limited movement of sculpin (Cunjak et al. 2005), it was necessary to ensure that salmon spawner density was high enough to increase the probability of observing evidence of egg consumption. We therefore specifically targeted sculpin from river habitats immediately downstream of redds excavated by salmon the previous autumn. Despite this concerted effort, we found no individuals that had isotope ratios suggestive of a marine food source. Isotopic change in response to a shift in diet is strongly related to growth in ectotherms such as fishes (Hesslein et al. 1993) and invertebrates (Fry and Arnold 1982). Since sculpin likely have limited somatic growth through the winter, the lack of evidence for marine food consumption in muscle tissue was not surprising. However, there was also no evidence in the Scorff River for a marine signature when analyzing isotope ratios of female sculpin gonads, which grow during winter and would be expected to have fast turnover rates similar to liver tissue (Perga and Gerdeaux 2005) and rapidly growing invertebrates. If there was preferential shunting of available nutrients to developing tissues, we predicted that female gonads might show evidence of marine food sources. Instead, female gonads had more negative $\delta^{13}C$ and more positive $\delta^{15}N$ than muscle, consistent with earlier work on sculpin with no access to marine food sources (Jardine et al. 2005). Differences in $\delta^{13}C$ between eggs and muscle are due to high lipid loads in the former tissue, while $\delta^{15}N$ differences are likely due to differences in amino acid content (Jardine et al. 2005). The lack of a marine isotopic signal in the tissues of sculpin suggests that while salmon eggs may have been consumed occasionally, they formed a small enough fraction of the diet as to avoid detection using isotope analysis. Invertebrates were not sampled concurrently with sculpin, so it is unknown whether eggs were being consumed by invertebrates at our Scorff River sites at times of high salmon egg deposition.

One of the greatest challenges in using SIA across systems is the confounding influence of varying baseline isotope ratios. Both $\delta^{13}C$ and $\delta^{15}N$ can vary considerably across the landscape due to natural and anthropogenic processes. Stable carbon ratios show a general trend of enrichment in ^{13}C from headwaters to lower reaches of rivers (Finlay 2001), owing to changes in dissolved inorganic carbon isotope ratios and CO_2 supply and demand (Finlay 2004). Human activities such as manure-based agriculture and forestry contribute to elevated baseline $\delta^{15}N$ in rivers and streams (Udy and Bunn 2001; Anderson and Cabana 2005). Taken together, we expected a progressive increase in both $\delta^{13}C$ and $\delta^{15}N$ from small undeveloped streams to more developed stretches of river downstream even in the absence of anadromous fish influence. While this is the case in the West River, Nova Scotia (our upstream reference site was a partially shaded stream with limited forestry and agricultural activity, whereas our downstream site was partially developed with considerable agricultural activity near an urban centre), N and P stream-water concentrations at the downstream site were low and similar to undeveloped systems (Harrington et al. 1998; Dodds and Welch 2000), suggesting that a high $\delta^{15}N$ baseline from agriculture is not a concern in this system. However, most coastal streams in Brittany, France drain intensive agricultural catchments, and dissolved N levels and baseline $\delta^{15}N$ are unusually high compared to most other areas. The lower reaches of the Scorff River have nitrate concentrations that exceed 20 mg/L (Giovanni 1996). As a consequence, the biota in the Scorff River is ^{15}N-enriched compared with the ocean (Charles et al. 2004), and $\delta^{15}N$ is therefore less effective as a tracer of marine inputs. In all of our systems, concerns about baseline $\delta^{15}N$ differences between sites prompted our use of temporal sampling and stable sulfur isotope analysis.

Temporal sampling was conducted by analyzing components of the food web before and after the arrival of blueback herring in the West River in June, sculpin in the Scorff River before and after the arrival of Atlantic salmon in December, and food web components in Doctor's Brook after the arrival of smelt in May. We assumed that if the differences in isotope ratios of short-lived organisms such as mayflies or actively growing tissues like female gonads were due to natural patterns and anthropogenic influences, they would remain apparent throughout the year, while differences caused by spawning anadromous fishes would be pronounced following their arrival but less so prior to their arrival. For sculpin, we found no evidence of temporal variations in C and N isotopes in relation to salmon egg availability. Conversely, the short-lived mayfly *Drunella* sp. showed a strong increase in stable isotope ratios at the downstream site in the West River that were not related to ontogenetic diet shifts, as individuals collected in the summer at both sites were smaller than those from the spring (T. Jardine, unpublished data). This suggests that a pulse of *Drunella* sp. emergence from the stream occurred prior to blueback herring spawning and the subsequent generation in the summer at the downstream site had access to marine food sources, resulting in elevated $\delta^{13}C$ and $\delta^{15}N$. *Drunella* sp. are also rapid colonizers (Miyake et al. 2003) and hence could drift to concentrated spawning areas to take advantage of available organic matter. In Doctor's Brook, our first sample occurred during the smelt spawning period, so we were unable to assess isotope ratios pre- and postspawning. Predatory stoneflies (Perlidae) in Doctor's Brook showed strong isotopic responses to smelt arrival at the downstream site, indicating direct consumption of eggs. This is not surprising given that eggs, a potential high-quality food resource, are in high abundance during this time period and are small enough in diameter (~1 mm, Scott and Crossman 1998) for consumption by most invertebrates. Blueback herring eggs also have a small diameter (~1 mm, Scott and Crossman 1998), making them an attractive food source for invertebrates, including stoneflies, in the West River; we have also observed elevated $\delta^{13}C$ and $\delta^{15}N$ in predatory stoneflies collected from Atlantic salmon egg incubation baskets in New Brunswick rivers (Cunjak, unpublished data), suggesting that even for patchy spawners such as salmon with larger eggs, a localized influence of marine organic matter inputs may be observed in smaller-bodied predators whose diet may consist entirely of eggs for an extended period of time (e.g., during winter).

Our second approach to assess the influence of marine inputs was to use stable sulfur isotopes. Sulfur isotope ratios differ markedly and consistently between freshwater and marine environments (Mekhtiyeva et al. 1976; Doucett et al. 1999) and seem to be less affected by land-use activities than $\delta^{13}C$ and $\delta^{15}N$. Benthic macroinvertebrates in rivers have known $\delta^{34}S$ values that range from –5‰ to 8‰ (Doucett et al. 1999; Wayland and Hobson 2001), while anadromous alosids have been shown to have $\delta^{34}S$ values that range from 13‰ to 15‰ (MacAvoy et al. 1998). *Drunella* sp. from the West River in the current study had $\delta^{34}S$ values that fell between these two groups, with samples from the downstream (spawning) site having values within the range of those observed previously in alosids (MacAvoy et al. 1998). While this observation and the observed increase in $\delta^{34}S$ from spring to summer in *Drunella* sp. at the downstream site support our hypothesis of marine inputs, the concurrent increase in $\delta^{34}S$ at the upstream site (9.8 ± 0.2‰ SD in spring to 12.5 ± 0.1‰ SD in summer), where no spawning occurs, exceeded the increase observed at the downstream site. This casts uncertainty on the degree of natural variability within a site, as our current knowledge of $\delta^{34}S$ patterns in freshwater biota is limited. Furthermore, sulfur inputs from sea spray and precipitation at coastal sites such as the downstream site in the West River could be entering the food web independently from anadromous fishes, leading to elevated $\delta^{34}S$ values (Novak et al. 2001). Trophic fractionation of sulfur is also highly variable among species, and the average $\delta^{34}S$ difference between an animal and its diet is 0.4 ± 0.5‰ SD (McCutchan et al. 2003). Seasonal changes may be related to changes in diet-tissue fractionation or inherent variation in $\delta^{34}S$ of algae as observed in $\delta^{13}C$ (France 1995b), but more research is required to draw any conclusions regarding sulfur isotope patterns.

Population sizes of anadromous fishes, including salmonids, alosids, and osmerids, are known to fluctuate considerably on an annual basis (Cunjak and Therrien 1998; Gresh et al. 2000; Chaput and Atkinson 2001; Acolas et al. 2006). These fluctuations would presumably influence isotope ratios in the receiving environment such that differences between spawning and reference sites would be more pronounced in years with higher spawner abundance. Consequently, $\delta^{15}N$ levels in sediments and consumers are now being promoted as surrogates for historic and current salmon escapement levels (Bilby et al. 2001; Finney et al. 2002, Gregory-Eaves et al. 2004). While we had reasonable estimates of spawner abundance in our salmon rivers (Catamaran Brook and Scorff River), we are unsure of population sizes of blueback herring in the West River and rainbow smelt in Doctor's Brook. Knowledge of changes in spawner biomass from year to year could assist in modeling the potential contribution of alosids and osmerids to freshwater energy budgets (Durbin et al. 1979). Similarly, detection of marine inputs using SIA rests largely on the timing of sampling. Inherent variability in primary producer isotope ratios (Finlay 2004; Figures 4 and 5) over time could lead to false conclusions regarding the presence or absence of marine inputs if sampling in time is limited.

Our results suggest that marine subsidies are likely present in Atlantic coast streams, but these pathways differ from Pacific coast systems in that, due to limited semelparity in the species present, they appear to be restricted to direct consumption of eggs by invertebrates and nutrient release from egg decomposition rather than through the incorporation of dissolved nutrients from carcass decay, particularly at the low population densities evident today. While bulk organic matter subsidies appear to be obtained from alosids (Durbin et al. 1979) and osmerids, contributions from Atlantic salmon at current population sizes may be more subtle, perhaps in the form of phosphorus additions as suggested by Nislow et al. (2004). Lyle and Elliott (1998) estimated minimal C, N, and P contributions (0.09–0.24%) by Atlantic salmon and sea trout to the current overall energy budgets of English rivers, which may be tempered further by energy export in the form of seaward migration by smolts. Given the historic range and abundance of anadromous fishes in eastern North America and Western Europe, and consequent declines over the past two centuries, organic matter and nutrient subsidies may be greatly reduced with consequences for the productivity of oligotrophic lakes, streams, and rivers. Future research is necessary to determine if productivity may be restored through management intervention such as fertilization (e.g., Ashley and Stockner 2003) to replace historic subsidies that are now reduced or absent. Stable isotope analysis should continue to assist in this regard by providing evidence that these subsidies are indeed being incorporated into freshwater food webs.

Acknowledgments

The authors would like to thank several individuals who assisted in collecting and analyzing samples for this project, including Charlie McInnis, Danielle Goff, Aaron Fraser, David Parsons, Alex Ready, Nicolas Jeannot, Dominique Huteau, Christine Paton, Mireille Savoie, Anne McGeachy, and Rick Doucett. Michelle Gray and Laura Weir supplied unpublished data. Funding was provided by the Canada Research Chairs Program. Two anonymous reviewers helped improve the quality of this manuscript.

References

Acolas, M. L., V. Véron, H. Jourdan, M. L. Bégout, M. R. Sabatié, and J. L. Baglinière. 2006. Upstream migration and reproductive patterns of a population of allis shad in a small river (l'Aulne, Brittany, France). ICES Journal of Marine Science 63:476–484.

Anderson, C., and G. Cabana. 2005. $\delta^{15}N$ in riverine food webs: effects of N inputs from agricultural watersheds. Canadian Journal of Fisheries and Aquatic Sciences 62:333–340.

Ashley, K. I., and J. G. Stockner. 2003. Protocol for applying limiting nutrients to inland waters. Pages 245–258 *in* J. Stockner, editor. Nutrients in salmonid ecosystems: sustaining production and biodiversity. American Fisheries Society, Symposium 34, Bethesda, Maryland.

Baglinière, J. L., and A. Champigneulle. 1986. Population estimates of juvenile Atlantic salmon, *Salmo salar*, as indexes of smolt production in the Scorff R, Brittany. Journal of Fish Biology 29:467–482.

Bilby, R. E., B. R. Fransen, and P. A. Bisson. 1996. Incorporation of nitrogen and carbon from spawning coho salmon into the trophic system of small streams: evidence from stable isotopes. Canadian Journal of Fisheries and Aquatic Sciences 53:164–173.

Bilby, R. E., B. R. Fransen, J. K. Walter, C. J. Cederholm, and W. J. Scarlett. 2001. Preliminary evaluation of the use of nitrogen stable isotope ratios to establish escapement levels for Pacific salmon. Fisheries 26:6–14.

Chaloner, D. T., K. M. Martin, M. S. Wipfli, P. H. Ostrom, and G. A. Lamberti. 2002. Marine carbon and nitrogen in southeastern Alaska stream food webs: evidence from artificial and natural streams. Canadian Journal of Fisheries and Aquatic Sciences 59:1257–1265.

Chaloner, D. T., G. A. Lamberti, R. W. Merritt, N. L. Mitchell, P. H. Ostrom, and M. S. Wipfli. 2004. Variation in responses to spawning Pacific salmon among three south-eastern Alaska streams. Freshwater Biology 49:587–599.

Chaput, G., and G. Atkinson. 2001. The gaspereau fisheries (*Alosa pseudoharengus* and *A. aestivalis*) of the Miramichi River with updates on the fishery of the Richibucto River of Gulf New Brunswick, 1997 to 2000. Canadian Science Advisory Secretariat, CSAS Research Document 2001/047, Ottawa.

Charles, K., J.-M. Roussel, and R. A. Cunjak. 2004. Estimating contribution of sympatric anadromous and freshwater resident brown trout to juvenile production. Marine and Freshwater Research 55:185–191.

Cunjak, R. A., D. Caissie, N. El-Jabi, P. Hardie, J. H. Conlon, T. L. Pollock, D. J. Giberson, and S. Komadina-Douthwright. 1993. The Catamaran Brook (New Brunswick) habitat research project: biological, physical and chemical conditions (1990–1992). Canadian Technical Report of Fisheries and Aquatic Sciences 1914.

Cunjak, R. A., J.-M. Roussel, M. A. Gray, J. P. Dietrich, D. F. Cartwright, K. R. Munkittrick, and T. D. Jardine. 2005. Using stable isotope analysis with telemetry or mark-recapture data to identify fish movement and foraging. Oecologia 144:636–646.

Cunjak, R. A., and J. Therrien. 1998. Inter-stage survival of wild juvenile Atlantic salmon, *Salmo salar* L. Fisheries Management and Ecology 5:209–223.

Dodds, W. K., E. Marti, J. L. Tank, J. Pontius, S. K. Hamilton, N. B. Grimm, W. B. Bowden, W. H. McDowell, B. J. Peterson, H. M. Valett, J. R. Webster, and S. Gregory. 2004. Carbon and nitrogen stoichiometry and nitrogen cycling rates in streams. Oecologia 140:458–467.

Dodds, W. K., V. H. Smith, and K. Lohman. 2002. Nitrogen and phosphorus relationships to benthic algal biomass in temperate streams. Canadian Journal of Fisheries and Aquatic Sciences 59:865–874.

Dodds, W. K., and E. B. Welch. 2000. Establishing nutrient criteria in streams. Journal of the North American Benthological Society 19:186–196.

Doucett, R. R. 1999. Food-web relationships in Catamaran Brook, New Brunswick, as revealed by stable-isotope analysis of carbon and nitrogen. Doctoral dissertation. University of Waterloo, Waterloo, Ontario.

Doucett, R. R., W. Hooper, and G. Power. 1999. Identification of anadromous and nonanadromous brook trout and their progeny in the Tabusintac River, New Brunswick, by means of multiple-stable-isotope analysis. Transactions of the American Fisheries Society 128:278–288.

Durbin, A. G., S. W. Nixon, and C. A. Oviatt. 1979. Effects of the spawning migration of the alewife, *Alosa pseudoharengus*, on freshwater ecosystems. Ecology 60:8–17.

Finlay, J. C. 2001. Stable-carbon-isotope ratios of river biota: implications for energy flow in lotic food webs. Ecology 82:1052–1064.

Finlay, J. C. 2004. Patterns and controls of lotic algal carbon isotope ratios. Limnology and Oceanography 49:850–861.

Finney, B. P., I. Gregory-Eaves, M. S. V. Douglas, and J. P. Smol. 2002. Fisheries productivity in the northeastern Pacific Ocean over the past 2,200 years. Nature (London) 416:729–733.

Foote, C. J., and G. S. Brown. 1998. Ecological relationship between freshwater sculpins (genus *Cottus*) and beach-spawning sockeye salmon (*Oncorhynchus nerka*) in Iliamna Lake, Alaska. Canadian Journal of Fisheries and Aquatic Sciences 55:1524–1533.

France, R. 1995a. Stable nitrogen isotopes in fish: literature synthesis on the influence of ecotonal coupling. Estuarine, Coastal and Shelf Science 41:737–742.

France, R. 1995b. Critical examination of stable isotope analysis as a means for tracing carbon pathways in stream ecosystems. Canadian Journal of Fisheries and Aquatic Sciences 52:651–656.

Fry, B., and C. Arnold. 1982. Rapid $^{13}C/^{12}C$ turnover during growth of brown shrimp (*Penaeus aztecus*). Oecologia 54:200–204.

Fry, B., and E. B. Sherr. 1984. $\delta^{13}C$ measurements as indicators of carbon flow in marine and freshwater ecosystems. Contributions in Marine Science 27:13–47.

Garman, G. C., and S. A. Macko. 1998. Contribution of marine-derived organic matter to an Atlantic coast, freshwater, tidal stream by anadromous clupeid fishes. Journal of the North American Benthological Society 17:277–285.

Giovanni, R. 1996. Dégradation de la qualité des eaux courantes en Bretagne: nitrates et cas de deux pesticides, l'atrazine et le lindane. Cybium 3(Supplement):143–162.

Gray, M. A., R. A. Cunjak, and K. R. Munkittrick. 2004. Site fidelity of slimy sculpin (*Cottus cognatus*): insights from stable carbon and nitrogen analysis. Canadian Journal of Fisheries and Aquatic Sciences 61:1717–1722.

Gray, M. A., and K. R. Munkittrick. 2005. An effects based assessment of slimy sculpin (*Cottus cognatus*) populations in agricultural regions of northwestern New Brunswick. Water Quality Research Journal of Canada 40:16–27.

Gregory-Eaves, I., B. P. Finney, M. S. V. Douglas, and J. P. Smol. 2004. Inferring sockeye salmon (*Oncorhynchus nerka*) population dynamics and water quality changes in a stained nursery lake over the past ~500 years. Canadian Journal of Fisheries and Aquatic Sciences 61:1235–1246.

Gresh, T. U., J. Lichatowich, and P. Schoonmaker. 2000. An estimation of historic and current levels of salmon production in the northeast Pacific ecosystem: evidence of a nutrient deficit in the freshwater system of the Pacific Northwest. Fisheries 25:15–21.

Gross, M. R., R. M. Coleman, and R. M. McDowall. 1988. Aquatic productivity and the evolution of diadromous fish migration. Science 239:1291–1293.

Harding, J. S., D. J. Hawke, R. N. Holdaway, and M. J. Winterbourn. 2004. Incorporation of marine-derived nutrients from petrel breeding colonies into stream food webs. Freshwater Biology 49:576–586.

Harrington, R. R., B. P. Kennedy, C. P. Chamberlain, J. D. Blum, and C. L. Folt. 1998. ^{15}N enrichment in agricultural catchments: field patterns and applications to tracking Atlantic salmon (*Salmo salar*). Chemical Geology 147:281–294.

Hesslein, R. H., K. H. Hallard, and P. Ramlal. 1993. Replacement of sulfur, carbon, and nitrogen in tissue of growing broad whitefish (*Coregonus nasus*) in response to a change in diet traced by δ^{34}S, δ^{13}C, and δ^{15}N. Canadian Journal of Fisheries and Aquatic Sciences 50:2071–2076.

Jamieson, I., N. R. Seymour, and R. P. Bancroft. 1982. Use of two habitats related to prey availability in a population of ospreys in northeastern Nova Scotia. Wilson Bulletin 94:557–564.

Jardine, T. D., M. A. Gray, S. M. McWilliam, and R. A. Cunjak. 2005. Stable isotope variability in tissues of temperate stream fishes. Transactions of the American Fisheries Society 134:1103–1110.

Jonsson, B., and N. Jonsson. 2003. Migratory Atlantic salmon as vectors for the transfer of energy and nutrients between freshwater and marine environments. Freshwater Biology 48:21–27.

Kelly, J. F. 2000. Stable isotopes of carbon and nitrogen in the study of avian and mammalian trophic ecology. Canadian Journal of Zoology 78:1–27.

Kline, T. C., Jr., J. J. Goering, O. A. Mathisen, P. H. Poe, and P. L. Parker. 1990. Recycling of elements transported upstream by runs of Pacific salmon: I. δ^{15}N and δ^{13}C evidence in Sashin Creek, southeastern Alaska. Canadian Journal of Fisheries and Aquatic Sciences 47:136–144.

Kline, T. C., Jr., J. J. Goering, O. A. Mathisen, P. H. Poe, P. L. Parker, and R. S. Scalan. 1993. Recycling of elements transported upstream by runs of Pacific salmon: II. δ^{15}N and δ^{13}C evidence in Kvichak River watershed, Bristol Bay, southeastern Alaska. Canadian Journal of Fisheries and Aquatic Sciences 50:2350–2365.

Lyle, A. A., and J. M. Elliott. 1998. Migratory salmonids as vectors of carbon, nitrogen and phosphorus between marine and freshwater environments in north-east England. The Science of the Total Environment 210/ 211:457–468.

MacAvoy, S. E., S. A. Macko, and G. C. Garman. 1998. Tracing marine biomass into tidal freshwater ecosystems using stable sulfur isotopes. Naturwissenschaften 85:544–546.

MacAvoy, S. E., S. A. Macko, S. P. McIninch, and G. C. Garman. 2000. Marine nutrient contributions to freshwater apex predators. Oecologia 122:568–573.

MacLaren Plansearch. 1983. Hydrotechnical study of the Antigonish area floodplain. Environment Canada/Nova Scotia Department of the Environment, Canada-Nova Scotia Flood Damage Reduction Program, Halifax.

McCutchan, J. H., Jr., W. M. Lewis, Jr., C. Kendall, and C. C. McGrath. 2003. Variation in trophic shift for stable isotope ratios of carbon, nitrogen, and sulfur. Oikos 102:378–390.

Mekhtiyeva, V. L., R. G. Pankina, and Y. Y. Gavrilov. 1976. Distributions and isotopic compositions of forms of sulfur in water animals and plants. Geochemistry International 13: 82–87.

Merritt, R. W., and K. W. Cummins. 1996. An introduction to the aquatic insects of North America. 3rd edition. Kendall/Hunt Publishing, Dubuque, Iowa.

Mitchell, S., A. Fraser, P. Hardie, and R. A. Cunjak. 2004. Electrofishing data summaries for Catamaran Brook, New Brunswick, 1990–2002. Canadian Data Report of Fisheries and Aquatic Sciences 1130.

Miyake, Y., T. Hiura, N. Kuhara, and S. Nakano. 2003. Succession in a stream invertebrate community: a transition in species dominance through colonization. Ecological Research 18:493–501.

Moyle, P. B. 1977. In defense of sculpins. Fisheries 2:20–23.

Naiman, R. J., J. M. Helfield, K. K. Bartz, D. C. Drake, and J. M. Honea. 2009. Pacific salmon, marine-derived nutrients and the dynamics of aquatic and riparian ecosystems. Pages 395–425 *in* A. J. Haro, K. L. Smith, R. A. Rulifson, C. M. Moffitt, R. J. Klauda, M. J. Dadswell, R. A. Cunjak, J. E. Cooper, K. L. Beal, and T. S. Avery, editors. Challenges for diadromous fishes in a dynamic global environment. American Fisheries Society, Symposium 69, Bethesda, Maryland.

Nislow, K. H., J. D. Armstrong, and S. McKelvey. 2004. Phosphorus flux due to Atlantic salmon (*Salmo salar*) in an oligotrophic upland stream: effects of management and demography. Canadian Journal of Fisheries and Aquatic Sciences 61:2401–2410.

Novak, M., S. H. Bottrell, and E. Prechova. 2001. Sulfur isotope inventories of atmospheric deposition, spruce forest floor and living Sphagnum along a NW-SE transect across Europe. Biogeochemistry 53:23–50.

Pastershank, G. M. 2001. Unifying ecosystem concepts and mercury biomagnification in an estuarine environment using stable isotopes (δ^{13}C and δ^{15}N). Doctoral dissertation. University of Ottawa, Ottawa.

Perga, M. E., and D. Gerdeaux. 2005. 'Are fish what they eat' all year round? Oecologia 144:598–606.

Petrosky, C. E., and T. F. Waters. 1975. Annual production by the slimy sculpin population in a small Minnesota trout stream. Transactions of the American Fisheries Society 104:237–244.

Polis, G. A., W. B. Anderson, and R. D. Holt. 1997.

Toward an integration of landscape and food web ecology: the dynamics of spatially subsidized food webs. Annual Review of Ecology and Systematics 28:289–316.

Scott, W. B., and E. J. Crossman. 1998. Freshwater fishes of Canada. Galt House Publications, Oakville, Ontario.

Tank, J. L., and W. K. Dodds. 2003. Nutrient limitation of epilithic and epixylic biofilms in ten North American streams. Freshwater Biology 48:1031–1049.

Udy, J. W., and S. E. Bunn. 2001. Elevated $\delta^{15}N$ values in aquatic plants from cleared catchments: why? Marine and Freshwater Research 52:347–351.

Wayland, M., and K. A. Hobson. 2001. Stable carbon, nitrogen, and sulfur isotope ratios in riparian food webs on rivers receiving sewage and pulp mill effluents. Canadian Journal of Zoology 79:5–15.

Wipfli, M. S., J. P. Hudson, D.T. Chaloner, and J. P. Caouette. 1999. Influence of salmon spawner densities on stream productivities in southeast Alaska. Canadian Journal of Fisheries and Aquatic Sciences 56:1600–1611.

American Fisheries Society Symposium 69:447–463, 2009

Spatial and Temporal Variability in the Trophic Ecology of Atlantic Salmon in the North Atlantic Inferred from Analyses of Stable Isotope Signatures

R. Niloshini Sinnatamby
Department of Biology, University of Waterloo
200 University Avenue West, Waterloo, Ontario N2L 3G1, Canada

J. Brian Dempson
Fisheries and Oceans Canada, Science Branch
Post Office Box 5667, St. John's, Newfoundland and Labrador A1C 5X1, Canada

Gerald Chaput
Fisheries and Oceans Canada, Science Branch
Post Office Box 5030, Moncton, New Brunswick E1C 9B6, Canada

Francois Caron
Ministère des Ressources naturelles et de la Faune
Direction de l'aménagement de la faune du Saguenay-Lac-Saint-Jean
3950, boul. harvey. 4e étage, Jonquière, Québec G7X 8L6, Canada

Eero Niemelä and Jaakko Erkinaro
Finnish Game and Fisheries Research Institute
Oulu Game and Fisheries Research Tutkijantie 2E, 90570 Oulu, Finland

Michael Power*
Department of Biology, University of Waterloo
200 University Avenue West, Waterloo, Ontario N2L 3G1, Canada

Abstract.—In many areas of the North Atlantic, populations of Atlantic salmon *Salmo salar* are now either in a state of decline or extirpated such that concern over the continued survival of the species has been given more attention in recent years despite large reductions in directed ocean fisheries. Previous investigations have established linkages between ocean climate conditions and variability in abundance or survival. However, one avenue not previously explored considers whether changes in marine food webs owing to ever increasing and unsustainable levels of exploitation on many marine species—the so-called "fishing down marine food webs" hypothesis—could influence survival and abundance of salmon as a result of shifts in trophic position or changes in energy flows. Since Atlantic salmon are opportunistic feeders during the marine life history phase, the species lends itself well to studies associated with marine environmental conditions and food web interactions. Here, we examine long-term variability in the trophic ecology of Atlantic salmon using analyses of stable isotope signatures of carbon and nitrogen ($\delta^{13}C$ and $\delta^{15}N$). Signatures were extracted from the marine growth portion of scales of maiden one-sea-winter fish. Data were obtained from nine Canadian and one north European river (Teno) covering periods extending over three to four decades. Significant differences in $\delta^{13}C$ and $\delta^{15}N$ signatures were found to exist among

* Corresponding author: m3power@sciborg.uwaterloo.ca

rivers, as well as among years within rivers. Trends over time in either $\delta^{13}C$ or $\delta^{15}N$ signatures were evident in only a few situations, thus providing little evidence of substantive changes in the trophic ecology of salmon in the North Atlantic. In addition, isotopic signatures were largely invariant in relation to variations in abundance or to various environmental measures characterizing ocean climate conditions in the North Atlantic.

Introduction

Atlantic salmon *Salmo salar* exhibit a complex life history that is influenced by factors encompassing various freshwater and marine habitats. Populations are characterized by annual fluctuations in survival and abundance, and at times, fluctuations can be quite severe and occur over wide geographic areas (Dempson et al. 1998). While studies have shown that the number of salmon returning to rivers is dependent upon smolt production (Jonsson et al. 1998) as well as marine exploitation rates (Crozier and Kennedy 1994, 1999), factors other than exploitation may also be responsible, especially in areas where directed marine fisheries on salmon have been greatly reduced or eliminated (Parrish et al. 1998; Walters and Ward 1998; Dempson et al. 2001, 2004; Jonsson and Jonsson 2004). Indeed, in many areas of the North Atlantic, populations of salmon are now either in a state of decline or extirpated despite large reductions in directed ocean fisheries (Parrish et al. 1998) and the introduction of other management strategies to conserve stocks (e.g., Crozier et al. 2004).

While the issue of Atlantic salmon survival is complicated by their complex life cycle requirements (Armstrong et al. 1998), there are various hypotheses regarding marine fish survival and production that may pertain to Atlantic salmon declines. One hypothesis focuses on oceanic climatic factors and their potential influence on population dynamics (Friedland et al. 1998; Attril and Power 2002; Jonsson and Jonsson 2003). Evidence exists for dramatic changes in ocean climate conditions in the northwest Atlantic, particularly during the early 1990s (e.g., Colbourne et al. 1997; Colbourne and Anderson 2003), prompting some to suggest that there has been a marine climate regime shift (Drinkwater 2000; Montevecchi et al. 2002). Linkages between ocean climate conditions and variability in abundance or survival have been established in various populations of Atlantic salmon (Friedland 1998; Friedland et al. 2000, 2003; Niemelä et al. 2004). In addition, Chaput et al. (2005) noted strong evidence for a phase shift in Atlantic salmon productivity in the early 1990s that has persisted to date. Nevertheless, as noted by Jonsson and Jonsson (2004), understanding the relationship between water temperature, growth rate, and sea age at maturation is still poor, while indirect effects of climate on marine food availability, for example, are not entirely evident (Jacobsen and Hansen 2000). Another hypothesis stresses the implications of trophic structure and anthropogenic disturbances of trophic structure that have led to shortened food chains. Indeed, recent investigations have proposed that marine food webs have changed dramatically owing to ever increasing and unsustainable levels of exploitation—the so-called "fishing down marine food webs" hypothesis (Pauly et al. 1998, 2001; Pauly and Palomares 2005)—and demonstrated measurable effects on the mean trophic level of Atlantic cod *Gadus morhua* and other Canadian east coast demersal fishes as a result of fishing (Pauly et al. 2001). Additional hypotheses associated with the decline of salmon have also been identified and discussed (e.g., O'Neil et al. 2000).

Atlantic salmon are considered to be opportunistic feeders during the marine life history phase, often targeting prey in the upper end of the size spectrum with a preference for fish over crustaceans should both be available (Hislop et al. 1993; Jacobsen and Hansen 2000). Thus, the species lends itself well to studies associated with marine environmental conditions and food web interactions. Accordingly, salmon populations may reflect marine food web changes through utilization of prey from lower trophic levels. Trophic level can be evaluated by an examination of stomach contents over time or through stable isotope analysis (SIA). While stomach contents provide a snapshot of recent dietary resource use, stable isotope analyses yield time integrated measures of energy assimilation since analyses are performed on body tissues built from diet assimilated over time (Fry and Sherr 1984). In addition, gut content data are known to underestimate reliance on soft-bodied prey (McCarthy et al. 2004) and are more influenced by temporal variation in diet than SIA (Grey 2001; Guiguer et al. 2002). Consequently, SIA has been increasingly used in ecological studies as a reliable means of inferring trophic status and the impacts of anthropogenic disturbance on trophic relationships (e.g., Vander Zanden et al. 1999).

Although fish muscle tissue is commonly analyzed for stable isotopes, studies have also used fish scale stable isotope values to evaluate diet and/or trophic position (Estep and Vigg 1985; Wainright et al. 1993; Perga and Gerdeaux 2003; Pruell et al. 2003; Bowman 2005; Estrada et al. 2005) as they are known to integrate dietary signatures effectively over the last season of growth (Hutchinson and Trueman 2006). Nitrogen stable isotope analysis provides a quantitative means to determine trophic level since nitrogen signatures from organism tissue are consistently 3–5‰ more enriched than dietary sources (Peterson and Fry 1987). In contrast, carbon stable isotopes are conserved up the food chain (DeNiro and Epstein 1978), owing to the slight 0.0–1.0‰ enrichment occurring between prey and consumer (Peterson and Fry 1987). Because ^{13}C is conserved during trophic transfer, but varies at the base of the food web, consumer tissue stable isotope signatures will also reflect dietary source information (Vander Zanden and Rasmussen 1999).

In light of the above, the aim of the present study was to assess spatial and temporal variability in the trophic ecology as inferred from analyses of stable isotope signatures of carbon and nitrogen ($\delta^{13}C$ and $\delta^{15}N$) of archived maiden one-sea-winter (1SW) Atlantic salmon scales from nine Canadian and one north European river. The discriminate use of 1SW salmon was intended to provide an integrated view of dietary conditions in the marine environment over 1 year. Specifically, data obtained from SIA were analyzed to determine whether there was evidence of changes in marine food webs that may have influenced salmon abundance. In addition, the time series of stable isotope data were assessed for significant trends or cycles in relation to environmental variables owing to the linkages between ocean climate conditions and variability in abundance and survival established by previous studies.

Methods

Atlantic salmon scales from nine Canadian rivers were obtained from historical collections maintained by annual monitoring programs managed by Fisheries and Oceans Canada (Newfoundland and Maritimes rivers), Ministère des Ressources naturelles et de la Faune, Québec (Rivière de la Trinité) (Figure 1; Table 1), or the Finnish Game and Fisheries Research Institute (River Teno/Tana) (Table 1). Rivers included were those for which longer time series of scale data sets were readily available and where salmon abundance data or an index of abundance was available. With the exception of the River Teno, all other stocks examined were from eastern Canada. River Teno data were specific to samples obtained from the River Pulmankijoki tributary, located approximately 70 km upstream from the sea. These samples were included as they provide an opportunity to contrast isotopic signatures of 1SW salmon from a very different geographic area of the North Atlantic. All scales were removed from individual fish, stored in envelopes, and subsequently aged to verify 1SW status. Scales from 10 randomly selected 1SW Atlantic salmon were retrieved per available year from the mid-portion of the annual run, except for a few years of low sample availability where fewer salmon were used.

Five to 15 scales per individual were cleaned by soaking in distilled water and gentle scraping with a scalpel to remove any surface residue. Under 25× magnification, the river-age portion of each scale was identified and separated from the marine-age portion using a scalpel (Power 1987). Scales were not treated with acid since a comparative assessment found that there was no significant influence of the scale inorganic fraction on the stable isotope signatures of nontreated Atlantic salmon scales (Sinnatamby et al. 2007). The marine-age portions of the scales were then dried and ground using a Retsch MM 2000 ball mill grinder (F. Kurt Retsch GmbH & Company, Haan, Germany) or cut using sterile laboratory scissors. Approximately 1.5 mg of the sample was used in the simultaneous analysis of stable carbon and nitrogen isotopes.

Stable isotope ratios are expressed here in conventional delta notation (δ) relative to Peedee Belemnite limestone for $\delta^{13}C$ (Craig 1957) and atmospheric nitrogen for $\delta^{15}N$ (Mariotti 1983). Enriched samples, which contain more of the heavier isotope

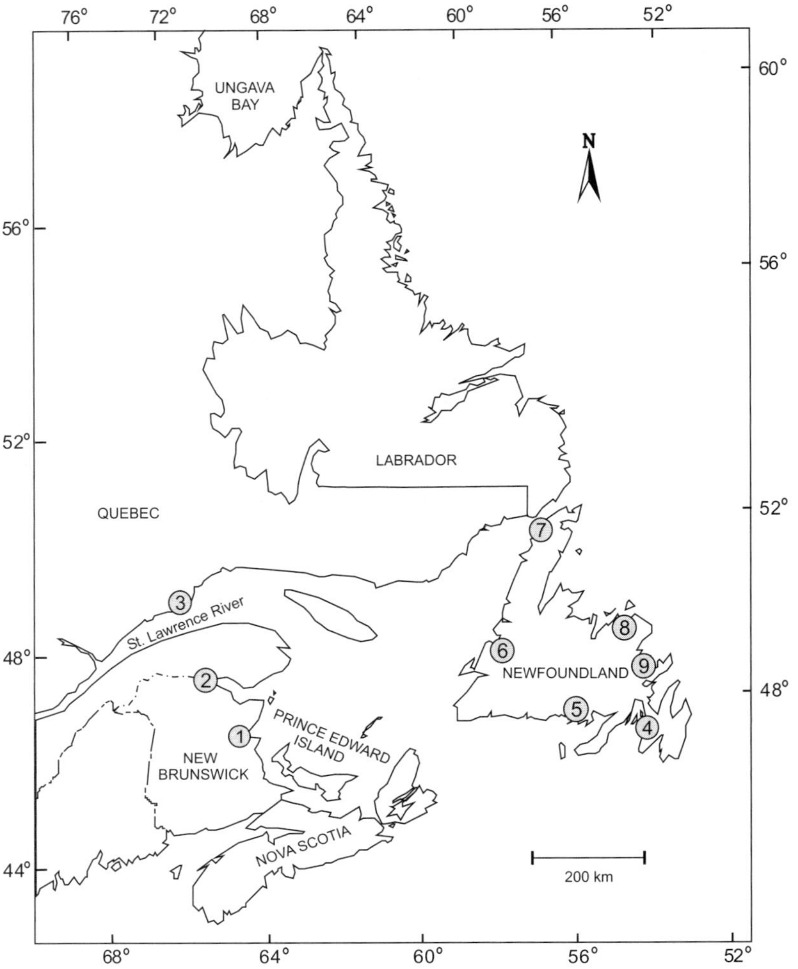

Figure 1.—Map of eastern Canada indicating the general location of the nine Canadian rivers included in this study. Sites are denoted by numbers as summarized in Table 1.

(^{13}C or ^{15}N) relative to the standard, are denoted by higher delta values. All isotopic analyses were performed at the Environmental Isotope Laboratory at the University of Waterloo on a Micromass VG Isochrom continuous-flow isotope ratio mass spectrometer connected to a Carlo Erba elemental analyzer with an analytical precision of ±0.2% ($\delta^{13}C$) and ±0.3% ($\delta^{15}N$). Precision was established through repeat analysis of International Atomic Energy Agency CH6 and N1 and N2 standards. Analytical accuracy was additionally assessed by repeat analysis of 1 in 10 samples.

Data normality and variance homogeneity were assessed prior to statistical analyses. Generalized linear models were used to assess differences among rivers, among years within rivers, and across rivers for each year. Linear regressions were used to detect and evaluate the significance of trends in mean $\delta^{13}C$ and $\delta^{15}N$ signatures over time for each site. Periodicity of mean $\delta^{13}C$ and $\delta^{15}N$ signatures was evaluated using randomization tests described by Manly (1991). In addition, randomization tests were used to examine stable isotope results for serial correlation following methods described by Manly (1991). In order to test both periodicity and serial correlation, only continuous years were used. Where only 1 or 2 years were missing from a series (approximately 4% of all observations), the data for those years were estimated using standard linear interpolation methods (Nelson 2003).

Table 1.—Mean fork length (cm) and river age (years), along with respective summary statistics of $\delta^{13}C$ and $\delta^{15}N$ obtained from analysing the marine growth section of one-sea-winter Atlantic salmon scales. Map location code is as given in Figure 1. N = sample size for isotope analyses. Mean, minimum (Min), and maximum (Max) values are presented in parts per mil (‰). SD denotes standard deviation.

Region	River	Map location code	Fork length	River age	N	$\delta^{13}C$ Mean	SD	Min	Max	$\delta^{15}N$ Mean	SD	Min	Max
Gulf of St. Lawrence													
	Miramichi	1	53.2	2.50	339	−16.10	0.32	−17.00	−15.32	11.32	0.46	10.10	13.04
	Restigouche	2	54.9	2.84	330	−15.95	0.31	−17.07	−15.18	11.54	0.49	10.03	12.64
	de la Trinité	3	54.7	3.10	247	−16.18	0.25	−17.12	−15.41	12.54	0.55	11.06	13.98
South Newfoundland													
	Northeast Placentia	4	53.5	3.10	194	−15.97	0.39	−17.22	−15.02	11.34	0.53	10.28	13.84
	Conne	5	51.6	3.14	196	−16.21	0.28	−16.75	−15.27	11.76	0.43	10.68	12.87
West Newfoundland													
	Humber	6	54.2	3.37	110	−15.84	0.32	−17.09	−14.92	11.50	0.56	10.44	13.47
	Western Arm Brook	7	53.7	3.70	294	−16.18	0.42	−17.44	−15.24	11.10	0.50	9.74	12.69
Northeast Newfoundland													
	Gander	8	52.8	3.75	257	−16.22	0.41	−17.29	−15.06	11.07	0.64	9.46	12.65
	Terra Nova	9	52.1	3.63	199	−15.96	0.31	−17.02	−15.15	11.18	0.62	9.71	13.80
Northern Europe													
	Teno		50.7	4.21	290	−16.99	0.43	−18.09	−15.99	11.87	0.50	10.45	13.31

Pearson's correlation coefficient and associated significance tests (Zar 1999) were used to find and test the significance of regression relationships between mean annual stable isotope signatures ($\delta^{13}C$, $\delta^{15}N$) and environmental variables as well as salmon abundance information. If first-order autocorrelation was present, the autocorrelative structure was incorporated into the analysis using the SAS Autoreg procedure (Brocklebank and Dickey 2003). Environmental variables selected for study of Canadian rivers included the following, all of which have been previously found to influence the abundance and/or growth of various marine species (e.g., Narayanan et al. 1995; Colbourne et al. 2002; Colbourne and Anderson 2003; Friedland et al. 2003; Colbourne 2004): the total Newfoundland ice cover from the general area 45–55°N during December to June (Colbourne et al. 1994; Narayanan et al. 1995), annual Station 27 temperature anomalies, an annual index of marine thermal habitat (Reddin and Friedland 1993), the areal extent of the cold intermediate layer defined as the area of water with temperatures less than 0°C overlaying the Newfoundland continental shelf (Colbourne 2004) from several standard monitoring sections in the northwest Atlantic (Bonavista and Flemish Cap transects), and the winter (December through March) index of the North Atlantic Oscillation index (NAOI) based on the difference in normalized pressures between Lisbon, Portugal and Stykkisholmur, Iceland (Hurrell 1995). For the River Teno, a border river between Finland and Norway that empties into the Barents Sea, environmental variables included in the correlation analysis included mean sea temperature (°C) calculated from average values integrated vertically from 0 to 50 m across all stations of the Kola section in the Barents Sea annually from 1975 to 2002 (Tereshchenko 1996; S. Prusov, Polar Research Institute of Marine Fisheries and Oceanography, unpublished data) and the NAOI. Salmon abundance information was obtained from published summaries (Niemelä et al. 2004; O'Connell et al. 2005; Chaput et al. 2006; Caron et al. 2007) or archived records maintained by the Gulf Fisheries Centre, Moncton, New Brunswick (Miramichi and Restigouche River data) and refer to numbers of returning individuals. Correlation analyses were performed using abundances of corresponding 1SW (de la Trinité, Teno) or small (Newfoundland, Maritimes) salmon where small salmon are predominately maiden 1SW fish.

The difference between stable isotope signatures of adjacent years was calculated and tallied in to 0.1‰ interval classes between –1.0 and 1.0 to assess the distribution of year-to-year changes around zero as an indicator of feeding stability. These differences were plotted and analyzed using the Shapiro-Wilk test for normality. Maximal type I error in all statistical test procedures was set at $\alpha = 0.05$.

Results

Mean annual $\delta^{13}C$ and $\delta^{15}N$ signatures of Atlantic salmon scales for all sites ranged from approximately –17.5‰ to –15‰ and 10–13‰, respectively (Table 1). Analysis of variance (ANOVA) indicated significant differences in $\delta^{13}C$ and $\delta^{15}N$ signatures among rivers with all years combined (Table 2). On average, the most enriched nitrogen values occurred in salmon from the Rivière de la Trinité where signatures exceeded those reported for the most depleted salmon found in the Gander River by 1.47‰. Salmon from the River Teno in northern Europe were the most depleted with respect to $\delta^{13}C$, whereas Humber River salmon were the least depleted with the average value being about 1.14‰ higher. A subsequent analysis showed that about 76% of the variation in $\delta^{13}C$ and 60% of the variation in $\delta^{15}N$ signatures could be explained by river, year, and a river–year interaction (Table 3). The significant interaction, however, indicates that

Table 2.—Analysis of variance results comparing $\delta^{13}C$ and $\delta^{15}N$ variation across all sampled rivers, including degrees of freedom (df), type III sum of squares (SS), F, and p-values (P). N = sample size for isotope analyses.

Source of variation	N	Model r^2	Type III SS	F	P
Carbon	2,456	0.44	239.16	26.57	<0.001
Nitrogen	2,456	0.40	448.55	180.70	<0.001

Table 3.—Analyis of variance results of $\delta^{13}C$ and $\delta^{15}N$ variation across years (1971–2005) and rivers including degrees of freedom (df), type IV sum of squares (SS), *F*, and *p*-values (*P*).

Isotope	Model r^2	Source of variation	df	Type IV SS	*F*	*P*
Carbon	0.76	River	9	173.55	327.16	<0.001
		Year	34	31.86	15.90	<0.001
		River × year	207	131.67	10.79	<0.001
Nitrogen	0.60	River	9	350.87	191.53	<0.001
		Year	34	31.17	4.50	<0.001
		River × year	207	149.72	3.55	<0.001

* Type IV SS used owing to missing cells.

differences among rivers depend on the year being analyzed. Owing to the complexity of the full model, river and year were further analyzed separately. ANOVA comparing $\delta^{13}C$ and $\delta^{15}N$ across years for individual rivers also revealed significant differences (Table 4). ANOVA comparing $\delta^{13}C$ and $\delta^{15}N$ across all rivers by individual year demonstrated significant differences among rivers in all years (maximum p = 0.043).

Annual trends in mean $\delta^{13}C$ signatures were evident in salmon scales originating from Gander River, Western Arm Brook, Conne River, Rivière de la Trinité, and River Teno (Figure 2). For Gander River, Western Arm Brook, Conne River, and Rivière de la Trinité salmon, annual carbon signatures demonstrated a significant positive trend over time (Table 5), while River Teno results indicated a negative relationship between $\delta^{13}C$ signatures and year when the analysis incorporated consideration of first-order autocorrelation that was present in most of the time series. Similarly, nitrogen signatures from Terra Nova River salmon demonstrated a sta-

Table 4.—Analysis of variance results comparing $\delta^{13}C$ and $\delta^{15}N$ across years for each site, including degrees of freedom (df), type III sum of squares (SS), *F*, and *p*-values (*P*).

		$\delta^{13}C$				$\delta^{15}N$			
Region	River	Model r^2	Type III SS	*F*	*P*	Model r^2	Type III SS	*F*	*P*
Gulf of St. Lawrence									
	Miramichi	0.50	17.66	9.28	<0.001	0.31	22.13	4.19	<0.001
	Restigouche	0.60	19.22	13.86	<0.001	0.36	27.51	5.12	<0.001
	de la Trinité	0.40	6.18	6.25	<0.001	0.24	18.02	2.99	<0.001
South Newfoundland									
	Conne	0.58	8.92	11.90	<0.001	0.34	12.51	4.53	<0.001
	Northeast Placentia	0.48	13.88	8.38	<0.001	0.38	21.00	5.64	<0.001
West Newfoundland									
	Humber	0.31	3.50	4.39	<0.001	0.27	9.22	3.71	<0.001
	Western Arm Brook	0.69	36.08	20.49	<0.001	0.29	21.65	3.80	<0.001
Northeast Newfoundland									
	Gander	0.52	22.44	9.71	<0.001	0.32	33.26	4.09	<0.001
	Terra Nova	0.48	8.83	8.54	<0.001	0.35	26.21	5.06	<0.001
Northern Europe									
	Teno	0.76	40.19	28.15	<0.001	0.47	34.10	7.82	<0.001

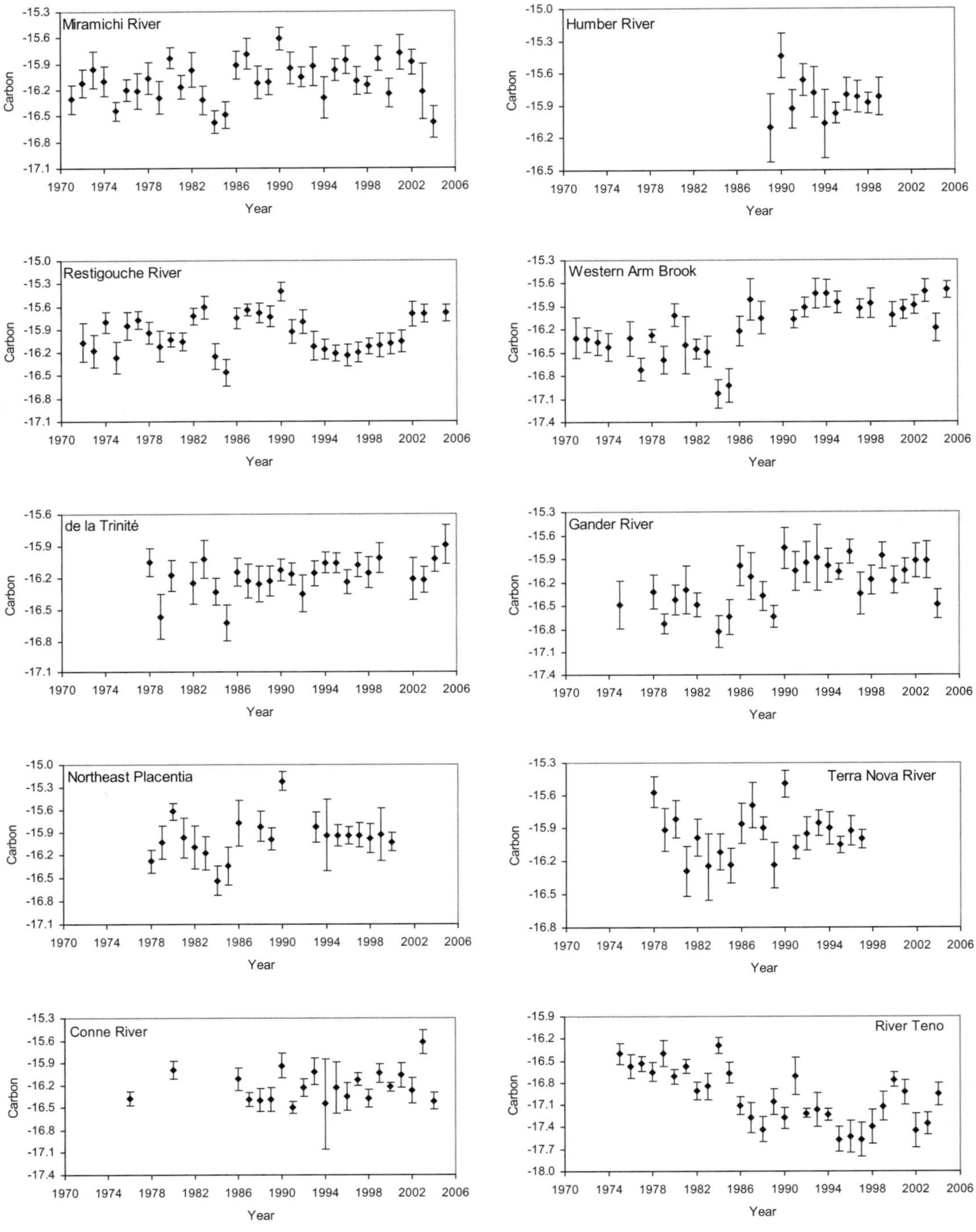

Figure 2.—Mean annual $\delta^{13}C$ signatures of the nine Canadian rivers and one Finnish river (Teno). Error bars indicate the 95% confidence interval.

Table 5.—Estimated regression coefficients and associated *p*-values (*P*) for regression models relating stable isotope values to year for each site. *N* = sample size for isotope analyses.

			$\delta^{13}C$		$\delta^{15}N$	
Region	River	*N*	Model r^2	*P*	Model r^2	*P*
Gulf of St. Lawrence						
	Miramichi	34	0.03	0.32	0.02	0.45
	Restigouche	33	0.00	0.75	0.10	0.08
	de la Trinité	25	0.31	0.00	0.00	0.87
South Newfoundland						
	Northeast Placentia	20	0.05	0.33	0.10	0.18
	Conne	19	0.40	0.01	0.33	0.01
West Newfoundland						
	Humber	11	0.03	0.61	0.01	0.79
	Western Arm Brook	30	0.26	0.00	0.03	0.36
Northeast Newfoundland						
	Gander	27	0.26	0.01	0.14	0.06
	Terra Nova	20	0.00	1.00	0.34	0.01
Northern Europe						
	Teno	30	0.274	0.004	0.012	0.564

tistically significant positive annual trend (Figure 3) and Conne River results indicated a negative relationship between $\delta^{15}N$ and year. For all other analyses, there were no consistent linear trends over time. Rather, SIA values often fluctuated (e.g., River Teno—nitrogen) without any consistent tendency to increase or decrease over the time series available.

Randomization tests for periodicity of mean annual carbon signatures demonstrated a statistically significant single cycle period in $\delta^{13}C$ obtained from Gander River, Western Arm Brook, and Teno River that were subsequently shown to have significant autocorrelative structure. Single period cycles in autocorrelated data must be treated with caution, as a time series with no periodic components but positive autocorrelation can yield false periodicity patterns (Manly 1991). Tests for periodicity of mean annual $\delta^{15}N$ signatures indicated significant 13-year periodicity in Gander River data and significant 7.5-year periodicity in Teno River data. Other periodicities were either insignificant or single long-cycle periodicities influenced by autocorrelative structure in the data. Overall, no consistent within and among river pattern of periodicity was demonstrated for either the $\delta^{13}C$ or $\delta^{15}N$ data.

Correlation analyses for evaluating the relationships between $\delta^{13}C$ and several environmental variables and salmon abundance produced no significant results. While the evaluation of linear relationships between $\delta^{15}N$ signatures with environmental variables and abundance generated some significant results, no consistent pattern emerged when comparing results across all sites (Table 6) and no single measure correlated with a large number of sites. Correlations of annual thermal habitat indices against $\delta^{15}N$ signatures for Canadian rivers were dominated by nonsignificant results (nine of nine rivers tested) using $p < 0.05$ as the judgment criterion. In contrast, the River Teno $\delta^{15}N$ data were significantly correlated ($P < 0.05$) with the vertically integrated (0–50 m) mean sea temperature (°C) calculated for the Kola section of the Barents Sea. Data for the Miramichi were significantly correlated with all considered thermal indicators (Table 6), whereas Conne and de la Trinité correlated only with the cold intermediate layer and no correlations were found between $\delta^{15}N$ and the considered thermal indicators for any of the other sites. Correlation analyses evaluating potential relationships between $\delta^{13}C$ and abundances data for each site yielded no significant results while correlation analyses between $\delta^{15}N$ and

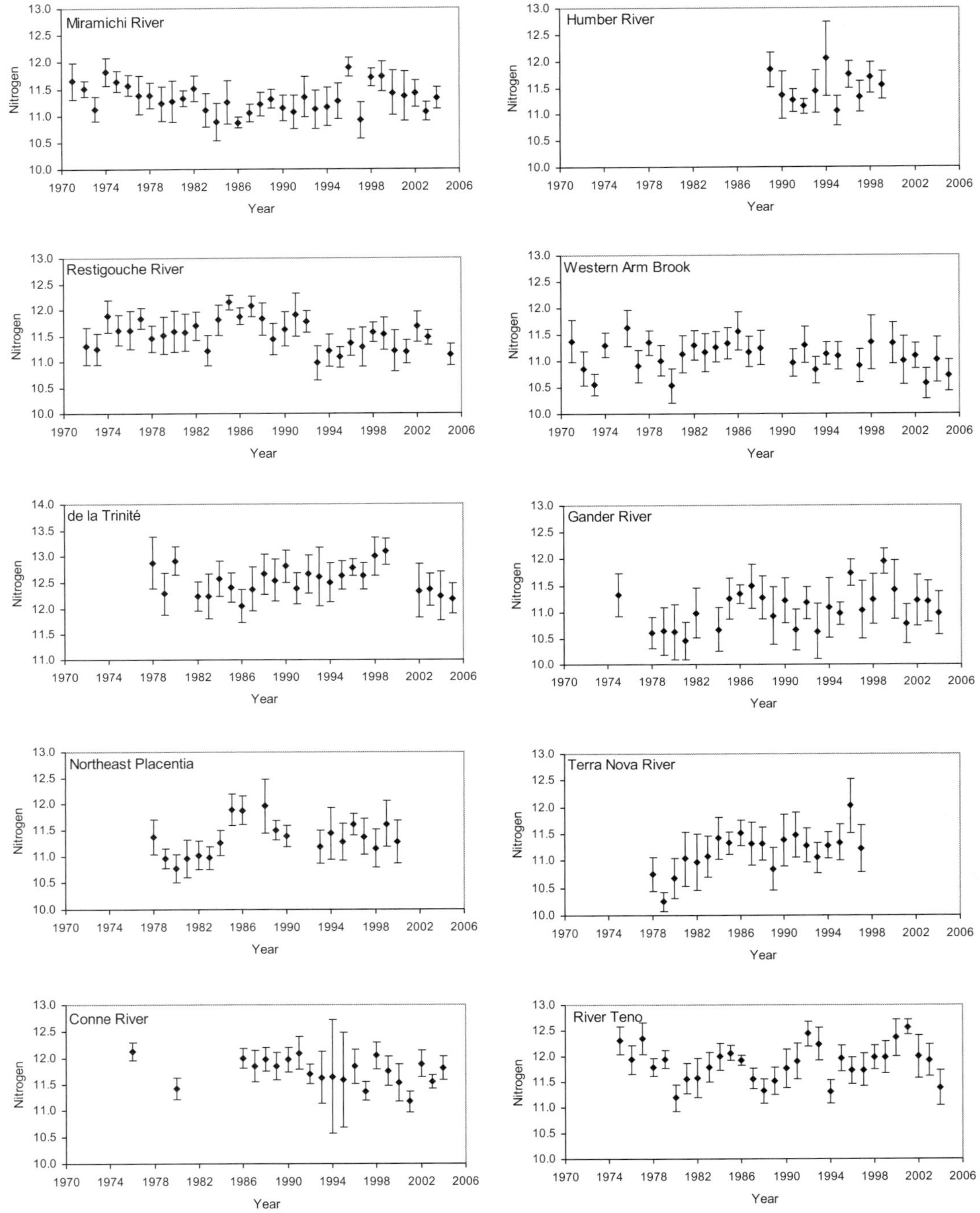

Figure 3.—Mean annual $\delta^{15}N$ signatures of the nine Canadian rivers and one Finnish river (Teno). Error bars indicate the 95% confidence interval.

Table 6.—Pearson's correlation analysis results of $\delta^{15}N$ signatures and environmental variables: Newfoundland ice cover (km^2); Station 27 located approximately 7 km east of the harbor of St. John's, Newfoundland; cold intermediate layer (CIL) at various transects; sea surface temperatures off the Kola Penninsula; and small salmon returns. Only significant results are reported. All other river-variable correlations were found to be nonsignificant.

	Environmental variate				
Site	Total Newfoundland ice cover	Station 27 temperature anomalies (0–175 m)	CIL at various transects	Annual Kola sea temperature (0–50 m)	Small salmon returns
Miramichi	0.37	0.40	0.46		
de la Trinité			0.46		
Conne			0.46		
Terra Nova					0.20
Teno				0.39	0.43

abundance data resulted in significant relationships for Terra Nova and Teno.

Distribution plots and tests for normality used to assess year to year feeding and trophic stability indicated mean annual shifts in both $\delta^{13}C$ and $\delta^{15}N$ ranging from –0.80‰ to 0.88‰, with means of 0.01‰ and 0.00‰, respectively (Figure 4). The distributions of annual changes in $\delta^{13}C$ and $\delta^{15}N$ signatures were both normal (Shapiro-Wilk W $P \geq 0.49$).

Discussion

Stable isotope signatures of $\delta^{13}C$ and $\delta^{15}N$ of 1SW salmon varied among rivers and among years within rivers and showed no consistency in pattern over time. As analyses were carried out on maiden 1SW salmon, they were not confounded by the inclusion of multiple sea-age-classes and variability in growth and reproductive strategies, as observed in some other studies (Berglund et al. 2001). Despite statistical differences among rivers, isotope signatures of salmon from Canadian rivers in the northwest Atlantic were generally similar to those from one salmon stock (Teno) from northern Europe. Similarities may relate to the utilization of a wide spectrum of related prey species and perhaps the broad spatial extent over which similar life history sea-age groups of salmon forage while at sea. Thus, results are consistent with reports concerning the mixing of stocks during marine feeding in areas such as the Labrador Sea (Reddin and Short 1991) and the Grand Banks (Reddin 1985).

Relationships between trophic status ($\delta^{15}N$), or energy source ($\delta^{13}C$), and salmon abundance were generally lacking. Furthermore, variations in stable isotope signatures did not vary consistently with any of the examined environmental indicators in a manner that would suggest the pervasive influence of any single environmental parameter on salmon survival. The $\delta^{15}N$ stable isotope ratios of primary producers at the base of food webs are known to vary spatially and to influence organisms at higher trophic levels (Vander Zanden and Rasmussen 1999; Jennings and Warr 2003). For example, in the North Sea, 51% and 77%, respectively, of the spatial variation in the $\delta^{15}N$ of demersal dab Limanda limanda and benthopelagic European whiting Merlangius merlangus was attributed to differences in $\delta^{15}N$ at the base of the food web and correlated to spatial differences in salinity and temperature (Jennings and Warr 2003). Estuarine and coastal variations in salinity and temperature will, therefore, have significant implications for base $\delta^{15}N$ and the trophic signatures of organisms feeding in inshore areas, as salmon smolts may do for some period of time before progressing to the open ocean (e.g., Dutil and Coutu 1988). In offshore regions, salinity and temperature evidence little variation over large areas, with the result that spatial comparisons of trophic position are subject to little bias (Jennings and Warr 2003). Observed variation in $\delta^{15}N$ among rivers, therefore, is likely influenced by the unique migratory and feeding experiences of each population, with gut contents of migrating smolts known to change as they move along salinity-temperature

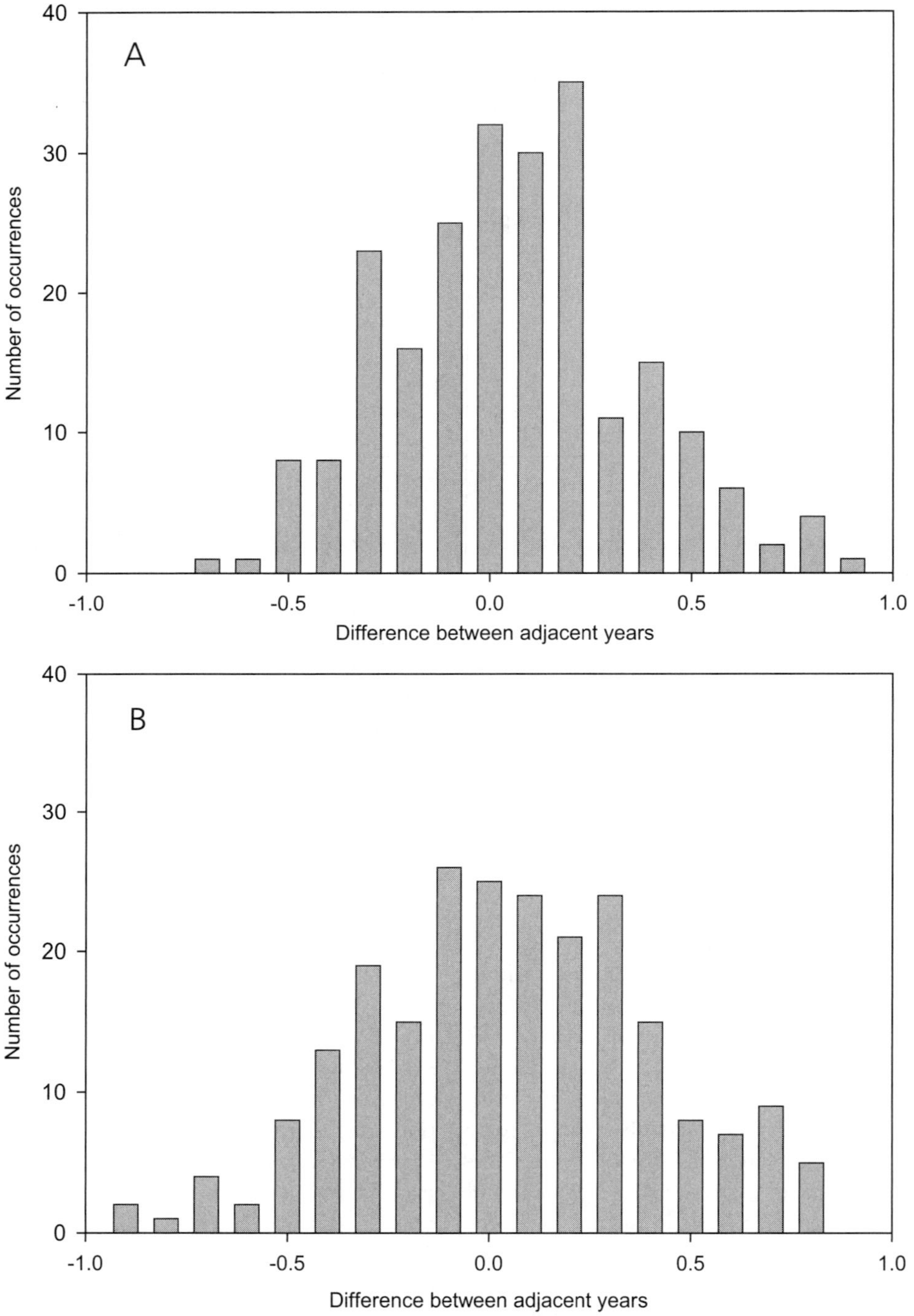

Figure 4.—Distribution plots indicating the spread of annual differences in (A) $\delta^{13}C$ and (B) $\delta^{15}N$ for all sites.

gradients from freshwater to coastal areas (Andreassen et al. 2001)

Comparisons of mean annual trends in $\delta^{13}C$ across all sites revealed a combination of significant (both positive and negative) trends also suggestive of river or stock-specific factors possibly related to differences in early marine feeding but no evidence of a larger spatial effect. Similarly, the lack of consistency among annual trends in nitrogen stable isotope signatures indicated that, in general, the mean trophic level of Atlantic salmon populations has not changed in recent decades. In particular, based on our inferences from analyses of isotopic signatures of scale samples of Atlantic salmon, trophic levels have not declined over time as might have been expected in light of documented fishing impacts on marine food webs (Pauly et al. 1998).

The lack of consistent correlations between carbon stable isotope signatures and the available environmental data suggests that environmental conditions have not affected the type or relative amounts of food sources available to the individual survivors included in our analysis. In contrast, other studies have often reported direct linkages between climate conditions and variations in salmon abundance or survival (e.g., Friedland 1998; Jonsson and Jonsson 2004; Condron et al. 2005; Beamish et al. 2009, this volume) rather than indirect linkages by way of trophic associations. Additionally, the low number of significant associations between $\delta^{15}N$ and environmental variables provided no convincing evidence that environmental conditions in the marine environment have affected the trophic level at which Atlantic salmon feed, at least 1SW salmon. Among Pacific salmon species hypothesized climate-induced changes in prey resource selection within or among trophic levels were similarly not reflected in pervasive year-to-year changes in measured stable isotopes (Kaeriyama et al. 2004). For Atlantic salmon, however, any effect of large-scale environmental change may have been dampened by variations in intra- and interspecific competition for prey. Likewise, the lack of significant correlations between $\delta^{13}C$ and associated abundance data suggest that minor variations in prey consumption patterns reflected in $\delta^{13}C$ signatures have not affected the eventual success and return rate of Atlantic salmon to their native rivers. The few significant correlations between $\delta^{15}N$ and abundance data that did exist provide little evidence of an association between trophic level and annual return rates. The observed stability and normality of year-to-year changes in trophic position and prey sourcing among all stocks suggests a level of compensation that allows successful individuals to maintain feeding patterns and trophic position regardless of climate variability or changes in fishing pressure.

Omnivorous predators in marine ecosystems (e.g., salmon) typically have broad diets and demonstrate an ability to rapidly switch feeding to more abundant prey items, often at different trophic levels, as they become available (Sissenwine et al. 1982; Overholtz et al. 2000; Kaeriyama et al. 2004). Indeed, Atlantic salmon are known to be opportunistic feeders targeting a wide array of invertebrate and fish prey species (Lear 1972; Jacobsen and Hansen 2001) such that Jacobsen and Hansen (2000) report that salmon abundance is unlikely to be affected by annual changes in the availability of any specific prey. Omnivory and opportunistic feeding should promote variability in resultant trophic signatures to the extent that variation in prey availability occurred, as has been documented for Pacific salmon where high plasticity in feeding strategies has been associated with dietary shifts in response to climate-induced (El Niño and La Niña) changes in ocean conditions (Kaeriyama et al. 2004).

Marine ecosystems in general are believed to be more open and contain fewer feeding specialists than their freshwater counterparts, with the result that marine food webs are more highly connected and species interactions are weak (Link 2002). Organisms within marine food webs, therefore, should typically display the feeding plasticity observed among Pacific salmon, unless the environmental changes are accompanied by increased food niche overlap that promotes more intense intra- and interspecific feeding competition. For example, variations in prey abundance induced by environmental changes may be quickly eliminated by the cumulative effect of opportunistic feeding by many species, yielding only minor changes in patterns of resource use by any single species. Unless persistent over time, such changes would manifest themselves as minor year-to-year fluctuations in stable isotope signatures, more or less about zero as was found here.

Changes in competitive conditions may be particularly applicable to the case of Atlantic salmon and contribute to the lack of significant results found here. Fish in this study consisted of maiden 1SW salmon that were sampled upon return to their native rivers. As a result, only survivors were sampled. By analyzing only survivors, this study is limited to those individuals that were able to compete and feed successfully, and it ignores those unsuccessful individuals that may have been influenced more heavily by changes in the marine ecosystem. It also ignores those stocks or components of stocks categorized by larger 2SW and 3SW fish. Larger 3SW salmon have been reported to be more piscivorous and feed on larger fishes than 1SW and 2SW salmon (Jacobsen and Hansen 2000) and consequently may be more likely to demonstrate changes in trophic feeding over time. Nonetheless, as with all analyses of changes in salmon abundance based on returning fish, stable isotope analysis cannot adequately account for critical changes in feeding opportunity or environmental risk that may ultimately be reflected in abundance. Thus our results would suggest that those salmon that survive to return to the river function trophically as they have always functioned over the period for which we have data. Changes in abundance not connected to increased fishing mortality, when coupled with our results from isotopic analyses, may suggest differences in individual feeding opportunities that might result in decreased ration and increased mortality for some individuals. Such a response would be consistent with documented fishing-induced changes in the food web (i.e., the fishing down the food web hypothesis), at least as it applies to prey species targeted by salmon, and depends on the weak connectivity between species that promotes omnivory in marine food webs and the intense intraspecific competition among conspecifics for available food resources.

Although the data presented here cannot infer the nature of competition among conspecifics that may lead to reduced abundance, the inconsistency in trends seen across all sites and the differences in mean signatures among sites suggests that local factors, particularly the early migration phase, may play a more significant role in the overall success of salmon populations. Signatures for many of the rivers included in this study were distinctive, with de la Trinité being notably $\delta^{15}N$ enriched by comparison with all other sites and the River Teno being notably $\delta^{13}C$ depleted by comparison to all other sites. Even within insular Newfoundland, there was a separation among rivers in $\delta^{15}N$ of about 1‰, a separation that may result from spatial variability in dissolved nitrate concentration (Coote and Yeats 1979). Most records indicate that time spent in nearshore environments is relatively short-lived but can be site-specific (e.g., Rikardsen et al. 2004). Nevertheless, Dutil and Coutu (1988) caught many postsmolts in the northern Gulf of St. Lawrence from August through October, suggesting that the role of this phase of the salmon life cycle on eventual survival may have been previously underestimated.

Acknowledgments

The authors would like to thank Neila Cochrane for graphical assistance, Dave Reddin and Eugene Colbourne for access to environmental data sets, and Geoff Veinott for commenting on an earlier version of the manuscript. Financial support was provided by an International Governance of High Seas Fisheries Canadian Science Program grant to JBD and MP and an NSERC Discovery Grant to MP.

References

Andreassen, P. M. R., M. B. Martinussen, N. A. Hvidsten, and S. O. Stefansson. 2001. Feeding and prey-selection of wild Atlantic salmon post-smolts. Journal of Fish Biology 58:1667–1679.

Armstrong, J. D., J. W. A. Grant, H. L. Forsgren, K. D. Fausch, R. M. DeGraaf, I. A. Fleming, T. D. Prowse, and I. J. Schlosser. 1998. The application of science to the management of Atlantic salmon (*Salmo salar*): integration across scales. Canadian Journal of Fisheries and Aquatic Sciences 55:303–311.

Attril, M. J., and M. Power. 2002. Climatic influences on a marine fish assemblage. Nature (London) 417:275–278.

Beamish, R. J., R. M. Sweeting and C. M. Neville. 2009. Changing climate and the need to change our thinking about the management of Pacific salmon. Pages 155–173 *in* A. J. Haro, K. L. Smith, R. A. Rulifson, C. M. Moffitt, R. J. Klauda, M. J. Dadswell, R. A. Cunjak, J. E. Cooper, K. L. Beal, and T. S. Avery, editors. Challenges for diadromous fishes in a dynamic global environment. American Fisheries Society, Symposium 69, Bethesda, Maryland.

Berglund, O., P. Larsson, and D. Broman. 2001. Organochlorine accumulation and stable isotope ratios in an Atlantic salmon (*Salmo salar*) population from the Baltic Sea. The Science of the Total Environment 281:141–151.

Bowman J. E. 2005. The use of stable isotopes to elucidate feeding relationships in key fish species in the

Bay of Quinte and Oneida Lake. Master's thesis. University of Waterloo, Ontario.

Brocklebank, J. C., and D. A. Dickey. 2003. SAS for forecasting time series. 2nd edition. SAS Institute, Cary, North Carolina.

Caron, F., D. Fournier, and V. Cauchon. 2007. Travaux de recherche sur le saumon des rivières Saint-Jean et de la Trinité en 2006. Ministère des Ressources naturelles et de la Faune, Direction de la recherché sur la faune, Jonquière, Québec.

Chaput, G., P. Cameron, D. Moore, D. Cairns, and P. LeBlanc. 2006. Stock status of Atlantic salmon (*Salmo salar* L.) from rivers of the Gulf Region, SFA 15 to 18. Canadian Science Advisory Secretariat, CSAS Research Document 2006/023, Ottawa.

Chaput, G., C. M. Legault, D. G. Reddin, F. Caron, and P. G. Amiro. 2005. Provision of catch advice taking account of non-stationarity in productivity of Atlantic salmon (*Salmo salar* L.) in the northwest Atlantic. ICES Journal of Marine Science 62:131–143.

Colbourne, E. B. 2004. Decadal changes in the ocean climate in Newfoundland and Labrador waters from the 1950s to the 1990s. Journal of the Northwest Atlantic Fishery Science 34:43–61.

Colbourne, E. B., and J. T. Anderson. 2003. Biological response in a changing ocean environment in Newfoundland waters during the latter decades of the 1990s. ICES Marine Science Symposia 219:169–181.

Colbourne, E. B., E. G. Dawe, D. G. Parsons, E. F. Murphy, W. R. Bowering, E. L. Dalley, J. T. Anderson, J. B. Dempson, D. Orr, D. E. Stansbury, and S. Epstein. 2002. A preliminary review of environmental-stock relationships for some species of marine organisms in NAFO waters of the northwest Atlantic. Northwest Atlantic Fisheries Organization, NAFO SCR Document 02/34, Dartmouth, Nova Scotia.

Colbourne, E., S. Narayanan, and S. Prinsenberg. 1994. Climatic changes and environmental conditions in the northwest Atlantic, 1970–1993. ICES Marine Science Symposia 198:311–322.

Colbourne, E., B. de Young, S. Narayanan, and J. Helbig. 1997. Comparison of hydrography and circulation on the Newfoundland shelf during 1990–1993 with the long-term mean. Canadian Journal of Fisheries and Aquatic Sciences 54(Supplement 1):68–80.

Condron, A., R. DeConto, R. S. Bradley, and F. Juanes. 2005. Multidecadal North Atlantic climate variability and its effect on North American salmon abundance. Geophysical Research Letters 32 L 23703:1–4.

Coote, A. R., and P. A. Yeats. 1979. Distribution of nutrients in the Gulf of St. Lawrence. Journal of the Fisheries Research Board of Canada 36:122–131.

Craig, H. 1957. Isotopic standards for carbon and oxygen and correction factors for mass-spectrometric analysis of carbon dioxide. Geochimica et Cosmochimica Acta 12(1–2):133–149.

Crozier, W. W., and G. J. A. Kennedy. 1994. Marine exploitation of Atlantic salmon (*Salmo salar* L.) from the River Bush, Northern Ireland. Fisheries Research 19:141–155.

Crozier, W. W., and G. J. A. Kennedy. 1999. Relationships between marine growth and marine survival of one sea winter Atlantic salmon, *Salmo salar* L., from the River Bush, Northern Ireland. Fisheries Management and Ecology 6:89–96.

Crozier, W. W., P-J. Schön, G. Chaput, E. C. E. Potter, N. Ó Maoiléidigh, and J. C. MacLean. 2004. Managing Atlantic salmon (*Salmo salar* L.) in the mixed stock environment: challenges and considerations. ICES Journal of Marine Science 61:1344–1358.

Dempson, J. B., D. G. Reddin, M. F. O'Connell, J. Helbig, C. E. Bourgeois, C. Mullins, T. R. Porter, G. Lilly, J. Carcadden, G. B. Stenson, and D. Kulka. 1998. Spatial and temporal variation in Atlantic salmon in the Newfoundland–Labrador region with emphasis on factors that may have contributed to low returns in 1997. Canadian Stock Assessment Secretariat, CSAS Research Document 98/114, Ottawa.

Dempson, J. B., C. J. Schwarz, D. G. Reddin, M. F. O'Connell, C. C. Mullins, and C. E. Bourgeois. 2001. Estimation of marine exploitation rates on Atlantic salmon (*Salmo salar* L.) stocks in Newfoundland, Canada. ICES Journal of Marine Science 58:331–341.

Dempson, J. B., M. F. O'Connell, and C. J. Schwarz. 2004. Spatial and temporal trends in abundance of Atlantic salmon, *Salmo salar*, in Newfoundland with emphasis on impacts of the 1992 closure of the commercial fishery. Fisheries Management and Ecology 11:387–402.

DeNiro, M. J., and S. Epstein. 1978. Influence of diet on distribution of carbon isotopes in animals. Geochimica et Cosmochimica Acta 42(5):495–506.

Drinkwater, K. F. 2000. Changes in ocean climate and its general effect on fisheries: examples from the northwest Atlantic. Pages 116–136 *in* D. Mills, editor. The ocean life of Atlantic salmon: environmental and biological factors influencing survival. Fishing News Books, Blackwell Scientific Publications, Oxford, UK.

Dutil, J. D., and J. M. Coutu. 1988. Early marine life of Atlantic salmon, *Salmo salar*, postsmolts in the northern Gulf of St. Lawrence. Fishery Bulletin 86(2):197–212.

Estep, M. L. F., and S. Vigg. 1985. Stable carbon and nitrogen isotope tracers of trophic dynamics in natural populations of the fisheries of the Lahontan Lake system, Nevada. Canadian Journal of Fisheries and Aquatic Sciences 42:1712–1719.

Estrada, J. A., M., Lutcavage, and S. R. Thorrold. 2005. Diet and trophic position of Atlantic bluefin tuna (*Thunnus thynnus*) inferred from stable carbon and nitrogen isotope analysis. Marine Biology 147:37.

Friedland, K. D. 1998. Ocean climate influences on critical Atlantic salmon (*Salmo salar*) life history events. Canadian Journal of Fisheries and Aquatic Sciences 55(Supplement):119–130.

Friedland, K. D., L. P. Hansen, and D. A. Dunkley. 1998. Marine temperatures experienced by postsmolts and the survival of Atlantic salmon, *Salmo salar* L., in the North Sea area. Fisheries Oceanography 7:22–34.

Friedland, K. D., L. P. Hansen, D. A. Dunkley, and J. C. MacLean. 2000. Linkage between ocean climate, post-smolt growth and survival of Atlantic salmon (*Salmo salar* L.) in the North Sea area. ICES Journal of Marine Science 57:419–429.

Friedland, K. D., D. G. Reddin, and M. Castonguay. 2003. Ocean thermal conditions in the post-smolt nursery of North American salmon. ICES Journal of Marine Science 60:343–355.

Fry, B., and E. B. Sherr. 1984. $\delta^{13}C$ measurements as indicators of carbon flow in marine and freshwater ecosystems. Contributions in Marine Science 27:13–47.

Grey, J. 2001. Ontongeny and dietary specialisation in brown trout (*Salmo trutta* L.) from Loch Ness, Scotland examined using stable isotopes of carbon and nitrogen. Ecology of Freshwater Fish 10:168–76.

Guiguer, K. R. R. A., J. D. Reist, M. Power, and J. A. Babaluk. 2002. Using stable isotopes to confirm the trophic ecology of Arctic charr morphotypes from Lake Hazen, Nunavut, Canada. Journal of Fish Biology 60:348–362.

Hislop, J. R. G., and R. G. J. Shelton. 1993. Marine predators and prey of Atlantic salmon. Pages 104–118 *in* D. Mills,editor. Salmon in the sea and new enhancement strategies. Fishing News Books, Blackwell Scientific Publications, Oxford, UK.

Hurrell, J. W. 1995. Decadal trends in the North Atlantic Oscillation: regional temperatures and precipitation. Science 269:677–679.

Hutchinson, J. J., and C. N. Trueman. 2006. Scale stable isotope analysis of collagen in fish scales: limitations set by scale architecture. Journal of Fish Biology 69:1874–1880.

Jacobsen, J. A., and L. P. Hansen. 2000. Feeding habitats of Atlantic salmon at different life stages at sea. Pages 170–192 *in* D. Mills, editor. The ocean life of Atlantic salmon: environmental and biological factors influencing survival. Fishing News Books, Blackwell Scientific Publications, Oxford, UK.

Jacobsen, J. A., and L. P. Hansen. 2001. Feeding habits of wild and escaped farmed Atlantic salmon, *Salmo salar* L., in the northeast Atlantic. ICES Journal of Marine Science 58:916–933.

Jennings, S., and K. J. Warr. 2003. Environmental correlates of large-scale spatial variation in the $\delta^{15}N$ of marine animals. Marine Biology 142:1131–1140.

Jonsson, N., and B. Jonsson. 2003. Energy allocation among developmental stages, age groups, and types of Atlantic salmon (*Salmo salar*) spawners. Canadian Journal of Fisheries and Aquatic Sciences 60:506–516.

Jonsson, B., and N. Jonsson. 2004. Factors affecting marine production of Atlantic salmon (*Salmo salar*). Canadian Journal of Fisheries and Aquatic Sciences 61:2369–2383.

Jonsson, N., B. Jonsson, and L. P. Hansen. 1998. The relative role of density-dependent and density-independent survival in the life cycle of Atlantic salmon *Salmo salar*. Journal of Animal Ecology 67:751–762.

Kaeriyama, M., Nakamura, M., Edpalina, J. R., Bower, H. Yamaguchi, R. V. Walker, and K. W. Myers. 2004. Change in feeding ecology and trophic dynamics of Pacific salmon (*Oncorhynchus* spp.) in the central Gulf of Alaska in relation to climate events. Fisheries Oceanography 13:197–207.

Lear, W. H. 1972. Food and feeding of Atlantic salmon in coastal areas and over oceanic depths. ICNAF Research Bulletin 9: 27–39.

Link, J. 2002. Does food web theory work for marine ecosystems? Marine Ecology Progress Series 230:1–9.

Manly, B. F. J. 1991. Randomization and Monte Carlo methods in biology. Chapman and Hall, New York.

Mariotti, A. 1983. Atmospheric nitrogen is a reliable standard for natural N-15 abundance measurements. Nature (London) 303(5919):685–687.

McCarthy, I. D., D. Fraser, S. Waldron, and C. E. Adams. 2004. A stable isotope analysis of trophic polymorphism among Arctic charr from Loch Ericht, Scotland. Journal of Fish Biology 65, 1435–40.

Montevecchi, W. A., D. K. Cairns, and R. A. Myers. 2002. Predation on marine-phase Atlantic salmon (*Salmo salar*) by gannets (*Morus bassanus*) in the northwest Atlantic. Canadian Journal of Fisheries and Aquatic Sciences 59:602–612.

Narayanan, S., J. Carascadden, J. B. Dempson, M. F. O'Connell, S. Prinsenberg, D. G. Reddin, and N. Shackell. 1995. Marine climate off Newfoundland and its influence on salmon (*Salmo salar*) and capelin (*Mallotus villosus*). Pages 461–171 *in* R. J. Beamish, editor. Climate change and northern fish populations. Canadian Special Publication of Fisheries and Aquatic Sciences 121:461–474.

Nelson, D. 2003. Dictionary of mathematics, 3rd edition. Penguin Books., London

Niemelä, E., J. Erkinaro, J. B. Dempson, M. Julkunen, A. Zubchenko, S. Prusov, M. A. Svenning, R. Ingvaldsen, M. Holm, and E. Hassinen. 2004. Temporal synchrony and variation in abundance of Atlantic salmon in two subarctic Barents Sea rivers: influence of oceanic conditions. Canadian Journal of Fisheries and Aquatic Sciences 61:2384–2391.

O'Connell, M. F., J. B. Dempson, D. G. Reddin, C. E. Bourgeois, T. R. Porter, N. M. Cochrane, and D. Caines. 2005. Status of Atlantic salmon (*Salmo salar* L.) stocks of insular Newfoundland (SFAs 3–14A), 2004. Canadian Science Advisory Secretariat, CSAS Research Document 2005/064, Ottawa.

O'Neil, S., J. Ritter, and K. Robichaud-LeBlanc. 2000. Proceedings of a workshop on research strategies into the causes of declining Atlantic salmon returns to North American rivers. Canadian Stock Assessment Secretariat, CSAS Proceedings Series 2000/018, Ottawa.

Overholtz, W. J., J. S. Link, and L. E. Suslowicz. 2000. The impact and implications of fish predation on pelagic fish and squid on the eastern USA shelf. ICES Journal of Marine Science 57:1147–1159.

Parrish, D. L., R. J. Behnke, S. R. Gephard, S. D. McCormick, and G. H. Reeves. 1998. Why aren't there more Atlantic salmon (*Salmo salar*)? Canadian Journal of Fisheries and Aquatic Sciences 55(Supplement 1):281–287. Pauly, D., V. Christensen, J. Dalsgaard, R. Froese, F. Torres Jr. 1998. Fishing down marine food webs. Science 279:860–863.

Pauly, D., and M-L. Palomares. 2005. Fishing down marine food web: it is far more pervasive than we thought. Bulletin of Marine Science 76:197–211.

Pauly, D., M. L. Palomares, R. Froese, P Sa-a, M. Vakily, D. Preikshot, and S. Wallace. 2001. Fishing down Canadian aquatic food webs. Canadian Journal of Fisheries and Aquatic Sciences 58:51–62.

Perga, M. E., and D. Gerdeaux. 2003. Using the δ^{13} Cand δ^{15}N of whitefish scales for retrospective ecological studies: changes in isotope signatures during the restoration of Lake Geneva, 1980–2001. Journal of Fish Biology 63:1197–1207.

Peterson, B. J., and B. Fry. 1987. Stable isotopes in ecosystem studies. Annual Review of Ecology and Systematics 18:293–320.

Power, G. 1987. Scales in the balance. Atlantic Salmon Journal 37:14–17.

Pruell, R. J., B. K. Taplin, and K. Cicchelli. 2003. Stable isotope ratios in archived striped bass scales suggest changes in trophic structure. Fisheries Management and Ecology 10:329–336.

Reddin, D. G. 1985. Atlantic salmon (*Salmo salar*) on and east of the Grand Bank. Journal of the Northwest Atlantic Fisheries Science 6:157–164.

Reddin, D. G., and K. D. Friedland. 1993. Marine environmental factors influencing the movement and survival of Atlantic salmon. Pages 79–103 *in* D. Mills, editor. Salmon in the sea and new enhancement strategies. Fishing News Books, Blackwell Scientific Publications, Oxford, UK.

Reddin, D. G., and P. B. Short. 1991. Postsmolt Atlantic salmon (*Salmo salar*) in the Labrador Sea. Canadian Journal of Fisheries and Aquatic Sciences 48:2–6.

Rikardsen, A. H., M. Haugland, P. A. Bjørn, B. Finstad, R. Knudsen, J. B. Dempson, J. C. Holst, N. A. Hvidsten, and M. Holm. 2004. Geographical differences in marine feeding of Atlantic salmon postsmolts in Norwegian fjords. Journal of Fish Biology 64:1655–1679.

Sinnatamby, R. N., J. E. Bowman, J. B. Dempson, and M. Power. 2007. An assessment of de-calcification procedures for δ^{13}C and δ^{15}N analysis of yellow perch, walleye, and Atlantic salmon scales. Journal of Fish Biology 70:1630–1635.

Sissenwine, M. P., B. E. Brown, J. E. Palmer, R. J. Essig, and W. Smith. 1982. Empirical examination of population interactions for the fishery resources off the northeastern USA. Canadian Special Publication of Fisheries and Aquatic Sciences 59:82–94.

Tereshchenko, V.V. 1996. Seasonal and year-to-year variations of temperature and salinity along the Kola meridian transect. International Council for the Exploration of the Seas, Council Meeting 1996/C:11, Copenhagen, Denmark.

Vander Zanden, M. J., and J. B. Rasmussen. 1999. Primary consumer delta δ^{13}C and δ^{15}N and the trophic position of aquatic consumers. Ecology 80:1395–1404.

Vander Zanden, M. J., J. M. Casselman, and J. B. Rasmussen. 1999. Stable isotope evidence for the food web consequences of species invasions in lakes. Nature (London) 401:464–467.

Wainright, South Carolina, M. J. Fogarty, R. C. Greenfield, and B. Fry. 1993. Long-term changes in the Georges Bank food web: trends in stable isotopic compositions of fish scales. Marine Biology 115:481–493.

Walters, C., and B. Ward. 1998. Is solar radiation responsible for declines in marine survival rates of anadromous salmonids that rear in small streams? Canadian Journal of Fisheries and Aquatic Sciences 55:2533–2538.

Zar, J. H. 1999. Biostatistical analysis. Prentice-Hall, Upper Saddle River, New Jersey.

American Fisheries Society Symposium 69:465–483, 2009

A Preliminary Investigation of the Effects of Anadromous Arctic Char on Food Web Structure and Nutrient Transport in Coastal Arctic Lakes

Heidi K. Swanson* and Karen A. Kidd
Canadian Rivers Institute, Department of Biology, University of New Brunswick
Saint John, New Brunswick E2L 4L5, Canada

Abstract.—While there has been much recent research conducted on the effects of semelparous anadromous (sea-run) fishes on nutrient cycling and productivity in freshwater lakes, we know very little about how iteroparous anadromous fishes may affect these processes and ecosystems. Although iteroparous sea-run fishes do not contribute mass postspawning carcass additions to freshwater systems, nutrients may still be transported from marine to freshwater environments through deposition of eggs, metabolic losses, and postspawning/overwintering mortality. As well, some iteroparous sea-run species spend significant time growing to juvenile and smolt stages in freshwater where they may affect top-down or bottom-up changes in food web structure. The goals of this study were to (1) conduct a preliminary investigation into the effect of sea-run Arctic char *Salvelinus alpinus* on food web structure in coastal lakes in the central Canadian Arctic, and (2) semiquantitatively assess the direction of net nutrient transport achieved by anadromous Arctic char during annual migrations to the sea. In 2006, fishes and macroinvertebrates were collected from four lakes near Hope Bay, Nunavut, and analyzed for stable C, N, and S isotopes to examine food web structure and anadromy. Preliminary results indicate that the presence of sea-run Arctic char may impact the $\delta^{15}N$-determined trophic position and condition of sympatric populations of lake trout *S. namaycush*. When stable S isotope results were applied to a two-source mixing model, marine food items accounted for 84–100% of the diet of mature, anadromous Arctic char. A semiquantitative assessment of char-mediated nutrient transport between marine and freshwater environments at Nauyuk Lake, Nunavut, indicated that masses of nutrients imported by char to freshwater systems are likely negligible in relation to ambient water chemistry. There was significant among-year variability in the magnitude and direction of nutrient transport, and the complex life history of anadromous Arctic char requires more research before detailed nutrient budgets can be calculated.

Introduction

It is well established that semelparous, anadromous Pacific salmon species transport a significant amount of nutrients from the marine environment to freshwater and riparian environments (e.g., Naiman et al. 2002 and 2009, this volume; Stockner 2003). Marine-derived nutrients from egg deposition, decomposing carcasses, and metabolic activity have been detected with stable isotope analysis in lakes, streams, and riparian zones (e.g., Kline et al. 1993; Reimchen et al. 2003; Bilby et al. 1996) and can influence lake productivity and recruitment of salmon (e.g., Schmidt et al. 1998), growth and biomass of stream invertebrates (Minakawa et al. 2002), and production and diversity of riparian plant communities (Helfield and Naiman 2001). The largest input of marine-derived nutrients from semelparous species to freshwater systems is from decomposing carcasses, and perhaps as a result, effects of iteroparous anadromous fishes on freshwater ecosystems have been largely unstudied.

Iteroparous anadromous fishes differ from their semelparous counterparts in that they may spawn multiple times and do not die en masse immediate-

* Corresponding author: heidikswanson@yahoo.ca

ly after spawning. It is therefore likely that marine transport of nutrients via iteroparous fishes will be relatively harder to detect and that the corresponding effects on freshwater ecosystems will be more subtle. Effects may still be ecologically relevant in systems that are oligotrophic (Nislow et al. 2004), however, because small inputs of marine-derived nutrients could comprise a relatively large portion of the overall nutrient budget.

Transport of marine-derived nutrients to freshwater by iteroparous anadromous fishes has been demonstrated in a few systems and may occur through egg deposition, metabolic breakdown of tissues, and postspawning or overwintering mortality. Durbin et al. (1979) found that migrations of alewife *Alosa pseudoharengus* (iteroparous and anadromous) resulted in a net import of nitrogen to small freshwater spawning habitats. They also found that streams with migratory alewife had higher microbial and invertebrate activity than streams without this species. Jonsson and Jonsson (2003) studied the effect of anadromous, iteroparous Atlantic salmon *Salmo salar* on the nutrient budget of a coastal river in Norway. They found that there was a net import of nutrients via migratory salmon that could account for as much as 17% of the total phosphorus budget of the river, even though there was low carcass deposition. Nislow et al. (2004) modeled the influence of anadromous Atlantic salmon on phosphorus fluxes in an oligotrophic river in Scotland. They found that the sign and magnitude of salmon-derived phosphorus fluxes depended on the number of returning Atlantic salmon from the sea, egg-smolt survival, and the retention of adult-derived P in the river. Finally, Jardine et al. (2009, this volume) found that benthic invertebrates from a stream that supports anadromous blueback herring *A. aestivalis* had $\delta^{13}C$, $\delta^{15}N$, and $\delta^{34}S$ ratios that indicated consumption of marine-derived organic matter.

To the best of our knowledge, no researchers have investigated anadromous Arctic char *Salvelinus alpinus* as a potential transport vector for marine-derived nutrients. Arctic char have the most northerly distribution of any salmonid species, are iteroparous, and have landlocked (or resident) and anadromous life histories that may occur in sympatry. Anadromous Arctic char occupy relatively cold, unproductive freshwater systems until they are 3–8 years old, at which time they make their first migration to sea (Johnson 1980). Summer marine residency lasts ~3–6 weeks each year, and fish usually require several sea migrations to reach sexual maturity at an age of approximately 10 years (Johnson 1980). In lakes in the central Canadian Arctic, fish do not usually migrate in the year they spawn and do not spawn every year once they reach sexual maturity. Once anadromous char begin migrations to sea, they are reliant on the marine environment for the majority of their nutriment and may increase their body weight by more than 50% during a single sea migration (Johnson 1980). Interannual variability in weight gain is high, however, and during a 5-year study at Nauyuk Lake, Nunavut, Canada, annual mean weight increase for migrating Arctic char varied from 12% to 42% (Gyselman 1984).

Arctic char may directly transfer nutrients obtained during sea migrations to freshwater systems through excretion of metabolic products, carcass deposition from postspawning and overwintering mortality, predation on first- and second-year migrants by sympatric species such as lake trout *Salvelinus namaycush*, and predation on/mortality of deposited eggs. Conversely, nutrients may leave the freshwater system via mortality (largely smolts) at sea and migrant straying. At Nauyuk Lake, Gyselman (1994) reported that fidelity to the natal freshwater system was size-dependent and often low. Using data from a detailed 5-year study, he estimated that ~30–50% of downstream migrants never returned, ~5% spent 1 year away from the system, and ~2% of migrants spent 2 years away from the system. These losses were apparently balanced, however, by immigration from other systems (Gyselman 1994).

In addition to functioning as a direct nutrient transport vector between marine and freshwater systems, anadromous populations of Arctic char may indirectly impact their freshwater habitats during presmolt freshwater residency periods. Little is currently known about how presmolt Arctic char affect the structure and function of lake food webs in the years before their first migration, but it appears that Arctic char can use all habitats and depth zones in lakes (Johnson 1980; Klemetsen et al. 2003) and that their presence may affect top-down or bottom-up changes in food web structure because they are both predator and prey. Also, lake trout and Arctic char are more likely to coexist at comparable densities when the population of Arctic char is anadromous

(Johnson 1980). Although sexually mature anadromous Arctic char do not usually feed in freshwater even if they are resting from spawning, they may require several sea migrations to reach sexual maturity, and until this happens, immature fish may feed on zooplankton, chironomid pupae, and surface insects after their return from the sea each fall (Rikardsen et al. 2003). In freshwater lakes in Norway, anadromous Arctic char have been shown to significantly alter the size and species composition of zooplankton communities because they preferentially feed on large-bodied cladocerans (Langeland 1982; Klemetsen et al. 2003). Based on current understanding, therefore, it is likely that anadromous Arctic char impact food web structure in the freshwater lakes that support them.

Stable carbon (C), sulfur (S), and nitrogen (N) isotope ratios can be used as continuous, time-integrated indicators of food source (C and S) and trophic position (N) in food webs (Minagawa and Wada 1984; Peterson and Fry 1987; Post 2002). Delta^{15}N (δ^{15}N) ratios increase ~3.4‰ per trophic transfer because of fractionation processes during protein synthesis, assimilation, and excretion (Post 2002). Delta^{13}C (δ^{13}C) and delta^{34}S (δ^{34}S) ratios do not change with trophic transfer and can be used to differentiate between littoral and pelagic carbon sources (C) (France 1995) and between freshwater and marine food sources (S) (Peterson and Fry 1987; Doucett et al. 1999).

The goals of this project were to (1) compare food web structure (using stable C and N isotope ratios) among lakes in the central Canadian Arctic that do and do not support anadromous Arctic char populations, (2) determine the importance of marine and freshwater food sources for sea-run Arctic char using a two-source mixing model with stable S isotope ratios as a tracer, and (3) semiquantitatively assess the direction of net nutrient transport achieved by anadromous Arctic char during annual return migrations to the sea using historical data from the Nauyuk Lake system, Nunavut.

Methods

Determination of Food Web Structure and Confirmation of Anadromy

In summer 2006, whole food web collections were made on four lakes located near Hope Bay, Nunavut (68°N/107°W; Figure 1). Our intent was to compare the food web structure of two lakes that contain sea-run Arctic char populations (Roberts and Glenn Lakes, connected to the sea by approximately 2.5 km of stream) to two lakes that do not (Patch Lake and Doris Lake; barriers to sea migration). Despite considerable effort and previous fisheries records indicating their presence (Golder Associates Ltd. 2002), however, no Arctic char were captured in Glenn Lake; Roberts Lake was the only system where Arctic char were captured. All lakes have populations of lake trout, lake whitefish *Coregonus clupeaformis*, cisco *C. artedi* and least cisco *C. sardinella*, and ninespine stickleback *Pungitius pungitius*. Glenn Lake also has a population of fourhorn sculpin *Myxocephalus quadricornis*. The lakes have similar morphometry and limnology (Miramar Hope Bay Ltd. 2005).

Invertebrates (opossum shrimp *Mysis relicta* [zooplanktivore], *Gammarus lacustris* [amphipod; scavenger], *Saduria entomon* [aquatic sow bug; predator on other invertebrates], and Chironomidae [midge larvae; detritivore]) were collected from each lake with an Eckman grab and kick nets. *Saduria entomon* and an amphipod species were also collected from Roberts Bay, the marine environment that Arctic char first enter when making sea migrations from the study lakes (Golder Associates Ltd. 2005). Approximately 10 g wet weight was collected for each taxon from each lake. Invertebrate samples were sorted to major taxon and frozen whole in lake water.

For fish, 8–10 individuals/species/lake were collected with minnow traps, a Smith-Root Type 12 backpack electrofisher, and sinking gill nets (mesh sizes ranging from 1.9 to 8.9 cm stretched mesh). All fish were measured and weighed and all fish except ninespine stickleback were dissected for dorsal muscle tissue; ninespine stickleback were frozen whole. Fish tissue was frozen immediately after processing and subsequently freeze-dried and ground to a fine powder along with the invertebrate samples. Subsamples of this powder were then analyzed for stable C and N isotope ratios. Amphipods and isopods collected from Roberts Bay and Roberts Lake (n = 3 for each species-location combination), Arctic char from Roberts Lake (n = 13), and a subset of lake trout from Doris, Glenn, and Patch lakes were also analyzed for stable S isotope ratios.

Analyses of stable C and N isotopes were performed on a Finnigan Mat Delta Plus continuous-

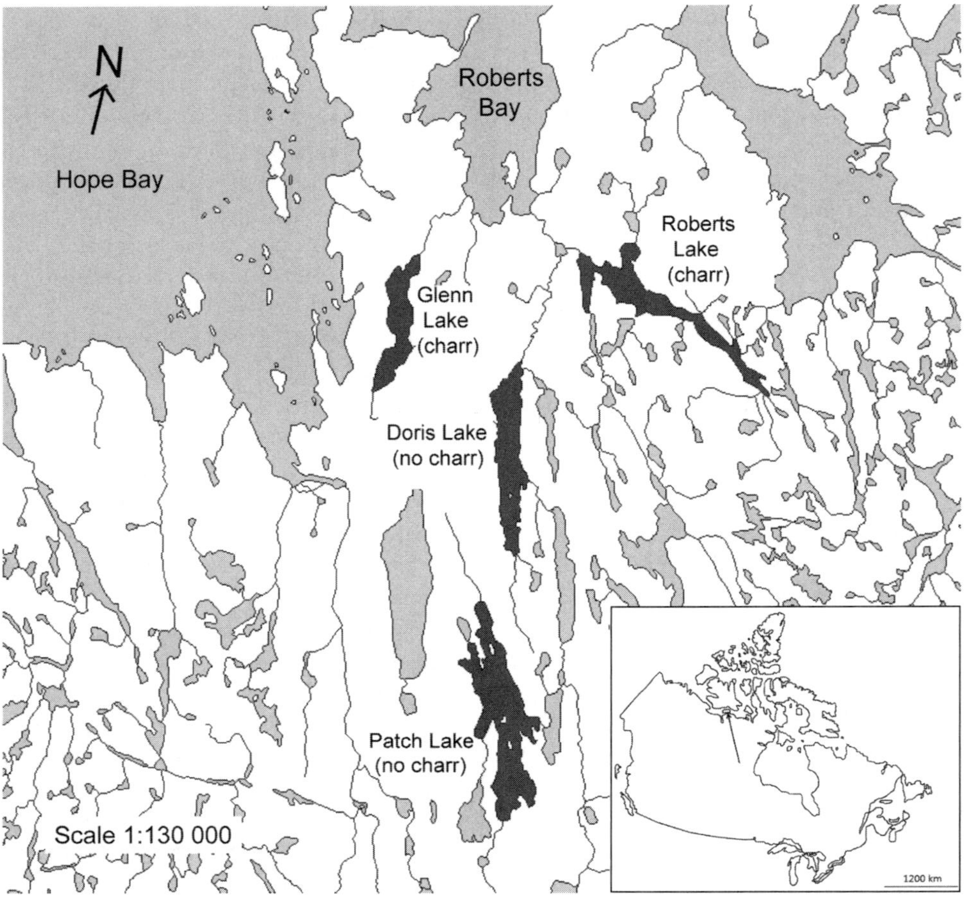

Figure 1.—Map of the Hope Bay study area.

flow isotope-ratio mass spectrometer connected to a Thermoquest NC2500 elemental analyzer (EA-CFIRMS) at the Stable Isotopes in Nature Laboratory at the University of New Brunswick. Carbon and nitrogen content were analyzed simultaneously with the elemental analyzer and converted to mole ratios (C:N). Analyses of stable S isotopes were performed at the Colorado Plateau Stable Isotope Laboratory at Northern Arizona University with a continuous-flow Costech Analytical ESC4010 elemental analyzer interfaced to a Thermo-Electron Delta Plus Advantage IRMS. Stable N, C, and S isotope ratios are expressed as parts per mil (‰) delta values ($\delta^{15}N$, $\delta^{13}C$, $\delta^{34}S$) from an international standard. International standards used for C, N, and S were Pee-Dee Belemnite, N_2 gas, and Vienna Canyon Diablo Troilite, respectively. Delta values were calculated using the following formula:

$$\delta^{15}N \text{ or } \delta^{13}C \text{ or } \delta^{34}S = [(R_{sample} - R_{standard})/R_{standard}] \times 1{,}000, \quad (1)$$

where $R = {}^{15}N/{}^{14}N$ or ${}^{13}C/{}^{12}C$ or ${}^{34}S/{}^{32}S$. At the Stable Isotopes in Nature Laboratory, isotopic, elemental, and internal standards were analyzed between every 15–20 samples for a total of 74 samples and 22 standards per run. Replicate analyses of internal laboratory standards (e.g., NIST 1577b, bovine liver) typically yielded standard deviations of ±0.15 ‰ for $\delta^{15}N$ and $\delta^{13}C$. The mean difference ± SD between duplicate subsamples was 0.19 ± 0.1‰ (n = 18) for $\delta^{15}N$ and 0.13 ± 0.08‰ (n = 18) for $\delta^{13}C$. At the Colorado Plateau Stable Isotope Laboratory, a total of 20 isotopic, elemental, and internal standards were analyzed per run. Replicate analyses of internal laboratory standards typically yielded standard deviations of

±0.3‰ or better, and the mean difference ± SD between duplicate subsamples was 0.25 ± 0.55‰ (n = 6).

Data Analyses

To correct for among-lake differences in $\delta^{15}N$ at the base of the food web (Cabana and Rasmussen 1996; Vander Zanden and Rasmussen 1999; Post 2002), we used a modified version of the model proposed by Cabana and Rasmussen (1996). Because there are no long-lived primary consumers in the study lakes, we assumed that dietary habits of the three main macroinvertebrates *S. entomon*, *G. lacustris*, and opossum shrimp were similar across systems and compared their $\delta^{15}N$ among lakes with an analysis of variance (ANOVA). Within species, among-lake differences in invertebrate $\delta^{15}N$ were less than 1.8‰ and the only species that displayed a significant difference in $\delta^{15}N$ among lakes was opossum shrimp (Figure 2; ANOVA, F = 11.82, df = 3,8, P = 0.0026). Lake-specific mean $\delta^{15}N$ for opossum shrimp was subtracted from fish $\delta^{15}N$ to create an adjusted $\delta^{15}N$ value for all resident and presmolt fish:

$$\text{Adjusted } \delta^{15}N\ (\delta^{15}N_{adj}) = \delta^{15}N_{fish} - \delta^{15}N_{baseline} \quad (2)$$

where $\delta^{15}N_{baseline}$ was lake-specific means for $\delta^{15}N$ in opossum shrimp. For sea-run fish, $\delta^{15}N_{baseline}$ was mean $\delta^{15}N$ calculated for all invertebrates collected from Roberts Bay.

Variation in lipid content among organisms has the potential to confound analyses of $\delta^{13}C$ because lipids are relatively depleted in ^{13}C (Post et al. 2007). We found that fish $\delta^{13}C$ was significantly related to C:N ratios (an indicator of lipid content, Post et al. 2007), even after the effects of lake and species were accounted for (general linear model, F = 7.17, df = 1,158, P = 0.0082). We therefore applied the normalization for aquatic organisms that Post et al. (2007) developed:

$$\delta^{13}C_{adjusted}\ (\delta_{adj}{}^{13}C) = \delta^{13}C - 3.32 + 0.99 \times C{:}N. \quad (3)$$

After applying this normalization, $\delta^{13}C$ ratios were not related to C:N ratios (GLM, F = 0.32, df = 1,158, P = 0.5713).

For the two lakes where both least cisco and lake cisco were caught, $\delta^{15}N_{adj}$, $\delta^{13}C_{adj}$, and fork

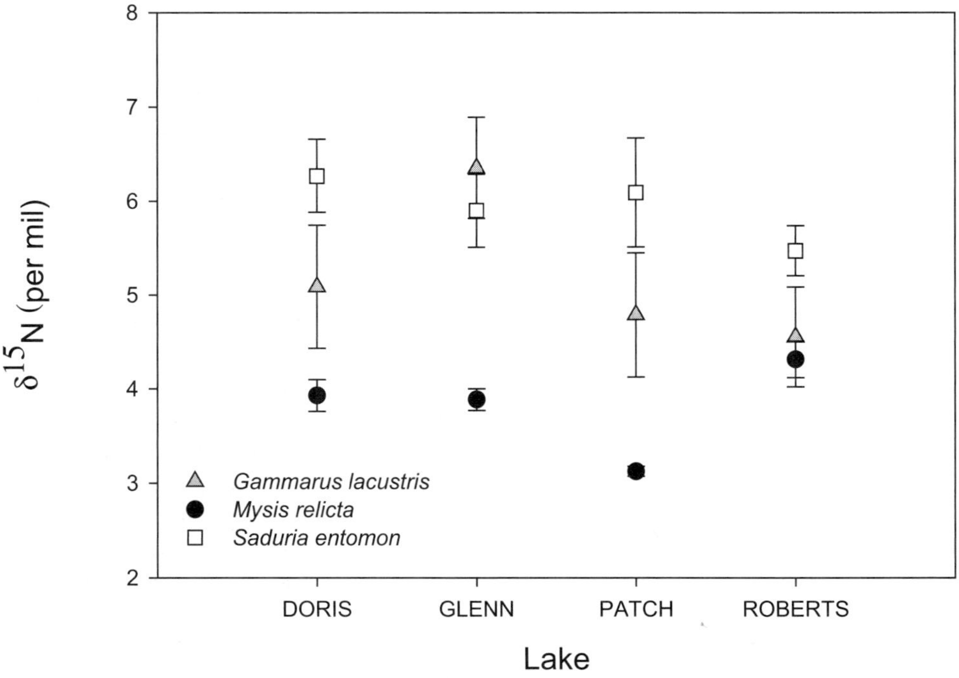

Figure 2.—Mean $\delta^{15}N$ ± SE for primary consumers. The only significant among-lake difference was for opossum shrimp (analysis of variance, F = 11.82, df = 3,8, P = 0.0026).

length were compared between the two species. No significant differences were found (t-test, $t < 4.24$, df = 1,8; 1,5, $P > 0.095$), and for all other analyses the two species were pooled.

After baseline and lipid correction, intraspecific $\delta^{15}N_{adj}$ and $\delta^{13}C_{adj}$ were compared among lakes with an analysis of covariance (ANCOVA) (SAS Institute Inc.); fish length was used as a covariate. Least-square mean (LSmean) $\delta^{15}N_{adj}$ and $\delta^{13}C_{adj}$ were generated at a specified body size for each species. These sizes represented the approximate modal length captured and were within the size range of captured individuals from each lake (650 mm for Arctic char and lake trout, 400 mm for lake whitefish, 270 mm for cisco, and 50 mm for ninespine stickleback). A posthoc Dunnett's test was used to identify differences in LSmeans among pairs of lakes, with Roberts Lake being set as the control (only lake with sea-run Arctic char). Adjusted $\delta^{15}N$, adjusted $\delta^{13}C$, and $\delta^{34}S$ were compared between anadromous and nonanadromous Arctic char with an ANCOVA or t-test, depending on whether fork length was a significant covariate in the model. C:N ratios were compared between anadromous and nonanadromous Arctic char with a t-test.

To facilitate comparisons with previous reports of Arctic char (e.g., Johnson 1989), condition factors were calculated for each species in each lake as

$$\text{Condition factor} = W \times 100/L^3, \quad (4)$$

where W = wet weight (g) and L = fork length (cm). Condition factors and C:N ratios for lake trout were compared among lakes with an ANOVA and a posthoc Dunnett's test was used to identify differences between Roberts Lake (control) and other lakes.

The contribution of marine and freshwater food sources to the diet of migratory and anadromous fish was quantified by using stable S isotope ratios in a two-source linear mixing model. Source proportions and 95% confidence intervals were estimated with the IsoError model developed by Phillips and Gregg (2001), which accounts for variability in the isotopic ratios of the sources as well as that of the mixture. *Saduria entomon* samples collected from Roberts Lake (freshwater environment; $\delta^{34}S = 3.38 \pm 0.36$) and Roberts Bay (marine environment; $\delta^{34}S = 18.33 \pm 0.23$) were used as the endpoints for this analysis.

Semiquantitative Estimates of Net Nutrient Movement in the Nauyuk Lake System, Nunavut

In the 1970s and 1980s, the Department of Fisheries and Oceans conducted an extensive study of ecology and life history characteristics of anadromous Arctic char at Nauyuk Lake, Nunavut (68°22′/107°35′) (Johnson 1989). Nauyuk Lake is approximately 27 km², has a maximum depth of 50 m, and is connected to Parry Bay by a 200-m stream. Immature and resting anadromous Arctic char migrate downstream to Parry Bay in June and July of every year and return upstream to Nauyuk Lake 3–6 weeks later (Johnson 1989). Mature, prespawning individuals do not migrate to sea in the year that they spawn, but move upstream from Nauyuk Lake to Willow Lake, where they spawn and overwinter (Figure 3). This results in spatial segregation of spawning and nonspawning individuals and means that prespawning fish spend at least 2 winters in freshwater. Willow Lake is a small (~2.5 km²), relatively shallow (maximum depth = 12 m) lake that is connected to Nauyuk Lake by a short stream and is reported to provide the only known spawning and rearing habitat for anadromous Arctic char in the Nauyuk Lake system (Johnson 1989).

A two-way counting fence operated from 1975 to 1980 on the stream that connects Willow Lake to Nauyuk Lake and from 1974 to 1979 on the stream that connects Nauyuk Lake to Parry Bay (Johnson 1980). From this study, biomass of pre- and postspawning fish moving in and out of Willow Lake and female gonad weight, postspawning mortality, and overwintering mass loss (Johnson 1980; Johnson 1989) were estimated. As well, increases in summer standing stock through sea migrations and overwintering biomass were estimated in Nauyuk Lake (Johnson 1989). Using this information, we estimated annual movement of nitrogen, phosphorus, and carbon into and out of Willow Lake. Because many of the estimates we used were published without an indication of error and there were many acknowledged difficulties with methods (Johnson 1989), our estimates of nutrient movement should be interpreted as semiquantitative. They are meant to provide an indication of the likely direction and order of magnitude of nutrient movement between the marine and freshwater systems. As well, this exercise is meant

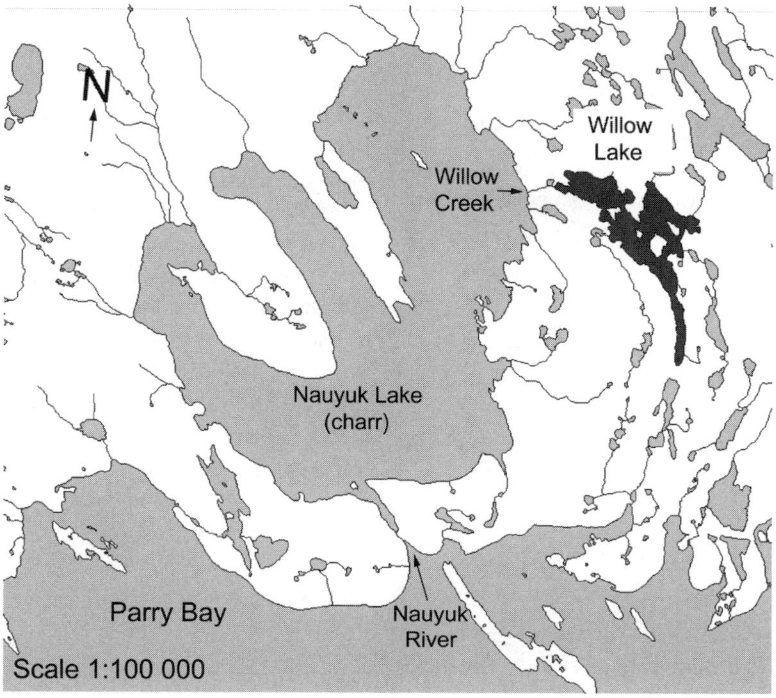

Figure 3.—Map of the Nauyuk Lake system.

to highlight areas where future research and refinements are needed.

In Willow Lake, we assumed that marine nutrients move into the system in the bodies of prespawners, are transferred into the lake via postspawning/overwintering mortality and egg deposition, and leave the system via out-migrating smolts:

$$X_{net} = [X_{adults\ (postspawning\ mortality)} + X_{eggs}] - X_{smolts} \quad \text{(Nislow et al. 2004)}, \qquad (5)$$

where X is an estimate of nitrogen (g), phosphorus (g), or carbon (g) that is derived by multiplying biomass (g) by concentration of the variable of interest (e.g., mg/g). We are conservatively omitting nutrients that may enter the lake through overwintering metabolic losses. Because mortality rates are sex-specific, nutrient inputs provided by postspawning and overwintering mortality (X_{adults}) were calculated separately for males and females and then summed (Johnson 1989). All nutrients entering Willow Lake due to adult mortalities were conservatively estimated to occur after spawning and overwintering. Annual inputs due to postspawning/overwintering mortality in females were calculated as

$X_{females\ (postspawning\ mortality)}$ = biomass of upstream-moving females in a given year (kg; calculated from Johnson 1989) – mass loss due to overwintering and spawning (kg; Johnson 1989) × annual mortality rate (Johnson 1989) × proportion of nutrient (nitrogen, phosphorus, or carbon; Lyle and Elliott 1998). (6)

Annual nutrient inputs from egg deposition by spawning females were calculated as

$X_{egg\ deposition}$ = biomass of upstream-moving females in a given year > 600 mm (kg; calculated from Johnson 1989) × portion of mass that is gonad (Johnson 1980) × proportion of nutrient (nitrogen, phosphorus, or carbon; Kamler 1992). (7)

Annual inputs due to postspawning/overwintering mortality in males were calculated as

X_{males} = biomass of upstream-moving males in a given year (kg; Johnson 1980) – mass loss due to overwintering (Johnson 1989) × mortality rate (Johnson 1989) × proportion of nutrient (nitrogen, phosphorus, or carbon; Lyle and Elliott 1998). (8)

Inputs due to the deposition of male gonadal products and nonprotein nitrogen were ignored. The export of nutrients was estimated as

$$X_{smolts} = \text{number of smolts (Johnson 1989; Department of Fisheries and Oceans, unpublished data)} \times \text{modal smolt weight (kg; calculated from Johnson 1980 and 1989)} \times \text{proportion of nutrient (nitrogen, phosphorus, carbon; Lyle and Elliott 1998)}. \quad (9)$$

Net masses of nutrients transported into Willow Lake were converted to concentrations by dividing by the estimated volume of the lake. Because detailed bathymetric data are not available, the volume of Willow Lake was estimated using the equation for a hyperbolic sinusoid:

$$\text{Volume} = 0.43 \times \text{area} \times \text{maximum depth (Post et al. 2000)}. \quad (10)$$

Values used for mortality rate, overwintering losses, gonad mass, and nutrient proportions are presented in Table 1.

In Nauyuk Lake, marine nutrients move into the system in the bodies of nonspawning migrants that have undergone growth at sea. These nutrients may be transferred into the lake via overwintering mortality and metabolic losses. In addition, there are nutrient inputs from postspawning mortality of fish moving between Willow Creek and Nauyuk River (Gyselman 1994). Freshwater-derived nutrients move out of the system in migrants (especially smolts because these fish contain the greatest proportion of freshwater nutrients) that perish at sea or stray to other systems. Due to methodological problems in operating the counting fence for small fish during downstream migrations (Johnson 1980; Johnson 1989), it was not possible for us to perform the same calculations as was done for Willow Lake. We did, however, compare estimates of summer increases in the standing stock of migratory fish to overwintering losses. For migrants greater than 400 mm, published estimates of increases in standing stock and overwintering losses were available (Johnson 1980). For fish less than 400 mm, we calculated these values using annual length–frequency plots of downstream and upstream migrants (Johnson 1989), weight–length relationships (Johnson 1980), and estimates of winter weight loss (Johnson 1980).

Results

Description of Anadromous Fish in the Study Lakes near Hope Bay, Nunavut

Previous fisheries records reported an anadromous population of Arctic char in Glenn Lake (Golder Associates Ltd. 2002). We did not capture any Arctic char in Glenn Lake in 2006, however, but captured two forms of lake trout—a "pale" form and a "normal" form. We suspect that previous reports of Arctic char in this system were actually misidentifications of pale-form lake trout. As a result, anadromous Arctic char were only collected from Roberts Lake.

Table 1.—Parameters used in the calculations for nutrients moving into Willow Lake and Nauyuk Lake via anadromous Arctic char. These constants were taken from the literature and applied, where appropriate, to equations (6)–(9). Where a range of values is presented, calculations were made at both the bottom and top of the range.

Parameter	Value(s) used	Reference
Proportion of prespawning female biomass that is gonad	0.10–0.13	Johnson 1980
Proportion of mass lost in Willow Lake overwintering	0.20–0.25	Johnson 1989
Annual postspawning female mortality rate in Willow Lake	0.12	Johnson 1989
Annual postspawning male mortality rate in Willow Lake	0.148	Johnson 1989
Proportion of mass lost in Nauyuk Lake overwintering	0.127	Johnson 1980
Proportion of N in fish flesh	0.03	Lyle and Elliott 1998
Proportion of C in fish flesh	0.135	Lyle and Elliott 1998
Proportion of P in fish flesh	0.0047	Lyle and Elliott 1998
Proportion of N in fish eggs	0.1211	Kamler 1992
Proportion of P in fish eggs	0.01	Kamler 1992

In Roberts Lake, 13 Arctic char were captured and analyzed for stable S, N, and C isotope ratios. All stable isotope ratios indicated that four of these fish were anadromous (Figures 4 and 5A, 5B) while the remaining nine had never migrated to sea and were either residents or presmolts. Anadromous Arctic char had significantly higher (or less negative) $\delta^{34}S$ (*t*-test, t = 24.22, df = 11, $P < 0.0001$; Figure 4), $\delta^{15}N_{adj}$ (ANCOVA, F = 70.82, df = 1,10, $P < 0.0001$; Figure 5A), and $\delta^{13}C_{adj}$ (*t*-test, t = 6.53, df = 11, $P < 0.0001$; Figure 5B) isotope ratios than nonanadromous Arctic char and had significantly higher C:N ratios than nonanadromous Arctic char (*t*-test, t = 9.28, df = 11, $P < 0.001$; Figure 6).

Anadromous Arctic char caught in Roberts Lake ranged from 460 to 670 mm fork length and from 8 to 11 years old (H. K. Swanson, unpublished data), whereas nonanadromous Arctic char caught in Roberts Lake ranged from 166 to 483 mm and from 3 to 9 years old (Swanson, unpublished data). Only the largest sea-run Arctic char (670 mm) had reached sexual maturity. Seven of the nonmigratory Arctic char were immature and two were maturing.

Two lake trout from Glenn Lake and one lake trout from Roberts Lake had $\delta^{34}S$, $\delta^{15}N$, and $\delta^{13}C$ ratios that supported use of brackish water environments (Swanson, unpublished data). These fish were excluded from the current analysis.

Food Web and Condition Factor Comparisons among Lakes

As reported above, anadromous Arctic char had higher $\delta^{15}N_{adj}$ than their resident and presmolt counterparts (Figure 5A). Resident lake trout in all lakes were approximately 2.5–5‰ enriched in $\delta^{15}N_{adj}$ compared to lake whitefish and cisco and approximately 4–6‰ enriched in $\delta^{15}N_{adj}$ compared to ninespine stickleback (Figure 5A). With anadromous Arctic char excluded, vertical food

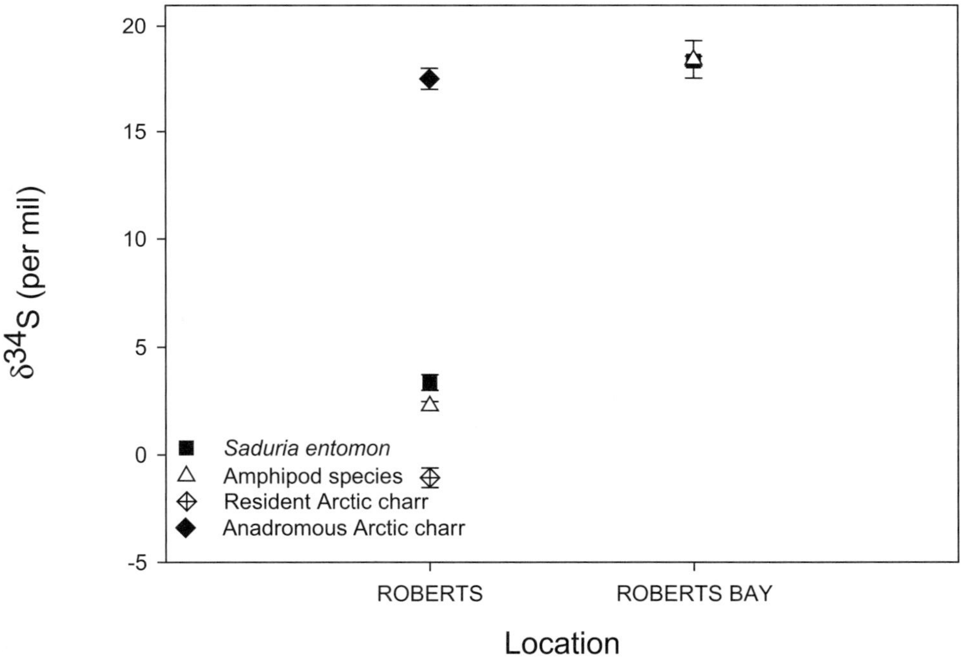

Figure 4.—Mean $\delta^{34}S \pm$ SE isotope ratios in lake trout, Arctic charr, and invertebrates from the study lakes and Roberts Bay. Anadromous Arctic charr had significantly higher $\delta^{34}S$ than resident/pre-smolt Arctic charr (*t*-test, t = 24.22, df = 11; $P < 0.0001$) and invertebrates from Roberts Bay had significantly higher $\delta^{34}S$ than invertebrates from Roberts Lake (*t*-test, $F > 316$, df = 1,4; 1,5, $P < 0.0001$). Mean $\delta^{34}S$ values were approximately coincident for lake trout and amphipod species in Roberts Lake and for amphipod species and *Saduria entomon* in Roberts Bay.

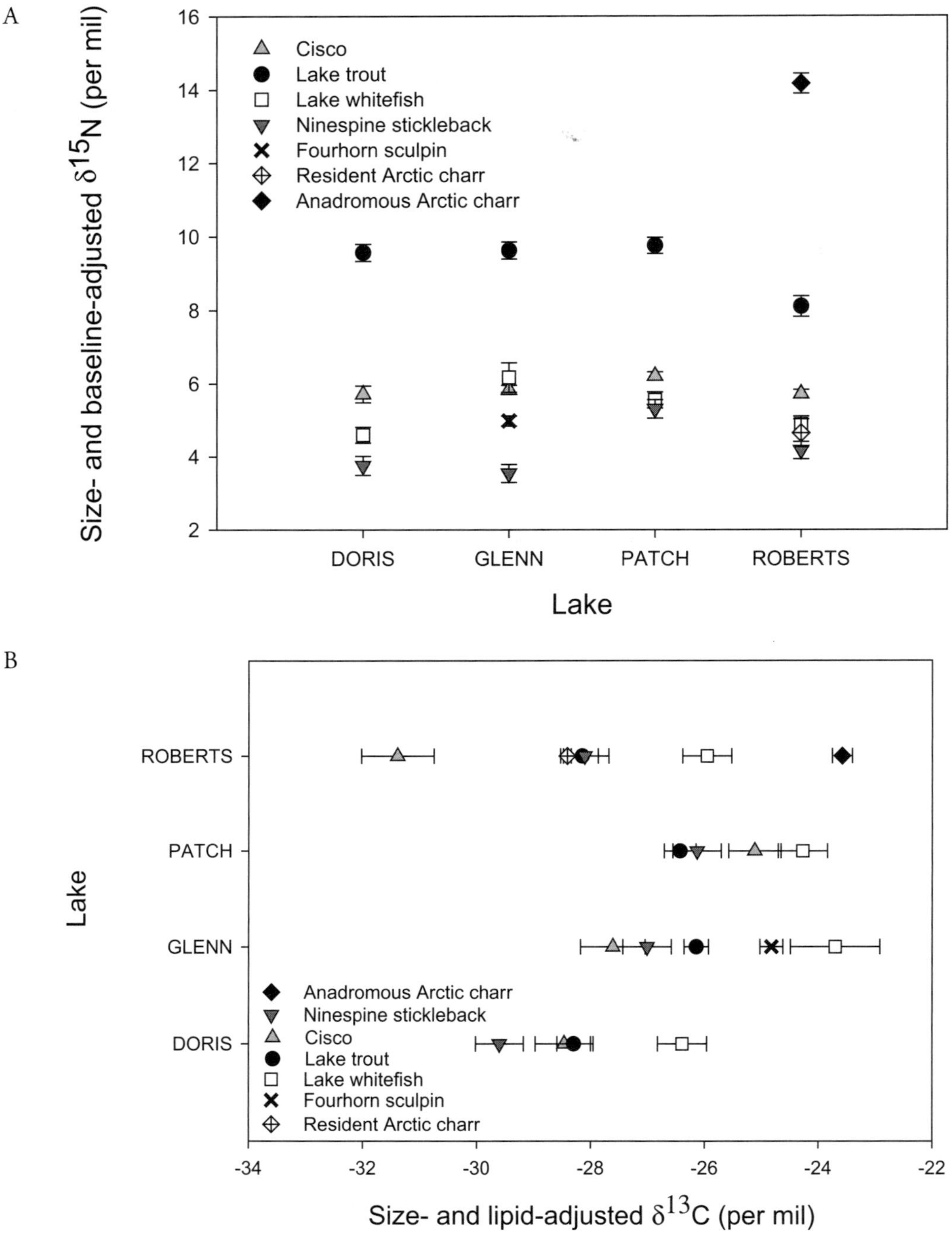

Figure 5.—(A) Least-squares mean $\delta^{15}N_{adj}$ ± SE at standardized body sizes (lake trout and anadromous Arctic charr = 650 mm; lake whitefish = 400 mm; cisco = 270 mm; ninespine stickleback = 50 mm; resident Arctic charr = 400 mm). In Roberts Lake, LSmean $\delta^{15}N_{adj}$ values were coincident for lake whitefish and resident/presmolt Arctic charr. (B) Least-squares mean $\delta^{13}C_{adj}$ ± SE at standardized body sizes (lake trout and anadromous Arctic charr = 650 mm; lake whitefish = 400 mm; cisco = 270 mm; ninespine stickleback = 50 mm; resident Arctic charr = 400 mm). In Roberts Lake, LSmean $\delta^{13}C_{adj}$ values were approximately coincident for ninespine stickleback and lake trout.

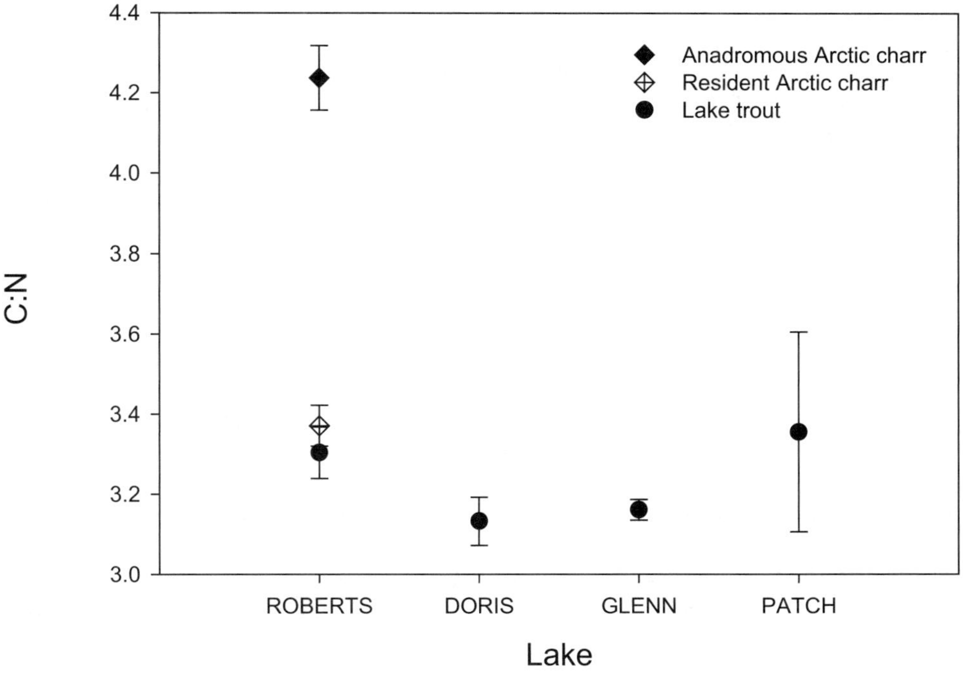

Figure 6.—Mean C:N ratios ± SE for lake trout and Arctic charr. Anadromous Arctic charr had significantly higher C:N ratios than resident/presmolt Arctic charr (*t*-test, $t = 9.28$, df = 11, $P < 0.001$). There were no significant differences among lakes in lake trout C:N ratios (analysis of variance, df = 3,42, $P = 0.20$).

web structure was very similar among lakes (Figure 5A) even though least-squares mean $\delta^{15}N_{adj}$ differed significantly among lakes for lake trout, lake whitefish, and ninespine stickleback (ANCOVA, $F > 5.81$, $P < 0.0032$). Posthoc Dunnett's tests (with Roberts Lake set as the control) applied to LSmeans indicated that lake trout in Roberts Lake had significantly lower $\delta^{15}N_{adj}$ than lake trout in Glenn, Patch, or Doris lakes (Dunnett's HSU test, $P < 0.0004$; Figure 5A). Resident/presmolt Arctic char in Roberts Lake had $\delta^{15}N_{adj}$ ratios that were intermediate between cisco and ninespine stickleback and coincident with lake whitefish (Figure 5A). Interestingly, LSmean $\delta^{15}N_{adj}$ for lake trout in Roberts Lake was approximately 3.2‰ higher than LSmean $\delta^{15}N_{adj}$ for resident Arctic char (Figure 5A). Lake whitefish in Glenn Lake had significantly higher $\delta^{15}N_{adj}$ than lake whitefish in Roberts Lake (Dunnett's HSU test, $P = 0.027$), and ninespine stickleback in Patch Lake had significantly higher $\delta^{15}N_{adj}$ than ninespine stickleback in Roberts Lake (Dunnett's HSU test, $P < 0.0001$; Figure 5A).

Overall variability in $\delta^{13}C_{adj}$ was higher in Roberts Lake than in any of the other lakes (Figure 5B). As mentioned above, anadromous Arctic char had less negative $\delta^{13}C_{adj}$ ratios than resident/presmolt Arctic char in Roberts Lake. Adjusted $\delta^{13}C$ ratios indicate that lake trout in Doris Lake are likely feeding on cisco, whereas lake trout in Patch Lake may be feeding more heavily on ninespine stickleback (Figure 5B). Resident lake trout in Glenn Lake appear to be integrating carbon signatures from a variety of prey sources, including ninespine stickleback, cisco, fourhorn sculpin, and perhaps lake whitefish. Resident lake trout in Roberts Lake are likely feeding most heavily on ninespine stickleback and resident/presmolt Arctic char (Figure 5B).

Least-squares mean $\delta^{13}C_{adj}$ differed significantly among lakes for lake trout, lake whitefish, cisco, and ninespine stickleback (ANCOVA, $F > 6.23$, $P < 0.0023$). Posthoc Dunnett's tests (with Roberts Lake set as the control) applied to LSmeans indicated that lake trout in Roberts lake were significantly more depleted in $\delta^{13}C_{adj}$ than lake trout in Glenn or Patch lakes (Dunnett's HSU test, $P < 0.048$; Figure 5B).

Mean condition factor in lake trout from Roberts Lake (1.159) was significantly higher than in lake trout from Doris (1.036), Patch (0.999), or Glenn lakes (0.936; ANOVA, $F = 9.88$, df = 3,42, $P < 0.001$). Mean C:N ratios in lake trout were not significantly different among lakes (ANOVA, df = 3,42, $P = 0.2$; Figure 6).

Estimation of Marine and Freshwater Food Sources in Anadromous Arctic Char

Delta^{34}S values for invertebrates were significantly higher in Roberts Bay than in Roberts Lake (*t*-test, $F > 316$, df = 1,4; 1,5, $P < 0.0001$; Figure 4). A two-source linear mixing model using δ^{34}S ratios from *Saduria entomon* as endpoints indicated that the mean proportion of marine-derived food sources in anadromous Arctic char diets was 0.95 (95% CI = 0.84–1) with freshwater food sources accounting for the remaining 0.05 (95% CI = 0–0.16). Results using amphipod species as endpoints were very similar, with the exception that the confidence interval was larger because fewer samples were analyzed (95% CI for marine-derived food sources = 0.75–0.1).

Estimates of Net Imports of Nutrients to Willow Lake, Nunavut

To arrive at estimates for net nutrient imports to Willow Lake, we calculated the biomass of fish moving into and out of Willow Lake (Table 2A, 2B) and applied the appropriate parameters for mortality rate, egg mass, and nutrient proportions described in equations (6)–(9) (Table 1). Wherever possible, we incorporated known error into the calculations. For example, we calculated gonad mass at 10% and 13% of female weight (Johnson 1980) and estimated overwintering mass loss at both 20% and 25% (Johnson 1980). We also incorporated our

Table 2A.—Annual estimates of marine-derived fish biomass moving into Willow Lake. These estimates were arrived at by multiplying the annual number of anadromous Arctic charr moving up Willow Creek by the sex-specific (and size-specific for females) mean mass. Estimates were made at the upper and lower 95% confidence limits for proportion of marine-derived biomass in mature and maturing anadromous Arctic char.

	Year	Assuming 100% marine-derived food sources (kg)	Assuming 84% marine-derived food sources (kg)	Reference
Biomass import via female prespawners (females > 600 mm)	1975	534	448	Johnson 1989[a]
	1976	1,510	1269	Johnson 1989[a]
	1977	290	244	Johnson 1989[a]
	1978	239	201	Johnson 1989[a]
	1979	154	129	Johnson 1989[a]
	1980	361	303	Johnson 1989[a]
Biomass import via female nonspawners (females < 600 mm)	1975	55	46	Johnson 1989[a]
	1976	101	85	Johnson 1989[a]
	1977	104	87	Johnson 1989[a]
	1978	144	121	Johnson 1989[a]
	1979	87	73	Johnson 1989[a]
	1980	142	119	Johnson 1989[a]
Biomass import via males	1975	120	101	Johnson 1989[a]
	1976	209	175	Johnson 1989[a]
	1977	142	119	Johnson 1989[a]
	1978	91	77	Johnson 1989[a]
	1979	63	53	Johnson 1989[a]
	1980	40	34	Johnson 1989[a]

[a] Calculations of error from published data were not possible.

Table 2B.—Annual estimates of freshwater-derived fish biomass moving out of Willow Lake via Arctic char smolts. The number of fish in the size range 180–240 mm that moved downstream in Willow Creek in a given year was multiplied by the modal smolt mass observed over the course of the entire study (Johnson 1980).

	Year	Assuming 100% freshwater-derived food sources (kg)	Reference
Biomass export via smolts	1975	29	Johnson 1989 and Department of Fisheries and Oceans, unpublished data[a]
	1976	21	Johnson 1989 and Department of Fisheries and Oceans, unpublished data[a]
	1977	21	Johnson 1989 and Department of Fisheries and Oceans, unpublished data[a]
	1978	25	Johnson 1989 and Department of Fisheries and Oceans, unpublished data[a]
	1979	25	Johnson 1989 and Department of Fisheries and Oceans, unpublished data[a]
	1980	82	Johnson 1989 and Department of Fisheries and Oceans, unpublished data[a]

[a] Calculations of error from published data were not possible.

results from the two-source mixing model and calculated marine-derived fish mass moving into Willow Lake as 100% of the observed biomass and 84% of the biomass (limits of the 95% confidence interval, see above). These estimates of error are reflected in the maximum and minimum values presented in Table 3.

Mean net annual imports of nitrogen ranged from 2.03 to 23.52 kg/year from 1975 to 1980, mean net annual imports of phosphorus ranged from 0.22 to 2.25 kg/year, and mean net annual imports of carbon ranged from –4.69 to 18.75 kg/year (Table 3). The large ranges were due to significant fluctuations in numbers of spawning females moving upstream into Willow Lake (largest input) and number of smolts moving downstream (output). When these net nutrient inputs were converted to concentrations based on the volume of the lake, mean loading concentrations for nitrogen varied from 0.28 to 1.82 μg/L between 1975 and 1980, mean loading concentrations for phosphorus varied from 0.01 to 0.17 μg/L, and mean loading concentrations for carbon varied from –0.36 to 1.45 μg/L (Table 3). These concentrations are two to three orders of magnitude lower than ambient concentrations measured in similarly-sized lakes in the same region (Miramar Hope Bay Ltd. 2005).

Net changes in standing stock of migrating fish in Nauyuk Lake ranged from –126 kg in 1974 to 2,307 kg in 1975 (Table 4). These estimates account for straying of fish out of the system and indicate that in some years, growth during the sea migration can compensate for overwintering losses and possibly overwintering mortality (not calculated here), but in other years, the marine subsidy is not adequate to maintain the standing stock of the population.

Discussion

Description of Arctic Char Captured in the Study Lakes

Previous studies have outlined the utility of stable C, N, and S signatures in distinguishing between anadromous and resident Arctic char (Doucett et al. 1999). Similar to the findings of Doucett et al. (1999), anadromous Arctic char had less negative $\delta^{13}C$ and higher $\delta^{34}S$ and $\delta^{15}N$ ratios relative to resident fish. Of the 13 Arctic char captured in Roberts Lake, four were anadromous and nine were residents or presmolts. The four anadromous Arctic char also had higher C:N ratios than the nine nonanadromous char, indicating that the former were higher in lipid. For the nine nonanadromous Arctic char,

Table 3.—Net mass and associated concentrations of nitrogen, phosphorus, and carbon deposited into Willow Lake via egg deposition and mortality of anadromous Arctic char. To arrive at estimates of net import, data presented in Tables 1, 2A, and 2B were applied to equation (6). Concentrations were calculated by dividing net imports by the calculated volume of the lake (equation 10).

	Year	Mean[a]	Max[a]	Min[a]		Year	Mean[a]	Max[a]	Min[a]
Net nitrogen imports (kg)	1975	7.87	9.67	6.24	Nitrogen concentration (μg/L)	1975	0.61	0.75	0.48
	1976	23.52	28.53	18.97		1976	1.82	2.21	1.47
	1977	4.55	5.58	3.62		1977	0.35	0.43	0.28
	1978	3.58	4.44	2.80		1978	0.28	0.34	0.22
	1979	2.03	2.59	1.53		1979	0.16	0.20	0.12
	1980	3.59	4.82	2.47		1980	0.28	0.37	0.19
Net phosphorus imports (kg)	1975	0.73	0.89	0.58	Phosphorus concentration (μg/L)	1975	0.06	0.07	0.04
	1976	2.25	2.71	1.83		1976	0.17	0.21	0.14
	1977	0.44	0.54	0.35		1977	0.03	0.04	0.03
	1978	0.33	0.42	0.26		1978	0.03	0.03	0.02
	1979	0.17	0.23	0.12		1979	0.01	0.02	0.01
	1980	0.22	0.33	0.11		1980	0.02	0.03	0.01
Net carbon imports (kg)	1975	4.66	5.69	3.66	Carbon concentration (μg/L)	1975	0.36	0.44	0.28
	1976	18.75	21.38	16.23		1976	1.45	1.66	1.26
	1977	3.74	4.55	2.98		1977	0.29	0.35	0.23
	1978	2.32	3.02	1.66		1978	0.18	0.23	0.13
	1979	0.28	0.73	–0.15		1979	0.02	0.06	–0.01
	1980	–4.69	–3.91	–5.43		1980	–0.36	–0.30	–0.42

[a] Mean, maximum, and minimum values reflect the range of parameters used in the calculations (e.g., proportion of marine biomass [0.84–1], proportion of female that is gonad [0.1–0.13; Table 1] and overwintering mass loss [0.2–0.25; Table 1]).

it is difficult to tell whether these were residents or presmolts, but it is likely that they represent a combination of both; there was an overlap in fork length between the anadromous and nonanadromous fish and the oldest nonanadromous fish was 9 years, which is older than the maximum smolt age reported by Johnson (1989; 8 years) for nearby Nauyuk Lake. The size of the one mature anadromous Arctic char (670 mm) captured is consistent with the findings of Johnson (1989), who found that female anadromous Arctic char did not mature until they reached a fork length of approximately 600 mm.

Using stable S isotope ratios of *Saduria entomon* collected from Roberts Bay (marine environment) and Roberts Lake (freshwater environment), the IsoError two-source mixing model (Phillips and Gregg 2001) estimated that marine food sources accounted for approximately 95% of the diet in anadromous Arctic char in our study lake. The 95% confidence interval extended from 84% to 100%. This agrees well with previous studies of stomach contents in anadromous Arctic char; once char begin migrating to sea, it seems that there is very little reliance on freshwater food sources (Klemetsen et al. 2003). In the future, it would be useful to perform this analysis on smaller anadromous Arctic char. The anadromous char we analyzed ranged in fork length from 460 to 670 mm. Fish of this size would have made at least one previous migration to sea (likely at least two) (Johnson 1980). It is therefore not surprising that they are primarily composed of marine-derived nutrients; most of their tissue would likely have turned over since the onset of sea migrations. It is also fairly certain that smolts (180–240 mm in size; Johnson 1989) are primarily composed of freshwater-derived nutrients because they have spent at least 3 years feeding in freshwater (Johnson 1989). In the size range of 240 to ~400 mm, however, it is likely that anadromous Arctic char are a more equal mixture of freshwater and marine-derived nutrients. Since fidelity to the natal system can be low in the smaller end of this

Table 4.—Annual net changes in the sea-going standing stock of anadromous Arctic char in Nauyuk Lake. These estimates were arrived at by comparing the number and length–frequency distribution of downstream migrants to that of upstream migrants in a given year. Overwintering losses were estimated by applying the constant presented in Table 1 to the annual overwintering standing stock in Nauyuk Lake (Johnson 1980, 1989).

	Year	Summer increase (kg)	Overwintering loss (kg)	Net change (kg)	Reference
Annual net change in standing stock in in fish > 400 mm	1974	3,589	–3,378	211	Johnson 1980
	1975	5,591	–3,153	2,438	Johnson 1980
	1976	1,844	–1,912	–68	Johnson 1980
	1977	2,284	–1,613	671	Johnson 1980
Annual net change in standing stock in fish < 400 mm	1974	–146.4	–75.4	–222	Calculated from Johnson 1980, 1989
	1975	8.5	–139.9	–131	Calculated from Johnson 1980, 1989
	1976	0.6	–58.7	–58	Calculated from Johnson 1980, 1989
	1977	–9.5	–35.5	–45	Calculated from Johnson 1980, 1989
Total annual net change in standing stock (kg)	1974	3,443	–3,453	–11	
	1975	5,599	–3,293	2,307	
	1976	1,845	–1,971	–126	
	1977	2,275	–1,648	626	

size range (Gyselman 1994), nutrient budgets need to account for straying of these fish. To determine mass of freshwater nutrients leaving the system, we need to establish the relative marine and freshwater-derived composition of these smaller anadromous fish.

Comparison of Food Web Structure and Condition

Glenn Lake did not support an anadromous Arctic char population as expected, so we compared the food web structure of one lake that contained anadromous Arctic char to three lakes that contained no Arctic char. Because of the low number of lakes and unbalanced design, our results must be considered as preliminary. We found that resident lake trout from Roberts Lake had significantly lower $\delta^{15}N_{adj}$ than lake trout from Glenn, Doris, or Patch lakes. There was some evidence that lake trout in Roberts Lake were feeding on nonanadromous Arctic char. Nonanadromous Arctic char had a $\delta^{15}N_{adj}$ ratio that was similar to lake whitefish and intermediate between ninespine stickleback and cisco. Stomach content analysis revealed that nonanadromous Arctic char were feeding primarily on detritus (Swanson, unpublished data); this may explain their relatively low $\delta^{15}N_{adj}$. The difference in LSmean $\delta^{15}N_{adj}$ between lake trout and nonanadromous Arctic char was 3.23‰. This is very close to the widely reported average value of 3.4‰ per trophic transfer (e.g., Post 2002). An almost identical difference in $\delta^{15}N_{adj}$ existed between lake trout and lake whitefish in Roberts Lake, but $\delta^{13}C_{adj}$ ratios suggest that lake trout in Roberts Lake were more likely feeding on ninespine stickleback and nonanadromous Arctic char; the difference in $\delta^{13}C_{adj}$ ratios between lake trout and nonanadromous Arctic char was only 0.26‰.

In addition to having lower $\delta^{15}N_{adj}$, lake trout from Roberts Lake had higher condition than those from Doris, Glenn, or Patch lakes. If lake trout in Roberts Lake are feeding on nonanadromous Arctic char, the higher condition could be due to feeding on relatively larger, higher-quality prey. Additional results from 2007 sampling will help determine whether the lower trophic position and higher condition observed in lake trout from Roberts Lake was a result of feeding on presmolt and resident Arctic char.

Overall variability in $\delta^{13}C_{adj}$ was highest in Roberts Lake and lowest in Patch Lake, even when anadromous Arctic char were omitted from the comparison. This may indicate use of a broader variety of habitats due to the presence of an additional species and the associated increased competition for food resources. Again, additional results from 2007 sampling are needed to verify whether this is a true result or an artifact of the small number of lakes sampled.

Estimates of Net Imports of Nutrients to Willow Lake, Nunavut

Very few studies have investigated net import of marine-derived nutrients to freshwater systems via iteroparous anadromous salmonids. This may be partly due to the complicated nature of performing such investigations. Lyle and Elliott (1998) found that carbon, nitrogen, and phosphorus transported from the marine environment into rivers by Atlantic salmon and brown trout *Salmo trutta* was in the range of metric tons but that this amount represented a tiny fraction of total annual river loads. In contrast, a study investigating the import of marine-derived nutrients to a river in Norway found that Atlantic salmon migration could account for up to 17% of the total river load of phosphorus (Jonsson and Jonsson 2003). The net import of nitrogen, carbon, and phosphorus to Willow Lake via spawning anadromous Arctic char varied widely between the years 1975–1980, but in general, these inputs were negligible in terms of water chemistry. Inputs were on the order of kilograms and, when converted to concentrations, were two to three orders of magnitude less than ambient concentrations in water from similarly sized lakes in the same region (Miramar Hope Bay Ltd 2005). It was interesting to note the net negative input of carbon in 1980. This was due to the relatively large number of smolts that left the system (and the dearth of prespawning adults). Net negative inputs of nutrients have been observed in other systems where the biomass of smolts exceeds the biomass of spawning adults (Nislow et al. 2004). It is possible that interannual variability in net negative and positive nutrient inputs may be interpreted as a reflection or consequence of partial migration or conditional life history strategies in iteroparous anadromous salmonids. It is common for some of these species (including Arctic char) to have sympatric resident and anadromous individuals that freely interbreed (Gyselman 1984). Alternate life history tactics may help populations adapt to variability in the environment (Secor 2007). In the context of this study, when conditions are poor at sea, the Arctic char population may be maintained by resident individuals in Nauyuk and Willow lakes and vice versa. We agree with Secor and Kerr (2009, this volume) that the consequences of life history diversity deserve further attention and argue that this diversity likely impacts the environments that support the fish as well as the populations themselves.

We could not perform the same detailed calculations on nutrient imports and exports for Nauyuk Lake as for Willow Lake because of greater uncertainty in a number of parameters. First, as mentioned above, we do not know the relative contribution of freshwater and marine food sources to fish in the size range of approximately 240–400 mm fork length. As a considerable number of these fish likely stray and/or perish at sea, this is an important piece of missing data. Also, we do no know how many of the downstream-migrating smolts in the Nauyuk River originated from Willow Lake and how many originated from Nauyuk Lake itself. Johnson (1989) reports that Willow Lake provides the only known spawning and rearing habitat for anadromous Arctic char in the Nauyuk system, but it seems unlikely that there would not be spawning in Nauyuk Lake. The number of smolt-sized fish migrating downstream in the Nauyuk River often exceeded the number migrating down Willow Creek during the study years (Johnson 1989; Gyselman and Broughton 1991), which also suggests that there is rearing in Nauyuk Lake.

Due to the uncertainties presented above, our budgeting of marine subsidies in Nauyuk Lake consisted of a comparison of the marine-derived increase in summer standing stock to overwintering losses. Although conceptually simple, these calculations did account for straying and mortalities at sea and reflected an extremely rare historical dataset; to achieve these estimates, we needed to know the number and size of all downstream and upstream-migrating fish in the study years. Our calculations revealed that marine-derived growth in the stock varied widely between years. In 1974 and 1976, increases in standing stock achieved during the sea migration would not be sufficient to maintain the population, especially considering that we did not include mortality estimates. Once again, this interannual fluctuation between positive and negative growth in the sea-going stock may be interpreted as a reflection or consequence of a partial migration life history strategy; in a year with high marine-derived growth (i.e., 1975), sea-going individuals may sustain the production of the population, whereas the existence of sympatric residents may protect the population against multiple years of poor growth at sea.

It is obvious that elucidating the effects of an iteroparous, low-fidelity anadromous fish species on freshwater ecosystems is considerably more difficult than for a semelparous, high-fidelity species. Anadromous Arctic char populations in Canada's north have complicated life histories, seem to function as metapopulations (Gyselman 1994), and thrive in systems that are not conducive to long-term field studies. This has precluded any detailed knowledge on their role in freshwater food webs and/or as a potential transport vector for nutrients. It is nevertheless likely that anadromous Arctic char have an effect, albeit more subtle than their semelparous counterparts, on the freshwater systems that support them. The results of this study are preliminary, but indicate that the presence of anadromous Arctic char may impact the trophic position and condition of sympatric populations of resident lake trout. It does not seem likely that Arctic char directly affect water chemistry through nutrient additions due to egg deposition and mortality, but many other potential effects have yet to be investigated. These include nutrient transport from the marine environment to terrestrial and riparian environments via predation of char by gulls, bears, and wolves (Johnson 1980; Swanson, unpublished data), and marine-driven increases in overall fish production. It has been shown in many previous studies that anadromous Arctic char are larger than resident Arctic char and that Arctic char are more likely to occur in sympatry with lake trout when the former are anadromous (e.g., Johnson 1980). This in itself indicates that marine subsidies are important in the structure and function of northern freshwater ecosystems and, in light of climate-induced changes to migration habitats used by anadromous Arctic char, illustrates the urgent need for further research in this area.

Acknowledgments

Funding for this project was provided by Miramar Mining Corporation, Natural Sciences and Engineering Research Council grants to K. A. Kidd (Collaborative Research and Development, Canada Research Chair, and Discovery) and to H. K. Swanson (Canada Graduate Scholarship), Golder Associates Ltd., Northern Scientific Training Program, Canadian Northern Studies Trust, and the Arctic Institute of North America. Field and laboratory assistance from K. Martell, M. Kullman, T. Antill, P. Emery, and K. Kristensen is gratefully acknowledged. Analytical advice and data provision from J. Babaluk and R. Wastle also contributed to the development of this manuscript.

References

Bilby, R. E., B. R. Fransen, and P. A. Bisson. 1996. Incorporation of nitrogen and carbon from spawning coho salmon into the trophic system of small streams: evidence from stable isotopes. Canadian Journal of Fisheries and Aquatic Sciences 53:164–173.

Cabana, G., and J. B. Rasmussen. 1996. Comparison of aquatic food chains using nitrogen isotopes. Ecology 93:10844–10847.

Doucett, R. R., M. Power, G. Power, F. Caron, and J. D. Reist. 1999. Evidence for anadromy in a southern relict population of Arctic charr from North America. Journal of Fish Biology 55:84–93.

Durbin, A. G., S. W. Nixon, and C. A. Oviatt. 1979. Effects of the spawning migration of the alewife, *Alosa pseudoharengus*, on freshwater ecosystems. Ecology 60:8–17.

France, R. L. 1995. Differentiation between littoral and pelagic food webs in lakes using stable carbon isotopes. Limnology and Oceanography 40:1310–1313.

Golder Associates Ltd. 2002. Aquatic baseline studies—Doris Hinge Project Data Compilation Report. 1995–2000. Prepared for Miramar Mining Inc., Edmonton, Alberta.

Golder Associates Ltd. 2005. Doris North Project "No Net Loss" Plan—revision 4. Prepared for Miramar Hope Bay Ltd, Edmonton, Alberta.

Gyselman, E. 1984. The seasonal movements of anadromous Arctic charr at Nauyuk Lake, Northwest Territories, Canada. Pages 575–578 *in* L. Johnson and B. L. Burns editors. Biology of the Arctic charr, Proceedings of the International Symposium on Arctic Charr. University of Manitoba Press, Winnipeg, Manitoba.

Gyselman, E. C. 1994. Fidelity of anadromous Arctic char (*Salvelinus alpinus*) to Nauyuk Lake, N.W.T., Canada. Canadian Journal of Fisheries and Aquatic Sciences 51:1927–1934.

Gyselman, E. C., and K. E. Broughton. 1991. Data on

the anadromous Arctic charr, (*Salvelinus alpinus*) of Nauyuk Lake, Northwest Territories, 1974–1978. Department of Fisheries and Oceans, Canadian Data Report of Fisheries and Aquatic Sciences No. 815, Winnipeg, Manitoba.

Helfield, J. M., and R. J. Naiman. 2001. Effects of salmon-derived nutrients on riparian forest growth and implications for stream habitat. Ecology 82:2403–2409.

Jardine, T. D., Roussel, J-M., Mitchell, S. C., and R. A. Cunjak. 2009. Pathways of marine nutrient and organic matter entry from anadromous fishes into Atlantic coast streams. Pages 427–445 *in* A. J. Haro, K. L. Smith, R. A. Rulifson, C. M. Moffitt, R. J. Klauda, M. J. Dadswell, R. A. Cunjak, J. E. Cooper, K. L. Beal, and T. S. Avery, editors. Challenges for diadromous fishes in a dynamic global environment. American Fisheries Society, Symposium 69, Bethesda, Maryland.

Johnson, L. 1980. The arctic charr, *Salvelinus alpinus*. Pages 15–98 *in* E. K. Balon, editor. Charrs: salmonid fishes of the genus *Salvelinus*. Dr. W. Junk bv Publishers, The Hague, Netherlands.

Johnson, L. 1989. The anadromous Arctic charr (*Salvelnius alpinus*) of Nauyuk Lake, N.W.T., Canada. Physiology and Ecology Japan Special Volume 1:201–227.

Jonsson, B., and N. Jonsson. 2003. Migratory Atlantic salmon as vectors for the transfer of energy and nutrients between freshwater and marine environments. Freshwater Biology 48:21–27.

Kamler, E. 1992. Early life history of fish: an energetics approach. Chapman and Hall, London.

Klemetsen, A., P. A. Amundsen, J. B. Dempson, B. Jonsson, N. Jonsson, M. F. O'Connell, and E. Mortensen. 2003. Atlantic salmon *Salmo salar* L., brown trout *Salmo trutta* L. and Arctic charr *Salvelinus alipnus* (L.): a review of aspects of their life histories. Ecology of freshwater fish 12:1–59.

Kline, T. C., J. J. Goering, O. A. Mathisen, P. Poe, P. L. Parker, and R. S. Scalan. 1993. Recycling of elements transported upstream by runs of Pacific salmon: II $\delta^{15}N$, $\delta^{13}C$ evidence in the Kvichak River watershed, Bristol Bay, southern Alaska. Canadian Journal of Fisheries and Aquatic Sciences 50:2350–2365.

Langeland, A. 1982. Interactions between zooplankton and fish in a fertilized lake. Holarctic Ecology 5:273–310.

Miramar Hope Bay Ltd. 2005. Final Environmental Impact Statement Doris North Project Nunavut, Canada. Submitted to Nunavut Impact Review Board. Miramar Hope Bay Ltd., Hope Bay, Nunavut, Canada.

Lyle, A. A., and J. M. Elliott. 1998. Migratory salmonids as vectors of carbon, nitrogen, and phosphorus between marine and freshwater environments in north-east England. Science of the Total Environment 210–211:457–468.

Minagawa, M., and E. Wada. 1984. Stepwise enrichment of ^{15}N along food chains: further evidence of the relation between $\delta^{15}N$ and animal age. Geochimica et Cosmochimica Acta 48:1135–1140.

Minakawa, N., R. I. Gara, and J. M. Honea. 2002. Increased individual growth rate and community biomass of stream insects associated with salmon carcasses. Journal of the North American Benthological Society 21:651–659.

Naiman, R. J., R. E. Bilby, D. E. Schindler, and J. M. Helfield. 2002. Pacific salmon, nutrients, and the dynamics of freshwater and riparian ecosystems. Ecosystems 5:399–417.

Naiman, R. J., J. M. Helfield, K. K. Bartz, D. C. Drake, and J. M. Honea. 2009. Pacific salmon, marine-derived nutrients and the dynamics of aquatic ecosystems. Pages 395–425 *in* A. J. Haro, K. L. Smith, R. A. Rulifson, C. M. Moffitt, R. J. Klauda, M. J. Dadswell, R. A. Cunjak, J. E. Cooper, K. L. Beal, and T. S. Avery, editors. Challenges for diadromous fishes in a dynamic global environment. American Fisheries Society, Symposium 69, Bethesda, Maryland.

Nislow, K. H., J. D. Armstrong, and S. McKelvey. 2004. Phosphorus flux due to Atlantic salmon (*Salmo salar*) in an oligotrophic upland stream: effects of management and demography. Canadian Journal of Fisheries and Aquatic Sciences 61:2401–2410.

Peterson, B. J., and B. Fry. 1987. Stable isotopes in ecosystem studies. Annual Review of Ecology and Systematics 18:293–320.

Phillips, D. L., and J. W. Gregg. 2001. Uncertainty in source partitioning using stable isotopes. Oecologia 127:171–179.

Post, D. M. 2002. Using stable isotopes to estimate trophic position: models, methods, and assumptions. Ecology 83:703–718.

Post, D. M., M. L. Pace, and N. G. J. Hairston. 2000. Ecosystem size determines food-chain length in lakes. Nature 405:1047–1049.

Post, D. M., C. A. Layman, D. A. Arrington, G. Takimoto, J. Quatrrochi, and C. G. Montana. 2007. Getting to the fat of the matter: models, methods and assumptions for dealing with lipids in stable isotope analysis. Oecologia 152:179–189.

Reimchen, T. E., D. D. Mathewson, M. D. Hocking, and J. Moran. 2003. Isotopic evidence for enrichment of salmon-derived nutrients in vegetation, soil, and insects in riparian zones in coastal British Columbia. Pages 59–69 *in* J. G. Stockner, editor. Nutrients in salmonid ecosystems: sustaining production and biodiversity. American Fisheries Society, Symposium 34, Bethesda, Maryland.

Rikardsen, A. H., P. A. Amundsen, and P-A. Bodin.

2003. Growth and diet of anadromous Arctic charr after their return to freshwater. Ecology of freshwater fish 12:74–80.

SAS Institute Inc. 2002. SAS v. 9.1.3. SAS Institute Inc., Cary, North Carolina.

Schmidt, D. C., S. R. Carlson, G. B. Kyle, and B. P. Finney. 1998. Influence of carcass-derived nutrients on sockeye salmon productivity of Karluk Lake, Alaska: importance in the assessment of an escapement goal. North American Journal of Fisheries Management 18:743–763.

Secor, D. H. 2007. The year-class phenomenon and the storage effect in marine fishes. Journal of Sea Research 57:91–103.

Secor, D. H., and L. H. Kerr. 2009. A lexicon of life cycle diversity in diadromous and other fishes. Pages 537–556 *in* A. J. Haro, K. L. Smith, R. A. Rulifson, C. M. Moffitt, R. J. Klauda, M. J. Dadswell, R. A. Cunjak, J. E. Cooper, K. L. Beal, and T. S. Avery, editors. Challenges for diadromous fishes in a dynamic global environment. American Fisheries Society, Symposium 69, Bethesda, Maryland.

Stockner, J. G., editor. 2003. Nutrients in salmonid ecosystems: sustaining production and biodiversity. American Fisheries Society, Symposium 34, Bethesda, Maryland.

Vander Zanden, M. J., and J. B. Rasmussen. 1999. Primary consumer $\delta^{13}C$ and $\delta^{15}N$ and the trophic position of aquatic consumers. Ecology 80:1395–1404.

American Fisheries Society Symposium 69:485–494, 2009

The Role of Anadromous Sea Lamprey in Nutrient and Material Transport between Marine and Freshwater Environments

Keith H. Nislow*
U.S. Forest Service Northern Research Station, 201 Holdsworth NRC, University of Massachusetts Amherst, Massachusetts 01003, USA

Boyd E. Kynard
Conte Anadromous Fish Research Center, Leetown Science Center, U.S. Geologic Survey One Migratory Way, Turners Falls, Massachusetts 01376, USA

Abstract.—The sea lamprey *Petromyzon marinus* is a widely distributed anadromous species spawning in coastal rivers and streams throughout the north Atlantic basin. In this paper, we review aspects of sea lamprey migration and ecology that relate to the transport of nutrients and materials to and from freshwater ecosystems and provide an example of a long-term study of a native wild population. Several aspects of lamprey life history (rapid growth in marine phase, many adults spawn in upper reaches of small oligotrophic rivers, all adults die after spawning) suggest that anadromous sea lampreys contribute marine-derived nutrients and materials (MDNM) to freshwater ecosystems. We used long-term (20 years) data on spawner abundance, along with literature-derived concentration values, to estimate the import of nutrients and materials to a spawning reach of the Fort River, a tributary of the Connecticut River in western Massachusetts, USA. Sea lamprey imported as much as 0.26 g of P per square meter of stream, as much as of 20% of the total annual P loading to a similar system where a full P budget has been developed. While the MDNM contribution of sea lamprey may be substantial, other aspects of their life history and habitat use may limit the overall magnitude and direction of lamprey influence on freshwater ecosystems. Spawning requirement for rocky substrate within a narrow size range may limit import at the watershed scale. In addition, marine survival rates of less than ~1% will result in a net export of nutrients and materials via out-migrating juveniles (transformers). While there is currently no information on survival rates in wild anadromous populations, the tight link between adult survival and prey/host fish populations observed in landlocked Great Lakes systems suggests that the ecological role of sea lamprey may be strongly related to the abundance of coastal marine fishes. Further research on adult survival, juvenile dispersal and distribution, and the paths of nutrient and material uptake in spawning streams are necessary to more fully evaluate the role of anadromous sea lamprey in the transport of MDNM.

Introduction

The sea lamprey *Petromyzon marinus* is a widely distributed anadromous species native to the North Atlantic and northern Mediterranean region (Maitland 2003). The species was highly valued as a food source during early European history (Dempson and Porter 1993) and continues to be in some European rivers today (Almeida et al. 2002). In addition to its food value, sea lamprey are harvested by biological supply companies for use in laboratories (Kircheis 2004). However, in the past half century, sea lamprey research and management in North America has been dominated by efforts to monitor and eradicate invasive landlocked populations from the Great Lakes region of the United States, where

* Corresponding author: knislow@fs.fed.us

they have had a devastating impact on valuable fisheries (Christie and Goddard 2003). In a Web-based literature search using the terms "sea lamprey ecology and management," more than 90% of the first 100 papers dealt with Great Lakes issues. Even in wild anadromous populations, fears of negative effects on commercial and game fishes have led some managers to cull prespawning lampreys in rivers, despite no evidence that these negative impacts occur. For example, until recently in the Sheepscot River, Maine, USA, it was the policy to exclude prespawning lampreys from the fishway at the lowermost dam in the system (Kircheis 2004).

More recently, given declines and local extirpations of native populations (Renaud 1997), due in large extent to the barriers to migration (Beamish and Northcote 1989), there has been an acknowledgment of the conservation importance of sea lamprey in its native range. Conservation of native sea lamprey is currently part of river management plans in both Europe (Maitland 2003) and North America (Connecticut River Atlantic Salmon Commission 1998). Interest in the conservation of native anadromous lamprey stems in part from their potential role in freshwater ecosystems. Anadromous fish can impact freshwater ecosystems in a number of different ways, but the role of Pacific salmonids in the transport of nutrients to and from stream ecosystems has received particular attention from ecologists (reviewed by Gende et al. 2002; Naiman et al. 2009, this volume). However, there is relatively little research on the role of Atlantic-basin anadromous fish in nutrient and material transport (Nislow et al. 2004; Saunders et al. 2006; Jardine et al. 2009, this volume). Further, while the focus has been on nutrient transport, anadromous fish transport other important materials in the course of their migratory life cycle. For example, in the north Atlantic region, mercury contamination of freshwaters is a major problem for freshwater fisheries (Mergler et al. 2007), but we have little information of the role of anadromous fish in Hg transport. In this paper, we assess the potential role of sea lamprey in freshwater ecosystems. We review lamprey life history and habitat-use characteristics that influence the direction and magnitude of nutrient and material transport and estimate potential import using long-term data on spawner abundance in a tributary of the Connecticut River, USA. Our goal is to establish a framework to evaluate the ecological role of sea lamprey, identifying important data gaps to help guide future research.

Factors Favoring Sea Lamprey as Important Nutrient and Material Importers

Life History Factors

Several life history attributes of anadromous sea lamprey point towards lamprey as significant importers of marine-derived nutrients and materials (MDNM). Adult growth rates in the marine environment are quite high, with adults increasing their body mass more than two orders of magnitude during an estimated 13–19 months at sea (Maitland 2003; Drevnick et al. 2006). This growth and sea age are similar to pink salmon *Oncorhynchus gorbuscha*, which are significant contributors of marine-derived nutrients to Pacific Northwest streams (Quinn 2005). Additionally, like all Pacific salmon (except for steelhead *O. mykiss* and cutthroat trout *O. clarkii*), sea lamprey are obligately semelparous, with all adults dying within a few days after spawning (Applegate 1950; Beamish 1980). This is in contrast to all other native Atlantic basin species in which some percentage of spawners survive to out-migrate, taking a considerable portion of their MNDM back out to the ocean (Lyle and Elliott 1998; Nislow et al. 2004).

In addition to the size, abundance, and fate of spawners, the timing of spawning and carcass availability may be key to the role of lamprey in nutrient and material transport. Studies in the Pacific Northwest (Gende et al. 2002) suggest that for some species in some systems, carcasses are present during a time when physical conditions (i.e., floods) prevent efficient uptake. Because lamprey generally spawn in late spring and die immediately after, carcasses are available at a time of the year characterized by relatively low flows, warm temperatures, and high fish growth rates. This makes the contribution of lamprey very well-timed to support instream production. Further, as a consequence of these conditions, carcasses decompose quickly. In the Fort River, western Massachusetts, USA, carcasses are essentially completely decomposed within 1–2 weeks after spawning (B. E. Kynard, personal observation). This is in strong contrast, for example, to decomposition rates observed for fall-spawning Atlantic salmon *Salmo salar* in Scottish Highland rivers (Nislow, unpublished data) where >50% of the carcass mass was still present months after placement in rearing streams in January. As a consequence, for the same nutrient and material total load (g) contained in spawners, loading rates (g/time) will be much greater for lamprey than for species with spawning timing and decomposition rates similar to Atlantic salmon (Figure 1).

Habitat Use Factors

In addition to life history characteristics, the types of spawning and juvenile rearing habitats used by anadromous species strongly influences their potential contribution to freshwater ecosystems. The ability of lamprey to access and use small, upland rivers makes it more likely that they will be important contributors of nutrients and materials. Total nutrient loading tends to be higher in larger, lowland rivers than in smaller upland rivers and streams (Lyle and Elliott 1998) due to greater residential, agricultural, and industrial use (Stockner et al. 2000; Nislow 2005) along with longitudinal increases in the storage capacity of rivers and floodplains (Naiman et al. 1987; Montgomery and Buffington 1997), so that inputs from migratory fish are likely to be a smaller proportion of total budgets in these downstream reaches. Further, in the north Atlantic basin, small rivers and streams tend to be most vulnerable to nutrient depletion associated with timber harvest, acidification, dams, and other anthropogenic factors, a phenomenon known as cultural oligotrophication (Stockner et al. 2000). As an illustration of this difference, Lyle and Elliott (1998) found that anadromous salmonids accounted for a very small fraction (<0.1%) of the total annual flux of nutrients in large river systems in the UK, while in a small oligotrophic Norwegian river, Atlantic salmon could account for as much as 18% of the total annual P budget (Jonsson and Jonsson 2003).

Differential use of rearing habitats within rivers may also influence the magnitude and direction of nutrient transport. Sea lamprey larvae and ammocoetes generally show a directed downstream movement from spawning areas with increasing age in

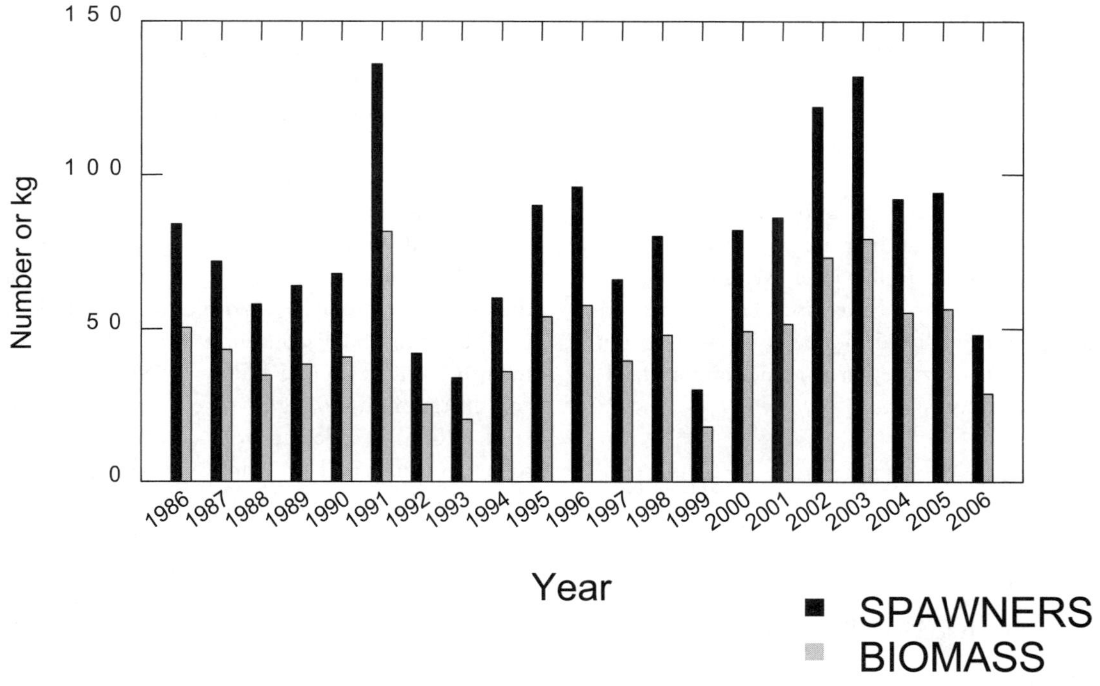

Figure 1.—Time series of estimated spawner abundance (numbers and biomass) for the upper spawning section of the Fort River, Massachusetts, USA from 1986 to 2006.

large river systems (Beamish and Potter 1975). In the Connecticut River, juvenile abundance tends to be very high in large tributary and main-stem habitats that are not appropriate for spawning (Kynard, unpublished data), an observation also made for Pacific lamprey *L. tridentata* (Torgerson and Close 2004). While some of this displacement is likely involuntary and associated with floods, an additional factor is the increase in productivity and growth potential of fish in downstream habitats, along with increased availability of fine-substrate (silt bed habitats) in lower-gradient reaches. As a consequence, a substantial percentage of the nutrients being exported by out-migrating juveniles are potentially not derived from the upstream spawning reach and does not subtract from net import by spawning adults.

Trophic Factors

One of the most striking aspects of sea lamprey life history is the major ontogenetic change in feeding strategy. As juveniles (ammocoetes), sea lamprey live in the stream bottom and filter-feed on diatoms and other algae, along with organic detritus (Maitland 2003). After developing into parasitic transformers, fish switch to feeding on coastal marine fishes and continue this foraging style as adults. This magnitude of change in trophic position (an increase of between two and three full trophic levels) is unique to north Atlantic anadromous fish and has important implications for mercury (Hg) accumulation and transport. Hg concentrations can increase by a factor of 10 with each increase in trophic level (Mergler et al. 2007). The combination of increased Hg concentration, along with increased biomass of spawning adults, suggests that lampreys may be net importers of Hg to stream ecosystems.

Case Study: Nutrient and Hg Import in a Wild Population in the Fort River, Western Massachusetts, USA

Study Site

The Fort River is a small (drainage area = 155 km^2) tributary of the Connecticut River draining into the main-stem Connecticut near the town of South Hadley, Massachusetts, USA. There are no barriers to fish migration in the Fort River, although access to the upper Connecticut and Fort rivers is regulated by Holyoke Dam ~20 river km downstream. Spawning occurs in several discrete sites where suitable gravel-cobble substrate is available (B. E. Kynard, unpublished data). The upper spawning site consists of a 300 m long (~5 m wide) section draining 39 km^2 of the upper Fort River watershed, which is largely an intact transitional hardwood forest (Fisher and Carpenter 1976). This river section that has been surveyed for the past 20 years by conducting a visual count of lamprey redds multiple times throughout the spawning period (late May through late June), depending on water temperature. To estimate spawner numbers, we multiplied the redd count by 2, as redds were generally used by single male–female pairs. This method yields a minimum estimate because it does not include a few males each year that build a late starter redd but do not spawn.

Calculating Nutrient and Hg Import

To calculate N and P import from spawning lampreys, we first multiplied spawner number by individual spawner mass to obtain total spawner biomass for a given year. This value is a slight underestimate of total adult lamprey biomass as those few adults that fail to find a mate and construct a redd also die on the spawning grounds. Individual spawner mass in the Connecticut River basin is 798 g (Stier and Kynard 1986) with very little variation (<2%) within and among years and sexes. We then multiplied total spawner biomass by standard N (0.03 g/g) and P concentrations (0.0047 g/g) for fish tissue (Lyle and Elliott 1998) and by Hg concentrations (400 ng/g) taken from Drevnick et al. (2006) to calculate total yearly import per square meter, based on a total spawning section area of 1,500 m^2. Finally, to give an estimate of the relative contribution of lamprey-derived nutrients, we compared lamprey-derived P to total annual P loading measured by Meyer and Likens (1979) in Bear Brook, New Hampshire, a stream with similar drainage are and land cover as the Fort River. As most MDNM from anadromous sea lamprey is imported, during 1 month, we also compared MDNM to monthly Bear Brook P loading.

Nutrient and Hg Import Results

Over the 20-year study period, an average of 40 redds (range, 15–68) were observed in the upper spawning site of the Fort River, translating into an average of 80 (range, 30–136) spawners and 46.7 kg total biomass (Figure 2). This biomass of spawners contributed an average of 0.15 g of P, 0.93 g of N, and 0.012 ng of Hg per square meter per year (Table 1). With respect to P, lamprey-derived contributions ranged from 5% to 20% of the total annual P loads in Bear Brook (Meyer and Likens 1979). However, given that almost all the lamprey-derived P loading occurs during 1 month (June), relative contribution for the month when carcasses are available is much higher, equal, or exceeding total Bear Brook P monthly loads in some years (Table 1.)

Factors Pointing Away from Sea Lamprey as Important Importers of MDNM

While consideration of several key life history, habitat, and trophic factors along with the results of a case study suggest an important role for sea lamprey as MDNM, other factors point in a different direction.

Life History Factors

The duration of freshwater residence can be an important determinant of the magnitude and direction of nutrient flux via anadromous fishes. In species such as pink salmon and chum salmon O. keta, the juvenile freshwater stage is very short (several months) with juveniles emigrating at very small sizes (Quinn 2005). As a consequence, nutrient export via juveniles is negligible and generation time is short, both of which increase net import via returning adults. In contrast, sea lamprey have an extended freshwater juvenile residence. Connecticut River sea lamprey generally spend ~4–5 years in freshwater (M. Horgan, Miami University, Oxford, Ohio, personal communication), which is similar to the residence time observed in another southerly river in the eastern Atlantic, the River Mondego in Portugal (Quintella et al. 2003). However, in a more northerly river, the St. John in New Brunswick, where colder water temperatures likely limit

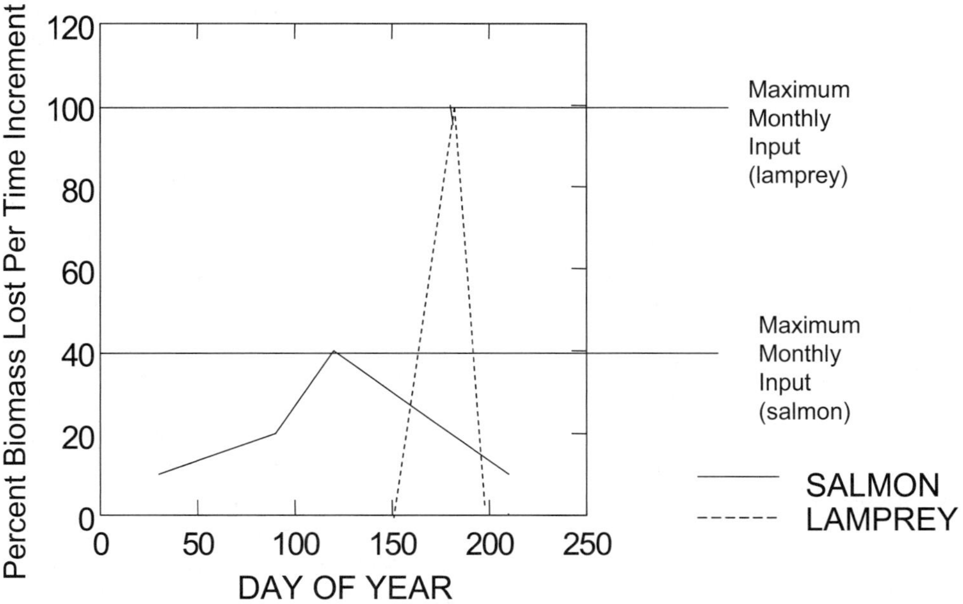

Figure 2.—Time course of carcass decomposition/disappearance of postspawned carcasses of sea lamprey (dashed line) compared to Atlantic salmon (solid line). Lines based on postspawning observations for lamprey in the Fort River, and on measured carcass mass loss rates for Atlantic salmon in a Scottish Highland stream.

Table 1.—Mean, minimum, and maximum annual transport of nitrogen, phosphorus, and mercury (Hg) from spawning sea lamprey to the upper spawning section of the Fort River, Masachusetts, USA based on a 20-year record of spawner abundance. Proportion of annual and monthly P load based on total annual loads measured at Bear Brook, New Hampshire (Meyer and Likens 1979).

	Mean	Minimum	Maximum
Hg (ng m^2)	0.012	0.005	0.022
Nitrogen (g m^2)	0.93	0.36	1.63
Phosphorus (g m^2)	0.15	0.06	0.26
Proportion of annual P load	0.12	0.05	0.20
Proportion of monthly P load	1.41	0.54	2.45

ammocoete growth, freshwater residence is 8–9 years (Beamish and Potter 1975).

Perhaps the key life history parameter determining whether lampreys represent a net gain or a net loss for freshwater ecosystems is the percentage of juveniles that survive to spawn (hereafter referred to as adult survival). The size ratio of adults to juveniles sets an approximate balance point for the direction of nutrient transfer with respect to adult survival. Lamprey increase their individual mass about two orders of magnitude while at sea. Assuming that proximate composition does not change markedly, adult survival of ~1% is the break-even point for nutrient and material transfer—below this value, lamprey represent a net loss to freshwater ecosystems. Unfortunately, we have essentially no information on adult survival rates in wild anadromous sea lamprey populations, and estimates from landlocked populations vary. In a widely used population model, transformer to spawner survival is assumed to be very high (50%) (Bence et al. 2003), which would make out-migrating juveniles largely inconsequential to the nutrient and material mass-balance. Other landlocked lamprey studies indicate significantly lower adult survival (minimum estimates of 4–8% based on tag recovery) (Stewart et al. 2003) which would, however, still result in substantial MDNM import if anadromous populations had similar adult survival rates. In general, however, survival rates for anadromous populations, which must contend with more extreme changes in habitats, as well as long and hazardous migrations, are considerably lower than landlocked populations (Jonsson and Jonsson 1993). In a 2-year study of a wild anadromous population of Pacific lamprey in Cedar Creek, Washington, USA (Stone et al. 2002), the population consisted of ~40 spawners and more than 3,000 juveniles (ammocoetes plus transformers). While the study did not measure survival per se, this ratio of adult to juvenile biomass is much closer to the break-even point for nutrient transport.

In the absence of data on adult survival in anadromous sea lamprey populations, it is worth considering several general aspects of sea lamprey life history and reproductive strategy that may influence whether they are a net nutrient/material gain or loss for freshwater. During transformation, sea lamprey shift to a parasitic existence and migrate to the sea. In many host–parasite systems, a very small percentage of individuals find suitable hosts (Anderson and May 1978). The reproductive strategies of spawning lamprey, which produce large numbers (hundreds of thousands) of small eggs, is consistent with overall high mortality rates characteristically experienced by parasites. These high fecundities may make it possible for populations to be sustained with low numbers of spawning adults, limiting potential nutrient and material import. Further, survival during parasitic life history stages is strongly dependent on host abundance and susceptibility (Anderson and May 1978). Evidence from landlocked populations indicates that spawner abundance is much more closely related to the abundance of important host fishes for newly transformed parasitic-phase lampreys than to the availability of spawning habitat (Young et al. 1996). Strong linkage between host abundance, adult survival, and nutrient and material import suggest that the role of sea lamprey may be dependent on the status of important marine and coastal host fishes. In the northwest Atlantic, anadromous and coastal clupeid fishes, which were likely important hosts for recently transformed parasitic-phase sea lampreys, have declined greatly as a consequence of river modification and overharvest (Saunders et al. 2006), potentially impacting the adult survival and ecological importance of anadromous sea lamprey. However, in the Connecticut River, lamprey abundance has increased while clupeids have decreased (Stier and Kynard 1986), suggesting that anadromous clupeids may not be the key posttransformer host species in this system.

Habitat Factors

Sea lamprey have fairly stringent habitat requirements for nest building and spawning. Maitland (2003) reported that nests are generally constructed of gravels between 9 and 60 mm, a size range that is similar to but narrower than that required by anadromous Atlantic salmon and brown trout (also known as sea trout) *Salmo trutta* (Armstrong et al. 2003). It is thought that the upper size limit of spawning substrate is set by limits on the ability of spawners to move large rocks. However, sea lamprey in the Connecticut River appear to be able to use somewhat larger substrate sizes, with male and female spawners often working together to move larger rocks (Kynard, personal observation) As a result of these habitat constraints, a relatively small percentage of the total river area, even for small rivers and streams, may be suitable for spawning. For example, in the Fort River, more than 90% of the redds are found in less than 5% of the total river length. This limitation will ultimately constrain the size of the spawning run and the magnitude of nutrient and material transport.

Trophic Factors

The potential for sea lamprey to be net importers of Hg to freshwater ecosystems is co-determined by (1) increased biomass of adults and (2) higher Hg concentration per unit biomass. The ontogenetic change in trophic level associated with the parasitic phase in sea lamprey should, given all else equal, result in increased Hg concentration. However, Drevnick et al. (2006) found that Hg concentrations in adult sea lamprey from the Connecticut River basin were highly variable and overlapped broadly with concentrations in ammocoetes. If this is the case generally, then Hg net import would be the result only of increases in net biomass of spawners and subject to the same considerations and constraints discussed in the previous two sections.

Lack of increased Hg concentration in adults, in spite of higher trophic position, may be the result of several factors. Hg methylation rates, which control the conversion of inorganic Hg to toxic, bioavailable methyl Hg, tend to be lower in marine environments due to high chloride ion concentrations (Fitzgerald et al. 2007) so that fish growing in marine environments have lower Hg concentrations than fish growing in freshwaters. As an example, Riget et al. (2000) found that Hg concentrations were 10–15× higher in lake-resident Arctic char *Salvelinus alpinus* than in co-occurring anadromous individuals. In addition, given all else equal, animals that grow slowly and inefficiently have been shown to accumulate more Hg per unit biomass than animals that grow quickly and efficiently, a phenomenon known as growth dilution (Simoneau et al. 2005; Karimi et al. 2007). Given the differences in growth rates and feeding ecology of ammocoetes (slow, inefficient growth and poor quality food resources) versus parasitic-phase (rapid, efficient growth on a high quality food source) lamprey, growth dilution may play an important role in determining relative Hg concentrations and the magnitude and direction of Hg transport in lamprey and anadromous fish in general.

Conclusions, Management Implications, and Research Recommendations

The unique ecology and life history of anadromous sea lamprey has important implications for the role of this species in aquatic ecosystems. With respect to nutrient and material transport, lamprey generally appear to use the right habitats (small, oligotrophic rivers and streams), at the right time (late spring, a time of high nutrient demand in north temperate systems) and do the right thing (die after spawning) to make them potentially important contributors of marine-derived nutrients to freshwater ecosystems in the north Atlantic region. Further, the ontogenetic shift in feeding ecology and trophic level, which is not observed in any other anadromous species, has major implications for material transport, specifically for substances such as Hg, which biomagnify along aquatic food chains. A more thorough understanding of the fate and effect of these nutrients and materials is necessary to evaluate ultimate impacts on freshwater ecosystems. For example, in the Pacific Northwest, salmon-derived MDNM is incorporated into fish and wildlife via direct consumption of carcasses, as well as indirectly via algal and invertebrate production (Kline et al. 1994). The relative importance of these two routes will strongly determine Hg transfer to higher trophic levels. Experimental additions of carcasses, coupled with isotope tracer experiments of the type used extensively in Pacific Northwest ecosystems (Bilby et al. 1998), would make a valuable contribution to this understanding.

While several attributes of sea lamprey point towards a significant ecological role other attributes suggest a lesser role under some circumstances. Perhaps most importantly, little information exists on adult survival rates in natural populations. However, the observed dependence of adult population size on the abundance of prey resources in landlocked populations has significant implications for management and conservation of anadromous populations. It suggests a strong connection between freshwater and coastal marine ecosystem health, as the ability of sea lamprey to import nutrients may depend strongly on the abundance of anadromous and coastal fishes. Many of these stocks have been declining for decades (Musick et al. 2000). Identifying key host species for anadromous sea lamprey, in the same way that bloater *Coregonus hoyi* was identified as a key species for Great Lakes populations, should help in this integrated conservation effort. In addition, it may be possible to use the frequency of lamprey marks on coastal fishes, which form at attachment and feeding sites on host fishes, to provide indices of lamprey abundance (Bence et al. 2003). Lamprey marks have been correlated to spawner abundance for landlocked Great Lakes populations (Smith et al. 1974).

One key question with respect to management and conservation is the extent to which population viability (an essential management target) coincides with ecological function. The high fecundity and parasitic life history of sea lamprey also suggest that population abundance sufficient to maintain population viability may be insufficient to fully achieve ecological contribution goals. Specifically, high adult fecundities may make it possible for relatively few adults to sustain local populations, but with either limited import or net export of nutrients and materials. This issue has been considered for Pacific salmonids, leading to recommendations that escapement levels be set high enough to ensure both population viability and adequate provision of marine-derived nutrients (Bilby et al. 2001). Similar considerations should apply to sea lamprey populations, particularly those populations that are either currently exploited (Almeida et al. 2002) or being considered for exploitation.

Stringent habitat requirements may limit the import of marine-derived nutrients and materials via wild anadromous sea lamprey at the watershed scale. These habitat restrictions may make it more appropriate to evaluate lamprey contributions at a smaller scale within river systems (i.e., the spawning reach—100 m to km). Given these smaller-scale considerations, it is crucial to understand within-river dispersal and movements of juvenile-stage lampreys in order to develop accurate mass-balance models of nutrient and material transport. Long-term demographic studies that incorporate these movements, coupled with experimental manipulations of carcasses and isotope tracer experiments, are critically needed to better understand the present and potential future role of lamprey in the transport of marine-derived nutrients and materials.

Acknowledgments

The authors would like to thank the conveners of ANACAT II, two anonymous reviewers, and R. Cunjak who organized this special session.

References

Almeida, P. R., Quintella, B. R., and N. M. Diaz. 2002. Movement of radio-tagged sea lamprey during the spawning migration in the River Mondego, Portugal. Hydrobiologia 483:1–8.

Anderson, R. M., and R. M. May. 1978. Regulation and stability of host-parasite interactions. Journal of Animal Ecology 47:219–249.

Applegate, V. C. 1950. Natural history of the sea lamprey, Petromyzon marinus, in Michigan. U.S. Fish and Wildlife Service Special Scientific Report: Fisheries 55.

Armstrong, J. D., P. S. Kemp, G. J. A. Kennedy, M. Ladle, and N. J. Milner. 2003. Habitat requirements of Atlantic salmon and brown trout in rivers and streams. Fisheries Research 62:143–170.

Beamish, F. W. H. 1980. Biology of the North Ameri-

anadromous sea lamprey, *Petromyzon marinus*. Canadian Journal of Fisheries and Aquatic Sciences 37:1924–1943.

Beamish, F. W. H., and I. C. Potter. 1975. The biology of the anadromous sea lamprey (*Petromyzon marinus*) in New Brunswick. Journal of Zoology 177:57–72.

Beamish, R. J., and T. G. Northcote. 1989. Extinction of a population of anadromous parasitic lamprey, *Lampetra tridentate*, upstream of an impassible dam. Canadian Journal of Fisheries and Aquatic Sciences 46:420–425.

Bence, J. R., R. A. Bergstedt, G. C. Christie, P. A. Cochran, M. P. Ebener, J. F. Koonce. M. A. Rutter, and W. D. Swink. 2003. Sea lamprey parasite-host interactions in the Great Lakes. Journal of Great Lakes Research 29(Supplement 1):252–283.

Bilby, R. E., B. R. Fransen, P. A. Bisson, and J. K. Walter. 1998. Response of juvenile coho salmon (*Oncorhynchus kisutch*) and steelhead (*Oncorhynchus mykiss*) to the addition of salmon carcasses to two streams in southwestern Washington, USA. Canadian Journal of Fisheries and Aquatic Sciences 55:1909–1918.

Bilby, R. E., Fransen, B. R., Walter, J. K., Cederholm, C. J., and W. J. Scarlett. 2001. Preliminary evaluation of the use of nitrogen stable isotope ratios to establish escapement levels for Pacific salmon. Fisheries 26:6–14.

Christie, G. C., and C. L. Goddard. 2003. Sea Lamprey International Symposium (SLIS II): advances in the integrated management of sea lamprey in the Great Lakes. Journal of Great Lakes Research 29(Supplement 1):1–14.

Connecticut River Atlantic Salmon Commission. 1998. Strategic plan for the restoration of Atlantic salmon to the Connecticut River. U.S. Fish & Wildlife Service. Available: www.fws.gov/r5src (July 2007).

Dempson, J. B., and T. R. Porter. 1993. Occurrence of sea lamprey, Petromyzon marinus, in a Newfoundland River, with additional records from the northwest Atlantic. Canadian Journal of Fisheries and Aquatic Sciences 50:1265–1269.

Drevnick, P. E., M. J. Horgan, T. P. Oris, and B. E. Kynard. 2006. Ontogenetic dynamics of mercury accumulation in the northwest Atlantic sea lamprey. Canadian Journal of Fisheries and Aquatic Sciences 63:1058–1066.

Fisher, S. G., and S. R. Carpenter. 1976. Ecosystem and macrophyte primary production of the Fort River, Massachusetts. Hydrobiologia 47:175–187.

Fitzgerald, W. F., C. H. Lamborg, and C. D. Hammerschmidt. 2007. Marine biogeochemical cycling of mercury. Chemical Reviews 107:641–662.

Gende, S. M., R. T. Edwards, M. F. Willson, and M. S. Wipfli. 2002. Pacific salmon in aquatic and terrestrial ecosystems. Bioscience 52:917–928.

Jardine, T. D., M. Roussel, S. C. Mitchell, and R. A. Cunjak. 2009. Detecting marine nutrients and organic matter inputs into multiple trophic levels in streams of Atlantic Canada and France. Pages 427–445 *in* A. J. Haro, K. L. Smith, R. A. Rulifson, C. M. Moffitt, R. J. Klauda, M. J. Dadswell, R. A. Cunjak, J. E. Cooper, K. L. Beal, and T. S. Avery, editors. Challenges for diadromous fishes in a dynamic global environment. American Fisheries Society, Symposium 69, Bethesda, Maryland.

Jonsson, B., and N. Jonsson. 1993. Partial migration in fishes: niche shift vs. sexual maturation. Reviews in Fish Biology and Fisheries 3:348–365.

Jonsson, B., and N. Jonsson. 2003. Migratory Atlantic salmon as vectors for the transfer of energy and nutrients between freshwater and marine environments. Freshwater Biology 48:21–27.

Karimi, R., C. Y. Chen, P. C. Pickhardt, N. S. Fisher, and C. L. Folt. 2007. Stoichiometric controls of mercury dilution by growth. Proceedings of the National Academy of Sciences of the United States of America 104:7477–7482.

Kircheis, F. W. 2004. Sea lamprey *Petromyzon marinus* L. 1758. F.W. Kircheis L.L.C., Carmel, Maine.

Kline, T. C., Jr., Goering, J. J., Mathisen, O. A., Poe, P. H., Parker, P. L., and R. S. Scanlan. 1994. Recycling of elements transported upstream by runs of Pacific salmon: II. $\delta^{15}N$ and $\delta^{13}C$ evidence in the Kvichak River watershed, Bristol Bay, southwestern Alaska. Canadian Journal of Fisheries and Aquatic Sciences 50:2350–2365.

Lyle, A. A., and J. M. Elliott. 1998. Migratory salmonids as vectors of carbon, nitrogen, and phosphorus between marine and freshwater environments in north-east England. The Science of the Total Environment 210/ 211:457–468.

Maitland, P. S. 2003. Ecology of the river, brook and sea lamprey. English Nature, Conserving Natura 2000 Rivers Ecology Series 5, Peterborough.

Mergler, D., H. A. Anderson, L. H. M. Chan, K. R. Mahaffey, M. Murray, M. Sakamoto, and A. H. Stern. 2007. Methylmercury exposure and health effects in humans: a worldwide concern. Ambio 36:3–11.

Meyer, J. L., and G. E. Likens. 1979. Transport and transformation of phosphorus in a forest stream ecosystem. Ecology 60:1255–1269.

Montgomery, D. R., and J. M. Buffington. 1997. Channel-reach morphology in mountain drainage basins. Geological Society of America Bulletin 109: 596–611.

Musick, J. A., M. M. Harbin, S. A. Berkeley, G. H. Burgess, and A. M. Eklund. 2000. Marine, estuarine, and diadromous fish stocks at risk of extinction in

North America (exclusive of Pacific salmonids). Fisheries 25:6–30.

Naiman, R. J., J. M. Melillo, M. A. Lock, T. E. Ford, and S. R. Reice. 1987. Longitudinal patterns of ecosystem processes and community structure in a subarctic river continuum. Ecology 68:1139–1156.

Naiman, R. J., J. M. Helfield, K. K. Bartz, D. C. Drake, and J. M. Honea. 2009. Pacific salmon, marine-derived nutrients, and the dynamics of aquatic and riparian ecosystems. Pages 395–425 *in* A. J. Haro, K. L. Smith, R. A. Rulifson, C. M. Moffitt, R. J. Klauda, M. J. Dadswell, R. A. Cunjak, J. E. Cooper, K. L. Beal, and T. S. Avery, editors. Challenges for diadromous fishes in a dynamic global environment. American Fisheries Society, Symposium 69, Bethesda, Maryland.

Nislow, K. H., J. D. Armstrong, and S. McKelvey. 2004. Transport of phosphorus by migratory Atlantic salmon: the role of management and demography. Canadian Journal of Fisheries and Aquatic Sciences 61:2401–2410.

Nislow, K. H. 2005. Forest change and stream fish habitat: lessons from 'Olde' and New England. Journal of Fish Biology 67:186–204.

Quinn, T. M. 2005. The behavior and ecology of Pacific salmon and trout. American Fisheries Society, Bethesda, Maryland and University of Washington Press, Seattle.

Quintella, B. R., N. O. Andrade, and P. R. Almeida. 2003. Distribution, larval stage duration and growth of the sea lamprey ammocoetes, *Petromyzon marinus* L., in a highly modified river basin. Ecology of Freshwater Fish 12:286–293.

Renaud, C. B. 1997. Conservation status of northern hemisphere lampreys (Petromyzontidae). Journal of Applied Ichthyology 13:143–148.

Riget, F., G. Asmund, and P. Aastrup. 2000. Mercury in Arctic char populations in Greenland. The Science of the Total Environment 245:161–172.

Saunders, R., M. Hachey, and C. W. Fay. 2006. Maine's diadromous fish community: past, present and implications for Atlantic salmon recovery. Fisheries 31:537–547.

Simoneau, M., M. Lucotte, S. Garceau, and D. Laliberte. 2005. Fish growth rates modulate mercury concentrations in walleye (*Sander vitreus*) from eastern Canadian lakes. Environmental Research 98:73–82.

Smith, B. R., J. J. Tibbles, and B. G. H. Johnson. 1974. Control of the sea lamprey, *Petromyzon marinus*, in Lake Superior, 1953–70. Great Lakes Fishery Commission, Technical Report 26, Ann Arbor, Michigan.

Stier, K., and B. E. Kynard. 1986. Abundance, size, and sex ratio of adult sea-run sea lampreys, *Petromyzon marinus*, in the Connecticut River. Fisheries Bulletin 84:476–480.

Stewart, T. J., J. R. Bence, R. A. Bergstedt, M. P. Ebener, F. Lupi, and M. A. Rutter, 2003. Optimizing the sea lamprey control program in the Great Lakes. Journal of Great Lakes Research 29(Supplement 1):783–893.

Stockner, J. G., E. Rydin, and P. Hyenstrad. 2000. Cultural oligotrophication: causes and consequences for fisheries resources. Fisheries 25:7–14.

Stone, J., j. Pirtle and S. Barndt. 2002. Evaluate habitat use and population dynamics of Pacific lamprey in Cedar Creek, Washington. U.S. Fish & Wildlife Service, Columbia River Fisheries Program, Vancouver, Washington.

Torgerson, C. E., and D. A. Close. 2004. Influence of habitat heterogeneity on the distribution of larval Pacific lamprey at two spatial scales. Freshwater Biology 49:614–630.

Young, R. J., G. C. Christie, R. B. McDonald, D. W. Cuddy, T. J. Morse, and N. R. Payne. 1996. Effects of habitat change in the St. Marys River and northern Lake Huron on sea lamprey (*Petromyzon marinus*) populations. Canadian Journal of Fisheries and Aquatic Sciences 53(Supplement 1):99–104.

American Fisheries Society Symposium 69:495–496, 2009

Population and Habitat Restoration Preamble

Alex Haro
S.O. Conte Anadromous Fish Research Laboratory, U.S. Geological Survey Biological Resources Discipline, Turners Falls, Massachusetts 01376, USA

Diadromous fish populations are particularly difficult to understand, model, and manage because they traverse multiple habitats that present not only environmental, ecological, reproductive, and physiological challenges, but also frequently convey them across multiple management jurisdictions. Our knowledge of population-level effects is also dependent on the quality and extent of biological, population, and demographic data. For some species, such as Pacific salmon, populations are routinely monitored, life cycles are fairly well understood, and population trends are, in general, well documented. However, for other species such as anguillid eels, our understanding of life history is incomplete, population-level data are meager, and the trends in abundance are less clear. Application of "one size fits all" models, or models that assume closed, nonmigratory populations, therefore have little application to highly migratory diadromous species with life history traits that vary both inter- and intraspecifically. Similarly, classic population metrics and models, most of which were originally developed for stocks of exploited marine fishes, may not apply easily to populations that migrate between diverse and often multijurisdictional environments.

The habitat of diadromous fishes also plays a significant, if not primary, role in the sustainability of their populations; without adequate habitat in marine, estuarine, and freshwater environments, natural populations are generally unviable. Diadromous fishes are uniquely dependent on multiple habitats and, equally important, connectivity between those habitats. As such, provision and maintenance of migratory corridors can be as significant as habitat itself in the sustainability of diadromous fishes. Often, particularly in freshwater, essential habitats can be drastically curtailed or altered by both natural and anthropogenic causes. Restriction of access to, or alteration of, historic habitat by dams is the classic example of how barriers to migration can interfere with critical life history cycles. Dams not only create a migratory barrier, but can also alter habitat adjacent to the dam so that it is unsuitable for reproduction or growth, in both upstream and downstream directions. In such cases, simple provision of passage (often only in the upstream direction) is not enough to ensure connectivity and population viability. However, our knowledge of the details of habitat and connectivity requirements for many species is often limited.

Under circumstances of habitat loss, increased natural and anthropogenic mortality, and genetic bottlenecks, populations of diadromous fishes are often targeted for restoration through supplementation of hatchery-reared fish. In some cases, compensation by culture techniques can be justified, especially if population sizes are critically low or habitats are extremely limited. However, the response of specific stocks of a species to artificial propagation will vary with life history characteristics, dynamics of the local genetic structure, aspects of fish health, and environmental productivity. At best, culture techniques should be viewed as short-term measures to sustain populations until adequate habitat and genetic diversity can be restored. Newly developed guidelines to minimize the adverse effects of fish culture suggest that stocks be supplemented by using progeny from local stocks (especially for species with strong homing behavior) and by carefully considering the effects of stocking on genetic population viability. However, these guidelines have yet to be implemented widely by fisheries managers and at a global scale.

The papers in this session focus on only some of the major aspects of population and habitat restoration of diadromous fishes. The variety of historic restoration techniques applied to salmonid populations and their relative effectiveness is critically reviewed by Jonsson and Jonsson (2009, this volume), who promote an adaptive approach to restoring populations and habitats and who document some of the pitfalls of hatchery management at various life history stages. Secor and Kerr (2009, this volume) recognize that diversity of life cycles in fish populations (i.e., individuals within populations that do not conform to a single life cycle pattern) can effectively offset environmental stochasticity and contribute to long-term persistence of populations. Castro-Santos et al. (2009, this volume) revisit the shortcomings in our knowledge of fishway design and performance—critical for successful migration and reproduction in diadromous fishes—and offer a new rationale for understanding correlates of fish passage performance. A review of a large-scale restoration of an entire fish community in the marshes of southern Iraq by Mohamed et al. (2009, this volume) illustrates the important biological role of large coastal marsh habitats for diadromous fishes and the significance of the maintenance of a natural flow regime in biodiversity and population sustainability. Finally, Woodland et al. (2009, this volume) evaluate how historic changes in water quality in the tidal freshwater Hudson River affect nursery habitat suitability for shortnose sturgeon, and identify future potential threats from diminished nursery habitat due to warming temperatures and increased benthic oxygen demand by invasive zebra mussels.

Our understanding of population and habitat restoration for diadromous fishes will continue to improve, as will applied methods and techniques to preserve sustainable and robust populations, for both exploited and unexploited species. However, at present, the rates of decline of some populations and destruction or alteration of critical supporting habitats often outpace our ability to apply the tools and resources we have at our disposal. Only a small fraction of anthropogenic migratory barriers for diadromous fish have been mitigated for provision of passage, and demand for hydroelectric development poses new threats of tidal and hydrokinetic generation to migrating fishes in both developed and developing countries. Clearly, the need for continued research and managements efforts for population and habitat restoration is critical and will remain so for the foreseeable future.

References

Castro-Santos, T., A. Cotel, and P. Webb. 2009. Fishway evaluations for better bioengineering—an integrative approach. Pages 557–575 *in* A. J. Haro, K. L. Smith, R. A. Rulifson, C. M. Moffitt, R. J. Klauda, M. J. Dadswell, R. A. Cunjak, J. E. Cooper, K. L. Beal, and T. S. Avery, editors. Challenges for diadromous fishes in a dynamic global environment. American Fisheries Society, Symposium 69, Bethesda, Maryland.

Jonsson, B., and N. Jonsson. 2009. Restoration and enhancement of salmonid populations and habitats with special reference to Atlantic salmon. Pages 497–535 *in* A. J. Haro, K. L. Smith, R. A. Rulifson, C. M. Moffitt, R. J. Klauda, M. J. Dadswell, R. A. Cunjak, J. E. Cooper, K. L. Beal, and T. S. Avery, editors. Challenges for diadromous fishes in a dynamic global environment. American Fisheries Society, Symposium 69, Bethesda, Maryland.

Mohamed, A. R. M., N. A. Hussain, S. S. Al-Noor, B. W. Coad, and F. M. Mutlak. 2009. Status of diadromous fish species in the restored East Hammar marash in southern Iraq. Pages 577–556 *in* A. J. Haro, K. L. Smith, R. A. Rulifson, C. M. Moffitt, R. J. Klauda, M. J. Dadswell, R. A. Cunjak, J. E. Cooper, K. L. Beal, and T. S. Avery, editors. Challenges for diadromous fishes in a dynamic global environment. American Fisheries Society, Symposium 69, Bethesda, Maryland.

Secor, D. H., and L. A. Kerr. 2009. Lexicon of life cycle diversity in diadromous and other fishes. Pages 537–556 *in* A. J. Haro, K. L. Smith, R. A. Rulifson, C. M. Moffitt, R. J. Klauda, M. J. Dadswell, R. A. Cunjak, J. E. Cooper, K. L. Beal, and T. S. Avery, editors. Challenges for diadromous fishes in a dynamic global environment. American Fisheries Society, Symposium 69, Bethesda, Maryland.

Woodland, R. J., D. H. Secor, and E. J. Niklitschek. 2009. Past and future habitat suitability for the Hudson River populations of shortnose sturgeon: a bioenergetic approach to modeling habitat suitability for an endangered species. Pages 589–604 *in* A. J. Haro, K. L. Smith, R. A. Rulifson, C. M. Moffitt, R. J. Klauda, M. J. Dadswell, R. A. Cunjak, J. E. Cooper, K. L. Beal, and T. S. Avery, editors. Challenges for diadromous fishes in a dynamic global environment. American Fisheries Society, Symposium 69, Bethesda, Maryland.

American Fisheries Society Symposium 69:497–535, 2009

Restoration and Enhancement of Salmonid Populations and Habitats with Special Reference to Atlantic Salmon

Bror Jonsson* and Nina Jonsson
Norwegian Institute for Nature Research, Gaustadalléen 21, Oslo N-0349, Norway

Abstract.—Populations of Atlantic salmon *Salmo salar* can be restored and enhanced through planting of green or eyed eggs (embryos) in rivers and by releasing fry, parr, smolts, or postsmolts. The success of the releases varies with time and site of release, broodstock origin, size and age of the fish, and rearing and release techniques applied. However, egg, fry or parr releases cannot be used for augmenting populations above the carrying capacity of the water course. To surpass the carrying capacity, the fish should be released as smolts or postsmolts. Smolts released in rivers during spring migrate to sea for feeding but return to the river of release for spawning. Atlantic salmon released at the postsmolt stage may return to the release site when adult, but thereafter, they may stray to any of a number of rivers for spawning. As a result of ecological interactions, released juvenile hatchery fish may partly displace, increase the mortality, and decrease the growth rate, adult size, reproductive output, biomass, and production of wild conspecifics through density-dependent mechanisms working in freshwater. Hatchery-reared Atlantic salmon are usually competitively inferior to wild conspecifics both during feeding and spawning in rivers, due to environmental impacts and genetic changes that occur during the juvenile rearing. Habitat restoration is preferred when restoring endangered, threatened, or weak populations. Degraded spawning habitats can be reconstructed, and poor freshwater quality can be mitigated. In regulated rivers, rapid fluctuations in water level should be avoided, and the migratory activity of the fish can be stimulated by increased water flow. Populations can also be enhanced by expanding the accessible nursery habitat by use of artificial fishways through human induced or natural migration hindrances. Adaptive management practice is useful when restoring and rehabilitating populations and habitats. More knowledge is needed about environmental and genetic influences on the phenotype of hatchery fish and how habitats constrain salmon production in rivers.

Introduction

Restoration ecology, or the study of renewing degraded, damaged, or destroyed ecosystems and populations, is a rapidly growing field, stimulated by new knowledge about population and community ecology, behavioral ecology, genetics, and evolution (Jordon, III et al. 1999; Van Andel and Aronson 2005). With the advancement of modern technology, the human ability to destroy habitats and use and overexploit populations have escalated at the same time as the awareness of our dependence of and responsibility for intact ecosystems have matured. Thus, researchers and laymen are searching for indicators of unspoiled ecosystems, and Atlantic salmon *Salmo salar* has become a symbol of clean, healthy aquatic ecosystems (Mills 1989).

Atlantic salmon forms anadromous populations. It spawns in freshwater. The offspring rear in rivers and lakes for 1–6 years before they migrate to sea as smolts, 10–30 cm in length. The postsmolts feed in the ocean for 1–4 years before attaining maturity and returning to their river of origin for spawning. An individual river may support one or more Atlantic salmon populations (Garcia de Leaniz et al. 2007).

Atlantic salmon is a very popular sporting species, and it is recognized as a delicacy. Therefore, the fishing pressure is high. The juvenile production in freshwater is recognized as the main limiting factor

* Corresponding author: bror.jonsson@nina.no

for the production of wild Atlantic salmon (Jonsson et al. 1998). Therefore, to increase the catch, populations have been enhanced by hatcheries for more than 150 years (Jonsson and Fleming 1993). In parallel with this activity, the salmon catch in the North Atlantic increased to a maximum in the mid-1970s. Since then, the catch has declined. This may be partly a result of reduced smolt production (Klemetsen et al. 2003; Jonsson and Jonsson 2004a; Quinn et al. 2006).

In freshwater, habitat destruction and alterations and introductions of exotic organisms have decimated salmon populations. For instance, acidification of Norwegian rivers has eradicated the Atlantic salmon in 25 water courses during the past century (Hesthagen and Hansen 1991). The monogene parasite *Gyrodactylus salaris*, accidentally introduced to Norway from the Baltic in the early 1970s, has since then decimated the juvenile Atlantic salmon production by between 80% and 90% in another 45 Norwegian rivers (Johnsen and Jensen 1991). These losses represent a reduction of about 50% in the Norwegian production of wild Atlantic salmon smolts (Hesthagen and Hansen 1991). Since the 1970s, the farming of Atlantic salmon has increased gradually and, with that, the escapement of fish from fish farms. The escapees may impact the wild Atlantic salmon populations negatively through ecological interactions, genetic introgressions, and the spreading of contagious diseases (Johnsen and Jensen 1994; Fleming et al. 2000; McGinnity et al. 2003; Jonsson and Jonsson 2006). Movements of hatchery fish between regions also increase the potential for the spreading of diseases such as furunculosis and proliferative kidney disease, killing salmon in the wild (Johnsen and Jensen 1994; Tops et al. 2006). Table 1 summarizes effects of hatchery salmon on wild conspecifics.

Survival and growth of Atlantic salmon at sea have also decreased with the growth of the salmon farming industry, and the abundances of marine parasites such as sea lice *Lepeophtheirus salmonis* and *Caligus* spp. have increased, with harmful effects on salmonids at sea (Heuch et al. 2005; Skilbrei and Wennevik 2006; Hvidsten et al. 2007). The decline in Atlantic salmon production since 1980 may be also related to climate change with warmer water (Friedland et al. 2000, 2005) and decreased food abundance at moderate latitudes (Beaugrand and Reid 2003; Kallio-Nyberg et al. 2006). With warmer climate, the annual growth rate of Atlantic salmon in freshwater has increased with reduced age and size at smolting as a consequence (Jonsson et al. 2005). Small fish are generally more vulnerable to predation at sea than larger conspecifics (Sundström et al. 2007). The abundances of some crustaceans in the northeast Atlantic have decreased since the 1980s, possibly with trophic effects mediated through the food chain, resulting in reduced marine salmon production (Helle and Pennington 1999; Beaugrand and Reid 2003). At least part of the Atlantic salmon population appears to feed farther north and in colder water than they did recently, resulting in slower growth (Jonsson and Jonsson 2004b). Fishing also reduces stock abundances, but it appears not to have contributed significantly to the recent stock decline in Atlantic salmon (Dempson et al. 2004). At present, the abundance of wild Atlantic salmon is low, and stock enhancements are very popular (Potter and Crozier 2000; Klemetsen et al. 2003). In the Pacific, coho salmon *Oncorhynchus kisutch* and Chinook salmon *O. tshawytscha* have undergone parallel decreases during the same period, probably for some of the same or similar reasons as Atlantic salmon (Noakes et al. 2000; Mote et al. 2003; Beamish et al. 2004b).

Here, we review common methods for restoring, rehabilitating, and enhancing Atlantic salmon populations and habitats. In particular, we focus on supportive breeding and effects of fish releases. During a long period, this has been the chief method for augmenting Atlantic salmon stocks. Furthermore, we discuss impacts of released hatchery salmon on wild populations and reasons why the success of hatchery fish often deviates from that of wild conspecifics. Then, we review possibilities for restoring and improving the Atlantic salmon habitat by various methods. We sum up by discussing the trade-off between restoring or stocking rivers and how adaptive management can be useful in this context and present some important research directions.

Population Restoration and Enhancement

Supportive Breeding

Supportive breeding involves the gathering of gametes artificially stripped and fertilized. The resulting progeny are reared in hatcheries and released at vari-

Table 1.—Effects of released hatchery-reared fish on wild populations.

Classes of interactions	Responses	Sources
Ecological competition	Parr emigration from river stretches at high fish densities.	McMichael et al. 1999, 2000; McGinnity et al. 2003; Weber and Fausch 2003
	Parr mortality at high fish densities.	Nickelson et al. 1986; Vincent 1987; Nielsen 1994; McGinnety et al. 1997, 2003
	Reduced growth at moderate fish densities.	Bohlin et al. 2002; Imre et al. 2005
Genetic interbreeding	Reduced reproductive success, offspring survival and production.	Fleming et al. 1996, 1997, 2000; McGinnity et al. 1997, 2003, 2004
Spreading of diseases and parasites	Furunculosis (*Aeromonas salmonicida*).	Johnsen and Jensen 1994; Glover et al. 2006
	Gyrodactylosis (*Gyrodactylus salaris*).	Johnsen and Jensen 1994; Bakke and Harris 1998; Peeler et al. 2006.
	Salmon lice (*Lepeophtheirus salmonis* and *Caligus* spp.).	McVicar 2004; Heuch et al. 2005; Hvidsten et al. 2007
	Proliferative kidney disease (*Tetracatculoides bryosalmonae*)	Tops et al. 2006

ous life stages (i.e., eggs, fry, older parr, smolts, or postsmolts). Releases of eggs or unfed fry (alevins) are often used where environmental conditions during spawning limit recruitment. If nursery areas limit population size, such as in many rivers regulated for hydropower purposes, releases of older juveniles may be more suitable. Economic costs of these release practices vary directly with the length of the hatchery rearing required. However, since juvenile survival in freshwater increases with the length of the hatchery rearing, this may at least partly compensate for the additional rearing costs.

Hatcheries are important tools in the supplementation and enhancement of yields for fisheries. Sea ranching operations, involving the release of hatchery juveniles, which return to the point of release as adults, are used to support recreational and professional fisheries. Supportive breeding is used in many rivers regulated for hydropower production where dams isolate the fish from upstream spawning grounds or water is channeled away from the river. Hatchery methods and technology have been much improved during recent years in parallel with the growth of the salmon farming industry, and massive development of hatchery programs for the above purposes has resulted in Atlantic salmon becoming one of the most intensely, artificially supplemented organisms in the world.

Why Are Fish Released?

There are periods in the life cycle of salmon with marked reductions in abundance because of population bottlenecks. Fish are released to escape the effect of such bottlenecks. Positive effects of stocking can be achieved if natural reproduction in the river is below its carrying capacity, if Atlantic salmon is released in habitats above the natural salmon producing stretches of rivers or in rivers where the spawning but not the juvenile rearing habitat is degraded.

Periods of high mortality occur when there is a marked ontogenetic shift in diet or habitat. Examples of such changes in diet are yolk to first feeding on small drifting invertebrates and then a diet change from invertebrates to fish. Examples of habitat change with fish age are open water in riffles to deeper pools in streams, pools in tributaries to the larger parent river, river to estuary, estuary to ocean, ocean to freshwater of the parent river, and river to spawning ground in the natal stream (Elliott 2001).

The resource limitation will affect the life stage most dependent on the resource, and a population bottleneck will occur. The chief bottlenecks occur in the early life stages such as the times of first feeding and smolting (Gibson 1993), although there may sometimes be population regulation later in life as well (Weatherley and Gill 1987; Shuter 1990). The bottlenecks may affect released hatchery fish as well as the local population. According to Sægrov et al. (2001), water discharge is a major factor influencing the carrying capacity of Atlantic salmonid parr. Based on studies from 11 rivers in western Norway, they found that the carrying capacity was inversely related to the natural logarithm of the water discharge between 2 and 70 m^3/s. Factors associated with high discharge constrain the parr production, especially during early summer when high water velocity restricts the area of available habitat.

Although the carrying capacity for salmon parr in a river changes from year to year, rivers can be characterized by an average carrying capacity with fluctuations around this average. The carrying capacity is largely determined by the variations in the physical and chemical conditions, the frequency of extreme events such as droughts and spates, the availability of food, and the density of other fish species and the density of different life stages of the same species (Elliott 2001). Thus, carrying capacity reflects the effect of all the environmental variables and density-dependent factors and sets the long-term maximum level of population density. As the population size approaches the carrying capacity of the area, emigration and mortality will increase (Einum and Nislow 2005). After that regulatory phase, mortality is influenced mainly by density-independent factors (Jonsson et al. 1998; Milner et al. 2003; Su et al. 2004), although exceptions exist where density-dependent mortality occurs at a later stage (Unwin 1997; Elliott and Hurley 1998). Thus, salmon rivers can be stocked to fill the available niches for salmon, and populations can be enhanced by releasing fish after the main periods of population regulation.

Salmon Stocking

Egg Planting

Salmon eggs (embryos) are placed in incubating boxes buried in the gravel bed of rivers or freely in the bottom substratum, imitating a natural salmon redd (Barlaup and Moen 2001; Johnson 2004). Newly fertilized eggs and eyed eggs are the two developmental stages usually used. (1) Newly fertilized eggs (green eggs) are planted between 24 and 48 h after the fertilization and water hardening. After that (but before the eye stage), the eggs are very sensitive to handling stress and easily killed if moved. (2) Eyed eggs are robust and tolerate substantial handling and are often used for planting (Wagner et al. 2006). The survival of planted eggs is variable (Table 2), and there is no systematic difference in survival between eggs planted in boxes or placed directly in the gravel substratum or whether they are buried as green or eyed eggs. The survival from fertilization to hatching is similar for the two methods (Kelly-Quinn et al. 1993). However, planting of eyed eggs is often preferred since this allows for proper veterinary health control of the spawners before the eggs are planted. Furthermore, when using eyed eggs, there is a less strict time constraint on the planting. Since newly fertilized eggs are very sensitive to movement after 48 h, movements of the substratum during freshets or spates may kill the eggs.

When eggs are planted directly into the gravel substratum, they are usually placed in areas where the salmon spawn. The selection of the site is critical to the survival of the eggs. Preferred spawning areas have variable particle size, and the nests have a few

Table 2.—Survival (%) of various developmental stages of hatchery-reared Atlantic salmon.

Stocked stage	Survival to	Survival %	References
Eggs	hatching	0–100	Reviewed in Barlaup and Moen 2001
Eggs	emergence	3.3–89	Reviewed in Barlaup and Moen 2001
Unfed fry	smolts	0.2–15	Rosseland 1975; reviewed in Fjellheim and Johnsen 2001
Fry	adults	0.7–5.9	Berg 1969; Hansen 1991
Smolts	adults	0–11.6	Reviewed in Finstad and Jonsson 2001

large stones in the centre of the egg pocket. These stones stabilize the redd and give a sheltered environment for the eggs. Usually, Atlantic salmon spawns between 500 and 1,000 eggs in each nest (Fleming 1996). The median particle size of the gravel substratum where salmon spawn is about 10% of their body length (Kondolf et al. 1993), and the eggs are often buried at 10–30 cm depth in the substratum, the deeper the larger the fish is (DeVries 1997). When mimicking natural redds, the egg pocket is covered by gravel washed free of fine sediment to allow proper oxygenation of the eggs. The survival of the eggs depends on the permeability of the gravel (Kondou et al. 2001), and there is negative correlation between the dissolved oxygen concentration and the mortality rate of the eggs (Malcolm et al. 2003). Furthermore, embryos developing at low dissolved oxygen concentrations are smaller at hatching than those developing under more favorable conditions (Youngson et al. 2004).

More commonly than planting eggs in artificial redds, they are planted in boxes or trays buried in the gravel substratum. Common devices are the Whitlock-Vibert box (Vibert 1949; Whitlock 1978), various types of perforated plastic boxes (Harris 1973; Scrivener 1988; Rubin 1995), and plastic trays (Raddum and Fjellheim 1995; Donaghy and Verspoor 2000). Common problems are a too high egg density in the boxes and that the eggs cluster together and become susceptible to bacterial and fungal infections, oxygen deficiencies, and sedimentation of fine-particulate material (Harshberger and Porter 1979; Chapman 1988; Scrivener 1988; Tabachek et al. 1993).

Under favorable conditions and properly done, the hatching success of planted eggs exceeds 90% (Humpesch 1985; Kelly-Quinn et al. 1993). A key to success is to provide suitable conditions such as proper gravel composition, burial depth, number of eggs per pocket, and hydrological conditions (Barlaup and Moen 2001). In most cases, egg plantings are more cost-effective than rearing and releasing hatchery fry, parr, or smolts. On the other hand, the survival is higher if the fish are released at a more developed stage (Coghlan and Ringler 2004; Johnson 2004).

Fry and Parr Stocking

Stocking of young salmon in rivers and lakes is a useful method if the habitat is spawning-site limited (Hyatt et al. 2005). Cultivation of Atlantic salmon commenced with the building of the first hatcheries in the 1850s. Initially, the fish were stocked as alevins and small fry, but older fish were released as the rearing technique improved. The stocking efforts were further stimulated by declining populations due to regulations of rivers for hydropower production and the continued acidification of rivers in northern countries, which started about 1875 (Hesthagen and Hansen 1991). River-owners' organizations, fishing societies, and management agencies ran hatcheries to enhance the river production and yield for fisheries, for conservation purposes to save populations at risk of extinction, or to re-establish populations that had been eradicated (Jonsson et al. 1999; Fleming and Peterson 2001). Often, the results of such releases were not evaluated, and most of the evaluations reported tend to be among the more successful ones.

The published results of fry and parr stocking vary. Particularly high survival of unfed Atlantic salmon fry was reported from releases in a tributary to the River Sandvikselva, south Norway (Rosseland 1975). There was no anadromous fish present prior to the release. A stocking density of two unfed fry per square meter gave about 0.3 smolt/m^2 (Table 1). This was a very productive stream, and a similar stocking density gave less than 8% of this when tested in tributaries of the River Vefsna, northern Norway (Johnsen et al. 1997b). In the latter case, the survival from smolts to returning adults was estimated at 2%. Later releases gave even poorer survival with 0.85% as the mean for a subsequent 6-year release period (Johnsen et al. 1997a).

There are also examples of parr releases giving excellent survival to adults. Hansen (1991) reported 2.3% survival to returning adults of 14 000 one-summer-old parr released in 1983. The fish were liberated above the natural Atlantic salmon producing area of the River Drammen, south Norway. When repeating the experiment in 1986, the survival of 50 000 one-summer-old parr released in the same area was 0.7%. The lower yield of the repeated release may be because the food base was exploited by the earlier release of parr or that the carrying capacity of the freshwater habitat was surpassed because too many fish were released. Releases of Atlantic salmon in rivers appear particularly successful if there is no other fish species present (Fjellheim and Johnsen 2001), but even with the presence of nonanadromous populations of salmonids such as brook char *Salvelinus fontinalis*, brown trout *S. trutta*, and rainbow trout *O. mykiss*, the parr survival of Atlantic salmon can be high (MacCrimmon 1954; Egglishaw and Shackley 1980; Kennedy and Strange 1986; Whalen and LaBar 1998; Jokikokko 1999; Jutila et al. 2003).

An example of an unfortunate result is the parr releases from the West-Norwegian River Teigdalselva, a tributary to the River Vosso (Fjellheim and Johnsen 2001). In total, 70,000 1-year-old parr were released giving almost no smolts. In this river, a large part of the water was directed to a hydropower station outside the catchment area. The carrying capacity for Atlantic salmon smolts had obviously decreased dramatically as a result of the decreased flow.

The success of the releases is influenced by the quality, size, and density of the stocked fish and time and place of stocking (Connor et al. 2004; Saltveit 2006). Jokikokko (1999) reported that both point and scatter stockings are suitable methods for supplementing Atlantic salmon parr in rivers. Letcher and Terrick (2001) reported that a wide range of developmental stages of Atlantic salmon fry will survive equally well and grow to a similar size when released in a natural system. On the other hand, even a small difference in introduction site and time can influence the migratory behavior of the fish (Pirhonen et al. 2003) and have long-term effects on body size, survival, and life history expressions (Letcher et al. 2004). Jokikokko and Jutila (2004) found that stocking of 1-year-old parr was economically more cost-effective than stocking of one-summer-old parr when restoring endangered stocks. A less cost-effective way of enhancement is the release of 2-year-old parr (Salminen et al. 2007). Large size at release has a positive effect on survival as also reported for masu salmon *O. masou* (Miyakoshi et al. 2003), but the economic cost of rearing one extra year is high. Therefore, there must be a good reason for an extra year of parr rearing before releasing the fish, such as avoidance of an effective local parr predator.

The release site influences the recapture rate. In some cases, stockings of Atlantic salmon parr in lakes have been very successful (Berg 1969; Pedley and Jones 1978; Pepper et al. 1992). However, the flow through of many lakes is small, and Hansen (1987) reported that one-summer-old parr released in Lake Storevatnet of the River Imsa, southwest Norway moved downstream over an extended period, compared with wild smolts produced in the downstream river. He attributed this to the low flow-through making it difficult for the fish to find the lake outlet. It is also possible that the River Imsa salmon are a river population not adapted to navigate through lakes. The outlet river is their original habitat to which they have been adapted through several thousand years. Further research is needed to establish whether difficulties in finding the lake outlet are an attribute of the habitat or a population-specific adaptation found in some Atlantic salmon populations.

Removing broodstock from rivers for hatchery rearing and release from an already depleted spawning stock may result in even further population decline (Saltveit 1998). In spite of extensive stocking of parr in the River Suldalslågen, southwest Norway, there has been a steady decline of Atlantic salmon in the river, where a large part of the water is channeled from the main river to the sea. Although about 50% of the smolts that leave the river are of stocked origin, most returning adults are naturally produced. In spite of this, broodstock was taken from the reduced number of returnees, leaving even fewer fish for reproduction. On this basis, the stocking program was advised to be discontinued. Thus, the results in various stocking programs vary, and poor results are reported in inferior habitats. There is little reason for stocking salmon at densities way above the carrying capacity of a system (Brännäs et al. 2004) because the density-dependent response is reduced parr survival and growth of the local fish (Imre et al. 2005). To enhance depleted populations, one may sample the broodstock in an abundant population nearby where the environmental conditions are similar, as the removing of spawners can be detrimental to the stock. Natural spawning is superior to fry or parr stocking in rivers with intact breeding grounds, and the stocking success is generally low in rivers where the density of naturally bred conspecifics is high (Crozier et al. 1997; Verspoor and Garcia de Leaniz 1997; Mowbray and Locke 1998).

Smolt Release

The productivity and size of the freshwater habitat constrain the sizes of Atlantic salmon populations (Jonsson et al. 1998), and the release of hatchery-reared smolts have been used to augment Atlantic salmon populations since the first part of the 20th century. Smolts have been released to compensate for habitat loss due to dam and impoundment building of streams and use of water for hydropower production. However, if the habitat is intact, the release of 1-year-old parr appears more cost effective than smolt releases (Jokikokko et al. 2006).

Hatchery smolts start their seaward migration immediately after release (Hansen and Jonsson 1985; Jonsson and Fleming 1993), and many return to the place of release when sexually mature (Hansen et al. 1993). To enhance the spawning population of a river, the smolts should be released in the river. Fish released in river estuaries stray more to other rivers and are delayed in their upstream spawning migration compared with smolts released higher upstream (Jonsson et al. 1994; Insulander and Ragnarsson 2001). The smolts should be released in spring, at the time of seaward migration for wild smolts in the same or similar, neighboring rivers. Smolts released at that time survive better and stray less frequently than fish released at other times of the year (Hansen and Jonsson 1989a, 1991a).

The survival of released hatchery salmon is often low (Table 3), usually less than half of that of wild smolts (Jonsson et al. 1991, 2003b; Jutila et al. 2003). Commonly, the recaptures of the adults are in the range of 0.5–3.0% (Finstad and Jonsson 2001), although recapture rates above 11% have been observed (Hansen and Jonsson 1990; Hansen et al. 1997). According to Moksness et al. (1998), the recapture rates should be above 10% to be economically profitable in sea ranching operations, which is only rarely obtained (Finstad and Jonsson 2001). The reduced survival of hatchery smolts may be partly caused by the artificial rearing conditions resulting in decreased smolt quality and poor handling and release procedures. Furthermore, the released fish may not be genetically adapted to the system of release or have a too small genetic variability (Ayllon et al. 2004; McGinnity et al. 2004; Garcia de Leaniz et al. 2007).

The yields of the releases differ depending on smolt size and age. Hansen and Jonsson (1989b) reported that 2-year-old smolts gave higher yields than 1 year olds. The yield of the 2 year olds varied between 125 and 1,050 kg/1,000 smolts released. A similar difference caused by smolt age was reported from smolt releases in southern Finland (Salminen et al. 2007). On the other hand, it is cheaper to produce 1- than 2-year-old smolts, which can make it economically more profitable to produce 1- than 2-year-old smolts as recently found in Norway (Jonsson et al. 2003b). Among similar-aged fish, large individuals appear to survive better than smaller ones as reported for masu salmon in Japan (Miyakoshi et al. 2003) and pink salmon *O. gorbuscha* in Alaska (Moss et al. 2005).

A large number of experiments have been performed to increase the survival of released hatchery smolts, and effects of the rearing and release methods have been tested (Finstad and Jonsson 2001). For instance, handling, transport, and anesthesia stress anadromous salmonids (Nikinmaa et al. 1983; Hansen and Jonsson 1988; Barton 2000), and stress-related cortisol surges can suppress the immunological capacity (Fries 1986; Iversen et al. 1998) and migratory activity (Specker and Schreck 1980). But even when handling and transport are kept at a minimum and no hatchery smolt is anesthetized within 2 weeks of release (Pickering et al. 1982), the survival rate of hatchery reared smolts is between one and two times lower than that of comparable groups of wild smolts (Jonsson et al. 2003b).

Time and place of release have been optimized through experimental releases during the 1980s and 1990s (Hansen and Jonsson 1986, 1989a, 1991a; Hansen et al. 1989; Jonsson et al. 1994). Experiments have been performed to adapt the parr to natural food items or predator training before release, but none of these have so far been successful in improving the sea survival substantially. Increased water level during the emigration period has a positive effect on the smolt survival (Hvidsten and Hansen 1988). Futhermore, physical exercise of the parr has proven to be positive for survival and growth in hatcheries (Jørgensen and Jobling 1993; Davidson 1997). So far, however, exercised Atlantic salmon have not significantly improved return rates to the river of release, although exercised fish strayed less to other rivers than unexercised fish (Skilbrei and Holm 1998).

Hatcheries tend to produce elevated levels of sexually mature male parr, which more often become freshwater resident than what immature parr do (Hansen et al. 1989). To increase their emigration rates to the same levels as those of immature smolts, their high steroid concentration can be decreased either through gonadal stripping or elevated

Table 3.—Changes occurring in hatcheries reducing the performance of released hatchery fish in nature.

Changes in	Changed character	Sources
Morphology	Body form and size	Taylor 1986; Swain et al. 1991; Fleming et al. 1994; Fleming and Einum 1997; Fiske et al. 2005; Von Cramon-Taubadel et al. 2005
	Distorted jaws	Fleming et al. 1994
	Fin damage	Höglund et al. 1997; Lellis and Barrows 1997; Ellis et al. 2002; Latremouille 2003
	Scale loss	MacLean et al. 2000; Lacroix and Knox 2005
	Adiposy	Rowe et al. 1991; Silverstein et al. 1999
Physiology and anadromy	Heart abnormity	Poppe et al. 2003; Seierstad et al. 2005
	Brain	Marchetti and Nevitt 2003; Lema et al. 2005
	Metabolic rate	Dunmall and Schreer 2003; Claireaux et al. 2005
	Smolting	Poole et al. 2003
	Hormone	Youngson and Webb 1992; McCormick et al. 2003
Life history characters	Growth rate	Jonsson et al. 1991a; Jonsson and Fleming 1993; Kistow 2004
	Survival	Piggins and Mills 1985; Jonsson and Fleming 1993; Jonsson et al. 1991a, 2003b; Kostow 2004; Saloniemi et al. 2004
	Smolt age	Økland et al. 1993; Yamamoto and Morita 2002; Jonsson et al. 2003b; Duston et al. 2005
	Age at maturity	Jonsson et al. 2003b; Kostow 2004; Patterson et al. 2004
	Reproductive output	Jonsson et al. 1996; Tamate and Maekawa 2000; Fleming et al. 2003; Quinn et al. 2004
	Longevity	Kostow 2004
Behavior	Time of river ascent	Jonsson et al. 1990b, 1994; Fleming et al. 1997; Skilbrei and Holm 1998
	Risk taking	Berejikian 1995; Fleming et al. 2002; Sundström et al. 2004
	Feeding behavior	Reiriz et al. 1998; Reinhardt 2001; Sundström and Johnsson 2001; Brown et al. 2003a, 2003b
	Aggressive behavior	Einum and Fleming 1997; Rhodes and Quinn 1998; Riley et al. 2005; Sundström et al. 2003; Yamamoto and Reinhardt 2003
	River movement	Jonsson et al. 1990a; Økland et al. 1995
	River stay	Jonsson et al. 1990a
	Straying to foreign rivers	Hansen et al. 1993; Jonsson et al. 2003a
	Predator recognition	Brown and Smith 1998; Mirza and Chivers 2000; Berejikian et al. 2003b; Vilhunen et al. 2005
	Refuge use	Griffiths and Armstrong 2002; Orpwood et al. 2004
	Swimming activity	McDonald et al. 1998; Claireaux et al. 2005
	Spawning time	Berejikian et al. 2003a
	Courting and spawning behavior	Fleming et al. 1996, 1997

water temperature during the winter after maturation (Berglund et al. 1991). Vaccines and chemical protection against contagious diseases and parasites such as sea lice have a positive effect on the survival of hatchery smolts in nature (Hvidsten et al. 2007). But even sea lice-protected smolts exhibit inferior survival to adulthood. Thus, improved handling and release strategies can increase the survival of released hatchery smolts but have so far not brought the survival rate up to a satisfactory level (Jonsson et al. 2003b).

Postsmolt Release

Survival of released smolts can be increased by increasing the size of the fish at release (Salminen et al. 1995), and the mortality may be particularly high due to predation during the first weeks at sea (Hvidsten and Møkkelgjerd 1987; Salminen et al. 1995; Dieperink et al. 2002). To avoid coastal smolt predators, postsmolts have also been released directly in the ocean after transportation in well-boats (Gunnerød et al. 1988; Heggberget et al. 1991). Furthermore, postsmolts have been retained in sea pens a few weeks during the first summer after smolting and then released. Both methods have given significantly higher recapture rates compared with fish released in rivers at the time of smolting (Eriksson and Eriksson 1991). A similar effect has been reported from coho and Chinook salmon, although the results vary among stocks (Linley 2001; Thrower and Joyce 2006). However, coastal released Atlantic salmon have exhibited a temporal delay in river ascent relative to the river-released fish and exhibited higher straying rates to other rivers compared with river-released fish (Hansen and Jonsson 1991a; Hansen et al. 1993; Jonsson et al. 1994).

Effects of Hatchery Salmon on the Local Wild Fish

Juvenile Competition

The results from experimental tests of feeding competition between wild and hatchery Atlantic salmon vary. Einum and Fleming (1997) reported that parr of hatchery Atlantic salmon dominated wild conspecifics in one-on-one challenges, with hybrids exhibiting an intermediate success. They related this to higher aggressiveness in hatchery than wild fish. A similar dominance of hatchery fish was reported by Rhodes and Quinn (1998) for coho salmon. Berejikian et al. (1999) found that juvenile coho salmon with cultured mothers won dominance challenges in a laboratory flume more frequently than parental half-sibs with wild mothers, suggesting that dominance may be a maternal effect. Riley et al. (2005), on the other hand, found no evidence that rearing environments caused higher aggression in cultured than in wild steelhead (anadromous rainbow trout) fry.

The higher aggressiveness observed in some hatchery populations can be modified by the environment. Fleming and Einum (1997) reported that hatchery parr were more aggressive in tank environments, contrasting the dominance of wild juveniles in stream-like environments. In brown trout, Höjsjö et al. (2004) found that the growth rate of dominant individuals relative to subordinates decreased with increased habitat complexity lending support to the hypothesis that habitat complexity favors wild salmonids in competition with hatchery reared conspecifics.

Prior residence influences the outcome of competition between wild and hatchery-reared fish (Reinhardt et al. 2001). In Atlantic salmon, it influences which individuals obtain territories (Cutts et al. 1999). In brown trout territory, owners are more likely to win contests, whether the fish are of wild or cultured origin (Sundström et al. 2003). A prior residence of 4 d motivated a stronger defense than a 2-d resident (Johnsson and Forser 2002). Furthermore, levels of aggression in juvenile Atlantic salmon are lowered by the presence of larger individuals (Adams et al. 2000; Peery et al. 2004). Thus, although hatchery parr may win feeding contests in tanks with slowly flowing water, the dominance can be reversed if intrinsic or extrinsic conditions change. Competition may result in increased emigration and mortality and decreased individual growth through density dependent mechanisms.

Displacement and Mortality in Freshwater

In rivers, hatchery parr may be displaced by wild conspecifics and vice versa, as found in experiments with rainbow trout (Table 3). Whether or not cultured fish dominate over wild conspecifics vary with the genetic background of the fish (Weber and Fausch 2003). McGinnity et al. (1997) reported that cultured Atlantic salmon fry outgrew and partly replaced wild conspecifics. The possible displacement may be linked to body size and density of fish. Weiss and Schmutz (1999) observed movement of resident brown trout from stocked stream sections. There are also examples where no effect of hatchery parr has been observed. For instance, Orpwood et al. (2004) reported that the ability of wild Atlantic salmon parr to find shelter in winter was unaffected by the presence of hatchery parr, even when the wild fish were outnumbered by four to one. Nickelson et al. (1986) found that the density of wild coho salmon juveniles was lower in streams stocked with hatchery fish than in unstocked streams, indicating that cultured fish replaced wild fish. The total density of juveniles had increased 1 year after stocking, but there was decreased production of juveniles in the next generation.

Weber and Fausch (2003) reported that at high density, hatchery rainbow trout were able to displace wild conspecifics from favorable stream positions when the hatchery fish were larger. At normal density, however, no consistent effect on emigration was found. In any case, it may be wise to delay the release of hatchery fish until after smolting of the local fish in the river to reduce potential interactions in freshwater.

There is little evidence of mortality effects of hatchery-reared Atlantic salmon parr on wild conspecifics, but experimental evidence from a number of other salmonid species indicates that density dependent mortality can result from releases of hatchery parr. Nielsen (1994) reported reduced production of wild coho salmon after hatchery coho salmon were stocked in a Californian river. Vincent (1987) found that densities of wild rainbow trout and brown trout increased after the stocking of adult hatchery rainbow trout ceased in two Montana streams, and Petrosky and Bjornn (1988) found that the mortality of wild rainbow and cutthroat trout *O. clarkii* increased at high, but not low stocking densities. In competition experiments with masu salmon in river enclosures, the hatchery fish survived in larger numbers than wild fish (Reinhardt et al. 2001). The mortality effect of released hatchery fish may be similar to that of adding wild fish, as reported by Bohlin et al. (2002) who tested effects of competition from hatchery on wild brown trout. Thus, in freshwater, density-dependent effects of cultured fish appear common among salmonid species and is probably also taking place in Atlantic salmon.

Releases of hatchery salmon may increase the mortality of competing species. Levin and Williams (2002) reported that the survival of wild Chinook salmon was negatively associated with releases of hatchery-reared steelhead in the Snake River, western USA, and similarly, Atlantic salmon releases influence the carrying capacity for brown trout as a result of competitive interactions (Heggenes et al. 1999; Harwood et al. 2001; Armstrong et al. 2003; Höjsjö et al. 2005). However, the effect of interspecific competition will probably be smaller than that of intraspecific competition between hatchery and wild Atlantic salmon. Although the ecological requirements of various species may be similar, they are less similar than those of hatchery and wild conspecifics (Harwood et al. 2002).

Growth

Density can influence the growth rate of salmonids (Brännäs et al. 2004). While density dependent displacement occurs at high population densities, density-dependent growth reduction can be noticeable even at low population densities (growth depensation) (Jenkins et al. 1999; Lobon-Cervia 2005). In addition to Atlantic salmon (Imre et al. 2005), growth depensation caused by released hatchery fish has been observed in brown trout and rainbow trout, and it probably occurs among stream-living salmonids in general (McMichael et al. 1997, 2000; Weiss and Schmutz 1999; Sundström et al. 2004). Bohlin et al. (2002) found that the addition of hatchery trout had a similar effect on growth rate of wild brown trout as increasing the density of wild conspecifics. For Chinook salmon, Weber and Fausch (2005) reported an even stronger negative effect on wild fish growth by adding hatchery fish than adding wild fish to the same density. In addition, releases of hatchery fish may influence growth rate of competing species, as found in experiments with brown trout and cutthroat trout (Shemai et al. 2007). But the negative interspecific effect on growth rate may be less than the intraspecific effect. An indirect consequence of the growth depensation may be decreased survival rate and impacts on other life history traits of the fish (Beamish et al. 2004a; Jonsson and Jonsson 2004b).

Other Life History Traits

The presence of hatchery salmon can contribute to the decline in adult body size of the fish in localities where they are released due to feeding competition. Hatchery practices together with fast juvenile growth in freshwater often results in younger age at maturity, as a phenotypic response (Salminen 1997; Quinn et al. 2001; Bates and McKeown 2003; Vøllestad et al. 2004; Scheuerell 2005). Furthermore, selective broodstock selection may alter the age at maturity of the fish as found for Chinook salmon (Unwin and Glova 1997).

With a decrease in juvenile growth rate and adult body size, egg size and fecundity may be altered (Unwin and Glova 1997). In Atlantic salmon, fast juvenile growth rate in freshwater, such as in hatcheries, reduces egg size and increases the fecundity of the fish as a plastic response of the phenotype, whereas the effect on egg size of growth rate variation at sea is minimal (Jonsson et al. 1996), relationships that also hold for masu salmon (Tamate and Maekawa 2000). Variation in growth rate, adult size, age at maturity, egg size, and fecundity influence competitive ability, reproductive success, and fitness of the fish with effects on biomass and production of fish in nature (Wertheimer et al. 2004).

Sea Survival

Released hatchery salmon survive less well than wild salmon at sea. In the Burrishoole, Ireland, smolt-to-adult survival of one sea-winter Atlantic salmon averaged 8% (2.9–12.6%) for wild fish and 2% (0.4–4.4%) for sea-ranched fish (Piggins and Mills 1985). In the River Imsa, the mean sea survival during 14 years of study was 8.9% for wild and 3.3% and 2.9% for cultured fish released as 1- and 2-year-old smolts, respectively (Jonsson et al. 2003b). In the Baltic Sea, the smolt to adult survival was 4.5 times higher in wild than in released hatchery Atlantic salmon (Saloniemi et al. 2004). It was reported that the difference in sea survival was more pronounced in low-survival years than in high-survival years. In good years, the larger size of hatchery smolts could compensate for their inferior performance, compared with wild smolts, but in poor survival years, wild smolts always exhibited higher survival. The estimated mean survival from smolts to adults of naturally produced steelhead was 5–6%, whereas that of hatchery populations was approximately 1%, and total egg to adult survival was 0.05% for wild fish and 0.56% for cultured fish (Kostow 2004). The 3–5 times higher sea survival of wild than hatchery-reared Atlantic salmon and steelhead released in rivers as smolts may be linked to more relaxed selection pressure in hatcheries than in nature and the phenotypic divergences of hatchery from wild fish (Jonsson and Fleming 1993; Reisenbichler and Rubin 1999; Ford 2002). Also, in other salmonids, the survival of released hatchery fish can be low as reported for released hatchery-reared anadromous brown trout in Denmark due to high mortality at sea (Hansen 2002; Ruzzante et al. 2004).

Spawning Competition and Reproductive Success

Returning adult hatchery Atlantic salmon enter rivers to spawn later in the season, move about more, and stay for a shorter time in the river than wild fish (Jonsson et al. 1990a; Økland et al. 1995). Upstream migrating hatchery salmon may not be heading for any particular spawning area. Many may move to the top of the river instead of entering the spawning grounds of wild fish lower downstream (Thorstad et al. 1998). Some released hatchery-reared Atlantic salmon spawn in the river they enter; others leave the river unspawned (Jonsson et al. 1990a). The spawning success of hatchery salmon may be reduced by their late river entry (Aarestrup et al. 2000).

On the spawning grounds, hatchery Atlantic salmon have been found competitively and reproductively inferior and injured more often than their wild counterparts (Jonsson et al. 1990a). Fleming et al. (1997) reported that the spawning success of male Atlantic salmon released as smolts was 51% of that of corresponding wild males from the same population, whereas there was no significant difference in reproductive success between wild and hatchery females. McGinnity et al. (2004) reported an overall lifetime success from fertilized egg to returning adult of nonnative Atlantic salmon to be 35% less than that of native and conspecifics released as smolts. Early survival was lower in offspring of hatchery than of wild fish; later, it was similar. Also, in other salmonids such as coho salmon, the reproductive success is higher for wild than for hatchery-produced fish (Fleming and Gross 1992, 1993; Berejikian et al. 1997).

The release of hatchery-reared adults is not an effective tool to rebuild a seriously depressed population (Carr et al. 2004). The reproductive success of hatchery fish, however, may increase with increasing time in nature. For instance, the reproductive success of sea-ranched salmon that have lived one year in nature is between that of wild and farmed Atlantic salmon coming directly from the net pens (Fleming et al. 1996, 1997). However, there is one example of high reproductive success of cultured fish. Dannewitz et al. (2004) found no significant difference in reproductive success between seventh-generation hatchery brown trout and wild-born brown trout in an experimental stream. Thus, hatchery fish may not always be an inferior competitor to wild fish on the spawning grounds.

The inferiority of hatchery fish is more pronounced in hatchery males than females, resulting in cross-breeding between hatchery females and wild males. In brown trout, hatchery-reared males seem to have lower reproductive success than wild males as found in an experimental stream, but no similar effect for females was reported (Dannewitz et al. 2004). Experimental evidence from Atlantic salmon suggests that the male fitness difference occurs because they arrive at the spawning grounds later than wild males, do not establish dominance hierarchies as effectively as wild males, court less, spawn with females in larger numbers, and partake in fewer spawnings, and they frequently fail to release sperm when the females release their eggs. On the spawning grounds, male hatchery Atlantic salmon are involved in more prolonged aggressive encounters, incur greater wounding, and have higher mortality than wild males originating from the same population (Fleming et al. 1996, 1997). The hatchery males ascend the spawning river later in the season, are less able to monopolize females, move about more in the river, and, after spawning, they leave the river earlier than wild fish originating from the same population. Hatchery salmon also return to sea without having spawned more often than wild salmon (Jonsson et al. 1990a).

Biomass and Production

Releases of salmon are meant to increase the productivity of habitats as found in the River Drammen by Hansen (1991), but hatchery production may decrease the productivity of the wild stock present. As a consequence of the spawning of hatchery salmon in the River Imsa, Fleming et al. (2000) found a 30% reduction in production of wild Atlantic salmon. Unwin and Glova (1997) found a 34% reduction in the production of wild Chinook salmon in a New Zealand river, probably due to density-dependent mortality caused by released hatchery fish. Furthermore, Nickelson (2003) reported decreased salmon production in Oregon coastal river basins and lakes where large numbers of cultured coho salmon smolts were released and recommended against such large releases in areas with high concentrations of wild fish. Chilcote (2003) maintained that removal rather than addition of hatchery fish may be the most effective strategy to improve productivity and resilience of steelhead. He found that populations consisting of equal numbers of cultured and wild fish produced 63% fewer recruits per spawner than one composed entirely of wild fish. In cases where fish releases result in a decrease rather than an increase in total population size, this may be due to a genetic change with the introduction of maladaptive traits or loss of genetic variation (Wang and Ryman 2001; Garcia de Leaniz et al. 2007) or an overexploitation of the food resources present with a resulting decrease in carrying capacity of the habitat. In some cases, there appear to be only minor effects of released cultured fish on the local wild populations, as reported by Hayes et al. (2004). When Goodman (2005) modeled the effects on natural spawning fitness in rivers where wild and cultured fish spawn together, he found potential, but not a certainty, for erosion of natural spawning fitness, a finding supported in the analysis of Naylor et al. (2005). Thus, there are variable results from an increased to decreased total production after releases of salmon, which are reasonable and depend on the environmental conditions where the fish are liberated. But most effects of releases of hatchery salmon seem negative.

Why Do Hatchery Salmon Often Perform Poorly in Nature?

The success of hatchery fish in nature is often low (e.g., Hjort and Schreck 1982; Swain et al. 1991; Fleming et al. 1994; Pelis and McCormick 2003; Kostow 2004; Von Cramon-Taubadel et al. 2005). Hatchery and wild conspecifics experience different environments before the release of the cultured fish. Hatchery salmon allocate more energy to protein growth and lipid deposition, and in association with this, several morphological changes occur (Fleming et al. 1994; Price 1999; Waples 1999). Hatchery tanks are space-restricted and simple; there is little seasonal change in environmental variables, high-quality food is readily available, and the fish are protected against predators and treated for some diseases. Furthermore, in hatcheries, salmon reproduce without having to compete for mates. On the other hand, hatchery fish are frequently disturbed by human treatment, and fish density is unnaturally high with the possibility of more social encounters, increased stress and aggression levels, and increased vulnerability to contagious diseases (Huntingford 2004).

Hatchery salmon deviate from wild salmon due to these differences in environments. The phenotype is both directly (plastic) and indirectly (genetic) influenced by the environment. Phenotypically plastic divergences are often shaped early in life. Von Cramon-Taubadel et al. (2005) found that the body form of Atlantic salmon parr grown from the eyed egg stage with a nonsibling group in a hatchery resembled the body shape of the nonsiblings more closely than the full siblings grown in their natal habitat. The morphological differences, however, are less pronounced after 1 year swimming freely in the sea (Fleming et al. 1994). Thus, some of the phenotypic differences caused by the rearing conditions disappear with time when the divergent groups are brought together in a common habitat.

Hatchery rearing also influences anatomic characters such as the development of the forebrain (telencephalon) of salmon and trout (Lema et al. 2005). It is found that cultured Pacific salmonids have smaller brains than wild conspecifics of similar size, but the reason is still unknown (Kihslinger and Nevitt 2006; Kihslinger et al. 2006). Furthermore, sensory organs such as the lateral system and eyes may be modified during hatchery rearing and influence the performance of hatchery fish in nature (Marchetti and Nevitt 2003; Anras and Lagardere 2004). Furthermore, it is found that brain gene expression profiles in Atlantic salmon are affected by rearing environment such as hatchery and river, as well as between reproductive tactics independent of rearing environment (Aubin-Horth et al. 2005).

Heart anatomy also differs between hatchery and wild salmonids. The normal shape of the salmonid ventricle is a triangular pyramid with the apex pointing caudoventrally. But Poppe et al. (2003) found that the hearts of hatchery-reared Atlantic salmon and rainbow trout were rounder than those in their wild counterparts and that the angle between the ventricular axis and the axis of the bulbus arteriosus was more acute in wild fish. Fish with abnormal heart morphology have higher mortality rate during stress-induced situations, and the cardiac output, heart rate and stroke volume, and active metabolic rate may be smaller (Dunmall and Schreer 2003; Claireaux et al. 2005).

Hatchery fish may be compromised in their ability to undergo smolting in terms of physiological changes needed to ionic regulation in marine waters. Lower gill $Na+$, $K+$ – ATPase activity, growth hormone, and plasma chloride levels of cultured than wild smolts was observed by Handeland et al. (2003), and survival on transfer to full-strength seawater at different temperatures indicates that wild Atlantic salmon smolts may tolerate the transfer better than cultured smolts. Handeland et al. (2003) concluded that the observed differences are genetic and associated with broodstock selection for rapid growth over several generations. On the other hand, such differences may well be phenotypic, linked to the seasonal development and size of the fish, as suggested by Ugedal et al. (1998), investigating seawater tolerance in cultured and wild smolts of brown trout. Hatchery Atlantic salmon smolts of the Irish Burrishoole stock had higher basal cortisol levels in April and May than wild smolts and did not exhibit the typical cortisol responses to capture stress. Similar differences were found in serum glucose levels, and cultured smolts had significantly higher concentrations of mucous cells in both skin and secondary gill lamellae, which may influence the subsequent marine survival (Poole et al. 2003).

Such phenotypic deviations results from (1) hatchery experiences, (2) developmental processes, and (3) physical damage incurred through hatchery rearing.

Hatchery Experiences

Cues sensed by fish influence behavioral traits (Brown et al. 2003) and differential juvenile experiences between hatchery and wild Atlantic salmon are likely to generate differences between them (Jonsson et al. 1990a; Huntingford 2004; Braithwaite and Salvanes 2005). For instance, early river experience influences the timing of the river entry for spawning (Jonsson et al. 1994; Skilbrei and Holm 1998), risk taking (Sundström et al. 2004), antipredator and feeding behavior (Reiriz et al. 1998; Brown and Laland 2001, 2002; Reinhardt 2001). Vilhunen et al.

(2005) reported that acquired predator recognition was socially transmitted from predator experienced to predator naïve conspecifics as found in experiments with Arctic char *Salvelinus alpinus*.

When released in nature, hatchery Atlantic salmon enter rivers to spawn (Jonsson et al. 1990a, 2003a; Clifford et al. 1998). However, their homing precision is less accurate than that of wild fish even when the two leave the river together as smolts (Jonsson et al. 2003a). Mean rates of straying of released hatchery versus wild Atlantic salmon of the River Imsa stock were estimated at 15% and 6%, respectively, and the more years the fish stayed away from the river, the larger was the straying rate. Both cultured and wild salmon strayed to many of the same rivers (ca. 80% of them drain into the fjord of the River Imsa within 60 km of the outlet).

Cues encountered by seaward migrating smolts influence the homing behavior of salmonids (Hansen et al. 1993; Dittman and Quinn 1996) and river ascent (Hansen and Jonsson 1994; Jonsson et al. 1994). Together, such observations indicate that differences in sensory stimulations between hatchery and wild salmon influence subsequent performance in nature. A more variable hatchery rearing environment might mitigate some of this difference between wild and hatchery salmon, as shown for hatchery-reared Atlantic cod *Gadus morhua* (Salvanes and Braithwaite 2006).

Developmental Processes

Developmental processes expressed by the phenotype are influenced by hatchery conditions. For instance, egg incubation temperature affects subsequent growth performance of the parr. In hatcheries, salmonid eggs are often incubated at elevated water temperature to induce early hatching and a prolonged first growing season. This gives the young fish a size advantage over similar-aged wild conspecifics when liberated in nature. This size advantage can influence the outcome of social encounters, with effects on other life history characters as previously explained.

Atlantic salmon parr are often faster growing in hatcheries than in nature owing to higher energy input and/or lower energy expenditure, with consequences for life history traits such as age and size at smolting (Økland et al. 1993), age at sexual maturity (Alm 1959; Vøllestad et al. 2004), and reproductive output (Jonsson et al. 1996). Fast-growing parr tend to smolt younger and smaller (Økland et al. 1993), but the size of hatchery smolts is variable and heavily dependent on smolt age (Jonsson et al. 2003b). Furthermore, high growth rate of female salmon in freshwater is associated with a relatively low growth increment at sea (Einum et al. 2002), and low growth increment at sea is associated with early age and small size at sexual maturity (Nicieza and Braña 1993; Jonsson and Jonsson 2004b). Gonadal mass and energy content increase with somatic mass in both sexes (Jonsson and Jonsson 2003), and as a reaction norm in Atlantic salmon, fast-growing parr tend to produce more and smaller eggs when they mature than if they grow more slowly (Jonsson et al. 1996; Fleming et al. 2003). In other species such as brown trout, coho salmon, and Chinook salmon, egg size and fecundity appear to be chiefly determined by the energy intake later in life and not flexibly dependent on the early, juvenile growth rate (Jonsson and Jonsson 1999; Quinn et al. 2004).

Lack of exercise in hatcheries may influence the hormone production of Atlantic salmon. Hatchery smolts challenged by a high current velocity are more active than the unchallenged smolts, probably because of elevated thyroxin level (Youngson and Webb 1992) with effects on the downstream smolt migration (Youngson et al. 1989; Iwata et al. 2003) and possibly the subsequent homing behavior (Dittman et al. 1996; Lema and Nevitt 2004). The hormone level can also be elevated if the hatchery smolts are retained for some time in so-called "imprinting ponds" with higher current velocity than experienced in hatcheries, before release (McCormick et al. 2003).

There is correlation between adiposy and maturation in salmonids (Rowe et al. 1991; Silverstein et al. 1999), and the lack of exercise in hatcheries influences lipid deposition, growth, swimming performance, and rate of fin healing, with possible effects on subsequent reproductive performance and success (Jørgensen and Jobling 1993). Male Chinook salmon reared in high-current velocity conditions started spawning 2.4 d earlier and defended their access to spawning females better than males reared in low-velocity tanks (Berejikian et al. 2003). Adult Atlantic salmon reared to smolting in high-velocity tanks enter freshwater for spawning more readily than those reared in a regular low-velocity environment (Skilbrei and Holm 1998). Patterson et al. (2004) reported effects of exercise on age at maturity, egg deposition rate, and egg survival in sockeye salmon *O. nerka*. Nonexercised females had delayed maturity, had lower egg deposition rates, and were more likely to die prior to ovulation and to exhibit poorer egg survival than exercised fish and wild spawners. Thus, lack of physical exercise by hatchery fish may diminish their success in nature relative to that of wild fish.

Physical Damages

Damage to the rayed fins of hatchery Atlantic salmon parr is primarily caused by aggressive encounters between fish with nipping of fins (Ellis et al. 2002), but may also result from abrasion on rough surfaces, nutritional deficiencies, and secondary bacterial infections (Höglund et al. 1997; Lellis and Barrows 1997; Latremouille 2003). The damaged or distorted jaws sometimes seen in hatchery salmon may also result from injuries in the tank environment, which hardly ever occur under natural conditions in rivers. While such damage incurred during culture can influence the performance of the fish and is therefore undesirable, it can be helpful when studying social interactions between groups of wild and hatchery fish (MacLean et al. 2000).

Genetic Diversity

Atlantic salmon segregate into distinct reproductive groups or local populations (Verspoor et al. 2007), and there is evidence of adaptive variation among the populations of Atlantic salmon (Hansen and Jonsson 1991b; Nislow et al. 2004; review in Garcia de Leaniz et al. 2007). In the hatcheries, however, salmon face new selection pressures, and the divergent phenotypic expression of hatchery relative to wild conspecifics can be influenced by natural selection in the hatchery conditions and artificial broodstock selection. Furthermore, hatchery populations may be influenced by genetic drift, inbreeding, and outbreeding depression.

The scale and extent of adaptive variations among salmon populations are poorly understood, but they depend probably on habitat heterogeneity, environmental stability, and the relative roles of selection and drift. As Garcia de Leaniz et al. (2007) maintained, maladaptation often results from phenotype–environment mismatch. To avoid this, one should act as if all populations are locally adapted. That means that one should minimize alterations to native populations and habitats to which populations may be adapted to and allow for population size to extend beyond the carrying capacity of the habitat to maintain genetic diversity and encourage competition and other sources of natural mortality required for natural or stabilizing selection.

Hatchery Selection

Population specific adaptations may be changed in hatcheries as artificial culture exposes fish to new selecting forces (Thorpe 2004). The genotypic change of cultured fish from their wild origin is a response to changed birth and/or death rates as a consequence of natural selection in the hatchery environment (Heath et al. 2003; Obedzinski and Letcher 2004). For instance, hatcheries appear to select for enhanced aggression in natural river environments, as found for Atlantic salmon (Einum and Fleming 1997), Chinook salmon (Wessel et al. 2006), coho salmon (Rhodes and Quinn 1998), masu salmon (Yamamoto and Reinhardt 2003), brown trout (Sundström et al. 2003), and rainbow trout (Riley et al. 2005). The higher aggressiveness may be linked to the high fish density in hatchery tanks. Glover et al. (2004) showed that the families of brown trout that survived best under conditions of abundant food were different from those that survived best on low rations.

Broodstock Selection

Farmed salmon selectively bred over several generations for production traits such as fast growth differ genetically from their wild origin when they are released in nature (Weber and Fausch 2003; McLean et al. 2005), with for example higher production rates of growth hormone (Fleming et al. 2002). The resulting fast growth is linked to enhanced appetite and greater risk taking (Fleming et al. 2002) and elevated standard metabolic rate (Metcalfe et al. 1995; Cutts et al. 2002; Lahti et al. 2002). Hybrid juveniles are often intermediate in character expression between hatchery and wild juveniles (McGinnity et al. 1997, 2003; Fleming et al. 2000). Thus, broodstock selection can cause correlated and unintended genetic changes. Hatchery salmon transported and released in new areas can deviate significantly from the local wild fish.

Selection for high growth rate, however, may reduce the aggressiveness of the fish. This was demonstrated in experiments with newly emerged brown trout fry (Hedenskog et al. 2002). Petersson and Järvi (2003) reported that wild juvenile brown trout were more aggressive than the offspring of sea-ranched brown trout and attacked novel objects sooner, a behavior that gives elevated dominance status (Sundström et al. 2004). Furthermore, Sundström et al. (2005) observed different responses of cultured and wild brown trout originating from the same stock, which may be caused by different selection regimes in the hatchery and nature (Huntingford and Adams 2005). In coho salmon, aggressiveness and growth rate are negatively correlated (Vøllestad and Quinn 2003), probably because the time spent on agonistic interactions reduces food consumption and/or increases the energy use. Thus, broodstock selection for production traits in hatcheries may counteract the selection for increased aggressiveness under hatchery conditions.

Selection response may also be obtained for a number of other traits such as sea survival and return rate (Jonasson et al. 1997), age at sexual maturity (Gjerde et al. 1994; Gjedrem 2000), disease and parasite resistance (Fjælestad et al. 1993; Gjøen et al. 1997; Kolstad et al. 2005), feeding efficiency (Kolstad et al. 2004), and low percentage of sexually mature male parr (Wild et al. 1994). However, many such changes may be undesirable when restoring salmon populations.

Genetic Drift, Inbreeding, and Outbreeding

Genetic changes of the populations should be avoided when restoring salmonid populations (Cross 2000), as this may result in increased rate of hybridization with related species or with endemic populations of the same species and give elevated rates of genetic drift (Wang and Ryman 2001). Furthermore, within-population genetic diversity may be eroded by stocking large numbers of genetically similar individuals into small populations (Yokota et al. 2003). To decrease such hazards, Harada et al. (1998) advised that one should use wild-born parents of both sexes for broodstocks. On the other hand, Riley et al. (2004) found few significant ecological effects of small-scale releases of hatchery coho and Chinook salmon even when the wild conspecifics occurred at low densities, indicating that genetic effects of hatchery salmon on wild populations are variable.

Inbreeding with loss of heterozygosity may occur in hatcheries with negative effects on individual and population performance when released in nature because of reduced adaptability and accumulation of genes with detrimental effects (Wang et al. 2001; Primmer et al. 2003; Tiira et al. 2006). Garant et al. (2005) reported an increased reproductive success of females with a higher number of mates, resulting in more outbred offspring. Furthermore, Ayllon et al. (2004) suggested that poor planting success of Atlantic salmon introduced to streams of the Kerguelen Island, Antarctica was due to a too low genetic variability of the broodstock. Thus, enhanced genetic diversity of released hatchery fish may increase their reproductive success in nature.

As a management measure, new nonnative alleles can be introduced to populations with low genetic variability (Tallmon et al. 2004; Hedrick 2005; Edmands 2007). On the other hand, adding new alleles may be detrimental to populations if it breaks coadapted gene complexes important for fitness traits in the local environment (outbreeding depression). Little is known about the effects of outbreeding in salmon, but it may decrease fitness (Fleming et al. 2000; McGinnity et al. 2004) as in a variety of other species (Edmands 1999; McClelland and Naish 2007). Because of this, several authors advocate habitat restoration, if possible, rather than supportive breeding when supporting threatened or endangered populations (Ford 2002; Dannewitz et al. 2004; Almodovar et al. 2006).

A sufficient amount of genetic variation is required for the persistence of self-recruiting populations, and genetic variation is important for the survival and success of cultivated smolts in natural ecosystems. If the cultivated population has been through a bottleneck and the genetic variation is low, due to a low effective population size, the cultivated fish may suffer from inbreeding depression, reducing the fitness considerably. Hence, the release may be more harmful than good for the augmented population.

Conclusion

With all these differences between hatchery and wild fish, what is the main reason for the low success of hatchery salmon in nature? Most probably, a large part of the released hatchery salmon die because of predation soon after being released. The performance of hatchery fish in nature is highly influenced by their early experiences in the hatchery environment. Hatchery practices affect both genotype and phenotype of the fish and may produce an inadequate behavior versus predators. Thus, the time period the fish spend in the hatchery may be minimized to offer the fish maximum experience in natural environments. This implies that the size of the managed population will be regulated by natural mechanisms and factors in the systems. Alternatively, the complexity of the hatchery environment should be increased to produce fish exhibiting adequate avoidance behavior against predators.

Habitat Restorations and Enhancement

For Atlantic salmon, less is known about habitat restoration than on effects and successes of fish releases, and much knowledge gained is from experiments and management operations on other salmon species. However, the abundance and growth can be augmented through habitat manipulations improving feeding opportunities and water quality (Lacroix 1996; Hesthagen et al. 1999), increasing the spawning habitat by introducing gravel (Avery 1996; House 1996; Scruton et al. 1997), increasing the productive area by constructing side channels (Pethon et al. 1998), removing blockages and constructing fishways (Saltveit 1989; Simenstad et al. 2005), and changing the flow regime (Armstrong et al. 2003). If the natural production of a river is to be restored, one should identify the constraints for salmon habitat use and relieve these constraints (Ebersole et al. 1997). One should keep in mind, however, that any river has a limited carrying capacity, although this may be increased to some degree by habitat improvements. To augment the population abundance, the easiest means may be to increase the exploitable area for the fish.

Fishways

Salmonid populations can be enhanced if the fish are allowed to colonize new habitats upstream of their original distribution area (Ritter 1997; Bryant et al. 1999). Such improvements may be accompanied by stocking programs to increase the colonization rate. The first step in colonizing new habitats is to provide free fish migration over natural and/or man-made barriers as waterfalls, dams, and turbines. Construction of fishways and nature-like bypasses similar to natural streams facilitates the upstream migration for salmon.

Often, however, fishways can delay or prevent the upstream migration of Atlantic salmon. The fish may have problems in locating entrances and successfully ascending them. To help avoid this, Katopodis (2005) presents a toolkit for fish passages. It is essential that entrances are designed and located properly to enable and to stimulate fish ascent (Clay 1995; Larinier 1998). Spawners search for the highest flow (Williams 1998) and if the fish need to leap, a downstream deep pool is needed where fish can initiate jumps. If care is not taken, the fish may be attracted towards impassable routes from turbine outlets or dams rather than to the bypass (Andrew and Geen 1960; Brayshaw 1967; Arnekleiv and Kraabøl 1996; Thorstad et al. 2003). For instance, upstream migrating salmon in the regulated River Tuloma, Kola Peninsula were rather reluctant to enter fish passes, occasionally even backing out after having entered the pass. The fish preferred to seek their way in strong current (i.e., the tailrace and spillway discharges below the Tuloma dam [Karppinen et al. 2002]). Upstream migrating Atlantic salmon in River Conon, northern Scotland were delayed and did not navigate through a series of four fish passes and an impoundment (Gowans et al. 2003). The proportion of the fish passing the individual obstructions ranged from 63% to 100%. No fish were lost when moving through the impoundment, whereas 63% of the approaching salmon passed each of the lifts. The fish were delayed for 1–41 d at a pool-and-overfall ladder and 1–52 d at a Borland fish lift. In the River Nidelva, south Norway, the salmon migrated quickly up to the tunnel outlet of the power station, but stayed on average 20 d (0–71 d) in the outlet area before continuing upstream (Thorstad et al. 2003). The size and design of the tunnel outlet, whether the outlet is submerged or not, and the slope of the tunnel appeared important for the salmon ascent.

Fishways can be size-selective. The Isohara fishway, close to the mouth of the regulated River Kemijoki, north Finland allowed one-sea-winter fish to pass (Laine et al. 1998). Larger, multi-sea-winter fish were observed close to the two alternative fish entrances without any attempt to enter. An inadequate fishway discharge together with poor attraction of the fish entrance was suspected to be the main reason for why they did not enter. After the entrance was replaced by a pool and a small waterfall, no more observations were made of salmon gathering close to the entrance, and the number and maximum size of the salmon increased in the fishway (Laine et al. 2002). The number of multi-sea-winter salmon passing through the fishway in the River Kemijoki was positively correlated with the mean tailwater level on the day of ascent and 1–3 d earlier. The tailwater level did not follow the river discharge but the seawater level, which was affected by the direction and velocity of the winds. Thus, the design of the entrance of the fishway is important, and a poor design is a migratory obstacle for the fish.

The colonization rate varies among species when new river stretches open for migratory fish. For example, the access of anadromous salmonids to the Margaret Creek water shed, southeast Alaska, previously blocked by a 7-m waterfall, was opened by a fish ladder constructed during 1989–1990. Pink salmon dominated numerically in the ladder during the subsequent 7-year study period (Bryant et al. 1999) and increased from 6,090 fish in 1991 to 39,499 fish in 1997. The number of sockeye and coho salmon passing the ladder ranged from 73 to 408 and 111–1,986, respectively. Few chum salmon *O. keta* and steelhead entered the ladder. Reasons for this may be that chum salmon does not ascend obstacles as easily as other anadromous species (Hale et al. 1985) and steelhead was less abundant in the watershed than other salmonids. Further examples of successful colonization of new areas by natural waterfalls by Atlantic salmon are described by Ritter (1997) from Torrent and Exploits rivers, west and east coast of insular Newfoundland and LaHave River, Nova Scotia.

Spawning Habitat Improvements

Gravelling can be a successful way of enhancing Atlantic salmon populations in rivers with reduced spawning opportunities (Merz and Setka 2004). For instance, gravel size, depth and compactness, and extent of suitable gravel available at the redd site affect density of redds in steelhead, in addition to water depth, velocity, and temperature (Orcutt et al. 1968). Spawning-bed enhancements have increased the survival and growth of Chinook salmon embryos in a regulated California stream (Merz et al. 2004). Salmon embryos planted in the improved spawning substratum exhibited higher rates of survival to the swim-up stage than embryos planted in the original spawning gravels. Furthermore, in the Mokelumne River, California Central Valley, 976 m^3 of clean river gravel (25–150 mm) was placed in berm and gravel bars along a 45-m enhancement site (Merz and Setka 2004). After gravel placement, the channel water velocities, intergravel permeability, and dissolved oxygen increased and the channel depth was reduced. Adult Chinook salmon began spawning at the previously unused site 2 months after gravel placement. However, in some cases, gravelling is not enough to create new salmonid spawning grounds (Zeh and Dönni 1994). In the River High-Rhine, Switzerland, washed gravel (grain size 16–50 mm) were introduced in an impounded section of the river to restore the spawning grounds for brown trout and Arctic grayling. Neither brown trout nor rainbow trout were observed spawning in the gravel beds. On the other hand, successful embryonic and larval development of grayling was observed, meaning that the new spawning area was suitable for this species.

Juvenile Habitat Improvements

Habitat preferences of salmonids vary with species and life stage and season (Heggenes et al. 1999). For instance, Atlantic salmon is distributed in faster-flowing habitats compared with brown trout, and often, but not always, they are associated with rivers with gravel bottom (Riley et al. 2006). The time as embryo in the bottom substratum and the transition from dependence on maternal yolk reserves to external feeding are critical periods. For instance, salmonid embryos are susceptible to fine-sediment infiltration during the incubation period (Julien and Bergerson 2006). In a field experiment, it was found that survival of pre-eyed, eyed, and hatched stages of Atlantic salmon were all negatively correlated with the percentage of fine sediment entering the incubation baskets. The pre-eyed and eyed stages were most strongly affected by silts and clays (<0.063 mm), although this size-class represented only a small fraction (0.03–0.41%) of the grain size inside. The hatched stage was most strongly correlated with the infiltration of medium sand (0.25–0.50 mm) material. On average, 66% of the implanted embryos survived to the pre-eyed stage of development compared to 63% for the eyed and 48% for the hatched stages of development.

At the commencement of external feeding, the fry are especially vulnerable to predators and adverse environmental conditions. The availability of slowly flowing habitats at the stream margins is crucial during the first month of independence (Armstrong and Nislow 2006). Atlantic salmon parr perform ontogenetic niche shifts, and later during summer, age-0 parr obtain high consumption rates over a wider range of current velocities (0.2–0.6 m/s) (Nislow et al. 1999), and the fish are often found in shallow riffle-chute habitats (Heggenes et al. 2002). Larger parr exploit even faster current velocities and greater water depths. During winter, the parr often hide in the gravel substratum or stay in deep and low-velocity habitats during daytime, but may seek more open feeding areas during night (Maki-Petays et al. 2004; Riley et al. 2006). High concentrations of fine sediment in the substratum degrades the habitat for steelhead parr (Suttle et al. 2004), and the same probably holds for juvenile Atlantic salmon in rivers, although there are differences in responses to substrate and cover among species (Sergeant and Beauchamp 2006). Thus, it appears important both to consider substrate conditions and current velocity and water level when restoring Atlantic salmon rivers (Armstrong et al. 2003; Hendry et al. 2003).

To manage salmon rivers well, it is important to protect existing high-quality habitats (Fullerton et al. 2006). For Pacific salmon, Roni et al. (2002) recommended that the restoration then should focus on connecting isolated high-quality fish habitats such as instream or off-channel habitats made inaccessible by culverts or other artificial obstructions. Removing small artificial barriers that hinder upstream migration of fish is a major task in riparian habitat restoration (O'Hanley and Tomberlin 2005). In cases where the juvenile habitat is highly degraded, such as in the case of hydropower developments, artificial fluvial habitat channels may be constructed to enhance the natural production of juvenile fish as successfully done in south-central Newfoundland (Enders et al. 2007). The introduction of boulder clusters in the river has been found to be another effective method of increasing the parr density of Atlantic salmon in rivers, as shown in Joe Farrell's Brook, Newfoundland (de Jong et al. 1997). Also, V-dams have proven to be effective in increasing juvenile density through the creation of a diverse pool habitat. Half-log covers increased the number of age-0 parr through an increased instream cover. The usefulness of the placement of boulder weirs appears however, to vary among species (Roni et al. 2006) and may be even more useful when restoring brown trout than Atlantic salmon rivers (Heggenes et al. 1999).

Atlantic salmon parr feed largely on larvae of aquatic insects (Lillehammer 1973), and the quality and quantity of the food are often viewed as important factors influencing the carrying capacity of salmon rivers (Jonsson et al. 1998). Presence of woody debris in streams is one factor influencing the abundance of insect larvae (Giannico and Hinch 2003; Milner and Gloyne-Phillips 2005); it provides greater surface areas for the growth of the prey species and shelter for the fish (Johnson et al. 2005). Furthermore, woody debris may give overhead cover that decreases predation risk and offer decreased contact between the fish. It also decreases the current velocity and thereby decreases the energetic costs of the fish in the streams (Crook and Robertson 1999). Presence of dead wood in streams is found to be profitable for the production in a number of salmonid species (Johnson et al. 2005; Fox and Bolton 2007). One way of providing woody debris in salmon rivers is to leave an effective riparian buffer zone along the banks where trees and other plants are allowed to grow undisturbed (Haberstock et al. 2000; Opperman and Merenlender 2004).

Liming of Acidified Rivers

Acidification of salmonid rivers represents a major threat to salmon production. Prominent physiological disturbances in fish exposed to acid water are failures in ionic regulation, acid–base regulation, circulation, and respiration, of which the first and last are held to be the primary causes of fish death in both acid and aluminum-rich water (McDonald 1983; Exley and Phillips 1988; Berntssen et al. 1997). Atlantic salmon are more sensitive to acid water than other naturally occurring salmonids in Scandinavia (Rosseland and Skogheim 1984), and the most sensitive stage is the smolt stage (Rosseland et al. 1986). Applications of crushed limestone in acidic rivers enhanced Atlantic salmon spawning habitat and improved the survival of juvenile salmonids (Staurnes et al. 1996). Liming of an acidic second-order stream, Fifteen Mile Brook, Canada, resulted in a twofold increase in the number of occasions where more than a few juvenile Atlantic salmon survived severe acidic episodes in the brook (Lacroix 1996). Densities of age-0 salmon were related to seasonal and interannual variability in pH, but they were always greater in the limed section than in the unlimed. Atlantic salmon consistently placed most of their redds (78%) in the limed section, and these were often on the limestone bar.

Water Level and Flow Regulation

Low river flow restricts the upstream migration of the fish. Low flow precludes salmon from entering small streams, and the effect is strongest for large fish early in the migration season (Jonsson et al. 1990b, 2007; Tetzlaff et al. 2005). A similar effect on migratory behavior is observed for brown trout, although the species is less sensitive than Atlantic salmon (Jonsson and Jonsson 2002). In the River Gudbrandsdalslågen, east Norway, Arnekleiv and Kraabøl (1996) found that ferox trout (large fish-eating brown trout) did not pass the outlet channel from the power station when the residual flow decreased below 20 m^3/s, and the upstream migration could be initiated by an artificial freshest of 60 m^3/s. Thus, increased water level at the time of migration may facilitate the upstream migration in rivers and streams suffering from low flow (Jonsson et al. 2007).

Changes in flow patterns due to impoundments or partial barriers may affect habitat and mortality of young fish living in the river. Sudden reduction in river flow may cause high mortality of juvenile Atlantic salmon and brown trout through stranding (Bradford 1997; Halleraker et al. 2003; Berland et al. 2004). The chance of stranding was higher when the water temperature was low (e.g., winter conditions < 4.5°C) compared with higher temperatures during late summer and early autumn. This is probably because of lower fish activity during the cold season and a substrate-seeking behavior. Stranding is not equal to mortality, as the fish can survive for several hours in the substrate after dewatering. A prolonged shut down procedure of the turbines in a regulated river decreased the stranding of salmon parr drastically under spring conditions. In an experiment with free-ranging parr and parr restricted to an area near the riverbank, Berland et al. (2004) found no stranding of free-ranging parr during rapid flow reductions during daytime. In the containment pen, the parr distributed themselves relatively evenly among the cells. They moved about more at changing than at stable flows, and most fish that were stranded were observed during rapid flow reduction at night. Thus, rapid reductions in water flow may cause increased mortality in salmon parr in shallow habitats if movements are restricted, and less abrupt water level fluctuations might be helpful to the fish.

Salmon Management and Research Tasks

Population or Habitat Restoration?

Restoration, rehabilitation, and enhancement of salmonid populations may draw on any of the techniques and methods mentioned earlier in this paper. The preferable approach will depend on the purpose of the activity, the status of the population, and the condition of the habitat. In a situation with weak or declining salmonid populations caused by increased mortality at sea due to climate change (Friedland et al. 2000; Jonsson and Jonsson 2004a), there may be a wish to (1) augment the bottlenecks of the juvenile production by restoring or rehabilitating the habitat, (2) increase the productive river area by building fishways to previously salmon-free habitats upstream, or (3) release fish to enhance the amount of seaward migration smolts.

The abundance of salmonid populations is variable with long-term trends upon which there are short-term fluctuations (Einum et al. 2003). From the 1980s onwards, the production of wild Atlantic salmon at sea has decreased gradually, viewed to be the beginning of a negative, long-term trend (Beaugrand and Reid 2003; Jonsson and Jonsson 2004a). Growth and survival rates have decreased at the same time that the populations exhibit younger sea-age at sexual maturity. The result is more relatively small one-sea-winter fish and fewer old, large adults (Jonsson and Jonsson 2004b; Boyland and Adams 2006). Since there is no sign of density dependent mortality in Atlantic salmon at sea (Jonsson et al. 1998; Niemelä et al. 2005), the population decline may be mitigated by an increased smolt output. Both habitat improvements and opening of new river and lake habitats for salmon are ways of expanding the carrying capacity of water courses. Furthermore, eggs may be planted if spawners are lacking, whereas parr releases can be successful if the spawning but not the juvenile rearing habitats are destroyed. Smolt release is a possible way of escaping a population regulatory bottleneck in freshwater, as is postsmolt release at sea. However, the latter may be avoided because of the risk of spreading the hatchery fish among rivers when they return to freshwater for spawning (Hansen and Jonsson 1991a).

Regulated rivers, characterized by rapid fluctuations in water flow and temperature, may offer poor juvenile rearing habitats favoring smolt releases before releases of younger fish. The marine mortality of hatchery Atlantic salmon released at the smolt stage is often very high owing to phenotypic changes caused by the juvenile rearing environment (Dannewitz et al. 2003), and inadequate behavior of hatchery fish towards predators (Brown and Laland 2001). In such cases, research efforts may be channeled to improving the sea-survival of the fish. Successful restoration and rehabilitation of salmon populations assume that the fish reproduce successfully. Reproductive success of released hatchery salmon is generally lower than that of wild conspecifics because of inadequate spawning behavior (Fleming et al. 1996, 1997, 2000). If a population is on the brink of extinction, the population may be saved by cryopreservation of spermatozoa in a gene bank (Gallant et al. 1993; Jodun et al. 2007) to be used for production and release of hatchery fish if and when the living conditions in the river are improved. There are also hatcheries serving as gene banks by keeping live fish (Bergan et al. 1991). This may be helpful in a critical situation, but with time, hatchery selection may gradually change such populations from their source populations. Hatchery preserved populations may also have a small genetic basis and be different from the original populations due to genetic drift, reducing their value as broodstock for supplementary releases. However, release of hatchery fish is the only known method to increase population sizes above the carrying capacity of the juvenile rearing habitat, when this is rehabilitated and enlarged to its maximum (e.g., by use of fishways).

Although stocking for population enhancement has been practiced by resource managers for more than 150 years, the general view now is that habitat restoration and rehabilitation are preferred over fish releases where these can be applied. Reasons are possible negative genetic effects on the local population and the risk of spreading contagious diseases from hatcheries. Hatchery salmon released in nature have in some cases led to loss of genetic variability due to a small effective population size of the stocked fish with introgression and adaptive divergences in wild populations (Crozier 1998; Utter 1998). Therefore, one has to balance the long-term risk of genetic change in natural fish populations against the immediate benefits of artificially increasing fish abundance (Tringali and Bert 1998). This is an often encountered trade-off for fisheries managers since genetics of weak populations are easily changed by massive releases because of the small size of the local gene pools. As long as there are wild spawners present and suitable opportunities for reproduction and juvenile rearing in rivers, it appears better to allow the fish to reproduce naturally than using spawners as broodstock for artificial rearing and release.

Adaptive Management

By using adaptive management practices (Miller et al. 1995), one can reduce uncertainty and improve the remedial actions in light of the results from a continued monitoring of the population in question (Figure 1). The establishment of restoration goals is often hindered by limited knowledge about the status of the population, the reason for long-term trends and short-term fluctuations, the role of specific components in the ecosystems exploited by the salmon, the vulnerability to loss of diversity, and the economic implications involved. Goal setting and the attainment of goals are also hindered because ecosystems are dynamic and populations and species evolve in response to selection pressures. Salmonid populations shift with changes in climate and human impacts, and they migrate among habitats. Consequently, even if we have a vision of a desirable future, it is difficult to foresee a precise pathway to its fulfillment. Management decisions are taken in the face of uncertainty, but by using the adaptive management technique, we gradually reduce uncertainty by a continued evaluation of the results used to improving the remedial actions. Furthermore, we have to view salmon as an integrated part of the ecosystem, meaning that they depend

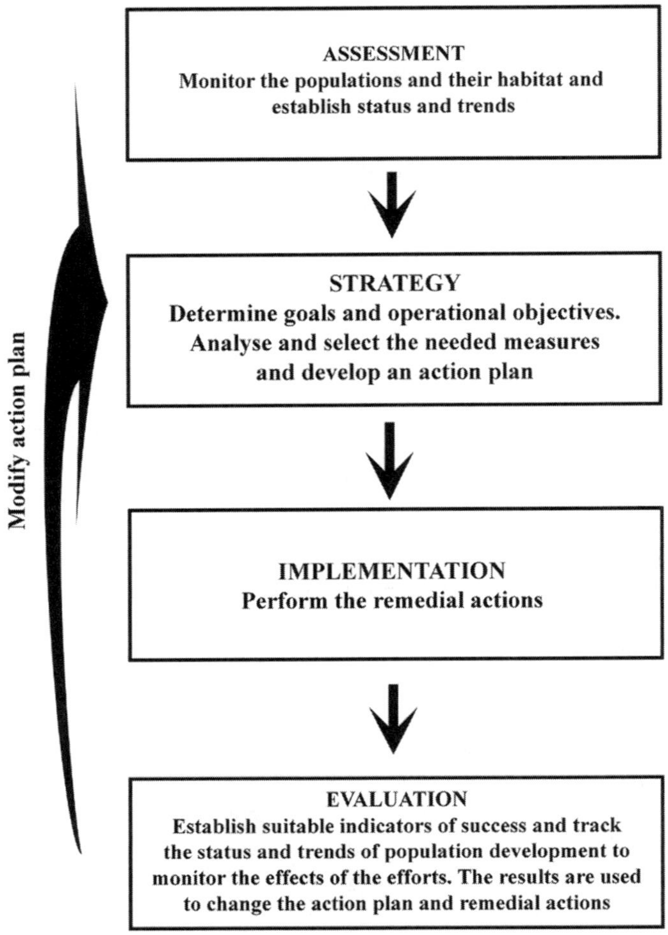

Figure 1.—Adaptive management of populations.

on the other components of their ecosystem. When restoring salmon populations and/or habitats, we therefore have to secure that the various needs for living in the habitat are fulfilled.

Density-dependent mechanisms constrain the parr production of freshwater habitats, as earlier explained. To manage Atlantic salmon population, one should monitor the habitats and densities of the age-groups of parr, establish status and trends, and compare the results with expectations based on experience from similar localities (Figure 1). Are parr densities significantly lower than found in good salmon rivers at similar latitudes? Are there too few young of the year or are older parr lacking? If the answer is yes to the first question, there may be a lack of spawners, the habitat may be unsuitable for spawning or early rearing, or the population is threatened by pollution (e.g., acidity, oxygen deficit, heavy metals) or a disease. These are all conditions that we may judge based on local knowledge, results of fishing efforts, observations on the spawning grounds, or experimental tests by use of, for instance, experiments such as planting eggs in incubation boxes buried in the spawning area or parr releases (see Population Restoration and Enhancement). If only older parr are lacking from the relevant habitats, they may have emigrated or an earlier year's recruitment has failed. At moderate and northern latitudes, there should at least be 1-year-old, if not older, parr, present. At southern localities, a large part of the smolt may be 1 year old, meaning that older parr can be naturally scarce.

Based on this assessment, one's goals and operational objectives can be set. One may wish to restore inferior salmon habitats or open new areas of a river for salmon above a previously impassable water fall. Furthermore, the spawning stock may be small and needs to be increased. If this is difficult, an alternative is egg planting at suitable sites in the river. If there is a lack of parr, releases of parr or smolt may be required. One can determine required measures and develop and implement the remedial action plan to restore or enhance salmon populations.

By annual assessments of the populations, it is possible to evaluate the effects of the efforts and use the results to change the management strategy and actions to further improve the results in relation to any environmental or population change that might occur. Thus, the adaptive management program functions as a long-term experiment for optimal production of salmon rivers.

Research Tasks

An important research task is to breed a more wild-like phenotype than the regular hatchery salmon. To do that, we need new knowledge on how hatchery conditions change the fitness of cultured fish in nature. We need more information on the plasticity of the genotype–environment interactions and how hatchery conditions influence the fitness in the wild of liberated fish. For instance, how does the water current velocity influence hormonal changes and what are the phenotypic expressions of the affected individuals (McCormick et al. 2003)?

Elevated egg incubation temperature affects the subsequent metabolic and growth rates of the fish with possible effects on age and size at smolting and sexual maturity (Jonsson et al. 2005), but we lack a quantitative understanding of how important egg incubation temperature is for the subsequent maximum growth rate, optimal temperature for growth and growth efficiency (Jonsson et al. 2001), and life history characters associated with growth through norms of reaction (Jonsson and Jonsson 2004a). There is also a need for further studies on causes for the insufficient antipredator behavior of hatchery salmon and how this behavior can best be changed by modifying rearing facilities. Early habitat complexity influences later behavioral performance in Atlantic cod (Salvanes and Braithwaite 2006), and there is reason to believe that a similar relationship may hold for Atlantic salmon.

There is a need for new knowledge on how the habitat constrains salmon production. What are the mechanisms of density regulation, and how are the environmental conditions influencing population abundance (Einum and Nislow 2005)? Is winter or summer flow most limiting for the survival of young fish in rivers? How can flow in a regulated river be modified to increase the smolt production? How is flow in fishways constraining adult fish size farther upstream? Furthermore, there is growing concern about ecosystem effects of hatchery salmon on communities and ecosystems (Pascual et al. 2002; Waknitz et al. 2003; Baxter et al. 2004). Salmon have the potential to reconstruct local food webs, and research is needed to evaluate this risk, especially when the species is spreads beyond its native range. Atlantic salmon is known as an ineffective colonizer (Naylor et al. 2005), but the current range represents colonization after the last glaciation period, meaning that there is a chance that releases in new areas may result in the establishment of self-sustaining populations.

Acknowledgments

We thank two anonymous reviewers and AlexHaro for helpful comments.

References

Aarestrup, K., N. Jepsen, G. Rasmussen, F. Økland, E. B. Thorstad, and G. Holdernsgaard. 2000. Prespawning migratory behaviour and spawning success of sea-ranched Atlantic salmon, *Salmo salar* L., in the River Gudenaa, Denmark. Fisheries Management and Ecology 7:387–400.

Adams, C., F. Huntingford, J. Turnbull, S. Arnott, and A. Bell. 2000. Size heterogeneity can reduce aggression and promote growth in Atlantic salmon parr. Aquaculture International 8:543–549.

Alm, G. 1959. Connection between maturity, size and age in fishes. Report of the Institute of Freshwater Research Drottningholm 40:5–145.

Almodovar, A., G. G. Nicola, B. Elvira, and J. L. Garcia-Marin. 2006. Introgression variability among Iberian brown trout evolutionary significant units: the influence of local management and environmental features. Freshwater Biology 51:1175–1187.

Andrew, F. J., and G. H. Geen. 1960. Sockeye and pink

salmon production in relation to proposed dams in the Fraser River system. Bulletin of the International Pacific Salmon Fisheries Commission 11:10–30.

Anras, M. L. B., and J. P. Lagardere. 2004. Domestication and behaviour in fish. Productions Animales 17:211–215.

Armstrong, J. D., P. S. Kemp, G. J. A. Kennedy, M. Ladle, and N. J. Milner. 2003. Habitat requirements of Atlantic salmon and brown trout in rivers and streams. Fisheries Research 62:143–170.

Armstrong, J. D., and K. H. Nislow. 2006. Critical habitat during the transition from maternal provisioning in freshwater fish, with emphasis on Atlantic salmon (*Salmo salar*) and brown trout (*Salmo trutta*). Journal of Zoology 269:403–413.

Arnekleiv, J. V., and M. Kraabøl. 1996. Migratory behaviour of adult fast-growing brown trout (*Salmo trutta* L.) in relation to water flow in a regulated Norwegian river. Regulated Rivers Research and Management 12:39–49.

Aubin-Horth, N., B. H. Letcher, and H. A. Hoffmann. 2005. Interaction of rearing environment and reproductive tactic on gene expression profiles in Atlantic salmon. Journal of Heredity 96:261–278.

Avery, E. L. 1996. Evaluations of sediment traps and artificial gravel riffles constructed to improve reproduction of trout in three Wisconsin streams. North American Journal of Fisheries Management 16:282–293.

Ayllon, F., P. Davaine, E. Beall, J. L. Martinez, and E. Garcia-Vazquez. 2004. Bottlenecks and genetic changes in Atlantic salmon (*Salmo salar* L.) stocks introduced in the Subarctic Kerguelen Islands. Aquaculture 237:103–116.

Bakke, T. A., and P. D. Harris. 1998. Diseases and parasites in wild Atlantic salmon (*Salmo salar*) populations. Canadian Journal of Fisheries and Aquatic Sciences 55(Supplement 1):247–266.

Barlaup, B. T., and V. Moen. 2001. Planting of salmonid eggs for stock enhancement—a review of the most commonly used methods. Nordic Journal of Freshwater Research 75:7–19.

Barton, B. A. 2000. Salmonid fishes differ in their cortisol and glucose responses to handling and transport stress. North American Journal of Aquaculture 62:12–18.

Bates, D. J., and B. A. McKeown. 2003. Growth in stream-stocked juvenile hatchery-reared coastal cutthroat trout (*Oncorhynchus clarki clarki*) and the implications for wild populations. Aquaculture 222:215–228.

Baxter, C., K. Fausch, M. Murakami, and P. Chapman. 2004. Fish invasion restructures stream and forest food webs by interrupting reciprocal prey subsidies. Ecology 85:2656–2663.

Beamish, R. J., C. Mahnken, and C. M. Neville. 2004a. Evidence that reduced early marine growth is associated with lower marine survival in coho salmon. Transactions of the American Fisheries Society 133:26–33.

Beamish, R. J., R. M. Sweeting, and C. M. Neville. 2004b. Improvement of juvenile Pacific salmon production in a regional ecosystem after the 1998 climate regime shift. Transactions of the American Fisheries Society 133:1163–1175.

Beaugrand, G., and P. C. Reid. 2003. Long-term changes in phytoplankton, zooplankton and salmon related to climate. Global Change Biology 9:801–817.

Berejikian, B. A. 1995. The effect of hatchery and wild ancestry and experience on the relative ability of steelhead trout fry (*Oncorhynchus mykiss*) to avoid a benthic predator. Canadian Journal of Fisheries and Aquatic Sciences 52:2476–2482.

Berejikian, B. A., W. T. Fairgrieve, P. Swanson, and E. P. Tezak. 2003a. Current velocity and injection of GnRHa affect reproductive behaviour and body composition of captively reared offspring of wild Chinook salmon (*Oncorhynchus tshawytscha*). Canadian Journal of Fisheries and Aquatic Sciences 60:690–699.

Berejikian, B. A., E. P. Tezak, and A. L. LaRae. 2003b. Innate and enhanced predator recognition in hatchery-reared chinook salmon. Environmental Biology of Fishes 67:241–251.

Berejikian, B. A., E. P. Tezak, S. L. Schroder, T. A. Flagg, and C. M. Knudsen. 1999. Competitive differences between newly emerged offspring of captive-reared and wild coho salmon. Transactions of the American Fisheries Society 128:832–839.

Berejikian, B. A., E. P. Tezak, S. L. Schroder, and C. M. Knudsen. 1997. Reproductive behavioural interactions between wild and captively reared coho salmon (*Oncorhynchus kisutch*). ICES Journal of Marine Science 54:1040–1050.

Berg, M. 1969. Stocking of salmon in lakes and ponds. Directorate of Nature Conservation, Fish and Fisheries Management, Trondheim, Norway.

Bergan, P. I., D. Gausen, and L. P. Hansen. 1991. Attempts to reduce the impact of reared Atlantic salmon on wild in Norway. Aquaculture 98:319–324.

Berglund, I., L. P. Hansen, H. Lundqvist, B. Jonsson, T. Eriksson, J. E. Thorpe, and L.-O. Eriksson. 1991. Effects of elevated winter temperature on seawater adaptability, sexual rematuration, and downstream migratory behaviour in mature male Atlantic salmon parr (*Salmo salar*). Canadian Journal of Fisheries and Aquatic Sciences 48:1041–1047.

Berland, G., T. Nickelsen, J. Heggenes, F. Økland, E. B. Thorstad, and J. Halleraker. 2004. Movements of wild Atlantic salmon parr in relation to peaking flows below a hydropower station. River Research and Applications 20:957–966.

Berntssen, M. H. G., F. Kroglund, B. O. Rosseland, and S. E. W. Bonga. 1997. Responses of skin mucous cells to aluminium exposure at low pH in Atlantic salmon (*Salmo salar*) smolts. Canadian Journal of Fisheries and Aquatic Sciences 54:1039–1045.

Bohlin, T., L. F. Sundström, J. I. Johnsson, J. Höjsjö, and J. Petterson. 2002. Density-dependent growth in brown trout: effects of introducing wild and hatchery fish. Journal of Animal Ecology 71:683–692.

Boyland, P., and C. E. Adams. 2006. The influence of broad scale climatic phenomena on long-term trends in Atlantic salmon population size: an example from the Foyle, Ireland. Journal of Fish Biology 68:276–283.

Bradford, M. J. 1997. An experimental study of stranding of juvenile salmonids on gravel ears and in sidechannels during rapid flow decreases. Regulated Rivers Research and Management 13:395–401.

Braithwaite, V. A., and A. G. V. Salvanes. 2005. Environmental variability in the early rearing environment generates behaviourally flexible cod: implications for rehabilitating populations. Proceedings of the Royal Society B 272:1107–1113.

Brännäs, E., S. Jonsson, and K. Brännäs. 2004. Density-dependent effects of prior residence and behavioural strategy on growth of stocked brown trout (*Salmo trutta*). Canadian Journal of Zoology 82:1638–1646.

Brayshaw, J. D. 1967. The effect of river discharge on inland fisheries. Pages 102–118 *in* P. G. Isaac, editor. River management. MacLaren, London.

Brown, C., T. Davidson, and K. Laland. 2003a. Environmental enrichment and prior experience of live prey improve foraging behaviour in hatchery-reared Atlantic salmon. Journal of Fish Biology 63(Supplement A):187–196.

Brown, C., and K. Laland. 2001. Social learning and life skills training for hatchery reared fish. Journal of Fish Biology 59:471–493.

Brown, C., and K. Laland. 2002. Social enhancement and social inhibition of foraging behaviour in hatchery-reared Atlantic salmon. Journal of Fish Biology 61:987–998.

Brown, C., K. Laland, and J. Krause. 2003. Learning in fishes: why are they smarter than you think? Fish and Fisheries 4:197–288.

Brown, C., A. Markula, and K. Laland. 2003b. Social learning of prey location in hatchery-reared Atlantic salmon. Journal of Fish Biology 63:738–745.

Brown, G. W., and R. J. F. Smith. 1998. Aquired predator recognition in juvenile rainbow trout (*Oncorhynchus mykiss*): conditioning hatchery-reared fish to recognize chemical cues of a predator. Canadian Journal of Fisheries and Aquatic Sciences 55:611–617.

Bryant, M. D., B. J. Frenette, and S. J. McCurdy. 1999. Colonization of a watershed by anadromous salmonids following the installation of a fish ladder in Margaret Creek, southeast Alaska. North American Journal of Fisheries Management 19:1129–1136.

Carr, J. W., F. Whoriskey, and P. O'Reilly. 2004. Efficacy of releasing captive reared broodstock into an imperilled wild Atlantic salmon population as a recovery strategy. Journal of Fish Biology 65(Supplement A):38–54.

Chapman, D. W. 1988. Critical review of variables used to define effects of fines in redds of large salmonids. Transactions of the American Fisheries Society 117:1–21.

Chilcote, M. W. 2003. Relationship between natural productivity and the frequency of wild fish in mixed spawning populations of wild and hatchery steelhead (*Oncorhynchus mykiss*). Canadian Journal of Fisheries and Aquatic Sciences 60:1057–1067.

Claireaux, G., D. J. McKenzie, A. G. Genge, A. Chatelier, J. Aubin, and A. P. Farrell. 2005. Linking swimming performance, cardiac pumping ability and cardiac anatomy in rainbow trout. Journal of Experimental Biology 208:1775–1784.

Clay, C. H. 1995. Design of fishways and other fish Facilities. Lewis Publisher, London.

Clifford, S. L., P. McGinnity, and A. Ferguson. 1998. Genetic changes in an Atlantic salmon population resulting from escaped juvenile farm salmon. Journal of Fish Biology 52:118–127.

Coghlan, S. M., and N. H. Ringler. 2004. Comparison of Atlantic salmon embryo and fry stocking in the salmon river, New York. North American Journal of Fisheries Management 24:1385–1397.

Connor, W. P., S. G. Smith, T. Andersen, S. M. Bradbury, D. C. Burum, E. E. Hockersmith, M. L. Schuck, G. W. Mendel, and R. M. Bugert. 2004. Postrelease performance of hatchery yearling and subyearling fall Chinook salmon released into the Snake River. North American Journal of Fisheries Management 24:545–560.

Crook, D. A., and A. I. Robertson. 1999. Relationships between riverine fish and wood debris: implications for lowland rivers. Marine and Freshwater Research 50:941–953.

Cross, T. F. 2000. Genetic implications of translocation and stocking of fish species, with particular reference to western Australia. Aquaculture Research 31:83–94.

Crozier, W. W. 1998. Genetic implications of hatchery rearing in Atlantic salmon: effects of rearing environment on genetic composition. Journal of Fish Biology 52:1014–1025.

Crozier, W. W., I. J. J. Moffett, and G. J. A. Kennedy. 1997. Comparative performance of native and nonnative strains of Atlantic salmon (*Salmo salar* L.) ranched from the River Bush, Northern Ireland. Fisheries Research 32:81–88.

Cutts, C. J., B. Brembs, N. B. Metcalfe, and A. C. Taylor. 1999. Prior residence, territory quality and life-history strategies in juvenile Atlantic salmon (*Salmo salar* L.). Journal of Fish Biology 55:784–794.

Cutts, C. J., N. B. Metcalfe, and A. C. Taylor. 2002. Fish may fight rather than feed in a novel environment: metabolic rate and feeding motivation in juvenile Atlantic salmon. Journal of Fish Biology 61:1540–1548.

Dannewitz, J., E. Petersson, J. Dahl, T. Prestegaard, A. C. Lof, and T. Järvi. 2004. Reproductive success of hatchery-produced and wild-born brown trout in an experimental stream. Journal of Applied Ecology 41:355–364.

Dannewitz, J., E. Petersson, T. Prestegaard, and T. Järvi. 2003. Effects of sea-ranching and family background on fitness traits in brown trout *Salmo trutta* reared under near-natural conditions. Journal of Applied Ecology 40:241–250.

Davidson, W. 1997. The effects of exercise training on teleost fish: a review of recent literature. Comparative Biochemistry and Physiology 117A:67–75.

de Jong, M. C. V., I. G. Cowx, and D. A. Scruton. 1997. An evaluation of instream habitat restoration techniques on salmonid populations in a Newfoundland stream. Regulated Rivers Research and Management 13:603–614.

Dempson, J. B., M. F. O'Connell, and C. J. Schwarz. 2004. Spatial and temporal trends in abundance of Atlantic salmon, *Salmo salar*, in Newfoundland with emphasis on impacts of the 1992 closure of the commercial fishery. Fisheries Management and Ecology 11:387–402.

DeVries, P. 1997. Riverine salmonid egg burial depth: review of published data and implications for scour studies. Canadian Journal of Fisheries and Aquatic Sciences 54:1685–1689.

Dieperink, C., B. D. Bak, L. F. Pedersen, M. I. Pedersen, and S. Pedersen. 2002. Predation on Atlantic salmon and sea trout during their first days as postsmolts. Journal of Fish Biology 61:848–852.

Dittman, A. H., and T. P. Quinn. 1996. Homing in Pacific salmon: mechanisms and ecological basis. Journal of Experimental Biology 199:83–91.

Dittman, A. H., T. P. Quinn, and G. A. Nevitt. 1996. Timing of imprinting to natural and artificial odors by coho salmon (*Oncorhynchus kisutch*). Canadian Journal of Fisheries and Aquatic Sciences 53:434–442.

Donaghy, M. J., and E. Verspoor. 2000. A new design of instream incubator for planting out and monitoring Atlantic salmon eggs. North American Journal of Fisheries Management 20:521–527.

Dunmall, K. M., and Schreer, J. F. 2003. A comparison of the swimming and cardiac performance of fanned and wild Atlantic salmon, *Salmo salar*, before and after stripping. Aquaculture 220:869–882.

Duston, J., T. Astatkie, and P. F. MacIsaac. 2005. Genetic influence of parr versus anadromous sires on the life histories of Atlantic salmon (*Salmo salar*). Canadian Journal of Fisheries and Aquatic Sciences 62:2067–2075.

Ebersole, J. L., W. J. Liss, and C. A. Frissell. 1997. Restoration of stream habitats in the western United States: restoration as reexpression of habitat capacity. Environmental Management 21:1–14.

Edmands, S. 1999. Heterosis and outbreeding depression in interpopulation crosses spanning a wide range of divergence. Evolution 53:1757–1768.

Edmands, S. 2007. Between a rock and a hard place: evaluating the relative risk of inbreeding and outbreeding for conservation and management. Molecular Ecology 16:463–475.

Egglishaw, H. J., and P. E. Shackley. 1980. Survival and growth of salmon, *Salmo salar* (L.) planted in a Scottish stream. Journal of Fish Biology 11:647–672.

Einum, S., and I. A. Fleming. 1997. Genetic divergence and interactions in the wild among native, farmed and hybrid Atlantic salmon. Journal of Fish Biology 50:634–651.

Einum, S., I. A. Fleming, I. M. Cote, and J. D. Reynolds. 2003. Population stability in salmon: effects of population size and female reproductive allocation. Journal of Animal Ecology 72:811–821.

Einum, S., and K. H. Nislow. 2005. Local-scale, density-dependent survival of mobile organisms in continuous habitats: an experimental test using Atlantic salmon. Oecologia 143:203–210.

Einum, S., E. B. Thorstad, and T. F. Næsje. 2002. Growth rate correlations across life-stages in female Atlantic salmon. Journal of Fish Biology 60:780–784.

Elliott, J. M. 2001. The relative role of density in the stock-recruitment relationship of salmonids. Pages 25–66 *in* E. Prevost, and G. Chaput, editors. Stock, recruitment and reference points: assessment and management of Atlantic salmon. INRA editions, Paris.

Elliott, J. M., and M. A. Hurley. 1998. Population regulation in adult, but not juveniles, resident trout (*Salmo trutta*) in a Lake District stream. Journal of Animal Ecology 67:280–286.

Ellis, T., B. North, A. P. Scott, N. R. Bromage, M. Porter, and D. Gadd. 2002. The relationships between stocking density and welfare in farmed rainbow trout. Journal of Fish Biology 61:493–531.

Enders, E. C., K. E. Smokorowski, C. J. Pennell, K. D. Clarke, B. Sellars, and D. A. Scruton. 2007. Habitat use and fish activity of landlocked Atlantic salmon and brook charr in a newly developed habitat compensation facility. Hydrobiologia 582:133–142.

Eriksson, T., and L. O. Eriksson. 1991. Spawning migratory behaviour of Baltic salmon (*Salmo salar*)—effects on straying frequency and time of river ascent. Aquaculture 98:79–87.

Exley, C., and M. J. Phillips. 1988. Acid rain: implications for the farming of salmonids. Pages 225–341 *in* J. F. Muir, and R. J. Robers, editors. Recent advances in aquaculture. Croom Helm, London.

Finstad, B., and N. Jonsson. 2001. Factors influencing the yield of smolt releases in Norway. Nordic Journal of Freshwater Research 75:37–55.

Fiske, P., R.A. Lund, and L.P. Hansen. 2005. Identifying fish farm escapees. Pages 659–680 *in* S. X. Cadrin, K. D. Friedland, and J. D. Waldman, editors. Stock identification methods. Elsevier Academic Press, Amsterdam.

Fjælestad, K. T., T. Gjedrem, and B. Gjerde. 1993. Genetic improvement of disease resistance in fish: an overview. Aquaculture 111:65–74.

Fjellheim, A., and B. O. Johnsen. 2001. Experiences from stocking salmonid fry and fingerlings in Norway. Nordic Journal of Freshwater Research 75:20–36.

Fleming, I. A. 1996. Reproductive strategies of Atlantic salmon: ecology and evolution. Reviews in Fish Biology and Fisheries 6:379–416.

Fleming, I. A., T. Augustsson, B. Finstad, J. L. Johnsson, and B. T. Björnsson. 2002. Effects of domestication on growth physiology and endocrinology of Atlantic salmon (*Salmo salar*). Canadian Journal of Fisheries and Aquatic Sciences 59:1323–1330.

Fleming, I. A., and S. Einum. 1997. Experimental tests of genetic divergence of farmed from wild Atlantic salmon due to domestication. ICES Journal of Marine Science 54:1051–1063.

Fleming, I. A., S. Einum, B. Jonsson, and N. Jonsson. 2003. Comment on "Rapid evolution of egg size on captive salmon." Science 302:59.

Fleming, I. A., and M. R. Gross. 1992. Reproductive behaviour of hatchery and wild coho salmon: does it differ? Aquaculture 103:1–21.

Fleming, I. A., and M. R. Gross. 1993. Breeding success of hatchery and wild coho salmon (*Oncorhynchus kisutch*) in competition. Ecological Applications 3:230–245.

Fleming, I. A., K. Hindar, I. B. Mjølnerød, B. Jonsson, T. Balestad, and A. Lamberg. 2000. Lifetime success and interactions of farmed Atlantic salmon invading a native population. Proceedings of the Royal Society of London Series B 267:1517–1523.

Fleming, I. A., B. Jonsson, and M. R. Gross. 1994. Phenotypic divergence of sea-ranched, farmed and wild salmon. Canadian Journal of Fisheries and Aquatic Sciences 51:2808–2824.

Fleming, I. A., B. Jonsson, M. R. Gross, and A. Lamberg. 1996. An experimental study of the reproductive behaviour and success of farmed and wild Atlantic salmon (*Salmo salar*). Journal of Applied Ecology 33:893–905.

Fleming, I. A., A. Lamberg, and B. Jonsson. 1997. Effects of early experience on the reproductive performance of Atlantic salmon. Behaviour Ecology 8:470–480.

Fleming, I. A., and E. Peterson. 2001. The ability of hatchery-reared salmonids to breed and contribute to the natural productivity of wild populations. Nordic Journal of Freshwater Research 75:71–98.

Ford, M. J. 2002. Selection in captivity during supportive breeding may reduce fitness in the wild. Conservation Biology 16:815–825.

Fox, M., and S. Bolton. 2007. A regional and geomorphic reference for quantities and volumes of instream wood in unmanaged forested basins of Washington State. North American Journal of Fisheries Management 27:342–359.

Friedland, K. D., G. Chaput, and J. C. MacLean. 2005. The emerging role of climate in post-smolt growth of Atlantic salmon. ICES Journal of Marine Science 62:1338–1349.

Friedland, K. D., L. P. Hansen, D. A. Dunkley, and J. C. MacLean. 2000. Linkage between ocean climate, post-smolt growth, and survival of Atlantic salmon (*Salmo salar* L.) in the North Sea area. ICES Journal of Marine Science 57:419–429.

Fries, C. R. 1986. Effects of environmental stressors and immunosuppressants on immunity of *Fundulus heteroclitus.* American Zoology 26:271–282.

Fullerton, A. H., T. J. Beechie, S. E. Baker, J. E. Hall, and K. A. Barnas. 2006. Regional patterns of riparian characteristics in the interior Columbia River basin, northwestern USA: applications for restoration planning. Landscape Ecology 21:1347–1360.

Gallant, R. K., G. F. Richardson, and M. A. McNiven. 1993. Comparison of different extenders for the cryopreservation of Atlantic salmon spermatozoa. Theriogenology 40:479–486.

Garant, D., J. D. Dodson, and L. Bernatchez. 2005. Offspring genetic diversity increases fitness of female Atlantic salmon (*Salmo salar*). Behavioral Ecology and Sociobiology 57:240–244.

Garcia de Leaniz, C., I. A. Fleming, S. Einum, E. Verspoor, W. C. Jordan, S. Consuegra, N. Aubin-Horth, D. Lajus, B. H. Letcher, A. F. Youngson, J. H. Webb, L. A. Vøllestad, B. Villanueva, A, Ferguson, and T. P. Quinn. 2007. A critical review of adaptive genetic variation in Atlantic salmon: implications for conservation. Biological Reviews 82:173–211.

Giannico, G. R., and S. G. Hinch. 2003. The effect of wood and temperature on juvenile coho salmon winter movement, growth, density and survival in side-channels. River Research and Applications 19:219–231.

Gibson, R. J. 1993. The Atlantic salmon in fresh water: spawning, rearing and production. Reviews in Fish Biology and Fisheries 3:39–73.

Gjedrem, T. 2000. Genetic improvement of cold-water fish species. Aquaculture Research 31:25–33.

Gjerde, B., H. Simianer, and T. Refstie. 1994. Estimates of genetic and phenotypic parameters for body-weight, growth-rate and sexual maturity in Atlantic salmon. Livestock Production Science 38:133–143.

Gjøen, H. M., T. Refstie, O. Ulla, and B. Gjerde. 1997. Genetic correlations between survival of Atlantic salmon in challenge and field tests. Aquaculture 158:277–288.

Glover, K. A., O. Bergh, H. Rudra, and Ø. Skaala. 2006. Juvenile growth and susceptibility to *Aeromonas salmonicida* subsp *salmonicida* in Atlantic salmon (*Salmo salar* L.) of farmed, hybrid and wild parentage. Aquaculture 254:72–81.

Glover, K. A., J. B. Taggart, Ø. Skaala, and A. J. Teale. 2004. A study of inadvertent domestication selection during start-feeding of brown trout families. Journal of Fish Biology 64:1168–1178.

Goodman, D. 2005. Selection equilibrium for hatchery and wild spawning fitness in integrated breeding programs. Canadian Journal of Fisheries and Aquatic Sciences 62:374–389.

Gowans, A. R. D., J. D. Armstrong, I. G. Priede, and S. Mckelvey. 2003. Movements of Atlantic salmon migrating upstream through a fish-pass complex in Scotland. Ecology of Freshwater Fish 12:177–189.

Griffiths, S. W., and Armstrong, J. D. 2002. Rearing conditions influence refuge use among over-wintering Atlantic salmon juveniles. Journal of Fish Biology 60:363–369.

Gunnerød, T. B., N. A. Hvidsten, and T. G. Heggberget. 1988. Open sea releases of Atlantic salmon smolts, *Salmo salar*, in central Norway, 1973–83. Canadian Journal of Fisheries and Aquatic Sciences 45:1340–1345.

Haberstock, A. E., H. G. Nichols, M. P. DesMeules, J. Wright, J. M. Christensen, and D. H. Hudnut. 2000. Method to identify effective riparian buffer widths for Atlantic salmon habitat protection. Journal of the American Water Resources Association 36:1271–1286.

Hale, S. S., T. E. McMahon, and P. C. Nelson. 1985. Habitat suitability index models and instream flow suitability curves: chum salmon. U.S. Fish and Wildlife Service Biological Report 82.

Halleraker, J. H., S. J. Saltveit, A. Harby, J. V. Arnekleiv, H. P. Fjeldstad, and B. Kohler. 2003. Factors influencing stranding of wild juvenile brown trout (*Salmo trutta*) during rapid and frequent flow decreases in an artificial stream. River Research and Applications 19:589–603.

Handeland, S. O., B. T. Bjørnsson, and A. M. Arnesen. 2003. Seawater adaptation and growth of post-smolt Atlantic salmon (*Salmo salar*) of wild and farmed strains. Aquaculture 220:367–384.

Hansen, L. P. 1987. Growth, migration and survival of lake reared juvenile anadromous Atlantic salmon *Salmo salar* L. Fauna Norvegica Serie A 8:29–34.

Hansen, L. P. 1991. Rehabilitation of the Atlantic salmon stock in the River Drammen, Norway. Pages 140–146 *in* D. Mills, editor. Strategies for the rehabilitation of salmon rivers. Proceedings of a Joint Conference held at the Linnean Society on Thursday 29 to Friday 30 November, 1990. The Atlantic Salmon Trust, The Institute of Fisheries Management, and The Linnean Society of London, London.

Hansen, L. P., and B. Jonsson. 1985. Downstream migration of hatchery reared smolts of Atlantic salmon (*Salmo salar* L.) in the River Imsa, Norway. Aquaculture 45:237–248.

Hansen, L. P., and B. Jonsson. 1986. Salmon ranching experiments in the River Imsa: effects of day and night release and seawater adaptation on recapture rates of adults. Report of Institute of Freshwater Research Drottningholm 63:47–51.

Hansen, L. P., and B. Jonsson. 1988. Salmon ranching experiments in the River Imsa: effects of dip-netting, transport and chlorobutanol anaesthesia on survival. Aquaculture 75:301–305.

Hansen, L. P., and B. Jonsson. 1989a. Salmon ranching experiments in the River Imsa: effects of timing of Atlantic salmon (*Salmo salar*) smolt migration on survival to adults. Aquaculture 82:367–373.

Hansen, L.P., and B. Jonsson. 1989b. Salmon ranching experiments in the River Imsa: returns of different stocks to the fishery and to River Imsa. Pages 445–452 *in* N. DePauw, E. Jaspers, H. Ackefors, and N. Wilkins, editors. Aquaculture: a biotechnology in progress. European Aquaculture Society, Bredene, Belgium.

Hansen, L. P., and B. Jonsson. 1990. Restocking the River Akerselva, Oslo with Atlantic salmon smolts *Salmo salar* L. of different stocks. Fauna Norvegica Series A 11:9–15.

Hansen, L. P., and B. Jonsson. 1991a. The effect of timing of Atlantic salmon smolt and post-smolt release on the distribution of adult return. Aquaculture 98:61–67.

Hansen, L. P., and B. Jonsson. 1991b. Evidence of a genetic component in seasonal return pattern of Atlantic salmon (*Salmo salar* L.). Journal of Fish Biology 38:251–258.

Hansen, L. P., and B. Jonsson. 1994. Homing in Atlantic salmon: effects of juvenile learning on transplanted post-spawners. Animal Behaviour 47:220–222.

Hansen, L. P., B. Jonsson, R.I.G. Morgan, and J. E. Thorpe. 1989. Influence of parr maturity on emigration of smolts of Atlantic salmon (*Salmo salar*). Canadian Journal of Fisheries and Aquatic Sciences 46:410–415.

Hansen, L. P., N. Jonsson, and B. Jonsson. 1993. Oceanic migration of homing Atlantic salmon. Animal Behaviour 45:927–941.

Hansen, L. P., D. G. Reddin, and R. A. Lund. 1997. The incidence of reared Atlantic salmon (*Salmo salar* L.) of fish farm origin at West Greenland. ICES Journal of Marine Science 54:152–155.

Hansen, M. M. 2002. Estimating the long-term effects of stocking domesticated trout into wild brown trout (*Salmo trutta*) populations: an approach using microsatellite DNA analysis of historical and contemporary samples. Molecular Ecology 11:1003–1015.

Harada, Y., M. Yokota, and M. Iizuka. 1998. Genetic risk of domestication in artificial fish stocking and its possible reduction. Research on Population Ecology 40:311–324.

Harris, G. S. 1973. A simple egg box planting technique for estimating the survival of eggs deposited in stream gravel. Journal of Fish Biology 5:85-88.

Harshberger, T. J., and P. E. Porter. 1979. Survival of brown trout eggs: two planting techniques compared. North American Journal of Fisheries Management 2:84–89.

Harwood, A. J., J. D. Armstrong, S. W. Griffiths, and N. B. Metcalfe. 2002. Sympatric association influences within-species dominance relations among juvenile Atlantic salmon and brown trout. Animal Behaviour 64:85–95.

Harwood, A. J., N. B. Metcalfe, J. D. Armstrong, and S. W. Griffiths. 2001. Spatial and temporal effects of interspecific competition between Atlantic salmon (*Salmo salar*) and brown trout (*Salmo trutta*) in winter. Canadian Journal of Fisheries and Aquatic Sciences 58:1133–1140.

Hayes, S. A., M. H. Bond, C. V. Hanson, and R. B. MacFarlane. 2004. Interactions between endangered wild and hatchery salmonids: can pitfalls of artificial propagation be avoided in small coastal streams? Journal of Fish Biology 65(Supplement A):101–121.

Heath, D. D., J. W. Heath, C. A. Bryden, R. M. Johnson, and C. W. Fox. 2003. Rapid evolution of egg size in captive salmon. Science 299:1738–1740.

Hedenskog, M., E. Peterson, and T. Järvi. 2002. Agonistic behaviour and growth in newly emerged brown trout (*Salmo trutta* L.) of sea ranched and wild origin. Aggressive Behavior 28:145–153.

Hedrick, P. 2005. 'Genetic restoration:' a more comprehensive perspective than 'genetic rescue.' Trends in Ecology and Evolution 20:109.

Heggberget, T. G., Hvidsten, N. A., T. B. Gunnerød, and P. I. Møkkelgjerd. 1991. Distribution of adult recaptures from hatchery-reared Atlantic salmon (*Salmo salar*) smolts released in and offshore of the river Surna, western Norway. Aquaculture 98:89–96.

Heggenes, J., J. L. Bagliniere, and R. A. Cunjak. 1999. Spatial niche variability for young Atlantic salmon (*Salmo salar*) and brown trout (*S. trutta*) in heterogeneous streams. Ecology of Freshwater Fish 8:1–21.

Heggenes, J., S. J. Saltveit, D. Bird, and R. Grew. 2002. Static habitat partitioning and dynamic selection by sympatric young Atlantic salmon and brown trout in southwest England streams. Journal of Fish Biology 60:72–86.

Helle, K., and M. Pennington. 1999. The relation of the spatial distribution of early juvenile cod (*Gadus morhua* L.) in the Barents Sea to zooplankton density and water flux during the period 1978–1984. ICES Journal of Marine Science 56:15–27.

Hendry, K., D. Cragg-Hine, M. O'Grady, H. Sambrook, and A. Stephen. 2003. Management of habitat for rehabilitation and enhancement of salmonid stocks. Fisheries Research 62:171–192.

Hesthagen, T., and L. P. Hansen. 1991. Estimates of the annual loss of Atlantic salmon, *Salmo salar* L., in Norway due to acidification. Aquaculture and Fisheries Management 22:85–91.

Hesthagen, T., J. Heggenes, B. M. Larsen, and T. Forseth. 1999. Effects on water chemistry and habitat on the density of young brown trout *Salmo trutta* in acidic streams. Water Air and Soil Pollution 112:85–106.

Heuch, P. A., P. A. Bjørn, B. Finstad, J. C. Holst, L. Asplin, and F. Nilsen. 2005. A review of the Norwegian 'national action plan against salmon lice on salmonids': the effect on wild salmonids. Aquaculture 246:79–92.

Hjort, R. C., and C. B. Schreck. 1982. Phenotypic differences among stocks of hatchery and wild coho salmon, *Oncorhynchus kisutch*, in Oregon, Washington, and California. Fisheries Bulletin 80:105–119.

Höglund, J., A. Alfjorden, and T. Nikkila. 1997. Infection of juvenile salmon *Salmo salar* with a Dermocystidium-like organism in Sweden. Diseases of Aquatic Organisms 30:171–176.

Höjsjö, J., J. D. Armstrong, and S. W. Griffiths. 2005. Sneaky feeding by salmon in sympatry with dominant brown trout. Animal Behaviour 69:1037–1041.

Höjsjö, J., J. I. Johnsson, and T. Bohlin. 2004. Habitat complexity reduces the growth of aggressive and dominant brown trout (*Salmo trutta*) relative to subordinates. Behavioral Ecology and Sociobiology 56:286–289.

House, R. 1996. An evaluation of stream restoration structures in a coastal Oregon stream, 1981–1993.

North American Journal of Fisheries Management 1996:272–281.

Humpesch, U. H. 1985. Inter- and intra-specific variation in hatching success and embryonic development of five species of salmonids and *Thymallus thymallus*. Archiv für Hydrobiology 104:129–144.

Huntingford, F. A. 2004. Implications of domestication and rearing conditions for the behaviour of cultivated fishes. Journal of Fish Biology 65(Supplement A):122–142.

Huntingford, F. A., and C. Adams. 2005. Behavioural syndromes in farmed fish: implications for production and welfare. Behaviour 142:1207–1221.

Hvidsten, N. A., and L. P. Hansen. 1988. Increased recapture rate of adult Atlantic salmon, *Salmo salar* L. stocked as smolts at high water discharge. Journal of Fish Biology 32:153–154.

Hvidsten, N. A., and P. I. Møkkelgjerd. 1987. Predation on salmon, *Salmo salar* L., in the estuary of the River Surna, Norway. Journal of Fish Biology 30:273–280.

Hvidsten, N. A., Finstad, B., Kroglund, F., Johnsen, B. O., Strand, R., Arnekleiv, J. V., and Bjørn, P. A. 2007. Does increased abundance of sea lice influence survival of wild Atlantic salmon post-smolts? Journal of Fish Biology 71:1639–1648.

Hyatt, K. D., K. L. Mathias, D. J. McQueen, B. Mercer, P. Milligan, and D. P. Rankin. 2005. Evaluation of hatchery versus wild sockeye salmon fry growth and survival in two British Columbia lakes. North American Journal of Fisheries Management 25:745–762.

Imre, I., J. W. A. Grant, and R. A. Cunjak. 2005. Density-dependent growth of young-of-the-year Atlantic salmon *Salmo salar* in Catamaran Brook, New Brunswick. Journal of Animal Ecology 74:508–516.

Insulander, C., and B. Ragnarsson. 2001. Homing pattern of Baltic salmon, *Salmo salar* L., from smolt released from two hatcheries in the river Dalälven, Sweden. Fisheries Management and Ecology 8:61–67.

Iversen, M., B. Finstad, and K. J. Nilssen. 1998. Recovery from loading and transport stress in Atlantic salmon (*Salmo salar*) smolts. Aquaculture 168:387–394.

Iwata, M., H. Tsuboi, T. Yamashita, A. Amemiya, H. Yamada, and H. Chiba. 2003. Function and trigger of thyroxin surge in migrating chum salmon *Oncorhynchus keta* fry. Aquaculture 222:315–329.

Jenkins, T. M., S. Diehl, K. W. Kratz, and S. D. Cooper. 1999. Effects of population density on individual growth of brown trout in streams. Ecology 80:941–956.

Jodun, W. A., K. King, P. Farrell, and W. Wayman. 2007. Metanol and egg yolk as cryoprotectants for Atlantic salmon spermatozoa. North American Journal of Fisheries Management 69:36–40.

Johnsen, B. O., and A. J. Jensen. 1991. The Gyrodactylus story in Norway. Aquaculture 98:289–302.

Johnsen, B. O., and A. J. Jensen. 1994. The spread of furunculosis in salmonids in Norwegian rivers. Journal of Fish Biology 45:47–55.

Johnsen, B. O., A. J. Jensen, J. I. Koksvik, and H. Reinertsen. 1997a. Predation of Atlantic salmon smolts based on stocking of juveniles in lakes. Water chemistry, zooplankton, benthic animals and fish in Lakes Øvre and Nedre Mosvasstjern, River Vefsna 1986–1994. Norwegian Institute for Nature Research, Assignment Report, Trondheim.

Johnsen, B. O., J. I. Koksvik, and A. J. Jensen. 1997b. Production of Atlantic salmon smolts based on stocking of juveniles in rivers. Benthic animals and fish in Klubbvasselva, River Vefsna 1987–1996. Norwegian Institute for Nature Research, Assignment Report, Trondheim.

Johnson. J. H. 2004. Comparative survival and growth of Atlantic salmon from egg stocking and fry releases. North American Journal of Fisheries Management 24:1409–1412.

Johnson, S. L., J. D. Rogers, M. F. Solazzi, and T E. Nickelson. 2005. Effects of an increase in large wood on abundance and survival of juvenile salmonids (*Oncorhynchus* spp.) in an Oregon stream. Canadian Journal of Fisheries and Aquatic Sciences 62:412–424.

Johnsson, J. I., and A. Forser. 2002. Residence duration influences the outcome of territorial conflicts in brown trout (*Salmo trutta*). Behavioral Ecology and Sociobiology 51:282–286.

Jokikokko, E. 1999. Density of brown trout, *Salmo trutta* L., and Atlantic salmon, *Salmo salar* L., parr after point and scatter stocking of fry. Fisheries Management and Ecology 6:475–486.

Jokikokko, E., and E. Jutila. 2004. Divergence in smolt production from the stocking of 1-summer-old and 1-year-old Atlantic salmon parr in a northern Baltic river. Journal of Applied Ichthyology 20:511–516.

Jokikokko, E., I. Kallio-Nyberg, I. Saloniemi, and E. Jutila. 2006. The survival of semi-wild, wild and hatchery-reared Atlantic salmon smolts of the Simojoki River in the Baltic Sea. Journal of Fish Biology 68:430–442.

Jonasson, J., B. Gjerde, and R. Gjedrem. 1997. Genetic parameters for return rate and body weight of sea ranched Atlantic salmon. Aquaculture 154:219–231.

Jonsson, B., and I. A. Fleming. 1993. Enhancement of wild salmon populations. Pages 209–238 *in* G. Sundnes, editor. Human impact on self-recruiting populations. The Royal Norwegian Society of Sciences and Letters Foundation, Tapir Publishers, Trondheim.

Jonsson, B., and N. Jonsson. 2004a. Factors affecting

marine production of Atlantic salmon (*Salmo salar*). Canadian Journal of Fisheries and Aquatic Sciences 61:2369–2383.

Jonsson, B., and N. Jonsson. 2006. Cultured salmon in nature: a review of their ecology and interactions with wild fish. ICES Journal of Marine Science 63:1162–1181.

Jonsson, B., T. Forseth, A.J. Jensen, and T. F. Næsje. 2001. Thermal performance of juvenile Atlantic salmon, *Salmo salar* L. Functional Ecology 15:701–711.

Jonsson, B., N. Jonsson, and L. P. Hansen. 1990a. Does juvenile experience affect migration and spawning of adult Atlantic salmon? Behavioral Ecology and Sociobiology 26:225–230.

Jonsson, B., N. Jonsson, and L. P. Hansen. 1991. Differences in life history and migratory behaviour between wild and hatchery reared Atlantic salmon in nature. Aquaculture 98:69–78.

Jonsson, B., N. Jonsson, and L. P. Hansen. 2003a. Atlantic salmon straying from the River Imsa. Journal of Fish Biology 62:641–657.

Jonsson, B., N. Jonsson, and L. P. Hansen. 2007. Factors affecting river entry of adult Atlantic salmon in a small river. Journal of Fish Biology 71:943–956.

Jonsson, B., R. S. Waples, and K. D. Friedland. 1999. Extinction considerations for diadromous fishes. ICES Journal of Marine Science 56:405–409.

Jonsson, N., L. P. Hansen, and B. Jonsson. 1994. Juvenile experience influences timing of adult river ascent in Atlantic salmon. Animal Behaviour 48:740–742.

Jonsson, N., and B. Jonsson. 1999. Trade-off between egg size and numbers in brown trout. Journal of Fish Biology 55:767–783.

Jonsson, N., and B. Jonsson. 2002. Migration of anadromous brown trout in a Norwegian river. Freshwater Biology 47:1391–1401.

Jonsson, N., and B. Jonsson. 2003. Energy allocation among developmental stages, age groups, and types of Atlantic salmon (*Salmo salar*) spawners. Canadian Journal of Fisheries and Aquatic Sciences 60:506–516.

Jonsson, N., and B. Jonsson. 2004b. Size and age at maturity of Atlantic salmon correlate with the North Atlantic Oscillation index (NAOI). Journal of Fish Biology 64:241–247.

Jonsson, N., B. Jonsson, and I. A. Fleming. 1996. Does early growth rate cause a phenotypically plastic response in egg production of Atlantic salmon? Functional Ecology 10:89–96.

Jonsson, N., B. Jonsson, and L. P. Hansen. 1990b. Partial segregation in the timing of migration of Atlantic salmon of different ages. Animal Behaviour 40:313–321.

Jonsson, N., B. Jonsson, and L. P. Hansen. 1998. The relative role of density-dependent and density-independent survival in the life cycle of Atlantic salmon *Salmo salar*. Journal of Animal Ecology 67:751–762.

Jonsson, N., B. Jonsson, and L. P. Hansen. 2003b. Marine survival and growth of wild and released hatchery reared Atlantic salmon. Journal of Applied Ecology 40:900–911.

Jonsson, N., B. Jonsson, and L. P. Hansen. 2005. Does climate during embryonic development influence parr growth and age of seaward migration in Atlantic salmon (*Salmo salar*) smolts? Canadian Journal of Fisheries and Aquatic Sciences 62:2502–2508.

Jordon, W. R., III, M. E. Gilpin, and J. D. Aber editors. 1999. Restoration ecology: a synthetic approach to ecological research. Cambridge University Press, Cambridge, UK.

Jørgensen, E. H., and M. Jobling. 1993. The effects of exercise on growth, food utilization and osmoregulatory capacity of juvenile Atlantic salmon, *Salmo salar*. Aquaculture 116:233–246.

Julien, H. P., and N. E. Bergerson. 2006. Effect of fine sediment infiltration during the incubation period on Atlantic salmon (*Salmo trutta*) embryo. Hydrobiologia 563:61–71.

Jutila, E., E. Jokikokko, and M. Julkunen. 2003. Management of Atlantic salmon in the Simojoki River, northern Gulf of Bothnia: effects of stocking and fishing regulation. Fisheries Research 64:5–17.

Kallio-Nyberg, I., E. Jutila, E. Jokikokko, and I. Saloniemi. 2006. Survival of reared Atlantic salmon and sea trout in relation to marine conditions of smolt year in the Baltic Sea. Fisheries Research 80:295–304.

Karppinen, P., T. S. Mäkinen, J. Erkinaro, V. V. Kostin, R. V. Sadkovskij, A. I. Lupandin, and M. Kaukoranta. 2002. Migratory and route-seeking behaviour of ascending Atlantic salmon in the regulated River Tuloma. Hydrobiologia 483:23–30.

Katopodis, C. 2005. Developing a toolkit for fish passage, ecological flow management and fish habitat works. Journal of Hydraulic Research 43:451–467.

Kelly-Quinn, M., D. Tierney, and J. J. Bracken. 1993. Survival of salmon, *Salmo salar*, eggs planted in upland streams. Aquaculture and Fisheries Management 23:791–796.

Kennedy, G. J. A., and C. D. Strange. 1986. The effects of intra- and inter-specific competition on the survival and growth of stocked juvenile Atlantic salmon, *Salmo salar* L., and resident trout, *Salmo trutta* L., in an upland stream. Journal of Fish Biology 28:479–489.

Kihslinger, R. L., S. C. Lema, and G. A. Nevitt. 2006. Environmental rearing conditions produce forebrain differences in wild Chinook salmon *Oncorhynchus tshawytscha*. Comparative Biochemistry and Physiology A 145:145–151.

Kihslinger, R. L., and G. A. Nevitt. 2006. Early rearing environment impacts cerebella growth in juvenile salmon. Journal of Experimental Biology 209:504–509.

Klemetsen, A., P. A. Amundsen, B. Dempson, B. Jonsson, N. Jonsson, M. O'Connel, and E. Mortensen. 2003. Atlantic salmon, brown trout and Arctic charr: a review of their life histories. Ecology of Freshwater Fish 12:1–59.

Kolstad, K., B. Grisdale-Helland, and B. Gjerde. 2004. Family differences in feed efficiency in Atlantic salmon (*Salmo salar*). Aquaculture 241:169–177.

Kolstad, K., P. A. Heuch, B. Gjerde, T. Gjedrem, and R. Salte. 2005. Genetic variation in resistance of Atlantic salmon (*Salmo salar*) to the salmon louse *Lepeophtheirus salmonis*. Aquaculture 247:145–151.

Kondolf, G. M., M. J. Sale, and M. G. Wolman. 1993. The size of salmonid spawning gravel. Water Resource and Research 29:2275–2285.

Kondou, T., N. Takeshita, A. Nakazono, and S. Kimura. 2001. Egg survival in a fluvial population of masu salmon in relation to intragravel conditions in spawning redds. Transactions of the American Fisheries Society 130:969–974.

Kostow, K. E. 2004. Differences in juvenile phenotypes and survival between hatchery stocks and a natural population provide evidence for modified selection due to captive breeding. Canadian Journal of Fisheries and Aquatic Sciences 61:577–589.

Lacroix, G. L. 1996. Long-term enhancement of habitat for salmonids in acidified running waters. Canadian Journal of Fisheries and Aquatic Sciences 53:283–294.

Lahti, K., H. Huuskonen, A. Laurila, and J. Piironen. 2002. Metabolic rate and aggressiveness between brown trout populations. Functional Ecology 16:167–174.

Laine, A., T. Jokivirta, and C. Katopodis. 2002. Atlantic salmon, Salmo salar L., and sea trout, *Salmo trutta* L., passage in a regulated northern river: fishway efficiency, fish entrance and environmental factors. Fisheries Management and Ecology 9:65–77.

Laine, A., R. Kamula, and J. Hooli. 1998. Fish and lamprey passage in a combined Denil and vertical slot fishway. Fisheries Management and Ecology 5:31–44.

Larinier, M. 1998. Upstream and downstream fish passage experience in France. Pages 127–145 *in* M. Jungwirth, S. Schmutz, and S. Weiss, editors. Fish migration and fish bypasses. Fishing News Book, University Press, Cambridge, UK.

Latremouille, D. N. 2003. Fin erosion in aquaculture and natural environments. Reviews in Fisheries Science 11:315–335.

Lellis, W. A., and F. T. Barrows. 1997. The effect of diet on dorsal fin erosion in steelhead trout (*Oncorhynchus mykiss*). Aquaculture 156:229–240.

Lema, S. C., M. J. Hodges, M. P Marchetti, and G. A. Nevitt. 2005. Proliferation zones in the salmon telencephalon and evidence for environmental influence on proliferation rate. Comparative Biochemistry and Physiology Part A 141:327–335.

Lema, S. C., and G. A. Nevitt. 2004. Evidence that thyroid hormone induces olfactory cellular proliferation in salmon during a sensitive period for imprinting. Journal of Experimental Biology 207:3317–3327.

Letcher, B. H., T. Dubreuli, M. J. O'Donnell, M. Obedzinski, K. Griswold, and K. H. Nislow. 2004. Long-term consequences of variation in timing and manner of fry introduction on juvenile Atlantic salmon (*Salmo salar*) growth, survival, and life-history expression. Canadian Journal of Fisheries and Aquatic Sciences 61:2288–2301.

Letcher, B. H., and T. D. Terrick. 2001. Effects of developmental stage at stocking on growth and survival of Atlantic salmon fry. North American Journal of Fisheries Management 21:102–110.

Levin, P. S., and J. G. Williams. 2002. Interspecific effects of artificially propagated fish: an additional conservation risk for salmon. Conservation Biology 16:1581–1587.

Lillehammer, A. 1973. An investigation of the food of one- to four-month-old salmon fry (*Salmo salar* L.) in the River Suldalslågen, west Norway. Norwegian Journal of Zoology 21:17–24.

Linley, T. J. 2001. Influence of short-term estuarine rearing on the ocean survival and size at return of coho salmon in southeastern Asia. North American Journal of Aquaculture 63:306–311.

Lobon-Cervia, J. 2005. Spatial and temporal variation in the influence of density dependence on growth of stream-living brown trout. Canadian Journal of Fisheries and Aquatic Sciences 62:1231–1242.

McClelland, E. K., and K. A. Naish. 2007. What is the fitness outcome of crossing unrelated fish populations? A meta-analysis and an evaluation of future research directions. Conservation Genetics 8:397–416.

McCormick, S. D., M. F. O'Dea, A. M. Moeckel, and B. T. Björnsson. 2003. Endocrine and physiological changes in Atlantic salmon smolts following hatchery release. Aquaculture 222:45–57.

MacCrimmon, H. R. 1954. Stream studies on planted Atlantic salmon. Journal of Fisheries Research Board Canada 11:362–403.

McDonald, D. G. 1983. The effects of H+ upon gills of freshwater fish. Canadian Journal of Fisheries and Aquatic Sciences 61:691–703.

McGinnity, P., P. Prodöhl, A. Ferguson, R. Hynes, N. O. Maoileidigh, N. Baker, D. Cotter, B. O'Hea, D. Cooke, G. Rogan, J. Taggart, and T. Cross. 2003.

Fitness reduction and potential extinction of wild populations of Atlantic salmon, *Salmo salar*, as result of interactions with escaped farmed salmon. Proceedings of the Royal Society of London Series B 270:2443–2450.

McGinnity, P., P. Prodöhl, N. O. Maoileidigh, R. Hynes, D. Cotter, N. Baker, B. O'Hea, and A. Ferguson. 2004. Differential lifetime success and performance of native and non-native Atlantic salmon examined under communal natural conditions. Journal of Fish Biology 65(Supplement A):173–187.

McGinnity, P., C. Stone, J. B. Taggart, D. Cooke, D. Cotter, R. Hynes, C. McCamley, T. Cross, and A. Ferguson. 1997. Genetic impact of escaped farmed Atlantic salmon (*Salmo salar* L.) on native populations: use of DNA profiling to assess freshwater performance of wild, farmed, and hybrid progeny in a natural river environment. ICES Journal of Marine Science 54:998–1008.

MacLean, A., N. B. Metcalfe, and D. Mitchell. 2000. Alternative comparative strategies in juvenile Atlantic salmon (*Salmo salar*): evidence from fin damage. Aquaculture 184:291–302.

McLean, J. E., P. Bentzen, and T. P. Quinn. 2005. Nonrandom, size- and timing-biased breeding in a hatchery population of steelhead trout. Conservation Biology 19:446–454.

Maki-Petays, A., J. Erkinaro, E. Niemelä, A. Huusko, and T. Muotka. 2004. Spatial distribution of juvenile Atlantic salmon (*Salmo salar*) in a subarctic river: size-specific changes in a strongly seasonal environment. Canadian Journal of Fisheries and Aquatic Sciences 61:2329–2338.

Malcolm, I. A., A. F. Youngson, and C. Soulsby. 2003. Survival of salmonid eggs in a degraded gravel-bed stream: effects of groundwater–surfacewater interactions. River Research and Applications 19:303–316.

McDonald, D. G., C. L. Milligan, W. J. McFarlane, S. Crooke, S. Currie, B. Hooke, R. B. Angus, B. L. Tufts, and K. Davidson. 1998. Condition and performance of juvenile Atlantic salmon (*Salmo salar*): effects of rearing practices on hatchery fish and comparison with wild fish. Canadian Journal of Fisheries and Aquatic Sciences 55:1208–1219.

McMichael, G. A., T. N. Pearsons, and S. A. Leider. 1999. Behavioral interactions among hatchery-reared juvenile steelhead smolts and wild *Oncorhynchus mykiss* in natural streams. North American Journal of Fisheries Management 19:948–956.

McMichael, G. A., T. N. Pearsons, and S. A. Leider. 2000. Minimizing ecological impacts of hatchery-reared juvenile steelhead trout on wild salmonids in a Yakima basin watershed. Pages 365–380 *in* E. E. Knudsen, C. R. Steward, D. D. MacDonald, J. E. Williams, and D. W. Reiser, editors. Sustainable fisheries management: Pacific salmon. Lewis Publishers, Boca Raton, Florida.

McVicar, A. H. 2004. Management actions in relation to the controversy about salmon lice infections in fish farms as a hazard to wild salmonid populations. Aquaculture Research 35:751–758.

McMichael, G. A., C. S. Sharpe, and T. N. Pearsons. 1997. Effects of residual hatchery-reared steelhead on growth of wild rainbow and spring Chinook salmon. Transactions of the American Fisheries Society 126:230–239.

Marchetti, M. P., and Nevitt, G. A. 2003. Effects of hatchery rearing on brain structures of rainbow trout, *Oncorhynchus mykiss*. Environmental Biology of Fishes 66:9–14.

Merz, J. E., and J. D. Setka. 2004. Evaluation of a spawning habitat enhancement site for Chinook salmon in a regulated California river. North American Journal of Fisheries Management 24:397–407.

Merz, J. E., J. D. Setka, G. B. Pasternack, and J. M. Wheaton. 2004. Predicting benefits of spawning-habitat rehabilitation to salmonid (*Oncorhynchus* spp.) fry production in a regulated California river. Canadian Journal of Fisheries and Aquatic Sciences 61:1433–1446.

Metcalfe, N. B., A. C. Taylor, and J. E. Thorpe. 1995. Metabolic rate, social status and life history strategies in Atlantic salmon. Animal Behaviour 49:431–436.

Miller, K., M. H. Allegretti, N. Johnson, and B. Jonsson. 1995. Measures for conservation of biodiversity and sustainable use of its components. Pages 915–1061 *in* V. H. Heywood, editor. Global biodiversity assessment. Cambridge University Press, Cambridge, UK.

Mills, D. 1989. Ecology and management of Atlantic salmon. Chapman and Hall, New York.

Milner, A. M., and I. T. Gloyne-Phillips. 2005. The role of riparian vegetation and woody debris in the development of macroinvertebrate assemblages in streams. River Research and Applications 21:403–420.

Milner, N. J., J. M. Elliott, J. D. Armstrong, R. Gardiner, J. S. Welton, and M. Ladle. 2003. The natural control of salmon and trout populations in streams. Fisheries Research 62:111–125.

Mirza, R. S., and D. P. Chivers. 2000. Predator-recognition training enhances survival of brook trout: evidence from laboratory and field-enclosure studies. Canadian Journal of Zoology 78:2198–2208.

Miyakoshi, Y., H. Hayano, M. Fujiwara, M. Nagata, and J. R. Irvine. 2003. Size-dependent smolt yield and overwinter survival of hatchery-reared masu salmon released in fall. North American Journal of Fisheries Management 23:264–269.

Moksness, E., R. Støle, and G. van der Meeren. 1998.

Profitability analysis of sea ranching with Atlantic salmon (*Salmo salar*), Arctic charr (*Salvelinus alpinus*), and European lobster (*Homarus gammarus*) in Norway. Bulletin of Marine Science 62:689–699.

Moss, J. H., D. A. Beauchamp, A. D. Cross, K. W. Myers, E. V. Farley, J. M. Murphy, and J. H. Helle. 2005. Evidence for size-selective mortality after the first summer of ocean growth by pink salmon. Transactions of the American Fisheries Society 134:1313–1322.

Mote, P. W., E. Parson, A. F. Hamlet, W. S. Keeton, D. Lettenmaier, E. L. Mantua, D. Peterson, D. L. Peterson, R. Slaughter, and A. K. Snover. 2003. Preparing for climatic change: the water, salmon, and forests of the Pacific Northwest. Climate Change 61:45–88.

Mowbray, F. K., and A. Locke. 1998. Biological characteristics of Atlantic salmon (*Salmo salar* L.) in the Nepisiguit River, New Brunswick, 1982–1996. Canadian Technical Report of Fisheries and Aquatic Sciences 2236:1–32.

Naylor, R., K. Hindar, I. A. Fleming, R. Goldburg, S. Williams, J. Volpe, F. Whoriskey, J. Eagle, D. Kelso, and M. Mangel. 2005. Fugitive salmon: assessing the risk of escaped fish from net-pen aquaculture. Bioscience 55:427–437.

Nicieza, A. G., and F. Braña. 1993. Relationships among smolt size, marine growth, and sea age at maturity of Atlantic salmon (*Salmo salar*) in northern Spain. Canadian Journal of Fisheries and Aquatic Sciences 50:1632–1640.

Nickelson, T. E. 2003. The influence of hatchery coho salmon (*Oncorhynchus kisutch*) on the productivity of wild coho salmon populations in Oregon coastal basins. Canadian Journal of Fisheries and Aquatic Sciences 60:1050–1056.

Nickelson, T. E., M. F. Solazzi, and S. L. Johnson. 1986. The influence of hatchery coho salmon (*Oncorhynchus kisutch*) on the productivity of wild coho salmon populations in Oregon coastal basins. Canadian Journal of Fisheries and Aquatic Sciences 60:1050–1056.

Nielsen, J. L. 1994. Invasive cohorts: impacts of hatchery-reared coho salmon on the trophic, developmental and genetic ecology of wild stocks. Pages 361–385 *in* D. J. Stouder, K. L. Fresh, and R. Feller, editors. Theory and application in fish feeding ecology. University of South Carolina Press, Columbia.

Niemelä, E., J. Erkinaro, M. Julkunen, and E. Hassinen. 2005. Is juvenile salmon abundance related to subsequent and preceding catches? Perspectives from a long-term monitoring programme. ICES Journal of Marine Science 62:1617–1629.

Nikinmaa, M., A. Soivio, T. Nakari, and S. Lindgren. 1983. Hauling stress in brown trout (*Salmo trutta*): physiological responses to transport in fresh water or salt water, and recovery in natural brackish water. Aquaculture 34:93–99.

Nislow, K. H., S. Einum, and C. L. Folt. 2004. Testing predictions of the critical period for survival concept using experiments with stocked Atlantic salmon. Journal of Fish Biology 65(Supplement A):188–200.

Nislow, K. H., C. L. Folt, and D. L. Parrish. 1999. Favourable foraging locations for young Atlantic salmon: application to habitat and populations restoration. Ecological Applications 9:1085–1099.

Noakes, D. J., R. J. Beamish, and M. L. Kent. 2000. On the decline of Pacific salmon and speculative links to salmon farming in British Columbia. Aquaculture 183:363–386.

Obedzinski, M., and B. H. Letcher. 2004. Variation in freshwater growth and development among five New England Atlantic salmon (*Salmo salar*) populations reared in a common environment. Canadian Journal of Fisheries and Aquatic Sciences 61:2314–2328.

O'Hanley, J. R., and D. Tomberlin. 2005. Optimizing the removal of small fish passage barriers. Environmental Modelling and Assessment 10:85–98.

Økland, F., T. G. Heggberget, and B. Jonsson. 1995. Migratory behaviour of wild and farmed Atlantic salmon (*Salmo salar*) during spawning. Journal of Fish Biology 46:1–7.

Økland, F., B. Jonsson, A. J. Jensen, and L. P. Hansen. 1993. Is there a threshold size regulating smolt size in brown trout and Atlantic salmon? Journal of Fish Biology 42:541–550.

Opperman, J. J., and A. M. Merenlender. 2004. The effectiveness of riparian restoration for improving instream fish habitat in four hardwood-dominated California streams. North American Journal of Fisheries Management 24:822–834.

Orcutt, D. R., B. R. Pulliam, and A. Arp. 1968. Characteristic of steelhead trout redds in Idaho streams. Transactions of the American Fisheries Societies 97:42–45.

Orpwood, J. E., S. W. Griffiths, and J. D. Armstrong. 2004. Effect of density on competition between wild and hatchery-reared Atlantic salmon for shelter in winter. Journal of Fish Biology 65(Supplement A):201–209.

Pascual, M., P. Macchi, J. Urbanski, F. Marcos, C. Rossi, M. Novara, and D. Dell'Arciprete. 2002. Evaluating potential effects of exotic freshwater fish from incomplete species presence–absence data. Biological Invasions 4:101–113.

Patterson, D. A., J. S. Macdonald, S. G. Hinch, M. C. Healey, and A. P. Farrell. 2004. The effect of exercise and captivity on energy partitioning, reproductive maturation and fertilization success in adult sockeye salmon. Journal of Fish Biology 64:1039–1059.

Pedley, R. B., and J. W. Jones. 1978. The comparative feeding behaviour of brown trout, *Salmo trutta* L. and Atlantic salmon, *Salmo salar* L., in Llyn Dwythwch, Wales. Journal of Fish Biology 12:239–256.

Peeler, E., M. Thrush, L. Paisley, and C. Rogers. 2006. An assessment of the risk of spreading the fish parasite *Gyrodactylus salaris* to uninfected territories in the European Union with the movement of live Atlantic salmon (*Salmo salar*) from coastal waters. Aquaculture 258:187–197.

Peery, C. A., T. C. Bjornn, and C. Bjornn. 2004. Interactions between natural and hatchery chinook salmon parr in a laboratory stream channel. Fisheries Research 66:311–324.

Pelis, R. M., and S. D. McCormick. 2003. Fin development in stream- and hatchery-reared Atlantic salmon. Aquaculture 220:525–536.

Pepper, V. A., T. Nicholls, and N. P. Oliver. 1992. An evaluation of the quality of fall-fingerling Atlantic salmon (*Salmo salar* L.) released to natural lacustrine nursery areas in Newfoundland, Canada. Pages 249–259 *in* V. Ilmavirta, and R. I. Jones, editors. The dynamics and use of lacustrine ecosystems, Helsinki. Springer, Dordrecht, Netherlands.

Petersson, E., and T. Järvi. 2003. Growth and social interactions of wild and sea-ranched brown trout and their hybrids. Journal of Fish Biology 63:673–686.

Petersson, E., D. J. Piggins, and C. P. R. Mills. 1985. Comparative aspects of the biology of naturally produced and hatchery-reared Atlantic salmon smolts (*Salmo salar* L.). Aquaculture 45:321–333.

Pethon, P., L. Lillehammer, and G. Barstad. 1998. Smolt production in the Førland Channel: The Salmon Enhancement Programme, River Suldalslågen. Report of Statkraft Engineering, Oslo.

Petrosky, C. E., and T. C. Bjornn. 1988. Response of wild rainbow (*Salmo gairdneri*) and cutthroat trout (*S. clarki*) to stocked rainbow trout in fertile and infertile stream. Canadian Journal of Fisheries and Aquatic Sciences 45:2087–2105.

Pickering, A. D., T. G. Pottinger, and P. Christie. 1982. Recovery of brown trout, *Salmo trutta* L., from acute handling stress: a time-course study. Journal of Fish Biology 20:229–244.

Piggins, D. J., and C. P. R. Mills. 1985. Comparative aspects of the biology of naturally produced and hatchery-reared Atlantic salmon smolts (*Salmo salar* L.). Aquaculture 45:321–333.

Pirhonen, J., P. Valkeajärvi, J. E. Thorpe, and A. Soivio. 2003. Effect of stocking time on yield and location of recapture in two forms of brown trout (*Salmo trutta*) when stocked in respect to migration activity. Aquaculture 222:189–201.

Poole, W. R., D. T. Nolan, T. Wevers, M. Dillane, D. Cotter, and O. Tully. 2003. An ecophysiological comparison of wild and hatchery-reared Atlantic salmon (*Salmo salar* L.) smolts from the Burrishole system, western Ireland. Aquaculture 222:301–314.

Poppe, T. T., R. Johansen, G. Gunnes, and B. Torud. 2003. Heart morphology in wild and farmed Atlantic salmon *Salmo salar* and rainbow trout *Oncorhynchus mykiss*. Diseases of Aquatic Organisms 57:103–108.

Potter, E. C. E., and W. W. Crozier,. 2000. A perspective on the marine survival of Atlantic salmon. Pages 19–36 *in* D. H. Mills, editor. The ocean life of Atlantic salmon: environmental and biological factors influencing survival. Fishing news Books, Blackwell Scientific Publications, Oxford, UK.

Price, E. O. 1999. Behavioural development in animals undergoing domestication. Applied Animal Behaviour Science 65:211–218.

Primmer, C. R., P. A. Landry, E. Ranta, J. Merilä, J. Piironen, K. Tiira, N. Peuhkuri, S. Pakkasmaa, and P. Eskelinen. 2003. Prediction of offspring fitness based on parental genetic diversity in endangered salmonid populations. Journal of Fish Biology 63:909–927.

Quinn, T. P., M. T. Kinnison, and M. J. Unwin. 2001. Evolution of chinook salmon (*Oncorhynchus tshawytscha*) populations in New Zealand: pattern, rate, and process. Genetica 112:493–513.

Quinn, T. P., P. McGinnity, and T. F. Cross. 2006. Long-term declines in body size and shifts in run timing of Atlantic salmon in Ireland. Journal of Fish Biology 68:1713–1730.

Quinn, T. P., L. A. Vøllestad, J. Peterson, and V. Gallucci. 2004. Influences of freshwater and marine growth on the egg size: egg number trade-off in coho and Chinook salmon. Transactions of the American Fisheries Society 133:55–65.

Raddum, G. G., and A. Fjellheim. 1995. Artificial deposition of eggs of Atlantic salmon (*Salmo salar*) in a regulated Norwegian river: hatching, dispersal and growth of fry. Regulated Rivers 10:169–180.

Reinhardt, U. G. 2001. Selection for surface feeding in farmed and sea-ranched masu salmon juveniles. Transactions of the American Fisheries Society 130:155–158.

Reinhardt, U. G., T. Yamamoto, and S. Nakano. 2001. Effects of body size and predator on intercohort competition in wild and domesticated juvenile salmon in a stream. Ecological Research 16:327–334.

Reiriz, L., A. G. Nicieza, and F. Braña. 1998. Prey selection by experienced and naïve juvenile Atlantic salmon. Journal of Fish Biology 53:100–114.

Reisenbichler, R. R., and S. P. Rubin. 1999. Genetic changes from artificial propagation of Pacific salmon affect the productivity and viability of supplemented populations. ICES Journal of Marine Science 56:459–466.

Rhodes, J. S., and T. P. Quinn. 1998. Factors affecting the outcome of territorial contests between hatchery and naturally reared coho salmon parr in the laboratory. Journal of Fish Biology 53:1220–1230.

Riley, S. C., H. J. Fuss, and L. L. VeClair. 2004. Ecological effects of hatchery-reared juvenile Chinook and coho salmon on wild juvenile salmonids in two Washington streams. North American Journal of Fisheries Management 24:506–517.

Riley, S. C., C. P. Tatara, and J. A. Scheurer. 2005. Aggression and feeding of hatchery-reared and naturally reared steelhead (*Oncorhynchus mykiss*) fry in a laboratory flume and a comparison with observations in natural streams. Canadian Journal of Fisheries and Aquatic Sciences 62:1400–1409.

Riley, W. D., M. J. Ives, M. G. Pawson, and D. L. Maxwell. 2006. Seasonal variation in habitat use by salmon, *Salmo salar*, trout, *Salmo trutta* and grayling, *Thymallus thymallus*, in a chalk stream. Fisheries Management and Ecology 13:221–236.

Ritter, J. A. 1997. The contribution of Atlantic salmon (*Salmo salar* L.) enhancement to a sustainable resource. ICES Journal of Marine Science 54:1177–1187.

Roni, P., T. J. Beechie, R. E. Bilby, F. E. Leonetti, M. M. Pollock, and G. R. Pess. 2002. A review of stream restoration techniques and a hierarchical strategy for prioritizing restoration in Pacific Northwest watersheds. North American Journal of Fisheries Management 22:1–20.

Roni, P., T. Bennett, S. Morley, G. R. Press, K. Hanson, D. Van Slyke, and P. Olmstead. 2006. Rehabilitation of bedrock stream channels: the effects of boulder weir placement on aquatic habitat and biota. River Research and Applications 22:967–980.

Rosseland, B. O., and O. K. Skogheim. 1984. A comparative study on salmonid fish species in acid aluminium-rich water. II. Physiological stress and mortality of one and two year old fish. Report of the Institute of Freshwater Research Drottningholm 61:186–194.

Rosseland, B. O., O. K. Skogheim, H. Abrahamsen, and D. Matzow. 1986. Limestone slurry reduces physiological stress and increases survival of Atlantic salmon. Canadian Journal of Fisheries and Aquatic Sciences 43:1888–1893.

Rosseland, L. 1975. Annual report 1974. Norwegian Directorate for Hunting and Fisheries Management, Trondheim.

Rowe, D. K., J. E. Thorpe, and A. M. Shanks. 1991. Role of fat stores in the maturation of male Atlantic salmon (*Salmo salar*) parr. Canadian Journal of Fisheries and Aquatic Sciences 48:405–413.

Rubin, J. F. 1995. Estimating the success of natural spawning of salmonids in streams. Journal of Fish Biology 46:603–622.

Ruzzante, D. E., M. M. Hansen, D. Meldrup, and K. M. Ebert. 2004. Stocking impact and migration pattern in an anadromous brown trout (*Salmo trutta*) complex: where have all the stocked spawning sea trout gone? Molecular Ecology 13:1433–1445.

Sægrov, H., K. Urdal, B. A. Hellen, S. Kålås, and S. J. Saltveit. 2001. Estimating carrying capacity and presmolt production of Atlantic salmon (*Salmo salar*) and anadromous brown trout (*Salmo trutta*) in west Norwegian rivers. Nordic Journal of Freshwater Research 75:99–108.

Salminen, M. 1997. Relationships between smolt size, postsmolt growth and sea age at maturity in Atlantic salmon ranched in the Baltic Sea. Journal of Applied Ichthyology 13:121–130.

Salminen, M., T. Alapassi, and E. Ikonen. 2007. The importance of stocking age in the enhancement of River Kymijoki salmon (*Salmo salar*). Journal of Applied Ichthyology 23:46–52.

Salminen, M., S. Kruikka, and E. Erkamo. 1995. Annual variability in survival of sea ranched Baltic salmon, *Salmo salar* L.: significance of smolt size and marine conditions. Fisheries Management and Ecology 2:171–184.

Saloniemi, I., E. Jokikokko, I. Kallio-Nyberg, E. Jutila, and P. Pasanen. 2004. Survival of reared and wild Atlantic salmon smolts: size matters more in bad years. ICES Journal of Marine Science 61:782–787.

Saltveit, S. J. 1989. An assessment of natural recruitment above the Sjurhaug waterfall, River Lærdalselva, Sogn og Fjordane. Report from Laboratory of Freshwater Ecology and Inlands Fisheries, University of Oslo, Oslo, Norway.

Saltveit, S. J. 1998. The effect of stocking Atlantic salmon, *Salmo salar*, in Norwegian rivers. Pages 22–33 *in* I. G. Cowx, editor. Stocking and introduction of fish. Fishing News Books, Blackwell Scientific Publications Publications, Oxford, UK.

Saltveit. S. J. 2006. The effects of stocking Atlantic salmon, *Salmo salar*, in a Norwegian regulated river. Fisheries Management and Ecology 13:197–205.

Salvanes, A. G. V., and V. A. Braithwaite. 2006. The need to understand the behaviour of fish reared for mariculture or restocking. ICES Journal of Marine Science 63:346–354.

Scheuerell, M. D. 2005. Influence of juvenile size on the age at maturity of individually marked wild Chinook salmon. Transactions of the American Fisheries Society 134:999–1004.

Scrivener, J. C. 1988. Two devices to assess incubation survival and emergence of salmonid fry in an estuary streambed. North American Journal of Fisheries Management 8:248–258.

Scruton, D. A., K. D. Clarke, T. C. Anderson, A. S. Hoddinott, M. C. Van Zyll de Joung, and K. A. Huston.

1997. Evaluation of habitat improvement and restoration initiatives for salmonids in Newfoundland, Canada. Canadian Manuscript Report of Fisheries and Aquatic Sciences 2413:1–35.

Seierstad, S. L. O., T. T. Poppe, E. O. Koppang, A. Svindland, G. Rosenlund, L. Fløyland, and S. Larsen. 2005. Influence of dietary composition on cardiac pathology in farmed Atlantic salmon, *Salmo salar*. Journal of Diseases 28:677–690.

Sergeant, C. J., and D. A. Beauchamp. 2006. Effects of physical habitat and ontogeny on lentic habitat preference of juvenile Chinook salmon. Transactions of the American Fisheries Society 135:1191–1204.

Shemai, B., R. Sallenave, and D. E. Cowley. 2007. Competition between hatchery-raised Rio Grande cutthroat trout and wild brown trout. North American Journal of Fisheries Management 27:315–325.

Shuter, B. J. 1990. Population-level indicators of stress. Pages 145–166 *in* S. M. Adams, editor. Biological indicators of stress in fish. American Fisheries Society, Bethesda, Maryland.

Silverstein, J. T., K. D. Shearer, W. W. Dickhoff, and E. M. Plisetskaya. 1999. Regulation of nutrient intake and energy balance in salmon. Aquaculture 177:161–169.

Simenstad, C., C. Tanner, C. Crandell, J. White, and J. Cordell. 2005. Challenges of habitat restoration in a heavily urbanized estuary: evaluating the investment. Journal of Coastal Research, Special Issue 40:6.23.

Skilbrei, O. T., and M. Holm. 1998. Effects of long-term exercise on survival, homing and straying of released Atlantic salmon smolts. Journal of Fish Biology 52:1083–1086.

Skilbrei, O. T., and V. Wennevik. 2006. Survival and growth of sea-ranched Atlantic salmon, *Salmo salar* L., treated against sea lice before release. ICES Journal of Marine Science 63:1317–1325.

Specker, J. L., and C. B. Schreck. 1980. Stress responses to transportation and fitness for marine survival in coho salmon (*Oncorhynchus kisutch*) smolts. Canadian Journal of Fisheries and Aquatic Sciences 37:765–769.

Staurnes, M., L. P. Hansen, K. Fugelli, and Ø. Haraldstad. 1996. Short-term exposure to acid water impairs osmoregulation, seawater tolerance, and subsequent marine survival of smolts of Atlantic salmon (*Salmo salar* L.). Canadian Journal of Fisheries and Aquatic Sciences 53:1695–1704.

Su, Z. M., R. M. Peterman, and S. L. Haeseker. 2004. Spatial, hierarchical Bayesian models for stock-recruitment analysis of pink salmon (*Oncorhynchus gorbuscha*). Canadian Journal of Fisheries and Aquatic Sciences 61:2471–2486.

Sundström, L. F., and J. I. Johansson. 2001. Experience and social environment influence the ability of young brown trout to forage on live novel prey. Animal Behaviour 61:249–255.

Sundström, L. F., M. Lohmus, and J. I. Johnsson. 2003. Investment in territorial defence depends on rearing environment in brown trout (*Salmo trutta*). Behavioral Ecology and Sociobiology 54:249–255.

Sundström, L. F., M. Lohmus, W. E. Tymchuk, and R. H. Devlin. 2007. Gene-environment interactions influence ecological consequences of transgenic animals. Proceedings of the National Academy of Sciences of the United States of America 104:2889–2894.

Sundström, L. F., E. Petersson, J. Höjsjö, J. I. Johnsson, and T. Järvi. 2004. Hatchery selection promotes boldness in newly hatched brown trout (*Salmo trutta*): implications for dominance. Behavioural Ecology 15:192–198.

Sundström, L. F., E. Peterson, J. L. Johnsson, J. Dannevitz, J. Höjsjö, and T. Järvi. 2005. Heart rate responses to predation risk in *Salmo trutta* are affected by the rearing environment. Journal of Fish Biology 67:1280–1286.

Suttle, K. B., M. E. Power, J. M. Levine, and C. McNeely. 2004. How fine sediment in river beds impairs growth and survival of juvenile salmonids. Ecological Applications 14:969–974.

Swain, D. P., B. E. Riddell, and C. B. Murray. 1991. Morphological differences between hatchery and wild populations of coho salmon (*Oncorhynchus kisutch*): environmental versus genetic origin. Canadian Journal of Fisheries and Aquatic Sciences 48:1783–1791.

Tabachek, J. L., M. J. Foster, C. E. Engel, and R. N. Olson. 1993. A system for the incubation of separate groups of salmonids eggs. Progressive Fish-Culturist 55:101–105.

Tallmon, D. A., G. Luikart, and R. S. Waples. 2004. The alluring simplicity and complex reality of genetic rescue. Trends in Ecology and Evolution 19:489–496.

Tamate, T., and K. Maekawa. 2000. Interpopulation variation in reproductive traits of female masu salmon, *Oncorhynchus masou*. Oikos 90:209–218.

Taylor, E. B. 1986. Differences in morphology between wild and hatchery populations of juvenile coho salmon. Progressive Fish-Culturist 48:171–176.

Tetzlaff, D., C. Soulsby, A. F. Youngson, C. Gibbins, P. J. Bacon, I. A. Malcolm, and S. Langan. 2005. Variability in stream discharge and temperature: a preliminary assessment of the implications for juvenile and spawning Atlantic salmon. Hydrology and Earth System Science 9:193–208.

Thorpe, J. E. 2004. Life history responses of fishes to culture. Journal of Fish Biology 65(Supplement A):263–285.

Thorstad, E. B., T. G. Heggberget, and F. Økland. 1998. Migratory behaviour of adult wild and escaped farmed Atlantic salmon, *Salmo salar* L., before, during and after spawning in a Norwegian river. Aquaculture Research 29:419–428.

Thorstad, E. B., F. Økland, F. Kroglund, and N. Jepsen. 2003. Upstream migration of Atlantic salmon on the River Nidelva, southern Norway. Fisheries Management and Ecology 10:139–146.

Thrower, F. J., and E. Joyce. 2006. The effects of stock and pre-release marine net-pen culture on survival to adulthood, age at maturity, and fisheries contribution for three stocks of Chinook salmon in Southeast Asia. North American Journal of Aquaculture 68:317–323.

Tiira, K., J. Piironen, and C. R. Primmer. 2006. Evidence of reduced genetic variation in severely deformed juvenile salmonids. Canadian Journal of Fisheries and Aquatic Sciences 63:2700–2707.

Tops, S., W. Lockwood, and B. Okamura. 2006. Temperature-driven proliferation of Tetracapsuloides bryosalonae in bryozoan hosts portends salmon declines. Diseases of Aquatic Organisms 70:227–236.

Tringali, M. D., and T. M. Bert. 1998. Risk of genetic effective population size should be an important consideration in fish stock enhancement programs. Bulletin of Marine Science 62:641–659.

Ugedal, O., B. Finstad, B. Damsgård, and A. Mortensen. 1998. Seawater tolerance and downstream migration in hatchery-reared and wild brown trout. Aquaculture 168:395–405.

Unwin, M. J. 1997. Fry-to-adult survival of natural and hatchery-produced chinook salmon *(Oncorhynchus tshawytscha)* from a common origin. Canadian Journal of Fisheries and Aquatic Sciences 54:1246–1254.

Unwin, M. J., and G. J. Glova. 1997. Changes in life history parameters in a natural spawning population of chinook salmon (*Oncorhynchus tshawytscha*) associated with releases of hatchery-reared fish. Canadian Journal of Fisheries and Aquatic Sciences 54:1235–1245.

Utter, F. 1998. Genetic problems of hatchery-reared progeny released into the wild, and how to deal with them. Bulletin of Marine Science 62:623–640.

Van Andel, J., and J. Aronson. 2005. Restoration ecology. The new frontier. Blackwell Scientific Publications Publishing, Oxford, UK.

Verspoor, E., and C. Garcia de Leaniz. 1997. Stocking success of Scottish Atlantic salmon in two Spanish rivers. Journal of Fish Biology 51:99–108.

Verspoor, E., L. Stradmeyer, and J. Nielsen, editors. 2007. The Atlantic salmon: genetics, conservation and management. Blackwell Scientific Publications Publishing, Oxford, UK.

Vibert, R. 1949. Du repeuplement en truites et saumons par enfouissement de "boites alevinage" garnies d'oeufs dans les graviers. Bulletin Francais de Pisciculture 153:125–150.

Vilhunen, S., H. Hirvonen, and M. M. Laakkonen. 2005. Less is more: social learning of predator recognition requires a low demonstrator to observer ration in Arctic charr (*Salvelinus alpinus*). Behavioral Ecology and Sociobiology 57:275–282.

Vincent, R. E. 1987. Effects of stocking catchable-size hatchery rainbow trout on two wild trout species in the Madison River and O'Dell Creek, Montana North American Journal of Fisheries Management 7:91–105.

Vøllestad, L. A., and T. P. Quinn. 2003. Trade-off between growth rate and aggression in juvenile coho salmon, *Oncorhynchus kisutch*. Animal Behaviour 66:561–568.

Vøllestad, L. A., J. Peterson, and T. P. Quinn. 2004. Effects of freshwater and marine growth rates on early maturity in male coho and Chinook salmon. Transactions of the American Fisheries Society 133:495–503.

Von Cramon-Taubadel, N., E. N. Ling, D. Cotter, and N. P. Wilkins. 2005. Determination of body shape variation in Irish hatchery-reared and wild Atlantic salmon. Journal of Fish Biology 66:1471–1482.

Wagner, E. J., R. E. Arndt, and R. Roubidoux. 2006. The effect of temperature changes and transport on cutthroat trout eggs soon after fertilization. North American Journal of Aquaculture 68:235–239.

Waknitz, F. W., R. N. Iwamoto, and M. S. Strom. 2003. Interactions of Atlantic salmon in the Pacific Northwest. 4. Impacts on local ecosystems. Fisheries Research 62:307–328.

Wang, J. L., and N. Ryman. 2001. Genetic effects of multiple generations of supportive breeding. Conservation Biology 15:1619–1631.

Wang, S. Z., J. Hard, and F. Utter. 2001. Salmonid inbreeding. Reviews in Fish Biology and Fisheries 11:301–319.

Waples, R. S. 1999. Dispelling some myths about hatcheries. Fisheries 24:12–21.

Weatherley, A. H., and H. S. Gill. 1987. The biology of fish growth. Academic Press, London.

Weber, E. D., and K. D. Fausch. 2003. Interactions between hatchery and wild salmonids in streams: differences in biology and evidence for competition. Canadian Journal of Fisheries and Aquatic Sciences 60:1018–1036.

Weber, E. D., and K. D. Fausch. 2005. Competition between hatchery-reared and wild juvenile Chinook salmon in enclosures in the Sacramento River, California. Transactions of the American Fisheries Society 134:44–58.

Weiss, S., and S. Schmutz. 1999. Performance of hatchery-reared brown trout and their effects on wild fish in two small Austrian streams. Transactions of the American Fisheries Society 128:302–316.

Wertheimer, A. C., W. R. Heard, J. M. Maselko, and W. W. Smoker. 2004. Relationship of size at return with environmental variation, hatchery production, and productivity of wild pink salmon in Prince William Sound, Alaska: does it matter? Reviews in Fish Biology and Fisheries 14:321–334.

Wessel, M., W. W. Smoker, R. M. Fagen, and J. Joyce. 2006. Variation of agonistic behaviour among juvenile Chinook salmon (*Oncorhynchus tshawytscha*) of hatchery, hybrid and wild origin. Canadian Journal of Fisheries and Aquatic Sciences 63:438–447.

Whalen, K. G., and G. W. LaBar. 1998. Survival and growth of unfed and fed Atlantic salmon fry stocked in a Vermont tributary of the Connecticut River. North American Journal of Fisheries Management 18:931–935.

Whitlock, D. 1978. The Whitlock Vibert box handbook. Federation of Fly Fishermen, Bozeman, Montana.

Wild, V., H. Simianer, H. M. Gjøen, and B. Gjerde. 1994. Genetic parameters and genotype x environment interaction for early sexual maturity in Atlantic salmon (*Salmo salar*). Aquaculture 128:51–65.

Williams, J. G. 1998. Fish passage in the Columbia River. Pages 180–191 *in* M. Jungwirth, S. Schmutz, and S. Weiss, editors. Fish migration and fish bypasses. Fishing News Books, University Press, Cambridge, UK.

Yamamoto, S., and K. Morita. 2002. Interpopulation comparison of size and age at smolting of white-spotted charr, *Salvelinus leucomaenis*. Ecology of Freshwater Fish 11:281–284.

Yamamoto, T., and U. G. Reinhardt. 2003. Dominance and predator avoidance in domesticated and wild masu salmon *Oncorhynchus masou*. Fisheries Science 69:88–94.

Yokota, M., Y. Harada, and M. Iizuka. 2003. Genetic drift in a hatchery and the maintenance of genetic diversity in hatchery-wild systems. Fisheries Science 69:101–109.

Youngson, A. F., L. P. Hansen, B. Jonsson, and T. F. Næsje. 1989. Effects of exogenous thyroxin or prior exposure to raised water-flow on the downstream movement of hatchery-reared Atlantic salmon smolts. Journal of Fish Biology 34:791–797.

Youngson, A. F., I. A. Malcolm, J. L. Thorley, P. J. Bacon, and C. Soulsby. 2004. Long-residence groundwater effects on incubating salmonid eggs: low hyporheic oxygen impairs embryo development. Canadian Journal of Fisheries and Aquatic Sciences 61:2278–2287.

Youngson, A. F., and J. H. Webb. 1992. The relationship between stream or river discharge and thyroid-hormone levels in wild adult Atlantic salmon (*Salmo salar* L.). Canadian Journal of Zoology 70:140–144.

Zeh, M., and W. Dönni. 1994. Restoration of spawning grounds for trout and grayling in the river High-Rhine. Aquatic Sciences 56:59–69.

American Fisheries Society Symposium 69:537–556, 2009

Lexicon of Life Cycle Diversity in Diadromous and Other Fishes

David H. Secor* and Lisa A. Kerr

University of Maryland Center for Environmental Science, Chesapeake Biological Laboratory Post Office Box 38, Solomons, Maryland 20688, USA

Abstract.—In the study of species life histories and the structure of diadromous populations, an emerging trend is the prevalence of life cycle diversity—that is, individuals within populations that do not conform to a single life cycle pattern. A rapid rise in publications documenting within-population variability in life cycles has resulted in the use of numerous terms and phrases. We argue that myriad terms specific to taxa, ecosystem types, and applications are in fact describing the same phenomenon—life cycle diversity. This phenomenon has been obscured by the use of multiple terms across applications, but also by the overuse of typologies (i.e., anadromy, catadromy) that fail to convey the extent of life cycle variations that underlay population, metapopulation, and species dynamics. To illustrate this, we review migration and habitat-use terms that have been used to describe life cycles and life cycle variation. Using a citation index (Cambridge Scientific Abstracts © Aquatic Sciences and Fisheries Abstracts), terms were tallied across taxonomic family, ecosystem, type of application, analytical approach, and country of study. Studies on life cycle diversity have increased threefold during the past 15 years, with a total of 336 papers identified in this review. Most of the 40 terms we identified described either sedentary or migratory lifetime behaviors. The sedentary-migratory dichotomy fits well with the phenomenon of partial migration, which has been commonly reported for birds and Salmonidae and is postulated to be the result of early life thresholds (switch-points). On the other hand, the lexicon supports alternate modes of migration, beyond the simple sedentary–migratory dichotomy. Here more elaborate causal mechanisms such as the entrainment hypothesis may have application. Diversity of life cycles in fish populations, whether due to partial migration, entrainment, or other mechanisms, is increasingly recognized as having the effect of offsetting environmental stochasticity and contributing to long-term persistence.

Introduction

Spatial management and population connectivity across freshwater, estuarine, and marine ecosystems is a priority theme within fisheries science and management. With the maturation of approaches that yield longitudinal migration data (i.e., electronic tags and otolith chemistry that record daily and seasonal migrations of individuals), recognition of life cycle diversity within populations is becoming increasingly prevalent. Many examples exist of alternate life cycles within populations, epitomized by resident males in Salmonidae (Groot and Margolis 1998) and "sea eels" in Anguillidae (Tsukamoto et al. 1998). Concepts explaining life cycle diversity include genetic polymorphism, density-dependent habitat selection, early life conditional tactics, and social transmission of learned migration behavior (Jonsson and Jonsson 1993; McQuinn 1997; Secor 1999; Corten 2002). Life cycle diversity can have consequences for mixed-stock fisheries (Fromentin and Powers 2005), classification of essential fish habitats (Kraus and Secor 2005), and spatial management tactics (Robichaud and Rose 2004). A key challenge presented by alternate life cycles is to identify commonalities across species in pattern, cause, and consequence. To aid in this, we present a brief review on theory underlying life cycle diversity

* Corresponding author: secor@cbl.umces.edu

in fishes and then examine the ways in which scientists have described life cycle diversity across a range of taxa, ecosystems, and problems.

Alternate Life Cycle Theory

In contrast to the fish literature, avian studies support a structured terminology associated with alternate life cycles, largely due to wide-scale adoption of a theory for partial migration. In partial migration, one portion of a population is migratory and the other portion is sedentary (Lack 1943; Berthold 2001). Partial migration is further subdivided into two types: (1) facultative, where migration behavior is only under environmental control; and (2) obligate, where migrations are under both genetic and environmental control (Terrill and Able 1988; Berthold 1996). In facultative migration, there typically is not a persistent migratory life cycle, but the population can respond (e.g., with irruptive or nomadic migrations) when triggered by large changes to the environment. Obligate partial migration is believed to be more prevalent among birds and is thought to be maintained through a conditional strategy: one based upon early life thresholds (Berthold 2001). Thus, the coexistence of sedentary and migratory tactics in a single genetic population is maintained by the combined effect of an individual's fitness and trade-offs associated with each behavior (Lundberg 1988).

This theory of partial migration (Table 1) has been adopted to describe the diversity of migration patterns observed in Salmonidae but has not been widely used outside this family. As in birds, the choice of alternative life cycles (resident versus migrant) within many salmonid populations is hypothesized to be under the control of a genetically defined developmental switch-point, which causes the individual to follow one or another conditional strategy (Thorpe 1989; Mangel 1994; Thorpe et al. 1998). For instance, individual growth rate relative to a growth threshold has been shown to be a cue that can either initiate or preclude migration depending upon species (Secor 1999). In addition to growth rate, partial migration has been postulated to be regulated by early physiological state and behavior (e.g., feeding hierarchies; Metcalfe and Thorpe 1992; Økland et al. 1993).

An alternative theory has been proposed to describe the return to diverse spawning habitats by Atlantic herring *Clupea harengus*. Atlantic herring populations show a diverse pattern of intrapopulation spawning habitat use, which has stimulated debate on population structure, life cycle closure, and alternate migration behaviors (Iles and Sinclair 1982; Sinclair 1988; McQuinn 1997). Strong circumstantial evidence has been documented for the role of learned behaviors in conserving multiple migration circuits within herring populations. The so-called "adopted migrant" or "entrainment" hypothesis (Table 1) postulates that behaviors associated with a particular life cycle circuit are learned by juveniles (aka subadults) from adults during periods of spatial overlap (McQuinn 1997; Corten 2002; Petitgas et al. 2007). Thus, a certain number of juveniles are entrained into an already established life cycle. Without association with older age-classes, juveniles can establish novel migration circuits through exploration, although the adoption of novel life circuits is expected to be much less efficient than the adoption of already established circuits. As populations decline, behavioral entrainment into minority groups is postulated to become less efficient, and entire circuits and associated patterns of habitat-use can be lost (Table 1). As a population expands, density-dependent habitat selection can promote exploration of new habitats and establishment of novel migration circuits by juveniles. The diverse patterns of winter, forage, and spawning habitat use in Norwegian, North Sea, and Scotian Shelf populations of Atlantic herring have been explained in terms of the entrainment hypothesis, and it could have more general application for other coastal species (Petitgas et al. 2007).

Partial migration and behavioral entrainment are complementary ideas in that they account for life cycle diversity through a polyphenic response rather than a genetic polymorphism. For instance, following an initial threshold switch-point, a juvenile undertakes a migratory rather than a sedentary behavior. Thereafter, at some point later in life, the individual behaviorally entrains into one of several migration circuits through its interaction with already entrained age-classes in common wintering or feeding habitats. Genetic polymorphism—the idea that different life cycles represent discrete subpopulations—can only be maintained through isolating mechanisms such as assortative mating. Such subpopulation structure has been associated with sedentary and migratory components of populations of

Table 1.—Description of two theories describing alternative life cycles within fish populations.

Partial migration hypothesis	Entrainment hypothesis
Background	
Explanation for migratory and sedentary forms of Atlantic salmon and other salmonids (Thorpe 1989).	Explanation for persistent homing to multiple spawning banks within the same population of Atlantic herring—also known as conservatism of life cycles (McQuinn 1997).
Theory related to frequency dependent mating success among migratory types for birds and salmon (Hutchings and Myers 1994; Berthold 2001) that commonly exhibit partial migration.	Theory related to schooling and learned behavior in animals (Helfman and Schultz 1984; Petitgas et al. 2007).
Concept of population life cycle organization	
Same spawning unit uses different nursery, feeding, and wintering habitats.	Different spawning units share nursery, feeding and wintering habitats.
Sedentary/migratory behaviors during first year of life result in divergent life cycles.	Overlap of juvenile and adult habitat permits opportunity for adopted migration behavior.
Dispersive contingent more likely to colonize new habitats or stray into adjacent populations.	Juveniles that do not adopt an adult behavior are vagrants and can colonize novel habitats.
Life cycle closure achieved through natal homing.	Life cycle closure achieved through entrainment.
Statement of hypotheses	
Adoption of migratory behavior occurs during first year of life, with subsequent conservatism of life cycle throughout subadult and adult period.	Conveyance and conservation of migration behavior occur through transmission of behaviors from repeat spawners (adults) to recruit spawners (juveniles).
Conditional responses to the larval/juvenile environment (thresholds and developmental switches) result in adoption and subsequent conservatism of life cycle behaviors. Conditional response thresholds can differ among populations.	Juveniles adopt adult life cycles after their distributions initially overlap.
Environmental factors during the first year of life will influence the frequency of life cycle pattern in a year-class. Overlapping generations will determine how climate and other conditions affect the frequency of life cycle patterns within populations.	The dominant spawning unit will influence the dominant life cycle pattern. Loss of a life cycle can only be recovered through recolonization.

Atlantic salmon *Salmo salar* and Atlantic cod *Gadus morhua*, but results have been inconsistent for the same study populations or for other populations of the same species (Verspoor and Cole 1989; Jonsson and Jonsson 1993; Ruzzante et al. 1997, 2000; Beacham et al. 2002; Klemetsen et al. 2003).

Life Cycle Typologies and Lexicon

In this paper, we suggest that discourse among fisheries ecologists on life cycle diversity has been hindered by terms used to describe this phenomenon. First, terms used to ascribe general life cycles at the species level (typologies) tend to obscure the phenomenon of life cycle diversity. On the other hand, terms used to describe life cycle diversity within species and populations are themselves diverse, which potentially limit cross-taxa and cross-system generalities because the dialogue is too restricted (e.g., straying salmon versus vagrant herring). Here, we examine patterns of term use with the intent to not be prescriptive or authoritarian, but rather to illustrate that across diverse taxa, ecosystems, methodologies, and applications, life cycle diversity is increasingly recognized as a common phenomenon, one deserving of dedicated investigation related to its general patterns, causes, and consequences.

At both the species and within species level, a literature search for the period 1975–2006 was conducted to identify terms used to describe life cycles and life cycle diversity of fish. Terms were analyzed for underlying patterns of usage associated with taxonomic group, ecosystem, discipline, analytical approach, study nationality, and precedence.

Methods

Terms and phrases that apply to alternate life cycles were searched using Cambridge Scientific Abstracts Aquatic Sciences and Fisheries Abstracts (ASFA). The database covers more than 4,500 periodicals and gray literature reports and was selected based upon its inclusion of journals pertinent to fish biology and ecology and fisheries science (http://www-md2.csa.com/factsheets/aquclust-set-c.php). It has substantial coverage across all aquatic, ecological, and evolutionary biology journals and includes high impact periodicals such as Science and Nature. Boolean operators were applied to both restrict and broaden searches of terms that occurred anywhere in the record (e.g., title, abstract, key words). Several spot checks conducted using the more inclusive ISI Web of Knowledge (http://isiwebofknowledge.com/database) yielded nearly identical search results.

For each record attributable to bony fishes (Osteichthyes), the abstract was read to determine whether usage was a species descriptor or used to describe individual behaviors within populations or species. For instance, the term "nondiadromous" was used only as a species or population descriptor rather than a reference to individuals that were sedentary within a diadromous population. For a minority of keywords, searches yielded too many records (>500) to be reviewed for term usage (e.g., "migrant," "migratory," and "anadromous") and were excluded. While this meant that our review was not absolutely all-inclusive, coverage was likely representative and captured the prevalence of life cycle studies and the numerous terms used. We tallied the incidence of terms by taxonomy, the ecosystems connected by migration, study type, methodological approach, and the country where the work was done. Ecosystems were broadly classified according to population (rather than groups or individuals within a population) into 10 categories. River–Coast represented most diadromous species that migrate between freshwater and marine ecosystems (e.g., Salmonidae and Anguillidae). Here, coast represents both neritic (continental shelf) and oceanic environments. River was broadly used to indicate all freshwater fluvial habitats, including streams and tidal freshwater. Dashed Ecosystem categories (e.g., Estuary–Coast) indicate species migrations between the two opposed ecosystems in either direction.

Applications were tallied according to a subjective decision based upon content of the reviewed abstracts. For instance, a study noting increased straying within a population of salmon augmented by hatchery releases was identified as a hatchery application. Conservation applications were driven by concerns of species extirpation, but also often included issues related to impact of invasive caged salmon on natural populations. Population structure applications were usually easily defined but, on occasion, overlapped with studies that only described patterns of behavior (Behavior category) or Conservation (e.g., when concerned with metapopulation structure). In such instances, if there was an effort to discriminate or otherwise quantify within-population groups, the record was assigned to population structure. The Evolution category was used to describe studies on mechanisms associated with alternate life cycles. The Habitat category pertained to those applications centered on altered habitats (e.g., river impoundments). Fisheries applied to papers that related to stock structure and its effect on population dynamics in exploited species. Oceanography dealt with climate and oceanographic forcing on key life cycle attributes. Ecosystem category included community dynamics and multispecies interactions.

Twelve broad categories of analytical approach were chosen, which emphasized the increased application of longitudinal methods in recent years. The Distribution category broadly included catch and survey data analysis and basic demographic characterizations of fish populations. Physiology represented laboratory and field measurements of energetic and physiological state. Tagging studies were broken down into a Conventional tag category, where tags were externally applied; an Electronic tag category, which includes microchip, acoustic, radio, datalogging ,and satellite tags; and a Natural tag category. Natural tags referred to meristics, morphometrics, parasites, and other tracers of environmental origin. Because otolith chemistry has become a dominant approach in uncovering patterns of life cycle diversity, it was treated separately from the Natural tag category. Other categories of approaches included Genetic studies and Review.

Papers were identified in terms of country of study site. Authors often use multiple terms related to alternate life cycles within the same paper. These instances were identified to ensure that individual papers were not redundantly tallied across classifications.

Results and Discussion

Life Cycle Typologies

At the species level, life cycle descriptions have generated two typological approaches based upon habitat use or migration (Table 2). The two typologies share their basis in the characterization of the salinity habitats utilized by species. Habitat-use types emphasize the area where fish spend most of their lives (e.g., freshwater, estuarine, and marine habitats), with qualifiers describing the degree of residency in these environments (e.g., resident, migrant, straggler, etc.; Whitfield 1999). Migration types characterize the movement of fish between habitats utilized for growth, breeding, or other purposes (Myers 1949; McDowall 1988). The migration typology is well accepted in the literature, as indicated by numerous citations (Table 2), with few synonyms in use. The two typologies have been deliberately developed with terms generalizing important seasonal, ontogenetic, and lifetime trends in habitat use. Substantial literature and scholarship support these typologies as valid ways of describing assemblages and communities across zoogeographical and geological scales.

Beyond species descriptors, the typologies do not accommodate the well-documented within-population diversity in life cycles of diadromous fishes. Additionally, the unique terminology used for diadromous fishes cannot be extended to characterize life-cycle patterns in marine and freshwater species. For instance, important migration cycles occur for marine fishes; yet, these populations are simply classified as marine or oceanodromous types because migrations do not traverse salinity zones.

Life Cycle Lexicon

For the period 1975–2006, the search identified 336 studies within which occurred 40 terms pertinent to life cycle diversity that were used in 401 instances (Figures 1 and 2). Zero to 10 papers appeared during the period 1975–1990, but over the past 15 years, publications have increased threefold to more than 30 papers per year in recent years. In contrast, during the same period, total papers across all fishes and other taxa reported by ASFA increased 80% from 829 in 1992 to 1489 in 2006. Prior to 2000, papers related to Salmonidae (predominately *Oncorhynchus* and *Salvelinus*) dominated the life cycle diversity literature, but in the most recent decade, papers constitute an increasing diversity of species and families (Figures 2 and 3), representing fishes from all major ecosystem types (Figure 4A).

Terms were grouped into four categories: (1) migration patterns and modalities, (2) migration process, (3) dispersive modes, and (4) retentive (sedentary) modes (Figure 1; Table 3). Here and elsewhere, we use the term mode to imply groupings or variants within populations. Modalities represent multiple groups. Within each category, repeated terms included migratory, type, ecotype, form, and history. Common process-oriented phrases included facultative, which specified alternate life cycles, and vagrant, which related to tests of the member-vagrant hypothesis on life-cycle evolution (Sinclair 1988). The most common term was stray, which has a long history of use in questions related to Salmonidae (Tables 3 and 4). Other commonly used terms that were predominately applied to anadromous salmon populations included ocean type, stream type, nonanadromous, and nonmigratory.

Terms that were searched, but not applied to life cycle diversity studies, included ecomorph, straggler, wanderer, dispersive, and facultative anadromy; with the exception of facultative anadromy, these terms did occur in our search but were applied at the species level or pertained to attributes unrelated to alternate life cycles. The absence of the terms ecomorph and facultative anadromy in the life cycle diversity literature was unexpected. Ecomorph would seem an apt term to describe morphological variation among sedentary and migratory forms in

Table 2.—Habitat use and migration typologies that describe fish life cycles at the species level. In the choice of terms to define, we considered the precedence and common usage of the term and included less utilized synonyms in parentheses. We used primary sources of term definitions and also used more recent review articles that synthesized terminology. The definitions appear verbatim as found in the bolded citations (however, in some of these cases terms where defined in illustrations, and in those instances, the definition was excerpted and summarized from figures and associated text). One paper from a peer-reviewed journal was chosen to illustrate the application of each term. The number of citations for each term, as determined by a Cambridge Scientific Abstracts Aquatic Sciences and Fisheries Abstracts (ASFA) term search of abstracts and titles of peer-reviewed journals (1975–2006), was recorded to evaluate usage of the term.

Term	Subcategory	Definition	Application	Number of citations (ASFA)
Habitat typology				
Marine resident		Species that utilize ocean habitat at all life stages (inferred definition; not formally defined).	Tsukamoto and Arai 2001	3
Oceanodromous		Truly migratory fishes that live and migrate wholly in the sea (**Myers 1949**).	Khalaf 2005	1
Marine stragglers (Marine adventitious visitor, marine visitors, local marine species, nondependent marine species, occasional marine visitor)		Marine species where only a small proportion of the overall population makes use of estuaries (Potter et al. 1986, 1990; Whitfield 1999).	Valesini et al. 1997	7
Marine migrant (Marine immigrant, seasonal marine migrant, dependent marine species, marine estuarine dependent, estuarine transient, temporary residents, marine estuarine-opportunist)		Marine species that make extensive use of estuaries during juvenile and/or adult life stages (**Whitfield 1999**).	Vorwerk et al. 2003	6
	Marine estuarine-opportunist	Marine species of teleost that enter estuaries regularly and in at least moderate numbers (**Potter 1990**).	Potter et al. 1997	7
	Marine estuarine-dependent	Marine species that enter estuaries in large numbers (**Potter et al. 1986**).	Laffaille et al. 2000	2
Estuarine resident (resident, true estuarine species, permanent resident, estuarine, inshore nonmigratory, truly estuarine species)		Truly estuarine resident species that spend their entier lives in the estuary (McHugh 1967; **Elliott and Dewailly 1995**; Whitfield 1999).	Knieb and Knowlton 1995	27

Table 2.—Continued.

Term	Subcategory	Definition	Application	Number of citations (ASFA)
Estuarine migrant		Fish species, usually of marine origin, that breed in estuaries but have a marine or freshwater aspect to their life cycle. Estuarine migrants often have marine or freshwater breeding populations (Whitfield 1999, 2005).	Whitfield 2005	1
Freshwater resident		Species that utilize freshwater habitat at all life stages (adopted term)	Secor et al. 2001	40
Freshwater straggler (freshwater adventitious, freshwater species, freshwater migratory species)		Freshwater fish species that sometimes enter estuaries when conditions are favorable (Elliott and Dewailly 1995; Whitfield 1999).	Whitfield 2005	1
Freshwater migrant (freshwater species, freshwater migratory species)		Freshwater species that are often in estuaries, retreating into catchment rivers when conditions become unfavorable (Whitfield 1999, 2005).	Whitfield 2005	0
Migration typology				
Potamodromous		Truly migratory fishes that live and migrate wholly within freshwater (Myers 1949).	Mallen-Cohen and Stuart 2003	18
Diadromous		Truly migratory fishes that migrate between the sea and freshwater (Myers 1949; McDowall 1988).	Tsukamoto et al. 2002	143
Anadromous		Diadromous fishes that spend most of their lives in the sea and that migrate to freshwater to breed (Myers 1949; McDowall 1988).	Wedemeyer et al. 1980	985
Semi-anadromous		Diadromous fishes that spend most of their lives in saline water and that migrate to, or almost to, freshwater for spawning (Cronin and Mansueti 1971).	Hoeksema and Potter 2006	7

Table 2.—Continued.

Term	Subcategory	Definition	Application	Number of citations (ASFA)
Catadromous		Diadromous fishes that spend most of their lives in freshwater and that migrate to the sea to breed (Myers 1949, McDowall 1988).	Sang et al. 1994	71
Semi-catadromous		Diadromous fishes that spend most of their lives in freshwater and that migrate to the estuary to breed (Whitfield 2005).	Tzeng et al. 2000	0
Amphidromous		Diadromous fishes whose migration from freshwater to the sea, and vice versa, is not for the purpose of breeding (Myers 1949; McDowell 2007).	Iguchi and Mizuno 1990	122
	Freshwater amphidromy	Diadromous fish whose return migration to freshwater from the sea occurs during the juvenile stage is not for the purpose of breeding (McDowell 1988; McDowall 1992).	No application found outside definition by McDowall	0
	Marine amphidromy	Diadromous fish whose return migration to the sea from freshwater occurs during the juvenile stage and is not for the purpose of breeding (McDowell 1988, 1992).	No application found ouside definition by McDowall	0

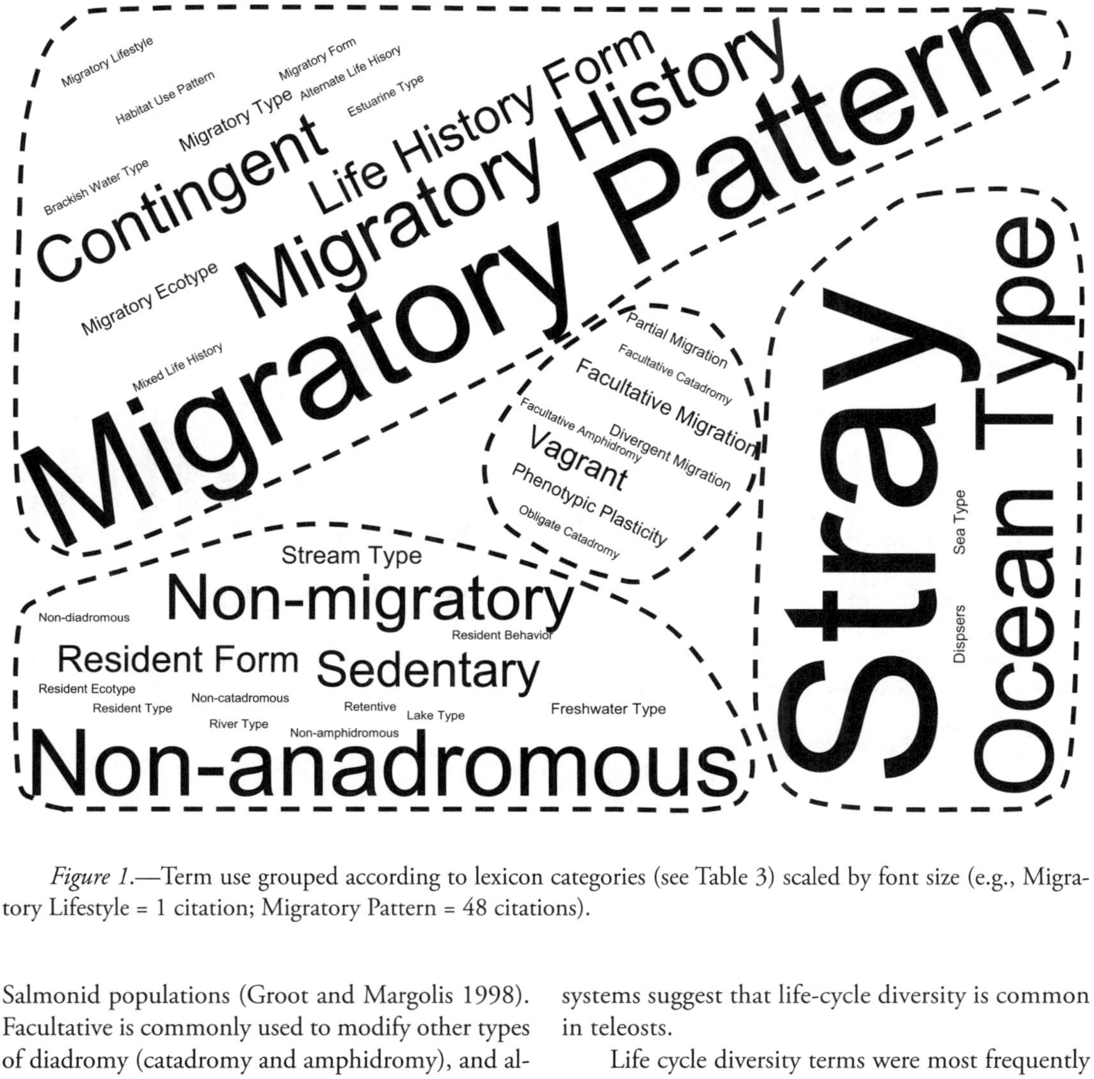

Figure 1.—Term use grouped according to lexicon categories (see Table 3) scaled by font size (e.g., Migratory Lifestyle = 1 citation; Migratory Pattern = 48 citations).

Salmonid populations (Groot and Margolis 1998). Facultative is commonly used to modify other types of diadromy (catadromy and amphidromy), and alternate life cycles are well known for many anadromous species.

Taxonomy and Applications

More than 25 species representing species from stream, lake, river, estuary, and coastal ecosystems were represented (Figure 4; Table 4). The lexicon was clearly dominated by Salmonidae, with most of these applications pertinent to anadromy (River–Coast ecosystem). Nondiadromous families include Centrarchidae, Cyprinidae, Esocidae, Gadidae, Mugilidae, Pleuronectidae, Pomatomidae, Scombridae, and Sparidae. Although alternate life cycles are best known among anadromous species, numerous studies across diverse species and ecosystems suggest that life-cycle diversity is common in teleosts.

Life cycle diversity terms were most frequently applied in Population structure, Behavior, Habitat, and Hatchery studies (Table 4). Application of these terms was also, although less frequently, used in the study of Physiology, Evolution, Fisheries, Conservation, and Pollution. Most studies were aimed at describing behaviors (n = 117), evaluating population structure (n = 104), or addressing habitat issues (n = 46). Otolith chemistry was a dominant approach in these applications, used in 41% of the Behavior studies, 26% of the Population structure studies, and 32% of the Habitat studies. Genetics was used in 41% of the Population structure studies. Conventional tagging and Distribution approaches were also commonly used across these applications. Hatchery applications mostly involved issues related to hatchery recruits straying into adjacent

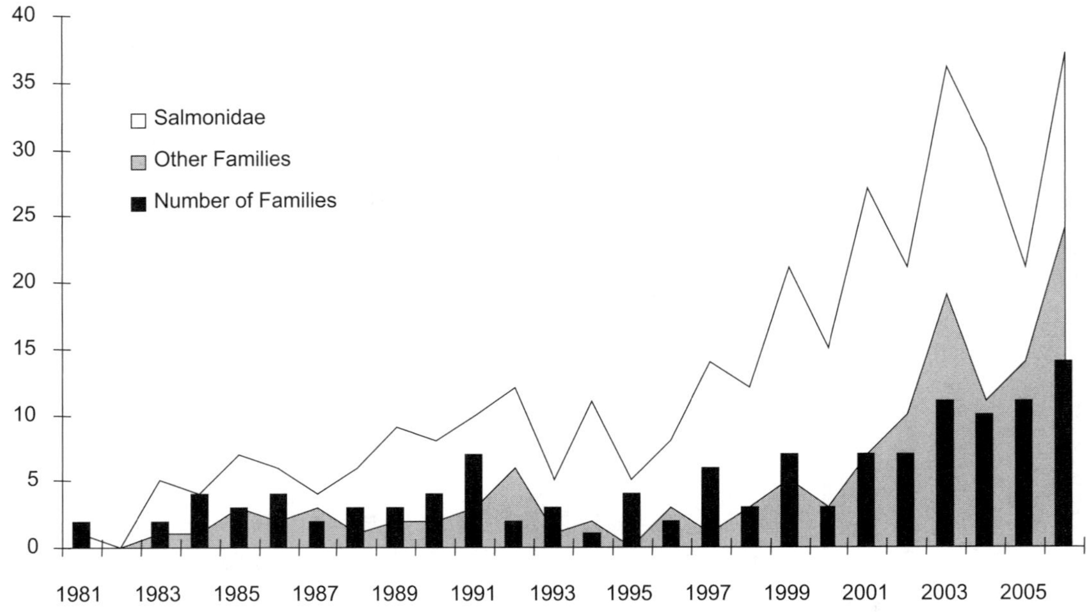

Figure 2.—Number of studies reviewed for life cycle lexicon from 1981 to 2006 using Cambridge Scientific Abstracts Aquatic Sciences and Fisheries Abstracts.

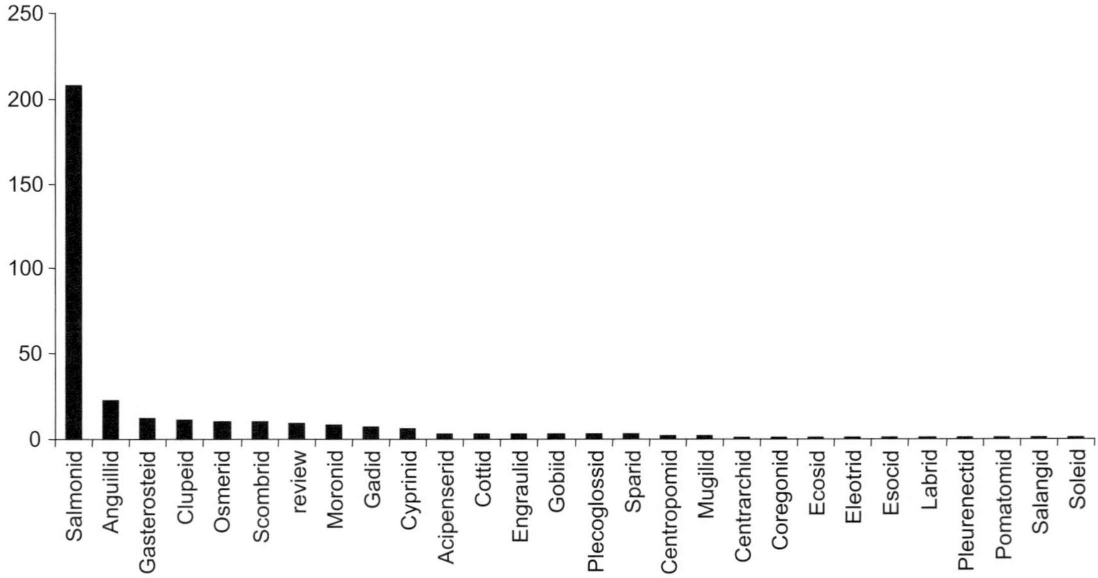

Figure 3.—Distribution of taxonomic families in life cycle lexicon.

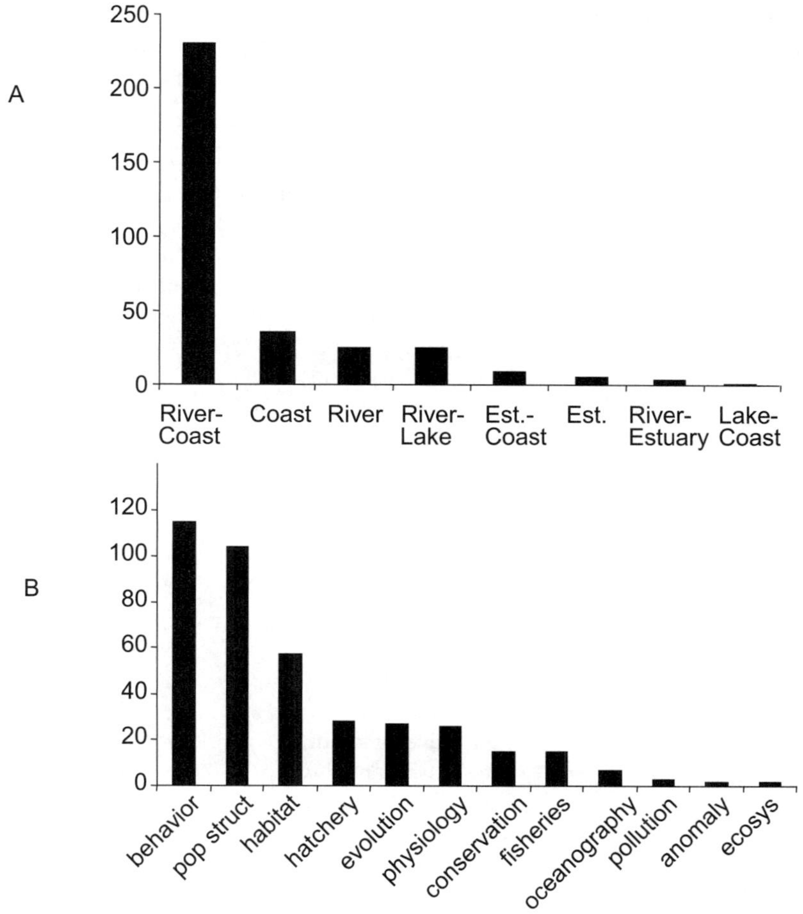

Figure 4.—Distribution of ecosystems (A) and applications (B) in life cycle lexicon. Est. = estuary. For details on categories, see Methods.

populations and principally involved Conventional tagging approaches.

Below are several examples of term use across applications:

- Nonanadromous, resident, migatory, and nonmigratory terms were collectively used in an examination of population structure of rainbow trout *Oncorhynchus mykiss* (Zimmerman and Reeves 2000). In this study, reproductive isolation between anadromous and nonanadromous life history forms was suggested by spawning surveys and otolith microchemistry.
- Facultative catadromy was used in a study of the migratory behavior of Japanese eel *Anguilla japonica* (Tsukamoto and Arai 2001). Otolith chemistry revealed remarkable flexibility in migratory pathways by Japanese eel in this behavioral study.
- Partial migration was used in the description of physiological differences between migratory and resident forms of brook trout *Salvelinus fontinalis* (Morinville and Rasmussen 2003). The study provided evidence for bioenergetic trade-offs in life history strategy. Migratory forms of brook trout exhibited higher consumptions rates prior to dispersal compared to resident forms; however, this gain in energy up take was balanced by higher metabolic costs.
- Straying was used in an examination of the significance of this life history tactic in Salmonidae from an evolutionary perspective (Thorpe 1994). The study highlighted the im-

Table 3.—Lexicon of terms and phrases used to describe life cycle diversity within species and populations. A total of 40 terms were distributed among four lexicons centered on migration pattern or process and degree of dispersive or retentive behaviors. More than 300 citations were distributed among terms. Alternate search strings are shown in parentheses.

Term grouping	Term or phrase	Citations
Patterns/modalities (13)	Migratory pattern (migration pattern)	48
	Migratory history (migration history)	26
	Contingent(s) (migratory contingents)	21
	Life history forms (___types, ___modes)	16
	Migratory ecotype (migration ecotype)	6
	Migratory type (migratory type)	6
	Habitat use pattern(s)	4
	Migratory form(s)	4
	Brackish water type(s)	3
	Alternate life history(ies)	2
	Mixed life history(ies)	2
	Estuarine type(s)	1
	Migratory lifestyle	1
Migration process (8)	Vagrants (member-vagrant)	10
	Facultative migration (___migrants)	7
	Phenotypic plasticity	6
	Divergent migration(s)	5
	Partial migration	5
	Facultative catadromy	4
	Facultative amphidromy	1
	Facultative anadromy	0
	Obligate catadromy	2
	Obligate anadromy/___amphidromy	0
Dispersive modes (4)	Stray(s)	70
	Ocean type(s)	35
	Sea type(s)	4
	Dispersers	1
Retentive modes (15)	Nonanadromous	30
	Nonmigratory	21
	Sedentary	16
	Resident form(s)	12
	Stream type(s)	8
	Freshwater type(s)	5
	Resident type(s)	4
	River type(s)	4
	Lake type(s)	3
	Nonamphidromous	2
	Resident behavior(s)	2
	Noncatadromous	1
	Nondiadromous	1
	Resident ecotype	1
	Retentive	1

Table 4.—Term usage in life cycle diversity lexicon. Number of times a term was used in a given application is shown for more than five citations for patterns of use across taxonomic family, ecosystem, application, approach, and study nationality. Migr = migration or migratory; LH = life history; Facult = facultative; P Plastic = phenotypic plasticity.

Term	Families	Ecosystem	Application	Approach	Country
Migr pattern	Salmonid (14) Acipenserid, Anguillid, Centropomid, Clupeid, Cottid, Cyprinid, Engraulid, Gadid, Gasterosteid, Labrid, Moronid, Mugillid, Osmerid, Pomatomid, Salangid, Scombrid, Sparid	River-Coast (20) Coast (12) Estuary-Coast, Ocean, River, River-Lake	Habitat (17) Pop Str (9) Fisheries (8) Behavior (6) Conservation, Ecosystem, Evoluation, Oceanography	Otol chem (16) Conv tag (8) Electronic (8) Acoustics, Distribution, Genetics, Model, Nat tag, Physiology, Review	USA (16) Canada (8) Japan (7) Australia, China, Finland, France, Germany, Lithuania, Norway Papua NG, Taiwan, UK, Yugoslavia
Migr history	Anguillid (7) Acipenserid, Centropomid, Cottid, Cyprinid, Engraulid, Gadid, Gasterosteid, Gobiid, Moronid, Osmerid, Salmonid, Scombrid	River-Coast (22) Estuary-Coast, Ocean, River-Estuary	Behavior (20) Habitat, Pop Str	Otol chem (26)	Japan (16) Canada, China, New Zealand, Papua NG, Russia, Spain, Taiwan, USA
Contingents	Anguillid (6) Clupeid, Esocid, Moronid, Pleuronectid, Salmonid, Scombrid, Sparid	River-Coast (8) Coast, Ocean, Review, River	Pop Str (15) Behavior, Ecosystem, Habitat	Otol chem (12) Conv tag, Distribution, Electronic, Genetics, Nat tag, Physiology	USA (8) Australia, Canada, France, New Zealand, Sweden, Taiwan, UK
LH forms	Salmonidae (16)	River-Coast (11) River-Lake	Conserv (6) Behavior, Evolution, Hatch, Pop Str	Distribution (6) Genetics (6) Electronic, Physiology Review	USA (8) Canada, Europe, Japan, Norway, Russia, Sweden

Table 4.—Continued.

Term	Families	Ecosystem	Application	Approach	Country
Migr ecotype	Salmonid (6)	River-Coast, River-Lake	Behavior, Evolution, Habitat, Pop Str	Behavior, Electronic, Genetics, Physiology, Review	Canada, Finland, Japan, Norway, USA
Migr type	Anguillid, Gadid, Mugillid, Salmonid, Sparid	Coast, River-Estuary, River-Coast	Behavior, Physiology, Pollution, Pop Str	Conv tag, Electronic, Genetics, Oto chem, Physiology	Canada Germany, Japan, Taiwan, UK
Vagrants	Anguillid, Clupeid, Gadid, Osmerid, Salmonid, Soleid	Coast, Estuary, River, River-Coast	Pop Str (6) Oceanography	Genetics (6) Distribution, Review	Canada, Mexico, Russia, Taiwan, UK, USA
Facult migr	Anguillid, Eleotrid	Estuary-Coast, River-Coast, River-Estuary, River-Lake	Behavior, Habitat	Otol chem, Review	Canada, France, Japan, New Zealand, Poland,
P plastic	Anguillid, Clupeid, Salmonid	River-Coast	Behavior, Evolution, Habitat	Distribution, Nat tag, Otol chem, Review	France, New Zealand, Norway, USA
Strays	Salmonid (60) Acipenserid, Clupeid, Cyprinid, Gadid, Moronid, Scombrid	River-Coast (57) Coast, River River-Lake	Hatch (23) Pop Str (16) Behavior (15) Conservation, Evolution, Fisheries, Habitat, Pollution	Conv tag (33) Genetics (16) Review (7) Distribution, Electronic, Electronic-otol chem, Genetics, Model, Nat tag, Physiology	USA (34) Canada (14) Norway (6) Argentina, Iceland, Netherlands, New Zealand, Spain, Sweden, UK
Ocean type	Salmonid (34) Gadid	River-Coast (34) Coast	Habitat (11) Behavior (7) Pop Str (6) Conservation, Evolution, Fisheries, Hat, Oceanography, Physiology	Distribution (16) Behavior Conv tab, Genetics, Model, Otol chem, Physiology, Review Distribution (16)	USA (26) Canada (7) New Zealand

Table 4.—Continued.

Term	Families	Ecosystem	Application	Approach	Country
Non-anadromous	Salmonid (29) Coregonid	River-Coast (29) River	Behavior (10) Pop Str (9) Evolution (6) Physiol	Genetics (10) Physiology (6) Behavior, Distribution, Nat tag, Otol chem	Canada (16) USA (8) Argentina, Finland, Japan, New Zealand, Norway, UK
Non-migr	Salmonid (14) Anguillid, Clupeid, Cyprinid, Osmerid, Sparid	River-Coast (15) Coast, River-Coast, River-Lake	Behavior (9) Physiology (7) Evolution, Hat, Pop Str	Physiology (8) Conv tag, Distribution, Nat tag, Otol chem, Review	Argentina, Australia, Canada, Denmark, France, Japan, Netherlands, New Zealand, Norway, Sweden, UK, USA
Sedentary	Anguillid, Centrarchid, Cottid, Cyprinid, Ecosid, Engraulid, Gadid, Osmerid, Salmonid	River (8) Coast, Lake-Coast, River-Coast, River-Lake	Pop Str (9) Behavior Evolution	Conv tag, Distribution, Electronic, Genetics, Nat tag, Review,	Belgium, Canada, Finland, France, Japan, Mexico, Switzerland, USA
Resident form	Salmonid (10) Osmerid	River-Coast (9) River-Lake	Behavior, Physiology, Pop str, Habitat	Distribution, Genetics, Otol chem, Physiology, Review	Canada, Japan, New Zealand, Norway, Sweden, USA
Stream type	Salmonid (8)	River, River-Coast	Habitat (5) Behavior	Conv tag Distribution,	Canada, USA, Argentina, New Zealand

portance of straying in the colonization of new habitat by modern salmon over the past 8,000–15,000 years. Implications included the need to minimize the probability of straying in released populations by ensuring that the homing process is engrained in smolts.

- Ocean-type and stream-type were applied in an oceanographic context in a study of variation in size of returning life history forms of Chinook salmon along the northeast Pacific coast. Here, Wells et al. (2006) found significant associations between adult sizes of differing migratory forms to Pacific Ocean teleclimatic indices.
- Zlokovitz and Secor (1999) described resident and migratory contingents of striped bass *Morone saxatilis* in the Hudson River. In this application, body burdens of polychlorinated bi-

phenyls in resident fish, identified through otolith chemistry, were far greater than migratory fish.

- Vagrant was used in the description of population structure of Atlantic salmon *Salmo salar* L. in a Canadian river system. Garant et al. (2000) provided genetic evidence of within-river population structure and established the importance of vagrants in population structuring and metapopulation persistence.
- Life history form was used to describe life history diversity documented in bull char *Salvelinus confluentus* in a conservation context (Nelson et al. 2002). A decline in the migratory form of bull char was documented in a Montana river drainage, with only the resident form present in several headwater streams. The absence of the migratory form of bull char was thought to be a factor in their current threatened status, and establishment of its stock was deemed an important conservation goal.

Approaches and Study Nationality

Contributing to the increased observations of life cycle diversity are longitudinal methods, which enable the spatial histories of fish to be hindcast over major portions of an individual's life cycle. Electronic tagging and otolith chemistry are particularly powerful longitudinal approaches. Although electronic tagging applications did not increase markedly during the past 30 years in the lexicon (ranging between zero and four studies per year), they represent perhaps the most precise method for elaborating patterns of habitat use and migration. Recent applications have also permitted spatial behaviors to be linked to demographic fates (Hightower et al. 2001). Terms associated with Electronic tagging studies include migratory pattern (8), sedentary, and strays (four each; Table 4). Otolith chemistry applications have increased notably during the past decade with 8–25 studies per year during 2003–2006. Migratory history, an apt term initially coined by Radtke and Morales-Nin (1989), was most commonly associated with these applications (26) followed by migratory pattern (16), contingent (12), and habitat-use pattern, facultative migration, and nonmigratory (four each) (Table 4).

The term stray was most often associated with genetic applications (16), conventional tagging studies (33), and review papers (7). Physiological applications were commonly associated with the sedentary terms nonmigratory (8) or nonanadromous (6). In distribution studies, life history form was the most commonly used term (Table 4).

Terms showed few obvious trends with study site nationality. As expected, terms typically applied to Salmonidae were used most frequently in northern nations (e.g., Canada, USA, UK, Japan, Norway, Sweden, Finland, and Russia). Other terms that were used more broadly across taxa, such as migratory pattern, migratory history, contingent, and sedentary, were applied broadly across nationalities and zoogeographical provinces (Table 4).

Implications

Both habitat-use and migration typologies emphasize salinity; yet, in some ecosystems such as the watersheds and neretic regions of eastern North America, salinity gradients are broad and thresholds between freshwater, estuarine, and marine difficult to discern. Here a migration of a "marine resident" into adjacent parts of estuaries may be no more significant to that fish's ecology than movements within the coastal environment. In contrast, Hawaiian freshwater gobies, alternatively termed amphidromous (McDowall 2007) or marine anadromous (Bell 2009, this volume) encounter dramatically different freshwater and marine environments with little gradient in between (Kinzie 1988). Clearly, in this instance, populations are dependent upon diadromous life cycles. Application of typologies across diverse ecosystems and taxa will necessarily be incomplete descriptions of life cycles.

Here, we argue that continued authoritative assertions for improved meanings within life cycle typologies will be of limited value in understanding the complex life cycles of diadromous species. As indicated previously, species-level descriptors can be useful shorthand but, if exclusively relied upon, will limit discourse on mechanisms and consequences of the complex and diverse life cycles of diadromous and other fishes. As evidence, there is little indication that authors have sought to modify species typologies to describe life cycle diversity. Terms such as facultative catadromy and facultative amphidromy rarely occur, and facultative anadromy was not found in our literature survey. Rather, the lexicon of terms used to describe within population diversity shows substantial variation across taxa, applications, and approaches.

The proliferation of terms in describing life cycle diversity is likely due to the maturation of sophisticated longitudinal approaches for tracking individual migration pathways within populations of diadromous, marine, and freshwater fishes. But it is also reflective of a lack of cross-taxa, cross-application, or cross-ecosystem integration. As a result, scientists working in diverse individual settings may not sufficiently have recognized the prevalence of life cycle diversity among fishes, nor have pursued a more fundamental understanding of causes and consequences of that diversity. Thus, discourse on life cycle diversity within species has been hampered by term proliferation. Still, general modalities in life cycles emerge from the lexicon.

The lexicon shows sedentary and migratory components being recognized in many populations of diadromous, estuarine, marine, and freshwater fishes. Earlier studies focused on the dichotomy of sedentary versus migratory behaviors within populations of Salmonidae, but more recent evidence exists for this dichotomy in diverse taxa. Indeed, nondiadromous species show patterns of migration that can be coarsely described as migratory and nonmigratory. For instance, some Atlantic bluefin tuna *Thunnus thynnus* complete their life cycle centered on the Mediterranean Sea, whereas others of this same population (spawning in the Mediterranean) undertake long migrations to the western Atlantic (Rooker et al. 2007). As we track individual migrations with increasing resolution, we are likely to discover ever increasing diversity and patterns. Here, the dichotomy between sedentary and migratory forms may break down. For instance, modalities in Chesapeake Bay striped bass include resident freshwater, estuarine, and marine forms. An epitome may be Atlantic herring, which utilize myriad local spawning habitats generation after generation without strong evidence of philopatry. In describing life cycle modes (variants), we have resurrected the term contingent (noun) in our work (Secor 1999), which has strong precedence (Hjort 1914) in describing life cycle diversity. The term also has advantages as not specifying ecosystems, species types, or implicit meanings pertaining to function or migration. We humbly suggest it as a neutral term.

Regardless of term usage, the sedentary–migratory dichotomy is consistent with the phenomenon of partial migration, which may be as general to fishes as it is to birds. Other modalities are likely and the entrainment hypothesis is an intriguing and untested theory that could have broad application to fish. Testing among these and other hypotheses has received very little treatment in fish studies, with the exception of very strong experimental research on Salmonidae (e.g., Nordeng 1983; Thorpe 1989; Metcalfe and Thorpe 1992), which supports the obligate partial migration hypothesis. While much science has focused on describing patterns of life cycle diversity, proportionately little scientific effort has been devoted to the causes and consequences of this diversity.

Diversity of life history tactics in fish populations is increasingly recognized as having the effect of offsetting environmental stochasticity and contributing to long-term persistence (Secor 2007). Contingents within populations should show some degree of independent response to environmental conditions because each contingent is responding to a different subset of those conditions (e.g., Kraus and Secor 2005; Fodrie and Mendoza 2006). Particular contingents can differ in their vulnerability to exploitation, habitat degradation, and climate change (Secor 1999, 2007; Cadrin and Secor, in press). Thus, diversity of life cycles within populations hedges against environmental uncertainty and contributes to population persistence.

Acknowledgments

This review was supported by National Science Foundation OCE-0324850. We thank Alex Haro and the Program Committee of the 2nd International Symposium on Diadromous Fishes for their kind invitation to participate and prepare this manuscript, and Kim Bell for a constructive review of this paper. The idea for the paper was stimulated by Katsumi Tsukamoto; in reference to his newly discovered sea eels, he asked D. S. many years ago, "What should we call these things? This phenomenon?" This is contribution 4254 of the University of Maryland Center for Environmental Science.

References

Beacham, T. D., J. Brattey, K. M. Miller , K. D. Le, and R. E. Withler. 2002. Multiple stock structure of Atlantic cod (*Gadus morhua*) off Newfoundland and

Labrador determined from genetic variation. ICES Journal of Marine Science 59(4):650–665.

Bell, K. N. I. 2009. What comes down must go up: migrations of Caribbean anadromous taxa. Pages 321–341 *in* A. J. Haro, K. L. Smith, R. A. Rulifson, C. M. Moffitt, R. J. Klauda, M. J. Dadswell, R. A. Cunjak, J. E. Cooper, K. L. Beal, and T. S. Avery, editors. Challenges for diadromous fishes in a dynamic global environment. American Fisheries Society, Symposium 69, Bethesda, Maryland.

Berthold, P. 1996. Control of bird migration. Chapman and Hall, London.

Berthold, P. 2001. Bird migration: a general survey. Oxford University press, New York.

Cadrin, S. X., and D. H. Secor. In press. Accounting for spatial population structure in stock assessment: past, present and future. In R. J. Beamish and R. J. Rothschild, editors. Future of fishery science in North America. Springer Science, 405 Fish & Fisheries Series, New York.

Corten, A. 2002. The role of conservatism in herring migrations. Reviews in Fish Biology and Fisheries 11:339–361.

Cronin, L. E., and A. J. Mansueti. 1971. The biology of the estuary. Pages 14–39 *in* P. A. Douglas and R. H. Stroud, editors. A symposium on the biological significance of estuaries. Sport Fishing Institute, Washington, D.C.

Elliott, M., and F. Dewailly. 1995. The structure and components of European estuarine fish assemblages. Netherlands Journal of Aquatic Ecology 29:397–417.

Fodrie, F. J., and G. Mendoza. 2006. Availability, usage and expected contribution of potential nursery habitats for the California halibut. Estuarine Coastal and Shelf Science 68:149–164.

Fromentin, J. M., and J. E. Powers. 2005. Atlantic bluefin tuna: population dynamics, ecology, fisheries and management. Fish and Fisheries 6:281–306.

Garant, D., J. J. Dodson, and L. Bernatchez. 2000. Ecological determinants and temporal stability of the within-river population structure in Atlantic salmon (*Salmo salar* L.) Molecular Ecology 9:615–628.

Groot C., and L. Margolis, editors. 1998. Pacific salmon life histories. UBC Press, Vancouver.

Helfman, G. S., and E. T. Schultz. 1984. Social transmission of behavioral traditions in a coral-reef fish. Animal Behaviour 32:379–384.

Hightower, J. E., J. R. Jackson, and K. H Pollock. 2001. Use of telemetry methods to estimate natural and fishing mortality of striped bass in Lake Gaston, North Carolina. Transactions of the American Fisheries Society 130:557–567.

Hjort, J. 1914. Fluctuations in the great fisheries of northern Europe. Rapports et Procès-Verbaux des Réunions du Conseil International pour l'Exploration de la Mer 20:1–228.

Hoeksema, S. D., and I. C. Potter. 2006. Diel, seasonal, regional and annual variations in the characteristics of the ichthyofauna of the upper reaches of a large Australian microtidal estuary. Estuarine Coastal and Shelf Science 67:503–520.

Hutchings, J. A. and R. A. Myers. 1994. The evolution of alternative mating strategies in variable environments. Evolutionary Ecology 8: 256–268.

Iguchi, K., and N. Mizuno. 1990. Diel changes of larval drift among amphidromous gobies in Japan, especially *Rhinogobius brunneus*. Journal of Fish Biology 37:255–264.

Iles, T. D., and M. Sinclair. 1982. Atlantic herring: stock discreteness and abundance. Science 215:627–633.

Jonsson, B., and N. Jonsson. 1993. Partial migration: niche shift versus sexual maturation in fishes. Reviews in Fish Biology and Fisheries 3:348–365.

Khalaf, M. A. 2005. Five additional records of fishes in the Gulf of Aqaba, including *Mola mola* (Forskael, 1775), new for the Red Sea. Zoology in the Middle East 34:45–52.

Kinzie, R. A., III. 1988. Habitat utilization by Hawaiian stream fishes with reference to community structure in oceanic island streams. Environmental Biology of Fish 22(3):179–192.

Klemetsen, A., P. A. Amundsen, J. B. Dempson, B. Jonsson, N. Jonsson, M. F. O'Connell, and E. Mortensen. 2003. Atlantic salmon *Salmo salar* L., brown trout *Salmo trutta* L. and Arctic charr *Salvelinus alpinus* (L.): a review of aspects of their life histories. Ecology of Freshwater Fish 12:1–59.

Knieb, R. T., and M. K. Knowlton. 1995. Stage-structured interactions between seasonal and permanent residents of an estuarine nekton community. Oecologia 103:425–434.

Kraus, R. T., and D. H. Secor. 2005. Application of the nursery-role hypothesis to an estuarine fish. Marine Ecology Progress Series 291:301–305.

Lack, D. 1943. The problem of partial migration. British Birds 37:122–130.

Laffaille, P., E. Feunteun, and J.-C. Lefeuvre. 2000. Composition of fish communities in a European macrotidal salt marsh (the Mont Saint-Michel Bay, France). Estuarine Coastal and Shelf Science 51:429–438.

Lundberg, P. 1988. The evolution of partial migration in birds. Trends in Ecology and Evolution 3:172–175.

Mallen-Cooper, M., and I. G. Stuart. 2003. Age, growth and non-flood recruitment of two potamodromous fishes in a large semi-arid/temperate river system. River Research and Applications 19:697–719.

Mangel, M. 1994. Climate change and salmonid life history variation. Deep Sea Research II 41:75–106.

McDowall, R. M. 1988. Diadromy in fishes: migrations between freshwater and marine environments. Timber Press, Portland, Oregon.

McDowall, R. M. 1992. Diadromy: origins and definitions of terminology. Copeia 1992(1):248–251.

McDowall, R. M. 2007. On amphidromy, a distinct form of diadromy in aquatic organisms. Fish and Fisheries 8:1–13.

McHugh, J. L. 1967. Estuarine nekton. Pages 581–619 *in* G. H. Lauff, editor. Estuaries. American Association for the Advancement of Science Special Publication 83.

McQuinn, I. 1997. Metapopulations and the Atlantic herring. Reviews in Fish Biology and Fisheries 7:297–329.

Metcalfe, N. B., and J. E. Thorpe. 1992. Early predictors of life-history events—the link between 1st feeding date, dominance and seaward migration in Atlantic salmon, *Salmo salar* L. Journal of Fish Biology 41(Supplement B):93–99.

Morinville, G. R., and J. B. Rasmussen. 2003. Early juvenile bioenergetic differences between anadromous and resident brook trout (*Salvelinus fontinalis*). Canadian Journal of Fisheries and Aquatic Sciences 60:401–410.

Myers, G. S. 1949. Usage of anadromous, catadromous and allied terms for migratory fishes. Copeia 2:89–97.

Nelson, M. L., T. E. McMahon, and R. F. Thurrow. 2002. Decline of the migratory form in bull charr, *Salvelinus confluentus*, and implications for conservation. Environmental Biology of Fishes 64:321–332.

Nordeng, H. 1983. Solution to the "char problem" based on Arctic char (*Salvelinus alpinus*) in Norway. Canadian Journal of Fisheries and Aquatic Sciences 40:1372–1387.

Økland, B., B. Jonsson, A. J. Jensen, and L. P. Hansen. 1993. Is there a threshold size regulating seaward migration of brown trout and Atlantic salmon? Journal of Fish. Biology 42:541–550.

Petitgas, P., G. Charrier, J. Coetzee, A. Corten, N. Lo, I. McQuinn, B. Planque, D. Secor, and G. Skaret. 2007. Report of the Workshop on Testing the Entrainment Hypothesis (WKTEST). ICES Living Resource Committee, ICES CM2007/LRC, Copenhagen, Denmark.

Potter, I. C., L. E. Beckley, A. K. Whitfield, and R. C. J. Lenanton. 1990. Comparisons between the roles played by estuaries in the life-cycles of fishes in temperate western Australia and southern Africa. Environmental Biology of Fishes 28:143–178.

Potter, I. C., P. N. Clandge, and K. M. Warwick. 1986. Consistency of seasonal changes in an estuarine fish assemblage. Marine Ecology Progress Series 32:217–228.

Potter, I. C., P. N. Claridge, G. A. Hyndes, and K. R. Clarke. 1997. Seasonal, annual and regional variations in ichthyofaunal composition in the inner Severn estuary and inner Bristol Channel. Journal of the Marine Biological Association of the United Kingdom 77:507–525.

Radtke, R. L., and B. Morales-Nin. 1989. Mediterranean juvenile bluefin tuna: life history patterns. Journal of Fish Biology 35:485–496.

Robichaud, D., and G. A. Rose. 2004. Migratory behaviour and range in Atlantic cod: inference from a century of tagging. Fish and Fisheries 5:185–214.

Rooker, J. R., J. R. Alverado Bremer, B. A. Block, H. Dewar,G. De Metrio, A. Corriero, R. T. Kraus, E. D. Prince, E. Rodrígues-Marín, and D. H. Secor. 2007. Life history and stock structure of Atlantic bluefin tuna (*Thunnus thynnus*). Reviews in Fisheries Science 15:265–310.

Ruzzante, D. E., C. T. Taggart, D. Cook, and S. V. Goddard. 1997. Genetic differentiation between inshore and offshore Atlantic cod (*Gadus morhua*) off Newfoundland: a test and evidence of temporal stability. Canadian Journal of Fisheries and Aquatic Sciences 54:2700–2708.

Ruzzante, D. E., J. S. Wroblewski, C. T. Taggart, R. K. Smedbol, D. Cook, and S. V. Goddard. 2000. Bay-scale population structure in coastal Atlantic cod in Labrador and Newfoundland. Canadian Journal of Fish Biology 56:431–447.

Sang, T. K., H. Y. Chang, C. T. Chen, and C. F. Hui. 1994. Population-structure of the Japanese eel, *Anguilla japonica*. Molecular Biology and Evolution 11:250–260.

Secor, D. H. 1999. Specifying divergent migrations in the concept of stock: the contingent hypothesis. Fisheries Research 43:13–34.

Secor, D. H. 2007. The year-class phenomenon and the storage effect in marine fishes. Journal of Sea Research 57:91–103.

Secor, D. H., J. R. Rooker, E. Zlokovitz, and V. S. Zdanowicz. 2001. Identification of riverine, estuarine, and coastal contingents of Hudson River striped bass based upon otolith elemental fingerprints. Marine Ecology Progress Series 211:245–253.

Sinclair, M. 1988. Marine populations: an essay on population regulation and speciation. University of Washington Press, Seattle.

Terrill, S. B., and K. P. Able. 1988. Bird migration terminology. Auk 105:205–206.

Thorpe, J. E. 1989. Developmental variation in salmonid populations. Journal of Fish Biology 5:295–303.

Thorpe, J. E. 1994. Significance of straying in salmonids and implications for ranching. Aquaculture and Fisheries Management 25:183–190.

Thorpe, J. E., M. Mangel, N. B. Metcalfe, and F. A. Huntingford. 1998. Modelling the proximate basis of salmonid life-history variation, with application to Atlantic salmon, *Salmo salar* L. Evolution and Ecology 12:581–599.

Tsukamoto, K., I. Nakai, and W. V. Tesch. 1998. Do all freshwater eels migrate? Nature 396:635–636.

Tsukamoto, K., J. Aoyama, and M. J. Miller. 2002. Migration, speciation, and the evolution of diadromy in anguillid eels. Canadian Journal of Fisheries and Aquatic Sciences 59:1989–1998.

Tsukamoto, K., and T. Arai. 2001. Facultative catadromy of the eel Anguilla japonica between freshwater and seawater habitats. Marine Ecology Progress Series 220:265–276.

Tzeng, W. N., C. H. Wang, H. Wichström, and M. Reizenstein. 2000. Occurrence of the semi-catadromous European eel *Anguilla anguilla* in the Baltic Sea. Marine Biology 137(1):93–98.

Valesini, F. J., I. C. Potter, M. E. Platell, and G. A. Hyndes. 1997. Ichthyofaunas of a temperate estuary and adjacent marine embayment. Implications regarding choice of nursery area and influence of environmental changes. Marine Biology 128:317–328.

Verspoor, E., and L. J. Cole. 1989. Genetically distinct sympatric populations of resident and anadromous Atlantic salmon, *Salmo salar*. Canadian Journal of Zoology 67:1453–1461.

Vorwerk, P. D., A. K. Whitfield, P. D. Cowley, and A. W. Paterson. 2003. The influence of selected environmental variables on fish assemblage structure in a range of southeast African estuaries. Environmental Biology of Fishes 66:237–247.

Wedemeyer, G. A., R. L. Saunders, and W. C. Clarke. 1980. Environmental-factors affecting smoltification and early marine survival of anadromous salmonids. Marine Fisheries Review 42:1–14.

Wells, B. K., C. B. Grimes, J. C. Field, and C. S. Reiss. 2006. Covariation between the average lengths of mature coho (*Oncorhynchus kisutch*) and Chinook salmon (*O. tshawytscha*) and the ocean environment. Fisheries Oceanography 15:67–79.

Whitfield, A. K. 1999. Ichthyofaunal assemblages in estuaries: a South African case study. Reviews in Fish Biology and Fisheries 9:151–186.

Whitfield, A. K. 2005. Preliminary documentation and assessment of fish diversity in sub-Saharan African estuaries. African Journal of Marine Science 27:307–324.

Zimmerman, C. E., and G. H. Reeves. 2000. Population structure of sympatric anadromous and nonanadromous *Oncorhynchus mykiss*: evidence from spawning surveys and otolith microchemistry. Canadian Journal of Fisheries and Aquatic Sciences 57:2152–2162.

Zlokovitz, E. R., and D. H. Secor. 1999. Effect of habitat use on PCB body burden in Hudson River striped bass (*Morone saxatilis*). Canadian Journal of Fisheries and Aquatic Sciences 56(Supplement 1):86–93.

American Fisheries Society Symposium 69:557–575, 2009

Fishway Evaluations for Better Bioengineering: An Integrative Approach

Theodore Castro-Santos*
S.O. Conte Anadromous Fish Research Center, U.S. Geological Survey, Leetown Science Center Post Office Box 796, One Migratory Way, Turners Falls, Massachusetts 01376, USA

Aline Cotel
Department of Civil and Environmental Engineering, University of Michigan Ann Arbor, Michigan 48109, USA

Paul Webb
School of Natural Resources and the Environment, University of Michigan Ann Arbor, Michigan 48109, USA

Abstract.—Effective fishway design requires extensive integration of biological and hydraulic data. Many relevant biological parameters remain poorly characterized, however, and the lack of adequate biological data has long been recognized as a central weakness in fish passage technology. This is of particular concern given the growing recognition of the importance of passing a broad diversity of species. Part of the reason for this weakness is the difficulty of identifying relevant biological, hydraulic, and other physical parameters. We propose that by both exploring questions suggested by current knowledge, and also by increasing the frequency and refining the methods with which fishways are evaluated, two results can be achieved: our understanding of design effectiveness can be improved, and research questions can be prioritized through adaptive management. We describe a framework and rationale for fishway evaluations that identifies several promising avenues of research. Understanding correlates of passage performance is increasingly important as fish passage needs expand on a global scale.

Background

Fishway engineers have long recognized the importance of fish swimming capacity to fishway design. Indeed, the recognized need to pass fish upstream of dams and other obstacles inspired many seminal studies on fish swimming performance, energetics, and biomechanics (Denil 1909, 1937; Stringham 1924; McLeod and Nemenyi 1940; Bainbridge 1960; Weaver 1963, 1965; Brett 1964, 1967; Beamish 1978). Close collaboration between biologists and fishway engineers reached a zenith in the 1950s and 1960s at the Columbia River Fisheries Research Laboratory (CRFRL). The CRFRL enabled researchers to gather valuable information on hydraulics, swimming performance, and physiological limits to passage on a scale that permitted testing of full-scale prototype fishways with live, actively migrating salmon, (Collins 1952, 1962; Collins and Elling 1960; Collins et al. 1963; Weaver 1963; 1965). Many of today's most common fishway designs were developed or modified at this facility.

By the late 1960s, however, close collaborations between fishway engineers and biologists became less frequent. Bioengineers continued to grapple with improving fishway performance (Haro et al. 1998, 2004; Bunt et al. 1999; Larinier and Travade 1999, 2002), and work on fish swimming performance and biomechanics proliferated (Lighthill 1971, 1977; Webb 1971a, 1971b, 1975; Brett 1972, 1973, 1995; Brett and Glass 1973; Wardle 1975; Brett and Groves 1979; Alexander 1983; Vi-

* Corresponding author: tcastro_santos@usgs.gov

deler and Wardle 1991), but the two areas of research progressed independently, having lost the core of integration that had existed previously.

At the time, there was a sense that sufficient information had been gathered to provide adequate passage (Powers et al. 1985; Bell 1991). Salmon passed structures in large numbers, and although fishway engineers recognized that knowledge gaps remained, it was unclear how important these factors were (Powers et al. 1985; Orsborn and Powers 1986; Orsborn 1987); specific recommendations for how to gather the needed data and how to prioritize research questions were lacking. Further, both managers and engineers were primarily concerned with salmonid passage. This taxonomic bias had deep cultural and economic roots (Clay 1995). Although the need for passage of other taxa has long been acknowledged, there is a growing consensus that failure to provide passage for a broad range of both diadromous and riverine species poses a serious threat to both freshwater and terrestrial ecosystems (Meyer et al. 2007). Part of this recognition comes from a growing awareness of the mobility of riverine organisms, and part comes from a body of literature that suggests that fishways often perform poorly. This is especially true for nonsalmonid species (Slatick and Basham 1985; Bunt et al. 1999, 2000, 2001; Bunt 2001; Boggs et al. 2004; Cooke et al. 2005), but recent data suggest that even salmon may not pass as well as previously thought (Naughton et al. 2005).

Why some fishways work better than others and why some species perform better than others in fishways is poorly understood. Although some efforts have been made to identify conditions conducive to passage (Monk et al. 1989; Bunt et al. 1999, 2001; Haro et al. 1999; Bunt 2001), a coherent framework and rationale for evaluations is lacking. Below, we propose a framework for evaluating fishways and explore how such a framework might lead to better integration of biology and engineering toward improved fish passage technology.

A Framework for Fishway Evaluations

Comprehensive evaluations of fishways are only possible if their purpose and expected product are clearly defined. Powers et al. (1985) described the objective of a fishway as follows:

> If an artificial barrier to upstream migration is created ... an alternate passage route must be provided for the anadromous and/or resident fish. The alternate route should not add to the level of stress (or even total energy expenditure) compared with the level of energy expenditure experienced by the fish under pre-[barrier] conditions.

This description, though reasonable, does not readily lend itself to quantification. Also, it misses the main point: persistence of populations above and below barriers depends on protecting fitness (i.e., spawning success). The "predammed condition" is almost always impossible to document, and cumulative effects of the structure itself, compromised riverine and marine habitats, and interests of commercial and recreational fisheries may direct managers to attempt to provide superior passage than existed before human alteration. Although such management policies can be controversial, passage is sometimes even provided in locations that contain natural obstacles.

The Ideal Fishway Dichotomy

A more generalized approach is to define the ideal fishway. This is more a conceptual than a realistic goal, but identifying ideal characteristics can serve as a heuristic exercise for establishing a rationale for evaluating fishway performance. The objective of an ideal fishway, we propose, is to make the dammed reach transparent to the movement of native species, allowing unfettered access to free-flowing reaches above and below the obstacle (note that this may include passage through reservoirs and other features associated with the obstacle). If this objective is accepted, then the ideal fishway must have the following characteristics: (1) any individual of any native species wishing to move upstream or downstream must be able to enter the fishway without experiencing any delay; (2) entry is immediately followed by successful passage, with (3) no temporal or energetic costs and (4) no stress, disease, injury, predation, or other fitness-relevant costs associated with passage.

This describes the ideal condition; it is also useful to contemplate the worst-case scenario. From the biological perspective, this would be immediate extinction of the local population. This extinction could result in several ways, for example, (1) by denying access to essential habitat, (2) by capturing and removing all individuals, or (3) by efficiently passing all individuals upstream (in the case of anadromous migrants) but then failing to provide adequate downstream passage for adults and/or juveniles.

A similar continuum can be defined from an operational perspective. Here, the objective is to design, build, and operate a structure with a minimum of cost. An ideally designed structure from this perspective would cost nothing to construct; would require no maintenance; would use no water, power, or other resources to operate; and would be free of licensing restrictions. In the absence of performance criteria, the ideal operational solution can thus be described as doing nothing. The worst-case scenario is essentially indeterminate, however, because it is always possible to commit more resources to solving a particular problem, usually with diminishing marginal gains as costs increase.

There is clearly an underlying dichotomy between biological and operational goals. However, it is important to recognize that success in one goal does not necessarily imply failure in the other. Three hypothetical scenarios illustrate some potential bioengineered outcomes (Figure 1). Consider the "obligate philopatriot," species A: an example might be a salmon population excluded from its spawning habitat by a dam. In the absence of fish passage, this population will quickly go extinct. On the other hand, modest efforts at providing passage can yield substantial benefits, and a biologically ideal situation might be approached without excessive costs from the operational perspective.

A second example ("opportunistic invader," species B) also suggests that good passage success may be attainable at reasonable cost. It is worth noting here, though, that doing nothing does not result in local extinction, merely a suboptimal condition for that species. An example of this might be the sea lamprey *Petromyzon marinus*, which persists in altered habitat but ascends certain fishways designed for salmon with seeming ease. The lamprey example is also interesting given the experience of the Laurentian Great Lakes. Here, the species is invasive and massive efforts to exterminate it have

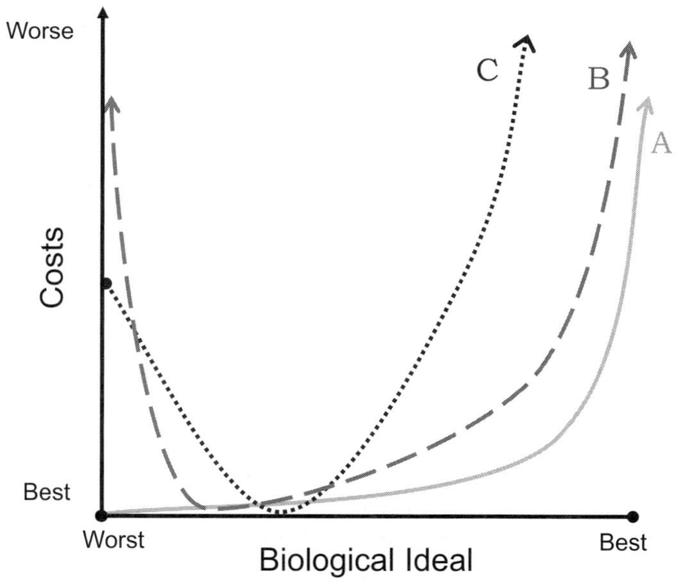

Figure 1.—A heuristic description of the intersection of operational ("Costs") and biological ideals. Three species are presented: species A is an obligate philopatriot (i.e., must return to specific spawning grounds to reproduce). Failure to provide fish passage quickly drives this population to extinction. Species B is an opportunistic invader. Passage provisions help this species, but populations persist without them, and extirpation may be difficult or impossible. Species C is a facultative philopatriot that benefits from fish passage but can also be harmed if passage provisions are not adequate.

failed. In this context, engineered solutions for extirpation, including construction of passage barriers, have proved unattainable, and emphasis has shifted to control (Madenjian et al. 2002; Sullivan et al. 2003).

A third example ("facultative philopatriot," species C) shows the dangers of unintended consequences; an example might be the American shad *Alosa sapidissima* in its native range. Although they often migrate long distances upriver, American shad are able to spawn in lower reaches. Thus, the widespread extirpations of Atlantic salmon *Salmo salar* that resulted from damming rivers in eastern North America merely restricted the available spawning habitat of American shad, and populations were maintained at stable but reduced levels. Provisions of upstream passage, however, have proved difficult (Foote 1976; Rideout et al. 1985; Slatick and Basham 1985; Monk et al. 1989; Larinier and Travade 2002; Sullivan 2004), and it appears that potential benefits of providing passage have not always been realized. Indeed, there is some concern that fishways, once constructed, can have negative effects on individuals and perhaps even populations (Foote 1976; Sullivan 2004). Although we know of no examples of extinctions of any species that have occurred as a result of mitigation measures, the potential for negative results is real and should be openly considered when evaluating performance of these structures.

Clearly, the goal of managers and fish passage engineers is to optimize designs with respect to both biological and operational ideals. Intrinsic incentives exist to identify and minimize factors associated with the costs of constructing and operating fishways. Optimization with respect to biological goals, however, is harder to characterize. Although fitness maximization is the desired outcome, it is unclear how this should be measured. The only biological parameter consistently included in fishway design is swimming performance. But is this an adequate predictor of passage success? How important is delay? What are the behavioral variables associated with passage success, and how might they be quantified? How do turbulence and other hydraulic complexities influence passage performance? How do energetics, stress, and disease associated with fish passage affect spawning success and subsequent survival? How do all of these translate into successful downstream migrations of juveniles and iteroparous adults? Hidden in these questions is the threat of uninformed compromise: how can a manager argue for more costly design criteria if s/he is unable to quantify benefits?

Rationale and Criteria for Fishway Evaluations

The proliferation of questions surrounding biological components of passage success quickly becomes overwhelming. This suggests that passage predictions from fundamental biological principles are likely to be imprecise. Instead, this complexity of potentially important and interacting biological factors argues for the importance of coordinated efforts to catalog performance of existing fishways. Further, there is a pressing need to characterize passage performance for a range of species and through a diversity of fishway types.

In the United States alone, there are thousands of engineered fishways; if we consider culverts as de facto fish passage structures, the number reaches into the millions. Despite the ubiquity of these structures, fishway evaluations are surprisingly rare and tend to focus on a small number of high-profile projects (Reback et al. 2004; Bernhardt et al. 2005; Palmer et al. 2005). This is a missed opportunity: collectively, these structures can be considered a functional experiment; the more information becomes available on how various taxa pass them, the easier it will become to identify and prioritize hypotheses as to what governs passage performance.

Information from evaluations is most useful if performance parameters are reported in consistent and biologically relevant terms. The more information provided the better. Thus, documentation of spawning fish upstream, or even window counts, indicate some level of success. This type of information can be misleading, however, and can mask significant problems if large numbers of individuals fail to enter, fail to pass, or otherwise incur fitness costs associated with the structure. We propose a suite of variables that might be quantified in fishway evaluations (Table 1). The list is not intended to be exhaustive, nor is it reasonable to expect that all identified variables will be quantified in each evaluation. Instead, it is intended as a framework from which managers can draw and assess how well fishways are being evaluated.

Note that many of the parameters identified

Table 1.—Biological indicators of fish passage success.

Biological variables	Units	Goal
Dependent variables		
Proportion guided to fishway entrance	P_g	Maximize P_g and dP_g/dt
Proportion entering fishway (P_e)	P_e	Maximize P_e and dP_e/dt
Proportion passing (P_p)		Maximize P_p per units
Height (H)	P_p/m	Height (dP_p/dH) and
Distance (D)	P_p/m	Distance (dP_p/dD)
Passage time ("delay," t)	h	Minimize t
Guidance	P_g/h	Maximize dP_g/dt
Entry	P_e/h	Maximize dP_e/dt
Passage	P_p/h	Maximize dP_p/dt
Failure (P_F)	P_F/h	Context-dependent[a]
Condition/health		Minimize fishway effects on condition
Energy content (E)	J/kg	Minimize ΔE
Change in E		
Per time (dE/dt)	J/kg/h	Minimize dE/Dt
Per height (dE/dH)	J/kg/m	Minimize dE/dH
Per distance (dE/dD)	J/kg/m	Minimize dE/dD
Stress indicators	[Cortisol] [Plasma ions] [Glucose] Sex steroids	Minimize fishway-induced stress
Injury	Wounds Scale loss	Minimize injuries
Proportion surviving	P_S	Maximize long-term survival, including spawning success of adults and migration success of juveniles (survival and postmigratory performance)
Covariates		
Species	Category	Identify species-specific differences; understand forcing mechanisms
Body length	m, cm, mm	Relate size to overall passage, hydraulic and other physical effects
Temperature (T)/water quality	°C, d°C, integral	Effect of temperature and water quality on passage performance and bioenergetics
Time (t)	h	Effects of residence time and seasonality on passage success, bioenergetics
Hydraulic, physical characteristics	See Table 2	Evaluate effects on all biological dependent variables

[a] Depending on importance of upstream habitat and risks of prolonged residence, it might be better to abandon fishway quickly (increase dP_F/dt) or retain fish to improve P_P (minimize dP_F/dt).

in Table 1 scale along biologically and structurally relevant dimensions. Thus proportion passing is useful information, but if it is presented as a continuous distribution with respect to fishway height and length, a much clearer understanding of where failure occurs emerges. Once the location of failure has been characterized, it becomes much easier to formulate hypotheses as to why.

Quantification along these axes will also allow for least-biased estimates of covariate effects on failure rates. Just as mortality rates act only on the surviving portion of a population, factors affecting failure rate in a fishway act on an ever-decreasing population of ascending fish (Castro-Santos and Haro 2006). The structural similarity between fish passage and mortality data means that well-established methods of survival analysis can be readily applied to fish passage studies. Successful ascenders constitute censored observations and will bias results unless explicitly accounted for. Survival analysis methods minimize this bias; they also allow failure rate to be modeled as either constant or varying with distance, enabling researchers to estimate both the magnitude and location of covariate effects. (Hosmer and Lemeshow 1999; Castro-Santos and Haro 2003; Haro et al. 2004; Castro-Santos 2005). Note that in these applications, the term "survival analysis" can be misleading, given that the dependent variables may not be directly linked to survival at all. Less ambiguous terms include "event-time analysis," "failure analysis," or perhaps most generally "analysis of censored data."

In addition to rates of passage and failure, some effort should be made to characterize delay incurred at the structure. The proportion of available fish entering a fishway changes over time; because the objective is to pass fish expeditiously through the desired route, rather than simply excluding them from less desirable routes, it is appropriate to report these data as probability functions that are continuous with respect to time, rather than multinomial responses (Castro-Santos and Haro 2003). Furthermore, fishway entry is actually a two-step process, each step having a discrete zone: (1) the guidance zone, where fish are guided to the fishway entrance; and (2) the entry zone, where fish are able to detect the structure and must choose to enter it. Performance within these two zones is best measured as time-based functions (i.e., guidance rate, the proportion of available fish arriving at the detection zone per unit time, and entry rate, the proportion entering per unit time. Once again, these data are prone to censoring, with individuals leaving the respective zones, entering alternate routes, and so forth. Having once entered, additional delays are incurred as fish ascend or descend the fishway, whether or not they ultimately pass the structure.

The amount of delay incurred at fishways is likely to be associated with fitness costs. These include energetic costs of passage, rates of injury, exposure to disease, and so forth. These factors, along with their consequent effects on survival and recruitment, remain poorly understood. We propose some variables to measure in Table 1 but encourage researchers to further explore the relationship between postpassage condition and spawning success (for adults) and successful seawater entry or adult return rates (for juveniles). For iteroparous species, postspawning survival is another feature that can have serious implications for population viability (Charnov and Schaffer 1973; Glebe and Leggett 1981). Migratory delay, both up-and downstream, holds considerable potential to affect this important life history stage (see also McCormick et al. 2009, this volume), and the cumulative effects of multiple barriers should be considered wherever they occur.

Many of the issues laid out in the preceding paragraphs have long been recognized (British Institution of Civil Engineers 1942; Powers et al. 1985; Orsborn 1987). Recent advances in animal tracking technology and analytical techniques, however, have now made it possible to address these questions directly. Passive integrated transponder telemetry promises to supplant visible external and coded wire tags for many mark–recapture experiments (Prentice et al. 1990; Skalski et al. 1998; Axel et al. 2005) and has also been used to provide fine-scale spatiotemporal data on behaviors associated with passage performance (Castro-Santos et al. 1996; Lucas et al. 1999; Sullivan 2004). Both radio and acoustic telemetry technology are being used to collect movement data over scales ranging from submeter to intercontinental. Telemetry of physiological and biomechanical processes (e.g., McKinley and Power 1992; Hinch et al. 1996; Lowe 1996; Lowe et al. 1998) also holds the potential to improve our understanding of fishway energetics. More work is needed in this area, though, especially with regard to contributions of anaerobic metabolism during passage through zones of high velocity.

In some cases, these technological advances have outstripped the statistical methodology needed to fully exploit them. There is an ongoing need for new and adapted techniques for gleaning as much information as possible from the data these technologies provide (e.g., Turchin 1998; Skalski et al. 2001; Castro-Santos and Haro 2003; Johnson et al. 2004; Castro-Santos 2006). This is particularly important with respect to guiding fish toward desirable passage routes and away from undesirable ones. Simply excluding fish from hazardous routes may incur added delays that offset the benefits of safer passage. Telemetry and hydroacoustic systems can provide a wealth of fine-scale data on three-dimensional movements, but effective methods for using these data to quantify attraction have yet to be demonstrated.

In addition to biological indicators of passage performance, concurrent reporting of engineering and hydraulic design specifications of the fishway under evaluation is also important. Here again, because we have limited information on hydraulic correlates of passage success, it makes sense to seek the most detailed descriptions possible. Current standard criteria for fish passage include maximum and mean flow velocity, total kinetic energy (or the related energy dissipation factor [EDF]), and various dimensions describing pools, chutes, weirs, and so forth. These, however, have proven to be poor predictors of passage performance, and more detail is needed. Recent work shows that eddy structure, vorticity, and circulation all affect fish behavior and may be important determinants of fishway performance (Crowder and Diplas 2002; Dabiri 2005; Tritico et al. 2007). We propose a suite of structural and hydraulic variables to quantify in Table 2. These criteria can ultimately be used as covariates to help explain biological performance measures.

Here again, technological advances hold great potential for improving our understanding of hydraulics, both in engineered fishways and in natural rivers. In particular, flow visualization using digital particle image velocimetry and computational fluid dynamics modeling can provide much more detailed descriptions of fishway hydraulics than were previously available (e.g., Khan 2006; Cea et al. 2007). These same technologies have also allowed significant advances in fish locomotion theory (Fish and Lauder 2006; Schultz and Webb 2002), with the intriguing possibility that hydraulic complexity and biomechanics could be linked to produce improved fishway designs.

Below, we pose several questions, based on current and developing knowledge, to serve as a starting point for future fish passage research. As more data become available, it will be possible, and indeed necessary, to refine these questions. In this way, understanding of bioengineering factors underlying fishway performance can progress. Ultimately, this should lead toward development of generalized predictive models to help direct management choices for site-specific fishway design. Developing questions may still function on the level of prototype fishways (e.g., distance between resting pools as predictors of passage performance). However, as hydraulic and biological details of fish passage emerge, the moment-to-moment experience of ascending fish will become more apparent. As these experiences become linked to performance measures, more basic biological questions can be prioritized and tested at the organismal level.

Some Questions Suggested by the Current Literature

Question 1: How Does Turbulence Structure Influence Swimming Performance?

Technical fishways are structures engineered using hydraulic principles and (sometimes) swimming performance estimates—generally, these are built of standard construction materials and make no effort to mimic a natural river. These fishways typically use turbulence to dissipate head, with design prescriptions including maximum EDF values. Turbulence may be defined in various ways, and it is likely that these affect conclusions regarding turbulence effects on fish swimming performance or other biologically relevant parameters. Most studies find turbulence increases costs of swimming performance, irrespective of how it is defined or induced in experimental systems (Powers et al. 1985; Orsborn 1987). Recent work by Enders et al. (2003, 2005) also seems to support the practice of limiting turbulence in fishways. On the other hand, Nikora et al. (2003) found negligible effects of turbulence created by wavy walls on swimming ability of inanga *Galaxius maculatus*. Other studies show that turbulence may be beneficial, certainly promoting resting opportunities and probably facilitating translocation through fishways. Thus Liao et al. (2003a, 2003b; Liao 2004) found that rainbow trout *Oncorhynchus mykiss* swimming in the presence of a controlled von Karman vortex street appeared to surf on the vortices, experiencing substantial reductions in activity of swimming muscles and presumably saving energy in the process. Webb (1998a) also described reduced swimming activity by river chub *Nocomis micropogon* and smallmouth bass *Micropterus dolo-*

Table 2.—Engineering and hydraulic descriptors of fishways. These variables should be considered when evaluating design effects on passage performance.

Variable	Units
Design	
Design type	Category
Dimensions	
Chute	m, m^2, m^3
Pool	m, m^2, m^3
Weir	m, m^2
Total length	m
Number of pools by pool type	Integer
Slope	m rise: m run or degrees from horizontal
Resting/turn pools	m, m^2, m^3
Dimensions	m, m^2, m^3
Elevation	m
Distance	m
Hydraulics	
Reynolds number	–
Froude number	–
Energy dissipation factor	J/m^3
Total kinetic energy	J/m^3
Flow velocity	
Minimum	m/s
Maximum	m/s
Mean	m/s
Turbulence	
SD of velocity ("RMS"; three-dimensional)	m/s
Vortices	
Size	m
Shedding frequency[a]	Hz
Rotation period	s
Vorticity	rad/s
Impulse: linear	kg m/s
Impulse: angular	kg m^2/s
Force or thrust	kg m/s^2
Circulation	m^2/s

[a] Shedding frequency can also be presented as period (seconds elapsed between vortices) or Strouhal number.

mieu swimming behind cylinders, and similar use of refuges is well known among stream species.

Evidently turbulence is not intrinsically costly. This fact raises the tantalizing possibility that turbulence structure might be controlled in such a way that hydraulic head is dissipated without reducing passage performance. Indeed, it may be possible to enhance passage through judicious control of turbulence structure. Before such applications could be realized, though, much more detail is needed to understand the correlations between turbulence structure and swimming performance, from mechanical, energetic, and behavioral perspectives.

One approach that might elucidate interactions of kinematics and energetics with respect to swimming in turbulent flow is to explore turbulence structure in natural environments (Crowder and Diplas 2002; Cotel et al. 2006; Tritico et al. 2007). Using developing technology like the portable small-scale Particle Image Velocimetry system (Tritico et al. 2007), researchers can obtain instantaneous measurements of velocity and vorticity at high spatial resolution. The detailed information on flow structure that this provides should help answer questions arising from the work described above: What ranges of size, period, and vorticity of eddies occur in the natural environment? How readily do fish pass flows with these various characteristics? How do fishways compare with natural rivers in this respect? Are there limits to design parameters for technical fishways? Depending on the answers, these questions could provide feedback for prioritization of design schemes.

Question 2: How Does Morphology Affect Fish Passage?

A common concern regarding fish passage is that fishways are often size-selective (Powers et al. 1985). This does not necessarily reflect swimming ability (Bunt et al. 2000), especially when it is the smaller individuals that pass (Sullivan 2004). Perhaps hydraulic structure somehow interacts with the morphology of the fish to produce selective passage. Several morphological characters besides body length affect swimming functions and performance and hence could be implicated. These include body depth, general shape (gibbose versus fusiform versus elongate), number and structure of fins, and so forth. Morphological form may vary to some extent within species, and variation is large among species. Aspects of such interspecific variation also should include consideration of musculoskeletal systems (Johnston 2001), sensory apparatus (Coombs and Montgomery 1999), and life history (Northcote et al. 1970). Advances in this area could result from comparative data on passage performance, but also from smaller scale information relating swimming capacity in controlled environments to controlled hydraulic conditions.

Question 3: Does Sensory Physiology Affect Passage Performance?

The lateral line system comprises an important sensory modality with which fish detect and process certain hydraulic stimuli. Lateral line morphology and associated neural processing vary with life history and habitat type, with fish inhabiting lentic and lotic environments having structures that are relatively more and less sensitive to high-frequency disturbances, respectively (Coombs et al. 1988; Mogdans and Bleckmann 2001; Engelmann et al. 2003). Could this factor be implicated in passage performance? Data from the Connecticut River (Sullivan 2004; T. Castro-Santos, personal observation) show a stark contrast between passage performance of Atlantic salmon *Salmo salar* and American shad *Alosa sapidissima*. Both species are strong swimmers (Beamish 1978; Peake et al. 1997; Castro-Santos 2005), but fishways and turbulent zones that salmon traverse with ease pose barriers to shad movements (Sullivan 2004; Castro-Santos and Haro 2005). Moreover, many shad do not abandon the fishways, but hold, sometimes for weeks, in the relatively quiescent waters of resting pools, where they gradually sicken and die. Why?

Engelmann et al. (2003) found that goldfish *Carassius auratus* were much more sensitive than trout to stimuli in moving water. Like the goldfish, clupeids possess a lateralis structure that appears to be designed to be sensitive to absolute velocity and small perturbations in flow. Could this heightened sensitivity explain the avoidance behaviors these fish appear to exhibit in the presence of turbulent flow? Perhaps the hydraulic environments common to salmon fishways are too "noisy" for American shad to pass. Larinier and Travade (2002) suggest that fishways of even moderate height seem to be problematic for shad. Given their swimming ability, it is unlikely that this is a result of fatigue. Could it be a stress response that triggers these fish to abandon their migrations?

Question 4: How Does Behavior Affect Fish Passage?

Connections between sensory physiology and behavior are also important in influencing entry rates. Flow velocity, acceleration, turbulence, and discharge, as well as temperature, pH, and CO_2 concentrations, have all been shown to influence rates of entry by fish into open-channel flumes (Collins 1952; Weaver 1963; Haro et al. 1998; Castro-Santos 2002, 2004). How and why these behaviors vary among species remains nearly unexplored, even though optimizing conditions for entry into fishways remains one of the most important and challenging problems of fish passage (Bunt 2001).

Other aspects of behavior may influence passage performance as well. In one fishway, passage rates of alewife *Alosa pseudoharengus* declined as density increased (Dominy 1973). How do social interactions like schooling behavior affect passage performance? What role might vision or other sensory modalities play in this? Similarly, the presence of conspecifics upstream has been shown to improve attraction and passage of sea lamprey (Sorensen and Vrieze 2003). Other chemical cues also influence homing ability and migratory behavior (Hasler 1971; Carruth et al. 2002; Sorensen and Stacey 2004), suggesting the potential exists to influence passage performance, especially among depleted populations.

The well-known biotic factors competition and predation might also influence passage performance. Fish passageways tend to concentrate fishes at greater densities than in adjacent natural habitats. In addition, preferred tracks through a fishway may be limited, such that fish compete for them. Predation both above and below dams remains a major concern surrounding fish passage and is one reason that passage delays should be minimized (Venditti et al. 2000; Agostinho et al. 2002; Laake et al. 2002; Aarestrup and Koed 2003). If fishway entrances act efficiently to attract target species, they might also prove attractive to predators, which could then intercept the fish or otherwise exclude them from the fishway. We know of no studies that evaluate this question, but the potential is troubling and worthy of examination.

One obvious way in which behavior can influence swimming performance is through volitional gait selection and swim speed optimization. Existing designs tend to be overly conservative with respect to flow velocity, in part because methods typically used to quantify swimming capacity are themselves intrinsically conservative. Most studies of fish locomotion have used enclosed chambers or respirometers (Brett 1964). By restricting the movement of study animals, these devices allow researchers to closely monitor and control flow velocity and minimize turbulence while offering a fixed frame of reference for the researcher. Such enclosed spaces reduce swimming performance, however, and fish that are allowed to swim volitionally in open-channel flow exhibit gait shifts and related behaviors that substantially expand their ability to traverse velocity barriers (Peake 2004; Peake and Farrell 2004, 2005; Tudorache et al. 2007). Furthermore, the speeds at which unrestrained fish actually swim and the ability to stage multiple attempts can dramatically affect distance of ascent, depending on how closely they approximate the distance-maximizing optimum (Castro-Santos 2004, 2005, 2006). Data from fish swimming in open-channel flow cast doubt on the one biological performance component that is consistently incorporated into fishway design. Characterizing the magnitude of error associated with the wealth of literature describing swimming capacity of constrained fish would be very helpful and could yield considerable benefits in improved design efficiencies.

Question 5: Do Fishways Affect Disease Transmission?

Because diadromous species are often under stress when they enter freshwater, they may be more susceptible to disease or other factors that reduce fitness. The example cited above (Dominy 1973), where alewife pass more slowly at higher densities, suggests that there may be greater risk of disease transmission under these conditions. Also, migrating fish often suffer injuries and scale loss, either within or outside the fishway. This fact, along with elevated densities and increased stress, will likely act to increase rates of disease transmission. Although the potential consequences seem significant (Foote 1976; Powers et al. 1985; Tierney and Farrell 2004) and extensive studies exist with respect to rates of injury associated with downstream passage (e.g., Reischel and Bjornn 2003; Boggs et al. 2004; Cada et al. 2006), we know of no studies examining fishway-induced injury and disease transmission.

Question 6: What Are the Energetic Consequences of Fish Passage?

One of the more exciting technologies to arise in recent years is in the field of physiological telemetry. Using radio tags that transmit rectified and integrated electromyograms (EMG tags), researchers have been able to identify zones of difficult passage within rivers and even to project the consequences of the energetic costs incurred there (Hinch et al. 1996; Hinch and Rand 1998; Rand and Hinch 1998; Hinch and Bratty 2000). These data have far-reaching implications, including the likely consequences of climate change (Rand et al. 2006). Although some concern remains regarding the reliability of the data this technology provides (Geist et al. 2002), its potential for fishway research is obvious.

Concerns over the energetic costs of fish passage and migration motivated several seminal studies (e.g.,

Idler and Clemens 1959; Brett 1962). These concerns are particularly acute for those species that do not feed during their migrations: a primary measure of a successful migration is that the animals arrive at their habitat with sufficient energy reserves to spawn successfully and, in the case of iteroparous migrants, to return to their feeding habitat. Indirect observations of fishway energetics have been performed by collecting fish above and below structures (e.g., Leonard and McCormick 1999), but direct observations are lacking. With improved data on activity levels of fish ascending fishways, it will become possible to measure their effect on energy depletion.

Energetics, while affecting fitness, is itself affected by environmental and behavioral factors. As mentioned above, turbulent flow affects swimming energetics, but mechanisms and magnitude of this effect are still ambiguous. Delay or residence time clearly can be expected to have an effect, but this will depend on where the fish spend most of their time. If fish exhibit anaerobically fuelled burst swimming behavior, this will have much higher costs than if they are able to swim at aerobically powered speeds (Boisclair and Tang 1993). It is worth noting that the EMG tags mentioned above do not accurately transmit data from anaerobic swimming behaviors, a limitation that may be overcome using alternate technologies (e.g., Lowe et al. 1998). Likewise, temperature will influence swimming energetics. This has implications both for climate change and for more immediate anthropogenic thermal impacts such as power plant effluents (Mustard et al. 1999). Energetics may also be associated with disease risk and reproductive physiology.

Question 7: How Do Hormonal Levels Influence Fish Passage?

Hormonal ontogeny is closely linked to migratory behavior throughout the Animalia (Dingle 1996). Among fish, this effect is particularly well documented for out-migrating juvenile salmonids (Hoar 1988; McCormick 1994). Although hormonal correlates of adult salmonid migrations have also been documented (Ueda et al. 1984; Leonard et al. 2001), we are aware of only one investigation that linked changes in circulating levels of hormones to upstream migratory behavior (Sato et al. 1997).

How and whether the physical environment within and around fishways influences hormonal levels of migratory fish is an open question. Stress in fish is known to affect the timing of reproduction, behavior during spawning, and the survival of offspring (Schreck et al. 2001). Some data exist to show that stress levels increase and sex steroids decrease in association with fishway residence and that this varies by fishway type (D. T. Lerner and S. D. McCormick, U.S. Geological Survey, S.O. Conte Anadromous Fish Research Center, personal communication)

Given growing concerns regarding endocrine disrupting compounds in the environment (McCormick 2009), the potential influence of these compounds on migratory motivation and performance is worthy of examination.

Despite the scope of topics covered in the preceding questions, this list is neither exhaustive nor comprehensive. As stated above, improved information on performance of existing fishways will at once expand and focus the list; and this result will be best attained if evaluations are conducted in ways that provide the greatest possible scope and detail on biologically relevant performance measures. Such data will improve measures of biological effectiveness (or lack thereof). This in turn will allow for better cost–benefit analysis and ultimately optimized bioengineered fishway designs.

It is also worth noting that the preceding sections have a pronounced bias toward improving upstream passage. Many of the same principles apply to downstream passage, though, and we in no way wish to diminish the importance of this issue. To the contrary, the concept of transparency requires that passage be facilitated in both directions, and improved downstream passage technology remains one of the most pressing needs in diadromous fishery management. Often, solutions for up- and downstream passage are considered separately. This is a practical matter that has to do with different flow requirements for up- and downstream fishways. In the same way, it is likely that improved passage for a range of species will only be achieved by providing more than one type of passage structure at a given location.

The Qualified Promise of Nature-Like Fishways

Inevitably, fish passage structures are designed on a site-specific basis, and engineers must rely on the best available knowledge in order to adapt the structures to local conditions. Insufficient information on behavior can lead to a dilemma: should the design be conservative and maintain low flow velocities? If so, there is a risk that the conditions will not stimulate the fish to move through. Conservative designs are often recommended by management agencies, despite the possibility of a reduced cost–benefit ratio. This trend troubled Orsborn (1987), and solutions remain elusive.

One approach to the one-size-fits-all problem of fishway technology is to design fishways that provide diverse hydraulic conditions. This is the rationale most often proposed for constructing so-called nature-like fishways. The theory underlying these designs is that if the fishway can adequately mimic its surrounding environment, including both dimensional diversity of substrate and spatially diverse hydraulic characteristics (called "stream simulation" in the culvert literature; Bates et al. 2003), then all native fauna (including invertebrates) should be able to pass it or even occupy it as habitat. This notion is intuitively appealing, and nature-like fishways are being widely constructed in Europe, with increasing support elsewhere (Jungwirth 1996; Eberstaller et al. 1998; Santos et al. 2005).

As with technical fishways, the enthusiasm to construct nature-like fishways has preceded and exceeded the willingness or ability to evaluate them (Elvira et al. 1998; FAO 2002). Available data indicate that they do sometimes meet their objectives of passing a diversity of species (Eberstaller et al. 1998; Santos et al. 2005). This success is not universal, however, even for salmonids (Aarestrup et al. 2003), and passage for some species has proved ineffective (Schmutz et al. 1998).

A further criticism of the nature-like approach is that these fishways tend to have very low slopes and so require extensive resources in land and materials to construct. They also can require a great deal of flow to operate (after all, it is difficult to simulate a natural river without passing a significant amount of the total discharge through the structure). Moreover, even by attempting to simulate the natural features of rivers questions remain as to the requirements of various fish species; the fact that it might look natural to a human does not eliminate this concern. Nevertheless, the nature-like approach explicitly recognizes the lack of supporting knowledge for effective technical fishway design and so holds real promise in the absence of reliable technical solutions. Perhaps the added costs associated with these structures will create an economic incentive to gather the information needed for more economical technical solutions (e.g., Stuart and Berghuis 2002).

Surrogate Species and the Big Picture

Perhaps the single greatest challenge in fish passage technology is the development of structures and design concepts that will pass a broad range of species. Of particular concern is the development of hydropower in regions with diverse and poorly understood fish stocks. Tremendous hydropower resources exist in countries like Brazil and China (Zhong and Power 1996; Anderson et al. 2009, this volume). Often, these same systems provide habitat for economically, sociologically, and ecologically important species, and planned and ongoing development may obstruct vital migratory corridors (Agostinho et al. 2002, 2005; Fernandes et al. 2004).

Biological data relevant to passage performance are insufficient even among the comparatively sparse native fauna in developed countries; this problem is even more acute in those regions currently undergoing the most rapid economic development. Because it is unrealistic to expect that all of the necessary data be acquired for the full diversity of taxa, managers are faced with the difficult task of identifying surrogate species that can be expected to be representative of local fauna. Although broad generalizations have been proposed linking body form to function and behavior (e.g., Webb 1998b; Webb and Gerstner 2000), this link is largely lacking with respect to fish passage. Even more troubling is the extent of intraspecific diversity of swimming performance; differences exist among individuals and among populations, leading some to question the reliability of even species-specific models (Northcote et al. 1970; Taylor and Mcphail 1986; Castro-Santos 2006; Castro-Santos and Haro 2006). Certainly, it is unrealistic to expect existing salmonid-based models to be of much value on a global scale (Santos et al. 2007).

Identification of surrogate species or trait characteristics associated with passage would be a major step toward improving fish passage worldwide.

In this paper, we have attempted to identify the biological variables most likely to lead to this goal. Further work on relations between hydraulics and morphology, sensory biology, life history, behavior, disease, energetics, and endocrinology can all help in the development of improved fishway technology. The most pressing question is how to prioritize these areas of research. By starting from the natural experiments that comprise existing structures while focusing on both biological and design objectives, practical solutions are likely to emerge.

References

Aarestrup, K., and A. Koed. 2003. Survival of migrating sea trout (*Salmo trutta*) and Atlantic salmon (*Salmo salar*) smolts negotiating weirs in small Danish rivers. Ecology of Freshwater Fish 12:169–176.

Aarestrup, K., M. C. Lucas, and J. A. Hansen. 2003. Efficiency of a nature-like bypass channel for sea trout (*Salmo trutta*) ascending a small Danish stream studied by PIT telemetry. Ecology of Freshwater Fish 12:160–168.

Agostinho, A. A., L. C. Gomes, D. R. Fernandez, and H. I. Suzuki. 2002. Efficiency of fish ladders for neotropical ichthyofauna. River Research & Applications 18:299–306.

Agostinho, A. A., S. M. Thomaz, and L. C. Gomes. 2005. Conservation of the biodiversity of Brazil's inland waters. Conservation Biology 19:646–652.

Alexander, R. M. 1983. The history of fish mechanics. Pages 1–35 *in* P. W. Webb and D. Weihs, editors. Fish biomechanics. Praeger, New York.

Anderson, E., M. Montoya, A. Soto, H. Flores, and M. McClain. 2009. Challenges and opportunities for co-management of a migratory fish (*Prochilodus nigricans*) in the Peruvian Amazon. Pages 741–756 *in* A. J. Haro, K. L. Smith, R. A. Rulifson, C. M. Moffitt, R. J. Klauda, M. J. Dadswell, R. A. Cunjak, J. E. Cooper, K. L. Beal, and T. S. Avery, editors. Challenges for diadromous fishes in a dynamic global environment. American Fisheries Society, Symposium 69, Bethesda, Maryland.

Axel, G. A., E. F. Prentice, and B. P. Sandford. 2005. PIT-tag detection system for large-diameter juvenile fish bypass pipes at Columbia River basin hydroelectric dams. North American Journal of Fisheries Management 25:646–651.

Bainbridge, R. 1960. Speed and stamina in three fish. Journal of Experimental Biology 37:129–153.

Bates, K., B. Barnard, B. Heiner, J. P. Klavas, and P. D. Powers. 2003. Design of road culverts for fish passage. Washington Department of Fish and Widlife, Olympia.

Beamish, F. W. H. 1978. Swimming capacity. Pages 101–187 *in* W. S. Hoar and D. J. Randall, editors. Fish physiology, volume VII. Locomotion. Academic Press, London.

Bell, M. C. 1991. Fisheries handbook of engineering requirements and biological criteria. U.S. Army Corps of Engineers, Portland, Oregon.

Bernhardt, E. S., M. A. Palmer, J. D. Allan, G. Alexander, K. Barnas, S. Brooks, J. Carr, S. Clayton, C. Dahm, J. Follstad-Shah, D. Galat, S. Gloss, P. Goodwin, D. Hart, B. Hassett, R. Jenkinson, S. Katz, G. M. Kondolf, P. S. Lake, R. Lave, J. L. Meyer, T. K. O'Donnell, L. Pagano, B. Powell, and E. Sudduth. 2005. Synthesizing US river restoration efforts. Science 308:636–637.

Boggs, C. T., M. L. Keefer, C. A. Peery, T. C. Bjornn, and L. C. Stuehrenberg. 2004. Fallback, reascension, and adjusted fishway escapement estimates for adult Chinook salmon and steelhead at Columbia and Snake River dams. Transactions of the American Fisheries Society 133:932–949.

Boisclair, D., and M. Tang. 1993. Empirical-analysis of the influence of swimming pattern on the net energetic cost of swimming in fishes. Journal of Fish Biology 42:169–183.

Brett, J. R. 1962. Some considerations in the study of respiratory metabolism in fish, particularly salmon. Journal of the Fisheries Research Board of Canada 19:1025–1038.

Brett, J. R. 1964. The respiratory metabolism and swimming performance of young sockeye salmon. Journal of the Fisheries Research Board of Canada 21:1183–1226.

Brett, J. R. 1967. Swimming performance of sockeye salmon in relation to fatigue time and temperature. Journal of the Fisheries Research Board of Canada 24:1731–1741.

Brett, J. R. 1972. Metabolic demand for oxygen in fish, particularly salmonids, and a comparison with other vertebrates. Respiration Physiology 14:151–170.

Brett, J. R. 1973. Energy expenditure of sockeye salmon (*Oncorhynchus nerka*) during sustained performance. Journal of the Fisheries Research Board of Canada 30:1799–1809.

Brett, J. R. 1995. Energetics. Pages 3–68 *in* C. Groot, L. Margolis, and W. C. Clarke, editors. Physiological ecology of Pacific salmon. UBC Press, Vancouver.

Brett, J. R., and N. R. Glass. 1973. Metabolic rates and critical swimming speeds of sockeye salmon (*Oncorhynchus nerka*) in relation to size and temperature. Journal of the Fisheries Research Board of Canada 30:379–387.

Brett, J. R., and T. D. D. Groves. 1979. Physiological energetics. Pages 279–352 *in* W. S. Hoar, D. J. Randall, and J. R. Brett, editors. Fish physiology. Volume 8: bioenergetics and growth. Academic Press, New York.

British Institution of Civil Engineers 1942. Report of the committee on fish-passes. William Clowes and Sons, London (UK) (Reprinted 1948).

Bunt, C. M. 2001. Fishway entrance modifications enhance fish attraction. Fisheries Management and Ecology 8:95–105.

Bunt, C. M., S. J. Cooke, and R. S. McKinley. 2000. Assessment of the Dunnville fishway for passage of walleyes from Lake Erie to the Grand River, Ontario. Journal of Great Lakes Research 26:482–488.

Bunt, C. M., C. Katopodis, and R. S. McKinley. 1999. Attraction and passage efficiency of white suckers and smallmouth bass by two Denil fishways. North American Journal of Fisheries Management 19:793–803.

Bunt, C. M., B. T. van Poorten, and L. Wong. 2001. Denil fishway utilization patterns and passage of several warmwater species relative to seasonal, thermal and hydraulic dynamics. Ecology of Freshwater Fish 10:212–219.

Cada, G., J. Loar, L. Garrison, R. Fisher, and D. Neitzel. 2006. Efforts to reduce mortality to hydroelectric turbine-passed fish: locating and quantifying damaging shear stresses. Environmental Management 37:898–906.

Carruth, L. L., R. E. Jones, and D. O. Norris. 2002. Cortisol and pacific salmon: a new look at the role of stress hormones in olfaction and home-stream migration. Integrative & Comparative Biology 42:574–581.

Castro-Santos, T. 2002. Swimming performance of upstream migrant fishes: new methods, new perspectives. Doctoral dissertation. University of Massachusetts, Amherst.

Castro-Santos, T. 2004. Quantifying the combined effects of attempt rate and swimming capacity on passage through velocity barriers. Canadian Journal of Fisheries and Aquatic Sciences 61:1602–1615.

Castro-Santos, T. 2005. Optimal swim speeds for traversing velocity barriers: an analysis of volitional high-speed swimming behavior of migratory fishes. Journal of Experimental Biology 208:421–432.

Castro-Santos, T. 2006. Modeling the effect of varying swim speeds on fish passage through velocity barriers. Transactions of the American Fisheries Society 135:1230–1237.

Castro-Santos, T., and A. Haro. 2003. Quantifying migratory delay: a new application of survival analysis methods. Canadian Journal of Fisheries and Aquatic Sciences 60:986–996.

Castro-Santos, T., and A. Haro. 2005. Turners Falls fish passage studies 2005: results from PIT and radio telemetry studies. U.S. Geological Survey, Biological Resource Discipline, S.O. Conte Anadromous Fish Research Center, Internal Report # 2005-04, Turner Falls, Massachusetts.

Castro-Santos, T., and A. Haro. 2006. Biomechanics and fisheries conservation. Pages 469–523 *in* R. E. Shadwick and G. V. Lauder, editors. Fish physiology. Volume 23: Fish biomechanics. Academic Press, New York.

Castro-Santos, T., A. Haro, and S. Walk. 1996. A passive integrated transponder (PIT) tagging system for monitoring fishways. Fisheries Research 28:253–261.

Cea, L., L. Pena, J. Puertas, M. E. Vazquez-Cendon, and E. Pena. 2007. Application of several depth-averaged turbulence models to simulate flow in vertical slot fishways. Journal of Hydraulic Engineering 133:160–172.

Charnov, E. L., and W. M. Schaffer. 1973. Life-history consequences of natural selection: Cole's result revisited. American Naturalist 107:791–793.

Clay, C. H. 1995. Design of fishways and other fish facilities. Lewis Publishers, Boca Raton, Florida.

Collins, G. B. 1952. Factors influencing the orientation of migrating anadromous fishes. Fisheries Bulletin 52:375–396.

Collins, G. B. 1962. Ability of salmonids to ascend high fishways. Transactions of the American Fisheries Society 91:1–7.

Collins, G. B., and C. H. Elling. 1960. Fishway research at the Fisheries-Engineering Research Laboratory. U.S. Fish and Wildlife Service Circular 98.

Collins, G. B., C. H. Elling, J. R. Gauley, and C. S. Thompson. 1963. Effect of fishway slope on performance and biochemistry of salmonids. Fisheries Bulletin 63:221–253.

Cooke, S. J., C. M. Bunt, S. J. Hamilton, C. A. Jennings, M. P. Pearson, M. S. Cooperman, and D. F. Markle. 2005. Threats, conservation strategies, and prognosis for suckers (Catostomidae) in North America: insights from regional case studies of a diverse family of non-game fishes. Biological Conservation 121:317–331.

Coombs, S., J. Janssen, and J. F. Webb. 1988. Diversity of lateral line systems: evolutionary and functional considerations. Pages 553–593 *in* J. Atema, R. R. Fay, A. N. Popper, and W. N. Tovolga, editors. Sensory biology of aquatic animals. Springer-Verlag, New York.

Coombs, S., and J. C. Montgomery. 1999. The enigmatic lateral line system. Pages 319–361 *in* R. R. Fay and A. N. Popper, editors. Comparative hearing: fish and amphibians. Springer, New York.

Cotel, A. J., P. W. Webb, and H. Tritico. 2006. Do brown trout choose locations with reduced turbulence? Transactions of the American Fisheries Society 135:610–619.

Crowder, D. W., and P. Diplas. 2002. Vorticity and circulation: spatial metrics for evaluating flow complexity in stream habitats. Canadian Journal of Fisheries and Aquatic Sciences 59:633–645.

Dabiri, J. O. 2005. On the estimation of swimming and flying forces from wake measurements. Journal of Experimental Biology 208:3519–3532.

Denil, G. 1909. Les eschelles a poissons et leru application aux barrages des Meuse et d'Ourthe. Annales des Travaux Publics de Belgique 2:1–152.

Denil, G. 1937. La mecanique du poisson de riviere: les capacites mecaniques de la truite et du saumon. Annales des Travaux Publics de Belgique 38:412–423.

Dingle, H. 1996. Migration: the biology of life on the move. Oxford University Press, New York.

Dominy, C. L. 1973. Effect of entrance-pool weir elevation and fish density on passage of alewives (*Alosa pseudoharengus*) in a pool and weir fishway. Transactions of the American Fisheries Society 102:398–404.

Eberstaller, J., M. Hinterhofer, and P. Parasiewicz. 1998. The effectiveness of two nature-like bypass channels in an upland Austrian river. Pages 363–383 *in* M. Jungwirth, S. Schmutz, and S. E. Weiss, editors. Fish migration and fish bypasses. Fishing News Books, Oxford, UK.

Elvira, B., G. G. Nicola, and A. Almodovar. 1998. A catalogue of fish passes at dams in Spain. Pages 203–207 *in* M. Jungwith, S. Schmutz, and S. Weiss, editors. Fish migration and fish bypasses. Fishing News Books, Oxford, UK.

Enders, E. C., Boisclair, D., and Roy, A.G. 2003. The effect of turbulence on the cost of swimming for juvenile Atlantic salmon (*Salmo salar*). Canadian Journal of Fisheries and Aquatic Sciences 60:1149–1160.

Enders, E. C., D. Boisclair, and A. G. Roy. 2005. A model of total swimming costs in turbulent flow for juvenile Atlantic salmon (*Salmo salar*). Canadian Journal of Fisheries and Aquatic Sciences 62:1079–1089.

Engelmann, J., S. Krother, H. Bleckmann, and J. Mogdans. 2003. Effects of running water on lateral line responses to moving objects. Brain, Behavior & Evolution 61:195–212.

FAO (Food and Agriculture Organization of the United Nations). 2002. Fish passes. Food and Agriculture Organization of the United Nations, Rome.

Fernandes, C. C., J. Podos, and J. G. Lundberg. 2004. Amazonian ecology: tributaries enhance the diversity of electric fishes. Science 305:1960–1962.

Fish, F. E., and G. V. Lauder. 2006. Passive and active flow control by swimming fishes and mammals. Annual Review of Fluid Mechanics 38:193–224.

Foote, P. S. 1976. Blood lactic acid levels and mortality of American shad (*Alosa sapidissima*) utilizing the Holyoke dam fishlift Massachusetts, 1974 and 1975. Master's thesis. University of Massachusetts, Amherst.

Geist, D. R., R. S. Brown, K. Lepla, and J. Chandler. 2002. Practical application of electromyogram radiotelemetry: the suitability of applying laboratory-acquired calibration data to field data. North American Journal of Fisheries Management 22:474–479.

Glebe, B. D., and W. C. Leggett. 1981. Latitudinal differences in energy allocation and use during the freshwater migrations of American shad (*Alosa sapidissima*) and their life history consequences. Canadian Journal of Fisheries and Aquatic Sciences 38:806–820.

Haro, A., T. Castro-Santos, J. Noreika, and M. Odeh. 2004. Swimming performance of upstream migrant fishes in open-channel flow: a new approach to predicting passage through velocity barriers. Canadian Journal of Fisheries and Aquatic Sciences 61:1590–1601.

Haro, A., M. Odeh, T. Castro-Santos, and J. Noreika. 1999. Effect of slope and headpond on passage of American shad and blueback herring through simple Denil and deepened Alaska steeppass fishways. North American Journal of Fisheries Management 19:51–58.

Haro, A., M. Odeh, J. Noreika, and T. Castro-Santos. 1998. Effect of water acceleration on downstream migratory behavior and passage of Atlantic salmon smolts and juvenile American shad at surface bypasses. Transactions of the American Fisheries Society 127:118–127.

Hasler, A. D. 1971. Orientation and fish migration. Pages 429–510 *in* W. S. Hoar and D. J. Randall, editors. Environmental relations and behavior. Academic Press, New York.

Hinch, S. G., and J. Bratty. 2000. Effects of swim speed and activity pattern on success of adult sockeye salmon migration through an area of difficult passage. Transactions of the American Fisheries Society 129:598–606.

Hinch, S. G., R. E. Diewert, T. J. Lissimore, A. M. J. Prince, M. C. Healey, and M. A. Henderson. 1996. Use of electromyogram telemetry to assess difficult passage areas for river-migrating adult sockeye salmon. Transactions of the American Fisheries Society 125:253–260.

Hinch, S. G., and P. S. Rand. 1998. Swim speeds and energy use of up-river migrating sockeye salmon (*Oncorhynchus nerka*): role of local environment and fish characteristics. Canadian Journal of Fisheries and Aquatic Sciences 55:1821–1831.

Hoar, W. S. 1988. The physiology of smolting salmonids. Pages 275–343 *in* W. S. Hoar and D. Randall, editors. Fish physiology. Volume XIB: The physiology of developing fish. Academic Press, New York.

Hosmer, D. W., and S. Lemeshow. 1999. Applied survival analysis. Wiley, New York.

Idler, D. R., and W. A. Clemens. 1959. The energy expenditures of Fraser River sockeye salmon during the spawning migration to Chilko and Stuart lakes. International Pacific Salmon Fisheries Commission, Vancouver.

Johnson, G. E., J. B. Hedgepeth, J. R. Skalski, and A. E. Giorgi. 2004. A Markov chain analysis of fish movements to determine entrainment zones. Fisheries Research 69:349–358.

Johnston, I. A. 2001. Genetic and environmental determinants of muscle growth patterns. Pages 141–186 *in* D. Randall and A. P. Farrell, editors. Fish physiology, volume 18: Muscle development and growth. Academic Press, London.

Jungwirth, M. 1996. Bypass channels at weirs as appropriate aids for fish migration in rhithral rivers. Regulated Rivers Research & Management 12:483–492.

Khan, L. A. 2006. A three-dimensional computational fluid dynamics (CFD) model analysis of free surface hydrodynamics and fish passage energetics in a vertical-slot fishway. North American Journal of Fisheries Management 26:255–267.

Laake, J. L., P. Browne, R. L. DeLong, and H. R. Huber. 2002. Pinniped diet composition: a comparison of estimation models. Fisheries Bulletin 100:434–447.

Larinier, M., and F. Travade. 1999. Downstream migration: problems and facilities [In French]. Bulletin Francais de la Peche et de la Pisciculture 364(Supplement):181–210.

Larinier, M., and F. Travade. 2002. The design of fishways for shad. Bulletin Francais de la Peche et de la Pisciculture 364(Supplement):135–146.

Leonard, J. B. K., M. Iwata, and Ueda, H. 2001. Seasonal changes of hormones and muscle enzymes in adult lacustrine masu (*Oncorhynchus masou*) and sockeye salmon (*O. nerka*). Fish Physiology and Biochemistry 25:153–163.

Leonard, J. B. K., and S. D. McCormick. 1999. Effects of migration distance on whole-body and tissue-specific energy use in American shad (*Alosa sapidissima*). Canadian Journal of Fisheries and Aquatic Sciences 56:1159–1171.

Liao, J. C. 2004. Neuromuscular control of trout swimming in a vortex street: implications for energy economy during the Karman gait. Journal of Experimental Biology 207:3495–3506.

Liao, J. C., D. N. Beal, G. V. Lauder, and M. S. Triantafyllou. 2003a. Fish exploiting vortices decrease muscle activity. Science 302:1566–1569.

Liao, J. C., D. N. Beal, G. V. Lauder, and M. S. Triantafyllou. 2003b. The Karman gait: novel body kinematics of rainbow trout swimming in a vortex street. Journal of Experimental Biology 206:1059–1073.

Lighthill, M. J. 1971. Large amplitude elongate body theory of fish locomotion. Proceedings of the Royal Society of London Series B 179:125–138.

Lighthill, M. J. 1977. Mathematical theories of fish swimming. Pages 131–144 *in* J. H. Steele, editor. Fisheries mathematics. Academic Press, London.

Lowe, C. G. 1996. Kinematics and critical swimming speed of juvenile scalloped hammerhead sharks. Journal of Experimental Biology 199:2605–2610.

Lowe, C. G., K. N. Holland, and T. G. Wolcott. 1998. A new acoustic tailbeat transmitter for fishes. Fisheries Research 36:275–283.

Lucas, M. C., T. Mercer, J. D. Armstrong, S. McGinty, and P. Rycroft. 1999. Use of a flat-bed passive integrated transponder antenna array to study the migration and behaviour of lowland river fishes at a fish pass. Fisheries Research 44:183–191.

Madenjian, C. P., G. L. Fahnenstiel, T. H. Johengen, T. F. Nalepa, H. A. Vanderploeg, G. W. Fleischer, P. J. Schneeberger, D. M. Benjamin, E. B. Smith, J. R. Bence, E. S. Rutherford, D. S. Lavis, D. M. Robertson, D. J. Jude, and M. P. Ebener. 2002. Dynamics of the Lake Michigan food web, 1970–2000. Canadian Journal of Fisheries and Aquatic Sciences 59:736–753.

McCormick, S. D. 1994. Ontogeny and evolution of salinity tolerance in anadromous salmonids: hormones and heterochrony. Estuaries 17:26–33.

McCormick, S. D., D. T. Lerner, M. Y. Monette, K. Nieves-Puigdoller, J. T. Kelly, and B. T. Björnsson. 2009. Taking it with you when you go: how perturbations to the freshwater environment, including temperature, dams, and contaminants, affect marine survival of salmon. Pages xxx–xxx *in* A. J. Haro, K. L. Smith, R. A. Rulifson, C. M. Moffitt, R. J. Klauda, M. J. Dadswell, R. A. Cunjak, J. E. Cooper, K. L. Beal, and T. S. Avery, editors. Challenges for diadromous fishes in a dynamic global environment. American Fisheries Society, Symposium 69, Bethesda, Maryland.

McKinley, R. S., and G. Power. 1992. Measurement of activity and oxygen consumption for adult lake sturgeon (*Acipenser fulvescens*) in the wild using radio-transmitted EMG signals. Pages 307–318 *in* I. G. Priede and S. M. Swift, editors. Wildlife telemetry: remote monitoring and tracking of animals. Horwood, New York.

McLeod, A. M., and P. Nemenyi. 1940. An investigation of fishways. State University of Iowa, Iowa City.

Meyer, J. L., D. L. Strayer, J. B. Wallace, S. L. Eggert, G. S. Helfman, and N. E. Leonard. 2007. The contribution of headwater streams to biodiversity in river networks. Journal of the American Water Resources Association 43:86–103.

Mogdans, J., and H. Bleckmann. 2001. The mechanosensory lateral line of jawed fishes. Pages 181–213 *in* B. G. Kapoor and T. J. Hara, editors. Sensory biology of jawed fishes: new insights. Science Publishers, Plymouth, UK.

Monk, B., D. Weaver, C. Thompson, and F. Ossiander. 1989. Effects of flow and weir design on the passage behavior of American shad and salmonids in an experimental fish ladder. North American Journal of Fisheries Management 9:60–67.

Mustard, J. F., M. A. Carney, and A. Sen. 1999. The use of satellite data to quantify thermal effluent impacts. Estuarine Coastal and Shelf Science 49:509–524.

Naughton, G. P., C. C. Caudill, M. L. Keefer, T. C. Bjornn, L. C. Stuehrenberg, and C. A. Peery. 2005. Late-season mortality during migration of radio-tagged adult sockeye salmon (*Oncorhynchus nerka*) in the Columbia River. Canadian Journal of Fisheries and Aquatic Sciences 62:30–47.

Nikora, V. I., J. Aberle, B. J. F. Biggs, I. G. Jowett, and J. R. E. Sykes. 2003. Effects of fish size, time-to-fatigue and turbulence on swimming performance: a case study of *Galaxias maculatus*. Journal of Fish Biology 63:1365–1382.

Northcote, T. G., S. N. Williscr, and H. Tsuyuki. 1970. Meristic and lactate dehydrogenase genotype differences in stream populations of rainbow trout below and above a waterfall. Journal of the Fisheries Research Board of Canada 27:1987–1995.

Orsborn, J. F. 1987. Fishways: historical assessment of design practices. Pages 122–130 *in* M. J. Dadswell, R. J. Klauda, C. M. Moffitt, R. L. Saunders, R. A. Rulifson, and J. E. Cooper, editors. Common strategies of anadromous and catadromous fishes. American Fisheries Society, Symposium 1, Bethesda, Maryland.

Orsborn, J. F., and P. D. Powers. 1986. Fishways: an assessment of their development and design. Part 3 of 4. Final project report. Bonneville Power Administration, BPA Report DOE/BP-36523-2, Portland, Oregon.

Palmer, M. A., E. S. Bernhardt, J. D. Allan, P. S. Lake, G. Alexander, S. Brooks, J. Carr, S. Clayton, C. N. Dahm, J. F. Shah, D. L. Galat, S. G. Loss, P. Goodwin, D. D. Hart, B. Hassett, R. Jenkinson, G. M. Kondolf, R. Lave, J. L. Meyer, T. K. O'Donnell, L. Pagano, and E. Sudduth. 2005. Standards for ecologically successful river restoration. Journal of Applied Ecology 42:208–217.

Peake, S. 2004. An evaluation of the use of critical swimming speed for determination of culvert water velocity criteria for smallmouth bass. Transactions of the American Fisheries Society 133:1472–1479.

Peake, S., R. S. McKinley, and D. A. Scruton. 1997. Swimming performance of various freshwater Newfoundland salmonids relative to habitat selection and fishway design. Journal of Fish Biology 51:710–723.

Peake, S. J., and A. P. Farrell. 2004. Locomotory behaviour and post-exercise physiology in relation to swimming speed, gait transition and metabolism in free-swimming smallmouth bass (*Micropterus dolomieu*). Journal of Experimental Biology 207:1563–1575.

Peake, S. J., and A. P. Farrell. 2005. Postexercise physiology and repeat performance behaviour of free-swimming smallmouth bass in an experimental raceway. Physiological and Biochemical Zoology 78:801–807.

Powers, P. D., J. F. Orsborn, T. W. Bumstead, S. Klinger-Kingsley, and W. C. Mih. 1985. Fishways: an assessment of their development and design. Bonneville Power Administration, U.S. Department of Energy No. DOE/BP-36523-4, Portland, Oregon..

Prentice, E. F., T. A. Flagg, and S. McCutcheon. 1990. Feasibility of using implantable passive integrated transponder (PIT) tags in salmonids. Pages 317–322 *in* N. C. Parker, A. E. Giorgi, R. C. Heidinger, D. B. Jester, Jr., E. D. Prince, and G. A. Winans, editors. Fish marking techniques. American Fisheries Society, Bethesda, Maryland

Rand, P. S., and S. G. Hinch. 1998. Swim speeds and energy use of upriver-migrating sockeye salmon (*Oncorhynchus nerka*): simulating metabolic power and assessing risk of energy depletion. Canadian Journal of Fisheries and Aquatic Sciences 55:1832–1841.

Rand, P. S., S. G. Hinch, J. Morrison, M. G. G. Foreman, M. J. MacNutt, J. S. Macdonald, M. C. Healey, A. P. Farrell, and D. A. Higgs. 2006. Effects of river discharge, temperature, and future climates on energetics and mortality of adult migrating Fraser River sockeye salmon. Transactions of the American Fisheries Society 135:655–667.

Reback, K. E., P. D. Brady, K. D. McLaughlin, and C. G. Milliken. 2004. A survey of anadromous fish passage in coastal Massachusetts (parts 1–4). Massachusetts Division of Marine Fisheries, Technical Report TR-15, Boston.

Reischel, T. S., and T. C. Bjornn. 2003. Influence of fishway placement on fallback of adult salmon at the Bonneville Dam on the Columbia River. North American Journal of Fisheries Management 23:1215–1224.

Rideout, S. G., L. M. Thorpe, and L. M. Cameron. 1985. Passage of American shad in an Ice Harbor style fishladder after flow pattern modifications. Pages 251–256 *in* F. W. Olson, R. G. White, and R. H. Hamre. Proceedings of the Symposium on Small Hydropower and Fisheries. American Fisheries Society, Bethesda, Maryland.

Santos, H. D. E., P. D. Pompeu, and C. B. Martinez. 2007. Swimming performance of the migratory neotropical fish *Leporinus reinhardti*

(Characiformes:Anostomidae). Neotropical Ichthyology 5:139–146.

Santos, J. M., M. T. Ferreira, F. N. Godinho, and J. Bochechas. 2005. Efficacy of a nature-like bypass channel in a Portuguese lowland river. Journal of Applied Ichthyology 21:381–388.

Sato, A., H. Ueda, M. Fukaya, M. Kaeriayama, Y. Zohar, A. Urano, and K. Yamauchi. 1997. Sexual differences in homing profiles and shortening of homing duration by gonadotropin-releasing hormone analog implantation in lacustrine sockeye salmon (*Oncorhynchus nerka*) in Lake Shikotsu. Zoological Science 14:1009–1014.

Schmutz, S., C. Giefing, and C. Wiesner. 1998. The efficiency of a nature-like bypass channel for pike-perch (*Stizostedion lucioperca*) in the Marchfeldkanalsystem. Hydrobiologia 372: 355–360.

Schreck, C. B., W. Contreras-Sanchez, and M. S. Fitzpatrick. 2001. Effects of stress on fish reproduction, gamete quality, and progeny. Aquaculture 197:3–24.

Schultz, W. W., and P. W. Webb. 2002. Power requirements of swimming: do new methods resolve old questions? Integrative & Comparative Biology 42:1018–1025.

Skalski, J. R., J. Lady, R. Townsend, A. E. Giorgi, J. R. Stevenson, C. M. Peven, and R. D. McDonald. 2001. Estimating in-river survival of migrating salmonid smolts using radiotelemetry. Canadian Journal of Fisheries and Aquatic Sciences 58:1987–1997.

Skalski, J. R., S. G. Smith, R. N. Iwamoto, J. G. Williams, and A. Hoffmann. 1998. Use of passive integrated transponder tags to estimate survival of migrant juvenile salmonids in the Snake and Columbia rivers. Canadian Journal of Fisheries and Aquatic Sciences 55:1484–1493.

Slatick, E., and L. R. Basham. 1985. The effect of Denil fishway length on passage of some nonsalmonid fishes. Marine Fisheries Review 47:83–85.

Sorensen, P. W., and N. E. Stacey. 2004. Brief review of fish pheromones and discussion of their possible uses in the control of non-indigenous teleost fishes. New Zealand Journal of Marine and Freshwater Research 38:399–417.

Sorensen, P. W., and L. A. Vrieze. 2003. The chemical ecology and potential application of the sea lamprey migratory pheromone. Journal of Great Lakes Research 29:66–84.

Stringham, E. 1924. The maximum speed of fresh-water fishes. American Naturalist 18:156–161.

Stuart, I. G., and A. P. Berghuis. 2002. Upstream passage of fish through a vertical-slot fishway in an Australian subtropical river. Fisheries Management and Ecology 9:111–122.

Sullivan, T. J. 2004. Evaluation of the Turners Falls fishway complex and potential improvements for passing adult American shad. Master's thesis. University of Massachusetts, Amherst.

Sullivan, W. P., G. C. Christie, F. C. Cornelius, M. F. Fodale, D. A. Johnson, J. F. Koonces, G. L. Larson, R. B. McDonald, K. M. Mullett, C. K. Murray, and P. A. Ryan. 2003. The sea lamprey in Lake Erie: a case history. Journal of Great Lakes Research 29:615–636.

Taylor, E. B., and J. D. Mcphail. 1986. Prolonged and burst swimming in anadromous and fresh-water threespine stickleback, *Gasterosteus aculeatus*. Canadian Journal of Zoology-Revue Canadienne de Zoologie 64:416–420.

Tierney, K. B., and A. P. Farrell. 2004. The relationships between fish health, metabolic rate, swimming performance and recovery in return-run sockeye salmon, *Oncorhynchus nerka* (Walbaum). Journal of Fish Diseases 27:663–671.

Tritico, H. M., A. J. Cotel, and J. N. Clarke. 2007. Development, testing and demonstration of a portable submersible miniature particle imaging velocimetry device. Measurement Science & Technology 18:2555–2562.

Tudorache, C., P. Viaenen, R. Blust, and G. De Boeck. 2007. Longer flumes increase critical swimming speeds by increasing burst and glide swimming duration in carp (*Cyprinus carpio*, L.). Journal of Fish Biology 71:1630–1638.

Turchin, P. 1998. Quantitative analysis of animal movement. Sinauer, Sunderland, Massachusetts.

Ueda, H., O. Hiroi, A. Hara, K. Yamauchi, and Y. Nagahama. 1984. Changes in serum concentrations of steroid-hormones, Thyroxine, and Vitellogenin during spawning migration of the chum salmon, *Oncorhynchus keta*. General and Comparative Endocrinology 53:203–211.

Venditti, D. A., D. W. Rondorf, and J. M. Kraut. 2000. Migratory behavior and forebay delay of radio-tagged juvenile fall Chinook salmon in a lower Snake River impoundment. North American Journal of Fisheries Management 20:41–52.

Videler, J. J., and C. S. Wardle. 1991. Fish swimming stride by stride: speed limits and endurance. Reviews in Fish Biology and Fisheries 1:23–40.

Wardle, C. S. 1975. Limit of fish swimming speed. Nature 255:725–727.

Weaver, C. R. 1963. Influence of water velocity upon orientation and performance of adult migrating salmonids. Fisheries Bulletin 63:97–121.

Weaver, C. R. 1965. Observations on the swimming ability of adult American shad (*Alosa sapidissima*). Transactions of the American Fisheries Society 94:382–385.

Webb, P. W. 1971a. The swimming energetics of trout: I.

Thrust and power output at cruising speeds. Journal of Experimental Biology 55:189–520.

Webb, P. W. 1971b. The swimming energetics of trout: II. Oxygen consumption and swimming capacity. Journal of Experimental Biology 55:521–540.

Webb, P. W. 1975. Hydrodynamics and energetics of fish propulsion. Bulletin of the Fisheries Research Board of Canada 190:1–158.

Webb, P. W. 1998a. Entrainment by river chub *Nocomis micropogon* and smallmouth bass *Micropterus dolomieu* on cylinders. Journal of Experimental Biology 201:2403–2412.

Webb, P. W. 1998b. Swimming. Pages 3–24 *in* D. H. Evans, editor. The physiology of fishes. CRC Press, Boca Raton, Florida.

Webb, P. W., and C. L. Gerstner. 2000. Fish swimming behaviour: predictions from physical principles. Pages 59–77 *in* P. Domenici and R. W. Blake, editors. Biomechanics in animal behaviour. BIOS Scientific Publishing, Oxford, UK.

Zhong, Y. G., and G. Power. 1996. Environmental impacts of hydroelectric projects on fish resources in China. Regulated Rivers Research & Management 12:81–98.

American Fisheries Society Symposium 69:577–588, 2009

Status of Diadromous Fish Species in the Restored East Hammar Marsh in Southern Iraq

ABDUL-RAZAK M. MOHAMED*
Department of Fisheries and Marine Resources, College of Agriculture Basrah University, Basrah, Iraq

NAJAH A. HUSSAIN AND SAJED S. AL-NOOR
Department of Biology, College of Science, Basrah University, Basrah, Iraq

BRIAN W. COAD
Canadian Museum of Nature, Box 3443, Station D, Ottawa, Ontario K1P 6P4, Canada

FALAH M. MUTLAK
Marine Science Centre, Basrah University, Basrah, Iraq

Abstract.—A total of 31 fish species belonging to 14 families were collected from the restored East Hammar marsh from October 2005 to September 2006 using seine net, electrofishing gear, fixed gill net, and dip net. Eleven marine diadromous species were caught, representing 35.5% of the total number of species, which is similar to the fish assemblage of the Shatt Al-Arab River during the 1970s and 1980s. Marine diadromous species in the East Hammar marsh were divided into a resident group, including only one species (greenback mullet *Liza subviridis*), a seasonal group represented by three species (hilsa *Tenualosa ilisha*, moustached thryssa *Thryssa mystax*, and dusky frillgoby *Bathygobius fuscus*), and an occasional group consisting of seven species (yellowfin seabream *Acanthopagrus latus*, spotted scat *Scatophagus argus*, sobaity seabream *Sparidentex hasta*, Klunzinger's mullet *L. klunzingeri*, mud skipper *Boleophthalmus dussumieri*, long billed half beak *Rhynchorhamphus georgii*, and ocellated sole *Brachirus orientalis*). These species were all derived from the Arabian Gulf. Marine fish comprised 15.8% of the total catch of which hilsa was the most abundant comprising 10.1%, followed by moustached thryssa (3.8%) and greenback mullet (1.6%), of the total catch. Water temperature showed positive correlations with both number and total catch of marine species. The marine assemblage in the East Hammar marsh consisted of small-sized (juvenile) individuals. Hilsa fed mostly on alga and diatoms; greenback mullet on diatoms, higher plants, and organic materials; Klunzinger's mullet on diatoms and organic materials; moustached thryssa on insects, shrimps, and fish; desky frillgoby on shrimps and fish; and yellowfin seabream on shrimps and insects. The presence of marine fish species has major effects on the seasonality and levels of ecological indices (richness, diversity, and evenness) of the fish assemblages in the East Hammar marsh. The restored East Hammar marsh has regained its original biological role as a nursery as well as a forage ground for marine diadromous fishes.

Introduction

The Tigris and Euphrates rivers and their tributaries are the main sources of inland freshwater in Iraq. The inland freshwater bodies cover between 600,000 and 700,000 ha, made up of natural lakes (39%), dams and reservoirs (13.3%), rivers and their branches (3.7%), and marshes (44%), in addition to coastline of approximately 50 km along the northwestern Arabian Gulf (FAO 1999). There are

* Corresponding author: abdul19532001@yahoo.com

more than 58 freshwater fish species in Iraqi inland waters, about a further 53 marine species penetrating estuarine and freshwater (Coad 1991; www.briancoad.com), and 125 fish species and five species of shrimps in the Iraqi marine waters (Mohamed et al. 2001).

In the 1950s and 1960s, the main elements of the marshlands economy in southern Iraq were based on their biological diversity: agriculture, livestock raising, birding, mat making, and fishing. Fishing was practiced widely in the marshes, and fish were a major food item for people inhabiting marshes. The estimated annual catch of fish in the Mesopotamian marshes in the early 1960s was 30,000 metric tons, of which 70% were Cyprinid species.

In 1990, the Food and Agriculture Organization of the United Nations estimated that the total inland catch of fish in Iraq was 23,600 metric tons, with more than 60% of this coming from the Mesopotamian marshes (Partow 2001). The marshlands also once served as a natural filter for waste and other pollutants transported by the Tigris and Euphrates rivers, thus protecting the Arabian Gulf, which has now become noticeably degraded along the coast of Kuwait (Al-Ghadban et al. 1999; Saeed et al. 1999). They were the permanent habitat for millions of birds and a flyway for millions more migrating between Siberia and Africa (Maltby 1994; Evans 2002). Several marine fish species of great economic importance in the Arabian Gulf are dependent on the estuarine systems and marshes for spawning, such as Clupeidae (Hussain et al. 1994), or for feeding, such as Mugilidae and Sparidae (Hussain et al. 1987; Hussain and Ahmed 1995). The penaeid shrimp, jinga shrimp *Metapenaeus affinis*, undertakes seasonal migrations between spawning grounds in the gulf and nursery and feeding grounds in the East Hammar (Mathews et al. 1986).

Concomitant factors, including the Iran–Iraq War in the 1980s; new hydrological projects in Turkey, Syria, and Iran; the constructions of drainage systems by diversions of major rivers surrounding the marsh areas; and drainage processes in the 1990s drained the southern marshlands and led to substantial loss of native aquatic flora and fauna well known for a long period as marsh biota of southern Iraq. In 2002, 85% of permanent marshes defined in 1973 had been environmentally destroyed. Only 3% of the central marshes and 14.5% of the East Hammar marshes remained, along with the largest natural expanse of marsh, the Hawizah, near the Iranian border (Richardson and Hussain 2006).

Since 2003, great efforts have been made to restore the marshes and revive the fishing industry, but major dams across the border in Turkey have reduced the river discharge significantly. As of August 2005, the marshes had recovered almost 40% of their former levels, according to the United Nations Environment Programme. Satellite imagery shows the southern marshlands now occupy about 3,500 km^2 after having dwindled to just 760 km^2.

Several taxonomic studies were previously published about freshwater and marine fishes of Iraq (Khalaf 1961; Mahdi 1962; Al-Nasiri and Shamsul-Houda 1975; Banister 1980; Al-Daham 1982; Coad 1991), and these refer to the marshes in passing. Other studies have focused on water quality, plankton and plant communities in the marshes before desiccation (Maulood et al. 1979; Pankow et al. 1979; Al-Saadi et al. 1981; Al-Saboonchi et al. 1982, 1986; Al-Zubaidy 1985; Mohamed and Barak 1988a), or on biological aspects of some of the freshwater fishes in the marshes (Al-Mukhtar 1982; Barak and Mohamed 1982, 1983; Naama 1982; Dawood 1986; Al-Sayab 1988; Jasim 1988; Mohamed and Barak 1988b; Al-Kanaani 1989; Al-Rudainy 1989; Mohamed and Ali 1994). A number of studies of the biological characteristics of some of the diadromous fish species in the Shatt Al-Arab River, Shatt Al-Basrah, and Iraqi marine waters have been undertaken (Hussain and Ali 1987; Hussain et al. 1987, 1989, 1999, 2001; Jabir and Faris 1989; Al-Daham et al. 1993; Younis 1995; Al-Noor 1998; Mohamed et al. 1998, 2000). Nevertheless, there is no published information on the species composition, fish ecology, and fisheries of the marshes before their draining, nor on its ecological role for marine fish species.

The specific objective of this research was to describe the composition, occurrence, abundance, and food habits of marine diadromous fish species in East Hammar marsh after restoration.

Study Site

The East Hammar marsh is situated south of the Euphrates, extending from near Nasiriyah in the west to the outskirts of Basrah on the Shatt al-Arab in the east (Figure 1). The marsh area historically consisted of 2,800 km^2 of permanent marsh and

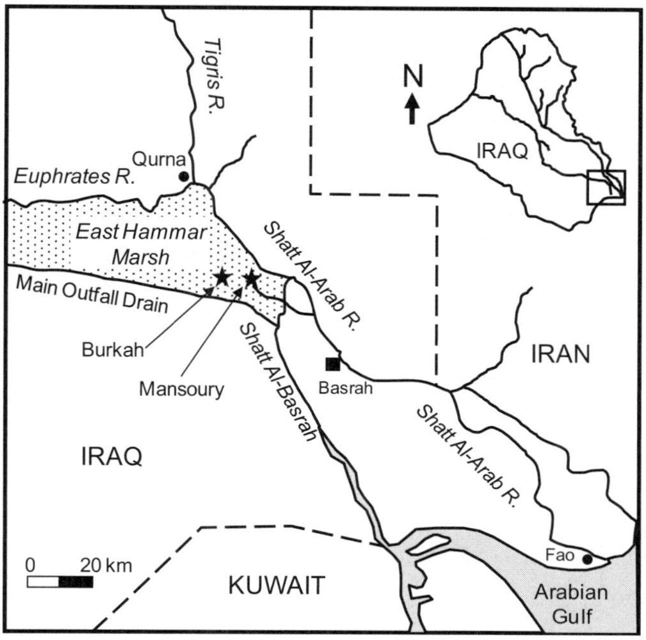

Figure 1.—Location of study area in the East Hammar marsh, south Iraq.

lake, expanding to more than 4,500 km^2 during periods of seasonal and temporary inundation. Hammar Lake (some authors used the term lake instead of marsh for Hammar), which used to dominate the marsh, is the largest water body on the lower Euphrates. In the late 1980s, it was approximately 120 km long and 25 km wide. Maximum water depth in the lake ranged from 1.8 to 3 m (Partow 2001).

During the 1990s, the East Hammar marsh completely dried out as a result of local drainage structures. These structures included two dams on the Euphrates, one at Nasiriyah and one at Medina to reduce water flow into the marsh area; sluice gates and dams placed on distributaries of the Euphrates to restrict water flow; and two canals and a pipeline that diverted water from the Euphrates and agricultural drainage water away from the marshlands.

In 2003 and 2004, the local Marsh Dwellers and Iraq's Ministry of Water Resources began to divert water back into the marshlands through opening sluice gates, breaching embankments, and other actions along the southern side (Garmat Ali marsh). Water from the Euphrates is returning from the western and northern sides of Hammar and from Shatt al-Arab through Garmat Ali marsh to the southeast (Figure 1).

Methods

The fishes were collected monthly from two selected sites (Mansoury: N 30° 40′ 32″, E 47° 37′ 21″; and Burkah: N 30° 40′ 22″, E 47° 33′ 03″) in the East Hammar (Figure 1) as part of a larger study to investigate the composition of the fish community in the three restored southern Iraqi marshes during the period from October 2005 to September 2006. The Mansoury site consisted of previously drained tidal channel habitat, while the Burkah site was previously drained tidal open-water habitat.

Sampling was carried out using seine net (20 m long with a 2.5 cm mesh), electrofishing gear, fixed gill nets (50–100 m long with 2.5–10 cm mesh size), and dip nets. Specimens were immediately transported to the laboratory on crushed ice. Water temperature and salinity were measured to determine the relationships of these two factors with the number of species and total catch of marine species. Fishes were identified to species by using Khalaf (1961), Mahdi (1962), and Beckman (1962). The total length (mm) and weight (g) of all fish captured were recorded. Three analytical methods were used to analyze stomach contents (i.e. numerical, volumetric, and frequency of occurrence; Windell

1971). The importance of food item was determined by using the index of relative importance of (Pinkas et al. 1971). The similarity among fish species based on their diet was calculated according to Jaccard similarity coefficient, using SPSS software (version 11, 2001) statistical package.

The analysis of the nature of the fish communities in the East Hammar marsh was carried out by the following methods and indices: relative abundance (Odum 1970), the diversity index (Shannon and Weaver 1949), the evenness index (Pielou 1977), and the richness index (Margalef 1968). Fish species were divided into three categories according to their occurrence in the monthly samples following Tyler (1971).

Results

Species Composition and Occurrence

The fish assemblage of East Hammar marsh consists of 31 species belonging to 14 families and is dominated by four species (Abu mullet *Liza abu*, crucian carp *Carassius carassius*, *Acanthobrama marmid*, and hilsa *Tenualosa ilisha*) that account for 80.4% of the total catch. The fish fauna was composed of 14 native, 6 alien, and 11 marine species. The marine species belonged to eight families, namely Mugilidae (greenback mullet *L. subviridis* and Klunzinger's mullet *L. klunzingeri*), Sparidae (yellowfin seabream *Acanthopagrus latus* and sobaity seabream *Sparidentex hasta*), Gobiidae (dusky frillgoby *Bathygobius fuscus* and mud skipper *Boelophthalmus dussumieri*), Clupeidae (hilsa), Engraulidae (moustached thryssa *Thryssa mystax*), Scatophagidae (spotted scat *Scatophagus argus*), Hemiramphidae (long billed half beak *Rhynchorhamphus georgii*) and Soleidae (ocellated sole *Brachirus orientalis*). The marine species comprised 35.5% of the total number of species. All marine species were collected from the Mansoury site and only four species, namely moustached thryssa, greenback mullet, dusky frillgoby, and yellowfin seabream, from Burkah site. The highest number of marine species (seven) was caught in July and the lowest (one) in November and December (Table 1). One marine species (greenback mullet) was considered as a resident species, three (hilsa, moustached thryssa, and dusky frillgoby) as seasonal species, and seven (yellowfin seabream, spotted scat, sobaity seabream, Klunzinger's mullet, mud skipper, long billed half beak, and ocellated sole) were occasional species.

Relative Abundance

Of the total catch of 16,199 fish of East Hammar marsh, 2,560 were marine fish (Table 1). The majority of marine fish (2,502 fish, 97.7%) were caught from the Mansoury site. The relative abundance of the marine fish fluctuated from 0.2% (two fish) in November to 55.1% (1,610 fish) in September (Table 1), with an overall value of 15.8% of the total catch. Hilsa was the most abundant marine species, comprising 10.1%, followed by moustached thryssa (3.8%) and greenback mullet (1.6%) of the total catch.

The water temperature fell to a minimum of 12.5°C in February and rose to a maximum of 29°C in July. Salinity changed from 1.2‰ in August to 2.0‰ in May and July (Figure 2). Water temperature showed positive correlations (0.713 and 0.407), with both the number and the total catch of marine species, respectively, while salinity exhibited weak negative and positive correlations (–0.107 and 0.195), with both the number and the total catch of marine species, respectively.

Ecological Indices

The monthly variations in ecological indices relevant to fish species composition in the East Hammar marsh during October 2005–September 2006 are given in Table 1. The diversity index (*H*) fluctuated from 1.07 in November to 2.01 in July, with an overall value of 1.53. The richness index (*D*) varied from 0.74 in December to 2.83 in July, with an overall value 1.76. The evenness index (*J*) ranged from 0.49 in November to 0.84 in December, with an overall value 0.60.

Length-Frequency Distributions

The overall length-frequency distributions of the most abundant marine fish species in the East Hammar marsh are illustrated in Figure 3. Hilsa was the most abundant marine species in the marshes and appeared in the catches. The lengths of Hilsa ranged from 3 to 13 cm, and a length of 6 cm dominated the catch. The second most abundant marine species was moustached thryssa, which was found in the catch from March to October. Lengths from

Table 1.—Monthly variations in relative abundance (%) of marine fish species caught in the Hammar marshes (October 2005–September 2006).

Fish species	Oct.	Nov.	Dec.	Jan.	Feb.	Mar.	Apr.	May	June	July	Aug.	Sep.	Total
Hilsa	18.7	–	–	–	–	–	0.06	0.1	–	15.7	1.4	38.6	10.1
Moustached thryssa	3.2	–	–	–	–	–	0.12	0.7	0.4	1.9	8.7	14.9	3.8
Greenback mullet	3.7	0.2	6.3	2.2	1.6	1.6	1.4	0.2	0.9	0.6	0.49	1.6	1.6
Dusky frillgoby	0.08	–	–	0.2	–	–	0.94	0.4	0.1	0.06	–	–	0.17
Yellowfin seabream	–	–	–	–	–	–	0.06	–	–	1.1	–	–	0.12
Klunzinger's mullet	–	–	–	–	–	–	–	–	–	0.18	–	–	0.02
Spotted cat	0.08	–	–	–	–	–	–	–	–	–	–	–	0.006
Sobaity seabream	–	–	–	–	–	–	–	–	–	–	0.1	–	0.006
Long billed half beak	–	–	–	–	–	–	0.06	–	–	–	–	–	0.006
Ocellated sole	–	–	–	–	–	–	–	–	–	0.06	–	–	0.006
Mudskipper	0.08	–	–	–	–	–	–	–	–	–	–	–	0.006
% of marine species	25.8	0.2	6.3	2.4	1.6	2.8	2.6	1.4	1.4	19.5	10.6	55.1	15.8
% of other species	74.2	99.8	93.7	97.6	98.4	98.4	97.4	98.6	98.6	80.5	89.4	44.9	84.2
No. of marine species	6	1	1	2	1	3	6	4	3	7	4	3	–
Catch of marine fish	307	2	50	23	14	37	44	17	21	326	109	1,610	2,560
Total catch	1,188	1,126	800	942	882	1,315	1,695	1,217	1,419	1,670	1,025	2,920	16,199
Diversity index	1.55	1.07	1.50	1.34	1.36	1.87	1.51	1.78	1.43	2.01	1.35	1.63	1.53
Richness index	2.12	1.14	0.74	1.61	1.33	2.23	2.15	1.83	1.65	2.83	1.73	1.76	1.76
Eveness index	0.56	0.49	0.84	0.538	0.59	0.66	0.53	0.67	0.56	0.65	0.53	0.60	0.60

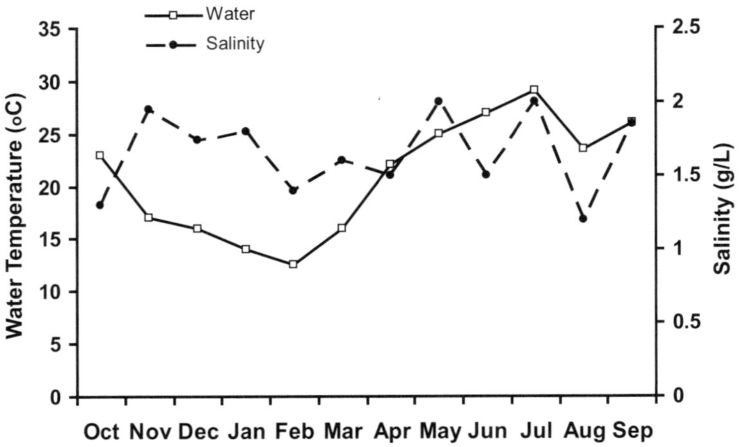

Figure 2.—Monthly variations in the water temperature and salinity in the East Hammar marsh.

6 to 19 cm were represented in the samples, and the dominant length-group was 10 cm. Greenback mullet was very common and regularly found in the catch throughout the year, except January. The length range of this species includes sizes from 7 to 22 cm, with fish of 11 cm dominating the catch.

Food Composition

The diet composition of some diadromous fish in the East Hammar marsh are given in Figure 4. Hilsa fed primarily on algae (40%) and diatoms (34%); other food items were detritus (8%), Copepoda (7%), Cladocera (5%), Rotifera (4%), and Ostracoda (2%). The food items of greenback mullet consisted of diatoms (39%), higher plants (37%), Copepoda (3%), and detritus (37%). Diatoms formed (37%) of the total food items of Klunzinger's mullet, followed by detritus (34%), higher plants (15%), algae (9%), Copepoda (3%), and Ostracoda (2%). Moustached thryssa preyed

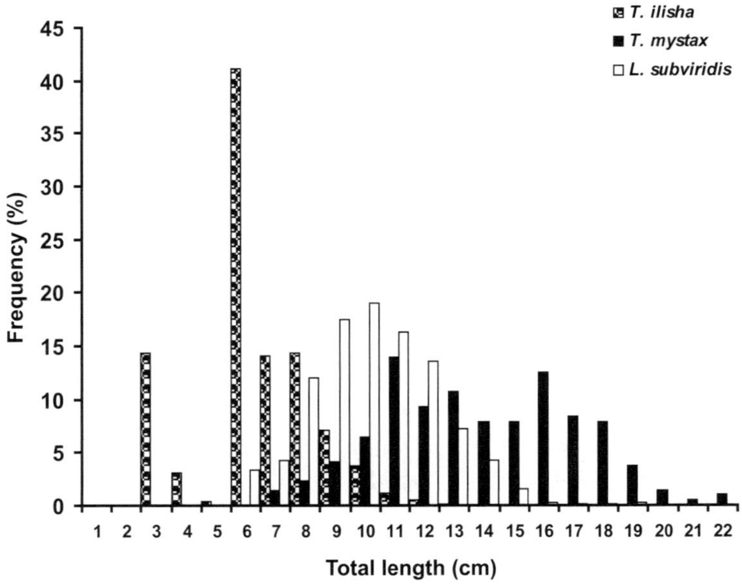

Figure 3.—The length frequencies of dominant diadromous fish in the East Hammar marshes.

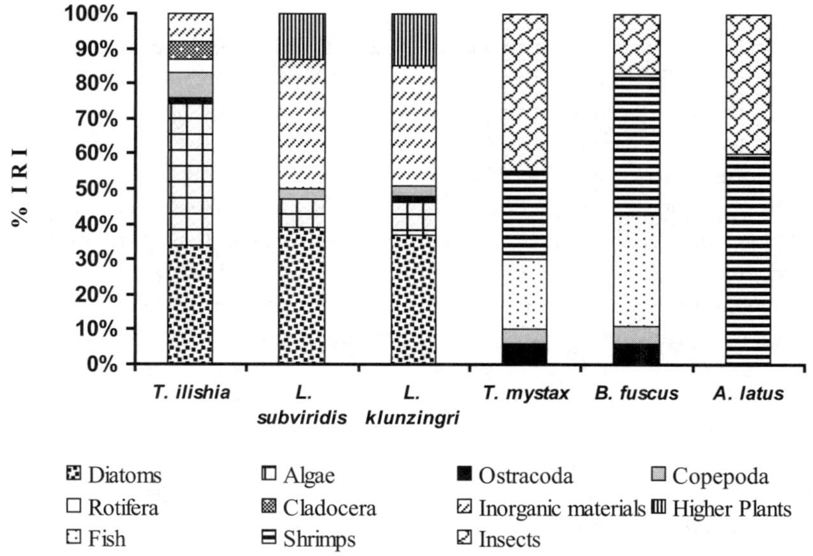

Figure 4.—Diet composition of some diadromous fish in the East Hammar marsh.

on insects (45%), shrimps (25%), fish (20%), Copepoda (6%), and Ostracoda (4%). Shrimps comprised 40% of the total food items of Dusky frillgoby, followed by fish (32%), insects (17%), Copepoda (6%), and Ostracoda (5%). Yellowfin seabream preyed mainly on shrimps (60%) and insects (40%).

Similarity dendrogram among fish species based on their diet is presented in Figure 5. Two main groups could be distinguished. Group I, consists of two subgroups: the first includes moustached thryssa and dusky frillgoby, which preyed mostly on shrimp, insects, and fish; and the second includes yellowfin seabream, which preyed on shrimp and insects. Group II, also consists of two subgroups: the first includes greenback mullet and Klunzinger's mullet, which fed mainly on diatoms, higher plants, and detritus; and the second included hilsa, whose diet consisted mostly of diatoms and algae.

Discussion

One of the obvious differences between the three restored southern marshes is that they represent different environments that receive their water from three separate sources, the Tigris, Euphrates, and Shatt Al-Arab rivers, each with a distinct water qual-

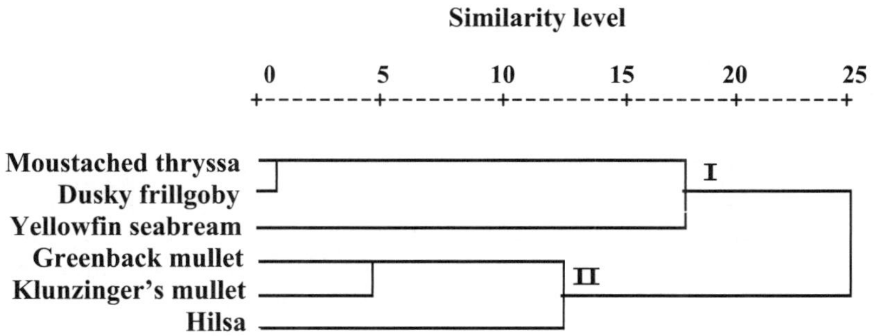

Figure 5.—Similarity dendrogram among some diadromous fish based on their diet in the East Hammar marsh.

ity (Rzóska et al. 1980; Richardson et al. 2005). The fish assemblage of East Hammar marsh, which receives water from both the Euphrates and the Shatt Al-Arab rivers, differs from other restored marshes by the occurrence of marine diadromous fish species, in addition to freshwater species both native and alien (Hussain et al. 2006). The dykes between the former East Hammar marsh and the Shatt Al-Arab River were demolished in late April 2003, and water has been flowing upstream due to the tidal action of Gulf waters forcing the waters of Shatt al-Arab back into the marsh. As a result, several diadromous species have returned to the Hammar marsh through the Shatt Al-Arab River for spawning, to nursery grounds, or for feeding (Richardson and Hussain 2006).

No species composition studies exist on this marsh before they were drained, but the occurrence of marine species in southern marshes was recorded by several authors (Al-Daham 1982, 1988; Al-Hassan and Naama 1986; Coad 1991). However, a few surveys were conducted in the upper reaches of the Shatt Al-Arab River, about 30 km from the marsh, before the draining of the marshes and indicated the presence of several marine species. Al-Nasiri and Shamsul-Hoda (1975) listed 12 marine species, forming 37.5% of the total species during 1973; Hussain et al. (1989) found 14 marine species, comprising 42.4% of the total species during 1982–1983; Hussain et al. (1997) collected seven marine species during 1992–1993; and Younis (2005) caught eight marine species during 2003. The present study indicated the presence of 11 marine species in East Hammar marsh. It can be seen that the proportion of marine species from the total species in the East Hammar was 35.5%, which is close to that of Shatt Al-Arab River during the 1970s and 1980s. Hussain and Taher (2007) suggested that during flood tide, a greater volume of water from Shatt Al-Arab River enters the Hammar marsh due to the semidiurnal tide, making the Hammar marsh mixed waters and not from the Euphrates River alone; consequently the environment of the Hammar is rather similar to Shatt Al-Arab to a certain degree.

The marine fish species recurring in the East Hammar marsh moved from the Shatt Al-Arab estuary via the Shatt Al-Arab River, a distance of 130 km, after the inundation of the marsh in April 2003 (Reiss and Farhan 2006). Moreover, Mohamed et al. (2006) reported that the Tigris fisheries at Qurna confluence, north of Basrah consisted of seven marine species during the spring and summer seasons of 2005, constituting 31.8% of the total number of species.

The marine species occurring in the East Hammar marsh were derived from the lower reaches of the Shatt Al-Arab estuary, northwest Arabian Gulf. However, the number of species invading the East Hammar marsh represents a small percentage of the total number of species in the fish assemblage of the Shatt Al-Arab estuary (Mohamed et al. 2001; Hussain et al. 2003). It seems that only the euryhaline species able to tolerate the low salinity of the marsh (1.2–2.0‰) have invaded. Al-Hassan and Hussain (1985) showed that hydrological parameters influenced the penetration of marine species from the gulf to the Shatt Al Arab River. Blaber et al. (1998) found that the juveniles of four species of Ilisha (Pristigasteridae) enter the upper reaches of Sarawak estuaries. These species were *Ilisha filigera*, *I. pristigastroides*, elongate ilisha *I. elongate*, and bigeye ilisha *I. megaloptera*. The last two were recorded from the Iraqi coastal water and Hammar marsh before desiccation during the 1990s. Utilization of salt marshes by marine fishes was reported in many Australian estuaries (Blaber and Blaber 1980; Blaber et al. 1985 and Morton et al. 1987).

It is also apparent that the relative abundance of marine diadromous species fluctuated temporarily. The low values coincided with low temperature (winter) and the high values with high temperature (summer), synchronizing with plankton blooms in the Hammar marsh (Reiss and Farhan 2006). Water temperature showed positive correlations with both the species number and total catch, indicating the environmental impact of this factor on migration and emigration of marine fish to the East Hammar marsh.

Only one marine species (greenback mullet) was considered a resident species and the other 10 species were considered seasonal or occasional species. These species enter into the East Hammar marsh in summer due to biological factors such as feeding, spawning, and protection from predators. The high abundance in late summer was due to an increase of Hilsa juveniles. Hilsa comprised 63.6% of the marine species and 10.1% of the total catch in the East Hammar marsh. Such a high number of juveniles confirms the importance of the marsh as a protected area for rearing young of the year of several species.

Examination of the stomach contents of the investigated fishes has shown that while the diet varied among the fish species, most of them depend on two or three major food items. In general, the diets of examined species were similar to that previously reviewed by Al-Daham et al. (1993), Hussain and Ali (1987), Hussain et al. (1987, 1999, 2001), and Mohamed et al. (1998, 2000) with certain differences in the percentages, which could be related to the difference in environments. Thick aquatic plants (*Ceratophyllum demersum*, *Potamogeton crispus*, and *Vallisneria spiralis*) could play a major role in Hammar marsh, offering extensive shelter from piscivorous predators and availability of food (insect, isopod, and amphipod). The same was noted previously by Lenanton (1982), that aquatic plants played a role as source of food (amphipod) and shelter. Blaber and Blaber (1980) identified that Australian salt marshes serve as nursery grounds for juveniles due to the absence of big predators and plenty of food. Subrahmanyan and Drake (1975) observed that most marine species in two tidal marshes in Florida were represented by juveniles and found a seasonal succession in the recruitment of juveniles.

The presence of marine fish species in the East Hammar marsh has also major effects on the seasonality and levels of ecological indices of the fish assemblages. There was a wide range in the richness index of the fish assemblages in the East Hammar marsh, leading to the conclusion that the marsh plays an important role as a nursery ground for several fish species. McErlean et al. (1973) considered a wide range of richness index as an indicator of the nursery function of an area. The marine diadromous fish species largely contributed to the high ecological indices obtained in the East Hammar marsh. These values were comparable to that of the Shatt Al- Arab River (Hussain et al. 1989), Shatt Al-Basrah (Al-Daham and Yousif 1990), and Khor Al-Zubair (Ali and Hussain 1990).

The aim of present study was to throw some light on the ecology of marine fishes in one of southern Iraq's marshes. However, the species composition within the Mesopotamian marshes requires detailed ongoing research.

Acknowledgments

We would like to express our thanks to the Canadian Iraqi Marsh Initiative (CIMI) and Iraq Nature for their support and help. Our thanks also go to the Canadian International Development Agency (CIDA) without whose financial support this work would never have been achieved.

References

Al-Daham, N. K. 1982. The ichthyofauna of Iraq: a check-list. Basrah Natural History Museum, Publication No. 4, Basrah, Iraq.

Al-Daham, N. K. 1988. Development of fisheries in the marshes, south of Iraq: obstacles and solutions. Arab Gulf Journal of Scientific Research 20(2):85–97.

Al-Daham, N. K., A. R. M. Mohamed, and A. Y. Al-Dibakel. 1993. Estuarine life of yellow fin sea bream *Acanthopagrus latus* in southern Iraq. Marina Mesopotamica 8(1):137–152.

Al-Daham, N. K., and A. Y. Yousif. 1990. Composition, seasonal and abundance of fishes in Shatt Al-Arab Canal, an estuary in southern Iraq. Estuarine Coastal and Shelf Science 31:411–421.

Al-Ghadban, A. N., T. Saeed, A. M. Al-Dousari, H. Al-Shemmari, and M. Al-Mutair. 1999. Preliminary assessment of the impact of draining of Iraqi marshes on Kuwait's northern marine environment. Part 1. Physical manipulation. Water Science & Technology 40(7):75–87.

Al-Hassan, A. J., and N. A. Hussain. 1985. Hydrological parameters influencing the penetration of Arabian Gulf fishes into the Shatt Al-Arab River, Iraq. Cybium 9(1):7–16.

Al-Hassan, L. A. J., and A. K. Naama. 1986. New record for some Arab Gulf fishes in freshwater systems of Iraq. Bulletin of the Basrah Natural History Museum 6:35–44, Basrah, Iraq.

Ali, T. S., and N. A. Hussain. 1990. Composition and seasonal fluctuations of intertidal fish assemblage in Khor Al-Zubair, northwest Arabian Gulf. Journal of Applied Ichthyology 6:24–36.

Al-Kanaani, S. M. 1989. Diet overlap among the common carp *Cyprinus carpio* L. and three native species in Al-Hammar marshes, southern Iraq. Master's thesis. Basrah University, Basrah, Iraq. [In Arabic]

Al-Mukhtar, M. A. 1982. Biological study for two freshwater fish, *Barbus luteus* (Heckel) and *Aspius vorax* (Heckel) in Hammar marsh, Basrah. Master's thesis. Basrah University, Basrah, Iraq. [In Arabic]

Al-Nasiri, S. K., and N. I. Shamsul-Houda. 1975. Survey of fish fauna of Shatt Al-Arab (from Abu Al-Khasib to Karmat Ali). Bulletin of the Basrah Natural History Museum 2:36–46, Basrah, Iraq.

Al-Noor, S. S. H. 1998. The reproductive biology of Tenualosa ilisha in Shatt Al-Arab and Iraqi marine waters. Doctoral thesis. Basrah University, Basrah, Iraq. [In Arabic]

Al-Rudainy, A. A. J. 1989. Morphometric study of four cyprinid fish and their relation to food in Hammar marsh, south Iraq. Master's thesis. Basrah University, Basrah, Iraq. [In Arabic]

Al-Saadi, H. A., S. R. Antoine, and A. K. M. Nurl Islam. 1981. Limnological investigation in Al-Hammar marsh area in southern Iraq. Nova Hedwigia 35:157–166 Stuttgart, Germany.

Al-Saboonchi, A. A., N. A. Barak, and A. R. M. Mohamed. 1986. Zooplankton of Garma marshes, Iraq. Journal of Biological Sciences Research 17(1):33–39.

Al-Saboonchi, A. A., A. R. M. Mohamed, and N. A. Barak. 1982. A study of phytoplankton in the Garma marshes, Iraq. Iraqi Journal of Marine Science 1:67–79, Basrah, Iraq.

Al-Sayab, A. A. A. 1988. Ecology and bioogy of Asian catfish, *Silurus triostegus* Heckel, 1843 in Hammar marsh, South Iraq. Master's thesis. Basrah University, Basrah, Iraq. [In Arabic]

Al-Zubaidy, A. J. M. 1985. Ecological study on the phytoplankton in some marshes near Qurna, South Iraq. Master's thesis. Master's thesis. Basrah University, Basrah, Iraq. [In Arabic]

Banister, K. E. 1980. The fishes of the Tigris and Euphrates rivers. In J. Rzóska, editor. Euphrates and Tigris Mesopotamian ecology and destiny. Monographiae Biologicae 38.

Barak, A. A., and A. R. M. Mohamed. 1982. Food habits of Cyprinid fish, *Barbus luteus* (Heckel). Iraqi Journal of Marine Science 1(1):59–67, Basrah, Iraq.

Barak, N. A. A., and A. R. M. Mohamed. 1983. Biological study of the cyprinid fish, *Barbus luteus* (Heckel) in Garma marshes. Journal of Biological Research 14(2):53–70.

Beckman, W. C. 1962. The fresh water fishes of Syria and their general biology and management. FAO Fisheries Biology Technical Paper 8.

Blaber, S. J. M., and T. G. Blaber. 1980. Factors affecting the distribution of juvenile estuarine and inshore fish. Journal of Fish Biology 17:143–162.

Blaber, S. J. M., J, Staunton-smith, D. A. Milton, G. Fry, T. Vanderelde, J. Pang, P. Wong, and O. Boon-Teck. 1998. The biology and life-history strategies of Ilisha (Teleostei: Pristigasteridae) in the coastal waters and estuaries of Sarawak. Estuarine and Coastal Shelf Science 47:499–511.

Blaber, S. J. M., J. W. Young, and M. C. Dunning. 1985. Community structure and zoogeographic affinities of the coastal fishes of the Dampier region of north-western Australia. Australian Journal of Marine and Freshwater Research 36:247–266.

Coad, B. W. 1991. Fishes of the Tigris–Euphrates basin: a critical-list. Canadian Museum of Nature, Syllogeus No. 68, Ottawa.

Dawood, A. H. 1986. Biology of common carp *Cyprinus carpio* L. in Al-Hammar Marsh, southern Iraq. Master's thesis. Basrah University, Basrah, Iraq. [In Arabic]

Evans, M. I. 2002. The ecosystem. Pages 201–219 *in* E. Nicholson and P. Clark, editors. The Iraqi marshlands: a human and environmental study. Politico's, London.

FAO (Food and Agriculture Organization of the United Nations). 1999. Fishery country profile. Food and Agriculture Organization of the United Nations, FID/CP/IRQ Rev. 2, Rome.

Hussain, N. A., and S. M. Ahmed. 1995. Seasonal composition, abundance and spatial distribution of ichthyoplankton in an estuarine subtropical part of the north western Arabian Gulf. Marine Research 4(2):135–146.

Hussain, N. A., and T. S. Ali. 1987. Some biological aspects of *Thryssa hamiltonii* and *Thryssa mystax* in Khor Al-Zubair, northwest Arab Gulf. Indian Journal of Fish of Fisheries 34(2):152–162.

Hussain, N. A., T. S. Ali, and K. D. Saud. 1989. Seasonal fluctuations and composition of fish assemblage in the Shatt Al-Arab River at Basrah, Iraq. Journal of Biological Sciences Research 20(1):139–150.

Hussain, N. A., H. A. Hamza, and K. D. Saud. 1987. Some biological aspect of the fresh water population of Shanag *Acanthopagrus latus* (Houttuyna) in Shatt Al-Arab River, Iraq. Marina Mesopotamica 2(1):29–40.

Hussain, N. A., M. K. Jabir, and U. H. Yousif. 1994. On the biology of sbour *Tenualosa ilisha* (Hamilton) in the Shatt Al-Arab river, south of Iraq, with notes on their distribution in the Arab Gulf. Marina Mesopotamica 9(1):115–139.

Hussain, N. A., A. R. M. Mohamed, S. S. Al-Noor, B. Coad, F. M. Mutlak, I. M. Al-Sudani, A. M. Mojer, A. J. Toman, and M. A. Abdad. 2006. Species composition, ecological indices, length frequencies and food habits of fish assemblages of the restored southern Iraqi marshes. Basrah University, Annual Report, Basrah, Iraq.

Hussain, N. A., A. R. M. Mohamed, J. H. Saleh, and F. M. Mutlak. 2001. The biology of juvenile and immature of *Acanthopagrus latus* in tidal pools of Khor Al-Zubair lagoon, Iraq. Marina Mesopotamica 16(1):59–68.

Hussain, N. A., A. R. M. Mohamed, U. H. Younis, and F. M. Mutlak. 1999. The biology of *Bathygobius fuscus* (Ruppell) at the intertidal mudflats of Khor

Al-Zubair lagoon, northwest Arabian Gulf. Marina Mesopotamica 14(1):119–132.

Hussain, N. A., F. M. Mutlak, and K. H. Younis. 2003. The small fish assemblages in the Shatt Al-Arab estuary, northwest Arabian Gulf. Marina Mesopotamica 18(2):179–188.

Hussain, N. A. and M. A. Taher. 2007. Effect of daily variations, diurnal fluctuations and tidal stage on water parameters of East Hammar marshland, southern Iraq. Marsh Bulletin 2(1):32–42.

Hussain, N. A., K. H. Younis, and U. H. Yousif. 1997. The composition of small fish assemblages in the river Shatt Al-Arab near Basrah, Iraq. Acta Hydrobiol 39:29–37.

Jabir, M. K., and A. A. Faris. 1989. Fecundity of sbour *Tenualosa ilisha* (Hamilton Buchana, 1822) in the Shatt Al-Arab rivers, Basrah. Marina Mesopotamica 4(2):281–296.

Jasim, A. A. W. 1988. The reproductive biology of Barbus sharpeyi in Hammar marsh, Iraq. Master's thesis. Basrah University, Basrah, Iraq. [In Arabic]

Khalaf, K. T. 1961. The marine and fresh water fishes of Iraq. Al-Rabitta Press, Baghdad, Iraq.

Lenanton, R. C. J. 1982. Alternative non-estuaries nursery habitats for some commercial and recreational important fish species of south-western Australia. Australian Journal of Marine and Freshwater Research 33:881–900.

Mahdi, N. 1962. Fishes of Iraq. Ministry of Education, Baghdad, Iraq.

Maltby, E., editor. 1994. An environmental and ecological study of the marshlands of Mesopotamia. University of Exeter, Wetland Ecosystems Research Group, Draft Consultative Bulletin, UK. Published by The AMAR Appeal Trust, London.

Margalef, R. 1968. Perspectives in ecology. University of Chicago Press, Chicago.

Mathews, C. P., J. M. Bishop, and S. Salman. 1986. Stocks of *Metapenaeus affinis* in Kuwait and Iraq waters. Kuwait Institute for Scientific Research, Final Report KISR 2231, Safat.

Maulood, B. K., G. C. F. Hinton, H. S. Kamees, F. A. K. Saleh, A. A. Shaban, and S. M. H. Al-Shahwani. 1979. An ecological survey of some aquatic ecosystems in southern Iraq. Tropical Ecology 20(1):27–40.

McErlean, A. J., S. G. O`Connor, J. A. Mihurky, and C. I. Gibson. 1973. Abundance, diversity and seasonal patterns of estuarine fish populations. Estuarine and Coastal Marine Science 1:19–36.

Mohamed, A. R. M., and T. S. Ali. 1994. The biological importance of Iraqi marshes in fish growth. Pages 205–215 *in* N. A. Hussain, editor. Ahwar of Iraq: an environmental approach. Basrah University, Marine Science Center Publication 18, Basrah, Iraq.

Mohamed, A. R. M., S. S. Al-Noor, and R. A. K. Faris. 2006. Analytical study of the fisheries in the lower reaches of the Tigris River, North Qurna, Basrah, Iraq. Iraqi Journal of Aquaculture 1:57–73, Basrah.

Mohamed, A. R. M., and N. A. Barak. 1988a. Growth and condition of a cyprinid fish, *Barbus sharpeyi* Gunther in Al-Hammar marsh, Barash, Iraq. Basrah Journal of Agricultural Science 2:18–25, Barash, Iraq.

Mohamed, A. R. M., and N. A. Barak. 1988b. Seasonal variations in some limnological features of the Garma Marshes. Basrah Journal of Agricultural Science 1:56–63, Barash, Iraq.

Mohamed, A. R. M., N. A. Hussain, and T. S. Ali. 2001. Estuarine components of the ichthyofauna of the Arabian Gulf. Marina Mesopotamica 16(2):209–224.

Mohamed, A. R. M., S. A. Hussein, and J. H. Saleh. 1998. The biology of green back grey mullet *Liza subviridis* in the northwest Arabian Gulf. Marina Mesopotamica 13(2):375–385.

Mohamed, A. R. M., S. A. Hussein, and J. H. Saleh. 2000. Food habits of *Liza carinata* in the Iraqi marine waters. Marina Mesopotamica 15(1):275–284.

Morton, R. M., B. R. Pollack, and J. P. Beumer. 1987. The occurrence and diet of fishes in a tidal inlet to a salt marsh in southern Moreton Bay, Queensland. Australian Journal of Ecology 12:217–237.

Naama, A. K. 1982. Study of some biological aspects of *Mugil dussumieri* and *Liza abu* in Al Hammar marsh. Master's thesis. Basrah University, Basrah, Iraq. [In Arabic]

Odum, W. A. 1970. Insidious alternation of the estuarine environment. Transactions of the American Fisheries Society 99:836–847.

Pankow, H., H. A. Al-Saadi, M. F. Huq, and R. A. M. Hadi. 1979. On the algae flora of the marshes near Qurna (southern Iraq). Willdenowata 8:493–506.

Partow, H. 2001. The Mesopotamian marshlands: demise of an ecosystem. United Nations Environment Programme, Early Warning and Assessment Technical Report, UNEP/DEWA/TR.01–3 Rev 1, Nairobi, Kenya.

Pielou, E. C. 1977. Mathematical ecology. John Wiely, New York.

Pinkas, L., M. S. Oliphant, and I. L. K. Iverson. 1971. Food habits of albacore, fin tuna and bonito in California waters. California Department of Fish and Game Fish Bulletin 152:1–105.

Reiss, P., and A. Farhan, editiors. 2006. Final report. Development Alternatives, Iraq Marshlands Restoration Program, Bethesda, Maryland.

Richardson, C. J., and N. A. Hussain. 2006. Restoring the Garden of Eden: an ecological assessment of the marshes of Iraq. BioScience 56(6):477–489.

Richardson, C. J., P. Reiss, N. A. Hussain, A. J. Alwash,

and D. J. Pool 2005. The restoration potential of the Mesopotamian marshes of Iraq. Science 307:1307–1311.

Rzóska, J., J. F. Talling, and K. E. Bannister. 1980. Euphrates and Tigris: Mesopotamian ecology and destiny. Dr. W. Junk, The Hague, Netherlands.

Saeed, T., N. Al-Ghandban, H. Al-Shemmari, M. Al-Mutair, and H. Al-Hashash. 1999. Preliminary assessment of the impact of draining of Iraqi marshes on Kuwait's northern marine environment, part II: sediment associated pollutants. Water Science and Technology 40:89–98.

Shannon, C. E., and W. Weaver. 1949. The mathematical theory of communication. University of Illinois Press, Urbane .

Subrahmanyam, C. B., and S. H. Drake. 1975. Studies on animals communities in two north Florida salt marshes. Bulletin of Marine Science 25(4):445–465.

Tyler, A. V. 1971. Periodic and resident component communities of the Atlantic fishes. Journal of the Fisheries Research Board of Canada 28(7):935–946.

Windell, J. T. 1971. Food analysis and rate of digestion. Pages 215–226 *in* W. E. Ricker, editor. Methods for assessment of fish production in fresh water. Blackwell Scientific Publications, Oxford, UK.

Younis, K. H. 1995. Occurrence and food habit of the juveniles of *Liza carinata* (valenciennes, 1836) in the Shatt Al-Arab river at Basrah, Iraq. Marina Mesoptamica 10(2):331–339.

Younis, K. H. 2005. Biological assessment of fish assemblage environment in Shatt Al-Arab River, Karmat Ali, Basrah. Doctoral thesis. Basrah University, Basrah, Iraq. [In Arabic]

American Fisheries Society Symposium 69:589–604, 2009

Past and Future Habitat Suitability for the Hudson River Population of Shortnose Sturgeon: A Bioenergetic Approach to Modeling Habitat Suitability for an Endangered Species

RYAN J. WOODLAND* AND DAVID H. SECOR
University of Maryland Center for Environmental Science, Chesapeake Biological Laboratory Post Office Box 38, 1 Williams Street, Solomons, Maryland 20688, USA

EDWIN J. NIKLITSCHEK
Universidad Austral de Chile, Portales 73, Coyhaique, CP 5950000, Region de Aisén, Chile

Abstract.—Diadromous species encounter highly variable water quality as they traverse freshwater, estuarine, and marine environments. The U.S. federally endangered shortnose sturgeon *Acipenser brevirostrum* is a diadromous estuarine resident species that relies heavily on tidal freshwater regions of estuaries as spawning, nursery, and foraging habitat. A recent recovery in abundance in the Hudson River shortnose sturgeon population coincided with an ecosystem shift in the tidal freshwater estuary from hypoxia to normoxia (dissolved oxygen > 4 mg/L) during the summer juvenile rearing period. Decades of persistent summertime hypoxia encompassing as much as 40% of shortnose sturgeon nursery habitat was followed by a sudden shift to normoxia (1970 to 1978) due to the U.S. Clean Water Act legislation. Here, we evaluate how past and present water quality in the tidal freshwater Hudson River affects nursery habitat suitability. Habitat suitability, as indexed by potential instantaneous growth rate, was estimated with an empirically derived bioenergetic growth model before (pre-1978: 20% and 40% dissolved oxygen [DO] saturation) and after (1988: 85% DO saturation) the shift in seasonal ecosystem oxygenation. Habitat suitability was then forecast in the context of regional climate change and potential zebra mussel *Dreissena polymorpha* oxygen demand. Results from this simulation study indicated that even moderate reductions in water quality can significantly lower habitat suitability, supporting the circumstantial association between improved water quality and shortnose sturgeon recovery. Although presently occurring at high abundance levels, Hudson River shortnose sturgeon in the future may encounter diminished nursery habitat due to warming temperatures and increased benthic oxygen demand by zebra mussels.

Introduction

Globally, most sturgeon species (Acipenseridae) are listed as either threatened or endangered (Birstein 1993). Sturgeons are particularly sensitive to anthropogenic habitat alteration (e.g., water quality, flow regulation) during early life phases (Secor and Gunderson 1998; Jager et al. 2002; Campbell and Goodman 2004; Coutant 2004; McAdam et al. 2005). Habitat degradation has been implicated as both a primary (NMFS 1998; McAdam et al. 2005; Paragamian et al. 2005) and contributing (Birstein 1993; NMFS 1998; Collins et al. 2000; Niklitschek and Secor 2005) factor to recruitment failure among sturgeon species. Lethal and sublethal environmental conditions are particularly important during the first year of life when recruitment bottlenecks can occur in restricted spawning and nursery habitats. By the same token, life table analysis demonstrated that positive population growth is expected to be

* Corresponding author: woodland@cbl.umces.edu

most sensitive to improvements in larval and juvenile habitats (Gross et al. 2002).

Despite the range-wide endangered status of shortnose sturgeon *Acipenser brevirostrum* along the east coast of North America, recent research on the Hudson River (New York, USA) population of shortnose sturgeon indicated that the population increased from c. 13,000 adults in 1979–1980 (Dovel et al. 1992) to 57,000 in 1997 (Bain et al. 2007). This corresponds to a fourfold increase over an 18-year period There is corroborating evidence that year-class strength built during the 1980s and peaked at 31,000–52,000 yearling recruits during the late 1980s and early 1990s (Woodland and Secor 2007). Therefore, the Hudson River is a unique case study: an example of robust population growth that can be examined in the context of promoting sturgeon recovery elsewhere.

Suitable nursery habitat is a useful index of the recruitment process because it integrates complex and interacting environmental factors that affect first-year growth and survival. A common means for modeling suitable habitat is empirical, based upon intensively surveyed regions. Abundances can be linked to multiple environmental factors through statistical models, providing prediction of responses under simulated environmental conditions (Crance 1986; Jager et al. 2002; Jensen et al. 2005; Austin 2007). Such methods are less feasible for highly mobile or rare species. Another principal approach uses ecophysiological responses to environmental conditions determined experimentally to predict the distribution and production of fish (Neill et al. 1994; Boisclair 2001; Niklitschek and Secor 2005). This method is intuitively appealing in that it incorporates fundamental biology into a predictive framework rather than applying a purely empirical approach. Still, it is subject to the assumption that fish can locate all suitable habitats. Therefore, such terms as potential growth, potential production, and potential habitat volume are commonly associated with this approach (Logerwell et al. 2001; Luo et al. 2001; Niklitschek and Secor 2005).

Here, we apply an ecophysiological bioenergetic modeling approach to examine the predicted suitability of nursery habitat for the upper Hudson River estuary. Historical water quality data indicate that prior to the 1970s, dissolved oxygen (DO) concentrations in a heavily polluted segment of the Hudson River estuary downstream of the Troy–Albany area (river kilometer [rkm] 235–175, Figure 1) dropped precipitously from springtime normoxia (6–7 mg/L) to c. 2 mg/L at the onset of summer (Figure 2) (20–25% DO saturation) before reaching a nadir in the early fall (<1 mg/L; Leslie et al. 1988). This c. 60-km segment, dubbed the "Albany Pool" (Boyle 1969), coincides with approximately 40% of the estimated nursery habitat for Hudson River age-0 shortnose sturgeon (Dovel et al. 1992; Haley 1999) and occurs directly downstream of spawning habitats. April–May spawning is believed to occur predominantly in the uppermost segment of the tidal freshwater Hudson River, below the Federal Dam in Troy (hereafter Troy dam) and 246 rkm from the river mouth (Dadswell et al. 1984). Larval and young-of-the-year juvenile nursery habitat in the Hudson River extends downstream from the spawning grounds and encompasses much of the tidal freshwater portion of the estuary (Bain 1997; Figure 1). Thus, the hypoxic zone below Albany may have functioned as a recruitment bottleneck, rendering much of the summertime nursery habitat unsuitable for juvenile shortnose sturgeon. The system recovered to summertime normoxia (>4 mg/L) between 1970 and 1978 (Figure 2) following national legislation that stipulated more stringent controls on pollutant discharge.

Although water quality in the tidal freshwater Hudson has improved substantially since the passage of the Federal Water Pollution Control Act Amendments of 1972 and 1977, future nursery conditions could be affected by more recent ecosystem changes. There is evidence that the invasive zebra mussel *Dreissena polymorpha* has already precipitated a 10% decline in DO saturation in bottom waters extending c. 40 km downstream of rkm 184 (Caraco et al. 2000). Caraco et al.'s analysis of water quality data (1986–1998) from this stretch of the Hudson River found that summertime (June–September) DO saturation declined from 95% to 85% following the colonization of the river by zebra mussels. This shift in water quality could not be explained through fluctuations in temperature, summer flow, organic loading, or wind stress (Caraco et al. 2000). Further, warming temperatures associated with global climate change could result in superoptimal water temperatures for shortnose sturgeon juveniles.

We used a bioenergetics model to both hindcast and forecast the effects of temperature and DO

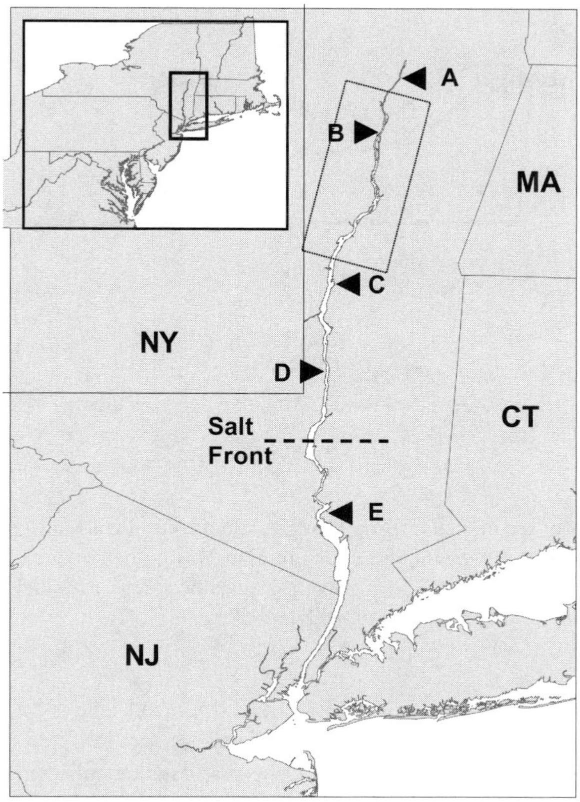

Figure 1.—Map of the tidal freshwater portion (river kilometer [rkm] 245–63) of the Hudson River, New York. The historical location of a large zone of summer hypoxia (Albany Pool), as indicated by a dotted rectangle, overlapped c. 40% of the tidal freshwater reach in this system. Letters mark the approximate locations of sampling stations used to generate long-term mean temperature estimates for the bioenergetics model: A—Green Island (rkm 248), B—Glenmont (rkm 215), C—Tivoli Bay (rkm 160), D—Poughkeepsie (rkm 120), and E-Haverstraw Bay (rkm 63).

on habitat suitability and availability. Habitat suitability was indexed by the potential instantaneous individual growth rate predicted for young-of-the-year shortnose sturgeon entering the system. Predictions of negative instantaneous growth rates for a given area were interpreted as dysfunctional nursery habitat. In our historical simulation, we investigated potential loss of nursery habitat in the presence of moderate (40% DO saturation) to severe (20% DO saturation) hypoxia during the summer months in the Albany Pool area. In the future simulation, we assumed moderate increases in temperature and decreases in DO due to the influence of global warming and zebra mussels, respectively. We evaluated the amplitude of nursery habitat degradation associated with an increase of temperature by 1–3°C and/or a decrease in DO (to 70% saturation) throughout the entire nursery region.

Methods

Habitat Data

Historical water temperature data were obtained for five independent sampling stations (Table 1) evenly spaced along the tidal freshwater portion of the Hudson River (Figure 1). Data were downloaded from the U.S. Geological Survey (Green Island station #1358000) and National Estuarine Research Reserve System Web sites (Tivoli Bay South), collected from the literature (Haverstraw Bay, Drisco et al. 2003), and obtained through individual cor-

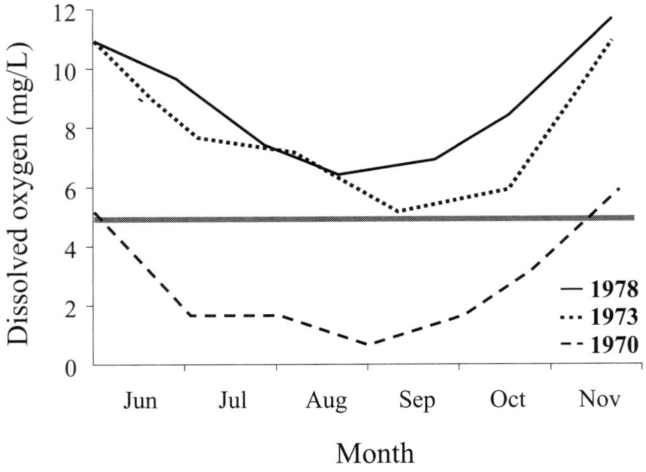

Figure 2.—Time series of spring–fall dissolved oxygen conditions in the Hudson River during the 1970s at Glenmont, New York (directly downstream of the Albany, New York). The horizontal thickened line (c. 5 mg/L) indicates an approximate threshold for negative growth in juvenile sturgeons under summer conditions (Secor and Niklitschek 2001). Data from Leslie et al. 1988 (Figure 72).

respondence (R. J. Alstadt, Poughkeepsie water treatment facility, personal communication; New York Department of Environmental Conservation, Glenmont station #13010142). Available data from 1987 to 2005 were used to generate single point-estimates of average summer temperature at the five primary water quality stations (Table 2). Data collection methods vary between stations: temperature data at Green Island is collected at 3 m (mid-depth), Glenmont data reflected surface conditions, Poughkeepsie data are collected just inside an intake pipe c. 2.4 m above the riverbed, and Haverstraw Bay (<3 m depth) and Tivoli Bay (0.7–2.1 m depth) data are collected using a handheld multiparameter probe in shallow areas. Ordinary least-squares linear regression was used to generate point-estimates

Table 1.—Sampling station site name, location (river kilometer), data attributes (datum collection frequency, total data included, data range [years]), and source of historical water temperature data from the Hudson River. Sources include the U.S. Geological Survey (USGS, www.usgs.gov), New York Department of Environmental Conservation (NYDEC), National Estuarine Research Reserve System (NERRS, www.nerrs.gov), Poughkeepsie water treatment facility (PWTF), and SUNY Marine Science Research Center (MSRC, Drisco et al. 2003).

		Temperature data			
Site	River km	Sampling frequency	Total data (*N*)	Data range	Source
Green Island	248	~monthly	73	1970–1985, 1988–1994	USGS
Glenmont	215	monthly	18	1993–2002	NYDEC
Tivoli Bay	160	0.5 h	7,616	1995–2000	NERRS
Poughkeepsie	120	daily	1,448	1987–2005	PWTF
Haverstraw Bay	37–63	biweekly	52[a]	1985–2002	MSRC

[a] Each Haverstraw Bay datum used in this analysis represents a biweekly mean calculated from 25 individual temperature measurements collected from the Haverstraw–Tappan Zee region of the Hudson River (see Drisco et al. 2003).

Table 2.—Nineteen-year time-series of observed and reconstructed (*) river temperature data from five sampling stations along the upper Hudson River estuary (river kilometer 248–63).

	Site				
Year	Green Island	Glenmont	Tivoli South	Poughkeepsie	Haverstraw
1987	20.8	21.2*	22.5*	21.7	22.6
1988	24.7	21.9*	23.2*	22.4	22.3
1989	14.7	21.4*	22.7*	21.9	22.7
1990	23.0	22.1*	23.3*	22.6	23.2
1991	23.0	22.3*	23.5*	22.8	21.5
1992	21.5	20.9*	22.3*	21.4	20.4
1993	21.4	21.3	23.1*	22.4	22.7
1994	22.6	21.8	23.0*	22.3	23.5
1995	21.6*	18.6	21.3	22.5	23.8
1996	20.7*	18.6	21.3	21.6	21.4
1997	21.6*	21.8	21.0	22.5	22.2
1998	22.2*	19.3	21.7	23.1	23.2
1999	21.6*	14.1	21.5	22.4	22.9
2000	20.2*	16.7	20.3	21.1	20.7
2001	22.2*	20.3	23.8*	23.1	23.1
2002	22.2*	19.4	23.7*	23.1	23.9
2003	21.4*	21.8*	23*	22.3	24.3*
2004	20.8*	21.1*	22.5*	21.7	23.7*
2005	22.9*	23.3*	24.3*	23.7	25.7*

of missing temperature data at some stations based on the continuous temperature data available from the Poughkeepsie water treatment facility (explanatory variable). The Poughkeepsie versus Green Island, Glenmont, and Tivoli Bay regressions ($R^2 >$ 0.62, *F*-stat = 144.2, $p \leq 0.01$) were restricted to data collected concurrently at a daily resolution to avoid the effect of short-term weather variability on temperature relationships between stations. The Poughkeepsie versus Haverstraw Bay temperature regression was based upon biweekly mean values; therefore, the relationship was less precise, $R^2 = 0.5$, although highly significant at $\alpha = 0.05$ (F-stat = 14.2, $p = 0.002$).

Observed and estimated temperature data from each station were averaged across months from July to September with summer water temperature calculated as the mean of the monthly averages. Incremental change in water temperature between sampling stations was assumed to follow a linear relationship with distance and was interpolated at c. 10-km intervals between stations. Each interpolated point was considered representative of the river conditions occurring in an entire river segment. The final result of these regressions was a spatially explicit reconstruction of average summer water temperature at 21 individual points and for 20 river segments along the upper Hudson River estuary. This reconstructed temperature record (Table 3) extended 185 km from an upstream boundary at Green Island (directly upstream of Troy dam, rkm 248) downstream to the salt front at Haverstraw Bay (rkm 63). The highly fragmented historical DO data precluded any attempt to reconstruct a spatially representative model of dissolved oxygen. Instead, we chose to simulate historical and future conditions through proscribed reductions in % DO in the area(s) of interest.

Habitat Availability and Suitability Index

Estimates of individual potential instantaneous daily growth rate for each river segment were generated using a bioenergetics model (Appendix 1) derived by Niklitschek (2001) through extensive laboratory and mesocosm studies. The model is parameterized with temperature, DO saturation, salinity, and fish weight, reflecting the observed and/or modeled de-

Table 3.—Nineteen year time-series of observed, reconstructed (*) and linearly interpolated river temperature data along the upper Hudson River estuary (river kilometer [rkm] 248–63). River km indicates the down-estuary boundary of each segment (excluding rkm 248, which represents the upstream boundary of the data set). Code identifiers indicate sampling stations (see key at bottom of table).

River km	1987	1988	1989	1990	1991	1992	1993	1994	1995	1996	1997	1998	1999	2000	2001	2002	2003	2004	2005
A 248	20.8	24.7	14.7	23.0	23.0	21.5	21.4	22.6	21.6*	20.7*	21.6*	22.2*	21.6*	20.2*	22.2*	22.2*	21.4*	20.8*	22.9*
239.8	20.9	24.0	16.4	22.8	22.8	21.3	21.4	22.4	20.9	20.2	21.6	21.5	19.7	19.4	21.8	21.5	21.5	20.8	23.0
231.3	21.0	23.3	18.0	22.5	22.6	21.2	21.4	22.2	20.1	19.7	21.7	20.7	17.8	18.5	21.3	20.8	21.6	20.9	23.1
222.6	21.1	22.6	19.7	22.3	22.5	21.0	21.3	22.0	19.3	19.1	21.7	20.0	15.9	17.6	20.8	20.1	21.7	21.0	23.2
B 215	21.2*	21.9*	21.4*	22.1*	22.3*	20.9*	21.3	21.8	18.6	18.6	21.8	19.3	14.1	16.7	20.3	19.4	21.8*	21.1*	23.3*
205.8	21.4	22.1	21.6	22.3	22.5	21.1	21.6	22.0	19.0	19.0	21.6	19.7	15.3	17.3	20.9	20.1	22.0	21.4	23.4
196.7	21.6	22.3	21.8	22.5	22.7	21.3	21.9	22.2	19.5	19.5	21.5	20.1	16.5	17.9	21.5	20.8	22.2	21.6	23.6
187.5	21.9	22.6	22.0	22.7	22.9	21.6	22.2	22.4	19.9	19.9	21.4	20.5	17.8	18.5	22.1	21.6	22.4	21.8	23.8
178.3	22.1	22.8	22.3	22.9	23.1	21.8	22.5	22.6	20.4	20.4	21.2	20.9	19.0	19.1	22.6	22.3	22.6	22.0	24.0
169.2	22.3	23.0	22.5	23.1	23.3	22.0	22.8	22.8	20.9	20.8	21.1	21.3	20.3	19.7	23.2	23.0	22.8	22.3	24.2
C 160	22.5*	23.2*	22.7*	23.3*	23.5*	22.3*	23.1*	23.0*	21.3	21.3	21.0	21.7	21.5	20.3	23.8*	23.7*	23.0*	22.5*	24.3*
150	22.3	23	22.5	23.1	23.3	22.1	22.9	22.9	21.6	21.4	21.4	22.1	21.7	20.5	23.6	23.6	22.9	22.3	24.2
140	22.1	22.8	22.3	22.9	23.1	21.8	22.8	22.7	21.9	21.5	21.7	22.4	22.0	20.7	23.4	23.4	22.7	22.1	24.0
130	21.9	22.6	22.1	22.7	22.9	21.6	22.6	22.5	22.2	21.5	22.1	22.7	22.2	20.9	23.3	23.2	22.5	21.9	23.9
D 120	21.7	22.4	21.9	22.6	22.8	21.4	22.4	22.3	22.5	21.6	22.5	23.1	22.4	21.1	23.1	23.1	22.3	21.7	23.7
110.5	21.9	22.4	22	22.7	22.6	21.3	22.4	22.5	22.7	21.6	22.4	23.1	22.5	21.1	23.1	23.2	22.6	22.0	24.1
101	22.0	22.4	22.2	22.8	22.4	21.1	22.5	22.7	22.9	21.6	22.4	23.1	22.6	21.0	23.1	23.4	23.0	22.3	24.4
91.5	22.2	22.4	22.3	22.9	22.1	20.9	22.5	22.9	23.1	21.5	22.3	23.1	22.6	20.9	23.1	23.5	23.3	22.7	24.7
82	22.3	22.3	22.4	23.0	21.9	20.8	22.6	23.1	23.3	21.5	22.3	23.1	22.7	20.8	23.1	23.6	23.6	23.0	25.0
72.5	22.5	22.3	22.6	23.1	21.7	20.6	22.6	23.3	23.5	21.4	22.2	23.2	22.8	20.8	23.1	23.8	24.0	23.4	25.3
E 63	22.6	22.3	22.7	23.2	21.5	20.4	22.7	23.5	23.8	21.4	22.2	23.2	22.9	20.7	23.1	23.9	24.3*	23.7*	25.7*

A—Green Island, B—Glenmont, C—Tivoli Bay, D—Poughkeepsie, E—Haverastraw Bay.

pendence of consumption, metabolism (routine and active), and excretion rates on these four parameters. Postprandial and egestion rates are modeled solely on temperature and DO saturation values in accordance with laboratory observations (Niklitschek 2001). The model was developed under maximum consumption conditions; therefore, growth estimates represent maximum potential rates. For the purposes of these simulations, we chose to predict $\hat{g}$ for a 5-g young-of-year shortnose sturgeon (hereafter "juvenile"), a size that approximates the weight of a 4–6 month old fish (Dadswell et al. 1984; Carson 1987). In all model runs, salinity was restricted to zero in accordance with observed low-salinity preferences of age-0 fish (Dadswell et al. 1984). We considered estimates of instantaneous growth rate at each river segment as an index of habitat suitability (HSI); therefore, increases or decreases in instantaneous growth rate estimates for each river segment were interpreted as an increase or decrease in HSI. It is important to note that there are several other definitions of HSI in the literature (e.g., Crance 1986; Bray 1996; Vadas and Orth 2001); for the purposes of this study, it is defined strictly as the modeled potential instantaneous growth rate for a particular river segment. The total available summertime nursery habitat (138 km) was calculated as the distance between Troy dam (rkm 246) and the mean seasonal location of the salt front (rkm 108, de Vries and Weiss 2001). The percent of available nursery habitat was calculated as the fraction of the total nursery habitat in which HSI values were predicted to be positive under a given scenario. Nursery habitats that did not support positive HSI values were considered unavailable to juveniles.

Scenario 1: Historical Perspective—Legacy of the Albany Pool

This scenario simulated the impact of the historically polluted Albany Pool area on the availability and suitability of nursery habitat (Table 4). A baseline model was populated with mean summer temperatures observed or reconstructed from 1987 to 2005 for each river segment. Dissolved oxygen saturation was fixed at 85% commensurate with recent summer (June–September) monitoring data (Caraco et al. 2000). For hindcast simulations, DO saturation was decremented for the area corresponding to the historical extent of the Albany Pool. Dissolved oxygen saturation was reduced to 40% under the moderate hypoxia scenario and 20% under the severe hypoxia scenario for all river segments between rkm 231 and 178 (DO remained fixed at 85% up- and downstream of the Albany Pool area). Temperatures used in the hindcast simulations were the same as those used in the baseline model. The model was run and the effect on HSI within the Albany Pool

Table 4.—Summary table of bioenergetic modeling simulations of habitat suitability for a 5-g juvenile shortnose surgeon in the upper Hudson River estuary. Temperature indicates range of years used for simulation and simulated change (e.g., +1°C), Salinity was held constant at 0 for all model runs. DO = dissolved oxygen.

		Model input	
Simulation	Scenario	DO (% sat)	Temperature
Historical: Albany Pool	Baseline run (current conditions)	85	1987–2005
	Moderate hypoxia	40	1987–2005
	Severe hyopoxia	20	1987–2005
Forecast: Climate change	Baseline run (85% DO)	85	1988
	1°C increase	85	1988 + 1°C
	3°C increase	85	1988 + 3°C
	Baseline run (70% DO)	70	1988
	1°C increase	70	1988 + 1°C
	3°C increase	70	1988 + 3°C

area was compared among the baseline and historical scenarios. We then examined the change (if any) in nursery habitat availability between scenarios.

Scenario 2: Future Perspective—Temperature and Dissolved Oxygen

We simulated the impact of future regional climate change due to a combination of increased temperature and decreased dissolved oxygen saturation (Table 4). Mean summer temperature from 1988 was selected as a baseline temperature regime and DO was fixed at 85%. This year was representative of the period 1987–2005: temperatures at Tivoli Bay, Poughkeepsie, and Haverstraw Bay deviated less than ±1°C from the 19-year average while the Green Island and Glenmont sites were within ±1.5 SD of the mean site-specific temperature. Also, 1988 produced the strongest year-class on record from 1980 to 1999 as hindcast from the 2004–2005 extant population (Woodland and Secor 2007) and is therefore assumed to represent appropriate physical conditions for juvenile shortnose sturgeon survival and growth.

In the first simulation series, temperature was increased by 1°C and then by 3°C for all river segments (DO fixed at 85%). Each increase was followed by a model run, the result of which was compared to baseline (1988) HSI values. We then simulated a concurrent decrease in dissolved oxygen saturation of 15%, representing increased oxygen demand from zebra mussels. Under this reduced oxygen simulation, DO saturation was fixed at 70% basin-wide while temperatures were again raised by 1°C and 3°C increments for each river segment. Patterns in estimated habitat suitability for all model runs were analyzed in a manner similar to the historical analysis.

Results

Habitat Suitability

Present-day summertime nursery habitat in the Hudson River appears to exceed the thermal growth optima for shortnose sturgeon by 2–3°C (Figure 3). The shape of the temperature-growth response is conserved under 85% and 70% DO conditions; yet, thermal conditions corresponding to maximum HSI differs between the two DO levels. At 85%

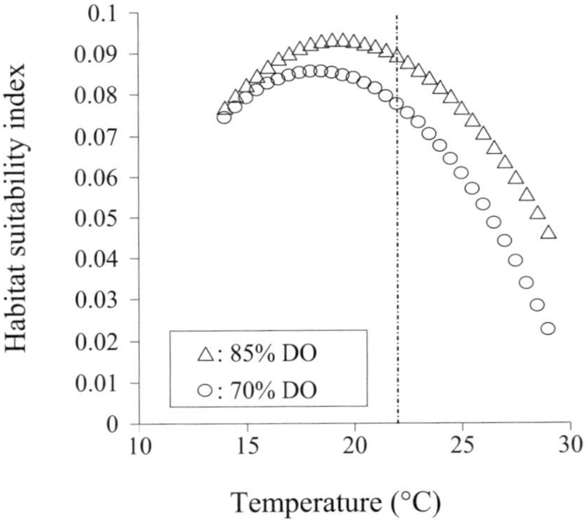

Figure 3.—Habitat suitability as indexed by modeled potential instantaneous growth rates for a 5-g juvenile shortnose sturgeon under the full spectrum of summer (July–September) water temperatures recorded from the tidal freshwater portion (river kilometer 248–63) of the Hudson River between 1987 and 2005. Mean water temp (21.9°C) over this period is denoted by the dashed line.

DO saturation HSImax (0.093/d) occurs at 19.5°C while HSImax (0.086/d) occurs at 18°C under 70% DO conditions (Figure 3). The average summer temperature of the upper Hudson River estuary from 1987 to 2005, 21.9°C, indicates a nursery-wide mean HSI of 0.089/d and 0.078/d under 85% and 70% DO saturation conditions, respectively. This corresponds to HSI values that are c. 9% less than optimal HSI values for either dissolved oxygen value (Figure 3).

Scenarios

Simulation 1: historical perspective—legacy of the Albany Pool.—The model predicted substantially reduced habitat suitability under moderate to severe hypoxia scenarios (Figure 4). The baseline model (85% DO) indicated a threefold higher average HSI value (0.091/d ± 0.0003 [±1 SD]) than that estimated under moderately hypoxic conditions (40% DO, HSI = 0.028/d ± 0.016) in the Albany Pool area. Overall habitat availability did not change under the moderate hypoxia scenario because HSI values remained positive throughout the river segments. The severe hypoxic stress scenario (20% DO) resulted in

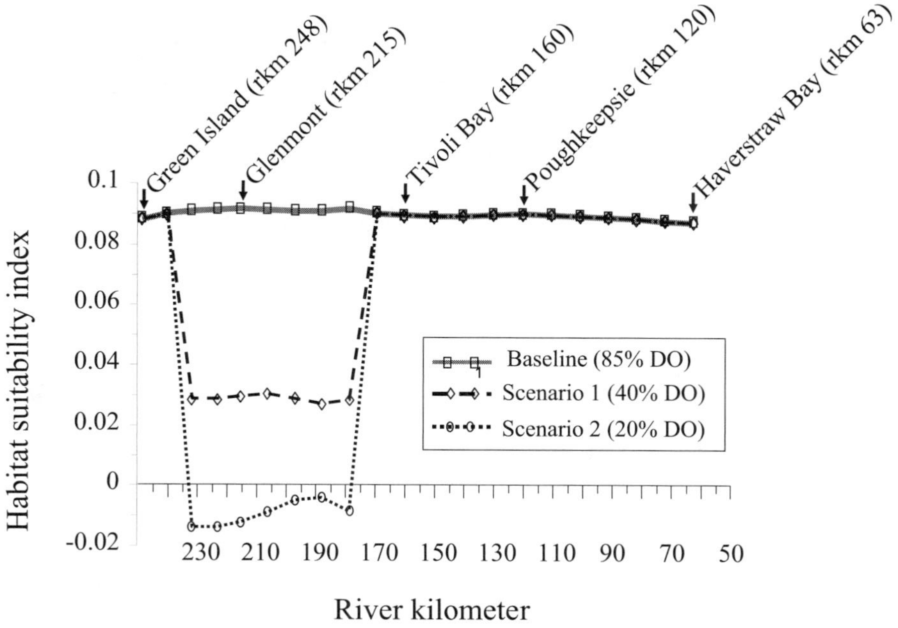

Figure 4.—Predicted habitat suitability is plotted against river kilometer (rkm) for a 5-g juvenile shortnose sturgeon under recent (baseline) and historical water quality simulations in the upper Hudson River estuary (tidal freshwater). Baseline conditions were modeled by fixing dissolved oxygen (DO) saturation at 85% for the entire estuary. Historical water quality scenarios were modeled by imposing moderate (40% DO) to severe (20% DO) seasonal summer hypoxia conditions on 53 km of the river downstream of Albany, New York (rkm 231–148).

a severe reduction in nursery habitat suitability with HSI declining to negative values in the Albany Pool area (–0.010/d ± 0.033). Under the severe hypoxia scenario, total habitat availability (185 km) was reduced to 55% of available habitats (Figure 4).

Simulation 2: Future scenarios—temperature and dissolved oxygen.—Habitat suitability index values declined in the presence of a 1°C increase in water temperature (Figure 5; % DO saturation was maintained at 85% across river segments). Mean HSI across all river segments was reduced to 0.086/d ± 0.0008, a 4% decline from average habitat suitability under baseline conditions (0.090/d ± 0.007). Raising water temperatures by a total of 3°C substantially reduced the average HSI, lowering the basin-wide mean suitability index to 0.077/d ± 0.011. Overall, the 3°C warming scenario reduced the average HSI by 14% from current temperature conditions. Habitat availability was unaltered.

When combined with a 15% decline in DO saturation, increasing water temperatures resulted in a significant decline in HSI values, both within and between climate change scenarios (Figure 5). The 70% DO scenario result was qualitatively similar to the 85% scenario: baseline HSI > 1°C HSI > 3°C HSI. In the 70% DO scenario, increasing temperature by 1°C reduced HSI to an average of 0.073/d ±0.010, 6% lower than from average baseline conditions (0.078/d ± 0.009). When temperature was increased by 3°C in concert with 70% DO, average HSI declined to 0.061/d ± 0.014. This represents a 22% reduction in average habitat suitability from the baseline 70% DO run. Habitat availability was unaltered. These results indicate that, over all river segments, habitat suitability is substantially less at 70% DO with a 1°C increase than under 85% DO and a 3°C increase.

Discussion

Assessing the current and future status of freshwater and estuarine nursery habitats is particularly impor-

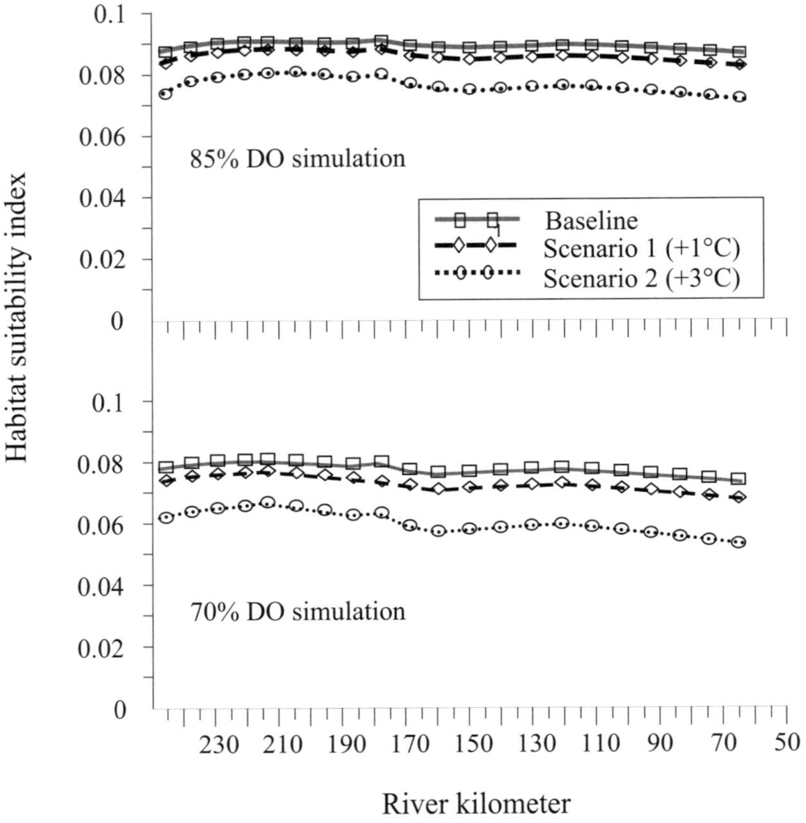

Figure 5.—Predicted habitat suitability is plotted against river kilometer for a 5-g juvenile shortnose sturgeon under a series of climate simulations on the Hudson River. Two baseline runs predict habitat suitability under average summer water temperatures (1988 data) and dissolved oxygen saturation conditions fixed at 85% (upper panel) and 70% (lower panel). The effect of two regional warming scenarios (+1°C and + 3°C above baseline temperatures) on habitat suitability are modeled under 85% (upper panel) and 70% (lower panel) dissolved oxygen saturation conditions.

tant to the management of diadromous species. A recent synthesis of available data concluded that 18 North American diadromous species (excluding Pacific salmonids) are at risk of extinction from habitat destruction or degradation (Musick et al. 2000). Risks to diadromous fishes arise in part from the influence of fixed migration corridors (e.g., riverine habitat) that can serve as natural constrictions and may force species to traverse or remain within a compromised environment. Human development of riverine and estuarine areas can directly influence local water quality (e.g., DO, temperature), hydrography, and sedimentation rates, processes that can substantially alter the physical and chemical habitat of diadromous fishes. In the case of anadromous species, this can limit the accessibility and availability of suitable spawning and nursery habitat (Maurice et al. 1987; Collins et al. 2000, Niklitschek and Secor 2005). Conversely, this suggests that these populations may respond favorably to restoration of these critical movement and nursery habitats. The potential benefit of nursery habitat restoration depends on the survival elasticity of the affected life stage(s) (Schaaf et al. 1987; Gross et al. 2002), the total area and potential contribution of the habitat to the parental stock (Beck et al. 2001; Dahlgren et al. 2006), and the emergent effects of habitat expansion on population resiliency (Kraus and Secor 2005). Nursery habitats may be limited to suitable patches within a single estuary-river system (i.e., distinct population segments) or it may include multiple estuarine systems in the case of a panmictic stock (Ray 1997). The life history of shortnose sturgeon relegates populations to specific estuaries wherein changes in habitat suitability affect local population dynamics, making this species a tractable choice for this type of analysis.

Habitat degradation is listed as a principal factor in the decline of shortnose sturgeon populations range-wide (NMFS 1998), a conclusion corroborated by results from this modeling exercise. Our analysis suggests nursery habitat degrades rapidly under realistic historical and future water quality scenarios. Laboratory studies have shown that juvenile and yearling shortnose sturgeon are unusually sensitive to hypoxia and demonstrate decreased routine metabolism, consumption, feeding metabolism, growth, and survival at DO levels less than 40% saturation (Jenkins et al. 1993; Niklitschek 2001; Secor and Niklitschek 2001; Campbell and Goodman 2004). Prolonged exposure to the level of pervasive hypoxia (<30% DO saturation) observed prior to the late 1970s would have been lethal to juveniles (Niklitschek 2001; Campbell and Goodman 2004). Also, there is evidence that larval dispersal downstream is extensive in northern populations (Kynard and Horgan 2002), a behavioral trait that would effectively increase the likelihood of exposure to summer hypoxia for late-spring spawned fish dispersing downriver (Figure 1).

In addition to losses from direct asphyxiation, reduced production associated with increased metabolic costs (Secor and Gunderson 1998) or synergistic interactions among stressors (i.e., "habitat squeeze" sensu Coutant 1985) may lead to indirect mortality. The potential impact of a temperature-hypoxia squeeze is apparent from the future scenario modeling (Figure 5) in which there was a greater net decline in suitability when increased temperature was coupled with a 15% reduction in DO (Figure 5). In the Hudson River, larvae and age-0 juveniles have been collected along channels and deeper habitats (3–23 m, Hoff et al. 1988), areas prone to hypoxia. A study of Atlantic croaker *Micropogonias undulatus* in the Neuse River estuary (North Carolina, USA) found that intermittent hypoxia restricted fish to resource-limited shallow habitats led to crowding that may have reduced growth rates and drastically reduced prey densities in preferred deeper habitats (Eby et al. 2005). Therefore, the return of normoxia during the critical summer growth and development period may have eliminated a substantial recruitment bottleneck to the Hudson River population. Further, improved habitat quality may be beneficial to all age-classes and stimulate population growth through increased production across life stages (Gross et al. 2002).

Our modeling efforts suggest that nursery habitat suitability for shortnose sturgeon is very responsive to changes in water quality. While empirical evidence is lacking to conclusively identify the factor or suite of factors that led to fourfold population growth in c. 20 years, nursery habitat suitability increased 62–111% and a potentially lethal dispersal barrier (Albany Pool) was eliminated just prior to the species' recovery. Results from the 20% DO scenario (historical simulation) suggest that shortnose sturgeon may have recovered 45% of the preindustrial total nursery habitat availability with the recovery of summertime normoxia. While hypoxia levels of 20% DO saturation have been reported in the Albany Pool area (Boyle 1969; Leslie et al. 1988), the actual magnitude of summertime hypoxia would have fluctuated seasonally and interannually due to variability in precipitation, wind stress, river depth, and anthropogenic inputs. Interestingly, the Albany Pool area, which was once the site of pervasive seasonal hypoxia, is predicted, based upon temperature and DO conditions, to be one of the most productive nursery areas (HSI = 0.089/d ± 0.001 [rkm 231–178]) in the river. Increasing habitat availability would increase potential recruitment and overall carrying capacity (Schaaf et al. 1993; Gallaway et al. 1999; Luo et al. 2001; Niklitschek and Secor 2005). In fact, the recent fourfold increase in abundance of adult shortnose sturgeon (Bain et al. 2007) may reflect a reversion in Hudson River carrying capacity to preindustrial conditions.

The accuracy of the simulations is necessarily dependent on the quality of the data available to the model and the parameterization of the model. Specific attributes of the temperature data that we used to reconstruct historical conditions in the Hudson River have the potential to bias our hindcast estimates of habitat suitability. These data were collected for a variety of water quality monitoring purposes and, as such, may not be truly representative of benthic conditions as experienced by shortnose sturgeon. Heterogeneity in the physical features of estuaries (e.g., width, depth, flow rate, bathymetry) can result in substantial spatial variability in vertical mixing, solar heating, and thermal stratification. Such issues are ameliorated in the freshwater Hudson River estuary, which shows little stratification in comparison to partially mixed estuaries like the Chesapeake Bay. Either over- or underestimating historical temperatures would bias our growth predictions; the directionality of the bias in this case would be an underestimation of potential growth rate in the face of exaggerated water temperatures. This is an important caveat that accompanies bioenergetic-based habitat modeling and must be taken into consideration when attempting to infer causality from model output. Second, the bioenergetic model applied in this analysis was parameterized through an extensive laboratory-based rearing study that utilized Savannah River broodstock. Research has shown that shortnose sturgeon display latitude-dependent growth trajectories in which fish from southern estuaries (e.g., Pee-Dee, Altamaha, Savannah) grow at an accelerated rate and reach a lower asymptotic size than northern fish (Dadswell et al. 1984 and references therein). Therefore, the bioenergetics of the Hudson River population may be sufficiently different from the Savannah River population. Still, we would expect Savannah progeny to be better adapted to warm conditions, suggesting a conservative response to increased temperature relative to northern stocks. In terms of dissolved oxygen tolerances, there is little reason to expect that Hudson River progeny shortnose sturgeon would be less sensitive to hypoxia than Savannah River progeny.

A change in weight-at-age of juvenile Hudson River shortnose sturgeon over time could introduce bias into our modeling results due to the sensitivity of the bioenergetic model to sturgeon weight. Similarly, if the historical bottleneck on juvenile survival occurred earlier or later in the year, our selection of 5 g as an appropriate size-at-age would be inaccurate. Changing the weight parameter in the model alters the specific values of individual estimates; yet the patterns of habitat suitability described here for a 5 g individual are qualitatively similar to those obtained for fish up to 18.3 g. Unfortunately, we do not have sufficient data on fish less than 5 g to evaluate scenario sensitivity to a smaller size-class.

This shortnose sturgeon case study indicates that efforts directed at recovering water quality can substantially increase the suitability and availability of spatially restricted nursery habitats. In light of the highly elastic survivorship rates during the young-of-the-year period (Gross et al. 2002), we believe that we have provided an important piece of circumstantial evidence linking the recovery of Hudson River shortnose sturgeon to the restoration of summertime normoxia. A similar situation occurred in the Delaware River (Delaware, USA), in which recent studies have linked restored water quality to the reproductive success of anadromous species. One study found the abundance of age-0 American shad *Alosa sapidissima* and striped bass *Morone saxatilis* increased 1,000-fold in a decade (1980s–1990s) following the return of normoxic conditions to a 40-km stretch of the river (Weisberg et al. 1996). In another study (Maurice et al. 1987), evidence of increased American shad egg and larvae abundance was attributed to the removal of a seasonal hypoxic zone that had prevented spawning aggregations of anadromous shad to reach the upstream nontidal spawning habitat. Interestingly, our simulation study indicated nonlinear bioenergetic responses due to the interaction of temperature and DO, suggesting that current regulations aimed at maintaining water quality standards may be insufficient to protect sensitive species or life stages under future climate changes. This uncertainty provides further incentive for fisheries managers to adopt a precautionary approach when considering proposals that develop or somehow alter known or potential habitat.

Acknowledgments

This work builds upon research that was supported by the Hudson River Foundation (RJW—thesis, EJN—dissertation), the Nature Conservancy, the National Science Foundation, and the Chilean Ministry of Planning and Cooperation (EJN—dissertation). We'd like to thank Randy J. Alstadt, Paul Lill, and Matthew Geho (PWTF); Timothy Hoffman (USGS); and William Andrews (NYDEC) for provision of temperature data and sampling information. Particular thanks to Alex Haro, Joseph Hightower, and an anonymous reviewer for providing constructive comments. Finally, thanks to Alex Haro and the organizational committee for accepting a submission of this manuscript. This is contribution 4253 of the University of Maryland Center for Environmental Science.

References

Austin, M. 2007. Species distribution models and ecological theory: a critical assessment and some possible new approaches. Ecological Modeling 200:1–19.

Bain, M. B. 1997. Atlantic and shortnose sturgeons of the Hudson River: common and divergent life history attributes. Environmental Biology of Fishes 48:347–358.

Bain, M. B., N. Haley, D. L. Peterson, K. K. Arend, K. E. Mills, and P. J. Sullivan. 2007. Recovery of a US endangered fish. PLoS ONE 2:e168, San Francisco.

Beck, M. W., K. L. Heck, K. W. Able, D. L. Childers, D. B. Eggleston, B. M. Gillanders, B. Halpern, C. G. Hays, K. Hoshino, T. J. Minello, R. J. Orth, P. F. Sheridan, and M. R. Weinstein. 2001. The identification, conservation, and management of estuarine and marine nurseries for fish and invertebrates. Bioscience 51:633–641.

Birstein, V. J. 1993. Sturgeons and paddlefishes: threatened fishes in need of conservation. Conservation Biology 7:773–787.

Boisclair, D. 2001. Fish habitat modeling: from conceptual framework to functional tools. Canadian Journal of Fisheries and Aquatic Sciences 58:1–9.

Boyle, R. H. 1969. The Hudson River: a natural and unnatural history. W. W. Norton & Company, New York.

Bray, K. E. 1996. Habitat models as tools for evaluating historic change in the St Marys river. Canadian Journal of Fisheries and Aquatic Sciences 53:88–98.

Campbell, J. G., and L. R. Goodman. 2004. Acute sensitivity of juvenile shortnose sturgeon to low dissolved oxygen concentrations. Transactions of the American Fisheries Society 133:772–776.

Caraco, N. F., J. J. Cole, S. E. G. Findlay, D. T. Fischer, G. G. Lampman, M. L. Pace, and D. L. Strayer. 2000. Dissolved oxygen declines in the Hudson River associated with the invasion of the zebra mussel (*Dreissena polymorpha*). Environmental Science & Technology 34:1204–1210.

Carson, D. M. 1987. Gut contents of juvenile shortnose sturgeon in the upper Hudson estuary. Copeia 3:796–802.

Collins, M. R., S. G. Rogers, T. I. J. Smith, and M. L. Moser. 2000. Primary factors affecting sturgeon populations in the southeastern United States: fishing mortality and degradation of essential habitats. Bulletin of Marine Science 66:917–928.

Coutant, C. C. 1985. Striped bass, temperature, and dissolved-oxygen: a speculative hypothesis for environmental risk. Transactions of the American Fisheries Society 114:31-61.

Coutant, C. C. 2004. A riparian habitat hypothesis for successful reproduction of white sturgeon. Reviews in Fisheries Science 12:23–73.

Crance, J. H. 1986. Habitat suitability index models and instream flow suitability curves: shortnose sturgeon. U.S. Fish and Wildlife Service Biological Report 82(10.129).

Dadswell, M. J., B. D. Taubert, T. S. Squiers, D. Marchette, and J. Buckley. 1984. Synopsis of biological data on shortnose sturgeon, *Acipenser brevirostrum* LeSueur 1818. NOAA Technical Report NMFS 14.

Dahlgren, C. P., G. T. Kellison, A. J. Adams, B. M. Gillanders, M. S. Kendall, C. A. Layman, J. A. Ley, I. Nagelkerken, and J. E. Serafy. 2006. Marine nurseries and effective juvenile habitats: concepts and applications. Marine Ecology Progress Series 312:291–295.

de Vries, M. P. and L. A. Weiss. 2001. Salt-front movement in the Hudson River estuary, New York—simulations by one-dimensional flow and solute transport models. U.S. Geological Survey, Water Resources Investigations Report 99–4024, Troy, New York.

Dovel, W. L., A. W. Pekovitch, and T. J. Berggren. 1992. Biology of the shortnose sturgeon (Acipenser brevirostrum LeSueur, 1818) in the Hudson River estuary, New York. Pages 187–216 *in* C. L. Smith, editor. Estuarine research in the 1980s. State University of New York Press, Albany.

Drisco, D. M., S. M. Moser, and D. O. Conover. 2003. Nearshore fish communities of the mid-Hudson River estuary, 1985-2002. Marine Sciences Research Center, Special Report #129, Ref. 03-01, Stony Brook, New York.

Eby, L. A., L. B. Crowder, C. M. McClellan, C. H. Peterson, and M. J. Powers. 2005. Habitat degradation from intermittent hypoxia: impacts on demersal fishes. Marine Ecology Progress Series 291:249–261.

Gallaway, B. J., J. G. Cole, R. Meyer, and P. Roscigno. 1999. Delineation of essential habitat for juvenile red snapper in the northwestern Gulf of Mexico. Transactions of the American Fisheries Society 128:713–726.

Gross, M. R., J. Repka, C. T. Robertson, D. H. Secor, and W. van Winkle. 2002. Sturgeon conservation: insights from elasticity analysis. Pages 13–30 *in* W. van Winkle, P. J. Anders, D. H. Secor, and D. A. Dixon, editors. Biology, management, and protection of North American sturgeon. American Fisheries Society, Symposium 28, Bethesda, Maryland.

Haley, N. J. 1999. Habitat characteristics and resource use patterns of sympatric sturgeons in the Hudson River estuary. Masters thesis. University of Massachusetts, Amherst.

Hoff, T. B., R. J. Klauda, and J. R. Young. 1988. Contribution to the biology of shortnose sturgeon in the Hudson River Estuary. Pages 171–189 *in* C. L. Smith, editor. Fisheries research in the Hudson River. State University of New York Press, Albany.

Jager, H. I., W. van Winkle, J. A. Chandler, K. B. Lepla, P. Bates, and T. D. Counihan. 2002. A simulation study of factors controlling white sturgeon recruitment in the Snake River. Pages 127–150 *in* W. van Winkle, P. J. Anders, D. H. Secor, and D. A. Dixon, editors. Biology, management, and protection of North American sturgeon. American Fisheries Society, Symposium 28, Bethesda, Maryland.

Jenkins, W. E., T. I. J. Smith, L. D. Heyward, and D. M. Knott. 1993. Tolerance of shortnose sturgeon, Acipenser brevirostrum, juveniles to different salinity and dissolved oxygen concentrations. Proceedings of the Annual Conference Southeastern Association of Fish and Wildlife Agencies 47: 476-484.

Jensen, O. P., R. Seppelt, T. J. Miller, and L. J. Bauer. 2005. Winter distribution of blue crab *Callinectes sapidus* in Chesapeake Bay: application and cross-validation of a two-stage generalized additive model. Marine Ecology Progress Series 299:239–255.

Kraus, R. T., and D. H. Secor. 2005. Application of the nursery-role hypothesis to an estuarine fish. Marine Ecology Progress Series 291:301–305.

Kynard, B., and M. Horgan. 2002. Ontogenetic behavior and migration of Atlantic sturgeon, *Acipenser oxyrinchus oxyrinchus*, and shortnose sturgeon, *A. brevirostrum*, with notes on social behavior. Environmental Biology of Fishes 63:137–150.

Leslie, J. A., K. A. Abood, E. A. Maikish, and P. J. Keeser. 1988. Recent dissolved oxygen trends in the Hudson River. Pages 287–303 *in* C. L. Smith, editor. Fisheries research in the Hudson River. State University of New York Press, Albany.

Logerwell, E. A., B. Lavaniegos, and P. E. Smith. 2001. Spatially-explicit bioenergetics of Pacific sardine in the Southern California Bight: are mesoscale eddies areas of exceptional prerecruit production? Progress in Oceanography 49:391–406.

Luo, J. G., K. J. Hartman, S. B. Brandt, C. F. Cerco, and T. H. Rippetoe. 2001. A spatially-explicit approach for estimating carrying capacity: an application for the Atlantic menhaden (*Brevoortia tyrannus*) in Chesapeake Bay. Estuaries 24:545–556.

McAdam, S. O., C. J. Walters, and C. Nistor. 2005. Linkages between white sturgeon recruitment and altered substrates in the Nechako River, Canada. Transactions of the American Fisheries Society 134:1448–1456.

Maurice, K. R., R. W. Blye, P. L. Harmon, and D. Lake. 1987. Increased spawning by American shad coincident with improved dissolved oxygen in the tidal Delaware River. Pages 79–88 *in* M. J. Dadswell, R. J. Klauda, C. M. Moffitt, and R. L. Saunders, editors. Common strategies of anadromous and catadromous fishes. American Fisheries Society, Symposium 1, Bethesda, Maryland.

Musick, J. A., M. M. Harbin, S. A. Berkeley, G. H. Burgess, A. M. Eklund, L. Findley, R. G. Gilmore, J. T. Golden, D. S. Ha, G. R. Huntsman, J. C. McGovern, S. J. Parker, S. G. Poss, E. Sala, T. W. Schmidt, G. R. Sedberry, H. Weeks, and S. G. Wright. 2000. Marine, estuarine, and diadromous fish stocks at risk of extinction in North America (exclusive of Pacific salmonids). Fisheries 25:6–30.

Neill, W. H., J. M. Miller, H. W. Vanderveer, and K. O. Winemiller. 1994. Ecophysiology of marine fish recruitment—a conceptual-framework for understanding interannual variability. Netherlands Journal of Sea Research 32:135–152.

Niklitschek, E. J. 2001. Bioenergetics modeling and assessment of suitable habitat for juvenile Atlantic and shortnose sturgeons (*Acipenser oxyrinchus* and *A. brevirostrum*) in the Chesapeake Bay. Doctoral dissertation. University of Maryland at College Park, College Park.

Niklitschek, E. J., and D. H. Secor. 2005. Modeling spatial and temporal variation of suitable nursery habitats for Atlantic sturgeon in the Chesapeake Bay. Estuarine Coastal and Shelf Science 64:135–148.

NMFS (National Marine Fisheries Service). 1998. Recovery plan for the shortnose sturgeon (*Acipenser brevirostrum*). Prepared by the Shortnose Sturgeon Recovery Team for the National Marine Fisheries Service, Silver Spring, Maryland.

Paragamian, V. L., R. C. P. Beamesderfer, and S. C. Ireland. 2005. Status, population dynamics, and future prospects of the endangered Kootenai River white sturgeon population with and without hatchery intervention Transactions of the American Fisheries Society 134:518–532.

Ray, G. C. 1997. Do the metapopulation dynamics of estuarine fishes influence the stability of shelf ecosystems? Bulletin of Marine Science 60:1040–1049.

Schaaf, W. E., D. S. Peters, L. Coston-Clements, and C. W. Krouse. 1987. Fish population responses to chronic and acute pollution: the influence of life history strategies. Estuaries 10:267–275.

Schaaf, W. E., D. S. Peters, L. Coston-clements, D. S. Vaughan, and C. W. Krouse. 1993. A simulation-model of how life-history strategies mediate pollution effects on fish populations. Estuaries 16:697–702.

Secor, D. H., and T. E. Gunderson. 1998. Effects of hypoxia and temperature on survival, growth, and respiration of juvenile Atlantic sturgeon, *Acipenser oxyrinchus*. Fishery Bulletin 96:603–613.

Secor, D. H., and E. J. Niklitschek. 2001. Sensitivity of

sturgeons to environmental hypoxia: a review of physiologial and ecological evidence. Pages 61–78 *in* Fish physiology, toxicology and water quality. Ecosystems Research Division, Athens, Georgia.

Vadas, R. L., and D. J. Orth. 2001. Formulation of habitat suitability models for stream fish guilds: do the standard methods work? Transactions of the American Fisheries Society 130:217–235.

Weisberg, S. B., P. Himchak, T. Baum, H. T. Wilson, and R. Allen. 1996. Temporal trends in abundance of fish in the tidal Delaware River. Estuaries 19:723–729.

Woodland, R. J., and D. H. Secor. 2007. Year-class strength and recovery of endangered shortnose sturgeon in the Hudson River, New York. Transactions of the American Fisheries Society 136:72–81.

Appendix I: Survival and Bioenergetics Model for Juvenile Shortnose Sturgeon (Modified from Niklitschek 2001)

Notes

1. Coefficient values from Niklitschek (2001) have been incorporated into the original equations, which have been simplified whenever possible.
2. Some coefficients of $f(\mathrm{DO,Sal})_{\mathrm{FC}}$ were re-estimated and formulas re-arranged to allow extrapolation down to 0% DO saturation.

Symbols Common to All Equations

W= fish weight (g); T= temperature (°C); DO = dissolved oxygen saturation (%); and Sal = salinity.

Growth

$$G = \mathrm{FC} - (\mathrm{RM} + \mathrm{SDA} + \mathrm{ACT}) - (F + U) \qquad (\mathrm{Kj/g/d})$$

Food Consumption (FC)

$$\mathrm{FC} = 1.23 \times W^{0.20} \times f(T)_{\mathrm{FC}} \times f(\mathrm{DO,Sal})_{\mathrm{FC}} \qquad (\mathrm{Kj/g/d})$$

$$f(T)_{\mathrm{FC}} = \frac{0.28 \times \exp[0.24 \times (T-6)]}{1 + 0.28 \times \{\exp[0.24 \times (T-6)] - 1\}}$$

$$f(\mathrm{DO,Sal})_{\mathrm{FC}} = \left| \begin{array}{ll} \mathrm{COK} + \dfrac{[1-\mathrm{COK}] \times \mathrm{KO}_i}{\mathrm{KO}_{\mathrm{max}}}, & \text{for} \quad \mathrm{DO} < \mathrm{CDO}_{\mathrm{crit}} \\ 1, \quad \text{otherwise} & \end{array} \right.$$

$$c_{\mathrm{FC}} = \frac{1}{\mathrm{CDO}_{\mathrm{crit}} - 30}$$

$$c_{\mathrm{FC}} = \frac{1}{\mathrm{CDO}_{\mathrm{crit}} - 30}$$

$$\mathrm{KO}_{\mathrm{max}} = (\mathrm{DO}_{\mathrm{crit}} - 30) \times e^{-c_{\mathrm{FC}} \times (\mathrm{DO}_{\mathrm{crit}} - 30)}$$

$$\mathrm{CDO}_{\mathrm{crit}} = 110.1 \times f(T)_{\mathrm{FC}}$$

$$\mathrm{COK} = 1 - 1.98 \times [f(\mathrm{Sal})_{\mathrm{RM}} \times f(T)_{\mathrm{RM}}]$$

Routine Metabolism

$$\mathrm{RM} = 0.42 \times W^{-0.196} \times f(T)_{\mathrm{RM}} \times f(\mathrm{DO})_{\mathrm{RM}} \times f(\mathrm{Sal})_{\mathrm{RM}} \qquad (\mathrm{Kj/g/d})$$

$$f(T)_{\mathrm{RM}} = e^{-0.085\times(T-6)}$$

$$f(\mathrm{DO})_{\mathrm{RM}} = \begin{vmatrix} \mathrm{COK}_{\mathrm{RM}} + \dfrac{(1-\mathrm{COK}_{\mathrm{RM}})\times \mathrm{KO}_i}{KO_{\max}}, & \text{for} \quad \mathrm{DO} < \mathrm{RDO}_{\mathrm{crit}} \\ 1, \quad \text{otherwise} & \end{vmatrix}$$

$$\mathrm{KO}_i = (\mathrm{DO}_i - 25)\times e^{[-c_{\mathrm{RM}}\times(\mathrm{DO}_1-25)]}$$

$$\mathrm{KO}_{\max} = \mathrm{KO}_i \text{ at } \mathrm{RDO}_{\mathrm{crit}}$$

$$\mathrm{COK}_{\mathrm{RM}} = e^{(-0.027\times T)}$$

$$\mathrm{RDO}_{\mathrm{crit}} = \frac{\log_e(T)}{0.051} + 25$$

$$c_{\mathrm{RM}} = \frac{1}{\mathrm{RDO}_{\mathrm{crit}} - 25}$$

$$f(\mathrm{Sal})_{\mathrm{RM}} = \frac{K_{\mathrm{sa}} + K_{\mathrm{sb}}}{1 + e^{2.52\times W^{-0.158}}}$$

$$K_{\mathrm{sa}} = e^{0.09\times(29-\mathrm{Sal})\times W^{-0.158}}$$

$$K_{\mathrm{sb}} = e^{0.11\times(\mathrm{Sal}-1)\times W^{-0.158}}$$

Postprandial Metabolism (SDA)

$$\mathrm{SDA} = [0.12 + 0.00099\times(\mathrm{DO}-79.6)]\times \mathrm{FC} \qquad (\mathrm{Kj/g/d/})$$

Active Metabolism (ACT)

$$\mathrm{ACT} = 0.35\times \mathrm{FC} \qquad (\mathrm{Kj/g/d})$$

Egestion (F)

$$F = 0.075\times \mathrm{FC}\times(T-18)^{-0.55}\times e^{\mathrm{CRO}} \qquad (\mathrm{Kj/g/d})$$

$$\mathrm{CRO} = -0.26\times p_{\mathrm{FC}} + 2.6\times p_{\mathrm{FC}}^{\ 2} - 0.011\times(\mathrm{DO}-76.7)$$

p_{FC} = actual proportion of FC consumed by the fish

Excretion (U)

$$U = \mathrm{RNE} + \mathrm{XNE} \qquad (\mathrm{Kj/g/d})$$

$$\mathrm{RNE} = 0.062\times W^{-0.29}\times \mathrm{RM}$$

$$\mathrm{XNE} = 0.039\times \mathrm{FC}$$

American Fisheries Society Symposium 69:605–607, 2009

Management and Governance of Diadromous Fisheries: Sociological, Economic, Political, and Ecological Considerations

Preamble

Ronald J. Klauda
Maryland Department of Natural Resources, Annapolis, Maryland 21401, USA

Katherine L. Smith
National Oceanic and Atmospheric Administration, Silver Spring, Maryland 20910, USA

Considerable research has focused on the biology and ecology of diadromous fishes since the 1986 symposium held in Boston, Massachusetts, USA (Dadswell et al. 1987), and a fairly complete presentation of new knowledge appears in Sections 1 through 5 of this publication. As a result of such efforts, our understanding of the basic ecology and life history of diadromous fishes has improved. Additionally, we have gained an understanding of the multiple stressors such as high mortality (e.g., overharvesting, turbine-related mortality, predation), as well as habitat degradation and loss (e.g., water pollution, channelization, dams and other barriers), affecting diadromous fishes in their freshwater and marine environments. Climatic changes and oceanic current shifts also pose an emerging threat to diadromous fishes, effects that are explored in Section 2.

Growing human populations and expanding global economies will only increase demands for energy, freshwater, and cheap protein, thus placing additional pressures on diadromous fish populations world-wide. Given the breadth of these threats, traditional single species management approaches are unlikely to adequately provide for sustainable diadromous fish populations. Since the 1986 symposium, interdisciplinary concepts such as adaptive management, place-based or watershed management, multispecies management, and ecosystem-based management have gained increased attention for addressing complex natural resource management issues such as those facing diadromous fishes (e.g., Pikitch et al. 2004).

Based on this growing body of literature, it is generally recognized that successful strategies to address threats to diadromous fishes and their habitats will require contributions from resource economists, sociologists, anthropologists, fishers, indigenous peoples, political leaders, and the public. Learning each others' vernaculars and potential contributions to resource management are critical steps in working together effectively in multidisciplinary teams to achieve sustainability. Recent publications have examined how a range of issues affect a particular species or region (e.g., Pacific salmon in Lackey et al. 2006). This section brings together examples of interdisciplinary efforts to improve management of diadromous fishes in various regions of the world.

The opening papers in this section provide a big-picture perspective on management needs to ensure sustainable diadromous fish populations. Lackey (2009, this volume) evaluates the long history of efforts to reverse the decline of Pacific salmon in western North America, providing policy lessons relevant to the potential recovery and sustainability of other diadromous fishes. Haggan et al. (2009, this volume) suggest that aboriginal people in the Pacific Northwest developed integrated ecosystem-based management of natural resources, including fish, long before European contact. Bevacqua et al. (2009, this volume) show how decision makers can use a multi-objective approach to identify trade-offs between conservation and harvest of European eels and to select effective management actions. The first paper by Blaber (2009a, this volume) explores the role of ecosystem-based approaches in the management of tropical diadromous fishes.

The other papers in this section provide case studies illustrating how components of these new integrated approaches can be incorporated in diadromous fish management strategies. The papers by Hammer (2009, this volume), Blaber (2009b, this volume), and Rochard et al. (2009, this volume) provide regional examples of linking the ecology of diadromous fishes with social, political, and economic dimensions. Local-level management issues associated with indigenous and artisanal/subsistence fisheries, including the use of traditional ecological knowledge and archaeological evidence, are illustrated in the papers by MacGregor et al. (2009, this volume), Anderson (2009, this volume), and Siddique (2009, this volume). The final papers in this section by Hill (2009, this volume), Trial (2009, this volume), and Willis (2009, this volume) discuss case studies of conflict resolution, cooperation, and collaboration in the management of important diadromous fishes.

Multidisciplinary efforts to improve the management of diadromous fishes have received considerable support in both the scientific and management communities. However, this remains an emerging field and its implementation will require decades of concerted effort. Achieving sustainable fisheries for diadromous fishes will be an ongoing and difficult challenge and probably never fully realized for some species in some regions of the world. Perhaps what is needed, at least in the near term, is not a complete abandonment of traditional fisheries management, but rather the adoption of a more holistic perspective that extends the successful components of conventional approaches into the complexity of the socioeconomic and political realms.

References

Anderson, E., M. Montoya, A. Soto, H. Flores, and M. McClain. 2009. Challenges and opportunities for co-management of a migratory fish (*Prochilodus nigricans*) in the Peruvian Amazon. Pages 741–756 *in* A. J. Haro, K. L. Smith, R. A. Rulifson, C. M. Moffitt, R. J. Klauda, M. J. Dadswell, R. A. Cunjak, J. E. Cooper, K. L. Beal, and T. S. Avery, editors. Challenges for diadromous fishes in a dynamic global environment. American Fisheries Society, Symposium 69, Bethesda, Maryland.

Bevacqua, D., P. Melià, A. J. Crivelli, M. Gatto, and G. A. De Leo. 2009. Assessing management plans for the recovery of the European eel: a need for multi-objective analyses. Pages 637–647 *in* A. J. Haro, K. L. Smith, R. A. Rulifson, C. M. Moffitt, R. J. Klauda, M. J. Dadswell, R. A. Cunjak, J. E. Cooper, K. L. Beal, and T. S. Avery, editors. Challenges for diadromous fishes in a dynamic global environment. American Fisheries Society, Symposium 69, Bethesda, Maryland.

Blaber, S. J. M. 2009a. Can the ecosystem approach improve management of tropical estuarine fisheries for diadromous species? Pages 649–661 *in* A. J. Haro, K. L. Smith, R. A. Rulifson, C. M. Moffitt, R. J. Klauda, M. J. Dadswell, R. A. Cunjak, J. E. Cooper, K. L. Beal, and T. S. Avery, editors. Challenges for diadromous fishes in a dynamic global environment. American Fisheries Society, Symposium 69, Bethesda, Maryland.

Blaber, S. J. M. 2009b. Socioeconomic and biopolitical linkages in the management of tropical shads. Pages 677–690 *in* A. J. Haro, K. L. Smith, R. A. Rulifson, C. M. Moffitt, R. J. Klauda, M. J. Dadswell, R. A. Cunjak, J. E. Cooper, K. L. Beal, and T. S. Avery, editors. Challenges for diadromous fishes in a dynamic global environment. American Fisheries Society, Symposium 69, Bethesda, Maryland.

Dadswell, M. J., R. J. Klauda, C. M. Moffitt, R. L. Saunders, R. A. Rulifson, and J. E. Cooper, editors. 1987. Common strategies of anadromous and catadromous fishes. American Fisheries Society, Symposium 1, Bethesda, Maryland.

Haggan, N., G. D. Jackson, and P. Lacroix. 2009. Salmon and eulachon in ecosystem space and time: a plea for wider collaboration and data integration. Pages 619–636 *in* A. J. Haro, K. L. Smith, R. A. Rulifson, C. M. Moffitt, R. J. Klauda, M. J. Dadswell, R. A. Cunjak, J. E. Cooper, K. L. Beal, and T. S. Avery, editors. Challenges for diadromous fishes in a dynamic global environment. American Fisheries Society, Symposium 69, Bethesda, Maryland.

Hammer, M. 2009. Whose fish? Managing Salmonidae and humans in complex social–ecological systems: examples from the Baltic Sea region. Pages 663–675 *in* A. J. Haro, K. L. Smith, R. A. Rulifson, C. M. Moffitt, R. J. Klauda, M. J. Dadswell, R. A. Cunjak, J. E. Cooper, K. L. Beal, and T. S. Avery, editors. Challenges for diadromous fishes in a dynamic global environment. American Fisheries Society, Symposium 69, Bethesda, Maryland.

Hill, A. K. 2009. The Santee River Basin Accord: restoring diadromous fish through prioritization of subbasins. Pages 769–782 *in* A. J. Haro, K. L. Smith, R. A. Rulifson, C. M. Moffitt, R. J. Klauda, M. J. Dadswell, R. A. Cunjak, J. E. Cooper, K. L. Beal, and T. S.

Avery, editors. Challenges for diadromous fishes in a dynamic global environment. American Fisheries Society, Symposium 69, Bethesda, Maryland.

Lackey, R. T. 2009. Challenges to sustaining diadromous fishes through 2100: lessons learned from western North America. Pages 609–617 *in* A. J. Haro, K. L. Smith, R. A. Rulifson, C. M. Moffitt, R. J. Klauda, M. J. Dadswell, R. A. Cunjak, J. E. Cooper, K. L. Beal, and T. S. Avery, editors. Challenges for diadromous fishes in a dynamic global environment. American Fisheries Society, Symposium 69, Bethesda, Maryland.

Lackey, R. T., D. H. Lach, and S. L. Duncan, editors. 2006. Salmon 2100: the future of wild Pacific salmon. American Fisheries Society, Bethesda, Maryland.

Pikitch, E. K., C. Santora, E. A. Babcock, A. Bakun, R. Bonfil, D. O. Conover, P. Dayton, P. Doukakis, D. Fluharty, B. Heneman, E. D. Houde, J. Link, P. A. Livingston, M. Mangel, M. K. McAllister, J. Pope, and K. J. Sainsbury. 2004. Ecosystem-based fishery management. Science 305(5682):346–347.

Rochard, E., P. Pellegrini, J. Marchal, M. Béguer, D. Ombredane, G. Lassalle, E. Menvielle, and J-L. Baglinière. 2009. Identification of diadromous fish species on which to focus river restoration: an example using an eco-anthropological approach (the Seine basin, France). Pages 691–711 *in* A. J. Haro, K. L. Smith, R. A. Rulifson, C. M. Moffitt, R. J. Klauda, M. J. Dadswell, R. A. Cunjak, J. E. Cooper, K. L. Beal, and T. S. Avery, editors. Challenges for diadromous fishes in a dynamic global environment. American Fisheries Society, Symposium 69, Bethesda, Maryland.

Siddique, M. A. L. 2009. Conservation of juvenile hilsa shad (jatka) in Bangladesh: need to address the livelihood fishers. Pages 757–768 *in* A. J. Haro, K. L. Smith, R. A. Rulifson, C. M. Moffitt, R. J. Klauda, M. J. Dadswell, R. A. Cunjak, J. E. Cooper, K. L. Beal, and T. S. Avery, editors. Challenges for diadromous fishes in a dynamic global environment. American Fisheries Society, Symposium 69, Bethesda, Maryland.

Trial, J. G. 2009. Antioxidants, irrigation, and Atlantic salmon. Pages 783–791 *in* A. J. Haro, K. L. Smith, R. A. Rulifson, C. M. Moffitt, R. J. Klauda, M. J. Dadswell, R. A. Cunjak, J. E. Cooper, K. L. Beal, and T. S. Avery, editors. Challenges for diadromous fishes in a dynamic global environment. American Fisheries Society, Symposium 69, Bethesda, Maryland.

Willis, T. V. 2009. How policy, politics, and science shaped a 25-year conflict over alewife in the St. Croix River, New Brunswick–Maine. Pages 793–811 *in* A. J. Haro, K. L. Smith, R. A. Rulifson, C. M. Moffitt, R. J. Klauda, M. J. Dadswell, R. A. Cunjak, J. E. Cooper, K. L. Beal, and T. S. Avery, editors. Challenges for diadromous fishes in a dynamic global environment. American Fisheries Society, Symposium 69, Bethesda, Maryland.

American Fisheries Society Symposium 69:609–617, 2009

Challenges to Sustaining Diadromous Fishes through 2100: Lessons Learned from Western North America

Robert T. Lackey*
National Health and Environmental Effects Research Laboratory
U.S. Environmental Protection Agency, Corvallis, Oregon 97333, USA

Abstract.—An evaluation of the history of efforts to reverse the long-term decline of Pacific salmon in western North America provides instructive policy lessons for the potential recovery of diadromous fishes throughout the world. From California to southern British Columbia, wild runs of Pacific salmon have universally declined and many have disappeared. Billions have been spent in so-far failed attempts to reverse the decline in response to the requirements of the U.S. Endangered Species Act, the Canadian Species at Risk Act, or other laws or policies. The annual expenditure of hundreds of millions of dollars continues, but a sustainable future for wild salmon in this region of North America remains elusive. Despite documented public support for restoring wild salmon, the long-term prognosis for a sustainable future appears problematic. After considering various policy options to increase the numbers of wild salmon and other diadromous species, the major lessons learned were (1) the rules of commerce, especially trends in international commerce and trade, tend to put relentless downward pressure on their numbers; (2) competition for critical natural resources, especially for high quality water, will continue to be great and will work to constrain the numbers of most diadromous species; (3) the aggregate demands of humans will continue to swell, thus tending to limit the available habitat for diadromous species; (4) individual and collective preferences directly determine the future of diadromous species, and there appears to be little indication of a massive and widespread change in those individual and collective preferences; (5) there is an almost insurmountable tendency on the part of elected and appointed officials to avoid explicitly confronting difficult, divisive ecological policy issues; and (6) there is a systematic, often subtle, pressure on fisheries scientists, managers, and analysts to avoid explicitly conveying unpleasant facts to the public, senior bureaucrats, and elected or appointed officials. Fisheries biologists and others continue to craft restoration plans, but an easy, effective approach has yet to emerge that will actually restore and sustain most runs of wild salmon in the region. For most diadromous fishes, restoration options exist that offer both ecological viability and appreciably reduced social disruption, but these options also tend to have more modest restoration objectives.

Introduction

If you read the majority of the other papers in this proceedings, you will learn about the poor state of many diadromous fish runs. There are the latest assessments about the causes of the declines. You can read well-informed judgments about the relative importance of this or that causal factor. Mostly, you will read explanations offered by biologists about their interpretations of the how and the why of the current state of diadromous fishes.

It is good stuff for a fisheries biologist and I feel right at home. It is an interesting story to be sure and one to which I am usually inclined to contribute my two cents worth. But it is time to shift to the "so what" question. What does all this science really mean? Is it relevant to the public in general and decision makers in particular? Does either group really care? Does the science make a difference to anyone but us? How are the myriad scientific facts relevant in public policy?

The steering committee knew about the Salm-

* Corresponding author: robert.lackey@oregonstate.edu

on 2100 Project and its no-nonsense analysis of the future of salmon in western North America (Lackey et al. 2006a). Specifically, of the nearly 1,400 historical populations of Pacific salmon (and steelhead *Oncorhynchus mykiss*) in California, Nevada, Oregon, Washington, Idaho, and southern British Columbia, 29% are estimated to have disappeared (Gustafson et al. 2007). Given current long-term trajectories, the future viability of many of the remaining populations looks bleak (Lackey et al. 2006b).

The committee's charge to me was blunt and unequivocal: offer answers to the "so what" questions. Look forward, predict, speculate, and consider the most likely future of diadromous fishes. And, while you are at it, identify what must change if the long-term, downward trajectory in wild diadromous fishes is to be reversed.

And how was the so-called future to be defined? "Use your own judgment" was the answer. Those of us who provide science to decision makers and politicians know that the typical time frame of interest may extend to next year, perhaps on some occasions it might be a few years, a salmon life cycle possibly, but surely no longer.

Let me switch from a decision maker's or politician's perspective to a policy analyst's perspective. We need to get real about a time context! A century is the correct time frame. One of the take-home lessons from the Salmon 2100 Project was that 2050 was just too soon to assess the efficacy of the various policy options. A single ocean cycle may last three or four decades. The changing hydrograph from a major dam may take half a century to show its full influence. A mature forest may take well over a century to fully recover from a major fire or an extensive logging effort.

Using 2100 as the time horizon means forecasting nine decades out and doing so publicly. Of course, most of us secretly have a forecast about the long-term future of our favorite species, but we keep it to ourselves; perhaps we might share it with a few close colleagues and such colleagues must be trusted and reliably discreet. Most of us are not willing to publicly speculate, much less publish, about diadromous fishes over the longer term.

At least in some ways, it is easier to forecast the long term. It does tend to dampen the confusion caused by year-to-year and decade-to-decade variations in ecological and social factors that often mask fundamental, underlying trends in fish populations.

I am well aware that it is easy to dispense generalities and feel-good platitudes, but I will try to be candid and frank. You may well argue with my take on science and policy, but I do not want to be Pollyannaish nor fatalistic. So forget both the temptation to be unduly optimistic and the tendency to be unnecessarily pessimistic. What follows is my stab at forthright realism.

Let me begin with my simple take-home message:

> In spite of abundant uncertainty about the relative importance of the various factors that drove the declines of most diadromous species, we fundamentally recognize and know the direct causes of nearly all long-term declines.

Most scientists tend to focus on what we do not know, but we know a lot. As we have examined at this symposium, and elsewhere, the causes have been, and often still are

- intense commercial, recreational, and subsistence fishing;
- freshwater, estuarine, and marine habitat alteration due to many human activities as well as changes in climate and shifts in ocean regimes;
- dams built and operated for electricity generation, flood control, irrigation, and other purposes;
- water withdrawals for agricultural, municipal, or commercial requirements;
- stream and river channel alteration, diking, and riparian corridor modifications;
- climatic changes caused at least in part by human actions;
- predation by marine mammals, birds, and other fish species often exacerbated by unintentionally concentrating diadromous fish or their predators (e.g., near dams, near hatchery release sites, or on islands created by discarded dredged material);
- competition, especially competition with exotic or nonnative fishes, many of which are better adapted to the highly altered aquatic environments we now have in many regions;
- introduction and spread of exotic diseases and

parasites, caused in some situations by aquacultural operations; and
- reduction in the annual replenishment of nutrients from decomposing, spawned-out diadromous fish.

In a few highly visible cases, supplemental stocking from hatcheries for sustaining high levels of fishing may have exacerbated the decline of naturally spawning runs.

It is a long list and it encompasses most of the entire human enterprise. In the background, of course, is the knowledge that changes in ocean and climatic conditions occur even in the absence of human activities or influences, and they have immense influence on diadromous fishes.

We can argue endlessly over details of the science or the relative importance of anthropogenic and natural causes, but consider my basic assertion that the primary cause of most long-term declines is anthropogenic and we pretty much know the key reasons why each decline happened. There are exceptions, but let us not quibble over these or lose sight of what we do know with reasonable assurance.

We know much more than we sometimes admit about the decline of diadromous fishes, especially about its trajectory. Based on what we found in the Salmon 2100 Project, I propose another statement of fact that applies to most other diadromous species:

> As we move forward in this century, in spite of ups and downs, good years and bad years, favorable and unfavorable ocean conditions, even newspaper headlines proclaiming some species have recovered, diadromous fishes have been on a long-term downward trend and many species are now at very low levels.

It is not easy to focus on the long term when most of us are thinking about this year's runs or perhaps living dangerously by speculating about next year's runs or those in 2011. 2100 seems way too far removed from the press of day-to-day business, but it provides the necessary policy perspective.

How can it be that the reasons for the declines are usually pretty well known, that in many cases they have been studied in grand detail, that typically the public appears to support reversing the long-term downward trends, yet the recovery prognosis is usually poor?

The answer is another simple policy statement of fact:

> Effecting any change in the long-term downward trend for wild diadromous fishes is impossible in the absence of shifts in the core policy drivers.

It is the core policy drivers, the root causes, that have determined the status of diadromous species and will continue to determine their status through this century. As we learned in the Salmon 2100 Project, habitat alteration, dams, water withdrawals, fishing, supplemental stocking from hatcheries, and many more causal agents are simply the way in which the core policy drivers are expressed.

No one is bent on eradicating salmon or other diadromous species, and we usually know the reasons for the declines. It is not sinister motives nor is it lack of knowledge, but rather it is that policy choices are made between conflicting and desirable alternatives. In real life, benefits come with costs. In short, ecological policy focused on the goal of restoring diadromous species is a zero-sum game (Lackey 2006). There are always winners and losers.

What then are the specific lessons I have learned from a 4-year involvement with the Salmon 2100 Project? These lessons could perhaps be more accurately called inconvenient truths. Whether called lessons learned or inconvenient truths, they drive ecological policy in general and will also define the future of wild diadromous species.

Lesson #1—The Marketplace Is Fundamental

My first lesson learned was that for wild salmon and other diadromous species, the rules of commerce, especially trends in international commerce and trade tend to work against increasing their numbers.

The rules of commerce over the long term are not fish-friendly. The drive for near-term low-cost production in free market economies is a widely professed approach to trade, both within nations and between nations. We can argue about whether so-called "free" markets are actually free; my purpose is not to argue either in favor of or against such a philosophy of commerce, but rather to note its impact on populations of diadromous fishes.

My presumption is that free trade and market-driven economies will continue to be a dominant government policy through this century. One consequence of such an approach to commerce is that noneconomic values, such as preserving declining or remnant diadromous fish runs, tend not to get weighted very heavily in decision making (Lackey 2005).

This tendency is not inherently good or bad, just a fact. We obtain our computers from wherever they can be manufactured most cheaply. We move our automobile assembly plants to wherever they can produce cars most inexpensively. We tend to produce electricity in the most cost-effective way. We obtain most of our wheat where it can be grown most productively and consistently. We obtain wood products where they can be grown and harvested most efficiently and sold at the lowest price. Yes, we have trade disputes, but most everyone seems to accept the basic value of free trade.

Our individual and collective choices are not entirely driven by cost, of course. Perception, personal preferences, and risk tolerance play important roles. Consider how we choose to generate electricity. Generating electricity from dams is usually inexpensive and does not produce greenhouse gas emissions, but it is not beneficial to diadromous species. Wind power is comparatively expensive but is greenhouse gas friendly. Unfortunately, it is not ideal for birds, nor for improving viewsheds. Depending on a person's perceptions, preferences, and risk tolerance, nuclear, coal, natural gas, solar, biofuels, and wave power all have their own strengths and weaknesses; thus, it is not merely cost that determines a market preference.

Considering the results of the Salmon 2100 Project, the benefits of public policies that favor free market economies are well recognized, but there are also consequences that are not all that favorable to diadromous species. How much more are people willing to pay for bread, for electricity, or for automobiles produced in ways that will help restore wild salmon or other diadromous species? Avoid the balderdash that bread, electricity, and automobiles can be produced just as cheaply in a salmon-friendly manner. As with all policy choices, there are winners and there are losers and we should make that point clear to all.

Searching for the nonexistent but ever politically tantalizing win–win salmon recovery solution often ends up frustrating everyone. Except for the most trivial policy aspects of salmon recovery, compromise is necessary to craft a proposed policy that is democratically possible. Thus, salmon policy winds up as the classic zero-sum game. Accepting this reality encourages serious discussion about how to best resolve what appears to be an intractable issue.

As I observe consumer behavior today and guess about the future, I do not see much willingness on most people's part to pay substantially more for salmon-friendly products. This lesson learned does not in itself discourage me, but failure to recognize its reality does.

Lesson #2—Competition for Scarce Resources Is Unyielding

My second lesson learned was that for wild salmon and other diadromous species, competition for critical natural resources, especially for high quality water, will continue to be great and will work against increasing their numbers.

You might argue that this lesson learned borders on the brilliantly obvious. It does. Perhaps that is why it is so often overlooked in policy analysis and forecasting. The region where we did the Salmon 2100 Project (California, Oregon, Washington, Idaho, and southern British Columbia) suffers from water shortages, especially shortages of high quality water. Unless the competition for scarce water bubbles into open political conflict, most people are oblivious of the severity of the challenge. Even with the myriad media stories about impending water scarcity, most written in a doom and gloom style, our seemingly insatiable demand for freshwater shows little sign of letting up.

I am not arguing that allocating water for salmon or other diadromous species is more important than allocating it for alternative uses, but consider the importance of these competing needs. As competition for scarce water continues and gets much more intense, how will advocates for diadromous species fare relative to advocates for competing priorities such as water for drinking, irrigation, manufacturing, generating electricity, or any of a thousand other needs?

The decades-long and continuing water war in the Klamath basin, along the California–Oregon border, gives us an indication of the future. Newspapers describe farmers defying law enforcement agents and illegally opening locked valves and releasing water to irrigate their fields. The television evening news shows streams choked with dying salmon caused by low water flows, poor water quality, and virulent diseases sustained by warm water. Lawyers from various competing interest groups duel in court over who will get how much water. Always, at the end of the day, every faction in the battle is dissatisfied with the result, feels their interest did not get a fair share of the water, and grapples for ways to be more politically effective in next year's battle.

Water is not the only resource in limited supply. Some argue that competition for high quality soil, especially for agricultural uses, is most crucial (Salonius 2007). Whether water or soil, as key natural resources become scarcer through this century, the individual and collective choices that permit long-term abundance of many diadromous fishes likely will become increasingly unacceptable to more and more people. For those of us who spend our working lives concerned with creatures with fins, no matter how economically, socially, or culturally important they might be to segments of society, it is a challenge to acknowledge that our focus of attention is just one of a set of multiple and competing priorities.

Lesson #3—The Human Population Exerts a Pervasive Influence

The third lesson I came to concede, albeit grudgingly, was that for wild salmon and other diadromous fishes, the aggregate resource demands of humans will continue to swell, thus tending to work against increasing the numbers of most diadromous species.

The most probable scenario for the human population trajectory through this century for places like the U.S. Pacific Northwest and British Columbia is upward, substantially upward. Any serious discussion about the future of diadromous species must consider the human population trajectory, but it is not fashionable to raise this issue.

Environmental advocacy groups avoid it like a pestilence, even though it dwarfs most of the human behaviors they are trying to modify. Even fish advocacy groups rarely mention it, much less take a clear policy position. It is the proverbial elephant in the room that few want to acknowledge. As one of my colleagues warned me during the planning of the Salmon 2100 Project,

> Bob, you are absolutely right, most people already know it, and that's exactly why you should let it rest. Back off. You'll leave the proponents of wild salmon restoration depressed. Worse, you'll have the rest of the audience wondering why you are pontificating on the intuitively obvious. And you run the risk of being attacked as a racist, nativist, xenophobe, cultural imperialist, sexist, or, at the least, an economic elitist.

Undoubtedly, very good advice. However, if society wishes to do anything meaningful about moving many diadromous species off their current downward trajectory, then something must be done about the unrelenting growth in the number of humans generally and especially in areas like the Pacific Northwest. It is not simply the number of people that causes problems for diadromous species, but it is also our individual and collective ecological footprint. Based on assessing our individual lifestyle choices, one car is no longer sufficient, nor is even two. Many families now have three cars, along with a three-car garage. To some of us, one house is apparently no longer adequate. We have a second home. Should any of us be expected to get by without a hot tub, central air conditioning, and three TVs?

I am not arguing that we necessarily ought to change any current public policy or our individual priorities. I am arguing that we should be candid and truthful about reality. A simple and inescapable fact is that the growth in the human population level that we can realistically anticipate through the rest of this century will create a serious barrier to achieving any significant long-term recovery of diadromous species. In regions such as the Pacific Northwest, what demographers like to call "fill-in" country, the consequences of a larger human population on diadromous species will be severe.

The latest demographic forecasts show a slowing of the world population through this century with a leveling off toward 2100. Yes, a leveling off is predicted, but at 9 billion people. As we found in the Salmon 2100 Project for the Pacific Northwest, however, there is a different story. It is largely one of past, current, and future immigration. Currently, Washington, Oregon, Idaho, and British Columbia are home to 15 million humans. Assuming a range of likely human reproductive rates, migration to the Pacific Northwest from elsewhere in Canada and the United States, and continuing immigration policy and patterns, by 2100, this region's human population will not be its present 15 million, but rather will be somewhere between 50 and 100 million, a quadrupling or more of the region's human population by the end of this century.

Visualize 50 or 100 million people in the Pacific Northwest, and their demands for housing, schools, tennis courts, football stadiums, expressways, planes, trains, automobiles, Starbucks, McDonalds, Wal-Mart, electricity, drinking water, pipelines, marinas, computers, DVDs, 12 screen movie theaters, ski resorts, golf courses, lush weed-free lawns, and ocean front hotels. For my Canadian colleagues, add to the list Tim Hortons, Canadian Tire, Zellers, and The Bay.

Visualize the western region of the state of Washington and southwestern corner of British Columbia in 2100 with its metropolis of Seavan. Seavan morphed into a truly great metropolis as smaller, discrete cities grew together. Seavan in 2100 stretches from Olympia in south Puget Sound northward through the once stand-alone cities of Tacoma and Seattle and on to Vancouver, east to Hope, and west to cover the southern half of Vancouver Island. Rather than the 6 million people back in 2009, Seavan in 2100 rivals present-day Mexico City and Tokyo with its 30 million inhabitants. Or think about the New York City—Boston corridor transplanted to the Pacific Northwest. It is within this context that salmon recovery strategies need to be developed if they have any chance of succeeding.

Regardless of the accuracy of my forecast, population issues are not easy ones to highlight without implying a preferred policy option. After all, there are understandable, strategic reasons why the big environmental groups, most groups in fact, stay clear of population issues these days, but a recognition of the expected overall increase in world population, coupled with the spectacular increases in certain "fill-in" regions such as western North America, must be at the core of any credible analysis of potential recovery strategies of diadromous species. Without such recognition, recovery strategies for salmon and other diadromous fishes are doomed to fail.

Lesson #4—Individuals Select from Among Desirable Alternatives

A straightforward lesson learned was that for wild salmon and other diadromous species, our individual and collective preferences directly determine their future, and substantial and pervasive changes must take place in these preferences if current long-term downward trends are to be reversed.

This lesson learned is perhaps the most obvious and arguably the most important. Among most fisheries scientists, it is easy to assume that charismatic and/or economically valuable diadromous species are near the top of the public's priorities. Just look at the polling results regarding restoring depleted salmon runs in the Pacific Northwest. Everyone supports salmon and especially wild salmon. But, the fact is that salmon recovery is only one of many priorities that individuals, when not forced to make a choice, profess to rank high. When forced to make a choice, salmon recovery drops substantially in importance as compared to other priorities. I have some difficulty conceiving of this, but that is the situation out there in the authentic world. Even my children, who I have had well over three decades to inculcate, regularly admonish me:

> Dad, get a life. Most people out here in the real world just don't care that much about restoring wild salmon. We have other things to worry about!

Society's collective behavior, not public opinion polls, not thick recovery plans, but people's individual and collective behavior, gives us the best indication.

Consider this example. In 1991, the first salmon "distinct population segment" in the Pacific Northwest was listed under terms of the U.S. Endangered Species Act. With this listing of salmon as a legally protected species, the policy debate shifted in California, Oregon, Washington, and Idaho away from restoring salmon runs in order to support fishing, to protecting wild salmon runs from extinction, two very different policy objectives. Starting with the first listing, followed by many others, protecting at-risk runs of wild salmon won out over maximizing fishing opportunities through stocking from hatcheries or through other efforts to put fish on the hook or fish in the net. The residents of the western United States apparently made a choice. Or did they?

Jump ahead 10 years to 2001. Just a decade after the first salmon listing, a severe drought, combined with ongoing electrical blackouts in California, prompted the Bonneville Power Administration to declare a power emergency, abandon previously agreed upon interagency salmon restoration commitments, and generate electricity at maximum capacity using water reserved to help salmon migrate later in the year.

In one of the most striking and recent barometers of competing societal priorities, air conditioners, electricity to be exact, won out over both wild and hatchery-bred salmon and with scant public opposition. No street protests. No legal challenges. No elected officials publicly pleading for salmon. No environmental group blanketing the Internet with calls to mobilize fax machines in defense of salmon. Near complete silence.

Over the past 150 years, each of us has made plenty of these kinds of choices: often contradictory and apparently inconsistent, and these choices roughly reflect our collective and relative priority for various diadromous species. These choices are trade-offs, and we continue to make them and they are a real measure of the relative importance of salmon and other diadromous species. That is not good or bad, just a fact, however unpleasant to some.

Now, I am not cheerleading for wild salmon or for any other diadromous species or for electricity or for property rights or for hatcheries or for having a McDonalds, Tim Hortons, and Starbucks on every corner, but it is naive to consider salmon recovery as anything but one often minor element in a constellation of competing, often mutually exclusive, wants, needs, and preferences.

Lesson #5—Policy Domestication Is Ubiquitous

Regarding the "users" of science and policy analysis (politicians and other decision makers), a sobering lesson learned was that there is an almost insurmountable tendency on the part of elected and appointed officials to avoid explicitly confronting difficult, divisive policy issues.

Even many of the participants in the Salmon 2100 Project (and they were selected in part for their ability to address complex science and policy issues in a direct and candid manner) had a tendency to slip into the mode of what political scientists call "domesticating" the policy issue. Policy domestication is the process of taking difficult, divisive policy issues off the table until a solution emerges or the problem disappears by solving itself (e.g., the species is extirpated). The most common indicators of domestication are funding more research or scientific reviews, holding more workshops and venues to get stakeholders involved through collaboration, forming more planning teams to assess policy options, and tweaking current regulations or policies. Such activities provide the illusion of substantive action.

It is easy to see why offering political actions to domesticate the salmon decline policy challenge, for example, is easier than developing explicit policies to reverse the decline. Reversing the long-term decline of salmon and other diadromous species requires changing at least some of the current political realities about the decline.

Few elected or appointed officials ever explicitly propose ways to change political realities about recovery of diadromous species. Instead, they suggest variations on existing policy options to revise the Endangered Species Act (U.S.), the Species at Risk Act (Canada), or similar legislation; protect more and/or different salmon habitat; create new hatchery practices; change K–12 education; and/or somehow transform people's attitudes.

Domesticating strategies are typically requests for extensions of practices already in place and they do not propose revolutionary approaches or challenge existing beliefs. They tacitly assume that at some future time, we will formulate and agree on a viable solution. In reality, the public may not even be sure what the problem is, much less know what possible solutions exist.

Lesson #6—Delusional Reality Is Tempting and Widespread

The final lesson learned was that there is a systemic, often subtle, pressure on fisheries scientists, managers, and analysts to avoid explicitly conveying unpleasant facts to the public, senior bureaucrats, and elected or appointed officials.

During the planning and implementation of the Salmon 2100 Project and the resulting book, the most fascinating aspect was the recurring suggestion, even a plea, to lighten up and be more optimistic and positive in assessing the future of wild salmon. The entire premise of the project was to be blunt, direct, and realistic and to avoid both pessimism and optimism. Outside reviewers (mostly scientists long active in salmon science, management, or policy) of the individual book chapters concluded that the results were realistic in content and conclusion, but at the same time, many of them encouraged us to abandon the blunt realism and forthright honesty in favor of a more encouraging sense of optimism.

Similarly, reacting to articles I had written, several reviewers suggested that if my objective in writing was to help save wild salmon (it was not), then the accurate, realistic message would leave proponents dejected. This common sentiment is captured by the following: You have to give those of us trying to restore wild salmon some hope of success.

Conversely, a few veterans of the salmon wars confessed their regret over the optimistic approach that they had taken during their careers in fisheries, and they endorsed the "tell it like it is" tactic. They felt that they had, especially early in their careers, given false hope about the effectiveness of fishways, hatcheries, and the ability of their agencies to manage mixed-stock fishing. I was left with a feeling that many professional fisheries scientists have been, and still are, subtly pressured by employers, funding organizations, and colleagues to spin fisheries science and policy realism to accentuate optimism.

Other reviewers took professional refuge in the reality that senior management or policy bureaucrats select and define the policy question to be addressed, thus constraining and bounding the research questions to be addressed. The result is often narrowly focused, reductionist, and constrained scientific information and assessments. Rarely are fisheries scientists empowered to provide big picture assessments of the future of salmon. Whether inadvertent or not, such information often misleads the public into endorsing false expectations of the likelihood of the recovery of wild salmon. For many of us, such implicit optimism is a healthy, rewarding way to go through life.

Is adopting unfounded professional optimism a harmless adaptive behavior of little import? After all, think positive slogans are a hallmark of many self-improvement programs. What is wrong is that optimism does not convey what is happening with wild salmon and it allows the public, elected officials, and fisheries managers to escape the torment of confronting triage. No one ever argues that you can have wild salmon everywhere they once were, but few are willing to be explicit about identifying those locations where the cost is very high and chance of success is very low. Triage, whether in medical emergencies or ecological crises, is difficult, but necessary.

Fisheries scientists should be realistic and avoid being either optimistic or pessimistic. This professional stance does not covertly argue in favor of an imperative to save wild salmon or any other diadromous species regardless of the cost to society, nor does it necessarily support a defeatist strategy. Such choices should be made by an informed public that is aware of the difficult trade-offs. Restoring diadromous species is only one of many competing, important priorities, and the public is entitled to be accurately informed about the long-term prospects of success.

In discussions about the future of salmon, it is easy to find comfort in debating the nuances of hatchery genetics, evolutionarily significant units, dam breaching, smolt barging, selective fishing regulations, predatory bird control, habitat restoration, and atmospheric and oceanic climate, and unintentionally mislead the public about the realities of the situation. As discomforting as it may be to disclose the future of wild salmon and other diadromous species relative to society's apparent values and preferences, our most useful contribution as fisheries scientists is providing information and assessments that are policy-relevant but policy-neutral, understandable to the public and decision makers, and scrupulously realistic about the future. Otherwise, we simply squander our professional credibility to become acolytes of delusion.

Conclusion

I have now presented my analysis of the most likely future of diadromous fishes, identified those policy factors that would have to change if the long-term, downward trajectory is to be reversed, and offered a few lessons learned from the Salmon 2100 Project. To some, my commentary may not be all that uplifting. A greater worry to me is that we will probably continue to spend billions of dollars in quick-fix restoration and management efforts that will in many cases be only marginally successful. One of my colleagues calls such expenditures "guilt money," a tax we willingly bear to alleviate our individual and collective remorse, money spent on activities not likely to achieve the stated purpose, but it makes us feel much better as we continue enjoying our current lifestyle.

As we found in the Salmon 2100 Project, there are options that we believe would work if implemented, but they each acknowledge people's priorities for what they are, not what some might want them to be. There were 23 different policy prescriptions developed as part of the project, each of which would work for recovering wild salmon runs (Lackey et al. 2006a) and, I would assert, for recovering most runs of diadromous fishes. We need to offer the public effective options for recovering diadromous species; there is no place for delusional reality, be it groundless optimism or needless pessimism.

In conclusion, I tender a challenge to those concerned with diadromous fisheries issues:

> Any policy or plan targeted to restore any diadromous species must incorporate the above lessons learned or that plan will fail. It will be added to an already long list of prior, noble, earnest, and failed fisheries management strategies.

Consider 2100, less than 10 decades away, only a few dozen generations of diadromous fishes beyond today's runs, just a few ocean oscillations from now, to a time when the world's population will likely be 50% greater and, in some regions, quadrupled. Even given all this, there are still sustainable management options that are likely to be ecologically viable and probably socially acceptable, but the range of options continues to dwindle. For professional fisheries experts—for fisheries scientists, technocrats, analysts, and managers, for those of us who are involved with diadromous fishes—it is a time for neither crippling pessimism nor delusional optimism; rather, it is a time for uncompromising ecological realism and honest, blunt, and unbiased assessments of the consequences of various policy choices to sustain diadromous fishes through 2100.

Acknowledgments

The Salmon 2100 Project, four years in the making, resulted in 23 policy alternative prescriptions developed by 33 senior salmon scientists, policy analysts, and wild salmon advocates. The contributions from these individuals in shaping my thinking about the future of diadromous species were immense. In addition, my two co-project leaders, Denise Lach and Sally Duncan, with their sociological and political science perspectives, worked diligently to broaden my provincial view of ecological policy, a view formed largely by spending my professional career surrounded by other biologists. This article benefited greatly from candid review by colleagues. Particularly helpful were the suggestions provided by Kenneth I. Ashley, Randy E. Bailey, David R. Bernard, Robert E. Bilby, Michael E. Colvin, Jeffrey J. Dose, Joseph M. Eilers, Peter F. Galbreath, Guillermo R. Giannico, Effie A. Greathouse, Michael J. Harte, David S. Hewlett, Gordon F. Hartman, Joan Hurley, Ronald J. Klauda, Arne R. Langston, William M. Lewis, Douglas F. Markle, T. Larry Marshall, Kim G. Mattson, Willis E. McConnaha, John H. Michael, Gary S. Morishima, Jeff Oveson, Derek Poon, Janine Salwasser, Benjamin B. Stout, David L. Trauger, Joan G. Trial, Webb Van Winkle, Katherine S. Whitehead, and Frederick G. Whoriskey. Some of these reviewers maintain different views than those expressed in this essay; thus, being acknowledged as a reviewer does not constitute an endorsement.

References

Gustafson, R. G., R. S. Waples, J. M. Myers, L. A. Weitkamp, G. J. Bryant, O. W. Johnson, and J. J. Hard. 2007. Pacific salmon extinctions: quantifying lost and remaining diversity. Conservation Biology 21(4):1009–1020.

Lackey, R. T. 2005. Economic growth and salmon recovery: an irreconcilable conflict? Fisheries 30(3):30–32.

Lackey, R. T. 2006. Axioms of ecological policy. Fisheries 31(6):286–290.

Lackey, R. T., D. H. Lach, and S. L. Duncan, editors. 2006a. Salmon 2100: the future of wild Pacific salmon. American Fisheries Society, Bethesda, Maryland.

Lackey, R. T., D. H. Lach, and S. L. Duncan. 2006b. Wild salmon in western North America: forecasting the most likely status in 2100. Pages 57–70 *in* R. T. Lackey, D. H. Lach, and S. L. Duncan, editors. Salmon 2100: the future of wild Pacific salmon. American Fisheries Society, Bethesda, Maryland.

Salonius, P. 2007. Will forestry follow agriculture toward unsustainable soil depletion. The Forestry Chronicle 83(3):375–377.

American Fisheries Society Symposium 69:619–636, 2009

Salmon and Eulachon in Ecosystem Space and Time: A Plea for Wider Collaboration and Data Integration

Nigel Haggan*
UBC Fisheries Centre, 2202 Main Mall, Vancouver, V6T 1Z4, Canada

George D. Jackson
Institute of Antarctic and Southern Ocean Studies, University of Tasmania Hobart, Tasmania 7001, Australia

Pacific Ocean Shelf Tracking Project, Vancouver Aquarium Post Office Box 3232 Vancouver British Columbia V68 3X8, Canada

Paul Lacroix
COIN Pacific
Post Office Box 3075 Stn CSC, R Hut, McKenzie Avenue Victoria, British Columbia V8W 3W2, Canada

Abstract.—Aboriginal people developed integrated ecosystem-based management long before European contact in the 1750s. Ecosystem knowledge contributed the lion's share of precontact wealth. Fisheries drove the early British Columbia economy, but now account for less than 0.5% of gross domestic product. Even thought West Coast research shows that precontact ecosystems could sustain many times current catch value, this still would not weigh heavily against other economic sectors. Single species management has failed to avert the depletion of many fisheries; hence, we now hear calls for ecosystem-based management as opposed to integrated management (used in reference to managing multiple sectors such as fisheries, farmed salmon, oil, and gas, as well as climate change). We suggest that reintegrating ecosystem-based and integrated management necessitates the cooperation of other ocean sectors in generating the information necessary to monitor and restore ecosystems while ensuring that their own operations are sustainable. Currently, there are a number of scientific initiatives, ocean and biological observing platforms, and high-powered models to help develop new management regimes. We consider how this new technology could help to understand the collapse of eulachon *Thaleichthys pacificus*. Eulachon are of great importance to Native peoples but could well be described as the forgotten anadromous fish of the research community. It is important that both industry and governments recognize the importance of maintaining the long-term viability of these important tools and invest appropriately to ensure sound ecosystem management practices into the future.

Introduction

> We believe that life supports life, that we are one with the animals of air, land and water. What you call "biodiversity" is only a part of it. (Chief Simon Lucas 2007)

Fisheries drove the early British Columbia economy. All fisheries, commercial, sport, aquaculture, and processing now account for less than 0.5% of gross domestic product (BC 2002). Commercial fisheries account for only 0.1%, a decline of 15% since 1984 (BC 2002). Within the fisheries sector, the relative importance of Pacific salmon and eulachon *Thaleichthys pacificus* mirrors the overall decline in fisheries value. Eulachon are of great importance to Aboriginal people (Drake and Wilson 1991; Hay and McCarter 2000; Hume 2007). Eulachon could equally be described as

* Corresponding author: nhaggan@gmail.com

the "forgotten anadromous fish" of the research community. Eulachon are not mentioned in the 1986 anadromous fish conference (Dadswell et al. 1987) and, what is even more remarkable, by other contributors to this volume, given their collapse throughout their range. For this reason, we consider how new tracking systems might provide some insight into the cause of decline in eulachon as well as insight into salmon and other species.

Balancing conservation of coastal ecosystems with immediate need for jobs and revenue requires a larger investment in knowledge acquisition than can be justified in our current economic assessment of fisheries value. How then can we justify the expenditure? This paper presents a vision for the harmonious use of ocean space where human activities that exploit living marine resources and other civil, industrial, and military sectors will together contribute to our knowledge base for integrated, ecosystem-based management, with subsequent benefits for all.

To do this, we do not look at the ecosystem at one point in time, but instead develop an integrated understanding of the extent of ecosystem change over long time periods and the natural and human drivers of that change. This is the vital context within which ecosystem-based management (EBM) efforts must operate, even though they also, of necessity, focus on the ecosystem as we see it today based on its very recent past. Broadening our temporal scope separates EBM from the struggle over who owns and manages the current system and requires us to frame questions over longer temporal scales and many different spatial scales. We now need to extend the concept of adaptive management to include modeling, seasonal, and real-time science in collaborative iteration with the maritime community (Figure 1). We see an urgent need to link relatively large-scale food web,

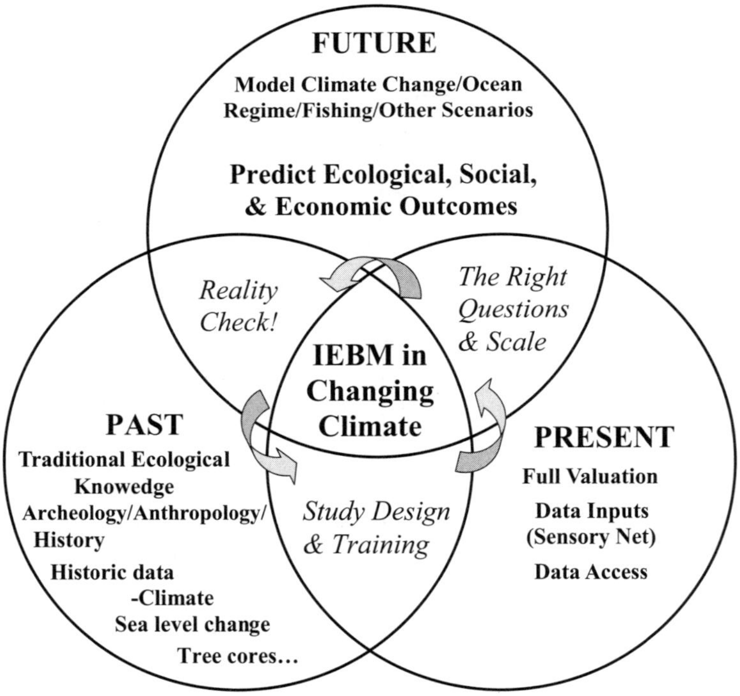

Figure 1.—Integrated ecosystem-based management (IEBM) concept: linking knowledge of the past, present research, and computer modeling facilitates prediction of ecosystem response to natural variability and human factors from fishing to climate change. Full valuation of present and future benefits of productive ecosystems is needed to assess the social, cultural, and economic consequences of different actions. Data inputs include traditional and local knowledge, single and multidisciplinary research, and anticipate a sensory net of ecological and ocean observation system assets. Present and future data volumes present a significant management and access challenge.

oceanographic, and climate models with the fine-scale spatial optimization capability of MARXAN[1] and geographic information system (GIS) platforms. We do not envisage one monster model for the entire West coast, but rather the capability to link whatever models are necessary to answer the questions at hand. This requires data that are standardized and permanently accessible. Here we envision a sensory net that links existing real-time data collection projects and anticipates collaboration between fisheries and ecological researchers, fishers, and other civil, industrial, and military users of ocean space in maintaining the net and data management systems.

Including all players in sensory net, model, and process construction builds social capital between collaborators and intellectual capital in the models and approaches (Haggan 2000). It also facilitates restoring depleted fish populations or reinvestment in natural capital, with benefits for coastal people and the community at large.

A Short History of British Columbia Ocean Space

Research indicates that Aboriginal people developed integrated ecosystem-based management (IEBM) several thousand years ago (Haggan et al. 2005). Advanced scientific knowledge of the biology, behavior, and genetics of Pacific salmon and other marine and terrestrial species contributed to a stable food supply and, in turn, cultural and economic wealth (Anderson 2005; Haggan et al. 2005, 2006; Turner 2005). For Aboriginal people, the commercial fishery was the new kid on the block. Aboriginal IEBM was replaced by single species management. By the 20th century, the breakdown of Canadian salmon catch of all species was 92% commercial, 4% recreational, and 4% Aboriginal (Pearse and Larkin 1992). The Canadian Aboriginal catch was also restricted to a stringently regulated food fishery, none of which might be sold. This contrasts with the United States where the Supreme Court Boldt decision returned 50% of salmon to the Washington State tribes.

By the 1990s, the new kids included salmon wars between Canada and the United States, a sport fishery taking increasing numbers of prime species and conservation organizations calling for reduced catch. Aboriginal people were also re-emerging as major players flowing from constitutional recognition of Aboriginal rights (Canada 1982).

The inadequacy of single species management under such complex conditions underpins calls for ecosystem-based management. The advent of farmed salmon, oil and gas, wind farms, natural gas pipelines, and so forth is behind the current call for integrated management. New geopolitical issues such as climate change and opening of the Northwest Passage make it critical that we reinvent integrated, ecosystem-based management, defined as the ability to understand and manage cumulative effects. As in the ancient past, our success will depend on two things: (1) the quality of knowledge and our ability to share it, and (2) our understanding of the full range of present and future benefits to humanity.

A new IEBM has to take into account three differences between the situation prior to European contact and the present day:

1. Precontact physical and cultural wealth derived from biota. By contrast, modern economies put more value on other resource sectors, with, in general, a shorter return on investment than most fish populations;
2. Apart from major river systems like the Fraser and Skeena,[2] precontact managment operated on a small scale. This is a very different model from today's coast-wide and international management, but new findings that many marine species are composed of numerous distinct subpopulations (e.g., Cury 1994; Hauser et al. 2002; Hutchinson et al. 2003; Prince 2003) point to the urgent need for management at a much finer spatial scale and a need for cross-scale linkages.
3. Precontact management was precautionary, based on respect for the intrinsic worth of salmon. There were also dire consequences for greed, waste, and disrespect, consisting of social sanctions, penalties, and even death for infra-

[1] MARXAN is a spatial optimization program used extensively in the design of marine reserves (Stewart et al. 2003) and identification of areas of high conservation utility (Arsron et al. 2002).

[2] Major river systems like the Fraser and Skeena required communication and negotiation to ensure that upstream tribes received sufficient salmon (Johnsen 2001), but the absence of major coastal interception fisheries made the type of coast-wide communication mediated through international salmon and halibut commissions unnecessary.

tion. This precautionary approach is reflected in the abundance of salmon at European contact, despite the existence of trap and weir technology capable of wiping out salmon runs many times over on all but the bigger rivers Anderson 1996; Johnsen 2001).

Changing Values

> The chiefs are instructed so that when they deliberate on the serious matters of the council, they are to consider the impact of their decisions on the seventh generation into the future. (Chief Oren Lyons: Opening Speech for UN Year of the Indigenous Peoples 1993[3])

The lion's share of precontact wealth came from the coastal and marine ecosystems. The ability to create this wealth demanded knowledge of the environment and ways to maintain and enhance resource productivity into the far future. This long view is common to Aboriginal peoples and many longstanding communities that depend on natural resources for their distinct identity and continuing existence. The long view is well articulated in the Seventh Generation principle of the Haudenosaunee or Six Nations (Clarkson et al. 1992). We note in passing that Thomas Jefferson and Benjamin Franklin drew on the Haudenosaunee "Great Law of Peace" for the principle of representative democracy in the U.S. Constitution (Grinde and Johansen 1990). From this perspective, catching all the fish to generate cash to invest in something else is illogical. It may also be illegal given that almost all resource management legislation requires consideration of future generations.

Today, we have many other sources of wealth: the information technology sector, service industries, pharmaceutical and bioengineering industries, entertainment, automotive, consumer electronics, construction, and so forth. Power and wealth tend to be measured in dollars. Fisheries are the most obvious source of dollars from marine ecosystems, but fisheries are a tiny fraction of 21st century economies (Table 1).

The small percentages in Table 1 explain why governments are reluctant to take fisheries seriously and why other investment opportunities such as offshore oil and gas and salmon farming appear so attractive to decision makers. Another key reason is that dollars grow faster than fish. Specifically, extinction is likely when the economic discount rate is more than twice the species population growth rate (Clark 1973a, 1973b). This accounts in large part for the depletion of ecosystems. Modern society tends to want money for a whole variety of reasons in less time than the ecosystem can produce fish to sell. It also accounts for government reluctance to invest in rebuilding depleted ecosystems (Sumaila 2004). These concerns can only be addressed through a thorough understanding of ecosystem response to natural variability, multiple human use, and climate change from the distant past to at least 100 years into the future. The problem is that the fisheries values in Table 1 do not justify the necessary computer models, data collection assets, collaborative research, and field studies. A full ecosystem valuation is needed.

[3] www.ratical.org/many_worlds/6Nations/OLatUNin92.html.

Table 1.—Fisheries as percentage of gross domestic product (GDP) of the United States, Canada, and British Columbia.

Country/region	% of GDP	Source (and remarks)
USA (2003)[a]	0.30	FAO country profiles (including forestry and and hunting)
Canada (2000)[b]	0.21	Fisheries and Oceans Department of Canada (commercial, aquaculture, and processing)
British Columbia (2001)[c]	0.50	BC government statistics (commercial, sport, aquaculture, and processing)

[a] www.fao.org/fi/fcp/en/usa/profile.htm.

[b] www.dfo-mpo.gc.ca/communic/statistics/oceans/economy/contribution/table3_4_e.htm.

[c] www.env.gov.bc.ca/omfd/reports/BC-Fisheries-Aquaculture-Sector-2002.pdf.

Full Ecosystem Valuation

The "total economic valuation" approach (Millennium Ecosystem Assessment 2003; National Research Council 2005) identifies five categories of value: (1) direct use (e.g., fisheries); (2) nonconsumptive use (e.g., eco-tourism); (3) indirect use (e.g. ecosystem services); (4) option value (e.g., the ability to enjoy something that you may not be aware of now or later in your lifetime); and (5) existence value, generally expressed as willingness to pay (e.g., to protect species).

The decline in fisheries and other natural resources has drawn increasing attention to new and more comprehensive ways to value the natural world. Costanza et al. (1997) valued the Earth's ecosystem services at $US33 trillion/year. While the approach has been criticized, it is generally agreed that ecosystem services (e.g., the air we breathe and the role of the coupled oceanic–atmospheric circulation in moderating climate) have value even when not traded in the market. New methods allow us to compute the ecological and social as well as the market value (Angelsen et al. 1994; Angelsen and Sumaila 1996; Sumaila et al. 2001) and to include the value to future generations (Sumaila 2004; Ainsworth and Sumaila 2005; Sumaila and Walters 2005).

Considering each category of value and reducing the discount rate from 7% to 3%, (i.e., taking a longer view of the value of renewable resources), U. R. Sumaila et al. (University of British Columbia, unpublished data) found that the net present value of U.S. fisheries increased from $33 to $879 billion. While this is an important step beyond the current >0.5% of GDP, it did not, or did not fully include, the higher value of waterside property: the percentage of waterside recreation and vacation sector value attributable to a view of the ocean as pristine, vast, mysterious, and providing the opportunity to see whales, birds, fish; and the physical and mental health benefits that marine ecosystems provide to stressed city dwellers.

While these values are the subject of vigorous debate, we believe that a full accounting of the natural benefits is essential to sustain the level of effort needed to understand ecosystems and to engage the support of the wider users of ocean space. Indeed, our models of ecosystems, ocean regime, and climate change can only acquire the necessary power, elegance, and sophistication through wider collaboration in model construction. The models also need a sensory net of data collection assets supported by government, industry, foundations, environmental nongovernment organizations (ENGOs), academia, and stable repositories of data in accessible formats.

Ocean management requires (1) the ability to understand and model ecosystem response at multiple temporal and spatial scales, (2) collaboration on the design of a sensory net greater than the sum of its parts, and (3) standards and accessibility for existing and new databases. Sustainability requires a new look at the interface between big science projects, traditionally the role of academia and government, and the intellectual property that gives business its competitive edge. Benefits could include (1) more timely and comprehensive environmental impact assessments, (2) ongoing assurance of safety and sustainability, (3) more efficient and cost-effective operation, and (4) new technology that can be marketed at home and abroad.

The Separation of Ecosystem-Based and Integrated Management

Aboriginal IEBM made sense because of the immediate link between ecosystems and wealth. Most of what we now count as wealth comes from other sources, each with its own specialized field of knowledge. This explains the separation of ecosystem-based management from integrated management in Canada's Oceans Act and Oceans Strategy (Haggan et al. 2004). What has been lost between the period prior to European contact and the present is the ability to understand and allow for cumulative effects. The ecosystem principles are good, but the game of integrated management pits the potential revenue from depleted ecosystems against the money-making potential of oil and gas and farmed salmon and weighs ecosystem health against the costs of proper treatment of urban and industrial waste.

This is a bleak picture. What it suggests is that even if we get the science right, even if we could attain the level of ecosystem knowledge of an hereditary chief of the precontact era, able to draw on many generations of knowledge and trained from infancy in the ecology and management of their territory, we would still have to contend with the perception that fisheries are trivial and ecosystem considerations are, at best, a barrier to development.

We also have to address a tradition that is deeply rooted in single-species management and unduly influenced by industry and government priorities (Finlayson 1994). The majority of fisheries science and management funds still go to single-species approaches. This is not entirely a bad thing. There is a great deal of excellent single-species work. What we are not doing with any consistency is relating these individual studies to an ecosystem framework that connects species, habitats, people, and environment. Figure 1 presents a concept that links computer models (food web, oceanographic, climate, and socioeconomic) to traditional and local knowledge and data collection. (Pitcher et al. 2007) present a formal ecosystem evaluation framework that can assess the extent of our knowledge or, as the title suggests, plumb "the depths of ignorance" to suggest where research and modeling effort may best be deployed.

Linking the Models

"Baseline shift" (Pauly 1995) describes a tendency of fishers and fisheries scientists to equate the amount of fish an ecosystem ought to produce with abundance at the start of their careers. Perception of productive potential thus ratchets down over the generations. MacGregor et al. (2009, this volume) illustrate baseline shift in American eel *Anguilla rostrata* with accounts of extremely high catch rates in indigenous and early fisheries. Eels were a mainstay of Aboriginal culture and economy and were once the most valuable fishery of Lake Ontario, but are now of deep conservation concern (MacGregor et al. 2009).

The "back to the future" project attempts to reverse this cognitive ratchet at a whole ecosystem level (Haggan 2000; Pitcher et al. 2005). The Coasts Under Stress project (Ommer et al. 2007) used Ecopath with Ecosim (Christensen and Pauly 1992; Walters et al. 1997) to model northern BC waters in collaboration with Aboriginal people, commercial fishers, and others (Ainsworth et al. 2002). Figure 2 compares the biomass of key model groups before European contact in the 1750s, the 1900s, the 1950s, and the present day (2000). Twenty-five-year fishing simulations with current gear types indicate that the 1750s ecosystem could sustain a catch 7–9 times higher than the actual 2000 catch, as reported by Department of Fisheries and Oceans Canada (DFO; Ainsworth 2006), and more than 20 times more (Haggan et al. 2007, Table 4.2) if fished in accordance with the Food and Agriculture Organization of the United Nations (FAO) Code of Conduct (FAO 1995).

The goal is to use benchmarks of past abundance, diversity, and trophic structure to set res-

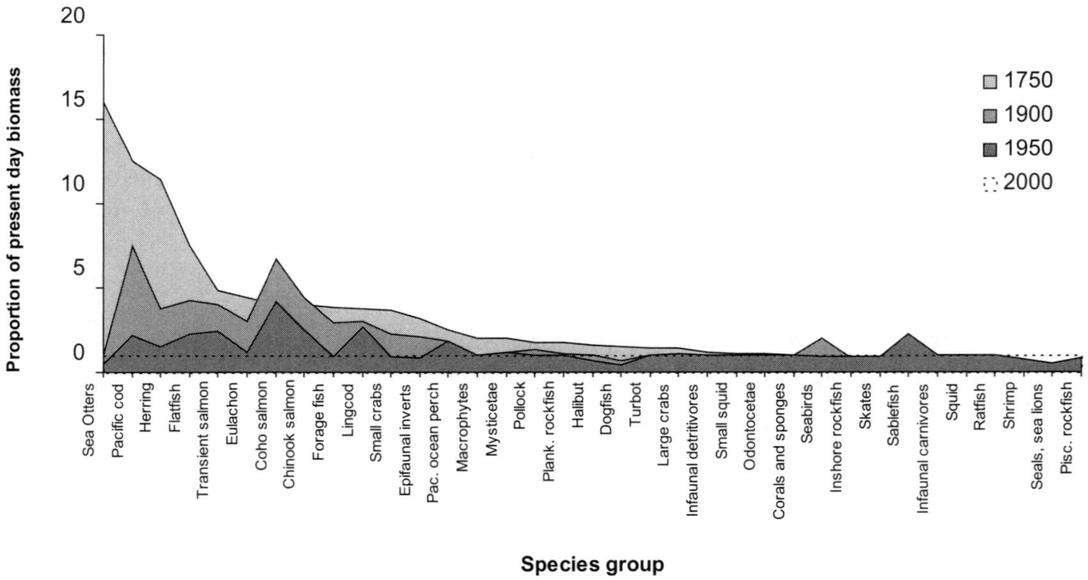

Figure 2.—Biomass of major marine species groups in northern British Columbia in the 1750s, 1900s, and 1950s compared to present day (dotted line), as modeled by Ainsworth (2006).

toration goals that relate to productive potential rather than present scarcity, not to restore some past golden age. Extinctions, exotic species, climate change, and possible (though unproven) irreversible effects of overfishing make this impossible. Accordingly, climate change modeling scenarios would adjust the benchmarks of Figure 2 to account for new species assemblages (e.g., warmer water ecosystems similar to those of Oregon or California). Rochard et al. (2009, this volume) explicitly consider future climate regimes in anadromous fish restoration in the Seine basin. Lassalle et al. (2009, this volume) model future distribution in a pan-European context.

Over the same period, the DFO Hecate Strait Ecosystem Project built a parallel model of Hecate Strait (HecStEP, no date). There is great potential to combine the regional food web models (e.g., Ainsworth et al. 2002; HecStEP, no date) with the spatial optimization capability of MARXAN used for conservation and marine use planning in British Columbia and the growing GIS databases of federal and provincial governments, Aboriginal people, and ENGOs. These models can in turn be driven by regional climate change models.

The Sensory Net

The sensory net consists of current and planned sensor arrays to track salmon and other species and anticipates linkage with a proliferating range of ocean observation systems (OOS). This linkage is critical. While the ecological arrays are visionary, they are subject to funding constraints related to our current assessment of fisheries values (Table 1). Ocean observation systems relate to the entire range of scientific enquiry, impact assessment, and development in the world's oceans and coupled ocean-atmospheric systems. The annual market for new OOS technology is approximately $2 billion (Douglas-Westwood 2006). Current ecological assets include, by geographic scale, the following:

The Victoria Experimental Network Under the Sea

The Victoria Experimental Network Under the Sea (VENUS) is the first of a new type of seafloor observatory (Tunnicliffe et al. 2003; Dewey et al. 2007). Instrument arrays are connected via cable to the shore and thus to unlimited power and two-way communication. This has three major advantages: (1) experiments are unconstrained by battery life; (2) instruments can be accessed anywhere at any time, not only to receive data, but to command and control; and (3) data and view events can be shared in real time with collaborators and students. The first VENUS array began operations in Saanich Inlet, British Columbia in February 2006. The second, commissioned in fall 2007, includes three seafloor instrument suites in the Strait of Georgia south of Vancouver. Receivers deployed across the mouth of the Fraser River will communicate information from tagged salmon, white sturgeon *Acipenser transmontanus*, or eulachon, as discussed below, in a daisy chain back to an acoustic modem cabled into the VENUS node. The capacity for higher frequency interrogation and notification once an emerging—or returning—cohort crosses the line lends immediacy to data return and may be of considerable value for alerting other detection systems or management agencies. The multitude of sensors also allows researchers to measure ocean parameters, including temperature, salinity, gas pressure, and plankton abundance, both at times of fish presence and absence. It would be possible, for example, to examine current records from the full Strait depth to determine the flow conditions at the times fish were crossing the receiver line.

VENUS provides simple graphical analyses for a quick inspection of conditions. Fish records can be displayed with a GoogleMap interface currently used to display instrument locations and metadata. The VENUS Web site (www.venus.uvic.ca) provides a convenient display and communication tool that is especially valuable for teaching. The Web site also allows researchers to perform data searches and multisensor integration using the Data Management and Archive System. The Data Management and Archive System is a highly sophisticated structure to deal with many types of ocean data: scalar, acoustic, and visual. The challenge for fish tracking is to develop a rapid response and alerting system and a data policy to integrate with partners such as the Pacific Ocean Shelf Tracking Project and the Ocean Tracking Network.

The Pacific Ocean Shelf Tracking Project

The Pacific Ocean Shelf Tracking Project (POST) mission is to "assess and explain the changing diversity, distribution and abundance of marine species, from the past to the present, and project future marine life" (Yarincik and O'Dor 2005). POST is funded by the Census of Marine Life (CoML, www.coml.org) and the Moore Foundation's Wild Salmon Ecosystems Initiative. POST is intended as a permanent continental-scale telemetry research platform. POST currently spans more than 2,500 km from Icy Strait in southeast Alaska to Oregon and extends almost 900 km up the U.S. Columbia–Snake River (Figure 3). Future plans anticipate continental shelf coverage from Baja California to the Bering Sea.

The acoustic frequencies selected allow the system to work seamlessly between freshwater and marine environments. Engineering standards applied in deployment and array geometry enable a more than 95% detection of fish passing the array. Surgical techniques have been standardized and refined to enable routine tagging of fish as small as 12.5 cm, with close to 100% survival. POST is now beginning to answer critical questions on salmon migration and survival:

- Where do salmon go?
- What do they do when they get there?
- How do they return to spawn in their home rivers?
- How do changes in the ocean environment affect their survival?

POST can also help to facilitate the evaluation of proposed networks of marine protected areas for salmon and other key species likely to be affected by climate change. The current situation of accelerating climate change is now well documented (e.g., Parmesan 2006). Although there are many unknowns, temperature alone may continue to push salmon populations north. This would severely limit the distribution of some species (e.g., steelhead *Oncorhynchus mykiss*,

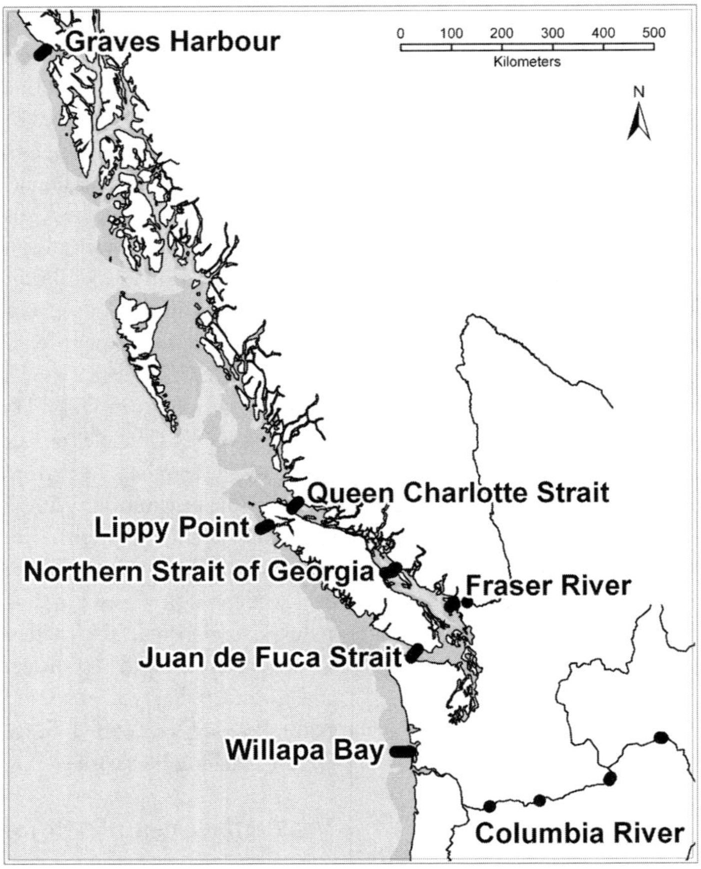

Figure 3.—The Pacific Ocean Salmon Tracking Project array configuration in 2007.

Welch et al. 1998) or even exclude others from the Pacific entirely (e.g., sockeye *O. nerka*, Welch et al. 1998). It is important that POST continues to collect baseline data on salmon movement, migration, and survival. A time series is needed to monitor how climate change and other anthropogenic influences might affect salmon populations.

Other species and globalization.—POST can provide insights into the ecology of nonsalmonids, which advance fisheries and ecosystem management and assist in conservation of important species. POST dramatically demonstrated that a white sturgeon tagged in the Sacramento River spent several months in the Fraser River (Welch et al. 2006). This finding has significant management implications as some sturgeon populations are protected while others are not. Key species for which POST can answer important questions include Pacific halibut *Hippoglossus stenolepis*, spiny dogfish *Squalus acanthias*, sablefish *Anoplopoma fimbria*, and invertebrates such as squid (e.g., Pecl et al. 2006; Stark et al. 2005), octopus, and Dungeness crab *Cancer magister*.

O'Dor and Gallardo (2005) anticipate that: "The POST arrays will be a coastal component of GOOS" (Global Ocean Observing System, www.ioc-goos.org). This is the critical linkage among tracking anadromous fishes and other aquatic species, ecosystem modeling, and the entire range of human activity in the oceans of the world.

The Ocean Tracking Network

POST is the flagship project in the Ocean Tracking Network (OTN) (www.oceantrackingnetwork.org), an international scientific endeavor to integrate animal movement and migration patterns with oceanographic data. The Ocean Tracking Network encompasses all seven continents and 14 ocean regions. Plans to deploy 5,000 receivers in 600 listening lines around the world will make it possible to track 1 million animals simultaneously. The volumes of data to be collected will be huge, requiring the development of new data structures and analysis tools. The Ocean Tracking Network will be working closely with both POST and ultimately the Ocean Biogeographic Information System (OBIS) for archiving, distributing, and analyzing the data.

The OTN will be able to track migrating animals, undertake global ocean physics modeling, and study how climate change will influence both. Conservation and resource management questions will determine where OTN lines are deployed and what species are tracked.

The OTN will drive acoustic technology forward to create new and innovative ways of tracking marine organisms. Step 1 will use miniature acoustic receivers or "business card" tags to transform predators such as sharks or seals into roving receiver platforms able to record interactions with the ecosystem. Step 2 will double tag predators with business card and geolocating archival tags to correlate individual movements of tracked animals to depth and temperature, giving us the clearest picture ever available of how organisms use the aquatic environment. Step 3 will develop new generation fast Communicating Histogram Archiving Transmitter (CHAT) tags (O'Dor et al. 2006) and new receivers to allow oceanographic and predator–prey interaction data to be downloaded as the organism crosses an acoustic line (Holland et al. 2001). Step 4 will combine the business card, archival geolocating, and fast CHAT tag functions in fully integrated tags (FITs). When the organism swims over the new generation listening line, the FIT will download location, oceanographic, and biological interaction data without need for recapture—essentially a complete underwater Argos system independent of satellites. This will ultimately lead to many more species being tagged with increasingly smaller tags and the acquisition of archival data at a much lower cost. For a more complete description of OTN, see Stokesbury et al. (2009, this volume).

Tracking eulachon: the next step.—Eulachon are suffering a severe decline and may have been extirpated from several rivers in the central coast of British Columbia (Hay and McCarter 2000), with profound impact on Aboriginal people (Hume 2007). Eulachon, the first species to enter the rivers in springtime when dried salmon and other food sources had run low or out, were referred to as "salvation fish" (Harrington 1967). For the same reason, eulachon are thought to be a keystone species based on the number of predators attending their arrival (Willson and Halupka 1995) and may be critical to the energetics of Steller's sea lion *Eumetopias jubatus* (Sigler et al. 2004). However, many aspects of their ecology remain unknown. Offshore abundance indicated by substantial bycatch in the BC shrimp trawl fishery is not reflected in numbers migrating into the rivers (Hay and McCarter 2000). Likewise, there appear to be interesting parallels between population levels of eulachon and salmon. Both salmon and eulachon appear to be doing better in Alaskan waters than in southern British Columbia. This suggests that there is likely an ocean survival problem with both groups of fish, with the impact not affecting Alaskan waters to the degree that it is in waters further south. POST is well placed to test hypotheses of eulachon survival in the ocean. By tagging individuals in the ocean, it would be possible to obtain a better indication of where individuals are moving and why there are low numbers returning to natal rivers.

Stokesbury et al. (2009) discuss how FITs attached to salmon sharks *Lamna ditropis* could track salmon throughout their ocean range, providing a major extension to POST and other arrays. Eulachon are important prey items of endangered beluga whales *Delphinapterus leucas* in Cook Inlet, Alaska and white sturgeon in the Fraser River system. It would be feasible to turn both belugas and sturgeon into roaming acoustic receivers by fitting them with OTN business card tags, providing a strong scientific framework with which to study trophic interactions of all three species. Central to any of these efforts is the ability to catch and tag sufficient numbers of eulachon.

Tagging eulachon at sea would then allow them to be counted over the POST array and as they move up the Fraser River. A more extensive array would enable tracking migration into other eulachon-bearing rivers such as the Nass and Skeena and even depleted systems such as the Bella Coola River and Rivers Inlet where their presence is revealed only by intensive field surveys (Moody 2008) and sporadic sightings by local people.

Ocean Gliders and Ocean Observing Systems

Ocean gliders provide a conceptual link between predators as mobile receptor platforms in OTN and the wider world of OOS. Ocean gliders are battery-powered, ocean-going robots that operate for periods up to 6 months in depths typically down to 1,000 m. Gliders use Global Positioning Systems and dead reckoning for positioning and can achieve speeds of 0.9 km/h by shifting buoyancy to descend and ascend at shallow angles. Ocean gliders surface to uplink data and receive updated mission instructions via satellite. Different sensor packages enable measurement of physical properties, location of marine mammals, and surveillance in areas where it is difficult to otherwise obtain information. In 2007, the Canadian Centre for Ocean Gliders, together with DFO, completed a project that demonstrated the capability of ocean gliders to provide mobile monitoring of the ocean under tough winter conditions (D. Gueret, Fisheries and Oceans Canada, personal communication). Key advantages of relative ease of operation, low capital and operations costs, extended mission duration, and flexibility must be balanced against limitations in speed, payload, and power.

Ocean Observing Systems

The global Ocean Observing Systems (OOS) market is conservatively projected to increase from $1.8 billion in 2006 to $2.2 billion by 2011, with governments as the main client and source of funding (Douglas-Westwood 2006). Key market drivers are the role of the oceans in climate change, environmental pollution, and industry needs to operate in an environment of increasing regulation. The complex nature of OOS, with multiple stakeholders and financing requirements, is a consequence of carrying out integrated management in the ocean. Douglas-Westwood (2006) identified 1,200 OOS applications in 500 programs currently underway. The United States is by far the largest player with 2002 expenditures of $750 billion, 50% generated by the oil and gas industry, with 33% attributed to the Navy.

These expenditures dwarf the largest sums allocated to fisheries and ecosystem considerations, underscoring the need for collaboration. Canada's fisheries budget actually decreased in recent years. Clearly, the future of IEBM lies in networking OOS systems with integrated physical, biological, environmental, and economic models, as contemplated in Figure 1.

Data Management, Accessibility, and Sustainability

The Sea Around Us Project—A Global Fisheries Database

Until the end of the 20th century, the FAO was the sole source of global fisheries data. The Sea Around

Us Project (SAUP), funded at the University of British Columbia Fisheries Centre by the Pew Charitable Trusts, was founded in 1999 to "provide an integrated analysis of the impacts of fisheries on marine ecosystems, and to devise policies that can mitigate and reverse harmful trends while ensuring the social and economic benefits of sustainable fisheries" (Pauly and Pitcher 2000). The SAUP database (www.seaaroundus.org) divides the world's oceans into 180,000 half-degree squares and applies a rule-based method to reassign FAO and other catch data to their ecosystems of origin (Watson et al. 2004). The Sea Around Us Project is extending FAO data to include discards and illegal, unregulated, and unreported catch. These figures are substantial (Pauly et al. 2002; Figure 1) but do not yet include a significant portion of artisanal fishery catch conservatively estimated at 30 million metric tons (Pauly 2006). The proportion of recreational catch included is also unclear.

The SAUP database is deep-linked to Fishbase (www.FishBase.org), the online global finfish database (Froese and Pauly 2007). The Sea Around Us Project, in partnership with the Oak Foundation, has recently launched SeaLifeBase (www.seaLifeBase.org), which currently includes more than 80,000 of an estimated 200,000 marine species. The SAUP database now includes or is linked to databases of cephalopods (www.cephbase.utmb.edu/), coral reefs, sea grass beds, estuaries, seamounts, and marine-protected areas. The SAUP database also processes data from the SeaWiFS satellite (http://oceancolor.gsfc.nasa.gov/SeaWiFS/) at a spatial resolution of ~6 min (11 km) to estimate the phytoplankton production using a model by Platt and Satyendranath (1988) that integrates primary production by depth based on chlorophyll pigment concentrations and photosynthetically active radiation (Hoepffner et al. 1999; Lai 2004).

These linkages facilitate the rapid construction of basic ecosystem models for any area, over and above those existing on the Ecopath with Ecosim Web site (www.ecopath.org). The Sea Around Us Project also contains a global database of exvessel fish prices (Sumaila et al. 2007), which, combined with the catch data, yields a database of catch value. It also includes a database of fishing agreements.

The entire SAUP edifice is sustained by the energy and vision of Dr. Daniel Pauly and the Pew grant. Individual pieces are sustainable, notably FishBase, which has evolved from a project of ICLARM to a consortium of eight institutions, including four museums. The risk is not limited to SAUP. The sheer volume of data that VENUS, POST, OTN, and affiliated networks will generate, the new database structures and analysis tools needed, and the 2010 CoML POST funding sunset call for sustaining partnerships. The Ocean Tracking Network is addressing this through more than 30 partnership agreements leveraging more than $160 million in cash and in-kind services, where partners have agreed to track, observe, maintain equipment, and upload data. The POST vision (O'Dor and Gallardo 2005) also anticipates the sustainability issue: "by 2010 POST will have tested and demonstrated continental-scale acoustic tracking....It will enter the accumulated migratory tracks in the Ocean Biogeographic Information System (OBIS)"

The Ocean Biological Information System—A Global Marine Biodiversity Database

The Ocean Biological Information System (www.iobis.org) is the CoML data depository. The Ocean Biological Information System is a Web-based, georeferenced catalog on marine species that currently houses more than 10 million records and incorporates marine data from other institutions and programs (Costello et al. 2007). As OBIS continues to grow and becomes more comprehensive, it will become an increasingly powerful resource on marine biodiversity. The Ocean Biological Information System will be a means to create linkages between programs and databases and will assimilate not only biological data on depth and location, but also distribution data such as that generated by POST and OTN.

Data Needs and Initiatives in the Pacific Northwest

While SAUP provides a useful model, the half degree or 30 × 30 nautical mile spatial resolution is too coarse for most EBM and management applications in Pacific West Coast waters. Access to data in standard formats is an issue for all EBM activities from whole ecosystem models to the GIS-based traditional-use studies of Aboriginal people. By their holistic nature, ecosystem models require a wide variety of biological, oceanographic, and socioeconomic information. Collaboration across disciplines is required to satisfy diverse data needs.

Centralized data repositories are critical for broad-swath modeling systems. They also hold many advantages, not the least of which is streamlining research and helping to overcome data sharing hurdles. This is especially true for modeling systems like Ecopath with Ecosim, Atlantis (Fulton et al. 2004), and other modeling platforms reviewed in Fulton et al. (2007). Thus, while both Ecopath with Ecosim and Atlantis are open source, freely available and flexible enough to represent many types of ecosystems and study areas, the question of gathering sufficient data for them to produce meaningful results remains.

Major ecosystem modeling projects such as Back to the Future (Haggan 2000; Pitcher et al. 2005), MARXAN spatial optimization analyses (Ardron 2002, 2003, 2005; Ardron et al. 2002), and GIS-based traditional-use studies by Aboriginal people all wrestle with data access[4] and standardization. These projects have each made huge investments of time in gathering and formatting data for comprehensive analyses. Because current data-sharing agreements usually preclude further sharing, other researchers have to do it all over again. This is unnecessarily costly and labor-intensive. It also means that issues addressed by one team may be missed by others and that data are not standardized across models, making comparisons of results more difficult. Many projects simply cannot get off the ground because they are denied access to data, resulting in a limited field of a few players who have had data access.

The field of ecosystem modeling and EBM in general would benefit greatly from a centralized and standard repository of oceanographic time series information. Data monitoring networks, such as POST, VENUS, and OTN, could contribute to such a collaboration, increasing the scientific dividends of the submarine acoustic monitoring network. Availability of highly resolved regional data is also conducive to the development of additional modeling platforms and capabilities to capitalize on the valuable site-specific information.

[4] For example, the georeferenced Pacific groundfish catch database collected by Canada's Department of Fisheries and Oceans in the 1990s has never, to our knowledge, been integrated into an ecosystem model due to stated concerns about identification of individual fishers and confusion about who actually owns and has rights to the data. The Fisheries and Oceans Department of Canada is still without a data-sharing policy, which would offer guidance in such situations.

The Cooperative Ocean Information Network for the Pacific.—The Cooperative Ocean Information Network for the Pacific (COINPac) was formed to address the issue of access and sustainability by creating a portal to all major databases and a mechanism to translate data from various formats. COINPac was successful in creating a BC user community forum, developing and reaching agreement on the portal and access mechanism. This forum, which represents a substantial investment in partnership building, has relocated to the University of Victoria. In early 2007, the Ocean Science and Technology Partnership (www.ostp-psto.ca), with COINPac as the BC partner, completed a strategic document that called for national and regional policy changes and investment to support the development of OOS (OSTP 2007). The report calls for a national strategy to facilitate and help coordinate the emerging OOS systems to maximize their value to society. Coordinating multiple ocean users as OOS stakeholders, using systems approaches to ocean observations, and concentrating on lessons learned elsewhere, were identified as key approaches. The north coast was identified as the location for an OOS in British Columbia.

The Pacific Marine Analysis and Research Association.—The Pacific Marine Analysis and Research Association (PacMARA, www.pacmara.org) grew out of a series of multi-stakeholder meetings held in 2002–2003, where users, practitioners, regulators, and advocates for the sea identified a number of common research needs. PacMARA seeks to develop and encourage the use of cross-disciplinary marine science in ecosystem-based decision making. PacMARA takes an impartial, nonadvocacy approach, believing that access to data, good science, and clear results are at the heart of sustainable oceans management.

It has been suggested that PacMARA could act as an honest broker between data sources and data users. Processed data sets (often compiled from a variety of sources, as described above) could be held by PacMARA and distributed to EBM modelers and users based upon agreed data protocols. Additionally, PacMARA could compile and relay back to data sources those corrections or issues that have arisen with regard to particular source data.

While the above arrangement would not represent the ideal of open Internet access often promoted by data-sharing advocates, it would nonetheless represent a significant step forward, allowing EBM research in British Columbia to proceed more broadly, based on standardized data sets, and allow for regular updates while adhering to protocols that would satisfy data sources.

Long-Term Sustainability—Global and Local

Even major biotech and health databases face problems of sustainability (Anon 2007). If health, which accounts for ~50% of tax dollars, cannot sustain the basic information needed to support research into human life and health, the problem of sustaining the sensory net and databases needed for EBM is even more acute. The risk that substantial collaborative projects such as SAUP or POST or even OTN might not survive the retirement of visionary leaders or the sunset of foundation funding is unacceptable. Nor is it reasonable that foundations should carry the entire cost. We hope that this brief overview of analytic tools and data collection and management assets will at least set the stage for a broader based collaboration between government, ENGOs, and industry to build sustaining partnerships.

The case for business support.—The information systems needed to understand the drivers of ecosystem abundance and response to fishing, climate change, and other factors can benefit multiple users of ocean space. There is therefore a cogent argument for the design of data acquisition assets to provide multiple data streams. This coordination is not cheap, but it is achievable through cost sharing between government and industry, provided that the benefits to operations as diverse as ports, national defense, marine navigation, ocean energy (renewable and hydrocarbon), wind farms, waste management, and so forth can be identified. We take the positive view that all sectors, new, old, large, and small, share a desire to maintain ecosystem health and keep options open for future generations. Collaboration and data sharing are much more likely to reveal effective ways to do this than campaigns against the presence of certain sectors.

The case for government support.—Coastal and marine ecosystems provide a wide range of public goods for which no market currently exists. Healthy marine ecosystems provide a wide range of natural resources and a growing number of pharmaceuticals and other marine products. Most Pacific salmon die after spawning, contributing nitrogen, phosphorus, and carbon to freshwater and forest ecosystems (Stockner 2003). The size of past salmon runs is reflected in the growth rings of riverside trees (Reimchen 2001; Naiman 2009, this volume), but it does not stop at the bank or lakeshore. At least 40 creatures from people to bears to insects transport nutrients from salmon carcasses into the forest (Watkinson 2001). We know this because marine nitrogen is a different isotope (^{15}N) from terrestrial nitrogen (^{14}N) and has been used to trace the influence of salmon carcasses up to 1 km away from the water's edge (Watkinson 2001). Gresh et al. (2000) estimated historic salmon escapements to the Pacific Northwest at 160,000–226,000 metric tons, versus 11.8–13.7 today, indicating that nutrient inputs are about 6–7% of pre-European contact levels. Out-migrating salmon smolts require significant amounts of nitrogen (Moore and Schindler 2004), so reduction in spawning runs may also impact freshwater and early marine survival.

Marine ecosystems also absorb, detoxify, and sequester significant amounts of industrial and human waste; support tourism, vacation, and outdoor recreation; enhance property values; and provide a source of psychological rejuvenation and inspiration for the arts (Peterson and Lubchenko 1997; National Research Council 2005). On a global scale, oceans sequester carbon, moderate climate, and provide half of the oxygen we breathe (Field et al. 1998). Given the significant impact of private investment and ownership on ecosystem health, there is a pressing case for government to protect and, where necessary, restore ecosystem services. Salzman (2005) outlines workable approaches.

Conclusion

While we need employment and prosperity today, we also need to ensure that the ecosystem goods and services that sustained rich Aboriginal societies and jump-started the BC economy continue into the far future. It is simplistic to cast this as a struggle between good (conservation) and evil (development). It is unhelpful to categorize ecosystem values as priceless or infinite (given future generations). It is, however, vitally important that we strive to understand each other's values and explore ways in which they can coexist.

Acknowledgments

The lead author gratefully acknowledges input on VENUS from Dr. Verena Tunnicliffe and helpful comments from Drs. Cameron Ainsworth, Rosemary Ommer, Rashid Sumaila, David Welch, and Mr Jeff Ardron. The senior author is also grateful to Dr. Ronald Klauda and the American Fisheries Society for covering expenses to attend the conference. We are grateful to the reviewers whose insightful comments strengthened the manuscript considerably.

References

Ainsworth, C. 2006. Strategic marine ecosystem restoration in northern British Columbia. Doctoral thesis. University of British Columbia, Vancouver.

Ainsworth, C., J. J. Heymans, T. J. Pitcher, and M. Vasconcelos. 2002. Ecosystem models of northern British Columbia for the time periods 2000, 1950, 1900 and 1750. University of British Columbia, Fisheries Centre Research Report 10(4), Vancouver.

Ainsworth, C., and U. R. Sumaila. 2005. Intergenerational valuation of fisheries resources can justify long-term conservation: a case study in Atlantic cod (*Gadus morhua*). Canadian Journal of Fisheries and Aquatic Sciences 62(5):1104–1110.

Anderson, E. N. 1996. Ecologies of the heart: emotion, belief, and the environment. Oxford University Press, New York.

Anderson, K. 2005. Tending the wild: Native American knowledge and the management of California's natural resources. University of California Press, Berkeley.

Angelsen, A., O. H. Fjeldstad, and U. R. Sumaila. 1994. Growth, development and sustainability. Development Matters 1(2):19–22.

Angelsen, A., and U. R. Sumaila. 1996. Hard methods for soft policies: environmental and social cost-benefit analysis. Forum for Development Studies 1:87–114.

Anon. 2007. The database revolution. Nature (London) 445(7125):229.

Ardron, J. 2003. BC Coast Information Team Marine Ecosystem spatial analysis, v. 1.2. Excerpted and revised from C. Rumsey, J. Ardron, K. Ciruna, T. Curtis, F. Doyle, Z. Ferdana, T. Hamilton, K. Heinemyer, P. Iachetti, R. Jeo, G. Kaise, D. Narver, R. Noss, D. Sizemor, A. Tautz, R. Tinger, and K. Vance-Borland, editors. An ecosystem analysis for Haida Gwaii, central coast, and north coast British Columbia. Available: www.livingoceans.org/files/PDF/mpa/CIT_marine_report1-2b.doc (November 2008).

Ardron, J. A. 2005. Protecting British Columbia's corals and sponges from bottom trawling. Living Oceans Society, Sointula, BC, Available: www.livingoceans.org/files/PDF/sustainable_fishing/Protecting_BCs_Corals_and_Sponges.pdf (November 2008).

Ardron, J. A. 2002. A recipe for determining benthic complexity: An indicator of species richness. Pages 169–176 *in* J. Breman, editor. Marine geography: GIS for the oceans and seas. ESRI Press, Redlands, California.

Ardron, J. A., J. Lash, and D. Haggerty. 2002. Modelling a network of marine protected areas for the central coast of British Columbia. Living Oceans Society, Sointula, BC. Available: www.livingoceans.org/files/PDF/mpa/LOS_MPA_model_v31_web.pdf (November 2008)

BC (British Columbia). 2002. British Columbia's fisheries and aquaculture sector British Columbia Ministry of Management Services, Victoria, BC. Available: www.env.gov.bc.ca/omfd/reports/BC-Fisheries-Aquaculture-Sector-2002.pdf (February 2008).

Canada. 1982. The Constitution Act. Enacted by the Canada Act, 1982. (UK), c. 11, Schedule B.

Christensen, V., and D. Pauly. 1992. ECOPATH II—a system for balancing steady-state ecosystem models and calculating network characteristics. Ecological Modeling 61:169–185.

Clark, C. W. 1973a. The economics of overexploitation. Pages 155–159 *in* L. G. Anderson, editor. Fisheries economics: collected essays. volume 1. International Library of Environmental Economics and Policy, Aldershot, UK.

Clark, C. W. 1973b. Profit maximization and the extinction of animal species. Journal of Political Economy 81(4):950–961.

Clarkson, L., L. V. Morrissette, and G. Régallet. 1992. Our responsibility to the seventh generation. International Institute for Sustainable Development, Winnipeg, Manitoba, Canada.

Costanza, R., R. d'Arge, R. de Groot, S. Farber, M. Grasso, B. Hannon, K. Limburg, S. Naeem, R. V. O'Neill, J. Paruelo, R. G. Raskin, P. Sutton, and M. van den Belt. 1997. The value of the world's ecosystem services and natural capital. Ecological Economics 25(1):3–15.

Costello M. J., K. Stocks, Y. Zhang, J. F. Grassle, and D. G. Fautin. 2007. About the Ocean Biogeographic Information System. Available: www.iobis.org (July 2007).

Cury, P. 1994. Obstinate nature: an ecology of individuals. Thoughts on reproductive behavior and biodiversity. Canadian Journal of Fisheries and Aquatic Sciences 51(7):1664.

Dadswell, M. J., R. J. Klauda, C. M. Moffit, R. L. Saunders, R. A. Rulifson, and J. E. Cooper, editors. 1987.

Common strategies in anadromous and catadromous fishes. American Fisheries Society, Symposium 1, Bethesda, Maryland.

Dewey, R., Round, R., Macoun, R., Vervynck, J., and Tunnicliffe, V. 2007. The VENUS cabled observatory: engineering meets science on the seafloor. In Proceedings of the OCEANS 2007 MTS/IEEE, Vancouver, Canada, October 2007. Available: http://venus.uvic.ca/resources/documents/Oceans07paper-070427-015.pdf (November 2008).

Douglas-Westwood. 2006. Global markets for ocean observation systems. Available: www.ostp-psto.ca/Files/Intro_and_OOS_Strategy.ppt (July 2007).

Drake, A., and L. Wilson, 1991. Eulachon: a fish to cure humanity. 32. University of British Columbia, UBC Museum of Anthropology, Museum Note No. 32, Vancouver.

FAO (Food and Agriculture Organization of the United Nations. 1995. Code of conduct for responsible fisheries. Food and Agriculture Organization of the United Nations, Rome.

Field, C. B., M. J. Behrenfeld, J. T. Randerson, and P. Falkowski. 1998. Primary production of the biosphere: integrating terrestrial and oceanic components. Science 281(5374):237.

Finlayson, A. C. 1994. Fishing for truth: a sociological analysis of northern cod stock assessments from 1977 to 1990. ISER Press, St. John's, Newfoundland, Canada.

Froese, R., and D. Pauly. 2007. FishBase. Available: www.fishbase.org (November 2008).

Fulton, E. A., T. Morato, and T. J. Pitcher. 2007. Modelling seamount ecosystems and their fisheries. In T. J. Pitcher, T. Morato, P. J. B. Hart, M. Clark, N. Haggan, and R. Santos, editors. Seamounts, ecology, fisheries and conservation. Blackwell Scientific Publications, Oxford, UK.

Fulton, E. A., J. S. Parslow, A. D. M. Smith, and C. R. Johnson. 2004. Biogeochemical marine ecosystem models II: the effect of physiological detail on model performance. Ecological Modelling 173(4):371.

Gresh, T., J. Lichatowich, and P. Schoonmaker. 2000. Salmon decline creates nutrient deficit in northwest streams. Fisheries 15(1):15–21.

Grinde, D. A., Jr. and B. A. Johansen. 1990. Exemplar of liberty: native America and the evolution of democracy. University of California Los Angeles, American Indian Studies Center, Los Angeles.

Haggan, N. 2000. Back to the future and creative justice: Recalling and restoring forgotten abundance in Canada's marine ecosystems. Pages 83–99 *in* H. Coward, R. E. Ommer, and T. J. Pitcher, editors. Just fish: ethics in the Canadian coastal fisheries. ISER Books, St. John's, Newfoundland, Canada.

Haggan, N., C. Ainsworth, T. J. Pitcher, and J. J. Heymans. 2007. Life in the fast food chain: où sont les poissons d'antan? Pages 51–74 *in* C. C. Parrish, N. Turner, and S. Solberg, editors. Resetting the kitchen table: Food security, culture, health and resilience in coastal communities. Nova Science Publishers, New York.

Haggan, N., A. Narcisse, U. R. Sumaila, C. S. Lucas, and N. Turner. 2005. Pacific ecosystems, past, present and future: integrating knowledge and values, anticipating climate change. Proceedings of the Proceedings of the Indigenous Peoples' Restoration Network Earth in Transition Session, Society for Ecological Restoration International Conference, Zaragoza, Spain, September 12–18, 2005. Available: www.ser.org/iprn/pdf/Nigel_Haggan.pdf (July 2007).

Haggan, N., N. J. Turner, J. Carpenter, J. T. Jones, C. Menzies, and Q. Mackie. 2006. 12,000+ years of change: linking traditional and modern ecosystem science in the Pacific Northwest. UBC Fisheries Centre, Working Paper #2006–02, Vancouver. Available: www.fisheries.ubc.ca/archive/publications/working/2006/2006-02.pdf (November 2008).

Haggan, N., T. J. Pitcher, and U. R. Sumaila. 2004. Back to the future in the Strait of Georgia. Proceedings of the 2003 Georgia Basin/Puget Sound Research Conference, Vancouver, BC.

Harrington, R. 1967. Eulachon and the grease trails of British Columbia. Canadian Geographic Journal (January):28–31.

Hauser, L., G. J. Adcock, P. J. Smith, J. H. Bernal Ramírez, and G. R. Carvalho. 2002. Loss of microsatellite diversity and low effective population size in an overexploited population of New Zealand snapper (*Pagrus auratus*). Proceedings of the National Academy of Sciences of the United States of America 99(18):11742.

Hay, D., and P. B. McCarter. 2000. Status of the eulachon, *Thaleichthys pacificus* in Canada. Department of Fisheries and Oceans Canada, Canadian Stock Assessment, Secretariat, Research Document 2000-145, Ottawa.

HecStEP. No date. HecStEP. HecStEP Strait Ecosystem Project. Available: Hecwww-sci.pac.dfo-mpo.gc.ca/sa-hecate/default.htm (February 2008).

Hoepffner, N., Z. Finenko, B. Sturm, and D. Larkin. 1999. Depth-integrated primary production in the eastern tropical and sub-tropical north Atlantic basin from ocean colour imagery. International Journal of Remote Sensing 20:1435–1456.

Holland, K. N., A. Bush, S. M. Kajiura, C. G. Meyer, B. M. Wetherbee, and C. G. Lowe. 2001. Five tags applied to a single species in a single location: the tiger shark experience. Pages 237–247 *in* J. R. Sibert, and J. Nielsen, editors. Electronic tagging and tracking in marine fisheries. Kluwer Academic Publishers, Dordrecht, The Netherlands.

Hume, M. 2007. Collapse of the eulachon and the loss of a culture. The Globe and Mail, Toronto (June 20).

Hutchinson, W. F., C. van Oosterhout, S. I. Rogers, and G. R. Carvalho. 2003. Temporal analysis of archived samples indicates marked genetic changes in declining North Sea cod (*Gadus morhua*). Proceedings of the Royal Society of London B 270:2125–2132.

Johnsen, D. B. 2001. Customary law, scientific knowledge, and fisheries management among northwest coast tribes. Environmental Law Journal 10(1):1–69.

Lai, S. 2004. Primary production methodology. In A global database on marine fisheries and ecosystems. University of British Columbia Fisheries Centre, Vancouver. Available: www.seaaroundus.org/primaryproduction/interpolation_method.htm (November 2008).

Lassalle, G., M. Beguer, E., Rochard, and L. Beaulaton 2009. Learning from the past to predict the future: responses of European diadromous fishes to climate change. Pages 175–193 *in* A. J. Haro, K. L. Smith, R. A. Rulifson, C. M. Moffitt, R. J. Klauda, M. J. Dadswell, R. A. Cunjak, J. E. Cooper, K. L. Beal, and T. S. Avery, editors. Challenges for diadromous fishes in a dynamic global environment. American Fisheries Society, Symposium 69, Bethesda, Maryland.

Lucas, S. 2007. Life supports life. Pages 61–70 *in* N. Haggan, B. Neis, and I. G. Baird, editors. Fishers' knowledge in fisheries science and management. Coastal Management Sourcebooks 4. UNESCO, Paris.

MacGregor, R. B., P. C. Thompson, A. Mathers, J. M. Casselman, J. M. Dettmers, and S. LaPan. 2009. Natural heritage, anthropogenic impacts, and biopolitical issues related to the status and sustainable management of American eel: a retrospective analysis and management perspective at the population level. Pages 713–740 *in* A. J. Haro, K. L. Smith, R. A. Rulifson, C. M. Moffitt, R. J. Klauda, M. J. Dadswell, R. A. Cunjak, J. E. Cooper, K. L. Beal, and T. S. Avery, editors. Challenges for diadromous fishes in a dynamic global environment. American Fisheries Society, Symposium 69, Bethesda, Maryland.

Millennium Ecosystem Assessment. 2003. Ecosystems and human well-being: a framework for assessment. Island Press, Washington, D.C.

Moody, M. 2008. Eulachon past and present. Master's thesis. University of British Columbia, Vancouver.

Moore, J. W., and D. E. Schindler. 2004. Nutrient export from freshwater ecosystems by anadromous sockeye salmon (*Oncorhynchus nerka*). Canadian Journal of Fisheries and Aquatic Sciences 16:1582–1589.

Naiman, R. J. 2009. Marine-derived nutrients and the dynamics of riverine ecosystems. Pages 395–425 *in* A. J. Haro, K. L. Smith, R. A. Rulifson, C. M. Moffitt, R. J. Klauda, M. J. Dadswell, R. A. Cunjak, J. E. Cooper, K. L. Beal, and T. S. Avery, editors. Challenges for diadromous fishes in a dynamic global environment. American Fisheries Society, Symposium 69, Bethesda, Maryland.

National Research Council. 2005. Valuing ecosystem services: toward better environmental decision-making. The National Academies Press, Washington, D.C.

O'Dor, R., and V. A. Gallardo. 2005. How to census marine life: ocean realm field projects. Scientia Marina 69(Supplement 1):181–199.

O'Dor, R., M. Stokesbury, and G. D. Jackson. 2006. Tracking marine species, taking the next steps. In Australian Society for Fish Biology 2006 Workshop Proceedings, 28-29 August 2006, Hobart, Tasmania.

Ommer, R. E. and the Coasts Under Stress Team. 2007. Coasts under stress: understanding restructuring and social-ecological health in coastal communities. McGill-Queens University Press, Montreal.

OSTP (Ocean Science and Technology Partnership). 2007. Prosperity through SMART ocean management: enabling Canada's ocean economic future through a new ocean science and technology strategy for Canada. Available: www.ostp-psto.ca/Files/SmartOceanMan-FEBRUARY2007.pdf (November 2008).

Parmesan, C. 2006. Ecological and evolutionary responses to recent climate change. Annual Review of Ecological and Evolutionary Systems 37:637–669.

Pauly, D. 2006. Major trends in small-scale marine fisheries, with emphasis on developing countries, and some implications for the social sciences. Maritime Studies 4(2):7–22.

Pauly, D. 1995. Anecdotes and the shifting baseline syndrome of fisheries. Trends in Ecology and Evolution 10:430.

Pauly, D., V. Christensen, S. Guénette, T. J. Pitcher, U. R. Sumaila, C. J. Walters, R. Watson, and D. Zeller. 2002. Toward sustainability in world fisheries. Nature 418:689–695.

Pauly, D., and T. J. Pitcher, editors. 2000. Methods for assessing the impact of fisheries on marine ecosystems of the North Atlantic. Fisheries Centre Research Reports 8(2).

Pearse, P. H., and P. Larkin. 1992. Managing salmon in the Fraser. Report to the Minister of Fisheries and Oceans on the Fraser River salmon investigation. Department of Fisheries and Oceans Canada, Vancouver.

Pecl, G. T., S. R. Tracey, J. M. Semmens, and G. D. Jackson. 2006. Use of acoustic telemetry for spatial management of southern calamary *Sepioteuthis australis*, a highly mobile inshore squid species. Marine Ecology Progress Series 328:1–15.

Peterson, C. H., and J. Lubchenko. 1997. Marine ecosystem services. Pages 177–194 *in* G. C. Daily, editor. Nature's services: societal dependence on natural ecosystems. Island Press, Washington, D.C.

Pitcher, T. J., C. Ainsworth, E. A. Buchary, W. L. Cheung, R. Forrest, N. Haggan, H. Lozano, T. Morato, and L. Morissette. 2005. Strategic management of marine ecosystems using whole-ecosystem simulation modelling: the 'back-to-the-future' policy approach. Pages 199–258 *in* E. Levner, I. Linkov, and J. Proth, editors. The strategic management of marine ecosystems. Springer, NATO Science Series: IV: Earth and Environmental Sciences, Dordrecht, Netherlands.

Pitcher, T. J., T. Morato, P. J. B. Hart, M. R. Clark, N. Haggan, and R. S. Santos. 2007. The depths of ignorance: an ecosystem evaluation framework for seamount ecology, fisheries and conservation. Pages 476–488 *in* T. J. Pitcher, T. Morato, P. J. B. Hart, M. R. Clark, N. Haggan, and R. S. Santos, editors. Seamounts: ecology, conservation and management. Blackwell Scientific Publications, Oxford, UK.

Platt, T., and S. Satyendranath. 1988. Oceanic primary production: estimation by remote sensing at local and regional scales. Science 241:1613–1620.

Prince, J. D. 2003. The barefoot ecologist goes fishing. Fish and Fisheries 4(4):359–371.

Reimchen, T. E. 2001. Salmon nutrients, nitrogen isotopes and coastal forests. Ecoforestry 16:13–17.

Rochard, E., P. Pellegrini, J. Marchal, M. Beguer, D. Ombredane, G. Lassalle, and E. Menvielle. 2009. Identification of diadromous fish species on which to focus river restoration: an example using an eco-anthropological approach (the Seine basin, France). Pages 691–711 *in* A. J. Haro, K. L. Smith, R. A. Rulifson, C. M. Moffitt, R. J. Klauda, M. J. Dadswell, R. A. Cunjak, J. E. Cooper, K. L. Beal, and T. S. Avery, editors. Challenges for diadromous fishes in a dynamic global environment. American Fisheries Society, Symposium 69, Bethesda, Maryland.

Salzman, J. 2005. Creating markets for ecosystem services: notes from the field. New York University Law Review 80(3):870–961.

Sigler, M. F., J. N. Womble, and J. J. Vollenweider. 2004. Availability to Steller sea lions (*Eumetopias jubatus*) of a seasonal prey resource: a prespawning aggregation of eulachon (*Thaleichthys pacificus*). Canadian Journal of Fisheries and Aquatic Sciences 61(8):1475–1484.

Stark, K. E., G. D. Jackson, and J. M. Lyle. 2005. Tracking arrow squid movements with an automated acoustic telemetry system. Marine Ecology Progress Series 299:167–177.

Stewart, R. R., T. Noyce, and H. P. Possingham. 2003. Opportunity cost of *ad hoc* marine reserve design decisions: an example from South Australia. Marine Ecology Progress Series 253:25–38.

Stockner, J. G., editor. 2003. Nutrients in salmon ecosystems: sustaining production and biodiversity. American Fisheries Society, Symposium 34, Bethesda, Maryland.

Stokesbury, M. J. W., R. K. O'Dor, G. D. Jackson, D. Pincock, and K. N. Holland. 2009. Tracking diadromous fishes at sea: the electronic future using hybrid acoustic and archival electronic tags. Pages 311–320 *in* A. J. Haro, K. L. Smith, R. A. Rulifson, C. M. Moffitt, R. J. Klauda, M. J. Dadswell, R. A. Cunjak, J. E. Cooper, K. L. Beal, and T. S. Avery, editors. Challenges for diadromous fishes in a dynamic global environment. American Fisheries Society, Symposium 69, Bethesda, Maryland.

Sumaila, U. R. 2004. Intergenerational cost benefit analysis and marine ecosystem restoration. Fish and Fisheries 5:329–343.

Sumaila, U. R., D. Marsden, R. Watson, and D. Pauly. 2007. Global ex-vessel fish price database: construction, spatial and temporal applications. Journal of Bioeconomics 9(1):39–51.

Sumaila, U. R., T. J. Pitcher, N. Haggan, and R.R. Jones, 2001. Evaluating the benefits from restored ecosystems: a back to the future approach. Pages 1–7 *in* R. S. Johnston, R. S. and A. L. Shriver, editors. Proceedings of the 10th International Conference of the International Institute of Fisheries Economics and Trade. Corvallis, Oregon, USA, July 2000. (CD-ROM).

Sumaila, U. R., and C. J. Walters. 2005. Intergenerational discounting: a new intuitive approach. Ecological Economics 52:135–142.

Tunnicliffe, V., R. Dewey, and D. Smith. 2003. Research plans for a mid-depth cabled seafloor observatory in western Canada. Oceanography 16:53–59.

Turner, N. J. 2005. The earth's blanket: traditional teachings for sustainable living. University of Washington Press, Seattle.

Walters, C. J., V. Christensen, and D. Pauly. 1997. Structuring dynamic models of exploited ecosystems from trophic mass-balance assessments. Reviews in Fish Biology and Fisheries 7:139–172.

Watkinson, S. 2001. The importance of salmon carcasses to watershed function. Master's thesis. University of British Columbia, Vancouver.

Watson, R., A. Kitchingman, A. Gelchu, and D. Pauly. 2004. Mapping global fisheries: sharpening our focus. Fish and Fisheries 5(2):168–177.

Welch, D. W., Y. Ishida, and K. Nagawasa. 1998. Thermal limits and ocean migration of sockeye salmon (*Oncorhynchus nerka*) long-term consequences of global warming. Canadian Journal of Fisheries and Aquatic Sciences 55:937–948.

Welch, D. W., Y. Ishida, K. Nagawasa, and J. P. Eveson. 1998. Thermal limits on the ocean distribution of steelhead trout (*Oncorhynchus mykiss*). North

Pacific Anadromous Fish Commission Bulletin 1:396–404.

Welch, D. W., S. Turo, and S. D. Batten. 2006. Large-scale marine and freshwater movements of white sturgeon. Transactions of the American Fisheries Society 135:140–143.

Willson, M. F., and K. C. Halupka. 1995. Anadromous fish as keystone species in vertebrate communities. Conservation Biology 9(3):489–497.

Yarincik, K., and R. O'Dor. 2005. The census of marine life: goals, scope and strategy. Scientia Marina 69(Supplement 1):201–208.

American Fisheries Society Symposium 69:637–647, 2009

Assessing Management Plans for the Recovery of the European Eel: A Need for Multi-Objective Analyses

DANIELE BEVACQUA*
Dipartimento di Scienze Ambientali, Università degli Studi di Parma
Parco Area delle Scienze 33A, I-43100 Parma, Italy

PACO MELIÀ
Dipartimento di Elettronica e Informazione, Politecnico di Milano
via Ponzio 34/5, I-20133 Milano, Italy

ALAIN J. CRIVELLI
Station Biologique de la Tour du Valat, Le Sambuc, F-13200 Arles, France

MARINO GATTO
Dipartimento di Elettronica e Informazione, Politecnico di Milano
via Ponzio 34/5, I-20133 Milano, Italy

GIULIO A. DE LEO
Dipartimento di Scienze Ambientali, Università degli Studi di Parma
Parco Area delle Scienze 33A, I-43100 Parma, Italy

Abstract.—The European eel *Anguilla anguilla* stock has been declining since the early 1970s and is currently considered to be outside safe biological limits. In June 2007, the Council of the European Union approved a regulation establishing measures for the recovery of the European eel stock. Each member state is required to develop eel management plans (EMPs) in order to achieve an escapement of at least 40% of the potential spawner biomass (with respect to undisturbed conditions) from each river basin. A reliable estimate of the potential spawner output of local stocks is crucial for the development of EMPs. Given the complexity of the eel life cycle, the use of mathematical models explicitly accounting for specific demographic traits and incorporating fundamental socioeconomic information is necessary to thoroughly assess the effectiveness of alternative management strategies. Here, using a case study approach, we show how mathematical modeling, based on sound field data, can contribute to the assessment of potential spawning stock and to the development of sound management plans. Then, we discuss how a multi-objective approach can be used to examine trade-offs between conservation and fishery goals and to help decision makers identify effective management policies.

Introduction

The European eel *Anguilla anguilla* is found and exploited in most European water bodies and a number of sites in northern Africa (Dekker 2000a). More than 25,000 people obtain a substantial income from eel fisheries (Moriarty and Dekker 1997). In recent decades, however, eel recruitment and eel catches have dramatically declined throughout the range of this species, which is presently considered outside safe biological limits (ICES 2005). The causes of its widespread decline are still poorly understood but most likely include changes in oceanic circulation (Castonguay et al. 1994; Knights 2003; Friedland et al. 2007), impact of new parasites (Lefebvre et al. 2002), habitat disruption, chemical contamination, and overfishing at different developmental stages (Dekker 2000b; Feunteun 2002; ICES 2005).

* Corresponding author: bevacqua@dsa.unipr.it

Devising suitable strategies for the recovery of the stock is made particularly difficult by the unique and complex life cycle of the European eel. The European eel is a catadromous amphihaline fish whose biological cycle is fairly well known in the continental phase but whose oceanic phase remains surrounded by much mystery. Spawning areas are believed to occur in the Sargasso Sea. Larvae (leptocephali) reach the North African and European continental shelf where they develop into glass eels (small, unpigmented eels) and then metamorphose to elvers (small, pigmented eels). Although eel catadromy may be facultative (Tsukamoto et al. 1998; Daverat et al. 2006), a significant proportion of glass eels colonize brackish and freshwater environments. There they gradually become yellow eels (larger, still immature, pigmented eels) and grow for 2–20 years until they attain the critical size triggering sexual maturation and metamorphosis into the silver stage. Silver eels begin a 5,000-km journey that brings them back to the spawning grounds where they eventually mate and die.

Although decline of eel catches began in the late 1960s and recruitment collapse became evident in the 1980s (ICES 2005), the first comprehensive restoration plans are only now being developed (Dekker 2008). Dekker (2008) provides an exhaustive description of the political and scientific process that contributed to developing these conservation plans. Briefly summarized, in 2003, the European Commission issued a "Proposal for a Community Action Plan for the Management of European Eel" (Commission of the European Communities 2003), further developed in a proposal for a "Council Regulation Establishing Measures for the Recovery of the Stock of European Eel" (Commission of the European Communities 2005). A revised version of the text was unanimously approved by the European Parliament and finally endorsed by the Council of the European Union in June 2007. Its main target is to permit "the escapement to the sea of at least 40% of the biomass of adult eel relative to the best estimate of the potential escapement from the river basin in the absence of human activities affecting the fishing area or the stock" (Commission of the European Communities 2005). Member states are required to provide an eel management plan (EMP) for each river basin, with the aim of achieving this target via locally implemented measures. Those member states that do not submit an EMP to the commission for approval by December 31, 2008 "shall either reduce the fishing effort by at least 50% relative to the average effort deployed from 2004 to 2006 or reduce the fishing effort to ensure a reduction of eel catches by at least 50% relative to the average catch from 2004 to 2006" (Commission of the European Communities 2005).

The regulation's target escapement of at least 40% of the potential adult eel biomass is not clearly defined. The regulation refers to the "absence of human activities affecting the fishing area or the stock," a pristine state that may be unrealistically difficult to determine due to lack of historical data. ICES (2005) recommends, when possible, the use of existing and scientifically reviewed historical data on eel abundance and glass eel recruitment to derive a reference point. However, historical data are often missing, and when present, they usually come from sites where eel exploitation has a long history. The longest European data sets on eel catches have been collected in Lake IJsselmeer (the Netherlands), Lough Neagh (northern Ireland), Baltic Sea, and Comacchio lagoons (northern Italy) (Moriarty and Dekker 1997). These local eel populations have been strongly affected by fishing activities in the past century and even before (Moriarty and Dekker 1997). Therefore, estimating the potential spawning stock in the absence of human activities is a very hard task even for these sites, as human pressure began far before the collection of data. Long-term data from unexploited systems are often missing. In any case, inferring the productive potential of exploited areas from data collected in unexploited areas (though with similar characteristics) could cause serious underestimates, since fisheries activities are likely to have developed where stock densities were higher while historically unexploited areas were probably the less productive ones (Dekker 2003). Where the fishery targets only silver eels and almost all individuals are caught at the outlet of the lagoons (like, for instance, at Comacchio), the potential spawning output can be easily estimated from historical silver eel catches. For the fisheries where yellow eels are also caught (for instance, at Lake IJsselmeer and Lough Neagh), or where the fraction of silver eels caught is not reliably known (for instance, in the Baltic sea), the potential spawning stock cannot be easily assessed.

In cases where few data are available, the development of suitable mathematical models is often the only way to examine the consequences of different management strategies (De Leo et al. 2009).

Here, we illustrate the process for developing a conceptual framework to assess different management policies via mathematical modeling supporting the decision process. We start with a review of existing efforts to estimate the potential spawner production, which is the key reference point of an EMP. According to ICES (2005) guidelines, we only consider models based on field data from well-studied local eel populations. Then, we show how demographic models have been used to assess the consequences of different management policies on the viability of eel populations and how socioeconomic information has been integrated into demographic models to evaluate the productivity and profitability of the fishery under different management scenarios. Finally, we discuss the potential contribution of multi-objective analysis supporting the identification of optimal management policies when decision makers are faced with contrasting objectives (typically, eel conservation and profitability of the fishery).

Bioeconomic Assessment of Eel Populations

A reliable estimate of the potential (pristine) spawner output of local stocks is the starting point for the development of EMPs in accordance to the regulation. However, conservation measures aimed at achieving the conservation target set by the regulation also affect, through possible limitations of fishing effort, the profitability of the fishery and, consequently, the social acceptability of a management plan. Therefore, the effectiveness of proposed EMPs must be evaluated not only from a conservation viewpoint, but also an economic one. To promote consensus among fishermen, decision makers should look for optimal fishing policies that achieve compromise between maximizing the viability of the stock and maximizing the profitability of the fishery. To this aim, it is crucial to have suitable tools to (1) estimate spawner output under undisturbed conditions, (2) predict the impact of different management policies on spawner output, and (3) estimate their influence on revenues.

Estimating Potential Spawner Outputs

Estimating potential spawner output (i.e., the biomass of mature silver eels that, in the absence of any fishing activity, would abandon a given site to begin their oceanic migration) provides a reference point to assess the impact of the fishery on the reproductive success of eel populations. Historically, most studies on eel population dynamics have been conducted in locations where commercial fisheries were present and fishing activities themselves provided the data used to develop models. These were usually aimed at assessing potential yields rather than spawner outputs. However, where the fishery targets only silver eels, estimating maximum yield also provides an assessment of potential spawner output. In contrast, where yellow eels are also exploited, estimating potential spawner output requires the development of demographic models, explicitly accounting for fishing mortality at all developmental stages. In the following, we briefly review the main studies that, in the past decades, provided reliable estimates of silver eel productions. Estimates are all expressed as a silver eel biomass per hectare in order to favor the comparison among results of different studies.

A first attempt to assess silver eel escapement through mathematical modeling was made by Rossi (1979) for the Comacchio lagoons and by Rossi and Cannas (1984) for the Porto Pino lagoons (southwestern Sardinia) through a simple life-table analysis. They provided an estimate of silver eel production before the recruitment drop of the 1980s, equal to 20 and 19 kg/ha at Comacchio and Porto Pino, respectively. Rossi (1979) reported also that silver eel production at Comacchio was much bigger in the period between the two world wars than in the 1900s, thanks to a water system configuration favoring juvenile recruitment.

Vøllestad and Jonsson (1988) used a long-term data series from the Imsa River (southwestern Norway) to develop an input–output model predicting total biomass and age distribution of silver eels from annual recruitment data. They estimated an overall yield of 3.51 kg/ha for the period 1975–1979. The mortality rate (assumed to be constant with age) was inversely correlated with the number of recruiting elvers, thus giving the first evidence for density dependence in eel mortality. Vøllestad and Jonsson's (1988) approach provides a powerful tool to predict yields at sites where elver recruitment and silver eel migration can be monitored and where the impact of commercial harvest is also reliably known. However, recruitment and silver eel migration cannot be readily measured in most eel fishing areas.

De Leo and Gatto (1995) developed the first model for the European eel, including a multiple classification of individuals by age and size. The model was based on data from the Valli di Comacchio lagoons (northern Italy) and accounted for inter-individual life history variability by means of a stochastic formulation. This represented a major improvement, as accounting for variability in the life cycle provides fundamental information about the uncertainty associated with harvest and the risk of stock decline. De Leo and Gatto (1995) estimated a silver yield of 6.15 kg/ha at Comacchio for the period 1989–1990. This figure is almost one-third of that estimated for the same lagoons by Rossi (1979) 15 years before, likely reflecting a recruitment drop in the 1980s. Later on, De Leo and Gatto (1996) applied their model to three data sets from the same sites, though from different periods (mid-1970s versus late 1980s) and revealed the dependence of prereproductive survival and mean body size at silvering upon eel density.

Dekker (2000), through a length-structured cohort analysis, estimated a potential silver eel production of about 4.4 kg/ha for Lake IJsselmeer. By using data from commercial catches, he calculated length-specific rates of total mortality, and under the assumption of constant natural mortality and silvering and escapement rates, he estimated fishing mortality. He argued that ceasing yellow eel exploitation in Lake IJsselmeer would lead to a many-fold increase in the adult eel population. He concluded that current, uncontrolled exploitation levels in the major eel fisheries might have negative consequences on the entire stock of European eel spawners.

Feunteun et al. (2000) used electrofishing and mark–recapture techniques in the Fremur catchment (northern France) to identify the relationship among silver eel dynamics, standing stock structure, and environmental factors such as flow, atmospheric pressure, rainfall, and lunar phase at a river basin scale. According to their study, silver eels represented almost 10% of the sedentary population in the catchment. However, only a small fraction of the silver eels (around 20%) effectively contributed to spawner output in the following migration period. In the Fremur catchment, where both natural and fishing mortality are low, the authors estimated an average spawner production of 1.3 kg/ha.

Rosell et al. (2005) examined data from tagging experiments and commercial sources at Lough Neagh (Northern Ireland) by means of cross-spectral analysis to investigate the relationship between explanatory variables (natural glass eel input, additional purchased glass eel input, mean water flow, and temperature) and response variables (yellow and silver eel yield). They found a significant dependence between yellow and silver eel yield on natural glass eel input 8 and 18 years before, probably reflecting the different life span of males and females. Their study revealed, despite the presence of commercial fisheries, a silver eel escapement of 2.5–3.5 kg/ha. This quite high estimate (considering Lough Neagh latitude and the fact that the local fishery exploits both yellow and silver eel) led the authors to consider the current management to be sustainable. Yet, the authors did not provide any assessment of the potential spawner production in undisturbed conditions.

Bevacqua et al. (2007) extended the approach proposed by De Leo and Gatto (1995) to account for interannual variability of glass eel recruitment and density-dependent juvenile survival while retaining the multiple classification of individuals by age and size and stochasticity in individual growth rates. They accounted for peculiarities of the eel life cycle such as delayed sex differentiation and sexual dimorphism in body growth through the model proposed by Melià et al. (2006) and described monthly variations in maturation rates with the model proposed by Bevacqua et al. (2006). By applying the model to a long-term data series from the Camargue lagoons (southern France), they estimated that current spawner escapement is far below the potential escapement (5.6 kg/ha) achievable in the absence of fishing at present recruitment levels.

Potential spawner escapement varies significantly among different sites, ranging between 1 and 20 kg/ha. Eel production is influenced by a number of environmental factors, such as temperature, salinity, and food availability, as well as by juvenile recruitment. Production is generally higher in brackish water bodies and at higher temperatures (up to 20 kg/ha before the recruitment drop of the past decades, around 6 kg/ha at present recruitment levels). In freshwater environments, lakes are usually more productive (about 4 kg/ha) than rivers (around 2–3 kg/ha). At sites where no historical data are available, gathering data on environmental parameters and present levels of juvenile recruitment is crucial to obtain preliminary estimates of potential production (through comparison with sites with similar conditions) and to provide the basis for the development of sound demographic models.

Assessing the Consequences of Different Management Scenarios on Spawner Output

Conservation of the European eel stock requires action in a number of fields, including structural measures to make rivers passable and to improve habitats, control of predators and parasites, improvement of water quality, and glass eel restocking. However, sustainable management of the fishery will certainly remain the central element of most EMPs due to their immediate consequences on spawner escapement. For this reason, we focus our analysis on management measures oriented to the regulation of the fishing effort. While it is easy to quantify the reduction in spawning output biomass due to silver eel fisheries, the consequences of yellow eel fishing on the spawning stock are not always easy to quantify. The impact of exploiting yellow eel was usually underestimated and often neglected, assuming strong compensatory density-dependent effects (ICES 2005). Recent works on different eel species, however, show that overfishing of yellow eels can dramatically impair spawner escapement.

Dekker (2000) established, through a length-structured cohort analysis, the historical impact of the well-documented fisheries of Lake IJsselmeer on both silver eel escapement and commercial catches. He concluded that yellow eel overexploitation in Lake IJsselmeer reduced female escapement to 0.14% and male escapement to 1.43% of pristine levels and that the fishing pressure on yellow eels precluded them from attaining a sufficient size to undergo sexual maturation and metamorphose into silver eels. Consequently, most of the catch was made up of yellow eels and spawner output was virtually absent.

Hoyle and Jellyman (2002) assessed the consequences of different management policies on yield and spawner biomass per recruit of two sympatric eel species in New Zealand, the New Zealand longfin eel *Anguilla dieffenbachii* and the shortfin eel *A. australis*. The local fishery targets both species, which are characterized by life cycles of different duration (the New Zealand longfin eel spends longer in freshwater and attains bigger sizes than the shortfin eel). They estimated that current exploitation rates have reduced the spawning per recruit of New Zealand longfin eel and shortfin eel females by about 95% and 40%, respectively. Then, they explored the effects of different decision variables, such as minimum legal weights and exploitation rates, on the spawning and fishing yield per recruit of both species, concluding that the two species require different management policies due to their different life cycles. Given the difficulty of developing different management policies for species that are hardly distinguished by fishermen, the authors suggested establishment of no-take reserves to protect New Zealand eels.

Doole (2005) applied a multiple-cohort bioeconomic model to the longfin eel fishery of the Waikato River (New Zealand) to investigate its optimal management and ascertain the appropriateness of current regulatory policies. He argued that using historical harvest data to calculate presently sustainable catches is inappropriate in light of the recent recruitment collapse. The author explored the consequences of management policies based on individual transferable quotas and the enforcement of protected areas on the status of the stock and the harvest. He argued that area closure and the spatial definition of harvest rights are attractive management options due to the territoriality of New Zealand longfin eels and that limiting the exploitation of older cohorts would increase yields. However, this last finding critically depends upon the specific spawner–recruitment relationship adopted. In fact, the author assumed that the number of juvenile eels entering the river in a given year and sustaining the local population depends on the abundance of local spawning stock 2 years before. Unfortunately, this assumption is not valid for local European eel populations, so that neither Doole's (2005) method nor his conclusions can be taken for granted in the management of European eel fisheries.

Bevacqua et al. (2007) evaluated the consequences of different management policies for the Camargue eel fishery. They estimated both silver eel escapement and harvest by local fishermen corresponding to different mesh sizes of the nets and different levels of fishing effort. By using realistic recruitment estimates, they could assess effective harvest and spawning output in absolute numbers and not only in terms of values per recruit.

All these studies clearly show that eels are particularly susceptible to overexploitation due to the singularity of their life cycle. In particular, downstream migration of silver eels facilitates their catch at particular places (e.g., bottlenecks). Also, their long lifespan is responsible for accumulation of high mortality rates as demonstrated by the major impact of fisheries on eel females and, in general, on long-living species (Dekker 2000; Jackson et al. 2001; Hoyle and Jellyman 2002). In addition, the absence of apparent stock–recruitment relationships at a local scale impairs the acceptance of sustainable management policies by fishermen communities.

Calculating Eel Fishery Profits

In addition to theoretical studies on the economics of fisheries (since the seminal works by Gordon 1954; Schaefer 1954), there are several studies regarding the exploitation of specific fish stocks in the literature (see, e.g., Myers et al. 1997; Orensanz et al. 1998; Kulmala et al. 2007). With regard to eels, however, examples of thorough bioeconomic analyses predicting the profitability of a fishery under different management scenarios are rare. While studies from other fisheries can provide useful information on the general guidelines to be followed to pursue sustainability in eel fisheries, the eel life cycle is so distinctive that general guidelines can be hardly applied to the development of specific policies for the management of eel stock. Here, we summarize the main results of the few bioeconomic analyses specifically focused on eel fisheries.

Gatto et al. (1982) assessed the profitability of different management strategies for the eel fishery of the Comacchio lagoons. The effort was traditionally exerted only on silver eels, which are fished by special devices called lavorieri, intercepting the entire flux of migrating fish. Gatto et al. (1982) concluded that extending the fishery to also target a fraction of yellow eels would allow fishermen to improve their gross economic return by about 10%. Almost 20 years later, De Leo and Gatto (2001) performed a stochastic bioeconomic analysis of eel fishing in the same lagoons aimed at optimizing the economic return from the Comacchio eel fishery. The authors explored the effect of extending the fishery to yellow eels and tested whether the decline of natural recruitment could be effectively supplemented by elver restocking. The authors analyzed different management policies in terms of fishing effort on yellow eels (defined as number of nets placed), net selectivity (mesh size), and restocking density. Net selectivity was expressed as a function of the fish size and the mesh size of the net. They explicitly considered harvesting costs, different selling prices for yellow and silver eels, and different discount rates to assess the optimal management policy for maximizing the average net economic benefit. They found that the highest profits could be obtained by fishing silver eels by lavorieri and fishing a fraction of yellow eels with 160 nets of 21-mm mesh. A stochastic approach allowed the authors to derive not only a point estimate of the economic benefit associated with the different management alternatives considered, but also the uncertainty of their estimates.

A Call for Multi-Objective Analyses

Fisheries managers must often cope with multiple, and possibly conflicting, objectives (Charles 1989; Hilborn 2007) such as maximizing catches, minimizing costs, minimizing bycatch, and maximizing spawning output per recruit. Considering several objectives at once provides a framework for the decision process, promotes a more appropriate role in the process for the analyst, and usually identifies a wider range of alternatives than those obtained by a single-objective analysis (Cohon 1978).

Multi-objective techniques represent an improvement with respect to traditional, single-objective approaches to planning problems (e.g., cost–benefit analysis) because they allow decision makers to address a number of objectives that cannot be reduced to a single dimension such as revenue (Meier and Munasinghe 1994). Multi-objective analysis can indeed help decision makers identify and highlight possible trade-offs among conflicting viewpoints. However, while it is widely agreed that the use of a multi-objective approach is highly desirable (Vaca-Rodriguez and Enríquez-Andrade 2006), substantial difficulties are encountered in identifying the ultimate goals of the different stakeholders and in providing a framework for the comparison between objectives. For these reasons, the use of multi-objective methods in fisheries research has been scarce over the past decades, although pioneering studies have been conducted since the early 1980s (e.g., Bishop et al. 1981; Charles 1989) and a few recent examples can also be found in the literature (e.g., Sylvia and Enríquez-Andrade 1994; Pan et al. 2001; Melià and Gatto 2005). Nevertheless, most efforts remain directed to the development of analytical tools to evaluate the impact of management strategies in a single-objective perspective.

The key concept of multiple-objective analysis is Pareto efficiency. An alternative (for instance, a fishing policy) is called Pareto-efficient when it is not possible to modify decision variables to improve a performance indicator (e.g., the viability of a fish stock) without worsening another performance indicator (e.g., the revenue of fishermen exploiting the stock). All other alternatives, for which there exists at least one feasible solution guaranteeing both higher viability and higher revenues, are called Pareto-dominated. The set of nondominated policies is called the Pareto boundary (or Pareto set) and represents the suite of alternatives among which the decision maker can reasonably choose. The Pareto optimal set and associated trade-offs supply a useful reference and important information to decision makers. Eventually, for any given problem, only one solution has to be selected by the decision makers. This solution is usually not the result of a formal maximization problem, but rather of a subjective evaluation of the relative importance of the objectives by the decision makers. Hence, it must be clear that the multi-objective approach concentrates on providing information to the decision makers regarding the range of effective choices and the consequences of different options rather than suggesting a single optimal solution (Gatto and De Leo 2000).

An example of applying this approach to eel management is provided in Bevacqua et al. (2007). They performed a Pareto analysis to identify the fishing policies providing the best compromise between two partially conflicting objectives in the management of the eel fishery of the Camargue lagoons: maximizing the escapement of silver eels towards the ocean and maximizing the harvest by commercial fishermen. Their results support the view that, at present, the Camargue eel fishery is inefficient with respect to the two objectives of maximizing spawner output and catch biomass. The main reason of such inefficiency is the use of highly selective fishing devices, which focus the fishing pressure on younger stages. In order to maximize the spawning output, the fishery would need to be closed, while yield maximization would require adopting a larger mesh size than currently used.

Yields or Profits?

Many local European eel fisheries are likely as inefficiently managed as the Camargue fishery. However, assuming the main goal of fishermen is the maximization of fishing yield is not always realistic. Indeed, informed fishermen are usually more interested in maximizing the revenues derived from selling their eel catches rather than maximizing yields (Hilborn 2007). Estimating the profitability of a fishery requires socioeconomic information on costs and revenues, which is often difficult to collect, especially for small-scale fisheries. Selling prices vary widely depending on a number of factors such as seasonal fluctuations in demand, provenance of the catch, and size of the fish. In most fish markets, bigger individuals are preferred to smaller ones (see, e.g., De Leo and Gatto 2001). In these cases, the fishing strategies adopted by fishermen cannot be correctly interpreted without explicitly incorporating the size price structure into the analysis. In the Camargue, for instance, the main cause of inefficiency of the fishery is the overexploitation of young yellow eels (Bevacqua et al. 2007), a practice that affects the size composition of the catch and, if the eel price per mass unit is related to the fish size or the maturation stage, can also affect profits. If we applied De Leo and Gatto's (2001) price structure to the analysis of the Camargue fishery, the inefficiency of current exploitation practices would become even more evident. In recent years, however, local market demand has shifted towards a marked preference for small-sized eels required by aquaculture. Therefore, the reduction of the overall harvestable catch caused by the use of a small mesh size might be compensated by the extra profit generated by selling the lucrative small eels to the aquaculture market. Despite the difficulty of gathering the necessary information, including fishing costs and revenues in the quantification of fishermen's objectives is of critical importance and can lead to very different management scenarios from those obtained by aiming to maximize yields. This approach is even more compelling when eels are exploited in a multispecies fishery where other species may also drive or influence the fishing strategy.

Final Considerations

The collapse of the European eel stock and glass eel recruitment requires immediate action to halt their decline. Although habitat considerations such as pollution and dams and hydropower stations certainly contribute to the decline, there is little doubt that, given the present level of recruitment, reducing fishing mortality is the most practical and effective short-term strategy to increase the spawning stock (ICES 2005). For some fisheries, this might simply require the reduction of the catch and consequently the profit—which, understandably, is strongly opposed by fishermen. Yet, recent analyses show that cases exist where the conservation target can be achieved without reductions in harvest (Bevacqua et al. 2007). Given the complexity of the eel life cycle and the duration of its continental life span, the only way to assess the efficacy of an EMP at the local level is to make use of demographic models of eel dynamics that allow the investigation of the effects of a large number of fishing alternatives (in terms of fishing effort, fishing gear, and length of the fishing season). The results can be surprising, as both spawning stock and catches can be substantially improved by using a suitable combination of fishing effort and mesh size, as demonstrated by Bevacqua et al. (2007).

As the European eel is a panmictic species, sustainable management strategies must have both a local and global scope. All bioeconomic analyses conducted on local eel populations have disregarded the existence of a global stock–recruitment relationship, considering spawner output as unrelated to recruitment. To date, the only attempts to assess the whole European eel stock and describe its dynamics have been conducted by Dekker (2000c) and Åström and Dekker (2007). Although these studies were chiefly focused on eel conservation rather than on the sustainability of the fishery from the fishermen's viewpoint, they represent a useful starting point for a comprehensive bioeconomic analysis of the European eel stock and its fishery.

Another often neglected aspect in the design of eel recovery plans is the inclusion of the economic component in fishery management. It is well known, in fact, that fishery dynamics are generally driven by economic forces occurring at the market level that try to match supply and demand (Pinnegar et al. 2006). A change in market price reflecting the consumer's willingness-to-pay to purchase eels at a given level of production can strongly affect fishermen's profits and consequently influence the set of optimal policies. For instance, in the late 1990s, the increasing Asian demand for glass eels on European and North American markets pushed selling prices to exceptionally high levels (up to €4,500/kg; Tesch 2003). In the following years, Japanese demand for glass eels was mostly satisfied on Asian markets, and glass eel prices in Europe began to decline. Market price fluctuations remain driven by the balance between Asian and local demand, which prizes small size eels for aquaculture (Allen et al. 2006). Given this global consideration, the European eel was included in Appendix II of the Convention on International Trade in Endangered Species in June 2007. This listing will likely have consequences on the international eel trade that will be important in shaping market scenarios and in producing cascading impacts on local fisheries. Moreover, after years of almost complete absence of regulation, European Union (EU) member states are now required by the end of 2008 to implement EMPs that will certainly affect thousands of small artisan fisheries scattered all around Europe. An effective enforcement of the EU regulation is difficult to achieve in such a fragmented situation, as free-riding incentives are always present.

Nevertheless, conflicts that are likely to arise between fishermen and policy makers as a result of the implementation of EMPs can be reduced and enforcement enhanced if the interests of fishermen are explicitly accounted for in the definition of the recovery plans. This implies that a detailed analysis of fishermen's preferences should be carried out. This task is far from trivial, as fishermen operating in small-scale fisheries affected by unpredictable environmental conditions do not always aim to maximize profits, but rather to minimize the variability of annual revenues (Chaboud 1995). Curiously enough, a similar attitude has been observed in small-scale African farmers who preferred to reject technological innovations when the potential increase in yield was associated with an increase in yield variance (Brossier 1989). In these cases, the use of stochastic models, explicitly accounting for the uncertainty in predictions, may allow decision makers to formulate risk-averse management policies that are more likely to be supported by fishermen.

Finally, when developing management plans, it should be remembered that small-scale fishing is not just a source of income but is often perceived also as a valuable "way of life" (Apostle et al. 1985). Factors such as sense of independence (i.e., being one's own boss), lack of options, socialization processes, cultural traditions, and so forth can indeed play a central role in defining fishermen's responses to regulations. Multi-objective methods can provide a way to explicitly account for several contrasting objectives that cannot be reduced to or evaluated in just monetary terms. Last but not least, multi-objective analysis also offers the further advantage of identifying a whole set of Pareto-efficient policies rather than just a single optimal policy simply expressed in economic terms. This approach provides decision makers with more opportunities to manage potential conflict among contrasting stakeholders while applying a rigorous and quantitative assessment of the conservation effectiveness of different fishing policies and conservation plans.

Acknowledgments

The authors thank the Joint ICES/EIFAC Working Group on Eels for stimulating exchanges of views about the management of the eel fisheries and two anonymous referees for their constructive comments to the manuscript draft. This work was supported by Fondation Sansouire (Tour du Valat, France) and the Italian Ministry of Research through Interlink project II04CE49G8 and PRIN project 2006054928.

References

Allen, M., R. Rosell, and D. Evans. 2006. Predicting catches for the Lough Neagh (Northern Ireland) eel fishery based on stock inputs, effort and environmental variables. Fisheries Management and Ecology 13(4):251–260.

Apostle, R., L. Kasdan, and A. Hanson. 1985. Work satisfaction and community attachment among fishermen in southwest Scotia. Canadian Journal of Fisheries and Aquatic Sciences 43:1187–1200.

Åström, M., and W. Dekker. 2007. When will the eel recover? A full life-cycle model. ICES Journal of Marine Science 64:1491–1498.

Bevacqua, D., P. Melià, A. J. Crivelli, G. A. De Leo, and M. Gatto. 2006. Timing and rate of sexual maturation of European eel in brackish and freshwater environments. Journal of Fish Biology 69(Supplement C):200–208.

Bevacqua, D., P. Melià, A. J. Crivelli, G. A. De Leo, and M. Gatto. 2007. Multi-objective assessment of conservation measures for the European eel (*Anguilla anguilla*): an application to the Camargue lagoons. ICES Journal of Marine Science 64:1483–1490.

Bishop, R., D. W. Bromley, and S. Langdon. 1981. Implementing multiobjective management of commercial fisheries: a strategy for policy-relevant research. In L. G. Anderson, editor. Economic analysis for fisheries management plans. Ann Arbor Science, Ann Arbor, Michigan.

Brossier, J. 1989. Risque et incertitude dans la gestion de l'exploitation agricole. Pages 25–26 *in* M. Edlin and P. Mileville, editors. Quelques principes methodologiques. Le risque en agriculture. Coll. A travers champs, ORSTOM, Paris.

Castonguay, M., P. V. Hodson, C. M. Couillard, M. J. Eckersley, J. D. Dutil, and G. Verreault. 1994. Why is recruitment of the American eel, *Anguilla rostrata,* declining in the St. Lawrence river and gulf? Canadian Journal of Fisheries and Aquatic Sciences 51(2):479–488.

Chaboud, C. 1995. Risques et incertitudes dans les peches: le point de vue de l'economiste. Pages 263–295 *in* F. Laloe, H. Rey, and J. L. Durand, editors. Questions sur la dynamique de l'exploitation halieutique. Colloques et séminaires, ORSTOM, Paris.

Charles, A. T. 1989. Bio-socio-economic fishery models. Labor dynamics and multi-objective management. Canadian Journal of Fisheries and Aquatic Sciences 46(8):1313–1322.

Cohon, J. L., 1978. Multiobjective programming and planning, 1st edition. Academic Press, New York.

Commission of the European Communities. 2003. Development of a community action plan for the management of European eel. Communication from the commission to the council and the European Parliament. Commission of the European Communities, COM (2003) 573 final, Brussels, Belgium.

Commission of the European Communities. 2005. Proposal for a council regulation establishing measures for the recovery of the stock of European eel (presented by the commission). Commission of the European Communities, COM (2005) 472 final, Brussels, Belgium.

Daverat, F., K. E. Limburg, I. Thibault, J. C. Shiao, J. D. Dodson, F. Caron, W. N. Tzeng, Y. Iizuka, and H. Wickström. 2006. Phenotypic plasticity of habitat use by three temperate eel species, *Anguilla anguilla,* A.japonica and A.rostrata. Marine Ecology Progress Series 308:231–241.

De Leo, G. A., and M. Gatto. 1995. A size and age-structured model of the European eel (*Anguilla anguilla*

L.). Canadian Journal of Fisheries and Aquatic Sciences 52(7):1351–1367.

De Leo, G. A., and M. Gatto. 1996. Trends in vital rates of the European eel: evidence for density dependence? Ecological Applications 6(4):1281–1294.

De Leo, G. A., and M. Gatto. 2001. A stochastic bioeconomic analysis of silver eel fisheries. Ecological Applications 11(1):281–294.

De Leo, G. A., P. Melià, A. J. Crivelli, and M. Gatto. 2009. Eel population modeling and its aplication to conservation management. Pages 327–345 *in* J. M. Casselman and D. K. Cairns, editors. Eels at the edge: science, status, and conservation concerns. American Fisheries Society, Symposium 58, Bethesda, Maryland.

Dekker, W. 2000a. Fractal geometry of the European eel stock. ICES Journal of Marine Science 57(1):109–121.

Dekker, W. 2000b. Impact of yellow eel exploitation on spawner production in Lake IJsselmeer, the Netherlands. Dana 12:25–40.

Dekker, W. 2000c. A procrustean assessment of the European eel stock. ICES Journal of Marine Science 57(4):938–947.

Dekker, W. 2003. On the distribution of the European eel (*Anguilla anguilla*) and its fisheries. Canadian Journal of Fisheries and Aquatic Sciences 60(7):787–799.

Dekker, W. 2008. Coming to grips with the eel stock slip-sliding away. Pages 335–355 *in* M. G. Schecter, N. J. Leonard, and W. Taylor, editors. International governance of fisheries ecosystems: learning from the past, finding solutions for the future. American Fisheries Society, Bethesda, Maryland.

Doole, G. J. 2005. Optimal management of the New Zealand longfin eel (*Anguilla dieffenbachii*). Australian Journal of Agricultural and Resource Economics 49(4):395–411.

Feunteun, E., A. Acou, P. Laffaille, and A. Legault. 2000. European eel (*Anguilla anguilla*): prediction of spawner escapement from continental population parameters. Canadian Journal of Fisheries and Aquatic Sciences 57(8):1627–1635.

Feunteun, E. 2002. Management and restoration of European eel population (*Anguilla anguilla*): an impossible bargain. Ecological Engineering 18:575–591.

Friedland, K., D., M. J. Miller, and B. Knights. 2007. Oceanic changes in the Sargasso Sea and declines in recruitment of the European eel. ICES Journal of Marine Science 64:519–530.

Gatto, M., and G. A. De Leo. 2000. Pricing biodiversity and ecosystem services: the never-ending story. Bioscience 50:347–355.

Gatto, M., E. Laniado, and R. Rossi. 1982. The management of eels in the Valli di Comacchio lagoon. Oceanologica Acta 5:303–307.

Gordon, H. S. 1954. The economic theory of a common property resource: the fishery. Journal of Political Economy 82:124–142.

Hilborn, R. 2007. Defining success in fisheries and conflicts in objectives. Marine Policy 31:153–158.

Hoyle, S. D., and D. J. Jellyman. 2002. Longfin eels need reserves: modelling the effects of commercial harvest on stocks of New Zealand eels. Marine and Freshwater Research 53(5):887–895.

ICES (International Council for the Exploration of the Sea). 2005. Report of the ICES/EIFAC Working Group on Eels. International Council for the Exploration of the Sea, ICES C.M. 2005/ I:01, Galway, Ireland.

Jackson, J. B. C., M. X. Kirby, W.H. Berger, K.A. Bjorndal, L. W. Botsford, B. J. Bourque, R. H. Bradbury, R. Cooke, J. Erlandson, J. A. Estes, T. P. Hughes, S. Kidwell, C. B. Lange, H. S. Lenihan, J. M. Pandolfi, C. H. Peterson, R. S. Steneck, M. J. Tegner, and R. R. Warner. 2001. Historical overfishing and the recent collapse of coastal ecosystems. Science 293:629–638.

Knights, B. 2003. A review of the possible impacts of long-term oceanic and climate changes and fishing mortality on recruitment of anguillid eels of the northern hemisphere. Science of the Total Environment 310(1–3):237–244.

Kulmala, S., H. Peltomaki, M. Lindroos, P. Soderkultalahti, and S. Kuikka. 2007. Individual transferable quotas in the Baltic Sea herring fishery: a socio-bioeconomic analysis. Fisheries Research 84(3):368–377.

Lefebvre, F., P. Contournet, F. Priour, O. Soulas, and A. J. Crivelli. 2002. Spatial and temporal variation in *Anguillicola crassus* counts: results of a 4 year survey of eels in Mediterranean lagoons. Diseases of Aquatic Organisms 50(3):181–188.

Meier, P., and M. Munasinghe. 1994. Incorporating environmental concerns into power sector decision making: a case study of Sri Lanka. World Bank, World Bank Environment Paper No. 6, Washington, D.C.

Melià, P., and M. Gatto. 2005. A stochastic bioeconomic model for the management of clam farming. Ecological Modelling 184(1):163–174.

Melià, P., D. Bevacqua, A. J. Crivelli, G. A. De Leo, J. Panfili, and M. Gatto. 2006. Age and growth of the European eel *Anguilla anguilla* in the Camargue lagoons. Journal of Fish Biology 68:876–890.

Moriarty, C., and W. Dekker. 1997. Management of the European eel. Fisheries Bulletin (Dublin) 15.

Myers, R. A., J. A. Hutchings, and N. J. Barrowman. 1997. Why do fish stocks collapse? The example of cod in Atlantic Canada. Ecological Applications 7(1):91–106.

Orensanz, J. M. L., J. Armstrong, D. Armstrong, and R.

Hilborn. 1998. Crustacean resources are vulnerable to serial depletion—the multifaceted decline of crab and shrimp fisheries in the Greater Gulf of Alaska. Reviews in Fish Biology and Fisheries 8(2):117–176.

Pan, M. L., P. S. Leung, and S. G. Pooley. 2001. A decision support model for fisheries management in Hawaii: a multilevel and multiobjective programming approach. North American Journal of Fisheries Management 21(2):293–309.

Pinnegar, J. K., T. P. Hutton, and V. Placenti. 2006. What relative seafood prices can tell us about the status of stocks. Fish and Fisheries 7(3):219–226.

Rosell, R., D. Evans, and M. Allen. 2005. The eel fishery in Lough Neagh, Northern Ireland—an example of sustainable management? Fisheries Management and Ecology 12(6):377–385.

Rossi, R. 1979. An estimate of the production of silver eel in the Valli di Comacchio (Po Delta) during 1974–1976. Bollettino di Zoologia 46:217–223.

Rossi, R., and A. Cannas. 1984. Eel fishing management in a hypersaline lagoon of southern Sardinia. Fisheries Research 2(4):285–298.

Schaefer, M. B. 1954. Some aspects of the dynamics of populations. Bulletin of the Inter-American Tropical Tuna Commission 1(2):26–56.

Sylvia, G., and R. R. Enríquez-Andrade. 1994. Multiobjective bioeconomic analysis: an application to the Pacific whiting fishery. Marine Resource Economics 9(4):311–328.

Tesch, F. W. 2003. The eel, 3rd edition. Blackwell Scientific Publications, Oxford, UK.

Tsukamoto, K., I. Nakai, and W. V. Tesch. 1998. Do all freshwater eels migrate? Nature 396(6712):635–636.

Vaca-Rodriguez, J. G., and R. R. Enríquez-Andrade. 2006. Analysis of the eastern Pacific yellowfin tuna fishery based on multiple management objectives. Ecological Modelling 191(2):275–290.

Vøllestad, L. A., and B. Jonsson. 1988. A 13-year study of the population dynamics of the European eel *Anguilla anguilla* in a Norwegian river: Evidence for density-dependent mortality, and development of a model for predicting yield. Journal of Animal Ecology 57(3):983–997.

American Fisheries Society Symposium 69:649–661, 2009

Can the Ecosystem Approach Improve Management of Tropical Estuarine Fisheries for Diadromous Species?

Stephen J. M. Blaber*
CSIRO Marine & Atmospheric Research
Post Office Box 120, Cleveland, Queensland 4163, Australia

Abstract.—The general acceptance that individual fisheries should be managed in the context of the ecosystem of which the target species are a part has led to the paradigm of ecosystem-based fisheries management (EBFM). Such management may be particularly suitable for diadromous fishes, which spend time in rivers, estuaries, and the sea, because their sustainability would benefit from a holistic approach. The extent to which this is possible and has been successful depends on a wide variety of biological, socioeconomic, and political factors. Fishing in all its forms is only one of a broad array of human activities in rivers, estuaries, and the coastal zone; hence, diadromous fisheries management requires not only integration across ecosystems, but also its inclusion in overall planning processes. Tropical diadromous fisheries in developing countries suffer from overfishing, lack of adequate human resources to manage the fisheries, insufficient scientific data for sound management and decision making, lack of monitoring capabilities, poor enforcement of laws, inefficient administration, and increasing conflicts between different types of users. Even in developed countries, many of these issues also have not been addressed. The extent to which an EBFM approach has been successful for tropical diadromous species is discussed using the various barramundi *Lates calcarifer*, mullet (Mugilidae), and tropical shad (*Tenualosa*) fisheries of Asia and Australia, taking into account issues of food security and the role of community-based management.

Introduction

Recognizing that fisheries not only affect the target organisms, but also the entire ecosystem of which they are a part (Hall 1999; Blaber et al. 2000; Barnes and Thomas 2005) has led to widespread acceptance throughout the world that fisheries should be managed in a broader context—often referred to as ecosystem-based fisheries management (EBFM) (Brodziak and Link 2002; Norse 2005). While intuitively logical, implementing EBFM requires a different approach from traditional single species management under precautionary principles. To be successful in rivers and estuaries, its implementation requires an integration of fisheries management with the myriad of other human activities that impact the coastal zone. Increasing scientific knowledge has assisted the shift in current thinking among biologists, fisheries managers, and legislators towards the management of fisheries in an ecosystem context (Elliott 2002), but such management requires a multidisciplinary approach. Importantly, however, practical outcomes need not only fisheries expertise, but also an organizational framework involving relatively complex financial, political, and administrative structures (Knecht 1997) that may be beyond the national capacities of many countries, particularly those in the developing world. Some developed countries have enacted legislation to try and ensure that fisheries become environmentally sustainable. For example, in Australia, the Environment Protection and Biodiversity Conservation Act 1999 requires that each fishery undergo ecological assessments to demonstrate that it is being managed in an ecologically sustainable manner and that export approval is contingent upon satisfactory assessment. Such assessments include the communities and habitats associated with or impacted by the target species (Scandol et al. 2005; Brewer et al. 2008). However, it

* Corresponding author: steve.blaber@csiro.au

is noteworthy that recently, the value of EBFM and the meaning of sustainability have been called into question by Longhurst (2006), particularly in relation to large-scale marine ecosystems. With regard to EBFM, Longhurst (2006) makes a number of important points: for example, that "concerns for the sustainability of 'holistic fishery systems' by economists and sociologists has drawn attention away from the sustainability crisis in the only really critical element of such systems—the fish themselves, about which these people know very little." Also that widening the discussion of EBFM to include such matters as the rights of stakeholders has not sharpened understanding of ecological realities and has led to increasing emphasis on institutional challenges rather than the science. Longhurst (2006) concludes that sustainable catches at mid-20th century levels are simply not attainable by conventional fish stock modeling, by EBFM, or by any other novel means and suggests that sustainability is mythic.

Diadromous species should particularly benefit from integrated management because they move to and from sea and river and spend time in estuaries. Quite apart from fishing pressure, they are subject to the impacts of a wide variety of human activities, including industry and agriculture, construction of dams, dredging, mangrove loss, residential development, and organic and inorganic pollution. In addition, in most countries, fisheries for diadromous species are managed by different agencies in the sea and in rivers. The extent to which EBFM and integrated management may be relevant to smaller-scale fisheries for tropical diadromous species in Australia and Asia is explored in the present paper against the background of various existing management approaches for tropical diadromous species, including the barramundi *Lates calcarifer*, mullet (Mugilidae), and tropical shad (*Tenualosa*) fisheries of Asia and Australia, taking into account issues of food security and the role of traditional or community-based management. A comprehensive table of tropical diadromous fishes is listed in Milton (in press).

Management of Diadromous Fisheries in Tropical Asia and Australia

Fisheries for diadromous species in tropical Asia and Australia fall into two categories: fisheries that target a diadromous species—these are relatively few, and multispecies fisheries in which diadromous species are part of the catch.

Target fisheries that are managed separately include the barramundi fisheries of Australia and Papua New Guinea, some of the *Tenualosa* spp. (tropical shad) fisheries of south and Southeast Asia, and the striped mullet *Mugil cephalus* fisheries of Australia. Most other mugilid species are not truly diadromous (McDowall 1988; Blaber 1987), but of those that are, species such as corsula *Rhinomugil corsula* and gold spot mullet *Liza parsia* in India, and sand grey mullet *Myxus elongatus* in Australia, form part of multispecies fisheries and are not managed separately. Similarly, most tropical *Anguilla* species are not managed separately, with the possible exception of speckled longfin eel *A. reinhardtii* in eastern Australia.

Target Fisheries

Barramundi.—The barramundi or sea bass is sparingly distributed throughout the estuaries and coastal waters of northern Australia, Papua New Guinea, and south and Southeast Asia. It has a complex life history and is a protandrous hermaphrodite that moves to coastal spawning grounds during the dry season when gonad maturation takes place in estuarine waters. Spawning takes place in the sea and between October and February, and as the wet season progresses, there is a tidal-based monthly cycle of spawning, with larvae moving into shallow coastal areas, back swamps, and freshwaters. The chief threats to maintenance of sustainable populations relate to habitat conservation and overfishing, primarily of the large females (Figure 1). In northern Australia, it is harvested by three different sectors: commercial, recreational, and traditional (subsistence) aboriginal fishers. It is a relatively small fishery with 1669 metric tons landed by the commercial fishery in 2004 (Zeller and Snape 2006), but it is well documented and carefully managed. It extends from the Ashburton River in West Australia to the Noosa River in Queensland and is described in detail by Lea et al. (1987), Quinn (1987), and Kailola et al. (1993).

Commercial fishing is carried out almost entirely using large-mesh monofilament gill nets in coastal waters and the tidal portions of estuaries across northern Australia. These commercial fisher-

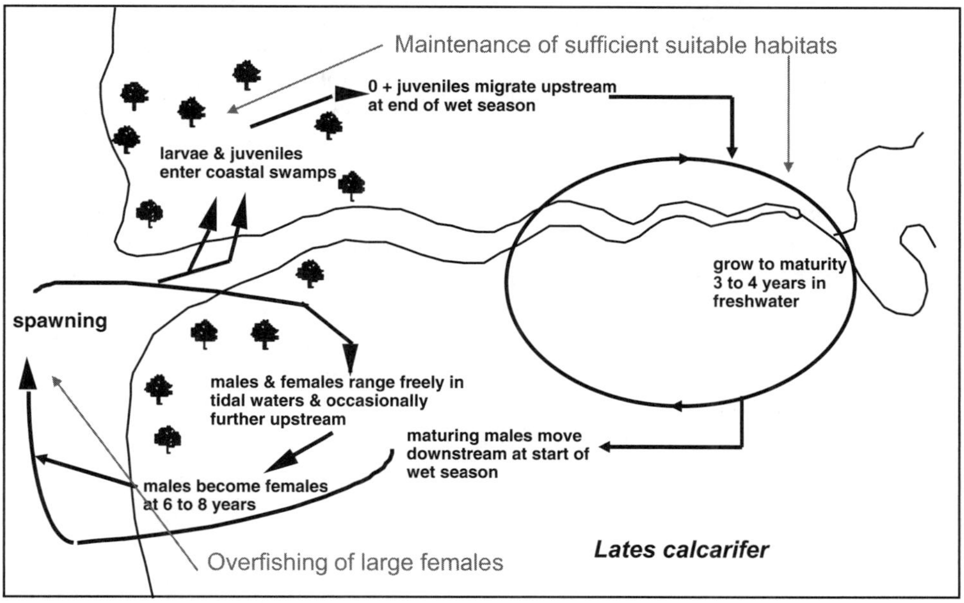

Figure 1.—Life cycle of barramundi with main management issues and areas for possible intervention shown in red (Modified from Blaber 2000).

ies are managed separately in the states of Queensland and West Australia and in the Northern Territory, but although the regulations vary, they are essentially similar, with upper and lower limits for both gill-net mesh sizes and lengths of fish retained. The need for careful management was reinforced by the discovery that genetically discrete stocks exist in different groups of estuaries, with more than 16 stocks in the Northern Territory alone (Salini and Shacklee 1988). Commercial fishing effort has been reduced by closing the freshwater areas of rivers in Queensland, Northern Territory, and West Australia to commercial fishing and restricting it to the lower reaches and coastal waters. There is a closed season from October to January, and about 22 estuaries are closed to commercial fishing in northern Queensland. Recreational fishing for barramundi is an increasingly popular leisure activity in Australia, particularly near population centers such as Darwin or Cairns. The increasing number of recreational fishers, together with improved access to remote areas, has increased the fishing effort in recent years, and this sector now takes as much barramundi as the commercial sector. The species forms the basis of a lucrative tourist industry in the Northern Territory, with the total monetary value of this industry plus the recreational catch exceeding that of the commercial fishery (Grey 1987), and hence, the Northern Territory government is gradually phasing out commercial barramundi perch fishing in favor of recreational fishing. It has instituted a buyback scheme for commercial licenses to try and increase the availability of fish for the recreational industry (Griffin 1995). In the Gulf of Carpentaria of Queensland, the commercial fishery is limited entry and each license holder is required to have a full-time commitment to the gill-net fishery for a minimum of 20 weeks a year.

Recreational fishers are subject to similar regulations with minimum fish sizes and bag limits enforced in Queensland and Northern Territory. Closed seasons also apply to recreational fishers.

The barramundi fisheries of northern Australia are regulated on a target species basis, but their management has essentially been brought into an EBFM framework through the Australian Government's Environment Protection and Biodiversity Conservation Act of 1999. Because of this act, each fishery, including those for barramundi, has to demonstrate that it is being managed in an ecologically sustainable manner. This has been achieved through the use of ecological risk assessment techniques

(Fletcher et al. 2005) that incorporate not only assessments of the impact of the fishery on the target species, but also on bycatch of the fishery and associated communities and habitats. Management for ecological sustainability is supported by fishery independent monitoring by the Queensland State Government, of both barramundi catch and bycatch, carried out on a regular basis in rivers that support commercial barramundi fishing. To date, such surveys indicate that the management regime is effective in ensuring that long-term recruitment and natural dynamics are largely unaffected by the fishery and that catch levels have remained steady at around 900 metric tons per annum (Zeller and Snape 2006).

In Papua New Guinea, barramundi is the most economically important species in the commercial component of the artisanal fisheries of Western Province, where it occurs on the coast and inland in most rivers, with the largest populations in the Fly River. Recent genetic research (Salini 2002) shows that, unlike the situation in northern Australia where there are multiple small genetic stocks of barramundi, there is just one large stock of barramundi living in the Fly River and associated coastal waters. They are the most valuable species in the Fly River and, in the past, have represented up to 80% of the catch in coastal waters. This fishery supports more than 400 small-scale fishers catching a few fish each. A similar artisanal fishery exists in the Fly River and its tributaries. Barramundi is the third most valuable export commercial fishery in Papua New Guinea. This export fishery produced a total of about 500 metric tons of barramundi in 2000, but the coastal artisanal fishery has continued to decline with recent catches in the coastal fishery only reaching 100 metric tons a year. Recent research by Milton et al. (1998) showed that reducing the numbers of juveniles caught may not be the best way to restore the fishery. Rather, the number of adult females surviving may be more critical. Fishery-independent catch data from monofilament gill nets and population simulations showed that the number of females surviving to maximum age, as well as their total egg production, increased as mesh sizes were reduced. Hence, reducing the mesh size to increase adult female escapement (but catch more juveniles) may be a better management strategy than increasing mesh sizes to improve recruitment.

An artisanal fishery in the Fly River is now expanding as part of community development initiatives by Ok Tedi Mining Limited. The coastal and lower Fly River fishery collapsed in the early 1990s, and while the middle Fly River fishery has, in the past few years, recorded catches of more than 300 metric tons, the coastal fishery produces only about another quarter of this. Several management measures, including a minimum size limit for barramundi (38 cm total length), were introduced after 1983 to reduce the catch of juvenile and immature fish in Western Province; none of these measures succeeded, probably because of a lack of will to enforce them (Kare 1995). A new barramundi management plan was gazetted into law in 2003. This plan took into account the research results from projects carried out by Ok Tedi Mining Limited, the National Fisheries Authority, and CSIRO Australia, together with commercial and artisanal fishery consultations, to produce a new set of regulations to try and ensure the long-term sustainability of the subsistence, artisanal, and commercial components of the fishery. The main points of the new law are that it stipulates a total allowable catch of 260 metric tons per annum for the Western Province; it prohibits the sale or export of barramundi with a total length of less than 36 cm, it prohibits the use of gill nets and beach seine nets with mesh sizes greater than 15 cm—in order to protect the larger spawning females, and it prohibits the use of gill and beach seine nets with mesh sizes between 6.35 and 12.7 cm during the peak periods of juvenile recruitment in March and April in coastal waters, and the use of gill nets greater than 12.7 cm is also prohibited during the peak spawning migration period in September and October. In addition, the main spawning and breeding grounds along the coast are closed to commercial fishing during the peak spawning periods in October and November.

The future sustainability of the Papua New Guinea barramundi fishery depends upon the implementation of these new regulations. Unfortunately, in the absence of their enforcement, the prognosis remains uncertain and EBFM is a distant horizon.

Fisheries for barramundi in south and Southeast Asia are unregulated and most are subject to high fishing pressure. Lake Chilka in India has a large barramundi fishery that has landed up to

700 metric tons per annum, but catches have been falling over the past decades (Jhingran 2002). Unfortunately, there is apparently little scope for new management regulations in the tradition oriented fishing practices in Lake Chilka (Kasim and James 1986).

Tropical shads (Tenualosa spp.).—Tropical shads of the genus *Tenualosa* consist of five species of considerable fisheries significance of which the most widespread and well studied is hilsa *T. ilisha*, found from north Sumatra in the east to Kuwait in the west. The closely related eastern hilsa *T. reevesii* occurs intermittently along the South China coast and far up the Yangtze, Pearl, and Qiantang rivers, and Laotian shad *T. thibaudeaui*, which only lives in the lower and middle Mekong system, is believed to be close to extinction. Once widespread, toli *T. toli* is now found only in the estuaries and adjacent coastal areas of Sarawak (north coast of Borneo), and longtail shad *T. macrura* lives in the coastal waters of Sumatra and Borneo (Blaber 2009, this volume). As with barramundi, the main threats to sustainability of *Tenualosa* come from habitat alterations and overfishing, particularly of spawning aggregations (Figures 2 and 3).

Most *Tenualosa* fisheries are not strongly regulated, and although hilsa and toli fisheries are subject to management controls in a number of countries (Blaber 2009; Siddique 2009, this volume), none form part of any ecosystem-based management regime. The biological, cultural, and fisheries importance of diadromous tropical shads has long been recognized in many Asian countries, but this has had little impact on overall government planning. Fisheries have seldom been integrated into broader management planning in south or Southeast Asia, and this has had severe impacts on several *Tenualosa* species that undertake long migrations to and from the sea. The building of dams for irrigation, hydropower, industry, and domestic water supply has interrupted the migration patterns of hilsa in Pakistan, India, and Bangladesh and caused massive declines in the fishery. For example, catches of the order of 200,000 metric tons per annum in Bangladesh are declining (Amin et al. 2004). Similarly dams on the Mekong have assisted the decline, almost to extinction, of Laotian shad.

The relatively successful attempts by the Malaysian state of Sarawak to conserve toli and its fishery represent the only example of the use of integrated planning in relation to tropical shads. Active collaboration between government and fishers led to new

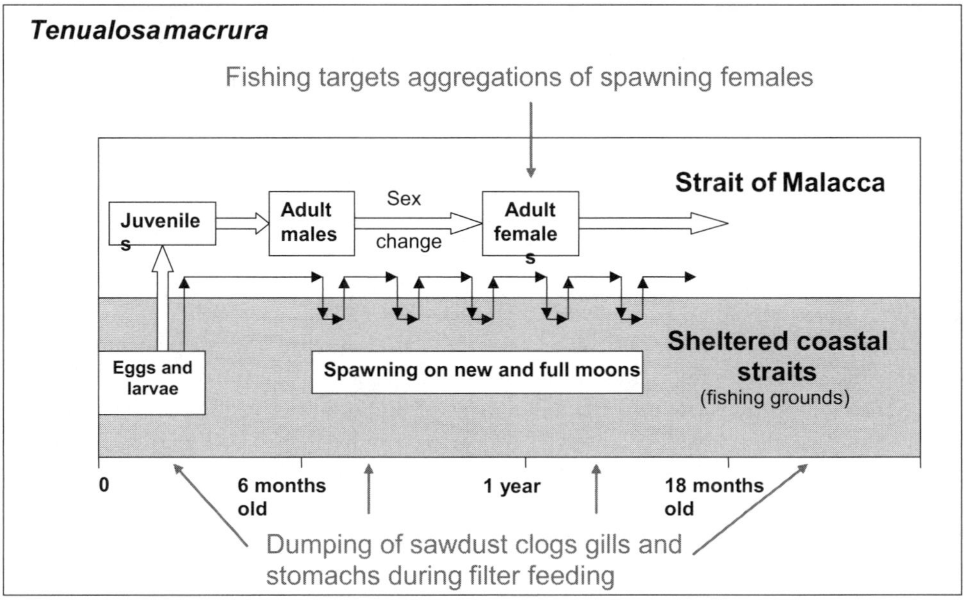

Figure 2.—Life cycle of longtail shad with main management issues and areas for possible intervention shown in gray text (Modified from Blaber et al. 1999).

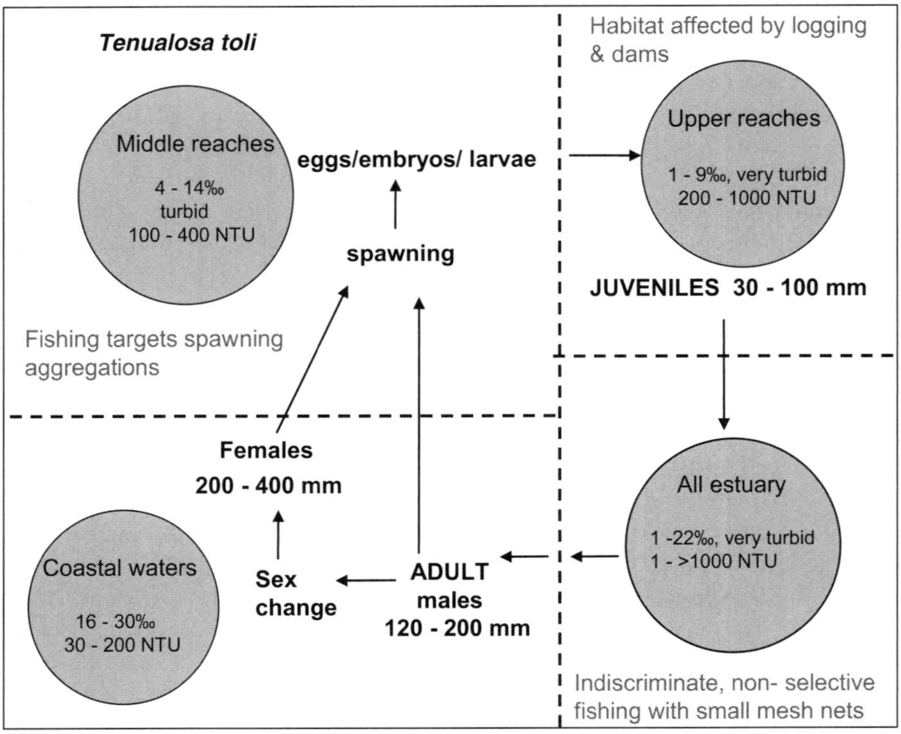

Figure 3.—Life-cycle of toli with main management issues and areas for possible intervention shown in gray text (Modified from Blaber 2000).

regulations that limited the number of fishers and imposed a closed season. These measures together with the gazetting of new national parks in the areas surrounding the upstream spawning grounds have apparently arrested the previously massive decline of the species (Blaber 2000).

Striped mullet.—Although most mullet species move between sea and estuary, only about 10 of the 70 or so species are truly catadromous and move into freshwater (Blaber 1987; McDowall 1988). Striped mullet is a cosmopolitan species and forms the basis of important commercial fisheries in many parts of the world. There are important fisheries in the Mediterranean, Mauritania, South Africa, Taiwan, Mexico, and Australia. However, Crosetti et al. (1993, 1994), Rossi et al. (1998), and Rocha-Olivares et al. (2000) have shown that populations from different oceans are highly isolated and estimates of gene flow rates extremely low. The species is catadromous, but not obligately so, with fry and juveniles also utilizing shallow coastal waters (Blaber 1987; Lenanton and Potter 1987) (Figure 4). Populations in most areas of the world appear to be sustainable at current harvest levels, but overfishing of spawning aggregations may occur (Figure 4). The largest fishery for striped mullet in South African waters is found at Lake St Lucia where both legal and illicit gill net fisheries operate (Mann 1995). Although striped mullet is not the primary targeted species, large numbers of these fish are captured by these operations since they are concentrated in North Lake and False Bay where adult prespawning striped mullet are abundant.

In Australia, the striped mullet fishery in Queensland and New South Wales is among the most important in Australia, with annual landings of more than 6,000 metric tons with a value of about $20 million (Bell et al. 2005). It is managed as a single stock and, as in many other parts of the world, the species makes pre- and postspawning migrations along the coast (Kesteven 1953; Wallace 1975), during which time they form large schools that become the target of commercial fisheries. In eastern Australia, there are two distinct commercial

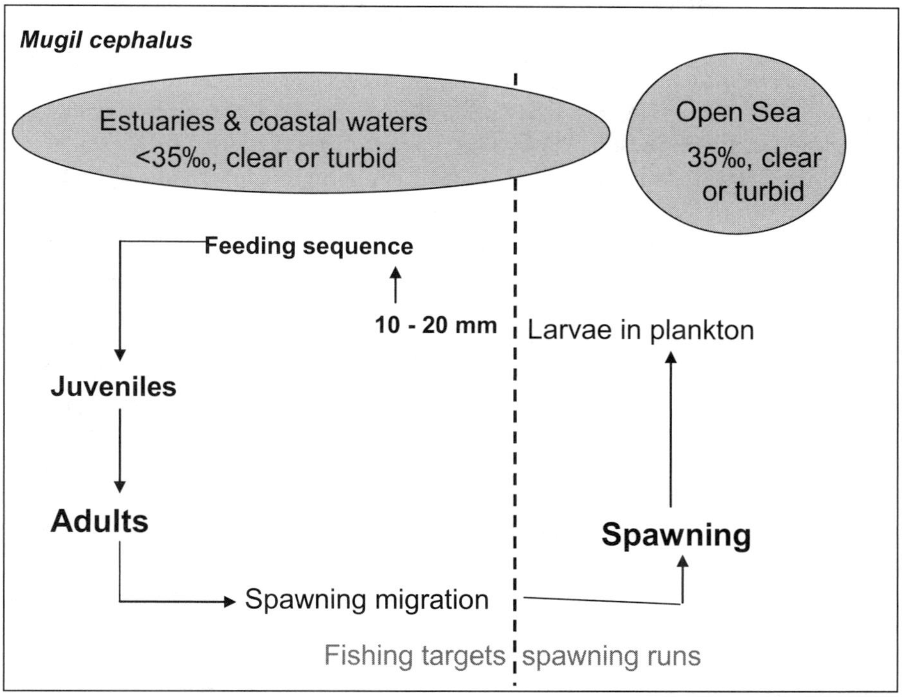

Figure 4.—Life-cycle of striped mullet with main management issues and areas for possible intervention shown in gray text (Modified from Blaber 2000).

fisheries: an ocean-beach sector that targets fish during the spring run using seine nets, and an estuarine sector that catches fish using gill nets and tunnel nets. The ocean-beach fishery contributes more than 70% of the total landings (Bell et al. 2005). In Queensland a limited-entry ocean-beach license regulates the targeting of the spawning run and allows license holders to fish from April to August each year. In New South Wales, licenses are restricted to fishers who can demonstrate a historical participation in the mullet fishery, and there are spatial and temporal closures. In both states, the minimum legal size for striped mullet is 30 cm total length. Current management concerns center on declines in recruitment

Declines in recruitment to the fishery and the occurrence of striped mullet in freshwater and estuarine environments has increased pressure for more effective overall management of the fishery, which, like the northern Australian barramundi fishery, has now been brought into an EBFM framework through the Australian Government's Environment Protection and Biodiversity Conservation Act of 1999. Because of this act, the mullet fisheries have to demonstrate that they are being managed in an ecologically sustainable manner. This incorporates not only assessments of the impact of the fishery on the target species, but also on any bycatch of the fishery and its associated communities and habitats.

Nontarget Fisheries

Quite a number of tropical diadromous species form part of multispecies coastal and inland fisheries in tropical Asia. These include a number of mullets, small clupeoids, and anguillids. Such fisheries are largely unmanaged. Tropical anguillids are an important component of some fisheries, notably Indian mottled eel *Anguilla bengalensis* in India and speckled longfin eel in eastern Australia and are of considerable commercial importance. Although some measures have been taken to mitigate the effects of dams (Thippeswamy and Hegde 1999) on the migration of eels, usually by providing fish ladders or bypasses (Stuart and Mallen-Cooper 1999), most management efforts in relation to anguillids have concerned the promotion of their aquaculture. Although tropical anguillids differ from temperate ones in that they have year-round spawning and recruitment (Arai et al. 2001, 2003; Pease et al. 2003), research on speckled longfin eel in eastern Australia indicates that eels reaching spawning condition are extremely vulnerable to overfishing, and minimum legal sizes would help maintain stocks (Collins 2000; W. Dekker, presented paper at the 2nd International Sympolsium on Diadromous Fishes, 2008). It is noteworthy that management of anguillids in temperate Europe and North America has even been difficult and not very successful and not involved EBFM in any significant way. Since the 1980s, a 90% decline in recruitment of European eel *Anguilla anguilla* has occurred across most of Europe (Dekker 2003a), and MacGregor et al. (2009, this volume) comment that North American eel management and regulatory systems, with few exceptions, are rudimentary. The massive declines in Europe have been ascribed to overfishing of all continental life stages, obstructions to upstream and downstream migrations, habitat loss, water quality, parasites, and xenobiotic contamination, which together have reduced spawner escapement from European inland waters to the sea (Feunteun 2002). Several authors have noted that restoration efforts, although substantial, have been relatively unsuccessful, mainly because it is necessary to work at a catchment scale and at a multiagency international scale (Feunteun 2002; Dekker 2003b; Lafaille et al. 2004; 2005).

Ecosystem or Integrated Management and Diadromous Fishes

The key point in relation to any ecosystem-based approach to management of tropical diadromous species is that the fish spend time in at least two, and usually three or more, ecosystems. Hence, any meaningful management plan is going to require the integration of the management of several ecosystems, ranging from the sea to rivers and including estuaries. This is a complex task at several levels: first, in relation to understanding the linkages between the key biological and physical processes; second, with regard to institutional and governance issues; and third, because any plan has to take account of other forms of management or human activity that impact the various ecosystems. These issues require fundamental changes in approach from single user management to multi-user management and coordinated interagency activities and decisions (Knecht 1997; Oakley 2001).

In terms of understanding linkages at the scientific level, the importance of a detailed multispecies approach was demonstrated by Ortiz and Wolff (2004) in their study of the holistic sustainability of integrated management regimes for the Caeté estuary in Brazil. A reductionist approach, which often happens in ecosystem modeling and integration, could not have delivered meaningful results. Understanding the influence of socioeconomics on a fishery is paramount in many developing countries where apart from food security, which is often an overriding issue, the behavior and cultures of fishers may be highly significant. The dynamics of fishing itself have often been ignored in designing management, particularly their different operating scales and the differing personal goals of individual fishers (Salas and Gaertner 2004).

Ecosystem-based management of fisheries cannot be successful unless it is undertaken in a broader context of integrating management with interests outside fisheries. The almost overwhelming problems of managing fisheries in the estuaries of tropical countries, in which all the diadromous species spend part of their life cycle, are well illustrated by the situation in Malaysia. In the mangrove areas of Malaysia, fisheries management is under the purview of the Department of Fisheries, while mangrove forests are managed by state forestry departments. Neither of these agencies are however responsible for the health of the water, which is the responsibility of the Department of the Environment. Such sectoral division of management responsibilities does not add up to effective management of the tropical estuarine/mangrove ecosystem (Chong 2006).

Integration of management has been relatively successful in Moreton Bay, a large subtropical mangrove fringed embayment adjacent to the city of Brisbane in Australia. It has extensive mangroves, sea grass meadows, and coral reefs that support an enormous diversity of life—10,000 turtles, more than 50,000 wading birds, and more than 700 different fish species and is the only place in the world where herds of dugongs graze on sea grasses within view of a major city. It also supports several commercial fisheries, such as those for penaeid prawns and the diadromous striped mullet. It was an obvious place for integrated and ecosystem-based management, and The Moreton Bay Waterways and Catchments Partnership between government and industry was established. The partnership set up an ecosystem health monitoring program in 2002 involving organizations that discharge nitrogen into Moreton Bay, including eight local councils and four major industrial companies. It is funded by councils on a per capita basis and from industry, based on their nitrogen discharge loads, to encourage improvement of wastewater treatment. An integrated research program is carried out by eight different organizations. Unfortunately, fisheries management has not yet become part of the consortium—fisheries and fisheries reserves are managed separately! However, it is hoped that the political and bureaucratic issues largely responsible for this will be overcome in the next few years.

The value of marine protected areas (MPAs) for the conservation and enhancement of fisheries sustainability has recently been seriously questioned, notably for offshore fisheries (Kaiser 2005; Longhurst 2006). However, for many smaller-scale fisheries, particularly in developing countries, MPAs may be the only practical administrative tool to effectively reconcile fisheries with conservation and to have had varying success. The problems in managing MPAs in Mexico (Bourillon et al. 2004) mirror the situation throughout the developing world: the Ministry of Environment, which manages MPAs does not have full authority over fisheries management, and there is a serious lack of interagency coordination and collaboration. Nevertheless, the declaration of MPAs has been important in bringing the problems faced by fisheries managers to the attention of governments and the public. The major problem in the role of MPAs for management of diadromous species is one of scale, but as Agardy (2005) states, "MPA networks are most successful in promoting large-scale conservation when their design is based on recognition of the interconnectivity of freshwater, coastal and marine ecosystems and habitats." Unfortunately, although there has been some progress in this respect in temperate northern hemisphere fisheries (Browman and Stergiou 2005), there are few examples of the use or value of MPAs for diadromous species in the tropics. An interesting exception is the use of fish conservation zones in the Mekong River in Laos. There is increasing evidence that these government supported community/comanagement based fish sanctuaries, usually in deepwater pools, are important places of concentration for many fish species, including highly migratory ones (Baird 2006). The declaration of MPAs in tropical developing countries is very often linked with various forms of community-based fisheries management (Pomeroy and Berkes 1997). Comanagement by government and fisher communities has been successful in some south Pacific countries, less so in south and Southeast Asia (Crawford et al. 2006), and again there are few examples where it has involved diadromous species. The recently concluded fisheries management plan for barramundi in Papua New Guinea is a possible exception: the development of the final plan that was gazetted into law involved the participation of many stakeholders, including fishers, fish processors, net sellers, and market traders, all of whom had input into the plan. Its ultimate success, however, rests not only on the general concurrence of the people, but also on the degree to which government is able to police its implementation. Lack of enforcement of fisheries laws, including those arrived at by community consultation and participation, remains a major problem in most developing countries.

In the context of benefits of an ecosystem approach, it is important in any examples to define the real or primary objectives of management. These can encompass one or all of a suite of objectives that may include maintenance of food security; fisher employment and income; conservation and maintenance of biodiversity; and fisheries sustainability at species, ecosystem, or cultural levels. Although many of these are intertwined, their outcomes will depend on the emphasis of the management, and not all stakeholders are likely to benefit equally. It is interesting that in the examples quoted in this paper, the main threats to the species are similar and relate mainly to habitat destruction and overfishing of spawning aggregations. Hence, it is perhaps these aspects that should be given attention in relation to any management prioritization. The latter threat—that of fishing spawning aggregations—may be most amenable to intervention, as illustrated by the actions taken in Sarawak for toli.

The management of tropical diadromous species is also a question of scale, regardless of the type of management. Diadromous species are migratory, and for management to be effective, it needs to be coordinated at local, provincial, national, and sometimes even international levels. This may be relatively straightforward if the management is a purely government function, but when comanagement of any sort is involved, it becomes increasingly difficult and complex. It is questionable whether comanagement can handle large-scale fisheries systems, with division of authority between community fisheries versus national fisheries and international regimes. If management is to be ecosystem based, then generally the ecosystems on which diadromous fisheries depend are usually too large to be within the control of a few communities. Therefore, unless management institutions can address the problems of resource access and sharing, there is likely to be alienation between fishers and managers (Viswanathan et al. 2003). The major problems and constraints to managing diadromous fisheries in developing countries are exemplified by those listed for Cambodia by Sour and Viseth (2004): overfishing, lack of adequate human resources to manage the fisheries, insufficient scientific data for sound management and decision making, lack of monitoring capabilities, poor enforcement of laws, inefficient administration, and increasing conflicts between different types of users.

The answers to the question forming the title of this paper "Can the ecosystem approach improve management of tropical estuarine fisheries for diadromous species?" may be summed up as (1) yes it can, and it is probably a very effective form of management; and (2) although it is broadly known why and how to do it, there are really no good examples of success because while some of the problems have been addressed in some countries, few if any have solved them all—even in developed countries!

Acknowledgments

Much of the material discussed in this paper is from research projects funded by the Australian Centre for International Agricultural Research, the Sarawak Government, and the Fisheries Research & Development Corporation of Australia. I am grateful to Thomas Hoff, Alan Whitfield, and an anonymous reviewer for their helpful comments on the paper and to the organizers of this symposium, particularly Ron Klauda, for providing the opportunity to present it.

References

Agardy, T. 2005. Global marine conservation policy versus site-level implementation: the mismatch of scale and its implications. Marine Ecology Progress Series 241:241–248.

Amin, S. M. N., M. A. Rahman, G. C. Haldar, M. A. Mazid, D. A. Milton, and S. J. M. Blaber. 2004. Stock assessment and management of *Tenualosa ilisha* in Bangladesh. Asian Fisheries Science 17:50–59.

Arai,T., D. Limbong, and K. Tsukamoto. 2003. Migration and recruitment of tropical glass eels to the mouth of the Poigar River, Sulawesi Island, Indonesia. Pages 37–46 *in* D. A. Dixon, editor. Biology, management, and protection of catadromous eels. American Fisheries Society, Symposium 33, Bethesda, Maryland.

Arai,T., D. Limbong, T. Otake, and K. Tsukamoto. 2001. Recruitment mechanisms of tropical eels *Anguilla* spp. and implications for the evolution of oceanic migration in the genus *Anguilla*. Marine Ecology Progress Series 216:253–264.

Baird, I. G. 2006. Strength in diversity: fish sanctuaries and deep-water pools in Lao PDR. Fisheries Management and Ecology 13: 1–8.

Barnes, P. W., and J. P. Thomas. 2005. Benthic habitats and the effects of fishing. American Fisheries Society, Symposium 41, Bethesda, Maryland.

Bell, P. A., M. F. O'Neill, G. M. Leigh, A. J. Courtney, and S. L. Peel. 2005. Stock assessment of the Queensland–New South Wales sea mullet fishery (*Mugil cephalus*). Queensland Department of Primary Industries and Fisheries Report, Brisbane, Australia.

Blaber, S. J. M. 2000. Tropical estuarine fishes. Ecology, exploitation and conservation. Blackwell Scientific Publications, Oxford, UK.

Blaber, S. J. M. 2009, Socioeconomic and biopolitical linkages in the management of tropical shads. This volume.

Blaber, S. J. M. 1987. Factors affecting recruitment and survival of Mugilidae in estuaries and coastal waters of southeastern Africa. Pages 507–518 *in* M. J. Dadswell, R. J. Klauda, C. M. Moffitt, R. L. Saunders, R. A. Rulifson, and J. E. Cooper, editors. Common strategies of anadromous and catadromous fishes. American Fisheries Society, Symposium 1, Bethesda, Maryland.

Blaber, S. J. M., D. P. Cyrus, J-J. Albaret, Chong Ving Ching, J. W. Day, M. Elliott, M. Fonseca, D. Hoss, J. Orensanz, I. C. Potter, and W. Silvert. 2000. Effects of fishing on the structure and functioning of estuarine and near shore ecosystems. ICES Journal of Marine Science 57:590–602.

Blaber, S. J. M., D. T. Brewer, D. A. Milton, G. S. Merta, D. Efizon, G. Fry, and T. Van der Velde. 1999. The life history of the protandrous tropical shad *Tenualosa macrura* (Alosinae: Clupeidae): fishery implications. Estuarine Coastal and Shelf Science 49:689–701.

Bourillon, L., J. Torre, A. H. Weaver, and A. Saenz. 2004. The role of marine protected areas in reconciling fisheries with conservation in the Gulf of California, Mexico. In Fourth World Fisheries Congress, abstract volume. May 2–6, 2004, Vancouver, British Columbia, Canada.

Brewer, D. T., D. S. Heales, S. P. Griffiths, M. Harwood, D. A. Milton, S. J. M. Blaber, P. N. Jones, and I. C. Stobutski. 2008. Moving towards sustainable bycatch populations in Australia's northern prawn fishery. Pages 1165–1176 *in* J. L. Nielsen, J. J. Dodson, K. Friedland, T. R. Hamon, J. Musick, and E. Verspoor, editors. Reconciling fisheries with conservation: proceedings of the Fourth Word Fisheries Congress. American Fisheries Society, Symposium 49, Bethesda, Maryland.

Brodziak, J., and J. Link. 2002. Ecosystem-based fishery management: what is it and how can we do it? Bulletin of Marine Science 70: 589–611.

Browman, H. I., and K. I. Stergiou. 2005. Politics and socioeconomics of ecosystem-based management of marine resources: introduction. Marine Ecology Progress Series 241:241–242.

Chong, V. C. 2006. Importance of coastal habitats in sustaining the fisheries industry. In National Fisheries Symposium on 'Advancing R & D Towards Fisheries Business Opportunities', 26–28 June 2006, Crown Plaza Riverside Hotel, Kuching, Sarawak, Malaysia.

Collins, A. 2000. Glass eel research in Australia. Pages 77–84 *in* Aquaculture and Fisheries Resources Management Proceedings. Taiwan Fisheries Research Institute, Conference Proceedings No. 4, Keelung.

Crawford, B., M. Kasmidi, F. Korompis, and R. B. Pollnac. 2006. Factors affecting progress in establishing community-based marine protected areas in Indonesia. Coastal Management 34:39–64.

Crosetti, D., J. C. Avise, F. Placidi, A. R. Rossi, and L. Sola. 1993. Geographic variability in the grey mullet *Mugil cephalus*: preliminary results of mtDNA and chromosome analyses. Aquaculture 111:95–101.

Crosetti, D., W. S. Nelson, and J. Avise. 1994. Pronounced genetic structure of mitochondrial DNA among populations of the circumglobally distributed grey mullet (*Mugil cephalus* Linaeus). Journal of Fish Biology 44:47–58.

Dekker, W. 2003a. Did lack of spawners cause the collapse of the European eel, *Anguilla anguilla*. Fisheries Management and Ecology 10:365–376.

Dekker, W. 2003b. On the distribution of the European eel (*Anguilla anguilla*) and its fisheries. Canadian Journal of Fisheries and Aquatic Sciences 60:787–799.

Elliott, M. 2002. An overview of the status, study and management of fishes in estuaries. Pages 555–576 *in* M. Elliott and K.L. Hemingway, editors. Fishes in estuaries. Blackwell Scientific Publications, Oxford, UK.

Feunteun, E. 2002. Management and restoration of European eel population (*Anguilla anguilla*): an impossible bargain. Ecological Engineering 18:575–591.

Fletcher, W. J., J. Chesson, K.J. Sainsbury, T. Hundloe, and M. Fisher. 2005. A flexible and practical framework for reporting on ecologically sustainable development for wild capture fisheries. Fisheries Research 71:175–183.

Grey, D. L. 1987. An overview of *Lates calcarifer* in Australia and Asia. ACIAR Proceedings 20:15–21.

Griffin, R. K. 1995. Recreational fishing surveys in the Northern Territory: 1978–1993. Pages 113–119 *in* D. A. Hancock, editor. Recreational fishing: what's the catch? Australian Government Publishing Service, Canberra.

Hall, S. J. 1999. The effects of fishing on marine ecosystems and communities. Blackwells, Oxford, UK.

Jhingran, V. G. 2002. Fish and fisheries of India. 3rd edition. Hindustan Publishing Corporation, New Delhi, India.

Kailola, P. J., M. J. Williams, P. C. Stewart, R. E. Reichelt, A. McNee, and C. Grieve, 1993. Australian fisheries resources. Bureau of Resource Sciences and the Fisheries Research and Development Corporation, Canberra, Australia.

Kaiser, M. J. 2005. Are marine protected areas a red herring or fisheries panacea? Canadian Journal of Fisheries and Aquatic Sciences 62:1194–1199.

Kare, B. 1995. A review of research on barramundi, reef fish, dugong, turtles and Spanish mackerel and their fisheries in the Torres Strait adjacent to Papua New Guinea. Science in New Guinea 21:43–56.

Kasim, H. M., and P.S.B.R. James. 1986. Distribution and fishery of *Lates calcarifer* in India. ACIAR Proceedings 20:109–115.

Kesteven, G. L. 1953. Further results of tagging sea mullet, *Mugil cephalus* Linneaus, on the eastern Australian coast. Australian Journal of Marine and Freshwater Research 4:251–306.

Knecht, R.W. 1997. Integrated coastal zone management for developing maritime countries. Pages 29–42 *in* B. U. Haq, editor. Coastal zone management imperative for maritime developing nations. Kluwer, Dordrecht, Netherlands.

Lafaille, P., A. Baisez, C. Rigaud, and E. Feunteun. 2004. Habitat preferences of different European eel size classes in a reclaimed marsh: a contribution to species and ecosystem conservation. Wetands 24:642–651.

Lafaille, P., A. Acou, J. Guillioueet, and A. Legault. 2005. Temporal changes in European eel, *Anguilla anguilla*, stocks in a small catchment after installation of fish passes. Fisheries Management and Ecology 12:123–129.

Lea, R., D. Grey, and R. Griffin. 1987. Utilisation and wildstock management of the barramundi (*Lates calcarifer*) in the Northern Territory. ACIAR Proceedings 20:82–86.

Lenanton, R. C. J., and I. C. Potter. 1987. Contribution of estuaries to commercial fisheries in temperate Western Australia and the concept of estuarine dependence. Estuaries 10:28–35.

Longhurst, A. 2006. The sustainability myth. Fisheries Research 81:107–112.

MacGregor, R. B., P. C. Thompson, A. Mathers, J. M. Casselman, J. M. Dettmers, and S. LaPan. 2009. Natural heritage, anthropogenic impacts, and biopolitical issues related to the status and sustainable management of American eel: a retrospective analysis and management perspective at the population level. Pages 713–740 *in* A. J. Haro, K. L. Smith, R. A. Rulifson, C. M. Moffitt, R. J. Klauda, M. J. Dadswell, R. A. Cunjak, J. E. Cooper, K. L. Beal, and T. S. Avery, editors. Challenges for diadromous fishes in a dynamic global environment. American Fisheries Society, Symposium 69, Bethesda, Maryland.

Mann, B. Q. 1995. Quantification of illicit fish harvesting in the Lake St Lucia Game Reserve, South Africa. Biological Conservation 74:107–113.

McDowall, R. M. 1988. Diadromy in fishes: migrations between freshwater and marine environments. Croom Helm, London.

Milton, D. A. In press. Diadromous fishes and ecological connectivity. Chapter 12 *in* I. Nagelkerken, editor. Ecological interactions among tropical coastal ecosystems. Springer-Verlag, Heidelberg, Germany.

Milton, D. A., D. Die, C. Tenakanai, and S. Swales. 1998. Selectivity for barramundi (*Lates calcarifer*) in the Fly River, Papua New Guinea: implications for managing gillnet fisheries of protandrous fishes. Marine and Freshwater Research 49:499–506.

Norse, E. A. 2005. Destructive fishing practices and evolution of the marine ecosystem-based management paradigm. Pages 101–114 *in* P. W. Barnes and J. P. Thomas, editors. Benthic habitats and the effects of fishing. American Fisheries Society, Symposium 41, Bethesda, Maryland.

Oakley, S. 2001. Issues, trends and integrated development in coastal zone management. Pages 1–8 *in* D. Chew, P. Sawai, and A. H. Sim, editors. Integrated coastal zone management. Sarawak Development Institute, Kuching, Sarawak, Malaysia.

Ortiz, M., and M. Wolff. 2004. Qualitative modelling for the Caeté mangrove estuary (north Brazil): a prelimninary approach to an integrated eco-social analysis. Estuarine coastal and shelf science 61:243–250.

Pease, B., V. Silberschneider, and T. Walford. 2003. Upstream migration by glass eels of two Anguilla species in the Hacking River, New South Wales, Australia. Pages 47–31 *in* D. A. Dixon, editor. Biology, management, and protection of catadromous eels. American Fisheries Society, Symposium 33, Bethesda, Maryland.

Pomeroy, R. S., and F. Berkes. 1997. Two to tango: the role of government in fisheries management. Marine Policy 21:465–480.

Rocha-Olivares, A., N. M. Garber, and K. C. Stuck. 2000. High genetic diversity, large inter-oceanic divergence and historical demography of the striped mullet. Journal of Fish Biology 57:1134–1149.

Rossi, A. R., M. Capula, D. Crosetti, L. Sola, and D. E. Campton. 1998. Allozyme variation in global populations of striped mullet, *Mugil cephalus* (Pisces: Mugilidae). Marine Biology 131:203–212.

Quinn, R. H. 1987. Analysis of fisheries logbook information from the Gulf of Carpentaria, Australia. ACIAR Proceedings 20:92–95.

Salas, S., and D. Gaertner. 2004. The behavioural dynamics of fishers: management implications. Fish and Fisheries 5:153–167.

Salini, J. P. 2002. Genetic structure and diversity of barramundi in Papua New Guinea and implications for managing the fishery. Journal of Fish Biology 61(Supplement A):267.

Salini, J. P., and J. B. Shaklee, J. B. 1988. Genetic structure of barramundi (*Lates calcarifer*) stocks from northern Australia. Australian Journal of Marine and Freshwater Research 39:317–329.

Scandol, J. P., M. G. Holloway, P. J. Gibbs, and K. L. Astles. 2005. Ecosystem-based management: an Australian perspective. Aquatic Living Resources 18:261–273.

Siddique, M. A. L. 2009. Conservation of juvenile hilsa (jatka) in Bangladesh: need to address the livelihood of fishers. Pages 757–768 *in* A. J. Haro, K. L. Smith, R. A. Rulifson, C. M. Moffitt, R. J. Klauda, M. J. Dadswell, R. A. Cunjak, J. E. Cooper, K. L. Beal, and T. S. Avery, editors. Challenges for diadromous fishes in a dynamic global environment. American Fisheries Society, Symposium 69, Bethesda, Maryland.

Sour, K., and H. Viseth. 2004. Fisheries and aquaculture in Cambodia's wetland environment. Pages 1–13 *in* M. Torell, A. M. Salamanca and B. D. Ratner, editors. Wetland management in Cambodia: sociological, ecological and policy perspectives. WorldFish Center Technical Report 64, Penang, Malaysia.

Stuart, I. G., and M. Mallen-Cooper. 1999. An assessment of the effectiveness of a vertical slot fishway for non-salmonid fish at a tidal barrier on a large tropical/subtropical river. Regulated Rivers Research and Management 15:575–590.

Thippeswamy, S., and S. N. Hegde. 1999. Ichthyofaunal diversity of the Sharavathi tail race hydroelectric power project area in the central western Ghats of India. Proceedings of the fourth Indian Fisheries Forum 1996:53–56.

Viswanathan, K. K., J. R. Nielsen, P. Degnbol, M. Ahmed, M. Hara, and N. M. Raja Abdullah. 2003. Fisheries co-management policy brief: findings from a worldwide study. Worldfish Center Policy Brief 2:1–26.

Wallace, J. H. 1975. The estuarine fishes of the east coast of South Africa. I. Species composition and length distribution in the estuarine and marine environments II. Seasonal abundance and migrations. Oceanographic Research Institute, Investigational Report 40, Durban, South Africa.

Zeller, B., and N. Snape. 2006. Ecological risk assessment of Queensland-managed fisheries in the Gulf of Carpentaria. Department of Primary Industries and Fisheries, Brisbane, Queensland, Australia.

American Fisheries Society Symposium 69:663–675, 2009

Whose Fish? Managing Salmonids and Humans in Complex Social–Ecological Systems: Examples from the Baltic Sea Region

Monica Hammer*
School of Life Sciences, Södertörn University
S 141 89 Huddinge, Sweden

Abstract.—Management of salmonids at local and regional levels in the Baltic Sea is best analyzed by viewing the Baltic region as a complex social–ecological system. The ecosystem approach, now in an implementation phase in the Baltic Sea region, provides a framework for understanding interdependencies between resources, ecosystems and society, and examining management trade-offs. This study focuses on application of the ecosystem approach to management of two salmonids: the Atlantic salmon *Salmo salar* fishery in the Baltic Sea region and the brown trout *S. trutta* recreational fishery in Stockholm archipelago, Sweden. These cases provide illustrative examples of some current trends and key challenges for sustainable salmonidae management. The Baltic Sea has low biodiversity and Atlantic salmon is one of the few commercially or recreationally important fish species. In this region, fish resources and the ecosystems they depend on are shared by nine coastal nations. The different salmon populations spawn in their native rivers but mix in offshore areas where they are harvested by national fishing fleets. In the Baltic, a number of factors have impacted wild salmon populations. Hydropower development in the Baltic coastal states has led to declines in wild salmon populations and, to compensate, annual stocked fish releases. In attractive coastal areas such as the Stockholm archipelago in Sweden, increased recreational fishing has led to a shift in user patterns from small-scale commercial fisheries by islanders towards a heterogeneous group of visiting recreational fishers. Efforts to reorganize fisheries management institutions in the Baltic Sea region indicate the need to consider the full range of user groups, develop coherent multilevel management institutions, and achieve a better understanding of the role of diverse fish populations in the ecosystem.

Introduction

Capture fisheries are characterized by interdependencies between ecosystems and society that are stronger than for most other food production systems. This may be particularly true for diadromous fishes that migrate between spatially separate ecosystems, during which time they are exposed to a variety of human activities and demands. To enhance sustainable resource use, management policies for marine systems and fisheries in the Baltic Sea region and elsewhere have evolved towards ecosystem approaches in recent years (SOU 2003; Hughes et al. 2005). Management priorities are thus shifting from target species towards the entire ecosystem and its broader range of services (Dale et al. 2000; Christensen et al. 2003; Pikitch et al. 2004). The ecosystem approach to management, as formulated by the United Nations (UN) Convention on Biodiversity (CBD), focuses on the resilience of ecosystems to sustain delivery of ecosystem goods and services for human well-being and emphasizes the capacity of society to cope with, adapt to, and shape change (Holling 1978; CBD 1998; Berkes et al. 2003; Folke 2006).

* Corresponding author: monica.hammer@sh.se

Ecosystem services (Daily 1997), such as fish, nutrient recirculation, and water flows are often characterized as common-pool resources at risk in modern societies (Hanna et al. 1996; Ostrom et al. 2002; Millennium Ecosystem Assessment 2005). In the Baltic Sea region, fish resources and the ecosystems they depend on are shared by nine coastal nations. For example, distinct Atlantic salmon *Salmo salar* populations spawn in their native rivers but mix in offshore areas where they are harvested by several national fishing fleets. Also, human activities have degraded local spawning grounds affecting salmon throughout the Baltic (ICES 2007). In many coastal regions along the Baltic Sea, there is a transition from traditional subsistence and commercial fisheries towards other activities, including increased recreational fisheries affecting fishing patterns (Hammer et al. 2003; Bruckmeier and Neuman 2005).

Controlling fishing pressure on salmon in the Baltic Sea and restoring its ecosystem depends to a large degree on establishment of institutional frameworks that align diverse economic interests with conservation. In particular, institutions that can interact across multiple scales to reinforce and balance each other are key (Young 2002; Hughes et al. 2005). The ecosystem approach provides a framework to address such interdependencies between a targeted resource, its supporting ecosystem, and the trade-offs made by a complex human society. In this paper, the development and challenges for salmonid management in the Baltic Sea are analyzed from an ecosystem management perspective. In particular, I focus on two examples that illustrate different management levels and important trends in Baltic Sea fisheries: the regional management of Atlantic salmon in the Baltic Sea and local brown trout *S. trutta* fisheries in Stockholm archipelago Sweden. First, I discuss ecosystem management approaches, outlining key elements of the ecosystem approach and providing recent examples of ecosystem management policy initiatives in the Baltic Sea region.

Ecosystem Management Approaches to Natural Resources Management

In its narrowest sense, ecosystem management extends managers' focus beyond target species to include the impacts of fishing on nontarget species or aquatic habitats. A broader view of ecosystem management includes both sustainability of harvestable resources and the recognition that these resources depend on a functioning ecosystem. Further, recognizing that fish generate a range of important ecological services contributing to the dynamics, structure, and functioning of complex ecosystems at multiple scales is also crucial (Daily 1997; Holmlund and Hammer 1999; MA assessment 2005).

The ecosystem approach as formulated by the UN CBD is described as "a strategy for the integrated management of land, water and living resources that promotes conservation and sustainable use in an equitable way" (CBD 2004). In this definition, the 12 principles and five guidelines to the ecosystem approach of the CBD stress a management focus on functional relationships and processes in ecosystems, including habitat fragmentation and underlying causes of biodiversity loss. Further, benefit sharing should be enhanced, in particular among stakeholders. Because change and uncertainties are inherent in social–ecological systems, adaptive management practices are recommended, including flexibility and learning. The CBD guidelines also stress that ecosystem management includes recognition of the diversity of social and cultural factors affecting natural resource use. Further, management actions should be carried out at the scale appropriate for the issue addressed, with decentralization to lowest level as appropriate. Finally, ecosystem management of natural resources calls for increased intersectorial communication and cooperation at a range of levels. Hence, stakeholder involvement, adaptive management and decentralization, as appropriate, are key to the successful and sustainable management of social–ecological systems in this framework (CBD 1998, 2004). In systems such as the Baltic Sea, characterized by interdependence of resource users and the existence of both local and regional management authorities, collaborative management approaches involving stakeholders is advocated (Pinkerton 1989; Wilson et al. 2003; Åqvist Almlöv and Hammer 2006).

The Baltic Sea, as a multi-use, multicultural region with a relatively enclosed and low-diversity ecosystem, provides an interesting case study for applying an ecosystem approach to management. A number of recent management initiatives based on ecosystem management principles such as The Helsinki Commission (HELCOM) Baltic Sea Action Plan adopted in November 2007, the European Union (EU) Common Fisheries Policy (CFP), and the EU Marine Strategy (Ask and Westerberg 2006; HELCOM 2007) have been established in this region. The overall goal of the HELCOM Baltic Sea Action Plan is to restore good ecological status in the Baltic marine environment by 2021. Specific goals of the plan include a Baltic Sea unaffected by eutrophication and undisturbed by hazardous substances, favorable conservation status of the Baltic Sea's biological diversity, and environmentally sound maritime activities in the Baltic Sea (HELCOM 2007). Examples of national-level ecosystem approaches include the Swedish national environmental goals. In particular, the National Swedish Marine Strategy, including fisheries, is based on the ecosystem approach as formulated by CBD (SOU 2003). Hence, the ecosystem management approach is currently in the process of being implemented in the Baltic Sea. The ecosystem approach puts new demands on local, national, as well as regional management institutions in the Baltic Sea. Specifically, the ecosystem approach, while focusing on ecosystem processes rather than individual ecosystem services, requires intersectorial cooperation, scale matching, and stakeholder involvement. The implications and main challenges for managing a resource, such as salmon, from an ecosystem approach is an interesting question that I explore in the following case studies.

The Baltic Sea Ecosystem and Its Fisheries

The Baltic Sea is a young brackish, nontidal relatively shallow estuary of about 413,000 km^2 with a drainage basin approximately four times larger than the sea itself (Figure 1). Fed by the drainage of more than 200 rivers, the Baltic Sea is connected to the Atlantic Ocean by narrow straits at its southwestern end. These biophysical conditions form a surface salinity gradient from south to north with 6–8 practical salinity units (psu) in the Baltic proper to 2–4 psu in the Bothnian Bay. This gradient structures the biotic community (Elmgren 1989). The ecosystem is composed of an assemblage of euryhaline marine and brackish estuarine species, freshwater species, and glacial marine or freshwater relicts (Segerstråle 1957). The species richness of the Baltic Sea is significantly lower than that of adjacent marine areas, and many species tend to occupy broader ecological niches in the Baltic than in their original habitats (Dahl 1973). Despite the low species richness, the structure and function of the Baltic Sea ecosystem is complex, with food chains up to five levels (Elmgren 1989; Hammer et al. 1993). Due to natural variations in temperature, salinity, and oxygen conditions in the Baltic Sea, fishes must tolerate constantly changing conditions.

Since at least the 1950s, human activities have played a dominant role in the degradation of the Baltic Sea ecosystem (Elmgren 1989; Ojaveer and Lehtonen 2001). In the nine coastal states, extensive human activities such as industrial production, agriculture, and forestry have caused heavy toxic pollution problems and severe eutrophication of the Baltic Sea. The nitrogen and phosphorous load has increased four- and eightfold, respectively, transforming the Baltic Sea from an oligotrophic to a eutrophic water body (Larsson et al. 1985; Österblom et al. 2007) and affecting primary production, water transparency, and oxygen conditions (Larsson et al. 1985; Jansson and Jansson 2002).

At the ecosystem level, the human activities outlined above have strongly contributed to changes in the food web where planktivorous fish such as Atlantic herring *Clupea harengus* and sprat *Sprattus sprattus* have increased, while heavily harvested species such as cod and wild salmon have decreased (Ojaveer and Lehtonen 2001). The gray seal *Halichoerus grypus*, a top predator in the Baltic Sea, population decreased from approximately 100,000 seals in the early 1900s to only a few thousand individuals in the 1970s. This decline resulted from pollution of toxic substances, seal hunting, and loss of prey species (Elmgren 1989; Folke et al. 1991). Efforts to restore the Baltic seals, in particular the gray seal, have resulted in an increase in individuals to around 20,000 today. This increase, however, has primarily affected the stationary gear (traps) salmon fishery as the seals eat and destroy the salmon in the traps (Westerberg et al. 2008).

In the Baltic Sea, the large-scale commercial fishery is dominated by three species in terms of volume: Atlantic herring, sprat, and Atlantic cod *Gadus morhua*. Atlantic salmon landings are considerably smaller but have been relatively important in economic value. Commercial fish landings for all species were fairly moderate (i.e., under 200,000

Figure 1. The Baltic Sea and its drainage basin.

metric tons) until the 1950s. In the following decades, with modern technology and increased fishing effort, landings increased to around 1 million metric tons by the 1980s (Ojaveer and Lehtonen 2001).

The International Council for the Exploration of the Seas assesses commercially important fish species, including salmon, yearly. Commercial fish catches for these species are regulated with yearly total allowable catches that were negotiated and divided between the coastal states within the International Baltic Sea Fishery Commission until 2005. Since 2005, all countries except Russia joined the EU and the responsibility for resource management shifted primarily to the EU CFP.

Although large-scale, offshore commercial fisheries dominate in the Baltic, there remains a declining small-scale coastal, commercial fishery. For example, along the Swedish Baltic coast, a small-scale coastal and archipelago fishery, often combined with agriculture, has a long tradition. Since the 1960s, this fishery has been marginalized by competition from the offshore commercial fishery. In particular, large catches in the offshore fisheries targeting the migrating species of salmon and cod have resulted in a decrease in local coastal fisheries and caused conflicts between these different fisher categories. In addition, the rapidly increasing recreational fishery also targets salmonid species such as Atlantic salmon and brown trout. Recreational fisheries have been actively supported in legislation in Finland and Sweden. Free fishing for the general public with hand gear in both public and private waters has been allowed since 1985 in Sweden and since 1995 in Finland. Partly as a result, the number of people engaged in recreational fishing in Sweden increased from about 1 million in 1963 to 2.3 million in 2000 (Bengtsson et al. 2000).

Salmon in the Baltic Sea

Among the first fish species to migrate into the Baltic Sea during the deglaciation some 15,000–9,000 years B.C. were Atlantic salmon and brown trout (Ojaveer and Lehtonen 2001). Salmon spawn in several rivers tributary to the Baltic Sea and are composed of numerous, genetically distinct local populations that home to separate rivers. Genetic differentiation, however, also exists within rivers (Ståhl, 1987). Atlantic salmon migrate hundreds of kilometers in the Baltic Sea, during which time they are exploited by both offshore and coastal commercial and recreational fisheries.

During the 20th century, most of the wild salmon populations in the Baltic Sea disappeared and the remaining ones were greatly depleted, mainly as a result of overfishing and freshwater habitat degradation (e.g., Karlsson and Karlström 1994). Hydropower development in the 1950s and 1960s, which limited both up and downstream migrations, also contributed to the decline of wild salmon (Karlsson and Karlström 1994).

In the early 20th century, there were more than 80 salmon-producing rivers in the Baltic Sea with an estimated smolt production of 8–10 million per year. Extensive stock and release programs started in the 1950s based on court decisions to compensate the commercial fishery for the losses of salmon due to hydropower development in Baltic rivers. This resulted in stocking of up to 90% of the salmon in the Baltic Sea (Petersson and Järvi 1999). The annual production of wild salmon in the Baltic has been estimated at only 1.2 million smolts in comparison to approximately 7 million stocked salmon per year in the early 2000s. Approximately 2 million of these fish are stocked in major Swedish Baltic rivers (Holmlund and Hammer 2004: Naturvårdsverket and Fiskeriverket 2005).

Until the early 1990s, catches in the offshore salmon fishery remained high and even increased because of the extensive stocked salmon releases. Offshore fisheries on mixed stocks dominated where most of the salmon were harvested using drift nets. Since this time there has been a decline of the total nominal salmon catch in the Baltic Sea from 5,636 metric tons in 1990 to 1,547 metric tons in 2003 (ICES 2006). The Swedish salmon catch was 719 metric tons in 2005. Of this, 50% was caught in the offshore fishery, 40% in the coastal fishery, and the remaining 10% in the rivers[1] (ICES 2006).

Major reasons for the changes in salmon fish catches and fishing profile during the past decade include a sharp decrease of offshore fisheries due to low profitability, partly a result of competition from reared Norwegian salmon (ICES 2006). In addition, because of high dioxin levels in salmon, there are national dietary advisories for human consumption of Baltic salmon and restrictions on salmon marketing in Sweden, Finland, Denmark, and Latvia (ICES 2006).

Another factor affecting the Baltic salmon populations is a disease called M74-syndrome that was discovered in 1974 and is believed to be connected to environmental pollution in the Baltic Sea (Vuori and Nikinmaa 2007). M74 causes thiamine deficiency and increased mortality in the yolk sack fry (maternally transmitted). M74 mortality has been extremely high periodically (Vuori and Nikinmaa 2007). Between 1992 and 1996 there was a more than 50% frequency (proportion females) among reared salmon (ICES 2006). There is also evidence that wild populations are affected by M74 during periods of high occurrence (Karlström, 1999).

Increases in the Baltic seal population during the past two decades has led to increased predation on coastal fish (Lunneryd et al. 2004). Seals interact with commercial fisheries, particularly in the northern Baltic Sea, by eating fish caught in passive gears. Kauppinen et al. (2005) recorded catch losses of salmon of at least 37% of the catch in numbers in the Bothnian Sea. In the Baltic main basin and Gulf of Bothnia, about 70% of the discards are estimated to be caused by seal damage (Anon. 2006). The phasing out of drift nets to protect the harbor porpoise *Phocoena phocoena* has led to additional salmon predation and resulting decline in the commercial salmon fishery (ICES 2007).

Hatchery-produced salmon are less genetically diverse compared to their wild counterparts. Weakening wild populations even further, enhancement operations may cause loss of biodiversity through

[1] River catches are not included in the Swedish total allowable catch.

hybridization, out-competing of local conspecifics, or increased rates of inbreeding (Waples 1991; Koljonen et al. 1999; Laikre et al. 2005, 2006). Also, salmon harvest quotas are set on the total populations, and the fishery exploits a mixture of reared and wild salmon, thus placing excessively high pressure on the wild salmon (Karlsson and Karlström 1994; Koljonen et al. 1999).

As indicated above, wild salmon in the Baltic Sea has been affected by a number of different factors. Some effects are directly related to fishing activities, such as overfishing of wild salmon. Other human activities, such as hydropower generation or pollution, result in changes in the supporting ecosystem, with effects on stocks. Some of these effects have been anticipated by management agencies while others have not. Ironically, the Water Court decisions to compensate the loss of wild salmon from hydropower development by enhancement fish stocking seems, in a longer time perspective, to have inadvertently reduced the resilience of the populations, creating instead one of the more serious threats to the wild salmon populations.

Other management trade-offs must be considered. For example, some fishing regulations, such as the time periods for closed seasons, change from year to year. In Finland, fishers claim that these variations cause problems in the planning of their long-term fishing activities (Salmi and Salmi 2005). The trade-off between salmon fisheries interests and conservation policies directed towards other species in the Baltic, in particular the gray seal and the harbor porpoise, must also be considered. Salmon catches have continued to decline and are expected to decrease further with the drift net ban in 2008 (ICES 2007).

Restoration Efforts for Salmon in the Baltic

To rehabilitate wild salmon populations, an international treaty, the Baltic Salmon Action Plan (SAP), was signed in 1997 by the Baltic Sea countries represented in the International Baltic Sea Fishery Commission. A main goal of the SAP was for wild salmon rivers to reach at least 50% of the potential production of out-migrating smolt and that 26 rivers have self-sustaining salmon populations by the year 2010. The SAP entails responsibility on all parties to rehabilitate wild salmon populations. Hence, this plan combined a regional management approach with national and local management responsibility.

In Sweden, the country with the longest Baltic coastline, 27 rivers historically harbored distinct salmon populations, but by the 1990s, natural reproduction occurred in only 14 rivers. Because 12 of the rivers in the SAP are situated within its boundaries, Sweden adopted a national salmon plan within the framework of the SAP in 1998. On a national level, Sweden established protected areas outside the wild salmon rivers and protection measures, including closures during the early summer period along the coast of the Bothnian Gulf and Bothnian Bay. As wild salmon have been shown to migrate back to the coast before entering rivers to spawn, the intention was to provide wild salmon time to traverse the coast before the start of the coastal fishery (Naturvårdsverket and Fiskeriverket 2005). In addition, in 2005, Sweden began marking stocked smolts by means of fin clipping to distinguish reared from wild salmon.

Wild salmon smolt production in the Baltic Sea has recovered somewhat in recent years. Since the SAP was adopted in 1997, wild smolt production in the entire Baltic Sea has increased about fourfold in the main basin and Gulf of Bothnia, and the current wild smolt production is now estimated to be around two-thirds of the estimated overall potential smolt production (Anon 2006). Lower mortality from the disease M74, lowered offshore fishery exploitation, and habitat restoration efforts also contributed to this recovery. The long-term success of the recovery, however, is still uncertain. In spite of the increased smolt production in rivers, postsmolt survival in the sea is declining. The reasons for this are still unclear (ICES 2007). There are also large variations between rivers. Most of the northernmost rivers are likely to reach the 50% objective while the southern stocks are more variable (ICES 2007). For example, the Mörrum and Emå rivers in the southern part of Sweden are more productive than the northern rivers due to a more suitable nutrient and climate situation. The Mörrum River is the strongest producer of wild salmon in the Baltic proper. In 2004, the densities were estimated as 100 young salmon per 100 m^2 in the Mörrum River and approximately 50 per 100 m^2 in the Emå River. In comparison, densities in the Ume-Vindeln River in northern Sweden are approximately 8 per 100 m^2.

The Mörrum River is projected to reach the target or even 75% of the objective while production in the nearby Emå River is unlikely to reach even 50% of the objective. As such, other factors, including the institutional setting, seem to be important. Thus, a vital component of the SAP is local management responsibility on a national and on a river scale.

On the local scale, involvement of stakeholders and balancing of different interests affect the likelihood of reaching the targets of the SAP. An illustrative example is the Ume-Vindeln River. The Vindeln River is one of Sweden's "national rivers" and stretches 500 km from the mountains to its confluence with the Ume River (a tributary of the Bothnian Bay). The production potential of salmon and trout in this river is large, due to suitable habitat, but spawning migration to this habitat is constrained by downstream hydropower dams. The national hydropower company, Vattenfall, has released fish above the hydropower station and thus the river system now contains a mixture of stocked and wild salmon and trout. (Fiskeriverket 2007). Also, in recent years, salmon in this river have been affected by M74 syndrome. As a result of these issues, the wild salmon population has been very weak in the past decade.

In the reaches of the Ume-Vindeln River above the hydropower plant, salmon fisheries are well organized, with the recreational fishery strongly regulated by catch limits during a restricted season. The river includes approximately 3,500 fishing rights owners organized in nearly 50 different collectively managed fishing management areas (FMA). An advisory council, Vindelälvens fiskeråd, was initiated in 1981, and 25 of the FMAs are members of the council. In the lower part of the river, below the hydropower plant, fisheries are less organized and rules vary. Fishing rights are also unclear for certain areas, and this uncertainty has led to conflicts between water owners and different categories of fishers, including the coastal commercial salmon fishery. To facilitate local management, a promising comanagement initiative was recently created, involving all stakeholders and in collaboration with the National Board of Fisheries. The long-term goal of this initiative is to allow natural increases in populations of river migrating brown trout and wild Atlantic salmon and to create a locally managed and attractive fishery based on these populations (Fiskeriverket 2007). The Ume-Vindeln River is projected to reach the 50% target for wild salmon (ICES 2007).

The Baltic Salmon Action Plan is in accordance with an ecosystem management approach in the sense that ecological boundaries for the resource are well matched with management boundaries. Addressing both the regional- and local (river)-scale issues is likely one reason for the plan's success thus far. Managers have recognized the variability in production among rivers. Suggested measures to further improve the Emå River salmon include restrictions in the offshore fishery along the migrating routes of this particular population (ICES 2007). Hence, a main lesson learned in the Baltic salmon case is the interdependence between different human as well as natural variations. The massive stocking of salmon to protect the commercial salmon fishery was successful only in the short run, but the decreased resilience of salmon populations to cope with disturbances clearly demonstrated the need for a broader view for sustainable salmon management. However, the difficulties for management agencies in balancing different human interests and activities are apparent and point to the importance of local management involving local stakeholders and creating coherent rules of use on multiple levels. This goal can be particularly difficult in situations where user groups and use patterns change. An illustrative example of this problem is described in the following example of recreational fisheries in Stockholm archipelago, Sweden.

Recreational Trout Fisheries in Stockholm Archipelago, Sweden

The Stockholm archipelago is part of the central Baltic archipelago stretching across the Baltic Sea from Stockholm through the Åland islands to the southwest Finland archipelago (Figure 1). Approximately 30,000 islands, islets, and skerries in Stockholm archipelago are home to 12,000 permanent residents, a large number of seasonal summer guests in approximately 50,000 summerhouses, as well as a large number of visiting tourists.

In Stockholm County, there are an estimated 400,000 recreational fishers (Nilsson 1990). In contrast, the number of licensed commercial fishers has declined from almost 3,000 to only 36 since the 1960s (Andersson 2003). The shift towards a recreational dominated fishery is related to a number of socioeconomic factors such as increased leisure time and transportation possibilities (Hammer et al. 2003). Commercial archipelago fisheries, generally a part-time occupation, became less profitable as the traditional way of making a living in the region from a combination of occupations declined. Economic incentives and subsidy systems for fisheries have also favored large-scale commercial operations. In addition, a compulsory licenses scheme for commercial fishing introduced in 1993 further decreased the number of small-scale commercial fishers. In 1985, a governmental decision that recreational fishing with hand held gear should also be free on private waters further diminished coastal commercial fisheries. A growing occupation in the archipelago is the fishing tourism industry, with professional guides arranging fishing trips. Currently, more than 20 fishing guide entrepreneurs are in operation (Jonson 2002). Sea migrating brown trout and Atlantic salmon are among the most highly attractive target species.

Fisheries Management in the Archipelago

Archipelago fisheries are affected by local, national, and international regulations. Since Sweden joined the EU in 1995, fisheries are regulated by international agreements within the framework of the EU's CFP. The Swedish Board of Fisheries has overall management responsibility and is responsible for monitoring fish stocks in coastal waters. It also provides subsidies to fishers. Statistics of commercial landings are combined with yearly net sampling by the Swedish National Board of Fisheries to estimate fish stock conditions (Ask and Westerberg 2006). However, catch statistics from private waters are largely lacking. Recreational catches (sport and household fishing) are estimated every fifth year in a national survey, providing a rough estimate.

At the regional level, the Stockholm County Administration Board (SCAB) is responsible for promoting sustainable use of fish resources, protecting the resource base, and, at the same time, increasing possibilities for both public and commercial fisheries within the archipelago. Since 1977, SCAB has also been responsible for authorizing all local requests for fish stocking (Holmlund and Hammer 2004). The Stockholm County Administration Board collaborates with Stockholm municipality in fish monitoring, especially regarding sea migrating brown trout enhancements. Together, the agencies perform monitoring activities in approximately 50 archipelago rivulets once per decade (Holmlund and Hammer 2004). Since the early 1990s, Stockholm municipality has been mandated to manage most of the sport fishery in the entire Stockholm archipelago, including the rivulets tributary to the archipelago.

A large number of interest groups and other organizations operate in the Stockholm archipelago, including more than 40 recreational fishing associations and three commercial fishing organizations (ÅqvistAlmlöv and Hammer 2006). Property rights constitute an important structuring factor for Stockholm archipelago fisheries. In nearshore waters in Sweden, fishing rights are tied to land ownership. In general, private waters extend 300 m from the coastline, in effect making most of the waters in the archipelago private. The trend towards converting more homes into summerhouses resulted in selling of smaller land parcels, and the associated fishing rights were not always regulated. Fishing is free for Swedish citizens in public waters. The 1985 regulation allowing free hand gear fishing on private waters also affected the commercial fishers, as they no longer had exclusive rights to their private or leased fishing waters. This change has caused conflict and debate where owners of waters held privately or in common have claimed that the incentives to conserve and manage their coastal fish resources were lost. A mixture of private, collectively owned, and state-owned water properties now exist in addition to properties where the current property rights are unclear (ÅqvistAlmlöv and Hammer 2006). The result is a heterogeneous group of fishing rights owners, with differing degrees of interest in the resource, ultimately leading to a mismatch of rights and interests between user groups and owners (ÅqvistAlmlöv and Hammer 2006).

Sea Migrating Brown Trout Enhancements

Both in terms of participants and catch, recreational fisheries now dominate in the archipelago. This trend has affected the demand for fish resources. Sea migrating brown trout is one of the most desired target species for sport fishers. To bridge the gap between human demand and the production of wild fish resources, large-scale enhancement of migrating brown trout is made primarily to support the growing recreational fishery in Stockholm archipelago. In Stockholm archipelago, stocking operations are made by several organizations but need to be approved by the Stockholm County fishery administration. Fish hatcheries have existed in Stockholm county since 1865, but the first indication of fish stocking in Stockholm archipelago dates from the 1940s (Holmlund 1996). After some experimental stocking in 1973 of sea trout smolts in Stockholm Stream (the outlet from Lake Mälaren to the sea) in downtown Stockholm, the yearly enhancements have increased from a few thousand in the 1970s–1980s to approximately 125,000 smolts between 1996 and 2000. Most of the released sea trout are 1–2 years old. Stockholm municipality has been responsible for the majority of migrating brown trout enhancements since 1973. The state owned hydropower company Vattenfall, mainly for goodwill reasons, has sponsored around half of the enhancements.

Sea migrating brown trout are released (generally delayed release) more or less regularly in more than 30 rivulets tributary to the archipelago with or without self-reproducing sea migrating brown trout populations. Habitats in most of these rivulets have been restored after several years of severe pollution, eutrophication, and destructive land-use changes. Rivulet enhancements have increased greatly since the late 1980s, as a direct consequence of the 1985 national regulation of free sportfishing with handheld gears in private waters (Holmlund and Hammer 2004). This regulation enabled the municipality to use governmental funds for releasing fish in both state-owned and private waters (Holmlund and Hammer 2004).

In the mid-1970s, there were 14 archipelago rivulets with spawning sea migrating brown trout. In 1995, sea migrating brown trout were found to spawn in 28 rivulets, as a result of habitat restoration and regular enhancements (Lovén and Ungsgård, 1999). Five to seven of these rivulets are believed to sustain genetically distinct sea trout populations with little or no history of fish enhancement. In the more than 20 remaining rivulets, the sea trout are of either wild or hatchery origin or a mixture of both (Holmlund and Hammer 2004).

The origin of the early sea trout enhancements included broodstock from more than 10 different locales along the Swedish east coast. Since the mid-1990s, however, one local archipelago population (Åva rivulet) has dominated as the source population (Holmlund and Hammer 2004). The Åva trout is considered to be stationary within the archipelago, and a comparatively large number of sea migrating brown trout return to the Åva rivulet to spawn, thus minimizing the risk for mixture of different populations (since it does not migrate far and thus does not mix with other populations).

Since 1989, municipal fishery managers have also been active in engaging local sportfishing associations in sea migrating brown trout enhancement efforts through education programs about habitat restoration and monitoring procedures in rivulets. This goal was primarily achieved by letting sportfishing associations become "godmothers" of a particular rivulet and its trout. For example, the sportfishing associations assist in fish monitoring and habitat restoration activities. The monitoring of sea migrating brown trout is partly made in collaboration with SCAB. The number of associations has varied over the years, but in general, 10–15 associations are active godmothers for some 20 rivulets.

In Stockholm archipelago, the transition from commercial towards recreational fisheries has shifted the focus from provisioning to cultural ecosystem services. The main user groups have changed from being stationary fishers living in the archipelago and fishing in their own waters that they supervised on a regular basis to a more heterogeneous group of fishers making occasional visits in the archipelago. It is worth noting that the commercial landings have to be reported to the authorities, while there is no such control of recreational fish catches. These altered fishing patterns and control over fish resources appear to have resulted in loss of monitoring and possibly knowledge of regulating and supporting services that might undermine sustainable development of fish populations. However, the initiative to collaborate with new user groups (sport fishers) is a step towards reorganizing management and rebuilding local knowledge. The example of the changing fishing patterns in Stockholm archipelago underlines the necessity to apply a broader, longer-term perspective to management. Acknowledging the possibility for changes in fishing patterns, the human community, or the broader ecosystem is necessary for long-term sustainable management.

An Ecosystem Approach to Salmonid Fishing

The ecosystem management approach illustrates many of the difficult management issues facing Baltic salmonid management. These issues include understanding the role of functioning ecosystems in supporting fish production. Also critical are management institutions that can act at a scale relevant to both ecological and socioeconomic borders and that are adaptive to change. This is particularly important for migrating species such as the anadromous salmon.

Present management regimes are, in many cases, not ecologically consistent with the resources they manage, creating a mismatch between the social and ecological system (Folke et al. 1998; Young 2002; Galaz 2005). The ecosystem approach attempts to overcome such mismatches by defining administrative areas according to ecological borders such as drainage basins. The salmon action plan attempts to align ecological and political boundaries by establishing appropriate management responsibilities on both the local river scale and the regional level. Aligning ecological and political boundaries, however, can be a complicated issue for migratory fishes, in particular for anadromous fish. As discussed by Carvalho and Hauser (1994), there are several definitions of stock, but in fisheries management, the term is often used to signify a group of fish exploited in a specific area or by a specific method. This definition does not necessarily reflect the biologically adequate substructure of the species. Fish species are rarely genetically homogeneous, but instead are often structured into groups of individuals that are isolated from each other by factors such as restricted dispersal and homing behavior. Hence, developing management strategies for species with different biological characteristics and requirements remains a challenge for sustainable management of coastal and marine ecosystems and their fish resources. This concern applies to management of Atlantic salmon in the Baltic, where genetically distinct populations occur in rivers (Ståhl 1987; Laikre et al. 2005).

Salmonids are dependent on a number of ecosystem attributes to sustain viable populations. In turn, salmon, like many other fishes, contribute to ecosystem functioning by performing an array of services and are fundamental for the maintenance of these ecosystems. Trade-offs between management activities directed towards effective fisheries and activities directed towards restoration of other species such as the harbor porpoise highlights the question of who has the primary right to the fish resource. This also encompasses the different ecosystem services performed by, for example, salmon and also other ecosystem services that salmon depend on during its life cycle. For example, a choice between salmon fisheries and seal conservation also implies a choice between allowing the salmon a role in an endangered food chain in the sea as prey for seals.

Wild salmon in the Baltic Sea have been negatively affected by a number of factors both in the ecosystem and in society. These issues, such as pollution due to industrial activities or degradation of habitats, often are not directly or even indirectly related to fisheries. A narrow scope of management, which emphasizes certain services, has resulted in adverse affects to other ecosystem services. An obvious example is the choice of hydropower development in Baltic rivers and compensatory fish stocking. Fish stocking has been a dominant management practice. As indicated by the dynamics of Baltic salmon populations, this practice appears to decrease buffer capacity and resilience of populations in the long term. These declines have ultimately led to alternative management policies. The Baltic Regional Action Plan, to some extent an ecosystem management plan for salmon, resulted in increased wild smolt production, although with large local variations. This action plan was key to coordinating local and regional actions to meet the needs of salmon during its life cycle. As suggested by Laikre and colleagues (2005), eventually terminating the stocking operations altogether would motivate stakeholders to determine the biological conditions of the wild populations, and develop biologically sustainable exploitation and management practices.

In any coupled human-natural system, the ecosystem, human-use patterns, and composition of user groups are likely to change over time. In Sweden, commercial fisheries are a small and diminishing sector. Nevertheless, Swedish national policies for regional and rural development point to the significance of fisheries as an important actor in reaching national environmental and sustainability goals. Intense stocking of trout in the Stockholm archipelago may prevent adoption of an adaptive management approach that focuses on sustaining ecosystem function and incorporating natural variability in management plans. From a public environmental awareness perspective, a recreational fishery that depends on understanding the natural variability of fish populations (rather than a constant supply of stocked fish) may reinforce ecosystem management approaches. In a large survey of recreational fishing in Sweden by the National Board of Fisheries (Bengtsson et al. 2000), more than 50% of the respondents claimed that wilderness experience was a main motivation for fishing. In heterogeneous ecosystems such as archipelago systems with a number of small, distinct fish populations, the role of local ecological knowledge and monitoring is particularly important for sustaining a resilient social–ecological system (Johannes 1978; Berkes et al. 2003). Also, there is an obvious difference between the type of knowledge needed to use a resource and that needed to sustain it in the long term. The initiative to collaborate with new user groups, including sports fisher organizations, in Stockholm archipelago is a step towards reorganizing management and rebuilding local knowledge.

Salmonid management in the Baltic involves a number of different user groups and ecosystem components. As such, sustainable management of salmonids requires an integrated ecosystem approach. Lessons learned regarding the interdependencies between a particular resource, such as salmon and its supporting ecosystem, and the complex human society provide a good point of departure for understanding the challenges to resilient management of social–ecological systems.

Acknowledgments

The author is grateful for constructive comments on this manuscript provided by Karin Limburg and an anonymous reviewer. The study was funded by the Baltic Sea Foundation, Sweden.

References

Andersson, H. C. 2003. Fiskar och fiskare i Stockholms län – år 2002. Stockholm County Administrative Board, Report 2003:19, Stockholm. [In Swedish)]

Anon. 2006. Report of the scientific, technical and economic committee for fisheries. Review of scientific advice for 2007. Brussels, 14–16 June 2006. Advice on deepwater resources and stocks in the Baltic Sea. Commission of the European Communities, Brussels, Belgium.

Åqvist Almlöv, M., and M. Hammer. 2006. Changing use patterns, changing feedback links: implications for reorganization of coastal fisheries management in the Stockholm archipelago, Sweden. Ecology and Society 11(2):3.

Ask, L., and H. Westerberg, editors. 2006. Fiskbestånd och miljö i hav och sötvatten. Resurs och miljööversikt 2006. National Board of Fisheries, Gothenburg, Sweden. [In Swedish]

Bengtsson, B., R. Lundgren, and G. Johansson. 2000. Fiske 2000. En undersökning om svenskarnas sport- och husbehovsfiske. National Board of Fisheries, Gothenburg, Sweden. [In Swedish]

Berkes, F., J. Colding, and C. Folke, editors. 2003. Navigating social–ecological systems building resilience for complexity and change. Cambridge University Press, Cambridge, UK.

Bruckmeier, K., and E. Neuman. 2005. Local fisheries management at the Swedish coast: Biological and social preconditions. Ambio 34:91–100.

Carvalho, G. R., and L. Hauser. 1994. Molecular genetics and the stock concept in fisheries. Reviews in Fish Biology and Fisheries 4:326–350.

CBD (Convention on Biodiversity). 1998. Report of the workshop on the ecosystem approach: Lilongwe, Malawi. 26–28 January, 1998. UN Convention on Biodiversity, UNEP/COP/4/Inf. 9, Montreal.

CBD (Convention on Bioviersity). 2004. The ecosystem approach. Secretariat of the Convention on Biodiversity, Montreal.

Christensen, N., A. Bartuska, J. Brown, S. Carpenter, S. d'Antonio, R. Francis, J. Franklin, J. MacMahon, R. Noss, D. Parsons, C. Peterson, M. Turner, and R. Woodmansee. 2003. The report of the Ecological Society of America on the scientific basis for ecosystem management. Ecological Applications 6:665–691.

Dahl, E. 1973. Ecological range of Baltic and North Sea species. Oikos(Supplement 15):85–90.

Daily, G., editor. 1997. Nature´s services. Societal dependence on natural ecosystems. Island Press, Washington, D.C.

Dale, V. H., S. Brown, R. A. Haeuber, N. T. Hobbs, N. Huntley, R. J. Naiman, W. E. Ribesame, M. G. Turner, and T. J. Valone. 2000. Ecological principles and guidelines for managing the use of land. Ecological Applications 10:639–670.

Elmgren, R. 1989. Understanding human impact on the Baltic ecosystem: changing views in recent decades. Ambio 30:222–231.

Fiskeriverket (National Board of Fisheries). 2007. Regional och lokal samförvaltning av fiske. Fiskeriverket, Göteborg. [In Swedish]

Folke, C. 2006. Resilience: the emergence of a perspective for social-ecological systems analyses. Global Environmental Change 16:253–267.

Folke, C., M. Hammer, and A. M. Jansson. 1991. Life-support value of ecosystems: a case study of the Baltic Sea region. Ecological Economics 3:123–137.

Folke, C., L. Pritchard, Jr., F. Berkes, J. Colding, and U. Svedin. 1998. The problem of fit between ecosystems and institutions. International Human Dimensions Program on Global Environmental Change, IHDP Working Paper No. 2, Bonn, Germany.

Galaz. V. 2005. Social-ecological resilience and social conflict: Institutions and strategic adaptation in Swedish water management. Ambio 34: 567–572.

Hammer, M., C. M. Holmlund, and M. Åqvist Almlöv. 2003. Social-ecological feedback links for ecosystem management: a case study of fisheries in the central Baltic Sea archipelago. Ocean & Coastal Management 46:527–545.

Hammer, M., A. M. Jansson, and B-O. Jansson. 1993. Diversity, change, and sustainability implications for fisheries. Ambio 22:97–105.

Hanna, S., C. Folke, and K.-G. Mäler. 1996. Rights to nature ecological, economic, cultural and political principles of institutions for the environment. Island Press, Washington, D.C.

Holling, C. S. 1978. Adaptive environmental assessment and management. John Wiley & Son, New York.

Holmlund, C. 1996. Fish stocking: consistent with sustainable management of natural resources? Incentives and effects of fish stocking in the county of Stockholm. Master's thesis. Stockholm University, Stockholm, Sweden.

Holmlund, C. M., and M. Hammer. 1999. Ecosystem services generated by fish populations. Ecological Economics 29:253–268.

Holmlund, C. M., and M. Hammer. 2004. Effects of fish stocking on ecosystem services: an overview and case study using the Stockholm archipelago. Environmental Management 33:799–820.

Hughes, T. P., D. R. Bellwood, C. Folke, R. S. Steneck, and J. Wilson. 2005. New paradigms for supporting the resilience of marine ecosystems. Trends in Ecology & Evolution 20:380–386.

ICES (International Council for Exploration of the Sea). 2006. Report of the Baltic salmon and trout assessment working group (WGBAST), 28 March-6 April 2006. The International Council for the Exploration of the Sea, ICES CM 2006/ACFM:21, Copenhagen, Denmark.

ICES (International Council for the Exploration of the Sea). 2007. Report of the Baltic salmon and trout assessment working group (WGBAST). The International Council for the Exploration of the Sea, Copenhagen, Denmark.

Jansson, B-O, and A. M. Jansson. 2002. The Baltic Sea: reversibly unstable or irreversibly stable? Pages 71–110 *in* L. Gunderson and L. Pritchard, editors. Resilience and the behavior of large-scale ecosystems. Scope 60. Island Press, Washington, D.C.

Johannes, R. E. 1978. Traditional marine conservation methods in Oceania and their demise. Annual Review of Ecology and Systematics 9:349–364.

Jonson, M. 2002. Fisketuristiskt företagande I Sverige. Fisketurismföretagare. Forshaga. Sweden. [In Swedish]

Karlsson, L., and Ö Karlström. 1994 The Baltic salmon (*Salmo salar* L.): its history, present situation and future. Dana 10:61–85.

Karlström, O. 1999. Development of the M74 syndrome in wild populations of Baltic salmon (*Salmo salar*) in Swedish rivers. Ambio 28:82–86.

Kauppinen, T., A. Siira, and P. Suuronen. 2005. Temporal and regional patterns in seal-induced catch and gear damage in the coastal trap-net fishery in the northern Baltic Sea: effect of netting material on damage. Fisheries Research 73:99–109.

Koljonen, M. L., H. Jansson, T. Paaver, O. Vasin, and J. Koskiniemi. 1999. Phylogeographica lineages and differentiation pattern of Atlantic salmon (*Salmo salar*) in the Baltic Sea with management implications. Canadian Journal of Fisheries and Aquaculture 56:1766–1780.

Laikre, L., S. Palm, and N. Ryman. 2005. Genetic population structure of fishes: Implications for coastal zone management. Ambio 2:111–119.

Laikre, L., A. Palmé, M. Josefsson, F. Utter, and N. Ryman. 2006. Release of alien populations in Sweden. Ambio 35:255–261.

Larsson, U., R. Elmgren, and F. Wulff. 1985. Eutrophication and the Baltic Sea:causes and consequences. Ambio 14:9–14.

Lovén, S. and Y Ungsgård. 1999. Fiskevårdsplan för treårsperioden 1999–2001. Länsstyrelsen i Stockholms Län. Stockholm. [In Swedish]

Lunneryd, S., S. Königson, and N. Sjöberg. 2004. By-catch of seals, harbour porpoises and birds in Swedish commercial fisheries. Finfo 2004:8. [In Swedish with English summary]

Millennium Ecosystem Assessment. 2005. Ecosystems and human well-being: synthesis. Island Press, Washington D.C.

Naturvårdsverket & Fiskeriverket. 2005. Förutsättningar för fisketurismens utveckling i Sverige. Fiskeriverket (National Board of Fisheries), Governmental Report, Göteborg, Sweden. [In Swedish]

Nilsson, M. 1990. Fritidsfiske—90. Statistiska centralbyrån. Fiskerietyrelsen. Örebro, Sweden. [In Swedish]

Ojaveer, E., and H. Lehtonen, 2001. Fish stocks in

the Baltic Sea: finite or infinite resource? Ambio 30:217–221.

Österblom, H., S. Hansson, U. Larsson, O. Hjerne, F. Wulff. R. Elmgren, and C. Folke. 2007. Human-induced trophic cascades and ecological regime shifts in the Baltic Sea. Ecosystems 10:877–889.

Ostrom, E., T. Dietz, N. Dolsak, D. C. Stern, S. Stonich, and E. U. Weber, editors. 2002. The drama of the commons. National Academy Press, Washington D.C.

Peterson, E., and T. Järvi. 1999. Odling, domesticiering och bevarandebiologi hos laxfiskar. Pages 51–79 *in* Fiskeriverket rapport. The Swedish National Board of Fisheries, Drottningholm.

Pikitch, E. K., C. Santora, E. A. Babcock, A. Bakun, R., Bonfil, D. O. Conover, P. Dayton, D. Fluharty, B. Heneman, E. D. Houde, J. Link, P. A. Livingston, M. Mangel, M. K. McAllister, J. Pope, and K. J. Sainsbury. 2004. Ecology: ecosystem-based fishery management. Science 305:346–347.

Pinkerton, E., editor. 1989. Co-operative management of local fisheries. University of British Columbia Press, Vancouver.

Salmi, J., and P. Salmi. 2005. Governance of the Baltic salmon fishery: fishers' views. International Council for the Exploration of the Seas, CM2005/V:29, Denmark, Copenhagen.

Segerstråle, S. G. 1957. Baltic Sea. In J. W. Hedepeth, editor. Treatise on marine ecology and paleoecology, volume 1. Ecology. Geological Society of America Memoir 67:751–800.

SOU (Statens Offentliga Utredningar). 2003. Havet – tid för en ny strategi. Statens offentliga utredningar, Stockholm, Sweden. [In Swedish with English summary]

Ståhl, G. 1987. Genetic population structure of Atlantic salmon. Pages 121–140 *in* N. Ryman, and F. Utter. Population genetics and fishery management. Washington Sea Grant Program/University of Washington Press, Seattle.

Vuori, K. A., and M. Nikinmaa. 2007. M74 syndrome in Baltic salmon and the possible role of oxidative stresses in its development: present knowledge and perspectives for future studies. Ambio 36:168–172.

Waples, R. S. 1991. Genetic interactions between hatchery and wild salmonids: lessons from the Pacific Northwest. Canadian Journal of Fisheries and Aquatic Sciences 48:124–133.

Westerberg, H. S.-G. Lunneryd, M. Wahlberg, and A. Fjälling. 2008. Reconciling fisheries activities with the conservation of seals through the development of new fishery: a case study from the Baltic fishery—grey seal conflict. Pages 1281–1291 *in* J. Nielsen, J. J. Dodson, K. Friedland, T. R. Hamon, J. Musick, and Eric Verspoor, editors. Reconciling Fisheries with Conservation: Proceedings of the Fourth World Fisheries Congress. American Fisheries Society, Symposium 49, Bethesda, Maryland.

Wilson, D. C., J. R. Nielsen, and P. Degnbol, editors. 2003. The fisheries co-management experience: challenges and prospects. Kluwer Academic Publishers, Dortrecht, Netherlands.

Young, O. 2002. The institutional dimensions of environmental change: fit, interplay, and scale. MIT Press, Cambridge, Massachusetts.

American Fisheries Society Symposium 69:677–690, 2009

Socioeconomic and Biopolitical Linkages in the Management of Tropical Shads

Stephen J. M. Blaber*
CSIRO Marine & Atmospheric Research
Post Office Box 120, Cleveland, Queensland 4163, Australia

Abstract.—The five species of shads (Clupeidae: *Tenualosa*) that form the basis of important fisheries in the estuaries and coastal waters of tropical Asia are subject to a variety of management regimes in the countries of the region. The most widespread species, hilsa *T. ilisha*, is managed differently in Bangladesh, India, Burma, Pakistan, and Kuwait. The fishery for Reeves shad *T. reevesii* along the south China coast and far up the Yangtze, Pearl, and Qiantang rivers has collapsed and the species is being cultured, while Laotian shad *T. thibaudeaui*, in the lower and middle Mekong system, was a component of co-managed multispecies fisheries, but is now almost extinct. The fishery for toli *T. toli* in the estuaries and coastal areas of Sarawak (Malaysian Borneo) is now successfully co-managed, while fisheries for Terubuk (also known as longtail shad) *T. macrura*, in the coastal waters of Sumatra are currently unmanaged. The degree to which agencies in the various countries have been successful in managing shad fisheries for sustainability is analyzed in relation to the socioeconomic and governance environments in which they operate. The key to the future of tropical shad fisheries may be determined by the extent to which fishery management plans can be integrated into overall coastal or river basin management plans and is discussed in relation to population pressures and widespread environmental changes, as well as biopolitical and social linkages that could facilitate actions leading to concrete measures to re-establish migration pathways for the fish.

Introduction

Tropical shads of the genus *Tenualosa* comprise five species of which the most widespread and well studied is hilsa *T. ilisha*, found from north Sumatra in the east to Kuwait in the west, and is the basis of important fisheries in Bangladesh, India, Burma, Pakistan, and Kuwait (Whitehead 1985; Al-Baz 2001; Sann Aung 2001; Blaber et al. 2003a). The closely related Reeves shad *T. reevesii* occurs intermittently along the south China coast and far up the Yangtze, Pearl, and Qiantang rivers (Wang 2003), and Laotian shad *T. thibaudeaui*, which only lives in the lower and middle Mekong system, is believed to be close to extinction (Warren 2001). Once widespread, toli *T. toli* is now found only in the estuaries and adjacent coastal areas of Sarawak (north coast of Borneo)(Blaber et al. 1996), and Terubuk (also known as longtail shad) *T. macrura*, lives in the coastal waters of Sumatra and Borneo (Blaber et al. 1999, 2005). All species are of considerable fisheries significance, both from cultural and economic points of view (Chew 2001; Fegan 2001; Blaber et al. 2003b), and indeed, it has been the cultural importance of these fishes in all the areas where they are fished, particularly in Bangladesh, Indonesia, and Sarawak, that has given much extra impetus and urgency to recent research aimed at ensuring their long-term survival. Despite the increasing amount of research however, the various fisheries for tropical shads (known as terubuk in Indonesia, terubok in Malaysia, hilsa in India and Bangladesh, palla in Pakistan, and Pba mak pang in the Mekong) exist in the context of very different socioeconomic and political environments, and hence, the types of fisheries management vary greatly. In this paper, the degree to which agencies in the various countries have been successful in managing shad fisheries for

* Corresponding author: steve.blader@csiro.au

sustainability is analyzed in relation to the socioeconomic and governance environments in which they operate.

Fisheries and Conservation Status of *Tenualosa* Species

Although a wide variety of nets and traps are used for catching tropical shads, the most widely used are monofilament gill nets and these are employed in all countries where *Tenualosa* are caught. The proliferation of monofilament gill nets in the 1970s, with their higher capture efficiency (Lilygreen and Meade 1982; Maki et al. 2006), has been the single most significant factor influencing the increase in catch rates of tropical shads and probably more latterly their decline. All *Tenualosa* have migrations associated with spawning and live mainly within the bounds of estuarine salinities or freshwater (Blaber et al. 2003b). Two species are the only clupeids known to change sex, and most species are fast-growing and short-lived (2–3 years). The relationships between life history strategies and how the various *Tenualosa* have withstood fishing pressure are discussed by Blaber et al. (2003b), and the predictability of the spawning behavior of shads is believed to be a major reason for their vulnerability to overfishing (Waldman 2003). The characteristics of *Tenualosa* fisheries in each country are summarized in Table 1.

Hilsa

Hilsa is the basis of very large fisheries in Myanmar, Bangladesh and India, totaling well over 200,000 metric tons per annum, mostly from the Bay of Bengal and its rivers and estuaries (Table 1). Until recently, populations appeared relatively secure, despite intense fishing pressure over most of its range. In this regard, it is similar to the very ecologically adaptable tropical shad of west Africa, bonga *Ethmalosa fimbriata*, which supports fisheries totaling about 150,000 metric tons per annum (Charles-Dominique and Albaret 2003). However, environmental changes as well as fishing pressure are now having a very significant impact on hilsa. As in temperate regions (Larinier 2001; Limburg et al. 2003), the building of dams has had a major impact (Pandey et al. 1996; FAO 2003). The construction of many dams on the Indus River in Pakistan have prevented upstream migrations of hilsa and resulted in a massive decline in catches throughout the Indus River and its delta (Ahmad 1997; Figure 1). The Farraka Barrage built on the Indian side of the border where part of the Ganges flows into Bangladesh has caused the Padma River in Bangladesh to nearly dry out for parts of the year, and hilsa no longer occur in the area (Pandey et al. 1996; Rahman and Haldar 1998), although, more recently, controlled releases of water from the barrage have improved the fishery in the Hooghly estuary on the Indian side of the border (Sinha and Khan 2001). Similarly, the canalization and damming of the Brahmaputra River in Bangladesh have prevented the upstream movement of hilsa into areas where they were previously common, and in the upper Brahmaputra in northeast India, the declines have been exacerbated by indiscriminate fishing, pollution, and heavy siltation (Sarma and Dutta 2000). As riverine landings of hilsa have decreased in Bangladesh and India, it has only been possible to maintain the high total landings by increasing catches from the marine waters of the Bay of Bengal.

The hilsa fisheries of Kuwait, Iraq, and Iran in the Arabian Gulf have also shown significant declines (Ye and Almatar 2003; FAO 2004; Figures 2 and 3), primarily as a result of environmental changes caused by a reduction in the spawning grounds in the Shatt el Arab waterway caused by the draining of 90% of the Iraqi marshlands prior to 2001 (Lawler 2005). The ongoing restoration of the marshlands and the reestablishment of the flow patterns in the Shatt el Arab are now vital to the future of the fishery (Mohamed et al. 2009).

Toli

The large declines in fisheries for toli can be attributed to a combination of overfishing and environmental degradation, and the following summary is taken from Blaber et al. (2003b). Until well into the 20th century, this species was widespread from Borneo westwards to Burma. Early literature indicates that it formed important fisheries in the Straits of Malacca for at least 500 years, only disappearing in the 1970s. This fishery appears to have steadily contracted but was still important in the 1930s, especially in the Rokan estuary and adjacent coastal area (Hardenberg 1934). Important fisheries also existed on the Malayan side of the straits but, by

Table 1.—Characteristics of tropical shad (*Tenualosa*) fisheries by country.

Species	Country	Fishery	Catch in metric tons (year)	Management	Status	Causes of declines if any	References
Hilsa	Kuwait	Gill nets	135 (2004)	Public Authority for Agriculture & Fisheries Resources—series of strong fishery regulations.	Large declines	Changes to Shatt el Arab waterways in Iraq	Al-Baz (2001); Ye and Almatar (2003)
	Iraq	Gill nets	<2,000 (2001)	Ministry of Agriculture, Fisheries Department regulations. Restoration of marshland spawning areas in progress.	Declining	Changes to Shatt el Arab waterways, dams, and massive pollution	Mohamed et al. (2001); FAO (2004)
	Iran	Traps, weirs, and gill nets in southern waters	400 (1997)	Iranian Fisheries Research Organization (IFRO) conducts research activities and is responsible for providing data and information for fisheries management and development (DBO).	Probably secure but heavily exploited	Overfishing and habitat changes, siltation of rivers	Coad et al. (2003); Marammazi et al. (2004)
	Pakistan	Gill net	200 (2004)	Ministry of Food, Agriculture & Livestock, Marine Fisheries Department plus Provincial Directorates of Fisheries.	Massive declines	Dams and overfishing of Indus	FAO (2003)
	India	Inland and marine gill nets	37,897 (31,765 in Orissa) (2003)	Union, state, and local government jurisdictions plus community. Many different local regulations.	Declining	Overfishing and habitat changes	Srinath et al. (2003); CMFRI (2004)
	Bangladesh	Inland and marine gill nets	255,000 (2004)	Multiple government agencies and community involvement.	Secure but signs of overfishing	Overfishing, including juveniles, dams, and habitat changes	Blaber et al. (2003a)
	Myanmar	Local and foreign purse seine and gill nets	42,000 (2000)	Department of Fisheries. Ban on hilsa fishing in rivers in spawning months of June, July, and August. Little enforcement.	Declining	Overfishing	Sann Aung (2001)
	Indonesia	Gill nets	<10 (2003)	Not separately managed.	Close to extinction	Unknown	
Toli	Malaysia (East)—Sarawak	Estuarine and coastal gill-net fishery	<100	Comanagement, state government, and fisheries.	Massive declines, now stabilized	Habitat loss and overfishing of spawning females following management, now secure?	Blaber et al. (1996, 2003b)

Table 1.—Continued.

Species	Country	Fishery	Catch in metric tons (year)	Management	Status	Causes of declines if any	References
Terubuk	Malaysia (East)—Sarawak	Coastal gill-net fishery	<1,000	Not managed separate from other coastal fishes.	Federal jurisdiction	Secure	Blaber et al. (2005)
	Indonesia	Gill-net fishery in Bengkalis region	<1,000	No effective management but within jurisdiction of local government (Riau Province and Bengkalis regency) and traditional fishers.	Declining	Pollution, habitat loss, and overfishing of spawning females	Brewer et al. (2001)
Reeves shad	China	Fishery closed since 1987	Close to zero	Chinese Ministery of Agriculture ban on harvest since 1987. Intensive aquaculture.	Almost extinct in wild	Habitat loss and overfishing	Wang (2003)
Laos shad	Laos	No separate fishery since 1980s	<1	Government and Mekong River Commission—effectively unmanaged.	Globally threatened	Habitat loss, dam construction, overfishing (introduction of monofilament gill nets)	Warren (2001)
	Cambodia	Bagnet fishery	<1	Government and Mekong River Commission—now part of community management program.	Globally threatened	Habitat loss and dam construction plus overfishing (introduction of monofilament gill nets)	Warren (2001); Rattner (2006)
	Thailand	No specific fishery	<1	Government and Mekong River Commission—effectively unmanaged.	Globally threatened	Habitat loss and dam construction plus overfishing (introduction of monofilament gill nets)	Warren (2001)
	Vietnam	No specific fishery	<1	Government and Mekong River Commission—effectively unmanaged.	Globally threatened	Habitat loss and dam construction plus overfishing (introduction of monofilament gill nets)	Warren (2001)

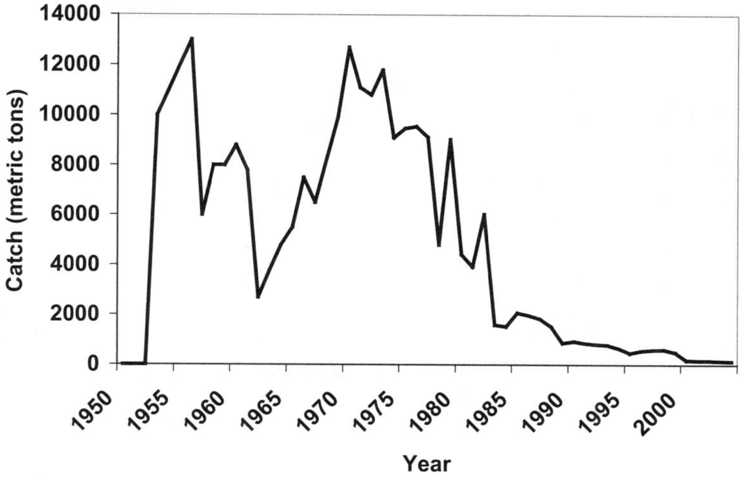

Figure 1.—Annual catches (1952–2004) of hilsa in Pakistan based on data from FAO statistical records (http://fao.org/figis/).

the 1920s, had been severely impacted by siltation of the estuaries and rivers caused by tin mining (Maxwell 1921). The impacts of this pollution are graphically described by Maxwell (1921): "These fish used to be taken in such numbers that the nets contained more than the boats could load. Within the last few years, the writer has, on several occasions, picked up these fish by hand in a dying condition apparently choked by silt in their attempt to ascend the rivers." It is also likely that in a similar way, the decline of toli in Borneo has at least been partially caused by the very extensive logging and agricultural development that has increased siltation in the rivers. Today, toli has completely disappeared from Burma where a large fishery existed prior to 1964 (S. Aung, Academy of Fisheries, Myanmar, personal communication) and is no longer found in the Straits of Malacca, now being restricted to the coast of Sarawak, where catches total less than 50 metric tons per annum.

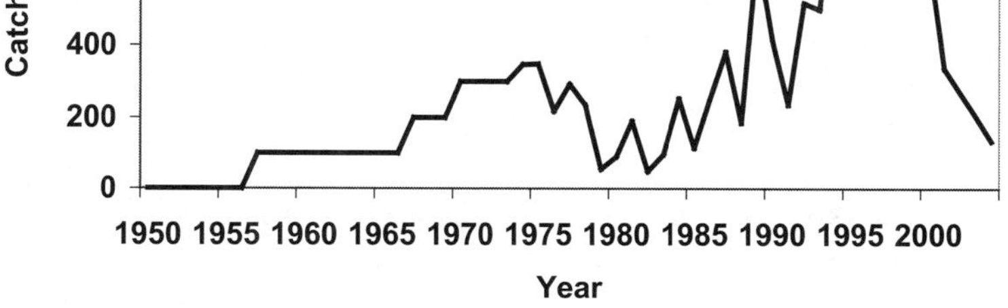

Figure 2.—Annual catches (1952–2004) of hilsa in Kuwait based on data from FAO statistical records (http://fao.org/figis/).

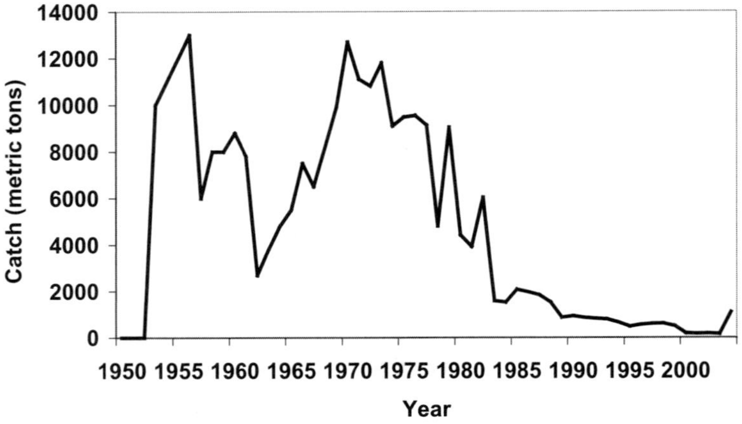

Figure 3.—Annual catches (1952–2004) of hilsa in Iraq based on data from FAO statistical records (http://fao.org/figis).

Terubuk

The contraction of the geographical range of Terubuk to the Sarawak coast and a small part of Sumatra and the large decline in its fishery mirror what has happened to three other species in the genus. The decline in catches of Terubuk in Sumatra is a primary result of fishers intensively targeting aggregations of spawning females, for most of these are caught before they have spawned (Brewer et al. 2001). Second, the overall deterioration of the estuarine and coastal environment caused by large-scale logging and oil industry developments (Adriman 1997; Feliatra 1997) may have influenced Terubuk abundance; for example, it has been shown that Terubuk in the sampling areas ingested large quantities of sawdust (Brewer et al. 2001).

The lunar periodicity of catches of Terubuk in Sarawak is apparently tied to the lunar spawning cycle (Blaber et al. 2005), as described for Terubuk in Sumatra (Blaber et al. 1999). However, although Terubuk in Sumatra and Sarawak both have a lunar spawning periodicity, those in Sumatra spawn throughout the year, whereas those in Sarawak have a definite seasonal cycle with a peak at the height of the northeast monsoon. From a fisheries management perspective, it is important to note that the Sarawak artisanal gill-net fishery largely ceases operations in January and February due to the adverse monsoon weather conditions, hence reducing fishing pressure on spawning Terubuk. For this reason, the management measures, including a closed season during July and August, recently introduced in Sarawak for toli may not be required for Terubuk. Terubuk populations in Sarawak appear to have withstood fishing pressure better than toli, perhaps because of the weather-driven closed season (Blaber et al. 2005). In contrast, toli spawn inside only a few Sarawak estuaries from May to November and the spawning aggregations are subject to heavy fishing pressure throughout the spawning period. The fishery for Terubuk in the sheltered waters of Bengkalis in Sumatra is less subject to adverse weather, and fishing occurs in all months. Catches have declined drastically in recent years (Blaber et al. 1999; Brewer et al. 2001), leading to a recommendation for a closed season.

Laotian Shad

The Pba Mak Pang, as it is known locally, was once one of the most abundant fish in the Mekong. Its migrations were "extensive and impressive and only exceeded in magnitude by a few other species" (Roberts 1993). It has undergone a widespread and dramatic decline in recent decades and has almost completely disappeared from the Thai and Laotian sections of the Mekong. It is now only caught in small numbers in the Mekong in Cambodia and in the Cambodian Great Lake (Tonle Sap), as well as parts of the Mekong delta in Vietnam. Effectively, there is now no separate fishery for Laotian shad and it is listed by the IUCN (International Union for Conservation of Nature) as globally threatened. Its decline has probably been due to a combination of a deterioration in the river ecosystem caused by deforestation and siltation, the introduction of monofilament gill nets in the 1960s, and the building of irrigation and hydropower dams (Warren 2001).

Reeves Shad

Reeves shad was an important commercial species in the Yangtze River system of China before the 1960s when annual catches were as high as 1,500 metric tons. However, by the 1980s, catches had declined to almost zero and the fishery was closed in 1987. In the Pearl River, the stock similarly declined and only 200 kg were taken in 1999 (Wang 2003). Currently, there is no separate fishery for Reeves shad. The reasons for the decline probably relate to habitat exclusion and changes as well as overfishing. The Chinese Ministry of Agriculture issued a ban on harvest of Reeves shad in 1987 and began research into aquaculture and induced breeding of the species (Wang 2003).

Factors Affecting Management of Shads and Their Success or Otherwise

Key biological attributes of *Tenualosa* that are particularly relevant to management and fisheries are their diadromy, their spawning behavior, and, in some species, their protandrous reproductive system (Blaber et al. 1996, 1999). In the socioeconomic context, most species occur in large estuary–river systems in developing countries, where large human populations are dependent on fish as a main source of protein. Although the role of socioeconomics and governance in fisheries management has received increasing attention of late (Wilson et al. 2003; Gray 2005), the issues and priorities in developing countries, such as food security, are usually somewhat different from those described for developed countries (Symes 2006). Another sociopolitical linkage aspect that has impeded management is collection of catch data and the accuracy of such data. This is particularly relevant in the large Bangladesh fishery for hilsa, where FAO (Food and Agriculture Organization of the United Nations) records suggest a steady increase in catches to beyond 200,000 metric tons (Figure 4), but anecdotal and market data suggest a declining catch. According to Fegan (2001), official catch data are based on calculated guesswork. Unfortunately, the data collection systems are inadequate and neither catch nor effort data are reliable enough for accurate scientific management planning. Catch data are not routinely collected from boats, landings, or markets, and changes in transport of fish make rail shipments, used as a proxy for catch data up to the 1960s, unusable. Additionally, collection of data on number of boats is spread over several agencies and probably severely underestimates effort (Blaber 2000).

Diadromy

The largest tropical shad fishery, that of hilsa in Bangladesh, exemplifies the fisheries management

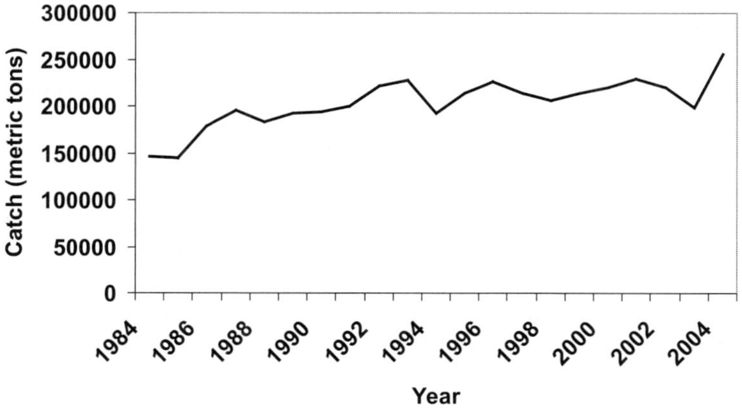

Figure 4.—Annual catches (1984–2004) of hilsa in Bangladesh based on data from FAO statistical records (http://fao.org/figis).

problems related to the roles of multiple government agencies together with long-standing socioeconomic and cultural traditions (Siddique 2008). The hilsa fishery consists of two main sectors: a marine open-water fishery in the Bay of Bengal and a riverine fishery throughout the large Meghna and Brahmaputra systems. The Department of Fisheries has jurisdiction over capture fisheries, but in reality, it is the Department of Land Revenue that collects taxes (via private tax agents) on fishing boats and fishing gears, the Department of Forestry that collects revenues and controls access to the extensive Sundarbans area where large amounts of hilsa are caught, and the Marine Department that has legislative control over boats in the Bay of Bengal sector. Finally, catch, effort, and marketing of hilsa is still very traditional and is run by auctioneers and middlemen (aratdars) via provision of credit to fishers. These fish traders are de facto controlling the sector for hilsa and are politically influential and generally resist any moves to implement management that might restrict their income by reducing catches (Fegan 2001). This transfer of responsibility for certain government functions to individuals means that about 10,000 fishing spots in Bangladesh's inland open-water fishery are allocated by leasing in an open auction, apparently a continuation of the revenue-generating system from colonial times. Those who purchase the leases in the auction can then fish the area and manage the fishery and establish their own security force that can close access to other potential fishers. The leaseholders do not fish the area themselves, but sell fishing rights to others, and hence the main beneficiaries of the fishery are not the fishers themselves but the middlemen. The system perpetuates itself because the real fishers (who are typically poor) owe money to middlemen and can never get out of a cycle of debt; they have little chance of buying fishing rights directly from the government (Pomeroy and Berkes 1997). As Khan (2004) states "coordinated management is essential for addressing the common issues in a cluster of waterbodies because resources, such as fish, are highly mobile and the complexity of different property rights and diversity of users has had a cumulative adverse effect." Recent government attempts to implement the New Fisheries Management Policy adopted in the late 1980s have not been successful, mainly due to legislative and budgetary constraints.

The effective management of the hilsa fisheries of the Arabian Gulf has been severely impacted by human conflicts. Kuwait, Iraq, and Iran have government agencies with laudable fisheries management strategies, but their attempts to manage for sustainability have been overwhelmed, first by the vast oil discharges in Kuwait in 1991, which harmed much of the coastal zone of the gulf, and second by the massive changes to the Shatt Al-Arab waterway where the previous Iraqi government drained the marshes, causing freshwater discharge to decline by more than 50%, and diverted rivers (Ye and Almatar 2003). This loss of the Euphrates delta habitats led to declines in diadromous species, particularly hilsa, whose juveniles were dependent on the 20,000 km2 of rivers and marshlands. The future health of hilsa populations in the northern Arabian Gulf is probably dependent on the ongoing restoration of the Shatt Al-Arab and the Iraqi wetlands (Mohamed et al. 2009).

The large-scale migrations of hilsa that took place in the Indus River of Pakistan have virtually ceased, and the fisheries collapsed due to the succession of dams along the river. That this situation occurred was not helped by the separation of marine and inland fisheries management in Pakistan, and the apparent lack of any integrated planning in relation to dam building (Ahmad 1997).

Food Security

Fish form 60% of the protein intake of the people of Bangladesh, and hilsa forms 30% of the total fish production (FAO 1987). The species is vital to the food security of millions of people both in Bangladesh and other countries of the Bay of Bengal. Quite apart from the complexity of any management and jurisdiction issues, the traditional role of hilsa in the culture of many of the peoples of the region is significant. Any management measures that do not take account of the cultural traditions and norms around hilsa fisheries are hence doomed to failure. For example, there have been numerous calls by researchers and managers to restrict the catching of juvenile hilsa in order to conserve stocks (Chong and Hossain 1997; Sarma and Dutta 2000; Haldar et al. 2001; Siddique 2009), but the fisheries for the juveniles are carried out largely by women and children who are among the most disadvantaged of the poor and are dependent on catching large numbers of these small fish for their livelihoods.

The people of the Mekong basin are similarly dependent on fish resources (Baran et al. 2003). Fish from capture fisheries constitute the dominant source of protein (75%) in the region. The populations of the countries of the lower Mekong basin have one of the highest rates of fish consumption in the world, with an average of 33.4 kg/person/year (Baran and Baird 2001), compared to the world average that is around 16 kg/person/year (FAO 2000). For instance, in Cambodia, fish constitute up to 65–75% of total protein in the diet (Guttman 1999) and access to this common property resource is particularly important for rural poor and landless people (Baran et al. 2003).

The Role of Traditional Management and Comanagement

The problems faced by fishing communities are not necessarily a result of the absence of management institutions, but rather of the inadequacy of those institutions to deal with recent developments (Viswanathan et al. 2003). There are countless examples of ways in which the regulation and access to fisheries were controlled before modern concepts of mainly top-down fisheries management were developed (Wilson et al. 2003). The involvement of the fishers themselves in management together with government—usually referred to as comanagement—is often seen as a solution to the failure of the top-down approach, but success has been variable. Viswanathan et al. (2003) define comanagement as "an arrangement where management responsibility is shared between the government and fishing communities and can be viewed as a set of institutional and organizational arrangements that define the cooperation between the fisheries managers and fishing communities."

Three *Tenualosa* species have been or are subject to various degrees of comanagement. Up until at least the 1940s, the Terubuk fishery in the Bengkalis area of Sumatra was managed in a traditional way by the regency and it was the regent who determined when the fishing season opened and closed and who had access to the fishery. Elaborate ceremonies surrounded the opening of the season in Bengkalis, which is still known as Kota Terubok (Terubuk City). The Terubuk fishery in the Bengkalis Strait historically supported a large fishery from at least the mid-19th century. A Dutch source (Gramberg 1880) records a thriving fishery at that time in Bengkalis, with 14–15 million fish a year taken in 1823, when a small boat could net 1,000–2,000 fish. The salted eggs were a delicacy valued like caviar. Salted eggs and salted fish were important items of commerce, collected by Chinese traders for the Singapore market. In 1876, there were 500 small fishing boats. The fishery continued to give good catches into the 1950s, but catches collapsed around the 1970s (Ahmad et al. 1995). Boats using much longer and deeper gill nets were taking only from 1 to 19 Terubuk per overnight trip in 1998 (Brewer et al. 2001) in the spawning aggregation areas. The collapse of the fishery has affected fishers and other stakeholders in the fishery and caused suspension of an important annual ceremony in Terubuk City (Bengkalis) that had some tourist potential. Following community concern and a coordinated research program (Blaber et al. 1999), steps are now being taken to try and manage the fishery at a local level. Comanagement is made more complex because many Terubuk boat captains are not free to choose what fish to target. The majority of boats that dock in Bengkalis are owned by tauke (fish traders) , who have fitted them with gear specialized for Terubuk. Most captains receive credit from tauke for fuel, ice, and crew food, sometimes for engine and gear repairs/replacements. In return, they are required to target Terubuk at spring tides and to sell them through the lender. In order to change the behavior of fishers, management must identify the taukes who control effort at each landing and include them in consultations.

The situation with regard to terubok in the estuaries of Sarawak (Malaysian Borneo) was somewhat similar. Fishers target the spawning females because of the high value of the roe, and by the early 1990s, catch levels had fallen drastically due to overfishing and environmental degradation. The cultural value of the terubok in Sarawak led the state government to institute a research and management program from 1993 onwards, with the express aims of conserving the species and sustaining the fishery. Once the few spawning sites had been identified (Blaber et al. 1996), the government moved quickly into a phase of consultation with the terubok fishers to institute a closed season, limit entry to the fishery (no new nontraditional fishers), and protect the areas adjacent to the spawning sites by declaring new national parks. The success of this strategy was due, first, to strong commitment from the highest levels of government and, second, to the successful involvement of fishers in decision making and their acceptance of the necessity for new controls. An additional important factor helping to sustain the fishery is that with increasing affluence and education, fewer young people are entering the fishing industry in Malaysia and the number of terubok fishers is declining.

The fate of Laotian shad in the Mekong, where it still survives in small numbers and forms a tiny part of the multispecies fishery, now rests largely with the outcomes of the new fisheries policy in Cambodia. The Cambodian government introduced a dramatic fisheries reform in 2001 that reduced the allocation of commercial fishing lots (defined fishing areas that could be owned or leased) in favor of local community access. Unfortunately, during the implementation of the policy, the freeing up of fishing lots and the recall of fisheries officers led to a management vacuum and effectively opened the fisheries to all, leading to intense exploitation and increased illegal fishing. Although the commercial lot system had been inequitable, the lot owners managed complex local systems of social organization complete with subleasing, user fees, and enforcement. The release of lots was intended to benefit the small-scale fishers and the poor, but the legal framework to support community-based management was absent (Rattner 2006). According to Rattner (2006), the implementing regulations to define the status, rights, and responsibilities of community fisheries organizations have still not been agreed on, and although it was possible to open fisheries to community access by decree, the work of building community-based management is proving a much greater challenge. Currently, the absence of good governance not only threatens the reforms, but also the sustainability of the fishery.

Is Integrated Management the Solution?

Despite successes in South Pacific and other small island states, community-based fisheries management has not been notably successful in South and Southeast Asia due to traditions of common property resources, social complexities, and variable government commitment (Crawford et al. 2006). There are very real difficulties in meshing community-based fisheries management with integrated coastal zone or floodplain management. For example, although many small-scale or traditional fishers have the detailed knowledge necessary to successfully manage and sustain their fisheries, a very careful (and sympathetic) government involvement is required if their ideas, and the needs of the fishery, are to be incorporated into planning on a broader scale.

Although all the tropical shad fisheries have been seriously affected by overfishing, there is no doubt that a suite of environmental changes have also negatively impacted their populations. The building of dams, logging of river catchments, and industrial and agricultural pollution have all played their part in reducing tropical shad numbers. Integrated planning would obviously have been beneficial in mitigating some of these effects, but whether maintenance of shad populations would have received priority over, for example, irrigation for agriculture (particularly rice) or industrial development, is questionable.

The sensitivity of a number of shad species, including hilsa, to changes in fishing mortality was analyzed by Boreman and Friedland (2003). They showed that given the life history characteristics of the species, relaxing fishing mortality would have relatively little effect on abundance, whereas such factors as habitat availability and unobstructed passage through river systems are paramount. This finding would apply particularly to hilsa, which has a rapid growth rate, early maturity, and high fecundity and is hence able to withstand relatively heavy fishing pressure. However, the probably unsustainable increase in fishing pressure on marine populations of hilsa in the Bay of Bengal, noted earlier, is significant in this context because it is increasing the mortality on the adult stage.

Although better fisheries management, or at least the idea of it, is becoming more widespread in the tropics, the outlook for most *Tenualosa* species is not very promising because of environmental changes. While the effects of, for example, logging, on catchments may slowly disappear if suitable mitigation measures, reafforestation, and stabilization take place, few dams are likely to be removed and their effects on migrations are drastic. The migratory nature of shads necessitates an integrated and usually intergovernmental approach. In some areas, the institutional structures needed to facilitate this exist; for example, the Mekong River Commission (MRC) coordinates many industrial, agricultural, and fisheries projects between Vietnam, Cambodia, Thailand, and Laos in the Mekong basin. Unfortunately, to date, dam building, hydroelectric power generation, flood mitigation, and irrigation have received priority, despite the strong emphasis of the MRC's Fisheries Programme on implementation, uptake, and impact of fisheries information into planning and development decisions in the basin (MRC 2006). Coordination is also complicated by the building of hydropower dams on the upstream part of the Mekong in China, which is not a member of the MRC (Sneddon and Fox 2006).

The science to underpin the actions needed to restore tropical diadromous populations is relatively well understood (Arthington et al. 2004), and much effort has been expended on modeling support systems for management (Baran 2002; Baran et al. 2003). However, the future of most tropical shad populations is going to depend not only on understanding the effects of dam building on river flow and fishes, but on building the necessary biopolitical and social linkages that will influence decision makers and facilitate actions (such as constructing fishways at dams) that will help to reestablish migration pathways for the fish. The effectiveness of such linkages was demonstrated by the success of the toli restoration program in Sarawak, which ultimately depended on political decisions.

Acknowledgments

I am very grateful to the American Fisheries Society and Ron Klauda in particular, for encouraging me to put this paper together, and to Alan Whitfield for his comments and suggestions. Many thanks to Eric Rochard and two anonymous reviewers for their helpful comments on the manuscript. Much of my research on shads was funded by the Australian Centre for International Agricultural Research as part of their fisheries program and by the Sarawak, Indonesian, and Bangladesh governments.

References

Adriman. 1997. The quality of Dumai coastal waters based on physico-chemical characteristics and community structure of macrozoobenthos (In Indonesian). Terubuk (Jurnal Perikanan dan Kelautan, Universitas Riau, Indonesia) 67:93–107.

Ahmad, M. 1997. Natural and human threats to biodiversity in the marine ecosystem of coastal Pakistan. Pages 319–332 *in* B. U. Haq, S. M. Haq, G. Kullenberg, and J. H. Stel, editors. Coastal zone management imperative for maritime developing nations. Kluwer, Dordrecht, Netherlands.

Ahmad, M., T. Dahril, and D. Efizon. 1995. Ecology and reproduction of Terubuk (*Alosa toli*) in the Bengkalis region, Riau. [In Indonesian]. Jurnal Perikanan dan Kelautan, Universitas Riau, Indonesia 1:2–19.

Al-Baz, A. 2001. The biology of *Tenualosa* ilisha in Kuwait and the status of the fishery. Pages 18–24 *in* S. J. M. Blaber, D. Brewer, D. Milton, and C. Baiano, editors. Proceedings of the International Terubok Conference, Sarawak, Malaysia. Sarawak Development Institute, Kuching, Sarawak, Malaysia.

Arthington, A. H., K. Lorenzen, B. J. Pusey, R. Abell, A. S. Halls, K. O. Winemiller, D. A. Arrington, and E. Baran. 2004. River fisheries: ecological basis for management and conservation. Pages 21–60 *in* R. L. Welcomme and T. Petr, editors. Proceedings of the Second International Symposium on the Management of Large Rivers for Fisheries, volume 1. Food and Agriculture Organization of the United Nations, Bangkok.

Baran E. 2002. Modelling the Mekong fisheries: what can be done? Pages 70–74 *in* S. Sverdrup-Jensen, editor. Fisheries in the lower Mekong basin: status and perspectives. Mekong River Commission, Phnom Penh, Cambodia.

Baran E., and I. Baird. 2001. Approaches and tools for sustainable management of fish resources in the Mekong River basin. Contribution to the International Symposium on Biodiversity Management and Sustainable Development in the Lancang-Mekong River basin. 4–7 December 2001, Xishuangbanna, Yunnan, China. WorldFish Center, Penang, Malaysia.

Baran E., I. Makin, and I. G. Baird. 2003. BayFish: a model of environmental factors driving fish production in the lower Mekong basin. A contribution to the Second International Symposium on Large Rivers for Fisheries. Phnom Penh, Cambodia, 11–14 February 2003. Worldfish Center, Penang, Malaysia.

Blaber, S. J. M. 2000. Tropical estuarine fishes: ecology, exploitation and conservation. Blackwell Scientific Publications, Oxford, UK.

Blaber, S. J. M., D. T. Brewer, D. A. Milton, G. S. Merta, D. Efizon, G. Fry, and T. Van der Velde. 1999. The life history of the protandrous tropical shad *Tenualosa* macrura (Alosinae: Clupeidae): fishery implications. Estuarine Coastal and Shelf Science 49:689–701.

Blaber, S. J. M., G. Fry, D. A. Milton, T. Van der Velde, J. Pang, Ong Boon-Teck, and P. Wong. 2005. The life history of *Tenualosa* macrura in Sarawak, further notes on protandry in the genus and management strategies. Fisheries Management and Ecology 12:1–10.

Blaber, S. J. M., D. A. Milton, D. T. Brewer, and J. P. Salini. 2003b. Biology, fisheries, and status of tropical shads *Tenualosa* spp. in south and Southeast Asia. Pages 49–58 *in* K. E. Limburg and J. R. Waldman, editors. Biodiversity, status, and conservation of the world's shads. American Fisheries Society, Symposium 35, Bethesda, Maryland.

Blaber, S. J. M., D. A. Milton, S. R. Chenery, and G. Fry. 2003a. New insights into the life history of *Tenualosa* ilisha and fishery implications. Pages 223–240 *in* K. E. Limburg and J. R. Waldman, editors. Biodiversity, status, and conservation of the world's shads. American Fisheries Society, Symposium 35, Bethesda, Maryland.

Blaber, S. J. M., D. A. Milton, J. Pang, P. Wong, B. Ong, L. Nyigo, and D. Lubim. 1996. The biology of the tropical shad *Tenualosa* toli from Sarawak: first evidence of protandry in the Clupeiformes? Environmental Biology of Fishes 46:225–242.

Boreman, J., and K. D. Friedland. 2003. Sensitivity of American shad to changes in fishing mortality. Pages 267–273 *in* K. E. Limburg and J. R. Waldman, editors. Biodiversity, status, and conservation of the world's shads. American Fisheries Society, Symposium 35, Bethesda, Maryland.

Brewer, D. T., S. J. M. Blaber, G. Fry, G. Sedana Merta, and D. Efizon. 2001. Sawdust ingestion by the tropical shad *Tenualosa* macrura (Teleostei: Clupeidae): implications for conservation and fisheries. Biological Conservation 97:239–249.

Charles-Dominique, E., and J.-J. Albaret. 2003. African shads, with emphasis on the West African shad Ethmalosa fimbriata. Pages 27–48 *in* K. E. Limburg and J. R. Waldman, editors. Biodiversity, status, and conservation of the world's shads. American Fisheries Society, Symposium 35, Bethesda, Maryland.

Chew, D. 2001. Socio-economics of the Terubok fishery in Sarawak. Pages 180–187 *in* S. J. M. Blaber, D. Brewer, D. Milton, and C. Baino, editors. Proceedings of the International Terubok Conference, Sarawak, Malaysia. Sarawak Development Institute, Kuching, Sarawak, Malaysia.

Chong, K-C., and M. Hossain. 1997. Toward sustainable fisheries in Bangladesh. Bay of Bengal News 2:14–16.

CMFRI (Central Marine Fisheries Research Institute). 2004. Estimated marine fish landings (in tonnes) in India during 2002 and 2003. Central Marine Fisheries Research Institute, Marine Fisheries Information Service Technical and Extension Series Number 179, Cochin, India.

Coad, B. W., N. A. Hussein, T.S. Ali, and K. E. Limburg. 2003. Middle Eastern shads. Pages 59–67 *in* K. E. Limburg and J. R. Waldman, editors. Biodiversity, status, and conservation of the world's shads. American Fisheries Society, Symposium 35, Bethesda, Maryland.

Crawford, B., M. Kasmidi, F. Korompis, and R. B. Pollnac. 2006. Factors affecting progress in establishing community-based marine protected areas in Indonesia. Coastal Management 34:39–64.

FAO (Food and Agriculture Organization of the United Nations). 1987. Hilsa investigations in Bangladesh. Food and Agriculture Organization of the United Nations, Bay of Bengal Programme, Report 36, Madras, India.

FAO (Food and Agriculture Organization of the United Nations). 2000. The state of world fisheries and aquaculture. Food and Agriculture Organization of the United Nations, Rome.

FAO (Food and Agriculture Organization of the United Nations). 2003. Information on fisheries management in the Islamic Republic of Pakistan. Available: www.fao.org/fi/fcp/en/PAK/body.htm (September 2006).

FAO (Food and Agriculture Organization of the United Nations). 2004. Information on fisheries management in Iraq. Available: www.fao.org/fi/fcp/en/IRQ/body.htm (September 2006).

Fegan, B. F. 2001. Socioeconomics of the Hilsa fishery in Bangladesh. Pages 188–191 *in* S. J. M. Blaber, D. T. Brewer, D. A. Milton D. A., and C. Baiano, editors. Proceedings of the International Terubok Conference, Sarawak, Malaysia. Sarawak Development Institute, Kuching, Sarawak, Malaysia.

Feliatra. 1997. Environmental parameters affecting the degradation of oil products in the Dumai region. [In Indonesian]. Terubuk (Jurnal Perikanan dan Kelautan, Universitas Riau, Indonesia) 69:2–14.

Gramberg, J. S. G. 1880. De visscherij en bezwering van troeboek. Indische Gids (Die Nederlands), tweede jaargang II:331–346 .

Gray, T. 2005. Participation in fisheries governance. Springer-Verlag, Dordrecht, Netherlands.

Guttman, H. 1999. Rice fields fisheries: a resource for Cambodia. NAGA 22:11–15.

Haldar G. C., M. A. Mazid, M. A. Rahman, and S. M. N. Amin. 2001. The present status of the hilsa (*Tenualosa* ilisha) fishery in Bangladesh. Pages 52–64 in S. J. M. Blaber, D. Brewer, D. Milton, and C. Baino, editors. Proceedings of the International Terubok Conference, Sarawak, Malaysia. Sarawak Development Institute, Kuching, Sarawak, Malaysia.

Hardenberg, J. D. F. 1934. The fish fauna of the Rokan mouth. Treubia 33:81–168.

Khan, A. K. M. 2004. Challenges in open water fisheries management in Bangladesh. Pages 1–12 *in* Y. Matsuda and T. Yamamoto, editors. What are responsible fisheries?: Proceedings of the Twelfth Biennial Conference of the International Institute of Fisheries Economics and Trade, July 21–30, 2004, Tokyo, Japan. International Institute of Fisheries Economics & Trade, Corvallis, Oregon.

Larinier, M. 2001. Environmental issues, dams and fish migration. Pages 45–89 *in* G. Marmulla, editor. Dams, fish and fisheries: opportunities, challenges and conflict resolution. Food and Agriculture Organization of the United Nations, Rome.

Lawler, A. 2005. Reviving Iraq's wetlands. Science 307:1186–1189.

Lilygreen, V., and B. Meade. 1982. Fish stocks under pressure from nylon monofilament gill-nets. Marine Pollution Bulletin 13:263–266.

Limburg, K. E., K. A. Hattala, and A. Kahnle. 2003. American shad in its native range. Pages 125–140 *in* K. E. Limburg and J. R. Waldman, editors. Biodiversity, status, and conservation of the world's shads. American Fisheries Society, Symposium 35, Bethesda, Maryland.

Maki, K. L., J. M. Hoenig, J.E. Olney, and D. M. Heisey. 2006. Comparing historical catch rates of American shad in multifilament and monofilament nets: a step toward setting restoration targets for Virginia stocks. North American Journal of Fisheries Management 26:282–288.

Marammazi, J. G., G. R. Eskandari, M.A. Al-Mukhtar, and B. H. Kiabi. 2004. Study of spawning season and spawning ground of Soboor (*Tenualosa* ilisha) during its migration in Khuzestan rivers. Iranian Journal of Fisheries Sciences 4:89–102.

Maxwell, C. N. 1921. Malayan fishes. Journal of the Straits Branch of the Royal Asiatic Society 84:1–194.

Mohamed, A. R. M., T. S. Ali, and N. A. Hussein. 2001. Stock assessment of hilsa shad *Tenualosa* ilisha in the Iraqi marine waters, northwest Arabian Gulf. Marina Mesopotamica 16:1–9.

Mohamed, A. R. M., N. A. Hussain, S. S. Al-Noor, B. W. Coad, and F. M. Mutlak. 2009. Status of diadromous fish species in the restored East Hammar marsh in southern Iraq. Pages 577–588 *in* A. J. Haro, K. L. Smith, R. A. Rulifson, C. M. Moffitt, R. J. Klauda, M. J. Dadswell, R. A. Cunjak, J. E. Cooper, K. L. Beal, and T. S. Avery, editors. Challenges for diadromous fishes in a dynamic global environment. American Fisheries Society, Symposium 69, Bethesda, Maryland.

MRC (Mekong River Commission). 2006. Annual Report. Mekong River Commission, Vientiane, Laos.

Pandey, B. L., B. N. Pandey, and A. Q. Siddiqui. 1996. Impact of Farakka Barrage on hilsa fishery of Padma River. Pages 171–173 *in* M. M. Joseph and C. V. Mohan, editors. Proceedings of the Third Indian Fisheries Forum, 11–14 October 1993, Pant Nagar, Utta Pradesh, India.

Pomeroy, R. S., and F. Berkes. 1997. Two to tango: the role of government in fisheries management. Marine Policy 21:465–480.

Rahman, M. A., and G. C. Haldar. 1998. Assessment of current hilsa resources in Bangladesh. Pages 20–27 *in* M. A. Mazid and S. J. M. Blaber editors. Hilsa Fisheries research in Bangladesh. Proceedings of a BFRI/ACIAR/CSIRO workshop, Dhaka, Bangladesh. Bangladesh Fisheries Research Institute, Mymensingh.

Rattner, B. 2006. Community management by decree? Lessons from Cambodia's fisheries reform. Society and Natural Resources 19:79–86.

Roberts, T. R. 1993. Artisanal fisheries and fish ecology below the Great Waterfalls of the Mekong River in southern Laos. Natural History Bulletin of the Siam Society 41:31–62.

Sann Aung. 2001. The present status of the hilsa (*Tenualosa ilisha*) fishery in Myanmar. Pages 65–76 *in* S. J. M. Blaber, D. Brewer, D. Milton, and C. Baiano, editors. Proceedings of the International Terubok Conference, Sarawak, Malaysia. Sarawak Development Institute, Kuching, Sarawak, Malaysia.

Sarma, D., and A. Dutta. 2000. Present status of *Tenualosa* ilisha (Ham.) in the Brahmaputra River. Pages 97–98 *in* A. G. Ponniah and U. K. Sarkar, editors. Fish biodiversity of north-east India 2:97–98. National Bureau for Fish Genetic Resources, NBFGR-NATP Publication 2, Lucknow, U.P., India.

Siddique, M. A. L. 2009. Conservation of juvenile hilsa (jatka) in Bangladesh: need to address the livelihood of fishers. Pages 757–768 *in* A. J. Haro, K. L. Smith, R. A. Rulifson, C. M. Moffitt, R. J. Klauda, M. J. Dadswell, R. A. Cunjak, J. E. Cooper, K. L. Beal, and T. S. Avery, editors. Challenges for diadromous fishes in a dynamic global environment. American Fisheries Society, Symposium 69, Bethesda, Maryland.

Sinha, M., and M. A. Khan. 2001. Impact of environmental aberrations on fisheries of the Ganga (Ganges) River. Aquatic Ecosystem Health & Management 4:493–504.

Sneddon, C., and C. Fox. 2006. Rethinking transboundary waters: a critical hydropolitics of the Mekong basin. Political Geography 25:181–202.

Srinath, M., S. Kuriakose, K. G. Mini, M. R. Beena, and S. K. Augustine. 2003. Status of exploited marine fishery resources of India. Central Marine Fisheries Research Institute, Kochi, India.

Symes, D. 2006. Fisheries governance: a coming of age for fisheries social science? Fisheries Research 81:113–117.

Viswanathan, K. K., J. R. Nielsen, P. Degnbol, M. Ahmed, M. Hara, and N. M. Raja Abdullah. 2003. Fisheries co-management policy brief: findings from a worldwide study. Worldfish Center Policy Brief 2:1–26.

Waldman, J. 2003. Introduction to the shads. Pages 3–9 *in* K. E. Limburg and J. R. Waldman, editors. Biodiversity, status, and conservation of the world's shads. American Fisheries Society, Symposium 35, Bethesda, Maryland.

Wang H.-P. 2003. Biology, population dynamics, and culture of Reeves Shad *Tenualosa* reevesii. Pages 77–83 *in* K. E. Limburg and J. R. Waldman, editors. Biodiversity, status, and conservation of the world's shads. American Fisheries Society, Symposium 35, Bethesda, Maryland.

Warren, T. J. 2001. Conservation status fishery and future for *Tenualosa* thibaudeaui (Durand, 1940) in the Mekong system. Pages 77–86 *in* S. J. M. Blaber, D. Brewer, D. Milton, and C. Baiano, editors. Proceedings of the International Terubok Conference, Sarawak, Malaysia. Sarawak Development Institute, Kuching, Sarawak, Malaysia.

Whitehead, P. J. P. 1985. FAO species catalogue 7, clupeoid fishes of the world. Part I—Chirocentridae, Clupeidae, and Pristigasteridae. FAO Fisheries Synopsis 125.

Wilson, D. G., J. R. Nielsen, and P. Degnbol. 2003. The fisheries co-management experience: accomplishments, challenges and prospects. Kluwer, Dordrecht, Netherlands.

Ye, Y., and S. Almatar. 2003. Toward integrated management of Kuwait's coastal resources. Pages 475–490 *in* B. Phillips, B. Megrey, and Y. Zhou, editors. Proceedings of the Third World Fisheries Congress: feeding the world with fish in the next millennium—the balance between production and environment. American Fisheries Society, Symposium 38, Bethesda, Maryland.

American Fisheries Society Symposium 69:691–711, 2009

Identification of Diadromous Fish Species on which to Focus River Restoration: An Example Using an Eco-Anthropological Approach (The Seine Basin, France)

Eric Rochard*
Cemagref, Estuarine Ecosystems and Diadromous Fish Research Unit
33612 Cestas cedex, France

Patricia Pellegrini
National Museum of Natural History, Eco-Anthropology Research Unit
75231 Paris cedex 05, France

Julie Marchal
INRA UMR 985 Écologie et santé des Ecosystèmes F- 35000 Rennes, France

Mélanie Béguer
Cemagref, Estuarine Ecosystems and Diadromous Fish Research Unit
33612 Cestas cedex, France

Dominique Ombredane
INRA UMR 985 Écologie et santé des Ecosystèmes F- 35000 Rennes, France

Géraldine Lassalle
Cemagref, Estuarine Ecosystems and Diadromous Fish Research Unit
33612 Cestas cedex, France

Erwan Menvielle
INRA UMR 985 Écologie et santé des Ecosystèmes F- 35000 Rennes, France

Jean-Luc Baglinière
INRA UMR 985 Écologie et santé des Ecosystèmes F- 35000 Rennes, France

Abstract.—The Seine River, like other large and developed European river basins, has lost most of its native diadromous fish populations since the Industrial Revolution. A restoration program has been planned to recover water quality and fish biodiversity in the Seine basin. In this prospective study, we used a dual approach (ecological and anthropological) to prioritize the 11 historic diadromous fish species for restoration. Using generalized additive models and geographic information systems, we assessed which species would have the highest probability of recovery and drew maps of their potential favorable habitats. Three fish appear to be good candidates as flagship species for a recovery plan: smelt *Osmerus eperlanus*, which has came back into the lower part of the basin; Allis shad *Alosa alosa*, a few individuals of which have recently been observed upstream from Paris, France; and brown trout *Salmo trutta*, which is limited to estuarine tributaries. Our anthropological survey revealed that for most citizens polled, these species have no particular significance. These citizens are not in favor of these fish species' recovery because it would lead to changes (e.g., to restore the

* Corresponding author: eric.rochard@cemagref.fr

upstream–downstream connectivity) and new constraints (e.g., rules of management). To address this potential conflict, we studied a conciliation process with the human inhabitants from one subbasin.

Introduction

Deforestation and the installation of mills during medieval times (Hoffmann 1996; 2005) modified the morphology and longitudinal connectivity of small rivers, while the industrial revolution (1850–1900 in Europe) modified larger rivers with weirs, dams, and reservoirs. Significant cumulative impacts have been observed due to (1) construction of dams for hydropower (Nicola et al. 1996; Larinier 2001), (2) channelization and dredging for shipping (de Groot 2002), (3) industrial and domestic pollution, (4) alterations of river flow regimes (Jager et al. 1997; Fernandez-Pasquier 1999), (5) intensification of agriculture, and (6) uncontrolled exploitation of fish. These pressures induced a long period of progressive degradation of the quality of European watersheds, leading to a decrease in diadromous fish (de Groot 2002). Most species in this group are now threatened and require strict protection (i.e., listed in Appendix IV of the Council Directive 92/43/EEC of 21 May on the conservation of natural habitats and of wild fauna and flora), including European sturgeon *Acipenser sturio*, Adriatic sturgeon *A. nacarii*, and houting *Coregonus oxyrinchus*. Other species are considered to be vulnerable and need species-specific management plans and special areas of conservation (i.e., Council of Europe 1992, Appendix IV), including sea lamprey *Petromyzon marinus*, European river lamprey *Lampetra fluviatilis*, Allis shad *Alosa alosa*, twaite shad *A. fallax*, and Atlantic salmon *Salmo salar*.

When concerned citizens objected to the unintentional consequences of these anthropogenic influences, progressive countries adopted laws to control human activities more strictly and reduce impacts on the aquatic environment. Since the 1980s, Europe has been developing species recovery plans, with most of the large river basins launching their own plans. In many cases (Rhine, Thames, Loire, Garonne, Dordogne), these plans focused on Atlantic salmon.

Since 2000, ambitious environmental rehabilitation plans have been implemented. One example is the European Water Framework Directive (European Parliament and Council of Europe 2000), which aims at achieving "good water status" for all waters by 2015. In this context, river basin authorities examine the possibilities of enhancing rare fish species or reintroducing extinct fish species, and diadromous species are frequently considered. Financial constraints forced the authorities to prioritize actions and identify cost-effective species for their basin, both in terms of ecological improvement and social satisfaction. Most large basins established global recovery plans targeting improvement of their ecosystems; however, these plans were mainly focused on flagship species. In Europe, the selection of these flagship species was primarily influenced by fisheries managers or fishermen's representatives, with a strong emphasis on the symbolic component associated with the species and its historic presence. Successful recovery plans are rare and those that are well documented have required at least several decades: 50 years of a fishing ban before a partial recovery of the white sturgeon *Acipenser transmontanus* population was achieved in the Sacramento–San Joaquin system in California, USA (Kohlhorst 1980), 40 years of a recovery plan for American shad *Alosa sapidissima* in the Susquehanna River basin, USA (St. Pierre 2003), and 20 years of a recovery plan for Atlantic salmon in the Couesnon River, France (Maisse and Baglinière 2001).

We are now facing global climate warming with ecological consequences (Walther et al. 2002). We must consider the expected changes in the environment as we assess which diadromous fish species could live in the basin in the future (Lassalle et al. 2009, this volume).

The objective of our study is to propose and test a process to identify a single diadromous fish species (or several species) that could be used as a flagship(s) to support population recovery actions in a river basin. We postulated that to accomplish this objective, present and future ecological potentialities should be combined with the wishes of the different water users in the basin. This study was carried out in the Seine River basin in the northwest region of France (Figure 1), a region that has been highly impacted by human activities for two centuries.

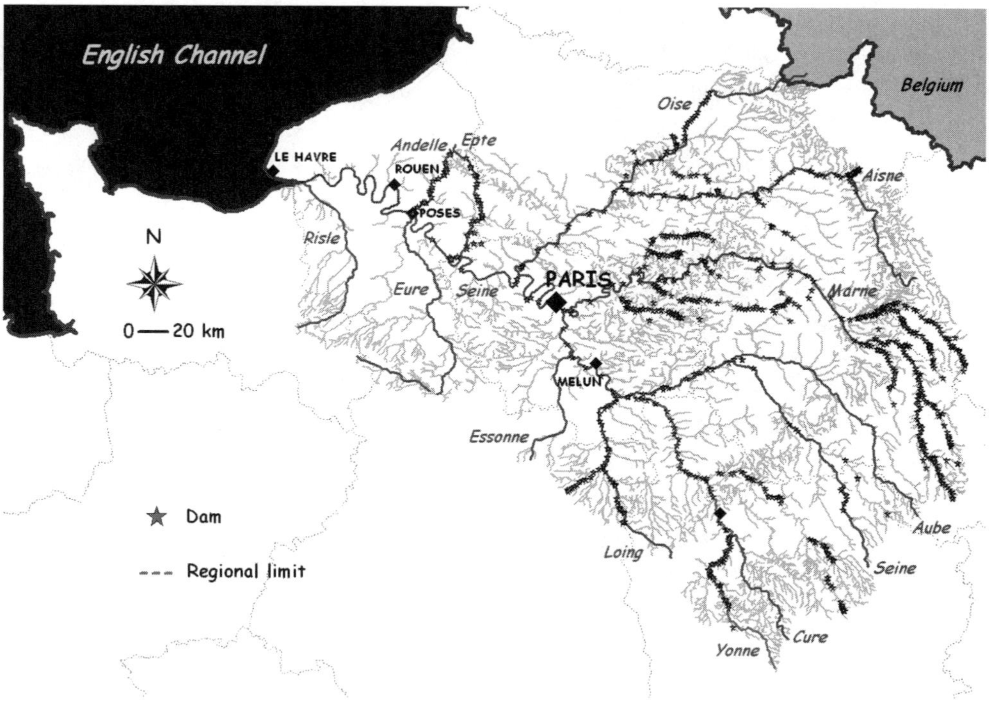

Figure 1.—The Seine River basin with tributaries and dams.

Study Site

The Seine River rises on the Langres plateau (elevation 470 m) and reaches the English Channel in Normandy after flowing 776 km through the plains of the Parisian basin (Figure 1). The climate in the Seine basin is oceanic temperate with a mean precipitation (1931–1960) of 750 mm/year (Ducharne et al. 2003) and a mean water flow of 650 m^3/s (maximum 1,350 m^3/s in March, minimum 274 m^3/s in September). The drainage area is 78,650 km^2, with 60% of the land use as intensive agriculture. This basin contains 40% of French economic activity and 26% of its human population (16 million inhabitants), mainly in Paris and its suburbs. Shipping is also an important activity in the Seine basin, with one harbor at the mouth (Le Havre) and another in the estuary (Rouen). We focused our analysis on three areas: the estuary, the main-stem river upstream to Melun, and two estuarine tributaries (Epte and Andelle). At the beginning of the study, Melun seemed to be the upper limit of the presence of the anadromous fish species in the Seine watershed.

History of the Diadromous Fish Community in the Seine Basin

In the 12th century, 11 species of diadromous fish were regularly observed in the Seine basin (Belliard et al. 1999): sea lamprey, European river lamprey, European eel *Anguilla anguilla* (L., 1758), European sturgeon, Allis shad, twaite shad, smelt *Osmerus eperlanus* (L., 1758), Atlantic salmon (L., 1758), brown trout *Salmo trutta*, thin-lipped mullet *Liza ramada* (Risso, 1826), and flounder *Platichthys flesus* (L., 1758). These populations were abundant and commercially exploited, as attested by global landings of 250 metric tons/year in the first half of the 19th century in Rouen (Figure 2; Euzenat et al. 1992).

However, the abundance of all diadromous fish species has decreased since 1850. Anadromous species requiring the most upstream spawning habitats exhibited the sharpest decreases, following the construction of several large dams on the main-stem Seine. Construction of the most downstream dam in Poses in 1885 essentially destroyed the Allis shad population (Roule 1920) and limited yearly cap-

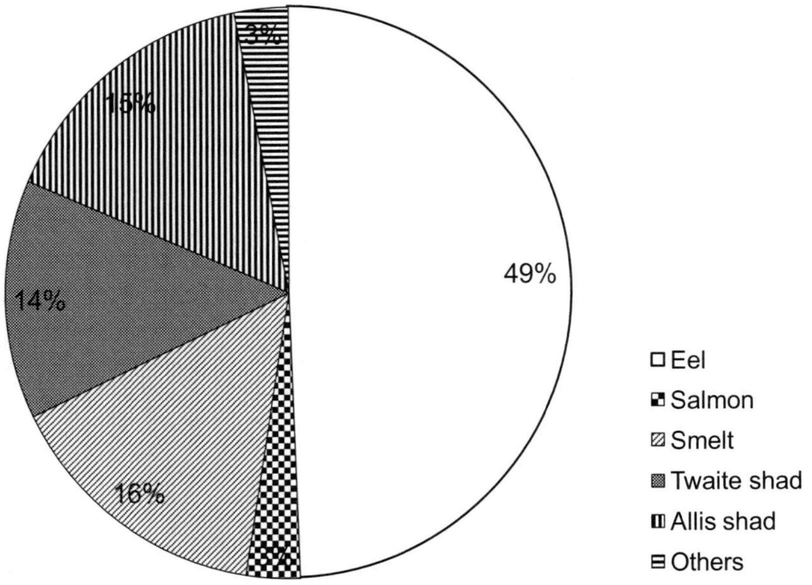

Figure 2.—Commercial landings of diadromous fish in Rouen in 1876–1894 (total quantity ≈ 250 metric tons).

tures of salmonids to some dozens of individuals (Euzenat et al. 1992). Species using only the lowest part of the basin (smelt, twaite shad) maintained their abundance up to 1950. They then decreased rapidly as water quality in the Seine became very poor in the 1970s (about 100 km of anoxic area downstream from the main waste water treatment plant in Paris; Belliard 1994). Euzenat et al. (1992) conducted a detailed analysis to document the historic distribution of Atlantic salmon in the Seine basin and called for implementation of a recovery plan. Although this plan is still supported (a workshop was held in late 2006) by the International Association for the Defense of the Atlantic Salmon, the plan has not been implemented.

Materials and Methods

We used a six-step process to identify which diadromous fish species could be used as flagship species in a recovery plan for the Seine basin (Table 1). (1) As we postulated that global change can drastically modify the environmental conditions in the basin, we decided to consider all diadromous species historically present in the region (western Europe) and not only those observed in the Seine basin. A species missing in the Seine due to unfavorable conditions could find favorable conditions in the future. However, only native fish diadromous species in the Seine have been taken into consideration to estimate the recovery potential. (2) For all these species, we established bioclimate envelope models using a data set from 197 European basins around 1900 and we estimated the favorability of the Seine for these species in 2000. We did the same estimation (3) in 2100, using the most likely climate change scenarios, (4) we mapped and qualified the suitable habitats and their accessibility, (5) we evaluated the interest in and importance of the fish species to stakeholders and other human inhabitants in the basin, and (6) we examined the technical feasibility of measures to enhance existing populations or reintroduce the species. Steps 1 through 4 can be considered as successive filters. They provided a group of species that could probably achieve their life cycles in the Seine basin. Step 5 evaluated the relationship(s) between fish species and people living in the study area. In Step 6, we calculated a score which is the sum of other columns. It helped us to summarize the global potentialities (species and habitat requirements) for improvement, technical feasibilities, and the cost of management actions required to enhance the species. However, we did not systematically select the species with the highest scores.

Table 1.—Successive steps in the process used to select flagship fish species. X indicates that the species (rows) satisfy criteria (columns). Criteria 1–4, 6 X: species fulfills conditions. Criterion 5 (0: no appropriation, 1: appropriation). Criteria 2 and 3 are satisfied according to the outputs of the bioclimate models. Criterion 6 is satisfied when both artificial breeding and fish passes are mastered. Score is the sum of Xs for one species.

Considered species	(1) Historically present	(2) Favorability of the basin in 2000	(3) Favorability of the basin in 2100	(4) Favorable and accessible habitats	(5) Significance for the inhabitants	(6) Technical feasibility	(7) Potential of recovery	Score
A	X	X	X	X	X	X	X	7
B	X							1
C	X	X	X	X		X		5
D	X					X		2
E	X	X	X		X			4
...	X	X		X		X		4

Large-Scale Approach

The list of the species historically present in the region (western Europe, from the Iberian Peninsula to Scandinavia) came from an extensive literature review in a database called "EuroDiad 1.0." For more details, see Béguer et al. (2007) and Lassalle et al. (2009).

To compare these species and determine which could be present in the Seine basin, we used generalized additive models (GAM; Hastie and Tibshirani 1990). GAM are one of the most appropriate methods to build bioclimate envelope models (Leathwick et al. 2006). These models relate, simultaneously, an observed species distribution to a set of variables describing the environment (Thuiller 2003). Béguer et al. (2007) and more recently Lassalle et al. (2009) established such models for European diadromous fish species. Using a data set from 41 basins (from the Iberian Peninsula to Scandinavia), Béguer et al. (2007) produced 10 single species models using: air temperature (all but one model), surface drainage area (six models), stream power of the river (four models), and net primary productivity of the area (three models) as predictors of species distribution circa 1900. These authors provided specific response curves that illustrate the influence of the variables selected in the models on the species. With a larger data set (197 basins in a larger area, from the Black Sea to Scandinavia), the second study built 22 single species models using longitude at the outlet (all models), air temperature (15 models), surface of the basin (13 models), precipitation (8 models), and source elevation (1 model) as predictors of species distribution at the same time (ca 1900). We initially used the same database, EuroDiad 1.0 (www.diadfish.org), and the same approach to establish new distribution models without data from the Seine basin. Then, using the measured values of the environmental variables required by the different models, we assessed the favorability of the Seine basin for each species ca 2000 and ca 2100. In both cases, we have neglected regional pressures to focus on the potential of the basin. We used different R software packages (R foundation core team 2005) for the model computations.

To assess and project species presence in 2000 and 2100, we used the CRU TS 2.1 data set and the TYN SC 2.03 data set, respectively (Mitchell et al. 2004), which are composed of monthly grids of modeled climate covering the global earth surface at 0.5 degree resolution, for 1970–1999 and 2001–2100, respectively.

Potential species distributions in 2100 were projected in the face of changing climate predicted by the HadCM3 global climatic model from a special report on greenhouse gas emission scenario (A2) (IPCC 2000). Temperatures and precipitation values used in the models were averaged over 30 years from 2070 to 2099 to smooth interannual variability. These two parameters have a real biological significance at the large-scale approach used. The climate model predicted a rise in annual temperature between 1°C and 7°C in Western Europe and a reduction in annual precipitation from 1% to 4%, mainly in the southern basins, including the Seine basin. Conversely, annual precipitation is predicted to increase in the same range for northern European basins. The other variables selected in the models to predict species distribution (longitude, surface of the basin, source elevation) remain constant for the basin.

To transform probability values for each model, we used the threshold (*T*), maximizing kappa statistics (Fielding and Bell 1997). Results were interpreted as (1) unlikely to be present in the Seine, (2) likely to be present in the Seine, or (3) impossible to get a reliable prediction.

Habitat Suitability and Connectivity

From a literature review, we listed the major habitat requirements for all diadromous fish species historically present (observations) or likely to be present (models) in the Seine basin (Table 2):

- temperature ranges (°C) and minimum concentration in dissolved oxygen (dissolved oxygen = O_2 in mg/L) for migration and reproduction;
- depth, current velocity, and substrate size in the spawning grounds and juveniles' growth habitats; and
- swimming, jumping, and crawling abilities of the species to overcome barriers.

Information on the ecological requirements of European river lamprey, thin-lipped mullet, and flounder is so scarce that we could not assess the suitability of habitats in the study area for them.

We considered concentrations of two water pollutants present in the lower part of the Seine River that could impact these species: the reduced form of nitrogen (nitrite = NO_2^- in mg/L) and some heavy metals (cadmium [Cd], copper [Cu], nickel [Ni], and zinc [Zn]). We defined five types of zones from the thresholds of these water pollutants used by French water agencies for water quality assessment (Anonyme 2003):

1. No risk zone ($NO_2^- \leq 0.5$ mg/L; Cd ≤ 0.7 mg/kg; Cu ≤ 1.9 mg/kg; Ni ≤ 1.6 mg/kg; Zn ≤ 124 mg/kg);
2. Potentially vulnerable zone ($0.5 < NO_2^-$ in spring ≤ 1 mg/L);
3. Potentially toxic zone ($0.7 <$ Cd ≤ 4.2 mg/kg; $1.9 <$ Cu ≤ 19 mg/kg; $1.6 <$ Ni ≤ 16 mg/kg; $124 <$ Zn ≤ 271 mg/kg);
4. Vulnerable zone (NO_2^- in spring > 1 mg/L); and
5. Toxic zone (Cd > 4.2 mg/kg; Cu > 19 mg/kg; Ni > 16 mg/kg; Zn > 271 mg/kg).

We used river flow and water quality data covering the period 2002–2004. Data were recorded at 43 monitoring stations distributed along the studied zone and were provided by the Seine Navigation Service and Regional Environment Direction. We characterized the morphodynamic units using Malavoi (1989) criteria and a classification scheme simplified by Haury et al. (1999). We found substrate size data for only the main-stem river below Poses dam and between Paris and Melun. We used the classification scheme proposed by Bain et al. (1985).

We located and mapped dams and barriers and assessed their crossing level (expressed by the difficulty level for passing the obstacle) from a gray literature review validated from field surveys in the Epte and Andelle tributaries.

We combined environmental data and biological requirements of the species into habitat suitability levels using MapInfo Professional 7.5 (Pitney Bowes MapInfo, New York) to interpolate linearly between stations. For each season and species, we established suitability maps for several ecological functions (migration, reproduction, growth).

Perception and Interest of the Human Inhabitants

We also carried out research in environmental anthropology. We questioned different categories of people (water managers, fish/fauna managers, river/special conservation area managers, fishermen associations, fishermen, residents, other water users) on how they understand and live in their environment and how they evaluate water quality and the role of fish. We developed interview questions adapted to different categories of people (Table 3). Items concerning fish appeared only at the end of the interviews so as not to guide the answers.

We also followed up the questionnaires with workshops and meetings with the commission in charge of managing diadromous species in the Seine basin (Copepomi [Comité de gestion des

Table 2.—Ecological requirements for diadromous species historically observed in the Seine basin (from Maisse and Costard 1977; Deniel 1981; Masson 1987; Boisneau et al. 1990; Baglinière and Maisse 1991; Möller and Scholz 1991; Euzenat et al. 1992; Mennesson-Boisneau et al. 1993; Gueguen and Prouzet 1994; Holmes and Lin 1994; Le-Teuff 1996; Pronier and Rochard 1998; Baglinière and Elie 2000; Dedual et al. 2000; Keith and Allardi 2001; Kelly and King 2001; Sabatié and Baglinière 2001; Jego et al. 2002; Gouneau 2003; Maitland 2003; Duhamel et al. 2004).

	Allis shad	Twaite shad	Smelt
	Adult migration		
Period	March–July	May–July	November–February
Temperature °C (min–max)	8–22	>10	0–20
O_2 dissolved (mg/L)	NA	NA	>5
Crossing barriers	swimming	swimming	swimming
	Reproduction		
Period	May–August	May–July	February–May
Location	Middle course of rivers, weak slope	Lower part of rivers, tidal river	Lower part of rivers and brooks including tidal part
Salinity	Freshwater	Freshwater	Freshwater
Facies	Running flats, running channels	Running flats, running channels	Running flats
Speed current (m/s)	0.9–2.0	0.5–1.5	0.3–2.0
Depth (m)	<3	1.5–3.0	Highly variable
Substrate size	Boulder–pebble	Sand–boulder	Boulder–pebble
Temperature °C (min-max)	>15 (optimum 18)	>15 (optimum 18)	4–12 (optimum 6–9)
	Juvenile growth and migration		
Location	Lower part of rivers and estuaries	Lower part of rivers and estuaries	Estuaries
O_2 dissolved (mg/L)	NA	≥4	> 3
Temperature °C (min–max)	NA	NA	<21.5
Period of migration	Summer–fall	Summer–fall	Summer–fall

	European sturgeon	European river lamprey	Sea lamprey
	Adult migration		
Period	May–June	October–December	March–May
Temperature °C (min–max)	>12	NA	10–16
O_2 dissolved (mg/L)	>3.5	NA	
Crossing obstacles	Swimming	Crawling, jumping	Crawling, jumping
	Reproduction		
Period	June	March–May	April–May
Location	Middle part of large basins	Upper part of rivers and brooks	Upper part of rivers and creeks
Salinity	Freshwater	Freshwater	Freshwater
Facies	Running channels	Running flats, running channels	Running flats, running channels
Speed current (m/s)	0.5–2.0	NA	> 0.4
Depth (m)	>5	NA	
Substrate size	Pebble–block	Sand–gravel	Gravel–pebble
Temperature °C (min–max)	15–20	NA	NA

Table 2.—Continued.

	European sturgeon	Europoean river lamprey	Sea lamprey
	Juvenile growth and migration		
Location	Mid-part of rivers and estuaries	Rivers and brooks	Rivers and creeks
Facies	NA	Pools	Pools
Substrate size	Silt–sand	Silt–sand	Silt–sand
Migration period	Fall	Fall–winter	Fall–winter

	Atlantic salmon	Sea trout
	Adult migration	
Period	October–March March–May June–July	May–January
Temperature °C (min–max)	4.5–24	6–23
O_2 dissolved (mg/L)	>6.0	>6.0
Crossing obstacles	Jumping	Jumping
	Reproduction	
Period	November–February	November–March
Location	Large rivers	Lower part of large basins, tributaries
Salinity	Freshwater	Freshwater
Facies	Riffles	Running flats, riffles
Speed current (m/s)	0.15–1.0	
Depth (m)	0.3–0.6	0.3–0.6
Substrate size	Gravel–boulder	Gravel–boulder
Temperature °C (min–max)	3–12	2–12
	Juvenile growth and migration	
Location	Upper part	Upper part
Salinity	Freshwater	Freshwater
Facies	Running flats, riffles	Running flats, riffles
Speed current (m/s)	0.5–0.7	0.2–0.5
O_2 dissolved (mg/L)s	>6	NA
Depth (m)	>0.3	0.1–0.4
Substrate size	Gravel–pebble Pebble–block	Gravel–boulder
Temperature °C (min–max)	13–16	18–23

poissons migrateurs]). Fifty ethnographic interviews with open answers were conducted between August 2005 and March 2006 in different locations, urban and rural.

Interviewees were not randomly selected (Table 4); instead, we chose to question people who were involved in the life of the Seine basin at different levels and for different reasons about their analysis of the situation, their actions, and their priority, and we did not question the common citizen. In addition to interviews, we spent time with anglers, we assisted at an electrofishing operation with fish technicians, and we attended workshops dealing with (1) sharing the Seine between various stakeholders,

Table 3.—Guidelines for interview questions.

For river residents
How long have you lived near the tributaries or near the Seine?
What are the advantages and inconveniences of this proximity?
What activities do you practice in relation to the river?
Have you noticed any evolution of the river during the past years and in what way?
What improvement or modification would you support?
Do you know which fish are in this river, and are there diadromous fish?
What kind of fish would you enjoy seeing in the river and more widely in the Seine? and among the diadromous fish?
For fishermen
Where are you from and how long have you practised fishing, here and in other places?
Are you involved in a fishing association and do you participate in its actions?
What do you think about environmental quality and the changes in quality?
Why do you say the quality is XXX?
Have you tried to do something and what?
What kind of fish do you catch and do you eat them?
What fish are you interested in and why?
Do you know and have you ever caught any diadromous fish?
What do they represent for you?
What kind of fish would you enjoy to seeing in this river, and among the diadromous fish species?
In your opinion, how might one facilitate their presence?

Table 4.—Categories and numbers of people interviewed during the study.

Category	Subcategory
Fisheries (n = 15)	
	3 commercial fishermen
	6 anglers
	4 representatives of recreational fishermen
	2 representatives of commercial fishermen
Riverine (n = 15)	
	2 representatives of river residents association
	3 technicians of river residents association
	3 river residents
	1 canoe-kayak rental company
	1 manufacturer
	2 mill owners
	1 hydropower producer
	2 fish passage technicians
Environmental managers (n = 11)	
Nature protection association (n = 7)	
Local authorities (n = 12)	

(2) the conservation of the natural environment in the Seine estuary, and (3) salmon in the Seine.

We also used various materials (e.g., Internet Web sites, reports and minutes of the commission meetings, scientific and press articles, archives) for a deeper understanding of the way fish and their aquatic environments are described and the place they are given in those segments of human society included in this research project.

We used a symmetric approach (Bloor 1982) and analyzed the scientific data, administrative data, and local knowledge in the same way. We recognize the necessity of addressing the influence of this study on the topic we are analyzing (Kilani 1995) as a way to question our own perception of things and the influence our research would have on people.

In one region (Andelle and Epte rivers), we identified a conflicting situation between mill owners who wanted to develop hydropower and anglers who wanted to improve habitat connectivity for fish. We chose to further explore this problem of a dam by addressing the conflict between two apparently opposing aims: a fish pass as a tool to manage diadromous fish population and development of hydroelectric power. We then added 10 more interviews conducted between January and March 2007. People were chosen according to their connection with this conflict.

We visited a wild salmon farm, a hydroelectric factory, and its fish pass. To address the question at a more global level, we followed three symposia dealing with renewable energy, biodiversity, and sustainable development. We also attended two workshops of the diadromous fish management committee of the Seine River to understand how committee members work, how they evaluate the situation, and what kind of management they focus on.

In order to understand whether or not there is a peculiar situation in the Seine basin, we attended a symposium about the management of the Rhône River and two workshops of the diadromous fish management committee of the Garonne River. We also met the president of the European river network.

We took this opportunity to examine whether a reconciliation approach using both ecological and sociological expertise could be helpful. We communicated with the two groups and organized a reconciliation meeting.

Results

Large-Scale Approach

Eleven diadromous fish species have been recorded historically in the Seine basin. Minimum species richness was observed in the 1970s with only four species (Table 5). Now, four species are encountered regularly (smelt, mullet, European eel, and flounder) while six others (Atlantic salmon, Allis shad, twaite shad, sea lamprey, and European river lamprey) remain rare and are not observed every year. Brown trout are encountered regularly in the estuarine tributaries (Duhamel et al. 2004; data from the Réseau Hydrobiologique et piscicole [ONEMA]; J.-L. Baglinière, unpublished data).

Distribution models were established for all species except two: houting and European eel. Data sets for these two species are too unbalanced to enable presence–absence modeling (houting is present only in a very limited number of basins; conversely, eel is present in almost all basins in the area). Even though we could not model the eel distribution, we can reasonably hypothesize that the Seine basin was favorable for European eel in 2000 and will remain favorable in 2100.

From these model predictions, the present Seine River environment is suitable for all species observed during the study period and even for one extant species, European sturgeon. Conversely, the models showed that the Seine environment is not suitable for fish species never observed there: Vimba *Vimba vimba* and Arctic char *Salvelinus alpinus*. Projections for 2100 revealed only one change: the disappearance of smelt, which, due to the increasing temperatures, would not find suitable conditions in the Seine to complete its life cycle.

Habitat Suitability

Lampreys.—The Epte and Andelle tributaries provided suitable habitats for adult reproduction (running flats with coarse size substrates) and for larval growth (pools with fine sand and silt). The upper part of the Seine presented good habitat for larval growth only. Temperature and oxygen were favorable for spring migration in all zones (temperature > 10°C). Nitrite concentrations in the main stem (>1 mg/L) were likely to induce poor conditions for juvenile migration to the sea. Indeed, nitrite uptake and accumulation in freshwater fish can modify

Table 5.—Presence (1) or absence (0) of diadromous fish species from the west European pool (from Béguer et al. 2007) recorded in the Seine basin at several time periods and according to the outputs from our models for 2000 and 2100 (projection using scenario A2). F indicates failure to get a sound estimate from our model.

	Observations					Modeling	
Species	1850	1900	1950	1970	2000	2000	2100
European river lamprey	1	1	1	1	1	1	1
Sea lamprey	1	1	1	0	1	1	1
European eel	1	1	1	1	1	F	F
European sturgeon	1	1	0	0	0	1	1
Allis shad	1	1	0	0	1	1	1
Twaite shad	1	1	1	0	1	1	1
Vimba	0	0	0	0	0	0	0
Smelt	1	1	1	0	1	1	0
Houting	0	0	0	0	0	F	F
Atlantic salmon	1	1	0	0	1	1	1
Brown trout	1	1	1	1	1	1	1
Arctic char	0	0	0	0	0	0	0
Thin-lipped mullet	1	1	1	0	1	1	1
Flounder	1	1	1	1	1	1	1
Species richness	11	11	8	4	10		

their hypoosmoregulatory ability (Jensen 2003). Furthermore, lampreys cannot presently complete the early stages of their life cycle because nitrite concentrations are usually higher in the substrate than in the water column in the study area, resulting in a strong decrease in their potential for recovery.

European eel.—All study areas presented suitable habitats for the inland growth of this species, which colonized the entire study area. However, (1) since this species is still present in most of the study areas, an improvement in the conditions for this species will only slightly extend its present distribution in the basin; and (2) the abundance trend of the European population is drastically decreasing (Dekker 2003) and management actions in progress at the continental scale (Commission des communautés européennes 2005) will have only delayed effects. For these reasons, this species is not a good candidate to go with the improvement in the basin.

Clupeids.—Deep-flowing habitats below the first dam (Poses) are suitable for twaite shad reproduction while spawning grounds for Allis shad with appropriate substrate size (gravel) are only found in the upper part of the main-stem Seine and in the Andelle (Figure 3). Barriers currently limit migratory fish access into the Andelle, but Allis shad were observed within the main stem near Melun in 2004 and above Poses dam in 2007 (J. Belliard, Cemagref Antony, unpublished data). Spring and summer temperatures are favorable for the reproduction of these two species, and oxygen concentrations are not limiting for juveniles (Möller and Scholz 1991) nor for upstream migration of adult twaite shad (Maes et al. 2008). Only twaite shad can currently complete its life cycle in the study area because the species can spawn in the estuarine and downstream parts of river. If improved fish passages were created at the main-stem Seine dams, Allis shad could also complete its life cycle.

Smelt.—The estuary and lower part of the Seine basin are favorable environments for migration (Figure 4) and reproduction of smelt, even though winter temperatures are below the optimum (8°C). Barriers prevent smelt from colonizing the Epte and Andelle tributaries where they could make use of suitable habitats (running flats, medium size substrate) and favorable environmental conditions for migration and reproduction in winter and spring. Smelt can currently complete its life cycle in the study area.

Salmonids.—Conditions encountered during winter and spring in the Seine estuary and in the Epte and Andelle tributaries fulfill sediment require-

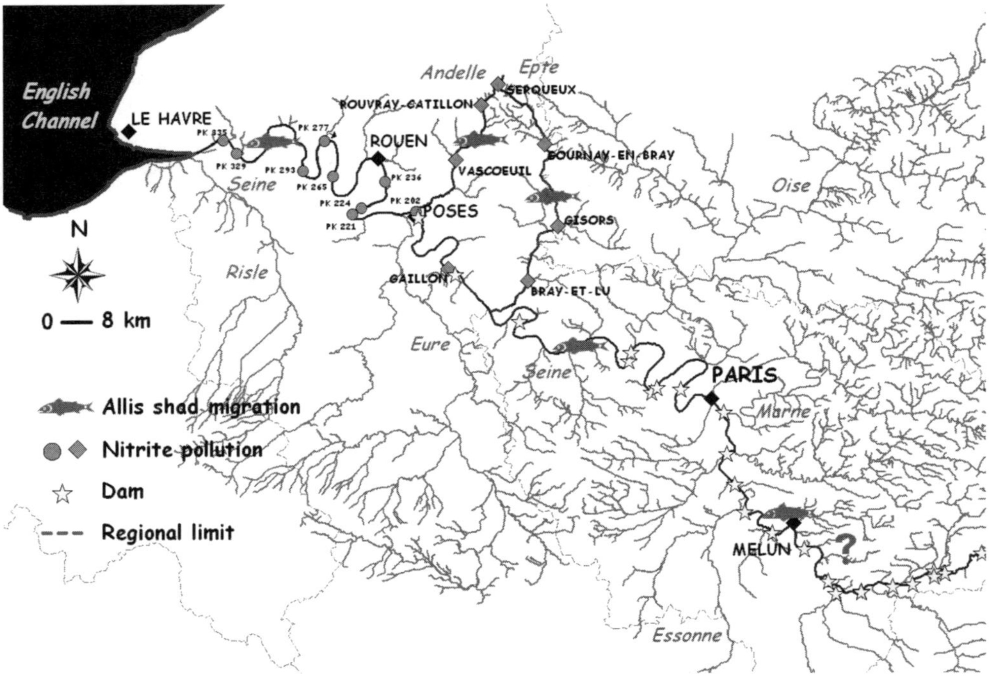

Figure 3.—Synthesis of migration possibilities for Allis shad.

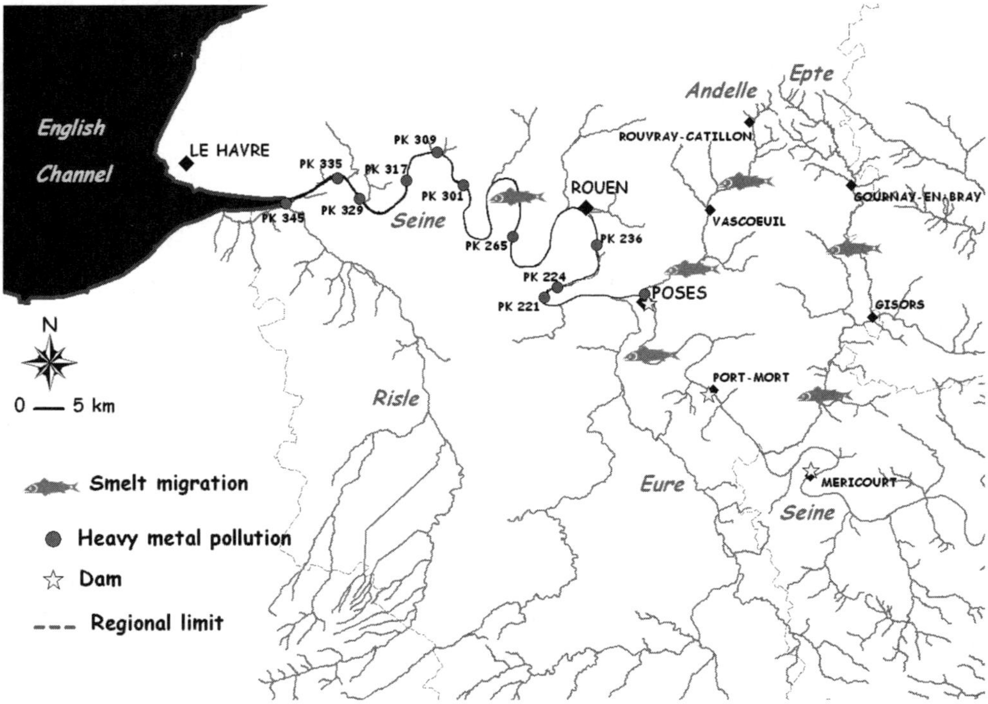

Figure 4.—Synthesis of migration possibilities for smelt.

ments for migration and reproduction of salmonids (temperature > 6°C, DO > 6 mg/L; Table 6). On the other hand, reproduction is not possible in fall because temperatures exceed 12°C. Cooler water during summer in the Epte and Andelle (<15°C) favor the survival of young of the year, even though nitrite concentrations (>1 mg/L) probably complicate their stay in freshwater. Indeed, salmonids are among the fish with high branchial Cl- uptake rates, which would make them more sensitive to nitrite concentrations (Jensen 2003). Moreover, low concentrations of dissolved oxygen (<5 mg/L) and high concentrations of heavy metals (copper, nickel, zinc) in fall 2003 and 2004 in the central part of the estuary certainly hinder the migration of spawning adults. Atlantic salmon and brown trout both have a high potential for recovery in the basin, but the habitats observed in the tributaries (mainly running flats) are much more suitable for sea trout (Figure 5) than for Atlantic salmon. Dams and barriers not equipped with effective fish passages restrict salmonid (mainly brown trout) colonization except in the lower part of the Andelle where dams are passable. Only sea trout can currently complete its life cycle in the study area.

Relationships between Fish and Human Inhabitants

The presence of three smelt on the blazon of the city of Caudebec en Caux in the Seine estuary suggests a long history between human inhabitants of the Seine estuary and this diadromous fish species. The actual history is more complicated. A royal decree of 1696 by Louis XIV, king of France, stipulated that these fish were in fact three salmon, acknowledging the preeminence of symbols over reality.

In comparison with other European rivers, we find far fewer organizations dealing with the river, its fauna, and its environment in the Seine water-

Table 6.—Parts of the study area (1: Seine estuary, 2: Epte River, 3: Andelle River, 4: upper Seine) where these diadromous fish species could complete parts of their life cycle in 2002, 2003 and 2004.

(A) Migration

Species	2002	2003	2004
Sea lamprey	1, 2, 3	1, 2, 3	1, 2, 3
European eel	1, 2, 3, 4	1, 2, 3, 4	1, 2, 3, 4
European sturgeon	1	1	1
Allis shad	1, 3, 4	1, 3, 4	1, 3, 4
Twaite shad	1	1	1
Smelt	1, 2, 3	1*, 2, 3	1*, 2, 3
Salmo sp.	1**, 3	1***, 3	1***, 3
Thin-lipped mullet	1	1	1
Flounder	1	1	1

* migration possible in winter and spring but not in fall, ** migration possible in winter, spring and fall but not in summer, *** migration possible in winter and spring but not in summer or fall.

(B) Growth of subadults

Species	2002	2003	2004
European eel	1, 2, 3, 4	1, 2, 3, 4	1, 2, 3, 4

(C) Reproduction

Species	2002	2003	2004
Sea lamprey	none	none	none
European sturgeon	4	4	4
Allis shad	4	4	4
Twaite shad	1	1	1
Smelt	2, 3	2, 3	2, 3
Salmo sp.	1, 2, 3	1, 2, 3	1, 2, 3

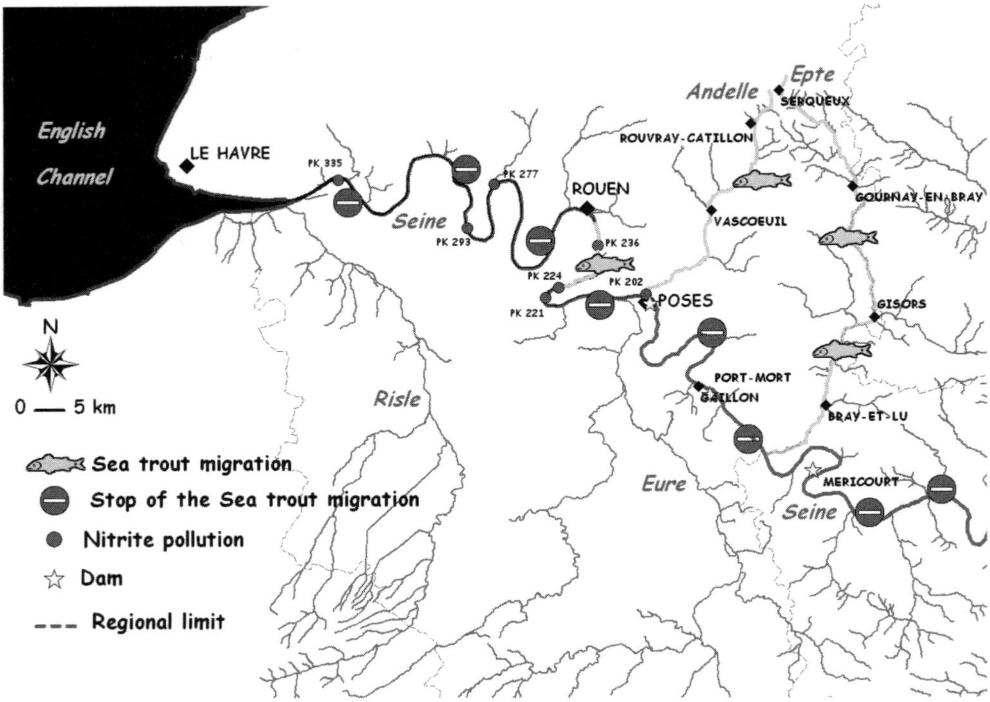

Figure 5.—Synthesis of reproduction possibilities for sea trout.

shed. For example, there was no technical (management) plan dealing with the conservation or enhancement of diadromous fish until 2007 (creation of Seinenormigr Association), while such plans exist for the Rhine, Loire, Garonne-Dordogne, Adour, and Rhône. This fact led us to conclude that the Seine, including its fauna and flora, are not appreciated by watershed inhabitants and the rest of society.

From our interviews, the upper part of the Seine (around Paris) is considered to be heavily polluted. People are unwilling to have any direct physical contact with the river water. Some local practices, like the cleaning of the Paris quays with hoses, emphasize this idea that the river carries waste. The water management structure in charge of the Paris region (SIAAP) works to reduce visible and olfactory pollution and to inform citizens of the evolution of the situation in the Seine ecosystem. On the SIAPP Web site (http://w1.siaap.fr) the water quality of the Seine is symbolized by different species of fish and we can find information about the organisms living in the turbid water. The altered, concrete, and rip-rap stabilized riverbanks do not help convince people that there is a relatively natural environment here, with animals and plants living below the surface of the river. However, in order to maintain the historic appearance of this part of Paris, it is forbidden to restore the riverbanks to a more natural state.

For most people interviewed (except anglers and fishermen), there is no obvious link between water quality and fish. However, most of them believed that some birds like the bittern *Botaurus stellaris* or the corn crake *Crex crex* were associated with the quality of the river environment. Fishing is limited to the lower part of the Seine (below Poses dam). For fishermen, the two major environmental problems are water pollution (mainly from agriculture) and obstacles to fish migration (hydroelectric dams on the main stem and water mills in the tributaries). These problems limit abundance of the species they target (trouts, northern pike *Esox lucius*, and whitefish). They appreciate eel and smelt, but, surprisingly, most of them do not desire the recovery of diadromous game species (salmonids). They are afraid of several possible negative consequences. They imagine that anglers would come from other regions, leading to an increase in the cost of fishing rights and an increase in fishing pressure. They also think that new, more restrictive management rules would result and that they would not be able to manage the river as they do now, especially the local arrangements concerning water levels and mills.

The reconciliation meeting we organized with people on both sides of the issue, supporters of mills and hydropower and supporters of fish, was successful. People who had not communicated before were able to discuss and exchange their viewpoints peacefully. They were able to find a solution to the problem of migration that was acceptable to all parties: leave the water mills open once a week instead of building expensive fish passage facilities. This meeting, proposed by our research group, which is not locally involved in the final decisions, provided them with an opportunity that they had not had via the official committees.

Technical Feasibility of Enhancement

There are techniques available to build effective fish passage for all of the diadromous species, but the cost of these facilities increases with the height of the barriers (dams and locks) and the width of the river. Stocking eggs, unfed fry, and alevins is feasible for salmonids if suitable habitats for juveniles are available. Artificial culture is also possible for clupeids, but techniques are not as well developed and controlled. For the other depleted diadromous fish species, there are no established culture methods.

In the vicinity of the Seine basin, we regularly observed brown trout and shads coming from other basins (Bresle River and Arques River for brown trout, Vire River and Orne River for Allis shad). Some individuals from these populations may stray and come back into the Seine basin, thereby strengthening the recovering populations. Results of the different steps in the process are summarized in Table 7.

Discussion

Recovery Conditions in the Future

The large-scale statistical models we used in this study are sound (Béguer et al. 2007) and can be used to establish projections for diadromous fish species recovery in the future, using at least one reasonable climate change scenario among many (Lassalle et al. 2009). Our study established that the historical status of the diadromous fish community is not a realistic reference for the future. Smelt, a well-known fish species and formerly among the most abundant in the Seine estuary, could probably not complete its life cycle in the Seine in this century. Conversely, Allis shad and thin-lipped mullet might encounter favorable environmental conditions in the future because they respond positively to increasing temperatures and because the Seine is in the northern part of their distribution range (Béguer et al. 2007; Lassalle et al. 2009).

Present Recovery Conditions

Suitable habitats for most of the diadromous fish species examined in this study exist within the Seine basin. However, environmental conditions also have to be favorable to make these suitable habitats functional. Such is not the case for the reproduction of lampreys. In certain locations and seasons, poor water quality (nitrites, heavy metals) can still be the limiting factor for the recovery of diadromous fish, owing to poor migration conditions.

Only those diadromous species using the lower part of the Seine basin, that have a short migration route within the basin, (smelt, twaite shad, thin-lipped mullet) or special colonization skills (eel), could easily reach suitable habitats. For the other species that must migrate further upstream to find suitable spawning habitats, barriers along the main -stem Seine and in the tributaries reduce spawning and nursery areas within the basin and represent the most important limiting factor for their recovery. The situation could be improved most easily on the Andelle, but many major improvements are needed on the main-stem Seine. It is presently possible to build effective passage for any migratory fish species given the extensive knowledge base on this topic (Larinier 2002; Porcher and Travade 2002). However, to construct fish passages requires money and political will. Unfortunately, local decision makers do not seem to support this idea.

Eco-Anthropological Approach

This dual approach is relatively new. The people we surveyed reacted very differently, depending on their link with fish. Fishermen and their representa-

Table 7.—Criteria used to select those diadromous fish species most likely to experience population recovery in the Seine basin. Criteria 1–4, 6 X: species fulfills conditions. Criterion 5. (0: no appropriation, 1: appropriation). Criteria 2 and 3 are satisfied according to the outputs of the bioclimate models, criterion 6 is satisfied when both artificial breeding and fish passes are mastered. Score is the sum of X for one species.

Considered species	(1) Historically present	(2) Favorability of the basin in 2000	(3) Favorability of the basin in 2100	(4) Favorable and accessible habitats	(5) Appropriation for the inhabitants	(6) Technical feasibility	(7) Potential of recovery	Score
European river lamprey	X	X	X					3
Sea lamprey	X	X	X					3
European eel	X	X	X	X	X			5
European sturgeon	X	X	X					3
Allis shad	X	X	X	X		X	X	6
Twaite shad	X	X	X	X			X	5
Vimba								0
Smelt	X	X		X	X		X	5
Houting						X		1
Atlantic salmon	X	X	X		X	X	X	6
Brown trout	X	X	X	X	X	X	X	7
Arctic char								
Thin-lipped mullet	X	X	X	X				4
Flounder	X	X	X	X				4

tives, as well as fisheries managers, were surprised to learn about the point of view of other citizens; indeed, in their mind it was clear that some species (i.e., salmonids) need much more consideration and management actions than others. Other citizens that were surveyed considered the process to be interesting. Enhancement of those species for which no costly management actions are needed only requires that citizens are kept better informed. In some cases, enhancement of a diadromous species will require that rules or practices be modified, possibly leading to conflicting situations (e.g., salmonids versus mills in the Andelle). Our experience demonstrated that with reconciliation, it is possible to find effective local solutions among groups that normally did not communicate. These solutions may require all groups to adapt to new rules. For example, the mill owners proposed opening the dam gates frequently during the upstream migration period of different fish species in the Andelle.

Interest of Stakeholders and Citizens

Our survey results suggested that citizens do not obviously establish a link between fish and river quality. The stronger link that exists between people and birds still needs to be established between people and fish. This issue must be addressed over the long term, with more communication and public education programs to help make fish more visible (e.g., observation room or Web camera inside fish passages).

Symbols and Species

Not all fish species have the same meaning for people. During the course of our study, several Atlantic salmon were captured in the lower part of the Seine and the first of these captures was widely publicized. Conversely, the capture of Allis shad upstream from Paris in 2004 ($N = 30$) and in 2007 ($N = 1$) was only mentioned a few times and not emphasized. This demonstrated that Atlantic salmon is generally perceived more as a symbol species than shads. Contrary to most of the other diadromous fish species, salmonids can be observed alive out of the water (because they jump and leap) and even in the water (because they reproduce in shallow areas of clear rivers). Their long-range migrations are well known. They also have several positive traits: they do not look like snakes (as do eels and lampreys), they are not linked with polluted areas (like mullet and flounder) or with food (like caviar from sturgeon). These traits all make salmonids useful to popularize restoration programs for them and other diadromous fish. During the period of the study, at least two poster advertising campaigns occurred, independently of our study, that used salmonids. The first promoted an outdoor clothing firm using two communication ideas. "The spring is back, it is time to practice sport," linked to the seasonal aspect of salmonid migration and the healthy and sporty aspect of these species swimming upstream and jumping over obstacles. The second poster promoted a Paris bus with this idea: "We clean the water which cleaned our bus." Here, salmonids were used as a symbol for pure, clear water. In both posters, fish are shown jumping out of the water.

Beyond this general perception, diadromous salmonids are fished for sport, and for anglers, they represent a kind of myth. For fish technicians, salmonid ecological requirements (Table 2), as well as their interest to anglers, make them an ideal umbrella species (Simberloff 1998; Collares-Pereira and Cowx 2004). However, because they can easily pass most of the small barriers, salmonids, with their great jumping ability, are no better indicators of habitat connectivity than shads. Moreover, because salmonids need good quality water for spawning, especially cool water, in the context of global warming, they may become less advantageous as a metaphor for future good water quality in the Seine basin.

Diadromous Flagship Species

Thirty years ago, most fish species (migratory or resident) had disappeared from the main stem of the Seine owing to poor water quality caused by polluted industrial discharges (Belliard 1994; Boet et al. 1999). However, improvements in water quality in the Seine in recent years have led to the recovery of a relatively rich diversity of resident fish returning from the tributaries (Burgeot et al. 1999; Rochard et al. 2007). This improvement was not followed by the recovery of diadromous fish, which were probably more affected by adverse anthropogenic influences than resident fish species because diadromous fish need good environmental conditions to successfully migrate to suitable spawning or growing habitats. This life history difference supports the importance of recovery plans for diadromous species as well as their use as flagship species.

An interesting case is the "Salmon 2000–Salmon 2020" program for the rehabilitation of the Rhine ecosystem (CIPR-IKSR 1999; www.iksr.org/index.php?id = 339). This program focuses on Atlantic salmon as a flagship species, but also includes management actions needed to enhance other species (brown trout, houting), to reintroduce others (European sturgeon, Allis shad) and also to reestablish habitat connectivity in the watershed. However, recently, two German regions (Nordrhein Westfalia and Hessen) participating in this program decided to replace Atlantic salmon with Allis shad as their symbolic species. This shift in focus reflects dissatisfaction with the slow progress in the recovery of Atlantic salmon. In the same way, Rhodanian shad *Alosa fallax* Rhodanensis has been preferred as a symbolic diadromous species for the restoration of the Rhône basin (Le Corre et al. 1997). These two situations emphasize the need to choose carefully the species that are used as symbolic or flagship species.

Improving biodiversity in a river basin implies that we consider more than one species (Ward 1998). Even if we have more interest in one species, a focus on the recovery of a single species does not appear to be a good strategy. A recent historical study of the diadromous fish communities in Maine rivers (USA) showed that coevolved diadromous fish may play an important role in key life history events for Atlantic salmon. Successful recovery of Atlantic salmon therefore requires the recovery of these coevolved diadromous fish species (shad, smelt, lamprey, etc.; Saunders et al. 2006).

In the Seine, we suggested smelt as a flagship species for the estuary, even if our model for 2100 forecasts its extinction in the future. Smelt will be used as a first step in the recovery of the river and will positively respond to the improvements in water quality and habitat connectivity that will also benefit other species. We propose brown trout as a flagship species for the tributaries of the lower basin. It will be used as a second step in the recovery and will also benefit from improvements in connectivity in the tributaries. Successful recovery of brown trout will require local reconciliation actions to defuse conflicting situations. And finally, we propose Allis shad as a flagship species for the middle and upper part of the Seine basin, where it will be used as a first step and will probably benefit from the increasing temperature and improvements in habitat connectivity along the main stem. However, more communication is required for better acceptance of the Allis shad as a flagship.

In conclusion, we suggest several paradigm changes for the conservation and management of diadromous fish. The historic fauna of a basin should not be considered as a reference for the future. Instead, long-term changes in environmental factors (especially temperature and hydrology) and the sensitivity of each diadromous fish species to projected environmental changes must be examined to establish a potential reference condition for the future. Citizens and decision makers must now be involved with fishermen and anglers in developing and implementing conservation and management plans, even though these groups do not necessarily share the same expectations and require more realistic and pragmatic choices. For financial and ecological reasons, the use of several flagship species is probably better than using only one species in developing plans for restoring good water and habitat quality throughout a large river basin. To be successful, people of different regions must be involved with a restoration plan and be committed to its success.

Acknowledgments

We thank the "Seine aval" scientific program for their financial support (grant O2/2005/03). We also thank Miran Aprahamian and William Richkus for their pertinent reviews and Ron Klauda for his very valuable help in the improvement of the manuscript.

References

Anonyme. 2003. Système d'évaluation de la qualité de l'eau des cours d'eau. Grilles d'évaluation SEQ-Eau. Ministère de l'Ecologie et du Développement Durable & Agences de l'eau, Paris.

Baglinière, J. L., and P. Elie. 2000. Les aloses (*Alosa alosa* et *Alosa fallax* spp.). Ecobiologie et variabilité des populations. Cemagref - Cemagref, Paris.

Baglinière, J. L., and G. Maisse, editors. 1991. La truite: biologie et écologie. INRA Editions, Paris.

Bain, M. B., J. T. Finn, and H. E. Booke. 1985. A quantitative method for sampling riverine microhabitats by electrofishing. North American Journal of Fisheries Management 5:499–506.

Béguer, M., L. Beaulaton, and E. Rochard. 2007. Distribution and richness of diadromous fish assemblages in western Europe: large scale explanatory factors. Ecology of Freshwater Fish 16:221–237.

Belliard, J. 1994. Le peuplement ichtyologique du bassin de la Seine. Rôle et signification des échelles temporelles et spatiales. Doctoral dissertation. Paris VI, Paris.

Belliard, J., R. Berrebi dit Thomas, and D. Monnier. 1999. Fish communities and river alteration in the Seine basin and nearby coastal streams. Hydrobiologia 400:155–166.

Bloor, D. 1982. Sociologique de la logique ou les limites de l'épistémologie. Pandore, Paris. Originally published as Knowledge and social imagery (1976, Chicago University Press, Chicago).

Boet, P., J. Belliard, R. Berrebi-dit-Thomas, and E. Tales. 1999. Multiple human impacts by the city of Paris on fish communities in the Seine river basin, France. Hydrobiologia 410:59–68.

Boisneau, P., C. Mennesson-Boisneau, and J. L. Baglinière. 1990. Description d'une frayère et comportement de reproduction de la grande alose (*Alosa alosa* L.) dans le cours supérieur de la Loire. Bulletin Français de la Pêche et de la Pisciculture 316:15–23.

Burgeot, T., C. Minier, G. Bocquené, F. Vincent, J. Cachot , V. Loizeau, A. Jaouen, P. Miramand, T. Guyot, P. Lesueur, E. Rochard, and P. Boet. 1999. Des organismes sous stress. Ifremer Editions, Fascicule n° 14 du Programme Scientifique Seine Aval, Issy-les-Moulineaux Cedex, France.

CIPR-IKSR. 1999. Le Rhin est-il redevenu un fleuve salmonicole? "Saumon 2000" Commission Internationale pour la protection du Rhin, Koblenz, Germany.

Collares-Pereira, M. J., and I. G. Cowx. 2004. The role of catchment scale environmental management in freshwater fish conservation. Fisheries Management and Ecology 11:303–312.

Commission des communautés européennes. 2005. Proposition de règlement du conseil instituant des mesures de reconstitution du stock d'anguille européenne. Commission des communautés européennes, COM(2005) 472 final 2005/0201 (CNS), Bruxelles, Belgium.

Council of Europe. 1992. Council Directive 92/43/EEC of 21 May 1992 on the conservation of natural habitats and of wild fauna and flora. Official Journal L 206, 22/07/1992, pp. 0007–0050. Available: http://eur-lex.europa.eu/LexUriServ/LexUriServ.do?uri=CELEX:31992L0043:EN:HTML (April 2008).

de Groot, S. J. 2002. A review of the past and present status of anadromous fish species in the Netherlands: is restocking the Rhine feasible? Hydrobiologia 478(1–3):205–218.

Dedual, M., I. D. Maxwell, J. W. Hayes, and R. R. Strickland. 2000. Distribution and movements of brown (*Salmo trutta*) and rainbow trout (*Oncorhynchus mykiss*) in Lake Otamangakau, central North Island, New Zealand. New Zealand Journal of Marine and Freshwater Research 34:615–327.

Dekker, W. 2003. On the distribution of the European eel (*Anguilla anguilla*) and its fisheries. Canadian Journal of Fisheries and Aquatic Sciences 60:787–799.

Deniel, C. 1981. Les poissons plats (téléostéens, pleuronectiformes) en baie de Douarnenez: reproduction, croissance et migration des Bothidae, Scophthalmidae, Pleuronectidae et Soleidae. Doctoral dissertation. Université de Bretagne Occidentale, Brest, France.

Ducharne, A., S. Théry, P. Viennot, E. Ledoux, E. Gomez, and M. Déqué. 2003. Influence du changement climatique sur l'hydrologie du basin de la Seine. VertigO, La revue en sciences de l'environnement sur le WEB 4:1–24.

Duhamel, S., N. Gouneau, T. Lefrançois, S. Mayot, Y. Perrot, E. Feunteun, and E. Parlier. 2004. Action 4.4.2: Etat de la ressource halieutique dans l'estuaire amont, Rapport final 2003. L'estuaire dulçaquicole, distribution longitudinale de l'ichtyofaune et comparaison de différents types d'habitats. Ifremer Editions, Programme Scientifique Seine Aval, Issy-les-Moulineaux Cedex, France.

European Parliament and Council of Europe. 2000. Directive 2000/60/EC of the European Parliament and of the Council of 23 October 2000 establishing a framework for Community action in the field of water policy. Official Journal L 327, 322.312.2000, pp. 2001–2073. Available: http://eur-lex.europa.eu/LexUriServ/LexUriServ.do?uri=CELEX:32000L0060:EN:NOT (April 2008).

Euzenat, G., C. Pénil, and J. Allardi. 1992. Migr'en Seine. Stratégie pour le retour du saumon en Seine. SIAAPP et Conseil Supérieur de la Pêche, Paris.

Fernandez-Pasquier, V. 1999. *Acipenser sturio* L. in the Guadalquivir River, Spain. Water regulation and fishery as factors in the stock decline from 1932 to 1967. Journal of Applied Ichthyology 15:133–135, 1999.

Fielding, A. H., and J. F. Bell. 1997. A review of methods for the assessment of predictor error in conservation presence/absence models. Environmental Conservation 24:38–49.

Gouneau, N. 2003. L'éperlan (*Osmerus eperlanus*) dans l'estuaire de la Seine: premiers éléments sur la biologie et l'écologie d'une espèce en réapparition depuis 2000. Diplôme D' Nospace Etudesupérieur Universitaire. Paul Sabatier, Toulouse, France.

Gueguen, J., and P. Prouzet, editors. 1994. Le saumon atlantique. Ifremer, Nantes, France.

Hastie, T. J., and R. J. Tibshirani. 1990. Generalized additive models. Chapman and Hall, London.

Haury, J., D. Ombredane, and J-L. Baglinière. 1999. The habitat of the brown trout (*Salmo trutta*) in running waters. Pages 37–89 *in* J-L. Baglinière, and G. Maisse, editors. Biology and ecology of the brown and sea trout. Springer-Praxis, Chichester, UK.

Hoffmann, R. C. 1996. Economic development and aquatic ecosystems in medieval Europe. American Historical Review:631–669.

Hoffmann, R. C. 2005. A brief history of aquatic resource use in medieval Europe. Helgolaender Meeresuntersuchungen 59:22–30.

Holmes, J. A., and P. Lin. 1994. Thermal niche of larval sea lamprey, *Petromyzon marinus*. Canadian Journal of Fisheries and Aquatic Sciences 51:253–262.

IPCC (Intergovernmental Panel on Climate Change). 2000. Emissions scenarios. Cambridge University Press, Cambridge, UK.

Jager, H. I., H. E. Cardwell, M. J. Sale, M. S. Bevelhimer, C. C. Coutant, and W. Van Winkle. 1997. Modelling the linkages between flow management and salmon recruitment in rivers. Ecological Modelling 103:171–191.

Jego, S., C. Gazeau, P. Jatteau, P. Elie, and E. Rochard. 2002. Les frayères potentielles de l'esturgeon européen *Acipenser sturio* L. 1758 dans le bassin Garonne-Dordogne. Méthodes d'investigation, état actuel et perspectives. Bulletin Français de la Pêche et de la Pisciculture 365/366:487–505.

Jensen, F. B. 2003. Nitrite disrupts multiple physiological functions in aquatic animals. Comparative Biochemistry and Physiology A (135):9–24.

Keith, P., and J. Allardi, editors. 2001. Atlas des poissons d'eau douce de France, volume 47. Muséum national d'Histoire naturelle, Paris.

Kelly, F. L., and J. J. King. 2001. A review of the ecology and distribution of three lamprey species, *Lampetra fluviatilis* (L.), *Lampetra planeri* (Bloch) and *Petromyzon marinus* (L.): a context for conservation and biodiversity considerations in Ireland. Proceedings of the Royal Irish Academy 101B(3):165–185.

Kilani, M. 1995. Les anthropologues et leur savoir: du terrain au texte. Pages 65–100 *in* J-M. Adam, M-J.

Borel, C. Calame, and M. Kilani, editors. Le discours anthropologique. Description, narration, savoir. Payot, Lausanne, Switzerland.

Kohlhorst, D. W. 1980. Recent trends in the white sturgeon population in California's Sacramento–San Joaquin estuary. California Fish and Game 66(4):210–219.

Larinier, M. 2001. Environmental issues, dams and fish migration. Pages 45–89 *in* G. Marmulla, editor. Dams, fish and fisheries: opportunities, challenges and conflict resolution. Food and Agriculture Organization of the United Nations, Rome.

Larinier, M. 2002. Biological factors to be taken into account in the design of fishways, the concept of obstruction of upstream migration. Bulletin Français de la Pêche et de la Pisciculture 364(Supplement):23–38.

Lassalle, G., M. Béguer, L. Beaulaton, and E. Rochard. 2009. Learning from the past to predict the future: responses of European diadromous fish to climate change. Pages xxx–xxx *in* A. J. Haro, K. L. Smith, R. A. Rulifson, C. M. Moffitt, R. J. Klauda, M. J. Dadswell, R. A. Cunjak, J. E. Cooper, K. L. Beal, and T. S. Avery, editors. Challenges for diadromous fishes in a dynamic global environment. American Fisheries Society, Symposium 69, Bethesda, Maryland.

Le Corre, M., J. L. Baglinière, R. Sabatié, J. Y. Menella, and D. Pont. 1997. Données récentes sur les caractéristiques morphologiques et biologiques de la population d'alose feinte du Rhône (*Alosa fallax rhodanensis* Roule, 1924). Bulletin Français de la Pêche et de la Pisciculture 346:527–545.

Le-Teuff, L. 1996. Premiers éléments de l'écologie de la lamproie marine (*Petromyzon marinus*) dans une rivière bretonne Le Scorff. Ecole Supérieure d'Ingénieurs et de Techniciens pour l'Agriculture, Rennes, France.

Leathwick, J. R., J. Elith, and T. Hastie. 2006. Comparative performance of generalized additive models and multivariate adaptive regression splines for statistical modelling of species distributions. Ecological Modelling 199(2):188–196.

Maes, J., M. Stevens, and J. Breine. 2008. Poor water quality constrains the distribution and movements of twaite shad *Alosa fallax fallax* (Lacépède, 1803) in the watershed of river Scheldt. Hydrobiologia 602(1):129–143.

Maisse, G., and J. L. Baglinière. 2001. Conservation et restauration des populations de saumon Atlantique en France. NRA Département HyFS-GIS GRISAM, SCRIBE INRA et UMR INRA-ENSA, Rennes, France.

Maisse, G., and J. Y. Costard. 1977. Etude préliminaire sur la protection des équilibres biologiques sur le bassin de la Dordogne en vue de la restauration du saumon. Atlantique. Ministère de la Culture et de l'Environnement, Paris.

Maitland, P. S. 2003. Ecology of the river, brook and sea lamprey. English Nature, Conserving Natura 2000 Rivers Ecology Series No. 5, Peterborough, UK.

Malavoi, J. R. 1989. Typologie des faciès d'écoulement ou unités morphodynamiques des cours d'eau à haute énergie. Bulletin Francais de la Peche et de la Pisciculture 315:189–210.

Masson, G. 1987. Biologie et écologie d'un poisson plat amphihalin, le flet (*Platichthys flesus flesus* Linné, 1758) dans l'environnement ligérien : distribution, démographie, place au sein des réseaux trophiques. Doctoral dissertation. Bretagne Occidentale, Brest, France.

Mennesson-Boisneau, C., P. Boisneau, and J. L. Baglinière. 1993. Migration de reproduction des aloses dans le bassin de la Loire. Incidence des obstacles. Bulletin Scientifique et Technique, Inra 27:23.

Mitchell, T. D., T. R. Carter, P. D. Jones, M. Hulme, and M. G. New. 2004. A comprehensive set of high-resolution grids of monthly climate for Europe and the globe: the observed record (1901–2000) and 16 scenarios (2001–2100). Tyndall Centre, Norwich, UK.

Möller, H., and U. Scholz. 1991. Avoidance of oxygen-poor zones by fish in the Elbe River. Journal of Applied Ichthyology 7:76–82.

Nicola, G. G., B. Elvira, and A. Almodovar. 1996. Dams and fish passage facilities in the large rivers of Spain: effects on migratory species. Archives of Hydrobiology 10(Supplement 113):375–379.

Porcher, J. P., and F. Travade. 2002. Fishways: biological basis, limits and legal considerations. Bulletin Français de la Pêche et de la Pisciculture 364(Supplement):9–20.

Pronier, O., and E. Rochard. 1998. Fonctionnement d'une population d'éperlan (*Osmerus eperlanus*, Osmériformes Osmeridae) située en limite méridionale de son aire de répartition, influence de la température. Bulletin Français de la Pêche et de la Pisciculture 350–51:479–497.

R foundation core team. 2005. R: a language and environment for statistical computing. R Foundation for Statistical Computing, Vienna, Austria.

Rochard, E., J. Marchal, P. Pellegrini, M. Béguer, D. Ombredane, C.Gazeau, J. L. Baglinière, E. Menvielle, and O. Croze. 2007. Identification éco-anthropologique d'espèces migratrices, emblématiques de la reconquête d'un milieu fortement anthropisé, la Seine. Programme Seine aval. No. 104. Cemagref Bordeaux, Bordeaux, France; Rennes Agrocampus, Rennes, France; et Muséum National d'Histoire Naturelle, Paris.

Roule, L. 1920. Les espèces d'aloses du bassin de la Seine. Bulletin du Muséum National d'Histoire Naturelle 26:610–611.

Sabatié, M. R., and J. L. Baglinière. 2001. Some ecobiological traits in morrocan shads; a cultural and socio-economic value interest which has disappeared. Bulletin Français de la Pêche et de la Pisciculture 362/363:903–917.

Saunders, R. L., M. A. Hachey, and C. W. Fay. 2006. Maines's diadromous fish community: past, present, and implications for Atlantic salmon recovery. Fisheries 31:537–547.

Simberloff, D. 1998. Flagships, umbrellas, and keystones: is single-species management passe in the landscape era? Biological Conservation 83(3):247–257.

St. Pierre, R. A. 2003. A case history: American shad restoration on the Susquehanna River. Pages 315–321 *in* K. E. Limburg and J. R. Waldman, editors. American Fisheries Society, Symposium 35, Bethesda, Maryland.

Thuiller, W. 2003. BIOMOD: optimizing predictions of species distributions and projecting potential future shifts under global change. Global Change Biology 9:1353–1362.

Walther, G. R., E. Post, P. Convey, A. Menzel, C. Parmesan, T. J. C. Beebee, J.-M. Fromentin, O. Hoegh-Guldberg, and F. Bairlein. 2002. Ecological responses to recent climate change. Nature (London) 416:389–395.

Ward, J. V. 1998. Riverine landscapes: biodiversity patterns, disturbance regimes and aquatic conservation. Biological Conservation 83(3):269–278.

American Fisheries Society Symposium 69:713–740, 2009

Natural Heritage, Anthropogenic Impacts, and Biopolitical Issues Related to the Status and Sustainable Management of American Eel: A Retrospective Analysis and Management Perspective at the Population Level

Rob MacGregor*
Ontario Ministry of Natural Resources
Post Office Box 7000, 300 Water Street - 5 North, Peterborough, Ontario K9J 8M5, Canada

John M. Casselman
Queen's University, Department of Biology
Biosciences Complex, 116 Barrie Street, Kingston, Ontario K7L 3N6, Canada

William A. Allen
Heritage One, 9 First Avenue, Box 85, Burk's Falls, Ontario P0A 1C0, Canada

Tim Haxton
Ontario Ministry of Natural Resources
300 Water Street - 4 South, Peterborough, Ontario K9J 8M5, Canada

John M. Dettmers
Great Lakes Fishery Commission
2100 Commonwealth Boulevard, Suite 100 Ann Arbor, Michigan 48105, USA

Alastair Mathers
Ontario Ministry of Natural Resources, RR 4, Picton, Ontario K0K 2T0, Canada

Steve LaPan
New York State Department of Environmental Conservation, Cape Vincent Fisheries Station
Post Office Box 292, Cape Vincent, New York 13618, USA

Thomas C. Pratt
Fisheries and Oceans Canada, Great Lakes Laboratory for Fisheries and Aquatic Sciences
1 Canal Drive, Sault Ste Marie, Ontario P6A 6W4, Canada

Peter Thompson
Fisheries and Oceans Canada, 520 Exmouth Street, Sarnia, Ontario N7T 8B1, Canada

Max Stanfield
Fisheries and Oceans Canada, 200 Kent Street, Ottawa, Ontario K1A 0E6, Canada

Lucian Marcogliese
30 Salem Road, Ameliasburgh, Ontario K0K 1A0, Canada

Jean-Denis Dutil
Pêches et Océans Canada, Institut Maurice-Lamontagne, C.P. Mont-Joli, Quebec G5H 3Z4, Canada

* Corresponding author: rob.macgregor@ontario.ca

Abstract.—We examine historical, archaeological, and current patterns in American eel *Anguilla rostrata* use, abundance, and distribution to improve understanding of current population-level status. Our research indicates that distribution and abundance has changed significantly in response to the cumulative impacts of fishing, turbine mortality, and major loss of freshwater habitat. The 1950–1970 peaks in dam construction and turbine mortalities, together with the unprecedented North American harvests in the 1970s, have lead to a perilous synergy of effects at the population level. Based on our findings, we call for coordinated conservation and management actions for American eel across North America. Preservation of life cycle diversity and coordinated conservation actions are required across the range to ensure continued and improved societal benefits, protect the legacy of cultural and natural heritage values, restore ecological services, and reinstate the benefits to biodiversity provided by this unique and important species. Finally, we describe key elements and recent progress in recovery planning.

Introduction

American eel *Anguilla rostrata*, once a widespread, abundant, and important species throughout all accessible freshwater and coastal habitats of eastern North America gradually has diminished in abundance and distribution over the past century, particularly in freshwater habitats. MacGregor et al. (2008) describe how the invisible collapse phenomenon (Post et al. 2002) applies to American eel, given its wide geographic range, diverse and localized fisheries, and paucity of coordinated management and stock assessment at the population level. In fact, eels appear to have declined or disappeared from many watersheds long before adequate records were kept. Consequently, biologists, particularly those new to the profession, often are surprised to learn that eels previously existed, were important, and were locally abundant in places like Lake Timiskaming in the upper Ottawa River watershed, northern Ontario. This loss of corporate memory, with respect to past abundance and distribution, has contributed to shifting baselines (Pauly 1995; Sáenz-Arroyo et al. 2005; Pinnegar and Engelhard 2008) in our understanding of American eel status, leading to difficulties in implementing adequate conservation measures.

While now a species of concern in some jurisdictions (COSEWIC 2006; OMNR 2007), shifting baselines and perspectives on the status and importance of eels clearly influences the management attention and conservation priority they receive (MacGregor et al. 2008). For instance, similar to the connections between humans and Pacific salmon (Lackey et al. 2006a), indigenous peoples and European settlers experienced long and close associations with the American eel in freshwater. Like Pacific salmon, eels took on spiritual status with some people in eastern North America (Prosper and Paulette 2003) and were clearly important for sustenance and commerce in North America for centuries. In contrast to Pacific salmon, however, the earlier importance and abundance of eels appear to have been forgotten by the mid-1900s. By that time, eels were viewed by some fisheries managers as a direct loss of nutrients in freshwater systems and as competitors with trout and salmon, and they were certainly in disfavor with many anglers (Smith and Saunders 1955; Eales 1968). These perceptions have limited efforts to conserve and manage American eel on a sustainable basis during the 20th century.

Markets for American eel recently have grown substantially, and its commercial value has generated renewed attention. At the same time, government and societal interests in protecting and restoring biodiversity, ecological services, and natural heritage values have increased, sparking considerable attention on this highly unique and historically important species. Others view eels as a bellwether species sending an integrated message about the state of the environment (Prosper and Paulette 2003; Hoag 2007; Zettler 2007). These interests have led to growing concern over the condition of the American eel population (Casselman 2003; Dekker et al. 2003; Commanda 2007; MacGregor et al. 2008).

Varying opinions regarding the status of American eel exist among the some 25 jurisdictions having responsibility for their management (ASMFC 2000, 2006; COSEWIC 2006; OMNR 2007; U.S. Office of the Federal Register 2007). Perceptions of eel status appear to be parochial in nature, depending in part on the location of a jurisdiction in relation to resource use and distance from the common source of recruitment, the Sargasso Sea. Given conflicting opinions regarding the population-level status of American eel, we re-establish indications of former abundance and distribution across North America by means of a comprehensive synthesis and review of historical and archaeological information. After examining changes in American eel abundance, we summarize the need for conservation action and coordinated binational management.

American Eel Life Cycle and Related Management Complexities

The American eel is a long-lived (Casselman 2003), semelparous, highly migratory, and apparently panmictic species (Avise et al. 1986; Wirth and Bernatchez 2003). Its complex life cycle is illustrated in Figure 1, and more detailed descriptions are available in Tesch (1977) and COSEWIC (2006). American eel are described as catadromous, but this life history strategy is facultative because some eels appear to spend their entire life cycle in marine or brackish environments (Lamson et al. 2006). Catadromy in anguillid eels apparently evolved from local migrations of tropical eels as a result of long-distance dispersal of leptocephali from tropical spawning sites (Tsukamoto et al. 2002). Nevertheless, freshwater occupancy clearly has been an important part of their population-level life history strategy. Many eels, particularly at the extremities of their range in Ontario and Québec watersheds of the St. Lawrence River/Great Lakes basin, spend almost two decades in freshwater (Casselman 2003), and virtually all are females. Impacts on the freshwater contingent may be especially important as they affect predominantly highly fecund females.

Given the panmictic nature of the eel life history, anthropogenic impacts accumulate over a wide geographic range on one common spawning stock. Significant harvests occur at all continental life stages (MacGregor et al. 2008), and mortalities and habitat losses have been pronounced across the range. Anthropogenic stressors are widespread and cumulative across much of the North American range, having population-level effects that must be considered in management.

Historical Importance of American Eel to Early Indigenous Peoples and European Settlers

Baseline understanding of the American eel has eroded over the 20th century, as information and

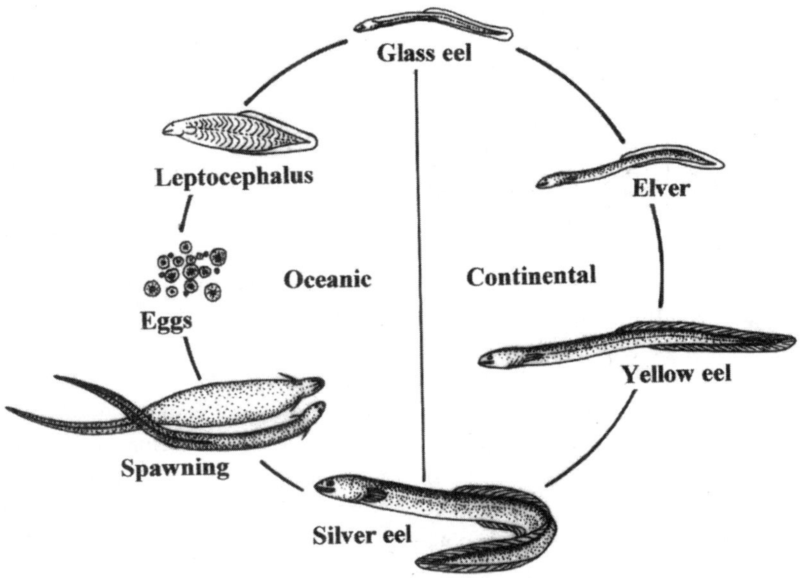

Figure 1.—American eel lifecycle.

knowledge have been lost or forgotten. Anecdotal information from an earlier time, and early eyewitness accounts gleaned from historical documents, archaeological records, and traditional ecological knowledge of indigenous peoples, provide valuable insight into the past distribution, abundance, and importance of American eel (Pauly 1995; Allen 2007a; Pinnegar and Englehard 2008). Here, we provide a comprehensive synthesis of such information for American eel, progressing through time to the 20th century.

Eels in Traditional Indigenous Lifeways

Eels have played a long and significant role in the lives of indigenous peoples in eastern North America. Junker-Anderson (1988) states that "most contemporary observers agree the eel was the single most important of the fish species for all of the Iroquoian peoples, with the possible exception of the Mohawks." Eels also were important in traditional Algonquin territory (Allen 2007b, 2007c). In precontact and early contact times, eels were so abundant, widely dispersed, easy to catch at certain times of year, highly nutritious, and easy to preserve that they were considered as gifts of Manitou (or Manidoo, the Good Spirit).

Indigenous peoples in eastern North America had many names for American eel and place names for important fishing areas, and this information provides valuable clues in improving our understanding of past abundance and distribution. The Mi'kmaq people called them Kat. The Algonquin words pimizi (McGregor 1994) and Ojibwe bimizi (Baraga 1878), derived from Algonquin pimide (McGregor 1994) or Ojibway bimide (Baraga 1880) meaning greasy, emphasize the high-calorie, high-fat content of eel as food. The Seneca, south of Lake Ontario, called the eel goda:noh (Bardeau 2002). Swatara Creek in Swatara Township, Pennsylvania is named after an Iroquoian or Susquehannock term meaning "where we feed on eels" (Dohne 2004). An Indian village at the site of Sunbury, Pennsylvania on the Susquhanna River was originally called Schahamoki, "the place of the eels" (Heckewelder 1823). Franklin Hough describes a place on the Onondaga River called Ongontenayea or Aquegontenayea, meaning "a place remarkable for eels," some 8 km downstream of the river outlet at Oneida Lake (Munsell 1861).

The Mi'kmaq had a deep connection with eels as a traditional and important food source (Prosper 2001; Prosper and Paulette 2002, 2003). Eels also were considered to have many spiritual qualities and to be a spiritual being; they were used as a ceremonial offering and found their place in traditional stories. Eels were offered in a ceremony called Apuknajit, feeding of the Grandfather (Marshall 1997), to give thanks to the spirits for allowing the people to survive through the most difficult time of the year. Seventeenth-century accounts indicate that "if they had roasted an Eel, they also believed that this would prevent them from catching [one] at a different time" (Denys 1672:480) and that frightened people threw eels on the fire to appease the devil (Thwaites 1896–1901). Such taboos indicate spiritual connections with eels. This spiritual connection continues today as evidenced by a 2007 letter to Fisheries and Oceans Canada by Elder Dr. William Commanda, Algonquin spiritual leader, in which he wrote,

> I am raising my voice concerning the plight of American eel. The eel has been of spiritual, nutritional and material importance to the Indigenous Peoples of the eastern seaboard, and to my ancestors of the Ottawa River Watershed, since time immemorial...I believe the eel spirit is intrinsic to the sacred Seven Fires Prophecy Wampum Belt that I have carried for the people for thirty six years. It is this prophecy that tells us that humanity is now at a cross roads, and that we need to regenerate our relationship with Mother Earth and each other. (Commanda 2007)

Indigenous peoples established close connections with eels in the St. Lawrence River, Lake Ontario, and Ottawa River watersheds. Before timber harvesting, in areas such as the Madawaska River watershed in Ontario, eel harvesting was part of an efficient food procurement process involving farming, fishing, hunting, and collecting (Allen 2004, 2007d). Le Jeune's 1634 description (Thwaites1896–1901) notes the Montagnais' reliance on eels from the St. Lawrence River and describes fishing techniques and preparation of eels. Father Louis Hennepin wrote that "the most considerable fishery...is that of the eels, which are very large" (Hennepin in Thwaites 1903:2:524). Fresh eels were the primary food during September and October. Eels were dried when they were abundant, providing sustenance from November through January. Smoked eels, being lightweight and highly nutritious (Casselman 2003), were commonly used as traveling food (Junker-Anderson 1988). Indigenous peoples often gave eels as gifts to early European travelers or used them in trade (Beauchamp 1916).

Twenty-one Ontario archaeological sites within the Lake Ontario–St. Lawrence River watershed yielded eel remains of precontact age between 500 and 1,000 years old (Pearce 1977; Junker-Andersen 1988; Fitzgerald 1990; Reed 1993; Finlayson 1998). In addition, two archaeological sites more than 4,000 years old on the Québec side of the Ottawa River yielded substantial eel remains (Clermont and Chapdelaine 1998; Clermont 1999; Clermont et al. 2003), and a complex of stone weirs and pools was documented in 2007 in Rapides des Allumettes (BkGg-29[1]) (W. A. Allen, Heritage One, unpublished data). At Pointe-du-Buisson Archaeological Park, near Montréal, two sites yielded eel bones dating from 920 to 940 A.D. (Courtemanche 2003) and 500–1000 A.D. in a second case (Cossette 1995). Aboriginals harvested eels as recently as the early 20th century within the upper Ottawa River watershed in Ontario's Mattawa and Montréal rivers (Allen, unpublished data).

In Virginia, traditional homeland of the Virginia Algonquins (Feest 1978), American eel remains were documented at several sites predating the arrival of Europeans (Whyte 1988, 2001, 2002; Whyte and Thompson 1989; Egloff et al. 1994; Clark et al. 2005). Some artifacts at a site in Delaware tested positively for American eel protein residues (Puseman and Cummings 2004). In New York State at the Bates site in Chenango County, a site carbon-dated to 1190 A.D. yielded eel remains (Ritchie and Funk 1973).

Indigenous peoples fished eels by using specialized eel splint baskets, weirs, and spears, particularly at night. Le Jeune described

> certain ones who will take three hundred in one night…After being well smoked, they are piled together in large packages, about a hundred in each. Here you have food up to the season of the snow, which brings them the Moose. (Thwaites 1896–1901:6:311–313)

De Quen noted that "eels are so abundant [at Onondaga] in the summer that a single fisherman can harpoon as many as a thousand in one night" (Thwaites 1896–1901:42:97). Father Louis Hennepin (Thwaites 1903:2:524) described the Iroquoians spearing an "infinite number" of eels by the light of a birch bark torch at night, being particularly successful because the "great white porpoises which pursue them make them fly towards the banks of the river where porpoises cannot follow."

Early explorers and missionaries described weirs as important eel fishing methods of indigenous peoples (Charlevoix 1761; Beauchamp 1916). The Jesuit Relation of 1634 describes the construction of weirs "capable of holding five or six hundred eels" (Thwaites 1896–1901:6:309). Stones were extended out at the sides to guide eels into a trap. A weir historically used by Mi'kmaqs to trap eels on the Mersey River, Nova Scotia is estimated to be 4,000 years old (Prosper and Paulette 2002). In fall on the Delaware River, the Lenape would catch thousands of downstream migrants in a single night, using a kind of basket weir or eel pot (Wade 2007). Archaeological study has been undertaken at several stone weirs that have a unique downstream "V" suggesting the ancient capture of downstream-migrating mature eels. These weirs include the 2,500-year-old Satucket River weir in Massachusetts (Watts and Watts 2004), 11 weirs on the Passaic River in New Jersey (Lutins 1992; Lutins and DeCondo 1999), and the Otselic River Weir in New York (E. McDowell-Loudan, State University of New York College at Cortland, personal communication) on a tributary of the Susquehanna River where remnants of more stone weirs are still visible in low water (Dohne 2004). In addition, sharpened wooden stakes in Maine's Sebasticiook Lake fish weir in the Kennebec River watershed, where eels were abundant, have been dated to between 5,800 and 1,700 years old (Miller et al. 2006).

Indigenous peoples used eels in multipurpose material and medicinal fashions. The Mi'kmaqs used eel skins as wrist and ankle straps to provide support and relief from rheumatism. Eel skins (kadaagel) were used as braces and bandages: "juniper balsam and eel skin make a good poultice for sprains" (Lacey

[1] This paper uses the standard "Borden" numbering system used across Canada at registered archaeological sites (see Garrad 1967).

1977; Prosper 2001) and to set broken bones (Allen, unpublished data). In New York, eel oil was collected by parboiling and was used to relieve ear aches (Allen, unpublished data). Eel skins also were used as hair braids; Champlain described indigenous girls attending dances "with a tuft of their hair behind tied up with eel skin which they arrange to serve as a band" (Champlain Society 1922–1936:4:312). Eel skins were used as bindings for sleds, moccasins, and clothing and for tying spears and harpoons on sticks (Prosper and Paulette 2003) since they tighten when dried. In Ontario, eel skins were used as bow grips, and fatty eel flesh was used to waterproof clothing (Allen, unpublished data).

In the United States, at least three of the New York State six nations—the Cayuga, Onondaga, and Tuscarora—had, at one time or another, an eel clan (Morgan 1877). Clans named after fish are rare so eel clan designations are a further indicator of the significance of this species (Casselman 2003). Tooker (1978) documented three of 14 Onondaga titles that indicated eel clans. Before 1800, the Mahican had a yellow eel clan (Barton 1797). In Indiana, in the upper Mississippi watershed within traditional eel range, an entire book has been written about the Eel River Tribe, a Wabash confederacy (Floyd 2007). Indigenous surnames include Chief Swimming Eel (Speck 1940) among the Scaghticoke on the New York/Connecticut border and Chief Eels from the Kawartha Lakes area of Ontario (Guillet 1957).

As Europeans moved west, many indigenous peoples moved outside the traditional native range of eels to the Lake Erie watershed where eel clan significance, over a few generations, appears to have dissolved into subclans. Nevertheless, indigenous peoples have maintained their connection with eels in many areas of Canada (Prosper 2001). While eel weirs have long been banned in many areas (Canada 1868a; Fraser 1912; Pulla 2003), indigenous peoples in the Maritimes region have continued their practice of spearing eels during both open water seasons and ice-covered periods. However, there is concern that as eels decline in abundance, the close connection and tradition associated with eels soon will be lost (Prosper 2001).

Importance of Eels to European Settlers

When European settlers arrived in North America, they did not keep harvest records, but historical anecdotes survive. For instance, Father Francois du Creux (1664) suggests "an almost unlimited supply" of eels. An account from 1652 to 1653 at Trois-Rivières notes that "the eel constitutes a manna exceeding all belief" (Thwaites 1896–1901:40:215–217); "a wonderful manna for this country, and one that costs nothing beyond the catching" (Thwaites 1896–1901:45:193). The eels kept well and were considered tastier than those in France because they swam in the vast waters of the St. Lawrence River rather than their muddy environments in France.

In the 1800s, eels of large size and fine quality were taken everywhere within the Gulf of St. Lawrence: "They are in reality a very valuable description of fish; they are very numerous, very prolific... They are in great esteem for the table, and the consumption in our large cities is considerable" (Canadian Naturalist and Geologist 1859:98). In the late 1800s, the lower St. Lawrence River eel fisheries were considered to be the most productive in the world, and their quality unsurpassed (The New York Times 1880a). Here, they deployed weirs, catching eels during tidal events in September and October; one weir was reported to have taken 3,000 eels at a single tide, but an average good catch in the late 1800s was considered to be 1,000–1,500 eels. In 1870, 100,000 eels were reported to have been taken at the mouth of Rivière Ouelle. Farmers and other individuals along the St. Lawrence harvested eels by spears, usually at low tide, often making as much profit as the profit derived from farming the land (The New York Times 1880a).

Early settlers harvested large quantities of eels in weirs well upstream on the west branch of the Delaware River at Schewakhawken (now Sunbury, Pennsylvania) (Heckewelder 1823; Munsell 1880). "Prodigious" quantities of eels were taken by weirs and hurdles in the St. Lawrence River "from Québec, as high as Trois Rivière...[where] eels come down from Lake Ontario...great quantities are taken during the time this fishery lasts, they are salted and barreled up like herrings" (Charlevoix 1761:1:261–262). Eels were in such large quantity that they were easily caught and stored at little expense and hence were important as a staple food item for the poor (Charlevoix 1761:1: 228).

The Canadian Sessional Papers document many thousands of barrels of eels harvested commercially in areas extending from Ontario to the Atlantic provinces. Many were pickled for export to

Europe or taken in "barges filled with eels" to New York City (The New York Times 1936), but significant quantities apparently were sold fresh locally. The 1880s Sessional Papers indicate many concerns over declines in other species, whereas eels remained abundant and eel fisheries were productive throughout Québec and the Maritime provinces.

The implementation of regulations can indicate conservation concern over fish stocks. However, few conservation concerns over eel status were raised in the early years, and regulations that were implemented regarding eel fisheries were related largely to minimizing conflicts with other fisheries. Charlevoix (1761:1:228) described early conflicts as a "great outcry" over "porpoise" fisheries below Québec City interfering with eel fisheries: "for the porpoises having finding themselves disturbed…have retired elsewhere, and the eels no longer finding those large fishes in their way, swim down the river without any hindrance." Hence, the success of eel fisheries was considered to be greatly reduced by the "porpoise" fishery. The Province of Prince Edward Island had fisheries legislation in existence as early as 1780: "An Act to regulate the Salmon, Salmon-trout and Eel fishery" (Prince Edward Island 1773–1834:1780). The then colony's conservation concerns appeared to focus on salmon and on resolving conflicts among the various fisheries. A report of a fishery overseer in Lower Canada in 1823, perhaps concerned with eel conservation, commented on a localized issue concerning the use of spears, torches, and sticks by indigenous people to kill eels (Canada 1823). However, this issue may have been as much an indication of conflict among users as a conservation issue.

Many of the Canadian eel fisheries regulatory measures introduced in the late 19th century seem to have focused not so much on eel conservation but in dealing with access and allocation conflicts in the face of abundance declines of more valuable species. The Fisheries Act (Canada 1886) included prohibitions on the obstruction of streams by fishing gear, but added, "provided that the use of weirs for catching eels exclusively, and the use of mill-dams for catching eels, shall be prevented only in the case where, and at times when they injure other fisheries, or by completely barring any passage, they deprive other weirs of a share in the run of eels."

Concerns were raised about eel fishing methods interfering with other fisheries, with spawning of other species (Canada 1868b), and about fishing for other species under the guise of fishing for eels. The Province of Prince Edward Island enacted a regulation to prevent fishing for eels from boats by means of torches during October, November, and December (Canada 1893), but the regulation was really related to effects on other fisheries, not specifically to eel conservation. In Nova Scotia, concerns were raised with eel fishing practices destroying young alewife *Alosa pseudoharengus*, and a regulation was implemented to limit certain eel fishing practices at the mouths of streams between July and November was instituted (Canada 1889a). In Upper Canada (now Ontario), on the Trent River (a tributary of Lake Ontario), an 1817 report to the colony's House of Assembly stated that "some unprincipled characters… [are] at full liberty to pursue their depredations by making weirs and dams across the river under the pretence of catching eels and whitefish, and do almost totally obstruct the passage of salmon up the same…and thereby cut off a very material source of supply to the back settlers" (Fraser 1912).

Similar conflicts occurred in the United States. For instance, the presidents of the Pennsylvania Commissioners of Fisheries and New York Fisheries, Game and Forest Commission agreed in 1895 that the eel weirs in the upper Delaware River "had to go" because many game fish were being caught that were stocked by Pennsylvania at "heavy expense" (The New York Times 1895), and concerns were raised over eel weirs illegally catching Atlantic salmon *Salmo salar* on the Beaverkill in Delaware County (The New York Times 1896). Conflicts with eel fisheries persisted at the turn of the century; while Pennsylvania had prohibited the use of eel weirs in the state, New York and New Jersey fishermen continued to take "wagonloads" from the Delaware River where eel weirs were still permitted (The New York Times 1901).

Early Commercial Harvests

Indigenous peoples used eels for trade, barter, and income (Supreme Court of Canada 1999). For instance, in the early1620s, Champlain reported the indigenous peoples to be hard bargainers in trade; his hungry men gave their coats and other possessions for eels and that he acquired 1,200 eels at the rate of 10 eels for one new beaver pelt (Champlain Society: 1922–1936). Specialized eel spears were included in a 1684 French inventory at Fort Frontenac

(Kingston, Ontario) (Preston 1958), and trade in eels was reported as early as 1770 (Schmalz 1991:96).

In the early 1800s, eel harvests generally were not recorded, but it is clear that eels were "taken in great numbers in Lake Ontario and the rivers Niagara and St. Lawrence" (Talbot 1824). During the mid-1800s to 1900, commercial harvest records began to be reported in Canada's Sessional Papers. At that time, eel prices (C$0.11–0.13/kg) were not as high as Atlantic salmon in Québec and the Atlantic provinces ($0.33–0.44/kg fresh) or as Ontario lake trout *Salvelinus namaycush* ($0.22/kg) and lake whitefish *Coregonus clupeaformis* ($0.18/kg). Although the quantities harvested commercially were quite small in comparison with other large fisheries, eels were always a valued commodity, in part because they were harvested regularly for sustenance. Quantities harvested by indigenous peoples and early European settlers for local uses were not reported, but clearly, eels were fished consistently by indigenous peoples for thousands of years and by early European settlers since contact. Indeed, eel fisheries were "highly productive, and enable[d] people to live when all else fails" (Thwaites 1896–1901:40:11–12).

Early commercial harvest records provide some insight into patterns of use and development of commercial eel fisheries. Ontario commercial eel landings were first reported in 1884 (Canada 1885) for Lake Ontario, the St. Lawrence River, and their watersheds. Ontario commercial eel fishing appears to have developed first in smaller tributary watersheds, likely as an extension of subsistence fishing practices. It is now clear that locally important eel harvests occurred in the inland waters of Ontario, particularly in the Ottawa and Trent River systems (Canada 1895, 1896,1889b).

Harvest records in Canada's Sessional Papers, spanning the late 1800s to 1900s, indicate that eels remained abundant and widely dispersed. Eel fisheries were locally important, providing stable exports and supplying local markets and family food. Eel fishing used relatively inexpensive techniques, including brush hurdles, weirs, spears, and pots. After a brief period in the early 1900s when eel prices rose to range from $0.13 to $0.20/kg and harvests in Ontario increased (Casselman 2003), harvests dropped to relatively low levels throughout most of Canada while prices fell by some 30% (Eales 1968) until the 1960s.

Twentieth Century Commercial Eel Harvests

Harvest data indicate that eels were abundant along the East Coast and in watersheds of North America at the beginning of the 20th century through to the mid-1900s, often accounting for more than 50% of total fish biomass (Smith and Saunders 1955; Ogden 1970; Lary et al. 1998). Trends in eel harvests for much of this period are described in detail by Eales (1968) and Casselman (2003).

Although long-term fisheries-independent data are lacking across much of the North American range, there is strong evidence of declining abundance (Haro et al. 2000; Casselman 2003; Dekker et al. 2003; Cairns and Casselman 2004; COSEWIC 2006; MacGregor et al. 2008). Early declines began in the Mississippi watershed where diminishing harvests were reported across 10 states between 1899 and 1922 (Coker 1929). In Canada, even into the mid-20th century, eels were considered to be underutilized (Eales 1968). However, harvests began to increase substantially after the 1950s. Total North American harvests rose from an average of 1,215 metric tons (mt) between 1950 and 1955 to an unprecedented peak of 2,915 mt in 1978 (Casselman and Marcogliese 2007). Lucrative glass eel fisheries also developed in North America at this time (Jessop 1987; ASMFC 2000). By the early 1990s, North American harvests had begun to plummet, and by 2004, eel harvests had declined to 840.4 mt. This decline occurred despite sustained high prices, well above the long-term mean (Casselman and Marcogliese 2007).

While there are temporal differences across regions in the onset of major declines in catches, overall trends in Ontario commercial harvests parallel those of Canada and the United States (Figures 2A–C). Between 1950 and 2003, Ontario commercial eel harvests averaged 80.1 mt (Figure 2A) but rose substantially in the 1970s to an unprecedented 228.2 mt. This harvest constituted 20% of total Canadian landings in that year. At the same time, North American harvests also increased dramatically in response to high prices and strong markets (Casselman and Marcogliese 2007), and concern grew over their sustainability (Kolenosky and Hendry 1982). In 1980, an experimental quota of 270 mt was implemented for Ontario's

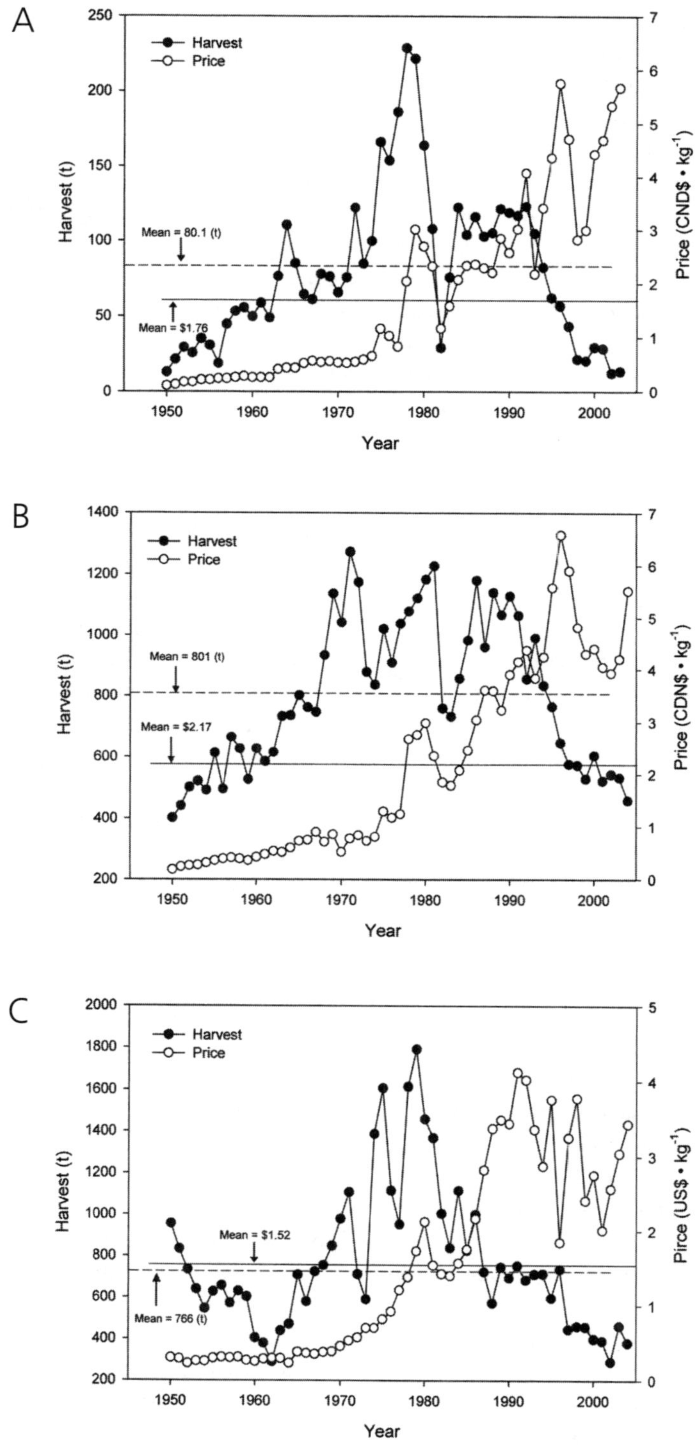

Figure 2.—Commercial eel harvest (metric tons) (solid circles) and weighted mean annual price/kg (Canadian dollars) (open circles) for (A) Ontario, 1950–2003; (B) Canada, 1950–2004; and (C) United States, 1950–2004. Mean harvest (metric tons) (broken line) and price/kg (solid line) are indicated. Ontario eel fishery was closed in 2004. Assembled from data from Casselman and Marcogliese (2007).

portion of the upper St. Lawrence River–Lake Ontario (USLR–LO; Kolenosky and Hendry 1982); this was a quota that was set when harvests were at record-high levels and was never approached. Ontario harvests declined substantially thereafter, in synchrony with strong harvest declines across North America (Figures 2A–C), as did the important silver eel harvests in Québec. Ontario fisheries-independent indices paralleled these trends (Casselman 2003). These declines, together with the dramatic reductions in recruitment to Lake Ontario, led to more widespread concern over the general status of American eel (Castonguay et al. 1994; ASMFC 2000; Casselman 2003). By the early 2000s, there were numerous calls for action (Casselman 2003; Dekker 2003; Dekker et al. 2003), and by 2004, the Ontario commercial fishery was closed.

Although effort data for commercial harvest statistics are universally lacking, if harvest and price are integrated across a broad geographic range, commercial catch data can be useful in detecting population-level shifts in abundance (Casselman and Marcogliese 2007). Prices during the early 20th century were low but began to rise in the late 1960s, particularly in the Atlantic provinces of Canada and some Atlantic coastal states (Casselman and Marcogliese 2007). By the late 1970s, both price and harvest had increased in Ontario and throughout Canada and the United States (Figures 2A–C), reflecting market demand from Europe and Asia (Loftus 1982; Casselman and Marcogliese 2007).

At that time, the American eel became the most valuable commercial fish species in Lake Ontario. However, despite sustained high prices in the latter half of the 20th century (well above the long-term mean), landings declined substantially, not only in Ontario, but throughout Canada and the United States (Figures 2A–C; Casselman and Marcogliese 2007).

Numerous issues may confound examination of trends in eel catches, depending on the density-dependent patterns of eel dispersal, geographic source of the data, and proximity to source of recruitment. Eels in the USLR/LO are at the extremity of the range. Edeline et al. (2007) describe density-dependent dispersal in juvenile European eel *Anguilla anguilla*, and it is probable that similar processes occur for American eel. If we assume that recruitment to the USLR/LO is density-dependent, as proposed by Casselman (2003), then changes in abundance at the range extremities would reflect population-level changes. Assuming panmixia and that recruitment to Ontario is influenced by density (Casselman 2003), and noting that Ontario and North American harvests are synchronous, Ontario eel indices broadly reflect overall species abundance in North America. The longest index of recruitment is from Ontario's eel ladder at the Moses-Saunders Dam on the St. Lawrence River. Since 1993, this index has shown a 99% decline in recruitment (Casselman 2003). Total annual eel passage, somewhat influenced by variable annual operating conditions, indicates only a very minor improvement in recruitment since 2003 (Figure 3). Assuming panmixia, Ontario's St. Lawrence River indices, at the extremities of the range, can be strong indicators of population-level change. These indices are among the most reliable since they are the longest and most consistently measured (Casselman et al. 1990). However, catch data across jurisdictions may be more or less sensitive to issues of hyperstability or hyperdepletion, depending on distance from the source of recruitment, and need to be interpreted carefully until appropriate population-level assessment programs are developed.

While diminishing abundance may be most pronounced at the extremities of the range in waters such as the USLR/LO, American eel appear to be in decline across North America (Petersen 1997; ASMFC 2000; Haro et al. 2000; Richkus and Whalen 2000; Casselman 2003). Key indicators of concern are (1) 99% decrease in eel recruitment to the USLR/LO (Casselman 2003); (2) indications that yellow eels are at historically low levels in the eastern United States (ASMFC 2006); (3) significant declines in silver eel harvests in Québec (Robitaille et al. 2003); and (4) declines in harvests in most of the commercial fisheries throughout North America, despite sustained high prices and strong demand in European and Asian markets. The general lack of long-term fishery-independent data series across the range has caused some to debate population status of American eel. In particular, the lack of reliable effort statistics is considered problematic as harvest trends can be driven by market conditions. While results may be misleading when effort is not taken into consideration, changes in commercial harvest can be interpreted much more precisely by taking price into consideration (Casselman and Marcogliese 2007).

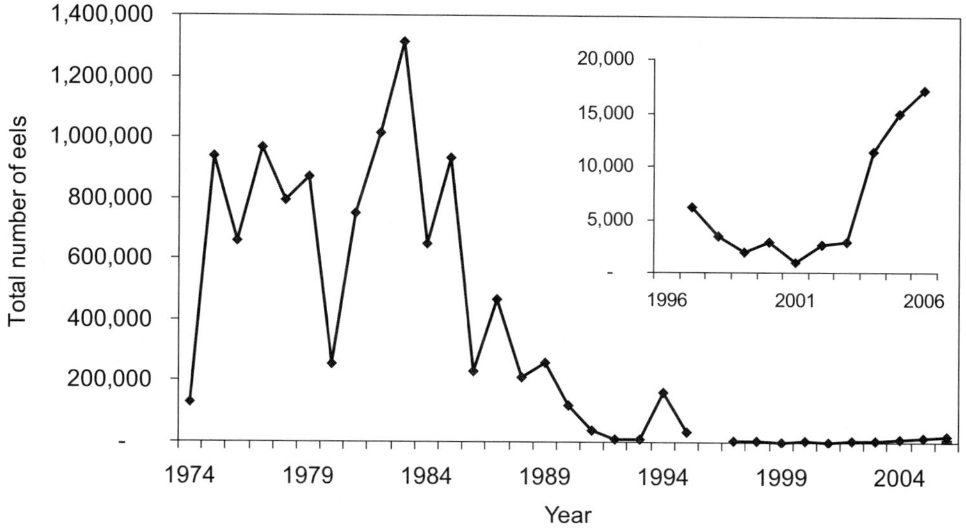

Figure 3.—Estimated number of eels annually ascending and exiting the Moses-Saunders eel ladder, Cornwall, Ontario, 1974 to 2005, and Moses-Saunders and New York Power Authority (Massena, New York) eel ladders combined, 2006 (inset 1997 to 2006) . Numbers of eels represents total passage, not a standardized index of abundance because of variability in operating conditions. Operation of Moses-Saunders eel ladder varied greatly among years (33 to 154 d), and two ladders were used in 2006.

Habitat Loss and Mortality Due to Barriers

Construction of barriers has reduced eel access to a large portion of formerly important freshwater habitat. From Maine to Florida, at least 15,115 dams restrict or prevent fish passage, leaving up to 84% of Atlantic coast river and stream habitat inaccessible (Busch et al. 1998). While this estimate assumes a worst-case scenario of 100% blockage at the first dam, a succession of barriers on a watershed will no doubt accumulate their effect. The majority of dams on Atlantic coastal rivers in the United States have no provision for fish passage (Haro et al. 2000). There also are numerous examples of lost habitat due to dams within the Mississippi River basin and other watersheds of the Gulf of Mexico. The situation is similar in Canada. For example, the St. Lawrence River basin, which constitutes about 19% of American eel freshwater habitat, contains some 8,411 dams (Verreault et al. 2004). These dams are estimated to block access to 12,140 km of eel habitat in the St. Lawrence River watershed (Verreault et al. 2004). There are 5,260 dams in Quebec watersheds alone that drain into the St. Lawrence River. In this watershed, as in much of North America, dam construction increased in the early 1900s and peaked between 1950 and 1970 (Figure 4).

Impacts of Hydroelectric Facilities and Other Barriers

Hydroelectric dams affect eels by forming barriers to upstream movement by young eels and subjecting silver eels to turbine mortality as they immigrate to the ocean (McCleave 2001; ICES 2003, 2006; Allen 2008). Young eels are able to use damp substrates to maneuver around obstructions (Haro et al. 2000), and locks, ditches, and marshes can provide limited passage. While some dams may be only partial obstructions, successive dams will have a cumulative impact on upstream passage of juvenile eels and will limit or completely restrict access to important rearing habitats. To illustrate, White and Knights (1997) found that dams on rivers had a greater effect on European eel than distance from the ocean, resulting in the upstream habitats being practically deserted.

As illustrated in early reports, dams have long affected American eel. Numerous late 19th and

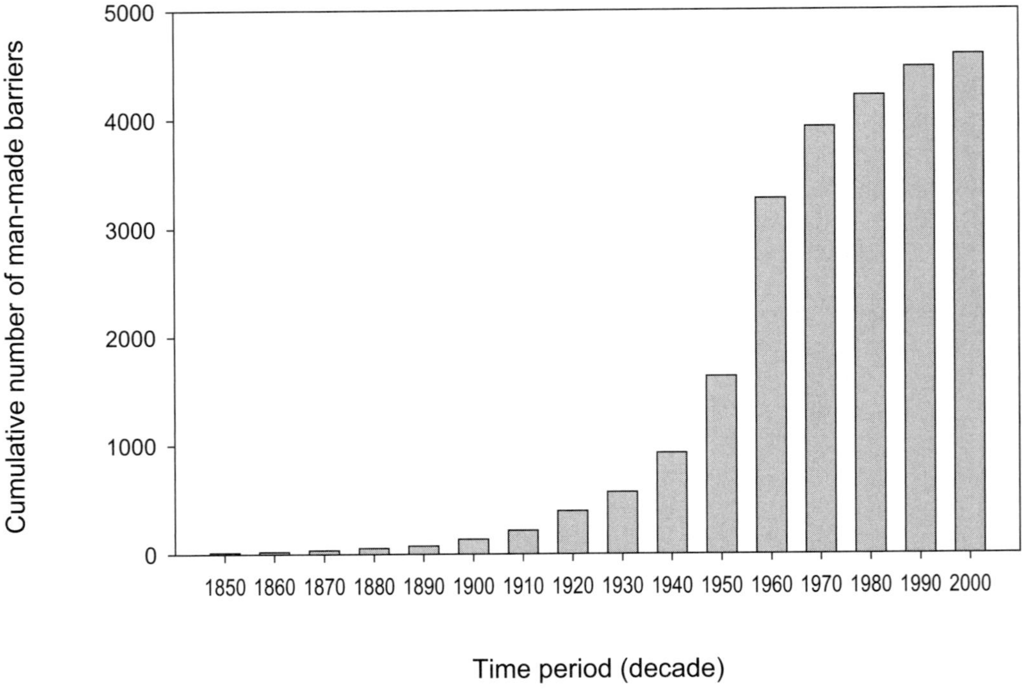

Figure 4.—Cumulative number of man-made barriers greater than 1 m in height in Quebec watersheds of the St. Lawrence River, plotted over time period (1850 to 2005). Data were obtained from the Centre d'expertise hydrique, Ministère du développement durable, de l'environnement et des parcs, Gouvernement du Québec.

early 20th century accounts reference the mortality of eels due to turbines and mill wheels in the eastern United States (The New York Times 1886, 1891, 1895, 1900, 1908) and in Michigan (Ballard 1948). Early 20th century accounts document mortality at mills within the Ottawa River watershed in Canada (Burnett 2007). Indeed, The New York Times (1906) reported that eels had become so abundant in the Connecticut River at a "local electric light plant" that they clogged the wheels and eventually stopped them. While these references imply the negative effects of eels on mill and power production, they also provide evidence of former eel abundance and indicate that this type of mortality problem has occurred for more than a century, likely increasing in significance as hydroelectric technology increased. As recently as the 1970s, eel mortalities due to turbines were so highly visible in the St. Lawrence River that Ontario Hydro annually hired contractors to pick up and dispose of metric tons of eel carcasses below the Ontario side of Moses-Saunders Dam (Verreault and Dumont 2003; J. M. Casselman, unpublished data).

Provisions for safe downstream passage of eels at hydroelectric facilities are rare, and turbine mortalities of downstream migrants can be high. Numerous successive facilities on a particular watershed will increase cumulative mortality. Donnil et al. (2001, in ICES 2003) estimated an average annual mortality of 92.7% for European eel in the River Rhine for a succession of 12 hydroelectric facilities. Cumulative turbine mortality at the two hydroelectric power facilities on the St. Lawrence River is calculated to be more than 40% (Desroches 1995; Normandeau Associates Inc. and Skalski 1998). Smaller turbines such as those on the Ottawa and Trent River systems likely induce higher eel mortality as it has been shown that eel survival increases with increasing turbine size (Larinier and Dartiguelongue 1989).

Eel size may also influence mortality rate. Haro et al. (2000) estimated turbine-induced injuries as high as 50% for small eels and up to 100% for large eels. Increased vulnerability of large eels is particularly troubling as females are larger and predominant in the upper reaches of watersheds. In a simulation model of the cumulative effect of 13 power stations on the Kennebec River in the United States, McCleave (2001) predicted that only 40% of the female eels would reach the ocean, assuming a 20% mortality rate at each facility, and this estimate did not consider fishing mortality. American eel travel far to reach the upper St. Lawrence River, Lake Ontario and associated inland waters, where they grow and mature for 12 or more years (Casselman, unpublished data). Eels from these waters are virtually all female and traditionally have produced the largest and most fecund females within the species' range (Casselman 2003). While estimates and modeling are still preliminary (COSEWIC 2006), these females undoubtedly made a substantial contribution to the reproductive potential for the entire species (Casselman 2003; Verreault and Dumont 2003). Given impacts on females in freshwater systems, the cumulative lost fecundity from North American freshwater watersheds due to dams cannot be overlooked.

Case Histories of Significant Habitat Loss Due to Barriers

Extensive habitat loss due to barriers has occurred throughout the American eel range. Examples spanning much of the North American range include the Richelieu River in Québec, the Oswego River–Oneida River system in Lake Ontario, and the Susquehanna River in Maryland. In the Richelieu River and Oswego–Oneida systems, harvests in the early 20th century regularly approached 70 and 100 mt, respectively, but now due to the construction of dams, eels in both systems are exceedingly rare (Adams and Hankinson 1928; Henke 1993; Axelsen 1997; Lary et al. 1998; Verdon et al. 2003; Verreault et al. 2004). In the Susquehanna system, eel harvests averaged some 393.4 mt annually prior to the construction of the Conowingo and three other main-stem dams in the 1920s. The commercial harvest since then has been zero, and it is estimated that there are 11 million fewer eels in the Susquehanna today than in the 1920s (MdDNR 1999).

To these examples, we add new evidence of dam-related habitat loss in inland Ontario. Archaeological, historical, and anecdotal information from the Ottawa and Trent rivers reveal surprising abundance of eels, a fact long since forgotten in the absence of readily accessible information. However, more recent commercial harvest and fishery-independent information illustrates the extent of habitat loss and reductions in eel distribution in these watersheds. Thus, these examples represent a pattern in habitat loss and shifting baselines that appears widespread across the North American range as eel diminished over the past century

The Ottawa River watershed is large, encompassing a drainage area of 146,000 km^2, representing about 12% of the St. Lawrence drainage area, including thousands of lakes in Ontario and Québec. An estimated 3,700 km^2 of suitable habitat was present within this system before extensive dam construction throughout the watershed (Verreault et al. 2004). Eels historically were abundant within this system (Small 1883) and penetrated the river as far inland as Lake Timiskaming (Purvis 1887), 580 km from its confluence with the St. Lawrence River. Archaeological evidence shows the presence and heavy use of American eel in the Ottawa River watershed by indigenous peoples, dating back some 4,000 years (Clermont and Chapdelaine 1998; Clermont 1999; Clermont et al. 2003). In the early 1900s, eels remained sufficiently abundant in the Ottawa River to support important commercial fisheries in Ontario and Quebec waters. Dymond (1939) notes that Quebec commercial eel harvests from the Ottawa River always exceeded those from Ontario; Quebec harvests ranged from 3.4 to 15.0 mt annually between 1930 and 1937.

This watershed has been heavily affected by fragmentation, first by the development of sawmills to support a burgeoning timber industry and later by numerous large hydroelectric installations (OMNR/Québec MNRF 1999; Haxton and Chubbuck 2002). The first hydroelectric dam was installed on the main branch of the Ottawa River at Chaudiere Falls (Ottawa/Hull) in the 1880s, and the ninth and final large dam on the main branch was constructed in 1964 at Carillon. Many more facilities were installed on tributaries within the watershed (Figure 5), making the system one of the most highly regulated watersheds in Canada (Telmer 1996).

American eel subsequently declined throughout the Ottawa River watershed. Historically abundant throughout the Mississippi River, a tributary of the Ottawa River, many were sampled during netting projects as late as the 1960s, but now are rarely

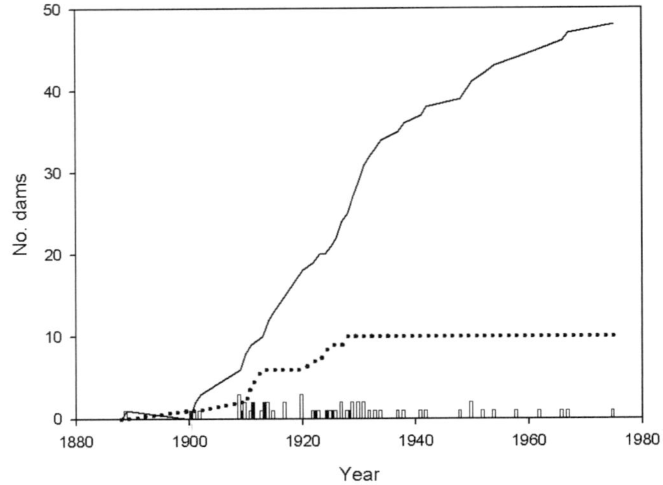

Figure 5.—Number of dams (open bar—Ottawa River; solid bar—Trent River) and cumulative number of dams (solid line—Ottawa River; dotted line—Trent River) constructed on the Ottawa River watershed or Trent River, 1880s to 1970s.

captured (Figure 6; Ontario Ministry of Natural Resures, unpublished data). The last American eel was reported from Algonquin Park in 1936 (Mandrak and Crossman 2003). Locks in the lower Ottawa River enable some immigration into the lower watershed. Eels are still found within the Ottawa River as far upstream as the Bonnechere River/Lac des Chats (260 km from the St. Lawrence River), but only as a remnant of their former abundance (Figure 7; T. Haxton, Ontario Ministry of Natural Resures, unpublished data). Four Ontario commercial fishing licenses remain in the lower Ottawa River, but prior to the Ontario eel closure in 2004, only a few hundred kilograms of eels were harvested

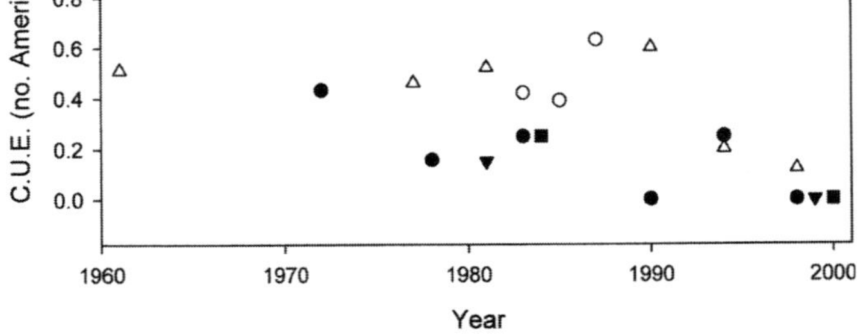

Figure 6.—Catch per unit effort (CPUE) of number of eels caught per night of trap netting in five lakes tributary to the St. Lawrence River or Ottawa River, Lanark County, 1961 to 2000. Fish-passage facilities do not exist at any dam. Data exist only for the years indicated.

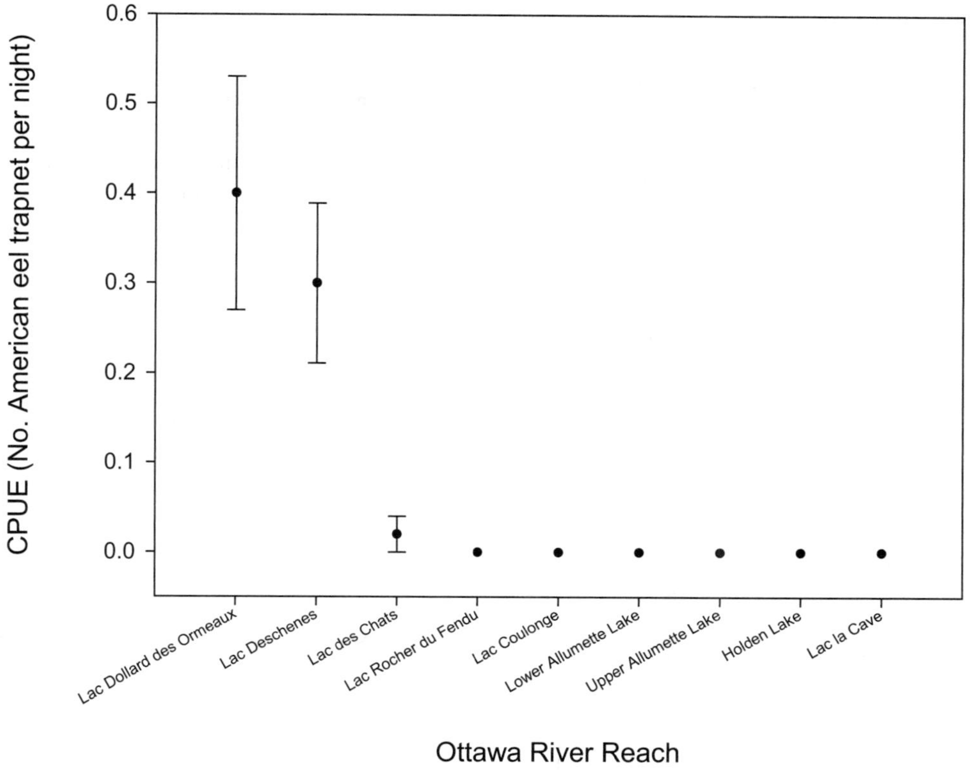

Figure 7.—Catch per unit effort (CPUE) of number of eels caught per night of trap netting in reaches of the Ottawa River, progressively farther upstream. Lac Dollard has the first upstream barrier, and the first reach is defined as below Lac Dollard des Ormeaux. Each additional barrier upstream defines the next reach as the area between dams. Vertical lines indicate 95% confidence limits.

annually (Ontario Ministry of Natural Resures, unpublished data).

Eels also were historically abundant in the Trent River watershed of Lake Ontario. The basin encompasses some 218 shallow and productive lakes in the Haliburton Highlands and Kawartha Lakes regions, which afford excellent eel habitat. As for the Ottawa River, there are historical and archaeological records of eel presence and use in the Trent River watershed. There is evidence that indigenous peoples had fish weirs in the system (Fraser 1912); stakes from an apparent fishing weir in Lovesick Lake was carbon-dated to more than 6,000 years old (Stevens 2004). Eel vertebrae were present in numerous test units at the significant precontact Late Woodland Gibson Site (BcGo-14) on the south shore of Chemong Lake near Peterborough, Ontario (Archaeological Services Inc. 2008). Copway (1847) noted that "Rice Lake contains quantities of the finest fish… bass, eels, etc. are also found in this lake." The Reverend Peter Jacobs (1857) described Rice Lake as one that abounds with "muskinoonj, bahs, eels." In 1817, fish weirs in the lower Trent River captured eels, whitefish, and other species (Fraser 1912). An advertisement in the Coburg Star (1838), marketing a mill property on the Otonabee River near Peterborough, Ontario provides further evidence that eels were abundant and valuable in the upper Trent River watershed. Among other selling points, the ad featured close to 50 £ sterling worth of eels that "were taken last season in the boxes fixed in the saw-mill."

In the late 1800s, commercial eel landings from counties surrounding the Trent River watershed ranged between 2.1 and 11.4 mt, annually (Canada 1885, 1886, 1887, 1888, 1889b, 1900). This system was first fragmented by mill dams, followed by the construction of the Trent-Severn

Waterway in the late 1800s. The waterway initially consisted of a series of dams and locks to aid navigation and was further altered between 1900 and 1928 with the addition of 14 hydroelectric facilities (Figure 5). None of the generating stations provide fishways, although a series of locks at each dam provides some upstream access. American eel have been virtually extirpated above the first barrier, and the last few were recorded by assessment gear in the Kawartha Lakes in the mid-1980s (Ontario Ministry of Natural Resurces, unpublished data).

Beginning in the early 20th century, eel abundance and distribution have declined substantially in response to decades of barrier construction across the North American range. Losses have been gradual over this period, leading to fading memories of the former presence and importance of this species. As noted earlier, perspectives on the importance of eels clearly shifted over the same period, leading to spotty record-keeping and low priority in fisheries management. Today, it often comes as a surprise that eels were even present, much less important, in many watersheds where they are now rare of absent.

While recent sharp declines in abundance have been detected in the late 20th century, it is clear that lost eel production began at a much earlier time. Major barrier construction began in the early 1900s and appears to have peaked in the 1950s-1960s. Given the long life span of eels in many freshwater systems and the unique panmictic life cycle, the cumulative effects of barrier construction would take a considerable amount of time to become apparent. The effects would manifest themselves through lost production from formerly accessible and productive habitat and would accumulate over many years at the population level. Thus, the accrued losses in spawner production from freshwater systems and effects on subsequent recruitment are particularly concerning.

Discussion

While the European Union has accepted that there are substantial declines in the European eel (EU 2007), there remains debate in North America regarding the status of American eel, despite the warnings by Casselman (2003) and Dekker et al. (2003). The evidence presented indicates that eels, particularly in freshwater habitats, have been severely marginalized due to anthropogenic impacts over the past century.

Population-Level Declines

Our extensive review of historical, traditional, and commercial harvest statistics indicates that American eels were formerly much more abundant in North America at the time of European contact. In more recent decades, commercial harvest statistics across the North American range have shown major declines despite sustained strong markets and high price. Long-term recruitment indices are rare, but as juvenile eel dispersal appears to be influenced by density, we are concerned that the recruitment indices in the USLR/LO (near the extremity of the range) may indicate population-level recruitment declines to continental waters.

Many fishery-independent indices for American eel across North American were not originally intended to target eels, are relatively recent, and often may not span an adequate time period, particularly in the case of recruitment indices. For instance, if only the most recent 15 years of the USLRL/LO recruitment index were to be examined, one would erroneously conclude stable conditions, whereas eel recruitment had actually declined by 99% since the late 1980s (Figure 3). However, fishery-independent indices for the USLR/LO extend back into the early 1980s, and these indices point to a substantial decline in yellow eel abundance that parallels the steep decline in recruitment to these waters. This information, combined with the observation that yellow eels are at nor near historic lows across the eastern seaboard of the United States (ASMFC 2006), matches the observed sharp declines in commercial harvests across North America (Figures 2A–C). While declining expectations and shifting baselines have clouded today's perceptions of American eel status, there is little doubt that American eel abundance has declined significantly over the past century, becoming particularly concerning over the past 20–30 years in many formerly significant components of its range.

Distribution Change

It is clear that eels formerly were distributed widely across all accessible freshwater habitats in eastern North America, ranging throughout watersheds and coastal habitats associated with the USLR/LO (including large systems such as the Ottawa River), eastern North America, and gulf states through to the upper Mississippi River (COSEWIC 2006; NatureServe 2007). We have described substantial, formerly highly productive areas where eels have been virtually extirpated and other areas where abundance obviously has declined significantly. These areas are all associated with the construction of numerous dams. This evidence suggests that cumulatively, dams have been a major cause in eel declines by reducing eel production and subsequent recruitment at the population level.

The extent of habitat loss has been substantial, with few large watersheds free of barriers. Adding to these concerns are observations that all of these waters formerly supported abundant production of predominantly large, highly fecund females. The examples of lost habitat documented herein appear symptomatic of a widespread problem across the North American range where significant lost eel production has occurred throughout the freshwater range, wherever major barriers have been constructed. Significant mortalities of predominantly female eels at hydroelectric facilities pose additional concern, especially when the mortalities are cumulative across a number of utilities on a single watershed (e.g. Ottawa and Trent River watersheds in Ontario).

Importance of Life Cycle Diversity and Panmixia to Management and Conservation

The science to date supports that American eel are panmictic (Avise et al. 1986; Wirth and Bernatchez 2003); therefore, it is particularly important to view the dynamics of American eel at the population level. For instance, if eels are panmictic, the consequences of lost female production in freshwater systems due to lost habitat combined with the costs of mortalities due to turbines and fishing would be additive and substantial on one common spawning stock. The protection of spawners and spawning habitats is particularly crucial in the management of long-lived species (Winemiller and Rose 1992) like eels. It is evident that spawner biomass of eels has declined over the past century. Under a model of panmixia and density-influenced dispersal of recruits, there would be direct negative feedback in terms of lower spawner biomass and reduced continental recruitment in the future (Casselman 2003). If geographic dispersal of recruits is influenced strongly by density, it would not be surprising to see stock indicators nearer to the source of recruitment remain apparently stable for a period of time while they collapsed in more distant waters.

Environmental variations also may induce recruitment vacillations in the American eel population (e.g., changes in oceanic productivity and currents, Beak 2001; Friedland et al. 2007; Bonhommeau et al. 2008). Cumulative anthropogenic effects, including overfishing, turbine mortalities, and lost habitat, may have destabalized the American eel population and increased its sensitivity to environmental conditions (Bonhommeau et al. 2008). Recruitment fluctuations due to environmental change highlight the need for improved escapement of spawners to boost reproductive output, hedging against environmentally induced periods of poor recruitment. Many diadromous fish populations use multiple modes of migration and habitats (McDowall 1996), and divergent migration behaviors or habitat use can reduce the overall variance of population responses to environmental change, thereby increasing stability and resilience. Termed "contingents" for eels and other species (Secor 1999; Jessop et al. 2002) in the concept of stock, multiple life cycles can induce a dampening effect on overall variance of population responses to environmental change, thus increasing stability and resilience.

Contingent structure has been shown for Japanese eel *Anguilla japonica*, European eel, and American eel (Tsukamoto et al. 1998; Tzeng et al. 2000; Tsukamoto and Arai 2001; Jessop et al. 2002; Morrison et al. 2003). In the case of American eel, at least two contingents can be visualized: one in which yellow eels predominately grow in freshwater; the other growing in marine habitats. These divergent life cycle pathways confer resiliency against spatial variation in mortality risk, and in the extreme case, if one contingent should fail, the other contingent would persist as an effective means of bet-hedging (Secor and Kerr 2009, this volume). Further, as eels mature at different rates across a wide variety of latitudes, and multiple cohorts from a widely diverse set of habitats apparently spawn together in any given year, overlapping generations can also contribute to population persistence. Here, strong recruitments are figuratively stored in the adult population (Secor 2007). At high latitudes in freshwater, eels primarily are long-lived, highly fecund females (Lary et al. 1998). Thus, protection and restoration of freshwater contingents will serve to improve American eel resiliency to further, seemingly inevitable anthropogenic disturbances and ensure sufficient population-level fecundity.

The evidence presented suggests that during the

past century the contribution of the freshwater contingent of American eels has been substantially marginalized by anthropogenic effects. As human population and development increase, impacts on both freshwater and marine systems will continue (Lackey et al. 2006a, 2006b). Our review suggests that it is imperative to restore and maintain the freshwater contingent to sustain resilience against future perturbations. The strictly marine segment of the population requires further study. While less understood, it is clear that the estuarine habitats can be particularly productive for eels and that the marine contingent may make important contributions to the overall population. Some of these habitats may not have been altered as much as freshwater ecosystems, and they are no less important to protect from overexploitation and further degradation. However, given that American eels extensively use all accessible freshwater habitats, it is ill-advised to discount the importance of freshwater eels on the premise that the marine contingent will sufficiently protect the species. This is especially important given the dominance of females in the freshwater contingent. Preserving only the most productive habitats may be a flawed management strategy (Kraus and Secor 2005). Rather, priority setting should be based on studies of habitat use that match the generation time (Secor 2007).

Constituent patterns of life cycle diversity within populations should be regarded as a portfolio that hedges against future environmental uncertainty (Secor and Kerr 2009). Conservation and recovery planning exercises need to recognize the diverse life cycle of American eel, the cumulative nature of impacts, and the importance of protecting and restoring their portfolio of habitat types. Contingent structure, proximity to the source of recruitment, and higher eel productivity in estuarine habitats may help explain differing views of American eel status. Edeline (2007) proposes that the decline in European eels could be less steep in saltwater than in freshwater habitats, suggesting that eel diadromy may be a conditional strategy and that anthropogenic changes are acting selectively on freshwater habitats.

Necessity for Coordinated Conservation and Management Actions

In the past, little attention was given to silver eel escapement in American eel management, apparently as a consequence of management perspectives described earlier, regional approaches, and absence of population-level assessment. Difficulties with fish stocks often occur when it is assumed that the number of spawners has little to do with subsequent recruitment (Clover 2006). Inadequate attention to spawning stocks is a common problem in fisheries science and management (Walters and Maguire 1996). In this context, the collapse of juvenile recruitment to the USLR/LO just a few years after the record-high North American harvests, and a decade or two after the peak in dam construction, is disconcerting.

Depensation was implied tentatively by Dekker (2008) for the European eel and may be an issue for the American eel. If the index of recruitment to the USLR/LO reflects existing or forthcoming continental declines, depensatory pressures may add significant hurdles to recovery efforts if actions are not undertaken soon Thus, future management initiatives should carefully consider the cumulative consequences of fishing and turbine mortalities on subsequent spawner biomass and recruitment. Strongly coordinated strategies to significantly improve silver eel escapement appear essential.

The impacts of significant widespread loss and fragmentation of freshwater habitat, primarily due to dam construction over the past 150 years, again are cumulative at the population level. While some American eel appear to complete their entire life cycle in brackish estuarine habitats (Lamson et al. 2006), it is clear that eels have made extensive use of freshwater habitat across their range, producing predominantly large females. Therefore, both habitats play critical roles, and both contingents merit protection.

The recently renewed interest in hydroelectric power, combined with growing eel markets, will add pressure on the population unless mitigation measures are implemented. It is clear to us that strong, coordinated conservation actions are required, with the ultimate objective of vastly improved silver eel escapement and recruitment. Further loss of eel habitat should be unacceptable, and existing levels of anthropogenic mortalities do not appear to be sustainable. Rather, improved access to formerly productive habitats is necessary, as are coordinated and strategic reductions in mortality due to fishing and turbines. Finally, given apparently reduced continental recruitment, harvests of glass eels and elvers need to be closely managed and coordinated.

Recovery Planning

The evidence presented herein suggests cause for concern at the population level and the need for immediate development and implementation of recovery plans. The absence of long-term stock assessment data should not be a deterrent to taking appropriate actions; rather, this situation requires adopting a precautionary approach until better information is available. Clearly, sharp increases in total anthropogenic mortality have occurred over the past century, accumulating in their effects at the population level. Largely as a result of the panmictic nature of eels, regional approaches to eel management have not been successful. For example, thousands of dams have been constructed over the past century with little consideration of their cumulative, population-level impacts on eels and other diadromous fishes. These structures were evaluated on a case by case basis over several decades; environmental impacts would only have been reviewed in the site-specific context. Similarly, eel fisheries have been managed independently across numerous jurisdictions, with little consideration of the population-level context.

More recently, some jurisdictions recognizing eel declines have undertaken some actions to protect American eel, but it is difficult for one or only a few jurisdictions to implement effective conservation measures given the panmictic nature of eels (MacGregor et al. 2008). As early as 1980, concerns over mortalities due to fishing and turbines prompted convening the North American Eel Symposium (Loftus 1982). While some expressed concern over increased mortalities (Kolenosky and Hendry 1982), others thought differently, assuming that (1) there was no reason why every silver eel should not be harvested since fisheries were localized and numerous areas fished either lightly or not at all, (2) escapement of adults from a few stocks could sustain the species, (3) silver eels represented a loss of organic material from freshwater since they spawned and died in the sea, and (4) silver eels would be lost from the fishery anyway (Eales 1968). With such conflicting opinions, there was little interest in developing a coordinated management plan for eels in North America at that time (Loftus 1982).

However, by 2005, the Committee on the Status of Endangered Wildlife in Canada reviewed the Canadian status of the American eel and recommended a designation of "special concern" (COSEWIC 2006). In Ontario, the Committee on the Status of Species at Risk in Ontario recommended a designation of "endangered" under Ontario's Endangered Species Act, and the species was officially listed on July 1, 2008 (OMNR 2007). In contrast, protection of American eel under species at risk legislation in the United States has not occurred to date. A recent petition to list American eel under the federal Endangered Species Act (Watts and Watts 2004) was unsuccessful (U.S. Office of the Federal Register 2007). American eel is not formally listed at this time as endangered or threatened in the U.S. under the federal Endangered Species Act, nor is it listed under the National Marine Fisheries Service's Species of Concern Program, a proactive conservation program for species that may become threatened or endangered.

Canada has announced the development of a national management plan to reduce anthropogenic mortality of American eel by 50% (DFO 2004). Further development of the plan and implementation of actions are underway (MacGregor et al. 2008). In the United States, the Atlantic States Marine Fisheries Commission has developed a management plan for American eel (ASMFC 2000), and fish passage plans have been developed to improve access for eels and other diadromous species to the important Susquehanna River watershed (PFBC 2007; MdDNR 2008; Native Fish Conservancy 2008). Binational efforts have begun to set the scene for recovery of the USLR/LO segment of the population under the auspices of the Great Lakes Fishery Commission.

A model developed by Beak (2001) was sensitive to the cumulative effects of anthropogenic mortality (fishing and turbines) on resident eels and to the effects of habitat exclusion by dams. The modeling exercise further suggests that the provision of high efficiency upstream passage facilities and mitigation of downstream passage at hydroelectric facilities could confer substantial benefits to egg production from Lake Ontario and St. Lawrence River eels (Beak 2001). In Canada, it now is considered critical that restoration efforts focus on substantially improved escapement of silver eels and that recovery plans be implemented using strategic watershed approaches involving improved access, reduced mortality, and increased escapement. The recently approved European Union regulation for European eel also has determined these approaches to be essential (EU 2007).

A range-wide approach for American eel recovery planning is warranted (Dannewitz et al. 2005;

MacGregor et al. 2008). While progress in both Canada and the United States has been made in moving away from uncoordinated regional approaches (ASMFC 2000; DFO 2007), the most appropriate model would be binational in scope (MacGregor et al. 2008). Formal discussions between Canada and the United States are underway to develop an effective memorandum of understanding among agencies that will establish coordinated binational management and science across the existing North American eel range. The adoption of a broad ecosystem approach covering a range of habitat and mortality issues will be an underlying theme.

Summary

> Any person stand[ing] on the banks of the Delaware...at this season of the year will see at what first sight appears to be a huge serpent moving up the stream near the shore. If the spectator wishes to see the end of this moving mass, he must remain...for three or four days...he will be surprised upon investigating... [that this] endless line of life...is made up of the diminutive eels. (The New York Times 1880b)

The formerly massive recruitment events reported in earlier times (The New York Times 1880b; Hartford Courant 1902) now rarely occur, and eel declines across North America are evident. We have shown that watershed by watershed, American eel gradually have declined or disappeared in numerous formerly productive waters, at times almost invisibly. Anthropogenic mortality and the harmful alteration and disruption of habitat have had cumulative, insidious, and substantial effects on the American eel population over the past century. In particular, the 1950–1970 peaks in dam construction and turbine mortalities, together with the unprecedented North American harvests in the 1970s, have lead to a perilous synergy of effects at the population level. Conservation and restoration efforts are long past due.

Overseas markets for American eel are expected to remain high in response to significant declines in European and Japanese eels and in response to new conservation measures implemented by the European Union (EU 2007). In addition, anthropogenic stresses will mount as human population growth continues. Under these circumstances, pressures to develop more renewable energy sources such as hydroelectric power can be expected to increase, as will pressures to maximize harvests of American eel. Difficulties in implementation of recovery measures will be exacerbated significantly if mitigation and conservation policies that address cumulative impacts are not executed soon.

While recovery efforts targeting full restoration to former abundance levels are likely unrealistic, timely efforts are required to halt the decline and prevent potential depensatory effects. The protection and restoration of freshwater and marine contingents are equally important to ensure population resilience and guard against future anthropogenic stressors. Conservation and recovery efforts are not only necessary to ensure long-term sustainability of the resource, but are also necessary to ensure preservation of the long-standing economic, ecological, cultural, and natural heritage values of this species. Measured, strategic steps across the North American range are required in a coordinated fashion.

Traditional ecological knowledge and anecdotal, historical, and archaeological information have been useful in establishing a baseline understanding of former eel abundance and distribution and will be very useful in recovery planning. Whatever the shape of the ultimate recovery prescription, it is clear that the recognition of rights and interests of indigenous peoples and stakeholders need to be incorporated appropriately. Finally, the circumstances of American eel should be viewed as symptomatic of the challenges faced by many diadromous fish species in North America; cumulative effects at the species level should not be overlooked.

Acknowledgments

We appreciate the suggestions from anonymous reviewers whose ideas greatly improved the organization and clarity of this manuscript. We thank Alex Haro, Katherine Smith, and the Program Committee of the 2nd International Symposium on Diadromous Fishes for their invitation to participate and prepare this manuscript. Finally, we thank David Secor for reviewing our initial considerations of life cycle diversity and contingent structure in American eel. The views in this manuscript are those of the authors and do not necessarily reflect official position of the associated agencies or organizations.

References

Adams, C. C., and T. L. Hankinson. 1928. The ecology and economics of Oneida Lake fish. Roosevelt Wildlife Annals 1:1–548.

Allen, W. A. 2004. Wa-nant-gi-tche-ang: canoe route to Lake Huron through southern Algonquia. Ontario Archaeology 73:38–68, Toronto.

Allen, W. A. 2007a. Dbaajmoowin: dialogue with the elders. Pages 18–24 *in* S. Weber and D. Harmon, editors. Protected areas in a changing world: proceedings of the 2007 George Wright Society Conference on Parks, Protected Areas, and Cultural Sites. The George Wright Society, Hancock, Michigan.

Allen, W. A. 2007b. American eel (Kat, Kinebikoinkosew, Pimizi): an open letter to Fisheries and Oceans Canada. Indigenous Cooperative on the environment. Available: www.icenetwork.ca/about_us/Americaneel01/American%20Eel%2C%20open%20letter.doc (March 2007).

Allen, W. A. 2007c. American eel in Algonquin Park. Ontario Ministry of Natural Resources, Algonquin Provincial Park, Algonquin Park Cultural Heritage Research Report, Whitney.

Allen, W. A. 2007d. Nineteenth century aboriginal farmers of the Madawaska River. Pages 56–75 *in* J. Molnar editor. 2007. Partners to the past: making connections in the Ottawa River Valley: proceedings of the 2005 Ontario Archaeological Society Symposium. Ontario Archaeological Society, Ottawa Chapter, Ottawa. Available www.ottawaoas.ca/Symposium/Allen.pdf (July 2008)

Allen, W. A. 2008. American eel: driving a ahift in power. Available: www.latornell.ca/pdf/2008sessions/2008_T2F_William_A_Allen_delivery.pdf (January 2009).

Archaeological Services Inc. 2008. Stage 3 archaeological assessment of the Gibson site (BcGo-14), Gibson Subdivision, Part Lot 18, Concession 10, Township of Smith-Ennismore-Lakefield, County of Peterborough. Archaeological Services Inc., Toronto.

ASMFC (Atlantic States Marine Fisheries Commission). 2000. Interstate fishery management plan for American eel (*Anguilla rostrata*). Atlantic States Marine Fisheries Commission, Fish Management Report No. 36, Washington, D.C.

ASMFC (Atlantic States Marine Fisheries Commission). 2006. Stock Assessment Report No. 06–01 of the Atlantic States Marine Fisheries Commission: Terms of References & Advisory Report to the American Eel Stock Assessment Peer Review, January 2006. Atlantic States Marine Fisheries Commission, Washington, D.C.

Avise, J. C., G. S. Helfman, N. C. Saunders, and L. S. Hales. 1986. Mitochondrial DNA differentiation and life history pattern in the North Atlantic eels. Proceedings of the National Academy of Sciences of the USA 83:4350–4354.

Axelsen, F. 1997. The status of the American eel (*Anguilla rostrata*) stock in Québec. Pages 121–131 *in* R. H. Peterson editor. The American eel in eastern Canada: stock status and management strategies. Canadian Technical Report of Fisheries and Aquatic Sciences 2196.

Ballard, R. 1948. Tales of early Niles. Niles Printing Co., Niles, Michigan.

Baraga, F. 1878. A dictionary of the Otchipwe language explained in english, part 1. English-Otchipwe. Beauchemin and Valois, Montréal.

Baraga, F. 1880. A dictionary of the Otchipwe language explained in English, part 2. Otchipwe-English. Beauchemin and Valois, Montréal.

Bardeau, P. E. W. 2002. Onondowa'Ga:' Gawe:no, new reference edition. Seneca Nation of Indians, Salamanaca, New York.

Barton, B .S. 1797. New views of the origin of the tribes and nations of America. John Bioren, Philadelphia.

Beak. 2001. The decline of American eel (*Anguilla rostrata*) in the Lake Ontario/St. Lawrence ecosystem: a modelling approach to identification of data gaps and research priorities. Beak International Incorporated, ref.: 21852.10. Available: www.glfc.org/lakecom/loc/eel.pdf (April 2009).

Beauchamp, W. M. 1916. Moravian journals relating to central New York, 1745–66. The Dehler Press, Syracuse, New York.

Bonhommeau, S., E. Chassot, B. Planque, E. Rivot, A. H. Knap, and O. Le Pape. 2008. Impact of climate change on eel populations of the Northern Hemisphere. Marine Ecology Progress Series 373:71–80.

Burnett, A. 2007. William (Bill) James Bridgeman—the last miller of Wakefield. Gatineau Valley Historical Society, Gatineau Valley Historical Society Newsletter 2007:04, Chelsea, Québec.

Busch, W. D. N., S. J. Lary, C. M. Castilone, and R. P. McDonald. 1998. Distribution and availability of Atlantic coast freshwater habitats for American eel (*Anguilla rostrata*). U.S. Fish and Wildlife Service, Administrative Report #98–2, Amherst, New York.

Cairns, D. K., and J. M. Casselman. 2004. Inaugural meeting of the Canadian Eel Science Working Group, 2–3 December 2003, Québec City. Fisheries and Oceans Canada, Canadian Science Advisory Secretariat, Proceedings Series 2004/017, Ottawa.

Canada. 1823. Canadian journals of House of Assembly. Appendix to the XXXII volume of the Journals of the House of Assembly of the province of Lower-Canada, third session, eleventh provincial Parliament.

Canada. 1868a. Statutes of Canada 31 Vic. c60:183.
Canada. 1868b. Sessional Papers of the Dominion of Canada, 1867–1868. Volume 9, pp. 43–113.
Canada. 1884. Sessional papers of the Dominion of Canada. No. 9, pp. 274–299.
Canada. 1885. Sessional papers of the Dominion of Canada. Volume 6, third session of the fifth Parliament, session 1885, pp. 218–219, 293.
Canada. 1886a. The revised statutes of Canada. The Fisheries Act. Volume II, p. 7.
Canada. 1886b. Sessional papers of the Dominion of Canada. Volume 9, fourth session of the fifth Parliament, session 1886, pp. 310–311, 321, 323.
Canada. 1887. Sessional papers of the Dominion of Canada. Volume 16, pp. 18–225.
Canada. 1888. Sessional papers of the Dominion of Canada. Volume 8, second session of the sixth Parliament, session 1888, p. 297.
Canada. 1889a. Consolidated orders in council o Canada: under the authority and direction of His Excellency the Governor-General in Council. Special Fishery Regulations, N.S. Chapter 96, p. 63.
Canada. 1889b. Sessional papers of the Dominion of Canada. Volume 8, third session of the sixth Parliament, session 1889, pp. 212–213.
Canada. 1893. Sessional papers of the Dominion of Canada. Volume 7, third session, pp. 10–24.
Canada. 1895. Sessional papers of the Dominion of Canada. Volume 8, fifth session of the seventh Parliament, session 1895, pp. 332–333.
Canada. 1896. Sessional papers of the Dominion of Canada. Volume 8, second session of the eighth Parliament session, pp. 204–205.
Canada. 1898. Sessional papers of the Dominion of Canada. Volume 9, third session of the eighth Parliament, session 1898, pp. 208–211.
Canada. 1900. Sessional papers of the Dominion of Canada. Volume 9, 5th session, pp. 225.
Perlet, H. M. 1859. Fisheries of the Gulf of St. Lawrence. Canadian Naturalist and Geologist 4(2):84, Montreal.
Casselman, J. M. 2003. Dynamics of resources of the American eel, *Anguilla rostrata*: declining abundance in the 1990s. Pages 255–274 *in* K. Aida, T. Tsukamoto, and K. Yamauchi, editors. Eel biology. Springer-Verlag, Tokyo.
Casselman, J. M., T. Penczak, L. Carl, R. H. K. Mann, and W. A. Wojtowich. 1990. An evaluation of fish sampling methodologies for large river systems. Polish Archives of Hydrobiology 37(4):521–551.
Casselman, J. M., and L. A. Marcogliese. 2007. Long-term changes in American eel (*Anguilla rostrata*) commercial harvest and price in relation to declining abundance. Prepared for Great Lakes Fishery Commission, Ontario Ministry of Natural Resources, and Department of Fisheries and Oceans by AFish-ESci Inc., Bath, Ontario.
Castonguay, M. P. V. Hodson, C. M. Couillard, M. J. Eckersley, J.-D. Dutil, and G. Verreault. 1994. Why is recruitment of the American eel, *Anguilla rostrata*, declining in the St Lawrence River and Gulf? Canadian Journal of Fisheries and Aquatic Sciences 51:479–488.
Champlain Society. 1922–1936. The works of Samuel de Champlain, volume IV. Available: http://link.library.utoronto.ca/champlain/search.cfm?lang=eng (January 2008).
Charlevoix, P. 1761. Journal of a voyage to North-America undertaken by order of the French king containing the geographical description and natural history of Canada, together with an account of the customs, characters, religion, manners, and traditions of the original inhabitants. Printed in two volumes for R. and J. Dodsley in Pall Mall, London.
Clark, W. E., J. T. Molderhauer, M .B. Barber and T. R. Whyte. 2005. The Bussard Rock site (44RN2): a late woodland village. Virginia Department of Historic Resources, Research Report Series 15, Richmond.
Clermont, N. 1999. The archaic occupation of the Ottawa Valley. Pages 43–53 *in* J-L. Pilon, editor. Ottawa Valley prehistory. Outaouais Historical Society, Hull.
Clermont, N. and C. Chapdelaine. 1998. Ile Morrison, Lieu sacre et atelier de l'Archaique dans l'Outaouais, Paleo-Québec 28. Musee Canadien Des Civilisations et Recherches amerindiennes au Québec, Montréal.
Clermont, N., C. Chapdelaine, and J. Cinq-Mars. 2003. Ile aux Allumettes L'Archaique superior dans L'Outaouais. Collection Paleo-Québec, Montreal.
Clover, C. 2006. The end of the line: how overfishing is changing the world and what we eat. The New York Press, New York.
Coburg Star. 1838. Valuable mill property and water power for sale. Coburg Star (May 9).
Coker, R. E. 1929. Studies of common fishes of the Mississippi River at Keokuk. Bulletin of the U.S. Bureau of Fisheries 45:172.
Commanda, W. 2007. Letter to Fisheries and Oceans Canada regarding American eel consultation with respect to listing eels under SARA. March 31:2007. Available: www.icenetwork.ca/about_us/gwcvictoria%20island/othergwcdomtar08 (February 2009).
Copway, G. 1847. The life, history and travels of Kah-Ge-Gah-Bowh: a young Indian chief of the Objebwa Nation. Weed and Parsons, Albany, New York.
COSEWIC (Committee on the Status of Endangered Wildlife in Canada). 2006. COSEWIC assessment and status report on the American eel *Anguilla rostrata* in Canada. Committee on the Status of Endangered Wildlife in Canada. Ottawa. Avail-

able: www.sararegistry.gc.ca/document/dspText_e.cfm?ocid=4251 (July 2008)

Cossette, É. 1995. Assemblages zooarchéologiques et stratégies de subsistances des groupes de chasseurs-pêcheurs du site Hector-Trudel (Québec) entre 500 et 1000 de notre ère. Thèse de doctorat en anthropologie. Université de Montréal, Montréal.

Courtemanche, M. 2003. Pratiques halieutiques à la station 4 de la Pointe-du-Buisson (BhFl-1) au Sylvicole Moyen tardif (920–940 AD). Mémoire de maîtrise. Université de Montréal, Montréal.

Dannewitz, J., G. E. Maes, L. Johansson, H. Wickstrom, F. A. M. Volckaert, and T. Jarvi. 2005. Panmixia in the European eel: a matter of time. Proceedings of the Royal Society London B 272:1129–1137.

Dekker, W., J. M. Casselman, D. K. Cairns, K. Tsukamoto, D. Jellyman, H. Lickers. 2003. Worldwide decline of eel resources necessitates immediate action: Québec declaration of concern. Fisheries 28(12):28–30.

Dekker, W. 2003. Status of the European eel stock and fisheries. Pages 237–254 *in* K. Aida, K. Tsukamoto, and K. Yamauchi, editors. Eel biology. Springer-Verlag, Tokyo.

Dekker, W. 2008. Coming to grips with the eel stock slip-sliding away. Pages 335–355 *in* M. G. Schechter, W. W. Taylor, and N. J. Leonard, editors. International governance of fisheries ecosystems: learning from the past, finding solutions for the future. American Fisheries Society, Bethesda, Maryland.

Denys, N. 1672. The description and natural history of the coasts of North America. Champlain Society, 1908. Available: http://link.library.utoronto.ca/champlain/item_record.cfm?Idno=9_96831&lang=eng&query=The%20AND%20description%20AND%20natural%20AND%20history%20AND%20of%20AND%20the%20AND%20coasts%20AND%20of%20AND%20North%20AND%20America&searchtype=Fulltext&startrow=21&Limit=All (February 2009).

Desroches, D. 1995. Suivi de la migration de l'anguille d'Amerique (*Anguilla rostrata*) au complexe Beauharnois, 1994. Rapport prepare par Milieu & Associes inc. Pour la vice-presidence Environnement, Hydro-Québec, Montréal.

DFO (Fisheries and Oceans Canada). 2004. Fisheries and Oceans Canada news release September 17, 2004, DFO and provinces release American eel management strategy. Available: www.dfo-mpo.gc.ca/media/npress-communique/2004/hq-ac79-eng.htm (July 2008).

DFO (Fisheries and Oceans Canada). 2007.American eel management plan. Available: www.dfo-mpo.gc.ca/communic/fish_man/consultations/eel-anguille_e.htm (July 2008).

Dohne, D. 2004. Are we seeing the end of American eel? The Patriot News (October 24).

Donnil, W., K-J. Maier, and H. Vicenti. 2001. Bestandsentwicklung des Aals (*Anguilla Anguilla*) im Hochrhein. Mitteilung Zur Fischeri, BUWAL, Bern, Germany.

du Creux, F. 1664. History of Canada or New France. Volume 1. The Champlain Society, Toronto.

Dymond, J. R. 1939. The Fishes of the Ottawa region. Contributions of the Royal Ontario Museum of Zoology 15, Toronto

Eales, J.G. 1968. The eel fisheries of eastern Canada. Fisheries. Research Board of Canada. Bulletin 166.

Edeline, E. 2007. Adaptive phenotypic plasticity of eel diadromy. Marine Ecology Progress Series 341:229–232.

Edeline, E., L. Beaulaton, R. Le Barh, and P. Elie. 2007. Dispersal in metamorphosing juvenile eel (*Anguilla anguilla*). Marine Ecology Progress Series 344:213–218.

Egloff, K.T. M., B. Barber, C. Reid and T. Whyte. 1994. Leggett Site (44HA23), Halifax County): a Dan River agricultural/riverine community. Archaeological Society of Virginia Quarterly Bulletin 49:3:89–120.

EU (European Union). 2007. Council Regulation (EC) No. 1100/2007 of 18 September 2007 establishing measures for the recovery of the stock of European eel. Official Journal of the European Union, 22.9.2007, L 248/17. Available: http://eur-lex.europa.eu/LexUriServ/site/en/oj/2007/l_248/l_24820070922en00170023.pdf (July 2008).

Feest, C. F. 1978. Virgina Algonquins. Pages 240–252 *in* W. C. Sturtevant, editor.

Handbook of North American Indians. Smithsonian Institution Press, Washington, D.C.

Finlayson, W. D. 1998. Iroquoian peoples of the land of rocks and water. A.D. 1000–1650: a study in settlement of archaeology. London Museum of Archaeology, Special Publication 1, 4 volumes, London.

Fitzgerald, W. 1990. Preliminary observations on the Ivan-Elliot (Ai-Ha-16) Village and the Raymond Reid (Ai-Ha-4) Hamlet, Wellington County, Ontario. KEWA 90:6:2–16. Ontario Archaeological Society, London Chapter, London.

Floyd, M. 2007. The Eel River Tribe of Indian: a Wabash confederacy. PageFee Publishing, Otsego, Michigan.

Fraser, A. 1912. Ninth Report of the Bureau of Archives for the Province of Ontario. Legislative Assembly of Ontario, Toronto.

Friedland, K. D. M. J. Miller, and B. Knights. 2007. Oceanic changes in the Sargasso Sea and declines in recruitment of the European eel. ICES Journal of Marine Science 64:519–530.

Garrad, C. 1967. The Borden Site designation scheme applied to Ontario. Ontario Archaeology 10:64–71.

Guillet, E. C. 1957. The Valley of the Trent. Champlain Society. Toronto.

Haro, A. J., W. Richkus, K. Whalen, A. Hoar, W.D. Busch, S. Lary, T. Bush, and D. Dixon. 2000. Population decline of the American eel: implications for research and management. Fisheries 25(9):7–16.

Hartford Courant. 1902. River full of eels. Hartford Courant (May 28).

Haxton, T., and D. Chubbuck. 2002. Review of the historical and existing natural environment and resource uses on the Ottawa River. Ontario Ministry of Natural Resources, Science and Information Branch, Southcentral Science and Information Section Technical Report Number 119, North Bay, Ontario.

Heckewelder, J. 1823. A narrative of the mission of the United Brethren among the Delaware and Mohegan Indians. W. E. Connelly, editors. The Burrows Brothers, Cleveland, Ohio.

Henke, J. 1993. Tales of Oneida Lake. North County Books. Utica, New York.

Hennepin, l. 1697. Nouvelle découvert d'un très grand pays situé dans l'Amérique entre le Nouveu Mexique et la mer Glaciale. Guillaume Broedelet, Utrecht.

Hoag, H. 2007. A slippery slope. The Globe and Mail (March 31): F7.

ICES (International Council for Exploration of the Sea). 2003. Report of the ICES/EIFAC Working Group on Eels. International Council for Exploration of the Sea, ICES CM 2003/ACFM:06, Copenhagen, Denmark.

ICES (International Council for Exploration of the Sea). 2006. Report of the 2006 session of the Joint EIFAC/ICES Working Group on Eels. Rome, 23–27 January 2006. Food and Agriculture Organization of the United Nations, European Inland Fisheries Advisory Commission, EIFAC Occasional Paper No. 38, Rome and International Council for Exploration of the Sea, ICES CM 2006/ACFM:16, Copenhagen, Denmark.

Jacobs, P. 1857. Journal of the Reverend Peter Jacobs, Jacobs Indian Wesleyan Missionary, from Rice Lake to the Hudson's Bay Territory and returning commencing May, 1852. The author, New York.

Jessop, B. M. 1987. Migrating American eels in Nova Scotia. Transactions of the American Fisheries Society 116:161–170.

Jessop, B. M., J. C. Shiao, Y. Iizuki, and W. N. Tzeng. 2002. Migratory behavior and habitat use by American eels, *Anguilla anguilla*, as revealed by otolith microchemistry, Marine Ecology Progress Series 233:217–229.

Junker-Anderson, C. 1988. The eel fisheries of the St. Lawrence Iroquoians. North American Archaeologist 9(2):97–121.

Kolenosky, D. P., and M. J. Hendry. 1982. The Canadian Lake Ontario fishery for American eel (*Anguilla rostrata*). Pages 8–16 *in* K. H. Loftus, editor. Proceedings of the 1980 North American Eel Conference. Ontario Ministry of Natural Resources, Fisheries Branch, Ontario Fish Technical Series No. 4, Toronto.

Kraus, R. T., and D. H. Secor. 2005. Application of the nursery-role hypothesis to a marine fish. Marine Ecology Progress Series 291:301–305.

Lacey, L. 1977. Micmac Indian medicine: traditional way of health. Formac Limited, Antigonish, Nova Scotia.

Lackey, R. T., D. H. Lach, and S. L. Duncan. 2006a. Introduction: the challenge of restoring wild salmon. Pages 1–11 *in* R. T. Lackey, D. H. Lach, and S. L. Duncan, editors. 2006. Salmon 2100: the future of wild Pacific salmon. American Fisheries Society, Bethesda, Maryland.

Lackey, R. T., D. H. Lach, and S. L. Duncan, editors. 2006b. Salmon 2100: the future of wild Pacific salmon. American Fisheries Society, Bethesda, Maryland.

Lamson, H. M., J.-C. Shiao, Y. Iizuka, W. -N. Tzeng, and D. K. Cairns. 2006. Movement patterns of American eels (*Anguilla rostrata*) between salt- and freshwater in a coastal watershed, based on otolith microchemistry. Marine Biology 149:1567–1576.

Larinier, M., and J. Dartiguelongue. 1989. La circulation des poisons migrateurs: le transit a travers les turbines des installations hydroelectriques. Bulletin Français de la peche et de la Piscicultur 312–313.

Lary, S. J., W-D Busch, C. M. Castigione, and R. MacDonald. 1998. Distribution and availability of Atlantic coast freshwater habitats for American eel (*Anguilla rostrata*). U.S. Fish and Wildlife Service, Draft Administrative Report 98–02, Washington, D.C.

Loftus, K. H, editor. 1982. Proceedings of the 1980 North American Eel Conference. Ontario Ministry of Natural Resources, Fisheries Branch, Ontario Fish Technical Series No. 4, Toronto.

Lutins, A. 1992. Prehistoric fisheries in Eastern North America. Master's thesis, State University of New York, Binghamton. Available: www.lutins.org/thesis/index.html (January 2009).

Lutins, A., and A. P. DeCondo. 1999. The Fair Lawn/Paterson Fish Weir. Bulletin of the Archaeological Society of New Jersey 54. Available: www.lutins.org/weir/index.html (January 2009).

MacGregor, R. B., A. Mathers, P. Thompson, J. M. Casselman, J. M. Dettmers, S. LaPan, T. C. Pratt, and W. A. Allen. 2008. Declines of American eel in North America: complexities associated with binational management. Pages 357–381 *in* M. G. Schechter, W. W. Taylor, and N. J. Leonard, editors. International governance of fisheries ecosystems: learning from the past, finding solutions for the future. American Fisheries Society, Bethesda, Maryland.

Mandrak, N. E., and E. J. Crossman. 2003. Fishes of Algonquin Park. Friends of Algonquin Park, Whitney, Ontario.

Marshall, M. 1997. Values, customs and traditions of the Mi'kmaq nation. Pages 51–63 *in* L. Choyce, editor. The Mi'kmaq anthology. Pottersfield Press, Lancetown Beach, Nova Scotia.

McCleave, J. D. 2001. Simulation of the impact of dams and fishing weirs on the reproductive potential of silver-phase American eels in the Kennebec River basin. North American Journal of Fisheries Management 21:592–605.

McDowall, R. M. 1996. Diadromy and the assembly and restoration of riverine fish communities: a downstream view. Canadian Journal of Fisheries and Aquatic Sciences 53(Supplement 1):213–236.

McGregor, E. 1994. Algonquin lexicon. Kitigan Zibi Education Council, Maniwaki, Québec.

MdDNR (Maryland Department of Natural Resources). 1999.Maryland waters: an evaluation of stream health. Available: www.dnr.state.md.us/streams/pubs/state_factsheet.pdf (July 2008).

MdDNR (Maryland Department of Natural Resources). 2008. The Renewed Bay Agreement. Available: http://dnrweb.dnr.state.md.us/bay/res_protect/c2k/index.asp (July 2008).

Miller, C. E., D. F. Belknap, and J. T. Kelly. 2006. Geoarchaeological setting of the Sebasticook Lake fish weir, Newport, Maine. The Geological Society of America. Available: http://gsa.confex.com/gsa/2006NE/finalprogram/abstract_100730.htm (July 2008).

Morgan, L. H. 1877. Ancient society or researches in the lines of human progress from savagery through barbarism to civilization. Henry Holt and Company, New York. [Reprinted, 1963, world Publishing Company, Cleveland, and 1976, Gordon Press, New York]

Morrison, W. E., D. H. Secor, and P. M. Piccoli. 2003. Estuarine habitat use by Hudson River American eels as determined by otolith strontium:calcium ratios. Pages 87–99 *in* D. A. Dixon, editor. Biology, management, and protection of catadromous eels. American Fisheries Society, Symposium, Bethesda, Maryland.

Munsell, J., 1861. Proceedings of the Commissioners of Indian Affairs appointed by law for the extinguishment of Indian titles in the state of New York published from the original manuscript in the library of the Albany Institute, volume 1. Munsell's Historical Series, No. X. Available: www.canadiana.org/ECO/PageView/58728/0059?id=11d141c04177adac (January 2008).

Munsell, W. W. 1880. The history of Delaware County, the town of Hancock. Available: www.dcnyhistory.org/books/munhank.html (January 2008).

Native Fish Conservancy. 2008. Migratory fish restoration and passage on the Susquehanna River. Available: www.nativefish.org/articles/Migratory_Fish_Restoration.php (January 2008).

NatureServe. 2007. Comprehensive report—*Anguilla rostrata*. Available: www.natureserve.org/explorer/servlet/NatureServe?searchName=ANGUILLA+ROSTRATA (July 2008).

The New York Times. 1880a. Catching canadian eel: the fisheries of the St. Lawrence River. The New York Times (October 18).

The New York Times. 1880b. Curiosities of eel life: some facts for fishermen and fish-eaters. The army of young eels marching up the Delaware—surmounting natural obstacles—some disputed points of natural history…"bobbing for eels and other ways of catching them." The New York Times (June 10).

The New York Times. 1886. A mill stopped by eels. New York Times (May 16).

The New York Times. 1888. An eel in a London Street. The New York Times (August 12).

The New York Times. 1891. Eels stopped the wheel. The New York Times (November 3).

The New York Times. 1895. Eels made riverhead dark. The New York Times (November 28).

The New York Times. 1896. Salmon caught in eel weir. New York Times (September 29).

The New York Times. 1900. Eels stop a water wheel. The New York Times (December 2).

The New York Times. 1901. Woodcock, bass and eels: vagaries of the game law as affecting the sportsmen of New York, New Jersey and Pennsylvania. The New York Times (September 1).

The New York Times. 1906. Eels stop an electric light plant. The New York Times (October 26).

The New York Times. 1908. Eels in Harriman pipes: three eels weighing 16 pounds, stop lighting plant on his estate. The New York Times (November 27).

The New York Times. 1936. The markets that feed New York. The New York Times (January 5).

Normandeau Associates Inc. and J. R. Skalski. 1998. Draft final report. Estimation of survival of American eel after passage through a turbine at the St. Lawrence-FDR power project. Prepared for New York Power Authority, White Plains, New York

Ogden, J. C. 1970. Relative abundance, food habits and age of American eel, *Anguilla rostrata*, in certain New Jersey streams. Transactions of the American Fisheries Society 99(1):54–59.

OMNR/Québec MNRF (Ontario Ministry of Natural Resources/Ministere des Ressources Naturelles et de la Faune du Québec). 1999. A strategic fisheries management framework for the Ottawa River. Ontario Ministry of Natural Resources, Pembroke.

OMNR (Ontario Ministry of Natural Resources). 2007. Endangered Species Act, 2007. Available: www.mnr.gov.on.ca/en/Business/Species/2ColumnSubPage/STEL01_131232.html (July 2008).

Pauly, D. 1995. Anecdotes and the shifting baseline syndrome in fisheries. Trends Ecology and Evolution 10:420.

Pearce, R. 1977. An eastern regional expression of the Pickering Branch. Master's thesis. Trent University, Peterborough, UK.

Petersen, R. H. 1997. The American eel in eastern Canada: stock status and management strategies. Proceedings of the Eel Workshop, January 13–14, 1997, Québec City, Québec, Canada. Canadian Technical Report of Fisheries and Aquatic Sciences 2196.

PFBC (Pennsylvania Fish and Boat Commission) 2007. Migratory fish restoration and passage on the Susquehanna River. Available: www.fish.state.pa.us/pafish/shad/migratory_fish.pdf (July 2008).

Pinnegar, J. K., and G. H. Engelhard. 2008. The 'shifting baseline' phenomenon: a global perspective. Reviews in Fish Biology and Fisheries 18:1–16.

Post, J. R., M. Sullivan, S. Cox, N. P. Lester, C. J. Walters, E. A. Parkinson, A. J. Paul, L. Jackson, and B. J. Shuter. 2002. Canada's recreational fisheries: the invisible collapse? Fisheries 27(1):6–17.

Preston, R. A. 1958. Royal Fort Frontenac. Champlain Society, Toronto. Available: http://link.library.utoronto.ca/champlain/item_record.cfm?Idno=9_96896&lang=eng&query=royal%20AND%20fort%20AND%20frontenac&searchtype=Fulltext&startrow=11&Limit=All (February 2009).

Prince Edward Island. 1773–1834. The acts of the General Assembly of Prince Edward Island. CAP. V. An act to regulate the salmon, salmon-trout and eel fishery. A.D. 1780.

Prosper, K. 2001. The Mi'Kmaq and Kat (American eel). Antigonish: a preliminary report of research results, phase 1. Social Research for Sustainable Fisheries, SSRF Research Report #4 in collaboration with the Paq'tnkek Fish and Wildlife Society, Antigonish, Nova Scotia. Available: www.stfx.ca/research/SRSF/researchreports1/FinalReports/Report4.pdf (January 2008).

Prosper, K., and M. J. Paulette. 2002. The Mi'Kmaq relationship with Kat. Fact Sheet 7. Social Research for Sustainable Fisheries in collaboration with the Paq'tnkek Fish and Wildlife Society, Antigonish, Nova Scotia. Available: http://faculty.msvu.ca/srsf/ResearchReports/FactSheets/Factsheet7.pdf (January 2008).

Prosper, K., and M. J. Paulette. 2003. Kat (American eel- *Anguilla rostrata*) life history. Pages 392–400 in N. Haggan, C. Brignall, and L. Wood, editors. Putting fishers's knowledge to work. University of British Columbia Fisheries Centre Research Reports 11(1). Available: www.fisheries.ubc.ca/archive/publications/reports/11-1/42_prosper_paulette.pdf (February 2009).

Pulla, S. P. 2003. Frank Speck and the Moisie River incident: anthropological advocacy and the question of Aboriginal fishing rights in Québec. Anthroplogica 45:129–145.

Purvis, F. 1887. Field notes of subdivision survey of the township of Hilliard, Field note book 1300. Ontario Ministry of Natural Resources, Peterborough.

Puseman, K., and L. S. Cummings. 2004. Protein residue analysis of artifacts from the Puncheon run site. Paleo Research Laboratories, Denver, Colorado.

Reed, P. L. 1993. The MacLeod site and a preliminary delineation of the Lake Ontario Iroquois. North and south: two views of the Black Creek–Lalond period. P. Ramsden editor. Copetown Press, Dundas, Ontario.

Richkus, W., and K. Whalen. 2000. Evidence for a decline in the abundance of American eel, *Anguilla rostrata* (LeSeur), in North America since the early 1980s. Dana 12:83–97.

Ritchie, W. A., and R. E. Funk. 1973. Aboriginal settlement patterns in the Northeast. New York State Museum Memoir 20.

Robitaille, J. A., P. Berube, S. Tremblay, and G. Verreault. 2003. Eel fishing in the Great Lakes/St. Lawrence River system during the 20th century: signs of overfishing. Pages 253–262 *in* D. A. Dixon, editor. Biology, management and protection of catadromous eels. American Fisheries Society, Symposium 33, Bethesda, Maryland.

Sáenz-Arroyo, A., C. M. Roberts, J. Torre, M. Cariño-Olvera, and R .R. Enríque-Andrade. 2005. Rapidly shifting environmental baselines among fishers of the Gulf of California. Proceedings of the Royal Society B 272:1957–1962.

Secor, D. H. 1999. Specifying divergent migrations in the concept of stock: the contingent hypothesis. Fisheries Research 43:13–34.

Secor, D. H. 2007. The year –class phenomenon and the storage effect in marine fishes. Journal of Sea Research 57:91–103.

Secor, D. H., and L. A. Kerr. 2009. Lexicon of life cycle diversity in diadromous and other fishes. Pages 537–556 *in* A. J. Haro, K. L. Smith, R. A. Rulifson, C. M. Moffitt, R. J. Klauda, M. J. Dadswell, R. A. Cunjak, J. E. Cooper, K. L. Beal, and T. S. Avery, editors. Challenges for diadromous fishes in a dynamic global environment. American Fisheries Society, Symposium 69, Bethesda, Maryland.

Schmalz, P. S. 1991. The Ojibwa of southern Ontario. University of Toronto Press, Toronto.

Small, H. B. 1883. Fishes of the Ottawa River. Ottawa Field Naturalists Club Transactions 4:31–49.

Smith, M. W., and J. W. Saunders. 1955. The American eel in certain fresh waters of the Maritime Provinces of Canada. Journal of the Fisheries Research Board of Canada 12(2):238–269.

Speck, F. G. 1940. Miscellaneous notes, Scatticook, two letters of Chief Swimming Eel to Speck, August 16, 1939 and 1940, concerning Indigenous social activities. American Philosophical Society Library. Available: www.amphilsoc.org/library/guides/indians/info/s.htm#scatticoo (December 2007).

Stevens, E. W. 2004. First Nations underwater archaeology of the Trent-Severn Waterway. Copy on file at Parks Canada, Ottawa.

Supreme Court of Canada. 1999. R. v. Marshall. Docket Number 26014. Available: http://csc.lexum.umontreal.ca/en/1999/1999rcs3-533/1999rcs3-533.html (February 2009).

Talbot, E. A. 1824. Five year's residence in the Canadas. 2 volumes. Longman, London.

Telmer, K. H. 1996. Biogeochemistry and water balance of the Ottawa River basin. Doctoral thesis. University of Ottawa, Ottawa.

Tesch, F. W. 1977. The eel: biology and management of anguillid eels. Chapman and Hall, London.

Thwaites, R. G. 1896–1901. The Jesuit relations and allied documents. 73 volumes. Burrows Bros., Cleveland, Ohio.

Thwaites, R. G. 1903. A new discovery of a vast country in America, by Father Louis Hennepin. A.C. McClurg and Company, Chicago.

Tooker, E. 1978. The league of the Iroquois: its history, politics, institution and ritual. Handbook of North American Indians 15:418–441. Smithsonian Institution, Washington, D.C.

Tsukamoto, K., I. Naka, and W. V. Tesch. 1998. Do all freshwater eels migrate? Nature 396:635–636.

Tsukamoto, K., and T. Arai. 2001. Facultative catadromy of the eel *Anguilla japonica* between freshwater and seawater habitats. Marine Ecology Progress Series 220:265–276.

Tsukamoto, K. J. Aoyma, and M. J. Miller. 2002. Migration, speciation, and the evolution of diadromy in anguillid eels. Canadian Journal of Fisheries and Aquatic Sciences 59(12):1989–1998.

Tzeng, W. H., C. H. Wang, H. Wickstrom, and M. Reizenstein. 2000. Occurrence of the semi-catadromous European eel *Anguilla anguilla* in the Baltic Sea. Marine Biology 137:93–98.

U.S. Office of the Federal Register. 2007. Endangered and threatened wildlife and plants: 12-month finding on a petition to list the American eel as threatened or endangered. Code of Federal Regulations, Title 50, Part 15. U.S. Government Printing Office, Washington, D.C.

Verdon, R., D. Desroches, and P. Dumont. 2003. Recruitment of American eels in the Richelieu river and Lake Champlain: provision of upstream passage as a regional-scale solution to a large-scale problem. Pages 125–138 *in* D. A. Dixon, editor. Biology, management and protection of catadromous eels. American Fisheries Society, Symposium 33, Bethesda, Maryland.

Verreault, G., and P. Dumont. 2003. An estimation of American eel escapement from the upper St. Lawrence River and Lake Ontario in 1996 and 1997. Pages 243–251 *in* D. A. Dixon, editor. Biology, management and protection of catadromous eels. American Fisheries Society, Symposium 33, Bethesda, Maryland

Verreault, G., P. Dumont, and Y. Mailot. 2004. Habitat losses and anthropogenic barriers as a cause of population decline for American eel (*Anguilla rostrata*) in the St. Lawrence watershed, Canada. International Council for the Exploration of the Sea, ICES CM 2004/S:04, Copenhagen, Denmark.

Wade, J. 2007. Indian autumn: the harvest, hunt and the big house ceremony. New Jersey Department of Environmental Protection, Washington Crossing State Park, The Nature Grapevine, Trenton. Available: http://liberty.state.nj.us/dep/parksandforests/parks/washington_crossing_grapevine.htm (January 2008).

Walters, C., and J.-J. Maguire. 1996. Lessons for stock assessment from the northern cod collapse. Fish Biology and Fisheries 6:125–137.

Watts, T .A., and D. H. Watts. 2004. Petition to list the American eel as an endangered species pursuant to the United States Endangered Species Act 16 U.S.C. ss 1531–1544. Available: www.glooskapandthefrog.org/ESA%20petition.htm (March 2008).

White, E. M., and B. Knights. 1997. Dynamics of upstream migration of the European eel *Anguilla anguilla* (L.), in the rivers of Severn and Avon, England, with special reference to the effects of man-made barriers. Fisheries Management Ecology 4:311–324.

Whyte, T. R. 1988. Fish remains recovered from a protohistoric village site on the Jackson River, Bath County, Virginia. Archaeological Society of Virginia, Quarterly Bulletin 43:4:167–171.

Whyte, T. R. 2001. Prehistoric archaeological record of freshwater fishes in the Roanoke River, Virginal and North Carolina. Banisteria 18: 24–30. Virginia Natural History Society, Hampden-Sydney.

Whyte, T. R. 2002. Archaeofaunal remains from the late prehistoric Mount Joy Site in Botetourt County,

Virginia. Banisteria 20: 45–52. Virginia Natural History Society.

Whyte, T .R., and, S.M. Thompson. 1989. A late woodland period Dan River and Page component village site on the upper James River. Archaeological investigations at the Bessemer Site (44B026). On file at the Virginia Department of Transportation, Richmond.

Winemiller, K. O., and K. A. Rose. 1992. Patterns of life-history diversification in North American fishes: implications for population regulation. Canadian Journal of Fisheries and Aquatic Sciences 49:2196–2218.

Wirth, T., and L. Bernatchez. 2003. Decline of North Atlantic eels: a fatal synergy? Proceedings of the Royal Society of London Series B 270:681–688.

Zettler, M. 2007. Algonquin elder seeks "at risk" status for the American eel. Pembroke Daily Observer (May 17):3, 14.

American Fisheries Society Symposium 69:741–756, 2009

Challenges and Opportunities for Co-Management of a Migratory Fish (*Prochilodus nigricans*) in the Peruvian Amazon

Elizabeth P. Anderson*,[1]
Global Water for Sustainability Program, Department of Environmental Studies
Florida International University, Miami, Florida 33199, USA

Mariana Montoya
Department of Geography and the Environment, University of Texas
1 University Station A3100, Austin, Texas 78712, USA

Aldo Soto and Hernán Flores
WWF-Peru Programme Office
Trinidad Morán 853, Lima 14 – Perú, Apartado Postal: 11-0205

Michael McClain
Global Water for Sustainability Program, Department of Environmental Studies
Florida International University, Miami, Florida 33199, USA

Abstract.—Fisheries co-management lies at the center of environmental, social, and economic issues in the lower Pastaza River basin in the Peruvian Amazon. In this remote region, the intermingling of black- and white-water systems creates unique aquatic habitats that harbor diverse assemblages of fishes, including *Prochilodus nigricans*, a migratory species of high ecological and socioeconomic importance throughout the Amazon. In the lower Pastaza basin, freshwater fishes provide a major source of food and income for indigenous communities such as the Kandozi. However, the basin's aquatic resources and the livelihoods of indigenous people are being compromised by environmentally unsound practices such as overfishing by unregulated commercial fishers and pollution from petroleum exploration and exploitation in the region. In response, there has been a concerted effort by indigenous communities, conservation organizations, and the Peruvian government to develop systems of fisheries co-management in recent years. Near Lake Rimachi, the largest lake in the Peruvian Amazon, an association of Kandozi fishermen was recently formed to oversee the management of the lake's aquatic resources. The association is working towards implementation of a newly created fisheries management plan, which includes provisions for co-management and protection of migratory species like *P. nigricans*. Although the challenges are many, the long-term objective of fisheries management around Lake Rimachi is to ensure sustainable livelihoods for indigenous communities while maintaining aquatic biodiversity. Here, we discuss the ecological and socioeconomic importance of *P. nigricans* and describe lessons learned from an ongoing process of fisheries co-management.

* Corresponding author: eanderson@fieldmuseum.org
[1] Current address: The Field Museum of Natural History, 1400 South Lake Shore Drive, Chicago, IL 60605, USA

Introduction

Tropical rivers of South America are heterogeneous systems that vary along longitudinal, lateral, and temporal scales and present an array of aquatic environments available to freshwater fishes. Migratory species, notably prochilodontid fishes, exploit this natural variation, moving between areas of distinct hydrology, soils, water quality, and vegetation at different periods in their life cycle (Winemiller and Jepsen 1998). Although not considered diadromous by the definition given by McDowall (1987), *Prochilodus* spp. exhibit complex spawning and feeding migrations spanning hundreds of kilometers and strongly influence distant ecosystems and human lives throughout the Amazon and Orinoco basins. Ecologically, *Prochilodus* play a primary role in the transfer of energy from basal resources to higher levels of the food web, and their foraging behavior shapes ecological processes in aquatic environments (Araujo-Lima et al. 1986; Flecker 1992, 1996; Forsberg et al. 1993; Winemiller 1996; Taylor et al. 2006). They are also considered one of the most important commercial and subsistence fishes in South America (Goulding 1981; Tello and Bayley 2001; Allan et al. 2005; Silva and Stewart 2006; Taylor et al. 2006; Junk et al. 2007).

Overfishing, water pollution, and dam construction threaten survival of inland water fishes worldwide, and in South America, *Prochilodus* may be particularly sensitive to the negative effects of human-induced changes to freshwater ecosystems (Allan et al. 2005). Migrations of *Prochilodus* are cued to seasonal fluctuations in water levels (Winemiller and Jepsen 1998). The presence of dams often results in hydrologic alterations and barriers to longitudinal movement that hinder the ability of *Prochilodus* and other migratory species to complete key phases of their life cycle (Ribeiro et al. 1995; Pringle et al. 2000). Some prochilodontids, such as *P. nigricans*, are known to move between white-water and black-water areas differing naturally in water chemistry (Carvalho de Lima and Araujo-Lima 2004). White-water systems are characterized by high turbidity, high oxygen, and high nutrient concentrations; conversely, black-water systems exhibit low turbidity, low oxygen, and low nutrient concentrations. Water pollution stemming from human activities like deforestation or petroleum exploitation can alter the chemistry of these systems and potentially disrupt fish movements or, in the worst cases, result in fish kills (Kimerling 2006). Overfishing of *Prochilodus* species is a pervasive problem in much of the Amazon and Orinoco basins and presents one of the greatest challenges to fisheries management. As a migratory fish, the same *Prochilodus* stocks are subject to exploitation by multiple fishing communities. Over the past 25 years, overfishing in the Orinoco drainage has substantially affected one prochilodontid species, *P. mariae*, evidenced by correlations of declining fish body mass with decreasing mesh size of fishing nets (Taylor et al. 2006). *Prochilodus* fisheries in other parts of tropical South America may mirror these conditions, but limited scientific information on their status is available (Tello and Bayley 2001).

The social and economic contexts of fisheries in South America, as well as the type of gear used in fishing, are all important factors that influence survival of *Prochilodus* spp. (Batista et al. 1998; Almeida et al. 2003). Published estimates place the potential yield of the inland food fishery of the entire Amazon River basin at 900,000 metric tons per year; in the 1990s, only about half of this potential yield was exploited by Amazonian fisheries (Bayley and Petrere 1989; Junk et al. 2007). In the Peruvian Amazon, the value of total fish capture (~80,000 metric tons/year actual yield) is estimated at US$80 million annually, shared between both commercial (25%) and subsistence (75%) based fisheries (Tello and Bayley 2001). *Prochilodus* represent a disproportionately high percentage of the number and biomass of fishes captured and sold at local and regional fish markets in the Amazon. In Manaus and Tefe, Brazil (middle Amazon), *Prochilodus* are the most important commercial species, representing 45% of market landings (Junk et al. 2007). The same is true in Iquitos and Pucallpa, Peru (upper Amazon), where *Prochilodus* spp. account for 45–55% of the commercial catch (Montreuil et al. 2001; Silva and Stewart 2006; Junk et al. 2007). These percentages are often higher for indigenous communities' fisheries; in parts of the remote northern Peruvian Amazon, *Prochilodus* represents 80% of fish capture and sale during certain times of the year (FECONAKADIP 2007). Because both artisanal and commercial fishermen target *Prochilodus*, competition for fisheries resources has resulted at times in conflicts between indigenous communities and commercial fleets over resource rights (McDaniel

1997; Almeida et al. 2003; Crampton et al. 2004b; FECONAKADIP 2007). In these situations, the response of both parties may be to increase fishing efforts or, in the case of artisanal fishermen, to adopt newer, more efficient fishing gear like gill nets. In this context, overfishing can quickly become a problem that threatens *Prochilodus* populations and the livelihoods of Amazonian people.

Studies of fishing activities and fish consumed by indigenous communities have received low priority, as most attempts to quantify importance of fishing in the Amazon have focused on commercial fisheries (Bayley 1981; Batista et al. 1998; Almeida et al. 2003). Nevertheless, indigenous fisheries deserve more attention for multiple reasons. First, indigenous communities are widespread throughout the Amazon over a large geographical area and most depend on fisheries for subsistence and income generation. Although preferences vary by region, *Prochilodus* are almost always among the most important species (Batista et al. 1998), and because *Prochilodus* are highly migratory, several different indigenous communities may harvest the same fish stocks. Second, indigenous fishermen typically harvest fish year-round and fisheries are often the primary or only source of livelihood for indigenous communities (FECONAKADIP 2007). Fishing for *Prochilodus* can intensify during migration and spawning periods or during low water, when many fish are concentrated in small areas (Batista et al. 1998; H. Flores, personal observation). Finally, although on an individual basis one indigenous community's effect on *Prochilodus* and other fish populations may be minimal, the collective influence of indigenous fishermen throughout the Amazon basin is substantial. Some authors have suggested that the amount of fish caught by subsistence and artisanal fishermen far exceeds that of commercial fisheries and may account for as much as 60% of fish landings in the Amazon (de Merona 1985; Batista et al. 1998; Maccord et al. 2007), but scant data on fish landings outside of regional markets make estimation of the quantity, value, and impact of these fisheries a challenge (Junk et al. 2007).

Effective management of *Prochilodus* and other fishes in the Amazon basin must consider the position of indigenous communities, both in terms of the pressure they exert on the resource and the integral role that fisheries play in their lives. To that end, indigenous communities are beginning to take part in fisheries co-management systems that involve a combination of traditional and modern approaches and offer a space for participation of multiple stakeholders in decision making (Dyer and McGoodwin 1994; Sen and Nielsen 1996; de Castro and McGrath 2003; Maccord et al. 2007). Fisheries co-management systems have emerged in response to several factors: failure of state-imposed restrictive regulations as a fisheries management strategy, more intense fishing pressures and more efficient gear, a desire by local communities to prevent outsiders from having access to resources (e.g., in lake settings), and a desire to regulate the distribution of benefits provided by fisheries to local communities (de Castro and McGrath 2003; Crampton et al. 2004a; Maccord et al. 2007). In some parts of Brazil and Peru, fisheries co-management systems allow rural residents and/or indigenous communities to share legal responsibility with government authorities in devising and enforcing fishing regulations (Almeida et al. 2003; Crampton et al. 2004a; FECONAKADIP 2007; Junk et al. 2007). The case of Mamirauá in Brazil provides an example: a zoned reserve was created to place fisheries resource rights in the hands of local residents with the aim of reconciling biodiversity conservation and economic growth and as a means for monitoring invasive, commercial fishing (Crampton et al. 2004b). Further, by increasing community organization, involvement in fisheries co-management systems has helped some rural and/or indigenous communities address other problems such as disease outbreaks or environmental degradation (McDaniel 1997).

The case of Lake Rimachi and the Kandozi people in the Peruvian Amazon is representative of the challenges facing management of *Prochilodus* and the opportunities for fisheries co-management in the Amazon. The Kandozi are an indigenous community that inhabits a landscape composed of interconnected black-water lakes adjacent to the turbulent, white-water Pastaza River in northern Peru. Freshwater fish provide the overwhelming majority of protein to the Kandozi diet, and the capture and sale of one species, *P. nigricans* (Spix and Agassiz 1829), provides the primary source of economic gain (FECONAKADIP 2007). Nevertheless, the long-term sustainability of Kandozi fisheries is at risk on the basis of internal threats such as overfishing within the community and external threats, including unregulated commercial fishing and petroleum exploration. Together with local authorities, the Kandozi have recently taken steps to address these threats through a system of fisheries co-management in the lower Pastaza River basin. In this article, we briefly discuss the ecology of *P. nigricans*, describe the threats and trends that affect this and other fish species in the lower Pastaza basin, and outline the approach to fisheries co-management being implemented in Lake Rimachi, Peru.

Ecology of *Prochilodus nigricans*

Prochilodus nigricans is a potamodromous fish species that inhabits a wide range of aquatic habitats over a broad spatial area of the Amazon basin. The species is found in the main-stem Amazon River, as well as in lowland areas of Amazonian drainages in Brazil, Colombia, Ecuador, Perú, and Bolivia (Silva and Stewart 2006). It is the most widely distributed of the *Prochilodus* spp. found in the Amazon and is locally known by several common names: boquichico in Peru, curimatá in Brazil, sábalo in Bolivia, and bocachico in Colombia and Ecuador (Araujo-Lima and Ruffino 2004). In the Peruvian Amazon, *P. nigricans* reach maturation at a total length of around 25 cm and attain a maximum total length of approximately 40 cm (Montreuil et al. 2001; Escobedo et al. 2005).

In neotropical rivers, many fish species are specialized consumers of detritus; *P. nigricans* falls within this feeding group (Yossa and Araujo-Lima 1998; Taylor et al. 2006). *Prochilodus nigricans* feeds primarily on amorphous detritus that contains algae and exhibits little to no seasonal variation in diet (Bowen 1983; Bowen et al. 1984; Yossa and Araujo-Lima 1998; Oliveira et al. 2006). On the basis of their feeding behavior, *Prochilodus* spp. act as conduits of energy in aquatic food webs. As detritivores, they facilitate transfer of energy from basal resources directly to species at higher trophic levels, thus increasing the ecological efficiency of piscivore production by means of short food chains (Winemiller 1996; Winemiller and Jepsen 1998). Further, bioturbation during feeding by *Prochilodus* spp. has been shown to increase nitrogen-fixing cyanobacteria and enhance downstream transport of particulate organic carbon (Flecker 1996; Taylor et al. 2006). Recent studies suggest a lack of functional redundancy for the ecological role of *Prochilodus*, even in rivers with diverse fish assemblages (Taylor et al. 2006).

Seasonal, long-range migrations in schools are a critical component of the life history of *P. nigricans*. Although exact migratory patterns vary in different basins, in general migrations occur in response to seasonal changes in water level, food availability, reproduction, and predation risk (Araujo-Lima and Ruffino 2004; Silva and Stewart 2006; Winemiller and Jepsen 1998). *Prochilodus nigricans* moves between black-water environments, such as floodplain lakes, and white-water systems, such as main-stem rivers draining the Andes Mountains. Reproductive migrations are thought to coincide with the onset of the wet season when water levels are rising, while migrations in response to other factors occur during the receding water season (Fernándes 1997; Araujo-Lima and Ruffino 2004).

The Lower Pastaza River Basin, Peru

The Pastaza River is a tributary of the Marañon River, draining an Andean subcatchment of the larger Amazon River basin. Shared between Ecuador and Peru, the Pastaza River basin encompasses an area of approximately 40,000 km^2 and spans an altitudinal gradient of more than 5,000 m. This gradient has created distinct environments in the basin: páramo ecosystems and glacier-covered volcanoes characterize headwater areas, high gradient streams drain cloud forests at middle elevations (1,000–2,500 meters above sea level), and meandering rivers wind through wetlands in the lowlands. While the upper and middle parts of the Pastaza River basin are home to more than 1 million people, wilderness areas of tropical rainforest comprise the sparsely populated lower two-thirds of the basin.

In Peru, the Pastaza River basin drains an extremely heterogeneous natural landscape, dominated primarily by wetland areas. Here, the white-water Pastaza River flows adjacent to or mixes with lentic and lotic black-water systems such as Lake Rimachi and the local rivers draining into the lake. With an area of 79 km^2, Lake Rimachi is considered the largest lake in the Peruvian Amazon and forms part of a series of interconnected black-water lakes, or cochas, that integrate the lower basin of the Pastaza River with two other rivers, the Chuinda and the Chapuli (Figure 1). The intermingling of black- and white-water systems, as well as the presence of floating mats of vegetation in the lake and flooded riparian forests (varzea), have produced an array of

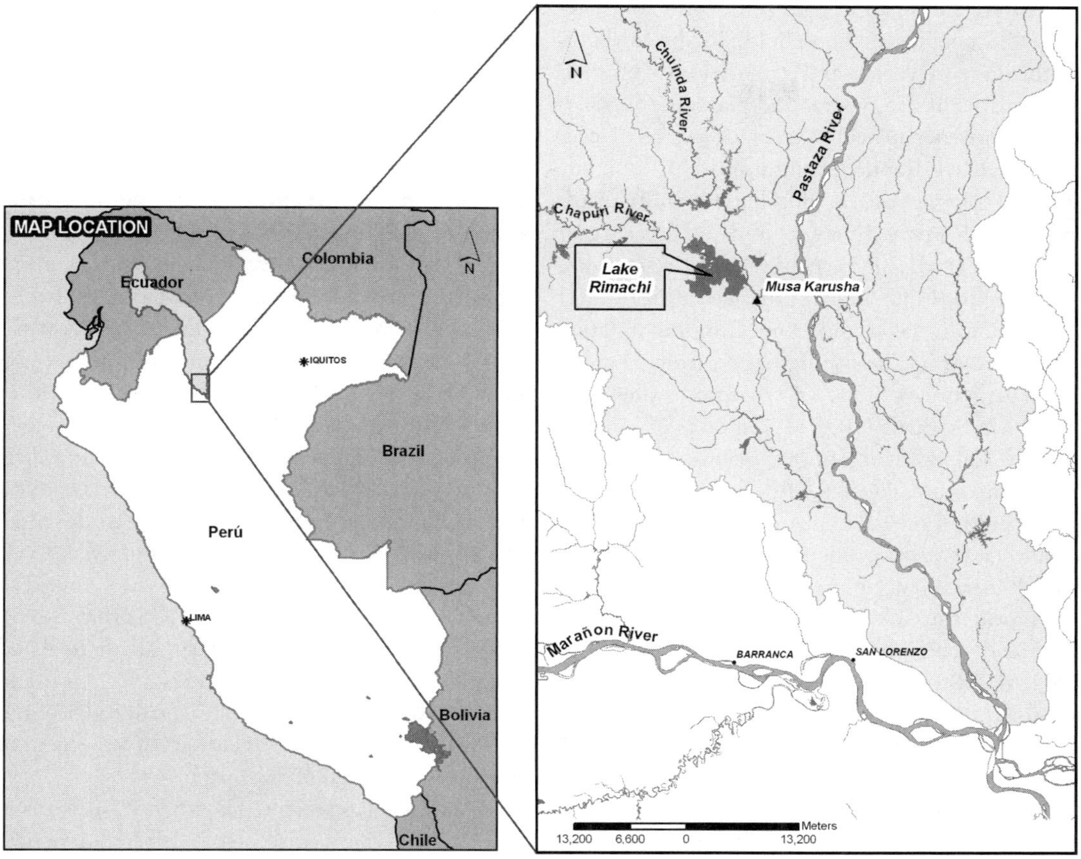

Figure 1. The Pastaza River basin is shared between Ecuador and Peru (inset). The Kandozi inhabit the area west of the Pastaza River around Lake Rimachi in the lower basin. The Kandozi village of Musa Karusha lies at the confluence of the lake near a side channel of the Pastaza River. Figure: C. Alvarez, WWF-Peru.

unique habitats for aquatic biota, especially fishes. Previous studies have reported 292 species of fish (33 families) from the lake and surrounding water bodies; however, these estimates are based on limited scientific information and the actual number of species may be much higher (Willink et al. 2005; FECONAKADIP 2007). A seasonal climate, with marked differences in rainfall between wet and dry seasons, characterizes the lower Pastaza River basin. Annual precipitation is estimated at 3,000 mm, and water levels in Lake Rimachi fluctuate as much as 7 m between the driest (December–February) and wettest (April–June) periods (IIAP-WWF 1999).

The lives and livelihoods of people in the Peruvian Pastaza are intimately linked to freshwater goods and services, especially fisheries. An estimated 11,600 people inhabit the Peruvian Pastaza, and most belong to one of four indigenous communities: Achuar, Kichwa, Shapra, or Kandozi. Settlements are grouped into small villages that average about 120 people and are typically located in riparian areas. In the absence of roads and other major infrastructure, waterways provide the major means of transportation and communication between villages and larger cities downstream. Fish constitute the primary source of protein for indigenous communities and fisheries drive local economies (Pinedo et al. 2000).

The importance of fisheries to indigenous communities is exemplified by the case of the Kandozi, who inhabit areas within and around Lake Rimachi. The Kandozi number about 2,500 people and are known as strong warriors that have resisted conversion by religious missionaries and enslavement by commercial resource exploitation operations. Most

Kandozi live in small villages (<25 households; average 7.5 people per household) in the lowlands of the Pastaza basin, and the majority of their diet consists of fish, yucca, and plantain (Surrallés 2007). Fisheries provide the main source of economic gain for Kandozi communities. As in many artisanal fisheries, the Kandozi harvest multiple fish species with many different kinds of fishing gear (Escobedo et al. 2005). Nevertheless, *P. nigricans* is the keystone of Kandozi fisheries and is primarily captured in lakes and connecting canals using gill nets. During the fishing period from October 2004 to February 2005, *P. nigricans* represented 78% of individuals (10,870) and 83% of biomass (10,520 kg) of fish captures by Kandozi fishermen, out of a total of 42 fish species (Escobedo et al. 2005); these data encompass just the time of year when *P. nigricans* typically spawn. Across the year, *P. nigricans* accounted for 51% of biomass of fish captures in Lake Rimachi between September 2005 and September 2006 (Figure 2). Data from 2004 to 2006 indicate that the catch per unit effort for *P. nigricans* in Lake Rimachi is significantly higher than that of other species, and capture of *P. nigricans* varies seasonally (Figures 3 and 4). When sold locally, average prices for *P. nigricans* were recently about 3.00 Peruvian soles per kilogram, just less than one U.S. dollar (Pacheco 2006).

Despite its remoteness, the Peruvian Pastaza and the area inhabited by the Kandozi has been subject to a series of human-induced threats over the past half century that have affected fishes in Lake Rimachi and surrounding environments. Petroleum production is among the most serious threats to fisheries resources and indigenous communities' livelihoods. The lower Pastaza River basin and surrounding areas are the site for major petroleum exploration and exploitation activities, and it is estimated that approximately 55% of the petroleum produced in Peru came from this region in 2008. Occasional oil spills present one form of water contamination, but a more chronic source of pollution is the failure to reinject production waters into oil wells, which leads to thermal and chemical water pollution as industrial waste waters are often released directly without treatment into rivers. Contamination events like oil spills have been linked to fish kills in the region in the past (Kimerling 2006), but it is still unclear how more chronic thermal and chemical pollution from by-waters affects fish and whether zones of degraded water quality adjacent to petroleum operations act as barriers to movement of migratory fishes like *P. nigricans*.

Additional stressors on fishes in the lower Pastaza basin have arisen from fisheries practices themselves. Commercial fishing fleets, operational

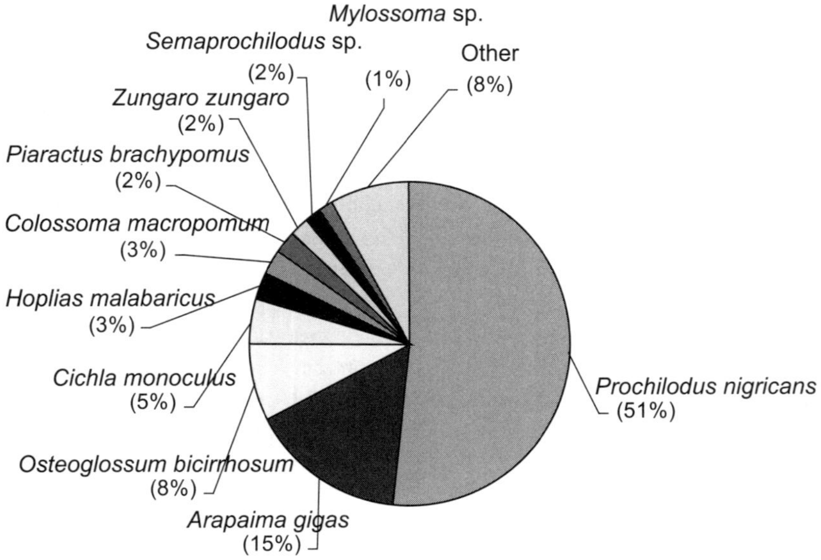

Figure 2.—Biomass of fish capture between September 2005 and September 2006 by species.

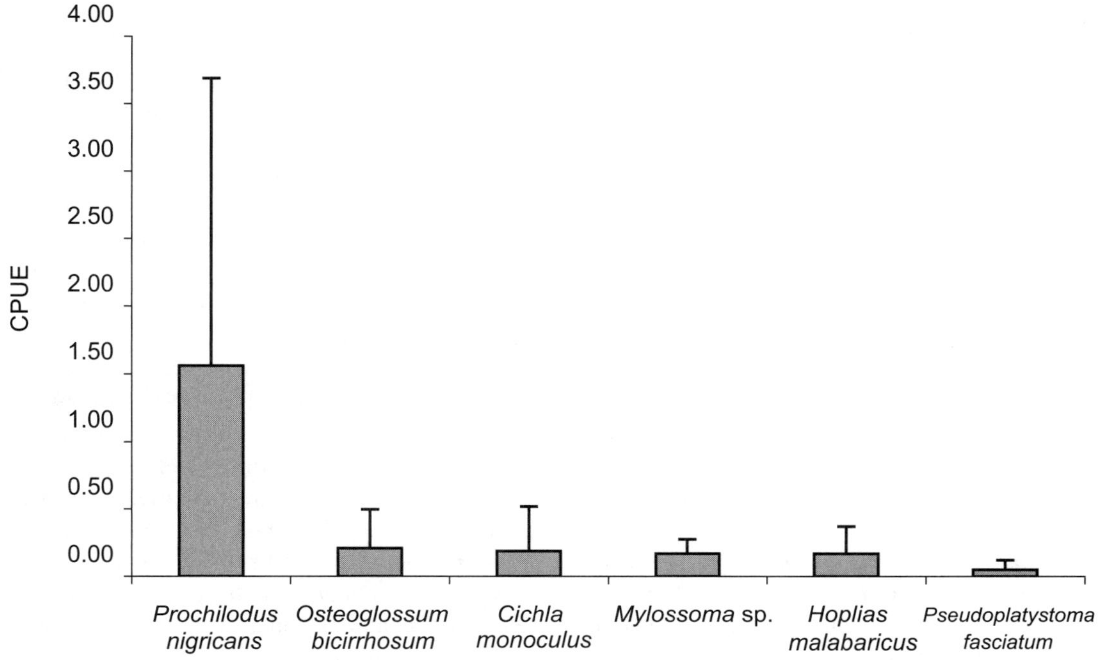

Figure 3.—Average catch per unit effort (CPUE) of the main commercial fish species in Lake Rimachi. Data cover the period from October 2004 to September 2006.

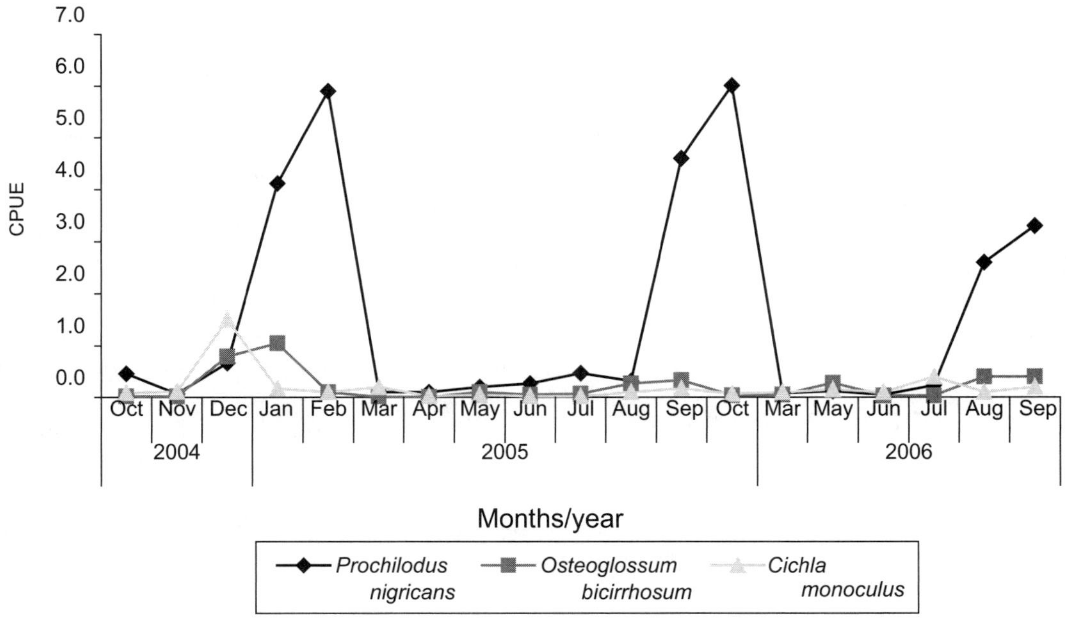

Figure 4.—Fluctuations across the year in catch per unit effort (CPUE) of three fish species in Lake Rimachi.

in Lake Rimachi since the 1970s, have been a significant source of mortality for fishes in the lower Pastaza basin. The Kandozi have perceived the commercial fleets' presence as a direct threat to their own fisheries-based livelihood. Consequently, conflicts have occurred in the past among Kandozi, commercial fishing fleets, and government fishing authorities over resource rights in Lake Rimachi. The presence of commercial fleets has also modernized fishing gear used in Lake Rimachi. Among the Kandozi, a shift from traditional fishing gear (e.g., hooks made from bird clavicle bones, wooden spears, plant-based fish poisons) to use of gill nets occurred in the mid-1990s (Bodmer et al. 2006). Changes have also occurred with respect to the primary species harvested from the lake. The focal species of fisheries in the past (1970–1990) was *Arapaima gigas*, locally known as paiche, but intense fishing pressures by commercial fleets led to depletion of *A. gigas* populations. The reduction in *A. gigas* populations and the adoption of net-based fishing techniques helped stimulate an increase in the importance of *P. nigricans* to both commercial and Kandozi fisheries. Over the past decade, the contribution of *P. nigricans* to local fisheries has only increased, in large part due to the increased efficiency of capture using nets with smaller mesh sizes. Intense exploitation of *P. nigricans* and the disproportionate contribution of this species to local and regional fisheries are topics of concern for the long-term sustainability of Kandozi fisheries. While these topics should not be overlooked, perhaps a more immediate threat to *P. nigricans* is a method practiced by both commercial and Kandozi fishermen that involves placing large gill nets (average 150 m length) in areas where migrants congregate to spawn (Pacheco 2006). Harvest of ripe individuals poses a direct threat to survival of *P. nigricans* populations in the lower Pastaza River basin.

Context for Fisheries Co-Management in Lake Rimachi

In this section, we discuss the Kandozi community's response to the threats that affect fisheries resources in Lake Rimachi and the current approach to co-management of *P. nigricans* and other economically valuable fish species. The Kandozi are one of the first indigenous communities in Peru to have legal responsibility to manage a freshwater body and its associated resources; understanding the present implications of this resource right for the Kandozi requires some appreciation of the historical context of fisheries management in Lake Rimachi. The Kandozi's response to fisheries management in Lake Rimachi has been a process-based approach, and here we provide details about that process.

Lake Rimachi has a long history of management interventions that relate to fisheries. In 1945, the Peruvian government established a Fishery Reserve (Reserva Pesquera) in Lake Rimachi as a measure to conserve fish stocks. Nevertheless, the reserve was subject to limited management and control on the part of the Peruvian Ministry of Fisheries, and commercial fishermen were permitted to exploit the lake's resources during the 1970s and 1980s. By 1990, the populations of several fishes, including *A. gigas*, had noticeably declined (Surrallés 2007). Today, information to support this decline is anecdotal, since fisheries records for 1970–1990 were lost in a fire. Declining fish populations alarmed the Kandozi community because of their fisheries-based livelihoods, and in August 1991, Lake Rimachi became the center of a confrontation between the Kandozi and the Ministry of Fisheries. Kandozi fishermen took control of the Fishery Reserve's base office of operations at the lake, with the intent to force the Ministry of Fisheries out of the region as they were permitting commercial fishermen to exploit the lake's resources and were also suspected to be involved in illegal commercialization of fishes. Since these initial conflicts in the early 1990s, Kandozi leaders have been trying to officially include Lake Rimachi in their territory. However, Peru's constitution does not allow titling of large bodies of water such as lakes or rivers. The Kandozi have thus had to explore alternative legal frameworks to enable them to have resource rights to Lake Rimachi and to have authority to monitor the entrance of commercial fishermen to the lake.

The case of Lake Rimachi lies at the heart of two important issues in the Peruvian Amazon: (1) management and conservation of inland water fishes, and (2) indigenous peoples' rights to ancestral territories and natural resources. The case has attracted the attention of two nongovernmental organizations, WWF-Peru and Racimos de Ungurahui, which specialize in environmental conservation and indigenous rights, respectively. In 2002, these organizations conducted an analysis of the legal and institutional frameworks in Peru that would support the Kandozi's rights to fisheries resources in Lake Rimachi. Several options were uncovered, including land title for the lake, land title for surrounding areas, or creation of a communal reserve. While the legal technicalities of these options differ, all require that the Kandozi prepare a fisheries management plan for Lake Rimachi and that this plan be approved by the Vice-Ministry of Fisheries (Ministerio de Pesquería: Ley No. 25977, D.S. 012–2001-PE and R.M. No. 147–2001-PE). To this end, a formal partnership among the Kandozi, WWF-Peru, and Racimos de Ungurahui was created in 2002 to work towards formalizing Kandozi fishing activities and protecting freshwater resources through the creation and implementation of a fisheries management plan for Lake Rimachi.

The development of a fisheries management plan for Lake Rimachi has been a multi-step process and an exercise for the Kandozi in navigating the Peruvian legal framework. According to the Reglamento de Ordenamiento Pesquero de la Amazonía Peruana (Ministerio de Pesquería R.M. No.147–2001-PE), the initiative to develop a fisheries management plan has to come from an organized group of fishermen. Therefore, the first step in the process for the Kandozi was to form an association of artisanal fishermen. In Peru, artisanal fishermen's associations are recognized through a formal process overseen by the Vice-Ministry of Fisheries that involves training and licensing of artisanal fishermen. Therefore, in November 2003, at the request of the Kandozi and WWF-Peru, the Vice-Ministry of Fisheries conducted a training course for ~150 Kandozi fishermen and subsequently issued certificates and fishing licenses. The training incorporated representatives from 28 Kandozi communities from the Chapuli, Chuinda, Huituyacu, and Ungurahui River basins (all subcatchments of the Pastaza), which primarily use resources from Lake Rimachi and surrounding bodies of water, and included information on Peruvian fisheries legislation, responsibilities of artisanal fishermen's associations, and basic concepts of fisheries management (Flores, personal communication). Following the training, Racimos de Urgurahui prepared a series of legal statutes based on Peruvian laws that were revised by the Kandozi and then presented to the Vice-Ministry of Fisheries. The purpose of these statutes was to legalize the authority of the Kandozi association of artisanal fishermen that was being created over activities in all bodies of water within Kandozi territories, including Lake Rimachi. Care was taken to ensure that the Vice-Ministry of Fisheries' proposed statutes did not interfere with the statutes of the Kandozi Federation, FECONAKADIP. The negotiation of the statutes by both Kandozi fishermen and Vice-Ministry of Fisheries merits discussion because these statutes were among the first of their kind in Peru and they could provide an innovative example of aquatic resource regulation by indigenous communities.

Yungani and the Lake Rimachi Fisheries Management Plan

The Kandozi association of artisanal fishermen, known as Yungani, was officially constituted in September 2004 following the processes outlined in Peru's General Fishing Law (Ministerio de Pesqueria Ley No 25977) and as a result of the technical assistance provided by the Vice-Ministry of Fisheries. Yungani counts 115 members, all Kandozi fishermen, and it is now legally responsible for the administration and control of fisheries in Lake Rimachi. Members are required to develop and implement activities that support co-management and conservation of freshwater resources of the lake and surrounding areas. These activities include managing fisheries within the lake and surrounding water bodies and signing agreements with other civil and governmental organizations that are compatible with Yungani's objectives (FECONAKADIP 2007). The establishment of Yungani has also resulted in the creation of two surveillance and control committees for fisheries in Lake Rimachi; these committees are an important component of artisanal fishermen's associations according to Peru's General Fishing Law. Each of these committees is composed of 10 Kandozi fishermen and the committees are specifically charged with enforcement of regulations that limit minimum mesh sizes of fishing nets (10 cm), prohibit the illegal use of toxic substances (e.g., barbasco, a plant-derived fish toxin used traditionally), and limit the minimum size of fish that can be captured according to Peruvian fishing policies (Reglamento de Ordenamiento Pesquero R.M. 147–2001-PE, art. 4 numeral 4.3). The committees do not control total harvest from the lake, but they do enforce no-fishing periods for certain species (*A. gigas*) and patrol areas that have been designated as no-take zones. All surveillance and control of fishing activities carried out by Yungani must be reported to the Regional Directorate of Fisheries Production (DIREPRO) and coordinated with the respective regional authorities. In essence, the role of Yungani is to guarantee the sustainable use of fisheries resources in Lake Rimachi (FECONAKADIP 2007).

Following the establishment of Yungani, the Kandozi developed a fisheries management plan according to guidelines set forth by Peruvian law (Reglamento de Ordenamiento pesquero de la Amazonia; R.M. N° 147–2001-PE) and with technical assistance from outside organizations such as WWF-Peru, FundAmazonía, and independent consultants. Both Kandozi fishermen and the international scientific community provided information that was incorporated into the plan. For example, Kandozi fishermen shared traditional ecological knowledge on fish species and the spatial and temporal variability of water conditions in Lake Rimachi and surrounding water bodies. With respect to *P. nigricans*, this included information on reproductive and migratory behavior, as well as designation of spawning areas. The Kandozi also provided information on valuable species from a socioeconomic perspective, fishing areas, fishing techniques, and objectives of fisheries management (Figure 5). During 2004–2005, scientific studies on water quality conditions and fish population dynamics were

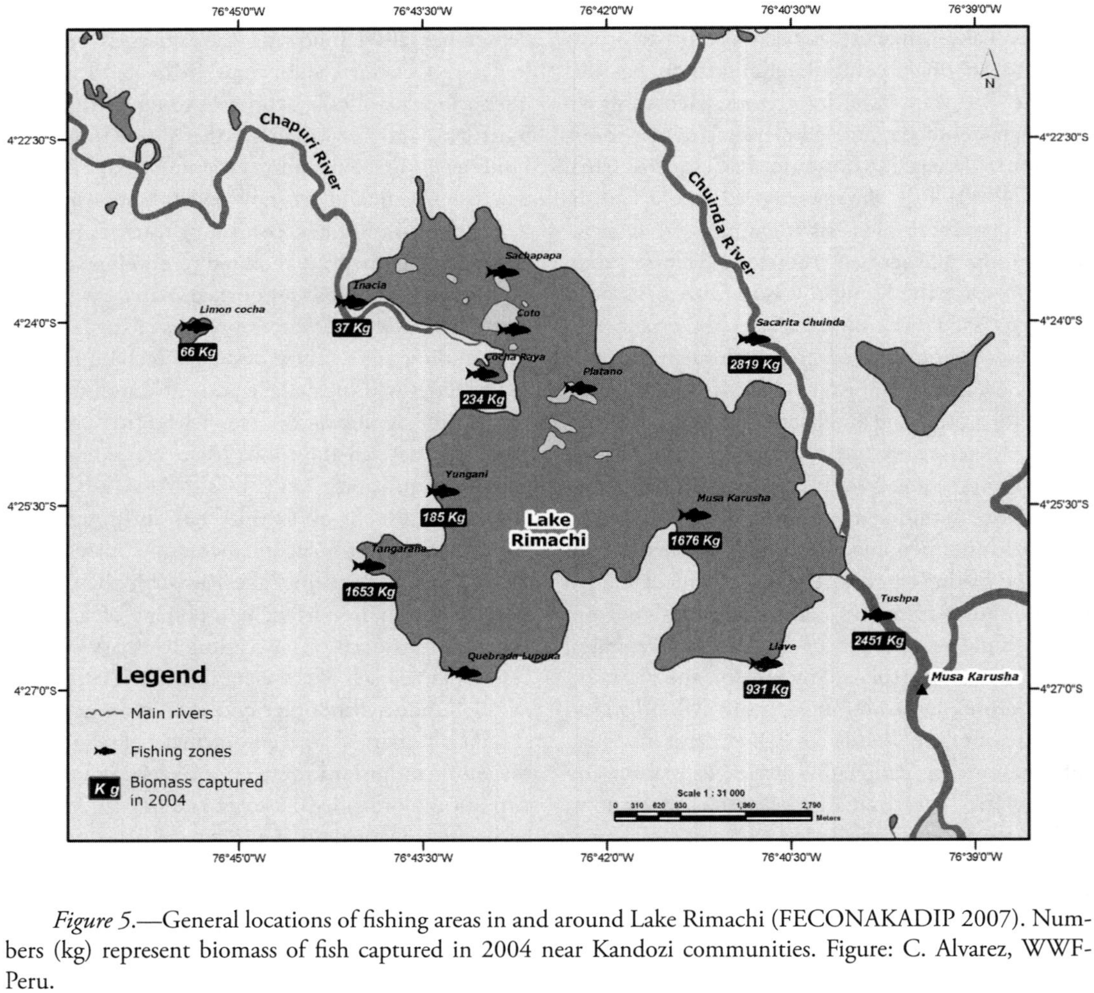

Figure 5.—General locations of fishing areas in and around Lake Rimachi (FECONAKADIP 2007). Numbers (kg) represent biomass of fish captured in 2004 near Kandozi communities. Figure: C. Alvarez, WWF-Peru.

conducted and their results incorporated into the development of the plan (Escobedo et al. 2005). In 2007, the Lake Rimachi fisheries management plan was completed and approved (see Table 1 for contents of the plan). The plan covers a 5-year period (2007–2011) and is implemented by Yungani in coordination with local authorities and with technical support from WWF-Peru.

The Lake Rimachi fisheries management plan focuses on the sustainable management of *P. nigricans* and three other key species. The plan recognizes the importance of *P. nigricans* to the livelihoods of the Kandozi and to the ecology of the lake and surrounding water bodies. The other species recognized in the plan include *Cichla monoculus*, *Hypothalamus edentatus*, and *Colossoma macropomum*, on the basis of their importance to the Kandozi fisheries. Specifically for *P. nigricans*, the plan limits the size of individuals that can be captured (>25 cm) and the stretched mesh size of fishing nets (10 cm). It also prohibits fishing of *P. nigricans* during spawning periods. These regulations represent a combination of general focus of local management systems on how to fish and more scientific management systems that focus on how much and what to fish (de Castro and McGrath 2003). In addition to the sustainable management of *P. nigricans* and the other three fish species, the Lake Rimachi plan also lists five more specific objectives: (1) to conserve freshwater resources and aquatic biodiversity in Lake Rimachi and surrounding water bodies, (2) to improve fisheries production in Lake Rimachi, (3) to establish a research program on the population ecology of species with high economic value like *P. nigricans*, (4) to transfer fisheries management tools applied in Lake Rimachi to other parts of the Peruvian Amazon, and (5) to strengthen Kandozi local organizations and secure their resource rights to Lake Rimachi (FECONAKADIP 2007).

Fisheries Co-Management as a Tool for Addressing Threats to *Prochilodus nigricans*

How has the formation of the Kandozi association of artisanal fishermen (Yungani) and the development and implementation of the fisheries management plan helped to address threats to *P. nigricans* and other fishes in the Peruvian Amazon? Because the fisheries management plan is still in its early phases of implementation, it is difficult to quantify its effects on fish populations and its overall success as a conservation tool. Nevertheless, several trends are evident that suggest that the plan is an appropriate management response to threats of overfishing and petroleum exploration and exploitation activities in the lower Pastaza River basin.

Table 1.—General contents of the Kandozi fisheries management plan (FECONAKADIP 2007).

1. A list of fish species to be managed and an ecological characterization of Lake Rimachi.
2. Detailed justification of the need for the management plan.
3. Objectives and expected results of the plan, including information about market prices and exploitation estimates.
4. Rights and obligations of fishermen from Kandozi communities, technical organizations, and state authorities that are part of the management plan.
5. Information about each fish species included in the plan.
6. Explanation of ongoing and future research and monitoring activities.
7. Fishing regulations and management techniques, including number of fishermen, fishing intensity, periods, quotas, and zones.
8. Description of resource use that explains how fish will be processed, transported, marketed, and sold.
9. Explanation of control and surveillance activities.
10. Fisheries Management Plan Evaluation.
11. Adaptive management techniques for adjusting fisheries production and management if necessary.
12. Description of conservation measures for Lake Rimachi.
13. References for further information.

The fisheries management plan directly addresses the threat of overfishing by providing regulations on fishing of *P. nigricans* and the other three key fish species and by giving Yungani the authority to enforce these regulations. Specifically, the plan controls fishing methods (e.g., size of nets and type of gear) and prohibits fishing during times of the year and in areas where these species are known to spawn. The times and locations of spawning were based on information gathered from Kandozi traditional ecological knowledge and from scientific studies (Escobedo et al. 2005; FECONAKADIP 2007). Areas of Lake Rimachi where juveniles are primarily found (especially *Colossoma macropomum*) are identified as no-fishing zones under the plan. Preliminary evidence suggests that these restrictions will have a substantial effect on sustainable fisheries management. In combination with the fisheries management plan, Yungani's control and surveillance committees also help to address threats of overfishing in Lake Rimachi through enforcement of general Peruvian standards that limit the minimum size of captured fishes and the minimum mesh size of fishing nets (Reglamento de Ordenamiento Pesquero; R.M. No. 147–2001-PE). For *P. nigricans*, the minimum mesh size of fishing nets is 10 cm, and when nets with smaller mesh are encountered, they are taken from the fisherman and given to DIREPRO. If one of the committees encounters a fisherman with fish that have been captured with illegal nets or during no-fishing periods, these resources are taken and distributed in the Kandozi community.

Sustainable management of fisheries resources by the Kandozi, formalized through the management plan, indirectly addresses threats stemming from overfishing and petroleum activities because it formally articulates the socioeconomic importance of *P. nigricans* and other fishes to Kandozi livelihoods and because it increases organization of the Kandozi community. The fishery based on *P. nigricans* in Lake Rimachi is a profitable activity that provides the main source of income for Kandozi families. The development of the fisheries management plan facilitated coordination between Kandozi fishermen and other local and regional authorities over the management of the lake and generated the interest of other municipalities to buy the fish from the Kandozi. Commercial fishermen from other municipalities are now aware that they must abide by the regulations for fishing established in the Kandozi fisheries management plan for Lake Rimachi. Regarding the unique management challenges posed by the migratory behavior of *P. nigricans*, commercial fishermen from downstream areas have recognized that they will benefit from the Kandozi's management practices, noting that the *P. nigricans* stocks upon which they depend for income most likely spawn in the area of Lake Rimachi. Further, within the Kandozi community, it can be argued that as members of Yungani, Kandozi fishermen have greater incentive to exercise control of their own fishing practices of *P. nigricans* and follow the regulations of the fisheries management plan (FECONAKADIP 2007).

With respect to petroleum activities, while the fisheries management plan cannot prevent oil spills and other forms of environmental contamination, it does put the Kandozi in a better position for negotiation over resource rights and for making claims in the case of contamination events. This advantage is important considering that the lower Pastaza River basin has been divided into lots and is being considered for exploration by multinational petroleum companies. In fact, one of these lots lies right over Lake Rimachi. A Brazilian company currently holds the concession to the Lake Rimachi lot, and they have already signed a contract with the Peruvian government that permits exploration and possible future exploitation of petroleum reserves in the area. Some believe that this contract was signed without adequate consultation with the Kandozi people, despite the fact that the Peruvian government signed the Convention 169 (concerning indigenous and tribal peoples) from the International Labor Organization that obligates the government to have a consultation process before deciding any administrative or legislative measure that could affect indigenous communities directly (Greenspan 2006). In the past, consultation processes between petroleum companies and indigenous communities over resource or land rights have suffered from a lack of transparency and petroleum companies have often been given unfair advantage (Greenspan 2006). One of the problems indigenous communities face in negotiation with petroleum companies is the lack of quantitative evidence that supports their socioeconomic dependence on the resources in question, in this case, fishes of Lake Rimachi. By establishing a direct link between the livelihoods of the Kandozi and the survival of *P. nigricans* and other fishes found in Lake Rimachi, the fisheries management plan advocates conservation of freshwater resources and makes a solid case for limiting petroleum activities in certain areas of the lower Pastaza River basin.

Challenges

The development and initial implementation of a fisheries co-management strategy and plan is only a part of the task of enabling the survival of *P. nigricans* and other species in Lake Rimachi while supporting the fisheries-based livelihoods of the Kandozi. Long-term implementation and adaptation of the plan presents an additional challenge, and thus the overall success of fisheries co-management in Lake Rimachi remains to be seen. Experiences from other parts of the Amazon basin have shown that community-based or co-management programs depend on the ability of local people to manage and defend natural resources and on appropriate and adequate financial, legal, and scientific support (Crampton et al. 2004a). Crampton et al. (2004a) list five provisos essential to successful local management of fisheries: (1) a single group has rights of access to the resource, (2) strong economic incentives for protecting fish stocks exist, (3) fishers have a general awareness of ecological and resource management concepts, (4) sound science or traditional ecological knowledge underpins management strategies, and (5) management includes efforts to protect habitats (e.g., like flooded forests).

Most of these conditions are manifest in the case of the Kandozi and fisheries co-management in Lake Rimachi. Nevertheless, several challenges for the long-term success of the program have been identified. First, enforcement of the regulations stipulated by the management plan, now being carried out by control and vigilance committees, depends on strong organization and leadership within Yungani, as well as the respect of other fishermen for the plan and the committees. Second, the maintenance and effectiveness of the Kandozi fishermen's association (Yungani), with minimal to no assistance from nongovernmental organizations (e.g., WWF-Peru and Racimos de Ungurahui), depends on sufficient economic support. It is proposed that these resources come from the sale of fish, but this would mean substantial changes to the way funds are currently used and managed by the Kandozi. Third, there is a need to form stronger links between the fishermen's association and FECONAKADIP (Kandozi Federation) and for outside collaborators (e.g., Ministry of Fisheries) to continue providing technical support to the association. Fourth, in the case of *P. nigricans*, management of migratory fish species is complicated by the fact that Lake Rimachi populations spend part of their life cycle outside of the area where the Kandozi fisheries management plan is applicable. Other authors have suggested that for migratory species, chains of fisheries management reserves may be an effective strategy on a regional scale (Crampton et al. 2004a); the possibility of establishing similar programs in areas near Lake Rimachi needs investigation. Fifth, more research is needed to evaluate management strategies from an ecological perspective to determine whether restrictions on net mesh sizes will actually prevent population declines of species like *P. nigricans*. Finally, a change in current Peruvian laws, increased pressure from commercial fishing operations or petroleum companies, or internal conflicts within the Kandozi community could shift the dynamics of the situation in Lake Rimachi to a scenario where the current fisheries co-management strategy might not apply. It is important to recognize these inherent uncertainties now, in case adaptations to fisheries co-management strategies in Lake Rimachi are needed in the future.

Conclusions

Here, we describe threats to an indigenous fishery in the Peruvian Amazon and the innovative steps that have been taken to manage those threats through fisheries co-management. The Kandozi example reinforces the importance of *P. nigricans* to indigenous communities and also shows that economics play a major role in how natural resources, particularly fisheries, are managed in the Amazon. Because *P. nigricans* is among the most important species for subsistence and artisanal fisheries for indigenous communities throughout the Amazon, some of the fishing regulations developed and implemented by the Kandozi for this species could potentially be exported to other parts of the basin.

In Lake Rimachi and among the Kandozi, a system for fisheries co-management emerged in response to increasing threats to the resources from commercial fisheries and the adoption of more efficient gear and more intense fishing practices. Thus, while the Kandozi depend on the fisheries for their livelihoods, increasing external and internal pressures have produced a situation where the long-term survival of *P. nigricans* and other fishes now depend on the Kandozi. Long-term sustainability of fisheries resources in Lake Rimachi will require continued involvement of the Kandozi community and local government authorities, regional recognition and respect for the Kandozi from other fishermen, and an adaptive approach to fisheries co-management.

Acknowledgments

The activities described in this article received support from a number of organizations including the Kandozi community, FECONAKADIP, the Regional Coordinator of Indigenous Peoples (CORPI), the Peruvian Ministry of Fisheries, WWF-Peru Programme Office, Racimos de Ungurahui, Fundación Amazonia, and the Global Water for Sustainability (GLOWS) Program. Technical assistance was provided by the Wildlife Conservation Society, the University of Kent-DICE, Florida International University, FundAmazonía (Luis Moya), and independent consultants (J. C. Pacheco). The MacArthur Foundation, DGIS, WWF-International, WWF-CH, and the U.S. Agency for International Development provided financial support for related technical studies and development of the fisheries management plan. Special thanks go to the organizers of the special session on Management and Governance of Diadromous Fishes at the 2007 Challenges for Diadromous Fishes in a Dynamic Global Environment Symposium. This manuscript was improved by comments from several reviewers.

References

Allan, J. D., R. Abell, Z. Hogan, C. Revenga, B. W. Taylor, R. L. Welcomme, and K. Winemiller. 2005. Overfishing of inland waters. Bioscience 55(12):1041–1051.

Almeida, O. T., K. Lorenzen, and D. G. McGrath. 2003. Commercial fishing in the Brazilian Amazon: regional differentiation in fleet characteristics and efficiency. Fisheries Management and Ecology 20:109–115.

Araujo-Lima, C. A. R. M., and M. L. Ruffino. 2004. Migratory fishes of the Brazilian Amazon. Pages 233–302 *in* J. Carolsfield, B. Harvey, C. Ross, and A. Baer, editors. Migratory fishes of South America. International Development Research Center and the World Bank, Washington, D.C.

Araujo-Lima, C. A. R. M., B. R. Forsberg, R. Victoria, and L. Martinelli. 1986. Energy sources for detritivorous fishes in the Amazon. Science 234:1256–1258.

Batista, V. S., A. J. Inhamuns, C. E. C. Freitas, and D. Freire-Brasil. 1998. Characterization of the fishery in river communities in the low-Solimões/high-Amazon region. Fisheries Management and Ecology 5:419–435.

Bayley, P. B. 1981. Fish yield from the Amazon in Brazil: comparisons with African river yields and management possibilities. Transactions of the American Fisheries Society 110:351–359.

Bayley, P. B., and M. Petrere. 1989. Amazon fisheries: assessment methods, current status, and management options. Pages 385–398 *in* D. P. Dodge, editor. Proceedings of the International Symposium on Large Rivers. Canadian Special Publication of Fisheries and Aquatic Sciences 106.

Bodmer, R., A. Escobedo, M. Inga, and D. Gallardo. 2006. Proyecto: conservación de fauna silvestre por las comunidades indígenas Kandozi del Chapuri. Technical Report prepared for WWF-Peru, Lima.

Bowen, S. H. 1983. Detrivory in neotropical fish communities. Environmental Biology of Fishes 9:137–144.

Bowen, S. H., A. A. Bonetto, and M. O. Ahlgren. 1984. Microorganisms and detritus in the diet of a typical neotropical riverine detritivore, *Prochilodus platensis* (Pisces:Prochilidontidae). Limnology and Oceanography 29:1120–1122.

Carvalho de Lima, A. C., and C. A. R. M. Araujo-Lima. 2004. The distributions of larval and juvenile fishes in Amazonian rivers of different nutrient status. Freshwater Biology 49:787–800.

Crampton, W. G. R., L. Castello, and J. P. Viana. 2004a. Fisheries in the Amazon Varzea: historical trends, current status, and factors affecting sustainability. Pages 76–98 *in* K. M. Silvius, R. E. Bodmer, and J. M. V. Fragoso, editors. People in nature: wildlife conservation in South and Central America. Columbia University Press, New York.

Crampton, W. G. R., J. P. Viana, L. Castello, and J. M. B. Damasceno. 2004b. Fisheries management in the Mamirauá Sustainable Development Reserve. Pages 99–122 *in* K. M. Silvius, R. E. Bodmer , and J. M. V. Fragoso, editors. People in nature: wildlife conservation in South and Central America. Columbia University Press, New York.

de Castro, F., and D. G. McGrath. 2003. Moving toward sustainability in the local management of floodplain lake fisheries in the Brazilian Amazon. Human Organization 62(2):123–133.

de Merona, B. 1985. Les peuplements de poissons et la peche dans le Bas Tocantins (Amazonia Bresilienne) avant fermerture du barrage de Tucurui. Verhanlungen der International Verein fur Limnologie 22:2698–2703.

Dyer, C. L., and J. R. McGoodwin, editors. 1994. Folk management in the world's fisheries: lessons for modern fisheries management. University Press of Colorado, Niwot.

Escobedo, A., L. Moya, and R. Bodmer. 2005. Iniciativas para el Manejo de los Recúrsos Naturales por los Kandozi del Lago Rimachi, Río Pastaza. Technical Report prepared for WWF-Peru, Iquitos.

FECONAKADIP (Federación de comunidades nativas Kandozi del Distrito de Pastaza and Asociación de Pescadores artesanales "Yungani" del Pueblo Kandozi cuenca del Pastaza). 2007. Programa de Manejo Pesquero de las Especies "boquichico, gamitana, tucunare, maparate." FECONAKADIP, San Lorenzo, Peru.

Fernándes, C. C. 1997. Lateral migration of fishes in Amazon floodplains. Ecology of Freshwater Fish 6:36–44.

Flecker, A. S. 1992. Fish trophic guilds and the organization of a neotropical stream: weak direct versus strong indirect effects. Ecology 73:927–940.

Flecker, A. S. 1996. Ecosystem engineering by a dominant detritivore in a diverse tropical stream. Ecology 77:1845–1854.

Forsberg, B. R., C. A. R. M. Araujo-Lima, L. A. Martinelli, R. L. Victoria, and J. A. Bonassi. 1993. Autotrophic carbon sources for fish of the central Amazon. Ecology 74:643–652.

Goulding, M. 1981. Man and fisheries on an Amazon frontier. Dr W Junk Publishers, Boston.

Greenspan, E. 2006. Strategies for improving government consultations with indigenous people in the Peruvian Amazon. Master's thesis. Harvard University, Cambridge, Massachusetts.

Junk, W. J., M. G. M. Soares, and P. B. Bayley. 2007. Freshwater fishes of the Amazon River basin: their biodiversity, fisheries, and habitats. Aquatic Ecosystem Health & Management 10(2):153–173.

Kimerling, J. 2006. Indigenous people and the oil frontier in Amazonia: the case of Ecuador, ChevronTexaco and Aguinda vs Texaco. International Law and Politics 38:413–664.

Maccord, P. F. L., R. A. M. Silvano, M. S. Ramires, M. Clauzet, and A. Begossi. 2007. Dynamics of artisanal fisheries in two Brazilian Amazonian reserves: implications of co-management. Hydrobiologia 583:365–376.

McDowall, R. M. 1987. Evolution and importance of diadromy: the occurrence and distribution of diadromy among fishes. Pages 1–13 *in* M. J. Dadswell, R. J. Klauda, C. M. Moffitt, R. L. Saunders, R. A. Rulifson, and J. E. Cooper, editors. Common strategies of anadromous and catadromous fishes. American Fisheries Society, Symposium 1, Bethesda, Maryland.

McDaniel, J. 1997. Communal fisheries management in the Peruvian Amazon. Human Organization 56(2):147–152.

Ministerio de Pesqueria. Reglamento y Ordenamiento Pesquero de la Amazonía Peruana, R.M. N° 147–2001-PE

Montreuil, V., A. Garcia, and R. Rodriguez. 2001. Biología reproductiva de boquichico, *Prochilodus nigricans*, en la Amazonia Peruana. Folia Amazónica 12(1–2):5–13.

Oliveira, A. C. B., M. G. M. Soares, L.A. Martinelli, and M. Z. Moreira. 2006. Carbon sources of fish in an Amazonian floodplain lake. Aquatic Sciences 68:229–238.

Pacheco, J.C. 2006. Proyecto desarrollo y fortalecimiento de la actividad pesquera de las comunidades indigenas del pueblo Kandozi. Technical Report prepared for WWF-Peru. Lima.

Pinedo, D., P. M. Summers, R. C. Smith, J. Saavedra, R. Zumaeta, and A. M. Almeyda. 2000. Community-based natural resource management as a non-linear process: a case study in the Peruvian Amazon Varzea. Instituto Bien Comun, Lima, Peru.

Pringle, C. M., M. C. Freeman, and B. J. Freeman. 2000. Regional effects of hydrologic alterations on riverine macrobiota in the new world: tropical-temperate comparisons. BioScience 50(9):807–823.

Ribeiro, M. C. L. B., M. Petrere, and A. A. Juras. 1995. Ecological integrity and fisheries ecology of the Araguaia–Tocantins River basin, Brazil. Regulated Rivers Research and Management 11:325–350.

Sen, S., and J. R. Nielsen. 1996. Fisheries co-management: a comparative analysis. Marine Policy 20:405–418.

Silva, E. A., and D. J. Stewart. 2006. Age structure, growth and survival rates of the commercial fish *Prochilodus nigricans* (bocachico) in northeastern Ecuador. Environmental Biology of Fishes 77:63–77.

Surrallés, A. 2007. Los Candozi. Pages 244–380 *in* F. Santos and F. Barclay, editors. Guía etnográfica de la alta Amazonía. Volume VI. Smithsonian Tropical Research Institute and Instituto Francés de Estudios Andinos, Lima, Peru.

Taylor, B. W., A. S. Flecker, and R. O. Hall. 2006. Loss of a harvested species disrupts carbon flow in a diverse tropical river. Science 313:833–836.

Tello, S. and P. Bayley. 2001. La pesquería comercial de Loreto con énfasis en el análisis de la relación entre captura y esfuerzo pesquero de la flota comercial de Iquitos, cuenca del Amazonas (Perú). Folia Amazónica 12:123–139.

Willink P.W., B. Chernoff, and J. McCullough, editors. 2005. A rapid biological assessment of the Aquatic ecosystems of the Pastaza River basin, Ecuador and

Perú. Conservation International, RAP Bulletin of Biological Assessment 33, Washington, D.C.

Winemiller, K. O. 1996. Factors driving spatial and temporal variation in aquatic floodplain food webs. Pages 298–312 *in* G. P. Polis and K. O. Winemiller, editors. Food webs: integration of patterns and dynamics. Chapman and Hall, New York.

Winemiller, K. O., and D. B. Jepsen. 1998. Effects of seasonality and fish movement on tropical river food webs. Journal of Fish Biology 53(Supplement A):267–296.

Yossa, M. I., and C.A.R.M. Araujo-Lima. 1998. Detrivory in two Amazonian fish species. Journal of Fish Biology 52:1141–1153.

American Fisheries Society Symposium 69:757–768, 2009

Conservation of Juvenile Hilsa (Jatka) in Bangladesh: Need to Address the Livelihood of Fishers

Mohammad Abdul Latif Siddique*,[1]
Marine Affairs Program, Dalhousie University, Halifax, Canada

Abstract.—Hilsa *Tenualosa ilisha* (Hamilton, 1822) is a major food fish in Bangladesh and is hailed in Bengali as *macher raja ilish*, meaning "hilsa shad—the king of fish." World catch of hilsa is approximately 0.3–0.4 million metric tons per year and 50–60% are caught in Bangladesh. Hilsa contributes 13–14% of total fish production and 1.25% of the gross domestic product in the country. The marine catch of hilsa increased threefold between 1984 and 2001, but inland production decreased by 19%. Dams lacking fish passage and indiscriminate catching of *jatka* (juvenile hilsa up to 23 cm long) in rivers are the major reasons for this declining trend in inland production. The livelihood of fishers is also vulnerable since 55% of hilsa and jatka fishers are engaged in inland fishing. Best management practices to control overfishing include conservation of jatka during their emigration from the spawning areas towards the sea to improve both the inland and sea production of hilsa. However, the livelihood of jatka fishers might be threatened during conservation measures because 60% of these fishers have no alternative income, not even subsistence agriculture. Therefore, it is as important to support these fishers (by providing appropriate alternative income generating activities) as it is to protect jatka, to ensure sustainable yields of hilsa in Bangladesh.

Introduction

Bangladesh is rich in fisheries resources that contribute 5.15% to the gross domestic product (GDP). Fish is the principal ingredient in people's diets, providing about 60% of their animal protein intake. Hilsa *Tenualosa ilisha* (Hamilton, 1822) is the national fish because of its importance. Hilsa is a major food fish in Bangladesh and is sacred according to Hindu mythology. It is hailed in Bengali as *macher raja ilish*, meaning "hilsa shad—the king of fish." Like American shad *Alosa sapidissima*, it is anadromous and grows and feeds mainly in marine waters. World catch of hilsa is approximately 0.3–0.4 million metric tons (mt) per year and 50–60% are caught in Bangladesh (DoF 2005). Hilsa contributes 13–14% of total fish production, 9.5% to total protein production and 1.25% to GDP in Bangladesh. The hilsa fishery employs approximately 447,000 fishers; 32% are engaged permanently and 68% are part time (DoF 2005). According to the report of the Fisheries Resources and Survey Systems (FRSS), the marine catch of hilsa increased threefold between 1984 and 2001, but inland production decreased by 19%. Halder et al. (2005) reported that the catch of hilsa from about 35 Bangladesh rivers was zero or almost zero during the past 25–30 years. The main problems are habitat destruction from pollution, siltation, construction of dams with no fish passage structures, minimum water discharges through some spawning rivers, and overfishing of juveniles. The livelihood of fishers is also vulnerable since 55% of hilsa fishers operate in inland waters.

Hilsa lives in the sea (Bay of Bengal) for most of its life but migrates upstream in some river systems in Bangladesh for spawning. There is little evidence of more than one genetic population of hilsa in Bangladesh or within the Bay of Bengal region (Salini et al. 2004). Younger adults may spawn in the tidal zone of rivers. During the start of the

* Corresponding author: latif@dal.ca
[1] Current address: Department of Fisheries, Matshya Bhaban, Ramna, Dhaka-1000, Bangladesh

southwest monsoon and associated flooding in all the rivers, hilsa starts its upstream spawning migration. There is some evidence that most hilsa populations in the Ganges and other large rivers are anadromous and migrate upstream to spawn, but there are also resident river populations that do not emigrate from freshwater to the sea to grow and mature (Rahman and Naevdal 2000). Similarly, there are marine populations that spawn in coastal areas.

The larvae and juveniles make their way downstream to the sea during a period of several months, feeding and growing along the way. During this stage in their life cycle, they are locally known as *jatka*. These juveniles are caught in large numbers using *current jal* (monofilament gill nets) and *ber jal* (seine nets) of small mesh size in rivers and in nearshore estuaries by artisanal fishers and sold in local markets. These juvenile hilsa are often mistakenly or intentionally called *chapilla* (Indian river shad *Gudusia chapra*), a different fish species. After growing for 1–2 years at sea, hilsa mature and reach a length of 32–55 cm prior to their spawning migration towards inland rivers, where their life cycle continues (Halder et al. 2005). During their upstream migration, brood fish are caught by fishers in tidal rivers using gill nets and seines, fishing methods so efficient that the potential spawners are heavily exploited. Hilsa are at risk from overfishing and also from natural barriers in their migratory routes associated with heavy siltation in inland rivers. During the upstream migration, the percentage of adult spawners is very high, with 80% of the harvested females ripe and ready to spawn (Halder et al. 2005).

Jatka are heavily caught during November and December in estuaries and from January to May in most rivers. The rate of jatka catch gradually increases each day due to increasing fishing effort through entry of many new fishers and weak enforcement of fishing regulations. A continuous indiscriminate harvest of jatka may adversely affect the annual total shad catch in the fishery (Amin et al. 2000). If these jatka were not harvested and instead grew to maturity, they would boost total production by an additional 0.2 million mt per year, double the present annual catch of hilsa. Therefore, conservation and management of this fish stock in Bangladesh has become very important both to optimize total production and ensure the livelihoods of fishers.

It is necessary to reduce the harvest of jatka in order to increase overall hilsa production. The government of Bangladesh has decided to protect jatka by declaring a ban on their harvest for a particular time of the year, with the sole purpose of increasing production. To ensure a target production of hilsa, the government first implemented a coordinated program to protect jatka in 2002–2003 (DoF 2004). Through participation of all concerned, this jatka protection program had a positive effect. Hilsa production increased by 56,807 mt in 2003–2004 and 76,831 mt in 2004–2005. The market value of 275,863 mt of hilsa caught in 2006 was about 41,370 million BDT (Bangladeshi taka; US$1 = BDT 70). To continue this success, a more detailed and strengthened conservation program was adopted and implemented in 2005–2006. The ban during the 2004–2005 and 2005–2006 seasons was effective because of Department of Fisheries (DoF) initiatives and increased enforcement by the coast guard service.

Both significant positive and negative effects have been associated with the conservation of juvenile hilsa. The populations of affected human communities in Bangladesh are polarized regarding the present and probable future of their livelihoods and the trend in hilsa catches. The fishers prevented from catching jatka under the new conservation program come from the poorer segments of society. Supporters see the conservation measures as an effective way to increase fish production, while those against it point to the socioeconomic losses to fishers and other issues like habitat restoration and the need for a regional management plan with at least neighboring countries that share the Bay of Bengal. Both arguments are equally important in the context of true sustainable management of hilsa in Bangladesh, actions that are required to meet the national demand to produce hilsa in an environmentally and socially sound way.

This study is an attempt to answer two questions: (1) Can the current conservation activities with juvenile hilsa be effectively implemented to minimize the declining trend in shad production without increasing socioeconomic hardships on fishers? and (2) What are the trade-offs between increasing hilsa production and providing livelihood subsidies for the fishers to achieve eco-friendly and sustainable hilsa management in Bangladesh? Although the current hilsa management strategy includes both coastal/marine and freshwater components, this paper will focus on the freshwater sector, with an emphasis on the fisher livelihood issues.

The Hilsa Fisheries in Bangladesh

Distribution of the Adult and Jatka Hilsa Shad Fisheries in Bangladesh

Adult and jatka hilsa shad are widely distributed in coastal, estuarine, and riverine waters of the Meghna, Padma, Jamuna, and other deltaic rivers of Bangladesh. Lengths of juvenile hilsa (jatka) can reach 15 cm (Hora and Nair 1940), although DoF defined jatka as being up to 23 cm long (Fish Act, 1950, amended in 2006). The exploitation of jatka has serious consequences on the overall production in the hilsa fishery and conservation of inland populations. Currently, the annual catch of jatka is estimated at more than 3,700 mt (DoF 2005). If 30% of the present catch of jatka was not harvested, these fish could mature and produce an additional value of 5–7.5 million BDT or US$1.2–1.8 million (Rahman and Naevdal 2000).

Trends in the Hilsa Catch

The FRSS of DoF compiled hilsa catch data from 1986–1987 to 2005–2006 (FRSS 1986–1987 to 2005–2006), with the harvest in 1986–1987 considered to be the base year for trend assessment. According to FRSS, the total hilsa catch has increased, ranging from 182,167 mt in 1990–1991 to 277,123 mt in 2005–2006 (Table 1). The catch in 2005–2006 was 57.5% higher than the base year, which could be a response to the jatka conservation program. A similar increasing trend also appeared in the marine catch of hilsa that ranged from 103,814 mt in 1986–1987 to 198,850 mt in 2005–2006. The marine catch of hilsa in 2005–2006 was 91.5% higher than the base year. However, an opposite trend was observed in the inland hilsa catch that ranged from 62,944 mt in 1989–1990 to 112,408 mt in 2002–2003, the year of sharp decline for all catch categories in Bangladesh. The highest catch occurred in 1989–1990. The overall inland hilsa catch decreased by 14.1% in 2005–2006, compared to the base year. This decreas-

Table 1.—Harvest of hilsa shad (Hamilton) from 1986 to 2006 in Bangladesh (data source: Fisheries Resources and Survey Systems, Department of Fisheries). mt = metric tons.

Year	Hilsa catch (mt)			% of hilsa in total capture fisheries
	Inland catch (mt)	Marine catch (mt)	Total (mt)	
1986–1987	91,167	103,814	194,981	30.1
1987–1988	78,551	104,950	183,501	28.2
1988–1989	81,651	110,311	191,952	29.2
1989–1990	112,408	113,943	226,351	34.1
1990–1991	66,809	115,358	182,167	26.6
1991–1992	68,356	120,106	188,462	26.0
1992–1993	74,715	123,115	197,830	25.3
1993–1994	71,370	121,161	192,531	23.3
1994–1995	84,420	129,115	213,535	25.0
1995–1996	80,625	126,660	207,285	23.6
1996–1997	83,230	131,204	214,434	24.5
1997–1998	81,634	124,105	205,739	23.2
1998–1999	73,809	140,710	214,519	22.4
1999–2000	79,165	140,367	219,532	21.9
2000–2001	75,060	154,654	229,714	21.5
2001–2002	68,250	152,343	220,593	20.0
2002–2003	62,944	136,088	199,032	17.4
2003–2004	71,001	184,838	255,839	21.6
2004–2005	77,499	198,363	275,862	20.7
2005–2006	78,273	198,850	277,123	19.3

ing trend in the inland catch is alarming (Halder et al. 2005; Islam 2006), and indiscriminate catching of jatka is the main explanation (Amin et al. 2000; Halder et al. 2005).

Possible Reasons for the Declining Trend in Hilsa Harvest

Dams without Fish Passage Structures

Reductions in the depth and discharge of rivers due to siltation and the construction of dams and obstacles have affected the spawning, feeding, and migration of hilsa. Amin et al. (2002) reported that until 1972, the upstream rivers were characterized by high hilsa abundance. Declines started gradually there but simultaneously increased downstream in the estuary and sea. The construction of the Farakka barrage (dam) in the Indian state of West Bengal blocked hilsa migrations, resulting in closure of the Goalondo Ghat landing station on the Padma River, once famous for its large shad landings.

Overfishing

Both adult and juvenile hilsa are normally exploited during the spawning migration. Jatka are usually caught using large- and medium-length beach seines and monofilament gill nets (Amin et al. 2000). According to Amin et al. (2000), the highest fishing mortality occurred on shad between 3.5 and 5.5 cm in length, associated with the dominant use of the large seine (mesh size 0.5 to 1.5 cm). Another peak in fishing mortality occurred on fish between 10.5 to 15.5 cm in length, attributed to the gill net (mesh size 1 to 2.5 cm). Amin et al. (2002) reported a rate of exploitation (E) of 0.66 on hilsa from Bangladesh waters. The E_{max} value was 0.59, indicating that hilsa were being overfished. Gulland (1971) stated that the yield is optimum when fishing mortality (F) is equal to natural mortality (M) and when E is not more than 0.50. If E is more than 0.50, a fish stock is generally considered to be overfished, which appears to be occurring with hilsa in Bangladesh.

Recruitment Pattern

Amin et al. (2002) reported that hilsa are recruited to the fishery in Bangladesh throughout the year, with one peak that produces 19.8% of the recruits, on average. This peak period (January to March) is the vulnerable season for juvenile shad (jatka) when they are also harvested intensively by the artisanal fishery.

Socioeconomic Profile of Hilsa Fishers

Millions of artisanal fishers are engaged throughout the year in the hilsa fishery in Bangladesh. The average family size is 7.2 persons per household in all the shad fishing areas except the Laxmipur district (Halder 2006), where the average family size is 6.1 per household. The people do not possess any land for crop cultivation so the rivers and hilsa are their only means of survival. Ignoring the intense heat of the sun, the lack of security and safety measures during monsoons and tidal waves, and having little or no food during fishing, these fishers struggle for their livelihood. Some fishers are happy with a catch of just one average-sized fish per day, as it provides them with money needed to feed their family or repay a boat loan. Most of the fishers are so poor that they are unable to upgrade their boats so they can fish in the marine environment. Most of these fishers are illiterate, and their children cannot attend school because they must help their fisher parents. The fishers who are prevented from catching jatka under the conservation program come from the poorest segments of the community. Their monthly income ranges from 1,200 to 1,500 BDT (US$17–20). Most of the adult and jatka Hilsa fishers live below the poverty line, and most work in teams as laborers/fishers. The wealthier people own the boats and nets. During the jatka season, the fishers are dependent on fishing for their livelihood and do not have any alternative sources of income to support their families. For these reasons, the jatka fishers need alternative sources of income, at least during the ban on juvenile harvest period.

Juvenile Conservation as a Tool for Sustainable Hilsa Management

The DoF, within the Ministry of Fisheries and Livestock in Bangladesh, developed an action plan for the conservation of juvenile hilsa. The action plan specified actions to protect jatka, developed an implementation strategy, ascertained the responsibilities of relevant agencies and target communities, and established a specific time frame for implementation. The action plan involved a variety of activities, including awareness meetings involving district ministers/public representatives, riverine rallies, awareness development through the public media, distribution of leaflets and posters about the need to protect jatka, enforcement of the Fish Protection and Conservation Act 1950 (amended in 2005), establishment of a hilsa sanctuary, and developing alternative livelihoods for jatka fishers (DoF 2006). Under the Fish Protection and Conservation Act 1950 (amended in 2005), harvesting of jatka is totally prohibited from November to May. This action plan has been implemented in 87 upazilas (subdistricts) of 20 districts where jatka catches are high.

Declaration of Hilsa Sanctuaries

Four sites in the coastal areas of Bangladesh have been declared hilsa sanctuaries (Table 2) under the Protection and Conservation of Fish Act, 1950, to achieve desired goals in hilsa management plans. Every year during the full moon of the Bengali month of Agrahayan (October 15–24), the hilsa harvest is banned for 10 d on all major spawning grounds. In these four sanctuaries, all types of fish harvesting are banned during certain periods of time (DoF 2006).

Effects of Conservation on Hilsa Production

According to the DoF report and data provided by FRSS, the conservation actions taken by the DoF have had positive effects on hilsa production, increasing by 28.5% and 38.6% in 2003–2004 and 2004–2005 respectively (Table 3). The DoF believes that protecting juvenile hilsa during the fishing ban period was achieved through strict enforcement of the 1950 Fish Act.

Effects of Conservation on the Livelihoods of Fishers

A baseline socioeconomic survey and impact assessment on hilsa fishers was conducted in December 2004 by Uniconsult International Limited (UCIL 2004) under the Fourth Fisheries Project of the DoF. One objective of the assessment was to gather information about the livelihood of hilsa fishers during and after the ban on harvesting jatka. According to the report, the ban was strictly enforced and controlled by regular coast guard participation.

The report also stated that the economic condition of the fishers during the initial fishing ban (2002–2003) was very poor. Their only source of income was totally stopped and strict supervision was carried out by the Bangladesh Naval Force and Coast Guards and DoF personnel to prevent fishing in three upazilas of the Chandpur district. The fishing community was informed before imposition of the ban and they expected compensation to maintain their livelihoods. However, the support provided to the fishers by the Bangladesh government was insufficient compared to their needs. The fishers lost their assets and access to fishing. The assistance was irregular and not adequate to compensate their losses or to provide enough food for their families.

Jatka fishers earn their livelihood by catching and selling juvenile hilsa and they have no other sources of income. That is why the Bangladesh government gave

Table 2.—List of hilsa shad sanctuaries (protected areas) in the major spawning and feeding grounds in the rivers of Bangladesh (source: DoF 2006).

Serial no.	Description of sanctuary	Fishing ban period
1.	From Shatnol of Chandpur district to Char Alexander of Laxmipur (100 km of lower Meghna estuary)	March and April
2.	Madanpur/Char Ilisha to Char Pial in Bhola district (90 km area of Shahbajpur river, a tributary of Meghna)	March and April
3.	Bheduria of Bhola district to Char Rustam of Patuakhali district (nearly 100-km area of Tentulia river)	March and April
4.	Whole 40-km stretch of Andharmanik river in Kalapara upazila of Patuakhali district.	November to January

Table 3.—Production performance of hilsa shad before and after implementation of the jatka conservation program in 2002–2003 (source: DoF 2006).

	Catch		Value[a] (BDT 150/kg) 1 US$= 70 BDT	
Financial year	Metric tons	Increase (%)	BDT in millions	Increase (%)
1999–2000	219,532	–	329.29	–
2000–2001	229,719	–	344.57	–
2001–2002	220,593	–	330.89	–
2002–2003	199,032	–9.0	298.54	–9.8
2003–2004	255,839	28.6	383.75	28.5
2004–2005	275,863	38.6	413.79	38.6

[a] Rough average value to get an idea about the economic return as a positive impact of jatka conservation.

special importance in 2006–2007 to providing alternative sources of income for jatka fishers during the "no catch period" (March and April in coastal sanctuaries and November to May in the rivers). A total of 30 million BDT was allocated for rehabilitation and to generate alternative incomes for fishers in the following districts: Chandpur, Laxmipur, Shariatpur, Barguna, Patuakhali, Bhola, Pirojpur, Munshiganj, Jhalakathi, Barisal, and Chittagong. Unfortunately, these subsidy programs were not sufficient to support the fishers during the period when the ban on fishing for hilsa was in effect. According to the DoF (2006) report, only US$4.90 were allocated per beneficiary during the closed season for jatka (Table 4). If a fisher spends this money to buy only rice, he can purchase around 12 kg of rice for his seven family members (average household members reported by UCIL 2004) during the 7-month conservation period (November to May). This situation is a hardship for the jatka fishers and their families. There is another problem associated with the uneven distribution of these meager government funds. The most money per beneficiary went to fishers in the Chittagong district and the lowest amount to the Bhola district (Table 4). The larger amount of money allocated for the Chittagong district appears to be politically motivated. Fishers in the Bhola district are the most vulnerable because it is a poor and remote coastal area.

Table 4.—District allocation of government funds allocated to jatka fishers during 2005–2006 (Modified from DoF 2006).

Serial no.	District	Funds allocated (in million BDT)	Number of beneficiaries (fishers)	Money/beneficiary (BDT/US$)
1	Jalakhathi	1.0	880	1,136/16.2
2	Bhola	5.0	27,448	182/2.6
3	Munshiganj	1.0	1,678	596/8.5
4	Barguna	3.0	6,539	459/6.5
5	Laxmipur	3.0	3,890	771/11.0
6	Patuakhali	2.0	2,423	825/34.6
7	Chandpur	6.0	22,066	272/3.8
8	Sariatpur	2.0	2,365	846/12.0
9	Pirojpur	2.0	8,979	223/3.1
10	Barisal	3.0	10,000	300/4.2
11	Chittagong	2.0	523	3,824/54.6
	Total	30.0	86,791	346/4.9

There are also other problems faced by the jatka fishers due to conservation actions. The fishing ban was announced by the government before it was implemented, which created problems for the fishers. Nongovernmental organizations (NGOs) working in the area stopped providing loans to fishing families before and during the fishing ban period. Only after the ban was lifted in June could the fishers apply for loans. Undoubtedly, the implementation of the fishing ban to protect jatka has had a direct negative impact on the livelihoods of the fishers, even though the positive impact of the ban on hilsa production is acknowledged by all. The goal of increasing hilsa production through imposition of the fishing ban can be realized with enforcement of the ban and the cooperation of all affected parties.

The Bangladesh government also launched a "Money for Work" program for jatka fishers during the closed season. Unfortunately the fishers could not take advantage of the benefits from this program because it was implemented in June, the start of the fishing season and not the appropriate time for implementing this type of government-assistance program. The appropriate time is the dry season (typically November to May), since this is also the fishing ban period and the fishers pass idle time. June is a prime fishing month so the fishers were not interested in the work (money for work) specified for them. Hence, the objectives of this government assistance program were not achieved.

This scenario has been changed. The fishers are now aware of the negative impacts of overfishing jatka, due to a positive DoF campaign. Awareness has increased among the jatka fishers and social consciousness levels have been raised. They agreed to refrain from jatka fishing during the ban period, but only if government assistance during the ban period was raised to a level that could support their families.

Conservation actions have also had immediate negative impacts on those who are not fishers but work in fishery-related activities like buying, selling, and transportation. During the ban on harvesting jatka, they depend on commerce in other fish species that are not banned. Some people try to carry on other professions like rickshaw pulling and various small businesses. These people are also affected by the ban, but not as severely as the hilsa fishers and they are fewer in numbers.

Necessity of Implementing Comprehensive Livelihood Programs for Jatka Fishers

From the above discussion on the impacts of the jatka conservation program, it is clear that these management measures alone cannot make the hilsa fishery sustainable, even though the species has experienced increased production. On the human side of this issue, the so-called rehabilitation or subsidy program (US$4.90/fisher/7 months) implemented by the government is not enough to support the fishers when they cannot legally catch jatka. This is a major reason why the jatka conservation program has not been accepted by fishers. The current conservation actions regarding juvenile hilsa might be able to reverse the declining trend in shad production, but only with increased socioeconomic costs to the fishers. It might be better for the fishers if a comprehensive rehabilitation program, including alternative income generating activities (AIGA), were added to the livelihood program. It is important that AIGA be selected based on an analysis of the jatka fishers' situations rather than being imposed by the government in a top-down system.

Situation Analysis of the Livelihoods of Jatka Fishers

A situation analysis of the jatka fishers is a bottom-up program to find the reasons for their declining living standard and also ways to minimize this decline to an acceptable level. In this paper, a participatory rural appraisal (PRA) tool (problem tree) is used to analyze the jatka fishers' situation (Figure 1).

From the problem tree analysis, there are three major reasons for the decreasing hilsa production. Although habitat destruction, large jatka harvest, and fewer adult fish available to spawn should be considered for the sustainable management of hilsa, only the issues associated with jatka conservation are discussed here. The root cause for overharvesting of jatka and hence reduced hilsa production is the extreme poverty of the jatka fishers and their nearly total dependence on this fishery for their livelihoods during certain times of the year (Islam 2006).

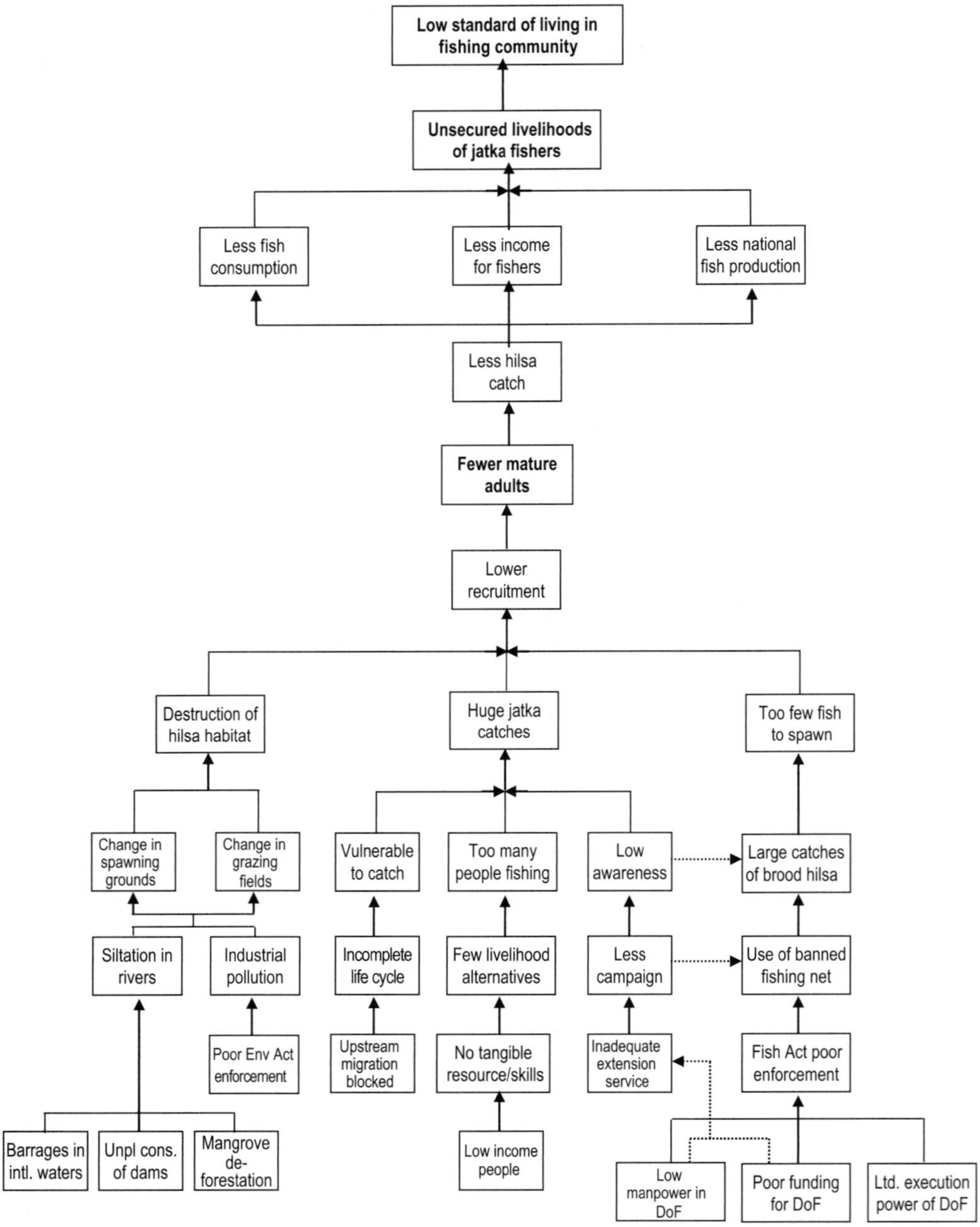

Figure 1.—Situational analysis of jatka fishers through problem tree, a participatory rural appraisal tool (modified from Islam 2006).

Alternative Income-Generating Activities Selection for Jatka Fishers

From the situation analysis of jatka fishers, it is now clear that conservation activities will not be sustainable unless they are combined with a comprehensive AIGA. Another PRA tool, "objective tree analysis" (Figure 2), can be used to find a way of selecting suitable alternative livelihood programs for the jatka fishers.

Jatka fishers do not fully understand hilsa biology and the contribution of juveniles to overall shad

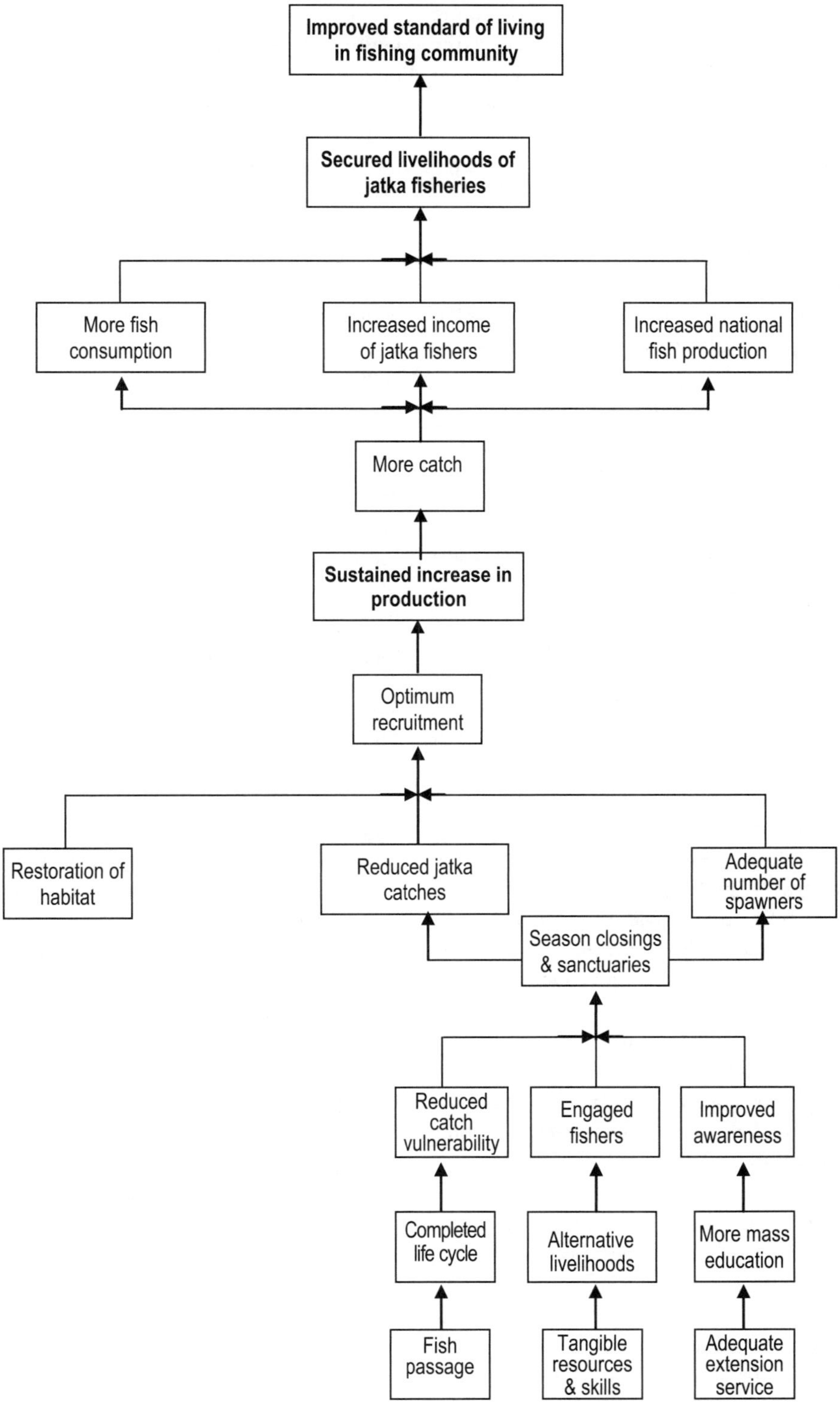

Figure 2.—Selection of alternative income-generating activities for jatka fishers through objective tree analysis.

production. This is due to both illiteracy among fishers and inadequate extension services of the DoF. Solution of these two problems is hampered by a technical manpower shortage and inadequate funding for the DoF.

During the closed season for jatka, the government initiated fisher assistance programs like Vulnerable Group Feeding, which provided 10 kg of rice per family, and the Money for Works program. Neither program was successful due to top-down selection of the program and inadequacies. Selection of the most appropriate assistance programs is very important and one of the prerequisites for successful fisheries conservation in a developing country like Bangladesh. Because the AIGA may not be equally acceptable to the beneficiaries and the government, objective tree analysis offers one option.

To develop an objective tree, the problems and their underlying causes are expressed as a positive goal. The objective tree shows the logical pathways and linkages between various objectives. The tree shows three main outputs: habitat restoration, reduced jatka harvests, and an adequate number of adult hilsa for spawning, which leads to the immediate objective of increased production through optimum recruitment of jatka. Reduced jatka harvests and the availability of adult hilsa are the result of conservation activities such as closing the jatka fishing season from November to May and in the coastal sanctuaries (Table 2). Reduced jatka harvests should result from the motivation of the fishers to not catch jatka during the conservation period because the fishers will have alternative income generating activities (Islam 2006).

In addition, adequate numbers of adult hilsa for spawning should result from improvements in a government-sponsored education program, which should motivate the fishers to not overexploit adult shad. Understanding the importance of adult fish will come from the education program for fishers and their family members. To motivate the fishers to conserve jatka stocks, awareness building among them is needed. In addition, the jatka fishers should be educated about hilsa biology so that they understand the contribution of jatka to overall stock production. It is possible to do this through the improvement of extension services of the DoF and by enforcement of Fish Act, 1950, tasks that the DoF is mandated to do. The highest priority is that the jatka fishers should be supported with AIGAs.

For any economic support program, a whole household approach will work best. In this approach, the husband, wife, and adult sons and daughters can all be associated with the program and reap maximum benefits. If one member of the household is unable to participate, another member can replace and support them. If one fails to meet an important deadline (e.g., missing the installment payment on microcredit, a useful program for the poor), other members of the household can be made liable for the failure. Special programs for the women in the fishers' families, such as making handicrafts with the support of a marketing facility, could be developed, since the women are also without work during the jatka fishing ban period. These programs would enable women to contribute directly to the family income and thereby empower the least empowered group, something required in a country like Bangladesh.

Risk and Flexibility with an Alternative Income Generating Activities Program

The major risks with an AIGA program are inadequate commitment of the wealthier villagers living around the jatka areas, in particular the money lenders. If the fishers develop AIGAs, with support from DoF or NGOs, in lieu of fishing for jatka, informal money lending businesses in the fishing village could be threatened. The money lenders could then discourage the fishers by trying to convince them that AIGA program support may not continue and then they will refuse to extend loans to the fishers. This threat from the money lenders could give the fishers a sense of insecurity. These possible risks could be avoided by mobilizing fishers' community based organizations (CBOs) and allocating the endowment fund well ahead of time. To form such CBOs, the jatka fishers should be organized first. Then, the money, which is now going to them as a grant or so-called "money for work," could be given to the CBOs as a revolving fund in a community bank to support jatka fishers during the harvest ban period.

The contributions of the Bangladesh Naval Force and Coast Guard to enforcement of Fish Act, 1950 along the coast and in the open ocean are of utmost importance to the management of hilsa because the DoF has very limited mobility in these marine areas. Withdrawal of Navy and Coast Guard support could occur and hinder the enforcement program as they have not been mandated to implement Fish Act, 1950. Their primary mission is surveillance for homeland security in coastal areas. To sustain participation of the Navy and Coast Guard in the jatka conservation program, a formal memorandum could be signed between the Ministry of Fisheries and Livestock and the Ministry of Defense.

Political interference may also be a threat to successful implementation of a jatka conservation program. Sometimes financial support may not go to the beneficiaries (in this case, the jatka fishers' families) due to interference from local elites and politicians (Islam 2006). Sometimes the attitudes of the local administrators and police may hinder program implementation due to their limited understanding of the assistance program. Orientation workshops that include all secondary stakeholders and make them fully aware of the jatka conservation program and AIGA program activities may alleviate these concerns. A good working liaison with local politicians could lessen and hopefully eliminate political interference.

The beneficiaries of the program (the jatka fishers and their families) may hesitate to adopt new livelihoods during the conservation period, even if alternative income activities are available. They may fear that abstaining from jatka fishing, a livelihood for which they have the required skills and experience, means that they must adopt alternative livelihoods for which they need to develop the requisite skills. Participatory analysis of the jatka fishers' livelihoods that generates several options may reduce this risk and help the fishers choose appropriate AIGAs that will minimize the number of new skills they must learn and master. Imposition of selected AIGAs may result in many jatka fishers rejecting the program.

Summary

The ban on the harvest of juvenile hilsa (jatka) in Bangladesh has increased the marine catch of adults—a positive effect. But, those opposed to these conservation measures point to the socioeconomic hardships inflicted on those fishers who depend on the jatka fishery for their livelihoods. Many of these artisanal fishers come from the poorer segments of society and have no alternative incomes, not even subsistence agriculture. Adequate compensation of the jatka fishers by the government is essential to the success of the conservation program. Stakeholder participation should be ensured. A community-based program to monitor and enforce the ban on jatka fishing should be implemented in all major rivers during juvenile hilsa out-migration. An interagency habitat restoration program focused on the spawning and nursery rivers should be developed and implemented. International funds may be needed. Population genetics studies are also needed to better understand the population structure of hilsa stocks that spawn in several Bangladesh rivers. Comprehensive studies should be conducted to understand how hilsa spawning migrations are being affected by dams. Hilsa is a migratory species and the stocks in the Bay of Bengal and surrounding waters are shared by several countries. Therefore, a collaborative, regional water management plan is needed to regulate water levels in the rivers during the dry season and maintain optimum water flows during the monsoons.

Acknowledgments

I am grateful to the American Fisheries Society's International Fisheries Section for providing travel funds that enabled me to attend and participate in the 2007 "Challenges for Diadromous Fishes in a Dynamic Global Environment" symposium in Halifax, Nova Scotia, Canada. I also thank Alex Haro who informed me about the symposium and Stephen Blaber, Md. Arshad Hossain, Ronald Klauda, and Katherine Smith for their pertinent reviews and valuable assistance in improving the manuscript.

References

Amin, S. M. N., M. A. Rahman, G. C. Haldar, M. A. Mazid, and D. Milton. 2002. Population dynamics and stock sssessment of hilsa shad, *Tenualosa ilisha* in Bangladesh. Asian Fisheries Science 15:123–128.

Amin, S. M. N., M. A. Rahman, G. C. Haldar, S. Nahar, S. Dewan, and M. A. Mazid. 2000. Population dynamics of jatka (juvenile hilsa) in the Meghna River, Bangladesh. Asian Fisheries Science 13:383–389.

DoF (Department of Fisheries). 2006. Annual Report 2006. Department of Fisheries, Dhaka, Bangladesh.

DoF (Department of Fisheries). 2005. Hilsa resource conservation and development management, a training manual. Department of Fisheries, Dhaka, Bangladesh.

DoF (Department of Fisheries). 2004. Hilsa fish conservation and development management implementation technique, a publication under Fourth Fisheries Project. Department of Fisheries, Dhaka, Bangladesh.

FRSS (Fisheries Resources and Survey Systems). 1986–1987 to 2005–2006. Fishery statistical yearbook of Bangladesh. Department of Fisheries, Dhaka, Bangladesh.

Gulland, J. A. 1971. Estimation of mortality rates. Annex to Arctic fisheries working group report. International Council for the Exploration of the Sea, C. M. Document 3, Copenhagen, Denmark.

Halder, G. C. 2006. Strategy report on introduction of selective fishing of hilsa: impact and benefits. Department of Fisheries, Fourth Fisheries Project, Dhaka, Bangladesh.

Halder, G. C., A. U. Ahmed, and A. Islam. 2005. Reclamation, conservation and development techniques of hilsa habitat, Jatio Matshya Pakkha 2005, a national fish fortnight publication. Department of Fisheries, Ministry of Fisheries and Livestock, Dhaka, Bangladesh.

Hora, S. L., and K. K. Nair. 1940. The jatka of eastern Bengal and its significance in the fishery of the so called *Hilsa ilisha* (Hamilton). Records of the Indian Museum 42(4):553–565.

Islam, A. 2006. Conservation of jatka stock for sustainable hilsa production. Individual assignment submitted to Copenhagen DC, Denmark for the fulfillment of the "project management course." Department of Fisheries, Dhaka, Bangladesh.

Rahman, M., and G. Naevdal. 2000. Population genetics studies of hilsa shad, *Tenualosa ilisha* (Hamilton), in Bangladesh waters: evidence for the existence of separate gene pools. Fisheries Management and Ecology 7:401–412.

Salini, J. P., D. A. Milton, M. J. Rahman, and M. G. Hussain. 2004. Allozyme and morphological variation throughout the geographic range of the tropical shad, hilsa *Tenualosa ilisha*. Fisheries Research 66:53–69.

UCIL (Uniconsult International Limited). 2004. A baseline socio-economic survey and impact assessment of hilsa fishers, under the Fourth Fisheries Project of Department of Fisheries, Uniconsult International Limited, Dhaka, Bangladesh.

American Fisheries Society Symposium 69:769–782, 2009

The Santee River Basin Accord: Restoring Diadromous Fish through Prioritization of Subbasins

Amanda K. Hill*
U.S. Fish and Wildlife Service, Ecological Services
Charleston Field Office, Charleston, South Carolina 29407, USA

Abstract.—This paper presents a case study of a collaborative, watershed approach among hydroelectric utilities and natural resource agencies to restore diadromous fish stocks in the Santee River basin, a large Atlantic coast river. The basin once supported large populations of diadromous fish prior to the era of dam construction. A series of hydroelectric dams now block fish spawning migrations from the majority of the Santee River basin. Three of these hydroelectric projects are currently at various stages of the Federal Energy Regulatory Commission's relicensing process for non-federal hydroelectric dam projects. South Carolina Electric & Gas and Duke Energy Carolinas LLC, in concert with the U.S. Fish and Wildlife Service, the South Carolina Department of Natural Resources, and the North Carolina Wildlife Resources Commission, have developed the Santee River Basin Accord to address diadromous fish restoration in the basin. The Santee River Basin Accord utilizes a watershed approach to prioritize the Santee River basin's subbasins and determine the subbasin with the greatest potential for successful passage, spawning, and recruitment of diadromous fish. The process used water quality and quantity, spawning habitat, historic occurrence, quantity of riverine habitat, and the number of dams that fish must pass to access these habitats as the prioritization criteria. Based on these criteria, the Broad River subbasin was chosen as the priority subbasin. The Santee River Basin Accord will fund a 10-year restoration program for this subbasin that includes a hatchery-based restoration program, scientific studies, monitoring, and fish passage at dams that currently impeded spawning migrations. This watershed approach represents a collaborative and innovative method in the southeastern United States to manage and restore diadromous fish populations within a large river basin.

Introduction

Hydroelectric dams impede migrations of diadromous fish, contributing to declines in their populations. In the United States, privately owned hydroelectric dams require a license to operate from the Federal Energy Regulatory Commission (FERC). FERC relicensing is a multiple year process requiring project owners to consult with federal and state natural resource agencies regarding the environmental impacts of their projects and to mitigate for continuing environmental impacts from project operations in accordance with United States federal laws. Under the U.S. Federal Power Act, the U.S. Fish and Wildlife Service (USFWS) and the National Marine Fisheries Service (NMFS) have mandatory conditioning authority to prescribe fish passage at dams as a requirement of a license. These licenses are issued for a period of 30–50 years. As such, the relicensing process offers a rare opportunity to restore diadromous fish to areas from which their populations have significantly declined or have previously been extirpated.

In an effort to address diadromous fish restoration from a watershed perspective, the USFWS, the South Carolina Department of Natural Resources (SCDNR), the North Carolina Wildlife Resources Commission (NCWRC) (together referred to as "the agencies"), Duke Energy Carolinas LLC (Duke Energy), and South Carolina Electric & Gas (SCE&G) developed the Santee River Basin Accord (Accord). The Santee River basin and its hydroelec-

* Corresponding author: amanda_hill@fws.gov

tric projects are complex, and many questions related to diadromous fish movements through several large reservoirs, including their specific migration routes, remain unanswered. The Santee River Basin Accord contains a 10-year action plan funded by the utilities that will monitor, study, augment populations, and provide fish passage at priority dams to restore diadromous fish populations within the basin. This approach is based on a watershed prioritization method that focuses efforts on the subbasin with the greatest potential for diadromous fish restoration. Criteria to prioritize the subbasins of the Santee River basin were identified by the USFWS in concert with the state natural resource agencies. These criteria include water quality and quantity, spawning habitat, historic occurrence, quantity of riverine habitat, and the number of dams fish must pass to access habitats. The prioritization of the Santee River basin's subbasins based on these criteria and the restoration actions planned for diadromous fish are the topic of this case study.

The Santee Basin

The Santee River basin is located along the South Atlantic coast in South Carolina (Figure 1). The Santee River basin consists of four subbasins: the Santee-Cooper, the Catawba-Wateree, the Broad, and the Saluda, together totaling 64,408 km^2. The major (i.e., at least third-order) rivers that form the three upper subbasins are the Broad River, the Saluda River, and the Catawba–Wateree River. These three rivers converge to form the Congaree River. The Congaree River flows into the Santee River, which then flows to the Atlantic Ocean. The Saluda, Broad, and Catawba rivers are located above the fall line zone, while the Wateree River, the lower Congaree River, and the Santee River lie below the fall line zone in the coastal plain. The fall line is a geologic feature along the Atlantic coast where the continental bedrock meets the alluvial coastal plain. Rivers and streams above the fall line are characterized by gravel and rock substrate while rivers below the zone are dominated by sandy substrates. In South Carolina, hydroelectric dams were constructed along the fall line zone because this provided the greatest relief in terms of "head" difference between the upstream reservoir and the turbines, providing the most energy in terms of electricity. The Santee–Cooper subbasin is the only one of the four subbasins located entirely within the coastal plain. This presents a unique situation considering that the majority of hydroelectric facilities are located at or above the fall line zone.

The Santee River basin once supported large populations of diadromous fish, including the American shad *Alosa sapidissima*, blueback herring *A. aestivalis*, shortnose sturgeon *Acipenser brevirostrum*, Atlantic sturgeon *A. oxyrinchus*, and the American eel *Anguilla rostrata*. Valuable commercial and subsistence fisheries were historically dependent on these species' spring upstream migrations (ASMFC 1999). The American shad has been documented ascending 45 km above the North Carolina and South Carolina state boundary (Mills 1826; Stevenson 1897, 1899), approximately 478 km from the Atlantic Ocean. American eel have been documented throughout the Santee basin and historically as far inland as Marion, North Carolina on the Catawba River (Jordan 1889). Shortnose sturgeon spawning adults are known to have migrated above the fall line zone (Mills 1826; Goode 1887).

In the late 18th century, tributaries in the headwaters of the Santee River basin were dammed to operate small grist mills. In the late 19th century, diversion dams were constructed along the fall line to allow for boat navigation around the rocky shoals that exist along this zone. In the 20th century, hydroelectric operations were added to these navigation canals, and hydropower dams were constructed along the main stem of rivers. Construction of hydroelectric dams in the lower portion of the Santee River basin impeded diadromous fish spawning migrations above the fall line. The reduction of spawning and maturation habitat in the Santee River basin is presumed to be a major contributor to the decline in the basin's diadromous fish stocks.

Federal Energy Regulatory Commission Relicensings

The Federal Energy Regulatory Commission is a U.S. government agency that licenses nonfederally owned and operated hydroelectric dam projects. Under Part I of the Federal Power Act, the FERC is responsible for determining the conditions of a license for the construction, maintenance, or continued operation of a hydroelectric project. The Federal Power Act identifies state and federal natu-

Figure 1.—Santee River basin with demarcation of the fall line zone. The fall line zone marks the geologic boundary between the alluvial coastal plain and the continental bedrock.

ral resource agencies as entities, whom the FERC must consult, for the protection of trust resources, including fish and wildlife. Under Section 18 of the Federal Power Act, the Department of the Interior and the Department of Commerce are provided mandatory conditioning authority to prescribe fish passage capabilities at FERC-licensed hydropower projects. This authority is delegated by the Department of the Interior to the USFWS and by the Department of Commerce to the National Marine Fisheries Service.

The FERC hydroelectric projects that impede

access to the majority of the Santee River basin are now within various stages of the relicensing process. Federal and state natural resource agencies actively participate in these processes, each lasting for a minimum of 5 years. Their management objectives for the basin are to improve water quality, enhance instream flows, and protect and restore aquatic communities, including diadromous stocks. Providing fish passage at these projects will allow diadromous fish to access riverine reaches that have been inaccessible for more than 100 years. Most hydroelectric projects in the headwaters of the Santee River basin were relicensed in the 1990s and contain Section 18 reservations for fish passage, put in place by the federal agencies, in the FERC licenses. A reservation of authority for fish passage allows for fish passage to be implemented at these dams when diadromous fish are provided access and passage at the downstream hydroelectric dam projects.

The agencies have established a goal of restoring diadromous fish populations and providing access to historic spawning and foraging habitats in the Santee River basin (USFWS 2001). The timing of relicensings, the project locations, and their common fisheries issues has resulted in an exceptional opportunity for the natural resource agencies and the utilities to work collaboratively to protect and enhance fisheries resources within the basin through a watershed approach rather than project by project, based on FERC relicensing timelines. The duration of these licenses further underscores the importance of the new licenses containing conditions for diadromous fish protection and restoration.

Subbasins

Santee–Cooper Subbasin

The first impediment to upstream migrating fish in the Santee River basin is the Santee–Cooper Hydroelectric Project at river kilometer (rkm) 142 on the Santee River, located in the Santee–Cooper subbasin within the Atlantic coastal plain (Figure 2). This complex project hydrologically connects the Santee River to the Cooper River, which was once just a tidal estuary. Both the Santee River and Cooper River were dammed during the 1942 construction of the Santee–Cooper Hydroelectric Project, creating two shallow coastal reservoirs connected by a canal. This complex of dams, canals, and reservoirs essentially forms the Santee–Cooper subbasin. The project increased in complexity in 1986 when the U.S. Army Corps of Engineers constructed an additional re-diversion canal connecting the lower reservoir to the Santee River, 59 km downstream of the Santee River Dam (USACE 1975). A fish lift was constructed with the hydroelectric dam on this canal. The annual alosine spring spawning migrations are monitored at the fish lift. Fish passage also occurs on the Cooper River through a navigation lock at the hydroelectric dam; however, the current monitoring system in the lock is based on an index of fish biomass rather than by specific species. As a result, only trends in passage of biomass can be determined. This prevents managers from deriving precise annual counts of alosines that are passed into the project. A settlement agreement for the relicensing for this project includes provisions for a new fish counter in the navigation lock capable of distinguishing alosine species from other species.

Catawba–Wateree Subbasin

The Catawba–Wateree subbasin consists of the Catawba River, which originates in the Blue Ridge Mountains in North Carolina and flows through South Carolina to form the Wateree River (Figure 3). The first obstruction on the Wateree River is the Wateree Hydroelectric Dam at rkm 122 (measured from its confluence with the Congaree River). The remaining 354 km of the Catawba River to its headwaters are impounded by a series of 11 dams and reservoirs operated as the Catawba–Wateree Hydroelectric Project by Duke Energy. The first of these dams was constructed in 1907 and the last in 1963. Diadromous fish are known to migrate to the Wateree Dam; however, it is presumed that years of hydropeaking operations and poor water quality from low dissolved oxygen concentrations in discharges have contributed to the declining migrations in the Wateree River.

Saluda River Subbasin

The Saluda subbasin has more than 354 km of river and contains more than 13 hydropower dams (Figure 4). The Saluda Hydroelectric Project, built

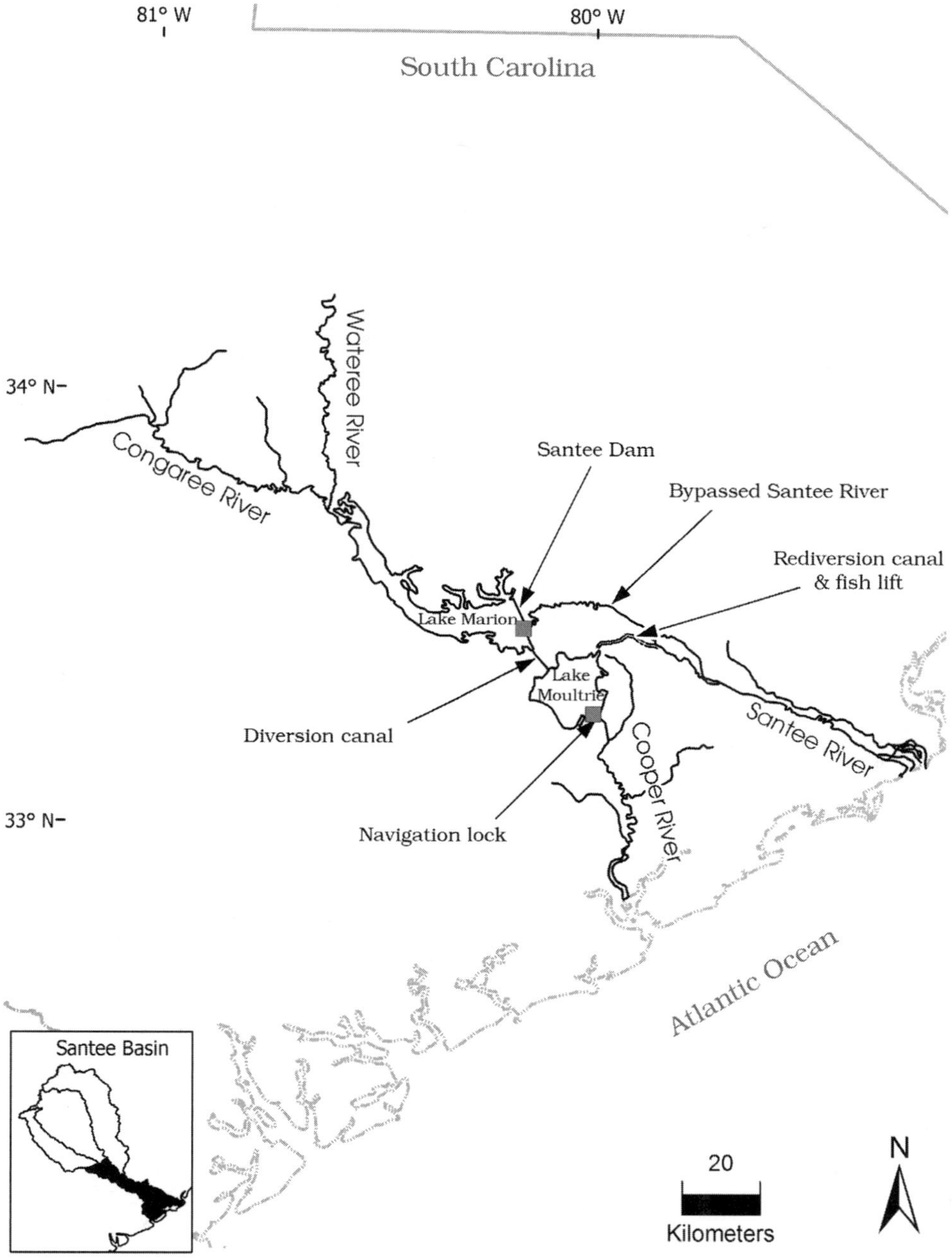

Figure 2.—Santee-Cooper subbasin, including the lower Santee River and Congaree River. Note: the Cooper River was a short estuarine river prior to diversion of the Santee River by the Santee-Cooper Project, a navigation and hydroelectric endeavor completed in 1943. Square dots indicate location of major hydroelectric dams.

in 1930 and owned and operated by SCE&G, impounds approximately 66 km of the Saluda River forming Lake Murray. The Lake Murray Dam is the first impediment to upstream fish migrations on the Saluda River. Below the Lake Murray Dam, there remains a 16-km reach of river to its confluence with the Broad River. This reach of river is characterized by bedrock and rocky shoal habitat.

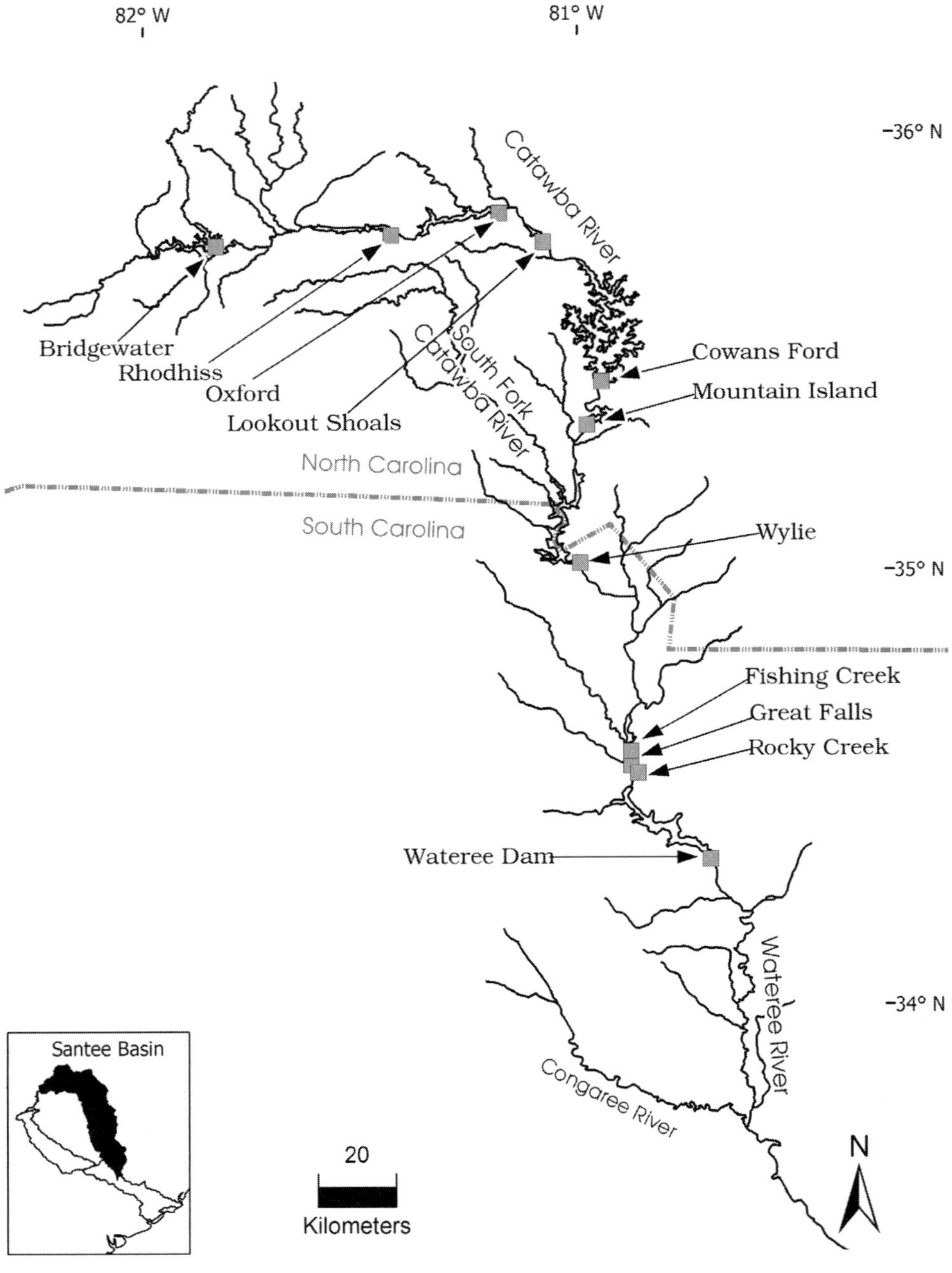

Figure 3.—The Catawba-Wateree subbasin. Square dots indicate location of major hydroelectric dams.

Hypolimnetic (cold, anoxic water) discharges from the project, along with past hydropeaking operations, have significantly reduced diadromous fish migrations in this reach (Kleinschmidt Associates 2007).

Broad River Subbasin

The headwaters of the Broad River subbasin originate in the Blue Ridge Mountains of North Carolina and flow through South Carolina to the confluence with

Figure 4.—The Saluda subbasin. Square dots indicate location of major hydroelectric dams.

the Saluda River and together form the Congaree River. The fourth-order Broad River contains seven hydroelectric projects and 19 small projects on the headwaters of its tributaries (Figure 5). The Broad River has three major tributaries: the Enoree, Tyger, and Pacolet rivers. The Columbia Diversion Dam at rkm 3.22 (measured from its confluence with the Congaree River) was constructed in 1824 and is the first impediment to upstream migrating fish on the Broad River. This project was relicensed by FERC in 2002 and was issued a fishway prescription. A vertical slot fishway was completed in 2006 and began

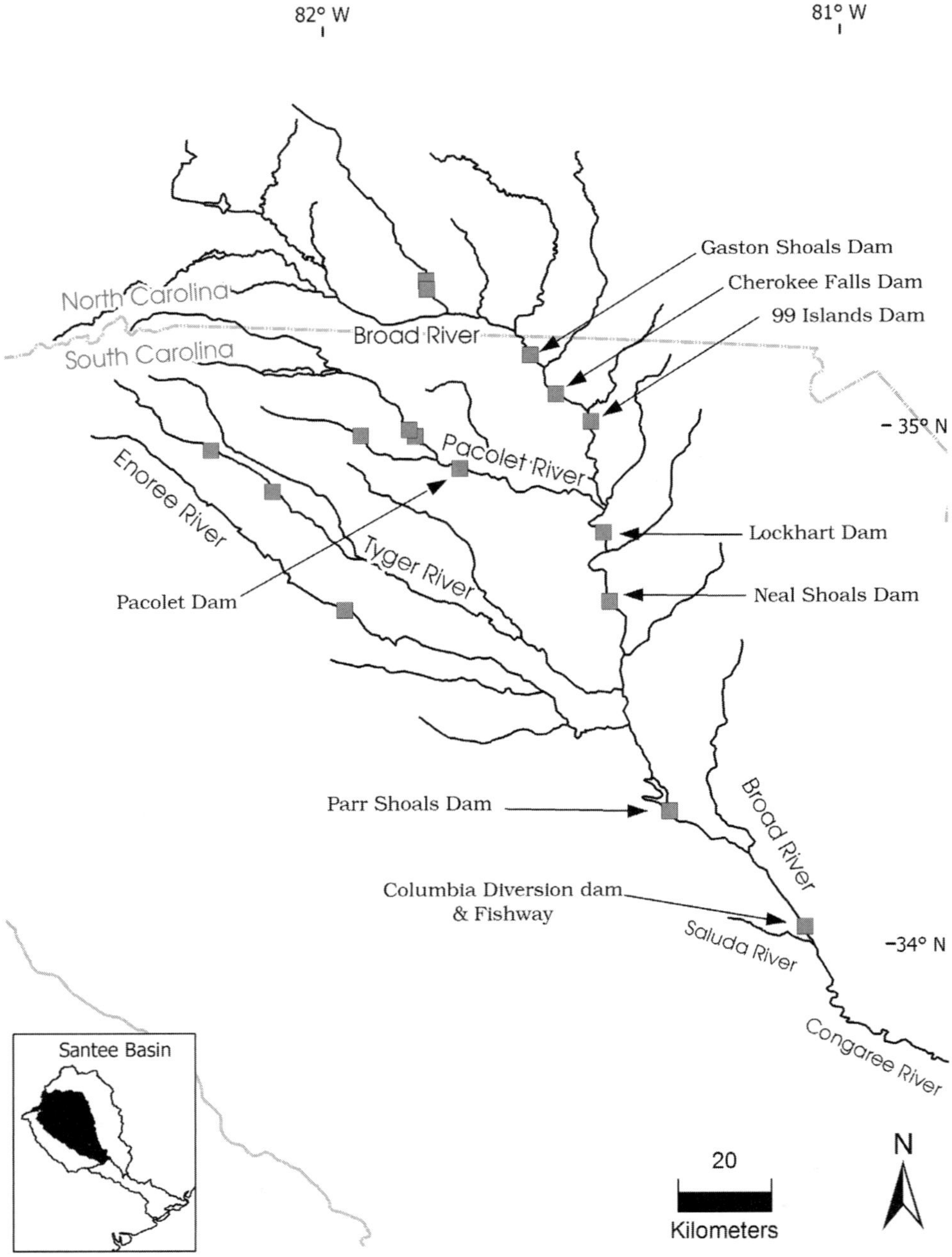

Figure 5.—The Broad River subbasin. Square dots indicate location of major hydroelectric dams.

operation for the first time during the spring 2007 spawning season. Passage beyond this dam provides access to 37 km of high-quality riverine habitat; this reach of river has been inaccessible to diadromous fish for 183 years. The next upstream impediment, which is owned and operated by SCE&G, the Parr Shoals Dam Hydroelectric Project, has a license that expires in 2020.

Subbasin Prioritization Methods

The three upper subbasins, the Saluda River basin, Broad River basin, and Catawba–Wateree basin, were prioritized by the agencies based on the quantity and quality of historic spawning habitat for anadromous fish, specifically for two alosine species, American shad and blueback herring. Other target diadromous fish include the American eel, Atlantic sturgeon, and shortnose sturgeon. American eel is a catadromous species and is characterized as a habitat generalist. The shortnose sturgeon is federally endangered and protected under the Endangered Species Act. Federal management of the shortnose and Atlantic sturgeons are under the purview of the NMFS. Because passage of Atlantic and shortnose sturgeon through fishways is considered experimental, restoration of sturgeon requires special considerations and consultation with the NMFS. For purposes of this accord, restoration through access to historic spawning habitats is focused on alosine species and American eel but includes restoration activities for sturgeon in the form of tagging studies to further understand their movements within the basin. Alosines are semelparous in the southern portion of their range (Leggett and Carscadden 1978) and entrainment and mortality studies at several of the hydroelectric projects have shown that high percentages of juveniles survive downstream passage through turbines (Normandeau Associates 2004; DTA 2005). Therefore, the agencies decided to focus restoration efforts on upstream migrations. American eel restoration is focused on determining the timing and location of elver migrations and providing for their upstream passage at dams. Methods for downstream passage of out-migrating adult silver eels is very site specific and will also be investigated and implemented at dams as part of the Accord.

The quantity of available habitat, its condition, and the practicality of accessing those habitats were the primary criteria used to assess the subbasins. Within each subbasin, reaches of historic riverine spawning habitat were identified as potential restoration areas. Higher priority was given to the subbasin containing the most unimpeded free-flowing riverine habitat. Habitat quality was assessed by evaluating the condition of the water quality, instream flows, substrate, and riparian habitat. Water quality parameters such as dissolved oxygen concentrations, temperatures, and impairments, as listed under Section 303(d) of the Clean Water Act, were evaluated. The agencies gave priority to the subbasin with the least amount of anthropogenic development adjacent to the rivers and the reservoirs; a subbasin with less than 25% development was considered as having low development, 25–50% was considered as medium development, and greater than 50% was considered highly developed. The USFWS assigned these designations based on project shoreline management plans, the amount of development surrounding project reservoirs, and regional populations within the subbasin. Experience in other river basins with multiple fish passage facilities indicates that some percentage of fish approaching each passage facility may not successfully pass upstream for reasons not fully understood. For this reason, the number of dams between currently accessible habitats and the priority habitats was assessed and priority was given to the subbasin with the fewest dams. Finally, the constructional and biological feasibility of providing passage at each dam was determined based on the size and characteristics of the dam, the upstream reservoir, and the amount of desirable upstream river habitat. Dams with a high head usually require a larger facility to pass fish upstream and thus can be very costly. Additionally, large reservoirs, as compared to run-of-river projects, are thought to be more difficult for spawning adults to traverse. Large reservoirs in this system have altered longer reaches of river than run-of-river projects, reducing the amount of available instream habitat. Therefore, priority was given to lower head dams, smaller reservoirs, and run-of-river projects.

Subbasin Prioritization Results

The Saluda subbasin offers the least amount of contiguous riverine reaches of the three subbasins. Upstream of Lake Murray are a series of 13 dams that effectively fragment the system, leaving only a few kilometers of riverine habitat between each of the project's reservoirs. Shoreline at Lake Murray is more than 60% developed, and contains areas of poor water quality due to phosphorus loading (Kleinschmidt Associates 2007). The reservoir experiences thermal stratification in the summer months, resulting in low dissolved oxygen concentrations. The project discharges cold, hypolimnetic flows due to the location of the penstocks deep in the reservoir. In 1998–1999, the five turbine units were modified with hub baffles to ameliorate the low dissolved oxygen releases caused from drawing water from the bottom of the reservoir. The project, however, contains a deep reservoir with a 65-m-high dam and continues to discharge flows that reduce the ambient temperature of the tailwaters. The Saluda subbasin ranks as the third priority subbasin due to the issues discussed above concerning habitat quality and quantity as well as the anticipated engineering difficulties and costs of fishway installation at the Lake Murray Dam.

The Catawba–Wateree system contains two reaches identified by the resource agencies as potential restoration areas for diadromous fish. These are the 3.2 km remaining of the Great Falls, an ecologically significant bedrock and boulder reach located on the fall line zone, and a 48 km tailwater reach below the Wylie Dam at the North and South Carolina border. The reaches are characterized by gravel, cobble, and boulder habitats. However, the system has suffered from poor water quality, including low dissolved oxygen concentrations, low instream flows from peaking operations having inadequate minimum flow requirements, and reaches listed as impaired under Section 303(d) of the Clean Water Act. Duke Energy has proposed enhancements for water quality improvements in the new FERC license for the Catawba–Wateree Hydroelectric Project scheduled for issuance in 2008 or 2009, which in time should rectify these impairments. In order for diadromous fish to access these reaches, however, they would need to pass beyond two dams to gain access to the Great Falls reach, and four dams to access the 48 km of riverine habitat below the Wylie Dam. Based on these considerations, the resource agencies ranked the Catawba–Wateree subbasin as the second priority of the three subbasins.

The Broad River subbasin was identified by the resource agencies as the highest priority for restoration. The Broad River subbasin contains the longest unimpeded reaches of high quality spawning habitat and the fewest number of impediments blocking its access. The installation and operation of the Columbia Diversion Dam fishway at rkm 3.2, resulting from a new FERC license with a fishway prescription issued in 2002, opened 37 km on the Broad River. Passage above the next impediment, Parr Shoals Dam, would provide access to an additional 293 km of high quality riverine and tributary habitat. The Enoree and Tyger rivers are major tributaries to the Broad River. The Enoree and Tyger rivers connect with the Broad River above the Parr Shoals Dam project and flow through the Sumter National Forest. The federally protected forest has protected these reaches from anthropogenic disturbances, resulting in good water quality conditions. These tributaries contain dams in their headwaters but, for the most part, are free-flowing rivers providing a combined total of approximately 243 km of high quality riverine spawning habitat. Additionally, there are 50 km of the Broad River from the Parr Shoals Dam to the next upstream project, the Neal Shoals Dam.

Santee River Basin Accord

Development

Three hydroelectric utilities, Duke Energy, SCE&G, and South Carolina Public Service Authority and four natural resource agencies, the USFWS, NMFS, South Carolina Department of Natural Resources, and North Carolina Wildlife Resources Commission began discussions in 2005 regarding a cooperative approach toward diadromous fish restoration and fish passage at dams within the Santee River basin. The utilities raised concerns that more information regarding diadromous fish migrations within the Santee River basin was needed than could be acquired in the time allocated for studies in the FERC relicensing process. The utilities expressed interest in a post-license study and enhancement program that would maximize the opportunity for successful mitigation at potential future fishway facilities and provide reassurance that expenditures were based on sound science. The utilities approached the natural resource agencies with a concept referred to as the Santee River Basin Accord. The intent of the Accord was to address diadromous fish issues within the Santee River basin through a collaborative process among the utilities and state and federal natural resource agencies. The USFWS, in concert with the SCDNR, NCWRC, and NMFS, responded in November 2006 with a proposal for a watershed approach to diadromous fish restoration that incorporates the utilities' desired study and enhancement program, along with fish passage, fry stocking and imprinting, and adult reintroductions.

The proposed watershed approach prioritized the three subbasins upstream of the Santee–Cooper subbasin based on criteria to identify the subbasin with the greatest potential for successful fish passage, spawning, and recruitment of diadromous fish. The intent behind the proposal was to combine funding from each utility that is party to the Accord and focus restoration efforts on the subbasin of highest priority. In exchange for the utilities' concurrence and support of such a plan, the USFWS agrees to phase fish passage capabilities at dams on the lower priority subbasins. The utilities recognized value in such a strategy in that it offers great potential for successful restoration and provides assurance that expenditures for fish passage facility construction are timely. From the USFWS perspective, consolidating funding from the utilities provides the opportunity to initiate restoration and enhancement activities that are outside the mandatory authorities of the Federal Power Act, for example, constructing a fish passage facility at a project in advance of its relicensing schedule and implementing a hatchery-based restoration program. The priority subbasin approach can also expedite restoration efforts by avoiding a schedule based solely on the FERC dam relicensing time frame. Depending on when a hydroelectric project was licensed within the 30–50 year range, there usually are temporal gaps between relicensing of projects within the same river system. This can result in a significant amount of time between opportunities to provide fish passage at consecutive projects on a river. The priority subbasin approach can obviate time constraints from licensing schedules and allow for a sequenced implementation of diadromous fish restoration.

Organization

Parties to the Santee River Basin Accord include two of the three utility companies from the 2005 discussions, Duke Energy and SCE&G, state natural resource agencies, including both the SCDNR and the NCWRC, and one federal agency, the USFWS, collectively referred to as the Cooperative Accord Partnership (CAP). The Accord is directed by a board consisting of representatives from each of the CAP members. The CAP board is the final decision-making entity. A technical committee consisting of biologists and other qualified professionals representing each of the CAP members will determine the scope and methodologies of the enhancement activities. The technical committee will analyze data and develop annual work plans to be reviewed by the CAP. CAP members have committed to participate in the implementation of the agreed upon enhancement and restoration activities in the Santee River basin for a minimum of 10 years. This collaborative approach is partly modeled on the Susquehanna River Anadromous Fish Restoration Committee developed for the Susquehanna River (St. Pierre 2003).

Restoration

From the USFWS perspective, restoration of diadromous stocks requires access to spawning and maturation habitats, adequate river flow and water quality, recruitment, and reduced mortality from ocean intercept fisheries. Through the relicensing processes, the USFWS is focusing on enhancing and restoring riverine habitats adversely affected by hydroelectric project operations. Major steps toward restoration from the relicensing processes include enhancing existing instream flows and water quality and providing safe and timely access to historic upstream spawning habitats. With these measures in place, the USFWS perceives relocation of adult spawners to upstream reaches and population augmentation through hatchery activities as additional methods to speed the recovery process. Hatchery-based restoration, including population augmentation, imprinting, and reintroductions, has successfully contributed to restoration of alosine populations in the Susquehanna, Potomac, and James rivers (Hendricks 1995, 2003; Olney et al. 003; St. Pierre 2003; Cummins 2004). The reestablishment of alosines upstream of a dam can be expedited by stocking cultured larvae for early imprinting so that in 4 to 5 years, spawning adults return to the specific reach. Restoration programs such as these generally are accomplished through the collaboration of partners. The timing of the relicensing processes within the Santee River basin offers a unique opportunity to combine funding efforts and develop a collaborative action-based restoration plan among utilities and natural resource agencies focused on the subbasin with the most potential for timely and successful restoration.

10-Year Action Plan

A 10-year action plan for the Broad River subbasin was proposed by the USFWS under the consultation of retired USFWS biologist Dick St. Pierre, in concert with the SCDNR and the NCWRC, and further developed by the parties to the Accord. The basic tenet of the plan is to rebuild diadromous fish stocks through a combination of enhancement activities and construction of fish passage facilities at dams. Some restoration efforts will be realized through new FERC licenses, such as the instream flows and water quality improvements. Enhancement activities of the 10-year action plan are prioritized based on those activities determined necessary to begin within years 1 through 5, and those that will be implemented in years 5 through 10.

The initial tasks of the plan are those associated with developing a hatchery-based restoration program. Culture and marking techniques for American shad and blueback herring will be developed and implemented at existing state and federal fish hatcheries. Alosine eggs will be collected by a combination of strip-spawning and collection of nonripe or green fish for hormone injections. As a starting goal, a target of 1 million shad eggs will be incubated by the second working year. In subsequent years, a goal of 10 million eggs will be incubated annually. Fry will be marked with oxytetracycline antibiotic and stocked at locations in the Broad River subbasin, as determined by the technical committee, and below the Wateree Dam on the Wateree River. Hatchery biologists are also investigating the use of genetic markers in lieu of otolith marking.

A trap and transport program for prespawned adults of American shad and blueback herring will be implemented to encourage natural spawning and to imprint progeny to historic upstream reaches. Juvenile alosine young of year will be collected within nursery zones during late summer and fall months for otolith analysis to determine the relative contribution of naturally produced versus hatchery produced shad and herring juveniles. In years two and eight, adult shad and herring will be collected, tagged, and monitored to determine movement patterns throughout the accessible portions of the basin. Beginning at year five, presence and abundance surveys will be implemented below dams that were previously selected for fry stocking during years one through five, consistent with the American shad return generation time of 4–5 years (Walburg and Nichols 1967; Loesch 1987). Adult shad will be collected for otolith analysis to determine the percentage of natural versus stocked returns.

American eel elvers will be monitored at the Parr Shoals Dam for 3 years to determine the optimal location for placement of an eel passage device. Elvers will be relocated upstream of the dam to access riverine, tributary, and reservoir habitats. A series of ramp/trap devices will be placed at each successive upstream dam. Following a 3-year sampling of elver migration and movements, a permanent or semipermanent eel ladder will be constructed at each dam. Out-migration studies will also be implemented at select dams to determine effective methods for downstream passage of adult silver American eels.

An existing tracking program utilizing acoustic telemetry for shortnose sturgeon will be continued through the Accord. Additional studies focused on shortnose sturgeon spawning habitat, juvenile maturation habitat, passage at dams, and food habits will be developed by the technical committee in consultation with NMFS and implemented through the Accord.

Restoration of diadromous fish to the Broad River subbasin will be approached by river reaches. Restoring fish to the 37-km reach above the Columbia Diversion Dam, utilizing the newly constructed fish passage facility and hatchery stocking efforts in the reach below Parr Shoals Dam, will constitute the first reach. The USFWS and SCDNR have established a restoration target number of 92,800 American shad and 464,000 blueback herring for the reach of the Broad River from the Columbia Diversion Dam to the Parr Shoals Dam. South Carolina Electric & Gas has agreed, as part of the Accord, to begin conceptual design plans for construction of a fishway at Parr Shoals Dam when 50% of the restoration target numbers for either species pass upstream at the Columbia Fishway. SCE&G agrees to construct the Parr Shoals Dam fishway before relicensing of the Parr Dam Shoals Project in 2020, provided the restoration target numbers are met for 3 of 5 years, but construction will begin no sooner than 2012. Construction of the fishway at the Parr Shoals Dam will commence when 75% of the restoration target numbers pass upstream of the Columbia Fishway. Providing fish passage capabilities at the Parr Shoals Project earlier than would have occurred based on the relicensing schedule will open 293 km of quality spawning habitat, potentially jumpstarting restoration in those reaches. The next phase will include passage at and stocking above Parr Shoals Dam within the Enoree and Tyger rivers, major tributaries to the Broad River. The USFWS and the SCDNR have established target restoration numbers of 128,150 American shad and 640,750 blueback herring for the reach of the Broad River between the Parr Shoals Dam and the next upstream project, the Neal Shoals Dam. Conceptual design plans for construction of a fishway at Neal Shoals Dam will begin when 50% of the target restoration numbers pass upstream at the Parr Shoals Dam fishway. Construction of the fishway at Neal Shoals Dam will commence when 75% of the target restoration numbers pass at the Parr Shoals Dam fishway. As part of the Accord, fishway construction at Parr Shoals Dam will begin no sooner, however, than 2016. It is assumed that it will be at least this long before stocked fry return as spawning adults or adequate numbers of fish access this reach.

Passage beyond the next four dams is outside the scope of the 10-year action plan but remains a goal of the USFWS. The next upstream dam on the Broad River after the Neal Shoals Dam (rkm 76) is Lockhart Dam (rkm 94). There are 18 km between Neal Shoals Dam and the Lockhart Dam, and above Lockhart Dam there are 35 km of the Broad River to the next dam, the Ninety-nine Islands Project. The Pacolet River joins with the Broad River above the Lockhart Dam Project and contains 37 km of high quality spawning habitat to its first impediment to passage, the Pacolet Dam. Accessing the Pacolet River and the Broad River up to the 99 Islands Project will provide an additional 92 km of habitat. Eleven kilometers upstream from the 99 Island Project is the Cherokee Falls Dam and six more kilometers further upstream is the Gaston Shoals Dam. Existing licenses for the Lockhart Hydroelectric Project, the 99 Islands Hydroelectric Project, the Cherokee Falls Hydroelectric Project, and the Gaston Shoals Hydroelectric Project contain Federal Power Act Section 18 reservations for fish passage. Passage above 99 Islands Dam, Cherokee Falls Dam, and Gaston Shoals Dam will provide access to an additional 134 km of the Broad River in North Carolina. The decision to seek passage beyond these dams to access these headwaters will most likely depend on the success of the downstream restoration efforts.

Restoration activities will be phased in the Catawba–Wateree subbasin, the second-priority subbasin. Cultured American shad larvae will be stocked on the Catawba–Wateree subbasin in the 48-km reach below the Wylie Dam and below the Wateree Dam, the first impediment on the system. Implementing the stocking program immediately in the Catawba–Wateree subbasin will continue to build stocks throughout the 10 years of the plan. In exchange for the utilities funding of the 10-year action plan and prioritizing of subbasins, the USFWS agrees to phase fish passage on the Catawba–Wateree Hydroelectric Project and to defer passage at the Lake Murray Dam. A fish passage facility will be constructed at the Wateree Dam in 2018, 10 years after the new FERC license issuance, allowing a minimum of two generations of alosines to return to the dam where they were stocked and imprinted. American eel elver monitoring will be implemented at the Wateree Dam and phased in 3- year periods sequentially at each of the dams of the Catawba–Wateree Hydroelectric Project.

Summary

Development of the Santee River Basin Accord is an attempt by hydroelectric utilities, the USFWS, SCDNR, and NCWRC to address diadromous fish enhancement and restoration in a large watershed through an interest-based, collaborative process. Working with the utility companies provides an opportunity to pursue restoration activities beyond the scope of the FERC relicensing process and to build cooperative relationships among private and government entities with vested interests in the watershed. The Santee River Basin Accord supports a unique watershed approach in the southeastern United States through focusing fish passage efforts, adult reintroductions, and fry augmentations in the subbasin with the greatest quantity and quality of instream habitat. Restoration efforts will continue on the lower-priority subbasins on a phased schedule.

Acknowledgments

The author would like to thank those individuals from Duke Energy Carolinas LLC, SCE&G, SCPSA, SCDNR, NCWRC, SCDHEC, NMFS, and USFWS who have participated in the development of the Santee River Basin Accord. I also thank Mark A. Cantrell of the U.S. Fish and Wildlife Service for his assistance and review of this manuscript. I especially want to thank Dick St. Pierre for his guidance during the development of this plan.

References

ASMFC (Atlantic States Marine Fisheries Commission). 1999. Amendment 1 to the Interstate Management Plan for Shad and River Herring. Fishery Management Report No. 35. Atlantic States Marine Fisheries Commission, Washington, D.C.

Cummins, J. D. 2004. The Potomac River American Shad Restoration Project Summary Report. Interstate Commission on the Potomac River, Rockville, Maryland.

DTA (Devine, Tarbell & Associates, Inc.). 2005. Fish entrainment evaluation. Final Report. Catawba-Wateree FERC License Application (FERC No. 2232). Duke Energy Carolinas, LLC, Charlotte, North Carlina.

Goode, G. B. 1887. The fisheries of South Carolina and Georgia. The fisheries and fishery industry of the United States. Pages 501–514 *in* Report of the U.S. Commission of Fish and Fisheries, Section 2. Government Printing Office, Washington, D.C.

Hendricks, M. L. 1995. The contribution of hatchery fish to the restoration of American shad in the Susquehanna River. Pages 329–336 *in* H. L. Schramm, Jr. and R. G. Piper, editors. Uses and effects of cultured fishes in aquatic ecosystems. American Fisheries Society, Symposium 15, Bethesda, Maryland.

Hendricks, M. L. 2003. Culture and transplant of alosines in North America. Pages 303–312 *in* K. E. Limburg and J. R. Waldman, editors. Biodiversity, status, and conservation of the world's shads. American Fisheries Society, Symposium 35, Bethesda, Maryland.

Jordan, D. S. 1889. Report of explorations made during1888 in the Alleghany region of Virginia, North Carolina, and Tennessee and in western Indiana, with an account of the fishes found in each of the river basins of those regions. The Bulletin of the United States Fish Commission, volume 8. Available from the National Oceanic and Atmospheric Administration, Asheville, North Carolina.

Kleinschmidt Associates LLC. 2007. Draft license application for Saluda Hydroelectric Project, FERC No. 516. Kleinschmidt Associates LLC, Lexington, South Carolina.

Leggett, W. C., and J. E. Carscadden. 1978. Latitudinal variation in reproductive characteristics of American shad (*Alosa sapidissima*): evidence for population-specific life history strategies in fish. Journal of the Fisheries Research Board Canada 35:1469–1478.

Loesch, J. G. 1987. Overview of life history aspects of anadromous alewife and blueback herring. Pages 89–103 *in* M. J. Dadswell, R. J. Klauda, C. M. Moffitt, R. L. Saunders, R. A. Rulifson, and J. E. Cooper, editors. Common strategies of anadromous and catadromous fishes. American Fisheries Society, Symposium 1, Bethesda, Maryland.

Mills, R. 1826. Statistics of South Carolina, including a view of the natural, civil, and military history, general and particular. Hurlbut and Lloyd.,Charleston, South Carolina.

Normandeau Associates. 2004. Analysis of entrainment and survival of target species at the Santee–Cooper Hydroelectric Project, South Carolina. South Carolina Public Service Authority FERC License Application, FERC No. 199, Moncks Corner, South Carolina.

Olney, J. E., and D. A. Hopler, Jr., T. P. Gunter, Jr., K. L. Maki, and J. M. Hoenig. 2003. Signs of recovery of American shad in the James River, Virginia. Pages 323–329 *in* K. E. Limburg and J. R. Waldman, editors. Biodiversity, status, and conservation of the world's shads. American Fisheries Society, Symposium 35, Bethesda, Maryland.

SCDHEC (South Carolina Department of Health and Environmental Control). 2006. State of South Carolina Integrated Report for 2006, part 1: Listing of impaired waters. South Carolina Department of Health and Environmental Control, Colunbia.

Stevenson, C. H. 1897. The restricted inland range of shad due to artificial obstructions and its effect on natural reproduction. Bulletin of the U.S. Fish Commission 17:265–271.

Stevenson, C. H. 1899. The shad fisheries of the Atlantic coast of the United States. Pages 103–269 *in* U.S. Commission of Fish and Fisheries. Report of the Commissioner 29. Government Printing Office, Washington, D.C.

St. Pierre, R. A. 2003. A case history: American shad restoration on the Susquehanna River. Pages 315–321 *in* K. E. Limburg and J. R. Waldman, editors. Biodiversity, status, and conservation of the world's shads. American Fisheries Society, Symposium 35, Bethesda, Maryland.

USACE (U.S. Army Corps of Engineers). 1975. Final environmental statement, Cooper River rediversion project, Charleston Harbor, South Carolina. U.S. Army Corps of Engineers, Charleston, South Carolina.

USFWS (U.S. Fish and Wildlife Service), National Marine Fisheries Service, and South Carolina Department of Natural Resources, 2001. Santee-Cooper Basin Diadromous Fish Passage Restoration Plan. U.S. Fish and Wildlife Service, Charleston, South Carolina.

Walburg, C. H., and P. R. Nichols. 1967. Biology and management of the American shad and status of the fisheries. Atlantic coast of the United States, 1960. U.S. Fish and Wildlife Service Special Scientific Report Fisheries 500.

American Fisheries Society Symposium 69:783–791, 2009

Antioxidants, Irrigation, and Atlantic Salmon

JOAN G. TRIAL*
Maine Department of Marine Resources, 650 State Street Bangor, Maine 04401, USA

Abstract.—Wild low bush blueberries, advertised as "nature's #1 antioxidant super fruit," require about 2.5 cm of water per 0.4 ha each week during late June and all of July when fruit is forming and maturing. To reduce annual variation in production, Maine's blueberry growers began rapidly expanding the acreage under irrigation in the 1990s. As a result of increased acreage and irrigation, average annual Maine blueberry production increased from 15.8 million kilograms in the 1980s to 32.1 million kilograms in 2006. Endangered Atlantic salmon *Salmo salar* inhabit the rivers and streams in the watersheds where blueberries are grown. Initially all irrigation water was drawn from surface sources and some withdrawals directly degraded salmon habitat. To address this concern, the Maine State Planning Office coordinated a collaborative planning process that included hydrologists, fisheries biologists, and agricultural scientists from state and federal agencies, nongovernmental organizations, and industry, as well as members of the public and industry management. Following this effort, the large growers within Atlantic salmon watersheds shifted irrigation sources to storage impoundments and groundwater. The process also helped moved the blueberry industry to acknowledge fish habitat as a legitimate competing water use, thus potentially setting the stage for their future acceptance of the concept that there are hydrologic and ecological limits on water use in Atlantic salmon watersheds.

Introduction

Washington County produces most of Maine's blueberries, and many of the remaining wild Atlantic salmon *Salmo salar* populations in the United States are found in five of its larger rivers. In 1995, the U.S. Fish and Wildlife Service (USFWS) and the National Oceanic and Atmospheric Administration's National Marine Fisheries Service (NMFS) proposed to list a distinct population segment (DPS) of Atlantic salmon populations in a portion of Maine, including Washington County, as a threatened species under the Endangered Species Act (ESA; 60 FR 50530). As Atlantic salmon populations continued to decline (Table 1; U.S. Atlantic Salmon Assessment Committee 2007) the Federal Services listed the Gulf of Maine (GOM) DPS as endangered, effective December 18, 2000 (61 FR 69459). During the listing process, land- and water-use practices were identified as threats to the quantity and quality of Atlantic salmon habitat throughout Maine. Forestry, agriculture, urbanization, flood control, water pollution, irrigation and other water withdrawals, and dams had potential to degrade spawning and rearing habitat, reduce habitat complexity and connectivity, degrade water quality, and alter ambient stream water temperatures. The threat to Atlantic salmon habitat from irrigation in Washington County has not lessened following listing, with irrigation volume increasing from 1,973,571 m^3 in 1995 (Maine Atlantic Salmon Task Force 1997) to 2,112,729 m^3 in 2006 (Maine Department of Environmental Protection 2006). The irrigation volume reported for 1995 is not representative of irrigation earlier in the 1990s because the practice was growing during the decade (Wild Blueberry Extension Office 2007a) and 1999 to 2002 was a drought period (Lombard 2004).

Wild blueberries, advertised as "nature's #1 antioxidant super fruit," are harvested from intensively managed naturally occurring plants. Farmers actively control insect pollinators and pests, plant diseases, and weeds; prune; mulch; fertilize; manipulate soil pH; and irrigate fields to increase yields. The fruit of

* Corresponding author: joan.trial@maine.gov

Table 1.—Estimated returns of adult Atlantic salmon in the core Gulf of Maine distinct population segment rivers from 1991 to 2006 with upper (U) and lower (L) confidence intervals (CI; 90% CI).

Year	L CI	Average	U CI
1991	235	294	367
1992	200	247	306
1993	222	264	315
1994	154	192	238
1995	131	162	200
1996	230	284	351
1997	130	165	208
1998	153	200	258
1999	137	175	221
2000	79	100	127
2001	90	103	120
2002	28	37	48
2003	62	76	96
2004	60	82	113
2005	44	71	111
2006	49	79	122

blueberry plants is approximately 85% water, and the quality and quantity of the wild blueberry harvest each year is directly related to the timing and volume of moisture provided either by precipitation or irrigation. Berry flavor, color, size, and firmness are related to water content. Harvest in a wet-year harvest can approach two to four times that of a dry year (without irrigation). The financial incentive for wild blueberry growers to irrigate contributed to average Maine blueberry production increasing from 15.8 million kilograms in the 1980s to 32.1 million kilograms from 1998 to 2006 (Wild Blueberry Extension Office 2007b). Demand for irrigation water is highest during late June, July, and August, with growers trying to ensure about 2.5 cm of water per 0.4 ha per week. During the same months, juvenile salmon survival is directly related to stream discharge (Gibson 1993).

The 1995 USFWS and NMFS proposal to list Atlantic salmon as a threatened species contained a special rule under Section 4(d) of the ESA, which allowed for a state plan, approved by USFWS and NMFS, to define the manner in which certain activities could be conducted without violating the ESA. The Maine Atlantic Salmon Conservation Plan (Maine Atlantic Salmon Task Force 1997) was developed in response to this special rule. It acknowledged that blueberry irrigation in Washington County had the potential to reduce the amount of Atlantic salmon habitat during the summer and that Maine had no comprehensive water-use policy. Starting in 1998, the Maine State Planning Office coordinated water-use planning, focusing on the Pleasant River, Narraguagus River, and Mopang Stream in the Machias River watershed to "ensure water withdrawals do not adversely affect Atlantic salmon" (Maine State Planning Office 2001). In this paper, I describe the information and process used to develop the Downeast Salmon Rivers Water Use Management Plan (Maine State Planning Office 2001) and give an overview of the effects of the planning on blueberry irrigation practices in the Pleasant River watershed, Washington County, Maine, the smallest of the three watersheds.

Site Description

The Pleasant River has a small watershed, approximately 326 km^2, and is approximately 45 km long from its source on Pleasant Mountain (elevation 418.8 m) to the head of tide in the town of Colombia Falls. The middle and upper portions of the watershed include significant sand and gravel aquifers. Annual discharge at the U.S. Geological Survey gauge upstream of the estuary for the period of record (1981–1991, 2001–2006) averages 127,588,714 m^3 (http://waterdata.usgs.gov/nwis). More than half (64.6%) of the approximately 1,120 100-m^2 units of mapped juvenile salmon rearing habitat are in the lower portion of the watershed (Table 2), with 75% of the rest upstream of a large heath. The majority of the approximately 17 km^2 of blueberry fields are in the watershed areas of river reaches containing juvenile salmon habitat.

The Process

The Maine State Planning Office contracted with the U.S. Army Corps of Engineers (ACE) to work with a Water Use Planning Steering Committee to prepare the water-use plan for the three rivers. ACE enlisted assistance from subcontractors Battelle and Horsley and Witten to prepare the plan and develop hydrologic models to assist planning. The steering committee included representatives from the forest industry, blueberry industry, state and federal natural resource and agriculture agen-

Table 2.—Distribution of surveyed Atlantic salmon habitat (100-m^2 units) in the Pleasant River, Washington County, Maine, from headwaters (reach 1) to head of tide (reach 5) and total units for the river. Habitat types in italics are considered suitable for juvenile rearing.

Habitat		Reach					
Category	Type	1	2	3	4	5	Total
Falls	*Falls*	0	0	0	9	3	12
Flat	Deadwater	686	81	3,637	41	1,583	6,027
	Glide	21	131	146	22	77	398
	Pool	7	13	10	30	16	75
Riffle	*High gradient*	26	3	0	30	51	110
	Low gradient	33	80	0	140	253	505
Run	*Run*	53	187	14	49	186	490
Totals	All types	825	495	3,807	320	2,169	7,616
	Rearing	112	270	14	228	493	1,116
	% of rearing	10	24	1	20	44	

cies, municipal governments, and local, regional, and national nongovernmental organizations (e.g., watershed councils, lake associations, recreational anglers, wildlife). The process was data intensive, requiring information on hydrology, geology, land use, stream geomorphology, Atlantic salmon habitat suitability at a range of hydrologic conditions, and irrigation. The steering committee was presented general information on hydrology, irrigation, and Atlantic salmon population and habitat, models of watershed hydrology, and potential irrigation scenarios for increases in production and irrigation. The steering committee enlisted the assistance of a flow technical team to evaluate the hydrologic assessments, input data, and models developed to assess the effects of alternative sources and volumes of irrigation on hydrology and Atlantic salmon habitat.

The Water Use Management Plan documents the science and the steering committee work to resolve water-use conflicts. Implicit in the goals of the plan is the importance of protecting Atlantic salmon habitat and populations. The plan included a variety of recommendations on potential water sources, best management practices, and water conservation measures and is meant to be a living watershed planning tool. It included a list of sources for irrigation water ranked based on the steering committees' understanding of their potential effects on Atlantic salmon habitat (Table 3). However, it does not address a method of allocating water among users nor hydrologic limits to irrigation. Implementing the recommendations of the plan was voluntary. Because discussions of sustainable water use and potential changes in regulations on surface water use were ongoing for the entire state, no changes in existing regulations were recommended by the steering committee.

The Science

The U.S. Geological Survey Maine Office conducted a record extension for the Pleasant River gauge in the lower river (Nielsen 1999). The state of Maine contracted Kleinschmidt Associates (1999) to conduct studies linking stream hydrology and Atlantic salmon habitat using the instreamflow incremental methodology (IFIM; Bovee et al. 1998). To conduct the IFIM, the river was segmented into five reaches based on hydrologic characteristics, river geomorphology, and the distribution of Atlantic salmon habitat (Figure 1). Horsley and Witten (2000) developed a simple deterministic model for predicting discharge at each reach using a ratio of the gauge discharge that was based on watershed area of the reach and the proportion of sand and gravel. With these three reports, the flow technical team developed a matrix to identify the months when natural median flows were less than flows that provided the maximum habitat for selected life stages (Table 4). The discharge that produced maximum habitat for juvenile rearing was the habitat protection

Table 3.—The list of alternative irrigation water sources developed for the Downeast Salmon Rivers Water Use Management Plan, Pleasant and Narraguagus Rivers, Mopang Stream (Maine State Planning Office 2001) prioritized within categories only.

Seasonal groundwater pumping (wells)	No hydraulic coupling to rivers or streams during the annual period of low flow.
	Hydraulic coupling to rivers and streams, affecting less than 25% of stream discharge, with flow augmentation.
	Hydraulically coupled wells without flow augmentation.
Withdrawals from lakes and ponds	Natural lakes and ponds that are hydraulically isolated from surrounding groundwater—maximum drawdown no more than what occurs naturally.
	Natural ponds and lakes that are hydraulically coupled to surrounding groundwater with seasonal outlets—maximum drawdown no more than that which naturally occurs.
	Natural or artificial ponds and lakes that are hydraulically coupled to surrounding groundwater with year round outlets—maintain environmentally protective flow in outlet stream.
Withdrawals from rivers and streams	In late winter and spring during annual period of high flow, as long as fish habitat is maintained.
	Skimming high flows during the low-flow season, as long as fish habitat is maintained.
	Direct pumping from streams during the annual low-flow period that reduce the flow below previously established protective environmental flows.
Storage alternatives	Existing lakes and ponds in which the long-term naturally occurring range of water levels is maintained during the low-flow season.
	Natural depressions, such as kettle holes, that will hold water.
	New constructed ponds filled with precipitation, surface runoff, and groundwater seepage alone.
	Existing lakes and ponds, manipulate long-term naturally occurring high or low water level to increase the stored volume.
	Impounded stream with adequate outflows are provided and all other environmental issues addressed.

standard carried forward to evaluate a variety of irrigation scenarios. Thus, in all but the wettest years, irrigation withdrawals with direct links to the river would be limited or unlikely after mid-June because natural flows will limit salmon populations (Gibson 1993). The standard was much higher than those developed based on USFWS New England aquatic base flow policies (Kulik 1990); however, it was not dynamic (Richter et al. 1997).

The major thrust of the irrigation modeling (Horsley and Witten 2000) was to predict changes in the base-flow contribution to surface water discharge at the downstream end of each of five reaches, for different types and distributions of water sources and volumes. The five reaches corresponded to those used in the IFIM study. The model was deterministic. For each reach, watershed area and recharge rates were multiplied for three surface geologies (glacial till, wetlands, and sand and gravel) and summed. The contribution to base flow for each reach was then summed sequentially from headwaters to mouth. Existing and future irrigation withdrawals were placed within reach watersheds and classed as direct, indirect, or impoundments. Direct withdrawals were modeled to remove the total irrigation volume from base flow, unless they occurred at high flows. Indirect withdrawals were modeled such that irrigation volume was removed from base flow over a 1-year period. Impoundments reduced base flow based on area and volume of any conservation releases. Irrigation demands were modeled based on a schedule of water demand and the area of crop and prune fields being irrigated provided by the industry. Varying levels of irrigation were simulated assuming 100%, 33%, and 25% of crop need was provided by irrigation.

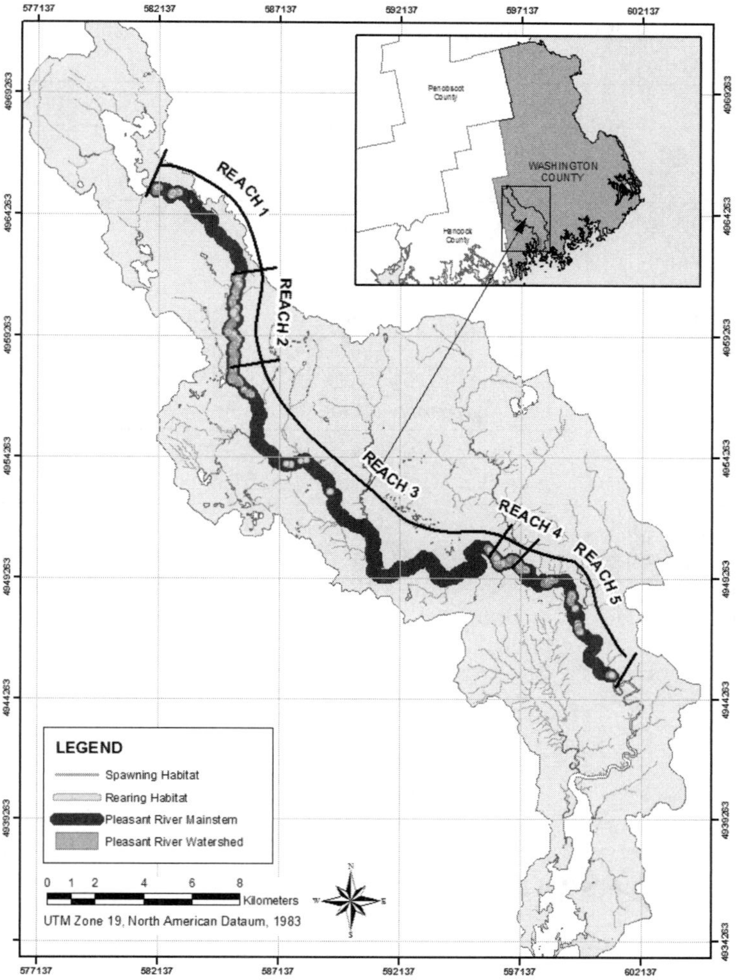

Figure 1.—Distribution of Atlantic salmon juvenile rearing and adult spawning habitat within reaches defined while conducting IFIM studies on the Pleasant River, Washington County, Maine.

Initial scenarios were based on the existing irrigation sources and fields being irrigated in the year prior to model development (1999). Future scenarios expanded irrigation to the entire 1999 production area and by annual increments of 5% to the year 2015. For the future scenarios, a variety of direct, indirect, and impoundment withdrawal sources were added to meet the water demand. These were placed in the reach watershed of the fields being irrigated. The model produced estimated monthly base-flow losses for each reach associated with irrigation scenarios. The effects on streamflow were predicted by subtracting base flow losses from historic monthly streamflow statistics for the reach based on the record extension.

The effects of the modeled scenarios on Atlantic salmon habitat were evaluated by graphing weighted usable habitat area linked with monthly flow statistics. These habitat-discharge duration curves were developed by determining the weighted usable habitat area available at the discharges in a flow duration curve and plotting weighted usable area versus the percent exceedence for these discharges. The curves were compared for historic streamflow statistics based on gauge data (Nielsen 1999) and for predicted flows for a variety of irrigation scenarios (Figure 2).

Table 4.—The discharge (m^3/s) at which maximum weighted usable area (WUA) of Atlantic salmon juvenile and spawning habitat occurs and the estimated median monthly discharge for reaches within the Pleasant River, Washington County, Maine.

Juvenile rearing: young of year and parr						
Reach	Max WUA	May median	Jun median	Jul median	Aug median	Sep median
1	0.5	1.1	0.6	0.3	0.2	0.2
2	0.6	1.7	0.9	0.5	0.4	0.4
3	1.7	4.6	2.4	1.4	1.0	1.0
4	1.7	4.6	2.4	1.4	1.0	1.0
5	1.1	7.2	3.7	2.1	1.6	1.6

Spawning and incubation								
Reach	Max WUA	Oct median	Nov median	Dec median	Jan median	Feb median	Mar median	Apr median
1	1.7	0.4	0.9	0.9	0.8	0.7	1.1	2.0
2	1.6	0.6	1.3	1.4	1.2	1.1	1.6	2.9
3	3.6	1.7	3.6	3.8	3.1	3.0	4.2	7.8
4	3.6	1.7	3.6	3.8	3.1	3.0	4.3	7.9
5	4.5	2.7	5.6	6.0	4.9	4.7	6.6	12.3

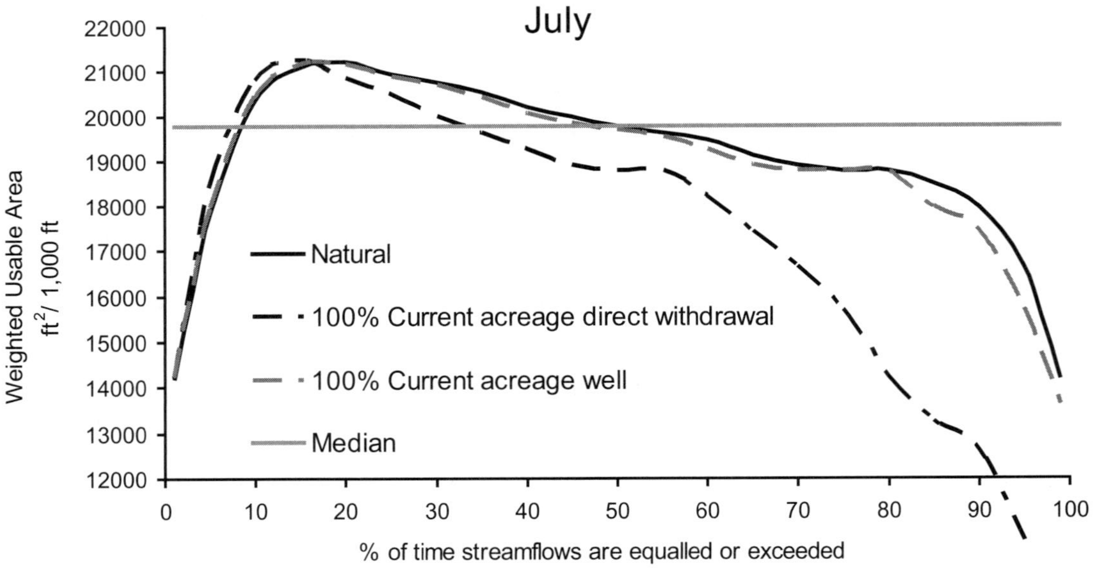

Figure 2.—July habitat (weighted usable area)-streamflow duration curve for reach four of the Pleasant River. Three irrigation scenarios for current acreage are compared: (1) no irrigation, flow statistics from the Pleasant River (solid black); (2) 100% of crop water demand met using direct stream and river withdrawals (broken black); or (3) 100% of the crop water demand met using wells that do not immediately reduce river base flow (broken gray). Horizontal gray line indicates habitat available at monthly median discharge.

Thus, in addition to recommendations for sources prioritized based on minimizing the effects of withdrawals on Atlantic salmon habitat, the planning process developed tools to evaluate the effects of changing irrigation strategies in the watersheds on hydrology and Atlantic salmon habitat. The watershed model was transferred to the State Planning Office, but responsibility for using the model was not assigned to an individual and no training was provided. Fisheries agencies are currently requesting that Maine regulatory agencies have growers evaluate cumulative effects of irrigation on the Pleasant River using the available watershed model.

Irrigation Changes

In 1995, blueberry fields in the Pleasant River watershed were irrigated from 16 surface water withdrawal sites: five streams, eight natural lakes, and three man-made ponds/reservoirs (U.S. Department of Agriculture 1996). The volume of irrigation water in 1995, a relatively dry year, was reported as approximately 1,603,526 m^3, with 60% of that coming directly from streams (Figure 3) in June, July, and August (Maine Atlantic Salmon Task Force 1997). These sources limited the grower's ability to irrigate because state regulations placed limits on withdrawals to protect recreational use on lakes and fish habitat in rivers. In developing the hydrologic model, Horsley and Witten (2000) estimated that 17.4 km^2 were in blueberry production and documented that 8.8 km^2 were being irrigated in 1999. The major grower in the watershed reported using 1,088,722 m^3 from sources in the Pleasant River watershed to irrigate blueberry fields in 1999.

That same year, in response to the ongoing planning process, the grower eliminated their largest upriver direct stream withdrawal in juvenile Atlantic salmon habitat. Further, this company decided to pursue wells as their primary irrigation source based on the ongoing planning, the preliminary ranked list of sources, and the presence of aquifers in the area. Wells were drilled and brought into the irrigation system in 2000, 2004, and 2005, and one is currently being evaluated for production use in 2008. As these wells became operational, the first surface water sources to be replaced were direct stream withdrawals. During the same period, two wells in an adjacent drainage that has no Atlantic salmon were brought into the company's irrigation system and used to irrigate blueberry fields within the Pleasant River watershed. In 2000, the one remaining direct stream withdrawal accounted for 30% of the irrigation water from Pleasant River watershed sources.

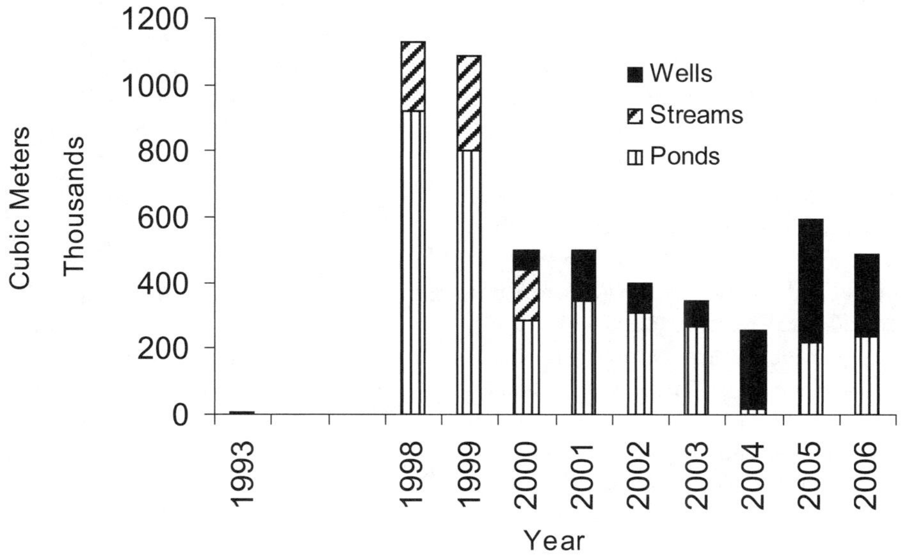

Figure 3.—Irrigation water volumes reported for pond, stream, and well sources within the Pleasant River watershed by the major blueberry grower. No data were available for 1994 to 1997.

No direct stream withdrawals have been used in the Pleasant River watershed since 2000. In 2006, the company reported using 484,643 m^3 for irrigation from Pleasant River watershed sources, with 51% of that from wells (Figure 3). In addition, approximately 65,000,000 m^3 of water from wells in an adjacent watershed were used to irrigate fields in the Pleasant River watershed.

Discussion

The major outcome of the planning process was a complete elimination of direct stream withdrawal in the Pleasant River watershed, the largest of which was within juvenile rearing habitat. The replacement sources were wells, which have less obvious effects on streamflow due to groundwater travel times. The water-use management plan recommended that wells be located such that withdrawals did not immediately affect streamflows and considered wells placed near a stream to have the same effect as a direct withdrawal. The preferred well should be located where withdrawal from the aquifer had no influence on streamflow and in the lower portions of the drainage. The blueberry company's consulting hydrologist found replacement sources; unfortunately, one has been identified as directly affecting streamflow (Dudley and Stewart 2006). Two of the wells used to irrigate blueberry fields in the watershed were located in an adjacent watershed. Groundwater and surface water boundaries do not always coincide, and these wells are very near the surface water divide. The wells have increased the company's ability to irrigate, creating new conflicts related to hydrologic and ecological limits on water use in Atlantic salmon watersheds that have yet to be resolved. Using the watershed model to assess the cumulative hydrologic effects of the changes in irrigation sources and volumes in the Pleasant River watershed could partly resolve these new conflicts.

Although Maine's water management issues are not on the scale of issues in the western United States, there was value in a collaborative approach (Poff et al. 2003). The planning process for the Conservation Plan and the Water Use Plan resulted in major blueberry growers altering irrigation sources in part to protect fish habitat and in part because the new sources provided more irrigation capacity than direct stream withdrawal. Through the process, the growers developed an appreciation for fish habitat as a legitimate competing water use, demonstrated by voluntarily eliminating direct stream withdrawals. This does not mean that irrigation conflicts in Atlantic salmon watersheds are entirely resolved. The industry has not yet accepted the concept that there are hydrologic and ecological limits on water use, especially in headwaters, and continues to increase the area being irrigated. The watershed irrigation model developed during the planning process could help address these limits. Further, newly enacted statewide water use standards are as protective as the seasonal limits agreed upon in this process. Fisheries agencies are lobbying for using the watershed model developed during this process to determine compliance with the new regulations and evaluating the cumulative effects of blueberry irrigation on fish habitat.

References

Bovee, K. D., B. L. Lamb, J. M. Bartholow, C. B. Stalnaker, J. Taylor, and J. Henriksen. 1998. Stream habitat analysis using the instream flow incremental methodology. U.S. Geological Survey, Biological Resources Division Information and Technology Report USGS/BRD-1998–0004, Reston, Virginia.

Dudley, R. W., and G. J. Stewart. 2006. Estimated effects of ground-water withdrawals on stream water levels of the Pleasant River near Crebo Flats, Maine, July 1 through September 30, 2005. U.S. Geological Survey, Scientific Investigations Report 2006–5268, Reston, Virginia. Available: http://pubs.water.usgs.gov/sir2006–5268 (May 2007).

Gibson, R. J. 1993. The Atlantic salmon in freshwater: spawning, rearing and production. Reviews in Fish Biology and Fisheries 3:39–73.

Horsley and Witten. 2000. Draft Hydrologic Aspects of the Water Use Management Plan for the Pleasant River, Downeast Maine. Horsley and Witten, Sandwich, Massachusetts.

Kleinschmidt Associates. 1999. Maine Atlantic Salmon Conservation Plan, report on instream flow studies. Volume 1. Pleasant River. www.maine.gov/dmr/searunfish/reports/MaineAtlanticSalmonConservationPlanReportonInstreamFlowStudies.pdf (April 2007).

Kulik, B. H. 1990. A method to refine the New England aquatic base flow policy. Rivers 1(1):8–22.

Lombard, P. J. 2004. Drought conditions in Maine, 1999–2002: a historical perspective. U.S. Geological Survey Water-Resources Investigations Report 03–4310. Available: http://pubs.usgs.gov/wri/wrir034310 (May 2007).

Maine Department of Environmental Protection. 2006. Water withdrawal reporting program 2004–2005. Annual Report. Maine Department of Environmental Protection, Document #DEPLW0752, Augusta.

Maine State Planning Office. 2001. Downeast Salmon Rivers Water Use Management Plan, Pleasant and Narraguagus Rivers, Mopang Stream. Maine State Planning Office, Augusta.

Maine Atlantic Salmon Task Force. 1997. Atlantic Salmon Conservation Plan for seven Maine rivers. State of Maine, Augusta.

Nielsen, J. P. 1999. Record extension and streamflow statistics for the Pleasant River, Maine. U.S. Geological Survey, Water-Resources Investigation Report 99–4078, Reston, Virginia. Available: http://me.water.usgs.gov/reports/finalreport.pdf (May 2007).

Poff, N. L., J. D. Allan, M. A. Palmer, D. D. Hart, B. D. Richter, A. H. Arthington, K. H. Rogers, J. L. Meyer, and J. A. Stanford. 2003. River flows and water wars: emerging science for environmental decision making. Frontiers in Ecology and the Environment 1(3):298–306.

Richter, B., J. Baumgartner, R. Wigington, and D. Braun. 1997. How much water does a river need? Freshwater Biology 37(1):231–249.

U.S. Atlantic Salmon Assessment Committee. 2007. Annual Report Of the U.S. Atlantic Salmon Assessment Committee. U.S. Atlantic Salmon Assessment Committee, Report No. 19–2006 Activities, Gloucester, Massachusetts. Available: www.nefsc.noaa.gov/USASAC (May 2007).

U.S. Department of Agriculture. 1996. Precipitation, runoff, irrigation water use and potential sources in the Narraguagus and Pleasant River watersheds, Maine. Natural Resources Conservation Service, Orono, Maine.

Wild Blueberry Extension Office. 2007a. Wild blueberry acres by counties. University of Maine Cooperative Extension, Orono.

Wild Blueberry Extension Office. 2007b. Wild blueberry crop statistics 1924–2006. University of Maine Cooperative Extension, Orono.

American Fisheries Society Symposium 69:793–811, 2009

How Policy, Politics, and Science Shaped a 25-Year Conflict over Alewife in the St. Croix River, New Brunswick–Maine

THEODORE V. WILLIS*
Aquatic Systems Group, University of Southern Maine
350 Commercial Street, Portland, Maine 04101, USA

Abstract.—Consensus-based research uses scientific inquiry as an unbiased tool to contrast opposing positions in resource conflicts. However, when applied inappropriately, science is more likely to polarize opposing parties. Opposing parties also may use a consensus-based research process to delay actions not in their interest. The recreational smallmouth bass (SMB) *Micropterus dolomieu* fishery in the St. Croix River, New Brunswick–Maine is an example of how consensus-based research failed to bring opposing sides in a natural resource conflict to a mutually agreeable resolution. Recreational fishing guides blamed high alewife *Alosa pseudoharengus* returns for the crash of a naturalized SMB fishery in the upper watershed and proposed an exclusion law in response. In 1995, Maine legislated that all fishways in the St. Croix River be closed to alewives, with a resulting decline from 2.6 million alewives in 1987 to 900 by 2002. Claims of insufficient data to describe alewife ecological impacts on SMB led to formation of a local, state, provincial, U.S., and Canadian stakeholders science committee charged with developing a study that would describe SMB and alewife interactions. The committee identified the question "has alewife presence contributed to poor SMB population performance" as of primary importance to moving forward to a resolution. To test this hypothesis, SMB condition and growth, and the diet habits and diet overlap of alewives and SMB, were used as indicators of competition. Results, presented here, indicate that growth did not decline in the presence of alewives, age-0 SMB condition did not decline when alewives were present, and diet overlap between the two species was low in three of four lakes examined. Thus, the available data did not indicate that alewives caused poor SMB population performance. The project results helped break down institutional barriers between Maine natural resource agencies. However, local recreational fishermen were not satisfied with the project process or results, claiming that their interests, as represented by the state freshwater fish and game agency, were not seriously considered. They opposed and nearly defeated a bill to reintroduce alewives to the St. Croix River in 2008, again citing insufficient data to justify the action.

Introduction

Natural resource conflicts may be based in a variety of misunderstandings or opposing viewpoints concerning the use of a resource (Odell et al. 2005). The perception that an institution failed to manage a resource fairly and efficiently is often the trigger for conflict among stakeholders (Bennett et al. 2001). Feelings of inequity or injustice by parties involved in an issue are likely to develop into conflict. Decisions that marginalize local stakeholders, under the assumption that communities are an obstacle to efficient or rational management, are often met with resistance (Agrawal and Gibson 1999). Communities that are not involved in the active management of natural resources may choose to use those resources destructively. More often, stakeholders who feel slighted by the management process will enter into a lengthy and costly appeal process, delaying implementation of management decisions (Germain et al. 2001). Allocation of natural resources may not be exclusive to one opposing group or another, but perception to the contrary by either side can further polarize a conflict, making

* Corresponding author: theowillis06@aim.com

trade-offs and resolution harder to reach (Grimble and Wellard 1997). Many interventions in natural resource management issues have failed because not enough attention was paid to the economic interests and objectives of stakeholders.

This manuscript describes a natural resource management conflict over alewife *Alosa pseudoharengus*, an anadromous herring that is native to the eastern seaboard of North America. The case study specifically deals with a northern population that spawns in the St. Croix River watershed, Passamaquoddy Bay, Gulf of Maine and the interaction of that population with a naturalized sport fish, smallmouth bass *Micropterus dolomieu*. The St. Croix River defines the border between New Brunswick, Canada and Maine, USA. Consequently, all conflicts that occur in the St. Croix are international by definition, even if local by nature. Stakeholders in this 20-year impasse include marine and freshwater resource management agencies from the state of Maine, the province of New Brunswick, the U.S. Fish and Wildlife Service, the Department of Fisheries and Oceans Canada, local fishing guides, and conservation groups. The fisheries issue also has relevance to the National Marine Fisheries Service because the alewife is a marine species that falls under the authority of the Magnuson-Stevens Fishery Conservation and Management Act. The International Joint Commission has jurisdiction over border disputes between the United States and Canada.

Despite the prevalence of international stakeholders, the fisheries issue in the St. Croix River region is specific to 21,945 ha of lake habitat and two Maine counties. The original complaints about alewives were made by eastern Maine freshwater fishing guides (EMFFG), who suspected that alewives were the cause behind a downturn in smallmouth bass (SMB) fishing in one upper watershed lake. The EMFFG successfully lobbied the Maine state legislature and pushed through a law in 1995, which excluded alewives from most of the St. Croix River drainage. The conflict today is between EMFFG, backed by the 1995 law that requires state of Maine natural resource agencies to prevent alewife from ascending fish ladders in the St. Croix River, and opposing local, national, and international conservation groups, state and national management agencies, and marine resource users.

What was essentially a local issue in 1995 is now seen as a contributing factor in the range-wide decline of the U.S. river herring fishery (including alewife and blueback herring *A. aestivalis*). Legislative decisions made in 1995 accounted for the economic needs of Washington County, Maine, where the median income is less than $26,000 per year. Little consideration was given to species, ecological, and geographic considerations surrounding the loss of more than 1.5 million pounds in alewife biomass.

In 2006, river herring were declared a "species of concern" by the National Marine Fisheries Service (USDOC 2006), a status that focuses research and conservation efforts to prevent a species from declining to the point where Endangered Species Act protection is needed. In 2007, the Atlantic States Marine Fisheries Commission started an amendment process to determine if closing the U.S. river herring fishery was warranted, based on declines in commercial landings since the 1990s (ASMFC 2007). Several Atlantic states, through their own management review processes, concluded that emergency closures of river herring fisheries were warranted (i.e., in Connecticut [2002], Massachusetts [2006], Rhode Island [2006], and North Carolina [2007]). The St. Croix fishery issue existed in a vacuum in 1995 but was very relevant to conservation of the species in 2008.

Alewife Management Issues in St. Croix River

Management actions that led to the St. Croix River impasse started almost two centuries earlier. The first dams were constructed on the St. Croix River around 1790; however, not until construction of the Union Dam in 1836 was there a drastic and long-term decline in anadromous fish returns (Perley 1852; Goode 1887). At about the same time, SMB were introduced to the upper St. Croix River watershed, where they have supported a significant recreational fishery since the mid-1900s (Warner 2005). A century of increasing industrialization and restricted fish passage followed the completion of Union Dam (Marshall 1976). In 1960, reliable fish passage was installed at Milltown Dam, the first dam without fish passage. Within a few years, alewife escapement numbers were around 10,000 (Marshall 1976). Pollution abatement in the 1970s and improvements to the Milltown fish passage facilities in 1981 likely led to the alewife population explosion of the late 1980s.

Alewife spawning escapement prior to 1981 was likely between 100,000 and 200,000 individuals. Between 1985 and 1988, the number of alewives returning to spawn increased exponentially to 2,624,700 fish (Figure 1). Beginning in 1988, parts of the St. Croix watershed were closed to alewives in response to EMFFG concern over the coincident decline of SMB in Spednic Lake. These closures, requested by Maine inland fisheries officials for research and management, were in effect permanent reductions in alewife spawning habitat, codified by passage of the 1995 legislation. A precipitous decline in the St. Croix alewife population resulted, with the population reaching a low of just 900 fish in 2002 (Figure 2).

The nature of interspecific interactions between alewives and SMB was central to the debate regarding what should be done about the alewife decline. Those opposed to reintroducing alewives claimed that the scientific data describing the interaction of alewives with organisms in the freshwater ecosystem was insufficient to justify the "experiment" of reintroducing alewives to the St. Croix watershed. The Maine legislature supported the alewife opposition in 2001 when conservation groups first attempted to repeal the 1995 law. It became apparent that intervention from organizations outside the state government would be necessary when a five-agency (Maine, New Brunswick, Canada, and U.S. federal) proposal to conduct staged alewife releases into St. Croix waters and study the effects on resident SMB populations failed shortly thereafter. The 1995 law prevented Maine fisheries agencies from participating in any release. A working group, with representatives from many of the aforementioned stakeholders, acted to fill the data gap by agreeing

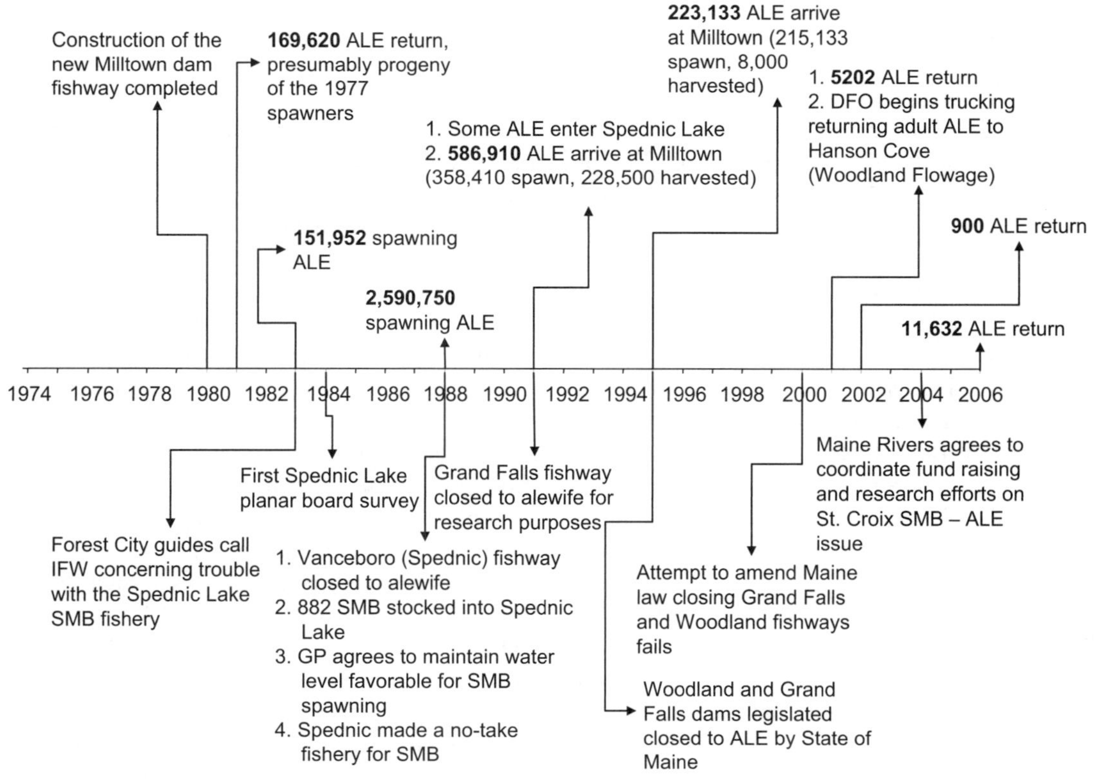

Figure 1.—Timeline of events in the management of alewives and alewife returns on the St. Croix River. Spawner escapement in any year consists mostly of fish spawned 4–5 years previous (SCIWC 2000). ALE = anadromous (sea-run) alewife, IFW = Maine Department of Inland Fisheries and Wildlife, SMB = smallmouth bass, GP = Georgia-Pacific Corporation, and DFO = Fisheries and Oceans Canada.

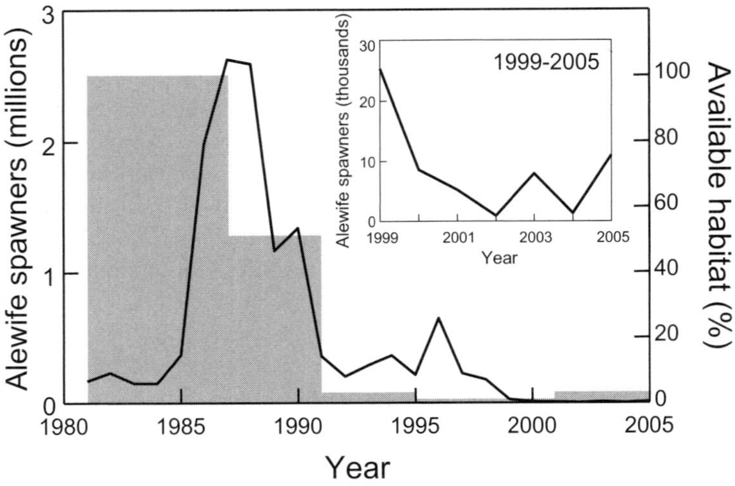

Figure 2.—Alewife spawning escapement (left axis, solid line) and available spawning habitat (right axis, shading) graphed against year in the St. Croix River system. The inset shows spawning escapement in number of alewives between 1999 and 2005. Spawning alewives were counted at Milltown Dam fishway by the St. Croix International Waterway Commission. The maximum spawning area was equal to Woodland Flowage, Grand Falls Flowage, Big Lake, and Spednic Lake. The minimum spawning area was equal to waters between Milltown Dam and Woodland Dam.

to support an independent research project organized by nongovernmental organizations. The role of science in the political landscape of the St. Croix issue became addressing how alewives and SMB interacted in freshwater.

Ecological studies of alewives in their native range, under natural conditions, are rare compared to investigation of the invasive land-locked form. Many of the concerns regarding anadromous alewife effects in freshwater ecosystems are derived from information about landlocked alewives in the Laurentian Great Lakes. Alewife opponents have cited early mortality syndrome (Fisher et al. 1996), water quality concerns (Brooks and Dodson 1965), competition, and predation (Bunnell et al. 2006) as evidence for keeping the St. Croix River fishways closed. Similarities between the two life histories do exist. However, a cohort of anadromous alewives are present in freshwater for 4–5 months in their first year of life, then for several weeks 4–6 years later when they return to spawn. Conversely, landlocked alewives are permanent freshwater residents.

The working group project was designed to investigate the potential negative impacts of anadromous alewives on SMB. The working group evolved into a science advisory committee that consisted of the U.S. Fish & Wildlife Service (USFWS), Maine Department of Inland Fisheries & Wildlife (MEIFW), Maine Department of Marine Resources (MEDMR), Fisheries and Oceans Canada (DFO), New Brunswick Department of Natural Resources (NBDNR), St. Croix International Waterway Commission (SCIWC), and Maine Rivers (MR). Each member of the working group represented a constituency of stake holders. USFWS represented U.S. national conservation interest, particularly those encompassed by the Anadromous Fish Act of 1965. MEDMR and DFO represented marine commercial and conservation interests; MEDMR and NBDNR represented local freshwater fishing interests. The role of MR and SCIWC was neutral and could be best described as facilitatory as far as securing funding and administering project activities (i.e., meetings, research, and publication). Focal questions researched by the project were chosen through the consensus process by the science advisory committee. Questions chosen in part reflected MEIFW SMB and alewife data that was available and had not been analyzed. The two primary questions were (1) do adult alewives eat age-0 SMB and/or compete with age-0 SMB for food, and (2) do age-0 alewives compete with age-0 SMB for food.

Methods

Alewife Effects on Smallmouth Bass Condition and Growth

Between 1989 and 2005, MEIFW collected scale samples and/or length–weight data from SMB in the lower St. Croix drainage, as well as from seven other lakes in eastern Maine (Figure 3). Alewives were absent from four lakes, and the alewife status of lakes in the St. Croix drainage changed through time (Table 1). Density data were not available for alewives or SMB in most study lakes. The exception was Woodland Flowage where alewife counts had been conducted at the Milltown Fishway since 1981. Densities ranged from an average high of 639 alewives/ha (259/acre; 1991–1994) to 0 (1995–2000) to a current average of 8 alewives/ha (3/acre; 2001–2005). All lakes were located in MEIFW resource management region C, including Washington County, Hancock County, and part of Penobscot County. Lakes ranged in size from 5 to 13 km^2. Six lakes were mesotrophic (http://pearl.maine.edu/), and the remainder were oligotrophic.

The condition analysis assessed the correlation between a low length–weight ratio for SMB at the end of their first year of growth and the presence of alewives. Condition data were collected by electrofishing (backpack or boat) of up to 50 age-0 SMB per year in early fall after growth for the year had ceased. One day of effort was expended per lake in a variety of locations; thus, 50 fish were not always caught. Fish were returned to the laboratory alive, sacrificed, measured, and weighed (0.01 g). Large fish (>85 mm) suspected of being age 1+ were aged by scale and excluded from the data set if found to be older than 0+. Condition of age-0 SMB was expressed as Fulton's condition factor (K_b) (Anderson and Neumann 1996) and percent of relative weight (W_r) (Murphy et al. 1990, 1991). W_r is a better metric than Kb for between-lake comparisons, although the calculation of W_r is more complicated (Murphy et al. 1991). Statistical analysis of condition data consisted of *t*-tests of alewife versus no alewife lakes and one-way analysis of variance (ANOVA) on the average annual condition of age-0 SMB in a lake (Systat 12, Systat Software, Inc., 2007). W_r was square root transformed and K_b was log (x + 1) transformed to approximate a normal distribution.

Woodland Flowage was omitted from the condition ANOVA and treated separately in its own analysis because it neither fell into the "alewife" nor the "no alewife" category. Differences in K_b associated with alewife spawning escapement were explored with a one-way ANOVA. Spawning escapement (i.e., the number of alewives either allowed to pass Woodland Dam or stocked into Woodland Flowage at Hanson Cove by DFO) and K_b were log (x + 1) transformed.

The growth analysis looked for a correlation between low SMB growth, between ages 1–3, and the presence of alewives. Growth data (i.e., back-calculated length from scales) were collected from age 1+ SMB in the spring before growth commenced for that year. Approximately 100 SMB were angled from each lake (Big Lake, Grand Falls Flowage and Woodland Flowage; see Table 1 for years of available data). One day of effort was expended per lake in a variety of locations; 100 fish were not always caught. Scales were aged, and distance between annuli and scale radius measured for fish up to age 5 only, in accordance with MEIFW protocols.

Statistical analysis of SMB growth data were performed on estimated length-at-age. Length-at-age was calculated using the scale back-calculation formula:

$$L_{ij} = 21 + \left(\frac{A_{ij}}{r_j}\right) \times (L_j - 21) \quad , \qquad (1)$$

where L_{ij} is the length-at-annuli (length-at-age) of fish j at annulus i, A_{ij} is the distance between annuli i and the focus of fish scale j, r_j is the radius length of fish scale j, and L_j is the length at capture of fish j. Growth was calculated by

$$G_{ji} = L_{ji} + 1 - L_{ji}, \qquad (2)$$

where G_{ij} is the growth of fish j at annuli i and L_{ji} is the estimated length of fish j at annuli i. Using this method, growth data were available as far back as 1984 for 5-year-old fish caught in Grand Falls Flowage in 1989, 1985 for fish caught in Woodland Flowage in 1990, and 1986 for fish caught in Big Lake in 1991.

Big Lake and Grand Falls and Woodland

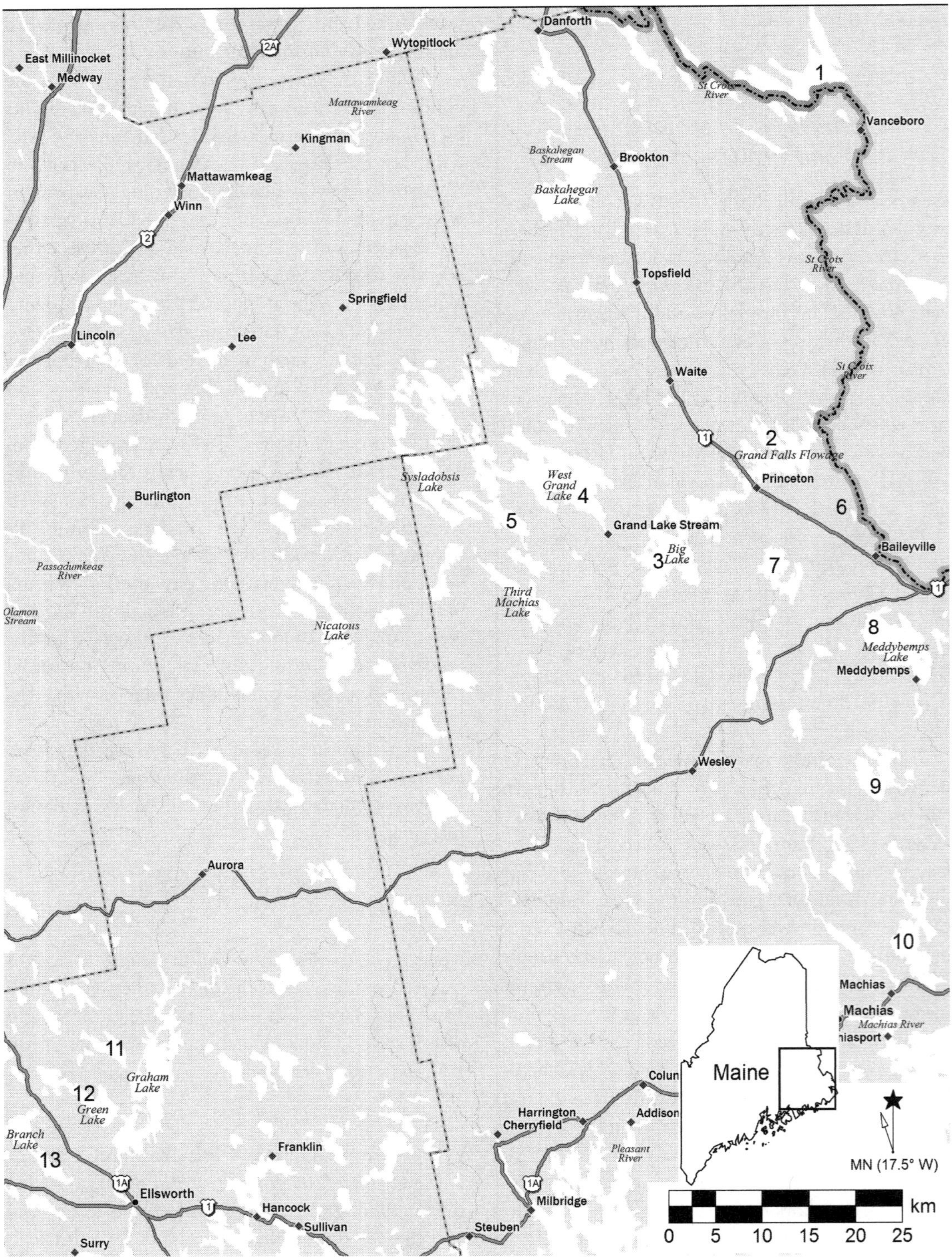

Figure 3.—Map of region with lakes sampled. 1. Spednic, 2. Grand Falls, 3. Big, 4. West Grand, 5. Pocumcus, 6. Woodland, 7. Pocomoonshine, 8. Meddybemps, 9. Cathance, 10. Gardner, 11. Beech Hill, 12. Green, 13. Branch.

Table 1.—Background information on lakes included in the smallmouth bass (SMB) condition, growth, and diet habit analyses (http://pearl.maine.edu). Adult SMB data were used for the growth analysis; age-0 SMB data were used in the condition analysis. Diet habit data were collected in 2005 from Big, Grand Falls, Pocumcus without alewives (ALE) and Cathance, Gardner, Meddybemps, Woodland with alewives. Trophic index is based on Secchi depth; the Chlorophyll-*a* trophic index is reported for Pocomoonshine Lake. Oligo = oligotrophic, and mes = mesotrophic.

Lake	Watershed	ALE present	Area (km^2)	Mean depth (m)	Trophic state	Years adult SMB sampled	Years age-0 SMB sampled
Beech Hill	Union	No	5.47	13.4	oligo	–	1999–2002
Big	St. Croix	Until 1990	41.7	3.7	meso	1991, 1993, 1995, 1997, 1999, 2001, 2003	1994–2005
Branch	Union	No	10.94	11.9	oligo	–	1995–2002
Cathance	Dennys	Yes	11.76	7.3	oligo	–	1995–2003
Gardner	East Machias	Yes	15.05	12.2	–	–	–
Grand Falls	St. Croix	Until 1990	27.08	3.0	meso	1989, 1993, 1995, 1997, 1999, 2001, 2003	1994–2005
Green	Union	No	12.1	13.4	oligo	–	1994–1998, 2000–2003
Meddybemps	Dennys	Yes	27.38	4.3	meso	–	1992, 1995–2003
Pocomoonshine	East Machias	Yes	9.97	4.3	meso	–	1992, 1995–2003
Pocumcus	St. Croix	No	8.95	7.6	–	–	–
West Grand	St. Croix	No	58.03	11.3	meso	–	1994–2005
Woodland	St. Croix	Yes (to 1994; 2001 to 2005), No (1995–2000)	4.75	4.6	–	1990, 1994, 1997, 2000–2004	1994–1995, 1997, 1999, 2001–2005

flowages underwent a series of sequential closures in the late 1980s and early 1990s so that each lake had scale samples collected from fish grown in the presence and the absence of alewives. Age-1 through age-3 SMB from each lake were tested separately to quantify variability in growth between alewife and no alewife years. Nonparametric methods were used because growth at age t was autocorrelated with growth at age $t - 1$. Statistical analysis was performed on Big Lake and Grand Falls using the Mann–Whitney U-test (Sokal and Rohlf 1995). Woodland Flowage had four distinct periods of alewife access between 1985 and 2005 (Table 1). Each time period was analyzed as a separate treatment in a 4×1 Kruskal-Wallis one-way nonparametric ANOVA for ages 1–3.

Comparison of Smallmouth Bass and Alewife Diet Habits

Seven lakes were chosen to conduct the diet analysis portion of the project. Four lakes with alewives and three lakes without were matched for size, average depth, productivity, and other limnological characteristics (Table 1). Alewives have free access to Cathance, Meddybemps, and Gardner lakes via Denil (Gardner and Meddybemps) or Alaskan steep-pass (Cathance and Meddybemps) fish ladders. Alewives were stocked into Woodland Flowage up to a density of 15 fish/ha (~7,100 alewives) by DFO, starting in 2001.

Adult alewife diets.—Adult alewife diets were collected to estimate what percentage of the diet consisted of fish, especially members of the family Centrarchidae. Adult alewives were targeted in Cathance, Meddybemps, and Gardner lakes and Woodland Flowage using nocturnal sets of stationary gear between June 25 and July 26, 2005. Trammel nets (1.8 m tall, 30.5 m long, 2.5-cm interior mesh, 30.4-cm outer mesh) were set three times between dusk and dawn for 30 min on at least four nights in each lake. Captured alewives were measured (mm) and weighed (g) on site. Diet samples were collected via gastric lavage (Hartleb and Moring 1995), flushing the gut contents into an 80-μm brass sieve. Diet contents were then washed into a 236.5-mL (8 oz.) plastic cup and preserved in a solution of 80% alcohol, 15% water ,and 5% polyethylene glycol (F13) (Warmington et al. 2000). In 24–36 h, samples were concentrated into a 90-mL plastic vial with 70% ethanol for long-term storage. Fifteen alewives were sacrificed and their stomachs removed to verify the effectiveness of gastric lavage. In all cases, gastric lavage cleared the foregut but left the contents of the hind gut largely intact. Hind gut contents were usually a single concentrated pellet of mostly digested material. Most of this material was not identifiable, so the gastric lavage protocol was considered effective with the stipulation that only the foregut was being sampled.

Diet contents were identified to family when possible, and in most cases, all organisms were counted. In cases where zooplankton were too numerous to count individually, three 2-mL subsamples were taken, all organisms of the focal taxa were counted, and the results averaged. Number of organisms was estimated by scaling the average number of organisms per 2-mL subsample up to the volume of fluid in the container prior to subsampling. Five representative diet items per fish from each category of diet items identified were measured to the nearest 0.01 mm with a stereo dissecting scope equipped with a 10× optical micrometer. Lengths of diet items were used to calculate an average length of prey ingested for each category of diet item identified. Diet item dry mass was calculated from length measurements using published formulas and conversions (Dumont et al. 1975; Rosen 1981; Culver et al. 1985; Yan and Mackie 1987; Benke et al. 1999; Johnston and Cunjak 1999; Nicol and Mackauer 1999; Dole-Olivier et al. 2000; Sabates et al. 2003).

Age-0 alewife and smallmouth bass diets.—Age-0 alewife and age-0 SMB bass diets were collected to assess degree of food resource overlap between the two species. Alewife and SMB age 0 were fished using a beach seine (1.2 m tall, 15.25 m long, 6.4-mm mesh). Lakes were sampled every 2 to 3 weeks between July 27 and September 9 during daylight hours (Table 2). Cathance Lake samples were analyzed in two periods, August 23 and September 8. All Gardner lake samples were combined; only five age-0 SMB were collected on September 7, so this date was combined with August 1. Age-0 SMB collected on August 18 were significantly longer than those collected on July 27 and August 3, which did not differ from each other; thus, the two earlier samples were combined. All Woodland dates were combined because length of SMB collected on three dates did not differ significantly.

Fish processing included lethal sampling of alewife and SMB age 0 in the field; subsequent data collection occurred in the laboratory. Fish caught in the field were immediately preserved in F13 preservative solution. Before diets were transferred to 70% ethanol for long-term storage, fish were removed from the preservative, measured to the nearest mm, and weighed to the nearest 0.1 g. The body cavity of the fish was opened and the stomach excised and rinsed into a 2-mL vial containing 70% ethanol. The goal was to excise the stomachs of 25 fish from each species for each sampling date. In some cases, the total catch of age-0 SMB or age-0 alewife was haphazardly subsampled to arrive at 25 diet sam-

Table 2.—Dates and location (lakes) of adult alewife, age-0 smallmouth bass, and age-0 alewife diet habit data collection.

Lake	Sample dates (2005)
Cathance	6/24, 7/6, 7/19, 7/26, 8/23, 8/8
Gardner	6/27, 7/11, 7/14, 7/18, 7/28, 8/17, 9/7
Meddybemps	6/23, 6/28, 7/8, 7/13, 7/20, 7/27, 8/3, 8/18, 9/6
Woodland	6/29, 6/30, 7/7, 7/12, 7/21, 8/1, 8/10, 8/19, 8/24, 9/9
Big	8/4, 8/24
Grand Falls	8/8, 8/9, 8/12, 8/30
Pocumcus	8/5, 8/31

ples. When less than 25 fish were available, diets were processed from all the fish collected. Diet processing, including estimating item dry mass, was the same as adult alewives.

Comparisons of diet habits were facilitated by calculating the index of relative importance (%IRI) for each species and then calculating Schoener's index for each sampling period. The IRI combines frequency of occurrence as percent occurrence (%O), diet category weight as percent weight (%W) and diet category numerical occurrence as percent of number (%N) (from Liao et al. 2002). Competition potential for the same food resources between age-0 alewife and age-0 SMB was assessed using Schoener's index (Schoener 1970). Schoener's index compares differences in the proportions of prey comparison units and sums across all units to arrive at an index value that is a measure of diet overlap between two species or groups of organisms. Schoener's index is calculated as

$$\alpha = 1 - 0.5 \times \left(\sum_{i=1}^{n} \left| P_{xi} - P_{yi} \right| \right), \qquad (3)$$

where P_{xi} is the proportion of food item (taxa) i in the diet of age-0 alewives, P_{yi} is the proportion of food item i in the diet of age-0 SMB, and n is the number of prey categories (Kahilainen and Ostbye 2006). Here, Schoener's index is expressed as a percentage (i.e., $\alpha \times 100 = \%\alpha$) and IRI values were substituted for P. A value of 0% indicates no diet overlap, and a value of 100% indicates complete diet overlap. It is assumed that fish diet overlap must be greater than or equal to 60% in order to be biologically significant, a value that has been widely accepted in scientific literature (Zaret and Rand 1971; Wallace 1981; Willis et al. 2002; Kahilainen and Ostbye 2006; Kahl and Radke 2006).

Results

Alewife Effects on Smallmouth Bass Condition and Growth

There was no difference in age-0 SMB condition between the groups of lakes with alewives present or alewives absent and no indication that high or low condition values from a subset of lakes affected the analysis for either W_r or K_b. Condition was significantly different between lakes for both W_r ($p < 0.001$, $F = 6.26$, $n = 85$) and K_b ($p < 0.001$, $F = 5.87$, $n = 85$). SMB from Grand Falls Flowage had the highest condition, followed by Cathance and Meddybemps lakes. Green Lake SMB had the fourth highest condition; SMB from West Grand Lake had the lowest condition.

Stocking rate (i.e., spawning escapement) did not correlate with increased or decreased SMB condition in Woodland Flowage. Linear regression of SMB log (K_b + 1) against alewife stocking rate was not statistically significant ($p = 0.1$, $F = 2.8$, $n = 9$). K_b was lowest in 1994, at a stocking level of 796 alewives/ha, followed by 2003, at a stocking level of 14 alewives/ha. K_b was highest in 2002, at 2 alewives/ha, followed by 2005 at 15 alewives/ha.

There was no evidence that the presence of alewives retarded SMB growth between the ages of 1 and 3. In Big Lake (age 2) and Grand Falls (age 1 and age 3) SMB growth was significantly higher ($p < 0.05$), coincident with years when alewives were allowed passage into the lake (Figure 4; Table 3). In Woodland Flowage, there was no statistically significant difference in growth between periods when the St. Croix River was open, when access to spawning area was reduced, when the river was closed to alewives, or when DFO stocked alewives. Age-1 SMB did have a higher average growth in Woodland for 2001–2005, during DFO stocking, but this was not statistically significant (Figure 4; Table 3).

Comparison of Smallmouth Bass and Alewife Diet Habits

Adult alewife diets.—Fish made up a very small proportion of the diet of the adult alewives collected, less than 0.28% of the diet. Out of 70 adult alewives sampled from three lakes, 10% had fish in their guts. Most fish prey that were found in alewife stomachs had the general body characteristics of either larval stage rainbow smelt *Osmerus mordax* or larval stage alewives. Adult alewives consumed a range of diet items once in freshwater, including pelagic plankton (Daphnidae, Polyphemidae, and Copepoda) (Table 4). Adult alewives also ate dipteran pupae and trichopteran larvae. Most adult alewives were caught in the cove nearest the lake outlet, presumably the point of access to the lake.

Age-0 alewife and smallmouth bass diets.—Zooplankton were the most frequently found diet item for both species for all lakes, but there was con-

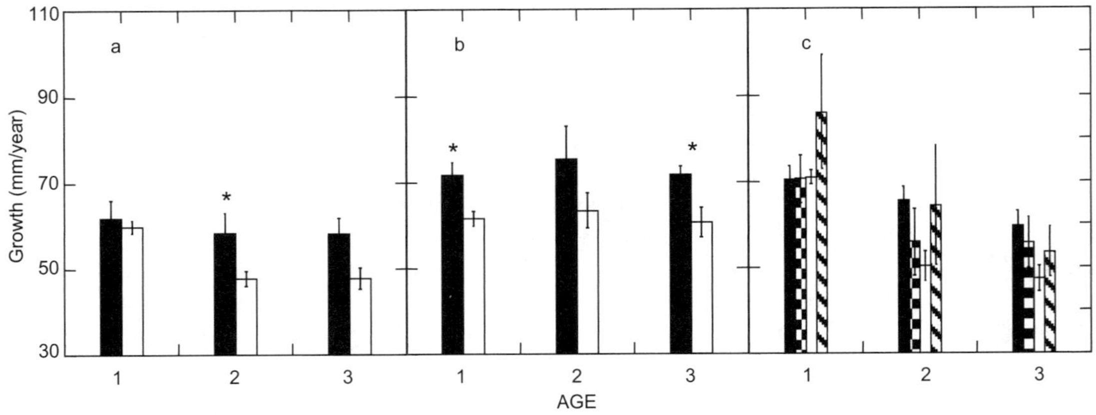

Figure 4.—Smallmouth bass growth (mm/year) for (A) Big Lake, (B) Grand Falls Flowage, and (C) Woodland Flowage. Black bars represent years in which alewives were present; white bars represent years when alewives were excluded. Error bars are 1 SE. Additional categories for Woodland included alewives reduced (block), when alewife spawning habitat was reduced to Woodland only, and alewives stocked (hash), when Fisheries and Oceans Canada trucked alewifes around Woodland Dam. * = $p < 0.05$.

siderable variation in age-0 diet by lake. Both age-0 alewives and SMB were caught in mixed schools, presumably feeding alongside each other. However, age-0 SMB, more than alewives, diversified their diets with other invertebrates (e.g., mayfly larvae, midge larvae, and adults of these invertebrates; Table 5). Of the four lakes sampled with sympatric populations of SMB and alewives, biologically significant diet overlap (Schoener's index > 60%) was detected only in Meddybemps Lake, on the first sample date (Table 6). Meddybemps Lake age-0 alewives foraged almost exclusively on cladocerans (Family Sididae) at the time of the first (98%) and second (93%) sampling. Age-0 SMB also ate Sididae (69%), along with other cladocerans (28%). By the second sampling, age-0 SMB in Meddybemps Lake were still eating Sididae (40%) and another zooplankter (Family Moinidae, 17%), but were also eating fish (41%).

Table 3.—Results of Kruskal-Wallis tests of smallmouth bass (SMB) growth in years with alewives present and years with alewives (ALE) absent for SMB of ages 1–3. Data were collected by Maine Department of Inland Fisheries & Wildlife between 1981 and 2004. Sample sizes, as number of years in which SMB scales were collected, appear below the p values. ALE reduced (1988–1994) and ALE stocked (2001–2005) refers to alewife management periods in Table 1. See Figure 1 for detailed explanation. * = significant test.

Lake	Age 1	Age 2	Age 3
Big	$p = 0.5$	$p = 0.02$*	$p = 0.06$
ALE present	5	4	3
ALE absent	11	12	12
Grand Falls	$p = 0.01$*	$p = 0.2$	$p = 0.03$*
ALE present	7	6	4
ALE absent	10	11	12
Woodland	$p = 0.2$	$p = 0.15$	$p = 0.14$
ALE present	6	5	3
ALE reduced	4	4	4
ALE absent	6	6	6
ALE stocked	3	3	3

Table 4.—Major diet items based on the %IRI (index of relative importance) method for adult alewives, sorted by lake. Diets were collected from alewives in the lake during the 2005 alewife spawning season. *n* = sample size.

Diet taxa	Gardner	Meddybemps	Woodland
Daphnidae	1.6	82.9	0.8
Polyphemidae	2.1	–	56.7
Diptera pupae	43.9	0.2	4.7
Trichoptera	10.6	0.1	26.7
Copepoda	13.6	5.2	0.1
Bosminidae	5.6	N/A	2.6
Sididae	5.6	–	–
other	17.1	10.6	8.3
n	38	21	11

The diet of allopatric age-0 SMB was similar to the diets of SMB that occurred with alewives. In all lakes, zooplankton made up a large proportion of the diet. Benthic invertebrates became more important in subsequent samples, as did fish prey (Table 7).

Discussion

Alewife Effects on Smallmouth Bass Condition and Growth

Analyses completed for this project did not indicate that a systematic decline occurred in SMB popu-

Table 5.—Major diet items, based on the %IRI (index of relative importance) method, for age-0 alewives and age-0 smallmouth bass in lakes with sympatric populations. Diets were collected from alewives in the lakes during the 2005 growing season. CA = Cathance, GA = Gardner, ME = Meddybemps, WO = Woodland. A = alewife, S = smallmouth bass. Numbers following A or S refer to sampling period. *n* = sample size.

Diet taxa	CA A1	CA A2	CA S1	CA S2	GA A	GA S	ME A1	ME A2	ME S1	ME S2	WO A1	WO S1
Sididae	0	0	1.2	24.1	0.3	53.1	97.8	93.0	69.3	40.0	0	1.0
Chaoborus	62.8	76.9	0	0.4	0.2	0	0	0	0	0	0	0
Chydoridae	0.6	0.1	68.8	0.1	36.8	2.6	0.1	0.6	0	0	0	0
Diptera adult	0.4	0	0.4	4.9	1.4	0.1	0	0	0	0	22.0	49.0
Bosminidae	2.3	0.5	0	0	56.9	0	0	0	0	0	0	0
Ephemeroptera	0.1	0.3	2.6	0.5	0	37.6	0	0	0	0.5	0	17.0
Fish	0	0	1.5	8.5	0	1.1	0	0	0	40.6	0	3.9
Polyphemidae	0	13.0	5.0	32.9	0.1	0	0	0.1	0	0	0	0
Simulidae	0	0	0	0	0	0	0	0	0	0	48.7	2.2
Trichoptera	0	0	0.2	0	0	0.1	0.1	0.1	0	0.5	21.7	17.1
Daphnidae	29.6	0.2	7.2	0.4	0	0	0	0	0	0	0	0.2
Zygoptera	0	0	0	27.9	0	0.1	0	0	0	0	0	0.3
Cladocera	0	0	0	0	0	0	0	0	27.6	0.5	0	0
Chironomidae	0.3	0	0.4	0	1.3	2.1	0.2	2.7	2.8	0.2	5.4	5.3
Moinidae	0	0	0	0	0	1.6	0	0	0	17.1	0	0
Holopedidae	0.2	8.1	0	0	1.3	0	0	0	0	0	0	0
Terrestrial insect	0	0	5.9	0	0	0	0	0	0	0.1	0	0
Plecoptera adult	0	0	5.4	0	0	0	0	0	0	0	0	0.3
Other	3.7	0.8	1.5	0.3	1.7	1.5	1.6	3.5	0.2	0.4	2.3	3.8
n	25	25	9	10	35	17	50	8	7	28	7	36

Table 6.—Schoener's index of diet overlap values for age-0 alewives (ALE) and age-0 smallmouth bass (SMB) and average length of fish sampled (SD). Data are presented as percent diet overlap between alewives and bass by sample period and lake. Values greater than 60% are considered biologically relevant.

Lake	First sampling	Species data	Second sampling	Species data
Cathance	8.7%	ALE: 58.7 mm (9.8) SMB: 49.4 mm (4.8)	14.1 %	ALE: 63.9 mm (12.1) SMB: 60.2 mm (3.7)
Gardner	4.3%	ALE: 44.1 mm (8.4) SMB: 52.6 mm (6.0)		
Meddybemps	69.8%	ALE: 33.9 mm (3.5) SMB: 40.1 mm (5.0)	40.4%	ALE: 35.6 mm (1.3) SMB: 62.6 mm (10.5)
Woodland	46.8%	ALE: 79.4 mm (8.4) SMB: 59.0 mm (7.4)		
Big		SMB: 45.0 mm (4.0)		SMB: 52.2 mm (5.1)
Grand Falls		SMB: 46.9 mm (10.7)		SMB: 54.7 mm (5.8)
Pocumcus		SMB: 42.6 mm (5.4)		SMB: 56.0 mm (7.3)

lation performance in the presence of anadromous alewives. The project predicted that size-structured interactions would inhibit age-0 SMB feeding, as reflected in growth, condition, and diet overlap measures (i.e., adult alewives would eat age-0 SMB and/or compete with age-0 SMB for food and age-0 alewives would compete with age-0 SMB for food). Interspecific competition between predator and prey fish can lead to a size-structured interaction (Bystrom et al. 1998; Olson et al. 1995) when prey impedes the growth of predator. Size-structured interactions are an acute problem for piscivorous fishes in northern latitudes, like St. Croix River SMB, because a minimum growth threshold (~60 mm) is required for overwinter survival (Shuter et al. 1980; Shuter and Post 1990; Curry et al. 2005). Reaching the minimum overwinter size is often associated with an ontogenetic shift to piscivory (Ludsin and DeVries 1997; Olson et al. 1998), which may in turn depend on the availability of smaller high-quality

Table 7.—Major diet items, based on the %IRI (index of relative importance) method, for age-0 smallmouth bass (SMB) in lakes without alewives. Diets were collected from SMB in the lakes during the 2005 growing season. GF = Grand Falls; Pocu = Pocumcus. Number following lake name refers to sampling period. *n* = sample size.

Diet taxa	Big 1	Big 2	GF 1	GF 2	Pocu 1	Pocu 2
Sididae	0	21.8	1.7	35.2	43.7	60.7
Ephemeroptera	0.1	20.2	5.6	58.6	0.1	1.7
Diptera pupae	62.2	0.2	0	0.8	3.1	0
Fish	1.1	0.9	0	1.4	37.4	19.3
Polyphemidae	0	5.4	40.7	0	7.3	1.9
Diptera adult	2.3	28.0	13.3	0.1	0	0.1
Daphnidae	29.0	0	0.5	0	0	0.1
Chironomidae	0	6.7	0.4	0.9	3.3	5.3
Hymenoptera	2.1	0.1	13.3	0	0	0
Trichoptera	0.1	0.5	11.3	1.1	0.1	1.0
Chydoridae	0	4.9	1.5	0.1	2.6	2.7
Hemiptera	0	0	8.9	0	0	1.5
Other	2.9	11.1	2.7	1.8	2.5	5.6
n	25	24	8	25	9	25

prey (Galarowicz and Wahl 2005). The results presented here do not suggest that alewives negatively affect age-0 SMB across the region as a whole or in a lake where alewives were intermittently introduced and removed.

There were, however, some limitations to the study. Existing data from MEIFW came from a variety of sampling efforts that were started in the late 1980s and early 1990s. In many cases, the sampling programs were a response to a previously identified problem. The methodology used reflected what was known in the fisheries field at the time. The depth of sampling also reflected the resources (hours and dollars) available to contribute to any one sampling need in light of regular management duties. Ancillary data that would have been useful in 2005 was often not available from years when samples were collected (e.g., annual spawning escapement, diet habits, prey abundance). For that part of the project executed in 2005, the number of sampling activities limited the depth of information that could be collected from any one water body, and the number of lakes sampled dictated how much time could be committed to collecting fish in different parts of the lake. For example, alewife spawning escapement numbers were not available for any specific water body other than Woodland Flowage, and compromises had to be made in sampling for adult alewife versus age-0 alewife versus age-0 SMB. The decision to cover a wide temporal and spatial breadth of lakes means that this study cannot suggest threshold stocking or escapement levels at which alewives and SMB interactions change or address the implications of diet overlap that might occur at the larval or black fry stage of development.

Despite these caveats, the available data yielded some insights into how the presence of alewives affected SMB population performance. Condition between groups of lakes with and without alewives was not significant, possibly due to significant differences in condition between lakes within groups. The lake with the highest condition was a no-alewife lake, followed by two alewife lakes. Individual lake characteristics appeared to be better determinants of condition than alewife presence or absence. Within the same lake, Woodland Flowage, annual variation was a stronger determinant of bass condition than alewife stocking rates. Negative effects of alewives on the growth of SMB for ages 1–3 were not apparent; there was no difference in bass growth whether alewives were present or absent. Conversely, growth was equal or greater for SMB in the presence of alewives in Big Lake and Grand Falls Flowage. Results of the current analysis do not support the presumed negative impacts of alewives on SMB population performance.

Comparison of Smallmouth Bass and Alewife Diet Habits

The adult and age-0 alewife diet analyses did not identify direct (predation) and/or indirect (size structured) interactions between alewife and SMB populations. Observations from Spednic Lake were the genesis for this research question. MEIFW and NBDNR observed low SMB survival between the black fry and fall age-0 stage, concurrent with increasing alewife returns and high age-0 alewife production (R. Jordan, Maine Inland Fisheries and Wildlife, personal communication). Diets of adult alewives indicated that fish prey made up an extremely small percentage of the diet. These results agreed with earlier results from Maine lakes (Jordan 1989). Based on the available data, one would conclude that adult alewives are not a significant predator of age-0 SMB.

Age-0 alewives and age-0 SMB diet data were examined for biologically significant diet overlap that might be indicative of a competitive interaction. The extent of diet overlap did not support this hypothesis for the lakes sampled. Schoener's index was less than 60% in most lakes, below the threshold for biologically significant interaction. The late June sampling of Meddybemps Lake produced the only biologically significant diet overlap of the study (69%). Zooplankton in the family Sididae were a common diet item of both species. Ironically, SMB and alewives have coexisted in Meddybemps Lake since 1877 (Warner 2005). Given the long history of sympatry between these two species, it is likely that Sididae were a shared diet item because they were the most prevalent food resource in the lake.

Late summer may represent the most obvious window for age-0 alewife and age-0 SMB to compete for food resources. In late summer, the decline in planktonic prey drives alewives into the littoral zone. Benthic food items (e.g., chironimids, ostracods, benthic zooplankton) become more prevalent in the guts of age-0 anadromous alewives (Vigerstad and Cobb 1978; Gregory et al. 1983; Jessop 1990;

Hanson and Curry 2005). However, overlap of bass and alewives in the littoral zone may be relatively short and not entirely detrimental. Some research indicates that the decline in planktonic prey is a cue for age-0 alewives to begin moving towards the ocean (Vigerstad and Cobb 1978; Yako et al. 2002). Although interactions between age-0 alewives and age-0 SMB could reduce bass growth if alewives out-compete bass for important intermediate prey resources, it is also possible for alewives spawned late in the season to become a prey resource themselves (Hanson and Curry 2005).

Alewife Management Issues in St. Croix River

The expectation of science in a dispute is that investigation will result in proof and an immediate resolution of the conflict. However, when science is brought in before the political process has identified the most important commonalities and objections, science can delay action on an issue (Oreskes 2004). Using scientific inquiry to help shape policy is considered, in some circles, a gross misrepresentation of how science and policy should intersect (Oreskes 2004). Two entities (people, scientists, agencies, etc.) interpreting the same set of results with incongruent a priori objectives will inevitably arrive at different conclusions (Oreskes 2004; Sarewitz 2004). Clarifying value disputes, reflected in study objectives agreed upon by the stakeholders, is considered a more appropriate, constructive, and less inflammatory use of scientific inquiry (Sarewitz 2004).

This project was not the first study to address the effects of alewives on freshwater ecosystems, but rather represented a second round of inquiry focused on anadromous alewife interactions with freshwater ecosystems. A 10-year study by MEIFW and MEDMR found that anadromous alewives had no statistically significant negative effects on freshwater game fish, forage fish, or water quality at a stocking rate of six fish per acre (14/ha; Kircheis et al. 2002). MEIFW studies conducted after the closure found that Spednic Lake adult alewives could have competed with age-0 SMB for prey when they were sympatric (Jordan 1988) and that excluding alewives from Spednic Lake resulted in recovery of the bass population (Figure 5; Smith 1988, 1998). Alewife proponents and opponents maintained that evidence contrary to their position lacked credibility. Neither of the two Spednic Lake studies nor Kircheis et al. (2002) addressed a widely agreed upon set of objectives. Supporting and opposing positions taken by the groups, and their associated actions, were based in strong commitments to collective goals and values (Sarewitz 2004). The alewife opposition claimed that no "adequate biological data about the impact on other species" was available, an argument that convinced the Maine legislators to maintain fishway closures in 2001. In response, DFO demonstrated their commitment to reestablishing alewives in the St. Croix River by commencing "trap & truck" operations to stock alewives in the first closed section of the river, Woodland Flowage.

Transparency and good faith negotiation are vital to the success of any collaborative effort, especially one as contentious as the St. Croix alewife issue. The 2001 legislative defeat forced alewife proponents to negotiate in earnest with alewife opponents. Negotiations prior to and immediately after the failure of the 2001 bill identified the most pertinent issues in the alewife–smallmouth bass debate for the St. Croix River stakeholders. Alewife proponents lead the efforts to form an ad hoc science committee to secure funding and to develop an implementation plan. Both opponents and proponents were represented on the science committee, either directly or through an agency (i.e., MEIFW represented the interests of the EMFFG). A contractor, naïve to the 20-year history of the St. Croix debate, was hired to execute the study. The implementation plan included studies of juvenile and adult SMB, anadromous alewives, landlocked alewives, and other freshwater species, along with a regional meta-analysis of data from tournament lakes where bass–alewife interactions might occur. The project, as it was originally conceived, was too big to complete in one year. The questions addressed here were deemed to be the most important and most achievable questions at the time. The study results did resolve MEDMR and MEIFW differences over alewife impacts, but focusing on SMB only, and not addressing what happened in Spednic Lake 20 years earlier, hardened the stance of the EMFFG.

The expectations from a stakeholder/science process should match the scope of the project. Unrealistic expectations will negatively affect the outcome because participants will be less willing to accept the results (Germain et al. 2001; Marshall et

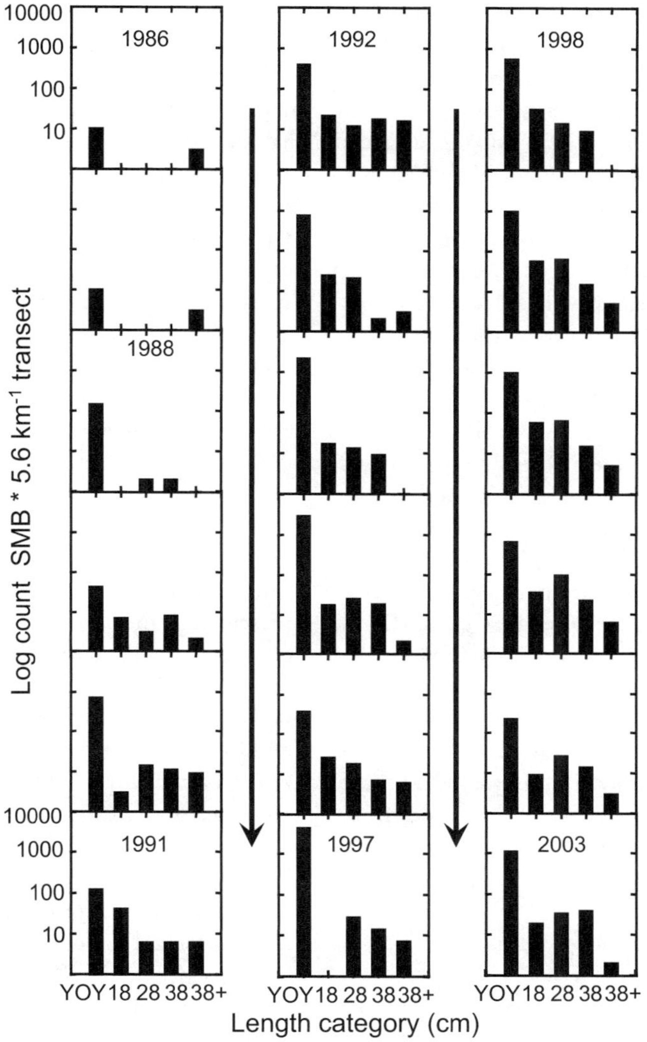

Figure 5.—Progressive time series of smallmouth bass (SMB) size-classes for Spednic Lake. Counts are averages per transect for at least three 5.6-km transects surveyed per year. Size-classes roughly correspond to alewife age-classes. YOY = young of year or age 0.

al. 2007). Despite what appeared to be an inclusive process, the EMFFG raised several more study objectives after data collection began in 2005. These included alewife–landlocked salmon interactions and whether alewives could negotiate the two lowest falls in the river. The EMFFG argued that these additional objectives warranted inclusion, but were ignored. Also, they claimed that the project sponsor, a conservation-focused nonprofit, biased the results towards justifying alewife reintroduction. The omitted research issues and bias claims were central to the EMFFG arguments to keep the fishways closed in 2008 when a second attempt to repeal the 1995 alewife exclusion legislation came before the Maine State Legislature. By attempting to change the terms of the St. Croix debate to include species in addition to SMB, the EMFFG may have been disingenuous in their participation. Conversely, the project organizers may have misinterpreted the EMFFG's interests in the process.

Investigation of Spednic Lake is an example of how scientific investigation can do as much to inflame a controversy as provide solutions. Even as of 2008, Spednic Lake was still the premiere example for alewife opponents concerning the harmful effects that alewives can have on sport fish. Year-class failures had been occurring for several years prior to 1984, when the scuba surveys began. Remediation was accomplished through stocking, alewife exclusion, new water control procedures, and changes in fishing regulations. For example, prior to 1988, water releases during bass spawning and rearing periods changed lake water levels by more than a meter in half the years between 1971 and 1986 (SCRPPT 1987), three times the threshold for negative impacts to SMB production (Neves 1975). Any of these could be credited with restoring the Spednic SMB population, but regional managers only credited excluding alewives (Smith 1988, 1998). Spednic Lake continues to be a quandary: the EMFFG do not consider studies that occur outside of the lake to effectively address their alewife issues; there are no plans to reintroduce alewives to the lake (nor does the EMFFG want them), and therefore, it is impossible to test the EMFFG's hypotheses about alewife–SMB interactions, and not enough data are available from the 1970–1985 period to provide definitive answers. The stigma on alewives from the Spednic Lake SMB crash may prevent the two sides from reaching a mutually agreeable solution that would reintroduce alewives to the St. Croix River.

Conclusions

This project was an important step in bringing the St. Croix controversy closer to resolution. The results challenged the position of alewife opponents within Maine natural resource management agencies, removing some of the institutional barriers to reintroduction. Failing to find negative impacts in MEIFW's data were largely responsible for changing MEIFW's position. The process of negotiation around a common table also helped reduce the rhetoric and facilitate compromise between natural resource agencies. The project, however, had a much smaller effect on the surrounding politics. "An Act to Restore Diadromous Fish in the St. Croix River" (LD 1957, 123rd Maine Legislature) as originally written, would have restored alewives to 52% or their 1987 spawning habitat if passed in 2008. Public hearings lasted 7 h. Maine Fishing Guides opposed the bill, citing the same reasons given in 2001. The Washington County, Maine legislative delegation (1of 35 senators, 5 of 153 representatives) aligned with the MEFFG before the public hearing. The Passamaquoddy Indian Nation, who have some jurisdiction over fisheries in the St. Croix River, also sided with the MEFFG. The final amended bill, brokered behind closed doors, opened the Woodland Flowage fishway in 2008, restoring alewives to 2.4% of their 1987 spawning habitat. The stipulation limiting further restoration was that the Grand Falls fishway remain closed until a study of the historical range of alewives could prove that alewives reached reservation lands above Grand Falls before Grand Falls Dam was built in 1913. Any further legislative action will be delayed until after a third round of studies is conducted on alewives in the St. Croix River.

Acknowledgments

With appreciation to the following organizations represented on the project's Scientific Advisory Committee: U.S. National Marine Fisheries Service, U.S. Fish & Wildlife Service, Maine Department of Inland Fisheries & Wildlife, Maine Department of Marine Resources, Canada Department of Fisheries & Oceans, New Brunswick Department of Natural Resource, and St. Croix International Waterway Commission, with data contributed by state and provincial agencies, where indicated. And with gratitude to the following entities for providing funding for these studies: Maine Outdoor Heritage Fund, Gulf of Maine Council on the Marine Environment/NOAA Habitat Restoration Partnership, National Fish and Wildlife Foundation, International Joint Commission, Passamaquoddy Tribe/U.S. Environmental Protection Agency, New Brunswick Wildlife Trust Fund, St. Croix International Waterway Commission, Norcross Wildlife Foundation, and Maine Rivers. I would also like to thank K. Wilson, N. Schalit, L. Sochasky, R. Bradford, P. Cronin, E. Hutchins, R. Jordan, S. Lary, P. Seymour, L. Sochasky, T. Squiers and G. Wippelhauser, W. MacDonald, C. Bohlen, L. Dwyer, K. Elowe, S. Fefer, S. Gephard, J. Kachmar, T. Kelsch, R. Lawrence, B. Townsend, J. Trial, F. Whoriskey, two anonymous reviewers, and—in memory—C. Fay.

References

Agrawal, A., and C. C. Gibson. 1999. Enchantment and disenchantment: the role of community in natural resource conservation. World Development 27(4):629–649.

Anderson, R. O., and R. M. Neumann. 1996. Length, weight, and associated structural indices. Pages 283–299 *in* B. R. Murphy, and D. W. E. Willis, editors. Fisheries techniques, 2nd edition. American Fisheries Society, Bethesda, Maryland.

ASMFC (Atlantic States Marine Fisheries Commission). 2007. Public information document for Amendment 2 to the Interstate Fishery Management Plan for shand and river herring. Available from Atlantic States Marine Fisheries Commission, Washington, D.C.

Benke, A. C., A. D. Huryn, L. A. Smock, and J. B. Wallace. 1999. Length–mass relationships for freshwater macroinvertebrates in North America with particular reference to the southeastern United States. Journal of the North American Benthological Society 18(3):308–343.

Bennett, E., A. Neiland, E. Anang, P. Bannerman, A. A. Rahman, S. Huq, S. Bhuiya, M. Day, M. Fulford-Gardiner, and W. Clerveaux. 2001. Towards a better understanding of conflict management in tropical fisheries: evidence from Ghana, Bangladesh and the Caribbean. Marine Policy 25(5):365–376.

Brooks, J. L., and S. I. Dodson. 1965. Predation body size and composition of plankton. Science 150(3692):28–35.

Bunnell, D. B., C. P. Madenjian, and R. M. Claramunt. 2006. Long-term changes of the Lake Michigan fish community following the reduction of exotic alewife (*Alosa pseudoharengus*). Canadian Journal of Fisheries and Aquatic Sciences 63(11):2434–2446.

Bystrom, P., L. Persson, and E. Wahlstrom. 1998. Competing predators and prey: juvenile bottlenecks in whole-lake experiments. Ecology 79(6):2153–2167.

Culver, D. A., M. M. Boucherle, D. J. Bean, and J. W. Fletcher. 1985. Biomass of fresh-water crustacean zooplankton from length weight regressions. Canadian Journal of Fisheries and Aquatic Sciences 42(8):1380–1390.

Curry, R. A., S. L. Currie, S. K. Arndt, and A. T. Bielak. 2005. Winter survival of age-0 smallmouth bass, *Micropterus dolomieu*, in north eastern lakes. Environmental Biology of Fishes 72(2):111–122.

Dole-Olivier, M. J., D. M. P. Galassi, P. Marmonier, and M. C. Des Chatelliers. 2000. The biology and ecology of lotic microcrustaceans. Freshwater Biology 44(1):63–91.

Dumont, H. J., I. Vandevelde, and S. Dumont. 1975. Dry weight estimate of biomass in a selection of cladocera, copepoda and rotifera from plankton, periphyton and benthos of continental waters. Oecologia 19(1):75–97.

Fisher, J. P., J. D. Fitzsimons, G. F. Combs, and J. M. Spitsbergen. 1996. Naturally occurring thiamine deficiency causing reproductive failure in Finger Lakes Atlantic salmon and Great Lakes lake trout. Transactions of the American Fisheries Society 125:167–178.

Galarowicz, T. L., and D. H. Wahl. 2005. Foraging by a young-of-the-year piscivore: the role of predator size, prey type, and density. Canadian Journal of Fisheries and Aquatic Sciences 62(10):2330–2342.

Germain, R. H., D. W. Floyd, and S. V. Stehman. 2001. Public perceptions of the USDA Forest Service public participation process. Forest Policy and Economics 3(3–4):113–124.

Goode, G. B. 1887. The fisheries and fishery industries of the United States, volumes 1 and 2. Government Printing Office, Washington, D.C.

Gregory, R. S., G. S. Brown, and G. R. Daborn. 1983. Food-habits of young anadromous alewives, *Alosa pseudoharengus*, in Lake Ainslie, Nova Scotia. Canadian Field-Naturalist 97(4):423–426.

Grimble, R., and K. Wellard. 1997. Stakeholder methodologies in natural resource management: a review of principles, contexts, experiences and opportunities. Agricultural Systems 55(2):173–193.

Hanson, S. D., and R. A. Curry. 2005. Effects of size structure on trophic interactions between age-0 smallmouth bass and juvenile anadromous alewives. Transactions of the American Fisheries Society 134:356–368.

Hartleb, C. F., and J. R. Moring. 1995. An improved gastric lavage device for removing stomach contents from live fish. Fisheries Research 24:261–265.

Jessop, B. M. 1990. Diel variation in density, length composition, and feeding-activity of juvenile alewife, *Alosa pseudoharengus* Wilson, and blueback herring, *A. aestivalis* Mitchill, at near-surface depths in a hydroelectric dam impoundment. Journal of Fish Biology 37(5):813–822.

Johnston, T. A., and R. A. Cunjak. 1999. Dry mass-length relationships for benthic insects: a review with new data from Catamaran Brook, New Brunswick, Canada. Freshwater Biology 41(4):653–674.

Jordan, R. M. 1988. Food habits of late summer young of year smallmouth bass in two eastern Maine lakes. Interim Summary Report No. 1 (F-404). Maine Department of Inland Fisheries and Wildlife, Interim Summary Report No. 1 (F-404), Augusta.

Jordan, R. M. 1989. Food habits of post-spawning adult anadromous alewives in freshwater. Maine Department of Inland Fisheries and Wildlife, Interim Summary Report No. 3 (F-409), Augusta.

Kahilainen, K., and K. Ostbye. 2006. Morphological differentiation and resource polymorphism in three sympatric whitefish *Coregonus lavaretus* (L.) forms in a subarctic lake. Journal of Fish Biology 68(1):63–79.

Kahl, U., and R. J. Radke. 2006. Habitat and food resource use of perch and roach in a deep mesotrophic reservoir: enough space to avoid competition? Ecology of Freshwater Fish 15(1):48–56.

Kircheis, F., J. G. Trial, D. P. Boucher, B. Mower, T. Squiers, N. Gray, M. O'Donnell, and J. Stahlnecker. 2002. Analysis of impacts related to the introduction of anadromous alewives into a small freshwater lake in central Maine, USA. Maine Department of Marine Resources, Maine Department of Inland Fisheries and Wildlife, Maine Department of Environmental Protection, Interagency Report Series No. 02-1, Augusta.

Liao, H., C. L. Pierce, and J. G. Larscheid. 2002. Diet dynamics of the adult piscivorous fish community in Spirit Lake, Iowa, USA 1995–1997. Ecology of Freshwater Fish 11(3):178–189.

Ludsin, S. A., and D. R. DeVries. 1997. First-year recruitment of largemouth bass: the interdependency of early life stages. Ecological Applications 7(3):1024–1038.

Marshall, K., R. White, and F. Anke. 2007. Conflicts between humans over wildlife management: on the diversity of stakeholder attitudes and implications for conflict management. Biodiversity and Conservation 16(11):3129–3146.

Marshall, T. 1976. Historical perspectives of resource development branch activities in restoring anadromous fishes to the St. Croix River, New Brunswick–Maine. Department of Fisheries and the Environment, Halifax, Nova Scotia.

Murphy, B. R., M. L. Brown, and T. A. Springer. 1990. Evaluation of the relative weight (W_r) index, with new applications to walleye. North American Journal of Fisheries Management 10:85–97.

Murphy, B. R., D. W. Willis, and T. A. Springer. 1991. The relative weight index in fisheries management—status and needs. Fisheries 16(2):30–38.

Neves, R. J. 1975. Factors affecting fry production of smallmouth bass (*Micropterus dolomieui*) in South Branch Lake, Maine. Transactions of the American Fisheries Society 104:83–87.

Nicol, C. M. Y., and M. Mackauer. 1999. The scaling of body size and mass in a host parasitoid association: influence of host species and stage. Entomologia Experimentalis et Applicata 90(1):83–92.

Odell, J., M. E. Mather, and R. M. Muth. 2005. A biosocial approach for analyzing environmental conflicts: a case study of horseshoe crab allocation. Bioscience 55(9):735–748.

Olson, M. H., S. R. Carpenter, P. Cunningham, S Gafny, B. R. Herwig, N. P. Nibbelink, T. Pellett, C. Storlie, A. S. Trebitz, and K. A. Wilson. 1998. Managing macrophytes to improve fish growth: a multi-lake experiment. Fisheries 23(2):6–12.

Olson, M. H., G. G. Mittelbach, and C. W. Osenberg. 1995. Competition between predator and prey—resource-based mechanisms and implications for stage-structured dynamics. Ecology 76(6):1758–1771.

Oreskes, N. 2004. Science and public policy: what's proof got to do with it? Environmental Science and Policy 7(5):369–383.

Perley, M. 1852. Reports on the sea and river fisheries of New Brunswick, 2nd edition. J. Simpson, Fredericton, New Brunswick, Canada.

Rosen, R. A. 1981. Length–dry weight relationships of some fresh-water zooplankton. Journal of Freshwater Ecology 1(2):225–229.

Sabates, A., S. Rossi, and E. Reyes. 2003. Lipid content in the early life stages of three mesopelagic fishes. Journal of Fish Biology 63(4):881–891.

Sarewitz, D. 2004. How science makes environmental controversies worse. Environmental Science and Policy 7(5):385–403.

Schoener, T. W. 1970. Nonsynchronous spatial overlap of lizards in patchy habitats. Ecology 51(3):408–418.

SCIWC (St. Croix International Waterway Commission). 2000. Anadromous alewives and freshwater smallmouth bass on the St. Croix system Maine and New Brunswick: responses to questions. St. Croix International Waterway Commission, St. Croix Steering Committee Report, Calais, Maine.

SCRPPT (St. Croix River Pilot Project Team). 1987. St. Croix River integrated data technical report. Inland Waters Directorate, Conservation and Protection, Atlantic Region, Dartmouth, Nova Scotia.

Shuter, B. J., J. A. Maclean, F. E. J. Fry, and H. A. Regier. 1980. Stochastic simulation of temperature effects on 1st-year survival of smallmouth bass. Transactions of the American Fisheries Society 109:1–34.

Shuter, B. J., and J. R. Post. 1990. Climate, population viability, and the zoogeography of temperate fishes. Transactions of the American Fisheries Society 119:314–336.

Smith, M. R. 1988. Status of smallmouth bass populations in Spednic Lake, Maine, Maine Department of Inland Fisheries and Wildlife, Summary Report No. 2, Augusta.

Smith, M. R. 1998. Status of the smallmouth bass populations in Spednic Lake, Maine, USA, and New Brunswick, Canada. Maine Department of Inland Fisheries and Wildlife, Augusta.

Sokal, R. R., and F. J. Rohlf. 1995. Biometry: the principles and practice of statistics in biological research, 3rd edition. Freeman, New York.

USDOC (U.S. Department of Commerce). 2006. Endangered and threatened species; Revision of species of concern list, candidate species definition, and candidate species list. Federal Register 71(200):61022–61025.

Vigerstad, T. J., and J. S. Cobb. 1978. Effects of predation by sea-run juvenile alewives (*Alosa pseudoharengus*) on zooplankton community at Hamilton Reservoir, Rhode Island. Estuaries 1(1):36–45.

Wallace, R. K. 1981. An assessment of diet-overlap indexes. Transactions of the American Fisheries Society 110(1):72–76.

Warmington, A. R., J. M. Wilkinson, and C. B. Riley. 2000. Evaluation of ethanol-based fixatives as a substitute for formalin in diagnostic clinical laboratories. The Journal of Histotechnology 23(4):299–308.

Warner, K. 2005. Smallmouth bass introductions in Maine: history and management implications. Fisheries 30(11):20–26.

Willis, D. W., C. P. Paukert, and B. G. Blackwell. 2002. Biology of white bass in eastern South Dakota glacial lakes. North American Journal of Fisheries Management 22(2):627–636.

Yako, L. A., M. E. Mather, and F. Juanes. 2002. Mechanisms for migration of anadromous herring: an ecological basis for effective conservation. Ecological Applications 12(2):521–534.

Yan, N. D., and G. L. Mackie. 1987. Improved estimation of the dry-weight of *Holopedium gibberum* (Crustacea, Cladocera) using clutch size, a body-fat index, and lake water total phosphorus concentration. Canadian Journal of Fisheries and Aquatic Sciences 44(2):382–389.

Zaret, T. M., and A. S. Rand. 1971. Competition in tropical stream fishes—support for competitive exclusion principle. Ecology 52(2):336–342.

American Fisheries Society Symposium 69:813–821, 2009

Concluding Remarks on the Halifax Diadromy Symposium

Robert M. McDowall
National Institute of Water and Atmospheric Research
Post Office Box 8602, Christchurch, New Zealand

Introduction

The Halifax symposium of 2007 was viewed as an opportunity to review progress in our understanding of the place of diadromy in fish and fisheries approximately 20 years after the 1986 meeting in Boston, also sponsored by the American Fisheries Society, and to address challenges for the future. Clearly, diadromous fishes, despite their relatively small number, are of profound importance across a wide range of values from economic to conservation, and this symposium, and others before it, such as at Hilo, Hawai'i and Bordeaux, France in 2005, reflect that importance.

International and Geographical Scope

The Halifax symposium was international, and although scientists attending from the United States and Canada were most abundant, there were people from, or discussing the fish species of, Albania, Argentina, Asia (by an Australian), Bangladesh, Chile, France, Iraq, Italy, Japan, the Netherlands, New Zealand, Norway, Peru, Russia, and Sweden. Some of the larger contingents from outside North America were from New Zealand and Japan, perhaps a reflection of the importance of diadromy to fish ecology and fisheries in these countries. Somewhat ironically, it was those from New Zealand and Japan who came the furthest to attend the Halifax meeting. Missing were papers on diadromous fishes from Africa, some parts of Asia, Australia, and some of Europe. The small islands of the tropical and subtropical Indo-West Pacific were only sparingly covered (Iida et al. 2009, this volume), as well as the Caribbean and Central America (Bell 2009, this volume) and the warm Neotropics. A particular and, I think, unfortunate absence was of biologists working on diadromous fishes across French Polynesia and the Mascarene Islands in the western Indian Ocean, who have a vigorous research program on the amphidromous fishes of these islands (Keith et al. 1999, 2002, 2006; Keith 2003; Berrebi et al. 2006; Hoareau et al. 2007).

I have real concerns about the lack of knowledge of diadromy (if any) in the great tropical rivers, like the Mekong, Zambezi, Congo, Amazon, Ganges, Nile, and so many more. At present, there is little indication of diadromous species in many of these rivers (McDowall 1988), but is this an accurate representation of what happens there? Similarly, we seem to know relatively little of the diadromous hilsa *Tenualosa ilisha*, though Siddique, from Bangladesh, provided some background to such species there (Siddique 2009, this volume).

Major conservation issues relate to the state of the Caspian and Black seas, and the Sea of Azov, but there was no reference to their diadromous fish and fisheries at the Halifax symposium—an unfortunate absence.

There were significant meetings on diadromy in other regions in 2005, one, held in Bordeaux, France, focused particularly on European species, though it did extend more widely, including papers on fish and fisheries on some of the French tropical islands. Another was in Hilo, Hawai'i, that was concerned particularly with the role of diadromous fishes in the fish communities in small tropical islands, like those in the Hawai'ian Archipelago. Interestingly, I think that only three people who had attended the meeting organized in Bordeaux in mid-2005 were also present in Halifax and only two who had been at the meeting on Hawai'ian (amphidromous) stream fishes, in Hilo, also in 2005 (Evenhuis and Fitzsimons 2007). There were more people attending in Halifax who had been at the Boston meeting 20 years ago than had been in Bordeaux or Hilo. The goal of making the Halifax meeting international was achieved, though representation was uneven, as it had been in Boston, Bordeaux, and Hawai'i. These differences in part reflect different goals of the meetings, different geographies, and different economics of diadromous fish and their fisheries. The Halifax symposium organizers generously supported attendance of many from outside of North America, helping to ensure an international perspective on diadromy. The Boston meeting made available a substantial body of information on diadromous fishes in a single volume (Dadswell et al. 1987), and this galvanized my book on diadromy (McDowall 1988). Additional volumes are emerging from the Bordeaux and Hawai'ian meetings (Dufour et al. 2008; Evenhuis and Fitzsimons 2007), as well as the further volume of which this summary paper is a part. Both of these diadromy meetings have resulted in published reviews, quite aside from what seems to me the primary value—bringing together workers from disparate nations with interests in disparate taxonomic groups but a common interest in exploring the nature and meaning of migrations across the marine–freshwater interface. The eventual results of interactions between people at these meetings are in some senses intangible but are nonetheless highly important. It is my expectation that the Halifax symposium will have established, or re-established, personal connections between fisheries biologists, galvanized continuing interactions, and energized new levels of research productivity.

Taxonomic Coverage

A concern at all meetings focused on diadromy has been the unevenness of taxonomic coverage, and this is interesting to examine. Table 1 shows (A) my estimate of the spread of diadromous species across the entire diversity of fishes (McDowall 1988); (B) percentage composition by family from column A; and the spread of effort at (C) the 2005 meeting in Bordeaux, (D) the 1986 Boston conference, and (E) Halifax in 2007.

Families Gobiidae, Clupeidae, Salmonidae, and Anguillidae are the families with the most diadromous species (Table 1). The papers and posters at the Halifax symposium covered all these groups but were focused mainly on salmonids and anguillid eels, with lesser reference to speciose diadromous groups such as clupeids, osmerid smelts, and percichthyid basses. Lampreys were little-mentioned though they are of economic value in some countries both as a food and as a pest, but they were heavily represented at the Bordeaux meeting. Sturgeons virtually disappeared below the horizon at this symposium, being mentioned only in the poster session (Simpson and Fox 2009, this volume) , despite the Acipenseridae being among the most severely threatened families of fish known (Birstein 1993). However, there has been a large number of recent symposia on threatened species and sturgeon, where this group was well represented. There was limited discussion of the species-rich, widespread, and diverse amphidromous gobioid fishes, especially those from the tropics and subtropics, but this probably reflects the fact that their migrations are not well understood. Among the most speciose of diadromous fishes are the sicydiine gobies, and the limited number of French fisheries biologists attending the meeting resulted in insufficient coverage of these species. Midori Iida et al. (2009) and Kim Bell (2009) told us something of the life history of an amphidromous temperate (Japanese) species of *Sicyopterus* and a Caribbean species of *Sicydium* and, in a sense, showed us how much there is to learn. Also, there was only minimal reference to the existence of diadromous migrations in some invertebrates, such as atyid and palaemonid shrimps and neritinid gastropods (McDowall 2009a; Greathouse and Compton 2009; both this volume). Both the 1986 Boston meeting and that in Bordeaux in 2005, reflected this taxonomic bias, though the Bordeaux meeting had notably more effort applied to lampreys.

Much of the research effort on diadromous fishes is driven by the economic importance of their fisheries. This fact probably partly explains the preceding anomalies as well as geographical locations of symposia—sadly, too much research is driven by economics rather than inquisitiveness, or even conservation agendas, despite the serious conservation status of some diadromous families, such as sturgeons (Birstein 1993) and galaxiids (McDowall 2006). Thus, it is likely that the spread of effort in all diadromy meetings reflects the economic importance of the various families of fishes. If the aim is to ensure an understanding of these economically important fishes and to ensure that information is available for management of their fisheries, then this is being achieved, at least for eels

Table 1.—Number of diadromous species, by family: A and B, derived from McDowall (1988), and their representation in papers and posters at C: Bordeaux, 2005, D: Boston 1986 (Anacat 1), and E: Halifax 2007 (Anacat 2)

	A	B	C	D	E
	Number of species in McDowall 1988	% by family in McDowall 1988	% by family at Bordeaux meeting	% by family included in Ancat 1	% by family included in Ancat 2
Acipenseridae	11.0	4.6	4.3	3.3	3.5
Anguillidae	15.0	6.6	37.2	31.1	13.8
Ariidae	1.0	0.4	0.0	0.0	0.0
Bovicthidae	1.0	0.4	0.0	0.0	0.0
Centropomidae	1.0	0.4	0.0	0.0	0.0
Clupeidae	32.0	14.1	12.1	12.2	15.5
Cottidae	6.0	2.6	0.0	0.0	0.0
Eleotridae	9.0	4.0	0.0	0.0	0.0
Engraulidae	2.0	0.9	0.0	0.0	0.0
Gadidae	1.0	0.4	0.0	0.0	0.0
Galaxiidae	9.0	4.0	0.0	3.3	0.0
Gasterosteidae	2.0	0.9	0.0	0.0	0.0
Gobiidae	44.0	19.4	1.7	1.1	0.0
Kuhliidae	3.0	1.3	0.0	0.0	0.0
Lutjanidae	1.0	0.4	0.0	0.0	0.0
Mugilidae	16.0	7.1	0.0	0.0	0.0
Mugiloididae	1.0	0.4	0.0	0.0	0.0
Osmeridae	7.0	3.1	0.0	6.7	3.5
Percichthyidae	5.0	2.2	0.0	1.1	3.5
Petromyzontiformes	9.0	4.0	13.6	3.3	0.0
Pleuronectidae	3.0	1.3	0.4	0.0	0.0
Retropinnidae	5.0	2.2	0.0	1.1	1.7
Rhyacichthyidae	1.0	0.4	0.0	0.0	0.0
Salangidae	11.0	4.9	0.0	0.0	0.0
Salmonidae	27.0	11.9	26.8	34.4	53.5
Scorpaenidae	1.0	0.4	0.0	0.0	0.0
Soleidae	1.0	0.4	0.0	0.0	0.0
Syngathidae	1.0	0.4	0.0	0.0	0.0
Terapontidae	1.0	0.4	0.0	0.0	0.0
"Technology"		0.0	4.3	2.2	5.2
	227.0	99.5	100.4	99.8	100.2

and salmonids. However, if the aim is to ensure conservation and have information available that makes this practicable, then there should be concerns, especially if there is a wish to develop a broad overarching perspective on the ecology and evolution of diadromy.

There was little placement of diadromy in an evolutionary context (but see Dodson 2009, this volume). This issue is important for determining the role of adaptiveness across the diversity of diadromous fishes (i.e., what is the adaptive role of diadromy, and to what extent is it simply an inherited trait?) There are complex patterns of presence/absence of diadromy, latitudinally, taxonomically, phylogenetically, and ontogenetically. Bradbury et al. (2009, this volume) discussed the place of dwarf morphs of an anadromous osmerid smelt, and Howland et al. (2009, this volume) explored ontogenetic variation in patterns of migration in the salmonid inconnu *Stenodus leucichthys*.

Regardless of the goals of meetings on diadromy in fishes, there continues to be a serious shortage of interest in sturgeons, although the workers in these groups may have been saturated by numerous previous meetings on threatened fishes. Work on clupeids had a strong cool-temperate bias in both North America and Europe and less focus on the tropical shads (though see Siddique 2009; Blaber 2009a, this volume). At all meetings, the relative effort applied to the most speciose family, Gobiidae, was greatly less than the richness of diadromous species in this family. Although goby fisheries have significant local interest in various tropical islands (the tismiche fisheries of the Caribbean, the ipon fisheries of the Philippines, and the bichique of tropical French islands), the family as a whole is poorly researched and little is known of life histories and ecologies, especially as regards the amphidromous sicydiine gobies. I suggest that this group needs a symposium of its own to bring these workers together and move their study forward.

Most of the papers and posters at Halifax had a species focus, and this was also true at the other, previous meetings. Various papers were involved with process in diadromy, looking at such issues as energetics, nutrient transport (Gregory-Eaves et al; Jardine et al. 2009; Naiman 2009; all this volume), habitat restoration, fish passage and connectivity (Castro-Santos et al. 2009, this volume). Some discussed environmental issues, such as climate change (Beamish et al. 2009; Lassalle et al. 2009; Miller et al. 2009; all this volume), which is likely to have growing importance if the predictions of climatologists are as accurate as we think they are. But rather less attention was given to comparative processes across the diversity of diadromous fishes (see Blaber 2009b, Dodson et al. 2009; Secor and Kerr 2009, all this volume), and of big-picture, overview issues for diadromous fishes generally—questions like evolution of diadromy, adaptiveness/cost-benefits of diadromous strategies, energetic, navigation, and homing.

Diadromy at Sea

Inherent in being diadromous is the fact that species spend a significant part of their lives at sea (if they did not, they would not be diadromous!), and little is known about the ecology of diadromous fishes as regards their place in marine populations/fish communities. An immense amount of work has been done on the lives of a few diadromous species of major economic importance, such as salmons at sea, both in the Atlantic and Pacific. And huge effort has been invested in trying to determine where anguillid eels spawn—and thus on just a narrow aspect of the life histories and migrations of eels. Discovering the spawning sites of Indo-Pacific eels is a problem of epic proportions that will not easily be solved, especially given that there are around 14 species compared with just the two rather better-known Atlantic eels.

There are prodigious problems involved with regard to the practicality and cost of elucidating the marine ecologies of diadromous species. Even for many diadromous species of economic importance, it is unrealistic to expect focused and definitive research, regardless of the monetary value of the product derived from these species, especially the galaxiid whitebait of southern lands or the sicydiine fisheries of the islands of the tropics. The marine lives of many diadromous species remain virtually unknown, and coming to grips with studies of this aspect of their ecology is always going to be challenging, owing to the high cost and logistical difficulties of deep-sea research. Some aspects of the marine ecology of diadromous species can be elucidated from study of the fish returning to freshwater (Rowe and Kelly 2009, this volume), but this is really only at the fringes of this great issue.

Freshwater Invasion Processes in Diadromous Fishes

An aspect of the ecology of diadromous fishes that is, as far as I know, virtually unstudied relates to the ecological processes involved in diadromous species taking their place in the preexisting resident fish communities of rivers and lakes as, year after year, they move out of the sea and into river ecosystems, although Swanson and Kidd (2009, this volume) addressed the place of Arctic char *Salvelinus alpi-*

nus in food webs in Arctic lakes. This is a distinct contrast with what might be regarded as analogous questions in the fish communities of tropical reefs, in which recruitment processes are well studied (Sale 1988, 1991).

Equally, little is known about the upstream/downstream patterns of species diversity, age and size structure, and abundance of diadromous fish species in freshwater ecosystems. Tsukamoto (2009, this volume) mentioned the question of migratory drives or instincts in diadromous fishes, a question about which we know very little, the idea of such drives at times being rejected as scarcely scientific (McDowall 1993) despite endogenous behavioral patterns clearly being an important part of the diadromous migration process.

The Changing Face of Research on Diadromy: New Technologies and New Ideas

A distinctive feature of the Halifax symposium was discussion of several new technologies associated with research on diadromous fishes. The use of daily growth rings for aging migrating diadromous fishes was not a part of the 1986 Boston meeting, though it had been in use for about a decade earlier (Panella 1971; Brothers 1987). At the 2007 Halifax symposium, we also heard of the application of otolith microchemistry for identifying biome shifts by diadromous fishes (Bradbury et al. 2009; Howland et al. 2009), DNA sequencing for understanding taxonomy and relationships of these fishes (Dodson et al. 2009), development of new tagging techniques for elucidating migration patterns (Jellyman and Bowen 2009; Stokesbury et al. 2009; both this volume), and application of new and sophisticated modeling techniques.

The past decade has seen the emergence of evidence that anguillid eels may eschew entry into freshwater, but rather spend their trophic lives feeding and growing in coastal seas before joining emigrants from freshwater on their long journey to oceanic spawning grounds. This parallels the facultativeness long recognized among salmonids, in which partial migration is a common phenomenon (Jonsson and Jonsson 1993).

There will always be problems with definitions of diadromous terminology, owing to the number of multidimensional continua associated with diadromy. But we face the choice of getting on with studying diadromy using the terminology we have or getting bogged down in the niceties of terminology and getting lost or becoming redundant. A key problem is the degree of facultativeness and within taxon/within population variation in diadromous life cycles. Sometimes we do not actually know whether to regard species as diadromous or not. I think we need to be inclusive rather than exclusive. In particular, acceptance of amphidromy as a distinctive form of diadromy varies geographically, though use of the term is routine by many fish biologists across the tropical islands of the Indian Ocean, in the Pacific and Caribbean, and in Japan, Australia, and New Zealand (Keith et al. 1999, 2002, 2006; Keith 2003; Fitzsimons et al. 2002; Tamada and Iwata 2005; McDowall 2007, 2009b).

There seemed to be less emphasis on issues of fish passage at the Halifax symposium, though this is surely not because the problems have been solved, or even understood in many instances, and the question of connectivity remains a crucial issue for the research, management, and conservation of diadromous fishes. In this area of concern, we need to understand the differing needs of different taxa and, in particular, to accommodate the needs of smaller species that can climb, in contrast with larger fish like salmonids and clupeids that can jump.

Connectivity is not just fish passage; it also deals with issues of community structuring and assembly. A particular focus of the 2005 meeting in Hilo, Hawai'i was the role of amphidromous fishes in aquatic ecology and conservation and the importance of connectivity, not just to fish passage, but also to stream community structure and ecology.

At times, it seems that the more we know about some diadromous fishes, the more complex the migratory scenarios become, and this is especially true of salmon; Secor's (1999) work with *Morone* bass seems to be pointing in the same direction.

Many diadromous fishes have ancient traditional importance to indigenous peoples all over the world, to which Haggan et al. (2009, this volume) alluded. Some of our research ought, in my view, to contribute to meeting the needs or aspirations of these people—perhaps highlighted during the week of the Halifax symposium, of a Canadian newspaper report of the virtual disappearance of the eulachon *Thaleichthys pacificus* from the rivers of the

Pacific coast of Canada and the implications for the native peoples living there. I think we need a global assessment of the importance of these diadromous species to traditional and contemporary communities of indigenous fishes. Siddique (2009) drew our attention to the human/fisheries issues that involve some diadromous fisheries. Management of diadromous fisheries is difficult enough when we are dealing with significant commercial/economic fisheries; it is much more difficult with fisheries that are artisanal/subsistence.

There are complex fisheries and resource management issues that relate to species-based approaches, as opposed to ecosystems approaches. We have tended to deal primarily with the species-level issues, though the paper by Rochard et al. (2009, this volume) explored the diadromous fish communities of the River Seine in France.

The place of diadromy in the application of techniques for the evaluation of habitat and fish community ecology has not been explored at any of the meetings on diadromy. Diadromy poses special problems owing to downstream–upstream gradients in community composition, age and size structures of the fish populations, and the application of tools such as indices of biotic integrity (McDowall 1993, 1998) that are being addressed at least in New Zealand (Joy and Death 2003; Leathwick et al. 2005; Eikaas et al. 2006).

Above all else, I came away from the Halifax symposium with a strong sense that diadromy is an important global phenomenon in fish ecology and conservation, that those of us interested in it have many common problems and issues and much to discuss, that we do not talk to each other enough, and that the opportunities to do so are too few and too far apart. It is more than 20 years since the Boston meeting; it would be unfortunate if it was another 20 years before we meet again, under the auspices of the American Fisheries Society or other interested fisheries organizations.

References

Beamish, R. J., R. M. Sweeting, and C. M. Neville. 2009. Planning the management of Pacific salmon in a changing climate. Pages 155–173 *in* A. J. Haro, K. L. Smith, R. A. Rulifson, C. M. Moffitt, R. J. Klauda, M. J. Dadswell, R. A. Cunjak, J. E. Cooper, K. L. Beal, and T. S. Avery, editors. Challenges for diadromous fishes in a dynamic global environment. American Fisheries Society, Symposium 69, Bethesda, Maryland.

Bell, K. N. I. 2009. What comes down must go up: the migration cycle of juvenile-return anadromous taxa. Pages 321–341 *in* A. J. Haro, K. L. Smith, R. A. Rulifson, C. M. Moffitt, R. J. Klauda, M. J. Dadswell, R. A. Cunjak, J. E. Cooper, K. L. Beal, and T. S. Avery, editors. Challenges for diadromous fishes in a dynamic global environment. American Fisheries Society, Symposium 69, Bethesda, Maryland.

Berrebi, P., T. Galewski, and P. Keith. 2006. *Sicyopterus lagocephalus*, a unique widespread taxon confirmed by mtDNA sequences. Molecular Phylogenetics and Evolution 40:903–904.

Birstein, V. J. 1993. Sturgeons and paddlefishes: threatened fishes in need of conservation. Conservation Biology 7:773–787.

Blaber, S. J. M. 2009a. Can the ecosystem approach improve management of tropical estuarine fisheries for diadromous species? Pages 649–661 *in* A. J. Haro, K. L. Smith, R. A. Rulifson, C. M. Moffitt, R. J. Klauda, M. J. Dadswell, R. A. Cunjak, J. E. Cooper, K. L. Beal, and T. S. Avery, editors. Challenges for diadromous fishes in a dynamic global environment. American Fisheries Society, Symposium 69, Bethesda, Maryland.

Blaber, S. J. M. 2009b. Socioeconomic and biopolitical linkages in the management of tropical shads. Pages 677–690 *in* A. J. Haro, K. L. Smith, R. A. Rulifson, C. M. Moffitt, R. J. Klauda, M. J. Dadswell, R. A. Cunjak, J. E. Cooper, K. L. Beal, and T. S. Avery, editors. Challenges for diadromous fishes in a dynamic global environment. American Fisheries Society, Symposium 69, Bethesda, Maryland.

Bradbury, I. R., M. W. Coulson, S. E. Campana, E. Baggs, and P. Bentzen. 2009. Postglacial recolonization and the loss of anadromy in euteleostean fishes. Pages 79–96 *in* A. J. Haro, K. L. Smith, R. A. Rulifson, C. M. Moffitt, R. J. Klauda, M. J. Dadswell, R. A. Cunjak, J. E. Cooper, K. L. Beal, and T. S. Avery, editors. Challenges for diadromous fishes in a dynamic global environment. American Fisheries Society, Symposium 69, Bethesda, Maryland.

Brothers, E. B. 1987. Methodological approaches to the examination of otoliths in aging studies. Pages 319–330 *in* R. C. Summerfelt and C. E. Hall, editors. The age and growth of fish. Iowa State University, Ames.

Castro-Santos, T., A. Cotel, and P. Webb. 2009. Fishway evaluations for better engineering: an integrative approach. Pages 557–575 *in* A. J. Haro, K. L. Smith, R. A. Rulifson, C. M. Moffitt, R. J. Klauda, M. J. Dadswell, R. A. Cunjak, J. E. Cooper, K. L. Beal, and T. S. Avery, editors. Challenges for diadromous fishes in a dynamic global environment. American Fisheries Society, Symposium 69, Bethesda, Maryland.

Dadswell, M. J., R. J. Klauda, C. M. Moffitt, R. L. Saunders, R. A. Rulifson, and J. E. Cooper, editors. 1987. Common strategies of anadromous and catadromous fishes. American Fisheries Society, Symposium 1, Bethesda, Maryland.

Dodson, J. J., J. Laroche, and F. Lecomte. 2009. Contrasting evolutionary pathways of anadromy in euteleostean fishes. Pages 63–77 *in* A. J. Haro, K. L. Smith, R. A. Rulifson, C. M. Moffitt, R. J. Klauda, M. J. Dadswell, R. A. Cunjak, J. E. Cooper, K. L. Beal, and T. S. Avery, editors. Challenges for diadromous fishes in a dynamic global environment. American Fisheries Society, Symposium 69, Bethesda, Maryland.

Dufour, S., E. Prévost, E. Rochard, and P. Williot, editors. 2008. Fish and diadromy in Europe (ecology, management, conservation): proceedings of the symposium held 29 March–1 April 2005, Bordeaux, France. Hydrobiologia 602.

Eikaas, H. S., A. D. Kliskey, and A. R. McIntosh. 2006. Analysis of patterns in diadromous fish distributions using GIS. Transactions in GIS 10:469–483.

Evenhuis, N. L. and M. J. Fitzsimons, editors. 2007. Biology of Hawai'ian streams and estuaries. Bishop Museum Bulletin in Cultural and Environmental Studies 3.

Fitzsimons, J. M., J. E. Parham and R. T. Nishimoto. 2002. Similarities in behavioural ecology among amphidromous and catadromous fishes in the oceanic islands of Hawai'i and Guam. Environmental Biology of Fishes 65:123–129.

Greathouse, E. A., and J. E. Compton. 2009. Beyond diadromous fishes: diadromous invertebrates are strong interactors too [poster abstract]. Page 939 *in* A. J. Haro, K. L. Smith, R. A. Rulifson, C. M. Moffitt, R. J. Klauda, M. J. Dadswell, R. A. Cunjak, J. E. Cooper, K. L. Beal, and T. S. Avery, editors. Challenges for diadromous fishes in a dynamic global environment. American Fisheries Society, Symposium 69, Bethesda, Maryland.

Gregory-Eaves, I., D. T. Selbie, J. N. Sweetman, B. P. Finney, and J. P. Smol. 2009. Tracking sockeye salmon population dynamics from lake sediment cores: a review and synthesis. Pages 379–393 *in* A. J. Haro, K. L. Smith, R. A. Rulifson, C. M. Moffitt, R. J. Klauda, M. J. Dadswell, R. A. Cunjak, J. E. Cooper, K. L. Beal, and T. S. Avery, editors. Challenges for diadromous fishes in a dynamic global environment. American Fisheries Society, Symposium 69, Bethesda, Maryland.

Haggan, N., G. D. Jackson, and P. Lacroix. 2009. Salmon and eulachon in ecosystem space and time: a plea for wider collaboration and data integration. Pages 619–636 *in* A. J. Haro, K. L. Smith, R. A. Rulifson, C. M. Moffitt, R. J. Klauda, M. J. Dadswell, R. A. Cunjak, J. E. Cooper, K. L. Beal, and T. S. Avery, editors. Challenges for diadromous fishes in a dynamic global environment. American Fisheries Society, Symposium 69, Bethesda, Maryland.

Hoareau, T. B., R. Lecomte-Finiger, H. P. Grondin, C. Conand, and P. Berrebi. 2007. Oceanic larval life of La Reunion 'bichiques,' amphidromous gobiid postlarvae. Marine Ecology Progress Series 333:303–308.

Howland, K. L., J. A. Babaluk, D. Chiperzak, R. F. Tallman, and G. Low. 2009. Variability in diadromous behavior of a northern Coregonid based on scanning proton microprobe analysis of otolith strontium. Pages 121–134 *in* A. J. Haro, K. L. Smith, R. A. Rulifson, C. M. Moffitt, R. J. Klauda, M. J. Dadswell, R. A. Cunjak, J. E. Cooper, K. L. Beal, and T. S. Avery, editors. Challenges for diadromous fishes in a dynamic global environment. American Fisheries Society, Symposium 69, Bethesda, Maryland.

Iida, M., S. Watanabe, and K. Tsukamoto. 2009. Life history characteristics of a Sicydiinae goby in Japan, compared with its relatives and other amphidromous fishes. Pages 355–373 *in* A. J. Haro, K. L. Smith, R. A. Rulifson, C. M. Moffitt, R. J. Klauda, M. J. Dadswell, R. A. Cunjak, J. E. Cooper, K. L. Beal, and T. S. Avery, editors. Challenges for diadromous fishes in a dynamic global environment. American Fisheries Society, Symposium 69, Bethesda, Maryland.

Jardine, T. D., J.-M. Roussel, S. Mitchell and R. A. Cunjak. 2009. Detecting marine nutrient and organic matter inputs into multiple trophic levels in streams of Atlantic Canada and France. Pages 427–445 *in* A. J. Haro, K. L. Smith, R. A. Rulifson, C. M. Moffitt, R. J. Klauda, M. J. Dadswell, R. A. Cunjak, J. E. Cooper, K. L. Beal, and T. S. Avery, editors. Challenges for diadromous fishes in a dynamic global environment. American Fisheries Society, Symposium 69, Bethesda, Maryland.

Jellyman, D. J., and M. M. Bowen. 2009. Modelling larval migration routes and spawning areas of anguillid eels of New Zealand and Australia. Pages 255–274 *in* A. J. Haro, K. L. Smith, R. A. Rulifson, C. M. Moffitt, R. J. Klauda, M. J. Dadswell, R. A. Cunjak, J. E. Cooper, K. L. Beal, and T. S. Avery, editors. Challenges for diadromous fishes in a dynamic global environment. American Fisheries Society, Symposium 69, Bethesda, Maryland.

Jonsson, B., and N. Jonsson. 1993. Partial migration: niche shift versus sexual maturation in fishes. Reviews in Fish Biology and Fisheries 3:348–365.

Joy, M. K., and R. G. Death. 2003. Assessing biological integrity using freshwater fish and decapod habitat selection functions. Environmental Management 32:747–759.

Keith, P. 2003. Biology and ecology of amphidromous Gobiidae of the Indo-Pacific and Caribbean regions. Journal of Fish Biology 63:831–847.

Keith, P., E. Vigneux, and P. Bosc. 1999. Atlas des poissons et des crustaces d'eau douce de la Reunion. Patrimoines Naturels 39, Paris.

Keith, P., E. Vigneux, and G. F. Marquet. 2002. Atlas des poissons et des crustaces d'eau douce de Polynesie Francaise. Patrimoines Naturels 55.

Keith, P., G. Marquet., P. Valade, P. Bosc, and E. Vigneux. 2006. Atlas des poissons et des Crustaces d'eau douces des Comoros, Mascareignes et Seychelles. Patrimoines Naturels 65, Paris.

Lassalle, G., M. Béguer, L. Beaulaton, and E. Rochard. 2009. Learning from the past to predict the future; responses of European diadromous fish to climate change. Pages 175–193 *in* A. J. Haro, K. L. Smith, R. A. Rulifson, C. M. Moffitt, R. J. Klauda, M. J. Dadswell, R. A. Cunjak, J. E. Cooper, K. L. Beal, and T. S. Avery, editors. Challenges for diadromous fishes in a dynamic global environment. American Fisheries Society, Symposium 69, Bethesda, Maryland.

Leathwick, J. R., D. Rowe, J. Richardson, J. Elith, and T. Hastie. 2005. Using multivariate adaptive regression spines to predict the distributions of New Zealand's freshwater diadromous fish. Freshwater Biology 50:2034–2052.

McDowall, R. M. 1988. Diadromy in fishes: migrations between freshwater and marine environments. Croom Helm, London.

McDowall, R. M. 1993. Implications of diadromy for the structuring and modelling of riverine fish communities in New Zealand. New Zealand Journal of Marine and Freshwater Research 27:453–462.

McDowall, R. M. 1998. Fighting the flow: downstream-upstream linkages in the ecology of diadromous fish faunas in West Coast New Zealand rivers. Freshwater Biology 40:111–122.

McDowall, R. M. 2006. Crying wolf, crying foul, or crying shame: alien salmonids and a biodiversity crisis in the southern cool-temperate galaxioid fishes. Reviews in Fish Biology and Fisheries 16:233–422.

McDowall, R. M. 2007. Hawai'ian stream fishes: the role of amphidromy in history, ecology and conservation biology. Bishop Museum Bulletin in Cultural and Environmental Studies 3: 3–9.

McDowall, R. M. 2009a. Early larval hatch: a strategy for safe downstream transport in amphidromous fishes. Reviews in Fish Biology and Fisheries 19:1–8.

McDowall, R. M. 2009b. Making the best of two worlds: diadromy in the evolution, ecology, and conservation of aquatic organisms. Pages 1–22 *in* A. J. Haro, K. L. Smith, R. A. Rulifson, C. M. Moffitt, R. J. Klauda, M. J. Dadswell, R. A. Cunjak, J. E. Cooper, K. L. Beal, and T. S. Avery, editors. Challenges for diadromous fishes in a dynamic global environment. American Fisheries Society, Symposium 69, Bethesda, Maryland.

Miller, M. J., S. Kimura, K. D. Friedland, B. Knights, H. Kim, D. J. Jellyman, and K. Tsukamoto. 2009. Review of ocean-atmospheric factors in the Atlantic and Pacific oceans influencing spawning and recruitment of anguillid eels. Pages 231–249 *in* A. J. Haro, K. L. Smith, R. A. Rulifson, C. M. Moffitt, R. J. Klauda, M. J. Dadswell, R. A. Cunjak, J. E. Cooper, K. L. Beal, and T. S. Avery, editors. Challenges for diadromous fishes in a dynamic global environment. American Fisheries Society, Symposium 69, Bethesda, Maryland.

Naiman, R. J., J. M. Helfield, K. K. Bartz, D. C. Drake, and J. M. Honea. 2009. Pacific salmon, marine-derived nutrients, and the characteristics of aquatic and riparian ecosystems. Pages 395–425 *in* A. J. Haro, K. L. Smith, R. A. Rulifson, C. M. Moffitt, R. J. Klauda, M. J. Dadswell, R. A. Cunjak, J. E. Cooper, K. L. Beal, and T. S. Avery, editors. Challenges for diadromous fishes in a dynamic global environment. American Fisheries Society, Symposium 69, Bethesda, Maryland.

Panella, G. 1971. Fish otoliths: daily growth layers and periodical patterns. Science 173:1124–1127.

Rochard, E., P. Pellegrini, J. Marchal, M. Béguer, D. Ombredane, G. Lassalle, E. Menvielle, and J.-L. Baglinere. 2009. Identification of diadromous fish species on which to focus river regulations: an eco-anthropological approach (the Seine basin, France). Pages 691–711 *in* A. J. Haro, K. L. Smith, R. A. Rulifson, C. M. Moffitt, R. J. Klauda, M. J. Dadswell, R. A. Cunjak, J. E. Cooper, K. L. Beal, and T. S. Avery, editors. Challenges for diadromous fishes in a dynamic global environment. American Fisheries Society, Symposium 69, Bethesda, Maryland.

Rowe, D. K., and G. Kelly. 2009. Duration of the oceanic phase for inanga whitebait (Galaxiidae) is inversely related to growth rate at sea. Pages 343–354 *in* A. J. Haro, K. L. Smith, R. A. Rulifson, C. M. Moffitt, R. J. Klauda, M. J. Dadswell, R. A. Cunjak, J. E. Cooper, K. L. Beal, and T. S. Avery, editors. Challenges for diadromous fishes in a dynamic global environment. American Fisheries Society, Symposium 69, Bethesda, Maryland.

Sale, P. F. 1988. Perception, pattern, chance and the structure of reef fish communities. Environmental Biology of Fishes 21:3–15.

Sale, P. F. 1991. Reef fish communities: open non-equilibrial systems. Pages 564–598 *in* P. F. Sale, editor. The ecology of fishes on coral reefs. Academic, New York.

Secor, D. H. 1999. Specifying divergent migrations in the concept of stock: the contingent hypothesis. Fisheries Research 43:13–34.

Secor, D. H., and L. A. Kerr. 2009. Lexicon of life cycle diversity in diadromous and other fishes. Pages 537–556 *in* A. J. Haro, K. L. Smith, R. A. Rulifson, C. M. Moffitt, R. J. Klauda, M. J. Dadswell, R. A. Cunjak, J. E. Cooper, K. L. Beal, and T. S. Avery, editors. Challenges for diadromous fishes in a dynamic global environment. American Fisheries Society, Symposium 69, Bethesda, Maryland.

Siddique, M. A. L. 2009. Conservation of juvenile hilsa (jatka) in Bangladesh: need to address the livelihood of fishers. Pages 757–768 *in* A. J. Haro, K. L. Smith, R. A. Rulifson, C. M. Moffitt, R. J. Klauda, M. J. Dadswell, R. A. Cunjak, J. E. Cooper, K. L. Beal, and T. S. Avery, editors. Challenges for diadromous fishes in a dynamic global environment. American Fisheries Society, Symposium 69, Bethesda, Maryland.

Stokesbury, M. J. W., M. J. Dadswell, K. N. Holland, G. D. Jackson, W. D. Bowen, and R. K. O'Dor. 2009. Tracking diadromous fishes at sea: the electronic future using hybrid acoustic and archival tags. Pages 311–320 *in* A. J. Haro, K. L. Smith, R. A. Rulifson, C. M. Moffitt, R. J. Klauda, M. J. Dadswell, R. A. Cunjak, J. E. Cooper, K. L. Beal, and T. S. Avery, editors. Challenges for diadromous fishes in a dynamic global environment. American Fisheries Society, Symposium 69, Bethesda, Maryland.

Swanson, H. K., and K. A. Kidd. 2009. A preliminary investigation of the effects of anadromous Arctic char on food web structure and nutrient transport in coastal Arctic lakes. Pages 465–483 *in* A. J. Haro, K. L. Smith, R. A. Rulifson, C. M. Moffitt, R. J. Klauda, M. J. Dadswell, R. A. Cunjak, J. E. Cooper, K. L. Beal, and T. S. Avery, editors. Challenges for diadromous fishes in a dynamic global environment. American Fisheries Society, Symposium 69, Bethesda, Maryland.

Tamada, K., and K. Iwata. 2005. Intra-specific variation of egg size, clutch size and larval survival related to maternal size in amphidromous Rhinogobius goby. Environmental Biology of Fishes 73:379–389.

Tsukamoto, K., M. J. Miller, A. Kotake, J. Aoyama, and K. Uchida. 2009. The origin of fish migration: the random escapement hypothesis. Pages 45–61 *in* A. J. Haro, K. L. Smith, R. A. Rulifson, C. M. Moffitt, R. J. Klauda, M. J. Dadswell, R. A. Cunjak, J. E. Cooper, K. L. Beal, and T. S. Avery, editors. Challenges for diadromous fishes in a dynamic global environment. American Fisheries Society, Symposium 69, Bethesda, Maryland.

American Fisheries Society Symposium 69:823–828, 2009

Changing Themes in the Study of Diadromous Fishes

Thomas P. Quinn*
School of Aquatic and Fishery Sciences
Box 355020, University of Washington, Seattle, Washington 98195, USA

Abstract.—In the two decades since the first symposium devoted to anadromous and catadromous fishes, the study of these fishes and their life history patterns has benefited from numerous advances in perspectives and techniques. Molecular ecology, fisheries oceanography, conservation biology, climate change, nutrient cycling, and landscape ecology are some of the many fields that have blossomed during this period, providing a wealth of insights for those studying migratory fishes. The purpose of this paper is to bridge the gap between the 1986 and 2007 symposia, noting some of the scientific advances, but also the fact that many populations of diadromous fishes have declined. These problems are reflected in the focus of the present symposium on conservation and socioeconomic aspects, as well as life history, diversity, and migration patterns. The enormity of the changes in the environment, both natural and anthropogenic, poses tremendous challenges for diadromous fishes and further emphasizes the need for linkage between research and conservation, using all the tools at the disposal of modern science and the development of new ones.

The conservation and management of migratory animals presents special challenges relative to more sedentary animals because they move between distinct habitats. Each species requires more than one critical habitat and safe passage between them. As pointed out in both scientific and popular literature, many migratory animals and indeed the phenomenon of their migrations are in jeopardy (Wilcove 2008). Whether they travel by air, land, or sea, migrants are vulnerable as well as fascinating. Fisheries traditionally take advantage of migration patterns to facilitate exploitation, and habitat degradation or passage barriers pose serious problems for migratory species. The regular migrations of diadromous fishes between freshwater and saltwater allow them to take full advantage of the differences in growing conditions and breeding habitats (Northcote 1978; Gross et al. 1988) in the two environments. However, diadromy is comparatively rare among fishes, being found in fewer than 1% of the species (McDowall 1988), perhaps because of the physiological challenges associated with life in both saltwater and freshwater, the demands of upstream migration, and the difficulties in adapting to such distinctly different ecological conditions.

* Corresponding author: tquinn@u.washington.edu

In 1986, a symposium in Boston was held to consider "common strategies" of these fishes, and its goal "was to identify key similarities or differences in life history patterns and to examine these to assist our understanding of diadromy" (Dadswell 1987). In the two decades since that symposium (Dadswell et al. 1987), tremendous synergies between technical advances and conceptual breakthroughs in diverse fields, including molecular biology (Carvalho et al. 1994), biotelemetry (Lacroix et al. 2004; Welch et al. 2004), analysis of stable isotopes and microchemistry (Ingram and Weber 1999; MacAvoy et al. 2000; Finney et al. 2002; Kennedy et al. 2002; Milton and Chenery 2003; Gillanders 2005; Harrod et al. 2005; Schindler et al. 2005), and ecotoxicology (Ewald et al. 1998), have greatly expanded our understanding of the life history and ecology of many diadromous fishes.

With respect to molecular biology, for example, the pivotal paper using mitochondrial DNA to show that rainbow trout *Oncorhynchus mykiss*, formerly *Salmo gairdneri*, were closely related to Pacific salmon *Oncorhynchus* spp. was published in 1986 by Thomas et al. Genetic techniques are now so fast and powerful that they can provide in-season management advice, as well as revealing patterns of relationships among populations (Habicht et al. 2007). It is also possible to take DNA from scales collected decades ago for analysis of long-term genetic changes (Nielsen et al. 1997) and parentage analysis to determine the lifetime reproductive success of individual fish (Seamons et al. 2007). Such applications were scarcely conceivable to most fisheries scientists in the mid-1980s, and no papers in the 1987 publication relied on genetics. A key journal, *Molecular Ecology*, did not publish its first issue until 1992. Another topic emphasized in the 2007 symposium, linkages with ecosystem energetics, reflects both the shift from single species to ecosystem perspectives and also the ways in which stable isotopes of elements (notably nitrogen and carbon) allow us to track "marine-derived nutrients" brought from the ocean to freshwater in the bodies of anadromous fishes. The existence of these isotopes was known, of course, but key papers applying this approach to salmon came after the 1986 meeting (Kline et al. 1990, 1993).

Beyond the importance of developing and emerging techniques in changing the face of research since the 1980s, there is the enormous increase in computing ability that we now enjoy. Indeed, many of us were getting our first personal computers in the mid-1980s, and their capacity was very limited. Now, we can run complex models, download huge amounts of data on climate and hydrology from Web sites, access maps and photographs from our offices, send large files electronically, and in so many ways expand our capacity for productive work. These tools are no substitute for ideas, but there have been many important conceptual advances and syntheses as well. The publication of two books (Roff 1992; Stearns 1992) spurred the application of life history theory to many topics in behavior and evolutionary ecology, including such topics of phenotypic plasticity and norms of reaction that are central to the life histories of many diadromous fishes (Hutchings 2004). The past two decades have also seen the fields of conservation biology and climate change grow from relative infancy to the forefront of ecological research. For perspective, the journal *Conservation Biology* was first published in 1987, and *Global Change Biology* began in 1995. The fields of fisheries and oceanography have grown together in many ways, reflected in part by the journal *Fisheries Oceanography*, first published in 1992, and many important papers linking climate, the ocean, and salmon have been published in the past decades (Beamish and Bouillon 1993; Mantua et al. 1997; Friedland et al. 2003). The concept of metapopulations (Hanski 1999) has come to prominence, as have the concepts of biodiversity, biocomplexity, ecosystem management, and many others. Some might feel that these are merely "old wine in new bottles," but there is little doubt that the study of diadromous fishes is changing rapidly, along with the rest of science and natural resource management.

Unfortunately, all the technological breakthroughs and conceptual advances have not always resulted in healthier fish populations. The past decades have seen declines of many fishes to the point where the list of threatened or endangered anadromous fishes under the U.S. Endangered Species Act includes four species of sturgeon, bull trout *Salvelinus confluentus*, and Atlantic salmon *Salmo salar* and two evolutionarily significant units (Waples 1991; Waples 1995) of sockeye salmon *Oncorhynchus nerka*, two of chum salmon *O. keta*, three of coho salmon *O. kisutch*, nine of Chinook salmon *O. tshawytscha*, and 11 of steelhead (nonadromous *O. mykiss*). The Committee on the Status of Endangered Wildlife in Canada (COSEWIC) is charged with assessing the status of species, and designatable units (DUs) within species, for possible listing under the Species at Risk Act (SARA). Atlantic salmon in the inner Bay of Fundy Atlantic are listed as Endangered under SARA, as are several DUs of white sturgeon *Acipenser transmontanus* and Atlantic whitefish *Coregonus huntsmani*. There are other diadromous species that COSEWIC has designated at some degree of risk but have no official protection under SARA at this time, including populations of coho, Chinook, and sockeye salmon; striped bass *Morone saxatilis*; and American eel *Anguilla rostrata*.

Given these increases in knowledge, new perspectives, and ongoing crises for diadromous fishes, it is appropriate that the 2007 symposium's stated goal was not only to "review the current state of scientific knowledge with respect to biology, ecology, and conservation of diadromous fishes," but also to "address how recent alterations to the environment and human activity have affected diadromous fishes with respect to their sustainability and role in aquatic ecosystems." The 1986 meeting had sessions devoted to habitats (freshwater, marine, and transition) and recruitment. In contrast, the 2007 meeting's sessions were on life history, climate change and human influences, ocean migration, ecosystem linkages, restoration of populations and habitats, and socioeconomic aspects, reflecting the explicit shift in focus towards conservation and the connection between life history and the human element in the management of diadromous fishes.

The 2007 symposium includes papers that reveal the expansion of new technologies such as those applied to tracking fish movements (Stokesbury et al. 2009, this volume), analysis of paleolimnology data from sediment cores in lakes (Gregory-Eaves et al. 2009, this volume), stable isotopes (Naiman et al. 2009, this volume), elemental analysis of otolith microchemistry (Howland et al. 2009, this volume), and the insights of molecular biology for fish phylogeny and the evolution of anadromy (Dodson et al. 2009, this volume). These diverse perspectives and techniques are also brought together to advance our understanding of many kinds of migratory fishes. For example, work on eels has drawn from the fields and techniques of fisheries oceanography, behavior, life history, electronics, chemistry, and more (e.g., Jellyman and Bowen 2009; Miller et al. 2009; Tsukamoto et al. 2009; all this volume).

In addition to the fascinating new information on diadromous fishes presented at the symposium, however, we were also reminded of the problems that many species face. Mixed stock fisheries are a common problem for management, especially when the fish migrate across governmental boundaries in coastal areas or enter international waters (e.g., Hammer 2009, this volume). These problems are all the more complex when applied to species such as anguillid eels, which can be fished and managed in specific areas but breed in common at a remote location. Thus, a degraded stream might continue to attract juveniles as long as they are produced by adults that survived in other rivers, unlike the river-specific stock and recruitment relationship typical of anadromous fishes. In addition to these general issues, the precise timing patterns that characterize many diadromous fishes only make them more vulnerable to fishing. The most obvious effect of fishing, the reduction in density, necessarily attracts the most attention, but there is also increasing awareness of the possible evolutionary effects of size-selective fishing (Conover and Munch 2002; Conover et al. 2005). The long-term effects of size-selective fishing will depend on the balance between genetic and environmental effects on growth and age at maturation in the population and the intensity of selection (Hard 2004). Likewise, many fisheries are selective with respect to migratory timing because of exploitation patterns (Consuegra et al. 2005) and management regulations, though this has attracted less scientific attention. Documenting such selection and then considering the possible long-term consequences will provide scientists working on such species with plenty of work in the future.

Fisheries management, however, involves much more than traditional scientific disciplines such as population dynamics and genetics, but also a thorough consideration of socioeconomic aspects of the people involved in the fisheries, as exemplified by the hilsa shad *Tenualosa ilisha* fishery in Bangladesh (Siddique 2009, this volume), and the complex facets of managing many tropical fisheries (Blaber 2009a, 2009b, both this volume). All of these processes will play out in the 21st century as we experience changing climate conditions (Beamish et al. 2009, this volume), posing even greater challenges for the fishes and those working on them. These challenges are exacerbated by the growing human population, with all the demands for water, space, and other resources, causing many scientists to take a pessimistic view of the prospects for Pacific salmon (Hartman et al. 2000; Lackey et al. 2006) and, by extension, other diadromous fishes. Without denying these problems and the inherent vulnerability of diadromous fishes, it is important to also recognize that diadromy confers some measure of resilience to their populations. The success of striped bass management, leading to the rebound of the species along the east coast of North America, and the increasing abundance of pink salmon *Oncorhynchus gorbuscha* in Puget Sound, in less than pristine habitat at the southern end of their range, should give us reason for hope if not optimism. I look forward to the next symposium on the exciting topic of diadromous fishes, sometime in the future, and I can only wonder what themes, techniques, and findings will be emphasized there.

References

Beamish, R. J., and D. R. Bouillon. 1993. Pacific salmon production trends in relation to climate. Canadian Journal of Fisheries and Aquatic Sciences 50:1002–1016.

Beamish, R. J., R. M. Sweeting, and C. M. Neville. 2009. Changing climate and the need to change our thinking about the management of Pacific salmon. Pages 155–173 *in* A. J. Haro, K. L. Smith, R. A. Rulifson, C. M. Moffitt, R. J. Klauda, M. J. Dadswell, R. A. Cunjak, J. E. Cooper, K. L. Beal, and T. S. Avery, editors. Challenges for diadromous fishes in a dynamic global environment. American Fisheries Society, Symposium 69, Bethesda, Maryland.

Blaber, S. J. M. 2009a. Can the ecosystem approach improve management of tropical estuarine fisheries for diadromous species? Pages 649–661 *in* A. J. Haro, K. L. Smith, R. A. Rulifson, C. M. Moffitt, R. J. Klauda, M. J. Dadswell, R. A. Cunjak, J. E. Cooper, K. L. Beal, and T. S. Avery, editors. Challenges for diadromous fishes in a dynamic global environment. American Fisheries Society, Symposium 69, Bethesda, Maryland.

Blaber, S. J. M. 2009b. Socioeconomic and biopolitical linkages in the management of tropical shads.Pages 677–690 *in* A. J. Haro, K. L. Smith, R. A. Rulifson, C. M. Moffitt, R. J. Klauda, M. J. Dadswell, R. A. Cunjak, J. E. Cooper, K. L. Beal, and T. S. Avery, editors. Challenges for diadromous fishes in a dynamic global environment. American Fisheries Society, Symposium 69, Bethesda, Maryland.

Carvalho, G. R., T. J. Pitcher, L. K. Park, P. Morgan, R. D. Ward, P. M. Grewe, L. Hauser, M. Ferguson, F. M. Utter, J. M. Wright, P. Bentzen, and R. Lincoln. 1994. Molecular genetics in fisheries. Reviews in Fish Biology and Fisheries 4:269–399.

Conover, D. O., S. A. Arnott, M. R. Walsh, and S. B. Munch. 2005. Darwinian fishery science: lessons from the Atlantic silverside (*Menidia menidia*). Canadian Journal of Fisheries and Aquatic Sciences 62:730–737.

Conover, D. O., and S. B. Munch. 2002. Sustaining fisheries yields over evolutionary time scales. Science 297:94–96.

Consuegra, S., C. García de Leaniz, A. Serdio, and E. Verspoor. 2005. Selective exploitation of early running fish may induce genetic and phenotypic changes in Atlantic salmon. Journal of Fish Biology 67(Supplement A):129–145.

Dadswell, M. J. 1987. Introduction. Pages xiii–xiv *in* M. J. Dadswell, R. J. Klauda, C. M. Moffitt, R. L. Saunders, R. A. Rulifson, and J. E. Cooper, editors. Common strategies of anadromous and catadromous fishes. American Fisheries Society, Symposium 1, Bethesda, Maryland.

Dadswell, M. J., R. J. Klauda, C. M. Moffitt, R. L. Saunders, R. A. Rulifson, and J. E. Cooper. 1987. Common strategies of anadromous and catadromous fishes. American Fisheries Society, Symposium No. 1, Bethesda, Maryland.

Dodson, J. J., J. Laroche, and F. Lecomte. 2009. Contrasting evolutionary pathways of anadromy in euteleostean fishes. Pages 63–77 *in* A. J. Haro, K. L. Smith, R. A. Rulifson, C. M. Moffitt, R. J. Klauda, M. J. Dadswell, R. A. Cunjak, J. E. Cooper, K. L. Beal, and T. S. Avery, editors. Challenges for diadromous fishes in a dynamic global environment. American Fisheries Society, Symposium 69, Bethesda, Maryland.

Ewald, G., P. Larsson, H. Linge, L. Okla, and N. Szarzi. 1998. Biotransport of organic pollutants to an inland Alaska lake by migrating sockeye salmon (*Oncorhynchus nerka*). Arctic 51:40–47.

Finney, B. P., I. Gregory-Eaves, M. S. V. Douglas, and J. P. Smol. 2002. Fisheries productivity in the northeastern Pacific Ocean over the past 2,200 years. Nature (London) 416:729–733.

Friedland, K. D., D. G. Reddin, J. R. McMenemy, and K. F. Drinkwater. 2003. Multidecadal trends in North American Atlantic salmon (*Salmo salar*) stocks and climate trends relevant to juvenile survival. Canadian Journal of Fisheries and Aquatic Sciences 60:563–583.

Gillanders, B. M. 2005. Otolith chemistry to determine movements of diadromous and freshwater fish. Aquatic Living Resources 18:291–300.

Gregory-Eaves, I., D. T. Selbie, J. N. Sweetman, B. P. Finney, and J. P. Smol. 2009. Tracking sockeye salmon population dynamics from lake sediment cores: a review and synthesis. Pages 379–393 *in* A. J. Haro, K. L. Smith, R. A. Rulifson, C. M. Moffitt, R. J. Klauda, M. J. Dadswell, R. A. Cunjak, J. E. Cooper, K. L. Beal, and T. S. Avery, editors. Challenges for diadromous fishes in a dynamic global environment. American Fisheries Society, Symposium 69, Bethesda, Maryland.

Gross, M. R., R. M. Coleman, and R. M. McDowall. 1988. Aquatic productivity and the evolution of diadromous fish migration. Science 239:1291–1293.

Habicht, C., L. W. Seeb, and J. E. Seeb. 2007. Genetic and ecological divergence defines population structure of sockeye salmon populations returning to Bristol Bay, Alaska, and provides a tool for admixture analysis. Transactions of the American Fisheries Society 136:82–94.

Hammer, M. 2009. Whose fish? Managing salmonidae and humans in complex social-ecological systems: examples from the Baltic Sea Region. Pages 663–675 *in* A. J. Haro, K. L. Smith, R. A. Rulifson, C. M. Moffitt, R. J. Klauda, M. J. Dadswell, R. A. Cunjak, J. E. Cooper, K. L. Beal, and T. S. Avery, editors. Challenges for diadromous fishes in a dynamic global environment. American Fisheries Society, Symposium 69, Bethesda, Maryland.

Hanski, I. 1999. Metapopulation ecology. Oxford University Press, Oxford, UK.

Hard, J. J. 2004. Evolution of Chinook salmon life history under size-selective harvest. Pages 315–337 *in* A. P. Hendry, and S. C. Stearns, editors. Evolution illuminated: salmon and their relatives. Oxford University Press, Oxford, UK.

Harrod, C., J. Grey, T. K. McCarthy, and M. Morrissey. 2005. Stable isotope analyses provide new insights into ecological plasticity in a mixohaline population of European eel. Oecologia 144:673–683.

Hartman, G. F., C. Groot, and T. G. Northcote. 2000. Science and management in sustainable salmonid fisheries: the ball is not in our court. Pages 31–50 *in* E. E. Knudsen, C. R. Steward, D. D. MacDonald, J. E. Williams, and D. W. Reiser, editors. Sustainable fisheries management: Pacific salmon. Lewis Publishers, Boca Raton, Florida.

Howland, K. L., J. A. Babaluk, D. Chiperzak, R. F. Tallman, and G. Low. 2009. Variability in diadromous behavior of a northern coregonid based on scanning proton microprobe analysis of otolith strontium. Pages 121–134 *in* A. J. Haro, K. L. Smith, R. A. Rulifson, C. M. Moffitt, R. J. Klauda, M. J. Dadswell, R. A. Cunjak, J. E. Cooper, K. L. Beal, and T. S. Avery, editors. Challenges for diadromous fishes in a dynamic global environment. American Fisheries Society, Symposium 69, Bethesda, Maryland.

Hutchings, J. A. 2004. Norms of reaction and phenotypic plasticity in salmonid life histories. Pages 154–174 *in* A. P. Hendry, and S. C. Stearns, editors. Evolution illuminated: salmon and their relatives. Oxford University Press, Oxford, UK.

Ingram, B. L., and P. K. Weber. 1999. Salmon origin in California's Sacramento–San Joaquin River system as determined by otolith strontium isotopic composition. Geology 27:851–854.

Jellyman, D. J. and M. M. Bowen. 2009. Modelling larval migration routes and spawning areas of temperate *Anguilla* spp. of New Zealand and Australia. Pages 255–274 *in* A. J. Haro, K. L. Smith, R. A. Rulifson, C. M. Moffitt, R. J. Klauda, M. J. Dadswell, R. A. Cunjak, J. E. Cooper, K. L. Beal, and T. S. Avery, editors. Challenges for diadromous fishes in a dynamic global environment. American Fisheries Society, Symposium 69, Bethesda, Maryland.

Kennedy, B. P., A. Klaue, J. D. Blum, C. L. Folt, and K. H. Nislow. 2002. Reconstructing the lives of fishes using Sr isotopes in otoliths. Canadian Journal of Fisheries and Aquatic Sciences 59:925–929.

Kline, T. C. Jr., J. J. Goering, O. A. Mathisen, and P. Poe. 1990. Recycling of elements transported upstream by runs of Pacific salmon: I. $\delta^{15}N$ and $\delta^{13}C$ evidence in Sashin Creek, southeastern Alaska. Canadian Journal of Fisheries and Aquatic Sciences 47:136–144.

Kline, T. C. Jr., J. J. Goering, O. A. Mathisen, P. Poe, P. L. Parker, R. S. Scalan. 1993. Recycling of elements transported upstream by runs of Pacific salmon: II. $\delta^{15}N$ and $\delta^{13}C$ evidence in the Kvichak River watershed, Bristol Bay, southern Alaska. Canadian Journal of Fisheries and Aquatic Sciences 50:2350–2365.

Lackey, R. T., D. H. Lach, and S. L. Duncan. 2006. Salmon 2100: the future of wild Pacific salmon. American Fisheries Society, Bethesda, Maryland.

Lacroix, G. L., P. McCurdy, and D. Knox. 2004. Migration of Atlantic salmon postsmolts in relation to habitat use in a coastal system. Transactions of the American Fisheries Society 133:1455–1471.

MacAvoy, S. E., S. A. Macko, S. P. McIntosh, and G. C. Garman. 2000. Marine nutrient contributions to freshwater apex predators. Oecologia 122:568–573.

Mantua, N. J., S. R. Hare, Y. Zhang, J. M. Wallace, and R. C. Francis. 1997. A Pacific interdecadal climate oscillation with impacts on salmon production. Bulletin of the American Meteorological Society 78:1069–1079.

McDowall, R. M. 1988. Diadromy in fishes. Timber Press, Portland, Oregon.

Miller, M. J., S. Kimura, K. D. Friedland, B. Knights, H. Kim, D. J. Jellyman, and K. Tsukamoto. 2009. Potential effects of ocean-atmospheric changes on recruitment of temperate anguillid eels in the Atlantic and Pacific oceans. Pages 231–249 *in* A. J. Haro, K. L. Smith, R. A. Rulifson, C. M. Moffitt, R. J. Klauda, M. J. Dadswell, R. A. Cunjak, J. E. Cooper, K. L. Beal, and T. S. Avery, editors. Challenges for diadromous fishes in a dynamic global environment. American Fisheries Society, Symposium 69, Bethesda, Maryland.

Milton, D. A., and S. R. Chenery. 2003. Movement patterns of the tropical shad hilsa (*Tenualosa ilisha*) inferred from transects of 87Sr/86Sr isotope rations in their otoliths. Canadian Journal of Fisheries and Aquatic Sciences 60:1376–1385.

Naiman, R. J., J. M. Helfield, K. K. Bartz, D. C. Drake, and J. M. Honea. 2009. Pacific salmon, marine-derived nutrients and the characteristics of aquatic and riparian ecosystems. Pages 395–425 *in* A. J. Haro, K. L. Smith, R. A. Rulifson, C. M. Moffitt, R. J. Klauda, M. J. Dadswell, R. A. Cunjak, J. E. Cooper, K. L. Beal, and T. S. Avery, editors. Challenges for diadromous fishes in a dynamic global environment. American Fisheries Society, Symposium 69, Bethesda, Maryland.

Nielsen, E. E., M. M. Hansen, and V. Loeschcke. 1997. Analysis of microsatellite DNA from old scale samples of Atlantic salmon *Salmo salar*: a comparison of genetic composition over 60 years. Molecular Ecology 6:487–492.

Northcote, T. G. 1978. Migratory strategies and production in freshwater fishes. Pages 326–359 *in* S. D. Gerking, editor. Ecology of freshwater fish production. Wiley, New York.

Roff, D. A. 1992. The evolution of life histories: theory and analysis. Chapman and Hall, New York.

Schindler, D. E., P. R. Leavitt, C. S. Brock, S. P. Johnson, and P. D. Quay. 2005. Marine-derived nutrients, commercial fisheries, and production of salmon and lake algae in Alaska. Ecology 86:3225–3231.

Seamons, T. R., P. Bentzen, and T. P. Quinn. 2007. DNA parentage analysis reveals inter-annual variation in selection: results from 19 consecutive brood years in steelhead trout. Evolutionary Ecology Research 9:409–431.

Siddique, M. A. L. 2009. Conservation of juvenile hilsa shad in Bangladesh: need to address the livelihood of fishers. Pages 757–768 *in* A. J. Haro, K. L. Smith, R. A. Rulifson, C. M. Moffitt, R. J. Klauda, M. J. Dadswell, R. A. Cunjak, J. E. Cooper, K. L. Beal, and T. S. Avery, editors. Challenges for diadromous fishes in a dynamic global environment. American Fisheries Society, Symposium 69, Bethesda, Maryland.

Stearns, S. C. 1992. The evolution of life histories. Oxford University Press, Oxford, UK.

Stokesbury, M. J. W., M. J. Dadswell, K. N. Holland, G. D. Jackson, W. D. Bowen, and R. K. O'Dor. 2009. Tracking of diadromous fishes at sea using hybrid acoustic and archival electronic tags. Pages 311–320 *in* A. J. Haro, K. L. Smith, R. A. Rulifson, C. M. Moffitt, R. J. Klauda, M. J. Dadswell, R. A. Cunjak, J. E. Cooper, K. L. Beal, and T. S. Avery, editors. Challenges for diadromous fishes in a dynamic global environment. American Fisheries Society, Symposium 69, Bethesda, Maryland.

Thomas, W. K., R. E. Withler, and A. T. Beckenbach. 1986. Mitochondrial DNA analysis of Pacific salmonid evolution. Canadian Journal of Zoology 64:1058–1064.

Tsukamoto, K., M. J. Miller, A. Kotake, J. Aoyama, and K. Uchida. 2009. The origin of fish migration: an evolutionary and behavioral approach. Pages 45–61 *in* A. J. Haro, K. L. Smith, R. A. Rulifson, C. M. Moffitt, R. J. Klauda, M. J. Dadswell, R. A. Cunjak, J. E. Cooper, K. L. Beal, and T. S. Avery, editors. Challenges for diadromous fishes in a dynamic global environment. American Fisheries Society, Symposium 69, Bethesda, Maryland.

Waples, R. S. 1991. Pacific salmon, *Oncorhynchus* spp., and the definition of "species" under the Endangered Species Act. Marine Fisheries Review 53(3):11–22.

Waples, R. S. 1995. Evolutionarily significant units and the conservation of biological diversity under the Endangered Species Act. Pages 8–27 *in* J. L. Nielsen, editor. Evolution and the aquatic ecosystem: defining unique units in population conservation. American Fisheries Society, Symposium 17, Bethesda, Maryland.

Welch, D. W., B. R. Ward, and S. D. Batten. 2004. Early ocean survival and marine movements of hatchery and wild steelhead trout (*Oncorhynchus mykiss*) determined by an acoustic array: Queen Charlotte Strait, British Columbia. Deep-Sea Research II 51:897–909.

Wilcove, D. S. 2008. No way home: the decline of the world's great animal migrations. Island Press, Washington, D.C.

American Fisheries Society Symposium 69:829, 2009

Poster Sessions
Preamble

John E. Cooper
Cooper Environmental Research
1444 County Route 23, Constantia, New York 13044, USA

Diadromy encompasses a wide range of life history characteristics, many of which are altered during the life cycle of the organism. Research is often focused into many narrow topics, but understanding the interactions between the environment and the organism requires a broad approach that incorporates the topics into a cohesive whole. It is hoped that this can be facilitated with the presentations of the following abstracts. The authors present new ideas on physiology, genetics, migration (in relation to ocean currents, barriers, pollutants, parasites, and diseases), growth in ocean and freshwater environments, behavior, design of fishways, analysis of stream networks, forming partnerships for stock restoration, and policy needs.

The underlying theme of these presentations (either expressed or implied) is the physical and chemical characteristics of habitat and how well we understand the responses of organisms to the unstable nature of the habitat caused by human activity. The organisms evolved with specific, or at least preferred, needs in habitat and must now adapt to altered conditions caused by migratory barriers, sedimentation, introduced chemical pollutants, and climate change. The effects of barriers, loss of historic substrate, introduced chemicals that can mask natural migratory cues, and changes in the open ocean are examined in the abstracts presented here, as well as possible solutions to some of these problems. However, the changes to specific habitats may be occurring faster than the pace of our understanding of what is needed for each species.

Advancements in technology have vastly improved our ability to collect fine-scale data that reveals migration movements and duration, growth habitats, and interactions of predator and prey, but can we utilize these tools quickly enough to restore the declining populations of organisms?

The primary authors represent nine countries and present information on more than 17 fish species: eels and salmonids were the more numerous, as in American Fisheries Society Symposium 1, *Common Strategies of Anadromous and Catadromous Fishes*, in 1986, but galaxids, ayu *Plecoglossus altivelis*, coregonids, smelt, lampreys, shads, sturgeons, and striped bass *Morone saxatilis* were also discussed. The presentation on diadromous invertebrates reminds us that diadromy is not limited to fish.

The 62 abstracts submitted for the meeting in Halifax required two separate sessions and were presented as they related to the paper sessions. Authors were given the choice of preparing extended abstracts for this publication, and these 24 abstracts, as well as 33 short abstracts, are presented here in the order of the respective paper session.

American Fisheries Society Symposium 69:831–834, 2009

Bay-Scale Population Differentiation among Anadromous Rainbow Smelt

Mark W. Coulson* and Paul Bentzen
Department of Biology and Marine Gene Probe Laboratory, Dalhousie University Halifax, Nova Scotia B3H 4J1, Canada

Spatial separation of spawning aggregations has long been recognized as a major determinant of population genetic structure. Isolation by distance (IBD) (Wright 1943) predicts that genetic differences accumulate along increasing geographic distances separating populations. However, local factors may also play a role in further structuring populations. Rainbow smelt *Osmerus mordax* are assumed to remain near the coast and do not undergo lengthy migrations, suggesting a potential for strong population structuring. It has been shown that regional forms may exhibit strikingly different degrees of genetic structure, such as that seen between Newfoundland and the mainland. While Newfoundland populations display a strong signal of IBD, mainland populations seemingly show little effect of distance on population structuring (Bradbury et al. 2006). Therefore, it is possible that structure among mainland samples may be influenced by more local effects (e.g., bay-scale). Factors such as variation in the hydrographic and temperature conditions encountered throughout their range may therefore be a more pronounced determinant of population structure than increasing distance alone. Indeed, member vagrancy has been proposed for smelt and other species whereby the spatial scale across which early life history stages are retained determines the scale at which structuring occurs (Iles and Sinclair 1982).

Anadromous runs of rainbow smelt were sampled from 22 populations by dip netting throughout much of their native range (Figure 1A). Approximately 50–100 individuals per population were sampled and genotyped for nine microsatellite loci (see Coulson et al. 2006 for loci characteristics and procedures). Basic descriptive microsatellite statistics were conducted as in Bradbury et al. (2006). Factorial correspondence analysis (FCA) was carried out to emphasize differences between populations based on their multilocus genotypes. Two approaches were taken to estimate the effects of distance on genetic differentiation. First, pairwise measures of F_{ST} were plotted against pairwise geographic distances, and second, an assignment test was carried out on all fish: the resultant proportion of misassigned fish to each location was plotted against the geographic distance of that location from the source (i.e., where the fish was sampled). Finally, a barrier analysis was conducted to determine the principal discontinuities in the spatial pattern of the genetic data.

Factorial correspondence analysis of all 22 populations resulted in four major groupings: Gulf of St. Lawrence, Atlantic coast, St. Mary's Bay, and inner Bay of Fundy (Figure 1B). Barrier analysis (Figure 1A) identified St. Mary's Bay as the first major discontinuity (A) based on pairwise F_{ST}. The second major barrier (B) was identified as separating the gulf region from the rest of the sites, and a third barrier (C) was identified separating the two bays sampled along the Atlantic coast (Musquodoboit Harbour [sites 9 and 10; Figure 1A] and St. Margaret's Bay [sites 6–8]). FCA was subsequently carried out on each of the above regions. Estimates of pairwise F_{ST} among the Gulf and Bras D'Or lakes populations revealed most comparisons to be nonsignificant; however, there was a tendency for populations within the Bras D' Or lakes (Eskasoni, Malagawatch, and Washabuck [sites 11–13]) to separate from those within the gulf, based on allele frequencies. Significant differentiation was detected between the two bays along the Atlantic coast, also suggesting within-bay differentiation. Finally, a large amount

* Corresponding author: mcoulson@dal.ca

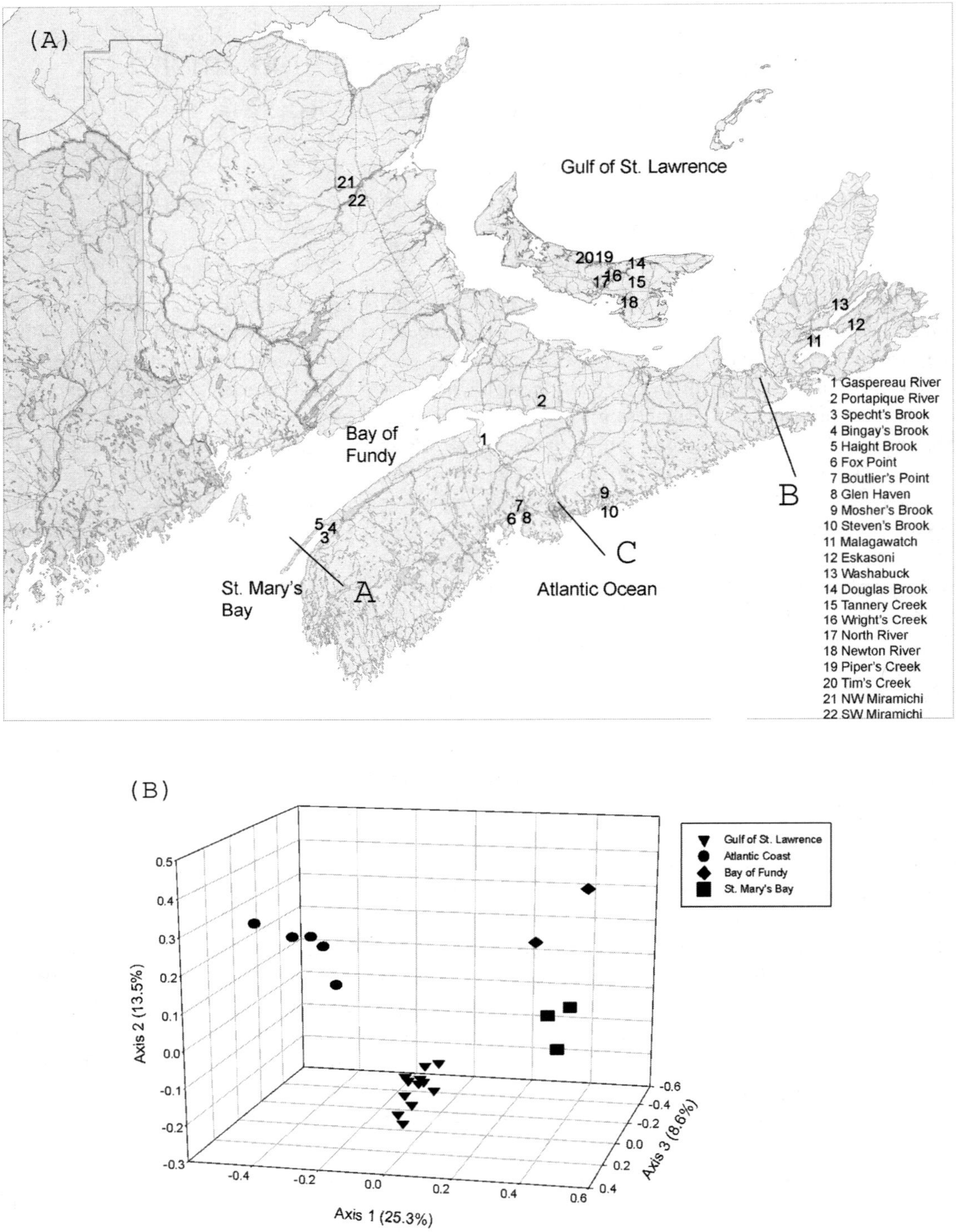

Figure 1.—(A) Map of eastern Canada, showing the 22 sampling locations included in the current study; (B) factorial correspondence analysis (FCA) plot showing the four major groupings based on multilocus genotypes.

of variation separated St. Mary's Bay from the Inner Bay of Fundy, while again showing separation between populations within each of these regions. For instance, within St. Mary's Bay, Haight Brook was separated out from Bingays and Spechts brooks, which are closer together geographically and on the opposite side of the bay.

The plot of genetic versus geographic distance revealed a weak effect of IBD (Mantel test, $r = 0.14$, $P = 0.045$; Figure 2A). By comparison, analysis of the proportion of misassigned fish at increasing spatial scales suggested a much stronger effect of geographic distance (Mantel test, $r = -0.46$, $P < 0.0001$; Figure 2B). Within the IBD plot, however, there was evidence of regional heterogeneity when comparing the four major areas. A very weak IBD was evidenced when comparing gulf populations, whereas high levels of genetic differentiation were observed at much smaller spatial scales for St. Margaret's Bay and St. Mary's Bay. This, taken to-

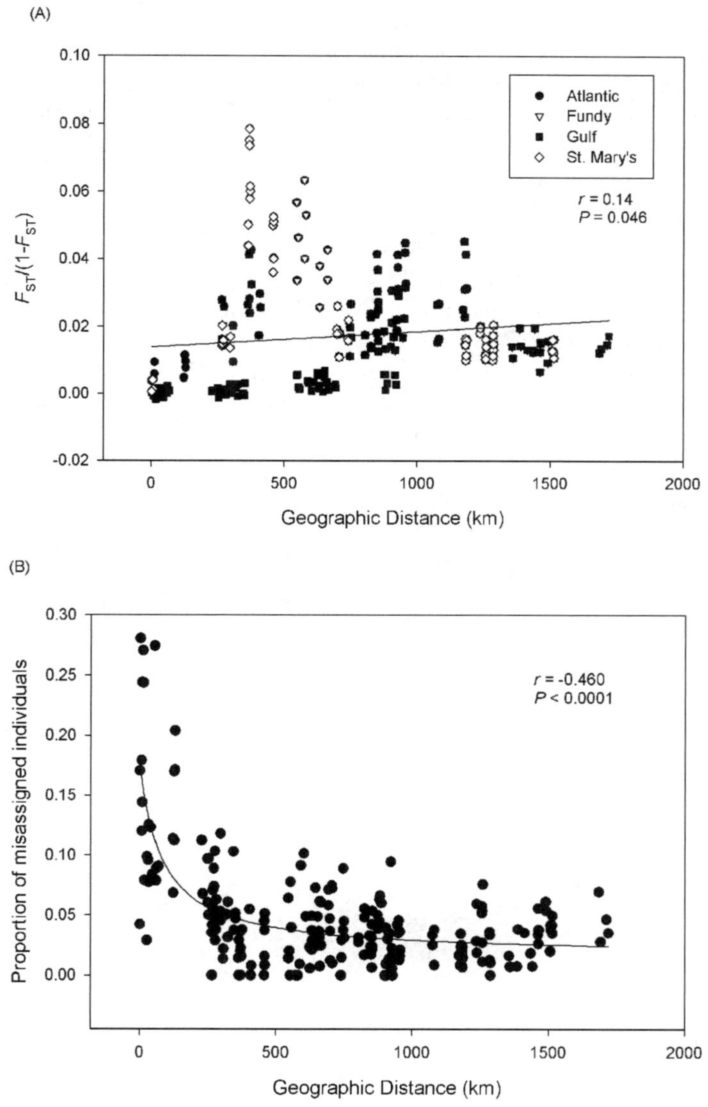

Figure 2.—(A) Plot of isolation by distance using pairwise comparisons of $F_{ST}/(1 - F_{ST})$ versus geographic distance, and (B) the proportion of misassigned individuals at increasing geographic scales.

gether with the results of the barrier analysis, suggests that populations within these two bays were the most discrete and resulted in greater levels of differentiation between them than would be expected on distance alone.

Smelt populations demonstrate evidence for restriction on dispersal at increased spatial scales despite a seemingly weak effect of overall IBD. The low power in resolving F_{ST}-based IBD, as well as nonsignificant pairwise tests of genetic differentiation in the face of geographical separation, based upon allele frequencies, may be due to the low amount of variation seen among F_{ST} values. Indeed, a negative relationship was seen between locus-specific variability and F_{ST}, suggesting decreased power of more variable loci to resolve population structure. This suggests that allele frequency information (assignment testing, FCA plots) offers a more powerful approach than average-based F_{ST} analyses in detecting population structure. Discrepancies between more traditional F_{ST}-based IBD and assignment-based IBD have been seen in other species (Castric and Bernatchez 2004).

Additionally, the weak F_{ST}-based IBD may be due, in part, to regional heterogeneity in IBD slope. For instance, compare the slopes of Gulf of St. Lawrence pairwise comparisons and that of St. Mary's Bay (Figure 2A). In addition, the high F_{ST} values for St. Mary's Bay centered around 400 km are all pairwise comparisons between these populations and those from St. Margaret's Bay. This suggests that there are unique factors involved with each of these bays that make them more discrete than would be expected based on distance alone.

The scale across which population structure is resolved within a species can vary across that species' distribution and is a result of interplay between the isolating effects of geographic distance, as well as more local effects. Therefore, knowledge of population structure and the factors determining the degree of connectivity among populations is crucial to their conservation and management efforts.

References

Bradbury, I. R., M. W. Coulson, S. E. Campana, and P. Bentzen. 2006. Morphological and genetic differentiation in anadromous smelt *Osmerus mordax* (Mitchill): disentangling the effects of geography and morphology on gene flow. Journal of Fish Biology 69:95–114.

Castric, V., and L. Bernatchez. 2004. Individual assignment test reveals differential restriction to dispersal between two salmonids despite no increase of genetic differences with distance. Molecular Ecology 13:1299–1312.

Coulson, M. W., I. G. Paterson, A. Green, R. Kepkay, and P. Bentzen. 2006. Characterization of di- and tetranucleotide microsatellite markers in rainbow smelt (*Osmerus mordax*). Molecular Ecology Notes 6:942–944.

Iles, T. D., and M. Sinclair. 1982. Atlantic herring—stock discreteness and abundance. Science 215:627–633.

Wright, S. 1943. Isolation by distance. Genetics 28:114–138.

American Fisheries Society Symposium 69:835–836, 2009

Population Structure of American Shad within and among Canadian Rivers

Daniel J. Hasselman*
Department of Biology, Dalhousie University, Halifax, Nova Scotia B3H4J1, Canada

Rod G. Bradford
Bedford Institute of Oceanography, Dartmouth, Nova Scotia, B2Y 4A2, Canada

Paul Bentzen
Department of Biology, Dalhousie University, Halifax, Nova Scotia B3H4J1, Canada

Knowledge of population structuring is essential to the management of diadromous fishes, within the context of fisheries that may target mixed stock assemblages, to maintain biodiversity, and in the context of species at risk. Fishery-based indices, or simple presence–absence, are not always adequate, or accurate, for resolving the population structure of highly migratory species. These considerations apply to American shad *Alosa sapidissima*, a widespread and seemingly loosely structured population-rich anadromous alosine of increasing conservation concern. American shad are philopatric and form reproductively discrete spawning runs that are genetically distinguishable (Bentzen et al. 1989). The largest spawning runs, within the Canadian portion of their range, are found in the Annapolis, Shubenacadie, Saint John, Miramichi, and St. Lawrence rivers (Chaput and Bradford 2003). However, shad have been reported from several additional drainages that are not known to support spawning runs (Leim 1924), and it is uncertain whether these accounts constitute genetically discrete spawning populations.

Genetic analyses can resolve population structure but may be compromised by stocking efforts. Canadian populations of American shad have not been subject to artificial stock transfers, unlike their U.S. counterparts, and present an opportunity to study population structuring in the absence of human-mediated gene flow, on a variety of spatial scales. Here, we examine broad and fine spatial scale population structuring among, and within, six drainages tributary to the Bay of Fundy using 13 polymorphic microsatellite loci.

Whole genomic DNA was extracted from tissue acquired from 50 to 100 live specimens captured above the head of tide in Nova Scotia: Annapolis River in 2004, Gaspereau River in 2004 and 2006, Kennetcook River in 2005, Shubenacadie River in 2004–2006, River Hebert in 2004–2005, and Saint John River, New Brunswick in 2003–2006. Eight sites were sampled from the Saint John River, and three sites were sampled from the Shubenacadie River between 2004 and 2006. Samples were genotyped across four dinucleotide (Aa14, Af6, Af13, Af20), four trinucleotide (Asa2, Asa4, Asa16, AsaB020), and five tetranucleotide (Asa8, Aps7C, AsaD042, AsaD429, AsaD029) microsatellite markers. Broad spatial scale analysis involved comparisons among drainages, while fine spatial scale analysis involved comparisons within drainages for the Saint John River and Shubenacadie River. Analyses were conducted using Bayesian clustering analysis (BAPS 4.1) and factorial correspondence analysis. Isolation by distance among drainages was assessed to determine whether an increasing degree of genetic differentiation with increasing geographic distance was evident. Geographic distances among sampled drainages were calculated following a counterclockwise migration route hypothesized for shad within the Bay of Fundy (Dadswell et al. 1987).

Bayesian clustering analysis identified five groups among drainages: (1) Saint John River 2003;

* Corresponding author: dhasselm@dal.ca

(2) Saint John River 2004–2006; (3) Gaspereau River 2004 and 2006; (4) Annapolis River 2004; and (5) Kennetcook River 2005, River Hebert 2004–2005, and Shubenacadie River 2004–2006. Shad collected in 2004 and 2006 from the Gaspereau River clustered together, but were independent from all other drainages, suggesting the existence of a previously unidentified genetically discrete spawning run. The three major spawning runs of the Bay of Fundy (Annapolis River, Shubenacadie River, and Saint John River) clustered independently of each other and supported their continued recognition as independent spawning populations. Saint John River shad collected in 2003 failed to group with those collected from 2004 to 2006. This discrepancy may be the result of temporal effects, as temporal variation had a greater effect on the genetic composition of shad in the Saint John River than on other drainages. The grouping of shad from the Shubenacadie River with the Kennetcook River and River Hebert suggests that the genetic composition of shad in these latter drainages is influenced by migration-gene flow from the Shubenacadie River.

Significant ($P < 0.05$) pairwise genetic differentiation (F_{ST}) was detected among sites in multiple years in the Shubenacadie River. However, the degree of differentiation among sites was not temporally stable. Significant genetic differentiation in 2004 was detected between the Nine Mile River and both the Shubenacadie and Stewiacke rivers. Shad genotypes from Nine Mile River and Shubenacadie River were not significantly different in 2005, whereas genotypes from both of these sites were significantly different from the Stewiacke River. Nonsignificant genetic differentiation was detected among all sites in all years within the Saint John River, regardless of the level of spatial hierarchical grouping.

A highly significant ($P < 0.01$) relationship was found between genetic differentiation and geographic distance among drainages, supporting the hypothesized counterclockwise migration route for shad in the Bay of Fundy (Dadswell et al. 1987).

We detected significant genetic structuring among shad spawning runs in the Bay of Fundy and evidence for subpopulation structure within the Shubenacadie River drainage. Shad from the Gaspereau River comprise a genetically distinguishable spawning run not recognized previously. Spawning runs maintain their genetic and geographic distinctiveness despite evidence of temporal variability in genetic composition. Strong evidence for isolation by distance among spawning runs is consistent with the hypothesis of a counterclockwise migration route for shad in the Bay of Fundy (Dadswell et al. 1987). Genetic tools provide a powerful means by which to resolve the spatial scale of population structure. Reports of shad in rivers other than known spawning runs are numerous. Resolving whether these reports constitute genetically discrete spawning runs is required for effective management of shad populations.

References

Bentzen, P., G. C. Brown, and W. C. Leggett. 1989. Mitochondrial DNA polymorphism, population structure, and life history variation in American shad (*Alosa sapidissima*). Canadian Journal of Fisheries and Aquatic Sciences 46:1446–1454.

Chaput, G., and R. G. Bradford. 2003. American shad (*Alosa sapidissima*) in Atlantic Canada. Canadian Science Advisory Secretariat, Research Document No. 2003/009, Ottawa.

Dadswell, M. J., G. D. Melvin, P. J. Williams, and D. E. Themelis. 1987. Influences of origin, life history, and chance on the Atlantic coast migration of American shad. Pages 313–330 *in* M. J. Dadswell, R. J. Klauda, C. M. Moffitt, R. L. Saunders, R. A. Rulifson, and J. E. Cooper, editors. Common strategies of anadromous and catadromous fishes. American Fisheries Society, Symposium 1, Bethesda, Maryland.

Leim, A. H. 1924. The life history of the shad (*Alosa sapidissima*) with special reference to the factors limiting its abundance. Contribution to Canadian Biology 2:163–284.

American Fisheries Society Symposium 69:837–840, 2009

A Comprehensive Approach to Understanding Diadromy at the Species Level: Learning from the Spawning Migration of Hickory Shad

JOSHUA G. MURAUSKAS*,[1] AND ROGER A. RULIFSON
East Carolina University, Institute for Coastal Science and Policy Flanagan Hall 250, Greenville, North Carolina 27858, USA

We consider diadromous fish migrations to include three principal components: demographics of the migrating population—who is migrating, external conditions, and movements—where and when is migration occurring, and, physical or internal changes—what changes do fish experience and why do they occur? Despite extensive work to characterize these components (Harden-Jones 1968; McKeown 1984; Smith 1985; McDowall 1988), synthesis of data at the species level is lacking. This is partially due to the narrow scope of many projects limited by objectives, time, resources, and sampling logistics. This process has led to piecemeal explanations of a multifaceted event for many diadromous fishes. The objective of this extended abstract is to briefly demonstrate how a broader approach to investigating diadromy is not only beneficial, but is needed to understand the whole picture.

Hickory shad *Alosa mediocris* was selected as a model to examine the collective changes experienced by an iteroparous fish during a spawning migration. This relatively unstudied fish belongs to a recreationally and commercially important family that likely accounts for a majority of diadromous fish along the eastern United States. We sampled fish of a particular run in two consecutive years at various points in their ascent in salt-, brackish, and freshwater in coastal North Carolina to describe our three components of diadromy. Time, location, and water quality data were collected at each capture site. Samples of blood, muscle tissue, gonads, stomach, otoliths, and scales were collected from each fish (Table 1). These data along with length, weight, and spatial-temporal data provided a detailed account of internal and external changes experienced by a population during the spawning migration.

One-hundred twenty-two females and 71 males were captured from the ocean (n = 41), the estuary (n = 113), and the river (n = 39). Mean total length (and weight) of males was 342 mm (0.590 kg) and of females was 366 mm (0.754 kg). Most fish were repeat spawners (~75%), and both sexes averaged 2.9 years of age, although females had a greater proportion in older age-classes. Fish were caught in coastal areas of the ocean in late January (water temperature = 7.4°C, salinity > 30 parts per thousand [ppt]), and entered the estuary by February (7.9°C, 5–10 ppt) in both years. Spawning at upriver locations occurred for about 2 to 3 weeks in early to mid-March (~13.0°C, 0 ppt). Postspawning fish, returning to the ocean, were captured in the estuary during April (>16.0°C, 3–5 ppt) in both years. Although migratory timing was similar between years, significantly colder water temperatures delayed freshwater entry in 2005 by at least 1 week. Regression analysis indicated that water temperature accounted for more than 70% of the variation in fish location. Feeding decreased dramatically early in migration, and energy reserves (indicated by mesentery fat, lipids, and condition factor) were depleted with time and distance upstream. An average 45% reduction in lipid content was observed in a subsample of females from 2004. Increases in the gonadosomatic index between the ocean and spawning grounds averaged nearly 60% for males and more than 300% for females. Histological examination of gonads suggests that batch spawning occurred over

* Corresponding author: joshmurauskas@gmail.com
[1] Current address: Public Utility District No. 1 of Douglas County, 1151 Valley Mall Parkway, East Wenatchee, Washington 98802, USA

Table 1.—Characterization of samples collected from hickory shad populations during the spring migrations of 2004 and 2005 to explore the three fundamental components of diadromy.

Data and sample	Results
Date and time of capture	Temporal description of migratory movements. Important for determining how populations move and how external and internal conditions change through time.
Capture location	Spatial description of migratory movements. Provides information regarding movements in relation to environmental conditions and time. Provides the ability to make spatial comparisons on population demographics, reproductive development, energy depletion, and feeding habits.
Water quality	Description of surrounding environment during migration. Important for determining how environmental cues and clues affect movements and internal physical conditions.
Length and weight	Physical description of migrating fish. Provides information for estimating health and reproductive potential of the population. Also used in indices for describing reproductive development and fish condition (*k*-factor).
Otoliths	Age of migrating fish. Provides information regarding population structure and year-class comparisons of size, bioenergetics, and reproductive development.
Scales	Age and spawning periodicity. In addition to results from age determination, provides information on iteroparity and its relationship with fish size and reproductive potential.
Stomach	Diet composition, feeding habits, and mesentery fat depletion during migration. Provides information regarding diet and energy acquisition prior to and energy use during migration. Implications on how diadromous populations affect local fauna during migrations and timing of movements.
Flesh	Stored energy (i.e. lipids) and metabolism during migration. Provides information regarding energy availability for movements and spawning, along with ability to examine how other internal and external conditions affect migration.
Blood[a]	Hormone levels during migration. Provides information regarding important changes in the endocrine system through time and space, such as reproductive and adaptive hormones (estrogen, cortisol). Allows comparison of these changes to observable traits and external conditions.
Gonads	Stage of reproductive development. Provides information regarding advancement through time and space, characterization of stages, and determination of spawning strategy. Allows comparisons of these changes to determine relationships between fish attributes, time/space, and other internal and external conditions.

[a] Results from blood samples of our research are not yet complete.

a several-week period. Significant correlations ($P < 0.05$) were found to exist between (1) movements and environmental conditions, bioenergetics, and reproductive factors, (2) environmental conditions and bioenergetics and reproductive factors, and (3) physical characteristics and bioenergetics and reproductive factors. For example, water temperature and salinity were positively correlated to fish location (distance inland). Location was positively correlated with reproductive potential and negatively correlated with stored energy. Reproductive potential was positively correlated with stored energy. Size was positively correlated to both stored energy and reproductive potential. These correlations demonstrate how external conditions, location, physical characteristics, bioenergetics, and reproductive development construct diadromy.

Our results provide, to our knowledge, the first all-encompassing description of diadromy by a particular species, illustrating the inner workings, relationships, and importance of each component. Beginning with the first element (who is migrating), sex ratios, age structure, fish size, and abundance largely determine the population dynamics (recruitment, reproductive potential, etc.) in the region. Undoubtedly, years with an overabundance of virgin spawners and high male to female ratios will contribute less in terms of reproductive potential and future recruitment. The second element (where and when) involves movements and timing, which ultimately exists to achieve optimal feeding, migration, spawning, postspawn survival, and, especially, fitness (Gross 1987). Unexpected changes in water temperature and salinity have substantial influence on fish physiology (particularly reproductive development, osmoregulatory costs, and metabolism) and the overall success of the migration (McKeown 1984; Smith 1985; McDowall 1988). The third element (what and why) of internal changes is related not only to overlapping physiological changes, but also to population characteristics and external conditions. Past research demonstrates powerful links between metabolism, reproductive potential, and endocrinology, along with water temperature, movements, and fish attributes (McKeown 1984; Smith 1985; McDowall 1988). These overlapping components and our ability to measure their levels over the migration of a population have confirmed our general hypotheses regarding diadromy (some depicted and explained in Figure 1) and provided details to describe the migration.

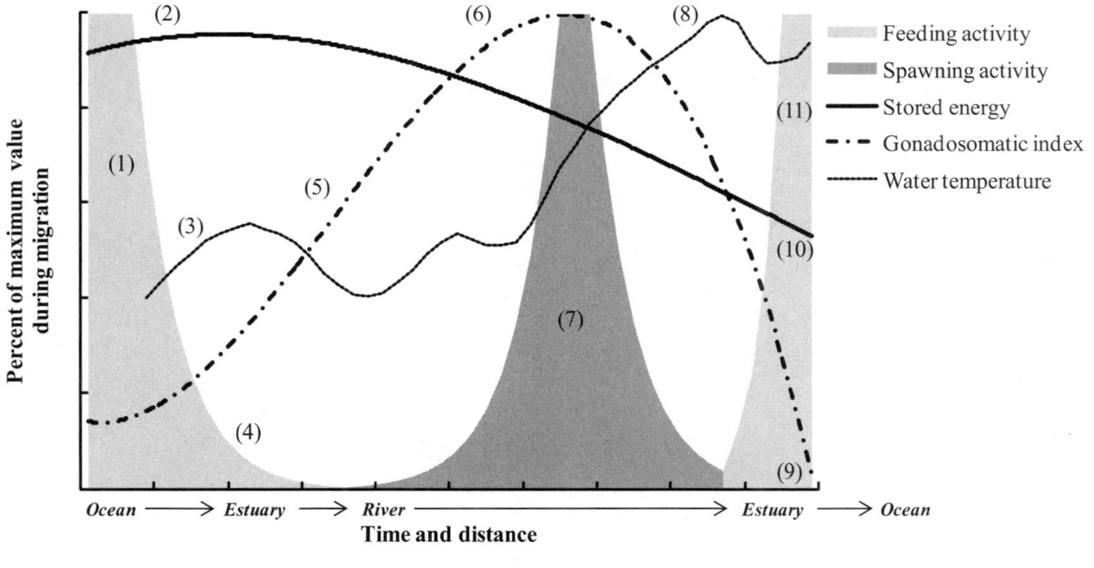

Figure 1.—Hypothetical model of internal changes of iteroparous, batch-spawning fish during anadromous migration. Fish feed intensely (1) preceding onshore movements, causing an increase of stored energy (2). Increasing water temperatures (3) initiates movement inland (*x*-axis). Feeding decreases significantly (4) as fish continue through the estuary. Gonads increase in size as they advance to latter stages of development (5). Reproductive potential approaches greatest values (6) just prior to the peak in spawning activity (7) that lasts for some period of time. Fish near their last spawning episode as fecundity decreases and temperatures rise (8). Reproductive potential decreases dramatically (9) and fish exit the spawning grounds with relatively little stored energy (10). Females reabsorb unspawned atretic oocytes and feeding resumes (11) as fish exit the estuary en route to the ocean.

Understanding the who, what, where, when, and why of diadromy has great biological, ecological, and management implications. Studies that simultaneously address these components (Table 1) will ultimately provide a more accurate description of the migration than otherwise possible. Examining diadromy as a whole explains better the importance of how the individual components function. Resulting data is useful for management through estimating effects of regulations and improving research questions. Broader knowledge will help to identify how climate change, habitat loss, and human interactions can and will influence diadromous fish populations. We acknowledge the drawbacks (e.g., time and resources) associated with a far-reaching approach to research; however, comprehensive results and explanations at the species level are lacking. This groundwork is needed to understand fish migration, identify uncertainties, and provide foundations for future research. Improved knowledge of the dynamic nature of diadromy is imperative to our contribution to maintaining these fish populations, especially in the face of a changing landscape.

References

Gross, M. R. 1987. Evolution of diadromy in fishes. Pages 14–25 *in* M. J. Dadswell, R. J. Klauda, C. M. Moffitt, R. L. Saunders, R. A. Rulifson, and J. E. Cooper, editors. Common strategies of anadromous and catadromous fishes. American Fisheries Society, Symposium 1, Bethesda, Maryland.

Harden-Jones, F. R. 1968. Fish migration. St. Martin's Press, New York.

McDowall, R. M. 1988. Diadromy in fishes: migrations between freshwater and marine environments. Timber Press, Portland, Oregon.

McKeown, B. A. 1984. Fish migration. Timber Press, Portland, Oregon.

Smith, R. J. F. 1985. The control of fish migration. Springer-Verlag, New York.

American Fisheries Society Symposium 69:841–844, 2009

Stream Water Creates a Discernable Odor Gradient that Migratory Juvenile American Eels May Follow Inshore

Peter W. Sorensen*,[1]
Graduate School of Oceanography, University of Rhode Island Narragansett, Rhode Island 02882, USA

Young, migrating anguillid eels face the monumental challenge of locating productive coastal and associated inland freshwater regions after they metamorphose over the continental shelf. Olfactory cues are efficacious in this challenge because of the distances they can travel, the quantity of information they can convey, and the remarkable sensitivity of fish olfactory systems. An example of the power of olfaction is found in adult sea lamprey *Petromyzon marinus*, which can locate spawning streams by detecting subpicomolar concentrations (grams in tens of billion liters; Sorensen et al. 2005) of sulfated steroids released by stream-resident larvae. Creutzberg (1961) appears to have been the first to suggest that innately recognized natural odors emanating from freshwater ecosystems guide young anguillid eels to natural freshwaters from the ocean. Testing various natural waters in a circular maze, he discovered that changes in water chemistry cause elvers to either surge forward with the water flow or reside on the bottom, mediating selective tidal transport. Cruetzberg (1961) also found that salinity gradients alone could not explain this phenomenon as charcoal filtration removed the attractive properties of freshwater. Later, Miles (1968) discovered that the attractive components found in stream water were biodegradable and recognized innately. Subsequently, Sorensen (1986) tested plant extracts and films, conspecific washings, and minerals on elver movement to demonstrate that the freshwater attractants are composed of multiple cues originating from microflora and fauna, which correlated with stream productivity. Although a few candidate compounds have been suggested, definitive biochemical identification has not been performed (Sola 1995). The possible role of freshwater attractants on the actual ability of recently metamorphosed anguillid eels to locate natural inland waters has not yet been described. Here, I describe a study that addressed the latter issue by testing the attractive properties of estuarine waters versus natural stream water, and the ions they contained.

Migrating young eels, starting to acquire pigmentation (i.e., glass eels and elvers), were captured by hand net as they entered Gilbert Stuart Brook in southern New England (USA) in the spring of 1982 and transported to the laboratory in the water they had been captured in. Test waters were collected simultaneously along a transect (A–G; Figure 1) running from Block Island (10 km offshore), up the Pettaquamscutt estuary to Gilbert Stuart Pond, the main source of freshwater for this system. Water samples were collected in polypropylene carboys on two occasions and permitted to reach room temperature (18°C ± 1°C; 4 h) before their attractive natures were tested (following established protocols; Sorensen 1986). Several laboratory waters (well water, tap water, and de-ionized water, to which various ions were added) were also tested to determine the role of inorganic ions on attraction and rheotaxis (tendency to swim up-current). Briefly, groups of 25–30 elvers were placed into a chamber located at the end of a Y-maze with 25-cm long arms through which test waters were introduced at 100 mL/min via two terminal chamber traps. Experiments lasted 10 min, after which the number and position of animals was noted in each arm trap. The experiment was then repeated, reversing the positions of the test waters. Most experiments were repeated a third and fourth time using a second group of elvers. Preferences between waters were evaluated by comparing

* Corresponding author: soren003@umn.edu
[1] Current address: Department of Fisheries, Wildlife and Conservation Biology, University of Minnesota, St. Paul, Minnesota 55108, USA

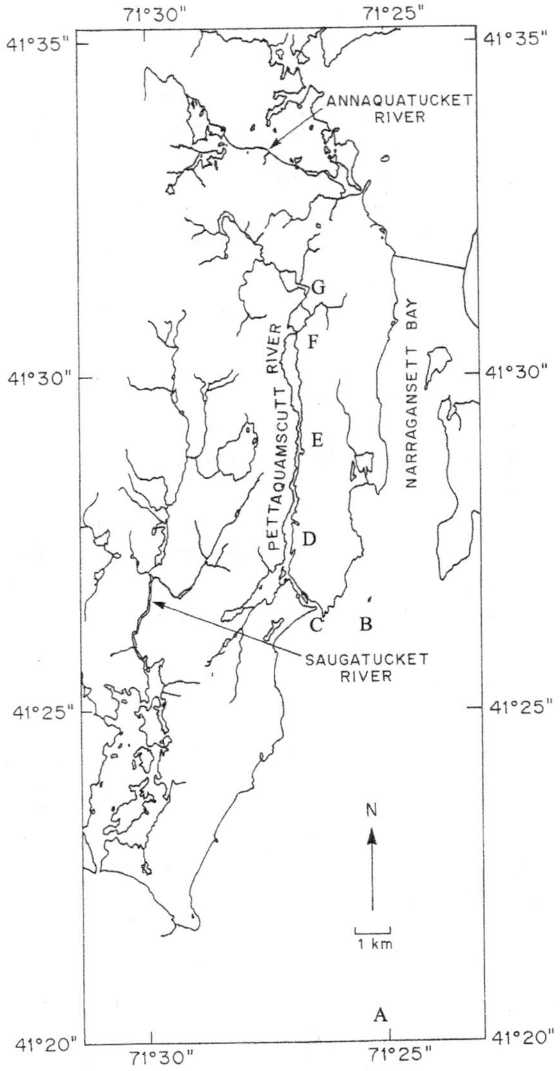

Figure 1.—Locations where water samples were collected for this study. Elvers were collected just downstream of position G (Gilbert Stuart Pond, Rhode Island).

the numbers of elvers found in each trap using two-tailed sign tests ($P < 0.05$), while rheotactic response was documented by noting the total number of animals that swam up-current into a trap.

When offered the choice of inland and offshore waters, elvers consistently chose water collected from the most inland site (Table 1). Thus, stream water was preferred to brackish estuarine waters ($P < 0.01$), estuarine waters were preferred to water collected at the mouth of the estuary ($P < 0.05$), and waters collected at the mouth of the estuary were preferred over waters collected offshore ($P < 0.01$). Water preference was inversely correlated with salinity; however, there was one exception: water from position C (higher salinity than expected owing to a variable plume structure) was preferred over water from position B. Only experiments employing natural stream water (versus laboratory water) caused the majority of elvers to swim up-current and choose a chamber. Inorganic ion content did not alter preferences, while high salinities (i.e., Ringer's and seawater) inhibited rheotaxis (Table 1).

This study demonstrates that migrating elvers have the ability to discern natural fresh waters at

Table 1.—Choices made by elvers offered paired natural and laboratory waters with differing salinities. Values are the number of elvers. $^{*}P < 0.05$; $^{**}P < 0.01$; N = majority did not choose; ppt = parts per thousand.

Paired water position		
Position G (0 ppt) 46**	Position F (15 ppt) 19	No choice 23
Position F (15 ppt) 37**	Position E (30 ppt) 5	No choice 50^{N}
Position E (30 ppt) 26	Position D (31 ppt) 28	No choice 40
Position D (31 ppt) 46**	Position C (32 ppt) 13	No choice 37
Position C (29 ppt) 31*	Position B (26 ppt) 16	No choice 52^{N}
Position B (31 ppt) 46**	Position A (32 ppt) 13	No choice 37
De-ionized (0 ppt) 14**	Position A (32 ppt) 33**	No choice 51^{N}
De-ionized (0 ppt) 14	Position G (0 ppt) 56**	No choice 29
De-ionized (0 ppt) 33	Artificial Pond (1 ppt) 32	No choice 13
De-ionized (0 ppt) 9	Ringer's (6 ppt) 7	No choice 62^{N}

very low concentrations (>10-fold dilutions of stream water), an ability first suggested by Creutzberg (1961). Whether and how this trait is translated into directed movement in the coastal waters cannot be discerned from these simple maze tests, which, none the less, suggest that elvers could be using a mixed strategy involving both chemokinesis and rheotaxis. Organic odorants derived from stream microflora and conspecifics (Sorensen 1986) are probably of greater importance in this process (indicated by weak responses to de-ionized water), but inorganic ions may have a role in modulating rheotaxis-selective tidal transport; high salinity flood tides would cause elvers to drift with the prevailing flow. This sensitive, olfactory-driven orientational strategy can explain both the presence of eels in moderately brackish offshore waters as well as inland freshwater (Edeline et al. 2004). It seems reasonable to speculate that one cause of recent declines in eel populations may be that the concentration of freshwater attractants may have become reduced with declining ecosystem integrity. Enormous strides have been made in natural products chemistry and olfactory physiology over the past decade and render this hypothesis testable.

References

Creutzberg, F. 1961. On the orientation of migrating elvers (*Anguilla vulgaris*) in a tidal area. Netherlands Journal of Sea Research 1:257–338.

Edeline, E., S. Dufour, and P. Elie. 2004. Role of glass eel salinity preference in the control of habitat selection and growth plasticity in *Anguilla anguilla*. Marine Ecology Progress Series 304:191–199.

Miles, S. G. 1968. Rheotaxis of elvers of the American eel (*Anguilla rostrata*) in the laboratory to water from different streams in Nova Scotia. Journal of the Fisheries Research Board of Canada 25:1591–1602.

Sola, C. 1995. Chemoattraction of upstream migrating glass eels, *Anguilla anguilla*, to earthy and green odorants. Environmental Biology of Fishes 43:179–185.

Sorensen, P. W. 1986. Origins of the freshwater attractant

of migrating elvers of the American eel, *Anguilla rostrata*. Environmental Biology of Fishes 17:185–200.

Sorensen, P. W., J. M. Fine, V. Dvornikovs, C. S. Jeffrey, F. Shao, J. Wang, L. A. Vrieze, K. R. Anderson, and T. R. Hoye. 2005. Mixture of new sulfated steroids functions as a migratory pheromone in the sea lamprey. Nature Chemical Biology 1:324–328.

American Fisheries Society Symposium 69:845–846, 2009

Partial Evolutionary Divergence of a Migratory Pheromone between Northern and Southern Hemisphere Lampreys

CINDY F. BAKER* AND MICHAEL STEWART
National Institute of Water and Atmospheric Research
Post Office Box 11 115, Hillcrest, Hamilton 3215, New Zealand

JARED M. FINE AND PETER W. SORENSEN
Department of Fisheries, Wildlife and Conservation Biology, University of Minnesota
200 Hodson Hall, 1980 Folwell Avenue, St Paul, Minnesota 55108, USA

Lampreys are ancient cartilaginous fish. Around half of the lamprey species spend their entire life in freshwater and are not parasitic while the other half have an anadromous life history. Larvae feed in freshwater streams before metamorphosing into a parasitic phase that migrates to the sea (or lake) to feed on fish or mammals before returning to freshwater streams to spawn. The pouched lamprey *Geotria australis* is the only species in the southern hemisphere family Geotriidae and is migratory and parasitic. Previous radio tracking of pouched lampreys indicated that adult fish were attracted to certain rivers and not others (National Institute of Water and Atmospheric Research, unpublished data).

Sorensen et al. (2005) have demonstrated that migratory adult sea lamprey *Petromyzon marinus* locate spawning streams using a pheromone mixture released by upstream resident larvae. This mixture is composed of three sterols (petromyzonamine disulfate [PADS], petromyzosterol disulfate [PSDS], and petromyzonol sulfate [PS]) that are detected at subpicomolar concentrations (1×10^{-13} M). Fine et al. (2004) discovered that the pheromone is not species-specific, as sea lampreys exhibited a behavioral attraction to water that held three other lamprey species. The present study examined sterol release by pouched lamprey as a first step in determining if native New Zealand lamprey could be selecting spawning streams based on the same pheromone mixture as northern hemisphere lampreys.

We examined sterol release by larvae of three species of northern hemisphere lampreys: sea lamprey, American brook lamprey *Lampetra appendix*, northern brook lamprey *Ichthyomyzon fossor*, and the southern hemisphere pouched lamprey.

One-liter tank water samples were extracted using a reverse-phase C18 solid phase extraction cartridge. PADS, PSDS, and PS were detected by high performance liquid chromatography and quantified with electrospray ionization mass spectrometry (following Fine et al. 2006). Samples were analyzed by flow injection analysis. Detection was by selected reaction monitoring for PADS and PSDS and selected ion monitoring (SIM) for PS. The fragment ions ($[M\text{-}H_2SO_4]^-$) for PADS and PSDS were analyzed at *m/z* (mass-to-charge ratio) 605.5 and 491.3, respectively. The pseudomolecular ion of *m/z* 473.5 ([M-Na]–) was detected for PS.

The three northern hemisphere species released PADS, PSDS and PS. All three standards were detected at low picomolar levels; however, only PS was detected in the tank water of pouched lamprey. Samples were concentrated fivefold further and analyzed by LC-MS, using SIM, to confirm results for pouched lamprey. PS was detected; however, no ions were observed for PADS or PSDS. Spiking samples with PADS and PSDS resulted in detection at the same levels as the standard alone.

The American brook and northern brook lampreys are not parasitic or migratory but release PADS, PSDS, and PS in similar quantities to the sea lamprey. Our results indicate that the pouched lamprey only releases PS, even though they have a migratory life history. Thus, it would appear that the production of these compounds as pheromones

* Corresponding author: c.baker@niwa.co.nz

does not reflect a migratory specialization and that the enzyme pathways responsible for their production may have been altered during the course of the evolution of these major lineages of this ancient fish.

References

Fine, J. M., L. A. Vrieze, and P. W. Sorensen. 2004. Evidence that petromyzontid lampreys employ a common migratory pheromone that is partially comprised of bile acids. Journal of Chemical Ecology 30(11):2091–2110.

Fine, J. M., S. P. Sisler, L. A. Vrieze, W. Swink, and P. W. Sorensen. 2006. A practical method for obtaining useful quantities of pheromones from sea lamprey and other fishes for identification and control. Journal of Great Lakes Research 32:832–838.

Sorensen, P. W., J. M. Fine, V. Dvornikovs, C. S. Jeffrey, F. Shao, J. Wang, L. A. Vrieze, K. R. Anderson, and T. R. Hoye. 2005. Mixture of new sulfated steroids functions as a migratory pheromone in the sea lamprey. Nature Chemical Biology 1:324–328.

American Fisheries Society Symposium 69:847–849, 2009

Spawning Areas of Anguillid Eels in the World

Jun Aoyama*
Ocean Research Institute, The University of Tokyo
1–15–1, Minamidai, Nakano, Tokyo 164–8639, Japan

Recent findings relevant to the spawning areas of the genus *Anguilla* are greatly changing our understanding of the migratory ecology and evolutionary history of these fascinating catadromous species. The anguillid larvae (leptocephali) of many sympatric tropical species are morphologically similar, so they have long been impossible to identify to species, and their spawning areas are virtually unknown. Progress in sequencing mitochondrial DNA (Minegishi et al. 2005) has enabled us to identify anguillid leptocephali, eggs, and newly hatched larvae to species. These techniques have been applied recently to determine the spawning areas and distribution of anguillid eel larvae in Indonesian waters and the Pacific and Indian oceans.

Small leptocephali of two tropical anguillid species, Celebes longfin eel *A. celebesensis* (12.3 mm total length [TL]) and *A. borneensis* (8.5 mm TL), were collected around Sulawesi Island, Indonesia (Aoyama et al. 2003). This indicated that *A. borneensis*, which is endemic to the east-central part of Borneo, spawns in the Celebes Sea and migrates back to their nearby freshwater growth habitats (Figure 1). This finding indicates that much shorter migrations of a few hundred kilometers are made by tropical eels to spawn in areas near their freshwater habitats. This clearly contrasts with the long distance migrations of their counterparts in temperate regions, such as European eel *A. anguilla*, American eel *A. rostrata*, and Japanese eel *A. japonica*, and suggests an evolutionary cline of migratory behavior in freshwater eels from tropical to temperate regions.

Interestingly, the small leptocephali of Celebes longfin eel were collected in two different, but neighboring areas, separated by the northern arm of Sulawesi Island: a 12.3-mm-TL specimen in the Celebes Sea and a 13-mm-TL specimen in Tomini Bay. This indicated that this species could have different spawning sites even in a narrow geographic range (Figure 1). Anguillid eels have been considered to be fully or mostly panmictic with single specific spawning sites, despite their wide geographic ranges (Dannewitz et al. 2005). However, it is also possible that Celebes longfin eel forms a single genetic population having multiple spawning sites or two clearly different genetic populations with respective spawning sites within a very small geographic range: these traits have never been found in anguillid eels. Gaining a greater understanding of Celebes longfin eel will give new insight into the ecology of the spawning migration of anguillid eels and the evolutionary processes associated with the development of these peculiar catadromous life histories.

In the North Pacific, newly hatched larvae (preleptocephali) of the Japanese eel were collected in 2005, and their spawning site was suggested to be limited in latitude and longitude near the Suruga Seamount (14°13.7 N, 142°53.0 E) of the West Mariana Ridge (Figure 1; Tsukamoto 2006). In contrast, the evidence from modern research cruises in the Atlantic Ocean indicated that the spawning areas of European and American eels were partly sympatric and widely extended in the Sargasso Sea (23°00.0–29°30.0 N, 52°00.0–74°00.0 W). Japanese eel leptocephali are known to be transported westward by the North Equatorial Current after hatching, and then they transfer to the northward flow of the Kuroshio Current to recruit to their freshwater growth habitats in East Asia while avoiding entrainment into the southward Mindanao Current (Figure 1). Japanese eel leptocephali may be shifted gradually northward in the North Equatorial Current by Ekman transport when they are in a shallow layer (roughly less than 70 m depth; Kimura et al. 1994). Diurnal vertical migration during the larval stage would provide adequate Ek-

* Corresponding author: jaoyama@ori.u-tokyo.ac.jp

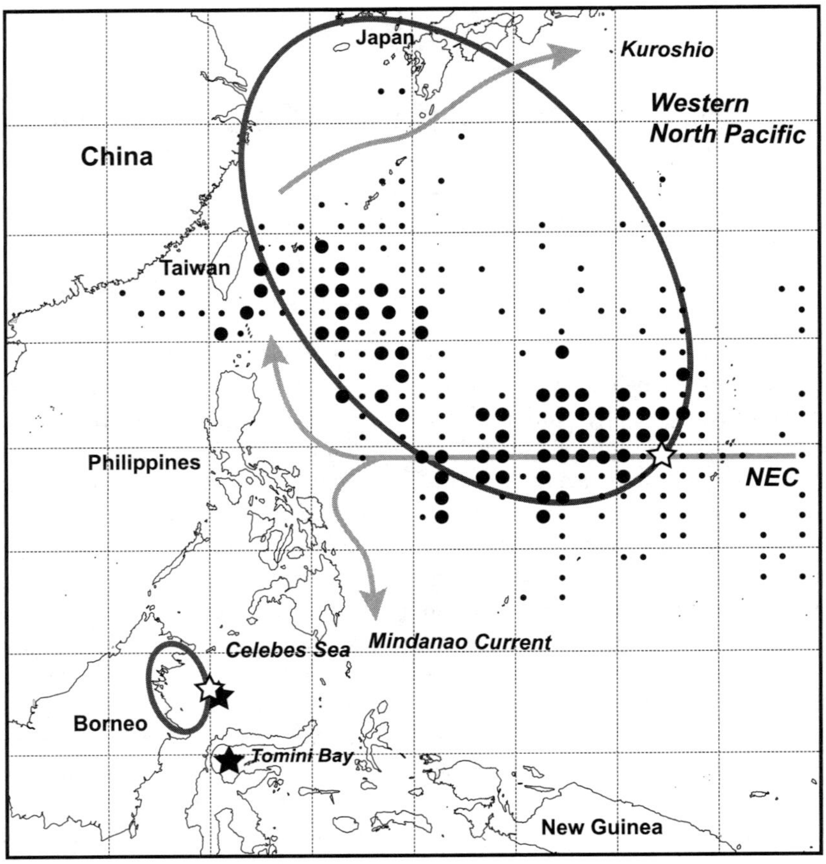

Figure 1.—Estimated spawning sites of anguillid eels in the western Pacific region. The white star in the Pacific indicates the spawning site of the Japanese eel, and dots show the distribution of larvae at one degree resolution (large dots—leptocephali present; small dots—leptocephali absent). The white star in the Celebes Sea indicates the spawning area of *Anguilla borneensis*. Solid stars in the Celebes Sea and Tomini Bay are the spawning areas of Celebes longfin eel. Ellipses in the Pacific and Celebes Sea show the diagrammatic migration loops for Japanese eel and *A. borneensis*, respectively. NEC is the North Equatorial Current.

man transport for their successful recruitment. This mechanism likely needs an exact spawning location that allows the Japanese eel leptocephali to transfer to the Kuroshio Current within their typical growth and ontogenetic behavioral patterns in relation to the velocity of the North Equatorial Current. Differences in ranges of the spawning areas between Pacific and Atlantic eels suggested that the Japanese eel has specific larval migratory strategies in relation to oceanographic conditions. Spawning over a wider area or changes in the oceanographic structure of the North Equatorial Current region would likely increase the number of their leptocephali transported toward the Mindanao Current that would bring larvae to areas outside the species range. This suggested that changes in ocean-atmospheric conditions, such as El Niño, could strongly affect the recruitment success of the Japanese eel.

Some anguillid leptocephali have been collected recently from the South Pacific and Indian oceans. Considering that less than 20 anguillid leptocephali have been reported from the South Pacific and western Indian oceans, these new collections could be very important historical records (Aoyama et al. 2007), and studies on their detailed ecological information are in progress.

All spawning areas of the anguillid eels confirmed to date are located in tropical or subtropical waters over the deep ocean and have similar physical conditions, as was suggested at the beginning of 19th century. However, recent findings on the spawning areas suggested some variation in which biological or ecological factors may strongly constrain anguillid eels to spawn in their species–population specific locations.

References

Aoyama, J., S. Wouthyzen, M. J. Miller, T. Inagaki, and K. Tsukamoto. 2003. Short-distance spawning migration of tropical freshwater eels. Biological Bulletin 204:104–108.

Aoyama, J., S. Wouthuyzen, M. J. Miller, Y. Minegishi, M. Kuroki, S. R. Suharti, T. Kawakami, K. O. Sumadhiharga, and K. Tsukamoto. 2007. Distribution of leptocephali of the freshwater eels, genus *Anguilla*, in the waters off west Sumatra in the Indian Ocean. Environmental Biology of Fishes 80:445–452.

Dannewitz, J., G. E. Maes, L. Johansson, H. Wickstrom, F. A. M. Volkaert, and T. Jarvi. 2005. Panmixia in the European eel: a matter of time. Proceedings of the Royal Society B 272:1129–1137.

Kimura, S., K. Tsukamoto, and T. Sugimoto. 1994. A model for the larval migration of the Japanese eel—roles of the trade winds and salinity front. Marine Biology 119:185–190.

Minegishi, Y., J. Aoyama, J. G. Inoue, M. Miya, M. Nishida, and K. Tsukamoto. 2005. Molecular phylogeny and evolution of the freshwater eels genus *Anguilla* based on the whole mitochondrial genome sequences. Molecular Phylogenetics and Evolution 34:134–146.

Tsukamoto, K. 2006. Spawning of eels near a seamount. Nature (London) 439:929.

American Fisheries Society Symposium 69:851–853, 2009

Migratory Characteristics as a Key Factor for Determining the Population Structure of a Diadromous Species: Eels Tend to Be Panmictic while Salmon Divide Populations with Ease

Yuki Minegishi*, Jun Aoyama, and Katsumi Tsukamoto
Ocean Research Institute, The University of Tokyo
1–15–1 Minamidai, Nakano-ku, Tokyo 164–8639, Japan

Eels and salmon are typical representative fishes of the diadromous life histories of catadromy and anadromy, respectively. Although the population structure of anadromous salmonid species has been well studied, those of catadromous anguillid eels, especially tropical species, are poorly known.

Anguillid eel population structures have been thought to be basically panmictic because they have species-specific spawning areas far offshore and the glass eels recruit randomly to freshwater habitats. However, we found multiple population structures for three anguillid species—giant mottled eel *Anguilla marmorata*, *A. bicolor*, and *A. australis*—by analyzing genetic variations of microsatellite (nuclear DNA) and control region sequences (mitochondrial DNA). Marbled eel has the widest geographic distribution among anguillid species, ranging from the western Indian Ocean to the eastern Pacific Ocean. The analyses for marbled eel (N = 449 individuals) showed significant differences (F_{ST} = 0.013–0.240 for microsatellite, 0.053–0.698 for the control region, $P < 0.05$) among 13 localities throughout their geographic range in different oceans (Madagascar, Réunion, Sumatra, Sulawesi, Philippines, Taiwan, Japan, Guam, Ambon, Papua New Guinea, New Caledonia, Fiji, and Tahiti), whereas no significant differences were observed among localities within the same ocean except for a few comparisons: Papua New Guinea and Tahiti in the South Pacific Ocean for microsatellite (F_{ST} = 0.020, $P < 0.05$), and Sumatra and Réunion in the Indian Ocean for the control region (F_{ST} = 0.563, $P < 0.001$). Large geographic groups were roughly clustered in the neighbor-joining tree based on the control region. We suggest that marbled eel was composed of four genetically different populations: North Pacific, South Pacific, Indian Ocean, and Mariana (Figure 1). Each population was considered to be fully panmictic or to have a totally connected metapopulation structure (Minegishi et al. 2008).

Anguilla bicolor and *A. australis* (Minegishi et al. 2008) have been classified as two subspecies based on their morphological differences, but this classification has been debated recently. Their subspecies have been described as *A. bicolor bicolor* in the Indian Ocean, *A. bicolor pacifica* in the Pacific Ocean, *A. australis australis* in Australia, and *A. australis schmidtii* in New Zealand. Our study revealed that *A. bicolor* (N = 145), from six representative localities in their species range (Madagascar, Réunion, Seychelles, Mayotte, Sumatra, and Philippines), showed clear divergence between the Indian Ocean (*A. bicolor bicolor*) and Pacific Ocean (*A. bicolor pacifica*) (F_{ST}= 0.018 and 0.697–0.706, $P < 0.001$). *Anguilla australis* (N = 73) from Australia and New Zealand showed slight but statistically significant differences (F_{ST} = 0.019 and 0.037, $P < 0.001$). These results support the subspecies designations that have long been accepted.

The population structure of anguillid eels found in our study corresponds to oceanic scales, and each population shows high genetic homogeneity despite their geographic distributions, ranging over thousands of kilometers. This suggests that they have unique migratory characteristics, such as spawning areas, migratory routes, or spawning periods, that differ from other populations.

Anadromous salmonids, unlike eels, breed principally in isolated streams and show clear ge-

* Corresponding author: minegisi@ori.u-tokyo.ac.jp

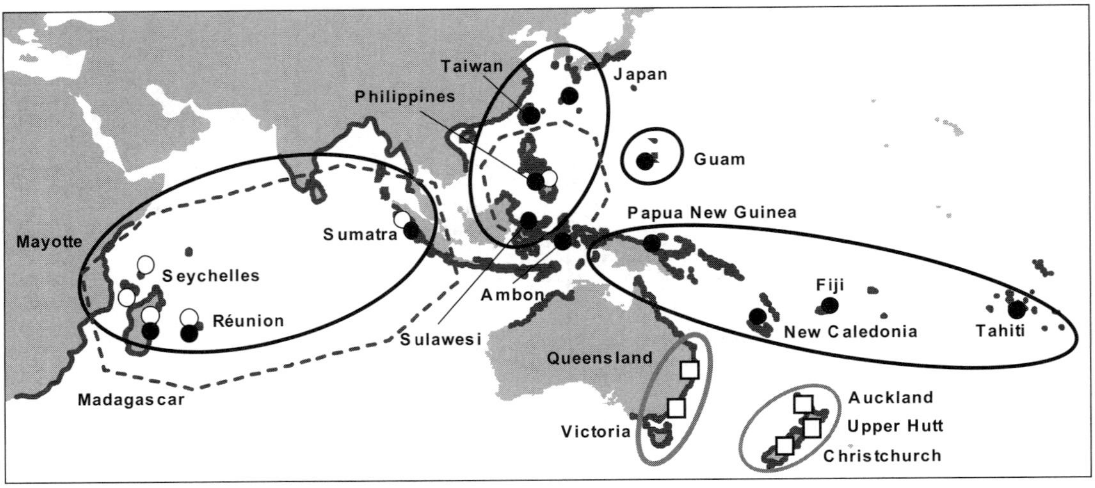

Figure 1.—Sampling localities and population structure of three anguillid species: filled circles and black lines are marbled eel, open circles and broken lines are *Anguilla bicolor*, and open squares and gray lines are shortfin eel. Bold lines along coasts show species distribution.

netic divergence at much smaller spatial scales than anguillid eels. For example, geographic differentiations were observed in brook trout *Salvelinus fontinalis* at a 100-km scale and in cutthroat trout *Oncorhynchus clarkii* at a 50-km scale (Castric et al. 2001; Wenburg and Bentzen 2001). In addition to geographic variations, clear ecological divergences were found in North Pacific pink salmon *O. gorbuscha*, depending on their year-class (Churikov and Gharrett 2002). Furthermore, rainbow trout *O. mykiss* shows genetic divergence between life history forms, with anadromous and resident types being found in the same river (Narum et al. 2004). This suggests that the highly diverged migratory ecology of salmonid fishes result in the various population subdivisions.

The population structures of diadromous fishes require that each population has a specific migration loop, which is the conceptual migration route connecting the spawning area and the growth habitat of a species or population, including particular life history parameters such as growth rates and timing of metamorphosis (Tsukamoto et al. 2002). Spatial or temporal shifts in one of these ecological characters can shift the migration loop and, thus, cause the subdivision of a population to result in speciation. However, the migration loops of catadromous anguillid eels appear to be comparatively stable and not easily shifted to subdivide populations, even over long evolutionary time scales. This is due to the long distance migrations of eels to a far offshore spawning area, the passive transport of leptocephali by ocean current systems, the random recruitment of glass eels, and the widely ranging ages of silver eels at downstream migration.

In contrast, anadromous salmon that have a spawning migration back to their natal rivers are easily divided into subpopulations because their migration loops can diverge depending on each stream, year-class, or migratory form, if homing is accurate and their life cycle is strict. Therefore, this suggests that migratory patterns strongly characterize the genetic population structure of diadromous fishes.

References

Castric, V., F. Bonney, and L. Bernatchez. 2001. Landscape structure and hierarchical genetic diversity in the brook charr, *Salvelinus fontinalis*. Evolution 55(5):1016–1028.

Churikov, D., and A. J. Gharrett. 2002. Comparative phylogeography of the two pink salmon broodlines: an analysis based on a mitochondrial DNA genealogy. Molecular Ecology 11(6):1077–1101.

Narum, S. R., C. Contor, A. Talbot, and M. S. Powell. 2004. Genetic divergence of sympatric resident and anadromous forms of *Oncorhynchus mykiss* in the Walla Walla River, USA. Journal of Fish Biology 65(2):471–488.

Tsukamoto, K., J. Aoyama, and M. J. Miller. 2002. Migration, speciation, and the evolution of diadromy in anguillid eels. Canadian Journal of Fisheries and Aquatic Sciences 59(12):1989–1998.

Wenburg, J. K., and P. Bentzen. 2001. Genetic and behavioral evidence for restricted gene flow among coastal cutthroat trout populations. Transactions of the American Fisheries Society 130(6):1049–1069.

American Fisheries Society Symposium 69:855–857, 2009

Climate Change May Have Affected Growth and Life History in Atlantic Salmon Juveniles over the Past 30 Years

Etienne Rivot*, Charle Perrier, Céline Jouanin, and Jean-Marc Roussel
Agrocampus Rennes, INRA, UMR 985 Ecologie et Santé des Ecosystèmes
65 rue de St Brieuc, 35042 Rennes, France

Pascal Riera
Station Biologique de Roscoff, UMR 7127 CNRS-INSUE Université Pierre et Marie Curie
Place Georges Tessier BP74, 29682 Roscoff, France

Jean-Pierre Porcher
ONEMA, Office National de l'Eau et des Milieux Aquatiques
84 rue de Rennes, 35510 Cesson Sévigné, France

Jean-Luc Baglinière
Agrocampus Rennes, INRA, UMR 985 Ecologie et Santé des Ecosystèmes
65 rue de St Brieuc, 35042 Rennes, France

Atlantic salmon stocks are declining in European and North American rivers (ICES 2006). French rivers are located at the southern edge of the species distribution in Europe, providing a good opportunity to study life history variants in a changing environment (local and global) where unfavorable conditions may occur rapidly. In these rivers, almost all juveniles smoltify at age 1 (Smolt1) or age 2 (Smolt2), and adults return to rivers to spawn after 1 or 2 sea winters (1SW or 2SW). In the present work, we focus on the age at smoltification as an integrated parameter of changes occurring during the juvenile period in rivers. More precisely, variations in the mean age at smoltification (MAS) were studied in 23 rivers of the Armoricain Massif (western France) belonging to three geographical regions: northern Brittany, southern Brittany, and lower Normandy.

Archived scales of 24,500 adult salmon caught by anglers from 1972 to 2005 were read to estimate smolt age and sea age of each individual (Baglinière 1985). Logistic regressions indicated that MAS decreased significantly over the study period whatever the geographical region and sea age. Moreover, the analysis of autocorrelation in the time series of residuals highlighted the existence of medium-term fluctuations that were synchronous across the rivers, suggesting a response to a global environmental forcing. The proportion of Smolt1 and average monthly air temperature showed significant increasing time trends over the period (Figure 1). Linear correlation between series of residuals (proportion of Smolt1 and air temperature after removing the linear trend) was positive and highly significant (R^2 = 0.55, P = 0.006) despite reduction of the degrees of freedom due to high autocorrelation in the series.

The hypothesis that the temporal changes in MAS results from modifications in juvenile growth conditions was investigated on the Scorff River in southern Brittany. We used archived scales of 720 adult salmon caught by anglers, juvenile salmon caught during annual electrofishing censuses, and complementary data from scientific surveys (adult and juvenile trapping), from 1972 to 2002. Back-calculations were performed to estimate length of juveniles at their first winter, according to the method proposed by Beall et al. (1992) with recommendations from Francis (1990). Back-calculated length of juveniles during their first winter showed a significant increase of 1.5 cm, on average, over the study period (Figure 2). Similarly, water tem-

* Corresponding author: etienne.rivot@agrocampus-rennes.fr

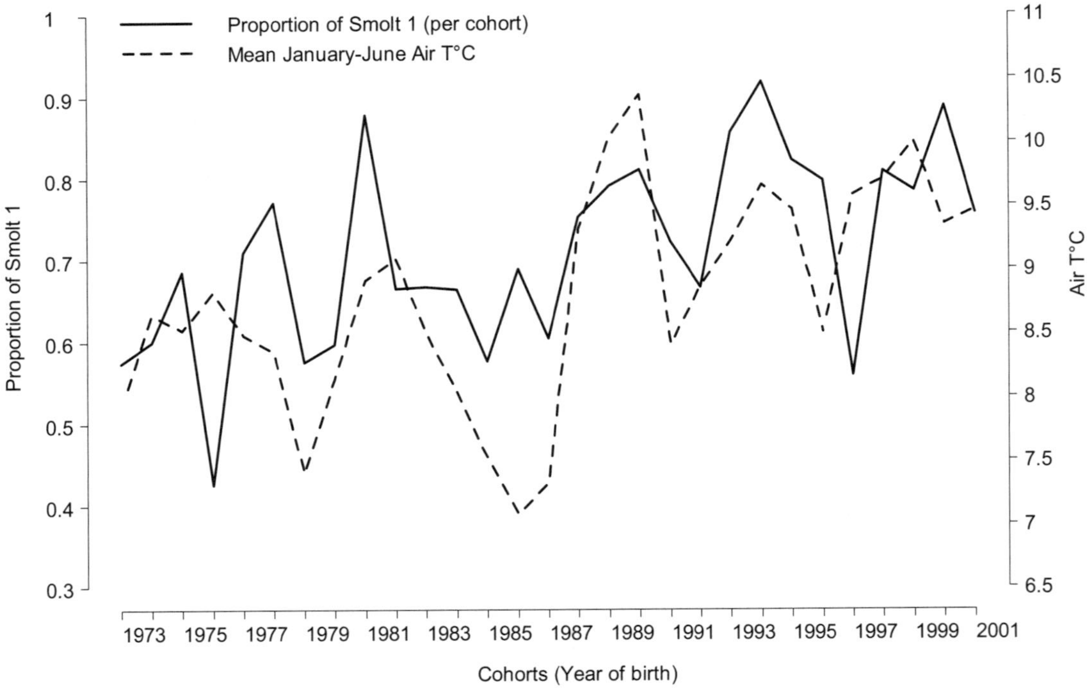

Figure 1.—Correlations between the proportions of age-1 smolt in Atlantic salmon populations in western France and mean air temperature during the spring growing period, from 1973 to 2001.

peratures during the March to June growth period correlated positively with juvenile body length.

Stable isotope analysis (C and N) were run on 162 samples of archived scales collected on juvenile Atlantic salmon caught in the Scorff River from 1977 to 2006. No specific trend was found for scale δ^{15}N, whereas scales became increasingly ^{13}C-enriched (+1.4‰ for δ^{13}C) over the period (Figure 2). Livestock production increased in the watershed until the mid-1990s, resulting in a remarkable enhancement of nitrate concentration (NO_3^-; 16 mg/L) in the river. A significant positive correlation (R^2 = 0.67, P = 0.004) was found between NO_3^- concentration and δ^{13}C in juvenile salmon scales, suggesting that scales became increasingly ^{13}C enriched as more dissolved nutrients were available in the river. This suggests that in the context of increasing carbon demand, fixation of dissolved atmospheric CO_2 (about –8‰ for δ^{13}C) by river primary producers increased over the study period, leading to more ^{13}C-enriched dissolved inorganic carbon (DIC). Under such hypothesis, the incorporation of ^{13}C-enriched DIC through photosynthesis could have led to higher δ^{13}C values from primary producers to higher trophic levels in the entire river food web, including insectivorous Atlantic salmon *Salmo salar* juveniles.

Taken in concert, results suggest that the decrease in MAS might be due to changes in juvenile growth performance in response to climate warming, increase in stream productivity, or both. We cannot rule out the alternative hypothesis that the decrease in MAS could be a response of populations to selective fishing on late-maturing and large fish.

References

Baglinière, J. L. 1985. La détermination de l'âge par scalimétrie chez le saumon atlantique dans son aire de répartition méridionale: utilisation pratique et difficultés de la méthode. Bulletin Français de la Pêche et de la Pisciculture 298:69–105.

Beall, E., P. Davaine, D. Bazin, and J. M. Blanc. 1992. Détermination d'un modèle de rétrocalcul pour l'estimation de la croissance de la truite de mer (*Salmo trutta*) à Kerguelen. Pages 193–198 *in* J. L. Baglinière, J. Castanet, F. Conand, and F. J.

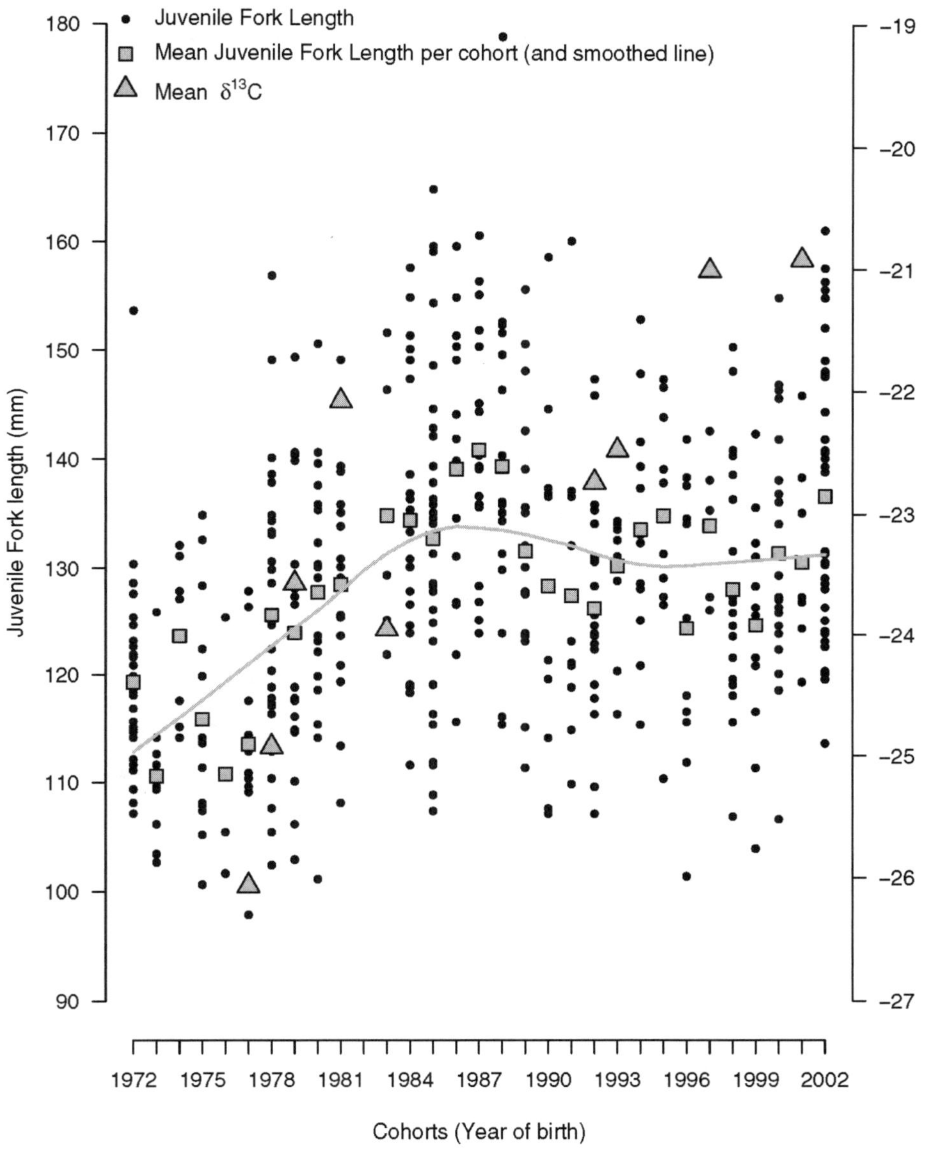

Figure 2.—Back-calculated length of juveniles during their first winter in the Scorff River (southern Brittany), and the increase in juvenile scale $\delta^{13}C$ (‰), from 1972 to 2002.

Meunier, editors. Tissus durs et âge individuel des vertébrés, Colloque national, Bondy, France. ORSTOM-INRA Editions, Colloques et Séminaires, Paris.

Francis, R. I. C .C. 1990. Back-calculation of fish length: a critical review. Journal of Fish Biology 36:883–902.

ICES (International Council for the Exploration of the Sea). 2006. Report of the working group on North Atlantic salmon (WGNAS). ICES CM 2006/ ACFM 23.

American Fisheries Society Symposium 69:859–862, 2009

The Spawning Habitat of Anadromous Rainbow Smelt: Trouble at the Tidal Interface

BRADFORD C. CHASE*
Massachusetts Division of Marine Fisheries
1213 Purchase Street, 3rd Floor, New Bedford, Massachusetts 02740, USA

Anadromous rainbow smelt *Osmerus mordax* depend on discrete spawning riffles near the tidal interface to deposit demersal, adhesive eggs. In many rivers, these locations are also centers of human development where watershed alterations have degraded water and substrate quality. The degradation of spawning habitat could reduce a fitness advantage in the life history of diadromous fish that depend on the integrity of freshwater habitats (Gross 1987). Smelt populations at the southern end of their range have declined sharply in the past 30 years, and in southern New England, some runs are now barely detectable. A monitoring project was conducted on the Gulf of Maine coast of Massachusetts, in response to growing concerns over the status of smelt, to document spatial occurrence and temporal use of smelt spawning habitat, identify influences on smelt spawning success, and develop recommendations on habitat restoration.

All freshwater drainages (N = 162) from the Cape Cod Canal to New Hampshire were surveyed to select spawning habitat monitoring stations. Sixty-four selected monitoring stations were visited twice each week from March 1 through May 31 for one to three seasons. Observations of deposited smelt eggs formed the basis for delineating smelt spawning habitat. Egg monitoring focused on the first riffle upstream of tidal influence. Once smelt eggs were identified, monitoring was expanded to record the upstream and downstream limits of egg deposition, resulting in spawning habitat measurements of river length and substrate area for each river. Water chemistry was measured at each station visit, and streamflow measurements and ichthyoplankton samples were collected at some stations.

* Corresponding author: brad.chase@state.ma.us

Smelt spawning habitat was identified at 45 locations (mapped by Global Positioning System coordinates reported by Chase 2006) in 30 river systems (Figure 1). Smelt spawning began at riffle habitat near the interface of saltwater and freshwater, with few exceptions. Spawning progressed upstream beyond the tidal influence as far as a kilometer and typically ceased at physical barriers to passage. The size of spawning habitats varied widely, although most were less than 200 m in stream length and less than 1,000 m^2 in substrate area. Only five locations provided 10,000–15,000 m^2 of spawning substrate.

he typical spawning period was from mid-March until mid-May. Egg deposition typically peaked in April and was intermittent in early March and late May. The temporal range when smelt eggs were present was March 3 to May 28. The spawning period of smelt in Massachusetts begins earlier and extends more than twice the duration as found in their northern range (McKenzie 1964; Trencia et al. 2005). The average water temperature at the onset of spawning runs was 5.3°C. However, water temperature, the onset of spawning, and the duration of the spawning period varied widely, with some dependence on the size of spawning habitat and seasonal weather. Smelt larvae were present in the tidal waters downstream of the spawning habitat from April 14 through May 31.

Water temperature, dissolved oxygen, pH, salinity, and specific conductivity were measured and compared to Massachusetts Surface Water Quality Standards. All water temperature and dissolved oxygen measurements were supportive of aquatic life, both favorably influenced by the cool air temperature and higher discharge during the spring freshet. Routine violations to the pH criterion (<6.5) were observed in most rivers in the South Coastal Basin, raising concerns about egg survival. Evidence of cul-

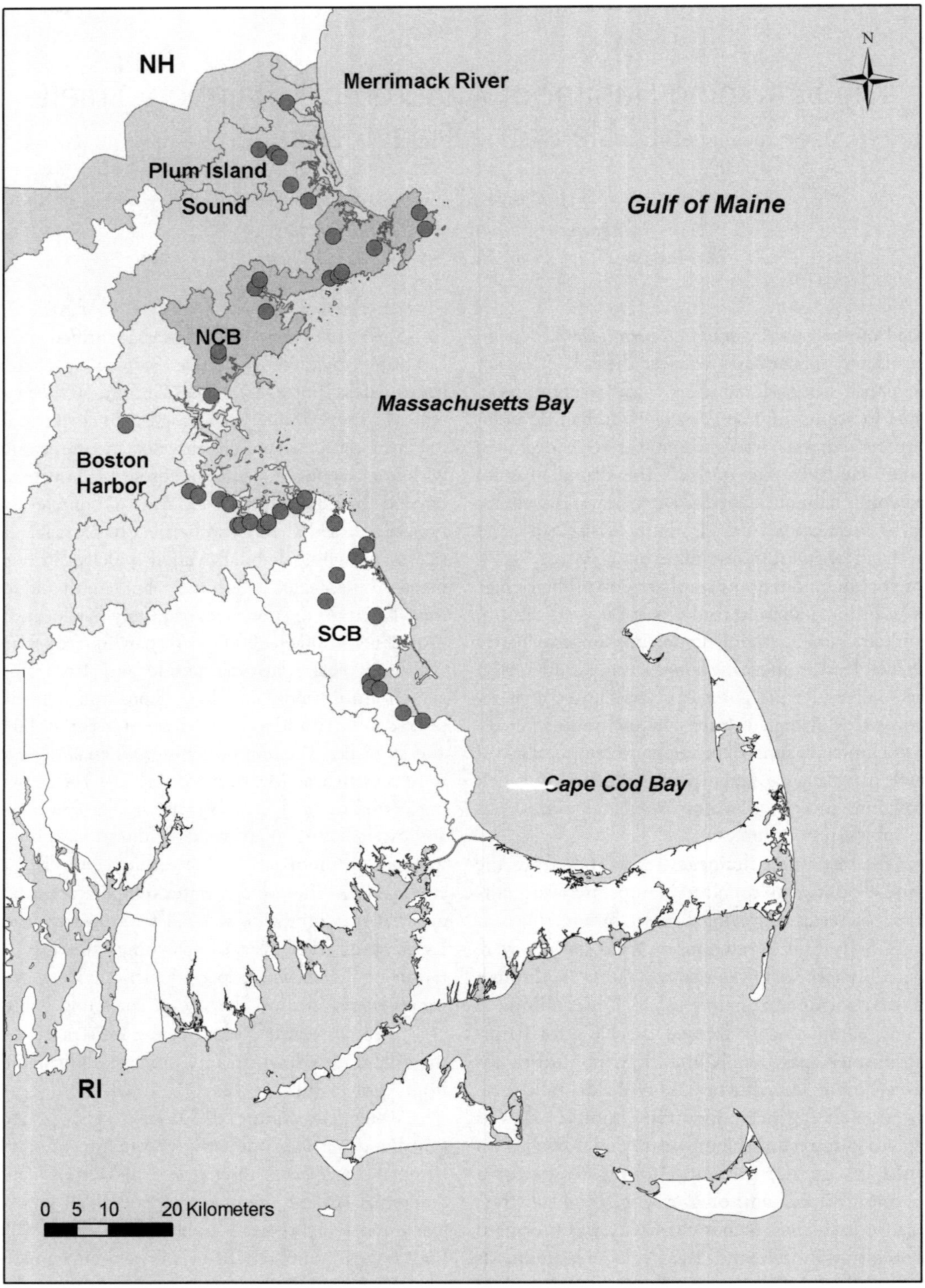

Figure 1.—Location of smelt spawning habitat on the Gulf of Maine coast of Massachusetts. Dots indicate the 45 locations designated as smelt runs in the study area. NCB = North Coastal basin; SCB = South Coastal basin.

tural eutrophication was found with high nutrient concentrations at spawning locations (subset of six rivers). All sampled rivers had total dissolved phosphorus (TDP) concentrations, and four of six rivers had nitrite + nitrate concentrations that exceeded predicted concentrations (6.4 μg/L TDP; 0.267 mg/L N+N; Sosiak 2002) that would cause nuisance periphyton biomass. The downstream limits of spawning routinely experienced tidal influence but were not typically exposed to the salt wedge. In locations where eggs received saline water, the cause was human alterations such as bridge constrictions, flood control structures, and dams.

Nine physical and chemical conditions were identified as degrading the stream habitat for migrating adults and egg survival: sedimentation, eutrophication, passage impediments, channel alterations, stream flow reduction, stormwater, tidal influence, vegetative buffer loss, and acidification. These conditions were ranked in each smelt run, resulting in the highest summed scores for sedimentation and eutrophication. The effects of eutrophication, in the form of excessive periphyton growth, on spawning substrata were nearly ubiquitous. The cumulative effect from all the negative influences is the reduction in the quantity and quality of spawning habitat. It is hypothesized that the degradation of smelt spawning habitat is causing lower egg survival and chronic reduction of smelt recruitment. Eutrophication may be the most significant threat because the periphyton growth can result in spawning substrate that is poorly suited for a life history depending on a demersal, adhesive egg with a long incubation period.

Most smelt spawning locations have been modified to accommodate transportation routes, stormwater drainage, and property development. Reduced presence of stream channelization, crossings, and passage impediments was highly favorable towards maintaining natural riffle habitat. A synopsis of monitoring provided a description of favorable smelt-spawning habitat: a deep-channel estuary that allows the salt wedge to meet a moderate gradient riffle at the tidal interface, unimpeded passage upstream to the freshwater zone with ample vegetative buffer and canopy, and an extended pool-riffle complex that limits egg crowding and provides shelter. This scenario was only approximated in the Fore River, the largest smelt run in the study area, because of common riparian and channel alterations. Only 1 of 45 spawning locations had no riparian or channel alterations. Other positive features recorded were substrate conditions that increased flow turbulence, velocity, and increased surface area for egg deposition. Riffle habitat with turbulent water velocity (0.5–0.8 m/s), coarse cobble (10–20 cm diameter), and aquatic moss *Fontinalis* spp. were suitable for adult attraction, egg attachment, and egg survival.

A large number of smelt-spawning locations were documented on the Gulf of Maine coast of Massachusetts. However, most locations suffered from a variety of stressors. The relation of spawning habitat degradation to smelt population decline remains unknown for this poorly assessed species. Robust smelt fisheries of the 1960s and 1970s were followed by sharply declining fisheries and smelt runs in the 1980s and 1990s, a period of extensive coastal development and improved industrial point source water quality but declining stormwater quality. Concerns for smelt have increased in New England, resulting in the 2004 designation of smelt as a species of concern under the Endangered Species Act review process based on decreased harvest records and distribution truncation (NOAA 2004). The degradation of spawning habitat is a likely contributor to the present status of smelt runs and fisheries. The deposition of demersal, adhesive eggs in freshwater riffles near the tidal interface is an important adaptation of smelt life history. This reproductive process may be at risk because of chronic degradation of spawning substrates in developed watersheds in Massachusetts. The documentation of smelt spawning habitat, in an accessible geographic information system database, will help resource managers protect these habitats.

References

Chase, B. C. 2006. Rainbow smelt (*Osmerus mordax*) spawning habitat on the Gulf of Maine coast of Massachusetts. Massachusetts Division of Marine Fisheries, Technical Report No. 30, Gloucester.

Gross, M. R. 1987. Evolution of diadromy in fishes. Pages 14–25 *in* M. J. Dadswell, R. J. Klauda, C. M. Moffitt, R. L. Saunders, R. A. Rulifson, and J. E. Cooper, editors. Common strategies of anadromous and catadromous fishes. American Fisheries Society, Symposium 1, Bethesda, Maryland.

McKenzie, R. A. 1964. Smelt life history and fishery in

the Miramichi River, New Brunswick. Bulletin of Fisheries Research Board of Canada No. 77.

NOAA (National Oceanic and Atmospheric Administration). 2004. Species of concern in the northeast region. National Marine Fisheries Service, Proactive Conservation Program, Silver Spring, Maryland.

Sosiak, A. 2002. Long-term response of periphyton and macrophytes to reduced municipal nutrient loading to the Bow River (Alberta, Canada). Canadian Journal of Fisheries and Aquatic Sciences 59:987–1001.

Trencia, G., G. Verreault, M. Legault, and V. Tremblay. 2005. L'eperlan arc-en-ciel (*Osmerus mordax*) anadrome du sud de l'estuaire du Saint-Laurent. Le Naturaliste Canadien 129(1):86–94.

American Fisheries Society Symposium 69:863–865, 2009

Differences in Accumulation Pattern of Organochlorine Compounds in Masu Salmon between Fluvial and Sea-Run Forms

Mayuko Oka*
Ocean Research Institute, The University of Tokyo
1–15–1 Minamidai, Nakano, Tokyo, Japan

Takaomi Arai
International Coastal Research Center, Ocean Research Institute, The University of Tokyo
2–106–1, Akahama, Otsuchi, Iwate, Japan

Nobuyuki Miyazaki
Ocean Research Institute, The University of Tokyo
1–15–1 Minamidai, Nakano, Tokyo, Japan

Masu salmon *Oncorhynchus masou masou* select one of two life histories after a year of life in the river: the sea-run form or the fluvial form (Kubo 1980). Since the masu salmon is semelparous, the sea-run form returns to the river for spawning after a year of life in the ocean. Marine organisms, such as masu salmon, that feed at higher trophic levels will have higher concentrations of organic pollutants due to bioaccumulation. Many of the masu salmon carcasses will be left on the river bottom after their spawning migration. Recently, these contaminated carcasses were of concern in transferring contaminants to the river sediments (Ewald et al. 1998; Krümmel et al. 2003), as well as transporting marine-origin nutrients. Pollution of the river sediments from masu salmon carcasses may lead to serious secondary pollution to the other organisms in the river. We determined the accumulation of organochlorine compounds in muscle and liver tissue of fluvial and sea-run masu salmon to understand the differences in contaminant accumulation between the two types of life styles. We also examined the possible source of their pollution.

Thirteen samples of fluvial masu salmon were collected by fishing from the Kogawa and Kumano rivers, located in Iwate Prefecture, northern Japan, and 10 samples of the sea-run masu salmon were collected by fixed fishing net from Otsuchi Bay in Iwate Prefecture. The pollutants analyzed were hexachlorocyclohexane, hexachlorobiphenyl (HCB), dieldrin, chlordane, DDT (dichlorodiphenyltrichloroethane), heptachlor, and mirex. Each 5-g sample was homogenized and extracted with 1:1 hexane–acetone solvent using the accelerated solvent extractor Dionex ASE 200. Cleanup preparation was done with gel permeation chromatography, Florisil column chromatography, and silica-gel column chromatography. Identification and quantification were done by gas chromatography-negative chemical ionization mass spectrometry (Agilent 5973N).

The concentrations of organochlorine compounds in liver and muscle tissue of the sea-run salmon were 8–83 times higher than that of fluvial salmon (Figure 1). The *p,p'*-DDE was the most predominant compound in liver and muscle tissue of the sea-run salmon and was predominant in the muscle tissue of the fluvial salmon. HCB was the most predominant compound in liver in fluvial salmon. Almost all contaminants analyzed were detected in the sea-run salmon, but many of the contaminants were below the detection limit in fluvial salmon. Only HCB, DDE, trans-chlordane, and trans-nonachlor were detected. The organochlorine pesticides, which were used mainly

* Corresponding author: delphi19_@hotmail.com

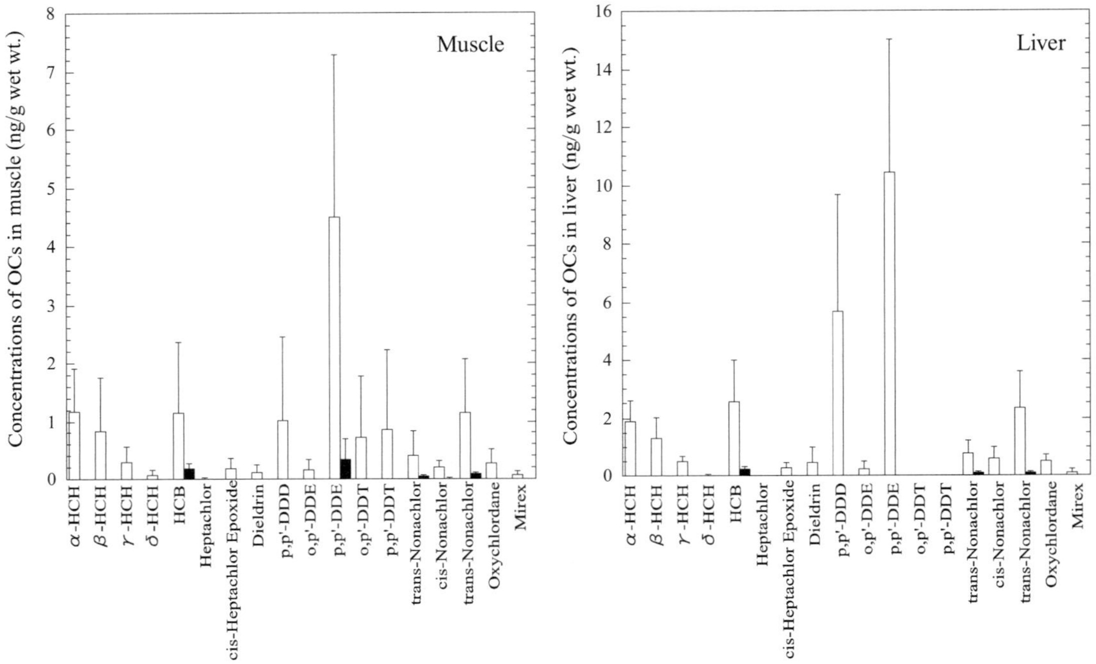

Figure 1.—Concentrations of organochlorines in liver and muscle of sea-run and fluvial forms. The white bars indicate sea-run form and black bars indicate fluvial form masu salmon.

on land, will ultimately accumulate in seawater via atmospheric deposition and biotransportation. The results show that the sea-run salmon are more contaminated than fluvial salmon due to differences in prey contamination level. Sea-run masu salmon have the potential to increase pollutants in the river sediments since the concentrations of the pollutants are much higher in the sea-run form. The upper river, which was originally free of heavy pollution, could receive considerable amounts of pollutants from the sea.

The trans-nonachlor/trans-chlordane (N:C) concentration ratio was used to estimate new input of pollutants or the distance from the pollution source (Iwata et al. 1993; de Brito et al. 2002). The N:C ratio in the environment is likely to be 0.15–0.45 under continuous usage of chlordanes, and the ratio will increase according to the distance or time from the pollution input.

The N:C ratio in the muscle of fluvial salmon was significantly lower than that of the sea-run salmon (Mann–Whitney *U*-test, $P < 0.05$, Table 1). This indicates that the source of the organic pollutants in the two forms could be different. The

Table 1.—Mean N/C ratio in muscle and liver tissue of masu salmon (mean N/C ratio ±standard deviation).

Form	Site	Mean N/C ratio
Fluvial	Muscle	2.4±1.4
	Liver	4.6±6.4
Sea-run	Muscle	12.0±12.9
	Liver	3.1±0.6

fluvial salmon was contaminated by chemicals of comparatively recent input. Although the fluvial salmon are not seriously affected by the sea-run origin pollutants at this point, further marine pollution may result in the sea-run salmon acting as vectors, increasing the contaminant load in fluvial fishes. Therefore, continuous monitoring and further research is essential.

References

de Brito, A. P. X., D. Ueno, S. Takahashi, and S. Tanabe. 2002. Organochlorine and butyltin residues in walleye pollock (*Theragra chalcogramma*) from Bering Sea, Gulf of Alaska and Japan Sea. Chemosphere 46(3):401–411.

Ewald, G., P. Larsson, H. Linge, L. Okla, and N. Szarzi. 1998. Biotransport of organic pollutants to an inland Alaska lake by migrating sockeye salmon (*Onchorhynchus nerka*). Arctic 51(1):40–47.

Iwata, H., S. Tanabe, N. Sakai, and R. Tatsukawa. 1993. Distribution of persistent organochlorines in the oceanic air and surface seawater and the role of ocean on their global transport and fate. Environmental Science and Technology 27:1080–1098.

Krümmel, E. M., R. W. MacDonald, L. E. Kimpe, I. Gregory-Eaves, M. J. Demers, J. P. Smol, B. Finney, and J. M. Blais. 2003. Delivery of pollutants by spawning salmon. Nature (London) 425:255–256.

Kubo, T. 1980. Studies on the life history of "masu" salmon (*Onchorhynchus masou*) in Hokkaido. Scientific Reports of the Hokkaido Salmon Hatchery 34:1–95. [In Japanese with English abstract.]

American Fisheries Society Symposium 69:867–870, 2009

A Contemporary Understanding of the Delaware River Atlantic Sturgeon: Survival in a Highly Impacted Aquatic Ecosystem

PHILIP C. SIMPSON* AND DEWAYNE A. FOX
Delaware State University, Department of Agriculture and Natural Resources
1200 North DuPont Highway, Dover, Delaware 19943, USA

The Delaware River and Bay, while once supporting the largest known population of Atlantic sturgeon *Acipenser oxyrinchus*, has long suffered under multiple anthropogenic stress sources. These include both point and nonpoint pollution sources, which resulted in considerable hypoxic zones and large-scale channel deepening projects. When coupled with these human-induced habitat alterations, historic harvest levels were not sustainable and quickly lead to the demise of the Delaware River stock. At present, there is still much uncertainty surrounding the status of Atlantic sturgeon in the Delaware River. We attempted to ascertain the contemporary population status of Atlantic sturgeon in the Delaware River, with a specific focus on the identification of spawning and juvenile riverine habitats.

Atlantic sturgeon were collected during 2005–2007 through independent gill-net sampling and a reward program established with local striped bass *Morone saxatilis* gill-net harvesters. Captured Atlantic sturgeon were measured, processed for both genetic and gonad (>1.3 m full length [FL]) samples, and, if large enough (>800 mm FL), implanted with an ultrasonic transmitter (Vemco V-16, 6-H). We monitored telemetered Atlantic sturgeon through both passive (Vemco Ltd. [VR2]) and active tracking. The majority of passive hydrophones were affixed to U.S. Coast Guard navigation aids with the remainder being moored independently. Habitat use comparisons were made using a combination of real-time water quality measurements, U.S. Geological Survey water quality data, and available sediment records (Sommerfield and Madsen 2003). Habitat selection was assessed using an individual-based chi-square model with a significance level of $P < 0.05$. We attempted to document the location of spawning sites with egg collection mats in areas where telemetry results and substrate data indicated that spawning may be occurring. Egg mats were inspected at least every third day for the presence of eggs.

To date, 32 Atlantic sturgeons have been implanted with transmitters resulting in a total of 184 active and 182,740 passive relocations. Telemetered juvenile Atlantic sturgeon began entering the Delaware estuary in early spring and then gradually moved into the river (Figure 1A, 1B). Juvenile Atlantic sturgeons are typically concentrated into one of three geographically constrained areas in the Delaware River during the summer months. Their summer movements were generally reduced compared to spring and fall where movements have exceeded 100 km/d. Telemetered juvenile Atlantic sturgeon selected deep water (>8 m) habitat but showed no statistical preference to substrate type (Figure 2). We have also documented that six telemetered juvenile Atlantic sturgeon have entered the Chesapeake and Delaware Canal, with at least three individuals passing completely through its length, transiting between the two bays. Our results indicate primarily unidirectional movements from east to west, although there is some evidence of peripheral movements into the canal throughout the spring and summer.

Only nine Atlantic sturgeons collected in 2006–2007 met the size criteria for being potentially mature (FL > 1.3 m; Huff 1975). The reproductive status of these individuals was later confirmed through histology, which indicated that three were ripe males. Movements of telemetered adult At-

* Corresponding author: philchristiansimpson@yahoo.com

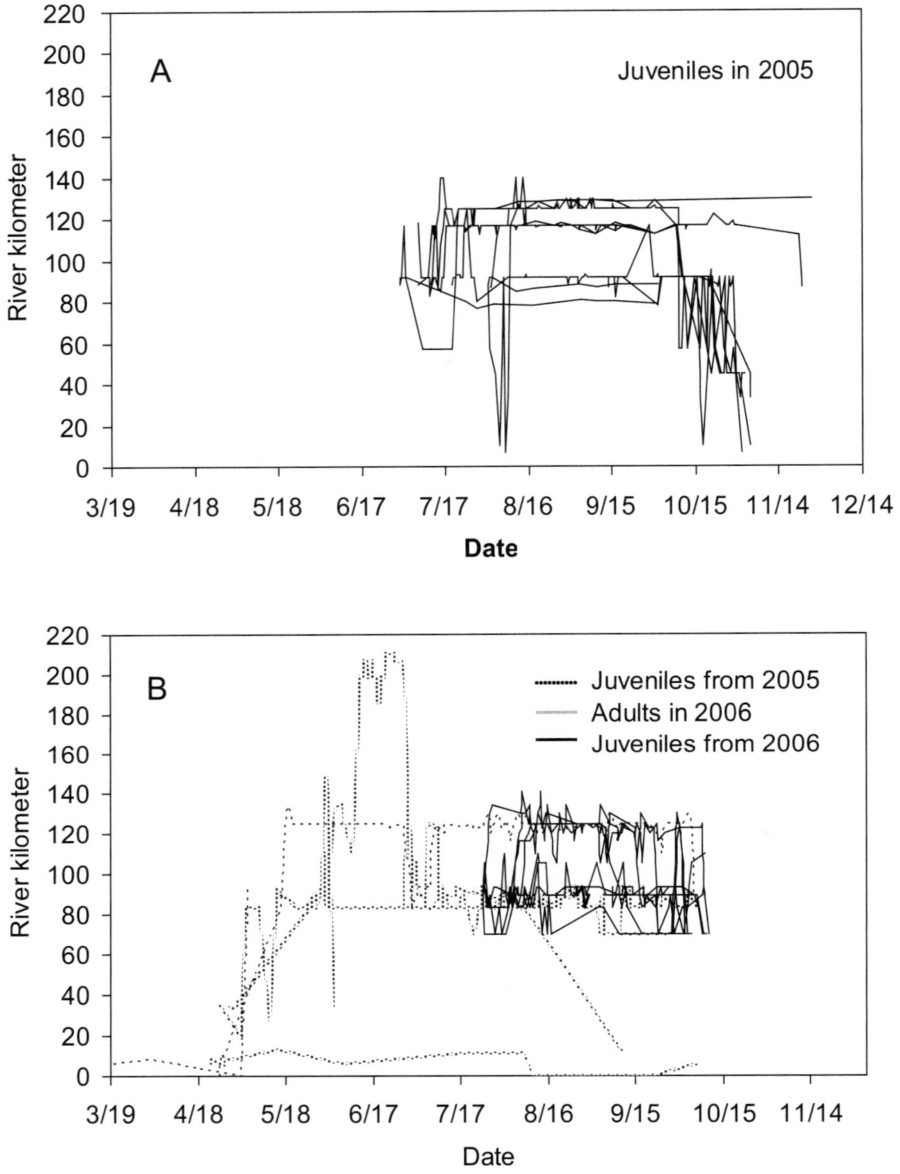

Figure 1.—Movement of juvenile sturgeon (A) in 2005 (N = 10), and (B) movement of sturgeon in 2006: returning juvenile sturgeon tagged in 2005 (N = 2), adult sturgeon (N = 5) in 2006, and juvenile sturgeon (N = 9) from 2006.

lantic sturgeon during 2006 were highly varied, with only two individuals entering the freshwater environment (>rkm [river kilometer] 130). Directional upriver movements occurred between early May and late June, with occupation of habitats near Philadelphia, Pennsylvania (rkm 163–186) and Trenton, New Jersey (rkm 199–211). Egg collection mats were deployed from June 14 to June 28, 2006, without success, between rkm 199 and 211. Telemetered adult sturgeon that entered the Delaware River remained in the estuary throughout the summer months following the likely spawning period.

Our telemetry results, coupled with the collection of ripe male Atlantic sturgeon, although not

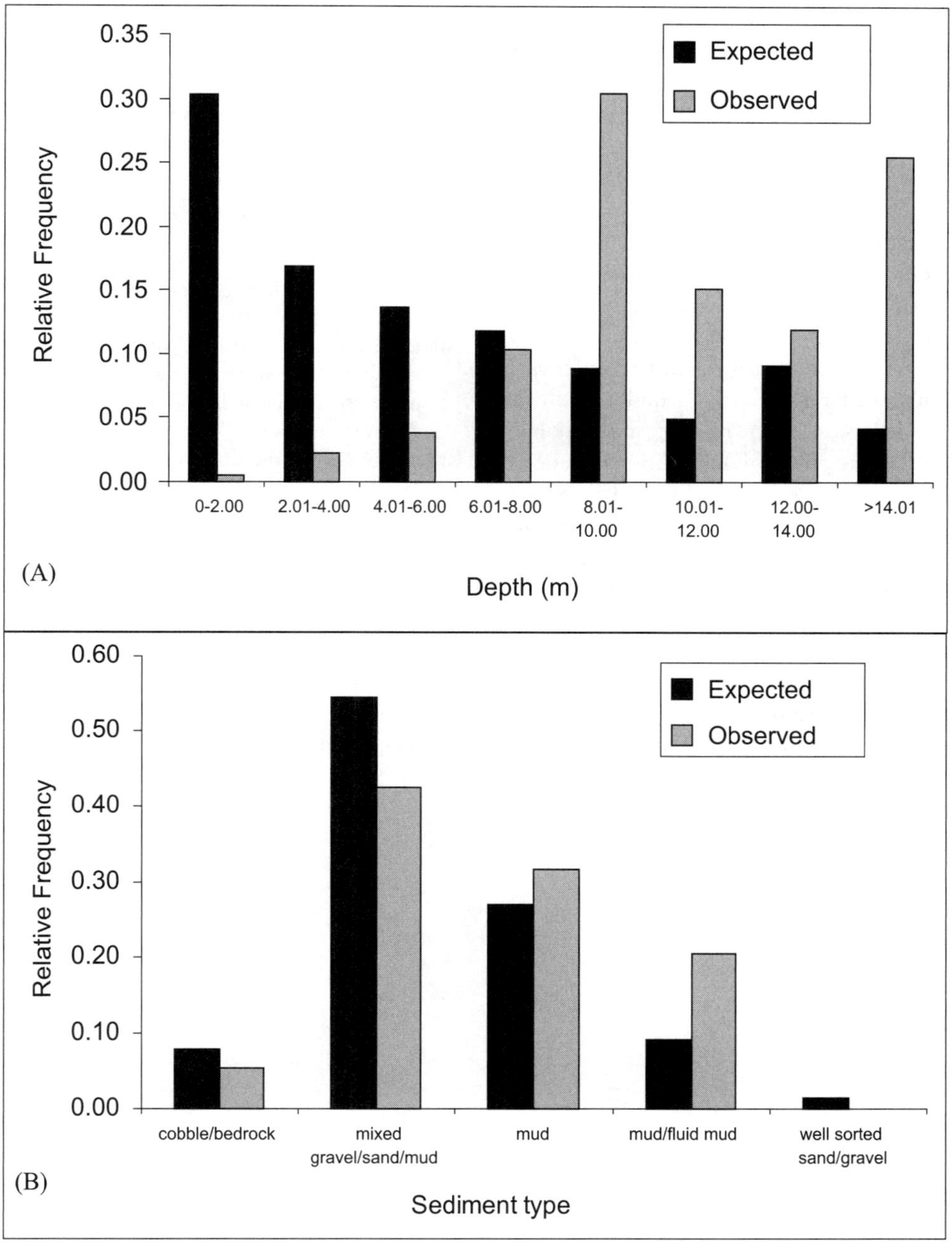

Figure 2.—(A) Expected and observed relative frequency of juvenile Atlantic sturgeon depth selection, and (B) expected and observed relative frequency of juvenile Atlantic sturgeon substrate selection.

conclusive, provide strong evidence to suggest that a remnant population of spawning Atlantic sturgeon exists in the Delaware River. These results are supported by analysis of genetic samples that indicate the presence of a Delaware River specific haplotype (T. King, U.S. Geological Survey, personal communication). Although the location of spawning habitat has not been confirmed, our telemetry results, together with information on substrate and water quality data, indicate that spawning may occur between mid-May and late June in freshwater-tidal reaches between north Philadelphia (rkm

176) and Trenton (rkm 211). These spawning areas are located much farther upriver than the historical spawning grounds (rkm 75–130; Ryder 1890; Cobb 1899) which may, in part, be due to increased saltwater intrusion. The effects of increased saltwater intrusion on Atlantic sturgeon in the Delaware River, although not well understood, may continue to play a role in the slow recovery of this population; yet, political and socioeconomic pressure to expand dredging efforts in the river is mounting.

Atlantic sturgeon bycatch mortality is known to occur in both the Delaware estuary and nearby continental shelf regions, which undoubtedly has played a role in slowing the recovery of this population. We became aware of an additional source of mortality during the course of this study: we documented numerous reports of adult, and juvenile, Atlantic sturgeon mortalities likely caused by boat propeller strikes. The Delaware River serves as one of the most important shipping ports in the United States and is the number one East Coast port for petroleum shipments. This strategic importance must be weighed against the costs to the ecological integrity of the Delaware River estuary. At present, there are numerous large-scale shipping and refining projects proposed for the Delaware River estuary. We argue that managers should apply a precautionary approach to the approval of any such projects and take into account how their potential impacts can alter the recovery for what was once the largest population of Atlantic sturgeon.

References

Sommerfield, C. K., and J. A. Madsen. 2003. Sedimentological and geophysical survey of the upper Delaware estuary. Final Report to the Delaware River Basin Commission, West Trenton, New Jersey.

Ryder, J. A. 1890. The sturgeon and sturgeon industries of the eastern coast of the United States, with an account of experiments bearing upon sturgeon culture. Bulletin of the U.S. Fish Commission 8:231–328.

Cobb, J. N. 1899. The sturgeon fishery of Delaware River and Bay. Commissioner's Report of the U.S. Commission of Fish and Fisheries. Government Printing Office, Washington, D.C.

Huff, J. A. 1975. Life history of Gulf of Mexico sturgeon, *Acipenser oxyrhynchus desotoi*, in the Suwannee River, Florida. Florida Marine Research Publications 16.

American Fisheries Society Symposium 69:871–874, 2009

European Eels from Deep Mediterranean Waters

Marco L. Bianchini*
Italian National Research Council, Istituto Biologia Agroambientale e Forestale (IBAF-CNR)
Monterotondo Scalo (RM), Italy

Gloria Vaggelli
Italian National Research Council, Istituto Geoscienze e Georisorse (IGG-CNR)
Torino, Italy

Roberto Cossio
Università di Torino, Dip. Scienze Mineralogiche e Petrologiche (DSMP)
Torino, Italy

Giovanni Battista Palmegiano and Francesco Gai
Italian National Research Council, Istituto Scienze Produzioni Animali (ISPA-CNR)
Torino, Italy

Luciana Sola and Anna Rita Rossi
Università di Roma "La Sapienza," Istituto Anatomia Comparata (DBAU)
Roma, Italy

Donatella Crosetti
Italian Institute for Applied Marine Research (ICRAM)
Roma, Italy

Giovanni Battista Giusto, Salvatore Gancitano, and Sergio Ragonese
Italian National Research Council
Istituto Ambiente Marino e Costiero (IAMC-CNR)–Mazara del Vallo (TP), Italy

It is generally agreed that the genus *Anguilla* is made up of 19 species, two of which, European eel *A. anguilla* and American eel *A. rostrata*, belong to the North Atlantic zone, and 17 to the Indo-Pacific area. All species share common problems in larval and reproductive migrations, metamorphosis, distribution overlaps, and meristic characters.

As a broad rule, subadult (yellow) eels live in fresh or brackish waters; after some years, as they start maturing (silvering), they migrate to the sea (catadromy), where they purportedly spawn. Leaf-like larvae (leptocephali) are transported passively back to coastal areas, where they metamorphose into glass eels and subsequently enter in-shore waters as elvers.

* Corresponding author: bradipo50@yahoo.com

After the supposition of a Mediterranean spawning ground, formulated in 1897, a comprehensive theory about the reproductive migratory behavior of the Atlantic eels, based on the reverse tracking of leptocephali, was proposed in 1922, suggesting that eels travel to the Sargasso Sea to spawn and that the European and American eels were separate taxa. An alternative theory was developed in 1959, according to which the eels from the western Atlantic were, in reality, the only parental stock for both Atlantic populations, Europe merely being the frontier of the distribution area. As a matter of fact, contrary to what has happened to other eel species, not a single Atlantic eel has ever been caught in the supposed spawning areas, nor in any other open-ocean water.

Therefore, while adult eel movements have

been studied extensively (Rossi et al. 1987), the genetic migration of European eels has not been completely explained: that adults return to the Sargasso Sea remains a hypothesis, supported by the inverse transport of leptocephali, but not based on data. In fact, the capture, and reporting, of mature or maturing eels at sea is an extremely rare event—a few reports in the past 100 years, as nearshore or as prey of other animals—and in the past 12 years, only one eel has been caught after more than 25,000 h of scientific trawling, encompassing a large part of the Mediterranean Sea (international Medits and Grund programs) on the upper shelf (L. Gil de Sola, Instituto Español de Oceanografia, Fuengirola, Spain, personal communication).

This paper reports the capture, for the first time, of silver eels in deep waters of the Strait of Sicily in the Central Mediterranean Sea. A total of seven eels were caught on four different occasions, in two locations 500 km apart, in the Strait of Sicily (Figure 1) by commercial bottom trawlers targeting red shrimps, in daylight, during early spring to early summer, at depths ranging from almost 300 to 650 m.

The eels were frozen and brought to the Italian National Research Council's Mazara laboratory, unluckily in less than perfect preservation condition. Eels were measured, sexed histologically, and X-rayed, and their otoliths were extracted. Otoliths were ground to thin sections and aged under a light microscope (Graynoth 1999). Thereafter, the sections were microanalyzed for Sr and Ca with quantitative X-ray mapping, using a Cambridge S360 SEM connected to an Inca Energy 200 Energy Dispersive Spectrometer. Samples of flesh were analyzed successively for their proximate composition; lipids were extracted and the fatty acids profile determined by gas chromatography. Total genomic DNA was extracted from other flesh samples, and mtDNA (cytochrome b) sequences were amplified via polymerase chain reaction; amplification products were finally sequenced and compared with GenBank records (Daemen et al. 2001).

The specimens, two males and five females (Table 1), were quite small (100–650 g), with ex-

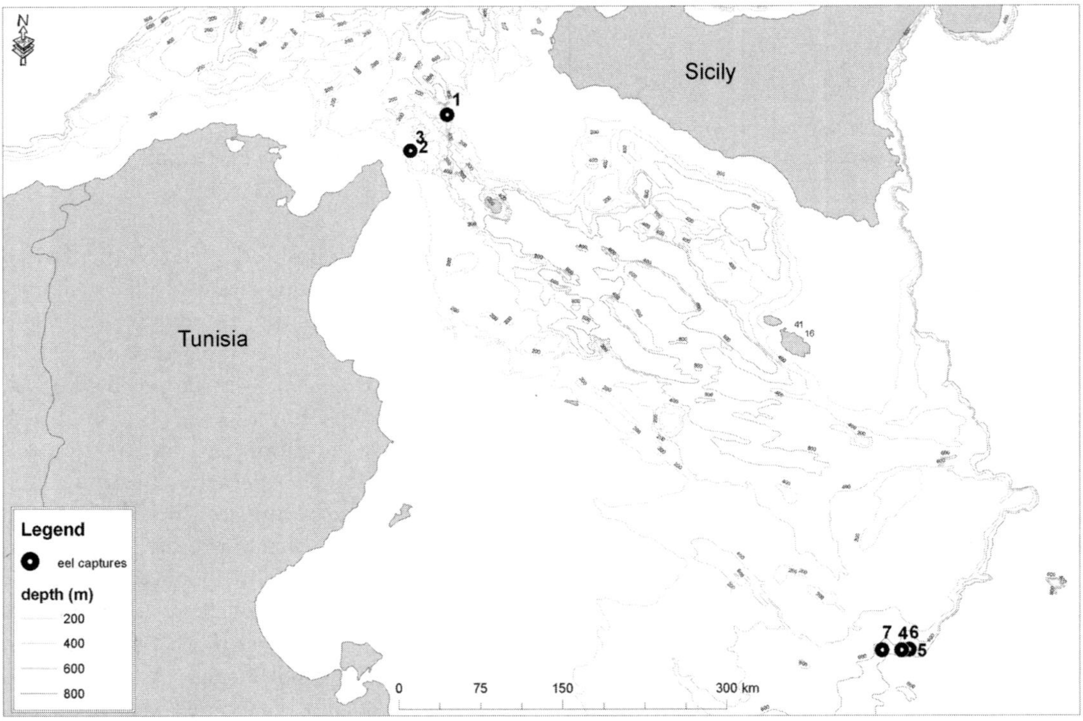

Figure 1.—Bathymetric map of the sites where seven European eels were collected in the Strait of Sicily.

ternal features identifying them as European eel, a fact confirmed by the genetic analyses (homologies greater than 99%) and the vertebral counts (low, but still well in the range accepted for this species). The morphological traits of the eels suggested an advanced maturation (e.g., ocular indexes up to 11.6; Pankhurst 1982), but the histological examination of the oocytes was not conclusive due to poor preservation. As expected, lipid content was high (30% d.m. of the muscle), and the fatty acid profile did not differ significantly from that of eels of similar status collected in "wild" freshwaters (Abrami et al. 1992). Aged otoliths resulted in ages from 4 to 11 years; the Sr:Ca ratio in the otoliths showed that all specimens had entered into freshwater (i.e., they were not "marine eels"); incidentally, some otolith Sr:Ca patterns suggested seasonal movements (broadly corresponding to the annuli pattern) or a life in the transition zone between fresh and brackish waters (Jessop et al. 2002).

This is the first report of European eels captured in the Mediterranean and the deepest record for the European eel; it is also the first time that Atlantic eels have been caught at sea in groups and not as isolated individuals. It is now possible to have a time of the year and a location to look for more animals to understand better the route taken by eels from the western Mediterranean during their spawning migration; thereafter, a broader and more articulated approach may produce results that are applicable to the whole basin.

References

Abrami, G., F. Natiello, P. Bronzi, D. J. McKenzie, L. Bolis, and E. Agradi. 1992. A comparison of highly

Table 1.—Collection and morphological data of seven European eels captured in deep waters in the Strait of Sicily.

	Individual eel						
Character	Eel 1	Eel 2	Eel 3	Eel 4	Eel 5	Eel 6	Eel 7
Total length (mm)	605	505	510	375	395	520	675
Ocular index	6.36	7.62	8.66	10.26	9.74	8.50	11.64
Number of vertebrae	111	109	112	–	–	–	–
Body weight (g)	384.4[a]	204.8[a]	258.3[a]	78.9	106.9	205.5	646.9
Gonad weight (g)	–	2.41	2.41	0	0.36	6.03	11.79
Gonadosomatic index	–	1.3	1.0	0.0	0.3	3.0	1.9
Color	silver	dark gray	silver	dark gray	silver	silver	silver
Sex	–	F	F	M	–	F	F
Maturity stage (five stages)	3	3	3	0	2	4	4
Age	4	9	6	6	2[a]	4	11
Sr:Ca ratio (min–max, ×1,000)	6.6–15.9	1.4–5.8[a]	6.8–16.8	2.1–6.1[a]	4.5–9.3	6.8–16.3	6.8–14.3
Catch date	Feb 2004	Mar 2004	Mar 2004	Jun 2004	Jun 2004	Jun 2004	Jun 2004
Location	N Cape Bon	E Cape Bon	E Cape Bon	Desert	Desert	Desert	Desert
Coordinates	N37°25' E11°37'	N37°12' E11°19'	N37°12' E11°19'	N33°52' E15°22'	N33°52' E15°22'	N33°52' E15°20'	N33°51' E15°10'
Depth (m)	220–330	450–510	450–510	660	660	640	640

[a] Denotes unreliable data due to poor preservation or possible artifacts.

unsaturated fatty acid levels in wild and farmed eels (*Anguilla anguilla*). Comparative Biochemistry and Physiology B 101:79–81.

Daemen, E., T. Cross, F. Ollevier, and F. A. Volckaert. 2001. Analysis of the genetic structure of European eel (*Anguilla anguilla*) using microsatellite DNA and mtDNA markers. Marine Biology 139(4):755–764.

Graynoth, E. 1999. Improved otolith preparation, aging and back-calculation techniques for New Zealand freshwater eels. Fisheries Research 42:137–146.

Jessop, B. M., J. C. Shiao, Y. Iizuka, and W. N. Tzeng. 2002. Migratory behaviour and habitat use by American eels *Anguilla rostrata* as revealed by otolith microchemistry. Marine Ecology Progress Series 233:217–229.

Pankhurst, N. W. 1982. Relation of visual changes to the onset of sexual maturation in the European eel *Anguilla anguilla* (L.). Journal of Fish Biology 21:127–140.

Rossi, R., M. Bianchini, A. Carrieri, and P. Franzoi. 1987. Observations on movements of yellow eels, *Anguilla anguilla* L., after displacement from coastal waters to sea. Journal of Fish Biology 31:155–164.

American Fisheries Society Symposium 69:875–877, 2009

Feeding and Prey of Pacific Lamprey in Coastal Waters of the Western North Pacific

Alexei M. Orlov
*Russian Federal Research Institute of Fisheries and Oceanography (VNIRO)
17, V. Krasnoselskaya Str., Moscow, 107140, Russia*

Richard J. Beamish
*Fisheries and Oceans Canada, Pacific Biological Station
Nanaimo, British Columbia V9T 6N7, Canada*

Andrei V. Vinnikov
*Kamchatka Research Institute of Fisheries and Oceanography (KamchatNIRO)
18, Naberezhnaya Str., Petropavlovsk-Kamchatsky, 683000, Russia*

Dmitry Pelenev
*Russian Federal Research Institute of Fisheries and Oceanography (VNIRO)
17, V. Krasnoselskaya Str., Moscow, 107140, Russia*

Pacific lamprey *Lampetra tridentata* is an endemic parasitic species widely distributed in the North Pacific Ocean. The known range of this species is from northern Hokkaido (Japan) and southern California (USA) to the northern Bering Sea. There is evidence that Pacific lamprey extended its range southward during recent years to central Honshu (Tochigi prefecture, Japan) and Mexico (Santo Domingo River).

Pacific lamprey plays an important and many-sided role in the marine and freshwater ecosystems of the North Pacific. On the one hand, it is consumed by a variety of aquatic animals, from river crayfishes to piscivorous birds, and marine mammals. On the other hand, Pacific lamprey is a parasite of many fishes, and even of whales, affecting commercial fish stocks, including those of Pacific herring *Clupea pallasii*, Pacific hake *Merluccius productus*, Pacific cod *Gadus macrocephalus*, walleye pollock *Theragra chalcogramma*, salmons *Oncorhynchus* spp., rockfishes *Sebastes* spp., Pacific halibut *Hippoglossus stenolepis*, Greenland halibut *Reinhardtius hippoglossoides*, Kamchatka flounder *Atheresthes evermanni*, and arrowtooth flounder *A. stomias*. Pacific lamprey became more abundant recently in the North Pacific, based on the number of marks on host species, and its effect on commercial fish stocks will notably increase if the population continues to increase. Although this species is of some interest for fisheries, it is currently fished in limited amounts in Oregon (USA) waters only.

The freshwater life of Pacific lamprey is well known, but data on the sea period of life, when Pacific lamprey is parasitic, is scarce (Novikov 1963; Abakumov 1964; Scott and Crossman 1973; and Beamish 1980). There is a large number of host fish species known for Pacific lamprey. However, its selectivity in respect to species, size, sex of prey, and the character of wounds (sizes and location on the body) are studied insufficiently.

In 2001–2004, data on Pacific lamprey hosts were collected in the northwestern Pacific waters of the northern Kuril Islands, southeastern Kamchatka, and in the western Bering Sea (Russia). The analysis was conducted according to Orlov and Vinnikov (2005). The following data were taken on fish, with lamprey marks, selected from the catch: total length, body weight, sex, number of marks, and mark type, position, diameter, and depth. The host body surface was divided into six sectors following Lennon (1954), with minor changes in numeration according to frequency of occurrence of marks: I—front part of back above lateral line, II—rear part of back above lateral line, III—front part of abdomen under lateral line, IV—rear part of abdomen under lateral line, V—head area, and VI—caudal area. The data allow us to characterize the selectivity of Pacific lamprey regarding host species, size, sex, and the properties of wounds. About 200 specimens of 11 fish species were analyzed.

The trawl catches showed that Pacific lamprey attacks fishes of a certain size range, body shape, flesh consistency, and affinity to the bottom. Prey of Pacific lamprey were large or moderate-sized fishes with fish-like body shape: salmons, Pacific cod, walleye pollock, Pacific hake, Pacific herring, lingcod *Ophiodon elongatus*, Atka mackerel *Pleurogrammus monopterygius*, rougheye rockfish *Sebastes aleutianus*, shortraker rockfish *S. borealis*, Pacific ocean perch *S. alutus*, sablefish *Anoplopoma fimbria*, Pacific and Greenland halibuts, and Kamchatka and arrowtooth flounders. Pacific lamprey did not attack skates (Rajidae), flatfishes (Pleuronectidae) lying on the bottom (with the exception of roughscale sole *Clidoderma asperrimum*), small sculpins (Cottidae), poachers (Agonidae), fishes with eel-like shape and slimy skin—eelpouts (Zoarcidae) and pricklebacks (Stichaeidae), and fishes with watery flesh—grenadiers (Macrouridae), and snailfishes (Liparidae).

Species composition of prey in the western Bering Sea and Pacific waters off the northern Kuril Islands and southeastern Kamchatka differed considerably. In the former area, Pacific lamprey attacked Greenland halibut most frequently (37.4% of total number of prey), followed by walleye pollock (28.4%), Pacific herring (15.3%), arrowtooth flounder (8.0%), Kamchatka flounder (4.0%), Pacific cod (3.4%), Pacific halibut (1.7%), Atka mackerel (0.6%), and Pacific ocean perch (0.6%). In the latter area, lamprey wounds were most frequently found in Pacific halibut (52.2%), followed by walleye pollock (17.4%), Kamchatka flounder (13.0%), Pacific cod (4.4%), Atka mackerel (4.4%), Pacific ocean perch (4.3%), and roughscale sole (4.3%).

Selectivity of lamprey attacks with regard to size, sex, body side, and body sectors was found in all host species. Mean total length of three host fishes was smaller than mean total length in trawl catches (walleye pollock [48.6 cm versus 51.5 cm], Kamchatka flounder [52.0 cm versus 56.5 cm] and Greenland halibut [46.6 cm versus 67.2 cm]), was similar in Pacific halibut (72.3 cm versus 73.4 cm) and arrowtooth flounder (55.3 cm versus 54.6 cm), and was larger in herring (34.5 cm versus 31.0 cm), and Pacific cod (63.2 cm versus 60.4 cm).

Male fish were attacked more frequently than female fish in Greenland halibut (100% versus 0%), walleye pollock (63% versus 37%), and Pacific cod (67% versus 33%), but the opposite situation was observed in Pacific herring (31% versus 69%), Pacific halibut (4% versus 96%), and Kamchatka flounder (25% versus 75%).

The majority of wounds was found on the blind side of the body of halibuts, on the right side of walleye pollock and Pacific herring, and on the left side of Pacific cod (Table 1). The belly section of walleye pollock, Greenland and Pacific halibuts, and arrowtooth and Kamchatka flounders were attacked more frequently by Pacific lamprey, but attacks were more frequent on the back section of Pacific herring and head area of Pacific cod (Table 1).

Maximum sizes and depth of wounds were observed in Greenland halibut, Pacific cod, and walleye pollock. Suck prints that characterized unsuccessful attacks were more frequently found in all halibut species (Table 2).

Table 1.—Position of Pacific lamprey marks (% of occurrence) on the more abundant prey caught in the northwestern Pacific, 2001–2004. Values in bold italics show highest occurrence.

	Side		Sector					
Species	Right	Left	I	II	III	IV	V	VI
Greenland halibut	4.4	***95.6***	18.9	22.2	24.4	***33.4***	1.1	0
Walleye pollock	***60.8***	39.2	10.0	28.3	***35.0***	21.7	0	5.0
Herring	96.9	3.4	***58.6***	0	31.1	0	6.9	3.4
Pacific halibut	3.6	***96.4***	21.4	21.4	***32.2***	14.3	10.7	0
Arrowtooth flounder	0	***100***	6.7	26.7	***33.3***	***33.3***	0.0	0
Kamchatka flounder	10	***90***	10	0	***60***	5	25	0
Pacific cod	37.5	***62.5***	12.5	12.5	12.5	25.0	***37.5***	0

Table 2.—Characteristics of Pacific lamprey marks on more abundant prey caught in the northwestern Pacific, 2001–2004 (na = not available).

Species	Imprint %	Wound %	Scar %	Fresh wound %	Diameter (mm)	Depth (mm)
Greenland halibut	10.3	89.7	54.9	45.1	22.3	na
Walleye pollock	1.7	98.3	20.7	79.3	21.0	6.0
Herring	0	100	0	100	13.7	4.5
Pacific halibut	10.7	89.3	4.0	96.0	10.9	2.6
Arrowtooth flounder	13.3	86.7	15.4	84.6	13.9	na
Kamchatka flounder	5.0	95.0	5.3	94.7	13.5	na
Pacific cod	0	100	25.0	75.0	24.4	6.0

References

Abakumov, V. A. 1964. On the sea life period of Pacific lamprey *Entosphenus tridentatus* (Richardson). Trudy VNIRO 49:253–256, Moscow. [In Russian.]

Beamish, R. J. 1980. Adult biology of the river lamprey (*Lampetra ayresi*) and Pacific lamprey (*Lampetra tridentata*) from the Pacific coast of Canada. Canadian Journal of Fisheries and Aquatic Sciences 37:1906–1923.

Lennon, R. E. 1954. Feeding mechanism of the sea lamprey and its effect on host fishes. Fish and Wildlife Service Fishery Bulletin 56(98):247–293.

Novikov, N. P. 1963. Events of attacks of Pacific lamprey *Enthosphenus tridentatus* (Gairdner) on halibuts and other fish of the Bering Sea. Voprosy Ikhtiologii 3(3):567–569 [In Russian.]

Orlov, A. M., and A. V. Vinnikov. 2005. Methodic aspects of studies on feeding behaviour of Pacific lamprey *Lampetra tridentata* during sea life period. Pages 404–409 *in* Fish behaviour. Proceedings of the International conference. 1–4 November, 2005, Borok, Russia. AQUAROS, Moscow. [In Russian.]

Scott, W. B., and E. J. Crossman. 1973. Freshwater fishes of Canada. Fisheries Research Board of Canada Bulletin 184.

American Fisheries Society Symposium 69:879–881, 2009

The Early Life History and Migration of Japanese Eel Larvae

Akira Shinoda* and Katsumi Tsukamoto
Ocean Research Institute, University of Tokyo
1–15–1 Minamidai, Nakano Tokyo 164–8639, Japan

The Japanese eel *Anguilla japonica* is a catadromous species that is widely distributed in the rivers and estuaries of East Asia. The spawning area of the Japanese eel has been determined to be in the North Equatorial Current near seamounts to the west of the Mariana Islands, 2,000–3,000 km away from their growth habitats (Tsukamoto 2006). The leptocephali of the Japanese eel are transported by surface flows in the North Equatorial and Kuroshio currents from the spawning area toward the coastal waters of East Asia. Therefore, long-term or short-term changes in hydrographic structure of the ocean have the potential to cause population fluctuations in the Japanese eel by influencing the timing or location of recruitment and the yearly recruitment success. The early life history parameters such as hatching date, duration of the marine migration, and timing and duration of metamorphosis may also have an effect on the location and timing of recruitment of each individual. We analyzed the otolith daily increments of glass eels collected throughout its range in East Asia during three fishing seasons to understand the key factors of the migration success of Japanese eel larvae.

Otoliths were analyzed from 307 glass eels collected from November 1997 to April 2000 in nine estuaries in China, Taiwan, and Japan. The average total length of the glass eels was 57.9 mm (range: 51.2–64.6 mm), and their pigmentation stages ranged from 5A–6A3 (mode: 5B). Sectioned sagittal otoliths were viewed and photographed with scanning electron microscopy, and thereafter, otolith increments were counted and their widths measured. We classified six stages (embryo, preleptocephalus, leptocephalus, metamorphosing larva, oceanic glass eel, and coastal glass eel) of the early life history of the Japanese eel, using morphological otolith characteristics of checks and patterns of change in increment width.

The average age at recruitment was 156 d (range: 98–227 d), and individual differences in age at recruitment within the samples were large (maximum: 84 d). These data suggest that larvae hatched in different months frequently recruit simultaneously to the same location. The duration of the leptocephalus stage varied widely from 59 to 170 d (mean = 106 d) and accounted for 68% of the average age at recruitment. The average duration of the other early life history stages were relatively shorter than the leptocephalus stage: 9 d for preleptocephali, 21 d for metamorphosing larvae, 21 d for oceanic glass eels, and 6 d for coastal glass eels. The large variation in ages at recruitment may originate from the long and variable duration of the leptocephalus stage. Drifter buoys released near the spawning area of the Japanese eel in 1998, and tracked by the ARGOS satellite, showed complicated trajectories (Tsukamoto et al. 2003), suggesting that the leptocephali were frequently mixed by the fine- or mesoscale eddies in the North Equatorial Current. Indeed, the ages of leptocephali had a 27-d age range in the North Equatorial Current, although they were collected from only one net tow (Tsukamoto et al. 1989).

Estimated hatching dates were from early May to mid-December, and the peak season of spawning was late August. The hatching dates were not correlated with ages at recruitment in 1998 and 1999 but were negatively correlated in 1997, so some relationships may differ among years as a result of annual fluctuations in oceanic conditions. In 1997, there was a strong El Niño event, and the glass eel catch of the Japanese eel was very low (Kimura et al. 2001). The southward shift of the spawning area associated with a southward movement of the salinity front during El Niño events

* Corresponding author: shinoda@ori.u-tokyo.ac.jp

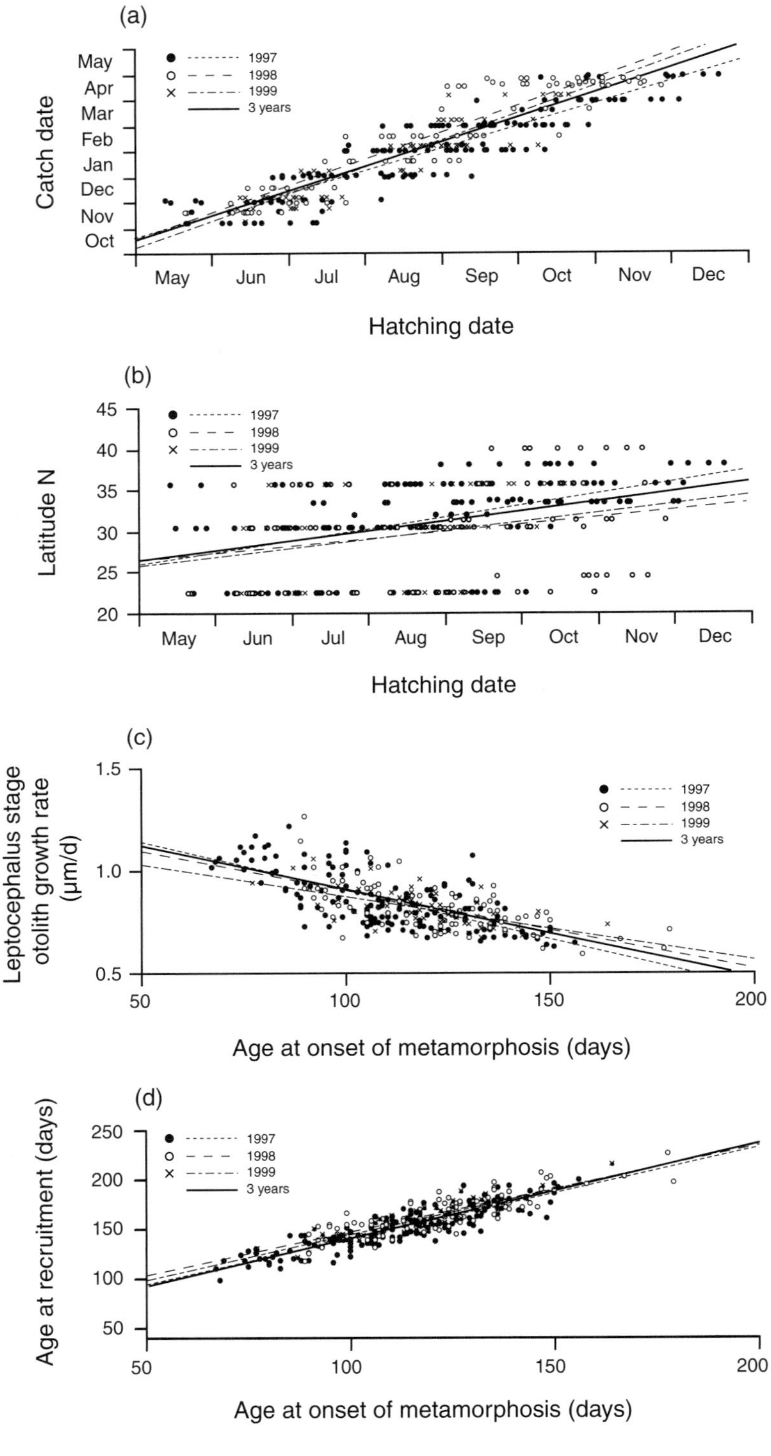

Figure 1.—Relationships between (a) hatching date and catch date (r^2 = 0.822, P < 0.0001), (b) hatching date and latitude (r^2 = 0.120, P < 0.0001), (c) age at onset of metamorphosis and otolith growth rate of leptocephalus stage (r^2 = 0.446, P < 0.0001), and (d) age at onset of metamorphosis and age at recruitment (r^2 = 0.768, P < 0.0001).

would increase larval transport into the Mindanao Current and reduce the recruitment of glass eels to their growth habitats in East Asia (Kimura and Tsukamoto 2006). Despite the changing oceanic conditions, other relationships were constant throughout the three fishing years. Hatching dates were correlated with sampling dates and with the latitudes of the sampling locations (Figure 1A, 1B), suggesting that the earlier they were hatched in the spawning area, the lower the latitude and the earlier in the fishing season they recruited to the estuaries. The age at metamorphosis was negatively correlated with otolith growth rate during the leptocephalus stage and positively correlated with age at recruitment (Figure 1C, 1D), suggesting that faster growing leptocephali metamorphose and recruit at younger ages. The leptocephali had various age, size, and growth histories when they arrived at the origin of the Kuroshio Current (where they would metamorphose to glass eels) because they were transported in complicated oceanic conditions. The leptocephali would serially metamorphose in the Kuroshio Current, beginning with the faster growing individuals, and would lose their buoyancy as they drop out of the current and recruit randomly to various areas of East Asia (drop-out recruitment hypothesis; Tsukamoto and Umezawa 1994). Thus, growth rate of the leptocephalus stage and timing of metamorphosis likely determine the latitude where recruitment takes place. The individual recruitment success of the Japanese eel would be increased by the plasticity in duration of the leptocephalus stage. The large variation in duration of the leptocephalus stage suggests that the Japanese eel has the ability to delay metamorphosis until they reach the origin of the Kuroshio Current. The long duration of leptocephalus stage and the ability to adjust its duration would play a crucial role in the risky oceanic migration. Therefore, the long and plastic duration of the leptocephalus stage may be a key factor for the recruitment mechanism of glass eels and the resultant recruitment success of the Japanese eel.

References

Kimura, S., and K. Tsukamoto. 2006. The salinity front in the North Equatorial Current: a landmark for the spawning migration of the Japanese eel (*Anguilla japonica*) related to the stock recruitment. Deep-Sea Research Part II Topical Studies in Oceanography 53:315–325.

Kimura, S., T. Inoue, and T. Sugimoto. 2001. Fluctuation in the distribution of low-salinity water in the North Equatorial Current and its effect on the larval transport of the Japanese eel. Fisheries Oceanography 10:51–60.

Tsukamoto, K., and A. Umezawa. 1994. Metamorphosis: a key factor of larval migration determining geographic distribution and speciation of eels. Pages 231–248 *in* Proceedings of 4th Indo-Pacific Conference, Bangkok, Thailand. Kasetart University, Faculty of Fisheries, Bangkok, Thailand.

Tsukamoto, K., A. Umezawa, O. Tabeta, N. Mochioka, and T. Kajihara. 1989. Age and birth date of *Anguilla japonica* leptocephali collected in September, 1986. Nippon Suisan Gakkaishi 55:1023–1028.

Tsukamoto, K., T. Otake, N. Mochioka, T. W. Lee, H. Fricke, T. Inagaki, J. Aoyama, S. Ishikawa, S. Kimura, M. J. Miller, H. Hasumoto, M. Oya, and Y. Suzuki. 2003. Seamounts, new moon and eel spawning: the search for the spawning site of the Japanese eel. Environmental Biology of Fishes 66:221–229.

Tsukamoto, K. 2006. Spawning of eels near a seamount. Nature(London) 439:929.

American Fisheries Society Symposium 69:883–885, 2009

First Microchemical Examination of Abnormal Otoliths of Amphidromous Ayu *Plecoglossus altivelis*

Tao Ma* and Mari Kuroki
Ocean Research Institute, The University of Tokyo
1–15–1 Minamidai, Nakano, Tokyo 164–8639, Japan

Rikizo Ishida
Fisheries Environment Institute
3–2-12 Kagurazaka, Shinjuku, Tokyo 162–0825, Japan

Katsumi Tsukamoto
Ocean Research Institute, the University of Tokyo
1–15–1 Minamidai, Nakano, Tokyo 164–8639, Japan

The ayu *Plecoglossus altivelis* is an annual fish that spawns in freshwater in autumn. The larvae drift downstream to overwinter in coastal waters and then migrate upstream as juveniles in spring (Tsukamoto et al. 1987). Landlocked populations are known from Biwa Lake (Azuma 1973) and several other lakes in Japan, in addition to the typical amphidromous populations. Juvenile ayu have been stocked extensively in many rivers as they are an important freshwater fishery species in Japan. However, some ayu are found to have abnormal otoliths, and little is known about their detailed morphology and microchemistry. We examined the morphology and microchemical concentrations of some common elements (Sr, Ca, Na, K, and S) of abnormal otoliths in ayu and consider the possible factors influencing otolith abnormality.

A total of 31 reared juvenile ayu (101.7–137.0 mm standard length) were sampled randomly from three local culture ponds in Wakayama Prefecture (WK), Kumamoto Prefecture (KM), and Biwa Lake (BL) during June 4–10, 2004. These ayu had been collected as larvae from the wild and then reared to a larger size in culture ponds for stocking into rivers. Sagittal otoliths were extracted from all specimens, observed using a dissecting microscope, and subsampled for life history transect and x-ray intensity map analyses using a wave-length dispersive x-ray electron microprobe (JEOL JXA-8900R).

Eight ayu (26%) were found to have abnormal otoliths, with differing abnormal rates from the three collection locations: 0% at WK, 45% at KM, and 30% at BL. Five ayu (16%) had the abnormal otolith only on one side, while the otolith on the other side was normal. The normal sagittal otoliths were oval with an elongated frontal side and had smooth surfaces with an opaque whitish color. The abnormal otoliths were more transparent, with irregular shapes, compared to the normal ones. Abnormal otoliths were divided into Type 1 (semi-abnormal) and Type 2 (fully abnormal; Figure 1). The Sr:Ca ratios of Type 1 showed two patterns: (1) a drastic change in Sr:Ca ratios between high values ($4.0 \times 10^{-3} – 12.0 \times 10^{-3}$) of the inner part, except for the core, and persistently low values ($<4.0 \times 10^{-3}$) of the outer part; and (2) a constant and low level in Sr:Ca ratio. The Sr:Ca ratio of both patterns in Type 1 otoliths were generally similar to the two patterns in normal otoliths that corresponded to the amphidromous and landlocked populations, but the ratios decreased more quickly to lower levels in outer parts of Type 1 otoliths than did the ratios in normal otoliths. The Sr:Ca ratios of Type 2 otoliths were lower throughout the life history transect. Furthermore, microchemistry analyses (Figure 2) showed that the concentration of some elements was different

* Corresponding author: tm0928@ori.u-tokyo.ac.jp

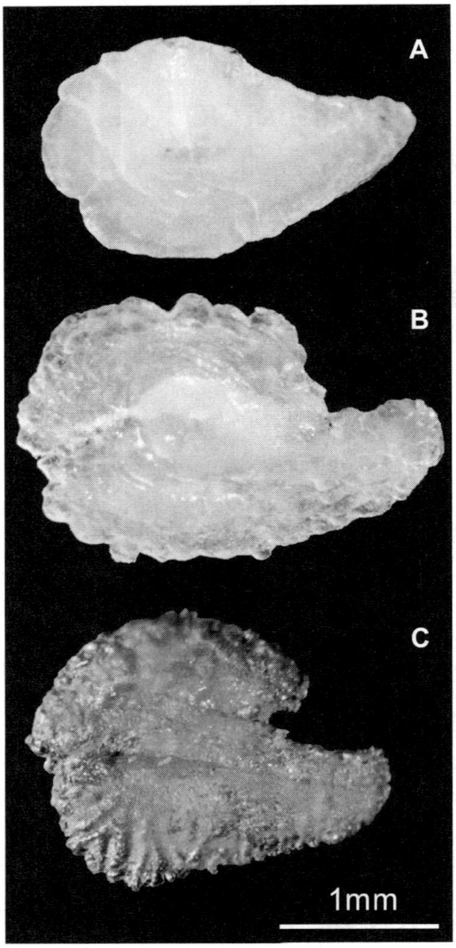

Figure 1.—Light microscope photos of sagittal otoliths (right side) of ayu. (A) Normal otolith (ayu of 101.7 mm standard length [SL]); (B) Type 1, semi-abnormal otolith (ayu of 127.0 mm SL); (C) Type 2, fully abnormal otolith (ayu of 116.2 mm SL).

between the normal and abnormal otoliths; Sr, Na, and K were much lower in the abnormal otoliths compared to the normal otoliths, regardless of whether the otolith was from the amphidromous or landlocked ayu. However, the concentration of Ca and S in the abnormal otoliths was not different from that of the normal otoliths.

This study examined the microchemistry of abnormal otoliths in ayu for the first time and showed that the abnormal formation is related to the deposition of some minor elements such as Sr, Na, and K. Comparison of Sr:Ca ratios between the normal and abnormal otoliths suggested that the abnormality may have occurred before the period when ayu ascend to rivers. High rates of abnormality were found in the KM (facing an inner sea) and BL (closed water) but not in the WK (facing the open ocean), suggesting that these relatively enclosed waters, where the larval ayu with abnormal otoliths were living, appear to be related to the abnormality. This is supported by the results of Kuroki et al. (2006), in which 12% of the 83 ayu examined from the Kurobe River (flowing into the semi-enclosed Toyama Bay and being heavily stocked by landlocked ayu from BL) were found to have abnormal otoliths. Physiological effects, as well as environmental conditions, may cause the abnormal otolith formation

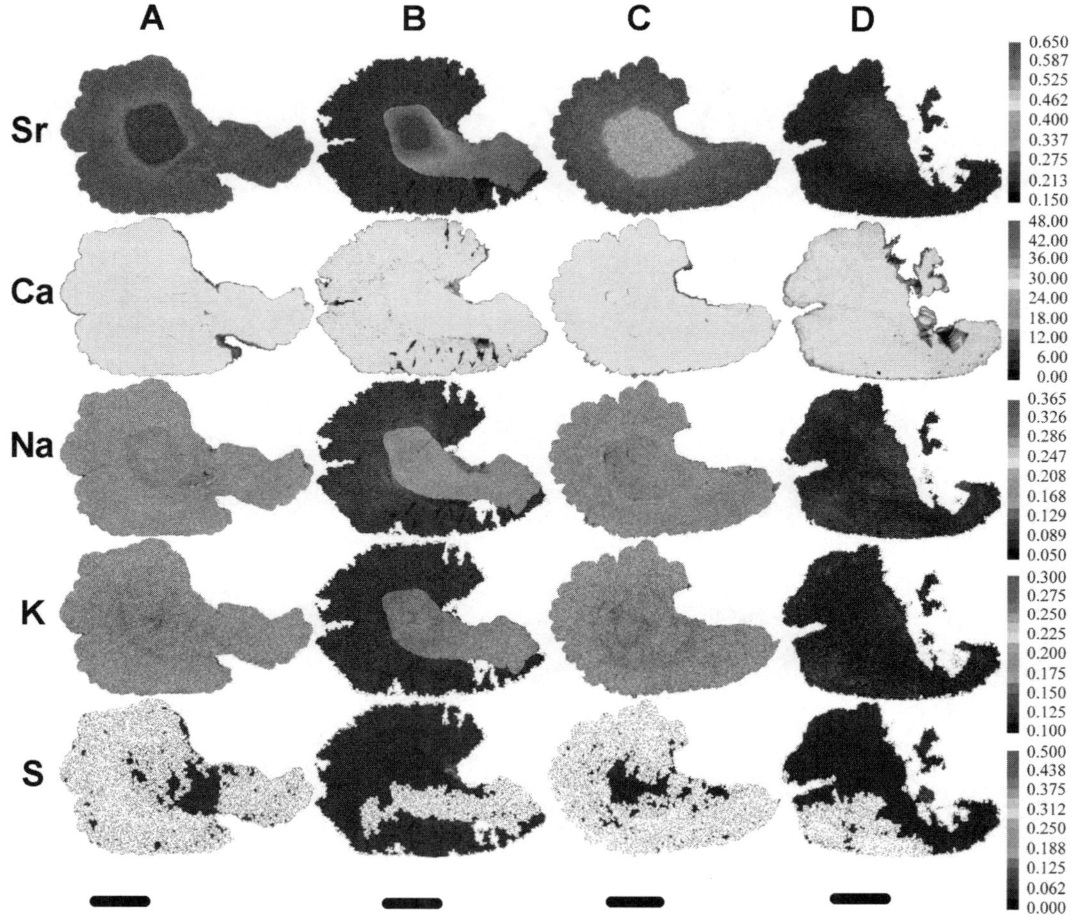

Figure 2.—X-ray intensity maps of element concentration (Sr, Ca, Na, K, and S) in otoliths of ayu. Higher concentrations are shown in light gray to gray, and lower concentrations are shown in dark gray to black. Column A shows a normal otolith of a landlocked ayu (107.1 mm standard length [SL]); B shows a Type 1, semi-abnormal otolith of a landlocked ayu (137.0 mm SL); C shows a normal otolith of an amphidromous ayu (104.0 mm SL); and D shows a Type 2, fully abnormal otolith of an amphidromous ayu (116.2 mm SL). Scale bars = 500 μm.

because some abnormality occurred in only one of the paired otoliths.

References

Azuma, M. 1973. Studies on the variability of the landlocked ayu-fish, *Plecoglossus altivelis* T. and S., in Lake Biwa IV. Considerations on the grouping and features of variability. Japanese Journal of Ecology 23:255–265.

Kuroki, M., T. Ma, R. Ishida, and K. Tsukamoto. 2006. Migratory history of wild and released ayu (*Plecoglossus altivelis*) in the Kurobe River, Japan. Coastal Marine Science 30:425–431.

Tsukamoto, K., R. Ishida, K. Naka, and T. Kajihara. 1987. Switching of size and migratory pattern in successive generations of landlocked ayu. Pages 492–506 *in* M. J. Dadswell, R. J. Klauda, C. M. Moffitt, R. L. Saunders, R. A. Rulifson, and J. E. Cooper, editors. Common strategies of anadromous and catadromous Fishes. American Fisheries Society, Symposium 1, Bethesda, Maryland.

American Fisheries Society Symposium 69:887–889, 2009

Migration Scales of Catadromous Eels: Diversity and Evolution of Larval Migration Based on Their Distribution, Morphology, and Early Life History

MARI KUROKI* AND JUN AOYAMA
Ocean Research Institute, University of Tokyo
Minamidai, Nakano, Tokyo 164–8639, Japan

HEEYONG KIM
National Fisheries Research and Development Institute
408–1 Sirang-ri, Gijang-eup, Gijang-kun, Busan 619–705, Korea

MICHAEL J. MILLER, SHINGO KIMURA, AND KATSUMI TSUKAMOTO
Ocean Research Institute, University of Tokyo
Minamidai, Nakano, Tokyo 164–8639, Japan

The long distance migrations of catadromous eels of the genus *Anguilla* consist of two different components, the larval migration toward their freshwater growth habitats and the adult migration back to the spawning area. These components have been poorly understood due to the critical lack of information about the spawning areas and migration routes. We examined the morphology and otolith microstructure of anguillid leptocephali to estimate the diversity of migration and its evolutionary origins. Leptocephali were examined from the Indo-Pacific region (12 species, $N = 759$), and glass eels (9 species, $N = 653$) were obtained from all over the world, including temperate and tropical eels. Tropical eels compose two-thirds of the genus *Anguilla*, but compared to temperate eels, information has been scarce in tropical eels, especially their leptocephalus larvae (Jespersen 1942).

Distributional analysis of leptocephali (Figure 1) suggested that most tropical species such as the Celebes longfin eel *A. celebesensis* and *A. borneensis* spawn near their growth habitats like in the Celebes Sea or in Tomini Bay in Indonesian waters, and their larval migrations are only short distances (small-scale migration; Aoyama et al. 2003). Other tropical species like the giant mottled eel *A. marmorata* and the Indian shortfin eel *A. bicolor pacifica* spawn in the open ocean, but inhabit wide regions of the tropics (middle-scale migration; Kuroki et al. 2006). These migratory distances contrast with that of temperate eel species, which show migrations of thousands of kilometers (large-scale migration; Tsukamoto et al. 2002).

Numerical simulations were made of the larval migration routes of the temperate Japanese eel *A. japonica* and the tropical giant mottled eel in the western North Pacific region, (Figure 2) based on their slightly different spawning areas west of the Mariana Islands. The simulated transport trajectories showed that the Japanese eel had an expected long distance passive transportation by the North Equatorial Current and northward into the Kuroshio Current (Figure 2A), whereas the giant mottled eel showed transport by the North Equatorial Current and then southward transport into the Mindanao Current (Figure 2B). The spawning sites of the tropical Celebes longfin eels and *A. borneensis* were based on the collection locations of small leptocephali, and simulated transport trajectories resulted in their larvae remaining near the spawning areas for about 100 d (Figure 2C, 2D). These results corresponded well with the mean duration the larval period of tropical and temperate species estimated from the otolith microstructure of their glass eels (108.5–194.4 d in four tropical species and 203.0–351.2 d in five temperate species) as well as their geographic distributions.

Morphological measurements of leptocephali

* Corresponding author: mari@ori.u-tokyo.ac.jp

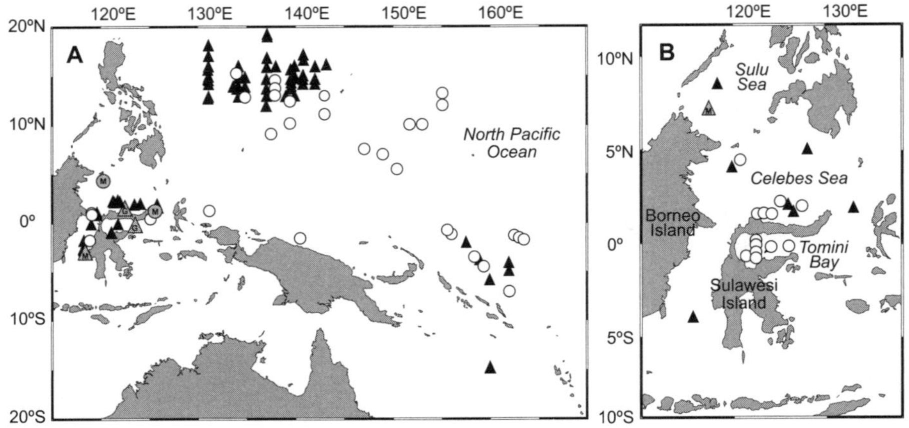

Figure 1.—Geographic distribution of anguillid leptocephali of tropical species collected from 1995 to 2004. A: giant mottled eels (triangles) and Indian shortfin eels (circles) that are distributed widely in the Indo-Pacific region (middle-scale migration), B: Celebes longfin eels (circles) and *Anguilla borneensis* (triangles) that are distributed only in Indonesian waters (small-scale migration). M = metamorphosing leptocephalus; G = oceanic glass eel.

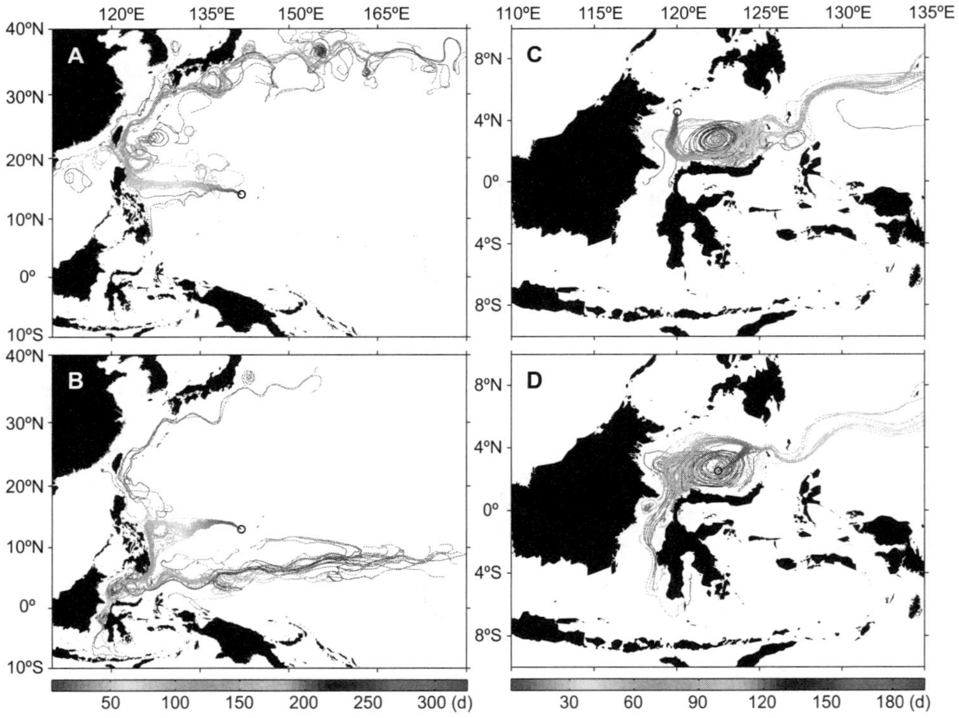

Figure 2.—Numerical simulations of four anguillid species of A: Japanese eel (spawning site/particle release point at 142.0°E, 14.0°N), B: giant mottled eel, (spawning site/particle release point at 142.0°E, 13.0°N) C: Celebes longfin eel (spawning site/particle release points at 120.0°E, 4.5°N), and D: *Anguilla borneensis* (spawning site/particle release points at 122.5°E, 3.5°N). The trajectories are shaded according to the days after release from each open circle, which were considered to be the spawning areas based on the collection of small leptocephali.

showed that the maximum total length and body depth at metamorphosis were shorter (about 50 mm) and deeper (20.7–22.9% of total length) in the typical tropical species than in temperate eels (about 60–75 mm and 17.6–18.9%, respectively), and those of other tropical eels that migrate at mid-scale distances were intermediate. These different body shapes might be an adaptation to facilitate the specific migrations of larvae.

Growth rates of larvae, estimated from otolith analysis, showed that the Celebes longfin eel and *A. borneensis*, which have small-scale migrations, grew faster (0.53–0.57 mm/d) than the giant mottled eel and Indian shortfin eel (0.40–0.53 mm/d) that have mid-scale migrations. Furthermore, temperate eels, such as the European eel *A. anguilla* and American eel *A. rostrata*, that have large-scale migrations (up to 8,000 km) showed slower growth rates (0.38 mm/d, Castonguay 1987) than did tropical eels. Leptocephali of tropical eels, with small-scale migrations, had shorter larval durations by growing faster and metamorphosing earlier; temperate species showed the opposite trends. These results strongly suggest that there are different scales of migration in anguillid eels.

These findings, and the recent molecular phylogenetic relationships of anguillid species, indicate that the spawning migrations of anguillid eels likely originated from short migrations in tropical regions and then gave rise to the surprisingly long distance migration in temperate species. The leptocephalus larva appears to have played an important role in the diversification of migration in the genus *Anguilla*.

References

Aoyama, J., S. Wouthuyzen, M. J. Miller, T. Inagaki, and K. Tsukamoto. 2003. Short-distance spawning migration of tropical freshwater eels. Biological Bulletin 204:104–108.

Castonguay, M. 1987. Growth of American and European eel leptocephali as revealed by otolith microstructure. Canadian Journal of Zoology 65:875–878.

Jespersen, P. 1942. Indo-Pacific leptocephalids of the genus *Anguilla*. Dana-Report 4:1–128.

Kuroki, M., J. Aoyama, M. J. Miller, S. Wouthuyzen, T. Arai, and K. Tsukamoto. 2006. Contrasting patterns of growth and migration of tropical anguillid leptocephali in the western Pacific and Indonesian seas. Marine Ecology Progress Series 309:233–246.

Tsukamoto, K., J. Aoyama, and M. J. Miller. 2002. Migration, speciation, and the evolution of diadromy in anguillid eels. Canadian Journal of Fisheries and Aquatic Sciences 59:1989–1998.

American Fisheries Society Symposium 69:891–894, 2009

Using Physiological Telemetry and Intervention Experiments to Examine the Maladaptive Shift in Fraser River's Late-Run Sockeye Salmon Spawning Migration

Scott G. Hinch*
Centre for Applied Conservation Research, Department of Forest Sciences, and Institute for Resources Environment and Sustainability, University of British Columbia Vancouver British Columbia V6T 1Z4, Canada

Anthony P. Farrell
Department of Zoology and Faculty of Agricultural Sciences, University of British Columbia Vancouver, British Columbia V6T 1Z4, Canada

Steven J. Cooke
Department of Biology and Institute of Environmental Science, Carleton University Ottawa, Ontario K1S 5B6, Canada

David A. Patterson
Fisheries and Oceans Canada, Cooperative Resource Management Institute, Simon Fraser University Burnaby, British Columbia V5A 1S6, Canada

Mike F. LaPointe
Pacific Salmon Commission 60 – 1155 Robson Street, Vancouver, British Columbia V6E 1B5, Canada.

David W. Welch
Kintama Research Corporation 10-1850 Northfield Road, Nanaimo British Columbia V9S 3B3, Canada

Karl K. English
LGL Environmental Research, 9768 Second Street, Sydney, British Columbia V8L 3Y8, Canada

Glenn T. Crossin
British Antarctic Survey, Cambridge University, Cambridge CB3 0ET, England,
and
Simon Fraser University, Burnaby, British Columbia V5A 1S6, Canada

Kristi Miller
Fisheries and Oceans Canada, Pacific Biological Station 3190 Hammond Road, Nanaimo, British Columbia V9T 6N7, Canada

Richard E. Thomson
Fisheries and Oceans Canada, Institute of Ocean Studies Post Office Box 6000, Sidney, British Columbia V8L 4B2, Canada

* Corresponding author: shinch@interchg.ubc.ca

Glen van der Kraak
Department of Integrative Biology, University of Guelph, Guelph, Ontario N1G 2W1, Canada

Ivan Olssen
Biology Department, Karlstad University, Sweden

Mark Shrimpton
Department of Biology, University of Northern British Columbia
Prince George, British Columbia V2N 4Z9, Canada

Michael S. Cooperman
National Research Council in residence at NOAA Fisheries, Northeast Fisheries Science Center
Maine Field Station, Orono, Maine 04473, USA

The Fraser River drains one-third of British Columbia as it flows to the Pacific Ocean and is the largest salmon-producing river in Canada. Sockeye salmon *Oncorhynchus nerka* are the most economically important species in the Fraser River system, and the population is composed of more than 150 stocks. It is managed as four groups (Early Stuart, Early Summer, Summer, Late-Run), based on timing of adult salmon returning to freshwater. Late-Run stocks were unique, historically, in that upon arrival to the Strait of Georgia (SOG), maturing adults paused for 3–6 weeks on their spawning migration prior to entering the Fraser River. However, since 1995, increasing proportions of Late-Run salmon have entered the Fraser River without delaying in SOG, despite early river entry causing high river migration mortality (60–96% per year compared to 20% per year, historically; Cooke et al. 2004). Mortality of early-timed migrants is thought to be caused by salmon encountering warmer river temperatures than normal-timed migrants, resulting in higher energetic costs, more severe infections (e.g., kidney parasite *Parvicapsula*; Wagner et al. 2006), and, in some years, the collapse of metabolic scope. The high mortality has caused a conservation crisis, including emergency listings of stocks, fisheries closures, and lost recreational and cultural fisheries (Cooke et al. 2004). We formed a research team, in 2001, involving ecologists, physiologists, oceanographers, geneticists, and resource managers, to explore the causes and consequences of this maladaptive shift in migration behavior.

Our initial efforts focused on identifying the environmental and physiological correlates of Late-Run sockeye migration behavior and fate (Cooke et al. 2006a, 2006b). The strongest environmental correlate of river entry timing is the level of surface water mixing occurring within SOG at the time of Late-Run sockeye arrival. Our data indicate that Late-Run sockeye that enter the SOG at times of low mixing encounter low salinity surface waters and are more likely to enter the Fraser River without delaying migration than are Late-Run sockeye that enter the SOG during well-mixed conditions (i.e, absence of low salinity surface water) (linear regression of mean river entry date and mixing strength, $N = 9$ years, $r^2 = 0.89$, $P < 0.00001$; R. Thomson, unpublished data). Exposure to low-salinity water may accelerate osmoregulatory preparedness for freshwater entry and thereby cause fish to enter the Fraser River prematurely—our salinity intervention study, described below, demonstrated that freshwater-exposed marine fish could survive for periods of time in saltwater. We used physiological biopsy telemetry (Cooke et al. 2005) to relate individual condition to the behavior and fate of Late-Run sockeye at various locations along marine and freshwater portions of the spawning migration. The biopsy involved collecting blood plasma for ions (indicators of osmoregulatory function and stress), reproductive and stress hormones (indicators of maturation level), gill tissue for Na^+ K^+ ATPase (osmoregulation status) and genomic analysis, somatic tissue for genomic analysis and stock identification using DNA, scales for stable isotopes, and microwave interrogation of somatic water (index of energy reserves). To date, we have biopsied more than 3,000 Late-Run sockeye, and the results indicate that sockeye that do not hold in the SOG are characterized by more variable osmoregulatory condition, lower energy reserves, and higher stress and reproductive hormone concentrations (Table 1).

Table 1.—Sex hormone concentrations (mean pg/mL ± 1 SD) for male and female Late-Run sockeye salmon that either did not pause in migration in the Strait of Georgia (SOG) and subsequently died in the Fraser River before reaching their natal stream or those did pause in SOG and successfully reached natal streams. Sample sizes were female—no pause, n = 6; female—pause, n = 11; male—no pause, n = 6; and male—pause, n = 4. Table is a simplified reproduction from Cooke et al. 2006.

Variable	No pause and died before reaching natal stream	Paused migration and reached natal stream	t-test P value
♀ Testosterone	13,566 ± 2,066	10,935 ± 472	0.125
♀ 11-ketotestosterone	1,016 ± 154	750 ± 26	0.038
♀ 17β-estradiol	6,029 ± 1,083	3,868 ± 659	0.019
♂ Testosterone	11,567 ± 1,211	10,229 ± 1,308	0.480
♂ 11-ketotestosterone	5,659 ± 787	4,786 ± 563	0.420
♂ 17β-estradiol	518 ± 170	395 ± 41	0.551

Genomic analysis (i.e., gene arrays) of the physiology samples involved the use of complementary DNA technology to construct comparisons of the relative strength of the up–down regulation of large numbers of individual genes between migration behavior groups in order to examine genome-wide changes in physiology associated with spawning migration and freshwater entry timing. Gene arrays have been applied to multiple tissue types (some from lethally sampled fish) to address specific hypotheses relating to entry timing, including gill (osmoregulation), brain and hypothalamus (reproductive maturation and sensory signaling), and liver and white muscle (energy and senescence). Gene arrays revealed striking physiological shifts, sometimes involving over 3,500 genes, occurring both in the marine and freshwater environments, but interestingly, major shift points (hence potential environmental triggers) were variable among tissues, as were the physiological relationships among individuals within sites. These data suggest a lack of cohesion between processes relating to freshwater preparedness (osmoregulation), reproductive maturation, and energy utilization. Importantly, in contrast with the environmental and other physiological indicators, the genomic analysis suggests that in 2006, there was not good evidence that freshwater entry timing was associated with osmoregulatory preparation or energetic shifts measured in coastal environments.

We developed two hypotheses of migration behavior, based on initial findings, for testing via intervention experiments during the 2006 migration: the accelerated sexual maturation hypothesis and the osmoregulation hypothesis. As a test of the maturation hypothesis, we collected 200 migrating sockeye salmon near the coastline of the Queen Charlotte Islands, approximately 700 km north of the Fraser River mouth, and injected sockeye salmon with time-released gonadotropin releasing hormone (GnRH, a chemical produced in the brain responsible for initiating a hormone cascade associated with maturation), testosterone (a sex hormone produced in gonads), or a combination of the two chemicals. Fish were biopsied, tagged, released at the site of capture, and tracked to freshwater entry and their ultimate fate. Telemetry tracking occurred in partnership with POST (the Pacific Ocean Shelf Telemetry program). Unfortunately, postrelease survival was low for all groups, including the control and sham injection groups; only 13 of the 200 sockeye were detected in SOG, and only 4 entered the Fraser River. We are currently investigating the causes for the poor survival.

Our second intervention experiment involved collecting 250 Late-Run sockeye salmon from SOG, transporting them to the Fisheries and Oceans Canada West Vancouver Laboratory, and placing them into five treatment groups: freshwater (FW, 0 parts per thousand [ppt] salinity), saltwater (SW, 28 ppt), iso-osmotic water (Iso, 14 ppt), FW+GnRH, and SW+GnRH. Sockeye salmon were held for 1 week, biopsied, telemetry tagged, and released to SOG. Data analyses are ongoing, but telemetry results indicate an effect from salinity; fish held in FW entered the Fraser River faster than did sockeye salmon held in either Iso or SW (Table 2). There was no obvious effect of hormone treatment on mi-

Table 2.—Time lag between post-treatment release to the Strait of Georgia (SOG) and first Fraser River detection for the five treatment groups of the osmoregulation intervention experiment. All sockeye salmon included in the experiment were members of the Adams River stock, and all treatments started with approximately 50 individuals. One iso-osmotic fish had a delay to river entry of more than 310 h and was excluded from calculations as an outlier. One-way analysis of variance of the three salinity-only treatments indicated that freshwater fish entered the river in significantly less time ($P = 0.0266$) than did iso-osmotic or saltwater-treated fish. SW = saltwater; FW = freshwater; GnRH = gonadotropin releasing hormone.

Treatment	Number released to SOG post-treatment	Number detected in Fraser River	Mean delay in hours ± SD
Saltwater	20	5	86.0 ± 19.6
Iso-osmotic	30	12	84.5 ± 45.6
Freshwater	31	7	49.8 ± 17.5
SW + GnRH	8	2	49.3 ± 0.3
FW + GnRH	16	1	109.0

gration behavior, but sample sizes were small since hormone-treated fish had lower survival than did nonhormone-treated fish.

References

Cooke, S. J., S. G. Hinch, G. T. Crossin, D. A. Patterson, K. K. English, M. Healey, J. M. Shrimpton, G. Van Der Kraak, and A. P. Farrell. 2006a. Mechanistic basis of individual mortality in Pacific salmon during spawning migrations. Ecology 87:1575–1586.

Cooke, S. J., S. G. Hinch, G. T. Crossin, D. A. Patterson, K. K. English, J. M. Shrimpton, G. Van Der Kraak, and A. P. Farrell. 2006b. Physiology of individual late-run Fraser River sockeye salmon (*Oncorhynchus nerka*) sampled in the ocean correlates with fate during spawning migration. Canadian Journal of Fisheries and Aquatic Sciences 63:1469–1480.

Cooke, S. J., G. T. Crossin, D. A. Patterson, K. K. English, S. G. Hinch, J. L. Young, R. F. Alexander, M. C. Healey, G. Van Der Kraak, and A. P. Farrell. 2005. Coupling non-invasive physiological assessments with telemetry to understand inter-individual variation in behaviour and survivorship of sockeye salmon: development and validation of a technique. Journal of Fish Biology 67:1342–1358.

Cooke, S. J., S. G. Hinch, A. P. Farrell, M. F. LaPointe, S. R. M. Jones, J. S. MacDonald, D. A. Patterson, M. Healey, and G. Van Der Kraak. 2004. Abnormal migration timing and high en route mortality of sockeye salmon in the Fraser River, British Columbia. Fisheries 29:22–33.

Wagner, G. N., L. J. Kuchel, A. Lotto, D. A. Patterson, J. M. Shrimpton, S. G. Hinch, and A. P. Farrell. 2006. Routine and active metabolic rates of migrating adult wild sockeye salmon (*Oncorhynchus nerka* Walbaum) in seawater and freshwater. Physiological and Biochemical Zoology 79:100–108.

American Fisheries Society Symposium 69:895–897, 2009

Modeling the Migratory Patterns and Habitat Use of Migratory Coregonids in the Mackenzie River System

Kimberly L. Howland*, M. VanGerwen-Toyne, and R. Tallman
Department of Fisheries and Oceans, Freshwater Institute
501 University Crescent, Winnipeg, Manitoba R3T 2N6, Canada

Diadromous coregonids are an important component of aboriginal fisheries in the Canadian north. Broad whitefish *Coregonus nasus*, inconnu *Stenodus leucichthys*, lake whitefish *C. clupeaformis*, and Arctic cisco *C. autumnalis* are the major fish species taken in subsistence fisheries in the Mackenzie River Valley, Northwest Territories (Stewart 1996), with broad whitefish and inconnu being of greater importance. All of these species are highly migratory, moving between coastal feeding areas and upstream spawning areas, and are likely targeted by multiple fisheries along their migration routes (Reist and Bond 1988; Dillinger et al. 1992; Chang-Kue and Jessop 1997; Howland et al. 2000).

Recent oil and gas developments—in particular, the proposed Mackenzie Valley natural gas pipeline—are expected to have impacts on these species, including the disruption of migration patterns, disturbance of spawning areas during pipeline construction and operation, and mortality from seismic activities.

It is important to try to limit impacts to fish populations in areas where fish are most vulnerable and congregate in large numbers, such as spawning grounds, overwintering areas, and migratory corridors. It is necessary to identify where these areas are located, and the critical times for their use by fish species, to mitigate and monitor such impacts. Although principal migratory corridors of the major harvested coregonids have been identified (Dillinger et al. 1992; Chang-Kue and Jessop 1997; Howland et al. 2000), knowledge of spawning and overwintering areas along the proposed pipeline route, and in production areas of the Mackenzie Delta, is limited.

We are using radio telemetry in this study to address the following questions: (1) where are critical spawning and overwintering areas for the key harvested populations of anadromous fish species (broad whitefish, inconnu, lake whitefish, and Arctic cisco) in the Mackenzie Valley, and (2) what is the timing of migration into spawning sites and actual time of spawning for key harvested fish populations?

This study was carried out between 2004 and 2007. Over this time, a total of 118 fish (66 broad whitefish, 30 inconnu, 20 lake whitefish, and 2 Arctic cisco) were surgically implanted with coded radio transmitters and released in the main-stem Mackenzie River (Figure 1). The battery life of transmitters ranged from 10 months to 4.5 years, allowing us the opportunity to follow movements of individual fish over consecutive years. Tracking was carried out in the region between Norman Wells and the Beaufort Sea coast (Figure 1) using a combination of aerial surveys with either a helicopter or fixed wing aircraft and strategically placed remote receiving towers.

We identified four reaches of the main-stem Mackenzie River as potential spawning areas, based on the congregation and staging of fish during the expected time of spawning (Figure 1). Overwintering areas were identified in the main-stem Mackenzie River, inner Mackenzie Delta lakes, and coastal Kittigazuit Bay-outer delta lakes areas. Broad whitefish began to move upstream from coastal overwintering sites by mid-July. Spawning appeared to occur in late October to early November followed by downstream movements to overwintering areas. Three and eight individuals were found in upstream areas on a biannual and annual basis, respectively. Two fish used the same overwintering area in consecutive years.

Specific spawning areas could not be identified due to weather problems with tracking during

* Corresponding author: howlandk@dfo-mpo.gc.ca

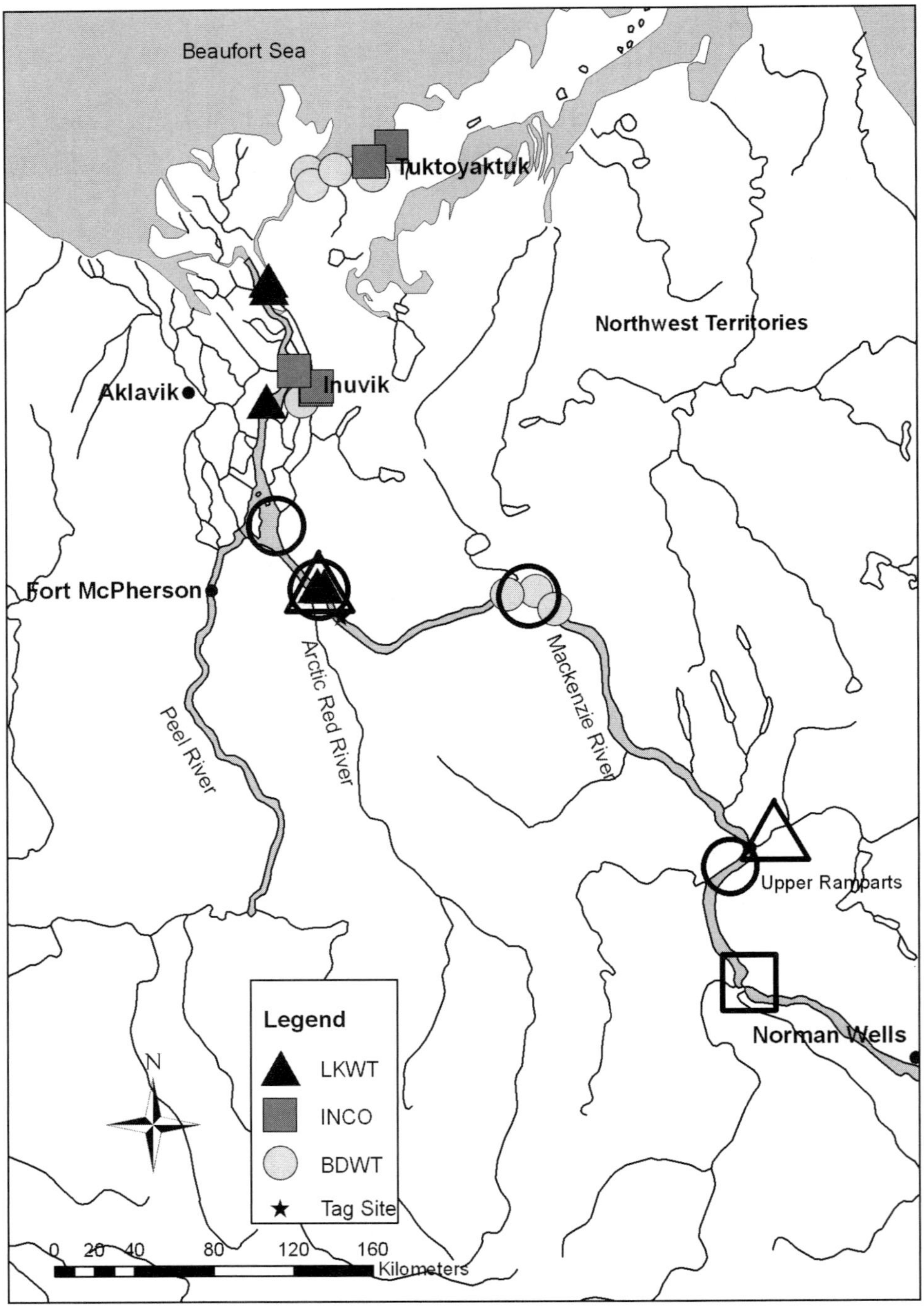

Figure 1.—Map of the lower Mackenzie River system indicating locations of potential spawning (open shapes) and overwintering (filled shapes) areas for broad whitefish (BDWT), inconnu (INCO), and lake whitefish (LKWT).

the expected spawning period in late September; however, patterns and timing of movements suggested that spawning areas for inconnu were located upstream of the Upper Ramparts (Figure 1). Overwintering areas were identified in the inner delta lakes and the Kittigazuit Bay–outer delta lakes area.

Inconnu began to move upstream from overwintering areas by mid-July and were found beyond the Upper Ramparts by August. Downstream movements towards overwintering areas were recorded in October. Four inconnu were found upstream in consecutive years, and three used the same overwintering areas in consecutive years.

Spawning areas were identified in the main-stem Mackenzie and the Hare Indian rivers (Figure 1). Overwintering areas were identified in the main-stem Mackenzie River and the inner Mackenzie Delta. Lake whitefish began to move upstream by late July. Spawning appeared to take place in late September–early October followed by downstream movement to overwintering areas. Four lake whitefish were found upstream in consecutive years.

Insufficient data were available to make any conclusions regarding this species.

Late June to early December is the most critical time for movements between spawning and overwintering sites for diadromous coregonids in the Mackenzie River system. Critical spawning habitat for diadromous coregonids is located in the main-stem Mackenzie River upstream of Point Separation and in the Hare Indian River. Critical overwintering areas include the Mackenzie Delta, Kittigazuit Bay, and outer delta lakes; however, some broad and lake whitefish ovewintered in the main-stem Mackenzie River, suggesting the possible existence of riverine populations. Some individuals, for all three species, were found in upstream areas in consecutive years, suggesting possible unsuccessful spawning, the existence of a riverine population, or annual spawning. Broad whitefish also showed evidence of biannual spawning suggesting that there is variability in spawning frequency. Some broad whitefish and inconnu returned to the same overwintering areas in consecutive years, suggesting that they may have fidelity to these locations.

Acknowledgments

Funding was provided by DFO MCII. Logistical support was provided by Polar Continental Shelf Project, the Gwich'in Renewable Resource Board, Gwichya Renewable Resource Council, Nihtat Renewable Resource Council, and the Fort Good Hope Renewable Resource Council. Thanks to the youth and community members from Tsiighetchic, Inuvik and Fort Good Hope, Zoya Pawlychyn, and Cam Barth for their help in the field.

References

Chang-Kue, K. T. J., and E. Jessop. 1997. Broad whitefish radio tagging studies in the lower Mackenzie River and adjacent coastal region, 1982–1993. Pages 117–146 *in* R. F. Tallman and J. D. Reist, editors. The proceedings of the broad whitefish workshop: the biology, traditional knowledge and scientific management of broad whitefish in the lower Mackenzie River. Canadian Technical Report Fisheries and Aquatic Sciences 2193.

Dillinger, R. E., Jr., T. P. Birt, and M. Green. 1992. Arctic cisco, Coregonus autumnalis, distribution, migration, and spawning in the Mackenzie River. Canadian Field-Naturalist 106:175–180.

Howland, K. L., R. F. Tallman, and W. M. Tonn. 2000. Migration patterns of freshwater and anadromous inconnu in the Mackenzie River system. Transactions of the American Fisheries Society 129:41–59.

Reist, J. D., and W. A. Bond. 1988. Life history characteristics of migratory coregonids of the lower Mackenzie River, Northwest Territories, Canada. Finnish Fisheries Research 9:133–144.

Stewart, D. B. 1996. A review of the status and harvests of fish stocks in the Gwich'in Settlement Area. Canadian Manuscript Report Fisheries Aquatic Science 2336.

American Fisheries Society Symposium 69:899–901, 2009

Rainbow Smelt Population Monitoring and Restoration on the Gulf of Maine Coast of Massachusetts

Bradford C. Chase*, Matthew H. Ayer, and Scott P. Elzey
Massachusetts Division of Marine Fisheries
1213 Purchase Street, 3rd Floor, New Bedford, Massachusetts 02740, USA

Rainbow smelt *Osmerus mordax* resources in Massachusetts have received little attention since the 1970s when sport fisheries attracted many anglers, resulting in large catches in numerous embayments. A recent survey of smelt spawning habitat found evidence that many smelt runs had declined since the 1980s (Chase 2006). Growing concern over the status of New England smelt resources prompted interest in assessing smelt populations and in developing restoration strategies. A fyke net survey of anadromous smelt on the Gulf of Maine coast of Massachusetts was initiated in 2004 to monitor annual spawning run demographics and to assist restoration experiments.

Fyke nets were set in six known smelt runs during the spring spawning seasons of 2005–2007. Four rivers were selected as annual population monitoring stations based on a range of expected run size and watershed characteristics. The Crane and North rivers, where recent restoration efforts improved habitat and documented a reestablished smelt run (Chase 2006), were selected to evaluate the methods and benefits of stocking smelt larvae. Nets were set in the intertidal zone below the downstream limit of smelt egg deposition. Three sets were made per week from March 7 to May 18 in each season. Captured fish were counted, measured, and released. Weekly samples of smelt were collected for aging and gametes in two rivers, Fore and Saugus, with the larger smelt runs. The collected smelt eggs were fertilized, incubated, marked with oxytetracycline (OTC), and stocked as yolk sac larvae. All smelt caught at the restoration sites will be inspected for marked otoliths. Analyses were conducted on size and age composition, and age comparisons were made to studies from the 1970s in the study area (Murawski and Cole 1978; Lawton et al. 1990).

The spawning run peak occurred between the last week of March and the first week of May, although smelt were captured during each week of sampling. The run onset was earlier and the duration longer than that in smelt runs at the northern end of their range (McKenzie 1964; Pettigrew et al. 2009). Each river displayed distinct characteristics of run peak and duration. The Fore River had a 10-week spawning season each year with a consistent peak during the second week of April. Contrary to expectations, the peak run in the Parker River (the most northern) occurred 3–4 weeks earlier than the peak in the Jones River (the most southern).

Fyke net catch rates varied widely among rivers, with the highest in the Fore River and the lowest in the two restoration rivers. Analyses of peak season catches (weeks 4–9) were made to assess the potential of using catch-per-unit-effort (CPUE) data as relative indices of abundance. The nominal CPUE data had high coefficients of variance, and transformed CPUE data also displayed high variance that could present difficulties for discerning annual trends in some rivers. Large year-to-year changes in CPUE were found with little evidence of annual synchrony among rivers. Stronger than average cohorts were evident in the Fore River for 2003 and in the Saugus River for 2005.

Fore River smelt scales were aged in 2005 (N = 274) and 2006 (N = 265), and age and length keys were applied to all measured smelt to construct age composition for each river. In both years, only ages 1–3 were represented in the Fore River age key. Smelt larger than the upper limit of the age

* Corresponding author: brad.chase@state.ma.us

key were few (<1% > 239 mm) and were assigned to age 4 based on the age data of Murawski and Cole (1978) and Lawton et al. (1990). The earlier studies on smelt in the study area found that age structure was dominated by age-2 smelt, and age 3 was the next most abundant age-class. Our catches had a reduced presence of older smelt and a higher proportion of age-1 smelt.

Mean total length by sex was tested (one-way analysis of variance) to determine if there were significant differences ($P < 0.05$) in length at age among rivers. For both years and sexes, the Jones River age-1 smelt were significantly shorter than age-1 smelt in the other rivers. All within-year comparisons showed that Saugus River age-1 smelt and Parker River age-2 smelt were longest among rivers, and Saugus River age-2 smelt were the shortest (Table 1), although not all comparisons were significantly different. Length data from previous Massachusetts smelt studies have similar mean lengths at age (Table 1), and the same traits of shorter Jones River age-1 smelt and longer Parker River age-2 smelt.

Annual survival rates (S) and instantaneous total mortality (Z) were calculated using the Chapman and Robson equation. The age-1 male cohort was excluded from mortality estimates because they are only partially recruited to the spawning run. Total annual mortality of age-2 smelt during 2005–2006 was high, ranging from 75% to 99%. The average annual survival for 2005–2006 data was 0.12, compared to 0.21 for the earlier studies in the Jones and Parker rivers. The precision of Chapman and Robson estimates declines as Z increases. This condition was found when few age 3, and no age 4, smelt were caught. Age structure was compared to the earlier Jones and Parker River data with chi-square contingency tables to test the hypothesis that age frequencies have the same distribution among years. All comparisons of past and present age structure were significantly different ($P < 0.05$). That age frequencies are independent is likely due to the large increase in age-1 smelt, and the decrease in age-3 smelt, in contemporary spawning runs.

More than 1 million smelt larvae, marked as embryos with 500 mg/L OTC, were stocked in the Crane River in both 2005 and 2006. Smelt marked with 500 mg/L in a 2006 controlled experiment failed to yield a consistent fluorescent mark in a blind test 1 year later. Smelt from that same experiment marked with 1,000 mg/L OTC passed the blind test. The 1.6 million smelt stocked in 2007 were marked with 500 mg/L OTC as larvae, and early results of a 2007 control experiment have found a stronger mark than that produced with embryo marking.

The analyses of catch data indicate that present

Table 1.—Smelt mortality and total length (mm) at age from fyke net catches and comparative data from earlier studies in Massachusetts (Murawski and Cole 1980; Lawton et al. 1990). The length data from 1974 to 1981 were only available as averages for all the years of the studies. Only age-1 and age-2 smelt were caught in the Jones River during 2005 (NA). S = annual survival rate; Z = instantaneous mortality.

River	Year	N	S	Z	Z (SE)	Age 1 M	Age 1 F	Age 2 M	Age 2 F
Parker River	1974	561	0.29	1.23	0.069				
Parker River	1975	1,358	0.11	2.18	0.075	141	140	188	197
Parker River	2005	482	0.04	3.30	0.262	139	139	190	198
Parker River	2006	123	0.25	1.39	0.042	137	139	190	192
Jones River	1979	13,093	0.20	1.62	0.017				
Jones River	1980	18,012	0.11	2.21	0.020				
Jones River	1981	5,298	0.33	1.10	0.016	132	130	184	190
Jones River	2005	449		NA		128	131	175	181
Jones River	2006	614	0.18	1.72	0.122	127	131	186	191
Fore River	2005	1,050	0.01	4.98	0.407	138	141	177	188
Fore River	2006	973	0.18	1.69	0.076	135	137	186	191
Saugus River	2005	141	0.04	3.11	0.560	141	144	172	183
Saugus River	2006	1,192	0.17	1.79	0.099	143	143	182	183

smelt runs have a truncated age distribution with a higher percentage of age-1 fish than previously documented. Mortality rates are higher than previously reported, as few smelt are reaching ages 3 and 4 in the present runs. These conditions support management concerns of declining smelt populations in Massachusetts.

The stressed condition of the Jones River smelt run raises questions related to the partial recruitment of age-1 smelt. Jones River age-1 smelt were smaller than that found in northern runs, and their proportion in the spawning run has increased. For short-lived species, spawning at an earlier age can be a response to environmental stress or a tactic to yield higher evolutionary fitness than remaining at sea an additional year to gain larger size (Gross 1987). Fyke net sampling and experimental stocking will continue and include examinations of the status of age-1 smelt, potential stressors related to age structure changes, and the influence of repeat spawning on population data.

References

Chase, B. C. 2006. Rainbow smelt (*Osmerus mordax*) spawning habitat on the Gulf of Maine coast of Massachusetts. Massachusetts Division of Marine Fisheries, Technical Report No. TR-30, Gloucester.

Gross, M. R. 1987. Evolution of diadromy in fishes. Pages 14–25 *in* M. J. Dadswell, R. J. Klauda, C. M. Moffitt, R. L. Saunders, R. A. Rulifson, and J. E. Cooper, editors. Common strategies of anadromous and catadromous fishes. American Fisheries Society, Symposium 1, Bethesda, Maryland.

McKenzie, R. A. 1964. Smelt life history and fishery in the Miramichi River, New Brunswick. Bulletin of the Fisheries Research Board of Canada 144.

Murawski, S. A., and C. F. Cole. 1978. Population dynamics of anadromous rainbow smelt, *Osmerus mordax*, in a Massachusetts estuary. Transactions of the American Fisheries Society 107:535–542.

Lawton, R., P. Brady, C. Sheehan, S. Correia, and M. Borgatti. 1990. Final report on spawning sea-run rainbow smelt (*Osmerus mordax*) in the Jones River and impact assessment of Pilgrim Station on the population, 1979–1981. Boston Edison. Massachusetts Division of Marine Fisheries, Pilgrim Nuclear Power Station Marine Environmental Monitoring Program Report Series No. 4, Sandwich.

Pettigrew, P. 2009. Suivi de la reproduction de l'eperlan arc-en-ciel dans la riviere Fouquette en 2008. Ministeres des Ressources naturelles et de la Faune, Direction generale du Bas-Saint-Laurent, Direction regionale de l'amenagement de la faune, Québec.

American Fisheries Society Symposium 69:903–906, 2009

Restoration of Diadromy in an Endangered Fish: Do Atlantic Whitefish Covet the Sea?

Adam M. Cook* and Paul Bentzen
Marine Gene Probe Laboratory, Dalhousie University, Halifax, Nova Scotia B3H 4J1, Canada

Diadromy is a tactic employed by many fish species around the globe. However, its significance as a strategy integral to species survival is often unknown. Numerous examples in the literature document cases where species, generally regarded as diadromous, harbor populations or cohorts that are freshwater residents and never descend rivers, or marine contingents that never fully ascend estuaries (Skaala and Nævdal 1989; Tzeng et al. 2000). As such, diadromy may be viewed as a plastic trait only displayed by some individuals or populations. Situations arise, occasionally, where the only remaining population of a diadromous species is found in a land-locked form. Prior to restoring diadromy to this population, information is required on their propensity for saltwater.

Atlantic whitefish *Coregonus huntsmani* is an endangered, endemic Nova Scotian species documented only in the Petite Riviere and the Tusket River. Historically, Atlantic whitefish in the Tusket River made diadromous upriver migrations in autumn. The form of diadromy used by this population was unknown as spawning locations were never identified, and unfortunately, this population was extirpated at some time during the 1970s or early 1980s (Edge and Gilhen 2001). Diadromy in the Petite Riviere population could not be quantified as the river has been dammed, with no fish passage, for the past 30–35 years (Bradford et al. 2004). The importance of diadromy to the population, and hence species recovery, is not known.

The aim of the present work was twofold: first, to determine the form of diadromy exploited by Atlantic whitefish by testing salinity tolerance of gametes, and second, to determine the salinity tolerance and preference of developing larvae and juveniles.

* Corresponding author: amcook@dal.ca

Three Atlantic whitefish males were stripped manually of milt. Three treatments were performed on each batch of milt, such that 1 μL of milt was activated in one drop of either freshwater, brackish water, or seawater, on a hanging drop microscope slide. Each treatment was replicated three times for each batch of milt. The mixture was observed and video recorded, while active, at 100× magnification.

Videos were analyzed to track the decay in sperm activity over time. Logistic decay models (equation [1] were fitted and compared (equation [2]; where Motility is the percent motility, min is the minimum sperm motility, max is the maximum sperm motility, T is time, and T_{50} is time at 50% motility. In equation (2), Δ represents the change in parameter estimate when combining data from both treatments, and X_1 is an indicator variable.

$$\text{Motility} = \min + \frac{\max - \min}{1 + \left(\dfrac{T}{T_{50}}\right)^{S}} \tag{1}$$

$$\text{Motility} = \left[\min + \left(\Delta \min X_1\right)\right] + \frac{\left[\max + \left(\Delta \max X_1\right)\right] - \left[\min + \left(\Delta \min X_1\right)\right]}{1 + T_{50} + \left\{\dfrac{T}{\left[T_{50} + \left(\Delta T_{50} X_1\right)\right]}\right\}^{\left[S + \left(\Delta S X_1\right)\right]}} \tag{2}$$

Salinity tolerance of Atlantic whitefish was determined on individuals ranging from premetamorphosis larvae to adults. Individuals were transferred from freshwater to full strength seawater (30 parts per thousand [ppt]) at a rate of 10 ppt/d and were reared at these salinities for 8–10 d.

An experiment to determine the effect of salinity on the growth of Atlantic whitefish was performed at salinity levels of 0, 15, and 30 ppt. Fish had an initial mean body length of 66 mm (SD = 12 mm) and were reared for 14 d in respective treatment salinities. Fish were fed three times daily and kept at a water temperature of 16°C.

Salinity preference was performed using juvenile Atlantic whitefish with a mean body length of 65 mm (SD = 9 mm). Ninety fish were used in these experiments with groups of 30 being acclimated to freshwater (0 ppt), brackish water (15 ppt), and full strength seawater (30 ppt) for 2 weeks prior to initiation. In each experiment, three salinity preference chambers were filled with freshwater, brackish water, or full strength seawater. The freshwater chamber was allowed to overflow to set up a stratified layer of freshwater across the surface of each chamber to allow for fish movement between chambers. Five fish from each acclimation salinity were introduced into each filled salinity chamber. Fish movements were monitored over 24 h. Experiments were performed twice at each acclimation salinity. Analyses were performed as analysis of variance on the change in the number of fish present in each chamber at 24 h, with the independent variable being acclimation salinity.

Atlantic whitefish sperm, activated in freshwater, were motile the longest time, with 50% motility occurring at 32 s (Figure 1). In comparison, time to 50% sperm motility in brackish water and saltwater was 1.72 and 1.52 s. Across all treatments, maximum sperm motility was similar with rates of 92–95%; however, percent motility decreased rapidly in brackish water and saltwater (Figure 1).

Atlantic whitefish are tolerant to full strength seawater prior to metamorphosis and retain this ability throughout life (data not shown). In a separate experiment, the growth of juvenile Atlantic whitefish was independent of salinity from 0 to 30 ppt ($P > 0.1$). Additionally, Atlantic whitefish juveniles prefer full strength seawater at a mean rate of 92.2% (SD = 5.0%), regardless of their acclimation salinity ($P < 0.01$; Figure 2). However, acclimation salinity did affect the time required to make a final salinity choice as individuals reared in freshwater required more than 5 h to choose seawater, whereas seawater-acclimated fish required less than 1 h.

Despite a long-standing lack of two-way access to the sea, the sole remaining population of Atlantic whitefish retains the attributes of an anadromous species. Although there are no records of directed movements, upstream or downstream, of Atlantic whitefish on the Petite Riviere, adults require freshwater to spawn and larvae and juveniles are tolerant to, and prefer, seawater. These results suggest that larvae, just after hatching, may make a downstream migration to feed and grow in estuarine or marine waters. This theory would account for the lack of information, either scientific or local, on the fish's downstream migration as the small (14 mm at hatch) larvae would likely go

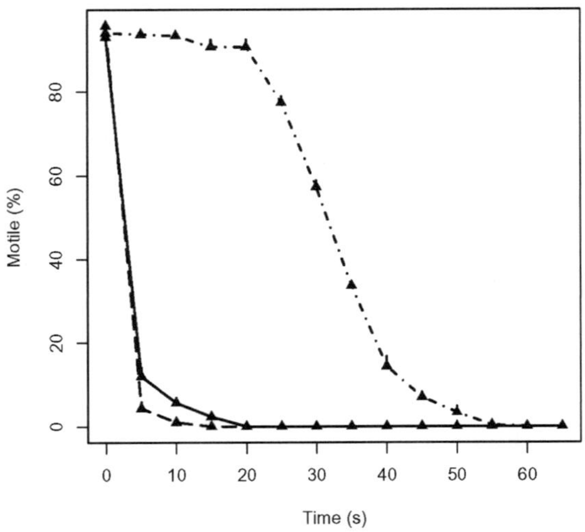

Figure 1.—Duration (s) of sperm motility of Atlantic whitefish milt activated in salinities of 0 (dash-dot), 15 (solid) and 30 parts per thousand (dash). Upper lines from symbols represent standard error.

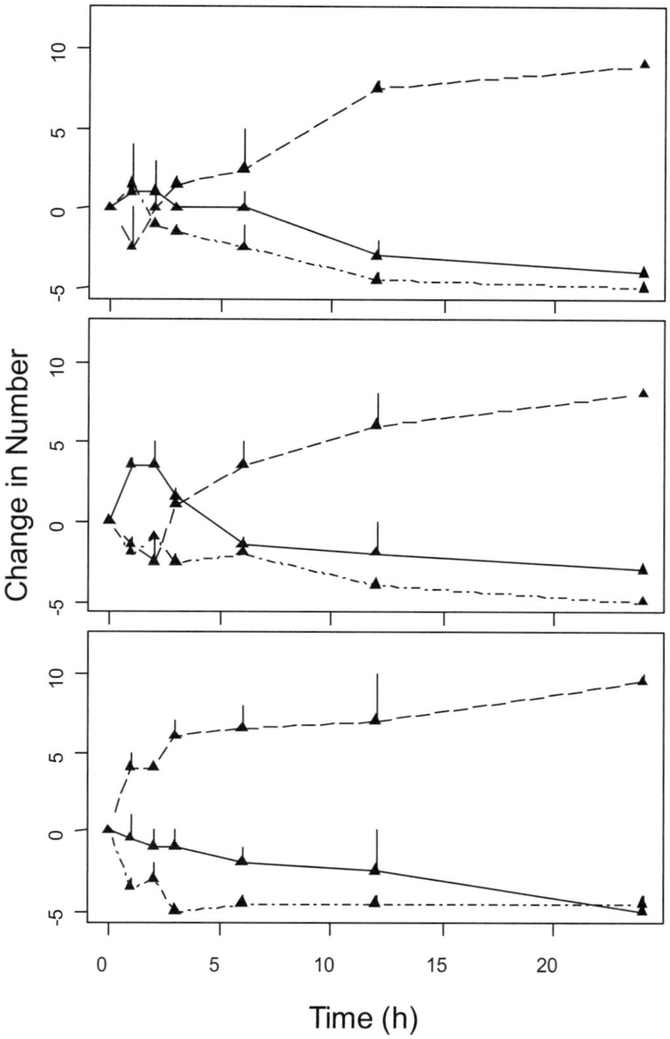

Figure 2.—Salinity preference of juvenile Atlantic whitefish acclimated to 0 (upper), 15 (middle) and 30 parts per thousand (ppt) (lower) salinity. Change in number represents the gain or loss of fish over time (hours) from the original five fish placed in each salinity chamber. which were set to 0 (dash-dot), 15 (solid). and 30 ppt (dash). Upper lines from symbols represent standard error.

undetected. Supporting evidence may be observed through the apparent low densities of young Atlantic whitefish in a land-locked form as only a single report of one age-1 juvenile Atlantic whitefish has ever been made (Hasselman et al. 2005). Testing this theory is vital to the Atlantic whitefish's viability: if significant proportions of larvae are leaving the Petite Riviere system and cannot return to spawn, they will not be recruited to the population, causing the species to head down an extinction vortex (Gilpin and Soulé 1986).

References

Bradford, R. G., D. Longard, and P. Longue. 2004. Status, trend and recovery considerations in support of an allowable harm assessment for Atlantic whitefish (*Coregonus huntsmani*). Department of Fisheries and Oceans Canada, Canadian Science Advisory Secretariate Research Document 2004/109, Ottawa.

Edge, T. A., and J. Gilhen. 2001. Updated status report on the endangered Atlantic whitefish, *Coregonus huntsmani*. Canadian Field Naturalist 115:635–651.

Gilpin, M. E., and M. E. Soulé. 1986. Minimum viable

populations: processes of species extinction. Pages 19–34 *in* M. E. Soulé, editor. Conservation biology: the science of scarcity and diversity. Sinauer, Sunderland, Massachusetts

Hasselman, D. J., P. Longue, and R. G. Bradford. 2005. First record of age 0+ Atlantic Whitefish, *Coregonus huntsmani*, from the wild. Canadian Field Naturalist 119:294–295.

Skaala, O., and G. Nævdal. 1989. Genetic differentiation between fresh-water resident and anadromous brown trout, *Salmo trutta*, within watercourses. Journal of Fish Biology 34:597–605.

Tzeng, W. N., C. H. Wang, H. Wickstrom, and M. Reizenstein. 2000. Occurrence of the semi-catadromous European eel *Anguilla anguilla* in the Baltic Sea. Marine Biology 137:93–98.

American Fisheries Society Symposium 69:907–909, 2009

An Evaluation of Nature-Like Fishways for Passage of Anadromous Alewife

ABIGAIL FRANKLIN*
Department of Natural Resources Conservation, University of Massachusetts Amherst, Massachusetts 01003, USA

ALEX HARO AND THEODORE CASTRO-SANTOS
S. O. Conte Anadromous Fish Research Center, U.S. Geological Survey Biological Resources Discipline, Turners Falls, Massachusetts 01376, USA

Nature-like fishways were designed in part as a response to poor or unknown passage efficiency of technical fishways, to address a desire to reconnect river corridors, and provide passage for all species occurring in a system (FAO 2002). The first nature-like fishways were developed in Austria, Germany, and France. A nature-like fishway typically consists of a wide, low-gradient channel with a 1:20 slope or less and a concave stream channel cross section. Natural cobble or boulder substrates are used to dissipate hydraulic kinetic energy and to reduce channel velocities to levels that allow fish to pass at sustainable (i.e., aerobic) swimming speeds. Nature-like fishways have been designed as bypass channels around dams, as roughened ramps constructed in association with a dam, and as stream restorations with complete dam removals. Fish are believed to find natural substrates more acceptable than concrete channels or channels with baffles in technical fishways (FAO 2002). The low velocities at the margins and hydraulic boundary layers of this fishway design allow for greater potential passage of small or weakly swimming species.

Evaluations of nature-like fishways in Europe have shown variable success. Nature-like fishways have been constructed in Canada and in the western and northeastern states in the United States, but have not been evaluated extensively for passage of North American species. The goal of this project was to evaluate a nature-like fishway design in the northeastern United States (Massachusetts) for passage of anadromous alewife *Alosa pseudoharengus*.

* Corresponding author: afranklin@nrc.umass.edu

Alewife are a relatively small (maximum fork length = 40 cm) anadromous, iteroparous species that range from northeastern Newfoundland to South Carolina (ASMFC 1999). Alewife migrate from the ocean to their natal river when water temperatures increase in the spring months. They spawn in lentic waters, and adults leave freshwater quickly after laying their eggs. The juveniles feed on zooplankton and migrate downstream to the ocean when water temperatures drop in autumn (Loesch 1987). Alewife populations have been declining since the industrial age due to overharvest, poor water quality, and blockages to spawning habitat. Movements of alewife may be blocked by a structure creating a vertical hydraulic drop of only 20–30 cm (ASMFC 1999).

The National Oceanic and Atmospheric Administration Habitat Restoration Program funded the removal of the Billington St. Dam on Town Brook in Plymouth, Massachusetts, in 2002, with the goal of increasing spawning habitat. The dam was replaced with a 30-m-long perturbation boulder-style nature-like fishway. Passive integrated transponder telemetry was used to monitor the movements of alewife through the fishway, using methods described in Castro-Santos et al. (1996). Four antennas were placed across the fishway, and 400 fish were tagged internally with 20-mm tags in seven separate batches from April 19, 2006 to May 15, 2006. Water temperature and level were recorded using underwater data loggers. Water temperatures ranged from 9.96°C to 23.58°C over the field season.

Data were collected from 175 (44% of total) tagged fish. Ninety-six percent of the fish ascended the fishway successfully (Figure 1). Further analysis

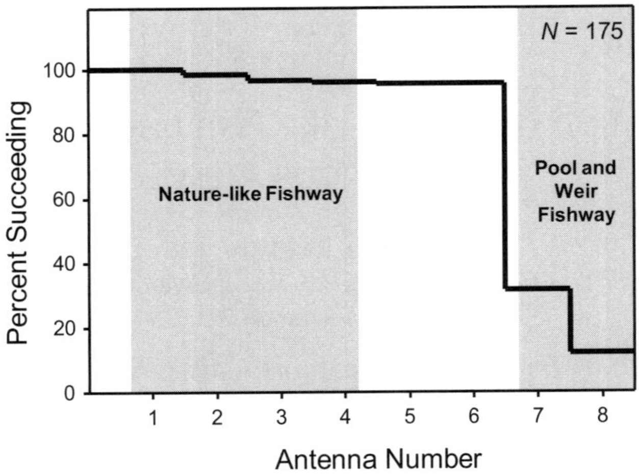

Figure 1.—Percentage of fish passing each antenna successfully at the passive integrated transponder antenna array at Town Brook. Antennas 1 through 4 cover the entire length of the nature-like fishway; other antennas cover the upstream reach (antennas 5 and 6) and a small pool-and-weir fishway (antennas 7 and 8).

is based on only103 fish due to an equipment malfunction that led to 5 d of lost data for 72 fish. The median transit time for the first attempt through the fishway was 10.9 min (range: 4.85–44.08 min; Figure 2). No relationship was present between transit time and fork length, sex, release group, or water level. Transit times decreased with increasing water temperature ($r^2 = -0.33$; $P < 0.01$).

Ninety-one fish (88.3% of total) were successful in ascending the entire fishway on their first attempt. Five of the remaining 12 fish that were unsuccessful on their first attempt, ascended on a subsequent attempt. Of a total of 160 attempts, 93.1% were made during daylight hours.

The nature-like fishway design at Town Brook passes alewife efficiently. Percent passage was high

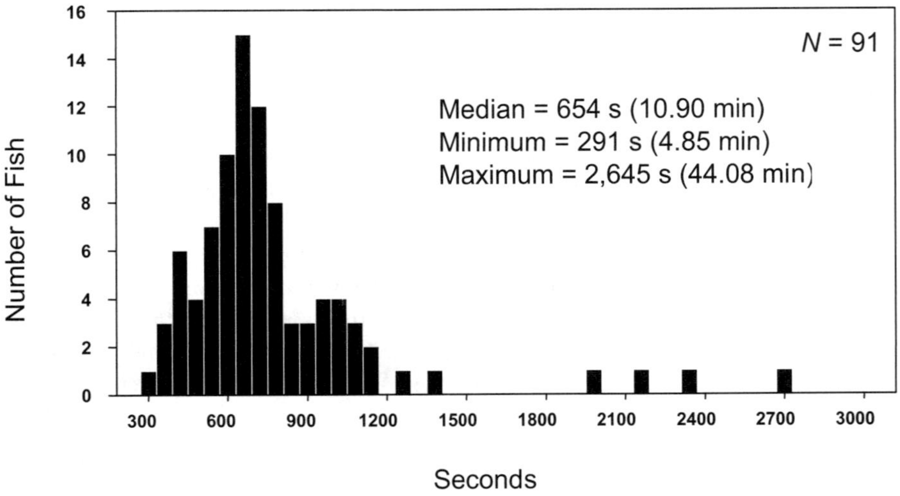

Figure 2.—Distribution of first attempt transit times through the nature-like fishway (antennas 1 through 4). Transit times are defined as the period between the last tag detection at antenna 1 and the first detection at antenna 4.

and 95% of fish ascended in less than 22 min. This nature-like fishway design shows promise as an appropriate passage structure for alewife at low head dam removals on small streams.

References

ASMFC (Atlantic States Marine Fisheries Commission). 1999. Amendment 1 to the Interstate Fishery Management Plan for Shad and River Herring. Atlantic States Marine Fisheries Commission, Fishery Management Report 35, Washington, D.C.

Castro-Santos, T., A. Haro, and S. Walk. 1996. A passive integrated transponder (PIT) tag system for monitoring fishways. Fisheries Research 28:253–261.

FAO (Food and Agriculture Organization of the United Nations). 2002. Fish passes—design, dimensions, and monitoring. Food and Agriculture Organization of the United Nations, Rome.

Loesch, J. G. 1987. Overview of life history of anadromous alewife and blueback herring in freshwater habitats. Pages 89–103 *in* M. J. Dadswell, R. K. Klauda, C. M. Moffitt, R. L. Saunders, R. A. Rulifson, and J. E. Cooper, editors. Common strategies of anadromous and catadromous fishes. American Fisheries Society, Symposium 1, Bethesda, Maryland.

American Fisheries Society Symposium 69:911–913, 2009

Why Has the Japanese Eel Not Declined to the Levels of the European and American Eels? A Possible Explanation from the Movement and Age Structure of Growth-Phase Eels

Kazuki Yokouchi*, Jun Aoyama, and Katsumi Tsukamoto
Ocean Research Institute, The University of Tokyo, Tokyo 164–8639, Japan

Catadromous temperate eel recruitment in the northern hemisphere has been declining drastically in some areas, but the degree of the decline appears to differ. Glass eel recruitment of the European eel *Anguilla anguilla* and American eel *A. rostrata* has dropped as much as 99% in some areas since the early 1980s, while the Japanese eel *A. japonica* has remained at levels as great as 20% of the levels of the early 1970s (Dekker 2003). The characteristics of the life cycle have an effect on the traits of the stock, and the demographic attributes during the growth phase of eels (e.g., sex, growth, and age at maturation) are important determinants for their life cycles that may differ among species. The objective of this study was to explore why the Japanese eel stock has not declined to the level of the European and American eel stocks by comparing the demography of the growth phase between Japanese eels and the Atlantic eels.

A total of 4,925 Japanese eels were collected by electrofishing, eel pots, set nets, longline, and traditional traps from nine prefectures of Japan (Ibaraki, Chiba, Shizuoka, Aichi, Tokushima, Kochi, Ohita, Miyazaki, and Kagoshima) from 1999 to 2004. Mean total length (±SD, mm) of females and males was 510 ± 118.2 and 420 ± 81.9, respectively (Figure 1). Mean age (±SD, years) of females and males was 5.0 ± 2.0 and 3.6 ± 1.7 (Figure 1; Table 1). Mean size-at-age growth rate (±SD, mm/year) of females (n = 4,077) and males (n = 1,038) was 96.3 ± 37.8 and 118.0 ± 63.5 (Table 1). The age of Japanese eels was less than that of European eels (mean age range: 8.5–20 years; Vøllestad 1992) and of American eels (range: 1–39 years; Morrison and Secor 2003) after eel stocks began to decline (European eel after 1985, American eel after 1990). The size-at-age growth rate of the Japanese eel was also faster than the Atlantic eels (28–60 mm/year; Vøllestad 1992; Moriarty 2003; Morrison and Secor 2003). These data on the age and growth of northern temperate anguillid eels suggest that the life cycle of the Japanese eel is shorter than that of European and American eels.

The movement patterns of eels are also important factors in fish stock assessment, and Japanese eels had restricted movement (81% of recaptured eels had moved less than 150 m from the release points) in the Nishijinden River in Shizuoka, Japan, inferred from a mark–recapture experiment from 2003 to 2004. This result was similar to previous studies of European and American eels (Oliveira 1997; Feunteun et al. 2003; Morrison and Secor 2003). Therefore, it was suggested that these three species have limited movements.

Although eels have been studied in freshwater and estuarine environments, it is not known if the causes for the declines of the temperate eels are changes in oceanic conditions, overfishing, or the loss or deterioration of growth habitats. However, since all temperate eels may experience similar effects of changing oceanic conditions during their larval dispersal by oceanic currents, some of the differences in the levels of declines of temperate eels may be due to factors during their yellow eel growth phase. European and American eels are characterized by a sedentary growth phase and a long life span, compared with Japanese eels. These characteristics would increase the time in which Atlantic eels would be more vulnerable to fishing pressure, and anthropogenic impacts on their growth habitats, before reaching maturation. It is hypothesized that this may be one of the reasons that the European and American eel stocks have declined more than the Japanese eel stock.

* Corresponding author: kazukiy@ori.u-tokyo.ac.jp

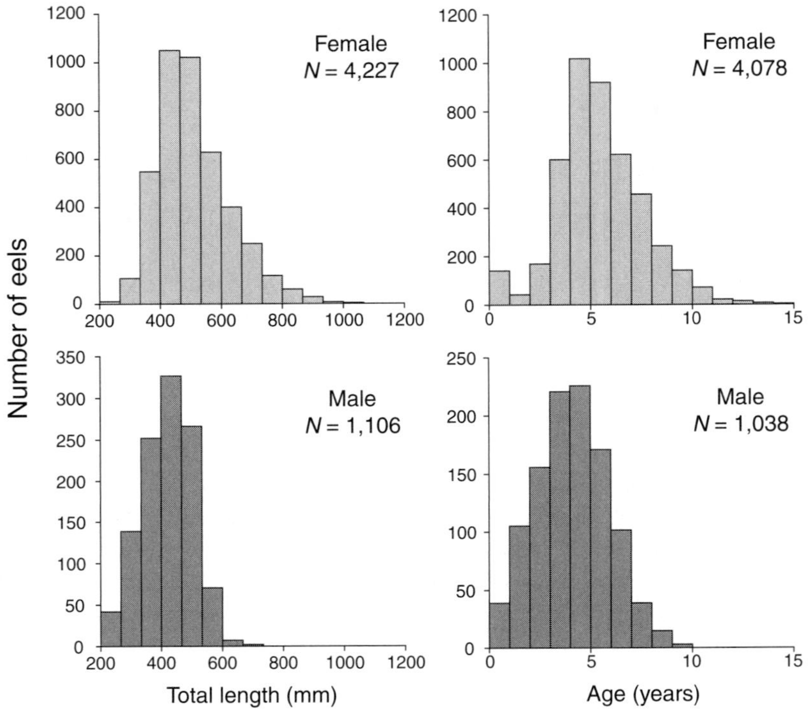

Figure 1.—Frequency histograms of total length and age for male and female Japanese eels.

Table 1.—Age and growth of temperate eels in the northern hemisphere. F = female; M = male.

Species	Location	Mean age (years)	Growth rate (mm/year)	Source
European eel			30.0–60.0	Moriarty (2003)
	Burrishoole	F: 20.0, M: 15.0		Vøllestad (1992)
	River Shannon	F: 16.1, M: 10.6		Vøllestad (1992)
	River Imsa	F: 7.9, M: 4.8		Vøllestad (1992)
American eel	Rhode Island river		29.9	Oliveira (1997)
	Hudson River freshwater site	17.0	34.0	Morrison and Secor (2003)
	Hudson River brackish-water site	8.0	80.0	Morrison and and Secor (2003)
Japanese eel	9 prefectures	F: 5.0, M: 3.6	F: 96.3, M: 118.0	This study

References

Dekker, W. 2003. Did lack of spawners cause the collapse of the European eel, *Anguilla anguilla*? Fisheries Management and Ecology 10:365–376.

Feunteun, E., P. Laffaille, T. Robinet, C. Briand, A. Baisez, J.-M. Oliver, and A. Acou. 2003. A review of upstream migration and movements in inland waters by anguillid eels: toward a general theory. Pages 191–213 *in* K. Aida, K. Tsukamoto, and K. Yamauchi, editors. Eel biology. Springer-Verlag, Tokyo.

Moriarty, C. 2003. The yellow eel. Pages 89–105 *in* K. Aida, K. Tsukamoto, and K. Yamauchi, editors. Eel biology. Springer-Verlag, Tokyo.

Morrison, W. E., and D. H. Secor. 2003. Demographic

attributes of yellow-phase American eels (*Anguilla rostrata*) in the Hudson River estuary. Canadian Journal of Fisheries and Aquatic Sciences 60:1487–1501.

Oliveira, K. 1997. Movements and growth rates of yellow-phase American eels in the Annaquatucket River, Rhode Island. Transactions of the American Fisheries Society 126:638–646.

Vøllestad, L. A. 1992. Geographic variation in age and length at metamorphosis of maturing European eel—environmental effects and phenotypic plasticity. Journal of Animal Ecology 61:41–48.

American Fisheries Society Symposium 69:915–917, 2009

Evaluating Small Barrier Removal and Passage Improvement Scenarios to Enhance Diadromous Fish Restoration in the Penobscot Basin, Maine

WESLEY S. PATRICK*,[1]
NOAA, National Marine Fisheries Service, Northeast Regional Office
Protected Resources Division
1 Blackburn Drive, Gloucester, Massachusetts 01930, USA

RORY SAUNDERS
NOAA, National Marine Fisheries Service, Northeast Regional Office
Protected Resources Division
17 Godfrey Drive-Suite 1, Orono, Maine 04473, USA

The Penobscot River Restoration Project (PRRP) offers some unique opportunities to reconnect 12 native diadromous species with historic habitats and restore associated ecological functions. In particular, alewives *Alosa pseudoharengus* are absent from the drainage with the exception of two tributaries of the Penobscot estuary: the Souadabscook Stream and the Orland River. While the PRRP will substantially improve passage conditions for diadromous fish in the Penobscot Basin, it is in no way a "silver bullet" for ecosystem restoration of the Penobscot Basin. There are roughly 100 small dams and other fish passage barriers in the Penobscot Basin that will prevent some diadromous species from gaining access to historic habitats after the implementation of the PRRP. Many other passage impediments will need to be addressed to realize the full potential of the PRRP. We assessed the restoration potential for alewives in the major subwatersheds using a novel, yet simple, geographic information system-based approach to help prioritize restoration efforts in the Penobscot basin.

We mapped the historical range of alewives in the Penobscot Basin (Figure 1) with ArcGIS 9.2 software and 1:24,000 scale data layers using a variety of sources, most notably Houston et al. (2007). We used four parameters to select streams that were likely to have supported historical alewife runs in subwatersheds, but for which no historical accounts were available: (1) presence of an adjacent stream that supported an alewife run, (2) presence of lake habitat, (3) river width leading to the lake habitat was greater than 50 ft, and (4) absence of a significant change in elevation that could create impassable falls within the tributary. We considered streams and lakes as being historical alewife habitat if these four conditions were met.

Subwatersheds were prioritized based on the acreage of available lake habitat that could have supported alewife spawning. We used a sensitivity analysis to determine the relative importance of each subwatershed given varying downstream impediments and associated passage efficiencies. The sensitivity analysis reduces the productivity of each unit of spawning habitat based on the passage efficiency. We simulated passage efficiency rates using a high (100%), medium (75%), and low (50%) range for each impediment. For example, given a "medium" passage scenario in a 1,000-acre lake upstream of five dams, the 1,000 acres would be reduced to 237 acres (i.e., 1,000 acres $\times$ 0.75^5 = 237.3 acres). The revised estimates of accessible lake habitat were then ranked from highest to lowest in terms of restoration potential, or the number of alewives that could be supported by the lake habitat.

* Corresponding author: wesley.patrick@noaa.gov
[1] Current address: NOAA's National Marine Fisheries Service, Office of Sustainable Fisheries, 1315 East-West Highway (SSMC3-F/SF3), Silver Spring, MD 20910, USA

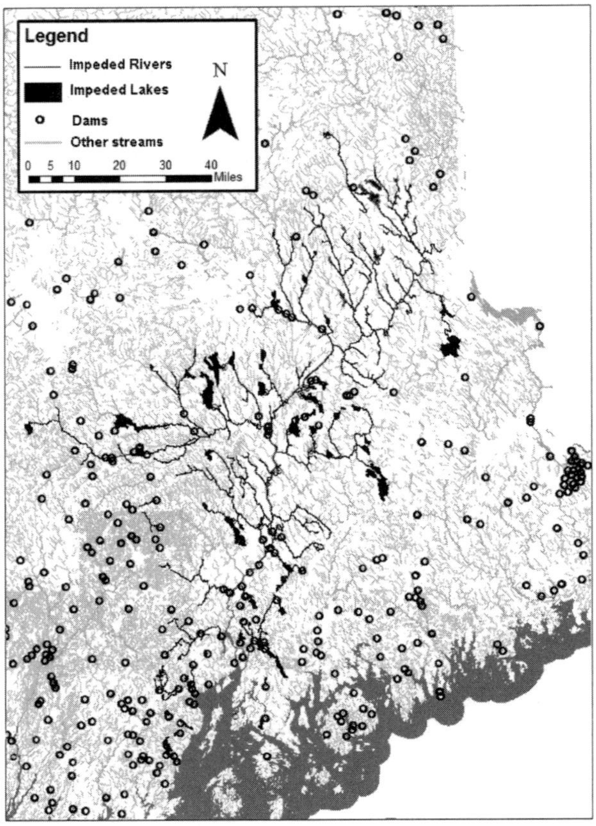

Figure 1.—A map of the impeded alewife habitat in the Penobscot Basin, Maine.

We estimated that there were roughly 90,000 acres of historically accessible lacustrine habitat in the Penobscot Basin. Adult returns of alewives could be as high as 21 million annually at full capacity (235 adult returns per acre; Flagg 1984). However, the vast majority of this lacustrine habitat is upstream of one or more dams, and these impediments have a negative effect on the number of annual adult returns. We examined the production potential of each major subwatershed in the Penobscot Basin to determine which subwatersheds should receive priority in terms of alewife restoration and fish passage improvements. We found that the Piscataquis, Mattawamkeag, and Passadumkeag drainages contain the most lacustrine habitat and, thus, the highest alewife production potential (Table 1). However, the production potential of upper tributaries (e.g., Mattawamkeag and Mattamiscontis) diminished substantially (Table 1) when passage efficiencies through multiple main-stem dams on the Penobscot River were taken into account. Subwatersheds that remained highly ranked were usually located in the lower (Orland, Sedgeunkedunk, Soudabscook, and Ducktrap) to middle (Piscataquis and Passadumkeag) portion of the Penobscot Basin, which contained large lake systems or few impediments or a combination of each.

Our mapping and modeling exercise represents the first attempt to quantify restoration potential of lacustrine habitats in the Penobscot Basin in a spatially explicit manner. A previous estimate of 14.5 million adults as the production potential (Flagg 1984) was lower than our estimate of 21 million adults. This difference is largely attributable to the absence of the West Branch and Cold Stream Pond in the estimate by Flagg (1984).

This approach only considers restoration of alewife, but the tool could become a multispecies model if other species with sufficient historic range data were included. We suggest that our

Table 1.—Alewife restoration potential for the top 10 subwatersheds of the Penobscot Basin and differences in ranking based on varying passage efficiency rates.

Subwatershed	Lake acres (total)	Potential alewife production	Ranking based on passage efficiency		
			100%	75%	50%
Piscataquis	22,637	5,319,695	1	1	3
Mattawamkeag	17,963	4,221,305	2	4	7
Passadumkeag	15,939	3,745,665	3	2	2
Pushaw	6,656	1,564,160	4	5	5
Orland	5,301	1,245,735	5	3	1
Mattamiscontis	3,324	781,140	6	8	*
Blackman	2,232	524,520	7	6	10
Sedgeunkedunk	1,476	346,860	8	7	8
Souadabscook	940	220,900	9	10	9
Ducktrap	814	191,290	10	9	4

* Did not rank in the top 10.

model is useful for understanding and managing the restoration potential of diadromous fish in the Penobscot Basin. Integration of our model into a broader ecosystem approach would require that barriers be physically evaluated, that site-specific passage efficiency be determined for each impediment, and that state and federal fisheries managers set geographically explicit recovery objectives.

References

Houston, R., K. Chadbourne, S. Lary, and B. Charry. 2007. Geographic distribution of diadromous fish in Maine [CD-ROM v1.0]. U.S. Fish & Wildlife Society and Maine Audubon Society, Falmouth, Maine.

Flagg, L. N. 1984. Penobscot River shad and alewife restoration potential. Maine Department of Marine Resources, Augusta.

American Fisheries Society Symposium 69:919–921, 2009

How Dammed Is Your Watershed: First Approximation of an Index to Relative Dammed-ness of U.S. Watersheds

MARK A. CANTRELL*
U.S. Fish and Wildlife Service, Ecological Services, Asheville Field Office
160 Zillicoa Street, Asheville, North Carolina 28801, USA

AMANDA K. HILL
U.S. Fish and Wildlife Service, Ecological Services, Charleston Field Office
176 Croghan Spur Road, Charleston, South Carolina 29407, USA

We describe an index of "dammed-ness" of waters of the United States, with an emphasis on the Atlantic coast of North America where there are many efforts to restore diadromous fishes. We compared the relative degree of dammed-ness at multiple scales, with specific examples where we are involved with diadromous fish restoration. This first approximation provides a quick means of ranking watersheds and evaluating dams as environmental stressors on diadromous fishes. Use of this dammed index can provide insights to relative conditions of subbasins when prioritizing diadromous fish restoration efforts, both within and among basins or regions. We offer additional metrics in the dammed index as a source for prioritization of restoration efforts, as identification of problem areas, and for understanding the seriousness of the challenges facing fishery managers on the Atlantic coast of the United States.

The dammed index is based on comprehensive data in the National Inventory of Dams, maintained by the U.S. Army Corps of Engineers (USACE 2005). We used a geographic information system to determine the dammed index, which lists and ranks watersheds classed by hydrologic unit codes (#d-HUCs) according to several metrics. The first group of metrics that we have developed are related to dam infrastructure: dam density, storage volume, and damless areas are summarized by two d-HUCs, four d-HUCs, six d-HUCs, and eight d-HUCs. Watersheds are ranked according to dam density and placed into five dammed categories: extreme, high (many, large), moderate (transitional), low, and few (or small). We also considered that as the combined storage volume of dams increased, so did the potential for streamflow regulation in the watershed. Similarly, as the total surface area of reservoirs within a HUC increase, so does the potential for thermal and chemical alteration, as well as potential reduction of streamflow by evaporative losses, in the watershed. These categories provide a framework for understanding potential watershed impacts to diadromous fish conservation at multiple scales. Dams are often cited as an impediment to diadromous fish restoration, usually because dams can block or slow fish spawning runs (Freeman et al. 2003). The dammed index provides a broad means of consideration of the storage capacity of dams as well as the relative position of these reservoirs within the subbasins. The dammed index can be used to evaluate diadromous fish restoration efforts among watersheds. The dammed index can be used to compare or guide diadromous fish restoration for subbasins within larger basins to prioritize restoration efforts (Hill 2009, this volume).

The dammed index furthers the work by Busch et al. (1998) to assess stream habitat for migratory fishes of Atlantic coast watersheds. Few subbasins in the United States are undammed. The rate of dammed-ness is generally highest in areas with highest rainfall, but not absent in arid regions. On a national level, the density of dams was highest for the New England, middle Atlantic, South Atlantic-Gulf, and Arkansas-White-Red River region (Figure 1). Subbasins along the Fall Line were the most dammed within the South Atlantic-Gulf region, consistently ranking as high in dam density, storage capacity, and surface area. Examples of dammed watersheds for the

* Corresponding author: mark_a_cantrell@fws.gov

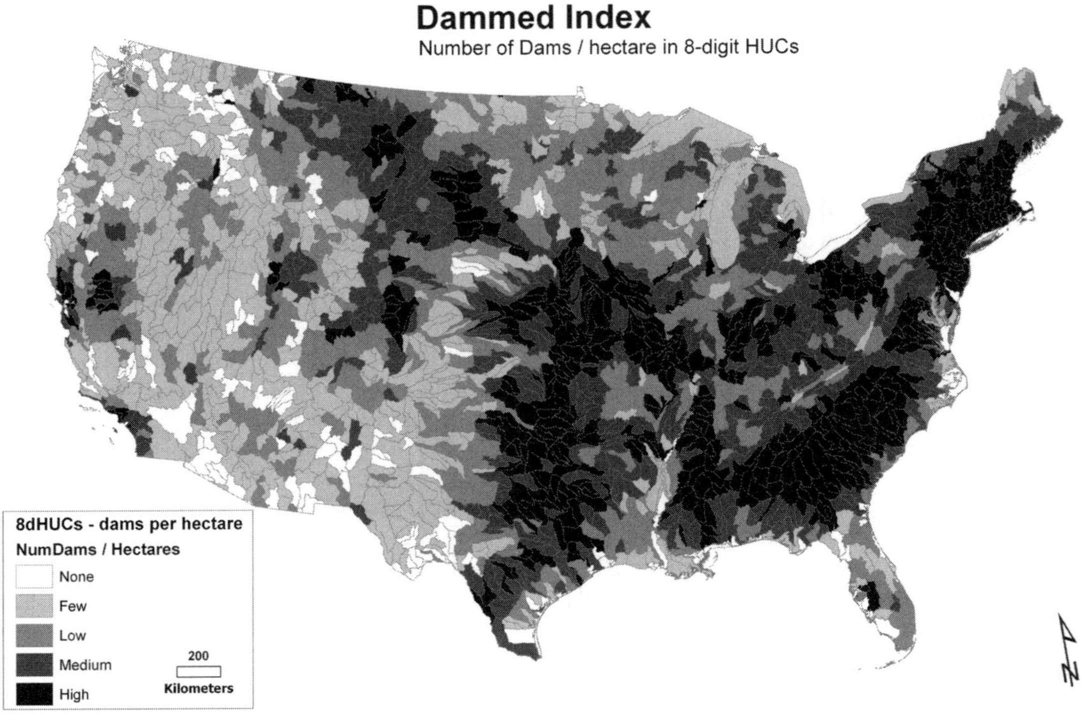

Figure 1.—Relative number of dams per hectare in the continental United States. Data from the National Inventory of Dams (USACE 2005).

South Atlantic are listed in Table 1. Further analyses, not completed, will examine the parameters of hydrology that are affected through the Indicators of Hydrologic Alteration (Richter et al. 1996) and the Hydroecological Integrity Assessment Process developed by the U.S. Geological Survey (Henriksen 2006). These additional detailed analyses could be assimilated as supplementary metrics to estimate ecological integrity at the reach or watershed scale relative to dammed-ness. Relative scores may be calculated for prior time frames and compared to understand the past changes in diadromous fisheries data because

Table 1.—Some additional dammed index metrics for 6-digit South Atlantic hydrologic unit code (HUC) watersheds related to metric 1 (dam density), metric 2 (storage volume of dams), and metric 3 (reservoir storage capacity). NID = National Inventory of Dams.

HUC	Watershed name	Basin size (km^2)	Number of dams	Dam density (dams/ (km^2)	Total NID storage (ha-m)	Storage per basin area (ha-m/km^2)	Total surface area (ha)	Surface area per basin area (surface ha/ watershed ha)
030601	Savannah	27,138	710	38.2	2,438,857	89.9	85,102,562	5.14%
030502	Edisto	20,583	453	45.4	607,565	29.5	63,996,884	5.09%
030501	Santee	39,909	1414	28.2	1,513,667	37.9	88,406,509	3.63%
030401/2	Pee Dee	36,806	1,044	35.3	158,797	4.3	13,367,504	0.59%
030300	Cape Fear	8,399	664	12.6	248,569	29.6	8,410,410	1.64%
030202	Neuse	14,587	380	38.4	142,825	9.8	4,651,586	0.52%
030101	Roanoke	25,359	321	79.0	721,103	28.4	22,164,760	1.43%

the data summarized by the dammed index include a time factor. The dammed index should be reviewed regularly in the future since efforts to restore diadromous fish may include dam removal, or additional dams may be constructed.

References

Busch, W., S. Larry, and C. Castiglione. 1998. Evaluating stream habitat for diadromous fish in Atlantic coast watersheds: a preliminary assessment. Habitat Hotline Atlantic 27:1–3. Atlantic States Marine Fish Commission, Washington, D.C.

Freeman, M. C., C. M. Pringle, E. A. Greathouse, and B. J. Freeman. 2003. Ecosystem-level consequences of migratory faunal depletion caused by dams. Pages 255–266 *in* K. E. Limburg and J. R. Waldman, editors. Biodiversity, status, and conservation of the world's shads. American Fisheries Society, Symposium 35, Bethesda, Maryland.

Henriksen, J. A., J. Heasley, J. G. Kennen, and S. Nieswand. 2006. Users' manual for the Hydroecological Integrity Assessment Process software (including the New Jersey Assessment Tools). U.S. Geological Survey, Fort Collins Science Center, Open-File Report 2006–1093, Fort Collins, Colorado.

Hill, A. K. 2009. The Santee River Basin Accord: restoring diadromous fish through prioritization of subbasins. Pages 769–782 *in* A. J. Haro, K. L. Smith, R. A. Rulifson, C. M. Moffitt, R. J. Klauda, M. J. Dadswell, R. A. Cunjak, J. E. Cooper, K. L. Beal, and T. S. Avery, editors. Challenges for diadromous fishes in a dynamic global environment. American Fisheries Society, Symposium 69, Bethesda, Maryland.

Richter, B. D., J. V. Baumgartner, J. Powell, and D. P. Braun. 1996. A method for assessing hydrologic alteration within ecosystems. Conservation Biology 10:1163–1174.

USACE (U.S. Army Corps of Engineers). 2005. National inventory of dams. Available: www.tec.army.mil/nid (March 2007).

POSTER ABSTRACTS

Greater Genetic Differentiation and Complex Migratory Behavior of Striped Bass in the Canadian Portion of the Species' Range

PAUL BENTZEN*, ROD G. BRADFORD, AND IAN G. PATERSON
Biology Department, Dalhousie University, Halifax, Nova Scotia B3H 4J1, Canada

Although genetic diversity commonly declines towards the periphery of species' ranges, the opposite pattern can occur, with heightened diversity in peripheral areas driven by diversifying selection, genetic drift, or a combination of the two factors. Peripheral populations assume particular importance for conservation in such cases because they may represent more distinctive combinations of adaptive and neutral genetic diversity than those in populations in the center of the range. Striped bass *Morone saxatilis*, a diadromous species widely distributed along the Atlantic coast, is a case in point. We used microsatellite markers and a mix of fin clips and archived scales to examine genetic variation among striped bass from eight rivers composed of five extant populations in Canada and the United States and a presumed extirpated population in Canada. We observed substantially greater differentiation among populations in the Canadian portion of the range than over a comparable spatial scale in the U.S. portion of the range. Bayesian clustering analyses also strongly suggested the existence of either a previously undocumented population or persistence of a population thought to be extirpated. Finally, the genetic analyses also revealed complex migratory behaviors, including the presence of striped bass from widespread populations more than 100 km upstream in the Saint John River during the summer months, and over-wintering of U.S.-origin striped bass in the lower portion of the Saint John River. These results demonstrate the genetic distinctiveness of Canadian striped bass populations and highlight the importance of conserving striped bass populations and habitat in Canada.

* Corresponding author: paul.bentzen@dal.ca

Buoyancy Regulation during the Freshwater–Marine Transition: Differential Use of Lipid by Hatchery and Wild Coho Salmon

LAURIE A. WEITKAMP*
National Oceanic and Atmospheric Administration, Northwest Fisheries Science Center Newport Research Station, Newport, Oregon 97365, USA

* Corresponding author: laurie.weitkamp@noaa.gov

One challenge to diadromy that has received little attention is buoyancy regulation as diadromous fishes move between freshwater and saltwater. Fish must adjust their density in order to avoid becoming positively or negatively buoyant as they change environments because of density differences between the two water types. I measured whole body densities, swim bladder volumes, and lipid levels of juvenile coho salmon *Oncorhynchus kisutch* collected from freshwater and marine environments to explore if, and how, fish are adjusting their buoyancy. These fish exhibited increased whole body density with increasingly dense environments, consistent with the need to avoid positive buoyancy. Most of the density increase was attributable to decreases in gas volume. However, hatchery coho salmon had significantly higher lipid levels than wild coho salmon and exhibited marked declines of both gas and lipid with increasing water density. These high lipid levels may interfere with hatchery fishes' ability to increase their density, and the resulting positive buoyancy may heighten their exposure to avian predators during the transition from freshwater to marine environments. In addition, lipid levels that vary with both water density and fish origin undoubtedly confound their value as an indicator of fish health, as is commonly practiced.

Diverse Life History Strategies amongst Brown Trout in a New Zealand Lowland River Catchment

ESBEN A. KRISTENSEN
Department of Zoology, University of Otago, Post Office Box 56, Dunedin, New Zealand
and
The National Environmental Research Institute, Post Office Box 314, DK-8600 Silkeborg, Denmark

GERARD P. CLOSS* AND RICKY OLLEY
Department of Zoology, University of Otago, Post Office Box 56, Dunedin, New Zealand

JONATHAN P. KIM
Department of Chemistry, University of Otago, Post Office Box 56, Dunedin, New Zealand

* Corresponding author: gerry.closs@stonebow.otago.ac.nz

Brown trout *Salmo trutta* are known to exhibit considerable diversity in life history, with resident and anadromous fish often occurring within the same catchment. Brown trout from mixed stocks were introduced to New Zealand in the 1860s. Resident and anadromous fish are known to occur in New Zealand, but little is known of the frequency or distribution of different life history strategies. We examined the proportion of resident and anadromous brown trout spawning in various tributaries within a large river catchment, the Taieri River, by analyzing trace element signatures of eggs collected from spawning redds. Across the Taieri River catchment, egg trace element signatures indicate that the middle Taieri Gorge is a significant barrier to brown trout upstream migration. No redds formed by anadromous trout were detected upstream of the Taieri Gorge whereas trace element signatures indicated spawning by resident and anadromous fish downstream of the Taieri Gorge. However, rather than forming discrete resident and anadromous stocks, trace element signatures varied continuously from resident fish through to anadromous sea-run fish, with estuarine fish dominating. Spawning by resident and migratory fish varied longitudinally within tributaries. In one steep tributary upstream of the middle Taieri Gorge, egg trace element signatures formed an upstream group most likely representing stream-resident fish and a downstream group representing migratory Taieri River fish. Below the middle Taieri Gorge, resident fish generally spawned upstream of anadromous fish in a 20-km-long tributary, whereas overlap in spawning by resident and anadromous fish was considerable in a shorter, low-gradient stream. Our results indicate considerable flexibility, and a diversity of life history strategies, amongst brown trout in a large lowland river system. Clearly defined resident and anadromous populations do not exist; instead, trace element signatures indicate a continuum of habitat use by adult fish from freshwater residency to riverine, estuarine, and marine migratory fish.

New Insights on American Eel Life History from Otolith Microchemistry

Brian M. Jessop*
Department of Fisheries and Oceans, Box 1006, Dartmouth, Nova Scotia B2Y 4A2, Canada

David K. Cairns
Department of Fisheries and Oceans
Box 1236, Charlottetown, Prince Edward Island C1A 7M8, Canada

Isabel Thibault
Département de Biologie, Université Laval, Québec, Québec G1K 7P4, Canada

Wann-Nian Tzeng
Institute of Fisheries Science, College of Life Science, National Taiwan University
Taipei, Taiwan 106, Republic of China

* Corresponding author: jessopb@mar.dfo-mpo.gc.ca

The temporal patterns in Sr:Ca concentration ratios in otoliths of American eels *Anguilla rostrata* consistently indicate variable patterns of residence in and migration between river, estuarine, and marine habitats. American eels are now considered to be facultatively catadromous. The diversity of habitat use by American eels has permitted them the most extensive geographic range of any diadromous species in the northwest Atlantic. Annual growth rate, based on Sr:Ca habitat determinations, increases with increasing habitat salinity (freshwater < estuarine < marine) and proportion of residence at higher salinity. Increased annual growth rate reduces the age at maturity. Brackish and marine waters may thus be important production areas. Most eels recruit to freshwater as elvers, but a proportion (12–25%) may recruit as juveniles. After entering freshwater, 23% to 100% of eels remain exclusively in freshwater. Exclusively freshwater residence increases with distance upstream from the river mouth. Most interhabitat migrants make only one such movement before their spawning migration. Interhabitat migration frequency may increase with latitude and decreasing freshwater productivity. Interhabitat migrations may produce otolith checks that can increase the difficulty of ageing. Otolith growth periods accurately indicate freshwater or saltwater habitat residency period when the otolith growth rates in each habitat are similar, but not if the growth rates differ greatly. Outlier otolith Sr:Ca values are not uncommon, but whether they represent a brief habitat transition is uncertain. Ignoring them potentially underestimates, while unnecessarily counting them overestimates, the frequency of interhabitat movement. The geographic extent and degree of periodic euryhaline migratory behavior by American eels remains to be determined. Development of a simple, comprehensive interhabitat migratory classification scheme would assist interpretation and comparison of otolith microchemistry studies.

Physiological Mechanisms of European Eel Upstream Migration Plasticity are Stage-Dependant

Hélène Imbert*, Rory Arrowsmith, Sylvie Dufour, and Pierre Elie
Cemagref, Unité EPBX, Estuarine Ecosystems and Diadromous Fish Research Unit
50 Avenue de Verdun - 33612 Cestas, France

* Corresponding author: helene.imbert@bordeaux.cemagref.fr

We developed a new experimental device to observe the impact of biometric parameters and thyroid status on consistent upstream locomotor behavior of elvers and yellow eels to identify physiological mechanisms in freshwater eel *Anguilla anguilla* migration. Our study reveals that the physiological release factors for freshwater eel migration may be stage-dependent. High thyroid gland activity, related to high body condition, may be the physiological release factor for migration for elvers. In contrast, for yellow eels, physiological stress may be the release factor and will increase deiodination activity in the weakest eels. The identification of predominant consistent locomotor behaviors (locomotor activity for elvers and inactivity for yellow eels) and the observation of alternative consistent behaviors within both elver and yellow eel stages demonstrate the existence of interstage and intrastage plasticity levels. Moreover, the observation of inconsistent behaviors provides evidence of an individual plasticity, which could reveal further possibilities of expressing different behaviors according to biotic and abiotic conditions. Thus, we assume that climate change and anthropogenic influences on eel physiological condition could affect eel behavioral plasticity in a stage-dependent manner and disturb the extent of continental colonization.

Silver Eel Tagging as a Method to Investigate the Migration Pattern in the Baltic Sea

Niklas B. Sjöberg*, Erik Petersson, Willem Dekker, and Håkan Wickström
Institute of Freshwater Research, Swedish National Board of Fisheries, Stockholm, Sweden

* Corresponding author: niklas.sjoberg@fiskeriverket.se

Silver eels have been tagged in the Baltic Sea since the beginning of the 20th century using Carlin and silver tags, with recapture rates up to 70%. We compiled data from 268 different experiments, covering the period 1903–2006, with a total of 9,370 recaptures. The current eel fishery in Swedish inland waters is based on restocked eels to a large extent. Practically all tagged eels were natural before 1955 as large-scale stocking was not practiced until the 1950s. We compared recapture rates, migration distance, speed and direction, and over-winter residence of the pre- and post-1955 period to analyze the vitality of restocked eels. The proportion of recaptured eels declined, more eels migrated in unexpected directions, and over-winter residence became more frequent after 1955. This indicates that vitality of the silver eels probably declined over the decades. This will need further research to determine if this is an autonomous development or related directly to restocking.

Homing and Ranging Concepts Are Compatible with Low Recapture Rate of Yellow European Eels

Hélène Imbert*, Patrick Lambert, and Christian Rigaud
Cemagref, Unité EPBX, Estuarine Ecosystems and Diadromous Fish Research Unit
50 Avenue de Verdun - 33612 Cestas Cedex, France

* Corresponding author: helene.imbert@cemagref.fr

We investigated if the home range concept could be compatible with upstream river colonization of European eel *Anguilla anguilla*. We have developed a 2-year mark–recapture program in order to follow eel movements on the 20-km-long Saucats River, France. We have observed recapture rates of 13% and 44%, respectively, of the 2,046 VIE (visible implant elastomer)-tagged eels (<240 mm) and the 229 PIT (passive integrated transponder)-tagged eels (>240 mm). More than 96% of the recaptured eels were located in their 50-m-long original marking site. In addition, the rate of recapture decreased as the time interval between the initial marking and recapture increased and was not explained by natural mortality. Our findings were in agreement with previous mark–recapture field data analysis. We have hypothesized that the catchment colonization process can be redefined as a combination of homing (movements within home range) and ranging (seeking a new home core) behaviors. We have developed an individual-based model that can reproduce the pattern observed in Saucats River by simulating these behaviors. We conclude that using "homing" and "ranging" concepts may provide an explanation of the apparent paradox between eel sedentary lifestyle and catchment colonization.

What Promotes Nondiadromous Recruitment in Migratory Galaxiids?

Andy S. Hicks* and Gerry P. Closs
Department of Zoology, University of Otago, Post Office Box 56, Dunedin, New Zealand

Bruno David
Department of Conservation, Raglan, New Zealand

Jon Waters
Department of Zoology, University of Otago, Post Office Box 56, Dunedin, New Zealand

Jonathan P. Kim
Department of Chemistry, University of Otago, Post Office Box 56, Dunedin, New Zealand

* Corresponding author: ashicks@gmail.com

Migratory galaxiids (whitebait) are characterized by having a distinct niche separation between the pelagic larvae and benthic adults. Whitebait species typically follow an amphidromous life history. Adult growth and reproduction occurs in freshwater, and marine larvae development is expected when populations have open access to the sea. Bet-hedging against unpredictable local conditions and increased colonizing ability are offered as adaptive benefits for a dispersing, marine larval phase. Our research has discovered whitebait populations that exhibit nondiadromous recruitment (NDR) despite having open access to the sea. The presence of freshwater lakes and wetlands, or slow-flowing sections of lowland rivers, appears to provide a pelagic environment that is suitable for larvae development and promotes NDR. These findings question the traditional rationale underlying the purpose of amphidromy. Pelagic larvae have not evolved specifically for dispersal, but rather a pelagic larval period provides small offspring with suitable rearing opportunities while requiring minimal maternal investment per offspring. The salinity of the pelagic environment is irrelevant. Oceanic dispersal may be common in river systems that lack a pelagic environment, but nondiadromous recruitment is likely when suitable conditions exist for larval growth prior to larvae reaching the sea. Management of whitebait metapopulations needs to incorporate an understanding of larvae ecology and appreciate the potential for localized recruitment dynamics in systems that promote retention.

The Anadromous Arctic Char of Quttinirpaaq National Park, Ellesmere Island, Canada, as Determined by Otolith Strontium Distribution

John A. Babaluk*, James D. Reist, and Richard J. Wastle
Fisheries and Oceans Canada, 501 University Crescent, Winnipeg, Manitoba R3T 2N6, Canada

JOHN (IAIN) L. CAMPBELL
Department of Physics, University of Guelph, Guelph, Ontario N1G 2W1, Canada

NORMAN M. HALDEN
Department of Geological Sciences, University of Manitoba, Winnipeg, Manitoba R3T 2N2, Canada

* Corresponding author: babalukj@dfo-mpo.gc.ca

Arctic char *Salvelinus alpinus* exhibit either lake-resident or anadromous life histories, with anadromy thought to be less prevalent at the northern and southern extremes of the distribution. Quttinirpaaq National Park, located at the northern end of Ellesmere Island in the Canadian High Arctic, has some of the most northern populations of the species. Effective conservation and management of char requires knowledge of life histories exhibited by individuals and populations. Otoliths, composed mainly of calcium carbonate, also incorporate trace elements such as strontium. Otolith strontium is derived mainly from waters that fish inhabit. Seawater contains ca. 80 times more strontium than does freshwater; thus, the differences in otolith composition can indicate past anadromous behavior. Scanning proton microprobe analysis was used to determine the strontium distribution in otoliths from char from all seven lakes in the park that empty directly into the sea. Five of the char populations exhibited anadromy. The two populations that did not exhibit anadromy were from lakes with outlets that had physical barriers, or their length and flows precluded anadromy due to energetic constraints. Where char had access to the sea via relatively short, low-gradient streams with no physical barriers, some component of the populations exhibited anadromy. One of the populations, located at 83°N, represents the most northern char population known to exhibit anadromy.

Seasonal Substrate and Water Flow Preference of Year-0 White, Green, Pallid, Shovelnose, Lake, Shortnose, and Gulf Sturgeons

ERIKA PARKER*, BOYD KYNARD, AND MARTIN HORGAN
U.S. Geological Survey, S. O. Conte Anadromous Fish Research Center
Turners Falls, Massachusetts 01376, USA

* Corresponding author: ehenyey@forwild.umass.edu

Capturing young sturgeons in the wild and determining habitat preference is difficult in small populations. Laboratory tests are sometimes the only option in this situation. We tested year-0 white sturgeon *Acipenser transmontanus*, green sturgeon *A. medirostris*, pallid sturgeon *Scaphirhynchus albus*, shovelnose sturgeon *S. platorynchus*, lake sturgeon *A. fulvescens*, shortnose sturgeon *A. brevirostrum*, and gulf sturgeon *A. oxyrinchus desotoi* for seasonal water velocity and substrate preference in a rectangular stream tank with half fast (mean = 9.3 cm/s) and half slow (mean = 3.0 cm/s) flow. Equal areas of cobble and sand substrate were present within each velocity. Fish were monitored for 8 min for the amount of time spent in each velocity-substrate combination and the point velocity where fish stopped. No species had seasonal differences in habitat preference. All species preferred the fastest available velocity. White and green sturgeons preferred cobble substrate; pallid and shovelnose sturgeon preferred sand substrate; and shortnose, lake, and gulf sturgeons had no strong substrate preference. These results suggest that wild juvenile sturgeons should be in areas of fast velocity during any season of their first year of life, but substrate preference varies by species. Water velocity and substrate type are likely to be important determinants of distribution of year-0 sturgeons, and anthropogenic alteration of river habitats could have a detrimental effect on this life stage.

Eel Growth under the Influence of Global Warming and Ecosystem Perturbation

FRANÇOISE DAVERAT* AND LAURENT BEAULATON
Cemagref, Centre de Bordeaux, 50 Avenue de Verdun, 33612 Cestas Cedex, France

RUSSEL POOLE
Marine Institute, Furnace Newport, County Mayo, Ireland

HÅKAN WICKSTRÖM
Swedish National Board of Fisheries, Institute of Freshwater Research, 178 93 Drottningholm, Sweden

GÉRALDINE LAMAISON
Cemagref, Centre de Bordeaux, 50 Avenue de Verdun, 33612 Cestas Cedex, France

JAN ANDERSSON
Swedish National Board of Fisheries, Institute of Freshwater Research, 178 93 Drottningholm, Sweden

GÉRALDINE LASSALLE
Cemagref, Centre de Bordeaux, 50 Avenue de Verdun, 33612 Cestas Cedex, France

* Corresponding author: francoise.daverat@cemagref.fr

The influence of a dynamic global environment on European eel *Anguilla anguilla* growth was investigated at a total of 144 sites over a wide area of the distribution from the south of Spain to Sweden. A model was calibrated with 56 length-at-age data sets (more than 13,000 eels) covering the distribution area from 1913 to 2005, obtained either from literature or from original data sets. Several types of effects were tested. Global effects, such as temperature and year of sampling, were tested to investigate human influence and impact of climate change. The effect of local environment inside a watershed (salinity, depth, connectivity, productivity) was also included in the model. Individual variability and sex differences were also taken into account. The model was further validated with mean length-at-age values obtained from 23 literature data sets.

Behavior at High Temperature: Can Physiology Explain Movement of Atlantic Salmon to Cool Water?

CINDY BREAU*, RICHARD A. CUNJAK, AND STEVE J. PEAKE
Canadian Rivers Institute, University of New Brunswick
Fredericton, New Brunswick E3B 6E1, Canada

* Corresponding author: cindy.breau@unb.ca

Wild salmonids experience water temperatures during the summer that approach their upper lethal limit. Juvenile Atlantic salmon *Salmo salar* of different age-classes exhibit different behavioral responses to high temperatures (>23°C) in the Little Southwest Miramichi River (New Brunswick, Canada). Age-1 and age-2 salmon cease feeding, abandon territorial behavior, and swim continuously in aggregations near coolwater refugia; however, age-0 fish continue to feed and defend territories. Movements of individually marked salmon prior, during, and after these temperature events were studied in the field using PIT (passive integrated transponder) technology. A laboratory experiment was conducted to determine (1) if the behavioral shift in older juvenile salmon occurs when basal metabolic rate, driven by increasing water temperature, approaches maximum metabolic rate; and (2) if anaerobic pathways were being recruited to provide energy to support vital processes. Behavior (feeding, startle response), oxygen consumption, circulating lactate and glucose levels, and muscle lactate and glycogen were measured in fish acclimated to temperatures between 16°C and 28°C. Results indicate that oxygen consumption of the age-2 salmon increased with temperature and reached a plateau at 24°C, a temperature that also corresponded to a significant increase in lactate levels and cessation of feeding. In contrast, oxygen consumption in age-0 fish did not plateau, feeding continued, and lactate did not increase, even at the highest temperatures tested. We therefore concluded that some behavioral responses of age-2 salmon to temperature stress may be employed in an effort to mitigate the physiological imbalances associated with an inability to support basal metabolism aerobically.

Biological Characteristics and Population Status of Anadromous Brook Trout 66 Years after an Initial Study in Moser River, Nova Scotia, Canada

John L. MacMillan* and Reginald J. Madden

Nova Scotia Department of Fisheries and Aquaculture, Pictou, Nova Scotia B0K 1H0, Canada

* Corresponding author: macmiljl@gov.ns.ca

Brook trout *Salvelinus fontinalis* are the most popular native sport fish in Nova Scotia. A decline in the annual catch of brook trout since the early 1980s may reflect a change in the resource. Nova Scotia's relatively short river systems are believed to facilitate anadromy in many brook trout populations, and these trout are important to the sport fishery. Live release traps and water quality measurements were used to assess brook trout population parameters and habitat in the same locations, as in historical studies conducted on the Moser River. The timing of anadromous brook trout migration was similar in 2006 to that in 1939 to 1941; downstream migration occurred in May, and most upstream migration occurred in July. However, a greater proportion of upstream migrating brook trout were detected in June 2006, compared to June of 1939 to 1941. The number of brook trout trapped per day in Mill Brook, a major tributary of Moser River, suggests that the median number of anadromous brook trout had declined to 1.5 in 2006, from 9.5 in 1940, and 18 in 1941. A significant difference was detected in the number of brook trout trapped per day in 2006 compared to 1940 and 1941 ($P < 0.05$). Relative stock density of brook trout indicated that large trout in the 2006 population was less than that of historical levels. The proportion of brook trout (fork length > 300 mm) captured while migrating upstream was 3% in 2006, significantly less than 26% in 1940 and 27% in 1941 ($P < 0.05$). The mean fork length of anadromous brook trout (221 mm, age 2; 276 mm, age 3) in 2006 was similar to mean fork length at age from 1939 to 1943. The oldest brook trout age was 4 years in 2006 and 6 years in 1939 to 1943. A maximum age of 4 years, in the 2006 study, was consistent with other recent studies on anadromous brook trout populations in Nova Scotia. Moser River mean pH was 6.5 in 1939 and 5.7 in 2006, during low-flow periods in August. Diatom inferred pH from a paleolimnological assessment of Mill Lake, located upstream of the Mill Brook trapping site, indicated an increase in acidity by a decline of 0.7 pH units from pH 6.8 in the 1930s. Diatom inferred Gran alkalinity changed from a pre-impact range of 4.1–7.2 mg/L to a present day 0.6 mg/L Water temperature was warmer in 1939 compared to that in 1940–1941 and 2006. Mortality of salmon and trout in the Moser River occurred during warm low flow conditions in 1937 and 1939. A warming trend occurred over the past 140 years, and the summers of 1937 and 1939 were two of the warmer on record for the Halifax region. Changes in environmental conditions and exploitation may have influenced the size and age structure of the Moser River anadromous brook trout population.

From Local to Global Climate Change: Effects on Southern European Atlantic Salmon

Francis Juanes*, America Valiente, Steve Gephard, and Eva Garcia-Vazquez
University of Massachusetts, Amherst, Massachusetts 01003, USA

* Corresponding author: juanes@forwild.umass.edu

Populations at the edge of the species distribution exhibit increased sensitivity to both local and global climatic effects compared to those inhabiting the central portion of the distribution. However, large-scale climate indices generally predict ecological processes better than local weather does. Southern European Atlantic salmon *Salmo salar*, which feed in West Greenland, are expected to be particularly affected by global warming. In northern Spanish rivers, the first salmon caught each year is called "campanu" and is highly prized in a public auction. Here, we demonstrate that salmon run timing has changed over the past 50 years, as inferred from delays in the date of capture of the campanu, correlated with both local and global temperature indices. We also demonstrate that weight decreases are associated with an increase in global temperatures and with changes in abundance of squid, an important marine prey. Decreased energetic reserves, as a consequence of reduced prey abundance and increasing temperatures, may lead to observed delays in return timing, likely a reflection of whole ecosystem-level effects, and will have important consequences for the species that interact with Atlantic salmon.

Behavior, Activity, and Movements of Juvenile Atlantic Salmon in Early Winter

Tommi Linnansaari* and Richard A. Cunjak
Canadian Rivers Institute, University of New Brunswick
Fredericton, New Brunswick E3B 6E1, Canada

* Corresponding author: tommi.linnansaari@unb.ca

The behavior of juvenile Atlantic salmon *Salmo salar* changes when the water temperature drops below 10°C; the fish remain inactive within the stream substrate during the daytime, becoming active only during the hours of darkness. This behavior may pose a threat to fish survival in high gradient nursery streams during the early part of the winter when subsurface ice formation takes place on the substrate material (i.e. anchor ice forms). Current climate change scenarios predict that the period when subsurface ice forms will be longer in future and would thus affect the freshwater production of juvenile Atlantic salmon. A total of 228 juvenile salmon (mean fork length 95 mm; 10.5 g) were tagged with 23-mm passive integrated transponders (PIT) in Catamaran Brook in northern New Brunswick during the winters of 2003 to 2005 to study how ice formations affect the behavior, activity, and movements of salmon parr. Fish were located by using a portable PIT-tag detector during multiple tracking surveys between November and January. Emigration prior to ice formation was relatively high (52–78 %) in all of the study sites. However, emigration during the subsurface ice events was lower (0–14 %), and only very few long-scale movements (>30 m) were observed. No salmon parr mortality was observed during subsurface ice events. The resulting home range during the subsurface ice events was small (median = 2.0 m^2) and not different from home range size prior to or after the subsurface ice events. Fish maintained mainly nocturnal activity during the subsurface ice events. During the hours of darkness, fish often followed the lateral growth of surface ice holding position just under the edge of the ice where the water velocity was, on average, 3.9 times slower than in the open-water column. The fish that entered open water remained close to ice formations that provided overhead cover. At dawn, a large proportion of fish returned to their initial home territory ("home stone"), but the proportion of fish returning to an exact initial location was significantly lower during the most extensive anchor ice events, as the fish were unable to locate the home stone. Under these conditions, the displaced fish chose a daytime position under ice formations that provided overhead cover close to their initial territory. Our results indicate that the small-sized juvenile Atlantic salmon exhibit a different behavioral strategy than many other rheophilic salmonid species during subsurface ice events and that subsurface ice production may not be as deleterious for juvenile salmon as thought previously.

An Unwelcome Guest? Preliminary Status of the American Eel Swim Bladder Parasite *Anguillicola crassus* in New Jersey Estuaries

MARK C. SULLIVAN* AND MARCY HOWLAND
The Richard Stockton College of New Jersey, Marine Science Program
Division of Natural Sciences and Mathematics, Pomona, New Jersey 08240, USA

KENNETH W. ABLE
Rutgers University Marine Field Station 800
c/o 132 Great Bay Boulevard, Tuckerton, New Jersey 08087, USA

* Corresponding author: mark.sullivan@stockton.edu

Recent evidence suggests that coastal ecosystems around the world are experiencing rising atmospheric temperatures. These physical changes may trigger corresponding increases in exotic pathogens and parasites that contribute to the mortality of catadromous species using estuaries as juvenile habitat. Young American eel *Anguilla rostrata* are exposed to the exotic swim bladder nematode *Anguillicola crassus* during the first few years of residency in North American estuaries. Given the changing climate of northeast ecosystems, updated information on infection rates of this parasite is needed, particularly for eels in the first few years of life. To address these issues, arrays of eel resettlement collectors were used to obtain eels less than 200 mm standard length along a polyhaline-to-freshwater gradient in two southern New Jersey watersheds from April to October 2006. Subsamples of eels were sacrificed to calculate prevalence and intensity of infection. Preliminary results indicate a higher infection rate (>80%) than reported previously from inlet mouth to freshwater habitats 30 km upriver. Individuals less than 80 mm standard length were infected, suggesting that young-of-the-year eels may be acquiring the parasite in the months after first feeding. Given these above-average rates of infection and small body sizes, the negative effect of infection on eel biology remains unclear. More research is needed to increase sample sizes and to better understand the timing and routes of new infections.

Investigation into the Role of Water Velocity on Fish Habitat Selection

Zoya A. Pawlychyn*
University of Manitoba, Winnipeg, Manitoba R3T 2N2, Canada
and
Department of Fisheries and Oceans, Winnipeg, Manitoba R3T 2N2, Canada

Mark Abrahams
University of Manitoba, Winnipeg, Manitoba R3T 2N2, Canada

Chris Katopodis and Ross Tallman
Department of Fisheries and Oceans, Winnipeg, Manitoba R3T 2N2, Canada

* Corresponding author: zapaqua@gmail.com

Pipeline development affects both terrestrial and aquatic environments. Roads need to be built to provide access for construction, operation, and maintenance of pipelines. The construction of roads requires clearing of forests and building culverts to allow for water crossings. The construction of culverts affects fish habitat by changing the physical characteristics of the water body. For example, increasing water velocity can affect migratory fish and their chances of survival before reaching the spawning grounds. Current research looking at the effects of culverts on fish migration has been focused on southern fish species. There is a need to understand how culverts will affect migratory fish in the arctic because of the development of a natural gas pipeline in the Mackenzie Valley, Northwest Territories, Canada. The objectives of this research are to understand the way fish use the natural flow variation in their habitat to migrate and to determine the physical and physiological swimming potential of broad whitefish *Coregonus nasus*. Water velocity profiling will provide information as to the natural variation present in the environment. Short-term gill netting will indicate the primary location of fish in the water column and allow for fish identification. Swimming potential will be assessed on wild fish using a swim tunnel; this will allow us to construct a swim curve that incorporates prolonged, sustained, and burst swimming speeds. We will be able to understand how fish use the water column for migration and how anthropogenic effects alter water velocity that may affect a fish's swimming ability and migration.

Dealing with Changing Environments: Emerging Diseases and Invasive Species in Changing Landscapes

Christine M. Moffitt*
U.S. Geological Survey, Idaho Cooperative Fish and Wildlife Research Unit
Department of Fish and Wildlife, University of Idaho, Moscow, Idaho 83844, USA

* Corresponding author: cmoffitt@uidaho.edu

Biologists and resource managers need tools to identify and prioritize the risks of altered environments on diadromous fishes within the landscape. Diadromous fishes are particularly vulnerable to direct and indirect consequences of human development since they use different environments and migration corridors to complete their life history in ocean and freshwater environments. Coastal communities continue to house the largest proportion of human populations of the world. Pollution from organic xenobiotic compounds and toxic metals can accumulate in tissues, causing endocrine disruption, tumors, susceptibility to diseases, and physiological impairment. Few studies consider large-scale or landscape-level approaches to understanding the extent of risks to species or habitats. Studies focus on mechanistic approaches and controlled laboratory assessments of single agents. Many field-oriented studies are retrospective, not prospective. Increased globalization and transportation networks pose increased risks of introduction of pathogens and other invasive species to aquatic systems. These factors need to be considered together in understanding and articulating risks. In this paper, I provide a framework approach to identify the potential and known hazards at the landscape level. This risk characterization can then be used to assess risk. Once these assessments are completed and the risks are communicated, managers can seek to prioritize monitoring, data needs, and management options.

The Nature of Dispersal in Atlantic Salmon: Source-Sink Reversals Revealed by Contemporary and Long-Term Genetic Analyses

Friso P. Palstra* and Daniel E. Ruzzante

Biology Department, Dalhousie University, 1355 Oxford Street, Halifax, Nova Scotia B3H 4J1, Canada

* Corresponding author: fpalstra@dal.ca

Migration forms an integral part of salmonid life-history, setting the stage for dispersal by which populations are interconnected. From a metapopulation point of view, source-sink dynamics characterize systems where migration occurs predominantly from large populations with high productivity into smaller populations, which depend on immigration for persistence. Such patterns have recently been reported for a number of salmonid species. If these dynamics indeed are common, they have important implications for management and conservation of (endangered) salmonid populations. Here, we examine population structure and connectivity in Atlantic salmon *Salmo salar* from Newfoundland and Labrador, a region where populations of this species are relatively pristine. We examine connectivity at several regional and temporal scales, using genetic variation (13 microsatellite loci) from samples (n = 1,346) collected from 20 rivers, to test the hypothesis that gene flow predominantly occurs from large into small populations. We reject this idea and find that the directionality of migration is affected by the temporal scale over which gene flow is assessed. While large populations tend to function as sources of migration over contemporary time scales, such patterns are often changed and even reversed over long-term evolutionary scales, where populations that are contemporarily (small) sinks appear to have functioned as sources in the past. Importantly, we find that patterns of population structure vary between different regions and are often compatible with demographic and life-history attributes. Assignment tests revealed no apparent sex-biased dispersal despite differences between rivers in sex ratios of anadromous runs. Regardless of directionality, migration between rivers within regions is common. Our findings caution against generalizations concerning the directionality of gene flow in Atlantic salmon and emphasize the need for detailed regional study, if such information is to be applied in conservation and management of salmonids.

American Shad Migrations Revisited: Otolith Chemistry, Natal Origins, and Mixed Stock Compositions

Benjamin D. Walther*

Southern Seas Ecology Laboratories, School of Earth and Environmental Science Darling Building, University of Adelaide, South Australia 5005, Australia

SIMON R. THORROLD
Biological Oceanography Department MS #50, Woods Hole Oceanographic Institution Woods Hole, Massachusetts 02543, USA

* Corresponding author: benjamin.walther@adelaide.edu.au

The migratory movements of American shad *Alosa sapidissima* are extensive and complex. Previous studies estimated oceanic movements and natal homing based on substantial tag-and-release efforts along the Atlantic coast. We revisited the patterns identified by these efforts using the geochemical compositions of otoliths to identify natal origin. Otoliths and water samples from the major spawning rivers from Florida to Quebec showed high separation of geochemical signatures among rivers, allowing confident assessments of origin based on Sr:Ca, Ba:Ca, $\delta^{18}O$, and ^{87}Sr:^{86}Sr ratios. Otolith chemistry offers advantages over traditional approaches in that all fish are tagged naturally from birth and the natal origin of migrants can be determined. These attributes allowed us to assess mixed-stock composition of juvenile American shad harvested in marine habitats in the upper Gulf of Maine and provided stage-specific information about the distribution and movements of immature fish. Mixed-stock compositions were dominated by only two to three rivers for all collections, and there was a distinct lack of southern populations in the mixtures. Estimates of mixed-stock compositions for these immature fish differ significantly from those reported for older age-classes. These patterns suggest that American shad may exhibit ontogenetic niche shifts in their migratory strategies after their transition to marine habitats.

Inshore Migration of Glass Eel Phase of Japanese Eel in Lake Hamana, Japan: Periodicity, Seasonality, and Recruitment Mechanisms

NOBUTO FUKUDA, JUN AOYAMA, AND KATSUMI TSUKAMOTO
Ocean Research Institute, University of Tokyo, Tokyo, Japan

* Corresponding author: bukuchan@ori.u-tokyo.ac.jp

Japanese glass eels *Anguilla japonica* must adapt to various environmental changes of great magnitude during inshore migration. Considerable information has been accumulated on this phase of the life history of catadromous eels, but the understanding of the details of inshore migration is still limited, so more research is needed. Glass eels were collected with an anchored plankton net from December 2003 to December 2005 to investigate the behavior of migrating glass eels near the entrance of Lake Hamana, Japan, where strong tidal flows occur. A total of 530 glass eels were collected during 629 sampling events. The results of the year-round collections showed that recruitment into Lake Hamana occurred mainly from November to May. Time-series auto-correlation analysis of the average daily catch in each season detected 32-d periodicity in the 2003 season and 30–32-d periodicity in the 2004 season. This suggests that the number of recruiting glass eels increases during new moons and fluctuates according to the lunar-cycle. The results of average hourly catch analyses showed a diel rhythm with glass eels appearing only at night and a tidal rhythm where the catch during flood tide is greater than during ebb tide. Glass eels might utilize flood tidal-stream selectively at night to migrate inshore. The hourly catch during flood tide throughout the recruitment season showed that a rapid increase of recruiting glass eels occurred 2 to 3 h after the beginning of flood tide and tended to be synchronized with the rise of water temperature caused by the inflow of warm water from outside of the lake. This environmental change during flood tide may be a cue for the ascending behavior of glass eels into the water column or could be a zeitgeber of the circa-tidal rhythm, as in the ascending behavior of American eels *Anguilla rostrata*. Inshore migration by glass eels, one of the most important migrations of catadromous eels, has been developed by adapting to various periodic environmental changes formed by the semi-permanent movements of the earth, sun, and moon.

Migration Patterns of American Eel Determined with Fin Tissue Stable Isotope and Otolith Microchemistry Analyses

MARIE CLÉMENT*
Department of Fisheries and Oceans Canada, Gulf Region, Moncton, New Brunswick E1C 9B6, Canada

ALYRE CHIASSON
Université de Moncton, Moncton, New Brunswick, Canada

GEOFF VEINOTT
Department of Fisheries and Oceans Canada
Newfoundland and Labrador Region, Newfoundland, Canada

DAVID K. CAIRNS
Department of Fisheries and Oceans Canada, Gulf Region, Moncton, New Brunswick E1C 9B6, Canada

* Corresponding author: clementm@dfo-mpo.gc.ca

Stable nitrogen and carbon isotope ratios were analyzed to determine migration patterns of American eels *Anguilla rostrata* in two low-productivity rivers of Fundy National Park, New Brunswick, Canada. Overall, eels captured during the summer showed an isotopic signature suggestive of feeding on freshwater (FW) organisms. Eels captured during the spring downstream migration and during the fall showed an isotopic signature indicative of feeding on saltwater (SW) organisms. The SW signature observed in the eels captured in the fall and in the spring implied a migration from SW to FW prior to overwintering in FW. Otolith microchemistry analysis was conducted on a subsample of the captured eels. The two analytical methods yielded similar conclusions regarding recent residency in FW and SW. However, otolith microchemistry analysis showed that eels between 4 and 13 years of age with Sr:Ca ratios suggestive of SW residency never entered the FW environment prior to capture. These results suggest that the otolith microchemistry analysis might not detect FW residency because of a slower growth rate of eels during the winter months.

The Oceanic Distribution of Alewives: An Examination of Seasonal and Interannual Patterns, and Bycatch Risk

JACOB KRITZER* AND PETER BLACK
Environmental Defense, Oceans Program, New York, New York 98115, USA

* Corresponding author: jkritzer@environmentaldefense.org

River herring are major prey for top predators in oceans, estuaries, and rivers, and support commercial and recreational fisheries. However, populations have declined in many areas. Multiple management measures are being implemented, including fish passage projects, habitat restoration, transplanting spawners to rejuvenate extirpated runs, and various fisheries management strategies, such as full moratoria in several states. Despite these efforts, many runs continue to decline. One hypothesis for the decline is that fish are harvested as bycatch by ocean trawl fleets, particularly in the Atlantic herring *Clupea harengus* fishery. Bycatch mortality would compromise not only the ecological and social roles of river herring, but also restoration investments in many rivers. As an initial assessment of bycatch risk, we mapped 2001 and 2006 data from the National Marine Fisheries Service (NMFS) independent trawl survey on the spatial and temporal distribution of alewives *Alosa pseudoharengus* during the oceanic phase and compared those patterns with locations of Atlantic herring landings to determine if alewives were aggregated where fishing effort was greatest. In winter, alewives were concentrated off Long Island in the northern part of herring Management Area 2, which is the source of the vast majority of winter herring landings. In the spring, alewives were distributed along the coast, presumably in preparation for spawning runs, but their distance offshore appeared to be temperature-dependent. Herring catch effort shifts from Area 2 to Area 1A (inshore Gulf of Maine) in the spring. Spring fishing effects on alewives are likely to vary among years, depending upon whether temperatures draw them inshore into Area 1A. In the fall, alewives are concentrated in the Gulf of Maine, where inshore–offshore distribution again appears to be temperature-dependent. The highest fall herring effort varies from year to year among Area 1A, Area 1B (offshore Gulf of Maine), and Area 3 (Georges Bank). Therefore, bycatch is likely to vary among years, depending upon the extent to which inshore–offshore patterns in effort overlap with temperature-driven changes in alewife distribution. Our next steps are to expand the analysis of fishery-independent data farther back in time, include an analysis of fishery-dependent bycatch data from the NMFS observer program, and recommend research and monitoring approaches to characterize and mitigate bycatch impacts better.

Habitat Use and Timing by Bull Trout in Marine Waters of Northern Puget Sound, Washington, USA

Reg Reisenbichler*, Steve Rubin, and Michael Hayes
U.S. Geological Survey, Western Fisheries Research Center, Seattle, Washington 98115, USA

Fred Goetz
U.S. Army Corps of Engineers, Seattle, Washington, USA

Mike Parsley
U.S. Geological Survey, Columbia River Research Laboratory, Cook, Washington, USA

* Corresponding author: reg_reisenbichler@usgs.gov

Bull trout *Salvelinus confluentus* and their habitats are protected because the species is listed as threatened under the U.S. Endangered Species Act; however, the scientific literature contains little information about the habitat used by anadromous bull trout during their residence in marine waters. We used stationary receivers located in the lower Skagit River and mobile tracking equipment in marine waters to study the timing and habitat use for bull trout in Skagit Bay during April to August 2006. Acoustic tags were implanted surgically in 23 bull trout that were captured by beach seining or hook and line in Puget Sound during April 2004 to June 2006 and in 20 bull trout that were captured by hook and line in the lower Skagit River during March 13 to April 3, 2006. The fish were 22 to 56 cm fork length at tagging. We detected fish moving past the stationary receivers toward Skagit Bay from March to May. Fish were detected returning to the river from mid-May to mid-August. While in Skagit Bay, fish typically resided within 300 m of the shoreline and at a depth less than 5 m. Each fish tended to remain in a limited area, generally no more than 1.5 km in length along the shore. We monitored four fish continually for 20 to 24 h, each on a different day at least 1 week apart. The only differences between diurnal and nocturnal habitat selection and behavior seemed to reflect changes in tidal height rather than solar position. We are assembling data on various habitat features such as bathymetric topography, geomorphic shoreline forms, macrophyte occurrence, and distance from man-made structures or shoreline modifications. These data will be used to characterize the marine habitat used by bull trout.

Dynamic Network Topology: Towards Developing a Stream Network Perspective

REBECCA L. FLITCROFT*, REEVES GORDON, AND KELLY BURNETT
U.S. Forest Service, Pacific Northwest Research Laboratory, Corvallis, Oregon 97330, USA

* Corresponding author: becky.flitcroft@orst.edu

As the broader discipline of ecology embraces dynamic processes as a fundamental element of ecosystems, aquatic ecologists are working to develop theory and techniques that allow for the analysis of dynamic stream processes and communities of organisms. The stream network is the basis upon which stream habitats are organized. Natural and anthropogenic disturbances in streams alter the configuration of stream habitats over seasons, decades, or centuries. Native aquatic species have developed different mechanisms for coping with the dynamic nature of habitats in stream networks. However, stream network structure complicates habitat connectivity for aquatic obligate species, due to limited directions of movement offered by stream channels, and the inherent directionality of water flow and other elements of the stream environment. A conceptual framework that includes the dynamic nature of streams within the context of the stream network may by useful for the study and conservation of aquatic species. Dynamic network topology is proposed as a conceptual framework that includes the location and juxtaposition of habitats in stream systems with the underlying structure of the stream network. Pacific salmon and trout were used to discuss the implications of dynamic network topology for aquatic obligate species. Ultimately, a perspective that includes the constraints and opportunities offered to stream species may help in the development of analytical tools and management recommendations that benefit native species. Aquatic obligate species that reside in streams offer ecologists an opportunity to explore new ways of thinking about movement and behavior by using the stream network as a critical organizing feature describing biotic diversity.

Evolutionary and Ecological Significance of Anadromy for Brook Char

R. ALLEN CURRY*, ERIC CHERNOFF, AND TIM D. JARDINE
Canadian Rivers Institute, University of New Brunswick
Fredericton, New Brunswick E3B 6E1, Canada

* Corresponding author: racurry@unb.ca

Early-life history characteristics of sea-run and resident salmonid fishes that coexist in sympatry are poorly understood. Stable isotope ratios ($\delta^{13}C$ and $\delta^{15}N$) were used to identify the progeny of resident and sea-run female brook char *Salvelinus fontinalis* in a tributary of the Miramichi River, New Brunswick, Canada to assess differences in characteristics of offspring of these two strategies following their emergence to the stream. Sea-run progeny captured in emergence traps were longer, were heavier, and had higher carbon to nitrogen ratios (C/N, an indicator of lipid content) than resident progeny but did not have higher body condition. Sea-run progeny captured in dip nets, and identified by their $\delta^{13}C$ and $\delta^{15}N$ ratio, had higher body condition than resident progeny after feeding and growth in the stream. The proportion of fish caught in dip nets at sites where the two forms coexisted was dominated by offspring of sea-run females (sea-run = 56%, resident = 19%, unknown = 25%). These analyses suggest that offspring of sea-run, female char benefit from maternally derived energy reserves gained at sea and thus have potential adaptive advantages over offspring of residents.

Beyond Diadromous Fishes: Diadromous Invertebrates Are Strong Interactors Too

EFFIE A. GREATHOUSE* AND JANA E. COMPTON
Western Ecology Division, National Health and Environmental Effects Research Laboratory
Office of Research and Development, U.S. Environmental Protection Agency
200 SW 35th Street, Corvallis, Oregon 97333, USA

* Corresponding author: effieg@gmail.com

Most research on diadromy focuses on temperate fishes, but a variety of diadromous invertebrates is common in the tropics. Invertebrates with diadromous life cycles include freshwater shrimps (e.g., Palaemonidae, Atyidae, Xiphocarididae), mitten crabs *Eriocheir* spp. and snails (Neritinidae). The form of diadromy displayed by shrimps and snails is freshwater amphidromy: adults feed and reproduce in freshwater, larvae passively drift downstream to saltwater, and juveniles migrate back upstream into freshwater. Functional roles of adult diadromous invertebrates are generally herbivory, detritivory, and omnivory. In insular and coastal tropical streams, diadromous shrimps and snails, together with herbivorous diadromous fishes (e.g., Gobiidae), play important roles as primary consumers and detritivores. When these fauna experience declines due to human activities such as dam building, there can be dramatic effects on primary producers and standing stocks of organic matter. Consequences of diadromous fauna loss encompass a wider variety of species interactions and biomes than the bottom-up effects recognized for northern-temperate systems (i.e., elimination of salmonid nutrient subsidies).

Quantification of Regional Differences in River Habitat Capacity for Atlantic Salmon in Eastern Canada—A Pilot Study

ROBERT G. RANDALL*
Fisheries and Oceans Canada, Canada Centre for Inland Waters
867 Lakeshore Road, Burlington, Ontario L7R 4A6, Canada

PETER G. AMIRO
Fisheries and Oceans Canada, Bedford Institute of Oceanography
Box 1006, Dartmouth, Nova Scotia B2Y 4A2, Canada

* Corresponding author: randallr@dfo-mpo.gc.ca

Regional and proximate factors affect habitat carrying capacity and the productivity of Atlantic salmon *Salmo salar* in eastern Canada. The efficacy of using area-production (area-P) relationships to determine and compare habitat capacity in different regions was investigated as a pilot study, using data from a subset of rivers ranging in latitude from 44°N (southwest Nova Scotia) to 53°N (southern Labrador). Average river fish yield data (1983 to 1987) were used as a proxy for salmon production, and tertiary watershed area (km^2) and fluvial habitat (ha) were used as measures of habitat supply. Area-P relationships were positive and significant, and habitat capacity was river-dependent, ranging between <1 and 17 fish per ha for small salmon and between <1 and 3 fish per ha for large salmon. Latitude, the occurrence of lacustrine rearing habitat in watersheds and air temperature were shown to affect habitat capacity in this pilot study. Results indicated that area-P relationships were useful for identifying and predicting regional patterns in salmon production in the different watersheds of eastern Canada. Area-P relationships can be used to account for both historic and emerging stressors on salmon, including temporal changes in habitat supply.

Genetic Evaluation of Restoration Success in American Shad Returning to the James River, Virginia, USA

Aaron W. Aunins*
Virginia Commonwealth University, 1000 West Cary Street, Richmond, Virginia 23284, USA

Tom P. Gunter
Virginia Department of Game and Inland Fisheries
4000 West Broad Street, Richmond, Virginia 23230, USA

John M. Epifanio
Center for Aquatic Ecology and Conservation, Illinois Natural History Survey
Champaign, Illinois 61820, USA

Bonnie L. Brown
Virginia Commonwealth University, 1000 West Cary Street, Richmond, Virginia 23284, USA

* Corresponding author: auninsaw@vcu.edu

The James River population of American shad *Alosa sapidissima* has been stocked annually since 1994 with millions of hatchery-reared, oxytetracycline tagged larvae derived from broodstock collected in the neighboring Pamunkey River. This hatchery restoration program was initiated in response to a severe population decline in the James River due to overfishing and habitat loss. The percentage of tagged hatchery recruits returning to the James River has declined from a high of 90% in 2002 to 45% in 2006. It is likely that the increasing proportion of recruits is predominantly composed of second generation hatchery-derived American shad. However, it also is possible that they are derived from the remnant James River population or perhaps from other East Coast populations known to contribute small numbers of individuals to Chesapeake Bay tributaries. We have begun a genetic evaluation of the restoration program using microsatellite-based mixed-stock analysis to investigate the origin of the increasing proportion of untagged recruits. Archived tissue samples are available from the donor and receiving prerestoration populations, yearly samples from returning recruits, broodstock, and samples of shad from other tributaries that have been shown to stray into the James River. Analysis of molecular variance results, using seven loci, suggest that although the prerestoration James and Pamunkey River populations displayed marginal differentiation ($\phi_{st} = 0.02$), the current James River population is now more similar to the prerestoration Pamunkey River ($\phi_{st} = 0.008$), presumably due to stocking. A preliminary mixed-stock analysis of a James River 2006 population sample using the Statistics Program for Analyzing Mixtures (SPAM) assigned 45% of individuals as Pamunkey River, 42% as native James River, and 13% as unassignable. This result may reflect the low level of differentiation between the James and Pamunkey rivers, but also may indicate the impact of strays from other river systems.

Conservation Concerns for Inanga *Galaxias maculatus* in Chile, South America

EVELYN HABIT*, PRISCILA PIEDRA, AND ROBERTO CIFUENTES
Environmental Science Centre EULA, Universidad de Concepción, Concepción, Chile

MARK BELK
Department of Integrative Biology, Brigham Young University, Utah 84602, USA

PEDRO VICTORIANO
Zoology Department, Universidad de Concepción, Chile

JUAN JOSE ORTIZ
Environmental Science Centre EULA, Universidad de Concepción, Concepción, Chile

DANIEL RUZZANTE
Department of Biology, Dalhousie University, Halifax, Nova Scotia B3H 4J1, Canada

VICTOR CUSSAC
Universidad Nacional de Comahue, Bariloche, Argentina

JORGE GONZÁLEZ
Environmental Science Centre EULA, Universidad de Concepción, Concepción, Chile

* Corresponding author: ehabit@udec.cl

Inanga *Galaxias maculatus* is among the most widely distributed of all freshwater fishes, and populations are exposed to several threats in New Zealand, Tasmania, Argentina, and Chile. Mitigating undesirable effects is difficult in Chile as essential information on distribution, abundance, and life history for inanga is scarce. We analyzed current distribution patterns based on sampling conducted from 28° to 54°S latitude in Chile. Comparison with previous records indicated loss or reduction in distribution of inanga, mainly in the extreme southern and northern areas. Predation of brown trout *Salmo trutta* on inanga in lacustrine populations of Chilean northern Patagonia was relatively high where native fishes were abundant. Lake populations exhibited significant differences in morphology based on the analysis of 19 populations using geometric morphometrics. Also, the genetic structure of inanga inhabiting lakes, river, and estuary of the Valdivia River basin showed significant differences, suggesting different life histories (landlocked versus diadromous populations). Conservation actions should be analyzed independently depending on the geographical area and life history characteristics. Major concerns are habitat fragmentation and exotic species. We suggest that habitat protection should be prioritized to maintain conditions where the undesirable effects of nonnative species are less likely to be realized.

The Atlantic Coast Fish Habitat Partnership: Forming a Coast-Wide Alliance to Protect and Restore Diadromous Fish Habitat, USA

JESSIE C. THOMAS*
Atlantic State Marine Fisheries Commission, 1444 Eye Street NW, Washington, D.C. 20005, USA

* Corresponding author: jthomas@asmfc.org

The Atlantic States Marine Fisheries Commission (ASMFC) was formed in 1942 by the 15 Atlantic coast states in recognition that fish do not adhere to political boundaries. The long-term vision of the ASMFC is working toward healthy, self-sustaining populations for all Atlantic coast fish species or successful restoration in progress by 2015. Fishery managers have recently come to realize that in order to restore sustainable fish stocks, essential habitat areas must be restored and protected as well. The ASMFC Interstate Fishery Management Program Policy Board charged the ASMFC Habitat Committee in May 2006 to work with the national and regional offices of the U.S. Fish and Wildlife Service to develop a pilot marine fish habitat partnership (FHP) under the auspices of the National Fish Habitat Action Plan (NFHAP). The resultant partnership between state and federal agencies, nongovernmental organizations, and others is now called the Atlantic Coast Fish Habitat Partnership (ACFHP). The initial focus of the ACFHP is on improving saltwater interface habitat for diadromous fish through region-specific habitat restoration and protection projects. The process of FHP formation is evolving due to the recent completion of the NFHAP. The ACFHP will serve as a model for future partnerships under the NFHAP.

Sustainability of American Eel in the St. Jones River, Delaware, USA: A Collaborative Approach towards Understanding Population Ecology

COLLETTE M. CAIRNS*, DEWAYNE A. FOX, LARISSA L. BAILEY, AND KYLE SHERTZER
Delaware State University, Department of Agriculture and Natural Resources
Dover, Delaware 19904, USA

* Corresponding author: colettemcairns@yahoo.com

There have been calls for more comprehensive data on eel *Anguilla* spp. ecology due to declines in abundance of eel populations. The American eel *A. rostrata* supports an important commercial fishery in the mid-Atlantic region, which accounted for 83% of the total U.S. landings in 2005. The American eel ranked fifth among commercial fisheries for pounds landed in Delaware, with 77% of landings destined for food and 23% for the bait market. We initiated a mark–recapture study with both fishery-dependent and fishery-independent components in the St. Jones River, Delaware, to better understand the population dynamics of American eel. All American eel captured (n = 5,961) were tagged in the dorsal musculature with a PIT (passive integrated transponder) tag and released. Higher salinity sites, which experienced the most commercial pressure, generally produced larger catches but with shorter (<350 mm) eels. There were 2,499 recaptures, of which 1,476 (59%) were from fishery-independent sampling, and 1,023 (41%) were from the commercial harvest. The high level of both commercial and directed sampling recapture rates suggests that American eels may be susceptible to over-harvest. American eels are subjected to other stressors, in addition to the commercial harvest, in the St. Jones River where a large proportion (42%) of the population have been infected by the nonnative parasite *Anguillicola crassus*. The combined effects of these anthropogenic impacts pose new challenges to American eels and fisheries professionals. We developed a more thorough understanding of American eel population dynamics by working closely with harvesters.

Challenges of Conservation and Management of Anadromous Fish in Relicensing Proceedings for Hydropower Projects, USA

MELANIE HARRIS* AND JANE HANNUKSELA

National Oceanic and Atmospheric Administration, National Marine Fisheries Service, Office of Habitat Conservation, F/HC2, 1315 East–West Highway, Silver Spring, Maryland 20910, USA

* Corresponding author: melanie.harris@noaa.gov

The U.S. National Oceanic and Atmospheric Administration's National Marine Fisheries Service (NMFS) and other federal agencies have statutory responsibility to provide resource management conditions for federal hydropower licenses issued by the Federal Energy Regulatory Commission (FERC). The National Marine Fisheries Service is the federal trustee for anadromous fish and ensures that these species receive adequate safeguards by providing appropriate protection, mitigation, enhancement, and fish passage measures for inclusion in licenses for the dams. Since FERC licenses have a term of 30–50 years, relicensings are once-in-a-lifetime opportunities to provide significant benefits to anadromous fish by making many miles of previously blocked upstream habitat accessible and improving habitat degraded by changes in stream flow and water quality. These hydropower relicensing proceedings often include parties such as federal, state, and local managers, tribes, licensees, landowners, environmental organizations, fishing groups, agricultural organizations, and rate payers, all with different objectives and interests. These differing interests result in potentially conflicting license conditions, and the challenge is in finding a way to address the concerns in a license. Another challenge NMFS faces is in making science-based conservation and management decisions with limited or incomplete data regarding habitat condition, fish movements and migrations in reservoirs and rivers, and engineering aspects of fish passage facilities. Working through these issues in the context of settlement negotiations can be an effective means to allow diverse interests to arrive at mutually agreeable license conditions. Adaptive management measures allow licensees to begin implementation of conservation measures and make modifications based on recommendations from NMFS and other interested parties as information becomes available. The National Marine Fisheries Service has helped to enhance significant amounts of habitat and improve access to thousands of river miles for anadromous fish through these actions.